PREFACE

The principal purpose of the Commentary is to provide a basic volume of knowledge and facts relating to building construction as it pertains to the regulations set forth in the 2015 *International Building Code*. The person who is serious about effectively designing, constructing and regulating buildings and structures will find the Commentary to be a reliable data source and reference to almost all components of the built environment.

As a follow-up to the *International Building Code*, we offer a companion document, the *International Building Code Commentary—Volume I*. Volume I covers Chapters 1 through 15 of the 2015 *International Building Code*. The basic appeal of the Commentary is thus: it provides in a small package and at reasonable cost thorough coverage of many issues likely to be dealt with when using the *International Building Code* — and then supplements that coverage with historical and technical background. Reference lists, information sources and bibliographies are also included.

Throughout all of this, effort has been made to keep the vast quantity of material accessible and its method of presentation useful. With a comprehensive yet concise summary of each section, the Commentary provides a convenient reference for regulations applicable to the construction of buildings and structures. In the chapters that follow, discussions focus on the full meaning and implications of the code text. Guidelines suggest the most effective method of application, and the consequences of not adhering to the code text. Illustrations are provided to aid understanding; they do not necessarily illustrate the only methods of achieving code compliance.

The format of the Commentary includes the full text of each section, table and figure in the code, followed immediately by the commentary applicable to that text. At the time of printing, the Commentary reflects the most up-to-date text of the 2015 *International Building Code*. As stated in the preface to the *International Building Code*, the content of sections in the code which begin with a letter designation (i.e., Section [F]307.1) are maintained by another code development committee. Each section's narrative includes a statement of its objective and intent, and usually includes a discussion about why the requirement commands the conditions set forth. Code text and commentary text are easily distinguished from each other. All code text is shown as it appears in the *International Building Code*, and all commentary is indented below the code text and begins with the symbol ❖.

Readers should note that the Commentary is to be used in conjunction with the *International Building Code* and not as a substitute for the code. The Commentary is advisory only; the code official alone possesses the authority and responsibility for interpreting the code.

Comments and recommendations are encouraged, for through your input, we can improve future editions. Please direct your comments to the Codes and Standards Development Department at the Chicago District Office.

The International Code Council would like to extend its thanks to the following individuals for their contributions to the technical content of this commentary:

Brian Black	Kevin Brinkman
John Woestman	Gregory Cahanin
David Cooper	Rebecca Quinn
Dave Collins	Joann Surmar
Vickie Lovell	James Milke
John Valiulis	Richard Walke
Marcelo Hirschler	Dave Frable
Jeff Hugo	Elley Klausbruckner

Dick Bukowski

S.I. CONVERSION CHART FOR ENGLISH UNITS

English Unit	Metric Equivalent
1 inch (in.)	25.4 millimeters (mm)
1 foot (ft.)	304.8 millimeters (mm)
1 square inch (in^2)	645.2 square millimeters (mm^2)
1 square foot (ft^2)	0.093 square meters (m^2)
1 cubic foot (ft^3)	0.028 cubic meters (m^3)
1 gallon (gal.)	.00379 cubic meters (m^3)
1 gallon (gal.)	3.79 liters (L)
1 pound (lb.)	0.454 kilograms (kg)
1 ton (T)	0.907 metric ton
1 ton (T)	907.2 kilograms (kg)
1 pound per square inch (psi)	6.894 kilo- Pascals (kPa)
1 pound per square inch (psi)	.0703 kilogram force per square centimeter
1 pound per square foot (psf)	47.88 Pascals (Pa)
1 pound per square foot (psf)	47.88 Newton per square meter (N/m^2)
1 foot candle	10.764 lux
1 inch water column	248.84 Pascal
1 degree	0.01745 radian
1 Btu/hr	0.2931 Watt
Temperature degrees Fahrenheit	1.8 x degrees C + 32.

TABLE OF CONTENTS

Chapter 1: Scope and Administration

General Comments

This chapter contains provisions for the application, enforcement and administration of subsequent requirements of the code. In addition to establishing the scope of the code, Chapter 1 identifies which buildings and structures come under its purview.

Chapter 1 is subdivided into two parts. Part 1 includes scope and application, Sections 101 and 102. Part 2 deals with administration and enforcement, Sections 103 through 116.

- Section 101 addresses the scope of the IBC and references the other *International Codes*® mentioned in the code.
- Section 102 establishes the applicability of the code and addresses existing structures.
- Section 103 establishes the department of building safety and the appointment of department personnel.
- Section 104 outlines the duties and authority of the building official with regard to permits, inspections and right of entry. It also establishes the authority of the building official to approve alternative materials, used materials and modifications.
- Section 105 states when permits are required and establishes procedures for the review of applications and the issuance of permits.
- Section 106 provides requirements for posting live loads greater than 50 pounds per square foot (psf) (2394 Pa).
- Section 107 describes the information that must be included on construction documents submitted with the application.
- Section 108 authorizes the building official to issue permits for temporary structures and uses.
- Section 109 establishes requirements for a fee schedule.
- Section 110 includes inspection duties of the building official or an inspection agency that has been approved by the building official.
- Section 111 details provisions for the issuance of certificates of occupancy.
- Section 112 gives the building official the authority to approve utility connections.
- Section 113 establishes the board of appeals and criteria for making applications for appeal.
- Section 114 addresses administrative provisions for violations, including provisions for unlawful acts, violation notices, prosecution and penalties.
- Section 115 describes procedures for stop work orders.
- Section 116 establishes the criteria for unsafe structures and equipment, and the procedures to be followed by the building official for abatement and notification to the responsible party.

Each state's building code enabling legislation, which is grounded within the police power of the state, is the source of all authority to enact building codes. In terms of how it is used, police power is the power of the state to legislate for the general welfare of its citizens. This power enables passage of such laws as building codes. If the state legislature has limited this power in any way, the municipality may not exceed these limitations. While the municipality may not further delegate its police power (e.g., by delegating the burden of determining code compliance to the building owner, contractor or architect), it may turn over the administration of the building code to a municipal official, such as a building official, provided that sufficient criteria are given to establish clearly the basis for decisions as to whether a proposed building conforms to the code.

Chapter 1 is largely concerned with maintaining "due process of law" in enforcing the building performance criteria contained in the body of the code. Only through careful observation of the administrative provisions can the building official reasonably hope to demonstrate that "equal protection under the law" has been provided. While it is generally assumed that the administration and enforcement section of a code is geared toward a building official, this is not entirely true. The provisions also establish the rights and privileges of the design professional, contractor and building owner. The position of the building official is merely to review the proposed and completed work and to determine if the construction conforms to the code requirements. The design professional is responsible for the design of a safe structure. The contractor is responsible for constructing the structure in compliance with the plans.

During the course of construction, the building official reviews the activity to ascertain that the spirit and intent of the law are being met and that the safety, health and welfare of the public will be protected. As a public servant, the building official enforces the code in an unbiased, proper manner. Every individual is guaranteed equal enforcement of the provisions of the code. Furthermore, design professionals, contractors and building owners have the right of due process for any requirement in the code.

Purpose

The building code, as with any other I-Code®, is intended to be adopted as a legally enforceable document to provide a reasonable level of safety, and protection of public health, general welfare and property. A building code cannot be effective without adequate provisions for its administration and enforcement. The official charged with the administration and enforcement of building regulations has a great responsibility, and with this responsibility goes authority. No matter how detailed the building code may be, the building official must, to some extent, exercise his or her own judgment in determining code compliance. The building official has the responsibility to establish that the homes in which the citizens of the community reside and the buildings in which they work are designed and constructed to be structurally stable with adequate means of egress, accessibility, light and ventilation, and to provide a minimum acceptable level of protection to life and property from fire.

Chapter 1 contains two parts. Part 1, Scope and Application, contains all issues related to the scope and intent of the code, as well as the applicability of this code relative to other standards and laws that might also be applicable on a given building project, such as federal or state. Part 2, Administration and Enforcement, contains all issues related to the duties and powers of the building official, the issuance of permits and certificates of occupancy, and other related operational items.

PART 1—SCOPE AND APPLICATION

SECTION 101
GENERAL

[A] 101.1 Title. These regulations shall be known as the *Building Code* of **[NAME OF JURISDICTION]**, hereinafter referred to as "this code."

❖ The purpose of this section is to identify the adopted regulations by inserting the name of the adopting jurisdiction into the code.

[A] 101.2 Scope. The provisions of this code shall apply to the construction, *alteration*, relocation, enlargement, replacement, *repair*, equipment, use and occupancy, location, maintenance, removal and demolition of every building or structure or any appurtenances connected or attached to such buildings or structures.

> **Exception:** Detached one- and two-family *dwellings* and multiple single-family *dwellings (townhouses)* not more than three *stories above grade plane* in height with a separate *means of egress*, and their accessory structures not more than three *stories above grade plane* in height, shall comply with the *International Residential Code*.

❖ This section establishes when the regulations contained in the code must be followed, whether all or in part. Something must happen (construction of a new building, modification to an existing one or allowing an existing building or structure to become unsafe) for the code to be applicable. While such activity may not be as significant as a new building, a fence is considered a structure and, therefore, its erection is within the scope of the code. The building code is not a maintenance document requiring periodic inspections that will, in turn, result in an enforcement action, although periodic inspections are addressed by the *International Fire Code*® (IFC®).

The exception indicates that detached one- and two-family dwellings and townhouses that are not more than three stories above grade and have separate means of egress are to comply with the *International Residential Code*® (IRC®). The definition of townhouse adds that an IRC townhouse must meet four criteria: 1. It is not more than three stories in height; 2. It has a separate means of egress; 3. Each unit extends from foundation to roof; and 4. There is open space on at least two sides.

This applies to all such structures, whether or not there are lot lines separating them, and also to accessory structures such as garages and pools. Accessory structures are also limited to not more than 3 stories in height. Such structures four stories or more in height are beyond the scope of the IRC and must comply with the provisions of the IBC and its referenced codes.

There are two exceptions in the IRC that allow for buildings otherwise required to be constructed in accordance with the IBC to be constructed in accordance with the IRC. These include live/work units (see Section 419) and small bed-and-breakfast style hotels where there are five or fewer guestrooms and the owner also lives in the hotel (see Section 310.5.2).

[A] 101.2.1 Appendices. Provisions in the appendices shall not apply unless specifically adopted.

❖ The provisions contained in Appendices A through M are not considered part of the code and are, therefore, not enforceable unless they are specifically included in the ordinance or other adopting law or regulation of the jurisdiction. See Section 1 of the sample legislation on page xix of the code for where the appendices to be adopted are to be specified in the adoption ordinance.

[A] 101.3 Intent. The purpose of this code is to establish the minimum requirements to provide a reasonable level of safety, public health and general welfare through structural strength, *means of egress* facilities, stability, sanitation, adequate light and ventilation, energy conservation, and safety to life and property from fire and other hazards attributed to the built environment and to provide a reasonable level of safety to fire fighters and emergency responders during emergency operations.

❖ The intent of the code is to establish regulations providing for the safety, health and general welfare of building occupants, as well as for fire fighters and emergency responders during building emergencies. The intent becomes important in the application of such sections as Sections 102, 104.11 and 114, as well as any enforcement-oriented interpretive action or judgment. Like any code, the written text is subject to interpretation. Interpretations should not be affected by economics or the potential impact on any party. The only considerations should be safety of the occupants, protection of occupant's health and welfare and emergency responder safety.

[A] 101.4 Referenced codes. The other codes listed in Sections 101.4.1 through 101.4.7 and referenced elsewhere in this code shall be considered part of the requirements of this code to the prescribed extent of each such reference.

❖ The International Code Council® (ICC®) promulgates a complete set of codes to regulate the built environment. These codes are coordinated with each other so as to avoid conflicting provisions. When the code is adopted by a jurisdiction, the codes that regulate a building's electrical, fuel gas, mechanical and plumbing systems are also included in the adoption and are considered a part of the code. The *International Property Maintenance Code®* (IPMC®), *International Existing Building Code®* (IEBC®) and the IFC are also referenced and enable the building official to address unsafe conditions in existing structures. Various other sections of the code also specifically refer to these codes. Note that these codes are listed in Chapter 35 and further identified by the specific year of issue. Only that edition of the code is legally adopted and any future editions are not enforceable. New editions of the *International Codes®* are issued concurrently and new editions of the referenced codes are adopted with each new edition of the code. Adoption is done in this manner so that there are not conflicting provisions in these codes.

[A] 101.4.1 Gas. The provisions of the *International Fuel Gas Code* shall apply to the installation of gas piping from the point of delivery, gas appliances and related accessories as covered in this code. These requirements apply to gas piping systems extending from the point of delivery to the inlet connections of appliances and the installation and operation of residential and commercial gas appliances and related accessories.

❖ The *International Fuel Gas Code®* (IFGC®) regulates gas piping and appliances and is adopted by reference from this section, as well as other sections in the code, as the enforceable document for regulating gas systems. This section also establishes the scope of the IFGC as extending from the point of delivery to the inlet connections of each gas appliance. The "point of delivery" is defined in the IFGC as the outlet of the service meter, regulator or shutoff valve.

[A] 101.4.2 Mechanical. The provisions of the *International Mechanical Code* shall apply to the installation, *alterations*, *repairs* and replacement of mechanical systems, including equipment, appliances, fixtures, fittings and/or appurtenances, including ventilating, heating, cooling, air-conditioning and refrigeration systems, incinerators and other energy-related systems.

❖ The *International Mechanical Code®* (IMC®) regulates all aspects of a building's mechanical system, including ventilating, heating, air-conditioning and refrigeration systems, incinerators and other energy-related systems, and is adopted by reference from this section, as well as other sections in this code, as the enforceable document for regulating these systems.

[A] 101.4.3 Plumbing. The provisions of the *International Plumbing Code* shall apply to the installation, *alteration*, *repair* and replacement of plumbing systems, including equipment, appliances, fixtures, fittings and appurtenances, and where connected to a water or sewage system and all aspects of a medical gas system. The provisions of the *International Private Sewage Disposal Code* shall apply to private sewage disposal systems.

❖ The *International Plumbing Code®* (IPC®) regulates the components of a building's plumbing system, including water supply and distribution piping; sanitary and storm drainage systems; the fixtures and appliances connected thereto; and medical gas and oxygen systems, and is adopted by reference from this section, as well as other sections in this code, as the enforceable document for regulating these systems. The *International Private Sewage Disposal Code®* (IPSDC®) is also adopted as the enforceable document for regulating on-site sewage disposal systems.

[A] 101.4.4 Property maintenance. The provisions of the *International Property Maintenance Code* shall apply to existing structures and premises; equipment and facilities; light, ventilation, space heating, sanitation, life and fire safety hazards; responsibilities of *owners*, operators and occupants; and occupancy of existing premises and structures.

❖ The applicability of the code to existing structures is set forth in the IEBC and is generally limited to new work or changes in use that occur in these buildings. The IPMC, however, is specifically intended to apply to existing structures and their premises, providing a jurisdiction with an enforceable document protecting occupant safety, public health and general welfare, including in buildings that were constructed prior to the adoption of the current building code.

[A] 101.4.5 Fire prevention. The provisions of the *International Fire Code* shall apply to matters affecting or relating to structures, processes and premises from the hazard of fire and explosion arising from the storage, handling or use of structures, materials or devices; from conditions hazardous to life, property or public welfare in the occupancy of structures or premises; and from the construction, extension, *repair, alteration* or removal of fire suppression, *automatic sprinkler systems* and alarm systems or fire hazards in the structure or on the premises from occupancy or operation.

❖ The IFC contains provisions which provide a reasonable level of safety for occupants from the hazards of fire and explosion that result from: materials, substances and operations that may be present in a structure; circumstances that endanger life, property or public welfare; and the modification or removal of fire suppression and alarm systems. Many of the provisions contained in the IBC, especially in Chapters 9 and 10, also appear in the IFC. So that all International Codes contain consistent provisions, only one development committee is responsible for considering proposed changes to such provisions. That committee is identified by a letter designation in brackets that appears at the beginning of affected sections. This is described more fully in the preface to the codes. The IFC also contains provisions that are specifically applicable to existing structures and uses and, like the IPMC, provides a jurisdiction with an enforceable document protecting occupant safety, public health and general welfare in all buildings.

[A] 101.4.6 Energy. The provisions of the *International Energy Conservation Code* shall apply to all matters governing the design and construction of buildings for energy efficiency.

❖ The *International Energy Conservation Code®* (IECC®) contains provisions for the efficient use of energy in buildings by regulating the design of building envelopes for thermal resistance and low air leakage, and the design and selection of mechanical systems for effective use of energy. The IECC® is adopted by reference in this section, as well as other sections in this code, as the enforceable document for regulating these systems.

[A] 101.4.7 Existing buildings. The provisions of the *International Existing Building Code* shall apply to matters governing the *repair, alteration*, change of occupancy, *addition* to and relocation of existing buildings.

❖ The *International Existing Building Code®* (IEBC®) is typically utilized when a building is undergoing some type of alteration, change of occupancy or addition. Maintenance of existing buildings is addressed in the IPMC and IFC. Three different options for compliance are provided within the IEBC.

In the 2012 IBC, Chapter 34 addressed existing buildings. This criterion was repeated in the IEBC as Chapter 4, Prescriptive Compliance Methods, and Chapter 14, Performance Compliance Methods. Now this information is only available in the IEBC.

SECTION 102 APPLICABILITY

[A] 102.1 General. Where there is a conflict between a general requirement and a specific requirement, the specific requirement shall be applicable. Where, in any specific case, different sections of this code specify different materials, methods of construction or other requirements, the most restrictive shall govern.

❖ In cases where the code establishes a specific requirement for a certain condition, that requirement is applicable even if it is less restrictive than a general requirement elsewhere in the code. As an example, the requirements contained in Section 402.8 for means of egress in a covered mall building would govern over any differing requirements located in Chapter 10, regardless of whether the requirements in Section 402.8 are more or less restrictive.

The most restrictive requirement is to apply where there may be different requirements in the code for a specific issue.

[A] 102.2 Other laws. The provisions of this code shall not be deemed to nullify any provisions of local, state or federal law.

❖ In some cases, other laws enacted by the jurisdiction or the state or federal government may be applicable to a condition that is also governed by a requirement in the code. In such circumstances, the requirements of the code are in addition to the other law that is still in effect, although the building official may not be responsible for its enforcement.

[A] 102.3 Application of references. References to chapter or section numbers, or to provisions not specifically identified by number, shall be construed to refer to such chapter, section or provision of this code.

❖ In a situation where the code may make reference to a chapter or section number or to another code provision without specifically identifying its location in the code, assume that the referenced section, chapter or provision is in the code and not in a referenced code or standard.

[A] 102.4 Referenced codes and standards. The codes and standards referenced in this code shall be considered part of the requirements of this code to the prescribed extent of each such reference and as further regulated in Sections 102.4.1 and 102.4.2.

❖ A referenced code, standard or portion thereof is an enforceable extension of the code as if the content of the standard were included in the body of the code. For example, Section 905.2 references NFPA 14 in its entirety for the installation of standpipe systems. In those cases when the code references only portions of a standard, the use and application of the referenced standard is limited to those portions that are specifically identified. For example, Section 412.4.6 requires that aircraft hangars must be provided with fire suppression systems as required in NFPA 409.

Section 412.4.6 cannot be construed to require compliance with NFPA 409 in its entirety. It is the intent of the code to be in harmony with the referenced standards. If conflicts occur because of scope or purpose, the code text governs.

[A] 102.4.1 Conflicts. Where conflicts occur between provisions of this code and referenced codes and standards, the provisions of this code shall apply.

❖ The use of referenced codes and standards to cover certain aspects of various occupancies and operations rather than write parallel or competing requirements into the code is a long-standing code development principle. Often, however, questions and potential conflicts in the use of referenced codes and standards can arise, which can lead to inconsistent enforcement of the code. In the code, several sections illustrate this concern, such as Section [F] 415.9.3.

> **Section [F] 415.9.3 Dry cleaning plants.** The construction and installation of dry cleaning plants shall be in accordance with the requirements of this code, the *International Mechanical Code*, the *International Plumbing Code* and NFPA 32. Dry cleaning solvents and systems shall be classified in accordance with the *International Fire Code*.

Based on this text, NFPA 32, Standard for Drycleaning Plants, 2011 edition, in Section 4.4.1.1 states, "General building and structure design and construction shall be in accordance with NFPA 5000, *Building Construction and Safety Code*®, except as modified herein." Since the extent of the reference to NFPA 32 in Section 415.9.3 includes "...construction...", it has happened that designers construed this to mean that the requirements for building construction of dry cleaning plants will be required to follow NFPA 5000 instead of the IBC.

Another example is in the IMC, which references ASHRAE 15 in Sections 1101.6 and 1108.1. ASHRAE 15 then references NFPA 54 (ANSI Z223.1), *National Fuel Gas Code*. This could lead code users to interpret the mechanical code to mean that the *National Fuel Gas Code* is applicable to specific situations rather than the IFGC.

In both cases, the reference is only applicable to the first referenced standard.

[A] 102.4.2 Provisions in referenced codes and standards. Where the extent of the reference to a referenced code or standard includes subject matter that is within the scope of this code or the International Codes listed in Section 101.4, the provisions of this code or the International Codes listed in Section 101.4, as applicable, shall take precedence over the provisions in the referenced code or standard.

❖ Section 102.4.2 expands upon the provisions of Section 102.4.1 by making it clear that, even if a referenced standard contains requirements that parallel the code (or the other referenced International Codes) in the standard's own duly referenced section(s), the provisions of the IBC (or the other referenced International Codes) will always take precedence. This proposed section does not intend to take the place of carefully scoped and referenced text for written standards for the International Codes but, rather, provides the policy underpinnings upon which sound code change proposals can be based.

[A] 102.5 Partial invalidity. In the event that any part or provision of this code is held to be illegal or void, this shall not have the effect of making void or illegal any of the other parts or provisions.

❖ Only invalid sections of the code (as established by the court of jurisdiction) can be set aside. This is essential to safeguard the application of the code text in situations where a provision is declared illegal or unconstitutional. This section preserves the legislative action that put the legal provisions in place.

[A] 102.6 Existing structures. The legal occupancy of any structure existing on the date of adoption of this code shall be permitted to continue without change, except as otherwise specifically provided in this code, the *International Existing Building Code*, the *International Property Maintenance Code* or the *International Fire Code*.

❖ An existing structure is generally "grandfathered" to be considered approved with code adoption, provided that the building meets a minimum level of safety. Frequently, the criteria for this level are the regulations (or code) under which the existing building was originally constructed. If there are no previous code criteria to apply, the building official must apply those provisions that are reasonably applicable to existing buildings. A specific level of safety in existing buildings is dictated by maintenance and hazard abatement provisions, as contained in this code, the IPMC and the IFC. These codes (see Sections 101.4.4 and 101.4.5) are applicable to existing buildings. Special attention should be paid to IFC Chapter 11, *Construction Requirements for Existing Buildings*. Additionally, IEBC (see Section 101.4.7) comprehensively identifies the pertinent requirements for existing buildings on which construction operations are intended or that undergo a change of occupancy.

[A] 102.6.1 Buildings not previously occupied. A building or portion of a building that has not been previously occupied or used for its intended purpose in accordance with the laws in existence at the time of its completion shall comply with the provisions of the *International Building Code* or *International Residential Code*, as applicable, for new construction or with any current permit for such occupancy.

❖ This section applies to any building that may have been completed but not occupied or used for its original intended purpose. The building remains a new structure in terms of code compliance until such time as it is occupied in whole or in part. Tenant buildouts are permitted to comply with the code adopted at the time of initial construction, unless that permit has expired. If the permit has expired, the tenant buildout must comply with new construction requirements, similar to alterations to existing buildings. See Section 105.5 regarding the expiration and extensions available for permits.

[A] 102.6.2 Buildings previously occupied. The legal occupancy of any building existing on the date of adoption of this code shall be permitted to continue without change, except as otherwise specifically provided in this code, the *International Fire Code* or *International Property Maintenance Code*, or as is deemed necessary by the *building official* for the general safety and welfare of the occupants and the public.

❖ This section allows for buildings that were legally occupied in whole or in part at the time the code was adopted to continue as is. There is a maintenance concern that is addressed by the requirement that the building comply with either the IFC or the IPMC. These codes ensure that life safety systems, such as means of egress pathways and fire protection systems, are kept in place and continue to be able to protect the life and safety of the inhabitants of these existing structures.

PART 2—ADMINISTRATION AND ENFORCEMENT

SECTION 103
DEPARTMENT OF BUILDING SAFETY

[A] 103.1 Creation of enforcement agency. The Department of Building Safety is hereby created and the official in charge thereof shall be known as the *building official*.

❖ This section creates the building department and describes its composition (see Section 110 for a discussion of the inspection duties of the department). Appendix A contains qualifications for the employees of the building department involved in the enforcement of the code. A jurisdiction can establish the qualifications outlined in Appendix A for its employees by specifically referencing Appendix A in the adopting ordinance.

The executive official in charge of the building department is named the "building official" by this section. In actuality, the person who is in charge of the department may hold a different title, such as building commissioner, building inspector or construction official. For the purpose of the code, that person is referred to as the "building official."

[A] 103.2 Appointment. The *building official* shall be appointed by the chief appointing authority of the jurisdiction.

❖ This section establishes the building official as an appointed position of the jurisdiction.

[A] 103.3 Deputies. In accordance with the prescribed procedures of this jurisdiction and with the concurrence of the appointing authority, the *building official* shall have the authority to appoint a deputy building official, the related technical officers, inspectors, plan examiners and other employees. Such employees shall have powers as delegated by the *building official*. For the maintenance of existing properties, see the *International Property Maintenance Code*.

❖ This section provides the building official with the authority to appoint other individuals to assist with the administration and enforcement of the code. These individuals would have the authority and responsibility as designated by the building official. Such appointments, however, may be exercised only with the authorization of the chief appointing authority.

SECTION 104
DUTIES AND POWERS OF BUILDING OFFICIAL

[A] 104.1 General. The *building official* is hereby authorized and directed to enforce the provisions of this code. The *building official* shall have the authority to render interpretations of this code and to adopt policies and procedures in order to clarify the application of its provisions. Such interpretations, policies and procedures shall be in compliance with the intent and purpose of this code. Such policies and procedures shall not have the effect of waiving requirements specifically provided for in this code.

❖ The duty of the building official is to enforce the code, and he or she is the "authority having jurisdiction" for all matters relating to the code and its enforcement. It is the duty of the building official to interpret the code and to determine compliance. Code compliance will not always be easy to determine and will require judgment and expertise, particularly when enforcing the provisions of Sections 104.10 and 104.11. In exercising this authority, however, the building official cannot set aside or ignore any provision of the code.

[A] 104.2 Applications and permits. The *building official* shall receive applications, review *construction documents* and issue *permits* for the erection, and *alteration*, demolition and moving of buildings and structures, inspect the premises for which such *permits* have been issued and enforce compliance with the provisions of this code.

❖ The code enforcement process is normally initiated with an application for a permit. The building official is responsible for processing applications and issuing permits for the construction or modification of buildings in accordance with the code.

[A] 104.2.1 Determination of substantially improved or substantially damaged existing buildings and structures in flood hazard areas. For applications for reconstruction, rehabilitation, *repair*, *alteration*, *addition* or other improvement of existing buildings or structures located in *flood hazard areas*, the *building official* shall determine if the proposed work constitutes substantial improvement or *repair* of *substantial damage*. Where the *building official* determines that the proposed work constitutes *substantial improvement* or *repair* of *substantial damage*, and where required by this code, the *building official* shall require the building to meet the requirements of Section 1612.

❖ "Substantial damage" and "Substantial improvement" are defined in Section 202 and in federal regulations (see 44 CFR 59.1, Definitions). Long-term reduction in exposure to flood hazards, including exposure of older buildings, is one of the purposes for regulating development in flood hazard areas. Existing buildings or structures located in flood hazard areas are to be brought into compliance with the flood-resistance provisions of Section 1612 when the cost of improvements or the cost of repair of damage equals or exceeds 50 percent of the market value of the building.

Applicants state the valuation of the proposed work as part of the information submitted to obtain a permit. If the proposed work will be performed on existing buildings or structures in flood hazard areas, including restoration of damage from any cause, this section requires the building official to determine the value of the proposed work. Guidance from the National Flood Insurance Program (NFIP) is in FEMA P-758, *Substantial Improvement/Substantial Damage Desk Reference*. This guidance advises that the value of the property owner's labor, as well as the value of donated labor and materials, must be included. For damaged buildings, the value of the proposed work is the value of work necessary to restore the building to its predamage condition, even if the applicant is proposing less work.

To make a determination about whether a proposed repair, reconstruction, rehabilitation, addition or improvement of a building or structure will constitute a substantial improvement or repair of substantial damage, the cost of the proposed work must be compared to the market value of the building or structure before the work is started or before the damage occurred. To determine market value, the building official may require the applicant to provide such information. Options for determining value are described in FEMA P-758. If the building official determines that the work is a substantial improvement or repair of substantial damage, the existing building must be brought into compliance with the flood-resistance provisions of Section 1612. See the IEBC for requirements for alterations.

[A] 104.3 Notices and orders. The *building official* shall issue necessary notices or orders to ensure compliance with this code.

❖ An important element of code enforcement is the necessary advisement of deficiencies and corrections, which is accomplished through written notices and orders. The building official is required to issue orders to abate illegal or unsafe conditions. Section 116.3 contains additional information for these notices.

[A] 104.4 Inspections. The *building official* shall make the required inspections, or the *building official* shall have the authority to accept reports of inspection by *approved agencies* or individuals. Reports of such inspections shall be in writing and be certified by a responsible officer of such *approved agency* or by the responsible individual. The *building official* is authorized to engage such expert opinion as deemed necessary to report upon unusual technical issues that arise, subject to the approval of the appointing authority.

❖ The building official is required to make inspections as necessary to determine compliance with the code, or to accept written reports of inspections by an approved agency. The inspection of the work in progress or accomplished is another significant element in determining code compliance. While a department does not have the resources to inspect every aspect of all work, the required inspections are those that are dictated by administrative rules and procedures based on many parameters, including available inspection resources. In order to expand the available resources for inspection purposes, the building official may approve an agency that, in his or her opinion, complies with the criteria set forth in Section 1703. When unusual, extraordinary or complex technical issues arise relative to building safety, the building official has the authority to seek the opinion and advice of experts. Since this usually involves the expenditure of funds, the approval of the jurisdiction's chief executive (or similar position) is required. A technical report from an expert requested by the building official can be used to assist in the approval process (also see Section 1704 for special inspection requirements).

[A] 104.5 Identification. The *building official* shall carry proper identification when inspecting structures or premises in the performance of duties under this code.

❖ This section requires the building official (including by definition all authorized designees) to carry identification in the course of conducting the duties of the position. This removes any question as to the purpose and authority of the inspector.

[A] 104.6 Right of entry. Where it is necessary to make an inspection to enforce the provisions of this code, or where the *building official* has reasonable cause to believe that there exists in a structure or upon a premises a condition that is contrary to or in violation of this code that makes the structure or premises unsafe, dangerous or hazardous, the *building official* is authorized to enter the structure or premises at reasonable times to inspect or to perform the duties imposed by this code, provided that if such structure or premises be occupied that credentials be presented to the occupant and entry requested. If such structure or premises is unoccupied, the *building official* shall first make a reasonable effort to locate the owner or other person having charge or control of the structure or premises and request entry. If entry is refused, the *building official* shall have recourse to the remedies provided by law to secure entry.

❖ The first part of this section establishes the right of the building official to enter the premises in order to make permit inspections required by Section 110.3. Permit application forms typically include a statement in the certification signed by the applicant (who is the owner or owner's agent) granting the building official the authority to enter areas covered by the permit in order to enforce code provisions related to the permit. The right to enter other structures or premises is more limited. First, to protect the right of privacy, the owner or occupant must grant the building official permission before an interior inspection of the property can be conducted. Permission is not required for inspections that can be accomplished from within the public right-of-way. Second, such access may be denied by the owner or occupant. Unless the inspector has reasonable cause to believe that a violation of the code exists, access may be unattainable. Third, building officials must present proper identification (see Section 104.5) and request admittance during reasonable hours—usually the normal business hours of the establishment—to be admitted. Fourth, inspections must be aimed at securing or determining compliance with the provisions and intent of the regulations that are specifically within the established scope of the building official's authority.

Searches to gather information for the purpose of enforcing other codes, ordinances or regulations are considered unreasonable and are prohibited by the Fourth Amendment to the U.S. Constitution. "Reasonable cause" in the context of this section must be distinguished from "probable cause," which is required to gain access to property in criminal cases. The burden of proof establishing reasonable cause may vary among jurisdictions. Usually, an inspector must show that the property is subject to inspection under the provisions of the code; that the interests of the public health, safety and welfare outweigh the individual's right to maintain privacy; and that such an inspection is required solely to determine compliance with the provisions of the code.

Many jurisdictions do not recognize the concept of an administrative warrant and may require the building official to prove probable cause in order to gain access upon refusal. This burden of proof is usually more substantial, often requiring the building official to stipulate in advance why access is needed (usually access is restricted to gathering evidence for seeking an indictment or making an arrest); what specific items or information is sought; its relevance to the case against the individual subject; how knowledge of the relevance of the information or items sought was obtained, and how the evidence sought will be used. In all such cases, the right to privacy must always be weighed against the right of the building official to conduct an inspection to verify that public health, safety and welfare are not in jeopardy. Such important and complex constitutional issues should be discussed with the jurisdiction's legal counsel. Jurisdictions should establish procedures for securing the necessary court orders when an inspection is deemed necessary following a refusal.

[A] 104.7 Department records. The *building official* shall keep official records of applications received, *permits* and certificates issued, fees collected, reports of inspections, and notices and orders issued. Such records shall be retained in the official records for the period required for retention of public records.

❖ In keeping with the need for efficient business practices, the building official must keep official records pertaining to permit applications, permits, fees collected, inspections, notices and orders issued. Such documentation provides a valuable resource of information if questions arise regarding the department's actions with respect to a building. The code does not require that construction documents be kept after the project is complete. It requires that other documents be kept for the length of time mandated by a jurisdiction's or its state's laws, or administrative rules for retaining public records.

Certain records related to buildings in flood hazard areas must be retained permanently in accordance with the community's agreement with the National Flood Insurance Program, including the lowest floor elevation information collected pursuant to Sections 110.3.3, 110.3.10.1, and 1612.5; determinations made when work on existing buildings is proposed to determine if the work constitutes substantial improvement or repair of substantial damage; and modifications granted pursuant to Section 104.10.1. Communities agree to allow inspection of these records upon request by FEMA or the NFIP state coordinating agency.

[A] 104.8 Liability. The *building official*, member of the board of appeals or employee charged with the enforcement of this code, while acting for the jurisdiction in good faith and without malice in the discharge of the duties required by this code or other pertinent law or ordinance, shall not thereby be civilly or criminally rendered liable personally and is hereby relieved from personal liability for any damage accruing to persons or property as a result of any act or by reason of an act or omission in the discharge of official duties.

❖ The building official, other department employees and members of the appeals board are not intended to be held liable, either civilly or criminally, for those actions performed in accordance with the code in a reasonable and lawful manner. The responsibility of the building official in this regard is subject to local, state and federal laws that may supersede this provision.

[A] 104.8.1 Legal defense. Any suit or criminal complaint instituted against an officer or employee because of an act performed by that officer or employee in the lawful discharge of duties and under the provisions of this code shall be defended by legal representatives of the jurisdiction until the final termination of the proceedings. The *building official* or any subordinate shall not be liable for cost in any action, suit or proceeding that is instituted in pursuance of the provisions of this code.

❖ This section establishes that building officials (or subordinates) must not be liable for costs in any legal action instituted in response to the performance of lawful duties. These costs are to be borne by the state, county or municipality. The best way to be certain that the building official's action is a "lawful duty" is always to cite the applicable code section on which the enforcement action is based.

[A] 104.9 Approved materials and equipment. Materials, equipment and devices *approved* by the *building official* shall be constructed and installed in accordance with such approval.

❖ The code is a compilation of criteria with which materials, equipment, devices and systems must comply to be suitable for a particular application. The building official has a duty to evaluate such materials, equipment, devices and systems for code compliance and, when compliance is determined, approve the same for use. The materials, equipment, devices and systems must be constructed and installed in compliance with, and all conditions and limitations considered as a basis for, that approval. For example, the manufacturer's instructions and recommendations are to be followed if the approval of the material was based even in part on those instructions and recommendations. The approval authority given to the building official is a significant responsibility and is a key to code compliance. The approval process is first technical and then administrative and must be approached as such. For example, if data to determine code compliance are required, such data should be in the form of test reports or engineering analysis and not simply taken from a sales brochure.

[A] 104.9.1 Used materials and equipment. The use of used materials that meet the requirements of this code for new materials is permitted. Used equipment and devices shall not be reused unless *approved* by the *building official*.

❖ Code criteria for materials and equipment have changed over the years. Evaluation of testing and materials technology has permitted the development of new criteria that old materials may not satisfy. As a result, used materials are required to be evaluated in the same manner as new materials. Used materials, equipment and devices must be equivalent to that required by the code if they are to be used again in a new installation.

[A] 104.10 Modifications. Where there are practical difficulties involved in carrying out the provisions of this code, the *building official* shall have the authority to grant modifications for individual cases, upon application of the *owner* or the owner's authorized agent, provided that the *building official* shall first find that special individual reason makes the strict letter of this code impractical, the modification is in compliance with the intent and purpose of this code and that such modification does not lessen health, *accessibility*, life and fire safety or structural requirements. The details of action granting modifications shall be recorded and entered in the files of the department of building safety.

❖ The building official may amend or make exceptions to the code as needed where strict compliance is impractical. Only the building official has authority to grant modifications. Consideration of a particular difficulty is to be based on the application of the owner and a demonstration that the intent of the code is accomplished. This section is not intended to permit setting aside or ignoring a code provision; rather, it is intended to provide acceptance of equivalent protection. Such modifications do not, however, extend to actions that are necessary to correct violations of the code. In other words, a code violation or the expense of correcting one cannot constitute a practical difficulty.

[A] 104.10.1 Flood hazard areas. The *building official* shall not grant modifications to any provision required in *flood hazard areas* as established by Section 1612.3 unless a determination has been made that:

1. A showing of good and sufficient cause that the unique characteristics of the size, configuration or topography of the site render the elevation standards of Section 1612 inappropriate.
2. A determination that failure to grant the variance would result in exceptional hardship by rendering the lot undevelopable.
3. A determination that the granting of a variance will not result in increased flood heights, additional threats to public safety, extraordinary public expense, cause fraud on or victimization of the public, or conflict with existing laws or ordinances.
4. A determination that the variance is the minimum necessary to afford relief, considering the flood hazard.
5. Submission to the applicant of written notice specifying the difference between the *design flood elevation* and the elevation to which the building is to be built, stating that the cost of flood insurance will be commensurate with the increased risk resulting from the reduced floor elevation, and stating that construction below the *design flood elevation* increases risks to life and property.

❖ Before granting a modification related to the flood-resistant provisions of the code, the code official must consider the listed factors. This determination is consistent with the requirements of the National Flood Insurance Program regulations. The community must consider these factors in order to grant variances to provide relief from selected provisions for flood-resistant construction. A record of the determination must be retained as part of the community's permanent records.

Granting modifications from these provisions may place people and property at significant risk. Therefore, code officials are cautioned to carefully evaluate the impacts, particularly the impact of modifications to the requirements for elevated buildings. The factors that must be evaluated are listed and include: impacts on the site; the applicant and other parties who may be affected, such as adjacent property owners; and the community as a whole. Floodplain development that is not undertaken in accordance with the flood-resistance provisions of the code will be exposed to increased flood damage.

Any modification granted must be the minimum necessary to afford relief. The code official must address each listed factor, especially the requirement to determine whether the failure to grant the modification would result in exceptional hardship. Modifications must be based solely on technical justifications and unique characteristics of a site, and not on the personal circumstances of an owner or applicant.

In guidance materials, FEMA cautions that financial hardship, inconvenience, aesthetic considerations, physical handicaps, personal preferences or the disapproval of one's neighbors do not qualify as exceptional hardships. Applicants sometimes request variances to the elevation requirements in order to improve access for the disabled and the elderly. Generally, variances of this nature should not be granted because these are personal circumstances that will change as the property changes ownership. Not only would persons of limited mobility be at risk, but a building that is below the required elevation would continue to be exposed to flood damage long after a personal need ends.

Code officials are cautioned that granting a modification under this section does not affect how the building will be rated for the purposes of federal flood insurance. Even if circumstances justify granting a modification related to the elevation of buildings, the rate used to calculate the cost of a federal flood insurance policy will be based on the risk to the building. Federal flood insurance, required by certain mortgage lenders, may be extremely expensive. Although an owner may not be required to purchase flood insurance, the requirement will be imposed on subsequent owners. The code official is to provide the applicant a written notice to this effect, along with the other cautions listed in this section.

[A] 104.11 Alternative materials, design and methods of construction and equipment. The provisions of this code are not intended to prevent the installation of any material or to prohibit any design or method of construction not specifically prescribed by this code, provided that any such alternative has been *approved*. An alternative material, design or method of construction shall be *approved* where the *building official* finds that the proposed design is satisfactory and complies with the intent of the provisions of this code, and that the material, method or work offered is, for the purpose intended, not less than the equivalent of that prescribed in this code in quality, strength, effectiveness, *fire resistance*, durability and safety. Where the alternative material, design or method of construction is not *approved*, the *building official* shall respond in writing, stating the reasons why the alternative was not *approved*.

❖ The code is not intended to inhibit innovative ideas or technological advances. A comprehensive regulatory document, such as a building code, cannot envision and then address all future innovations in the industry. As a result, a performance code must be applicable to and provide a basis for the approval of an increasing number of newly developed, innovative materials, systems and methods for which no code text or referenced standards yet exist. The fact that a material, product or method of construction is not addressed in the code is not an indication that such material, product or method is intended to be prohibited. The building official is expected to apply sound technical judgment in accepting materials, systems or methods that, while not anticipated by the drafters of the current code text, can be demonstrated to offer equivalent performance. The code regulates new and innovative construction practices while addressing the relative safety of building occupants. The building official is responsible for determining if a requested alternative provides the equivalent level of protection of public health, safety and welfare as required by the code. In order to ensure effective communication and due process of law, if an alternative is not approved, the building official should state in writing the reasons for the disapproval. This is similar to when a permit is rejected in Section 105.3.1.

[A] 104.11.1 Research reports. Supporting data, where necessary to assist in the approval of materials or assemblies not specifically provided for in this code, shall consist of valid research reports from *approved* sources.

❖ When an alternative material or method is proposed for construction, it is incumbent upon the building official to determine whether this alternative is, in fact, an equivalent to the methods prescribed by the code. Reports providing evidence of this equivalency are required to be supplied by an approved source, meaning a source that the building official finds to be reliable and accurate. The ICC Evaluation Service is an example of an agency that provides research reports for alternative materials and methods.

[A] 104.11.2 Tests. Whenever there is insufficient evidence of compliance with the provisions of this code, or evidence that a material or method does not conform to the requirements of this code, or in order to substantiate claims for alternative materials or methods, the *building official* shall have the authority to require tests as evidence of compliance to be made at no expense to the jurisdiction. Test methods shall be as specified in this code or by other recognized test standards. In the absence of recognized and accepted test methods, the *building official* shall approve the testing procedures. Tests shall be performed by an *approved agency*. Reports of such tests shall be retained by the *building official* for the period required for retention of public records.

❖ To provide the basis on which the building official can make a decision regarding an alternative material or method, sufficient technical data, test reports and documentation must be provided for evaluation. If evidence satisfactory to the building official indicates that the alternative material or construction method is equivalent to that required by the code, he or she may approve it. Any such approval cannot have the effect of waiving any requirements of the code. The burden of proof of equivalence lies with the applicant who proposes the use of alternative materials or methods.

The building official must require the submission of appropriate information and data to assist in the determination of equivalency. This information must be submitted before a permit can be issued. The type of information required includes test data in accordance with referenced standards, evidence of compliance with the referenced standard specifications and design calculations. A research report issued by an authoritative agency is particularly useful in providing the building official with the technical basis for evaluation and approval of new and innovative materials and methods of construction. The use of authoritative research reports can greatly assist the building official by reducing the time-consuming engineering analysis necessary to review these materials and methods. Failure to substantiate adequately a request for the use of an alternative is a valid reason for the building official to deny a request. Any tests submitted in support of an application must have been performed by an agency approved by the building official based on evidence that the agency has the technical expertise, test equipment and quality assurance to properly conduct and report the necessary testing. The test reports submitted to the building official must be retained in accordance with the requirements of Section 104.7.

SECTION 105 PERMITS

[A] 105.1 Required. Any *owner* or owner's authorized agent who intends to construct, enlarge, alter, *repair*, move, demolish or change the occupancy of a building or structure, or to erect, install, enlarge, alter, *repair*, remove, convert or replace any electrical, gas, mechanical or plumbing system, the installation of which is regulated by this code, or to cause any such work to be performed, shall first make application to the *building official* and obtain the required *permit*.

❖ This section contains the administrative rules governing the issuance, suspension, revocation or modification of building permits. It also establishes how and by whom the application for a building permit is to be made, how it is to be processed, fees and what information it must contain or have attached to it.

In general, a permit is required for all activities that are regulated by the code or its referenced codes (see Section 101.4), and these activities cannot begin until the permit is issued, unless the activity is specifically exempted by Section 105.2. Only the owner or a person authorized by the owner can apply for the permit. Note that this section indicates a need for a permit for a change in occupancy, even if no work is contemplated. Although the occupancy of a building or portion thereof may change and the new activity is still classified in the same group, different code provisions may be applicable. The means of egress, structural loads and light and ventilation provisions are examples of requirements that are occupancy sensitive. The purpose of the permit is to cause the work to be reviewed, approved and inspected to determine compliance with the code.

[A] 105.1.1 Annual permit. Instead of an individual *permit* for each *alteration* to an already *approved* electrical, gas, mechanical or plumbing installation, the *building official* is authorized to issue an annual *permit* upon application therefor to any person, firm or corporation regularly employing one or more qualified tradepersons in the building, structure or on the premises owned or operated by the applicant for the *permit*.

❖ In some instances, such as large buildings or industrial facilities, the repair, replacement or alteration of electrical, gas, mechanical or plumbing systems occurs on a frequent basis, and this section allows the building official to issue an annual permit for this work. This relieves both the building department and the owners of such facilities from the burden of filing and processing individual applications for this activity; however, there are restrictions on who is entitled to these permits. They can be issued only for work on a previously approved installation and only to an individual or corporation that employs persons specifically qualified in the trade for which the permit is issued. If tradespeople who perform the work involved are required to be licensed in the jurisdiction, then only those persons would be permitted to perform the work. If trade licensing is not required, then the building official needs to review and approve the qualifications of the persons who will be performing the work. The annual permit can apply only to the individual property that is owned or operated by the applicant.

[A] 105.1.2 Annual permit records. The person to whom an annual *permit* is issued shall keep a detailed record of *alterations* made under such annual *permit*. The *building official* shall have access to such records at all times or such records shall be filed with the *building official* as designated.

❖ The work performed in accordance with an annual permit must be inspected by the building official, so it is necessary to know the location of such work and when it was performed. This can be accomplished by having records of the work available to the building official either at the premises or in the official's office, as determined by the official.

[A] 105.2 Work exempt from permit. Exemptions from *permit* requirements of this code shall not be deemed to grant authorization for any work to be done in any manner in violation of the provisions of this code or any other laws or ordinances of this jurisdiction. *Permits* shall not be required for the following:

Building:

1. One-story detached accessory structures used as tool and storage sheds, playhouses and similar uses, provided the floor area is not greater than 120 square feet (11 m^2).
2. Fences not over 7 feet (2134 mm) high.
3. Oil derricks.
4. Retaining walls that are not over 4 feet (1219 mm) in height measured from the bottom of the footing to the top of the wall, unless supporting a surcharge or impounding Class I, II or IIIA liquids.
5. Water tanks supported directly on grade if the capacity is not greater than 5,000 gallons (18 925 L) and the ratio of height to diameter or width is not greater than 2:1.
6. Sidewalks and driveways not more than 30 inches (762 mm) above adjacent grade, and not over any basement or *story* below and are not part of an *accessible route*.
7. Painting, papering, tiling, carpeting, cabinets, counter tops and similar finish work.
8. Temporary motion picture, television and theater stage sets and scenery.
9. Prefabricated *swimming pools* accessory to a Group R-3 occupancy that are less than 24 inches (610 mm) deep, are not greater than 5,000 gallons (18 925 L) and are installed entirely above ground.
10. Shade cloth structures constructed for nursery or agricultural purposes, not including service systems.
11. Swings and other playground equipment accessory to detached one- and two-family *dwellings*.
12. Window awnings in Group R-3 and U occupancies, supported by an exterior wall that do not project more than 54 inches (1372 mm) from the *exterior wall* and do not require additional support.
13. Nonfixed and movable fixtures, cases, racks, counters and partitions not over 5 feet 9 inches (1753 mm) in height.

Electrical:

Repairs and maintenance: Minor repair work, including the replacement of lamps or the connection of *approved* portable electrical equipment to *approved* permanently installed receptacles.

Radio and television transmitting stations: The provisions of this code shall not apply to electrical equipment used for radio and television transmissions, but do apply to equipment and wiring for a power supply and the installations of towers and antennas.

Temporary testing systems: A *permit* shall not be required for the installation of any temporary system required for the testing or servicing of electrical equipment or apparatus.

Gas:

1. Portable heating appliance.
2. Replacement of any minor part that does not alter approval of equipment or make such equipment unsafe.

Mechanical:

1. Portable heating appliance.
2. Portable ventilation equipment.
3. Portable cooling unit.
4. Steam, hot or chilled water piping within any heating or cooling equipment regulated by this code.
5. Replacement of any part that does not alter its approval or make it unsafe.
6. Portable evaporative cooler.
7. Self-contained refrigeration system containing 10 pounds (4.54 kg) or less of refrigerant and actuated by motors of 1 horsepower (0.75 kW) or less.

Plumbing:

1. The stopping of leaks in drains, water, soil, waste or vent pipe, provided, however, that if any concealed trap, drain pipe, water, soil, waste or vent pipe becomes defective and it becomes necessary to remove and replace the same with new material, such work shall be considered as new work and a *permit* shall be obtained and inspection made as provided in this code.
2. The clearing of stoppages or the repairing of leaks in pipes, valves or fixtures and the removal and reinstallation of water closets, provided such repairs do not involve or require the replacement or rearrangement of valves, pipes or fixtures.

❖ Section 105.1 essentially requires a permit for any activity involving work on a building, its systems and other structures. This section lists those activities that are permitted to take place without first obtaining a permit from the building department. Note that in some cases, such as Items 9, 10, 11 and 12, the work is exempt only for certain occupancies. It is further the intent of the code that even though work may be exempted for permit purposes, it must still comply with the code and the owner is responsible for proper and safe construction for all work being done. Work exempted by the codes adopted by reference in Section 101.4 is also included here. However, even if a permit is not required, construction must not violate any code provisions. For example: If you replace a sink faucet, you don't need a permit, but the faucet would still have to meet material standard and water flow requirements in Chapter 6 of the IPC.

In flood hazard areas, work exempt from a permit must still be undertaken in ways that minimize flood damage. Accessory structures below the design flood elevation must be anchored to prevent flotation, have flood openings, be made of flood damage-resistant materials. Equipment and electrical service must also be elevated above the design flood elevation. Water tanks on grade must be anchored to prevent flotation, collapse or lateral movement. Additional descriptions of how the listed activities should be performed in order to meet the intent are found in the commentary for Appendix G.

[A] 105.2.1 Emergency repairs. Where equipment replacements and repairs must be performed in an emergency situation, the *permit* application shall be submitted within the next working business day to the *building official*.

❖ This section recognizes that in some cases, emergency replacement and repair work must be done as quickly as possible, so it is not practical to take the necessary time to apply for and obtain approval. A permit for the work must be obtained the next day that the building department is open for business. Any work performed before the permit is issued must be done in accordance with the code and corrected if not approved by the building official. For example, if a concealed trap failed on a Sunday, the plumber could replace the trap at that time, but he would have to apply for a permit on Monday and have the repair pass an inspection.

[A] 105.2.2 Repairs. Application or notice to the *building official* is not required for ordinary *repairs* to structures, replacement of lamps or the connection of *approved* portable electrical equipment to *approved* permanently installed receptacles. Such *repairs* shall not include the cutting away of any wall, partition or portion thereof, the removal or cutting of any structural beam or load-bearing support, or the removal or change of any required *means of egress*, or rearrangement of parts of a structure affecting the egress requirements; nor shall ordinary repairs include *addition* to, *alteration* of, replacement or relocation of any standpipe, water supply, sewer, drainage, drain leader, gas, soil, waste, vent or similar piping, electric wiring or mechanical or other work affecting public health or general safety.

❖ This section distinguishes between what might be termed by some as repairs but are in fact alterations, wherein the code is to be applicable, and ordinary repairs, which are maintenance activities that do not require a permit.

[A] 105.2.3 Public service agencies. A *permit* shall not be required for the installation, *alteration* or repair of generation, transmission, distribution or metering or other related equipment that is under the ownership and control of public service agencies by established right.

❖ Utilities that supply electricity, gas, water, telephone, television cable, etc., do not require permits for work involving the transmission lines and metering equipment that they own and control; that is, to their point of delivery. Utilities are typically regulated by other laws that give them specific rights and authority in this area. Any equipment or appliances installed or serviced by such agencies that are not owned by them and under their full control are not exempt from a permit.

[A] 105.3 Application for permit. To obtain a *permit*, the applicant shall first file an application therefor in writing on a form furnished by the department of building safety for that purpose. Such application shall:

1. Identify and describe the work to be covered by the *permit* for which application is made.
2. Describe the land on which the proposed work is to be done by legal description, street address or similar description that will readily identify and definitely locate the proposed building or work.
3. Indicate the use and occupancy for which the proposed work is intended.
4. Be accompanied by *construction documents* and other information as required in Section 107.
5. State the valuation of the proposed work.
6. Be signed by the applicant, or the applicant's authorized agent.
7. Give such other data and information as required by the *building official*.

❖ This section requires that a written application for a permit be filed on forms provided by the building department and details the information required on the application. Permit forms will typically have sufficient space to write a very brief description of the work to be accomplished, which is sufficient for only small jobs. For larger projects, the description will be augmented by construction documents as indicated in Item 4. As required by Section 105.1, the applicant must be the owner of the property or an authorized agent of the owner, such as an engineer, architect, contractor, tenant or other. The applicant must sign the application, and permit forms typically include a statement that if the applicant is not the owner, he or she has permission from the owner to make the application.

[A] 105.3.1 Action on application. The *building official* shall examine or cause to be examined applications for *permits* and amendments thereto within a reasonable time after filing. If the application or the *construction documents* do not conform to the requirements of pertinent laws, the *building official* shall reject such application in writing, stating the reasons therefor. If the *building official* is satisfied that the proposed work conforms to the requirements of this code and laws and ordinances applicable thereto, the *building official* shall issue a *permit* therefor as soon as practicable.

❖ This section requires the building official to act with reasonable speed on a permit application. In some instances, this time period is set by state or local law. The building official must refuse to issue a permit when the application and accompanying documents do not conform to the code. In order to ensure effective communication and due process of law, the reasons for denial of an application for a permit are required to be in writing. Once the building official determines that the work described conforms to the code and other applicable laws, the permit must be issued upon payment of the fees required by Section 109.

[A] 105.3.2 Time limitation of application. An application for a *permit* for any proposed work shall be deemed to have been abandoned 180 days after the date of filing, unless such application has been pursued in good faith or a *permit* has been issued; except that the *building official* is authorized to grant one or more extensions of time for additional periods not exceeding 90 days each. The extension shall be requested in writing and justifiable cause demonstrated.

❖ Typically, an application for a permit is submitted and goes through a review process that ends with the issuance of a permit. If a permit has not been issued within 180 days after the date of filing, the application is considered abandoned, unless the applicant was diligent in efforts to obtain the permit. The building official has the authority to extend this time limitation (in increments of 90 days), provided there is reasonable cause. This would cover delays beyond the applicant's control, such as prerequisite permits or approvals from other authorities within the jurisdiction or state. The intent of this section is to limit the time between the review process and the issuance of a permit.

[A] 105.4 Validity of permit. The issuance or granting of a *permit* shall not be construed to be a *permit* for, or an approval of, any violation of any of the provisions of this code or of any other ordinance of the jurisdiction. *Permits* presuming to give authority to violate or cancel the provisions of this code or other ordinances of the jurisdiction shall not be valid. The issuance of a *permit* based on *construction documents* and other data shall not prevent the *building official* from requiring the correction of errors in the *construction documents* and other data. The *building official* is authorized to prevent occupancy or use of a structure where in violation of this code or of any other ordinances of this jurisdiction.

❖ This section states the fundamental premise that the permit is only a license to proceed with the work. It is not a license to violate, cancel or set aside any provisions of the code. This is significant because it means that despite any errors or oversights in the approval process, the permit applicant, not the building official, is responsible for code compliance. Also, the permit can be suspended or revoked in accordance with Section 105.6.

[A] 105.5 Expiration. Every *permit* issued shall become invalid unless the work on the site authorized by such *permit* is commenced within 180 days after its issuance, or if the work authorized on the site by such *permit* is suspended or abandoned for a period of 180 days after the time the work is commenced. The *building official* is authorized to grant, in writing, one or more extensions of time, for periods not more than 180 days each. The extension shall be requested in writing and justifiable cause demonstrated.

❖ The permit becomes invalid under two distinct situations—both based on a 180-day period. The first situation is when no work was initiated 180 days from issuance of a permit. The second situation is when the authorized work has stopped for 180 days. The person who was issued the permit should be notified, in writing, that the permit is invalid and what steps must be taken to reinstate it and restart the work. The building official has the authority to extend this time limitation (in increments of 180 days), provided the extension is requested in writing and there is reasonable cause, which typically includes events beyond the permit holder's control.

[A] 105.6 Suspension or revocation. The *building official* is authorized to suspend or revoke a *permit* issued under the provisions of this code wherever the *permit* is issued in error or on the basis of incorrect, inaccurate or incomplete information, or in violation of any ordinance or regulation or any of the provisions of this code.

❖ A permit is a license to proceed with the work. The building official, however, can suspend or revoke permits shown to be based, all or in part, on any false statement or misrepresentation of fact. A permit can also be suspended or revoked if it was issued in error, such as an omitted prerequisite approval or code violation indicated on the construction documents. An applicant may subsequently apply for a reinstatement of the permit with the appropriate corrections or modifications made to the application and construction documents.

[A] 105.7 Placement of permit. The building *permit* or copy shall be kept on the site of the work until the completion of the project.

❖ The permit, or copy thereof, is to be kept on the job site until the work is complete, and made available to the building official or representative to conveniently make required entries thereon.

SECTION 106
FLOOR AND ROOF DESIGN LOADS

[A] 106.1 Live loads posted. In commercial or industrial buildings, for each floor or portion thereof designed for *live loads* exceeding 50 psf (2.40 kN/m^2), such design *live loads* shall be conspicuously posted by the owner or the owner's authorized agent in that part of each *story* in which they apply, using durable signs. It shall be unlawful to remove or deface such notices.

❖ This section requires that live loads be posted for most occupancies, since many of the live loads specified in Table 1607.1 exceed 50 pounds per square foot (psf) (2.40 kN/m^2). Where part of the floor is designed for 50 psf (2.40 kN/m^2) or less and part for more than 50 psf (2.40 kN/m^2), the live loads are required to be posted for those portions more than 50 psf (2.40 kN/m^2). The code requires that the posting be done in the part where it applies. For example, an assembly area such as a restaurant would need to have the live load posted in the dining room.

This live load posting gives the building department easy access to the information for field verification. It also serves as a notice of the loading restriction that is stated in Section 106.3.

[A] 106.2 Issuance of certificate of occupancy. A certificate of occupancy required by Section 111 shall not be issued until the floor load signs, required by Section 106.1, have been installed.

❖ The design live load signs required by Section 106.1 need to be in place prior to the occupancy of the building for reference purposes. They serve as a record of the structural design loads for future reference, particularly when a change in occupancy is contemplated.

[A] 106.3 Restrictions on loading. It shall be unlawful to place, or cause or permit to be placed, on any floor or roof of a building, structure or portion thereof, a load greater than is permitted by this code.

❖ The design live load signs required by Section 106.1 need to be in place prior to the occupancy of the building for reference purposes. They serve as a record of the structural design loads for future reference, particularly when a change in occupancy is contemplated.

SECTION 107
SUBMITTAL DOCUMENTS

[A] 107.1 General. Submittal documents consisting of *construction documents*, statement of *special inspections*, geotechnical report and other data shall be submitted in two or more sets with each *permit* application. The *construction documents* shall be prepared by a *registered design professional* where required by the statutes of the jurisdiction in which the project is to be constructed. Where special conditions exist, the *building official* is authorized to require additional *construction documents* to be prepared by a *registered design professional*.

Exception: The *building official* is authorized to waive the submission of *construction documents* and other data not required to be prepared by a *registered design professional* if it is found that the nature of the work applied for is such that review of *construction documents* is not necessary to obtain compliance with this code.

❖ This section establishes the requirement to provide the building official with construction drawings, specifications and other documents that describe the structure or system for which a permit is sought (see Section 202 for a complete definition). It describes the information that must be included in the documents, who must prepare them and procedures for approving them.

A detailed description of the work for which an application is made must be submitted. When the work can be briefly described on the application form and the services of a registered design professional are not required, the building official may utilize judgment in determining the need for detailed documents. An example of work that may not involve the submission of detailed construction documents is the replacement of an existing 60-amp electrical service with a 200-amp service. Other sections of the code also contain specific requirements for construction documents, such as Sections 1603, 1901.5, 2111.2, 2207.2 and 3103.2. These provisions are intended to reflect the minimum scope of information needed to determine code compliance. Although this section specifies that "one or more" sets of construction documents be submitted, note that Section 106.3.1 requires one set of approved documents be retained by the building official and one set be returned to the applicant, essentially requiring at least two sets of construction documents. The building official should establish a consistent policy of the number of sets required by the jurisdiction and make this information readily available to applicants.

This section also requires the building official to determine that any state professional registration laws be complied with as they apply to the preparation of construction documents.

[A] 107.2 Construction documents. *Construction documents* shall be in accordance with Sections 107.2.1 through 107.2.6.

❖ This section provides instructions regarding the information and form of construction documents.

[A] 107.2.1 Information on construction documents. *Construction documents* shall be dimensioned and drawn upon suitable material. Electronic media documents are permitted to be submitted where *approved* by the *building official*. *Construction documents* shall be of sufficient clarity to indicate the location, nature and extent of the work proposed and show in detail that it will conform to the provisions of this code and relevant laws, ordinances, rules and regulations, as determined by the *building official*.

❖ The construction documents are required to be of a quality and detail such that the building official can determine whether the work conforms to the code and other applicable laws and regulations. General statements on the documents, such as "all work must comply with the *International Building Code*®," are not an acceptable substitute for showing the required information. The following subsections and sections in other chapters indicated in the commentary to Sections 107.2.2 through 107.2.6 specify the detailed information that must be shown on the submitted documents. Where specifically allowed by the building official, documents can be submitted in electronic form.

[A] 107.2.2 Fire protection system shop drawings. Shop drawings for the *fire protection system(s)* shall be submitted to indicate conformance to this code and the *construction documents* and shall be *approved* prior to the start of system installation. Shop drawings shall contain all information as required by the referenced installation standards in Chapter 9.

❖ Since the fire protection contractor(s) may not be selected at the time a permit is issued for construction of a building, detailed shop drawings for fire protection systems are not available. Because they provide the information necessary to determine code compliance, as specified in the appropriate referenced standard in Chapter 9, they must be submitted and approved by the building official before the contractor can begin installing the system. For example, the professional responsible for the design of an automatic sprinkler system should determine that the water supply is adequate, but will not be able to prepare a final set of hydraulic calculations if the specific materials and pipe sizes, lengths and arrangements have not been identified. Once the installing contractor is selected, specific hydraulic calculations can be prepared. Factors, such as classification of the hazard, amount of water supply available and the density or concentration to be achieved by the system, are to be included with the submission of the shop drawings. Specific data sheets identifying sprinklers, pipe dimensions, power requirements for smoke detectors, etc., should also be included with the submission.

[A] 107.2.3 Means of egress. The *construction documents* shall show in sufficient detail the location, construction, size and character of all portions of the *means of egress* including the path of the *exit discharge* to the *public way* in compliance with the provisions of this code. In other than occupancies in Groups R-2, R-3, and I-1, the *construction documents* shall designate the number of occupants to be accommodated on every floor, and in all rooms and spaces.

❖ The complete means of egress system is required to be indicated on the plans to allow the building official to initiate a review and identify pertinent code requirements for each component. Additionally, requiring such information to be reflected in the construction documents requires the designer not only to become familiar with the code, but also to be aware of egress principles, concepts and purposes. The need to ensure that the means of egress leads to a public way is also a consideration during the plan review. Such an evaluation cannot be made without the inclusion of a site plan, as required by Section 107.2.5.

Information essential for determining the required capacity (see Section 1005) and number (see Sections 1006) of egress components from a space must be provided. The designer must be aware of the occupancy of a space and properly identify that information, along with its resultant occupant load, on the construction documents. In occupancies in Groups I-1, R-2 and R-3, the occupant load can be readily determined with little difference in the number so that the designation of the occupant load on the construction documents is not required.

The exit discharge path to the public way must also be shown on the construction documents. The exit discharge path to the public way is an important component of the means of egress system for all buildings or structures. The exit discharge path needs to be delineated on the submitted and approved plans to ensure the path is reviewed for compliance with the provisions of the code. This will also provide an historical reference once the building is occupied to ensure the exit discharge path is maintained as intended for the life of the building or structure unless modifications are approved.

[A] 107.2.4 Exterior wall envelope. *Construction documents* for all buildings shall describe the *exterior wall envelope* in sufficient detail to determine compliance with this code. The *construction documents* shall provide details of the *exterior wall envelope* as required, including flashing, intersections with dissimilar materials, corners, end details, control joints, intersections at roof, eaves or parapets, means of drainage, water-resistive membrane and details around openings.

The *construction documents* shall include manufacturer's installation instructions that provide supporting documentation that the proposed penetration and opening details described in the *construction documents* maintain the weather resistance of the *exterior wall envelope*. The supporting documentation shall fully describe the *exterior wall* system that was tested, where applicable, as well as the test procedure used.

❖ This section specifically identifies details of exterior wall construction that are critical to the weather resistance of the wall and requires those details to be provided on the construction documents. Where the weather resistance of the exterior wall assembly is based on tests, the submitted documentation is to describe the details of the wall envelope and the test procedure that was used. This provides the building official with the information necessary to determine code compliance.

[A] 107.2.5 Site plan. The *construction documents* submitted with the application for *permit* shall be accompanied by a site plan showing to scale the size and location of new construction and existing structures on the site, distances from *lot lines*, the established street grades and the proposed finished grades and, as applicable, *flood hazard areas*, *floodways*, and *design flood elevations*; and it shall be drawn in accordance with an accurate boundary line survey. In the case of demolition, the site plan shall show construction to be demolished and the location and size of existing structures and construction that are to remain on the site or plot. The *building official* is authorized to waive or modify the requirement for a site plan where the application for *permit* is for *alteration* or *repair* or where otherwise warranted.

❖ Certain code requirements are dependent on the structure's location on the lot (see Sections 506.3, 507, 705, 1027 and 1206), the topography of the site (see Sections 1104, 1107.4, 1107.7.4 and 1804.4), and whether the site has flood hazard areas (see Sections 1612 and 1804.4). As a result, a scaled site plan containing the data listed in this section is required to permit review for compliance. The building official can waive the requirement for a site plan when it is not required to determine code compliance, such as for work involving only interior alterations or repairs.

[A] 107.2.5.1 Design flood elevations. Where *design flood elevations* are not specified, they shall be established in accordance with Section 1612.3.1.

❖ Some Flood Insurance Rate Maps (FIRMs) prepared by FEMA show mapped special flood hazard areas that do not have either flood elevations or floodway designations (floodways are areas along riverine bodies of water where the water will be deeper and flow faster during flooding conditions). Section 1612.3 gives the authority to the code official to require use of data which may be obtained from other sources, or to require the applicant to develop flood hazard data.

[A] 107.2.6 Structural information. The *construction documents* shall provide the information specified in Section 1603.

❖ The purpose of this reference to Section 1603 is as a reminder that there are requirements for structural information to be part of the construction documents. Section 1603 requires the design professional to provide appropriate structural details, criteria and design load data for verifying compliance with the provisions of Chapters 16 through 23. See the commentary to Section 1603 for additional information.

[A] 107.3 Examination of documents. The *building official* shall examine or cause to be examined the accompanying submittal documents and shall ascertain by such examinations whether the construction indicated and described is in accordance with the requirements of this code and other pertinent laws or ordinances.

❖ The requirements of this section are related to those found in Section 105.3.1 regarding the action of the building official in response to a permit application. The building official can delegate review of the construction documents to subordinates as provided for in Section 103.3.

[A] 107.3.1 Approval of construction documents. When the *building official* issues a *permit*, the *construction documents* shall be *approved*, in writing or by stamp, as "Reviewed for Code Compliance." One set of *construction documents* so reviewed shall be retained by the *building official*. The other set shall be returned to the applicant, shall be kept at the site of work and shall be open to inspection by the *building official* or a duly authorized representative.

❖ The building official must stamp or otherwise endorse as "Reviewed for Code Compliance" the construction documents on which the permit is based. One set of approved construction documents must be kept on the construction site to serve as the basis for all subsequent inspections. To avoid confusion, the construction documents on the site must be the documents that were approved and stamped. This is because inspections are to be performed with regard to the approved documents, not the code itself. Additionally, the contractor cannot determine compliance with the approved construction documents unless they are readily available. If the approved construction documents are not available, the inspection should be postponed and work on the project halted.

[A] 107.3.2 Previous approvals. This code shall not require changes in the *construction documents*, construction or designated occupancy of a structure for which a lawful *permit* has been heretofore issued or otherwise lawfully authorized, and the construction of which has been pursued in good faith within 180 days after the effective date of this code and has not been abandoned.

❖ If a permit is issued and construction proceeds at a normal pace and a new edition of the code is adopted by the legislative body, requiring that the building be constructed to conform to the new code is unreasonable. This section provides for the continuity of permits issued under previous codes, as long as such permits are being "actively prosecuted" subsequent to the effective date of the ordinance adopting this edition of the code.

[A] 107.3.3 Phased approval. The *building official* is authorized to issue a *permit* for the construction of foundations or any other part of a building or structure before the *construction documents* for the whole building or structure have been submitted, provided that adequate information and detailed statements have been filed complying with pertinent requirements of this code. The holder of such *permit* for the foundation or other parts of a building or structure shall proceed at the holder's own risk with the building operation and without assurance that a *permit* for the entire structure will be granted.

❖ The building official has the authority to issue a partial permit to allow for the practice of "fast tracking" a job. Any construction under a partial permit is "at the holder's own risk" and "without assurance that a permit for the entire structure will be granted." The building official is under no obligation to accept work or issue a complete permit in violation of the code, ordinances or statutes simply because a partial permit had been issued. Fast tracking puts an unusual administrative and technical burden on the building official. The purpose is to proceed with construction while the design continues for other aspects of the work. Coordinating and correlating the code aspects into the project in phases requires attention to detail and project tracking so that all code issues are addressed. The coordination of these submittals is the responsibility of the registered design professional in responsible charge described in Section 107.3.4.

[A] 107.3.4 Design professional in responsible charge. Where it is required that documents be prepared by a *registered design professional*, the *building official* shall be authorized to require the *owner* or the owner's authorized agent to engage and designate on the building *permit* application a *registered design professional* who shall act as the *registered* *design professional in responsible charge*. If the circumstances require, the *owner* or the owner's authorized agent shall designate a substitute *registered design professional in responsible charge* who shall perform the duties required of the original *registered design professional in responsible charge*. The *building official* shall be notified in writing by the *owner* or the owner's authorized agent if the *registered design professional in responsible charge* is changed or is unable to continue to perform the duties.

The *registered design professional in responsible charge* shall be responsible for reviewing and coordinating submittal documents prepared by others, including phased and deferred submittal items, for compatibility with the design of the building.

❖ At the time of permit application and at various intervals during a project, the code requires detailed technical information to be submitted to the building official. This will vary depending on the complexity of the project, but typically includes the construction documents with supporting information, applications utilizing the phased approval procedure in Section 107.3.3 and reports from engineers, inspectors and testing agencies required in Chapter 17. Since these documents and reports are prepared by numerous individuals, firms and agencies, it is necessary to have a single person charged with responsibility for coordinating their submittal to the building official. This person is the point of contact for the building official for all information relating to the project. Otherwise, the building official could waste time and effort attempting to locate the source of accurate information when trying to resolve an issue such as a discrepancy in plans submitted by different designers. The requirement that the owner or their representative engage a person to act as the design professional in responsible charge is applicable to projects where the construction documents are required by law to be prepared by a registered design professional (see Section 107.1) and where required by the building official. The person employed by the owner to act as the design professional in responsible charge must be identified on the permit application, but the owner can change the designated person at any time during the course of the review process or work, provided the building official is so notified in writing.

[A] 107.3.4.1 Deferred submittals. Deferral of any submittal items shall have the prior approval of the *building official*. The *registered design professional in responsible charge* shall list the deferred submittals on the *construction documents* for review by the *building official*.

Documents for deferred submittal items shall be submitted to the *registered design professional in responsible charge* who shall review them and forward them to the *building official* with a notation indicating that the deferred submittal documents have been reviewed and found to be in general conformance to the design of the building. The deferred submittal items shall not be installed until the deferred submittal documents have been *approved* by the *building official*.

❖ "Deferred submittals" is defined in Chapter 2. Often, especially on larger projects, details of certain building parts are not available at the time of permit issuance because they have not yet been designed; for example, exterior cladding, prefabricated items such as trusses and stairs and the components of fire protection systems (see Section 107.2.2). The design professional in responsible charge must identify on the construction documents the items to be included in any deferred submittals. Documents required for the approval of deferred items must be reviewed by the design professional in responsible charge for compatibility with the design of the building, forwarded to the building official with a notation that this is the case and approved by the building official before installation of the items. Sufficient time must be allowed for the approval process. Note that deferred submittals differ from the phased permits described in Section 107.3.3 in that they occur after the permit for the building is issued and are not for work covered by separate permits.

[A] 107.4 Amended construction documents. Work shall be installed in accordance with the *approved construction documents*, and any changes made during construction that are not in compliance with the *approved construction documents* shall be resubmitted for approval as an amended set of *construction documents*.

❖ Any amendments to the approved construction documents must be filed before constructing the amended item. In the broadest sense, amendments include all addenda, change orders, revised drawings and marked-up shop drawings. Building officials should maintain a policy that all amendments be submitted for review. Otherwise, a significant amendment may not be submitted because of misinterpretation, resulting in an activity that is not approved and that causes a needless delay in obtaining approval of the finished work.

[A] 107.5 Retention of construction documents. One set of *approved construction documents* shall be retained by the *building official* for a period of not less than 180 days from date of completion of the permitted work, or as required by state or local laws.

❖ A set of the approved construction documents must be kept by the building official as may be required by state or local laws, but for a period of not less than 180 days after the work is complete. Questions regarding an item shown on the approved documents may arise in the period immediately following completion of the work and the documents should be available for review. See Section 104.7 for requirements to retain other records that are generated as a result of the work.

SECTION 108
TEMPORARY STRUCTURES AND USES

[A] 108.1 General. The *building official* is authorized to issue a *permit* for temporary structures and temporary uses. Such *permits* shall be limited as to time of service, but shall not be permitted for more than 180 days. The *building official* is authorized to grant extensions for demonstrated cause.

❖ In the course of construction or other activities, structures that have a limited service life are often necessary. This section contains the administrative provisions that permit such temporary structures without full compliance with the code requirements for permanently occupied structures. This section should not be confused with the scope of Section 3103, which regulates temporary structures larger than 120 square feet (11 m^2) in area.

This section allows the building official to issue permits for temporary structures or uses. The applicant must specify the time period desired for the temporary structure or use, but the approval period cannot exceed 180 days. Structures or uses that are temporary but are anticipated to be in existence for more than 180 days are required to conform to code requirements for permanent structures and uses. The section also authorizes the building official to grant extensions to this time period if the applicant can provide a valid reason for the extension, which typically includes circumstances beyond the applicant's control. This provision is not intended to be used to circumvent the 180-day limitation.

[A] 108.2 Conformance. Temporary structures and uses shall comply with the requirements in Section 3103.

❖ By a reference to Section 3103, this indicates that structures that will be permitted for a period of 180 days or less will comply with the IFC and the IBC. IBC provisions (see Section 3103.1.1) include those dealing with structural strength, fire safety, means of egress, accessibility, light, ventilation and sanitation requirements. These categories of the code must be complied with, despite the fact that the structure will be removed or the use discontinued. These criteria are essential for measuring the safety of any structure or use, temporary or permanent; therefore, the application of these criteria to a temporary structure cannot be waived.

"Structural strength" refers to the ability of the temporary structure to resist anticipated live, environmental and dead loads (see Chapter 16). It also applies to anticipated live and dead loads imposed by a temporary use in an existing structure.

"Fire safety" provisions are those required by Chapters 7, 8 and 9, invoked by virtue of the structure's size, use or location on the property.

"Means of egress" refers to full compliance with Chapter 10.

"Accessibility" refers to full compliance with Chapter 11 for making buildings accessible to physically disabled persons, a requirement that is repeated in Section 1103.1.

"Light, ventilation and sanitary" requirements are those imposed by Chapter 12 of the code or applicable sections of the IPC or IMC.

If temporary structures are permitted in flood hazard areas established by Section 1612, a certain level of conformance is appropriate in order to minimize the likelihood of increasing flood heights or flood damage. Communities that participate in the NFIP must take the following measures into consideration: 1. Anchoring should be required to prevent flotation and movement during conditions of the base flood; 2. In A Zones, walled and roofed structures with floors below the elevation required by Section 1612 should have flood openings to minimize hydrostatic loads (see ASCE 24); 3. Portions of structures that are below the elevation required by Section 1612 must be constructed of flood damage-resistant materials; 4. If the structures have utility service or contain equipment, those elements should be elevated above the DFE; and 5. Placement in floodways and flood hazard areas subject to high-velocity wave action (V Zones) should be avoided.

[A] 108.3 Temporary power. The *building official* is authorized to give permission to temporarily supply and use power in part of an electric installation before such installation has been fully completed and the final certificate of completion has been issued. The part covered by the temporary certificate shall comply with the requirements specified for temporary lighting, heat or power in NFPA 70.

❖ Commonly, the electrical service on most construction sites is installed and energized long before all of the wiring is completed. This procedure allows the power supply to be increased as construction demands. However, temporary permission is not intended to waive the requirements set forth in NFPA 70. Construction power from the permanent wiring of the building does not require the installation of temporary ground-fault circuit-interrupter (GFCI) protection or the assured equipment grounding program, because the building wiring installed as required by the code should be as safe for use during construction as it would be for use after completion of the building.

[A] 108.4 Termination of approval. The *building official* is authorized to terminate such *permit* for a temporary structure or use and to order the temporary structure or use to be discontinued.

❖ This section provides the building official with the necessary authority to terminate the permit for a temporary structure or use. The building official can order that a temporary structure be removed or a temporary use be discontinued if conditions of the permit have been violated or the structure or use poses an imminent hazard to the public, in which case the provisions of Section 116 become applicable. This text is important because it allows the building official to act quickly when time is of the essence in order to protect public health, safety and welfare.

SECTION 109
FEES

[A] 109.1 Payment of fees. A *permit* shall not be valid until the fees prescribed by law have been paid, nor shall an amendment to a *permit* be released until the additional fee, if any, has been paid.

❖ The code anticipates that jurisdictions will establish their own fee schedules. It is the intent that the fees collected by the department for building permit issuance, plan review and inspection be adequate to cover the costs to the department in these areas. If the department has additional duties, then its budget will need to be supplemented from the general fund. This section requires that all fees be paid prior to permit issuance or release of an amendment to a permit. Since department operations are intended to be supported by fees paid by the user of department activities, it is important that these fees are received before incurring any expense. This philosophy has resulted in some departments having fees paid prior to the performance of two areas of work: plan review and inspection.

[A] 109.2 Schedule of permit fees. On buildings, structures, electrical, gas, mechanical, and plumbing systems or *alterations* requiring a *permit*, a fee for each *permit* shall be paid as required, in accordance with the schedule as established by the applicable governing authority.

❖ The jurisdiction inserts its desired fee schedule at this location. The fees are established by law, such as in an ordinance adopting the code (see page xv of the code for a sample), a separate ordinance or legally promulgated regulation, as required by state or local law. Fee schedules are often based on a valuation of the work to be performed. This concept is based on the proposition that the valuation of a project is related to the amount of work to be expended in plan review, inspections and administering the permit, plus an excess to cover department overhead.

To assist jurisdictions in establishing uniformity in fees, building evaluation data are published twice each year in ICC's *Building Safety Journal*.

[A] 109.3 Building permit valuations. The applicant for a *permit* shall provide an estimated *permit* value at time of application. *Permit* valuations shall include total value of work, including materials and labor, for which the *permit* is being issued, such as electrical, gas, mechanical, plumbing equipment and permanent systems. If, in the opinion of the *building official*, the valuation is underestimated on the application, the *permit* shall be denied, unless the applicant can show detailed estimates to meet the approval of the *building official*. Final building *permit* valuation shall be set by the *building official*.

❖ As indicated in Section 109.2, jurisdictions usually base their fees on the total value of the work being performed. This section requires the applicant to provide this figure, including materials and labor, for work for which the permit is sought. If the building official believes that the value provided by the applicant is underestimated, the permit is to be denied unless the applicant can substantiate the value by providing detailed estimates of the work to the satisfaction of the building official. For the construction of new buildings, the building valuation data referred to in Section 109.2 can be used by the building official as a yardstick against which to compare the applicant's estimate.

[A] 109.4 Work commencing before permit issuance. Any person who commences any work on a building, structure, electrical, gas, mechanical or plumbing system before obtaining the necessary *permits* shall be subject to a fee established by the *building official* that shall be in addition to the required *permit* fees.

❖ The building official will incur certain costs (e.g., inspection time and administrative) when investigating and citing a person who has commenced work without having obtained a permit. The building official is, therefore, entitled to recover these costs by establishing a fee, in addition to that collected when the required permit is issued, to be imposed on the responsible party. Note that this is not a penalty, as described in Section 114.4, for which the person can also be liable.

[A] 109.5 Related fees. The payment of the fee for the construction, *alteration*, removal or demolition for work done in connection to or concurrently with the work authorized by a building *permit* shall not relieve the applicant or holder of the *permit* from the payment of other fees that are prescribed by law.

❖ The fees for a building permit may be in addition to other fees required by the jurisdiction or others for related items, such as sewer connections, water service taps, driveways and signs. It cannot be construed that the building permit fee includes these other items.

[A] 109.6 Refunds. The *building official* is authorized to establish a refund policy.

❖ This section allows for a refund of fees, which may be full or partial, typically resulting from the revocation, abandonment or discontinuance of a building project for which a permit has been issued and fees have been collected. The refund of fees should be related to the cost of enforcement services not provided because of the termination of the project. The building official, when authorizing a fee refund, is authorizing the disbursement of public funds; therefore, the request for a refund must be in writing and for good cause.

SECTION 110
INSPECTIONS

[A] 110.1 General. Construction or work for which a *permit* is required shall be subject to inspection by the *building official* and such construction or work shall remain accessible and exposed for inspection purposes until *approved*. Approval as a result of an inspection shall not be construed to be an approval of a violation of the provisions of this code or of other ordinances of the jurisdiction. Inspections presuming to give authority to violate or cancel the provisions of this code or of other ordinances of the jurisdiction shall not be valid. It shall be the duty of the *owner* or the owner's authorized agent to cause the work to remain accessible and exposed for inspection purposes. Neither the *building official* nor the jurisdiction shall be liable for expense entailed in the removal or replacement of any material required to allow inspection.

❖ The inspection function is one of the more important aspects of building department operations. This section authorizes the building official to inspect the work for which a permit has been issued and requires that the work to be inspected remains accessible to the building official until inspected and approved. Any expense incurred in removing or replacing material that conceals an item to be inspected is not the responsibility of the building official or the jurisdiction. As with the issuance of permits (see Section 105.4), approval as a result of an inspection is not a license to violate the code and an approval in violation of the code does not relieve the applicant from complying with the code and is not valid.

[A] 110.2 Preliminary inspection. Before issuing a *permit*, the *building official* is authorized to examine or cause to be examined buildings, structures and sites for which an application has been filed.

❖ The building official is granted authority to inspect the site before permit issuance. This may be necessary to verify existing conditions that impact the plan review and permit approval. This section provides the building official with the right-of-entry authority that otherwise does not occur until after the permit is issued (see Section 104.6).

[A] 110.3 Required inspections. The *building official*, upon notification, shall make the inspections set forth in Sections 110.3.1 through 110.3.10.

❖ The building official is required to verify that the building is constructed in accordance with the approved construction documents. It is the responsibility of the permit holder to notify the building official when the item is ready for inspection. The inspections that are necessary to provide such verification are listed in the following sections, with the caveat in Section 110.3.8 that inspections in addition to those listed here may be required depending on the work involved.

[A] 110.3.1 Footing and foundation inspection. Footing and foundation inspections shall be made after excavations for footings are complete and any required reinforcing steel is in place. For concrete foundations, any required forms shall be in place prior to inspection. Materials for the foundation shall be on the job, except where concrete is ready mixed in accordance with ASTM C94, the concrete need not be on the job.

❖ It is necessary for the building official to inspect the soil upon which the footing or foundation is to be placed. This inspection also includes any reinforcing steel, concrete forms and materials to be used in the foundation, except for ready-mixed concrete that is prepared off site.

[A] 110.3.2 Concrete slab and under-floor inspection. Concrete slab and under-floor inspections shall be made after in-slab or under-floor reinforcing steel and building service equipment, conduit, piping accessories and other ancillary equipment items are in place, but before any concrete is placed or floor sheathing installed, including the subfloor.

❖ The building official must be able to inspect the soil and any required under-slab drainage, waterproofing or dampproofing material, as well as reinforcing steel, conduit, piping and other service equipment embedded in or installed below a slab prior to placing the concrete. Similarly, items installed below a floor system other than concrete must be inspected before they are concealed by the floor sheathing or subfloor.

[A] 110.3.3 Lowest floor elevation. In *flood hazard areas*, upon placement of the lowest floor, including the *basement*, and prior to further vertical construction, the elevation certification required in Section 1612.5 shall be submitted to the *building official*.

❖ Where a structure is located in a flood hazard area, as established in Section 1612.5, the building official must be provided with surveyed documentation of specific elevations on buildings depending on flood zone: 1. The lowest floor elevation for structures located in flood hazard areas not subject to high-velocity wave action (called A Zones); or 2. The elevation of the lowest horizontal structural member for structures located in coastal high-hazard areas (called V Zones). This certification is the first of two such certifications. This certification must be submitted after the lowest floor is established and prior to any additional construction proceeding above this level, so that errors in the elevation can be corrected. Section 110.3.10.1 requires the second certification of elevations just prior to the final inspection.

Most communities use the Elevation Certificate form developed by FEMA, FEMA Form 086-0-33 (insurance agents are required to use this form to write NFIP flood insurance policies). The Elevation Certificate is also used to record information that is useful during final inspections, including flood openings, garage floor elevations, and the elevation of equipment that serves buildings. Section 104.7 requires the building official to maintain a copy of this certification in the department's permanent official records.

[A] 110.3.4 Frame inspection. Framing inspections shall be made after the roof deck or sheathing, all framing, *fireblocking* and bracing are in place and pipes, chimneys and vents to be concealed are complete and the rough electrical, plumbing, heating wires, pipes and ducts are *approved*.

❖ This section requires that the building official be able to inspect the framing members, such as studs, joists, rafters and girders and other items, such as vents and chimneys, that will be concealed by wall construction. Rough electrical work, plumbing, heating wires, pipes and ducts must have already been approved in accordance with the applicable codes prior to this inspection.

[A] 110.3.5 Lath, gypsum board and gypsum panel product inspection. Lath, gypsum board and gypsum panel product inspections shall be made after lathing, gypsum board and gypsum panel products, interior and exterior, are in place, but before any plastering is applied or gypsum board and gypsum panel product joints and fasteners are taped and finished.

Exception: Gypsum board and gypsum panel products that are not part of a fire-resistance-rated assembly or a shear assembly.

❖ In order to verify that lath, gypsum board or gypsum wallboard products are properly attached to framing members, it is necessary for the building official to be able to conduct an inspection before the plaster or joint finish material is applied. This is required only for gypsum board or gypsum panel products that are part of either a fire-resistant assembly or a shear wall. See the definitions for gypsum board and gypsum panel products.

[A] 110.3.6 Fire- and smoke-resistant penetrations. Protection of joints and penetrations in *fire-resistance-rated* assemblies, *smoke barriers* and smoke partitions shall not be concealed from view until inspected and *approved.*

❖ The building official must have an opportunity to inspect joint protection required by Section 715 and penetration protection required by Section 714 for fire-resistance-rated assemblies, smoke barriers and smoke partitions before they become concealed from view.

[A] 110.3.7 Energy efficiency inspections. Inspections shall be made to determine compliance with Chapter 13 and shall include, but not be limited to, inspections for: envelope insulation R- and U-values, fenestration U-value, duct system R-value, and HVAC and water-heating equipment efficiency.

❖ Items installed in a building that are required by the IECC to comply with certain criteria, such as insulation material, windows, HVAC and water-heating equipment, must be inspected and approved.

[A] 110.3.8 Other inspections. In addition to the inspections specified in Sections 110.3.1 through 110.3.7, the *building official* is authorized to make or require other inspections of any construction work to ascertain compliance with the provisions of this code and other laws that are enforced by the department of building safety.

❖ Any item regulated by the code is subject to inspection by the building official to determine compliance with the applicable code provision, and no list can include all items in a given building. This section, therefore, gives the building official the authority to inspect any regulated items.

[A] 110.3.9 Special inspections. For *special inspections*, see Chapter 17.

❖ Special inspections are to be provided by the owner for the types of work required in Section 1704. The building official is to approve special inspectors and verify that the required special inspections have been conducted. See the commentary to Section 1704 for a complete discussion of this topic.

[A] 110.3.10 Final inspection. The final inspection shall be made after all work required by the building *permit* is completed.

❖ Special inspections are to be provided by the owner for the types of work required in Section 1704. The building official is to approve special inspectors and verify that the required special inspections have been conducted. See the commentary to Section 1704 for a complete discussion of this topic.

[A] 110.3.10.1 Flood hazard documentation. If located in a *flood hazard area*, documentation of the elevation of the lowest floor as required in Section 1612.5 shall be submitted to the *building official* prior to the final inspection.

❖ The lowest floor inspection called for in Section 110.3.3 of the code requires submission of documentation of elevations upon placement of the lowest floor and prior to further vertical construction. The purpose for submission at that time is to confirm compliance at a point during construction when insufficient elevation can be corrected most readily. The purpose of submission of elevation information when construction is completed is to confirm compliance. Work that is performed subsequent to the placement of the lowest floor may alter the reference level that is deemed the lowest floor. Building owners must provide this "as-built" documentation when they obtain federal flood insurance policies from the NFIP. Documentation of the "as-built" lowest floor elevations is required to be obtained and maintained by communities that participate in the NFIP. A building for which the community does not have this documentation is, by federal regulation, considered to be in violation of the minimum NFIP requirements (see definition of "violation" in 44 C.F.R. §59.2).

[A] 110.4 Inspection agencies. The *building official* is authorized to accept reports of *approved* inspection agencies, provided such agencies satisfy the requirements as to qualifications and reliability.

❖ As an alternative to the building official conducting the inspection, he or she is permitted to accept inspections of and reports by approved inspection agencies. Appropriate criteria on which to base approval of inspection agencies can be found in Section 1703.

[A] 110.5 Inspection requests. It shall be the duty of the holder of the building *permit* or their duly authorized agent to notify the *building official* when work is ready for inspection.

It shall be the duty of the *permit* holder to provide access to and means for inspections of such work that are required by this code.

❖ It is the responsibility of the permit holder or other authorized person, such as the contractor performing the work, to arrange for the required inspections when completed work is ready and to allow for sufficient time for the building official to schedule a visit to the site to prevent work from being concealed prior to being inspected. Access to the work to be inspected must be provided, including any special means such as a ladder.

[A] 110.6 Approval required. Work shall not be done beyond the point indicated in each successive inspection without first obtaining the approval of the *building official*. The *building official*, upon notification, shall make the requested inspections and shall either indicate the portion of the construction that is satisfactory as completed, or notify the *permit* holder or his or her agent wherein the same fails to comply with this code. Any portions that do not comply shall be corrected and such portion shall not be covered or concealed until authorized by the *building official*.

❖ This section establishes that work cannot progress beyond the point of a required inspection without the building official's approval. Upon making the inspection, the building official must either approve the completed work or notify the permit holder or other responsible party of that which does not comply with the code. Approvals and notices of noncompliance must be in writing, as required by Section 104.4, to avoid any misunderstanding as to what is required. Any item not approved cannot be concealed until it has been corrected and approved by the building official.

SECTION 111
CERTIFICATE OF OCCUPANCY

[A] 111.1 Use and occupancy. A building or structure shall not be used or occupied, and a change in the existing use or occupancy classification of a building or structure or portion thereof shall not be made, until the *building official* has issued a certificate of occupancy therefor as provided herein. Issuance of a certificate of occupancy shall not be construed as an approval of a violation of the provisions of this code or of other ordinances of the jurisdiction.

Exception: Certificates of occupancy are not required for work exempt from *permits* in accordance with Section 105.2.

❖ This section establishes that a new building or structure cannot be occupied until a certificate of occupancy is issued by the building official, which reflects the conclusion of the work allowed by the building permit. Also, no change in occupancy or the use of an existing building is permitted without first obtaining a certificate of occupancy for the new use.

The tool that the building official uses to control the uses and occupancies of various buildings and structures within the jurisdiction is the certificate of occupancy. It is unlawful to use or occupy a building or structure unless a certificate of occupancy has been issued. Its issuance does not relieve the building owner from the responsibility for correcting any code violation that may exist.

The exception simply states that when work is not under the monitor of the building department, there is no need to deal with a certificate of occupancy.

[A] 111.2 Certificate issued. After the *building official* inspects the building or structure and does not find violations of the provisions of this code or other laws that are enforced by the department of building safety, the *building official* shall issue a certificate of occupancy that contains the following:

1. The building *permit* number.
2. The address of the structure.
3. The name and address of the *owner* or the owner's authorized agent.
4. A description of that portion of the structure for which the certificate is issued.
5. A statement that the described portion of the structure has been inspected for compliance with the requirements of this code for the occupancy and division of occupancy and the use for which the proposed occupancy is classified.
6. The name of the *building official*.
7. The edition of the code under which the *permit* was issued.
8. The use and occupancy, in accordance with the provisions of Chapter 3.
9. The type of construction as defined in Chapter 6.
10. The design *occupant load*.
11. If an *automatic sprinkler system* is provided, whether the sprinkler system is required.
12. Any special stipulations and conditions of the building *permit*.

❖ The building official is required to issue a certificate of occupancy after a successful final inspection has been completed and all deficiencies and violations have been resolved. This section lists the information that must be included on the certificate. This information is useful to both the building official and the owner because it indicates the criteria under which the structure was evaluated and approved at the time the certificate was issued. This is important when applying the IEBC or IFC to existing buildings.

[A] 111.3 Temporary occupancy. The *building official* is authorized to issue a temporary certificate of occupancy before the completion of the entire work covered by the *permit*, provided that such portion or portions shall be occupied safely. The *building official* shall set a time period during which the temporary certificate of occupancy is valid.

❖ The building official is permitted to issue a temporary certificate of occupancy for all or a portion of a building prior to the completion of all work. Such certification is to be issued only when the building or portion in question can be safely occupied prior to full completion. The certification is intended to acknowledge that some building features may not be completed even though the building is safe for occupancy, or that a portion of the building can be safely occupied while work continues in another area. This provision precludes the occupancy of a building or structure that does not contain all of the required fire protection systems and means of egress. Temporary certificates should be issued only when incidental construction remains, such as site work and interior work that is not regulated by the code and exterior decoration not necessary to the integrity of the building envelope. The building official should view the issuance of a temporary certificate of occupancy as substantial an act as the issuance of the final certificate. Indeed, the issuance of a temporary certificate of occupancy offers a greater potential for conflict because once the building or structure is occupied, it is very difficult to remove the occupants through legal means. The certificate must specify the time period for which it is valid.

[A] 111.4 Revocation. The *building official* is authorized to, in writing, suspend or revoke a certificate of occupancy or completion issued under the provisions of this code wherever the certificate is issued in error, or on the basis of incorrect information supplied, or where it is determined that the building or structure or portion thereof is in violation of any ordinance or regulation or any of the provisions of this code.

❖ This section is needed to give the building official the authority to revoke a certificate of occupancy for the reasons indicated in the code text. The building official may also suspend the certificate of occupancy until all of the code violations are corrected.

SECTION 112
SERVICE UTILITIES

[A] 112.1 Connection of service utilities. A person shall not make connections from a utility, source of energy, fuel or power to any building or system that is regulated by this code for which a *permit* is required, until released by the *building official*.

❖ This section establishes the authority of the building official to approve utility connections to a building for items such as water, sewer, electricity, gas and steam, and to require their disconnection when hazardous conditions or emergencies exist.

The approval of the building official is required before a connection can be made from a utility to a building system that is regulated by the code, including those referenced in Section 101.4. This includes utilities supplying water, sewer, electricity, gas and steam services. For the protection of building occupants, including workers, such systems must have had final inspection approvals, except as allowed by Section 112.2 for temporary connections.

[A] 112.2 Temporary connection. The *building official* shall have the authority to authorize the temporary connection of the building or system to the utility, source of energy, fuel or power.

❖ The building official is permitted to issue temporary authorization to make connections to the public utility system prior to the completion of all work. This acknowledges that, because of seasonal limitations, time constraints or the need for testing or partial operation of equipment, some building systems may be safely connected even though the building is not suitable for final occupancy. The temporary connection and utilization of connected equipment should be approved when the requesting permit holder has demonstrated to the building official's satisfaction that public health, safety and welfare will not be endangered.

[A] 112.3 Authority to disconnect service utilities. The *building official* shall have the authority to authorize disconnection of utility service to the building, structure or system regulated by this code and the referenced codes and standards set forth in Section 101.4 in case of emergency where necessary to eliminate an immediate hazard to life or property or where such utility connection has been made without the approval required by Section 112.1 or 112.2. The *building official* shall notify the serving utility, and wherever possible the *owner* and occupant of the building, structure or service system of the decision to disconnect prior to taking such action. If not notified prior to disconnecting, the *owner* or occupant of the building, structure or service system shall be notified in writing, as soon as practical thereafter.

❖ Disconnection of one or more of a building's utility services is the most radical method of hazard abatement available to the building official and should be reserved for cases in which all other lesser remedies have proven ineffective. Such an action must be preceded by written notice to the utility and the owner and occupants of the building. Disconnection must be accomplished within the time frame established by the building official in the notice. When the hazard to the public health, safety or welfare is so imminent as to mandate immediate disconnection, the building official has the authority and even the obligation to cause disconnection without notice. In such cases, the owner or occupants must be given written notice as soon as possible.

SECTION 113
BOARD OF APPEALS

[A] 113.1 General. In order to hear and decide appeals of orders, decisions or determinations made by the *building official* relative to the application and interpretation of this code, there shall be and is hereby created a board of appeals. The board of appeals shall be appointed by the applicable governing authority and shall hold office at its pleasure. The board shall adopt rules of procedure for conducting its business.

❖ This section provides an aggrieved party with a material interest in the decision of the building official a process to appeal such a decision before a board of appeals. This provides a forum, other than the court of jurisdiction, in which to review the building official's actions.

This section literally allows any person to appeal a decision of the building official. In practice, this section has been interpreted to permit appeals only by those aggrieved parties with a material or definitive interest in the decision of the building official. An aggrieved party may not appeal a code requirement per se. The intent of the appeal process is not to waive or set aside a code requirement; rather, it is intended to provide a means of reviewing a building official's decision on an interpretation or application of the code or to review the equivalency of protection to the code requirements. The members of the appeals board are appointed by the "governing body" of the jurisdiction, typically a council or administrator, such as a mayor or city manager, and remain members until removed from office. The board must establish procedures for electing a chairperson, scheduling and conducting meetings and administration. Note that Appendix B contains complete, detailed requirements for creating an appeals board, including number of members, qualifications and administrative procedures. Jurisdictions desiring to utilize these requirements must include Appendix B in their adopting ordinance.

[A] 113.2 Limitations on authority. An application for appeal shall be based on a claim that the true intent of this code or the rules legally adopted thereunder have been incorrectly interpreted, the provisions of this code do not fully apply or an equally good or better form of construction is proposed. The board shall not have authority to waive requirements of this code.

❖ This section establishes the grounds for an appeal, which claims that the building official has misinterpreted or misapplied a code provision. The board is not allowed to set aside any of the technical requirements of the code. It is, however, allowed to consider alternative methods of compliance with the technical requirements (see Section 104.11).

[A] 113.3 Qualifications. The board of appeals shall consist of members who are qualified by experience and training to pass on matters pertaining to building construction and are not employees of the jurisdiction.

❖ It is important that the decisions of the appeals board are based purely on the technical merits involved in an appeal. It is not the place for policy or political deliberations. The members of the appeals board are, therefore, expected to have experience in building construction matters. Appendix B provides more detailed qualifications for appeals board members and can be adopted by jurisdictions desiring that level of expertise.

SECTION 114
VIOLATIONS

[A] 114.1 Unlawful acts. It shall be unlawful for any person, firm or corporation to erect, construct, alter, extend, *repair*, move, remove, demolish or occupy any building, structure or equipment regulated by this code, or cause same to be done, in conflict with or in violation of any of the provisions of this code.

❖ Violations of the code are prohibited and form the basis for all citations and correction notices.

[A] 114.2 Notice of violation. The *building official* is authorized to serve a notice of violation or order on the person responsible for the erection, construction, *alteration*, extension, *repair*, moving, removal, demolition or occupancy of a building or structure in violation of the provisions of this code, or in violation of a *permit* or certificate issued under the provisions of this code. Such order shall direct the discontinuance of the illegal action or condition and the abatement of the violation.

❖ The building official is required to notify the person responsible for the erection or use of a building found to be in violation of the code. The section that is allegedly being violated must be cited so that the responsible party can respond to the notice.

[A] 114.3 Prosecution of violation. If the notice of violation is not complied with promptly, the *building official* is authorized to request the legal counsel of the jurisdiction to institute the appropriate proceeding at law or in equity to restrain, correct or abate such violation, or to require the removal or termination of the unlawful occupancy of the building or structure in violation of the provisions of this code or of the order or direction made pursuant thereto.

❖ The building official must pursue, through the use of legal counsel of the jurisdiction, legal means to correct the violation. This is not optional.

Any extensions of time, so that the violations may be corrected voluntarily, must be for a reasonable and valid cause, otherwise the building official may be subject to criticism for "arbitrary and capricious" actions. In general, it is better to have a standard time limitation for correction of violations. Departures from this standard must be for a clear and reasonable purpose, usually stated in writing by the violator.

[A] 114.4 Violation penalties. Any person who violates a provision of this code or fails to comply with any of the requirements thereof or who erects, constructs, alters or repairs a building or structure in violation of the *approved construction documents* or directive of the *building official*, or of a *permit* or certificate issued under the provisions of this code, shall be subject to penalties as prescribed by law.

❖ Penalties for violating provisions of the code are typically contained in state law, particularly if the code is adopted at that level, and the building department must follow those procedures. If there is no such procedure already in effect, one must be established with the aid of legal counsel.

SECTION 115 STOP WORK ORDER

[A] 115.1 Authority. Where the *building official* finds any work regulated by this code being performed in a manner either contrary to the provisions of this code or dangerous or unsafe, the *building official* is authorized to issue a stop work order.

❖ Whenever the building official finds any work regulated by this code being performed in a manner that is contrary to the provisions of this code, dangerous or unsafe, the building official is authorized to issue a stop work order.

This section provides for the suspension of work for which a permit was issued, pending the removal or correction of a severe violation or unsafe condition identified by the building official.

Normally, correction notices, issued in accordance with Section 110.6, are used to inform the permit holder of code violations. Stop work orders are issued when enforcement can be accomplished no other way or when a dangerous condition exists.

[A] 115.2 Issuance. The stop work order shall be in writing and shall be given to the *owner* of the property involved, the owner's authorized agent or the person performing the work. Upon issuance of a stop work order, the cited work shall immediately cease. The stop work order shall state the reason for the order and the conditions under which the cited work will be permitted to resume.

❖ Upon receipt of a violation notice from the building official, all construction activities identified in the notice must immediately cease, except as expressly permitted to correct the violation.

[A] 115.3 Unlawful continuance. Any person who shall continue any work after having been served with a stop work order, except such work as that person is directed to perform to remove a violation or unsafe condition, shall be subject to penalties as prescribed by law.

❖ This section states that the work in violation must terminate and that all other work, except that which is necessary to correct the violation or unsafe condition, must cease as well. As determined by the municipality or state, a penalty may be assessed for failure to comply with this section.

SECTION 116 UNSAFE STRUCTURES AND EQUIPMENT

[A] 116.1 Conditions. Structures or existing equipment that are or hereafter become unsafe, insanitary or deficient because of inadequate *means of egress* facilities, inadequate light and ventilation, or that constitute a fire hazard, or are otherwise dangerous to human life or the public welfare, or that involve illegal or improper occupancy or inadequate maintenance, shall be deemed an unsafe condition. Unsafe structures shall be taken down and removed or made safe, as the *building official* deems necessary and as provided for in this section. A vacant structure that is not secured against entry shall be deemed unsafe.

❖ This section describes the responsibility of the building official to investigate reports of unsafe structures and equipment and provides criteria for such determination.

Unsafe structures are defined as buildings or structures that are insanitary; are deficient in light, ventilation or adequate exit facilities; constitute a fire hazard; or are otherwise dangerous to human life.

This section establishes that unsafe buildings can result from illegal or improper occupancies. For example, prima facie evidence of an unsafe structure is an unsecured (open at door or window) vacant building. All unsafe buildings must either be demolished or made safe and secure as deemed appropriate by the building official.

[A] 116.2 Record. The *building official* shall cause a report to be filed on an unsafe condition. The report shall state the occupancy of the structure and the nature of the unsafe condition.

❖ The building official must file a report on each investigation of unsafe conditions, stating the occupancy of the structure and the nature of the unsafe condition. This report provides the basis for the notice described in Section 116.3.

[A] 116.3 Notice. If an unsafe condition is found, the *building official* shall serve on the *owner*, agent or person in control of the structure, a written notice that describes the condition deemed unsafe and specifies the required repairs or improvements to be made to abate the unsafe condition, or that requires the unsafe structure to be demolished within a stipulated time. Such notice shall require the person thus notified to declare immediately to the *building official* acceptance or rejection of the terms of the order.

❖ When a building is found to be unsafe, this information must be provided to the building owner or agent so that they have the opportunity to fix the problem or tear down the structure. The notice should include a time frame for when the items need to be addressed. After the building owner or agent receives the report (see Section 116.2), they must inform the building official how they will address the issues, or stipulate why they disagree with the findings.

[A] 116.4 Method of service. Such notice shall be deemed properly served if a copy thereof is (a) delivered to the *owner* personally; (b) sent by certified or registered mail addressed to the *owner* at the last known address with the return receipt requested; or (c) delivered in any other manner as prescribed by local law. If the certified or registered letter is returned showing that the letter was not delivered, a copy thereof shall be posted in a conspicuous place in or about the structure affected by such notice. Service of such notice in the foregoing manner upon the owner's agent or upon the person responsible for the structure shall constitute service of notice upon the *owner*.

❖ The notice must be delivered to the owner in person, by certified mail or in some other lawful manner, such as delivery to a specified agent of the owner. If the owner or agent cannot be located, additional procedures are established, including posting the unsafe notice on the premises in question. Such action may be considered the equivalent of personal notice. However, it may or may not be deemed by the courts as representing a "good faith" effort to notify. In addition to complying with this section, therefore, public notice through the use of newspapers and other postings in a prominent location at the government center should be used.

[A] 116.5 Restoration. Where the structure or equipment determined to be unsafe by the *building official* is restored to a safe condition, to the extent that repairs, *alterations* or *additions* are made or a change of occupancy occurs during the restoration of the structure, such *repairs*, *alterations*, *additions* and change of occupancy shall comply with the requirements of Section 105.2.2 and the *International Existing Building Code*.

❖ This section provides that unsafe structures may be restored to a safe condition. This means that the cause of the unsafe structure notice can be abated without the structure being required to comply fully with the provisions for new construction. Any work done to eliminate the unsafe condition, as well as any change in occupancy that may occur, must comply with the code.

Bibliography

The following resource materials were used in the preparation of the commentary for this chapter of the code.

Legal Aspects of Code Administration. Country Club Hills, IL: International Code Council, 2002.

ASTM C94/C94M-13, *Specification for Ready-Mixed Concrete*, West Conshohocken, PA: ASTM International, 2013.

NFPA 14-13, *Standpipe and Hose Systems*. Quincy, MA: National Fire Protection Association, 2013.

NFPA 70-14, *National Electrical Code*. Quincy, MA: National Fire Protection Association, 2014.

NFPA 409-11, *Standard for Aircraft Hangars*. Quincy, MA: National Fire Protection Association, 2011.

Chapter 2:
Definitions

General Comments

Nearly all terms defined in the code are listed and defined in Chapter 2. While many terms are used primarily in one chapter or another, the vast majority of terms are used outside of their chapter of primary significance. An example would be the term "fire barrier." The requirements for the construction of a fire barrier are found in Section 707, but fire barriers are required to provide fire-resistive separations by such diverse Sections as 508.4 for separated mixed occupancies, and 1022 for the enclosure of interior exit stairways.

There are some terms which have definitions that are specifically limited to the use of the term in a specific chapter or section of the code. This chapter specifically states where this is the case. There are other terms that have more than one definition. For example, "basement" is defined for general application, but is also defined differently as it applies to flood plain hazard regulation in Section 1612.

This chapter only provides a reference to definitions which, in Chapter 19, are amended versions of definitions in a referenced standard.

Purpose

Codes, by their very nature, are technical documents. As such, literally every word, term and punctuation mark can add to or change the meaning of the intended result. This is even more so with a performance-based code where the desired result often takes on more importance than the specific words. Furthermore, the code, with its broad scope of applicability, includes terms inherent in a variety of construction disciplines. These terms often have multiple meanings depending on the context or discipline being used at the time. For these reasons, it is necessary to maintain a consensus on the specific meaning of terms contained in the code. Chapter 2 performs this function by stating clearly what specific terms mean for the purpose of the code.

SECTION 201
GENERAL

❖ This section contains language and provisions that are supplemental to the use of Chapter 2. It gives guidance to the use of the defined words relevant to tense, gender and plurality. Finally, this section provides direction on how to apply terms that are not defined in the code.

201.1 Scope. Unless otherwise expressly stated, the following words and terms shall, for the purposes of this code, have the meanings shown in this chapter.

❖ The use of words and terms in the code is governed by the provisions of this section. This includes code-defined terms as well as those terms that are not defined in the code.

201.2 Interchangeability. Words used in the present tense include the future; words stated in the masculine gender include the feminine and neuter; the singular number includes the plural and the plural, the singular.

❖ While the definitions contained in Chapter 2 are to be taken literally, gender and tense are interchangeable.

201.3 Terms defined in other codes. Where terms are not defined in this code and are defined in the *International Energy Conservation Code*, *International Fuel Gas Code*, *International Fire Code*, *International Mechanical Code* or *International Plumbing Code,* such terms shall have the meanings ascribed to them as in those codes.

❖ Definitions that are applicable in other *International Codes*® (I-Codes®) are applicable everywhere the term is used in the code. Definitions of terms can help in the understanding and application of code requirements.

201.4 Terms not defined. Where terms are not defined through the methods authorized by this section, such terms shall have ordinarily accepted meanings such as the context implies.

❖ Words or terms not defined within the I-Code series are intended to be applied based on their "ordinarily accepted meanings." The intent of this statement is that a dictionary definition may suffice, provided it is in context. Often, construction terms used throughout the code are not specifically defined in the code or even in a dictionary. In such a case, the definitions contained in the referenced standards (see Chapter 35) and published textbooks on the subject in question are good resources.

SECTION 202
DEFINITIONS

24-HOUR BASIS. The actual time that a person is an occupant within a facility for the purpose of receiving care. It shall not include a facility that is open for 24 hours and is capable of providing care to someone visiting the facility during any segment of the 24 hours.

❖ Care offered on a 24-hour basis is used to differentiate groups and levels of protection between institutional facilities that typically house patients for more than a day, such as hospitals, detoxification facilities, foster care and nursing homes, from other care facilities that keep patients for only part of a day, such as day cares, clinics, day surgery centers and outpatient facilities. To better understand how these concepts work together, see the definitions for "Ambulatory care facility," "Custodial care," "Personal care," "Medical care" and "Incapable of self-preservation." Facilities that have patients/residents/customers who typically stay for 24 hours or more are considered to be providing care on a 24-hour care basis. However, a facility that operates 24 hours a day, such as a day care or an urgent care facility, would not be considered as providing care on a 24-hour basis if the clients did not stay 24 hours, but instead were in and out of the facility similar to one that closed for the night.

[BS] AAC MASONRY. *Masonry* made of autoclaved aerated concrete (AAC) units, manufactured without internal reinforcement and bonded together using thin- or thick-bed *mortar*.

❖ This definition establishes that the requirements of Chapter 21 apply to masonry units manufactured from autoclaved aerated concrete (AAC). AAC masonry units are low-density cementitious products first introduced into the 2005 edition of the *Building Requirements Code for Masonry Structures* (TMS 402/ACI 530/ASCE 5) and *Specifications for Masonry Structures* (TMS 602/ACI 530.1/ASCE 6), although its use has been prevalent in other countries for several decades. Besides being a relatively lightweight product, AAC is also considered to provide good thermal and acoustic insulation. AAC masonry units are bonded together using a thin-bed polymer mortar specifically manufactured for use with AAC masonry.

ACCESSIBLE. A *site*, *building*, *facility* or portion thereof that complies with Chapter 11.

❖ This definition identifies the fundamental concept of Chapter 11. Accessibility is deemed to be accomplished if a building, site or facility complies with the applicable provisions of Chapter 11 and ICC A117.1. It is not the intent of the code to accommodate fully every type and range of disability, as it would not be feasible to do so. The extent to which the code requires accessible features in the various occupancies covered by Chapter 11 (scoping) and the characteristics those features are required to meet through reference to ICC A117.1 (technical requirements) establish that which the code considers accessible.

There are elements that are related to accessibility, but also have requirements that are generally applicable for public safety. Examples are audible and visible alarms in Chapter 9; ramps, doors, protruding objects and accessible means of egress in Chapter 10; and elevator requirements in Chapter 30. These items have been "mainstreamed" into the code.

ACCESSIBLE MEANS OF EGRESS. A continuous and unobstructed way of egress travel from any *accessible* point in a *building* or *facility* to a *public way*.

❖ Accessible means of egress requirements are needed to provide those persons with physical disabilities or mobility impairments a means of egress to exit the building. Because of physical limitations, some occupants may need assistance to exit a building. See Section 1009 for requirements establishing areas where people can safely wait for assisted rescue. Chapter 4 of the IFC also includes requirements in the fire safety and evacuation plans for specific planning to address occupants who may need assistance in evacuation during emergencies. In addition, Chapter 9 of the code includes requirements for emergency evacuation notification for persons with hearing and vision disabilities.

The accessible means of egress requirements may not be the same route as that required for ingress into the building (see Sections 1104 and 1105). For example, a two-story building requires one accessible route to connect all accessible spaces within the building. The accessible route to the second level is typically by an elevator. During a fire emergency, persons with mobility impairments on the second level would be moving to the exit stairways for assisted rescue, not back to the way they came onto the level, via the elevator.

ACCESSIBLE ROUTE. A continuous, unobstructed path that complies with Chapter 11.

❖ There are typically more physical barriers in the built environment to people with a mobility impairment than in any other category of disability. An accessible route enables a person with a mobility impairment to approach and utilize a facility's accessible fixtures and features. While there are a variety of mobility devices, the design and construction of an accessible route is based predominantly on provisions necessary for accessibility to a person using a wheelchair. Accessible routes are required for both ingress and egress (see Sections 1009 and 1104).

An accessible route must also be safe and usable by people with other disabilities and those without disabilities. Therefore, requirements are set forth in consideration of those needs. For example, there are restrictions on objects that protrude into a circulation path in consideration of a person with a visual impairment as well as consideration of the possibility of smoke limiting visibility during an emergency.

ACCESSIBLE UNIT. A *dwelling unit* or *sleeping unit* that complies with this code and the provisions for Accessible units in ICC A117.1.

❖ There are three levels of accessibility described in the code pertaining to dwelling units: Accessible units (always spelled with a capital "A"), Type A units and Type B units. Accessible units are required to be constructed as fully accessible, meaning all required features are present at first occupancy. Unlike Type A and Type B units, Accessible units have no features left as adaptable. Accessible units provide a "higher" level of accessibility than Type A and Type B units and are mandated in all Group I occupancies (as a percentage), in Group R-1 occupancies (per Table 1107.6.1.1), in most Group R-2 congregate living (per Table 1107.6.1.1) and in Group R-4 (at least one unit). The technical criteria for Accessible dwelling units are identified in Section 1002 of the 2003 and 2009 ICC A117.1, whereas in the 1998 ICC A117.1, they were spread throughout Chapters 1 through 9. Also see the commentary for the definitions of "Dwelling unit" and "Sleeping unit," and Section 1107.2.

ACCREDITATION BODY. An *approved*, third-party organization that is independent of the grading and inspection agencies, and the lumber mills, and that initially accredits and subsequently monitors, on a continuing basis, the competency and performance of a grading or inspection agency related to carrying out specific tasks.

❖ The process of determining the grade of lumber not only includes the actual method of applying the grade stamp to the product, but also the certification of the grading agency and its methods of quality control, its rules and the work of its agents. For example, an independent third-party quality control program meeting the accreditation requirements of the American Lumber Standard Committee, Inc. (ALSC) or an equivalent process is required for all softwood lumber that is to be graded for use in construction in the United States.

[A] ADDITION. An extension or increase in floor area or height of a building or structure.

❖ This term is used to describe the condition when the floor area or height of an existing building or structure is increased. This term is only applicable to existing buildings, never new ones. This would include additional floor area that is added within an existing building, such as adding a new mezzanine. [See Section 101.4.7 for a reference to the *International Existing Building Code®* (IEBC®).]

[BS] ADHERED MASONRY VENEER. *Veneer* secured and supported through the adhesion of an *approved* bonding material applied to an *approved backing*.

❖ This type of masonry veneer relies on the backing surface for both vertical and lateral load resistance. The components of adhered masonry veneer construction generally include the masonry veneer, the adhering material and the backing to which the veneer is attached. It should be noted that the term "approve" means components are subject to approval by the building official or the authority having jurisdiction, in accordance with Section 104.

[BS] ADOBE CONSTRUCTION. Construction in which the exterior *load-bearing* and *nonload-bearing walls* and partitions are of unfired clay *masonry units*, and floors, roofs and interior framing are wholly or partly of wood or other *approved* materials.

Adobe, stabilized. Unfired clay *masonry units* to which admixtures, such as emulsified asphalt, are added during the manufacturing process to limit the units' water absorption so as to increase their durability.

Adobe, unstabilized. Unfired clay *masonry units* that do not meet the definition of "Adobe, stabilized."

❖ Adobe masonry was popular in the southwest United States due to the availability of soil for units, the frequent exposure to intense sunlight to dry the units, the thermal mass provided by the completed adobe structure and the low cost of this form of construction. This form of construction has relatively low strength, a lack of formalized design procedures and labor-intensive manufacture of units and construction of the building; thus, it has not been used as much in recent years.

Two types of adobe masonry, stabilized and unstabilized, are briefly described. Prescriptive design requirements for adobe masonry are contained in Section 2109.3.

In stabilized adobe, admixtures are used to produce more durable units (see Section 2109.3.2).

Unstabilized adobe does not contain stabilizers in the soil and is, therefore, not as durable as stabilized adobe (see Section 2109.3.1).

[F] AEROSOL. A product that is dispensed from an *aerosol container* by a propellant. Aerosol products shall be classified by means of the calculation of their chemical heats of combustion and shall be designated Level 1, Level 2 or Level 3.

Level 1 aerosol products. Those with a total chemical heat of combustion that is less than or equal to 8,600 British thermal units per pound (Btu/lb) (20 kJ/g).

Level 2 aerosol products. Those with a total chemical heat of combustion that is greater than 8,600 Btu/lb (20 kJ/g), but less than or equal to 13,000 Btu/lb (30 kJ/g).

Level 3 aerosol products. Those with a total chemical heat of combustion that is greater than 13,000 Btu/lb (30 kJ/g).

❖ The intent of the code is to regulate those aerosols that contain a flammable propellant, such as butane, isobutane or propane. An aerosol product such as whipped cream is a water-based material with a nonflammable propellant (nitrous oxide) and would, therefore, not be regulated as a hazardous material. The contents of the aerosol container may be dispensed in the form of a mist spray, foam, gel or aerated powder.

Because of the wide range of flammability of aerosol products, a classification system was established to determine the required level of fire protection. Categories are defined according to the aerosol's chemical heat of combustion expressed in Btus per pound (Btu/lb). Aerosol category classifications of Levels 1, 2 and 3 are used to avoid confusion with flammable liquid classifications.

Examples of Level 1 aerosol products are shaving gel, whipped cream and air fresheners. Level 1 aerosols are not regulated as a hazardous material and are essentially exempt from the requirements of Sections 307 and 414. Examples of Level 2 aerosols include some hair sprays and insect repellents. Level 3 aerosols include carburetor cleaner and other petroleum-based aerosols.

While aerosols are defined as hazardous materials, note that they are not listed in Table 307.1(1) or 307.1(2) as having a maximum allowable quantity per control area. As stated in Item 12 in Section 307.1.1, a building or structure used for aerosol storage is classified as Group S-1, provided the requirements of the IFC are satisfied. Therefore, the Group H classification is not utilized since the design must satisfy the IFC in order to be in compliance.

[F] AEROSOL CONTAINER. A metal can or a glass or plastic bottle designed to dispense an aerosol.

❖ All design criteria for the aerosol container, including the maximum size and minimum strength, are set by the U.S. Department of Transportation (DOTn 49 CFR) and addressed in Section 5104 of the IFC.

[BS] AGGREGATE. In roofing, crushed stone, crushed slag or water-worn gravel used for surfacing for *roof coverings*.

❖ Aggregate is gravel, stone or slag used as a roof surfacing to provide a walking surface and protection to the roof covering. An aggregate typically is not used as a ballast material based on its small size.

AGRICULTURAL BUILDING. A structure designed and constructed to house farm implements, hay, grain, poultry, livestock or other horticultural products. This structure shall not be a place of human habitation or a place of employment where agricultural products are processed, treated or packaged, nor shall it be a place used by the public.

❖ This definition is needed for the proper application of the utility and miscellaneous occupancy group and Appendix C provisions. The use of the building is quite restricted such that buildings that include habitable or public spaces are not agricultural buildings by definition.

AIR-IMPERMEABLE INSULATION. An insulation having an air permeance equal to or less than 0.02 l/s × m2 at 75 pa pressure differential tested in accordance with ASTM E2178 or ASTM E283.

❖ In buildings with conditioned spaces, an air-impermeable layer may be required to prevent the movement of air through the building's thermal envelope. In some cases insulation will serve that purpose, provided it is insulation as defined here. Air leakage, also known as air infiltration, is also specifically addressed for both commercial and residential buildings under the *International Energy Conservation Code*® (IECC®).

AIR-INFLATED STRUCTURE. A structure that uses air-pressurized membrane beams, arches or other elements to enclose space. Occupants of such a structure do not occupy the pressurized area used to support the structure.

❖ This type of membrane structure is characterized by multiple layers arranged such that air-pressurized membrane beams, arches or similar elements are formed. These elements are pressurized with air and form the membrane structure. Note that the occupants of the structure are not subjected to the pressurized areas, because the pressurization is in the structural elements, not within the space used by the occupants.

AIR-SUPPORTED STRUCTURE. A structure wherein the shape of the structure is attained by air pressure and occupants of the structure are within the elevated pressure area. Air-supported structures are of two basic types:

Double skin. Similar to a single skin, but with an attached liner that is separated from the outer skin and provides an airspace which serves for insulation, acoustic, aesthetic or similar purposes.

Single skin. Where there is only the single outer skin and the air pressure is directly against that skin.

❖ An air-supported structure identifies those membrane structures that are completely pressurized for the purposes of supporting the membrane covering. Most "domed" sports arenas use air pressure within the structure to support the membrane covering. The membrane covering can consist of one layer or multiple layers; thus, air-supported structures are classified as either "single skin" or "double skin."

A double-skin, air-supported structure contains multiple layers of membrane sheathing. The membranes are usually separated by enough distance to allow for pressurized air or other materials to be inserted between the plies. The pressurized air or other materials usually serve to increase the insulating and acoustical properties.

A single-skin, air-supported structure consists of just one membrane covering that is directly supported by the interior pressurized air. No other membranes are provided for insulating or acoustical purposes. If the membrane covering consists of several laminated plies, such an arrangement is still considered a single-skin, air-supported structure.

AISLE. An unenclosed *exit access* component that defines and provides a path of egress travel.

❖ Aisles and aisle accessways are both utilized as part of the means of egress in facilities where tables, seats, displays or other furniture may limit the path of travel. The aisle accessways lead to the main aisles that lead to the exits from the space and building [see Commentary Figure 202(1)]. While both may result in a confined path of travel, an aisle is an unenclosed component, while a corridor would be an enclosed component of the means of egress (see Sections 1018 and 1029 for requirements for aisles).

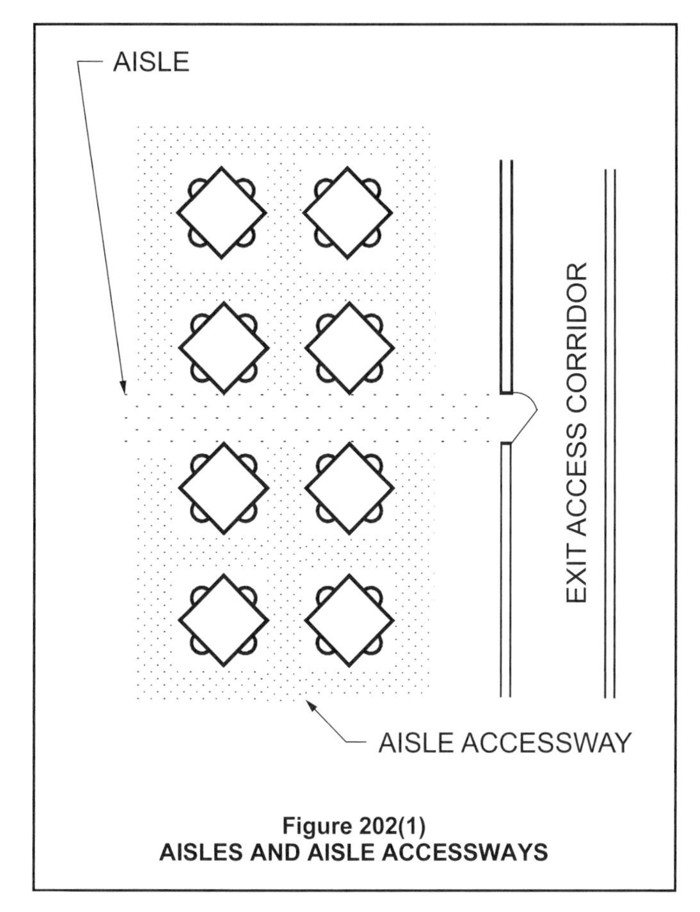

Figure 202(1)
AISLES AND AISLE ACCESSWAYS

AISLE ACCESSWAY. That portion of an *exit access* that leads to an *aisle*.

❖ As illustrated in Commentary Figure 202(1), an aisle accessway is intended for one-way travel or limited two-way travel. The space between tables, seats, displays or other furniture (i.e., aisle accessway) utilized for means of egress will lead to a main aisle (see Sections 1018 and 1029 for requirements for aisle accessways).

[F] ALARM NOTIFICATION APPLIANCE. A *fire alarm system* component such as a bell, horn, speaker, light or text display that provides audible, tactile or visible outputs, or any combination thereof.

❖ The code requires that fire alarm systems be equipped with approved alarm notification appliances so that in an emergency, the fire alarm system will notify the occupants of the need for evacuation or implementation of the fire emergency plan. Alarm notification devices required by the code are of two general types: visible and audible. Except for emergency voice/alarm communication systems, once the fire alarm system has been activated, all visible and audible communication alarms are required to activate. Emergency voice/alarm communication systems are special signaling systems that are activated selectively in response to specific emergency conditions but have the capability to be activated throughout the building if necessary.

[F] ALARM SIGNAL. A signal indicating an emergency requiring immediate action, such as a signal indicative of fire.

❖ This is a general term for all types of supervisory and trouble signals. An example would be a supervisory (tamper) switch on a sprinkler control valve. It could also be the response to a specific device that is not part of the alarm notification system but that causes a specific function such as a smoke detector for elevator recall. The activation of the device does not necessarily indicate that there is a fire; however, the level of protection may have been compromised (see the definition of "Fire alarm signal").

[F] ALARM VERIFICATION FEATURE. A feature of *automatic* fire detection and alarm systems to reduce unwanted alarms wherein *smoke detectors* report alarm conditions for a minimum period of time, or confirm alarm conditions within a given time period, after being *automatically* reset, in order to be accepted as a valid alarm-initiation signal.

❖ False fire (evacuation) alarms are a nuisance. For this reason, the code specifies that alarms activated by smoke detectors are not to be sounded until the alarm signal is verified by cross-zoned detectors in a single protected area or by system features that will retard the alarm until the signal is determined to be valid. Valid alarm initiation signals can be determined by detectors that report alarm conditions for a minimum period of time or that, after being reset, continue to report an alarm condition. The alarm verification feature may not retard signal activation for a period of more than 60 seconds and must not apply to alarm-initiating devices other than smoke detectors (which may be connected to the same circuit). Alarm verification is not the same as presignal features that delay an alarm signal for more than 1 minute and that are allowed only where specifically permitted by the authority having jurisdiction.

ALLOWABLE STRESS DESIGN. A method of proportioning structural members, such that elastically computed stresses produced in the members by *nominal loads* do not exceed *specified* allowable stresses (also called "working stress design").

❖ This definition describes the allowable stress design (ASD) method, which is one of the design approaches recognized under the code in Section 1604.1. In this approach, the computed stresses determined from the unfactored, or nominal, loads (see the definition for "Nominal loads") cannot exceed allowable stresses, which provide a factor of safety. The material chapters of the code specify either the allowable stresses for a given material or, in some cases, provide allowable capacities of specific assemblies, such as the wood structural panel shear wall capacities specified in Chapter 23.

[A] ALTERATION. Any construction or renovation to an *existing structure* other than *repair* or *addition.*

❖ The code utilizes this term to reflect construction operations intended for an existing building, but not within the scope of an addition or repair (see the definitions of "Addition" and "Repair"). (See Section 101.4.7 for a reference to the IEBC.)

ALTERNATING TREAD DEVICE. A device that has a series of steps between 50 and 70 degrees (0.87 and 1.22 rad) from horizontal, usually attached to a center support rail in an alternating manner so that the user does not have both feet on the same level at the same time.

❖ An alternating tread device is commonly used in areas that would otherwise be provided with a ladder where there is not adequate space for a full stairway. Where these devices are permitted is specifically listed in the code (i.e., Section 1006.2.2.1). The device is used extensively in industrial facilities for worker access to platforms or equipment. Requirements are found with stairways in Section 1011.

AMBULATORY CARE FACILITY. Buildings or portions thereof used to provide medical, surgical, psychiatric, nursing or similar care on a less than 24-hour basis to individuals who are rendered *incapable of self-preservation* by the services provided.

❖ The code provides different requirements for outpatient clinics, ambulatory care facilities and hospitals. Ambulatory care facilities, while still classified as a Group B occupancy, have additional standards above those of an outpatient clinic because its patients are temporarily unable to respond to emergencies due to treatment processes (see commentary, Section 422). Ambulatory care facilities include day surgery centers and similar facilities where patients may receive fairly intensive treatment, but do not stay at the facility more than a few hours. If patients are receiving care on a 24-hour basis, such facilities would be defined as hospitals (see definition of "24-hour basis").

ANCHOR BUILDING. An exterior perimeter building of a group other than H having direct access to a *covered or open mall building* but having required *means of egress* independent of the mall.

❖ A key to understanding what distinguishes an anchor building from a tenant space is that anchor buildings are typically retail establishments (Group M), although this may not always be the case. The anchor building is typically some facility that, by its nature, draws a considerable number of people. The tenants in the adjoining covered or open mall building then seek to capitalize on this traffic generated by the anchor building. The scale or size of the building is not a primary factor in determining whether it is an anchor building; rather, its function is such that it draws people to the site in sufficient numbers so that other facilities can benefit from being located in the same facility [see Commentary Figure 202(2)].

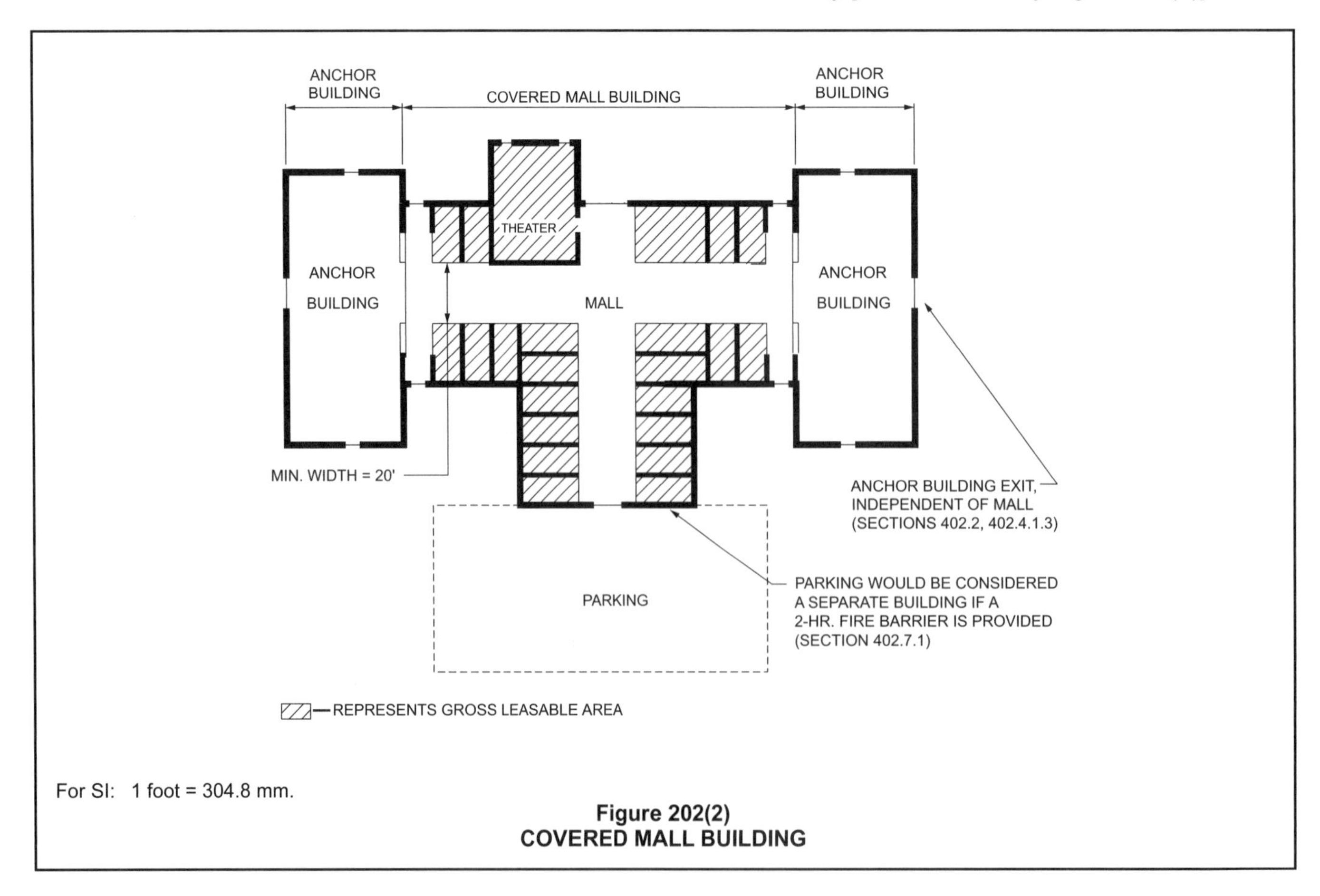

Figure 202(2)
COVERED MALL BUILDING

Generally, the anchor building will have its own identity and there is a high probability that it will have separate management and its own hours of operation. This will necessitate a means of egress that does not rely on the mall being available for its patrons to enter or leave. Similarly, the means of egress from the covered or open mall building cannot rely on the anchor building being open and available for patrons to exit through. Therefore, egress facilities for the anchor building must be independent of those for the covered or open mall building. An anchor building is a separate building from the covered or open mall building and must comply with the provisions of the code for its own identity, except as modified by Section 402 (see Sections 402.4 and 402.4.2.2.).

[BS] ANCHORED MASONRY VENEER. *Veneer* secured with *approved* mechanical fasteners to an *approved backing.*

❖ This type of masonry veneer is generally supported from below and anchored to the sheathing, studs or other structural portion of the wall. Veneers provide little, if any, strength to the wall and are, therefore, considered to be nonstructural. Anchored masonry veneer is unique in that it is usually supported from below by a footing, lintels or shelf angles. Anchored masonry veneer must not, however, support loads other than its own dead loads, wind loads and seismic loads resulting from the dead load of the wall.

ANNULAR SPACE. The opening around the penetrating item.

❖ The annular space is the space created between the outer surface of a penetrating item and the construction penetrated. If left unfilled, the space can provide a means of free passage for smoke, fire and products of combustion [see Commentary Figure 202(3)] for an example of an annular space]. Sections 714.3.1, 714.3.2, 714.4.1.1 and 714.4.1.2 contain the requirements for protection of annular spaces. Tested and listed through-penetration firestop systems (see Sections 714.3.1.2 and 714.4.1.1.2), which are commonly used to seal these annular spaces, generally specify maximum and minimum allowable annular spaces, neither of which must be exceeded in order to ensure that the intended fire-resistance rating is achieved. The minimum and maximum annular space for a penetrating item is frequently not equal, as penetrating items are often located eccentrically in the hole (i.e., not perfectly centered).

[F] ANNUNCIATOR. A unit containing one or more indicator lamps, alphanumeric displays or other equivalent means in which each indication provides status information about a circuit, condition or location.

❖ This refers to the panel that displays the status of the monitored fire protection systems and devices. It is not the fire alarm control unit, though the control panel may include an annunciator.

[A] APPROVED. Acceptable to the *building official.*

❖ As related to the process of acceptance of building installations, including materials, equipment and construction systems, this definition identifies where the ultimate authority rests. Whenever this term is used, it intends that only the enforcing authority can accept a specific installation or component as complying with the code. For the code and the *International Residential Code®* (IRC®), the "building official" is identified as the person responsible for administering its provisions. For the *International Fire Code®* (IFC®), the "fire code official" is identified as the person responsible for administering IFC provisions. For the *International Energy Conservation Code®* (IECC®), *International Fuel Gas Code®* (IFGC®), *International Green Construction Code®* (IgCC®), *International Mechanical Code®* (IMC®), *International Plumbing Code®* (IPC®), *International Property Maintenance Code®* (IPMC®), *International Swimming Pool and Spa Code®* (ISPSC®) and *International Wildland-Urban Interface Code®* (IWUIC®), the "code official" is identified as the person responsible.

[A] APPROVED AGENCY. An established and recognized agency that is regularly engaged in conducting tests or furnishing inspection services, where such agency has been *approved* by the *building official.*

❖ Third-party testing or inspections may be needed for elements within the built environment. The basis for the building official's approval of any agency for a particular activity may include, but is not necessarily limited to, the capacity and capability of the agency to perform the work in accordance with Section 1705 and other applicable sections. This is typically done through a review of the résumés and references of the agency and its personnel. For this code, the building official is identified as the person responsible for approval.

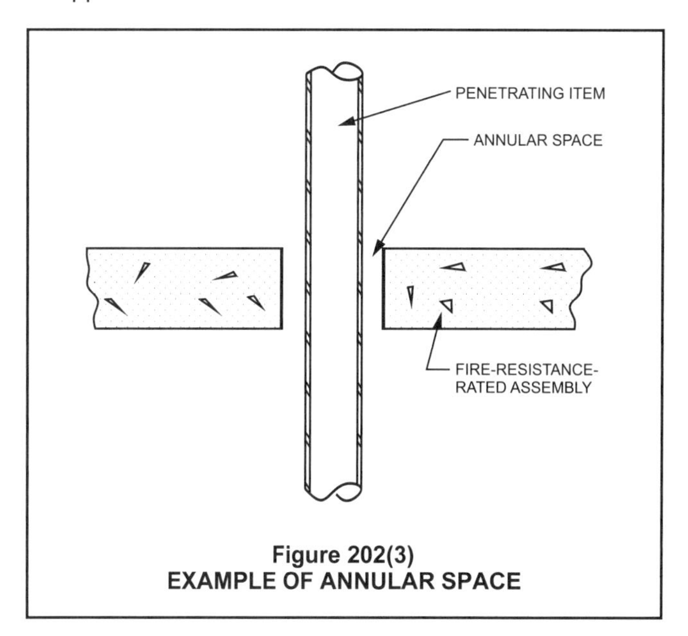

Figure 202(3)
EXAMPLE OF ANNULAR SPACE

[BS] APPROVED FABRICATOR. An established and qualified person, firm or corporation *approved* by the *building official* pursuant to Chapter 17 of this code.

❖ An approved fabricator is one who has received approval to perform work without a code-required special inspection. The approval is based upon review of the fabricator's written procedural and quality-control manuals, and periodic auditing of fabrication practices by an approved special inspection agency.

[A] APPROVED SOURCE. An independent person, firm or corporation, *approved* by the *building official*, who is competent and experienced in the application of engineering principles to materials, methods or systems analyses.

❖ The building official sometimes needs to rely on evaluation reports, analyses or other types of reports that purport to validate the use of a material, system or method as complying with the code. This definition establishes that the building official needs to rely on independent, competent individuals or agencies as the source of these reports.

[BS] AREA (for masonry).

Gross cross-sectional. The *area* delineated by the out-to-out *specified* dimensions of *masonry* in the plane under consideration.

Net cross-sectional. The *area* of *masonry units*, grout and *mortar* crossed by the plane under consideration based on out-to-out *specified* dimensions.

❖ Various areas are used for structural calculations in Chapter 21. It is important to use the appropriate areas required by the code, since they may give dramatically different results.

The gross cross-sectional area of the masonry is the specified masonry width (thickness) multiplied by the specified length, as illustrated in Commentary Figure 202(4). While subtraction of core areas of the masonry unit is not required, the space between wythes must be discounted in noncomposite walls. Empirical compressive stress design is based on the gross cross-sectional area of the masonry.

The net cross-sectional area encompasses the area of units, grout and mortar contained within the plane under consideration. For ungrouted masonry, this area is sometimes equal to the bedded area, or more often to the minimum specified area of the face shells. For grouted masonry, this also includes that area of cores, cells or spaces filled with grout.

AREA, BUILDING. The area included within surrounding *exterior walls* (or *exterior walls* and *fire walls*) exclusive of vent *shafts* and *courts*. Areas of the building not provided with surrounding walls shall be included in the building area if such areas are included within the horizontal projection of the roof or floor above.

❖ Allowable building areas (as established by the provisions of Chapter 5 and Table 506.2) are a function of the potential fire hazard and the level of fire endurance of the building's structural elements, as defined by the types of construction in Chapter 6. A building area is the "footprint" of the building; that is, the area measured within the perimeter formed by the inside surface of the exterior walls. This excludes spaces that are inside this perimeter and open to the outside atmosphere at the top, such as open shafts and courts (see Section 1206). When a portion of the building has no exterior walls, the area regulated by Chapter 5 is defined by the projection of the roof or floor above [see Commentary Figure 202(5)]. The roof overhang on portions of a building where there are exterior enclosure walls does not add to the building area because the area is defined by exterior walls.

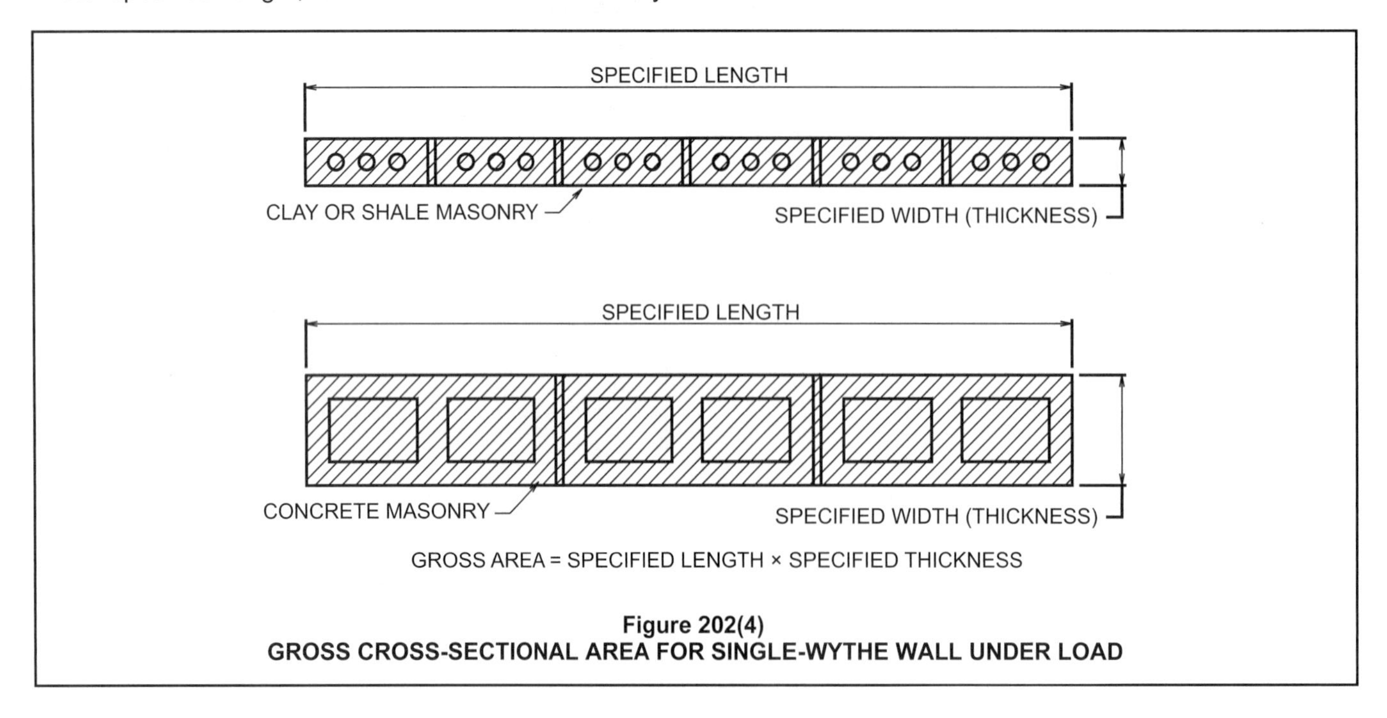

Figure 202(4)
GROSS CROSS-SECTIONAL AREA FOR SINGLE-WYTHE WALL UNDER LOAD

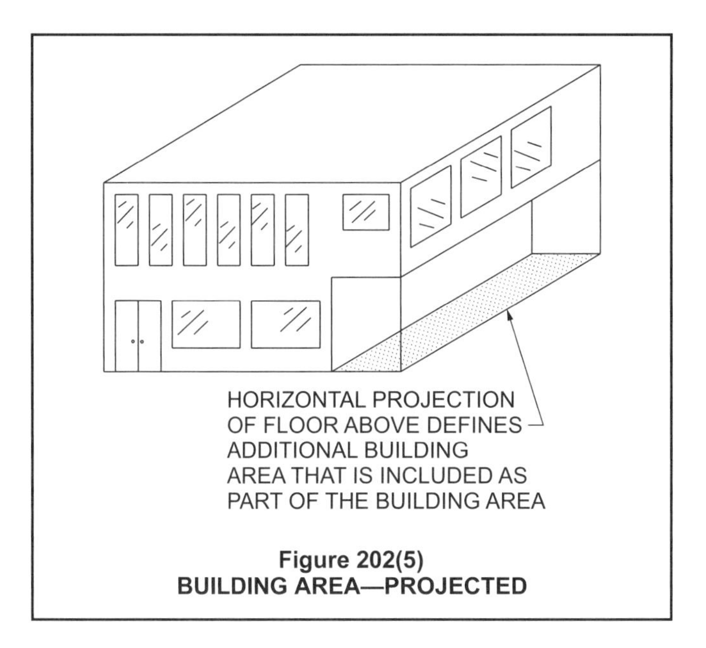

Figure 202(5)
BUILDING AREA—PROJECTED

AREA OF REFUGE. An area where persons unable to use *stairways* can remain temporarily to await instructions or assistance during emergency evacuation.

❖ The area of refuge is a temporary waiting area used during emergency evacuations for persons who are unable to exit the building using the stairways. The fire safety plans (in accordance with IFC Section 404) include the locations of areas of refuge so that the fire department will know where people may be waiting for rescue assistance. See Section 1009 for where areas of refuge are required at stairways and elevators. Areas of refuge have requirements for separation, size, signage, instructional information and two-way communication systems.

AREA OF SPORT ACTIVITY. That portion of an indoor or outdoor space where the play or practice of a sport occurs.

❖ The broad term, "area of sports activity," addresses indoor and outdoor courts, fields and other sport areas. Examples are basketball and tennis courts; practice areas for dance or gymnastics; baseball, soccer and football fields; skating rinks; running tracks; or skateboard parks. The application of the phrase "portion where the play or practice of a sport occurs," varies depending on the sports. Football fields include the playing field boundary lines, the end zones and the space between the boundary lines and safety border. Players may run or be pushed into this safety zone during play. In football, this safety zone is used as part of the playing field, and is therefore included in the area of sports activity. See the commentary for Section 1110 regarding the general accessible route allowances for areas of sports activity. Some of the areas listed under the recreational facilities addressed in Section 1110 (i.e., bowling lanes, exercise equipment facilities, golf courses, miniature golf, and pools) are considered areas of sports activity; however, they have additional requirements.

AREAWAY. A subsurface space adjacent to a building open at the top or protected at the top by a grating or *guard*.

❖ Areaways are often constructed to provide access to below-grade building services, including transformers, ventilation shafts and pipe tunnels.

ASSEMBLY SEATING, MULTILEVEL. See "Multilevel assembly seating."

ATRIUM. An opening connecting two or more *stories* other than enclosed *stairways*, elevators, hoistways, escalators, plumbing, electrical, air-conditioning or other equipment, which is closed at the top and not defined as a mall. *Stories*, as used in this definition, do not include balconies within assembly groups or *mezzanines* that comply with Section 505.

❖ The definition identifies that an atrium is a floor opening or a series of floor openings that connects the environments of adjacent stories. The definition of "Atrium" excludes enclosed stairways, elevators, hoistways and other similar openings in order to clarify that those elements would not fall under the purview as to what is considered an atrium, and therefore, the associated requirements found in Section 404 would not apply. What this does not preclude is the inclusion of elevators and open stairways within atriums. Such elements would need to be entirely within the atrium to meet the separation requirements found in Section 404.6. Building features, such as stairways, elevators, hoistways, escalators, plumbing, electrical, air conditioning or other equipment openings, are required to be enclosed in fire-resistance-rated shafts in accordance with Sections 712 and 713. Atriums are specified in Section 712.1.7 to address openings in horizontal assemblies. An atrium is not defined by size or use. A series of floor openings that are enclosed with exterior walls, yet open at the roof, would be considered a court and would be exempt from the requirements of Section 404. Balconies associated with assembly occupancies and mezzanines are not considered individual stories that would contribute to the classification of a space as an atrium.

ATTIC. The space between the ceiling beams of the top *story* and the roof rafters.

❖ The definition of "Attic" identifies the specific portion of a building or structure for the purpose of determining the applicability of requirements that are specific to attics, such as ventilation (see Section 1203) and draftstopping (see Section 718). Additionally, the code has access requirements (see Section 1209) and uniformly distributed live load requirements (see Table 1607.1) for attics. An attic is considered the space or area located immediately below the roof sheathing within the roof framing system of a building. Pitched roof systems, such as gabled, hip, saw-toothed or curved roofs, all create spaces between the roof sheathing and ceiling membrane, which are considered attics.

[F] AUDIBLE ALARM NOTIFICATION APPLIANCE. A notification appliance that alerts by the sense of hearing.

❖ Audible alarms that are part of a fire alarm system must be loud enough to be heard in every occupied space of a building. Section 907.5.2.1.1 prescribes the minimum sound pressure level for all audible alarm notification appliances depending on the occupancy of the building and the function of the space.

AUTOCLAVED AERATED CONCRETE (AAC). Low density cementitious product of calcium silicate hydrates, whose material specifications are defined in ASTM C1386.

❖ Autoclaved aerated concrete (AAC) is a relatively lightweight concrete product which is manufactured in panels that can be cut into smaller blocks to use as masonry units. The referenced material standard establishes the minimum compressive strength based on the class of the AAC. The definition of "AAC masonry" given above establishes that requirements in Chapter 21 apply to masonry units that consist of AAC.

[F] AUTOMATIC. As applied to fire protection devices, a device or system providing an emergency function without the necessity for human intervention and activated as a result of a predetermined temperature rise, rate of temperature rise or combustion products.

❖ This term, when used in conjunction with fire protection systems or devices, means that the system or device will perform its intended function without a person being present or performing any task in its control or operation. The device or system has the inherent capability to detect a developing fire condition and perform some predetermined function. Automatic devices and systems operate completely without human presence or intervention.

[F] AUTOMATIC FIRE-EXTINGUISHING SYSTEM. An *approved* system of devices and equipment which *automatically* detects a fire and discharges an *approved* fire-extinguishing agent onto or in the area of a fire.

❖ This term is the generic name for all types of automatic fire-extinguishing systems, including the most common type—the automatic sprinkler system. See Section 904 for requirements for particular alternative automatic fire-extinguishing systems, such as wet-chemical, dry-chemical, foam, carbon dioxide, Halon and clean-agent systems.

[F] AUTOMATIC SMOKE DETECTION SYSTEM. A *fire alarm system* that has initiation devices that utilize *smoke detectors* for protection of an area such as a room or space with detectors to provide early warning of fire.

❖ Chapter 9 provides requirements for various automatic protection systems including: "automatic fire detection system," "automatic sprinkler system," "automatic fire alarm system" and "automatic smoke detection system." Automatic smoke detection systems are required for various occupancies as specified in Section 907. They are required to increase the likelihood that fire is detected and occupants of the building are given an early warning.

[F] AUTOMATIC SPRINKLER SYSTEM. An *automatic sprinkler system*, for fire protection purposes, is an integrated system of underground and overhead piping designed in accordance with fire protection engineering standards. The system includes a suitable water supply. The portion of the system above the ground is a network of specially sized or hydraulically designed piping installed in a structure or area, generally overhead, and to which *automatic* sprinklers are connected in a systematic pattern. The system is usually activated by heat from a fire and discharges water over the fire area.

❖ An automatic sprinkler system is one type of automatic fire-extinguishing system. Automatic sprinkler systems are the most common, and their life safety attributes are widely recognized. The code specifies three types of automatic sprinkler systems: one installed in accordance with NFPA 13, one in accordance with NFPA 13R and the other in accordance with NFPA 13D. To be considered for most code design alternatives, a building's automatic sprinkler system must be installed throughout in accordance with NFPA 13 (see Section 903.3.1.1).

In a fire, sprinklers automatically open and discharge water onto the fire in a spray pattern that is designed to contain or extinguish the fire. Originally, automatic sprinkler systems were developed just for the protection of buildings and their contents. Because of the development and improvements in sprinkler head response time and water distribution, however, automatic sprinkler systems are now also considered life safety systems. Proper operation of an automatic sprinkler system requires careful selection of the sprinkler heads so that water in sufficient quantity at adequate pressure and properly distributed will be available to suppress the fire. Note that the context of the use of the term "fire area" in the last sentence of the definition is to refer to the area in which the fire is occurring, not in the context of the defined term "fire area."

There are many different types of automatic sprinkler systems—wet pipe, dry pipe, preaction, antifreeze and various combinations. Sprinklers can be pendant, upright or sidewall and can be designed for standard or extended coverage. Additional information can be found in NFPA 13.

[F] AUTOMATIC WATER MIST SYSTEM. A system consisting of a water supply, a pressure source, and a distribution piping system with attached nozzles, which, at or above a minimum operating pressure, defined by its listing, discharges water in fine droplets meeting the requirements of NFPA 750 for the purpose of the control, suppression or extinguishment of a fire. Such systems include wet-pipe, dry-pipe and pre-action types. The systems are designed as engineered, pre-engineered, local-application or total flooding systems.

❖ The code recognizes water mist systems as an alternative extinguishing system to automatic sprinkler systems. However, no exceptions, reductions, or "tradeoffs" for water mist systems are granted or permitted by the code because such systems are not considered to be equivalent to automatic sprinkler systems. Automatic water mist systems have been approved by FM Global for occupancies similar to Light Hazard (as defined by NFPA 13) and listed by UL for occupancies similar to Ordinary Hazard Group I (as defined by NFPA 13). These listings permit automatic water mist systems to be installed as the primary suppression system in a variety of occupancy classifications.

Water mist systems are used for special applications in which creating a heat-absorbent vapor consisting of water droplets with a size of less than 1000 microns at the discharge nozzle is the primary extinguishing method. These systems are typically used where water damage may be an issue, or where water supplies are limited. The droplet size can be controlled by adjusting the discharge pressure through the nozzle. By creating a mist, an equal volume of water will create a larger total surface area exposed to the fire and thus better facilitate the absorption of heat, allowing more water droplets to turn to steam more quickly and thus more effectively cool the room.

[F] AVERAGE AMBIENT SOUND LEVEL. The root mean square, A-weighted sound pressure level measured over a 24-hour period, or the time any person is present, whichever time period is less.

❖ The ambient noise that can be expected depends on the occupancy of the building. To attract the attention of the occupants, the audible alarm devices must be heard above the ambient noise in the space. For this reason, the alarm devices must have minimum sound pressure levels above the average ambient sound level. Section 907.5.2.1.1 prescribes the minimum sound pressure levels for the audible alarm notification appliances for all occupancy conditions.

Although it is possible to measure the ambient sound within an occupied space, the alarm notification devices are usually designed and installed before buildings are occupied, thus it is typically a careful analysis of the types of uses within a space that will determine the average ambient sound level. If, after the building is occupied, the alarm notification devices are below expected audibility, a field measurement may be necessary to determine whether the design assumptions are correct.

AWNING. An architectural projection that provides weather protection, identity or decoration and is partially or wholly supported by the building to which it is attached. An awning is comprised of a lightweight *frame structure* over which a covering is attached.

❖ Similar to a canopy, an awning typically provides weather protection, signage or decoration. Its distinguishing characteristic is the lightweight frame structure. It is also supported, at least in part, by the building from which it projects. See Section 3105 for general requirements and Section 3202 for encroachment requirements. This definition helps to clarify the applicability of roof loads in Section 1607.12.4. Also see the definitions for "Retractable awning," "Canopy" and "Marquee."

BACKING. The wall or surface to which the *veneer* is secured.

❖ The backing is the portion of a structure that provides support for the exterior veneer. The backing also typically resists lateral and transverse loads imposed by the veneer and may be load bearing or nonload bearing.

BALANCED DOOR. A door equipped with double-pivoted hardware so designed as to cause a semicounterbalanced swing action when opening.

❖ Balanced doors are commonly used to decrease the force necessary to open the door or to reduce the length of the door swing. Balanced doors typically reduce the clear opening width more than normally hinged doors [see Commentary Figure 202(6) and Section 1010.1.10.2].

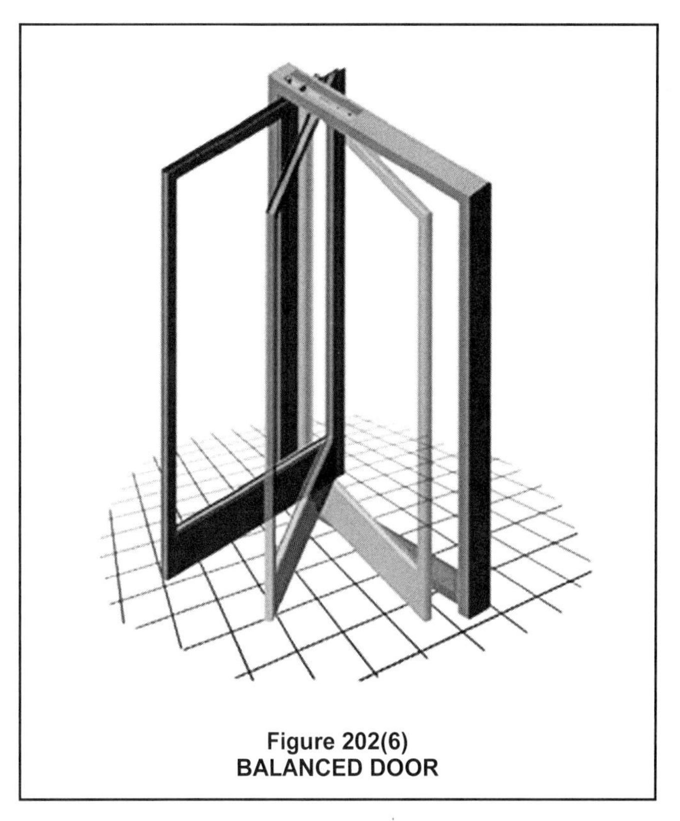

Figure 202(6)
BALANCED DOOR

[F] BALED COTTON. A natural seed fiber wrapped in and secured with industry accepted materials, usually consisting of burlap, woven polypropylene, polyethylene or cotton or sheet polyethylene, and secured with steel, synthetic or wire bands or wire; also includes linters (lint removed from the cottonseed) and motes (residual materials from the ginning process).

❖ This definition of standard "Baled cotton" is being included only to distinguish it from "Baled cotton, densely packed" (see the commentary to the definition of "Baled cotton, densely packed"). The Joint Cotton Industry Bale Packaging Committee (JCIBPC) represents all parts of the cotton industry and sets standards and specifications for packaging of cotton bales, including bale density. The JCIBPC specifications for baling of cotton require that all cotton bales be secured with fixed-length wire bands, polyester plastic strapping or cold-rolled, high-tensile steel strapping, then covered in fully coated woven polyolefin, polyethylene film or burlap.

[F] BALED COTTON, DENSELY PACKED. Cotton made into banded bales with a packing density of not less than 22 pounds per cubic foot (360 kg/m^3), and dimensions complying with the following: a length of 55 inches (1397 mm), a width of 21 inches (533.4 mm) and a height of 27.6 to 35.4 inches (701 to 899 mm).

❖ Currently, over 99 percent of all U.S. cotton is pressed and stored as densely packed baled cotton, with bales meeting the weight and dimension requirements of ISO 8115. One reason that the cotton industry has chosen to use such bales is because they are very difficult to ignite, which allows the industry to transport them without being labeled as "flammable solids" or "dangerous goods" by national or international transport authorities. This definition is intended to be used to distinguish such bales from other combustible fibers.

In order to counteract some erroneous information regarding the combustibility characteristics of densely packed cotton bales, flammability research was conducted on baled cotton. The research demonstrated that densely packed baled cotton meeting the size and weight requirements of ISO 8115 is not a hazardous material. In view of that data, the U.S. Department of Transportation (U.S. Coast Guard), the United Nations (UN) and the International Maritime Organization (IMO) have all removed baled cotton from the list of hazardous materials and from the list of flammable solids, provided the cotton bales are the densely packed type that meet the standard noted above. The research conclusions were:

1. Standard cotton fiber "passed" the Department of Transportation's spontaneous combustion test: the cotton did not exceed the oven temperature and was not classified as self-heating.
2. Cotton, as densely packed baled cotton, did not cause sustained smoldering propagation: an electric heater placed within the bales was unable to cause sustained smoldering propagation, because of the lack of oxygen inside the densely packed bale.
3. Cotton, as densely packed baled cotton, was exposed to ignition from a cigarette and a match and performed very well: no propagating combustion with either.
4. Cotton, as densely packed baled cotton, was exposed to ignition from the gas burner source in ASTM E1590 (also known as California Technical Bulletin 129) of 12 L/min of propane gas for 180 seconds and passed all the criteria, including mass loss of less than 1.36 kg (3 pounds), heat release rate less than 100 kW and total heat release of less than 25 MJ in the first 10 minutes of the test.

[BS] BALLAST. In roofing, ballast comes in the form of large stones or paver systems or light-weight interlocking paver systems and is used to provide uplift resistance for roofing systems that are not adhered or mechanically attached to the *roof deck.*

❖ Ballast is a material, usually stone or concrete, that is used specifically to provide uplift resistance for roof coverings that are not otherwise attached to the structure.

[F] BARRICADE. A structure that consists of a combination of walls, floor and roof, which is designed to withstand the rapid release of energy in an *explosion* and which is fully confined, partially vented or fully vented; or other effective method of shielding from explosive materials by a natural or artificial barrier.

Artificial barricade. An artificial mound or revetment a minimum thickness of 3 feet (914 mm).

Natural barricade. Natural features of the ground, such as hills, or timber of sufficient density that the surrounding exposures that require protection cannot be seen from the magazine or building containing explosives when the trees are bare of leaves.

❖ Barricade means effectively screening a building containing explosives by means of natural or artifical barrier from a magazine, another building, a railway or a highway. When suitable natural features do not exist to protect adjacent people and structures from flying debris and blast effects if an explosion in a magazine occurs, an artificial barrier must be constructed. For further information, see Chapter 56 of the IFC.

[BS] BASE FLOOD. The *flood* having a 1-percent chance of being equaled or exceeded in any given year.

❖ This term is used to define the land area along a body of water that is subject to flooding and within which flood-resistant design and construction requirements are applied. Typically, an authority, such as FEMA or a state or local jurisdiction, prepares flood hazard maps. In doing so, the base flood is derived by applying hydrologic models or by examining regional or local flood history to determine the 1-percent annual chance event, also referred to as the "100-year flood." The 1-percent annual chance event has been the standard used to regulate flood hazard areas for more than 40 years.

[BS] BASE FLOOD ELEVATION. The elevation of the *base flood*, including wave height, relative to the National

Geodetic Vertical Datum (NGVD), North American Vertical Datum (NAVD) or other datum specified on the *Flood Insurance Rate Map* (FIRM).

❖ The base flood elevation is the height to which floodwaters are predicted to rise during passage or occurrence of the base flood. It is determined using commonly accepted computer models that estimate hydrologic and hydraulic conditions to determine the 1-percent annual chance (base) flood. Along rivers and streams, statistical methods and computer models may be used to estimate runoff and to develop flood elevations. The models take into consideration watershed characteristics, and the shape and nature of the flood plain, including natural ground contours and the presence of buildings, bridges and culverts. Along coastal areas, base flood elevations may be developed using models that take into account offshore bathymetry, historical storms and typical wind patterns. In many coastal areas, the base flood elevation includes wave heights.

[BS] BASEMENT (for flood loads). The portion of a building having its floor subgrade (below ground level) on all sides. This definition of "Basement" is limited in application to the provisions of Section 1612.

❖ ASCE 24 specifies that lowest floors, including the floors of basements that are below grade on all sides, must be elevated to or above the elevations required by ASCE 24, which vary based on the occupancy category and flood zone. Buildings that are specifically designed to be dry floodproofed in accordance with ASCE 24 may have their lowest floors (including basements) below the design flood elevation. In terms of National Flood Insurance Program (NFIP) flood insurance, buildings with basements (below grade on all sides) are subject to higher premium rates.

BASEMENT. A *story* that is not a *story above grade plane* (see "*Story above grade plane*"). This definition of "Basement" does not apply to the provisions of Section 1612 for flood *loads*.

❖ Unlike previous editions of the I-Codes, where a story was defined as a basement if any portion of the story was below grade, a basement is now defined as a story that has its floor surface below the adjoining ground level and that does not qualify as a story above grade plane (see the commentary to the definition of "Story above grade plane"). Commentary Figure 202(7) illustrates the application of the definition of "Story above grade plane." Since a basement is not a story above grade, it does not contribute to the height of the building for the purpose of applying the allowable building height in stories from Table 504.4. This definition of "Basement" applies to all sections of the code except for flood loads.

Basements in buildings that are located in flood hazard areas and subject to flood loads are defined differently than this general definition [see the definition of "Basement (for flood loads)"].

BEARING WALL STRUCTURE. A building or other structure in which vertical *loads* from floors and roofs are primarily supported by walls.

❖ This definition describes the structure types that are covered under Section 1615.4.

[BS] BED JOINT. The horizontal layer of *mortar* on which a *masonry unit* is laid.

❖ This is a horizontal mortar joint [see Commentary Figure 202(8)] that separates a course of masonry units from the ones above and supports the weight of the masonry. Unlike the head or collar joint, it is easily closed when the masonry unit is placed. For a masonry unit in the typical (stretcher) orientation, the bed joint faces are the top and bottom, while the bed surface of the masonry unit is the underside. A special type of bed joint is the base-course joint or starting joint placed over foundations.

BLEACHERS. Tiered seating supported on a dedicated structural system and two or more rows high and is not a building element (see "*Grandstand*").

❖ Bleachers, folding and telescopic seating, and grandstands are essentially unique forms of tiered seating that are supported on a dedicated structural system. All types are addressed in ICC 300, *Standard on Bleach-*

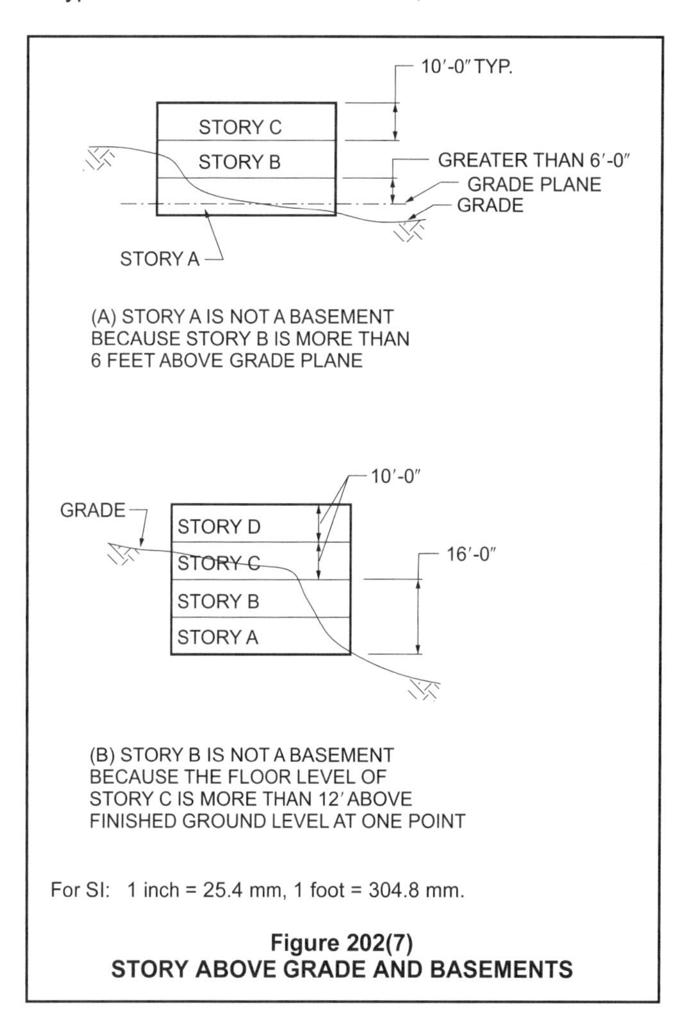

Figure 202(7)
STORY ABOVE GRADE AND BASEMENTS

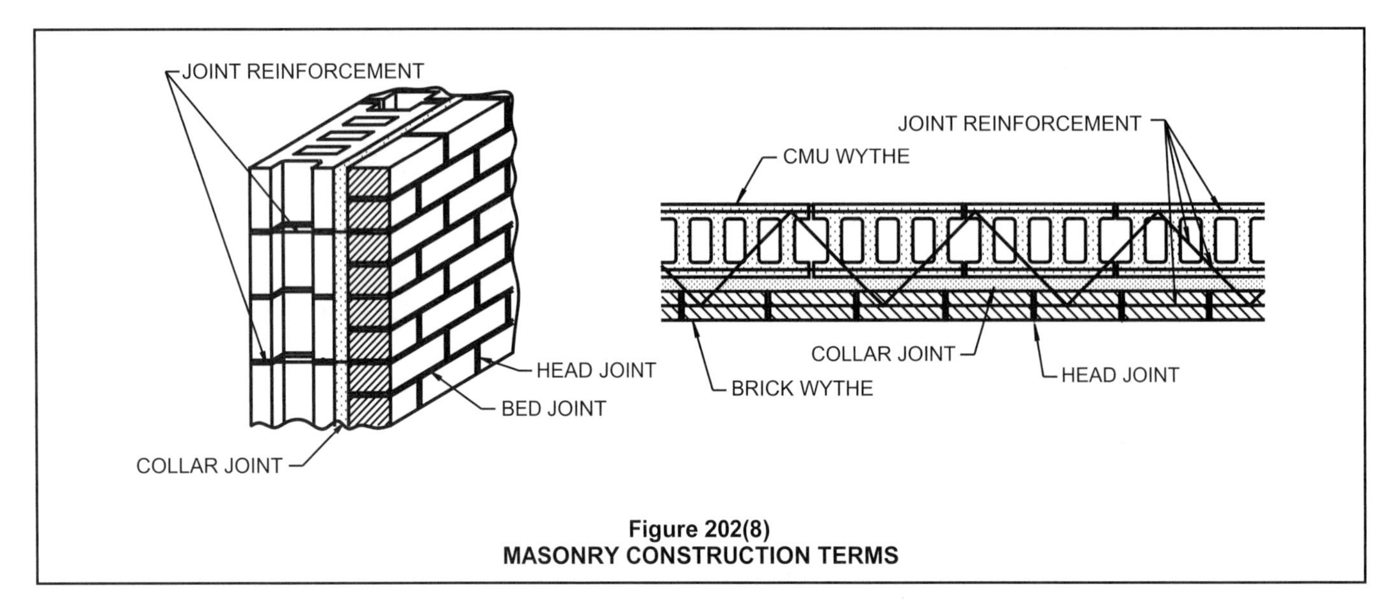

Figure 202(8)
MASONRY CONSTRUCTION TERMS

ers, Folding and Telescopic Seating and Grandstands, the safety standard for these types of seating arrangements (see Section 1029.1.1). Bleachers can have backrests or just seatboards. "Building element" is a defined term that is primarily used in conjunction with the structural elements regulated in Table 601. Bleachers have a separate structural system and are not considered a building element of the building or structure in which they are located. An individual bench seat directly attached to a floor system is not a bleacher. The terms "bleacher" and "grandstand" are basically interchangeable. There is no cutoff in size or number of seats that separates bleachers and grandstands.

BOARDING HOUSE. A building arranged or used for lodging for compensation, with or without meals, and not occupied as a single-family unit.

❖ A boarding house is a structure housing boarders in which the occupants are provided sleeping accommodations or meals and accommodations for a fee. The individual rooms used usually do not contain all of the permanent living provisions of a dwelling unit (e.g., permanent cooking facilities). Most often, the term "boarding house" describes a facility that is primarily for transient occupants; however, these facilities might also be used for nontransient purposes. Depending on the extent of transiency, a boarding house could be classified as Group R-1 when an occupant typically stays for not more than 30 days or Group R-2 when the length of stay is greater than 30 days [see Section 310 and Commentary Figure 202(9)]. Boarding houses are distinct from Lodging houses. Lodging houses allow transient guests within a residence that is also occupied by the owner or primary residents of the dwelling unit. Lodging houses are classified as Group R-3.

[F] BOILING POINT. The temperature at which the vapor pressure of a *liquid* equals the atmospheric pressure of 14.7 pounds per square inch (psia) (101 kPa) or 760 mm of mercury. Where an accurate boiling point is unavailable for the material in question, or for mixtures which do not have a constant boiling point, for the purposes of this classification, the 20-percent evaporated point of a distillation performed in accordance with ASTM D86 shall be used as the boiling point of the *liquid*.

❖ The boiling point of a liquid is significant in determining the appropriate division for Class I flammable liquids. Temperatures above the established boiling point for a given liquid would result in the atmospheric pressure no longer being able to keep the liquid in a liquid state. Liquids with low boiling points present a greater fire hazard because of the increased vapor pressure at normal ambient temperatures.

[BS] BRACED WALL LINE. A straight line through the building plan that represents the location of the lateral resistance provided by the wall bracing.

❖ A braced wall line consists of conventionally framed braced wall panels that provide lateral support for buildings constructed in accordance with the prescriptive conventional construction provisions of Section 2308.

[BS] BRACED WALL PANEL. A full-height section of wall constructed to resist in-plane shear loads through interaction of framing members, sheathing material and anchors. The panel's length meets the requirements of its particular bracing method and contributes toward the total amount of bracing required along its *braced wall line*.

❖ A braced wall panel is a segment of a braced wall line that is constructed in accordance with one of the bracing methods that are permitted in Section 2308. Braced wall panels are discussed in Section 2308.6. Braced wall panels are similar to shear walls but are designed by prescriptive rules rather than engineered like shear walls.

BREAKOUT. For revolving doors, a process whereby wings or door panels can be pushed open manually for *means of egress* travel.

❖ In addition to the swinging doors in the immediate area, revolving doors have a breakout feature as an additional safety requirement. The panels in the door can be operated manually to collapse or fold in the direction of egress during an emergency. This should increase the number of people that could exit per minute compared to using the revolving door in the standard manner.

[BS] BRICK.

❖ Brick is composed of masonry units that are generally prismatic (rectangular) in shape.

Calcium silicate (sand lime brick). A pressed and subsequently autoclaved unit that consists of sand and lime, with or without the inclusion of other materials.

❖ This solid brick unit is made principally from high-silica sand and lime.

Clay or shale. A solid or hollow *masonry unit* of clay or shale, usually formed into a rectangular *prism*, then burned or fired in a kiln; brick is a ceramic product.

❖ These masonry units are manufactured from surface clay, shale or fire clay. Different manufacturing processes and physical properties are associated with each material. Surface clays are found in sedimentary layers near the surface. Shales are clays subjected to geologic pressure, resulting in a solid state similar to slate. Fire clays are mined from deeper layers, resulting in more uniform properties. These units are formed into the desired shape by extrusion, molding or pressing. They are then fired in a kiln to increase their strength and durability.

Concrete. A concrete *masonry unit* made from Portland cement, water, and suitable aggregates, with or without the inclusion of other materials.

❖ Concrete brick units are made from a zero-slump mix of Portland cement (and possibly other cementitious materials), aggregates, water and admixtures. These units are solid or have a shallow depression called a "frog." Slump brick, for example, is a decorative concrete brick with bulged sides resulting from the consistency of the mix and the manufacturing process.

[A] BUILDING. Any structure used or intended for supporting or sheltering any use or occupancy.

❖ The code uses this term to identify those structures that provide shelter for a function or activity. See the definition for "Area, building" for situations when a single structure may be two or more "Buildings" created by fire walls.

BUILDING AREA. See "Area, building."

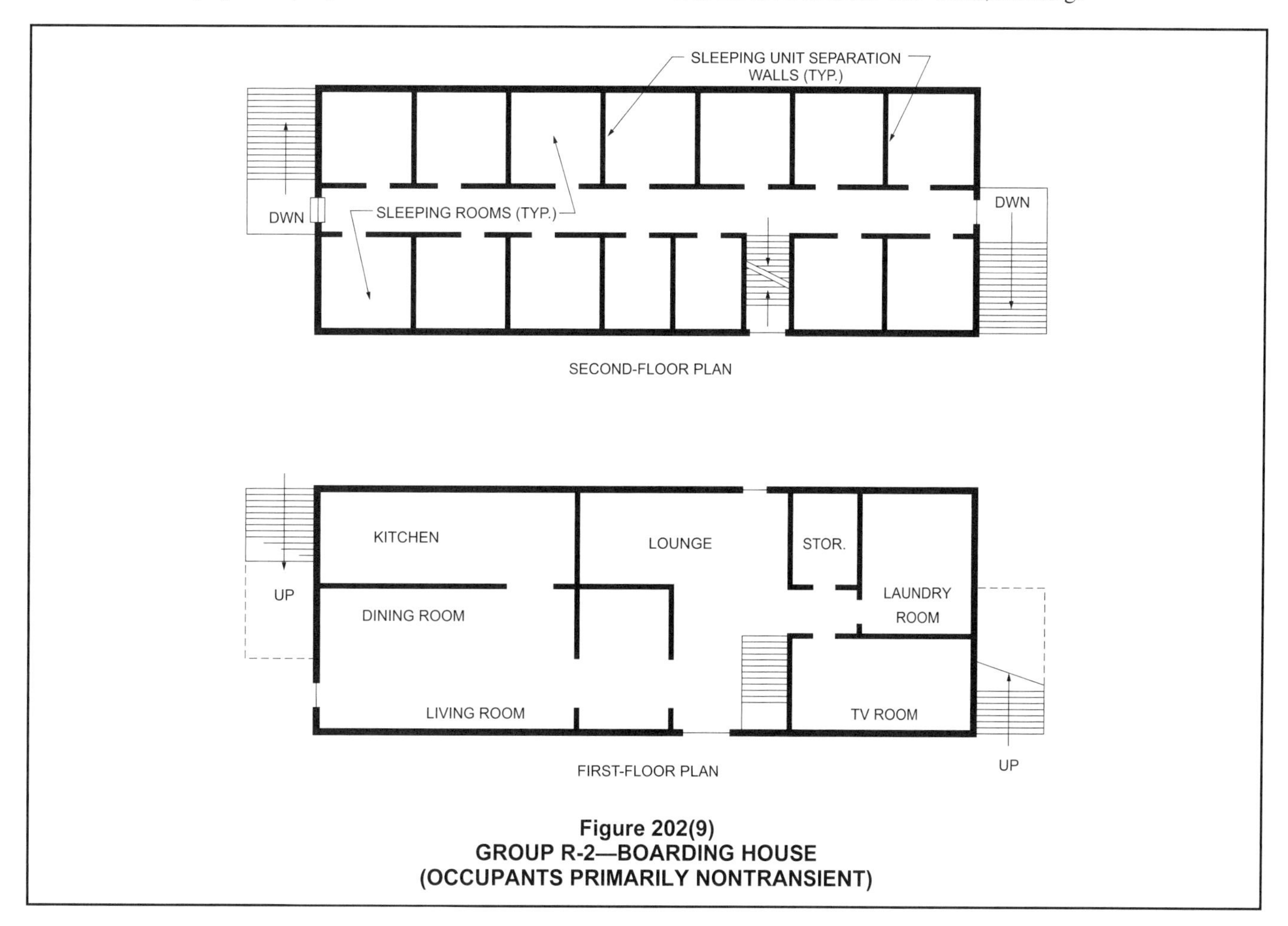

Figure 202(9)
GROUP R-2—BOARDING HOUSE (OCCUPANTS PRIMARILY NONTRANSIENT)

BUILDING ELEMENT. A fundamental component of building construction, listed in Table 601, which may or may not be of fire-resistance-rated construction and is constructed of materials based on the building type of construction.

❖ Building elements include primary structural members, secondary structural members, exterior walls and interior partitions. The fire-resistance ratings of the elements are found in Tables 601 and 602.

BUILDING HEIGHT. See "Height, building."

BUILDING-INTEGRATED PHOTOVOLTAIC (BIPV) PRODUCT. A building product that incorporates photovoltaic modules and functions as a component of the building envelope.

❖ This definition clarifies which photovoltaic products must comply with the roof covering fire classifications required in Section 1505.

BUILDING LINE. The line established by law, beyond which a building shall not extend, except as specifically provided by law.

❖ This term defines the limitations or boundaries for construction of a building. This line is typically established by a zoning statute or rights-of-way dedication and is not specified in the code.

[A] BUILDING OFFICIAL. The officer or other designated authority charged with the administration and enforcement of this code, or a duly authorized representative.

❖ The statutory power to enforce the code is normally vested in a building department of a state, county or municipality that has a designated enforcement officer termed the "building official" (see Section 103.1).

For the code and IRC, the "building official" is identified as the designated enforcement officer. For the IFC, the "fire code official" is identified as the designated enforcement officer. For other I-Codes, the "code official" is identified as the designated enforcement officer.

[BS] BUILT-UP ROOF COVERING. Two or more layers of felt cemented together and surfaced with a cap sheet, mineral *aggregate*, smooth coating or similar surfacing material.

❖ Because of their low melting points and self-healing characteristics, built-up roofs are typically constructed of coal tar membranes that are commonly installed on lower slopes and dead-level roofs.

CABLE-RESTRAINED, AIR-SUPPORTED STRUCTURE. A structure in which the uplift is resisted by cables or webbings which are anchored to either foundations or dead men. Reinforcing cable or webbing is attached by various methods to the membrane or is an integral part of the membrane. This is not a cable-supported structure.

❖ This definition establishes a variation of the air-supported membrane structure. A single-skin or double-skin membrane is still pressurized by interior air, but does not have the strength or tearing resistance against the internal air pressure. A "fishnet" system of cable wires is placed over the exterior surface of the membrane or is integrally woven into the membrane material. The "fishnet" system is then tied to exterior walls of the building or other structural supports to resist the internal air pressure.

CANOPY. A permanent structure or architectural projection of rigid construction over which a covering is attached that provides weather protection, identity or decoration. A canopy is permitted to be structurally independent or supported by attachment to a building on one or more sides.

❖ A canopy can be either an architectural projection from a building, or it can be an independent structure. An example of the former is typically found covering an entrance walkway in front of a hotel or apartment building, or perhaps a fancy restaurant (see the figures in Chapter 31). An example of the latter is a canopy built over fuel pumps at a gasoline station. This definition distinguishes a canopy from an awning, which consists of a lightweight frame that is supported by the building from which it projects. In doing so, the definition clarifies the applicability of roof loads in Section 1607.12.4 (see Section 3105 for general requirements and Section 3202 for encroachment requirements).

[F] CARBON DIOXIDE EXTINGUISHING SYSTEMS. A system supplying carbon dioxide (CO_2) from a pressurized vessel through fixed pipes and nozzles. The system includes a manual- or *automatic*-actuating mechanism.

❖ Carbon dioxide (CO_2) extinguishing systems are useful in extinguishing fires in specific hazards or equipment in occupancies where an inert electrically nonconductive medium is essential or desirable and where cleanup of other extinguishing agents, such as dry-chemical residue, presents a problem. The system works by displacing the oxygen in an enclosed area by flooding the space with CO_2. To effectively flood the enclosure, automatic door and window closers and control dampers for the mechanical ventilation system must be installed.

These types of gaseous extinguishing systems have some inherent disadvantages that should be considered before selection. Because the oxygen is being displaced, occupants should not be in the space for a period after discharge, depending on the concentration of CO_2 to be achieved. Additionally, the discharge rate can result in a rapid increase in pressure within the space where the system is discharged. Where water is not a desired means of suppression, CO_2 and other gaseous suppression systems can be very effective. NFPA 12 contains minimum requirements for the design, installation, testing, inspection, approval, operation and maintenance of carbon dioxide extinguishing systems.

CARE SUITE. In Group I-2 occupancies, a group of treatment rooms, care recipient sleeping rooms and the support rooms or spaces and circulation space within the suite where staff are in attendance for supervision of all care recipients within the suite, and the suite is in compliance with the requirements of Section 407.4.4.

❖ Care suites are designed to allow for a group of rooms to function as a unit in the treatment and care of patients. Suites are recognized to be an effective tool to provide flexibility in reaching an exit access due to functional considerations. Use of suites is a particularly useful tool at intensive care units and emergency departments in patient treatment areas. Suites allow staff to have clear and unobstructed supervision of patients/care recipients in specific treatment and sleeping rooms through the elimination of corridor width or rating requirements. The term is not intended to apply to day rooms or business sections of the hospital. This term is only applicable to suites of patient rooms in Group I-2 occupancies, and should not be confused with similar layouts in other parts of the hospital or within other occupancies that may be referred to as a "suite." Care suites are to meet the requirements of Section 407.

[BS] CAST STONE. A building stone manufactured from Portland cement concrete precast and used as a *trim*, *veneer* or facing on or in buildings or structures.

❖ Cast stone is a simulated stone precast from Portland cement concrete. This material is typically used for veneer, but can also be used in other applications.

[F] CEILING LIMIT. The maximum concentration of an air-borne contaminant to which one may be exposed. The ceiling limits utilized are those published in DOL 29 CFR Part 1910.1000. The ceiling Recommended Exposure Limit (REL-C) concentrations published by the U.S. National Institute for Occupational Safety and Health (NIOSH), Threshold Limit Value—Ceiling (TLV-C) concentrations published by the American Conference of Governmental Industrial Hygienists (ACGIH), Ceiling Workplace Environmental Exposure Level (WEEL-Ceiling) Guides published by the American Industrial Hygiene Association (AIHA), and other *approved*, consistent measures are allowed as surrogates for hazardous substances not lissZted in DOL 29 CFR Part 1910.1000.

❖ This term is used in Section 908.3 and indicates the threshold at which a gas detection system is required for highly toxic and toxic materials. It represents the maximum level of exposure for employees or occupants to hazardous air contaminants during any part of a normal workday. DOL 29 CFR Part 1910.1000 provides acceptable ceiling limits of contamination for various substances.

CEILING RADIATION DAMPER. A *listed* device installed in a ceiling membrane of a fire-resistance-rated floor/ceiling or roof/ceiling assembly to limit *automatically* the radiative heat transfer through an air inlet/outlet opening. Ceiling radiation dampers include air terminal units, ceiling dampers and ceiling air diffusers.

❖ See the commentary to the definition of "Damper."

CELL (Group I-3 occupancy). A room within a housing unit in a detention or correctional facility used to confine inmates or prisoners.

❖ Cells are the smallest unit in the portion of a Group I-3 prison or jail for the housing of inmates. In contrast to other occupancies, many of the provisions in Section 408 refer to cells. Requirements for Group I-3 occupancies in other chapters may refer to cells that are parts of housing units as "sleeping units."

[BS] CELL (masonry). A void space having a gross cross-sectional *area* greater than $1^1/_2$ square inches (967 mm^2).

❖ This term defines a large intentional void within a masonry unit. Grout and reinforcing steel are often placed in cells to form reinforced masonry.

CELL TIER. Levels of *cells* vertically stacked above one another within a *housing unit*.

❖ A tier is one level of cells. There can be multiple tiers within a single housing unit as well as more than one cell tier within a story [see the commentary for "Cell (Group I-3 occupancy)"].

[BS] CEMENT PLASTER. A mixture of Portland or blended cement, Portland cement or blended cement and hydrated lime, masonry cement or plastic cement and aggregate and other *approved* materials as specified in this code.

❖ Cement plaster (often referred to as "stucco") is a cementitious-based plaster material with excellent water-resistant properties. It is the only type of plaster that is permitted by the code to be used as an exterior wall covering and as a base coat and finish coat. Cement plaster can also be used as an interior finish wall covering, but is required by the code to be used in all interior wet areas such as toilet rooms, showers, saunas, steam rooms, indoor swimming pools or any other area that will be exposed to excessive amounts of moisture or humidity for prolonged periods of time.

CERAMIC FIBER BLANKET. A high-temperature *mineral wool* insulation material made of alumina-silica ceramic or calcium magnesium silicate soluble fibers and weighing 4 to 10 pounds per cubic foot (pcf) (64 to 160 kg/m^3).

❖ This form of insulation is used in conjunction with the provisions of Section 722.2.1.3.1, where joints in precast walls must be insulated (protected) in order to maintain the fire-resistance integrity of the precast wall panel (see Code Figure 722.2.1.3.1).

CERTIFICATE OF COMPLIANCE. A certificate stating that materials and products meet specified standards or that work was done in compliance with *approved construction documents*.

❖ A certificate of compliance is a document issued by a supplier of materials and products that certifies they meet the specified requirements.

[A] CHANGE OF OCCUPANCY. A change in the purpose or level of activity within a building that involves a change in application of the requirements of this code.

❖ This term describes the condition when an existing building or structure is used for a different use or the same use with an increase in the level of activity (see Section 101.4.7 for a reference to the IEBC). This term is only applicable to existing buildings, never new ones. For example, Group B includes both beauty parlors and post offices. If a beauty shop moved into an old post office, while remaining a Group B building, it would be a change in the level of activity; therefore, this would be considered a change of occupancy.

[M] CHIMNEY. A primarily vertical structure containing one or more flues, for the purpose of carrying gaseous products of combustion and air from a fuel-burning appliance to the outdoor atmosphere.

❖ The IFGC and the IMC address the installation of chimneys and venting systems that are required to convey products of combustion from fuel-burning appliances to the atmosphere. The code regulates the construction of masonry chimneys in Section 2113. Chimneys differ from metal vents in the materials from which they are constructed and the type of appliance they are designed to serve. Chimneys can vent much hotter flue gases than metal vents.

Factory-built chimney. A *listed* and *labeled chimney* composed of factory-made components, assembled in the field in accordance with manufacturer's instructions and the conditions of the listing.

❖ A factory-built chimney is a manufactured, listed and labeled chimney that has been tested by an approved agency to determine its performance characteristics. Factory-built chimneys are manufactured in two basic designs: a double-wall insulated design, or a triple-wall air-cooled design. Both designs use stainless steel inner liners to resist the corrosive effects of combustion products.

Masonry chimney. A field-constructed *chimney* composed of solid masonry units, bricks, stones, or concrete.

❖ Masonry chimneys can have one or more flues within them, and are field constructed of brick, stone, concrete or fire-clay materials. Masonry chimneys can stand alone or be part of a masonry fireplace.

Metal chimney. A field-constructed *chimney* of metal.

❖ A metal chimney is an unlisted chimney constructed and installed in accordance with NFPA 211 and is sometimes referred to as a 'smokestack.' Metal chimneys are typically field constructed and installed in industrial structures.

[M] CHIMNEY TYPES.

❖ Provisions for several types of chimneys are contained in Chapter 21, as described below. The construction of the chimney is affected by the type of appliance and the temperature of the flue gases that are generated.

High-heat appliance type. An *approved* chimney for removing the products of combustion from fuel-burning, high-heat appliances producing combustion gases in excess of 2000°F (1093°C) measured at the appliance flue outlet (see Section 2113.11.3).

❖ High-heat chimneys are used in industrial applications, such as incinerators, kilns and blast furnaces. Section 2113.11.3 contains requirements for the construction and installation of chimneys for high-heat appliances.

Low-heat appliance type. An *approved* chimney for removing the products of combustion from fuel-burning, low-heat appliances producing combustion gases not in excess of 1000°F (538°C) under normal operating conditions, but capable of producing combustion gases of 1400°F (760°C) during intermittent forces firing for periods up to 1 hour. Temperatures shall be measured at the appliance flue outlet.

❖ Most domestic fuel-burning appliances are low-heat appliances. Low-heat appliances include solid fuel-burning appliances, such as room heaters and wood stoves. Section 2113 contains requirements for the construction and installation of chimneys for low-heat appliances.

Masonry type. A field-constructed chimney of solid *masonry units* or stones.

❖ Masonry chimneys can have one or more flues and are field constructed of masonry units, stone, concrete and fired-clay materials. Masonry chimneys can stand alone or be part of a masonry fireplace. Section 2113 contains requirements for the construction and installation of masonry chimneys.

Most masonry chimneys require a chimney liner that is resistant to heat and the corrosive action of the products of combustion. Chimney liners are generally made of fired-clay tile, refractory brick, poured-in-place refractory materials or stainless steel.

Medium-heat appliance type. An *approved* chimney for removing the products of combustion from fuel-burning, medium-heat appliances producing combustion gases not exceeding 2000°F (1093°C) measured at the appliance flue outlet (see Section 2113.11.2).

❖ Some examples of medium-heat appliances are annealing furnaces, galvanizing furnaces, pulp dryers and charcoal furnaces. Section 2113.11.2 contains requirements for the construction and installation of chimneys for medium-heat appliances.

CIRCULATION PATH. An exterior or interior way of passage from one place to another for pedestrians.

❖ Examples of circulation paths include sidewalks, walkways, corridors, aisles, courtyards, ramps, stairways and landings. While a stairway is never part of an accessible route, it can be part of a general circulation path for ambulatory persons. Not all floor areas are circulation paths. What part of a floor is not a circulation path will be subjective. For example, if a drinking fountain is moved into an alcove, it has been moved off the circulation path. So, while still being over a floor, it is no longer considered a protruding object.

[F] CLEAN AGENT. Electrically nonconducting, volatile or gaseous fire extinguishant that does not leave a residue upon vaporation.

❖ The two categories of clean agents are halocarbon compounds and inert gas agents. Halocarbon compounds include bromine, carbon, chloride, fluorine, hydrogen and iodine. Halocarbon compounds suppress the fire through a combination of breaking the chemical chain reaction of the fire, reducing the ambient oxygen supporting the fire and reducing the ambient temperature of the fire origin to reduce the propagation of fire. The clean agents that are inert gas agents contain primary components consisting of helium, neon or argon, or a combination of all three. Inert gases work by reducing the oxygen concentration around the fire origin to a level that does not support combustion (see commentary, Section 904.10).

[E] CLIMATE ZONE. A geographical region that has been assigned climatic criteria as specified in Chapters 3CE and 3RE of the *International Energy Conservation Code*.

❖ Each region is assigned a climate zone. The IECC provides a list of each county in the United States with the designated climate zone for each. These designations are also mapped. Climate zones for other regions are described. Climate zone is used to determine the requirements for a building's thermal envelope including insulation, fenestration and thermal barriers.

CLINIC, OUTPATIENT. Buildings or portions thereof used to provide *medical care* on less than a 24-hour basis to persons who are not rendered *incapable of self-preservation* by the services provided.

❖ Outpatient clinics generally consist of doctors' offices where various medical services can be provided. These clinics typically function during normal business hours (i.e., less than 24 hours) and, unlike ambulatory care facilities, the patients are generally ambulatory and capable of self-preservation. This definition clarifies the difference between ambulatory surgery centers (ambulatory care facilities) and the typical doctor's office. In many cities, outpatient clinics are open at all hours to be available to people who work a variety of shifts. The term "urgent care" is often used to describe such facilities. An outpatient facility that is open 24/7 may still be classified as a Group B occupancy, provided all patients are outpatients and individual patients are not treated for periods in excess of 24 hours. The latter would describe a Group I-2 hospital (see the commentaries to Sections 407 and 423).

[F] CLOSED SYSTEM. The *use* of a *solid* or *liquid hazardous material* involving a closed vessel or system that remains closed during normal operations where vapors emitted by the product are not liberated outside of the vessel or system and the product is not exposed to the atmosphere during normal operations; and all *uses* of *compressed gases*. Examples of closed systems for *solids* and *liquids* include product conveyed through a piping system into a closed vessel, system or piece of equipment.

❖ The difference between a closed system and an open system is whether the hazardous material involved in the process is exposed to the atmosphere. While not specific in the definition, certain gases are also allowed in closed systems, as indicated in Tables 307.7(1) and 307.7(2). Materials in closed or open systems are assumed to be "in use" as opposed to "in storage." Gases are always assumed to be in closed systems, since they would be immediately dispersed in an open system if exposed to the atmosphere without some means of containment (see the definition of "Open system").

[BS] COASTAL A ZONE. Area within a *special flood hazard area*, landward of a V zone or landward of an open coast without mapped *coastal high-hazard areas*. In a coastal A zone, the principal source of flooding must be astronomical tides, storm surges, seiches or tsunamis, not riverine flooding. During the base flood conditions, the potential for breaking wave height shall be greater than or equal to $1^1/_2$ feet (457 mm). The inland limit of the coastal A zone is (a) the Limit of Moderate Wave Action if delineated on a FIRM, or (b) designated by the authority having jurisdiction.

❖ For years, post-disaster investigations have identified wave-related damage in flood zones other than coastal high-hazard areas, especially in flood hazard areas that are immediately inland of coastal high-hazard areas (Zone V). If a Coastal A Zone is shown on the Flood Insurance Rate Map by the delineation of the inland extent of the $1^1/_2$-foot wave, or if a community designates such a zone by other means (such as a certain distance inland of the Zone V boundary), then ASCE 24 (referenced standard), requires buildings and structures in Coastal A Zones to be designed and constructed in accordance with the requirements for coastal high-hazard areas.

[BS] COASTAL HIGH-HAZARD AREA. Area within the *special flood hazard area* extending from offshore to the inland limit of a primary dune along an open coast and any other area that is subject to high-velocity wave action from storms or seismic sources, and shown on a Flood Insurance Rate Map (FIRM) or other flood hazard map as velocity Zone V, VO, VE or V1-30.

❖ Some coastal and lake shorelines experience flooding that includes wind-driven waves. The presence of waves and the potential for local scour must be taken into consideration when determining flood loads. Coastal high-hazard areas are specifically designated on FEMA's flood hazard area maps and commonly are referred to as "V Zones." In these areas, wave heights are predicted to be greater than or equal to 3 feet (914 mm) or the wave runup elevations are predicted to reach 3 feet (914 mm) or more above grade.

[BS] COLLAR JOINT. Vertical longitudinal space between *wythes* of *masonry* or between *masonry wythe* and backup construction that is permitted to be filled with *mortar* or grout.

❖ A collar joint is a filled space between masonry wythes [see Commentary Figure 202(8)]. Care is necessary for proper construction of collar joints, particularly where solid filling is required.

[BS] COLLECTOR. A horizontal *diaphragm* element parallel and in line with the applied force that collects and transfers *diaphragm* shear forces to the vertical elements of the lateral force-resisting system or distributes forces within the *diaphragm*, or both.

❖ The collector is an element of the diaphragm that transfers the diaphragm shear from the point of application into the adjoining vertical lateral force-resisting elements, usually the shear walls.

COMBINATION FIRE/SMOKE DAMPER. A *listed* device installed in ducts and air transfer openings designed to close *automatically* upon the detection of heat and resist the passage of flame and smoke. The device is installed to operate *automatically*, controlled by a smoke detection system, and where required, is capable of being positioned from a *fire command center*

❖ A combination damper is used when the code requires not only a fire damper but also a smoke damper designed to limit the passage of smoke from one side of fire-resistance-rated construction to the other. Fire and smoke dampers are required at duct penetrations of shafts in accordance with Section 717.5.3. Both fire and smoke dampers are required at duct and air transfer openings in fire walls utilized as horizontal exits in accordance with Section 717.5.1.1. Subject to the exceptions listed, both shall also be required at duct penetrations of fire barrier walls utilized as horizontal exits in accordance with Section 717.5.2.1. The combination fire/smoke damper must be actuated automatically by a smoke detection system. Where the damper is part of a smoke control system in Section 909, it shall also be controlled from the fire command center.

[F] COMBUSTIBLE DUST. Finely divided *solid* material that is 420 microns or less in diameter and which, when dispersed in air in the proper proportions, could be ignited by a flame, spark or other source of ignition. Combustible dust will pass through a U.S. No. 40 standard sieve.

❖ Combustible dusts are combustible solids in a finely divided state that are suspended in the air. An explosion hazard exists when the concentration of the combustible dust is within the explosive limits and exposed to an ignition source of sufficient energy and duration to initiate self-sustained combustion. A review of the occupancy classification for Group H-2 in Section 307.4 indicates that combustible dusts are classified in that occupancy group. To apply this portion of Section 307.4, a determination of the deflagration hazard of the manufactured, generated or used combustible dust is required. This is typically determined by an engineering analysis based on the combustible dust dispersion and proportion. When the analysis determines that this dispersion and proportion of combustible dust can be ignited by an ignition source, then an occupancy classification of Group H-2 would be appropriate. Combustible dust that, as a material, does not rise to the defined level of hazard in a particular building, would not cause the building or portion thereof housing the hazard to be classified in Group H-2, but rather in the occupancy group that is most appropriate for the particular operation.

The original tabular maximum allowable quantity (MAQ) per control area for combustible dust included in the legacy building and fire codes was deleted because of its questionable value given the complexities of dust explosion hazards. In the 2012 edition of the code, a row for combustible dust was added to Table 307.1(1) along with a new Note q. The note reinforces the fact that determining a theoretical maximum allowable quantity of combustible dust and the potential for a dust explosion requires a thorough evaluation and technical report based on the provisions of Section 414.1.3. Such determination is complex and requires evaluation far beyond the simple 1 pound per 1,000 cubic feet (16 g/m^3) maximum allowable quantity previously used by the legacy codes. Critical factors, such as particle size, material density, humidity and oxygen concentration, play a major role in the evaluation of the dust hazard and are much too complex to be simply addressed.

[F] COMBUSTIBLE FIBERS. Readily ignitable and free-burning materials in a fibrous or shredded form, such as cocoa fiber, cloth, cotton, excelsior, hay, hemp, henequen, istle, jute, kapok, oakum, rags, sisal, Spanish moss, straw, tow, wastepaper, certain synthetic fibers or other like materials. This definition does not include densely packed baled cotton.

❖ Operations involving combustible fibers are typically associated with paper milling, recycling, cloth manufacturing, carpet and textile mills and agricultural operations, among others. The primary hazards associated with such operations involve the abundance of materials and their ready ignitability. Many organic fibers are prone to spontaneous ignition if improperly dried and kept in areas without sufficient ventilation. Densely packed baled cotton is a special type of combustible fiber that, based on its weight and dimension requirements, is not easily ignitable and is not a hazardous material.

[F] COMBUSTIBLE LIQUID. A *liquid* having a closed cup *flash point* at or above 100°F (38°C). Combustible liquids shall be subdivided as follows:

Class II. *Liquids* having a closed cup *flash point* at or above 100°F (38°C) and below 140°F (60°C).

Class IIIA. *Liquids* having a closed cup *flash point* at or above 140°F (60°C) and below 200°F (93°C).

Class IIIB. *Liquids* having a closed cup *flash point* at or above 200°F (93°C).

The category of combustible liquids does not include *compressed gases* or *cryogenic fluids*.

❖ Combustible liquids differ from flammable liquids in that the closed cup flash point of all combustible liquids is at or above 100°F (38°C) (see the definition of "Flash point"). There are three categories of combustible liquids. The range of their closed cup flash point dictates the class of combustible liquid. The flash point range of 100°F (38°C) to 140°F (60°C) for Class II liquids is based on a possible indoor ambient temperature exceeding 100°F (38°C). Only a moderate degree of heating would be required to bring the liquid to its flash point in this type of condition. Class III liquids, which have flash points higher than 140°F (38°C), would require a significant heat source besides ambient temperature conditions to reach their flash point (see the definition of "Flammable liquid"). Class IIIA has a closed cup flash point range of 140°F (93°C). Class IIIB has a closed cup flash point at or above 200°F (93°C). Combustible liquids are primarily considered Group H-2 materials except for Class II and IIIA liquids that are considered Group H-3 when used or stored in normally closed containers or systems pressurized at less than 15 psig (103.4 kPa). Motor oil is a typical example of a Class IIIB combustible liquid. Note that Class IIIB liquids are not regulated to be classified as Group H per Table 307.1(1). While cryogenic fluids and compressed gases may be combustible, they are to be regulated separately from combustible liquids.

COMMERCIAL MOTOR VEHICLE. A motor vehicle used to transport passengers or property where the motor vehicle:

1. Has a gross vehicle weight rating of 10,000 pounds (4540 kg) or more; or
2. Is designed to transport 16 or more passengers, including the driver.

❖ This definition provides the necessary clarification as to what constitutes a "commercial vehicle." This term has often been misinterpreted in previous editions. These criteria are from the DOT regulations 49 CFR 390.5, and correlate with Section 1607.7. Where vehicles of this size are present in buildings greater than 5,000 square feet (464 m^2) in area, the code requires an automatic sprinkler system be provided in repair garages, in commercial parking garages and in Group S-1 storage occupancies (see Section 903.2).

COMMON PATH OF EGRESS TRAVEL. That portion of the *exit access* travel distance measured from the most remote point within a *story* to that point where the occupants have separate access to two *exits* or *exit access* doorways.

❖ The common path of egress travel is a concept used to refine travel distance criteria. A common path of travel is the route an occupant will travel where the one way in is also the one way out, similar to a dead-end corridor or single-exit suite. Once occupants reach a point where two different routes are available, and the two different routes continue to two separate exits, then common path of travel is finished. The length of a common path of egress travel is limited so that the means of egress path of travel provides a choice before the occupant has traveled an excessive distance (see Section 1006). This reduces the possibility that, although the exits are remote from one another, a single fire condition will render both paths unavailable. The common path of egress travel is part of the overall exit access travel distance. To be compliant, the path of egress must meet criteria for both common path of egress travel and exit access travel distance.

COMMON USE. Interior or exterior *circulation paths*, rooms, spaces or elements that are not for public use and are made available for the shared use of two or more people.

❖ Some buildings include areas that are restricted to employees only or where public access is limited. Common-use spaces may be part of employee work areas but do not include public-use spaces. Any space that is shared by two or more persons, such as copy areas, break rooms, toilet rooms or circulation paths, are common use areas. A grade school classroom would be another example of a common use space (see also the commentary for the definition of "Public-use areas" and "Employee work area").

[F] COMPRESSED GAS. A material, or mixture of materials, that:

1. Is a gas at 68°F (20°C) or less at 14.7 pounds per square inch atmosphere (psia) (101 kPa) of pressure; and
2. Has a *boiling point* of 68°F (20°C) or less at 14.7 psia (101 kPa) which is either liquefied, nonliquefied or in solution, except those gases which have no other health- or physical-hazard properties are not considered to be compressed until the pressure in the packaging exceeds 41 psia (282 kPa) at 68°F (20°C).

The states of a compressed gas are categorized as follows:

1. Nonliquefied compressed gases are gases, other than those in solution, which are in a packaging under the charged pressure and are entirely gaseous at a temperature of 68°F (20°C).
2. Liquefied compressed gases are gases that, in a packaging under the charged pressure, are partially *liquid* at a temperature of 68°F (20°C).
3. Compressed gases in solution are nonliquefied gases that are dissolved in a solvent.

4. Compressed gas mixtures consist of a mixture of two or more compressed gases contained in a packaging, the hazard properties of which are represented by the properties of the mixture as a whole.

❖ This term refers to all types of gases that are under pressure at normal room or outdoor temperatures inside their containers, including, but not limited to, flammable, nonflammable, highly toxic, toxic, cryogenic and liquefied gases. The vapor pressure limitations provide the distinction between a liquid and a gas. Gases are materials that boil at a temperature of 68°F (20°C) or less at a pressure of 14.7 psia (101.3 kPa). Liquefied and nonliquefied compressed gases are determined by the state of the gas at a temperature of 68°F (20°C). Nonliquefied gases are entirely gaseous, while liquefied gases are partially liquid.

[BS] CONCRETE.

Carbonate aggregate. Concrete made with aggregates consisting mainly of calcium or magnesium carbonate, such as limestone or dolomite, and containing 40 percent or less quartz, chert or flint.

❖ Concrete made with this type of aggregate is listed in Tables 721.1(1) and 721.1(2) (Item 4) with prescriptive details regarding concrete fire-resistance ratings. This type of aggregate is also indicated for concrete assemblies that require calculations to determine the fire-resistance rating in accordance with Section 722.2 (i.e., Table 722.2.1.1). Concrete with this aggregate provides a lower fire resistance than others.

Cellular. A lightweight insulating concrete made by mixing a preformed foam with Portland cement slurry and having a dry unit weight of approximately 30 pcf (480 kg/m^3).

❖ Concrete made with this type of aggregate is listed in Table 721.1(3) with prescriptive details regarding concrete fire-resistance ratings. This type of aggregate is also indicated for concrete assemblies that require calculations to determine the fire-resistance rating in accordance with Section 722.2.1 [i.e., Table 722.2.1.2(1), Note a]. Concrete with this aggregate provides a higher fire resistance than others.

Lightweight aggregate. Concrete made with aggregates of expanded clay, shale, slag or slate or sintered fly ash or any natural lightweight aggregate meeting ASTM C330 and possessing equivalent fire-resistance properties and weighing 85 to 115 pcf (1360 to 1840 kg/m^3).

❖ Concrete made with this type of aggregate is listed in Tables 721.1(1), 721.1(2) and 721.1(3) with prescriptive details regarding concrete fire-resistance ratings. This type of aggregate is also indicated for concrete assemblies that require calculations to determine the fire-resistance rating in accordance with Section 722.2.1 (i.e., Table 722.2.1.1) (also see Section 702.1). Concrete with this aggregate provides a higher fire resistance than others.

Perlite. A lightweight insulating concrete having a dry unit weight of approximately 30 pcf (480 kg/m^3) made with perlite concrete aggregate. Perlite aggregate is produced from a volcanic rock which, when heated, expands to form a glass-like material of cellular structure.

❖ Concrete made with this type of aggregate is listed in Tables 721.1(1), 721.1(2) and 721.1(3) with prescriptive details regarding concrete fire-resistance ratings. This type of aggregate is also indicated for concrete assemblies that require calculations to determine the fire-resistance rating in accordance with Section 722.2.1 (i.e., Section 722.2.1.1.2). Concrete with this aggregate provides a higher fire resistance than others.

Sand-lightweight. Concrete made with a combination of expanded clay, shale, slag, slate, sintered fly ash, or any natural lightweight aggregate meeting ASTM C330 and possessing equivalent fire-resistance properties and natural sand. Its unit weight is generally between 105 and 120 pcf (1680 and 1920 kg/m^3).

❖ Concrete made with this type of aggregate is listed in Tables 721.1(1), 721.1(2) and 721.1(3) with prescriptive details regarding concrete fire-resistance ratings. This type of aggregate is also indicated for concrete assemblies that require calculations to determine the fire-resistance rating in accordance with Section 722.2.1 (i.e., Table 722.2.1.1) (also see Section 702.1). Concrete with this aggregate provides an average fire resistance compared to others.

Siliceous aggregate. Concrete made with normal-weight aggregates consisting mainly of silica or compounds other than calcium or magnesium carbonate, which contains more than 40-percent quartz, chert or flint.

❖ Concrete made with this type of aggregate is listed in Tables 721.1(1), 721.1(2) and 721.1(3) with prescriptive details regarding concrete fire-resistance ratings. This type of aggregate is also indicated for concrete assemblies that require calculations to determine the fire-resistance rating in accordance with Section 722.2.1 (i.e., Table 722.2.1.1). Concrete with this aggregate provides the lowest fire resistance.

Vermiculite. A light weight insulating concrete made with vermiculite concrete aggregate which is laminated micaceous material produced by expanding the ore at high temperatures. When added to a Portland cement slurry the resulting concrete has a dry unit weight of approximately 30 pcf (480 kg/m^3).

❖ Concrete made with this type of aggregate is listed in Tables 721.1(1), 721.1(2) and 720.1(3) with prescriptive details regarding concrete fire-resistance ratings. This type of aggregate is also indicated for concrete assemblies that require calculations to determine the fire-resistance rating in accordance with Section 722.2.1 [i.e., Table 722.2.1.2(1), Note a]. Concrete with this aggregate provides a higher fire resistance than others.

CONGREGATE LIVING FACILITIES. A building or part thereof that contains *sleeping units* where residents share bathroom or kitchen facilities, or both.

❖ Congregate living facilities are those pertaining to group housing (i.e., dormitories, fraternities, convents) that combine individual sleeping quarters with communal facilities for food, care, sanitation and recreation. The number of occupants in the facility determines the appropriate occupancy classification. There are two thesholds: 10 and 16. A congregate living facility with 16 or fewer nontransient residents falls in the R-3 classification. For above 16 nontransient residents, the classification is R-2. For transient residents, if there are 10 or fewer in the facility, it is also in the R-3 classification. If over 10 transient residents, it is an R-1 occupancy.

[F] CONSTANTLY ATTENDED LOCATION. A designated location at a facility staffed by trained personnel on a continuous basis where alarm or supervisory signals are monitored and facilities are provided for notification of the fire department or other emergency services.

❖ These locations are intended to receive trouble, supervisory and fire alarm signals transmitted by the fire protection equipment installed in a protected facility. It is the intent of Chapter 9 to have both an approved location and personnel who are acceptable to the code official responsible for actions taken when the fire protection system requires attention. The term "constantly attended" implies 24-hour surveillance of the system, at the designated location.

[A] CONSTRUCTION DOCUMENTS. Written, graphic and pictorial documents prepared or assembled for describing the design, location and physical characteristics of the elements of a project necessary for obtaining a building *permit*.

❖ To determine whether proposed construction is in compliance with code requirements, it is necessary that sufficient information be submitted to the building official for review. This typically consists of drawings (such as floor plans, elevations, sections and details), specifications and product information describing the proposed work.

CONSTRUCTION TYPES. See Section 602.

Type I. See Section 602.2.

Type II. See Section 602.2.

Type III. See Section 602.3.

Type IV. See Section 602.4.

Type V. See Section 602.5.

❖ Construction types are "defined" by the materials and fire-resistance rating of the construction as specified by the provisions in Chapter 6.

[F] CONTINUOUS GAS DETECTION SYSTEM. A gas detection system where the analytical instrument is maintained in continuous operation and sampling is performed without interruption. Analysis is allowed to be performed on a cyclical basis at intervals not to exceed 30 minutes.

❖ This term refers to a system that is capable of constantly monitoring the presence of highly toxic or toxic compressed gases at or below the permissible exposure limit (PEL) for the gas. A continuous gas detection system will provide notification of a leak or rupture in a compressed gas cylinder or tank in a storage or use condition.

[F] CONTROL AREA. Spaces within a building where quantities of *hazardous materials* not exceeding the maximum allowable quantities per control area are stored, dispensed, *used* or handled. See the definition of "Outdoor control area" in the *International Fire Code*.

❖ Control areas provide an alternative method for the use and storage of hazardous materials without classifying the building or structure as a high-hazard occupancy (Group H). This concept is based on regulating the allowable quantities of hazardous materials per control area, rather than per building area, by giving credit for further compartmentation through the use of fire barriers and horizontal assemblies having a minimum fire-resistance rating of not less than 1 hour. The maximum quantities of hazardous materials within a given control area cannot exceed the amounts for a given material listed in either Table 307.1(1) or 307.1(2) (see commentary, Section 414.2). Control areas are not limited to within buildings. A storage area that is exposed to the elements (wind, rain, snow, etc.) also cannot exceed the maximum allowable quantity.

CONTROLLED LOW-STRENGTH MATERIAL. A self-compacted, cementitious material used primarily as a backfill in place of compacted fill.

❖ The definition provided is from ACI 229R. This type of material is known by many "local" names (e.g., flowable fill) and is commonly used instead of a compacted backfill. Requirements for its use under the code are outlined in Section 1804.7.

CONVENTIONAL LIGHT-FRAME CONSTRUCTION. A type of construction whose primary structural elements are formed by a system of repetitive wood-framing members. See Section 2308 for conventional light-frame construction provisions.

❖ Conventional light-frame construction is one of the three methods for designing wood buildings or structures that are recognized in Chapter 23. Section 2308 provides prescriptive provisions and specifications for this method. Since it is a prescriptive approach, conventional construction is necessarily limited in applicability (see Section 2308.2).

CORNICE. A projecting horizontal molded element located at or near the top of an architectural feature.

❖ This definition facilitates the applicability of roof loads that are listed in Table 1607.1.

CORRIDOR. An enclosed *exit access* component that defines and provides a path of egress travel.

❖ Corridors are regulated in the code because they serve as principal elements of travel in many means of egress systems within buildings. Typically, corridors have walls that extend from the floor to the ceiling. They need not extend above the ceiling or have doors in their openings unless a fire-resistance rating is required (see Section 1020).

While both aisles and corridors may result in a confined path of travel, an aisle is an unenclosed or partially closed component, while a corridor would be an enclosed component of the means of egress. The enclosed character of the corridor restricts the sensory perception of the user. A fire located on the other side of the corridor wall, for example, may not be as readily seen, heard or smelled by the occupants traveling through the egress corridor. The code does not specifically state what is considered "enclosed" when corridors are not fire-resistance rated. When an egress path is bounded by partial-height walls, such as work-station partitions in an office, issues would be if the walls provided a confined path of travel and limited fire recognition in adjacent spaces by restricting line of sight, hearing and smell.

CORRIDOR, OPEN-ENDED. See "Open-ended corridor."

CORRIDOR DAMPER. A *listed* device intended for use where air ducts penetrate or terminate at horizontal openings in the ceilings of fire-resistance-rated corridors, where the corridor ceiling is permitted to be constructed as required for the corridor walls.

❖ Under certain conditions, the code allows the corridor enclosure to include ceiling constructed as a fire-rated assembly equivalent to that required for the corridor walls (fire partitions). Under most conditions, this would allow the fire partitions to be continuous only to the fire-rated ceiling assembly, rather than to the structure above. When ducts penetrate through or terminate at these ceiling assemblies, a corridor damper is required. Corridor damper requirements are included in Section 717.

[BS] CORROSION RESISTANCE. The ability of a material to withstand deterioration of its surface or its properties when exposed to its environment.

❖ There are different environments that contain different types of materials to which building construction materials are exposed. "Corrosion resistance" is not an absolute term; it is relative to the building material and where it is being used. For instance, some plastic polymers might resist corrosion when used in an exterior environment, but might not be resistant to corrosion from certain chemical gases that could be present in a laboratory.

[F] CORROSIVE. A chemical that causes visible destruction of, or irreversible alterations in, living tissue by chemical action at the point of contact. A chemical shall be considered corrosive if, when tested on the intact skin of albino rabbits by the method described in DOTn 49 CFR, Part 173.137, such chemical destroys or changes irreversibly the structure of the tissue at the point of contact following an exposure period of 4 hours. This term does not refer to action on inanimate surfaces.

❖ This definition is derived from DOL 29 CFR, Part 1910.1200. While corrosive materials do not present a fire, explosion or reactivity hazard, they do pose a handling and storage problem. Corrosive materials, therefore, are primarily considered a health hazard and are classified as Group H-4 material. Many corrosive chemicals are also strong oxidizing agents that require classification as a multiple hazard in accordance with Section 307.8.

COURT. An open, uncovered space, unobstructed to the sky, bounded on three or more sides by exterior building walls or other enclosing devices.

❖ Though not specifically identified in the definition, the provisions in the code for courts (see Section 1206) are only applicable to those areas created by the arrangement of exterior walls and used to provide natural light or ventilation (see Section 1206.1 and the definition of "Yard." See also the definition of "Egress court" for courts that are utilized for exit discharge.).

COVERED MALL BUILDING. A single building enclosing a number of tenants and occupants, such as retail stores, drinking and dining establishments, entertainment and amusement facilities, passenger transportation terminals, offices and other similar uses wherein two or more tenants have a main entrance into one or more malls. *Anchor buildings* shall not be considered as a part of the covered mall building. The term "covered mall building" shall include *open mall buildings* as defined below.

❖ The covered mall building is the entire area of the building (area of mall plus gross leasable area), excluding the anchor buildings. Passenger transportation terminals frequently are developed as wide concourses with small shops along the sides. For this reason, passenger transportation facilities are included. Transportation facilities used for freight or other purposes are not to be considered a covered mall building [see Commentary Figure 202(2)]. The term "covered mall building" also includes open mall buildings. Unless noted otherwise, open mall buildings shall comply with all the provisions for a covered mall building.

Mall. A roofed or covered common pedestrian area within a *covered mall building* that serves as access for two or more tenants and not to exceed three levels that are open to each other. The term "mall" shall include open malls as defined below.

❖ The mall is an interior, climate-controlled pedestrian way that is open to the tenant spaces within the mall building and typically connects to the anchor buildings. The term "mall" shall also include open malls. Unless noted otherwise, open malls must comply with all the provisions for malls.

Open mall. An unroofed common pedestrian way serving a number of tenants not exceeding three levels. Circulation at levels above grade shall be permitted to include open exterior balconies leading to *exits* discharging at grade.

❖ The open mall is an uncovered common pedestrian walk that is open to the sky above and to tenant spaces within the open mall building, and typically connects to the anchor buildings. The size of the openings to the sky is in accordance with Section 402.4.3. Unless noted otherwise, open malls must comply with all the provisions for malls.

Open mall building. Several structures housing a number of tenants, such as retail stores, drinking and dining establishments, entertainment and amusement facilities, offices, and other similar uses, wherein two or more tenants have a main entrance into one or more open malls. *Anchor buildings* are not considered as a part of the open mall building.

❖ The open mall building includes all of the buildings housing a number of tenants wherein two or more tenants have a main entrance into one or more open malls. Because the open mall is characterized by there not being a roof connecting one side of the pedestrian mall to the other, the covered mall "building" may actually be a collection of separate buildings which all rely on a shared pedestrian concourse for egress. Similar to the covered mall building, the open mall "building" does not include the anchor buildings. Unless noted otherwise, open mall buildings have to comply with all the provisions for covered mall buildings.

[BS] CRIPPLE WALL. A framed stud wall extending from the top of the foundation to the underside of floor framing for the lowest occupied floor level.

❖ Cripple walls are built on the top of footings or foundation walls. They can typically be found along the top of stepped foundation walls where the grade adjoining the structure changes height. Cripple walls must be properly braced to resist lateral forces. They are often treated the same as a first-story wall. Provisions for the bracing of cripple walls are located in Section 2308.6.6.

[F] CRITICAL CIRCUIT. A circuit that requires continuous operation to ensure safety of the structure and occupants.

❖ The purpose of this definition is to clarify the applicability of the provisions of Section 2702.3. Critical circuits are those electrical circuits supplying power to systems and equipment that are vital to the safety of building occupants and to the operational continuity of safety systems such as fire alarm systems, security systems, emergency communication systems and similar systems identified throughout the code, most notably in sections pertaining to emergency power. Although the term "critical circuit" is used widely throughout many industry standards, there is no specific definition for the term. As such, this definition was created based on definitions found in NFPA 70 and is similar to the definition of "Critical Operations Power Systems (COPS)" in that standard. See the commentary to Section 2702.3.

[BS] CROSS-LAMINATED TIMBER. A prefabricated engineered wood product consisting of not less than three layers of solid-sawn lumber or *structural composite lumber* where the adjacent layers are cross oriented and bonded with structural adhesive to form a solid wood element.

❖ This definition enables the code user to correctly apply the material standards in Section 2303 for these wood products.

[F] CRYOGENIC FLUID. A *liquid* having a *boiling point* lower than -150°F (-101°C) at 14.7 pounds per square inch atmosphere (psia) (an absolute pressure of 101 kPa).

❖ Cryogenic fluids present a hazard because they are extremely cold. Should a spill occur, their extremely cold temperature affects other compounds exposed to the spilled cryogenic fluid. Cryogenic fluids may be flammable or nonflammable. However, nonflammable cryogenics may possess properties that cause them to support combustion or react severely with other materials. The code is only intended to classify flammable or oxidizing cryogenic fluids as a hazardous material.

CUSTODIAL CARE. Assistance with day-to-day living tasks; such as assistance with cooking, taking medication, bathing, using toilet facilities and other tasks of daily living. Custodial care includes persons receiving care who have the ability to respond to emergency situations and evacuate at a slower rate and/or who have mental and psychiatric complications.

❖ Care facilities are used by patients of varying acuity seeking a broad spectrum of available support services. These facilities span a wide range of occupancy types including Groups E, I and R. There are three types of care defined in the codes: personal, custodial, and medical.

- Personal care is on one end of the care spectrum. It occurs in Group E for child daycare services for persons over $2^1/_2$ years of age. Occupants are supervised but do not need custodial or medical care.
- Custodial care occurs in Groups I-1, I-4 and R4, where occupants may be elderly or impaired, or require adult or child daycare of any age. Care recipients may need daily living assistance such as cooking, cleaning, bathing, or with taking medications. Persons who receive custodial care may or may not require assistance with evacuation depending on the occupancy and/or the "condition" in the occupancy. See also the commentary to Section 308.3 for Group I-1/R4 and Section 308.6 for Group I-4.
- Medical care occurs in Group I-2 on the opposite end of the spectrum, where care recipients are incapable of self-preservation. They may be completely bedridden, meaning bed movement

may be required during emergencies, and may be dependent on life support systems such as medical gases and emergency power to maintain life. This level of acuity is not allowed in custodial care or personal care.

There are two defining aspects of custodial care, which further differentiate it from medical care. The first is the evacuation capability of custodial care recipients. Custodial care recipients' evacuation capabilities are limited by the occupancy classification criteria or the occupancy condition in which care occurs. Groups I-1/R-4 Condition 1 only include occupants with the ability to self-evacuate. Groups I-1/R-4 Condition 2 include limited assistance with evacuation. Group R-4 also assumes that occupants may not be able to respond on their own during emergencies. The second differentiating aspect is that I-1 and R-4 custodial care recipients also participate in fire drills per the IFC, versus Group I-2 medical care which implements defend-in-place strategies during emergencies.

The level of care provided describes the condition and capabilities of an occupant which then indicates the appropriate standards for protection systems, both active and passive. See also the definition of "24-hour basis," "Ambulatory care facility," "Detoxification facility," "Foster care facility," "Group home," "Hospitals and psychiatric hospitals," "Medical care," "Nursing home," "Personal care services," and "Incapable of self-preservation."

[BS] DALLE GLASS. A decorative composite glazing material made of individual pieces of glass that are embedded in a cast matrix of concrete or epoxy.

❖ Dalle glass is a form of "stained" or "decorative" glass, and is more commonly known as "faceted" glass. Dalle glass is an outdated term and is rarely used. It can be prefabricated in panels that are set into framed openings, or it can be constructed in place, similar to the way a mason would construct a single wythe masonry wall. The actual glazing can be heavy pieces of flat or textured glass, irregular-shaped chunks of glass, glass shapes or translucent minerals cut to form geometric patterns or artistic scenes. Knowing that Dalle glass or faceted glass is a type of decorative glass is necessary because decorative glass is not required to be safety glazed in hazardous locations as required in Section 2406.4 (see Sections 2406.4.1, 2406.4.2 and 2406.4.3 for specific exceptions).

DAMPER. See "*Ceiling radiation damper*," "*Combination fire/smoke damper*," "*Corridor damper*," "*Fire damper*" and "*Smoke damper*."

❖ Dampers are used primarily in heating, ventilating and air-conditioning (HVAC) duct systems that pass through fire-resistance-rated walls, floors and ceilings. Dampers may also be required in rated walls independent of HVAC duct systems. Dampers are provided to maintain the fire-resistance rating of the penetrated assembly, to provide a barrier to resist smoke migration, or both. Damper types include fire dampers, smoke dampers, combination fire/smoke dampers, ceiling radiation dampers and corridor dampers.

[BS] DANGEROUS. Any building, structure or portion thereof that meets any of the conditions described below shall be deemed dangerous:

1. The building or structure has collapsed, has partially collapsed, has moved off its foundation or lacks the necessary support of the ground.
2. There exists a significant risk of collapse, detachment or dislodgment of any portion, member, appurtenance or ornamentation of the building or structure under service loads.

❖ This definition describes two conditions that are considered to be dangerous in buildings in regards to support, loading and the ability of the structure to resist loads

[F] DAY BOX. A portable magazine designed to hold explosive materials constructed in accordance with the requirements for a Type 3 magazine as defined and classified in Chapter 56 of the *International Fire Code*.

❖ A day box is an explosive magazine that is listed in Note e of Table 307.1(1). Where used, a day box will allow an increase in the maximum allowable quantities of hazardous materials.

[BS] DEAD LOAD. The weight of materials of construction incorporated into the building, including but not limited to walls, floors, roofs, ceilings, *stairways*, built-in partitions, finishes, cladding and other similarly incorporated architectural and structural items, and the weight of fixed service equipment, such as cranes, plumbing stacks and risers, electrical feeders, heating, ventilating and air-conditioning systems and *automatic sprinkler systems*.

❖ The definition of "Dead load" identifies the type of items that must be accounted for (also see Section 1607.12.3.1 for inclusion of items on landscaped roofs). This definition is necessary to distinguish dead loads from other loads, and for use in load combinations as specified by Section 1605. Dead loads are considered permanent in the load combinations. The nominal dead load is to be determined in accordance with Section 1606.

The weights of service equipment, such as plumbing stacks and risers; heating, ventilating and air-conditioning (HVAC) equipment; elevators and elevator machinery; fire protection systems and similar fixed equipment are to be included in the dead load. For the most part, tracking the weights of each utility system is not practical and the structural design is therefore based on a dead load allowance for these items. At times, the actual weight of equipment to be installed is unknown during the design phase of a building because the supplier of the equipment has yet to be determined. The structural design must often proceed based on an estimated equipment dead load. For additional comments on dead load estimates, see the commentary to Section 1606.2.

[BS] DECORATIVE GLASS. A carved, leaded or *Dalle glass* or glazing material whose purpose is decorative or artistic, not functional; whose coloring, texture or other design qualities or components cannot be removed without destroying the glazing material and whose surface, or assembly into which it is incorporated, is divided into segments.

❖ This definition is needed to understand and properly apply exceptions to the requirements for safety glazing. Decorative glass is excluded at specific hazardous locations, and thus is not required to be safety glazing (see Sections 2406.4.1, 2406.4.2 and 2406.4.3).

[F] DECORATIVE MATERIALS. All materials applied over the building *interior finish* for decorative, acoustical or other effect including, but not limited to, curtains, draperies, fabrics and streamers; and all other materials utilized for decorative effect including, but not limited to, bulletin boards, artwork, posters, photographs, batting, cloth, cotton, hay, stalks, straw, vines, leaves, trees, moss and similar items, foam plastics and materials containing foam plastics. Decorative materials do not include wall coverings, ceiling coverings, floor coverings, ordinary window shades, *interior finish* and materials 0.025 inch (0.64 mm) or less in thickness applied directly to and adhering tightly to a substrate.

❖ The significance of this definition is to provide information as to what items are not regulated as decorative materials in the application of code requirements. While any dictionary would define floor coverings, window shades and wall paper as being "decorative" in a building interior, they are not considered decorative materials for the flame-resistance testing to which the code requirements are intended to apply.

[BS] DEEP FOUNDATION. A deep foundation is a foundation element that does not satisfy the definition of a *shallow foundation*.

❖ Chapter 18 identifies two major classifications for foundations; either shallow or deep. This definition clarifies that foundations not meeting the definition of "Shallow foundation" are, by default, considered to be deep foundations.

DEFEND-IN-PLACE. A method of emergency response that engages building components and trained staff to provide occupant safety during an emergency. Emergency response involves remaining in place, relocating within the building, or both, without evacuating the building.

❖ Nursing homes, hospitals and separated ambulatory care facilities may use defend-in-place methods as part of their strategy to deal with some emergency situations. Reasons why this may be a necessary option include situations such as a surgery or other procedure in process, emergency life monitoring/support equipment needed, or weather outside that could be hazardous to patient health. Defend-in-place, or protect in place, is a concept that has long been employed as the preferred method of fire response in nursing homes and hospitals due to the fragile nature of the occupants. Occupants in these settings are often dependent on the building infrastructure (i.e., medical gases, life support) and immediate evacuation would place their lives at risk.

These types of facilities are subdivided into different smoke compartments with refuge areas that can accommodate all building occupants. The IFC requires all staff to have training on the fire and safety evacuation plans and practice drills. The idea is for staff to move patients and guests to areas that are away from the area where there may be smoke, flames or other safety hazards, but not necessarily outside of the building. Typically, movement is into an adjacent smoke compartment. Patients or guests already in the smoke compartments away from the hazard are defended in place. Many hospital systems are on emergency or standby power to keep patient care in operation during an emergency.

These types of facilities will also have full evacuation plans for other types of emergencies, such as floods. See the IFC for specific requirements for fire and safety evacuation plans. These plans must be developed and reviewed with the fire department.

[A] DEFERRED SUBMITTAL. Those portions of the design that are not submitted at the time of the application and that are to be submitted to the *building official* within a specified period.

❖ Submittal documents are required at the time of permit application (See Section 107). If a building is "fast-track" or is a shell building, a complete set of construction documents for the building may not be available at the time of the initial permit application. So that a complete set of documents is available for the building department review and records, some of the construction drawings may be provided at a later time. A list of pending documents, specifying when they will be provided, must be included in the initial submittal and approved by the building official.

[F] DEFLAGRATION. An exothermic reaction, such as the extremely rapid oxidation of a flammable dust or vapor in air, in which the reaction progresses through the unburned material at a rate less than the velocity of sound. A deflagration can have an explosive effect.

❖ Materials that present a deflagration hazard usually burn very rapidly with the release of energy from a chemical reaction in the form of intense heat. Confined deflagration hazards under pressure can result in an explosion. Most hazardous materials that pose a severe deflagration hazard are classified as Group H-2 in accordance with Section 307.4 (see the definition of "Detonation").

[F] DELUGE SYSTEM. A sprinkler system employing open sprinklers attached to a piping system connected to a water supply through a valve that is opened by the operation of a detection system installed in the same areas as the sprinklers. When this valve opens, water flows into the piping system and discharges from all sprinklers attached thereto.

❖ A deluge system applies large quantities of water or foam throughout the protected area by means of a system of open sprinklers. In a fire, the system is activated by a fire detection system that makes it possible to apply water to a fire more quickly and to cover a larger area than with a conventional automatic sprinkler system, which depends on sprinklers being activated individually as the fire spreads. As the definition indicates, the sprinklers are open. There is no fusible link, so when water is admitted into the system by the fire detection system, it flows through the piping and is immediately discharged through the sprinkler heads.

Deluge systems are particularly beneficial in hazardous areas where the fuel loads (combustible contents) are of such a nature that fire may grow with exceptional rapidity and possibly flash ahead of the operations of conventional automatic sprinklers.

[BS] DESIGN DISPLACEMENT. See Section 1905.1.1.

❖ This definition is a modification of ACI 318 text and is therefore located in Section 1905.

[BS] DESIGN EARTHQUAKE GROUND MOTION. The earthquake ground motion that buildings and structures are specifically proportioned to resist in Section 1613.

❖ A structure must be designed for the level of ground motion corresponding to the design earthquake. This is determined in Section 1613.3.4 as two-thirds of the maximum earthquake ground motions with adjustments to account for the site's soil profile.

[BS] DESIGN FLOOD. The *flood* associated with the greater of the following two areas:

1. Area with a flood plain subject to a 1-percent or greater chance of *flooding* in any year.
2. Area designated as a *flood hazard area* on a community's flood hazard map, or otherwise legally designated.

❖ The design flood is either the base flood or another flood based on other criteria. A state or local jurisdiction may choose to prepare and adopt flood hazard maps that show flood hazard areas that are not on maps prepared by FEMA. These may be areas that FEMA did not study, or areas that were studied with different criteria. For example, as a general rule, FEMA is concerned primarily with inland flooding sources that have a drainage area of 1 square mile (2.590 km^2) or more. For another example, some communities elect to prepare flood hazard maps based on the assumption that the upland watershed is built out to existing zoning, often called "ultimate development," and sometimes the "flood of record" is the basis for regulation.

[BS] DESIGN FLOOD ELEVATION. The elevation of the "*design flood*," including wave height, relative to the datum specified on the community's legally designated flood hazard map. In areas designated as Zone AO, the *design flood elevation* shall be the elevation of the highest existing grade of the building's perimeter plus the depth number (in feet) specified on the flood hazard map. In areas designated as Zone AO where a depth number is not specified on the map, the depth number shall be taken as being equal to 2 feet (610 mm).

❖ The design flood elevation is the height to which floodwaters are predicted to rise during passage or occurrence of the design flood. Unless a community has adopted another flood hazard map or otherwise designated the design flood, the design flood elevation is the base flood elevation shown on the Flood Insurance Rate Maps (FIRM). The datum specified on the flood hazard map is important because it may differ from the datum used locally for other purposes. Also, the datum shown on the FIRM may change over time, as FEMA updates and revises the maps.

Some flood hazard maps have areas denoted as AO Zones, identifying areas subject to sheet flow flooding, shallow flooding or ponding. Specific instruction for these zones is provided in the code because the maps specify a flood depth above grade rather than a height above datum.

FEMA refers to some areas subject to shallow flooding and/or unpredictable flow paths as AH Zones. AH Zones have specified base flood elevations and they are treated the same as the other A Zones with base flood elevations.

[A] DESIGN PROFESSIONAL, REGISTERED. See "Registered design professional."

[A] DESIGN PROFESSIONAL IN RESPONSIBLE CHARGE, REGISTERED. See "Registered design professional in responsible charge."

[BS] DESIGN STRENGTH. The product of the nominal strength and a *resistance factor* (or strength reduction factor).

❖ This definition is needed to apply the strength design requirements in the code. The design strength is the nominal strength multiplied by a resistance or strength reduction factor that is less than one. The design strength and corresponding strength reduction factors are specified in the applicable material chapter of the code or a standard that is referenced therein.

[BS] DESIGNATED SEISMIC SYSTEM. Those nonstructural components that require design in accordance with Chapter 13 of ASCE 7 and for which the component importance factor, I_p, is greater than 1 in accordance with Section 13.1.3 of ASCE 7.

❖ Designated seismic systems are those architectural, electrical and mechanical components and systems in the referenced ASCE 7 provisions that are assigned a component importance factor of 1.5. The importance factor is an indication that, compared with nonstructural components with the typical factor of 1.0, there is a higher level of importance placed on the system so that it will remain operational during and after an earthquake. This definition is important in correctly applying the provisions for special inspection and testing for designated seismic systems.

[F] DETACHED BUILDING. A separate single-*story* building, without a basement or crawl space, used for the storage or *use* of *hazardous materials* and located an *approved* distance from all structures.

❖ The term is used to define the type of structure the code recognizes for the use and storage of hazardous materials in excess of the maximum allowable quantities. While the definition addresses all hazardous materials, a detached storage building is only required for Group H-1, H-2 and H-3 structures as indicated in Sections 415.6.2 and 415.7, and Table 415.6.2. The location of the structure may be regulated by Section 415.6.1 based on the characteristics of the materials contained in the building.

[BS] DETAILED PLAIN CONCRETE STRUCTURAL WALL. See Section 1905.1.1

❖ This definition is a modification of ACI 318 text and is therefore located in Section 1905.

DETECTABLE WARNING. A standardized surface feature built in or applied to walking surfaces or other elements to warn visually impaired persons of hazards on a *circulation path*.

❖ A detectable warning is a change in texture that is detectable by a person with a vision impairment. Technical specifications for a detectable warning surface are set forth in ICC A117.1. Detectable warnings are only required for transit platform edges so that users can be confident of the warning that is intended to be communicated and not be confused by multiple, different surfaces intended to convey the same warning (see commentary, Section 1109.10).

[F] DETECTOR, HEAT. A fire detector that senses heat—either abnormally high temperature or rate of rise, or both.

❖ In a fire, heat is released that causes the temperature in a room or space to increase. Automatic fire detectors that sense abnormally high temperature or rate of temperature rise are known as heat detectors. These include fixed temperature detectors, rate compensation detectors and rate-of-rise detectors. The code requires all automatic fire detectors to be smoke detectors, except that heat detectors tested and approved in accordance with NFPA 72 may be used as an alternative to smoke detectors in rooms and spaces where, during normal operation, products of combustion are present in sufficient quantity to actuate a smoke detector.

[F] DETONATION. An exothermic reaction characterized by the presence of a shock wave in the material which establishes and maintains the reaction. The reaction zone progresses through the material at a rate greater than the velocity of sound. The principal heating mechanism is one of shock compression. Detonations have an explosive effect.

❖ Detonations are distinguished from deflagrations (which are produced by explosive gases, dusts, vapors and mists) by the speed with which they propagate a blast effect. Detonations occur much faster than deflagrations, since they propagate a combustion zone at a velocity greater than the speed of sound. Deflagrations propagate a combustion zone at a velocity less than the speed of sound. The speed of sound is approximately 1,100 feet per second (336 m/s) at sea level. Both detonations and deflagrations may produce explosive results when they occur in a confined space. Materials that are considered a detonation hazard are classified as Group H-1 materials in accordance with Section 307.3.

DETOXIFICATION FACILITIES. Facilities that provide treatment for substance abuse, serving care recipients who are *incapable of self-preservation* or who are harmful to themselves or others.

❖ Persons in detoxification facilities may be physically incapable of self-preservation, or they may be confined within an area of a building for care or security purposes. See the commentary to Sections 308 and 407, Group I-2.

Care facilities are used by patients of varying acuity seeking a broad spectrum of available support services. These facilities span a wide range of occupancy types including Groups E, I and R. The level of care provided describes the condition and capabilities of an occupant, which, in turn, indicates appropriate standards for protection systems, both passive and active. See also the definitions for "24-hour care," "Ambulatory care facility," "Group home," "Hospitals and psychiatric hospitals," "Incapable of self-preservation," "Medical care," "Nursing home" and "Personal care services."

[BS] DIAPHRAGM. A horizontal or sloped system acting to transmit lateral forces to vertical elements of the lateral force-resisting system. When the term "diaphragm" is used, it shall include horizontal bracing systems.

❖ Floor and roof diaphragms act to transfer the lateral forces, such as wind or seismic loads, to the vertical-resisting elements (e.g., shear walls, braced frames, moment frames, etc.), supporting them at their perimeter or intermittent locations.

Diaphragm, blocked. In *light-frame construction*, a diaphragm in which all sheathing edges not occurring on a framing member are supported on and fastened to blocking.

❖ Unblocked diaphragms can resist low and moderate in-plane shear forces. Diaphragm sheathing may be applied with the long dimension of the sheathing either perpendicular or parallel to the main framing members. When there is no blocking under the edge of the sheathing that is not supported by the main framing member, the diaphragm is considered to be unblocked. Tables in Chapter 23 provide different values for stapled structural sheathing when it is blocked and when it is not blocked. The ANSI/AWC *Special Design Provisions for Wind and Seismic* (SDPWS) contains separate tables for nailed diaphragms that are blocked and unblocked.

Diaphragm boundary. In *light-frame construction,* a location where shear is transferred into or out of the diaphragm sheathing. Transfer is either to a boundary element or to another force-resisting element.

❖ Diaphragm boundary is typically the connection between the floor or roof sheathing and the band board surrounding the diaphragm.

Diaphragm chord. A diaphragm boundary element perpendicular to the applied load that is assumed to take axial stresses due to the diaphragm moment.

❖ A diaphragm acts as a deep horizontal beam. The chords of the beam are the elements at the boundary of the diaphragm that are perpendicular to the direction of the applied load.

Diaphragm, unblocked. A diaphragm that has edge nailing at supporting members only. Blocking between supporting structural members at panel edges is not included. Diaphragm panels are field nailed to supporting members.

❖ Unblocked diaphragms can resist low and moderate shear forces. Diaphragm sheathing may be applied with the long dimension of the sheathing either perpendicular or parallel to the main framing members. When the edge of the sheathing is not supported by the main framing member, the diaphragm is considered to be unblocked. Various tables in Chapter 23 provide different values for structural sheathing when it is blocked and when it is not blocked.

DIMENSIONS (for Chapter 21).

❖ Different dimensions are used to designate sizes of masonry units and masonry elements. The terms denote the common meanings of various types of dimensions used in Chapter 21.

Nominal. The *specified dimension* plus an allowance for the *joints* with which the units are to be laid. Nominal dimensions are usually stated in whole numbers. Thickness is given first, followed by height and then length.

❖ The nominal dimensions of a masonry unit are the specified dimensions, plus the specified thickness of one mortar joint. Nominal dimensions are used for architectural layout of masonry structures. Commentary Figure 202(10) shows nominal dimensions for a specific concrete masonry unit.

Specified. Dimensions specified for the manufacture or construction of a unit, *joint* or element.

❖ The specified dimensions are prescribed in the construction documents. Commentary Figure 202(10) shows specified and nominal dimensions for a concrete masonry unit.

DIRECT ACCESS. A path of travel from a space to an immediately adjacent space through an opening in the common wall between the two spaces.

❖ This term is related to the interior exit stairway associated with the lobby required for fire service access elevators (FSAE) and occupant evacuation elevators (OEE). In order for FSAE and OEE to provide the protection and function necessary, a stairway needs to be located with the elevators themselves. See the commentary for Sections 3007.6.1 and 3008.6.1.

[F] DISPENSING. The pouring or transferring of any material from a container, tank or similar vessel, whereby vapors, dusts, fumes, mists or gases are liberated to the atmosphere.

❖ This term refers to a specific operation whereby the act of transferring a material occurs and has a hazard associated with the liberation of the material in the forms listed in the definition. It is not "handling" and should not be confused with that term (see the definitions of "Closed system" and "Handling").

DOOR, BALANCED. See "Balanced door."

DOOR, LOW-ENERGY POWER-OPERATED. See "Low-energy power-operated door."

DOOR, POWER-ASSISTED. See "Power-assisted door."

DOOR, POWER-OPERATED. See "Power-operated door."

DOORWAY, EXIT ACCESS. See "Exit access doorway."

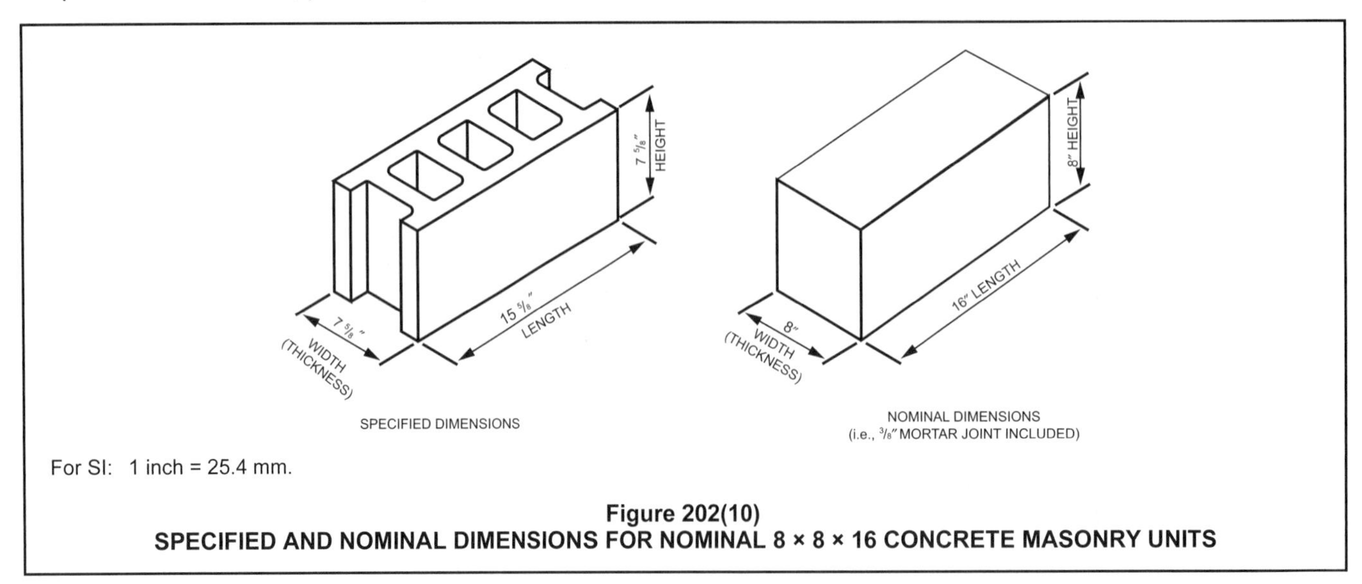

Figure 202(10)
SPECIFIED AND NOMINAL DIMENSIONS FOR NOMINAL 8 × 8 × 16 CONCRETE MASONRY UNITS

DORMITORY. A space in a building where group sleeping accommodations are provided in one room, or in a series of closely associated rooms, for persons not members of the same family group, under joint occupancy and single management, as in college dormitories or fraternity houses.

❖ Dormitories typically consist of a large room serving as a community sleeping room or many smaller rooms grouped together and serving as private or semiprivate sleeping rooms (sleeping units). A typical setting for dormitories is on college campuses. However, sleeping areas of a fire station and similar lodging facilities for occupants not of the same family group are also considered dormitories. Dormitories most often are not the permanent residence of the occupants. They are typically occupied only for a designated period of time, such as a school year. Though limited, the period of occupancy is usually more than 30 days, which provides the occupant with a familiarity of the structure such that the occupancy is not considered transient. A dormitory is classified as Group R-2 (see Section 310.4).

Structures containing a dormitory often have a cafeteria or central eating area and common recreational areas. When such conditions exist, the structure must comply with the mixed occupancy provisions of the code [see Section 508 and Commentary Figure 202(11)].

DRAFTSTOP. A material, device or construction installed to restrict the movement of air within open spaces of concealed areas of building components such as crawl spaces, floor/ceiling assemblies, roof/ceiling assemblies and *attics*.

❖ Draftstopping is required in concealed combustible spaces to limit the movement of air, smoke and other products of combustion. Draftstopping materials are permitted to be combustible based on the rationale that a large and thick enough combustible material will act as a hindrance against the free movement of air, of flame/fire and of the products of combustion [see Commentary Figures 202(12) and 202(13) for typical draftstopping applications (also see Section 718)]. Although the term "draftstopping" would seem to imply that its primary purpose is to hinder the circulation of air within the space, its intended purpose is to stop the movement of fire and products of combustion, as evidenced by the fact that draftstopping can be omitted in some cases when appropriate automatic fire sprinkler protection is installed (see Sections 718.3 and 718.4).

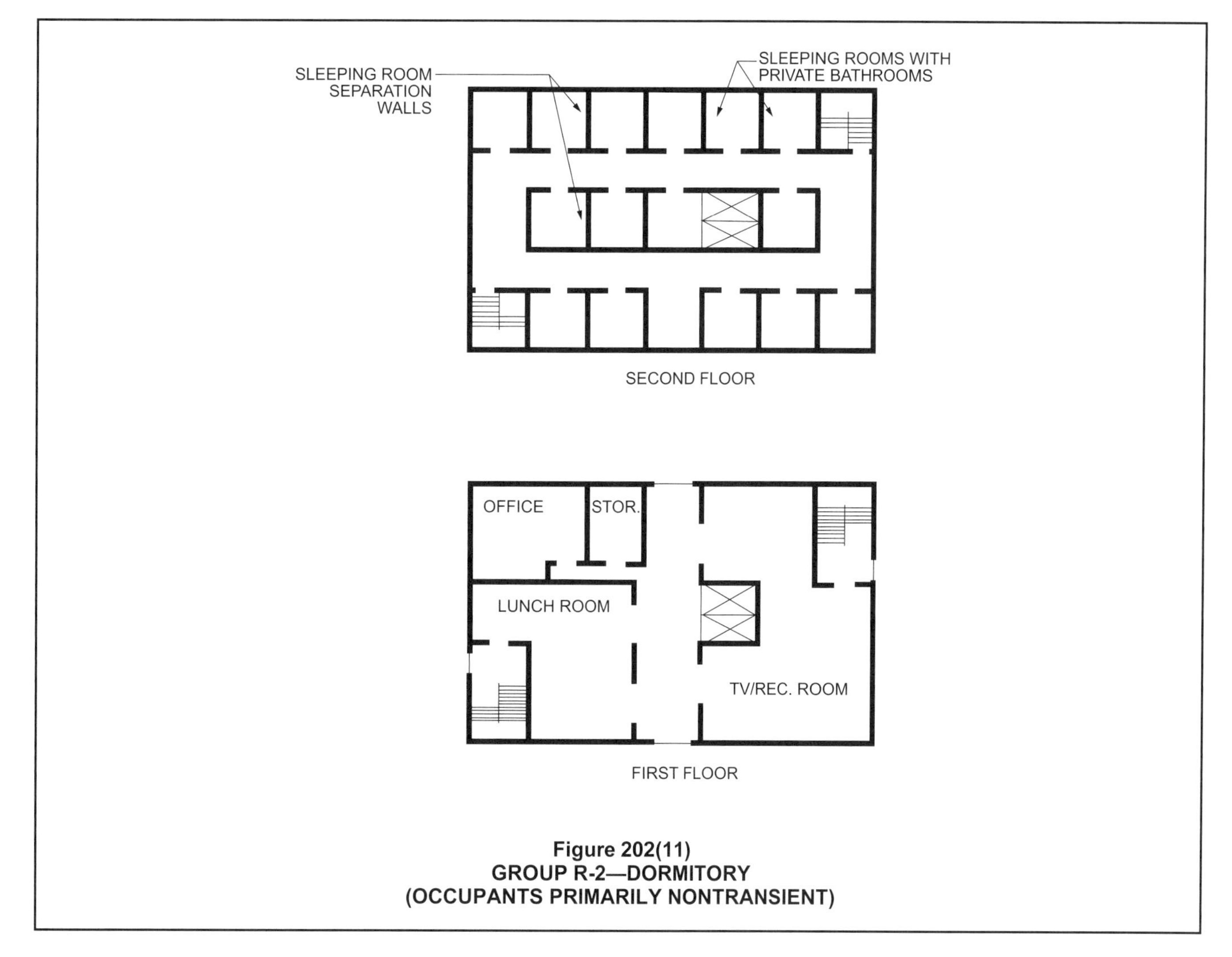

Figure 202(11)
GROUP R-2—DORMITORY
(OCCUPANTS PRIMARILY NONTRANSIENT)

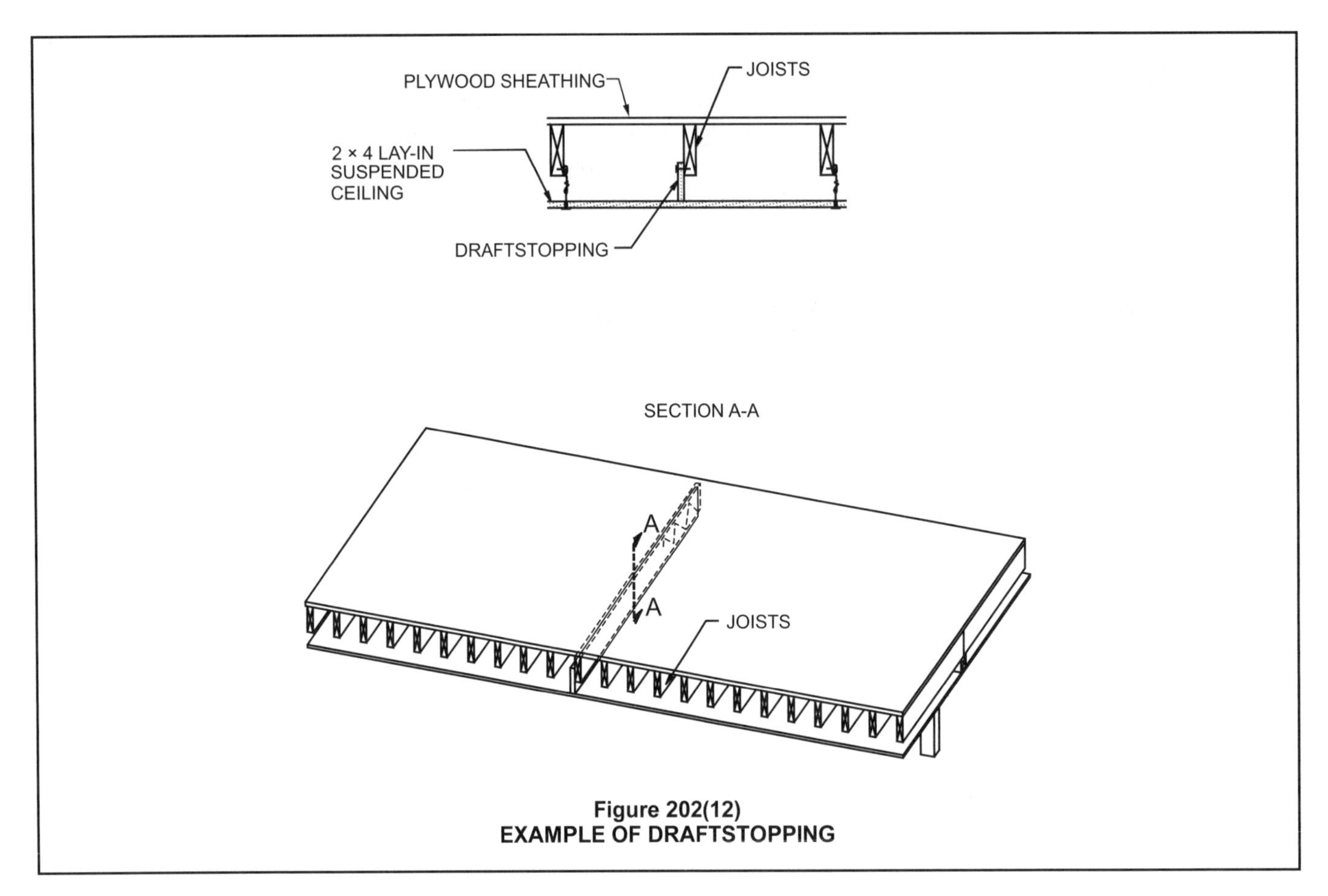

Figure 202(12)
EXAMPLE OF DRAFTSTOPPING

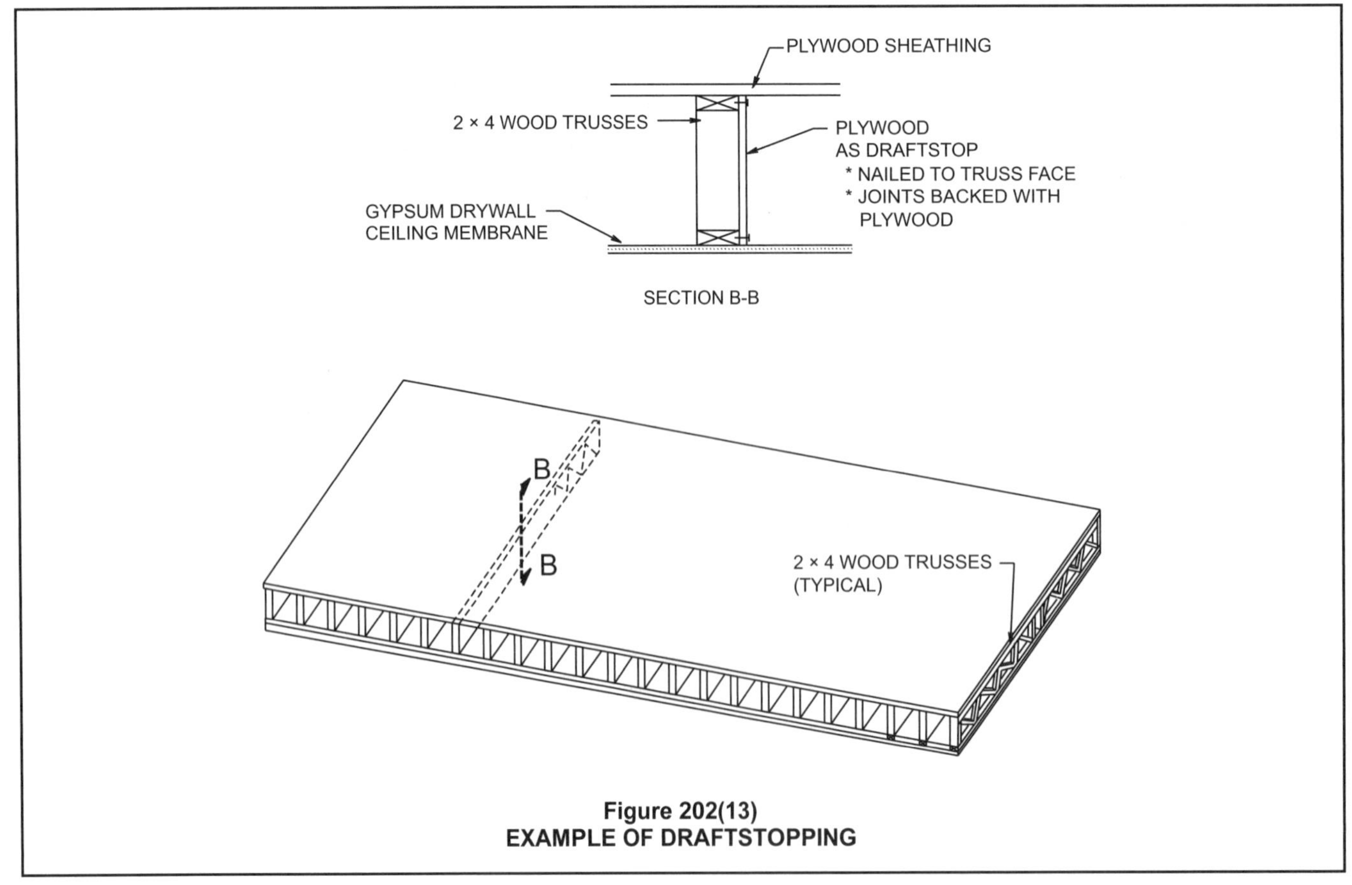

Figure 202(13)
EXAMPLE OF DRAFTSTOPPING

[BS] DRAG STRUT. See "Collector."

❖ The term "drag strut" is another term for "collector."

[BS] DRILLED SHAFT. A cast-in-place deep foundation element constructed by drilling a hole (with or without permanent casing) into soil or rock and filling it with fluid concrete.

❖ This definition clarifies a term that refers to a specific type of deep foundation element (see the requirements of Section 1810.3.9.5).

Socketed drilled shaft. A drilled shaft with a permanent pipe or tube casing that extends down to bedrock and an uncased socket drilled into the bedrock.

❖ This definition clarifies a term that refers to a specific type of deep foundation element (see the requirements of Sections 1810.3.9.6 and 1810.4.9).

[F] DRY-CHEMICAL EXTINGUISHING AGENT. A powder composed of small particles, usually of sodium bicarbonate, potassium bicarbonate, urea-potassium-based bicarbonate, potassium chloride or monoammonium phosphate, with added particulate material supplemented by special treatment to provide resistance to packing, resistance to moisture absorption (caking) and the proper flow capabilities.

❖ A dry-chemical system extinguishes a fire by placing a chemical barrier between the fire and oxygen, which acts to smother a fire. This system is best known for protecting commercial ranges, commercial fryers and exhaust hoods. Wet-chemical extinguishing systems, however, are more commonly used for new installations in commercial cooking equipment.

The type of dry chemical to be used in the extinguishing system is a function of the hazard expected. The type of dry chemical used in a system must not be changed unless it has been proven changeable by a testing laboratory, is recommended by the manufacturer of the equipment and is acceptable to the fire code official for the hazard expected. Additional guidance on the use of various dry-chemical agents can be found in NFPA 17, which gives minimum requirements for the design, installation, testing, inspection, approval, operation and maintenance of dry-chemical extinguishing systems.

[BS] DRY FLOODPROOFING. A combination of design modifications that results in a building or structure, including the attendant utilities and equipment and sanitary facilities, being water tight with walls substantially impermeable to the passage of water and with structural components having the capacity to resist *loads* as identified in ASCE 7.

❖ All residential buildings and most nonresidential buildings in flood hazard areas are elevated so that the lowest floors (including basements) are at or above the design flood elevation. Under certain circumstances, nonresidential buildings (and nonresidential portions of mixed-use buildings) may be designed and constructed to incorporate dry floodproofing modifications (structural and nonstructural additions, changes, adjustments or other measures) that are designed to keep water out, and to withstand hydrostatic and hydrodynamic loads.

DWELLING. A building that contains one or two *dwelling units* used, intended or designed to be used, rented, leased, let or hired out to be occupied for living purposes.

❖ Dwellings are buildings intended to serve as residences for one or two families. Dwellings can be owner occupied or rented. The term "dwelling," which refers to the building itself, is defined to distinguish it from the term "dwelling unit," which is a single living unit within a building. It is important to recognize that the code is not intended to regulate detached one- and two-family dwellings and townhouses that are regulated by the IRC (see Section 101.2). See also the definition for "Townhouse."

DWELLING UNIT. A single unit providing complete, independent living facilities for one or more persons, including permanent provisions for living, sleeping, eating, cooking and sanitation.

❖ A dwelling unit, as stated, is a residential unit that contains all of the necessary facilities for independent living. This provides a single, independent unit that serves a single family or single group of individuals. This terminology is used throughout the code for the determination of the application of various provisions. A dwelling unit is also distinguished from a sleeping unit, which does not have all of the features of a dwelling unit and must comply with a different set of requirements (see the definition for "Sleeping unit"). A building containing one or more dwelling units is a "dwelling" (see the definitions for "Dwelling" and "Townhouse"). A building containing three or more dwelling units is regulated as a Group R-2 occupancy. The most common term used for such a building is an apartment house or condominium. To be considered a Group R-3 occupancy, the structure must have one or two dwelling units, or be subdivided by fire walls between every unit or every two units.

DWELLING UNIT OR SLEEPING UNIT, MULTI-STORY. See "Multistory unit."

EGRESS COURT. A *court* or *yard* which provides access to a *public way* for one or more *exits*.

❖ The egress court requirements address situations where the exit discharge portion of the means of egress passes through confined areas near the building and therefore faces a hazard not normally found in the exit discharge (see Section 1028.4).

ELECTRICAL CIRCUIT PROTECTIVE SYSTEM. A specific construction of devices, materials, or coatings installed as a fire-resistive barrier system applied to electrical system components, such as cable trays, conduits and other raceways, open run cables and conductors, cables, and conductors.

❖ Electrical circuit protective systems are used primarily for fire protection of wiring and cables that serve smokeproof enclosure ventilation systems and fire service access elevators. These systems are an alternative to conventional rated construction such as fire barriers and horizontal assemblies.

[F] ELEVATOR GROUP. A grouping of elevators in a building located adjacent or directly across from one another that responds to common hall call buttons.

❖ This definition clarifies the application of the emergency voice/alarm communication system requirements in Section 907.5.2.2 as to the locations, called "paging zones," where system speakers are required (see commentary, Section 907.5.2.2).

[F] EMERGENCY ALARM SYSTEM. A system to provide indication and warning of emergency situations involving *hazardous materials*.

❖ Because of the potentially volatile nature of hazardous materials, an emergency alarm system is required outside of interior building rooms or areas containing hazardous materials in excess of the maximum allowable quantities permitted in Tables 307.1(1) and 307.1(2). The intent of the emergency alarm, upon actuation by an alarm-initiating device, is to alert the occupants to an emergency condition involving hazardous materials. The initiation of the emergency alarm can be by manual or automatic means depending on the hazard and the specific requirements for the type of hazard (see Section 908).

[F] EMERGENCY CONTROL STATION. An *approved* location on the premises where signals from emergency equipment are received and which is staffed by trained personnel.

❖ This definition identifies the room or area located in the hazardous production materials (HPM) facility that is utilized for the purpose of receiving various alarms and signals. The smoke detectors located in the building's recirculation ventilation ducts, the gas-monitoring/detection system and the telephone/fire protective signaling systems located outside of HPM storage rooms are all required to be connected to the emergency control station. The location of the emergency control station must be approved by the building official. An approved location should be based on personnel being able to adequately monitor the necessary alarms and signals, and on the fire department being able to gain access quickly when responding to emergency situations. Additionally, the room must be occupied by persons who are trained to respond to the various alarms and signals in the appropriate fashion.

EMERGENCY ESCAPE AND RESCUE OPENING. An operable window, door or other similar device that provides for a means of escape and access for rescue in the event of an emergency.

❖ These are commonly windows that are sized and located such that they can be used to exit a building directly from a basement or bedroom during an emergency condition. The openings are also used by emergency personnel to rescue the occupants in a building (see Section 1030). Windows are never considered to be exit or exit access components for purposes of meeting minimum number of exit requirements. An emergency escape and rescue opening could be a type of door, such as a basement door with direct access to an exterior stairway or a door to a balcony. Bulkhead style cellar doors could also be evaluated as possible emergency escape and rescue openings.

[F] EMERGENCY POWER SYSTEM. A source of automatic electric power of a required capacity and duration to operate required life safety, fire alarm, detection and ventilation systems in the event of a failure of the primary power. Emergency power systems are required for electrical loads where interruption of the primary power could result in loss of human life or serious injuries.

❖ Section 2702 of the code mandates the installation of emergency power systems and standby power systems for various buildings and occupancies in accordance with the IFC and NFPA 70, 110 and 111. This definition is intended to provide clarity for the fire code official as to exactly what systems are considered to be emergency power systems and is consistent with definitions in NFPA 110 and NFPA 111. A primary difference between emergency and standby power systems is that emergency power systems are essentially life safety systems (e.g., exit signage and egress illumination), whereas standby power systems focus more on the continued operation of critical equipment in a building (e.g., elevators, fire pumps). Another difference between the two is the time limit within which the power supply activates. More specifically, emergency power is available in 10 seconds after primary power fails, whereas standby power is available in 60 seconds. Both types of systems are required to operate for a minimum of two hours under full design load unless otherwise indicated in Section 2702. Emergency power systems are required for the following systems or features, among others:

1. Emergency voice/alarm communication systems (24-hour duration).
2. Emergency alarm systems.
3. Exit signs.
4. Means of egress illumination.
5. HPM facility equipment.

See also the commentary to the definition of "Standby power system" and Section 2702.

[F] EMERGENCY VOICE/ALARM COMMUNICATIONS. Dedicated manual or *automatic* facilities for originating and distributing voice instructions, as well as alert and evacuation signals pertaining to a fire emergency, to the occupants of a building.

❖ An emergency voice/alarm communication system is a special feature of fire alarm systems in buildings with special evacuation considerations, such as a high-rise building or a large assembly space. Emergency voice/alarm communication systems automatically communicate a fire emergency message to all occupants of a building on a general or selective basis. Such systems also enable the fire service to manually transmit voice instructions to the building occupants about a fire emergency condition and the action to be taken for evacuation or movement to another area of the building. Although most systems use prerecorded messages, some now use computer synthesized voices to communicate messages that allow customized messages unique to the facility.

EMPLOYEE WORK AREA. All or any portion of a space used only by employees and only for work. *Corridors*, toilet rooms, kitchenettes and break rooms are not employee work areas.

❖ An employee work area is different in an office versus on a factory line. An employee work area will most likely expand past the station or desk where an employee performs his or her job. An employee work area could include common use spaces, but not public use spaces. Depending on the duties of the employee, it may also include copy areas, stock rooms, filing areas, an assembly line, etc. (see also the commentary for the definitions of "Common use" and "Public-use areas").

Note that not all employee-only areas are considered part of employee work areas (i.e., bathrooms, corridors, breakrooms).

[BS] ENGINEERED WOOD RIM BOARD. A full-depth structural composite lumber, wood structural panel, structural glued laminated timber or prefabricated wood I-joist member designed to transfer horizontal (shear) and vertical (compression) loads, provide attachment for diaphragm sheathing, siding and exterior deck ledgers, and provide lateral support at the ends of floor or roof joists or rafters.

❖ Engineered rim board is a key structural element in many engineered wood floor applications where both structural load path through the perimeter member and dimensional change compatibility are design considerations. Section 2303.1.13 provides the compliance standards for this product.

ENTRANCE, PUBLIC. See "Public entrance."

ENTRANCE, RESTRICTED. See "Restricted entrance."

ENTRANCE, SERVICE. See "Service entrance."

EQUIPMENT PLATFORM. An unoccupied, elevated platform used exclusively for mechanical systems or industrial process equipment, including the associated elevated walkways, stairways, alternating tread devices and ladders necessary to access the platform (see Section 505.3).

❖ A distinction is made between equipment platforms and mezzanines by way of this definition. Equipment platforms, covered in Section 505.3, are unoccupied and used exclusively for housing equipment and providing access thereto, and are not subject to the requirements for mezzanines. Their purpose could also be to allow access for maintenance, repair or modification of elevated or very large equipment. Equipment platforms allow efficient use of high bay areas by locating infrequently accessed equipment or processes overhead without the occupant load or increasing the hazard to occupants in the room. Elevated floor areas that do not meet this definition would be subject to the requirements for mezzanines.

ESSENTIAL FACILITIES. Buildings and other structures that are intended to remain operational in the event of extreme environmental loading from *flood*, wind, snow or earthquakes.

❖ This definition is needed to facilitate identification of Risk Category IV buildings under Section 1604.5. This is critical in determining the design earthquake, flood, snow and wind load criteria that apply to the facility.

[F] EXHAUSTED ENCLOSURE. An appliance or piece of equipment that consists of a top, a back and two sides providing a means of local exhaust for capturing gases, fumes, vapors and mists. Such enclosures include laboratory hoods, exhaust fume hoods and similar appliances and equipment used to locally retain and exhaust the gases, fumes, vapors and mists that could be released. Rooms or areas provided with general *ventilation*, in themselves, are not exhausted enclosures.

❖ Exhausted enclosures, such as laboratory hoods or exhaust fume hoods, are utilized to contain hazardous fumes and vapors. The use of an approved exhausted enclosure may allow an increase in the maximum allowable quantities per control area of a given hazardous material. Exhausted enclosures are typically utilized when highly toxic or toxic compressed gases are involved. The exhausted enclosures are required to be of noncombustible construction, have specific ventilation criteria and be protected by an approved fire-extinguishing system.

EXISTING STRUCTURE. A structure erected prior to the date of adoption of the appropriate code, or one for which a legal building *permit* has been issued. For application of provisions in *flood hazard areas*, an existing structure is any building or structure for which the start of construction commenced before the effective date of the community's first flood plain management code, ordinance or standard.

❖ This definition clarifies which buildings and structures would be considered as existing for the application of the flood provisions of the IEBC (see Section 101.4.7). More specifically, they are defined as buildings that were previously built or that have valid permits. Buildings without permits would not be legally considered existing as far as receiving any of the exceptions afforded to buildings undergoing repairs and alterations, thus full compliance with the code for new buildings would be required. For the application of flood hazard area requirements, buildings and structures are considered to be "existing" if construction started before the local jurisdiction adopted its first regulation governing development in flood hazard areas. This distinction is important because all new construction, for which the start of construction is after the date of the first regulation, must be in full compliance with the regulation, including all subsequent alterations, improvements, repairs and additions.

EXIT. That portion of a *means of egress* system between the *exit access* and the *exit discharge* or *public way*. Exit components include exterior exit doors at the *level of exit discharge*, *interior exit stairways* and *ramps*, *exit passageways*, *exterior exit stairways* and *ramps* and *horizontal exits*.

❖ Exits are the critical element of the means of egress system that building occupants travel through to reach the exterior at the level of exit discharge. Exit stairways and ramps from upper and lower stories must be separated from adjacent areas with fire-resistance-rated construction. The fire-resistance-rated construction serves as a barrier between the fire and the means of egress and protects occupants while they travel through the exit. Separation by fire-resistance-rated construction is not required, however, where the exit leads directly to the exterior at the level of exit discharge (e.g., exterior door at grade). Commentary Figure 202(14) illustrates three different types of exits: interior exit stairway, exterior exit stairway and exterior exit door.

A horizontal exit, while not discharging to the outside, does discharge to another building or refuge area. The door to the refuge area is through a fire wall or fire barrier (see the definition for "Exit, horizontal" and Section 1026).

EXIT ACCESS. That portion of a *means of egress* system that leads from any occupied portion of a building or structure to an *exit*.

❖ The exit access portion of the means of egress consists of all floor areas that lead from usable spaces within the building to the exit or exits serving that floor area. Crawl spaces and concealed attic and roof spaces are not considered to be part of the exit access. As shown in Commentary Figure 202(15), the exit access begins at the furthest points within each room or space and ends at the entrance to the exit.

EXIT ACCESS DOORWAY. A door or access point along the path of egress travel from an occupied room, area or space where the path of egress enters an intervening room, *corridor*, *exit access stairway* or *ramp*.

❖ Exit access doorways are used to design many critical aspects of the means of egress, including arrangement, number, separation, opening protection and exit sign placement. The term "doorway" has traditionally been limited to those situations where an

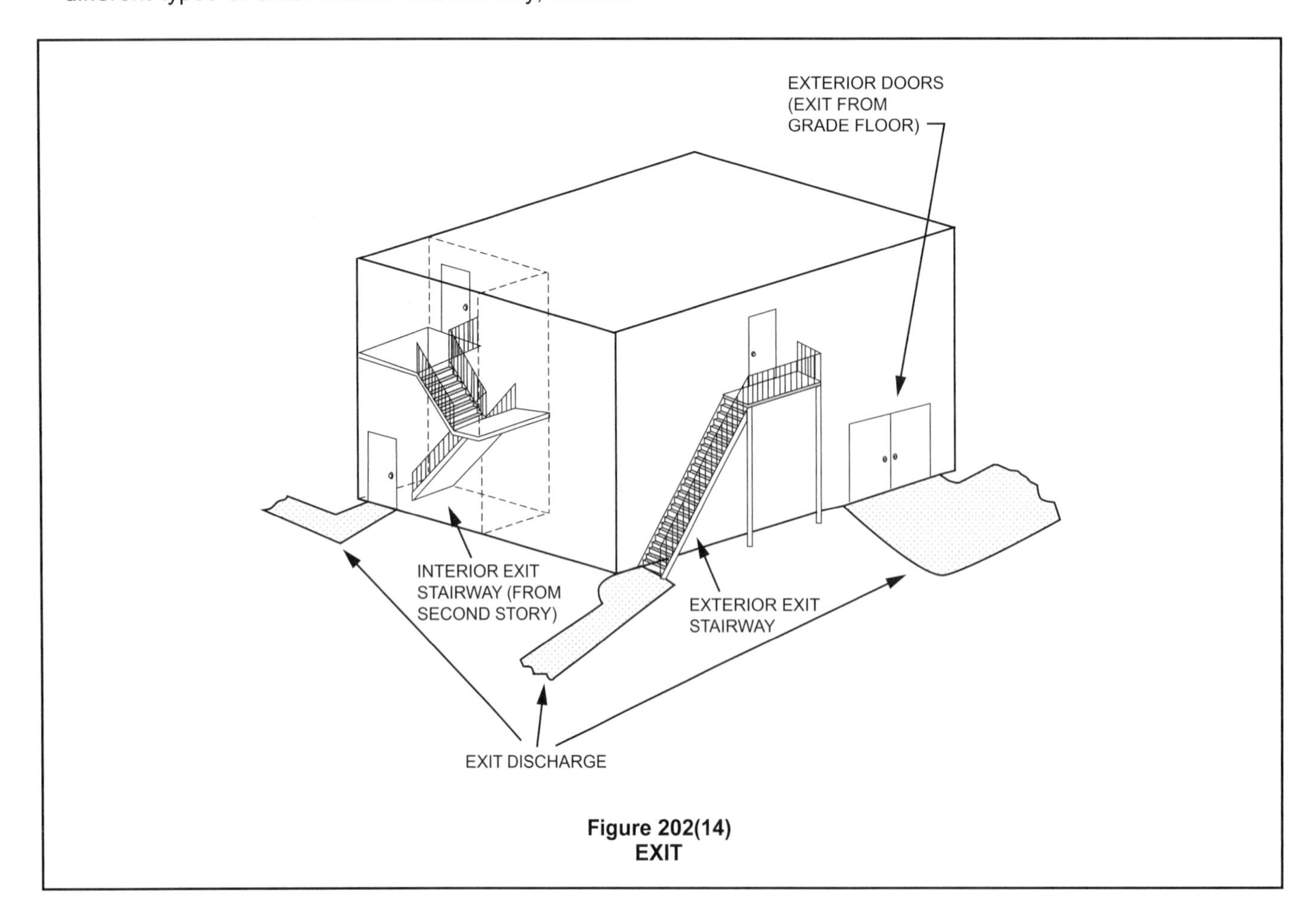

Figure 202(14)
EXIT

actual opening, either with or without a door, is present. With "access point," the term "exit access doorway" is inclusive of specific points in the means of egress, which may not include a "door," such as when an unenclosed exit access stairway is used in the egress path.

EXIT ACCESS RAMP. A *ramp* within the exit access portion of the means of egress system.

❖ Unenclosed ramps may serve as part of the route for exit access when floor levels have changes in elevation. For the limited situations where exit access ramps can be used between stories, see Sections 1006.3 and 1019. Exit access elements are included in the travel distance requirements unless specifically exempted (e.g., open parking garages, outdoor facilities) (see Section 1017.3.1).

EXIT ACCESS STAIRWAY. A *stairway* with the exit access portion of the means of egress system.

❖ Unenclosed steps and stairways may serve as part of the route for exit access when floor levels have changes in elevation. For the limited situations where exit access stairways can be used between stories, see Sections 1006.3 and 1019. Exit access elements are included in the travel distance requirements unless specifically exempted (e.g., open parking garages, outdoor facilities) (see Section 1017.3.1).

EXIT DISCHARGE. That portion of a *means of egress* system between the termination of an *exit* and a *public way*.

❖ The exit discharge will typically begin when the building occupants reach the exterior at or very near grade level. It provides occupants with a path of travel away from the building. All components between the building and the public way are considered to be the exit discharge, regardless of the distance. In areas of sloping terrain, it is possible to have steps or stairs in the exit discharge leading to the public way. The exit discharge is part of the means of egress and, therefore, its components are subject to the requirements of the code [see Commentary Figures 202(14) and 202(16) and Section 1028].

EXIT DISCHARGE, LEVEL OF. The *story* at the point at which an *exit* terminates and an *exit discharge* begins.

❖ The term is intended to describe the story where the transition from exit to exit discharge occurs. At this level, the occupant needs only to move in a substantially horizontal path to move along exit discharge [see Commentary Figure 202(17)]. Since the level is a volume rather than a horizontal plane, exterior exit steps may be part of the exit discharge when they provide access to the level that is closest to grade.

EXIT, HORIZONTAL. See "Horizontal exit."

EXIT PASSAGEWAY. An *exit* component that is separated from other interior spaces of a building or structure by fire-resistance-rated construction and opening protectives, and

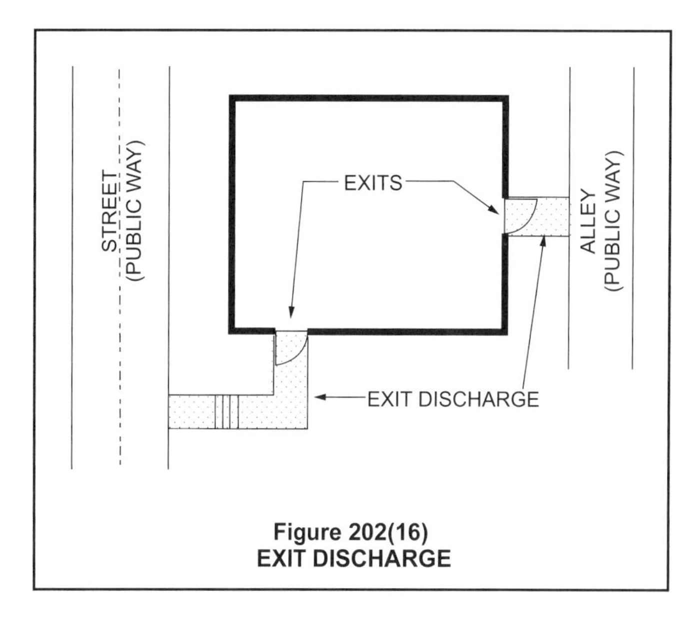

Figure 202(16)
EXIT DISCHARGE

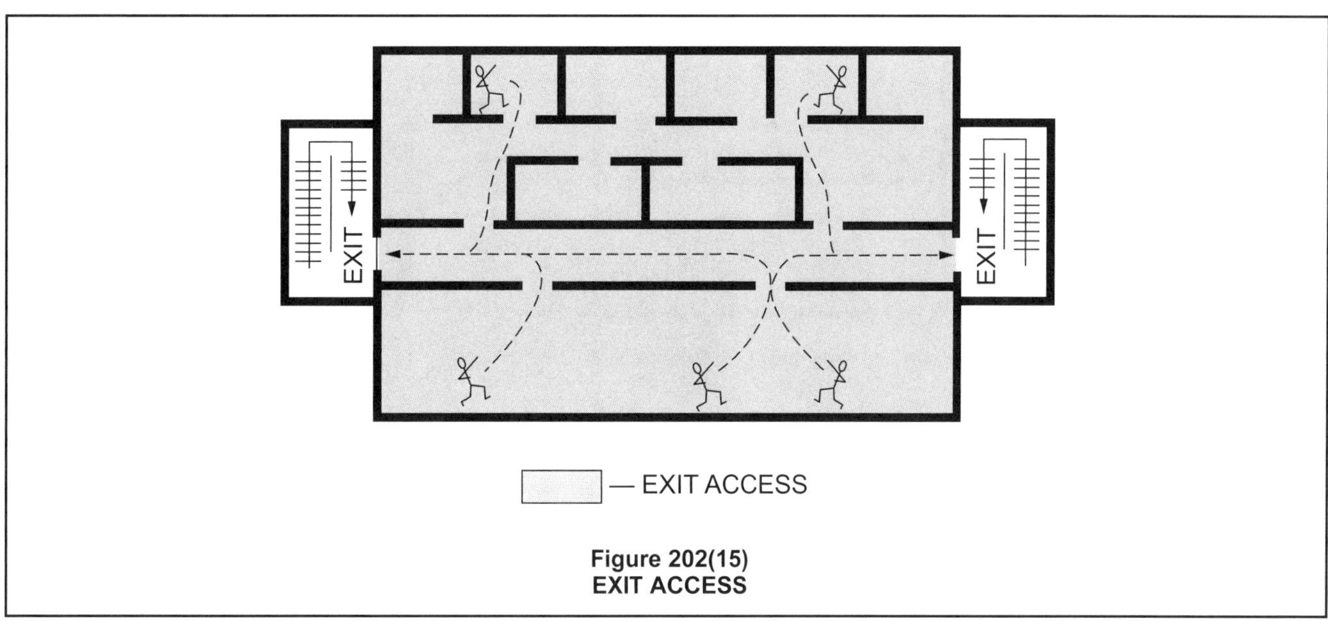

Figure 202(15)
EXIT ACCESS

provides for a protected path of egress travel in a horizontal direction to an *exit* or to the *exit discharge*.

❖ This term refers to a horizontal portion of the means of egress that serves as an exit element. Since an exit passageway is considered an exit element, it must be protected and separated as required by the code for exits (see Section 1024). Exit passageways between a vertical exit enclosure and an exterior exit door are typically found on the level of exit discharge to provide a protected path from a centrally located exit stairway to the exit discharge. In taller buildings that reduce floor sizes as they move up (sometimes called a wedding cake building), exit passageways may be utilized at "transfer floors" as stairway locations shift to move the vertical shafts out as the floor size increases. Exit passageways that lead to an exterior exit door are commonly used in malls to satisfy the travel distance in buildings having a large floor area.

EXPANDED VINYL WALL COVERING. Wall covering consisting of a woven textile backing, an expanded vinyl base coat layer and a nonexpanded vinyl skin coat. The expanded base coat layer is a homogeneous vinyl layer that contains a blowing agent. During processing, the blowing agent decomposes, causing this layer to expand by forming closed cells. The total thickness of the wall covering is approximately 0.055 inch to 0.070 inch (1.4 mm to 1.78 mm).

❖ Expanded vinyl wall coverings are manufactured with a textile backing, an expanded vinyl base coat layer consisting of closed cells and a vinyl skin coat. Expanded vinyl wall coverings are subject to the requirements of Section 803.7.

[F] EXPLOSION. An effect produced by the sudden violent expansion of gases, which may be accompanied by a shock wave or disruption, or both, of enclosing materials or structures. An explosion could result from any of the following:

1. Chemical changes such as rapid oxidation, *deflagration* or *detonation*, decomposition of molecules and runaway polymerization (usually *detonations*).
2. Physical changes such as pressure tank ruptures.
3. Atomic changes (nuclear fission or fusion).

❖ Materials that pose a threat of explosion are classified as Group H-1 when present in quantities exceeding the maximum allowable quantities in Table 307.1(1), and are required to be kept in a detached storage building meeting the requirements of Section 415.7 of the code and Chapter 56 of the IFC.

[F] EXPLOSIVE. A chemical compound, mixture or device, the primary or common purpose of which is to function by explosion. The term includes, but is not limited to, dynamite, black powder, pellet powder, initiating explosives, detonators, safety fuses, squibs, detonating cord, igniter cord, igniters and display fireworks, 1.3G.

The term "explosive" includes any material determined to be within the scope of USC Title 18: Chapter 40 and also includes any material classified as an explosive other than consumer fireworks, 1.4G by the *hazardous materials* regulations of DOTn 49 CFR Parts 100-185.

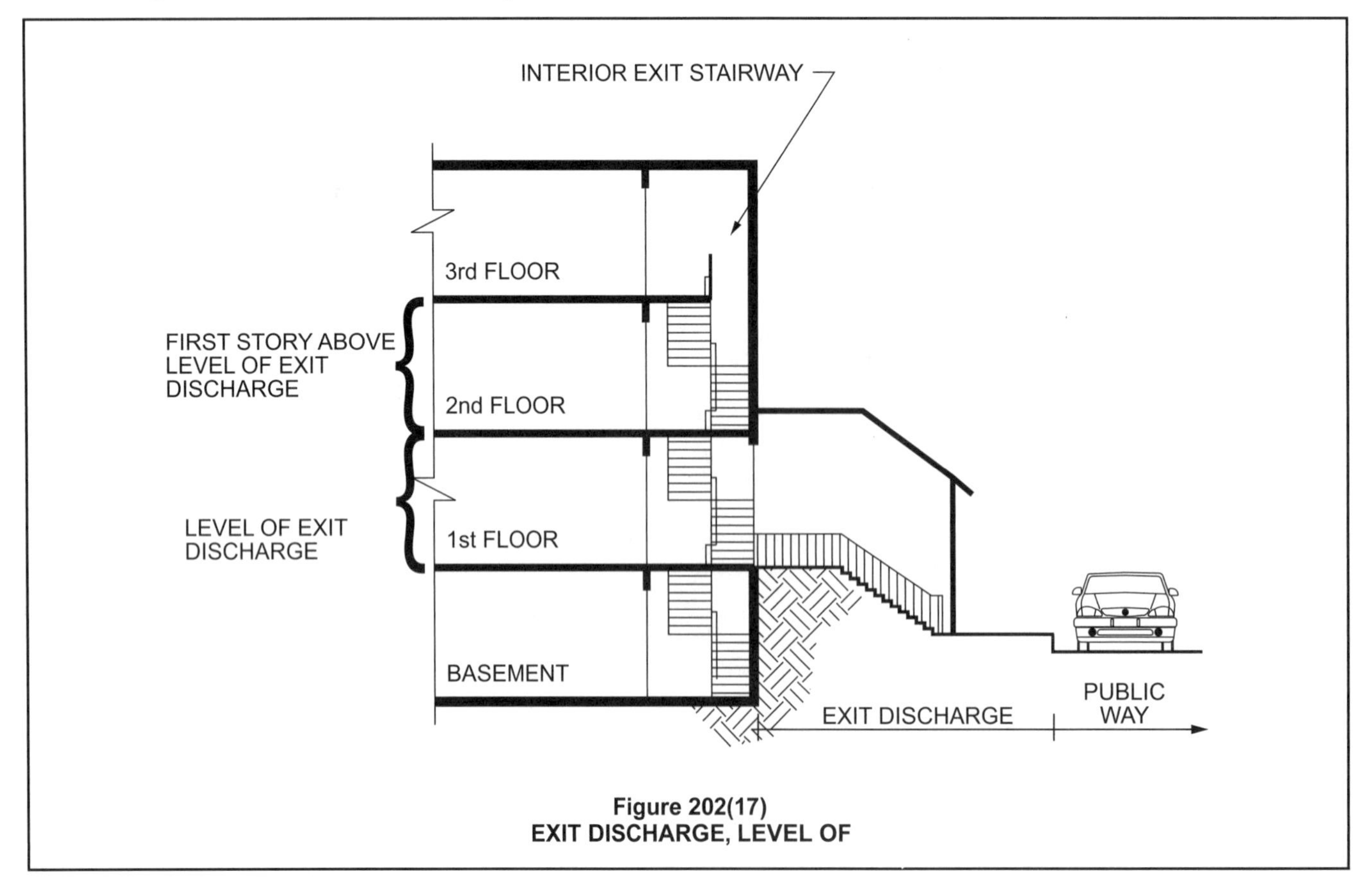

Figure 202(17)
EXIT DISCHARGE, LEVEL OF

High explosive. Explosive material, such as dynamite, which can be caused to detonate by means of a No. 8 test blasting cap when unconfined.

Low explosive. Explosive material that will burn or deflagrate when ignited. It is characterized by a rate of reaction that is less than the speed of sound. Examples of low explosives include, but are not limited to, black powder; safety fuse; igniters; igniter cord; fuse lighters; fireworks, 1.3G and propellants, 1.3C.

Mass-detonating explosives. Division 1.1, 1.2 and 1.5 explosives alone or in combination, or loaded into various types of ammunition or containers, most of which can be expected to explode virtually instantaneously when a small portion is subjected to fire, severe concussion, impact, the impulse of an initiating agent or the effect of a considerable discharge of energy from without. Materials that react in this manner represent a mass explosion hazard. Such an explosive will normally cause severe structural damage to adjacent objects. Explosive propagation could occur immediately to other items of ammunition and explosives stored sufficiently close to and not adequately protected from the initially exploding pile with a time interval short enough so that two or more quantities must be considered as one for quantity-distance purposes.

UN/DOTn Class 1 explosives. The former classification system used by DOTn included the terms "high" and "low" explosives as defined herein. The following terms further define explosives under the current system applied by DOTn for all explosive materials defined as hazard Class 1 materials. Compatibility group letters are used in concert with the division to specify further limitations on each division noted (i.e., the letter G identifies the material as a pyrotechnic substance or article containing a pyrotechnic substance and similar materials).

Division 1.1. Explosives that have a mass explosion hazard. A mass explosion is one which affects almost the entire load instantaneously.

Division 1.2. Explosives that have a projection hazard but not a mass explosion hazard.

Division 1.3. Explosives that have a fire hazard and either a minor blast hazard or a minor projection hazard or both, but not a mass explosion hazard.

Division 1.4. Explosives that pose a minor explosion hazard. The explosive effects are largely confined to the package and no projection of fragments of appreciable size or range is to be expected. An external fire must not cause virtually instantaneous explosion of almost the entire contents of the package.

Division 1.5. Very insensitive explosives. This division is comprised of substances that have a mass explosion hazard, but that are so insensitive there is very little probability of initiation or of transition from burning to *detonation* under normal conditions of transport.

Division 1.6. Extremely insensitive articles which do not have a mass explosion hazard. This division is comprised of articles that contain only extremely insensitive detonating substances and which demonstrate a negligible probability of accidental initiation or propagation.

❖ Explosives either detonate or deflagrate when initiated by heat, shock or electric current. While these materials are normally designed and intended to be initiated by detonators under controlled conditions, heat, shock or electric current from uncontrolled sources may initiate these materials to produce an explosion. DOTn classifies explosives in six classes according to the degree of hazard posed by the material. The most dangerous of these materials is capable of almost simultaneous detonation of all of the material in a single load or store. The least-sensitive explosives produce blasts limited to the packages in which they are transported. This definition of explosives includes materials such as detonators, blasting agents and water gels. Examples of these materials are listed in DOTy 27 CFR, Part 55.23.

Explosive materials are subdivided into high, low, mass-detonating and UN/DOTn Class 1 explosives. High explosives and mass-detonating explosives are typically classified as Group H-1 and present a detonation hazard. Low explosives more commonly are classified as Group H-2, as they tend to deflagrate or burn upon ignition. Mass-detonating devices present a greater threat to adjacent objects and structures. The IFC, therefore, contains provisions in the form of Table 5605.3 to deal with the separation distances for mass explosion hazards.

The definitions cited in this section are consistent with DOTn 49 CFR, Part 173.50. The hazards of this group of materials vary with the nature of the material, with some explosives being very sensitive and others less sensitive. Some explosives detonate, others deflagrate and the hazards of others are limited to intense burning. The classification system was designed to correlate with the system of classification developed under recommendations of the UN, wherein all explosive materials are placed into a hazard class of Class 1. This class is further divided into six divisions: Divisions 1.1 through 1.6.

EXTERIOR EXIT RAMP. An *exit* component that serves to meet one or more *means of egress* design requirements, such as required number of *exits* or *exit access* travel distance, and is open to *yards*, *courts* or *public ways*.

❖ Requirements for an exterior exit ramp are different from those for an interior exit ramp. Exterior exit ramps are typically outside of the building, provide exits from levels above or below the level of exit discharge, and are exposed to at least some elements of weather. The protection requirements, therefore, are for the exterior walls between the building and the ramp, rather than interior walls. For the ramp requirements, see Section 1012. For specifics for exterior exit ramps, see Section 1027

EXTERIOR EXIT STAIRWAY. An *exit* component that serves to meet one or more *means of egress* design requirements, such as required number of *exits* or *exit access* travel distance, and is open to *yards*, *courts* or *public ways*.

❖ Requirements for an exterior exit stairway are different from those for an interior exit stairway. Exterior exit stairways are typically outside of the building, provide exits from above or below the level of exit discharge, and are exposed to at least some elements of weather. The protection requirements, therefore, are for the exterior walls between the building and the stairway, rather than interior walls. For the stairway requirements, see Section 1011. For specifics for exterior exit stairways, see Section 1027.

EXTERIOR INSULATION AND FINISH SYSTEMS (EIFS). EIFS are nonstructural, nonload-bearing, *exterior wall* cladding systems that consist of an insulation board attached either adhesively or mechanically, or both, to the substrate; an integrally reinforced base coat and a textured protective finish coat.

❖ An exterior insulation and finish system (EIFS) is an exterior cladding that is specifically addressed in Section 1408. A definition is necessary in order to ensure the proper application of these requirements.

EXTERIOR INSULATION AND FINISH SYSTEMS (EIFS) WITH DRAINAGE. An EIFS that incorporates a means of drainage applied over a *water-resistive barrier*.

❖ Although EIFS and EIFS with drainage are somewhat similar, the presence of the water-resistive barrier makes it necessary to treat this material differently in code regulations.

EXTERIOR SURFACES. Weather-exposed surfaces.

❖ Exterior surfaces are the vertical and horizontal surfaces located on the outside of a building or structure and are subject to weather conditions, such as wind, changes in temperature, humidity, moisture and dampness. The commentary for the definition of "Weather-exposed surfaces" allows for specific exceptions to areas on the exterior of a building or structure that will not be exposed to direct contact with water, such as rain, sleet and frozen or melting snow.

EXTERIOR WALL. A wall, bearing or nonbearing, that is used as an enclosing wall for a building, other than a *fire wall*, and that has a slope of 60 degrees (1.05 rad) or greater with the horizontal plane.

❖ An exterior wall is defined as an exterior element that encloses a structure and that has a slope equal to or greater than 60 degrees (1.05 rad) from the horizontal plane. Exterior enclosing elements with slopes less than this are generally subjected to more severe weather exposure than vertical surfaces and thus may experience a greater amount of water intrusion. These sloped surfaces, which may include elements such as inset windowsills, sloped parapets and other architectural elements, should be designed to resist water penetration in a manner similar to a roof.

EXTERIOR WALL COVERING. A material or assembly of materials applied on the exterior side of *exterior walls* for the purpose of providing a weather-resisting barrier, insulation or for aesthetics, including but not limited to, *veneers*, siding, *exterior insulation and finish systems*, architectural *trim* and embellishments such as *cornices*, soffits, facias, gutters and leaders.

❖ Materials such as wood, masonry, metal, concrete, structural glass and plastics are used for exterior wall coverings. It should be noted that exterior wall coverings that are combustible are to meet the requirements for combustible materials found in Section 1406.2.

EXTERIOR WALL ENVELOPE. A system or assembly of *exterior wall* components, including *exterior wall* finish materials, that provides protection of the building structural members, including framing and sheathing materials, and conditioned interior space, from the detrimental effects of the exterior environment.

❖ This definition is needed for understanding the proper application of the weather protection requirements in Section 1403.2.

F RATING. The time period that the *through-penetration firestop system* limits the spread of fire through the penetration when tested in accordance with ASTM E814 or UL 1479.

❖ See the definition of "Through-penetration firestop system." The F rating is determined not only from a fire endurance test, but also includes a test for mechanical integrity of the through-penetration seal after fire exposure, known as the "hose stream test." The F rating indicates that the firestop system is capable of stopping the fire, flame and hot gases from passing through the assembly at the penetration.

FABRIC PARTITION. A partition consisting of a finished surface made of fabric, without a continuous rigid backing, that is directly attached to a framing system in which the vertical framing members are spaced greater than 4 feet (1219 mm) on center.

❖ This definition identifies which assemblies are considered fabric partitions for the purpose of applying the specific load criteria that is given in Section 1607.14.1. They are typically for uses such as office furniture panel systems, open floor plans in offices and similar occupancies.

[BS] FABRICATED ITEM. Structural, load-bearing or lateral load-resisting members of assemblies consisting of materials assembled prior to installation in a building or structure, or subjected to operations such as heat treatment, thermal cutting, cold working or reforming after manufacture and prior to installation in a building or structure. Materials produced in accordance with standards referenced by this code, such as rolled structural steel shapes, steel reinforcing bars, *masonry units* and *wood structural panels*, or in accordance with a referenced standard that provides requirements for quality control done under the supervision of a third-party quality control agency, are not "fabricated items."

❖ The term "fabricated item" can easily be misinterpreted to encompass a number of items for which the code does not intend special inspections; therefore, the term is defined to clarify the intent of the code (see Section 1704). Common construction items are often fabricated under standards that are referenced in the code and many of these provide for quality assurance by third-party supervision. Since special inspections are analogous to such quality assurance programs, in these cases it is not warranted to also require special inspections.

[F] FABRICATION AREA. An area within a semiconductor fabrication facility and related research and development areas in which there are processes using hazardous production materials. Such areas are allowed to include ancillary rooms or areas such as dressing rooms and offices that are directly related to the fabrication area processes.

❖ This definition describes the basic component of an HPM facility. The code uses this definition to provide certain material limitations on both a quantity and density basis, and to require enclosure of the fabrication areas with fire barrier assemblies. The fabrication area of an HPM facility is the area where the hazardous materials are actively handled and processed. The fabrication area includes accessory rooms and spaces, such as work stations and employee dressing rooms.

[A] FACILITY. All or any portion of buildings, structures, *site* improvements, elements and pedestrian or vehicular routes located on a *site*.

❖ This term is intentionally broad and includes all portions within a site and all aspects of that site that contain features required to be accessible. This includes parking areas, exterior walkways leading to accessible features, recreational facilities such as playgrounds and picnic areas, as well as any structures on the site (see also the commentary to the definition of "Site").

[BS] FACTORED LOAD. The product of a *nominal load* and a *load factor*.

❖ This definition explains the term "factored loads" so that the loads to be applied in strength design are clear. Factored loads are determined for strength design for load and resistance factor design (LRFD) by multiplying nominal load (see the definition of "Nominal loads") by a load factor, as in the load combinations of Section 1605.2. While this definition is clear for most loading, the exception would be the wind load, W, and earthquake load, E, both of which are computed directly at a strength level without the need for applying a load factor. This is evidenced by a load factor of 1.0 applied to earthquake and wind loads in the strength load combinations of Section 1605.2.

FENESTRATION. Skylights, roof windows, vertical windows (fixed or moveable), opaque doors, glazed doors, glazed block and combination opaque/glazed doors. Fenestration includes products with glass and nonglass glazing materials.

❖ The term "fenestration" refers to both opaque and glazed doors and the light transmitting areas of a wall or roof. It primarily means windows and skylights. The IECC sets performance requirements for fenestration by establishing separate requirements that differ from wall and roof requirements based on the type of fenestration. For certain categories of buildings, the total fenestration area is limited.

[BS] FIBER-CEMENT (BACKER BOARD, SIDING, SOFFIT, TRIM AND UNDERLAYMENT) PRODUCTS. Manufactured thin section composites of hydraulic cementitious matrices and discrete nonasbestos fibers.

❖ This definition enables the code user to apply the correct material requirements for fiber cement products in Chapters 14, 23 and 25.

[BS] FIBER-REINFORCED POLYMER. A polymeric composite material consisting of reinforcement fibers, such as glass, impregnated with a fiber-binding polymer which is then molded and hardened. Fiber-reinforced polymers are permitted to contain cores laminated between fiber-reinforced polymer facings.

❖ Fiber-reinforced polymer is used in building construction for purposes such as light-transmitting sheet panels, structural forms for concrete, sandwich panel construction and the replication of historic building ornamentation. The purpose of this definition is to describe the composition of this material in order to properly identify the element that is to comply with Section 2613.

[BS] FIBERBOARD. A fibrous, homogeneous panel made from lignocellulosic fibers (usually wood or cane) and having a density of less than 31 pounds per cubic foot (pcf) (497 kg/m^3) but more than 10 pcf (160 kg/m^3).

❖ Fiberboard is used primarily as an insulating board and for decorative purposes, but may also be used as wall or roof sheathing. The cellulosic components of fiberboard are broken down to individual fibers and molded to create the bond between the fibers. Other ingredients may be added during processing to provide or improve certain properties, such as strength and water resistance, in addition to surface finishes for decorative products. Fiberboard is used in all locations where panels are desirable, including wall sheathing, insulation of walls and roofs, roof decking, doors and interior finish.

[BS] FIELD NAILING. See "Nailing, field."

FIRE ALARM BOX, MANUAL. See "Manual fire alarm box."

[F] FIRE ALARM CONTROL UNIT. A system component that receives inputs from *automatic* and manual *fire* *alarm* devices and may be capable of supplying power to detection devices and transponders or off-premises transmitters. The control unit may be capable of providing a transfer of power to the notification appliances and transfer of condition to relays or devices.

❖ The fire alarm control unit (panel) acts as a point where all signals initiated within the protected building are received before the signal is transmitted to a constantly attended location. As the name implies, it also contains controls to test and manually activate or silence systems.

[F] FIRE ALARM SIGNAL. A signal initiated by a *fire alarm-initiating device* such as a *manual fire alarm box*, *automatic fire detector*, waterflow switch or other device whose activation is indicative of the presence of a fire or fire signature.

❖ This signal is transmitted to a fire alarm control unit as a warning that requires immediate action. The personnel at the constantly attended location are trained to immediately respond to a fire alarm signal, which indicates the presence of a fire. A fire alarm signal assumes an actual fire has been detected (see the definition of "Alarm signal"). The fire alarm signal is not the signal used to notify the occupants of an emergency condition. Such an action would involve the audible alarm, visual alarm or emergency voice/alarm notification appliances.

[F] FIRE ALARM SYSTEM. A system or portion of a combination system consisting of components and circuits arranged to monitor and annunciate the status of *fire alarm* or *supervisory signal-initiating devices* and to initiate the appropriate response to those signals.

❖ Fire alarm systems are installed in buildings to limit fire casualties and property losses by notifying the occupants of the building, the local fire department or both of an emergency condition. The alarm notification appliances associated with fire alarm systems are intended to be evacuation alarms. All fire alarm systems must be designed and installed to comply with NFPA 72. The term is among the most generic used in the code. It does not necessarily imply an automatic or manual system, nor does it identify what type of notification, if any, should be provided. The definition indicates that an appropriate response must be provided but does not indicate what that response must be. The appropriate responses are identified within the respective sections of Section 907.

FIRE AREA. The aggregate floor area enclosed and bounded by *fire walls*, *fire barriers*, *exterior walls* or *horizontal assemblies* of a building. Areas of the building not provided with surrounding walls shall be included in the fire area if such areas are included within the horizontal projection of the roof or floor next above.

❖ This term is used to describe a specific and controlled area within a building that may consist of a portion of the floor area within a single story, one entire story or the combined floor area of several stories, depending on how these areas are enclosed and separated from other floor areas. Where a fire barrier with a fire-resistance rating in accordance with Section 707.3.10 divides the floor area of a one-story building, the floor area on each side of the wall would constitute a separate fire area. If a horizontal assembly separating the two stories in a two-story building is fire-resistance rated in accordance with Section 711.2.4, each story would be a separate fire area. In cases where mezzanines are present, the floor area of the mezzanine is included in the fire area calculations, even though the area of the mezzanine does not contribute to the building area calculations. See the commentary to Sections 707.3.10 and 711.2.4 for further information.

Note that fire walls are one way of creating fire areas but are typically used to create separate buildings.

FIRE BARRIER. A fire-resistance-rated wall assembly of materials designed to restrict the spread of fire in which continuity is maintained.

❖ The term represents wall assemblies with a fire-resistance rating that are constructed in accordance with Section 707. Even though the definition applies to walls, horizontal assemblies also can be fire barriers. See the definition of "Horizontal assembly" and the requirements in Section 711 that would apply to floor and roof assemblies designed to restrict the spread of fire. See Commentary Figure 202(18) for examples of fire barriers and Commentary Figure 202(29) for examples of horizontal assemblies.

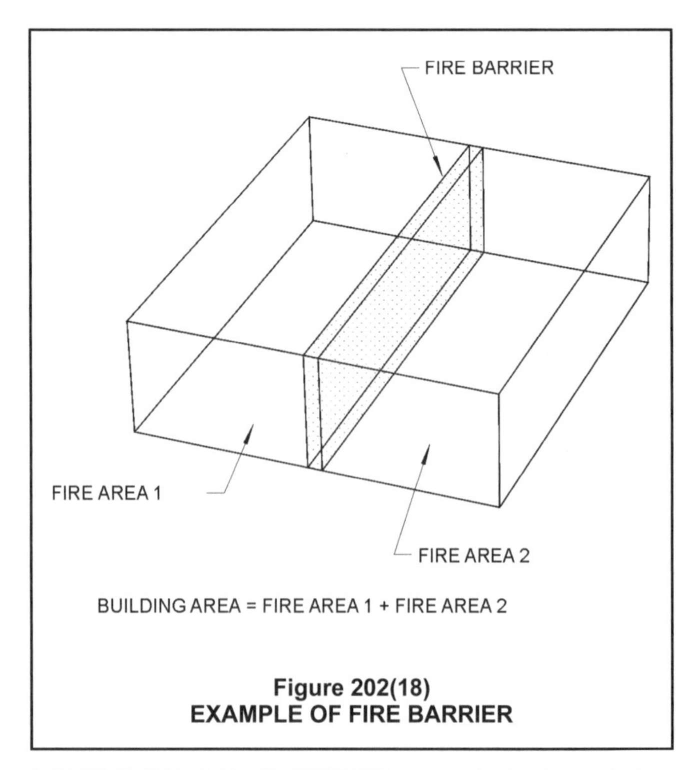

Figure 202(18)
EXAMPLE OF FIRE BARRIER

[F] FIRE COMMAND CENTER. The principal attended or unattended location where the status of detection, alarm communications and control systems is displayed, and from which the systems can be manually controlled.

❖ Fire command centers are communication centers where dedicated manual and automatic facilities are located for the origination, control and transmission of information and instruction pertaining to a fire emergency to the occupants (including fire department personnel) of the building. Fire command centers must provide facilities for the control and display of the status of all fire protection (detection, signaling, etc.) systems. These stations must be located in secure areas as approved by the authority having jurisdiction. Often this is a location near the primary building entrance. Fire command centers also may be combined with other building operations and security facilities when permitted by the authority having jurisdiction; however, operating controls for use by the fire department must be clearly marked.

FIRE DAMPER. A *listed* device installed in ducts and air transfer openings designed to close *automatically* upon detection of heat and resist the passage of flame. Fire dampers are classified for use in either static systems that will *automatically* shut down in the event of a fire, or in dynamic systems that continue to operate during a fire. A dynamic fire damper is tested and rated for closure under elevated temperature airflow.

❖ See the commentary for all dampers with the definition of "Damper."

[F] FIRE DETECTOR, AUTOMATIC. A device designed to detect the presence of a fire signature and to initiate action.

❖ Automatic fire detectors include all approved devices designed to detect the presence of a fire and automatically initiate emergency action. These include smoke-sensing fire detectors, heat-sensing fire detectors, flame-sensing fire detectors, gas-sensing fire detectors and other fire detectors that operate on other principles as approved by the fire code official. Automatic fire detectors must be selected based on the type and size of fire to be detected and the response required. The automatic fire detector sends a signal to a processing unit to initiate some predetermined action. The processing unit may be internal to the device, as is the case with single-station smoke detectors, or it may be an external unit, as in the case of a fire alarm control unit. Automatic fire detectors must be approved, installed and tested to comply with the code and NFPA 72.

FIRE DOOR. The door component of a *fire door assembly*.

❖ A fire door is the primary component of a fire door assembly. The fire protection rating assigned to a tested fire door is only valid if the door is installed in a labeled frame with appropriate hardware. Installation requirements are contained in the referenced standard, NFPA 80. Door ratings are expressed in minutes or hours. Field modification of doors are primarily limited to the mounting of listed hardware.

FIRE DOOR ASSEMBLY. Any combination of a *fire door*, frame, hardware and other accessories that together provide a specific degree of fire protection to the opening.

❖ Fire door assemblies, (door, frame and hardware) are required to be tested using the appropriate standard and then installed in accordance with NFPA 80. Side-hinged doors, hardware and frames are often manufactured separately with manufacturers and listing agencies defining acceptable combinations of assembly components that have been tested together.

FIRE DOOR ASSEMBLY, FLOOR. See "Floor fire door assembly."

FIRE EXIT HARDWARE. *Panic hardware* that is *listed* for use on *fire door assemblies*.

❖ Where a door that is required to be of fire-resistance-rated construction also has panic hardware, the hardware is required to be listed for use on the fire door. Thus, fire door hardware has been tested to function properly when exposed to the effects of a fire (see the definition for "Panic hardware" and Section 1010.1.10).

[F] FIRE LANE. A road or other passageway developed to allow the passage of fire apparatus. A fire lane is not necessarily intended for vehicular traffic other than fire apparatus.

❖ The term "fire lane" is synonymous with the term "fire apparatus access road," both being a road that provides access from a fire station to a building, or portion thereof. However, it should be noted that the driving surface is not necessarily the same as that provided for a public road. The driving surface must be a surface that can be shown to adequately support the load of anticipated emergency vehicles.

FIRE PARTITION. A vertical assembly of materials designed to restrict the spread of fire in which openings are protected.

❖ Fire partitions are used as wall assemblies to separate adjacent tenant spaces in covered and open mall buildings, to separate dwelling units, to separate sleeping rooms, to enclose corridors and to enclose elevator lobbies. Section 708 establishes the construction requirements for fire partitions. The fire-resistance ratings, continuity requirements and opening protective requirements for fire partitions are usually less restrictive than those for fire barriers.

FIRE PROTECTION RATING. The period of time that an opening protective will maintain the ability to confine a fire as determined by tests specified in Section 715. Ratings are stated in hours or minutes.

❖ The term "fire protection rating" applies to the fire performance of an opening protective, such as a fire door, determined through tests performed in accordance with NFPA 252 or UL 10C.

[F] FIRE PROTECTION SYSTEM. *Approved* devices, equipment and systems or combinations of systems used to detect a fire, activate an alarm, extinguish or control a fire, control or manage smoke and products of a fire or any combination thereof.

❖ A fire protection system is any approved device or equipment, used singly or in combination, manually or automatically, that is intended to detect a fire, notify the building occupants of a fire or suppress the fire. Fire protection systems include fire suppression systems, standpipe systems, fire alarm systems, fire detection systems, smoke control systems and smoke vents. All fire protection systems must be approved by the fire code official and tested in accordance with the referenced standards and Section 901.6.

FIRE-RATED GLAZING. Glazing with either a *fire protection rating* or a *fire-resistance rating*.

❖ Fire-rated glazing can be of two types—fire-resistance rating or fire protection rating. Fire-resistance-rated glazing is tested to ASTM E119 or UL 263, may be used in walls and is not considered as an opening. Fire-protection-rated glazing is tested to NFPA 257 or UL 9, and is usually in doors or limited size window openings in fire barriers and fire partitions

FIRE RESISTANCE. That property of materials or their assemblies that prevents or retards the passage of excessive heat, hot gases or flames under conditions of use.

❖ All materials offer some degree of fire resistance. A sheet of plywood has a low level of fire resistance as compared to a concrete block.The fire resistance of a material or an assembly is evaluated by testing performed in accordance with ASTM E119. Tested materials will be assigned a fire-resistance rating consistent with the demonstrated performance.

FIRE-RESISTANCE RATING. The period of time a building element, component or assembly maintains the ability to confine a fire, continues to perform a given structural function, or both, as determined by the tests, or the methods based on tests, prescribed in Section 703.

❖ This refers to the period of time a building element, component or assembly maintains the ability to confine a fire, continues to perform a given structural function, or both, as determined by tests or the methods based on tests prescribed in Section 703.

The fire-resistance rating is developed using standardized test methods (e.g., ASTM E119). Assemblies rated under these tests are deemed to be able to perform their function for a specified period of time under specific fire conditions (standard time-temperature curve).

The fire-resistance rating is not intended to be a prediction of the actual length of time that an assembly will perform its intended function under actual fire conditions. Although the time-temperature curves of standardized fire test methods are usually selected to approximate at least some real-life fire conditions, the very wide range of actual fire conditions makes the listed fire-resistance rating more of a nominal, comparative index than a predictor of fire-resistance time in any given fire incident.

FIRE-RESISTANT JOINT SYSTEM. An assemblage of specific materials or products that are designed, tested and fire-resistance rated in accordance with either ASTM E1966 or UL 2079 to resist for a prescribed period of time the passage of fire through *joints* made in or between fire-resistance-rated assemblies.

❖ In order to maintain the fire-resistant integrity of fire-resistance-rated assemblies, joints that occur within an assembly or between adjacent assemblies must be protected through an installation that has been tested in accordance with ASTM E1966 or UL 2079. Some common examples of applications where a fire-resistant joint system would be required are expansion joints in fire-resistance-rated floors or walls and the junction between fire-resistance-rated floors and walls [see Commentary Figure 202(19) for examples]. The regular joints that occur within a uniform assembly are most often tested as part of fire testing (e.g., in accordance with ASTM E119) for that entire assembly. The required details for these joints are specified in the listings for the underlying assembly. These joints do not need additional testing in accordance with ASTM E1966 or UL 2079. Examples of such joints are the joints between individual sheets of gypsum board in a gypsum-sheathed stud wall. Consequently, other than the joints covering or filling the gaps within an assembly, the need for ASTM E1966 or UL 2079 tested joint systems is usually for the joints between dissimilar assemblies or adjacent assemblies.

[F] FIRE SAFETY FUNCTIONS. Building and fire control functions that are intended to increase the level of life safety for occupants or to control the spread of harmful effects of fire.

❖ In many cases, automatic fire detectors are installed even in buildings not required to have a fire alarm system. These fire detectors perform specific functions such as releasing door hold-open devices, activating elevator recall, smoke damper activation or air distribution system shutdown (see Section 907.3).

FIRE SEPARATION DISTANCE. The distance measured from the building face to one of the following:

1. The closest interior *lot line*.
2. To the centerline of a street, an alley or *public way*.
3. To an imaginary line between two buildings on the lot.

The distance shall be measured at right angles from the face of the wall.

❖ Fire separation distance is the distance from the exterior wall of the building to one of the three following locations, measured perpendicular to the exterior wall face: an interior lot line [see Commentary Figure 202(20)]; the centerline of a street or public way [see Commentary Figure 202(21)]; or an imaginary line between two buildings on the same property [see Commentary Figure 202(22)]. The imaginary line can be located anywhere between the two buildings; it is the designer's choice, but, once established, the location of the line applies to both buildings and cannot be revised.

The distance can vary with irregular-shaped lots and buildings, as shown in Commentary Figures 202(20) and 202(21). When applying the exterior wall requirements of Table 602, the required exterior wall fire-resistance rating might vary along a building side; for example, where the lot line is not parallel to the exterior wall.

FIRE WALL. A fire-resistance-rated wall having protected openings, which restricts the spread of fire and extends continuously from the foundation to or through the roof, with sufficient structural stability under fire conditions to allow collapse of construction on either side without collapse of the wall.

❖ Fire walls must meet the construction requirements in Section 706. The requirements for fire walls are much more restrictive than fire barriers or fire partitions. The material constituting the fire wall must be noncombustible in all construction types except Type V. The vertical and horizontal continuity requirements are much more restrictive, as are those for the opening protectives. A fire wall, unlike the fire barrier and fire partition, must be constructed so it will remain in place if the construction on either side of it collapses. However, the fire wall is not required to remain in place if construction on both sides of it collapses (i.e., the fire wall is not required to be a free-standing cantilever wall). Fire walls are used to divide a structure into separate buildings (see the definition of "Areas, building"). To be considered separate buildings, the division must be vertical. The code applies the term "fire wall" to vertically constructed assemblies only and not to horizontal assemblies.

FIRE WINDOW ASSEMBLY. A window constructed and glazed to give protection against the passage of fire.

❖ Fire windows are "opening protectives" and contain glazing (see Section 716.6). They are required to be tested in accordance with NFPA 257 or UL 9 and are then to be installed in accordance with NFPA 80.

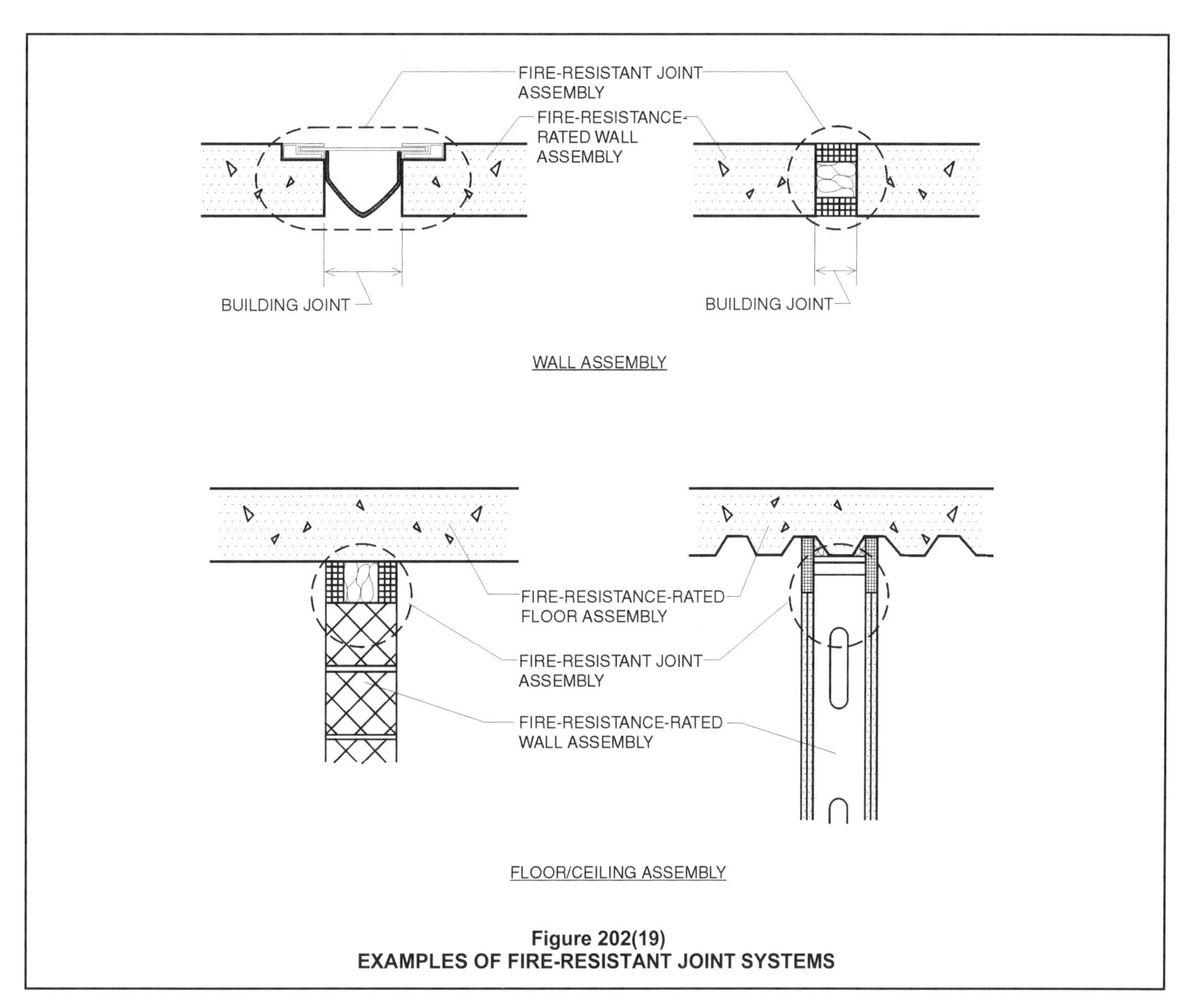

Figure 202(19)
EXAMPLES OF FIRE-RESISTANT JOINT SYSTEMS

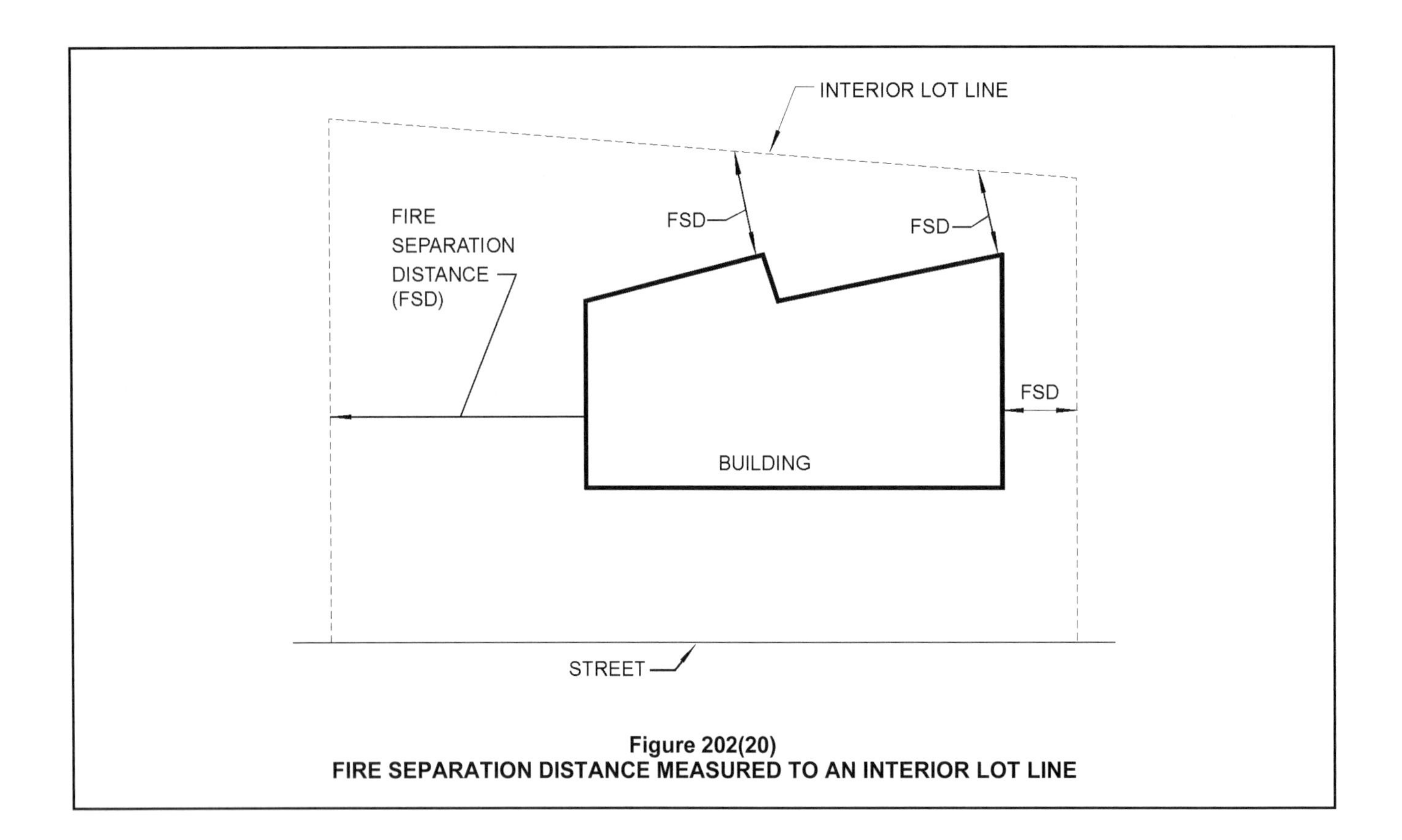

Figure 202(20)
FIRE SEPARATION DISTANCE MEASURED TO AN INTERIOR LOT LINE

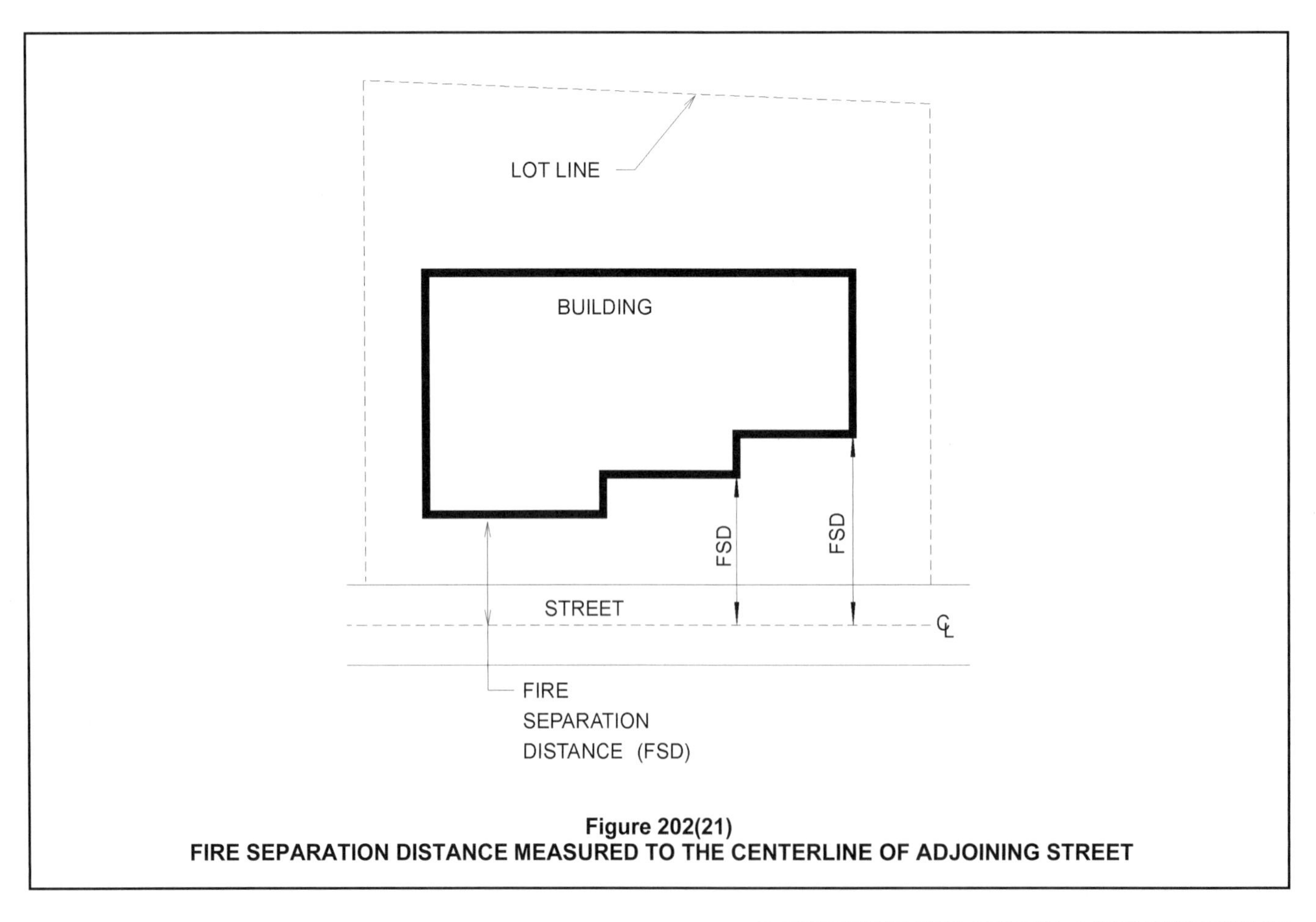

Figure 202(21)
FIRE SEPARATION DISTANCE MEASURED TO THE CENTERLINE OF ADJOINING STREET

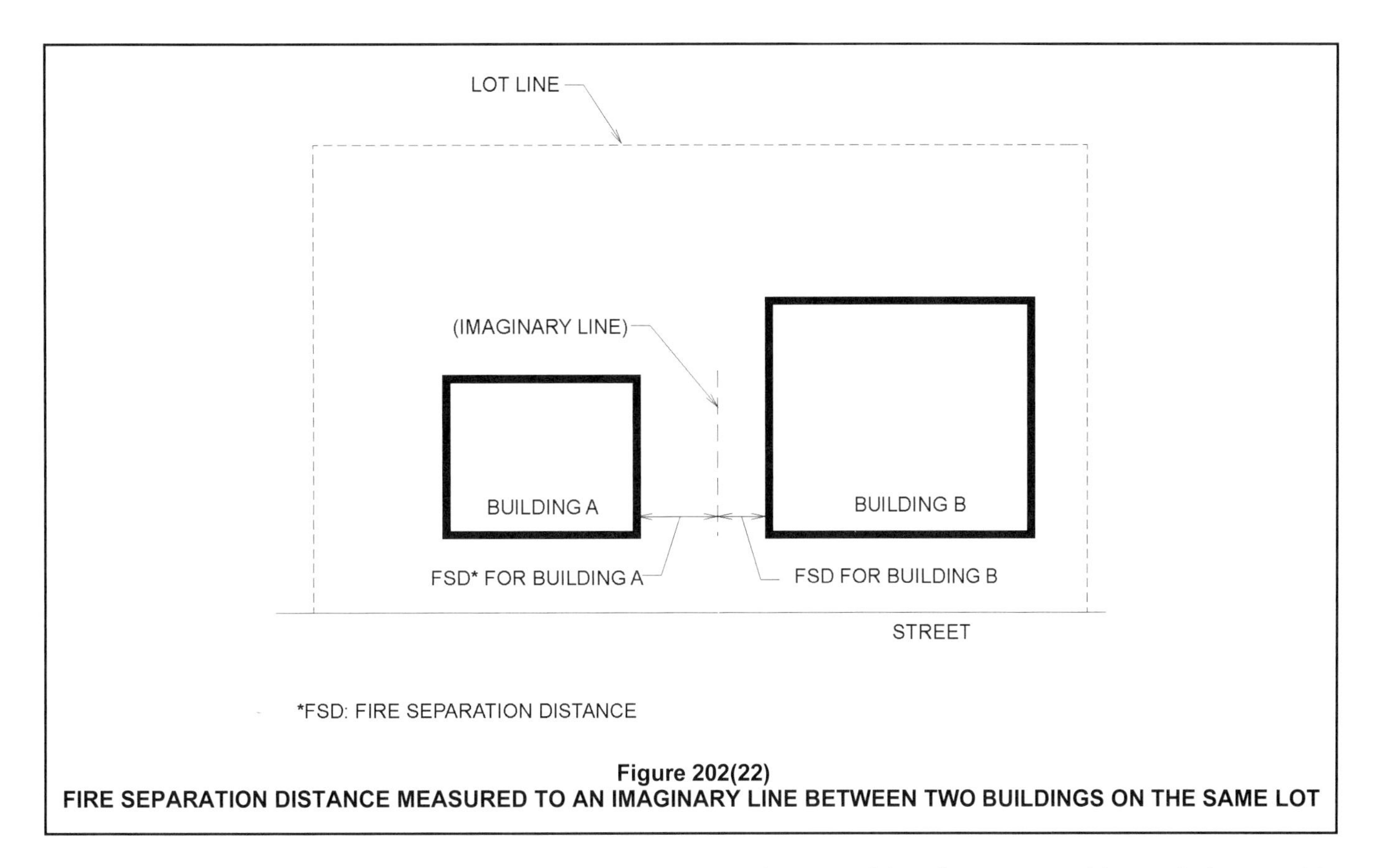

Figure 202(22)
FIRE SEPARATION DISTANCE MEASURED TO AN IMAGINARY LINE BETWEEN TWO BUILDINGS ON THE SAME LOT

FIREBLOCKING. Building materials, or materials *approved* for use as fireblocking, installed to resist the free passage of flame to other areas of the building through concealed spaces.

❖ Fireblocking is required to hinder the concealed spread of flame, heat and other products of combustion within hollow spaces inside of walls or floor/ceiling assemblies. This is done by periodically subdividing that space, as indicated in Section 718.2, using construction materials that have some resistance to fire and by sealing the openings around penetrations through those materials.

Some fireblocking materials are permitted to be combustible based on the rationale that a substantial combustible material will provide a barrier adequate to perform the intended function (also see Section 718).

[M] FIREPLACE. A hearth and fire chamber or similar prepared place in which a fire may be made and which is built in conjunction with a chimney.

❖ Requirements for masonry fireplaces are contained in Section 2111. Note that a chimney must be provided, as that is the means of conveying the products of combustion to the exterior of the building.

FIREPLACE THROAT. The opening between the top of the firebox and the smoke chamber.

❖ This definition is necessary for proper understanding of the code criteria for location and minimum cross-sectional area. This criterion is based on many years of successful performance and is needed to provide proper construction requirements (see Section 2111.8).

FIRESTOP, MEMBRANE-PENETRATION. See "Membrane-penetration firestop."

FIRESTOP, PENETRATION. See "Penetration firestop."

FIRESTOP SYSTEM, THROUGH-PENETRATION. See "Through-penetration firestop system."

[F] FIREWORKS. Any composition or device for the purpose of producing a visible or audible effect for entertainment purposes by combustion, *deflagration* or *detonation* that meets the definition of 1.4G fireworks or 1.3G fireworks.

Fireworks, 1.3G. Large fireworks devices, which are explosive materials, intended for use in fireworks displays and designed to produce audible or visible effects by combustion, *deflagration* or *detonation*. Such 1.3G fireworks include, but are not limited to, firecrackers containing more than 130 milligrams (2 grains) of explosive composition, aerial shells containing more than 40 grams of pyrotechnic composition, and other display pieces which exceed the limits for classification as 1.4G fireworks. Such 1.3G fireworks are also described as fireworks, UN0335 by the DOTn.

Fireworks, 1.4G. Small fireworks devices containing restricted amounts of pyrotechnic composition designed primarily to produce visible or audible effects by combustion. Such 1.4G fireworks which comply with the construction, chemical composition and labeling regulations of the DOTn for fireworks, UN0336, and the U.S. Consumer Product Safety Commission (CPSC) as set forth in CPSC 16 CFR: Parts 1500 and 1507, are not explosive materials for the purpose of this code.

❖ Any device containing an explosive material that produces an audible or visible effect through combustion, deflagration, detonation or explosion is considered a firework. Fireworks are divided into two categories, 1.4G and 1.3G, based on the amount of pyrotechnic composition present.

The definitions of "Fireworks, 1.3G" and "Fireworks, 1.4G" are derived from the DOTn 49 CFR classification system for transporting explosives, and from NFPA 1124. The amount of pyrotechnic composition is the distinguishing factor between the two types of fireworks (see the commentary to the definition of "Pyrotechnic composition"). Fireworks that contain a limited amount of pyrotechnic composition are classified as 1.4G fireworks. Such fireworks represent a physical hazard (Group H-3), while display fireworks represent a detonation hazard (Group H-1).

The requirements for storage, display and labeling depend on the correct application of the definition of "Fireworks, 1.4G." This definition reflects the construction, chemical composition and labeling requirements of the U.S. Consumer Product Safety Commission (CPSC) found in 16 CFR, Parts 1500 and 1507. Also, 1.4G fireworks are not considered to be explosives in accordance with the provisions of Chapter 56 of the IFC.

FIXED BASE OPERATOR (FBO). A commercial business granted the right by the airport sponsor to operate on an airport and provide aeronautical services, such as fueling, hangaring, tie-down and parking, aircraft rental, aircraft maintenance and flight instruction.

❖ Fixed base operator (FBO) is an aviation industry term used to describe a firm that is permanently based at an airport and provides a variety of aircraft services. It is used in conjunction with Section 412.4.6 in the context of determining the appropriate level of fire suppression required in various aircraft hangars at an airport (see commentary, Section 412.4.6).

FIXED SEATING. Furniture or fixture designed and installed for the use of sitting and secured in place including bench-type seats and seats with or without backs or arm rests.

❖ Fixed seating is secured to the floor or is a part of a seating system. Some of the varieties are: bench seating such as in bleachers, which can come with or without back rests; theater seating with arms for each viewer and seats that flip up; lecture halls with individual seats with tablet arms that are fixed or retractable; pew-type seating in courtrooms or places of religious worship; or booths in a restaurant. To allow appropriate egress from these spaces, the occupant load tables allow for the individual seats to be counted to determine occupant load for the space. Space between the seats (i.e., aisle accessways) and aisles leading to the exits for these types of seating is addressed in Section 1029. Given their unique issues, bleachers, folding and telescopic seating and grandstands are referenced in ICC 300 in Section 1029.1.1.

FLAME SPREAD. The propagation of flame over a surface.

❖ The rate at which flames travel along the surface of a combustible finish material directly impacts the speed with which a fire spreads within a room or space and is, therefore, regulated by Chapter 8.

FLAME SPREAD INDEX. A comparative measure, expressed as a dimensionless number, derived from visual measurements of the spread of flame versus time for a material tested in accordance with ASTM E84 or UL 723.

❖ The ASTM E84 and UL 723 test methods render measurements of surface flame spread (and smoke density) in comparison with test results obtained by using select red oak as a control material. Red oak is used as a control material for furnace calibration because it is a fairly uniform grade of lumber that is readily available nationally, is uniform in thickness and moisture content, and generally gives consistent and reproducible results. The results of this test simply provide a relative understanding of flame spread potential. The flame spread index is sometimes abbreviated as FSI.

[F] FLAMMABLE GAS. A material that is a gas at 68°F (20°C) or less at 14.7 pounds per square inch atmosphere (psia) (101 kPa) of pressure [a material that has a *boiling point* of 68°F (20°C) or less at 14.7 psia (101 kPa)] which:

1. Is ignitable at 14.7 psia (101 kPa) when in a mixture of 13 percent or less by volume with air; or
2. Has a flammable range at 14.7 psia (101 kPa) with air of at least 12 percent, regardless of the lower limit.

The limits specified shall be determined at 14.7 psi (101 kPa) of pressure and a temperature of 68°F (20°C) in accordance with ASTM E681.

❖ This term essentially refers to any type of compressed gas that burns in normal concentrations of oxygen in the air (see the definition of "Compressed gas"). The definition is consistent with the provisions of ASTM E681.

[F] FLAMMABLE LIQUEFIED GAS. A liquefied compressed gas which, under a charged pressure, is partially liquid at a temperature of 68°F (20°C) and which is flammable.

❖ This term essentially refers to any type of liquefied compressed gas that burns in normal concentrations of oxygen in the air (see the definition of "Compressed gas").

[F] FLAMMABLE LIQUID. A *liquid* having a closed cup *flash point* below 100°F (38°C). Flammable liquids are further categorized into a group known as Class I liquids. The Class I category is subdivided as follows:

Class IA. *Liquids* having a *flash point* below 73°F (23°C) and a *boiling point* below 100°F (38°C).

Class IB. *Liquids* having a *flash point* below 73°F (23°C) and a *boiling point* at or above 100°F (38°C).

Class IC. *Liquids* having a *flash point* at or above 73°F (23°C) and below 100°F (38°C). The category of flammable liquids does not include *compressed gases* or *cryogenic fluids.*

❖ While all flammable liquids have a closed cup flash point less than 100°F (38°C), the further classification of the Class I liquid is dependent on the boiling point (see the definition of "Boiling point"). The 100°F (38°C) flash point limitation for flammable liquids assumes possible indoor ambient temperature conditions of 100°F (38°C).

[F] FLAMMABLE MATERIAL. A material capable of being readily ignited from common sources of heat or at a temperature of 600°F (316°C) or less.

❖ Many standardized tests, such as those contained in ASTM E136 and NFPA 701, have been developed to assess the flammability and fire hazards of materials. Both of these test standards include objective criteria for evaluating the combustibility of different materials, however, great care must be taken in conducting and evaluating the results of such tests.

[F] FLAMMABLE SOLID. A *solid*, other than a blasting agent or *explosive*, that is capable of causing fire through friction, absorption or moisture, spontaneous chemical change, or retained heat from manufacturing or processing, or which has an ignition temperature below 212°F (100°C) or which burns so vigorously and persistently when ignited as to create a serious hazard. A chemical shall be considered a flammable *solid* as determined in accordance with the test method of CPSC 16 CFR; Part 1500.44, if it ignites and burns with a self-sustained flame at a rate greater than 0.1 inch (2.5 mm) per second along its major axis.

❖ Flammable solids are combustible materials that ignite easily and burn rapidly. Solids that may cause a fire due to friction are considered flammable solids as are metal powders that can be readily ignited. Examples of flammable solids include nitrocellulose and combustible metals, such as magnesium and titanium. For further discussion of this material definition, see the commentary to it in Chapter 2 of the *International Fire Code® Commentary.*

[F] FLAMMABLE VAPORS OR FUMES. The concentration of flammable constituents in air that exceeds 25 percent of their *lower flammable limit (LFL).*

❖ Vapors or fumes are considered to be flammable when they exceed 25 percent of their LFL. The LFL of a given vapor in air is the concentration at which flame propagation could occur in the presence of an ignition source (see the definition of "Lower flammable limit").

[F] FLASH POINT. The minimum temperature in degrees Fahrenheit at which a *liquid* will give off sufficient vapors to form an ignitable mixture with air near the surface or in the container, but will not sustain combustion. The flash point of a *liquid* shall be determined by appropriate test procedure and apparatus as specified in ASTM D56, ASTM D93 or ASTM D3278.

❖ The flash point is the characteristic used in the classification of flammable and combustible liquids. The Tag Closed Tester (ASTM D56), the Pensky-Martens Closed Cup Tester (ASTM D93) and the Small Scale Closed-Cup Apparatus (ASTM D3278) are the referenced test procedures for determining the flash points of liquids. The applicability of the respective test method is dependent on the viscosity of the test liquid and the expected flash point.

FLIGHT. A continuous run of rectangular treads, *winders* or combination thereof from one landing to another.

❖ Two points of clarification for stairways have been addressed by the definition of "Flight." First, a flight is made up of the treads and risers that occur between landings. Therefore, a stairway connecting two stories that includes an intermediate landing consists of two flights. Secondly, the inclusion of winders within a stairway does not create multiple flights. Winders are simply treads within a flight and are often combined with rectangular treads within the same flight.

[BS] FLOOD or FLOODING. A general and temporary condition of partial or complete inundation of normally dry land from:

1. The overflow of inland or tidal waters.
2. The unusual and rapid accumulation or runoff of surface waters from any source.

❖ The term "flood" is broadly defined because the condition may occur near all types and sizes of bodies of water. Along streams and riverine areas, flooding results from the accumulation of rainfall runoff that drains from upland watersheds. Along coasts and the shorelines of large lakes, flooding is caused by wind-driven surges and waves that push water onshore, often augmented by tidal influences.

[BS] FLOOD DAMAGE-RESISTANT MATERIALS. Any construction material capable of withstanding direct and prolonged contact with floodwaters without sustaining any damage that requires more than cosmetic *repair*.

❖ A building or structure located within a flood hazard area will have certain structural and nonstructural elements below the design flood elevation, even when the lowest floor is elevated in compliance with the requirements of Section 1612. To minimize damage and facilitate cleanup, the materials used in those elements are to be resistant to damage by floodwater. In coastal high-hazard areas (commonly called "V zones"), walls that are designed to break away during the design flood may be wetted during floods of lesser magnitude. When flood damage-resistant materials are used, the owners are not faced with significant repairs. In riverine and inland coastal flood hazard areas (commonly called "A zones"), enclosures may be allowed if designed and constructed with openings that allow floodwater to enter and exit without causing structural damage. To minimize costs and facilitate cleanup, materials used for those enclosures are to resist flood damage. Prolonged contact is generally considered to be partial or total inundation for at least 72 hours. In some instances, materials that are not flood damage-resistant materials, such as wiring for fire alarms and emergency lighting, are allowed below the flood elevation, if specifically required to address life safety and electrical code requirements for parking vehicles, building access and storage areas. For further guidance, refer to FEMA TB #2, *Flood Damage-Resistant Materials Requirements for Buildings Located in Special Flood Hazard Areas*.

FLOOD, DESIGN. See "Design flood."

FLOOD ELEVATION, DESIGN. See "Design flood elevation."

[BS] FLOOD HAZARD AREA. The greater of the following two areas:

1. The area within a flood plain subject to a 1-percent or greater chance of *flooding* in any year.
2. The area designated as a flood hazard area on a community's flood hazard map, or otherwise legally designated.

❖ FEMA prepares Flood Insurance Rate Maps (FIRMs) which delineate the land area that is subject to inundation by the 1-percent annual chance flood. Some states and local jurisdictions develop and adopt maps of flood hazard areas that are more extensive than the areas shown on FEMA's maps. For the purpose of the code, the flood hazard area within which the requirements are to be applied is the greater of the two delineated areas.

FLOOD HAZARD AREAS, SPECIAL. See "Special flood hazard area."

[BS] FLOOD HAZARD AREA SUBJECT TO HIGH-VELOCITY WAVE ACTION. Area within the *flood hazard area* that is subject to high-velocity wave action, and shown on a Flood Insurance Rate Map (FIRM) or other flood hazard map as Zone V, VO, VE or V1-30.

❖ Some coastal and lake shorelines experience flooding that includes wind-driven waves. Flood hazard areas that are anticipated to have high-velocity wave action, which changes anticipated flood loads, are specifically designated on FEMA's flood hazard area maps and commonly are referred to as "V Zones." In these areas, wave heights are predicted to be greater than or equal to 3 feet (914 mm), or the wave runup elevations are predicted to reach 3 feet (914 mm) or more above grade.

[BS] FLOOD INSURANCE RATE MAP (FIRM). An official map of a community on which the Federal Emergency Management Agency (FEMA) has delineated both the *special flood hazard areas* and the risk premium zones applicable to the community.

❖ The FIRM shows the flood hazard areas along bodies of water where some level of assessment or study by FEMA has determined that there is a risk of flooding. In addition to showing the extent of flood hazard areas, base flood elevations are shown where studies using detailed methods were conducted and floodways may be delineated. Users are cautioned that FIRMs may not show all bodies of water or all areas prone to flooding, including areas with local drainage problems.

[BS] FLOOD INSURANCE STUDY. The official report provided by the Federal Emergency Management Agency containing the Flood Insurance Rate Map (FIRM), the Flood Boundary and Floodway Map (FBFM), the water surface elevation of the *base flood* and supporting technical data.

❖ A Flood Insurance Study is prepared for each local jurisdiction for which FEMA has evaluated flood hazards. The report summarizes information that describes and supports the flood hazard area maps, including narratives of records of past floods, descriptions of the methodologies used, floodway data tables with additional information for waterways for which floodways have been designated, and tables with additional information for shoreline transects.

[BS] FLOODWAY. The channel of the river, creek or other watercourse and the adjacent land areas that must be reserved in order to discharge the *base flood* without cumulatively increasing the water surface elevation more than a designated height.

❖ FEMA delineates floodways along most rivers and streams that are studied using detailed methods. The floodway is the area that must be kept clear of encroachments, such as fill and buildings, in order to allow a base flood to pass without increasing the water surface more than a designated height. The designated height is found by referencing the floodway data table in the Flood Insurance Study. In general, floodways have faster flow velocities and deeper water than in adjacent floodway fringe areas.

FLOOR AREA, GROSS. The floor area within the inside perimeter of the *exterior walls* of the building under consideration, exclusive of vent *shafts* and *courts*, without deduction for *corridors*, *stairways*, *ramps*, closets, the thickness of interior walls, columns or other features. The floor area of a building, or portion thereof, not provided with surrounding *exterior walls* shall be the usable area under the horizontal projection of the roof or floor above. The gross floor area shall not include *shafts* with no openings or interior *courts*.

❖ Gross floor area is that area measured within the perimeter formed by the inside surface of the exterior walls. The area of all occupiable and nonoccupiable spaces, including mechanical and elevator shafts, toilet rooms, closets and mechanical equipment rooms, are included in the gross floor area. This area could also include any covered porches, carports or other exterior space intended to be used as part of the building's occupiable space. Both gross and net floor areas are used for the determination of occupant load in accordance with Table 1004.1.2.

FLOOR AREA, NET. The actual occupied area not including unoccupied accessory areas such as *corridors*, *stairways*, *ramps*, toilet rooms, mechanical rooms and closets.

❖ This area is intended to be only the room areas that are used for specific occupancy purposes and does not include circulation areas, such as corridors, ramps or stairways, or service and utility spaces, such as toilet rooms and mechanical and electrical equipment rooms. Net floor area is typically measured between inside faces of walls within a room. Floor area, net and gross, is utilized in Table 1004.1.2 to determine occupant load for a space.

FLOOR FIRE DOOR ASSEMBLY. A combination of a *fire door*, a frame, hardware and other accessories installed in a horizontal plane, which together provide a specific degree of fire protection to a through-opening in a fire-resistance-rated floor (see Section 712.1.13.1).

❖ Floor fire door assemblies are required to be tested in accordance with NFPA 288 and are used to protect openings in fire-resistance-rated floors. They are one alternative for protecting a floor opening, such as an access opening to mechanical equipment. See the commentary to Section 712.1.13.1 for additional information on floor fire door assemblies.

[F] FOAM-EXTINGUISHING SYSTEM. A special system discharging a foam made from concentrates, either mechanically or chemically, over the area to be protected.

❖ Foam-extinguishing systems must be of an approved type and installed and tested to comply with NFPA 11, 11A and 16. All foams are intended to exclude oxygen from the fire, cool the area of the fire and insulate adjoining surfaces from heat caused by fires. Foam systems are commonly used to extinguish flammable or combustible liquid fires (see commentary, Section 904.7). While water applied by an automatic sprinkler system can only act horizontally upon the surface that it reaches, foam-extinguishing agents have the ability to act vertically in addition to horizontally. Further, unlike gaseous extinguishing agents, foam does not dissipate rapidly where there is no confined space. Thus, foam systems are also used where there is a need to fill a nonconfined space with extinguishing material as in the case of certain industrial applications.

FOAM PLASTIC INSULATION. A plastic that is intentionally expanded by the use of a foaming agent to produce a reduced-density plastic containing voids consisting of open or closed cells distributed throughout the plastic for thermal insulating or acoustical purposes and that has a density less than 20 pounds per cubic foot (pcf) (320 kg/m^3).

❖ Foam plastic insulation is plastic that is imbedded with gas.

[BS] FOLDING AND TELESCOPIC SEATING. Tiered seating having an overall shape and size that is capable of being reduced for purposes of moving or storing and is not a building element.

❖ Bleachers, folding and telescopic seating and grandstands are essentially unique forms of tiered seating that are supported on a dedicated structural system. All types are addressed in ICC 300, the safety standard for these types of seating arrangements. Folding and telescopic seating is commonly used in gymnasiums and sports arenas where the seating can be configured in a variety of ways for various types of events. "Building element" is a defined term that is primarily used in conjunction with the structural elements regulated in Table 601. While telescopic seating may be attached to a wall, the system when pulled out or folded includes its main support system. Such seating is not considered a building element of the building or structure in which it is located (see Section 1029.1.1).

FOOD COURT. A public seating area located in the *mall* that serves adjacent food preparation tenant spaces.

❖ Typical mall building layouts include a central gathering area for food and drink consumption. These areas are usually located in the mall itself and chairs are provided for the public's use to consume the food and drink. This public area is usually surrounded by numerous tenant spaces where food is prepared and sold over the counter. A separate design occupant load is required to be calculated in accordance with Section 402.8.2.4.

FOSTER CARE FACILITIES. Facilities that provide care to more than five children, $2^1/_2$ years of age or less.

❖ Foster care facilities are group homes where children live, not day care facilities. By being under $2^1/_2$ years of age, children are assumed not capable of self-preservation, thus there is a need for higher levels of active and passive protection in the building (see the commentary to Group I-2 in Section 308.4). Group homes with children over $2^1/_2$ years of age would be classified as Group R-4 or I-1, depending on the number of children housed in the facility. The "more than five children" is intended to clarify that a foster care family would not be considered a Group I-2 facility. See also the definitions for "24-hour care," "Custodial care," "Group home," "Personal care" and "Incapable of self-preservation."

[BS] FOUNDATION PIER (for Chapter 21). An isolated vertical foundation member whose horizontal dimension measured at right angles to its thickness does not exceed three times its thickness and whose height is equal to or less than four times its thickness.

❖ This definition is intended specifically for masonry construction. Masonry foundation piers are similar to masonry columns, except that they are shorter and do not need to comply with the prescriptive reinforcement detailing requirements for masonry columns.

FRAME STRUCTURE. A building or other structure in which vertical *loads* from floors and roofs are primarily supported by columns.

❖ This definition describes the structure types that are covered under Section 1615.3.

GABLE. The triangular portion of a wall beneath the end of a dual-slope, pitched, or mono-slope roof or portion thereof and above the top plates of the story or level of the ceiling below.

❖ This definition enables the code user to correctly apply sheathing requirements found in Sections 2304 and 2308.

[F] GAS CABINET. A fully enclosed, ventilated noncombustible enclosure used to provide an isolated environment for *compressed gas* cylinders in storage or *use*. Doors and access ports for exchanging cylinders and accessing pressure-regulating controls are allowed to be included.

❖ Gas cabinets are used to provide adequate control for escaping gas in the event of a leaking cylinder of compressed gases. Gas cabinets are commonly used when dealing with highly toxic and toxic compressed gases. Sections 5003.8.6 and 6004.1.2 of the IFC provide additional construction and ventilation requirements for gas cabinets.

[F] GAS ROOM. A separately ventilated, fully enclosed room in which only *compressed gases* and associated equipment and supplies are stored or *used.*

❖ Gas rooms are used exclusively for the storage or use of hazardous gases in excess of the maximum allowable quantities permitted by Tables 307.1(1) and 307.1(2). Gas rooms are commonly used as an alternative storage area for HPM gases in a Group H-5 facility.

[F] GASEOUS HYDROGEN SYSTEM. An assembly of piping, devices and apparatus designed to generate, store, contain, distribute or transport a nontoxic, gaseous hydrogen-containing mixture having not less than 95-percent hydrogen gas by volume and not more than 1-percent oxygen by volume. Gaseous hydrogen systems consist of items such as *compressed gas* containers, reactors and appurtenances, including pressure regulators, pressure relief devices, manifolds, pumps, compressors and interconnecting piping and tubing and controls.

❖ This term includes the source of hydrogen and all piping and devices between the source and the equipment being used. The gas in a hydrogen system is above the upper flammable limit (UFL) and is therefore "too rich" to burn. Any leakage, however, can quickly create conditions that will be explosive under ambient conditions.

GLASS FIBERBOARD. Fibrous glass roof insulation consisting of inorganic glass fibers formed into rigid boards using a binder. The board has a top surface faced with asphalt and kraft reinforced with glass fiber.

❖ Depending on the type and location of insulation in walls, floors and roofs, the insulation may impact the fire-resistance rating (see Sections 703.2 and 721.1). Glass fiber insulation is specifically listed in Table 722.6.2(5).

GRADE FLOOR OPENING. A window or other opening located such that the sill height of the opening is not more than 44 inches (1118 mm) above or below the finished ground level adjacent to the opening.

❖ Openings used for emergency escape or rescue are clearly easier to use the closer they are to grade. This definition specifies that the maximum sill height above the exterior adjacent grade must be no more than 44 inches (1118 mm) to qualify as a grade floor opening (see Section 1030.2).

[BS] GRADE (LUMBER). The classification of lumber in regard to strength and utility in accordance with American Softwood Lumber Standard DOC PS 20 and the grading rules of an *approved* lumber rules-writing agency.

❖ The grade identifies the ability of a particular piece of lumber to resist applied loads. The mark is applied to solid sawn pieces of wood and includes the species, grade and whether it was finished (surfaced) green or dry. The species and grade designations on the grade mark, when used in conjunction with the design value in the NDS, provide all the information necessary to determine what load the piece of lumber is capable of holding. The ALSC provides facsimile sheets for agencies accredited to grade lumber. Although rare, fake grade stamps can be found in the marketplace.

GRADE PLANE. A reference plane representing the average of finished ground level adjoining the building at *exterior walls*. Where the finished ground level slopes away from the *exterior walls*, the reference plane shall be established by the lowest points within the area between the building and the *lot line* or, where the *lot line* is more than 6 feet (1829 mm) from the building, between the building and a point 6 feet (1829 mm) from the building.

❖ This term is used in the definitions of "Basement" and "Story above grade plane." It is critical in determining the height of a building and the number of stories, which are regulated by Chapter 5. Since the finished ground surface adjacent to the building may vary (depending on site conditions), the mean average taken at various points around the building constitutes the grade plane. One method of determining the grade plane elevation is illustrated in Commentary Figure 202(23), where the ground slopes uniformly along the length of each exterior wall.

Where a site has a more complex slope, a more detailed calculation that takes into account the various segments of the perimeter walls must be taken. Commentary Figure 202(24) shows an example of a complex finished grade. A full calculation will show the grade plane to be at an elevation of 498.64 feet (151 986 mm). If a calculation is done based on just the four extreme corners, grade plane would be thought to be 495.5 feet (151 029 mm), an error of more than 3 feet (914 mm).

Situations may arise where the ground adjacent to the building slopes away from the building because of site or landscaping considerations. In this case, the lowest finished ground level at any point between the building's exterior wall and a point 6 feet (1829 mm) from the building [or the lot line, if closer than 6 feet (1829 mm)] comes under consideration. These points are used to determine the elevation of the grade plane as illustrated in Commentary Figures 202(25) and 202(26).

In the context of the code, the term "grade" means the finished ground level at the exterior walls. While the grade plane is a hypothetical horizontal plane derived as indicated above, the grade is that which actually exists or is intended to exist at the completion of site work. The only situation where the grade plane and the grade are identical is when the site is perfectly level for a distance of 6 feet (1829 mm) from all exterior walls.

GRADE PLANE, STORY ABOVE. See "Story above grade plane."

GRANDSTAND. Tiered seating supported on a dedicated structural system and two or more rows high and is not a building element (see "*Bleachers*").

❖ Bleachers, folding and telescopic seating and grandstands are essentially unique forms of tiered seating supported on a dedicated structural system. All types are addressed in the safety standard for these types of seating arrangements, ICC 300. Grandstands can be found at a county fair ground, along a parade route or within indoor facilities. Examples are sports arenas and public auditoriums, as well as places of religious worship and gallery-type lecture halls. "Building element" is a defined term that is primarily used in conjunction with the structural elements regulated in Table 601. Grandstands have a separate structural system. Individual bench seats directly attached to a floor system are not a grandstand. The terms "bleacher" and "grandstand" are basically interchangeable. There is no cutoff in size or number of seats that separates bleachers and grandstands (see Section 1029.1.1).

GROSS LEASABLE AREA. The total floor area designed for tenant occupancy and exclusive use. The area of tenant occupancy is measured from the centerlines of joint partitions to the outside of the tenant walls. All tenant areas, including areas used for storage, shall be included in calculating gross leasable area.

❖ The gross leasable area represents the aggregate area available in a covered or open mall building for tenant occupancy. It does not include the area of the mall, unless portions of it are leased for the purposes of setting up separate tenant spaces (kiosks). The area is used to determine the design occupant load in accordance with Section 402.8.2.1.

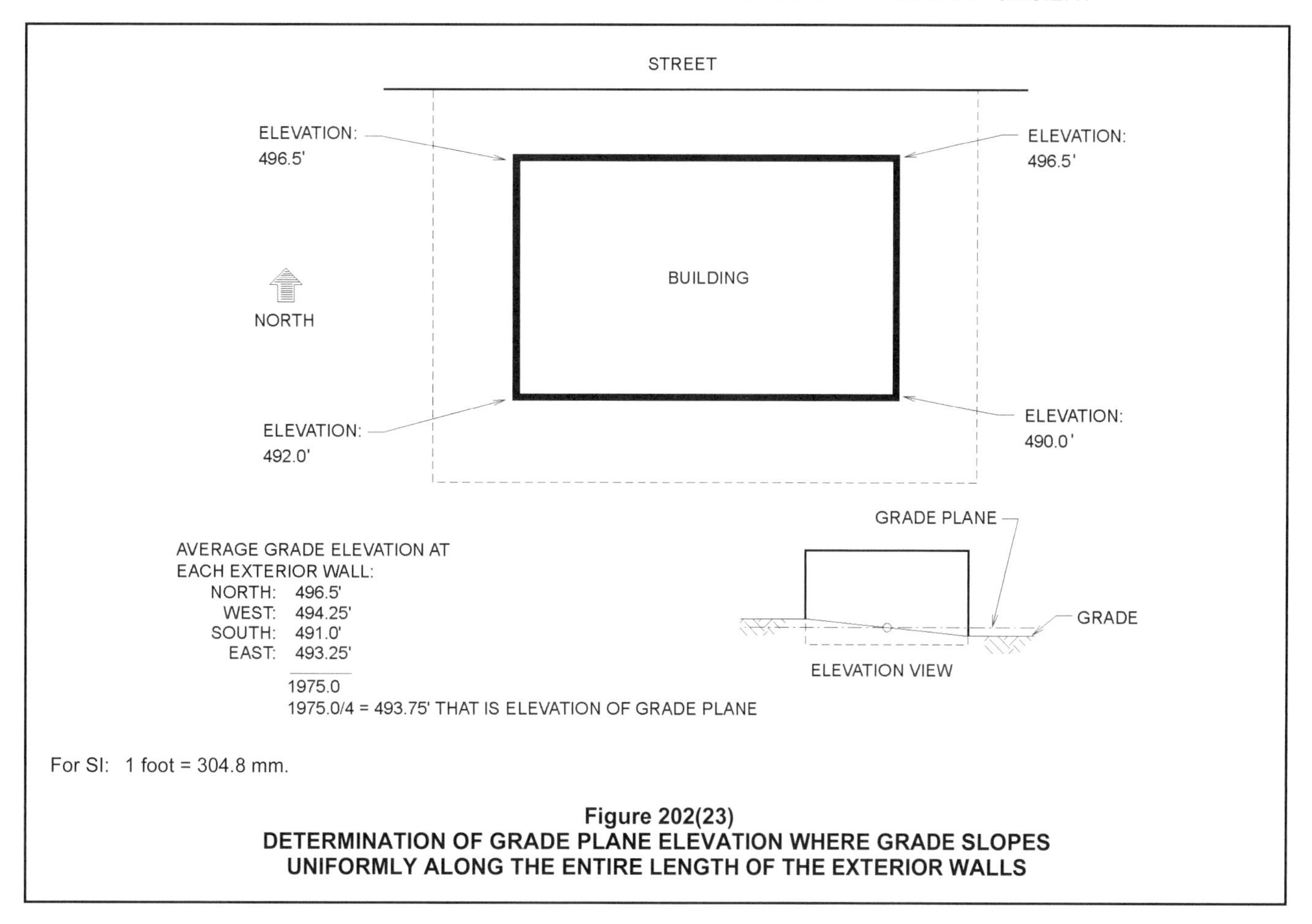

Figure 202(23)
DETERMINATION OF GRADE PLANE ELEVATION WHERE GRADE SLOPES UNIFORMLY ALONG THE ENTIRE LENGTH OF THE EXTERIOR WALLS

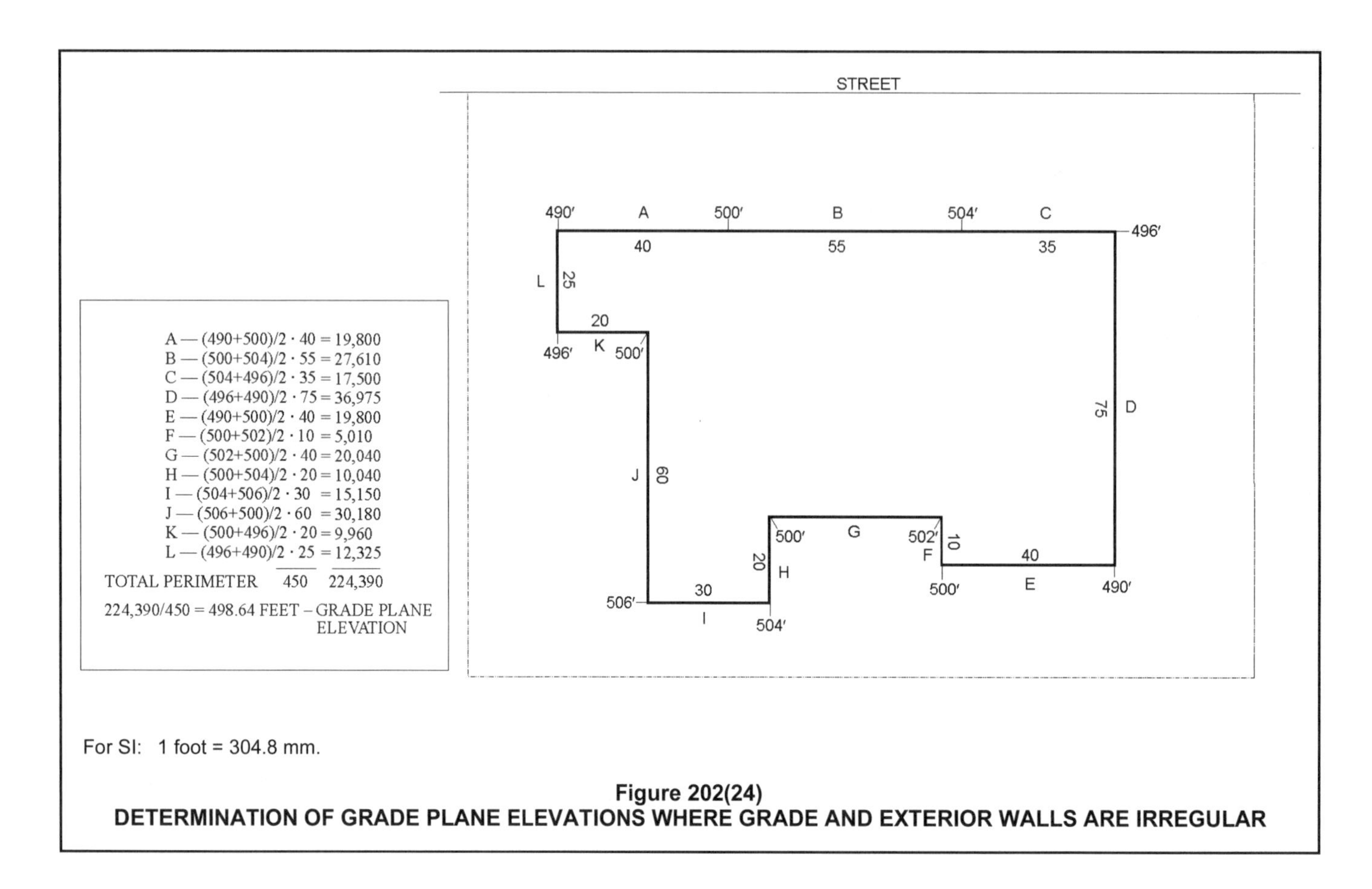

For SI: 1 foot = 304.8 mm.

Figure 202(24)
DETERMINATION OF GRADE PLANE ELEVATIONS WHERE GRADE AND EXTERIOR WALLS ARE IRREGULAR

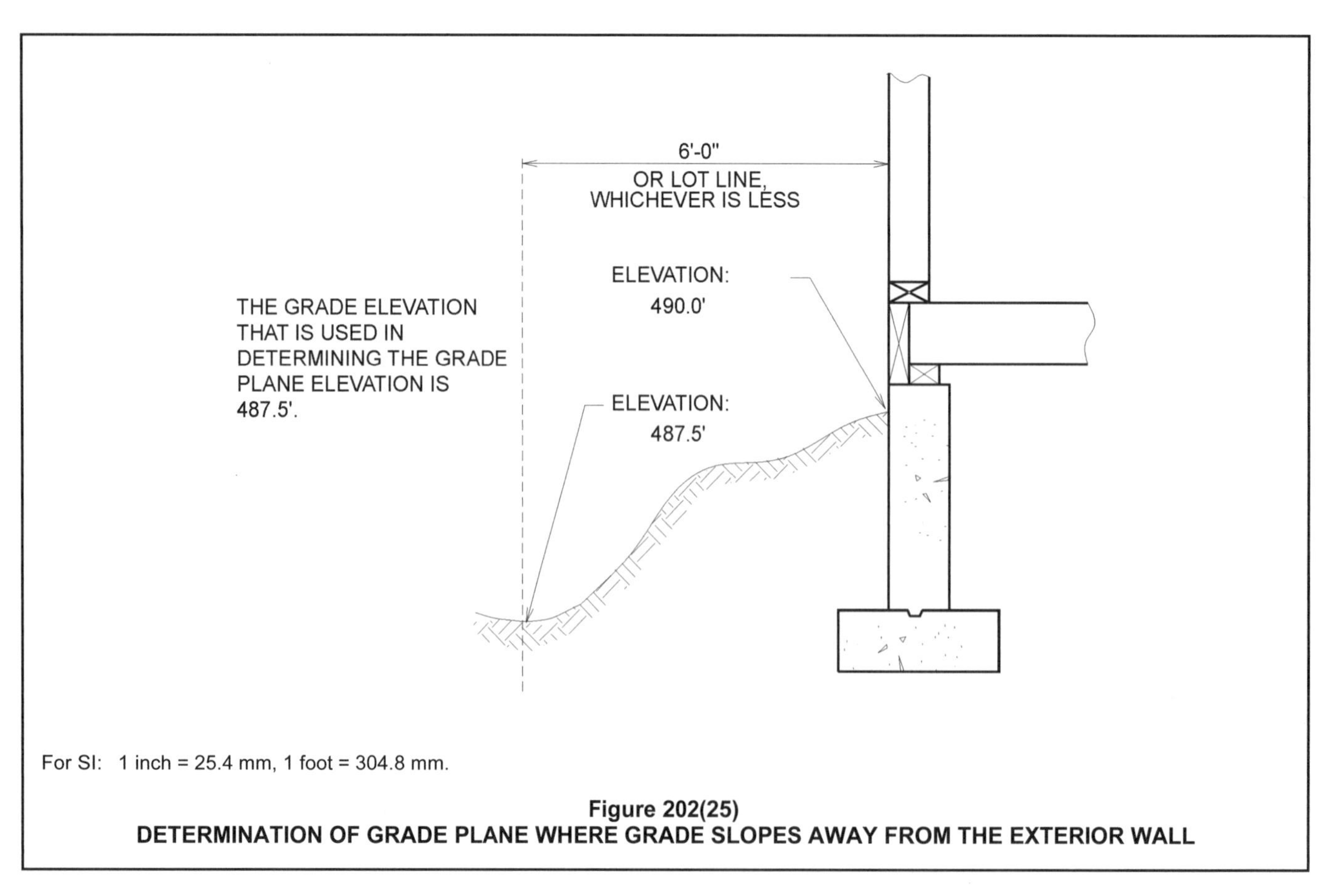

For SI: 1 inch = 25.4 mm, 1 foot = 304.8 mm.

Figure 202(25)
DETERMINATION OF GRADE PLANE WHERE GRADE SLOPES AWAY FROM THE EXTERIOR WALL

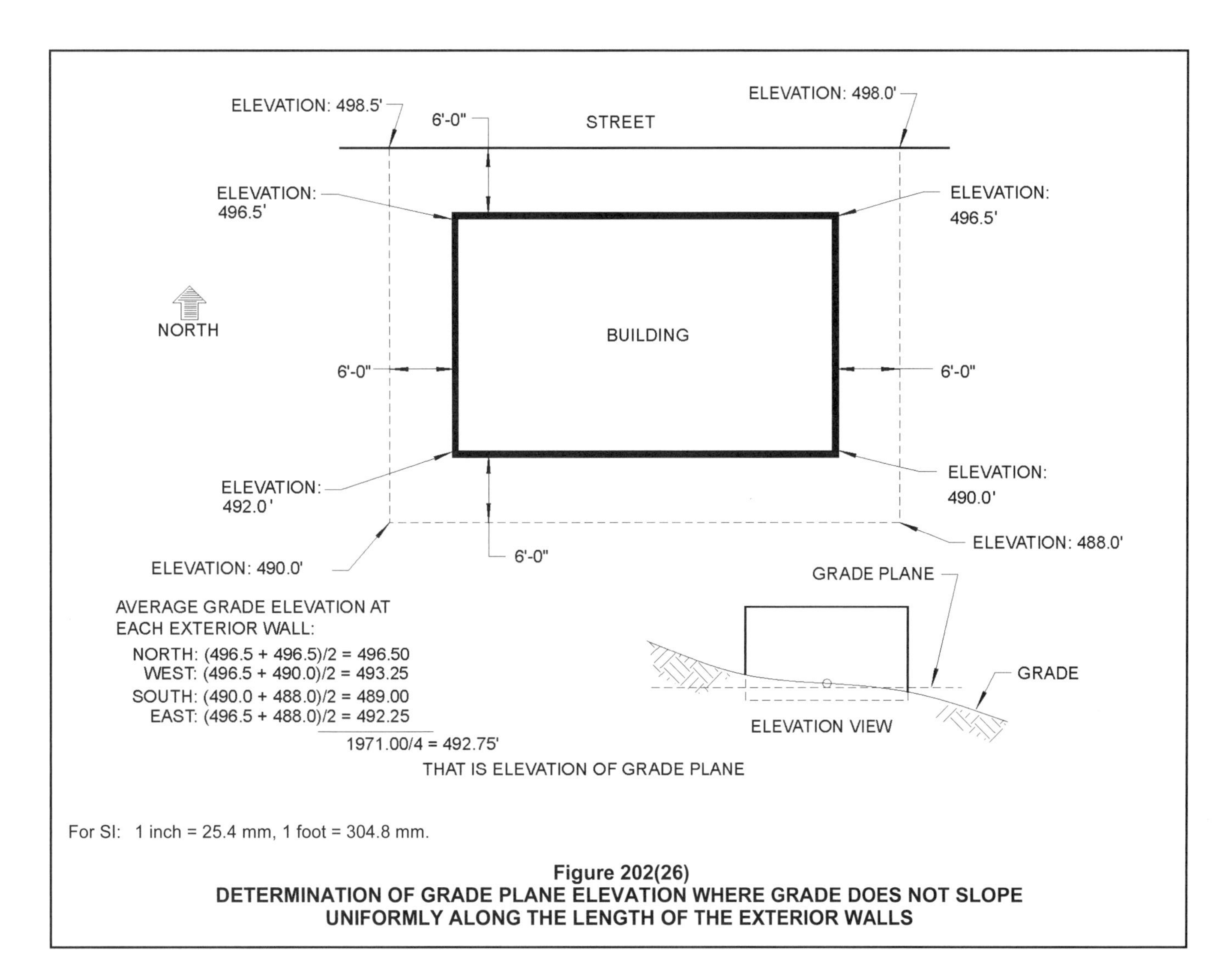

Figure 202(26)
DETERMINATION OF GRADE PLANE ELEVATION WHERE GRADE DOES NOT SLOPE UNIFORMLY ALONG THE LENGTH OF THE EXTERIOR WALLS

GROUP HOME. A facility for social rehabilitation, substance abuse or mental health problems that contains a group housing arrangement that provides *custodial care* but does not provide medical care.

❖ The term "group home" is listed under Group I-1 and R-4 occupancies. See the commentary under "Custodial care." The number of occupants would differentiate which occupancy classification is applicable to the facility. Residents live in a supervised living arrangement. Facilities can be for persons with developmental, emotional, mental or physical difficulties or for persons recovering from drug or alcohol abuse. The intent of these facilities is to promote recovery, community integration and improved quality of life. While there may be security restrictions, occupants are not restrained (see Section 308, Group I-3 for restrained conditions). The focus is to increase an individual's capacity to be successful in living, working, learning and social environments to the best of their ability. See also the commentary for "24-hour basis," "Foster care facility," "Personal care services," "Custodial care," "Nursing home" and "Incapable of self-preservation."

[BS] GUARD. A building component or a system of building components located at or near the open sides of elevated walking surfaces that minimizes the possibility of a fall from the walking surface to a lower level.

❖ Guards are sometimes mistakenly referred to as "guardrails." In actuality, the guard consists of the entire vertical portion of the barrier, not just the top rail (see commentary for "Handrail" and Section 1015). The purpose of guards is to minimize the potential for falls at dropoffs adjacent to walking surfaces. Loading requirements for guards are located in Section 1607.8.

GUEST ROOM. A room used or intended to be used by one or more guests for living or sleeping purposes.

❖ Lodging houses with five or fewer guest rooms are allowed by the code as R-3 occupancies. They are also permitted under the IRC. See the definition of "Lodging house" and Section 310.5.

GYPSUM BOARD. The generic name for a family of sheet products consisting of a noncombustible core primarily of gypsum with paper surfacing. Gypsum wallboard, gypsum sheathing, gypsum base for gypsum veneer plaster, exterior gypsum soffit board, predecorated gypsum board and water-resistant gypsum backing board complying with the standards listed in Tables 2506.2, 2507.2 and Chapter 35 are types of gypsum board.

❖ Gypsum board is the most commonly used material for interior wall covering. Gypsum board is also used for exterior sheathing, plaster lath and ceiling covering. Because it is installed in sheet form, it is less labor intensive and generally considered more cost effective than other wall and ceiling materials, such as plaster. Gypsum board requires a minimal amount of finishing and will readily accept paint, wallpaper, vinyl fabric, special textured paint and similar surface finish materials.

Gypsum board will be subject to severe failure when placed in direct contact with water or continuous moisture. For this reason, the code does not allow gypsum board to be used in wet areas unless it is provided with a finish material impervious to moisture.

[BS] GYPSUM PANEL PRODUCT. The general name for a family of sheet products consisting essentially of gypsum.

❖ This definition enables the code user to correctly apply the requirements for these products in Chapter 25. The wording, extracted from ASTM C11, was created by the gypsum manufacturing industry to describe gypsum sheet products that are manufactured either unfaced or with facing other than paper.

[BS] GYPSUM PLASTER. A mixture of calcined gypsum or calcined gypsum and lime and aggregate and other *approved* materials as specified in this code.

❖ Gypsum plaster is available in several forms and is designed for a variety of applications, including those where fire protection of building components, construction of fire-resistance-rated assemblies within the building or control of sound transmission is needed. Gypsum plaster can be applied using a two- or three-coat method, depending upon the backing or lathing system.

Gypsum plaster provides a hard, smooth finish surface that will readily receive most decorative finishes, such as paint, vinyl wall fabric, wallpaper and textured paint. Various aggregates can be added to the plaster composition to achieve different surface textures, from a smooth troweled to a heavily textured finish.

However, the code does not permit the use of gypsum plaster at exterior locations or at wet areas on the interior of the building.

[BS] GYPSUM VENEER PLASTER. *Gypsum plaster* applied to an *approved* base in one or more coats normally not exceeding $^{1}/_{4}$ inch (6.4 mm) in total thickness.

❖ Gypsum veneer plaster is specifically designed for use as a one-coat plaster veneer finish surface. Gypsum veneer plaster must be applied over a solid base such as gypsum board lath, gypsum base for veneer plaster, concrete or concrete unit masonry. Although gypsum veneer plaster is designed to be applied in one coat, it can be applied in two coats, as long as the maximum thickness does not exceed $^{1}/_{4}$ inch (6.4 mm).

Gypsum veneer plaster provides a hard, smooth finish surface, much like three-coat gypsum plaster, that will readily receive a variety of decorative finishes. Like gypsum plaster, the code does not permit the use of gypsum veneer plaster at exterior locations or at wet areas on the interior of the building.

HABITABLE SPACE. A space in a building for living, sleeping, eating or cooking. Bathrooms, toilet rooms, closets, halls, storage or utility spaces and similar areas are not considered habitable spaces.

❖ These spaces are normally considered inhabited in the course of residential living and provide the four basic characteristics associated with it: living, sleeping, eating and cooking. All habitable spaces are considered occupiable spaces, though other occupiable spaces, such as halls or utility rooms, are not considered habitable (see the definition of "Occupiable space").

[F] HALOGENATED EXTINGUISHING SYSTEM. A fire-extinguishing system using one or more atoms of an element from the halogen chemical series: fluorine, chlorine, bromine and iodine.

❖ Halon is a colorless, odorless gas that inhibits the chemical reaction of fire. Halon extinguishing systems are useful in occupancies such as computer rooms, where an electrically nonconductive medium is essential or desirable and where cleanup of other extinguishing agents presents a problem. The Halon extinguishing system must be of an approved type and installed and tested to comply with NFPA 12A.

Halon extinguishing agents have been identified as a source of emissions resulting in the depletion of the stratospheric ozone layer. For this reason, production of new supplies of Halon has been phased out. Alternative gaseous extinguishing agents, such as clean agents, have been developed as alternatives to Halon.

[F] HANDLING. The deliberate transport by any means to a point of storage or *use*.

❖ The term "handling" pertains to the transporting or movement of hazardous materials within a building. Handling presents a level of hazard less than that of use or dispensing operations but greater than storage. Material is handled only when it is transported from one point to another; it is the act of conveyance. The definition provides the means to determine proper controls necessary to provide safety in the transport mode. Specific handling requirements for various hazardous materials are contained in the IFC.

[BS] HANDRAIL. A horizontal or sloping rail intended for grasping by the hand for guidance or support.

❖ Handrails are provided along walking surfaces that lead from one elevation to another, such as ramps and stairways. Handrails may be any shape in cross section provided that they can be gripped by hand for support and guidance and for checking possible falls on the adjacent walking surface. In addition to being necessary in normal day-to-day use, handrails are especially needed in times of emergency when the pace of egress travel is hurried and the probability for occupant instability while traveling along the sloped or stepped walking surface is greater. Handrails, by themselves, are not intended to be used in place of guards to limit falls at drop-offs. Where guards and handrails are used together, the handrail is a separate element typically attached to the inside surface of the guard. The top guard cannot be used as a required handrail, except within dwelling units where the height is restricted to that of a handrail (see Section 1014). See the commentary for "Guard." For loading on handrails, see Section 1607.8.

HARDBOARD. A fibrous-felted, homogeneous panel made from lignocellulosic fibers consolidated under heat and pressure in a hot press to a density not less than 31 pcf (497 kg/m^3).

❖ Hardboard is used for various interior applications, as well as siding applications. Other ingredients may be added during processing to provide or improve properties, such as strength, water resistance and general utility.

HARDWARE. See "Fire exit hardware" and "Panic hardware."

[F] HAZARDOUS MATERIALS. Those chemicals or substances that are *physical hazards* or *health hazards* as classified in Section 307 and the *International Fire Code*, whether the materials are in usable or waste condition.

❖ The term "hazardous materials" refers to those materials that present either a physical or health hazard. A specific listing of hazardous materials is indicated in Sections 307.3, 307.4, 307.5 and 307.6. An occupancy containing greater than the maximum allowable quantities per control area of these materials as indicated in Table 307.1(1) or 307.1(2) is classified in one of the four high-hazard occupancy classifications. Section 307.1.1 provides 14 instances where, under specific conditions of handling, use, storage or packaging, the presence of one or more hazardous substances does not result in an occupancy being classified one of the Group H occupancies.

[F] HAZARDOUS PRODUCTION MATERIAL (HPM). A *solid*, *liquid* or gas associated with semiconductor manufacturing that has a degree-of-hazard rating in health, flammability or instability of Class 3 or 4 as ranked by NFPA 704 and which is *used* directly in research, laboratory or production processes which have as their end product materials that are not hazardous.

❖ This definition identifies those specific materials that can be contained within an HPM facility. The restriction in the definition for only hazardous materials with a Class 3 or 4 rating is not intended to exclude materials that are less hazardous, but to clarify that materials of the indicated higher ranking are still permitted in an HPM facility without classifying the building as Group H. NFPA 704 is referenced in order to establish the degree of hazard ratings for all materials as related to health, flammability and instability risks.

[BS] HEAD JOINT. Vertical *mortar joint* placed between *masonry units* within the *wythe* at the time the *masonry units* are laid.

❖ Vertically oriented joints between masonry units are head joints [see Commentary Figure 202(8)].

[F] HEALTH HAZARD. A classification of a chemical for which there is statistically significant evidence that acute or chronic health effects are capable of occurring in exposed persons. The term "health hazard" includes chemicals that are *toxic* or *highly toxic*, and *corrosive*.

❖ Materials that present risks to people from handling or exposure are considered health hazards. Examples of these types of materials are indicated in Section 307.6. Buildings and structures containing materials that present a health hazard in excess of the maximum allowable quantities would be classified as Group H-4. Materials that present a health hazard may also present a physical hazard (see the definition of "Physical hazard") and must comply with the requirements of the code applicable to both hazards.

HEAT DETECTOR. See "Detector, heat."

HEIGHT, BUILDING. The vertical distance from *grade plane* to the average height of the highest roof surface.

❖ This definition establishes the two points of measurement that determine the height of a building. This measurement is used to determine compliance with the building height limitations of Sections 503.1 and 504 and Tables 504.3 and 504.4, which limit building height both in terms of the number of stories and the number of feet between the two points of measurement.

The lower point of measurement is the grade plane (see the definition of "Grade plane"). The upper point of measurement is the roof surface of the building, with consideration given to sloped roofs (such as a hip or gable roof). In the case of sloped roofs, the average height would be used as the upper point of measurement, rather than the eave line or the ridge line. The average height of the roof is the mid-height between the roof eave and the roof ridge, regardless of the shape of the roof.

This definition also indicates that building height is measured to the highest roof surface. In the case of a building with multiple roof levels, the highest of the various roof levels must be used to determine the building height. If the highest of the various roof levels is a sloped roof, then the average height of that sloped roof must be used. The average height of multiple roof levels is not to be used to determine the building height. Where structures are divided into multiple buildings by fire walls, building height is determinable for each building separately.

The distance that a building extends above ground also determines the relative hazards of that building.

Simply stated, a taller building presents relatively greater safety hazards than a shorter building for several reasons, including fire service access and time for occupant egress. The code specifically defines how building height is measured to enable various code requirements, such as type of construction and fire suppression, to be consistent with those relative hazards [see Commentary Figure 202(27) for the computation of building height in terms of feet and stories].

The term "height" is also used frequently in the code for other limitations related to, and sometimes not related to, "building height." For example, Section 1509 limits the height of a penthouse above the top of the roof. Since a "Penthouse" is defined as a structure that is built above the roof of a building, it is above the point to which "Building height" is measured. Therefore a penthouse would not affect the measurement of building height and can be located above the maximum allowed roof height provided it complies with the limitations of Section 1509. Other provisions such as Sections 1013 and 1406 specify requirements based on height, but such height is usually measured from a location other than grade plane and is not intended to be building height.

HELICAL PILE. Manufactured steel deep foundation element consisting of a central shaft and one or more helical bearing plates. A helical pile is installed by rotating it into the ground. Each helical bearing plate is formed into a screw thread with a uniform defined pitch.

❖ This definition clarifies a term that refers to a specific type of deep foundation element (see the design requirements in Section 1810.3.1.5).

HELIPAD. A structural surface that is used for the landing, taking off, taxiing and parking of helicopters.

❖ This definition provides a specific term that refers to the portion of a structure that is subject to the helicopter live loads in Section 1607.6.

HELIPORT. An area of land or water or a structural surface that is used, or intended for use, for the landing and taking off of helicopters, and any appurtenant areas that are used, or intended for use, for heliport buildings or other heliport facilities.

❖ A heliport includes not only the immediate landing and take-off pad, but also all other adjacent service areas. The fueling, maintenance, repairs or storage of helicopters may be done within or outside of a building or structure. These outside areas or enclosed spaces are considered as part of the heliport.

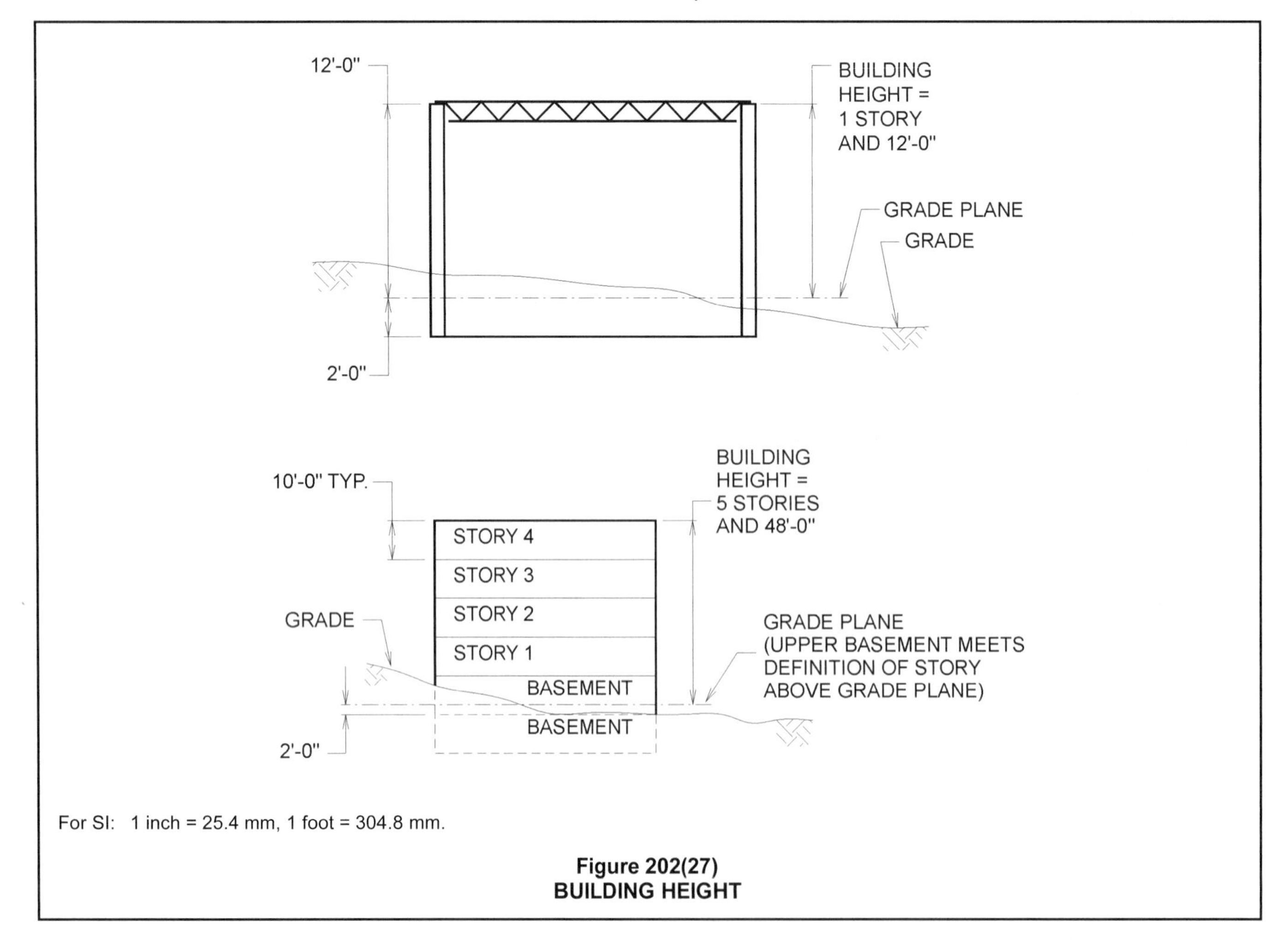

For SI: 1 inch = 25.4 mm, 1 foot = 304.8 mm.

Figure 202(27)
BUILDING HEIGHT

HELISTOP. The same as "heliport," except that no fueling, defueling, maintenance, repairs or storage of helicopters is permitted.

❖ A helistop, by definition, is limited only to the immediate landing and take-off pad. Examples of helistops would be the pad located on top of a hospital for the unloading of emergency room patients, a pad for discharging commuters outside of an office building or the pad used to load and unload tourists at a sightseeing attraction.

HIGH-PRESSURE DECORATIVE EXTERIOR-GRADE COMPACT LAMINATE (HPL). Panels consisting of layers of cellulose fibrous material impregnated with thermosetting resins and bonded together by a high-pressure process to form a homogeneous nonporous core suitable for exterior use.

❖ HPL is an exterior finish material. While in common use in Europe, HPL is finding expanded use elsewhere. The definition is based on the International Standard EN 428. Section 1409 specifies the requirements and uses for HPL.

HIGH-PRESSURE DECORATIVE EXTERIOR-GRADE COMPACT LAMINATE (HPL) SYSTEM. An *exterior wall covering* fabricated using HPL in a specific assembly including *joints*, seams, attachments, substrate, framing and other details as appropriate to a particular design.

❖ HPL systems are intended for exterior application for buildings. The definition is based on the International Standard EN 428. Section 1409 specifies the requirements and uses for HPL systems.

HIGH-RISE BUILDING. A building with an occupied floor located more than 75 feet (22 860 mm) above the lowest level of fire department vehicle access.

❖ Determining what qualifies as a high-rise building is a fairly unique measurement of height and is not based on the definition of "Building height." The critical measurement is from the lowest ground location where a fire department will be able to set its fire-fighting equipment to a floor level of occupied floors as shown in Commentary Figure 202(28). It is not a measurement from grade plane to top of the building. The basis of the measurement is analyzing the capability of fighting a fire and rescuing occupants from the outside the building. Once past a height of 75 feet (22 860 mm) above ground level, ground-based fire fighting will not be sufficient. High-rise buildings must comply with the requirements of Section 403.

[F] HIGHLY TOXIC. A material which produces a lethal dose or lethal concentration that falls within any of the following categories:

1. A chemical that has a median lethal dose (LD_{50}) of 50 milligrams or less per kilogram of body weight when administered orally to albino rats weighing between 200 and 300 grams each.
2. A chemical that has a median lethal dose (LD_{50}) of 200 milligrams or less per kilogram of body weight when administered by continuous contact for 24 hours (or less if death occurs within 24 hours) with the bare skin of albino rabbits weighing between 2 and 3 kilograms each.
3. A chemical that has a median lethal concentration (LC_{50}) in air of 200 parts per million by volume or less of gas or vapor, or 2 milligrams per liter or less of mist, fume or dust, when administered by continuous inhalation for 1 hour (or less if death occurs within 1 hour) to albino rats weighing between 200 and 300 grams each.

Mixtures of these materials with ordinary materials, such as water, might not warrant classification as *highly toxic.* While this system is basically simple in application, any hazard evaluation that is required for the precise categorization of this type of material shall be performed by experienced, technically competent persons.

❖ The definition is derived from DOL 29 CFR, Part 1910.1200. These materials are considered danger-

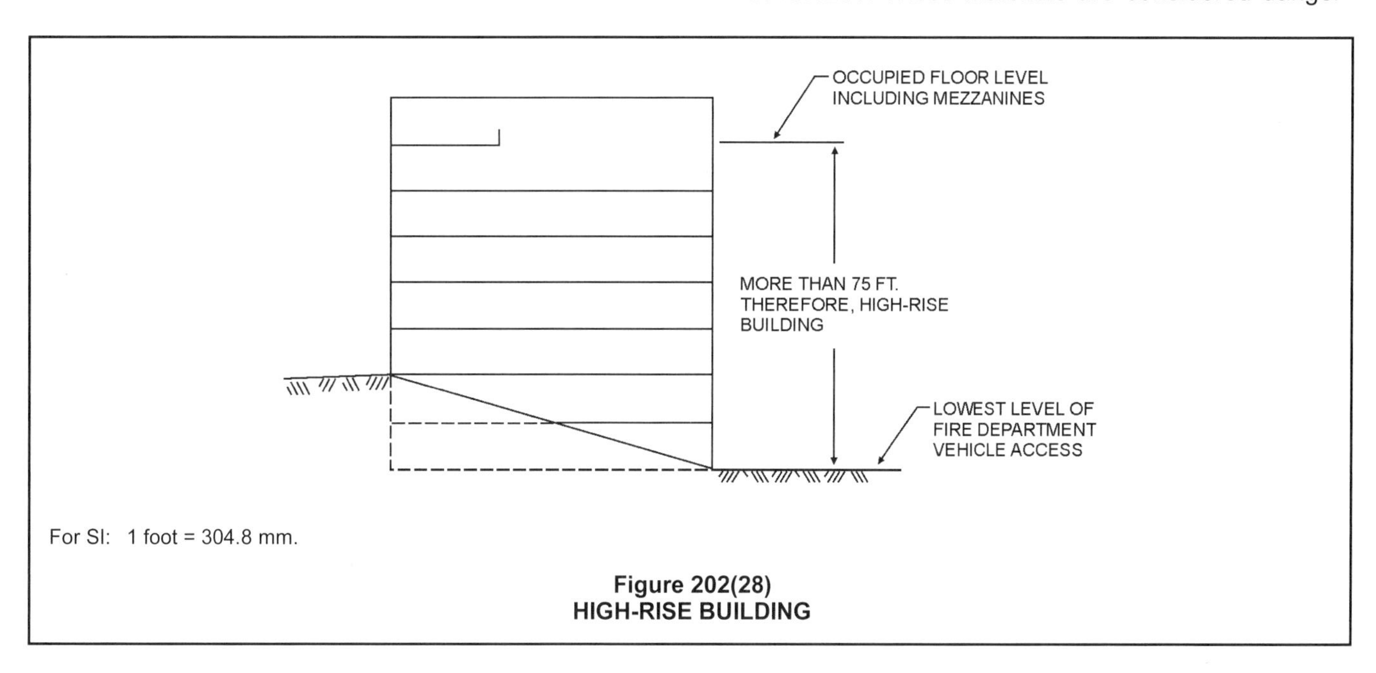

Figure 202(28)
HIGH-RISE BUILDING

ous to humans when inhaled, absorbed or injected through the skin or ingested orally. Highly toxic materials present a health hazard and are subsequently listed as Group H-4 in Section 307.6. Examples of highly toxic materials include gases such as arsine, fluorine and hydrogen cyanide, liquid acrylic acid and calcium cyanide in solid form.

Mixtures of these materials with ordinary materials, such as water, might not warrant a highly toxic classification. While this system is basically simple in application, any hazard evaluation that is required for the precise categorization of this type of material is to be performed by experienced, technically competent persons. For further discussion of this material definition, see the commentary to it in Chapter 2 of the *International Fire Code and Commentary*.

[A] HISTORIC BUILDINGS. Buildings that are listed in or eligible for listing in the National Register of Historic Places, or designated as historic under an appropriate state or local law.

❖ Buildings technically considered historic must be designated as such through a federal, state or local law. In addition, there are buildings that have been reviewed for eligibility to be listed as a national historic building. Those listed as eligible for national listing also are considered historic for the purposes of the code. Buildings that are within a historic district are not necessarily, themselves, historic buildings. The determination of their designation as historic would depend on the specifics of the listing of the historic area. The IEBC provides specific provisions applying to historic buildings.

[BF] HORIZONTAL ASSEMBLY. A fire-resistance-rated floor or *roof assembly* of materials designed to restrict the spread of fire in which continuity is maintained.

❖ A horizontal assembly is a component for completing compartmentation. Horizontal assemblies have all openings and penetrations protected equal to the rating for the fire-resistance-rated floor or roof assembly [see Commentary Figure 202(29)].

HORIZONTAL EXIT. An *exit* component consisting of fire-resistance-rated construction and opening protectives intended to compartmentalize portions of a building thereby creating refuge areas that afford safety from the fire and smoke from the area of fire origin.

❖ This term refers to a fire-resistance-rated wall that subdivides a structure into multiple compartments and provides an effective barrier to protect occupants from a fire condition within one of the compartments. After occupants pass through a horizontal exit, they must be provided not only with sufficient space to gather but also with access to another exit, such as an exterior door or exit stairway, through which they can exit the building. Commentary Figure 202(30) depicts the exits serving a single building that is subdivided with a fire-resistance-rated wall (see Section 1026).

HOSPITALS AND PSYCHIATRIC HOSPITALS. Facilities that provide care or treatment for the medical, psychiatric, obstetrical, or surgical treatment of care recipients who are *incapable of self-preservation*.

❖ Persons in hospital facilities may be physically incapable of self-preservation, or at least extremely limited in their ability to evacuate. In psychiatric hospitals, they may be confined within an area of a building for care or security purposes. In consideration of an occupant's health, as well as safety, hospitals and nursing homes at least partially rely on defend-in-place strategies. See the commentary for Sections 308 and 407, Group I-2.

Care facilities are used by patients of varying acuity seeking a broad spectrum of available support services. These facilities span a wide range of occupancy types including Groups E, I and R. The level of

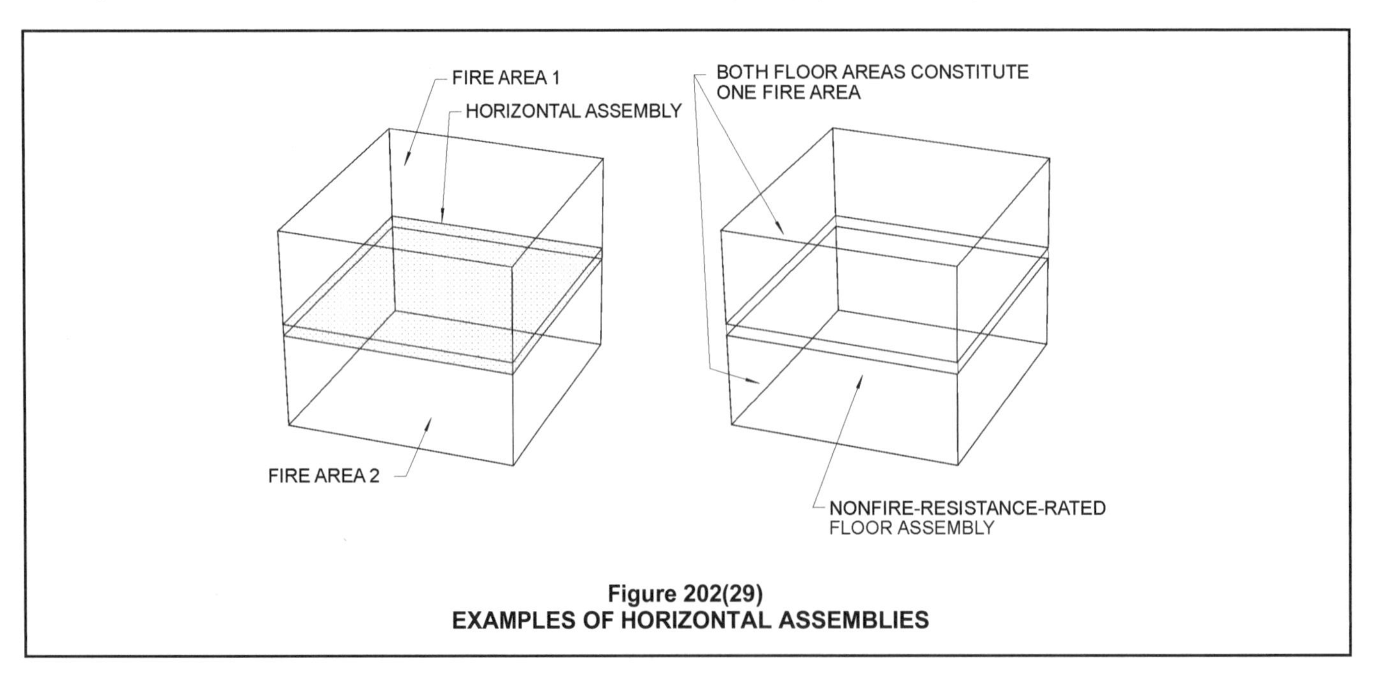

Figure 202(29)
EXAMPLES OF HORIZONTAL ASSEMBLIES

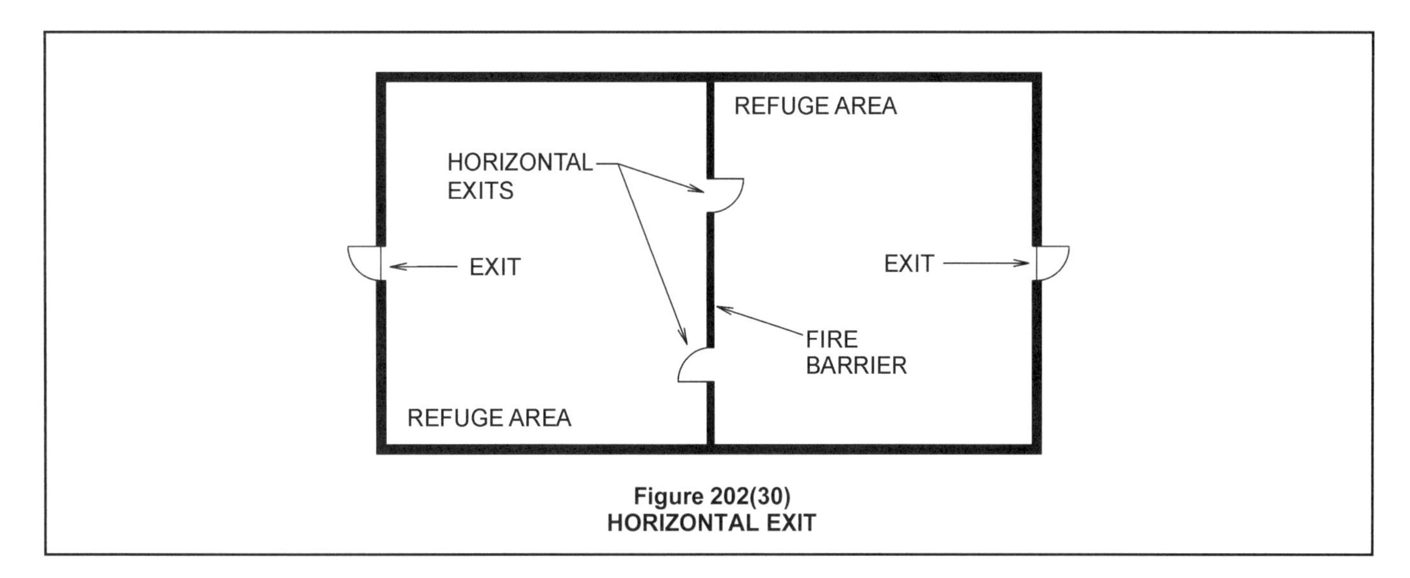

Figure 202(30)
HORIZONTAL EXIT

care provided describes the condition and capabilities of an occupant, which then indicates appropriate standards for protection systems, both passive and active. See also the definitions for "24-hour basis," "Care suites," "Custodial care," "Detoxification facilities," "Medical care," "Nursing homes" and "Incapable of self-preservation."

HOUSING UNIT. A *dormitory* or a group of *cells* with a common dayroom in Group I-3.

❖ A housing unit is a key delineation of space within a Group I-3 occupancy. Often inmates are restricted to their housing unit and not allowed to move freely outside of the unit. A housing unit can include more than one tier of cells. Because a housing unit is often designed as a multiple-level area with a shared day room, it will typically share a common atmosphere (see Section 408.5 for the protection of vertical openings). See also the commentaries for "Cell" and "Cell tier."

[F] HPM ROOM. A room used in conjunction with or serving a Group H-5 occupancy, where *HPM* is stored or *used* and which is classified as a Group H-2, H-3 or H-4 occupancy.

❖ An HPM room in a Group H-5 facility is utilized for the storage and use of hazardous production materials in excess of the maximum allowable quantities permitted in Table 307.1(1) or 307.1(2). The rooms are, therefore, considered a Group H-2, H-3 or H-4 occupancy depending on the type of hazardous material.

[BS] **HURRICANE-PRONE REGIONS.** Areas vulnerable to hurricanes defined as:

1. The U. S. Atlantic Ocean and Gulf of Mexico coasts where the ultimate design wind speed, V_{ult}, for Risk Category buildings is greater than 115 mph (51.4 m/s);
2. Hawaii, Puerto Rico, Guam, Virgin Islands and American Samoa.

❖ This definition identifies the areas where hurricane-force winds are expected.

[F] HYDROGEN FUEL GAS ROOM. A room or space that is intended exclusively to house a *gaseous hydrogen system*.

❖ This term refers to an enclosed space used exclusively for a gaseous hydrogen system that requires construction and protection unique to the hazards associated with this use. The room itself may be considered as an incidental accessory occupancy or a Group H occupancy, depending on the amount of hydrogen in such rooms. The definition itself should not be interpreted to prevent hydrogen piping systems from serving distributed hydrogen-using equipment and appliances located elsewhere on site or in the building. However, the amount of hydrogen within such piping needs to be evaluated with respect to the maximum allowable quantities in Table 307.1(1).

[BS] ICE-SENSITIVE STRUCTURE. A structure for which the effect of an atmospheric ice *load* governs the design of a structure or portion thereof. This includes, but is not limited to, lattice structures, guyed masts, overhead lines, light suspension and cable-stayed bridges, aerial cable systems (e.g., for ski lifts or logging operations), amusement rides, open catwalks and platforms, flagpoles and signs.

❖ This definition describes the type of structures that must consider atmospheric ice loading (see Section 1614).

[F] IMMEDIATELY DANGEROUS TO LIFE AND HEALTH (IDLH). The concentration of air-borne contaminants which poses a threat of death, immediate or delayed permanent adverse health effects, or effects that could prevent escape from such an environment. This contaminant concentration level is established by the National Institute of Occupational Safety and Health (NIOSH) based on both toxicity and flammability. It generally is expressed in parts per million by volume (ppmv/v) or milligrams per cubic meter (mg/m^3). If adequate data do not exist for precise establishment of IDLH concentrations, an independent certified industrial hygienist, industrial toxicologist, appropriate regulatory agency or other source *approved* by the *building official* shall make such determination.

❖ The definition of "Immediately dangerous to life and health (IDLH)" is the minimum concentration of an airborne contaminant, such as a highly toxic compressed gas, that a person could be exposed to before the risk of permanent adverse side effects. IDLH is important in determining the design of treatment systems for highly toxic or toxic compressed gases.

[BS] IMPACT LOAD. The *load* resulting from moving machinery, elevators, craneways, vehicles and other similar forces and kinetic *loads*, pressure and possible surcharge from fixed or moving *loads*.

❖ This definition identifies the scope of the type of loading addressed in Section 1607.9. The effect of impact loads on a structure can be significantly higher than the weight of the impacting elements because of their movement or vibration.

INCAPABLE OF SELF-PRESERVATION. Persons who, because of age, physical limitations, mental limitations, chemical dependency or medical treatment, cannot respond as an individual to an emergency situation.

❖ Patients/residents of nursing homes and hospitals may be in situations where they are unable to self-evacuate due to physical limitations arising from, for example, medication, surgery, injury or connection to medical equipment.

Care facilities are used by patients of varying acuity seeking a broad spectrum of available support services. These facilities span a wide range of occupancy types including Groups E, I and R. There are three types of care defined in the codes: personal, custodial and medical. The level of care provided describes the condition and capabilities of an occupant, which then indicate appropriate standards for protection systems, both active and passive. See also the definitions for "24-hour basis," "Custodial care," "Detoxification facilities," "Foster care facility," "Group home," "Hospitals and psychiatric hospitals," "Medical care," "Nursing home" and "Personal care services."

[F] INCOMPATIBLE MATERIALS. Materials that, when mixed, have the potential to react in a manner that generates heat, fumes, gases or byproducts which are hazardous to life or property.

❖ These materials, whether in storage or in use, constitute a dangerous chemical combination. Determination of which chemicals in combination present a hazard is a difficult situation for the building official. Material Safety Data Sheets (MSDS) alone may not provide all of the necessary information. When in doubt, the building official should seek additional information from the manufacturer of the chemicals involved, the building owner or from experts who are knowledgeable in industrial hygiene or chemistry.

[F] INERT GAS. A gas that is capable of reacting with other materials only under abnormal conditions such as high temperatures, pressures and similar extrinsic physical forces. Within the context of the code, inert gases do not exhibit either physical or health hazard properties as defined (other than acting as a simple asphyxiant) or hazard properties other than those of a *compressed gas*. Some of the more common inert gases include argon, helium, krypton, neon, nitrogen and xenon.

❖ Inert gases do not react readily with other materials under normal temperature and pressure (NTP), but it is possible for a reaction to occur. For example, even nitrogen combines with some of the more active metals such as lithium and magnesium to form nitrides, and at high temperatures it will also combine with oxygen and other elements. The formation of compounds utilizing inert gases is also possible. Xenon combines with fluorine to form various fluorides, and with oxygen to form oxides. The compounds formed are crystalline solids. As indicated in Table 307.1(1), there are no maximum allowable quantities specified for inert gases. As such, inert gases are not regulated by the code as hazardous materials with respect to a potential Group H occupancy classification.

[F] INITIATING DEVICE. A system component that originates transmission of a change-of-state condition, such as in a *smoke detector*, *manual fire alarm box* or supervisory switch.

❖ All fire protection systems consist of devices, which upon use or actuation will initiate the intended operation. A manual fire alarm box, for example, upon actuation will transmit a fire alarm signal. In the case of a single-station device, the initiating device and the notification appliance are one and the same.

INTENDED TO BE OCCUPIED AS A RESIDENCE. This refers to a *dwelling unit* or *sleeping unit* that can or will be used all or part of the time as the occupant's place of abode.

❖ A unit that is a person's home, rather than a unit used for a more transient nature, is a place of abode. Fair housing regulations do not include a 30-day criteria for transient/nontransient, similar to what has been traditionally used by the building codes (see the commentary to Section 1107 for additional information); therefore, beach homes, timeshares and extended-stay hotels may be included.

INTERIOR EXIT RAMP. An *exit* component that serves to meet one or more *means of egress* design requirements, such as required number of *exits* or *exit access* travel distance, and provides for a protected path of egress travel to the *exit discharge* or *public way*.

❖ To qualify as an interior exit ramp, the ramp must be enclosed with a fire-rated enclosure in order to provide a protected path between the exit access and exit discharge. This enclosure must extend directly to the exterior at grade; extend through an exit passageway to grade; or comply with one of the allowances for exit discharge through a lobby, vestibule or horizontal exit. Travel distance is measured to the entrance to the enclosure for the interior exit ramp. Ramps that are utilized by occupants for evacuation but do not meet the provisions for exits are considered exit access elements (see "Exit access ramp"). For exterior exit ramp requirements, see Section 1027.

INTERIOR EXIT STAIRWAY. An *exit* component that serves to meet one or more *means of egress* design requirements, such as required number of *exits* or *exit access* travel distance, and provides for a protected path of egress travel to the *exit discharge* or *public way*.

❖ To qualify as an interior exit stairway, the stairway must be enclosed with a fire-rated enclosure in order to provide a protected path between the exit access and exit discharge. This enclosure must extend directly to the exterior at grade; extend through an exit passageway to grade; or comply with one of the allowances for exit discharge through a lobby, vestibule or horizontal exit. Travel distance is measured to the entrance to the enclosure for the interior exit stairway. Stairways that are utilized by occupants for evacuation but do not meet the provisions for exits are considered exit access elements (see "Exit access stairway"). For exterior exit stairway requirements, see Section 1027.

INTERIOR FINISH. Interior finish includes *interior wall and ceiling finish* and *interior floor finish*.

❖ This is a general term that addresses all exposed surfaces, which includes walls, ceilings and floors. Interior finish material is exposed to the interior space enclosed by these building elements.

INTERIOR FLOOR FINISH. The exposed floor surfaces of buildings including coverings applied over a finished floor or *stair*, including risers.

❖ This definition clarifies which part of the interior finish is considered the floor. Floors are treated differently than walls and ceilings as they pose a minimal fire hazard in comparison to that potentially created by the interior finish on walls and ceilings (see commentary, Section 804). Materials that are installed above the structural floor element and exposed to the room or space are considered floor finish materials. Such materials are, therefore, subject to the requirements of Section 804. This section also restricts the combustibility and ability to propagate a fire based upon radiant exposure levels.

INTERIOR FLOOR-WALL BASE. *Interior floor finish trim* used to provide a functional or decorative border at the intersection of walls and floors.

❖ This definition, which addresses interior floor-wall base trim materials, provides an understanding and clarification of these types of products versus other interior trim materials. In many cases, floor-covering material is just seamlessly turned up or used at the intersection of the floor and the wall, thus becoming the floor-wall base trim. Because of their location at the floor line, floor-wall base materials are not likely to be involved in a fire until the floor covering is also involved, usually at room flashover (see also commentary, Section 806.6).

INTERIOR SURFACES. Surfaces other than weather exposed surfaces.

❖ Interior surfaces are all surfaces, exposed and unexposed to view, on the inside of a building or structure that are not subject to damage or failure of their intended purpose because of the effects of uncontrollable weather conditions, such as rain, sleet, snow and wind. Interior surfaces are generally exposed to mechanically controlled climatic conditions through air-conditioning, heating and humidity control. Because of these controls, interior surfaces require far fewer restrictions than exterior surfaces. Therefore, the quantities and types of materials and finishes permitted for use on interior surfaces are greatly increased.

INTERIOR WALL AND CEILING FINISH. The exposed *interior surfaces* of buildings, including but not limited to: fixed or movable walls and partitions; toilet room privacy partitions; columns; ceilings; and interior wainscoting, paneling or other finish applied structurally or for decoration, acoustical correction, surface insulation, structural fire resistance or similar purposes, but not including *trim*.

❖ A material that is applied to ceilings as well as walls, columns, partitions (including the privacy partitions in bathrooms that could pose a significant threat in larger bathrooms if unrated) and other vertical interior surfaces whether fixed or movable. The application of this material may be for structural, decorative, acoustical, structural fire resistance and other similar reasons. Trim, such as baseboard, door or window casing, is not considered interior wall and ceiling finish. Interior wall and ceiling finish is regulated by Section 803.

[BS] INTERLAYMENT. A layer of felt or nonbituminous saturated felt not less than 18 inches (457 mm) wide, shingled between each course of a wood-shake *roof covering*.

❖ According to Section 1507.9.5, interlayment is required to comply with ASTM D226, Type I, which is commonly referred to as No. 15 asphalt felt.

INTUMESCENT FIRE-RESISTANT COATINGS. Thin film liquid mixture applied to substrates by brush, roller, spray or trowel which expands into a protective foamed layer to provide fire-resistant protection of the substrates when exposed to flame or intense heat.

❖ Section 1705.15 requires special inspections for intumescent fire-resistant coatings in accordance with AWCI 12-B, based on the fire-resistance design. Because AWCI 12-B does not define or describe these materials, this definition facilitates correct application of the special inspections.

[BS] JOINT. The opening in or between adjacent assemblies that is created due to building tolerances, or is designed to allow independent movement of the building in any plane caused by thermal, seismic, wind or any other loading.

❖ This term defines the void created when two assemblies meet and an open space occurs between the assemblies [see Commentary Figure 202(19) for examples of building joints].

[A] JURISDICTION. The governmental unit that has adopted this code under due legislative authority.

❖ The governmental unit such as a town, township, county, or state that has the legal authority under state statues to adopt a building code.

L RATING. The air leakage rating of a *through penetration firestop system* or a fire-resistant *joint* system when tested in accordance with UL 1479 or UL 2079, respectively.

❖ The term "L rating" is a companion rating similar to F ratings and T ratings. The term is already widely used in the industry. Air leakage for breaches of smoke barriers can be limited by adherence to the requirements in Section 714.4. Only through-penetrations and joints in smoke barriers are required to meet the limits on air leakage.

[A] LABEL. An identification applied on a product by the manufacturer that contains the name of the manufacturer, the function and performance characteristics of the product or material and the name and identification of an *approved agency*, and that indicates that the representative sample of the product or material has been tested and evaluated by an *approved agency* (see Section 1703.5, "Manufacturer's designation" and "Mark").

❖ A label provides verification of testing and inspection of materials, products or assemblies (see commentary Section 1703.5). See also the commentary for definitions of "Listed," "Mark," "Labeled" and "Manufacturer's designation."

[A] LABELED. Equipment, materials or products to which has been affixed a *label*, seal, symbol or other identifying *mark* of a nationally recognized testing laboratory, *approved* agency or other organization concerned with product evaluation that maintains periodic inspection of the production of the above-labeled items and whose labeling indicates either that the equipment, material or product meets identified standards or has been tested and found suitable for a specified purpose.

❖ The term is an adjective applied to equipment, materials and products that has been tested or otherwise determined to meet the intended purpose or meet a standard. The label is that of a laboratory or other agency qualified to do the evaluations (see commentary, Section 1703.5). See also the commentary for definitions of "Listed," "Mark," "Label" and "Manufacturer's designation."

LEVEL OF EXIT DISCHARGE. See "Exit discharge, level of."

LIGHT-DIFFUSING SYSTEM. Construction consisting in whole or in part of lenses, panels, grids or baffles made with light-transmitting plastics positioned below independently mounted electrical light sources, skylights or light-transmitting plastic roof panels. Lenses, panels, grids and baffles that are part of an electrical fixture shall not be considered as a light-diffusing system.

❖ Heat-resistant plastics, such as polystyrene and urea, are frequently used for light diffusion.

LIGHT-FRAME CONSTRUCTION. A type of construction whose vertical and horizontal structural elements are primarily formed by a system of repetitive wood or cold-formed steel framing members.

❖ The code uses the term "light frame" to distinguish this unique type of framing system from other structural systems. The structural integrity of light-frame construction is dependent upon numerous connections or frequent bracing. Other framing systems or terms commonly used in the building industry that are considered as light-frame construction include: "stick built," "platform frame," "western frame" and "balloon frame." Section 2211 pertains to light-frame cold-formed steel construction. Section 2308 defines a specific subcategory of light-frame construction called "conventional light-frame construction," which is limited to wood materials.

LIGHT-TRANSMITTING PLASTIC ROOF PANELS. Structural plastic panels other than skylights that are fastened to structural members, or panels or sheathing and that are used as light-transmitting media in the plane of the roof.

❖ Plastic roof panels are primarily used in areas where light transmission rather than visual clarity is the goal (e.g., factory and industrial applications). Even though they are used as light-transmitting media, they must support the loads and forces encountered, just like roof surfaces. Typical applications include factories, warehouses and greenhouses where glass breakage is a practical concern.

LIGHT-TRANSMITTING PLASTIC WALL PANELS. Plastic materials that are fastened to structural members, or to structural panels or sheathing, and that are used as light-transmitting media in *exterior walls*.

❖ This term is only applicable to plastic wall panels allowing the transmission of light. Even though they are used as light-transmitting media, they must support the loads and forces encountered, just like other wall surfaces. Foam core insulated panels are regulated by Section 2603 and are not considered plastic wall panels.

[BS] LIMIT OF MODERATE WAVE ACTION. Line shown on FIRMs to indicate the inland limit of the $1^1/_2$-foot (457 mm) breaking wave height during the base flood.

❖ FEMA shows a Limit of Moderate Wave Action line to delineate the inland extent of the Coastal A Zone (see commentary for Coastal A Zone). If this line is on the FIRM, then the code, by reference to ASCE 7 and ASCE 24, imposes requirements that apply in areas with wave action.

[BS] LIMIT STATE. A condition beyond which a structure or member becomes unfit for service and is judged to be no longer useful for its intended function (serviceability limit state) or to be unsafe (strength limit state).

❖ This definition is needed for a clear understanding of the load and resistance factor design (LRFD) methodology.

[F] LIQUID. A material that has a melting point that is equal to or less than 68°F (20°C) and a *boiling point* that is greater than 68°F (20°C) at 14.7 pounds per square inch absolute (psia) (101 kPa). When not otherwise identified, the term "liquid" includes both *flammable* and *combustible liquids.*

❖ This definition specifies the criteria to establish when material is considered a liquid based on its melting and boiling points. When the term "liquid" is referred to, it is intended to include both flammable and combustible liquids.

[F] LIQUID STORAGE ROOM. A room classified as a Group H-3 occupancy used for the storage of *flammable* or *combustible liquids* in a closed condition.

❖ Liquid storage rooms are utilized exclusively for the storage of flammable and combustible liquids in closed containers in excess of the maximum allowable quantities permitted by Tables 307.1(1) and 307.1(2). The storage room itself is considered a Group H-3 occupancy in accordance with Section 307.5.

[F] LIQUID USE, DISPENSING AND MIXING ROOM. A room in which Class I, II and IIIA *flammable* or *combustible liquids* are *used,* dispensed or mixed in open containers.

❖ This term refers to all nonstorage rooms utilized exclusively for flammable and combustible liquids other than Class IIIB liquids. Class IIIB liquids have a flash point in excess of 200°F (93°C) and are not considered hazardous.

[A] LISTED. Equipment, materials, products or services included in a list published by an organization acceptable to the *building* official and concerned with evaluation of products or services that maintains periodic inspection of production of listed equipment or materials or periodic evaluation of services and whose listing states either that the equipment, material, product or service meets identified standards or has been tested and found suitable for a specified purpose.

❖ When a product is listed and labeled, it indicates that it has been tested for conformance to an applicable standard and is subject to a third-party inspection quality assurance (QA) program. The QA verifies that the minimum level of quality required by the appropriate standard is maintained. Labeling provides a readily available source of information that is useful for field inspection of installed products. The label identifies the product or material and provides other information that can be further investigated if there is any question as to its suitability for the specific installation. The labeling agency performing the third-party inspection must be approved by the building official, and the basis for this approval may include, but is not limited to, the capacity and capability of the agency to perform the specific testing and inspection. See also the commentary for definitions of "Mark," "Label," "Labeled" and "Manufacturer's designation."

The question is often asked whether the listing of a product can be voided or violated. The use of a listing mark applied to a product is authorized by the listing agency and is a "statement" by the product manufacturer that the product, as manufactured, met all appropriate requirements (such as the criteria contained in a test standard) at the time of manufacture and shipment to a point-of-use or point-of-sale. After that point in time, any alteration or modification makes it impossible for the building official or the listing agency to determine if the product continues to meet the criteria by which its listing was originally attained. Listed products are subject to the review and approval of the building official. Where the building official determines that a field modification or alteration to the product is significant enough to call its impact on the listing into question, an evaluation in the field by representatives of the listing agency may be required in order to verify the continued compliance of the product with the original listing criteria. Such field evaluations or tests would be the responsibility of the permit holder and conducted at no cost to the jurisdiction (see Section 104.11.2).

LIVE/WORK UNIT. A *dwelling unit* or *sleeping unit* in which a significant portion of the space includes a nonresidential use that is operated by the tenant.

❖ Live/work dwelling units incorporate both living space and nonresidential use spaces. Live/work units are growing in popularity in urban centers to bring a mix of uses to neighborhoods. The provisions of Section 419 address building code issues, but local zoning regulations may also provide limits on use or location.

[BS] LIVE LOAD. A *load* produced by the use and occupancy of the building or other structure that does not include construction or environmental *loads* such as wind load, snow load, rain load, earthquake load, flood load or *dead load.*

❖ This definition identifies the scope of the type of loading included in Section 1607. Generally, live loads are not environmental loads or dead loads, but are transient in nature and will vary in magnitude over the life of a structure.

[BS] LIVE LOAD, ROOF. A *load* on a roof produced:

1. During maintenance by workers, equipment and materials;
2. During the life of the structure by movable objects such as planters or other similar small decorative appurtenances that are not occupancy related; or
3. By the use and occupancy of the roof such as for roof gardens or assembly areas.

❖ This definition is needed for the proper application of the load combinations in Chapter 16. This definition clarifies what is considered a roof live load, as well as clarifying (implicitly) that snow loads are not considered roof live loads.

[BS] LOAD AND RESISTANCE FACTOR DESIGN (LRFD). A method of proportioning structural members and their connections using load and *resistance factors* such that no applicable *limit state* is reached when the structure is subjected to appropriate *load* combinations. The term "LRFD" is used in the design of steel and wood structures.

❖ This definition describes the load and resistance factor design (LRFD) method, which is one of the design approaches recognized under the code in Section 1604.1. It is needed for the proper application of the steel design requirements in Chapter 22 and the wood design requirements in Chapter 23.

[BS] LOAD EFFECTS. Forces and deformations produced in structural members by the applied *loads*.

❖ This definition is needed to properly apply the structural load requirements in Chapter 16. "Load effects" is a collective term used to refer to the internal member forces and member deformations that result from the applied loads.

[BS] LOAD FACTOR. A factor that accounts for deviations of the actual *load* from the *nominal load*, for uncertainties in the analysis that transforms the *load* into a *load effect*, and for the probability that more than one extreme *load* will occur simultaneously.

❖ This definition clarifies the application of the LRFD load combinations in Chapter 16.

[BS] LOADS. Forces or other actions that result from the weight of building materials, occupants and their possessions, environmental effects, differential movement and restrained dimensional changes. Permanent loads are those loads in which variations over time are rare or of small magnitude, such as *dead loads*. All other loads are variable loads (see "*Nominal loads*").

❖ This definition is needed for the proper application of the structural load requirements in Chapter 16. It also includes a definition of "Permanent loads" that serves as the distinction between variable loads and permanent loads. This is important in Section 1605.1, which requires all load combinations to be investigated with one or more of the variable loads taken as zero.

The distinction made between variable loads and permanent loads is fairly straightforward. Dead load is specifically mentioned because it is an ideal example of a load that does not vary over time. But dead load is not necessarily the only load that could be considered permanent. An additional complication is the definition of certain loads that may include both permanent and variable components. An indicator of this is provided in the load combinations in Section 1605. For instance, in the ASD basic load combinations given by Equations 16-11, 16-13 and 16-14, the variable loads are reduced by a factor of 0.75. Based on the application of the reduction factor in these load combinations, live load, L, roof live load, L_r, snow load, S, rain load, R, wind load, W, and earthquake load, E, are explicitly treated as variable loads. Since they are not reduced in these load combinations, pressure of a well-defined fluid, F, and lateral earth pressure, H, could be considered, at least in part, permanent loads.

LODGING HOUSE. A one-family dwelling where one or more occupants are primarily permanent in nature and rent is paid for guest rooms.

❖ The code establishes a lodging house as a Group R-3 occupancy where there are five or fewer guest rooms. This definition provides a distinction from Group R-1 occupancies where the occupants are expected to be transient. For a lodging house, there are one or more occupants who are permanent; this is their home.

[A] LOT. A portion or parcel of land considered as a unit.

❖ A lot is a legally recorded parcel of land, the boundaries of which are described on a deed. When code requirements are based on some element of a lot (such as yard area or lot line location), it is the physical attributes of the parcel of land that the code is addressing, not issues of ownership. Adjacent lots owned by the same party are treated as if they were owned by different parties because ownership can change at any time. However, a group of platted lots or subdivision lots could be joined together and "considered as a unit" for the purposes of the code. For example, a collection of platted lots could be used as a single building lot for the construction of a covered mall and its associated anchor buildings. Local jurisdictions may require for taxing or other purposes that the lots be legally joined, or merged, as well.

A condominium form of building ownership, whether a residential or commercial condominium, does not create separate lots (i.e., parcels of land) and such unit owners are treated as separate tenants, not separate lot owners. The lines separating one part of a condominium from another are not lot lines but lines indicating the limits of ownership. As such, walls constructed on lines separating condominium ownership would not need to be fire (or party) walls.

Legal property lines do not always constitute site boundaries (e.g., malls, condominiums, townhouses). A site could contain multiple legal "lot" divisions.

[A] LOT LINE. A line dividing one lot from another, or from a street or any public place.

❖ Lot lines are legally recorded divisions between two adjacent land parcels or lots. They are the reference point for the location of buildings for exterior separation and other code purposes (see the definition of "Lot" above).

LOW-ENERGY POWER-OPERATED DOOR. Swinging door which opens automatically upon an action by a pedestrian such as pressing a push plate or waving a hand in front of a sensor. The door closes automatically, and operates with decreased forces and decreased speeds (see "Power-assisted door" and "Power-operated door").

❖ There are basically three different types of doors that provide some type of power assistance for entry: low-energy power-operated doors, power-assisted doors and power-operated doors. The low-energy power-operated door is typically a side-swinging door that also operates as a manual door. However, the door has the additional feature of automatic operation when a person pushes on a plate or sensor located on a wall or post near the door [see Commentary Figures 202(31), 202(32) and 202(33)]. The low-energy power-assisted door and power-assisted door both are operated by the user touching something; therefore, they both must comply with BMHA156.19.

[F] LOWER FLAMMABLE LIMIT (LFL). The minimum concentration of vapor in air at which propagation of flame will occur in the presence of an ignition source. The LFL is sometimes referred to as "LEL" or "lower explosive limit."

❖ When the vapor-to-air ratio is somewhere between the LFL and the UFL, fires and explosions can occur upon introduction of an ignition source. The UFL is the maximum vapor-to-air concentration above which propagation of flame will not occur. If a vapor-to-air mixture is below the LFL, it is described as being "too lean" to burn, and if it is above the UFL, it is "too rich" to burn. For further discussion of this material definition, see the commentary to it in Chapter 2 of the *International Fire Code Commentary*.

Figure 202(31)
LOW-ENERGY POWER-OPERATED DOOR—SINGLE LEAF WITH REMOTE PLATE

LOWEST FLOOR. The floor of the lowest enclosed area, including *basement*, but excluding any unfinished or flood-resistant enclosure, usable solely for vehicle parking, building access or limited storage provided that such enclosure is not built so as to render the structure in violation of Section 1612.

❖ The lowest floor is the most important reference point when designing and constructing a building or structure in a flood hazard area. The term is specifically defined to include basements, which are any areas that are below grade on all sides. For elevated buildings, compliance with the flood-resistant design and

Figure 202(32)
LOW-ENERGY POWER-OPERATED DOOR—DUAL ACTION, DOUBLE LEAF SWINGING TYPE

Figure 202(33)
LOW-ENERGY POWER-OPERATED DOOR—DOUBLE LEAF SLIDING TYPE

construction provisions of the code is determined by an elevation survey that documents that the lowest floor, including the basement, is at or above the design flood elevation. It is important to note that enclosures may be built below elevated buildings provided they meet both the design criteria and use limitations (see ASCE 24). If compliant with the design criteria and use limitations, then enclosures are not considered the lowest floor. This distinction is also important because the premium rate used to determine the cost of a federal flood insurance policy is dependent on the elevation of the lowest floor, along with other factors.

[BS] MAIN WINDFORCE-RESISTING SYSTEM. An assemblage of structural elements assigned to provide support and stability for the overall structure. The system generally receives wind loading from more than one surface

❖ This definition identifies the structural components that are to comply with special inspection requirements. The main windforce-resisting system is the global structural system designed to resist wind loads on the building or structure.

MALL BUILDING, COVERED and MALL BUILDING, OPEN. See "Covered mall building."

[F] MANUAL FIRE ALARM BOX. A manually operated device used to initiate an *alarm signal*.

❖ Manual fire alarm boxes are commonly known as pull stations. Manual fire alarm boxes include all manual devices used to activate a manual fire alarm system and have many configurations, depending on the manufacturer. All manual fire alarm devices, however, must be approved and installed in accordance with NFPA 72 for the particular application. Manual fire alarm boxes may be combined in guard tour boxes.

[A] MANUFACTURER'S DESIGNATION. An identification applied on a product by the manufacturer indicating that a product or material complies with a specified standard or set of rules (see "*Label*" and "*Mark*").

❖ This represents terminology for a manufacturer's self-certification that a product complies with a given standard (see commentary, Section 1703.4). See also the commentary for definitions of "Listed," "Mark," "Label" and "Labeled."

[A] MARK. An identification applied on a product by the manufacturer indicating the name of the manufacturer and the function of a product or material (see "*Label*" and "Manufacturer's designation").

❖ A mark represents the manufacturer's identification placed on a product, stating who made the product and describing its function. There is, however, no certification of compliance to any particular standard and no third-party quality control (see commentary, Section 1703.4). Also see the commentary for definitions of "Listed," "Label," "Labeled" and "Manufacturer's designation."

MARQUEE. A *canopy* that has a top surface which is sloped less than 25 degrees from the horizontal and is located less than 10 feet (3048 mm) from operable openings above or adjacent to the level of the marquee.

❖ Marquees, unlike awnings, are fixed, permanent structures that justify sufficiently different requirements from those for other projections (see Section 3106 for code requirements for marquees). This definition also facilitates application of live loads in Table 1607.1

[BS] MASONRY. A built-up construction or combination of building units or materials of clay, shale, concrete, glass, gypsum, stone or other *approved* units bonded together with or without *mortar* or grout or other accepted methods of joining.

❖ The materials (other than gypsum) and elements constructed as stated in this definition are considered masonry construction and are regulated by Chapter 21. This term identifies the building elements of plain (unreinforced) masonry, reinforced masonry, grouted masonry, glass unit masonry and masonry veneer.

Glass unit masonry. Masonry composed of glass units bonded by *mortar*.

❖ Glass unit masonry consists of nonclad-bearing assemblies designed in accordance with Section 2110.

Plain masonry. Masonry in which the tensile resistance of the masonry is taken into consideration and the effects of stresses in reinforcement are neglected.

❖ Plain masonry has historically been referred to as "unreinforced masonry." Since such masonry may actually contain some reinforcement, however, the term "unreinforced" has fallen out of favor. When reinforcement is contained in plain masonry, its contribution to the strength of the system is required to be ignored. The bond between the masonry units and mortar is critical in the performance of plain masonry.

Reinforced masonry. Masonry construction in which reinforcement acting in conjunction with the masonry is used to resist forces.

❖ Reinforced masonry contains reinforcement (currently limited to steel reinforcement) and is designed considering the tensile strength of that reinforcement. Not all masonry containing reinforcement is considered reinforced masonry. Some plain masonry contains reinforcement (usually to reduce the size of any cracks that may form), but the contribution of that reinforcement is required to be neglected.

Solid masonry. Masonry consisting of solid masonry units laid contiguously with the *joints* between the units filled with *mortar*.

❖ This term describes single- or multiwythe walls composed of solid masonry units, including the thickness of the collar joint if it is filled with mortar or grout.

Unreinforced (plain) masonry. Masonry in which the tensile resistance of masonry is taken into consideration and the resistance of the reinforcing steel, if present, is neglected.

❖ See the commentary for "Plain masonry."

[BS] MASONRY UNIT. *Brick*, tile, stone, glass block or concrete block conforming to the requirements specified in Section 2103.

❖ Masonry units are natural stone units or manufactured units of fired clay, shale, cementitious materials or glass.

Hollow. A masonry unit whose net cross-sectional *area* in any plane parallel to the load-bearing surface is less than 75 percent of its gross cross-sectional *area* measured in the same plane.

❖ Hollow masonry units are those having a specified net cross-sectional area less than 75 percent of their corresponding gross cross-sectional area. Where the specified net cross-sectional area is equal to or greater than 75 percent of the gross cross-sectional area, the unit is considered to be solid.

Solid. A masonry unit whose net cross-sectional *area* in every plane parallel to the load-bearing surface is 75 percent or more of its gross cross-sectional *area* measured in the same plane.

❖ Solid masonry units have a specified net cross-sectional area 75 percent or greater of their corresponding gross cross-sectional area. Where the specified net cross-sectional area is less than 75 percent of the gross cross-sectional area, the unit is considered to be hollow.

MASTIC FIRE-RESISTANT COATINGS. Liquid mixture applied to a substrate by brush, roller, spray or trowel that provides fire-resistant protection of a substrate when exposed to flame or intense heat.

❖ Section 1705.15 requires special inspections for mastic fire-resistant coatings in accordance with AWCI 12-B, based on the fire-resistance design. Because AWCI 12-B does not define or describe these materials, this definition facilitates correct application of the special inspections.

MEANS OF EGRESS. A continuous and unobstructed path of vertical and horizontal egress travel from any occupied portion of a building or structure to a *public way*. A means of egress consists of three separate and distinct parts: the *exit access*, the *exit* and the *exit discharge*.

❖ The means of egress is the path traveled by building occupants to leave the building and the site on which it is located. It includes all interior and exterior elements that the occupants must utilize as they make their way from every room and usable space within the building to a public way such as a street or alley. The elements that make up the means of egress create the lifeline that occupants utilize to travel out of the structure and to a safe distance from the structure. The means of egress provisions strive to provide a reasonable level of life safety in every structure. The means of egress provisions are subdivided into three distinct portions (see the definitions of "Exit access," "Exit" and "Exit discharge").

MECHANICAL-ACCESS OPEN PARKING GARAGES. *Open parking garages* employing parking machines, lifts, elevators or other mechanical devices for vehicles moving from and to street level and in which public occupancy is prohibited above the street level.

❖ These types of garages are constructed with most of the same attributes as other garages. They have numerous parking levels that house motor vehicles; however, public access to these upper levels is not permitted. Employees take control of the vehicles at the street level and drive into some sort of a vertical mechanical device, which conveys them to the upper parking levels. These structures are still provided with the required openings of Section 406.5.2.

MECHANICAL EQUIPMENT SCREEN. A rooftop structure, not covered by a roof, used to aesthetically conceal plumbing, electrical or mechanical equipment from view.

❖ These are used mainly for appearance and to conceal roof-mounted equipment from view.

MEDICAL CARE. Care involving medical or surgical procedures, nursing or for psychiatric purposes.

❖ Persons who need medical care are likely to be incapable of self-preservation or at least extremely limited in their ability to evacuate. In consideration of occupant health as well as safety, hospitals and nursing homes at least partially rely on defend-in-place strategies. See also the commentary for Section 308, Group I-2.

Care facilities are used by patients of varying acuity seeking a broad spectrum of available support services. These facilities span a wide range of occupancy types including Groups E, I and R. There are three types of care defined in the codes: personal, custodial and medical. On the lower end of the care spectrum (i.e., personal care) is when occupants are supervised but do not need custodial or medical care. Where occupants may be elderly or impaired (i.e., custodial care), they may need occasional daily living assistance such as cooking and cleaning. Persons who receive custodial care may or may not require assistance with evacuation depending on the occupancy and/or the "condition" of the occupancy. On the opposite end of the care spectrum, persons receiving care may be completely bedridden and dependent on medical gases and emergency power to maintain life (i.e., medical care). The level of care provided describes the condition and capabilities of an occupant, which then indicate appropriate standards for protection systems, both active and passive. See also the definitions for "24-hour basis," "Custodial care," "Detoxification facility," "Foster care facility," "Group home," "Hospitals and psychiatric hospitals," "Nursing home," "Personal care services," and "Incapable of self-preservation."

MEMBRANE-COVERED CABLE STRUCTURE. A nonpressurized structure in which a mast and cable system provides support and tension to the membrane weather barrier and the membrane imparts stability to the structure.

❖ This definition identifies a structure in which the membrane fabric is draped over a cable framework. These structures do not involve any interior air pressurization. The cable framework system is supported by a mast or other tower system. The cables are usually pretensioned to provide structural support for the rest of the exterior structure.

MEMBRANE-COVERED FRAME STRUCTURE. A nonpressurized building wherein the structure is composed of a rigid framework to support a tensioned membrane which provides the weather barrier.

❖ This type of membrane structure is the same as the membrane-covered cable structure, except that a trussed framework constructed of structural shapes provides the support instead of pretensioned cables. Again, no air pressurization is required or provided.

MEMBRANE PENETRATION. A breach in one side of a floor-ceiling, roof-ceiling or wall assembly to accommodate an item installed into or passing through the breach.

❖ This term refers to the situation where a penetration is made of a single layer of a fire-resistance-rated assembly. Sections 714.3.2 and 714.4.2 establish criteria where the penetration of a single membrane of an assembly is permitted while still considering the assembly to have the required fire-resistance rating [see Commentary Figure 202(34) for an example of a single membrane penetration].

Where a penetration is made completely through a fire-resistance-rated wall or floor/ceiling assembly, it must be protected in accordance with Section 714.3 or 714.4.1, as applicable for the assembly penetrated.

MEMBRANE-PENETRATION FIRESTOP. A material, device or construction installed to resist for a prescribed time period the passage of flame and heat through openings in a protective membrane in order to accommodate cables, cable trays, conduit, tubing, pipes or similar items.

❖ Sections 714.3.2 and 714.4.2 both refer to protection methods for membrane penetrations of fire-resistance-rated walls, floor/ceiling assemblies and roof/ceiling assemblies.

MEMBRANE-PENETRATION FIRESTOP SYSTEM. An assemblage consisting of a fire-resistance-rated floor-ceiling, roof-ceiling or wall assembly, one or more penetrating items installed into or passing through the breach in one side of the assembly and the materials or devices, or both, installed to resist the spread of fire into the assembly for a prescribed period of time.

❖ Membrane-penetration firestop systems maintain the required protection from the spread of fire, passage of hot gases and transfer of heat. A firestop usually refers to a single item, where a system is a combination of elements. See also the commentary to the definition of "Through-penetration firestop system."

MERCHANDISE PAD. A merchandise pad is an area for display of merchandise surrounded by *aisles*, permanent fix-

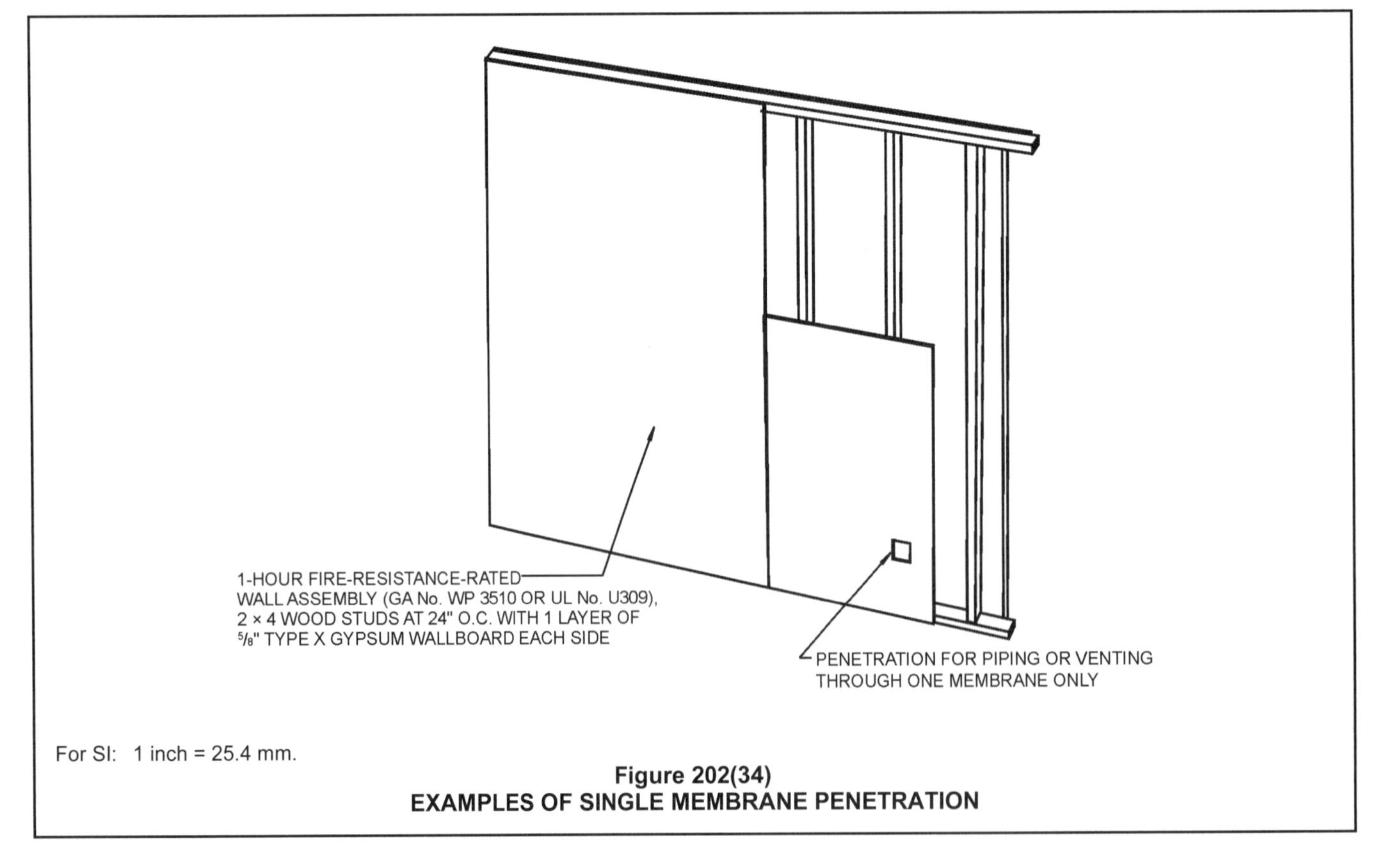

Figure 202(34)
EXAMPLES OF SINGLE MEMBRANE PENETRATION

tures or walls. Merchandise pads contain elements such as nonfixed and moveable fixtures, cases, racks, counters and partitions as indicated in Section 105.2 from which customers browse or shop.

❖ Merchandise pads would most likely be found in large stores with changing displays of clothes or furniture. This is not a raised display-only area. These areas allow customers to move between displays or racks. In regard to means of egress, merchandise pads could be considered analogous to areas of fixed seating or groups of tables. The aisle accessways are within the merchandise pad and lead to the aisles on the outside edges of the merchandise pads. Not all stores will contain merchandise pads (e.g., a typical grocery store with fixed shelves and aisles).

METAL COMPOSITE MATERIAL (MCM). A factory-manufactured panel consisting of metal skins bonded to both faces of a solid plastic core.

❖ These types of panels are sandwich construction composed of thin metal (usually aluminum or steel) sheets covering a solid plastic core. The panels typically have thicknesses that do not exceed $^1/_4$ inch (6.4 mm). Metal composite material (MCM) panels are different from other types of sandwich panels in that they do not contain foam plastic cores and are not typically intended to provide thermal insulation.

METAL COMPOSITE MATERIAL (MCM) SYSTEM. An *exterior wall covering* fabricated using MCM in a specific assembly including *joints,* seams, attachments, substrate, framing and other details as appropriate to a particular design.

❖ The provisions of Section 1407 provide the components that are typically part of a metal composite material (MCM) installation. The components of the exterior wall-covering system consist of framing members for the attachment and support of the MCM panels; the types of joints and seams used to maintain the weather resistance of the system; and the means for attaching the entire system to the building substrate or structural frame.

[BS] METAL ROOF PANEL. An interlocking metal sheet having a minimum installed weather exposure of 3 square feet (0.279 m^2) per sheet.

❖ There are two general categories of metal roofing systems: architectural metal roofing and structural metal roofing. Architectural metal roofs are generally watershedding roof systems and structural metal roofs have hydrostatic (water barrier) characteristics. The difference between a "metal roof panel" and a "metal roof shingle" is the weather exposure (i.e., that portion of the roofing exposed) (see Section 1507.4).

[BS] METAL ROOF SHINGLE. An interlocking metal sheet having an installed weather exposure less than 3 square feet (0.279 m^2) per sheet.

❖ See the commentary for the definition of "Metal roof panel."

MEZZANINE. An intermediate level or levels between the floor and ceiling of any *story* and in accordance with Section 505.

❖ A common design feature in factories, warehouses and mercantile buildings is an intermediate loft, or platform, between the story levels of a building. This type of feature, or mezzanine, can be found in buildings of all occupancies. The code must deal with whether this intermediate level is another story of the building or whether it can simply be treated as part of the story in which it is contained. The basic rule is that the intermediate level must be less than one-third of the area of the story below (of the room in which it is located) in order to be considered a mezzanine. Requirements for mezzanines are found in Section 505.

[BS] MICROPILE. A micropile is a bored, grouted-in-place *deep foundation* element that develops its load-carrying capacity by means of a bond zone in soil, bedrock or a combination of soil and bedrock.

❖ This definition clarifies a term that refers to a specific type of deep foundation element. The term "micropile" is a reference to its relative size compared to conventional deep foundation elements. They are also referred to as "minipiles" or "pin piles."

MINERAL BOARD. A rigid felted thermal insulation board consisting of either felted *mineral fiber* or cellular beads of expanded aggregate formed into flat rectangular units.

❖ This form of insulation is listed in Table 721.1(2), Items 15-1.9, 15-1.10 and 15-1.11.

MINERAL FIBER. Insulation composed principally of fibers manufactured from rock, slag or glass, with or without binders.

❖ This term provides a definition for a type of insulation that is commonly used or accepted within rated assemblies [see Table 721.1(2), Item 15-1.11] or as a fireblocking material (see Section 718.2.1).

MINERAL WOOL. Synthetic vitreous fiber insulation made by melting predominately igneous rock or furnace slag, and other inorganic materials, and then physically forming the melt into fibers.

❖ This term provides a definition for a type of insulation that is commonly used or accepted within rated assemblies [see Table 721.1(2), Item 15-1.15] or as a fireblocking material (see Section 718.2.1).

[BS] MODIFIED BITUMEN ROOF COVERING. One or more layers of polymer-modified asphalt sheets. The sheet materials shall be fully adhered or mechanically attached to the substrate or held in place with an *approved* ballast layer.

❖ These are composite sheets consisting of copolymer-modified bitumen, often reinforced and sometimes surfaced with various types of films, foils and mats.

[BS] MORTAR. A mixture consisting of cementitious materials, fine aggregates, water, with or without admixtures, that is used to construct unit masonry assemblies.

❖ Mortar is the material that bonds masonry units and accessories together and compensates for dimensional variations of the units. Both the plastic and hardened properties of mortar are important for strong, durable, water-tight construction. Material requirements and referenced standards for several permitted mortar types are given in Section 2103.2.

[BS] MORTAR, SURFACE-BONDING. A mixture to bond concrete *masonry units* that contains hydraulic cement, glass fiber reinforcement with or without inorganic fillers or organic modifiers and water.

❖ This mortar is a packaged, dry, combined material permitted for use in the surface bonding of concrete masonry units that have not been prefaced, coated or painted. Masonry units are stacked without mortar joints and surface-bonding mortar is then applied to both sides of the wall surface, creating a structural element.

MULTILEVEL ASSEMBLY SEATING. Seating that is arranged in distinct levels where each level is comprised of either multiple rows, or a single row of box seats accessed from a separate level.

❖ Assembly rooms may include a sloped seating arrangement (i.e., either ramped or stepped) to improve the viewing of the event for the occupants. These spaces can be single- or multiple-level arrangements. For example, for an auditorium with a sloped floor, the entire main floor is a single level. A level can be a balcony or a separate section of seating in an arena or stadium, such as skyboxes. The upper-seating bowl in the coliseum is a separate level, as is the loge in the theater. However, it is not the intent of this provision that each row of seats be considered a separate level.

[F] MULTIPLE-STATION ALARM DEVICE. Two or more single-station alarm devices that can be interconnected such that actuation of one causes all integral or separate audible alarms to operate. A multiple-station alarm device can consist of one single-station alarm device having connections to other detectors or to a *manual fire alarm box.*

❖ This definition refers to a combination of similar or different types of alarm devices that could be interconnected. The actuation of any two devices, whether a smoke detector or manual fire alarm box, will activate the required audible alarms at all interconnected devices.

[F] MULTIPLE-STATION SMOKE ALARM. Two or more single-station alarm devices that are capable of interconnection such that actuation of one causes the appropriate *alarm signal* to operate in all interconnected alarms.

❖ In occupancies with sleeping areas, occupants must be notified of a fire so that they can promptly evacuate the premises. In accordance with the requirements of NFPA 72, multiple-station smoke alarms are self-contained, smoke-activated alarm devices built in accordance with UL 217 that can be interconnected with other devices so that all integral or separate alarms will operate when any one device is activated.

MULTISTORY UNIT. A *dwelling unit* or *sleeping unit* with *habitable space* located on more than one *story*.

❖ A multistory dwelling or sleeping unit has living, sleeping, eating, cooking or bathroom space on more than one floor level within the unit (see Section 1107.7.2). A residence with only a garage underneath or an unfinished basement would not be a multistory unit.

[BS] NAILING, BOUNDARY. A special nailing pattern required by design at the boundaries of *diaphragms*.

❖ When designing a diaphragm or a shear wall, the sheathing for those elements must be attached to provide the design resistance to the applied load. Shear capacity tables for wood structural panels specify the nail size and spacing to be used along the edge and in the field of each panel. Boundary nailing is the required nailing pattern for panels located along the edges of diaphragms and at other boundaries where the stresses are typically high.

[BS] NAILING, EDGE. A special nailing pattern required by design at the edges of each panel within the assembly of a *diaphragm* or *shear wall*.

❖ When designing a diaphragm or shear panel, the sheathing for those elements must be attached in a fashion that will provide resistance to the design load. Shear capacity tables for wood structural panels used as sheathing specify fastener size and spacing. Edge nailing occurs at diaphragm or shear wall edges. In some cases, such as at blocked diaphragm boundaries, the edge of the panel occurs at the diaphragm boundary, in which case the edge must be boundary nailed.

[BS] NAILING, FIELD. Nailing required between the sheathing panels and framing members at locations other than *boundary nailing* and *edge nailing*.

❖ When designing a diaphragm or shear panel, the sheathing for those elements must be attached in a fashion that will provide resistance to the applied load. Shear capacity tables for wood structural panels used as sheathing specify fastener size and spacing. For example, Note b in Table 2306.2(1) indicates that the maximum spacing for fasteners along intermediate framing members must be 12 inches (305 mm) on center.

[BS] NATURALLY DURABLE WOOD. The heartwood of the following species except for the occasional piece with corner sapwood, provided 90 percent or more of the width of each side on which it occurs is heartwood.

❖ Because of their natural ability to resist deterioration, the harder portions of some species of wood are considered to be naturally durable. The code specifies that "occasional" sapwood is permitted if heartwood constitutes 90 percent of each side.

Decay resistant. Redwood, cedar, black locust and black walnut.

❖ Redwood, cedar, black locust and black walnut lumber are known to resist deterioration due to the action of microbes that enter the wood fibers. The code defines these species of lumber as being decay resistant.

Termite resistant. Redwood, Alaska yellow cedar, Eastern red cedar and Western red cedar.

❖ Alaskan yellow cedar, redwood, Eastern red cedar and Western red cedar are considered to be resistant to infestation by termites and are thus listed as naturally durable.

[BS] NOMINAL LOADS. The magnitudes of the *loads* specified in Chapter 16 (dead, live, soil, wind, snow, rain, *flood* and earthquake).

❖ This definition explains the term "nominal loads" so that the loads to be applied are clear. For the most part, the loads determined in Chapter 16 are considered nominal loads, which are used directly in allowable stress design (ASD) or multiplied by a load factor for use in strength or load and resistance factor design (LRFD). The exceptions are earthquake-load effect, *E*, as well as the wind load, *W*, both of which are considered strength level loads.

[BS] NOMINAL SIZE (LUMBER). The commercial size designation of width and depth, in standard sawn lumber and glued-laminated lumber *grades*; somewhat larger than the standard net size of dressed lumber, in accordance with DOCPS 20 for sawn lumber and with the AWC NDS for glued-laminated lumber.

❖ Unless specifically required by the code, all dimensions listed for wood member sizes are nominal, not actual. U.S. Department of Commerce (DOC) PS 20 specifies the required minimum dimension of lumber for each stated nominal size. The process used to smooth and finish the surface of the lumber and to dry the wood removes a certain thickness of wood on each side of the piece; therefore, a 2 by 4 is approximately $1^1/_2$ inches by $3^1/_2$ inches (38 mm by 89 mm). The actual dimension of the lumber, otherwise known as dressed size, is limited to ensure consistency in lumber sizes, which is essential to the satisfactory construction of a wood-frame building.

NONCOMBUSTIBLE MEMBRANE STRUCTURE. A membrane structure in which the membrane and all component parts of the structure are noncombustible.

❖ Any membrane structure that is constructed of all noncombustible materials is classified by this definition. To qualify as noncombustible, the materials, including the membrane itself, must be tested in accordance with ASTM E136 and must satisfy the criteria in Section 703.4.

[BS] NONSTRUCTURAL CONCRETE. Any element made of plain or reinforced concrete that is not part of a structural system required to transfer either gravity or lateral loads to the ground.

❖ This definition enables the code user to correctly apply the durability requirements of Section 1904.

[F] NORMAL TEMPERATURE AND PRESSURE (NTP). A temperature of 70°F (21°C) and a pressure of 1 atmosphere [14.7 psia (101 kPa)].

❖ This term refers to the standard room temperature at atmospheric pressure. Atmospheric pressure results from the weight of air elevated above the earth's surface. At sea level, the atmosphere exerts a pressure of 14.7 pounds per square inch (psi) (101 kPa). The properties of commercially available compressed gases are indicated at their normal temperature and pressure (NTP).

NOSING. The leading edge of treads of *stairs* and of landings at the top of *stairway flights*.

❖ The front edge of the tread that is exposed to the user's foot provides the visual clue for the placement of the foot in both ascent and descent. The nosings of a stair are a reference point for the measurement of the tread depth and riser height. The line connecting the nosings serves as the reference for the measurement of handrail and guard heights as well as headroom. The code ensures their uniformity by limiting the projection of the tread and landing nosings that results in a stairway that is easy to use. If too large, they are a tripping hazard when walking up a stair. If too small in relation to tread depth, the effective tread depth required for heel clearance in descent is minimized. The code provides limits for both minimum and maximum nosing projections and establishes a minimum tread depth when no projection is required [see Commentary Figures 1011.5.2 and 1009.5.5(1)]. An exception to these limits exists in the requirements for both alternating tread devices and ships ladders where an exaggerated projected tread depth, required to provide for reasonable foot room, is unique to the steeper gradient and functional use of these devices.

NOTIFICATION ZONE. See "Zone, notification."

[F] NUISANCE ALARM. An alarm caused by mechanical failure, malfunction, improper installation or lack of proper maintenance, or an alarm activated by a cause that cannot be determined.

❖ A nuisance alarm is essentially any alarm that occurs as a result of a condition that does not arise during the normal operation of the equipment. A nuisance alarm is not the same as a false alarm. A person who intentionally initiates an alarm by using a manual pull station or a person who accidentally initiates a smoke detector is not initiating a nuisance alarm. A nuisance alarm is, by nature, a factor of the system itself.

NURSING HOMES. Facilities that provide care, including both intermediate care facilities and skilled nursing facilities where any of the persons are *incapable of self-preservation*.

❖ Persons in nursing homes may be physically incapable of self-preservation or at least extremely limited in their ability to evacuate. In dementia wards they may be confined within an area of a building for care or security purposes. In consideration of an occupant's health as well as safety, hospitals and nursing homes at least partially rely on defend-in-place strategies. See the commentary for Sections 308 and 407, Group I-2.

Care facilities are used by patients of varying acuity seeking a broad spectrum of available support services. These facilities span a wide range of occupancy types including Groups E, I and R. The level of care provided describes the condition and capabilities of an occupant which then indicates appropriate standards for protection systems, both passive and active. See also the definitions for "24-hour basis," "Care suites," "Custodial care," "Detoxification facilities," "Hospitals and psychiatric hospitals," "Medical care" and "Incapable of self-preservation."

OCCUPANT LOAD. The number of persons for which the *means of egress* of a building or portion thereof is designed.

❖ In addition to the limitation on the maximum occupant load for a space, the code also requires the determination of the occupant load that is to be utilized for the design of the means of egress system. The number for the floor area per occupant (occupant load factor) in Table 1004.1.2 reflects common and traditional occupant densities based on the empirical data for the density of similar spaces. This occupant load is also utilized to determine the required number of plumbing fixtures (see Chapter 29) and when automatic sprinkler systems or fire alarm and detection systems are required (see Chapter 9).

OCCUPIABLE SPACE. A room or enclosed space designed for human occupancy in which individuals congregate for amusement, educational or similar purposes or in which occupants are engaged at labor, and which is equipped with *means of egress* and light and *ventilation* facilities meeting the requirements of this code.

❖ Occupiable spaces are those areas designed for human occupancy. It applies to both residential and nonresidential spaces alike. Most spaces in a building are occupiable spaces. Based on the nature of the occupancy, various code sections apply. All habitable spaces are also considered occupiable (see the definition of "Habitable space"). However, all occupiable spaces are not habitable. Additionally, some spaces are neither habitable nor occupiable. The code identifies crawl spaces, attics, penthouses and elevated platforms (mechanical or industrial equipment) as unoccupied spaces. Since the code generally states how these spaces must be accessed, but does not specifically require means of egress, they would not be occupiable spaces. If access is limited to maintenance and service personnel, it is likely that a space is not occupiable.

OPEN-ENDED CORRIDOR. An interior corridor that is open on each end and connects to an exterior *stairway* or *ramp* at each end with no intervening doors or separation from the corridor.

❖ Breezeway configurations are common in hotels and apartment buildings, especially in areas where being open to the outside is considered an amenity. By essentially being open to the outside, the intent and level of safety is similar to an exterior egress balcony (See Section 1027.6, Exception 4 for requirements)

OPEN PARKING GARAGE. A structure or portion of a structure with the openings as described in Section 406.5.2 on two or more sides that is used for the parking or storage of private motor vehicles as described in Section 406.5.3.

❖ Open parking garages are defined as having uniformly distributed openings on no less than two sides totaling no less than 40 percent of the building perimeter. The aggregate area of the openings is to be a minimum of 20 percent of the total wall area of all perimeter walls (see Commentary Figure 406.5.2).

[F] OPEN SYSTEM. The *use* of a *solid* or *liquid hazardous material* involving a vessel or system that is continuously open to the atmosphere during normal operations and where vapors are liberated, or the product is exposed to the atmosphere during normal operations. Examples of open systems for *solids* and *liquids* include dispensing from or into open beakers or containers, dip tank and plating tank operations.

❖ See the commentary to the definition of "Closed system."

[F] OPERATING BUILDING. A building occupied in conjunction with the manufacture, transportation or *use* of explosive materials. Operating buildings are separated from one another with the use of intraplant or intraline distances.

❖ Magazines are used for the storage of explosive materials. Manufacturing or operating buildings used for the storage of explosives are not magazines and are not intended to be used for storage, although at times there may be storage incidental to the manufacturing function. This definition is included here to clarify this difference.

[BS] ORDINARY PRECAST STRUCTURAL WALL. See Section 1905.1.1.

❖ This definition is a modification of ACI 318 text and is therefore located in Section 1905.

[BS] ORDINARY REINFORCED CONCRETE STRUCTURAL WALL. See Section 1905.1.1.

❖ This definition is a modification of ACI 318 text and is therefore located in Section 1905.

[BS] ORDINARY STRUCTURAL PLAIN CONCRETE WALL. See Section 1905.1.1.

❖ This definition is a modification of ACI 318 text and is therefore located in Section 1905.

[F] ORGANIC PEROXIDE. An organic compound that contains the bivalent -O-O- structure and which may be considered to be a structural derivative of hydrogen peroxide where one or both of the hydrogen atoms have been replaced by an organic radical. Organic peroxides can pose an *explosion* hazard (*detonation* or *deflagration*) or they can be shock sensitive. They can also decompose into various unstable compounds over an extended period of time.

Class I. Those formulations that are capable of *deflagration* but not *detonation*.

Class II. Those formulations that burn very rapidly and that pose a moderate reactivity hazard.

Class III. Those formulations that burn rapidly and that pose a moderate reactivity hazard.

Class IV. Those formulations that burn in the same manner as ordinary combustibles and that pose a minimal reactivity hazard.

Class V. Those formulations that burn with less intensity than ordinary combustibles or do not sustain combustion and that pose no reactivity hazard.

Unclassified detonable. Organic peroxides that are capable of *detonation*. These peroxides pose an extremely high *explosion* hazard through rapid explosive decomposition.

❖ The chemical structure of organic peroxides differs from that of hydrogen peroxide (an oxidizer) in that an organic radical replaces the hydrogen atoms. Organic chemicals are all carbon based. As a result, organic peroxides pose varying degrees of fire or explosion hazards, in addition to their oxidizing properties. The classification of organic peroxides is based on the provisions of NFPA 432 (formerly NFPA 43B). Proper material classification of organic peroxides is essential to determining the appropriate occupancy classification of the structure. Examples of organic peroxides include acetyl cyclohexane, sulfonyl peroxide and benzoyl peroxide. The actual class of these materials is dependent on the percentage of concentration by weight. Most organic peroxides are available as liquids, pastes or solids in a powder form. For further discussion of this material definition, see the commentary to it in Chapter 2 of the *International Fire Code Commentary*.

[BS] ORTHOGONAL. To be in two horizontal directions, at 90 degrees (1.57 rad) to each other.

❖ This definition applies to the alternative seismic design category determination permitted under Section 1613.3.5.1. It makes it clear that the directions (axes) referred to are in the horizontal plane.

[BS] OTHER STRUCTURES (for Chapters 16-23). Structures, other than buildings, for which *loads* are specified in Chapter 16.

❖ This definition is needed for application of the minimum load requirements to structures other than buildings that are included in Chapter 16. Note that buildings are defined as "Structures used to support or shelter an occupancy."

OUTPATIENT CLINIC. See "Clinic, outpatient."

[A] OWNER. Any person, agent, operator, entity, firm or corporation having any legal or equitable interest in the property; or recorded in the official records of the state, county or municipality as holding an interest or title to the property; or otherwise having possession or control of the property, including the guardian of the estate of any such person, and the executor or administrator of the estate of such person if ordered to take possession of real property by a court.

❖ This term defines the person or other legal entity responsible for a building and its compliance with code requirements.

[F] OXIDIZER. A material that readily yields oxygen or other *oxidizing gas*, or that readily reacts to promote or initiate combustion of combustible materials and, if heated or contaminated, can result in vigorous self-sustained decomposition.

Class 4. An oxidizer that can undergo an explosive reaction due to contamination or exposure to thermal or physical shock and that causes a severe increase in the burning rate of combustible materials with which it comes into contact. Additionally, the oxidizer causes a severe increase in the burning rate and can cause spontaneous ignition of combustibles.

Class 3. An oxidizer that causes a severe increase in the burning rate of combustible materials with which it comes in contact.

Class 2. An oxidizer that will cause a moderate increase in the burning rate of combustible materials with which it comes in contact.

Class 1. An oxidizer that does not moderately increase the burning rate of combustible materials.

❖ The classification of oxidizers is based on the provisions of NFPA 430. Oxidizers, whether a solid, liquid or gas, yield oxygen or another oxidizing gas during a chemical reaction or readily react to oxidize combustibles. The rate of reaction varies with the class of oxidizer. Specific classification of oxidizers is important because of the varying degree of hazard. Examples of oxidizers include liquid hydrogen peroxide, nitric acid, sulfuric acid and solids such as sodium chlorite, chromic acid and calcium hypochlorite. Many commercially available swimming pool chemicals are indicative of Class 2 or 3 oxidizers. For further discussion of this material definition, see the commentary to it in Chapter 2 of the *International Fire Code Commentary*.

[F] OXIDIZING GAS. A gas that can support and accelerate combustion of other materials more than air does.

❖ Oxidizers sometimes yield oxidizing gases during a chemical reaction. These gases are capable of supporting and accelerating the combustion of other materials. Examples of oxidizing gases include bromine, chlorine and fluorine.

[BS] PANEL (PART OF A STRUCTURE). The section of a floor, wall or roof comprised between the supporting frame of two adjacent rows of columns and girders or column bands of floor or roof construction.

❖ This definition is needed to apply the structural load requirements in Chapter 16.

PANIC HARDWARE. A door-latching assembly incorporating a device that releases the latch upon the application of a force in the direction of egress travel. See "Fire exit hardware."

❖ Panic hardware is commonly used in educational and assembly-type spaces where the number of occupants who would use a doorway during a short time frame in an emergency is high in relation to an occupancy with a less dense occupant load, such as an office building. The hardware is required so that the door can be easily opened during an emergency when pressure on a door from a crush of people could render normal hardware inoperable. Not all types of panic hardware are permitted on doors required to be fire-protection rated (see the definition for "Fire exit hardware" and Section 1010.1.10).

[BS] PARTICLEBOARD. A generic term for a panel primarily composed of cellulosic materials (usually wood), generally in the form of discrete pieces or particles, as distinguished from fibers. The cellulosic material is combined with synthetic resin or other suitable bonding system by a process in which the interparticle bond is created by the bonding system under heat and pressure.

❖ Particleboard is one of the family of wood-based products that can be used as wood panels. The definition describes the characteristics unique to particleboard as compared to other types of wood-based composite panels. Note that particleboard is not considered wood structural panel. Particleboard intended for use as wood structural panels is subject to the requirements of Section 2303.1.8, which references ANSI A208.1. Particleboard intended for use in shear panels is regulated by the requirements of Section 2306.3, which references the SDPWS.

PENETRATION FIRESTOP. A through-penetration firestop or a *membrane-penetration firestop*.

❖ See the commentaries for the definitions of "Membrane-penetration firestop" and "Through-penetration firestop system."

PENTHOUSE. An enclosed, unoccupied rooftop structure used for sheltering mechanical and electrical equipment, tanks, elevators and related machinery, and vertical *shaft* openings.

❖ Any enclosed structure that is located above the surrounding roof surfaces can be considered a penthouse as long as it meets the criteria within Section 1509.2. By complying with these requirements, the penthouse is considered to not contribute to the height of the building either in number of stories or feet above grade plane. If the penthouse does not meet these requirements, it must be considered as an additional story of the building.

[BS] PERFORMANCE CATEGORY. A designation of wood structural panels as related to the panel performance used in Chapter 23.

❖ The term "performance category" reflects the latest versions of the DOC PS 1 and PS 2 standards, which use terminologies of bond classification to reference glue type and performance categories to reference the thicknesses tolerance consistent with nominal panel thicknesses in the code. The performance category value is the "nominal panel thickness" or "panel thickness" (see Section 2303.1.4).

[A] PERMIT. An official document or certificate issued by the *building official* that authorizes performance of a specified activity.

❖ The permit constitutes a license issued by the building official to proceed with specific activity, such as construction of a building, in accordance with all applicable laws. For the code and the IRC, the "building official" is identified as the person responsible for issuing the permit. For the IFC, the "fire code official" is identified as the person responsible. For other I-Codes, the "code official" is identified as the person responsible.

[A] PERSON. An individual, heirs, executors, administrators or assigns, and also includes a firm, partnership or corporation, its or their successors or assigns, or the agent of any of the aforesaid.

❖ Corporations and other organizations listed in the definition are treated as persons under the law. Also, when the code provides for a penalty (see Section 114.4), the definition makes it clear that the individuals responsible for administering the activities of these various organizations are subject to these penalties.

PERSONAL CARE SERVICE. The care of persons who do not require *medical care*. Personal care involves responsibility for the safety of the persons while inside the building

❖ Persons who need personal care may need supervision, but they are capable of self-preservation (see the commentary for Section 305, Group E day care).

Care facilities are used by patients of varying acuity seeking a broad spectrum of available support services. These facilities span a wide range of occupancy types including Groups E, I and R. There are three types of care defined in the codes: personal, custodial and medical. On the lower end of the care spectrum (i.e., personal care) is when occupants are supervised but do not need custodial or medical care. Where occupants may be elderly or impaired (i.e., custodial care) they may need occasional daily living assistance such as cooking and cleaning. While occupants may take longer to evacuate than average, they are capable of self-preservation. On the opposite end of the care spectrum, persons receiving care may be completely bedridden and dependent on medical gases and emergency power to maintain life (i.e., medical care). The level of care provided describes the condition and capabilities of an occupant which then indicates appropriate standards for protection systems and building. See also the definitions for "24-hour basis," "Custodial care," "Detoxification facility," "Foster care facility," "Group home," "Hospitals and psychiatric hospitals," "Medical care," "Nursing home," and "Incapable of self-preservation."

PHOTOLUMINESCENT. Having the property of emitting light that continues for a length of time after excitation by visible or invisible light has been removed.

❖ Examples of photoluminescent material are paint and tape that are charged by exposure to light. When the lights are turned off, the product will "glow" in the dark. Products utilized to meet the requirements for luminous egress path markings in high-rise buildings (see Sections 403.5.5 and 1025) or exit signs (see Section 1013.5) may be photoluminescent or self-luminous. A variety of materials can comply with the referenced standards for egress path markings, including ASTM E2072 and UL 1994; and for signs, UL 924.

PHOTOVOLTAIC MODULE. A complete, environmentally protected unit consisting of solar cells, optics and other components, exclusive of tracker, designed to generate DC power when exposed to sunlight.

❖ This definition provides clarification of requirements for photovoltaic modules that appear in Sections 1505 and 1510.

PHOTOVOLTAIC PANEL. A collection of modules mechanically fastened together, wired and designed to provide a field-installable unit.

❖ This definition provides clarification of roof requirements for photovoltaic panels that appear in Sections 1505 and 1510.

PHOTOVOLTAIC PANEL SYSTEM. A system that incorporates discrete photovoltaic panels, that converts solar radiation into electricity, including rack support systems.

❖ This definition provides clarification of requirements for photovoltaic panel systems in Chapter 15.

PHOTOVOLTAIC SHINGLES. A *roof covering* resembling shingles that incorporates photovoltaic modules.

❖ This definition provides clarification of roof-covering requirements that appear in Section 1507.17.

[F] PHYSICAL HAZARD. A chemical for which there is evidence that it is a *combustible liquid*, *cryogenic fluid*, *explosive*, flammable (*solid*, *liquid* or gas), *organic peroxide* (*solid* or *liquid*), *oxidizer* (*solid* or *liquid*), *oxidizing gas*, *pyrophoric* (*solid*, *liquid* or gas), *unstable (reactive) material* (*solid*, *liquid* or gas) or *water-reactive material* (*solid* or *liquid*).

❖ Those materials that present a detonation hazard, a deflagration hazard or that readily support combustion are considered physical hazards. Examples of the types of materials that present a physical hazard are included in the definition. Buildings and structures containing materials that present a physical hazard in excess of the maximum allowable quantity would be classified in Group H-1, H-2 or H-3. Materials that present a physical hazard may also present a health hazard (see the definition of "Health hazard").

[F] PHYSIOLOGICAL WARNING THRESHOLD LEVEL. A concentration of air-borne contaminants, normally expressed in parts per million (ppm) or milligrams per cubic meter (mg/m^3), that represents the concentration at which persons can sense the presence of the contaminant due to odor, irritation or other quick-acting physiological response. When used in conjunction with the permissible exposure limit (PEL) the physiological warning threshold levels are those consistent with the classification system used to establish the PEL. See the definition of "Permissible exposure limit (PEL)" in the *International Fire Code*.

❖ The term "physiological warning properties" is not defined. From a practical standpoint, the physiological warning properties are represented by a concentration of a contaminant that allows the average individual to sense its presence by a body warning signal including, but not limited to, odor, irritating effects such as stinging sensations, coughing, scratchy feeling in the throat, running of the eyes or nose and similar signals.

There may be a wide variability reported for some of the more common threshold levels including that of olfactory perception. Variations that may be encountered are due to a number of factors including the methods used in their determination, the population exposed and others. The requirements for gas detection established in the code are tied to the PEL, and there are several methods for determining the PEL inherent in the definition of that term. The intent of this definition is to link the determination of the physiological warning threshold level to the data used to determine the PEL.

For example, the PEL as established by DOL 29 CFR, Part 1910.1000 is primarily based on data developed by the American Conference of Governmental Industrial Hygienists (ACGIH) called threshold limit values (TLVs), as referenced in the definition of PEL found in Chapter 2 of the IFC. To substantiate the TLVs (PELs), the ACGIH publishes the *Documentation of the Threshold Limit Values* (TLVs®) and *Biological Exposure Indices* (BEIs®) where the user is provided with data used in their establishment. The significant commercially available toxic and highly toxic gases with published TLVs are listed by ACGIH, and perception thresholds are provided.

These warning properties are considered, as evidenced by the documentation, when the TLV and, hence, the PEL is established. It is appropriate that the data used in the base documents be used as the basis for determining the threshold level when such data are available. The use of data from other sources may be used in the absence of data within the system used for the establishment of the PEL, but where such data have been considered in determining the PEL, such data should take precedent.

By providing a definition for "Physiological warning threshold level" and guidance as to how it is to be applied, the code user is given guidance that carries out the intent of the provisions for gas detection that have been established in the code. See the commentary to Section 6004.2.2.10 of the IFC for further discussion of gas detection.

PLACE OF RELIGIOUS WORSHIP. See "Religious worship, place of."

PLASTIC, APPROVED. Any thermoplastic, thermosetting or reinforced thermosetting plastic material that conforms to combustibility classifications specified in the section applicable to the application and plastic type.

❖ This term applies to plastic that has properties that conform to the code requirements for its intended use.

PLASTIC COMPOSITE. A generic designation that refers to wood/plastic composites and plastic lumber.

❖ Plastic composites can be plastic lumber or wood/plastic composite lumber. These components are typically used for exterior decking and railing systems. See also the definitions for "Plastic lumber" and "Wood/plastic composite."

PLASTIC GLAZING. Plastic materials that are glazed or set in frame or sash and not held by mechanical fasteners that pass through the glazing material.

❖ In lieu of glass, plastic is permitted to be used as a glazing material. Plastic glazing is required to meet the same load requirements as glass.

[BS] PLASTIC LUMBER. A manufactured product made primarily of plastic materials (filled or unfilled) which is generally rectangular in cross section.

❖ Plastic lumber is manufactured of plastic and a uses a variety of materials to provide strength and stiffness. These components are typically used for exterior decking and railing systems. Chapter 26, which regulates plastic materials, contains requirements for plastic lumber used as exterior decking components, such as deck boards, stair treads, handrails and guardrails.

PLATFORM. A raised area within a building used for worship, the presentation of music, plays or other entertainment; the head table for special guests; the raised area for lecturers and speakers; boxing and wrestling rings; theater-in-the-round *stages*; and similar purposes wherein, other than horizontal sliding curtains, there are no overhead hanging curtains, drops, scenery or stage effects other than lighting and sound. A temporary platform is one installed for not more than 30 days.

❖ Platforms are raised areas that are used for public performances and presentations that, except for lighting, do not incorporate any overhead hanging curtains, drops, scenery or stage effects. Additionally, while it is not specified, the fuel load on platforms (e.g., a podium for a lecturer or a boxing ring) is anticipated to be low.

Thus, since the fuel load on platforms is ordinarily low and there is no fuel load overhead in areas that would be difficult to access, the code requirements for platforms are less stringent than for stages.

Since many platforms are installed and used on a temporary basis, this definition specifically defines temporary use as 30 days or less. Temporary use is less likely to accumulate additional fuel loads, storage or waste materials on, beneath or above the platform. Platforms are distinct from equipment platforms (see Section 505.3). A platform addressed by this definition is typically not considered a mezzanine (see Section 505.2).

POLYPROPYLENE SIDING. A shaped material, made principally from polypropylene homopolymer, or copolymer, which in some cases contains fillers or reinforcements, that is used to clad *exterior walls* of buildings.

❖ Polypropylene siding is an exterior finish product that is available in many textures, colors and profiles. These products come from various manufacturers with specific installation requirements with which compliance is very important. Specification and installation requirements for polypropylene siding are discussed in Sections 1404.12 and 1405.18. Proper installation will result in expected product performance at varying temperatures by allowing for the expansion and contraction of the siding materials and its components.

[BS] PORCELAIN TILE. Tile that conforms to the requirements of ANSI 137.1.3 for ceramic tile having an absorption of 0.5 percent or less in accordance with ANSI 137.4.1–Class Table and ANSI 137.1.6.1 Allowable Properties by Tile Type–Table 10.

❖ Porcelain tile is one of a myriad of types of ceramic tile. Its unique characteristics and extremely low absorption rate requires it be dealt with differently from other materials, especially when applied as an exterior adhered veneer. These materials fall outside the scope of TMS 402/ACI 530/ASCE 6. General installation limitations are found in Section 1405.10.2.

[BS] POSITIVE ROOF DRAINAGE. The drainage condition in which consideration has been made for all loading deflections of the *roof deck*, and additional slope has been provided to ensure drainage of the roof within 48 hours of precipitation.

❖ The primary purpose of positive roof drainage is to allow for prompt roof drainage, thereby preventing the roof from being damaged or adversely impacting the structural support due to the additional load.

POWER-ASSISTED DOOR. Swinging door which opens by reduced pushing or pulling force on the door-operating hardware. The door closes automatically after the pushing or pulling force is released and functions with decreased forces.

See "Low-energy power-operated door" and "Power-operated door."

❖ There are basically three different types of doors that provide some type of power assistance for entry: low-energy power-operated doors, power-assisted doors and power-operated doors. The power-assisted door is typically a side-swinging door that has the additional feature of powered assistance to move the door to the open position. When a door has power assistance, the force or effort it takes to open the door while it is being pushed or pulled is reduced as long as a user maintains pressure on the hardware. When the hardware is released, the door will move to the closed position. Power-assist doors are typically be used when a door is an unusual size or weight. The low-energy power-assisted door and power-assisted door both are operated by the user touching something; therefore they both must comply with BMHA156.19.

POWER-OPERATED DOOR. Swinging, sliding, or folding door which opens automatically when approached by a pedestrian or opens automatically upon an action by a pedestrian. The door closes automatically and includes provisions such as presence sensors to prevent entrapment. See "Low energy power-operated door" and "Power-assisted door."

❖ There are basically three different types of doors that provide some type of power assistance for entry: low-energy power-operated doors, power-assisted doors and power-operated doors. The power-operated door can be a sliding, hinged or side-swinging door that operates automatically by either a motion sensor or sensor mat when someone approaches the door. Power-operated doors are most commonly installed at the busy entrances of commercial buildings. As a door with hands-free operation, the power-operated door must comply with BMHA156.10.

[BS] PREFABRICATED WOOD I-JOIST. Structural member manufactured using sawn or structural composite lumber flanges and wood structural panel webs bonded together with exterior exposure adhesives, which forms an "I" cross-sectional shape.

❖ Wood I-joists are structural members typically used in floor and roof construction manufactured out of sawn or structural composite lumber flanges and structural panel webs, bonded together with exterior adhesives forming an "I" cross section. This definition explains which engineered wood products are considered prefabricated wood I-joists in order to clarify the requirements in Section 2303.1.2.

[BS] PRESTRESSED MASONRY. *Masonry* in which internal stresses have been introduced to counteract potential tensile stresses in *masonry* resulting from applied *loads*.

❖ The definition provides an understanding of the term "prestressed masonry," which is used to define particular types of shear wall systems recognized under the Masonry Standards Joint Committee (MSJC) Code (a.k.a. TMS 402/ACI 530/ASCE 5 – Building Code Requirements for Masonry Structures).

PRIMARY STRUCTURAL FRAME. The primary structural frame shall include all of the following structural members:

1. The columns.
2. Structural members having direct connections to the columns, including girders, beams, trusses and spandrels.
3. Members of the floor construction and roof construction having direct connections to the columns.
4. Bracing members that are essential to the vertical stability of the primary structural frame under gravity loading shall be considered part of the primary structural frame whether or not the bracing member carries gravity *loads*.

❖ The primary structural frame and secondary members must meet different standards of design and protection as specified in Chapters 6 and 7. The definitions of these two terms spell out which elements of a structure's framing system are part of the primary structural framing system essential to carrying the gravity loads of the building. Such elements are generally required to have greater fire-resistance protection (see commentary, Table 601 and Section 704).

PRIVATE GARAGE. A building or portion of a building in which motor vehicles used by the tenants of the building or buildings on the premises are stored or kept, without provisions for repairing or servicing such vehicles for profit.

❖ In Section 406, the code regulates two types of garages: private and public. Public garages are further broken into open or enclosed garages. Private garages are limited in size by Section 406.3 and can be accessory to either residential or nonresidential uses. The definition is intended to help distinguish private garages from public garages. Carports are also addressed in Section 406.3

PROSCENIUM WALL. The wall that separates the *stage* from the auditorium or assembly seating area.

❖ A proscenium wall separates the stage area from the audience. The performance on the stage is viewed through a large opening known as the main proscenium opening. Where the stage height is greater than 50 feet (15 240 mm), a proscenium wall of approved construction is required by Section 410.3.4.

PSYCHIATRIC HOSPITALS. See "Hospitals."

PUBLIC ENTRANCE. An entrance that is not a *service entrance* or a *restricted entrance*.

❖ A public entrance is one that provides access for the general public or employees, other than the service entrance or a restricted entrance (see the commentaries for the definitions of "Service entrance" and "Restricted entrance" and to Section 1105.1).

Entrances may be locked for security purposes, but if the door serves as an entrance for employees or residents with keys, that door would still be considered a public entrance. A sports facility where there is control at the entrance for ticket holders only would still be considered a public entrance

PUBLIC-USE AREAS. Interior or exterior rooms or spaces that are made available to the general public.

❖ This term is utilized to describe all interior and exterior spaces or rooms that may be occupied by the general public for any amount of time. Spaces that are utilized by the general public may be located in facilities that are publicly or privately owned. Examples include the lobby in an office building, a high-school gymnasium with assembly seating, an open-air stadium, a multipurpose room, an exposition hall, a restaurant dining room, a health club, etc. (see also the commentaries for the definitions of "Common use" and "Employee work area").

[A] PUBLIC WAY. A street, alley or other parcel of land open to the outside air leading to a street, that has been deeded, dedicated or otherwise permanently appropriated to the public for public use and which has a clear width and height of not less than 10 feet (3048 mm).

❖ Public ways serve a variety of purposes in the code, including the determination of the allowable area of a building (see Section 506.2); the use as an open space for unlimited area buildings (see Section 507); the measurement of fire separation distance (see Section 202); as well as many provisions for the means of egress (Chapter 10).

[F] PYROPHORIC. A chemical with an auto-ignition temperature in air, at or below a temperature of 130°F (54.4°C).

❖ The definition is derived from DOL 29 CFR, Part 1910.1200. Pyrophoric materials, whether in a gas, liquid or solid form, are capable of spontaneous ignition at low temperatures. Examples of pyrophoric materials include silane and phosphine gas; liquid diethylaluminum chloride; and inert solids, such as cesium, plutonium, potassium and robidium. For further discussion of this material definition, see the commentary to it in Chapter 2 of the *International Fire Code Commentary*.

[F] PYROTECHNIC COMPOSITION. A chemical mixture that produces visible light displays or sounds through a self-propagating, heat-releasing chemical reaction which is initiated by ignition.

❖ Pyrotechnic composition consists of those chemical components, including oxidizers, that cause fireworks to make noise or display light when ignited. The definition is derived from NFPA 1124. The amount of pyrotechnic composition is the determining factor in whether the storage area for consumer fireworks is classified as Group H-3. The pyrotechnic content of consumer fireworks is contained within a significant amount of packaging and nonexplosive materials used in their manufacture, which constitute the bulk of the weight of the fireworks devices.

RADIANT BARRIER. A material having a low-emittance surface of 0.1 or less installed in building assemblies.

❖ This definition enables the code user to correctly apply the material standards for these products in Section 1509.

RAMP. A walking surface that has a running slope steeper than one unit vertical in 20 units horizontal (5-percent slope).

❖ This definition is needed to determine the threshold at which the ramp requirements apply to a walking surface. Walking surfaces steeper than specified in the definition are subject to the ramp requirements in Sections 1012 and 1029.

RAMP-ACCESS OPEN PARKING GARAGES. *Open parking garages* employing a series of continuously rising floors or a series of interconnecting ramps between floors permitting the movement of vehicles under their own power from and to the street level.

❖ These types of garages employ vehicular ramps or sloped tiers that connect all of the levels of the structure. Whether the vehicles are self-parked or employee parked is immaterial.

RAMP, EXIT ACCESS. See "Exit access ramp."

RAMP, EXTERIOR EXIT. See "Exterior exit ramp."

RAMP, INTERIOR EXIT. See "Interior exit ramp."

[A] RECORD DRAWINGS. Drawings ("as builts") that document the location of all devices, appliances, wiring sequences, wiring methods and connections of the components of a *fire alarm system* as installed.

❖ To verify that the system has been installed to comply with the code and applicable referenced standards, complete as-built drawings of the fire alarm system must be available on site for review.

REFLECTIVE PLASTIC CORE INSULATION. An insulation material packaged in rolls, that is less than $^1/_2$ inch (12.7 mm) thick, with not less than one exterior low-emittance surface (0.1 or less) and a core material containing voids or cells.

❖ This is a product distinct from "foam plastic insulation." Its unique properties dictate the need for a different testing procedure as specified in Section 2614.

[A] REGISTERED DESIGN PROFESSIONAL. An individual who is registered or licensed to practice their respective design profession as defined by the statutory requirements of the professional registration laws of the state or *jurisdiction* in which the project is to be constructed.

❖ Legal qualifications for engineers and architects are established by the state having jurisdiction. Licensing and registration of engineers and architects are accomplished by written or oral examinations offered by states, or by reciprocity (licensing in other states).

[A] REGISTERED DESIGN PROFESSIONAL IN RESPONSIBLE CHARGE. A *registered design professional* engaged by the owner or the owner's authorized agent to review and coordinate certain aspects of the project, as determined by the *building official*, for compatibility with the design of the building or structure, including submittal documents prepared by others, deferred submittal documents and phased submittal documents.

❖ This definition refers to the registered design professional named by the owner or the owner's representative where required by Section 107.3.4, which states that the owner must designate a registered design profession when required by the laws applicable to the jurisdiction in which its building is constructed. The role of the registered design professional in responsible charge includes the review and coordination of the following items for compatibility with a project's design requirements:

1. Submittal documents prepared by others;
2. Deferred submittal documents;
3. Phases submittal documents; and
4. Special inspection reports.

RELIGIOUS WORSHIP, PLACE OF. A building or portion thereof intended for the performance of religious services.

❖ This term has been added to the code for the purpose of making the code more broadly applicable to the worship facilities of all religions. Major religions for the world include Christianity, Islam, Hinduism, Buddhism and Judaism, which use different terms to describe the main space used for religious services. The intent in the code is for the same application for all similar types of religious facilities. The term also makes it clear that it defines the room or sanctuary for the performance of religious worship services and not retreat complexes, rectories, convents and classroom or office areas.

[A] REPAIR. The reconstruction or renewal of any part of an existing building for the purpose of its maintenance or to correct damage.

❖ As indicated in Section 105.2.2, the repair of an item typically does not require a permit. This definition makes it clear that repair is limited to work on the item, and does not include complete or substantial replacement or other new work.

[EB] REROOFING. The process of recovering or replacing an existing *roof covering*. See "Roof recover" and "Roof replacement."

❖ This term refers to the process of covering an existing roof system with a new roofing system (see Section 1511).

RESIDENTIAL AIRCRAFT HANGAR. An accessory building less than 2,000 square feet (186 m^2) and 20 feet (6096 mm) in *building height* constructed on a one- or two-family property where aircraft are stored. Such use will be considered as a residential accessory use incidental to the dwelling.

❖ A residential aircraft hangar is considered an accessory or auxiliary structure to a residential house similar to any other shed or detached garage. One- or two-family dwellings are required to be constructed in accordance with the IRC (see Section 101.2). Although that code is also applicable to accessory structures associated with one- or two-family dwellings, it is clear that Section 412.5 of the code was intended to control aircraft hangars since the IRC is silent on the subject.

A residential aircraft hangar is limited to 2,000 square feet (186 m^2) in area and 20 feet (6096 mm) in height. The hangar must be integral to a residential home on the same property. As such, the hangar would be considered the same as the private garage used to store the home's motor vehicles.

[BS] RESISTANCE FACTOR. A factor that accounts for deviations of the actual strength from the *nominal strength* and the manner and consequences of failure (also called "strength reduction factor").

❖ This definition is needed to apply material strength adjustments by way of the resistance factor that is specified in structural material chapters and referenced structural standards.

[BS] RESTRICTED ENTRANCE. An entrance that is made available for *common use* on a controlled basis, but not public use, and that is not a *service entrance*.

❖ The key to this provision is that the entrance has a controlled access or some type of limiting basis. This may be an entrance for jurors only at a courthouse, visitors only at a jail or employees only at a factory. A sports facility where there is control at the entrance for ticket holders only, or a building with locked entrances, is not typically considered a restricted entrance (see the commentaries for the definitions of "Public entrance" and "Service entrance," and to Section 1105.1.3).

RETRACTABLE AWNING. A retractable *awning* is a cover with a frame that retracts against a building or other structure to which it is entirely supported.

❖ The key to this definition is that it provides for an element of building construction that is popular in residential construction. Retractable awnings are a subcategory of awnings, and therefore are included in the regulations in Section 3105.

[BS] RISK CATEGORY. A categorization of buildings and other structures for determination of *flood*, wind, snow, ice and earthquake *loads* based on the risk associated with unacceptable performance.

❖ Risk category is a classification given to all buildings in Section 1604.5 based on the nature of the occupancy. This classification serves as a threshold for earthquake, flood, snow and wind load requirements (also see the definition of "Essential facilities" and the commentary to Section 1604.5).

[BS] RISK-TARGETED MAXIMUM CONSIDERED EARTHQUAKE (MCE_R) GROUND MOTION RESPONSE ACCELERATIONS. The most severe earthquake effects considered by this code, determined for the orientation that results in the largest maximum response to horizontal ground motions and with adjustment for targeted risk.

❖ This term refers to the mapped earthquake ground motions that are provided in Figures 1613.3.1(1) through 1613.3.1(8) for a Site Class B (rock) soil profile. It reflects the maximum level of earthquake ground shaking that is considered reasonable for the design of new structures.

[BS] ROOF ASSEMBLY (For application to Chapter 15 only). A system designed to provide weather protection and resistance to design *loads*. The system consists of a *roof covering* and *roof deck* or a single component serving as both the roof covering and the *roof deck*. A roof assembly includes the *roof deck*, *vapor retarder*, substrate or thermal barrier, insulation, *vapor retarder* and *roof covering*.

❖ With respect to its application as it relates to Chapter 15, a roof assembly is an assembly of interacting roof components (including the roof deck) designed to weatherproof and, normally, to insulate a building's top surface.

[BS] ROOF COVERING. The covering applied to the *roof deck* for weather resistance, fire classification or appearance.

❖ This definition identifies the specific membrane of the entire roof system that provides weather protection and any required resistance to exterior fire exposure. The code has specific performance and prescriptive requirements for this covering to provide a durable, weather-resistant surface for the entire structure. A roof covering is considered the membrane that provides the weather-resistance and fire-performance characteristics required by the code.

ROOF COVERING SYSTEM. See "Roof assembly."

[BS] ROOF DECK. The flat or sloped surface constructed on top of the *exterior walls* of a building or other supports for the purpose of enclosing the *story* below, or sheltering an area, to protect it from the elements, not including its supporting members or vertical supports.

❖ A roof deck is the structural surface to which the roofing and waterproofing system (including insulation) is applied.

ROOF DRAINAGE, POSITIVE. See "Positive roof drainage."

[EB] ROOF RECOVER. The process of installing an additional *roof covering* over a prepared existing *roof covering* without removing the existing *roof covering*.

❖ This term refers to the process of covering an existing roof system with a new roofing system.

[EB] ROOF REPAIR. Reconstruction or renewal of any part of an existing roof for the purposes of its maintenance.

❖ Roofs should be maintained for the purpose of protection. If a section is damaged, then it should be repaired immediately.

[EB] ROOF REPLACEMENT. The process of removing the existing *roof covering*, repairing any damaged substrate and installing a new *roof covering*.

❖ This definition refers to the process of covering an existing roof system with a new roofing system.

ROOF VENTILATION. The natural or mechanical process of supplying conditioned or unconditioned air to, or removing such air from, *attics*, cathedral ceilings or other enclosed spaces over which a *roof assembly* is installed.

❖ Ventilation of the attic prevents moisture condensation on cold surfaces and, therefore, will prevent dry rot on the bottom surface of shingles or wood roof decks.

ROOFTOP STRUCTURE. A structure erected on top of the *roof deck* or on top of any part of a building.

❖ This definition includes all appurtenances constructed and located above the surrounding roof surfaces. These items (water tanks or cooling towers) or architectural features (spires or cupolas) are regulated by specific code provisions.

[BS] RUNNING BOND. The placement of *masonry units* such that *head joints* in successive courses are horizontally offset at least one-quarter the unit length.

❖ Commentary Figure 202(35) illustrates the required overlap for running bonds. The minimum overlap is necessary to provide strength between units when masonry spans horizontally. Masonry not laid in running bond is often referred to as "stack bond" construction.

SALLYPORT. A security vestibule with two or more doors or gates where the intended purpose is to prevent continuous and unobstructed passage by allowing the release of only one door or gate at a time.

❖ While sallyports are not unique to Group 1-3 occupancies, they are essential to the control and safe movement of inmates in a Group 1-3 facility. Section 408.3.7 allows sallyports to be in the means of egress system. See the commentary for that section for an additional description of sallyport design and use.

SCISSOR STAIRWAY. Two interlocking *stairways* providing two separate paths of egress located within one *exit* enclosure.

❖ A scissor or interlocking stairway is sometimes used in high-rise buildings or to increase exit capacity of a stairwell enclosure. In this configuration, two independent stairway paths are located within the same exit enclosure and may or may not be visually open to one another. When interlocking stairways are separated from each other with compliant fire barriers and horizontal assemblies, they are not considered scissor stairways (see Section 1007.1.1).

[BS] SCUPPER. An opening in a wall or parapet that allows water to drain from a roof.

❖ These devices are for the purpose of allowing overflowing water to drain off the roof. These devices are commonly larger than the roof drain.

SECONDARY MEMBERS. The following structural members shall be considered secondary members and not part of the *primary structural frame*:

1. Structural members not having direct connections to the columns.
2. Members of the floor construction and roof construction not having direct connections to the columns.
3. Bracing members other than those that are part of the *primary structural frame*.

❖ This term works in conjunction with the term "primary structural frame" to distinguish the level of fire-resistance protection needed by various elements of a building's structure. These requirements are found in Table 601 and Sections 704.

[BS] SEISMIC DESIGN CATEGORY. A classification assigned to a structure based on its *risk category* and the severity of the *design earthquake ground motion* at the site.

❖ The seismic design category serves as a trigger mechanism for many seismic requirements, including the following:

1. Permissible seismic force-resisting systems.
2. Limitations on height.
3. Consideration of structural irregularities.
4. The types of lateral-force analysis that may be used.
5. The need for additional special inspections.

The first step in the design of a structure for earthquake forces is the determination of its seismic design category. Each building is assigned the more restrictive seismic design category classification depending upon the site soil profile, the mapped spectral response accelerations at the site and the nature of the uses that occupy the building, using both Tables 1613.3.5(1) and 1613.3.5(2). The following outlines the procedure used to determine a structure's seismic design category.

Step 1: Determine the mapped maximum considered earthquake spectral response acceleration at short periods, S_s, and at 1-second period, S_l, for the site location from Figures 1613.3.1(1) through 1613.3.1(8).

Step 2: Determine the (soil) site class in accordance with Section 1613.3.2.

Step 3: Determine the site coefficients, F_a and F_v, from Tables 1613.3.3(1) and 1613.3.3(2), respectively.

Step 4: Determine the 5-percent damped design spectral response acceleration at short

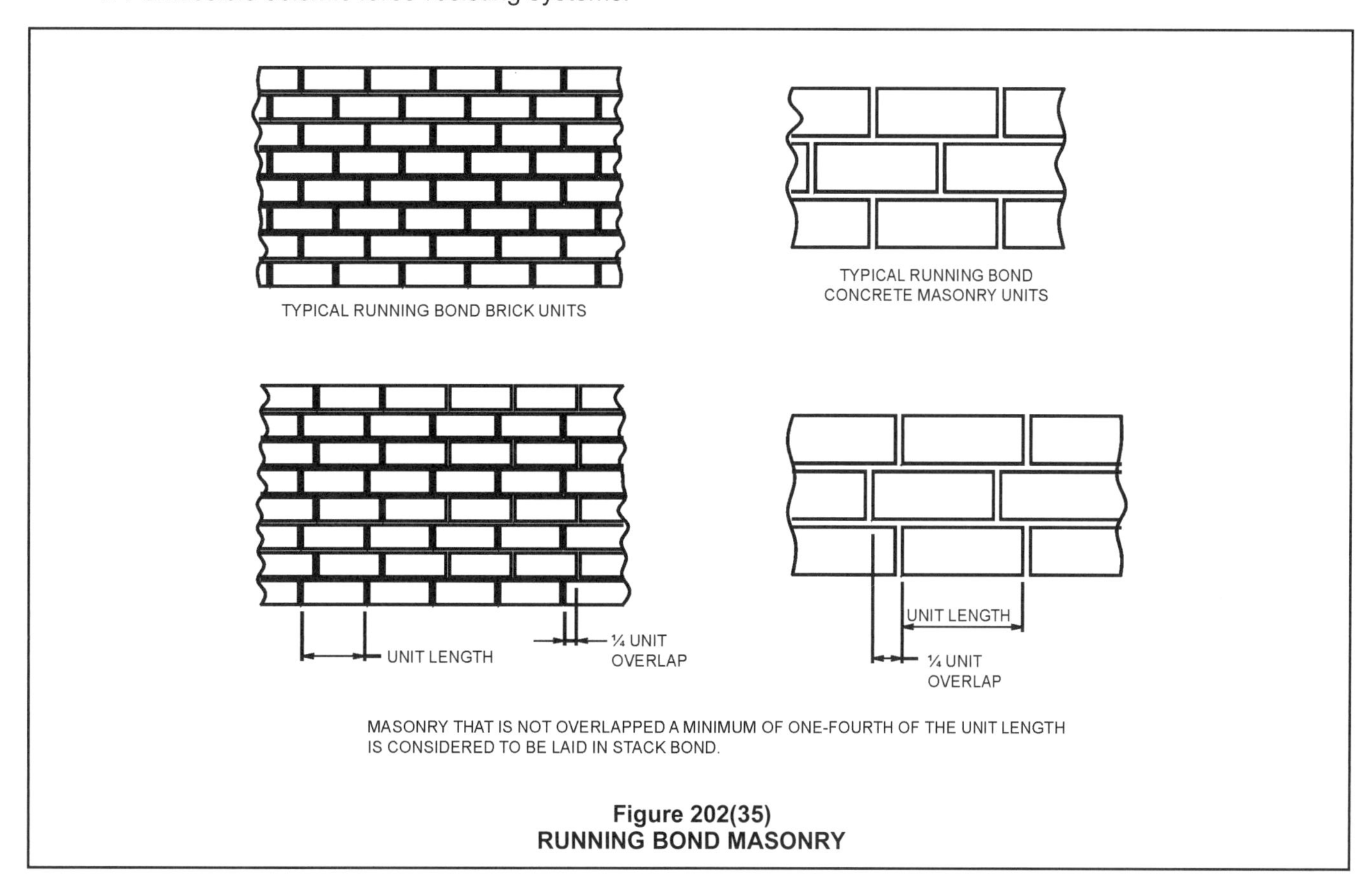

Figure 202(35)
RUNNING BOND MASONRY

periods, S_{DS}, and at 1-second period, S_{D1}, as follows:

$$S_{DS} = (^2/_3)(F_a)(S_s)$$

$$S_{D1} = (^2/_3)(F_v)(S_1)$$

Step 5: Determine the seismic design category as prescribed by Tables 1613.3.5(1) and 1613.3.5(2).

The highest of the seismic design categories from the two tables is the category assigned to the building, unless Section 1613.3.5.1 is applicable. For example, if the seismic design category from Table 1613.3.5(1) is D and from Table 1613.3.5(2) it is C, then the building would be assigned to Seismic Design Category D.

[BS] SEISMIC FORCE-RESISTING SYSTEM. That part of the structural system that has been considered in the design to provide the required resistance to the prescribed seismic forces.

❖ This definition is needed for the correct application of special inspections of seismic force-resisting systems. Unless classified as Seismic Design Category A, the type of seismic-force-resisting system must be selected to conform to Section 12.2 of ASCE 7. Providing a level of detailing that is appropriate for the selected system is critical to complying with the code's earthquake provisions.

SELF-CLOSING. As applied to a *fire door* or other opening protective, means equipped with an device that will ensure closing after having been opened.

❖ A self-closing opening protective refers to a fire or smoke door assembly equipped with a listed closer for doors that must be maintained in the normally closed position. When the door is opened and released, the self-closing feature returns the door to the closed position. It is important to distinguish between the terms "self-closing" and "automatic closing" because they are not interchangeable. "Automatic closing" refers to an opening protective that is normally in the open position (see Section 716.5.9.2). Opening protectives with automatic closers are often held open and then returned to the closed position upon activation of fire detectors or smoke detectors or loss of power, which automatically releases the hold-open device and allows the door to close.

SELF-LUMINOUS. Illuminated by a self-contained power source, other than batteries, and operated independently of external power sources.

❖ Self-luminous products do not need an outside light source to charge them like photoluminescent materials do. Products utilized to meet the requirements for luminous egress path markings in high-rise buildings (see Sections 403.5.5 and 1025) or exit signs (see Section 1013.5) may be photoluminescent or self-luminous. A variety of materials can comply with the referenced standards for egress path markings (UL 1994 and ASTM E2072) and for signs (UL 924).

SELF-PRESERVATION, INCAPABLE OF. See "Incapable of self-preservation."

SELF-SERVICE STORAGE FACILITY. Real property designed and used for the purpose of renting or leasing individual storage spaces to customers for the purpose of storing and removing personal property on a self-service basis.

❖ A portion or space within these facilities can be rented by persons to store personal property. Movement of items into and out of the space is handled by the individual.

[F] SERVICE CORRIDOR. A fully enclosed passage used for transporting *HPM* and purposes other than required *means of egress*.

❖ Though HPM facility occupants may be exposed to limited HPM quantities during the course of their employment, their means of egress are protected from the HPM hazards by confining the HPM being transferred to its own passageway. A service corridor is only required when the HPM must be carried from a storage room or external area to a fabrication area through a passageway.

SERVICE ENTRANCE. An entrance intended primarily for delivery of goods or services.

❖ This entrance is utilized primarily for accepting or sending deliveries of goods and services. Often this entrance is directly associated with a loading dock, and is not considered a public or restricted entrance (see the commentary for the definitions of "Service entrance" and "Public entrance" and to Section 1105.1.5).

SHAFT. An enclosed space extending through one or more *stories* of a building, connecting vertical openings in successive floors, or floors and roof.

❖ Shafts are successive openings in the floors. A shaft is required to be enclosed with fire-resistance-rated assemblies to help prevent the vertical spread of fire and resist the spread of products of combustion from story to story. Provisions for vertical shafts are found in Section 713. These provisions are applicable to vertical openings required to be enclosed in fire-resistant shaft construction, such as for elevators and enclosed exit stairways.

SHAFT ENCLOSURE. The walls or construction forming the boundaries of a *shaft*.

❖ Fire-resistance-rated walls forming shaft enclosures must be constructed as fire barrier walls in accordance with Section 707. Usually, vertical ducts penetrating multiple floors or buildings are required to be in shaft enclosures. Unlike a common chase wall, a shaft wall must be fire-resistance rated from the room side and from the shaft side.

[BS] SHALLOW FOUNDATION. A shallow foundation is an individual or strip footing, a mat foundation, a slab-on-grade foundation or a similar foundation element.

❖ Chapter 18 identifies two major classifications for foundations; either shallow or deep. This definition clarifies the type of foundations that are considered to be shallow foundations. All others are, by default, considered to be deep foundations.

[BS] SHEAR WALL (for Chapter 23). A wall designed to resist lateral forces parallel to the plane of a wall.

❖ Shear walls are the supporting elements for the lateral forces that are applied to a floor or roof diaphragm. They function structurally to provide a path for these loads to the foundation. Typically in wood framing these walls are framed with studs and have wood structural panel sheathing attached to the outside using specific nailing patterns to provide the necessary strength and stiffness. They also may have a tie-down attached at either or both ends of the wall to prevent overturning. Guidance for design of wood shear walls can be found in APA Research Report 154.

Shear wall, perforated. A wood structural panel sheathed wall with openings, that has not been specifically designed and detailed for force transfer around openings.

❖ This term refers to shear walls that are designed by using the provisions of ANSI/AWC *Special Design Provisions for Wind and Seismic* (SDPWS) as an alternative to the traditional segmented shear wall design methodology. Aspect ratio limits apply to shear wall segments within the perforated shear wall as described by the definition of "Perforated shear wall segment." This method of shear wall design utilizes empirically determined reductions of wood structural panel shear wall capacities that are based on the maximum opening height, as well as the percentage of a particular shear wall that qualifies as a perforated shear wall segment. This method recognizes the strength and stiffness of sheathed areas above and below openings without the need to specifically design and detail for force transfer around the openings. It is not expected that sheathed wall areas above and below openings behave as coupling beams acting end to end, but rather that they provide local restraint at their ends. As a consequence, significantly reduced capacities are attributed to interior perforated shear wall segments with limited overturning restraint. Further background on the development of provisions for perforated shear walls and example problems are provided in the National Earthquake Hazards Reduction Program (NEHRP) *Recommended Provisions for Seismic Regulations for New Buildings and Other Structures*, Part 2 Commentary (FEMA 450-2). Other resources and articles on perforated shear wall design are available from the American Wood Council at www.awc.org.

Shear wall segment, perforated. A section of shear wall with full-height sheathing that meets the height-to-width ratio limits of Section 4.3.4 of AWC SDPWS.

❖ This term refers to a fully sheathed portion of a perforated shear wall that also meets the limit on height-to-width ratio. Only these portions of the perforated shear wall are considered in determining the percentage of full-height sheathing.

[BS] SHINGLE FASHION. A method of installing roof or wall coverings, water-resistive barriers, flashing or other building components such that upper layers of material are placed overlapping lower layers of material to provide for drainage via gravity and moisture control.

❖ This term is used to describe the required method of applying moisture control layers such as roof underlayment and water-resistive barriers to the building. The intent is to direct the user to place upper layers of material lapping over lower layers of material, in the fashion of placing roof shingles, so moisture is provided with a clear path to drain down and away from the building.

[BS] SINGLE-PLY MEMBRANE. A roofing membrane that is field applied using one layer of membrane material (either homogeneous or composite) rather than multiple layers.

❖ This is a flexible or semiflexible roof covering or waterproofing layer, whose primary function is the exclusion of water.

[F] SINGLE-STATION SMOKE ALARM. An assembly incorporating the detector, the control equipment and the alarm-sounding device in one unit, operated from a power supply either in the unit or obtained at the point of installation.

❖ A single-station smoke alarm is a self-contained alarm device that detects visible or invisible particles of combustion. Its function is to detect a fire in the immediate area of the detector location. Single-station smoke alarms are individual units with the capability to stand alone. Where single-station smoke alarms are interconnected with other single-station devices, they would be considered a multiple-station smoke alarm system. Single-station smoke alarms are not capable of notifying or controlling any other fire protection equipment or systems. They may be battery powered, directly connected to the building power supply, or a combination of both. Single-station smoke alarms must be built to comply with UL 217, and are to be installed as required by Section 907.2.11.

SITE. A parcel of land bounded by a *lot line* or a designated portion of a public right-of-way.

❖ A site, for purposes of accessibility requirements, is the same as that which is considered in the application of other code requirements. The property within the boundaries of the site is under the control of the owner. The owner can be held responsible for code compliance of the site and all facilities on it. Legal property lines do not always constitute site boundaries (e.g., malls, condominiums, townhouses). A site could contain multiple legal "lot" divisions.

[BS] SITE CLASS. A classification assigned to a site based on the types of soils present and their engineering properties as defined in Section 1613.3.2.

❖ This definition is necessary because the earthquake load on a structure is greatly affected by the soil at the site. There are six site classes in the provisions that are referenced in Section 1613.3.2.

[BS] SITE COEFFICIENTS. The values of F_a and F_v indicated in Tables 1613.3.3(1) and 1613.3.3(2), respectively.

❖ The site coefficients F_a and F_v are used to adjust the mapped ground motions to account for the characteristics of the site's soil profile when determining the soil-adjusted earthquake spectral response accelerations in Section 1613.3.3. These coefficients are functions of the soil (site class), as well as the seismicity at the site of the structure.

SITE-FABRICATED STRETCH SYSTEM. A system, fabricated on site and intended for acoustical, tackable or aesthetic purposes, that is composed of three elements:

1. A frame (constructed of plastic, wood, metal or other material) used to hold fabric in place;
2. A core material (infill, with the correct properties for the application); and
3. An outside layer, composed of a textile, fabric or vinyl, that is stretched taut and held in place by tension or mechanical fasteners via the frame.

❖ Site-fabricated stretch systems are interior finish materials that are pulled taut across walls or ceilings with a frame that holds a fabric and core. These systems are now being used extensively because they can stretch to cover decorative walls and ceilings with unusual looks and shapes. The systems consist of three parts: a fabric (or vinyl), a frame and an infill core material. This type of product is not exclusive to any particular manufacturer. It is important to point out that these materials are not curtains or drapes because they are not free hanging (see commentary, Section 803.10).

SKYLIGHT, UNIT. A factory-assembled, glazed fenestration unit, containing one panel of glazing material that allows for natural lighting through an opening in the *roof assembly* while preserving the weather-resistant barrier of the roof.

❖ This is a specific type of sloped glazing assembly that is factory assembled. The code contains specific provisions that are appropriate for this type of building component. Factory-assembled units, as opposed to site-built skylights, can be designed, tested and rated as one component that incorporates both glazing and framing, if applicable. The individual components of site-built glazing must be designed to resist the design loads of the codes individually, and are not usually rated as an assembly.

SKYLIGHTS AND SLOPED GLAZING. Glass or other transparent or translucent glazing material installed at a slope of 15 degrees (0.26 rad) or more from vertical. Glazing material in skylights, including *unit skylights*, *tubular daylighting devices*, solariums, *sunrooms*, roofs and sloped walls, are included in this definition.

❖ The code regulates skylights and sloped glazing since their failure could result in injury and building damage (see Section 2405 for the code requirements). Fenestration is a key element in a building's thermal envelope.

SLEEPING UNIT. A room or space in which people sleep, which can also include permanent provisions for living, eating, and either sanitation or kitchen facilities but not both. Such rooms and spaces that are also part of a *dwelling unit* are not sleeping units.

❖ This definition is included to coordinate the *Fair Housing Act Guidelines* with the code. The definition for "Sleeping unit" clarifies the differences between sleeping units and dwelling units. In addition, using the term "sleeping unit" for spaces where people sleep will replace a multitude of other terms (i.e., patient room, cell, guestroom) so that there is consistent application across occupancies. Some examples of sleeping units are hotel guest rooms; dormitories; bedrooms in boarding houses; patient sleeping rooms in hospitals, nursing homes or assisted living facilities; or housing cells in a jail. Another example would be a studio apartment with a kitchenette (i.e., countertop microwave, sink, refrigerator). Since the cooking arrangements are not the traditional permanent appliances (i.e., a cooktop, range or oven), this configuration would be considered a sleeping unit, and not a dwelling unit. As defined in the code, a "Dwelling unit" must contain permanent facilities for living, sleeping, eating, cooking and sanitation.

[F] SMOKE ALARM. A single- or multiple-station alarm responsive to smoke. See "Multiple-station smoke alarm" and "Single-station smoke alarm."

❖ This is a general term that applies to both single- and multiple-station smoke alarms that are not part of an automatic fire detection system. It is the generic term for any device that both detects the products of combustion and initiates an alarm signal for occupant notification.

SMOKE BARRIER. A continuous membrane, either vertical or horizontal, such as a wall, floor or ceiling assembly, that is designed and constructed to restrict the movement of smoke.

❖ A smoke barrier is a fire-resistance-rated assembly that is different from a fire partition, fire barrier or fire wall. Smoke barriers include walls and floor/ceiling assemblies that are constructed with a 1-hour fire-resistance rating and are one of the components in a smoke compartment. In Group I-2 and I-3 occupancies, the smoke barriers are intended to create adjacent smoke compartments to which building occupants can be safely and promptly relocated during a fire, thus preventing the need to have complete and immediate egress from the building. For these occupancies, complete egress from the building would not be practical in most cases, due to restrictions on the mobility of the occupants. To maintain tenability in the adjacent smoke compartment, the smoke barrier is therefore intended to resist the spread of fire and hinder the movement of smoke. Smoke barriers are also used to compartment a building into separate smoke control zones when using the provisions of Section 909. The construction requirements for a smoke barrier provide resistance to the transmission of smoke [see Commentary Figure 202(36) and Section 709].

SMOKE COMPARTMENT. A space within a building enclosed by *smoke barriers* on all sides, including the top and bottom.

❖ Smoke compartments create spaces that protect occupants from the products of combustion produced by a fire in an adjacent smoke compartment and restrict smoke to the compartment of fire origin.

SMOKE DAMPER. A *listed* device installed in ducts and air transfer openings designed to resist the passage of smoke. The device is installed to operate *automatically*, controlled by a smoke detection system, and where required, is capable of being positioned from a *fire command center*.

❖ See the commentary for all dampers under the definition of "Damper."

[F] SMOKE DETECTOR. A *listed* device that senses visible or invisible particles of combustion.

❖ These devices are considered early-warning devices and have saved many people from smoke inhalation and burns. Smoke detectors have a wide range of uses, from sophisticated fire detection systems for industrial and commercial uses to residential. A smoke detector is a device, typically listed in accordance with UL 268, that activates a fire alarm system. These system smoke detectors contain only the components required to detect the products of combustion and activate a fire alarm system and are, therefore, different from single- and multiple-station smoke alarms.

Smoke detectors typically consist of two types: ionization and photoelectric. An ionization detector contains a small amount of radioactive material that ionizes the air in a sensing chamber and causes a current to flow through the air between two charged electrodes. When smoke enters the chamber, the particles cause a reduction in the current. When the level of conductance decreases to a preset level, the detector responds with an alarm.

A photoelectric smoke detector consists primarily of a light source, a light beam and a photosensitive device. When smoke particles enter the light beam, they reduce the light intensity in the photosensitive device. When obscuration reaches a preset level, the detector initiates an alarm.

SMOKE-DEVELOPED INDEX. A comparative measure, expressed as a dimensionless number, derived from measurements of smoke obscuration versus time for a material tested in accordance with ASTM E84.

❖ The ASTM E84 test method of measuring the density of smoke emitted from combustible materials determines the smoke-developed index. This value is only comparative and provides only a relative understanding of the smoke generation potential of a material. The smoke-developed index is sometimes abbreviated as SDI.

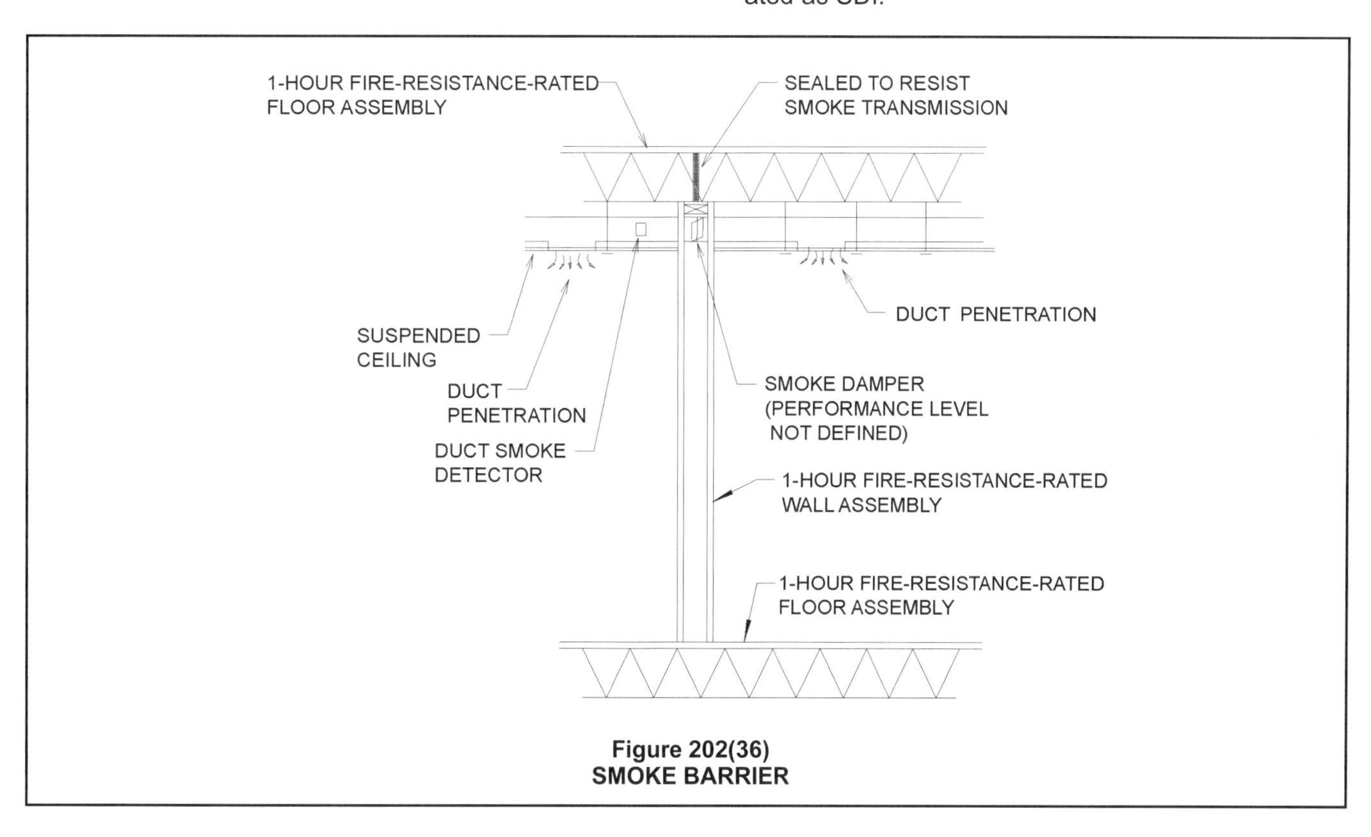

Figure 202(36)
SMOKE BARRIER

SMOKE-PROTECTED ASSEMBLY SEATING. Seating served by *means of egress* that is not subject to smoke accumulation within or under a structure.

❖ An example of smoke-protected assembly seating is an open outdoor grandstand or an indoor arena with a smoke control system. The code has less stringent requirements for certain aspects of smoke-protected assembly seating than for seating that is not smoke protected, since occupants are subject to less hazard from the accumulation of smoke and fumes during a fire event. For example, an assembly dead-end aisle is permitted to be longer for a smoke-protected assembly area. For smoke control system requirements, see Section 909.

SMOKEPROOF ENCLOSURE. An *exit stairway* designed and constructed so that the movement of the products of combustion produced by a fire occurring in any part of the building into the enclosure is limited.

❖ A smokeproof enclosure is intended to provide an effective barrier to the entry of smoke into an exit stairway, thereby offering an additional level of protection for occupants of high-rise and underground structures.

[F] SOLID. A material that has a melting point, decomposes or sublimes at a temperature greater than 68°F (20°C).

❖ The temperature at which a solid melts is the melting point. A material will begin to melt as heat is added to it and, thus, eventually change to a liquid state.

SPECIAL AMUSEMENT BUILDING. A special amusement building is any temporary or permanent building or portion thereof that is occupied for amusement, entertainment or educational purposes and that contains a device or system that conveys passengers or provides a walkway along, around or over a course in any direction so arranged that the *means of egress* path is not readily apparent due to visual or audio distractions or is intentionally confounded or is not readily available because of the nature of the attraction or mode of conveyance through the building or structure.

❖ In general, a special amusement building is a building or portion thereof in which people gather (thus, an assembly occupancy) and in which egress is either not readily apparent due to distractions, is intentionally confounded (e.g., a maze) or is not readily available. The definition includes all such facilities, including portable and temporary structures. The hazard associated with such buildings is not related to the permanence or length of use; therefore, seasonal uses (such as haunted houses at Halloween) and portable uses (carnival rides) are included if they meet the criteria in the definition.

[BS] SPECIAL FLOOD HAZARD AREA. The land area subject to flood hazards and shown on a *Flood Insurance Rate Map* or other flood hazard map as Zone A, AE, A1-30, A99, AR, AO, AH, V, VO, VE or V1-30.

❖ This term is used in the definition of "Flood insurance study" and in Section 1612.3, in which the flood hazard area is established. The special flood hazard area shown on maps prepared by FEMA is the area associated with the base flood.

[BS] SPECIAL INSPECTION. Inspection of construction requiring the expertise of an *approved special inspector* in order to ensure compliance with this code and the *approved construction documents*.

❖ This category of inspection is intended to apply to those material installations that require a special level of knowledge and attention. For example, special inspections are required for the installation of high-strength bolts, welded connections, concrete reinforcement, prestressed concrete, fabrication of laminated wood structural elements and pile installations to comply with the contract documents and the standards under which they are assembled.

Continuous special inspection. Special inspection by the *special inspector* who is present when and where the work to be inspected is being performed.

❖ Continuous special inspection is the constant monitoring by a special inspector of specific tasks. These inspections must be carried out continuously over the duration of the particular tasks. "Full-time" should not be construed as requiring the special inspector to be present when these tasks are not being performed.

Periodic special inspection. Special inspection by the *special inspector* who is intermittently present where the work to be inspected has been or is being performed.

❖ Periodic special inspection allows intermittent monitoring of specified tasks designated in the statement of special inspections as requiring periodic inspection. To avoid misunderstandings, the frequency or timing of these inspections should be established before a permit is issued.

[BS] SPECIAL INSPECTOR. A qualified person employed or retained by an *approved* agency and *approved* by the *building official* as having the competence necessary to inspect a particular type of construction requiring *special inspection*.

❖ This definition facilitates the application of Chapter 17.

[BS] SPECIAL STRUCTURAL WALL. See Section 1905.1.1.

❖ This definition is a modification of ACI 318 text and is therefore located in Section 1905.

[BS] SPECIFIED COMPRESSIVE STRENGTH OF MASONRY, f'_m. Minimum compressive strength, expressed as force per unit of net cross-sectional area, required of the *masonry* used in construction by the *approved construction documents*, and upon which the project design is based. Whenever the quantity f'_m is under the radical sign, the square root of numerical value only is intended and the result has units of pounds per square inch (psi) (MPa).

❖ Engineered design of structural masonry is based on the specified compressive strength of the masonry, f'_m. This specified strength is required to be shown on the approved construction documents. Strength of the constructed masonry, determined by the unit strength method or prism strength method, is required to equal or exceed the specified compressive strength of masonry.

[BS] SPLICE. The result of a factory and/or field method of joining or connecting two or more lengths of a *fire-resistant joint system* into a continuous entity.

❖ In many instances, the actual length of a joint required to be protected by a fire-resistant joint system exceeds the length of the joint system, especially when a prefabricated system is used. When two or more lengths of a joint system are necessary to protect a joint, the ends of each section must be joined by a splice to maintain the fire-resistive integrity of the assembly. Splices may be either factory or field installed, but in either case, the splice must be tested in accordance with UL 2079.

SPORT ACTIVITY, AREA OF. See "Area of sport activity."

SPRAYED FIRE-RESISTANT MATERIALS. Cementitious or fibrous materials that are sprayed to provide fire-resistant protection of the substrates.

❖ The cementitious or fibrous material is pneumatically projected onto a surface such that the density, thickness and cohesion/adhesion of the material will provide fire resistance to the surface.

STAGE. A space within a building utilized for entertainment or presentations, which includes overhead hanging curtains, drops, scenery or stage effects other than lighting and sound.

❖ The building feature known as a "stage" is defined as the area or space where the performers are located. The presentation area is usually elevated to provide clear sightlines for the audience. The stage area also includes the wing areas and backstage areas where curtains and scenery are placed.

STAIR. A change in elevation, consisting of one or more risers.

❖ All steps, even a single step, are defined as a stair. This makes the stair requirements applicable to all steps unless specifically exempt in the code.

STAIRWAY. One or more *flights* of *stairs*, either exterior or interior, with the necessary landings and platforms connecting them, to form a continuous and uninterrupted passage from one level to another.

❖ It is important to note that this definition characterizes a stairway as connecting one level to another. The term "level" is not to be confused with "story." Steps that connect two levels, one of which is not considered a "story" of the structure, would be considered a stairway. For example, a set of steps between the basement level in an areaway and the outside ground level would be considered a stairway. A series of steps between the floor of a story and a mezzanine within that story would also be considered a stairway (see definitions for "Flight," "Exit access stairway," "Exterior exit stairway," "Interior exit stairway" and Sections 1011, 1019, 1023 and 1027).

STAIRWAY, EXIT ACCESS. See "Exit access stairway."

STAIRWAY, EXTERIOR EXIT. See "Exterior exit stairway."

STAIRWAY, INTERIOR EXIT. See "Interior exit stairway."

STAIRWAY, SCISSOR. See "Scissor stairway."

STAIRWAY, SPIRAL. A *stairway* having a closed circular form in its plan view with uniform section-shaped treads attached to and radiating from a minimum-diameter supporting column.

❖ Spiral stairways are permitted as part of a means of egress in limited circumstances given in Section 1011.10. Spiral staircases could be used for supplemental/convenience stairways in other locations. Spiral stairways are commonly used where a small number of occupants use the stairway and the floor space for the stair is very limited. Spiral stairways are typically supported by a center pole. Requirements are found with stairways in Section 1011.10.

[F] STANDBY POWER SYSTEM. A source of automatic electric power of a required capacity and duration to operate required building, hazardous materials or ventilation systems in the event of a failure of the primary power. Standby power systems are required for electrical loads where interruption of the primary power could create hazards or hamper rescue or fire-fighting operations.

❖ Section 2702 of the code mandates the installation of emergency power systems and standby power systems for various buildings and occupancies in accordance with the IFC and NFPA 70, 110 and 111. This definition is consistent with definitions in NFPA 110 and NFPA 111 and is intended to provide clarity for the fire code official as to which systems are considered to be standby power systems. A primary difference between emergency and standby power systems is that emergency power systems are essentially life safety systems (e.g., exit signage and egress illumination), whereas standby power systems focus more on the continued operation of critical equipment in a building (e.g., elevators, fire pumps). Another difference between the two is the time limit within which the power supply activates. More specifically, emergency power is available in 10 seconds after primary power fails, whereas standby power is available in 60 seconds. Both types of system are required to operate for a minimum of two hours under full design load unless otherwise indicated in Section 2702. Standby power systems are required for the following systems or features, among others:

1. Elevator and platform lift operation.
2. Emergency responder radio coverage systems.
3. Horizontal sliding exit doors.

4. Smoke control systems.

5. High-rise building fire command centers.

See also the commentary to Section 202, the definition of "Emergency power system" and Section 2702.

[F] STANDPIPE SYSTEM, CLASSES OF. Standpipe classes are as follows:

❖ A standpipe system is typically an arrangement of vertical piping located in exit stairways that allows fire-fighting personnel to connect hand-carried hoses at each level to manually extinguish fires. Section 905 and NFPA 14 recognize three different classes of standpipe systems. For a further discussion of standpipe classes and types, see the commentary to Section 905.

Class I system. A system providing $2^1/_2$-inch (64 mm) hose connections to supply water for use by fire departments and those trained in handling heavy fire streams.

❖ A Class I standpipe system is intended for use by trained fire service personnel as a readily available water source for manual fire-fighting operations. A Class I standpipe system is equipped with only $2^1/_2$-inch (64 mm) hose connections to allow the fire service to attach the appropriate hose and nozzles. A Class I standpipe system is not equipped with hose stations, which include a cabinet, hose and nozzle.

Class II system. A system providing $1^1/_2$-inch (38 mm) hose stations to supply water for use primarily by the building occupants or by the fire department during initial response.

❖ A Class II standpipe system is intended for use by building occupants or by the fire department for manual suppression. The hose stations defined in NFPA 14 as part of the Class II standpipe system include a hose rack, hose nozzle, hose and hose connection. The intent of providing the hose is for use by properly trained personnel. Occupant-use hose stations should only be provided where they can be used by people who have been properly trained in the use of the hose and nozzle.

Class III system. A system providing $1^1/_2$-inch (38 mm) hose stations to supply water for use by building occupants and $2^1/_2$-inch (64 mm) hose connections to supply a larger volume of water for use by fire departments and those trained in handling heavy fire streams.

❖ A Class III standpipe system is intended for use by building occupants as well as trained fire service personnel. The $1^1/_2$-inch (38 mm) hose station is for use by the building occupants or fire department for manual fire suppression and the $2^1/_2$-inch (64 mm) hose connection is intended for use primarily by fire service personnel or those who have received training in the use of the larger hoses. Class III systems allow the fire department to select the types of hose necessary based on the fire hazard present. If the fire is effectively controlled by an automatic sprinkler system, the smaller hose size may be all that is necessary for fire department mop-up operations.

[F] STANDPIPE, TYPES OF. Standpipe types are as follows:

❖ Section 905 recognizes five types of standpipe systems. The use of each type of system depends on specific occupancy conditions and the presence of an automatic sprinkler system. For a further discussion of standpipe classes and types, see the commentary to Section 905.

Automatic dry. A dry standpipe system, normally filled with pressurized air, that is arranged through the use of a device, such as dry pipe valve, to admit water into the system piping *automatically* upon the opening of a hose valve. The water supply for an *automatic* dry standpipe system shall be capable of supplying the system demand.

❖ A typical automatic dry standpipe system has an automatic water supply retained by a dry pipe valve. The dry pipe valve clapper is kept in place by air placed in the standpipe system under pressure. Once a standpipe hose valve is opened, the air is released from the system, allowing water to fill the system through the dry pipe valve. This system is traditionally used in areas where the temperature falls below 40°F (4°C), where a wet system could freeze and possibly burst the pipe or simply be unavailable when needed.

Automatic wet. A wet standpipe system that has a water supply that is capable of supplying the system demand *automatically.*

❖ An automatic wet standpipe system is used in locations where the entire system would remain above 40°F (4°C). Because the system is pressurized with water, an immediate release of water occurs when a hose connection valve is opened. This is the most generally preferred type of standpipe but it is not necessarily the required type unless so stipulated.

Manual dry. A dry standpipe system that does not have a permanent water supply attached to the system. Manual dry standpipe systems require water from a fire department pumper to be pumped into the system through the fire department connection in order to meet the system demand.

❖ A manual dry standpipe system is filled with water only when the fire service is present. Typically, the fire service connects the discharge from a water source, such as a pumper truck, to the fire department connection of a manual dry standpipe system. When the fire service has suppressed the fire and is preparing to leave, the system is drained of the remaining water. Manual dry standpipe systems are commonly installed in open parking structures.

Manual wet. A wet standpipe system connected to a water supply for the purpose of maintaining water within the system but does not have a water supply capable of delivering the system demand attached to the system. Manual-wet standpipe systems require water from a fire department pumper (or the like) to be pumped into the system in order to meet the system demand.

❖ A manual wet standpipe system is connected to an automatic water supply, but the supply is not capable of providing the system demand. The manual wet system could be one that is connected with the sprinkler system such that it is capable of supplying the demand for the sprinkler system but not for the standpipe. The standpipe system demand is met when the fire service provides additional water through the fire department connection from the discharge of a water source, such as a pumper truck.

Semiautomatic dry. A dry standpipe system that is arranged through the use of a device, such as a deluge valve, to admit water into the system piping upon activation of a remote control device located at a hose connection. A remote control activation device shall be provided at each hose connection. The water supply for a semiautomatic dry standpipe system shall be capable of supplying the system demand.

❖ This type of dry standpipe is a special design that uses a solenoid-activated valve to retain the automatic water supply. Once the standpipe hose valve is opened, a signal is sent to the deluge valve retaining the automatic water supply to allow water to fill the system. This kind of system is used in areas where the temperature falls below 40°F (4°C), where a wet system would otherwise freeze. As such, there is no semiautomatic wet system type.

START OF CONSTRUCTION. The date of issuance for new construction and *substantial improvements* to *existing structures,* provided the actual start of construction, *repair*, reconstruction, rehabilitation, *addition*, placement or other improvement is within 180 days after the date of issuance. The actual start of construction means the first placement of permanent construction of a building (including a manufactured home) on a site, such as the pouring of a slab or footings, installation of pilings or construction of columns.

Permanent construction does not include land preparation (such as clearing, excavation, grading or filling), the installation of streets or walkways, excavation for a *basement*, footings, piers or foundations, the erection of temporary forms or the installation of accessory buildings such as garages or sheds not occupied as *dwelling units* or not part of the main building. For a *substantial improvement*, the actual "start of construction" means the first *alteration* of any wall, ceiling, floor or other structural part of a building, whether or not that *alteration* affects the external dimensions of the building.

❖ This term is used in the definition of "Existing structure," and is also applicable to substantial improvements of existing buildings and structures.

[BS] STEEL CONSTRUCTION, COLD-FORMED. That type of construction made up entirely or in part of *steel structural members* cold formed to shape from sheet or strip steel such as *roof deck*, floor and wall panels, studs, floor joists, roof joists and other structural elements.

❖ This definition is necessary to distinguish structural members of cold-formed steel from other types of steel elements. A distinguishing characteristic is that the forming operation takes place at room temperature without the addition of heat. Specific references to AISI standards and specifications applicable to cold-formed steel construction are provided in Sections 2210 and 2211.

[BS] STEEL ELEMENT, STRUCTURAL. Any *steel structural member* of a building or structure consisting of rolled shapes, pipe, hollow structural sections, plates, bars, sheets, rods or steel castings other than cold-formed steel or steel joist members.

❖ This definition facilitates application of steel requirements, primarily in Chapter 22, by enabling the code user to distinguish structural steel elements from other types of steel elements. See the references to the AISC standards and specifications for structural steel that are provided in Section 2205.

[BS] STEEL JOIST. Any *steel structural member* of a building or structure made of hot-rolled or cold-formed solid or open-web sections, or riveted or welded bars, strip or sheet steel members, or slotted and expanded, or otherwise deformed rolled sections.

❖ This definition is necessary to distinguish steel joists from other types of steel elements. References to SJI specifications are provided in Section 2207.

STEEP SLOPE. A roof slope greater than two units vertical in 12 units horizontal (17-percent slope).

❖ This is the general criterion for roof slope that is used throughout the code. Slope requirements for specific roof covering materials are specified in Chapter 15.

[BS] STONE MASONRY. *Masonry* composed of field, quarried or *cast stone* units bonded by *mortar*.

❖ Stone masonry is composed of natural marble, limestone, granite, sandstone and slate for building purposes. Ashlar stone is further distinguished as coursed or random. Rubble stone masonry is further distinguished as coursed, random or rough.

[F] STORAGE, HAZARDOUS MATERIALS. The keeping, retention or leaving of hazardous materials in closed containers, tanks, cylinders, or similar vessels; or vessels supplying operations through closed connections to the vessel.

❖ This term refers to all hazardous materials that are essentially being stored in a static condition. The material is considered in use once it is placed into action by either handling or transport or in a closed or open system.

[BS] STORAGE RACKS. Cold-formed or hot-rolled steel structural members which are formed into steel storage racks, including pallet storage racks, movable-shelf racks, rack-supported systems, automated storage and retrieval systems (stacker racks), push-back racks, pallet-flow racks, case-flow racks, pick modules and rack-supported platforms. Other types of racks, such as drive-in or drive-through racks, cantilever racks, portable racks or racks made of materials other than steel, are not considered storage racks for the purpose of this code.

❖ This definition of storage rack is drawn from RMI MH 16.1 in order to facilitate application of code requirements for steel storage racks in Chapters 17 and 22.

[BS] STORM SHELTER. A building, structure or portions thereof, constructed in accordance with ICC 500 and designated for use during a severe wind storm event, such as a hurricane or tornado.

Community storm shelter. A storm shelter not defined as a "Residential storm shelter."

Residential storm shelter. A storm shelter serving occupants of *dwelling units* and having an *occupant load* not exceeding 16 persons.

❖ The significance of the definitions is to distinguish between residential shelters and community shelters. Residential shelters are somewhat more basic than community shelters. A residential shelter allows for less floor area per person than that required for a community shelter because the occupants are presumed to be more familiar with each other. A shelter that serves a group of dwelling units but where the capacity needs to be more than 16 persons would be considered a community storm shelter.

STORY. That portion of a building included between the upper surface of a floor and the upper surface of the floor or roof next above (see "*Basement*," "*Building height,*" "*Grade plane" and "Mezzanine*"). A story is measured as the vertical distance from top to top of two successive tiers of beams or finished floor surfaces and, for the topmost story, from the top of the floor finish to the top of the ceiling joists or, where there is not a ceiling, to the top of the roof rafters.

❖ All levels in a building that conform to this description are stories, including basements. A mezzanine is considered part of the story in which it is located. See Chapter 5 for code requirements regarding limitations on the number of stories in a building as a function of the type of construction.

STORY ABOVE GRADE PLANE. Any *story* having its finished floor surface entirely above *grade plane*, or in which the finished surface of the floor next above is:

1. More than 6 feet (1829 mm) above *grade plane*; or
2. More than 12 feet (3658 mm) above the finished ground level at any point.

❖ The determination of the allowed height of a building under Section 504 is based on the number of stories above grade plane (see definitions of "Grade plane," "Height, building" and "Basement"). The code establishes by this definition which stories of a building are those above grade plane. Clearly, it includes those stories that are fully above grade plane. It also includes stories that may be partially below finished ground level, but the finished floor level is more than 6 feet (1829 mm) above grade plane. It also includes those floor levels that, due to an irregular terrain, have a finished floor level more than 12 feet (3658 mm) above finished ground level at any point surrounding the building. Any building level not qualifying as a story above grade plane is, by definition, a basement. See the commentaries for the definitions of "Grade plane," "Height, building" and "Basement" and Commentary Figures 202(7) and 202(23) through (27).

[BS] STRENGTH (For Chapter 21).

❖ In masonry design under Chapter 21 of the code, the term "strength" is used in both general and specific senses. In the general sense, the strength of a member is its capacity to resist internal forces and moments. In the specific sense, strength is further categorized by type. The tensile strength of a member, for instance, refers to how much tensile force the member can support.

In the context of strength design, the force resulting from factored design actions is referred to as the "required strength." An approximation to the "minimum expected" strength of the member is referred to as the "nominal strength" (see the commentary to "Nominal strength"). This nominal strength is then multiplied by a strength reduction factor to account for material, design and construction variabilities to determine the design strength. The design strength must equal or exceed the required strength.

In the context of allowable stress design (ASD), the force resulting from unfactored design loads is referred to as the "applied (actual) force." The anticipated strength would then be reduced by appropriate safety factors to either an allowable stress or an allowable strength (for example, for anchor bolts). This allowable stress or strength is required to equal or exceed the applied (actual) stress or force.

"Design strength," "Nominal strength" and "Required strength" are defined in more detail immediately below. Allowable stresses and strengths are defined in the *Building Code for Masonry Structures* (TMS 402/ACI 530/ASCE 5).

Strength can also refer to the load at which a test specimen fails (for example, prism strength).

Design strength. Nominal strength multiplied by a strength reduction factor.

❖ The strength design procedures of Section 2108 use the term "design strength" to indicate a realistic capacity of a member considering material, design and construction variabilities. The design strength is obtained by multiplying the "nominal strength" by a strength reduction factor. The design strength must equal or exceed the "required strength."

Nominal strength. Strength of a member or cross section calculated in accordance with these provisions before application of any strength-reduction factors.

❖ The strength design procedures of Section 2108 use the term "nominal strength" to refer to the capacity of a masonry member, which is determined based on the assumptions contained in Section 2108. It is sometimes referred to as the "expected" strength of the member; however, this is a misnomer. The nominal strength would equal the expected strength if: the masonry member were constructed of materials complying exactly with the minimum material requirements; design equations were perfect; and construction tolerances were zero. Since material strengths commonly exceed minimum requirements, expected strength is often much higher than nominal strength. For instance, Grade 60 reinforcement often has a yield strength of around 66,000 pounds per square inch (psi) (455 MPa), even though the minimum specified yield strength requirement is 60,000 psi (414 MPa). Therefore, the nominal strength is not the expected strength of the member, although it can be grossly classified as the minimum expected strength. Expected strength design is not included in Chapter 21 and is currently beyond the scope of the code. To determine the design strength, the nominal strength is multiplied by a strength reduction factor to account for material, design and construction variability. The design strength is required to equal or exceed the required strength.

Required strength. Strength of a member or cross section required to resist *factored loads*.

❖ In the strength design of masonry (see Section 2108), the required strength is that which corresponds to the factored design loads on the structure. The design strength (nominal strength times the appropriate strength reduction factor) must equal or exceed the required strength.

[BS] STRENGTH (for Chapter 16).

Nominal strength. The capacity of a structure or member to resist the effects of *loads*, as determined by computations using *specified* material strengths and dimensions and equations derived from accepted principles of structural mechanics or by field tests or laboratory tests of scaled models, allowing for modeling effects and differences between laboratory and field conditions.

❖ This definition is needed to apply the structural analysis and structural material design requirements in Chapter 16 and the structural materials requirements in Chapters 19 through 23.

Required strength. Strength of a member, cross section or connection required to resist *factored loads* or related internal moments and forces in such combinations as stipulated by these provisions.

❖ This definition is needed to apply the structural analysis requirements in Chapter 16 and the structural material design requirements in Chapters 19 through 23.

Strength design. A method of proportioning structural members such that the computed forces produced in the members by *factored loads* do not exceed the member design strength [also called "*load and resistance factor design*" (LRFD)]. The term "strength design" is used in the design of concrete and *masonry* structural elements.

❖ This definition describes the strength design method, which is one of the design approaches recognized under the code in Section 1604.1. It is needed to apply the structural analysis requirements in Chapter 16 and the structural material design requirements in Chapters 19 through 23.

[BS] STRUCTURAL COMPOSITE LUMBER. Structural member manufactured using wood elements bonded together with exterior adhesives. Examples of structural composite lumber are:

Laminated strand lumber (LSL). A composite of wood strand elements with wood fibers primarily oriented along the length of the member, where the least dimension of the wood strand elements is 0.10 inch (2.54 mm) or less and their average lengths not less than 150 times the least dimension of the wood strand elements.

Laminated veneer lumber (LVL). A composite of wood *veneer* sheet elements with wood fibers primarily oriented along the length of the member, where the *veneer* element thicknesses are 0.25 inches (6.4 mm) or less.

Oriented strand lumber (OSL). A composite of wood strand elements with wood fibers primarily oriented along the length of the member, where the least dimension of the wood strand elements is 0.10 inches (2.54 mm) or less and their average lengths not less than 75 times and less than 150 times the least dimension of the strand elements.

Parallel strand lumber (PSL). A composite of wood strand elements with wood fibers primarily oriented along the length of the member where the least dimension of the wood strand elements is 0.25 inches (6.4 mm) or less and their average lengths not less than 300 times the least dimension of the wood strand elements.

❖ This definition identifies which engineered wood products are considered structural composite lumber. This facilitates the application of the requirements of Section 2303.1.10, which references ASTM D5456 for determination of structural capacities. The four types are: laminated strand lumber (LSL), laminated veneer lumber (LVL), oriented strand lumber (OSL) and parallel strand lumber (PSL). LSL, OSL and PSL are a composite of wood strand elements with wood fibers primarily oriented along the length of the member. What distinguishes between the three is the least dimension of the wood strand elements and their average lengths. Note that in OSL, the limit on the least dimension of the wood strands is the same as LSL. The average length limit on the wood strand is what distinguishes OSL from LSL. LVL is made of composite wood veneer sheet elements with wood fibers primarily oriented along the length of the member where the thicknesses of the veneers are 0.25 inches (6.4 mm) or less.

[BS] STRUCTURAL GLUED-LAMINATED TIMBER. An engineered, stress-rated product of a timber laminating plant, comprised of assemblies of specially selected and prepared wood laminations in which the grain of all laminations is approximately parallel longitudinally and the laminations are bonded with adhesives.

❖ Structural wood members can be produced by laminating smaller pieces of lumber together in a specific way using glues to create high-performing structural members. Typically, these 2-inch-wide (51 mm) members are arranged in a specific order so that their strength characteristics are used most efficiently. For example, the outer laminations are for tension and compression while the inner laminations are for horizontal shear. Although these members are made of smaller pieces of lumber, they act together to create sizes that are included in the requirements for heavy timber framing members. Section 2303.1.3 references ANSI/APA A 190.1, which specifies the method of manufacturing these members, and the means of identifying their performance capabilities. ASTM D3737 covers procedures for establishing their structural properties.

[BS] STRUCTURAL OBSERVATION. The visual observation of the structural system by a *registered design professional* for general conformance to the *approved construction documents*.

❖ Where required based on wind or seismic hazards, the registered design professional must visit the site to visually determine general conformance to the approved construction documents. Structural observation is an additional tier of inspections and does not substitute for the inspections required by Sections 110 or 1705; or other sections of the code.

[A] STRUCTURE. That which is built or constructed.

❖ This definition is intentionally broad so as to include within its scope, and therefore the scope of the code (see Section 101.2), everything that is built as an improvement to real property. See also the definitions for "Building" and "Area, building" for the difference between a building and structure.

SUBSTANTIAL DAMAGE. Damage of any origin sustained by a structure whereby the cost of restoring the structure to its before-damaged condition would equal or exceed 50 percent of the market value of the structure before the damage occurred.

❖ This term is used in the definition of "Substantial improvement." The work necessary to repair a substantially damaged building qualifies as substantial improvement. A building is substantially damaged if the cost of restoring damage to pre-damage condition equals or exceeds 50 percent of the market value of the structure before the damage occurred. Buildings that are determined to be substantially damaged are required to be brought into compliance with the flood resistant design requirements for new buildings and structures. The market rate for owner labor, volunteer labor, and donated or discounted supplies, if any, must be estimated when determining the cost of repairing or restoring damage (see Section 104.2.1). A substantial damage determination must be made regardless of what causes damage. Buildings may sustain substantial damage due to flood, fire, wind, earthquake, deterioration and other causes. For more guidance, see FEMA P-758.

SUBSTANTIAL IMPROVEMENT. Any *repair*, reconstruction, rehabilitation, *alteration*, *addition* or other improvement of a building or structure, the cost of which equals or exceeds 50 percent of the market value of the structure before the improvement or repair is started. If the structure has sustained *substantial damage*, any *repairs* are considered substantial improvement regardless of the actual *repair* work performed. The term does not, however, include either:

1. Any project for improvement of a building required to correct existing health, sanitary or safety code violations identified by the *building official* and that are the minimum necessary to assure safe living conditions.
2. Any *alteration* of a historic structure provided that the *alteration* will not preclude the structure's continued designation as a historic structure.

❖ One of the long-range objectives of regulating flood hazard areas is to reduce the exposure of older buildings that were built in flood hazard areas before local jurisdictions adopted flood hazard area maps and regulations. Section 105.3 directs the applicant to state the valuation of the proposed work as part of the information submitted to obtain a permit. To make a determination (see Section 104.2.1) as to whether a proposed repair, reconstruction, rehabilitation, addition or improvement constitutes substantial improvement or repair of substantial damage, the building official compares the cost of the proposed work to the market value of the building or structure before the work is started. In order to determine market value, the building official may use the assessed valuation of the building (adjusted as specified by the appropriate appraisal authority) or may require the applicant to provide such information as allowed under Section 105.3. For additional guidance, refer to FEMA P-758 and FEMA P-784, Substantial Damage Estimator (user's manual, workbook and tool).

[BS] SUBSTANTIAL STRUCTURAL DAMAGE. A condition where one or both of the following apply:

1. The vertical elements of the lateral force-resisting system have suffered damage such that the lateral load-carrying capacity of any *story* in any horizontal direction has been reduced by more than 33 percent from its predamage condition.

2. The capacity of any vertical component carrying gravity load, or any group of such components, that supports more than 30 percent of the total area of the structure's floors and roofs has been reduced more than 20 percent from its predamage condition and the remaining capacity of such affected elements, with respect to all dead and *live loads*, is less than 75 percent of that required by this code for new buildings of similar structure, purpose and location.

❖ This definition gives the specific parameters that determine whether a building has sustained substantial structural damage. The IEBC provides direction on how such damage must be addressed.

[E] SUNROOM. A one-*story* structure attached to a building with a glazing area in excess of 40 percent of the gross area of the structure's *exterior walls* and roof.

❖ This terminology is provided in order to address separate requirements for sunrooms with regard to ventilation of adjoining spaces (see Section 1203.5.1.1). Defined in previous editions as "sunroom addition," sunrooms can be either an addition or part of the original construction. Provisions in Chapter 12 are applicable regardless of whether the sunroom is an addition to an existing structure or is part of the original design and construction. Sunrooms that are provided with "thermal isolation" are also addressed in the *International Energy Conservation Code.*

[F] SUPERVISING STATION. A facility that receives signals and at which personnel are in attendance at all times to respond to these signals.

❖ The supervising station is the location where all fire protection-system-related signals are sent and where trained personnel are present to respond to an emergency. The supervising station may be an approved central station, a remote supervising station, a proprietary supervising station or other constantly attended location approved by the fire code official. Each type of supervising station must comply with the applicable specific provisions described in NFPA 72.

[F] SUPERVISORY SERVICE. The service required to monitor performance of guard tours and the operative condition of fixed suppression systems or other systems for the protection of life and property.

❖ The supervisory service is responsible for maintaining the integrity of the fire protection system by notifying the supervising station of a change in protection system status.

Guard tours are recognized as a nonrequired (voluntary) system. If a guard tour is provided, the signals from that system can be transmitted through the supervisory service to the supervision station. Guard tours are not a required part of a fire alarm system.

[F] SUPERVISORY SIGNAL. A signal indicating the need of action in connection with the supervision of guard tours, the fire suppression systems or equipment or the maintenance features of related systems.

❖ Activation of a supervisory signal-initiating device transmits a signal indicating that a change in the status of the fire protection system has occurred and that action must be taken. These signals are the basis for the actions taken by the attendant at the supervising station. These signals do not indicate an emergency condition but indicate that a portion of the system is not functioning in the manner in which it should and that if the condition is not corrected it could impair the ability of the fire protection system to perform properly. A supervisory signal is also a part of the nonrequired guard tour system.

[F] SUPERVISORY SIGNAL-INITIATING DEVICE. An initiation device, such as a valve supervisory switch, water-level indicator or low-air pressure switch on a dry-pipe sprinkler system, whose change of state signals an off-normal condition and its restoration to normal of a fire protection or life safety system, or a need for action in connection with guard tours, fire suppression systems or equipment or maintenance features of related systems.

❖ The supervisory signal-initiating device detects a change in protection system status. Examples of a supervisory signal-initiating device include a flow switch to detect movement of water through the system and a tamper switch to detect when someone shuts off a water control valve.

[BS] SUSCEPTIBLE BAY. A roof or portion thereof with:

1. A slope less than $^1/_4$-inch per foot (0.0208 rad); or
2. On which water is impounded, in whole or in part, and the secondary drainage system is functional but the primary drainage system is blocked.

A roof surface with a slope of $^1/_4$-inch per foot (0.0208 rad) or greater towards points of free drainage is not a susceptible bay.

❖ This term is needed to identify portions of roofs that must be investigated for the possibility of progressive deflections due to ponding.

SWIMMING POOL. Any structure intended for swimming, recreational bathing or wading that contains water over 24 inches (610 mm) deep. This includes in-ground, aboveground and on-ground pools; hot tubs; spas and fixed-in-place wading pools.

❖ Section 3109 provides reference to the ISPSC, which provides requirements for these facilities.

T RATING. The time period that the *penetration firestop system*, including the penetrating item, limits the maximum temperature rise to 325°F (163°C) above its initial temperature through the penetration on the nonfire side when tested in accordance with ASTM E814 or UL 1479.

❖ See the definition of "Through-penetration firestop system." Some through penetrations having a high thermal conductivity (e.g., metal pipes, cables) will not be able to achieve a T rating without some method to prevent the heat from being transmitted from the fire side to the nonfire side. For metallic pipes, insulation must be applied to the through penetrant for some distance on one or, more often, both sides of the fire-resistance-rated assembly. Thus, when a T rating is desired, the seemingly unusual situation can arise of needing to insulate items that would normally not need insulation, such as drain pipes or conduit. Some plastic pipes are also unable to achieve a T rating without some insulation, particularly those used and, therefore, tested as "open" systems, where hot gases from the test furnace are able to flow within the plastic pipe once the pipe burns through. The type and thickness of insulation required, as well as the length over which it must be applied, are all indicated in the through-penetration firestop system listings.

Some highly intumescent firestop products can close (choke down) a plastic pipe early enough during a fire test to prevent hot gases from heating up the pipe from inside, thus allowing plastic pipe through-penetration firestop systems to be developed that have a T rating without the need for any insulation.

To avoid the need for a T rating altogether, through penetrations that pass through floors can be contained within a chase wall (see Section 714.4.1.2 exceptions). It is particularly important for the designer to plan ahead to have through penetrations concealed within a chase wall or located in a shaft in cases where adding insulation to the through penetrant is an undesirable or unacceptable option.

In the case of cable through penetrations, the need for heat dissipation would usually make the application of additional insulating material unacceptable. The few cable through-penetration firestop systems that do have a T rating achieve this by enclosing the cables within some protective enclosure for a distance from the rated assembly, thus allowing any heat transmitted through the cables during fire exposure to be dissipated, resulting in the exposed section of cabling to be cool enough for the duration of the fire test exposure to meet the T rating requirement [see Commentary Figure 202(37)].

Overall, due to the unavoidable difficulty in having a conductive or semiconductive through penetrant obtain a T rating, less than 5 percent of all tested and listed through-penetration firestop systems have a T rating. This is because the T rating indicates that the firestop system is not only capable of stopping the fire, flame and hot gases from passing through the assembly at the penetration as an F rating does, but it also must limit the temperature transfer to the unexposed side of the assembly.

TECHNICAL PRODUCTION AREA. Open elevated areas or spaces intended for entertainment technicians to walk on and occupy for servicing and operating entertainment technology systems and equipment. Galleries, including fly and lighting galleries, gridirons, catwalks, and similar areas are designed for these purposes.

❖ The definition replaces the dated terms and definitions of fly gallery and gridiron. A technical production area encompasses all areas, regardless of their traditional name, used to support entertainment technology from above the performance area. Such areas are sometimes called the backstage. Technical production areas may also be used in venues without stages or platforms, such as sports arenas. Technical production areas are regulated in Section 410.

TENSILE MEMBRANE STRUCTURE. A membrane structure having a shape that is determined by tension in the membrane and the geometry of the support structure. Typically, the structure consists of both flexible elements (e.g., membrane and cables), nonflexible elements (e.g., struts, masts, beams and arches) and the anchorage (e.g., supports and foundations). This includes frame-supported tensile membrane structures.

❖ This definition is associated the requirements in Section 3102.1 and 3102.1.1 for the construction of such structures. These types of structures have specific features and are regulated by ASCE 55. The definition serves as a clarification as to which structures must comply with ASCE 55. See commentary for Section 3102.1.1.

TENT. A structure, enclosure or shelter, with or without sidewalls or drops, constructed of fabric or pliable material supported in any manner except by air or the contents it protects.

❖ Tents can be temporary or permanent structures. When permanent, they are considered membrane-covered structures and are regulated by Section 3102. When erected as temporary enclosures, they are regulated by Section 3103.

[E] THERMAL ISOLATION. A separation of conditioned spaces, between a *sunroom* and a *dwelling unit*, consisting of existing or new walls, doors or windows.

❖ This terminology is needed for the same reason provided in the definition of "Sunroom." A sunroom with thermal isolation according to Section 1203.5 can be a source of natural ventilation.

THERMOPLASTIC MATERIAL. A plastic material that is capable of being repeatedly softened by increase of temperature and hardened by decrease of temperature.

❖ Thermoplastic materials become plastic and deform easily at higher temperatures. Indeed, many of these materials have limited construction value at high temperatures. Examples of thermoplastics are cellulose acetate, nylon, polycarbonates, polyethylene (three different densities), methyl methacrylate (plexiglass), polypropylene, polybutylene, polystyrene, acrylonitrile-butadiene-styrene (ABS), polytetrafluro ethylene (Teflon®), polyvinyl chloride (PVC), chlorinated polyvinyl chloride (CPVC) and polyethylene terephthalate (polyester). Note there are both thermosetting and thermoplastic polyesters. These materials have high thermal expansion coefficients and, while they may burn slowly in a fire, they decompose rapidly to yield a heavy black smoke and acrid, sometimes very toxic, fumes depending on the specific material and burning conditions.

THERMOSETTING MATERIAL. A plastic material that is capable of being changed into a substantially nonreformable product when cured.

❖ This term is the opposite of a thermoplastic material. Common thermosetting plastic includes epoxy, melamine, urea, polyester and polyurethane.

THROUGH PENETRATION. A breach in both sides of a floor, floor-ceiling or wall assembly to accommodate an item passing through the breaches.

❖ This term is different from a single "membrane penetration." It defines the situation where an opening (a breach in both membranes of an assembly) is created to accommodate an object that passes completely through an assembly from one side to the other [see Commentary Figure 202(37) for an example of a through penetration]. It is important to realize that this term applies only to an item that passes all the way through the assembly; back-to-back membrane penetrations, such as outlet boxes on each side of a wall assembly, should not be considered a through penetration.

THROUGH-PENETRATION FIRESTOP SYSTEM. An assemblage consisting of a fire-resistance-rated floor, floor-ceiling, or wall assembly, one or more penetrating items pass-

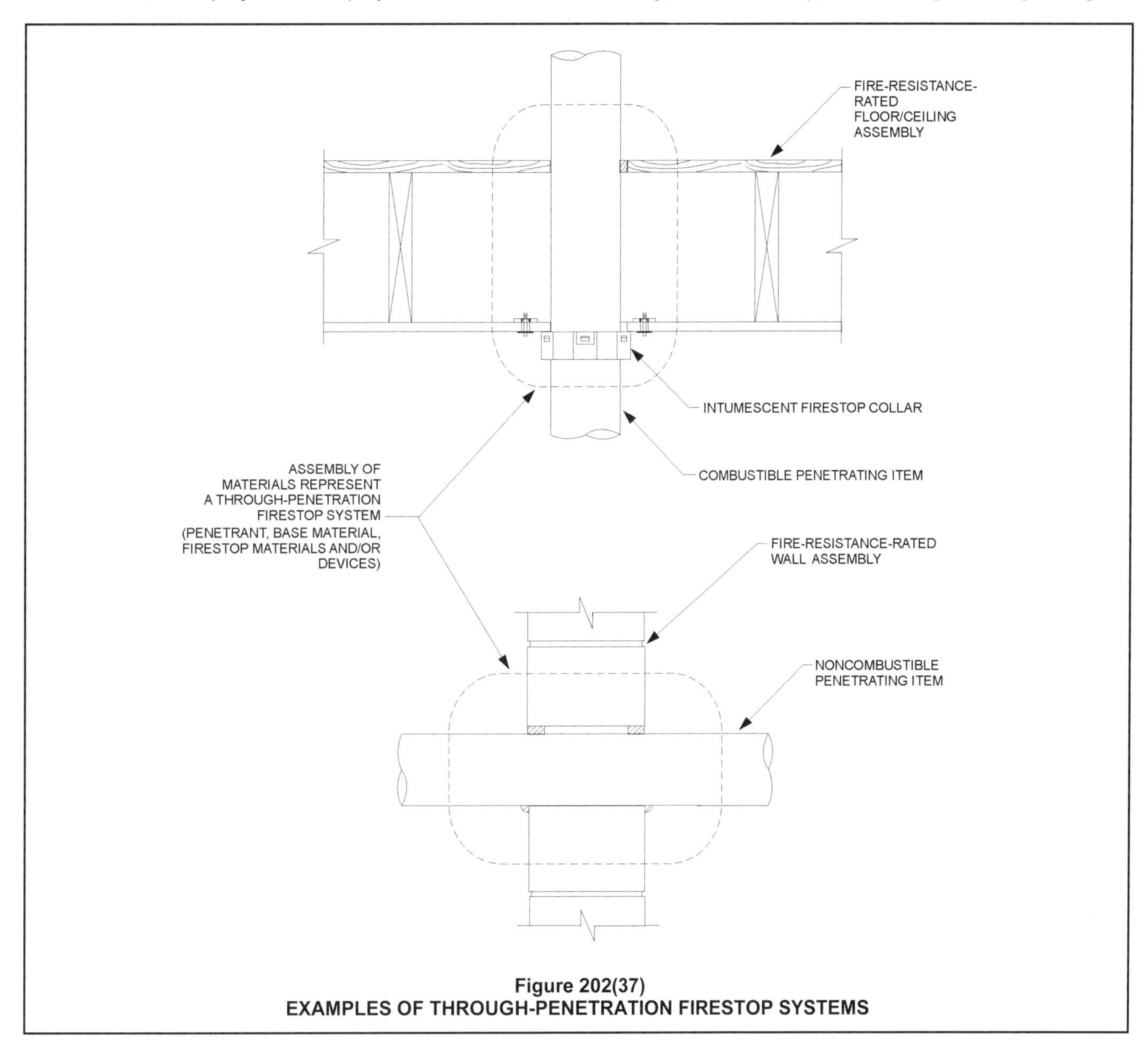

Figure 202(37)
EXAMPLES OF THROUGH-PENETRATION FIRESTOP SYSTEMS

ing through the breaches in both sides of the assembly and the materials or devices, or both, installed to resist the spread of fire through the assembly for a prescribed period of time.

❖ One method of protection for penetrations through fire walls, fire barriers, fire partitions and fire-resistance-rated floor/ceiling assemblies is to provide a through-penetration firestop system (see Sections 714.3.1.2 and 714.4.1.2). Through-penetration firestop systems maintain the required protection from the spread of fire, passage of hot gases and transfer of heat. The protection is often provided by an intumescent material. Upon exposure to high temperatures, this material expands as much as eight to 10 times its original volume, forming a high-strength char.

This is one of several types of through-penetration firestop systems available. This definition is based on information from three sources: ASTM E814; a compilation of definitions from ASTM International standards and the *Fire Resistance Directory* by Underwriters Laboratories Inc.

[BS] TIE-DOWN (HOLD-DOWN). A device used to resist uplift of the chords of *shear walls.*

❖ A shear wall resists lateral forces that are typically applied at the top of and parallel to the wall from the floor or roof diaphragm. This load will rack the wall and attempt to turn it over. If the strength of the sheathing and framing is adequate to resist the load and appropriately attached, the sheathing will resist the tendency to rack the wall. To prevent the wall from overturning, the tension chord must be adequately anchored to the foundation or structure below. Anchor bolts in the sill plate are not typically designed to resist these overturning loads. Hold-down devices are often necessary at the ends of the shear wall to resist overturning from high lateral loads due to wind loads or seismic ground motion.

[BS] TIE, WALL. Metal connector that connects *wythes* of *masonry* walls together.

❖ Ties are used to connect adjacent masonry wythes and are subject to requirements for strength, durability and installation. Ties are either adjustable or non-adjustable.

[BS] TILE, STRUCTURAL CLAY. A hollow *masonry unit* composed of burned clay, shale, fire clay or mixture thereof, and having parallel *cells.*

❖ These clay masonry units are produced as end tiles and side tiles, and differ from clay brick by having required cells with thinner webs between them.

[F] TIRES, BULK STORAGE OF. Storage of tires where the area available for storage exceeds 20,000 cubic feet (566 m^3).

❖ This definition describes a storage space that is larger than what would be found in most typical mercantile and storage occupancies. Because of its size and the volume of combustible material it would house, it poses an extraordinary hazard for fire protection.

The volume is based on the legacy code definition, which was based on 10,000 passenger vehicle tires weighing an average of 25 pounds (11 kg) each, rather than the volume of the stored tires. Assuming a 24-inch by 24-inch space (610 mm by 610 mm) for an average passenger vehicle tire and a 6-inch (152 mm) thickness, the result is 20,000 cubic feet (566 m^3):

$$10{,}000 \text{ tires} \times 2 \text{ ft} \times 2 \text{ ft} \times 0.5 \text{ ft} = 20{,}000 \text{ ft}^3$$

The 20,000 cubic feet (566 m^3) represents the actual volume of stored materials based on an equivalent height and area for passenger vehicle tires as shown in the calculation above and does not include circulation area or other portions of the building. Rather, it focuses on how much of the material is present. Although the definition uses the term "area" rather than "volume," it is the volume that becomes the threshold consideration. Still, the area where the tires are stored implies the footprint used for storage. It is not the intent to apply this to areas outside of those used for bulk tire storage.

Buildings used for the bulk storage of tires are classified as Group S-1 occupancies in accordance with Section 311.2. All Group S-1 occupancies, regardless of square footage, are required by Section 903.2.9.2 to be equipped with an NFPA 13 automatic sprinkler system if used for the bulk storage of tires. Chapter 34 of the IFC also requires that bulk tire storage buildings be further designed to comply with NFPA 13, and Chapter 32 of the IFC includes additional requirements for high-piled rubber tire storage as a high-hazard commodity (see commentary, Chapters 32 and 34 of the IFC).

[A] TOWNHOUSE. A single-family *dwelling unit* constructed in a group of three or more attached units in which each unit extends from the foundation to roof and with open space on at least two sides.

❖ This specific configuration of construction is called different things in different parts of the country, such as a rowhouse. A townhouse structure that meets the following four criteria is not regulated by the code, but is regulated by the IRC. Those criteria are:

1. Each unit extends from foundation to roof with no vertical overlap of any parts of adjoining units;
2. Each unit must have open space on at least two sides (either two opposite or two adjoining sides);
3. Each unit must have a separate means of egress; and
4. The building must not exceed three stories above grade plane.

❖ If all of these criteria are met, then according to the exception to Section 101.2, the structure is within scope of the IRC. (It should also be noted that townhouses within the IRC must be separated by a wall or walls meeting specific criteria.) If a structure does not meet these four criteria, it will need to be regulated under the code and will either be classified as a Group R-2 or Group R-3 structure, depending on how the units are separated. A building containing three or more dwelling units is regulated as a Group R-2 occupancy. To be considered a Group R-3, the structure must have one or two dwelling units, or be subdivided by fire walls between every unit or every two units (see Section 310 and the definitions for “Area, building,” “Dwelling” and “Dwelling unit”). Finally, the definition of “Townhouse” is not dependent on the presence of individual lots. A townhouse structure could be built with any number of attached units on the same lot, or it could be developed such that a property line lies at each common wall separating two units (see definition for “Lot”).

[F] TOXIC. A chemical falling within any of the following categories:

1. A chemical that has a median lethal dose (LD_{50}) of more than 50 milligrams per kilogram, but not more than 500 milligrams per kilogram of body weight when administered orally to albino rats weighing between 200 and 300 grams each.
2. A chemical that has a median lethal dose (LD_{50}) of more than 200 milligrams per kilogram, but not more than 1,000 milligrams per kilogram of body weight when administered by continuous contact for 24 hours (or less if death occurs within 24 hours) with the bare skin of albino rabbits weighing between 2 and 3 kilograms each.
3. A chemical that has a median lethal concentration (LC_{50}) in air of more than 200 parts per million, but not more than 2,000 parts per million by volume of gas or vapor, or more than 2 milligrams per liter but not more than 20 milligrams per liter of mist, fume or dust, when administered by continuous inhalation for 1 hour (or less if death occurs within 1 hour) to albino rats weighing between 200 and 300 grams each.

❖ The definition is derived from DOL 29 CFR; Part 1910.1200. These materials are considered dangerous to humans when either inhaled, absorbed or injected through the skin or when orally ingested. Toxic materials differ from highly toxic materials with regard to the specified median lethal dose or concentration of a given chemical. Toxic materials present a health hazard and are subsequently listed as a Group H-4 material in Section 307.6.

TRANSIENT. Occupancy of a *dwelling unit* or *sleeping unit* for not more than 30 days.

❖ The intent of this definition is to establish a time parameter to differentiate between transient and nontransient as listed under Groups R-1 and R-2. Real estate law often dictates that a lease must be created after 30 days, and 30-day time periods are typically how extended-stay hotels and motels rent to people. Such a time period gives the occupant time to be familiar with the surroundings and, therefore, become more accustomed to any hazards of the built environment than an overnight guest or a guest who stays for just a few days. Since nontransient occupancies do not have the same level of protection in the code as transient occupancies, it is important to determine what makes an occupancy transient so as to provide consistency in enforcement.

Since the requirements for Type B units are tied to the facilities that are intended to be occupied as a residence under both the Group R-1 and Group R-2, this definition does not have a detrimental effect on matching the Fair Housing Act provisions.

TRANSIENT AIRCRAFT. Aircraft based at another location and that is at the transient location for not more than 90 days.

❖ Transient aircraft are those that are merely visiting an airport as compared to those that are based at that location. The definition is used in conjunction with Section 412.4.6 to establish the level of fire suppression needed in various aircraft hangars. Fixed-base operators, especially at larger airports, will have distinct hangars that are used for repair and maintenance of aircraft. The hangars used by transient aircraft are primarily a storage place for aircraft based at another location. This better identifies the intent of this type of aircraft hangar. Most frequently, the owner who wants to develop an aircraft hangar that fits the Group II category will do no “major maintenance” and will only store airplanes in the hangar (see commentary, Section 412.4.6).

[BS] TREATED WOOD. Wood products that are conditioned to enhance fire-retardant or preservative properties.

❖ There are two types of treated wood: fire-retardant-treated wood and preservative-treated wood. Wood is treated to reduce its ability to propagate flame or resist damage caused by fungus or insects.

Fire-retardant-treated wood. Wood products that, when impregnated with chemicals by a pressure process or other means during manufacture, exhibit reduced surface-burning characteristics and resist propagation of fire.

❖ The ability of the wood to extinguish itself once the source of ignition is consumed or removed is an important characteristic of this material. Section 2303.2 requires testing in accordance with ASTM E84 or UL 723. The section requires the test to be continued an additional 20 minutes beyond the 10 minutes required to establish the flame spread. According to Section 2303.2, there can be no significant progressive combustion during this period.

Preservative-treated wood. Wood products that, conditioned with chemicals by a pressure process or other means, exhibit reduced susceptibility to damage by fungi, insects or marine borers.

❖ Wood that is exposed to high levels of moisture or heat is susceptible to decay both from fungus and other organisms as well as insect attack. The damage caused by decay or insects can reduce the performance of wood members below the minimum level required by the code. Section 2304.11 identifies the locations where the use of naturally durable or preservative-treated wood is required. It is important to note that preservative treatment requires a pressure-treatment process. Painting, coating or other surface treatment does not necessarily produce preservative-treated wood that will perform as required. The treatment process uses pressure to achieve the depth of penetration of preservative into the wood that is needed to verify that the wood will be resistant to decay and insects over time. Surface treatments may be washed away by rain or ground water, or may chip or peel. Preservative-treated wood will not reduce susceptibility to all insects, only those that actually eat the wood.

The American Wood Preservers' Association (AWPA) standards are prescribed in Section 2303.1.9. These include water-borne, as well as oil-borne, materials for treatment, testing and quality control of various wood and wood-based products. The American Lumber Standard Committee (ALSC) also accredits third-party inspection agencies, in a manner similar to their grading rules certification, for the quality control of preservative-treated wood products. Facsimiles of accredited agencies' quality marks for treatment are available from ALSC.

TRIM. Picture molds, chair rails, baseboards, *handrails*, door and window frames and similar decorative or protective materials used in fixed applications.

❖ As interior trim, this material is usually combustible and permanently affixed. Trim is primarily located around door and window openings (frames), around walls at floors (baseboard) and on walls (chair rail). Interior trim material should only constitute 10 percent of the wall or ceiling area (see Section 806.5).

[F] TROUBLE SIGNAL. A signal initiated by the *fire alarm system* or device indicative of a fault in a monitored circuit or component.

❖ This type of signal indicates that there has been an abnormal change in the normal status of the fire detection system or devices and that a response is required to determine the nature of the fault condition. The trouble signal is only associated with electronic portions of a fire protection system. Physical conditions such as a closed valve are monitored electronically and would report as a supervisory signal rather than a trouble signal. A valve supervisory switch, or "tamper switch," for example, would perform such a function.

[BS] TUBULAR DAYLIGHTING DEVICE (TDD). A non-operable *fenestration* unit primarily designed to transmit daylight from a roof surface to an interior ceiling via a tubular conduit. The basic unit consists of an exterior glazed weathering surface, a light-transmitting tube with a reflective interior surface, and an interior-sealing device such as a translucent ceiling panel. The unit can be factory assembled, or field-assembled from a manufactured kit.

❖ This definition helps to differentiate tubular daylighting devices (TDDs) from unit skylights, both of which are subject to similar requirements in Chapter 24. These devices are typically field assembled from a manufactured kit, unlike a unit skylight which is typically shipped as a factory-assembled unit.

24-HOUR BASIS. See "24-hour basis" located preceding "AAC masonry."

TYPE A UNIT. A *dwelling unit* or *sleeping unit* designed and constructed for accessibility in accordance with this code and the provisions for *Type A units* in ICC A117.1.

❖ There are three levels of accessibility described in the code pertaining to dwelling units. Accessible units (always spelled with a capital "A"), Type A units and Type B units. A Type A unit has some elements that are constructed for accessibility [e.g., 32-inch (813 mm) clear width doors with maneuvering clearances] and some elements that are constructed as adaptable (e.g., blocking for future installation of grab bars). A Type A dwelling unit is designed and constructed to provide accessibility for wheelchair users throughout the unit, and, as such, is considered more accessible than a Type B dwelling unit. Because some adaptable features are permitted in the standard, the Type A unit is considered less wheelchair-user friendly than an Accessible unit. Type A units are only scoped in large apartment buildings of Group R-2 occupancies. The technical requirements for the interior of Type A units are in Section 1003 of ICC A117.1 (see commentary, Section 1107.2).

TYPE B UNIT. A *dwelling unit* or *sleeping unit* designed and constructed for accessibility in accordance with this code and the provisions for *Type B units* in ICC A117.1, consistent with the design and construction requirements of the federal Fair Housing Act.

❖ There are three levels of accessibility described in the code pertaining to dwelling units. Accessible units (always spelled with a capital "A"), Type A units and Type B units. A Type B dwelling or sleeping unit is designed and constructed to provide a minimal level of accessibility, and, as such, is considered less accessible than either an Accessible unit or a Type A unit. The requirements for Type B units are intended to be consistent with the Fair Housing Amendments Act (FHA). Type B units are scoped in many types of housing. The technical requirements for the interior of Type B units are in Section 1004 of ICC A117.1 (see commentary, Section 1107.2 and the definition for "Intended to be occupied as a residence").

[BS] UNDERLAYMENT. One or more layers of felt, sheathing paper, nonbituminous saturated felt or other *approved* material over which a steep-slope *roof covering* is applied.

❖ According to Section 1507.2.3, underlayment is required to comply with ASTM D226 or ASTM D4869,

Type I, which is commonly referred to as No. 15 asphalt felt.

UNIT SKYLIGHT. See "Skylight, unit."

[F] UNSTABLE (REACTIVE) MATERIAL. A material, other than an explosive, which in the pure state or as commercially produced, will vigorously polymerize, decompose, condense or become self-reactive and undergo other violent chemical changes, including *explosion*, when exposed to heat, friction or shock, or in the absence of an inhibitor, or in the presence of contaminants, or in contact with *incompatible materials*. Unstable (reactive) materials are subdivided as follows:

Class 4. Materials that in themselves are readily capable of *detonation* or explosive decomposition or explosive reaction at *normal temperatures and pressures*. This class includes materials that are sensitive to mechanical or localized thermal shock at *normal temperatures and pressures*.

Class 3. Materials that in themselves are capable of *detonation* or of explosive decomposition or explosive reaction but which require a strong initiating source or which must be heated under confinement before initiation. This class includes materials that are sensitive to thermal or mechanical shock at elevated temperatures and pressures.

Class 2. Materials that in themselves are normally unstable and readily undergo violent chemical change but do not detonate. This class includes materials that can undergo chemical change with rapid release of energy at *normal temperatures and pressures*, and that can undergo violent chemical change at elevated temperatures and pressures.

Class 1. Materials that in themselves are normally stable but which can become unstable at elevated temperatures and pressure.

❖ The classification of unstable (reactive) materials is based on provisions in NFPA 704. The different classes of unstable (reactive) materials reflect the degree of susceptibility of the materials to release energy. Unstable (reactive) materials polymerize, decompose or become self-reactive when exposed to heat, air, moisture, pressure or shock. Separation from incompatible materials is essential to minimizing the hazards. Examples of unstable (reactive) materials include nitromethane, perchloric acid, sodium perchlorate, vinyl acetate and acetic acid.

[F] USE (MATERIAL). Placing a material into action, including *solids, liquids* and gases.

❖ This definition describes the active utilization mode of a material as opposed to the inactive nature of storage or the limited movement or transport involved in handling. This mode tends to be more hazardous in that the material is exposed to human or mechanical contact rather than being confined to a closed container, thus increasing the potential for spills or other releases of liquids, vapors, gases or solids.

VAPOR PERMEABLE MEMBRANE. The property of having a moisture vapor permeance rating of 5 perms (2.9 × 10-10 kg/Pa × s × m^2) or greater, when tested in accordance with the desiccant method using Procedure A of ASTM E96. A vapor permeable material permits the passage of moisture vapor.

❖ Greater demands on the building envelope due to energy considerations now dictate the need for an outer membrane that reduces wind infiltration. The membranes used in this application may need to allow vapor to pass through it, given that a vapor barrier would be needed on the inside of the wall and would be undesirable on the outside of the wall. In such cases, a vapor-permeable membrane would be used. Membranes with a permeance rating lower than 10 perms are considered vapor retarders.

VAPOR RETARDER CLASS. A measure of a material or assembly's ability to limit the amount of moisture that passes through that material or assembly. Vapor retarder class shall be defined using the desiccant method of ASTM E96 as follows:

Class I: 0.1 perm or less.

Class II: 0.1 < perm ≤ 1.0 perm.

Class III: 1.0 < perm ≤ 10 perm.

❖ Vapor retarders are used to limit moisture intrusion into a building. The definition establishes three classes of vapor retarders based on the amount of moisture that can pass through in a given time period.

Class I: 0.1 perm or less.

Class II: 0.1 < perm ≤ 1.0 perm.

Class III: 1.0 < perm ≤ 10 perm.

VEGETATIVE ROOF. An assembly of interacting components designed to waterproof and normally insulate a building's top surface that includes, by design, vegetation and related landscape elements.

❖ Vegetative roofs are a specific design element used in the IgCC to provide a cool roof or to limit heat island impacts of a building. For the IBC, their effects on the roofing requirements and structural design are addressed in Chapters 15 and 16, respectively.

VEHICLE BARRIER. A component or a system of components, near open sides of a garage floor or ramp or building walls that act as restraints for vehicles.

❖ This definition explains vehicle barriers in order to clarify the minimum design loads that must be applied to these components in accordance with Section 1607.8.3.

VEHICULAR GATE. A gate that is intended for use at a vehicular entrance or exit to a facility, building or portion thereof, and that is not intended for use by pedestrian traffic.

❖ This definition provides the scope for the particular type of gates addressed by Section 3110. They are, specifically, those gates associated with vehicles at a building or facility as a opposed to a parking lot or access road. In addition, it clarifies that these gates are not used by people.

VENEER. A facing attached to a wall for the purpose of providing ornamentation, protection or insulation, but not counted as adding strength to the wall.

❖ Veneers are wall facings or claddings of various materials that are used for environmental protection or ornamentation on the exterior of walls. Veneers are nonstructural in that they do not carry any load other than their own weight.

[M] VENTILATION. The natural or mechanical process of supplying conditioned or unconditioned air to, or removing such air from, any space.

❖ Ventilation is the process of moving air to or from building spaces. This definition of ventilation requirements is used in the code to establish minimum levels of air movement within a building for the purposes of providing a healthful interior environment. Ventilation would include both natural (openable exterior windows and doors for wind movement) and mechanical (forced air with mechanical equipment) methods.

VINYL SIDING. A shaped material, made principally from rigid polyvinyl chloride (PVC), that is used as an *exterior wall covering*.

❖ Vinyl siding is discussed in Sections 1404.9 and 1405.14. This material offers a siding product that does not require regular painting. It is offered in a variety of colors and textures. Some vinyl siding is manufactured with some relief on the face of the product to emulate wood grain. It is important when applying the material to use an attachment method that allows the product to move as it expands and contracts. Static connections will cause the material to crack and warp.

[F] VISIBLE ALARM NOTIFICATION APPLIANCE. A notification appliance that alerts by the sense of sight.

❖ Visible alarm notification appliances are located anywhere an occupant notification system is required, where occupants may be hearing impaired and in sleeping accommodations of Group I-1 and R-1 occupancies. These alarm notification devices must be located and oriented so that they will display alarm signals throughout the required space. Visible alarms, when provided, are typically installed in the public and common areas of buildings (see commentary, Section 907.5.2.3).

WALKWAY, PEDESTRIAN. A walkway used exclusively as a pedestrian trafficway.

❖ A pedestrian walkway is an enclosed passageway external to, and not considered part of, the buildings it connects. Intended only for pedestrian use, it can be at grade, below grade or elevated above grade.

[BS] WALL (for Chapter 21). A vertical element with a horizontal length-to-thickness ratio greater than three, used to enclose space.

❖ Masonry walls, as addressed in Chapter 21, typically enclose space and therefore are generally required to be designed and installed for weather resistance, durability and in addition to adequate strength. The given dimensional requirements differentiate masonry walls from masonry columns.

Cavity wall. A wall built of *masonry units* or of concrete, or a combination of these materials, arranged to provide an airspace within the wall, and in which the inner and outer parts of the wall are tied together with metal ties.

❖ Cavity walls are made up of solid or hollow masonry units separated by a continuous airspace or cavity. This continuous airspace adds insulating value and acts as a barrier to moisture when detailed with flashing and weep holes. In many cavity walls, thermal insulation is placed between the masonry wythes to further enhance thermal efficiency.

Dry-stacked, surface-bonded wall. A wall built of concrete *masonry units* where the units are stacked dry, without *mortar* on the bed or *head joints*, and where both sides of the wall are coated with a surface-bonding *mortar*.

❖ Although this type of masonry wall is dry stacked, a leveling course set in a full bed of mortar is often used. Dry-stacked walls are also required to be placed in a running bond pattern (see the commentary to the definition of "Mortar, surface-bonding").

Parapet wall. The part of any wall entirely above the roof line.

❖ These portions of masonry walls project above the roof. A parapet wall is exposed to weather on both sides and is laterally unsupported at the top. Parapets often have copings.

[BS] WALL, LOAD-BEARING. Any wall meeting either of the following classifications:

1. Any metal or wood stud wall that supports more than 100 pounds per linear foot (1459 N/m) of vertical load in addition to its own weight.
2. Any *masonry* or concrete wall that supports more than 200 pounds per linear foot (2919 N/m) of vertical load in addition to its own weight.

❖ This definition is necessary since the structural requirements and fire-resistance-rating requirements in the code vary for nonload-bearing walls and load-bearing walls. The term "load-bearing walls" is intended to refer to wall elements that support part of the structural framework of a building.

[BS] WALL, NONLOAD-BEARING. Any wall that is not a *load-bearing wall*.

❖ This definition is necessary since the structural requirements and fire-resistance-rating requirements in the code vary for nonload-bearing walls and load-bearing walls. Nonload-bearing walls do not support any portion of the building or structure except the weight of the wall itself.

[F] WATER-REACTIVE MATERIAL. A material that explodes; violently reacts; produces *flammable*, *toxic* or other hazardous gases; or evolves enough heat to cause autoignition or ignition of combustibles upon exposure to water or moisture. Water-reactive materials are subdivided as follows:

Class 3. Materials that react explosively with water without requiring heat or confinement.

Class 2. Materials that react violently with water or have the ability to boil water. Materials that produce *flammable*, *toxic* or other hazardous gases or evolve enough heat to cause autoignition or ignition of combustibles upon exposure to water or moisture.

Class 1. Materials that react with water with some release of energy, but not violently.

❖ These materials liberate significant quantities of heat when reacting with water. Combustible water-reactive materials are capable of self-ignition. Even noncombustible water-reactive materials present a hazard because of the heat liberated during their reaction with water, which is sufficient to ignite surrounding combustible materials. While a definition for Class 1 water-reactive materials is provided for informational purposes, the maximum allowable quantities of these materials in accordance with Table 307.1(1) is not limited. The descriptions of each of the subdivisions is consistent with the approach used for the determination of water hazards in NFPA 704.

WATER-RESISTIVE BARRIER. A material behind an *exterior wall covering* that is intended to resist liquid water that has penetrated behind the exterior covering from further intruding into the *exterior wall* assembly.

❖ Protection of the building envelope from moisture intrusion is a primary concern. The ability of the water-resistive barrier to provide weather resistance and maintain the integrity of the building envelope is key to controlling water-based problems such as mold, decay and deterioration of a structure. Water-resistive barriers are discussed in Section 1404.2.

WEATHER-EXPOSED SURFACES. Surfaces of walls, ceilings, floors, roofs, soffits and similar surfaces exposed to the weather except the following:

1. Ceilings and roof soffits enclosed by walls, fascia, bulkheads or beams that extend not less than 12 inches (305 mm) below such ceiling or roof soffits.
2. Walls or portions of walls beneath an unenclosed roof area, where located a horizontal distance from an open exterior opening equal to not less than twice the height of the opening.
3. Ceiling and roof soffits located a minimum horizontal distance of 10 feet (3048 mm) from the outer edges of the ceiling or roof soffits.

❖ Outside surfaces of a building or structure are generally exposed to elements of the weather. Therefore, it is necessary to take special precautions to protect them from damage. The most common way to protect exterior surfaces from damage caused by rain, sleet, wind, ice and snow or water from other sources is by the use of building materials that are designed, manufactured and installed to resist or withstand the extremes produced by weather. There are many building materials other than those identified in Chapter 25 that provide excellent protection from weather damage, and are permitted by the I-Codes for use on weather-exposed surfaces. Requirements for other approved materials, such as, but not limited to, brick, concrete, precast concrete panels, concrete unit masonry, wood, vinyl, aluminum, glass and steel, are specified in other chapters of the code.

There are isolated surfaces located on the exterior of a building or structure that are not directly exposed to or in contact with the damaging effects of the weather. These areas are identified in Items 1, 2 and 3 above. Although these surfaces are located on the exterior of the building or structure, they are provided with a permissible amount of protection and are not directly exposed to the weather with regard to moisture. This must be taken into consideration when wall and ceiling covering materials are being selected. It is not required to use waterproof materials at these locations, but materials approved by the code and the product manufacturer for exterior applications must be used.

[F] WET-CHEMICAL EXTINGUISHING SYSTEM. A solution of water and potassium-carbonate-based chemical, potassium-acetate-based chemical or a combination thereof, forming an extinguishing agent.

❖ This extinguishing agent is a suitable alternative to the use of a dry chemical, especially when protecting commercial kitchen range hoods. There is less cleanup time after system discharge. Wet chemical solutions are considered to be relatively harmless and normally have no lasting effect on the skin or respiratory system. These solutions may produce temporary irritation, which is usually mild and disappears when contact is eliminated. These systems must be preengineered and labeled. NFPA 17A applies to the design, installation, operation, testing and maintenance of wet-chemical extinguishing systems.

WHEELCHAIR SPACE. A space for a single wheelchair and its occupant.

❖ A wheelchair space is a designated space for a person to be stationary in his or her wheelchair as part of a fixed assembly seating configuration. The wheelchair space must be sized in accordance with ICC A117.1 (see Section 1108.2.4 for wheelchair space dispersion requirements for multilevel assembly seating). ICC A117.1 provides additional dispersion criteria within a level (i.e., side-to-side, front-to-back and by type), depending on the number of seats provided in the assembly seating space.

[BS] WIND-BORNE DEBRIS REGION. Areas within hurricane-prone regions located:

1. Within 1 mile (1.61 km) of the coastal mean high water line where the ultimate design wind speed, V_{ult}, is 130 mph (58 m/s) or greater; or
2. In areas where the ultimate design wind speed is 140 mph (63.6 m/s) or greater; or Hawaii.

For *Risk Category* II buildings and structures and *Risk Category* III buildings and structures, except health care facilities, the wind-borne debris region shall be based on Figure 1609.3.(1). For *Risk Category* IV buildings and structures and *Risk Category* III health care facilities, the wind-borne debris region shall be based on Figure 1609.3(2).

❖ This definition identifies those areas that require consideration of the impact of wind-borne debris on the building envelope (see the commentary to Sections 1609.1 and 1609.1.2 for information on the effects of openings in the building envelope). Note that Section 1609.3 makes Figure 1609.3.1(2) applicable to both Risk Category III and IV structures for obtaining the ultimate design wind speed. However, this definition of wind-borne debris regions provides somewhat of an exception by allowing Figure 1609.3.1(1) to be used for establishing the wind-borne debris regions that are applicable to Risk Category III structures, other than health care facilities. This results in wind-borne debris regions applicable to Risk Category III (other than health care facilities) that are consistent with the areas that were defined in the prior editions of the code. At the same time, it should be noted that the wind-borne debris region established by Figure 1609.3.1(2) for Risk Category IV, as well as health care facilities that are Risk Category III, is an expansion of the wind-borne debris region defined in prior editions of the code.

WINDFORCE-RESISTING SYSTEM, MAIN. See "Main windforce-resisting system."

[BS] WIND SPEED, V_{ult}. Ultimate design wind speeds.

❖ This term is needed to identify the mapped wind speed in Section 1609, which is one of the primary criteria for determination of wind loading.

[BS] WIND SPEED, V_{asd}. Nominal design wind speeds.

❖ This term is used to identify wind speeds that are converted from the ultimate (mapped) wind speeds in accordance with Section 1609.3.1. Nominal wind speed is used as a wind criteria, rather than the ultimate wind speed, for requirements in Chapters 14, 15 and 23, for example.

WINDER. A tread with nonparallel edges.

❖ Winders are used as components of stairways that change direction, just as "fliers" (rectangular treads) are components in straight stairways. A winder performs the same function as a tread, but its shape allows the additional function of a gradual turning of the stairway direction. The tread depth of a winder at the walkline and the minimum tread depth at the narrow end control the turn made by each winder. Winders are not landings. Winder treads are limited to curved or spiral stairways with all groups but are all stairways within dwelling units (see Section 1011.5.3).

[BS] WIRE BACKING. Horizontal strands of tautened wire attached to surfaces of vertical supports which, when covered with the building paper, provide a *backing* for cement plaster

❖ Where metal lath attached to exterior wall studs is used as backing for a plaster base coat, additional reinforcing is necessary to stiffen the wall framing system and minimize unwanted movement to the wall itself. Plaster is a fairly rigid material and excessive movement will cause it to crack. Eighteen-gage metal wire attached horizontally to the face of the wall studs and covered with building paper will provide the additional stiffness required to minimize cracking in the finish surface (see Section 2510.5 for additional commentary on the requirements for wire backing).

[F] WIRELESS PROTECTION SYSTEM. A system or a part of a system that can transmit and receive signals without the aid of wire.

❖ These systems use radio frequency transmitting devices that comply with the special requirements for supervision of low-power wireless systems in NFPA 72. Wireless devices have the advantage of flexibility in positioning. Consequently, portable wireless notification devices are frequently used in existing facilities where visual devices are not present throughout.

[BS] WOOD/PLASTIC COMPOSITE. A composite material made primarily from wood or cellulose-based materials and plastic.

❖ Wood/Plastic composite is manufactured of wood and cellulosic materials to provide strength and stiffness. These components are typically used for exterior decking and railing systems. Chapter 26, which regulates plastic materials, contains requirements for plastic lumber used as exterior decking components, such as deck boards, stair treads, handrails and guardrails.

[BS] WOOD SHEAR PANEL. A wood floor, roof or wall component sheathed to act as a *shear wall* or *diaphragm*.

❖ Wood shear panels are horizontal (diaphragm) or vertical (wood shear wall) panels, which transmit forces in the plane of the sheathing caused by lateral loads transmitted from one component of the structure to another. Wood shear panels usually consist of wood framing (studs or joists), either singly or doubly sheathed. The code has design provisions for wood shear panels that are sheathed with 4-foot by 8-foot (1219 mm by 2438 mm) wood structural panels.

[BS] WOOD STRUCTURAL PANEL. A panel manufactured from *veneers*, wood strands or wafers or a combination of *veneer* and wood strands or wafers bonded together with waterproof synthetic resins or other suitable bonding systems. Examples of wood structural panels are:

❖ The code defines wood structural panels in terms of the materials commonly used in their manufacture.

Wood structural panels intended for structural use must comply with DOC PS 1, DOC PS 2 or ANSI/APA PRP 210 in accordance with Section 2303.1.5. These documents specify the required structural performance characteristics of wood structural panels and detail the requirements for the third-party label each panel is required to bear (see Section 2303.1.5 for a more detailed discussion).

Composite panels. A wood structural panel that is comprised of wood *veneer* and reconstituted wood-based material and bonded together with waterproof adhesive;

❖ Composite panels are structural panels that are made up of a combination of wood veneers and wood-based materials. The wood veneer usually forms the two outer layers and may also be used in the core of the panel. These layers of veneer and wood flakes or strands, called "furnish," may be cross laminated.

Oriented strand board (OSB). A mat-formed wood structural panel comprised of thin rectangular wood strands arranged in cross-aligned layers with surface layers normally arranged in the long panel direction and bonded with waterproof adhesive; or

❖ Oriented strand board (OSB) is one of three types of wood structural panel. OSB panels are fabricated out of multiple layers made up of wood flakes or strands called "furnish." Like plywood, these layers of furnish are oriented at 90 degrees (1.57 rad) to each other. This gives OSB panels properties very similar to those of plywood.

Plywood. A wood structural panel comprised of plies of wood *veneer* arranged in cross-aligned layers. The plies are bonded with waterproof adhesive that cures on application of heat and pressure.

❖ Plywood is one of three types of wood structural panel. It is manufactured by gluing together three or more cross-laminated layers of wood veneer. The code does not differentiate between OSB and plywood, since manufacturing is done according to either the DOC PS 1 standard for construction and industrial plywood or for OSB, DOC PS 2 performance standard for wood-based structural panels.

[F] WORKSTATION. A defined space or an independent principal piece of equipment using *HPM* within a *fabrication area* where a specific function, laboratory procedure or research activity occurs. *Approved* or *listed hazardous materials storage cabinets*, *flammable liquid* storage cabinets or *gas cabinets* serving a workstation are included as part of the workstation. A workstation is allowed to contain *ventilation* equipment, fire protection devices, detection devices, electrical devices and other processing and scientific equipment.

❖ Workstations further subdivide a fabrication area and provide relatively self-contained, specialized areas where HPM processes are conducted. Workstation controls limit the quantity of materials and impose limitations on the design of these processes to include, but not be limited to, protection by local exhaust; sprinklers; automatic and emergency shutoffs; construction materials and HPM compatibility. Excess materials are prohibited and must be contained in storage rooms designed to accommodate such hazards.

[BS] WYTHE. Each continuous, vertical section of a wall, one *masonry unit* in thickness.

❖ Sometimes referred to as a "leaf" or "tier," each wythe is one thickness of a masonry unit.

YARD. An open space, other than a *court*, unobstructed from the ground to the sky, except where specifically provided by this code, on the lot on which a building is situated.

❖ This definition is used, similar to the definition of "Court," to establish the applicability of code requirements when yards are utilized for natural light or natural ventilation purposes (see Section 1206.1). Whereas a court is bounded on three or more sides with the building or structure, a yard is bounded on two or fewer sides by the building or structure (see also the definition of "Egress court").

[F] ZONE. A defined area within the protected premises. A zone can define an area from which a signal can be received, an area to which a signal can be sent or an area in which a form of control can be executed.

❖ Zoning a system is important to emergency personnel in locating a fire. When an alarm is designated to a specific zone, it allows the fire service to immediately respond to the area where the fire is in progress instead of searching the entire building for the origin of an alarm.

[F] ZONE, NOTIFICATION. An area within a building or facility covered by notification appliances which are activated simultaneously.

❖ This definition is provided to clarify the code by making a clear distinction between fire alarm system initiation device zones required by Section 907.6.4 and the zones that may be designed into occupant notification device systems in a building. The term is used primarily in the exceptions for sprinkler systems found in the manual fire alarm system requirements in Section 907.2 and its subsections. Note that the code does not require audible and visible occupant notification device systems to be zoned. If such zones are provided, it is a matter of the system design engineer's judgment. The voice paging component in high-rise building emergency voice/alarm communication systems is, however, required to be zoned in accordance with Section 907.5.2.2.

Bibliography

The following resource materials were used in the preparation of the commentary for this chapter of the code.

24 CFR, *Fair Housing Accessibility Guidelines* (FHAG). Washington, DC: US Department of Housing and Urban Development, 1991.

ACI 318-14, *Building Code Requirements for Structural Concrete*. Farmington Hills, MI: American Concrete Institute, 2014.

ANSI/APA A 190.1-12, *Structural Glued-Laminated Timber*. Tacoma, WA: APA-Engineered Wood Association, 2012.

APA Research Report 154, *Wood Structural Panel Shear Wall*. Tacoma, WA: APAC Engineered Wood Association.

ASCE 7-10, *Minimum Design Loads for Buildings and Other Structures with Supplement No.1* (also ASCE 7-88, ASCE 7-93, ASCE 7-95). Reston, VA: American Society of Civil Engineers, 2010.

ASTM C11-13 *Standard Terminology Related to Gypsum and Related Building Materials and Systems*. West Conshohocken, PA: ASTM International, 2013.

ASTM D56-05 (2010), *Test Method for Flash Point by Tag Closed Tester*. West Conshohocken, PA: ASTM International, 2005 (2010).

ASTM D93-12, *Test Methods for Flash Point by Pensky-Martens Closed-Cup Tester*. West Conshohocken, PA: ASTM International, 2012.

ASTM D226/D226M-09, *Specification for Asphalt-Saturated Organic Felt Used in Roofing and Waterproofing*. West Conshohocken, PA: ASTM International, 2009.

ASTM D3278-96 (2011), *Test Methods for Flash Point of Liquids by Small-Scale Closed-Cup Apparatus*. West Conshohocken, PA: ASTM International, 2011.

ASTM D3737-12, *Practice for Establishing Allowable Properties for Structural Glued-Laminated Timber* (Glulam). West Conshohocken, PA: ASTM International, 2012.

ASTM E119-12A, *Standard Test Method for Fire Tests of Building Construction and Materials*. West Conshohocken, PA: ASTM International, 2012A.

ASTM E136-12, *Test Method for Behavior of Materials in a Vertical Tube Furnace at 750°C*. West Conshohocken, PA: ASTM International, 2012.

ASTM E1966-07A (2011), *Test Method for Fire-resistant Joint Systems*. West Conshohocken, PA: ASTM International, 2007A (2011).

ASTM E2072-10, *Standard Specification for Photoluminescent (Phosphorescent) Safety Markings*. West Conshohocken, PA: ASTM International, 2010.

AWCI 12-B-04, *Technical Manual 12-B Standard Practice for the Testing and Inspection of Field-Applied Thin-Film Intumescent Fire-Resistive Materials; an Annoted Guide, Second Edition*. Falls Church, VA: The Association of the Wall and Ceiling Industries International, 2004.

AWC SDPWS-15, *Special Design Provisions for Wind and Seismic*. Leesburg, VA: American Wood Council, 2015.

CPA (ANSI) A208.1-09, *Particleboard*. Leesburg, VA, Composite Panel Association, 2009.

DOC PS 1-09, *Structural Plywood*, Washington, DC, MD: US Department of Commerce, 2010.

DOC PS 2-10, *Performance Standard for Wood-Based Structural-Use Panels*. Gaithersburg, MD: US Department of Commerce, 2010.

DOL 29 CFR, Part 1910.1000-09, *Air Contaminants*. Washington, DC: US Department of Labor, 2009.

DOL 29 CFR, Part 1910.1200-07, *Principal Emergency Response and Preparedness, Requirements and Guidance—Hazard Communication*. Washington, DC: US Department of Labor, 2007.

FEMA 450-2-04, NEHRP *Recommended Provisions for Seismic Regulations for New Buildings and Other Structures, Part 2: Commentary*. Washington, DC: Federal Emergency Management Agency, 2004.

FEMA FIA-TB #2, *Flood Damage-resistant Materials Requirements for Buildings Located in Special Flood Hazard Areas*. Washington, DC: Federal Emergency Management Agency, 2008.

FEMA P-758-09, *Substantial Improvement/Substantial Damage Desk Reference*. Washington, DC: Federal Emergency Management Agency, 2009.

FEMA P-784-09, *Substantial Damage Estimator*. Washington, DC: Federal Emergency Management Agency, 2009.

ICC A117.1-09, *Accessible and Usable Buildings and Facilities*. Washington, DC: International Code Council, 2011.

ISO 8115-86, *Cotton Bales Dimensions and Density*. Switzerland International Organization for Standardization, 1986.

NFPA 11-10, *Standard for Low-Expansion Foam*. Quincy, MA: National Fire Protection Association, 2010.

NFPA 11A-99, *Standard For Medium- And High-Expansion Foam Systems*. Quincy, MA: National Fire Protection Association, 1999.

NFPA 12A-09, *Standard for Halon 1301 Fire-Extinguishing Systems*. Quincy, MA: National Fire Protection Association, 2009.

NFPA 13-13, *Installation of Sprinkler Systems*. Quincy, MA: National Fire Protection Association, 2013

NFPA 13D-13, *Standard for the Installation of Sprinkler Systems in One- and Two-Family Dwellings and Manufactured Homes*. Quincy, MA: National Fire Protection Association, 2013.

NFPA 13R-13, *Standard for the Installation of Sprinkler Systems in Low Rise Residential Occupancies.* Quincy, MA: National Fire Protection Association, 2013.

NFPA 14-13, *Standard for the Installation of Standpipe and Hose Systems*. Quincy, MA: National Fire Protection Association, 2013.

NFPA 17A-13, *Standard for Wet-Chemical Extinguishing Systems*. Quincy, MA: National Fire Protection Association, 2013.

NFPA 72-13, *National Fire Alarm Code and Signaling Code*. Quincy, MA: National Fire Protection Association, 2013.

NFPA 80-13, *Standard for Fire Doors and Other Opening Protectives*. Quincy, MA: National Fire Protection Association, 2013.

NFPA 252-12, *Standard Methods of Fire Tests of Door Assemblies*. Quincy, MA: National Fire Protection Association, 2012.

NFPA 257-12, *Standard On Fire Test For Window And Glass Block Assemblies*. Quincy, MA: National Fire Protection Association, 2012.

NFPA 701-10, *Standard Method of Fire Tests for Flame- Propagation of Textiles and Films*. Quincy, MA: National Fire Protection Association, 2010.

NFPA 704-12, *Standard System for the Identification of the Hazards of Materials for Emergency Response*. Quincy, MA: National Fire Protection Association, 2012.

NFPA 1124-06, *Code for the Manufacture, Transportation, Storage of Fireworks and Pyrotechnic Articles*. Quincy, MA: National Fire Protection Association, 2006.

TMS 402/ACI 530/ASCE 5-13, *Building Code Requirements for Masonry Structures*. Farmington Hills, MI: American Concrete Institute; Reston, VA: American Society of Civil Engineers; Longmont, CO: The Masonry Society, 2013.

TMS 602/ACI 530.1/ASCE 6-13, *Specifications for Masonry Structures*. Farmington Hills, MI: American Concrete Institute; Reston, VA: American Society of Civil Engineers; Longmont, CO: The Masonry Society, 2013.

UL 9-09, Fire Tests of Window Assemblies. Northbrook, IL: Underwriters Laboratories Inc., 2009.

UL 10C-09, *Positive Pressure Fire Tests of Door Assemblies.* Northbrook, IL: Underwriters Laboratories Inc., 2009.

UL 217-06, *Single- and Multiple-Station Smoke Alarms - with Revisions through April 2002*. Northbrook, IL: Underwriters Laboratories Inc., 2006.

UL 924-06, *Standard for Safety of Emergency Lighting and Power Equipment with Revisions through February 2011*. Northbrook, IL: Underwriters Laboratories Inc., 2005.

UL 1994-04, *Standard for Luminous Egress Path Marking Systems with Revisions through November 2010.* Northbrook, IL: Underwriters Laboratories Inc., 2005.

UL 2079-04, *Tests for Fire Resistance for Building Joint Systems with Revisions through December 2012.* Northbrook, IL: Underwriters Laboratories Inc., 2004.

Chapter 3: Use and Occupancy Classification

General Comments

Chapter 3 provides for the classification of buildings, structures and parts thereof based on the purpose or purposes for which they are used.

Section 302 identifies the occupancy groups into which all buildings, structures and parts thereof must be classified.

Sections 303 through 312 identify the occupancy characteristics of each group classification. In some sections, specific group classifications having requirements in common are collectively organized such that one term applies to all. For example, Groups A-1, A-2, A-3, A-4 and A-5 are individual groups. The general term Group A, however, includes each of these individual groups. For this reason, each specific assembly group classification is included in Section 303.

Definitions play a key role in determining the occupancy classification. All definitions are located in Chapter 2. This chapter lists key definitions for classification of occupancies.

In the early years of building code development, the essence of regulatory safeguards from fire was to provide a reasonable level of protection to property. The idea was that if property was adequately protected from fire, then the building occupants would also be protected.

From this outlook on fire safety, the concept of equivalent risk has evolved in the code. This concept maintains that, in part, an acceptable level of risk against the damages of fire respective to a particular occupancy type (group) can be achieved by limiting the height and area of buildings containing such occupancies according to the building's construction type (i.e., its relative fire endurance).

The concept of equivalent risk involves three interdependent considerations: 1. The level of fire hazard associated with the specific occupancy of the facility; 2. The reduction of fire hazard by limiting the floor areas and the height of the building based on the fuel load (combustible contents and burnable building components); and 3. The level of overall fire resistance provided by the type of construction used for the building.

The interdependence of these fire safety considerations can be seen by first looking at Tables 601 and 602, which show the fire-resistance ratings of the principal structural elements comprising a building in relation to the five classifications for types of construction. Type I construction is the classification that generally requires the highest fire-resistance ratings for structural elements, whereas Type V construction, which is designated as a combustible type of construction, generally requires the least amount of fire-resistance-rated structural elements. If one then looks at Tables 504.3, 504.4 and 506.2, the relationship among group classification, allowable heights and areas and types of construction becomes apparent. Respective to each group classification, the greater the fire-resistance rating of structural elements, as represented by the type of construction, the greater the floor area and height allowances. The greater the potential fire hazards indicated as a function of the group, the lesser the height and area allowances for a particular construction type.

As a result of extensive research and advancements in fire technology, today's building codes are more comprehensive and complex regulatory instruments than they were in the earlier years of code development. While the principle of equivalent risk remains an important component in building codes, perspectives have changed and life safety is now the paramount fire issue. Even so, occupancy classification still plays a key part in organizing and prescribing the appropriate protection measures. As such, threshold requirements for fire protection and means of egress systems are based on occupancy classification (see Chapters 9 and 10).

Other sections of the code also contain requirements respective to the classification of building groups. For example, Section 705 addresses requirements for exterior wall fire-resistance ratings that are tied to the occupancy classification of a building and Section 803.9 contains interior finish requirements that are dependent upon the occupancy classification

Purpose

The purpose of this chapter is to classify a building, structure or part thereof into a group based on the specific purpose for which it is designed or occupied. Throughout the code, group classifications are considered a fundamental principle in organizing and prescribing the appropriate features of construction and occupant safety requirements for buildings, especially general building limitations, means of egress, fire protection systems and interior finishes.

SECTION 301
GENERAL

301.1 Scope. The provisions of this chapter shall control the classification of all buildings and structures as to use and occupancy.

❖ As used throughout the code, the classification of an occupancy into a group is established by the requirements of this chapter. The purpose of these provisions is to provide rational criteria for the classification of various occupancies into groups based on their relative fire hazard and life safety properties. This is necessary because the code utilizes group classification as a fundamental principle for differentiating requirements in other parts of the code related to fire and life safety protection.

SECTION 302
CLASSIFICATION

302.1 General. Structures or portions of structures shall be classified with respect to occupancy in one or more of the groups listed in this section. A room or space that is intended to be occupied at different times for different purposes shall comply with all of the requirements that are applicable to each of the purposes for which the room or space will be occupied. Structures with multiple occupancies or uses shall comply with Section 508. Where a structure is proposed for a purpose that is not specifically provided for in this code, such structure shall be classified in the group that the occupancy most nearly resembles, according to the fire safety and relative hazard involved.

1. Assembly (see Section 303): Groups A-1, A-2, A-3, A-4 and A-5.
2. Business (see Section 304): Group B.
3. Educational (see Section 305): Group E.
4. Factory and Industrial (see Section 306): Groups F-1 and F-2.
5. High Hazard (see Section 307): Groups H-1, H-2, H-3, H-4 and H-5.
6. Institutional (see Section 308): Groups I-1, I-2, I-3 and I-4.
7. Mercantile (see Section 309): Group M.
8. Residential (see Section 310): Groups R-1, R-2, R-3 and R-4.
9. Storage (see Section 311): Groups S-1 and S-2.
10. Utility and Miscellaneous (see Section 312): Group U.

❖ This section requires all structures to be classified in one or more of the groups listed according to the structure's purpose and function (i.e., its occupancy). By organizing occupancies with similar fire hazard and life safety properties into groups, the code has incorporated the means to differentiate occupancies such that various fire protection and life safety requirements can be rationally organized and applied. Each specific group has an individual classification. Each represents a different characteristic and level of fire hazard that requires special code provisions to lessen the associated risks. There are some group classifications that are very closely related to other specific groups and, therefore, are collectively referred to as a single group (e.g., Group F applies to Groups F-1 and F-2). In these cases, there are requirements within the code that are common to each specific group classification. These common requirements are applicable based on the reference to the collective classification. For example, the automatic sprinkler system requirement of Section 903.2.8 applies to each specific group classification (R-1, R-2, R-3 and R-4) listed under the term "Group R." Although many requirements applicable to a general occupancy classification are the same for all of the subclassifications within the occupancy group, there are enough differences to warrant the division of the general category into two or more specific classifications.

Example: Both a restaurant (Group A-2) and a church (Group A-3) are included in Group A, but they have different specific group classifications. Both Groups A-2 and A-3 are subject to the same travel distance limitations (see Table 1017.1) and corridor fire-resistance ratings (see Table 1020.1), but have different thresholds for when automatic sprinkler systems are required (see Section 903).

Buildings that contain more than one occupancy group are mixed occupancy buildings. Buildings with mixed occupancies must comply with one of the design options contained in Section 508. Options established in Section 508 include the regulation of the mixed-occupancy conditions as accessory occupancies, nonseparated occupancies or separated occupancies.

Occasionally, a building or space is intended to be occupied for completely different purposes at different times. For instance, a church hall might be used as a day care center during weekdays and as a reception hall for weddings and other similar events at other times. In these cases, the code provisions for each occupancy must be satisfied.

In cases where a structure has a purpose that is not specifically identified within any particular occupancy classification, that structure is to be classified in the group that it most closely resembles. Before an accurate classification can be made, however, a detailed description of the activities or processes taking place inside the building, the occupant load and the materials and equipment used and stored therein must be submitted to the building official. The building official must then compare this information to the various occupancy classifications, determine which one the building most closely resembles and classify the building as such.

Example: A designer presents the building official with a building needing an occupancy group classification. The building official is informed that the building is to be used as an indoor shooting gallery, open to the public but used mostly by police officers. After reviewing the code, the official cannot find a specific reference to a shooting gallery in Sections 303 through 312 or in the associated tables. The building official asks the designer for additional information about the activities to be conducted in the building and is told that there will be a small sign-in booth, patron waiting/viewing area and the actual shooting area. Based on this information, the building official can determine that the most logical classification of the building is Group A-3, assembly. This classification is based on the fact that the building is used for the congregation of people for recreation. A shooting gallery is similar in many respects to a bowling center, which is classified as Group A-3 (see Commentary Figure 302.1).

SECTION 303
ASSEMBLY GROUP A

303.1 Assembly Group A. Assembly Group A occupancy includes, among others, the use of a building or structure, or a portion thereof, for the gathering of persons for purposes such as civic, social or religious functions; recreation, food or drink consumption or awaiting transportation.

❖ Because of the arrangement and density of the occupant load associated with occupancies classified in the Group A assembly category, the potential for multiple fatalities and injuries from fire is comparatively high. For example, no other use listed in Section 302.1 contemplates occupant loads as dense as 5 square feet (0.46 m^2) per person (see Table 1004.1.1). Darkened spaces in theaters, nightclubs and similar spaces serve to increase hazards. In sudden emergencies, the congestion caused by large numbers of people rushing to exits can cause panic conditions. For these and many other reasons, there is a relatively high degree of hazard to life safety in assembly facilities. The relative hazards of assembly occupancies are reflected in the height and area limitations of Tables 504.3, 504.4 and 506.2, which are, in comparison, generally more restrictive than for buildings in other group classifications.

A room or space with an occupant load of 50 or more persons should not be automatically classified as Group A. However, if a room or space is used for assembly purposes (i.e., gathering of persons for purposes such as civic, social or religious functions; recreation, food or drink consumption; or awaiting transportation per Section 303.1) and the occupant load is 50 or more, Group A is likely to be the appropriate designation. Other uses can have an occupant load of more than 50 in a space or room—for example, a large office space, a grocery store or the main floor of a major retail business—but these are not assembly occupancies.

There are five specific assembly group classifications, Groups A-1 through A-5, described in this section. Where used in the code, the general term "Group A" is intended to include all five classifications.

The fundamental characteristics of all assembly occupancies are identified in this section. Structures that are designed or occupied for assembly purposes must be placed in one of the assembly group classifications. There are buildings and spaces which are used for assembly purposes, but are not classified as assembly occupancies. The "exceptions" to this rule include small assembly buildings, tenant spaces and assembly spaces in mixed-use buildings. These exceptions to the Group A classification are addressed in Sections 303.1.1 through 303.1.4.

303.1.1 Small buildings and tenant spaces. A building or tenant space used for assembly purposes with an *occupant load* of less than 50 persons shall be classified as a Group B occupancy.

❖ There are often small establishments that typically serve food and have a few seats that technically meet the definition of an assembly Group A occupancy but due to the low occupant load pose a lower risk than a typical assembly occupancy. These types of buildings

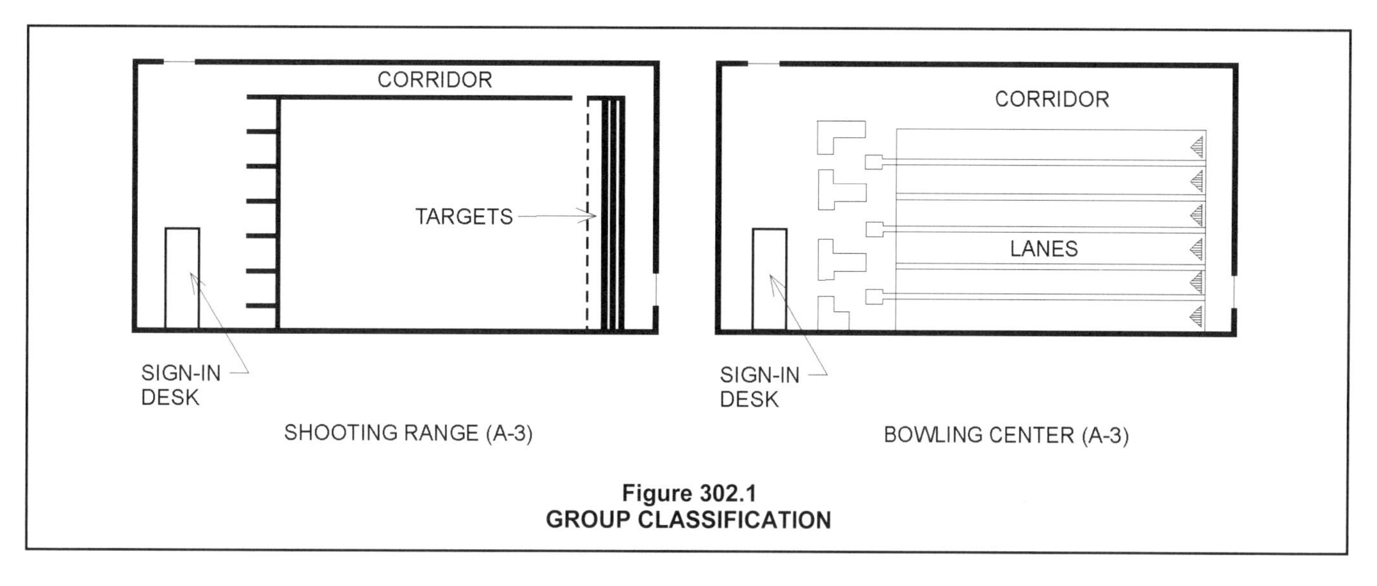

Figure 302.1
GROUP CLASSIFICATION

and tenant spaces are to be considered as Group B occupancies when the occupant load is determined to be less than 50 persons. Examples of this include small "fast food" establishments and small "mom-and-pop" cafes or coffee shops.

303.1.2 Small assembly spaces. The following rooms and spaces shall not be classified as Assembly occupancies:

1. A room or space used for assembly purposes with an *occupant load* of less than 50 persons and accessory to another occupancy shall be classified as a Group B occupancy or as part of that occupancy.
2. A room or space used for assembly purposes that is less than 750 square feet (70 m^2) in area and accessory to another occupancy shall be classified as a Group B occupancy or as part of that occupancy.

❖ Assembly rooms or spaces within larger buildings that house other uses may be classified as other than Group A depending on occupant load or the size of the space. Where the occupant load of the assembly space is less than 50, or where the floor area of the space used for assembly purposes is less than 750 square feet (65 m^2), a classification of other than Group A is permitted. In both cases, the purpose of the assembly space must be accessory to the principal occupancy of the structure (i.e., the activities in the assembly space are subordinate and secondary to the primary occupancy). If either the maximum occupant load or floor area limit requirement is satisfied and the purpose of the assembly space is accessory to the principal occupancy, the space is permitted to be classified as either a Group B occupancy or as part of the principal occupancy. In either case, Section 508.2 does not apply to this section; these assembly spaces (individually or in aggregation) are not required to be less than 10 percent of the area of the story on which they are located (IBC Interpretation No. 20-04).

The allowances given to assembly spaces in buildings containing multiple uses reflect a practical code consideration that permits a mixed-use condition to exist without requiring compliance with the provisions for mixed occupancies (see Section 508). Although the term "accessory" is used in describing the relationship of the uses, the intent of the term here is that the use of the space is related to, or part of, the main use of the space. These exceptions are not limited by the accessory use requirements found in Section 508.2.

Example 1: An office building, classified as a Group B occupancy, has a conference room used for staff meetings with an occupant load of 40 [see Commentary Figure 303.1.2(1)]. The occupancy classification of a conference room is generally considered a Group A-3. Since the occupant load of the conference room is less than 50 and its function is clearly accessory to the business area, the room is permitted to be classified the same as the main occupancy, Group B.

Example 2: A 749-square-foot (70 m^2) assembly area is located adjacent to a mercantile floor area of 5,000 square feet (465 m^2) [see Commentary Figure 303.1.2(2)]. Although the assembly use area occupies 15 percent of the 5,000-square foot (465 m^2) floor area, it does not exceed 750 square feet (70 m^2) and is not considered a Group A occupancy, but rather is classified as part of the Group M occupancy.

303.1.3 Associated with Group E occupancies. A room or space used for assembly purposes that is associated with a Group E occupancy is not considered a separate occupancy.

❖ A typical educational facility for students in the 12th grade and below invariably contains many types of assembly spaces other than classrooms, including auditoriums, cafeterias, gymnasiums and libraries. These assembly spaces in a Group E building are not intended to be regulated as separate Group A occupancies, regardless of their floor area, but rather an extension of the Group E classification. It is worth mentioning, for these assembly functions to be considered part of the primary Group E occupancy, the assembly functions must be ancillary and supportive to the educational operation of the building. Otherwise, they would be classified into the appropriate Group A occupancy based upon their specific function. These assembly spaces, where classified as a portion of the Group E occupancy, are still considered as assembly in nature and must comply with assembly space requirements specified for accessibility and means of egress. However, often these school facilities are used for other functions such as a meeting of a community service organization or a community crafts fair. These types of uses fall outside of the intent of this section, and therefore such assembly spaces would need to be classified as a Group A.

303.1.4 Accessory to places of religious worship. Accessory religious educational rooms and religious auditoriums with *occupant loads* of less than 100 per room or space are not considered separate occupancies.

❖ "Places of religious worship" are listed as Group A-3 occupancies. In addition to the worship hall, it is common for these facilities to contain smaller rooms used

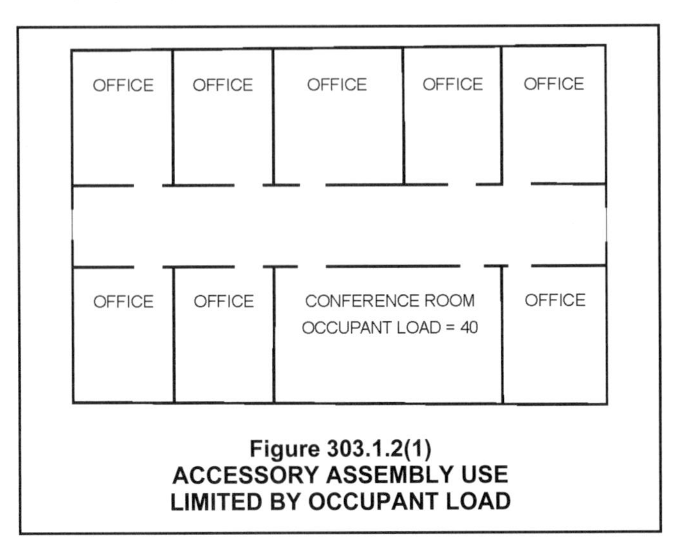

Figure 303.1.2(1)
ACCESSORY ASSEMBLY USE LIMITED BY OCCUPANT LOAD

for educational activities. This provision allows such spaces to be considered as part of the Group A-3 classification rather than create a mixed-occupancy condition. For example, classrooms are normally classified as Group E if occupied by persons of ages through the 12th grade, or as Group B if the education is provided to adults. These types of classrooms could be considered as a part of the Group A-3 occupancy under the provisions of this section. Where such rooms are used at certain times for other than a religious auditorium or for religious education, Section 302 requires that the requirements of each occupancy be applied.

303.2 Assembly Group A-1. Group A-1 occupancy includes assembly uses, usually with fixed seating, intended for the production and viewing of the performing arts or motion pictures including, but not limited to:

Motion picture theaters
Symphony and concert halls
Television and radio studios admitting an audience
Theaters

❖ Some of the characteristics of Group A-1 occupancies are large, concentrated occupant loads, low lighting levels, above-normal sound levels and a moderate fuel load.

Group A-1 is characterized by two basic types of activities. The first type is one in which the facility is occupied for the production and viewing of theatrical or operatic performances. Facilities of this type ordinarily have fixed seating; a permanent raised stage; a proscenium wall and curtain; fixed or portable scenery drops; lighting devices; dressing rooms; mechanical appliances; or other theatrical accessories and equipment [see Commentary Figure 303.2(1)].

The second type is one in which the structure is primarily occupied for the viewing of motion pictures. Facilities of this type ordinarily have fixed seating, no

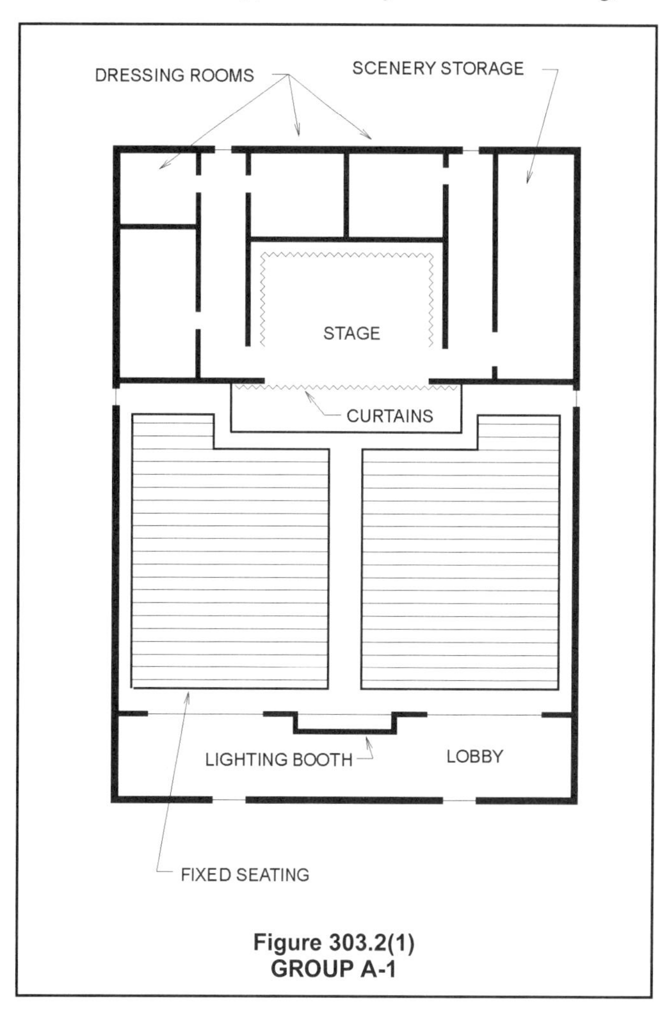

Figure 303.2(1)
GROUP A-1

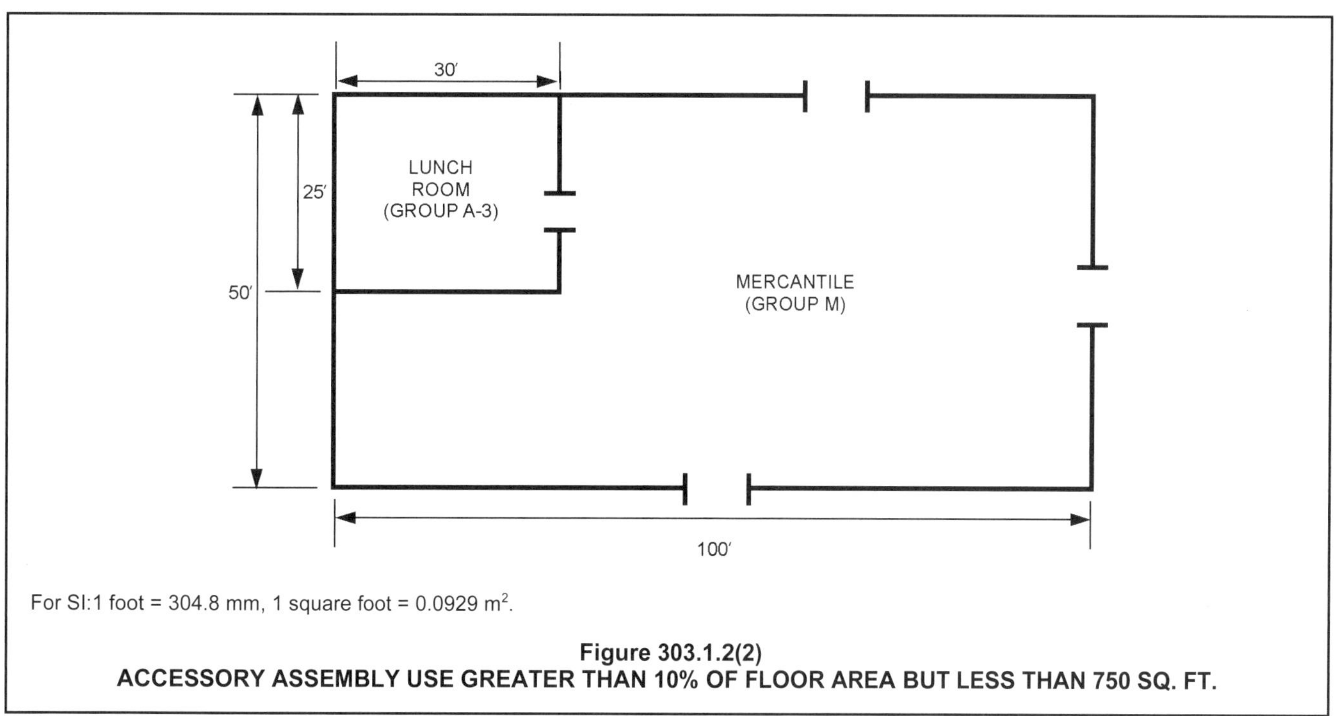

Figure 303.1.2(2)
ACCESSORY ASSEMBLY USE GREATER THAN 10% OF FLOOR AREA BUT LESS THAN 750 SQ. FT.

stage, a viewing screen, motion picture projection booths and equipment [see Commentary Figure 303.2(2)].

Group A-1 presents a significant potential life safety hazard because of the large occupant loads and the concentration of people within confined spaces. The means of egress is an important factor in the design of such facilities. Theaters for the performing arts that require stages are considered particularly hazardous because of the amount of combustibles such as curtains, drops, scenery, construction materials and other accessories normally associated with stage operation. As such, special protection requirements applicable to stages and platforms are provided in Section 410.

303.3 Assembly Group A-2. Group A-2 occupancy includes assembly uses intended for food and/or drink consumption including, but not limited to:

Banquet halls
Casinos (gaming areas)
Nightclubs
Restaurants, cafeterias and similar dining facilities (including associated commercial kitchens)
Taverns and bars

❖ Group A-2 includes occupancies in which people congregate in high densities for social entertainment, including drinking and dancing (e.g., nightclubs, banquet halls, cabarets) and food and drink consumption (e.g., restaurants). The uniqueness of these occupancies is characterized by some or all of the following:

- Low lighting levels;
- Entertainment by a live band or recorded music generating above-normal sound levels;
- No theatrical stage accessories;
- Later-than-average operating hours;
- Tables and seating arranged or positioned so as to create ill-defined aisles;
- A specific area designated for dancing;
- Service facilities for alcoholic beverages and food; and
- High occupant load density.

The fire records are very clear in identifying that the characteristics listed above often cause a delayed awareness of a fire situation and confuse occupants regarding the appropriate response, resulting in an increased egress time and sometimes panic. Together, these factors may result in extensive life and property losses. These characteristics are only advisory in determining whether Group A-2 is the appropriate classification. Often there are additional characteristics that are unique to a project, which also must be taken into consideration when a classification is made.

Not all restaurants have all of these characteristics. Most fast food restaurants will only have two or three of these, yet they are appropriately classified as Group A-2.

Example: The Downtown Club, a popular local nightclub/dance hall, features a different band every weekend [see Commentary Figure 303.3]. It is equipped with a bar and basic kitchen facilities so that beverages and appetizers can be served. There is a platform for a band to perform, a dance floor in front of the platform and numerous cocktail tables and chairs. The tables and chairs are not fixed, resulting in an arrangement with no distinct aisles. When the band

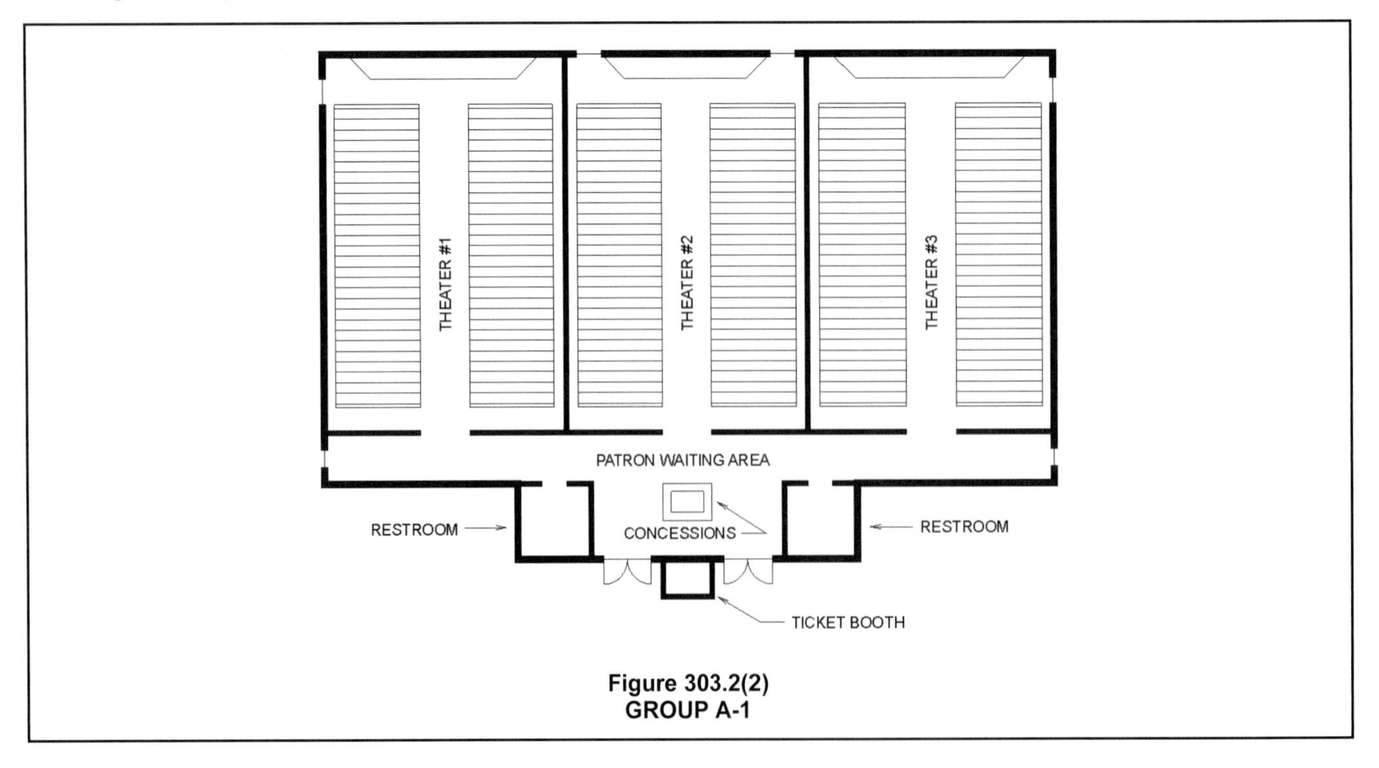

Figure 303.2(2)
GROUP A-1

performs, the house lights are dimmed and spotlights are keyed in on the performers. The club is equipped with a sound system that is used at loud levels. The club is open until 3:00 a.m.—the latest time the local jurisdiction will allow.

From this description of the Downtown Club, one can readily see that the appropriate classification is Group A-2. Sometimes, however, it is not this easy to determine the appropriate classification. In such cases, the building official must seek additional information regarding the functions of the building and each area within the building.

Two of the specific uses listed as Group A-2 occupancies are typically not considered as facilities primarily used for food and/or drink consumption, however their classification as such has been deemed appropriate for varying reasons. The placement of casino gaming areas in the Group A-2 classification is because they share many hazard characteristics with nightclubs and, to some extent, the other uses in the category. The presence of distracting lights, sounds and decorations, along with the potential for alcohol consumption, create an assembly environment that is best addressed under the Group A-2 provisions. The classification is specific to the gaming areas of a casino, therefore other related uses such as administrative, storage and lodging areas are to be individually classified based upon their specific use.

Commercial kitchens, when associated with a Group A-2 dining and drinking establishment, are also classified as Group A-2. Although commercial kitchens do not pose the same conditions and concerns as the other uses classified as Group A-2, their classification as such recognizes the relationship that exists between the dining and cooking areas.

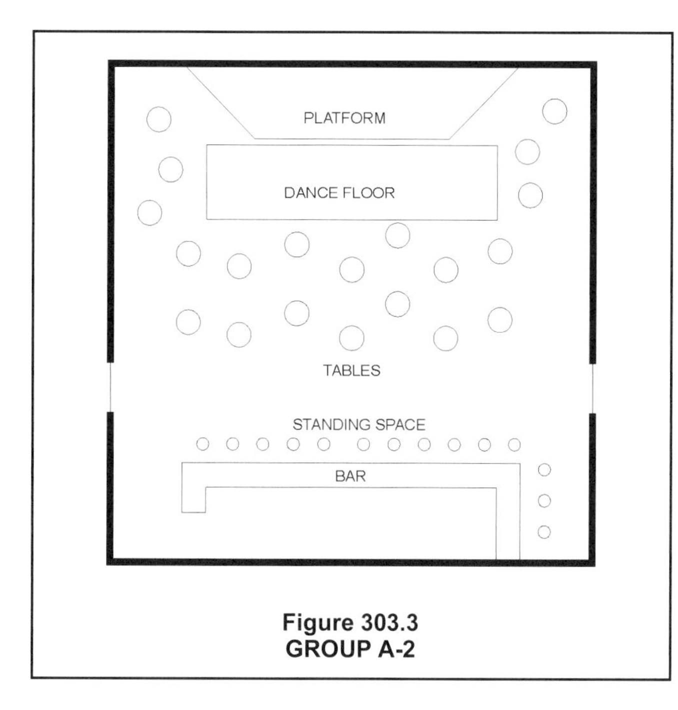

Figure 303.3
GROUP A-2

Providing a physical fire-resistive separation between the kitchen and the dining area that it serves is often found to be impractical. Assigning a single occupancy classification for both the commercial kitchen and its associated dining area eliminates any potential for an unnecessary mixed-occupancy condition. This single classification approach can also be applied where the occupant load of the dining area is below 50, allowing for a Group B classification. Under such circumstances, the kitchen would be considered an extension of the Group B dining facility. See also commercial kitchens in Group F-1.

303.4 Assembly Group A-3. Group A-3 occupancy includes assembly uses intended for worship, recreation or amusement and other assembly uses not classified elsewhere in Group A including, but not limited to:

Amusement arcades
Art galleries
Bowling alleys
Community halls
Courtrooms
Dance halls (not including food or drink consumption)
Exhibition halls
Funeral parlors
Gymnasiums (without spectator seating)
Indoor *swimming pools* (without spectator seating)
Indoor tennis courts (without spectator seating)
Lecture halls
Libraries
Museums
Places of religious worship
Pool and billiard parlors
Waiting areas in transportation terminals

❖ Structures in which people assemble for the purpose of social activities (such as entertainment, recreation and amusement) that are not classified in Group A-1, A-2, A-4 or A-5 are to be classified in Group A-3. Some of the uses listed as examples in Group A-3 are more appropriately classified as Group A-4 or A-5 where there is spectator seating provided. An indoor tennis court where no spectator seating is provided is an A-3. Add spectator seating and it becomes an A-4 occupancy. Take away the roof and now it is a Group A-5 occupancy. Exhibition halls, libraries, dance halls (not including food and drink), places of religious worship, museums, gymnasiums, recreation centers, health clubs, fellowship halls, indoor shooting galleries, bowling centers and billiard halls are among the facilities often classified in Group A-3. Also, since they most nearly resemble this occupancy classification, public and private spaces used for assembly are often classified in Group A-3. These include large courtrooms, meeting rooms and conference centers.

To more fully understand the occupancy classification of places of religious worship, see the commentary for the defined term in Section 202 and Section 303.1.4.

Schools (Group E) and colleges (Group B) may have spaces that have an occupant load of more than

50 in a room, but are ancillary to the place of education and used only for programs directly associated with the school. See the commentaries for Groups E and B for additional information (Sections 304 and 305). For college buildings, similar to other office buildings, if there are spaces with occupant loads of greater than 50 (such as cafeterias or lecture halls), by the character of the space and the level of fire hazard they would be appropriately classified as Group A-2 or A-3, respectively (see Section 302.1).

The fire hazard in terms of combustible contents (fuel load) in structures classified in Group A-3 is most often expected to be moderate to low. Since structures classified in Group A-3 vary widely as to the purposes for which they are used, the range of fuel load varies widely. For example, the fuel load in a library or an exhibition hall usually is considerably greater than that normally found in a gymnasium. While the code specifically addresses kitchens associated with Group A-2 restaurants and similar dining facilities, it is silent regarding accessory kitchens one might find associated with community centers or a fellowship hall of a place of religious worship. These kitchens should be considered carefully by the building official. Each facility and its use will be unique. The intensity and frequency of the use of the space for serving of food as well as the type of cooking equipment installed are important factors. The occasional use of the fellowship hall for a fundraising dinner where food is brought in and warmed is one end of the spectrum and may warrant keeping a Group A-3 classification. But when the same hall becomes a daily charity "soup" kitchen, then Group A-2 classification is more appropriate.

303.5 Assembly Group A-4. Group A-4 occupancy includes assembly uses intended for viewing of indoor sporting events and activities with spectator seating including, but not limited to:

Arenas
Skating rinks
Swimming pools
Tennis courts

❖ Structures provided with spectator seating in which people assemble to watch an indoor sporting event are to be classified as Group A-4. Arenas, skating rinks, swimming pools and tennis courts are among the facilities often classified as Group A-4. The list of four uses offers examples of A-4 occupancies and is not an exhaustive list. The distinguishing factor between Group A-4 and A-5 structures is whether the event is indoors or outdoors. Group A-4 facilities are limited to indoor structures only. The distinguishing factor between Group A-4 and Group A-3 facilities is the presence of a defined seating area. While Group A-3 facilities are indoors (e.g., tennis courts, swimming pools), they typically do not have a defined seating area in which to view the event. Only facilities that are both indoors and have a defined seating area are to be classified as Group A-4.

303.6 Assembly Group A-5. Group A-5 occupancy includes assembly uses intended for participation in or viewing outdoor activities including, but not limited to:

Amusement park structures
Bleachers
Grandstands
Stadiums

❖ Structures classified in Group A-5 are outdoor facilities where people assemble to view or participate in social and recreational activities (e.g., stadiums, grandstands, bleachers, coliseums). In order to qualify as an outdoor facility, the structure must be one where the products of combustion are freely and rapidly vented to the atmosphere (i.e., a structure without enclosures that would prevent the free movement of smoke from the occupied area to the outside). Any recreation facility that has exterior walls that enclose the facility and a roof that fully covers the area would not be classified in Group A-5, but rather in Group A-3 or A-4 depending on whether a seating area has been provided. In the case of a structure with a retractable roof, the more stringent occupancy classification (i.e., Group A-4) would be required. The list of four uses offers examples of A-5 occupancies and is not an exhaustive list.

Since occupancies classified in Group A-5 are primarily viewing and sports participation areas, the fuel load associated with them is very low (i.e., the structure itself and seats). Since the fuel load present is relatively low and the expectation is that smoke will be quickly evacuated from the structure, the relative fire hazard of occupancies classified in Group A-5 is expected to be low. The life safety hazard from panic that might occur in an emergency, however, is a serious concern. Hence, the capability of large crowds to exit the structure quickly and orderly during emergencies is an important design consideration (see Section 1029).

Both Group A-4 and A-5 occupancies will include a variety of uses that support the viewing of sports and similar activities. There will likely be luxury seating suites, locker rooms, toilet facilities and press boxes, which are clearly part of the overall uses of the facility. There will also be offices, food concession stands and merchandise stands, which by their use are different occupancies, but are probably within the accessory occupancy limits established in Section 508.2. Because of the multitiered design of most Group A-4 and A-5 occupancies, the limit for accessory occupancies to account for less than 10 percent of the story will need to be creatively applied. There may be full-fledged restaurants that are in the same building and open to guests, not just limited to those attending an event. A Group A-2 occupancy designation is likely the most appropriate classification and the mixed-occupancy conditions would most typically be addressed under the accessory occupancy provisions of Section 508.2.

SECTION 304
BUSINESS GROUP B

304.1 Business Group B. Business Group B occupancy includes, among others, the use of a building or structure, or a portion thereof, for office, professional or service-type transactions, including storage of records and accounts. Business occupancies shall include, but not be limited to, the following:

Airport traffic control towers
Ambulatory care facilities
Animal hospitals, kennels and pounds
Banks
Barber and beauty shops
Car wash
Civic administration
Clinic, outpatient
Dry cleaning and laundries: pick-up and delivery stations and self-service
Educational occupancies for students above the 12th grade
Electronic data processing
Food processing establishments and commercial kitchens not associated with restaurants, cafeterias and similar dining facilities not more than 2,500 square feet (232 m^2) in area.
Laboratories: testing and research
Motor vehicle showrooms
Post offices
Print shops
Professional services (architects, attorneys, dentists, physicians, engineers, etc.)
Radio and television stations
Telephone exchanges
Training and skill development not in a school or academic program (this shall include, but not be limited to, tutoring centers, martial arts studios, gymnastics and similar uses regardless of the ages served, and where not classified as a Group A occupancy).

❖ The risks to life safety in the business occupancy classification are relatively low. Exposure to the potential effects of fire is limited because business-type facilities most often have low fuel loads, are normally occupied only during the daytime and, with some exceptions, are usually occupied for a set number of hours. The occupants, because of the nature of the use, are typically alert, ambulatory, conscious, aware of their surroundings and generally familiar with the building's features, particularly the means of egress. Historically, this occupancy has one of the better fire safety records for the protection of life and property.

This section identifies the general characteristics and lists examples of occupancies that are classified in Group B. Note that the description recognizes the need for limited storage spaces that are incidental to office occupancies. Accessory storage classified as a Group B occupancy is limited by Section 311.1.1 to be no more than 10 percent of a given story of a building.

Colleges (educational occupancies for students above the 12th grade) may have spaces that have an occupant load of more than 50 in a room, but are ancillary to the place of education and used only for programs directly associated with training and education (Section 303.4). For college buildings, similar to other office buildings, if there are spaces with occupant loads greater than 50, such as cafeterias or lecture halls, by the character of the space and the level of fire hazard they would be appropriately classified as Group A-2 or A-3, respectively (see Section 302.1). Where lecture facilities for large groups (i.e., occupant load of 50 or more) are located within the same building as classrooms with an occupant load less than 50, the building is a mixed occupancy (Groups A-3 and B) and is subject to the provisions of Section 508

While civic administration covers a broad range of state and local government buildings, many such buildings will have a variety of uses and need to be considered under mixed occupancy provisions. Frequently, police stations will include jails or holding cells. Fire stations will be a mix of offices, parking and maintenance facilities for the fire engines, and living spaces for the fire fighters. Often a meeting room that is open to the public is also included. This type of facility is a mix of Group A, B, R and S occupancies.

Ambulatory care facilities are those used to provide medical, or similar care, on less than a 24-hour basis to patients who are rendered incapable of self-preservation (see Section 202). Frequently called "day surgery centers" or "ambulatory surgical centers," ambulatory care facilities perform procedures that render care recipients (patients) temporarily incapable of self-preservation due to the use of nerve blocks, sedation or anesthesia. Because of the condition of the care recipients, the need for medical staff to stabilize the patients before evacuation and the use of medical gases such as oxygen and nitrous oxide, these types of facilities pose greater fire and life safety hazards than other business occupancies. Accordingly, additional fire protection and means of egress requirements specific to ambulatory care are provided in Section 422.

Facilities that provide medical services for inpatient care where the care recipients (patients) stay for more than 24 hours would be classified as Group I-2. Buildings used as sleep clinics would be classified as Group B since these spaces are not typical dwelling or sleeping units where people live, the occupants are assumed to be capable of self-preservation and the occupants are not living in a supervised environment. Although the patients in a sleep clinic may be sleeping, they can be easily awakened and alerted to an emergency as compared to the patients at an ambulatory care facility.

The code distinguishes, based on size, between food processing operations that are not directly associated with a restaurant. Those 2,500 square feet

(232 m^2) or smaller fall under the Group B classification. A small catering business would fall under this classification. Other food-related businesses which do not provide a space for their product to be eaten on site, such as a take-out pizza store or a neighborhood bakery, would also be classified as a Group B occupancy. A commercial kitchen or food processing facility larger than 2,500 square feet (232 m^2) would be classified as a Group F occupancy.

Training and skill development is classified as a Group B occupancy due to the similarity in use of spaces between education above the 12th grade and professional consultation. Often unions provide training facilities for their members so they can keep up with new materials and updates of regulations. Other facilities can provide one-on-one tutoring such as remedial reading or math skills for students. Those receiving the training or skill development can include those whose ages are typically associated with grades 12 or earlier. The determination of the appropriate classification requires the building official to consider whether the training is given as part of a traditional educational program. Examples provided by the code allow a range of size in the numbers receiving the training from one-on-one tutoring to a large class of children learning martial arts or ballet. The presence of children does not automatically mean a classification as a Group E. Where the occupant load of a training classroom or space exceeds 50, a Group A classification may be appropriate for the space, especially if the space is to be used for different activities at different times. If the training room is used for a martial arts competition with spectators on an evening or weekend, then a Group A designation should also be considered.

304.2 Definitions. The following terms are defined in Chapter 2:

AMBULATORY CARE FACILITY.

CLINIC, OUTPATIENT.

❖ This section lists terms that are specifically associated with the subject matter of this section. It is important to emphasize that these terms are not exclusively related to this section but may or may not also be applicable where the term is used elsewhere in the code.

Definitions of terms can help in the understanding and application of the code requirements. The purpose for including a list within this chapter is to provide more convenient access to terms which may have a specific or limited application within this chapter. For the complete definition and associated commentary, refer back to Chapter 2. Terms that are italicized provide a visual identification throughout the code that a definition exists for that term. The use and application of all defined terms are set forth in Section 201.

SECTION 305
EDUCATIONAL GROUP E

305.1 Educational Group E. Educational Group E occupancy includes, among others, the use of a building or structure, or a portion thereof, by six or more persons at any one time for educational purposes through the 12th grade.

❖ The risks to life safety in this occupancy vary with the composition of the facilities and also with the ages of the occupants. In general, children require more safeguards than do older, more mature persons.

This section identifies the criteria for classification of a building in Group E. The two fundamental characteristics of a Group E facility are as follows:

1. The facility is occupied by more than five persons (excluding the instructor); and
2. The purpose of the facility is for educating persons at the 12th-grade level and below, but not including more than five occupants $2^1/_2$ years of age or less.

It is common for a school to also have gymnasiums (Group A-3), auditoriums (Group A-1), libraries (Group A-3) and offices (Group B). Storage rooms might be either a Group S-1 occupancy or, if less than 10 percent of a story, they can be classified as Group E (see Section 311.1.1). When this occurs, the building is considered as a mixed occupancy condition and is subject to the provisions of Section 508. In accordance with Section 303.1.3, assembly spaces, such as the gymnasium, auditorium, library and cafeteria, do not have to be considered separate occupancies if used for school purposes (see commentary, Section 303.1.3). For such assembly functions to be considered part of the primary Group E occupancy, the assembly functions must be ancillary and supportive to the educational operation of the building.

Occupancies used for the education of persons above the 12th grade level are not included in Group E. These facilities are occupied by adults who are not expected to require special supervision, direction or instruction in a fire or other emergency.

305.1.1 Accessory to places of religious worship. Religious educational rooms and religious auditoriums, which are accessory to *places of religious worship* in accordance with Section 303.1.4 and have *occupant loads* of less than 100 per room or space, shall be classified as Group A-3 occupancies.

❖ In places of religious worship, worship halls, religious educational rooms and religious auditoriums are often all provided in the same building complex. Such religious educational rooms and auditoriums are not to be considered separate occupancies (i.e., Group E) (see commentary, Section 303.1.4). Where such rooms are used at certain times for other than a religious auditorium or for religious education, Section 302 requires that the requirements of each occupancy be applied.

305.2 Group E, day care facilities. This group includes buildings and structures or portions thereof occupied by more than five children older than $2^1/_2$ years of age who receive educational, supervision or *personal care services* for fewer than 24 hours per day.

❖ Group E day care occupancies include facilities intended to be used for the care and supervision of more than five children older than $2^1/_2$ years of age where individual care is for a period of less than 24 hours per day. Day care centers are a special concern since they are generally occupied by preschool children who are less capable of responding to an emergency. The hazards found in a day care center are far greater than in normal educational facilities, not so much because of the occupant or fuel load, but because of the inability of the occupants to respond.

Children $2^1/_2$ years of age or less usually are not able to recognize an emergency situation, may not respond appropriately or simply may not be able to egress without assistance. Thus, facilities that have more than five children $2^1/_2$ years of age or less are classified as child care facilities and considered to be Group I-4 (see Section 308.6) unless the provisions of Sections 308.6.1 through 308.6.4 allow for a different classification.

Locations where child care may be provided that would not be considered Group E or I-2 are addressed in Sections 305.2.1 through 305.2.3.

305.2.1 Within places of religious worship. Rooms and spaces within *places of religious worship* providing such day care during religious functions shall be classified as part of the primary occupancy.

❖ Cry rooms and other types of child care areas within places of worship that are used for this purpose during a religious function need not be classified as Group E day care facilities. Such rooms and spaces may take on the classification of the primary occupancy, which in most cases would be Group A-3. The limited occupant load makes the need for classification as a Group E occupancy unnecessary. If the child care areas are used when there is not a service or other religious function going on simultaneously, such as for child day care during the week, the religious facility would be a mixed use building (see commentary, Section 303.5.2 or 308.6).

305.2.2 Five or fewer children. A facility having five or fewer children receiving such day care shall be classified as part of the primary occupancy.

❖ Where a child care facility has no more than five children receiving care at any one time, the classification of the main occupancy may extend to the child care use. The limited number of occupants requiring care services does not warrant classification as a separate and distinct occupancy from that of the major use.

305.2.3 Five or fewer children in a dwelling unit. A facility such as the above within a *dwelling unit* and having five or fewer children receiving such day care shall be classified as a Group R-3 occupancy or shall comply with the *International Residential Code*.

❖ Where child care services are performed within a single-family dwelling or within a dwelling unit of a two-family dwelling, residential provisions are applicable provided the number of children receiving care does not exceed five. The facility may be classified under the code as a Group R-3 occupancy or may be regulated under the provisions of the *International Residential Code*® (IRC®). Where this use occurs within a dwelling unit of a Group R-2 multifamily building, it is expected that the child care facility be considered as an extension of the Group R-2 classification as addressed in Section 305.2.2.

SECTION 306
FACTORY GROUP F

306.1 Factory Industrial Group F. Factory Industrial Group F occupancy includes, among others, the use of a building or structure, or a portion thereof, for assembling, disassembling, fabricating, finishing, manufacturing, packaging, repair or processing operations that are not classified as a Group H hazardous or Group S storage occupancy.

❖ The purpose of this section is to identify the characteristics of occupancies that are classified in factory and industrial occupancies and to differentiate Groups F-1 and F-2.

Because of the vast number of diverse manufacturing and processing operations in the industrial community, it is more practical to classify such facilities by their level of hazard rather than their function. In industrial facilities, experience has shown that the loss of life or property is most directly related to fire hazards, particularly the fuel load contributed by the materials being fabricated, assembled or processed.

Statistics show that property losses are comparatively high in factory and industrial occupancies, but the record of fatalities and injuries from fire has been remarkably low. This excellent life safety record can, in part, be attributed to fire protection requirements of the code.

This section requires that all structures used for fabricating, finishing, manufacturing, packaging, assembling or processing products or materials are to be classified in either Group F-1 (moderate hazard) or F-2 (low hazard). These classifications are based on the relative level of hazard for the types of materials that are fabricated, assembled or processed. Where the products and materials in a factory present an extreme fire, explosion or health hazard, such facilities are classified in Group H (see Section 307). It should be noted that the term "Group F" is not a specific occupancy, but is a term that collectively applies to Groups F-1 and F-2.

306.2 Moderate-hazard factory industrial, Group F-1. Factory industrial uses that are not classified as Factory Industrial F-2 Low Hazard shall be classified as F-1 Moderate Hazard and shall include, but not be limited to, the following:

Aircraft (manufacturing, not to include repair)
Appliances
Athletic equipment
Automobiles and other motor vehicles
Bakeries
Beverages: over 16-percent alcohol content
Bicycles
Boats
Brooms or brushes
Business machines
Cameras and photo equipment
Canvas or similar fabric
Carpets and rugs (includes cleaning)
Clothing
Construction and agricultural machinery
Disinfectants
Dry cleaning and dyeing
Electric generation plants
Electronics
Engines (including rebuilding)
Food processing establishments and commercial kitchens not associated with restaurants, cafeterias and similar dining facilities more than 2,500 square feet (232 m^2) in area.
Furniture
Hemp products
Jute products
Laundries
Leather products
Machinery
Metals
Millwork (sash and door)
Motion pictures and television filming (without spectators)
Musical instruments
Optical goods
Paper mills or products
Photographic film
Plastic products
Printing or publishing
Recreational vehicles
Refuse incineration
Shoes
Soaps and detergents
Textiles
Tobacco
Trailers
Upholstering
Wood; distillation
Woodworking (cabinet)

❖ Structures classified in Group F-1 (moderate hazard) are occupied for the purpose of fabrication, finishing, manufacturing, packaging, assembly or processing of materials that are combustible or that use combustible products in the production process. Food processing facilities and commercial kitchens that are 2,500 square feet (232 m^2) or smaller in area fall under the Group B occupancy category, see Section 304.1.

306.3 Low-hazard factory industrial, Group F-2. Factory industrial uses that involve the fabrication or manufacturing of noncombustible materials that during finishing, packing or processing do not involve a significant fire hazard shall be classified as F-2 occupancies and shall include, but not be limited to, the following:

Beverages: up to and including 16-percent alcohol content

Brick and masonry
Ceramic products
Foundries
Glass products
Gypsum
Ice
Metal products (fabrication and assembly)

❖ Structures classified in Group F-2 (low hazard) are occupied for the purpose of fabrication, manufacturing or processing of noncombustible materials. It is acceptable for noncombustible products to be packaged in a combustible material, provided that the fuel load contributed by the packaging is negligible when compared to the amount of noncombustible product.

The use of a significant amount of combustible material to package or finish a noncombustible product, however, will result in a Group F-1 (moderate-hazard factory and industrial) classification.

To distinguish when the presence of combustible packaging constitutes a significant fuel load, possibly requiring the reclassification of the building or structure as Group F-1, a reasonable guideline to follow is the "single thickness" rule, which is when a noncombustible product is put in one layer of packaging material.

Examples of acceptable conditions in Group F-2 include:

- Vehicle engines placed on wood pallets for transportation after assembly;
- Washing machines in corrugated cardboard boxes; and
- Soft-drink glass bottles packaged in pressed paper boxes.

Occupancies involving noncombustible items packaged in more than one layer of combustible packaging material are most appropriately classified in Group F-1.

Typical examples of packaging that would result in a Group F-1 classification include:

- Chinaware wrapped in corrugated paper and placed in cardboard boxes;
- Glassware set in expanded foam forms and placed in cardboard boxes; and
- Fuel filters individually packed in pressed paper boxes, placed by the gross in a cardboard box and stacked on a pallet for transportation.

Factories and industrial facilities often have offices and areas where large quantities of materials are kept in the same building as manufacturing operations, fabrication processes and assembly processes. The stock areas are classified as either Group S-1 or S-2, depending on the combustibility of the materials stored. Areas used for offices are classified in Group B. Where these and other combinations of occupancies occur, the building is subject to the mixed occupancy provisions in Section 508.

SECTION 307 HIGH-HAZARD GROUP H

[F] 307.1 High-hazard Group H. High-hazard Group H occupancy includes, among others, the use of a building or structure, or a portion thereof, that involves the manufacturing, processing, generation or storage of materials that constitute a physical or health hazard in quantities in excess of those allowed in *control areas* complying with Section 414, based on the maximum allowable quantity limits for *control areas* set forth in Tables 307.1(1) and 307.1(2). Hazardous occupancies are classified in Groups H-1, H-2, H-3, H-4 and H-5 and shall be in accordance with this section, the requirements of Section 415 and the *International Fire Code*. Hazardous materials stored, or used on top of roofs or canopies, shall be classified as outdoor storage or use and shall comply with the *International Fire Code*.

❖ This section identifies the various types of facilities contained in the high-hazard occupancy. This occupancy classification relates to those facilities where operations or the storage of materials are deemed to be extremely hazardous to life and property, especially when they involve the use of significant amounts of highly combustible, flammable or explosive materials, regardless of their composition (i.e., solids, liquids, gases or dust). Although they are not explosive or highly flammable, other hazardous materials, such as corrosive liquids, highly toxic materials and poisonous gases, still present an extreme hazard to life. Many materials possess multiple hazards, whether physical or health related.

There is a wide range of high-hazard operations in the industrial community; therefore, it is more practical to categorize such facilities in terms of the degree of hazard they present, rather than attempt to define a facility in terms of its function. This method is similar to that used to categorize factory (see Section 306) and storage (see Section 311) occupancies.

Group H is handled as a separate classification because it represents an unusually high degree of hazard that is not found in the other occupancies. It is important to isolate those industrial or storage operations that pose the greatest dangers to life and property and to reduce such hazards by providing systems or elements of protection through the regulatory provisions of building codes.

There are numerous provisions and exceptions throughout the code that cannot be used when one or more Group H occupancies are present.

Operations that, because of the materials utilized or stored, cause a building or portion of a building to be classified as a high-hazard occupancy are identified in this section. While buildings classified as Group H may not have a large occupant load, the unstable chemical properties of the materials contained on the premises constitute an above-average fuel load and serve as a potential danger to the surrounding area.

The dangers created by the high-hazard materials require special consideration for the abatement of the danger. The classification of a material as high hazard is based on information derived from National Fire Protection Association (NFPA) standards and the Code of Federal Regulations (DOL 29 CFR).

The wide range of materials utilized or stored in buildings creates an equally wide range of hazards to the occupants of the building, the building proper and the surrounding area. Since these hazards range from explosive to corrosive conditions, the high-hazard occupancy has been broken into four subclassifications: Groups H-1 through H-4. A fifth category, Group H-5, is used to represent structures that contain hazardous production material (HPM) facilities. Each of these subclassifications addresses materials that have similar characteristics, and the protection requirements attempt to address the hazard involved. These subclassifications are defined by the properties of the materials involved with only occasional reference to specific materials. This performance-based criterion may involve additional research to identify a hazard, but it is the only way to remain current in a rapidly changing field. Material Safety Data Sheets (MSDS) will be a major source for information.

Additional information on hazardous materials can be found in Section 415 as well as the commentary to the *International Fire Code*® (IFC®).

Section 307.1 acknowledges that a building is not classified as a high-hazard occupancy unless the maximum allowable quantities per control area as prescribed in Tables 307.1(1) and 307.1(2) are exceeded, subject to the applicable control area provisions of Section 414.2. The maximum quantity limitations per control area prescribed in Tables 307.1(1) and 307.1(2) have been determined to be relatively safe when maintained in accordance with the IFC. Therefore, a building containing less than the maximum allowable quantities specified in Tables 307.1(1) and 307.1(2) would not be classified as a Group H occupancy but rather as the occupancy group it most nearly resembles. The materials in these tables, which are listed in Section 307.2 are defined in detail in Section 202.

Section 414.2 establishes the control area concept for regulating hazardous materials. This concept would allow the maximum allowable quantities of hazardous materials per control area in Tables 307.1(1) and 307.1(2) to be exceeded within a given building without classifying the building as a high-hazard occupancy by utilizing a multiple control area approach. The permitted number of control areas, maximum percentage of allowable quantities of hazardous materials per control area and degree of fire separation between control areas are regulated by Section 414.2 (see commentary, Section 414.2).

Section 307.1 also clarifies that hazardous materials outside of the building envelope should be classified as outdoor storage. As such, hazardous material quantities on roofs or canopies are not included in evaluating the occupancy classification of a building or structure. Canopies used to support gaseous hydrogen systems must comply with Section 406.7.2.1.

[F] 307.1.1 Uses other than Group H. An occupancy that stores, uses or handles hazardous materials as described in one or more of the following items shall not be classified as Group H, but shall be classified as the occupancy that it most nearly resembles.

1. Buildings and structures occupied for the application of flammable finishes, provided that such buildings or areas conform to the requirements of Section 416 and the *International Fire Code*.
2. Wholesale and retail sales and storage of flammable and combustible liquids in mercantile occupancies conforming to the *International Fire Code*.
3. Closed piping system containing flammable or combustible liquids or gases utilized for the operation of machinery or equipment.
4. Cleaning establishments that utilize combustible liquid solvents having a flash point of 140°F (60°C) or higher in closed systems employing equipment *listed* by an *approved* testing agency, provided that this occupancy is separated from all other areas of the building by 1-hour *fire barriers* constructed in accordance with Section 707 or 1-hour *horizontal assemblies* constructed in accordance with Section 711, or both.
5. Cleaning establishments that utilize a liquid solvent having a flash point at or above 200°F (93°C).
6. Liquor stores and distributors without bulk storage.
7. Refrigeration systems.
8. The storage or utilization of materials for agricultural purposes on the premises.
9. Stationary batteries utilized for facility emergency power, uninterruptable power supply or telecommunication facilities, provided that the batteries are provided with safety venting caps and *ventilation* is provided in accordance with the *International Mechanical Code*.
10. Corrosive personal or household products in their original packaging used in retail display.
11. Commonly used corrosive building materials.
12. Buildings and structures occupied for aerosol storage shall be classified as Group S-1, provided that such buildings conform to the requirements of the *International Fire Code*.
13. Display and storage of nonflammable solid and nonflammable or noncombustible liquid hazardous materials in quantities not exceeding the maximum allowable quantity per *control area* in Group M or S occupancies complying with Section 414.2.5.
14. The storage of black powder, smokeless propellant and small arms primers in Groups M and R-3 and special industrial explosive devices in Groups B, F, M and S, provided such storage conforms to the quantity limits and requirements prescribed in the *International Fire Code*.

❖ Section 307.1.1 provides 14 cases where facilities would not be classified as Group H because of the specific type of material, how it is expected to be used or stored, or both; the building's construction and use; the packaging of materials; the quantity of materials; or the precautions taken to prevent fire. Even if a high-hazard material meets one of these 14 cases, its storage and use must comply with the applicable provisions of Section 414 and the IFC.

Item 1 exempts spray painting and similar operations within buildings from being classified as a high-hazard occupancy. This item requires that all such operations, as well as the handling of flammable finishes, are in accordance with the provisions of Section 416 and the IFC. Therefore, an adequately protected typical paint spray booth in a factory (Group F-1) would not result in a high-hazard occupancy classification for either the building or the paint spray area.

Item 2 relies on the provisions of Section 5704.3.4.1 of the IFC to regulate the storage of flammable and combustible liquids for wholesale and retail sales and storage in mercantile occupancies. The overall permitted amount of flammable and combustible liquids is dependent on the class of liquid, storage arrangement, container size and level of sprinkler protection. For nonsprinklered buildings, the maximum allowable quantity per control area permitted by Table 5704.3.4.1 of the IFC is 1,600 gallons (6057 L) of Class IB, IC, II and IIIA liquids with a maximum of 60 gallons (227 L) of Class IA liquids. Depending on storage and ceiling heights, buildings equipped with a sprinkler system with a minimum design density for an Ordinary Hazard Group 2 occupancy may have an aggregate total of 7,500 gallons (28 391 L) of Class IB, IC, II, and IIIA liquids with a maximum of 60 gallons (227 L) of Class IA liquids. The quantities of Class IB, IC, II and IIA liquids could be further increased depending on the potential storage conditions and enhanced degree of sprinkler protection (see Section 5704.3.4.1 of the IFC for additional design information). Again, it should be noted that, despite the increased quantities far exceeding the base quantity limitations of Table 307.1(1), compliance with this item would result in the building not being classified as a Group H occupancy.

Item 3 exempts from Group H classification closed systems that are used exclusively for the operation of machinery or equipment. The closed piping systems, which are essentially not open to the atmosphere, keep flammable or combustible liquids from direct exposure to external sources of ignition and prevent the users from coming in direct contact with liquids or harmful vapors. This item would include systems such as oil-burning equipment, piping for diesel fuel generators and LP-gas cylinders for use in forklift trucks.

Item 4 allows cleaning establishments that utilize a closed system for all combustible liquid solvents with a flash point at or above 140°F (60°C) to be classified as something other than Group H. The reference to using equipment listed by an approved testing laboratory does not mean that the entire system needs to be approved, but rather the individual pieces of equipment. As with any mechanical equipment or appliance, it should bear the label of an approved agency and be installed in accordance with the manufacturer's installation instructions [see the *International Mechanical Code*® (IMC®)].

Item 5 covers cleaning establishments that use solvents that have very high flash points [at least 200°F (93°C)] and that are exceedingly difficult to ignite. Such liquids can be used openly, but with due care.

Item 6 exempts all retail liquor stores and liquor distribution facilities from the high-hazard occupancy classification, even though most of the contents are considered combustible liquids. The item takes into account that alcoholic beverages are packaged in individual containers of limited size.

Item 7 refers to refrigeration systems that utilize refrigerants that may be flammable or toxic. Refrigeration systems do not alter the occupancy classification of the building, provided they are installed in accordance with the IMC. The IMC has specific limitations on the quantity and type of refrigerants that can be used, depending on the occupancy classification of the building.

Item 8 addresses materials that are used for agricultural purposes, such as fertilizers, pesticides, and fungicides, when used on the premises. Agricultural materials stored for direct or immediate use are not usually of a quantity large enough to constitute a large fuel load or an exceptionally hazardous condition. A group H classification is not appropriate for these situations.

Item 9 addresses battery storage rooms when used as part of an operating system, such as for providing standby power. The batteries used in installations of this type do not represent a significant health, safety or fire hazard. The electrolyte and battery casing contribute little fuel load to a fire. The release of hydrogen gas during the operation of battery systems is minimal. Ventilation in accordance with the IMC will disperse the small amounts of liberated hydrogen. This item also assumes that rooms containing stationary storage battery systems are in compliance with Section 608 of the IFC.

Without Item 10, certain products that technically are corrosive could cause grocery stores and other mercantile occupancies to be inappropriately classified as Group H-4. This item allows the maximum allowable quantity per control area in Table 307.1(2) for corrosives to be exceeded in the retail display area. This would include such things as bleaches, detergents and other household cleaning supplies in normal-size containers.

Item 11 exempts the storage or manufacture of commonly used building materials, such as Portland cement, from being inappropriately classified as Group H.

Item 12 exempts from a Group H classification those buildings and structures used for the storage of aerosol products, provided they are protected in accordance with the provisions of NFPA 30B and the IFC. The aerosol storage requirements in the IFC, referred to in this item, are based on the provisions of NFPA 30B. Compliance with the item exempts buildings from the Group H classification, provided the storage of aerosol products complies with applicable separation, storage limitations and sprinkler design requirements specified in the IFC and NFPA 30B.

Item 13 permits certain products found in mercantile and storage occupancies, which may be composed of hazardous materials, to exceed the maximum allowable quantity per control area of Tables 307.1(1) and 307.1(2). The products, however, must be composed of nonflammable solids or liquids that are nonflammable or noncombustible. Materials could include swimming pool chemicals, which are typically Class 2 or 3 oxidizers or industrial corrosive cleaning agents (see commentary, Section 414.2.5).

Item 14 permits the base maximum allowable quantity per control area of black powder, smokeless propellant and small arms primers in Group M and R-3 occupancies to be exceeded, provided the material is stored in accordance with Chapter 33 of the IFC. The requirements are based on the provisions in NFPA 495. Similarly, special industrial explosive devices are found in a number of occupancies other than Group H (Groups B, F, M and S). Storage of these devices in accordance with the IFC is not required to have a high-hazard occupancy classification. Power drivers are commonly used in the construction industry, and there are stocks of these materials maintained for sale and use by the trade. The automotive airbag industry has evolved with the use of these devices, and they are located in automotive dealerships and personal use vehicles throughout society. The IFC currently exempts up to 50 pounds (23 kg) of these materials from regulation under Chapter 56 (explosives).

TABLE 307.1(1). See page 3-18.

❖ The maximum allowable quantities (MAQ) of high-hazard materials allowed in each control area before having to classify a part of the (or the entire) building as a high-hazard occupancy are given in the table. This table is referenced in Section 307.1. The materials listed in this table (see Chapter 2 for definitions of the materials) are classified according to their specific occupancy in Sections 307.3 through 307.5. This table only contains materials applicable to Groups H-1, H-2 and H-3. The maximum allowable quantities per control area for Group H-4 materials are listed in Table 307.1(2).

The presence of any one of the materials listed in Table 307.1(1) in an amount greater than allowed requires that the building or area in which the material is contained be classified as a Group H, high-hazard occupancy.

If a building or area contains only the materials listed in either Table 307.1(1) or 307.1(2) in the maximum allowable quantity per control area or less, then that building or area would not be classified as a Group H, high-hazard occupancy. The possible increase in overall danger that might exist should this occur because of the storage and use of incompatible materials is an issue that the code does not specifically address. In such situations, the building official can seek the advice of chemical engineers, fire protection engineers, fire service personnel or other experts in the use of hazardous materials. Based on their advice, the building official can deem the building a high-hazard occupancy.

Table 307.7(1) is subdivided based on whether the material is in storage or in use in a closed or open system. Definitions of both closed and open systems are found in Section 202. Within these subdivisions, the appropriate maximum allowable quantity per control area is listed in accordance with the physical state (solid, liquid or gas) of the material. A column for gas in open systems is not indicated because hazardous gaseous materials should not be allowed in a system that is continuously open to the atmosphere. While hazardous materials within a closed or open system are considered to be in use, Note b clearly indicates that the aggregate quantity of hazardous materials in use and storage within a given control area should not exceed the quantity listed in Table 307.7(1) for storage. Without Note c, many common alcoholic beverages and household products containing a negligible amount of a hazardous material could result in a Group M occupancy being classified as a high hazard. Note c recognizes the reduced hazard of the materials based on their water miscibility and limited container size.

Notes d and e of Table 307.1(1) are significant in that, for certain materials, the maximum allowable amount may be increased due to the use of approved hazardous material storage cabinets, or where the building is fully protected by an automatic sprinkler system, or both. The notes are intended to be cumulative in that up to four times the base maximum quantity may be allowed per control area, if both sprinklered and in cabinets, without classifying the building as Group H. While the use of cabinets is not always a feasible or practical method of storage, they do provide additional protection to warrant an increase if provided. Construction requirements for hazardous material storage cabinets are contained in the IFC. Note that the use of day boxes, gas cabinets, exhausted enclosures or listed safety cans would allow the same increase as for cabinets. Listed safety cans, which are primarily intended for flammable and combustible liquids, must be in compliance with UL 30 when used to increase the maximum allowable quantities permitted by Table 307.1(1).

While classified as a hazardous material, the code recognizes the relative hazard of Class IIIB liquids as compared to that of other flammable and combustible liquids by establishing a base maximum allowable quantity per control area of 13,200 gallons (49 962 L). As indicated in Note f, the quantity of Class I oxidizers and Class IIIB liquids would not be limited, provided the building is fully sprinklered in accordance with NFPA 13. Since any building that exceeds this maximum amount would be required to be classified as Group H and these buildings are required to be sprinklered, the maximum allowable amount would then be unlimited. As such, a Group H classification would not be warranted. The hazard presented by Class I oxidizers is that they slightly increase the burning rate of combustible materials that they may come into contact with during a fire. Class IIIB combustible liquids have flash points at or above 200°F (93°C). Motor oil is a typical example of a Class IIIB combustible liquid.

Note g recognizes that the hazard presented by certain materials is such that they may be stored or used only inside buildings that are fully sprinklered.

Note h clarifies for the user that while there is a combination maximum allowable quantity for flammable liquids, no individual class of liquid (Class IA, IB or IC) may exceed its own individual maximum allowable quantity.

Note i is a specific exception for inside storage tanks of combustible liquids that are connected to a fuel-oil piping system in accordance with Section 603.3.2 of the IFC. This exception applies to most oil-fired stationary equipment, whether in industrial, commercial or residential occupancies. NFPA 31 and NFPA 37 provide further guidance on the type of installations this exception is intended to permit. This exception would permit fuel-oil storage tanks containing a maximum of 660 gallons (2498 L) of combustible liquids within a building without being classified as a Group H-3 occupancy. This quantity limitation could be further increased to 3,000 gallons (11 356 L) for combustible liquids stored in protected above-ground tanks in rooms protected by an automatic sprinkler system complying with NFPA 13.

Note k permits a larger amount of Class 3 oxidizers in a building when used for maintenance and sanitation purposes. An example is the use of 51 percent hydrogen peroxide (H_2O_2) during industrial-scale food production operations. In a diluted form, 51 percent hydrogen peroxide is used for disinfecting piping systems and machinery used for manufacturing and packaging dairy products, such as ice cream, cheese and milk, to effectively sterilize all surfaces that are in contact with these milk products. The method used to store the oxidizers is subject to the evaluation and approval of the building official. Note k also provides consistency with Note k of Table 5003.1.1(1) of the IFC.

Note l clarifies that the 125 pounds (57 kg) of storage permitted for consumer fireworks represents the net weight of the pyrotechnic composition of the fireworks in a nonsprinklered building. This amount represents approximately $12^1/_2$ shipping cases (less than $1^1/_2$ pallet loads) of fireworks in a nonsprinklered storage condition. In cases where the net weight of the pyrotechnic composition of the fireworks is unknown, 25 percent of the gross weight of the fireworks is to be used. The gross weight is to include the weight of the packaging.

Note n provides an exception when the amount of hazardous material in storage and display in Group M and S occupancies meet the requirements of Section 414.2.5.

Note o clarifies that densely packed baled cotton is not considered a hazardous material when meeting the size and weight requirements of ISO 8115 and, as such, is not subject to the maximum allowable quantity per control area specified for combustible fibers.

Note p clarifies that vehicles with closed fuel systems should be treated no differently than machinery or equipment when considering the allowable quantities of materials within a building. This note also clarifies that the fuels contained within the fuel tanks of vehicles or motorized equipment are not to be considered when calculating the aggregate quantity of hazardous materials within a control area of a building. For example, when evaluating a parking garage with several hundred cars parked inside, the fuel tanks of vehicles are not counted. When motorized equipment, such as a floor buffer or forklift, is used, those fuels are not included as long as other code requirements are satisfied. This note also exempts alcohol-based hand rubs when contained in dispensers complying with the IFC.

Note q directs code users to Section 414.1.3 in an effort to determine the appropriateness of a Group H-2 classification for facilities where combustible dust is manufactured, generated or used. The presence of combustible dust may or may not result in a high-hazard condition. In order to determine if a potential dust hazard exists that would warrant a Group H-2 occupancy classification, the conditions must be evaluated and a report submitted to the building official that sets forth a determination of the degree of hazard and recommended safeguards, including the appropriate occupancy classification.

TABLE 307.1(2). See page 3-20.

❖ Table 307.1(2), similar to Table 307.1(1), specifies the maximum quantities of hazardous materials, liquids or chemicals allowed per control area before having to classify building, or a portion of a building, as a high-hazard occupancy. Table 307.1(2), as referenced in Section 307.1, contains materials classified as Group H-4 in accordance with Section 307.6. While the materials listed in this table are considered health hazards, some materials may also possess physical hazard characteristics more indicative of materials classified as Group H-1, H-2 or H-3.

The maximum allowable quantities per control area listed in Table 307.1(2) are indicative of industry practice and assume the materials are properly stored and handled in accordance with the IFC. Group H-4 materials, while indeed hazardous, are primarily considered a handling problem and do not possess the same fire, explosion or reactivity hazard associated with other hazardous materials. The base maximum allowable quantity per control area of 810 cubic feet (23 m^3) for gases that are either corrosive or toxic is based on a standard-size chlorine cylinder. The use of 150 pounds (68 kg) as the baseline quantity for liquefied corrosive and toxic gases is intended to be consistent with the philosophical approach to the same maximum quantity permitted for liquefied oxidizing gases in Table 307.1(1). The 150-pound (68 kg) limitation allows a single cylinder of chlorine, which could be considered both a corrosive and oxidizing gas, to not result in either a Group H-3 or H-4 occupancy classification.

Note b clearly indicates that the aggregate quantity of hazardous materials in use and storage, within a given control area, cannot exceed the quantity listed in the table for storage.

Without Note c, many common household products, such as household bleach and window cleaners, containing a negligible amount of a hazardous material could result in a Group M occupancy being classified as a high hazard. Note c recognizes the reduced hazard of the materials based on their water miscibility and limited container size.

Notes d and e are identical to Notes d and e to Table 307.1(1) and allow up to four times the maximum allowed quantities. See the commentary for Notes d and e of Table 307.1(1).

Note f provides an exception when the amount of hazardous material in storage and display in Group M and S occupancies meets the requirements of Section 414.2.5.

Note g is significant in that, for certain materials, their hazard is so great that their maximum allowable quantity per control area may be stored in the building only when approved exhausted enclosures or gas cabinets are utilized.

TABLE 307.1(1)
MAXIMUM ALLOWABLE QUANTITY PER CONTROL AREA OF HAZARDOUS MATERIALS POSING A PHYSICAL HAZARD[a, j, m, n, p]

MATERIAL	CLASS	GROUP WHEN THE MAXIMUM ALLOWABLE QUANTITY IS EXCEEDED	STORAGE[b]			USE-CLOSED SYSTEMS[b]			USE-OPEN SYSTEMS[b]	
			Solid pounds (cubic feet)	Liquid gallons (pounds)	Gas cubic feet at NTP	Solid pounds (cubic feet)	Liquid gallons (pounds)	Gas cubic feet at NTP	Solid pounds (cubic feet)	Liquid gallons (pounds)
Combustible dust	NA	H-2	See Note q	NA	NA	See Note q	NA	NA	See Note q	NA
Combustible fiber[q]	Loose Baled[o]	H-3	(100) (1,000)	NA	NA	(100) (1,000)	NA	NA	(20) (200)	NA
Combustible liquid[c, i]	II IIIA IIIB	H-2 or H-3 H-2 or H-3 NA	NA	120[d, e] 330[d, e] 13,200[e, f]	NA	NA	120[d] 330[d] 13,200[f]	NA	NA	30[d] 80[d] 3,300[f]
Consumer fireworks	1.4G	H-3	125[e, l]	NA	NA	NA	NA	NA	NA	NA
Cryogenic flammable	NA	H-2	NA	45[d]	NA	NA	45[d]	NA	NA	10[d]
Cryogenic inert	NA	NA	NA	NA	NL	NA	NA	NL	NA	NA
Cryogenic oxidizing	NA	H-3	NA	45[d]	NA	NA	45[d]	NA	NA	10[d]
Explosives	Division 1.1 Division 1.2 Division 1.3 Division 1.4 Division 1.4G Division 1.5 Division 1.6	H-1 H-1 H-1 or H-2 H-3 H-3 H-1 H-1	1[e, g] 1[e, g] 5[e, g] 50[e, g] 125[d, e, l] 1[e, g] 1[e, g]	(1)[e, g] (1)[e, g] (5)[e, g] (50)[e, g] NA (1)[e, g] NA	NA	0.25[g] 0.25[g] 1[g] 50[g] NA 0.25[g] NA	(0.25)[g] (0.25)[g] (1)[g] (50)[g] NA (0.25)[g] NA	NA	0.25[g] 0.25[g] 1[g] NA NA 0.25[g] NA	(0.25)[g] (0.25)[g] (1)[g] NA NA (0.25)[g] NA
Flammable gas	Gaseous Liquefied	H-2	NA	NA (150)[d, e]	1,000[d, e] NA	NA	NA (150)[d, e]	1,000[d, e] NA	NA	NA
Flammable liquid[c]	IA IB and IC	H-2 or H-3	NA	30[d, e] 120[d, e]	NA	NA	30[d] 120[d]	NA	NA	10[d] 30[d]
Flammable liquid, combination (IA, IB, IC)	NA	H-2 or H-3	NA	120[d, e, h]	NA	NA	120[d, h]	NA	NA	30[d, h]

(continued)

TABLE 307.1(1)—continued
MAXIMUM ALLOWABLE QUANTITY PER CONTROL AREA OF HAZARDOUS MATERIALS POSING A PHYSICAL HAZARD[a, j, m, n, p]

MATERIAL	CLASS	GROUP WHEN THE MAXIMUM ALLOWABLE QUANTITY IS EXCEEDED	STORAGE[b]			USE-CLOSED SYSTEMS[b]			USE-OPEN SYSTEMS[b]	
			Solid pounds (cubic feet)	Liquid gallons (pounds)	Gas cubic feet at NTP	Solid pounds (cubic feet)	Liquid gallons (pounds)	Gas cubic feet at NTP	Solid pounds (cubic feet)	Liquid gallons (pounds)
Flammable solid	NA	H-3	125[d, e]	NA	NA	125[d]	NA	NA	25[d]	NA
Inert gas	Gaseous	NA	NA	NA	NL	NA	NA	NL	NA	NA
	Liquefied	NA	NA	NA	NL	NA	NA	NL	NA	NA
Organic peroxide	UD	H-1	1[e, g]	(1)[e, g]	NA	0.25[g]	(0.25)[g]	NA	0.25[g]	(0.25)[g]
	I	H-2	5[d, e]	(5)[d, e]		1[d]	(1)[d]		1[d]	(1)[d]
	II	H-3	50[d, e]	(50)[d, e]		50[d]	(50)[d]		10[d]	(10)[d]
	III	H-3	125[d, e]	(125)[d, e]		125[d]	(125)[d]		25[d]	(25)[d]
	IV	NA	NL	NL		NL	NL		NL	NL
	V	NA	NL	NL		NL	NL		NL	NL
Oxidizer	4	H-1	1[g]	(1)[e, g]	NA	0.25[g]	(0.25)[g]	NA	0.25[g]	(0.25)[g]
	3[k]	H-2 or H-3	10[d, e]	(10)[d, e]		2[d]	(2)[d]		2[d]	(2)[d]
	2	H-3	250[d, e]	(250)[d, e]		250[d]	(250)[d]		50[d]	(50)[d]
	1	NA	4,000[e, f]	(4,000)[e, f]		4,000[f]	(4,000)[f]		1,000[f]	(1,000)[f]
Oxidizing gas	Gaseous	H-3	NA	NA	1,500[d, e]	NA	NA	1,500[d, e]	NA	NA
	Liquefied			(150)[d, e]	NA		(150)[d, e]	NA		
Pyrophoric	NA	H-2	4[e, g]	(4)[e, g]	50[e, g]	1[g]	(1)[g]	10[e, g]	0	0
Unstable (reactive)	4	H-1	1[e, g]	(1)[e, g]	10[e, g]	0.25[g]	(0.25)[g]	2[e, g]	0.25[g]	(0.25)[g]
	3	H-1 or H-2	5[d, e]	(5)[d, e]	50[d, e]	1[d]	(1)[d]	10[d, e]	1[d]	(1)[d]
	2	H-3	50[d, e]	(50)[d, e]	750[d, e]	50[d]	(50)[d]	750[d, e]	10[d]	(10)[d]
	1	NA	NL	NL	NL	NL	NL	NL	NL	NL
Water reactive	3	H-2	5[d, e]	(5)[d, e]	NA	5[d]	(5)[d]	NA	1[d]	(1)[d]
	2	H-3	50[d, e]	(50)[d, e]		50[d]	(50)[d]		10[d]	(10)[d]
	1	NA	NL	NL		NL	NL		NL	NL

For SI: 1 cubic foot = 0.028 m^3, 1 pound = 0.454 kg, 1 gallon = 3.785 L.

NL = Not Limited; NA = Not Applicable; UD = Unclassified Detonable.

a. For use of control areas, see Section 414.2.

b. The aggregate quantity in use and storage shall not exceed the quantity listed for storage.

c. The quantities of alcoholic beverages in retail and wholesale sales occupancies shall not be limited provided the liquids are packaged in individual containers not exceeding 1.3 gallons. In retail and wholesale sales occupancies, the quantities of medicines, foodstuffs or consumer products, and cosmetics containing not more than 50 percent by volume of water-miscible liquids with the remainder of the solutions not being flammable, shall not be limited, provided that such materials are packaged in individual containers not exceeding 1.3 gallons.

d. Maximum allowable quantities shall be increased 100 percent in buildings equipped throughout with an *automatic sprinkler system* in accordance with Section 903.3.1.1. Where Note e also applies, the increase for both notes shall be applied accumulatively.

e. Maximum allowable quantities shall be increased 100 percent when stored in approved storage cabinets, day boxes, gas cabinets, gas rooms or exhausted enclosures or in *listed* safety cans in accordance with Section 5003.9.10 of the *International Fire Code*. Where Note d also applies, the increase for both notes shall be applied accumulatively.

f. Quantities shall not be limited in a building equipped throughout with an *automatic sprinkler system* in accordance with Section 903.3.1.1.

g. Allowed only in buildings equipped throughout with an *automatic sprinkler system* in accordance with Section 903.3.1.1.

h. Containing not more than the maximum allowable quantity per *control area* of Class IA, IB or IC flammable liquids.

i. The maximum allowable quantity shall not apply to fuel oil storage complying with Section 603.3.2 of the *International Fire Code*.

j. Quantities in parenthesis indicate quantity units in parenthesis at the head of each column.

k. A maximum quantity of 200 pounds of solid or 20 gallons of liquid Class 3 oxidizers is allowed when such materials are necessary for maintenance purposes, operation or sanitation of equipment when the storage containers and the manner of storage are approved.

l. Net weight of the pyrotechnic composition of the fireworks. Where the net weight of the pyrotechnic composition of the fireworks is not known, 25 percent of the gross weight of the fireworks, including packaging, shall be used.

m. For gallons of liquids, divide the amount in pounds by 10 in accordance with Section 5003.1.2 of the *International Fire Code*.

n. For storage and display quantities in Group M and storage quantities in Group S occupancies complying with Section 414.2.5, see Tables 414.2.5(1) and 414.2.5(2).

o. Densely packed baled cotton that complies with the packing requirements of ISO 8115 shall not be included in this material class.

p. The following shall not be included in determining the maximum allowable quantities:
   1. Liquid or gaseous fuel in fuel tanks on vehicles.
   2. Liquid or gaseous fuel in fuel tanks on motorized equipment operated in accordance with the *International Fire Code*.
   3. Gaseous fuels in piping systems and fixed appliances regulated by the *International Fuel Gas Code*.
   4. Liquid fuels in piping systems and fixed appliances regulated by the *International Mechanical Code*.
   5. Alcohol-based hand rubs classified as Class I or II liquids in dispensers that are installed in accordance with Sections 5705.5 and 5705.5.1 of the *International Fire Code*. The location of the alcohol-based hand rub (ABHR) dispensers shall be provided in the construction documents.

q. Where manufactured, generated or used in such a manner that the concentration and conditions create a fire or explosion hazard based on information prepared in accordance with Section 414.1.3.

[F] TABLE 307.1(2)
MAXIMUM ALLOWABLE QUANTITY PER CONTROL AREA OF HAZARDOUS MATERIAL POSING A HEALTH HAZARD[a, c, f, h, i]

MATERIAL	STORAGE[b]			USE-CLOSED SYSTEMS[b]			USE-OPEN SYSTEMS[b]	
	Solid pounds[d, f]	Liquid gallons (pounds)[d, f]	Gas cubic feet at NTP (pounds)[d]	Solid pounds[d]	Liquid gallons (pounds)[e, d]	Gas cubic feet at NTP (pounds)[d]	Solid pounds[d]	Liquid gallons (pounds)[d]
Corrosives	5,000	500	Gaseous 810[e, f] Liquefied (150)	5,000	500	Gaseous 810[e] Liquefied (150)	1,000	100
Highly Toxic	10	(10)	Gaseous 20[g] Liquefied (4)[g]	10	(10)	Gaseous 20[g] Liquefied (4)[g]	3	(3)
Toxic	500	(500)	Gaseous 810[e] Liquefied (150)[e]	500	(500)	Gaseous 810[e] Liquefied (150)[e]	125	(125)

For SI: 1 cubic foot = 0.028 m^3, 1 pound = 0.454 kg, 1 gallon = 3.785 L.

a. For use of control areas, see Section 414.2.

b. The aggregate quantity in use and storage shall not exceed the quantity listed for storage.

c. In retail and wholesale sales occupancies, the quantities of medicines, foodstuffs or consumer products, and cosmetics containing not more than 50 percent by volume of water-miscible liquids and with the remainder of the solutions not being flammable, shall not be limited, provided that such materials are packaged in individual containers not exceeding 1.3 gallons.

d. Maximum allowable quantities shall be increased 100 percent in buildings equipped throughout with an *approved automatic sprinkler system* in accordance with Section 903.3.1.1. Where Note e also applies, the increase for both notes shall be applied accumulatively.

e. Maximum allowable quantities shall be increased 100 percent where stored in approved storage cabinets, gas cabinets or exhausted enclosures as specified in the *International Fire Code*. Where Note d also applies, the increase for both notes shall be applied accumulatively.

f. For storage and display quantities in Group M and storage quantities in Group S occupancies complying with Section 414.2.5, see Tables 414.2.5(1) and 414.2.5(2).

g. Allowed only where stored in approved exhausted gas cabinets or exhausted enclosures as specified in the *International Fire Code*.

h. Quantities in parenthesis indicate quantity units in parenthesis at the head of each column.

i. For gallons of liquids, divide the amount in pounds by 10 in accordance with Section 5003.1.2 of the *International Fire Code*.

[F] 307.1.2 Hazardous materials. Hazardous materials in any quantity shall conform to the requirements of this code, including Section 414, and the *International Fire Code*.

❖ The use of high-hazard materials must be regulated in accordance with Sections 414 and 415 as well as the applicable requirements of the IFC. While the building may be exempt from a high-hazard occupancy classification (i.e., Group H-1, H-2, H-3, H-4 or H-5), any potential hazard with regard to the use of storage of any hazardous material, regardless of quantity, must be abated.

[F] 307.2 Definitions. The following terms are defined in Chapter 2:

AEROSOL

Level 1 aerosol products.

Level 2 aerosol products.

Level 3 aerosol products.

AEROSOL CONTAINER.

BALED COTTON.

BALED COTTON, DENSELY PACKED.

BARRICADE.

Artificial barricade.

Natural barricade.

BOILING POINT.

CLOSED SYSTEM.

COMBUSTIBLE DUST.

COMBUSTIBLE FIBERS.

COMBUSTIBLE LIQUID.

Class II.

Class IIIA.

Class IIIB.

COMPRESSED GAS.

CONTROL AREA.

CORROSIVE.

CRYOGENIC FLUID.

DAY BOX.

DEFLAGRATION.

DETONATION.

DISPENSING.

EXPLOSION.

EXPLOSIVE.

High explosive.

Low explosive.

Mass-detonating explosives.

UN/DOTn Class 1 explosives.

Division 1.1.

Division 1.2.

Division 1.3.

Division 1.4.

Division 1.5.

Division 1.6.

FIREWORKS.

Fireworks, 1.3G.

Fireworks, 1.4G.

FLAMMABLE GAS.

FLAMMABLE LIQUEFIED GAS.

FLAMMABLE LIQUID.

Class IA.

Class IB.

Class IC.

FLAMMABLE MATERIAL.

FLAMMABLE SOLID.

FLASH POINT.

HANDLING.

HAZARDOUS MATERIALS.

HEALTH HAZARD.

HIGHLY TOXIC.

INCOMPATIBLE MATERIALS.

INERT GAS.

OPEN SYSTEM.

OPERATING BUILDING.

ORGANIC PEROXIDE.

Class I.

Class II.

Class III.

Class IV.

Class V.

Unclassified detonable.

OXIDIZER.

Class 4.

Class 3.

Class 2.

Class 1.

OXIDIZING GAS.

PHYSICAL HAZARD.

PYROPHORIC.

PYROTECHNIC COMPOSITION.

TOXIC.

UNSTABLE (REACTIVE) MATERIAL.

Class 4.

Class 3.

Class 2.

Class 1.

WATER-REACTIVE MATERIAL.

Class 3.

Class 2.

Class 1.

❖ This section lists terms that are specifically associated with the subject matter of this section. It is important to emphasize that these terms are not exclusively related to this section but may or may not also be applicable where the term is used elsewhere in the code.

Definitions of terms can help in the understanding and application of the code requirements. The purpose for including a list within this chapter is to provide more convenient access to terms which may have a specific or limited application within this chapter. For the complete definition and associated commentary, refer back to Chapter 2. Terms that are italicized provide a visual identification throughout the code that a definition exists for that term. The use and application of all defined terms are set forth in Section 201.

[F] 307.3 High-hazard Group H-1. Buildings and structures containing materials that pose a detonation hazard shall be classified as Group H-1. Such materials shall include, but not be limited to, the following:

Detonable pyrophoric materials

Explosives:

Division 1.1

Division 1.2

Division 1.3

Division 1.4

Division 1.5

Division 1.6

Organic peroxides, unclassified detonable

Oxidizers, Class 4

Unstable (reactive) materials, Class 3 detonable and Class 4

❖ The contents of occupancies in Group H-1 present a detonation hazard. Examples of materials that create this hazard are listed in the section. The definitions for Group H-1 materials are listed in Section 307.2 and defined in Chapter 2. Because of the explosion hazard potential associated with Group H-1 materials, occupancies in Group H-1, which exceed the maximum allowable quantity per control area indicated in Table 307.1(1), are required to be located in detached one-story buildings without basements (see commentary, Sections 415.6.2, 415.7 and 508.3). Group H-1 occupancies cannot be located in a mixed occupancy building.

[F] 307.3.1 Occupancies containing explosives not classified as H-1. The following occupancies containing explosive materials shall be classified as follows:

1. Division 1.3 explosive materials that are used and maintained in a form where either confinement or configuration will not elevate the hazard from a mass fire

to mass explosion hazard shall be allowed in H-2 occupancies.

2. Articles, including articles packaged for shipment, that are not regulated as a Division 1.4 explosive under Bureau of Alcohol, Tobacco, Firearms and Explosives regulations, or unpackaged articles used in process operations that do not propagate a detonation or deflagration between articles shall be allowed in H-3 occupancies.

❖ There are certain explosive materials that pose a hazard level less than that anticipated for a Group H-1 occupancy. A Group H-2 classification is permitted for Division 1.3 explosive materials used or maintained under conditions where the hazard level will not rise from that of a mass fire hazard to a mass explosion hazard. A Group H-3 occupancy classification is permitted for packaged and unpackaged articles not regulated as Division 1.4 explosives by the Bureau of Alcohol, Tobacco and Firearms, as well as unpackaged articles used in process operations, provided there is no concern regarding the propagation of a detonation or deflagration between the articles during process operations.

[F] 307.4 High-hazard Group H-2. Buildings and structures containing materials that pose a deflagration hazard or a hazard from accelerated burning shall be classified as Group H-2. Such materials shall include, but not be limited to, the following:

Class I, II or IIIA flammable or combustible liquids that are used or stored in normally open containers or systems, or in closed containers or systems pressurized at more than 15 pounds per square inch gauge (103.4 kPa).
Combustible dusts where manufactured, generated or used in such a manner that the concentration and conditions create a fire or explosion hazard based on information prepared in accordance with Section 414.1.3.
Cryogenic fluids, flammable.
Flammable gases.
Organic peroxides, Class I.
Oxidizers, Class 3, that are used or stored in normally open containers or systems, or in closed containers or systems pressurized at more than 15 pounds per square inch gauge (103 kPa).
Pyrophoric liquids, solids and gases, nondetonable.
Unstable (reactive) materials, Class 3, nondetonable.
Water-reactive materials, Class 3.

❖ The contents of occupancies in Group H-2 present a deflagration or accelerated burning hazard. Examples of materials that create this hazard are listed. The definitions for Group H-2 materials are contained in Section 202. Because of the severe fire or reactivity hazard associated with these types of materials, proper classification is essential in determining the applicable requirements with regard to the mitigation of these hazards.

[F] 307.5 High-hazard Group H-3. Buildings and structures containing materials that readily support combustion or that pose a physical hazard shall be classified as Group H-3. Such materials shall include, but not be limited to, the following:

Class I, II or IIIA flammable or combustible liquids that are used or stored in normally closed containers or systems pressurized at 15 pounds per square inch gauge (103.4 kPa) or less.
Combustible fibers, other than densely packed baled cotton, where manufactured, generated or used in such a manner that the concentration and conditions create a fire or explosion hazard based on information prepared in accordance with Section 414.1.3.
Consumer fireworks, 1.4G (Class C, Common)
Cryogenic fluids, oxidizing
Flammable solids
Organic peroxides, Class II and III
Oxidizers, Class 2
Oxidizers, Class 3, that are used or stored in normally closed containers or systems pressurized at 15 pounds per square inch gauge (103 kPa) or less
Oxidizing gases
Unstable (reactive) materials, Class 2
Water-reactive materials, Class 2

❖ The contents of occupancies in Group H-3 present a hazard inasmuch as they contain materials that readily support combustion or that present a physical hazard. Examples of materials that create this hazard are listed. The definitions for Group H-3 materials are contained in Section 202. While Group H-3 materials are generally less of a fire or reactivity hazard than Group H-2 materials, they still present a greater physical hazard than materials not currently regulated as high hazard.

[F] 307.6 High-hazard Group H-4. Buildings and structures containing materials that are health hazards shall be classified as Group H-4. Such materials shall include, but not be limited to, the following:

Corrosives
Highly toxic materials
Toxic materials

❖ The contents of occupancies in Group H-4 present a hazard inasmuch as they contain materials that are health hazards. Examples of these hazards are listed in this section. The definitions for Group H-4 materials are contained in Section 202. While reference is made to chemicals that cause these hazards, the data sheets for these chemicals, which are furnished by the applicant, will need considerable subjective evaluation. Some materials falling into the category of health hazard may also present a physical hazard and would, therefore, require the structure to be designed for multiple hazards in accordance with Section 307.8.

[F] 307.7 High-hazard Group H-5. Semiconductor fabrication facilities and comparable research and development areas in which hazardous production materials (HPM) are used and the aggregate quantity of materials is in excess of those listed in Tables 307.1(1) and 307.1(2) shall be classified as Group H-5. Such facilities and areas shall be designed and constructed in accordance with Section 415.10.

❖ HPM includes flammable liquids and gases, corrosives, oxidizers and, in many instances, highly toxic materials (see the definition for "Hazardous production material" in Section 202). In determining the applicable requirements of other sections of the code, HPM facilities are considered to be Group H-5 occupancies. It is intended that the quantities of materials permitted in Table 415.11.1.1.1 will take precedence over Tables 307.1(1) and 307.1(2).

[F] 307.8 Multiple hazards. Buildings and structures containing a material or materials representing hazards that are classified in one or more of Groups H-1, H-2, H-3 and H-4 shall conform to the code requirements for each of the occupancies so classified.

❖ If materials are present that possess characteristics of more than one Group H, high-hazard occupancy, then the structure must be designed to protect against the hazards of each relevant high-hazard occupancy classification. For example, a material could be classified as both a Class 2 oxidizer (Group H-3) and a corrosive (Group H-4). If the given quantity exceeded the maximum allowable quantity per control area individually for both a Class 2 oxidizer and a corrosive, the structure is required to conform to the applicable requirements of both Groups H-3 and H-4.

SECTION 308
INSTITUTIONAL GROUP I

308.1 Institutional Group I. Institutional Group I occupancy includes, among others, the use of a building or structure, or a portion thereof, in which care or supervision is provided to persons who are or are not capable of self-preservation without physical assistance or in which persons are detained for penal or correctional purposes or in which the liberty of the occupants is restricted. Institutional occupancies shall be classified as Group I-1, I-2, I-3 or I-4.

❖ Institutional occupancies are composed of two basic types. The first includes Groups I-1, I-2 and I-4 and relates to facilities where personal care, custodial care or medical care is provided for people who due to age, physical limitations, diseases, mental disabilities or other infirmities need a supervised environment (see the commentary for the definitions of "Personal care," "Custodial care" and "Medical care"). This includes persons who are ambulatory and are capable of self-preservation as well as those who are restricted in their mobility or are totally immobile to the extent that they are incapable of self-preservation and therefore may need assistance to evacuate during an emergency situation, such as a fire. The IFC also addresses the idea of a defend-in-place protection option for hospitals and nursing homes (see the commentary for the definition of "Incapable of self-preservation" and the IFC Chapter 4 Commentary). The second type, Group I-3, relates primarily to detention and correctional facilities. Since security is the major operational consideration in these kinds of facilities, the occupants (inmates) are under some form of supervision and restraint and may be rendered incapable of self-preservation without direct intervention from staff in emergency situations due to locked cells and exits.

The degree of hazards in each type of institutional facility identified in this section varies respective to each kind of occupancy. The code addresses each occupancy separately and the regulatory provisions throughout the code provide the proper means of protection so as to produce an acceptable level of safety to life and property.

Groups I-1, I-2 and I-3 are further divided into "conditions" relative to unique aspects of the respective occupancies. Groups I-1 and R-4 are closely related and are primarily distinguished by the number of persons residing in the facility (see Section 308.3).

Another of the distinguishing characteristics between the different Group I occupancies and other occupancies is when care is provided for a length of time exceeding 24 hours. The intent is that these criteria are not specific to the hours of operation of the facility, but the length of time that care is provided for the patients, residents or those in day care. For example, an outpatient clinic that is open 24 hours a day is a Group B occupancy provided care recipients are treated as outpatients and there are no inpatients who would stay at the facility 24 hours or longer. Another example would be a "day care" facility that is open 24 hours to serve workers who work any shift and need to have children in "day care" while they work. Provided that individual children receive care for less than 24 hours, the occupancy would be classified as a Group I-4 or possible a Group E.

Each individual facility will have unique characteristics or a combination of characteristics which should be considered when classifying it to one occupancy and condition or another. For example, some of the newer care facilities are offering a combination of care levels to allow for persons to age in place within the same complex. A facility could easily have a mix of occupancies such as one wing providing full-time nursing care (Group I-2, Condition 1); a second wing with assisted living care for residents with dementia who may need some direct physical contact from staff to react to an emergency (Group I-1, Condition 2); and yet a third wing with custodial care where residents are capable of responding to an emergency on their own (Group I-1, Condition 1). Health care and custodial care facilities are subject to many state and local regulations. Such regulations may factor in which IBC occupancy classification is appropriate. Commentary Figure 308.1 provides a summary of care facility classifications.

24-HOUR CARE					
Type of care	**Capability of patients receiving care**	**Types of facilities**	**1-5 Occupants**	**6-16 Occupants**	**More than 16 occupants**
Custodial	Capable of responding to emergency situation and complete building evacuation without assistance	Alcohol & drug center; assisted living; congregate care; group home; halfway house; residential board and care; social rehabilitation	R-3[a] Sec: 308.3.4	R-4, Condition 1 Sec. 308.3.3 & 310.6.1	I-1 Condition 1 Sec. 308.3.1
Custodial	Any residents who require limited verbal or physical assistance while responding to emergency situation	Alcohol & drug center; assisted living; congregate care; group home; halfway house; residential board and care; social rehabilitation	R-3[a] Sec: 308.3.4	R-4, Condition 2 Sec. 308.3.3 & 310.6.2	I-1 Condition 2 Sec. 308.3.2
Medical	Incapable of self-preservation	Nursing homes; foster care facilities; facilities providing nursing and medical care but without emergency, surgery or obstetric services or inpatient stabilization for psychiatric or detoxification.	R-3[a] Sec: 308.4.2	I-2 Condition 1 Sec. 308.4.1.1	I-2 Condition 1 Sec. 308.4.1.1
		Hospitals; facilities providing nursing and medical care including emergency, surgery or obstetric services or inpatient stabilization for psychiatric or detoxification.	R-3[a] Sec: 308.4.2	I-2 Condition 2 Sec. 308.4.1.2	I-2 Condition 2 Sec. 308.4.1.2
LESS THAN 24-HOUR CARE					
Type of Care or Service	**Age and Capability of occupants**	**Types of Facilities**	**1-5 In a dwelling unit**	**1-5 occupants**	**6 or more occupants**
Medical	Any age—Capable of self-preservation	Outpatient clinic; doctor's office	NA	B Sec, 304.1	B Sec. 304.1
Medical, surgical, psychiatric, nursing	Any age—Rendered incapable of self-preservation	Ambulatory care facility	NA	B Sec. 304.1	B Sec. 304.1
Educational, supervisory or personal care services	Older than $2^1/_2$ years and 12th grade or younger	Day care, Child care	R-3 Sec, 305.2.3	Same of primary occupancy of Building Sec. 305.2.1 & 305.2.2	E[b] Sec. 305.2
Custodial care	$2^1/_2$ years or less and older where incapable of self-preservation	Day care, Adult care	R-3 Sec. 308.6.4	Same of primary occupancy of Building Sec. 308.6.2 & 308.6.3	I-4[b,c] Sec, 308.6

a. This is the option of complying with the *International Residential Code* providing an NFPA 13D automatic sprinkler system installed in accordance with IBC Section 903.3.1.3.
b. Rooms or spaces within places of religious worship; classified as part of primary occupancy—usually A-3.
c. Floors in day care facilities where more than five but no more than 100 infants and toddlers (i.e., children $2^1/_2$ years or younger in age) can be classified as Group E provided rooms used as such are on the level of exit discharge and have exit door directly to exterior.

Figure 308.1
OCCUPANCY CLASSIFICATION OF CARE FACILITIES

308.2 Definitions. The following terms are defined in Chapter 2:

24-HOUR BASIS.

CUSTODIAL CARE.

DETOXIFICATION FACILITIES.

FOSTER CARE FACILITIES.

HOSPITALS AND PSYCHIATRIC HOSPITALS.

INCAPABLE OF SELF-PRESERVATION.

MEDICAL CARE.

NURSING HOMES.

❖ This section lists terms that are specifically associated with the subject matter of this section. It is important to emphasize that these terms are not exclusively related to this section but may or may not also be applicable where the term is used elsewhere in the code.

Definitions of terms can help in the understanding and application of the code requirements. The purpose for including a list within this chapter is to provide more convenient access to terms which may have a specific or limited application within this chapter. For the complete definition and associated commentary, refer back to Chapter 2. Terms that are italicized provide a visual identification throughout the code that a definition exists for that term. The use and application of all defined terms are set forth in Section 201.

308.3 Institutional Group I-1. Institutional Group I-1 occupancy shall include buildings, structures or portions thereof for more than 16 persons, excluding staff, who reside on a 24-hour basis in a supervised environment and receive custodial care. Buildings of Group I-1 shall be classified as one of the occupancy conditions specified in Section 308.3.1 or 308.3.2. This group shall include, but not be limited to, the following:

Alcohol and drug centers
Assisted living facilities
Congregate care facilities
Group homes
Halfway houses
Residential board and care facilities
Social rehabilitation facilities

❖ Groups I-4 and R-4 are similar facilities that differ only by the number of residents receiving care. Groups I-1 and R-4 occupancies are based on three characterizations described in the occupancy classification: custodial care is provided; there is 24-hour-a-day supervision; and they are either Condition 1 or Condition 2. The difference is the number of persons receiving car and residing in such facilities. Group I-1 has more than 16 residents while Group R-4 has six to 16 persons. Note that Group I-1 and R-4 occupancies are limited facilities where custodial care is provided and not where medical care is provided. See the commentary to Section 202 Definitions, "Custodial care" and "Medical care." Groups I-1 and R-4 occupancies list the same eight generic uses as examples which fall under the Group I-1/R-4 umbrella. Of these eight, only "Group home" is defined (see commentary in Chapter 2). Some of these terms may be used in state and local regulations of care facilities. Caution should be taken before assuming that a state-defined "assisted living" facility should be classified under the IBC as a Group I-1 or R-4.

Both Groups I-1 and R-4 include "conditions" to cover the variety of acuity and ability levels of custodial care recipients. Group I-1/R-4, Condition 1 match requirements for previous editions of the code for Group I-1 and R-4, before conditions were included. The intent of the conditions was to address concerns that some residents may need limited assistance or verbal direction to evacuate. The building protection offered for Group I-1/R-4 in previous editions of the code is maintained in Condition 1. Some additional requirements were added for Condition 2. Note that this is custodial care. Where nursing care is provided, the facility is a Group I-2, Condition 1. The Condition 1 care recipients may be slower during evacuation, but all are capable of emergency evacuation without any physical assistance from others. However, they require minor verbal cues from others during emergencies, as might be expected in the general population. Condition 2 custodial care recipients are also slower to evacuate and include any care recipients who may require limited assistance during evacuation. Group I-1/R4, Condition 2 integrates additional protection features, such as smoke barriers to subdivide the building as well as increased automatic sprinkler requirements.

In Group I-1/R-4, Condition 2 facilities, assistance with evacuation can occur because of care recipients' physical or mental limitations, or both. The Condition 2 assistance with evacuation includes help getting out of bed and into a wheelchair or to a walker, or help initiating ambulation. It includes continued physical assistance getting out of the building from a sleeping room, apartment, or other rooms during an emergency. Assistance with evacuation includes assisting persons who may have resistance or confusion in response to an alarm, or require help with instructions. It can also include help for persons with short periods of impaired consciousness intermittently due to medications or illness. Custodial care Group I-1/R-4, Condition 2 evacuation assistance does not include moving occupants in beds or stretchers during emergencies, as is allowed in Group I-2 medical care.

How individual state licensing agencies name, classify and regulate many of the uses listed in Groups I-1 and R-4 vary significantly from state to state and may not correlate with the IBC classifications. It is for this reason that the Groups I-1 and R-4 list of uses is included under the general occupancy classifications and not under each "condition." The building permit applicant should confirm how the specific state licensing regulations correlate to the code's care type, occupancy, condition, evacuation capability, and number of persons receiving care. The permit application drawings should identify the five criteria, while specifically noting that the state licensing regulations limit occupants to only include Condition 1 criteria, or allow Condition 2 criteria. Most assisted living facilities and many residential board and care facilities will be classified as Group I-1, Condition 1 or Group R-4, Condition 1. Generally, almost all specially designated Alzheimer's/memory care facilities providing custodial care will be classified as Group I-1, Condition 2 or Group R-4, Condition 2, due to the inability of some residents to recognize how to respond to an emergency situation. Note that nursing facilities with specialized dementia wings that provide medical care would be classified as Group I-2, Condition 1. Also, it is important to keep in mind that facilities that may be classified initially as Group I-1, Condition 1 (capable of self-preservation) or Group R-4, Condition 1 can very easily need to be reclassified as a Group I-1/R-4, Condition 2 or as a Group I-2, Condition 1 if the abilities of the persons receiving care change over time. Therefore, it is essential for the proponents of a new facility to provide to the building official information regarding the full range of patients and residents expected at a facility both initially and over time.

The occupant load for occupancy classification purposes refers to the number of care recipients only. The number of guests or staff is not included. Note however, that the number of guests and staff is included for means of egress purposes.

For clarification purposes, a dormitory or apartment complex that houses only elderly people and has a nonmedically trained live-in manager is not classified as an institutional occupancy but rather as a residential occupancy (see Section 310). A critical phrase in the code to consider when evaluating this type of facility is "live in a supervised residential environment." Such dormitories or apartment complexes may contain features such as special emergency call switches that are located in each dwelling unit and monitored by health center staff. These emergency call switches are a convenience and do not necessarily indicate infirmity of the care recipients.

308.3.1 Condition 1. This occupancy condition shall include buildings in which all persons receiving custodial care who, without any assistance, are capable of responding to an emergency situation to complete building evacuation.

❖ See the commentary to Section 308.3.

308.3.2 Condition 2. This occupancy condition shall include buildings in which there are any persons receiving custodial care who require limited verbal or physical assistance while responding to an emergency situation to complete building evacuation.

❖ See the commentary to Section 308.3.

308.3.3 Six to 16 persons receiving custodial care. A facility housing not fewer than six and not more than 16 persons receiving custodial care shall be classified as Group R-4.

❖ Any building that has the characteristics of a Group I-1 occupancy but has more than five and not more than 16 persons receiving custodial care is classified as Group R-4 (see Section 310.6). Ninety- eight percent of households in the U.S. have less than 16 occupants, thus the limit of 16 is considered appropriate for a residential occupancy. Similar to Group I-1, Group R-4 is also divided into Conditions 1 and 2.

308.3.4 Five or fewer persons receiving custodial care. A facility with five or fewer persons receiving custodial care shall be classified as Group R-3 or shall comply with the *International Residential Code* provided an *automatic sprinkler system* is installed in accordance with Section 903.3.1.3 or Section P2904 of the *International Residential Code*.

❖ Any building that has the characteristics of a Group I-1 occupancy (Condition 1 or 2, or both) but has five or fewer persons receiving custodial care is classified as Group R-3 (see Section 310.5) or may be constructed in accordance with the *International Residential Code®* (IRC®) (see Section 310.5.1). When the code allows compliance in accordance with the IRC, the only requirements that would apply would be those of the IRC, including the installation of automatic sprinkler protection. The intent is to allow persons to be cared for in a residential, or home, environment, often by family members. Please note similar provisions for Group E occupancies as well as Groups I-2 and I-4.

308.4 Institutional Group I-2. Institutional Group I-2 occupancy shall include buildings and structures used for *medical care* on a 24-hour basis for more than five persons who are *incapable of self-preservation*. This group shall include, but not be limited to, the following:

Foster care facilities
Detoxification facilities
Hospitals
Nursing homes
Psychiatric hospitals

❖ An occupancy classified in Group I-2 is characterized by three conditions: it is a health care facility where the level of care offered is medical care; there is 24-hour-a-day medical supervision for the individuals receiving care; and patients/residents require physical assistance by staff or others to reach safety in an emergency situation (see the definitions for "Custodial care," "Medical care," "24-hour basis" and the five facility examples listed). Where a facility offers medical care instead of custodial care, it is assumed that residents may not be capable of self-preservation. This assessment of the level of care provided needs to be taken with caution, and reliance on other state and federal guidelines and associated regulations may be necessary. Also, it is important to keep in mind that facilities that may be classified initially as

Group I-1, Condition 1 (capable of self-preservation) or Group R-4, Condition 1 can very easily need to be reclassified as a Group I-1/R-4, Condition 2 or as a Group I-2, Condition 1 if the abilities of the persons receiving care change over time. Therefore, it is essential for the proponents of a new facility to provide to the building official information regarding the full range of patients and residents expected at a facility both initially and over time.

Due to the diversification of how medical care is provided in the five characteristic occupancies currently specified in the IBC for Group I-2 occupancies, the Group I-2 occupancy has been split into two basic conditions: Condition 1, nursing homes and foster care; and Condition 2, short-term care (hospitals). Although both of these subsets are based on medical treatment and are occupancies within which the occupants are protected with a defend-in-place method of safety, changes in the delivery of care in the two different conditions has changed in the past 10 to 20 years. Some examples of these changes include:

- Within hospitals, there has been a general increase in the floor area per patient due to the increase in diagnostic equipment and the movement toward single-occupant patient rooms.
- Within nursing homes, there has been a trend to provide more residential-type accommodations, such as group/suite living, gathering areas and cooking facilities in residential areas.

The most common examples of facilities classified in Group I-2 are hospitals (Condition 2) and nursing homes (Condition 1) [see Commentary Figures 308.4(1) and 308.4(2)]. Other facilities included are detoxification facilities, foster care facilities and psychiatric hospitals. How state licensing agencies name, classify and regulate many of the uses listed in Group I-2 varies from state to state and may not correlate with the IBC classifications. It is for this reason that the Group I-2 list of uses is included under the general occupancy classification and not under each "condition."

The benefit to the "condition" concept is that a majority of code requirements will still apply to all Group I-2 occupancies as well as allow the provision of specific code requirements that apply to different levels of care and acuity that are found in different facility types, thus allowing the establishment of specific code requirements that are based on the operation of the facility. See Commentary Figure 308.1.

It is not uncommon to find dining rooms (Group A-2), staff offices (Group B), gift shops (Group M), laundries (Group F) and other nonmedically related areas in buildings otherwise classified as Group I-2. Where such other occupancies occur, the building is considered as a mixed occupancy and subject to the provisions of Section 508. In addition to the general requirements contained in this section, Section 407 contains specific requirements for Group I-2.

308.4.1 Occupancy conditions. Buildings of Group I-2 shall be classified as one of the occupancy conditions specified in Section 308.4.1.1 or 308.4.1.2.

❖ Sections 308.4.1.1 and 308.4.1.2 provide the distinction between the Condition 1 and Condition 2 for the Group I-2 occupancy. Section 407 provides many requirements which apply to both conditions under the Group I-2 occupancy.

308.4.1.1 Condition 1. This occupancy condition shall include facilities that provide nursing and medical care but do

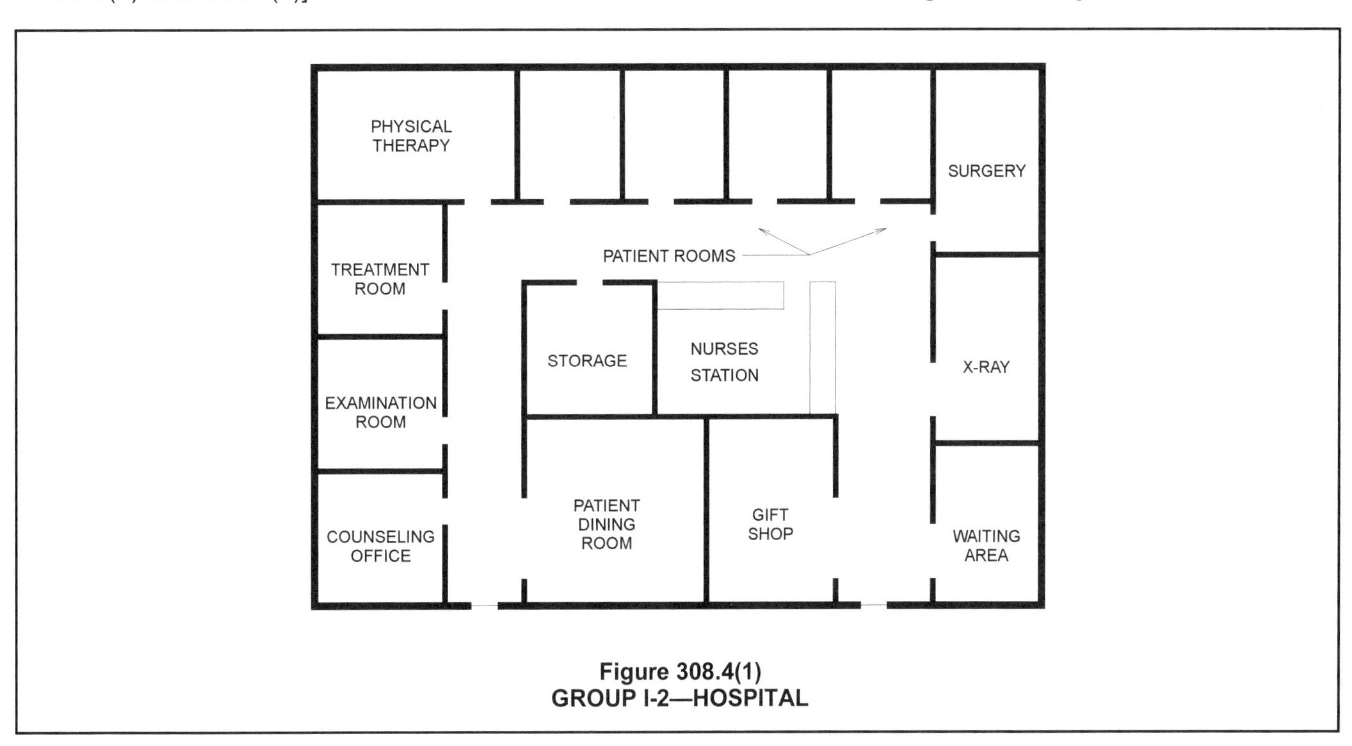

Figure 308.4(1)
GROUP I-2—HOSPITAL

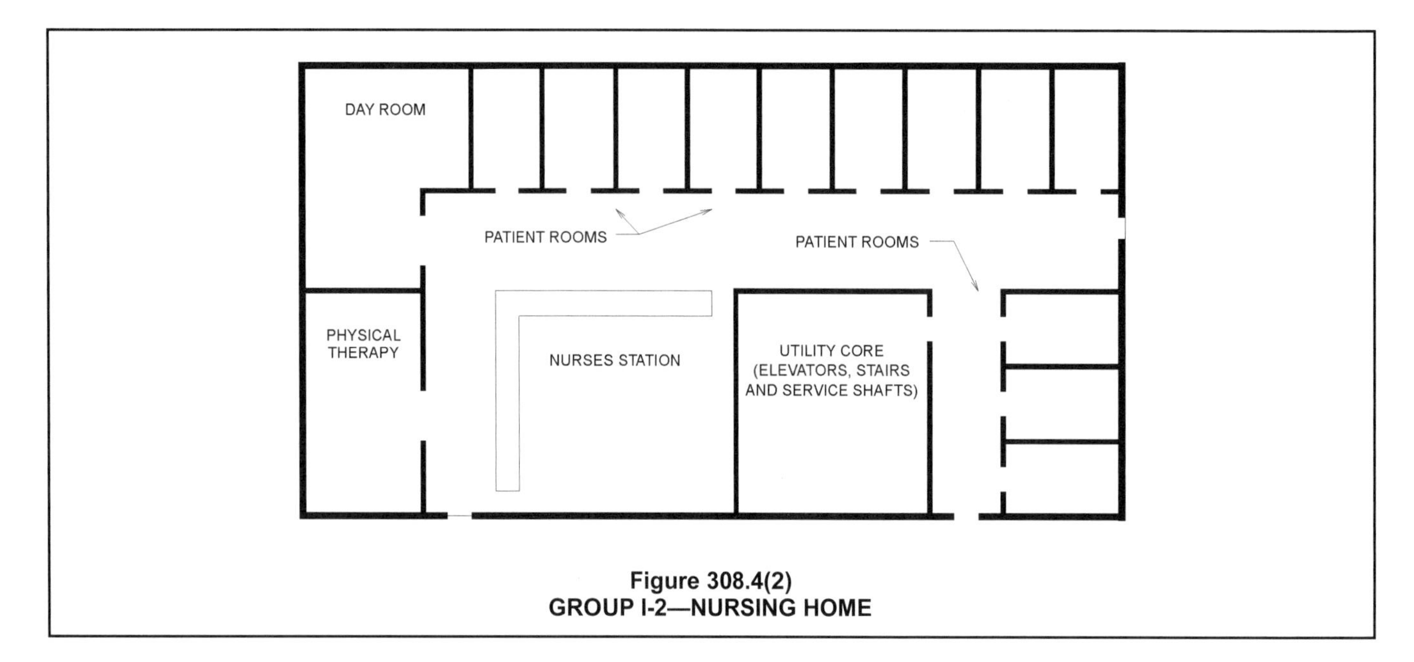

Figure 308.4(2)
GROUP I-2—NURSING HOME

not provide emergency care, surgery, obstetrics or in-patient stabilization units for psychiatric or detoxification, including but not limited to nursing homes and foster care facilities.

❖ The principal use in this category is nursing homes, which are typically facilities providing long-term medical care but not the types of care typically found in hospitals. Foster care facilities, by the Chapter 2 definition, are where children up to the age of $2^1/_2$ receive care. Foster care for more than five infants and toddlers would also be classified as Group I-2, Condition 2. Foster care facilities that provide care for five or more children older than age $2^1/_2$ would typically be Group I-1/R-4, Condition 1 based on supervised living and capability of the residents.

308.4.1.2 Condition 2. This occupancy condition shall include facilities that provide nursing and medical care and could provide emergency care, surgery, obstetrics or in-patient stabilization units for psychiatric or detoxification, including but not limited to hospitals.

❖ Hospitals and psychiatric hospitals are both included in the Condition 2 category of the Group I-2 occupancy. Treatment is usually for periods longer than 24 hours and provides medical care not typically available in long-term nursing care facilities.

308.4.2 Five or fewer persons receiving medical care. A facility with five or fewer persons receiving medical care shall be classified as Group R-3 or shall comply with the *International Residential Code* provided an *automatic sprinkler system* is installed in accordance with Section 903.3.1.3 or Section P2904 of the *International Residential Code*.

❖ Any facility that has the characteristics of a Group I-2 occupancy but does not have more than five persons receiving care at any one time is to be classified as a Group R-3 occupancy (see Section 310.5). As an option, the facility may be designed and constructed under the provisions of the IRC, provided the building has a sprinkler system (see Section 310.5.1). The intent is to allow persons to be cared for in a residential, or home, environment, often under the care of family members. The persons receiving the care do not need to be capable of self-preservation. The sprinkler system is to comply with the requirements for an NFPA 13D system or those of Section P2904 of the IRC. Please note similar provisions for Group E occupancies as well as Groups I-1 (see Section 308.3.4) and I-4 (See Section 308.6.4). See Commentary Figure 308.1.

308.5 Institutional Group I-3. Institutional Group I-3 occupancy shall include buildings and structures that are inhabited by more than five persons who are under restraint or security. A Group I-3 facility is occupied by persons who are generally *incapable of self-preservation* due to security measures not under the occupants' control. This group shall include, but not be limited to, the following:

Correctional centers
Detention centers
Jails
Prerelease centers
Prisons
Reformatories

Buildings of Group I-3 shall be classified as one of the occupancy conditions specified in Sections 308.5.1 through 308.5.5 (see Section 408.1).

❖ An occupancy classified in Group I-3 is characterized by three conditions: it is a location where persons are under restraint or where security is closely supervised; there are more than five such persons; and they are not capable of self-preservation because the conditions of confinement are not under their control (i.e., they require assistance by the facilities' staff to reach safety in an emergency situation). For occupancy classification purposes, the provision refers to the number of persons being secured or restrained only. The number of guests or staff is not included. Please note, however, that the number of guests and staff is included for means of egress purposes.

Buildings that have these characteristics but that contain no more than five persons who are being secured or restrained are to be classified based upon the function to which they are associated. For example, the small holding cell in a Group B police station having only the one cell would simply be classified as a portion of the Group B occupancy. Regardless of the occupancy classification, the means of egress provisions for places of restraint are still applicable (see Chapter 10).

It is recognized that not all Group I-3 occupancies have the same level of restraint. Thus, to distinguish these different levels, the code defines five different conditions of occupancy based on the degree of access to the exit discharge.

The five occupancy conditions are summarized in Commentary Figure 308.5

308.5.1 Condition 1. This occupancy condition shall include buildings in which free movement is allowed from sleeping areas, and other spaces where access or occupancy is permitted, to the exterior via *means of egress* without restraint. A Condition 1 facility is permitted to be constructed as Group R.

❖ Condition 1 areas are those where the secured persons have unrestrained access to the exterior of the building. As such, a key or remote control release device is not needed for any occupant to reach the exterior of the building (exit discharge) at any time. These types of buildings are referred to as low-security facilities. A work-release center is a typical Condition 1 facility (see Commentary Figure 308.5.1).

Because of the lack of restraint associated with a Condition 1 building, it resembles a residential use more than a detention facility and, therefore, is permitted to be classified in Group R (see Section 310).

308.5.2 Condition 2. This occupancy condition shall include buildings in which free movement is allowed from sleeping areas and any other occupied *smoke compartment* to one or more other *smoke compartments*. Egress to the exterior is impeded by locked *exits*.

❖ Condition 2 areas are those in which the movement of the persons under restraint is not controlled within the exterior walls of the building (i.e., the occupants have unrestrained access within the building). As such, there is free movement by such persons between smoke compartments (as created by smoke barriers); however, the restrained persons must rely on someone else to allow them to exit the building to the area of discharge. This is illustrated in Commentary Figure 308.5.2.

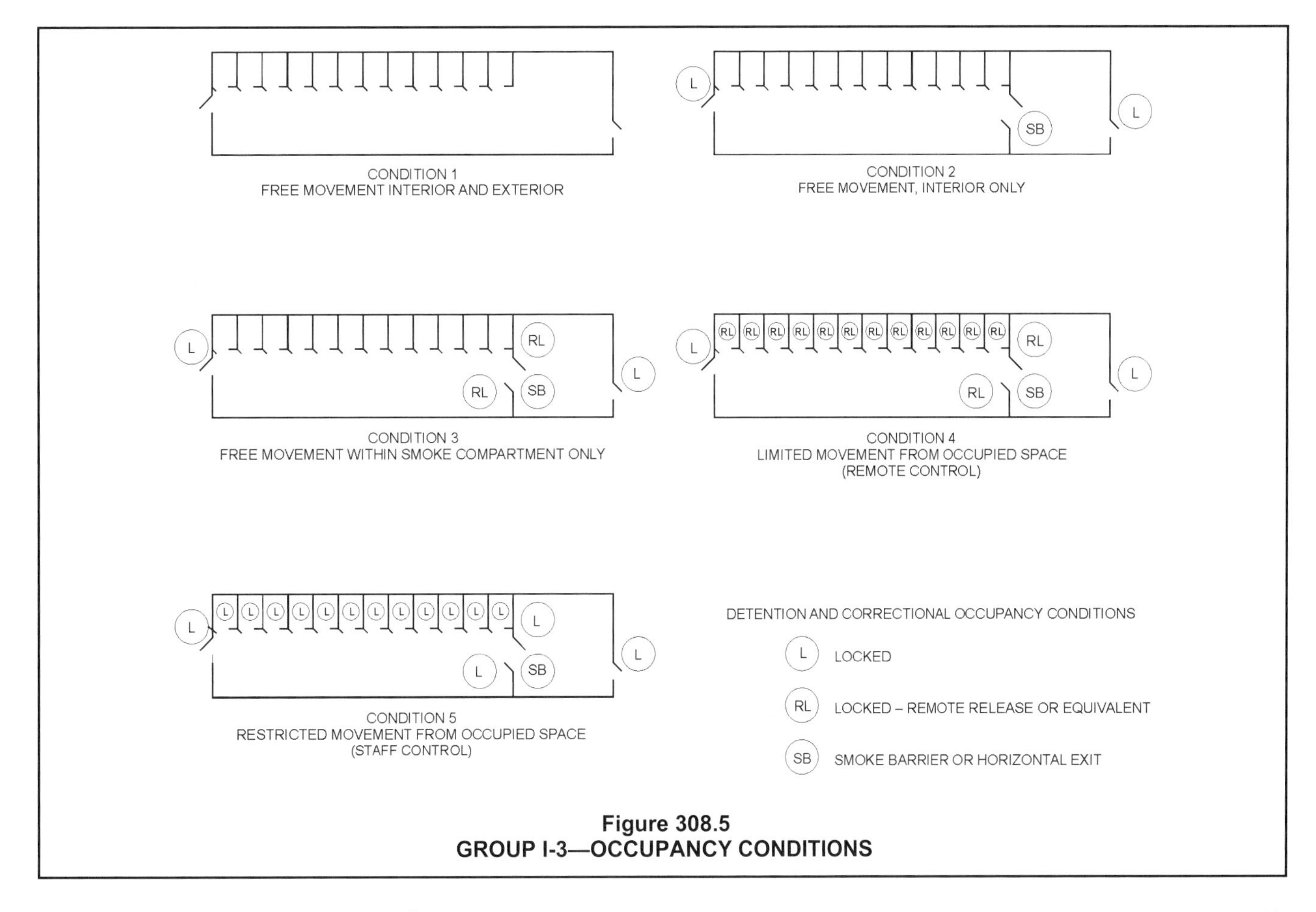

Figure 308.5
GROUP I-3—OCCUPANCY CONDITIONS

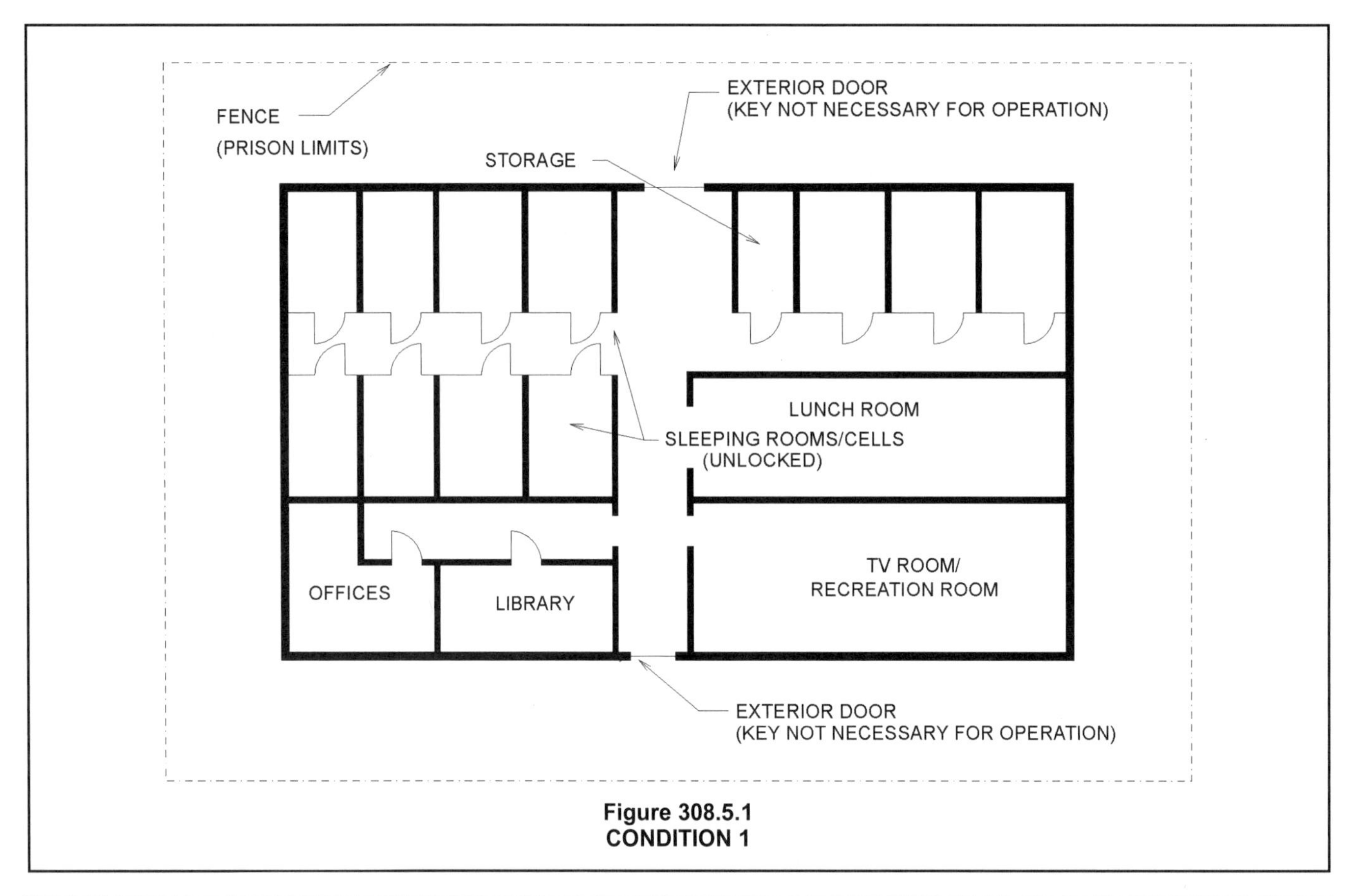

Figure 308.5.1
CONDITION 1

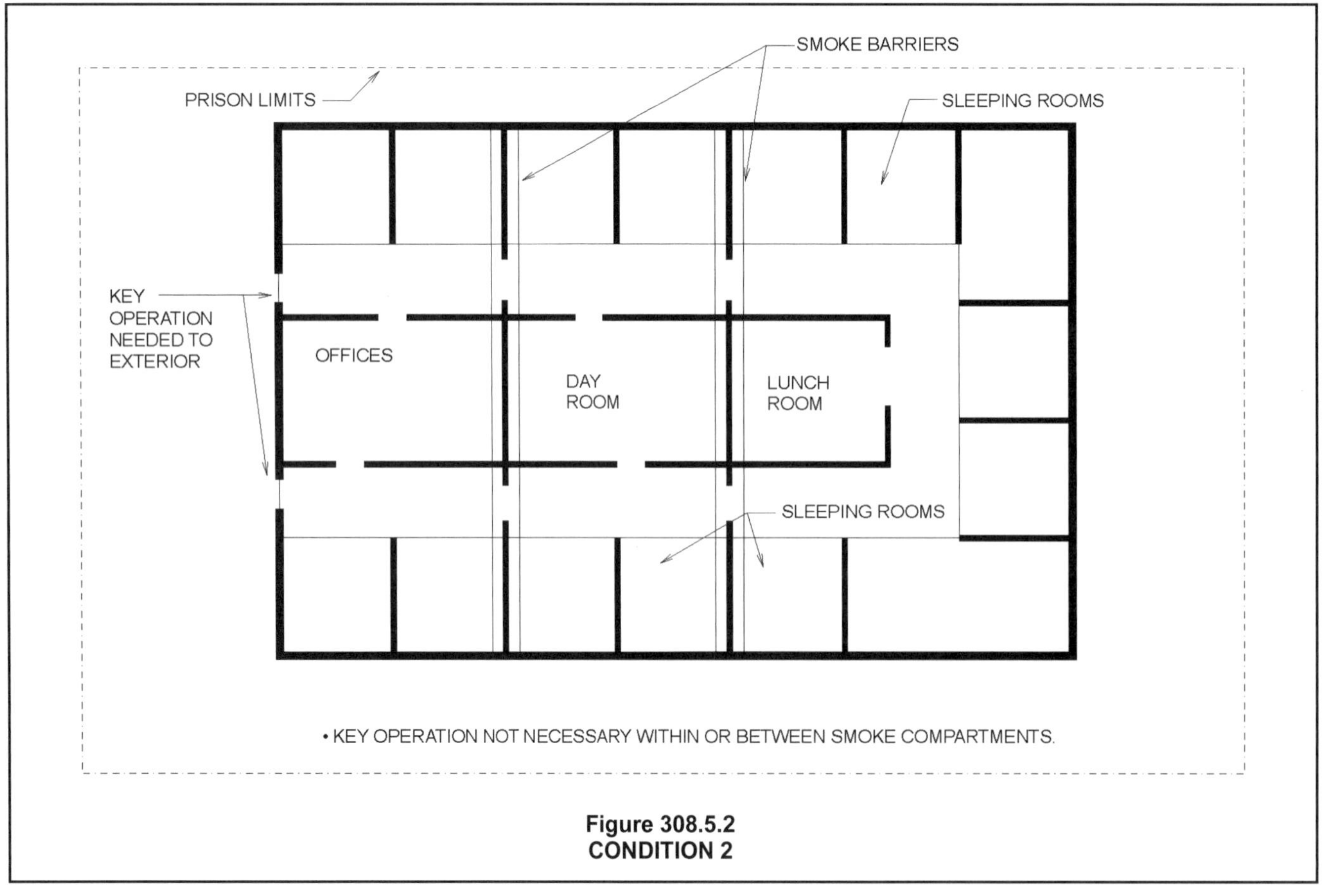

Figure 308.5.2
CONDITION 2

308.5.3 Condition 3. This occupancy condition shall include buildings in which free movement is allowed within individual *smoke compartments*, such as within a residential unit comprised of individual *sleeping units* and group activity spaces, where egress is impeded by remote-controlled release of *means of egress* from such a *smoke compartment* to another *smoke compartment*.

❖ Condition 3 areas are those in which free movement by the persons under restraint is permitted within an individual smoke compartment. However, movement of such persons from one smoke compartment (as created by smoke barriers) to another smoke compartment and from within the building to the exterior (exit discharge) is controlled by remote release locking devices. As such, the restrained persons in the facility are dependent on the staff for their release from each smoke compartment or to the exterior (exit discharge). This condition is illustrated in Commentary Figure 308.5.3.

308.5.4 Condition 4. This occupancy condition shall include buildings in which free movement is restricted from an occupied space. Remote-controlled release is provided to permit movement from *sleeping units*, activity spaces and other occupied areas within the *smoke compartment* to other *smoke compartments*.

❖ Condition 4 areas are those in which the movement of restrained persons from any room or space within a smoke compartment (as created by smoke barriers) to another smoke compartment or to the exterior (exit discharge) is controlled by remote-release locking devices. Any movement within the facility requires activation of a remote-control lock system to release the designated area (see Commentary Figure 308.5.4). The persons being restrained or secured within a Condition 4 area must rely on an activation system in the event of an emergency in order to evacuate the area.

Condition 4 facilities most often are penal facilities where the restrained persons are considered relatively safe to handle in large groups. As such, many persons can be released simultaneously from their individual sleeping areas when they need to travel to dining areas or move to another area.

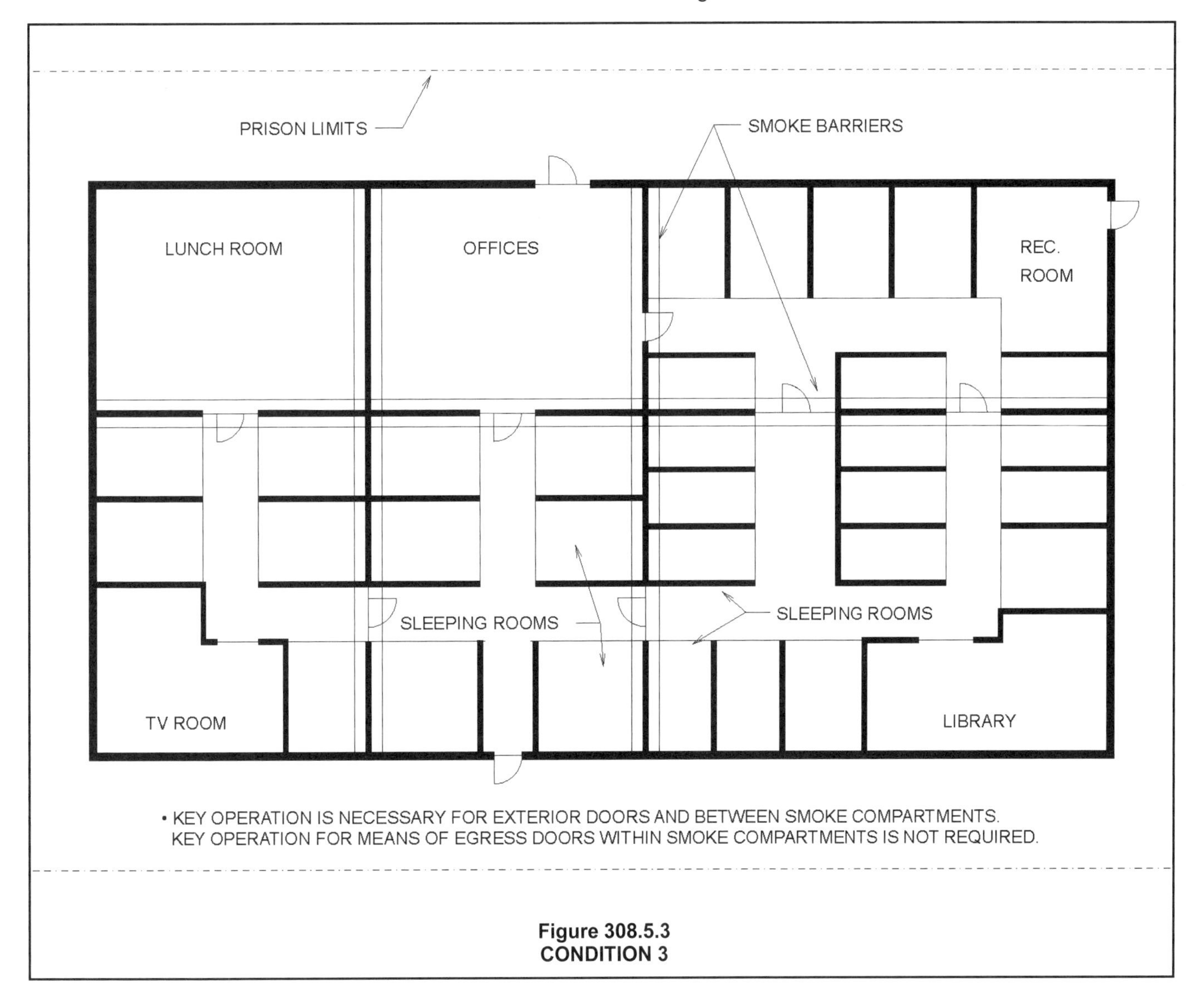

Figure 308.5.3
CONDITION 3

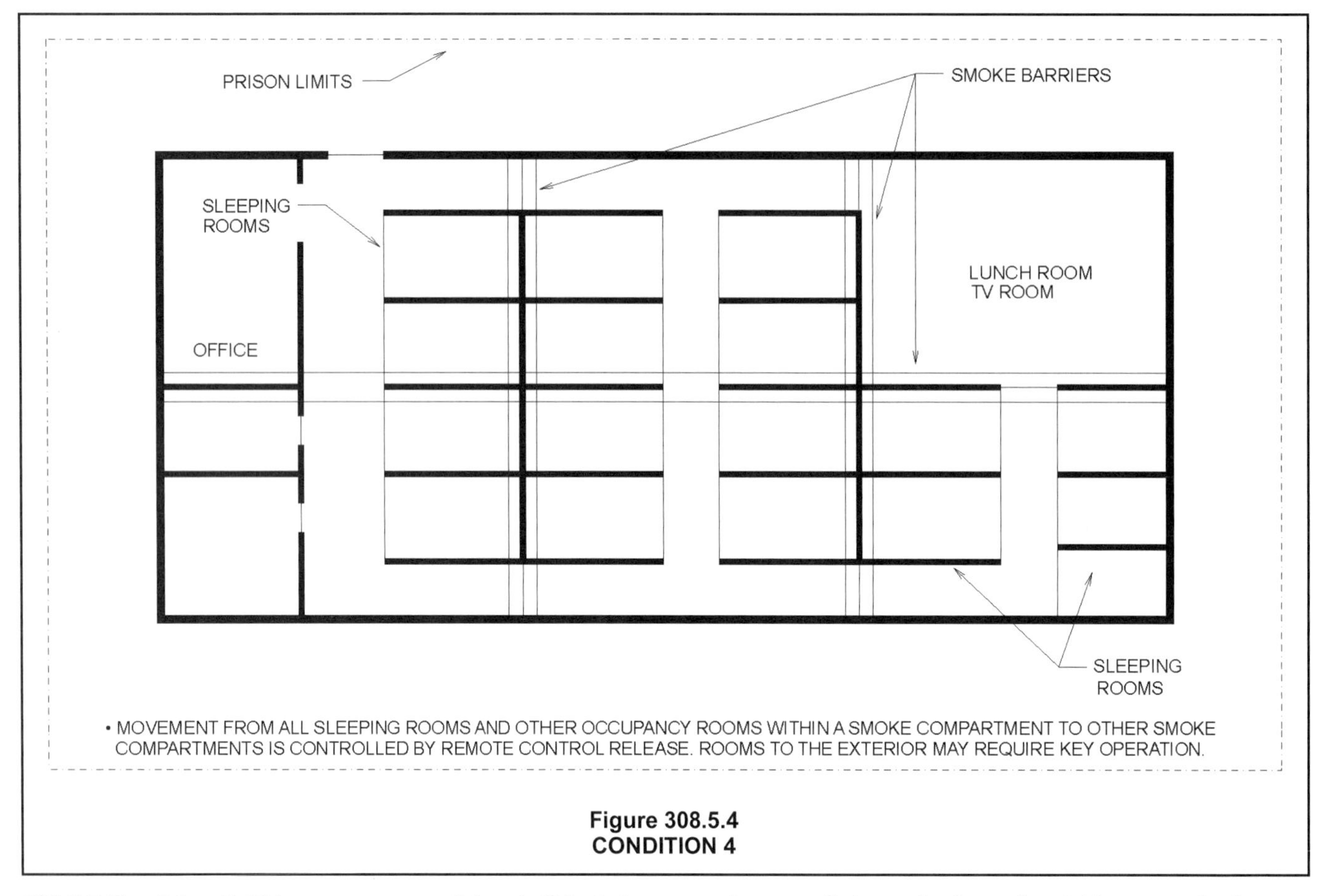

Figure 308.5.4
CONDITION 4

308.5.5 Condition 5. This occupancy condition shall include buildings in which free movement is restricted from an occupied space. Staff-controlled manual release is provided to permit movement from *sleeping units*, activity spaces and other occupied areas within the *smoke compartment* to other *smoke compartments*.

❖ Condition 5 areas are those in which the persons being secured or restrained are not allowed free movement from a room or space within a smoke compartment (as created by smoke barriers) to another smoke compartment or to the exterior (exit discharge) unless the locking device controlling their area of confinement is manually released by a staff member. Once an occupant is released from an individual space, a staff member is responsible for unlocking all doors from that location to the next smoke compartment. This is the most restrictive occupancy condition, as each secured person must be released on an individual basis and escorted to other areas.

Condition 5 facilities are most often used for maximum security or solitary confinement areas where the persons are considered to be dangerous to others, including staff members, and cannot safely be handled in large groups (see Commentary Figure 308.5.5).

308.6 Institutional Group I-4, day care facilities. Institutional Group I-4 occupancy shall include buildings and structures occupied by more than five persons of any age who receive *custodial care* for fewer than 24 hours per day by persons other than parents or guardians, relatives by blood, marriage or adoption, and in a place other than the home of the person cared for. This group shall include, but not be limited to, the following:

Adult day care
Child day care

❖ Facilities that contain provisions for the custodial care of more than five persons of any age are classified as Group I-4 (see definitions of "Personal care" and "Custodial care"). Group I-4 facilities are less restrictive in some requirements (e.g., height and area) than the other Group I occupancies. Group I-4 facilities are intended to be used to provide care for less than 24 hours. Day care facilities are not intended to be a full-time residence for the people receiving care (see Sections 308.3 and 308.4). Staff members are assumed not to be related to individuals in day care facilities. The premise of the provisions is that the numbers receiving care are exclusive of staff. The care recipients in a Group I-4 occupancy are not expected to respond to an emergency without physical assistance from others. Group I-4 occupancies include both adult day care and child day care.

Adult care facilities are assumed to be for people other than children who require some type of custodial (i.e., nonmedical) care. A facility where adults gather for social activities, such as a community center or a YMCA, is not an adult care facility (Group I) and would be regulated under other provisions of the code (Group A-3 or B). The classification of Group I-4 for an adult day care facility does not apply to facilities that provide personal care services for adults who are capable of responding to an emergency unassisted. In that case, the facility is simply classified into the occupancy group it most resembles. A facility providing a similar degree of custodial care for infants and toddlers on less than a 24-hour-per-day basis would be considered as a Group I-4 day care facility (see Sections 308.5.1 and 305.2).

308.6.1 Classification as Group E. A child day care facility that provides care for more than five but not more than 100 children $2^1/_2$ years or less of age, where the rooms in which the children are cared for are located on a *level of exit discharge* serving such rooms and each of these child care rooms has an *exit* door directly to the exterior, shall be classified as Group E.

❖ Children $2^1/_2$ years of age or less (i.e., infants and toddlers) are not typically capable of independently responding to an emergency and must be led or carried to safety. Under such circumstances, the infants and toddlers are considered nonambulatory. Therefore, a Group I-4 classification is given to those facilities where six or more infants and toddlers receive custodial care for less than 24 hours per day. A similar condition is found in foster care facilities (Group I-2, Condition 1), where infants and toddlers stay for extended periods of time. The distinguishing factor between the two occupancies is the amount of time the facility provides care for each individual. Group I-2 facilities provide care on a 24-hour basis while in Group I-4 facilities, individual care must be less than 24 hours. It is also assumed that medical care is not present in Group I-4 facilities.

A child care facility in which the number of infants and toddlers is greater than five but not more than 100 is permitted to be classified as Group E, provided the infants and toddlers are all located in rooms on the level of exit discharge that serves such rooms and all of the rooms have exit doors directly to the exterior. This exception is only applicable to rooms and spaces used for child care and is not intended to apply to accessory spaces such as restrooms, offices and kitchens. Many day care facilities primarily catering to those under primary school age tend to divide the children into three general categories based on state laws and regulations. These include infant, toddler and preschool.

Some variations do occur in that larger day care facilities will have transition rooms for mobile infants or pre-K oriented rooms for those entering kindergarten. But basically there is a mixture of children $2^1/_2$ years or less and older children. The older children can automatically be in a facility classified as a Group E occupancy, but for the younger children the exception as discussed above would need to be applied to classify the entire occupancy as Group E. The total number of children can exceed 100 and the Group E

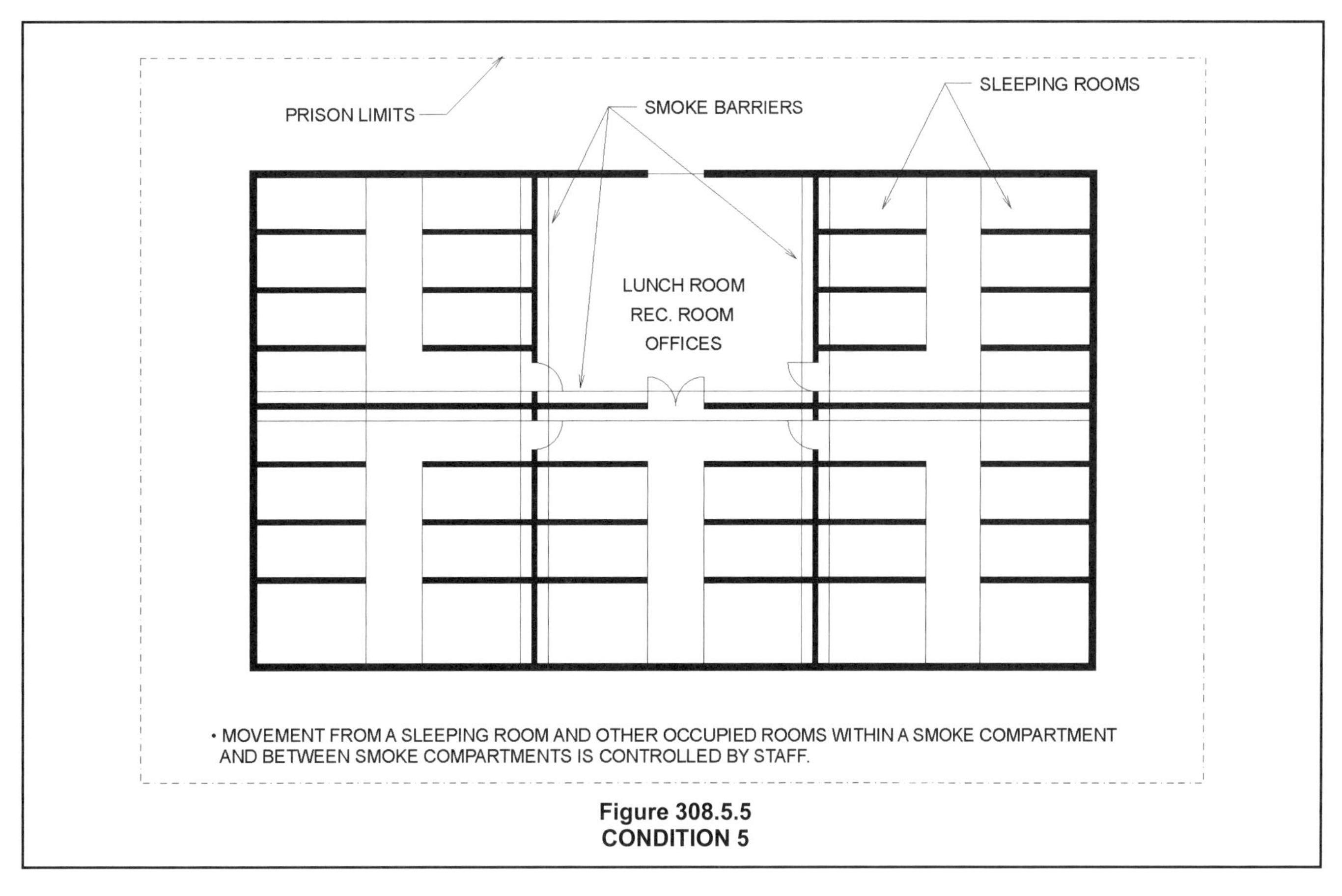

Figure 308.5.5
CONDITION 5

classification is retained, provided that the number of children $2^1/_2$ years or less is limited to 100 or fewer. The infant and toddler rooms would need to have exits directly to the outside on the level of exit discharge. If the exception is not applied, the entire facility would need to be classified as Group I-4 or a mixed occupancy classification would be necessary.

By permitting the facility to be classified as Group E, the building would not be required to be sprinklered unless the fire area was greater than 12,000 square feet (115 m^2). A Group I-4 facility would be required to be sprinklered regardless of the area. But as a Group E occupancy, panic hardware would be required in rooms and spaces exceeding 50 occupants.

See Commentary Figure 308.1.

308.6.2 Within a place of religious worship. Rooms and spaces within *places of religious worship* providing such care during religious functions shall be classified as part of the primary occupancy.

❖ Group I-4 provisions do not apply to places of religious worship simply providing care services during worship and related religious functions. If the space is used at other times as a day care facility, then it would be classified as Group I-4 or E as applicable.

308.6.3 Five or fewer persons receiving care. A facility having five or fewer persons receiving *custodial care* shall be classified as part of the primary occupancy.

❖ Where five or fewer persons receive custodial care in a facility other than a dwelling unit, the classification of the care area is to be consistent with that of the primary occupancy. The limited number of care recipients reduces the hazard level to the point that classification as a Group I-4 occupancy is not warranted.

308.6.4 Five or fewer persons receiving care in a dwelling unit. A facility such as the above within a *dwelling unit* and having five or fewer persons receiving *custodial care* shall be classified as a Group R-3 occupancy or shall comply with the *International Residential Code.*

❖ Buildings that have five or fewer persons receiving custodial care within a dwelling unit are to be classified as Group R-3, or shall be constructed in accordance with the IRC. The assumption is that this type of activity is possible in a residential environment where one or more family members require the high level of care regulated by Section 308.6. Please note similar provisions for Group E occupancies as well as Groups I-1 and I-2.

SECTION 309
MERCANTILE GROUP M

309.1 Mercantile Group M. Mercantile Group M occupancy includes, among others, the use of a building or structure or a portion thereof for the display and sale of merchandise, and involves stocks of goods, wares or merchandise incidental to such purposes and accessible to the public. Mercantile occupancies shall include, but not be limited to, the following:

Department stores
Drug stores
Markets
Motor fuel-dispensing facilities
Retail or wholesale stores
Sales rooms

❖ The characteristics of occupancies classified in Group M are contained in this section. Because mercantile occupancies normally involve the display and sale of large quantities of combustible merchandise, the fuel load in such facilities can be relatively high, potentially exposing the occupants (customers and sales personnel) to a high degree of fire hazard. Mercantile operations often attract large crowds (particularly in large department stores and covered and open malls and especially during weekends and holidays). There are two factors that alleviate the risks to life safety: the occupant load normally has a low to moderate density, and the occupants are alert, mobile and able to respond in an emergency situation. The degree of openness and the organization of the retail display found in most mercantile occupancies are generally orderly and do not present an unusual difficulty for occupant evacuation.

Listed here are general descriptions of the kinds of occupancies that are classified in Group M. Mercantile buildings most often have both a moderate occupant load and a high fuel load, which is in the form of furnishings and the goods being displayed, stored and sold [see Commentary Figure 309.1(1)].

The key characteristics that differentiate occupancies classified in Group M from those classified in Group B (see Section 304) are the larger quantity of goods or merchandise available for sale and the lack of familiarity of the occupants with the building, particularly its means of egress. To be classified in Group M, the goods that are on display must be accessible to the public. If a patron sees an item for sale, then that item is generally available for purchase at that time (i.e., there is a large stock of goods). If a store allows people to see the merchandise but it is not available on the premises, such as an automobile showroom, then the occupancy classification of business (Group B) should be considered. A mercantile building is open to the public, many of whom may not be regular visitors. A business building, however, is primarily occupied by regular employees who are familiar with the building arrangement and, most importantly, the exits. This awareness of the building and the exits can be an important factor in a fire emergency.

Where storage rooms are limited to a maximum of 100 square feet (9.3 m^2) in area, Section 311.1.1 allows such rooms to still be considered part of the primary occupancy, provided that the aggregate of such rooms does not exceed 10 percent of the story. Therefore, many mercantile stores with stockroom areas will be considered a mixed use building for height and area, construction type and sprinkler requirements. Note that Table 508.4 does not require a fire-resistance rating on the separation between Groups M and S-1.

Automotive, fleet-vehicle, marine and self-service fuel-dispensing facilities, as defined in the IFC, are classified in the mercantile occupancy, as are the convenience stores often associated with such occupancies [see Commentary Figure 309.1(2)]. Quick-lube, tune-up, muffler and tire shops are not included in this classification. Those facilities that typically conduct automotive service and repair work are treated as a repair garage (Group S-1, also defined in the IFC).

Simply because a building containing a mercantile-

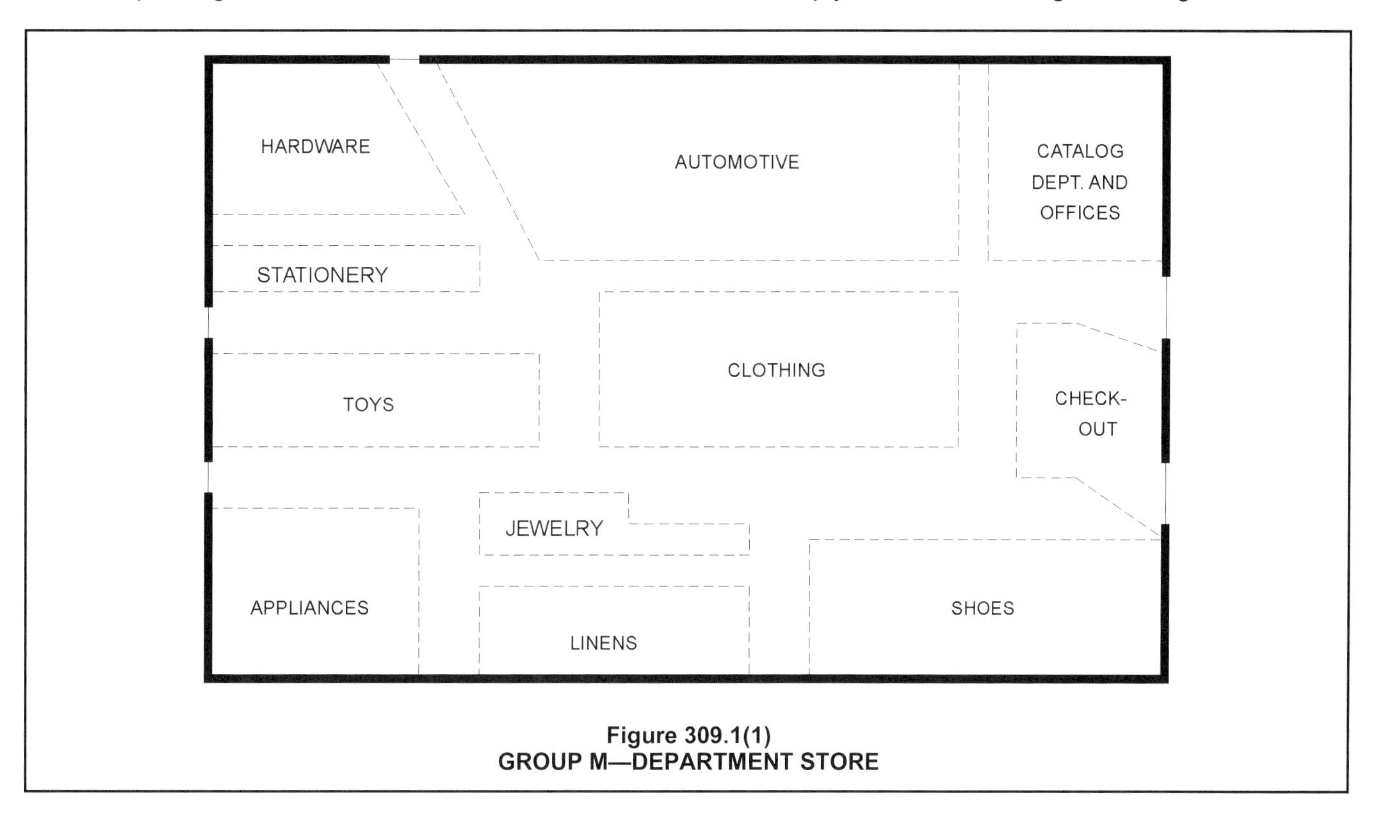

Figure 309.1(1)
GROUP M—DEPARTMENT STORE

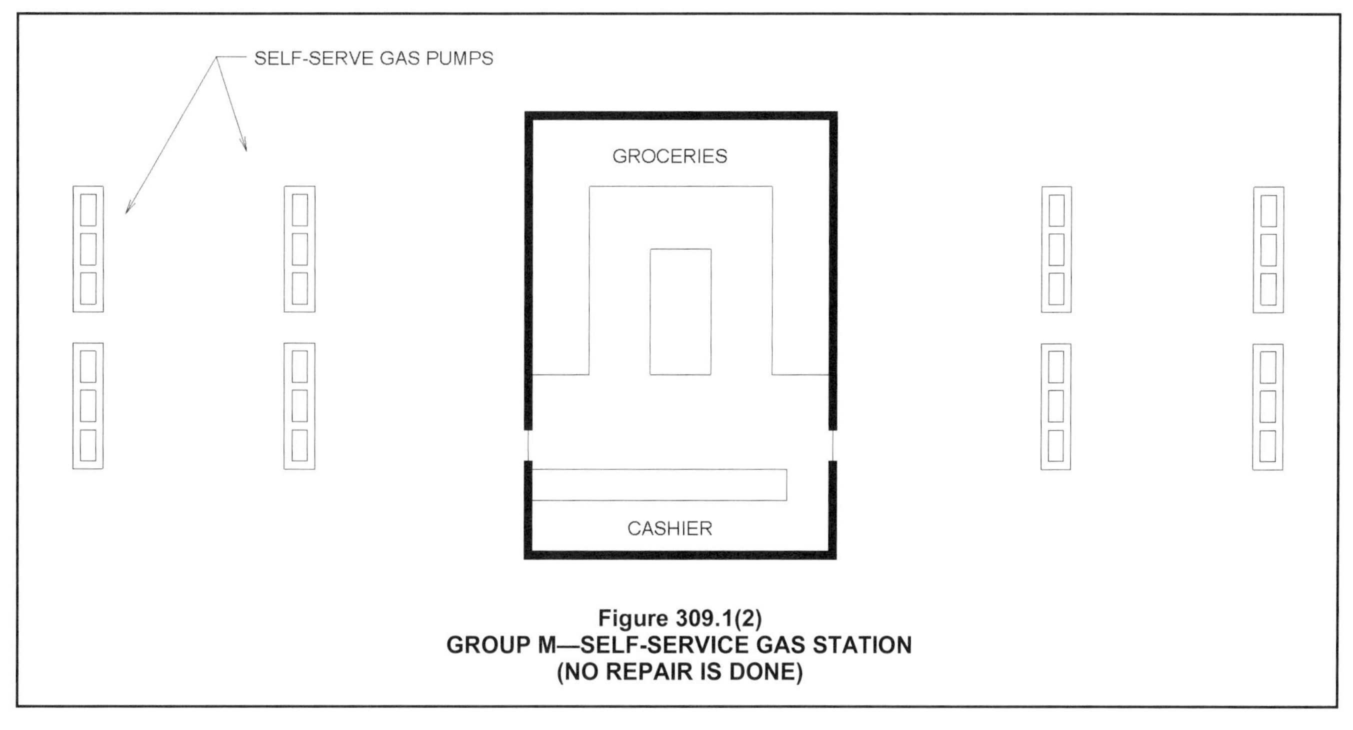

Figure 309.1(2)
GROUP M—SELF-SERVICE GAS STATION
(NO REPAIR IS DONE)

type occupancy has a dense occupant load does not necessitate the need to classify the building as an assembly occupancy, unless the activity includes an assembly-type area where purchasing of goods is a group activity versus individual shoppers independently considering and purchasing merchandise. For example, a building in which auction sales occur may have a highly concentrated occupant load where the sales occur, but Section 309.1 describes mercantile occupancies as "the use of a building or structure or portion thereof for the display and sale of merchandise and involves stocks of goods, wares or merchandise incidental to such purposes and accessible to the public." In an auction, however, the activity is dominated by an assembly use of the space as people gather to conduct and participate in the auction. As such, auction spaces need to be assigned a Group A-3 occupancy (IBC Interpretation No. 38-03). The presence of highly concentrated occupant loads does not in itself mandate an assembly use classification unless the activity is assembly in nature versus large numbers of people pursuing individual activities of acquisition.

309.2 Quantity of hazardous materials. The aggregate quantity of nonflammable solid and nonflammable or noncombustible liquid hazardous materials stored or displayed in a single *control area* of a Group M occupancy shall not exceed the quantities in Table 414.2.5(1).

❖ This section addresses an exception for control areas of mercantile occupancies containing certain nonflammable or noncombustible materials and health hazard gases that are stored in accordance with Table 414.2.5(1). This section allows Group H-4 materials, which pose a health hazard rather than a physical hazard, as well as limited Group H-2 and H-3 materials to be stored in both the retail display and stock areas of mercantile occupancies in excess of the maximum allowable quantity per control area of Tables 307.1(1) and 307.1(2) without classifying the building as Group H. To correctly classify a building where products that have hazardous properties are stored and sold, the code user must also be aware of the provisions contained in Section 307.1, Notes c and n of Table 307.1(1) and Notes b and c of Table 307.1(2). These provisions give the quantity limitations for specific high-hazard products in mercantile display areas, including medicines, foodstuffs, cosmetics and alcoholic beverages.

Without this option, many mercantile occupancies could technically be classified as Group H. The increased quantities of certain hazardous materials are based on the recognition that, while there is limited risk in mercantile occupancies, the packaging and storage arrangements can be controlled. For further information on the storage limitations required for these types of materials in mercantile occupancies, see Section 2703.11 of the IFC.

SECTION 310 RESIDENTIAL GROUP R

310.1 Residential Group R. Residential Group R includes, among others, the use of a building or structure, or a portion thereof, for sleeping purposes when not classified as an Institutional Group I or when not regulated by the *International Residential Code*.

❖ Residential occupancies represent some of the highest fire safety risks of any of the occupancies listed in Chapter 3. There are several reasons for this condition.

Structures in the residential occupancy house the widest range of occupant types, i.e., infants to the aged, for the longest periods of time. As such, residential occupancies are more susceptible to the careless acts of the occupants; therefore, the consequences of exposure to the effects of fire are the most serious.

Most residential occupants are asleep approximately one-third of every 24-hour period. When sleeping, they are not likely to become immediately aware of a developing fire. Also, if awakened from sleep by the presence of fire, the residents often may not immediately react in a rational manner and delay their evacuation.

The fuel load in residential occupancies is often quite high, both in quantity and variety. Also, in the construction of residential buildings, it is common to use extensive amounts of combustible materials.

Another portion of the fire problem in residential occupancies relates to the occupants' lack of vigilance in the prevention of fire hazards. In their own domicile or residence, people tend to relax and are often prone to allow fire hazards to go unabated; thus, in residential occupancies, fire hazards tend to accrue over an extended period of time and go unnoticed or are ignored.

Most of the nation's fire problems occur in Group R buildings and, in particular, one- and two-family dwellings, which account for more than 80 percent of all deaths from fire in residential occupancies and about two-thirds of all fire fatalities in all occupancies. One- and two-family dwellings also account for more than 80 percent of residential property losses from fire and more than one-half of all property losses from fire.

Because of the relatively high fire risk and potential for loss of life in buildings classified in Groups R-1 (hotels and motels) and R-2 (apartments and dormitories), the code has stringent provisions for the protection of life in these occupancies. Group R-3 occupancies, however, are not generally considered to be in the same domain and, thus, are not subject to the same level of regulatory control as is provided in other occupancies.

Because of the growing trend to care for people in a residential environment, residential care/assisted living facilities are also classified as Group R. Specifically, these facilities are classified as Group R-4. Mainstreaming people who are recovering from alcohol or drug addiction and people who are developmentally disabled is reported to have therapeutic and social benefits. A residential environment often fosters this mainstreaming.

A building or part of a building is considered to be a residential occupancy if it is intended to be used for sleeping accommodations (including assisted living facilities) and is not an institutional occupancy. Institutional occupancies are similar to residential occupancies in many ways. However, they differ from each other in that institutional occupants are in a supervised environment, and, in the case of Groups I-2 and I-3 occupancies, are under some form of restraint or physical limitation that makes them incapable of complete self-preservation. The number of these occupants who are under supervision or are incapable of self-preservation is one distinguishing factor for being classified as an institutional or residential occupancy.

The term Group R refers collectively to the four individual residential occupancy classifications: Groups R-1, R-2, R-3 and R-4. These classifications are differentiated in the code based on the following criteria: 1. Whether the occupants are transient or nontransient in nature; 2. The type and number of dwelling units or sleeping units contained in a single building; and 3. The number of occupants in the facility.

310.2 Definitions. The following terms are defined in Chapter 2:

BOARDING HOUSE.

CONGREGATE LIVING FACILITIES.

DORMITORY.

GROUP HOME.

GUEST ROOM.

LODGING HOUSE.

PERSONAL CARE SERVICE.

TRANSIENT.

❖ This section lists terms that are specifically associated with the subject matter of this section. It is important to emphasize that these terms are not exclusively related to this section but may or may not also be applicable where the term is used elsewhere in the code.

Definitions of terms can help in the understanding and application of the code requirements. The purpose for including a list within this chapter is to provide more convenient access to terms which may have a specific or limited application within this chapter. For the complete definition and associated commentary, refer back to Chapter 2. Terms that are italicized provide a visual identification throughout the code that a definition exists for that term. The use and application of all defined terms are set forth in Section 201.

310.3 Residential Group R-1. Residential Group R-1 occupancies containing *sleeping units* where the occupants are primarily *transient* in nature, including:

Boarding houses (*transient*) with more than 10 occupants
Congregate living facilities (*transient*) with more than 10 occupants
Hotels (*transient*)
Motels (*transient*)

❖ The key characteristic of Group R-1 that differentiates it from other Group R occupancies is that the occupants are considered transient in nature (i.e., those whose length of stay is not more than 30 days). There is an expectation that the occupants are not as familiar with the building as those residents in nontransient facilities such as apartment buildings and single-family dwellings. If occupants are unfamiliar with their surroundings, they may not recognize potential hazards or be able to use the means of egress effectively.

The most common building types classified in Group R-1 are hotels, motels and boarding houses. Facilities classified as Group R-1 occupancies may include dwelling units, sleeping units, or a combination of both. Group R-1 occupancies do not typically have cooking facilities in the unit. When a unit is not equipped with cooking facilities, it does not meet the definition of a "dwelling unit" in Section 202. When this occurs, such units are treated as sleeping units for the application of code provisions (see Commentary Figure 310.3). A recent trend in development is the construction of "extended-stay hotels." While these units may have all of the characteristics of a typical dwelling unit (i.e., cooking, living, sleeping, eating, sanitation), the length of stay is still typically not more than 30 days. As such, these buildings would still be classified as Group R-1. If the length of stay is more than 30 days, these buildings would be classified as Group R-2. If a hotel offers rooms for short-term housing (i.e., more than 30 days), the facility must comply with the provisions for both Groups R-1 and R-2 (see Section 302.1).

Other occupancies are often found in buildings classified in Group R-1. These occupancies include nightclubs (Group A-2), restaurants (Group A-2), gift shops (Group M), business offices (Group B), health clubs (Group A-3) and storage facilities (Group S-1). When this occurs, the building is a mixed occupancy and is subject to the provisions of Section 508.

Transient congregate living facilities and boarding houses with 10 or fewer occupants can be constructed to the standards of Group R-3 occupancies rather than the general category of Group R-1. The primary intent of this provision is to permit bed-and-breakfast-type facilities to be established in existing single-family (one-family) structures. In comparison to the provision under Group R-2 which permits congregate living facilities with fewer than 16 nontransient occupants to be built as a Group R-3, the Group R-3 "transient" facility is limited to 10 or fewer occupants in reflection of the limited number of occupants.

See Section 310.5.2 for lodging houses with five or fewer rooms to rent. While these are transient in nature and potentially have more than 10 occupants, these types of bed-and-breakfast type facilities are specifically listed under Group R-3.

310.4 Residential Group R-2. Residential Group R-2 occupancies containing *sleeping units* or more than two *dwelling units* where the occupants are primarily permanent in nature, including:

Apartment houses
Boarding houses (nontransient) with more than 16 occupants
Congregate living facilities (nontransient) with more than 16 occupants
Convents
Dormitories
Fraternities and sororities
Hotels (nontransient)
Live/work units
Monasteries
Motels (nontransient)
Vacation timeshare properties

❖ The length of the occupants' stay plus the arrangement of the facilities provided are the basic factors that differentiate occupancies classified in Group R-2 from other occupancies in Group R. The occupants of facilities or areas classified in Group R-2 are primarily nontransient, capable of self-preservation and share their means of egress in whole or in part with other occupants outside of their sleeping area or dwelling unit. Building types ordinarily classified in Group R-2 include apartments, boarding houses (when the occupants are not transient) and nontransient congregate living facilities, such as dormitories, where there are more than 16 occupants [see Commentary Figures 310.4(1) and 310.4(2)].

Individual dwelling units in Group R-2 are either rented by tenants or owned by the occupants. The code does not make a distinction between either type of tenancy. Residential condominiums are treated in the code the same as Group R-2 apartments. Such condominiums are based on shared ownership of a building and related facilities. While an individual

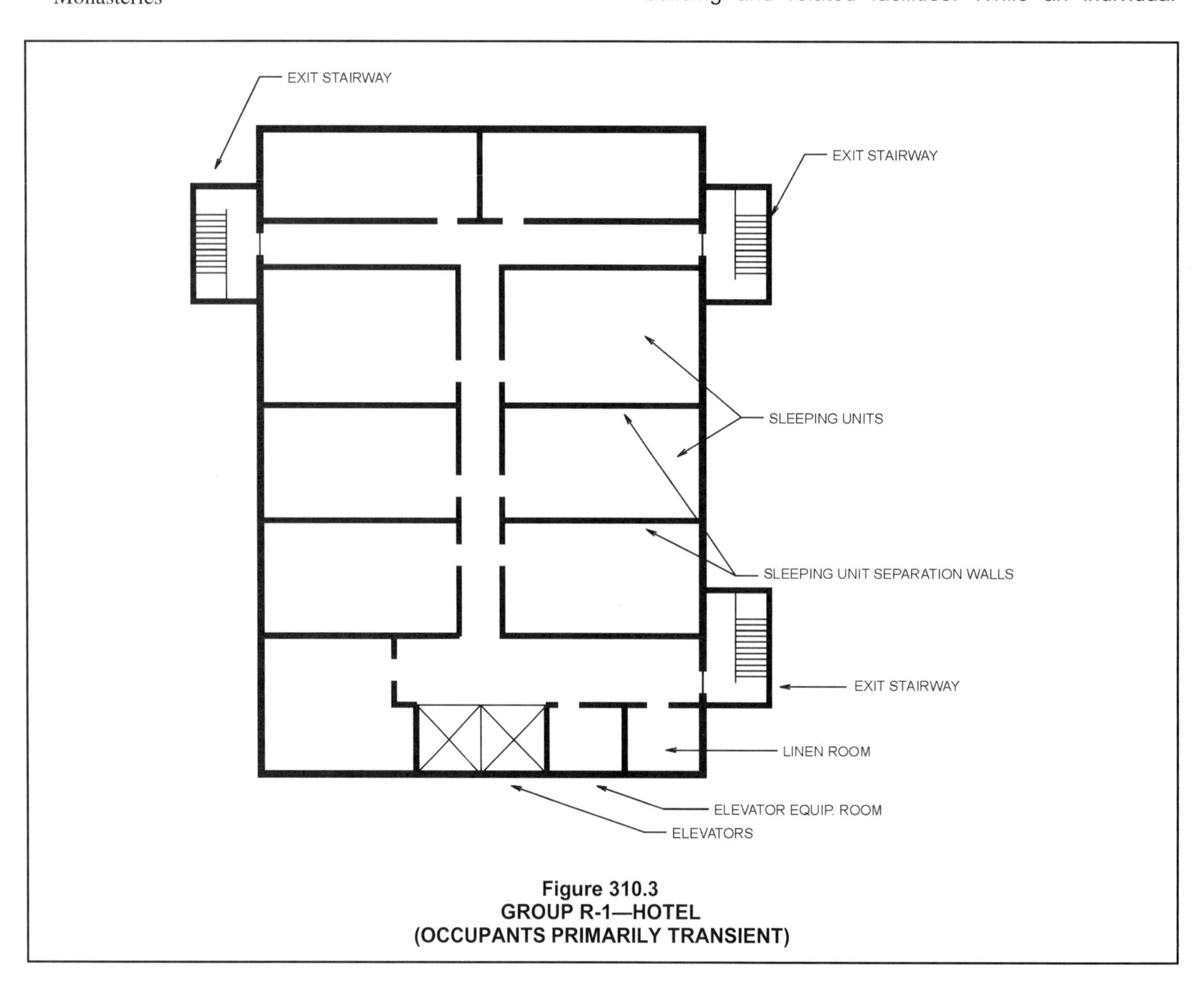

Figure 310.3
GROUP R-1—HOTEL
(OCCUPANTS PRIMARILY TRANSIENT)

owner will have exclusive rights to a certain unit, the building, the lot the building sits upon, parking, common recreational facilities and similar features are owned in common by all the owners of individual dwelling units. In most cases condominiums do not establish separate lots and the walls between units are not setting on lot lines. Another type of shared ownership is referred to as a "co-op," short for cooperative. Occasionally a condominium will establish actual lots and lot lines distinguishing individual ownership. When the dwelling unit is located on a separate parcel of land, lot lines defining the parcel exist and the requirements for fire separation must be met.

Dormitories are generally associated with university or college campuses for use as student housing, but dormitory uses continue to evolve. Many dormitories are now being built as housing for elderly people who wish to live with other people their own age and who do not need 24-hour-a-day medical supervision. The only difference between the dormitory that has just been described and the dormitory found on a college campus is the age of its occupants. If the elderly people must have 24-hour-a-day medical supervision (i.e., a nurse or doctor on the premises), the building is no longer considered a residential occupancy but an institutional occupancy and would have to comply with the applicable provisions of the code for the appropriate Group I occupancy.

Similar to Group R-1, individual rooms in dormitories are sleeping units. When college classes are not in session, the rooms in dormitories are sometimes rented out for periods of less than 30 days to convention attendees and other visitors. When dormitories undergo this type of transient use, they more closely resemble Group R-1. A style of dormitory design which is being used more frequently is one where there are groups of sleeping units around a common living space, including shared bathrooms and sometimes a kitchen or kitchenette. At this time, the code does not indicate if these groups of rooms should be addressed as separate sleeping units or as a dwelling unit, except for when counting units for determining the number of Accessible units required (see Section 1107.6.2.3.1).

Buildings containing dormitories often contain other occupancies, such as cafeterias or dining rooms (Group A-2), recreation rooms (Group A-3), offices (Group B) and study and meeting rooms (Group A-3). When this occurs, the building is considered a mixed occupancy and is subject to the provisions of Section 508 [see Commentary Figure 310.4(2)].

Included in the listing of Group R-2 are live/work units. A live/work unit is a dwelling unit or sleeping unit in which a significant portion of the space includes a nonresidential use operated by the tenant. Reflecting a growing trend in urban neighborhoods and the reuse of existing buildings, live/work units must comply with the provisions of Section 419.

The intent of the congregate living facility reference is to better define when a congregate living facility is operating as a single-family home. Blended families are now commonplace and not necessarily defined strictly by blood or marriage. Small boarding houses, convents, dormitories, fraternities, sororities, monasteries and nontransient hotels and motels may be small enough to operate as a single-family unit and would be permitted to be constructed as Group R-3 occupancies as intended by the code. The threshold of 16 persons is consistent with the results of the most recent census, which showed that 98 percent of households in the U.S. that identified themselves as a single family have less than 16 occupants. The 16-occupant limit is also consistent with the limits of an NFPA 13D sprinkler system.

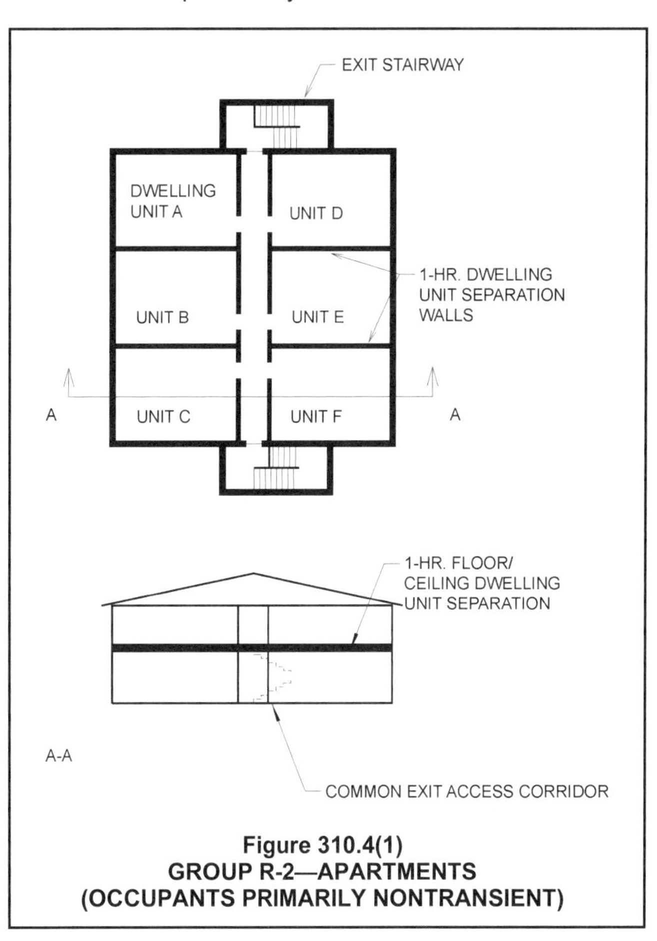

Figure 310.4(1)
GROUP R-2—APARTMENTS (OCCUPANTS PRIMARILY NONTRANSIENT)

310.5 Residential Group R-3. Residential Group R-3 occupancies where the occupants are primarily permanent in nature and not classified as Group R-1, R-2, R-4 or I, including:

Buildings that do not contain more than two *dwelling units*

Boarding houses (nontransient) with 16 or fewer occupants

Boarding houses (*transient*) with 10 or fewer occupants

Care facilities that provide accommodations for five or fewer persons receiving care

Congregate living facilities (nontransient) with 16 or fewer occupants

Congregate living facilities (*transient*) with 10 or fewer occupants

Lodging houses with five or fewer *guest rooms*

❖ Group R-3 facilities include all detached one- and two-family dwellings and multiple (three or more) single-family dwellings (townhouses) more than three stories in height. Those buildings three or less stories in height are not classified as Group R-3 and are regulated by the IRC. Each pair of dwelling units in multiple single-family dwellings greater than three stories in height must be separated by fire walls (see Section 706) or by two exterior walls (see Table 602) in order to be classified as Group R-3. (Duplexes, buildings with two dwelling units, must be detached from other structures in order to be regulated by the IRC.) A duplex attached to another duplex would be required to comply with the code and be classified as Group R-2 or R-3, depending on the presence of fire walls.

Buildings that are classified as Group R-3, while limited in height, are not limited in the allowable area as indicated in Table 503.

Buildings that are one- and two-family dwellings and multiple single-family dwellings less than three stories in height and that contain another occupancy (e.g., Groups B, M, I-4) must be regulated as a mixed occupancy in accordance with this code and are not permitted to be regulated by the provisions of the IRC [see Commentary Figures 310.5(1) and 310.5(2)]. However, some mixed-use dwelling units may qualify as live/work units under Section 419 and be classified as a Group R-2 occupancy.

In addition, boarding houses and congregate living facilities with no more than 16 nontransient occupants or no more than 10 transient occupants are to be classified as Group R-3.

Group R-3 occupancies include small care facilities where care is provided to five or fewer persons. The intent is to allow persons to be cared for in a residential, or home, environment, often by family members, in a manner that is typical within a single-family-type home. Allowances for the Group R-3 classification of smaller care facilities are established in Sections 305.2.3 (Group E), 308.3.4 (Group I-1), 308.4.2 (Group I-2) and 308.6.4 (Group I-4). Because the intent is to accommodate persons who might otherwise be in other group occupancies including Group I-2, the Group R-3 care facility is not limited to just persons who are capable of self-preservation. The

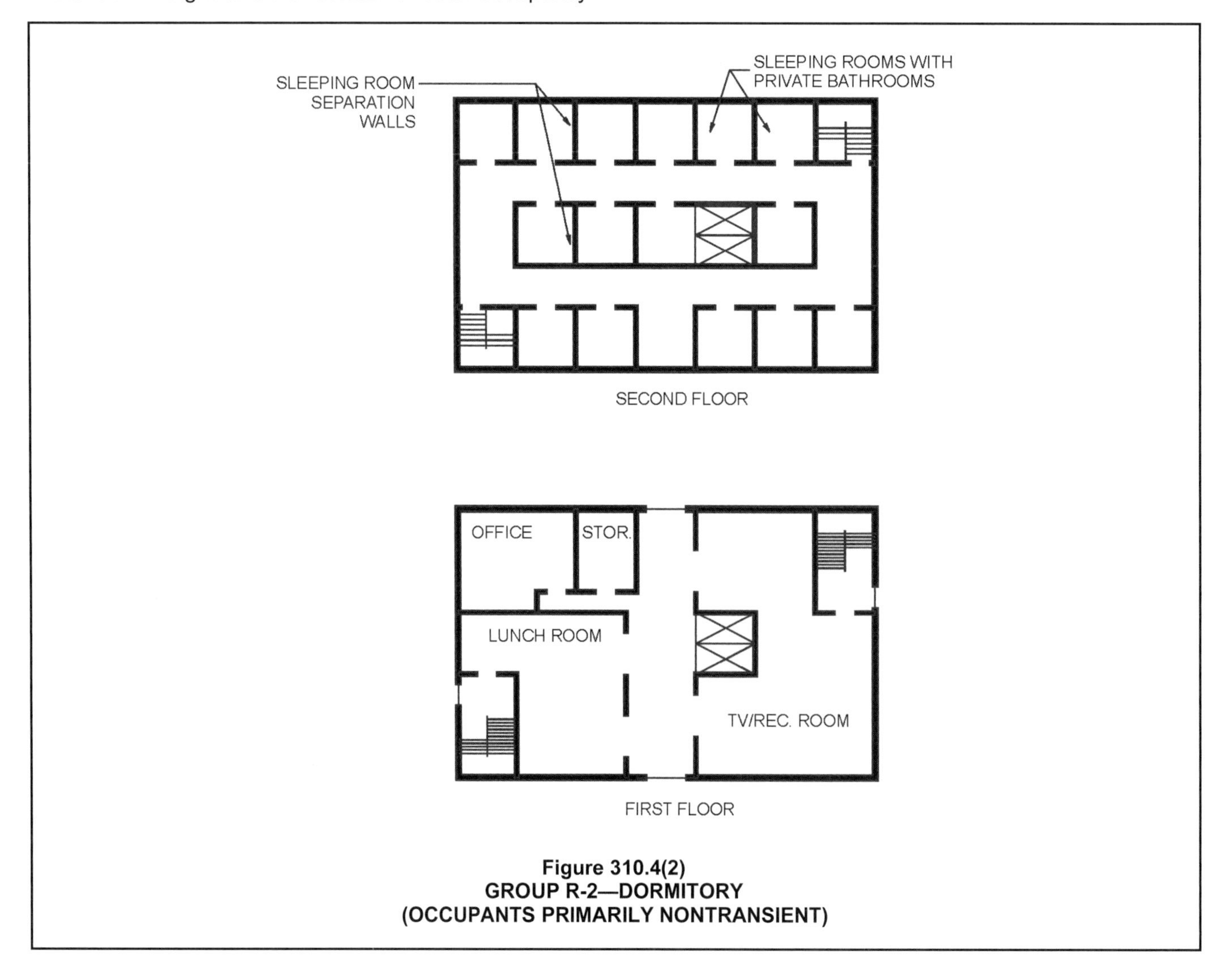

Figure 310.4(2)
GROUP R-2—DORMITORY
(OCCUPANTS PRIMARILY NONTRANSIENT)

only limit is the number receiving care, not the total number of occupants in the dwelling unit. If two people are receiving care, and the rest of the family is four people, the total of six occupants of the dwelling does not move this out of being a Group R-3 occupancy. See also Section 310.5.1.

A lodging house with five or fewer guest rooms can be classified as a Group R-3 occupancy or, under Section 310.5.2, can be constructed under the provisions of the IRC. The definition of lodging house allows the rental of guest rooms to transients, provided that there are one or more occupants who are permanent in nature. While Section 310.5.2 requires owner occupancy of the dwelling unit in order for it to be built in compliance with the IRC, there is no owner occupancy requirement for lodging houses that comply with Group R-3 requirements. The broad intent of the lodging house provisions is to allow bed and breakfast and similar facilities under the Group R-3 category even though transient housing generally falls under the Group R-1 classification. The limit is set to the number of guest rooms. There is no specified limit on the number of guests, the number of residents, or the combination of both.

310.5.1 Care facilities within a dwelling. Care facilities for five or fewer persons receiving care that are within a single-family dwelling are permitted to comply with the *International Residential Code* provided an *automatic sprinkler system* is installed in accordance with Section 903.3.1.3 or Section P2904 of the *International Residential Code*.

❖ Section 310.5 already states that care facilities that accommodate five or fewer persons receiving care can be classified as a Group R-3 occupancy. Sections 305.2.3 (Group E), 308.3.4 (Group I-1), 308.4.2 (Group I-2) and 308.6.4 (Group I-4) each state the option of providing the care of five or fewer persons within a structure regulated under the IRC. The intent is to allow persons to be cared for in a residential, or home, environment, often by family members, in a manner that is typical within a single-family-type home. As stated for those care facilities allowed within the R-3 occupancy, the persons receiving care in a building designed according to the IRC are not limited to those who are capable of self-preservation. Similar to Sections 308.3.4 and 308.4.2, this section specifies that any such IRC-regulated facility be provided with a sprinkler system.

310.5.2 Lodging houses. Owner-occupied *lodging houses* with five or fewer *guest rooms* shall be permitted to be constructed in accordance with the *International Residential Code*.

❖ This section allows bed-and-breakfast type hotels that are both owner occupied and have five or fewer rooms to rent to be constructed under the IRC. There are no occupant load limitations for this option as there are for boarding houses or congregate residences. See commentary for Section 310.5.

310.6 Residential Group R-4. Residential Group R-4 occupancy shall include buildings, structures or portions thereof for more than five but not more than 16 persons, excluding staff, who reside on a 24-hour basis in a supervised residential environment and receive *custodial care*. Buildings of Group R-4 shall be classified as one of the occupancy conditions specified in Section 310.6.1 or 310.6.2. The persons

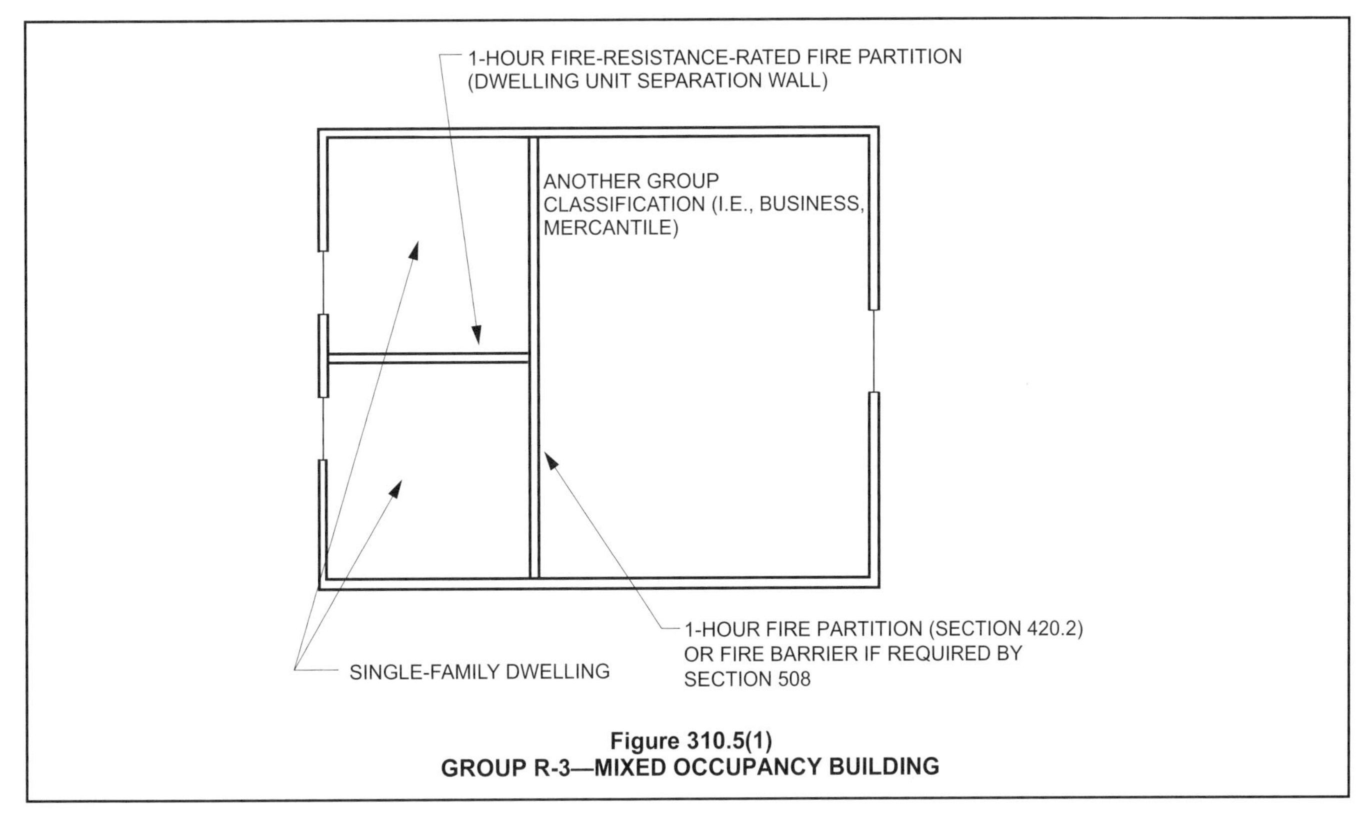

Figure 310.5(1)
GROUP R-3—MIXED OCCUPANCY BUILDING

receiving care are capable of self-preservation. This group shall include, but not be limited to, the following:

Alcohol and drug centers
Assisted living facilities
Congregate care facilities
Group homes
Halfway houses
Residential board and care facilities
Social rehabilitation facilities

Group R-4 occupancies shall meet the requirements for construction as defined for Group R-3, except as otherwise provided for in this code.

❖ Where five to 16 residents live in a supervised environment and receive custodial care, such a facility is classified as Group R-4. Ninety-eight percent of households in the U.S. that identified themselves as a single-family household have less than 16 occupants. The 16-occupant limit is also consistent with the limits of an NFPA 13D sprinkler system. Thus the limit of 16 residents was established as an appropriate limit considering that this facility will operate similarly to a single-family home. Under federal housing laws regarding nondiscrimination, families cannot be determined by blood or marriage. If a Group R-4 occupancy is expanded or allowed to have more than 16 care recipients, the facility needs to be reclassified as a Group I-1 occupancy. The number of persons used in the determination includes those who receive care but does not include staff.

Similar to Group I-1, a Group R-4 occupancy is also one of two "conditions." In a Condition 1 facility, care recipients may be slower during evacuation but are capable of self-preservation. In a Condition 2 facility, care recipients may require limited assistance with evacuation during emergency situations. See the commentary in Section 308.3 for Group I-1, for a further detailed explanation of both Group I-1 and R-4 custodial care occupancies. Also see Section 202, Definitions, Custodial care.

Group R-4 facilities must satisfy the construction requirements of Group R-3. Facilities with five or fewer persons receiving care will be either a Group R-3 occupancy (see Section 310.5), or can be built under the IRC (see Section 310.5.1).

See Commentary Figure 308.1.

310.6.1 Condition 1. This occupancy condition shall include buildings in which all persons receiving custodial care, without any assistance, are capable of responding to an emergency situation to complete building evacuation.

❖ See the commentary to Section 310.6.

310.6.2 Condition 2. This occupancy condition shall include buildings in which there are any persons receiving custodial care who require limited verbal or physical assistance while responding to an emergency situation to complete building evacuation.

❖ See the commentary to Section 310.6.

SECTION 311 STORAGE GROUP S

311.1 Storage Group S. Storage Group S occupancy includes, among others, the use of a building or structure, or a portion thereof, for storage that is not classified as a hazardous occupancy.

❖ This section requires that all structures (or parts thereof) designed or occupied for the storage of moderate- and low-hazard materials are to be classified in either Group S-1 (moderate hazard) or S-2 (low hazard).

Life safety problems in structures used for storage of moderate- and low-hazard materials are minimal

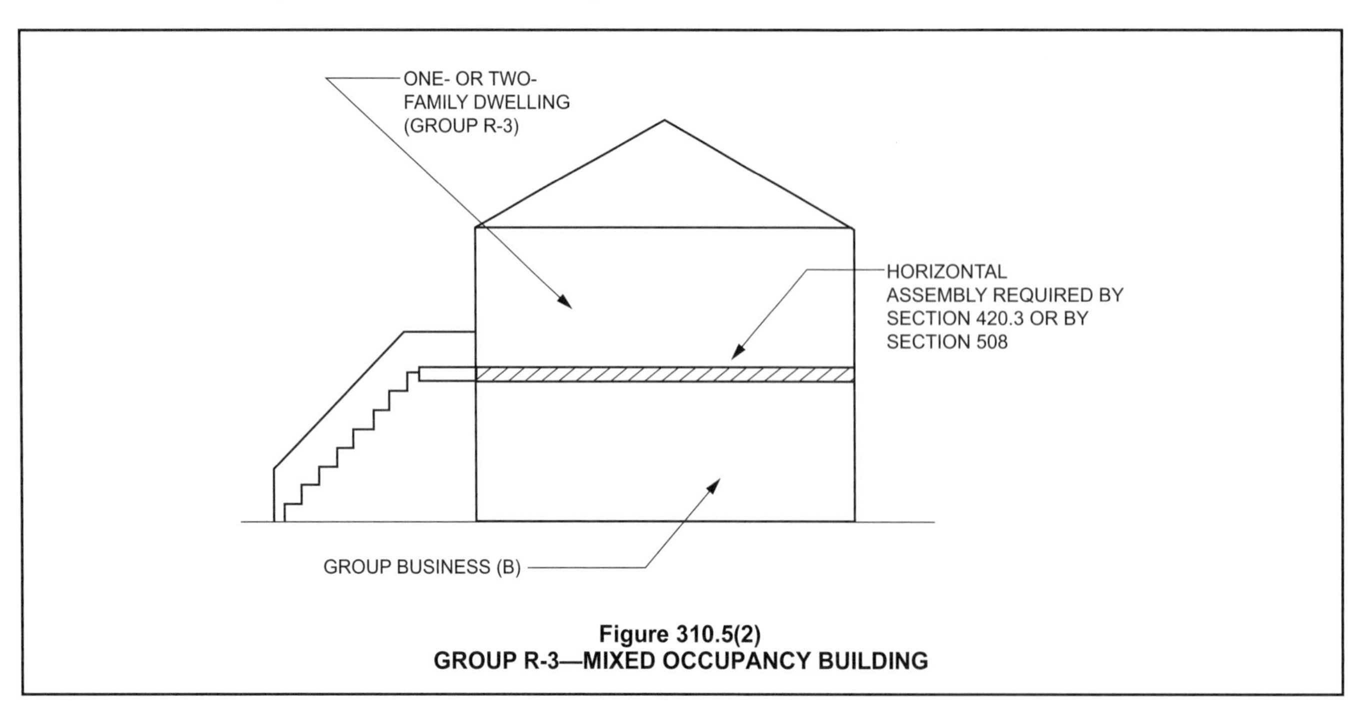

Figure 310.5(2)
GROUP R-3—MIXED OCCUPANCY BUILDING

because the number of people involved in a storage operation is usually small and normal work patterns require occupants to be dispersed throughout the facility.

Problems of fire safety, particularly as they relate to the protection of stored contents, are directly associated with the amount and combustibility of the materials (including packaging) that are housed on the premises.

Storage facilities typically contain significant amounts of combustible or noncombustible materials that are kept in a common area. Because of the combustion, flammability or explosive characteristics of certain materials (see Section 307), a structure or portion thereof that is used to store high-hazard materials exceeding the maximum allowable quantities (MAQ), or that does not meet one of the identified uses in Section 307.1.1, cannot be classified as Group S. Such a structure is to be classified as Group H, high-hazard use, and is to comply with Section 307.

Hazardous materials may be located in Group S occupancy buildings, provided the amount of materials in each control area does not exceed the MAQ specified in Tables 307.1(1) and 307.1(2). Control areas must comply with Section 414. Storage occupancies consist of two basic types: Groups S-1 and S-2, which are based on the properties of the materials being stored. The distinction between Groups S-1 and S-2 is similar to that between Groups F-1 and F-2, as outlined in Section 306.

311.1.1 Accessory storage spaces. A room or space used for storage purposes that is less than 100 square feet (9.3 m^2) in area and accessory to another occupancy shall be classified as part of that occupancy. The aggregate area of such rooms or spaces shall not exceed the allowable area limits of Section 508.2.

❖ This provision allows small storage rooms to be classified in the same occupancy group as the primary occupancy of a space. There are two limits: each space is limited to 100 square feet (9.3 m^2); and the aggregate of such spaces on any story is limited to 10 percent of the floor area. These small storage spaces could occur in any other occupancy, such as a Group B office building, a Group E classroom or a Group M retail store. Individual spaces in excess of 100 square feet (9.3 m^2) would be classified as a Group S occupancy and the provisions of mixed occupancies (Section 508) would apply. Section 508.2 limits accessory occupancies on any story to 10 percent of the area of that story. The area of storage spaces would need to be aggregated to determine if the spaces can be classified as part of the main occupancy or if they need to be classified as a Group S occupancy. As a separated mixed use building, Table 508.4 does not require a separation between Group S-1 and Group F-1, B or M. Table 509 for incidental uses would require storage rooms in Group I-2 and ambulatory care facilities with an area over 100 square feet (9.3 m^2) to be separated by fire barrier or horizontal assemblies with a fire-resistance rating of at least 1 hour.

311.2 Moderate-hazard storage, Group S-1. Storage Group S-1 occupancies are buildings occupied for storage uses that are not classified as Group S-2, including, but not limited to, storage of the following:

Aerosols, Levels 2 and 3
Aircraft hangar (storage and repair)
Bags: cloth, burlap and paper
Bamboos and rattan
Baskets
Belting: canvas and leather
Books and paper in rolls or packs
Boots and shoes
Buttons, including cloth covered, pearl or bone
Cardboard and cardboard boxes
Clothing, woolen wearing apparel
Cordage
Dry boat storage (indoor)
Furniture
Furs
Glues, mucilage, pastes and size
Grains
Horns and combs, other than celluloid
Leather
Linoleum
Lumber
Motor vehicle repair garages complying with the maximum allowable quantities of hazardous materials listed in Table 307.1(1) (see Section 406.8)
Photo engravings
Resilient flooring
Silks
Soaps
Sugar
Tires, bulk storage of
Tobacco, cigars, cigarettes and snuff
Upholstery and mattresses
Wax candles

❖ Buildings in which combustible materials are stored and that burn with ease are classified in Group S-1, moderate-hazard storage occupancies. Examples of the kinds of materials that, when stored, are representative of occupancies classified in Group S-1 are also listed in this section.

As defined by the IFC, a repair garage is any structure used for servicing or repairing motor vehicles. Therefore, regardless of the extent of work done (e.g., quick lube, tune-up, muffler and tire shops, painting, body work, engine overhaul), repair garages are classified as Group S-1 (see Commentary Figure 311.2) and must be in compliance with Section 406.8. In addition, to avoid a Group H classification, the amounts of hazardous materials in the garage must be less than the maximum allowable quantity per control area permitted in Tables 307.1(1) and 307.1(2).

Aircraft hangars for storage, repair or both would be classified as Group S-1. This classification correlates with the actual use of such hangars, which very frequently would include some level of repair work, and also works with the requirements of NFPA 409. Aircraft hangers accessory to one- and two-family structures are classified as Group U occupancies.

311.3 Low-hazard storage, Group S-2. Storage Group S-2 occupancies include, among others, buildings used for the storage of noncombustible materials such as products on wood pallets or in paper cartons with or without single thickness divisions; or in paper wrappings. Such products are permitted to have a negligible amount of plastic *trim*, such as knobs, handles or film wrapping. Group S-2 storage uses shall include, but not be limited to, storage of the following:

Asbestos
Beverages up to and including 16-percent alcohol in metal, glass or ceramic containers
Cement in bags
Chalk and crayons
Dairy products in nonwaxed coated paper containers
Dry cell batteries
Electrical coils
Electrical motors
Empty cans
Food products
Foods in noncombustible containers
Fresh fruits and vegetables in nonplastic trays or containers
Frozen foods
Glass
Glass bottles, empty or filled with noncombustible liquids
Gypsum board
Inert pigments
Ivory
Meats
Metal cabinets
Metal desks with plastic tops and *trim*
Metal parts
Metals
Mirrors
Oil-filled and other types of distribution transformers
Parking garages, open or enclosed
Porcelain and pottery
Stoves
Talc and soapstones
Washers and dryers

❖ Buildings in which noncombustible materials are stored are classified as Group S-2, low-hazard storage occupancies (see Commentary Figure 311.3). It is acceptable for stored noncombustible products to be packaged in combustible materials as long as the quantity of packaging is kept to an insignificant level.

As seen in Group F-1 and F-2 classifications, it is important to be able to distinguish when the presence of combustible packaging constitutes a significant fuel load. As such, a fuel load might require the building to be classified in Group S-1, moderate-hazard storage. A simple guideline to follow is the "single thickness" rule, which is when a noncombustible product is put in one layer of packaging material.

Examples of materials qualified for storage in Group S-2 storage facilities are as follows:

- Vehicle engines placed on wood pallets for transportation after assembly;
- Washing machines in corrugated cardboard boxes; and
- Soft-drink glass bottles packaged in pressed paper boxes.

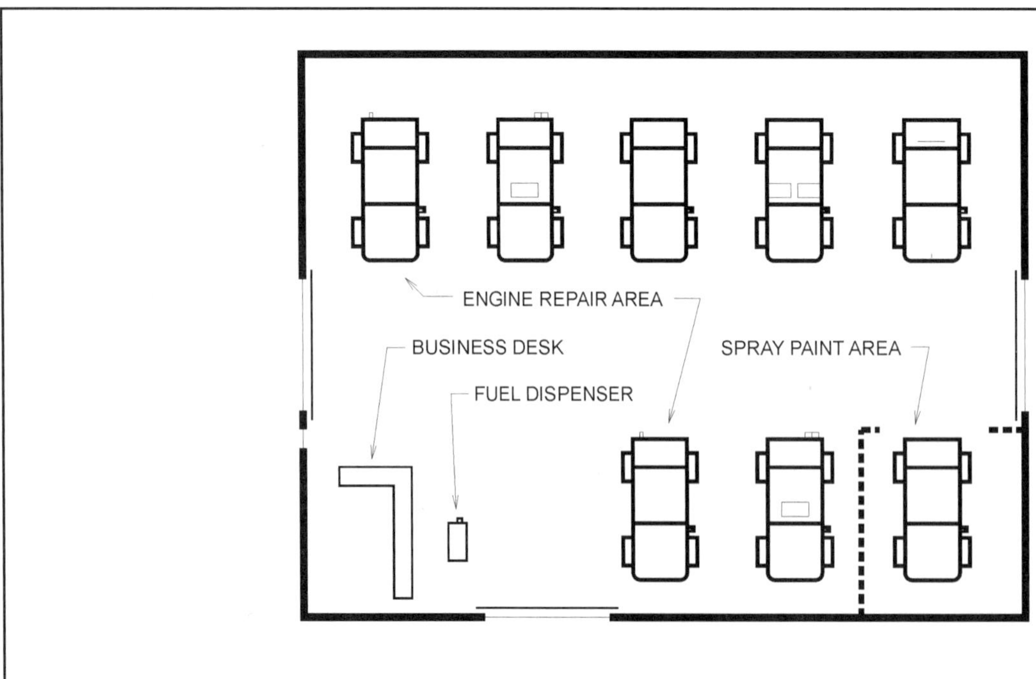

Figure 311.2
GROUP S-1—REPAIR GARAGES

Structures used to store noncombustible materials packaged in more than one layer of combustible packaging material are to be classified in Group S-1.

Examples of materials that, because of packaging, do not qualify for classification in Group S-2 are:

- Chinaware wrapped in corrugated paper and placed in cardboard boxes;
- Glassware set in expanded foam forms and placed in a cardboard box; and
- Fuel filters individually packed in pressed paper boxes, placed by the gross in a cardboard box and then stacked on a wood pallet for transportation.

An area of the IFC that is often related to Group S occupancies is Chapter 23, which regulates high-piled combustible storage [storage over 12 feet (3658 mm) in height or 6 feet (1829 mm) if the material is considered high hazard]. Chapter 23 of the IFC is focused not only on the type of materials being stored but also the height and configuration of such storage. It is important to note that not all Group S occupancies will contain high-piled storage and that high-piled storage is not limited to Group S occupancies. High-piled storage can be found in occupancies such as Groups H and F.

Open and enclosed parking garages are classified as Group S-2 occupancies as long as no repair activities as discussed in the commentary to Section 311.2 occur in such buildings. A garage in a fire station, for example, that undertakes maintenance and repairs limited to cleaning, hose change, water fill, fire equipment upgrades or wheel removal for repair off premises would not constitute the same hazard associated with repair garages and would be appropriately classified as a Group S-2 classification.

SECTION 312
UTILITY AND MISCELLANEOUS GROUP U

312.1 General. Buildings and structures of an accessory character and miscellaneous structures not classified in any specific occupancy shall be constructed, equipped and maintained to conform to the requirements of this code commensurate with the fire and life hazard incidental to their occupancy. Group U shall include, but not be limited to, the following:

Agricultural buildings
Aircraft hangars, accessory to a one- or two-family residence (see Section 412.5)
Barns
Carports
Fences more than 6 feet (1829 mm) in height
Grain silos, accessory to a residential occupancy
Greenhouses
Livestock shelters
Private garages
Retaining walls
Sheds
Stables
Tanks
Towers

❖ This section identifies the characteristics of occupancies classified in Group U. Structures that are classified in Group U are typically accessory to another building or structure and are not more appropriately classified in another occupancy. Miscellaneous storage buildings accessory to detached one- and two-family dwellings and multiple single-family dwellings (townhouses) not more than three stories in height, however, are intended to be designed and built in accordance with the !RC (see Section 101.2).

Structures classified as Group U, such as fences, equipment, foundations and retaining walls, are

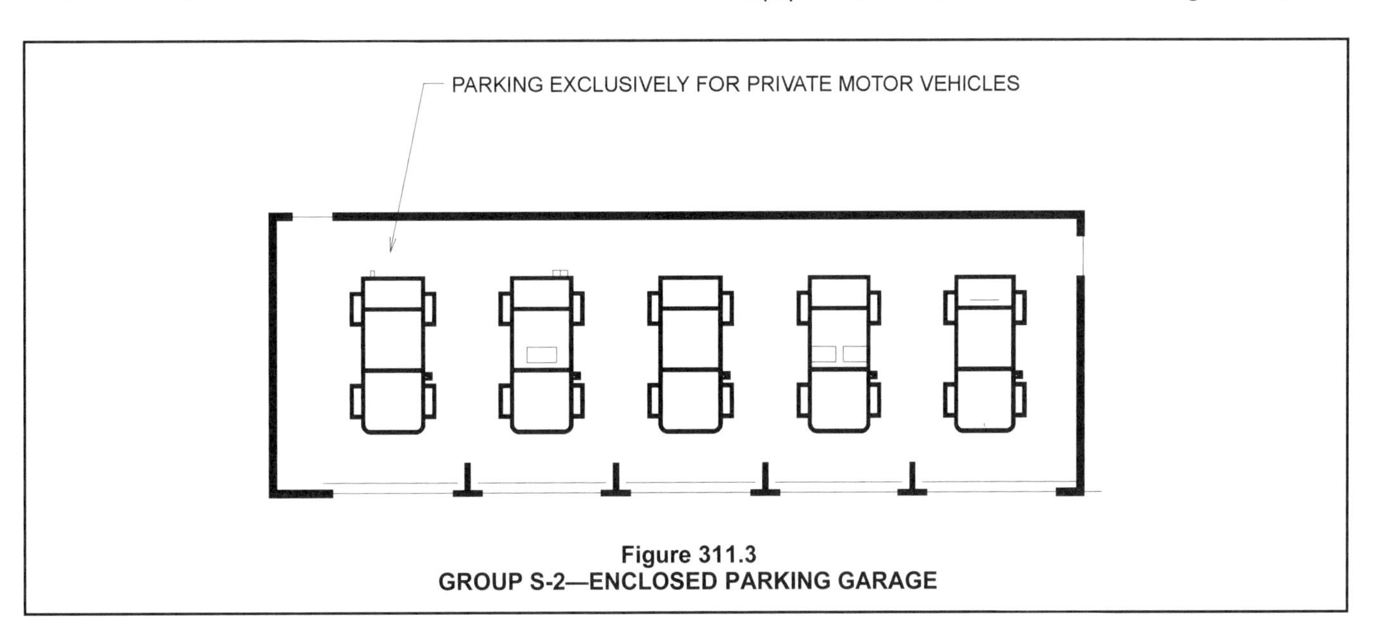

Figure 311.3
GROUP S-2—ENCLOSED PARKING GARAGE

somewhat outside the primary scope of the code (i.e., means of egress, fire resistance). They are not usually considered to be habitable or occupiable. Nevertheless, many code provisions do apply and need to be enforced (e.g., structural design and material performance).

Structures housing accessory equipment that is part of a utility or communications system are often classified as Group U occupancies when there is no intent that these structures be occupied except for servicing and maintaining the equipment within the structure. A pumphouse for a water or sewage system or an equipment building at the base of a telecommunications tower are examples of such buildings.

Group U occupancies are subject to the same structural loadings as other occupancies, such as snow loads. Section 312.1 establishes that occupancies classified as utility and miscellaneous structures shall be constructed, equipped and maintained to conform to the code requirements that are commensurate with the fire and life hazards incidental to their occupancy. The structural design requirements for roofs are the minimum deemed necessary to withstand such elements. Allowing construction of a building with an accessory occupancy that could reasonably be expected to collapse under the snow loads known to prevail in a certain area is not in the best interest of public safety.

Greenhouses are one of the listed uses in the Group U classification. Greenhouse is not a defined term in the code. Therefore, greenhouses can fall into categories other than Group U depending on their specific use. Greenhouse can be classified as business (Group B), education (Group E), and mercantile (Group M). The building code does not presently address these use conditions for greenhouses specifically, and the interpretation of how to apply these provisions can be inconsistent. However, there are specific considerations applicable in determining whether a greenhouse should be in another use group.

1. Is the greenhouse intended to have more than occasional access by the general public for business or education or retail purposes; and
2. Are those purposes more accurately reflected by Group B, E or M?

If the answer to one or more of these questions is yes, the greenhouse may be required to be constructed for a purpose in addition to the cultivation, protection and maintenance of plants. The greenhouse may be subject to the Group U provisions and the fire, egress, and accessibility requirements of another applicable use group, depending on the specific conditions for use.

The code provides Appendix C - Agricultural Buildings which, if adopted, provides alternative standards for Group U structures including egress, height and area limits, exits and mixed occupancy requirements. The building height and area limits in Chapter 5 which apply to Group U greenhouses are more restrictive than Groups S-1 and S-2, which have higher safety and hazard risks than the agricultural building specified in Appendix C. The height and area limits in Appendix C may more realistically reflect the relative hazards and occupancy level for Group U greenhouses. Even where Appendix C is not adopted, it may serve as a useful guide where larger greenhouse structures are proposed.

Bibliography

The following resource materials were used in the preparation of the commentary for this chapter of the code.

DOL 29 CFR; Part 1910-09, *Occupational Safety and Health Standards*. Washington, DC: United States Department of Labor, 2007.

ISO 8115-86, *Cotton Bales—Dimensions and Density*. Geneva 20, Switzerland: International Organization for Standardization, 1986.

NFPA 13-13, *Installation of Sprinkler Systems*. Quincy, MA: National Fire Protection Association, 2013.

NFPA 30B-15, *Manufacture and Storage of Aerosol Products*. Quincy, MA: National Fire Protection Association, 2015.

NFPA 31-11, *Installation of Oil-burning Equipment*. Quincy, MA: National Fire Protection Association, 2011.

NFPA 37-06, *Installation and Use of Stationary Combustion Engines and Gas Turbines*. Quincy, MA: National Fire Protection Association, 2006.

NFPA 495-13, *Explosive Materials Code*. Quincy, MA: National Fire Protection Association, 2013.

UL 30-95, *Metal Safety Cans—with Revisions through December 2004*. Northbrook, IL: Underwriters Laboratories, Inc., 1995.

Chapter 4: Special Detailed Requirements Based on Use and Occupancy

General Comments

The provisions of Chapter 4 are supplemental to the remainder of the code. Chapter 4 contains provisions that may alter requirements found elsewhere in the code; however, the general requirements of the code still apply unless modified within the chapter. For example, the height and area limitations established in Chapter 5 apply to all special occupancies unless Chapter 4 contains height and area limitations. In this case, the limitations in Chapter 4 supersede those in other sections. An example of this is the height and area limitations given in Section 406.5.4, which supersede the limitations given in Tables 504.3, 504.4 and 506.2 for open parking garages.

The *International Fire Code*® (IFC®) contains provisions applicable to the storage, handling and use of hazardous substances, materials or devices and, therefore, must also be complied with when addressing such occupancies as those involving flammable and combustible liquids. Similarly, the *International Mechanical Code*® (IMC®) and the *International Plumbing Code*® (IPC®) include provisions for specific applications, such as hazardous exhaust systems and hazardous material piping.

In some instances, it may not be necessary to apply the provisions of Chapter 4. For example, if a covered mall building complies with the provisions of the code for the occupancies within the building, then Section 402 does not apply. However, other sections that address a use, process or operation must be applied to that specific occupancy, such as Sections 410, 411 and 414.

Purpose

The purpose of Chapter 4 is to combine in one chapter the provisions of the code applicable to special uses and occupancies. Hazardous materials and operations may occur in more than one group; therefore, the applicable provisions for the specific hazardous occupancy or operation apply to multiple groups. Also, while the provisions for all structures are interrelated to form an overall protection system, by providing requirements for specific occupancies in Chapter 4 the package of protection features is more easily identified.

Chapter 4 contains the requirements for protecting special uses and occupancies. The provisions in this chapter reflect those occupancies and groups that require special consideration and are not addressed elsewhere in the code. The chapter includes requirements for buildings and conditions that apply to one or more groups, such as high-rise buildings or atriums. Special uses may also imply specific occupancies and operations, such as for Groups H-1, H-2, H-3, H-4 and H-5, application of flammable finishes and combustible storage; or for a specific occupancy within a much larger occupancy, such as covered mall buildings, motor-vehicle-related occupancies, special amusement buildings and aircraft-related occupancies. Finally, in order that the overall package of protection features can be easily understood, occupancies, such as Groups I-2 and I-3, live/work units and underground buildings, are addressed.

SECTION 401 SCOPE

401.1 Detailed use and occupancy requirements. In addition to the occupancy and construction requirements in this code, the provisions of this chapter apply to the special uses and occupancies described herein.

❖ This section provides guidance on how Chapter 4 is to be applied with respect to other sections of the code. Section 401.1 indicates that all other provisions of the code apply except as modified by Chapter 4.

The requirements contained in Chapter 4 are intended to apply to special uses and occupancies, as well as special construction features as defined in various sections in this chapter. These requirements are applicable in addition to other chapters of the code.

SECTION 402 COVERED MALL AND OPEN MALL BUILDINGS

402.1 Applicability. The provisions of this section shall apply to buildings or structures defined herein as *covered or open mall buildings* not exceeding three floor levels at any point nor more than three *stories above grade plane*. Except as specifically required by this section, *covered and open mall buildings* shall meet applicable provisions of this code.

Exceptions:

1. Foyers and lobbies of Groups B, R-1 and R-2 are not required to comply with this section.
2. Buildings need not comply with the provisions of this section where they totally comply with other applicable provisions of this code.

❖ This section primarily addresses shopping centers with a maximum of three levels that consist of one or more sizeable anchor businesses, typically department stores, known as anchor buildings, and numerous smaller retail stores, all of which are interconnected by means of a mall or common pedestrian way. See Commentary Figure 402.1. The mall may be covered, providing a climate-controlled environment, or open to the sky. The complex may include movie theaters, bowling lanes, ice arenas, offices, and dining and drinking establishments. Historically, anchor buildings were large department stores, and malls remain attached to department stores, but under this section anchor buildings can contain any occupancy with the exception of a Group H. The complex may also include single- or multiple-level buildings, with a vast majority of shopping centers being one or two levels, and with an occasional mezzanine. Many malls contain large food courts with a variety of food vendors surrounding a common seating area.

This section addresses the special considerations associated with covered or open mall buildings, including construction, egress and fire protection systems. It does not, in general, apply to the anchor buildings around the perimeter of the covered or open mall building. To be considered an anchor building for purposes of applying the code, the building must have complete egress facilities, including the required number and capacity of exits, independent of the covered or open mall building.

Originally, the provisions of this section applied to the typical covered mall building: a one- to three-level structure consisting primarily of retail space and a covered pedestrian way. More recently, other types of covered mall buildings, such as airport passenger terminals and office centers, have also been constructed in accordance with this section.

Malls with an unroofed common pedestrian way, or open malls, are becoming common in the "sun belt" areas of the country and in similar climates around the world. The code includes open malls within the broader definition of covered mall building and allows the same benefits of the covered mall provisions, because an open-to-the-sky mall provides equivalent or better life safety and property protection. The key to the open mall concept is to have everything a covered mall building would have, except for the roof

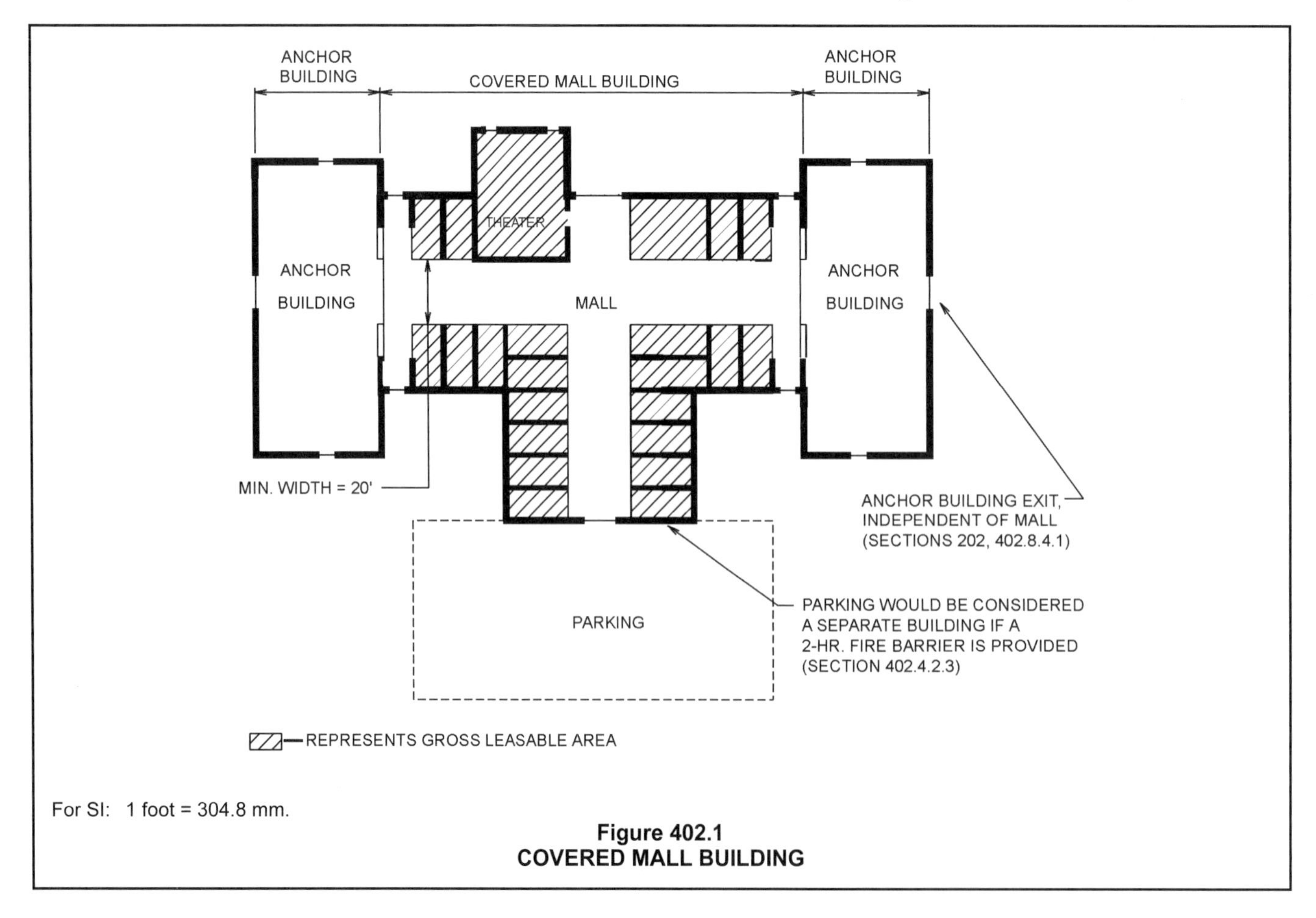

Figure 402.1
COVERED MALL BUILDING

over the mall area. Without a roof over the mall area and with required openings from the grade level to the sky above, natural ventilation is provided and mechanical smoke control is no longer necessary in the mall area and adjoining tenant spaces.

This section is not intended to apply to the foyers and lobbies of business and residential occupancies that may contain retail spaces, small restaurants and offices unrelated to the use of upper stories. If these lobbies have multiple levels open to each other, the provisions of Section 404, Atriums, may apply. In all such cases, the provisions for mixed occupancies found in Section 508 would apply. This section is also not intended to apply to large-footprint retail buildings, which may include smaller retail spaces, such as banks, florists or coffee stands, around the perimeter of a central sales area. These buildings do not provide the intended concept of a central pedestrian mall that is distinct from the surrounding tenant spaces.

Covered or open mall buildings that comply with all other applicable provisions of the code need not comply with the provisions of this section; however, covered or open mall buildings that are designed and constructed to comply with the provisions of this section must also comply with the provisions of the code that are not otherwise modified in this section. For example, the egress provisions in Section 402.8 are applicable to covered or open mall buildings designed in accordance with this section. The provisions in Chapter 10 related to door swing and stairways are also applicable, however, since Section 402 contains no similar provisions.

402.1.1 Open space. A *covered mall building* and attached *anchor buildings* and parking garages shall be surrounded on all sides by a permanent open space or not less than 60 feet (18 288 mm). An *open mall building* and *anchor buildings* and parking garages adjoining the perimeter line shall be surrounded on all sides by a permanent open space of not less than 60 feet (18 288 mm).

Exception: The permanent open space of 60 feet (18 288 mm) shall be permitted to be reduced to not less than 40 feet (12 192 mm), provided the following requirements are met:

1. The reduced open space shall not be allowed for more than 75 percent of the perimeter of the *covered or open mall building* and *anchor buildings*;
2. The *exterior wall* facing the reduced open space shall have a *fire-resistance rating* of not less than 3 hours;
3. Openings in the *exterior wall* facing the reduced open space shall have opening protectives with a *fire protection rating* of not less than 3 hours; and
4. Group E, H, I or R occupancies are not located within the *covered or open mall building* or *anchor buildings*.

❖ One fundamental aspect of the covered mall building concept is that significant open space be provided around the perimeter of the entire building. The open space serves two key roles: separation of these buildings from other buildings and ample space on all sides for fire-fighting operations. The required minimum open space of 60 feet (18 288 mm) must also be provided at any anchor buildings and parking garages attached to the covered mall building. In all cases the open space can occur either within public ways surrounding the site, or by yards provided on the lot between the covered mall building and the lot lines, or a combination of yards and public ways. Where a yard is used to achieve the open space, it must be on the same lot as the building receiving the benefit. With respect to the 60-foot (18 288 mm) requirement, the entire width of the public way can be included. The open space must be provided in all directions around the perimeter of the building, not just measured at right angles to the building. A similar degree of open space is required to surround open mall buildings and any adjacent anchor buildings and parking garages.

The exception allows covered or open mall buildings to reduce the required open space around the building in the same manner as unlimited area buildings regulated in Section 507. Where permitted in Section 507, unlimited area buildings require an open space around the building of 60 feet (18 288 mm) to prevent exposure fire spread. Section 507.2.1 permits five of the nine categories of unlimited area buildings to have a reduced open space of only 40 feet (12 192 mm). This reduction is permitted in these locations when the exterior wall in the required space area is rated for 3 hours and does not include more than 75 percent of the building perimeter. A significant majority of the uses described in Section 507 are found in the definition of a covered or open mall building. Therefore, it can be assumed that the fire load within a mall is similar to the fire load permitted in a code-compliant unlimited area building. Since the intent of the open space requirement is to prevent fire spread from one building to another and the fire loads are similar, allowing covered or open mall buildings to be subject to the same separation requirements as other unlimited area buildings does not reduce the level of fire protection afforded. The fourth requirement of the exception aligns this section with the occupancies permitted to use the unlimited area provisions of Section 507.2.1. Group E, H, I and R occupancies are not prohibited from being within a covered or open mall building or an anchor store, but the exception would prohibit the use of reduced open space if one of these occupancies was a part of the covered or open mall building or anchor building.

402.1.2 Open mall building perimeter line. For the purpose of this code, a perimeter line shall be established. The perimeter line shall encircle all buildings and structures that comprise the *open mall building* and shall encompass any open-air interior walkways, open-air courtyards or similar open-air spaces. The perimeter line shall define the extent of the *open mall building*. *Anchor buildings* and parking structures shall be outside of the perimeter line and are not considered as part of the *open mall building*.

❖ Where the intent is to apply the open mall building provisions of Section 402, it is necessary to establish a building perimeter line that can be used for the application of various special requirements, such as those regarding travel distance in Section 402.8.5. As shown in Commentary Figure 402.1.2, the perimeter line encloses all portions of the open mall building, including open-air walkways and spaces, but does not enclose any anchor buildings or parking garages adjacent to the open mall building. The designer has the discretion as to the relative location of the perimeter line, similar to the discretion they have to place an assumed imaginary line between two buildings located on the same lot. But even with that discretion, it is anticipated that the perimeter line will, in essence, enclose the same space as exterior walls would in a covered mall building.

402.2 Definitions. The following terms are defined in Chapter 2:

ANCHOR BUILDING.

COVERED MALL BUILDING.

Mall.

Open mall.

Open mall building.

FOOD COURT.

GROSS LEASABLE AREA.

❖ This section lists terms that are specifically associated with the subject matter of Section 402. It is important to emphasize that these terms are not exclusively related to this chapter but may or may not also be applicable where the term is used elsewhere in the code.

Definitions of terms can help in the understanding and application of the code requirements. The purpose for including a list within this chapter is to provide more convenient access to terms which may have a specific or limited application within this chapter. For the complete definition and associated commentary, refer back to Chapter 2. Terms that are italicized provide a visual identification throughout the code that a definition exists for that term. The use and application of all defined terms are set forth in Section 201.

402.3 Lease plan. Each *owner* of a *covered mall building* or of an *open mall building* shall provide both the building and fire departments with a lease plan showing the location of each occupancy and its *exits* after the certificate of occupancy has been issued. No modifications or changes in occupancy or use shall be made from that shown on the lease plan without prior approval of the *building official*.

❖ The required lease plan is permitted to be submitted after the certificate of occupancy has been issued because many times the developer does not have the information at the time of construction. The location of tenant separations may not be known until leases are negotiated with prospective tenants. This may occur after the mall has opened.

During initial construction, it is anticipated that the building department will require tenant improvements to be submitted through the building permit process. After tenant spaces are prepared for occupancy and the lease plan is developed, subsequent modifications and changes must be submitted for approval

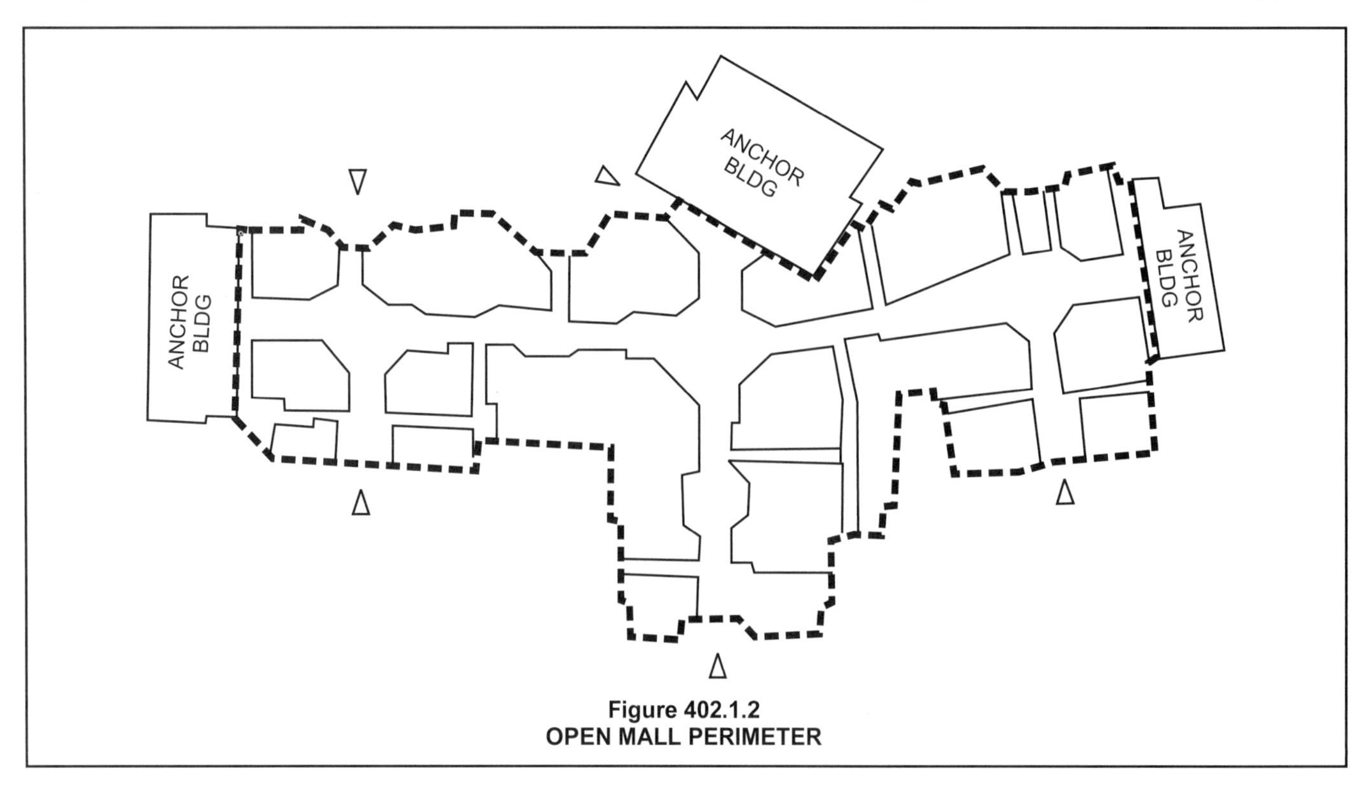

Figure 402.1.2
OPEN MALL PERIMETER

prior to commencing construction or changing the use. It is important that the fire department receives copies of current lease plans, since not only does this help the fire department while performing fire prevention inspections, but also the lease plans assist in fire department response to an emergency.

402.4 Construction. The construction of *covered and open mall buildings*, *anchor buildings* and parking garages associated with a *mall* building shall comply with Sections 402.4.1 through 402.4.3.

❖ A number of construction-related provisions are specific to covered and open mall buildings, anchor buildings and associated parking garages. Such buildings are regulated in this section for allowable floor area and height, type of construction, and fire-resistance-rated separations. The various requirements in these code sections are illustrated in Commentary Figure 402.4.

402.4.1 Area and types of construction. The *building area* and type of construction of *covered mall* or *open mall buildings*, *anchor buildings* and parking garages shall comply with this section.

❖ Specific construction requirements and area allowances are provided for mall buildings, anchor buildings and parking garages in Sections 402.4.1.1, 402.4.1.2 and 402.4.1.3, respectively.

402.4.1.1 Covered and open mall buildings. The *building area* of any *covered mall* or *open mall building* shall not be limited provided the *covered mall* or *open mall building* does not exceed three floor levels at any point nor three *stories above grade plane*, and is of Type I, II, III or IV construction.

❖ Covered or open mall buildings are considered to be special types of unlimited area buildings and are exempt from the allowable area limitations of Section 506 where they are limited to not more than three stories above grade plane. It should be noted that the height limitations in feet, specified in Table 504.3 based on the type of construction classification, are applicable to covered or open mall buildings. The allowance of an unlimited area is also based on the restriction of construction types to noncombustible (Types I and II), noncombustible/combustible (Type III) and heavy timber (Type IV). It assumes an effective automatic sprinkler system is provided in accordance with Section 402.5.

402.4.1.2 Anchor buildings. The *building area* and *building height* of any *anchor building* shall be based on the type of

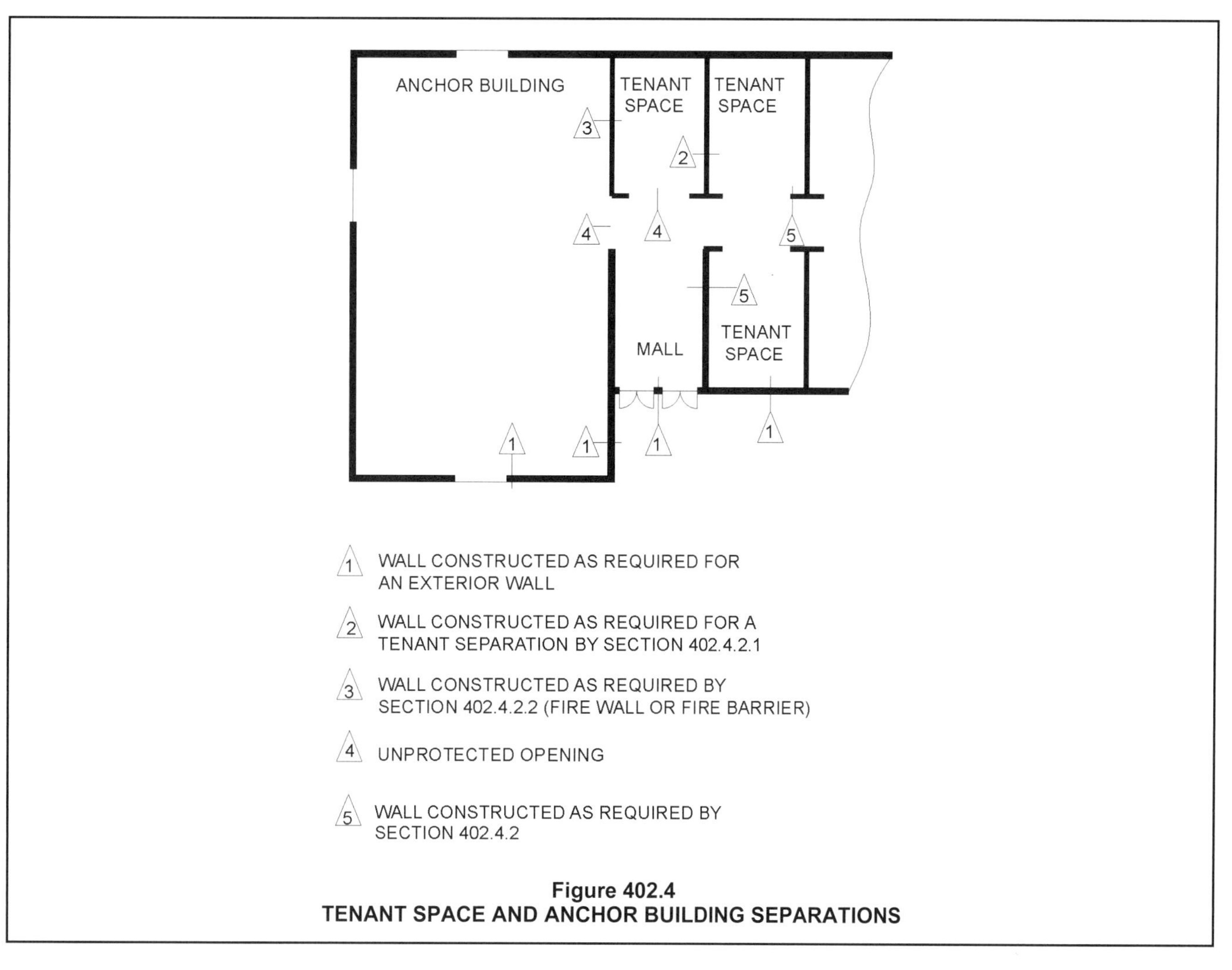

Figure 402.4
TENANT SPACE AND ANCHOR BUILDING SEPARATIONS

construction as required by Section 503 as modified by Sections 504 and 506.

Exception: The *building area* of any *anchor building* shall not be limited provided the *anchor building* is not more than three *stories above grade plane*, and is of Type I, II, III or IV construction.

❖ Anchor buildings, although attached to the mall building, can be treated as separate buildings with respect to building height, building area and construction type. For example, where a mall is limited to three stories, the attached anchor building built of Type I-A construction could be a high-rise building.

According to the exception, where the anchor building is limited to the same limits as the mall building, the anchor building can also be of unlimited area.

402.4.1.3 Parking garage. The *building area* and *building height* of any parking garage, open or enclosed, shall be based on the type of construction as required by Sections 406.5 and 406.6, respectively.

❖ This section provides a reminder that the construction of parking garages is regulated by the general provisions found in Sections 406.5 and 406.6 and that they are not regulated by the covered mall building's type of construction. It also clarifies that the garage is not included in the unlimited area allowance given to the mall building and anchor stores.

402.4.2 Fire-resistance-rated separation. Fire-resistance-rated separation is not required between tenant spaces and the *mall*. Fire-resistance-rated separation is not required between a *food court* and adjacent tenant spaces or the *mall*.

❖ From an operational point of view, separating the tenant space from the mall is not practical in that customer flow would be restricted and, therefore, detract from the merchandising purpose of covered or open malls. Historical experience and reliability data on automatic sprinkler performance indicate that a separation between the tenant space and the mall is not warranted.

402.4.2.1 Tenant separations. Each tenant space shall be separated from other tenant spaces by a *fire partition* complying with Section 708. A tenant separation wall is not required between any tenant space and the *mall*.

❖ Covered mall buildings are essentially large unlimited area mixed-occupancy buildings. Rather than complying with Section 508.3 or 508.4, the building has a series of protections including the tenant separations specified in this section. In order to limit the spread of smoke, tenant separation walls are required to be fire partitions (see Section 708) having a fire-resistance rating of at least 1 hour and extending from the floor to the underside of the ceiling (see Commentary Figure 402.4). Extending tenant separations to the floor slab or roof deck above is not always practical or possible because of operation of the heating, ventilating and air-conditioning (HVAC) system. The effectiveness of the automatic sprinkler system is also a reason for not requiring tenant separations to extend above the ceiling, including attic spaces (see Section 708.4, Exception 4).

The open mall "building" provides a unique question regarding separation of tenant spaces. The open mall building may likely be a collection of separate structures treated as one open mall "building." The extent of the open mall building will be established by the perimeter line. Where a separate structure within an open mall contains multiple tenant spaces, those spaces need to separated as required in this section.

Within the perimeter line, the walls between the tenant spaces and the mall will be "interior walls" to the whole open mall building complex, but they are also exterior walls as they separate interior from exterior environments. As part of a complex of buildings, the exterior walls of each individual tenant space facing the mall will not need to meet the wall and opening protection of Section 705, but do need to comply with other exterior wall provisions of Chapter 14. In most cases where there are separate "buildings" within the perimeter line, there will be a walkway—or pedestrian mall—between the buildings. Even where these are off the main mall, these are still considered tenant walls facing the mall. The walls on the exterior face of the open mall "building," those adjoining the perimeter line, would be subject to the provisions of Section 705.

402.4.2.2 Anchor building separation. An *anchor building* shall be separated from the *covered or open mall building* by *fire walls* complying with Section 706.

Exceptions:

1. *Anchor buildings* of not more than three *stories above grade plane* that have an occupancy classification the same as that permitted for tenants of the *mall building* shall be separated by 2-hour fire-resistance-rated *fire barriers* complying with Section 707.
2. The exterior walls of *anchor buildings* separated from an *open mall building* by an *open mall* shall comply with Table 602.

❖ In general, anchor buildings are typically viewed as being separate from the covered or open mall building (see commentary, Sections 402.1 and 402.4.1). As separate buildings, the code requires that a fire wall complying with Section 706 be constructed.

The first exception to this section may be used in situations where the occupancy and size limitations of the anchor building are the same as what would be permitted for any other tenant within the covered mall building. In those situations, the code permits the separation to be constructed as a fire barrier instead of as a fire wall. The exception would not be applicable if the anchor building was over three stories above grade plane or if it contained an occupancy that was not normally permitted within the mall. An example would be a high-rise office building or a hotel that was connected with the covered mall building (see Commentary Figure 402.4 and commentary, Section 402.4.2.2.1).

For an open mall building where the anchor building is actually a separate building, the application of a fire wall may not be appropriate, but since it is considered a separate structure even under the covered mall scenario, it is appropriate to have the anchor building provide exterior wall protection set in Table 602. This allowance is set forth in Exception 2.

A growing trend in the development of mall buildings and anchor buildings is the desire of the operators of the businesses in an anchor building to own the structure of the anchor building as well as a distinct parcel of land on which the anchor building sits. The original intent of the mall provisions was to apply to one building, or a collection of buildings, located on one lot under the control of one owner. Various provisions assume a single "lot" for the covered mall and associated anchor buildings and parking structures. If the mall and associated buildings are managed as a single complex, strict interpretation would eliminate the application of these provisions to such a complex. Application of Section 402 provisions to a situation where the mall and anchor buildings are separate ownership lots would be problematic, requiring special legal and technical considerations and treatment by the authority having jurisdiction.

402.4.2.2.1 Openings between anchor building and mall. Except for the separation between Group R-1 *sleeping units* and the *mall*, openings between *anchor buildings* of Type IA, IB, IIA or IIB construction and the *mall* need not be protected.

❖ As with tenant separations, openings between the anchor building and the pedestrian area of the mall need not be protected. Such separations would defeat the merchandising purpose of covered malls by restricting customer flow and visual access. While the openings may be unprotected, the separations between the anchor building and all tenant spaces in the covered mall building are to be constructed as required in Section 402.4.2.2 for a fire wall (see commentary, Section 706) or fire barrier (see commentary, Section 707) since the anchor building is considered a separate building (see Commentary Figure 402.4).

402.4.2.3 Parking garages. An attached garage for the storage of passenger vehicles having a capacity of not more than nine persons and *open parking garages* shall be considered as a separate building where it is separated from the *covered or open mall building* or *anchor building* by not less than 2-hour *fire barriers* constructed in accordance with Section 707 or *horizontal assemblies* constructed in accordance with Section 711, or both.

Parking garages, open or enclosed, which are separated from *covered mall buildings*, *open mall buildings* or *anchor buildings*, shall comply with the provisions of Table 602.

Pedestrian walkways and tunnels that connect garages to *mall* buildings or *anchor buildings* shall be constructed in accordance with Section 3104.

❖ Buildings adjacent to mall buildings are typically considered separate buildings provided they are separated by exterior walls or fire walls. In recognition of the limited hazard presented by attached garages and open parking garages, this section permits such structures to be considered separate buildings, if fire barriers and horizontal assemblies having a fire-resistance rating of at least 2 hours are provided. See the commentary to Section 707 relative to the construction of fire barriers and Section 711 relative to the construction of horizontal assemblies.

The 2-hour fire-resistance rating is consistent with the requirements of Table 508.4 for Groups M and S-2. If the 2-hour fire barrier or horizontal assembly is not provided, the parking structure can be considered part of the covered mall building and be limited. Table 602, regulating exterior wall protection, is applicable for those parking garages that are not attached to the covered or open mall building or anchor buildings. See the commentary to Section 3104 for the requirements related to any pedestrian walkways or tunnels which connect the garages to mall buildings or anchor stores.

402.4.3 Open mall construction. Floor assemblies in, and *roof assemblies* over, the *open mall* of an *open mall building* shall be open to the atmosphere for not less than 20 feet (9096 mm), measured perpendicular from the face of the tenant spaces on the lowest level, from edge of balcony to edge of balcony on upper floors and from edge of roof line to edge of roof line. The openings within, or the unroofed area of, an *open mall* shall extend from the lowest/grade level of the open mall through the entire *roof assembly*. Balconies on upper levels of the *mall* shall not project into the required width of the opening.

❖ An open mall requires a minimum open area width of 20 feet (9096 mm) between opposing structures from the lowest grade level to the sky above. This dimension aligns with Section 402.8.1.1, minimum width for egress. The width of the opening is measured perpendicular from the face of the tenant spaces, essentially across the pedestrian mall. This is similar to the mall width for covered malls. The opening, or the unroofed area, must extend from the lowest grade level of the common open mall to the sky above. This will provide natural ventilation to all levels. Balconies on the upper levels are permitted but may not project into the required 20-foot (9096 mm) width. Pedestrian walkways and overhangs at the end locations of the open mall are permitted (see Commentary Figure 402.4.3).

402.4.3.1 Pedestrian walkways. *Pedestrian walkways* connecting balconies in an *open mall* shall be located not less than 20 feet (9096 mm) from any other *pedestrian walkway*.

❖ Where pedestrian walkways are provided, they must be spatially separated in order to maintain the expected openness of the open mall area. The minimum required distance of 20 feet (9096 mm) is consistent with the minimum open space required between various buildings that make up an open mall building.

[F] 402.5 Automatic sprinkler system. *Covered* and *open mall buildings* and buildings connected shall be equipped throughout with an *automatic sprinkler system* in accordance with Section 903.3.1.1, which shall comply with all of the following:

1. The *automatic sprinkler system* shall be complete and operative throughout occupied space in the *mall building* prior to occupancy of any of the tenant spaces. Unoccupied tenant spaces shall be similarly protected unless provided with *approved* alternative protection.
2. Sprinkler protection for the *mall* of a *covered mall building* shall be independent from that provided for tenant spaces or *anchor buildings*.
3. Sprinkler protection for the tenant spaces of an *open mall building* shall be independent from that provided for *anchor buildings*.
4. Sprinkler protection shall be provided beneath exterior circulation balconies located adjacent to an *open mall*.
5. Where tenant spaces are supplied by the same system, they shall be independently controlled.

   Exception: An *automatic sprinkler system* shall not be required in spaces or areas of *open parking garages* separated from the *covered or open mall building* in accordance with Section 402.4.2.3 and constructed in accordance with Section 406.5.

❖ The covered or open mall building and connected buildings, such as anchor buildings, must be protected with an automatic sprinkler system to protect life and property effectively. As has been discussed throughout the section, numerous allowances (such

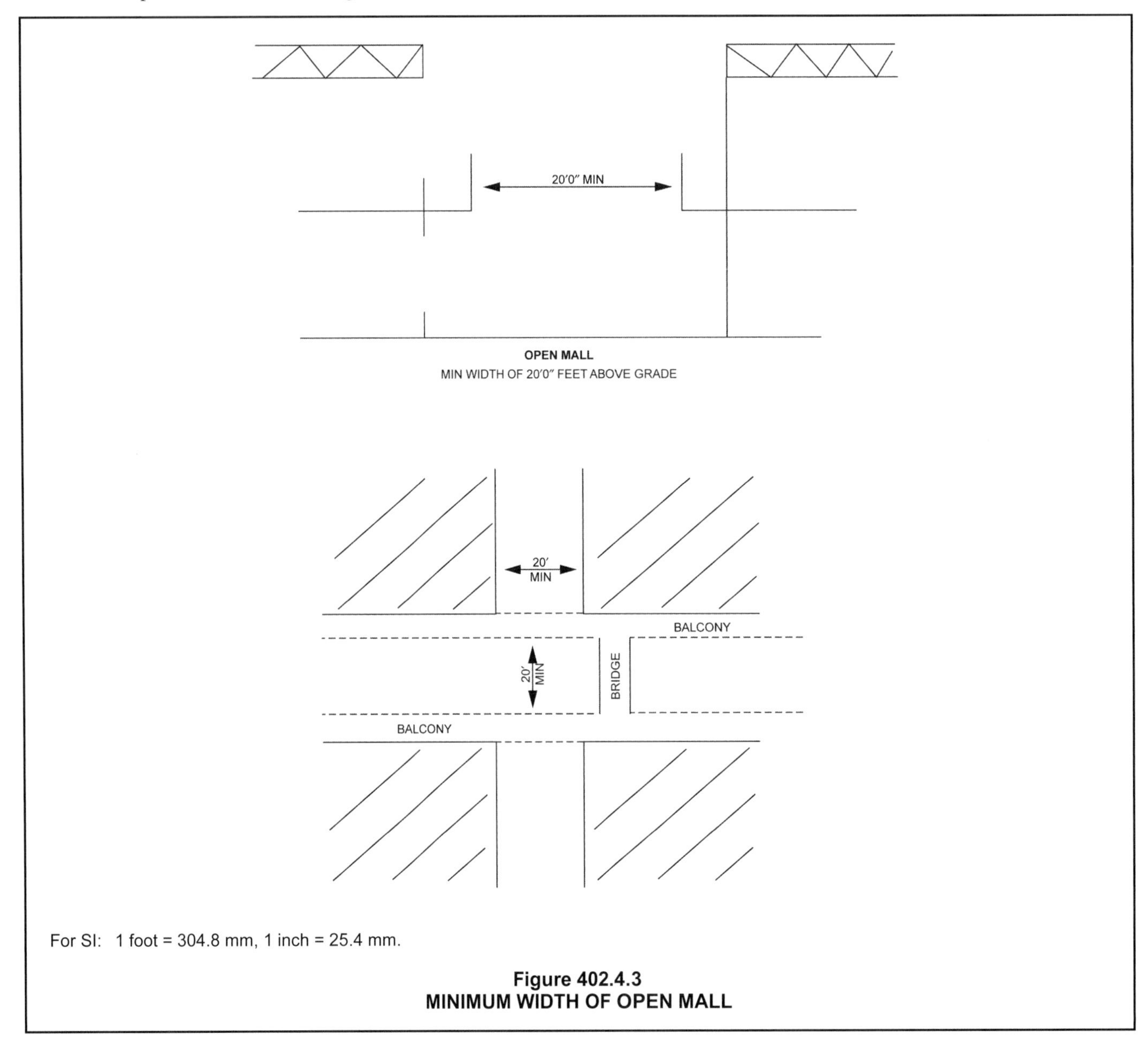

For SI: 1 foot = 304.8 mm, 1 inch = 25.4 mm.

Figure 402.4.3
MINIMUM WIDTH OF OPEN MALL

as reduced tenant separations and elimination of area limitations) are based on the effectiveness of the automatic sprinkler system.

The sprinkler system is to be designed, installed, tested and maintained in accordance with Chapter 9, the IFC and NFPA 13. Additionally, the system must be installed such that any portion serving tenant spaces in a covered mall building may be shut down independently without affecting the operation of the systems protecting the mall area. This special feature is in recognition of the frequent need to shut down the system so that changes can be made to it as a result of tenant improvements and modifications.

In an open mall building, the sprinkler systems for the tenant spaces are to be separate from those provided for the anchor buildings. It is necessary that sprinkler protection for open mall buildings be extended to the underneath side of pedestrian walkways and exterior balconies used for circulation and egress purposes. Although the open mall itself requires no sprinkler protection, it is important that those areas below overhead walkways be provided with sprinklers.

Section 909.12.3 requires operation of the sprinkler system to activate automatically the mechanical smoke control system (where an automatic control system is utilized). It is imperative that the zoning of the sprinkler system match the zoning of the smoke control system. This is necessary so that the area where water flow has occurred will also be the area from which smoke is removed. Smoke control is only required where an atrium that connects three stories is present (see Section 402.7.2).

The exception clarifies that sprinkler protection need not be extended to complying open parking garages that are adequately separated from the covered mall building, open mall building or anchor store. The allowance is consistent with the automatic sprinkler provisions of Section 903.2 where no sprinkler protection is required in open parking garages based on occupancy classification (Group S-2).

402.6 Interior finishes and features. Interior finishes within the *mall* and installations within the *mall* shall comply with Sections 402.6.1 through 402.6.4.

❖ This section addresses one general fire safety feature (interior finishes) and three specific features commonly found in covered mall buildings. Such features are also addressed where applicable within the perimeter line of an open mall building.

402.6.1 Interior finish. *Interior wall* and *ceiling finishes* within the *mall* of a *covered mall building* and within the *exits* of *covered or open mall buildings* shall have a minimum *flame spread index* and smoke-developed index of Class B in accordance with Chapter 8. *Interior floor finishes* shall meet the requirements of Section 804.

❖ Although Section 804 addresses the wall and ceiling finish requirements for mall building tenant spaces and anchor buildings based on the occupancies involved, Section 804 does not specifically state the requirements for interior finish within the mall of a covered mall building. In addition, it does not specify the interior wall and ceiling finish requirements for exits of open or covered mall buildings. Since a covered mall building is actually a mixed-use building, the most restrictive occupancy requirement should most likely apply. The use of the mall has gone beyond its original intent of providing a wide-open walking space for people. Mall spaces have become locations to assemble to watch a dance recital, attend a fashion show, or for the local radio station to run a promotion. It has become clear that malls are assembly spaces themselves, not taking into account the tenant assembly spaces that open into them. Therefore, the mall spaces and exits are regulated for interior finishes in a manner similar to Group A requirements. The wall and ceiling finishes must have a minimum flame spread index and smoke-developed index of Class B.

402.6.2 Kiosks. Kiosks and similar structures (temporary or permanent) located within the *mall* of a *covered mall building* or within the perimeter line of an *open mall building* shall meet the following requirements:

1. Combustible kiosks or other structures shall not be located within a *covered* or *open mall* unless constructed of any of the following materials:
   - 1.1. *Fire-retardant-treated* wood complying with Section 2303.2.
   - 1.2. Foam plastics having a maximum heat release rate not greater than 100 kW (105 Btu/h) when tested in accordance with the exhibit booth protocol in UL 1975 or when tested in accordance with NFPA 289 using the 20 kW ignition source.
   - 1.3. Aluminum composite material (ACM) meeting the requirements of Class A *interior finish* in accordance with Chapter 8 when tested as an assembly in the maximum thickness intended.
2. Kiosks or similar structures located within the *mall* shall be provided with *approved automatic sprinkler system* and detection devices.
3. The horizontal separation between kiosks or groupings thereof and other structures within the *mall* shall be not less than 20 feet (6096 mm).
4. Each kiosk or similar structure or groupings thereof shall have an area not greater than 300 square feet (28 m^2).

❖ Other potential sources of combustibles within the mall area are kiosks and similar structures. As with the restrictions of Section 402.6.4 on plastic signs, in order to maintain the mall as a viable means of egress, the amount of combustibles within the mall must be minimized. The restriction on kiosk construction and location is also intended to minimize the potential for fire spread through the mall area. These restrictions apply to both permanent and temporary structures and are applicable to both covered and open mall buildings.

Kiosks and similar structures are permitted to be of noncombustible construction or may be of combustible construction where the provisions of Item 1 are followed. The code allows combustible kiosks to be made of fire-retardant-treated wood or certain foam plastics and aluminum composites that provide performance comparable to that of the fire-retardant-treated wood. As such, the amount of combustibles within the mall and the potential for fire spread through the mall are minimized.

If the kiosk or similar structure has a cover, the automatic sprinkler system within the covered mall building will not be able to control effectively a fire within the kiosk (or similar structure) in the early stages of development. Suppression and detection devices must, therefore, be installed within such structures when there is anything that shields the mall sprinkler system from areas within the kiosk.

In order to minimize a fire exposure hazard, kiosks and similar structures must be located at least 20 feet (6096 mm) from each other or be situated in groups that are appropriately separated from other kiosks. Although the structure itself is often noncombustible or of the limited types of acceptable combustible materials, it is recognized that combustibles may be displayed within the structure and, therefore, the separation minimizes the potential for fire spread from kiosk to kiosk. If a series of kiosks do not have 20 feet (6096 mm) separating them, they can be treated as a group and would be limited to 300 square feet (28 m^2), as a group. The open space between individual kiosks in the group can be excluded from the aggregate area for determining the area of the group (see IBC Interpretation No. 11-07). Within a group of kiosks treated as one, there can be multiple "tenants" or one, and there are no specifics about the relative size of individual kiosks within the group (see IBC Interpretation Nos. 13-07 and 14-07). As required by Section 402.8.1.1, the kiosk must also be located at least 10 feet (3048 mm) from any projection of a tenant space.

To further control the use of kiosks and similar structures within the mall, their size is restricted to 300 square feet (28 m^2) or less. This area restriction attempts to minimize the potential for a significant fire within the mall area. A kiosk larger than 300 square feet (28 m^2) must be considered as a tenant space.

402.6.3 Children's play structures. Children's play structures located within the *mall* of a *covered mall building* or within the perimeter line of an *open mall building* shall comply with Section 424. The horizontal separation between children's play structures, kiosks and similar structures within the *mall* shall be not less than 20 feet (6096 mm).

❖ A common attraction found within covered and open mall buildings is one or more dedicated play areas for children. These attractions include such things as jungle gyms, playhouses and activity areas. If these structures exceed 10 feet (3048 mm) in height and 150 square feet (14 m^2) in area, then the provisions of Section 424 must be followed. Children's play structures have to be separated from each other, as well as from kiosks and similar structures, by at least 20 feet (6096 mm) in order to limit the aggregate combustible fire loading created by such structures.

402.6.4 Plastic signs. Plastic signs affixed to the storefront of any tenant space facing a *mall* or *open mall* shall be limited as specified in Sections 402.6.4.1 through 402.6.4.5.

❖ In order that the mall can be used as an exit access, combustible materials in the mall must be restricted; therefore, the size, area, location and amount of exposed plastic signs are restricted. This section restricts only plastic signs that are along the wall separating the tenant space from the mall (see Commentary Figure 402.6.4). The limit applies to both covered and open mall buildings. While the potential hazards of a plastic sign are somewhat dissipated in an open air mall, the pedestrian malls within an open mall are still considered space within a mall and are used for egress from the complex, therefore appropriate protections are needed. If the sign housing and face panels are noncombustible or glass, there are no limitations. Light-transmitting plastic interior signs not located in the mall, such as those located within an individual tenant space, must comply with the requirements of Chapter 26.

402.6.4.1 Area. Plastic signs shall be not more than 20 percent of the wall area facing the *mall*.

❖ The signs located on the wall facing the mall must not exceed 20 percent of the wall area. The restrictions on size, coupled with the restrictions on location and height, are intended to prevent fire spread via the plastic signs.

402.6.4.2 Height and width. Plastic signs shall be not greater than 36 inches (914 mm) in height, except that where the sign is vertical, the height shall be not greater than 96 inches (2438 mm) and the width shall be not greater than 36 inches (914 mm).

❖ Signs that are wider than they are high (horizontal panels) must not be more than 36 inches (914 mm) in height. The width of signs is controlled by the maximum area (see Section 402.6.4.1) and the location restrictions with respect to adjacent tenant spaces (see Section 402.6.4.3). Vertical signs must not be more than 36 inches (914 mm) in width and 96 inches (2438 mm) in height and are also subject to the area and location restrictions of other sections. The restrictions are intended to minimize horizontal and vertical fire spread through the mall area via the signs.

402.6.4.3 Location. Plastic signs shall be located not less than 18 inches (457 mm) from adjacent tenants.

❖ To minimize the potential for fire spread, plastic signs must be located at least 18 inches (457 mm) from adjacent tenant spaces.

402.6.4.4 Plastics other than foam plastics. Plastics other than foam plastics used in signs shall be light- transmitting plastics complying with Section 2606.4 or shall have a self-ignition temperature of 650°F (343°C) or greater when tested in accordance with ASTM D1929, and a *flame spread index* not greater than 75 and smoke-developed index not greater than 450 when tested in the manner intended for use in accordance with ASTM E84 or UL 723 or meet the acceptance criteria of Section 803.1.2.1 when tested in accordance with NFPA 286.

❖ To minimize the potential fire hazard, light-transmitting plastic signs located in a mall must comply with the specific material requirements of Section 2606.4 (see commentary, Section 2606.4) or must meet one of the alternative test methods that are permitted. Additionally, the backs and edges of the sign must be encased to reduce the likelihood of ignition and fire spread.

402.6.4.4.1 Encasement. Edges and backs of plastic signs in the *mall* shall be fully encased in metal.

❖ In order that the amount of exposed plastic materials in the mall is minimized, as well as to reduce the potential for ignition of a plastic panel, the edges and backs of plastic signs must be encased in metal. To conceptually understand this requirement, consider how easily a piece of paper ignites when a flame is held to its edge versus being held in the middle of the paper.

402.6.4.5 Foam plastics. Foam plastics used in signs shall have flame-retardant characteristics such that the sign has a maximum heat-release rate of 150 kilowatts when tested in accordance with UL 1975 or when tested in accordance with NFPA 289 using the 20 kW ignition source, and the foam plastics shall have the physical characteristics specified in this section. Foam plastics used in signs installed in accordance with Section 402.6.4 shall not be required to comply with the *flame spread* and smoke-developed indices specified in Section 2603.3.

❖ Foam plastics that are used in signs located in a mall must comply with the provisions of Sections 402.6.4.5.1 and 402.6.4.5.2. The provisions of the code include specific material requirements for the foam plastics, such as the allowable heat-release rate, density and thickness permitted.

402.6.4.5.1 Density. The density of foam plastics used in signs shall be not less than 20 pounds per cubic foot (pcf) (320 kg/ m^3).

❖ Foam plastics used in signs are limited to a minimum density of no less than 20 pounds per cubic foot (pcf) (320 kg/m^3). This limitation lessens the likelihood that the sign will add a significant fuel load to the tenant space entrance.

402.6.4.5.2 Thickness. The thickness of foam plastic signs shall not be greater than $^1/_2$ inch (12.7 mm).

❖ The thickness of the foam plastic sign is also limited to lessen the likelihood that the sign will add a significant fuel load to the tenant space entrance.

[F] 402.7 Emergency systems. *Covered and open mall buildings*, *anchor buildings* and associated parking garages

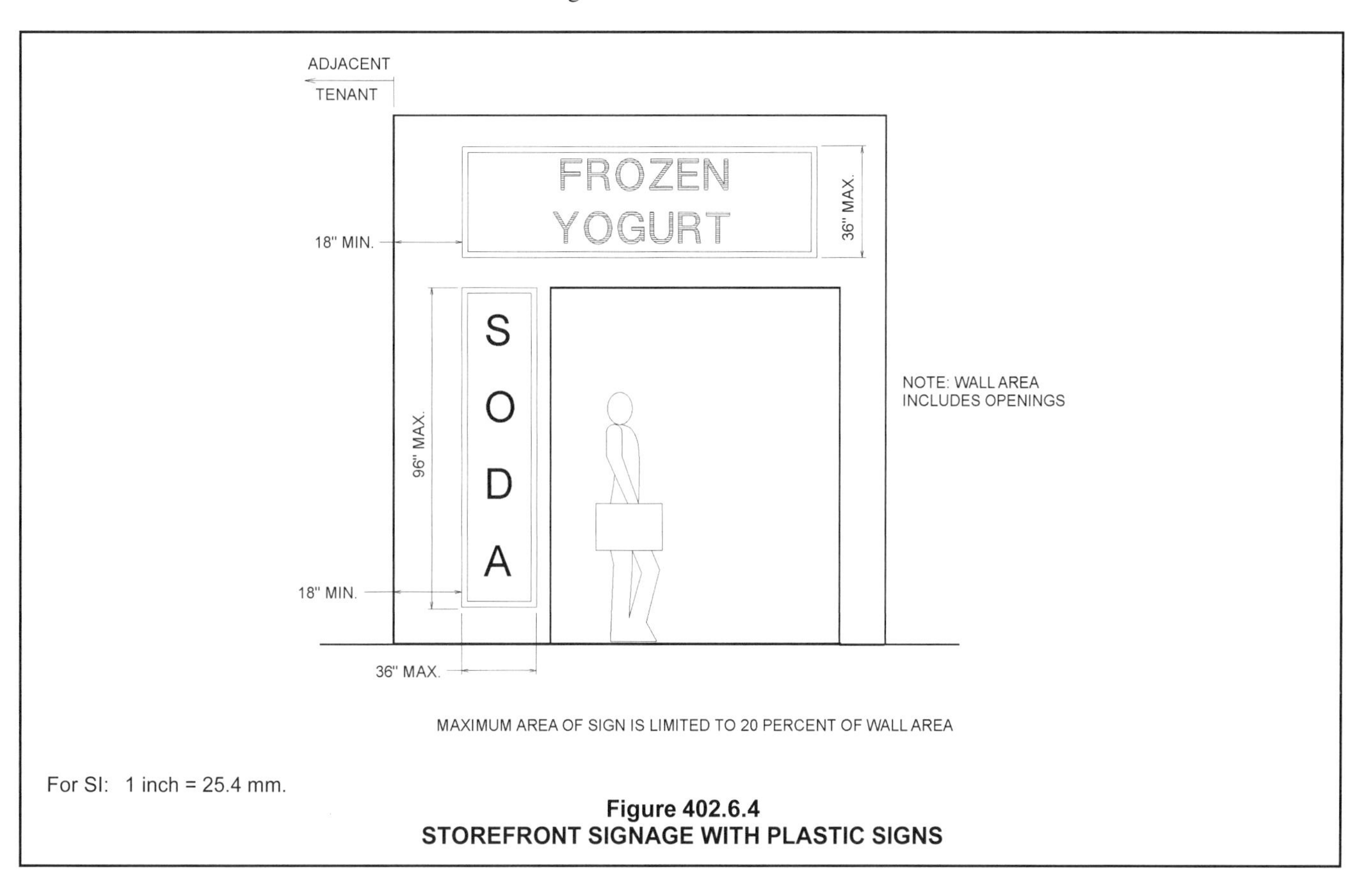

Figure 402.6.4
STOREFRONT SIGNAGE WITH PLASTIC SIGNS

shall be provided with emergency systems complying with Sections 402.7.1 through 402.7.5.

❖ The special allowances for covered and open mall buildings are available in great part due to the presence of several emergency fire protection systems within the building. In addition to the automatic sprinkler system required by Section 402.5, this section identifies those systems and methods by addressing standpipe systems, smoke control, standby power, emergency voice/alarm communication and fire department access to equipment.

[F] 402.7.1 Standpipe system. *Covered and open mall buildings* shall be equipped throughout with a standpipe system as required by Section 905.3.3.

❖ This section provides a reminder to the code user and a direct reference to the section that regulates the need for standpipes or hose connections for covered or open mall buildings (see commentary, Section 905.3.3).

[F] 402.7.2 Smoke control. Where a *covered mall building* contains an *atrium*, a smoke control system shall be provided in accordance with Section 404.5.

Exception: A smoke control system is not required in *covered mall buildings* where an *atrium* connects only two stories.

❖ A smoke control system must be provided for all covered mall buildings that contain an atrium which connects more than two stories. An "atrium" is defined (Section 202) as an opening connecting two or more stories other than enclosed stairways, elevators, hoistways, escalators, plumbing, electrical, air-conditioning or other equipment, which is closed at the top. Malls inherently contain unenclosed floor openings that may allow for the migration of smoke and hot gases resulting from a fire. See Sections 404.5 and 909 for a discussion of the requirements of the smoke control system, noting the exceptions listed for floor openings. Open malls and adjoining tenant spaces are not required to have a smoke control system because they are open at the top and would not meet the definition of an "atrium." However, individual tenant spaces that contain an atrium and are part of an open mall "building" are not exempted from the smoke control requirement.

[F] 402.7.3 Emergency power. *Covered mall buildings* greater than 50,000 square feet (4645 m^2) in area and *open mall buildings* greater than 50,000 square feet (4645 m^2) within the established perimeter line shall be provided with emergency power that is capable of operating the *emergency voice/alarm communication system* in accordance with Section 2702.

❖ Covered and open mall buildings of a substantial size are required to be provided with emergency power only for the emergency voice/alarm communication system (see Section 2702.2.3). For means of egress illumination, see Section 1008.

[F] 402.7.4 Emergency voice/alarm communication system. Where the total floor area is greater than 50,000 square feet (4645 m^2) within either a *covered mall building* or within the perimeter line of an *open mall building*, an *emergency voice/alarm communication system* shall be provided.

Emergency voice/alarm communication systems serving a *mall*, required or otherwise, shall be accessible to the fire department. The systems shall be provided in accordance with Section 907.5.2.2.

❖ An emergency voice/alarm communication system is required to allow the fire department control and communication during an emergency situation. Such a system is required when the covered or open mall building exceeds a floor area of 50,000 square feet (4645 m^2) (see commentary, Section 907.5.2.2). For open mall buildings, the threshold of 50,000 square feet (4645 m^2) is based on the floor area within the building's perimeter line as described in Section 402.1.2.

[F] 402.7.5 Fire department access to equipment. Rooms or areas containing controls for air-conditioning systems, *automatic fire-extinguishing systems, automatic sprinkler systems* or other detection, suppression or control elements shall be identified for use by the fire department.

❖ Consideration should be given to fire department access during an emergency. Fire department personnel may need to either determine that the fire protection systems are functioning properly or override the automatic features to manually activate or shut down a particular system. For this reason, the fire department must have access to controls for the air-conditioning and fire protection systems. The controls are to be clearly identified so that fire department personnel can properly operate them.

402.8 Means of egress. *Covered mall buildings, open mall buildings* and each tenant space within a mall building shall be provided with *means of egress* as required by this section and this code. Where there is a conflict between the requirements of this code and the requirements of Sections 402.8.1 through 402.8.8, the requirements of Sections 402.8.1 through 402.8.8 shall apply.

❖ Covered mall buildings and open mall buildings, as well as each individual tenant space within the covered or open mall building, are required to have means of egress that comply with other provisions of the code—especially Chapter 10. Keep in mind that the requirements of Section 402.8 will supersede some of the provisions of Chapter 10. In order to comply, travel through the mall area is considered as exit access and must comply with Section 402.8.5. Travel distance within a tenant space need only be measured to the entrance of the mall from the space (see Commentary Figure 402.8). As such, two distinct criteria must be met with regard to exit access travel distance: travel within the mall and travel within the tenant space.

402.8.1 Mall width. For the purpose of providing required egress, *malls* are permitted to be considered as *corridors* but need not comply with the requirements of Section 1005.1 of this code where the width of the *mall* is as specified in this section.

❖ Pedestrian malls that serve as exit access routes for tenant space occupants are not required to comply with Section 1005.1. The design capacity of the mall width is already established by Section 402.8.1.1. No further egress width calculations are required by the code for mall corridors.

402.8.1.1 Minimum width. The aggregate clear egress width of the *mall* in either a *covered or open mall building* shall be not less than 20 feet (6096 mm). The *mall* width shall be sufficient to accommodate the *occupant load* served. No portion of the minimum required aggregate egress width shall be less than 10 feet (3048 mm) measured to a height of 8 feet (2438 mm) between any projection of a tenant space bordering the *mall* and the nearest kiosk, vending machine, bench, display opening, *food court* or other obstruction to *means of egress* travel.

❖ The minimum 20-foot (6096 mm) clear egress width of a mall is based on the need to provide adequate access to exits or, for open malls, the "building perimeter line," which is the point where people transition from being within the open mall "building" to outside of the "building." Together with the automatic sprinkler system, the physical separation further reduces the need for a separation between tenant spaces and the mall.

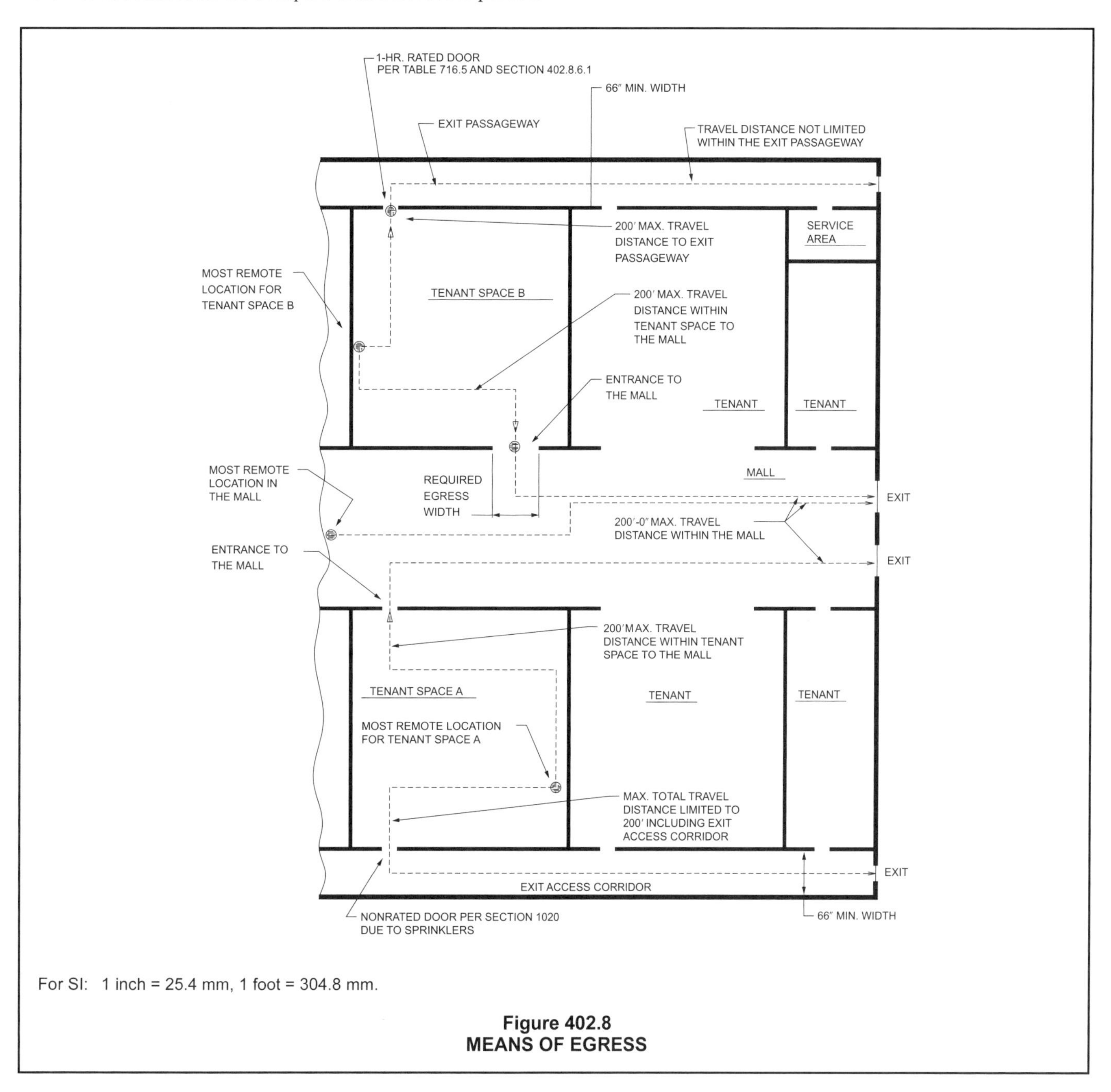

For SI: 1 inch = 25.4 mm, 1 foot = 304.8 mm.

Figure 402.8
MEANS OF EGRESS

In order that an aggregate clear width of 20 feet (6096 mm) is always provided in the mall, a minimum 10-foot (3048 mm) clear and unobstructed space is to be maintained to a height of 8 feet (2438 mm) in front of, adjacent to and parallel to the storefronts (tenant spaces). The requirement applies to kiosks, vending machines, benches, small stands, merchandise displays and any other potential obstruction to egress (see Commentary Figure 402.8.1.1).

402.8.2 Determination of occupant load. The *occupant load* permitted in any individual tenant space in a *covered or open mall building* shall be determined as required by this code. *Means of egress* requirements for individual tenant spaces shall be based on the *occupant load* thus determined.

❖ Since the tenant spaces of covered or open mall buildings can be used for varied occupancies, the design occupant loads will also vary. Although each tenant space contributes to the gross leasable area of Section 402.8.2.1, each must have its own occupant load calculated along with the applicable means of egress requirements. For example, a restaurant with its own dining area may occupy a tenant space. The design occupant load for this space is calculated in accordance with Table 1004.1.2 based on the assembly occupancy of the dining areas and on the commercial kitchen occupancy of the food preparation areas. Door swing direction and panic and fire exit hardware requirements would be based on this occupant load.

402.8.2.1 Occupant formula. In determining required *means of egress* of the *mall*, the number of occupants for whom *means of egress* are to be provided shall be based on *gross leasable area* of the *covered or open mall building* (excluding *anchor buildings*) and the *occupant load* factor as determined by Equation 4-1.

$OLF = (0.00007)(GLA) + 25$ **(Equation 4-1)**

where:

OLF = The *occupant load* factor (square feet per person).

GLA = The *gross leasable area* (square feet).

Exception: Tenant spaces attached to a *covered or open mall building* but with a *means of egress* system that is totally independent of the *open mall* of an *open mall building* or of a *covered mall building* shall not be considered as *gross leasable area* for determining the required *means of egress* for the *mall building*.

❖ The capacity of the exits serving the covered and open mall buildings must satisfy the calculated occupant load based on the gross leasable area. The occupant load factors (OLF) were determined empirically by surveying more than 270 covered mall shopping centers, studying mercantile occupancy parking requirements [5.0 cars per 1,000 square feet (93 m^2) of gross leasable area] and observing the number of occupants per vehicle during peak seasons (4.0 per car).

The formula used for determining occupant load results in a smooth transition between the occupant load and the gross leasable area with a range of 30 to 50.

For example, if the gross leasable area of the covered or open mall building is 400,000 square feet (37 160 m^2), the calculated OLF is determined as follows:

$OLF = (0.00007)(400{,}000) + 25$

OLF = 53 square feet (4.9 m^2) per person

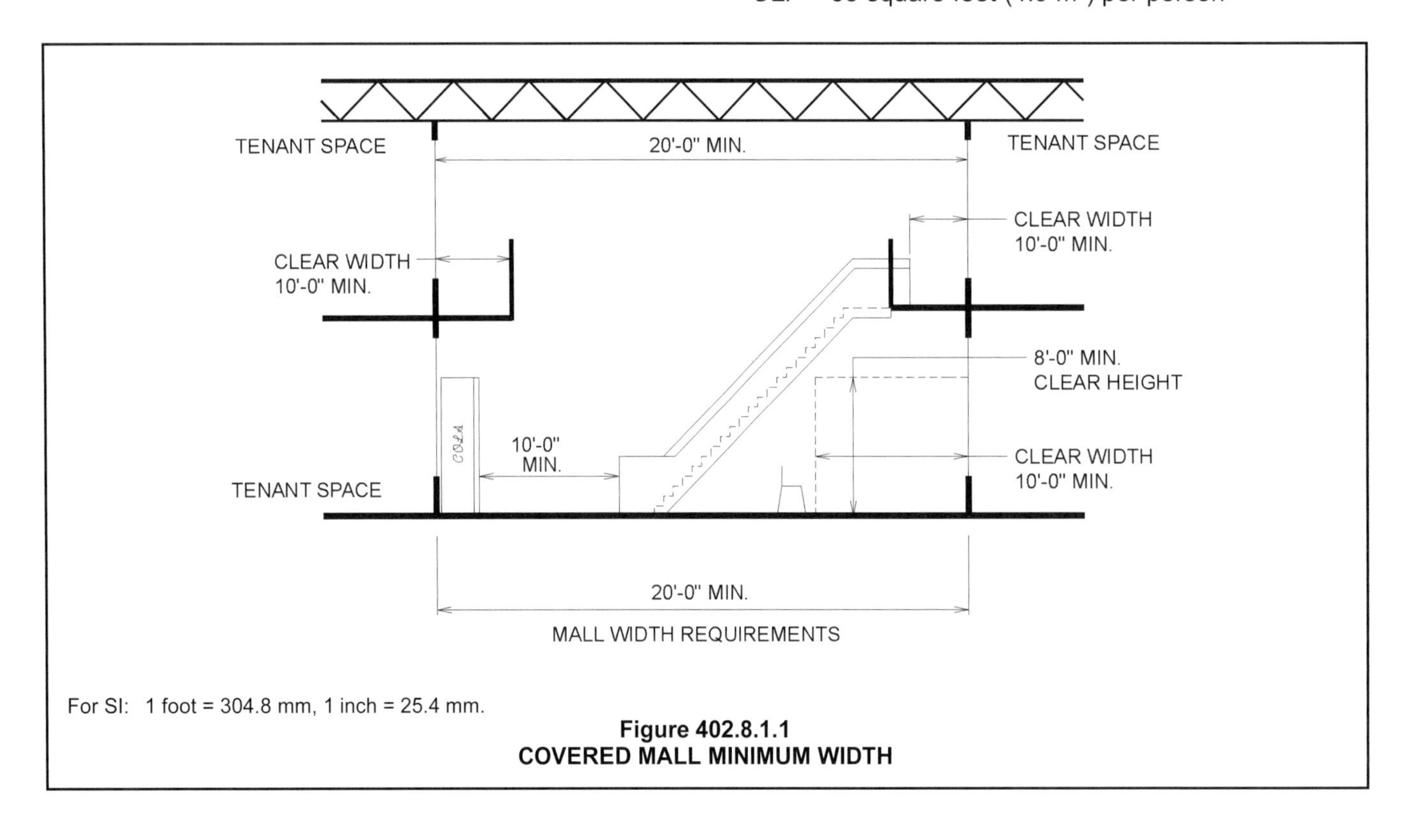

For SI: 1 foot = 304.8 mm, 1 inch = 25.4 mm.

Figure 402.8.1.1
COVERED MALL MINIMUM WIDTH

Since the design occupant load factor cannot exceed 50 in accordance with Section 402.8.2.2, the required OLF would be 50 square feet (4.6 m^2) per person.

If the gross leasable area of the mall building is 55,000 square feet (5109 m^2), the calculated OLF is determined as follows:

OLF = (0.00007) (55,000) + 25

OLF = 29 square feet (2.7 m^2) per person

Since the OLF need not be less than 30 in accordance with Section 402.8.2.2, the required occupant load factor would be 30 square feet (2.8 m^2) per person. The code would permit the use of an occupant load factor of 29, since it would result in a higher design occupant load and, consequently, increased width of exits.

402.8.2.2 OLF range. The *occupant load* factor (*OLF*) is not required to be less than 30 and shall not exceed 50.

❖ Although not mandatory, an OLF of less than 30 may be used at the designer's discretion (see Section 1004.2). In no case should an OLF greater than 50 be used, since it would result in too small a design occupant load. It should be noted that Table 1004.1.2 assigns an OLF of 60 to mercantile occupancies. The 60 OLF would apply to stores not located in a mall. The assumption behind a lower OLF in malls is that malls attract larger numbers of shoppers than standalone stores. A lower OLF means more people in a given space. This results in greater width for egress elements such as corridors or more egress elements such as exit doors.

402.8.2.3 Anchor buildings. The *occupant load* of *anchor buildings* opening into the *mall* shall not be included in computing the total number of occupants for the *mall*.

❖ Although anchor buildings are allowed access to the mall portions of the covered or open mall building, the occupant load of the mall does not have to be increased for this access. People may certainly discharge from the anchor building into the mall on their own accord; however, the anchor building is required to provide sufficient exits for its occupant load at locations other than the mall.

402.8.2.4 Food courts. The *occupant load* of a *food court* shall be determined in accordance with Section 1004. For the purposes of determining the *means of egress* requirements for the *mall*, the *food court occupant load* shall be added to the *occupant load* of the *covered or open mall building* as calculated above.

❖ The occupant load for food courts must be determined in accordance with Section 1004. Typically, tables and chairs are provided for the public's use in a food court. As such, the design occupant load for a food court is usually calculated at a rate of one person for every 15 square feet (1.4 m^2) of net floor space for an unconcentrated assembly occupancy. The occupant loads for all food courts are added to the mall occupant load calculated in accordance with Section 402.8.2.1 and Equation 4-1. Food courts are not considered part of the gross leasable area of the mall; therefore, the occupant loads of the food courts must be calculated separately and then added to the overall covered or open mall building occupant load.

402.8.3 Number of means of egress. Wherever the distance of travel to the *mall* from any location within a tenant space used by persons other than employees is greater than 75 feet (22 860 mm) or the tenant space has an *occupant load* of 50 or more, no fewer than two *means of egress* shall be provided.

❖ This section applies within each tenant space. It dictates the minimum number of paths of travel an occupant is to have available to avoid a fire incident in the occupied tenant space. While providing multiple egress doorways from every tenant space is unrealistic, a point does exist where alternative egress paths must be provided, based on the number of occupants at risk, the distance occupants must travel to reach the mall and the relative hazards associated with the occupancy of the space.

The 75-foot (22 860 mm) travel distance and the 49-person occupant load limitations represent an empirical judgment of the risks associated with a single means of egress from a tenant space based on the inherent risks associated with the occupancy (such as occupant mobility, occupant familiarity with the mall, occupant response and the fire growth rate). The requirement for two exits for 50 or more occupants applies to all occupants of the tenant space, including employees. The 75-foot (22 860 mm) travel distance does not apply to employees. Employee-only areas of the tenant space can require travel greater than 75 feet (22 860 mm) to reach the mall or other exit. A second means of egress is not required in spaces with an occupant load less than 50.

402.8.4 Arrangements of means of egress. Assembly occupancies with an *occupant load* of 500 or more located within a *covered mall building* shall be so located such that their entrance will be immediately adjacent to a principal entrance to the *mall* and shall have not less than one-half of their required *means of egress* opening directly to the exterior of the *covered mall building*. Assembly occupancies located within the perimeter line of an *open mall building* shall be permitted to have their main *exit* open to the *open mall*.

❖ Tenant spaces which are assembly occupancies with an occupant load of 500 or more must be carefully located to minimize the hazards of egress. Movie theaters, nightclubs and large restaurants must be located on an exterior wall of the covered mall building and adjacent to the mall's exits if their occupant loads exceed 499. A maximum of 50 percent of the means of egress from these assembly occupancies is permitted to discharge into the mall. Other means of egress from these tenant spaces must discharge directly to the exterior with a capacity of at least one-half the occupant load. For open mall buildings, any assembly occupancy located with the building perimeter line is permitted to have its main exits open directly into the mall due to the decreased hazard level of an open, unroofed, mall compared to an enclosed and covered mall.

402.8.4.1 Anchor building means of egress. Required *means of egress* for *anchor buildings* shall be provided independently from the *mall means of egress* system. The *occupant load* of *anchor buildings* opening into the *mall* shall not be included in determining *means of egress* requirements for the *mall*. The path of egress travel of *malls* shall not exit through *anchor buildings*. *Malls* terminating at an *anchor building* where no other *means of egress* has been provided shall be considered as a dead-end *mall*.

❖ As indicated in the definition of anchor building (see Section 202), the required means of egress for an anchor building must be provided independently of the mall and the means of egress from the mall. Since independent means of egress are provided, the occupant load of anchor buildings is not included in determining the means of egress requirements for the mall. If independent means of egress are not provided, the space cannot be considered an anchor building and is to be treated as any other tenant space of the mall. As such it would need to be included in the gross leasable area under Section 402.8.2.1.

402.8.5 Distance to exits. Within each individual tenant space in a *covered or open mall building*, the distance of travel from any point to an *exit* or entrance to the *mall* shall be not greater than 200 feet (60 960 mm).

The distance of travel from any point within a *mall* of a *covered mall building* to an *exit* shall be not greater than 200 feet (60 960 mm). The maximum distance of travel from any point within an *open mall* to the perimeter line of the *open mall building* shall be not greater than 200 feet (60 960 mm).

❖ The maximum permissible travel distance from any point in the tenant space to the mall is 200 feet (60 960 mm). A similar 200-foot (60 960 mm) limitation is applied from any point within the mall in a covered mall building to an exit, as well as from any point within the mall of an open mall building to the perimeter line. As such, the mall is treated like an aisle in a large store, the large store being the mall building. Although the distance is less than that permitted in Group M occupancies with an automatic sprinkler system, it must be recognized that this travel distance is measured from a point within the mall and not from the most remote point within the covered or open mall building (i.e., within a tenant space). Because of the uncertainty with respect to tenant improvements, it is more reliable to regulate travel distance within the mall instead of throughout the covered or open mall building.

402.8.6 Access to exits. Where more than one *exit* is required, they shall be so arranged that it is possible to travel in either direction from any point in a *mall* of a *covered mall building* to separate *exits* or from any point in an *open mall* of an *open mall building* to two separate locations on the perimeter line, provided neither location is an exterior wall of an *anchor building* or parking garage. The width of an *exit passageway* or *corridor* from a *mall* shall be not less than 66 inches (1676 mm).

Exception: Access to exits is permitted by way of a dead-end *mall* that does not exceed a length equal to twice the width of the *mall* measured at the narrowest location within the dead-end portion of the *mall*.

❖ In order to accommodate the many occupants anticipated in a covered or open mall building, exits are to be distributed equally throughout the mall. If exits were congregated in certain areas, egress would be compounded by the convergence of a large number of people to one or more areas of the building. In an open mall building, the distribution of exits must occur at separate locations on the perimeter line that are open and provide continued egress to the public way. The corridors and exit access passageways from a mall must be at least 66 inches (1676 mm) wide.

The maximum dead end permitted in a mall is twice the width of the mall. For a mall with a minimum width of 20 feet (6096 mm), dead ends are permitted to be 40 feet (12 192 mm) or less in length. The allowance for dead ends is in recognition of the reduced hazard represented by a dead end in a mall and because the relatively large minimum width provides alternative paths of travel. Additionally, dead ends are not as critical in a mall, since the volume of the space and the automatic sprinkler system minimize the potential for the space to become untenable in a fire condition.

402.8.6.1 Exit passageways. Where *exit passageways* provide a secondary *means of egress* from a tenant space, doorways to the *exit passageway* shall be protected by 1-hour *fire door assemblies* that are self- or automatic-closing by smoke detection in accordance with Section 716.5.9.3.

❖ If exit passageways must be provided because of either travel distance limitations or the number of means of egress required from a tenant space, the passageways must be enclosed as required by the code. Additionally, the doors opening from the tenant spaces into the exit enclosures must be not less than 1-hour rated. These doors must be self-closing or automatic-closing by smoke detection to provide the necessary protection as the occupants of a tenant space flee a fire situation.

402.8.7 Service areas fronting on exit passageways. Mechanical rooms, electrical rooms, building service areas and service elevators are permitted to open directly into *exit passageways*, provided the *exit passageway* is separated from such rooms with not less than 1-hour *fire barriers* constructed in accordance with Section 707 or *horizontal assemblies* constructed in accordance with Section 711, or both. The *fire protection rating* of openings in the *fire barriers* shall be not less than 1 hour.

❖ In a covered mall building, it is necessary to provide for services to the tenant spaces that are maintained by the mall management (e.g., water, electricity, telephone and fire protection). These services must be located in a common space controlled by the mall management and, therefore, cannot be located within the tenant spaces. Frequently, these services are logically located with direct access to exit passageways at the rear of the tenant spaces. For open mall buildings, the use of exit passageways will probably not occur as frequently, but where they do, they also need to comply with this section.

Exit passageways are generally treated similarly to exit stairways in that only openings from normally occupied spaces are permitted. This would prohibit doors or utility penetrations to such mechanical/electrical rooms. In the case of malls, the code allows an exception to that general rule, provided that the fire-resistance rating of the exit enclosure is maintained by appropriate opening protection, such as fire doors, fire dampers and through-penetration firestopping.

402.8.8 Security grilles and doors. Horizontal sliding or vertical security grilles or doors that are a part of a required *means of egress* shall conform to the following:

1. Doors and grilles shall remain in the full open position during the period of occupancy by the general public.
2. Doors or grilles shall not be brought to the closed position when there are 10 or more persons occupying spaces served by a single *exit* or 50 or more persons occupying spaces served by more than one *exit*.
3. The doors or grilles shall be openable from within without the use of any special knowledge or effort where the space is occupied.
4. Where two or more *exits* are required, not more than one-half of the *exits* shall be permitted to include either a horizontal sliding or vertical rolling grille or door.

❖ When improperly installed or when their use is not properly supervised, security grilles and doors can prohibit or delay egress beyond an acceptable time period. Such security devices are common in covered mall buildings. In every case, the grille must be openable to permit egress from the inside without the use of tools, keys or special knowledge or effort when the grille spans across a means of egress. For example, during business hours in a mercantile use, the grille must remain in its full, open position. The grille may be taken to a partially closed position when it provides the sole means of egress and no more than nine persons occupy the space. If a second means of egress is required, the grille may not be closed, as long as more than 49 persons occupy the space.

Because security grilles represent such a threat to prompt egress, the number of occupants exposed to the risk of a lack of supervision of the devices is limited. Except where a single means of egress is permitted, alternative means of egress must be available and not equipped with such devices. Security grilles must not be used on more than 50 percent of the exits.

SECTION 403
HIGH-RISE BUILDINGS

403.1 Applicability. *High-rise buildings* shall comply with Sections 403.2 through 403.6.

Exception: The provisions of Sections 403.2 through 403.6 shall not apply to the following buildings and structures:

1. Airport traffic control towers in accordance with Section 412.3.
2. *Open parking garages* in accordance with Section 406.5.
3. The portion of a building containing a Group A-5 occupancy in accordance with Section 303.6.
4. Special industrial occupancies in accordance with Section 503.1.1.
5. Buildings with:
   - 5.1. A Group H-1 occupancy;
   - 5.2. A Group H-2 occupancy in accordance with Section 415.8, 415.9.2, 415.9.3 or 426.1; or,
   - 5.3. A Group H-3 occupancy in accordance with Section 415.8.

❖ "High-rise buildings" are defined in Chapter 2 as buildings with occupied floors located more than 75 feet (22 860 mm) above the lowest level of fire department vehicle access [see Commentary Figure 403.1(1)]. Such buildings require special consideration relative to fire protection and occupant safety because of difficulties associated with smoke movement (stack effect), egress time, access by fire department personnel and perceived vulnerability to terrorist attack. This section contains provisions for high-rise buildings to address these special considerations.

While most of the provisions in Section 403 apply to all high-rise buildings, additional elevator requirements apply to high-rise buildings in excess of 120 feet (36 576 mm). Further standards apply to high-rise buildings greater than 420 feet (128 m). These "super high-rises" must comply with enclosure structural integrity requirements (see Section 403.2.3), increased bond strength for sprayed fire-resistant materials (see Section 403.2.4), additional sprinkler risers (see Section 403.3.1) and have an additional exit stairway (see Section 403.5.2). In addition, these taller high-rise buildings are not allowed to be designed with the reduced fire-resistance ratings permitted in Section 403.2.1.

The provisions of Section 403 apply to all high-rise buildings except those identified in the five exceptions.

Exception 1 addresses airport control towers and is based on the limited fuel load and the limited number of persons occupying the tower (see Section 412.3).

The intent of Exception 2 is exempt places of outdoor assembly (Group A-5) stadiums, but not other significant facilities which might be attached to the stadium. Stadiums which reach a 75 foot occupied floor level height are generally going to be large facilities at major universities or occupied by professional sports teams. Such stadiums will have accessory spaces such as concession stands. However, the trend for elaborate skyboxes and large, multilevel press boxes should be addressed by the high-rise building standards. Major attached facilities such as office towers and residential buildings are not exempted even though attached to a Group A-5 structure.

Stand-alone open parking garages are exempted by Exception 3 because of the free ventilation to the outside that exists in such structures.

Low-hazard special industrial occupancies, which meet the criteria of Section 503.1.1, are exempt from the high-rise provisions.

Finally, Exception 5 addresses buildings with Group H-1, H-2 or H-3 occupancies. H-1 occupancies must be in detached buildings and are limited to single-story buildings (see Section 415.7 and Table 504.4). Where Groups H-2 and H-3 are required to be in detached buildings in accordance with Section 415.8, high-rise building standards do not apply. Detached buildings, by definition, are limited to a single story. Finally, H-2 occupancy buildings that are liquefied petroleum gas facilities, dry cleaning plants, grain processing facilities or facilities where combustible dusts are generated or stored are not subject to Section 403. Other H-2 and H-3 occupancies could be in a high-rise building and are subject to Section 403 requirements.

High-rise buildings must comply with all provisions of the code. Section 403 provides additional requirements for high-rise development.

- Section 403.2 specifies additional construction requirements, as well as some requirement reductions, applicable to constructing a high-rise building.
- Section 403.3 states the basic requirement for all high-rise buildings that an automatic sprinkler system be provided throughout the building. Specific standards unique to high-rise development are also provided.
- Section 403.4 specifies the various emergency detection and response systems that are required in a high-rise building.
- Section 403.5 provides additional means of egress system requirements for the occupants of a high-rise building.
- Section 403.6 provides elevator-related requirements for these structures.

The provisions are applicable to all buildings when the highest occupied floor is more than 75 feet (22 860 mm) above the lowest level of fire department vehicle access [see Commentary Figure 403.1(1)]. The lowest level of fire department vehicle access refers to the lowest ground level at which the fire department vehicle could be staged at the exterior of the building for carrying out fire-fighting operations. The definition of a "High-rise building" comes from the International Symposium on Fire Safety in High-Rise Buildings sponsored by the General Services Administration (GSA). The definition developed at the symposium is as follows:

- "A high-rise building is one in which emergency evacuation is not practical and in which fire must

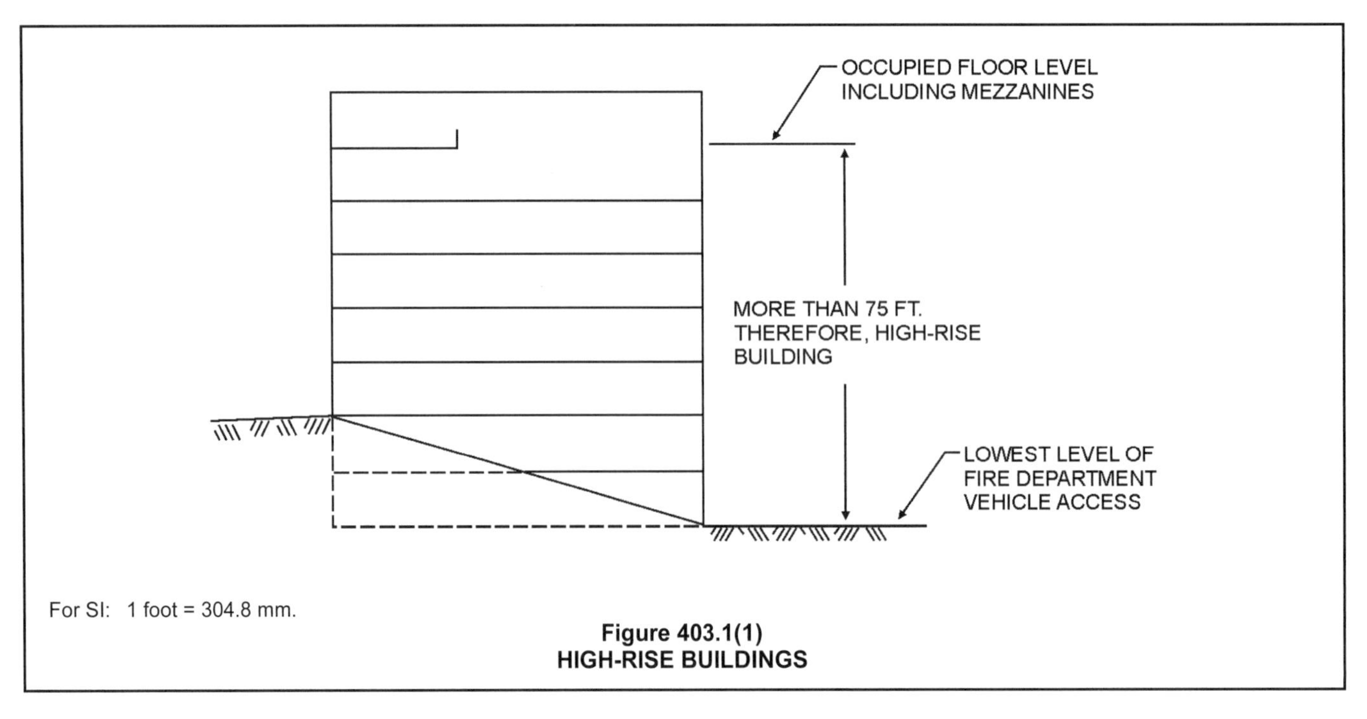

Figure 403.1(1)
HIGH-RISE BUILDINGS

be fought internally because of height." The usual characteristics of such a building are:

- Beyond the reach of fire department equipment;
- Poses a potential for significant stack effect; and
- Requires unreasonable evacuation time.

The 75-foot (22 860 mm) height threshold was determined from the effective reach of a 100-foot (30 480 mm) aerial apparatus based on the assumptions that the building will be set back from the curb, that access might be restricted by such things as parked cars and that obstructions, such as utility lines, would be present. The applicability of this section, however, is not based on the availability of such apparatus within the community.

Stack effect is illustrated in Commentary Figure 403.1(2), while typical evacuation times for buildings are shown in Commentary Figure 403.1(3).

403.2 Construction. The construction of *high-rise buildings* shall comply with the provisions of Sections 403.2.1 through 403.2.4.

❖ Section 403.2 includes subsections allowing the reduction of fire-resistance rating in some high-rise buildings, requiring minimum structural integrity for exit and elevator enclosures and setting bond strength for sprayed fire-resistant materials.

403.2.1 Reduction in fire-resistance rating. The *fire-resistance-rating* reductions listed in Sections 403.2.1.1 and 403.2.1.2 shall be allowed in buildings that have sprinkler control valves equipped with supervisory initiating devices and water-flow initiating devices for each floor.

❖ Since the overall reliability of the sprinkler system is greatly improved by the required control valves and water-flow devices, certain code modifications are allowed.

Sprinkler control valves with supervisory initiating devices and water-flow initiating devices must be provided for each floor to realize the reduction in rating. Annunciation by floor is intended to allow rapid identification of the fire location by the fire department. Sprinkler control valves on each floor are intended to permit servicing activated systems without impairing the water supply to large portions of the building. It is, therefore, necessary to equip the valves with supervisory initiating devices. NFPA 13 does not require annunciation per floor, since some systems may take on configurations that would not lend themselves to this type of zoning. In accordance with Section 102.4, the criteria for this section takes precedence over the permissive language of NFPA 13; therefore, in order to use the reductions allowed by this section, the sprinkler system must comply with these features. Without these sprinkler system enhancements, no reductions are permissible.

403.2.1.1 Type of construction. The following reductions in the minimum *fire-resistance rating* of the building elements in Table 601 shall be permitted as follows:

1. For buildings not greater than 420 feet (128 000 mm) in *building height*, the *fire-resistance rating* of the building elements in Type IA construction shall be permitted to be reduced to the minimum *fire-resistance ratings* for the building elements in Type IB.

   Exception: The required *fire-resistance rating* of columns supporting floors shall not be reduced.

2. In other than Group F-1, M and S-1 occupancies, the *fire-resistance rating* of the building elements in Type IB construction shall be permitted to be reduced to the *fire-resistance ratings* in Type IIA.
3. The *building height* and *building area* limitations of a building containing building elements with reduced *fire-resistance ratings* shall be permitted to be the same as the building without such reductions.

❖ This section specifies the allowed reductions of fire-resistance rating of building elements specified in Table 601, without this being considered a reduction in the construction type.

Item 1 allows high-rise buildings that must be of Type IA construction for building height and building area purposes to have building elements which comply with the fire-resistance rating of the Type IIB construction type as shown in Table 601. This reduction is not applicable to buildings with a height in excess of 420 feet (128 m) and does not apply to columns which are supporting floors in any building classified as a high-rise building.

Item 2 allows high-rise buildings that must be of Type IB construction for building height and area purposes to have elements that comply with the fire-resistance rating of Type IIA construction as shown in Table 601. This reduction is not allowed if the building contains Group F-1, M or S-1 occupancies. This reduction is not allowed for these moderate-hazard buildings because of their customary higher fuel loads.

Item 3 states that even though the fire-resistance ratings have been reduced to the equivalent of the next lower construction type, the construction type of the building is not considered reduced. A Type IB building with 1-hour fire-resistant protection of the structural frame is still a Type IB building.

In determining the minimum allowable type of construction for a high-rise building, Tables 504.3, 504.4 and 506.2 are applied in the usual fashion (see Sections 504 and 506). Item 3 reminds the code user that even though reductions are taken, the allowable area for the building is that of the "original" type of construction and is not reduced because building elements have reduced fire protection. Section 403.2.1.1 is applied once the minimum type of construction has been determined.

For example, an eight-story high-rise office building is to be constructed and equipped with an automatic sprinkler system in accordance with Sections 403.2.1 and 403.3. The minimum allowable construction type is Type IB. In accordance with Section 403.2.1.1, however, the required fire-resistance ratings for the structural elements would be determined using the column for Type IIA construction in Tables 601 and 602. It should be noted that Tables 504.3, 504.4 and 506.2 do not permit the use of unprotected noncombustible construction in a high-rise building.

403.2.1.2 Shaft enclosures. For buildings not greater than 420 feet (128 000 mm) in *building height*, the required *fire-resistance rating* of the *fire barriers* enclosing vertical *shafts*, other than *interior exit stairway* and elevator hoistway enclosures, is permitted to be reduced to 1 hour where automatic sprinklers are installed within the *shafts* at the top and at alternate floor levels.

❖ As with previous sections, based on the effectiveness and reliability of a complete automatic sprinkler system in accordance with Sections 403.2.1 and 403.3, the fire-resistance rating of shaft enclosures may be reduced to 1 hour, provided the sprinklers are also installed within the shafts at the top and at alternate floor levels. This reduction is not applicable to shafts within buildings with a building height in excess of 420 feet (128 m).

The reduced fire-resistance rating also does not apply to shafts which are either interior exit stairway enclosures or elevator hoistways (see also Section 403.2.3 for requirements applicable to these shafts.) The reduction does not apply to exit stairway and elevator shafts since they may continue to be used for egress or for fire-fighter access during a fire; therefore, the integrity of the shaft must be maintained. If, for some reason, the fire penetrates the 1-hour shaft, it will affect conditions on other floors, hence, the need to maintain the protection for interior exit stairway and elevator shafts. By the time the fire penetrates a 1-hour fire-resistance-rated shaft, the affected floors should be evacuated.

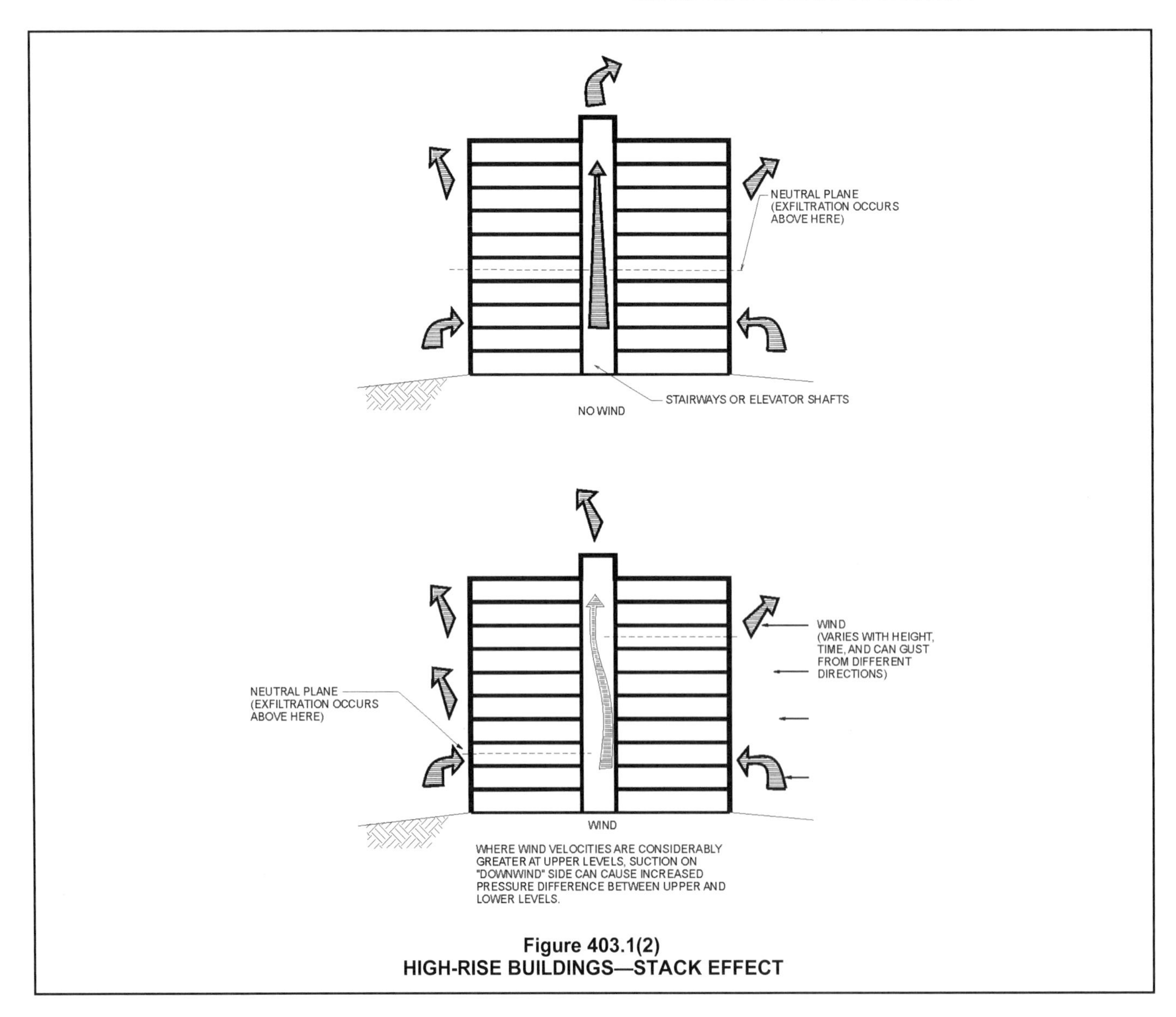

Figure 403.1(2)
HIGH-RISE BUILDINGS—STACK EFFECT

403.2.2 Seismic considerations. For seismic considerations, see Chapter 16.

❖ A reminder is provided in this section of the code that high-rise buildings are required to be designed for the effects of seismic activity. The reminder was primarily placed in Section 403.2.2, in addition to general design for seismic forces, to note that mechanical and electrical components are required to be designed for forces determined by Section 1613. This would also require that elevators be designed to resist the effects of the seismic forces. These elements are critical for the safety of high-rise buildings so that all of the required fire protection systems and the electrical components that secure them will be designed for the applicable seismic loads.

403.2.3 Structural integrity of interior exit stairways and elevator hoistway enclosures. For *high-rise buildings* of *Risk Category* III or IV in accordance with Section 1604.5, and for all buildings that are more than 420 feet (128 000 mm) in *building height*, enclosures for *interior exit stairways* and elevator hoistway enclosures shall comply with Sections 403.2.3.1 through 403.2.3.4.

❖ This section specifies a minimum standard for the integrity of the walls of interior exit stairways and elevator hoistway enclosures. The requirement applies to all high-rise buildings with a building height in excess of 420 feet (128 m). It also applies to any high-rise building, regardless of height, that is either a Risk Category Class III or IV as specified in Section 1604.5. Risk Category Class IV are those uses that are considered to be essential in that their continuous use is needed, particularly in response to disasters. Risk Category Class III buildings include those occupancies that have relatively large numbers of occupants, where there is an elevated life safety hazard, or where the occupants' ability to respond to an emergency is limited. Also see the commentary for Sections 3007 and 3008

403.2.3.1 Wall assembly. The wall assemblies making up the enclosures for *interior exit stairways* and elevator hoistway enclosures shall meet or exceed Soft Body Impact Classification Level 2 as measured by the test method described in ASTM C1629/C1629M.

❖ The requirement applies to all wall assemblies enclosing interior exit stairways, as well as elevator hoistway enclosures in the buildings specified in Section 403.2.3. The requirement applies regardless of the height or location of the particular enclosure. The standard applies to exit passageways that connect the enclosures of interior exit stairways to the exterior of the building, as well as those that connect an "offset" in an interior exit stairway. While this section of the code doesn't specifically include these exit passageways, Section 1024 requires that the level of protection in an exit be maintained until discharge.

The code relies on ASTM C1629 and C1629M for determining compliance with this requirement. The standard was developed specifically to test gypsum and fiber-reinforced cement panels; however, it can readily be used to test the impact resistance of other board and panel materials. For concrete and

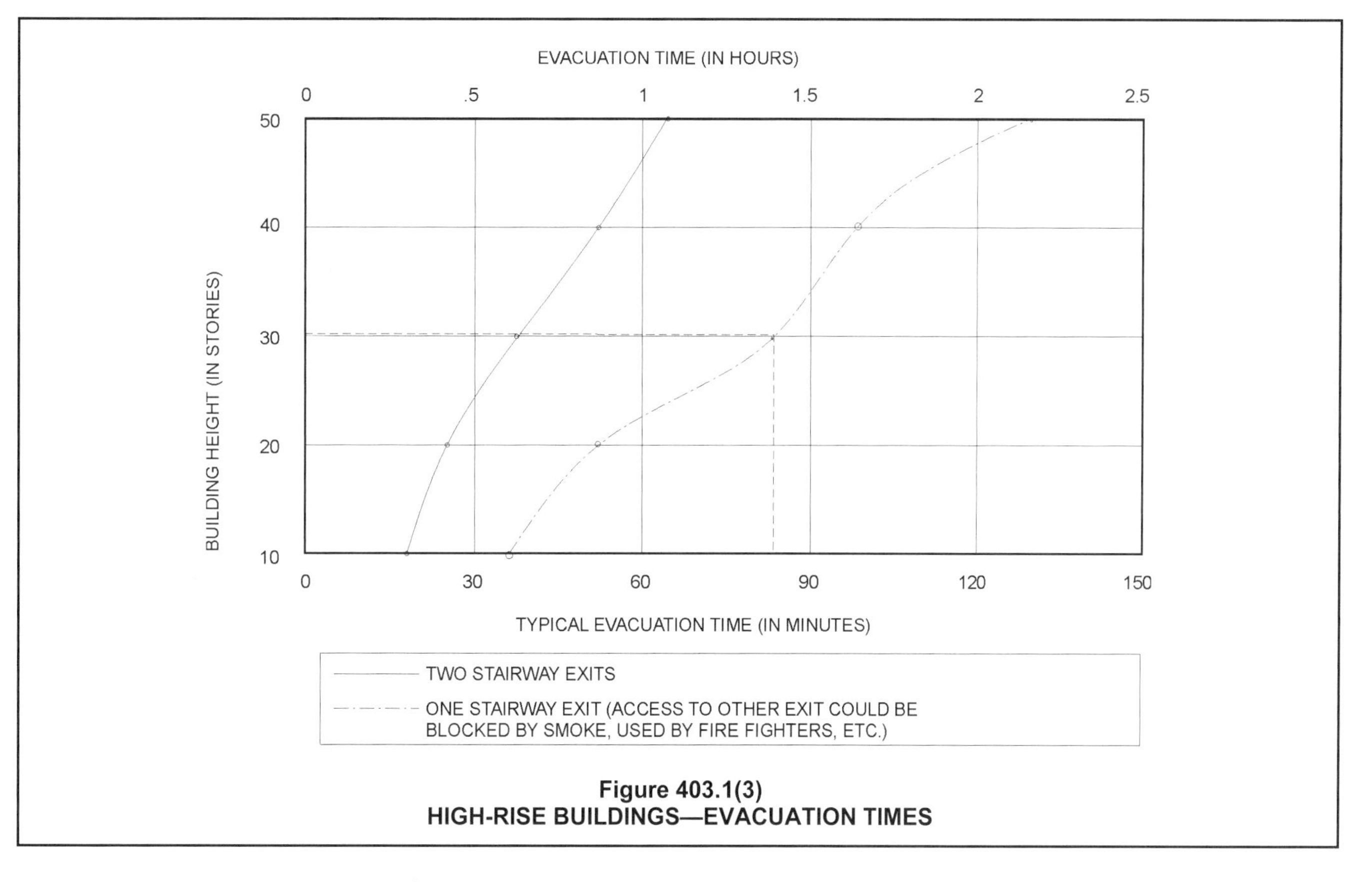

Figure 403.1(3)
HIGH-RISE BUILDINGS—EVACUATION TIMES

masonry walls, see Section 403.2.3.3. The code requires the entire wall assembly to withstand an impact of 195 pound force (867 N), as measured by the ASTM C1629, Soft Body Impact Test. The test method used in ASTM C1629 is conducted in accordance with the ASTM E695 test method, which covers the measurement of the relative resistance of wall, floor and roof construction to impact loading.

403.2.3.2 Wall assembly materials. The face of the wall assemblies making up the enclosures for *interior exit stairways* and elevator hoistway enclosures that are not exposed to the interior of the enclosures for *interior exit stairways* or elevator hoistway enclosure shall be constructed in accordance with one of the following methods:

1. The wall assembly shall incorporate no fewer than two layers of impact-resistant construction board each of which meets or exceeds Hard Body Impact Classification Level 2 as measured by the test method described in ASTM C1629/C1629M.
2. The wall assembly shall incorporate no fewer than one layer of impact-resistant construction material that meets or exceeds Hard Body Impact Classification Level 3 as measured by the test method described in ASTM C1629/C1629M.
3. The wall assembly incorporates multiple layers of any material, tested in tandem, that meets or exceeds Hard Body Impact Classification Level 3 as measured by the test method described in ASTM C1629/C1629M.

❖ Section 403.2.3.2 requires that the face of the wall assembly that is not exposed to the shaft enclosure containing the interior exit stairway or elevators (or exit passageway)—the outside face—be protected by a material or materials that comply with a level of impact resistance. To comply with the requirement, at least two layers of Level 2 material or one layer of Level 3 material must be incorporated into the system. Level 2 material must withstand a Hard Body impact of 100 lbf (445 N) to comply with the standard. Level 3 material must withstand a Hard Body impact of 150 lbf (667 N). The code also permits the use of a system composed of multiple layers of different materials, provided the composite system can comply with a Level 3 Hard Body test.

403.2.3.3 Concrete and masonry walls. Concrete or masonry walls shall be deemed to satisfy the requirements of Sections 403.2.3.1 and 403.2.3.2.

❖ Concrete and masonry walls used for such enclosures are deemed to be adequate to meet the intent of this requirement without any testing under ASTM C1629.

403.2.3.4 Other wall assemblies. Any other wall assembly that provides impact resistance equivalent to that required by Sections 403.2.3.1 and 403.2.3.2 for Hard Body Impact Classification Level 3, as measured by the test method described in ASTM C1629/C1629M, shall be permitted.

❖ This section allows other assemblies to be tested and approved for application in these walls. If a designer wished to include materials or a combination of materials not specified in Section 403.2.3.2 or 403.2.3.3, and wished to have the assembly tested, it would be acceptable as a wall for these enclosures if it complies with the Level 3 Hard Body test.

403.2.4 Sprayed fire-resistant materials (SFRM). The bond strength of the SFRM installed throughout the building shall be in accordance with Table 403.2.4.

❖ For buildings other than high-rise buildings, the minimum bond strength required for sprayed fire-resistant materials (SFRM) required by the code is 150 psf (667 N) when tested in accordance with ASTM E736. Recommendation 6 of the National Institute of Standards and Technology (NIST) investigation of the World Trade Center (WTC) Report called for improvement of the in-place performance of SFRM. For high-rise buildings with a building height in the 75- to 420-foot (22.9 to 128 m) range, SFRM must have a minimum bond strength of 430 psf (1913 N) wherever applied in the building. For high-rise buildings over 420 feet (128 m), the bond strength must be at least 1,000 psf (4448 N) wherever SFRM is applied in the building. While the NIST report was based on the specific events of the WTC fire and collapse, events far less dramatic can dislodge SFRM, such as elevator movement, building sway or maintenance activities. Using greater bond strengths will increase the probability that the protection will stay in place. These factors should provide for a longer time of safety.

TABLE 403.2.4
MINIMUM BOND STRENGTH

HEIGHT OF BUILDING[a]	SFRM MINIMUM BOND STRENGTH
Up to 420 feet	430 psf
Greater than 420 feet	1,000 psf

For SI: 1 foot = 304.8 mm, 1 pound per square foot (psf) = 0.0479 kW/m^2.
a. Above the lowest level of fire department vehicle access.

❖ See the commentary to Section 403.2.4.

[F] 403.3 Automatic sprinkler system. Buildings and structures shall be equipped throughout with an *automatic sprinkler system* in accordance with Section 903.3.1.1 and a secondary water supply where required by Section 403.3.3.

Exception: An *automatic sprinkler system* shall not be required in spaces or areas of:

1. *Open parking garages* in accordance with Section 406.5.
2. Telecommunications equipment buildings used exclusively for telecommunications equipment, associated electrical power distribution equipment,

batteries and standby engines, provided that those spaces or areas are equipped throughout with an automatic fire detection system in accordance with Section 907.2 and are separated from the remainder of the building by not less than 1-hour *fire barriers* constructed in accordance with Section 707 or not less than 2-hour *horizontal assemblies* constructed in accordance with Section 711, or both.

❖ Because of the difficulties associated with manual suppression of a fire in a high-rise building, all high-rise buildings regulated by this section are required to be protected throughout with an automatic sprinkler system. The sprinkler system is to be in accordance with Section 903.3.1.1, which requires compliance with NFPA 13.

In order to increase the reliability of the sprinkler system should an earthquake disable the primary water supply, a secondary water supply is required for buildings at sites with the specified seismic design category listed in Section 403.3.3. The secondary water supply must be equal to the hydraulically calculated sprinkler demand and must have a duration of no less than 30 minutes.

Exception 1 addresses where a high-rise building contains an open parking garage. The exception is limited where Section 903.2 dictates the extent of the sprinkler system coverage. Where the occupancy of the upper stories of a high-rise building is a Group B, then the open parking garage portion of the building need not be protected with an automatic sprinkler system. However, if the high-rise building contains Group R occupancies such as an R-1 hotel or an R-2 apartment/condominium occupancy, then Section 903.2.8 requires a sprinkler system throughout the building, including the parking garage. See the commentary to Section 406.3 for a discussion of the fire hazards associated with open parking garages.

Exception 2 addresses where a telecommunications equipment facility is part of a high-rise building. The automatic sprinkler protection can be eliminated from certain areas if an automatic sprinkler fire detection system is provided and the telecommunication equipment areas are separated from other building areas with fire-resistance-rated construction of the specified ratings. This same exception is provided in Section 903.2, but is restated here to emphasize its application to all buildings, including high-rise buildings. This exception is based on the need to maintain uninterrupted operation of telecommunications equipment, which frequently includes emergency telephone and similar communication services. The presence of an automatic sprinkler system or other automatic fire suppression system may be detrimental to public safety should disruption of emergency services occur.

[F] 403.3.1 Number of sprinkler risers and system design. Each sprinkler system zone in buildings that are more than 420 feet (128 000 mm) in *building height* shall be supplied by no fewer than two risers. Each riser shall supply sprinklers on alternate floors. If more than two risers are provided for a zone, sprinklers on adjacent floors shall not be supplied from the same riser.

❖ For all high-rise buildings with a building height in excess of 420 feet (128 m), this section requires the sprinkler system to be designed with additional risers for each sprinkler zone. This redundancy provides increased reliability of the automatic sprinkler system. At least two risers must be provided for each sprinkler zone. Recommendation 12 of the NIST WTC report called for the redundancy of active fire suppression systems to be increased to accommodate the greater risks associated with increasing building height and population. Providing two risers in each zone allows that if one riser is taken out of service, the other will be able to supply sprinklers on the floors above and below. This will impede any fire spread and allow the fire department time to respond and extinguish the fire.

[F] 403.3.1.1 Riser location. Sprinkler risers shall be placed in *interior exit stairways* and ramps that are remotely located in accordance with Section 1007.1.

❖ This section requires sprinkler risers to be located in the interior exit stairways of buildings. Such interior exit stairways must be separated in accordance with the requirements of Section 1007.1. This separation reduces the possibility that one incident could incapacitate both risers and additionally provides the protection of the rated enclosures. While the section references the separation requirements of Chapter 10, Section 403.5.1 can result in an even greater separation of the exit enclosures in these taller high-rise buildings. Also, Section 403.2.3 regarding the integrity of the interior exit stairways is intended to make sure that the walls of the enclosures protecting the risers meet a minimum impact standard.

[F] 403.3.2 Water supply to required fire pumps. In buildings that are more than 420 feet (128 000 mm) in *building height*, required fire pumps shall be supplied by connections to no fewer than two water mains located in different streets. Separate supply piping shall be provided between each connection to the water main and the pumps. Each connection and the supply piping between the connection and the pumps shall be sized to supply the flow and pressure required for the pumps to operate.

Exception: Two connections to the same main shall be permitted provided the main is valved such that an interruption can be isolated so that the water supply will continue without interruption through no fewer than one of the connections.

❖ Fire pumps are installed in sprinkler and standpipe systems to pressurize the water supply for the minimum required sprinkler and standpipe operation. Fire pumps are only "required" to meet the system needs. Therefore, whether a particular high-rise building includes fire pumps will depend on interaction of the height of the building, the local water system and the designs of the sprinkler and standpipe systems in the building (see Section 913 of the IFC for more information on fire pumps).

Where the systems require a fire pump and the building is more than 420 feet (128 m) in height, this section requires the fire pumps to be supplied by two water mains located in separate streets. Having two connections will greatly reduce the possibility of the loss of water due to a main break, given the valving of a public water system. Each connection must be adequate to provide the flow and pressure needed for the fire pumps to operate. The exception is a performance-based provision that is not tied to any specific configuration.

[F]403.3.3 Secondary water supply. An automatic secondary on-site water supply having a capacity not less than the hydraulically calculated sprinkler demand, including the hose stream requirement, shall be provided for *high-rise buildings* assigned to Seismic Design Category C, D, E or F as determined by Section 1613. An additional fire pump shall not be required for the secondary water supply unless needed to provide the minimum design intake pressure at the suction side of the fire pump supplying the *automatic sprinkler system*. The secondary water supply shall have a duration of not less than 30 minutes.

❖ The intent of this section is that a secondary water supply be provided on the high-rise building site in order to provide a high level of functional reliability for the fire protection systems if a seismic event disables the primary water supply for high-rise buildings assigned these seismic design categories. The categories are described in Section 1613.

The text's specific wording that the secondary supply be on site rather than to the site would preclude the use, for example, of a secondary connection to the municipal supply to achieve compliance with this requirement.

The required amount of water is equal to the hydraulically calculated sprinkler demand plus hose stream demand for a minimum 30-minute period, depending on the appropriate occupancy hazard classification in NFPA 13.

Note that the beginning of Section 403.3.3 requires that the secondary water supply be automatic; in other words, switchover to the secondary water source cannot occur manually. This is consistent with the definitions of "Automatic sprinkler system" and "Classes of standpipe systems" in that both systems are required to be connected to a reliable water supply.

Generally, this section does not automatically require a second fire pump but, if necessary, an additional pump may be required. A second pump would be necessary if the secondary water supply does not provide the necessary water pressure for intake into the primary fire pump.

[F] 403.3.4 Fire pump room. Fire pumps shall be located in rooms protected in accordance with Section 913.2.1.

❖ Where fire pumps are provided, they must be properly isolated from other portions of the high-rise building with fire-resistance-rated construction in accordance with Section 913.2.1. The separation must consist of minimum 2-hour fire barriers or horizontal assemblies, or both. Such separation is not required where the fire pumps are physically separated from the building in accordance with NFPA 20.

[F] 403.4 Emergency systems. The detection, alarm and emergency systems of *high-rise buildings* shall comply with Sections 403.4.1 through 403.4.8.

❖ Section 403.4 lists the requirements for detection and emergency response systems applicable to high-rise buildings. These systems detect smoke and heat in these buildings, and provide systems to notify occupants of the hazards of, and proper response to, the fire or other event. Systems and facilities are provided for emergency responders to assist the occupants and fight a fire. Finally, the section provides requirements for standby and emergency power to keep vital systems operating in case of power loss, either during an event or power loss to the building.

[F] 403.4.1 Smoke detection. Smoke detection shall be provided in accordance with Section 907.2.13.1.

❖ Automatic smoke detectors are required in all high-rise buildings in locations so that a fire will be detected in its early stages of development. Smoke detectors must be installed in rooms that are not typically occupied, such as mechanical equipment rooms, elevator equipment rooms and similar spaces. Smoke detectors are also required in elevator lobbies. Finally, smoke detectors are required at various locations in the HVAC ducts. See Section 907.2.13 for the specific requirements for high-rise buildings.

[F] 403.4.2 Fire alarm system. A *fire alarm* system shall be provided in accordance with Section 907.2.13.

❖ Fire alarm systems must be provided in high-rise buildings in accordance with the provisions of Section 907.2.13. Although a fire alarm system is not specifically listed as a requirement in Section 907.2.13, this section does require an automatic smoke detection system. Section 907.2.13.1.1 requires that the smoke detectors be connected to an automatic fire alarm system. Therefore, regardless of the occupancy category or categories in the high-rise building, a fire alarm system must comply with the code and NFPA 72, as applicable.

[F] 403.4.3 Standpipe system. A *high-rise building* shall be equipped with a standpipe system as required by Section 905.3.

❖ Section 905.3 provides the applicable provisions for standpipes in buildings, including high-rise buildings. At a minimum, a Class I standpipe system is required based upon Exception 1 to Section 905.3.1.

[F] 403.4.4 Emergency voice/alarm communication system. An *emergency voice/alarm communication system* shall be provided in accordance with Section 907.5.2.2.

❖ By definition, one characteristic of high-rise buildings is longer evacuation times. As such, the traditional fire alarm system, which usually results in simultaneous total building evacuation, is not practical. Therefore, the alarm communication system should be able to:

- Direct the occupants of the fire zone to an area of refuge or exit;
- Notify the fire department of the existence of the fire; and
- Sound no alarms outside the fire zone until deemed desirable.

The alarm signal to the fire zone may include continuous sounding devices (e.g., bells, horns, chimes), as well as voice direction, which may momentarily silence the continuous sounding devices so as to be heard clearly. In accordance with Section 907.5.2.3, visible alarm notification appliances are required in the common and public areas of the high-rise buildings. Visible alarm appliances are also required in certain dwelling units and sleeping units, such as hotel guest rooms. The message of the voice/alarm is to be predetermined but needs not be a recorded message. It would typically indicate that a fire has been reported at a specific location and that occupants should await further instructions or evacuate in accordance with the building's fire safety and evacuation plan. The voice/alarm signal will usually commence with a 3- to 10-second alert signal followed by the message. See the commentary to Section 907.5.2.2 for further requirements of the emergency voice/alarm communication system.

[F] 403.4.5 Emergency responder radio coverage. Emergency responder radio coverage shall be provided in accordance with Section 510 of the *International Fire Code*.

❖ High-rise buildings have posed a challenge to the traditional communication systems used by the fire service for fire-to-ground communications. Therefore, emergency responder radio coverage complying with Section 510 of the IFC must be provided. The coverage is needed to coordinate with the fire department radio system to allow fireground officers to remain in communication with fire fighters working in various areas of a building.

As facilities grow larger and more complex, they become more challenging for effective fire response. Therefore, more fire protection features must be provided within the building. While modeling and other techniques may provide a good prediction as to whether a building will interfere with radio communications, the reality is that it is unknown if a building will need to have an enhanced radio system installed until after the building is constructed. The presumption is that high-rise buildings will more likely than not need to be provided with an emergency responder system. Section 510.1 of the IFC offers three exceptions that may be considered by the building and fire code officials.

[F] 403.4.6 Fire command. A *fire command center* complying with Section 911 shall be provided in a location *approved* by the fire department.

❖ Fireground operations usually involve establishing an incident command post where the fire official can observe what is happening, control arriving personnel and equipment, and direct resources and fire-fighting operations effectively. Because of the difficulties in controlling a fire in a high-rise building, a separate room (enclosed in 1-hour-rated construction) within the building must be established as a fire command center. The room must be provided at a location that is acceptable to the fire department, usually along the front of the building or near the main entrance. The room must contain equipment necessary to monitor or control fire protection and other building service systems (see Section 911 for further information).

403.4.7 Smoke removal. To facilitate smoke removal in post-fire salvage and overhaul operations, buildings and structures shall be equipped with natural or mechanical *ventilation* for removal of products of combustion in accordance with one of the following:

1. Easily identifiable, manually operable windows or panels shall be distributed around the perimeter of each floor at not more than 50-foot (15 240 mm) intervals. The area of operable windows or panels shall be not less than 40 square feet (3.7 m^2) per 50 linear feet (15 240 mm) of perimeter.

   Exceptions:

   1. In Group R-1 occupancies, each *sleeping unit* or suite having an *exterior wall* shall be permitted to be provided with 2 square feet (0.19 m^2) of venting area in lieu of the area specified in Item 1.
   2. Windows shall be permitted to be fixed provided that glazing can be cleared by fire fighters.

2. Mechanical air-handling equipment providing one exhaust air change every 15 minutes for the area involved. Return and exhaust air shall be moved directly to the outside without recirculation to other portions of the building.
3. Any other *approved* design that will produce equivalent results.

❖ Section 403.4.7 requires new high-rise buildings to be provided with a smoke removal system to exhaust products of combustion following a fire incident. The intent of the requirement as clearly specified in the code is to facilitate the removal of smoke during the salvage and overhaul of a building after a fire has ended. “Overhaul and salvage operations” are generally understood to include searching the fire scene to detect and extinguish hidden fires or “hot spots.” They also include controlling additional losses, stabilizing the incident scene by providing for fire-fighter safety and securing the structure. The smoke removal system is not a smoke control system similar to those required for atriums or underground buildings. The smoke removal system in a high-rise building is not intended for use during a fire event. It is not intended to serve any health and life safety function, and, therefore, is only required to operate after, and not during, a fire event. In comparison, a smoke control system for an atrium is a life safety function and is designed to operate during a fire event to control smoke that is generated by the fire and keep it from migrating to other portions of the structure. This is not a smoke control system required elsewhere in the code; therefore, Section 909 is not referenced and is not applicable.

A smoke removal system can be a whole building system or a floor-by-floor localized system. It can also be provided by either natural means or a mechanical system. When a mechanical ventilation system is utilized, the controls for the smoke removal system are allowed to be independent of the building's HVAC system, or it may be an integral part of that or other ventilation systems present in the building. The technical requirements for smoke removal design for high-rise buildings are included in Section 403.4.7. Two primary options for accomplishing post-fire smoke removal are provided.

The first option is a natural ventilation system. This design option relies on reasonably spaced and sized openings to allow air movement through a story. This is accomplished by providing operable windows or panels of a minimum size and spacing around the exterior of each story.

There are two exceptions to this design option. Exception 1 applies to Group R-1 occupancy hotels. It permits each guestroom or suite to have a single 2-square-foot (0.19 m^2) operable window or panel in lieu of the 40 square feet (3.7 m^2). Exception 2 allows fixed windows to be used in lieu of operable windows or panels, if the glazing can be cleared by fire fighters. Traditional tempered glass may be an appropriate type of glazing for this exception provided it is not coated, or has an applied film that modifies its natural breaking characteristics. Designers should work with the building official and the fire department to determine what type of glazing might meet the intent of this exception.

The second option is a mechanical ventilation system. This design option requires a mechanical air-handling system capable of exhausting the equivalent of one air change every 15 minutes. The code requires both the air being exhausted from the building and any return air that might be coming from the fire salvage areas to be exhausted outside of the building. When using the mechanical ventilation option, the design will most likely result in the use of dampers to zone floors and to introduce fresh air.

Since the smoke removal system is meant to be used after a fire for cleanup and recovery and not during a fire event, it is not necessary to connect this system to standby or emergency power.

In Item 3 of this section, a third design option is given. It simply states that any system that can provide results similar to the prescribed natural or mechanical systems in Item 1 or 2 can be used when such alternative has been approved by the building official. This provision restates the general code allowance for alternative methods permitted by Section 104.11.

[F] 403.4.8 Standby and emergency power. A standby power system complying with Section 2702 and Section 3003 shall be provided for the standby power loads specified in Section 403.4.8.2. An emergency power system complying with Section 2702 shall be provided for the emergency power loads specified in Section 403.4.8.3.

❖ Standby and emergency power are required to increase the probability that the fire protection systems and elevators will continue to function in the event of failure of normal building service. By referencing Chapter 27, these provisions pick up the requirements of not only the *National Electrical Code* (NEC), but also NFPA 110 and 111. Section 403.4.8.3 lists three systems that are considered standby power loads and must be supplied with standby power. Section 403.4.8.4 lists six systems that must be supplied with emergency power.

[F] 403.4.8.1 Equipment room. If the standby or emergency power system includes a generator set inside a building, the system shall be located in a separate room enclosed with 2-hour *fire barriers* constructed in accordance with Section 707 or *horizontal assemblies* constructed in accordance with Section 711, or both. System supervision with manual start and transfer features shall be provided at the *fire command center.*

Exception: In Group I-2, Condition 2, manual start and transfer features for the critical branch of the emergency power are not required to be provided at the *fire command center.*

❖ The section requires that, where generators are used to provide standby power and/or emergency power, such generators be protected in a room of not less than 2-hour construction. The walls of the room are to be constructed as fire barriers. Since Section 2702.1.4 requires these systems to be supplied with power for 2 hours, for high-rise buildings it was thought to be essential that the generator be protected for that length of time. The purpose of the enclosure is to decrease the probability that a fire can affect both the normal and standby power systems. It is important to note that this requirement is more restrictive than the provisions found in the NEC (NFPA 70). NFPA 70 would only require either the 2-hour fire protection or a sprinkler system, but not both. Because high-rise buildings are required to be sprinkler protected, these standby systems end up with both the sprinkler protection and the separation.

The controls of the systems are to be located in the building's fire command center. The exception allows hospitals to place the controls for critical branch emergency power in a location more appropriate for the hospital's chief engineer or other authorized person to control.

[F] 403.4.8.2 Fuel line piping protection. Fuel lines supplying a generator set inside a building shall be separated from areas of the building other than the room the generator is located in by an approved method or assembly that has a fire-resistance rating of not less than 2 hours. Where the building is protected throughout with an automatic sprinkler system installed in accordance with Section 903.3.1.1 or 903.3.1.2, the required fire-resistance rating shall be reduced to 1 hour.

❖ This section adds to the protection of standby and emergency power systems. Section 403.4.8.1 requires that an on-site generator set be in an enclosure with not less than 2 hours of protection. If the generators are fuel dependent, the fuel lines delivering the fuel should also be protected.

[F] 403.4.8.3 Standby power loads. The following are classified as standby power loads:

1. Power and lighting for the *fire command center* required by Section 403.4.6.
2. *Ventilation* and automatic fire detection equipment for *smokeproof enclosures*.
3. Elevators.
4. Where elevators are provided in a *high-rise building* for *accessible means of egress*, fire service access or occupant self-evacuation, the standby power system shall also comply with Sections 1009.4, 3007 or 3008, as applicable.

❖ The standby power system must be adequate to provide power to the fire command center (Section 403.4.6), smokeproof enclosures (Section 403.5.4) and elevators.

Elevators in a high-rise building must be provided with standby power. Additional requirements are found in Section 1009.4 for elevators used as accessible means of egress; Section 3007 for fire service access elevators required by Section 403.6.1; and Section 3008 for occupant evacuation elevators provided under Section 403.6.2. These additional provisions may add to the standby power load.

Where standby power is required for elevators, Section 3003 describes how it is to be provided. Under Section 3003, an arrangement must be made such that any elevator may be connected to the standby power with one elevator designated as the primary recipient of power once the standby system is activated. The primary elevator must be capable of serving all floors of a building. In extremely tall buildings, it may be necessary to designate multiple elevators as being primary recipients in order to permit usage of both local and express elevators. Based on language in the NEC, the capacity of the standby system must be such that all equipment that has to be operational at the same time will be able to function. The system need not, however, be capable of supplying the load for all connected equipment simultaneously if automatic load shedding is provided. For example, if multiple elevators are connected to the system, the standby power system need only be capable of supplying one elevator such that the elevator provides access to all floors and that central controls restrict operation of other elevators at the same time. The intent of the provisions of Section 3003 is not only to keep one elevator in operation, but also to make sure that all elevators at least initially make it to the designated level and allow passengers to exit the elevators (see commentary, Section 3003).

Note that the requirements for elevators designated as fire service access elevators or elevators used for occupant evacuation have different and generally more restrictive requirements for standby power (see commentary, Sections 3007.8 and 3008.8).

[F] 403.4.8.4 Emergency power loads. The following are classified as emergency power loads:

1. Exit signs and *means of egress* illumination required by Chapter 10.
2. Elevator car lighting.
3. *Emergency voice/alarm communications systems*.
4. Automatic fire detection systems.
5. *Fire alarm* systems.
6. Electrically powered fire pumps.

❖ The section simply lists those systems that must be connected to the emergency power system. The illumination of exit signs and the means of egress are required by Sections 1008 and 1013. Other systems are required by the provisions of Section 403. These systems must be provided power within 10 seconds of loss of regular power.

403.5 Means of egress and evacuation. The *means of egress* in *high-rise buildings* shall comply with Sections 403.5.1 through 403.5.6.

❖ Section 403.5 lists the requirements for means of egress systems applicable to high-rise buildings. Because of the size and height of high-rise buildings, they will contain large occupant loads which, when a building is evacuated, have longer evacuation times than low-rise buildings. Therefore, additional standards for egress systems are imposed for high-rise buildings. These include greater separation of the exit stairways in all high-rise buildings, requiring most stairways to be smokeproof enclosures, and for the tall high-rise buildings over 420 feet (128 m), an extra stairway, in addition to those provided based on the occupant load, is required. Luminous egress path markings are required for both vertical and horizontal exit enclosures. Section 403.5.3 addresses the practice of locking doors of an interior exit stairway.

403.5.1 Remoteness of interior exit stairways. Required *interior exit stairways* shall be separated by a distance not less than 30 feet (9144 mm) or not less than one-fourth of the length of the maximum overall diagonal dimension of the building or area to be served, whichever is less. The distance shall be measured in a straight line between the nearest points of the enclosure surrounding the *interior exit stairways*. In buildings with three or more *interior exit stairways,* no fewer than two of the *interior exit stairways* shall comply with this section. Interlocking or *scissor stairs* shall be counted as one *interior exit stairway*.

❖ This section requires that two of the interior exit stairways be separated from each other by a minimum distance. Requiring a minimum separation, even as little as 30 feet (9144 mm), reduces the probability that a fire or other event that damages one interior exit stairway enclosure will also cause the second interior exit stairway to be unusable. This separation requirement must be met, as well as the separation standards of Section 1007.1.1. The separation is to be measured between the enclosures surrounding the interior exit stairways. If the design of a building under Section 1007 results in two interior exit stairways being separated by 30 feet (9144 mm) or more, then Section 403.5.1 is also satisfied. While Section 1007 requires that two means of egress be separated by at least one-half or one-third of the diagonal measurement of the area served, the distance is measured between the doors into those interior exit stairways. Depending on the size and configuration of the building, two interior exit stairways could directly adjoin each other and still comply with Section 1007. For high-rise buildings, Section 403.5.1 requires that the interior exit stairway enclosures be completely separated at all points by the distance prescribed (see Commentary Figure 403.5.1).

Section 403.5.1 applies to all high-rise buildings subject to Section 403, regardless of the height of the building above 75 feet (22 860 mm). The requirement is either 30 feet (9144 mm), or one-fourth of the diagonal measurement of the area served, whichever is less. This is the same diagonal that is measured for compliance with Section 1007. For any building with a diagonal exceeding 120 feet (36 576 mm), the enclosures of interior exit stairways must be at least 30 feet (9144 mm) apart at their closest points. Only where the diagonal is less than 120 feet (36 576 mm) will the one-fourth of the diagonal come into play. For example, a building with a 100-foot (30 480 mm) maximum diagonal would only need 25 feet (7620 mm) between the two interior exit stairway enclosures. In larger buildings, with an occupant load per story in excess of 500, a third stairway is required. Also in buildings in excess of 420 feet (128 m), an additional stairway is required. The separation

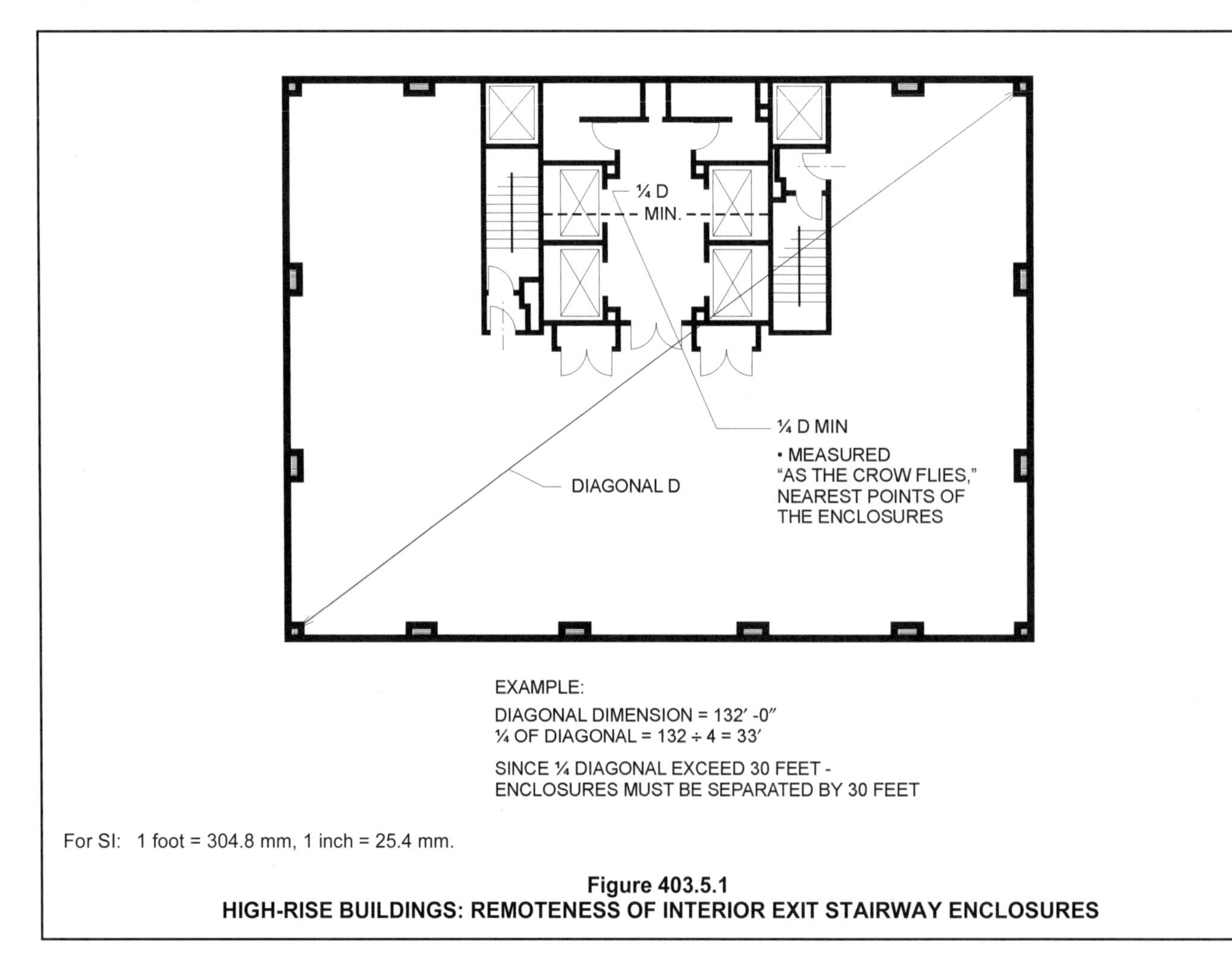

For SI: 1 foot = 304.8 mm, 1 inch = 25.4 mm.

Figure 403.5.1
HIGH-RISE BUILDINGS: REMOTENESS OF INTERIOR EXIT STAIRWAY ENCLOSURES

required by Section 403.5.1 only applies to two of the interior exit stairways. Any required third or fourth stairway could be directly adjacent to one of the first two interior exit stairways.

Interlocking interior exit stairways (two interior exit stairways that "wrap" around each other as they descend through the building) or scissor stairways (two stairways in the same shaft enclosure) have to be considered as one interior exit stairway for the purposes of this section. Clearly a scissor stairway does not create two independent interior exit stairways, and therefore would not meet the needs of this requirement. Similarly, interlocking interior exit stairways often share a common enclosure wall with an interior exit stairway on each side. Obviously, two interior exit stairways sharing a common enclosure wall would not meet the minimum separation requirement.

403.5.2 Additional interior exit stairway. For buildings other than Group R-2 that are more than 420 feet (128 000 mm) in *building height*, one additional *interior exit stairway* meeting the requirements of Sections 1011 and 1023 shall be provided in addition to the minimum number of *exits* required by Section 1006.3. The total width of any combination of remaining *interior exit stairways* with one *interior exit stairway* removed shall be not less than the total width required by Section 1005.1. *Scissor stairways* shall not be considered the additional *interior exit stairway* required by this section.

Exception: An additional *interior exit stairway* shall not be required to be installed in buildings having elevators used for occupant self-evacuation in accordance with Section 3008.

❖ For buildings in excess of 420 feet (128 m) in height, this provision requires an extra stairway to be provided in addition to the number of stairways (means of egress) required based on occupant load of the building as well as travel distance (see Sections 1004 and 1017). This requirement does not apply to high-rise buildings that are occupied as Group R-2 apartments or condominiums, but would apply to R-1 hotels. If a building is a mixed-occupancy structure, the building official should consider which occupancies are located above the 420-foot (128 m) level.

The intent of this provision is to accommodate a simultaneous occurrence of evacuation of the occupants of a high-rise building and fire-fighting operations. The traditional approach to emergencies in a high-rise building is to allow people on a floor affected by a fire to move up or down in the building to another story that is relatively safe from the effects of the fire. Fire fighters will usually set up a staging point one or two stories below the fire and commandeer one of the two stairways to move up to the fire floor. If for some reason it becomes necessary to evacuate the building during active fire fighting, the capacity of the means of egress system can be cut in half. This section implements, in part, Recommendation 17 of the NIST report. The report states that buildings should be designed to accommodate timely full-building evacuation of occupants when required by building-specific or large-scale emergencies, such as widespread power outages, major earthquakes, tornadoes, hurricanes without sufficient advance warning, fires, explosions or terrorist attacks.

The code requires that the extra stairway be sized so that the loss of any one stairway will still result in the remaining stairways being of sufficient size to accommodate the occupant load served by the stairways. For example, in a high-rise office building that has a calculated occupant load of 250 on each story, two stairways each designed to accommodate 125 occupants will be the minimum requirement for interior exit stairways in accordance with Chapter 10. The minimum stairway width of 44 inches (1118 mm) is sufficient to accommodate 146 people, or slightly more than needed for this building. In this building a third interior exit stairway would therefore need to be provided that is also at least 44 inches (1118 m) in width. The result is that if the fire fighters need to commandeer stairway 1, stairways 2 and 3, which each have a capacity to serve 146 people, will still be large enough to meet the occupant load requirements for the means of egress system. In larger buildings with occupant loads in excess of 500, a third stairway is required by Section 1006.2.1.1. In such buildings, Section 403.5.2 would require a fourth exit stairway.

The "additional" stairway has to comply with all the same requirements of any stairway in a high-rise building, including a design complying with smoke-proof enclosures required by Section 403.5.4.

The additional stairway is not affected by the remoteness requirement of Section 403.5.1. That section applies to the first two exit stairway enclosures required in a building. Additional stairways, whether required for occupant load, travel distance or this section, do not also need to comply with the remoteness requirement of Section 403.5.1.

Finally, the exception to this section permits the installation of occupant evacuation elevators as an alternative to providing the additional exit stairway (see also the commentary to Sections 403.6.2 and 3008). The intent behind the exception is recognition that the reason behind the requirement for an additional stairway is the loss of egress capacity under the unusual circumstance of simultaneous total building evacuation and fire fighting. Recognizing that occupant evacuation elevators provide an alternative way to meet a portion of the evacuation needs in these circumstances, the code allows the elevators as a substitute for the additional capacity provided by a third or fourth stairway. Where occupant evacuation elevators are proposed, the requirements of Section 3008 apply to all passenger elevators for general public use in the building (see commentary, Section 3008).

403.5.3 Stairway door operation. *Stairway* doors other than the *exit discharge* doors shall be permitted to be locked from the *stairway* side. *Stairway* doors that are locked from the *stairway* side shall be capable of being unlocked simultaneously without unlatching upon a signal from the *fire command center.*

❖ Section 1010.1.9.11 requires that all egress doors for interior stairways be readily openable from both sides. It is often desirable to control movement of people within a building and to provide additional security from external threats. This section permits locking of stairway doors from the stair side when all doors are capable of being simultaneously unlocked. Since high-rise buildings are difficult to evacuate and people are often relocated to another floor level during an emergency, access from the stairway to a floor could be essential in a fire or other emergency. Therefore, all stairway doors that are to be locked from the stairway side must have the capability of being unlocked on a signal from the fire command center. The unlocking of the door must not negate the latching feature, which is essential to the operation of the door as a fire door. Section 403.4.8.2, Item 1, by its reference to Section 403.4.6, requires the locking feature to be connected to the standby power system. When the door is unlocked during an emergency, it should not automatically relock on closure. Electrically powered locks should be designed such that when power to the locking device is interrupted, the lock is released. This is intended to enable doors to be operable from the inside of the stairway, and not locked, if power to the lock is interrupted. The building official should review the emergency release operation of stairway doors to determine that they remain unlocked.

403.5.3.1 Stairway communication system. A telephone or other two-way communications system connected to an *approved constantly attended station* shall be provided at not less than every fifth floor in each *stairway* where the doors to the *stairway* are locked.

❖ If the stairway doors are locked to restrict reentry as permitted in Section 403.5.3, a two-way communication system must be provided at no less than every fifth floor and must be connected to the standby power system. This system is required to be connected to a constantly attended location, which could be within the building, or to a central station that monitors fire alarms and is manned 24 hours a day, 7 days a week. Use of the fire command center is not recommended, since it may not be constantly attended. The system will permit occupants in the stairway to notify the attended location that the stairway doors need to be unlocked to access another floor or because conditions in the stairway prevent its continued use.

403.5.4 Smokeproof enclosures. Every required *interior exit stairway* serving floors more than 75 feet (22 860 mm) above the lowest level of fire department vehicle access shall be a *smokeproof enclosure* in accordance with Sections 909.20 and 1023.11.

❖ This section serves as a reminder that the smokeproof enclosure provisions of Section 1023.11 are applicable to high-rise buildings. Section 909.20 provides the standards on the construction of a smokeproof enclosure. It is important to understand that this requirement is only applicable in the portions of the building where the stack effect is more of a factor and the resulting spread of smoke will be the greatest. For example, if the high-rise building consists of a tower portion with a larger area on the bottom that is only a couple of stories in height, these requirements would only be applicable to stairways in the tower portion that serve floors that are more than 75 feet (22 860 mm) above the level where fire department vehicles would be located. The requirement would not apply to stairways in a low-rise portion of a building not exceeding 75 feet (22 860 mm), nor to stairways within the tower that do not serve floors more than 75 feet (22 860 mm) above the lowest level of fire department access.

403.5.5 Luminous egress path markings. Luminous egress path markings shall be provided in accordance with Section 1025.

❖ This section requires the provision of luminous egress path markings in all high-rise buildings containing any Group A, B, E, I, M or R-1 occupancy. The specific requirements are found in Section 1025. The requirements apply only to exit enclosures, vertical or horizontal. The markings are not required in the exit access or exit discharge portions of the means of egress system, nor at a horizontal exit.

403.5.6 Emergency escape and rescue. Emergency escape and rescue openings specified in Section 1030 are not required.

❖ This section of the code reinforces other tradeoffs or allowances permitted in the code for buildings equipped with automatic sprinkler systems. Previous editions of Section 1030 required emergency escape and rescue windows for windows located four stories or less, but then exempted high-rise buildings. Section 1030 now only requires the windows in R-3 buildings and single-exit R-2 occupancy buildings, neither of which is likely to occur in a high-rise building. This section is a leftover reference from previous editions of the code.

403.6 Elevators. Elevator installation and operation in *high-rise buildings* shall comply with Chapter 30 and Sections 403.6.1 and 403.6.2.

❖ This section explicitly states that elevator installation and operation are to be in accordance with Chapter 30, which contains provisions for the design, construction, installation, operation and maintenance of elevators. While a vital purpose of the elevator requirements in high-rise buildings has always been for the use of the fire department, this role has been enhanced (see Section 403.6.1). Elevators have been one method of providing an accessible means of egress under the requirements of Section 1009, but have never been considered part of the means of egress system for the nondisabled occupants of a building or for those with disabilities to self-evacuate. However, elevators are now allowed to augment the egress system for all occupants in buildings over 420 feet (128 m) in height. The allowance for self-evacuation is based on occupants using the elevators prior to the elevators being recalled. This is much different from the requirements of Section 1009, under which fire department personnel use the elevators to assist in the rescue of occupants while those elevators are on Phase II emergency operation.

Chapter 30 provides important standards for the size of elevators, allowing ambulance stretchers to easily fit into at least one elevator serving high-rise buildings. Protection of the hoistway and machine room is also addressed.

403.6.1 Fire service access elevator. In buildings with an occupied floor more than 120 feet (36 576 mm) above the lowest level of fire department vehicle access, no fewer than two fire service access elevators, or all elevators, whichever is less, shall be provided in accordance with Section 3007. Each fire service access elevator shall have a capacity of not less than 3,500 pounds (1588 kg) and shall comply with Section 3002.4.

❖ For buildings with occupied floors in excess of 120 feet (36 576 mm) above the lowest level of fire department vehicle access, at least two of the elevators must comply with the fire service access elevator requirements, which are detailed in Section 3007 (see Commentary Figure 403.6.1). The requirements in Section 3007 are another result of NIST research following the World Trade Center fire and collapse. The fire service access provisions work with previously enacted emergency operation requirements for elevators, allowing elevators that comply with Section 3007 to remain in service for trained fire fighters to reach the upper levels of a building within a reasonable amount of time and to stage their fire-fighting operations at a level below the actual fire. These elevators must be provided with a lobby of a minimum size which is protected by at least 1-hour fire-resistance-rated smoke barriers. The lobby must be directly connected to one of the interior exit stairways and that stairway must contain a standpipe as required by Section 905. These lobbies may also serve as the area of refuge required as part of an

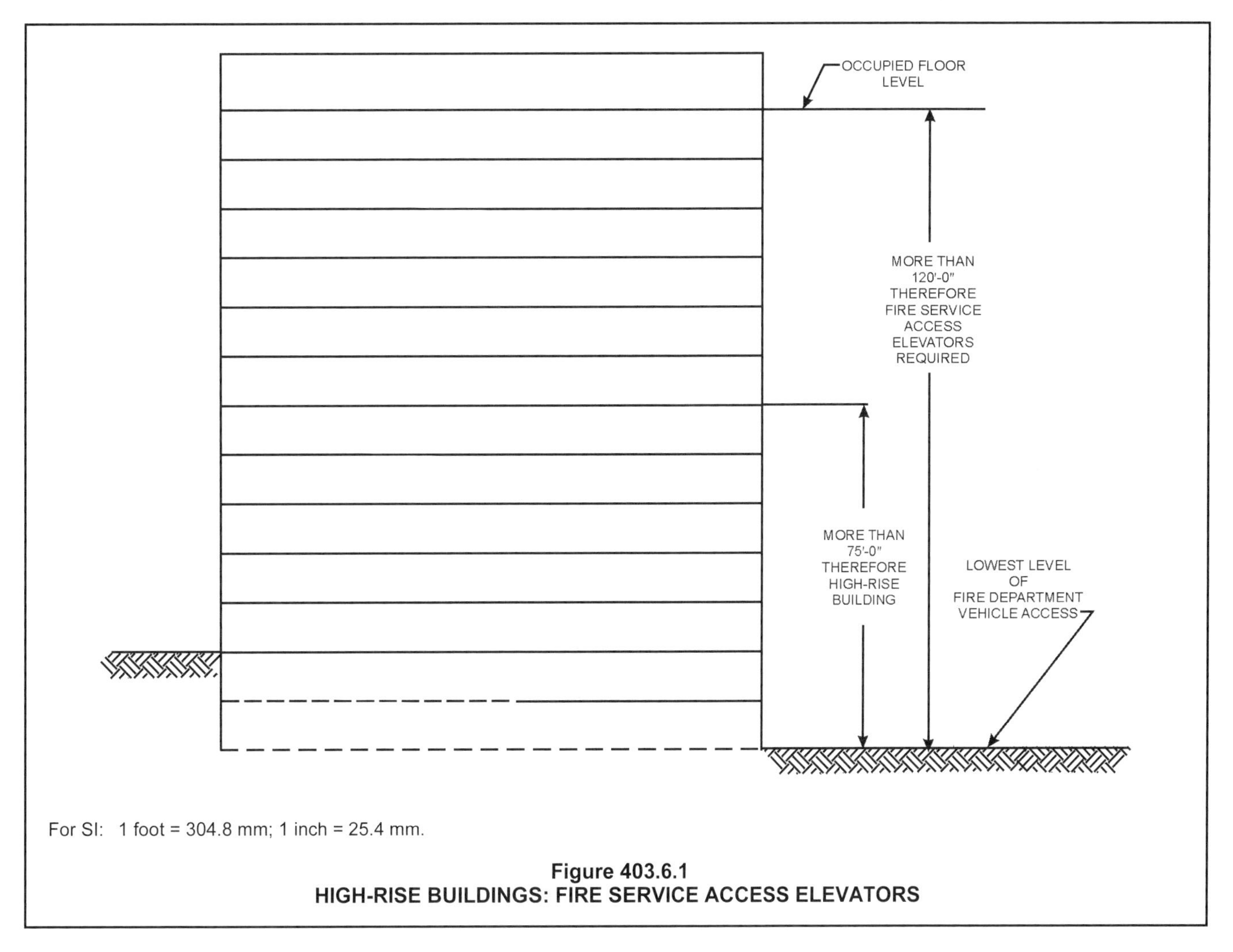

For SI: 1 foot = 304.8 mm; 1 inch = 25.4 mm.

Figure 403.6.1
HIGH-RISE BUILDINGS: FIRE SERVICE ACCESS ELEVATORS

accessible means of egress. People waiting for assistance in egress will need to be helped to a safe location before one lobby or another is used as the staging location for fire fighting.

The need for at least two elevators complying with Section 3007 is based on past experience showing that on many occasions elevators are not available because of various reasons, including problems in operation, routine maintenance, modernization programs and EMS operations in the building prior to fire fighter arrival. A minimum of two fire service elevators provided with all of the benefits afforded by such elevators better ensures that there will be a fire service access elevator available for the fire fighters' use in the performance of their duties.

The requirement for a minimum capacity of 3,500 pounds (1588 kg) for fire service access elevators allows for design flexibility in the selection of elevator cars while still providing the necessary capacity and size for emergency responder needs. Finally, Section 3002.4 is referenced to ensure that fire service access elevators will be large enough to accommodate an ambulance stretcher.

403.6.2 Occupant evacuation elevators. Where installed in accordance with Section 3008, passenger elevators for general public use shall be permitted to be used for occupant self-evacuation.

❖ Section 3008 provides the requirements for occupant evacuation elevators. When part of a fire safety and evacuation plan, occupant evacuation elevators are allowed in a building. A key element of applying Section 3008 is that all passenger elevators for general public use in the building must comply. Similar to fire service access elevators, occupant evacuation elevators must be provided with a lobby constructed of 1-hour fire-resistance-rated smoke barriers. Occupant evacuation elevators also require signs and other indicators, as well as a communication system connected to the fire command center to help occupants use the elevators. Where occupant evacuation elevators are provided, their requirements make for a substantially improved accessible means of egress over the requirements of Section 1009, which only allow for fire-fighter-assisted evacuation.

For any high-rise building, elevators may be designed for occupant self-evacuation. In fact, Section 3008 could be applied to any building as long as the design complies with all of the requirements.

Elevators are still prohibited by Section 1003.7 from serving as a means of egress for the general population of a building. However, in high-rise buildings in excess of 420 feet (128 m) in height, Section 403.5.2 requires there be an exit stairway provided in addition to those needed for egress capacity or travel distance purposes. This additional stairway is not required when the elevators in the building comply with Section 3008 and allow occupant self-evacuation.

SECTION 404
ATRIUMS

404.1 General. In other than Group H occupancies, and where permitted by Section 712.1.7, the provisions of Sections 404.1 through 404.10 shall apply to buildings or structures containing vertical openings defined as "Atriums."

❖ Unprotected vertical openings are often identified as the factors responsible for fire spread in incidents involving fire fatalities or extensive property damage. Section 404 addresses one method for protection of these specific building features in lieu of providing a complete floor separation. As noted in this section, the atrium provisions are primarily one of 18 options for addressing vertical openings as set forth in Section 712. Atriums are regulated by this section and are considered acceptable when in compliance with the criteria of this section. Additionally, the code permits other types of vertical openings that are not addressed by this section. For example, Section 712 permits openings within residential dwelling units to be unprotected. Other unprotected vertical openings that are permitted by the code include covered malls (see Section 402); communicating spaces in buildings of Group I-3 (see Section 408.5); mezzanines (see Section 505); escalator or supplemental stairway openings (see Section 712.1.3); open parking garages (see Section 406.5); enclosed parking garages (see Section 406.6); and as otherwise noted in Section 712.

An atrium is a space within a building that extends vertically and connects two or more stories. Atriums are not to be considered unprotected vertical openings; rather, the vertical openings are protected by means other than enclosure of the shaft or a complete floor assembly. This section does not apply to spaces that comply with Section 712 based on the itemized options, other than Section 712.1.7. In other words, if compliance with Section 712 is achieved by applying one of the options other than Section 712.1.7 that permits an atrium, then compliance with Section 404 is not required. Likewise, if the provisions of Section 404 have been complied with, then the other options listed in Section 712 do not need to be addressed. Spaces that are separated from the atrium by shaft enclosure assemblies complying with the provisions of Section 713 are not considered as part of the atrium. Section 404 applies only to the spaces that are contained within the atrium. Atriums are permitted in all buildings except Group H.

The options listed in Section 712 are all considered legitimate methods of appropriately addressing vertical openings within buildings, but were not developed with specific consideration of the application of multiple options in the same building/space. In most cases this will be appropriate but can become complicated where, for example, an atrium is open to a space that is applying Section 712.1.3 for escalator openings. More discussion of this particular scenario is found in the commentary to Section 404.6.

404.1.1 Definition. The following term is defined in Chapter 2:

ATRIUM.

❖ This section defines a term that is specifically associated with the subject matter of Section 404. It is important to emphasize that this term is not exclusively related to this chapter but may or may not also be applicable where the term is used elsewhere in the code.

Definitions of terms can help in the understanding and application of the code requirements. The purpose for including a definition in this chapter is to provide more convenient access to a term that may have a specific or limited application within this chapter. For the complete definition and associated commentary, refer back to Chapter 2. Terms that are italicized provide a visual identification throughout the code that a definition exists for that term. The use and application of all defined terms are set forth in Section 201.

404.2 Use. The floor of the *atrium* shall not be used for other than low fire hazard uses and only *approved* materials and decorations in accordance with the *International Fire Code* shall be used in the *atrium* space.

Exception: The *atrium* floor area is permitted to be used for any *approved* use where the individual space is provided with an *automatic sprinkler system* in accordance with Section 903.3.1.1.

❖ Because an automatic sprinkler system at the ceiling of an atrium may not be effective for a fire on the floor of the atrium due to the ceiling height or obstructions to the sprinkler discharge, the use and activities of the floor level and the types of materials in the atrium space must be controlled. This section applies to all atriums regardless of their height or area. Low fire-hazard uses would limit the atrium floor to such functions as pedestrian walk-through areas, security desks and reception areas. Storage areas, fabrication areas and office areas would not be low fire-hazard uses. Chapter 8 regulates the use of decorative materials and furnishings.

If the floor area is equipped with an automatic sprinkler system that can provide the required protection, then its use is not restricted. The exception stipulates that such areas must be equipped with an automatic sprinkler system as is required throughout the remainder of the atrium.

[F] 404.3 Automatic sprinkler protection. An *approved automatic sprinkler system* shall be installed throughout the entire building.

Exceptions:

1. That area of a building adjacent to or above the *atrium* need not be sprinklered provided that portion of the building is separated from the *atrium* portion by not less than 2-hour *fire barriers* constructed in accordance with Section 707 or *horizontal assemblies* constructed in accordance with Section 711, or both.
2. Where the ceiling of the *atrium* is more than 55 feet (16 764 mm) above the floor, sprinkler protection at the ceiling of the *atrium* is not required.

❖ One means of controlling the spread of fire and smoke through vertical openings is to control and extinguish the fire as early as possible. Therefore, all floor areas that are connected by the atrium are to be protected with an approved sprinkler system, including the atrium space itself. The system is to be designed, installed, tested and maintained in accordance with the provisions of Section 903.3 and the IFC. Since Chapter 9 requires that the sprinkler system be supervised, the reliability of the sprinkler system is improved. Exception 1 clarifies that since the atrium protection is one of the various options for protecting vertical openings, typically an alternative to a shaft enclosure, the sprinkler system is not required in areas that are separated from the atrium by 2-hour fire barrier walls or horizontal assemblies that would otherwise be required for a shaft enclosure. Such fire barrier walls must conform to Section 707, while the horizontal assemblies must comply with Section 711.

Exception 2 permits the required sprinkler system to be deleted from the ceiling areas of atriums where the vertical distance between the atrium floor and atrium ceiling is greater than 55 feet (16 764 mm). A ceiling height of more than 55 feet (16 764 mm) is the height at which the system is no longer effective and installing such systems provides little benefit. This exception does not alter the use limitations of the atrium as stated in Section 404.2, nor does it exempt any adjacent floor areas with lower ceiling heights that are included in the atrium boundary in accordance with Exception 1. It is important to note that if a smoke control system is required by Section 404.5, the smoke control design should not take sprinkler activation into account for design fires that are located in areas not protected with sprinklers or in areas where sprinklers are installed but, due to their location, may not be able to control a fire.

[F] 404.4 Fire alarm system. A *fire alarm* system shall be provided in accordance with Section 907.2.14.

❖ Section 907.2.14 of the code contains a requirement for a fire alarm system in atriums connecting more than two stories. In addition, when such atriums are located in Group A, E and M occupancies, an emergency voice communication system is required.

404.5 Smoke control. A smoke control system shall be installed in accordance with Section 909.

Exception: In other than Group I-2, and Group I-1, Condition 2, smoke control is not required for *atriums* that connect only two *stories*.

❖ In order to prevent the migration of smoke throughout interconnected levels of a building via the atrium, a mechanical smoke control system is to be installed in all atriums connecting more than two stories and in Group I-2 and Group I-1, Condition 2 buildings where atriums connect two or more stories. The smoke control system is to be designed and installed in accordance with the provisions of Section 909. See Commentary Figure 404.6(1) for examples of when smoke control is and is not required. For spaces such as atriums, the primary method of smoke control, in accordance with Section 909, is the exhaust method. Pressure differences in such a large space will be difficult to achieve and the area of fire origin when undertaking the pressurization method is not required to be tenable in accordance with Section 909.6. The airflow method is impractical with such large spaces. The airflow method is sometimes used in combination with the exhaust methods to protect openings into the atrium (see commentary, Section 909). The smoke control system for the atrium is required to be connected to a standby source of power in accordance with Sections 404.7 and 909.11.

Section 402.7.2 for covered mall buildings also references Section 404.5 for smoke control provisions when a covered mall building contains an atrium. It should be noted that when a covered mall building contains an atrium, all areas adjacent to the atrium without separation would need to be addressed when designing the smoke control system as required by Section 404.6, Exception 3.

As discussed in the commentary to Section 404.1, the atrium provisions are basically one of many options provided in the vertical opening requirements in Section 712. Therefore, if another option in Section 712 is chosen, such openings would not be considered atriums and smoke control would not be required.

404.6 Enclosure of atriums. *Atrium* spaces shall be separated from adjacent spaces by a 1-hour *fire barrier* constructed in accordance with Section 707 or a *horizontal assembly* constructed in accordance with Section 711, or both.

Exceptions:

1. A *fire barrier* is not required where a glass wall forming a smoke partition is provided. The glass wall shall comply with all of the following:
    1.1. Automatic sprinklers are provided along both sides of the separation wall and doors, or on the room side only if there is not a walkway on the *atrium* side. The sprinklers shall be located between 4 inches and 12 inches (102 mm and 305 mm) away from the glass and at intervals along the glass not greater than 6 feet (1829 mm). The sprinkler system shall be designed so that the entire surface of the glass is wet upon activation of the sprinkler system without obstruction;

    1.2. The glass wall shall be installed in a gasketed frame in a manner that the framing system deflects without breaking (loading) the glass before the sprinkler system operates; and

    1.3. Where glass doors are provided in the glass wall, they shall be either *self-closing* or automatic-closing.
2. A *fire barrier* is not required where a glass-block wall assembly complying with Section 2110 and having a $^{3}/_{4}$-hour *fire protection rating* is provided.
3. A *fire barrier* is not required between the *atrium* and the adjoining spaces of any three floors of the *atrium* provided such spaces are accounted for in the design of the smoke control system.

❖ One of the basic premises of atrium requirements is that an engineered smoke control system combined with an automatic fire sprinkler system that is properly supervised provide an adequate alternative to the fire-resistance rating of a shaft enclosure. It is also recognized that some form of a boundary is required to assist the smoke control system in containing smoke to just the atrium area. The basic requirement, therefore, is that the atrium space be separated from adjacent areas by fire barriers and horizontal assemblies having a fire-resistance rating of at least 1 hour.

Also, openings in the wall are required to be protected, in accordance with Section 707.6. In accordance with Section 713.4, shafts are required to have a fire-resistance rating of at least 2 hours if connecting more than three stories, and 1 hour when connecting two or three stories. The basis for the 1-hour requirement in Section 404.6 is that an automatic sprinkler system can be substituted for the 1 hour of fire resistance of a shaft enclosure. The allowance is consistent with the 1-hour fire-resistance-rating reduction permitted in high-rise buildings (see Section 403.2.1.2).

In lieu of a 1-hour fire-resistance-rated fire barrier separation, Exception 1 allows adjacent spaces to be separated by glass walls where automatic sprinklers have been installed to protect the glass. The sprinklers are to be located so as to wet the entire surface of the glass wall. If there is a floor surface on each side of the wall, both sides of the glass must be protected. The glass must be in a gasketed frame such that the framing system can deflect without breaking the glass. Although this exception does not address obstructions or other window treatments, consideration must be given to locating such items to avoid interference with the required sprinkler heads. Without specific test evidence, curtain rods, traverse rods, curtains and draperies must be located at least 12 inches (305 mm) from the window surface [see Commentary Figure 404.6(1)]. The sprinkler system required for Exception 1 is not intended to be a deluge system. Instead, it is intended to protect the glazing material from breakage as a result of thermal shock. It is not necessary to activate all the sprinklers along the glazing material to provide such protection as long as the entire surface of the glazed panel is designed such that it can be wetted by the sprinkler system.

Any doors through the required 1-hour fire barrier wall must be $^{3}/_{4}$-hour rated in accordance with Table 716.5. However, where glass doors occur within the complying glass wall system established in Exception 1, such doors need not be fire-protection rated but they must be either self-closing or automatic-closing.

Exception 2 allows a glass-block wall assembly conforming to Section 2110 in lieu of a 1-hour fire barrier. It is important to note that these glass-block assemblies do not require the sprinkler protection that is required by Exception 1.

Exception 3 recognizes the desire to have at least some floors open to the atrium, and permits a maximum of three. The three-floor restriction is consistent with the basic premise that the life safety hazard becomes significant when more than three floors are open and is also consistent with the allowances for covered mall buildings. It should be noted that the three floor levels may be at any height and need not be consecutive floor levels [see Commentary Figure 404.6(2)]. The exception also states that the smoke control design must account for these spaces. This particular reference to the smoke control design does not require that the 6-foot-high (1829 mm) layer required by Section 909.8.1 be maintained in these spaces. Instead it is saying that if a smoke control system is required by Section 404.5, such spaces must be accounted for in terms of the hazard they pose to the atrium and to smoke migrating to other adjacent spaces on other stories open to the atrium. Essentially these spaces have simply increased the possible design fires that may send smoke into the atrium, thus threatening to send smoke throughout the building and other adjoining spaces. If the atrium smoke control system is not designed to handle fires in these areas, the system may become overpowered. This exception can also permit a two-story atrium to have adjacent areas open to the atrium. In this case, since there is no required smoke control system, obviously there is no need to account for the added space (see IBC Interpretation No. 54-07).

As discussed in the commentary to Section 404.1, the atrium requirements are basically one of many options for protecting vertical openings through floors as established in the requirements of Section 712. More specifically, Section 712 has multiple options that are all legitimate approaches to addressing vertical openings. If there are portions of the building applying other options in Section 712 that are not separated from the atrium, they can be considered part of the atrium themselves and be subject to the smoke control layer height of 6 feet (1829 mm) or could be considered as adjacent spaces and need to be accounted for when designing the smoke control system, as discussed above. It should be noted that if the space is considered part of the atrium itself, the other options in Section 712 need not be considered as the atrium provisions are already being applied.

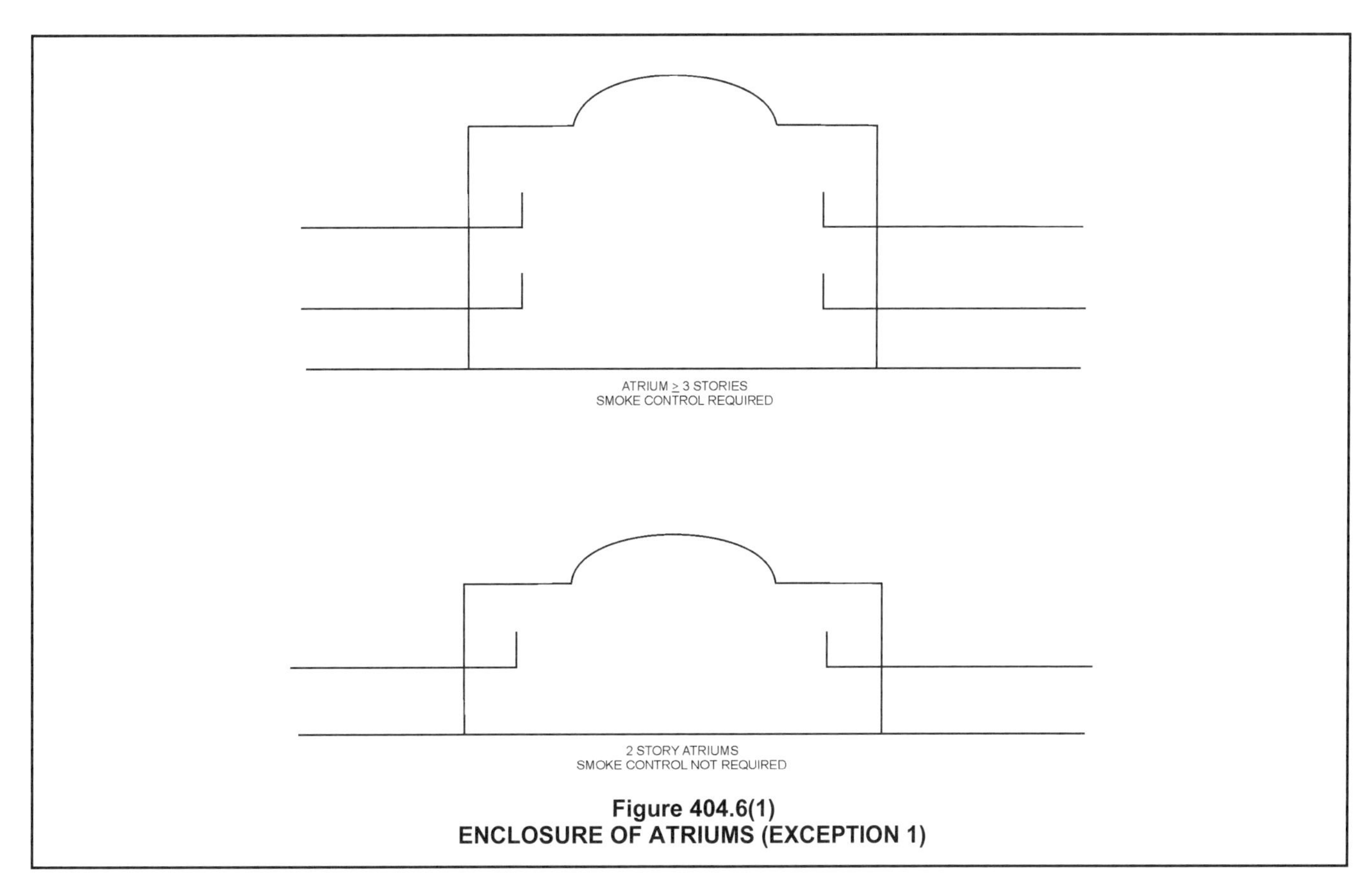

Figure 404.6(1)
ENCLOSURE OF ATRIUMS (EXCEPTION 1)

[F] 404.7 Standby power. Equipment required to provide smoke control shall be provided with standby power in accordance with Section 909.11.

❖ To enhance the reliability of the required smoke control system, this section of the code mandates that the mechanical smoke control system must be provided with standby power. The reference to Section 909.11 allows the atrium's equipment to receive its primary power from the building's power system. A secondary standby power source is required should the building's primary system fail.

404.8 Interior finish. The *interior finish* of walls and ceilings of the *atrium* shall be not less than Class B with no reduction in class for sprinkler protection.

❖ Although Chapter 8 contains provisions governing the use of interior finishes, trim and decorative materials, specific limitations are provided for an atrium. Similar to the use limitations contained in Section 404.2, a minimum interior finish classification is specified to limit the fuel load within the atrium spaces. At least a Class B finish is mandated for the walls and ceilings of the atrium. This finish classification limits the material's flame spread index to a range of 26 through 75 and a smoke-developed index rating of 450 or less. The sprinkler reductions shown in Table 803.11 are not applicable to the interior finishes of an atrium space.

404.9 Exit access travel distance. *Exit access* travel distance for areas open to an *atrium* shall comply with the requirements of this section.

❖ The means of egress from buildings containing atriums are required to conform to Chapter 10, the provisions of this section which address exit travel distance and Section 404.10.

404.9.1 Egress not through the atrium. Where required access to the *exits* is not through the *atrium*, *exit access* travel distance shall comply with Section 1017.

❖ Exit access travel which does not go through the atrium is determined in the usual method found in Chapter 10.

404.9.2 Exit access travel distance at the level of exit discharge. Where the path of egress travel is through an *atrium* space, *exit access* travel distance at the *level of exit discharge* shall be determined in accordance with Section 1017.

❖ This section indicates that if you have exit access travel across the floor of the atrium, essentially the lowest level, then this travel should also be determined in accordance with Chapter 10. However, where the egress travel is on upper levels, it is limited according to Section 404.9.3.

404.9.3 Exit access travel distance at other than the level of exit discharge. Where the path of egress travel is not at the *level of exit discharge* from the *atrium*, that portion of the total permitted *exit access* travel distance that occurs within the *atrium* shall be not greater than 200 feet (60 960 mm).

❖ On upper stories of an atrium, exit access travel through the atrium is limited to 200 feet (60 960 mm). An example is a hotel designed to have the doors to all guest rooms located on balconies in the atrium. The exit access travel from the room door to an exit enclosure (or to an area of the building not within the atrium) is limited to 200 feet (60 960 mm). That limit applies on each floor open to the atrium with the

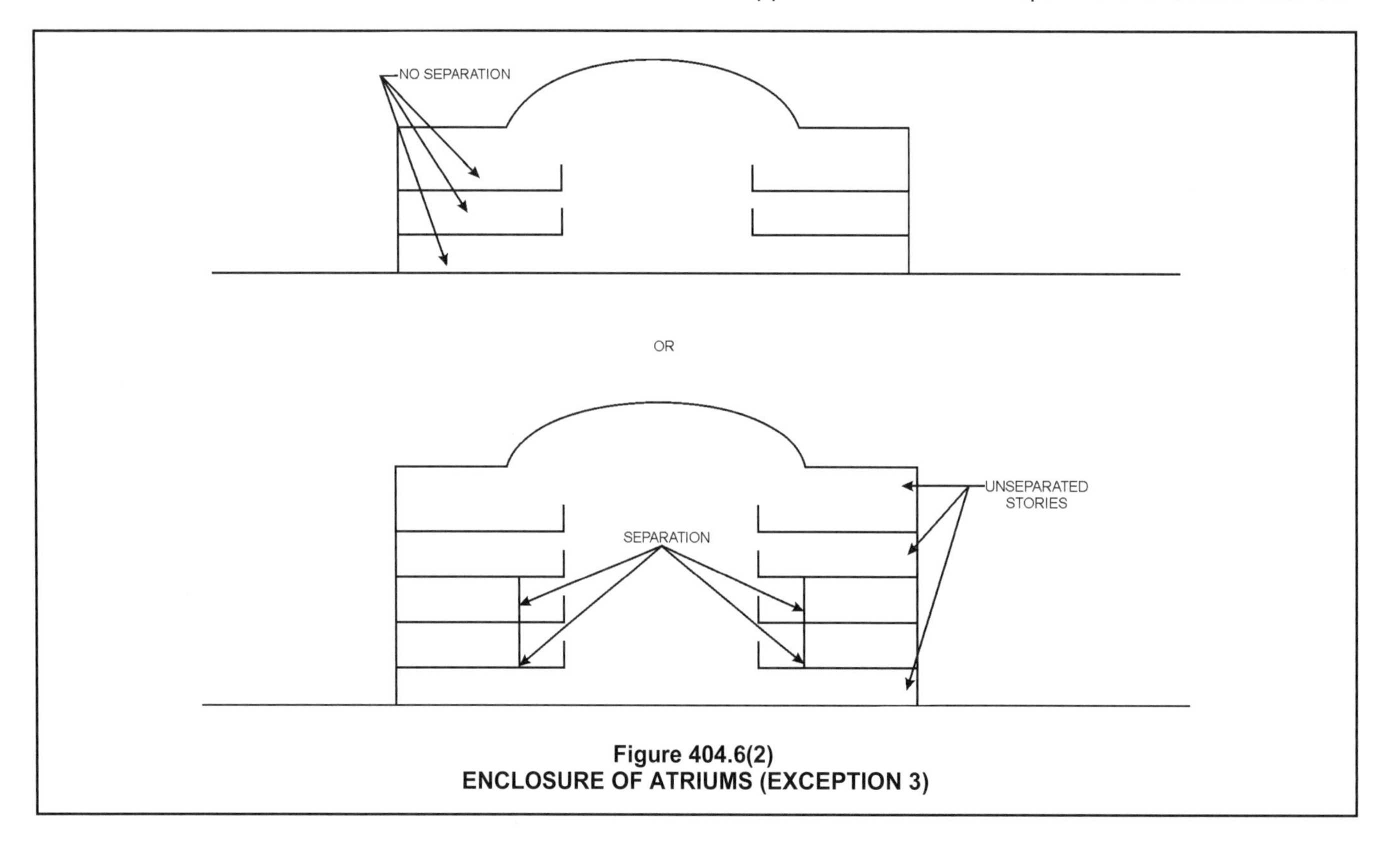

Figure 404.6(2)
ENCLOSURE OF ATRIUMS (EXCEPTION 3)

exception of the level of exit discharge, which is usually the floor of the atrium.

404.10 Interior exit stairways. A maximum of 50 percent of *interior exit stairways* are permitted to egress through an *atrium* on the *level of exit discharge* in accordance with Section 1028.

❖ Section 1028 of the code permits 50 percent of interior exit stairways to terminate on the level of exit discharge such that building occupants travel across the floor of that level before reaching an exterior exit door. The intent of this section is to clearly allow the same option for a building with an atrium. It allows the path of egress to be through the atrium.

SECTION 405 UNDERGROUND BUILDINGS

405.1 General. The provisions of Sections 405.2 through 405.9 apply to building spaces having a floor level used for human occupancy more than 30 feet (9144 mm) below the finished floor of the lowest *level of exit discharge*.

Exceptions: The provisions of Section 405 are not applicable to the following buildings or portions of buildings:

1. One- and two-family *dwellings*, sprinklered in accordance with Section 903.3.1.3.
2. Parking garages provided with *automatic sprinkler systems* in compliance with Section 405.3.
3. Fixed guideway transit systems.
4. *Grandstands*, *bleachers*, stadiums, arenas and similar facilities.
5. Where the lowest *story* is the only *story* that would qualify the building as an underground building and has an area not greater than 1,500 square feet (139 m^2) and has an *occupant load* less than 10.
6. Pumping stations and other similar mechanical spaces intended only for limited periodic use by service or maintenance personnel.

❖ An underground building presents a unique hazard to life safety. Because of its isolation and inaccessibility, occupants within the structure and fire fighters attempting to locate and suppress a fire are presented with a unique fire protection challenge.

Underground buildings that require the occupants of the lowest floor level to travel upward for more than 30 feet (9144 mm) to reach the finished floor of the lowest level of exit discharge present a significant hazard to the occupants. As such, Section 405 is applicable to buildings with a floor level more than 30 feet (9144 mm) below the finished floor surface of the lowest level of exit discharge (see Commentary Figure 405.1).

To egress the structure, occupants must travel in an upward direction. The direction of occupant travel is the same as the direction that the products of combustion travel. As such, the occupants are potentially exposed to the products of combustion along the entire means of egress.

Fire fighters are also confronted by constant exposure to the products of combustion. Beginning their descent above the actual location of the fire source, fire fighters encounter an increasing amount of smoke, heat and flame as they attempt to locate and extinguish the fire source. These extreme conditions could significantly hinder the effectiveness of the fire department if not offset by appropriate fire protection requirements. The requirements for underground buildings are, in some ways, similar to those for high-rise buildings. Both types of structures present an unusual hazard, since they are virtually inaccessible to exterior fire department suppression and rescue operations with the increased potential to trap occupants inside. To counteract these hazards, such structures are required by Section 405.2 to be built of noncombustible, fire-resistance-rated construction.

Additionally, they are required by Section 405.3 to be equipped with an automatic sprinkler system and a smoke control system in accordance with Section 405.5. Standby and emergency power systems are also required in these structures by Sections 405.8 and 405.9.

Structures regulated by Section 405 are also subject to all other applicable code provisions. Additionally, underground buildings to which this section does not apply are still subject to all other code provisions, including fire suppression (see Section 903); standpipe systems (see Section 905); and fire alarm and detection (see Section 907).

There are six exceptions to the applicability of Section 405. These exceptions are in consideration of specific types of structures to which the requirements of this section are impractical, unnecessary or to which alternative provisions apply.

405.2 Construction requirements. The underground portion of the building shall be of Type I construction.

❖ Two of the key goals with underground construction are to limit the fuel load and to increase the fire resistance of underground buildings. To reasonably ensure that the building elements will remain structurally sound during exposure to fire, all structural elements of the underground portions are required to be of noncombustible, protected construction. Therefore, any portion of the structure located below grade level is required to be of Type I construction (see Commentary Figure 405.1).

Portions of the structure above grade, however, are permitted to be of any type of construction.

Any portion of the structure above ground level must conform to the height and area limitations of Tables 504.3, 504.4 and 506.2. These tables do not regulate the maximum depth of an underground structure; therefore, the height, area and depth of most Type I structures are not limited.

[F] 405.3 Automatic sprinkler system. The highest *level of exit discharge* serving the underground portions of the building and all levels below shall be equipped with an *automatic* *sprinkler system* installed in accordance with Section 903.3.1.1. Water-flow switches and control valves shall be supervised in accordance with Section 903.4.

❖ One of the most effective preventive measures to fire growth is the installation of an automatic sprinkler system. Because of the unique conditions for occupant egress and fire department access in the underground portion of a building, automatic sprinkler protection is required. The level of exit discharge and all floor levels below are required to be sprinklered throughout in accordance with Section 903.3.1.1. This section does permit a portion of a building to extend above the level of exit discharge and not be equipped with an automatic sprinkler system. If, however, another code section (see Section 403.3 or 903) requires an automatic sprinkler system in the above-ground portion, such a requirement would still be applicable. Note that a smoke control system is required in accordance with Section 405.5. Automatic sprinkler systems are essential elements of any smoke control system. Without suppression, the size of the fire or the resulting products of combustion will rapidly overwhelm most mechanical smoke control systems. The pressurization system requirements in Section 909 are based upon the use of sprinklers. Pressure differences needed to contain the smoke to the area of origin would need to be higher in an nonsprinklered building.

405.4 Compartmentation. Compartmentation shall be in accordance with Sections 405.4.1 through 405.4.3.

❖ Compartmentation is a key element in the egress and fire access plan for floor areas in an underground building. Subdivision into separate compartments through the use of smoke barriers (see Section 709) permits occupants to travel horizontally to escape the fire condition and provides a staging area for the fire service (see Commentary Figure 405.4).

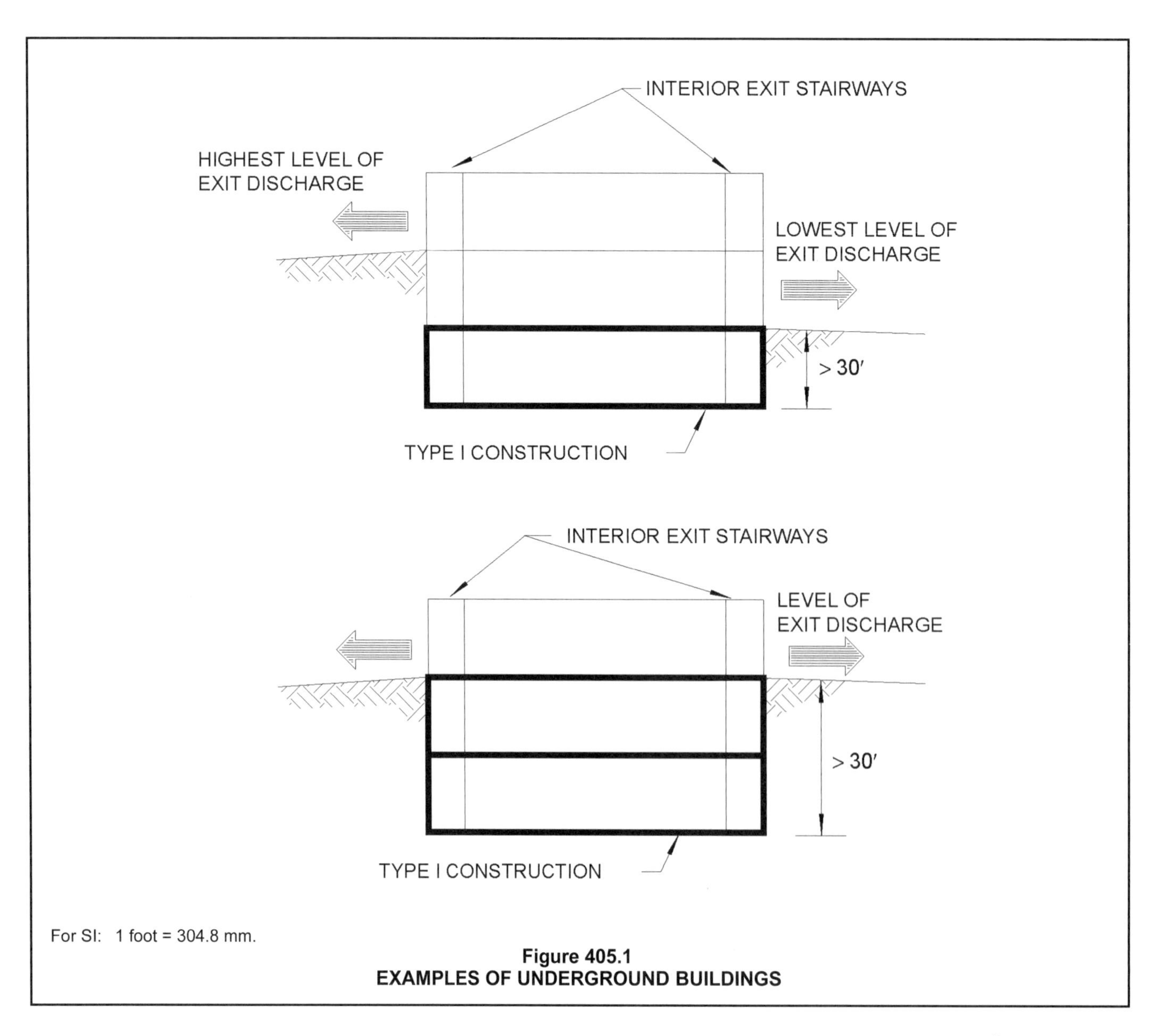

Figure 405.1
EXAMPLES OF UNDERGROUND BUILDINGS

405.4.1 Number of compartments. A building having a floor level more than 60 feet (18 288 mm) below the finished floor of the lowest *level of exit discharge* shall be divided into no fewer than two compartments of approximately equal size. Such compartmentation shall extend through the highest *level of exit discharge* serving the underground portions of the building and all levels below.

Exception: The lowest *story* need not be compartmented where the area is not greater than 1,500 square feet (139 m^2) and has an *occupant load* of less than 10.

❖ The 60-foot (18 288 mm) threshold is based on establishing a reasonable limitation on the required vertical travel distance for the occupants and fire service before the added protection of compartmentation is beneficial.

It is important to realize that the code requires, at a minimum, two compartments of approximately equal size. The maximum size of each compartment is not limited and is, therefore, a design decision subject to the approval of the building official.

An exception is permitted at the lowest level of the underground building as long as that level is relatively small in area [less than or equal to 1,500 square feet (139 m^2)] and serves a low occupant load (less than 10), since an evacuation to the next higher level

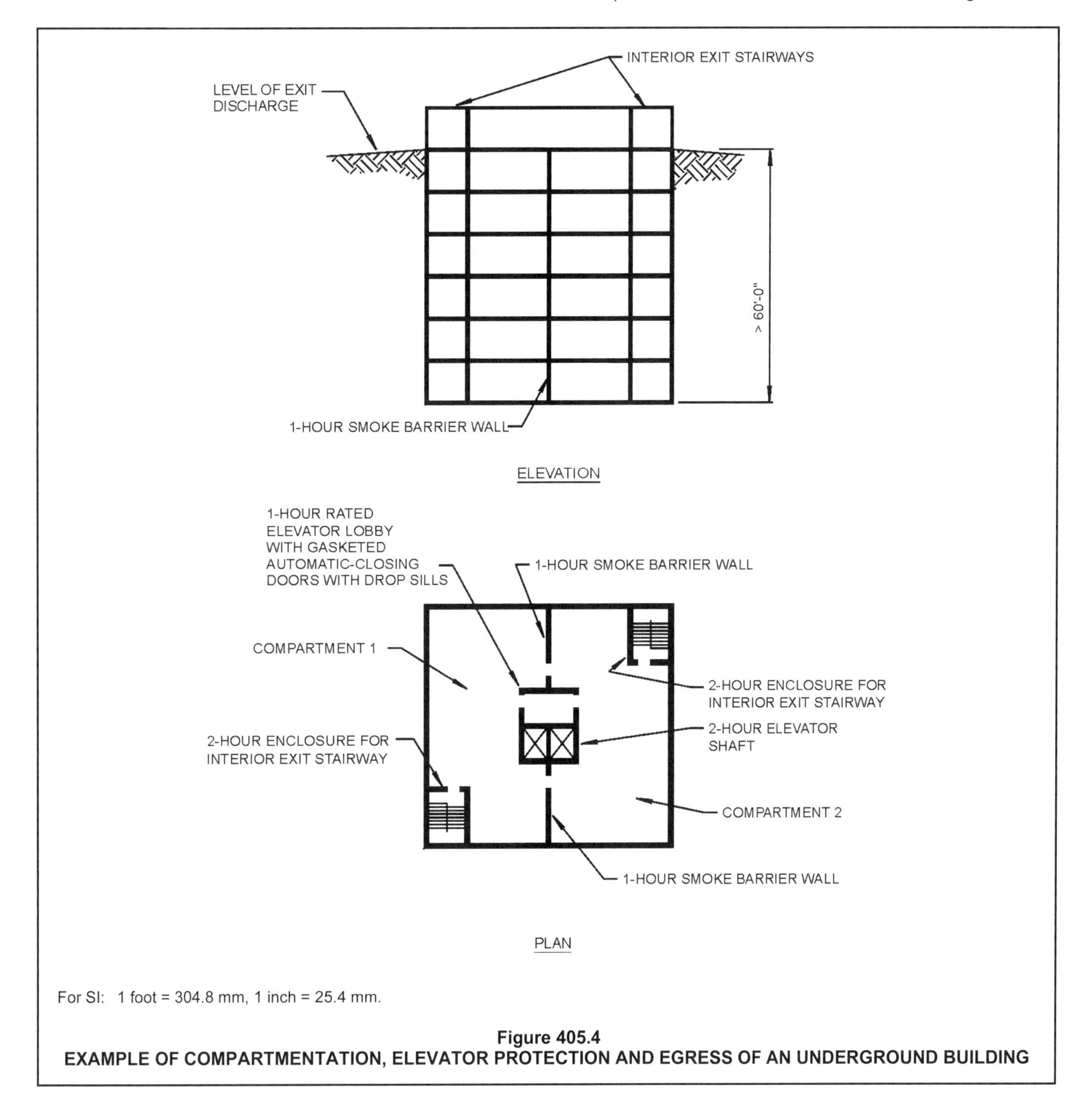

For SI: 1 foot = 304.8 mm, 1 inch = 25.4 mm.

Figure 405.4
EXAMPLE OF COMPARTMENTATION, ELEVATOR PROTECTION AND EGRESS OF AN UNDERGROUND BUILDING

would not be expected to adversely affect that level. Note that this exception coordinates with Exception 5 to Section 405.1.

It should be noted that smoke control would be required for all underground buildings in accordance with Section 405.5 that fall under the scope of Section 405.1. Smoke control systems, which are required to be designed in accordance with Section 909, would generally require the use of smoke barriers to create compartments to achieve the design goals stated in Section 405.5.1. Therefore, even when compartmentation is not required for a building that has a floor level less than 60 feet (18 288 mm) below the finished floor of the level of exit discharge, compartmentation will probably still be utilized within a building to achieve the goals of Section 909. These compartments may be utilized in a different manner than as mandated in Section 405.4.1. In other words, such compartments may not span vertically from the finished floor of the level of exit discharge to the lowest level of the underground building. The compartments may be designated by story (see Commentary Figure 405.4.1).

The provisions for smoke control and compartmentation found in Section 405 were historically focused on smoke exhaust versus smoke control. More specifically, smoke control is for the safety of the occupants only, whereas smoke exhaust tends to focus on the needs of the fire department in managing smoke in underground conditions. The compartmentation requirements in Section 405.4 were meant to work with the smoke exhaust requirements and, at the very least, provide passive smoke management for the building for both the occupants and the fire department. As noted, the requirements of Section 909 as required by Section 405.5 have a different focus and will result in many compartments within the building.

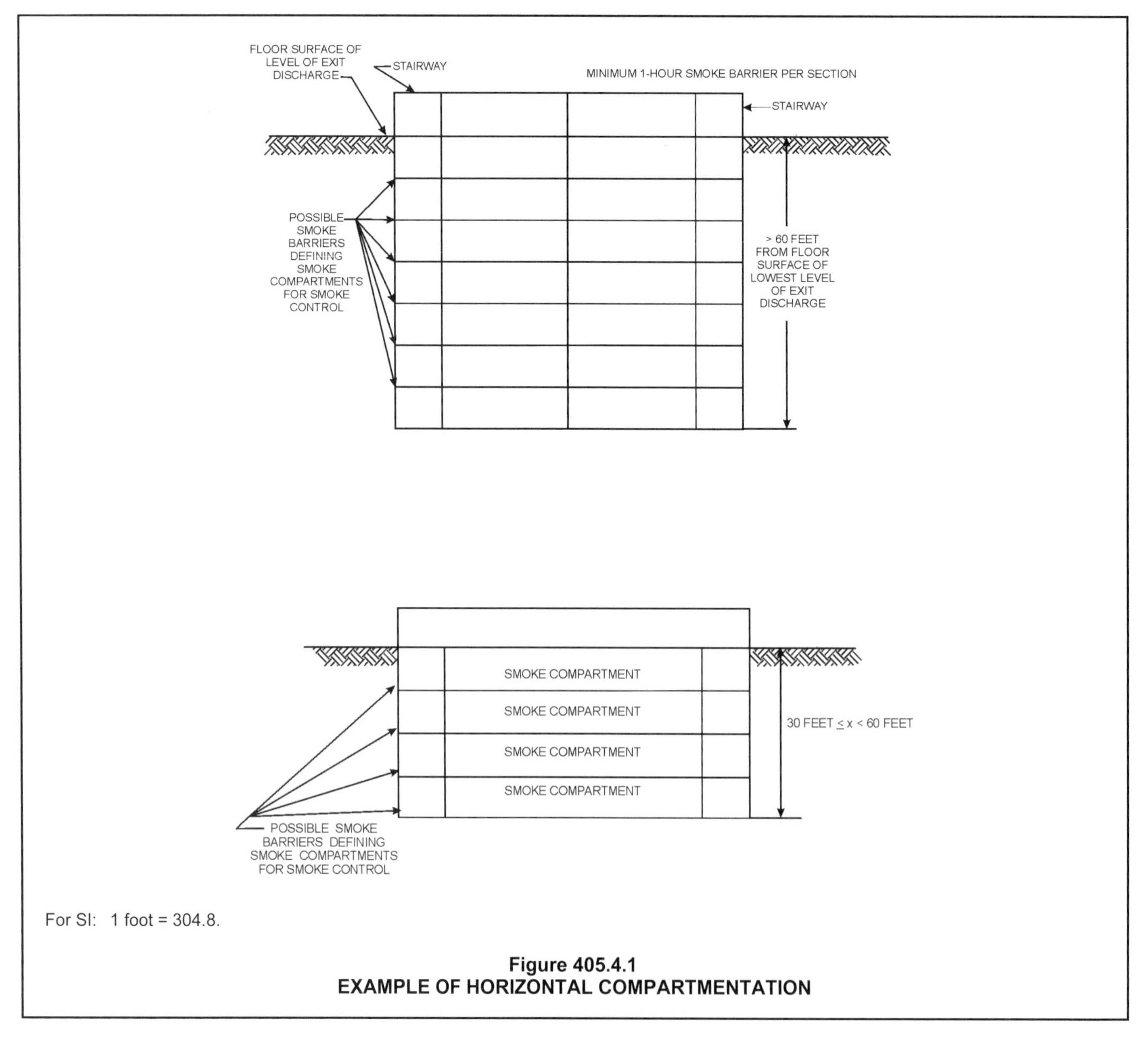

Figure 405.4.1
EXAMPLE OF HORIZONTAL COMPARTMENTATION

405.4.2 Smoke barrier penetration. The compartments shall be separated from each other by a *smoke barrier* in accordance with Section 709. Penetrations between the two compartments shall be limited to plumbing and electrical piping and conduit that are firestopped in accordance with Section 714. Doorways shall be protected by *fire door assemblies* that are automatic-closing by smoke detection in accordance with Section 716.5.9.3 and are installed in accordance with NFPA 105 and Section 716.5.3. Where provided, each compartment shall have an air supply and an exhaust system independent of the other compartments.

❖ The smoke barrier walls and floor/ceiling assemblies that create the compartments must be fire-resistance rated for 1-hour and must meet the requirements of Section 709. Penetrations through the smoke barrier wall are limited to plumbing, piping and penetrations that are vital to the fire protection systems (i.e., sprinkler piping and electrical raceways). Special fire-resistance-rated doors are required to maintain the intended compartmentation, including gasketing and drop sills. Separate air distribution systems are required for each compartment. Components of the air distribution system are not permitted to penetrate the smoke barrier walls. These requirements are intended to help provide for the independence of the compartments separated by the smoke barrier walls by minimizing the penetrations therein. Note also that Section 711.2 addresses horizontal assemblies that are used as smoke barriers.

The separate air distribution systems required by this section are focused on the smoke compartments specifically required by this section and would not necessarily require separate systems for each compartment as created by the requirements of Section 405.5.

405.4.3 Elevators. Where elevators are provided, each compartment shall have direct access to an elevator. Where an elevator serves more than one compartment, an elevator lobby shall be provided and shall be separated from each compartment by a *smoke barrier* in accordance with Section 709. Doors shall be gasketed, have a drop sill and be automatic-closing by smoke detection in accordance with Section 716.5.9.3.

❖ Elevators are permitted to serve more than one compartment when properly protected. When different compartments utilize a common elevator, a smoke barrier consisting of a 1-hour fire-resistance-rated elevator lobby is required. This lobby provides additional separation between adjacent compartments, thus helping to create their independence from the immediate effects of fire (see Commentary Figure 405.4). Essentially, the elevator lobby as prescribed in this section creates another smoke compartment within the building.

405.5 Smoke control system. A smoke control system shall be provided in accordance with Sections 405.5.1 and 405.5.2.

❖ A smoke control system is required for all underground buildings. The smoke control system is an integral part of the required fire protection systems for underground buildings and is focused upon the safety of occupants while exiting the building during a fire. In the case where the building contains both underground and above-ground portions, such specific details and requirements are listed in Sections 405.5.1 and 405.5.2.

405.5.1 Control system. A smoke control system is required to control the migration of products of combustion in accordance with Section 909 and the provisions of this section. Smoke control shall restrict movement of smoke to the general area of fire origin and maintain *means of egress* in a usable condition.

❖ The general design requirements, special inspection and test requirements, and applicable smoke control methods are contained within Section 909. The performance goal of the smoke control system is to contain the smoke and hot gases generated by a fire condition to the immediate area of origin. "Containing to the immediate area of origin" is usually taken to mean the compartment of origin. This allows the building occupants to access their required exits and evacuate the building before smoke movement traps them below grade. Generally the type of system will be a pressurization system where the smoke is managed by maintaining pressure differences across smoke barriers. It could be possible that a different type of system such as an exhaust system is required in larger more open spaces such as an atrium. It is important to note that the systems in Section 909 are intended for the protection of occupants and do not necessarily provide longer term capacity for the fire department. In addition, such systems are not intended as a method of smoke exhaust for salvage and overhaul after a fire. It is recognized that overhaul in underground buildings after a fire is far more complex because the structure is entirely underground, but as Section 909 is currently written such systems are not specifically designed with this role in mind.

As discussed in the commentary to Section 405.4.1, the smoke control system required by Section 405.5 will likely result in many smoke compartments in order to meet the design criteria of Section 909.

405.5.2 Compartment smoke control system. Where compartmentation is required, each compartment shall have an independent smoke control system. The system shall be automatically activated and capable of manual operation in accordance with Sections 907.2.18 and 907.2.19.

❖ The compartmentation referred to in this section is that discussed in Section 405.4.1, which requires that the building be separated from the story of level of exit discharge to the lowest level in the building. It is not the intent that each smoke compartment (smoke zone) as created through the application of Section 405.5.1 would be required to have independent smoke control systems (see commentary, Section 405.4.2).

[F] 405.6 Fire alarm systems. A *fire alarm* system shall be provided where required by Sections 907.2.18 and 907.2.19.

❖ The ability to communicate and offer warning of a fire scenario can increase the time available for egress from the building. Underground buildings with a floor level greater than 60 feet (18 288 mm) below the finished floor of the level of exit discharge are, therefore, required to be provided with a manual fire alarm system. An emergency voice/alarm communication system is required as part of this system.

405.7 Means of egress. *Means of egress* shall be in accordance with Sections 405.7.1 and 405.7.2.

❖ This section is simply the introduction for means of egress requirements in underground buildings. Egress from underground buildings can be more challenging than buildings above grade, thus additional requirements are necessary.

405.7.1 Number of exits. Each floor level shall be provided with no fewer than two *exits*. Where compartmentation is required by Section 405.4, each compartment shall have no fewer than one *exit* and shall also have no fewer than one *exit access* doorway into the adjoining compartment.

❖ The means of egress from underground buildings is an integral part of the safety precautions necessary to abate the hazards of people being located more than 30 feet (9144 mm) below the level of exit discharge. Sections 405.7.1 and 405.7.2 provide the minimum number of exits and requirements for smokeproof enclosures. These provisions create safe and usable elements that the building occupants can enter to flee the immediate effects of a fire located well below grade.

The arrangement of exits and exit access doors for an underground building is regulated by this section (see Commentary Figure 405.4).

Based on the requirements of this section and Section 405.4, an underground building having a floor level more than 30 feet (9144 mm) below the level of exit discharge requires a minimum of two exits, and cannot qualify as a single-exit building. Additional exits may be required in accordance with Section 1006.2. Underground buildings requiring compartmentation in accordance with Section 405.4.1 must have an exit in each and every compartment plus a doorway into the adjoining compartment.

405.7.2 Smokeproof enclosure. Every required *stairway* serving floor levels more than 30 feet (9144 mm) below the finished floor of its *level of exit discharge* shall comply with the requirements for a *smokeproof enclosure* as provided in Section 1023.11.

❖ In order for interior exit stairways to provide the necessary means of egress, all stairways of an underground building must be constructed as smokeproof enclosures. The requirements for smokeproof enclosures are contained in Section 1023.11, which in turn references Section 909.20, and include a 2-hour enclosure along with limited access via a vestibule where the mechanical ventilation alternative described in Section 909.20 is provided. Another alternative is a pressurized stairway. The requirements for pressurized stairways are also listed in Section 909.20. Pressurized interior exit stairways do not require a vestibule, but pressure differences as prescribed in Section 909.20.5 need to be provided and demonstrated through testing. These provisions for a smokeproof enclosure are intended to maintain a protected path of travel for building occupants to egress the underground building and also allow the fire department to enter for rescue and fire-fighting operations.

[F] 405.8 Standby and emergency power. A standby power system complying with Section 2702 shall be provided for the standby power loads specified in Section 405.8.1. An emergency power system complying with Section 2702 shall be provided for the emergency power loads specified in Section 405.8.2.

❖ All underground buildings regulated by this section are required to be provided with standby and emergency power systems to increase the probability that critical systems will be operational in the event of a loss of normal power supply. The purpose of the standby and emergency power systems is to provide an alternative means of supplying power to selected building systems should the normal electrical source fail. The systems which must be provided with standby power are specified in Section 405.8.1. The systems required to be provided with emergency power are specified in Section 405.8.2.

[F] 405.8.1 Standby power loads. The following loads are classified as standby power loads:

1. Smoke control system.
2. *Ventilation* and automatic fire detection equipment for *smokeproof enclosures*.
3. Fire pumps.
4. Elevators, as required in Section 3003.

❖ The standby power system is required to supply electrical power to equipment that is essential to emergency egress, fire-fighting and rescue operations. See the commentary in Sections 403.4.8.3 and 3003 regarding how elevators operate under standby power.

[F] 405.8.2 Emergency power loads. The following loads are classified as emergency power loads:

1. *Emergency voice/alarm communications systems*.
2. *Fire alarm* systems.
3. Automatic fire detection systems.
4. Elevator car lighting.
5. *Means of egress* and exit sign illumination as required by Chapter 10.

❖ The emergency power system is required to supply electrical power to equipment that is essential to detecting and warning others of a fire condition and to provide for illuminated evacuation of the underground building. Because of the critical nature of the systems and the potential for panic in the areas covered, the loads must be picked up within 10 seconds after failure of the normal power supply. This maximum time is consistent with the requirements of NFPA 70.

[F] 405.9 Standpipe system. The underground building shall be equipped throughout with a standpipe system in accordance with Section 905.

❖ Just as high-rise buildings or other large buildings with special features are required to be provided with a standpipe system, so are underground buildings. A Class I automatic wet or manual wet standpipe system must be provided. Standpipe systems allow fire department or other trained personnel to fight fires at basement levels where it is impractical to run supply lines from fire department trucks or outside hydrants. A quick and convenient water source for fire department use is essential to containing an underground fire.

SECTION 406 MOTOR-VEHICLE-RELATED OCCUPANCIES

❖ Included in this section of the code are the special use and occupancy requirements for those buildings and structures that house motor vehicles. Corresponding requirements for aircraft-related occupancies can be found in Section 412. By definition, all structures that provide services or are used for the storage or parking of motor vehicles are regulated by these provisions. These requirements are applicable regardless of the type of fuel source used to power the vehicles' motors, the number of vehicles present, whether the vehicles are privately or commercially owned and whether they are used strictly for passengers or for freight. Typically, fire hazards for motor-vehicle occupancies are low because of the limited amounts of combustibles present along with the steel frame and metal clad bodies. Motor vehicle structures normally have sufficient space around each vehicle for parking purposes, which serves to further limit the fuel load. The design occupant loads are usually low and pose little if any life safety risks.

This section contains requirements for five different motor-vehicle occupancies along with general requirements for parking garages. Section 406.3 addresses private garages and carports, which are classified as a Group U occupancies. All other parking garages are considered to be public parking garages and divided into either open parking garages or enclosed parking garages. These Group S-2 structures represent a slightly higher hazard than private garages or carports. As such, the heights and areas of these structures along with their type of construction classification are controlled. Requirements are also provided for those buildings and structures that provide service, care and repair of the vehicles. Stations where motor fuels are sold and dispensed are classified as Group M occupancies. The requirements for such facilities are contained in Section 406.7. Garages that provide repair services for vehicles also represent a slightly higher hazard; therefore, Section 406.8 identifies those additional code requirements necessary to abate the hazards of repair garages and their Group S-1 occupancy classification.

Although not specifically referenced in the code, both Section 304 of the *International Mechanical Code*® (IMC®) and Section 305 of the *International Fuel Gas Code*® (IFGC®) address the location of mechanical equipment in both private and public garages.

406.1 General. Motor-vehicle-related occupancies shall comply with Sections 406.1 through 406.8.

❖ The special provisions for motor vehicle related occupancies address private garages and carports, public parking (both open and unenclosed) garages, motor fuel-dispensing facilities and repair garages.

406.2 Definitions. The following terms are defined in Chapter 2:

MECHANICAL-ACCESS OPEN PARKING GARAGES.

OPEN PARKING GARAGE.

PRIVATE GARAGE.

RAMP-ACCESS OPEN PARKING GARAGES.

❖ This section lists terms that are specifically associated with the subject matter of Section 406. It is important to emphasize that these terms are not exclusively related to this chapter but may or may not also be applicable where the term is used elsewhere in the code.

Definitions of terms can help in the understanding and application of the code requirements. The purpose for including a list within this chapter is to provide more convenient access to terms which may have a specific or limited application within this chapter. For the complete definition and associated commentary, refer back to Chapter 2. Terms that are italicized provide a visual identification throughout the code that a definition exists for that term. The use and application of all defined terms are set forth in Section 201.

406.3 Private garages and carports. Private garages and carports shall comply with Sections 406.3.1 through 406.3.6.

❖ This section regulates private garages and carports. Such garages are frequently attached to residential-type occupancies. However, it is not mandatory that a private garage be accessory to a Group R occupancy. The code defines private garages (See Section 202). The definition states that a private garage does not accommodate repairing or servicing vehicles for a profit. But a person using the garage to service their own vehicle is not prohibited.

406.3.1 Classification. Private garages and carports shall be classified as Group U occupancies. Each private garage shall be not greater than 1,000 square feet (93 m^2) in area. Multiple private garages are permitted in a building where each private garage is separated from the other private garages by 1-hour *fire barriers* in accordance with Section 707, or 1-hour *horizontal assemblies* in accordance with Section 711, or both.

❖ Private garages and carports are usually considered accessory to the building they serve. Section 312 classifies these utility and miscellaneous uses as Group U. This section limits an individual private garage to 1,000 square feet (93 m^2) in area. However, the code recognizes the concept of having multiple private garages in the same structure. Where such occurs, there is a minimum 1-hour fire barrier between individual private garages (and a 1-hour horizontal assembly where private garages are on multiple levels). However, there is no upper limit on the number of private garages that could be in any one structure. Section 406.3.4 specifies how private garages are to be separated from other occupancies.

406.3.2 Clear height. In private garages and carports, the clear height in vehicle and pedestrian traffic areas shall be not less than 7 feet (2134 mm). Vehicle and pedestrian areas accommodating van-accessible parking shall comply with Section 1106.5.

❖ The clear height is only required for vehicle and pedestrian traffic areas. If a portion of the garage was used for storage, as one typically finds for a garage attached to a residence, the ceiling height could be lower in such areas. Van parking, where provided in accordance with Section 1106.5, is required to have a minimum ceiling height of 98 inches (8 feet 2 inches) as specified in the ICC A117.1 Accessibility standard.

406.3.3 Garage floor surfaces. Garage floor surfaces shall be of *approved* noncombustible material. The area of floor used for parking of automobiles or other vehicles shall be sloped to facilitate the movement of liquids to a drain or toward the main vehicle entry doorway.

❖ The use of a combustible floor finish that could absorb flammable and combustible liquids commonly used and stored within garages and carports presents a potential safety hazard to the occupants of an attached building. The floors of private garages must be sloped toward the main vehicle doors. Carports have an exception allowing use of asphalt paving, which is not allowed for private garages because they are enclosed resulting in the described potential safety hazards.

406.3.4 Separation. For other than private garages adjacent to dwelling units, the separation of private garages from other occupancies shall comply with Section 508. Separation of private garages from *dwelling units* shall comply with Sections 406.3.4.1 through 406.3.4.3.

❖ The section directs code users to places for separation requirements between private garages and other occupancies. The separations other than from dwelling units are treated no differently than other occupancy separations and therefore must comply with Section 508. Separation between private garages and dwelling units is specified in the three following sections. Please note that individual private garages in the same building are separated by 1-hour fire barriers as specified in Section 406.3.1.

406.3.4.1 Dwelling unit separation. The private garage shall be separated from the *dwelling unit* and its *attic* area by means of gypsum board, not less than $^1/_2$ inch (12.7 mm) in thickness, applied to the garage side. Garages beneath habitable rooms shall be separated from all habitable rooms above by not less than a $^5/_8$-inch (15.9 mm) Type X gypsum board or equivalent and $^1/_2$-inch (12.7 mm) gypsum board applied to structures supporting the separation from habitable rooms above the garage. Door openings between a private garage and the *dwelling unit* shall be equipped with either solid wood doors or solid or honeycomb core steel doors not less than $1^3/_8$ inches (34.9 mm) in thickness, or doors in compliance with Section 716.5.3 with a fire protection rating of not less than 20 minutes. Doors shall be *self-closing* and self-latching.

❖ When a private garage is attached to a dwelling unit, the adjacent areas (including attics) are to be separated to provide a minimum level of protection. This separation is to be constructed of not less than $^1/_2$-inch (12.7 mm) gypsum wallboard applied to the garage side. In a location where the garage is beneath any type of habitable space, the separation requirement is increased by requiring $^5/_8$-inch (15.9 mm) Type X gypsum board or some other material that would provide an equivalent level of protection. Bearing walls and other structural elements that support the horizontal separation shall be protected with minimum $^1/_2$-inch (12.7 mm) gypsum board. Doors opening within the adjacent wall are required to be protected with a minimum $1^3/_8$-inch-thick (34 mm) solid core wood or $1^3/_8$-inch-thick (34 mm) solid or honeycomb steel door. Although the $1^3/_8$-inch-thick (34 mm) solid core wood and honeycomb steel doors are not listed as fire doors, they still provide some protection. Alternatively, doors with a 20-minute fire protection rating in accordance with Section 716.5.3 may be used. All door openings from the garage, whether solid core wood, honeycomb steel or 20-minute rated, are prohibited from opening directly into a bedroom or any other room used for sleeping purposes. It is important to note that this separation requirement, including the protection and the limitation on the openings, is only applicable to garages and not to carports.

406.3.4.2 Openings prohibited. Openings from a private garage directly into a room used for sleeping purposes shall not be permitted.

❖ All openings, including doors, are strictly prohibited between a garage and any room used for sleeping purposes. Even the doors specified in Section 406.3.4.2. are prohibited. Openings are not prohibited for carports.

406.3.4.3 Ducts. Ducts in a private garage and ducts penetrating the walls or ceilings separating the *dwelling unit* from the garage, including its *attic* area, shall be constructed of sheet steel of not less than 0.019 inch (0.48 mm) in thickness and shall have no openings into the garage.

❖ Ducts in the walls and ceilings that do not penetrate into the garage are already protected by the gypsum wallboard; therefore, additional separation is not necessary. If the ducts do penetrate through the separation, then the specified protection must be provided.

406.3.5 Carports. Carports shall be open on at least two sides. Carport floor surfaces shall be of an *approved* noncombustible material. Carports not open on at least two sides shall be considered a garage and shall comply with the requirements for private garages.

Exception: Asphalt surfaces shall be permitted at ground level in carports.

The area of floor used for parking of automobiles or other vehicles shall be sloped to facilitate the movement of liquids to a drain or toward the main vehicle entry doorway.

❖ Carports are covered structures, either free-standing or attached to the side of a building, for the purposes of providing shelter for motor vehicles. Carports must be open on at least two sides. This allows a fire condition to vent quickly to the outside. This further prevents a fire from going undetected and usually limits the amount of other incidental storage that may be included in a fully enclosed garage. Attached carports that are not open on at least two sides are considered a garage and must then be separated from the adjoining dwelling or other occupancy. Where a garage door is provided on a carport, that side should not be considered an open side.

The use of a combustible floor finish that could absorb flammable and combustible liquids commonly used and stored within carports presents a potential safety hazard to the occupants of an attached building. Recognizing that a carport is open on at least two sides and thus has a lower hazard than an enclosed private garage, the exception allows asphalt paving on a ground-level carport. Regardless of the surfacing material, the floor needs to be sloped to a drain or vehicle entry. This permits any spilled material to be moved out of the carport and not accumulate.

406.3.5.1 Carport separation. A separation is not required between a Group R-3 and U carport, provided the carport is entirely open on two or more sides and there are not enclosed areas above.

❖ A roofed structure such as a carport, open on two or more sides without any enclosed uses above, poses no special hazard to the occupants of the dwelling. If a fire were to start under the carport roof, the smoke, hot gases and flames would be able to escape out the open sides, providing the structure with an adequate amount of protection. See the commentary to Section 406.3.3 for additional related discussion.

406.3.6 Automatic garage door openers. Automatic garage door openers, where provided, shall be *listed* in accordance with UL 325.

❖ This section provides a requirement that automatic door openers comply with UL 325. This provision is also found in the *International Residential Code*® (IRC®).

406.4 Public parking garages. Parking garages, other than *private garages*, shall be classified as public parking garages and shall comply with the provisions of Sections 406.4.2 through 406.4.8 and shall be classified as either an *open parking garage* or an enclosed parking garage. *Open parking garages* shall also comply with Section 406.5. Enclosed parking garages shall also comply with Section 406.6. See Section 510 for special provisions for parking garages.

❖ There are fundamentally two types of parking garages regulated by the IBC: private garages and public garages. Private garages are addressed by Section 406.3 and are defined in Section 202. Those structures that fall outside of the scope of Section 406.3 and the Section 202 definition of "Private garage" are considered as public parking garages. The primary difference between private and public garages is the size of the facility, rather than the use. Strictly limited in permissible height and area, private parking garages are typically not commercial in nature. They serve only a specific tenant or building and are not open for public use. It is important to note that there is no implication that public parking garages must be open to the public, as they are only considered public in comparison to private garages.

Public parking garages are considered to be storage occupancies (Group S-2). In addition to the provisions of Sections 406.4, 406.5 and 406.6, public parking garages must also comply with the code provisions for Group S occupancies.

The overall building fire loading in public parking garages is typically low because of the considerable amount of metal in vehicles, which absorbs heat, and the average weight of combustibles per square foot being relatively low [approximately 2 pounds per square foot (9.8 kg/m^2 psf)]. Still, a vehicle fire may be quite extensive.

Public parking garages are divided into one of two categories. Those parking garages that contain sufficient clear openings in their exterior walls that meet the requirements of Section 406.5.2 can be classified as open parking garages. All other public parking garages must be considered as enclosed parking garages and must comply with Section 406.6. Special height and area provisions for parking garages located above or below other uses and occupancies are contained in Section 510.

406.4.1 Clear height. The clear height of each floor level in vehicle and pedestrian traffic areas shall be not less than 7 feet (2134 mm). Vehicle and pedestrian areas accommodating van-accessible parking shall comply with Section 1106.5.

❖ A clear height of 7 feet (2134 mm) is required for all parking garages. This is an exception to the 7-foot 6-inch (2286 mm) minimum height required in Chapter 10 for the means of egress system. This minimum height permits free and unobstructed egress around the vehicles and exit access areas. Van-accessible parking areas and the vehicular route to them must comply with Section 1106.5.

406.4.2 Guards. Guards shall be provided in accordance with Section 1015. Guards serving as *vehicle barriers* shall comply with Sections 406.4.3 and 1015.

❖ For the same reasons as identified in the commentary to Section 1015, guards are required around all vertical openings in the floors and roofs of parking garages where the vertical distance between adjacent levels exceeds 30 inches (762 mm). As a separate requirement, vehicle barriers are required at specified locations where the vertical distance between adjacent levels exceeds 12 inches (305 mm). Guards that serve a dual purpose of guard and vehicle barrier systems must comply with both Sections 1015 and 406.4.3.

406.4.3 Vehicle barriers. *Vehicle barriers* not less than 2 feet 9 inches (835 mm) in height shall be placed where the vertical distance from the floor of a drive lane or parking space to the ground or surface directly below is greater than 1 foot (305 mm). *Vehicle barriers* shall comply with the loading requirements of Section 1607.8.3.

Exception: *Vehicle barriers* are not required in vehicle storage compartments in a mechanical access parking garage.

❖ This section requires vehicle barriers at locations where a 1-foot (305 mm) dropoff exists adjacent to parking spaces or drive lanes. Vehicle barriers are to be designed in accordance with Section 1607.8.3. The vehicle barriers are required to resist a single load of 6,000 pounds (2722 kg). The minimum height of the barriers is 2 feet 9 inches (835 mm), in recognition of the extensive number of vehicles with higher profiles on the road today. Barriers are not required in areas where the employees of the garage park the vehicles, such as mechanical access garages.

406.4.4 Ramps. Vehicle ramps shall not be considered as required *exits* unless pedestrian facilities are provided. Vehicle ramps that are utilized for vertical circulation as well as for parking shall not exceed a slope of 1:15 (6.67 percent).

❖ Since the vehicular ramps of parking garages are open to all levels, they are directly exposed to the effects of smoke and hot gases. In addition, the vehicle ramps are often sloped at a rate greater than 1:12 (8-percent slope). These ramps, therefore, cannot be counted as part of the required exits from each level or tier. Certainly occupants can travel from their vehicles on sloped parking levels to access the required exit stairways. However, occupants cannot continuously travel down the vehicle ramps to reach the exit discharge unless the slope is 1:15 or less and sidewalks or other protected walking surfaces are provided.

406.4.5 Floor surface. Parking surfaces shall be of concrete or similar noncombustible and nonabsorbent materials.

The area of floor used for parking of automobiles or other vehicles shall be sloped to facilitate the movement of liquids to a drain or toward the main vehicle entry doorway.

Exceptions:

1. Asphalt parking surfaces shall be permitted at ground level.
2. Floors of Group S-2 parking garages shall not be required to have a sloped surface.

❖ To avoid acquiring a buildup of flammable liquids on the floor of parking garages, the floor finish is required to be noncombustible and nonabsorbent. Because of pollution concerns, however, garage floors are not automatically required to be drained into the building drainage system. This is consistent with requirements of the Environmental Protection Agency (EPA) for storm water drainage, as well as the philosophy of hazardous material handling (i.e., localize spills and treat them). If floor drains are provided, however, they must be installed in accordance with the IPC.

Exception 1 recognizes that many states have allowed the use of asphalt paving surfaces at grade levels of parking garages with no record of fire hazards. This exception does not allow asphalt paving to be used on stories above or below grade. Although not specified, the floor must be positively sloped at some rate, such as 1 in 48 (2-percent slope), to prevent the accumulation of any spilled flammable and combustible liquids and their vapors. This minimizes the risk of the vapors building up to a point where a fire condition could result.

Exception 2 was introduced into the code based on the recognition that many parking garages are constructed of prefabricated materials which prove difficult to achieve the required slope when assembled on site. The exception, however, waives the slope requirement for all Group S-2 parking facilities, not just those of prefabricated materials.

406.4.6 Mixed occupancy separation. Parking garages shall be separated from other occupancies in accordance with Section 508.1.

❖ If a building or structure consists not only of a Group S-2 parking garage but also other occupancies, then the building is treated as any other mixed occupancy. The building designer must choose one of the options available in Section 508 to address the mixed use and occupancy issues. Keep in mind that Sections 510.2, 510.3, 510.4, 510.7, 510.8 and 510.9 contain special provisions for mixed separations when the parking garages are located above or below other groups.

406.4.7 Special hazards. Connection of a parking garage with any room in which there is a fuel-fired appliance shall be by means of a vestibule providing a two-doorway separation.

Exception: A single door shall be allowed provided the sources of ignition in the appliance are not less than 18 inches (457 mm) above the floor.

❖ As part of the special use and occupancy requirements for parking garages, all possible ignition sources must be controlled and isolated. Specifically, all heating equipment must be located in rooms that are separated from the main areas where the vehicles are parked. Doors connecting the heating equipment rooms and the main parking area must be done with a vestibule or airlock arrangement such that one must pass through two doors prior to entering the other room. Again, this is done to minimize the possibility of any spilled flammable liquids and the resulting vapors from coming in contact with the ignition sources of the heating equipment.

An allowance is made if the heating equipment is located at least 18 inches (457 mm) above the floor of the separated room. In such a case, the vestibule/airlock arrangement with a double-door system is not required and access can be done with just a single door. If this exception is used, it must be understood by the building and fire officials that these special stipulations and conditions are part of the certificate of occupancy. Please note that there are more specific standards regarding the installation of mechanical equipment in public and private garages in Section 304 of the IMC and Section 305 of the IFGC.

406.4.8 Attached to rooms. Openings from a parking garage directly into a room used for sleeping purposes shall not be permitted.

❖ Just as a private garage cannot have openings into bedrooms (see Section 406.3.4.2), public parking garages are likewise limited. The risks of a fire and smoke quickly spreading into adjacent areas where occupants may be sleeping are too great.

406.5 Open parking garages. *Open parking garages* shall comply with Sections 406.5.1 through 406.5.11.

❖ The code establishes four types of garages: private, enclosed parking, repair garages and open parking. With the exception of private garages, all such structures are classified as Group S. Unlike other storage occupancies in which the fuel load is evenly distributed, garages represent a different fire hazard.

Because of generally fast-burning upholstery and gasoline or diesel fuel content, fires in individual vehicles can be quite extensive. Because of both the overall combustible loading within a parking garage, which yields a low combustible fuel load [approximately 2 pounds per square foot (psf) (9.8 kg/m^2)], and the considerable amount of metal in vehicles, which absorbs heat, the overall fire hazards are low. Section 406.5 provides requirements that are unique to open parking garages, while Section 406.3 addresses private garages and Section 406.6 addresses enclosed parking garages. Because of the permanently open exterior walls of open parking garages, which permit the dissipation of heated gases, special provisions are made for heights and areas of such structures.

Open parking garages are classified as Group S-2 and all of the provisions for this occupancy group are applicable except as modified herein.

406.5.1 Construction. *Open parking garages* shall be of Type I, II or IV construction. *Open parking garages* shall meet the design requirements of Chapter 16. For *vehicle barriers*, see Section 406.4.3.

❖ One of the basic premises of the provisions for open parking garages is that the overall fuel load is low and that the average fuel load per square foot is low. The construction of open parking garages, therefore, is restricted to Type I, II or IV, so that the combustible loading of cars is not exceeded by that of the structure.

406.5.2 Openings. For natural *ventilation* purposes, the exterior side of the structure shall have uniformly distributed openings on two or more sides. The area of such openings in *exterior walls* on a tier shall be not less than 20 percent of the total perimeter wall area of each tier. The aggregate length of the openings considered to be providing natural *ventilation* shall be not less than 40 percent of the perimeter of the tier. Interior walls shall be not less than 20 percent open with uniformly distributed openings.

Exception: Openings are not required to be distributed over 40 percent of the building perimeter where the required openings are uniformly distributed over two opposing sides of the building.

❖ Key attributes of an open parking garage, based on the fire records of such facilities, include the exterior wall openings and ventilation of the structure. This section requires that 40 percent of the building perimeter has openings that are uniformly distributed on no less than two sides of the structure. In addition to providing for adequate distribution, the section also requires that a minimum of 20 percent of the total perimeter wall area at each level must be open. The openings in the exterior wall must be free so that natural ventilation will occur without interior wall obstructions.

In instances where the open parking garage is provided with openings on two opposite sides of the structure, thereby providing cross ventilation, the 40-percent criterion does not apply. However, the openings must still meet the 20-percent criterion of the total perimeter wall area.

In every case, interior walls are to have distributed openings totaling 20 percent of the wall area to allow ventilation of all spaces (see Commentary Figure 406.5.2).

406.5.2.1 Openings below grade. Where openings below grade provide required natural *ventilation*, the outside horizontal clear space shall be one and one-half times the depth of the opening. The width of the horizontal clear space shall be maintained from grade down to the bottom of the lowest required opening.

❖ Because permanently open exterior walls provide sufficient natural ventilation and permit the dissipation of heated gases, open parking garages are viewed as a relatively low hazard. However, there are situations where the required openings are located below the surrounding grade, which makes it more difficult to provide the necessary openness required for good performance. A clear horizontal space as described by this section must be provided adjacent to the garage's exterior openings to allow for adequate air movement through the opening. As the distance of the openings below the adjoining ground level increases, the minimum required exterior horizontal clear space also increases proportionally. The dimensional requirements are based on the provisions of Section 1203.5.1.2. The extent of the required clear space allows for adequate exterior open space to meet the intent and dynamics of natural ventilation requirements for open parking garages.

406.5.3 Uses. Mixed uses shall be allowed in the same building as an *open parking garage* subject to the provisions of Sections 402.4.2.3, 406.5.11, 508.1, 510.3, 510.4 and 510.7.

❖ Open parking garages can be constructed as part of a mixed-occupancy building. The building cannot be designed to take advantage of the height and area increases allowed by Sections 406.5.4.1 and 406.5.5. The code provides references to the placement of open parking garages as part of a covered or open mall building. It also provides specific references to three of the special building provisions in Section 510 where open parking garages may be located above or below other buildings. There is also a reference to the mixed-occupancy provisions of Section 508. There is also a reference to Section 406.5.11, where there are specific prohibitions for certain activities within an open parking garage.

406.5.4 Area and height. Area and height of *open parking garages* shall be limited as set forth in Chapter 5 for Group S-2 occupancies and as further provided for in Section 508.1.

❖ When an open parking garage is located within a building containing other occupancies, the code requires that the height and area limitations for the open parking garage be determined by using the normal provisions found in Chapter 5 for a Group S-2 occupancy. This essentially means that the garage would receive the same allowable areas of an enclosed parking garage or a storage building that was classified in the same occupancy.

406.5.4.1 Single use. Where the *open parking garage* is used exclusively for the parking or storage of private motor vehicles, with no other uses in the building, the area and height shall be permitted to comply with Table 406.5.4, along with increases allowed by Section 406.5.5.

Exception: The grade-level tier is permitted to contain an office, waiting and toilet rooms having a total combined area of not more than 1,000 square feet (93 m^2). Such area need not be separated from the *open parking garage*.

In *open parking garages* having a spiral or sloping floor, the horizontal projection of the structure at any cross section

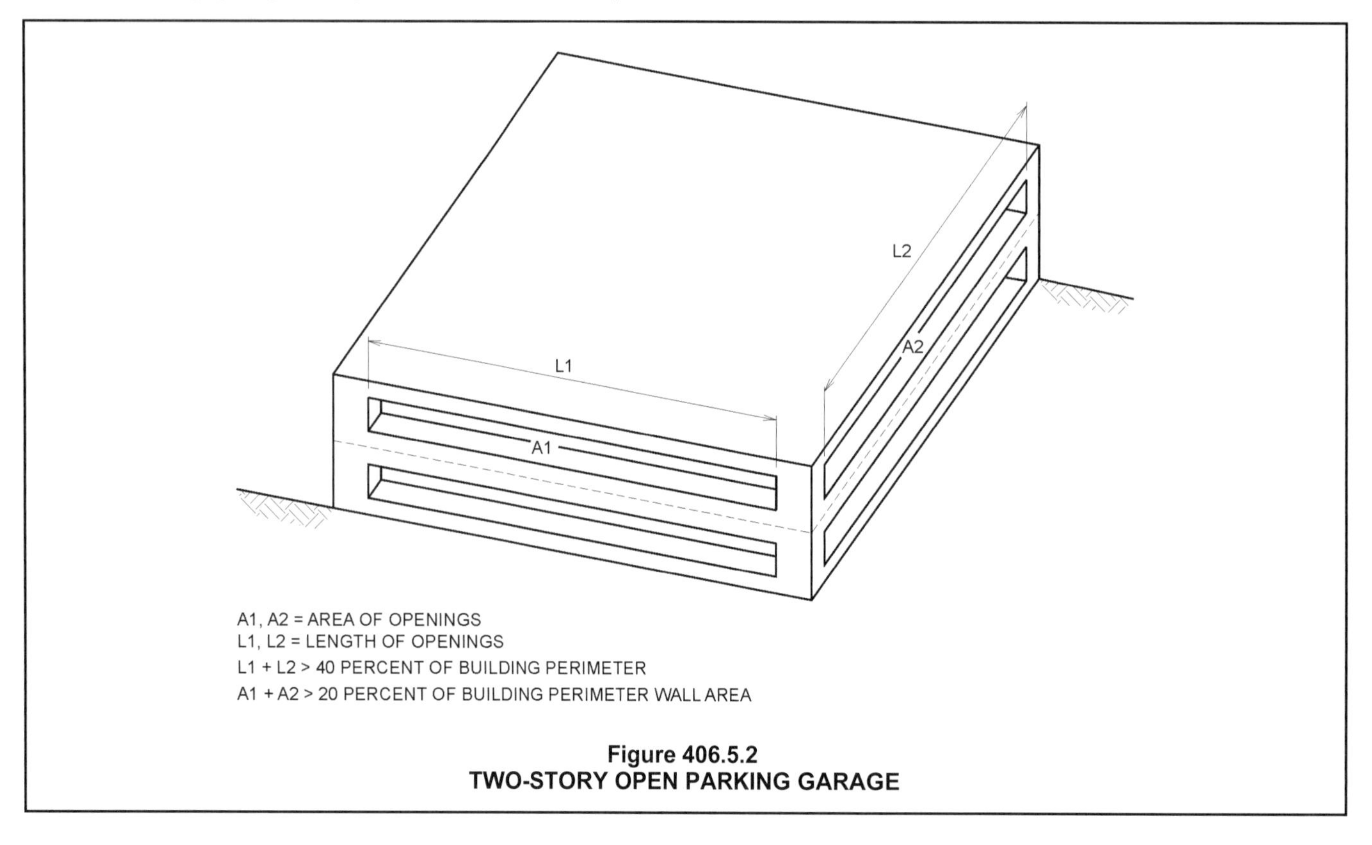

Figure 406.5.2
TWO-STORY OPEN PARKING GARAGE

shall not exceed the allowable area per parking tier. In the case of an *open parking garage* having a continuous spiral floor, each 9 feet 6 inches (2896 mm) of height, or portion thereof, shall be considered a tier.

The clear height of a parking tier shall be not less than 7 feet (2134 mm), except that a lower clear height is permitted in mechanical-access *open parking garages* where *approved* by the *building official*.

❖ When an open parking garage is located in a building that is only used as a parking garage, the type of hazard that is created by the occupancy is greatly reduced. Because of the openness requirements in Section 406.5.2 and the dissipation of heated gases, as well as the overall low fuel load, the height and area limitations for single-use open parking garages are not the same as for other buildings of Group S-2. The height requirements are based on actual fire experience and full-scale fire tests conducted in the United States, Great Britain, Japan and Switzerland. For these single-use buildings that comply with the other provisions of this section, Table 406.5.4 replaces Tables 504.3, 504.4 and 506.2 for the determination of heights and areas. Although these open parking garages are considered as being a single occupancy, the exception permits a limited amount of support-type spaces to be included in the structure.

TABLE 406.5.4. See below.

❖ Table 406.5.4 is used in the same manner as Tables 504.3, 504.4 and 506.2. For a given construction type, the height and area limitations can be determined from the table. For example, an open parking garage of Type IIB construction may be a maximum of eight tiers in height for a ramp access garage. The area would be limited to 50,000 square feet (4545 m^2) per tier.

The height restrictions in the table are measured as defined in Section 202. In this instance, parking may be permitted on the highest level of the structure, which also constitutes the roof of the structure. The areas provided are per tier.

406.5.5 Area and height increases. The allowable area and height of *open parking garages* shall be increased in accordance with the provisions of this section. Garages with sides open on three-fourths of the building's perimeter are permitted to be increased by 25 percent in area and one tier in height. Garages with sides open around the entire building's perimeter are permitted to be increased by 50 percent in area and one tier in height. For a side to be considered open under the above provisions, the total area of openings along the side shall not be less than 50 percent of the interior area of the side at each tier and such openings shall be equally distributed along the length of the tier. For purposes of calculating the interior area of the side, the height shall not exceed 7 feet (2134 mm).

Allowable tier areas in Table 406.5.4 shall be increased for *open parking garages* constructed to heights less than the table maximum. The gross tier area of the garage shall not exceed that permitted for the higher structure. No fewer than three sides of each such larger tier shall have continuous horizontal openings not less than 30 inches (762 mm) in clear height extending for not less than 80 percent of the length of the sides and no part of such larger tier shall be more than 200 feet (60 960 mm) horizontally from such an opening. In addition, each such opening shall face a street or *yard* accessible to a street with a width of not less than 30 feet (9144 mm) for the full length of the opening, and standpipes shall be provided in each such tier.

Open parking garages of Type II construction, with all sides open, shall be unlimited in allowable area where the *building height* does not exceed 75 feet (22 860 mm). For a side to be considered open, the total area of openings along the side shall be not less than 50 percent of the interior area of the side at each tier and such openings shall be equally distributed along the length of the tier. For purposes of calculating the interior area of the side, the height shall not exceed 7 feet (2134 mm). All portions of tiers shall be within 200 feet (60 960 mm) horizontally from such openings or other natural *ventilation* openings as defined in Section 406.5.2. These openings shall be permitted to be provided in *courts* with a minimum dimension of 20 feet (6096 mm) for the full width of the openings.

❖ The area modifications permitted by Section 406.5.5 for open building perimeters, which are similar to Section 506.3, are also applicable to the area restrictions in Table 406.5.4.

Open parking garages of Type IA construction are unlimited in area and height. This is consistent with Type I construction in Tables 504.3, 504.4 and 506.2.

Open parking garages of Type IB construction are

TABLE 406.5.4
OPEN PARKING GARAGES AREA AND HEIGHT

TYPE OF CONSTRUCTION	AREA PER TIER (square feet)	HEIGHT (in tiers)		
		Ramp access	Mechanical access	
			Automatic sprinkler system	
			No	Yes
IA	Unlimited	Unlimited	Unlimited	Unlimited
IB	Unlimited	12 tiers	12 tiers	18 tiers
IIA	50,000	10 tiers	10 tiers	15 tiers
IIB	50,000	8 tiers	8 tiers	12 tiers
IV	50,000	4 tiers	4 tiers	4 tiers

For SI: 1 square foot = 0.0929 m^2.

also unlimited in area, but are limited in height to 12 tiers (or 18 tiers for a mechanical-access garage which is sprinkler protected). If a Type IB garage meets either of the open perimeter criteria below, it can be one tier higher than allowed in the table.

Where the building is of Type II or IV construction and 75 percent of the perimeter is open, it is permitted to be increased 25 percent in area per tier [or from 50,000 square feet (4645 m^2) to 62,500 square feet (5806 m^2)] and one tier in height above the limits for the type of construction in Table 406.5.4.

Where the building is of Type II or IV construction and is open for the entire perimeter, a 50-percent increase in area is permitted, as well as one tier in height. This results in the same number of tiers as the 75-percent open perimeter, but in this case the area of each tier can be as large as 75,000 square feet (6968 m^2).

In addition to complying with the opening criteria found in Section 406.5.2, the openings for an increased area and/or height open parking garage must be at least equal to 50 percent of the interior area of each tier and must be at least 30 inches (762 mm) in vertical dimension. They must be found on 80 percent of the perimeter sides considered open and each such opening shall front on a yard or public way with a minimum dimension of 30 feet (9144 mm). Each part of enlarged tiers needs to be within 200 feet (60 960 mm) of such openings.

For Type II garages, there is another design option. If all sides are open, the area per tier is unlimited provided the building height does not exceed 75 feet (22 860 mm) above grade plane. As in the other options, the openings must be at least 50 percent of the interior of each tier, but in this case the openings must be evenly distributed around the perimeter. All portions of each tier must be within 200 feet (60 960 mm) of such openings, and, where necessary, some of the openings needed for this ventilation may be located in courts, provided the courts have a least dimension of 20 feet (6096 mm).

When calculating the interior area of the side of a tier, the maximum height to be used in the calculation is 7 feet (2134 mm) as shown in Commentary Figure 406.5.2. This method allows for consistent application of the provision regardless of the wall height.

406.5.6 Fire separation distance. *Exterior walls* and openings in *exterior walls* shall comply with Tables 601 and 602. The distance to an adjacent *lot line* shall be determined in accordance with Table 602 and Section 705.

❖ The required fire separation is intended to provide sufficient open space adjacent to the exterior wall openings to allow for free ventilation and to reduce the probability for fire spread to adjacent structures. The exterior wall openings in an open parking garage are not subject to the opening limitations with a fire separation distance greater than 10 feet (3048 mm) (see Table 705.8, Note g).

406.5.7 Means of egress. Where persons other than parking attendants are permitted, *open parking garages* shall meet the *means of egress* requirements of Chapter 10. Where no persons other than parking attendants are permitted, there shall be no fewer than two *exit stairways.* Each *exit stairway* shall be not less than 36 inches (914 mm) in width. Lifts shall be permitted to be installed for use of employees only, provided they are completely enclosed by noncombustible materials.

❖ Typical means of egress elements are required for open parking garages based on a design occupant load of one person per 200 square feet (19 m^2) of gross floor area (see Table 1004.1.2). For mechanical-access open parking garages or for attendant-only ramp-access parking garages, the code would permit two exit stairs, each 36 inches (914 mm) wide.

[F] 406.5.8 Standpipe system. An *open parking garage* shall be equipped with a standpipe system as required by Section 905.3.

❖ Depending on the height of an open parking garage, Section 905.3.1 may require a standpipe system. The standpipe system would aid the fire department in on-site fire-fighting operations, since Table 406.5.4 permits buildings that may not allow for conventional hoses and apparatus to be carried from ground operations to the fire floor.

406.5.9 Enclosure of vertical openings. Enclosure shall not be required for vertical openings except as specified in Section 406.5.7.

❖ Based on the use of open parking garages requiring convenient and frequent access between floor levels, enclosure of most vertical openings is not required or desired due to the opening requirements of Section 406.5.2. This would include the exit access stairways as permitted by Section 1019.3, Exception 6. Where a designer chooses to provide an enclosure—for example, around an elevator hoistway—since the enclosure does not need to be rated, there is no requirement that a lobby be provided for the elevator (see IBC Interpretation No. 36-07).

406.5.10 Ventilation. *Ventilation*, other than the percentage of openings specified in Section 406.5.2, shall not be required.

❖ Mechanical ventilation systems are not required in a structure that is inherently open to the exterior atmosphere.

406.5.11 Prohibitions. The following uses and alterations are not permitted:

1. Vehicle repair work.
2. Parking of buses, trucks and similar vehicles.
3. Partial or complete closing of required openings in exterior walls by tarpaulins or any other means.
4. Dispensing of fuel.

❖ This section reinforces the concepts and requirements of Section 406.5.2 and the definition of "Open parking garage." Any other use or alteration of an open parking garage, which is designed to take advantage of the increased heights and areas, is specifically prohibited.

406.6 Enclosed parking garages. Enclosed parking garages shall comply with Sections 406.6.1 through 406.6.3.

❖ Enclosed parking garages are considered to be storage occupancies. In addition to the provisions of Sections 406.4 and 406.6, enclosed parking garages must also comply with the code provisions for Group S-2 low-hazard storage occupancies.

The overall building fire load in enclosed parking garages is typically low because the considerable amount of metal in vehicles absorbs heat and the average weight of combustibles per square foot is relatively low [approximately 2 psf (9.8 kg/m^2)]. Still, a vehicle fire may be quite extensive.

406.6.1 Heights and areas. Enclosed vehicle parking garages and portions thereof that do not meet the definition of *open parking garages* shall be limited to the allowable heights and areas specified in Sections 504 and 506 as modified by Section 507. Roof parking is permitted.

❖ Special height and area limitations allowed for open parking garages in Table 406.5.4 are not provided for enclosed parking garages. The typical allowable building heights and areas of Chapter 5 are to be used for these Group S-2 structures. It should be noted that Section 510 also contains special provisions for enclosed parking garages located beneath other groups and occupancies. See that section for further discussion of these special building arrangements.

406.6.2 Ventilation. A mechanical *ventilation* system shall be provided in accordance with the *International Mechanical Code*.

❖ Enclosed parking garages are required to be ventilated in accordance with the provisions of the IMC. If motor vehicles will be operated for a period of time exceeding 10 seconds, the ventilation return air for the enclosed parking garage ventilation system must be exhausted (see Section 404 of the IMC).

[F] 406.6.3 Automatic sprinkler system. An enclosed parking garage shall be equipped with an *automatic sprinkler system* in accordance with Section 903.2.10.

❖ A building containing a Group S-2 enclosed parking garage is required to be sprinklered throughout where the fire area of the garage exceeds 12,000 square feet (1115 m^2) or where the garage is located under occupancies other than Group R-3. A full discussion is provided in the commentary to Section 903.2.10.

406.7 Motor fuel-dispensing facilities. Motor fuel-dispensing facilities shall comply with the *International Fire Code* and Sections 406.7.1 and 406.7.2.

❖ Motor fuel-dispensing facilities are those areas where motor fuels are stored, sold or otherwise dispensed to vehicles. These buildings and structures are classified as Group M. Motor fuel-dispensing facilities must be constructed in accordance with the applicable requirements of Chapter 23 of the IFC. Specific details are provided for the special hazards relative to fuel-dispensing systems and areas.

406.7.1 Vehicle fueling pad. The vehicle shall be fueled on noncoated concrete or other *approved* paving material having a resistance not exceeding 1 megohm as determined by the methodology in EN 1081.

❖ The expansion of the use of hydrogen as a vehicle fuel necessitates that the code more clearly address the hazard of potential electrostatic discharge around fueling facilities. The preferred material for fueling pads is concrete, but the referenced standard provides alternative materials to achieve an antistatic performance of the pad.

406.7.2 Canopies. Canopies under which fuels are dispensed shall have a clear, unobstructed height of not less than 13 feet 6 inches (4115 mm) to the lowest projecting element in the vehicle drive-through area. Canopies and their supports over pumps shall be of noncombustible materials, *fire-retardant-treated wood* complying with Chapter 23, wood of Type IV sizes or of construction providing 1-hour *fire resistance*. Combustible materials used in or on a *canopy* shall comply with one of the following:

1. Shielded from the pumps by a noncombustible element of the *canopy*, or wood of Type IV sizes;
2. Plastics covered by aluminum facing having a thickness of not less than 0.010 inch (0.30 mm) or corrosion-resistant steel having a base metal thickness of not less than 0.016 inch (0.41 mm). The plastic shall have a *flame spread index* of 25 or less and a smoke-developed index of 450 or less when tested in the form intended for use in accordance with ASTM E84 or UL 723 and a self-ignition temperature of 650°F (343°C) or greater when tested in accordance with ASTM D1929; or
3. Panels constructed of light-transmitting plastic materials shall be permitted to be installed in *canopies* erected over motor vehicle fuel-dispensing station fuel dispensers, provided the panels are located not less than 10 feet (3048 mm) from any building on the same *lot* and face *yards* or streets not less than 40 feet (12 192 mm) in width on the other sides. The aggregate areas of plastics shall be not greater than 1,000 square feet (93 m^2). The maximum area of any individual panel shall be not greater than 100 square feet (9.3 m^2).

❖ Almost all motor fuel-dispensing facilities are provided with a flat canopy-style roof structure that is column supported. Obviously these structures must be designed to meet the applicable wind and snow loads of Chapter 16 for life safety purposes. For fire safety purposes, the canopies must pose little fire risk and be located above any immediate fire hazards from the fuel-dispensing areas. Limitations are provided for all combustible materials used in the canopy construction.

406.7.2.1 Canopies used to support gaseous hydrogen systems. *Canopies* that are used to shelter dispensing operations where flammable compressed gases are located on the roof of the *canopy* shall be in accordance with the following:

1. The *canopy* shall meet or exceed Type I construction requirements.
2. Operations located under *canopies* shall be limited to refueling only.
3. The *canopy* shall be constructed in a manner that prevents the accumulation of hydrogen gas.

❖ This provision is part of the overall package of code requirements related to the gaseous hydrogen concept of fuel. These additional requirements are necessary to abate the inherent hazards while permitting the use of weather shelters. Not only must the canopies be of noncombustible materials but they must also have the fire-resistance ratings specified in Table 601 for either Type IA or IB construction.

406.8 Repair garages. Repair garages shall be constructed in accordance with the *International Fire Code* and Sections 406.8.1 through 406.8.6. This occupancy shall not include motor fuel-dispensing facilities, as regulated in Section 406.7.

❖ Repair garages are those in which provisions are made for the care, repair and painting of vehicles. Repair garages have an inherently higher fire hazard represented by numerous vehicles in some state of repair, with possible spray painting. As such, these buildings or structures are classified as Group S-1. Repair garages must also meet the applicable requirements of Chapter 23 of the IFC and Section 416 if there are paint-spraying operations.

All motor vehicle work that involves repairs or reconstruction of damaged or nonfunctioning vehicles is considered a higher hazard. These types of facilities provide a variety of services, including tune-ups, oil changes, engine or transmission overhauls and body work, and are all considered as repair garages. Additional protection and detection systems are required to match the increased hazards.

406.8.1 Mixed uses. Mixed uses shall be allowed in the same building as a repair garage subject to the provisions of Section 508.1.

❖ Repair garages usually have integral office and administration areas or are part of an automobile dealership. To minimize the hazards and functions housed in the repair areas, the code reminds the user that these spaces are considered as different occupancies and must therefore comply with the general code requirements for mixed uses. The actual requirements will depend on the size of the spaces and how the designer chooses to deal with the different uses.

406.8.2 Ventilation. Repair garages shall be mechanically ventilated in accordance with the *International Mechanical Code*. The *ventilation* system shall be controlled at the entrance to the garage.

❖ Repair garages are required to be ventilated in accordance with the provisions of the IMC. If motor vehicles will be operated for a period of time exceeding 10 seconds, the ventilation return air for the repair garage ventilation system must be exhausted (see Sections 403 and 404 of the IMC).

406.8.3 Floor surface. Repair garage floors shall be of concrete or similar noncombustible and nonabsorbent materials.

Exception: Slip-resistant, nonabsorbent, *interior floor finishes* having a critical radiant flux not more than 0.45 W/cm^2, as determined by NFPA 253, shall be permitted.

❖ To avoid a buildup of flammable liquids on the floor of repair garages, the floor finish is in general required to be noncombustible and nonabsorbent. Because of pollution concerns, however, garage floors are not automatically required to be drained into the building drainage system. This is consistent with requirements of the EPA for storm water drainage, as well as the philosophy of hazardous material handling (i.e., localize spills and treat them). If floor drains are provided, however, they must be installed in accordance with the IPC.

406.8.4 Heating equipment. Heating equipment shall be installed in accordance with the *International Mechanical Code*.

❖ As part of the special use and occupancy requirements for repair garages, all possible ignition sources must be controlled and isolated. Specifically, all heating equipment must be installed in accordance with the IMC, which contains provisions that address items such as the elevation of the appliance above the floor and the use of overhead unit heaters.

[F] 406.8.5 Gas detection system. Repair garages used for the repair of vehicles fueled by nonodorized gases such as hydrogen and nonodorized LNG, shall be provided with a flammable gas detection system.

❖ Some gases contain additives, such as ethyl mercaptan, that produce pungent odors for easy sensory detection and recognition. If the repair garage repairs vehicles equipped with fuel systems that do not use these odorized gases, a gas detection system must be installed to replace the lost sensory detection capability.

[F] 406.8.5.1 System design. The flammable gas detection system shall be *listed* or *approved* and shall be calibrated to the types of fuels or gases used by vehicles to be repaired. The gas detection system shall be designed to activate when the level of flammable gas exceeds 25 percent of the lower flammable limit (LFL). Gas detection shall be provided in lubrication or chassis service pits of repair garages used for repairing nonodorized LNG-fueled vehicles.

❖ The flammable gas detection system is designed to produce an alarm or signal when exposed to different concentrations of gases or vapor. The gas detector or gas sensor is a critical part of the system for the detection of the different gases. Section 2311.7.2.1 of the IFC requires gas detection equipment to be listed in accordance with UL 2075 and include an indication of the different gases it will detect. Under UL 2075, a set of flammable gases and concentrations (PPM) is developed for each detector or sensor and the manufacturer is required to provide information as to what gases and concentrations the device is designed to detect. Tests under the standard then verify the performance of each detector or sensor for each gas it is designed to detect. The gases that the equipment will detect may be shown in the manufacturer's instructions rather than on the product. This section will require quick-lube-type facilities that change oil and lubricate vehicles to install gas detection systems in the pit area if they service vehicles that are equipped with LNG fuel systems using nonodorized LNG.

[F] 406.8.5.1.1 Gas detection system components. Gas detection system control units shall be *listed* and *labeled* in accordance with UL 864 or UL 2017. Gas detectors shall be *listed* and *labeled* in accordance with UL 2075 for use with the gases and vapors being detected.

❖ This section mandates that gas detection system components are listed in accordance with nationally recognized safety standards. These standards include a comprehensive set of construction and performance requirements that are used to evaluate and list gas detection system control units and gas detectors.

[F] 406.8.5.2 Operation. Activation of the gas detection system shall result in all of the following:

1. Initiation of distinct audible and visual alarm signals in the repair garage.
2. Deactivation of all heating systems located in the repair garage.
3. Activation of the mechanical *ventilation* system, where the system is interlocked with gas detection.

❖ Once the required flammable gas detection system identifies a gas level exceeding 25 percent of the lower explosive limit (LEL), it must do three things: alert the occupants of the emergency, cut power to all possible ignition sources and dilute the concentration level with additional ventilation air.

[F] 406.8.5.3 Failure of the gas detection system. Failure of the gas detection system shall result in the deactivation of the heating system, activation of the mechanical *ventilation* system where the system is interlocked with the gas detection system and cause a trouble signal to sound in an *approved* location.

❖ The required flammable gas detection system must include a default mode of operation. If the system malfunctions or loses power, it must in effect do the same operational requirements referenced in Section 406.8.5.2. This redundancy will lessen the likelihood of an explosive fire risk for the specialized motor fuels.

[F] 406.8.6 Automatic sprinkler system. A repair garage shall be equipped with an *automatic sprinkler system* in accordance with Section 903.2.9.1.

❖ Section 903.2.9.1 requires buildings where repair garages are located to be provided with an automatic sprinkler system where any one of four listed conditions is present. The sprinkler threshold is reduced in those repair garages where work is performed on trucks or buses. See the commentary on Section 903.2.9.1 for further information.

SECTION 407
GROUP I-2

407.1 General. Occupancies in Group I-2 shall comply with the provisions of Sections 407.1 through 407.10 and other applicable provisions of this code.

❖ The provisions contained in Section 407 reflect requirements of previous model codes, full-scale fire tests and historical fire experience. Features unique to health care facilities, such as a low fuel load, the presence of staff, operating practices and procedures and the regulatory process, were also included in developing the provisions. One of the concerns addressed by this section is the need to provide an acceptable level of protection without interfering with the operation of the facility. To that end, various provisions of Section 407 are designed to soften the rigid barrier between rooms and corridors. For example, Section 407.2.5 allows a nursing home housing unit to be open to the corridor under certain conditions. Such designs allow nursing homes to be a little less institutional while still providing a safe environment for the residents and staff of the home.

Group I-2 has been divided into two conditions (see Section 308.4). Group I-2, Condition 1 occupancies are primarily nursing homes where both nursing and medical care is provided, but services such as surgery, obstetrics and emergency care are not provided. Care recipients in a Group I-2, Condition 1 facility are usually staying for long periods. Group I-2, Condition 2 occupancies are primarily hospitals which provide a wide range of medical treatment and services. Patients generally stay over a night or two, but stays are generally short term. See also the definitions in Chapter 2 of the terms "24-hour basis," "Medical care," "Custodial care" and "Personal care."

Underlying the protection requirements for Group I-2 is a "defend-in-place" philosophy based on the difficulties associated with evacuating such facilities.

The first level of protection established is the individual room (typically a patient sleeping room). Horizontal evacuation to an adjacent smoke compartment provides the second level of protection. If necessary, the third level of protection involves evacuation of the floor or entire building.

This section has been created not only because of

considerations unique to health care facilities, but also to facilitate the consideration of the entire package of protection features. The basic protection features include early detection, fire containment, horizontal evacuation and fire extinguishment. Additionally, Chapter 4 of the IFC contains criteria for emergency planning and preparedness.

As previously noted, the provisions of this section are based in part on a relatively low fuel load. Certain areas within Group I-2, however, will have a higher concentration of combustibles. These areas, as well as other areas that represent a specific hazard within Group I-2, are addressed in Section 509 as incidental uses. Because of the nature of the activity, the incidental use area is required to be protected with an automatic sprinkler system, separated with a fire barrier and horizontal assembly, or both (see Table 509).

The provisions of Section 407 apply to both Group I-2 conditions unless the section specifically refers to one or the other. The requirements of Section 407 are intended to be additional special use and occupancy provisions for facilities that provide medical care for more than five persons for longer than 24 hours. For medical care outside of a Group I occupancy, please also see Section 422 for ambulatory care facilities. Other requirements for Group I-2 occupancies are provided throughout the code.

407.2 Corridors continuity and separation. *Corridors* in occupancies in Group I-2 shall be continuous to the *exits* and shall be separated from other areas in accordance with Section 407.3 except spaces conforming to Sections 407.2.1 through 407.2.6.

❖ Since the occupants of areas classified as Group I-2 may be incapable of self-preservation, the means of egress serving such areas must facilitate the evacuation of such persons, as well as other people with mobility impairments. Accordingly, corridors serving Group I-2 occupancies are required to be continuous to the exits and, except as provided for in Sections 407.2.1 through 407.2.6, separated from other use areas. As such, corridors in Group I-2 occupancies are intended to provide a protected path of travel that, once entered, leads directly and continuously to the exits (see also commentary, Sections 407.4.1 and 1020.6). Note that a door through a horizontal exit constructed with a fire barrier or fire wall would constitute an exit (see Section 1026); however, moving through a smoke barrier into another smoke compartment is not a horizontal exit and therefore corridor continuity must be maintained on both sides of the smoke barrier.

To address the various operational requirements of occupancies classified as Group I-2, the code includes provisions that both permit various areas to be open to corridors and address care suite arrangements for the exit access portions of the facilities. Standards for areas open to corridors are covered primarily in this section. The design of care suites is regulated by provisions found in Section 407.4.4.

For the ease of operations of Group I-2 occupancies, it is often desirable to have certain areas open to exit access corridors. Recognizing these needs and that these occupancies are equipped throughout with an automatic sprinkler system, the code permits specific areas to be open to corridors, provided they are properly arranged and equipped with compensating protection features (e.g., automatic fire detection). Sections 407.2.1 through 407.2.6 identify those areas and the protection features required to maintain the safety of the corridor.

In addition to practical reasons, such as in the case of care providers' stations, it is also desirable to designate certain areas being open to the corridor for treatment purposes. In many instances, the care recipient is encouraged to leave the sleeping room for both physical as well as social benefit. So that care recipients can be encouraged to socialize and benefit from group activities and social interaction, the areas need to be open to the corridor.

Waiting areas and other areas open to the corridor also serve a useful purpose during a fire emergency. If evacuation becomes necessary, care recipients usually will be relocated to an adjacent smoke compartment in which the areas open to the corridor will serve as a refuge area. Section 407.5.1 contains provisions for the net area that must be available for the relocation of care recipients. Waiting areas provide a space in which care recipients can be relocated such that treatment can continue during a fire emergency.

407.2.1 Waiting and similar areas. Waiting areas and similar spaces constructed as required for *corridors* shall be permitted to be open to a *corridor*, only where all of the following criteria are met:

1. The spaces are not occupied as care recipient's sleeping rooms, treatment rooms, incidental uses in accordance with Section 509, or hazardous uses.
2. The open space is protected by an automatic fire detection system installed in accordance with Section 907.
3. The *corridors* onto which the spaces open, in the same *smoke compartment*, are protected by an automatic fire detection system installed in accordance with Section 907, or the *smoke compartment* in which the spaces are located is equipped throughout with quick-response sprinklers in accordance with Section 903.3.2.
4. The space is arranged so as not to obstruct access to the required *exits*.

❖ In the design of health care facilities, it is desirable to have small waiting areas and similar spaces open to the corridor. Such seating areas or gathering spaces provide care recipients and others a place to sit outside of the sleeping rooms or to wait near special care areas for their turn in a treatment room. In consideration of the additional protection features specified, small waiting areas and similar spaces are permitted to be open to corridors in Group I-2 occupancies within each smoke compartment, provided they are constructed to the corridor standards for

Group I-2 occupancies, as specified in Section 407.3.

Care recipient sleeping rooms, treatment rooms, and hazardous and incidental use areas cannot be open to the corridors under this section. Under Section 508, all Group H hazardous occupancies must be separated from other occupancies, therefore it follows that such can not be open to the corridors in Group I-2 facilities. This configuration is commonly found in a typical ward situation.

While Item 1 restricts treatment facilities in general from being open to the corridor, Section 407.2.3 specifically allows treatment areas for psychiatric wards to be open to the corridor. Further, the broad restriction that sleeping rooms not be open to the corridor is relaxed under Section 407.2.5, which relaxes the relationship between nursing home housing units and the corridors. Areas similar to waiting rooms could include common visiting areas, reception or information stations and patient check-in stations. General hospital office areas or support spaces beyond those associated with waiting or reception areas must be separated from the corridor (see Commentary Figure 407.2.1) unless permitted by other subsections.

To reduce the likelihood that a fire within such an open space could develop beyond the incipient stage, thereby jeopardizing the integrity of the corridor, the area is to be equipped with an automatic fire detection system (see Section 907). The detectors are to be located to provide the appropriate coverage to the space.

There is no size limit set by the code for these spaces. Due to the unlimited size of the space and, consequently, the potential for a significant number of people in the area, an automatic fire detection system is also required in all corridors open to the spaces within the smoke compartment, or the smoke compartment in which such spaces are located must be protected throughout with quick-response sprinklers.

Waiting areas and similar spaces open to corridors must be arranged so as to not obstruct access to exits. There must be a clear path of travel of the minimum required width for the corridor, which should be maintained at all times. As such, furniture layouts must be arranged in a manner that obstruction of the exits does not occur under normal use.

407.2.2 Care providers' stations. Spaces for care providers', supervisory staff, doctors' and nurses' charting, communications and related clerical areas shall be permitted to be open to the *corridor*, where such spaces are constructed as required for *corridors*.

❖ Stations for care providers need to be located to provide quick access to care recipients and to permit visual or audible monitoring of the sleeping rooms. For these and other practical reasons, care providers' stations are permitted to be open to the corridor, provided the walls surrounding the area are constructed as required for corridor walls (see Commentary Figure 407.2.2). Except as required in Section 407.8, the additional protection of automatic fire detection is not required because of the reduced risk to life safety and attendance by the staff (see Section 407.8). This sec-

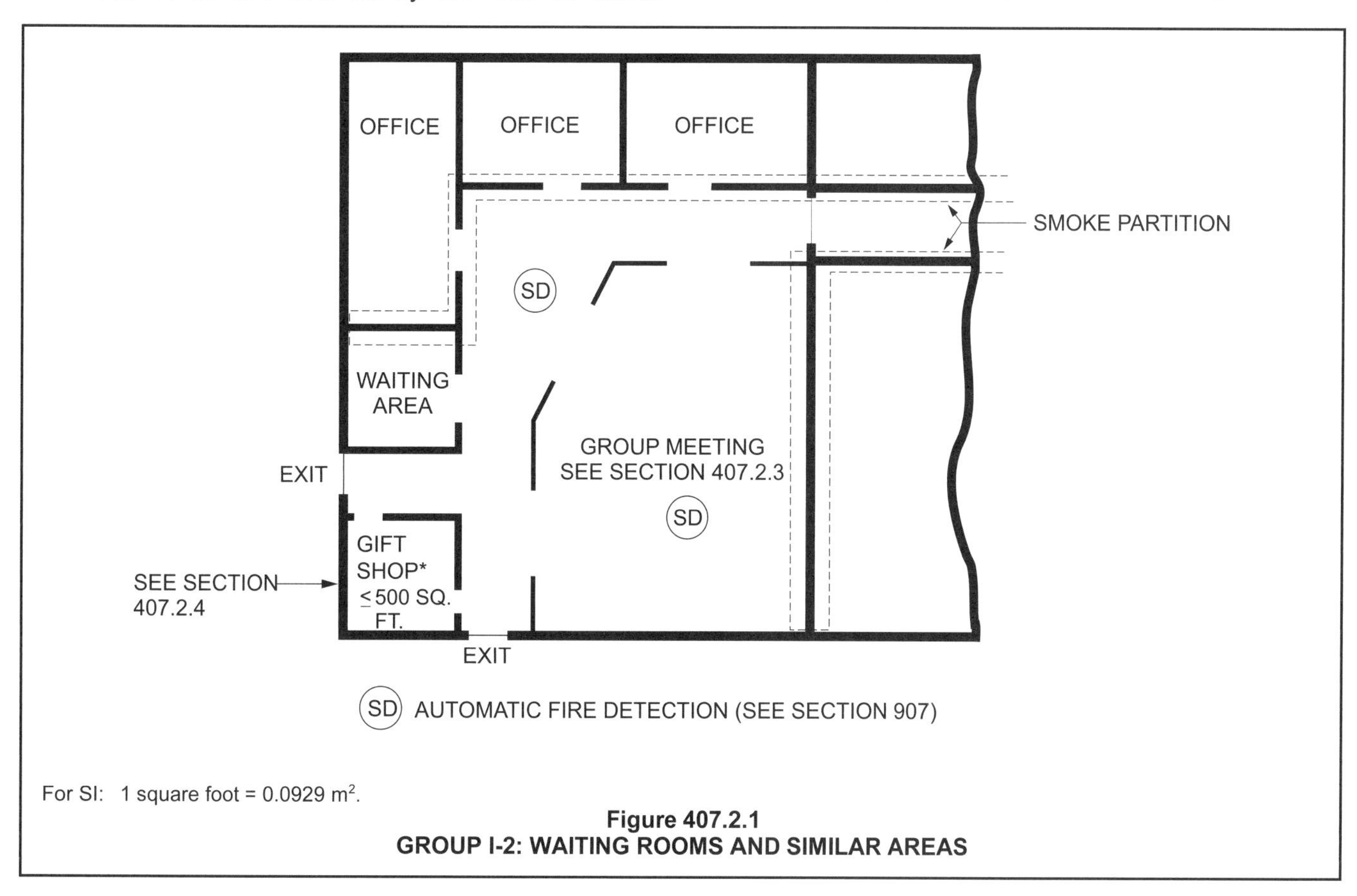

For SI: 1 square foot = 0.0929 m^2.

Figure 407.2.1
GROUP I-2: WAITING ROOMS AND SIMILAR AREAS

tion is not intended to apply to nurses' offices, administrative supply storage areas, drug distribution stations or similar areas with a higher fuel load and from which continual observation of the nursing unit is not essential.

407.2.3 Psychiatric treatment areas. Areas wherein psychiatric care recipients who are not capable of self-preservation are housed, or group meeting or multipurpose therapeutic spaces other than incidental uses in accordance with Section 509, under continuous supervision by facility staff, shall be permitted to be open to the *corridor*, where the following criteria are met:

1. Each area does not exceed 1,500 square feet (140 m^2).
2. The area is located to permit supervision by the facility staff.
3. The area is arranged so as not to obstruct any access to the required *exits*.
4. The area is equipped with an automatic fire detection system installed in accordance with Section 907.2.
5. Not more than one such space is permitted in any one *smoke compartment*.
6. The walls and ceilings of the space are constructed as required for *corridors*.

❖ This section applies to all areas that house psychiatric care recipients who are not physically capable of self-preservation because the area is secured (see Section 1010.1.9.6).

To encourage care recipients of psychiatric facilities to participate in group meetings and therapeutic activities, treatment areas may be open to the corridor, provided that the area does not exceed 1,500 square feet (140 m^2) and only one such area is provided within a smoke compartment. Other protection features that must be provided to permit the area to be open include: staff supervision, unobstructed access to exits and automatic fire detection in accordance with Section 907.2 within the area. The group meeting room, while not separated from the corridor, must be separated from other areas of the unit with construction as required for a corridor.

407.2.4 Gift shops. Gift shops and associated storage that are less than 500 square feet (455 m^2) in area shall be permitted to be open to the *corridor* where such spaces are constructed as required for *corridors*.

❖ Gift shops are not permitted to be open to a corridor in occupancies classified as Group I-2 unless they are less than 500 square feet (46 m^2) in area (see Commentary Figure 407.2.1). The section allows any associated storage to be included in the 500-square-foot (46.5 m^2) limitation since the associated fire hazard in both areas would be the same. If the gift shop and associated storage were greater than 500 square feet (46.5 m^2) combined, but the gift shop itself was less than 500 square feet (46.5m^2), an option would be to leave the gift shop open to the corridor and separate the storage from the gift shop and other areas with construction as required for corridors.

407.2.5 Nursing home housing units. In Group I-2, Condition 1, occupancies, in areas where nursing home residents are housed, shared living spaces, group meeting or multipurpose therapeutic spaces shall be permitted to be open to the *corridor*, where all of the following criteria are met:

1. The walls and ceilings of the space are constructed as required for *corridors*.

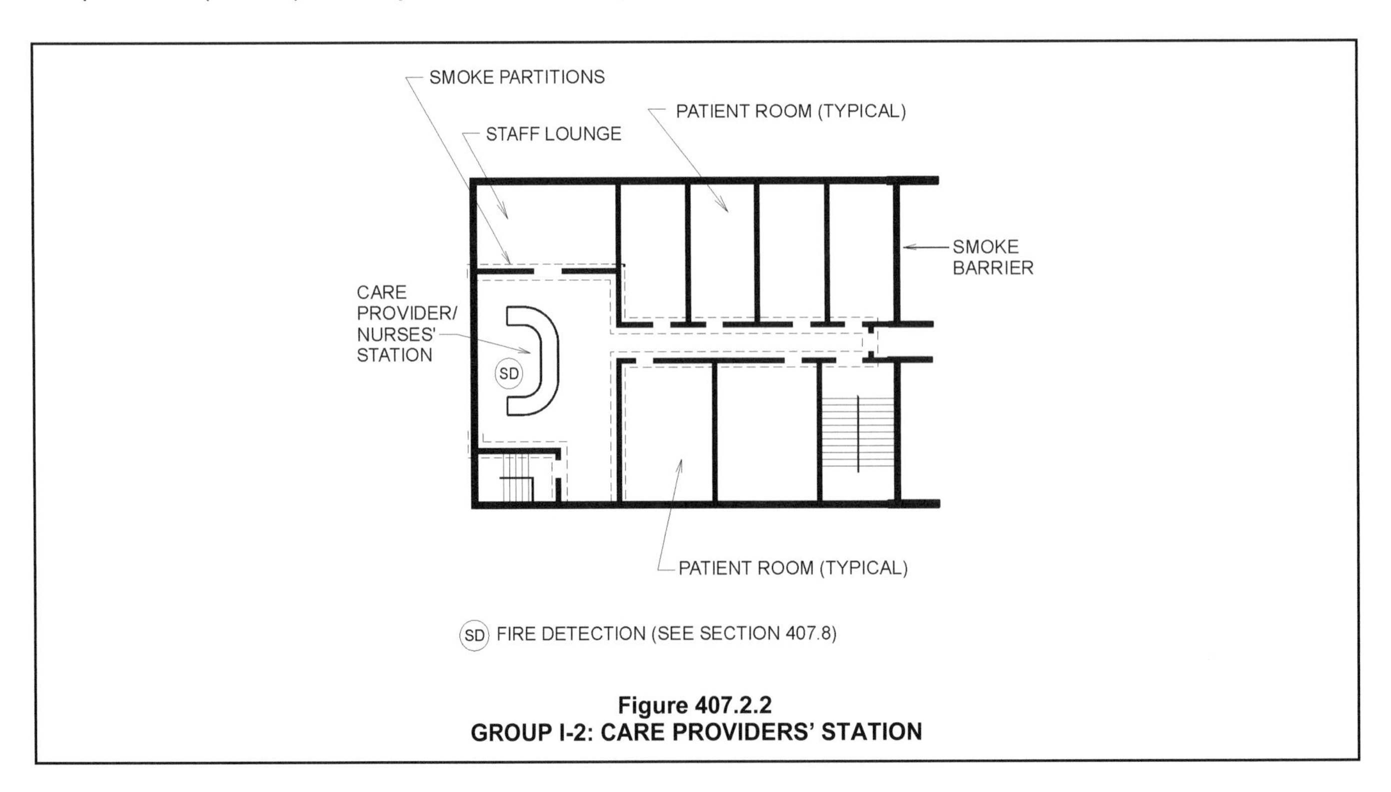

Figure 407.2.2
GROUP I-2: CARE PROVIDERS' STATION

2. The spaces are not occupied as resident sleeping rooms, treatment rooms, incidental uses in accordance with Section 509, or hazardous uses.
3. The open space is protected by an automatic fire detection system installed in accordance with Section 907.
4. The *corridors* onto which the spaces open, in the same *smoke compartment*, are protected by an automatic fire detection system installed in accordance with Section 907, or the *smoke compartment* in which the spaces are located is equipped throughout with quick-response sprinklers in accordance with Section 903.3.2.
5. The space is arranged so as not to obstruct access to the required *exits*.

❖ In nursing home occupancies, residents are encouraged to spend time outside of their rooms. Way finding and orientation problems are common in nursing home residents, and research has shown that direct visibility to a desired location is more effective for cuing than signage. Therefore, having a variety of shared living spaces open to the corridor encourages socialization, encourages interaction, and is important to resident well-being. Further, being able to preview activities that are occurring helps to encourage joining and allows reluctant participants to join at their own pace. Finally, a more open plan allows staff to see residents more easily throughout the course of the day. Commentary Figure 407.2.5 shows examples of typical areas open to a corridor that would be allowed by this provision. The protections used for other areas allowed open to the corridor in other parts of Section 407.2 are repeated here.

407.2.6 Nursing home cooking facilities. In Group I-2, Condition 1, occupancies, rooms or spaces that contain a cooking facility with domestic cooking appliances shall be permitted to be open to the corridor where all of the following criteria are met:

1. The number of care recipients housed in the smoke compartment is not greater than 30.
2. The number of care recipients served by the cooking facility is not greater than 30.
3. Only one cooking facility area is permitted in a smoke compartment.
4. The types of domestic cooking appliances permitted are limited to ovens, cooktops, ranges, warmers and microwaves.
5. The corridor is a clearly identified space delineated by construction or floor pattern, material or color.
6. The space containing the domestic cooking facility shall be arranged so as not to obstruct access to the required exit.
7. A domestic cooking hood installed and constructed in accordance with Section 505 of the *International Mechanical Code* is provided over the cooktop or range.
8. The domestic cooking hood provided over the cooktop or range shall be equipped with an automatic fire-extinguishing system of a type recognized for protection of domestic cooking equipment. Preengineered automatic extinguishing systems shall be tested in accordance with UL 300A and *listed* and *labeled* for the intended application. The system shall be installed in accordance with this code, its listing and the manufacturer's instructions.
9. A manual actuation device for the hood suppression system shall be installed in accordance with Sections 904.12.1 and 904.12.2.
10. An interlock device shall be provided such that upon activation of the hood suppression system, the power or fuel supply to the cooktop or range will be turned off.
11. A shut-off for the fuel and electrical power supply to the cooking equipment shall be provided in a location that is accessible only to staff.

Figure 407.2.5
NURSING HOME HOUSING UNIT

12. A timer shall be provided that automatically deactivates the cooking appliances within a period of not more than 120 minutes.
13. A portable fire extinguisher shall be installed in accordance with Section 906 of the *International Fire Code*.

❖ As nursing homes move away from institutional models, it is critical to have a functioning kitchen that can serve as the hearth of the home. Instead of a large, centralized, institutional kitchen where all meals are prepared and delivered to a central dining room or the resident's room, the new "household model" nursing home uses de-centralized kitchens and small dining areas to create the feeling and focus of home. For persons with dementia, it is particularly important to have spaces that look familiar, like the kitchen in their former home, to increase their understanding and ability to function at their highest level.

Allowing kitchens that serve a small, defined group of residents to be open to common spaces and, in some instances, corridors, is critically important to enhancing the feeling and memories of home for older adults. This allows residents to see and smell the food being prepared, which can enhance their appetites and evoke positive memories. Some residents, based on their abilities and cognition level, may even be able to participate in food preparation activities such as stirring, measuring ingredients, peeling vegetables, or folding towels. This becomes a social activity, where they can easily converse with the staff member cooking, as well as a way for the resident to maintain their functional abilities and to feel that they are still an important, contributing member of society. Commentary Figure 407.2.6(1) provides a picture of an open kitchen area. Commentary Figure 407.2.6(2) provides an example floor plan of a nursing home housing unit with an open kitchen.

It is recognized that unattended cooking is a leading cause of fires. Therefore, the provisions of this section are aimed at limiting the potential of such fires and, in the event of such a fire, to limit the area affected by the fire. Item 11 is key to limiting the use of these kitchens to when the staff is aware of the activity and authorizes it by turning on the fuel sources. In addition to the 13 safeguards provided in this section, all new nursing home are fully protected with an automatic sprinkler system.

Figure 407.2.6(1)
NURSING HOME WITH OPEN-PLAN KITCHEN

407.3 Corridor wall construction. *Corridor* walls shall be constructed as smoke partitions in accordance with Section 710.

❖ Corridor walls are required to form a barrier to limit smoke transfer in accordance with provisions for smoke partitions in Section 710, thus providing protection for exit access routes, from separate areas containing combustible materials that could produce smoke and to provide adequate separation of the care recipient sleeping rooms. In a building protected throughout with an automatic sprinkler system, the probability that a fire will develop that is life threatening to persons outside the room of origin is reduced; therefore, corridor walls need only be able to resist the passage of smoke.

The corridor wall is not required to have a fire-resistance rating (Section 710.3), but the wall is to terminate at the underside of the floor or roof deck above or to the underside of the ceiling membrane if it is capable of resisting the passage of smoke (Section 710.4). The walls must be of materials consistent with the building's type of construction classification.

Note that the provisions of Section 509 must be considered in determining the fire-resistance rating of any corridor walls that also enclose incidental uses where fire barriers are required (Section 509.4.1) or construction capable of resisting the passage of smoke (Section 509.4.2). Note that construction capable of resisting the passage of smoke does not have all the same requirements as smoke partitions as required in Section 710 (see Commentary Figure 407.3).

407.3.1 Corridor doors. *Corridor* doors, other than those in a wall required to be rated by Section 509.4 or for the enclosure of a vertical opening or an *exit*, shall not have a required *fire protection rating* and shall not be required to be equipped with *self-closing* or automatic-closing devices, but shall provide an effective barrier to limit the transfer of smoke and shall be equipped with positive latching. Roller latches are not permitted. Other doors shall conform to Section 716.5.

❖ As with corridor wall construction (see commentary, Section 407.3), the door assemblies need only be capable of resisting the passage of smoke. Corridor doors in occupancies classified as Group I-2 are not required to be self-closing or automatic-closing except where required by other sections, such as for doors serving incidental uses as required in Section 509, or doors in shaft walls that enclose a vertical opening or exit stairway or ramp. This provision is primarily intended to apply to care recipient sleeping room corridor doors. Doors in walls that separate incidental uses in accordance with Section 509 are required to be self-closing to minimize the possibility that corridors will be affected by a fire originating in those adjacent areas. Self-closing or automatic-closing doors in the shaft walls of vertical openings and exit stairways and ramps are still required to avoid a breech in vertical opening protection that would enable rapid fire spread from story to story and for adequate protection of the exits. It is important to recognize that this section contains a specific set of requirements for the corridor doors in this occupancy and that the provisions of this section must be followed versus the provisions of Section 710.5.2 for doors in smoke partitions (see code and commentary, Section 102.1). Full-scale fire tests and historical experience indicate that the automatic sprinkler system will provide adequate protection to persons outside of the room of origin. It is also expected that staff will be adequately trained and will close the doors to care recipient sleeping rooms within the affected areas, or will evacuate patients to an adjacent smoke compartment. The presence of the automatic sprinkler system, particularly one equipped with quick-response heads, increases the probability that staff will be able to close the door to the room of origin and thereby limit the size of the fire. Roller latches are not permitted on these doors, as they are not regarded as providing positive latching that will allow the door to act as a reliable smoke protective.

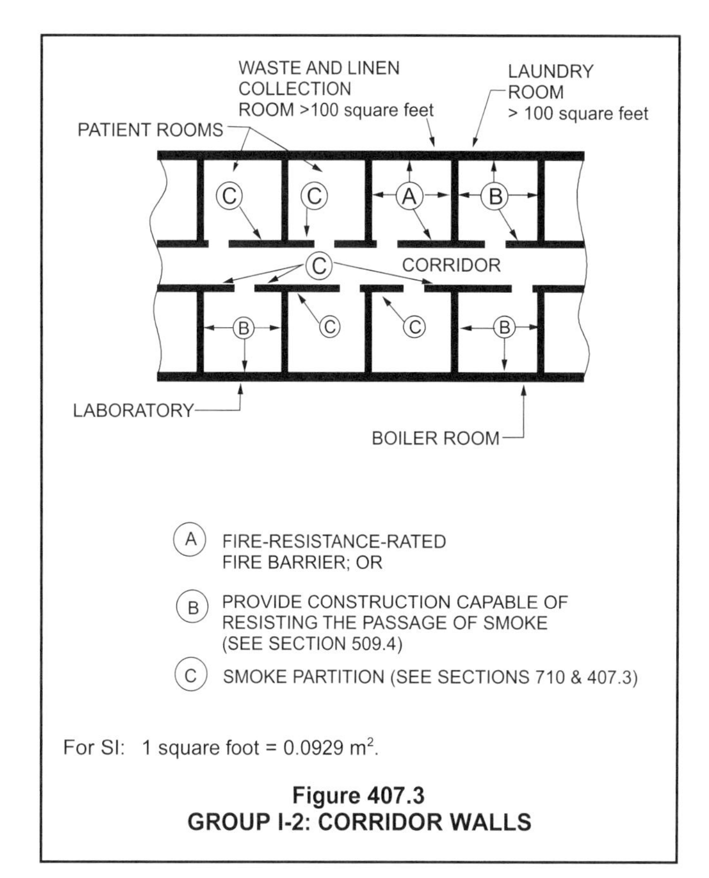

For SI: 1 square foot = 0.0929 m^2.

Figure 407.3
GROUP I-2: CORRIDOR WALLS

13 Resident Nursing Household
w/ open cooking facility

Figure 407.2.6(2)
NURSING HOME HOUSING UNIT FLOOR PLAN

407.4 Means of egress. Group I-2 occupancies shall be provided with means of egress complying with Chapter 10 and Sections 407.4.1 through 407.4.4. The fire safety and evacuation plans provided in accordance with Section 1001.4 shall identify the building components necessary to support a *defend-in-place* emergency response in accordance with Sections 404 and 408 of the *International Fire Code*.

❖ In addition to required compliance with Chapter 10, Group I-2 occupancies must also comply with special provisions regarding direct access to a corridor (Section 407.4.1), travel distance (Section 407.4.2), projections into corridors (Section 407.4.3) and care suites (Section 407.4.4).

Defend-in-place is a concept that has long been employed as the preferred method of fire response in nursing homes and hospitals due to the fragile nature of the occupants. Occupants in this setting are often dependent upon the building infrastructure, active and passive fire protection and staff training and response since immediate evacuation would place their lives at risk. The infrastructure system typically includes life-support systems such as medical gases, emergency power and environmental controls that rely on continued building operation. A combination of active and passive fire protection typically includes fire and smoke detection, fire sprinkler systems, fire and smoke barriers and dampers along with protective openings. Integral to the staff training and response is the fire safety and evacuation plan. Health care facilities typically have a facility-wide plan that describes the overall safety program and details the specific measures for a facility-wide response. In addition to this plan, facilities will also have department- or unit-specific safety plans. These unit-specific plans detail specific actions that will need to be taken to protect the various patients, staff and visitors within the area such as what actions are to be taken immediately, like who is responsible to close what doors, remove allowable equipment from the corridor, where and when to meet to discuss patient needs and to prepare for possible horizontal evacuation.

Previous versions of this code and legacy codes have created a tried and tested set of requirements to support the defend-in-place concept, such as smoke compartments and areas of refuge. However, previous codes have not specifically described the concept of occupants remaining within a building during an emergency, which has led to confusion and misapplication during design and enforcement. This section identifies Group I-2 as a location where this type of emergency response is permitted. The codes governing hospitals, nursing homes and other Group I-2 classes are designed to support the defend-in-place use. While the code has been silent on the underlying concept, the defend-in-place strategy has been the commonly accepted practice in these occupancies (see IFC Sections 403 and 404 for additional information on defend-in-place).

407.4.1 Direct access to a corridor. Habitable rooms in Group I-2 occupancies shall have an *exit access* door leading directly to a *corridor*.

Exceptions:

1. Rooms with *exit* doors opening directly to the outside at ground level.
2. Rooms arranged as *care suites* complying with Section 407.4.4.

❖ Direct access to the corridor system from a patient sleeping room or treatment room is a key component to staff access and patient movement. The term "habitable rooms" is not intended to include individual bathrooms, closets and similar spaces, or briefly occupied spaces such as control rooms in radiology and small storage/supply rooms. Habitable areas would include staff areas within the patient treatment and sleeping areas (e.g., nutrition rooms, clean/dirty linen rooms, staff lounge, staff work areas).

The exceptions would allow direct access to the outside rather than needing to go back into the corridor, or access through a portion of a care suite before accessing the corridor.

407.4.1.1 Locking devices. Locking devices that restrict access to a care recipient's room from the *corridor* and that are operable only by staff from the *corridor* side shall not restrict the *means of egress* from the care recipient's room.

Exceptions:

1. This section shall not apply to rooms in psychiatric treatment and similar care areas.
2. Locking arrangements in accordance with Section 1010.1.9.6.

❖ Locking devices are permitted on care recipient room doors, provided egress from the care recipient room is not restricted. Such locking devices enable care recipients in long-term care facilities to secure their rooms and staff to secure unoccupied rooms. In such cases, it is expected that the staff still has access, via key or other opening device.

The need to lock corridor doors to restrain care recipients in psychiatric facilities, however, is recognized. When such locking is necessary, provisions must be made for continuous supervision and prompt release. Additional guidance on locking arrangements, applying to occupancies in Group I-3, can be found in Sections 408.4 and 1010.1.9.10. A decision needs to be made to determine if the area being considered more closely represents an occupancy in Group I-2 or I-3 (e.g., a holding room in the emergency room of the hospital for patients who may be violent). In occupancies in Group I-3, it is generally assumed that the occupants are capable of self-preservation once the doors are unlocked. This may not be the case in some psychiatric facilities and, therefore, the provision applicable to occupancies in Group I-2 may be more appropriate.

Similar protection may also be necessary in certain portions of other occupancies in Group I-2 where the movement of the care recipients must be controlled (e.g., care recipients with Alzheimer's disease). For special locking arrangements for suites or floors where elopement may be an issue, see Section 1010.1.9.6. It is not the intent of this section to prohibit other locking options that will allow free access in accordance with Sections 1010.1.9.7, 1010.1.9.8 and 1010.1.9.9.

407.4.2 Distance of travel. The distance of travel between any point in a Group I-2 occupancy sleeping room, not located in a *care suite*, and an *exit access* door in that room shall be not greater than 50 feet (15 240 mm).

❖ This section limits the overall travel distance within an individual patient sleeping room to the exit access door from that room. This is a separate concept from the travel distance within a care suite (see Section 407.4.4) and must be considered for each patient sleeping room individually.

407.4.3 Projections in nursing home corridors. In Group I-2, Condition 1, occupancies, where the corridor width is a minimum of 96 inches (2440 mm), projections shall be permitted for furniture where all of the following criteria are met:

1. The furniture is attached to the floor or to the wall.
2. The furniture does not reduce the clear width of the corridor to less than 72 inches (1830 mm) except where other encroachments are permitted in accordance with Section 1005.7.
3. The furniture is positioned on only one side of the *corridor*.
4. Each arrangement of furniture is 50 square feet (4.6 m^2) maximum in area.
5. Furniture arrangements are separated by 10 feet (3048 mm) minimum.
6. Placement of furniture is considered as part of the fire and safety plans in accordance with Section 1001.4.

❖ Many nursing homes have long corridors that residents must traverse. Residents who are physically unable to traverse the distance without being able to rest periodically have little recourse but use a wheelchair, an outcome counter to maintaining their ambulatory skills. The provisions of this section are aimed at maintaining the corridor's function by keeping furniture secure in place and only allowing limited installations. The spacing of the furniture also supports the therapeutic use of the furniture by residents moving from one location to another farther down the corridor.

407.4.4 Group I-2 care suites. *Care suites* in Group I-2 shall comply with Sections 407.4.4.1 through 407.4.4.4 and either Section 407.4.4.5 or 407.4.4.6.

❖ Care suites are typically either groups of patient sleeping rooms or groups of care rooms and associated support/staff spaces that operate as a unit. Care suites are recognized to be an effective tool to provide some flexibility, due to functional considerations, in providing appropriate patient care and allowing for the ability to reach an exit access. Use of suites is a particularly useful tool at intensive care units and emergency departments in patient treatment areas. The ability to have a full visual wall system or open plan not only allows for patient privacy and improved patient care but also functions as extremely beneficial during any type of emergency situation, including defending-in-place and evacuation.

Patient treatment and sleeping rooms are permitted to be arranged as suites (see Commentary Figure 407.4.4 for an example plan of a patient suite). The figure illustrates a patient suite of more than 1,000 square feet (93 m^2) where two remote egress doors are required. The criteria in this section recognize the low patient-to-staff ratio of these facilities where the staff is directly responsible for the safety of the patients in the event of a fire.

Care suites allow for the elimination of the smoke partition between the patient rooms and the staff areas. By considering the connected space an intervening room rather than a corridor, these provisions can be applied.

While not prohibited in nursing homes, suite configurations are more commonly found in hospitals.

The use of the term "care suite" is not intended to prohibit other groups of rooms that are not patient care areas, such as records departments or the business offices found in hospitals.

407.4.4.1 Exit access through care suites. *Exit* access from all other portions of a building not classified as a *care suite* shall not pass through a *care suite*. In a *care suite* required to have more than one *exit*, one *exit access* is permitted to pass through an adjacent *care suite* provided all of the other requirements of Sections 407.4 and 1016.2 are satisfied.

❖ Suites are intended to serve specific functions for patient treatment or sleeping. This section clarifies that the suite should not be used as part of the exit access of other areas within the building. It is intended to prevent corridors from discharging into a maze of intervening rooms prior to entering the exit. However, the second sentence would allow for a group of suites to work together as long as the overall exit access travel distance of 200 feet (60 960 mm) was met (see Table 1017.2); the next suite could meet the egress through intervening spaces as addressed in Section 1016.2; and both suites complied with the provisions in this section.

407.4.4.2 Separation. *Care suites* shall be separated from other portions of the building, including other care suites, by a smoke partition complying with Section 710.

❖ This section clarifies the type of partition that should be used to separate the suite from the corridor, adjacent suite or any other portion of the building. This section requires separation by a smoke partition, similar to the construction of a corridor wall in Group I-2 (see Section 407.3).

407.4.4.3 Access to corridor. Movement from habitable rooms shall not require passage through more than three doors and 100 feet (30 480 mm) distance of travel within the suite.

Exception: The distance of travel shall be permitted to be increased to 125 feet (38 100 mm) where an automatic smoke detection system is provided throughout the *care suite* and installed in accordance with NFPA 72.

❖ In previous editions of the code, travel distance was controlled within suites by limiting the number of intervening rooms. The approach was confusing and led to inconsistent application of the care suite provisions. Travel distance within the suite is limited to 100 feet (30 480 mm) and allows for the movement of a patient through a maximum of three doors. The three doors are within the care suite, thus the door from the care suite to the corridor would be in addition to the three allowed within the suite. The distance can be expanded to 125 feet (38 100 mm) within the care suite as provided by the exception. Table 1017.2 limits the total travel distance within Group I-2 occupancies to 200 feet (60 960 mm). Therefore, if you have 125 feet (38 100 mm) of travel within the suite, you only have 75 feet (22 860 mm) outside of the care suite before your travel must get you to an exit. The Section 407.4.2 limit of a maximum 50-foot (15 240 mm) travel distance within a patient sleeping room to an exit access door does not apply to patient rooms within care suites.

407.4.4.4 Doors within care suites. Doors in care suites serving habitable rooms shall be permitted to comply with one of the following:

1. Manually operated horizontal sliding doors permitted in accordance with Exception 9 to Section 1010.1.2.
2. Power-operated doors permitted in accordance with Exception 7 to Section 1010.1.2.
3. Means of egress doors complying with Section 1010.

❖ Within care suites, patient rooms and treatment rooms are generally not required by the code to have doors. However, for clinical needs (infection control, privacy, noise control, etc.), doors are commonly required to patient or treatment rooms within care suites. This section clarifies what types of doors are permitted within a suite. Considered in the allowances is that staff will be assisting in evacuation and will be familiar with door operation.

407.4.4.5 Care suites containing sleeping room areas. Sleeping rooms shall be permitted to be grouped into care suites where one of the following criteria is met:

1. The *care suite* is not used as an *exit access* for more than eight care recipient beds.

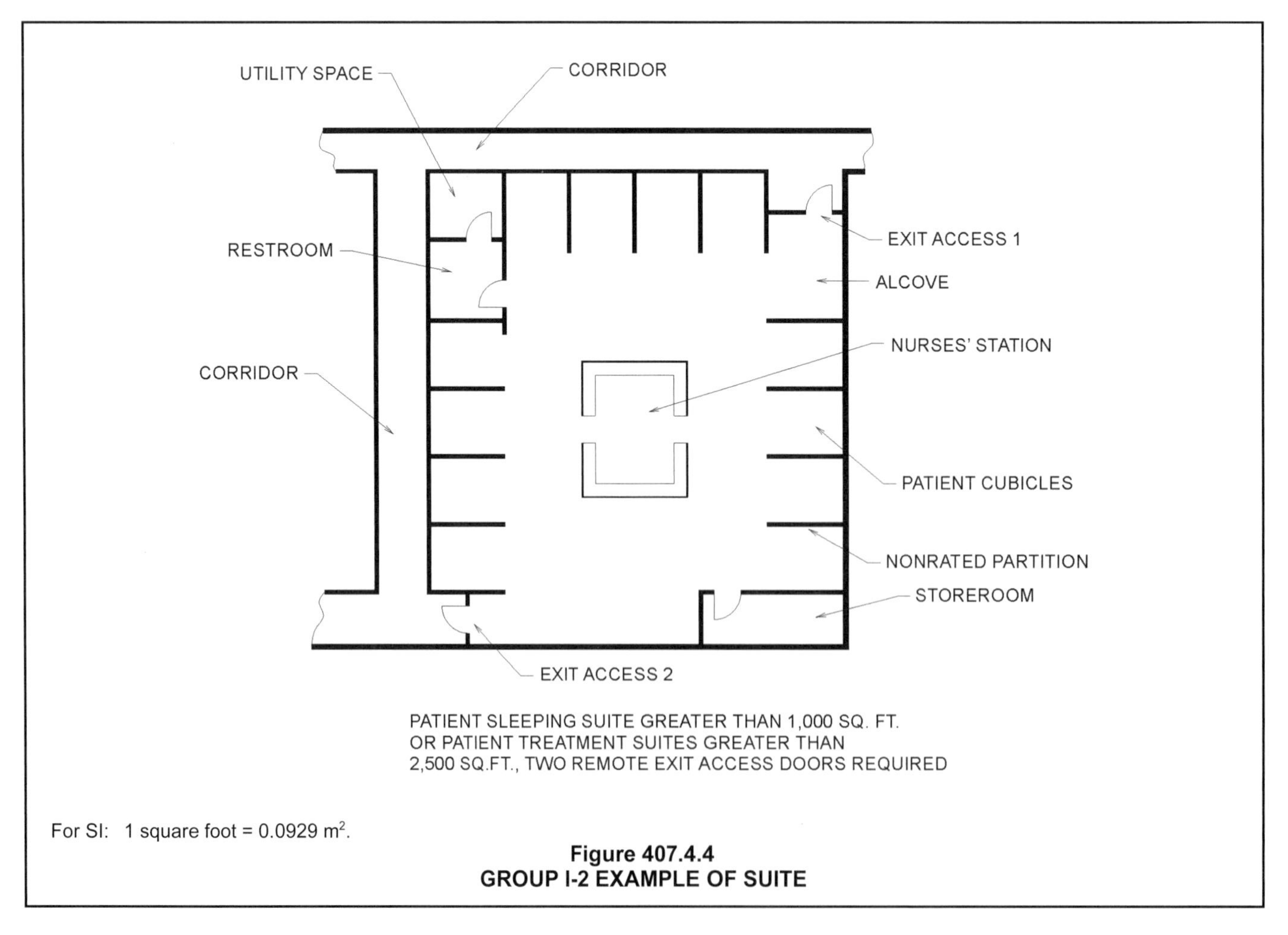

For SI: 1 square foot = 0.0929 m^2.

Figure 407.4.4
GROUP I-2 EXAMPLE OF SUITE

2. The arrangement of the *care suite* allows for direct and constant visual supervision into the sleeping rooms by care providers.
3. An automatic smoke detection system is provided in the sleeping rooms and installed in accordance with NFPA 72.

❖ In addition to complying with Sections 407.4.4 through 407.4.4.4, care suites containing patient sleeping rooms must comply with Sections 407.4.4.5, 407.4.4.5.1 and 407.4.4.5.2. The three items of this section are options. The suite can be limited to eight beds. It can have more than eight beds provided the suite is arranged such that each patient sleeping room is within direct and constant visual supervision or an automatic smoke detection system is provided. The smoke detection option is provided where direct supervision of patients by staff is not possible. Smoke detection in the patient room provides equivalent early detection of a fire. Section 407.4.4.5.1 will provide the overall size limit to the care suite.

407.4.4.5.1 Area. *Care suites* containing sleeping rooms shall be not greater than 7,500 square feet (696 m^2) in area.

Exception: *Care suites* containing sleeping rooms shall be permitted to be not greater than 10,000 square feet (929 m^2) in area where an automatic smoke detection system is provided throughout the *care suite* and installed in accordance with NFPA 72.

❖ The combined area of the patient sleeping rooms, the staff areas and any associated spaces (e.g., clean room, nutrition center, storage room) and any intervening rooms are permitted to have a total area of 7,500 square feet (696 m^2). The care suite perimeter is defined by its separations from other care suites, other portions of the occupancy and from corridors in accordance with earlier provisions of Section 407. Increasing the area to 10,000 square feet (929 m^2) allows inclusion of more staff and support spaces within the suite. Operationally, this is a key factor because the staff will not need to leave the suite as often in order to retrieve supplies or to access the staff toilet, which will improve the response time of the staff during a medical emergency, or a fire/safety situation.

407.4.4.5.2 Exit access. Any sleeping room, or any *care suite* that contains sleeping rooms, of more than 1,000 square feet (93 m^2) shall have no fewer than two *exit access* doors from the *care suite* located in accordance with Section 1007.

❖ A care suite that contains a patient sleeping room and an area of 1,000 square feet (93 m^2) or less can be provided with only one exit access door to a corridor, provided that the 100-foot (30 480 mm) maximum travel distance (see Section 407.4.4.3) is met. For suites that contain patient sleeping rooms and have an area of between 1,000 square feet (93 m^2) and 10,000 square feet (929 m^2), there must be at least two exit access doors to corridors, or one access door to a corridor and one to an adjoining care suite as allowed by Section 407.4.4.1.

Because this section allows a single exit access door from suites smaller than 1,000 square feet (93 m^2), and Section 407.4.4.3 allows a maximum travel within suite of either 100 feet (30 480 mm) or 125 feet (38 100 mm), there could be a configuration of a suite where the common path of travel within the suite exceeds the limits found in Table 1006.2.1. For rooms within suites, a greater common path of travel is allowed, provided the maximum travel distance established in Section 407.4.4.3 is met.

407.4.4.6 Care suites not containing sleeping rooms. Areas not containing sleeping rooms, but only treatment areas and the associated rooms, spaces or circulation space, shall be permitted to be grouped into *care suites* and shall conform to the limitations in Sections 407.4.4.6.1 and 407.4.4.6.2.

❖ According to the definition for "Care suite" in Chapter 2, suites covered by these provisions are limited to areas that include patient sleeping rooms or patient treatment rooms. This group of sections addresses suites for patient treatment rooms. Common examples include diagnostic or therapy areas where the function of treatment requires adjacency. This could also include inpatient or outpatient surgery suites, emergency treatment area suites or areas such as recovery rooms or dialysis treatment centers. Patients may be anesthetized, but staff/patient ratios are high and staff is trained in proper evacuation procedures. In addition to complying with Sections 407.4.4 through 407.4.4.4, care suites that do not contain patient sleeping rooms must comply with Sections 407.4.4.6, 407.4.4.6.1 and 407.4.4.6.2.

407.4.4.6.1 Area. *Care suites* of rooms, other than sleeping rooms, shall have an area not greater than 12,500 square feet (1161 m^2).

Exception: *Care suites* not containing sleeping rooms shall be permitted to be not greater than 15,000 square feet (1394 m^2) in area where an automatic smoke detection system is provided throughout the *care suite* in accordance with Section 907.

❖ A 10,000-square-foot (929 m^2) limitation for care suites not containing sleeping rooms was in the legacy codes before sprinkler protection was required in Group I-2 occupancies. Designs to address current diagnostic and treatment methods have proven difficult to achieve within the previous 10,000-square-foot (929 m^2) requirement. Sprinkler protection provides an additional level of life safety to building occupants, which justifies the area increase to 12,500 square feet (1161 m^2). Additionally, providing an automatic smoke detection system throughout a care suite provides an added level of life safety, which can be used to increase the area to 15,000 square feet (1394 m^2). Sprinkler protection and smoke detection are very effective measures of providing life safety to building occupants. Providing these additional levels of life safety will allow for more functional design to better meet increased space requirements for the area of a care suite not containing sleeping rooms.

407.4.4.6.2 Exit access. *Care suites*, other than sleeping rooms, with an area of more than 2,500 square feet (232 m^2) shall have no fewer than two *exit access* doors from the *care suite* located in accordance with Section 1007.1.

❖ A suite that contains a patient treatment room and an area of 2,500 square feet (232 m^2) or less can have only one exit access door to a corridor, provided that the maximum travel distance within the suite does not exceed the 100-foot (30 480 mm) exit access limit set by Section 407.4.4.3 is met. For suites that contain patient treatment rooms and have an area of between 2,500 square feet (232 m^2) and 15,000 square feet (1394 m^2), there must be at least two exit access doors to the corridor, or one access door to the corridor and one to an adjoining care suite as allowed by Section 407.4.4.1. These doors must meet the separation requirements in Section 1007. In a two-exit suite, at least one of the exits must be within the travel distance of every point in the suite as specified in Sections 407.4.4.3.

Because this section allows a single-exit access door from suites smaller than 2,500 square feet (232 m^2), and Section 407.4.4.3 allows a maximum travel within suite of either 100 feet (30 480 mm) or 125 feet (38 100 mm), there could be a configuration of a suite where the common path of travel within the suite exceeds the limits found in Table 1006.2.1. For rooms within suites, a greater common path of travel is allowed, provided the maximum travel distance established in Section 407.4.4.3 is met.

407.5 Smoke barriers. *Smoke barriers* shall be provided to subdivide every *story* used by persons receiving care, treatment or sleeping and to divide other *stories* with an *occupant load* of 50 or more persons, into no fewer than two *smoke compartments*. Such stories shall be divided into *smoke compartments* with an area of not more than 22,500 square feet (2092 m^2) in Group I-2, Condition 1, and not more than 40,000 square feet (3716 m^2) in Group I-2, Condition 2, and the distance of travel from any point in a *smoke compartment* to a *smoke barrier* door shall be not greater than 200 feet (60 960 mm). The *smoke barrier* shall be in accordance with Section 709.

❖ One of the basic premises of this section is that vertical evacuation of care recipients in Group I-2 occupancies is extremely difficult where patients are incapable of self-preservation or require ongoing care. It is essential, therefore, to provide a refuge area on every story used by care recipients for sleeping, care or treatment. The refuge area is intended to provide a protected area to which care recipients may be relocated if it becomes necessary to evacuate them from their original rooms. For this reason, each story used by care recipients for sleeping, care or treatment and all other stories with an occupant load of 50 or more must be divided into at least two smoke compartments. Each smoke compartment is limited to an area of no more than 22,500 square feet (2092 m^2) for nursing homes (Group I-2, Condition 1) and 40,000 square feet (3716 m^2) for hospitals (Group I-2, Condition 2) (see the definition of "Smoke compartment" in Section 202). The travel distance from any point in a smoke compartment to a smoke barrier door is limited to 200 feet (60 960 mm). This is distinct from the travel distance to exits. This distance is limiting the distance that care providers and care recipients must travel until reaching another smoke compartment and the refuge area it contains. Travel to an exit enclosure or exit door is not a substitute for travel to the smoke barrier and a separate compartment (see Commentary Figure 407.5). Every occupied space must be able to meet both exit access travel distance to an exit and to the entrance to an adjacent smoke compartment. The maximum travel distance to an exit is also 200 feet (60 960 mm).

Smoke barriers used to divide one smoke compartment from the next are to be constructed in accordance with Section 709 and with horizontal assemblies as specified in Section 407.5.3. The smoke barrier is different from the smoke partitions required for corridor walls in Section 407.3. Among other things, a smoke barrier is required to have a fire-resistance rating of 1 hour and, except as other-

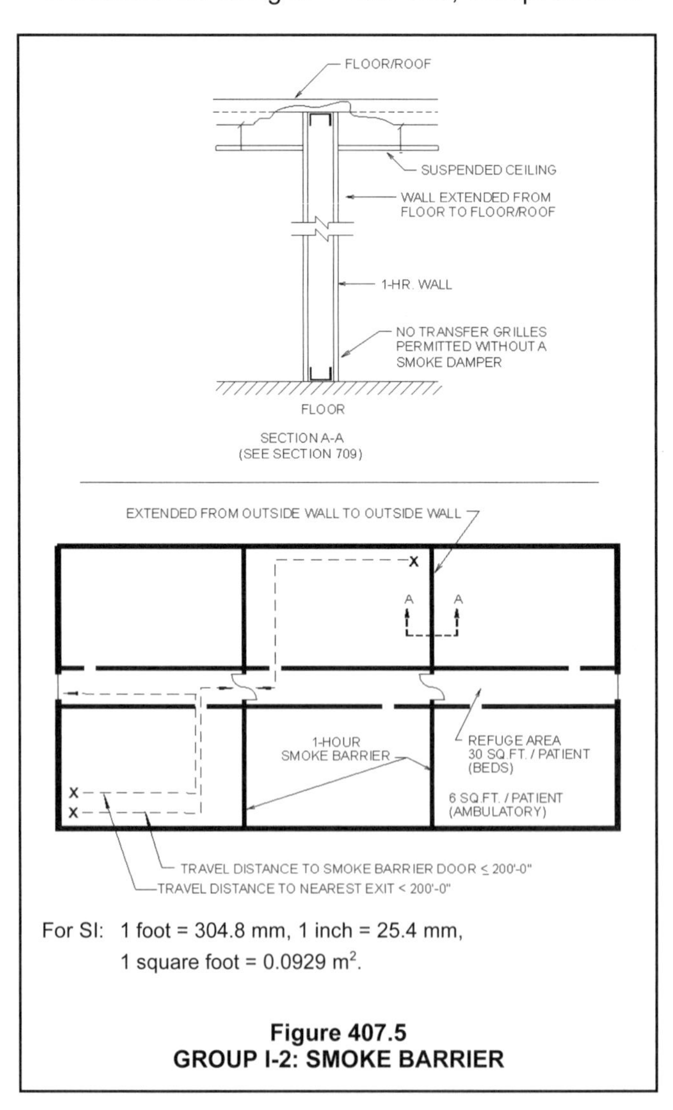

For SI: 1 foot = 304.8 mm, 1 inch = 25.4 mm, 1 square foot = 0.0929 m^2.

Figure 407.5
GROUP I-2: SMOKE BARRIER

wise provided for in the Exception 1 to Section 709.4, provide a continuous separation through all concealed spaces to resist the passage of fire and smoke (see Commentary Figure 407.5). The maximum area of the smoke compartment provides two protection features. First, the number of persons who are exposed to smoke and fumes will be limited. Typical designs indicate that the number of persons exposed will usually be less than 100 and, more likely, 50 or less. Second, the overall dimensions of the smoke compartment must result in a maximum travel distance to the smoke barrier door of 200 feet (60 960 mm) for all design configurations.

407.5.1 Refuge area. Refuge areas shall be provided within each *smoke compartment*. The size of the refuge area shall accommodate the occupants and care recipients from the adjoining *smoke compartment*. Where a *smoke compartment* is adjoined by two or more *smoke compartments*, the minimum area of the refuge area shall accommodate the largest *occupant load* of the adjoining compartments. The size of the refuge area shall provide the following:

1. Not less than 30 net square feet (2.8 m^2) for each care recipient confined to bed or stretcher.
2. Not less than 6 square feet (0.56 m^2) for each ambulatory care recipient not confined to bed or stretcher and for other occupants.

Areas or spaces permitted to be included in the calculation of refuge area are *corridors*, sleeping areas, treatment rooms, lounge or dining areas and other low-hazard areas.

❖ To provide adequate space for occupants who might be relocated to an adjacent smoke compartment, open space for refuge areas must be available. In the determination of the adequacy of the refuge areas, minimum net areas per occupant are required. In determining the area available, it should be noted that the facility may still need to function as a medical care facility even while the care recipients are relocated. On floors housing nonambulatory patients (i.e., care recipient sleeping floors), a minimum of 30 net square feet (2.8 m^2) per patient is required to be provided in low-hazard areas, such as waiting areas, corridors or patient rooms. This is to permit the relocation of the care recipient on a bed or stretcher to an area of safety on the same floor. The 30 net square feet (2.8 m^2) is intended only to apply to the number of care recipients and not the calculated occupant load as determined using Table 1004.1.2.

For floors classified as Group I-2 that are not used for care recipient sleeping rooms (i.e., treatment floors), the net area required is 6 square feet (0.56 m^2) per occupant. In this instance, the area required is based on the calculated occupant load as determined by Table 1004.1.2. The 6-square-foot (0.56 m^2) minimum is based on an assumption that some occupants will be using walkers, wheelchairs and possibly a few litters for care recipients who are being transported through the area at the time of the fire.

The refuge area must be able to accommodate all of the occupants and care recipients from the adjoining smoke compartment based on the factors described above. If a smoke compartment containing a refuge area is adjoined by multiple smoke compartments, the minimum required size of the refuge area is to be based upon the largest of the occupant loads of the adjoining compartments. It is assumed that the refuge area will only need to serve one adjacent smoke compartment.

When there are multiple smoke compartments on the same level, each smoke compartment must have open space for the largest number of evacuees. Space for people already in the smoke compartment should be considered in the design, but do not have to be considered for determining the refuge area square footage requirements. Remember, the idea is for smoke compartments to be part of the "defend-in-place" scenario commonly used in hospitals and nursing homes (see commentary, Section 407.1).

Refuge areas required by this section are distinct from areas of refuge and exterior areas of assisted rescue required by Section 1009 for accessible means of egress.

407.5.2 Independent egress. A *means of egress* shall be provided from each *smoke compartment* created by *smoke barriers* without having to return through the smoke compartment from which *means of egress* originated.

❖ To prevent creation of a dead-end smoke compartment, exits are to be arranged so as to permit access without returning through a smoke compartment from which egress originated. This section does not require an exit from within each smoke compartment. See Commentary Figures 407.5.2(1) and 407.5.2(2) for acceptable and unacceptable egress arrangements.

407.5.3 Horizontal assemblies. *Horizontal assemblies* supporting *smoke barriers* required by this section shall be designed to resist the movement of smoke. Elevator lobbies shall be in accordance with Section 3006.2.

❖ In order to maintain the integrity of a smoke compartment, the horizontal assemblies supporting the smoke barrier walls must also be designed to resist the passage of smoke. With the combination of these horizontal assemblies and the fire barriers, a complete smoke-protected compartment is achieved. Provisions of Section 711.2.4.4 address openings and penetrations of each horizontal assembly providing a barrier to smoke. Elevators and other large openings will have to be in a shaft enclosure. A lobby complying with Section 3006.2 will need to be provided for such elevators.

[F] 407.6 Automatic sprinkler system. *Smoke compartments* containing sleeping rooms shall be equipped throughout with an *automatic sprinkler* system in accordance with Sections 903.3.1.1 and 903.3.2.

❖ Recognizing the effectiveness of an automatic sprinkler system, the reference to Section 903.3.1.1 clarifies that the suppression system required by Section 903.2.6 must be a sprinkler system in accordance with the code and NFPA 13. Furthermore, in response to recent proven technology, smoke compartments containing patient rooms are required to be equipped throughout with quick-response or residential sprinklers as set forth in Item 1 in Section 903.3.2.

Most fire injuries in hospitals begin with the ignition of clothing on a person, a mattress, a pillow, bedding or linen. The Building and Fire Research Laboratory of NIST conducted fire tests in patient rooms using quick-response sprinklers and smoke detectors. The test results were published in a July 1993 report entitled, "Measurement of Room Conditions and Response of Sprinklers and Smoke Detectors During a Simulated Two-bed Hospital Patient Room Fire." The report concluded that quick-response sprinklers actuated before any life would be threatened by a fire in a care recipient's (patient's) room.

Prior to the development of quick-response sprinkler technology, smoke detection in care recipient rooms classified as Group I-2 occupancies was typically required, due to the slower response time of standard sprinklers. While properly operating standard sprinklers are effective, the extent of fire growth and smoke production that can occur before sprinkler activation creates the need for early warning to

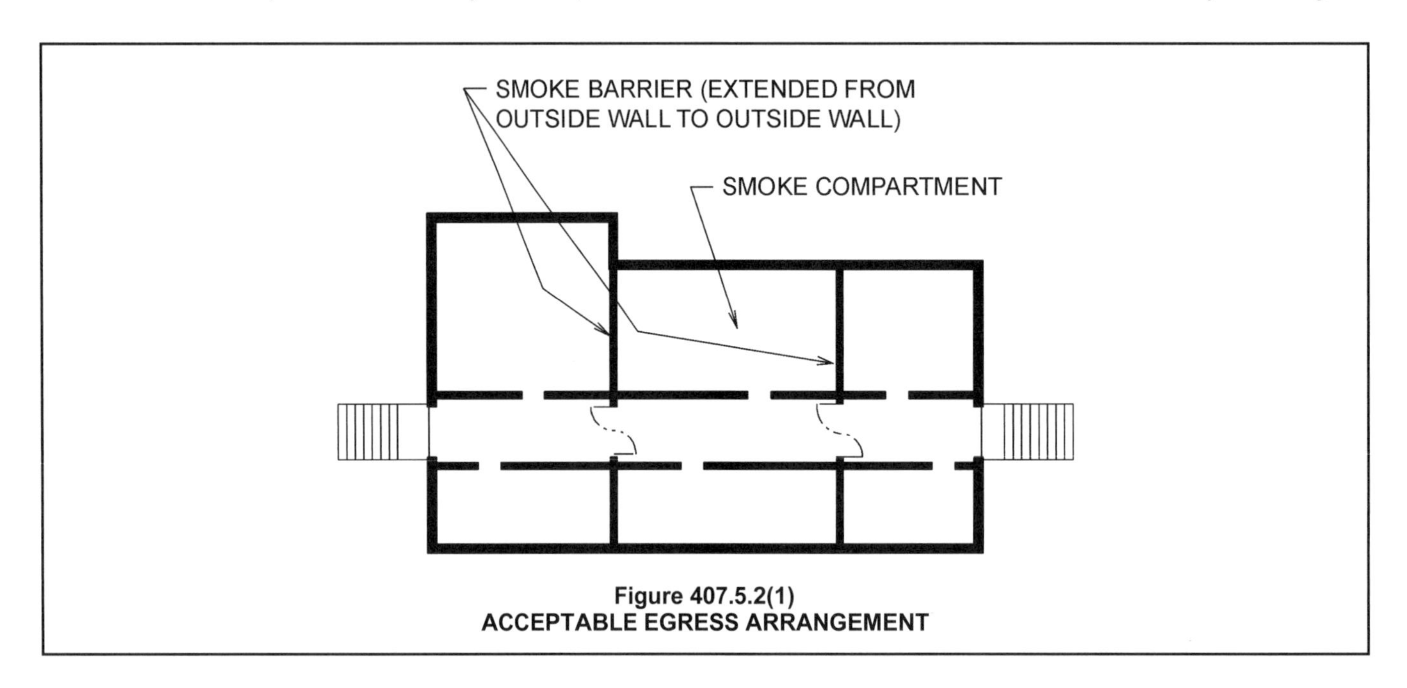

Figure 407.5.2(1)
ACCEPTABLE EGRESS ARRANGEMENT

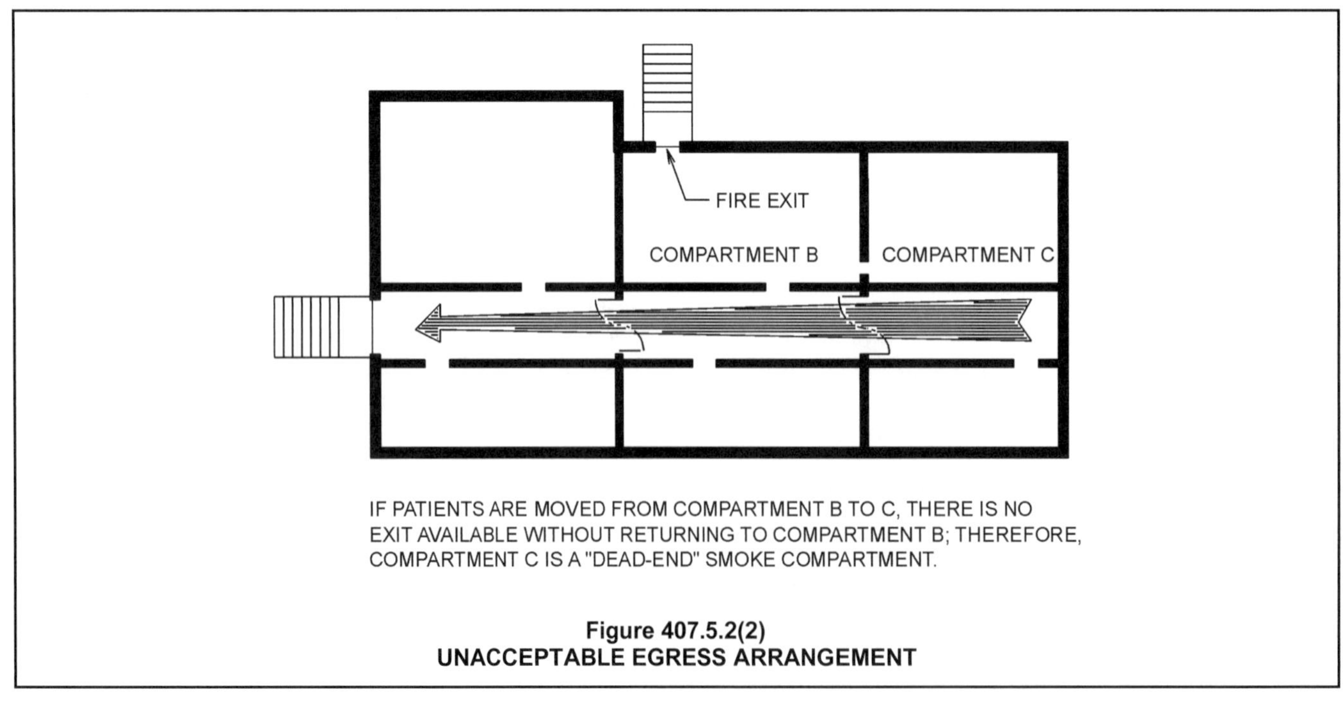

Figure 407.5.2(2)
UNACCEPTABLE EGRESS ARRANGEMENT

enable faster response by the care providers and initiation of egress that is critical in occupancies containing persons incapable of self-preservation. With quick-response or residential sprinklers installed throughout the smoke compartment, and additional protection provided as indicated for spaces open to corridors, incidental uses and certain specific health care facilities, smoke detectors in care recipient sleeping rooms are not required in Group I-2 occupancies.

[F] 407.7 Fire alarm system. A *fire alarm* system shall be provided in accordance with Section 907.2.6.

❖ Fire alarms are essential to this occupancy where many of those receiving care and treatment cannot leave a fire incident area without the assistance of staff. The alarm system is a key part of the safeguards to protect care recipients. Standards for this alarm system are found in Sections 907.2.6 and 907.2.6.2.

[F] 407.8 Automatic fire detection. *Corridors* in Group I-2, Condition 1, occupancies, long-term care facilities, *detoxification facilities* and spaces permitted to be open to the *corridors* by Section 407.2 shall be equipped with an automatic fire detection system.

Group I-2, Condition 2, occupancies shall be equipped with smoke detection as required in Section 407.2.

Exceptions:

1. *Corridor* smoke detection is not required where sleeping rooms are provided with *smoke detectors* that comply with UL 268. Such detectors shall provide a visual display on the *corridor* side of each sleeping room and an audible and visual alarm at the care provider's station attending each unit.
2. *Corridor* smoke detection is not required where sleeping room doors are equipped with automatic door-closing devices with integral *smoke detectors* on the unit sides installed in accordance with their listing, provided that the integral detectors perform the required alerting function.

❖ Automatic fire detection is required in corridors and areas open to corridors in nursing homes, Group I-2, Condition 1, occupancies and long-term care and detoxification facilities. In recognition of quick-response sprinkler technology and the fact that the sprinkler system is electronically supervised and doors to sleeping rooms are supervised by staff on a continual basis when in the open position, it is now believed that sleeping room smoke detectors are not required for adequate fire safety in sleeping rooms. In nursing homes, long-term care facilities and detoxification facilities, however, some redundancy is appropriate because such facilities typically have less control over furnishings and personal items and, thereby, result in a less predictable and usually higher fire hazard load than other Group I-2 occupancies. Also, there is generally less staff supervision in these facilities than in other health care facilities and, thus, less control over patient smoking and other fire causes. To provide additional protection against fires spreading from the room of origin, therefore, automatic fire detection is required in corridors of nursing homes, long-term care facilities and detoxification facilities. It should be noted that fire detection is not required in corridors of other Group I-2 occupancies except where otherwise specifically required in the code (see Sections 407.2.1 through 407.2.4, 907.2.6 and 907.2.6.2). Similarly, since areas open to the corridor very often are the room of fire origin and because such areas are not required by the code to be under visual supervision by staff, some redundancy to protection by the sprinkler system is requested. Accordingly, all areas open to corridors must be protected by an automatic fire detection system. This requirement provides an additional level of protection against sprinkler system failures or lapses in staff supervision. There are two exceptions to the requirement for an automatic fire detection system in corridors of nursing homes, long-term care facilities and detoxification facilities. In both exceptions, the alternative methods of protection specifically provide an equivalent level of safety to what is required in care recipient sleeping rooms.

Exception 1 requires smoke detectors to be located in the care recipient's sleeping room and to activate both a visual display on the corridor side of the patient room and a visual and audible alarm at the care providers' station serving or attending the room. Detectors complying with UL 268 are intended for open area protection and for connection to a normal power supply or as part of a fire alarm system.

This exception, however, is specifically designed so as not to require the detectors to activate the building's fire alarm system where approved sleeping room smoke detectors are installed and where visual and audible alarms are provided. This is in response to the concern over unwanted alarms. It should be noted that the required alarm signals will not necessarily indicate to staff that a fire emergency exists. The nursing/ care provider call system may typically be used to identify numerous conditions within the room.

Exception 2 addresses the situation where smoke detectors are incorporated within automatic door-closing devices. The units are acceptable as long as the required alarm functions are still provided. Such units are usually listed as combination door closer and hold-open devices.

407.9 Secured yards. Grounds are permitted to be fenced and gates therein are permitted to be equipped with locks, provided that safe dispersal areas having 30 net square feet ($2.8 m^2$) for bed and stretcher care recipients and 6 net square feet ($0.56 m^2$) for ambulatory care recipients and other occupants are located between the building and the fence. Such provided safe dispersal areas shall be located not less than 50 feet (15 240 mm) from the building they serve.

❖ Group I-2 occupancies, particularly those specializing in the treatment of mental disabilities such as

Alzheimer's, often provide a secured yard for care recipient use. During an emergency, it may be difficult to safeguard care recipients if they are allowed to have direct access to a public way. This section essentially creates a safe area where people may move from the building without having unrestricted access to the public way.

407.10 Electrical systems. In Group I-2 occupancies, the essential electrical system for electrical components, equipment and systems shall be designed and constructed in accordance with the provisions of Chapter 27 and NFPA 99.

❖ Currently, emergency power systems are required to comply with NFPA 99 by the Center for Medicare/Medicaid Services (CMS) in order for a facility to receive federal reimbursement funds. This section's mandate for compliance with NFPA 99 will ensure the required power system is provided in Group I-2 occupancies. A reference to Chapter 27 will comprehensively address electrical systems, including references to NFPA 70, 110 and 111.

SECTION 408 GROUP I-3

408.1 General. Occupancies in Group I-3 shall comply with the provisions of Sections 408.1 through 408.11 and other applicable provisions of this code (see Section 308.5).

❖ The provisions of Section 408 address the unique features of detention and correctional occupancies in Group I-3. The provisions are based on full-scale fire tests, fire experience (in particular several multiple-death fires that occurred between 1974 and 1979) and the provisions for occupancies in Group I-2. With respect to evacuation, occupancies in Groups I-2 and I-3 are similar in that the occupants are typically not capable of self-preservation and, therefore, staff must assist in evacuation.

The general approach is a defend-in-place philosophy based on the difficulties associated with evacuation, especially in medium- and high-security facilities. The first level of protection is established as the room that is typically a sleeping room or cell. Horizontal evacuation to an adjacent smoke compartment provides the second level of protection, and vertical evacuation is the third level of protection. The evacuation difficulty associated with occupancies in Group I-3 is not occupant mobility; rather, it is the need to maintain security and, in some instances, to keep the residents segregated. There are also instances, such as protected witness facilities, in which the identity of the occupant must not be revealed to other occupants or the public.

It is recognized that the broad classification of Group I-3 includes a variety of locking arrangements based on the various levels of security required. As such, Section 308.5 contains definitions for five different occupancy conditions that are intended to represent the various locking arrangements and different levels of security (see commentary, Section 308.5). Regardless of the security level, fire protection and life safety features must be provided in order to achieve an acceptable level of protection without interfering with the operation of the facility and the need to maintain security.

This section has been developed out of considerations unique to detention and correctional facilities and to facilitate the consideration of the entire package of protection features. The basic protection features provided include early detection, fire containment, evacuation and fire extinguishment. Section 403.8.3 of the IFC contains provisions for emergency preparedness, including staff training, staffing requirements and the need to be able to obtain and identify emergency keys. Additional consideration should be given to restricting the fuel load within resident housing areas. It should also be noted that Section 509 contains provisions for incidental uses within Group I-3. The incidental uses represent different degrees of hazard than are associated with the hazards of the primary building area—in this case, the housing of occupants under physical restraint. The different hazard is typically related to the basic nature of the contents or the quantity of combustibles.

The provisions of Section 408 apply to all occupancies in Group I-3 and those portions of occupancies that are considered Group I-3, except Condition 1 facilities (see Commentary Figure 408.1). As defined in Section 308.5.1, Condition 1 facilities permit free egress even to the exterior and, therefore, such occupancies are subject to the same provisions applicable to Group R.

The elimination of Condition 1 from the provisions of Section 408 may result in some interesting applications of the code. For example, the corridor separation requirements in Section 1020.1 for occupancies in Group R are more restrictive than the corridor requirements for occupancies in Group I-3, Condition 2. This may result in a request to consider a facility as Condition 2 even though the building is really Condition 1. Such a request could be considered acceptable only if the additional requirements in the code and in the IFC for Condition 2 are met, including automatic fire suppression and staffing (see Section 403.8.3.2 of the IFC).

408.1.1 Definitions. The following terms are defined in Chapter 2:

CELL.

CELL TIER.

HOUSING UNIT.

SALLYPORT.

❖ This section lists terms that are specifically associated with the subject matter of Section 408. It is important to emphasize that these terms are not exclusively related to this chapter but may or may not also be applicable where the term is used elsewhere in the code.

Definitions of terms can help in the understanding and application of the code requirements. The purpose for including a list within this chapter is to provide more convenient access to terms which may have a specific or limited application within this chapter. For the complete definition and associated commentary, refer back to Chapter 2. Terms that are italicized provide a visual identification throughout the code that a definition exists for that term. The use and application of all defined terms are set forth in Section 201.

408.2 Other occupancies. Buildings or portions of buildings in Group I-3 occupancies where security operations necessitate the locking of required *means of egress* shall be permitted to be classified as a different occupancy. Occupancies classified as other than Group I-3 shall meet the applicable requirements of this code for that occupancy where provisions are made for the release of occupants at all times.

Means of egress from detention and correctional occupancies that traverse other use areas shall, as a minimum, conform to requirements for detention and correctional occupancies.

Exception: It is permissible to exit through a *horizontal exit* into other contiguous occupancies that do not conform to detention and correctional occupancy egress provisions but that do comply with requirements set forth in the appropriate occupancy, as long as the occupancy is not a Group H use.

❖ In accordance with the provisions of Section 508, portions of an occupancy in Group I-3 may be classified as a separate occupancy and meet the provisions of the code for that occupancy. Since the area may be used by the residents, however, a need may exist for the other occupancy to also contain security provisions, such as the locking of egress doors. This section specifically permits such a condition as long as arrangements have been made for release of the occupants within these areas at any time they are occupied. Acceptable methods include having either

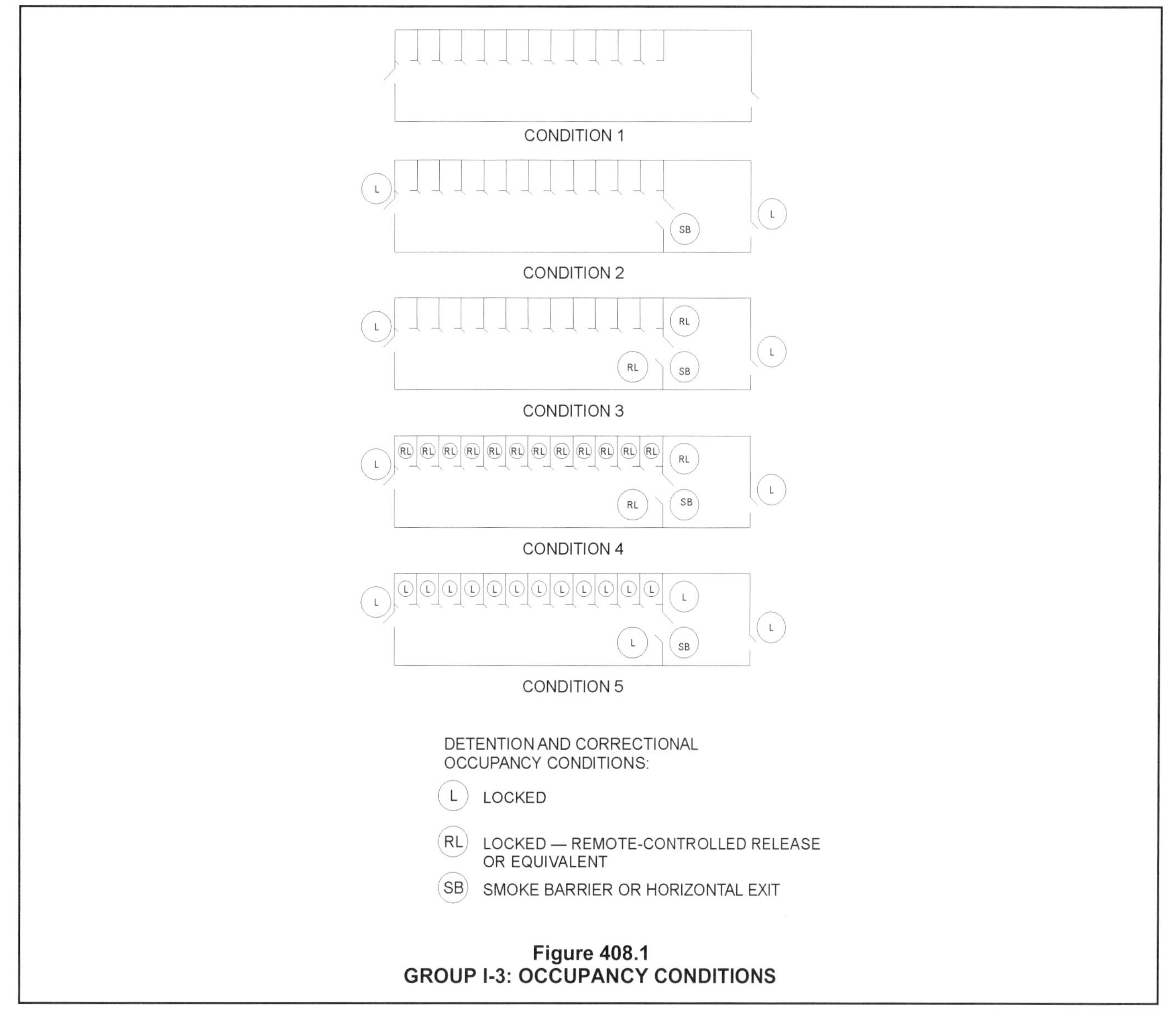

Figure 408.1
GROUP I-3: OCCUPANCY CONDITIONS

staff operate the locks or remote release of the locks, similar to that which is provided in the housing areas.

Although it is indicated that the section applies to portions of occupancies in Group I-3, consideration should be given to allow the necessary security to be provided in separate buildings that are part of a detention or correctional facility. Applications of this section should be restricted to the buildings that must be secure because they are occupied by residents, and arrangements for quick release of the locks must be provided at all times the building is occupied. If residents are to be permitted into an area or building without staff supervision, the residents should be able to initiate their own evacuation.

The means of egress provisions from this section and Chapter 10 that are applicable to Group I-3 are also applicable to the path through different occupancies. For example, if the means of egress from a cell block wing traverses through a Group A area, then the requirements of the code for the Group I-3 egress are also applicable to the path through the assembly occupancy.

The exception in the code permits egress through a different occupancy, other than Group H, that is separated from the Group I-3 areas by either a fire wall or fire barrier. A horizontal exit complying with Section 1026 must be provided in this wall or barrier. This separation voids the requirements that Group I-3 egress provisions are also applicable to the different occupancies. The reason for this, of course, is that Group I-3 occupants have fled the immediate effects of the fire condition in the original detention or correctional areas and are now harbored in a refuge area.

408.3 Means of egress. Except as modified or as provided for in this section, the *means of egress* provisions of Chapter 10 shall apply.

❖ Because of the need to provide security in occupancies in Group I-3, many of the means of egress requirements of Chapter 10 have been modified for Group I-3. Except as modified in Section 408, however, all other provisions of Chapter 10 are applicable.

408.3.1 Door width. Doors to resident *sleeping units* shall have a clear width of not less than 28 inches (711 mm).

❖ Except for sleeping units that are provided to meet accessible unit provisions of Section 1107.5.5, sleeping unit doors may be a minimum of 28 inches (711 mm) in width. This section is not intended to negate the need to provide a 32-inch (813 mm) door if the room is intended to serve as part of an accessible route. In an occupancy in Group I-3, most sleeping units and even sections of the prison need not be accessible (see Sections 1103.2.14 and 1107.5.5).

408.3.2 Sliding doors. Where doors in a *means of egress* are of the horizontal-sliding type, the force to slide the door to its fully open position shall be not greater than 50 pounds (220 N) with a perpendicular force against the door of 50 pounds (220 N).

❖ Swinging doors in occupancies in Group I-3 may not be acceptable for security reasons. If the door swings into a room, the door can easily be blocked and staff access can be prevented. If the door swings out, staff does not have control of the door and the residents can use the door as a means to overcome a staff member. Therefore, horizontal sliding doors are used quite extensively and are permitted in the means of egress. The doors must be capable of being opened with a force of 50 pounds (220 N) even if a 50-pound (220 N) force is being applied perpendicular to the door. The maximum force permitted to open the door exceeds the restrictions in Section 1010.1.3 in recognition of the physical capabilities of staff members.

408.3.3 Guard tower doors. A hatch or trap door not less than 16 square feet (610 m^2) in area through the floor and having dimensions of not less than 2 feet (610 mm) in any direction shall be permitted to be used as a portion of the *means of egress* from guard towers.

❖ This provision allows the use of trap doors in the floor of an observation point as both the access and the means of egress from the tower. In order to provide the 360-degree visibility necessary for guard observation stations, the size of the base of each such elevated station must be kept to a minimum. These towers are usually limited to prison/jail staff. The trap doors work in conjunction with the spiral stairway and ship's ladders provisions of the next two sections.

408.3.4 Spiral stairways. *Spiral stairways* that conform to the requirements of Section 1011.10 are permitted for access to and between staff locations.

❖ Recognizing the physical capabilities of the staff, spiral stairways are permitted for access to and between staff locations. In multiple story facilities, spiral stairways are often used between staff control areas on different levels. As such, staff can move between the areas without entering adjacent housing areas and without losing the space that would be required for a traditional stairway. Spiral stairways provide an option for vertical circulation in guard towers and similar observation areas where the floor area is limited and the desire is to have the least amount of obstruction of floor area possible.

408.3.5 Ship's ladders. Ship's ladders shall be permitted for egress from control rooms or elevated facility observation rooms in accordance with Section 1011.15.

❖ Ship's ladders are another alternative to spiral stairways for providing vertical paths of access and egress within a Group I-3 occupancy. These are intended for access to locations used by prison/jail staff. The design standard for ship's ladders is found in Section 1011.15. In that section, the code limits any space accessed by a ship's ladder to a maximum of 250 square feet (23 m^2) and a maximum of three occupants (see commentary, Section 1011.15).

408.3.6 Exit discharge. *Exits* are permitted to discharge into a fenced or walled courtyard. Enclosed *yards* or *courts* shall be of a size to accommodate all occupants, be located not less than 50 feet (15 240 mm) from the building and have an area of not less than 15 square feet (1.4 m^2) per person.

❖ For security purposes, exits from areas in Group I-3 do not normally provide access to a public way. In some instances, the building is located in a complex that has perimeter walls preventing access to a public way but that permit occupants to move away from the building. In other instances or in complexes where the mixing of residents must be controlled, one or more yards or courts are provided for exit discharge. Such arrangements are acceptable, provided that the occupants have an area that is at least 15 square feet (1.4 m^2) per person and is located at least 50 feet (15 240 mm) from the building.

This 50-foot (15 240 mm) distance enables residents to move away from the building and should have adequate spatial separation—especially since the building is required to be protected with an automatic fire suppression system in accordance with Section 903.2.6. The 15-square-foot (1.4 m^2) measurement is provided so that the residents have adequate space in which to stand and to move around, since they may need to remain in the yard or court for some time.

408.3.7 Sallyports. A *sallyport* shall be permitted in a *means of egress* where there are provisions for continuous and unobstructed passage through the *sallyport* during an emergency egress condition.

❖ A sallyport is a compartment established for security purposes that restricts the movement of individuals. A sallyport consists of two or more security doors that do not normally operate simultaneously. One door will not normally open until the other doors are closed and locked. Sallyports are provided for security purposes to control movement and passage of residents from one area to another. The sallyport is to be arranged such that during an emergency, the doors may be opened simultaneously so as to permit free and unobstructed movement from the area. In many instances, the sallyport doors contain a manual release that can be operated by staff during an emergency to override the normal electric operation.

Although not addressed by this section, if the sallyport is in the path that the fire department or fire brigade will use to bring fire hoses, additional consideration may need to be given to permit the hose to go through the sallyport. An alternative would be to provide a standpipe connection on the housing-area side of the sallyport.

408.3.8 Interior exit stairway and ramp construction. One *interior exit stairway* or *ramp* in each building shall be permitted to have glazing installed in doors and interior walls at each landing level providing access to the *interior exit stairway or ramp*, provided that the following conditions are met:

1. The *interior exit stairway or ramp* shall not serve more than four floor levels.
2. *Exit* doors shall be not less than $^3/_4$-hour *fire door assemblies* complying with Section 716.5
3. The total area of glazing at each floor level shall not exceed 5,000 square inches (3.2 m^2) and individual panels of glazing shall not exceed 1,296 square inches (0.84 m^2).
4. The glazing shall be protected on both sides by an *automatic sprinkler system*. The sprinkler system shall be designed to wet completely the entire surface of any glazing affected by fire when actuated.
5. The glazing shall be in a gasketed frame and installed in such a manner that the framing system will deflect without breaking (loading) the glass before the sprinkler system operates.
6. Obstructions, such as curtain rods, drapery traverse rods, curtains, drapes or similar materials shall not be installed between the automatic sprinklers and the glazing.

❖ In addressing the security needs in Group I-3, the limitation on openings in the interior exit stairways given in Sections 1023.2, 1023.4 and 1023.5 is modified to facilitate visibility into the exit.

An interior exit stairway that is open to view in a correctional facility is useful and necessary for two reasons. First, the movement of the inmates is observable, thus reducing the potential for concealment, physical attacks on other inmates and other undesirable activities that could otherwise take place in enclosed spaces that are not observable.

Second, the movement of inmates who are under physical restraint is observable from the exterior, thus increasing the level of safety for correctional officers. The alternative of closed-circuit television within the interior exit stairway is not as desirable because the potential for malfunction of or intentional damage to the equipment makes this method a less reliable form of observation.

The conditions specified that permit glazing require full enclosure of the floor opening so that there is no direct communication between stories. These criteria provide a measure of fire integrity, if somewhat less than that otherwise required of an interior exit stairway.

408.4 Locks. Egress doors are permitted to be locked in accordance with the applicable use condition. Doors from a refuge area to the outside are permitted to be locked with a key in lieu of locking methods described in Section 408.4.1. The keys to unlock the exterior doors shall be available at all times and the locks shall be operable from both sides of the door.

❖ The locking arrangements for egress doors depend on the occupancy condition (1, 2, 3, 4 or 5) given in

Section 308.5. Except for Condition 1 buildings, the doors to the exterior from an occupancy in Group I-3 may be locked and need not be remotely controlled. The door locks are to be arranged so that they can be released from the exterior of the building, as well as the interior. This feature accomplishes two purposes. First, since the door can be unlocked from the outside, staff need not enter the housing area in which the fire may be located to release the locks. Second, many facilities do not permit staff members in the housing area to carry the keys for the exterior doors; therefore, there may be no benefit in requiring the door lock to be arranged to permit operation from the interior, although such an arrangement is not prohibited.

The provisions of Section 403.8.3.4 of the IFC concerning the marking of keys applies to both exterior and interior door locks. For this reason, it is usually desirable to minimize the number of keys required for an emergency; therefore, the exterior doors will most likely be keyed alike. Unfortunately, many facilities do not take this into consideration and the number of emergency keys becomes excessive to the extent that it is virtually impossible to identify each key by sight and touch.

408.4.1 Remote release. Remote release of locks on doors in a *means of egress* shall be provided with reliable means of operation, remote from the resident living areas, to release locks on all required doors. In Occupancy Condition 3 or 4, the arrangement, accessibility and security of the release mechanisms required for egress shall be such that with the minimum available staff at any time, the lock mechanisms are capable of being released within 2 minutes.

Exception: Provisions for remote locking and unlocking of occupied rooms in Occupancy Condition 4 are not required provided that not more than 10 locks are necessary to be unlocked in order to move occupants from one smoke compartment to a refuge area within 3 minutes. The opening of necessary locks shall be accomplished with not more than two separate keys.

❖ The provisions of this section do not mandate remote release locks; rather, the facility determines the level and manner in which security is to be provided. Door-locking arrangements fall into one of the conditions defined in Section 308.5. If remote release locks are provided, the means of operating them must be external to the resident housing area so the staff is not required to enter the housing area to release the locks. Clearly, in Conditions 3 and 4, excessive delay in releasing locks puts the occupants at additional risk. Locks in Conditions 3 and 4 must be able to be released promptly (within 2 minutes) with the minimum staff that will be available. A control center is typically provided for two or more housing areas from which the locks are controlled. In evaluating the use and arrangement of the remote release, consideration must be given to the intended staffing levels at different times of the day, as well as the locations of the remote release.

The condition definitions of Section 308.5 are intended to relate to the time necessary to evacuate the residents to a refuge area, such as a smoke compartment; therefore, in Condition 4, manual release locks may be used only when no more than 10 locks have to be unlocked in order to move all occupants from one smoke compartment to another and provided that this can be accomplished within 3 minutes. The time period of 3 minutes is presumed to be a reasonable staff response time to a fire emergency. So that the time to release the locks is kept to a minimum, it cannot take more than two separate keys to release the 10 locks. Again, the intended staff levels of the facility need to be considered (see Commentary Figure 408.4.1).

[F] 408.4.2 Power-operated doors and locks. Power-operated sliding doors or power-operated locks for swinging

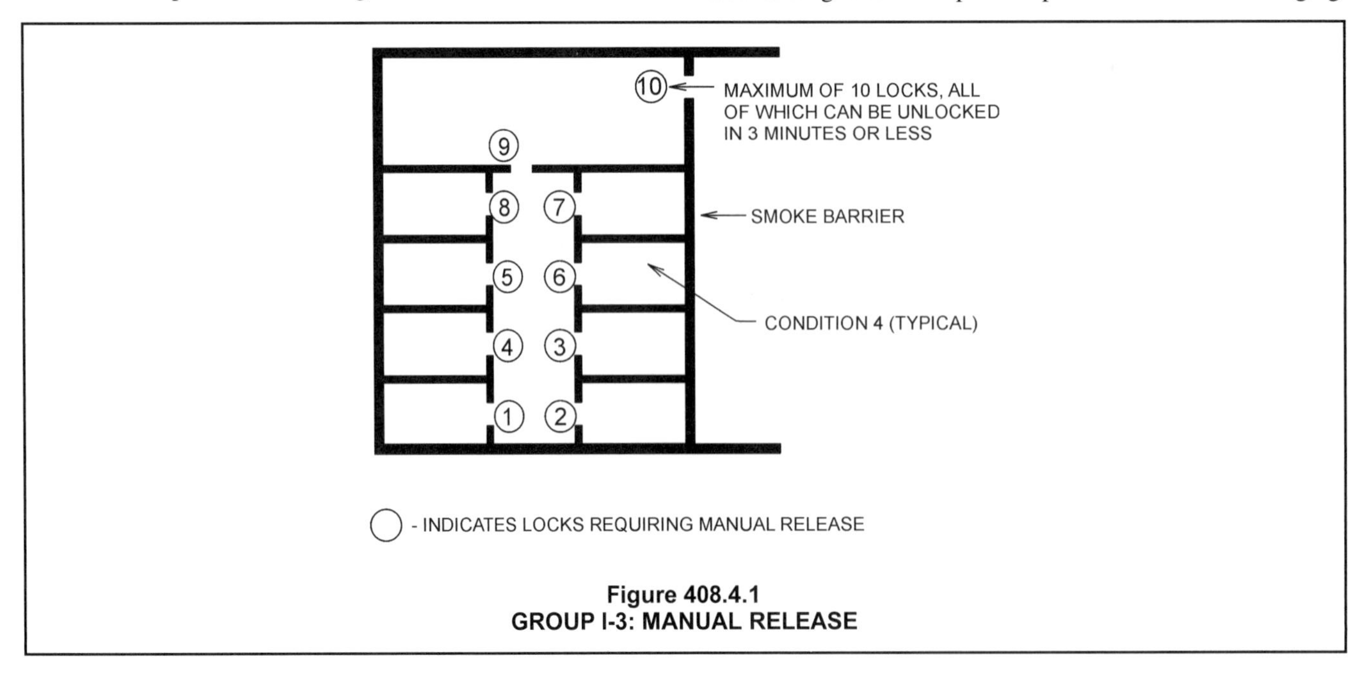

Figure 408.4.1
GROUP I-3: MANUAL RELEASE

doors shall be operable by a manual release mechanism at the door. Emergency power shall be provided for the doors and locks in accordance with Section 2702.

Exceptions:

1. Emergency power is not required in facilities with 10 or fewer locks complying with the exception to Section 408.4.1.
2. Emergency power is not required where remote mechanical operating releases are provided.

❖ To increase the likelihood that the doors in the means of egress can be operated even during a power failure, two alternative locking arrangements are required wherever the doors or locks are electrically operated. First, a manual release mechanism is to be provided at the door. Also, a remote backup (second) mechanical operating door release system must be provided or the electrical door control system must be provided with an emergency power source. If provided, the mechanical operating door release system must be capable of being operated from outside the residential housing area.

This section does not apply to manually operated doors and locks that meet the requirements of the exception to Section 408.4.1.

408.4.3 Redundant operation. Remote release, mechanically operated sliding doors or remote release, mechanically operated locks shall be provided with a mechanically operated release mechanism at each door, or shall be provided with a redundant remote release control.

❖ If a mechanically operated remote release mechanism is provided as the remote release system, either a second, redundant system is required or each remote release door or lock is to be provided with a mechanically operated device at the door. The redundant operation is essential to increase the likelihood that the doors will continue to operate in the event of failure of the primary locking system. If provided, the redundant mechanical operating door release system must also be operable from outside the residential housing area.

This provision does not apply to power-operated doors and locks that are required to have alternative systems in accordance with Section 408.4.2.

408.4.4 Relock capability. Doors remotely unlocked under emergency conditions shall not automatically relock when closed unless specific action is taken at the remote location to enable doors to relock.

❖ So as to maintain means of egress doors in an open position during an emergency condition, once a door in a means of egress is unlocked via an emergency unlocking system, the door must not automatically relock when closed unless a specific action is taken at the remote location. The normal operation of such doors and locks is typically arranged so that the door will automatically relock upon closure.

Many electronic-locking systems operate such that the door automatically relocks upon closure for security reasons. To comply with the provisions of this section, however, a specific switch is usually provided that serves as an emergency control and does not permit the door to relock without operating it. This section does not require that doors automatically release upon activation of the fire alarm system.

408.5 Protection of vertical openings. Any vertical opening shall be protected by a *shaft enclosure* in accordance with Section 713, or shall be in accordance with Section 408.5.1.

❖ This section requires that vertical openings be protected in the manner established in Section 713. The section allows compliance with Section 408.5.1 as an alternative within a housing unit. If the provisions of Section 408.5.1 for floor openings are used, then Section 408.5.2 regarding plumbing chases is also available for use in the design.

408.5.1 Floor openings. Openings in floors within a *housing unit* are permitted without a *shaft enclosure*, provided all of the following conditions are met:

1. The entire normally occupied areas so interconnected are open and unobstructed so as to enable observation of the areas by supervisory personnel;
2. *Means of egress* capacity is sufficient for all occupants from all interconnected *cell tiers* and areas;
3. The height difference between the floor levels of the highest and lowest *cell tiers* shall not exceed 23 feet (7010 mm); and
4. Egress from any portion of the *cell tier* to an *exit* or *exit access* door shall not require travel on more than one additional floor level within the *housing unit*.

❖ This section is essentially an exception to the protection of vertical openings required by Section 712. It allows openings between the various floor levels and cell tiers within a single housing unit. This provision applies regardless of whether the communicating floor levels are stories or mezzanines as long as they are within the same housing unit [see Commentary Figures 408.5(1) and (2)].

The following is a discussion of the conditions that must be met:

1. The areas must be sufficiently open and unobstructed so that a fire in one part will be immediately obvious to the occupants and supervisory personnel in the area. The intent of the provision is not to require open cell fronts; rather, it is to enable the staff to observe events occurring on all levels from one location. This arrangement is usually desirable from a security standpoint [see Commentary Figure 408.5(1)].
2. The exit capacity required is determined based on the occupant load of all floor levels exiting simultaneously. It is not unusual for most of the residents to be on the main level in the common space during periods when they are not required to be in the sleeping units (cells);

therefore, this level must be able to handle the potential occupant load of all cell tier levels.

3. The height differential between the floor levels of highest and lowest cell tiers must not exceed 23 feet (7010 mm). The limitation of 23 feet (7010 mm) expands the general code limitations on unprotected floor openings beyond two stories, because in Group I-3 it is not uncommon to have more than two interconnected levels in this open arrangement; however, the effective limitation in number of cell tiers is three. As with atriums (see Section 404), an automatic sprinkler system is required to control the spread of smoke and fire through vertical openings. Beyond the specified dimension of 23 feet (7010 mm), effective unobstructed sight becomes less attainable [see Commentary Figure 408.5(2)].

4. To minimize the potential that a fire in the day room would block access to all exits, the exits must be arranged such that it is possible to reach an exit on each level, or from the level next above or below. This would permit, for example, a three-tiered housing unit to have exits or exit access doors out of the housing unit, which are located on the lowest level and mid-level only; or on the lowest level and the top level only. In these two scenarios, the residents have access to an exit either from the level they are on or on the next level directly above or below.

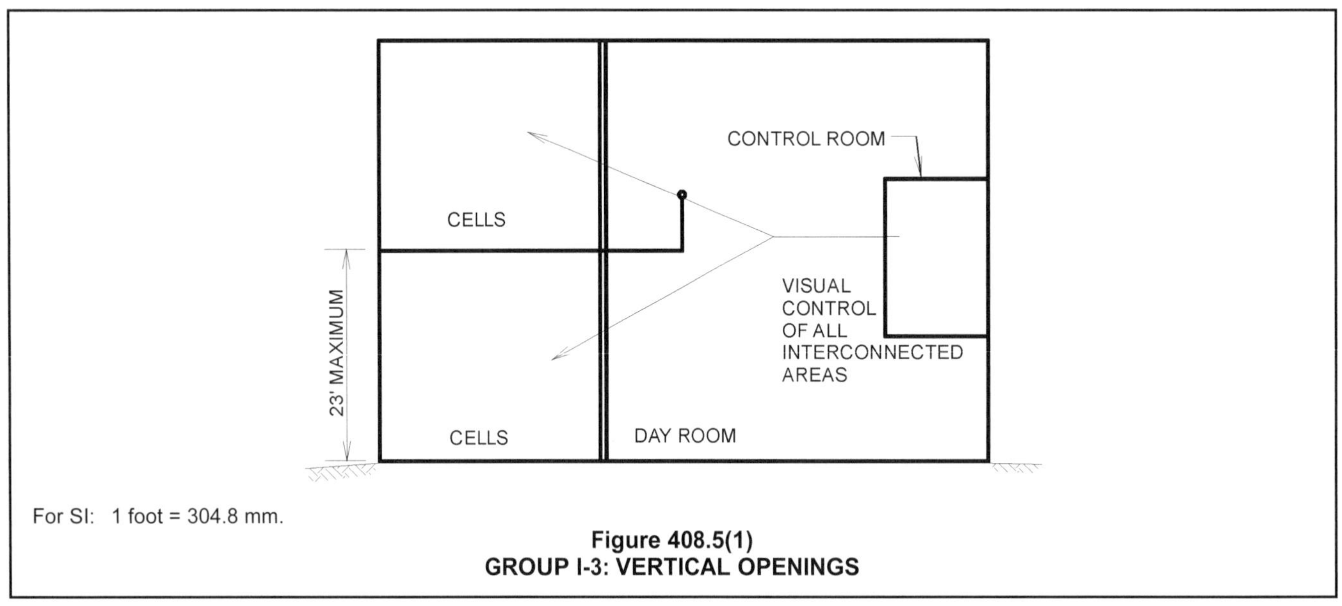

Figure 408.5(1)
GROUP I-3: VERTICAL OPENINGS

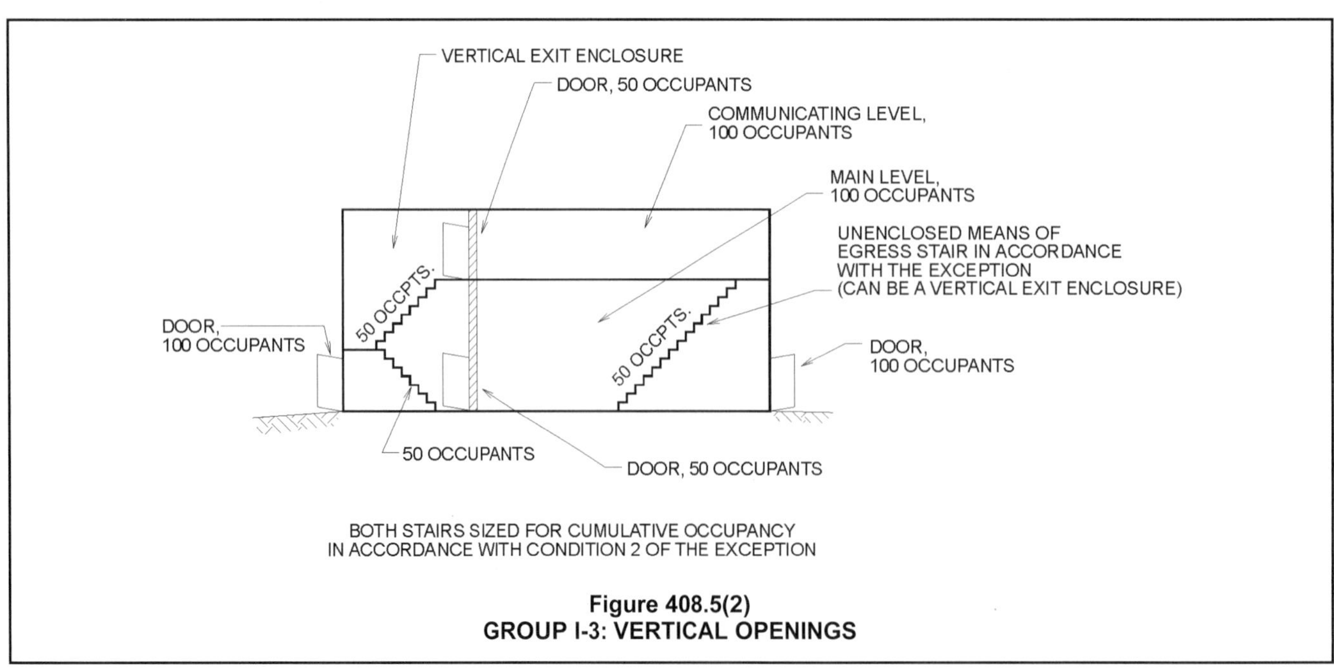

Figure 408.5(2)
GROUP I-3: VERTICAL OPENINGS

408.5.2 Shaft openings in communicating floor levels. Where a floor opening is permitted between communicating floor levels of a *housing unit* in accordance with Section 408.5.1, plumbing chases serving vertically staked individual *cells* contained with the *housing unit* shall be permitted without a *shaft enclosure*.

❖ Where a housing unit is designed in compliance with Section 408.5.1, this section allows plumbing chases within the same housing unit to also be unprotected where they penetrate through the floors separating cell tiers. If opening protection is provided within a housing unit under the provisions of Section 712, the provisions of this section cannot be used in a building design.

408.6 Smoke barrier. Occupancies in Group I-3 shall have *smoke barriers* complying with Sections 408.7 and 709 to divide every *story* occupied by residents for sleeping, or any other *story* having an *occupant load* of 50 or more persons, into no fewer than two *smoke compartments*.

Exception: Spaces having a direct *exit* to one of the following, provided that the locking arrangement of the doors involved complies with the requirements for doors at the *smoke barrier* for the use condition involved:

1. A *public way*.
2. A building separated from the resident housing area by a 2-hour fire-resistance-rated assembly or 50 feet (15 240 mm) of open space.
3. A secured *yard* or *court* having a holding space 50 feet (15 240 mm) from the housing area that provides 6 square feet (0.56 m^2) or more of refuge area per occupant, including residents, staff and visitors.

❖ One of the basic premises of this section is that evacuation of a Group I-3 facility is impractical because of the need to maintain security. The security measures may also result in a delayed evacuation as compared to residential occupancies in which free egress is allowed. Therefore, it is essential to provide a refuge area to which residents can be relocated when it becomes necessary to evacuate them from the housing units. For this reason, each cell tier and each floor with an occupant load of 50 or more is required to be divided into at least two smoke compartments using smoke barriers that comply with Sections 408.8 and 709 (see commentary, Section 709). It should be noted that Section 709 provides several exceptions to smoke barriers used in Group I-3 occupancies (e.g., construction, doors and opening protectives). A further exception allowing increased glazing is permitted by the provisions of Section 408.7.

The exception provides three alternatives that are common-sense options to movement into a smoke compartment. If the particular occupancy condition affords free, unobstructed access with no delay to the exterior, to a separate building or through a horizontal exit, then defend-in-place provisions are not needed.

408.6.1 Smoke compartments. The number of residents in any *smoke compartment* shall be not more than 200. The distance of travel to a door in a *smoke barrier* from any room door required as exit access shall be not greater than 150 feet (45 720 mm). The distance of travel to a door in a *smoke barrier* from any point in a room shall be not greater than 200 feet (60 960 mm).

❖ This section defines the maximum allowable size of smoke compartments by limiting the number of residents in any single smoke compartment to 200, the maximum door-to-door distance within the compartment to 150 feet (45 720 mm) and the maximum total travel distance to 200 feet (60 960 mm). These limitations are intended to enable evacuation of a smoke compartment for all of the occupants (see Commentary Figure 408.6.1).

408.6.2 Refuge area. Not less than 6 net square feet (0.56 m^2) per occupant shall be provided on each side of each *smoke barrier* for the total number of occupants in adjoining *smoke compartments*. This space shall be readily available wherever the occupants are moved across the *smoke barrier* in a fire emergency.

❖ To provide adequate space for the residents evacuated to an adjacent smoke compartment, an additional 6 square feet (0.56 m^2) per person must be readily available for the total number of occupants in the adjoining smoke compartment (see Commentary Figure 408.6.2). The 6-square-foot (0.56 m^2) criterion enables the occupants to be evacuated to the adjacent smoke compartment without delay from crowding.

In a fire emergency, any action that must be taken to unlock the space will result in a delay in moving occupants into the refuge area. As such, the code requires the refuge area to be readily available in order to preclude consideration (as a refuge area) of a space that may be normally locked and, therefore, not immediately usable.

408.6.3 Independent egress. A *means of egress* shall be provided from each *smoke compartment* created by *smoke barriers* without having to return through the *smoke compartment* from which *means of egress* originates.

❖ To prevent creating a dead-end smoke compartment, exits are to be arranged so as to permit access without returning through a compartment from which egress originated. An exit is not required from within each smoke compartment [see Commentary Figures 407.5.2(1) and (2) for acceptable and unacceptable exit arrangements].

408.7 Security glazing. In occupancies in Group I-3, windows and doors in 1-hour *fire barriers* constructed in accordance with Section 707, *fire partitions* constructed in accordance with Section 708 and *smoke barriers* constructed in accordance with Section 709 shall be permitted to have security glazing installed provided that the following conditions are met.

1. Individual panels of glazing shall not exceed 1,296 square inches (0.84 m^2).

2. The glazing shall be protected on both sides by an *automatic sprinkler system*. The sprinkler system shall be designed to, when actuated, wet completely the entire surface of any glazing affected by fire.
3. The glazing shall be in a gasketed frame and installed in such a manner that the framing system will deflect without breaking (loading) the glass before the sprinkler system operates.
4. Obstructions, such as curtain rods, drapery traverse rods, curtains, drapes or similar materials shall not be installed between the automatic sprinklers and the glazing.

❖ This section allows sprinkler-protected glazing in walls in a Group I-3 facility provided the wall is rated for 1 hour or less. As with many provisions in Section 408, this enhances the security of a facility by provid-

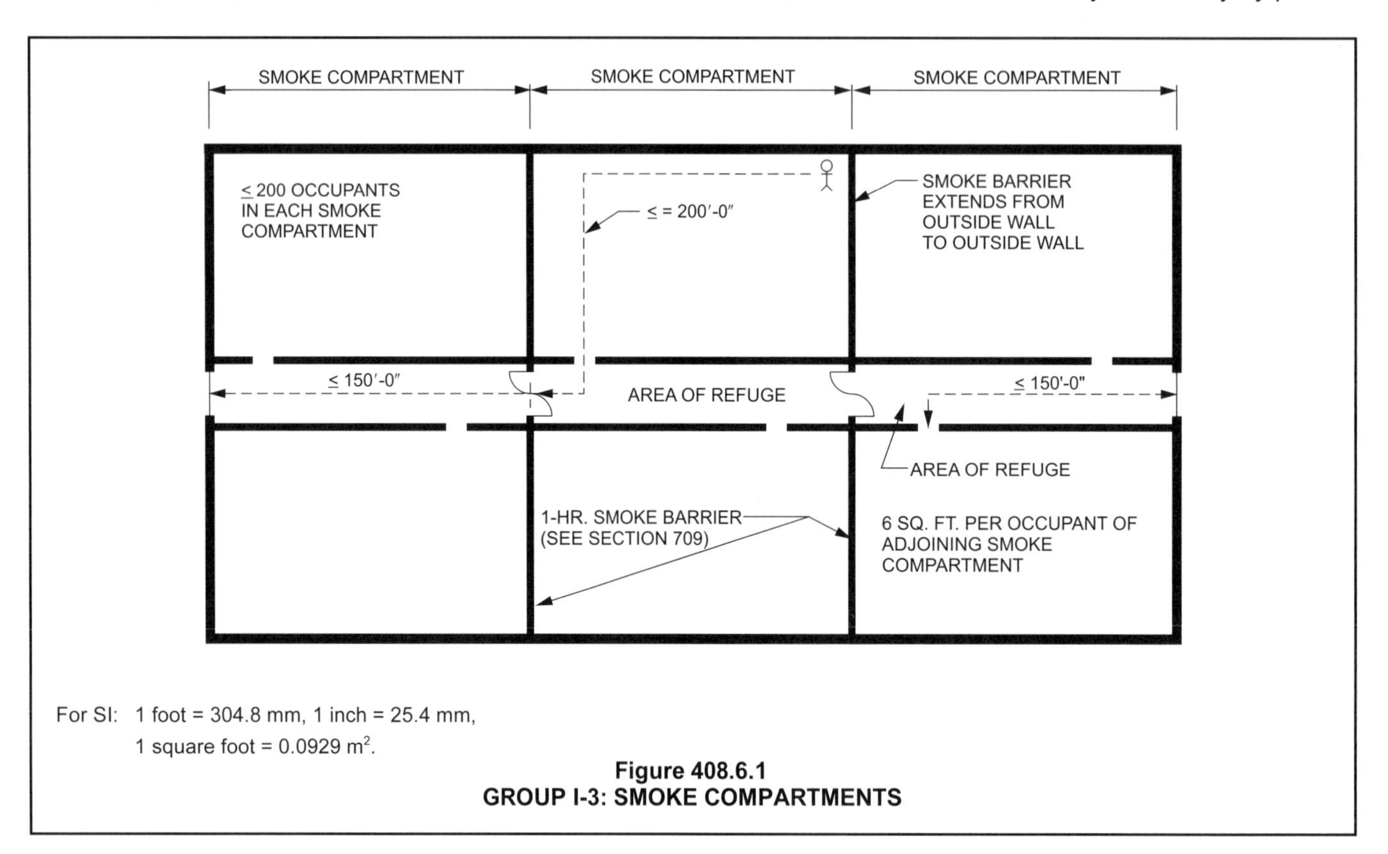

For SI: 1 foot = 304.8 mm, 1 inch = 25.4 mm, 1 square foot = 0.0929 m^2.

Figure 408.6.1
GROUP I-3: SMOKE COMPARTMENTS

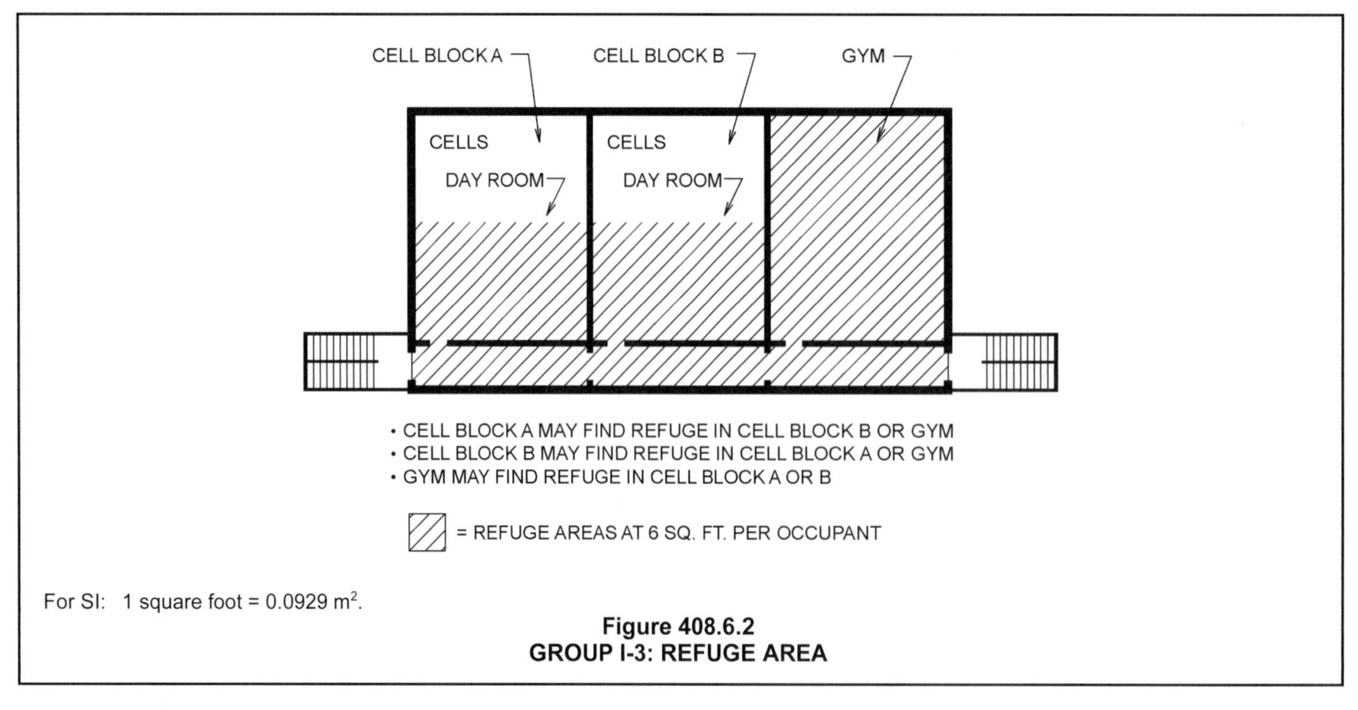

For SI: 1 square foot = 0.0929 m^2.

Figure 408.6.2
GROUP I-3: REFUGE AREA

ing more opportunities for staff to monitor the activities of residents

408.8 Subdivision of resident housing areas. Sleeping areas and any contiguous day room, group activity space or other common spaces where residents are housed shall be separated from other spaces in accordance with Sections 408.8.1 through 408.8.4.

❖ Resident housing areas usually consist of sleeping areas and other contiguous common area spaces, such as day rooms or activity areas onto which sleeping areas open. The separation required by this section is intended to separate housing areas, where the occupants in those areas may be asleep, from other activity areas since people asleep are more vulnerable to a fire emergency. The separation requirements are based on the relative evacuation difficulty as determined by the occupancy condition. Sections 408.8.1 through 408.8.4 give the separation requirements for the various occupancy conditions for which separation is necessary.

408.8.1 Occupancy Conditions 3 and 4. Each sleeping area in Occupancy Conditions 3 and 4 shall be separated from the adjacent common spaces by a smoke-tight partition where the distance of travel from the sleeping area through the common space to the *corridor* exceeds 50 feet (15 240 mm).

❖ A smoke-tight partition is required between the sleeping areas and common areas in Conditions 3 and 4 when the distance of travel from the sleeping area through the common area to an exit access corridor exceeds 50 feet (15 240 mm). This requirement recognizes the need to protect sleeping areas from non-sleeping areas due to the increased hazards associated with common areas, delayed recognition of fire hazards by sleeping persons and the restrictions on movement for these two occupancy conditions. Moreover, release from the sleeping area may be delayed and as such, evacuation with extended travel through a common space may not be safe. In such cases, the sleeping area will be a refuge area until the fire is controlled and the occupants in the sleeping area can be safely evacuated.

It should be noted that the provisions contained in Section 1016.2 are applicable to this circumstance, and as such, sleeping areas, including those in Group I-3 occupancies, are precluded from egressing through another sleeping area. Additionally, egress is not allowed from sleeping areas through toilet rooms.

408.8.2 Occupancy Condition 5. Each sleeping area in Occupancy Condition 5 shall be separated from adjacent sleeping areas, *corridors* and common spaces by a smoke-tight partition. Additionally, common spaces shall be separated from the *corridor* by a smoke-tight partition.

❖ Given that movement from an occupied space in Condition 5 is even more restricted than for the other conditions, smoke-tight partitions are required to be installed between areas, sleeping areas, common spaces and the exit access corridor. Protection is provided because the staff-controlled release from the sleeping area is not as immediate as in other conditions.

408.8.3 Openings in room face. The aggregate area of openings in a solid sleeping room face in Occupancy Conditions 2, 3, 4 and 5 shall not exceed 120 square inches ($0.77 m^2$). The aggregate area shall include all openings including door undercuts, food passes and grilles. Openings shall be not more than 36 inches (914 mm) above the floor. In Occupancy Condition 5, the openings shall be closeable from the room side.

❖ Food pass-throughs, door undercuts and grilles are a functional necessity; therefore, openings are limited in size and close to the floor. The openings are required to be close to the floor to minimize the potential for smoke to pass through the opening. By placing the opening close to the floor, smoke will not pass through a smoke-tight barrier initially; rather, there will be some time delay as the smoke layer descends to 36 inches (914 mm) above the floor before smoke is transmitted through such openings.

408.8.4 Smoke-tight doors. Doors in openings in partitions required to be smoke tight by Section 408.8 shall be substantial doors, of construction that will resist the passage of smoke. Latches and door closures are not required on *cell* doors.

❖ If the partition is required to be capable of resisting the passage of smoke, door openings within the enclosure are also to be protected with an assembly that is capable of resisting the passage of smoke. As with other sections of the code that deal with doors and walls capable of resisting the passage of smoke, application of the section requires judgment by the building official. The intent is a solid door surface without louvered openings. Door closers and latches are not required. It is anticipated that the security lock will secure the door in the closed position. A door closer can interfere with the removal of the occupants not only in an emergency but also during normal operations.

408.9 Windowless buildings. For the purposes of this section, a windowless building or portion of a building is one with nonopenable windows, windows not readily breakable or without windows. Windowless buildings shall be provided with an engineered smoke control system to provide a tenable environment for exiting from the *smoke compartment* in the area of fire origin in accordance with Section 909 for each windowless *smoke compartment*.

❖ An engineered smoke control system is required for smoke compartments in which there are no openings through which the products of combustion can be vented. The intent of the system is to provide a tenable environment during the period it takes the occupants to egress from a smoke compartment that is the area of fire origin. As it often takes longer to evacuate a Group I-3 facility due to the locking and restricted egress paths, the maintenance of a tenable environment is essential. As defined for this section only, a windowless building is a building or portion thereof with only nonoperable windows, windows that are not readily breakable or without windows. The intent of this section is that staff must have some means to ventilate the products of combustion; therefore, if the window cannot be broken by items readily available to the staff, the area is considered windowless.

[F] 408.10 Fire alarm system. A *fire alarm* system shall be provided in accordance with Section 907.2.6.3.

❖ Fire alarm systems are required in Group I-3 facilities. See the commentary to Section 907.2.6.3 for detailed requirements for this system.

[F] 408.11 Automatic sprinkler system. Group I-3 occupancies shall be equipped throughout with an *automatic sprinkler system* in accordance with Section 903.2.6.

❖ Automatic sprinkler systems are required in all Group I-3 facilities. See the commentary to Section 903.2.6 for detailed requirements for this system.

SECTION 409 MOTION PICTURE PROJECTION ROOMS

409.1 General. The provisions of Sections 409.1 through 409.5 shall apply to rooms in which ribbon-type cellulose acetate or other safety film is utilized in conjunction with electric arc, xenon or other light-source projection equipment that develops hazardous gases, dust or radiation. Where cellulose nitrate film is utilized or stored, such rooms shall comply with NFPA 40.

❖ Cellulose acetate or other safety film used in the presence of potential ignition sources, including projectors or related equipment, creates a potential hazard. The provisions of this section are intended to minimize the potential for exposure of the audience and other occupants to the hazards associated with the presence of ignition sources in close proximity to the fuel load.

Provisions are established for projection rooms in recognition of the hazards attendant to the projection of ribbon-type cellulose acetate or other safety film in such spaces.

Safety film is typically made of cellulose acetate, polyester or triacetate film. In the early 1950s, safety film began to replace cellulose nitrate film. Although the use and manufacture of cellulose nitrate film has virtually ceased in the United States, large quantities of such film still exist, primarily for archival purposes. Cellulose nitrate film presents a serious hazard because degradation under external heat below ignition temperatures causes a chemical reaction. This chemical reaction releases heat of sufficient quantity to raise the substance to ignition temperatures, resulting in spontaneous combustion. After ignition, rapid burning and the production of toxic and flammable gases takes place. If such film is used, the provisions of NPFA 40 are applicable as are the provisions of Section 306 of the IFC.

The provisions of this section apply only where the previously mentioned hazards exist. The use of ribbon-type cellulose acetate or other safety film in conjunction with electric arc, xenon or other light-source projection equipment that develops hazardous gases, dust or radiation invokes these provisions. The provisions of this section do not apply where nonprofessional projection equipment is used, except that statements specific to cellulose nitrate film apply if such film is present regardless of the nature of the equipment. The safety film used by the motion picture industry since 1951 has fire hazard characteristics similar to ordinary paper of the same thickness and form.

If cellulose nitrate film is used, the provisions of NFPA 40 apply. NFPA 40 contains minimum requirements to provide a reasonable level of protection for the storage and handling of cellulose nitrate film. The standard does not contain provisions for the manufacture of cellulose nitrate film, since it has not been manufactured in the United States since 1951.

409.1.1 Projection room required. Every motion picture machine projecting film as mentioned within the scope of this section shall be enclosed in a projection room. Appurtenant electrical equipment, such as rheostats, transformers and generators, shall be within the projection room or in an adjacent room of equivalent construction.

❖ The projection room is to be enclosed in accordance with Section 409.2 to isolate the spaces housing the potential hazard. To minimize the potential for arcs or sparks to come in contact with the film, electrical equipment, such as rheostats, transformers and generators, should be located in a separate room. This section permits such equipment to be in the same room, but it should be arranged to minimize the potential for arcs or sparks to come in contact with the film, reducing the potential for ignition. Additional requirements for electrical equipment permitted in the projection room and requirements for nonprofessional equipment are given in NFPA 70.

409.2 Construction of projection rooms. Every projection room shall be of permanent construction consistent with the construction requirements for the type of building in which the projection room is located. Openings are not required to be protected.

The room shall have a floor area of not less than 80 square feet (7.44 m^2) for a single machine and not less than 40 square feet (3.7 m^2) for each additional machine. Each motion picture projector, floodlight, spotlight or similar piece of equipment shall have a clear working space of not less than 30 inches by 30 inches (762 mm by 762 mm) on each side and at the rear thereof, but only one such space shall be required between two adjacent projectors. The projection room and the rooms appurtenant thereto shall have a ceiling height of not less than 7 feet 6 inches (2286 mm). The aggregate of openings for projection equipment shall not exceed 25 percent of the area of the wall between the projection room and the auditorium. Openings shall be provided with glass or other *approved* material, so as to close completely the opening.

❖ The projection room is to be constructed as required by the construction type of the building. Since the room dimensions are small, ventilation is provided to reduce heat buildup and accumulation of dust particles. A minimum fire-resistance rating is not required unless by other provisions of the code. Limited openings are permitted and are not required to be protected so as to facilitate the intended function of projecting the image onto the auditorium screen. Openings to the auditorium must not exceed 25 percent of the area of the wall between the projection room and the auditorium. Glass or other approved material must be provided that will completely close the opening and prevent a sharing of atmospheres with the audience or seating area.

In order to minimize the buildup of heat in any area, adequate floor space must be available. The space is required to be at least 80 square feet (7.44 m^2) in size for a single projector and an additional 40 square feet (3.7 m^2) for each additional projector. The room must be of adequate size so that not less than 30 inches by 30 inches (762 mm by 762 mm) of clear working space is available on each side and behind the projectors, floodlights, spotlights and similar equipment. For adjacent equipment, a clear space of 30 inches (762 mm) is adequate.

Because the projection room is considered an occupiable room, the ceiling height must not be less than 7 feet, 6 inches (2286 mm) (see commentary, Section 1208.2).

409.3 Projection room and equipment ventilation. *Ventilation* shall be provided in accordance with the *International Mechanical Code*.

❖ All projection rooms and the equipment located within the rooms must be properly ventilated to abate the hazards of any hot gases and smoke generated by the equipment. Further, any intense heat given off by the equipment must be directly exhausted to the exterior rather than be absorbed by the building's ventilation system. Sections 502.11.1 and 502.11.2 of the IMC contain requirements for projectors provided with an exhaust discharge and those without an exhaust connection.

409.3.1 Supply air. Each projection room shall be provided with adequate air supply inlets so arranged as to provide well-distributed air throughout the room. Air inlet ducts shall provide an amount of air equivalent to the amount of air being exhausted by projection equipment. Air is permitted to be taken from the outside; from adjacent spaces within the building, provided the volume and infiltration rate is sufficient; or from the building air-conditioning system, provided it is so arranged as to provide sufficient air when other systems are not in operation.

❖ Separate supply (see Section 409.3.1) and exhaust (see Section 409.3.2) requirements are provided for projection rooms. Exhaust rates are specified in Section 502.11.2 of the IMC for projectors without exhaust connection.

Sufficient air must be introduced into each projection room to allow the effects of hot gases, smoke and heat to be diluted and to provide a makeup air source for the exhausted air. Although the code does not prescriptively state where the air supply inlets are to be located, the intent is to create a well-ventilated room. Several air supply inlets should be provided so that the entire room has adequate air changes. The amount of air supplied to each projection room must be approximately equal to the volume of air that is exhausted according to the IMC. Makeup supply air can be obtained from any source as long as sufficient quantities are provided at all times that the exhaust system is operating.

409.3.2 Exhaust air. Projection rooms are permitted to be exhausted through the lamp exhaust system. The lamp exhaust system shall be positively interconnected with the lamp so that the lamp will not operate unless there is the required airflow. Exhaust air ducts shall terminate at the exterior of the building in such a location that the exhaust air cannot be readily recirculated into any air supply system. The projection room *ventilation* system is permitted to also serve appurtenant rooms, such as the generator and rewind rooms.

❖ Another method of providing exhaust is to exhaust the room itself. This method can only be used when the projectors do not have an integral lamp exhaust system. Exhaust rates for this type of system are specified in the IMC based on the type of projector. The exhaust ducts for the projection room must be terminated at the exterior of the building and cannot be combined with any other exhaust or return system present in the building. The intent of this requirement is so that the hot gases, smoke and heat will not be allowed to be disbursed elsewhere in the building.

409.3.3 Projection machines. Each projection machine shall be provided with an exhaust duct that will draw air from each lamp and exhaust it directly to the outside of the building. The lamp exhaust is permitted to serve to exhaust air from the projection room to provide room air circulation. Such ducts shall be of rigid materials, except for a flexible connector *approved* for the purpose. The projection lamp or projection room exhaust system, or both, is permitted to be combined but shall not be interconnected with any other exhaust or return system, or both, within the building.

❖ Projectors with an integral lamp exhaust system use air to cool the equipment during operation. The amount of air exhausted with these systems is part of the manufacturer's installation instructions. These integral exhaust systems must be directly connected to the mechanical exhaust systems. Further, the lamp exhaust system must be interconnected to the projector's power supply such that when the projector is operating, its exhaust system will be automatically activated.

409.4 Lighting control. Provisions shall be made for control of the auditorium lighting and the *means of egress* lighting systems of theaters from inside the projection room and from not less than one other convenient point in the building.

❖ The projection room is usually an attended location, but it is also a likely area for a fire to originate. Lighting controls that provide adequate normal and emergency auditorium lighting must be provided in the projection room. Upon detection of a fire anywhere in the building, the operator can immediately provide adequate illumination for the occupants egressing the auditorium. A second location for the lighting controls is also required somewhere else in the building, so that entry into the projection room is not required to provide the necessary illumination, as it could be the location of fire origin. This second control is required to be conveniently located for the staff yet inaccessible to unauthorized persons.

409.5 Miscellaneous equipment. Each projection room shall be provided with rewind and film storage facilities.

❖ In order to keep all operations related to projection activities in the protected enclosure, rewind and film storage facilities are to be located in the projection room or in an adjacent room in accordance with Section 409.1.1.

SECTION 410 STAGES, PLATFORMS AND TECHNICAL PRODUCTION AREAS

410.1 Applicability. The provisions of Sections 410.1 through 410.8 shall apply to all parts of buildings and structures that contain *stages* or *platforms* and similar appurtenances as herein defined.

❖ Historically, most significant theater fires originated on the stage. The 1903 Iroquois Theater fire in Chicago serves as a vivid example of a stage fire and its potentially tragic effects—602 people lost their lives. Hazards associated with stages include: combustible scenery and lighting suspended overhead; scenic elements, contents and acoustical treatment on the back and sides of the stage; technical production areas including workshops, scene docks and dressing rooms located around the stage perimeter; and storage areas and property rooms located underneath the stage.

The protection requirements set forth in this section are intended to limit the threat from stage fires to an audience and reduce the likelihood of a large fire in the stage area. These provisions include construction restrictions, automatic sprinkler systems, ventilation, separation of the stage from the audience and compartmentalization of appurtenant rooms to the stage. Special allowances are provided for galleries, gridirons, catwalks and other technical production areas.

Based on historical events and the expertise of a broad range of professionals with experience in theater design, operations and fire events, a set of principles guides the requirements for theaters. These principles are summarized as follows:

1. The fire hazards of stages are not necessarily a function of type (legitimate, regular or thrust) but rather area and height. Accordingly, a stage area in excess of 1,000 square feet (93 m^2) or a height in excess of 50 feet (15 240 mm) is the threshold that represents a significant potential for fuel load due to scenery and drops.
2. Stages that exceed 1,000 square feet (93 m^2) or the 50-foot (15 240 mm) height threshold require a means of emergency ventilation.
3. Stages that exceed 1,000 square feet (93 m^2) and the 50-foot (15 240 mm) height threshold require automatic sprinkler protection.
4. Stages and platforms are similar to floor construction; therefore, the type of construction should be consistent with that of the floor construction in the building.
5. Separation of the stage from the seating area, dressing rooms, property rooms and similar spaces is critical in order to provide a degree of fire containment.

Section 410.3 contains provisions for the construction of the stage and gallery. Stage openings, decorations, equipment and scenery are addressed in Sections 410.3.3 through 410.3.6. Stage ventilation is required in accordance with Section 410.3.7. Platforms must be constructed in accordance with Section 410.4. The construction of auxiliary stage spaces is addressed in Section 410.5. Special means of egress requirements for stages, platforms and technical production areas are established in Section 410.6. Sections 410.7 and 410.8 contain the requirements for automatic sprinkler systems and standpipes, respectively.

All parts of a building or structure that contain a stage, platform or similar appurtenance are required to comply with the provisions of this section.

410.2 Definitions. The following terms are defined in Chapter 2:

PLATFORM.

PROSCENIUM WALL.

STAGE.

TECHNICAL PRODUCTION AREA.

❖ This section lists terms that are specifically associated with the subject matter of Section 410. It is important to emphasize that these terms are not exclusively related to this chapter but may or may not also be applicable where the term is used elsewhere in the code.

Definitions of terms can help in the understanding and application of the code requirements. The purpose for including a list within this chapter is to provide more convenient access to terms which may have a specific or limited application within this chapter. For the complete definition and associated commentary, refer back to Chapter 2. Terms that are italicized provide a visual identification throughout the code that a definition exists for that term. The use and application of all defined terms are set forth in Section 201.

410.3 Stages. *Stage* construction shall comply with Sections 410.3.1 through 410.3.7.

❖ This section indicates the scope of the code provisions that apply to stage construction.

410.3.1 Stage construction. *Stages* shall be constructed of materials as required for floors for the type of construction of the building in which such *stages* are located.

Exception: *Stages* need not be constructed of the same materials as required for the type of construction provided the construction complies with one of the following:

1. *Stages* of Type IIB or IV construction with a nominal 2-inch (51 mm) wood deck, provided that the *stage* is separated from other areas in accordance with Section 410.3.4.
2. In buildings of Type IIA, IIIA and VA construction, a fire-resistance-rated floor is not required, provided the space below the *stage* is equipped with an *automatic sprinkler system or fire-extinguishing system* in accordance with Section 903 or 904.
3. In all types of construction, the finished floor shall be constructed of wood or *approved* noncombustible materials. Openings through *stage* floors shall be equipped with tight-fitting, solid wood trap doors with *approved* safety locks.

❖ Stages, most simply, are floor construction, and as such, must conform to the requirements for all other floors in the building so that the building's designated construction classification will be consistent throughout.

Since wood is the material of choice for stage floors because it allows for nailing of sets and props and provides the best acoustics for dance and other performances, the code allows three exceptions to the requirement for stage construction.

Exception 1 permits stages in buildings of any construction classification to be constructed with a nominal 2-inch (51 mm) wood deck supported by either unprotected noncombustible construction (Type IIB) or heavy timber construction (Type IV) where the stage is separated from the audience by a proscenium wall built in accordance with Section 410.3.4.

Under Exception 2, stage floors are not required to be fire-resistance rated in buildings of Types IIA, IIIA and VA where the space below the stage is equipped with an automatic sprinkler system or fire-extinguishing system installed in accordance with Section 903 or 904. In this case, the requirement for fire-resistance-rated floors is offset by the protection supported by the automatic fire-extinguishing system.

Exception 3 recognizes that stages are predominantly constructed with a finished floor of wood and identifies wood as an acceptable finish material for all construction classifications. It also permits openings in stage floors to be protected with tight-fitting solid wood doors with safety locks to secure them. The normal operation of stage trap doors does not typically allow them to have a fire-resistance rating.

410.3.1.1 Stage height and area. *Stage* areas shall be measured to include the entire performance area and adjacent backstage and support areas not separated from the performance area by fire-resistance-rated construction. *Stage* height shall be measured from the lowest point on the *stage* floor to the highest point of the roof or floor deck above the *stage*.

❖ As previously stated, many of the code's requirements for stages and platforms are triggered based on the volume of the stage or platform space. All of the wing areas and backstage areas are included as the stage area, unless they are separated with fire-resistance-rated construction. Stage height is not the height of the opening in the proscenium wall, but the height from the stage floor to the highest horizontal assembly that encloses the stage space. It will typically include the fly gallery and fly loft above the stage.

410.3.2 Technical production areas: galleries, gridirons and catwalks. Beams designed only for the attachment of portable or fixed theater equipment, gridirons, galleries and catwalks shall be constructed of *approved* materials consistent with the requirements for the type of construction of the building; and a *fire-resistance rating* shall not be required. These areas shall not be considered to be floors, *stories*, *mezzanines* or levels in applying this code.

Exception: Floors of fly galleries and catwalks shall be constructed of any *approved* material.

❖ This section identifies the components of the technical production areas located above the stage that are used to conceal movable scenery from the audience and where the operation of such scenery and other stage effects is controlled.

The normal use and operation of a stage requires that equipment, normally rigging, be installed and rearranged as production requirements vary. This equipment is normally installed by clamping or welding to the structural framing over and around the stage, making protecting the framing by encasement or membranes infeasible; hence, a fire-resistance rating is not required for these elements. In order to control the fuel load in this area, however, the code requires that all of the elements must be of approved materials consistent with the requirements of the building's construction type. The exception to this requirement is that in all cases, the floors of fly galleries and catwalks may be constructed of any approved material regardless of the building's construction type. Such spaces are generally limited in area, access is restricted to authorized personnel and they require floor materials that will deaden sound so as not to disrupt the performance below. These auxiliary areas do not have to meet the code's requirements as another story, mezzanine level or equipment platforms.

410.3.3 Exterior stage doors. Where protection of openings is required, exterior *exit* doors shall be protected with *fire door assemblies* that comply with Section 716. Exterior openings that are located on the *stage* for *means of egress* or loading and unloading purposes, and that are likely to be open during occupancy of the theater, shall be constructed with vestibules to prevent air drafts into the auditorium.

❖ If exterior opening protectives are required, exit discharge doors directly from the stage to the outside must be fire door assemblies in accordance with Section 716.

Since one of the major concerns is containing a stage fire to the stage area, any exterior opening from the stage that is likely to be opened during a performance is to be constructed with a vestibule to prevent air drafts into the auditorium. This vestibule requirement applies to all exterior openings that are likely to be open regardless of their intended use.

410.3.4 Proscenium wall. Where the *stage* height is greater than 50 feet (15 240 mm), all portions of the *stage* shall be completely separated from the seating area by a proscenium wall with not less than a 2-hour *fire-resistance rating* extending continuously from the foundation to the roof.

❖ The protection afforded by a 2-hour fire separation and opening protection (see Section 410.3.5) is required where the stage height is greater than 50 feet (15 240 mm). Stages with a height greater than 50 feet (15 240 mm) permit multiple settings and large amounts of scenery and scenic elements in dense configurations. The height may reduce the effectiveness of the suppression system and the multiple settings hung over the stage may further obstruct the suppression system and impede access to a fire originating high above the stage. Stages with a height less than 50 feet (15 240 mm) do not require separation from the audience since they represent a limited fuel load potential caused by scenery, drops and curtains.

410.3.5 Proscenium curtain. Where a proscenium wall is required to have a *fire-resistance rating*, the *stage* opening shall be provided with a fire curtain complying with NFPA 80, horizontal sliding doors complying with Section 716.5.2 having a fire protection rating of at least 1 hour, or an *approved* water curtain complying with Section 903.3.1.1 or, in facilities not utilizing the provisions of smoke-protected assembly seating in accordance with Section 1029.6.2, a smoke control system complying with Section 909 or natural *ventilation* designed to maintain the smoke level not less than 6 feet (1829 mm) above the floor of the *means of egress*.

❖ Fuel loads in a stage with a height greater than 50 feet (15 240 mm) most often are high and the resulting fire is severe. In such cases, to permit the audience to evacuate the seating area without being threatened directly or indirectly by a stage fire, this section requires the proscenium opening be protected. The code specifies three options: 1. A fire curtain (complying with NFPA 80); 2. Horizontal sliding doors which are fire door assemblies complying with Section 716.5.2; or 3. An approved water curtain. In lieu of directly protecting the opening, the theater design can provide either a smoke control system or natural ventilation designed to prevent the accumulation of smoke within the first 6 feet (1829 mm) above the floor. These smoke control or ventilation options are only permissible if a smoke-protected assembly seating design in accordance with Section 1029.6.2 is not utilized.

In accordance with NFPA 80, a fire curtain must be capable of preventing the passage of flame or smoke, thereby providing a certain level of protection for the audience when a stage fire occurs. It is notable that the fire curtain is not required to provide the same level of protection as is typically required for a fire door assembly. Also, since one of the goals is to protect against panic, flame and smoke must not be capable of passing through to the unexposed side of the fire curtain. The level of protection for the fire curtain is based upon evacuation time of the audience. Fire curtains are designed to provide 20 minutes of protection to allow the audience to evacuate.

NFPA 80 requirements for the fire curtain include, but are not limited to: testing of the fabric (NFPA 701, Test Method 2), a fire curtain sample assembly subjected to the standard fire test as specified in NFPA 251, activation (emergency and manual) and rate of closing speed. The speed of descent must be slowed for the last 8 feet (2438 mm) of travel in order to allow individuals on the stage to move away from the curtain and to avoid injury from being hit by the batten in the bottom pocket of the curtain. It should be noted that only stages with a height greater than 50 feet (15 240 mm) are required to have a proscenium wall and curtain.

Horizontal sliding fire door assemblies can also be used to protect the opening. These doors are tested in accordance with either NFPA 252 or UL 10B. For this installation, the door must have a fire protection rating of not less than one hour. This is a longer protection period than required by the fire curtain option.

Water curtains conforming with Section 903.3.1.1 are viewed as providing an equivalent degree of protection to that of a fire curtain. This, in part, recognizes that emergency ventilation is also required in accordance with Section 410.3.7, which will assist in controlling smoke movement. The duration of protection is, of course, continuous as long as the sprinkler system is in operation.

Lastly, today's assembly facilities, such as arenas and theaters in the round, make the use of a traditional fire curtain impractical. As such, a smoke control system or a natural ventilation system may be utilized in lieu of methods directly protecting the stage opening, provided the special allowances for smoke-protected assembly seating are not being applied.

410.3.6 Scenery. Combustible materials used in sets and scenery shall meet the fire propagation performance criteria of Test Method 1 or Test Method 2, as appropriate, of NFPA 701, in accordance with Section 806 and the *International Fire Code*. Foam plastics and materials containing foam plastics shall comply with Section 2603 and the *International Fire Code*.

❖ This section recognizes that scenery and sets are decorative materials. As such, combustible materials used in sets and scenery must meet the fire propagation performance criteria in accordance with the provisions of NFPA 701, Section 806 and the IFC. Section 806 references Section 2604 for specific requirements for foam plastics used as trim. It should also be noted that Section 2603 addresses the uses of foam plastics in insulation. The materials when tested in accordance with NFPA 701 must comply with the performance criteria of one of the two identified test methods. The historic "small-scale test" is no longer accepted.

410.3.7 Stage ventilation. Emergency *ventilation* shall be provided for *stages* larger than 1,000 square feet (93 m^2) in floor area, or with a *stage* height greater than 50 feet (15 240 mm). Such *ventilation* shall comply with Section 410.3.7.1 or 410.3.7.2.

❖ In addition to an automatic sprinkler system, emergency ventilation is one of the key fire protection components for stages that contain large fuel loads. Stages with large areas or heights permit multiple settings and large amounts of scenery and scenic elements in dense configurations. Increased height also permits multiple settings to be hung over the stage, possibly reducing the effectiveness of the suppression system or impeding access to a fire originating high above the stage. Based on these factors and the potential hazard presented by them, stages larger than 1,000 square feet (93 m^2) in area or with a height greater than 50 feet (15 240 mm) are required to be equipped with emergency ventilation to control smoke movement and minimize the potential for fire and smoke to spread to the seating area.

410.3.7.1 Roof vents. Two or more vents constructed to open automatically by *approved* heat-activated devices and with an aggregate clear opening area of not less than 5 percent of the area of the *stage* shall be located near the center and above the highest part of the *stage* area. Supplemental means shall be provided for manual operation of the ventilator. Curbs shall be provided as required for skylights in Section 2610.2. Vents shall be *labeled*.

❖ Where vents are used to ventilate a stage, the code requires that there be a minimum of two vents provided for redundancy. Those vents are to be located near the center and above the highest part of the stage, since this is the point at which smoke and hot gases are likely to accumulate. The minimum aggregate clear opening area of the vents is required to be 5 percent of the floor area of the stage. The vents must open automatically by approved heat-activated devices, often fusible links, and must also be operable by a supplemental manual means. The manual means serves as a backup to the automatic detectors and permits the vent to be opened as a precaution prior to achieving the heat required to activate the heat device. The vents are to be provided with curbs as required for skylights in Section 2610.2. In addition, the vents must be labeled by an approved agency (see Section 1703.5 for further labeling requirements).

[F] 410.3.7.2 Smoke control. Smoke control in accordance with Section 909 shall be provided to maintain the smoke layer interface not less than 6 feet (1829 mm) above the highest level of the assembly seating or above the top of the proscenium opening where a proscenium wall is provided in compliance with Section 410.3.4.

❖ In addition to roof vents, another option for emergency ventilation (i.e., smoke control) of stages is provided in this section. This section, in concert with the provisions contained in Section 909, provides the criteria for smoke control for stages larger than 1,000 square feet (93 m^2) in floor area or with a stage height greater than 50 feet (15 240 mm). In such cases, the smoke layer interface must be maintained 6 feet (1829 mm) above the highest level of the assembly seating or above the top of the proscenium opening when one is provided. The smoke layer is maintained 6 feet (1829 mm) above the proscenium opening in order to prevent smoke from entering into the audience area.

410.4 Platform construction. Permanent *platforms* shall be constructed of materials as required for the type of construction of the building in which the permanent *platform* is located. Permanent *platforms* are permitted to be constructed of *fire-retardant-treated wood* for Types I, II and IV construction where the *platforms* are not more than 30 inches (762 mm) above the main floor, and not more than one-third of the room floor area and not more than 3,000 square feet (279 m^2) in area. Where the space beneath the permanent *platform* is used for storage or any purpose other than equipment, wiring or plumbing, the floor assembly shall be not less than 1-hour fire-resistance-rated construction. Where the space beneath the permanent *platform* is used only for equipment, wiring or plumbing, the underside of the permanent *platform* need not be protected.

❖ This section establishes the scope of the code requirements for platform construction. A permanent platform is basically raised floor construction. As such, the construction must be consistent with that of all the floors in the building.

Permanent platforms can be of fire-retardant-treated wood in limited applications. In buildings of Type I, II or IV construction, platforms can be of fire-retardant-treated wood if they are no more than 30 inches (762 mm) in height, are no more than one-third of the floor area of the room and are 3,000 square feet (279 m^2) or less in area. A platform that meets these three size limits poses a small fire risk relative to the normal fuel load in the room. As such, the platform deck and its supporting construction can be of fire-retardant-treated wood materials. If the platform exceeds any one of the three size limits, then it must be constructed of materials consistent with the building's type of construction classification. A basic premise of Section 410.4 is that the platform must not significantly increase the fire hazard of the space or building. As such, the space beneath platforms is regulated to appropriately abate the risk. Where the space beneath a permanent platform is utilized for any purpose other than electrical wiring or plumbing, the platform is to have a 1-hour fire-resistance rating. The purpose of the fire-resistance-rating requirement is to provide some structural integrity to the platform should a fire occur in the concealed space. In the case of permanent platforms where the space beneath the platform is only used for plumbing or electrical wiring, no further protection is required since the fire risk will be minimal.

410.4.1 Temporary platforms. *Platforms* installed for a period of not more than 30 days are permitted to be constructed of any materials permitted by this code. The space between the floor and the *platform* above shall only be used for plumbing and electrical wiring to *platform* equipment.

❖ Temporary platforms may be constructed of any material regardless of the building's construction classification due to their limited time of use and their normally limited size and fuel load. Because temporary platforms are permitted to be constructed of any approved material, the space beneath a temporary platform may never be used for any purpose other than electrical wiring or plumbing to the platform equipment. Such lines serving other areas may not be located beneath a temporary platform.

410.5 Dressing and appurtenant rooms. Dressing and appurtenant rooms shall comply with Sections 410.5.1 and 410.5.2.

❖ Because stages are open to the viewing audience and typically contain a substantial fuel load, it is essential to contain fires in rooms around the stage to the room of origin; therefore, this section contains provisions for the separation of such areas from the stage and from one another.

410.5.1 Separation from stage. The *stage* shall be separated from dressing rooms, scene docks, property rooms, workshops, storerooms and compartments appurtenant to the *stage* and other parts of the building by *fire barriers* constructed in accordance with Section 707 or *horizontal assemblies* constructed in accordance with Section 711, or both. The *fire-resistance rating* shall be not less than 2 hours for *stage* heights greater than 50 feet (15 240 mm) and not less than 1 hour for *stage* heights of 50 feet (15 240 mm) or less.

❖ Separation of the stage from appurtenant rooms provides a significant level of fire containment. Such containment is fundamental to limiting the spread of fire from adjacent spaces to the stage area, as well as the growth of fires in the stage area itself. The 2-hour fire-resistance rating for stages with a height greater than 50 feet (15 240 mm), such as large theatrical stages, acknowledges the significant potential for a large fuel load and is consistent with the 2-hour rating of Section 410.3.4. Although stages with a height of 50 feet (15 240 mm) or less represent a limited potential fuel load, a 1-hour fire-resistance rating will provide an additional level of compartmentation in the event a fire originates at the stage (which may not be sprinklered if the stage complies with Section 410.7, Exception 2). Such stages are likely to be found in educational occupancies or meeting halls.

410.5.2 Separation from each other. Dressing rooms, scene docks, property rooms, workshops, storerooms and compartments appurtenant to the *stage* shall be separated from each other by not less than 1-hour *fire barriers* constructed in accordance with Section 707 or *horizontal assemblies* constructed in accordance with Section 711, or both.

❖ As an additional level of protection, rooms and compartments appurtenant to the stage must be separated from one another by an approved 1-hour fire-resistance-rated fire barrier and horizontal assemblies. Rooms appurtenant to the stage (e.g., property rooms and activities with a history of fire incidents, such as scenery workshops) often have very high fuel loads. This additional level of fire containment is intended to minimize fire growth and limit fires in these areas to the room of origin.

410.6 Means of egress. Except as modified or as provided for in this section, the provisions of Chapter 10 shall apply.

❖ The general means of egress requirements in Chapter 10 are applicable to stages, platforms and technical production areas except where modified by this section. The modifications provided in this section provide adjustments to the Chapter 10 general egress provisions because of the uniqueness of stage areas.

410.6.1 Arrangement. Where two or more *exits* or *exit access doorways* from the *stage* are required in accordance with Section 1006.2, no fewer than one *exit* or *exit access doorway* shall be provided on each side of a *stage*.

❖ Because of the relative fire hazards associated with stages, this section requires that a minimum of one approved means of egress be provided from each side of the stage if at least two means of egress are required based on the provisions of Section 1006.2. This is intended to provide the occupants of a stage ready access to evacuate the stage area in the event of a fire. Where the occupant load of the stage is such that only one means of egress is required by Section 1006.2, then it is not necessary to provide an exit or exit access doorway on both sides of the stage.

410.6.2 Stairway and ramp enclosure. *Exit access stairways* and *ramps* serving a *stage* or *platform* are not required to be enclosed. *Exit access stairways* and ramps serving *technical production areas* are not required to be enclosed.

❖ Exit access stairways that serve a stage, platform or technical production areas do not require enclosure due to the nature of the activities involved. These stairways are required to meet the stairway provisions in Section 1011, including handrail and guard requirements where applicable.

410.6.3 Technical production areas. *Technical production areas* shall be provided with means of egress and means of escape in accordance with Sections 410.6.3.1 through 410.6.3.5.

❖ A variety of special means of egress provisions are provided for technical production areas that take into account the unique characteristics of such areas. Defined in Section 202, technical production areas include all technical support areas regardless of their traditional name (fly gallery, gridiron). The provisions of Section 410.6.3 reflect the special allowances for minimum number of means of egress, maximum travel distance, allowable exit access components and minimum travel path width. It should be noted that these provisions are also applicable to those "technical production areas" that are not necessarily associated with a stage or platform, such as those that are present at a sports arena or stadium.

410.6.3.1 Number of means of egress. No fewer than one *means of egress* shall be provided from *technical production areas*.

❖ Fly galleries, gridirons and other technical production areas are ordinarily occupied by a few persons at any one time who are likely to be very familiar with the stage operations. As such, these areas need only be provided with one approved means of egress.

410.6.3.2 Exit access travel distance. The *exit access* travel distance shall be not greater than 300 feet (91 440 mm) for buildings without a sprinkler system and 400 feet (121 900 mm) for buildings equipped throughout with an *automatic sprinkler system* in accordance with Section 903.3.1.1.

❖ The exit access travel distance limitations reflect the limited hazard presented by technical production areas, the limited number of persons affected, and the necessary extent of such areas that would cause such distances to be reasonable.

410.6.3.3 Two means of egress. Where two *means of egress* are required, the *common path of travel* shall be not greater than 100 feet (30 480 mm).

Exception: A means of escape to a roof in place of a second *means of egress* is permitted.

❖ In those cases where a second means of egress from a technical production area is required, typically where the travel distance limits of Section 410.6.3.2 cannot be met, it is necessary to limit the common path of travel to 100 feet (30 480 mm). This ensures that limited occupant travel will be necessary before the occupants have a choice of two paths that lead to the two required means of egress. The exception permits one of the means of egress to lead to the roof where at least two means of egress are provided.

410.6.3.4 Path of egress travel. The following *exit access* components are permitted where serving *technical production areas:*

1. *Stairways.*
2. *Ramps.*
3. *Spiral stairways.*
4. Catwalks.
5. *Alternating tread devices.*
6. Permanent ladders.

❖ Due to the reduced hazards previously described, a variety of exit access components may be utilized throughout the means of egress serving technical production areas. In addition to stairways and ramps, the egress path may include spiral stairways as addressed in Section 1011.10, alternating tread devices as described in Section 1011.14, and permanent ladders as addressed in Section 1011.16 and the IMC, Section 306.5.

410.6.3.5 Width. The path of egress travel within and from technical support areas shall be not less than 22 inches (559 mm).

❖ Since the means of egress from a technical production area will not be used by the public nor will it serve a significant occupant load, it is considered acceptable that a reduced egress width be provided. The required minimum width of 22 inches (559 mm) is applicable to all of the exit access components serving only the technical production areas, including stairways, ramps, spiral stairways, alternating tread devices and permanent ladders, where provided.

[F] 410.7 Automatic sprinkler system. *Stages* shall be equipped with an *automatic sprinkler system* in accordance with Section 903.3.1.1. Sprinklers shall be installed under the roof and gridiron and under all catwalks and galleries over the *stage*. Sprinklers shall be installed in dressing rooms, performer lounges, shops and storerooms accessory to such *stages*.

Exceptions:

1. Sprinklers are not required under *stage* areas less than 4 feet (1219 mm) in clear height that are utilized exclusively for storage of tables and chairs, provided the concealed space is separated from the adjacent spaces by Type X gypsum board not less than $^{5}/_{8}$-inch (15.9 mm) in thickness.
2. Sprinklers are not required for *stages* 1,000 square feet (93 m^2) or less in area and 50 feet (15 240 mm) or less in height where curtains, scenery or other combustible hangings are not retractable vertically. Combustible hangings shall be limited to a single main curtain, borders, legs and a single backdrop.

3. Sprinklers are not required within portable orchestra enclosures on *stages*.

❖ Stages contain significant quantities of combustible materials stored in, around and above the stage that are located in close proximity to large quantities of lighting equipment (i.e., scenery and lighting above the stage). There also is scenery on the sides and rear of the stage; shops located along the back and sides of the stage; and storage, props, trap doors and lifts under the stage floor. This combination of fuel load and ignition sources increases the potential for a fire. As such, stages and technical production areas, such as dressing rooms, workshops and storerooms, are required to be protected with an automatic sprinkler system.

The references to Chapter 9 indicate that the sprinkler system may be designed in accordance with NFPA 13 and Section 903.3.1.1 or, if not more than 20 sprinklers are required on any single connection, a limited area sprinkler system may be used (see Section 903.3.8).

Exception 1 applies to areas less than 4 feet (1219 mm) in clear height under stages that are used only for the storage of tables and chairs. Because of the limited fuel load present, such areas are not required to be protected with sprinklers, provided that the concealed space is separated from all adjacent spaces by no less than $^5/_8$-inch (15.9 mm) Type X gypsum board. This level of separation is intended to provide fire containment in the absence of sprinkler protection. This arrangement is often found in educational buildings where the room is used as a multipurpose room.

Exception 2 recognizes that stages 1,000 square feet (93 m^2) or less in area and 50 feet (15 240 mm) or less in height represent a limited potential for combustibles that does not warrant the requirements for an automatic sprinkler system. The code further limits the amounts of combustible materials by not allowing any vertical retractable curtains, hangings and similar scenery elements. It should be noted, however, that although sprinkler protection is not required, the requirements of Section 410.5 still apply.

Exception 3 acknowledges the limited hazards associated with portable orchestra enclosures. These elements are temporary in nature and are intended to improve the acoustics of the stage performances. These temporary enclosures do not lend themselves to temporary sprinkler heads; therefore, none are required.

[F] 410.8 Standpipes. Standpipe systems shall be provided in accordance with Section 905.

❖ Because of the potentially large fuel load and three-dimensional aspect of the fire hazard associated with stages greater than 1,000 square feet (93 m^2) in area, a Class III wet standpipe system is required on each side of such stages. The standpipes are required to be equipped with both a $1^1/_2$-inch (38 mm) and $2^1/_2$-inch (64 mm) hose connection. The required design criteria for the standpipe system, including hoses and cabinets, is specified in Section 905.3.4.

SECTION 411
SPECIAL AMUSEMENT BUILDINGS

411.1 General. *Special amusement buildings* having an *occupant load* of 50 or more shall comply with the requirements for the appropriate Group A occupancy and Sections 411.1 through 411.8. *Special amusement buildings* having an *occupant load* of less than 50 shall comply with the requirements for a Group B occupancy and Sections 411.1 through 411.8.

Exception: *Special amusement buildings* or portions thereof that are without walls or a roof and constructed to prevent the accumulation of smoke need not comply with this section.

For flammable *decorative materials*, see the *International Fire Code*.

❖ A special amusement building is one in which the egress is not readily apparent, is intentionally confounded or is not readily available (see Section 411.2). This section addresses the hazards associated with such use by requiring automatic fire detection (see Section 411.3), automatic sprinkler protection (see Section 411.4), alarm requirements (see Section 411.5), an emergency voice/alarm communication system (see Section 411.6) and specific means of egress lighting and marking (see Section 411.7). Additionally, only interior finish materials that meet the most stringent flame spread classification, Class A, are permitted in special amusement buildings (see Section 411.8).

In addition to the provisions of this section, other applicable requirements, such as means of egress (i.e., occupant load and travel distance) and building construction (i.e., type of construction, and fire-resistance ratings), apply in accordance with the appropriate assembly group classification (see Section 303.1).

Special amusement buildings are considered assembly uses based on the provisions of Section 303.1. Section 303.1 further specifies that a Group A classification is only applicable when the building's design occupant load is 50 or more. The provisions of Section 411 apply in addition to the other requirements for the appropriate assembly use, usually Group A-1 or A-3. Very small special amusement buildings are not required to be classified as Group A. If the design occupant load is less than 50, then the building is classified as Group B. This acknowledges the lesser hazards associated with smaller groups of people. Although smaller buildings are classified as Group B, they must still meet the requirements of this section as special amusement buildings.

Section 411 does not apply to a facility that is constructed to permit the free and immediate ventilation of the products of combustion to the outside. Such free and immediate ventilation addresses the hazard associated with many special amusement buildings relative to the rapid accumulation of smoke in a building in which the egress is not readily apparent, confusing or not readily available.

All flammable decorative materials used in special amusement buildings are required to follow the provisions of the IFC. The use of flammable and combustible materials in these types of buildings must be strictly regulated to limit the fire hazards to the building occupants (refer to Chapter 8 of the IFC for those restrictions).

411.2 Definition. The following term is defined in Chapter 2:

SPECIAL AMUSEMENT BUILDING.

❖ This section lists a term specifically associated with the subject matter of Section 411. It is important to emphasize that this term is not exclusively related to this chapter but may or may not also be applicable where the term is used elsewhere in the code.

Definitions of terms can help in the understanding and application of the code requirements. The purpose for including the above term within this chapter is to provide more convenient access to a term that may have a specific or limited application within this chapter. For the complete definition and associated commentary, refer back to Chapter 2. Terms that are italicized provide a visual identification throughout the code that a definition exists for that term. The use and application of all defined terms are set forth in Section 201.

[F] 411.3 Automatic fire detection. *Special amusement buildings* shall be equipped with an automatic fire detection system in accordance with Section 907.

❖ The automatic fire detection system is required to provide early warning of fire and must comply with Section 907. The detection system is required regardless of the presence of staff in the building.

[F] 411.4 Automatic sprinkler system. *Special amusement buildings* shall be equipped throughout with an *automatic sprinkler system* in accordance with Section 903.3.1.1. Where the *special amusement building* is temporary, the sprinkler water supply shall be of an *approved* temporary means.

Exception: Automatic sprinklers are not required where the total floor area of a temporary *special amusement building* is less than 1,000 square feet (93 m^2) and the exit access travel distance from any point to an exit is less than 50 feet (15 240 mm).

❖ One protection strategy to minimize the potential hazard to occupants is to control fire development. As such, special amusement buildings are required to be protected with an automatic sprinkler system. If the building is small [less than 1,000 square feet (93 m^2)] and the travel distance to exits is short [less than 50 feet (15 240 mm)] and only used on a temporary basis (such as at Halloween), automatic sprinklers are not required. In such buildings, it is anticipated that automatic fire detection and the resulting alarm (see Section 411.5) will provide additional egress time for the limited number of occupants.

Since many special amusement buildings are portable or temporary, it is not practical to require a permanent, automatic sprinkler water supply as required in Chapter 9. Instead, the building official may allow a reliable, temporary water supply. There are other unique design considerations for the sprinkler system, such as drainage and pipe and sprinkler support, which may be necessary because of the movement of the structures from one location to another.

[F] 411.5 Alarm. Actuation of a single *smoke detector*, the *automatic sprinkler system* or other automatic fire detection device shall immediately sound an alarm at the building at a *constantly attended location* from which emergency action can be initiated including the capability of manual initiation of requirements in Section 907.2.12.2.

❖ Upon activation of either the automatic fire detection or the automatic sprinkler systems, an alarm is required to be sounded at a constantly attended location. The staff at the location is expected to be capable of then providing the required egress illumination, stopping the conflicting or confusing sounds and distractions and activating the exit marking required by Section 411.7. It is also anticipated that the staff would be capable of preventing additional people from entering the building.

[F] 411.6 Emergency voice/alarm communications system. An *emergency voice/alarm communications system* shall be provided in accordance with Sections 907.2.12 and 907.5.2.2, which is also permitted to serve as a public address system and shall be audible throughout the entire *special amusement building*.

❖ An integral part of the fire protection systems required for special amusement buildings is an emergency voice/alarm communications system. This system can serve as a public address system to alert the building occupants of the fire emergency and provide them with the proper emergency instructions. The system must be installed in accordance with NFPA 72 and must be heard throughout the entire special amusement building. Upon activation, the system must sound an alert tone followed by the necessary voice instructions. These instructions can save valuable time in directing the building occupants to quickly and safely egress.

411.7 Exit marking. Exit signs shall be installed at the required *exit* or *exit access doorways* of amusement buildings in accordance with this section and Section 1013. *Approved* directional exit markings shall also be provided. Where mirrors, mazes or other designs are utilized that disguise the path of egress travel such that they are not apparent, *approved* and *listed* low-level exit signs that comply with Section 1013.5, and directional path markings *listed* in accordance with UL 1994, shall be provided and located not more than 8 inches (203 mm) above the walking surface and on or near the path of egress travel. Such markings shall become visible in an emergency. The directional exit marking shall be activated by the automatic fire detection system and the *automatic sprinkler system* in accordance with Section 907.2.12.2.

❖ During normal operation of a special amusement building, the illumination and marking of the egress path may not be adequate to allow for prompt egress from the building. This section clearly reminds the reader that exit signs must be provided at the required exit or exit access doors, and that signs must comply with the requirements of Section 1013. As special amusement spaces quite often consist of low-level lighting and/or spaces and features that can confuse a person's orientation, exit signs are required regardless of the number of required means of egress. Where obstructions and confusion may still exist because of the nature of the facility, low-level exit signs and directional path markings (listed in accordance with UL 1994) must be provided and located not higher than 8 inches (203 mm) above the walking surface in order to direct the occupants toward the exits. In an emergency, activation of the automatic fire detection system or automatic sprinkler system must activate the directional exit markings to become visible.

411.7.1 Photoluminescent exit signs. Where *photoluminescent exit* signs are installed, activating light source and viewing distance shall be in accordance with the listing and markings of the signs.

❖ Photo luminescent exit signs operate due to exposure to specific sources of light as indicated in their listing and labeling. In some situations, not all types of photo luminescent exit signs can be used and, as such, the normal lighting levels in the area where such exit signs are to be installed must be assessed to verify compatibilty.

411.8 Interior finish. The *interior finish* shall be Class A in accordance with Section 803.1.

❖ All interior finish materials must be tested for surface-burning performance in accordance with ASTM E84 or UL 723. Due to the potential for fire to spread quickly in the relatively confined spaces in these structures, only Class A materials are permitted to be used as interior finish in a special amusement building. These special amusement buildings are not permitted the one classification reduction that would normally be allowed by Table 803.11 in a sprinklered building.

SECTION 412
AIRCRAFT-RELATED OCCUPANCIES

412.1 General. Aircraft-related occupancies shall comply with Sections 412.1 through 412.8 and the *International Fire Code*.

❖ Section 412 provides specific details for the construction of the full range of aircraft-related occupancies from residential aircraft hangars to those handling large commercial aircraft, as well as helistops and heliports. This section contains code requirements for some very specialized occupancies dealing with aircraft. Aircraft pose some of the same hazards associated with motor vehicles; therefore, some of the same requirements are applicable for life and fire safety issues. The unique issues surrounding aircraft traffic control towers are addressed in Section 412.3.

Sections 412.4 and 412.5 are composed of code provisions that regulate both commercial and residential aircraft hangars. The overall building fire load for hangars is typically low, because the considerable amounts of metal in aircraft absorb heat and provide limited combustibility to sustain a fire. An aircraft fire, however, can be quite severe because of the flammability of the fuel contained in its tanks. Provisions for regulating the exterior walls, basements and floor surfaces of commercial aircraft hangars are specified along with separation and fire suppression requirements.

Section 412.6 addresses specialized hangars that are used for the painting of aircraft with flammable materials. Just as Section 416 contains limitations and requirements for the application of flammable finishes, so does Section 412.6 for aircraft paint hangars. Use and occupancy requirements commensurate with the hazards are provided, including construction, fire protection systems and necessary ventilation.

Section 412.7 addresses the unique and very large facilities where aircraft are manufactured. The size of such buildings where both commercial and military aircraft are assembled have such large footprints that traditional exit access travel distances are unrealistic.

Lastly, Section 412.8 identifies those code requirements applicable to helicopter landing and service ports. Although the code relies heavily on the requirements of NFPA 418 for heliports and helistops located on the roofs of buildings and structures, there are provisions applicable for all locations. This section contains requirements for size limitations, design and means of egress.

There are additional requirements applicable to these occupancies in Chapter 11 of the IFC, Aviation Facilities, as well as other chapters of the IFC.

412.2 Definitions. The following terms are defined in Chapter 2:

FIXED BASE OPERATOR (FBO).

HELIPORT.

HELISTOP.

RESIDENTIAL AIRCRAFT HANGAR.

TRANSIENT AIRCRAFT.

❖ This section lists terms that are specifically associated with the subject matter of Section 412. It is important to emphasize that these terms are not exclusively related to this chapter but may or may not also be applicable where the term is used elsewhere in the code.

Definitions of terms can help in the understanding

and application of the code requirements. The purpose for including a list within this chapter is to provide more convenient access to terms that may have a specific or limited application within this chapter. For the complete definition and associated commentary, refer back to Chapter 2. Terms that are italicized provide a visual identification throughout the code that a definition exists for that term. The use and application of all defined terms are set forth in Section 201.

412.3 Airport traffic control towers. The provisions of Sections 412.3.1 through 412.3.8 shall apply to airport traffic control towers occupied only for the following uses:

1. Airport traffic control cab.
2. Electrical and mechanical equipment rooms.
3. Airport terminal radar and electronics rooms.
4. Office spaces incidental to the tower operation.
5. Lounges for employees, including sanitary facilities.

❖ Section 412.3 addresses airport traffic control towers. Although these structures do not house aircraft, they are included in this section for consistency. These structures pose unique hazards to occupants because of their height and the fact that they may have limited routes of escape. This section contains requirements governing the permitted types of construction and necessary egress along with the needed fire protection systems. This section applies only to airport traffic control towers that are limited to uses exclusively related to air traffic control purposes and the uses listed in this section. If the building contains uses not listed in this section, then the building must comply with the balance of the IBC and this section (412) cannot be applied.

Airport traffic control towers are structures designed for highly specific functions. These functions include the housing of vital electronic equipment, providing an elevated structure for electronic communication, such as radar, and providing an observation area that allows an unobstructed view of the ground and airspace for air traffic controllers. This functional configuration creates special hazards, including limited means of egress, limited fire-fighting access to upper floors and vulnerability to exposure fires. The intent of the code is to restrict the fuel load by limiting the structure construction to primarily noncombustible materials; minimizing combustible contents and potential ignition sources by limiting the use of the structure; providing automatic fire detection and sprinkler systems, and adequate and reliable egress. Some provisions of Section 412 only apply where the control tower is restricted to a single means of egress.

Regardless of the height of control towers, those towers regulated under Section 412 are not considered high-rise buildings, and therefore they are exempt from Section 403.

Standby or emergency power is not specified in Section 412 for aircraft control towers. However, many of the systems in the tower such as means of egress illumination, smokeproof enclosures, the elevator and the fire detection system will need standby or emergency power as specified in those provisions.

412.3.1 Type of construction. Airport traffic control towers shall be constructed to comply with the height limitations of Table 412.3.1.

❖ The height of aircraft control towers is limited as specified in Table 412.3.1. The height limitations of Section 504 do not apply to aircraft control towers.

Table 412.3.1. See below.

❖ Table 412.3.1 functions the same as Table 504.3. Height of control towers is based on feet only and not on stories, and the actual height is measured from grade plane to the finished floor of the cab or the highest occupied level. "Cab" is the term of art for control towers and applies to the room at the top of a control tower where the controllers do their work. The number of stories is not a criterion, since many of the towers do not have occupied stories for the entire height between the ground and the occupiable level at the top. If the type of construction is known, the allowable height is given by the corresponding figure in the second column. If the height is known, the permitted construction types are determined from the first column.

TABLE 412.3.1
HEIGHT LIMITATIONS FOR AIRPORT TRAFFIC CONTROL TOWERS

TYPE OF CONSTRUCTION	HEIGHT[a] (feet)
IA	Unlimited
IB	240
IIA	100
IIB	85
IIIA	65

For SI: 1 foot = 304.8 mm, 1 square foot = 0.0929 m^2.
a. Height to be measured from grade plane to cab floor.

412.3.2 Stairways. Stairways in airport traffic control towers shall be in accordance with Section 1011. Stairways shall be smokeproof enclosures complying with one of the alternatives provided in Section 909.20.

Exception: Stairways in airport traffic control towers are not required to comply with Section 1011.12.

❖ In general, stairways are to comply with the stairway provisions of Chapter 10. Specifically, they are required to be within smokeproof enclosures. The only stairway provision that does not apply to control tower stairways is Section 1011.12, which would require a stairway to the roof. The whole point of control towers is to have a 360-degree view. Pushing a stairway to the roof through the cab of the control tower would defeat the purpose.

412.3.3 Exit access. From observation levels, airport traffic control towers shall be permitted to have a single means of exit access for a distance of travel not greater than 100 feet

(30 480 mm). Exit access stairways from the observation level need not be enclosed.

❖ Again, the intent of the control tower is visibility of the aircraft and the airfield. It is essential to not have construction, such as a stairway enclosure, interrupting vision. Therefore, from the cab of the tower (the observation level) to an exit enclosure, a single exit access travel distance of 100 feet is allowed and would include the travel down an unenclosed stairway. Where the tower has two exit stairways, providing only a single means of exit access should not be needed.

412.3.4 Number of exits. Not less than one *exit stairway* shall be permitted for airport traffic control towers of any height provided that the *occupant load* per floor is not greater than 15 and the area per floor does not exceed 1,500 square feet (140 m^2).

❖ Where a tower has two exit stairways, this section does not apply. The code recognizes that the benefit of a second exit in a tower-type structure is greatly reduced when the two exits would be in very close proximity, as would occur for towers with a small cab. Larger cabs that one would expect at larger airports such as O'Hare in Chicago and Los Angeles International would more easily be able to be provided with two exit stairways.

Therefore, for towers with a small cab (observation level) with limited area per floor and limited number of occupants, only one exit is required per floor provided that the occupant load of each floor is 15 or less. The occupant load restriction of 15 persons is based on the calculated occupant load for a business use (1:100), with a maximum tower floor area of 1,500 square feet (139 m^2) (see Table 1004.1.2).

412.3.4.1 Interior finish. Where an airport traffic control tower is provided with only one exit stairway, interior wall and ceiling finishes shall be either Class A or Class B.

❖ Typically, rooms and spaces in most occupancies can have Class C finishes. Where egress is limited to one stairway as permitted in Section 412.3.4, the finished surfaces of the control tower cab and other interior spaces must have safer finishes.

[F] 412.3.5 Automatic fire detection systems. Airport traffic control towers shall be provided with an automatic fire detection system installed in accordance with Section 907.2.

❖ To ensure early fire detection and to alert the occupants to egress the building during the incipient stage of a fire event, an automatic smoke detection system is required in accordance with Section 907.2.22. Section 907.2.22.1 applies specifically to towers where sprinklers are provided and there are at least two means of egress. Where there is only a single means of egress, Section 907.2.22.2 specifies the coverage of the fire detection systems. Section 907.2.22.2 also applies to towers that are not sprinkler protected. Only towers of a height of less than 35 feet (10 668 mm) would be without sprinklers (see Section 412.3.6).

412.3.6 Automatic sprinkler system. Where an occupied floor is located more than 35 feet (10 668 mm) above the lowest level of fire department vehicle access, airport traffic control towers shall be equipped with an *automatic sprinkler system* in accordance with Section 903.3.1.1.

❖ Sprinkler systems are required for all towers being constructed under Section 412 once there is an occupied floor above 35 feet (10 668 mm). This height was selected because shorter towers should be reachable by ladders typically found at rural fire stations and smaller airports. These are also the types of airports that would have the shorter towers.

412.3.7 Elevator protection. Wires or cables that provide normal or standby power, control signals, communication with the car, lighting, heating, air conditioning, *ventilation* and fire detecting systems to elevators shall be protected by construction having a *fire-resistance rating* of not less than 1 hour, or shall be circuit integrity cable having a fire-resistance rating of not less than 1 hour.

❖ This provision is intended to increase the probability of a functioning elevator to aid fire fighters in the event of a fire and to increase the probability that the facility can be rapidly returned to service after a minor fire incident.

412.3.7.1 Elevators for occupant evacuation. Where provided in addition to an exit stairway, occupant evacuation elevators shall be in accordance with Section 3008.

❖ This section provides a simple reference to the provisions governing occupant evacuation elevators. Such elevators are not required for these towers, but could be provided to improve egress options.

412.3.8 Accessibility. Airport traffic control towers need not be *accessible* as specified in the provisions of Chapter 11.

❖ Airport traffic control towers are not required to have an accessible route to the cab and the floor immediately below the cab (see Section 1104.4, Exception 3). Other areas of the control tower are typically employee work stations, which must be connected by an accessible route so that such stations can be approached, entered and exited (see Sections 1103.2.2 and 1104.3.1).

412.4 Aircraft hangars. Aircraft hangars shall be in accordance with Sections 412.4.1 through 412.4.6.

❖ The requirements of Section 412.4 address commercial (or nonresidential) aircraft hangars. It should be noted that most commercial aircraft hangars will not be limited in height in terms of feet (see Section 504.1) or in area (see Section 507) based on the presence of an automatic sprinkler system. All commercial aircraft hangars, however, must be regulated in regard to exterior wall fire-resistance ratings, basement limitations, floor surface construction requirements, heating equipment separation and finishing restrictions. All of these provisions serve to abate the hazards associated with large aircraft and their integral fuel tanks to acceptable fire safety levels.

412.4.1 Exterior walls. *Exterior walls* located less than 30 feet (9144 mm) from *lot lines* or a *public way* shall have a *fire-resistance rating* not less than 2 hours.

❖ To abate the hazards of a fire condition from spreading from a commercial aircraft hangar to adjacent buildings and structures, the code requires exterior walls that are located less than 30 feet (9144 mm) from lot lines or public ways to have fire-resistance-rated construction of not less than 2 hours. Fire-resistance-rated exterior walls permit the fire department additional time and protection as it attempts to take control of a fire situation in a hangar located less than 30 feet (9144 mm) from lot lines or public ways.

412.4.2 Basements. Where hangars have *basements,* floors over *basements* shall be of Type IA construction and shall be made tight against seepage of water, oil or vapors. There shall be no opening or communication between *basements* and the hangar. Access to *basements* shall be from outside only.

❖ As part of the hangar requirements, the use and separation of any basement levels are rigidly controlled. A fire in the main level of a hangar could pose a severe hazard not only to the occupants of basements but also to the fire department itself. The floor/ceiling assembly located between the hangar and the first basement level below must be of Type IA construction. With a floor of 2-hour fire-resistance-rated construction along with the supporting construction, a fire on the main level should be contained. Not only is the floor construction required to be rated, but it must be properly sealed or otherwise made waterproof. This requirement prevents the chances of a fire or other liquids and vapors from seeping into the floor construction and leaking into the basement level. Further, the code prohibits any openings or access between the main level of the hangar and the basement level. This would include fire-resistance-rated shafts. Any opening into the basement has to be made from the exterior of the structure via areaway construction.

412.4.3 Floor surface. Floors shall be graded and drained to prevent water or fuel from remaining on the floor. Floor drains shall discharge through an oil separator to the sewer or to an outside vented sump.

Exception: Aircraft hangars with individual lease spaces not exceeding 2,000 square feet (186 m^2) each in which servicing, repairing or washing is not conducted and fuel is not dispensed shall have floors that are graded toward the door, but shall not require a separator.

❖ These provisions go hand in hand with the waterproof requirements of Section 412.4.2. The floors of all hangars must be positively sloped to prevent standing liquids. This requirement is not only for personnel safety but also to minimize the effects of any spilled flammable liquids. Floor drains, if provided, must discharge their contents into an oil separator or an outside sump. This prevents the flammable liquids from entering the jurisdiction's sewer system and causing additional hazards.

An exception has been included for those commercial aircraft hangars that are divided into individual tenant or lease spaces. Hangars with tenant or lease spaces of 2,000 square feet (186 m^2) or less in area do not pose the same overall hazard. The likelihood of all the aircraft being used at the same time is remote; therefore, the hazard of ponding water or spilled fuel is remote. The floor surfaces of small aircraft hangar tenant or lease spaces only need to be sloped toward the main exterior wall openings. It is imperative that no other servicing, repair work or aircraft washing is done and that no fuel dispensing can occur in these small tenant lease spaces.

412.4.4 Heating equipment. Heating equipment shall be placed in another room separated by 2-hour *fire barriers* constructed in accordance with Section 707 or *horizontal assemblies* constructed in accordance with Section 711, or both. Entrance shall be from the outside or by means of a vestibule providing a two-doorway separation.

Exceptions:

1. Unit heaters and vented infrared radiant heating equipment suspended not less than 10 feet (3048 mm) above the upper surface of wings or engine enclosures of the highest aircraft that are permitted to be housed in the hangar need not be located in a separate room provided they are mounted not less than 8 feet (2438 mm) above the floor in shops, offices and other sections of the hangar communicating with storage or service areas.
2. Entrance to the separated room shall be permitted by a single interior door provided the sources of ignition in the appliances are not less than 18 inches (457 mm) above the floor.

❖ As part of the special use and occupancy requirements for commercial aircraft hangars, all possible ignition sources must be controlled and isolated. Specifically, all heating equipment must be located in rooms that are separated from the main areas where the aircraft are parked. This separation (both fire barrier walls and horizontal assemblies) must be 2-hour fire-resistance-rated construction. Although not explicitly stated, all openings through the rated walls must be protected. Doors connecting the heating equipment rooms and the main hangar area must be done with a vestibule or airlock arrangement such that one must pass through two doors prior to entering the other room. Again, this is done to minimize the possibility of any spilled flammable liquids and the resulting vapors from coming in contact with the ignition sources of the heating equipment.

Two exceptions to the separation requirements are provided. Unit heaters or vented infrared radiant heating equipment that are carefully located high above not only the floor surfaces but also the fuel tanks and engine compartments of the aircraft pose little risk. An allowance is also made if the heating equipment is located at least 18 inches (457 mm) above the floor of the separated room. In such a case, the vestibule/airlock arrangement with a double door system is not required and can be done with just a single door. If either exception is used, care must be taken by the building and fire officials that these special stipulations and conditions are part of the certificate of occupancy.

412.4.5 Finishing. The process of "doping," involving use of a volatile flammable solvent, or of painting, shall be carried on in a separate *detached building* equipped with *automatic fire-extinguishing equipment* in accordance with Section 903.

❖ Any application of spraying or "doping" of flammable finishes or solvent treatments to aircraft is prohibited within the hangar. These types of operations must be done in a separate detached building that is provided with an automatic fire-extinguishing system. Although not specifically stated here, the intent of the code is to treat those types of buildings as aircraft paint hangars in accordance with Section 412.6.

Doping is a type of lacquer used to protect, waterproof and make taut cloth surfaces of airplane wings. It is used on lighter-than-air, ultra-light and some light aircraft. It is essentially painting on fabric. Doping is not used on metallic surfaces; however, the use of flammable paints is also addressed in this section. When flammable finishes are applied, the process must occur in a separate building not attached to the hangar. Because the code text refers to Section 903, the intent is for an automatic sprinkler system to be installed, unless otherwise approved.

[F] 412.4.6 Fire suppression. Aircraft hangars shall be provided with a fire suppression system designed in accordance with NFPA 409, based upon the classification for the hangar given in Table 412.4.6.

Exception: Where a *fixed base operator* has separate repair facilities on site, Group II hangars operated by a *fixed base operator* used for storage of *transient aircraft* only shall have a fire suppression system, but the system is exempt from foam requirements.

❖ To minimize the fire hazards associated with aircraft hangars, most hangars are required to be protected with a fire suppression system. Where required, the fire suppression system must be designed and installed in accordance with NFPA 409, which requires fire suppression based on the type and construction of, and the activities in, a given hangar. In the standard, the suppression requirements are broken down based on three categories: Group I, Group II and Group III. Table 412.4.6 designates which group designation applies to various sizes of fire areas within a hangar and the type of construction. For example, a hangar that is 28,000 square feet (2601 m^2) in Type IIB construction is a Group II hangar. Group I and II hangars are required to have fire suppression as specified in NFPA 409. In general, Group III hangars are exempt from providing fire suppression unless one or more of the hazardous operations listed in Section 412.4.6.1 occur within the hanger. In these situations, fire suppression based on the appropriate portion of the standard for either Group I or II is required.

The exception would not require a foam system for Group II hangers if the hangar is essentially a parking garage for transient aircraft. The exception would likely only be applicable at a larger airport facility that has multiple hangars and separate hangars for repair operations.

TABLE 412.4.6. See below.

❖ As discussed, Table 412.4.6 simply determines the hangar classification to which the fire suppression system must be designed in accordance with NFPA 409. This is based upon the construction type of each building and its floor area. Table 412.4.6, Note a indi-

[F] TABLE 412.4.6
HANGAR FIRE SUPPRESSION REQUIREMENTS[a,b,c]

MAXIMUM SINGLE FIRE AREA (square feet)	TYPE OF CONSTRUCTION								
	IA	IB	IIA	IIB	IIIA	IIIB	IV	VA	VB
≥ 40,001	Group I	Group I	Group I	Group I	Group I	Group I	Group I	Group I	Group I
40,000	Group II	Group II	Group II	Group II	Group II	Group II	Group II	Group II	Group II
30,000	Group III	Group II	Group II	Group II	Group II	Group II	Group II	Group II	Group II
20,000	Group III	Group III	Group II	Group II	Group II	Group II	Group II	Group II	Group II
15,000	Group III	Group III	Group III	Group II	Group III	Group II	Group III	Group II	Group II
12,000	Group III	Group III	Group III	Group III	Group III	Group III	Group III	Group II	Group II
8,000	Group III	Group III	Group III	Group III	Group III	Group III	Group III	Group III	Group II
5,000	Group III	Group III	Group III	Group III	Group III	Group III	Group III	Group III	Group III

For SI: 1 foot = 304.8 mm, 1 square foot = 0.0929 m^2.

a. Aircraft hangars with a door height greater than 28 feet shall be provided with fire suppression for a Group I hangar regardless of maximum fire area.

b. Groups shall be as classified in accordance with NFPA 409.

c. Membrane structures complying with Section 3102 shall be classified as a Group IV hangar.

cates that regardless of size or construction type, any hangar with a door opening greater than 28 feet high (8534 mm) requires that hangar to have fire suppression system required for a Group I. Note c provides a Group IV designation for any hangar located in a membrane structure.

[F] 412.4.6.1 Hazardous operations. Any Group III aircraft hangar according to Table 412.4.6 that contains hazardous operations including, but not limited to, the following shall be provided with a Group I or II fire suppression system in accordance with NFPA 409 as applicable:

1. Doping.
2. Hot work including, but not limited to, welding, torch cutting and torch soldering.
3. Fuel transfer.
4. Fuel tank repair or maintenance not including defueled tanks in accordance with NFPA 409, inerted tanks or tanks that have never been fueled.
5. Spray finishing operations.
6. Total fuel capacity of all aircraft within the unsprinklered single *fire area* in excess of 1,600 gallons (6057 L).
7. Total fuel capacity of all aircraft within the maximum single *fire area* in excess of 7,500 gallons (28 390 L) for a hangar with an *automatic sprinkler system* in accordance with Section 903.3.1.1.

❖ Any of the operations listed in Section 412.6.1 which are occurring in a Group III hangar will require, under NFPA 409, some level of fire suppression. Doping is clarified in Section 412.4.5.

[F] 412.4.6.2 Separation of maximum single fire areas. Maximum single *fire areas* established in accordance with hangar classification and construction type in Table 412.4.6 shall be separated by 2-hour *fire walls* constructed in accordance with Section 706. In determining the maximum single *fire area* as set forth in Table 412.4.6, ancillary uses that are separated from aircraft servicing areas by a *fire barrier* of not less than 1 hour, constructed in accordance with Section 707, shall not be included in the area.

❖ The classification of aircraft hangars for fire suppression purposes is based on the hangar's type of construction and fire area size. Fire area size is based on the aggregate floor area bounded by exterior walls and 2-hour fire walls where provided. For the purposes of hangar classification, ancillary uses located within the fire area are not required to be included in the fire area size provided they are separated from the aircraft servicing area by minimum 1-hour fire barriers.

Many times there are ancillary areas associated with an aircraft hangar, such as offices, maintenance shops and storage rooms. Since the fire suppression requirements of NFPA 409 are primarily for the protection of aircraft within the storage and servicing area, inclusion of the floor area of the ancillary spaces into the fire suppression criteria is considered unnecessary. The fire protection requirements in the ancillary areas are considered to be less extensive than those required for the aircraft servicing and storage areas. Therefore, their inclusion in the application of Table 412.4.6 for fire area size is not required where a limited degree of fire separation is provided.

In order to be exempted from the fire area calculation within the aircraft hangar, it is necessary that the ancillary areas be separated from the aircraft storage and servicing areas by minimum 1-hour fire barriers. The 1-hour requirement intends to provide an acceptable fire separation without the creation of additional fire areas that would require separation by 2-hour fire walls.

412.5 Residential aircraft hangars. *Residential aircraft hangars* shall comply with Sections 412.5.1 through 412.5.5.

❖ This section of the code contains provisions that account for small, limited-size aircraft hangars that are truly accessory and auxiliary to a dwelling unit. Housing developments located along or adjacent to small-scale airports are the most obvious application of these code requirements. As part of the scoping requirement of this section, the hangar must meet all of the criteria listed in its definition in Section 202. A hangar that exceeds the limitations must then meet the provisions of Section 412.4. Included in the code requirements are fire separation from the adjacent residence; adequate means of egress; smoke detection and alarms; independent mechanical and plumbing systems; and height and area limitations.

412.5.1 Fire separation. A hangar shall not be attached to a *dwelling* unless separated by a *fire barrier* having a *fire-resistance rating* of not less than 1 hour. Such separation shall be continuous from the foundation to the underside of the roof and unpierced except for doors leading to the *dwelling unit*. Doors into the *dwelling unit* shall be equipped with *self-closing* devices and conform to the requirements of Section 716 with a noncombustible raised sill not less than 4 inches (102 mm) in height. Openings from a hangar directly into a room used for sleeping purposes shall not be permitted.

❖ A residential aircraft hangar can either be a standalone detached structure or attached to the residential dwelling. If the hangar is in close proximity or attached to the dwelling, it must then be separated with 1-hour fire-resistance-rated fire barriers. This fire separation must be continuous from the floor slab up to the roof sheathing. The only openings permitted in the fire separation are normal access doors. Windows or other vent openings are prohibited, as are any openings between the hangar and the bedrooms of the dwelling. The doors must have a $^3/_4$-hour opening protective fire protection rating in accordance with Table 716.5. A 4-inch-high (102 mm) noncombustible step is required at the door along with a self-closing device. All of these fire separation requirements serve to isolate the fire hazards associated with the hangar from the occupants of the dwelling, similar to a private garage.

412.5.2 Egress. A hangar shall provide two *means of egress*. One of the doors into the dwelling shall be considered as meeting only one of the two *means of egress*.

❖ The hangar must be provided with two separate and remotely located means of egress. This redundancy provides persons with an alternative means of escaping a fire in the hangar. One of the means of egress can be the access door into the adjacent dwelling unit. The other means of egress could be the aircraft entrance door subject to the requirements of Sections 1010.1.1 and 1010.1.2 regarding door size and swing.

[F] 412.5.3 Smoke alarms. *Smoke alarms* shall be provided within the hangar in accordance with Section 907.2.21.

❖ A smoke alarm is required in the hangar space in accordance with Section 907.2.21. Similar to the smoke alarms required in each and every bedroom, in the immediate vicinity of the bedrooms and in each and every story of the dwelling, the hangar smoke alarm must be interconnected such that one alarm will activate all other alarms. An early detection and warning alert is essential to abate the hazards of an aircraft hangar located adjacent to a residence.

412.5.4 Independent systems. Electrical, mechanical and plumbing drain, waste and vent (DWV) systems installed within the hangar shall be independent of the systems installed within the dwelling. Building sewer lines shall be permitted to be connected outside the structures.

Exception: *Smoke detector* wiring and feed for electrical subpanels in the hangar.

❖ To maintain the fire separation requirements of Section 412.5.1, the only openings permitted are normal access doors. Likewise, the mechanical and plumbing systems for the hangar must be independent of the systems within the residential house. The plumbing drain or waste line could discharge into its own building sewer line that then connects to the house's building sewer line. This connection must be done outside of both the house and the hangar. Electrical wiring serving as the feed for the subpanels in the hangar is permitted to penetrate the fire separation along with the wiring for the smoke alarm required by Section 412.5.3 and the necessary interconnection with the other interior alarms.

412.5.5 Height and area limits. *Residential aircraft hangars* shall be not greater than 2,000 square feet (186 m^2) in area and 20 feet (6096 mm) in *building height*.

❖ As stated in the definition in Section 202, the residential aircraft hangar is limited to 2,000 square feet (186 m^2) in area and 20 feet (6096 mm) in building height. These limits control the fire hazard associated with such a use. The type and number of aircraft stored in the hangar are not limited. Since the hangar is considered part of the residential use, it can be constructed of any materials that are permitted for the house, including wood-frame construction.

[F] 412.6 Aircraft paint hangars. Aircraft painting operations where flammable liquids are used in excess of the maximum allowable quantities per *control area* listed in Table 307.1(1) shall be conducted in an aircraft paint hangar that complies with the provisions of Sections 412.6.1 through 412.6.6.

❖ This section provides requirements for aircraft-related structures that exceed normal hangar storage purposes. The painting or cleaning of all aircraft with flammable liquids must be carefully controlled in an aircraft paint hangar. To determine the applicability of these requirements, the building owner must provide a complete list of all flammable liquids intended to be used in the building along with their anticipated quantities. If the amounts exceed the maximum allowable quantities per control area, then the building must be classified as an aircraft paint hangar. See Table 307.1(1) and Section 414.2 for further discussion of the maximum allowable quantities per control area of hazardous materials.

[F] 412.6.1 Occupancy group. Aircraft paint hangars shall be classified as Group H-2. Aircraft paint hangars shall comply with the applicable requirements of this code and the *International Fire Code* for such occupancy.

❖ Similar to any other building containing hazardous materials in excess of the maximum allowable quantities per control area, the building or structure must be classified as Group H. Because of the flammable liquids present, the aircraft paint hangar is classified as Group H-2. Based on the equivalent risk theory, the requirements of Section 412.6 and other applicable portions of the code and the IFC must be followed for this Group H-2 occupancy.

412.6.2 Construction. The aircraft paint hangar shall be of Type I or II construction.

❖ All aircraft paint hangars must be constructed as a Type I or II building. Special allowable height modifications are provided in Section 504.1 for aircraft paint hangars along with special area limitations in Section 507.10. If these modifications are not applicable, then the height and area limitations of Tables 504.3, 504.4 and 506.2 for a Group H-2 structure must be used. The requirement for noncombustible construction limits the fuel load that is added to the occupancy contents by the structure.

[F] 412.6.3 Operations. Only those flammable liquids necessary for painting operations shall be permitted in quantities less than the maximum allowable quantities per *control area* in Table 307.1(1). Spray equipment cleaning operations shall be conducted in a liquid use, dispensing and mixing room.

❖ To lessen the likelihood of a fire in the actual painting or aircraft cleaning areas, only flammable liquids necessary for those operations are permitted. The cleaning and maintenance of spray equipment is further limited to a liquid use, dispensing and mixing room. This requirement for separation and segregation of the different operations reduces the fire risks associated with the painting and cleaning services.

[F] 412.6.4 Storage. Storage of flammable liquids shall be in a liquid storage room.

❖ The storage of all flammable liquids must be limited to a liquid storage room, which is further defined in Section 202. The applicable requirements of Section 415 and the IFC must be followed for liquid storage rooms to reduce the fire risks to the rest of the hangar operation.

[F] 412.6.5 Fire suppression. Aircraft paint hangars shall be provided with fire suppression as required by NFPA 409.

❖ To minimize the fire hazards associated with aircraft paint hangars, all such buildings are required to be protected with a fire suppression system. This requirement is applicable regardless of the height or area of the hangar or the types and quantities of aircraft being cleaned or painted. The fire suppression system must be designed and installed in accordance with referenced standard NFPA 409. This standard contains specific requirements for the suppression systems needed to properly protect paint hangars.

[F] 412.6.6 Ventilation. Aircraft paint hangars shall be provided with *ventilation* as required in the *International Mechanical Code*.

❖ Integral to the requirements for aircraft paint hangars are the ventilation provisions. Hazardous exhaust systems for controlling the overspray of painting and aircraft cleaning operations are required. These systems along with the necessary ventilation of the occupiable spaces must be in accordance with the IMC.

412.7 Aircraft manufacturing facilities. In buildings used for the manufacturing of aircraft, exit access travel distances indicated in Section 1017.1 shall be increased in accordance with the following:

1. The building shall be of Type I or II construction.
2. Exit access travel distance shall not exceed the distances given in Table 412.7.

❖ Aircraft, both domestic and military, are assembled within huge facilities. What is being assembled is huge. For example, the tail height of a Boeing 747 is over 63 feet (19 202 mm) tall. The assembly facilities for such aircraft frequently contain multiple planes at various stages of assembly. The traditional exit access travel distances are simply not reasonable for these facilities. The only way to comply with Chapter 10 travel distances would be to create exit passageways below the assembly floor. This would require occupants to behave in a counterintuitive fashion, moving into an underground enclosure where they would not be able to judge where the incident is occurring rather than away from the incident to an open area. Modeling of designs complying with this provision demonstrate the safety of these provisions for this specific use.

TABLE 412.7. See below.

❖ See commentary for Section 412.7

412.7.1 Ancillary areas. Rooms, areas and spaces ancillary to the primary manufacturing area shall be permitted to egress through such area having a minimum height as indicated in Table 412.7. Exit access travel distance within the ancillary room, area or space shall not exceed that indicated in Table 1017.2 based on the occupancy classification of that ancillary area. Total exit access travel distance shall not exceed that indicated in Table 412.7.

❖ It is common to have various rooms surrounding the main assembly floor where support functions are provided, perhaps storage of some parts or assembly of smaller units that are then installed into the aircraft being assembled on the main floor. These spaces are part and parcel to the function of the aircraft manufacturing/assembly process. This section allows these spaces to essentially use the main assembly space as an intervening room in the exit access pathway. The travel distance within such rooms is going to comply with the distances in Table 1017.2. Total travel distance starting within the room until an exit is reached is limited to those found in Table 412.7.

[F] 412.8 Heliports and helistops. *Heliports* and *helistops* shall be permitted to be erected on buildings or other locations where they are constructed in accordance with Sections 412.8.1 through 412.8.5.

❖ This section contains special use and occupancy requirements for a very unique and specialized aircraft occupancy—those related to helicopters. Because of the limited space requirements necessary for the landing and taking off of helicopters, these areas are more likely to be incorporated into other buildings and structures, such as hospitals or large office buildings, than facilities for fixed-wing aircraft.

TABLE 412.7
AIRCRAFT MANUFACTURING EXIT ACCESS TRAVEL DISTANCE

HEIGHT (feet) [b]	MANUFACTURING AREA (sq. ft.) [a]					
	≥ 150,000	≥ 200,000	≥ 250,000	≥ 500,000	≥ 750,000	≥ 1,000,000
≥ 25	400	450	500	500	500	500
≥ 50	400	500	600	700	700	700
≥ 75	400	500	700	850	1,000	1,000
≥ 100	400	500	750	1,000	1,250	1,500

For SI: 1 foot = 304.8 mm.
a. Contiguous floor area of the aircraft manufacturing facility having the indicated height.
b. Minimum height from finished floor to bottom of ceiling or roof slab or deck.

Included in these requirements are provisions for minimum clearance sizes, structural use and design, means of egress and a referenced standard for rooftop locations. Heliports and helistops pose less fire hazard than the storage of these aircraft in hangars, but increased life safety risks.

Heliports and helistops are permitted to be located anywhere as long as they meet the requirements of Section 412.8. Certainly, federal, state and local governments may have restrictions on the locations of heliports and helistops for general aviation purposes. However, the code only addresses those fire and life safety hazards associated with heliport and helistop locations to other buildings and structures and the means of egress from the same.

[F] 412.8.1 Size. The landing area for helicopters less than 3,500 pounds (1588 kg) shall be not less than 20 feet (6096 mm) in length and width. The landing area shall be surrounded on all sides by a clear area having a minimum average width at roof level of 15 feet (4572 mm) but with no width less than 5 feet (1524 mm).

❖ The landing pad for small helicopters [less than 3,500 pounds (1588 kg) in weight] must be a minimum 20-foot-diameter (6096 mm) circle. In addition, a concentric circle must be provided around the landing pad, providing a clear area with an average width of 15 feet (4572 mm) but with the least dimension of not less than 5 feet (1524 mm). This additional clear space provides an increased landing area during windy conditions when pinpoint landing is not possible. Further, this clear area maintains the necessary separation between the rotating blades and all adjacent construction.

[F] 412.8.2 Design. Helicopter landing areas and the supports thereof on the roof of a building shall be noncombustible construction. Landing areas shall be designed to confine any flammable liquid spillage to the landing area itself and provisions shall be made to drain such spillage away from any *exit* or *stairway* serving the helicopter landing area or from a structure housing such *exit* or *stairway*. For structural design requirements, see Section 1607.6.

❖ Landing areas (helistops) located on the roofs of buildings must be of noncombustible materials, including the supporting construction. This requirement is necessary to provide a structurally sound support for the additional weight of the helicopter and its loads. This section refers to Section 1607.6 for further structural design requirements. Section 1607.6 requires that the building designer must account for the increased roof loads, including impact loads in the structural design.

Rooftop landing areas must be sloped or diked to prevent any spillage of flammable fuel from the helicopter from entering the building. This is most important since penthouse doors or exit stairways could allow spilled hazardous materials and vapors to enter the building, which would pose an unacceptable fire hazard.

[F] 412.8.3 Means of egress. The *means of egress* from *heliports* and *helistops* shall comply with the provisions of Chapter 10. Landing areas located on buildings or structures shall have two or more *means of egress*. For landing areas less than 60 feet (18 288 mm) in length or less than 2,000 square feet (186 m^2) in area, the second *means of egress* is permitted to be a fire escape, *alternating tread device* or ladder leading to the floor below.

❖ As with all means of egress, the required egress paths from heliports and helistops must be in accordance with Chapter 10. Rooftop landing areas must be provided with at least two remotely located means of egress so that the helicopter occupants have redundant means to leave the landing area and enter the building. For very small landing areas, the code would permit one exit into the building while the other could take the form of a fire escape, alternating tread device or ladder to the next lower floor level.

[F] 412.8.4 Rooftop heliports and helistops. Rooftop *heliports* and *helistops* shall comply with NFPA 418.

❖ In addition to the specific requirements of Section 412.8, rooftop heliports and helistops must comply with referenced standard NFPA 418. That standard provides further life safety and fire safety requirements associated with rooftop landing areas.

[F] 412.8.5 Standpipe system. In buildings equipped with a standpipe system, the standpipe shall extend to the roof level in accordance with Section 905.3.6.

❖ In order to assist with fire suppression activities, a standpipe system is required in those buildings with a rooftop helistop or heliport. Section 905.3.6 requires that that the standpipe system be extended to the roof level on which the helistop or heliport is located. Section 2007.5 of the IFC provides additional information in regard to the placement of standpipe connections.

SECTION 413
COMBUSTIBLE STORAGE

413.1 General. High-piled stock or rack storage in any occupancy group shall comply with the *International Fire Code.*

❖ This section alerts the code user to the specific high-piled combustible storage requirements contained in Chapter 32 of the IFC. High-piled storage presents a hazard above that of normal combustible storage. By increasing the height of the storage, the ability for a fire to grow and thrive is increased dramatically. Therefore, such storage requires special consideration with regard to arrangement and fire protection design features.

Chapter 32 of the IFC provides requirements for the high-piled storage of combustible materials regardless of the occupancy classification. High-piled storage can occur in many different occupancies, but is most typical in Group M, S and F occupancies.

High-piled storage of combustible materials includes solid-piled, palletized, shelf or rack storage where the top of storage is in excess of 12 feet (3658 mm) in height or 6 feet (1829 mm) for high-hazard commodities. Commodity classifications for all types of products, as well as fire protection requirements, are provided in Chapter 32 of the IFC and the high-piled storage provisions of NFPA 13.

413.2 Attic, under-floor and concealed spaces. *Attic*, under-floor and concealed spaces used for storage of combustible materials shall be protected on the storage side as required for 1-hour fire-resistance-rated construction. Openings shall be protected by assemblies that are *self-closing* and are of noncombustible construction or solid wood core not less than $1^3/_4$ inch (45 mm) in thickness.

Exception: Neither fire-resistance-rated construction nor open protectives are required in any of the following locations:

1. Areas protected by *approved automatic sprinkler systems*.
2. Group R-3 and U occupancies.

❖ The severity of a potential fire hazard increases when combustibles are located within concealed spaces and similar areas that provide limited access to manual fire fighting. The areas typically have low supervision and, therefore, there is increased potential for a fire to develop and spread undetected through the building. This section regulates the minimum level of separation required between storage areas and the main occupiable area in nonsprinklered buildings. Since the intent is to protect against a fire in the storage area endangering the other occupied areas of the building, the required 1-hour fire-resistance rating need only be achieved from the storage side. While any access openings in the 1-hour fire-resistant construction need not be rated, they must be self-closing and of either noncombustible construction or a minimum $1^3/_4$-inch (44 mm) thickness of solid wood core.

Item 1 of the exception exempts the storage area from being separated by 1-hour fire-resistance-rated construction, provided the area is protected by an approved automatic sprinkler system. This exception only requires the sprinkler system in the attic, underfloor or concealed space. Complete sprinkler protection throughout the building is not required in order to be in compliance with the exception.

Item 2 of the exception clarifies that storage in residential occupancies classified as Group R-3 and Group U utility structures are exempt from the separation requirement.

SECTION 414 HAZARDOUS MATERIALS

[F] 414.1 General. The provisions of Sections 414.1 through 414.6 shall apply to buildings and structures occupied for the manufacturing, processing, dispensing, use or storage of hazardous materials.

❖ This section, along with Sections 307 (High-hazard Group H) and 415 (Groups H-1, H-2, H-3, H-4 and H-5) and the IFC, are intended to be companion provisions for the treatment of occupancies that contain hazardous materials. Any building or structure utilizing hazardous materials, regardless of quantity, is to comply with all of the applicable provisions of both the code and the IFC. This section also contains design alternatives for the use and storage of hazardous materials without classifying the building as a high-hazard Group H occupancy through the use of control areas (Section 414.2) or the mercantile display option (Section 414.2.5). While Section 414 contains general construction-related requirements for high-hazard occupancies, they are not indicative of a specific Group H occupancy classification but are dictated by hazardous material requirements in the IFC. Construction-related provisions for specific Group H occupancies are contained in Section 415.

The provisions of Section 414 apply to the use and storage of hazardous materials whether or not the building is classified as Group H. Requirements for specific materials are contained in the IFC.

[F] 414.1.1 Other provisions. Buildings and structures with an occupancy in Group H shall comply with this section and the applicable provisions of Section 415 and the *International Fire Code*.

❖ Section 415 is referenced for specific provisions applicable to occupancies classified as Groups H-1, H-2, H-3, H-4 and H-5. Regardless of the actual quantity of hazardous materials present, the use and storage of all such materials are required to comply with the applicable provisions of the IFC.

[F] 414.1.2 Materials. The safe design of hazardous material occupancies is material dependent. Individual material requirements are also found in Sections 307 and 415, and in the *International Mechanical Code* and the *International Fire Code*.

❖ This section emphasizes that high-hazard occupancies are different than other occupancies in that they are material dependent. This section alerts the code user to companion provisions in both the IMC and the IFC. Section 307 contains specific parameters for when a high-hazard occupancy classification is warranted. Section 415 contains specific building requirements dependent on the actual Group H occupancy classification for the building or area.

[F] 414.1.2.1 Aerosols. Level 2 and 3 aerosol products shall be stored and displayed in accordance with the *International Fire Code*. See Section 311.2 and the *International Fire Code* for occupancy group requirements.

❖ Where Level 2 and 3 aerosol products are stored or displayed in accordance with Chapter 51 of the IFC, they may be classified as a Group S-1 occupancy as stated in Section 311.2. Section 307.1.1, Item 12, specifically exempts aerosol storage from being classified as a Group H occupancy when in compliance with the IFC. The protection required by the IFC is important so that the hazards created by these aerosols are addressed. The reference to the IFC will also address those locations where limited quantities of aerosol products are allowed in other occupancies.

[F] 414.1.3 Information required. A report shall be submitted to the *building official* identifying the maximum expected quantities of hazardous materials to be stored, used in a *closed system* and used in an *open system*, and subdivided to separately address hazardous material classification categories based on Tables 307.1(1) and 307.1(2). The methods of protection from such hazards, including but not limited to *control areas*, fire protection systems and Group H occupancies shall be indicated in the report and on the *construction documents*. The opinion and report shall be prepared by a qualified person, firm or corporation *approved* by the *building official* and provided without charge to the enforcing agency.

For buildings and structures with an occupancy in Group H, separate floor plans shall be submitted identifying the locations of anticipated contents and processes so as to reflect the nature of each occupied portion of every building and structure.

❖ A detailed plan addressing storage of hazardous materials, as well as their use in both closed and open systems, must be prepared and submitted to the building official. Such a plan is essential for assisting fire department and other emergency response personnel in hazardous materials situations. A report, such as a *Hazardous Materials Management Plan* (HMMP), as indicated in the IFC, or other approved plan, should be submitted to aid fire department personnel in the building design preplanning phase.

[F] 414.2 Control areas. *Control areas* shall comply with Sections 414.2.1 through 414.2.5 and the *International Fire Code*.

❖ As defined in Section 202, control areas are spaces within a building where quantities of hazardous materials not exceeding the maximum allowable quantities per control area are stored, dispensed, used or handled.

This section, in conjunction with the maximum allowable quantity tables in Section 307, utilizes a limited density concept for hazardous materials through the use of control areas. The intent of the control area concept is to provide an alternative method for the handling of hazardous materials without classifying the occupancy as Group H. In order to not be considered Group H, the amount of hazardous materials within any single control area bounded by fire barriers, horizontal assemblies, fire walls and exterior walls cannot exceed the maximum allowable quantity for a specific material listed in Table 307.1(1) or 307.1(2) (see Commentary Figure 414.2). A control area may be an entire building or a portion thereof. Note that when an entire building is the control area, the entire maximum allowable quantity of material from Table 307.1(1) or 307.1(2) located on any story is subject to the limitations of Table 414.2.2 (see IFC Interpretation Numbers 51-07 and 52-07).

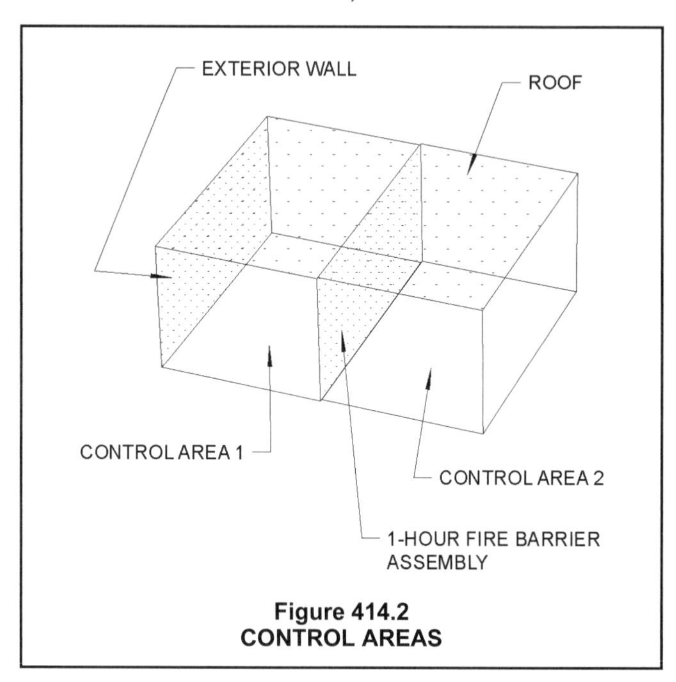

Figure 414.2
CONTROL AREAS

[F] 414.2.1 Construction requirements. *Control areas* shall be separated from each other by *fire barriers* constructed in accordance with Section 707 or *horizontal assemblies* constructed in accordance with Section 711, or both.

❖ Control areas are compartments of a building surrounded by fire barrier walls and fire-resistance-rated horizontal assemblies. If there are no fire barriers or fire-resistance-rated horizontal assemblies, the entire building is a single control area, for the purpose of applying these code provisions. Therefore, if more than the permitted maximum allowable quantities of Table 307.1(1) or 307.1(2) are anticipated in the building, additional control areas with minimum 1-hour fire barrier wall construction (2 hours where more than three stories) must be provided in order to not warrant a high-hazard occupancy classification. The provisions for required fire barriers also minimize the possibility of simultaneous involvement of multiple control areas due to a single fire condition. A fire in a single control area would involve only the amount of hazardous materials as limited by the maximum allowable quantities.

[F] 414.2.2 Percentage of maximum allowable quantities. The percentage of maximum allowable quantities of hazardous materials per *control area* permitted at each floor level within a building shall be in accordance with Table 414.2.2.

❖ Table 414.2.2 specifies the percentage of maximum allowable quantities of hazardous materials per control area dependent on the location of a given floor level with respect to grade. The noted percentages are a percentage of the maximum allowable quantities of hazardous materials permitted per control area in accordance with Tables 307.1(1) and 307.1(2).

For example, Table 307.1(1) would allow 240 gallons (908 L) of Class IB flammable liquid (in a storage condition) per control area in a fully sprinklered building [see Table 307.1(1), Note d]. Table 414.2.2, in turn, would allow 75 percent of the maximum allowable quantity per control area for control areas located on the second floor level above grade. As such, 180 gallons (681 L) of Class IB flammable liquids per control area could be stored on the second floor of a fully sprinklered building without classifying the building as a Group H-3 high-hazard occupancy.

TABLE 414.2.2. See below.

❖ The purpose of Table 414.2.2 is to establish the maximum allowable quantity of hazardous materials permitted in a building without classifying the building as a high-hazard occupancy based on the use of control areas. The number of control areas and permitted quantities of hazardous materials per control area are reduced when stored or used above the first floor level. This table also sets forth the minimum vertical fire barrier assemblies between adjacent control areas on the same floor. For floor levels above the third floor, a minimum 2-hour fire barrier is required from adjacent areas to aid fire department response due to the additional time necessary for them to access the hazardous material storage areas on upper floors. The required fire-resistance rating of the horizontal assembly above the control area is dictated by the required fire-resistance rating of the enclosure walls of the control area in order to maintain continuity and integrity. Special attention needs to be given to the rating of the floor of the control area, especially for levels below the fourth level. Section 414.2.4 requires that the floor of the control area and its supporting construction have a minimum 2-hour fire-resistance rating. In general, this will not affect floors above the third level since the continuity provisions of Section 707.5 when combined with the 2-hour separation requirement from Table 414.2.2 will ensure the floor that supports the fire barrier has an equivalent rating. However, in situations where Table 414.2.2 only requires the walls separating control areas to be of 1-hour fire-resistance-rated construction, Section 414.2.4 still would require a 2-hour fire-resistance-rated floor. This could greatly affect the design of a two-story building that was originally intended to be of Type IIB construction (see commentary, Section 414.2.4).

The percentage of quantities of hazardous materials per control area per floor area is intended to be cumulative. A two-story building, therefore, could contain three control areas with each having 75 percent of the maximum allowable quantity on the second floor in addition to the four control areas containing 100 percent each of the maximum allowable quantity on the first floor. This condition would require that adequate fire-resistance-rated separation be provided.

Note a clarifies that the maximum allowable quantity of hazardous materials per control area is based on Tables 307.1(1) and 307.1(2). The maximum permitted amount includes the increases allowed by an automatic sprinkler system in accordance with NFPA 13, approved hazardous material storage cabinets or both where applicable.

Note b clarifies the fire barrier separation needed to establish the boundaries of the control area that include not only the vertical wall assemblies but also the floor/ceiling assemblies in order to be adequately separated from all adjacent interior spaces.

Example: Determine the maximum amount of Class IB flammable liquids that can be stored within a single-story, 10,000-square-foot (929 m^2) nonsprinklered Group F-1 occupancy (see Commentary Figure 414.2.2) of Type IIB construction without classifying the storage area as Group H-2. Based on a maximum allowable quantity of 120 gallons (454 L) for Class IB flammable liquids from Table 307.1(1), a maximum of 120 gallons (454 L) can be stored in

[F] TABLE 414.2.2
DESIGN AND NUMBER OF CONTROL AREAS

FLOOR LEVEL		PERCENTAGE OF THE MAXIMUM ALLOWABLE QUANTITY PER CONTROL AREA[a]	NUMBER OF CONTROL AREAS PER FLOOR	FIRE-RESISTANCE RATING FOR FIRE BARRIERS IN HOURS[b]
Above grade plane	Higher than 9	5	1	2
	7-9	5	2	2
	6	12.5	2	2
	5	12.5	2	2
	4	12.5	2	2
	3	50	2	1
	2	75	3	1
	1	100	4	1
Below grade plane	1	75	3	1
	2	50	2	1
	Lower than 2	Not Allowed	Not Allowed	Not Allowed

a. Percentages shall be of the maximum allowable quantity per control area shown in Tables 307.1(1) and 307.1(2), with all increases allowed in the notes to those tables.

b. Separation shall include fire barriers and horizontal assemblies as necessary to provide separation from other portions of the building.

each of the four control areas. Therefore, while the building may actually contain a total of 480 gallons (1817 L), a maximum of 120 gallons (454 L) is permitted in each control area that is separated from all adjacent control areas by minimum 1-hour fire barriers constructed in accordance with Section 707. The building, in this case, could still be classified as Group F-1. An automatic fire suppression system would not be required, since the 12,000-square-foot (1115 m^2) threshold for suppression of Group F-1 fire areas is not exceeded (Section 903.2.4) and there are no control areas containing hazardous materials that exceed the maximum allowable quantities. Notes d and e of Table 307.1(1) would allow the base quantity of Class IB flammable liquids to be increased 100 percent in buildings protected with an automatic sprinkler system or when the material is stored in approved hazardous material storage cabinets. In this example, this would result in increasing the maximum allowable quantity of Class IB flammable liquids by a factor of two. Therefore, the building could now contain a total of 960 gallons (3634 L) with a maximum of 240 gallons (908 L) in each of the four control areas, separated as required by the code, and still maintain a Group F-1 classification. If both an automatic sprinkler system and hazardous material storage cabinets are used to protect Class IB flammable liquids, then the base quantity of Table 307.1(1) could be increased by a factor of four. The building in this example, therefore, with both sprinkler protection and approved cabinets, could contain a total of 1,920 gallons (7267 L) with a maximum of 480 gallons (1817 L) in each of the four control areas, separated as required by the code, and still maintain a Group F-1 classification. The allowable increase in the maximum allowable quantities is offset by the additional level or levels of protection. The use of control areas provides a tradeoff based on building compartmentation. Fire protection (automatic sprinkler systems) and controlled storage through the use of approved hazardous material storage cabinets also adds a degree of protection, justifying the increased allowable quantities.

[F] 414.2.3 Number. The maximum number of *control areas* within a building shall be in accordance with Table 414.2.2.

❖ The maximum quantity of hazardous materials permitted in a building without classifying the building as a Group H, high-hazard occupancy is regulated per control area and not per building area. The quantity limitation for the entire building would be established based on the number of permitted control areas on each floor of the building in accordance with Table 414.2.2. Based on the table, the first floor could contain four control areas with up to 100 percent of the maximum allowable quantity of hazardous materials per control area. For example, a single control area in a nonsprinklered building could contain up to 30 gallons (113 L) of Class IA flammable liquids, 125 pounds (57 kg) of Class III organic peroxides, 250 pounds (113 kg) of Class 2 oxidizers and 500 gallons (1893 L) of corrosive liquids based on the maximum allowable quantities of Tables 307.1(1) and 307.1(2). Those quantities could be contained in each of four different control areas, provided that all control areas are separated from each other with minimum 1-hour fire barriers and horizontal assemblies. Please note that in order to have more control areas per floor than indicated in Table 414.2.2, a fire wall in accordance with Section 706 would be required in order to create separate buildings within the total structure.

[F] 414.2.4 Fire-resistance-rating requirements. The required *fire-resistance rating* for *fire barriers* shall be in accordance with Table 414.2.2. The floor assembly of the

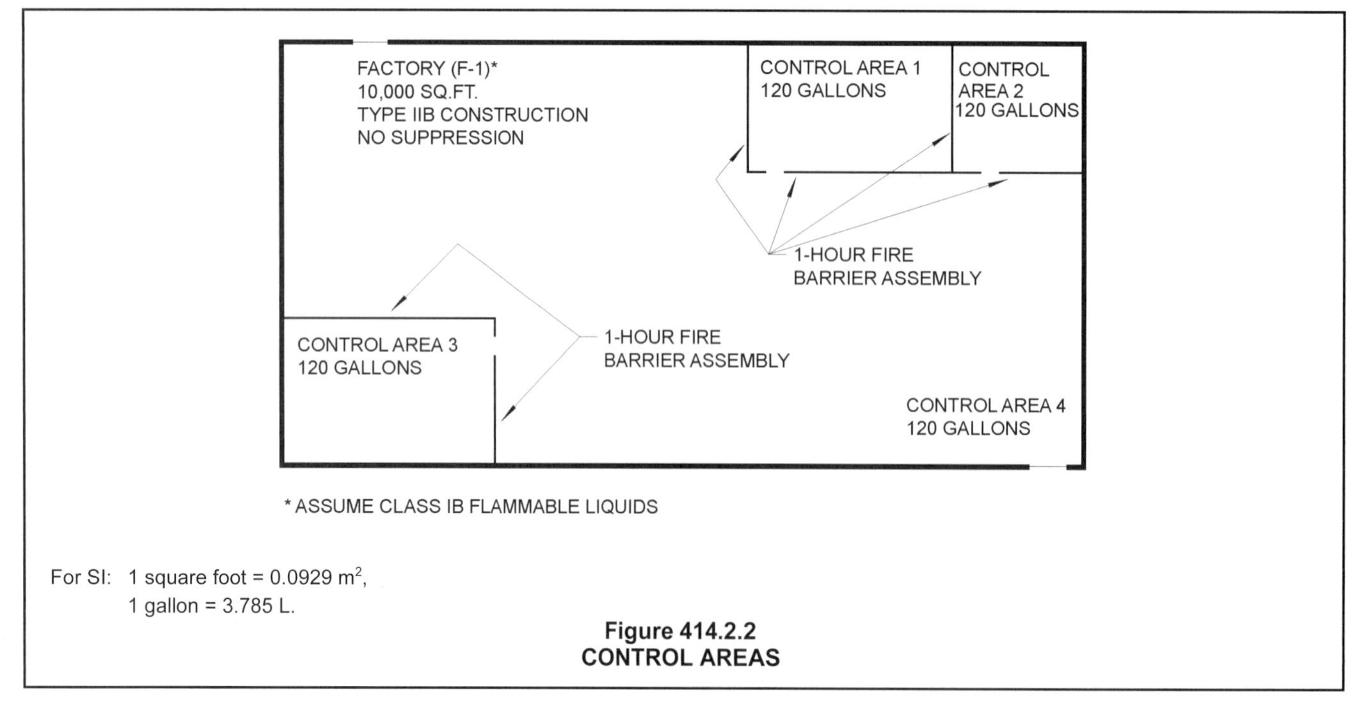

Figure 414.2.2
CONTROL AREAS

control area and the construction supporting the floor of the *control area* shall have a *fire-resistance rating* of not less than 2 hours.

Exception: The floor assembly of the *control area* and the construction supporting the floor of the *control area* are allowed to be 1-hour fire-resistance rated in buildings of Types IIA, IIIA and VA construction, provided that both of the following conditions exist:

1. The building is equipped throughout with an *automatic sprinkler system* in accordance with Section 903.3.1.1; and
2. The building is three or fewer *stories above grade plane*.

❖ The fire separation requirements for control areas, both horizontal and vertical, are dependent on their location in a building in accordance with Table 414.2.2. The amount of hazardous materials per control area, as well as the number of control areas per floor, are reduced if stored or used above the first floor.

Where the control area is located above the first floor, the floor assembly and all supporting construction for the control area would require a minimum 2-hour fire-resistance rating. The required 2-hour fire-resistance rating of the floor construction only refers to the floor of the control area. The increased fire-resistance rating and reduced quantities are intended to aid fire department personnel. The use of control areas on upper floors provides an alternative method for multistory research and laboratory-type facilities that may need to use a limited amount of hazardous materials throughout various portions of the building. Without control areas, the maximum allowable quantity for a hazardous material would be limited to a single building area regardless of the overall size or height of the building. For example, if control areas are not utilized, a 50,000-square-foot (4645 m^2) single-story building would be limited to the same quantity of hazardous materials as a two-story building with 5,000 square feet (464 m^2) per floor.

Buildings of Type IIA, IIIA and VA construction are required to have floor construction with a minimum fire-resistance rating of 1-hour as indicated in Table 601. The exception recognizes the combination of a 1-hour horizontal assembly in conjunction with sprinkler protection as a reasonable alternative for the noted construction types. The three-story limitation is consistent with the fire-resistant-rating requirements for fire barriers in Table 414.2.2.

[F] 414.2.5 Hazardous material in Group M display and storage areas and in Group S storage areas. The aggregate quantity of nonflammable solid and nonflammable or noncombustible liquid hazardous materials permitted within a single *control area* of a Group M display and storage area, a Group S storage area or an outdoor *control area* is permitted to exceed the maximum allowable quantities per *control area* specified in Tables 307.1(1) and 307.1(2) without classifying the building or use as a Group H occupancy, provided that the materials are displayed and stored in accordance with the *International Fire Code* and quantities do not exceed the maximum allowable specified in Table 414.2.5(1).

In Group M occupancy wholesale and retail sales uses, indoor storage of flammable and combustible liquids shall not exceed the maximum allowable quantities per *control area* as indicated in Table 414.2.5(2), provided that the materials are displayed and stored in accordance with the *International Fire Code.*

The maximum quantity of aerosol products in Group M occupancy retail display areas, storage areas adjacent to retail display areas and retail storage areas shall be in accordance with the *International Fire Code*.

❖ This section addresses an option for control areas containing certain nonflammable or noncombustible hazardous materials that are stored in mercantile and storage occupancies, including outdoor control areas. This option would allow Group H-4 materials, which present a health hazard rather than a physical hazard, as well as limited Group H-2 and H-3 materials, such as oxidizers, to be stored in both retail display and stock areas of regulated mercantile occupancies and in storage-related occupancies in excess of the maximum allowable quantities of Tables 307.1(1) and 307.1(2) without classifying the building as Group H. Without this option, many mercantile and storage occupancies could be classified technically as Group H. The increased quantities of certain hazardous materials are based on the recognition that while there is limited risk in these occupancies, the packaging and storage arrangements can be controlled. For further information on the storage limitations required for these types of materials, see Section 5003.11.3 of the IFC.

This section, in conjunction with Table 414.2.5(1), establishes the maximum quantity of the indicated hazardous materials permitted within a single control area of a mercantile occupancy. As indicated in Table 414.2.5(1), this section only applies to certain nonflammable solids and nonflammable or noncombustible liquids. Please note that this option is not applicable to mercantile and storage occupancies containing hazardous materials other than those indicated in Table 414.2.5(1).

This section also addresses Group M occupancies utilized for the wholesale and retail sale of flammable and combustible liquids. Group M occupancies must be able to display flammable and combustible liquids for sale to the public. The maximum allowable quantities of flammable and combustible liquids per control area can exceed the limitations of Table 307.1(1), provided they are in compliance with the amounts in Table 414.2.5(2) and displayed and stored in accordance with Section 5704.3.6 of the IFC. Section 307.1.1, Item 2 also addresses this design alternative to a Group H occupancy classification for flammable and combustible liquids in mercantile occupancies.

The retail sales of aerosol products requires compliance with Chapter 51 of the IFC and the applicable provisions of NFPA 30B.

TABLE 414.2.5(1). See below.

❖ Table 414.2.5(1) lists the hazardous materials eligible for the mercantile and storage occupancy option and the corresponding maximum permitted quantities per control area depending on the extent of protection provided. The permitted quantities of each listed material are independent of each other, as well as the various classes or physical state of a specific material. For example, a given control area could contain up to the permitted maximum quantity of Class 2 solid oxidizers, Class 3 solid oxidizers and Class 2 liquid oxidizers, in addition to the permitted quantities of corrosive materials.

Notes b and c would allow the listed maximum quantity in Table 414.2.5(1) to be increased because of the use of sprinklers, approved hazardous materials storage cabinets, or both. The notes are intended to be cumulative in that up to four times the listed amount may be allowed per control area, if the building is fully sprinkler protected and approved cabinets are utilized, without classifying the building as Group H.

Note d simply refers to Table 414.2.2 for the design and permitted number of control areas.

The 100-percent increase in maximum quantities for outdoor control areas permitted by Note f is based on the reduced exposure hazard to the building and its occupants. The increase encourages exterior storage applications without mandating sprinkler protection or approved hazardous material storage cabinets.

Notes g and h recognize that Class 2 and 3 solid

[F] TABLE 414.2.5(1)
MAXIMUM ALLOWABLE QUANTITY PER INDOOR AND OUTDOOR CONTROL AREA IN GROUP M AND S OCCUPANCIES NONFLAMMABLE SOLIDS AND NONFLAMMABLE AND NONCOMBUSTIBLE LIQUIDS[d,e,f]

CONDITION		MAXIMUM ALLOWABLE QUANTITY PER CONTROL AREA	
Material[a]	Class	Solids pounds	Liquids gallons
A. Health-hazard materials—nonflammable and noncombustible solids and liquids			
1. Corrosives[b,c]	Not Applicable	9,750	975
2. Highly toxics	Not Applicable	20[b,c]	2[b,c]
3. Toxics[b,c]	Not Applicable	1,000	100
B. Physical-hazard materials—nonflammable and noncombustible solids and liquids			
1. Oxidizers[b,c]	4	Not Allowed	Not Allowed
	3	1,150[g]	115
	2	2,250[h]	225
	1	18,000[i,j]	1,800[i,j]
2. Unstable (reactives)[b,c]	4	Not Allowed	Not Allowed
	3	550	55
	2	1,150	115
	1	Not Limited	Not Limited
3. Water reactives	3[b,c]	550	55
	2[b,c]	1,150	115
	1	Not Limited	Not Limited

For SI: 1 pound = 0.454 kg, 1 gallon = 3.785 L.

a. Hazard categories are as specified in the *International Fire Code.*

b. Maximum allowable quantities shall be increased 100 percent in buildings that are sprinklered in accordance with Section 903.3.1.1. When Note c also applies, the increase for both notes shall be applied accumulatively.

c. Maximum allowable quantities shall be increased 100 percent when stored in approved storage cabinets, in accordance with the *International Fire Code.* When Note b also applies, the increase for both notes shall be applied accumulatively.

d. See Table 414.2.2 for design and number of control areas.

e. Allowable quantities for other hazardous material categories shall be in accordance with Section 307.

f. Maximum quantities shall be increased 100 percent in outdoor control areas.

g. Maximum amounts shall be increased to 2,250 pounds when individual packages are in the original sealed containers from the manufacturer or packager and do not exceed 10 pounds each.

h. Maximum amounts shall be increased to 4,500 pounds when individual packages are in the original sealed containers from the manufacturer or packager and do not exceed 10 pounds each.

i. The permitted quantities shall not be limited in a building equipped throughout with an automatic sprinkler system in accordance with Section 903.3.1.1.

j. Quantities are unlimited in an outdoor control area.

oxidizers include several disinfectants that are commonly used in recreational, potable and wastewater treatment. Without these exceptions, the tabular maximum allowable quantities allowed in Group M and S occupancies would not be sufficient to sustain trade demand during times of peak usage. Because small containers of these materials have not been involved in losses, the exceptions permit additional containers of 10 pounds (5 kg) or less. Note that Section 5003.11.3.6 of the IFC limits the tabular quantities to individual containers of 100 pounds (45 kg) or less, whereas these exceptions give the retailer/wholesaler the option of increasing quantities on the shelves when the packaging sizes are reduced and limited to 10 pounds (5 kg) or less.

Note i recognizes the inherently higher level of protection and safety afforded by a sprinkler system and that, by definition, the only hazard presented by Class 1 oxidizers is that they slightly increase the burning rate of combustible materials with which they may come into contact during a fire. Materials with such properties present nowhere near the level of hazard of many ordinary commodities that might be found in a Group M or S occupancy, such as foam plastics. To put this matter into perspective, Class 1 oxidizers are materials with a degree of hazard similar to that of toilet bowl cleaner crystals. Note i also correlates with Table 307.1(1), Note f.

Note j further recognizes the lesser hazard of Class 1 oxidizers and the inherent safety of storing hazardous materials outdoors by allowing quantities to be unlimited in outdoor control areas. Note j also correlates with Table 5003.1.1(3) of the IFC.

TABLE 414.2.5(2). See below.

❖ Table 414.2.5(2) provides the maximum allowable quantity of flammable and combustible liquids per control area in mercantile occupancies. The limitations are based on the type of flammable and combustible liquid in the control area, the type of storage (such as rack or palletized) and the level of sprinkler protection.

The easier the flammable or combustible liquid is to ignite, the more restrictive is the quantity of liquid allowed. As such, the table severely limits the quantity of Class IA liquids. Class IIIB liquids, on the other hand, given their flash points above 200°F (93°C), are unlimited in a building with an automatic sprinkler system.

The automatic sprinkler system must be designed in accordance with NFPA 13. Flammable and combustible liquids displayed on shelves of 6 feet (1829 mm) or less are treated as an Ordinary Hazard Group 2 occupancy in accordance with NFPA 13, which would require a minimum sprinkler density of 0.20 gallon per minute (0.72 L/min) per square foot over the most remote 1,500-square-foot (140 m^2) area.

Because the flammable and combustible liquid is more exposed in individual packaging, the sprinkler system can provide better fire control. If the flammable and combustible liquids are displayed or stored in cartons, pallets or racks, the minimum sprinkler density is 0.21 gallons per minute (0.79 L/min) per square foot over the most remote 1,500-square-foot (140 m^2) area. This type of display or storage is limited to a maximum height of 4 feet, 6 inches (1372 mm). This type of packaging is more difficult for the fire sprinkler system to handle, so a greater density is required.

To allow a larger quantity and increased storage height of flammable and combustible liquids, the mercantile occupancy would require an enhanced sprinkler system in accordance with the applicable provisions of the IFC and NFPA 30. These provisions would require the sprinkler system to have a greater design density and operating area to adequately protect most storage conditions. For additional guidance, see Section 5704.3.6 of the IFC.

[F] TABLE 414.2.5(2)
MAXIMUM ALLOWABLE QUANTITY OF FLAMMABLE AND COMBUSTIBLE LIQUIDS IN WHOLESALE AND RETAIL SALES OCCUPANCIES PER CONTROL AREA[a]

TYPE OF LIQUID	MAXIMUM ALLOWABLE QUANTITY PER CONTROL AREA (gallons)		
	Sprinklered in accordance with note b densities and arrangements	Sprinklered in accordance with Tables 5704.3.6.3(4) through 5704.3.6.3(8) and 5704.3.7.5.1 of the *International Fire Code*	Nonsprinklered
Class IA	60	60	30
Class IB, IC, II and IIIA	7,500[c]	15,000[c]	1,600
Class IIIB	Unlimited	Unlimited	13,200

For SI: 1 foot = 304.8 mm, 1 square foot = 0.0929 m^2, 1 gallon = 3.785 L, 1 gallon per minute per square foot = 40.75 L/min/m^2.

a. Control areas shall be separated from each other by not less than a 1-hour fire barrier wall.

b. To be considered as sprinklered, a building shall be equipped throughout with an approved automatic sprinkler system with a design providing minimum densities as follows:

1. For uncartoned commodities on shelves 6 feet or less in height where the ceiling height does not exceed 18 feet, quantities are those permitted with a minimum sprinkler design density of Ordinary Hazard Group 2.
2. For cartoned, palletized or racked commodities where storage is 4 feet 6 inches or less in height and where the ceiling height does not exceed 18 feet, quantities are those permitted with a minimum sprinkler design density of 0.21 gallon per minute per square foot over the most remote 1,500-square-foot area.

c. Where wholesale and retail sales or storage areas exceed 50,000 square feet in area, the maximum allowable quantities are allowed to be increased by 2 percent for each 1,000 square feet of area in excess of 50,000 square feet, up to a maximum of 100 percent of the table amounts. A control area separation is not required. The cumulative amounts, including amounts attained by having an additional control area, shall not exceed 30,000 gallons.

[F] 414.3 Ventilation. Rooms, areas or spaces in which explosive, corrosive, combustible, flammable or highly toxic dusts, mists, fumes, vapors or gases are or may be emitted due to the processing, use, handling or storage of materials shall be mechanically ventilated where required by this code, the *International Fire Code* or the *International Mechanical Code*.

Emissions generated at workstations shall be confined to the area in which they are generated as specified in the *International Fire Code* and the *International Mechanical Code*.

❖ This section requires mechanical ventilation for occupancies utilizing hazardous materials when required by the IFC. The specific ventilation requirements are required to be in accordance with the applicable provisions in the IMC.

With regard to ducts conveying hazardous exhaust, the intent of this section is to minimize the potential for spreading hazardous exhaust to other parts of the building as a result of duct leakage or failure. In the event of a duct fire or explosion, other areas of the building could be jeopardized. Section 414.3 requires the exhaust flow to be maintained at concentrations below the lower flammability limit (LFL) of the contaminant. In all cases, ducts conveying hazardous exhaust must not extend into or through other ducts or plenum spaces, unless the flammable vapor-air mixtures are less than 25 percent of the LFL of the vapor being generated. The reference to work stations addresses spaces within a Group H-5 occupancy that utilize hazardous production materials within a fabrication area. The required control of emissions at work stations is intended to prevent exposure of the work station operator to hazardous fumes or vapors and to prevent hazardous concentrations of the materials used in the manufacturing process.

The exhaust system required by this section must be provided with an emergency manual shutoff control (kill switch) that will permit the exhaust system to be shut down without requiring personnel to enter the storage room. In the event of an emergency, it would be desirable to shut off an exhaust fan that is a source of ignition or is exhausting significant quantities of hazardous substances as a result of leaking storage containers. Under some circumstances, continued operation of an exhaust system could increase the level of hazard or cause the spread of fire. To prevent tampering and unauthorized use of the shutoff control, such controls must be of the type that requires a seal to be broken before they can be actuated. The control must be clearly identified as to its purpose.

[F] 414.4 Hazardous material systems. Systems involving hazardous materials shall be suitable for the intended application. Controls shall be designed to prevent materials from entering or leaving process or reaction systems at other than the intended time, rate or path. Automatic controls, where provided, shall be designed to be fail safe.

❖ Process-type systems involving the use of hazardous materials generally involve many design variables. Many times, the building official may not be aware of the potentially dangerous chemical combination or use of incompatible materials that might be inherent in a given system. This section simply requires that all systems involving the use of hazardous materials should be designed to prevent an unwanted mixture of hazardous materials from occurring.

[F] 414.5 Inside storage, dispensing and use. The inside storage, dispensing and use of hazardous materials shall be in accordance with Sections 414.5.1 through 414.5.3 of this code and the *International Fire Code*.

❖ Sections 414.5.1 through 414.5.5 contain construction-related items for the inside storage and use of hazardous materials. The applicability of the provisions of Section 414.5 is dependent on the specific hazardous material requirements in the IFC. It is important to note that these provisions are not solely applicable to Group H occupancies.

[F] 414.5.1 Explosion control. Explosion control shall be provided in accordance with the *International Fire Code* as required by Table 414.5.1 where quantities of hazardous materials specified in that table exceed the maximum allowable quantities in Table 307.1(1) or where a structure, room or space is occupied for purposes involving explosion hazards as required by Section 415 or the *International Fire Code*.

❖ It is usually impractical to design a building to withstand the pressure created by an explosion; therefore, this section mandates an explosion relief system for all structures, rooms or spaces with uses involving explosion hazards. Explosions may result from the overpressurization of a containing structure, by physical/chemical means or by a chemical reaction. During an explosion, a sudden release of a high-pressure gas occurs and the energy is dissipated in the form of a shock wave.

All structures, rooms or spaces with those special uses listed in Table 414.5.1 involving explosion hazards must be equipped with some method of explosion control as required by Section 415 or the material-specific requirements in the IFC. Table 414.5.1 also specifies when explosion control is required based on certain materials where the quantities of hazardous materials involved exceed the maximum allowable quantities in Table 307.1(1). Similarly, Section 911 of the IFC recognizes explosion (deflagration) venting and explosion (deflagration) prevention systems as acceptable methods of explosion control, where appropriate. The use of barricades or other explosion protective devices, such as magazines, may be permitted as the means of explosion control where indicated in the IFC as an acceptable alternative and approved by the building official.

TABLE 414.5.1. See below.

❖ This table designates when some methods of explosion control are required for specific material or special use conditions. The applicability of this table assumes either the quantities of hazardous materials involved exceed the maximum allowable quantities in Table 307.1(1), listed in the table under "Hazard Category," or special explosion hazards exist as listed under "Special Uses." Section 911 of the IFC provides design criteria for explosion (deflagration) venting. Explosion prevention (suppression) systems, where utilized, must comply with NFPA 69. Barricade construction must be designed and installed in accordance with NFPA 495. As indicated in Table 414.5.1, Note b, the IFC provides additional guidance as to the applicability and design criteria for explosion control methods.

[F] 414.5.2 Emergency or standby power. Where required by the *International Fire Code* or this code, mechanical ventilation, treatment systems, temperature control, alarm, detection or other electrically operated systems shall be provided with emergency or standby power in accordance with Section 2702. For storage and use areas for highly toxic or toxic

[F] TABLE 414.5.1
EXPLOSION CONTROL REQUIREMENTS[a, h]

MATERIAL	CLASS	EXPLOSION CONTROL METHODS	
		Barricade construction	Explosion (deflagration) venting or explosion (deflagration) prevention systems[b]
HAZARD CATEGORY			
Combustible dusts[c]	—	Not Required	Required
Cryogenic flammables	—	Not Required	Required
Explosives	Division 1.1 Division 1.2 Division 1.3 Division 1.4 Division 1.5 Division 1.6	Required Required Not Required Not Required Required Required	Not Required Not Required Required Required Not Required Not Required
Flammable gas	Gaseous Liquefied	Not Required Not Required	Required Required
Flammable liquid	IA[d] IB[e]	Not Required Not Required	Required Required
Organic peroxides	U I	Required Required	Not Permitted Not Permitted
Oxidizer liquids and solids	4	Required	Not Permitted
Pyrophoric gas	—	Not Required	Required
Unstable (reactive)	4 3 Detonable 3 Nondetonable	Required Required Not Required	Not Permitted Not Permitted Required
Water-reactive liquids and solids	3 2[g]	Not Required Not Required	Required Required
SPECIAL USES			
Acetylene generator rooms	—	Not Required	Required
Grain processing	—	Not Required	Required
Liquefied petroleum gas-distribution facilities	—	Not Required	Required
Where explosion hazards exist[f]	Detonation Deflagration	Required Not Required	Not Permitted Required

a. See Section 414.1.3.
b. See the *International Fire Code*.
c. As generated during manufacturing or processing.
d. Storage or use.
e. In open use or dispensing.
f. Rooms containing dispensing and use of hazardous materials when an explosive environment can occur because of the characteristics or nature of the hazardous materials or as a result of the dispensing or use process.
g. A method of explosion control shall be provided when Class 2 water-reactive materials can form potentially explosive mixtures.
h. Explosion venting is not required for Group H-5 fabrication areas complying with Section 415.11.1 and the *International Fire Code*.

materials, see Sections 6004.2.2.8 and 6004.3.4.2 of the *International Fire Code*.

❖ A backup emergency power source is considered essential for required systems monitoring hazardous materials. Therefore, when limit controls, detection systems or mechanical ventilation are required for a specific hazardous material, an emergency electrical system or standby power system is required.

[F] 414.5.2.1 Exempt applications. Emergency or standby power is not required for the mechanical ventilation systems provided for any of the following:

1. Storage of Class IB and IC flammable and combustible liquids in closed containers not exceeding 6.5 gallons (25 L) capacity.
2. Storage of Class 1 and 2 oxidizers.
3. Storage of Class II, III, IV and V organic peroxides.
4. Storage of asphyxiant, irritant and radioactive gases.

❖ Exemption 1 correlates with industry treatment of portable container storage. Notably, storage of small, closed containers does not pose a risk that warrants ventilation for these materials. FM Data Sheet 7-29, "Flammable and Combustible Liquid Storage in Portable Containers," does not require mechanical ventilation for flammable liquids in closed containers of not greater than 6.5 gallons (2290 L) individual capacity, with a flash point of not greater than 100°F (38°C) and a boiling point equal to or greater than 100°F (38°C). NFPA 30, *Flammable and Combustible Liquids Code*, also recognizes that closed container storage does not pose a risk that warrants ventilation (ventilation is required if there is open dispensing). These materials are in sealed containers in storage. Any loss of power would require an immediate cessation of operations, which would eliminate spill risk. By limiting the container size, the potential for accidental spills is significantly reduced.

Exemptions 2 and 3 address low-hazard oxidizers and organic peroxides that do not present a severe fire or reactivity hazard. Highly toxic and toxic materials (see Exception 5) must conform to applicable requirements of Chapter 60 of the IFC. For example, emergency power may be required for treatment systems utilized to process the accidental release of highly toxic or toxic compressed gases caused by a leak or rupture in storage cylinders or tanks. Without emergency power, all required monitoring systems, including the treatment system for neutralizing potential leaking gas, would be rendered inoperative if a power failure or other electrical system failure occurred.

Exemption 4 exempts storage areas for asphyxiant, irritant or radioactive gases because, unlike the requirements for hazard categories that use the Maximum Allowable Quantity (MAQ) per control area as a trigger threshold, the requirement for ventilation in storage areas containing asphyxiant, irritant and radioactive gases is not quantity based. The construction of compressed gas containers is robust compared to the containers used for other materials that may be of glass, plastic or paper. The integrity of the containers alone represents a major safeguard against likely failure. While leakage from containers is a consideration, the need for the reestablishment of power to the ventilation system within a 60-second period is not warranted given the fact that the requirement could be imposed for insignificant quantities of the gas, and given the fact that occupancy of a storage area during power outage is not the norm.

Exception 6 recognizes the use of an engineered system that is designed to always fail in the appropriate design mode without human intervention in lieu of the emergency power system. The intent of the exception is to permit alternative systems that are not subject to power interruptions. The exception, as noted, does not apply to detection and alarm systems, but addresses those systems essential to the removal of hazardous fumes and vapors from potentially occupied areas.

[F] 414.5.2.2 Fail-safe engineered systems. Standby power for mechanical ventilation, treatment systems and temperature control systems shall not be required where an approved fail-safe engineered system is installed.

❖ This section recognizes the use of an engineered system that is designed to always fail in the appropriate design mode without human intervention in lieu of the emergency power system. The intent of this section is to permit alternative systems that are not subject to power interruptions. This section does not apply to detection and alarm systems, but addresses those systems essential to the removal of hazardous fumes and vapors from potentially occupied areas.

[F] 414.5.3 Spill control, drainage and containment. Rooms, buildings or areas occupied for the storage of solid and liquid hazardous materials shall be provided with a means to control spillage and to contain or drain off spillage and fire protection water discharged in the storage area where required in the *International Fire Code*. The methods of spill control shall be in accordance with the *International Fire Code*.

❖ This section references the IFC for material-specific occupancies containing materials in excess of the maximum allowable quantities, which would require some method of spill control, drainage and containment. The specific provisions for providing adequate spill control, drainage and containment, when required, are located in Section 5004.2 of the IFC.

[F] 414.6 Outdoor storage, dispensing and use. The outdoor storage, dispensing and use of hazardous materials shall be in accordance with the *International Fire Code*.

❖ This section requires the outdoor storage, dispensing and use of hazardous materials to be in accordance with the provisions of the IFC, regardless of the quantity. Certain provisions in the IFC, however, are only applicable when the maximum allowable quantities are exceeded. In general, the permitted quantity per outdoor control area exceeds that permitted for the inside storage or use of the same material. The outdoor storage or use of hazardous materials in excess of the maximum allowable quantities does not result in a high-hazard occupancy classification but simply dictates the need for additional requirements.

[F] 414.6.1 Weather protection. Where weather protection is provided for sheltering outdoor hazardous material storage or use areas, such areas shall be considered outdoor storage or use when the weather protection structure complies with Sections 414.6.1.1 through 414.6.1.3.

❖ This section provides the minimum construction requirements for outdoor storage areas of hazardous materials that require protection from the elements. The need for weather protection is dependent on the specific material requirements in the IFC. For ready access to materials, many such structures are commonly constructed as an attached canopy. Depending on the hazardous material involved, additional protection, such as fire suppression of the outside storage area, fire-resistance-rated exterior wall construction or a limitation on exterior wall openings, may be necessary.

Structures not in compliance with the provisions of this section would be regulated as inside storage, and as such, would be considered a Group H, high-hazard occupancy if the maximum allowable quantities of a specific hazardous material were exceeded.

[F] 414.6.1.1 Walls. Walls shall not obstruct more than one side of the structure.

Exception: Walls shall be permitted to obstruct portions of multiple sides of the structure, provided that the obstructed area is not greater than 25 percent of the structure's perimeter.

❖ The structure should be sufficiently open to allow for adequate cross ventilation. The intent of this section is to allow either one full side of a weather protection structure to be obstructed or, through the exception, to permit portions of multiple sides to be obstructed, provided the obstructed perimeter does not exceed 25 percent of the total perimeter.

[F] 414.6.1.2 Separation distance. The distance from the structure to buildings, *lot lines*, *public ways* or *means of egress* to a *public way* shall be not less than the distance required for an outside hazardous material storage or use area without weather protection.

❖ The structure is required to be located with respect to lot lines, public ways or the means of egress to the public way as required for the specific hazardous material provisions in the IFC.

[F] 414.6.1.3 Noncombustible construction. The overhead structure shall be of *approved* noncombustible construction with a maximum area of 1,500 square feet (140 m^2).

Exception: The maximum area is permitted to be increased as provided by Section 506.

❖ The overhead structure is required to be of noncombustible construction to eliminate the possibility of adding to the fuel load in a fire condition. The exception permits the area of the overhead structure to exceed the 1,500-square-foot (139 m^2) area limitation if either open building frontage or an automatic sprinkler system is provided.

SECTION 415
GROUPS H-1, H-2, H-3, H-4 AND H-5

[F] 415.1 Scope. The provisions of Sections 415.1 through 415.11 shall apply to the storage and use of hazardous materials in excess of the maximum allowable quantities per *control area* listed in Section 307.1. Buildings and structures with an occupancy in Group H shall also comply with the applicable provisions of Section 414 and the *International Fire Code*.

❖ This section establishes the application of Section 415 and references the IFC for additional specific hazardous material requirements. Section 415 is only applicable when the maximum allowable quantity of a hazardous material listed in either Table 307.1(1) or 307.1(2) is exceeded. The provisions of Section 414, however, are applicable wherever hazardous materials are stored or used, regardless of quantity.

[F] 415.2 Definitions. The following terms are defined in Chapter 2:

CONTINUOUS GAS DETECTION SYSTEM.

DETACHED BUILDING.

EMERGENCY CONTROL STATION.

EXHAUSTED ENCLOSURE.

FABRICATION AREA.

FLAMMABLE VAPORS OR FUMES.

GAS CABINET.

GASROOM.

HAZARDOUS PRODUCTION MATERIAL (HPM).

HPM FLAMMABLE LIQUID.

HPM ROOM.

IMMEDIATELY DANGEROUS TO LIFE AND HEALTH (IDLH).

LIQUID.

LIQUID STORAGE ROOM.

LIQUID USE, DISPENSING AND MIXING ROOM.

LOWER FLAMMABLE LIMIT (LFL).

NORMAL TEMPERATURE AND PRESSURE (NTP).

PHYSIOLOGICAL WARNING THRESHOLD LEVEL.

SERVICE CORRIDOR.

SOLID.

STORAGE, HAZARDOUS MATERIALS.

USE (MATERIAL).

WORKSTATION.

❖ This section lists terms that are specifically associated with the subject matter of Section 415. It is important to emphasize that these terms are not exclusively related to this chapter but may or may not also be applicable where the term is used elsewhere in the code.

Definitions of terms can help in the understanding and application of the code requirements. The purpose for including a list within this chapter is to provide more convenient access to terms which may have a specific or limited application within this chapter. For the complete definition and associated commentary, refer back to Chapter 2. Terms that are italicized provide a visual identification throughout the code that a definition exists for that term. The use and application of all defined terms are set forth in Section 201.

[F] 415.3 Automatic fire detection systems. Group H occupancies shall be provided with an automatic fire detection system in accordance with Section 907.2.

❖ This section references Section 907.2.5 regarding the installation of automatic fire detection systems in Group H occupancies. Specifically required are automatic smoke detection systems when highly toxic gases, organic peroxides and oxidizers are present. See also Chapters 60, 62 and 63 of the IFC commentary. Section 907.2.5 also requires a manual fire alarm system in Group H-5 occupancies. This system shall activate the occupant notification system as specified in Section 907.5.

[F] 415.4 Automatic sprinkler system. Group H occupancies shall be equipped throughout with an *automatic sprinkler system* in accordance with Section 903.2.5.

❖ An automatic sprinkler system is required throughout buildings or portions of buildings classified as Group H occupancies as established in Section 903.2.5. In other than Group H-5 occupancies, the sprinkler system required by this section is only required in those areas that have the Group H classification. For example, in an 8,000-square-foot (743.2 m^2) Group F-1 factory with a 600-square-foot (55.7 m^2) Group H-3 storage room, only the storage room would require an automatic sprinkler system.

[F] 415.5 Emergency alarms. Emergency alarms for the detection and notification of an emergency condition in Group H occupancies shall be provided as set forth herein.

❖ An emergency alarm is required in all areas utilized for the storage, dispensing, use and handling of hazardous materials in accordance with Sections 415.5.1 through 415.5.4. This section assumes the area in question utilizes hazardous materials in excess of the maximum allowable quantities indicated in Tables 307.1(1) and 307.1(2) and, therefore, the building is classified as a Group H occupancy.

[F] 415.5.1 Storage. An approved manual emergency alarm system shall be provided in buildings, rooms or areas used for storage of hazardous materials. Emergency alarm-initiating devices shall be installed outside of each interior exit or exit access door of storage buildings, rooms or areas. Activation of an emergency alarm-initiating device shall sound a local alarm to alert occupants of an emergency situation involving hazardous materials.

❖ A manual pull station or other emergency signal device approved by the building official must be provided outside of each egress door to hazardous material storage areas. Activation of the device is intended to warn the building occupants of a potential dangerous condition within the hazardous material storage area. The alarm signal required by this section is intended only to be a local alarm. A complete evacuation alarm system, excluding a local trouble alarm located within the immediate high-hazard area, is not required.

[F] 415.5.2 Dispensing, use and handling. Where hazardous materials having a hazard ranking of 3 or 4 in accordance with NFPA 704 are transported through corridors, interior exit stairways or ramps, or exit passageways, there shall be an emergency telephone system, a local manual alarm station or an approved alarm-initiating device at not more than 150-foot (45 720 mm) intervals and at each exit and exit access doorway throughout the transport route. The signal shall be relayed to an approved central, proprietary or remote station service or constantly attended on-site location and shall initiate a local audible alarm.

❖ This section requires access to an approved supervised signaling device along the transport route when hazardous materials must be transported through corridors, interior exit stairways or exit passageways. A spill or other incident involving hazardous materials within a corridor or exit may render it unusable for egress. A loud audible alarm, which relays a signal to a remote station, allows a quick response by emergency responders.

[F] 415.5.3 Supervision. Emergency alarm systems shall be supervised by an approved central, proprietary or remote station service or shall initiate an audible and visual signal at a constantly attended on-site location.

❖ This section requires an approved method of electrical supervision for the emergency alarm systems for the hazardous material storage and use areas required by Sections 415.5.1 and 415.3.2. The method of supervision should also be in compliance with the applicable provisions of NFPA 72.

[F] 415.5.4 Emergency alarm systems. *Emergency alarm systems* shall be provided with emergency power in accordance with Section 2702.

❖ This section requires an approved emergency power system for emergency alarm systems for the hazardous material storage and use areas required by Sections 415.5.1 and 415.5.2. The system is required to be in accordance with Section 2702. Technical requirements are found in the International Fire Code, NFPA 70, NFPA 110 and NFPA 111 as stated in Section 2702.1.2.

[F] 415.6 Fire separation distance. Group H occupancies shall be located on property in accordance with the other provisions of this chapter. In Groups H-2 and H-3, not less than 25 percent of the perimeter wall of the occupancy shall be an *exterior wall*.

Exceptions:

1. *Liquid use, dispensing and mixing rooms* having a floor area of not more than 500 square feet (46.5 m^2) need not be located on the outer perimeter of the building where they are in accordance with the *International Fire Code* and NFPA 30.
2. *Liquid storage rooms* having a floor area of not more than 1,000 square feet (93 m^2) need not be located on the outer perimeter where they are in accordance with the *International Fire Code* and NFPA 30.
3. Spray paint booths that comply with the *International Fire Code* need not be located on the outer perimeter.

❖ This section specifies the location of Group H storage areas within a building. In order to provide adequate access for fire-fighting operations and venting of the products of combustion, Group H-2 and H-3 storage areas within a building must be located along an exterior wall.

Exception 1 recognizes the use of inside storage rooms that are utilized for operations involving flammable and combustible liquids. These types of rooms, which have no perimeter access, are permitted by the IFC and NFPA 30. The size of the room as well as the quantity of liquids, however, are restricted due to the lack of perimeter access and ventilation options.

Exception 2 is similar to Exception 1 except that a larger room area is permitted since the flammable and combustible liquids are in a static storage condition.

Spray paint booths are typically power-ventilated structures within a building that have their own enclosure separate from the exterior wall construction. The location and fire separation requirements for the spray paint booth must be in accordance with Section 416 and NFPA 33.

[F] 415.6.1 Group H occupancy minimum fire separation distance. Regardless of any other provisions, buildings containing Group H occupancies shall be set back to the minimum *fire separation distance* as set forth in Sections 415.6.1.1 through 415.6.1.4. Distances shall be measured from the walls enclosing the occupancy to *lot lines,* including those on a public way. Distances to assumed *lot lines* established for the purpose of determining exterior wall and opening protection are not to be used to establish the minimum *fire separation distance* for buildings on sites where explosives are manufactured or used when separation is provided in accordance with the quantity distance tables specified for explosive materials in the *International Fire Code*.

❖ Due to the potentially volatile nature of hazardous materials, specific setback requirements are necessary for Group H occupancies. These provisions take precedence over provisions in the code that may specify a minimum fire separation distance based on building construction type and exposure (see Table 602). The listed conditions are dependent on the type of materials that are indicative of the specified Group H occupancies, the size of the hazardous material storage area and whether a detached building is required by Table 415.6.2.

When dealing with explosives, it is important to note that the base paragraph makes a distinction between the assumed lot lines (see Section 705.3) that are used for determining exterior wall and opening protection and those that are used for the separation of buildings, based on fire separation distance, where explosives are involved. Where explosives are involved, the separation distances are measured between structures and not to some imaginary lot line that is assumed to be between them. It is reasonable to make this distinction since the separation distances for explosives far exceed the normal exterior wall and opening separation requirements.

[F] 415.6.1.1 Group H-1. Group H-1 occupancies shall be set back not less than 75 feet (22 860 mm) and not less than required by the *International Fire Code*.

Exception: Fireworks manufacturing buildings separated in accordance with NFPA 1124.

❖ Unless a greater distance is required by the IFC, a minimum setback of 75 feet (22 860 mm) is required between the walls enclosing a Group H-1 occupancy and all lot lines. The exception recognizes that fireworks manufacturing buildings pose unique hazards due to the potential volume (net weight) of fireworks in any single building. NFPA 1124 specifies the minimum separation distances between all process buildings, public highways and other inhabited buildings.

[F] 415.6.1.2 Group H-2. Group H-2 occupancies shall be set back not less than 30 feet (9144 mm) where the area of the occupancy is greater than 1,000 square feet (93 m^2) and it is not required to be located in a *detached building*.

❖ Unless required to be in a detached building as regulated by Section 415.6.2, all Group H-2 occupancies that exceed 1,000 square feet (93 m^2) in floor area are to have at least 30 feet (9144 mm) between the walls enclosing the Group H-2 occupancy and the lot lines surrounding the site.

[F] 415.6.1.3 Groups H-2 and H-3. Group H-2 and H-3 occupancies shall be set back not less than 50 feet (15 240 mm) where a *detached building* is required (see Table 415.6.2).

❖ Where Table 415.6.2 requires a detached building due to the quantity of hazardous materials involved, the Group H-2 and H-3 occupancies must be set back at least 50 feet (15 240 mm) from surrounding lot lines.

[F] 415.6.1.4 Explosive materials. Group H-2 and H-3 occupancies containing materials with explosive characteristics shall be separated as required by the *International Fire Code*. Where separations are not specified, the distances required shall be determined by a technical report issued in accordance with Section 414.1.3.

❖ The physical separation requirements for Group H-2 and H-3 occupancies that contain explosive materials are not regulated by the IBC, but rather by Chapter 56 in the IFC. In those situations where the IFC does not provide minimum separation requirements, it is necessary to do a specific analysis of the conditions and develop a technical report as described in Section 414.1.3.

[F] 415.6.2 Detached buildings for Group H-1, H-2 or H-3 occupancy. The storage or use of hazardous materials in excess of those amounts listed in Table 415.6.2 shall be in accordance with the applicable provisions of Sections 415.7 and 415.8.

❖ Table 415.6.2 requires detached buildings for some of the materials found in Groups H-2 and H-3 due to the large quantities involved. Even though these materials are less of a hazard when compared to those found in Group H-1, they do create a significant hazard in sufficiently large quantities

TABLE 415.6.2. See below.

❖ Table 415.6.2 establishes when hazardous materials must be stored in detached structures. The need for

[F] TABLE 415.6.2
DETACHED BUILDING REQUIRED

A DETACHED BUILDING IS REQUIRED WHEN THE QUANTITY OF MATERIAL EXCEEDS THAT LISTED HEREIN			
Material	**Class**	**Solids and Liquids (tons)[a, b]**	**Gases (cubic feet)[a, b]**
Explosives	Division 1.1 Division 1.2 Division 1.3 Division 1.4 Division 1.4[c] Division 1.5 Division 1.6	Maximum Allowable Quantity Maximum Allowable Quantity Maximum Allowable Quantity Maximum Allowable Quantity 1 Maximum Allowable Quantity Maximum Allowable Quantity	Not Applicable
Oxidizers	Class 4	Maximum Allowable Quantity	Maximum Allowable Quantity
Unstable (reactives) detonable	Class 3 or 4	Maximum Allowable Quantity	Maximum Allowable Quantity
Oxidizer, liquids and solids	Class 3 Class 2	1,200 2,000	Not Applicable Not Applicable
Organic peroxides	Detonable Class I Class II Class III	Maximum Allowable Quantity Maximum Allowable Quantity 25 50	Not Applicable Not Applicable Not Applicable Not Applicable
Unstable (reactives) nondetonable	Class 3 Class 2	1 25	2,000 10,000
Water reactives	Class 3 Class 2	1 25	Not Applicable Not Applicable
Pyrophoric gases	Not Applicable	Not Applicable	2,000

For SI: 1 ton = 906 kg, 1 cubic foot = 0.02832 m^3, 1 pound = 0.454 kg.

a. For materials that are detonable, the distance to other buildings or lot lines shall be in accordance with Chapter 56 of the *International Fire Code* based on trinitrotoluene (TNT) equivalence of the material. For materials classified as explosives, see Chapter 56 of the *International Fire Code*.

b. "Maximum Allowable Quantity" means the maximum allowable quantity per control area set forth in Table 307.1(1).

c. Limited to Division 1.4 materials and articles, including articles packaged for shipment, that are not regulated as an explosive under Bureau of Alcohol, Tobacco, Firearms and Explosives (BATF) regulations or unpackaged articles used in process operations that do not propagate a detonation or deflagration between articles, provided the net explosive weight of individual articles does not exceed 1 pound.

detached storage is a function of the type, physical state and quantity of material.

[F] 415.6.2.1 Wall and opening protection. Where a *detached building* is required by Table 415.6.2, there are no requirements for wall and opening protection based on *fire separation distance*.

❖ Detached buildings are required to be adequately separated from lot lines and other important buildings. As such, additional exposure protection for exterior walls is not needed.

[F] 415.7 Special provisions for Group H-1 occupancies. Group H-1 occupancies shall be in detached buildings used for no other purpose. Roofs shall be of lightweight construction with suitable thermal insulation to prevent sensitive material from reaching its decomposition temperature. Group H-1 occupancies containing materials that are in themselves both physical and health hazards in quantities exceeding the maximum allowable quantities per *control area* in Table 307.1(2) shall comply with requirements for both Group H-1 and H-4 occupancies.

❖ Due to the explosion hazard potential associated with Group H-1 materials, Group H-1 occupancies are required to be in separate detached structures. The limitation of one story is based on the need to exit a building with a detonation hazard as soon as possible. Exiting from a second story or basement, even within an interior exit stairway, may not offer sufficient protection where an explosion hazard exists. The one-story limitation is also intended to limit the volume of detonable materials that could be stored in any one structure.

Group H-1 occupancies that contain materials that present a health as well as a detonation hazard must also comply with the applicable requirements for a Group H-4 occupancy (see commentary, Section 307.8).

[F] 415.7.1 Floors in storage rooms. Floors in storage areas for organic peroxides, pyrophoric materials and unstable (reactive) materials shall be of liquid-tight, noncombustible construction.

❖ Noncombustible floors are required to prevent the structure from contributing to a fire scenario. The floors are also required to be liquid tight to prevent the spread of hazardous materials to areas outside the storage room.

[F] 415.8 Special provisions for Group H-2 and H-3 occupancies. Group H-2 and H-3 occupancies containing quantities of hazardous materials in excess of those set forth in Table 415.6.2 shall be in *detached buildings* used for manufacturing, processing, dispensing, use or storage of hazardous materials. Materials listed for Group H-1 occupancies in Section 307.3 are permitted to be located within Group H-2 or H-3 *detached buildings* provided the amount of materials per *control area* do not exceed the maximum allowed quantity specified in Table 307.1(1).

❖ This section, in conjunction with Table 415.6.2, specifies when a Group H-2 or H-3 occupancy must be in a detached structure. Detached structures used for the storage of hazardous materials are not intended to be mixed-use occupancies. In addition to the materials listed in Table 415.6.2, along with those addressed in Section 415.8.2, it is acceptable to locate materials listed in Section 307.3 in such detached structures provided they do not exceed the maximum allowable quantities per control area based on Table 307.1(1).

[F] 415.8.1 Multiple hazards. Group H-2 or H-3 occupancies containing materials that are in themselves both physical and health hazards in quantities exceeding the maximum allowable quantities per *control area* in Table 307.1(2) shall comply with requirements for Group H-2, H-3 or H-4 occupancies as applicable.

❖ Group H-2 and H-3 occupancies that contain materials that pose a health as well as a physical hazard must also comply with the applicable requirements for a Group H-4 occupancy (see commentary, Section 307.8).

[F] 415.8.2 Separation of incompatible materials. Hazardous materials other than those listed in Table 415.6.2 shall be allowed in manufacturing, processing, dispensing, use or storage areas when separated from incompatible materials in accordance with the provisions of the *International Fire Code*.

❖ In addition to those hazardous materials specifically listed in Table 415.6.2, it is acceptable to use or store other types of hazardous materials in a detached building, provided the incompatible materials are isolated from each other in accordance with the IFC.

[F] 415.8.3 Water reactives. Group H-2 and H-3 occupancies containing water-reactive materials shall be resistant to water penetration. Piping for conveying liquids shall not be over or through areas containing water reactives, unless isolated by *approved* liquid-tight construction.

> **Exception:** Fire protection piping shall be permitted over or through areas containing water reactives without isolating it with liquid-tight construction.

❖ The design and construction of buildings used for storing water-reactive materials must be such that water will not be permitted to come in contact with the stored materials. The building materials should be designed to resist the passage of flowing water. Piping conveying liquids, other than sprinkler system piping, is prohibited in any area containing water-reactive materials.

[F] 415.8.4 Floors in storage rooms. Floors in storage areas for organic peroxides, oxidizers, pyrophoric materials, unstable (reactive) materials and water-reactive solids and liquids shall be of liquid-tight, noncombustible construction.

❖ Noncombustible floors are required to prevent the structure from contributing to a fire scenario. The floors are also required to be liquid tight to prevent the spread of hazardous materials to areas outside the storage room.

[F] 415.8.5 Waterproof room. Rooms or areas used for the storage of water-reactive solids and liquids shall be constructed in a manner that resists the penetration of water through the use of waterproof materials. Piping carrying water for other than *approved automatic sprinkler systems* shall not be within such rooms or areas.

❖ Similar to Section 415.8.5, storage rooms containing water-reactive materials must be waterproofed. Though water piping may not be run into or through such rooms, the code recognizes that automatic sprinkler systems are a more regulated type of water piping system and have a low leakage and failure rate when properly installed and maintained.

[F] 415.9 Group H-2. Occupancies in Group H-2 shall be constructed in accordance with Sections 415.9.1 through 415.9.3 and the *International Fire Code*.

❖ Sections 415.9.1 through 415.9.3 contain specific construction requirements for occupancies that contain Group H-2 materials in excess of the maximum allowable quantities. Such occupancies are also required to comply with all material-specific related provisions in the IFC. In addition, see the commentary on Section 426.

[F] 415.9.1 Flammable and combustible liquids. The storage, handling, processing and transporting of flammable and combustible liquids in Group H-2 and H-3 occupancies shall be in accordance with Sections 415.9.1.1 through 415.9.1.9, the *International Mechanical Code* and the *International Fire Code*.

❖ Although Section 415.9.1 and its subsections are part of Section 415.9, these sections apply to both Group H-2 and H-3 occupancies. The storage of flammable and combustible liquids constitutes a special hazard because of the potential fire severity and ease of ignition related to such liquids. The flash point is the most important criterion of the hazards associated with flammable and combustible liquids because it represents the lowest temperature at which sufficient vapor will be given off to form an ignitable or flammable mixture with air. Although it is the flammable vapors that burn or explode, flammable liquids (by definition) are normally stored above their flash point. Since flammable and combustible liquids are found in virtually every industrial plant and many other occupancies, minimum requirements for the storage of such liquids are set forth in the provisions of Chapter 57 of the IFC and NFPA 30.

Additionally, the provisions of Sections 415.9.1.1 through 415.9.1.9 permit the inside tank storage of flammable and combustible liquids. It is intended that all of these provisions, where applicable, be complied with in order to allow the installation of large tanks of flammable and combustible liquids within the building. These provisions were developed in part due to the cost of complying with environmental considerations regarding the installation of underground storage tanks. In downtown areas, above-ground inside storage tanks are commonly used in conjunction with emergency power generators. While not specifically defined, these provisions are intended for tanks proposed for fixed installation as opposed to containers and portable tanks, such as 55-gallon (208 L) drums.

[F] 415.9.1.1 Mixed occupancies. Where the storage tank area is located in a building of two or more occupancies and the quantity of liquid exceeds the maximum allowable quantity for one *control area*, the use shall be completely separated from adjacent occupancies in accordance with the requirements of Section 508.4.

❖ This section requires mandatory fire-resistance-rated separation of the Group H storage tank area from adjacent areas containing other groups. By referencing Section 508.4 for separated mixed occupancies, the storage tank area would require a minimum fire barrier construction, both horizontally and vertically, whenever a mixed-occupancy condition occurs.

[F] 415.9.1.1.1 Height exception. Where storage tanks are located within a building no more than one story above grade plane, the height limitation of Section 504 shall not apply for Group H.

❖ Group H storage tank areas may be located on an upper floor in a multiple-story building subject to the limitations of Section 504. As long as the storage tank area is located within a building no more than one story, the height limitations of Tables 504.3 and 504.4, with respect to both height in feet and number of stories, would not apply to the Group H fire area. The height limitation for the building would be dictated by the other occupancy groups in the building and the actual construction type. The height exception assumes the storage tank area complies with all of the applicable provisions of Section 415.9.1 and Chapter 57 of the IFC.

[F] 415.9.1.2 Tank protection. Storage tanks shall be noncombustible and protected from physical damage. *Fire barriers* or *horizontal assemblies* or both around the storage tanks shall be permitted as the method of protection from physical damage.

❖ Storage tanks shall be noncombustible and protected from physical damage. Fire barriers, horizontal assemblies or both around the storage tanks shall be permitted as the method of protection from physical damage.

The intent of this section is to reduce the potential for industrial accidents involving storage tanks by providing a physical barrier, such as a pipe bollard or barricade. This section would also allow the fire barrier or horizontal assembly that is enclosing the storage tank area to qualify as the means of protection from physical damage. Any method of physical protection, other than a fire barrier wall assembly, must be approved by the building official.

[F] 415.9.1.3 Tanks. Storage tanks shall be *approved* tanks conforming to the requirements of the *International Fire Code*.

❖ The design, construction and installation of all storage tanks for flammable and combustible liquids must comply with the applicable provisions of Chapter 57 of the IFC.

[F] 415.9.1.4 Leakage containment. A liquid-tight containment area compatible with the stored liquid shall be provided. The method of spill control, drainage control and secondary containment shall be in accordance with the *International Fire Code*.

Exception: Rooms where only double-wall storage tanks conforming to Section 415.9.1.3 are used to store Class I, II and IIIA flammable and combustible liquids shall not be required to have a leakage containment area.

❖ In order to prevent the spread of flammable and combustible liquids to adjacent nonstorage areas in the event of a leak in a storage tank, an adequately sized containment area is required. The sizing of the containment area should be designed to contain a spill from the largest vessel plus the fire protection water for the required duration. It is also important that the drainage system be sufficient to drain not only the flammable or combustible liquid but also the drainage water from the sprinkler system as well as hose streams. Since the outer wall of a double-walled storage tank acts as the secondary means of containment, the exception would allow the omission of the leakage containment area for all flammable and combustible liquids.

[F] 415.9.1.5 Leakage alarm. An *approved* automatic alarm shall be provided to indicate a leak in a storage tank and room. The alarm shall sound an audible signal, 15 dBa above the ambient sound level, at every point of entry into the room in which the leaking storage tank is located. An *approved* sign shall be posted on every entry door to the tank storage room indicating the potential hazard of the interior room environment, or the sign shall state: WARNING, WHEN ALARM SOUNDS, THE ENVIRONMENT WITHIN THE ROOM MAY BE HAZARDOUS. The leakage alarm shall also be supervised in accordance with Chapter 9 to transmit a trouble signal.

❖ This section requires an alarm system to indicate a leak in the storage tank. The alarm must be supervised, as well as sound a local signal. Automatic electrical supervision of the leakage alarm is required in accordance with Section 901.6.2

[F] 415.9.1.6 Tank vent. Storage tank vents for Class I, II or IIIA liquids shall terminate to the outdoor air in accordance with the *International Fire Code*.

❖ Section 5704.2.7.3.3 of the IFC specifies the proper location of the tank vent to enable flammable vapors to be released to the outside air and dissipate without creating a hazard. It also requires the tank vent to terminate a minimum of 12 feet (3658 mm) above grade and 5 feet (1524 mm) from building openings and lot lines to reduce the possibility of ignitable concentrations of the vapors collecting near the discharge end of the vent pipe, entering building openings or exposing adjoining property.

[F] 415.9.1.7 Room ventilation. Storage tank areas storing Class I, II or IIIA liquids shall be provided with mechanical *ventilation*. The mechanical *ventilation* system shall be in accordance with the *International Mechanical Code* and the *International Fire Code*.

❖ This section requires mechanical ventilation for storage tank areas in order to maintain any vapor buildup at safe levels until the hazard of a spill or leak can be abated. The ventilation requirements for flammable or combustible liquid storage tanks must be in accordance with the applicable provisions of the IMC and the IFC.

[F] 415.9.1.8 Explosion venting. Where Class I liquids are being stored, explosion venting shall be provided in accordance with the *International Fire Code*.

❖ Class I flammable liquids, when stored in sufficient quantities, can readily produce vapors within the explosive limitations. Explosion venting must be in accordance with the provisions of Sections 414.5.1 and 911 and Chapter 57 of the IFC.

[F] 415.9.1.9 Tank openings other than vents. Tank openings other than vents from tanks inside buildings shall be designed to ensure that liquids or vapor concentrations are not released inside the building.

❖ Additional provisions are necessary to minimize the possibility of an accidental release of flammable or combustible liquids or their vapors via any connection to the tank other than a vent. This includes requirements for liquid-tight openings below the liquid level in the tank, overflow protection and vapor recovery connections as indicated in Section 5704.2.7.5 of the IFC.

[F] 415.9.2 Liquefied petroleum gas facilities. The construction and installation of liquefied petroleum gas facilities shall be in accordance with the requirements of this code, the *International Fire Code*, the *International Mechanical Code*, the *International Fuel Gas Code* and NFPA 58.

❖ The key to this section is the reference to NFPA 58 for design and construction of liquefied petroleum gas facilities. NFPA 58 provides a comprehensive set of construction requirements for LP-gas distribution facilities, as well as bulk plants, and industrial plants in which LP-gas systems, storage systems, vaporizers, mixing systems and similar activities are involved.

[F] 415.9.3 Dry cleaning plants. The construction and installation of dry cleaning plants shall be in accordance with the requirements of this code, the *International Mechanical Code*, the *International Plumbing Code* and NFPA 32. Dry cleaning solvents and systems shall be classified in accordance with the *International Fire Code*.

❖ Dry cleaning plants that do not qualify under Item 4 or 5 of Section 307.1.1 and in which the maximum allowable quantity per control area of dry cleaning solvents exceeds the values in Table 307.1(1) would be classified in Group H and would be required to comply with the applicable provisions of the IPC, the IMC and NFPA 32. It should be noted that Chapter 21 of the IFC also contains provisions for dry cleaning plants and references NFPA 32 for their maintenance. NFPA 32 contains provisions for the prevention and control of fire and explosion hazards associated with dry cleaning operations and for the protection of employees and the public. This standard addresses such issues as location, construction, building services, processes and equipment, and fire control.

[F] 415.10 Groups H-3 and H-4. Groups H-3 and H-4 shall be constructed in accordance with the applicable provisions of this code and the *International Fire Code*.

❖ This section contains provisions that are applicable to all Group H-3 and H-4 occupancies in addition to material-specific requirements in the IFC. These requirements assume that the Group H-3 and H-4 occupancies contain more than the maximum allowable quantities of hazardous materials per control area permitted by Tables 307.1(1) and 307.1(2). Sections 307.5 and 307.6 contain a list of those materials that could be present in Group H-3 and H-4 occupancies.

[F] 415.10.1 Flammable and combustible liquids. The storage, handling, processing and transporting of flammable and combustible liquids in Group H-3 occupancies shall be in accordance with Section 415.9.1.

❖ This section refers back to Section 415.9.1, where the standards for the handling, storage and processing of these liquids are provided for both Group H-3 and H-2 occupancies.

[F] 415.10.2 Gas rooms. Where gas rooms are provided, such rooms shall be separated from other areas by not less than 1-hour *fire barriers* constructed in accordance with Section 707 or *horizontal assemblies* constructed in accordance with Section 711, or both.

❖ In order to restrict the potential fire involvement of a storage room of hazardous gases within a Group H-3 or H-4 occupancy, a 1-hour fire-resistance-rated fire barrier, horizontal assemblies or both, are required to separate gas rooms from other adjacent areas. The required fire-resistance-rated 1-hour barrier separation is not due to a mixed-occupancy condition but rather further compartmentation in a Group H-3 or H-4 occupancy.

[F] 415.10.3 Floors in storage rooms. Floors in storage areas for corrosive liquids and highly toxic or toxic materials shall be of liquid-tight, noncombustible construction.

❖ Storage rooms of corrosive liquids and highly toxic and toxic materials would be classified as a Group H-4 occupancy. Noncombustible floors are required to prevent the structure from contributing to a fire scenario. The floors are also required to be liquid tight to prevent the spread of liquids from a spill to areas outside the storage room.

[F] 415.10.4 Separation—highly toxic solids and liquids. Highly toxic solids and liquids not stored in *approved* hazardous materials storage cabinets shall be isolated from other hazardous materials storage by not less than 1-hour *fire barriers* constructed in accordance with Section 707 or *horizontal assemblies* constructed in accordance with Section 711, or both.

❖ The potential involvement of materials that may be incompatible in storage situations with highly toxic solids and liquids must be addressed. The required degree of separation, either by approved cabinets or 1-hour fire-resistance-rated fire barrier and horizontal assembly construction, is also intended to restrict the highly toxic materials from other combustibles.

[F] 415.11 Group H-5. In addition to the requirements set forth elsewhere in this code, Group H-5 shall comply with the provisions of Sections 415.11.1 through 415.11.11 and the *International Fire Code.*

❖ This section contains the requirements for Group H-5 facilities that utilize HPM. Section 415.11 is intended to be a design option for buildings that utilize hazardous materials in excess of the maximum allowable quantities permitted by Tables 307.1(1) and 307.1(2). HPM facilities are considered unique high-hazard occupancies. The HPM occupancy classification assumes that while the manufacturing process involves the use of hazardous materials, the end product by itself is not hazardous.

The provisions of Section 415.11 resulted from nonuniform regulation of semiconductor fabrication facilities and are based on code changes submitted by groups associated with the semiconductor industry. These industries include electronic high-tech industries, such as semiconductor microchip fabrication, as well as the floppy disk and telecommunication industries. Additionally, there are other comparable activities involving similar technology that also use high-hazard materials. HPM include flammable liquids and gases, corrosives, oxidizers and, in many instances, highly toxic materials. The HPM may be a solid, liquid or gas with a degree of hazard rating in health, flammability or reactivity of Class 3 or 4 in accordance with NFPA 704, which is used directly in research, laboratory or production processes.

Structures that contain HPM facilities are required to meet the provisions of Section 415.11 as well as any other code provisions for a Group H-5 occupancy classification unless specifically modified in Section 415.11. Additionally, Chapter 27 of the IFC provides complementary fire safety controls that reduce the risk of hazard to an acceptable level in existing HPM buildings and fabrication areas.

[F] 415.11.1 Fabrication areas. *Fabrication areas* shall comply with Sections 415.11.1.1 through 415.11.1.8.

❖ The fabrication area of an HPM facility is where the hazardous materials are actively handled and processed. The fabrication area includes accessory rooms and spaces such as workstations and employee dressing rooms.

[F] 415.11.1.1 Hazardous materials. Hazardous materials and hazardous production materials (HPM) shall comply with Sections 415.11.1.1.1 and 415.11.1.1.2.

❖ Sections 415.11.1.1.1 and 415.11.1.1.2 regulate the quantities and densities of HPM for each fabrication area of a Group H-5 facility.

[F] 415.11.1.1.1 Aggregate quantities. The aggregate quantities of hazardous materials stored and used in a single *fabrication area* shall not exceed the quantities set forth in Table 415.11.1.1.1.

Exception: The quantity limitations for any hazard category in Table 415.11.1.1.1 shall not apply where the *fabrication area* contains quantities of hazardous materials not exceeding the maximum allowable quantities per *control area* established by Tables 307.1(1) and 307.1(2).

❖ This section regulates the total amount of hazardous materials, whether in use or storage, within a single fabrication area based on the density/quantity of material specified in Table 415.11.1.1.1. The exception permits a fabrication area to have a total quantity of HPM of either the quantity specified in Table 415.11.1.1.1 or the maximum allowable quantities per control area specified in Table 307.1(1) or 307.1(2), whichever is greater. For example, a small fabrication area may have a quantity in accordance with Tables 307.1(1) and 307.1(2) even though it may exceed the quantities specified in Table 415.11.1.1.1. Conversely, a large fabrication area may have an aggregate quantity that exceeds those specified in Tables 307.1(1) and 307.1(2) if it does not exceed the quantities specified in Table 415.11.1.1.1.

It should be reiterated that this section refers to an aggregate quantity of hazardous materials that are in use and storage within a single fabrication area. Section 415.11.1.1.1 further limits the amount of HPM that is being stored within a fabrication area.

TABLE 415.11.1.1.1. See page 4-116.

❖ Table 415.11.1.1.1 controls the permitted densities/quantities of hazardous materials in a single fabrication area. In accordance with Section 415.11.1.1.1, the quantities permitted in Tables 307.1(1) and 307.1(2) may be exceeded, provided the quantities in Table 415.11.1.1.1 are maintained subject to the storage limitations of Section 415.11.1.1.2. Conversely, the maximum quantities of Table 415.11.1.1.1 may be exceeded, provided the maximum allowable quantities of Tables 307.1(1) and 307.1(2) are not exceeded.

In accordance with Note a, hazardous materials in piping need not be included in evaluating the permitted quantity. Note b establishes that a specific quantity limitation, rather than a calculated density, is required for certain hazardous materials.

[F] 415.11.1.1.2 Hazardous production materials. The maximum quantities of hazardous production materials (HPM) stored in a single *fabrication area* shall not exceed the maximum allowable quantities per *control area* established by Tables 307.1(1) and 307.1(2).

❖ This section provides the overall limit of HPM that can be stored within a single fabrication area based on Tables 307.1(1) and 307.1(2). Any storage in excess of the permitted maximum allowable quantities would be required to be stored in an HPM storage room, liquid storage room or gas room. The aggregate quantities of hazardous materials in both a storage and use condition within a single fabrication area are regulated by Section 415.11.1.1.1.

[F] TABLE 415.11.1.1.1
QUANTITY LIMITS FOR HAZARDOUS MATERIALS IN A SINGLE FABRICATION AREA IN GROUP H-5[a]

HAZARD CATEGORY		SOLIDS (pounds per square foot)	LIQUIDS (gallons per square foot)	GAS (cubic feet @ NTP/square foot)
PHYSICAL-HAZARD MATERIALS				
Combustible dust		Note b	Not Applicable	Not Applicable
Combustible fiber	Loose Baled	Note b Notes b, c	Not Applicable	Not Applicable
Combustible liquid Combination Class	II IIIA IIIB I, II and IIIA	Not Applicable	0.01 0.02 Not Limited 0.04	Not Applicable
Cryogenic gas	Flammable Oxidizing	Not Applicable	Not Applicable	Note d 1.25
Explosives		Note b	Note b	Note b
Flammable gas	Gaseous Liquefied	Not Applicable	Not Applicable	Note d Note d
Flammable liquid Combination Class Combination Class	IA IB IC IA, IB and IC I, II and IIIA	Not Applicable	0.0025 0.025 0.025 0.025 0.04	Not Applicable
Flammable solid		0.001	Not Applicable	Not Applicable
Organic peroxide	Unclassified detonable Class I Class II Class III Class IV Class V	Note b Note b 0.025 0.1 Not Limited Not Limited	Not Applicable	Not Applicable
Oxidizing gas Combination of gaseous and liquefied	Gaseous Liquefied	Not Applicable	Not Applicable	1.25 1.25 1.25
Oxidizer Combination Class	Class 4 Class 3 Class 2 Class 1 1, 2, 3	Note b 0.003 0.003 0.003 0.003	Note b 0.03 0.03 0.03 0.03	Not Applicable
Pyrophoric materials		0.01	0.00125	Notes d and e
Unstable (reactive)	Class 4 Class 3 Class 2 Class 1	Note b 0.025 0.1 Not Limited	Note b 0.0025 0.01 Not Limited	Note b Note b Note b Not Limited
Water reactive	Class 3 Class 2 Class 1	Note b 0.25 Not Limited	0.00125 0.025 Not Limited	Not Applicable
HEALTH-HAZARD MATERIALS				
Corrosives		Not Limited	Not Limited	Not Limited
Highly toxic		Not Limited	Not Limited	Note d
Toxics		Not Limited	Not Limited	Note d

For SI: 1 pound per square foot = 4.882 kg/m^2, 1 gallon per square foot = 40.7 L/m^2, 1 cubic foot @ NTP/square foot = 0.305 m^3 @ NTP/m^2, 1 cubic foot = 0.02832 m^3.

a. Hazardous materials within piping shall not be included in the calculated quantities.

b. Quantity of hazardous materials in a single fabrication shall not exceed the maximum allowable quantities per control area in Tables 307.1(1) and 307.1(2).

c. Densely packed baled cotton that complies with the packing requirements of ISO 8115 shall not be included in this material class.

d. The aggregate quantity of flammable, pyrophoric, toxic and highly toxic gases shall not exceed 9,000 cubic feet at NTP.

e. The aggregate quantity of pyrophoric gases in the building shall not exceed the amounts set forth in Table 415.6.2.

[F] 415.11.1.2 Separation. *Fabrication areas*, whose sizes are limited by the quantity of hazardous materials allowed by Table 415.11.1.1.1, shall be separated from each other, from *corridors* and from other parts of the building by not less than 1-hour *fire barriers* constructed in accordance with Section 707 or *horizontal assemblies* constructed in accordance with Section 711, or both.

Exceptions:

1. Doors within such *fire barrier* walls, including doors to *corridors*, shall be only *self-closing fire door assemblies* having a *fire protection rating* of not less than $^3/_4$ hour.
2. Windows between *fabrication areas* and *corridors* are permitted to be fixed glazing *listed* and labeled for a *fire protection rating* of not less than $^3/_4$ hour in accordance with Section 716.

❖ Fabrication areas must be separated from other parts of the building, egress corridors and other fabrication areas by fire barriers, horizontal assemblies, or both, having a fire-resistance rating of at least 1 hour. While the potential fire exposure could warrant a higher fire-resistance rating, the additional protection features indicative of a Group H-5 facility result in a 1-hour fire-resistance rating being acceptable. The fire barriers must be constructed in accordance with Section 707. The horizontal assemblies must be constructed in accordance with Section 711. One reason for separating one fabrication area from another is the limitation of HPM quantities in a single fabrication area. If the need exists for quantities in excess of those permitted by Section 415.11.1.1.1, the area can be divided into two fabrication areas with 1-hour fire barriers and horizontal assemblies.

Exception 1 clarifies that all doors within the fire barrier walls that comprise the fabrication area must be self-closing and have a fire protection rating of $^3/_4$ hour; therefore, if a corridor wall was also part of the fabrication area, a $^3/_4$-hour opening protective would be required for the door opening. Openings in rated corridors that are not part of a fabrication area typically are required to only have a 20-minute fire protection rating. The doors are required to be self-closing since fabrication areas are "clean rooms" under positive pressure.

The $^3/_4$-hour requirement of Exception 2 will permit the use of fire protection-rated glazing or wired glass windows in accordance with Section 716.6. The use of windows is often desirable to facilitate the security of activities in the fabrication area as well as emergency response in the event of an accident.

[F] 415.11.1.3 Location of occupied levels. Occupied levels of *fabrication areas* shall be located at or above the first *story above grade plane*.

❖ Due to the potential extensive use of hazardous materials, the occupiable levels of fabrication areas are not permitted in basements or other areas below grade. Maintaining fabrication areas at or above grade aids in the egress of the occupants and access for fire department operations.

[F] 415.11.1.4 Floors. Except for surfacing, floors within *fabrication areas* shall be of noncombustible construction.

Openings through floors of *fabrication areas* are permitted to be unprotected where the interconnected levels are used solely for mechanical equipment directly related to such *fabrication areas* (see also Section 415.11.1.5).

Floors forming a part of an occupancy separation shall be liquid tight.

❖ Fabrication area floors are required to be noncombustible. The intent of the requirement for noncombustible floors is to prevent the structure from being involved in the fire from a liquid that may seep into the floor system. The requirement is not intended to apply to floor-covering material.

This section permits the interconnection of the operation floor and a mechanical floor, provided that the mechanical floor is essentially unoccupied. The intent is to prevent leakage past a required separation and to allow multiple-level fabrication areas where the space above or below the operation level is used for mechanical equipment, ducts and similar service equipment. Where a rated separation is required, the floors must be liquid tight. Whereas the hazards often involve liquids, floors are required to be liquid tight to prevent the spread of liquids from a spill to areas outside the fabrication area. "Liquid tight" requires that floors be sealed where they intersect with walls. This can be accomplished by a cove base of the same material as the floor.

[F] 415.11.1.5 Shafts and openings through floors. Elevator hoistways, vent *shafts* and other openings through floors shall be enclosed where required by Sections 712 and 713. Mechanical, duct and piping penetrations within a *fabrication area* shall not extend through more than two floors. The *annular space* around penetrations for cables, cable trays, tubing, piping, conduit or ducts shall be sealed at the floor level to restrict the movement of air. The *fabrication area*, including the areas through which the ductwork and piping extend, shall be considered a single conditioned environment.

❖ Section 712 specifies the conditions when a shaft enclosure or some other application of addressing vertical openings in multistory buildings is required. This section provides the requirements for specific opening penetrations through a floor within a fabrication area. The maximum number of stories that are allowed to be interconnected is three, since penetrations are limited to two floors. The penetrations for duct and piping are to have the annular space protected. Where such penetrations exist, the fire separation required by Section 415.11.1.2 is to separate the entire interconnected volume.

[F] 415.11.1.6 Ventilation. Mechanical exhaust *ventilation* at the rate of not less than 1 cubic foot per minute per square foot [0.0051 $m^3/(s \cdot m^2)$] of floor area shall be provided throughout the portions of the *fabrication area* where HPM are used or stored. The exhaust air duct system of one *fabrication area* shall not connect to another duct system outside that *fabrication area* within the building.

A *ventilation* system shall be provided to capture and exhaust gases, fumes and vapors at workstations.

Two or more operations at a workstation shall not be connected to the same exhaust system where either one or the combination of the substances removed could constitute a fire, explosion or hazardous chemical reaction within the exhaust duct system.

Exhaust ducts penetrating *fire barriers* constructed in accordance with Section 707 or *horizontal assemblies* constructed in accordance with Section 711 shall be contained in a *shaft* of equivalent fire-resistance- rated construction. Exhaust ducts shall not penetrate *fire walls*.

Fire dampers shall not be installed in exhaust ducts.

❖ A major design factor is the required minimum ventilation. The minimum rate is 1 cubic foot per minute (cfm) per square foot (0.0051 m^3/s x m^2), or about eight air changes per hour for an 8-foot-high (2438 mm) fabrication area. Typically, at least 24 air changes per hour are provided for operational needs and where unclassified electrical systems are to be used. In order to prevent the spread of gases or fumes, the exhaust air duct system of any fabrication area must not be connected to any other ventilation system. With regard to workstations, the required ventilation system is intended to prevent exposure of the work station operator to hazardous fumes or vapors and prevent hazardous concentrations of materials used in the manufacturing process. Additionally, ventilation systems must be designed to reduce the possibility of a chemical reaction occurring between two or more HPM or a material's and exhaust system's components. Where two or more chemicals may react dangerously if exhausted through a common duct, they must be removed separately. NFPA 491, which is included in the NFPA Fire Protection Guide on Hazardous Materials, is a good first reference when trying to determine if any two agents are reactive.

Ventilation should not be interrupted by a fire damper when a fire or other emergency occurs involving a workstation. This helps reduce the likelihood that hazardous combustion byproducts or hazardous concentrations of HPM are not forced back into the workstation or clean room. Continuous ventilation through a duct enclosed in a fire-resistance-rated shaft is required. Fire walls define separations between buildings. Ducts must never penetrate a barrier common to another building or occupancy. This reduces the likelihood of tampering with or interrupting the duct integrity.

[F] 415.11.1.7 Transporting hazardous production materials to fabrication areas. HPM shall be transported to *fabrication areas* through enclosed piping or tubing systems that comply with Section 415.11.6, through *service corridors* complying with Section 415.11.3, or in *corridors* as permitted in the exception to Section 415.11.2. The handling or transporting of HPM within *service corridors* shall comply with the *International Fire Code*.

❖ Except as permitted in the exception to Section 415.11.2 for existing facilities, all HPM must be transported to the fabrication area by service corridors or piping (see Sections 415.11.4 and 415.11.6). This reduces the chance that an egress route will be affected by an accidental leak or spill. Chapter 27 of the IFC contains additional requirements for the transport of HPM through all egress components as well as service corridors.

[F] 415.11.1.8 Electrical. Electrical equipment and devices within the *fabrication area* shall comply with NFPA 70. The requirements for hazardous locations need not be applied where the average air change is at least four times that set forth in Section 415.11.1.6 and where the number of air changes at any location is not less than three times that required by Section 415.11.1.6. The use of recirculated air shall be permitted.

❖ Consistent with Chapter 27, electrical installations are required to comply with NFPA 70. This section specifically permits the dilution of concentrations of hazardous areas such that the hazardous location provisions of NFPA 70 need not apply. This is because some of the equipment used in HPM facilities cannot comply with the hazardous location provisions of NFPA 70. The nature of the fabrication process also dictates the use of the dilution concept. Typically, such facilities have a ventilation system designed to provide at least 24 air changes per hour.

At least 3 cfm per square foot (0.132 $L/s/m^2$) or 24 air changes in an 8-foot-high (2438 mm) fabrication area are required at all locations in order not to apply the hazardous location provisions of NFPA 70. Also, an average rate of 4 cfm per square foot (0.176 $L/s/m^2$) or 32 air changes in an 8-foot-high (2438 mm) fabrication area are required if the hazardous location provisions of NFPA 70 are not applied.

[F] 415.11.1.8.1 Workstations. Workstations shall not be energized without adequate exhaust *ventilation*. See Section 415.11.1.6 for workstation exhaust *ventilation* requirements.

❖ Accidental exposure to flammable fumes or vapors can occur at workstations where hazardous materials are used. Either a mechanical or electrical interlock must be provided to engage the required exhaust ventilation system before HPM enters the workstation. This reduces the likelihood of gas or vapor exposure.

[F] 415.11.2 Corridors. *Corridors* shall comply with Chapter 10 and shall be separated from *fabrication areas* as specified in Section 415.11.1.2. *Corridors* shall not contain HPM and shall not be used for transporting such materials except through closed piping systems as provided in Section 415.11.6.4

Exception: Where existing *fabrication areas* are altered or modified, HPM is allowed to be transported in existing *corridors,* subject to the following conditions:

1. Nonproduction HPM is allowed to be transported in *corridors* if utilized for maintenance, lab work and testing.
2. Where existing *fabrication areas* are altered or modified, HPM is allowed to be transported in existing *corridors*, subject to the following conditions:
   - 2.1. Corridors. *Corridors* adjacent to the *fabrication area* where the alteration work is to be done shall comply with Section 1020 for a length determined as follows:
     - 2.1.1. The length of the common wall of the *corridor* and the *fabrication area*; and
     - 2.1.2. For the distance along the *corridor* to the point of entry of HPM into the *corridor* serving that *fabrication area.*
   - 2.2. *Emergency alarm system.* There shall be an emergency telephone system, a local manual alarm station or other *approved* alarm-initiating device within *corridors* at not more than 150-foot (45 720 mm) intervals and at each *exit* and doorway. The signal shall be relayed to an *approved* central, proprietary or remote station service or the emergency control station and shall also initiate a local audible alarm.
   - 2.3. Pass-throughs. *Self-closing* doors having a *fire protection rating* of not less than 1 hour shall separate pass-throughs from existing *corridors.* Pass-throughs shall be constructed as required for the *corridors* and protected by an *approved automatic sprinkler system.*

❖ Corridors are required to comply with the requirements of Section 1020 and must be separated from fabrication areas in accordance with Section 415.11.1.2. Additionally, typical egress corridors in new buildings are not to be used for transporting HPM to the fabrication area. In accordance with Section 415.11.1.7, HPM must be transported by service corridors or piping (see Sections 415.11.3 and 415.11.6, and Commentary Figure 415.11.2).

Figure 415.11.2. See page 4-120.

❖ Exception 1 allows corridors to be used for the movement of HPM that is only used for maintenance purposes, laboratory work and testing activities. The second exception addresses HPM facilities that existed before the adoption and enforcement of this section. It permits the transport of HPM in egress corridors in existing buildings under the conditions specified in Items 2.1, 2.2 and 2.3. When alterations are made to a fabrication area, such corridors must be upgraded. Chapter 27 of the IFC provides additional requirements for the storage, handling and transporting of HPM in existing HPM facilities. Item 2.1 of the exception requires that corridors be upgraded to a 1-hour fire-resistance rating for the entire length of the wall common to the fabrication area. Additionally, the entire length of the corridor that is used for transporting HPM to the fabrication area being renovated must also be upgraded to have a 1-hour fire-resistance rating. Item 2.2 requires a means to notify trained personnel of an emergency condition in an egress corridor. A local alarm is required to alert the occupants of a potential hazardous condition. The intent of Item 2.3 is to permit a "pass-through" for providing HPM to a fabrication area. A pass-through, such as a storage cabinet, is used to store and receive HPM for the fabrication area. The pass-through must be separated from the egress corridor by 1-hour fire-resistance-rated construction, including a 1-hour-rated self-closing fire door, and be protected by an automatic sprinkler system.

[F] 415.11.3 Service corridors. *Service corridors* within a Group H-5 occupancy shall comply with Sections 415.11.3.1 through 415.11.3.4.

❖ This section regulates the corridor that may be used for the transport of HPM in a Group H-5 facility of new construction. For the transport of HPM via piping systems, see Section 415.11.6.

It is also clarified that a potential separation due to the mixed occupancy provisions of Section 508.4 does not apply, since service corridors are considered part of the Group H-5 occupancy.

[F] 415.11.3.1 Use conditions. *Service corridors* shall be separated from *corridors* as required by Section 415.11.1.2. *Service corridors* shall not be used as a required *corridor.*

❖ Service corridors are used for transporting HPM in lieu of piping. The separation between a service corridor and an egress corridor must be in accordance with Section 415.11.1.2. As such, a minimum 1-hour fire separation is required. Service corridors are also not intended to be used as a portion of the required means of egress.

[F] 415.11.3.2 Mechanical ventilation. Service corridors shall be mechanically ventilated as required by Section 415.11.1.6 or at not less than six air changes per hour.

❖ Since the same HPM that may be present in the fabrication area may be transported in the service corridors, an adequate means of ventilation is required. This section requires a ventilation rate of either 1 cfm per square foot (0.044 L/s/m^2) of floor area in accordance with Section 415.11.1.6 or six air changes per hour, whichever is greater depending on the size and volume of the service corridor.

[F] 415.11.3.3 Means of egress. The distance of travel from any point in a *service corridor* to an *exit*, *exit access corridor* or door into a *fabrication area* shall be not greater than 75 feet (22 860 mm). Dead ends shall be not greater than 4 feet (1219 mm) in length. There shall be not less than two *exits*, and not more than one-half of the required *means of egress* shall require travel into a *fabrication area*. Doors from *service corridors* shall swing in the direction of egress travel and shall be *self-closing*.

❖ The exit access travel distance is limited to 75 feet (22 860 mm), except that the travel distance may be measured from a door to a fabrication area. In order to minimize the potential for an occupant to be trapped in a service corridor, dead ends must not exceed 4 feet (1219 mm). As such, the occupant would most likely be intimate with the source of the problem or located in an adjacent fabrication area from which egress is provided via exit access, not service, corridors.

An occupant at any location in a service corridor is required to have access to at least two means of egress. No more than 50 percent of the means of egress may be through a fabrication area that, in most cases, will be one of the means of egress. This assumes that, should an incident or problem occur, it will be in either the service corridor or the fabrication

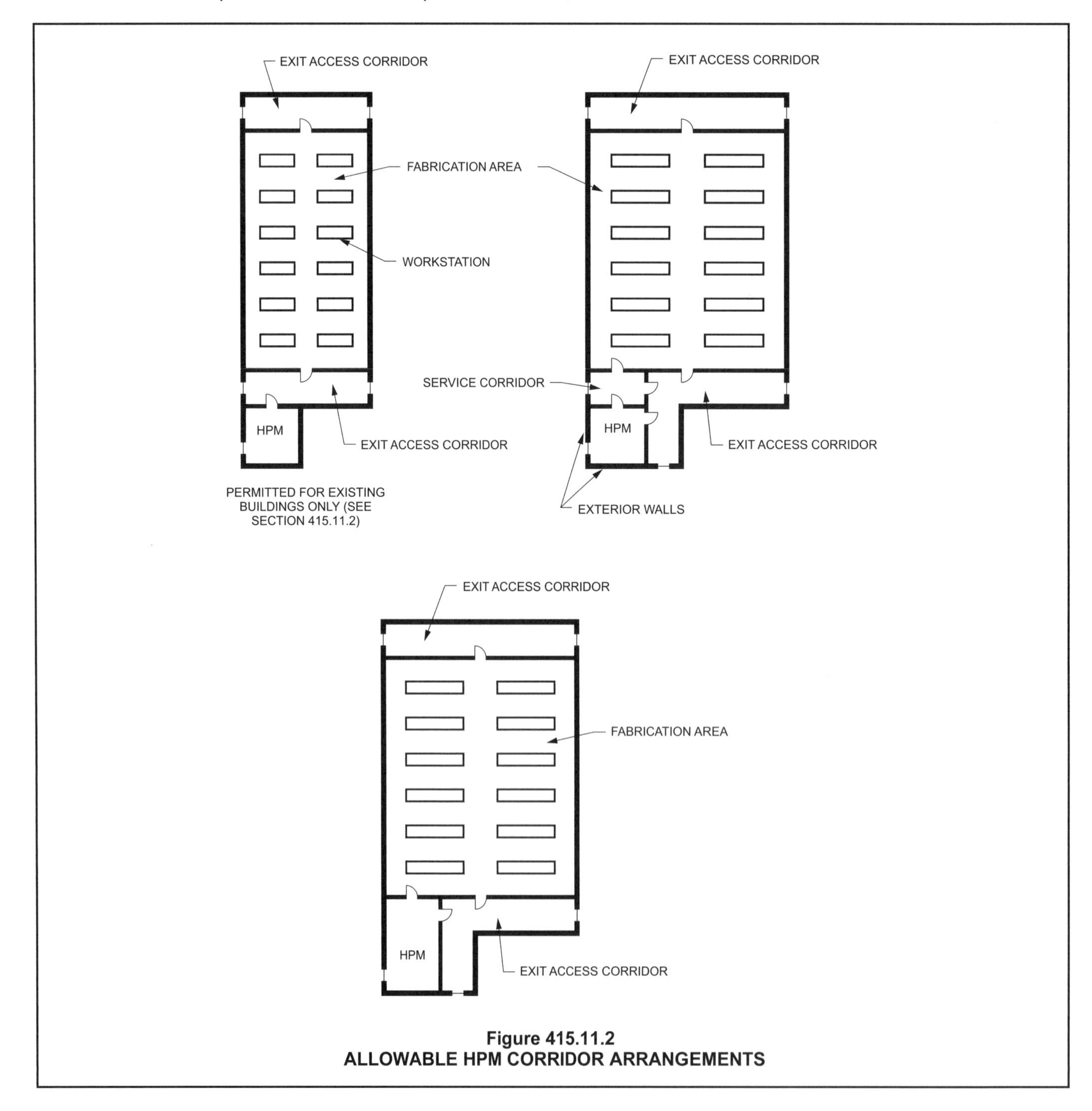

Figure 415.11.2
ALLOWABLE HPM CORRIDOR ARRANGEMENTS

area—not both. Because of the potential for a hazardous condition, all doors to service corridors must be self-closing and swing in the direction of egress travel. Smoke-detector-actuated hold-open devices are not acceptable because smoke detectors may not respond to hazardous conditions that do not involve products of combustion.

[F] 415.11.3.4 Minimum width. The clear width of a *service corridor* shall be not less than 5 feet (1524 mm), or 33 inches (838 mm) wider than the widest cart or truck used in the *service corridor*, whichever is greater.

❖ The minimum width requirement for the service corridor is to provide adequate clearance to get around the device used to transport the HPM in the event it is involved in an emergency incident in the service corridor. The size of the transport carts and trucks depends on the type and quantity of materials utilized in the facility.

[F] 415.11.3.5 Emergency alarm system. *Emergency alarm systems* shall be provided in accordance with this section and Sections 415.5.1 and 415.5.2. The maximum allowable quantity per *control area* provisions shall not apply to *emergency alarm systems* required for HPM.

❖ This section requires an emergency alarm system in all areas where HPM is transported or stored. The applicability of either Section 415.5.1 or 415.5.2 depends on whether the HPM material is in a storage or use condition. This section also clarifies that the requirement for an emergency alarm system in a Group H-5 facility in the locations identified in Sections 415.11.3.5.1 through 415.11.3.5.3 is not dependent on whether the maximum allowable quantities per control area of Tables 307.1(1) or 307.1(2) are exceeded.

[F] 415.11.3.5.1 Service corridors. An *emergency alarm system* shall be provided in *service corridors,* with no fewer than one alarm device in each *service corridor*.

❖ An emergency telephone system or manual alarm pull station is required in service corridors. Such devices must initiate an alarm at the emergency control station as well as activate a local audible device.

[F] 415.11.3.5.2 Corridors and interior exit stairways and ramps. Emergency alarms for *corridors, interior exit stairways* and *ramps* and *exit passageways* shall comply with Section 415.5.2.

❖ Since HPM would not be in exit access corridors or interior exit stairways unless they were being transported to another approved area, the emergency alarm requirements of Section 415.5.2 for dispensing, use and handling must be complied with.

[F] 415.11.3.5.3 Liquid storage rooms, HPM rooms and gas rooms. Emergency alarms for liquid storage rooms, HPM rooms and gas rooms shall comply with Section 415.5.1.

❖ This section requires compliance with the emergency alarm requirements of Section 415.5.1 for hazardous materials in a storage condition. This section addresses storage areas, which by their designation contain HPM in quantities greater than those listed in Table 307.1(1) or 307.1(2).

[F] 415.11.3.5.4 Alarm-initiating devices. An *approved* emergency telephone system, local alarm manual pull stations, or other *approved* alarm-initiating devices are allowed to be used as emergency alarm-initiating devices.

❖ This section classifies what constitutes an approved alarm-initiating device (see commentary, Section 415.5).

[F] 415.11.3.5.5 Alarm signals. Activation of the *emergency alarm system* shall sound a local alarm and transmit a signal to the emergency control station.

❖ The alarm signal is required to be transmitted to the emergency control station in order to notify trained personnel of an emergency condition. A local alarm is required to alert the occupants of a potentially hazardous condition.

[F] 415.11.4 Storage of hazardous production materials. Storage of hazardous production materials (HPM) in *fabrication areas* shall be within *approved* or *listed* storage cabinets or gas cabinets or within a workstation. The storage of HPM in quantities greater than those listed in Section 5004.2 of the *International Fire Code* shall be in liquid storage rooms, HPM rooms or gas rooms as appropriate for the materials stored. The storage of other hazardous materials shall be in accordance with other applicable provisions of this code and the *International Fire Code*.

❖ Section 415.11.4 addresses general storage provisions for HPM in a Group H-5 facility.

Even though the amount of HPM in a fabrication area is controlled, it still must be within approved cabinets or a workstation. This requirement is intended to limit the exposure to occupants of the fabrication area to only the material in use within that area.

The larger amounts of HPM typically stored in separate areas present a hazard comparable to other Group H facilities; therefore, such storage rooms need to meet similar code requirements. Storage rooms containing HPM in quantities greater than permitted by Section 2704.2 of the IFC are required to comply with the applicable provisions of Sections 415.11.5.1 through 415.11.5.9, depending on the state of the material.

It should be noted that the HPM quantity limitations per fabrication area are essentially identical to the limitations of Section 415.11.1.1.2.

[F] 415.11.5 HPM rooms, gas rooms, liquid storage room construction. HPM rooms, gas rooms and liquid shall be constructed in accordance with Sections 415.11.5.1 through 415.11.5.9.

❖ Sections 415.11.5.1 through 415.11.5.9 selectively address construction requirements for two types of storage rooms: those classified as HPM rooms or gas rooms, and those utilized as liquid storage rooms. The size and separation of such rooms is dependent on the type of materials stored.

[F] 415.11.5.1 HPM rooms and gas rooms. HPM rooms and gas rooms shall be separated from other areas by *fire barriers* constructed in accordance with Section 707 or *horizontal assemblies* constructed in accordance with Section 711, or both. The *fire-resistance rating* shall be not less than 2 hours where the area is 300 square feet (27.9 m^2) or more and not less than 1 hour where the area is less than 300 square feet (27.9 m^2).

❖ The amount of hazardous material in an HPM room or gas room is only limited by room size; therefore, since storage rooms in excess of 300 square feet (27.9 m^2) are not limited by size or quantity of hazardous material, construction of fire barriers and horizontal assemblies of not less than 2-hour fire-resistance rating is required. It should be noted that this section assumes the storage areas have at least one exterior wall with a minimum 30-foot (9144 mm) fire separation distance (see Section 415.11.5.4). If the room has an area of less than 300 square feet (27.9 m^2), the construction can be rated for 1 hour.

[F] 415.11.5.2 Liquid storage rooms. Liquid storage rooms shall be constructed in accordance with the following requirements:

1. Rooms greater than 500 square feet (46.5 m^2) in area, shall have no fewer than one exterior door *approved* for fire department access.
2. Rooms shall be separated from other areas by *fire barriers* constructed in accordance with Section 707 or *horizontal assemblies* constructed in accordance with Section 711, or both. The *fire-resistance rating* shall be not less than 1 hour for rooms up to 150 square feet (13.9 m^2) in area and not less than 2 hours where the room is more than 150 square feet (13.9 m^2) in area.
3. Shelving, racks and wainscotting in such areas shall be of noncombustible construction or wood of not less than 1-inch (25 mm) nominal thickness or fire-retardant-treated wood complying with Section 2303.2.
4. Rooms used for the storage of Class I flammable liquids shall not be located in a *basement*.

❖ This section assumes storage rooms are utilized for the storage of flammable and combustible liquids in closed containers. The size of the rooms is severely restricted when only 1-hour fire barriers and 1-hour horizontal assemblies are provided to limit the size of a potential fire. For rooms greater than 150 square feet (14 m^2) in area, a minimum 2-hour-rated construction is required because the size of the room or quantity of hazardous materials is not limited. Due to the need for readily available access in a fire scenario, all liquid storage rooms in excess of 500 square feet (46.5 m^2) must have at least one exterior door that provides access to the room. In accordance with Section 415.11.5.4, all liquid storage rooms, regardless of size, must have at least one exterior wall.

The construction limitations on the shelving racks and wainscotting are intended to limit their potential involvement in a fire scenario. Due to the volatile nature of Class I flammable liquids and the need for quick fire department access, they must be located in liquid storage rooms at or above grade.

[F] 415.11.5.3 Floors. Except for surfacing, floors of HPM rooms and liquid storage rooms shall be of noncombustible liquid-tight construction. Raised grating over floors shall be of noncombustible materials.

❖ Similar to the floor construction of fabrication areas, the floors of HPM rooms and liquid storage rooms are required to be noncombustible and liquid tight (see commentary, Section 415.11.1.4).

[F] 415.11.5.4 Location. Where HPM rooms, liquid storage rooms and gas rooms are provided, they shall have no fewer than one *exterior wall* and such wall shall be not less than 30 feet (9144 mm) from *lot lines*, including *lot lines* adjacent to *public ways*.

❖ Access from the exterior for fire-fighting operations is essential for storage rooms of hazardous materials. These rooms by their designation contain more than the maximum allowable quantities per control area listed in Tables 307.1(1) and 307.1(2). The 30-foot (9144 mm) limitation is based in part on the fact that the exterior wall need not have a fire-resistance rating depending on the construction type of the building (see Table 602). Such walls may also be used for explosion venting when such venting is required.

[F] 415.11.5.5 Explosion control. Explosion control shall be provided where required by Section 414.5.1.

❖ This section requires a method of explosion control only when an explosion hazard exists. Not all storage rooms are required to have explosion control. Section 414.5.1 requires a method of explosion control, such as explosion venting or barricade construction, depending on the specific hazardous material involved in accordance with the IFC. Table 414.5.1 provides guidance as to when explosion control is required for various hazardous materials.

[F] 415.11.5.6 Exits. Where two *exits* are required from HPM rooms, liquid storage rooms and gas rooms, one shall be directly to the outside of the building.

❖ Depending on the common path of egress travel available in accordance with Section 1006.2.1, two exits may be required from all HPM rooms, liquid storage rooms and gas rooms. For example, a liquid storage room containing Class IB flammable liquids in normally closed containers would be considered a Group H-3 occupancy. As such, a maximum common path of travel of 25 feet (7620 mm) would be permitted before a second exit, one of which was directly to the exterior, is required. Storage rooms are required to be located along an exterior wall. In addition to the requirements of this section, all liquid storage rooms in excess of 500 square feet (46 m^2) are required to have at least one exterior door regardless of travel distance.

[F] 415.11.5.7 Doors. Doors in a *fire barrier* wall, including doors to *corridors*, shall be *self-closing fire door assemblies* having a *fire protection rating* of not less than $^3/_4$ hour.

❖ Doors that are in an interior wall that comprises part of the storage room enclosure are required to have a minimum $^3/_4$-hour fire protection rating. The actual rating is dependent on the fire-resistance rating of the fire barrier separation required. While a 1-hour fire barrier would permit a door with a $^3/_4$-hour fire protection rating, a 2-hour fire barrier would require a door to have a $1^1/_2$-hour fire protection rating in accordance with Table 716.5.

[F] 415.11.5.8 Ventilation. Mechanical exhaust ventilation shall be provided in liquid storage rooms, HPM rooms and gas rooms at the rate of not less than 1 cubic foot per minute per square foot (0.044 L/s/m^2) of floor area or six air changes per hour.

Exhaust ventilation for gas rooms shall be designed to operate at a negative pressure in relation to the surrounding areas and direct the exhaust ventilation to an exhaust system.

❖ Similar to fabrication areas and service corridors, mechanical exhaust ventilation is required in liquid storage rooms, HPM rooms and gas rooms (see commentary, Sections 415.11.1.6 and 415.11.3.2).

[F] 415.11.5.9 Emergency alarm system. An *approved emergency alarm system* shall be provided for HPM rooms, liquid storage rooms and gas rooms.

Emergency alarm-initiating devices shall be installed outside of each interior *exit* door of such rooms.

Activation of an emergency alarm-initiating device shall sound a local alarm and transmit a signal to the emergency control station.

An *approved* emergency telephone system, local alarm manual pull stations or other *approved* alarm-initiating devices are allowed to be used as emergency alarm-initiating devices.

❖ An emergency alarm system is required for all storage rooms containing hazardous materials in excess of the quantities permitted in Tables 307.1(1) and 307.1(2). See the commentary to Section 415.11.3.5 for similar emergency alarm requirements.

[F] 415.11.6 Piping and tubing. Hazardous production materials piping and tubing shall comply with this section and ASME B31.3.

❖ This section addresses the design requirements for all piping and tubing utilized to transport HPM throughout a Group H-5 facility, including where it can be located within the building.

The nature of the HPM involved requires strict control of both the materials used in piping and the joint methods used to avoid leakage. ASME B31.3 provides criteria for piping installation applicable to HPM facilities.

[F] 415.11.6.1 HPM having a health-hazard ranking of 3 or 4. Systems supplying HPM liquids or gases having a health-hazard ranking of 3 or 4 shall be welded throughout, except for connections, to the systems that are within a ventilated enclosure if the material is a gas, or an *approved* method of drainage or containment is provided for the connections if the material is a liquid.

❖ The primary purpose of this section is to minimize the potential for a leak of HPM. The use of mechanical compression-type fittings or other nonwelded joints in areas of health hazard 3 or 4 must be limited to areas where any leaks will be vented and the materials exhausted. If the material is a liquid, then a method of drainage or containment is required to minimize the hazard.

[F] 415.11.6.2 Location in service corridors. Hazardous production materials supply piping or tubing in *service corridors* shall be exposed to view.

❖ Concealment of piping in service corridors must be avoided to provide a means of monitoring for leaks.

[F] 415.11.6.3 Excess flow control. Where HPM gases or liquids are carried in pressurized piping above 15 pounds per square inch gauge (psig) (103.4 kPa), excess flow control shall be provided. Where the piping originates from within a liquid storage room, HPM room or gas room, the excess flow control shall be located within the liquid storage room, HPM room or gas room. Where the piping originates from a bulk source, the excess flow control shall be located as close to the bulk source as practical.

❖ Excess flow control valves regulate the flow of hazardous materials within the piping system. The valves are designed to shut off in the event the predetermined flow is exceeded.

[F] 415.11.6.4 Installations in corridors and above other occupancies. The installation of HPM piping and tubing within the space defined by the walls of *corridors* and the floor or roof above, or in concealed spaces above other occupancies, shall be in accordance with Sections 415.11.6.1 through 415.11.6.3 and the following conditions:

1. Automatic sprinklers shall be installed within the space unless the space is less than 6 inches (152 mm) in the least dimension.
2. *Ventilation* not less than six air changes per hour shall be provided. The space shall not be used to convey air from any other area.
3. Where the piping or tubing is used to transport HPM liquids, a receptor shall be installed below such piping or tubing. The receptor shall be designed to collect any discharge or leakage and drain it to an *approved* location. The 1-hour enclosure shall not be used as part of the receptor.
4. HPM supply piping and tubing and nonmetallic waste lines shall be separated from the corridor and from occupancies other than Group H-5 by fire barriers or by an approved method or assembly that has a fire-resistance rating of not less than 1 hour. Access openings into the enclosure shall be protected by approved fire-protection-rated assemblies.

5. Readily accessible manual or automatic remotely activated fail-safe emergency shutoff valves shall be installed on piping and tubing other than waste lines at the following locations:
   5.1. At branch connections into the *fabrication area*.
   5.2. At entries into *corridors*.

Exception: Transverse crossings of the *corridors* by supply piping that is enclosed within a ferrous pipe or tube for the width of the *corridor* need not comply with Items 1 through 5.

❖ The installation of HPM piping in the space above an egress corridor or another occupancy, as well as the cavity of the egress corridor wall, present a potential source of hazard to the building's occupants. The five requirements provided for such installations serve to mitigate the hazard by suppression, ventilation, containment and ignition control. In order to address the potential hazards, the containment must be a fire barrier with at least a 1-hour fire-resistance rating, drainage receptors, excess flow control and shutoff valves.

When the piping traverses a corridor, the use of a coaxial enclosed pipe around the HPM piping is considered acceptable for providing the required separation and containment of a potential leak. The assumption is that the open ends of that pipe are in an HPM facility and, therefore, a leak into the outer casing can be monitored. If the adjacent areas that contain the open ends are not in an HPM facility, then the outer-jacket method cannot be used.

[F] 415.11.6.5 Identification. Piping, tubing and HPM waste lines shall be identified in accordance with ANSI A13.1 to indicate the material being transported.

❖ To facilitate monitoring, detecting and controlling any possible leaks, all piping, tubing and HPM waste lines must be identified in accordance with ANSI A13.1. This standard contains criteria for properly identifying the piping system, including labeling and frequency or distribution of the signs or labels.

[F] 415.11.7 Continuous gas detection systems. A *continuous gas detection system* shall be provided for HPM gases where the physiological warning threshold level of the gas is at a higher level than the accepted permissible exposure limit (PEL) for the gas and for flammable gases in accordance with Sections 415.11.7.1 and 415.11.7.2.

❖ A gas detection system in the room or area utilized for the storage or use of HPM gases provides early notification of a leak that is occurring before the escaping gas reaches hazardous concentration levels.

[F] 415.11.7.1 Where required. A *continuous gas detection system* shall be provided in the areas identified in Sections 415.11.7.1.1 through 415.11.7.1.4.

❖ Sections 415.11.7.1.1 through 415.11.7.1.4 prescribe the locations in a Group H-5 facility when a gas detection system is required.

[F] 415.11.7.1.1 Fabrication areas. A *continuous gas detection system* shall be provided in *fabrication areas* where gas is used in the *fabrication area*.

❖ All fabrication areas that utilize HPM gases must have a gas detection system. It should be noted that gas detection is often installed within workstations as a means of early detection of leaks. Such detection is generally not acceptable as an alternative to gas detection for the fabrication area, since a leak may occur remote from the workstation.

[F] 415.11.7.1.2 HPM rooms. A *continuous gas detection system* shall be provided in HPM rooms where gas is used in the room.

❖ HPM rooms, which by definition contain more than the quantities of hazardous materials per control area permitted by Tables 307.1(1) and 307.1(2), are required to have a gas detection system.

[F] 415.11.7.1.3 Gas cabinets, exhausted enclosures and gas rooms. A *continuous gas detection system* shall be provided in gas cabinets and exhausted enclosures. A *continuous gas detection system* shall be provided in gas rooms where gases are not located in gas cabinets or exhausted enclosures.

❖ In the potential event of a leaking cylinder of a hazardous gas, gas cabinets, exhausted enclosures and gas rooms must have a gas detection system.

[F] 415.11.7.1.4 Corridors. Where gases are transported in piping placed within the space defined by the walls of a *corridor* and the floor or roof above the *corridor*, a *continuous gas detection system* shall be provided where piping is located and in the *corridor*.

Exception: A *continuous gas detection system* is not required for occasional transverse crossings of the *corridors* by supply piping that is enclosed in a ferrous pipe or tube for the width of the *corridor*.

❖ In addition to the requirements of Section 415.11.6.4 for HPM piping and tubing in an egress corridor, a gas detection system is required for early notification of a potential leak of an HPM gas.

The exception is similar to the exception in Section 415.11.6.4. In essence, the gas detection system required by this section as well as the additional provisions in Section 415.11.6.4 are not required if the exception is met (see commentary, Section 415.11.6.4).

[F] 415.11.7.2 Gas detection system operation. The *continuous gas detection system* shall be capable of monitoring the room, area or equipment in which the gas is located at or below all the following gas concentrations:

1. Immediately dangerous to life and health (IDLH) values where the monitoring point is within an exhausted enclosure, ventilated enclosure or gas cabinet.
2. Permissible exposure limit (PEL) levels where the monitoring point is in an area outside an exhausted enclosure, ventilated enclosure or gas cabinet.

3. For flammable gases, the monitoring detection threshold level shall be vapor concentrations in excess of 25 percent of the lower flammable limit (LFL) where the monitoring is within or outside an exhausted enclosure, ventilated enclosure or gas cabinet.
4. Except as noted in this section, monitoring for highly toxic and toxic gases shall also comply with Chapter 60 of the *International Fire Code*.

❖ This section requires gas detection systems to be capable of sensing a leak at or below the permissible exposure limit (PEL). This exposure limit regulated by OSHA to prevent adverse health effects is the breathing zone exposure limit for employees over an 8-hour time weighted average. In most cases, gas detection in the semiconductor industry is conducted in an exhausted enclosure, ventilated enclosure or gas cabinet and not in the breathing zone of the employee; is designed to detect and alert employees of leaks inside exhausted enclosures, ventilated enclosures or gas cabinets; and is not intended to estimate potential employee breathing zone exposures. The semiconductor industry addressed this by codifying Section 10.9 of NFPA 318 to differentiate gas detection set points in exhausted enclosures (set at the IDLH) with gas detection when the monitoring point is in an area outside an exhausted enclosure, ventilated enclosure or gas cabinet. This section is consistent with the provisions of Section 10.9 of NFPA 318 guidelines that are much more relevant to the type of monitoring performed in semiconductor manufacturing (inside exhausted enclosures, ventilated enclosures or gas cabinets). Monitoring in the semiconductor industry is designed to detect and alert employees of leaks inside exhausted enclosures, ventilated enclosures and gas cabinets, and is not intended to estimate potential employee breathing zone exposures. Therefore, set points are not required or recommended to be set at occupational exposure limits (e.g., TLVs or PELs). Additionally, the 25 percent LFL is consistent with both Section 510.2 of the IMC and Section 10.9 of NFPA 318. Section Chapter 60 of the IFC contains additional requirements for the monitoring of highly toxic and toxic compressed gases.

[F] 415.11.7.2.1 Alarms. The gas detection system shall initiate a local alarm and transmit a signal to the emergency control station when a short-term hazard condition is detected. The alarm shall be both visual and audible and shall provide warning both inside and outside the area where the gas is detected. The audible alarm shall be distinct from all other alarms.

❖ The required local alarm is intended to alert occupants to a hazardous condition in the vicinity of where the HPM gases are being stored or used. The alarm is not intended to be an evacuation alarm; however, it is required to be monitored to hasten emergency personnel response.

[F] 415.11.7.2.2 Shutoff of gas supply. The gas detection system shall automatically close the shutoff valve at the source on gas supply piping and tubing related to the system being monitored for which gas is detected when a short-term hazard condition is detected. Automatic closure of shutoff valves shall comply with the following:

1. Where the gas detection sampling point initiating the gas detection system alarm is within a gas cabinet or exhausted enclosure, the shutoff valve in the gas cabinet or exhausted enclosure for the specific gas detected shall automatically close.
2. Where the gas detection sampling point initiating the gas detection system alarm is within a room and compressed gas containers are not in gas cabinets or an exhausted enclosure, the shutoff valves on all gas lines for the specific gas detected shall automatically close.
3. Where the gas detection sampling point initiating the gas detection system alarm is within a piping distribution manifold enclosure, the shutoff valve supplying the manifold for the compressed gas container of the specific gas detected shall automatically close.

Exception: Where the gas detection sampling point initiating the gas detection system alarm is at the use location or within a gas valve enclosure of a branch line downstream of a piping distribution manifold, the shutoff valve for the branch line located in the piping distribution manifold enclosure shall automatically close.

❖ Where gas detection systems are required, automatic emergency shutoff valves are required to stop the flow of hazardous material from possibly deteriorating further in an emergency.

[F] 415.11.8 Manual fire alarm system. An *approved* manual *fire alarm* system shall be provided throughout buildings containing Group H-5. Activation of the alarm system shall initiate a local alarm and transmit a signal to the emergency control station. The *fire alarm* system shall be designed and installed in accordance with Section 907.

❖ Due to the type and potential quantities of hazardous materials that could be present in a Group H-5 facility, a manual means of activating an evacuation alarm is essential for the safety of the occupants in an emergency situation. The alarm signal must be transmitted to the emergency control station. The fire alarm must also be in accordance with the applicable provisions of Section 907 and NFPA 72.

[F] 415.11.9 Emergency control station. An emergency control station shall be provided in accordance with Sections 415.11.9.1 through 415.11.9.3.

❖ Due to the extent of hazardous materials permitted in an HPM facility, an approved on-site location is needed where the alarm signals from emergency equipment can be received.

[F] 415.11.9.1 Location. The emergency control station shall be located on the premises at an *approved* location outside the *fabrication area*.

❖ The emergency control station must be located on-site at a location approved by the building official. The emergency control station must not be located in an area where hazardous materials are stored, used or transported, such as a fabrication area.

[F] 415.11.9.2 Staffing. Trained personnel shall continuously staff the emergency control station.

❖ Appropriate response by trained personnel is essential to any emergency event involving HPM materials.

[F] 415.11.9.3 Signals. The emergency control station shall receive signals from emergency equipment and alarm and detection systems. Such emergency equipment and alarm and detection systems shall include, but not be limited to, the following where such equipment or systems are required to be provided either in this chapter or elsewhere in this code:

1. *Automatic sprinkler system* alarm and monitoring systems.
2. Manual *fire alarm* systems.
3. *Emergency alarm systems.*
4. *Continuous gas detection systems.*
5. Smoke detection systems.
6. Emergency power system.
7. Automatic detection and alarm systems for pyrophoric liquids and Class 3 water-reactive liquids required in Section 2705.2.3.4 of the *International Fire Code.*
8. Exhaust *ventilation* flow alarm devices for pyrophoric liquids and Class 3 water-reactive liquids cabinet exhaust *ventilation* systems required in Section 2705.2.3.4 of the *International Fire Code.*

❖ This section specifies the types of systems that are to be monitored by the emergency control station.

[F] 415.11.10 Emergency power system. An emergency power system shall be provided in Group H-5 occupancies in accordance with Section 2702. The emergency power system shall supply power automatically to the electrical systems specified in Section 415.11.10.1 when the normal electrical supply system is interrupted.

❖ A backup emergency power source is considered essential for systems that are monitoring and protecting hazardous materials in a Group H-5 occupancy. Without an emergency power system, all required electrical controls or equipment monitoring hazardous materials would be rendered inoperative if a power failure or other electrical system failure were to occur.

[F] 415.11.10.1 Required electrical systems. Emergency power shall be provided for electrically operated equipment and connected control circuits for the following systems:

1. HPM exhaust *ventilation* systems.
2. HPM gas cabinet *ventilation* systems.
3. HPM exhausted enclosure *ventilation* systems.
4. HPM gas room *ventilation* systems.
5. HPM gas detection systems.
6. *Emergency alarm systems.*
7. Manual and automatic *fire alarm* systems.
8. *Automatic sprinkler system* monitoring and alarm systems.
9. Automatic alarm and detection systems for pyrophoric liquids and Class 3 water-reactive liquids required in Section 2705.2.3.4 of the *International Fire Code.*
10. Flow alarm switches for pyrophoric liquids and Class 3 water-reactive liquids cabinet exhaust *ventilation* systems required in Section 2705.2.3.4 of the *International Fire Code.*
11. Electrically operated systems required elsewhere in this code or in the *International Fire Code* applicable to the use, storage or handling of HPM.

❖ This section specifies the types of systems within a Group H-5 occupancy that are required to be connected to an approved emergency power system. As indicated in Section 2702, all emergency power systems must be installed in accordance with the applicable requirements of NFPA 70, 110 and 111.

[F] 415.11.10.2 Exhaust ventilation systems. Exhaust *ventilation* systems are allowed to be designed to operate at not less than one-half the normal fan speed on the emergency power system where it is demonstrated that the level of exhaust will maintain a safe atmosphere.

❖ Emergency power for exhaust ventilation is required to prevent hazardous concentrations of HPM fumes or vapors in areas such as workstations or fabrication areas. Fans for exhaust ventilation draw a considerable amount of current when operating. Running exhaust fans at a reduced speed may be desirable when it will not endanger the operator or result in a hazardous condition. However, an exhaust fan must not be run at a speed less than 50 percent of its rating, even if a slower speed will not produce a serious hazard.

[F] 415.11.11 Automatic sprinkler system protection in exhaust ducts for HPM. An *approved automatic sprinkler system* shall be provided in exhaust ducts conveying gases, vapors, fumes, mists or dusts generated from HPM in accordance with Sections 415.11.11.1 through 415.10.11.3 and the *International Mechanical Code.*

❖ This section prescribes the sprinkler system requirements for exhaust ducts for HPM. The requirements depend on the construction materials of the exhaust duct.

An exhaust duct for HPM materials could convey flammable and combustible gases, fumes, vapors or ducts. To provide protection against the spread of fire within the exhaust system and to prevent a duct fire from involving the building, sprinkler protection is required in the exhaust duct. The use of an extinguishing agent other than water, based on agent compatibility, would be subject to local approval. Section 510 of the IMC contains additional requirements for protecting hazardous exhaust systems.

[F] 415.11.11.1 Metallic and noncombustible nonmetallic exhaust ducts. An *approved automatic sprinkler system* shall be provided in metallic and noncombustible nonmetallic exhaust ducts where all of the following conditions apply:

1. Where the largest cross-sectional diameter is equal to or greater than 10 inches (254 mm).
2. The ducts are within the building.
3. The ducts are conveying flammable gases, vapors or fumes.

❖ Sprinklers are required within each individual duct when all three of the following conditions exist:

1. Cross-sectional diameter at the widest point is equal to or exceeds 10 inches (254 mm).
2. Ducts are located within the building.
3. Ducts convey gases or vapors within the flammable range.

Commentary Figure 415.11.11.1 illustrates how to measure the cross-sectional diameter of various duct shapes. Provisions of this section require the square and rounded ducts to be protected by automatic sprinklers. The round or elliptical ducts depicted on the right side of the diagram are not required to be protected.

[F] 415.11.11.2 Combustible nonmetallic exhaust ducts. *Automatic sprinkler system* protection shall be provided in combustible nonmetallic exhaust ducts where the largest cross-sectional diameter of the duct is equal to or greater than 10 inches (254 mm).

Exception: Ducts need not be provided with automatic sprinkler protection as follows:

1. Ducts *listed* or *approved* for applications without *automatic sprinkler system* protection.
2. Ducts not more than 12 feet (3658 mm) in length installed below ceiling level.

❖ Galvanized steel is not an appropriate duct material for all substances handled in the broad category of hazardous exhaust systems. A duct material compatible with the exhaust must be selected, and factors such as corrosiveness, abrasion resistance, chemical resistance and operating temperatures must be taken into account. Section 510.8 of the IMC contains additional provisions for duct construction that is part of a hazardous exhaust system. As such, automatic sprinkler protection is also required for all combustible nonmetallic ducts with a cross-sectional diameter equal to or greater than 10 inches (254 mm).

Exception 1 states that automatic sprinklers are not required where the risk to people or property is limited, such as when nonmetallic ducts approved for installation without sprinklers are used. Exception 2 states that when ducts do not exceed 12 feet (3658 mm) in length and are installed exposed below ceiling level, sprinklers may be omitted. The limited duct length and exposure reduces the potential of a concealed fire hazard.

[F] 415.11.11.3 Automatic sprinkler locations. Sprinkler systems shall be installed at 12-foot (3658 mm) intervals in

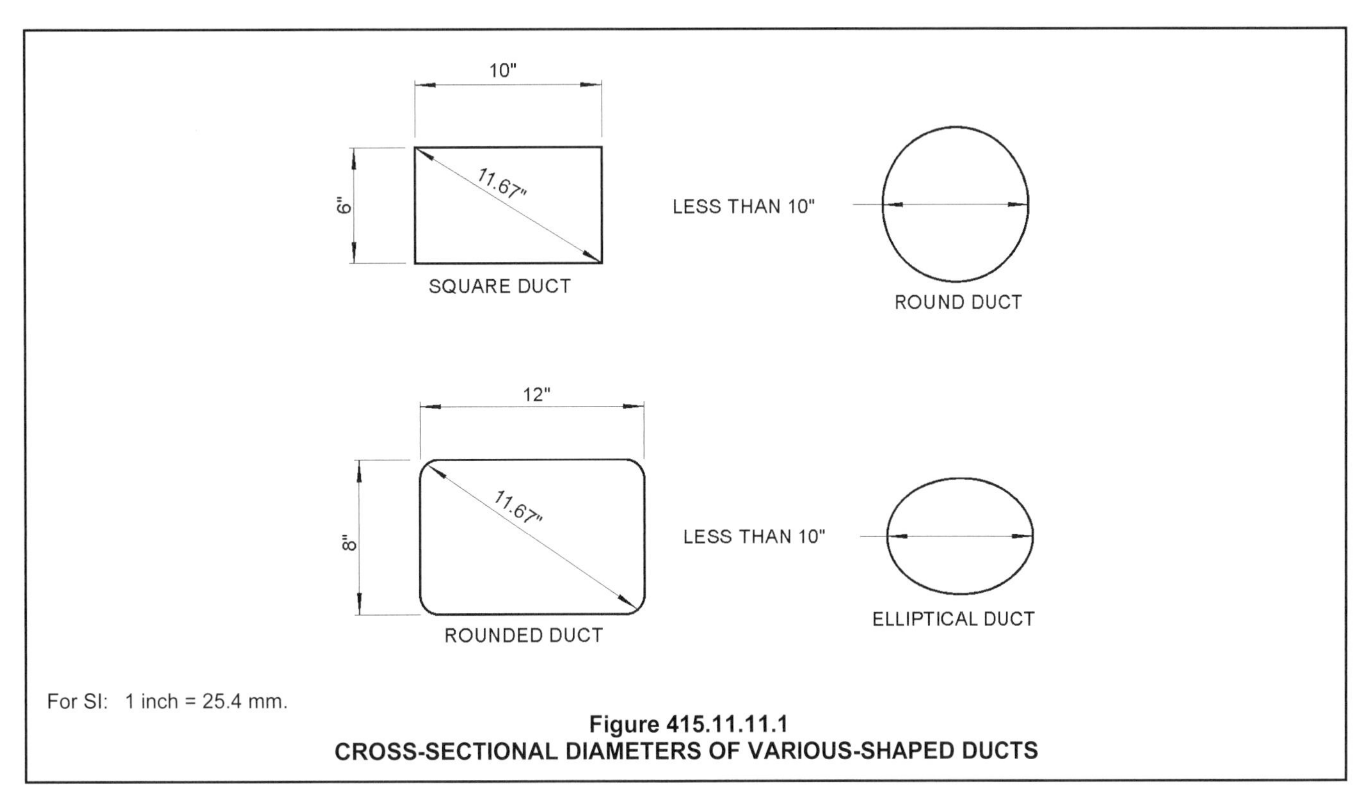

Figure 415.11.11.1
CROSS-SECTIONAL DIAMETERS OF VARIOUS-SHAPED DUCTS

horizontal ducts and at changes in direction. In vertical ducts, sprinklers shall be installed at the top and at alternate floor levels.

❖ Adequate sprinkler coverage needs to be maintained to prevent the spread of fire within the exhaust system. All sprinkler system components must also be in compliance with the applicable provisions of NFPA 13.

SECTION 416
APPLICATION OF FLAMMABLE FINISHES

[F] 416.1 General. The provisions of this section shall apply to the construction, installation and use of buildings and structures, or parts thereof, for the application of flammable finishes. Such construction and equipment shall comply with the *International Fire Code*.

❖ The principal hazards associated with paint spraying and spray booths originate from the presence of flammable liquids or powders and their vapors or mists. The purpose of this section is to provide requirements that address the hazards associated with spray applications and dipping or coating applications involving flammable paints, varnishes and lacquers. The requirements listed include such areas as ventilation, automatic sprinklers, control of ignition sources and proper operation of the equipment. In addition to the provisions of this section, the provisions of the IMC and Chapter 24 of the IFC also apply.

The provisions of this section apply to the indoor use of spray applications and dipping or coating applications involving flammable paints, varnishes and lacquers. Outdoor applications involving flammable paints, varnishes and lacquers are not covered, since overspray deposits are not likely to create hazardous conditions, and flammable vapor-air mixtures are minimized because of atmospheric dilution. Safeguards, such as ventilation, ignition control, material storage and waste disposal (see Chapter 24 of the IFC), should still apply. This reference to the IFC and its provisions regarding flammable finishes covers the application of flammable or combustible materials when applied as a spray by compressed air; via "airless" or "hydraulic atomization" steam; electrostatically; or by any other means in continuous or intermittent processes. They also cover the application of combustible powders when applied by powder spray guns, electrostatic powder spray guns, fluidized beds or electrostatic fluidized beds. The IFC and its referenced standards contain provisions relating to the location of spray areas, ignition sources, ventilation, liquid storage and handling, protection, operation, maintenance and training. The provisions found in Chapter 24 of the IFC also apply to: processes in which articles or materials are passed through the contents of tanks, vats or containers of flammable or combustible liquids, including dipping; roll; flow and curtain coating; finishing; treating; cleaning; and similar processes. Requirements for the location of dipping and coating processes, ventilation, equipment construction, liquid storage and handling, protection, operation, maintenance and training are also provided.

[F] 416.2 Spray rooms. Spray rooms shall be enclosed with not less than 1-hour *fire barriers* constructed in accordance with Section 707 or *horizontal assemblies* constructed in accordance with Section 711, or both. Floors shall be waterproofed and drained in an *approved* manner.

❖ A spray room is a power-ventilated, fully enclosed room used exclusively for open spraying of flammable and combustible materials. The entire spray room is considered the spray area. The primary difference between a spray room and a spray booth is that spray booths are partially open. A spray room is to be enclosed with fire barriers and horizontal assemblies having a fire-resistance rating of at least 1 hour. Waterproof floors are to be arranged to drain to the outside of the building, to internal drains or to other suitable places. Properly designed and guarded drains or scuppers of sufficient number and size to dispose of all surplus water likely to be discharged by automatic sprinklers must be provided.

[F] 416.2.1 Surfaces. The interior surfaces of spray rooms shall be smooth and shall be so constructed to permit the free passage of exhaust air from all parts of the interior and to facilitate washing and cleaning, and shall be so designed to confine residues within the room. Aluminum shall not be used.

❖ Rough, corrugated or uneven surfaces are difficult to clean. Periodic cleaning of interior surfaces reduces the fire hazard posed by the accumulation of flammable or combustible coatings. Due to the physical properties of aluminum, it is unsuitable for cleaning and scraping of overspray residue.

[F] 416.2.2 Ventilation. Mechanical *ventilation* and interlocks with the spraying operation shall be in accordance with the *International Mechanical Code*.

❖ The provisions addressing ventilation in spray rooms where flammable finishes are applied are found in Section 502.7 of the IMC. For effective ventilation, the system must run continuously during the spraying operation phase and the drying phase of the process when vapors are generated. The required interlock between the ventilation and spraying equipment is intended to reduce the likelihood that the operator will fail to activate the ventilation system prior to use of the spray equipment.

[F] 416.3 Spraying spaces. Spraying spaces shall be ventilated with an exhaust system to prevent the accumulation of flammable mist or vapors in accordance with the *International Mechanical Code*. Where such spaces are not separately enclosed, noncombustible spray curtains shall be provided to restrict the spread of flammable vapors.

❖ The objective of ventilation is to remove flammable vapors and mists so as to minimize the potential for a flash fire or explosion. The spray area is required to be ventilated in accordance with the provisions of the IMC. When the spray process is not separated from the other operations or areas, a noncombustible spray curtain must be provided to restrict the spread of flammable vapors. The IFC requires ventilation systems and enclosures to be interlocked with the spraying equipment to ensure that fans operate and doors remain closed when in use.

[F] 416.3.1 Surfaces. The interior surfaces of spraying spaces shall be smooth and continuous without edges; shall be so constructed to permit the free passage of exhaust air from all parts of the interior and to facilitate washing and cleaning; and shall be so designed to confine residues within the spraying space. Aluminum shall not be used.

❖ See commentary, Section 416.2.1.

[F] 416.4 Spray booths. Spray booths shall be designed, constructed and operated in accordance with the *International Fire Code*.

❖ Detailed requirements for the design and construction of spray booths are given in Section 2404.3.2 of the IFC. The IFC also addresses requirements for ventilation of adjacent areas, fire protection, fire extinguishers, control of ignition sources, housekeeping during operations, lighting and operation control through interlocks.

[F] 416.5 Fire protection. An *automatic sprinkler system* or *fire-extinguishing system* shall be provided in all spray, dip and immersing spaces and storage rooms and shall be installed in accordance with Chapter 9.

❖ Spray application operations are to be protected with an automatic sprinkler system or fire-extinguishing system. While an automatic sprinkler system is the most desirable type of suppression system, this section would allow the use of other effective extinguishing agents, such as dry chemical or foam. Because of the ease of ignition of flammable liquids, specifically spray paint applications of a flammable liquid, an automatic fire sprinkler system or fire-extinguishing system is required in rooms in which spray painting is conducted, as well as rooms in which flammable materials are used for painting, brushing, dipping or mixing on a regular basis. Dry sprinklers should not be used for the protection of spray operations, except possibly at the exhaust duct penetrations to the outside. Wet-pipe, preaction or deluge systems should be used so that water is placed on the fire in the shortest possible time. If the entire building is not protected with an automatic sprinkler system, then the system protecting the spray booth or room may be a limited area sprinkler system in accordance with Section 903.3.8. Locating sprinklers in paint spray booths presents an unusual problem, since the sprinkler may become clogged with paint. The most satisfactory solution is to locate the sprinklers in areas in a way that the paint spray will most likely not reach the sprinkler. Even in the most ideal locations, the sprinkler will need to be cleaned very frequently. A coating of grease with a low melting point, such as petroleum jelly, motor oil or a soft neutral soap, will facilitate cleaning of the sprinkler. Other alternatives that are commonly used include polyethylene or cellophane bags with a thickness of 0.003 inches (0.076 mm) or less or thin paper bags to protect the sprinklers.

Additional guidance on fire protection requirements for the application of flammable finishes can be found in Chapter 24 of the IFC.

SECTION 417
DRYING ROOMS

[F] 417.1 General. A drying room or dry kiln installed within a building shall be constructed entirely of *approved* noncombustible materials or assemblies of such materials regulated by the *approved* rules or as required in the general and specific sections of this chapter for special occupancies and where applicable to the general requirements of the *International Mechanical Code*.

❖ This section establishes specific safety requirements for drying rooms and dry kilns used in conjunction with the drying (accelerated seasoning) of lumber, and dryers or dehydrators used to reduce the moisture content of agricultural products. Included are drying operations for other combustible materials, such as certain building materials, textiles and fabrics. Drying operations associated with the application of flammable finishes are regulated by Section 416. The hazards associated with drying rooms and dry kilns relate to the volume of readily combustible materials being processed at temperatures conducive to their ignition in the event of a system malfunction.

The materials or assemblies used in the construction of drying rooms or dry kilns located within buildings must be noncombustible, regardless of the type of construction of the building itself (see commentary, Section 703.5 for a discussion of noncombustibility criteria) when required by Chapter 4 or the International Mechanical Code. The use of noncombustible materials of construction for drying rooms and dry kilns provides for a measure of confinement of a fire occurring within and prevents the room or kiln from contributing to the fuel load exposed to an unwanted fire.

[F] 417.2 Piping clearance. Overhead heating pipes shall have a clearance of not less than 2 inches (51 mm) from combustible contents in the dryer.

❖ Piped steam and hot-water indirect heating systems used in drying rooms or dry kilns can operate at temperatures of up to several hundred degrees Fahrenheit and are considered a safe and appropriate method of providing heat to the drying processes. A minimum clearance of 2 inches (51 mm) between hydronic piping and adjacent combustible contents in the drying room or dry kiln is required to reduce the likelihood of contact ignition of the combustible contents. Though hydronic piping is frequently installed overhead, the clearance is required regardless of the pipes' location within the drying room or kiln.

[F] 417.3 Insulation. Where the operating temperature of the dryer is 175°F (79°C) or more, metal enclosures shall be insulated from adjacent combustible materials by not less than 12 inches (305 mm) of airspace, or the metal walls shall be lined with $^1/_4$-inch (6.4 mm) insulating mill board or other *approved* equivalent insulation.

❖ The insulation of metal drying room and kiln enclosures from adjacent combustible materials is required when the drying apparatus operates at or above 175°F (79°C). Insulation can be accomplished by the use of airspace or by lining the metal dryer walls with insulating material. The insulation will reduce the possibility that combustible materials in proximity to the drying room or dry kiln will be ignited by heat transfer from it. The temperature of 175°F (79°C) has been determined to be the highest allowable temperature without any clearance to combustibles.

[F] 417.4 Fire protection. Drying rooms designed for high-hazard materials and processes, including special occupancies as provided for in Chapter 4, shall be protected by an *approved automatic fire-extinguishing system* complying with the provisions of Chapter 9.

❖ The possibility of ignition increases in a drying room utilized in conjunction with hazardous materials, such as combustible fibers and other finely divided materials. The installation of an approved fire-extinguishing system in accordance with Chapter 9 will contain a fire within the drying room until the fire department (or industrial fire brigade) arrives. In sprinklered buildings, fire suppression can be easily and economically provided by installing sprinklers within the drying room or area. Wet-pipe, preaction or deluge systems should be used so that water is placed on the fire in the shortest possible time. Note that dryers used in conjunction with plywood, veneer and composite board mill operations are required by the IFC to be protected by an approved deluge water-spray system. See the commentary for Chapter 28 of the IFC for further information on lumber and woodworking facilities. In nonsprinklered or partially sprinklered buildings, the system protecting the drying room may be an approved, limited area sprinkler system in accordance with Section 903.3.8. An alternative fire-extinguishing system consisting of dry chemical, carbon dioxide, clean agent or other approved nonwater-based fire-extinguishing system in accordance with Section 904 can provide an equivalent level of protection, depending on the nature of the drying room contents. See the commentary to Section 904 for a discussion of the various alternative fire-extinguishing systems available.

SECTION 418
ORGANIC COATINGS

❖ The manufacture of organic coatings encompasses operations that produce decorative and protective coatings for architectural uses, industrial products and other specialized purposes. Requirements of this chapter address the hazards associated with the manufacture of solvent-based organic coatings. Water-based materials are exempt from these requirements. These provisions are consistent with Chapter 29 of the IFC, which contains more detailed requirements for the manufacture of organic coatings. NFPA 35 also provides additional guidance on the maintenance of facilities utilized for the manufacture of organic coatings.

[F] 418.1 Building features. Manufacturing of organic coatings shall be done only in buildings that do not have pits or *basements*.

❖ Basements, pits and depressed first-floor construction are prohibited because of the tendency for hazardous material vapors to accumulate in low areas and the difficulty in fighting fires in basements and pits in such occupancies.

[F] 418.2 Location. Organic coating manufacturing operations and operations incidental to or connected therewith shall not be located in buildings having other occupancies.

❖ Operations and activities closely related to the primary occupancy are necessary for the efficient, continuous and safe performance of the manufacture of organic coatings. Administration, storage, shipping and receiving, as well as other related but not indispensable operations, should be located in separate buildings.

[F] 418.3 Process mills. Mills operating with close clearances and that process flammable and heat-sensitive materials, such as nitrocellulose, shall be located in a *detached building* or noncombustible structure.

❖ Milling of heat-sensitive materials, such as nitrocellulose, is an extraordinary hazard. Such operations must be located in separate, single-purpose buildings away from other uses and high-hazard operations. Pebble mills pose a special vapor ignition hazard caused by static electricity. Both the grinding material and inner lining of these mills are made of materials with good insulating characteristics. Because static electricity is produced during milling, it has nowhere to go. Generally, the inside of this type of mill is either partially or totally inerted using nitrogen or carbon dioxide gas to prevent ignition.

[F] 418.4 Tank storage. Storage areas for flammable and combustible liquid tanks inside of structures shall be located at or above grade and shall be separated from the processing area by not less than 2-hour *fire barriers* constructed in accordance with Section 707 or *horizontal assemblies* constructed in accordance with Section 711, or both.

❖ Tank storage located below grade is prohibited. Basements located under grade-level storage areas should be discouraged. Below-grade flammable liquid fires are extremely difficult to fight. Similarly, above-grade spills will flow to lower floors, possibly resulting in spill fires on more than one building level. Tank storage of raw materials must be confined to locations at or above grade level. Tank storage rooms must be separated from process and other storage areas by fire barriers and horizontal assemblies that are a minimum of 2-hour fire-resistance rated. If possible, these rooms should be able to be accessed on at least one side from the outside. Cutoff rooms with access from two or even three sides are considered ideal.

[F] 418.5 Nitrocellulose storage. Nitrocellulose storage shall be located on a detached pad or in a separate structure or a room enclosed with not less than 2-hour *fire barriers* constructed in accordance with Section 707 or *horizontal assemblies* constructed in accordance with Section 711, or both.

❖ Once ignited, nitrocellulose will continue to burn even in the absence of oxygen. Extra precautions must be used to prevent ignition and fire spread. Therefore, nitrocellulose storage should be in either a separate detached structure, on a detached exterior pad or fully enclosed by a construction of minimum 2-hour fire-resistance rating complying with fire barrier and horizontal assembly requirements.

[F] 418.6 Finished products. Storage rooms for finished products that are flammable or combustible liquids shall be separated from the processing area by not less than 2-hour *fire barriers* constructed in accordance with Section 707 or *horizontal assemblies* constructed in accordance with Section 711, or both.

❖ Finished products, which are also classified as flammable or combustible liquids, must also comply with the applicable provisions of Chapter 57 of the IFC and NFPA 30. As a minimum, a 2-hour fire-resistance-rated construction consisting of horizontal assemblies and fire barriers with approved opening protectives is required between the processing area and the storage area to eliminate mutual involvement in a single fire scenario. A higher degree of fire separation may be required depending on the building's occupancy classification.

SECTION 419 LIVE/WORK UNITS

419.1 General. A *live/work unit* shall comply with Sections 419.1 through 419.9.

Exception: Dwelling or sleeping units that include an office that is less than 10 percent of the area of the *dwelling unit* are permitted to be classified as *dwelling units* with accessory occupancies in accordance with Section 508.2.

❖ These provisions allow a live/work unit that includes both living and working environments to be considered a single Group R-2 dwelling unit for application of the code. Several limitations and specific requirements are applied to both the living and work portions of the unit. Prior to the adoption of these provisions, the code and the IRC did not allow residential live/work units in a form that is typically desirable for community development. This concept has become increasingly popular, allowing design and construction of a public business with employees working within a residence, allowing the public to enter the work area of the unit to acquire service. Examples of live/work commercial functions are artists' studios, beauty parlors, nail salons and chiropractors' offices. It is important to note that live/work is specifically not to apply to an in-home office. The exception to Section 419.1 is intended to address these small home offices which involve less than 10 percent of the dwelling's floor area.

These concepts reflect an era of planning which created a community where residents could walk to all needed services such as the typical corner commercial store. Examples of this form of planning can be found in many older cities, as well as many "planned communities." Live/work units began to reemerge in the 1990s through a development style known as "Traditional Neighborhood Design" (TND). More recently, adaptive reuse in many older urban structures in city centers incorporated the same live/work tools to provide a variety of business offerings combined with residential unit types.

Historically, building codes did not have to deal with many live/work issues because zoning codes generally precluded a mixing of uses within a neighborhood, much less within a building. However, recent planning trends have been adopted in many jurisdictions, encouraging mixing of commercial and residential uses, not just in neighborhoods, but also in buildings, and even within unit types, such as the live/work unit commonly found in TND projects.

The provisions for live/work units apply to the code criteria based on the Group R-2 provisions for construction. The occupant loads will be determined by the "function" of each space in accordance with Table 1004.1.2.

Section 419 allows mixed use within the dwelling unit or sleeping unit without classification as a mixed-occupancy condition as long as the unit meets the limits within this section. It is then to be classified as a Group R-2 occupancy. Special features that are common within a dwelling unit and are likely within the live/work unit are addressed in order to clearly delineate the means for designing a live/work unit.

419.1.1 Limitations. The following shall apply to all live/work areas:

1. The *live/work unit* is permitted to be not greater than 3,000 square feet (279 m^2) in area;
2. The nonresidential area is permitted to be not more than 50 percent of the area of each *live/work unit*;
3. The nonresidential area function shall be limited to the first or main floor only of the *live/work unit*; and
4. Not more than five nonresidential workers or employees are allowed to occupy the nonresidential area at any one time.

❖ These provisions were meant to apply strictly to small businesses associated with dwelling and sleeping units. In fact, the intent is that the main occupancy of the building is residential, with some business activity within the building. The code limits the nonresidential aspect to a maximum of 50 percent of the area of each unit. In addition, the total area of the live/work unit is limited to 3,000 square feet (279 m^2). Therefore, the total area of the work unit would be a maximum of 1,500 square feet (139 m^2). Since a nonresidential use is being located in a dwelling unit or sleeping unit, the nonresidential area is limited to the first or main floor. Therefore, those coming to the place of business do not need to enter the residential portion of the building. Finally, in keeping with the intent that these are small occupancies and that such occupancies could not create unnecessary life safety concerns, the number of nonresidential workers (employees) is limited to five. This limit of five is not the limit on the number of occupants that can be located within the work area, but simply the number of workers from outside the household that can be there on a regular basis. The 1,500 square-foot (139 m^2) limit on area would limit the number of occupants based upon the occupant load factors.

419.2 Occupancies. *Live/work units* shall be classified as a Group R-2 occupancy. Separation requirements found in Sections 420 and 508 shall not apply within the *live/work unit* where the *live/work unit* is in compliance with Section 419. Nonresidential uses that would otherwise be classified as either a Group H or S occupancy shall not be permitted in a *live/work unit.*

Exception: Storage shall be permitted in the *live/work unit* provided the aggregate area of storage in the nonresidential portion of the *live/work unit* shall be limited to 10 percent of the space dedicated to nonresidential activities.

❖ The entire live/work unit is to be classified as Group R-2 regardless of the types of business being conducted. This exempts such units from the requirements for separation in Sections 420 and 508 within the unit, but would still require the separation between each live/work unit.

The provisions that prohibit uses that would otherwise be classified as either Group H or S occupancies intend to avoid the accumulation of excessive and dangerous fire loads in residential related occupancies. This section would not prohibit the occupancy from containing the maximum allowable quantities of hazardous materials per control area within a building. Control areas are regulated in Section 414 (see the commentary for that section).

Storage is a potential fire load in any business. To ensure that it does not become a large fire hazard, it is limited to 10 percent of the nonresidential portion of the live/work unit. That would be a maximum of 150 square feet (14 m^2); about the size of a large closet.

419.3 Means of egress. Except as modified by this section, the *means of egress* components for a *live/work unit* shall be designed in accordance with Chapter 10 for the function served.

❖ This section requires compliance with Chapter 10 for means of egress unless the general requirements are modified by the following two subsections. The means of egress for the nonresidential portion of the live/work unit is regulated based upon the specific function of the space rather than the egress requirements for a Group R-2 occupancy. In many instances, this will require an occupancy classification determination for the nonresidential portion for means of egress purposes without assigning such a classification to the function itself. For example, if the nonresidential portion of the live/work unit is a retail sales use, that portion of the unit would be regulated as a Group M occupancy for the means of egress provisions in Chapter 10.

419.3.1 Egress capacity. The egress capacity for each element of the *live/work unit* shall be based on the occupant load for the function served in accordance with Table 1004.1.2.

❖ The egress capacity must be based upon the actual use of the space. Therefore if you had a mercantile type use in the live/work unit, the egress capacity must be based on 60 square feet (5.6 m^2) per person of the gross area of the mercantile space. If a 3,000-square-foot (279 m^2) unit is equally divided for mercantile and residential use, in such a case the calculated occupant load would be determined by dividing the area by the square feet per person: 1,500 square feet (139 m^2)/60 square feet (5.6 m^2) per person = 25 occupants. In addition, the capacity for residential use is 200 square feet (18.6 m^2) per person (gross area), which would be 1,500 square feet (139 m^2)/200 square feet (18.6 m^2) per person = 8 occupants.

419.3.2 Spiral stairways. *Spiral stairways* that conform to the requirements of Section 1011.10 shall be permitted.

❖ Spiral staircases are allowed in dwelling units and this section simply emphasizes this allowance with a specific reference to the design criteria in Section 1011.10. There is no limit on the use of spiral stairways regardless of the floor area of the nonresidential uses in the live/work unit.

419.4 Vertical openings. Floor openings between floor levels of a *live/work unit* are permitted without enclosure.

❖ Within a multistory unit, openings through one or more floors created by stairways and other vertical elements need not be enclosed. This is consistent with the allowance for unconcealed vertical openings within individual residential dwelling units and sleeping units (see Sections 712.1.2 and 1009.3, Exception 2).

[F] 419.5 Fire protection. The *live/work unit* shall be provided with a monitored *fire alarm* system where required by Section 907.2.9 and an *automatic sprinkler system* in accordance with Section 903.2.8.

❖ Since the unit is considered as Group R-2, the entire building would be required to be sprinklered in accordance with Section 903.3.1.1 (NFPA 13) or 903.3.1.2 (NFPA 13R).

This section requires the installation of a fire alarm system as required for a Group R-2 occupancy. Section 907.2.9 would only require a fire alarm system in certain cases. The requirements for a manual fire alarm are based on the location of the Group R-2 dwelling unit (three or more stories above the lowest level of exit discharge or one story below the highest level of exit discharge) and the number of units (more than 16). If a building has a sprinkler system and would require a manual system, the manual aspect is no longer required. Instead, the sprinkler system is required to be tied to the notification appliances and activate the system upon water flow.

419.6 Structural. Floors within a *live/work unit* shall be designed for the live loads in Table 1607.1, based on the function within the space.

❖ Since live/work units may have structural loads not normally anticipated by Group R-2 occupancies, the code specifically requires structural design of floor live loads to be addressed in accordance with Table 1607.1, based on what is actually occurring in the space. For instance, if the nonresidential activity is a business, there may be equipment such as computers, files or a large copy machine that requires loading based on office loads of 100 psf (4788 Pa).

419.7 Accessibility. Accessibility shall be designed in accordance with Chapter 11 for the function served.

❖ Accessibility to and within the live/work unit must be designed and constructed in compliance with Chapter 11. The business/work area on the first floor must be fully accessible. In accordance with Section 1107.6.2.1, the dwelling unit portion must be evaluated separately. If the structure has four or more units, or the site has more than 20 units, Type A unit or Type B unit requirements may be applicable. The exceptions under Section 1107.7 are applicable. The most common application will typically be the multistory dwelling unit (see Section 1107.7.2).

Section 1107.3 requires that rooms and spaces available to the public be accessible. Accessible spaces that may be utilized by the public in accordance with Section 1107.3 include other spaces such as kitchens, living and dining areas and any exterior spaces including patios, terraces and balconies that may be on the same level as the entrance to the unit. These may not be "available to the public" unless they are part of the function of the nonresidential activity. Such spaces that are strictly part of the residential function of the live/work unit are only required to be accessible to the degree that such units are required to be made accessible.

419.8 Ventilation. The applicable *ventilation* requirements of the *International Mechanical Code* shall apply to each area within the *live/work unit* for the function within that space.

❖ Similar to egress and structural requirements, the use of the space must be looked at individually to ensure ventilation being provided fits the use. For instance, Table 403.3.1.1 of the IMC would require a florist shop to have 15 cubic feet (0.42 m^3) per minute of outdoor air per person. With a possible floor area of 1,500 square feet (139 m^2) for the work area and a minimum of eight persons per 1,000 square feet (92.9 m^2) as required by Table 403.3.1.1, the cfm required would be as follows:

(8 occupants/1,000 square feet) x 1,500 square feet = 12 occupants

12 occupants x 15 cfm = 180 cfm of outside air ventilation

A two-bedroom dwelling unit would typically require 105 cfm (15 cfm per occupant x 7 occupants — 200 square feet per occupant in accordance with Table 1004.1.2). However, it is more typical that designs of residential dwelling units include the provision of fresh air by the natural ventilation option.

419.9 Plumbing facilities. The nonresidential area of the *live/work unit* shall be provided with minimum plumbing facilities as specified by Chapter 29, based on the function of the nonresidential area. Where the nonresidential area of the *live/work unit* is required to be *accessible* by Section 1103.2.13, the plumbing fixtures specified by Chapter 29 shall be *accessible*.

❖ In a manner consistent with the application of means of egress provisions to live/work units, plumbing facilities as specified by Chapter 29 are to be provided for the nonresidential use based on the function of the nonresidential area. Based on the example of a retail sales use as set forth in the commentary to Section 419.3, plumbing fixture requirements for the nonresidential retail sales area would be based on a Group M classification. However, the Group M classification would not be specifically assigned to the live/work unit. As previously stated in Section 419.7, the accessibility requirements for the nonresidential area would be based on the function of that area

SECTION 420
GROUPS I-1, R-1, R-2, R-3 AND R-4

420.1 General. Occupancies in Groups I-1, R-1, R-2, R-3 and R-4 shall comply with the provisions of Sections 420.1 through 420.6 and other applicable provisions of this code.

❖ The nature of occupancies in Groups I-1, R-1, R-2, R-3 and R-4 is such that some level of protection against fire is needed for occupants in dwelling units and sleeping units. There remains a high frequency of fires where people live. These occupancies need to be provided automatic sprinkler system protection. Requiring a minimum fire resistance to the construction separating both units and residential areas from nonresidential areas provides an extra level of protection in people's homes from occurrences in their neighbors' homes. These separations are required between live/work units as provided in Section 419, but separations within each live/work unit between residential and nonresidential uses are not required.

420.2 Separation walls. Walls separating *dwelling units* in the same building, walls separating *sleeping units* in the same building and walls separating *dwelling* or *sleeping units* from other occupancies contiguous to them in the same building shall be constructed as *fire partitions* in accordance with Section 708.

❖ The sleeping units or dwelling units that are in a single building are required to be separated by fire partitions complying with Section 708, or horizontal fire-resistance-rated assemblies complying with Section 711. Fire partitions are the least robust of all fire-resistance-rated walls called out in the code. These occupancies all require smoke alarms in the sleeping areas of the units, and occupancies in Groups I-1 and R-1 also require general fire alarms. In addition, these occupancies are all required to be sprinklered. The reason for the nominal fire-resistant separation is to account for the fact that individuals asleep in these units could respond more slowly to a fire; therefore, some amount of fire resistance is deemed necessary to protect these occupants.

Section 708.3 requires fire partitions to be not less than 1-hour fire-resistance rated. If the building's sprinkler protection is provided by a system complying with NFPA 13, the rating can be reduced to 30 minutes. Section 708.4 requires 1-hour-rated fire partitions to be supported by construction that has the same rating or better; however, this requirement is waived for these separation walls in buildings of Type II B, IIIB and VB construction.

This requirement also applies to walls that separate these Group R and I occupancies from other occupancies in the building. Even if those other areas are regulated under the accessory occupancies or nonseparated occupancies option of mixed occupancies contained in Sections 508.2 or 508.3 respectively, these partitions and horizontal assemblies are still required.

420.3 Horizontal separation. Floor assemblies separating *dwelling units* in the same buildings, floor assemblies separating *sleeping units* in the same building and floor assemblies separating *dwelling* or *sleeping units* from other occupancies contiguous to them in the same building shall be constructed as *horizontal assemblies* in accordance with Section 711.

❖ See the commentary to Section 420.2. Section 711.2 requires floor assemblies providing this separation to be not less than 1-hour fire-resistance rated. If the building's sprinkler protection is provided by a system complying with NFPA 13, the rating can be reduced to 30 minutes. Section 708.4 requires 1-hour-rated fire partitions to be supported by construction that has the same rating or better; however, this requirement is waived for these floor assemblies in buildings of Type IIB, IIIB and VB construction.

420.4 Smoke barriers in Group I-1, Condition 2. Smoke barriers shall be provided in Group I-1, Condition 2, to subdivide every story used by persons receiving care, treatment or sleeping and to provide other stories with an occupant load of 50 or more persons, into no fewer than two smoke compartments. Such stories shall be divided into smoke compartments with an area of not more than 22,500 square feet (2092 m^2) and the distance of travel from any point in a smoke compartment to a smoke barrier door shall not exceed 200 feet (60 960 mm). The smoke barrier shall be in accordance with Section 709.

❖ Group I-1, Condition 2 allows occupants that may need assistance with evacuation (see Section 308.3.2). Smoke barriers and the associated compartmentalization is a key protective feature utilized for this occupant type. Requirements mostly match Group I-2 smoke barrier criteria provided in Section 407 with the following noted differences: The refuge area is 15 feet (4572 mm) for each care recipient in a Group I-1, Condition 2 versus the 30 square feet (2.8 m^2) required in Group I-2. There are no bed or stretcher occupants in Group I-1, Condition 2, meaning occupants can be moved during emergencies without having to be moved in a bed. The Group I-1, Condition 2 areas permitted to be included in the calculation do not include sleeping areas (sleeping room) as permitted in Group I-2. Sleeping areas are included in Group I-2 because of defend-in-place strategies during emergencies. Group I-1, Condition 2 occupants utilize smoke compartments for a staged evacuation in accordance with the IFC but are not supposed to stay in their rooms during emergencies.

420.4.1 Refuge area. Refuge areas shall be provided within each smoke compartment. The size of the refuge area shall accommodate the occupants and care recipients from the adjoining smoke compartment. Where a smoke compartment is adjoined by two or more smoke compartments, the minimum area of the refuge area shall accommodate the largest occupant load of the adjoining compartments. The size of the refuge area shall provide the following:

1. Not less than 15 net square feet (1.4 m^2) for each care recipient.
2. Not less than 6 net square feet (0.56 m^2) for other occupants.

Areas or spaces permitted to be included in the calculation of the refuge area are corridors, lounge or dining areas and other low-hazard areas.

❖ Under this provision, refuge areas are only required for Group I-1, Condition 2 occupancies. They are not required for other Group I-1 occupancies or the Group R occupancies. Similar refuge areas are required for Group I-2 occupancies (Section 407) and

Group I-3 occupancies (Section 408.) See also the commentary for Section 420.4.

[F] 420.5 Automatic sprinkler system. Group R occupancies shall be equipped throughout with an *automatic sprinkler system* in accordance with Section 903.2.8. Group I-1 occupancies shall be equipped throughout with an *automatic sprinkler system* in accordance with Section 903.2.6. Quick-response or residential automatic sprinklers shall be installed in accordance with Section 903.3.2.

❖ The provisions requiring automatic sprinkler protection for all buildings containing Group I or R occupancies are referenced here. In addition, reference is made to Section 903.3.2 regarding the required use of quick-response or residential automatic sprinklers.

[F] 420.6 Fire alarm systems and smoke alarms. Fire alarm systems and smoke alarms shall be provided in Group I-1, R-1, R-2 and R-4 occupancies in accordance with Sections 907.2.6, 907.2.8, 907.2.9 and 907.2.10, respectively. Single- or multiple- station smoke alarms shall be provided in Groups I-1, R-2, R-3 and R-4 in accordance with Section 907.2.11.

❖ References to Sections 907.2.6, 907.2.8, 907.2.9 and 907.2.10 are a reminder of the requirements for fire alarm systems and smoke alarms in Group I-1, R-1 and R-2 occupancies. It is also noted that Section 907.2.11 sets forth the requirements for single-station and multiple-station smoke alarms.

SECTION 421
HYDROGEN FUEL GAS ROOMS

[F] 421.1 General. Where required by the *International Fire Code*, hydrogen fuel gas rooms shall be designed and constructed in accordance with Sections 421.1 through 421.7.

❖ This section is simply stating that all hydrogen cutoff rooms are to be constructed in accordance with the provisions contained in Section 421. Hydrogen cutoff rooms were created to address the increasing and emerging concepts of fuel cells that use hydrogen and actually generate hydrogen onsite to run the fuel cells. IFC Section 2309.3.2.3 requires that generation, compression, storage and dispensing equipment related to hydrogen be located in one of three places inside buildings. The first is a hydrogen cutoff room in accordance with this section. The second is outside a cutoff room where the gaseous hydrogen system is listed and labeled for indoor installation and installed in accordance with the manufacturer's instructions. The third is in a dedicated hydrogen fuel dispensing area having an aggregate hydrogen delivery capacity no greater than 12 standard cubic feet (0.34 m^3) per minute. The provisions of Section 421 address construction-related issues for hydrogen cutoff rooms such as location, fire-resistant separation and ventilation, and safety features such as gas detection and explosion control.

[F] 421.2 Definitions. The following terms are defined in Chapter 2:

GASEOUS HYDROGEN SYSTEM.

HYDROGEN FUEL GAS ROOM.

❖ This section lists terms that are specifically associated with the subject matter of Section 421. It is important to emphasize that these terms are not exclusively related to this chapter but may or may not also be applicable where the term is used elsewhere in the code.

Definitions of terms can help in the understanding and application of the code requirements. The purpose for including a list within this chapter is to provide more convenient access to terms that may have a specific or limited application within this chapter. For the complete definition and associated commentary, refer back to Chapter 2. Terms that are italicized provide a visual identification throughout the code that a definition exists for that term. The use and application of all defined terms are set forth in Section 201.

[F] 421.3 Location. Hydrogen fuel gas rooms shall not be located below grade.

❖ Restrictions against installation of hydrogen fuel gas rooms below grade are similar to those restricting the location of flammable and combustible liquids in basements. Explosion hazards are the primary concern, and placement of materials that have an ability to cause an explosion in below-grade spaces is not appropriate. Such spaces are more difficult to evacuate, create a fire explosion hazard to the structure above and are very difficult for the fire department to address.

[F] 421.4 Design and construction. Hydrogen fuel gas rooms not classified as Group H shall be separated from other areas of the building in accordance with Section 509.1.

❖ Hydrogen fuel gas rooms are required to be separated by not less than 1-hour fire-resistance-rated fire barriers and horizontal assemblies, which is consistent with the requirements in Table 509 for incidental uses. In addition, the classification of the space will affect the separation requirements. Hydrogen fuel gas rooms can contain any amount of hydrogen, but if the maximum allowable quantities are exceeded, the room can no longer be considered an incidental use and instead must be classified as a Group H-2 occupancy. A Group H-2 must then be addressed as a separated occupancy. In all cases, separation in accordance with Table 508.4 would apply in addition to the applicable occupancy-specific requirements.

[F] 421.4.1 Pressure control. Hydrogen fuel gas rooms shall be provided with a ventilation system designed to maintain the room at a negative pressure in relation to surrounding rooms and spaces.

❖ This section requires hydrogen fuel gas rooms to be maintained at a negative pressure with regard to surrounding spaces to provide some level of protection against a flammable mixture being attained in other parts of the building.

[F] 421.4.2 Windows. Operable windows in interior walls shall not be permitted. Fixed windows shall be permitted where in accordance with Section 716.

❖ Operable windows are prohibited to further reduce the likelihood of allowing hydrogen from escaping into the room and entering other portions of the building that may not be properly ventilated. An operable window could inadvertently be left in the open or partially open position and go unnoticed. Fixed window openings must meet the requirements of Section 716 to ensure that the requirements for opening protectives are met. More specifically, the proper fire protection ratings for openings in fire barriers are required.

[F] 421.5 Exhaust ventilation. Hydrogen fuel gas rooms shall be provided with mechanical exhaust ventilation in accordance with the applicable provisions of Section 502.16.1 of the *International Mechanical Code*.

❖ The purpose of this section is to prevent a dangerous accumulation of flammable gas in the room through the use of an exhaust ventilation system. The *Sourcebook for Hydrogen Applications* recommends ventilation at the rate of 1 cfm per square foot [0.00508 $m^3/(s \times m^2)$] of floor area, which is consistent with the requirements in Chapter 5 of the IMC.

[F] 421.6 Gas detection system. Hydrogen fuel gas rooms shall be provided with an approved flammable gas detection system in accordance with Sections 421.6.1 through 421.6.4.

❖ Some gases contain additives that produce pungent odors for easy recognition. Systems using nonodorized gases, such as hydrogen and liquid natural gas, must utilize gas detection systems to detect leaks. This section specifically requires such detection due to the hazards associated with a buildup of hydrogen at hazardous levels within a building.

[F] 421.6.1 System design. The flammable gas detection system shall be listed for use with hydrogen and any other flammable gases used in the hydrogen fuel gas room. The gas detection system shall be designed to activate when the level of flammable gas exceeds 25 percent of the lower flammability limit (LFL) for the gas or mixtures present at their anticipated temperature and pressure.

❖ The detection system must initiate the operations specified in Section 421.6.2 at any time that the flammable gas concentration exceeds one-fourth of the concentration necessary to support combustion. Early detection of the presence of a flammable gas will allow adequate mitigation procedures to be taken. Hydrogen fires are not normally extinguished until the supply of hydrogen has been shut off because of the danger of reignition or explosion. A gas detection system in the room or space housing a gaseous hydrogen system results in early notification of a leak that is occurring before the escaping gas reaches a hazardous concentration.

[F] 421.6.2 Gas detection system components. Gas detection system control units shall be listed and labeled in accordance with UL 864 or UL 2017. Gas detectors shall be listed and labeled in accordance with UL 2075 for use with the gases and vapors being detected.

❖ This section provides that control units and gas detectors used in these systems be listed and labeled equipment. The section provides the specific standard under which the equipment needs to be listed and labeled.

[F] 421.6.3 Operation. Activation of the gas detection system shall result in all of the following:

1. Initiation of distinct audible and visual alarm signals both inside and outside of the hydrogen fuel gas room.
2. Activation of the mechanical exhaust ventilation system.

❖ The detection system must activate the mechanical ventilation system that is required by Section 421.5, in addition to causing alarms to activate. Note that the mechanical ventilation alternative in Section 421.4.1.1 is continuously in operation.

The required local alarm is intended to alert the occupants to an emerging hazardous condition in the vicinity. The monitor control equipment must also initiate operation of the mechanical ventilation system in the event of a leak or rupture in the gaseous hydrogen system to prevent an accumulation of flammable gas.

[F] 421.6.4 Failure of the gas detection system. Failure of the gas detection system shall result in activation of the mechanical exhaust ventilation system, cessation of hydrogen generation and the sounding of a trouble signal in an approved location.

❖ Systems must be designed to be self-monitoring and fail-safe in that all safety systems are activated to alert any occupants that a problem exists and to prevent more hydrogen from being generated by any appliances in the room when hazardous conditions cannot be monitored.

[F] 421.7 Explosion control. Explosion control shall be provided where required by Section 414.5.1.

❖ The requirements of this section are intended to address the circumstance resulting from a catastrophic failure of the cutoff room. These requirements are the final safeguard in case safety features such as interlocked doors, ventilation and gas detection systems should fail. An ignited hydrogen mixture produces large quantities of heat, causing a rapid expansion of the surrounding air. This can cause a pressure increase in a confined space and a catastrophic failure. Explosion control methods are identified in Section 911 of the IFC to prevent such a catastrophic failure. The explosion control requirements for hydrogen are consistent with the requirements in NFPA 50A and the Sourcebook for Hydrogen Applications.

[F] 421.8 Standby power. Mechanical *ventilation* and gas detection systems shall be provided with a standby power system in accordance with Section 2702.

❖ The ventilation system and gas detection system are life safety systems and, therefore, must be dependable. Both safety systems must remain active in the event of a power failure of the primary power supply. Hydrogen is a colorless, odorless gas; a release might go undetected if detection systems are not functioning. The accumulation of hydrogen in an unventilated area can lead to mixtures in the flammable range if safety systems and mechanical ventilation systems are not in operation. Chapter 50 of the IFC addresses emergency and standby power requirements for emergency systems. It also allows an exception to the requirement for systems that are fail-safe (see IFC Section 5004.7, Exception 4). This exception may be used in cutoff rooms where hydrogen is generated, but not stored. Any storage of hydrogen within the cutoff room would not qualify for the exception because, in the event of a power failure, there will be no way to detect or ventilate a release from a storage vessel.

SECTION 422
AMBULATORY CARE FACIILITIES

422.1 General. Occupancies classified as *ambulatory care facilities* shall comply with the provisions of Sections 422.1 through 422.5 and other applicable provisions of this code.

❖ Complex outpatient surgeries outside of the hospital are now commonplace. They are performed in facilities often called "day surgery centers" or "ambulatory surgical centers" because patients are able to walk in and walk out the same day. Procedures render patients temporarily incapable of self-preservation by application of nerve blocks, sedation or anesthesia. Patients in these facilities typically recover quickly. A definition of "Ambulatory care facility" is provided in Section 202.

The code identifies medical care Group I occupancies as having 24-hour stay. Without a 24-hour stay, these surgery centers are classified as Group B. Strictly regulating such occupancies as a typical Group B occupancy is considered inappropriate, as this would allow the rendering of an unlimited number of people incapable of self-preservation with no more protection than a business office. These types of facilities contain distinctly different hazards to life and safety than other business occupancies, such as:

1. Patients incapable of self-preservation require rescue by other occupants or fire personnel.
2. Medical staff must stabilize the patient prior to evacuation; therefore, staff may require evacuation as well.
3. Use of oxidizing medical gases, such as oxygen and nitrous oxide.
4. Prevalence of surgical fires.

In the past, there was a movement to classify ambulatory care facilities as Group I-2 occupancies. This was determined to be a poor fit, because these are not hospitals. Federal and state jurisdictions have recognized that there is a middle ground between Groups B and I-2. These requirements provide a scaled approach to protection. The occupancy classification is a Group B, but with some enhanced safety features focused on the concern with occupants being incapable of self-preservation on a temporary basis. The enhanced requirements are based on the concepts in the regulation of the Group I-2 occupancy requirements found in Section 407.

422.2 Separation. *Ambulatory care facilities* where the potential for four or more care recipients are to be *incapable of self-preservation* at any time, whether rendered incapable by staff or staff accepted responsibility for a care recipient already incapable, shall be separated from adjacent spaces, *corridors* or tenants with a *fire partition* installed in accordance with Section 708.

❖ In a multitenant or mixed-occupancy building where there are other uses present in addition to an ambulatory care facility, a degree of fire separation between the ambulatory care facility and those nonrelated spaces is required where the ambulatory care facility is intended to have at least four care recipients incapable of self-preservation at any one time. The minimum separation, a fire partition complying with Section 708, is viewed as an important tool in isolating the ambulatory care portion of building from fire hazards that may occur in other portions of the building. Where a building is considered wholly as an ambulatory care facility, or if the ambulatory care facility occupies the entire story of a multistory building, there is no requirement for a fire partition to be constructed.

422.3 Smoke compartments. Where the aggregate area of one or more *ambulatory care facilities* is greater than 10,000 square feet (929 m^2) on one *story*, the *story* shall be provided with a *smoke barrier* to subdivide the *story* into no fewer than two *smoke compartments*. The area of any one such *smoke compartment* shall be not greater than 22,500 square feet (2092 m^2). The distance of travel from any point in a *smoke compartment* to a *smoke barrier* door shall be not greater than 200 feet (60 960 mm). The *smoke barrier* shall be installed in accordance with Section 709 with the exception that *smoke barriers* shall be continuous from outside wall to an outside wall, a floor to a floor, or from a *smoke barrier* to a *smoke barrier* or a combination thereof.

❖ In larger facilities, the creation of smoke compartments is required to allow a protect-in-place environment. These compartments allow staff a safer environment to stabilize the patients before evacuation, and protection for fire personnel who may have to evacuate both patients and staff. The maximum size of a smoke compartment is limited to 22,500 square feet (2090 m^2), consistent with the maximum size allowed for Group I-2, Condition 1 occupancies. It is important to note that each floor must be divided into at least two smoke compartments if the total floor area exceeds 10,000 square feet (929 m^2) on each story. Conversely, smoke compartments are not required for ambulatory care facilities with a floor area that does not exceed 10,000 square feet (929 m^2) on each story. The 10,000-square-foot (929 m^2) threshold is based on a story-by-story basis and is based on the aggregate floor areas where multiple ambulatory care facilities are present. The travel distance of 200 feet (60 960 mm) to a door providing egress from the smoke compartment is the same requirement as for Group I-2 occupancies.

422.3.1 Means of egress. Where ambulatory care facilities require smoke compartmentation in accordance with Section 422.3, the fire safety evacuation plans provided in accordance with Section 1001.4 shall identify the building components necessary to support a *defend-in-place* emergency response in accordance with Sections 404 and 408 of the *International Fire Code*.

❖ Defend-in-place, or protect in place, is a concept that has long been employed as the preferred method of fire response in hospitals due to the fragile nature of the occupants. Occupants in this setting, and similarly in ambulatory care facilities, are often dependent on the building infrastructure given that immediate evacuation would place their lives at risk. This infrastructure typically includes life support systems such as medical gases, emergency power and environmental controls that rely on continued building operation. Previous versions of this code have created a tried and tested set of requirements to support this concept, such as smoke compartmentation and areas of refuge. However, previous codes have not specifically described the concept of occupants remaining within a building during a fire emergency, leading to confusion and misapplication during design and enforcement. The concept is predicated on the use of smoke compartmentation.

422.3.2 Refuge area. Not less than 30 net square feet (2.8 m^2) for each nonambulatory care recipient shall be provided within the aggregate area of *corridors,* care recipient rooms, treatment rooms, lounge or dining areas and other low-hazard areas within each *smoke compartment.* Each occupant of an *ambulatory care facility* shall be provided with access to a refuge area without passing through or utilizing adjacent tenant spaces.

❖ This requirement is very similar to that found in Section 407.5.1 for Group I-2 occupancies. The purpose is to provide adequate space for patients, on beds, who might be relocated to an adjacent smoke compartment. Since the facility may still be in operation during the fire, space must be provided in corridors, patient rooms, treatment rooms, lounge or dining areas and other low-hazard areas for both the patients being relocated and the patients being treated in that smoke compartment. Section 407.5.1 acknowledges that there may be floors that do not house patients who are confined to beds or litter. It is also important that access to a refuge area be provided without requiring the occupant to travel through any other tenant space.

422.3.3 Independent egress. A *means of egress* shall be provided from each *smoke compartment* created by smoke barriers without having to return through the *smoke compartment* from which *means of egress* originated.

❖ This section is identical to Section 407.5.2 and requires that occupants should be able to exit the building without having to re-enter a smoke compartment from where they started. Although the strategy is to protect in place, occupants should not have to travel back through a smoke compartment to ultimately exit the building. This prevents the creation of a dead-end smoke compartment [see Commentary Figures 407.5, 407.5.2(1) and 407.5.2(2)].

[F] 422.4 Automatic sprinkler systems. *Automatic sprinkler systems* shall be provided for *ambulatory care facilities* in accordance with Section 903.2.2.

❖ Section 422.6 is part of the package of enhanced Group B requirements to accommodate the risk to occupants that is higher than the typical business occupancy but does not warrant classification as a Group I-2 occupancy. The actual sprinkler requirements are located in Section 903.2.2. The specific requirements apply when there are four or more care recipients incapable of self-preservation or if one or more care recipients are located on a story other than the level of exit discharge. The entire fire area meeting either one of the criteria must be sprinklered.

[F] 422.5 Fire alarm systems. A *fire alarm* system shall be provided for *ambulatory care facilities* in accordance with Section 907.2.2.

❖ As with the sprinkler requirements, the alarm requirements are actually located in Chapter 9. More specifically, Section 907.2.2.1 requires a manual fire alarm system in any fire area containing a Group B ambulatory care facility. Furthermore, an automatic smoke detection system is required within ambulatory care facility and public use spaces, such as lobbies and corridors, except where all areas of the building are protected by an automatic sprinkler system. In addition, the sprinkler system is required to activate the occupant notification appliances upon sprinkler water flow (see commentary, Section 907.2.2.1). The scope of the smoke detection was purposely narrowed to limit the amount of retrofit measures that a building owner would be required to undertake if a new ambulatory care tenant was established in an existing building.

SECTION 423
STORM SHELTERS

423.1 General. In addition to other applicable requirements in this code, storm shelters shall be constructed in accordance with ICC 500.

❖ Standard ICC 500 provides requirements for the design and construction of shelters to protect people from the violent winds of hurricanes and tornadoes. The standard includes special requirements for structural design, including wind loads that are considerably higher than the wind loads required by Chapter 16 for all structures.

Residential storm shelters are defined as shelters for dwelling units and with an occupant load of 16 or fewer. Commercial storm shelters are for anything else, including shelter for larger groups of homes and apartment buildings, as well as other occupancies.

Wind loads for storm shelters will be based upon wind speed contour maps developed specially for this standard. The wind load design requirements are relatively severe when compared to the wind speed maps in Chapter 16. Contour maps for wind speeds in hurricane prone regions were determined based upon a 10,000-year mean return period. The map shows 200 mph (89 m/s) wind speeds on the coast of Florida and the Carolinas, and wind speeds higher than 200 mph (89 m/s) in some locations. These are wind speeds associated with a Category 5 hurricane. Shelter design wind speeds in the central part of the United States (a region called "tornado alley") are as high as 250 mph (112 m/s).

Such high wind speeds, of course, produce flying debris, turning construction materials into deadly missiles. The standard contains specific test methods and pass-fail criteria for window and doors protection from flying debris.

ICC 500 addresses nonstructural issues, as well. Storm shelters for hurricanes will be required to house people for 24 hours. Tornado shelters will be required to house people for 2 hours. The standard addresses minimum requirements for ventilation air, sanitation facilities, potable water supply, lighting and other minimal power needs.

It should be noted that the entrances and exits to commercial storm shelters will be required to be accessible. In addition, the occupant load requirements are such that some wheelchair space will be required.

423.1.1 Scope. This section applies to the construction of storm shelters constructed as separate detached buildings or constructed as safe rooms within buildings for the purpose of providing safe refuge from storms that produce high winds, such as tornados and hurricanes. Such structures shall be designated to be hurricane shelters, tornado shelters, or combined hurricane and tornado shelters.

❖ Most storm shelters are safe rooms within a bigger facility. Shelters can be used for other purposes during normal building operation. Examples are gymnasiums and classrooms in a school, and locker rooms and bathrooms in a fire station. The purpose of the storm shelter is to provide refuge for people during a storm. The standard does not address the use of the shelter as a post-storm recovery facility, although it may well be used for that purpose. Sections 423.3 and 423.4 specify two instances where storm shelters are required in buildings constructed in compliance with the code.

423.2 Definitions. The following terms are defined in Chapter 2:

STORM SHELTER.

Community storm shelter.

Residential storm shelter.

❖ This section lists terms that are specifically associated with the subject matter of Section 423. It is important to emphasize that these terms are not exclusively related to this chapter but may or may not also be applicable where the term is used elsewhere in the code.

Definitions of terms can help in the understanding and application of the code requirements. The purpose for including a list within this chapter is to provide more convenient access to terms that may have a specific or limited application within this chapter. For the complete definition and associated commentary, refer back to Chapter 2. Terms that are italicized provide a visual identification throughout the code that a definition exists for that term. The use and application of all defined terms are set forth in Section 201.

423.3 Critical emergency operations. In areas where the shelter design wind speed for tornados in accordance with Figure 304.2(1) of ICC 500 is 250 MPH, 911 call stations, emergency operation centers and fire, rescue, ambulance and police stations shall have a storm shelter constructed in accordance with ICC 500.

Exception: Buildings meeting the requirements for shelter design in ICC 500.

❖ Figure 304.2(1) of ICC 500 is replicated as Commentary Figure 423.3 for this commentary. The figure outlines all or part of 23 states where the design speed for tornadoes is 250 mph (112 m/s), thus requiring that these emergency service facilities be provided with a storm shelter. The types of facilities that are required to have shelters are locations that house emergency responders or systems that a community would want operational after a tornado event.

The code offers the option of providing a shelter within the building, or constructing the whole building as a shelter. The advantage of constructing the whole building as a shelter is that, should a tornado hit the building, the building has a better chance of surviving the storm and therefore being ready to provide emergency services to the community immediately after the storm passes.

The code does not specify the capacity of the shelter if it is not the whole building. While the first thought of capacity should be protecting the occupants of the building containing the shelter, thought should be given to people in adjacent and nearby buildings that may not have their own shelter. Once the number of occupants to be accommodated is decided, ICC 500 provides details for the design and construction of the shelter.

423.4 Group E occupancies. In areas where the shelter design wind speed for tornados is 250 MPH in accordance with Figure 304.2(1) of ICC 500, all Group E occupancies with an aggregate occupant load of 50 or more shall have a storm shelter constructed in accordance with ICC 500. The shelter shall be capable of housing the total occupant load of the Group E occupancy.

Exceptions:

1. Group E day care facilities.
2. Group E occupancies accessory to places of religious worship.
3. Buildings meeting the requirements for shelter design in ICC 500.

❖ See the commentary for Section 423.3. For Group E occupancies such as grade, middle and high schools, the code requires the provision of a storm shelter large enough to accommodate the total occupant load of the Group E occupancy. The option exists to build the whole school as a shelter (Exception 3). Schools that include an auditorium or gymnasium where large numbers will gather for sporting events or other school activities must have shelters capable of accommodating those occupants. It is not the intent to require the shelter to be designed for the total occupant load of the building that is used for means of egress. Where there is a situation where the classrooms would be empty if the gymnasium was full, such as an all-school assembly, or after-hours sporting event, the storm shelter can be designed to accommodate that occupant load. The best option may be to construct the auditorium or gymnasium as the shelter. While the text says "a" storm shelter, there is nothing to prevent creating multiple shelters in diverse locations within the facility to accommodate all occupants. Some schools use the bathrooms throughout the school as shelters to lessen the time and travel for students to access those shelters. Three E occupancies are exempt from the requirement: 1. Those with an occupant load of less than 50; 2. Day care facilities (see Sections 308.6.1 and 305.2); and 3. Group E occupancies accessory to a place of religious worship (see Sections 303.1.4 and 305.2.1).

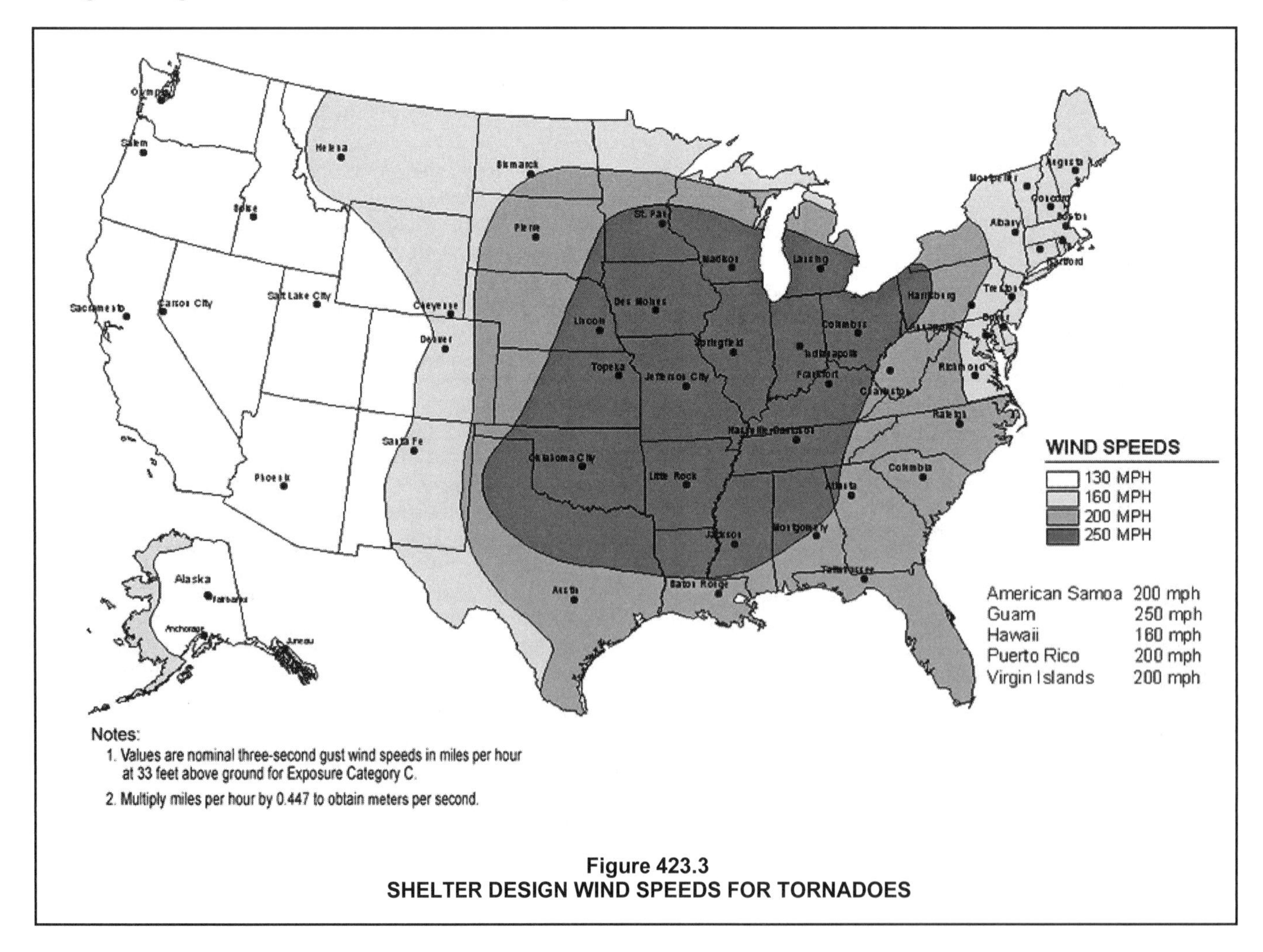

Figure 423.3
SHELTER DESIGN WIND SPEEDS FOR TORNADOES

SECTION 424 CHILDREN'S PLAY STRUCTURES

424.1 Children's play structures. Children's play structures installed inside all occupancies covered by this code that exceed 10 feet (3048 mm) in height and 150 square feet (14 m^2) in area shall comply with Sections 424.2 through 424.5.

❖ Play structures for children's activities have been regulated for some time by the code where such structures are located within covered mall buildings. The primary concern, consistent with other structures located within a covered mall building, is the combustibility of such play structures. Due to the potential fire hazards, the regulations are now applicable where such structures are located within any building regulated by this code, regardless of occupancy classification. The current provisions are essentially identical to those that historically regulated children's play structures located within covered mall buildings. If these structures exceed 10 feet (3048 mm) in height and 150 square feet (14 m^2) in area, then the provisions of Sections 424.2 through 424.5 must be followed.

424.2 Materials. Children's play structures shall be constructed of noncombustible materials or of combustible materials that comply with the following:

1. *Fire-retardant-treated* wood complying with Section 2303.2.
2. Light-transmitting plastics complying with Section 2606.
3. Foam plastics (including the pipe foam used in soft-contained play equipment structures) having a maximum heat-release rate not greater than 100 kilowatts when tested in accordance with UL 1975 or when tested in accordance with NFPA 289, using the 20 kW ignition source.
4. Aluminum composite material (ACM) meeting the requirements of Class A *interior finish* in accordance with Chapter 8 when tested as an assembly in the maximum thickness intended for use.
5. Textiles and films complying with the fire propagation performance criteria contained in Test Method 1 or Test Method 2, as appropriate, of NFPA 701.
6. Plastic materials used to construct rigid components of soft-contained play equipment structures (such as tubes, windows, panels, junction boxes, pipes, slides and decks) exhibiting a peak rate of heat release not exceeding 400 kW/ m^2 when tested in accordance with ASTM E1354 at an incident heat flux of 50 kW/m^2 in the horizontal orientation at a thickness of 6 mm.
7. Ball pool balls, used in soft-contained play equipment structures, having a maximum heat-release rate not greater than 100 kilowatts when tested in accordance with UL 1975 or when tested in accordance with NFPA 289, using the 20 kW ignition source. The minimum specimen test size shall be 36 inches by 36 inches (914 mm by 914 mm) by an average of 21 inches (533 mm) deep, and the balls shall be held in a box constructed of galvanized steel poultry netting wire mesh.
8. Foam plastics shall be covered by a fabric, coating or film meeting the fire propagation performance criteria contained in Test Method 1 or Test Method 2, as appropriate, of NFPA 701.
9. The floor covering placed under the children's play structure shall exhibit a Class I interior floor finish classification, as described in Section 804, when tested in accordance with NFPA 253.

❖ The amount of combustible materials associated with children's play structures is rigidly limited. The provisions of Section 424.2 provide nine items of combustible materials that can be used to construct play equipment. These limitations are consistent with those previously applicable only to children's play structures located in covered mall buildings. The materials when tested in accordance with NFPA 701 must comply with the performance criteria of one of the two identified test methods. The historic "small-scale test" is no longer accepted.

[F] 424.3 Fire protection. Children's play structures shall be provided with the same level of *approved* fire suppression and detection devices required for other structures in the same occupancy.

❖ Children's play structures within a building must be provided with fire suppression and detection devices that provide an equal or better degree of fire protection than that required for the occupancy in which the play structures are located.

424.4 Separation. Children's play structures shall have a horizontal separation from building walls, partitions and from elements of the *means of egress* of not less than 5 feet (1524 mm). Children's playground structures shall have a horizontal separation from other children's play structures of not less than 20 feet (6090 mm).

❖ Multiple children's play structures must be isolated from each other by a minimum distance of 20 feet (6090 mm). This separation reduces the concentration of combustible materials within a single area. A minimum separation of 5 feet (1524 mm) is also required between any play structure and portions of the building's construction, including means of egress elements, to allow for emergency personnel access and to reduce fire spread potential.

424.5 Area limits. Children's play structures shall be not greater than 300 square feet (28 m^2) in area, unless a special investigation, acceptable to the building official, has demonstrated adequate fire safety.

❖ Play structures are limited to a 300-square-foot (28 m^2) area limitation unless adequate fire safety is proven to the building official by way of a special investigation.

SECTION 425
HYPERBARIC FACILITIES

425.1 Hyperbaric facilities. Hyperbaric facilities shall meet the requirements contained in Chapter 20 of NFPA 99.

❖ For these unique treatment facilities, Chapter 20 of NFPA 99 provides appropriate design criteria. While hyperbaric facilities are often located in a hospital facility (Group I-2, Condition 2), they can also be found at independent medical facilities.

SECTION [F] 426
COMBUSTIBLE DUSTS, GRAIN PROCESSING AND STORAGE

426.1 Combustible dusts, grain processing and storage. The provisions of Sections 426.1.1 through 426.1.7 shall apply to buildings in which materials that produce combustible dusts are stored or handled. Buildings that store or handle combustible dusts shall comply with the applicable provisions of NFPA 61, NFPA 85, NFPA 120, NFPA 484, NFPA 654, NFPA 655 and NFPA 664 and the *International Fire Code*.

❖ Combustible dusts can generate high-pressure gas by combustion in air. The pressure created can destroy process equipment and structures. While there is a minimum concentration of dust required, there is no reliable upper concentration limitation beyond which combustion will not occur. The ignition source is often the process equipment itself. The provisions of this section primarily address the construction of a building that contains this hazard. The construction requirements are primarily intended to reduce the exposure hazard should a fire or explosion occur. To minimize the impact of the explosion on the structure, explosion control in accordance with Section 414.5.1 and the IFC is required. The provisions of this section apply to all buildings in which combustible dusts and particles are of sufficient quantity to be readily ignited and subject to explosion. The concentration that would be required to constitute such a hazard is dependent on the particle size. Additional guidance on the relative fire risk associated with various combustible dusts can be found in the referenced standards. In addition to the provisions of this section, compliance with the provisions of the IFC and the referenced standards is essential to minimize potential ignition and fire and explosion hazards. Chapter 22 of the IFC contains specific safety precautions for combustible dust-producing operations and references additional standards for explosion protection based on the type of process involved. Essentially, each of the referenced standards prescribes reasonable requirements for safety to life and property from fire and explosion. The standards also minimize the resulting damage should a fire or explosion occur. More specifically, they contain provisions for construction, ventilation, explosion venting, equipment, heating devices, dust control and fire protection; supplemental requirements related to electrical wiring and equipment; provisions concerning protection from sparks; and regulations addressing cutting and welding, smoking and signage.

[F] 426.1.1 Type of construction and height exceptions. Buildings shall be constructed in compliance with the height, number of stories and area limitations specified in Sections 504 and 506; except that where erected of Type I or II construction, the heights and areas of grain elevators and similar structures shall be unlimited, and where of Type IV construction, the maximum building height shall be 65 feet (19 812 mm) and except further that, in isolated areas, the maximum building height of Type IV structures shall be increased to 85 feet (25 908 mm).

❖ The construction of buildings in which combustible dusts can be readily ignitable and subject to an explosion hazard is restricted to the height and area limits in Tables 504.3, 504.4 and 506.2 for Group H-2. The limitation on construction is intended to reduce the available fuel source that could ignite and serve as the source of a dust explosion. Grain elevators and similar structures that require a height in excess of that permitted in Table 504.3 may be of unlimited height if the structures are of Type I or II construction; or such structures may be 65 feet (19 812 mm) if they are of Type IV construction. If the structure is of Type IV construction and located so as not to constitute an explosion hazard, the allowable height may be increased to 85 feet (25 908 mm). Although the buildings are required to have an automatic sprinkler system in accordance with Section 903.2.5, the speed at which the combustion process or explosion could occur may reduce the effectiveness of the sprinkler system. Additional guidance on methods to prevent explosions can be found in Section 911 of the IFC and applicable referenced standards, such as NFPA 69, for explosion prevention systems.

[F] 426.1.2 Grinding rooms. Every room or space occupied for grinding or other operations that produce combustible dusts in such a manner that the room or space is classified as a Group H-2 occupancy shall be enclosed with fire barriers constructed in accordance with Section 707 or horizontal assemblies constructed in accordance with Section 711, or both. The fire-resistance rating of the enclosure shall be not less than 2 hours where the area is not more than 3,000 square feet (279 m^2), and not less than 4 hours where the area is greater than 3,000 square feet (279 m^2).

❖ Grinding rooms are to be separated from remaining parts of the building by construction having a fire-resistance rating of at least 2 hours. If the area of the room is greater than 3,000 square feet (279 m^2), the fire-resistance rating of the enclosure must be at least 4 hours. Not only is it the intent to protect the room from an explosion fire but also, assuming the explosion venting results in minimal structural damage, for the enclosure to contain the resulting fire, if any, to the room of origin. Without venting, it is doubtful that the enclosure could withstand the forces exerted on the wall by an explosion.

[F] 426.1.3 Conveyors. Conveyors, chutes, piping and similar equipment passing through the enclosures of rooms or spaces shall be constructed dirt tight and vapor tight, and be of *approved* noncombustible materials complying with Chapter 30.

❖ The intent of this section is to restrict conveyors that pass through walls so that they do not serve as an ignition source or avenue of fire spread to adjacent areas. These conveyors and similar equipment must be constructed of noncombustible materials and be dirt and vapor tight.

[F] 426.1.4 Explosion control. Explosion control shall be provided as specified in the *International Fire Code,* or spaces shall be equipped with the equivalent mechanical *ventilation* complying with the *International Mechanical Code.*

❖ The pressure exerted by a combustible dust explosion typically ranges from 13 to 89 psi (89 to 614 kPa). It is impractical to construct a building that will withstand such pressures; therefore, either a means of explosion control must be provided in accordance with Section 911 of the IFC, or an equivalent mechanical ventilation system must be provided. The mechanical ventilation system must be in accordance with the IMC and must be designed to control the dust concentration below hazardous levels. The mechanical ventilation is dependent on the concentration at which a dust explosion could occur. Additional guidance on the relative fire risk associated with various combustible dusts can be found in the referenced standards.

[F] 426.1.5 Grain elevators. Grain elevators, malt houses and buildings for similar occupancies shall not be located within 30 feet (9144 mm) of interior *lot lines* or structures on the same *lot*, except where erected along a railroad right-of-way.

❖ The 30-foot (9144 mm) separation between grain elevators (or similar structures with regard to lot lines) and other structures is intended to reduce the exposure hazard. It is intended not only to reduce the damage to the adjacent structure but also to minimize the potential that a fire in the adjacent structure would affect the grain elevator. Grain elevators are singled out because of the additional height permitted in Section 415.8.1.1 and the typical openness of the structure. The 30-foot (9144 mm) separation is not required if the grain elevator is located along a railroad right-of-way. The railroad right-of-way will offer some separation, except when a car is being loaded from the grain elevator, in which case the 30-foot (9144 mm) criterion may not be practical.

[F] 426.1.6 Coal pockets. Coal pockets located less than 30 feet (9144 mm) from interior *lot lines* or from structures on the same *lot shall* be constructed of not less than Type IB construction. Where more than 30 feet (9144 mm) from interior *lot lines*, or where erected along a railroad right-of-way, the minimum type of construction of such structures not more than 65 feet (19 812 mm) in *building height* shall be Type IV.

❖ Like grain elevators, coal pockets are typically open structures and there is a need to provide separation between the coal pocket and adjacent structures to reduce exposure fire risk. Type IB construction is the required minimum if a coal pocket is within 30 feet (9144 mm) of another structure or lot line. Like grain elevators, if the separation is in excess of 30 feet (9144 mm), Type IV construction is permitted to a maximum building height of 65 feet (19 812 mm).

[F] 426.1.7 Tire rebuilding. Buffing operations shall be located in a room separated from the remainder of the building housing the tire rebuilding or tire recapping operation by a 1-hour *fire barrier.*

Exception: Buffing operations are not required to be separated where all of the following conditions are met:

1. Buffing operations are equipped with an *approved* continuous automatic water-spray system directed at the point of cutting action;
2. Buffing machines are connected to particle-collecting systems providing a minimum air movement of 1,500 cubic feet per minute (cfm) (0.71 m^3/s) in volume and 4,500 feet per minute (fpm) (23 m/s) in-line velocity; and
3. The collecting system shall discharge the rubber particles to an *approved* outdoor noncombustible or fire-resistant container, which is emptied at frequent intervals to prevent overflow.

❖ Tire rebuilding plants are classified in Group F-1 occupancies in accordance with Section 306.2. Depending on an evaluation of the actual hazards presented by a given operation, which could include grinding, buffing or gluing of tires or tire components, the plant or portions of it might be classified as Group H. This section specifies that the buffing operations must be separated from other operations by a 1-hour fire barrier. The intent is to keep the higher hazard operations separate, thereby reducing the potential for a rapidly spreading fire.

The exception recognizes that meeting the three outlined conditions will afford protection equivalent to the 1-hour fire barrier required by this section.

Bibliography

The following resource materials were used in the preparation of the commentary for this chapter of the code.

ANSI A13.1-07, *Scheme for the Identification of Piping Systems*. New York: American National Standards Institute, 2007.

ASME B31.3-12, *Process Piping*. New York: American National Standards Institute, 2012.

ASTM C1628-06, *Specification for Joints for Concrete Gravity Flow Sewer Pipe, Using Rubber Gaskets*. West Conshohocken, PA: ASTM International, 2006.

ASTM C1629/C1629M-06, (2011) *Standard Classification for Abuse-resistant Nondecorated Interior Gypsum Panel Products and Fiber-reinforced Cement Panels*. West Conshohocken, PA: ASTM International, 2011.

ASTM E84-2013A, *Test Methods for Surface-burning Characteristics of Building Materials*. West Conshohocken, PA: ASTM International, 2013.

ASTM E119-2012A, *Test Methods for Fire Tests of Building Construction and Materials*. West Conshohocken, PA: ASTM International, 2012.

ASTM E695-09, *Method for Measuring Relative Resistance of Wall, Floor, and Roof Construction to Impact Loading*. West Conshohocken, PA: ASTM International, 2009.

ASTM E736-00 (2011), *Test Method for Cohesion/Adhesion of Sprayed Fire-resistive Materials Applied to Structural Members*. West Conshohocken, PA: ASTM International, 2011.

Bain, A., J. Barclay, T. Bose, F. Edeskuty, M. Farlie, J. Hansel, R. Hay and M. Swain (et al). *Sourcebook for Hydrogen Application*. Golden, CO: Hydrogen Research Institute and National Renewable Energy Laboratory, 1998.

Boring, Delbert F. and others. *Fire Protection Through Modern Building Codes*, 5th ed. Washington, DC: American Iron and Steel Institute, 1981.

Final Report of the National Construction Safety Team on the Collapses of the World Trade Center Towers. Washington, DC: National Institute of Standards and Technology, 2005.

Fire Protection Handbook, 20th ed. Quincy, MA: National Fire Protection Association, 2008.

Fothergill, John W. and John H. Klote. *Design of Smoke Control Systems for Buildings*. Atlanta: American Society for Heating, Refrigerating and Air-Conditioning Engineers, Inc., 1983.

ICC/ANSI A117.1-09, *Accessible and Usable Buildings and Facilities*. Washington, DC: International Code Council, 2010.

ICC 500-14, *ICC/NSSA Standard on the Design and Construction of Storm Shelters*. Washington, DC: International Code Council, 2014.

NFPA 13-13, *Installation of Sprinkler Systems*. Quincy, MA: National Fire Protection Association, 2013.

NFPA 13-D-13, *Installation of Sprinkler Systems in One- and Two-family Dwellings and Manufactured Homes*. Quincy, MA: National Fire Protection Association, 2013.

NFPA 13R-13, *Installation of Sprinkler Systems in Residential Occupancies up to and Including Four Stories in Height*. Quincy, MA: National Fire Protection Association, 2013.

NFPA 30-12, *Flammable and Combustible Liquids Code*. Quincy, MA: National Fire Protection Association, 2011.

NFPA 30A-15, *Code for Motor Fuel-dispensing Facilities and Repair Garages*. Quincy, MA: National Fire Protection Association, 2015.

NFPA 30B-15, *Manufacture and Storage of Aerosol Products*. Quincy, MA: National Fire Protection Association, 2015.

NFPA 32-11, *Dry Cleaning Plants*. Quincy, MA: National Fire Protection Association, 2011.

NFPA 33-15, *Spray Application Using Flammable and Combustible Materials*. Quincy, MA: National Fire Protection Association, 2015.

NFPA 34-15, *Dipping and Coating Processes Using Flammable or Combustible Liquids*. Quincy, MA: National Fire Protection Association, 2015.

NFPA 35-11, *Manufacture of Organic Coatings*. Quincy, MA: National Fire Protection Association, 2011.

NFPA 40-11, *Storage and Handling of Cellulose Nitrate Film*. Quincy, MA: National Fire Protection Association, 2011.

NFPA 50A-99, *Gaseous Hydrogen Systems at Consumer Sites*. Quincy, MA: National Fire Protection Association, 1999.

NFPA 58-14, *Liquefied Petroleum Gas Code*. Quincy, MA: National Fire Protection Association, 2014.

NFPA 61-13, *Prevention of Fires and Dust Explosions in Agricultural Food Products Facilities*. Quincy, MA: National Fire Protection Association, 2013.

NFPA 68-07, *Explosion Protection by Deflagration Venting*. Quincy, MA: National Fire Protection Association, 2007.

NFPA 69-14, *Explosion Prevention Systems*. Quincy, MA: National Fire Protection Association, 2014.

NFPA 70-14, *National Electrical Code*. Quincy, MA: National Fire Protection Association, 2014.

NFPA 72-13, *National Fire Alarm Code*. Quincy, MA: National Fire Protection Association, 2013.

NFPA 80-13, *Fire Doors and Other Opening Protectives*. Quincy, MA: National Fire Protection Association, 2013.

NFPA 99-15, *Standard for Health Care Facilities*. Quincy, MA: National Fire Protection Association, 2015.

NFPA 110-13, *Emergency and Standby Power Systems*. Quincy, MA: National Fire Protection Association, 2013.

NFPA 111-13, *Stored Electrical Energy Emergency and Standby Power Systems*. Quincy, MA: National Fire Protection Association, 2013.

NFPA 120-15, *Coal Preparation Plants*. Quincy, MA: National Fire Protection Association, 2015.

NFPA 251-06, *Standard Methods of Tests of Building Endurance of Building Construction and Materials*. Quincy, MA: National Fire Protection Association, 2006.

NFPA 318-15, *Standard for the Protection of Semiconductor Fabrication Facilities*. Quincy, MA: National Fire Protection Association, 2015.

NFPA 409-11, *Aircraft Hangars*. Quincy, MA: National Fire Protection Association, 2011.

NFPA 418-11, *Standard for Heliports*. Quincy, MA: National Fire Protection Association, 2011.

NFPA 484-15, *Combustible Metals.* Quincy, MA: National Fire Protection Association, 2015.

NFPA 495-13, *Explosive Materials Code*. Quincy, MA: National Fire Protection Association, 2013.

NFPA 654-13, *Prevention of Fire and Dust Explosions from the Manufacturing, Processing and Handling of Combustible Particulate Solids*. Quincy, MA: National Fire Protection Association, 2013.

NFPA 655-12, *Prevention of Sulfur Fires and Explosions*. Quincy, MA: National Fire Protection Association, 2011.

NFPA 664-12, *Prevention of Fires and Explosions in Wood Processing and Woodworking Facilities*. Quincy, MA: National Fire Protection Association, 2011.

NFPA 701-10, *Standard Methods of Fire Tests for Flame-propagation of Textiles and Films*. Quincy, MA: National Fire Protection Association, 2010.

NFPA 704-12, *Flammable and Combustible Liquids Code*. Quincy, MA: National Fire Protection Association, 2011.

NFPA 1124-06, *Manufacture, Transportation and Storage of Fireworks and Pyrotechnic Articles*. Quincy, MA: National Fire Protection Association, 2006.

NFPA, *Fire Protection Guide on Hazardous Materials*. Quincy, MA: National Fire Protection Association, 2002.

NIST IR 5240-93, *Measurement of Room Conditions and Response of Sprinklers and Smoke Detectors During a Simulated Two-bed Hospital Patient Room Fire*. Gaithersburg, MD: National Institute of Standards and Technology, July 1993.

UL 217-06, *Single- and Multiple-station Smoke Detectors with Revisions through August 2005*. Northbrook, IL: Underwriters Laboratories Inc., 2006.

UL 268-09, *Smoke Detectors for Fire Protection Signaling*. Northbrook, IL: Underwriters Laboratories Inc., 2009.

UL 325-02, *Door, Drapery, Gate, Louver and Window Operations and Systems—with Revisions through February 2006*. Northbrook, IL: Underwriters Laboratories Inc., 2002.

UL 1994-04, *Standard for Luminous Egress Path Marking Systems, with Revisions through February of 2005*. Northbrook, IL: Underwriters Laboratories Inc., 2005.

UL 2075-2013, *Standard for Gas and Vapor Detectors and Sensors*. Northbrook, IL: Underwriters Laboratories Inc., 2013.

Chapter 5: General Building Heights and Areas

General Comments

Chapter 5 contains the provisions that regulate the minimum type of construction of a building, including allowable heights and areas; mezzanines; unlimited area structures; mixed occupancies; and special provisions. The occupancy (or occupancies) and type of construction of a building are of key importance to the application of these provisions.

Section 501.1 gives the scope of Chapter 5. It specifies that the provisions of Chapter 5 apply not only to new construction, but also to existing building additions.

Section 502 lists terms that are related to the use and application of Chapter 5 and other provisions of the code related to height and area. The definitions are found in Section 202.

Section 503 addresses the general allowable height and area limitations of buildings. New to the 2015 Codes, this section points to separate sections for limitations on building height and building area. The single Table 503 is replaced by three tables—Table 504.3 for building height in feet, Table 504.4 for building height in stories, and Table 506.2 for building area.

Section 504 addresses the height limits of buildings in both height in feet and number of stories. Table 504.3 provides the height limits, in feet, based upon type of construction, and Table 504.4 provides the maximum allowable number of stories above grade based upon type of construction. Both tables give allowable heights or number of stories with and without sprinkler systems.

Section 505 provides the criteria and regulations for mezzanines and industrial equipment platforms.

Section 506 addresses area limits for a building. Table 506.2 provides tabular areas for buildings based upon the type of construction and occupancy classification for buildings with and without sprinkler systems. The allowable area is then calculated by adding a street frontage increase to these values.

Section 507 includes requirements for buildings with no area limitation. These provisions are commonly used for very large retail sales buildings, industrial warehouses and factories.

Section 508 provides requirements for buildings with multiple occupancies.

Section 509 addresses incidental uses such as waste and linen collection rooms, laundry rooms and incinerator rooms.

Section 510 contains special options for buildings that modify the general limitations of a building's allowable height and/or area.

Chapter 5 is an important chapter in the code because many other code requirements depend on the establishment of the minimum required type of construction for a building. A building has one minimum required construction type and must meet or exceed those requirements, with certain exceptions for buildings conforming to the special provisions of Section 510. Without the correct determination of the minimum type of construction, misapplication of the code is probable.

Sections 504 and 506 provide the foundation for setting thresholds for building size based on the building's use and the materials with which it is constructed. The aim is to reduce risk of injury to an acceptable level for building occupants by limiting fire load and fire hazards relating to height, area and occupancy. In accordance with Tables 504.3, 504.4 and 506.2, buildings of higher construction types [noncombustible materials and a higher degree of fire-resistance rating of elements (see Chapter 6)] can be larger and taller than buildings of lower construction types (combustible and unprotected with fire-resistance ratings).

Purpose

Following occupancy classification of uses based on Chapter 3, the code review for the design of any building continues with Chapter 5, with the determination of the minimum required type of construction based on the use and size of the building. Misapplication of Chapter 5 can result in a multitude of related errors in subsequent code application, since many code requirements depend on the type of construction that is required for the building.

The main purpose of Chapter 5 is to regulate the size of structures based on the specific hazards associated with their occupancy and the materials of which they are constructed. Chapter 5 also provides for adjustments to the allowable area and height based on the presence of fire protection systems, which offer protection for both building occupants and responding fire service personnel in the event of a fire.

SECTION 501 GENERAL

501.1 Scope. The provisions of this chapter control the height and area of structures hereafter erected and *additions* to existing structures.

❖ Chapter 5 is applicable to all new buildings as well as any existing structures that are to be enlarged. Allowable building height and building area are evaluated on the basis of occupancy classification, type of construction, location on the property relative to lot lines and other structures on the same lot, and the presence of an automatic sprinkler system throughout the building.

In the case of additions to existing buildings that are not separated by fire walls, for purposes of determining allowable building area and height, the designer and plan reviewer must evaluate the entire building, including the addition, as if it were a new structure. For instance, if an existing Type IIB building is to have an addition, the aggregate area of the modified building (existing plus the addition) must be within the limits established by Table 506.2 for the allowable area, Table 504.3 for the allowable height in feet, and Table 504.4 for the allowable number of stories above grade of a Type IIB building. The tables take into account area and height limitations for nonsprinklered or sprinklered buildings.

[F] 501.2 Address identification. New and existing buildings shall be provided with *approved* address identification. The address identification shall be legible and placed in a position that is visible from the street or road fronting the property. Address identification characters shall contrast with their background. Address numbers shall be Arabic numbers or alphabetical letters. Numbers shall not be spelled out. Each character shall be a minimum of 4 inches (102 mm) high with a minimum stroke width of $^1/_2$ inch (12.7 mm). Where required by the fire *code official*, address identification shall be provided in additional approved locations to facilitate emergency response. Where access is by means of a private road and the building address cannot be viewed from the public way, a monument, pole or other approved sign or means shall be used to identify the structure. Address identification shall be maintained.

❖ Identifying buildings, both new and existing, during an emergency (i.e., fire department, ambulances, medical, police) is greatly aided by the proper placement of address identification. The size and color criteria are intended to aid visibility from the street. When several structures are remotely located on a site or set back into a property, or at locations where multiple addresses are provided (e.g., strip malls) where the address is not readily visible from the public way, an approved method of identification will also be required, which will have characters posted in a location that will help in an emergency. The fire code official has the authority to require that address numbers be located in all locations deemed necessary to properly identify the building by street address. The primary concern is for emergency personnel to locate the building without going through a lengthy search procedure. In the case of a strip mall, identification would be provided for the backs of buildings that face alleys or roads since the emergency response unit may often be directed to the back entrance. The address numbers must be maintained in a readily visible condition to provide for continuous identification. This would include the repainting of faded numbers or the trimming of trees or other vegetation that may be obscuring visibility of the address.

SECTION 502 DEFINITIONS

502.1 Definitions. The following terms are defined in Chapter 2:

AREA, BUILDING.

BASEMENT.

EQUIPMENT PLATFORM.

GRADE PLANE.

HEIGHT, BUILDING.

MEZZANINE.

❖ This section lists terms that are specifically associated with the subject matter of this chapter. It is important to emphasize that these terms are not exclusively related to this chapter but may or may not also be applicable where the term is used elsewhere in the code. For example, the defined term "basement" in most portions of the code applies to stories that do not meet the definition of a story above grade. However, for the application of flood plain regulations in Section 1612, a different definition of basement applies.

Definitions of terms can help in the understanding and application of the code requirements. The purpose for including a list within this chapter is to provide more convenient access to terms which may have a specific or limited application within this chapter. For the complete definition and associated commentary, refer back to Section 202. Terms that are italicized provide a visual identification throughout the code that a definition exists for that term. The use and application of all defined terms are set forth in Section 201.

SECTION 503 GENERAL BUILDING HEIGHT AND AREA LIMITATIONS

503.1 General. Unless otherwise specifically modified in Chapter 4 and this chapter, *building height*, number of stories and *building area* shall not exceed the limits specified in Sections 504 and 506 based on the type of construction as determined by Section 602 and the occupancies as determined by Section 302 except as modified hereafter. *Building height*, number of stories and *building* area provisions shall be applied independently. Each portion of a building separated by one or more *fire walls* complying with Section 706 shall be considered to be a separate building.

❖ The provisions for governing the height and area of buildings on the basis of occupancy classification and type of construction are established in this section. This section sends the user to Sections 504 and 506 to establish allowable heights, in feet and in stories, and allowable building area, based upon the type of construction. All buildings are subject to these limitations unless more specific code requirements for a building type provide for different height or area limitations. For instance:

- Section 507 allows certain buildings to be unlimited in area due to lack of exposure, low hazard level, construction type, the presence of fire safety systems such as a sprinkler system, or a combination of these characteristics.
- Section 510 allows certain buildings with additional safeguards to adjust the heights and areas allowed by Sections 504 and 506. These are the only provisions that allow a mixture of construction types. In many cases, these provisions involve a distinct portion of the structure being used for parking.
- Where more than one occupancy is present within one building, the provisions of Section 508 must be used in conjunction with Sections 504 and 506 to determine the appropriate construction type for the occupancies involved.

Table 601 is used in conjunction with this chapter to determine acceptable risk and fire safety levels for a building. Classification by occupancy, in accordance with the descriptions in Chapter 3, can be considered as establishing the level of "risk" associated with the use of a building. The various construction types, described in Chapter 6 and Table 601, can be thought of as various levels of safety in regard to fire resistance. Tables 504.3, 504.4 and 506.2 become a set of risk/safety matrices that sets a minimum level of safety (construction type) in accordance with the risk (the occupancy classification).

Fire walls are useful when a single building exceeds the allowable area limitation of Tables 504.3, 504.4 and 506.2. When fire walls (see Section 706) are used in a structure, multiple buildings are created. Each building created by the fire walls would be permitted to have its own occupancy classification and type of construction. Since multiple buildings would be created, each building would be evaluated separately in accordance with the height and area limitations of this Chapter (see commentary, Section 706 and the definition for "Area, building").

503.1.1 Special industrial occupancies. Buildings and structures designed to house special industrial processes that require large areas and unusual *building heights* to accommodate craneways or special machinery and equipment, including, among others, rolling mills; structural metal fabrication shops and foundries; or the production and distribution of electric, gas or steam power, shall be exempt from the *building height*, number of stories and *building area* limitations specified in Sections 504 and 506.

❖ This section provides an exemption from the limits of Sections 504 and 506. The occupancies that may use this exemption are quite limited. The exemption is only applicable when large areas or unusual heights beyond that permitted by Tables 504.3, 504.4 and 506.2 are necessary to accommodate the specific low-hazard or moderate-hazard manufacturing process.

It is the responsibility of the building official to determine when the application of this section is appropriate. The mere cost impact of the application of Tables 504.3, 504.4 and 506.2 does not dictate an exemption. The building official, in assessing the proposed construction, may wish to compare what protection features are being proposed to those features of facilities similar to those described in this section.

503.1.2 Buildings on same lot. Two or more buildings on the same lot shall be regulated as separate buildings or shall be considered as portions of one building where the *building height*, number of stories of each building and the aggregate *building area* of the buildings are within the limitations specified in Sections 504 and 506. The provisions of this code applicable to the aggregate building shall be applicable to each building.

❖ Ordinarily, two buildings on the same lot are considered independently for compliance with the requirements of the code. Section 705.3 requires that when two buildings are on the same lot, an "imaginary" line is assumed between the buildings. This is used to determine the appropriate exterior wall fire-resistance ratings in Table 602, exterior wall requirements of Section 705, projection requirements of Section 705.2 and opening protective ratings and requirements set forth in Table 705.8 and Section 716 (also see the commentary for the definition of "Fire separation distance"). The primary purpose of Section 503.1.2 is to eliminate the application of the provisions of Section 705.3 when two separate buildings can be regulated as one larger building on the lot. In other words, if the two buildings under consideration were actually connected (making one building) and could meet the area and height limitations for one building based on construction type, the connecting interior wall would not be required to have been rated. Therefore, it is inconsistent to require the protection of the facing exterior walls simply because there is not a physical connection between two portions of the same building.

For example, in Commentary Figure 503.1.2(1), Buildings A and B are being constructed on the same lot at the same time. If they are considered to be separate buildings, then the code requirements for exterior wall and opening protection, as well as projections and roof coverings, would apply to each building on the basis of the placement of an imaginary lot line between the two buildings.

Under this section, however, if they are considered to be one building (for minimum type of construction purposes), then there is no need for protection of the facing exterior walls and the code requirements for exterior wall and openings do not apply to the walls between the two buildings. These walls would have to meet exterior wall requirements for the type of construction specified in Table 601 for exterior bearing walls only. If they were nonbearing walls, then the required fire-resistance rating would be zero.

Although facing exterior walls between two buildings on the same lot are not subject to fire-resistance rating and opening size limitations by using this provision (except as previously indicated for bearing walls), the remainder of the exterior walls in both buildings must comply with any applicable exterior wall and opening provisions, including Tables 601, 602, 705.8, 716.5 and 716.6. Three of the walls in Building A and three of the walls in Building B, therefore, would still be required to meet the applicable requirements, and the exemption would only apply to the two walls (one from each building) that face each other.

A special circumstance occurs when a large structure is divided by fire walls into two or more buildings. In such a case, the exception provided by Section 503.1.2 would not be applicable because it is likely that a fire wall has been established because the overall structure exceeds the allowable area for a single building. If the configuration of the buildings results in a court, an imaginary line should be established between the two buildings in order to determine exterior wall and opening protection requirements [see Commentary Figure 503.1.2(2)].

503.1.3 Type I construction. Buildings of Type I construction permitted to be of unlimited tabular *building heights and areas* are not subject to the special requirements that allow unlimited area buildings in Section 507 or unlimited *building height* in Sections 503.1.1 and 504.3 or increased *building heights and areas* for other types of construction.

❖ Buildings of Type I construction permitted by Sections 504 and 506 to be of unlimited height and area do not need to comply with the provisions of Section 503.1.1 or the exception in Section 504.3, which allow height and area increases. While most buildings of Type I construction are permitted to be unlimited in area (see Section 506.2), they are not required to comply with any of the provisions in Section 507 for unlimited area buildings. The requirements in Section 507 address special circumstances that permit a building to be unlimited in area that would not otherwise be allowed to be unlimited in area. As there are no limitations to the size of these structures, the application of these sections would be superfluous. These buildings may be of unlimited size based on their type of construction alone.

Conversely, high-rise buildings are required to be protected by an automatic sprinkler system in accordance with Section 403.3, and this section should not be construed to be a release from that requirement if a building fits the definition of "High-rise building."

Also note that a special industrial occupancy building, in accordance with Section 503.1.1, is permitted to be unlimited in height. However, this does not

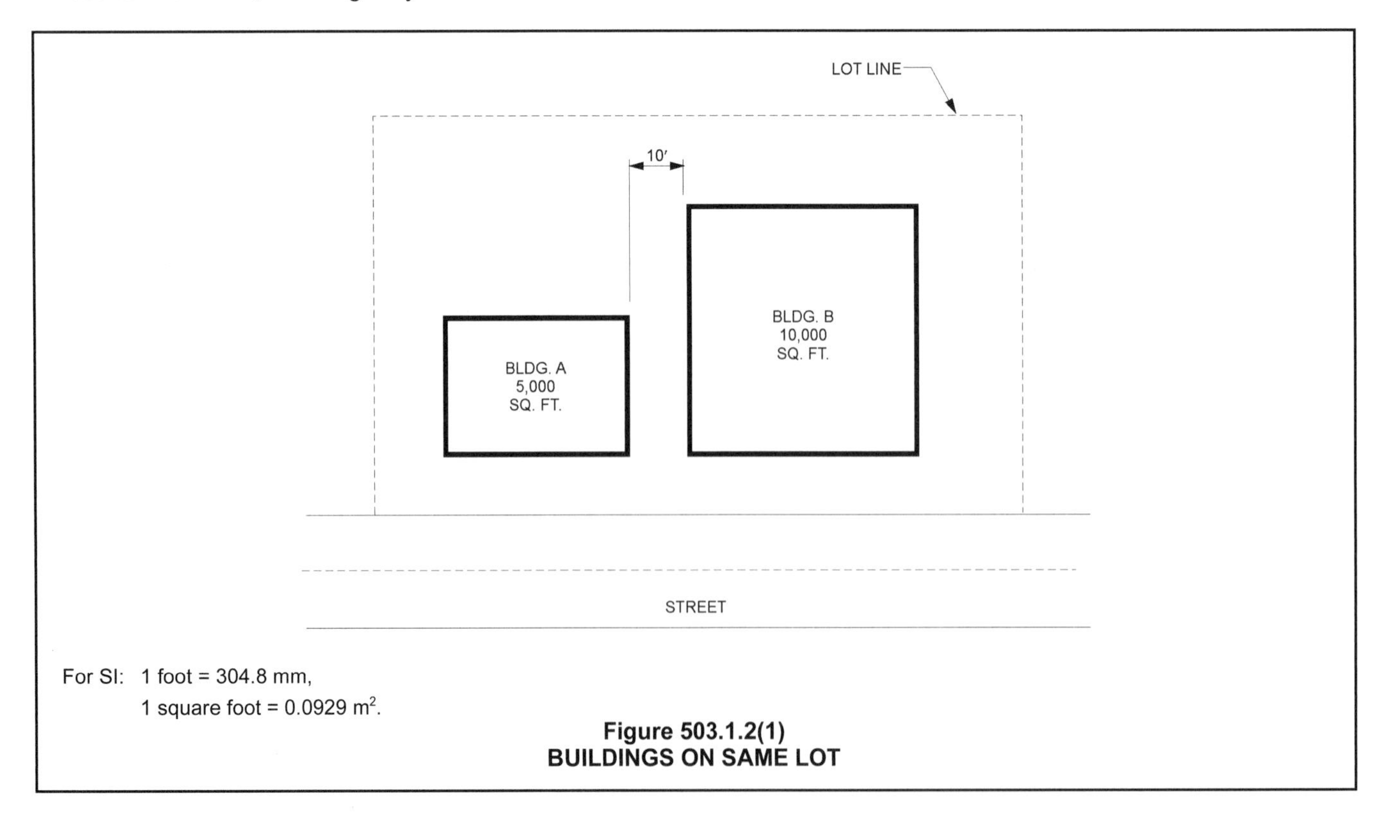

For SI: 1 foot = 304.8 mm,
1 square foot = 0.0929 m^2.

Figure 503.1.2(1)
BUILDINGS ON SAME LOT

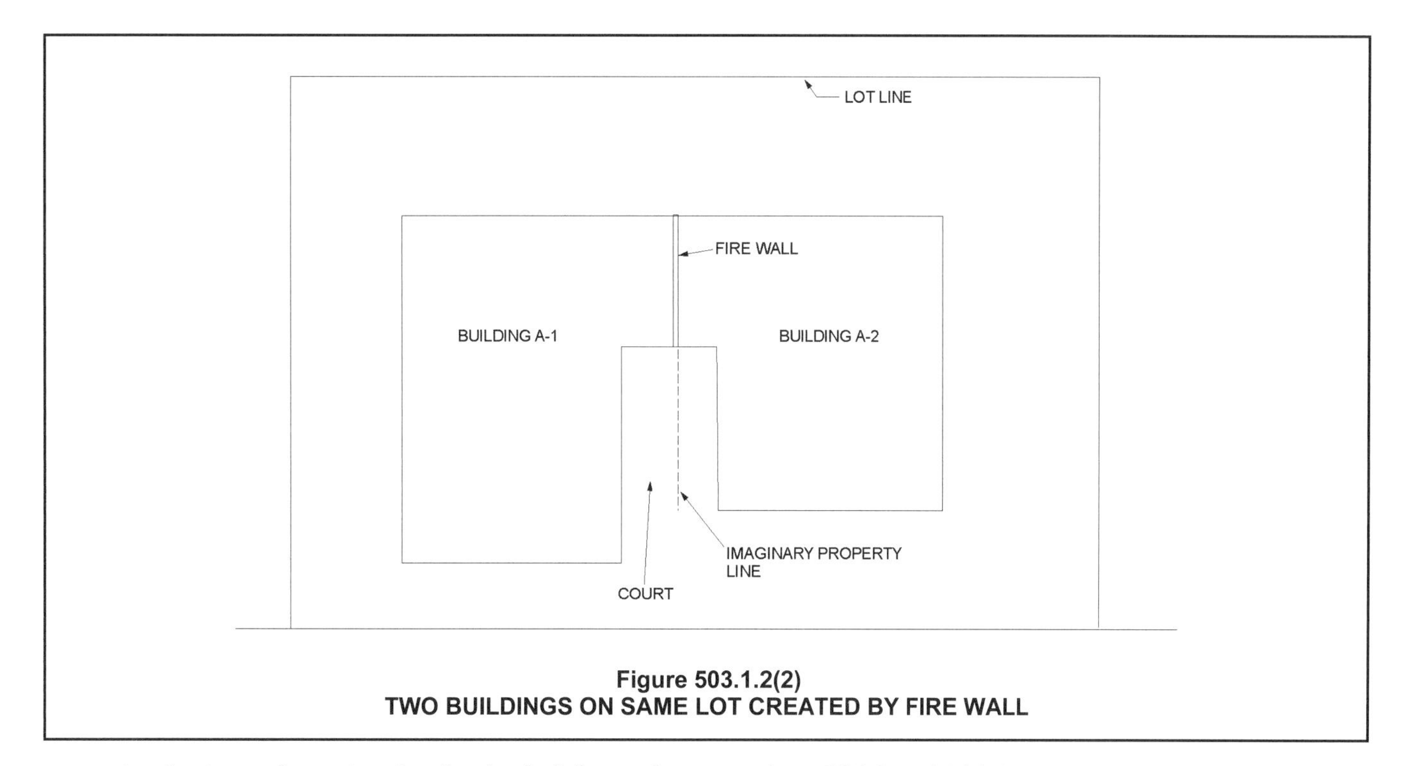

Figure 503.1.2(2)
TWO BUILDINGS ON SAME LOT CREATED BY FIRE WALL

require the type of construction for the building to be Type I. Any type of construction is permitted for special industrial occupancy buildings. Similarly, noncombustible roof structures are permitted to be unlimited in height on all buildings, including those of Type I construction. See also Sections 504.1.1 and 506.1.1 for additional options for unlimited area buildings.

SECTION 504
BUILDING HEIGHT AND NUMBER OF STORIES

504.1 General. The height, in feet, and the number of stories of a building shall be determined based on the type of construction, occupancy classification and whether there is an *automatic sprinkler system* installed throughout the building.

Exception: The *building height* of one-*story* aircraft hangars, aircraft paint hangars and buildings used for the manufacturing of aircraft shall not be limited where the building is provided with an *automatic sprinkler system* or *automatic fire-extinguishing system* in accordance with Chapter 9 and is entirely surrounded by *public ways* or *yards* not less in width than one and one-half times the *building height*.

❖ This section is a fairly straightforward statement regarding those attributes of a building that impact the allowable height of a building. These attributes are: the type of construction, as described in Chapter 6; the occupancy classification of the building, as described in Chapter 3; and the presence, or lack thereof, of an automatic sprinkler system for the entire building. This section is simply a charging statement for what is to come. The actual limitations on numbers of stories and feet come later, in Sections 504.3 and 504.4.

The exception permits fully suppressed aircraft hangars to exceed the height limits of Table 503 as long as they have the specified open area surrounding the building. The exception is necessary to accommodate the size of large aircraft within the building, and the hazard is mitigated by the requirement for suppression and the very large open space provided by the yards and public ways. While the code refers to fire-extinguishing requirements in Chapter 9, the detailed suppression requirements for aircraft hangars are specified in Section 412.4.6. Aircraft paint hangars that are Group H-2 may be unlimited in area as well, under certain conditions (see commentary, Section 507.10).

504.1.1 Unlimited area buildings. The height of unlimited area buildings shall be designed in accordance with Section 507.

❖ Section 507 allows certain buildings to be unlimited in area due to lack of exposure, low hazard level, construction type, the presence of fire safety systems or a combination of these characteristics. This section is a pointer to Section 507 for determination of the allowable height of buildings when the building is being built, using the specific provisions of the unlimited area buildings option. See also Sections 503.1.3 and 506.1.1 for additional options for unlimited area buildings.

504.1.2 Special provisions. The special provisions of Section 510 permit the use of special conditions that are exempt from, or modify, the specific requirements of this chapter regarding the allowable heights of buildings based on the occupancy classification and type of construction, provided the special condition complies with the provisions specified in Section 510.

❖ Section 510 allows certain buildings with additional safeguards to adjust the height in feet and stories allowed by Sections 504.3 and 504.4. Many of the provisions allow a mixture of construction types, and most involve a distinct portion of the structure being used for parking (see also Section 506.1.2).

504.2 Mixed occupancy. In a building containing mixed occupancies in accordance with Section 508, no individual occupancy shall exceed the height and number of story limits specified in this section for the applicable occupancies.

❖ This section is a pointer to Section 508 regarding how to apply the height limitations of Sections 504.3 and 504.4 when a building contains two or more occupancies. The actual allowable height of the building depends on the specific provisions of the applicable portion of Section 508. That is, are the occupancies separated or nonseparated? Or, is one of the occupancies accessory to a main occupancy?

For example: A two-story, nonsprinklered building of 5B construction is proposed to have a restaurant on the first floor and a business on the second floor. Per Table 504.4, the number of stories for Group A-2 is limited to one story and for Group B, to two stories. The building could not be built without a separation between the floors because the nonseparated option must comply with the most restrictive height limitation. Options that could be investigated include equipping the building with sprinklers, separating the first and second floors or choosing a higher construction type.

504.3 Height in feet. The maximum height, in feet, of a building shall not exceed the limits specified in Table 504.3.

Exception: Towers, spires, steeples and other roof structures shall be constructed of materials consistent with the required type of construction of the building except where other construction is permitted by Section 1510.2.5. Such structures shall not be used for habitation or storage. The structures shall be unlimited in height where of noncombustible materials and shall not extend more than 20 feet (6096 mm) above the allowable building height where of combustible materials (see Chapter 15 for additional requirements).

❖ Table 504.3 provides limitations on the height, in feet, of a building based on the occupancy classification, the type of construction and whether the building is equipped throughout with a sprinkler system. The height in feet is the distance from grade plane to the average height of the highest roof surface. See the commentary in Section 202 on "Height, building."

Example: An architect is planning an office building to be constructed of Type VB construction without a sprinkler system. What is the maximum height, in feet, above grade plane to which this building can be constructed?

Answer: The building can be constructed to a maximum height above grade plane of 40 feet (12 192 mm). The building is Occupancy Classification B. Therefore, to find this number, one would go to the row with Occupancy Classifications A, B, E, F, M, S, and U and find the row aligned with "NS." Following to the right on that row, the number 40 can be found under the column TYPE V and B. If a taller building is desired, the options to increase the allowable building height would be to either sprinkler the building, or choose a higher type of construction.

The exception relates to certain roof structures that may exceed the height limitations of Table 504.3 in accordance with this section and Chapter 15. Structures that are used for habitation or storage must be considered to be an additional story and must be included when considering the height limitations in stories in Table 504.4 as well as the height limits in feet in Table 504.3. Please note that Section 1510, Rooftop Structures, contains additional restrictions for structures of a certain size or use. Roof structures must be consistent with the type of construction for the supporting building and, in some cases, must have ratings equivalent to those required for exterior walls. Section 1510.2.5 contains exceptions for penthouse-type structures, mechanical equipment enclosures and screens when not in proximity to a property line. The height for noncombustible roof structures is unlimited. Combustible roof structures are permitted to extend up to 20 feet (6096 mm) above the allowable height set in Table 504.3.

TABLE 504.3. See page 5-7.

❖ The use of Table 504.3 is explained in the example in Section 504.3.

Regarding the Notes:

In this table, "UL" indicates where the height in feet is unlimited; however, it is important to also look at the limits for height in stories and area in Tables 504.4 and 506.2, respectively. The notes also contain some very important information for compliance with the code. In particular, "S" indicates an automatic sprinkler system installed in accordance with Section 903.3.1.1, which is a sprinkler system installed in accordance with NFPA 13 (as opposed to NFPA 13R or 13D), with some exempt locations as specified. "S13R" refers to buildings equipped with a sprinkler system installed in accordance with Section 903.3.1.2, which is a sprinkler system installed in accordance with NFPA 13R, with some exempt locations as specified. NFPA 13R is a standard for installation of sprinkler systems in low-rise residential and Group I-1 occupancies. NFPA 13 contains some provisions that are more rigorous than NFPA 13R.

Therefore, a residential building with a sprinkler system installed in accordance with Section 903.3.1.1 (NFPA 13) has a higher limitation for height than a residential building with sprinklers installed in accordance with Section 903.3.1.2 (NFPA 13R). Since there are no specific height limitations for buildings permitted to use an NFPA 13D sprinkler system, the height limitations for a nonsprinklered building should be used for this option.

Note a: Chapters 4 and 5 contain provisions for special detailed requirements based on use and occupancy, some of which include exceptions to these height limitations. An example is open parking garages, Section 406.5. Open parking garages have their own area and height limitations stated in Table 406.5.4. Similarly, this chapter contains some provisions that are exceptions to the height limitations contained in this section. An example is Section 507, which contains different height limits for unlimited area buildings.

Note b: Section 903.2 specifies thresholds for occupancies that require sprinklers, regardless of height and area or type of construction. For instance, a fire area containing a Group A-2 occupancy greater than 5,000 square feet (465 m^2), or an occupant load greater than 100, or one that is located above the level of exit discharge must be equipped with a sprinkler system, as well as all of the intervening floors from the level of exit discharge to the floor containing the A-2 fire area. If there were floors above this occupancy that were not sprinklered, then the building would be designed based on "NS," Nonsprinklered, because the building is not equipped throughout with a sprinkler system.

Note c: This is a specific mention of an item in Section 903.2 as already mentioned in Note b. All new construction in Group H occupancies is required to be equipped with a sprinkler system. Therefore the "NS" row will never apply.

Note d: There are some applications where "NS" values are given, but do not apply to new

TABLE 504.3[a]
ALLOWABLE BUILDING HEIGHT IN FEET ABOVE GRADE PLANE

<table>
<tr><th rowspan="3">OCCUPANCY CLASSIFICATION</th><th colspan="10">TYPE OF CONSTRUCTION</th></tr>
<tr><th rowspan="2">SEE FOOTNOTES</th><th colspan="2">TYPE I</th><th colspan="2">TYPE II</th><th colspan="2">TYPE III</th><th>TYPE IV</th><th colspan="2">TYPE V</th></tr>
<tr><th>A</th><th>B</th><th>A</th><th>B</th><th>A</th><th>B</th><th>HT</th><th>A</th><th>B</th></tr>
<tr><td rowspan="2">A, B, E, F, M, S, U</td><td>NS[b]</td><td>UL</td><td>160</td><td>65</td><td>55</td><td>65</td><td>55</td><td>65</td><td>50</td><td>40</td></tr>
<tr><td>S</td><td>UL</td><td>180</td><td>85</td><td>75</td><td>85</td><td>75</td><td>85</td><td>70</td><td>60</td></tr>
<tr><td rowspan="2">H-1, H-2, H-3, H-5</td><td>NS[c, d]</td><td rowspan="2">UL</td><td rowspan="2">160</td><td rowspan="2">65</td><td rowspan="2">55</td><td rowspan="2">65</td><td rowspan="2">55</td><td rowspan="2">65</td><td rowspan="2">50</td><td rowspan="2">40</td></tr>
<tr><td>S</td></tr>
<tr><td rowspan="2">H-4</td><td>NS[c, d]</td><td>UL</td><td>160</td><td>65</td><td>55</td><td>65</td><td>55</td><td>65</td><td>50</td><td>40</td></tr>
<tr><td>S</td><td>UL</td><td>180</td><td>85</td><td>75</td><td>85</td><td>75</td><td>85</td><td>70</td><td>60</td></tr>
<tr><td rowspan="2">I-1 Condition 1, I-3</td><td>NS[d, e]</td><td>UL</td><td>160</td><td>65</td><td>55</td><td>65</td><td>55</td><td>65</td><td>50</td><td>40</td></tr>
<tr><td>S</td><td>UL</td><td>180</td><td>85</td><td>75</td><td>85</td><td>75</td><td>85</td><td>70</td><td>60</td></tr>
<tr><td rowspan="2">I-1 Condition 2, I-2</td><td>NS[d, f, e]</td><td>UL</td><td>160</td><td>65</td><td rowspan="2">55</td><td rowspan="2">65</td><td rowspan="2">55</td><td rowspan="2">65</td><td rowspan="2">50</td><td rowspan="2">40</td></tr>
<tr><td>S</td><td>UL</td><td>180</td><td>85</td></tr>
<tr><td rowspan="2">I-4</td><td>NS[d, g]</td><td>UL</td><td>160</td><td>65</td><td>55</td><td>65</td><td>55</td><td>65</td><td>50</td><td>40</td></tr>
<tr><td>S</td><td>UL</td><td>180</td><td>85</td><td>75</td><td>85</td><td>75</td><td>85</td><td>70</td><td>60</td></tr>
<tr><td rowspan="3">R</td><td>NS[d, h]</td><td>UL</td><td>160</td><td>65</td><td>55</td><td>65</td><td>55</td><td>65</td><td>50</td><td>40</td></tr>
<tr><td>S13R</td><td>60</td><td>60</td><td>60</td><td>60</td><td>60</td><td>60</td><td>60</td><td>60</td><td>60</td></tr>
<tr><td>S</td><td>UL</td><td>180</td><td>85</td><td>75</td><td>85</td><td>75</td><td>85</td><td>70</td><td>60</td></tr>
</table>

For SI: 1 foot = 304.8 mm.

Note: UL = Unlimited; NS = Buildings not equipped throughout with an automatic sprinkler system; S = Buildings equipped throughout with an automatic sprinkler system installed in accordance with Section 903.3.1.1; S13R = Buildings equipped throughout with an automatic sprinkler system installed in accordance with Section 903.3.1.2.

a. See Chapters 4 and 5 for specific exceptions to the allowable height in this chapter.

b. See Section 903.2 for the minimum thresholds for protection by an automatic sprinkler system for specific occupancies.

c. New Group H occupancies are required to be protected by an automatic sprinkler system in accordance with Section 903.2.5.

d. The NS value is only for use in evaluation of existing building height in accordance with the *International Existing Building Code*.

e. New Group I-1 and I-3 occupancies are required to be protected by an automatic sprinkler system in accordance with Section 903.2.6. For new Group I-1 occupancies, Condition 1, see Exception 1 of Section 903.2.6.

f. New and existing Group I-2 occupancies are required to be protected by an automatic sprinkler system in accordance with Section 903.2.6 and Section 1103.5 of the *International Fire Code*.

g. For new Group I-4 occupancies, see Exceptions 2 and 3 of Section 903.2.6.

h. New Group R occupancies are required to be protected by an automatic sprinkler system in accordance with Section 903.2.8.

construction. They might, however, apply to existing construction, applying the evaluation provisions of the *International Existing Building Code®* (IEBC®).

Notes e, f, g, and h: Similar to the discussion on Note c, there are occupancies that are required to have sprinkler systems installed, and therefore the only time the values in the "NS" row might apply are for existing buildings, where the IEBC applies. Also, in the case of Note g, there are areas where the code does not require sprinklers in Group I-4. In those cases, the "NS" row would be the appropriate row to use in determining height limitations.

504.4 Number of stories. The maximum number of stories of a building shall not exceed the limits specified in Table 504.4.

❖ Table 504.4 provides height limitations, in stories, of a building based upon the occupancy classification, the type of construction, and whether the building is equipped throughout with an automatic sprinkler system. The height is in number of stories above grade plane. See the commentary in Section 202 for the definition of "Story above grade plane."

Example: An architect is planning an office building to be constructed of Type VB construction without a sprinkler system. What is the maximum height, in stories, above grade plane to which this building can be constructed?

Answer: The building can be constructed to a maximum height of two stories above grade plane. The building is Occupancy Classification B. To find this number, one would go to the row with Occupancy Classification "B" and find the row aligned, with "NS." Following to the right on that row, the number 2 can be found under the column TYPE V and B. If a taller building is desired, the options to increase the allowable building height would be to either sprinkler the building or choose a higher type of construction.

TABLE 504.4. See page 5-9.

❖ The use of Table 504.4 is explained in the example in Section 504.4. See the commentary provided for Table 504.3.

SECTION 505 MEZZANINES AND EQUIPMENT PLATFORMS

505.1 General. *Mezzanines* shall comply with Section 505.2. *Equipment platforms* shall comply with Section 505.3.

❖ Special provisions are established for two types of elevated floor areas: mezzanines and equipment platforms. Each of the elements has its own unique characteristics, requirements and allowances. Section 505.2 addresses mezzanines. Special provisions for equipment platforms are located in Section 505.3.

505.2 Mezzanines. A *mezzanine* or *mezzanines* in compliance with Section 505.2 shall be considered a portion of the *story* below. Such *mezzanines* shall not contribute to either the *building area* or number of *stories* as regulated by Section 503.1. The area of the *mezzanine* shall be included in determining the *fire area*. The clear height above and below the *mezzanine* floor construction shall be not less than 7 feet (2134 mm).

❖ Although mezzanines provide an additional or intermediate useable floor level in a building, they are not considered an additional story as long as they comply with the requirements of Section 505.2. Building height and area limitations are intended to offset the inherent fire hazard associated with specific occupancies and with materials and features of a specific construction type. Because of a mezzanine's restricted size and its required openness to the room or space in which it is located, a mezzanine is not considered to contribute significantly to a building's inherent fire hazard. Therefore, the area of a mezzanine is not considered when applying the provisions of Section 506.2 for building area limitations, and mezzanines are not considered in determining the height in stories of a building as regulated by Table 504.4. The occupant and fuel loads of the mezzanine should be taken into consideration, however, when determining the necessity for fire protection systems. As such, the area of the mezzanine is to be included in the calculation of the size of the fire area for sprinkler thresholds (see the commentary for the definition of "Fire area" in Section 202).

This section does not include any special requirements for the construction of a mezzanine or for fire-resistance ratings; therefore, the mezzanine is to be constructed of materials consistent with the construction type of the building. Required fire-resistance ratings are determined on the basis of Table 601 for the appropriate construction type.

Mezzanines are required to have a ceiling height of not less than 7 feet (2134 mm), and the ceiling height below the mezzanine must also be not less than 7 feet (2134 mm). Even habitable rooms located in mezzanines may have a ceiling height of 7 feet (2134 mm), in accordance with Exception 3 to Section 1208.2.

TABLE 504.4[a, b]
ALLOWABLE NUMBER OF STORIES ABOVE GRADE PLANE

OCCUPANCY CLASSIFICATION	TYPE OF CONSTRUCTION									
	SEE FOOTNOTES	TYPE I		TYPE II		TYPE III		TYPE IV	TYPE V	
		A	B	A	B	A	B	HT	A	B
A-1	NS	UL	5	3	2	3	2	3	2	1
	S	UL	6	4	3	4	3	4	3	2
A-2	NS	UL	11	3	2	3	2	3	2	1
	S	UL	12	4	3	4	3	4	3	2
A-3	NS	UL	11	3	2	3	2	3	2	1
	S	UL	12	4	3	4	3	4	3	2
A-4	NS	UL	11	3	2	3	2	3	2	1
	S	UL	12	4	3	4	3	4	3	2
A-5	NS	UL	UL	UL	UL	UL	UL	UL	UL	UL
	S	UL	UL	UL	UL	UL	UL	UL	UL	UL
B	NS	UL	11	5	3	5	3	5	3	2
	S	UL	12	6	4	6	4	6	4	3
E	NS	UL	5	3	2	3	2	3	1	1
	S	UL	6	4	3	4	3	4	2	2
F-1	NS	UL	11	4	2	3	2	4	2	1
	S	UL	12	5	3	4	3	5	3	2
F-2	NS	UL	11	5	3	4	3	5	3	2
	S	UL	12	6	4	5	4	6	4	3
H-1	NS[c, d]	1	1	1	1	1	1	1	1	NP
	S									
H-2	NS[c, d]	UL	3	2	1	2	1	2	1	1
	S									
H-3	NS[c, d]	UL	6	4	2	4	2	4	2	1
	S									
H-4	NS[c, d]	UL	7	5	3	5	3	5	3	2
	S	UL	8	6	4	6	4	6	4	3
H-5	NS[c, d]	4	4	3	3	3	3	3	3	2
	S									
I-1 Condition 1	NS[d, e]	UL	9	4	3	4	3	4	3	2
	S	UL	10	5	4	5	4	5	4	3
I-1 Condition 2	NS[d, e]	UL	9	4	3	4	3	4	3	2
	S	UL	10	5						
I-2	NS[d, f]	UL	4	2	1	1	NP	1	1	NP
	S	UL	5	3						
I-3	NS[d, e]	UL	4	2	1	2	1	2	2	1
	S	UL	5	3	2	3	2	3	3	2
I-4	NS[d, g]	UL	5	3	2	3	2	3	1	1
	S	UL	6	4	3	4	3	4	2	2
M	NS	UL	11	4	2	4	2	4	3	1
	S	UL	12	5	3	5	3	5	4	2

(continued)

TABLE 504.4[a, b]—continued
ALLOWABLE NUMBER OF STORIES ABOVE GRADE PLANE

OCCUPANCY CLASSIFICATION	TYPE OF CONSTRUCTION									
	SEE FOOTNOTES	TYPE I		TYPE II		TYPE III		TYPE IV	TYPE V	
		A	B	A	B	A	B	HT	A	B
R-1	NS[d, h]	UL	11	4	4	4	4	4	3	2
	S13R	4	4						4	3
	S	UL	12	5	5	5	5	5	4	3
R-2	NS[d, h]	UL	11	4	4	4	4	4	3	2
	S13R	4	4	4					4	3
	S	UL	12	5	5	5	5	5	4	3
R-3	NS[d, h]	UL	11	4	4	4	4	4	3	3
	S13R	4	4						4	4
	S	UL	12	5	5	5	5	5	4	4
R-4	NS[d, h]	UL	11	4	4	4	4	4	3	2
	S13R	4	4						4	3
	S	UL	12	5	5	5	5	5	4	3
S-1	NS	UL	11	4	2	3	2	4	3	1
	S	UL	12	5	3	4	3	5	4	2
S-2	NS	UL	11	5	3	4	3	4	4	2
	S	UL	12	6	4	5	4	5	5	3
U	NS	UL	5	4	2	3	2	4	2	1
	S	UL	6	5	3	4	3	5	3	2

Note: UL = Unlimited; NP = Not Permitted; NS = Buildings not equipped throughout with an automatic sprinkler system; S = Buildings equipped throughout with an automatic sprinkler system installed in accordance with Section 903.3.1.1; S13R = Buildings equipped throughout with an automatic sprinkler system installed in accordance with Section 903.3.1.2.

a. See Chapters 4 and 5 for specific exceptions to the allowable height in this chapter.

b. See Section 903.2 for the minimum thresholds for protection by an automatic sprinkler system for specific occupancies.

c. New Group H occupancies are required to be protected by an automatic sprinkler system in accordance with Section 903.2.5.

d. The NS value is only for use in evaluation of existing building height in accordance with the *International Existing Building Code*.

e. New Group I-1 and I-3 occupancies are required to be protected by an automatic sprinkler system in accordance with Section 903.2.6. For new Group I-1 occupancies, Condition 1, see Exception 1 of Section 903.2.6.

f. New and existing Group I-2 occupancies are required to be protected by an automatic sprinkler system in accordance with Section 903.2.6 and Section 1103.5 of the *International Fire Code*.

g. For new Group I-4 occupancies, see Exceptions 2 and 3 of Section 903.2.6.

h. New Group R occupancies are required to be protected by an automatic sprinkler system in accordance with Section 903.2.8.

505.2.1 Area limitation. The aggregate area of a *mezzanine* or *mezzanines* within a room shall be not greater than one-third of the floor area of that room or space in which they are located. The enclosed portion of a room shall not be included in a determination of the floor area of the room in which the *mezzanine* is located. In determining the allowable *mezzanine* area, the area of the *mezzanine* shall not be included in the floor area of the room.

Where a room contains both a *mezzanine* and an *equipment platform*, the aggregate area of the two raised floor levels shall be not greater than two-thirds of the floor area of that room or space in which they are located.

Exceptions:

1. The aggregate area of *mezzanines* in buildings and structures of Type I or II construction for special industrial occupancies in accordance with Section 503.1.1 shall be not greater than two-thirds of the floor area of the room.
2. The aggregate area of *mezzanines* in buildings and structures of Type I or II construction shall be not greater than one-half of the floor area of the room in buildings and structures equipped throughout with an *approved automatic sprinkler system* in accordance with Section 903.3.1.1 and an *approved emergency voice/alarm communication system* in accordance with Section 907.5.2.2.

❖ So as not to contribute significantly to a building's inherent fire hazard, a mezzanine is restricted to a maximum of one-third of the area of the room with which it shares a common atmosphere. The area may consist of multiple mezzanines open to the same room at the same or different levels, provided that the aggregate area does not exceed the one-third limitation. (This determination is based on the gross floor area of the mezzanines.) If the area limitation is exceeded, the provisions of this section do not apply and the mezzanine level is considered an additional story.

In determining the allowable area of the mezzanine, any enclosed spaces in the room in which the mezzanine is contained are not to be included in calculating the room size. Although the mezzanine area is included in the calculation of fire area size, it is not included in the area of the room when computing the allowable mezzanine area. For example, a room contains 5,000 square feet (465 m^2), 500 square feet (46 m^2) of which are enclosed and not part of the common atmosphere with the mezzanine. A mezzanine may be provided in the room such that the area of the mezzanine is not more than 1,500 square feet (139 m^2) [5,000 square feet — 500 square feet = 4,500 square feet x $^1/_3$ = 1,500 square feet (139 m^2)]. The mezzanine level must be open, except that enclosed spaces are permitted for mezzanines in accordance with the exceptions to Section 505.2.3. Even if the mezzanine is enclosed, the allowable area of the mezzanine is always determined assuming the space is open to the room in which it is located.

By definition, in special industrial occupancies (see commentary, Section 503.1.1) the inherent fire hazard is very low. Therefore, mezzanines in such occupancies located in buildings of Type I or II construction are permitted to constitute up to two-thirds of the area of the room in which they are located, in accordance with Exception 1. The limitation on construction types further reduces the fire hazard associated with such occupancies.

Exception 2 provides additional allowable area for mezzanines in Type I or II construction in any occupancy, when sprinklers and an approved emergency voice/alarm communication system are installed. The tradeoff for additional area for installation of a sprinkler system recognizes the additional time that sprinklers provide for occupant evacuation. The emergency voice/alarm communication system is also seen as a necessary feature of this tradeoff as it enables occupants of a building to become aware of a fire that could start in a remote part of the mezzanine, prompting evacuation before smoke spread could cause a problem for occupants. Experience has shown that occupants react more readily when given voice instructions than when given a general alarm.

Equipment platforms, as defined in Section 202, are not considered mezzanines (see Section 505.3 for provisions addressing equipment platforms). Where one or more mezzanines and one or more equipment platforms are located within the same room, the aggregate area of such elevated spaces cannot exceed more than two-thirds of the floor area of the room in which they are located. In addition, the maximum floor area for any portion determined to be a mezzanine is limited based upon the provisions of this section.

505.2.2 Means of egress. The *means of egress* for *mezzanines* shall comply with the applicable provisions of Chapter 10.

❖ A mezzanine can be likened to a room or space when considering means of egress (see Section 1006.2). As with rooms, if the occupant load of the mezzanine exceeds the limitation of Table 1006.2.1 for the specific use of the space, at least two independent means of egress must be provided for the mezzanine. For example, if a mezzanine containing office areas (Group B) has an occupant load exceeding 49, a second means of egress from the mezzanine is required. Additionally, if a mezzanine has one means of egress and it is by means of an exit access stairway to the floor below, the common path of travel distance from the most remote point on the mezzanine to the bottom of the stairway may not exceed 75 feet (22 860 mm) in accordance with Section 1006.2.1 [where the occupant load is 30 or less, a common path of egress travel is allowed to increase to 100 feet (30 480 mm)]. If the travel distance to the bottom of the exit access stairway exceeds the limits of Table 1006.2.1, then a second means of egress must be provided from the mezzanine. This assumes that at the bottom of the stairway there are at least two separate travel paths available that lead to at least two exits. If the room below has only one exit, the travel distance from the most remote point on the mezzanine is measured to the exit access door of the room.

Because mezzanines that comply with Section 505.2 are considered a portion of the story in which they are located, the exit access stairway leading from it is not required to be enclosed. Per Section 1019.2, stairways that are contained within a single story are not required to be enclosed.

When two exit access stairways are required, they are to be located remote from each other in accordance with Section 1007.1.1. The measurement for separation is to the top of the closest riser of the two open exit access stairways (see Section 1007.1.1.1, Item 2), and the separation must be maintained for the entire length of the open exit access stairways as they move down to the room below (see Section 1007.1.3).

The occupant load of the mezzanine is added to the room or space below, and the required means of egress width for that room is determined accordingly (see Section 1004.1.1.2). For example, if a room (Group B) has an occupant load of 45 and a mezzanine (also Group B) has an occupant load of 15, the total occupant load for the space served is 60. Therefore, the room must have two exit access doors or exits in accordance with Table 1006.2.1. The mezzanine itself, however, needs only one means of egress provided the common path of travel provisions of Table 1006.2.1 are met.

Alternatively, if exit access travel distance cannot

be met, an exit providing direct access to the outside, an exterior exit stairway or an enclosed interior exit stairway can be utilized to provide the required means of egress from the mezzanine.

The requirements for accessible means of egress are not in any way intended to be affected by these provisions. If two means of egress are required from the mezzanine and an accessible route is required to the mezzanine, then two accessible means of egress must be provided from the mezzanine (see Section 1009.1). If an accessible route is not required to the mezzanine (see Sections 1104.3 and 1104.4) or the mezzanine is not required to be accessible (see Section 1103.2), then accessible means of egress are not required.

505.2.3 Openness. A *mezzanine* shall be open and unobstructed to the room in which such *mezzanine* is located except for walls not more than 42 inches (1067 mm) in height, columns and posts.

Exceptions:

1. *Mezzanines* or portions thereof are not required to be open to the room in which the *mezzanines* are located, provided that the *occupant load* of the aggregate area of the enclosed space is not greater than 10.
2. A *mezzanine* having two or more exits or access to exits is not required to be open to the room in which the *mezzanine* is located.
3. *Mezzanines* or portions thereof are not required to be open to the room in which the *mezzanines* are located, provided that the aggregate floor area of the enclosed space is not greater than 10 percent of the *mezzanine* area.
4. In industrial facilities, *mezzanines* used for control equipment are permitted to be glazed on all sides.
5. In occupancies other than Groups H and I, that are no more than two *stories* above *grade plane* and equipped throughout with an *automatic sprinkler system* in accordance with Section 903.3.1.1, a *mezzanine* having two or more *means of egress* shall not be required to be open to the room in which the *mezzanine* is located.

❖ A mezzanine presents a unique fire threat to the occupant. If a mezzanine is closed off from the larger room, an undetected fire could develop such that it would jeopardize or eliminate the opportunity for occupant escape. The initial requirement is for the mezzanine to be open to the room below so that there will be the same atmosphere as the room below. This should make fire recognition quicker. The 42-inch (1067 mm) height for the perimeter walls is to allow walls that meet the height requirements for guards (see Section 1015.3). The columns and posts should be limited to those that support the roof or floor above the mezzanine.

The exceptions address situations where the hazard is reduced. Exception 1 would allow for a small mezzanine, or a small portion of a larger mezzanine, to be enclosed where the occupant load for that enclosed area is 10 or less. Occupant load is calculated in accordance with Section 1004.1 for the use or function of the mezzanine space. Similarly, Exception 3 permits the enclosure of a limited portion of a mezzanine.

Exception 2 permits enclosure of the mezzanine based on at least two means of egress being provided. Since mezzanines are elevated, means of egress is typically by either open exit access stairways or enclosed interior exit stairways. Exception 4 addresses industrial facilities, where enclosure may be necessary for noise reduction or atmospheric control.

Exception 5 permits enclosure of a mezzanine in a building that is fully protected by a sprinkler system and limited to two stories. While Exception 5 has more restrictions than Exception 2, the allowances for enclosure of the mezzanine are essentially the same where there are two means of egress provided.

505.3 Equipment platforms. *Equipment platforms* in buildings shall not be considered as a portion of the floor below. Such *equipment platforms* shall not contribute to either the *building area* or the number of *stories* as regulated by Section 503.1. The area of the *equipment platform* shall not be included in determining the *fire area* in accordance with Section 903. *Equipment platforms* shall not be a part of any *mezzanine* and such platforms and the walkways, *stairs*, *alternating tread devices* and ladders providing access to an *equipment platform* shall not serve as a part of the *means of egress* from the building.

❖ "Equipment platform" is defined in Section 202 as an unoccupied, elevated platform used exclusively for supporting mechanical systems or industrial process equipment and providing access to that equipment. If an elevated platform does not meet all the conditions of this definition, then it must be considered either a mezzanine or another story.

Equipment platforms are treated as part of the equipment they support (within the limitations of the subsections to this section), and do not contribute in any way to the area of the building, the number of stories, the area of any mezzanine or any fire area. If equipment platforms are located in the same room as a mezzanine, however, the aggregate area of the platforms and mezzanines is limited by Sections 505.2.1 and 505.3.1.

The definition of "Equipment platform" includes the associated elevated walkways, stairs, alternating tread devices and ladders necessary to access the platform. Any egress elements that serve an equipment platform are not permitted to serve as a means of egress for occupants from other areas of the building. Because they are not used for general means of egress, the elements used to access these platforms could be something other than a stairway or ramp. For example, a permanent ladder could be used to access an equipment platform.

505.3.1 Area limitation. The aggregate area of all *equipment platforms* within a room shall be not greater than two-thirds of the area of the room in which they are located. Where an *equipment platform* is located in the same room as a *mezzanine*, the area of the *mezzanine* shall be determined by Section 505.2.1 and the combined aggregate area of the *equipment platforms* and *mezzanines* shall be not greater than two-thirds of the room in which they are located.

❖ In determining the allowable area of an equipment platform, neither the enclosed spaces of the room in which the equipment platform is located nor the area of the equipment platform itself is to be included in calculating the room size. Whereas mezzanines are limited to one-third of the area of the room in which they are located, equipment platforms are permitted to be two-thirds of the area of the room. Whereas the area of mezzanines is included in the calculation of the fire area (see Section 505.1), the area of equipment platforms is not included as part of the fire area. The area of mezzanines and equipment platforms, however, is summed when they are in the same room, and the aggregate area is limited to two-thirds of the area of the room.

505.3.2 Automatic sprinkler system. Where located in a building that is required to be protected by an *automatic sprinkler system*, *equipment platforms* shall be fully protected by sprinklers above and below the platform, where required by the standards referenced in Section 903.3.

❖ In buildings or spaces that are required to be protected with an automatic fire sprinkler system, sprinkler protection above and below the equipment platform is needed so the equipment platform will not obstruct sprinkler coverage or delay sprinkler activation if a fire develops below the platform. This section should not be construed to require sprinkler protection for equipment platforms where such a system is not otherwise required.

505.3.3 Guards. *Equipment platforms* shall have *guards* where required by Section 1015.2.

❖ Guards are required for equipment platforms in the same manner that they are required at open walking surfaces in other parts of the building, and are subject to the same requirements for height, design load and configuration. Exception 3 to Section 1015.4 allows, in areas where electrical, mechanical and plumbing systems or equipment are accessed and used, guard balusters to be spaced to prevent the passage of a 21-inch (533 mm) sphere.

SECTION 506
BUILDING AREA

506.1 General. The floor area of a building shall be determined based on the type of construction, occupancy classification, whether there is an automatic sprinkler system installed throughout the building and the amount of building frontage on public way or open space.

❖ This section is a fairly straightforward statement regarding which attributes of a building impact the allowable area of a building. These attributes are the type of construction, as described in Chapter 6; the occupancy classification of the building, as described in Chapter 3; the presence, or lack thereof, of an automatic sprinkler system for the entire building; and the amount of the building that has frontage on an open space or public way. This section is simply a charging statement for what is to come. The limitations on numbers of stories and feet are found in Sections 504.3 and 504.4.

506.1.1 Unlimited area buildings. Unlimited area buildings shall be designed in accordance with Section 507.

❖ Section 507 allows certain buildings to be unlimited in area due to lack of exposure, low hazard level, construction type, the presence of fire safety systems or a combination of these characteristics. Under these specific provisions, Section 506.2 does not apply unless referenced. See also Section 503.1.3 for additional options for unlimited area buildings.

506.1.2 Special provisions. The special provisions of Section 510 permit the use of special conditions that are exempt from, or modify, the specific requirements of this chapter regarding the allowable areas of buildings based on the occupancy classification and type of construction, provided the special condition complies with the provisions specified in Section 510.

❖ Section 510 allows certain buildings with additional safeguards to adjust the heights and areas allowed by Section 506.2. Many of the provisions allow a mixture of construction types, and most involve a distinct portion of the structure being used for parking.

506.1.3 Basements. Basements need not be included in the total allowable floor area of a building provided the total area of such basements does not exceed the area permitted for a one-story above grade plane building.

❖ Assuming the basement is the same footprint as the building, the area of a single basement is not required to be counted as part of the total building area when evaluating total allowable area in accordance with Section 506.2. Determining the maximum area for the basement can include the frontage increase in Section 506.3. Where there are multiple basements, the area of the basements is combined in order to determine if the basements will be considered to count toward the limits for the total permitted building area. Therefore, if the allowable area of a single-story building is 50,000 square feet (4645 m^2) in accordance with Sections 506.2 and 506.3, the total aggregate area of all stories below grade plane must not exceed 50,000 square feet (4645 m^2). If the aggregate area exceeds that allowed for a single-story building, the basement levels must be counted as contributing to the total allowable area for the building.

506.2 Allowable area determination. The allowable area of a building shall be determined in accordance with the applicable provisions of Sections 506.2.1 through 506.2.4 and Section 506.3.

❖ The process for determining the allowable area of a building is provided in Sections 506.2.1 through 506.2.4 and Section 506.3. At the core of the requirements is Table 506.2, which provides tabular area values that are plugged into the applicable equations in Sections 506.2.1 through 506.2.4. The tabular values are established based upon the type of construction, the occupancy classification and whether the building is equipped throughout with an automatic sprinkler system. The allowable area values are then determined based upon these tabular values, the number of stories and the amount of frontage the building has on open space or a public way.

TABLE 506.2. See page 5-15.

❖ The use of Table 506.2 is explained in the example in Section 506.2. Regarding the notes, see the commentary provided on the notes for Table 504.3.

506.2.1 Single-occupancy, one-story buildings. The allowable area of a single-occupancy building with no more than one story above grade plane shall be determined in accordance with Equation 5-1:

$$A_a = A_t + (NS \times I_f) \qquad \textbf{(Equation 5-1)}$$

where:

A_a = Allowable area (square feet).

A_t = Tabular allowable area factor (NS, S1, or S13R value, as applicable) in accordance with Table 506.2.

NS = Tabular allowable area factor in accordance with Table 506.2 for nonsprinklered building (regardless of whether the building is sprinklered).

I_f = Area factor increase due to frontage (percent) as calculated in accordance with Section 506.3.

❖ This section gives the basic formula for determining the allowable area for a one-story, single-occupancy building.

Example 1: A one-story motel, planned to be Type IIIB construction. As required by the code, the building will be equipped throughout with an NFPA 13R sprinkler system. The building has only 25 percent of its perimeter facing a public way or open space. What is the allowable area?

Answer: The allowable area for this building is 16,000 square feet (1486 m^2).

Explanation: The occupancy classification for a motel is R-1. Under the column Type III, Column B, in the row for R-1, S13R, the value for A_t is 16,000, and NS = 16,000.

In Equation 5-1, I_f = 0. (See commentary on Section 506.3 for a detailed explanation of I_f)

Therefore, in Equation 5-1: A_a = 16,000 + (16,000 x 0) = 16,000.

Observe that, for residential occupancies, there is no advantage to having a sprinkler system installed in accordance with NFPA 13R versus having no sprinkler system at all. However, for this same building, the allowable area would be 64,000 square feet (5946 m^2) if an NFPA 13 sprinkler system were installed.

Example 2: A one-story retail store building is proposed to be built 21,000 square feet (1951 m^2) in area. Like Example 1, the building has only 25 percent of its perimeter facing a public way or open space. What is the minimum type of construction allowed for this building? If an NFPA 13 sprinkler system was to be installed, what would be the minimum type of construction allowed?

Answer: The minimum type of construction for this building without a sprinkler system would be Type IIA. With an NFPA 13 sprinkler system, the minimum type of construction would be Type VB.

The area factor increase due to frontage, I_f = 0. (For discussion on this factor, see Section 506.3.) Therefore, Equation 5-1 becomes $A_a = A_t$. That is, the allowable area equals the tabular area when there is no increase for frontage. The occupancy classification for a retail store is M. Therefore, entering Table 506.2 for occupancy classification M, in the row labeled "NS" for nonsprinklered buildings, note that the tabular area for Type IIA construction is 21,500 square feet (1997 m^2), which is greater than the area of 21,000 square feet (1951 m^2) proposed. Therefore, Type IIA construction would be required.

If the building was provided with an NFPA 13 sprinkler system, the construction type could be VB. Again, entering the table at occupancy classification M, but in the row "SI" for one-story sprinklered buildings, note that the tabular area for Type VB construction is 36,000 square feet (3344 m^2), which is greater than the proposed 21,000 square feet (1951 m^2).

TABLE 506.2[a, b]
ALLOWABLE AREA FACTOR (A_t = NS, S1, S13R, or SM, as applicable) IN SQUARE FEET

OCCUPANCY CLASSIFICATION	SEE FOOTNOTES	TYPE OF CONSTRUCTION								
		TYPE I		TYPE II		TYPE III		TYPE IV	TYPE V	
		A	B	A	B	A	B	HT	A	B
A-1	NS	UL	UL	15,500	8,500	14,000	8,500	15,000	11,500	5,500
	S1	UL	UL	62,000	34,000	56,000	34,000	60,000	46,000	22,000
	SM	UL	UL	46,500	25,500	42,000	25,500	45,000	34,500	16,500
A-2	NS	UL	UL	15,500	9,500	14,000	9,500	15,000	11,500	6,000
	S1	UL	UL	62,000	38,000	56,000	38,000	60,000	46,000	24,000
	SM	UL	UL	46,500	28,500	42,000	28,500	45,000	34,500	18,000
A-3	NS	UL	UL	15,500	9,500	14,000	9,500	15,000	11,500	6,000
	S1	UL	UL	62,000	38,000	56,000	38,000	60,000	46,000	24,000
	SM	UL	UL	46,500	28,500	42,000	28,500	45,000	34,500	18,000
A-4	NS	UL	UL	15,500	9,500	14,000	9,500	15,000	11,500	6,000
	S1	UL	UL	62,000	38,000	56,000	38,000	60,000	46,000	24,000
	SM	UL	UL	46,500	28,500	42,000	28,500	45,000	34,500	18,000
A-5	NS									
	S1	UL	UL	UL	UL	UL	UL	UL	UL	UL
	SM									
B	NS	UL	UL	37,500	23,000	28,500	19,000	36,000	18,000	9,000
	S1	UL	UL	150,000	92,000	114,000	76,000	144,000	72,000	36,000
	SM	UL	UL	112,500	69,000	85,500	57,000	108,000	54,000	27,000
E	NS	UL	UL	26,500	14,500	23,500	14,500	25,500	18,500	9,500
	S1	UL	UL	106,000	58,000	94,000	58,000	102,000	74,000	38,000
	SM	UL	UL	79,500	43,500	70,500	43,500	76,500	55,500	28,500
F-1	NS	UL	UL	25,000	15,500	19,000	12,000	33,500	14,000	8,500
	S1	UL	UL	100,000	62,000	76,000	48,000	134,000	56,000	34,000
	SM	UL	UL	75,000	46,500	57,000	36,000	100,500	42,000	25,500
F-2	NS	UL	UL	37,500	23,000	28,500	18,000	50,500	21,000	13,000
	S1	UL	UL	150,000	92,000	114,000	72,000	202,000	84,000	52,000
	SM	UL	UL	112,500	69,000	85,500	54,000	151,500	63,000	39,000
H-1	NS[c]	21,000	16,500	11,000	7,000	9.500	7,000	10,500	7,500	NP
	S1									
H-2	NS[c]									
	S1	21,000	16,500	11,000	7,000	9.500	7,000	10,500	7,500	3,000
	SM									
H-3	NS[c]									
	S1	UL	60,000	26,500	14,000	17,500	13,000	25,500	10,000	5,000
	SM									
H-4	NS[c, d]	UL	UL	37,500	17,500	28,500	17,500	36,000	18,000	6,500
	S1	UL	UL	150,000	70,000	114,000	70,000	144,000	72,000	26,000
	SM	UL	UL	112,500	52,500	85,500	52,500	108,000	54,000	19,500
H-5	NS[c, d]	UL	UL	37,500	23,000	28,500	19,000	36,000	18,000	9,000
	S1	UL	UL	150,000	92,000	114,000	76,000	144,000	72,000	36,000
	SM	UL	UL	112,500	69,000	85,500	57,000	108000	54,000	27,000

(continued)

TABLE 506.2[a, b]—continued
ALLOWABLE AREA FACTOR (A_t = NS, S1, S13R, or SM, as applicable) IN SQUARE FEET

OCCUPANCY CLASSIFICATION	SEE FOOTNOTES	TYPE OF CONSTRUCTION								
		TYPE I		TYPE II		TYPE III		TYPE IV	TYPE V	
		A	B	A	B	A	B	HT	A	B
I-1	NS[d, e]	UL	55,000	19,000	10,000	16,500	10,000	18,000	10,500	4,500
	S1	UL	220,000	76,000	40,000	66,000	40,000	72,000	42,000	18,000
	SM	UL	165,000	57,000	30,000	49,500	30,000	54,000	31,500	13,500
I-2	NS[d, f]	UL	UL	15,000	11,000	12,000	NP	12,000	9,500	NP
	S1	UL	UL	60,000	44,000	48,000	NP	48,000	38,000	NP
	SM	UL	UL	45,000	33,000	36,000	NP	36,000	28,500	NP
I-3	NS[d, e]	UL	UL	15,000	10,000	10,500	7,500	12,000	7,500	5,000
	S1	UL	UL	45,000	40,000	42,000	30,000	48,000	30,000	20,000
	SM	UL	UL	45,000	30,000	31,500	22,500	36,000	22,500	15,000
I-4	NS[d, g]	UL	60.500	26,500	13,000	23,500	13,000	25,500	18,500	9,000
	S1	UL	121,000	106,000	52,000	94,000	52,000	102,000	74,000	36,000
	SM	UL	181,500	79,500	39,000	70,500	39,000	76,500	55,500	27,000
M	NS	UL	UL	21,500	12,500	18,500	12,500	20,500	14,000	9,000
	S1	UL	UL	86,000	50,000	74,000	50,000	82,000	56,000	36,000
	SM	UL	UL	64,500	37,500	55,500	37,500	61,500	42,000	27,000
R-1	NS[d, h] S13R	UL	UL	24,000	16,000	24,000	16,000	20,500	12,000	7,000
	S1	UL	UL	96,000	64,000	96,000	64,000	82,000	48,000	28,000
	SM	UL	UL	72,000	48,000	72,000	48,000	61,500	36,000	21,000
R-2	NS[d, h] S13R	UL	UL	24,000	16,000	24,000	16,000	20,500	12,000	7,000
	S1	UL	UL	96,000	64,000	96,000	64,000	82,000	48,000	28,000
	SM	UL	UL	72,000	48,000	72,000	48,000	61,500	36,000	21,000
R-3	NS[d, h] S13R S1 SM	UL	UL	UL	UL	UL	UL	UL	UL	UL
R-4	NS[d, h] S13R	UL	UL	24,000	16,000	24,000	16,000	20,500	12,000	7,000
	S1	UL	UL	96,000	64,000	96,000	64,000	82,000	48,000	28,000
	SM	UL	UL	72,000	48,000	72,000	48,000	61,500	36,000	21,000
S-1	NS	UL	48,000	26,000	17,500	26,000	17,500	25,500	14,000	9,000
	S1	UL	192,000	104,000	70,000	104,000	70,000	102,000	56,000	36,000
	SM	UL	144,000	78,000	52,500	78,000	52,500	76,500	42,000	27,000
S-2	NS	UL	79,000	39,000	26,000	39,000	26,000	38,500	21,000	13,500
	S1	UL	316,000	156,000	104,000	156,000	104,000	154,000	84,000	54,000
	SM	UL	237,000	117,000	78,000	117,000	78,000	115,500	63,000	40,500
U	NS	UL	35,500	19,000	8,500	14,000	8,500	18,000	9,000	5,500
	S1	UL	142,000	76,000	34,000	56,000	34,000	72,000	36,000	22,000
	SM	UL	106,500	57,000	25,500	42,000	25,500	54,000	27,000	16,500

(continued)

TABLE 506.2[a,b]—continued
ALLOWABLE AREA FACTOR (A_t = NS, S1, S13R, or SM, as applicable) IN SQUARE FEET

Note: UL = Unlimited; NP = Not permitted;

For SI: 1 square foot = 0.0929 m^2.

NS = Buildings not equipped throughout with an automatic sprinkler system; S1 = Buildings a maximum of one story above grade plane equipped throughout with an automatic sprinkler system installed in accordance with Section 903.3.1.1; SM = Buildings two or more stories above grade plane equipped throughout with an automatic sprinkler system installed in accordance with Section 903.3.1.1; S13R = Buildings equipped throughout with an automatic sprinkler system installed in accordance with Section 903.3.1.2.

a. See Chapters 4 and 5 for specific exceptions to the allowable height in this chapter.
b. See Section 903.2 for the minimum thresholds for protection by an automatic sprinkler system for specific occupancies.
c. New Group H occupancies are required to be protected by an automatic sprinkler system in accordance with Section 903.2.5.
d. The NS value is only for use in evaluation of existing building area in accordance with the *International Existing Building Code*.
e. New Group I-1 and I-3 occupancies are required to be protected by an automatic sprinkler system in accordance with Section 903.2.6. For new Group I-1 occupancies, Condition 1, see Exception 1 of Section 903.2.6.
f. New and existing Group I-2 occupancies are required to be protected by an automatic sprinkler system in accordance with Section 903.2.6 and Section 1103.5 of the *International Fire Code*.
g. New Group I-4 occupancies see Exceptions 2 and 3 of Section 903.2.6.
h. New Group R occupancies are required to be protected by an automatic sprinkler system in accordance with Section 903.2.8.

506.2.2 Mixed-occupancy, one-story buildings. The allowable area of a mixed-occupancy building with no more than one story above grade plane shall be determined in accordance with the applicable provisions of Section 508.1 based on Equation 5-1 for each applicable occupancy.

❖ The allowable area of a building with more than one occupancy is dependent upon the option used in Section 508 for the multiple occupancies. The commentary for Section 508 will delve into this in some detail. The code provides three basic options for mixed occupancies: accessory occupancies, nonseparated occupancies and separated occupancies. The availability of the options depends upon the area of each story desired for each different occupancy and the relationship of one occupancy to the other.

Example: A single-story business office of 23,000 square feet (2137 m^2) contains a lunch room that is 2,000 square feet (186 m^2) in area. The main occupancy is Group B and the accessory occupancy is Group A-2. Since the Group A-2 occupancy has a floor area of less than 10 percent of the overall floor area of the single story, this can be viewed as an accessory occupancy and there is no fire separation requirement between the Group A-2 and B occupancies. For purposes of evaluating the area limitation, the building is evaluated solely as Group B.

In contrast to the accessory occupancy provisions illustrated above, the fundamental concept for nonseparated occupancies is that the allowable building areas are based upon the most restrictive requirements of each of the occupancies in the mixed occupancy building.

Example: A single-story Type IIB building contains both Group B and M occupancies and is being evaluated under the "nonseparated occupancies" provisions of Section 508.3. The Group B portion is 1,000 square feet (93 m^2) in area while the Group M portion is 10,000 square feet (929 m^2) in area. The building does not have a sprinkler system and does not qualify for any frontage increase. With the exception of Type I construction, the area limitation of Section 503.1 is more restrictive for Group M than for Group B.

Group M: A_a = 12,500 + 0 = 12,500 square feet

Therefore, for purposes of evaluating the area limitation of Table 503, the building is evaluated solely as Group M.

Since the total area is 11,000 square feet (1022 m^2), the building can be constructed as a Type IIB building, or another evaluation:

Actual area (Group M)/Allowable area (Group M) <1.0

11,000/12,500 = 0.88 which is < 1.0; therefore OK

The separated occupancies concept requires the different occupancies to be separated in accordance with Table 508.4. The sum of the ratios of the actual floor area of each separated occupancy contained therein, as compared to the area allowed per story by Section 506.2, must not exceed one (see Section 508.4.2).

Example: A single-story Type IIB building contains separated Group B and M occupancies. The Group B portion is 1,000 square feet (93 m^2) in area while the Group M portion is 10,000 square feet (929 m^2) in area. The building does not have a sprinkler system and does not qualify for any frontage increase.

$$\frac{\text{Actual area (B)}}{\text{Allowable area (B)}} + \frac{\text{Actual area of (M)}}{\text{Allowable area of (M)}} \leq 1.0$$

$$\frac{1,000}{23,000} + \frac{10,000}{12,500}$$

0.04 + 0.8 = 0.84 < 1.0; therefore OK

506.2.2.1 Group H-2 or H-3 mixed occupancies. For a building containing Group H-2 or H-3 occupancies, the allowable area shall be determined in accordance with Section 508.4.2, with the sprinkler system increase applicable only to the portions of the building not classified as Group H-2 or H-3.

❖ Section 506.2.2.1 allows Group H-2 and H-3 occupancies to be placed in a building where the increase for a sprinkler system given in Table 506.2 would be applied, but with the limitation that the increase for automatic sprinklers would not apply to the Group H-2 or H-3 areas. These Group H-2 or H-3 areas would be required to be separated from the other uses, given that Sections 508.2.4, 508.3.3 and 508.4.4 specifically require separation of these occupancies from other occupancies. The required separation would be as given for separated occupancies in Section 508.4.4, and the allowable area would be based upon the sum of the ratios of the actual area of each occupancy to the allowable area per story (see commentary, Section 508.4.2). In the case of the Group H-2 or H-3 areas, the allowable area per story would be based upon the tabular area given in Table 506.2 for H-2 or H-3 as applicable. This would have the effect of limiting the allowable areas of the entire building when a Group H-2 or H-3 area is being included.

Example: It is desired to construct a single-story Group F-1 factory using Type IIIB construction, with a 3,500-square-foot (325 m^2) Group H-2 area. An automatic sprinkler system will be used throughout the building. The building does not qualify for any frontage increase.

Applying Equation 5-1 to each occupancy:

Allowable area of Group F-1 = 48,000 square feet

Allowable area of Group H-2 = 7,000 square feet

Allowable area of the Group F-1 area with an NFPA 13 automatic sprinkler system = 48,000 square feet (4459 m^2). So, without a Group H-2 area, the building could be 48,000 square feet (4459 m^2) in area.

However, with the Group H-2 area present:

$$\frac{\text{Actual area of F-1}}{\text{Allowable area F-1}} + \frac{\text{Actual area of H-2}}{\text{Allowable area of H-2}}$$

= Actual Area F-1/48000 + 3,500/7,000 $\leq$ 1

Solving this equation for the actual area of Group F-1, it is determined that the maximum area of Group F-1 that could be allowed would now be 24,000 square feet (1672 m^2). Thus, while the high hazard is allowed, it does restrict the size of the host occupancy.

506.2.3 Single-occupancy, multistory buildings. The allowable area of a single-occupancy building with more than one story above grade plane shall be determined in accordance with Equation 5-2:

$$A_a = [A_t + (NS \times I_f)] \times S_a \qquad \textbf{(Equation 5-2)}$$

where:

A_a = Allowable area (square feet).

A_t = Tabular allowable area factor (NS, S13R or SM value, as applicable) in accordance with Table 506.2.

NS = Tabular allowable area factor in accordance with Table 506.2 for a nonsprinklered building (regardless of whether the building is sprinklered).

I_f = Area factor increase due to frontage (percent) as calculated in accordance with Section 506.3.

S_a = Actual number of building stories above grade plane, not to exceed three. For buildings equipped throughout with an automatic sprinkler system installed in accordance with Section 903.3.1.2, use the actual number of building stories above grade plane, not to exceed four.

No individual story shall exceed the allowable area (A_a) as determined by Equation 5-2 using the value of S_a = 1.

❖ Just as in Section 506.2.1, this section provides limitations on the area of a building based on occupancy classification, type of construction and whether the building is equipped throughout with a sprinkler system. However, in this case, the building is more than one story, so the equation for allowable area is a little more complicated. Basically, the code requirements are as follows: The total allowable area of a multistory building is based upon the tabular area of Table 506.2 multiplied by the perimeter frontage increase and multiplied by the total number of stories, but no more than three. Finally, no one story can be greater than the allowable area of a one-story building. This does not mean that buildings are limited to three stories in height. This means that the total allowable area of a building three or more stories in height is the same, regardless of the number of stories above three. Some examples will illustrate.

Example 1: What is the allowable area of a two-story office building, Type IIB construction, equipped throughout with an NFPA 13 sprinkler system, with no increase for frontage?

Answer: Total allowable area is 138,000 square feet (12 820 m^2). Maximum area of each story is 69,000 square feet (6410 m^2). For Equation 5-2: A_t = 69,000, NS = 23,000, I_f = 0, and S_a = 2. Therefore:

$A_a = [A_t + (NS \times I_f)] \times S_a$ = [69,000 + (23,000 x 0)] x 2 = 69,000 x 2 = 138,000. Again maximum area of each story is 69,000 square feet (6410 m^2).

Example 2: What is the allowable area of a three-story office building, constructed like the building in Example 1?

Answer: Total allowable area is 207,000 square feet (19 230 m^2). Maximum area of each story is 69,000 square feet (6410 m^2). For Equation 5-2, all is the same as in Example 1, except S_a = 3. Therefore, A_a = 69,000 x 3 = 207,000 square feet (19 230 m^2) total.

Example 3: What is the allowable area of a five-story office building, constructed like the building in Examples 1 and 2?

Answer: Total allowable area is 207,000 square feet (19 230 m^2). Maximum area of each story is 69,000 square feet (6410 m^2). For Equation 5-2 all is the same as in Example 2. The maximum that S_a can be is 3. The maximum area of each story is 69,000 square feet (6410 m^2); however, the average area of all five stories must be 41,400 square feet (3846 m^2) to be within the allowable area of the code. The area of each story can vary, as long as the total area does not exceed 207,000 square feet (19 230 m^2). For example: the first story could be 69,000 square feet (6410 m^2), the second story 50,000 square feet (4645 m^2), and the top three stories could be 29,333 square feet (2725 m^2) for a total of 206,999 square feet (19 230 m^2).

Example 4: What is the allowable area of a four-story apartment building, Type IIIB construction, equipped throughout with NFPA 13R sprinklers, with no allowable frontage increase?

Answer: Total allowable area is 64,000 square feet (5946 m^2). Maximum area of each story is 16,000 square feet (1486 m^2). Entering the Equation 5-2: A_t = 16,000, I_f = 0, NS = 16,000, S_a = 4. Therefore:

A_a = [16,000 + (16,000 x 0)] x 4 = 64,000 square feet (5946 m^2). The code increases the maximum number of stories as the base for the building equipped with an NFPA 13R system to four rather than three because NFPA 13R is specifically written to be capable of dealing effectively with buildings up to a maximum of four stories. Note that this is a limitation, and that an NFPA13 system would be required for a building over four stories in height.

506.2.4 Mixed-occupancy, multistory buildings. Each story of a mixed-occupancy building with more than one story above grade plane shall individually comply with the applicable requirements of Section 508.1. For buildings with more than three stories above grade plane, the total building area shall be such that the aggregate sum of the ratios of the actual area of each story divided by the allowable area of such stories, determined in accordance with Equation 5-3 based on the applicable provisions of Section 508.1, shall not exceed three.

$$A_a = [A_t + (NS \times I_f)] \quad \textbf{(Equation 5-3)}$$

where:

A_a = Allowable area (square feet).

A_t = Tabular allowable area factor (NS, S13R or SM value, as applicable) in accordance with Table 506.2.

NS = Tabular allowable area factor in accordance with Table 506.2 for a nonsprinklered building (regardless of whether the building is sprinklered).

I_f = Area factor increase due to frontage (percent) as calculated in accordance with Section 506.3.

Exception: For buildings designed as separated occupancies under Section 508.4 and equipped throughout with an *automatic sprinkler system* installed in accordance with Section 903.3.1.2, the total building area shall be such that the aggregate sum of the ratios of the actual area of each story divided by the allowable area of such stories determined in accordance with Equation 5-3 based on the applicable provisions of Section 508.1, shall not exceed four.

❖ This section addresses allowable area for multiple story buildings that have mixed occupancy. This can seem to be a little counterintuitive. The allowable area of a single story of a building depends upon the method of mixed occupancy being employed.

When buildings containing mixed occupancies are two or more stories above grade plane, the same concept applied to Section 506.2.2 is applied to each story of the building. Once a mixed-occupancy building exceeds three stories in height, the maximum allowed area for the total building is set at three times the sum of the ratios. The following examples illustrate the application of the total area limitations for each building as well as showing the variety of ways in which the provisions of Section 508, Mixed Use and Occupancy, can be used in the design of the building.

Example 1: Nonsprinklered, two-story Type IIA building.

First story: Group B occupancy—
3,000 square feet (278 m^2)
Group S-2 occupancy—
23,000 square feet (2137 m^2)

Second story: Group A-3 occupancy—
2,000 square feet (186 m^2)
Group B occupancy—
24,000 square feet (2230 m^2)

The first story qualifies for the nonseparated occupancy design method since the total square footage [26,000 square feet (2415 m^2)] is less than that permitted for the most restrictive occupancy [Group B-37,500 square feet (3484 m^2)].

The second story is larger than that permitted for a Group A-3 occupancy; however, since the Group A-3 occupancy is less than 10 percent of the area of the second story, it qualifies as an accessory occupancy if subsidiary to the Group B occupancy.

In this instance, each story individually qualifies based on the applicable mixed-occupancy provision.

Example 2: Sprinklered, four-story Type IIB building, no frontage increase.

First story: Group B occupancy—20,000 square feet (1858 m^2)
Group M occupancy—25,000 square feet (3159 m^2)

Second story: Group A-3 occupancy—4,000 square feet (372 m^2)
Group B occupancy—41,000 square feet (3809 m^2)

Third story: Group B occupancy—15,000 square feet (1394 m^2)
Group S-1 occupancy—15,000 square feet (1394 m^2)
Group F-1 occupancy—15,000 square feet (1394 m^2)

Fourth story: Group B occupancy—10,000 square feet (929 m^2)
Group S-2 occupancy—35,000 square feet (3252 m^2)

The first story does not qualify as an accessory or nonseparated occupancy. As separated occupancies, the ratio on the story is 0.96 [(20,000/69,000) + (25,000/37,500)].

The second story is larger than that permitted for a Group A-3 occupancy; however, since the Group A-3 occupancy is less than 10 percent of the area of the second story, it qualifies as an accessory occupancy if subsidiary to the Group B occupancy. The ratio on the story is 0.65 (45,000/69,000).

The third story qualifies as a nonseparated occupancy, since the allowable area per story for the most restrictive occupancy [Group F-1; 46,500 square feet (4320 m^2)] is larger than the actual area on any individual story. The ratio of the floor is 0.97 (45,000/46,500).

The fourth floor qualifies as a nonseparated occupancy, since the allowable area per floor for the most restrictive occupancy [Group B; 69,000 square feet (6410 m^2)] is larger than the actual area on any individual story. The ratio of the story is 0.65 (45,000/69,000).

The aggregate sum of the ratios is 3.23 (0.96 + 0.65 + 0.97 + 0.65). Since the aggregate sum of the ratios exceeds three, Type IIB construction would not be permitted without further modification. Options would be to increase the construction type or increase the open frontage.

506.2.4.1 Group H-2 or H-3 mixed occupancies. For a building containing Group H-2 or H-3 occupancies, the allowable area shall be determined in accordance with Section 508.4.2, with the sprinkler system increase applicable only to the portions of the building not classified as Group H-2 or H-3.

❖ This section is a reminder that Group H-2 or H-3 occupancies are allowed in mixed occupancy buildings, but that there is no advantage for sprinklers. Therefore, Table 506.2 has the same area factors for nonsprinklered and sprinklered buildings for H-2 or H-3 construction. See the commentary to Section 506.2.2.1.

506.3 Frontage increase. Every building shall adjoin or have access to a public way to receive an area factor increase based on frontage. Area factor increase shall be determined in accordance with Sections 506.3.1 through 506.3.3.

❖ The allowable area of a building is allowed to be increased when it has a certain amount of frontage on streets (public ways) or open spaces, since this provides access to the structure by fire service personnel, a temporary refuge area for occupants as they leave the building in a fire emergency and a reduced exposure to and from adjacent structures. Sections 506.3.1 through 506.3.3 describe how this increase for frontage is determined.

506.3.1 Minimum percentage of perimeter. To qualify for an area factor increase based on frontage, a building shall have not less than 25 percent of its perimeter on a public way or open space. Such open space shall be either on the same lot or dedicated for public use and shall be accessed from a street or approved fire lane.

❖ There is no requirement in the code that buildings have at least 25 percent of their perimeter on a public way or open space. However, in order to qualify for an area increase, a building must have more than 25 percent of its perimeter on a public way or open space having a minimum width of at least 20 feet (6096 mm) (see Section 506.3.2). When the calculations are done, the maximum percent increase for a fully open perimeter (full frontage—the entire perimeter fronts on a public way or open space) is 75 percent. Width is measured at right angles to the perimeter walls as set forth in Section 506.3.2, but open frontage is not the same as fire separation distance. Fire separation distance, as defined in Section 202, is measured to the centerline of public ways or to an imaginary line between buildings on the same lot. Open space can include the total width of the public way as well as the total open space between buildings on the same lot. "Public way" is a defined term.

If a structure is divided into two or more buildings by fire walls complying with Section 706, the area modifications allowed by Section 506 must be determined based on each separate building within the structure. This especially comes into play in determining increases allowable based on frontage. The fire wall is essentially the perimeter wall for that side of the building and must be included in determining the P in Equation 5-5. Since there is another building on the other side of the fire wall, this portion of the perimeter is not considered fronting on a public way or yard and, therefore, is not included in F [see Commentary Figure 506.3.1(1)].

This section requires that an open space that is not a public way be on the same lot or dedicated for public use, and it must have access from a street or an approved fire lane in order to contribute to the frontage increase.

The requirement that the open space be on the same lot is so that the owner or the jurisdiction can control the space that is assumed to be open for purposes of the area increase. One cannot encumber a neighbor's property with a requirement that the space will always remain unoccupied.

Any part of the perimeter that is not accessible to the fire department by means of a street or fire lane cannot be considered open for the purposes of this section. For instance, if the back side of a building on a narrow lot cannot be reached by means of a fire lane on one side of the building (and there is no alley or street at the back), that portion of the perimeter is not considered open for purposes of frontage increase, even if there is actual open space exceeding 20 feet (6096 mm) in width. See Commentary Figure 506.3.1(2) as an illustration of this limitation.

This section does not require that a fire lane or street extend immediately adjacent to every portion of the perimeter that is considered open for purposes of the increase. Rather, access by a fire lane must be provided up to the open side such that fire department personnel can approach the side and pull hoses across the open area to fight a fire, and no corner of the building will impede the use of hoses and equipment on that side of the building. The following examples demonstrate this point.

Example 1: In Commentary Figure 506.3.1(2), the south and east side of the building facing the street can be considered open perimeter (frontage). The north side of the building cannot be considered open perimeter for purposes of the increase, since it is not accessible from the street or a fire lane. Even though the 20-foot-wide yard can be included in open frontage because it does not provide a fire lane to the rear of the building, the 100-foot yard cannot be included.

Example 2: In Commentary Figure 506.3.1(3), all sides of the building are considered open perimeter (frontage) for purposes of the increase. Access up to each side of the building is provided by means of a fire lane or street.

Section 503 of the *International Fire Code*® (IFC®) specifies that access roads extend to within 150 feet (45 720 mm) of all portions of the exterior walls of a building. However, there are exceptions that would permit the omission of such roads under certain circumstances. One such exception is for buildings equipped throughout with an automatic sprinkler system where the fire code official is permitted to extend the 150-foot (45 720 mm) limit.

The IFC also stipulates that the access roads must be at least 20 feet (6096 mm) in unobstructed width, although it also gives the building official authority to require greater widths if necessary for effective firefighting operations. The type of surface necessary for the approved fire lane is determined by the local building official with input from the fire department, but the road must be capable of supporting the imposed loads of fire apparatus and be surfaced so as to provide all-weather driving capabilities.

BUILDING B
BUILDING A ONLY HAS 3 SIDES OPEN
BUILDING A
FIRE WALL
STREET

For SI: 1 inch = 25.4 mm, 1 foot = 304.8 mm.

Figure 506.3.1(1)
FRONTAGE INCREASE—FIRE WALL

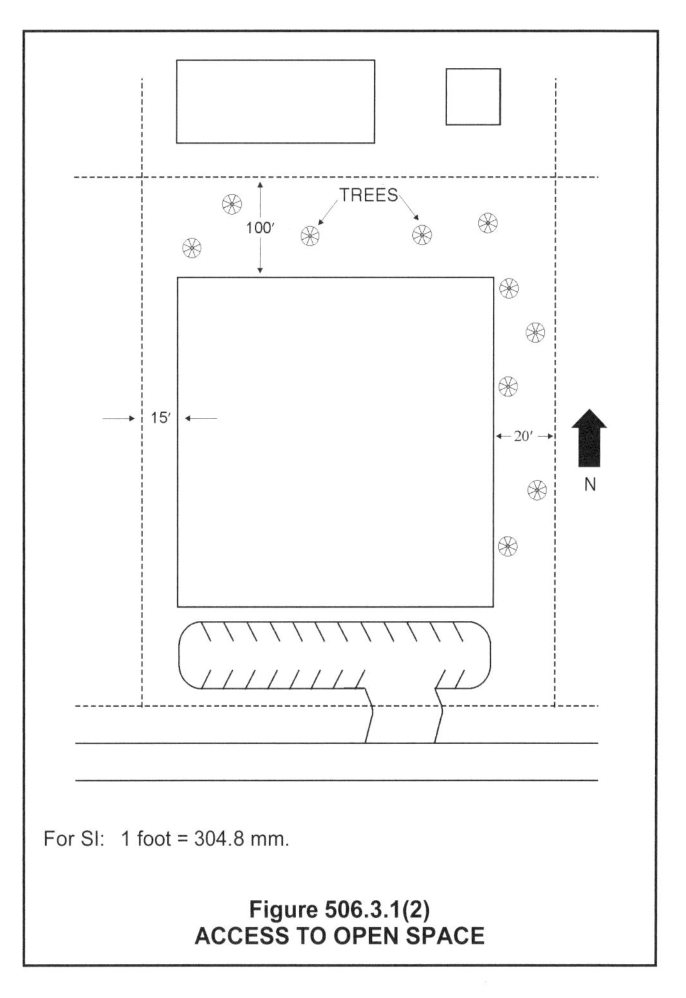

For SI: 1 foot = 304.8 mm.

Figure 506.3.1(2)
ACCESS TO OPEN SPACE

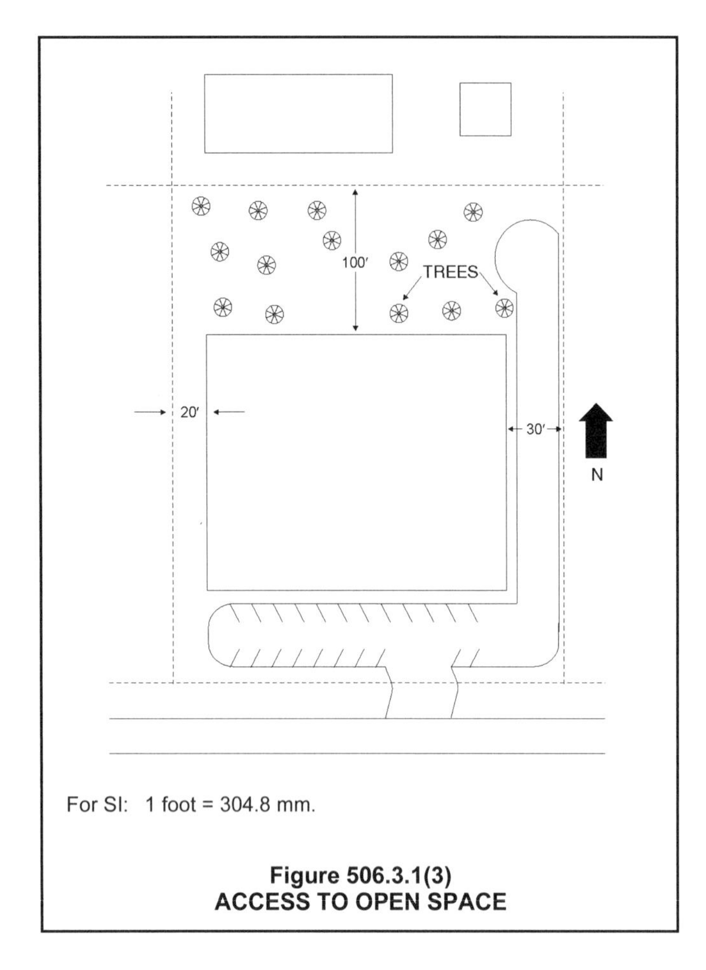

Figure 506.3.1(3)
ACCESS TO OPEN SPACE

506.3.2 Minimum frontage distance. To qualify for an area factor increase based on frontage, the public way or open space adjacent to the building perimeter shall have a minimum distance (*W*) of 20 feet (6096 mm) measured at right angles from the building face to any of the following:

1. The closest interior lot line.
2. The entire width of a street, alley or public way.
3. The exterior face of an adjacent building on the same property.

Where the value of *W* is greater than 30 feet (9144 mm), a value of 30 feet (9144 mm) shall be used in calculating the building area increase based on frontage, regardless of the actual width of the public way or open space. Where the value of *W* varies along the perimeter of the building, the calculation performed in accordance with Equation 5-5 shall be based on the weighted average calculated in accordance with Equation 5-4.

$$W = (L_1 \times w_1 + L_2 \times w_2 + L_3 \times w_3 \ldots)/F \qquad \textbf{(Equation 5-4)}$$

where:

W (Width: weighted average) = Calculated width of public way or open space (feet).

L_n = Length of a portion of the exterior perimeter wall.

w_n = Width (≥ 20 feet) of a public way or open space associated with that portion of the exterior perimeter wall.

F = Building perimeter that fronts on a public way or open space having a width of 20 feet (6096 mm) or more.

Exception: Where a building meets the requirements of Section 507, as applicable, except for compliance with the minimum 60-foot (18 288 mm) *public way* or *yard* requirement, and the value of *W* is greater than 30 feet (9144 mm), the value of *W* shall not exceed 60 feet (18 288 mm).

❖ The amount of area increase (that is, the value of I_f in Section 506.3.3) will vary depending on the minimum width of the open space used in the frontage increase calculation. The general requirement is that the value of *W* must equal at least 20 feet (6096 mm) and cannot exceed 30 feet (9144 mm). The value of *W* is the weighted average of the portions of the wall and each open space when the width of the open space is between 20 and 30 feet (6096 and 9144 mm). Except as provided in the exception, the width of open space used in this calculation cannot exceed 30 feet (9144 mm) even when part of the open space is wider than 30 feet (9144 mm). See the following two examples illustrating this requirement. Example 1 also illustrates the use of Equation 5-4 in the determination of the weighted average *W*.

Example 1: In Commentary Figure 506.3.2(1), the value *W* = [(200 feet x 25 feet) + (100 feet x 20 feet) + (200 feet x 30 feet)]/500 feet = 26 feet (7925 mm).

Example 2: In Commentary Figure 506.3.2(2), the value of *W* would be 30 feet (9144 mm), since the minimum width of the open space that qualifies as open frontage is greater than 30 feet (9144 mm) on all

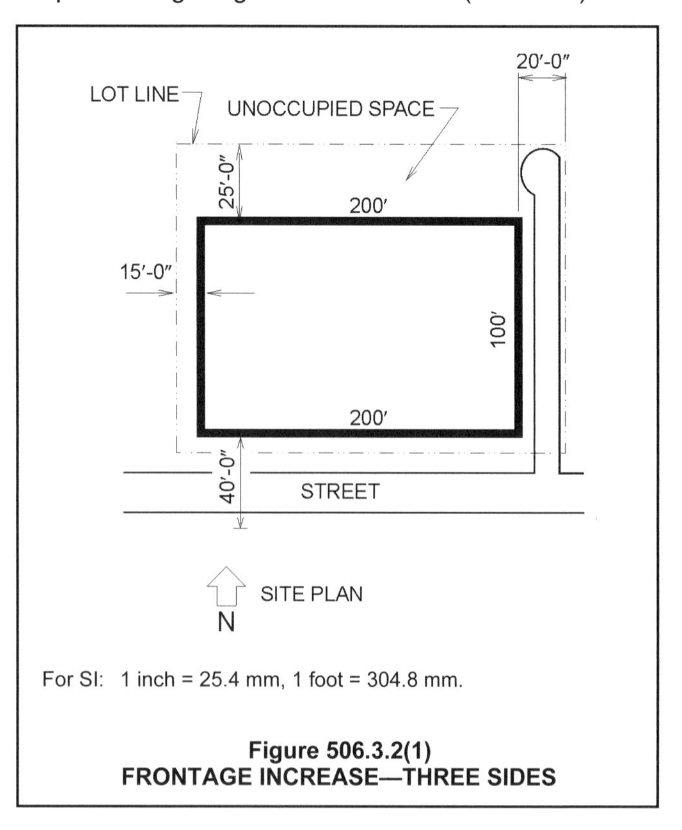

Figure 506.3.2(1)
FRONTAGE INCREASE—THREE SIDES

sides of this building. The reason the value of *W* must be taken at 30 feet (9144 mm) and not 50 feet (15 240 mm) is that this section sets an upper limit on the value of the *W*/30 term in Equation 5-5.

Typically, when multiple buildings are located on the same lot, an imaginary lot line must be established somewhere between the buildings in order to determine the fire separation distance. However, the value of *W* is defined as the width of the public way or open space, and is not dependent upon the fire separation distance. Therefore, for purposes of determining the value of *W* between buildings on the same lot, the entire distance between the buildings is permitted to be used, not solely the fire separation distance. Similarly, the entire width of a public way is used and not just the distance to the centerline of the public way. Such measurements are to be taken at right angles from the building.

The exception states that for certain buildings, the value of *W* divided by 30 is permitted to have a maximum value of 2.0. This exception applies to buildings that would be allowed to be unlimited in area in accordance with Section 507, save for the fact that the open area of 60 feet (18 288 mm) required by Section 507 is not met. Therefore, the weighted average of *W* would be calculated between 20 feet (6096 mm) and 60 feet (18 288 mm).

506.3.3 Amount of increase. The area factor increase based on frontage shall be determined in accordance with Equation 5-5:

$$I_f = [F/P - 0.25]W/30 \qquad \textbf{(Equation 5-5)}$$

where:

I_f = Area factor increase due to frontage.

F = Building perimeter that fronts on a *public way* or open space having minimum distance of 20 feet (6096 mm).

P = Perimeter of entire building (feet).

W = Width of *public way* or open space (feet) in accordance with Section 506.3.2.

❖ The terms that are used in Equation 5-5 have been discussed in the commentary to Sections 506.3.1 and 506.3.2. One additional example to demonstrate the application:

Example 1: Refer to Commentary Figure 506.3.3. In Equation 5-5, the term would have the following values:

F = 500 feet (15 240 mm), since all sides of the building front on a public way or open space having 20 feet (6096 mm) minimum open width.

P = 500 feet (15 240 mm), the length of the entire perimeter.

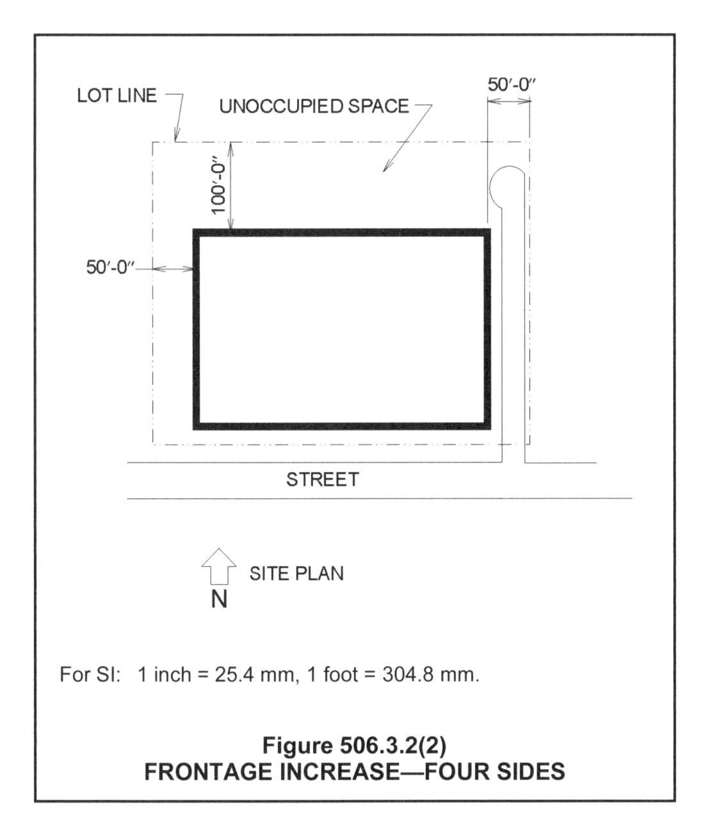

Figure 506.3.2(2)
FRONTAGE INCREASE—FOUR SIDES

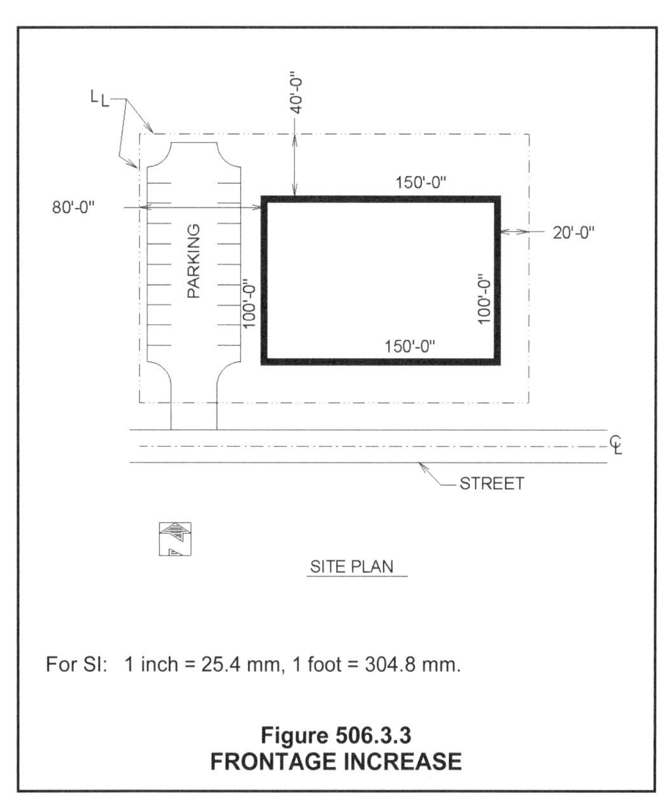

Figure 506.3.3
FRONTAGE INCREASE

W = The weighted average of the widths of the public way or open space = [(20 ft.)(100 ft.) + (30 ft.)(150 ft.) + (30 ft.)(100 ft.) + (30 ft.)(150 ft.)]/500 ft. = 28 feet (8534 mm).

Note in the above computation for W that, where the actual width exceeds 30 feet (9144 mm), a value of 30 feet (9144 mm) is used as required by Section 506.3.2.

Equation 5-5, $I_f = [(F/P) - 0.25](W/30)$, becomes:

$I_f = [(500/500) - 0.25](28/30) = (.75)(.93) = 0.70$

Therefore, the term "I_f" equals 0.70 when using Equation 5-1 or 5-2 for determining the allowable area per story (A_a).

Example 2: The building shown on Commentary Figure 506.3.3 for Example 1 is proposed to be a Mercantile Occupancy, Type VB construction, one story, equipped throughout with an NFPA 13 sprinkler system. The allowable area, using Equation 5-1:

$A_a = A_t + (NS \times I_f) = 36{,}000 + (9{,}000 \times 0.7) = 42{,}300$ square feet (3930 m^2).

SECTION 507 UNLIMITED AREA BUILDINGS

507.1 General. The area of buildings of the occupancies and configurations specified in Sections 507.1 through 507.12 shall not be limited. Basements not more than one story below grade plane shall be permitted.

❖ This section addresses circumstances under which a building would be allowed to be constructed unlimited in building area. Depending upon the height (in stories) of the structure and the occupancy classification, the code presents different circumstances that would allow an unlimited area building. There is a common thread among the requirements for these facilities:

1. Except for the option in Section 507.3, the building is equipped throughout with an automatic sprinkler system.
2. The building is surrounded by increased open space, usually 60 feet (18 288 mm) in width (see also Section 507.2.1).
3. The buildings are limited to one or two stories above grade plane, with a single-story basement.
4. The building only houses low-hazard or moderate-hazard occupancies.

Historically, structures constructed under the provisions for unlimited area buildings have performed quite well in regard to fire and life safety. In general, only those occupancies and configurations that are specifically identified in Section 507 are subject to the unlimited area allowance.

507.1.1 Accessory occupancies. Accessory occupancies shall be permitted in unlimited area buildings in accordance with the provisions of Section 508.2, otherwise the requirements of Sections 507.3 through 507.13 shall be applied, where applicable.

❖ This section clarifies that the accessory use provisions of Section 508.2 can also be applied to unlimited area buildings. This allows uses not specifically addressed in Sections 507.3 through 507.13 to be located in any of these buildings, provided they meet the requirements of accessory occupancies according to Section 508.2. The provision is distinct from Section 507.4.1, which addresses Group A-1 and A-2 occupancies in unlimited area buildings under Section 507.4. The provisions of Section 507.4.1 are for additional primary uses in these unlimited area buildings. A large office building could have an employee cafeteria, which is a Group A-2 occupancy, as an accessory, provided it was less than 10 percent of the floor area and didn't exceed the allowable area of Sections 503 and 506. Such accessory space would not need to be separated as required for the larger spaces allowed under Section 507.4.1.

507.2 Measurement of open spaces. Where Sections 507.3 through 507.13 require buildings to be surrounded and adjoined by *public ways* and *yards*, those open spaces shall be determined as follows:

1. Yards shall be measured from the building perimeter in all directions to the closest interior *lot lines* or to the exterior face of an opposing building located on the same *lot*, as applicable.
2. Where the building fronts on a *public way*, the entire width of the *public way* shall be used.

❖ In all cases, the open space can occur within public ways surrounding the site, by yards provided on the lot between the building and the lot lines, or a combination of yards and public ways. Where a yard is used to achieve the open space, it must be on the same lot as the building receiving the benefit. With respect to the 60 feet (18 288 mm), the code specifies that the entire width of the public way can be included. (Fire separation distance, which is a different requirement, requires measurement to the centerline of the public way.) Unlike the street frontage increase in Section 506.3.2, the open space must be provided in all directions around the perimeter of the building, not just measured at right angles to the building (see Commentary Figure 507.2). Two unlimited area buildings on the same lot must be separated by 60 feet (18 288 mm) [or 40 feet (12 192 mm) if Section 507.2.1 is used] unless they are treated as a single building under the provisions of Section 503.1.2.

The open space located on the private property does not need to be dedicated to the public or publicly owned, but can be the location of parking, landscaping, roadways and other minor accessory features (tanks, generators, trash dumpster enclosures) (see IBC Interpretation No. 20-03). However, the yard cannot be occupied by any exterior use that is essentially a continuation of use of the building. For example, many big box retailers will have an adjoining lawn and garden merchandise area; or a lumber supply area that is only partially enclosed by walls, fencing and roof covering. This type of use would need to be considered part of the unlimited area building, and the 60 feet (18 288 mm) of open space provided beyond this area (see IBC Interpretation No. 03-05).

The open areas serve two key roles: separation of these buildings from other buildings and ample space on all sides for fire-fighting operations.

Please note that Section 507.2.1 permits a reduction in the open space in exchange for increased wall and opening protection for those unlimited area buildings allowed by Sections 507.3, 507.4, 507.5, 507.6 and 507.12.

507.2.1 Reduced open space. The *public ways* or *yards* of 60 feet (18 288 mm) in width required in Sections 507.3, 507.4, 507.5, 507.6 and 507.12 shall be permitted to be reduced to not less than 40 feet (12 192 mm) in width provided all of the following requirements are met:

1. The reduced width shall not be allowed for more than 75 percent of the perimeter of the building.
2. The *exterior walls* facing the reduced width shall have a *fire-resistance rating* of not less than 3 hours.
3. Openings in the *exterior walls* facing the reduced width shall have opening protectives with a *fire protection rating* of not less than 3 hours.

❖ The minimum width of the open space surrounding those unlimited area buildings specified in Sections 507.3, 507.4, 507.5, 507.6 and 507.12 may be reduced if the exterior walls are protected with minimum 3-hour fire-resistance-rated construction and the openings are protected with 3-hour rated assemblies. All three conditions of this section must be met in order to reduce the open space to a minimum of 40 feet (12 192 mm). The criteria and standards for wall assemblies and opening protection are contained in Chapter 7. The rated walls and protected openings are required only for those portions of the walls that do not face at least 60 feet (18 288 mm) of unoccupied space or public way.

507.3 Nonsprinklered, one-story buildings. The area of a Group F-2 or S-2 building no more than one story in height shall not be limited where the building is surrounded and adjoined by *public ways* or *yards* not less than 60 feet (18 288 mm) in width.

❖ By definition, occupancies of Groups F-2 and S-2 are not permitted to contain significant amounts of combustible materials (see Sections 306.3 and 311.3); therefore, because the fire load of the contents is lower, the hazard is lower. The type of construction is not restricted, and sprinklers are not required for unlimited area buildings containing Group F-2 and S-2 occupancies, provided the building is no more than one story above grade plane. See the commentary in Section 507.2 regarding the required open space. Section 507.2.1 can be applied to buildings built under this section.

507.4 Sprinklered, one-story buildings. The area of a Group A-4 building no more than one *story above grade plane* of other than Type V construction, or the area of a Group B, F, M or S building no more than one story above grade plane of any construction type, shall not be limited

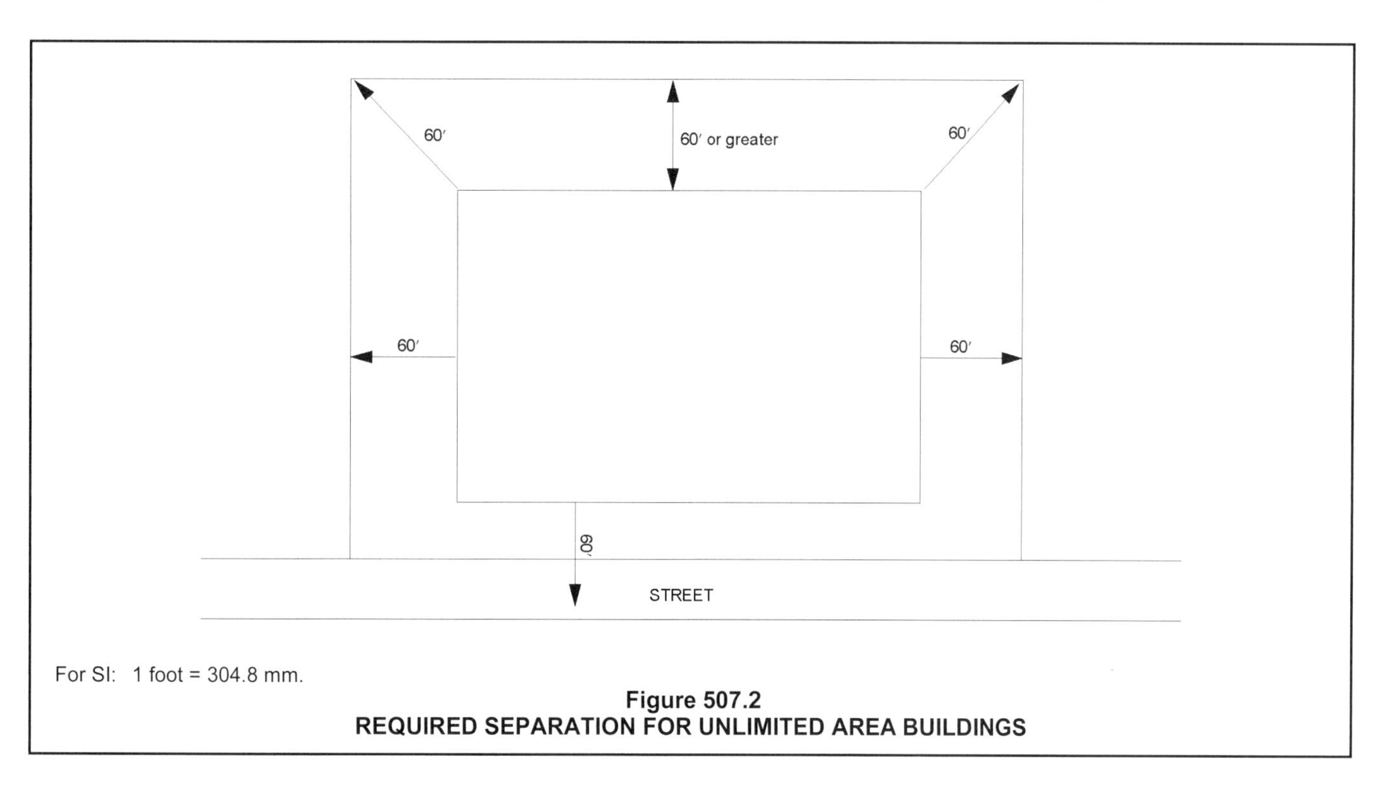

For SI: 1 foot = 304.8 mm.

Figure 507.2
REQUIRED SEPARATION FOR UNLIMITED AREA BUILDINGS

where the building is provided with an *automatic sprinkler system* throughout in accordance with Section 903.3.1.1 and is surrounded and adjoined by *public ways* or *yards* not less than 60 feet (18 288 mm) in width.

Exceptions:

1. Buildings and structures of Type I or II construction for rack storage facilities that do not have access by the public shall not be limited in height, provided that such buildings conform to the requirements of Sections 507.4 and 903.3.1.1 and Chapter 32 of the *International Fire Code.*
2. The *automatic sprinkler system* shall not be required in areas occupied for indoor participant sports, such as tennis, skating, swimming and equestrian activities in occupancies in Group A-4, provided that both of the following criteria are met:
   2.1. *Exit* doors directly to the outside are provided for occupants of the participant sports areas.
   2.2. The building is equipped with a *fire alarm system* with *manual fire alarm boxes* installed in accordance with Section 907.

❖ Because of their excellent record in controlling and preventing fires, the installation of a sprinkler system throughout single-story buildings of the listed groups, individually or in combination, permits them to be unlimited in area (IBC Interpretation No. 44-06). This also applies to two-story above-grade-plane buildings of these groups, with the exception of Group A-4 (see Section 507.5). See the commentary in Section 507.2 regarding the required open space. Section 507.2.1 can be applied to buildings built under Section 507.4.

The life-safety hazards in buildings of these occupancies, because of the typical activities of the occupants and their level of awareness, are considered low enough that larger building areas and increased fire loads can be tolerated. For Groups B, F, M and S, this section may be applied to buildings of all construction types and is not limited to noncombustible construction. For indoor arenas and other Group A-4 buildings, the unlimited area provision can be applied to buildings of Type I, II, III and IV construction only. This reflects the size of the occupant load in these types of facilities. The sprinkler system is required to be designed and installed in accordance with NFPA 13 (see Section 903.3.1.1). The height in feet of the unlimited area building cannot exceed the limits established in Section 504.3 for the occupancy and type of construction. See Section 507.8 for similar buildings that include Group H-5 facilities.

Exception 1 specifically permits Type I and II buildings with rack storage to be a single-story building, but unlimited in the height in feet, if they are surrounded by 60 feet (18 288 mm) of open space, do not permit access by the public, have an NFPA 13 sprinkler system and comply with the *International Fire Code*® (IFC®) Chapter 32. These standards contain provisions for storage configuration, in-rack sprinkler coverage and system requirements that offset the hazard introduced by increased height.

Exception 2 eliminates providing a sprinkler system in certain areas of unlimited area Group A-4 occupancies where a lack of fuel loading and excessive ceiling heights would reduce the need for, and effectiveness of, the system. Group A-4 occupancies frequently have indoor participant sports areas such as tennis courts, skating rinks, swimming pools, baseball fields, basketball courts and equestrian arenas, with spectator seating usually situated around the perimeter of the sports field or area. These types of indoor recreation areas often require very large, open areas with such high ceilings that the installation of an automatic sprinkler system in the immediate participant sport area would be ineffective. The potential for significant fire involvement in such an area is generally quite low because of the low fuel load; therefore, sprinkler coverage is unnecessary for the playing field in most of these buildings. These areas are, therefore, exempt from the suppression requirement of this section, provided the conditions regarding exiting and the required fire alarm system are met. The use of the sports area for other uses, such as a weekend crafts fair or other activity not specifically listed, would be in violation of the exception and should not be permitted. When an indoor arena or sports facility is built without suppression in the participant sport area, occupants of the playing field or participant sport area must be able to exit the building directly from the playing area, without having to pass through other parts of the building. This eliminates the hazard of having to pass through higher fuel load areas such as locker rooms or concession areas. The building must also be equipped with a manual alarm system that complies with Section 907. This manual alarm system provides an additional and acceptable level of life safety in spite of the omission of sprinklers over the playing field area. This exception only applies to Group A-4 occupancies that are contained within an unlimited area building.

Omission of sprinkler coverage is permitted in the unlimited area building for the participant sport area only. All other areas are required to be equipped with an automatic sprinkler system. This includes all other rooms and spaces in the building, such as the spectator seating areas, locker rooms, restaurants, lounges, shops, arcades, skyboxes and storage areas.

507.4.1 Mixed occupancy buildings with Groups A-1 and A-2. Group A-1 and A-2 occupancies of other than Type V construction shall be permitted within mixed occupancy buildings of unlimited area complying with Section 507.4, provided all of the following criteria are met:

1. Group A-1 and A-2 occupancies are separated from other occupancies as required for separated occupancies in Section 508.4.4 with no reduction allowed in the *fire-resistance rating* of the separation based upon the installation of an *automatic sprinkler system.*

2. Each area of the portions of the building used for Group A-1 or A-2 occupancies shall not exceed the maximum allowable area permitted for such occupancies in Section 503.1.
3. *Exit* doors from Group A-1 and A-2 occupancies shall discharge directly to the exterior of the building.

❖ This section allows Group A-1 and A-2 occupancies in mixed occupancy, single-story unlimited area buildings under limited conditions. These are considered as additional primary use occupancies allowed in buildings regulated by Section 507.4. For accessory uses, see the commentary for Section 507.1.1. A typical example of a practical application of this would be the construction of a strip mall that is mainly for retail stores, but might contain a restaurant or a movie theater, or both. Group A-1 or A-2 buildings would not be permitted as stand-alone unlimited area buildings. Similar to the requirement in Section 507.4 for Group A-4 buildings, unlimited area buildings that contain a Group A-1 or A-2 occupancy are not permitted to be built of Type V construction.

507.5 Two-story buildings. The area of a Group B, F, M or S building no more than two *stories above grade plane* shall not be limited where the building is equipped throughout with an *automatic sprinkler system* in accordance with Section 903.3.1.1 and is surrounded and adjoined by *public ways* or *yards* not less than 60 feet (18 288 mm) in width.

❖ The rationale for allowing unlimited area two-story buildings in these four occupancy groups is the same as for one-story structures of these uses. The type of construction is not restricted. The number of stories are those above grade plane. Group A is not included in the occupancies that can be located within a two-story, unlimited area building because of the hazards associated with higher occupant loads. Even though limited Group A occupancies can be located in one-story unlimited area buildings, such occupancies cannot be located on either story of a two-story, unlimited area building unless they are accessory occupancies (see Section 507.1.1). The provisions of Section 507.4.1 do not apply to two-story buildings. See the commentary to Section 507.2 regarding the required open space. Section 507.2.1 can be applied to buildings built under Section 507.5.

507.6 Group A-3 buildings of Type II construction. The area of a Group A-3 building no more than one *story above grade plane*, used as a *place of religious worship*, community hall, dance hall, exhibition hall, gymnasium, lecture hall, indoor *swimming pool* or tennis court of Type II construction, shall not be limited provided all of the following criteria are met:

1. The building shall not have a *stage* other than a *platform*.
2. The building shall be equipped throughout with an *automatic sprinkler system* in accordance with Section 903.3.1.1.
3. The building shall be surrounded and adjoined by *public ways* or *yards* not less than 60 feet (18 288 mm) in width.

❖ Assembly buildings pose a greater risk in general, primarily because of their relatively large occupant loads and the density of such loads; therefore, the unlimited area provisions are more restrictive for assembly occupancies. Limited types of Group A-3 buildings are allowed as unlimited area buildings, provided each is limited to one story above grade plane. In the case of the Group A-3 occupancies listed, the construction classifications allowed to be unlimited area include Type II, which are noncombustible structures, and Type III and IV (see Section 507.7), which are combustible structures. The types of Group A-3 occupancies listed—churches, dance halls, community halls, exhibition halls, gymnasiums, lecture halls, indoor swimming pools and tennis courts—are typically well-lit, open spaces, containing a low fuel load. Libraries and museums are excluded, as they generally have a higher fuel load, as are bowling alleys and pool halls that are often poorly lit and have higher levels of activity. The dance halls included in Group A-3 are not to be confused with nightclubs, which are Group A-2 occupancies.

The Group A-3 buildings listed in this section must have 60 feet (18 288 mm) of open space surrounding the structure. Similar to the occupancies referenced in Sections 507.3 and 507.4, a sprinkler system must be provided and is required to be designed and installed in accordance with NFPA 13. Group A-3 buildings are further restricted, however, in that the building must not have a stage (other than a platform). These requirements aid in the exiting of the building in an emergency. See the commentary to Section 507.2 regarding the required open space. Section 507.2.1 can be applied to buildings built under Section 507.6.

507.7 Group A-3 buildings of Type III and IV construction. The area of a Group A-3 building of Type III or IV construction, with no more than one *story above grade plane* and used as a *place of religious worship*, community hall, dance hall, exhibition hall, gymnasium, lecture hall, indoor *swimming pool* or tennis court, shall not be limited provided all of the following criteria are met:

1. The building shall not have a *stage* other than a *platform*.
2. The building shall be equipped throughout with an *automatic sprinkler system* in accordance with Section 903.3.1.1.
3. The assembly floor shall be located at or within 21 inches (533 mm) of street or grade level and all *exits* are provided with ramps complying with Section 1012 to the street or grade level.

4. The building shall be surrounded and adjoined by *public ways* or *yards* not less than 60 feet (18 288 mm) in width.

❖ The requirements for Group A-3 buildings of Type III or IV construction are similar to the provisions for buildings of Type II construction. Due to the permitted use of combustible materials in Group A-3 buildings of Type III and IV construction, these buildings are further restricted in that the assembly floor level of the building must not be higher than 21 inches (533 mm) above street or grade level. These requirements aid in the exiting of the building occupants in an emergency. In order to further assist in the evacuation of the building in an emergency, assembly floors which are not at grade level must be provided with ramps (i.e., no stairs) which lead to grade level. Note that the 21-inch (533 mm) height limitation for the assembly floor would prohibit the installation of a mezzanine, balcony or most raised-floor surfaces in the building. See the commentary to Section 507.2 regarding the required open space. Please note that Section 507.2.1 can not be applied to buildings built under Section 507.7.

507.8 Group H-2, H-3 and H-4 occupancies. Group H-2, H-3 and H-4 occupancies shall be permitted in unlimited area buildings containing Group F or S occupancies in accordance with Sections 507.4 and 507.5 and the provisions of Sections 507.8.1 through 507.8.4.

❖ Group F and S buildings, single story, with an NFPA 13 sprinkler and a 60-foot open perimeter can include some high-hazard areas. This section sets forth limits for Group H-2, H-3 and H-4 occupancies in these unlimited area buildings. Group F facilities often inherently contain a certain amount of Group H-2, H-3 and H-4 occupancies when hazardous materials are needed for the industrial processes. Similarly, it is reasonable to provide for limited quantities of hazardous material storage in a storage (Group S) occupancy. See the commentary to Section 507.2 regarding the required open space. Please note that Section 507.2.1 can not be applied to buildings built under Section 507.8.

507.8.1 Allowable area. The aggregate floor area of Group H occupancies located in an unlimited area building shall not exceed 10 percent of the area of the building or the area limitations for the Group H occupancies as specified in Section 506 based on the perimeter of each Group H floor area that fronts on a *public way* or open space.

❖ Multiple high-hazard (Group H-2, H-3 and H-4) occupancies are permitted within a manufacturing or storage building constructed under the unlimited area building provisions; however, the aggregate floor area of all such occupancies cannot be more than 10 percent of the floor area of the entire building. It is important that the higher hazards be limited in amount in order to maintain the high degree of fire and life safety anticipated in an unlimited area building. In addition, the aggregate floor area of Group H occupancies cannot exceed area limitations established by Section 506 with any permitted frontage increases as calculated by Section 506.3.

507.8.1.1 Located within the building. The aggregate floor area of Group H occupancies not located at the perimeter of the building shall not exceed 25 percent of the area limitations for the Group H occupancies as specified in Section 506.

❖ For those Group H occupancies that are not located at the perimeter of the building, an additional limitation is placed on their allowable area. If the high-hazard occupancy is surrounded on all sides by an unlimited area building, it is difficult for fire department personnel to locate and access that area. Therefore, Group H-2, H-3 and H-4 occupancies that are not located at the perimeter of the building are limited to 25 percent of the area limitation specified in Section 506 for the building type of construction and occupancy classification, as shown in Commentary Figure 507.8.1.1. With this option, a frontage increase per Section 506.3 is not permitted.

507.8.1.1.1 Liquid use, dispensing and mixing rooms. Liquid use, dispensing and mixing rooms having a floor area of not more than 500 square feet (46.5 m^2) need not be located on the outer perimeter of the building where they are in accordance with the *International Fire Code* and NFPA 30.

❖ Identical language is found in Exception 1 to Section 415.6, allowing small liquid use, dispensing and mixing rooms to be located within the building's interior rather than at its perimeter.

507.8.1.1.2 Liquid storage rooms. Liquid storage rooms having a floor area of not more than 1,000 square feet (93 m^2) need not be located on the outer perimeter where they are in accordance with the *International Fire Code* and NFPA 30.

❖ Liquid storage rooms no more than 1,000 square feet (93 m^2) are permitted to be located within the building. This allowance is also reflected in Section 415.6, Exception 2.

507.8.1.1.3 Spray paint booths. Spray paint booths that comply with the *International Fire Code* need not be located on the outer perimeter.

❖ Spray paint booths are regulated independently by the IFC and NFPA 33. There is no requirement that they be located on a building's perimeter. This language is also provided in Exception 3 of Section 415.6. It should be noted that an approved spray booth is not typically classified as a Group H occupancy, but rather simply regulated as an appliance (see definition of "Spray booth" in Section 202) located within the general building occupancy. Therefore, the allowable floor area for one or more spray booths in an unlimited area building would be regulated by the provisions of IFC Section 2404, not by this section. In addition, the floor area of the spray booths would not be included in the aggregate floor area of the Group H occupancies in the unlimited area building for the purposes of applying the allowable area limitations of Sections 507.8.1 and 507.8.1.1.

507.8.2 Located on building perimeter. Except as provided for in Section 507.8.1.1, Group H occupancies shall be located on the perimeter of the building. In Group H-2 and H-3 occupancies, not less than 25 percent of the perimeter of such occupancies shall be an *exterior wall*.

❖ Unless in compliance with Section 507.8.1.1, Group H occupancies are to be located on the building's perimeter. More ready access to the high-hazard occupancy from the exterior of the building provides the fire department with an opportunity to respond more effectively to an incident. As set forth in Section 507.8.1, Group H-2, H-3 and H-4 occupancies that are located on the perimeter of an unlimited area building are permitted to be a maximum of 10 percent of the total building area, or the maximum area permitted by Section 506 for the Group H occupancy, whichever is less.

The increase permitted in Section 506.3 is to be based on the open frontage of Group H-2, H-3 and H-4 floor areas. In Commentary Figure 507.8.2(1), the Group H-2 floor area is located in the corner of the Group F-1 building, so that two walls of the perimeter of the Group H-2 floor area front an unoccupied open space. In accordance with Section 506.2.1, the area increase due to frontage (I_f) is calculated to be 25 percent, in accordance with the figure. Applying this factor in Equation 5-1, the allowable area for the Group H-2 occupancy is [7,000 square feet + (7000 square feet x 0.25)] = 8,750 square feet (813 m^2). This is less than 10 percent of the total area per story [i.e., 15,000 square feet (1394 m^2)], so 8,750 square feet (813 m^2) is the maximum allowable area for the Group H-2 occupancy.

Locating the Group H-2 floor area further away from the perimeter of the Group F-1 building, as occurs in Commentary Figure 507.8.2(2), qualifies the Group H-2 floor area for an even greater allowable area increase [see Commentary Figure 507.8.2(2)].

The mandate that Group H-2 and H-3 occupancies have a minimum of 25 percent of their perimeter as an exterior wall is based on the provisions of Section 415.6. Additionally, the three exceptions to Section 415.6 are the basis for those of Sections 507.8.1.1.1, 507.8.1.1.2 and 507.8.1.1.3. Please refer to the commentary to Section 415.6 for further information.

507.8.3 Occupancy separations. Group H occupancies shall be separated from the remainder of the unlimited area building and from each other in accordance with Table 508.4.

❖ In order to maintain the necessary fire resistance between the high-hazard areas and the remainder of the unlimited area building, the required occupancy separations established in Table 508.4 for separated occupancies must be provided.

507.8.4 Height limitations. For two-*story,* unlimited area buildings, Group H occupancies shall not be located more than one *story above grade plane* unless permitted based on the allowable height and number of *stories* and feet as specified in Section 504 based on the type of construction of the unlimited area building.

❖ This section notes that the Group H occupancy must be on the first story above grade unless Table 504.4 allows for such occupancies on the second story above grade. This will depend on the type of construction. All the other size limitations still apply.

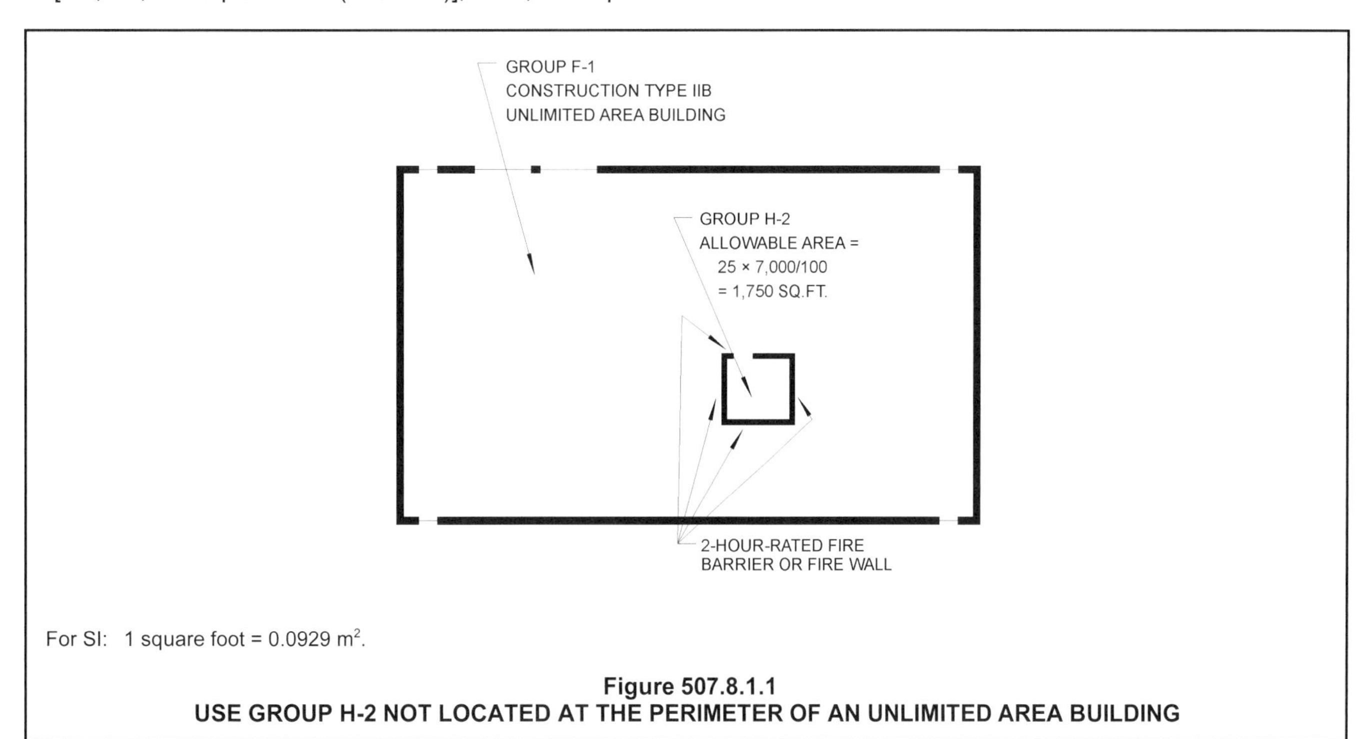

For SI: 1 square foot = 0.0929 m^2.

Figure 507.8.1.1
USE GROUP H-2 NOT LOCATED AT THE PERIMETER OF AN UNLIMITED AREA BUILDING

507.9 Unlimited mixed occupancy buildings with Group H-5. The area of a Group B, F, H-5, M or S building no more than two *stories above grade plane* shall not be limited where the building is equipped throughout with an *automatic sprinkler system* in accordance with Section 903.3.1.1, and is surrounded and adjoined by *public ways* or *yards* not less than 60 feet (18 288 mm) in width, provided all of the following criteria are met:

1. Buildings containing Group H-5 occupancy shall be of Type I or II construction.

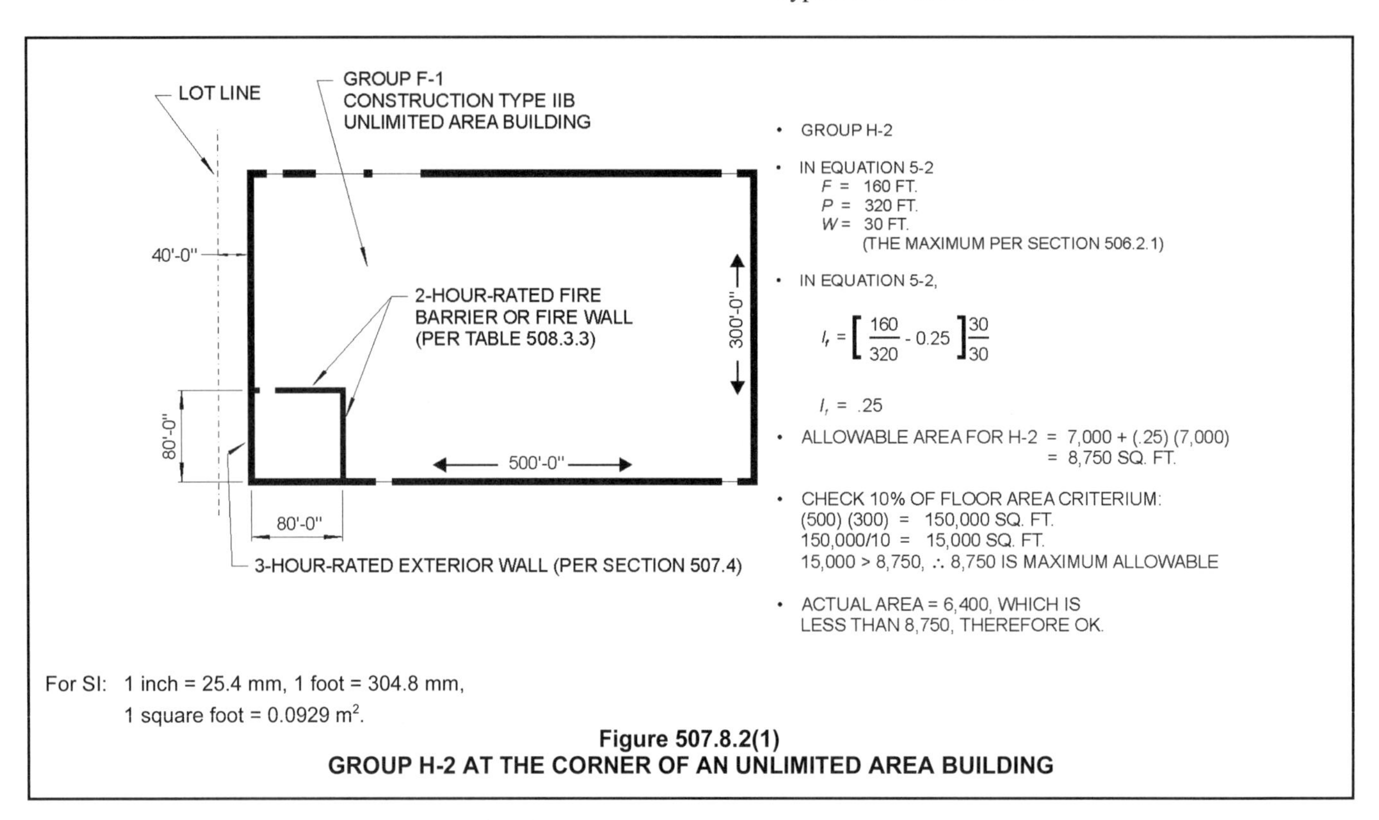

For SI: 1 inch = 25.4 mm, 1 foot = 304.8 mm, 1 square foot = 0.0929 m^2.

Figure 507.8.2(1)
GROUP H-2 AT THE CORNER OF AN UNLIMITED AREA BUILDING

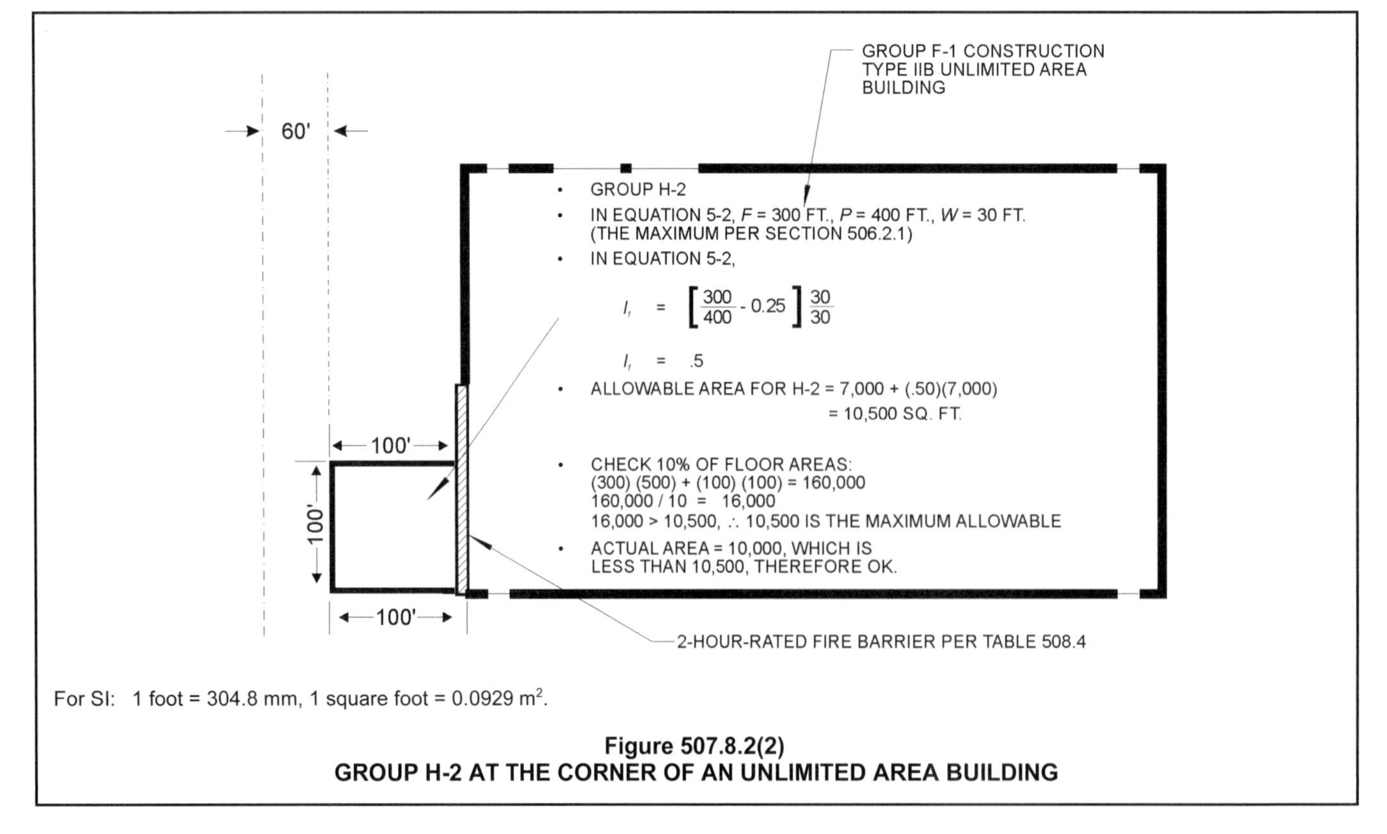

For SI: 1 foot = 304.8 mm, 1 square foot = 0.0929 m^2.

Figure 507.8.2(2)
GROUP H-2 AT THE CORNER OF AN UNLIMITED AREA BUILDING

2. Each area used for Group H-5 occupancy shall be separated from other occupancies as required in Sections 415.11 and 508.4.

3. Each area used for Group H-5 occupancy shall not exceed the maximum allowable area permitted for such occupancies in Section 503.1 including modifications of Section 506.

Exception: Where the Group H-5 occupancy exceeds the maximum allowable area, the Group H-5 shall be subdivided into areas that are separated by 2-hour fire barriers.

❖ H-5 occupancy has a relative hazard based on allowable area per Table 506.2 that is equal to or better than B, F, M, and S occupancies. Adding H-5 occupancy to the occupancies that are allowed to have unlimited area in accordance with Section 507.5 would be consistent with the permitted level of hazard and mitigation established by this section. The code-required mitigating features of H-5 occupancy have been demonstrated to be effective since the introduction of the semiconductor fabrication facility occupancy in the model codes. The exception permits larger H-5 facilities than under previous codes, where 2-hour fire barriers are provided, separating areas of the building.

507.10 Aircraft paint hangar. The area of a Group H-2 aircraft paint hangar no more than one *story above grade plane* shall not be limited where such aircraft paint hangar complies with the provisions of Section 412.6 and is surrounded and adjoined by *public ways* or *yards* not less in width than one and one-half times the *building height*.

❖ Because of their specialized nature and the required size of these facilities, one-story Group H-2 aircraft paint hangars are not limited in area if they meet all of the requirements of Section 412.6. This section requires that the building be of Type I or II construction and have a fire suppression system in accordance with NFPA 409. This standard requires a foam-water suppression system in the paint areas and water sprinklers in all accessory areas. The criteria for the foam-water system depends on the size (and classification in accordance with NFPA 409) of the hangar. Other requirements of Section 412.6 include compartmentalized storage and use of flammable liquids, compliance with the IFC for spray application of flammable liquids and compliance with the *International Mechanical Code*® (IMC®) for ventilation of flammable-finish application areas (see commentary, Section 412.6.6).

In addition to compliance with Section 412.6, the hangar must be surrounded by open space. However, instead of the 60-foot (18 288 mm) width required for other buildings, the open space surrounding the hangar must be at least one and one-half times the height of the building. Therefore, if the hangar was 30 feet (9144 mm) high, there would need to be at least 45 feet (13 716 mm) [30 feet (9144 mm) x 1.5] of open space surrounding the building.

Note that one-story paint hangars would also be permitted to exceed the height limit (in feet) of Table 504.3 if the exception to Section 504.1 is satisfied. Note, further, that Section 507.2.1 cannot be applied to buildings built under Section 507.10.

507.11 Group E buildings. The area of a Group E building no more than one *story above grade plane*, of Type II, IIIA or IV construction, shall not be limited provided all of the following criteria are met:

1. Each classroom shall have not less than two *means of egress*, with one of the *means of egress* being a direct *exit* to the outside of the building complying with Section 1022.

2. The building is equipped throughout with an *automatic sprinkler system* in accordance with Section 903.3.1.1.

3. The building is surrounded and adjoined by *public ways* or *yards* not less than 60 feet (18 288 mm) in width.

❖ This section permits Group E structures to be built as unlimited area buildings. A direct exit to the outside from each classroom must be provided in addition to another means of egress. It is clear in the wording that to fulfill this requirement, students must be able to egress the classroom directly to the outside without intervening corridors, exit passageways or exit enclosures. The unlimited area school building must also be fully sprinklered in accordance with NFPA 13, have open space on all sides of at least 60 feet (18 288 mm) and be no more than one story above grade plane. See the commentary to Section 507.2 regarding the required open space. Please note that Section 507.2.1 cannot be applied to buildings built under Section 507.11.

Where associated assembly spaces are included as part of the Group E occupancy classification in accordance with Section 303.1.3, they would also be allowed to be unlimited in area under this section. The type of construction is limited to Type II, IIIA or IV.

507.12 Motion picture theaters. In buildings of Type II construction, the area of a motion picture theater located on the first *story above grade plane* shall not be limited where the building is provided with an *automatic sprinkler system* throughout in accordance with Section 903.3.1.1 and is surrounded and adjoined by *public ways* or *yards* not less than 60 feet (18 288 mm) in width.

❖ Motion picture theaters located in buildings of Type II construction that are no more than one story above grade plane may be unlimited in area by virtue of being of noncombustible construction, and having low fire loading and sprinkler protection in accordance with NFPA 13. The building must also be surrounded by at least 60 feet (18 288 mm) of open space. See the commentary to Section 507.2 regarding the required open space. Section 507.2.1 can be applied to buildings built under Section 507.12.

507.13 Covered and open mall buildings and anchor buildings. The area of *covered and open mall buildings* and *anchor buildings* not exceeding three *stories* in height that comply with Section 402 shall not be limited.

❖ This section is simply included as a reference to Section 402 for covered malls. The definition of "Covered mall" includes open malls as well. An open mall is very similar to a covered mall but the central pedestrian mall area is open to the sky. If an open mall complies with the provisions for covered malls under Section 402, it is allowed to be of unlimited area. See the commentary to Section 507.2 regarding the required open space. While Section 507.2.1 does not include Section 507.13, Section 402.1.1 permits a similar reduction in open space.

SECTION 508 MIXED USE AND OCCUPANCY

508.1 General. Each portion of a building shall be individually classified in accordance with Section 302.1. Where a building contains more than one occupancy group, the building or portion thereof shall comply with the applicable provisions of Section 508.2, 508.3 or 508.4, or a combination of these sections.

Exceptions:

1. Occupancies separated in accordance with Section 510.
2. Where required by Table 415.6.2, areas of Group H-1, H-2 and H-3 occupancies shall be located in a *detached building* or structure.
3. Uses within *live/work units*, complying with Section 419, are not considered separate occupancies.

❖ Very frequently, buildings are designed for multiple uses, and typically that will result in the building having more than one occupancy classification. This section describes the provisions governing the condition when a building contains more than one of the occupancies identified in Sections 303 through 312. For example, a facility that appears to be used solely for storage often has a small office area for bookkeeping and other purposes. Because the office would be classified as a business occupancy, the building would be identified as a mixed occupancy, storage (Group S-1) and business (Group B).

The section contains three major parts:

1. Accessory occupancies (see Section 508.2). This portion of the section addresses situations where a building contains a main occupancy with minor areas of a building classified as one or more different occupancies that can be treated for allowable area purposes as part of the main occupancy.
2. Mixed occupancies that are not separated from each other (see Section 508.3) since the building is regulated for several of the worst-case conditions of each occupancy involved. This section addresses circumstances where the building contains more than one different occupancy classification in which certain requirements for each occupancy are addressed on a building-wide basis.
3. Mixed occupancies where occupancies of different hazard levels are separated (see Section 508.4). This section addresses circumstances where the building contains more than one occupancy classification but in which occupancies with different hazard levels are separated from each other by fire-resistance-rated construction.

The final sentence of this section reemphasizes the choice for the designer to use just the provisions of one of the three parts, or options, and not to use the other options. But the code also allows a mixture of the options in different portions of a building. For example, a building could be designed to comply with only Section 508.3 for nonseparated mixed occupancies and not comply with any of the provisions of either Section 508.2 or 508.4. A different example could be to use the provisions of Section 508.3 on the first story of a building which perhaps had three or four different occupancies, but then separate the upper stories of the building from the first story and treat the upper stories under Section 508.4 for separated occupancies. A final example would be a building that was predominately one use but had small accessory areas on various stories. For this building the best option might be Section 508.2 regarding accessory occupancies.

These options depend on the limitations of their use, as stated in the respective sections of the code. Basically, the availability of the options depends on the size of the areas desired for each different occupancy, and the relationship of one occupancy to the other.

Exception 1 provides a fourth set of options for mixed occupancies by referring to Section 510, which describes various special circumstances, including where parking garages are located above or below other uses.

Exception 2 reminds the user that including Group H-1, H-2 or H-3 in the same building with other occupancies when quantities exceed certain amounts is prohibited by Section 415.5.2 because of the nature of the hazard for high quantities of flammable, combustible and explosive materials.

Exception 3 exempts uses within a live/work unit from needing to comply with Section 508 provided the uses are within the limitations of Section 419. Live/work units provide the option for as much as 50 percent of nonresidential uses within what is classified as a Group R-2 dwelling unit.

Note that Section 509 regulates incidental uses that, when they occur as part of an occupancy, require additional protection or separation from the balance of the building. Incidental uses are part of the occupancy and the presence of an incidental use in a building does not establish a mixed occupancy condition and as such is not required to comply with Section 508.

508.2 Accessory occupancies. Accessory occupancies are those occupancies that are ancillary to the main occupancy of the building or portion thereof. Accessory occupancies shall comply with the provisions of Sections 508.2.1 through 508.2.4.

❖ Buildings often have rooms or spaces with an occupancy classification that is different from, but accessory to, the principal occupancy classification of the building. When such accessory areas are limited in size, they will not ordinarily represent a significantly different life safety hazard. This principle does not apply where otherwise indicated in Section 508.2.4 for areas classified as Group H, I-1 or R.

The accessory occupancy must be ancillary to the principal purpose for which the structure is occupied. This means that the purpose and function of the area is subordinate and secondary to the structure's primary function. As such, the activities that occur in accessory use areas are necessary for the principal occupancy to properly function and would not otherwise reasonably exist apart from the principal occupancy. See also Section 311, Small Storage Spaces.

508.2.1 Occupancy classification. Accessory occupancies shall be individually classified in accordance with Section 302.1. The requirements of this code shall apply to each portion of the building based on the occupancy classification of that space.

❖ Under Section 508.2.1, each accessory use is to be classified in accordance with Section 302 in the appropriate occupancy classification. Code requirements such as means of egress, the provision of sprinkler protection and structural load are to be determined for this occupancy as if it were a main occupancy of the building. For example, an accessory lunchroom located in a business office regulated as an accessory occupancy would need to be protected with an automatic sprinkler system if the lunchroom's occupant load exceeds 100. In addition, any means of egress doors serving the lunchroom, from the lunchroom to the exterior exit doors, would need to be provided with panic hardware. The lunchroom would need to comply with all other code requirements applicable to a Group A-2 use.

Please note that Sections 508.2.2 and 508.2.3 state that height and area of the building are based on the main occupancy. Therefore, for determining the construction type for the total building, the building's floor area and height should be compared to the limits of Tables 504.3, 504.4 and 506.2 to determine the appropriate construction type. See Section 508.2.3 regarding the area of the accessory occupancy itself and see Section 508.2.2 regarding the height of the accessory occupancy itself.

When applying Chapter 9, many of the requirements for automatic sprinkler systems are based on the size of the fire area in which the occupancy is located. For instance, a Group A-2 occupancy requires a sprinkler system when the fire area is greater than 5,000 square feet (465 m^2) or when the fire area has an occupant load of 100 or more. By understanding the definition of a "Fire area," if there is no rated separation of the accessory occupancy from the main occupancy, the fire area will be everything between fire barriers, fire walls or exterior walls, and therefore could, and most likely will, contain both the accessory occupancy area as well as the main occupancy.

Continuing with this example, suppose that a single-story business office of 23,000 square feet (2137 m^2) contains a lunchroom that is 2,000 square feet (186 m^2) and an occupant load of 134. The main occupancy is a Group B, and the accessory occupancy is a Group A-2. Because the Group A-2 occupancy has a floor area of less than 10 percent of the overall floor area of the single story, it could be regulated as an accessory occupancy, and there is no fire separation requirement between the Group A-2 and Group B. A Group A-2 occupancy is required to be provided with an automatic sprinkler system when it is either in excess of 5,000 square feet (465 m^2) or 100 occupants. Since this fire area is 23,000 square feet (2137 m^2), contains a Group A-2 occupancy with an occupant load of over 100 and is located within a fire area of greater than 5,000 square feet (465 m^2), this fire area would need to be provided with an automatic sprinkler system throughout based on either of the two thresholds.

508.2.2 Allowable building height. The allowable height and number of stories of the building containing accessory occupancies shall be in accordance with Section 504 for the main occupancy of the building.

❖ See the commentary to Section 508.2.1 for accessory occupancies.

508.2.3 Allowable building area. The allowable area of the building shall be based on the applicable provisions of Section 506 for the main occupancy of the building. Aggregate accessory occupancies shall not occupy more than 10 percent of the floor area of the story in which they are located and shall not exceed the tabular values for nonsprinklered buildings in Table 506.2 for each such accessory occupancy.

❖ For accessory occupancies, the area of the building containing the accessory occupancy is based on the area of the main occupancy because accessory occupancies must be limited in size and pose a limited increased degree of hazard. The aggregate area within a story devoted to the occupancies that are designated as accessory occupancies must not be greater than 10 percent of the area of that story [see Commentary Figure 508.2.3(1)].

The area of the portion of the building devoted to an accessory occupancy must also be less than the tabular building area for nonsprinklered buildings given in Table 506.2, based on the group classification that most nearly resembles the accessory occupancy under consideration. Area increases for street frontage based on the provisions of Section 506 are not allowed [see Commentary Figure 508.2.3(2)].

See also the commentary to Section 508.2.1.

508.2.4 Separation of occupancies. No separation is required between accessory occupancies and the main occupancy.

Exceptions:

1. Group H-2, H-3, H-4 and H-5 occupancies shall be separated from all other occupancies in accordance with Section 508.4.
2. Group I-1, R-1, R-2 and R-3 *dwelling units* and *sleeping units* shall be separated from other *dwelling* or *sleeping units* and from accessory occupancies contiguous to them in accordance with the requirements of Section 420.

❖ When a designer is using Section 508.2, most accessory occupancies need not be separated from the main occupancy. Please note that even where an accessory occupancy is not separated from the main occupancy under the provisions of Section 508.2.4, there may be other provisions of the code that might require a fire-resistance-rated separation. For instance, if it was desired not to provide a sprinkler system in the fire area discussed in the commentary example in Section 508.2.1, then a fire barrier would need to be installed to make the fire area containing the Group A-2 occupancy smaller than the threshold amount of 5,000 square feet (465 m^2) and below the 100-occupant threshold. Or, in some cases there could be a wall between two different occupancies that is required by another section of the code to be fire-resistance rated (e.g., a separation wall is provided that also serves as a corridor wall that is required to be fire-resistance rated).

Exception 1 addresses the need to separate Group H-2, H-3, H-4 and H-5 occupancies because of the higher risk posed by these high-hazard occupancies. Please notice that Group H-1 occupancies are not mentioned because these occupancies are always required to be located in a separate building (see Section 415.7).

Exception 2 states that the separation between dwelling units and sleeping units and any accessory occupancies must still comply with Section 420. For example, a common lounge or manager's office that otherwise might qualify as an accessory occupancy within an apartment building must still be separated from adjoining dwelling units.

Those rooms and spaces listed in Table 509 are incidental uses and are not to be considered as accessory occupancies, but rather must be separated or protected as specified in the table.

508.3 Nonseparated occupancies. Buildings or portions of buildings that comply with the provisions of this section shall be considered as nonseparated occupancies.

❖ This section describes the second option a designer may choose to apply to a building that contains more than one occupancy classification. Except with respect to occupancies classified in Group H, this

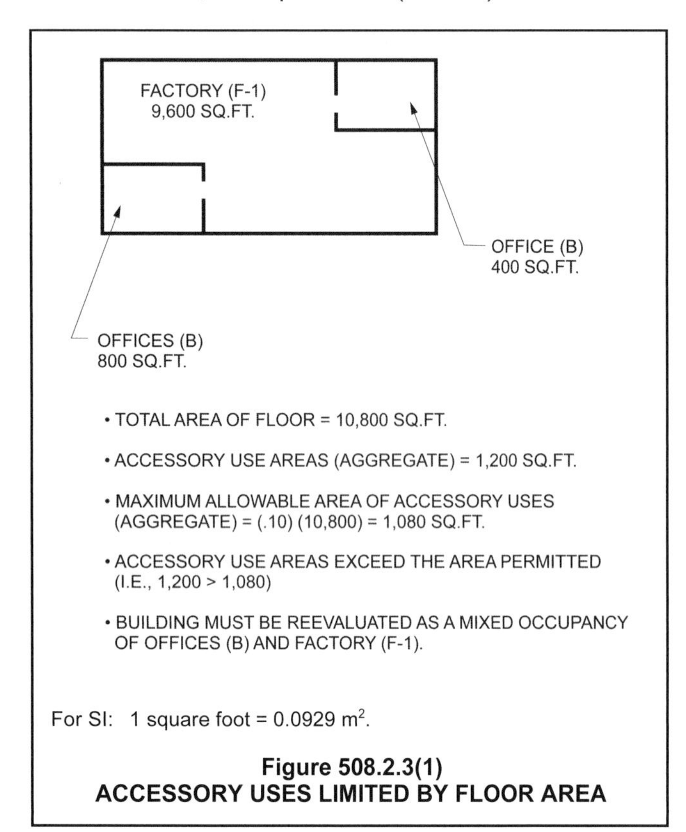

For SI: 1 square foot = 0.0929 m^2.

Figure 508.2.3(1)
ACCESSORY USES LIMITED BY FLOOR AREA

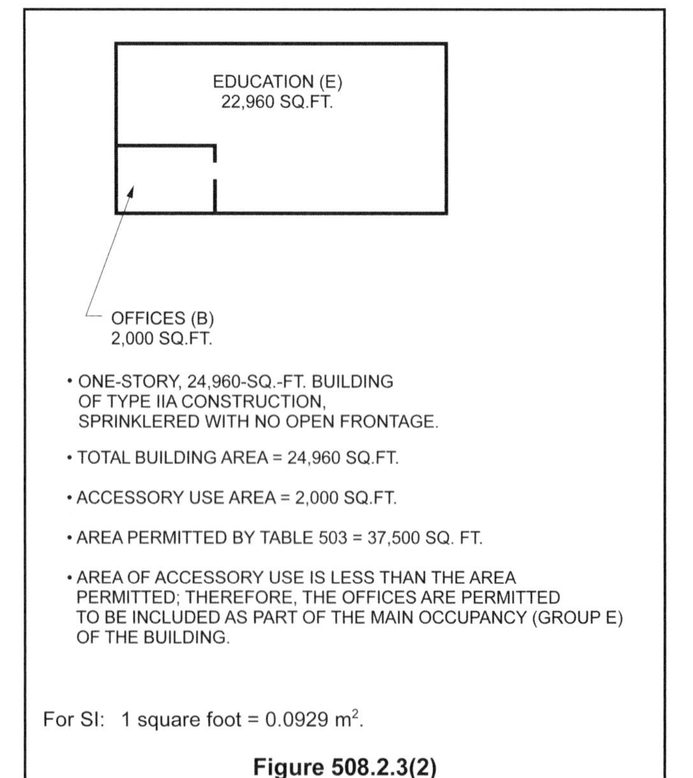

For SI: 1 square foot = 0.0929 m^2.

Figure 508.2.3(2)
ACCESSORY USES LIMITED BY TABULAR AREA

option is similar to the accessory occupancy option (see Section 508.2) in that there is no requirement for various occupancies to be physically separated by any type of fire-resistance-rated assembly (see Commentary Figure 508.3).

The principle behind nonseparated occupancies is that the different occupancies within the same building do not have to be separated by fire-resistance-rated assemblies if the building complies throughout with the more restrictive code requirements for minimum construction type and fire protection systems (Chapter 9) applicable to the occupancies in the building. If the building is also a high-rise building, the most restrictive provisions of Section 403 applicable to the specific occupancies in the building will also apply. Although each occupancy is separately classified as to its group, a fire-resistance-rated assembly is not required by the nonseparated occupancies option. A designer may choose to physically separate the occupancies; however, a fire-resistance rating of these separations would not be required by this section.

There are four basic steps to follow when applying the nonseparated occupancies concept:

Step 1: Determine which occupancy group classifications are present in the building.

Step 2: Determine the minimum type of construction based on the height and area of the building for each occupancy in accordance with Section 503 as if each occupancy occurred throughout the building. Compare the requirements for each occupancy and then apply the requirement for the highest type of construction to the entire building (see Section 508.3.2).

Step 3: Apply the most restrictive provisions found in Chapter 9 throughout the building containing nonseparated occupancies. For example, if the two occupancies in a 6,000-square-foot (557 m^2) building are Group B and Group A-2, even though Section 903 does not require an automatic sprinkler system in nonhigh-rise Group B buildings, it does require sprinklers in a Group A-2 building of this size. For this example, the building with these nonseparated occupancies would need to be fully sprinklered.

For a high-rise building, also apply the most restrictive provisions contained in Section 403 related to the occupancies present (see Section 508.3.1).

Step 4: Apply all other requirements of the code, except for Section 403 and Chapter 9, to each occupancy individually based on the specific occupancy of each space (e.g., means of egress).

508.3.1 Occupancy classification. Nonseparated occupancies shall be individually classified in accordance with Section 302.1. The requirements of this code shall apply to each portion of the building based on the occupancy classification of that space. In addition, the most restrictive provisions of Chapter 9 that apply to the nonseparated occupancies shall apply to the total nonseparated occupancy area. Where nonseparated occupancies occur in a *high-rise building*, the most restrictive requirements of Section 403 that apply to the nonseparated occupancies shall apply throughout the *high-rise building*.

❖ The nonseparated occupancies option requires that each occupancy area be considered a different occupancy for purposes of application of code requirements and therefore requires that all code issues related to that occupancy be considered separately

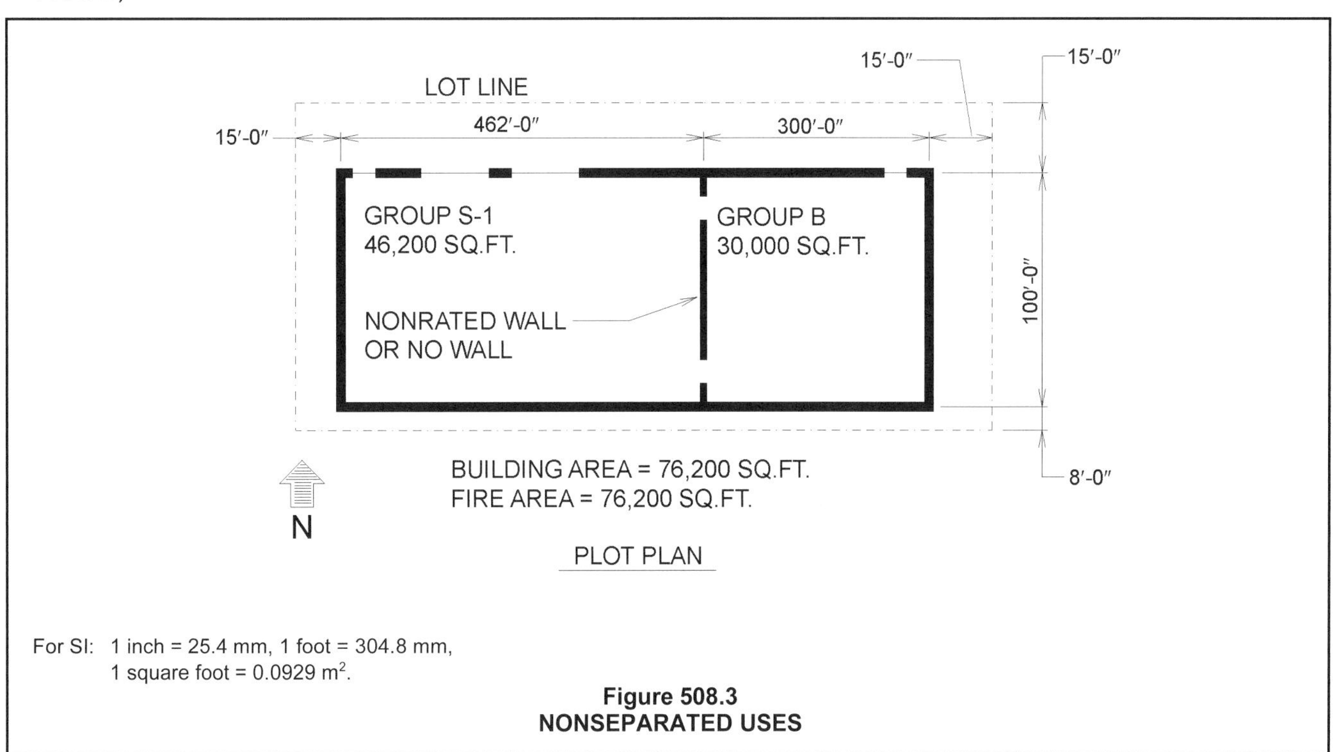

For SI: 1 inch = 25.4 mm, 1 foot = 304.8 mm, 1 square foot = 0.0929 m^2.

Figure 508.3
NONSEPARATED USES

for the areas of the building where the occupancy is located. The exception to this is for the provisions of Chapter 9 and Section 403, which must be applied to the whole building containing the nonseparated occupancies.

The principle behind the nonseparated occupancies concept is that the different occupancies within the same building do not have to be separated by fire-resistance-rated assemblies if the building complies throughout with the more restrictive code requirements for construction, fire protection systems and, where applicable, requirements for high-rise buildings. Although each occupancy group is separately classified as to its group, a fire-resistance-rated assembly is not required under the nonseparated occupancies option. A designer may choose to physically separate the occupancies; however, a fire-resistance rating of these separations would not be required by this section, as stated in Section 508.3.3.

For example, if a building contains both Groups B and M, each portion of the building must comply with the code requirements for its respective group classification. The occupant load for the area classified as Group B is based on the code requirements applicable to the business occupancy. Similarly, the occupant load for the Group M area is based on requirements applicable to the mercantile occupancy. Live loads for the Group B area are applied to the area of business occupancy. Group M live loads are applied to the area of mercantile occupancy. Exterior wall ratings based on fire separation distance (see Table 602) apply to exterior walls of the business occupancy for Group B buildings; the same is true for walls in Group M occupancies.

It is important to note the threshold requirements for fire protection systems are contained in Chapter 9. In some cases they are simply based on the occupancy, and in other cases they are based on height or area criteria. Also, the requirements to be met in some cases only apply to the fire area in which a given occupancy is contained, while in others they apply to the entire building.

For example, if the business occupancy as described in Commentary Figure 508.3 required a fire alarm system with manual fire alarm boxes, then one would be required for the entire building—even though Chapter 9 does not require a fire alarm system for a storage occupancy. Note that when determining if a fire alarm system is required and the threshold for the requirement is based on occupant load, it is based on the occupant load of the individual occupancy and not for all the occupancies in the nonseparated building (see commentary, Section 907.2).

508.3.2 Allowable building area and height. The allowable *building area and height* of the building or portion thereof shall be based on the most restrictive allowances for the occupancy groups under consideration for the type of construction of the building in accordance with Section 503.1.

❖ The fundamental concept underlying the nonseparated occupancies option is that the allowable building heights and allowable building areas are based on the most restrictive requirements of Tables 504.3, 504.4 and 506.2 applicable to each of the occupancy groups in the mixed occupancy building.

For example, the tabular area for Group S-1 in Table 506.2 is less than the corresponding tabular area for Group B. Therefore, Group S-1 results in a requirement for a higher type of construction for buildings of equal size and will determine the minimum construction type of the building. If the building were, say, Type IIB construction with a sprinkler system, then the allowable area for the building would be 70,000 square feet (6503 m^2), based on the allowable area of Group S-1 occupancies for Type IIB construction in Section 506. Area increases for frontage could be applied. The same methodology of applying the most restrictive criteria of Table 506.2 is used to evaluate the maximum permitted height in stories and feet above grade plane based upon Section 504.

508.3.3 Separation. No separation is required between nonseparated occupancies.

Exceptions:

1. Group H-2, H-3, H-4 and H-5 occupancies shall be separated from all other occupancies in accordance with Section 508.4.
2. Group I-1, R-1, R-2 and R-3 *dwelling units* and *sleeping units* shall be separated from other *dwelling* or *sleeping units* and from other occupancies contiguous to them in accordance with the requirements of Section 420.

❖ As is indicated by the name of this mixed occupancy option, "nonseparated occupancies," the occupancy groups within the building need not be separated from each other by fire-resistance-rated construction. It must be noted that other portions of the code might require fire separation assemblies, and where the separations are required by other provisions, they must be provided even in a building designed under the nonseparated mixed occupancy option. For instance, if it was desired not to provide a sprinkler system in the fire area discussed in the commentary example in Section 508.2.1, then a fire barrier would need to be installed to make the fire area containing the Group A-2 occupancy smaller than the threshold amount of 5,000 square feet (465 m^2) and the occupant load less than 100. Or, in some cases there could be a wall between two different occupancies that is required by another section of the code to be fire-resistance rated (e.g., a separation wall is provided that also serves as a corridor wall that is required to be fire-resistance rated).

When one or more of the occupancies is a Group H-2, H-3, H-4 or H-5 high-hazard occupancy, the nonseparated occupancies option could not be applied for the Group H portions of the building (Exception 1). Any Group H occupancies would need to be separated as required by Section 508.4 for separated occupancies.

Exception 2 states that the horizontal assemblies and walls required to separate dwelling units and sleeping units from each other and from other occupancies need to comply with Section 420 even if the nonseparated occupancies provisions are applied.

Example: A building contains a 46,200-square-foot (4292 m^2) general warehouse and an accompanying 30,000-square-foot (2787 m^2) regional dispatch office for a total area of 76,200 square feet (7079 m^2). The building is a one-story, metal building. Assume that there is no allowable area increase for frontage. In applying the four-step procedure for nonseparated uses, the following holds true:

Step 1: The warehouse is classified in Group S-1 and the offices are classified in Group B.

Step 2: The minimum type of construction is based on the lesser allowable area of the building for each occupancy classification. First, determine if any increases can be obtained (no frontage increase). Note, in Step 3, the building is required to be sprinklered. For the purpose of analysis, assume the building is of Type IIB construction. The Group B requirements would allow the building to be 92,000 square feet (8547 m^2) (A_t = 92,000 square feet from Table 506.2, with no frontage increases). The Group S-1 requirements, however, only allow a building of Type IIB construction to be 70,000 square feet (6503 m^2). The most restrictive is the Group S-1 requirement; therefore, it governs. Hence, Type IIB construction would not allow adequate area for the S-1 occupancy. Since the most restrictive allowable area is required to be considered, a higher construction type would be required. The minimum type of construction needs to be Type IIA.

Step 3: The fire area containing the occupancy classified in Group S-1 is more than 12,000 square feet (1115 m^2); thus, in accordance with Section 903.2.9, this building must be sprinklered. The area classified as Group B ordinarily is not required to be equipped with sprinklers by Section 903. However, since the Group S-1 provision does require sprinkler protection, it is more restrictive than the requirement for Group B under Chapter 9. Therefore, the entire building must be sprinklered.

Step 4: The occupant load of each area should be calculated separately (based on the occupancy of that area) using Section 1004.1 and Table 1004.1.1.

Business: = 30,000 square feet (2787 m^2) divided by 100 square feet (9 m^2) per person = 300 people.

Storage: = 46,200 square feet (4292 m^2) divided by 500 (warehouses) square feet (46 m^2) per person = 93 people.

Total occupant load: 393 people

The evaluation of means of egress in accordance with Chapter 10 is then based on these occupant loads.

The load-bearing members in the north, east and west exterior walls of the building must provide a minimum 1-hour fire-resistance rating as required for Type IIA construction in Table 601. The nonload-bearing walls are also required to have a 1-hour fire-resistance rating because of the 15-foot (4572 mm) fire separation distance (see Table 602).

The fire separation distance on the south side of the mixed occupancy building is 8 feet (2438 mm). Table 602 requires the Group S-1 and B portions of this wall to have a 1-hour fire-resistance rating. The ratings required by this table apply to both load-bearing and nonload-bearing exterior walls. The substitution of an automatic sprinkler system in lieu of 1-hour fire-resistance-rated construction (as indicated in Note d of Table 602) does not apply to exterior walls; therefore, all exterior walls (whether load-bearing or nonload-bearing) in this building are required to have a 1-hour fire-resistance rating.

This analysis must be continued for other aspects of code compliance for each respective occupancy.

508.4 Separated occupancies. Buildings or portions of buildings that comply with the provisions of this section shall be considered as separated occupancies.

❖ This section describes the third option, separated occupancies, that a designer may choose to apply when constructing a building that contains more than one occupancy classification. This third option differs from the first two alternatives (i.e., accessory occupancies and nonseparated occupancies) (see Sections 508.2 and 508.3) in four ways:

1. Occupancies that are to be evaluated as separated occupancies must be separated in accordance with Table 508.4. When the table requires a separation, the occupancies must be separated completely, both horizontally and vertically, by the provision of fire-resistance-rated fire barriers and horizontal assemblies (see Sections 707 and 711). It should be noted that Table 508.4 indicates "N" at the intersection of some occupancies. The "N" indicates that no rated separation is required. This is allowed where the adjacent occupancies pose similar levels of risk. Separations, where provided, do not necessarily create separate fire areas. Table 508.4 is not intended to be used for determining the required fire-

resistance rating for creating multiple fire areas. The minimum separations for fire area purposes are established in Table 707.3.9.

2. The determination of the minimum type of construction is based on both the height of each occupancy relative to the grade plane and the areas of each occupancy relative to the total area per story.
3. Allowable building area is determined based on the occupancies on each story through a comparison of the allowed area for each occupancy category and the actual floor area proposed for each occupancy.
4. In comparison to the nonseparated occupancies provisions in Section 508.3 where the most restrictive of the requirements of Chapter 9 apply throughout the building, under separated occupancies, requirements for such systems as sprinkler systems, alarms and standpipes are determined based on the areas containing each occupancy. It must be noted that the determination of sprinkler requirements in Section 903 is based on the occupancies present. The threshold for providing a sprinkler system is often based on the size of a fire area. In the cases where Table 508.4 does not require a separation, the result is that certain fire areas will contain more than one occupancy. Therefore, when determining where sprinklers are required, the total size of the fire area needs to be considered, not just the portion containing each occupancy. For example, a proposed building is 16,000 square feet (1486 m^2), there is 8,000 square feet (743 m^2) of Group M occupancy and 8,000 square feet (743 m^2) of Group B. According to Table 508.4, there does not need to be a separation between these two occupancies. Individually, neither is large enough to require a sprinkler system. However, if no separation is provided, the fire area is 16,000 square feet (1486 m^2), and in accordance with Section 903.2.7 a fire area containing a Group M occupancy and exceeding 12,000 square feet (1115 m^2) must be sprinkler protected. [Fire areas (defined in Section 202) are used in Section 903.2 to determine if sprinklers are required. Fire areas are a distinct requirement from separation of occupancies and the separation of occupancies by a fire-resistance-rated wall may not always create a separate fire area. Fire areas can be delineated by fire walls and fire barriers per Chapter 7.]

There are four basic steps to follow when using the separated occupancies option:

Step 1: Determine which occupancies are present in the building (see Section 508.4.1).

Step 2: Separate the occupancies in accordance with Table 508.4 with fire barrier walls and horizotal assemblies in accordance with Sections 707 and 711 (see Section 508.4.4).

Step 3: Apply all code requirements for each separated space individually based on the occupancy or occupancies present (i.e., design occupant load, means of egress elements, exterior wall requirements and fire protection). For the application of code provisions, each separated space is taken into consideration individually. Again, it must be remembered that Table 508.4 may not require a separation and, therefore, code requirements based on fire area must be addressed for the total fire area even though it contains more than one occupancy.

Step 4: Determine the minimum type of construction of a building based on the building height limitations of Sections 503 and 504 and the building area limitations of Sections 503 and 506 (see Sections 508.4.3 and 508.4.2, respectively).

Part A: Determine the minimum type of construction required based on the height of each occupancy relative to the grade plane (see Section 508.4.3). This needs to be evaluated for both height in feet and stories above grade plane.

Part B: Determine the minimum type of construction based on a weighted average of areas occupied by the various occupancies (see Section. 508.4.2).

TABLE 508.4. See page 5-39.

❖ The purpose of Table 508.4 is to set forth the fire-resistance rating required for fire barrier walls and horizontal assemblies used to separate occupancies. The fire-resistance rating of the separation between different occupancies is based on the relative anticipated fire severity of the occupancies.

"N" in this table is present where a separation is not required between the uses because the hazards of the two uses are the same. "NP" in this table is used where an occupancy is required to be sprinklered, so the nonsprinklered option is not permitted. "S" is for occupancies that are equipped throughout with an NFPA 13 sprinkler system. "NS" is for occupancies that are either not sprinklered or are sprinklered with and NFPA 13R or NFPA 13D system.

Note a: A reference is made to Section 420 for provisions requiring the separation of dwelling units and sleeping units in Group I-1, R-1, R-2, R-3 and R-4 occupancies (see commentary, Section 420).

Note b: The fire-resistance rating of spaces used solely for private or pleasure vehicles may be reduced by 1 hour but never less than 1 hour.

Note c: See the commentary for Section 406.3.4 for separation for private garages and carports.

Note d: Occupancies in the same classification do not need to be separated. A Group H-3 occupancy does not need to be separated from a different Group H-3 use.

Note e: See Section 422.2 for specific requirements for ambulatory care facilities that are required to be separated.

508.4.1 Occupancy classification. Separated occupancies shall be individually classified in accordance with Section 302.1. Each separated space shall comply with this code based on the occupancy classification of that portion of the building.

❖ In the separated occupancies option, occupancies that are to be evaluated as separated occupancies must be separated in accordance with Table 508.4. When a separation is required, occupancies must be separated completely, both horizontally and vertically, with fire barriers and horizontal assemblies (see Sections 707 and 711).

Except for certain occupancies, it is the designer's prerogative to use the accessory occupancies option, nonseparated occupancies option or the separated occupancies option when establishing a mixed occupancy building. It is also possible to apply both or all three options within different portions or different stories of a building. Where a mixture of options is used, the design documentation needs to clearly show how the requirements of each option are applied in each portion of the building.

508.4.2 Allowable building area. In each *story*, the *building area* shall be such that the sum of the ratios of the actual *building area* of each separated occupancy divided by the allowable *building* area of each separated occupancy shall not exceed 1.

❖ For each story, it must be determined that the sum of the ratios of the actual floor area of each separated occupancy, respective to the most restrictive occupancy contained therein as compared to the areas per story allowed by Section 506 for each respective occupancy, does not exceed one. In the evaluation of allowable area, intervening fire barriers between different fire areas containing the same occupancy are not a consideration. In determining the floor area per occupancy, all areas of the same occupancy are added together irrespective of the presence of multiple fire areas.

In determining the allowable areas for each occupancy, the frontage increase can be applied; thus, the allowable areas are intended to include the increases permitted for open perimeter. For determination of the allowable perimeter increase, use the entire building perimeter—not the occupancy perimeters.

508.4.3 Allowable height. Each separated occupancy shall comply with the *building height* limitations based on the type of construction of the building in accordance with Section 503.1.

Exception: Special provisions of Section 510 shall permit occupancies at *building heights* other than provided in Section 503.1.

❖ The allowable height is occupancy dependent. As long as individual occupancies meet the height limitations based upon a measurement from grade plane,

TABLE 508.4
REQUIRED SEPARATION OF OCCUPANCIES (HOURS)

OCCUPANCY	A, E		I-1[a], I-3, I-4		I-2		R[a]		F-2, S-2[b], U		B[e], F-1, M, S-1		H-1		H-2		H-3, H-4		H-5	
	S	NS	S	NS	S	NS	S	NS	S	NS	S	NS	S	NS	S	NS	S	NS	S	NS
A, E	N	N	1	2	2	NP	1	2	N	1	1	2	NP	NP	3	4	2	3	2	NP
I-1[a], I-3, I-4	—	—	N	N	2	NP	1	NP	1	2	1	2	NP	NP	3	NP	2	NP	2	NP
I-2	—	—	—	—	N	N	2	NP	2	NP	2	NP	NP	NP	3	NP	2	NP	2	NP
R[a]	—	—	—	—	—	—	N	N	1[c]	2[c]	1	2	NP	NP	3	NP	2	NP	2	NP
F-2, S-2[b], U	—	—	—	—	—	—	—	—	N	N	1	2	NP	NP	3	4	2	3	2	NP
B[e], F-1, M, S-1	—	—	—	—	—	—	—	—	—	—	N	N	NP	NP	2	3	1	2	1	NP
H-1	—	—	—	—	—	—	—	—	—	—	—	—	N	NP	NP	NP	NP	NP	NP	NP
H-2	—	—	—	—	—	—	—	—	—	—	—	—	—	—	N	NP	1	NP	1	NP
H-3, H-4	—	—	—	—	—	—	—	—	—	—	—	—	—	—	—	—	1[d]	NP	1	NP
H-5	—	—	—	—	—	—	—	—	—	—	—	—	—	—	—	—	—	—	N	NP

S = Buildings equipped throughout with an automatic sprinkler system installed in accordance with Section 903.3.1.1.

NS = Buildings not equipped throughout with an automatic sprinkler system installed in accordance with Section 903.3.1.1.

N = No separation requirement.

NP = Not permitted.

a. See Section 420.

b. The required separation from areas used only for private or pleasure vehicles shall be reduced by 1 hour but not to less than 1 hour.

c. See Section 406.3.4.

d. Separation is not required between occupancies of the same classification.

e. See Section 422.2 for ambulatory care facilities.

then the building complies. For example, a building of Type IIB construction with no increases for sprinklers containing a Group B occupancy and a Group F-1 occupancy could be three stories. The only limitation in this case, in terms of building height, would be that the Group F-1 occupancy could not be located any higher than the second story.

508.4.4 Separation. Individual occupancies shall be separated from adjacent occupancies in accordance with Table 508.4.

❖ Table 508.4 provides the required separation between areas containing the separated occupancies. Separations must be of fire-resistance-rated construction for 1, 2 or 3 hours as specified in the cells of the table. Generally, the required rating will be higher if a building is not fully protected by an automatic sprinkler system. The table states "NP" in certain cells. These occur where occupancies are not permitted to be in a mixed occupancy building, or where one or both occupancies cannot be in a nonsprinklered building. Other cells contain the letter "N." For the adjoining occupancies represented by each such cell, no separation is required. Although certain adjacent occupancies are not required to be physically separated, they are still evaluated under the separated option, but note that, where code requirements are based on established fire areas, occupancies not separated must be considered as sharing the same fire area.

For example, a completely sprinklered building of Type VB construction contains areas devoted to business and assembly occupancies. The designer has chosen the separated occupancies option and has completely separated the areas containing the two different occupancies by fire barriers (complying with Section 707) and horizontal assemblies (complying with Section 711) having a minimum 1-hour fire-resistance rating in accordance with Table 508.4. This is found by consulting the box that intersects with "A, E" in the first column with "B, F-1, M, S-1" in the first row and, because the entire building is sprinklered, the fire-resistance rating of the fire barrier walls and horizontal assemblies is required to be 1 hour, as indicated in the column designated "S." Had the building not been sprinklered, the required rating would have been 2 hours, as indicated in the column designated "NS."

Please note that Table 508.4 contains groupings of some of the occupancies, including:

- A, E
- I-1, I-3, I-4
- R
- F-2, S-2, U
- B, F-1, M, S-1
- H-3, H-4

These are occupancies that share the same level of hazard with respect to fire safety. It is possible, therefore, to have two occupancies that comply with these separated use provisions that require no separation between them. For instance, a mixed occupancy of Groups B and M would not be required to have a separation between them, but the provisions for calculation of the sum of the ratios of actual areas to allowable areas would still be applied to this circumstance.

508.4.4.1 Construction. Required separations shall be *fire barriers* constructed in accordance with Section 707 or *horizontal assemblies* constructed in accordance with Section 711, or both, so as to completely separate adjacent occupancies.

❖ This section is merely a reference to the appropriate provisions in Chapter 7 for fire-resistance-rated assemblies that are required to meet the provisions for adequate separation of occupancies.

The following three examples illustrate the application of the separated occupancies option.

Example 1: A four-story building contains a retail store on the first story, a lecture hall with fixed seats on the second story and offices on the top two stories. Each story is 10,500 square feet (975 m^2). A sprinkler system is not provided throughout the building. Assume that there is no open perimeter increase [see Commentary Figure 508.4.4.1(1)]. In applying the four-step procedure for separated uses, the following holds true:

Step 1: First story is Group M for the entire story. Second story is Group A-3 for the entire story. Third and fourth stories are Group B for both stories.

Step 2: The separation requirements in accordance with Table 508.4 require a 2-hour fire-resistance rating for the horizontal assembly between the Group M occupancy and the Group A-3 occupancy. Likewise, a 2-hour fire-resistance-rated horizontal assembly is required between the Group A-3 occupancy and the Group B occupancy. Each occupancy is separated horizontally with an approved fire-resistance-rated assembly. There is no rating required for the horizontal assembly between the third and fourth stories because they are in the same occupancy.

Step 3: The occupancy-specific requirements need to be addressed. The building is essentially broken into the following three areas based on the separations provided to achieve the separated occupancy requirements:

Group M area = 10,500 square feet (975 m^2).

Group A-3 area = 10,500 square feet (975 m^2).

Group B area = 21,000 square feet (1951 m^2) [combined area for two stories, each 10,500 square feet (975 m^2)].

Because of the construction required in this example for each separation, these occupancy areas may also be separate fire areas as determined by Table 707.3.10.

Since the story containing the Group M occupancy is also a fire area that is less than 12,000 square feet (1115 m^2) and is located below the fourth story, sprinklers are not required for the Group M occupancy.

The story containing the Group A-3 occupancy is also a fire area that is less than 12,000 square feet (1115 m^2), but it is located on the second story and above the level of exit discharge. According to Section 903.2.1.3, a Group A-3 occupancy fire area in this location must be provided with a sprinkler system. In addition, this would mean that the Group M occupancy would be required to be sprinklered in accordance with Section 903.2.1, which requires that all stories between the Group A-3 occupancy and the level of exit discharge be sprinklered.

The third and fourth stories containing the Group B occupancy do not require sprinklers because neither the size nor height of these stories triggers any threshold in Section 903.

Additionally, provisions in other parts of the code for each occupancy area (e.g., means of egress, exterior wall requirements, interior finishes) should be used based on the occupancy located on each story. For example, the design occupant load of the area containing the Group M occupancy is based on the mercantile occupant load factor found in Table 1004.1.1 while the area containing the Group A-3 occupancy is based upon the number of fixed seats in the lecture hall in accordance with Section 1004.4.

Step 4:

Part A: Review for the minimum types of construction based on the height limitations for each of the three occupancies as follows [see Commentary Figure 508.4.4.1(2)]:

1. Group M, one story in height; the minimum type of construction is Type VA [two stories and 40 feet (12 192 mm) in accordance with Section 504].
2. Group A-3, two stories in height; the minimum type of construction is Type VA [two stories and 50 feet (15 240 mm) in accordance with Section 504].
3. Group B, four stories in height; the minimum type of construction is Type IIIA [five stories and 65 feet (19 812 mm) in accordance with Section 504].

The highest type of construction for any one of the occupancies governs for the entire building; thus, the minimum allowable type of construction permitted for the building based on height is Type IIIA. The designer is permitted to choose any one of the construction types other than Types IIB, IIIB, VA and VB for the design of the building. The separated occupancy option presumes that each occupancy is restricted to the story (or stories) for which it is designated. For this example, the first story is separated and designated for the Group M occupancy, and the Group M occupancy can only occur on that story. In comparison, the nonseparated

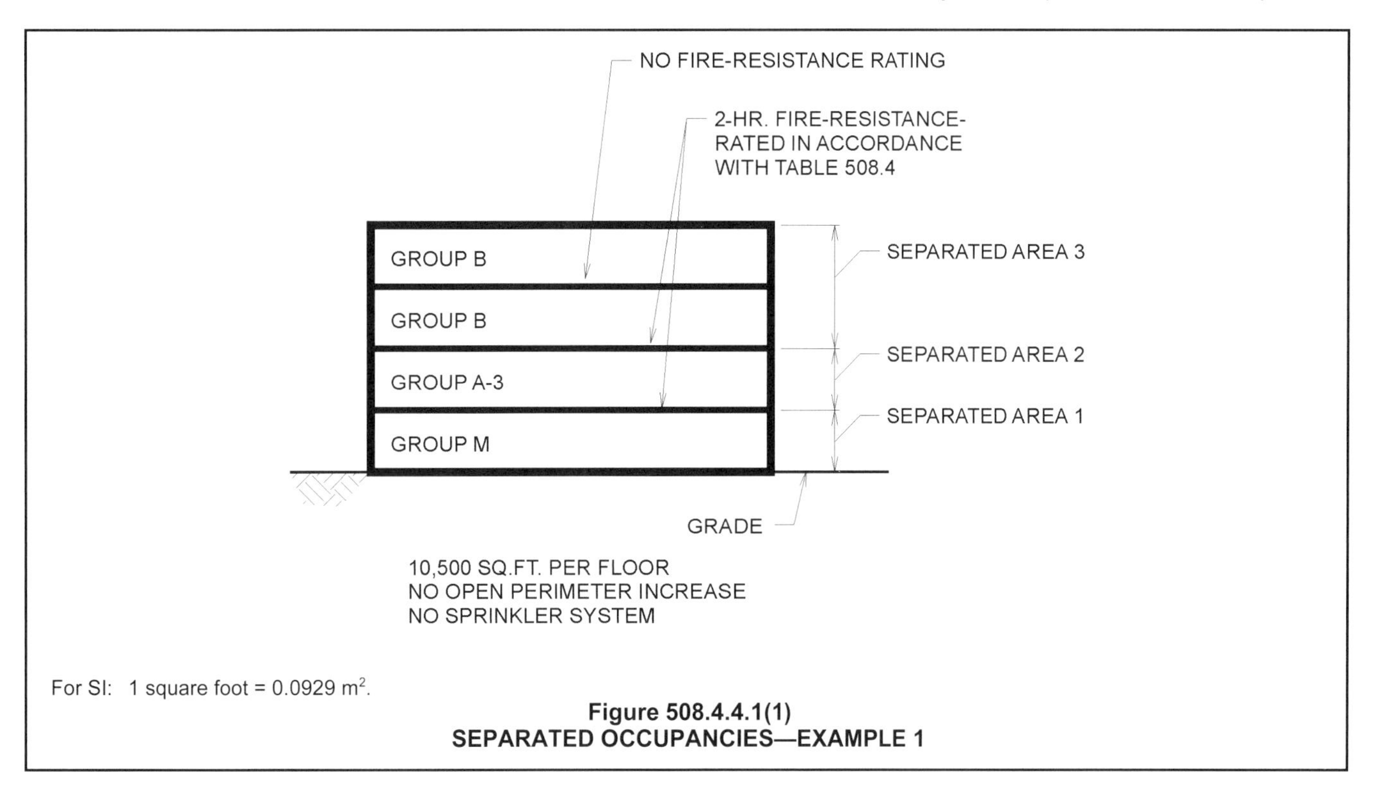

Figure 508.4.4.1(1)
SEPARATED OCCUPANCIES—EXAMPLE 1

option would assume that the Group M occupancy could occur on any of the four stories.

If the building were completely sprinklered, the allowable heights of all occupancies would be higher in accordance with Section 504. Accordingly, under Section 504, the minimum type of construction allowed for the building would be Type VA. In addition, the minimum fire-resistance rating of each floor/ceiling assembly would need to be only 1 hour, in accordance with Table 508.4.

Part B: Since each story has only one occupancy, the maximum allowable area for a story of the building can be taken directly from Table 506.2. Based on the allowable construction type determined, Type IIIB construction is acceptable for Groups B and M, but it is not acceptable for Group A-3. Based on the given area of 10,500 square feet (975 m^2) and without any area increases, the allowable types of construction permitted for Group A-3 based on area are Types IA, IB, IIA, IIIA, IV and VA. Although the Group A-3 story and the story below will be required to be sprinkler protected, because the whole building is not protected, area increases for sprinklers are not allowed in this example.

From Part A it was determined that Type VA construction is not permitted for the height of the building due to Group B; therefore, the allowable types of construction for the building based on height and area are Types IA, IB, IIA, IIIA and IV.

Example 2: Consider a 76,200-square-foot (7079 m^2) building with a fire barrier wall separating a 46,200-square-foot (4292 m^2) warehouse from a 30,000-square-foot (2787 m^2) office space [see Commentary Figure 508.4.4.1(3)].

Step 1: The warehouse occupancy (which is also a fire area) is classified in Group S-2. The office occupancy (which is also a fire area) is classified in Group B.

Step 2: A 2-hour fire barrier is required in accordance with Table 508.4 for nonsprinklered buildings (1 hour for sprinklered buildings) between the two occupancies and extends continuously through all concealed spaces to the underside of the roof deck in accordance with Section 707.

Step 3: The separation required by Table 508.4 also creates two fire areas since it requires the same level of fire separation as Table 707.3.10 which is used to determine the minimum required fire area separation. The Group B area is one story and 30,000 square feet (2787 m^2). A sprinkler system is not required by the code for a one-story Group B fire area (unless it is an ambulatory care center). The Group S-2 fire area is 46,200 square feet (4292 m^2), which is also not required to have a sprinkler system. However, the designer wants to be able to utilize Type IIB construction. The tabular area for nonsprinklered buildings from Table 506.2 for Group S-2 occupancies within a Type IIB building is only 26,000 feet (2416 m^2); therefore, a sprinkler system will need to be provided to allow an increase in allowable area. For a single-story building, the allowable area for a Group S-2 building in Type IIB would then be 104,000 square feet (9662 m^2). To allow this increase, the entire building including both occupancies would need to be protected throughout with an NFPA 13 sprinkler system.

Further, evaluate each occupancy area in the same manner for compliance with all provisions of the code.

Step 4:

Part A: The proposed building is only one story in height. Based on height, any type of construction is acceptable for the occupancies under consideration.

Part B: Make a trial evaluation of the building assuming it to be Type IIB construction:

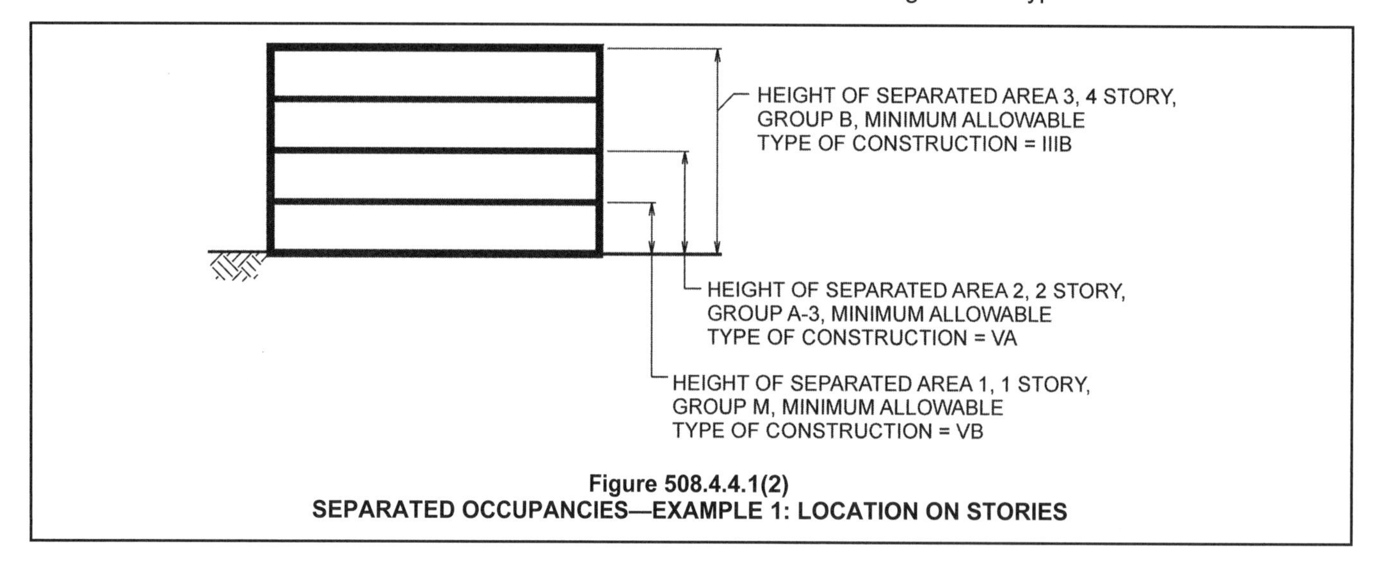

Figure 508.4.4.1(2)
SEPARATED OCCUPANCIES—EXAMPLE 1: LOCATION ON STORIES

Actual area of Group S-2 = 46,200 square feet (4292 m^2) and actual area of Group B = 30,000 square feet (2787 m^2).

Group S-2 maximum allowable area, = 104,000 square feet (9662 m^2).

The maximum allowable area for Group B = 92,000 square feet (8547 m^2).

Using the sum of the ratios approach, 46,200/104,000 + 30,000/92,000 = 0.77 which does not exceed 1.0.

Therefore, Type IIB construction is an acceptable minimum type of construction. When the factor arrived at is substantially below 1.0, it indicates that the building might qualify as a lower type of construction or as nonseparated occupancies.

The building is deemed to comply with the code after each of the code requirements is satisfied, as determined by Steps 1 through 4.

In the case of multistory buildings, an evaluation of the allowable area must be performed for every story of the building that contains separated uses.

Example 3: Consider a 76,200-square-foot (7079 m^2) building identical to that used in Example 2, except that the warehouse is classified in Group S-1. The Group S-1 warehouse is 46,000 square feet (4292 m^2) and the Group B office space is 30,000-square-foot (2787 m^2) [see Commentary Figure 508.4.4.1(4)].

Step 1: The warehouse is classified as Group S-1. The office is classified as Group B.

Step 2: Table 508.4 indicates that there is no separation requirement between Groups B and S-1, since they are in the same grouping of occupancies in the table.

Step 3: As there is no separation between the occupancies, there is only one fire area including both occupancies. The fire area for the occupancy of Groups B and S-1 is the entire area of the one-story building, or 76,200 square feet (7079 m^2). In accordance with Section 903.2.9, since the fire area containing Group S-1 is greater than 12,000 square feet (1115 m^2) the entire building is required to be sprinklered. Note that even if the uses were separated, Section 903.2.9 requires sprinklers throughout all buildings containing a Group S-1 fire area greater than 12,000 square feet (1115 m^2).

Step 4:

Part A: Based on the one-story height, any type of construction is acceptable for the occupancies under consideration.

Part B: Make a trial evaluation of the building assuming it to be Type IIB construction.

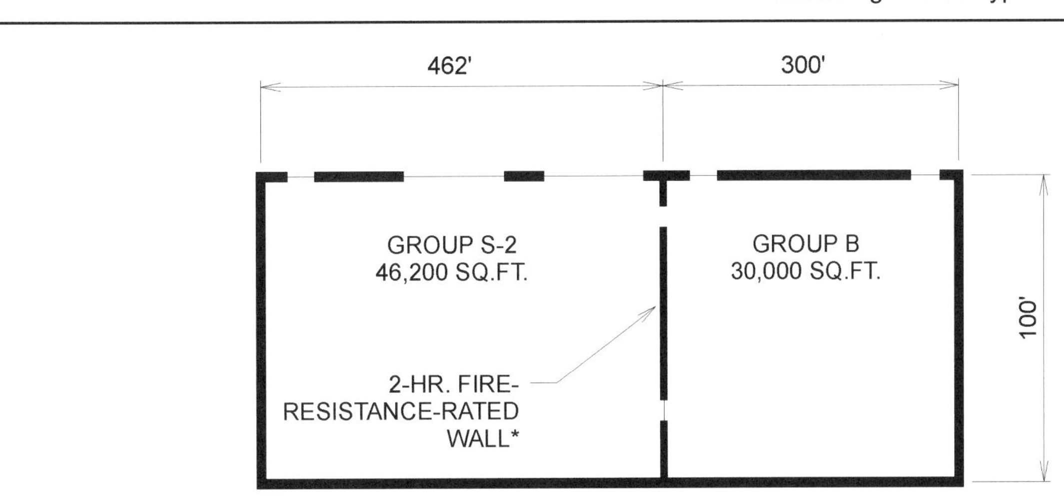

For SI: 1 foot = 304.8 mm, 1 square foot = 0.0929 m^2.

Figure 508.4.4.1(3)
SEPARATED OCCUPANCIES—EXAMPLE 2

Actual area of Group S-1 = 46,200 square feet (4292 m^2) and actual area of Group B = 30,000 square feet (2787 m^2).

Group S-1 maximum allowable area, = 70,000 square feet (6503 m^2).

The maximum allowable area for Group B including allowable increases = 92,000 square feet (8547 m^2). Using the sum of the ratios approach, 46,200/70,000 + 30,000/92,000 = 0.99 which does not exceed 1.0.

Note that, in this case, the separated use option requires no separation between these particular uses, and the Type IIB classification is acceptable. Comparing this to the example for nonseparated uses, with the same dimensions of each occupancy, the Type IIB construction classification is not acceptable, and a Type IIA construction classification is required.

Since Section 508.3 (nonseparated occupancies) makes a reference to applying the most restrictive requirements of Section 403 and Chapter 9, there may be the assumption that under the separated occupancies option, requirements can be applied separately to each occupancy in all cases. This is not completely true. There are sections of Chapter 9, such as Section 903.2.6, 903.2.8 or 903.2.9, that require sprinklers throughout the entire building when a particular occupancy is present. Some sections in Chapter 9, however, do permit separate application of requirements, even though the term "fire area" is not used. For example, a warehouse building with an office would be categorized as Groups B (office) and S-1 (warehouse). If the separated occupancies option is selected and the office has an occupant load of at least 500 people, then a fire alarm system is only required in the Group B area and not the Group S-1 area (see Section 907.2). It should be noted that alarm requirements must address the path of egress for occupants and may require notification appliances along those paths.

With respect to the high-rise provisions of Section 403, the requirements generally apply throughout the building and are not specific to fire areas. There are limited provisions in Section 403 that are specific to one occupancy or another. Regardless of the option selection in a mixed occupancy building, requirements in Section 403 that are based on occupancy will need to be applied throughout the building.

Another approach available to the designer of structures that contain more than one occupancy requires that the occupancies are completely separated by fire walls (see Section 706), thus creating separate buildings for each use. This is the simplest option to apply and analyze. When a fire wall is utilized, for code application purposes, two or more separate and independent buildings (see the definition of "Area, building" in Section 202) are created, and each building is reviewed individually.

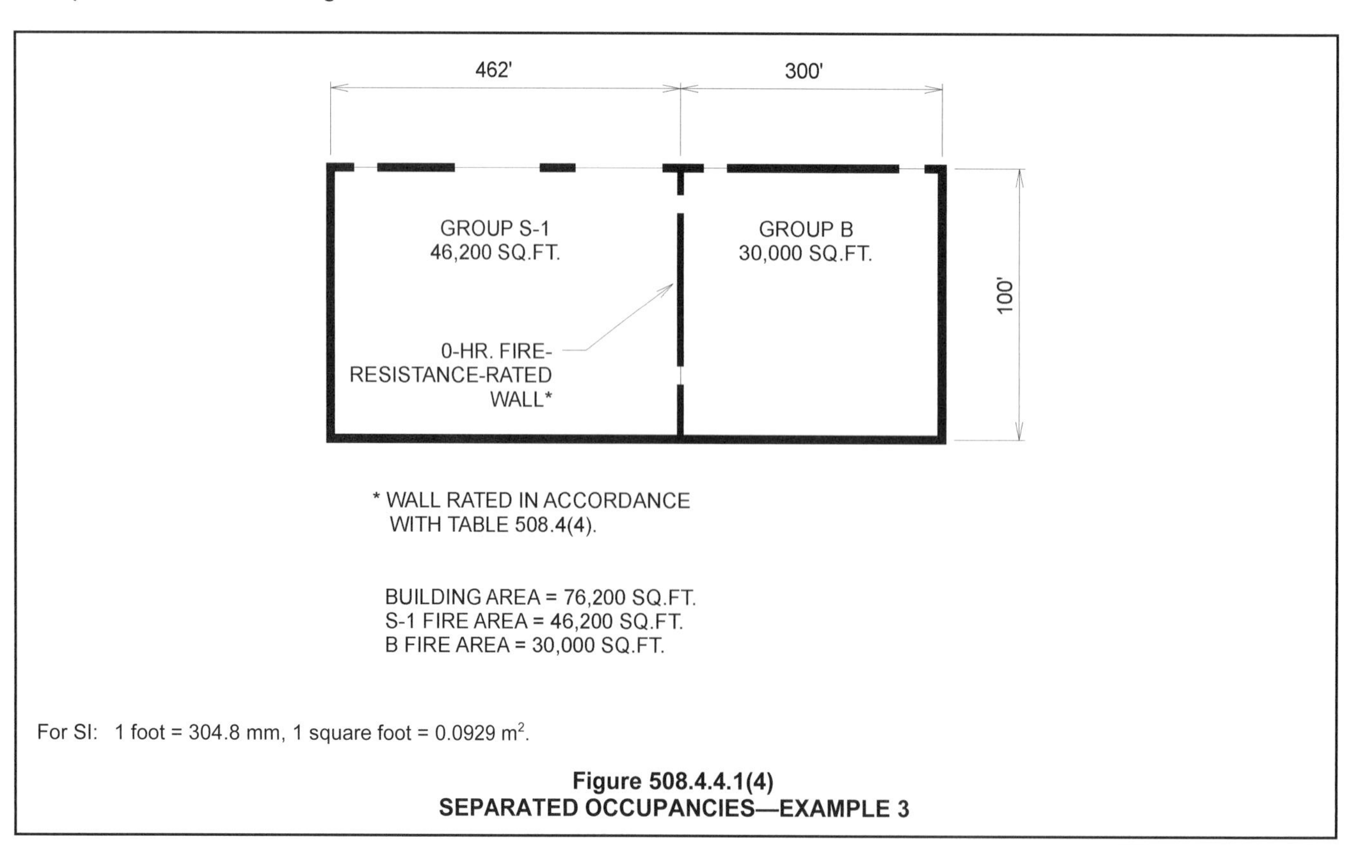

For SI: 1 foot = 304.8 mm, 1 square foot = 0.0929 m^2.

Figure 508.4.4.1(4)
SEPARATED OCCUPANCIES—EXAMPLE 3

SECTION 509
INCIDENTAL USES

509.1 General Incidental uses located within single occupancy or mixed occupancy buildings shall comply with the provisions of this section. Incidental uses are ancillary functions associated with a given occupancy that generally pose a greater level of risk to that occupancy and are limited to those uses listed in Table 509.

> **Exception:** Incidental uses within and serving a *dwelling unit* are not required to comply with this section.

❖ Incidental uses are rooms or areas that constitute special hazards or risks to life safety. Such spaces often pose risks that are not typically addressed by the provisions for the general occupancy groups under consideration. However, such rooms or areas may functionally be an extension of the primary use. Only those rooms or areas found in Table 509 are to be regulated as incidental uses. Incidental uses can be located within both single-occupancy and mixed-occupancy buildings. The concern is that those areas designated as incidental uses pose a risk to the remainder of the building, and as such, some degree of protection is required. The protection requirements, however, are not applicable to incidental uses that are located within and serve a dwelling unit. Incidental uses are not required to also comply with the accessory use provisions of Section 508.2.

TABLE 509. See page 5-46.

❖ Table 509 identifies incidental uses and the required separation or other protection to be provided. Where a fire-resistance-rated separation is required, the incidental use must be separated from other portions of the building with fire barriers that comply with Section 707 and/or horizontal assemblies complying with Section 711. Where Table 509 permits protection by an automatic sprinkler system without fire barriers, the walls enclosing the incidental use must comply with Section 509.4.2.

509.2 Occupancy classification. Incidental uses shall not be individually classified in accordance with Section 302.1. Incidental uses shall be included in the building occupancies within which they are located.

❖ This provision expressly states that incidental uses are not considered as separate and distinct occupancy classifications, but rather are classified the same as the building occupancies in which they are located. As an example, a 200-square-foot (18.5 m^2) laundry room in an apartment building would simply be classified as a part of the Group R-2 occupancy. Similarly, a waste and linen collection room in a nursing home would be classified as a portion of the Group I-2 occupancy.

509.3 Area limitations. Incidental uses shall not occupy more than 10 percent of the *building area* of the *story* in which they are located.

❖ The floor area limitation of 10 percent for incidental uses emphasizes the ancillary nature of such rooms and areas. Each incidental use is limited to a maximum of 10 percent of the floor area of the story in which it is located. Where there are two or more tenants located on the same story, the 10 percent limitation is presumably based upon the floor area of each individual tenant space rather than that of the entire story. The application of the limit on a tenant-by-tenant basis is consistent with the concept of incidental uses typically being ancillary only to a portion of the building, the specific tenant occupancy.

509.4 Separation and protection. The incidental uses listed in Table 509 shall be separated from the remainder of the building or equipped with an *automatic sprinkler system*, or both, in accordance with the provisions of that table.

❖ In addition to identifying those rooms or areas that warrant regulation as incidental uses, Table 509 also indicates the required degree of protection or separation. The requirements identified in Table 509 vary depending on the incidental use. In some cases, a specific type of separation and/or protection is required, while in others there is an option. As indicated by Section 509.4.2.1, the requirement for an automatic sprinkler system for the majority of those listed applies only to the incidental use room or area, not the entire building. If a building is required to be provided with a sprinkler system for other reasons, then any sprinkler requirement for an incidental use is also met.

509.4.1 Separation. Where Table 509 specifies a fire-resistance-rated separation, the incidental uses shall be separated from the remainder of the *building* by a *fire barrier* constructed in accordance with Section 707 or a *horizontal assembly* constructed in accordance with Section 711, or both. Construction supporting 1-hour *fire barriers* or *horizontal assemblies* used for incidental use separations in buildings of Type IIB, IIIB and VB construction is not required to be fire-resistance rated unless required by other sections of this code.

❖ Where a fire-resistance rated separation is required, the incidental use must be separated from other portions of the building with fire barriers that comply with Section 707, horizontal assemblies complying with Section 711, or both. Where Table 509 permits protection by an automatic sprinkler system without fire barriers, the construction enclosing the incidental use must resist the passage of smoke in accordance with Section 509.4.2. Where the construction surrounding an incidental use is only required to be 1 hour, and the building is of nonrated construction (Type IIB, IIIB or VB), the rated construction does not need to be supported by 1-hour fire-resistance-rated construction. In all other instances, the construction supporting incidental occupancy separations must be supported by construction with at least the same rating as the separations.

509.4.2 Protection. Where Table 509 permits an *automatic sprinkler system* without a *fire barrier*, the incidental uses shall be separated from the remainder of the building by construction capable of resisting the passage of smoke. The walls shall extend from the top of the foundation or floor assembly below to the underside of the ceiling that is a component of a fire-resistance-rated floor assembly or roof assembly above or to the underside of the floor or roof sheathing, deck or slab above. Doors shall be self- or automatic-closing upon detection of smoke in accordance with Section 716.5.9.3. Doors shall not have air transfer openings and shall not be undercut in excess of the clearance permitted in accordance with NFPA 80. Walls surrounding the incidental use shall not have air transfer openings unless provided with smoke dampers in accordance with Section 710.8.

❖ Where Table 509 permits protection by an automatic sprinkler system without fire barriers, the construction enclosing the incidental use must resist the passage of smoke. While this section can be viewed as a performance standard, construction details for resisting the passage of smoke are provided in this section. Although the section specifically states that air transfer openings must be provided with smoke dampers, it is silent with respect to ducts. If ducts are penetrating this separation, the arrangement of the duct system should be analyzed to determine if it will allow smoke to pass through the wall and not restrict it to the incidental use.

The wall construction described here is not required to be a smoke barrier conforming to Section 709 or a smoke partition conforming to Section 710.

509.4.2.1 Protection limitation. Where an *automatic sprinkler system* is provided in accordance with Table 509, only the space occupied by the incidental use need be equipped with such a system.

❖ The point of this section is that the sprinkler system stipulated in Table 509 is required for the incidental use only. In general, the nature of these incidental uses is such that they are small areas, not often frequented by building occupants, in which a fire could get underway and go unnoticed for a longer time than in a part of the building that is constantly occupied.

SECTION 510
SPECIAL PROVISIONS

510.1 General. The provisions in Sections 510.2 through 510.9 shall permit the use of special conditions that are exempt from, or modify, the specific requirements of this

TABLE 509
INCIDENTAL USES

ROOM OR AREA	SEPARATION AND/OR PROTECTION
Furnace room where any piece of equipment is over 400,000 Btu per hour input	1 hour or provide automatic sprinkler system
Rooms with boilers where the largest piece of equipment is over 15 psi and 10 horsepower	1 hour or provide automatic sprinkler system
Refrigerant machinery room	1 hour or provide automatic sprinkler system
Hydrogen fuel gas rooms, not classified as Group H	1 hour in Group B, F, M, S and U occupancies; 2 hours in Group A, E, I and R occupancies.
Incinerator rooms	2 hours and provide automatic sprinkler system
Paint shops, not classified as Group H, located in occupancies other than Group F	2 hours; or 1 hour and provide automatic sprinkler system
In Group E occupancies, laboratories and vocational shops not classified as Group H	1 hour or provide automatic sprinkler system
In Group I-2 occupancies, laboratories not classified as Group H	1 hour and provide automatic sprinkler system
In ambulatory care facilities, laboratories not classified as Group H	1 hour or provide automatic sprinkler system
Laundry rooms over 100 square feet	1 hour or provide automatic sprinkler system
In Group I-2, laundry rooms over 100 square feet	1 hour
Group I-3 cells and Group I-2 patient rooms equipped with padded surfaces	1 hour
In Group I-2, physical plant maintenance shops	1 hour
In ambulatory care facilities or Group I-2 occupancies, waste and linen collection rooms with containers that have an aggregate volume of 10 cubic feet or greater	1 hour
In other than ambulatory care facilities and Group I-2 occupancies, waste and linen collection rooms over 100 square feet	1 hour or provide automatic sprinkler system
In ambulatory care facilities or Group I-2 occupancies, storage rooms greater than 100 square feet	1 hour
Stationary storage battery systems having a liquid electrolyte capacity of more than 50 gallons for flooded lead-acid, nickel cadmium or VRLA, or more than 1,000 pounds for lithium-ion and lithium metal polymer used for facility standby power, emergency power or uninterruptable power supplies	1 hour in Group B, F, M, S and U occupancies; 2 hours in Group A, E, I and R occupancies.

For SI: 1 square foot = 0.0929 m^2, 1 pound per square inch (psi) = 6.9 kPa, 1 British thermal unit (Btu) per hour = 0.293 watts, 1 horsepower = 746 watts, 1 gallon = 3.785 L, 1 cubic foot = 0.0283 m^3.

chapter regarding the allowable *building heights and areas* of buildings based on the occupancy classification and type of construction, provided the special condition complies with the provisions specified in this section for such condition and other applicable requirements of this code. The provisions of Sections 510.2 through 510.8 are to be considered independent and separate from each other.

❖ The subsections of Section 510 are exceptions to the general height and area limitations of Chapter 5. Most of the subsections address attached parking structures and contain conditions by which these can be attached to buildings without creating hardship in regard to allowable area and height for the building. These allowances are not always related to buildings containing parking below but can involve other uses and occupancies. Also many of these scenarios will be applied such that there may be multiple buildings above the parking garage.

510.2 Horizontal building separation allowance. A building shall be considered as separate and distinct buildings for the purpose of determining area limitations, continuity of *fire walls*, limitation of number of *stories* and type of construction where all of the following conditions are met:

1. The buildings are separated with a *horizontal assembly* having a *fire-resistance rating* of not less than 3 hours.
2. The building below the *horizontal assembly* is of Type IA construction.
3. *Shaft*, *stairway*, *ramp* and escalator enclosures through the *horizontal assembly* shall have not less than a 2-hour *fire-resistance rating* with opening protectives in accordance with Section 716.5.

   Exception: Where the enclosure walls below the *horizontal assembly* have not less than a 3-hour *fire-resistance rating* with opening protectives in accordance with Section 716.5, the enclosure walls extending above the *horizontal assembly* shall be permitted to have a 1-hour *fire-resistance rating*, provided:

   1. The building above the *horizontal assembly* is not required to be of Type I construction;
   2. The enclosure connects fewer than four *stories*; and
   3. The enclosure opening protectives above the *horizontal assembly* have a *fire protection rating* of not less than 1 hour.
4. The building or buildings above the *horizontal assembly* shall be permitted to have multiple Group A occupancy uses, each with an *occupant load* of less 300, or Group B, M, R or S occupancies.
5. The building below the *horizontal assembly* shall be protected throughout by an *approved automatic sprinkler system* in accordance with Section 903.3.1.1, and shall be permitted to be any occupancy allowed by this code except Group H.
6. The maximum *building height* in feet (mm) shall not exceed the limits set forth in Section 504.3 for the building having the smaller allowable height as measured from the *grade plane*.

❖ Section 510.2 essentially allows a 3-hour fire-resistance-rated horizontal assembly to create separate buildings similar to the concept used for fire walls (see Commentary Figure 510.2). However, consideration as separated buildings by this method is only applicable to a limited number of specified conditions. This allowance provides an extensive benefit for height and area in these structures. Buildings constructed under this section are frequently referred to as "pedestal," "podium" or "platform" buildings. It should be noted that multiple buildings may be located above the horizontal assembly. Structures built under this section are considered to be distinct buildings above and below the 3-hour fire-resistance-rated horizontal assembly. As distinct buildings, they are individually evaluated with respect to allowable building area, the number of stories and the type of construction. In addition, if a fire wall is needed to address building area issues in the upper building or buildings, the fire wall construction can stop at the 3-hour fire-resistance-rated horizontal assembly and does not need to extend to the foundation. However, other building systems and requirements must be evaluated using the total structure. For example, if the upper building is apartments and the lower building an open parking garage, both buildings will need to be protected by an automatic sprinkler system because of the requirement for buildings occupied by Group R occupancies (see Section 903.2.8).

There are six conditions that set the limits of this design:

1. Separation of upper and lower buildings by a 3-hour fire-resistance-rated horizontal assembly.
2. The building below the horizontal assembly is of Type IA construction. As such it is allowed to be of unlimited area in accordance with Section 506.
3. All openings through the 3-hour fire-resistance-rated horizontal assembly are to be protected by a 2-hour shaft (the shaft openings having $1^1/_2$-hour protectives, see Table 716.5).
4. The uses in the upper building or buildings are limited to Group A occupancy uses where each Group A area has an occupant load of less than 300, or Group B, M, R or S occupancies. The limit of 300 occupants in Group A occupancies is also reflective of the history of these provisions from the legacy codes, and the intent of restricting the assembly spaces to smaller businesses so that the assembly use does not dominate the upper building. This same limit for assembly is not applied to the lower building.
5. The building below is allowed to be any occupancy or any combination of occupancies except Group H, provided that the lower

building is equipped throughout with an NFPA 13 sprinkler system. While historic versions of Section 510.2 limited the use of the lower building to parking, the provision has evolved to reflect a mixed occupancy setting of urban neighborhoods.

6. The height of the combined buildings above and below the horizontal assembly is limited to the number of feet above grade plane allowed by Section 504.3 for the type of construction of the upper building. However, the charging language of Section 510.2 does not restrict the number of stories to the entire structure but only to that which is above the horizontal assembly. Thus, a Type VA building, protected by a NFPA 13R sprinkler system and containing a Group R-2 occupancy, can have four stories above the horizontal assembly, provided the overall height of both buildings does not exceed 60 feet (18 288 mm) above grade plane.

This is one of the rare circumstances where there could be two different construction types in a single structure without being separated by a fire wall. It is possible that following the conventional provisions for mixed occupancies in Sections 508.3 and 508.4 would be less restrictive than the conditions of this section. Compliance with the general provisions of Section 508.3 or 508.4 is permissible, and this section should be viewed as an alternative means of compliance with Section 508.3 or 508.4.

There is one exception within the six items: the exception to Item 3 indicates the conditions by which the shaft construction protecting openings in the horizontal separation may be 1-hour rated above the horizontal separation. This allowance is only for openings connecting a maximum of four stories. This would include the connection to the area below the horizontal assembly (see Commentary Figure 510.2).

A common example of a building constructed under the provisions of Section 510.2 is a Type IA building that contains parking at and below grade level, with up to four stories of Group R-2 apartments in a Type VA, wood-frame constructed building above the separation (see Commentary Figure 510.2). In many city neighborhoods where zoning laws encourage a mixture of uses, the first story is often occupied with retail shops, service businesses and small restaurants. The height of this structure measured in feet is limited by the Type VA upper structure to a maximum of 70 feet (21 336 mm) above grade plane. But the number of stories of the Type VA portion of the building is determined by starting at the horizontal separation between the construction types. See also Section 510.9 for where there are multiple buildings above the horizontal separation.

510.3 Group S-2 enclosed parking garage with Group S-2 open parking garage above. A Group S-2 enclosed parking garage with not more than one *story* above *grade plane* and located below a Group S-2 *open parking garage* shall be classified as a separate and distinct building for the purpose of determining the type of construction where all of the following conditions are met:

1. The allowable area of the building shall be such that the sum of the ratios of the actual area divided by the allowable area for each separate occupancy shall not exceed 1.
2. The Group S-2 enclosed parking garage is of Type I or II construction and is at least equal to the *fire-resistance* requirements of the Group S-2 *open parking garage*.

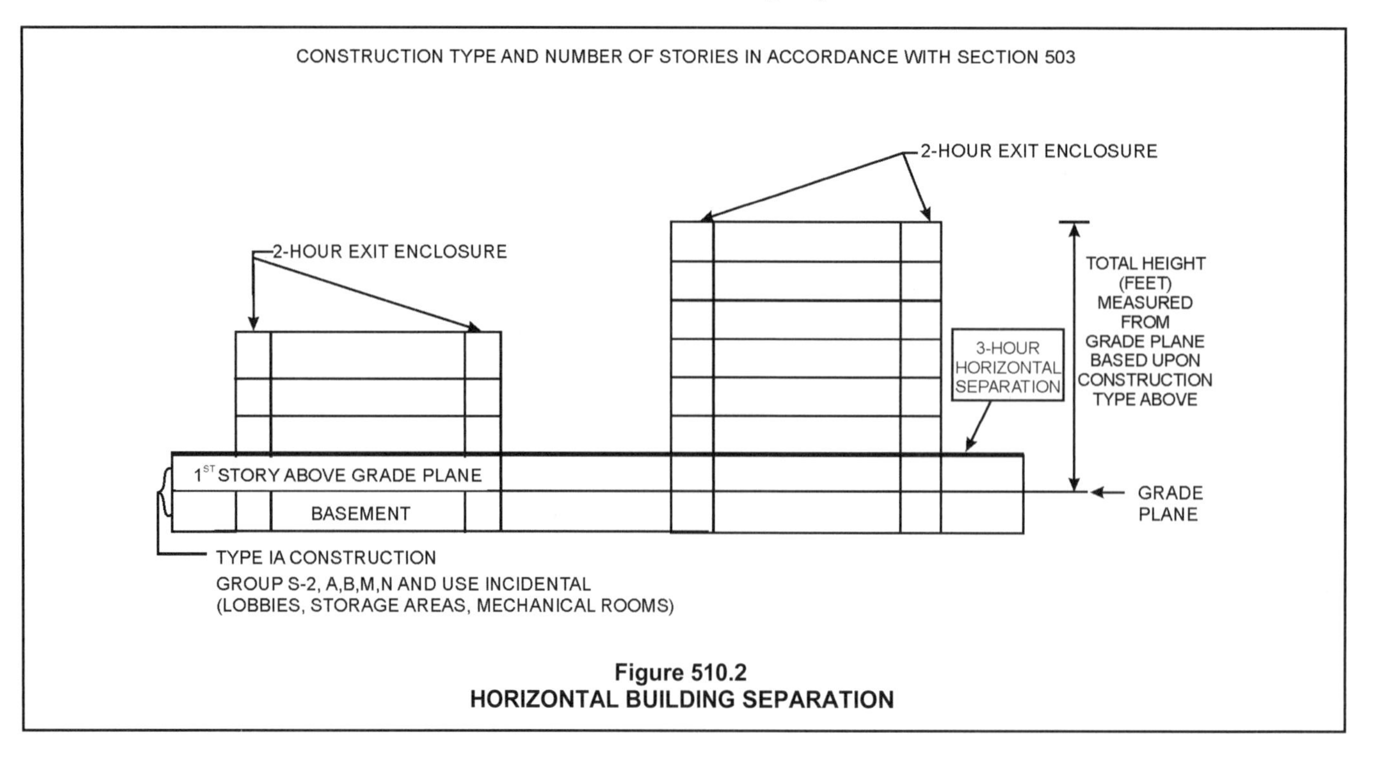

Figure 510.2
HORIZONTAL BUILDING SEPARATION

3. The height and the number of tiers of the Group S-2 *open parking garage* shall be limited as specified in Table 406.5.4.
4. The floor assembly separating the Group S-2 enclosed parking garage and Group S-2 *open parking garage* shall be protected as required for the floor assembly of the Group S-2 enclosed parking garage. Openings between the Group S-2 enclosed parking garage and Group S-2 *open parking garage*, except *exit* openings, shall not be required to be protected.
5. The Group S-2 enclosed parking garage is used exclusively for the parking or storage of private motor vehicles, but shall be permitted to contain an office, waiting room and toilet room having a total area of not more than 1,000 square feet (93 m^2) and mechanical equipment rooms incidental to the operation of the building.

❖ Parking garages of both types, enclosed and open, are addressed in Section 406. Special height and area allowances are given in Section 406.5.4 for open parking structures. However, these special height and area provisions are not applicable if any level of the parking garage does not meet the definition of "Open parking garage" by not having the requisite clear open area to the exterior. This would normally preclude having parking levels below grade in an open parking garage.

Section 510.3 contains provisions that would allow an open parking structure to take advantage of the special height and area limits for open parking structures in Section 406.5.4 while incorporating, in the same building, enclosed parking levels below grade. This is an alternative to treating the whole building as an enclosed parking garage.

There are five conditions listed that must be met in order to use this alternative:

1. Appropriate increases in accordance with Section 506 are to be considered for each portion, and for purposes of frontage increase the same measurement of open perimeter may apply to both the upper and lower garages.
2. The enclosed parking structure below must meet or exceed the fire resistance of the open parking structure above. At a minimum, the enclosed parking structure must be of Type IIB construction (all noncombustible materials, without fire-resistance-rated protection). However, if the open parking structure above is Type VA, IIIA, IIA or IA, the enclosed parking structure would also have to meet or exceed the required ratings of the building elements from Table 601 for the construction type of the open parking structure.
3. Both the entire height of the building and the number of tiers are limited by Table 406.5.4. Therefore, even if the first story above grade is part of the enclosed parking garage portion of the building, the height of the structure above grade plane could not exceed what would be permitted by Table 406.5.4 if the entire structure above grade plane were part of the open parking garage portion.
4. Protection of the floor assembly of the Group S-2 building depends on the construction type and Table 601 for the S-2 enclosed parking garage.
5. Certain accessory uses are permitted to be present when taking advantage of these alternative provisions, such as a small office and waiting area.

See also Section 510.9 for where there are multiple buildings above the horizontal separation.

510.4 Parking beneath Group R. Where a maximum one *story above grade plane* Group S-2 parking garage, enclosed or open, or combination thereof, of Type I construction or open of Type IV construction, with grade entrance, is provided under a building of Group R, the number of *stories* to be used in determining the minimum type of construction shall be measured from the floor above such a parking area. The floor assembly between the parking garage and the Group R above shall comply with the type of construction required for the parking garage and shall also provide a *fire-resistance rating* not less than the mixed occupancy separation required in Section 508.4.

❖ This section permits an extra story (above the limits of Table 504.4), based on construction type, for Group R buildings with parking on the first story (see Commentary Figure 510.4). There are several conditions that must be met: the parking must be limited to one story above grade; must be Type IV (if open) or I (open or enclosed) construction; the entrance to the garage must be at grade; and a 1-hour horizontal assembly must be provided between the parking and the Group R occupancy in accordance with Table 508.4. The limitation of Table 504.3 for height above grade plane (in feet) is not changed under this circumstance. It should be noted that all buildings containing Group R occupancies are required to be sprinklered in the code. Therefore, the separation required would only be 1 hour, but the parking area must also be sprinklered.

For instance, a fully sprinklered Group R-2 building of Type VB construction would normally be permitted to be three stories above grade (Table 504.4 for R-2, "S"). If it meets the conditions of this section for parking on the first story above grade, then it could actually be four stories above grade. However, the building height in feet cannot exceed 60 feet (18 288 mm) above grade plane in accordance with Table 504.3 for Type VB construction. Commentary Figure 510.4 provides an example of a Group R-1 building over enclosed parking.

510.5 Group R-1 and R-2 buildings of Type IIIA construction. The height limitation for buildings of Type IIIA construction in Groups R-1 and R-2 shall be increased to six *stories* and 75 feet (22 860 mm) where the first floor assembly above the *basement* has a *fire-resistance rating* of not less than 3 hours and the floor area is subdivided by 2-hour fire-resistance-rated *fire walls* into areas of not more than 3,000 square feet (279 m^2).

❖ This section contains special provisions for increasing the height of Type IIIA Group R-1 and R-2 buildings based upon increases in fire resistance and compartmentation. More specifically, the higher rating would apply to the floor structure of the first story above grade (3 hours), and the fire walls (2 hours) subdividing the building into floor areas of not more than 3,000 square feet (279 m^2). The fire walls must extend from foundation to roof. In addition, the fire walls must comply with the requirements in Section 706.

It should be noted that this section is independent of the fire area requirements of Section 901.7, but such separations could certainly be considered as creating fire areas or separate buildings. In either case, all buildings containing a Group R-1 or R-2 fire area would require sprinklers in accordance with Section 903.2.8.

This section constitutes a trade-off of building area for extra height, and depends on the materials and rating requirements for Type IIIA construction, as well as the extra fire resistance of the first story.

See Commentary Figure 510.5 for an example of a Type IIIA building height increase for a Group R-2 occupancy.

510.6 Group R-1 and R-2 buildings of Type IIA construction. The height limitation for buildings of Type IIA construction in Groups R-1 and R-2 shall be increased to nine *stories* and 100 feet (30 480 mm) where the building is separated by not less than 50 feet (15 240 mm) from any other building on the *lot* and from *lot lines*, the *exits* are segregated in an area enclosed by a 2-hour fire-resistance-rated *fire wall* and the first floor assembly has a *fire-resistance rating* of not less than $1^1/_2$ hours.

❖ This section contains special provisions for increasing the height of Type IIA, Group R-1 and R-2 buildings. The higher $1^1/_2$-hour rating would apply to the floor assembly of the first story above grade, and the exits are required to be enclosed with fire walls extending from foundation to roof, otherwise meeting all the requirements of Section 706. Such fire walls would create separate fire areas or buildings for the purposes of the application of the rest of the code.

A separation distance of at least 50 feet (15 240 mm) from other buildings on the same lot or property lines is also required to use this alternative. This section constitutes a tradeoff of extra protection for exits for extra building height, and depends also on the materials and rating requirements for Type IIA construction, as well as the extra fire-resistance rating protecting the first story from the basement. The entire building must be sprinklered in accordance with Section 903.2.8 with an NFPA 13 sprinkler system. See Commentary Figure 510.6 for an example of a Type IIA building height increase for a Group R-2 occupancy.

Note that such buildings will likely be considered high-rise buildings and, as such, also must comply with Section 403.

510.7 Open parking garage beneath Groups A, I, B, M and R. *Open parking garages* constructed under Groups A, I, B, M and R shall not exceed the height and area limitations

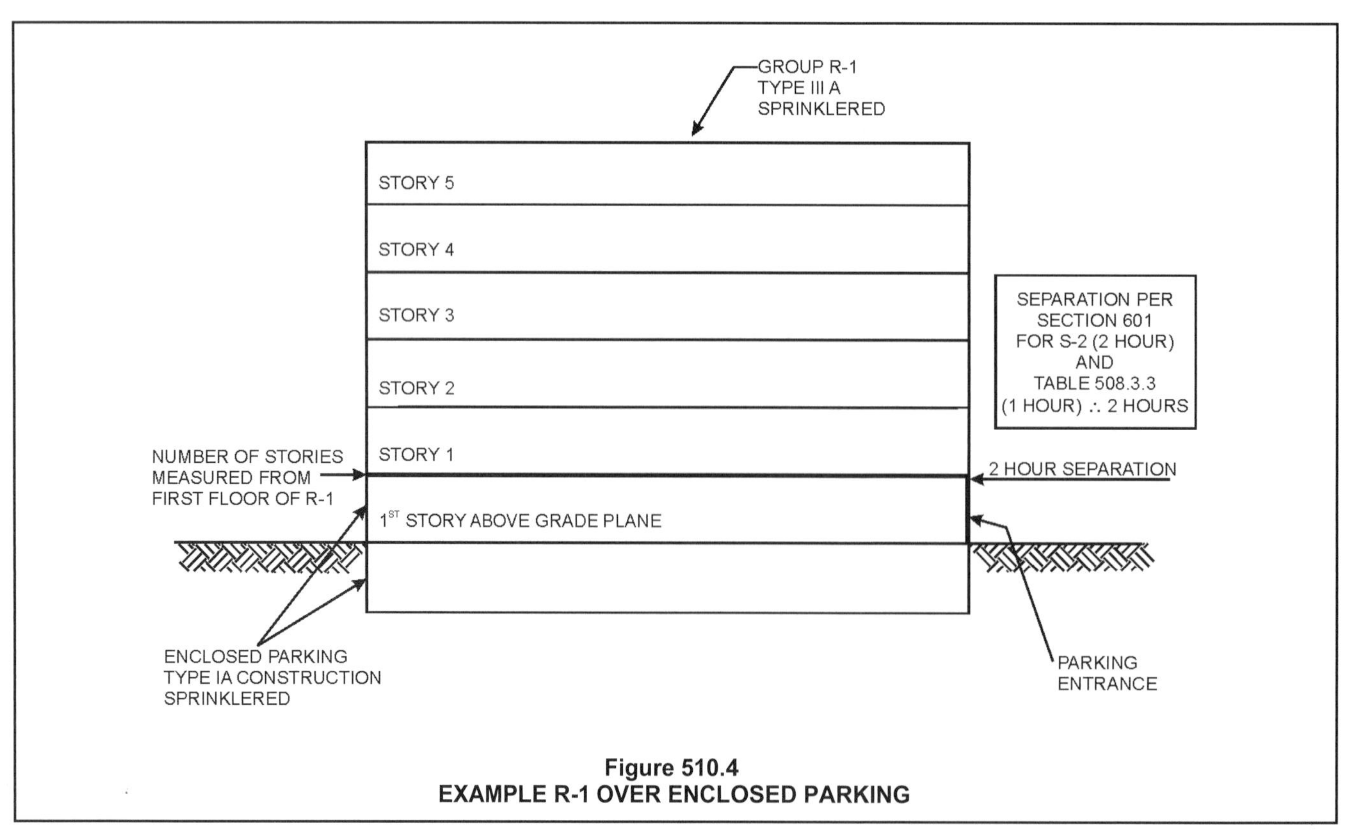

Figure 510.4
EXAMPLE R-1 OVER ENCLOSED PARKING

permitted under Section 406.5. The height and area of the portion of the building above the *open parking garage* shall not exceed the limitations in Section 503 for the upper occupancy. The height, in both feet and *stories*, of the portion of the building above the *open parking garage* shall be measured from *grade plane* and shall include both the *open parking garage* and the portion of the building above the parking garage.

❖ This section addresses a special mixed-use condition, and is another circumstance wherein a building can be designated with two different types of construction for determining height and area. This provision is only to be applied when an open parking structure (Group S-2) is to be constructed below a Group A, I, B, M or R occupancy (see Commentary Figure 510.7). If an open parking structure is located below any other occupancy group, Section 508 must be applied, as for any other mixed-use condition. In accordance with Section 510.1, this section is an alternative to the general mixed-occupancy provisions in Section 508 that can be applied where advantageous to the design of buildings with open parking on the lower levels.

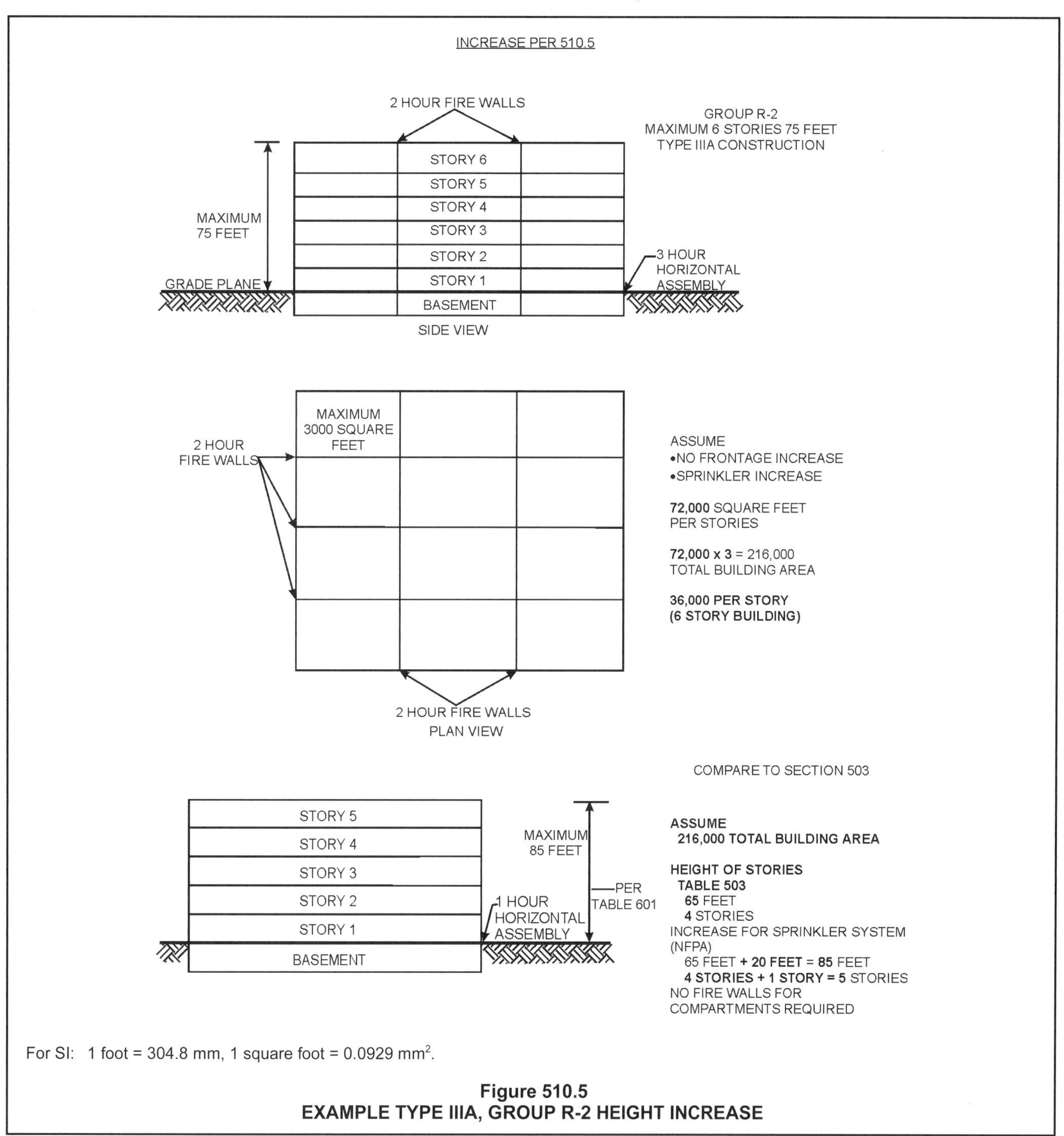

For SI: 1 foot = 304.8 mm, 1 square foot = 0.0929 mm^2.

Figure 510.5
EXAMPLE TYPE IIIA, GROUP R-2 HEIGHT INCREASE

In the application of Section 510.7, there are two criteria that must be met for code compliance (see Commentary Figure 510.7):

1. The height (measured both in feet and stories) and area of the open parking structure comprising a part of a mixed-use group building must not exceed the limitations for open parking structures permitted in Section 406.5.4 and Table 406.5.4.
2. The allowable height of the occupancy located above the open parking structure is to be determined in accordance with Section 504. The height is the vertical distance (measured both in feet and stories) from the grade plane to the top of the average height of the highest roof surface, in accordance with the definition of "Height, building" in Section 202.

510.7.1 Fire separation. *Fire barriers* constructed in accordance with Section 707 or *horizontal assemblies* constructed in accordance with Section 711 between the parking occupancy and the upper occupancy shall correspond to the required *fire-resistance rating* prescribed in Table 508.4 for the uses involved. The type of construction shall apply to each occupancy individually, except that structural members, including main bracing within the open parking structure, which is necessary to support the upper occupancy, shall be protected with the more restrictive fire-resistance-rated assemblies of the groups involved as shown in Table 601. *Means of egress* for the upper occupancy shall conform to Chapter 10 and shall be separated from the parking occupancy by *fire barriers* having not less than a 2-hour *fire-resistance rating* as required by Section 707 with *self-closing* doors complying with Section 716 or *horizontal assemblies* having not less than a 2-hour *fire-resistance rating* as required by Section 711, with *self-closing* doors complying

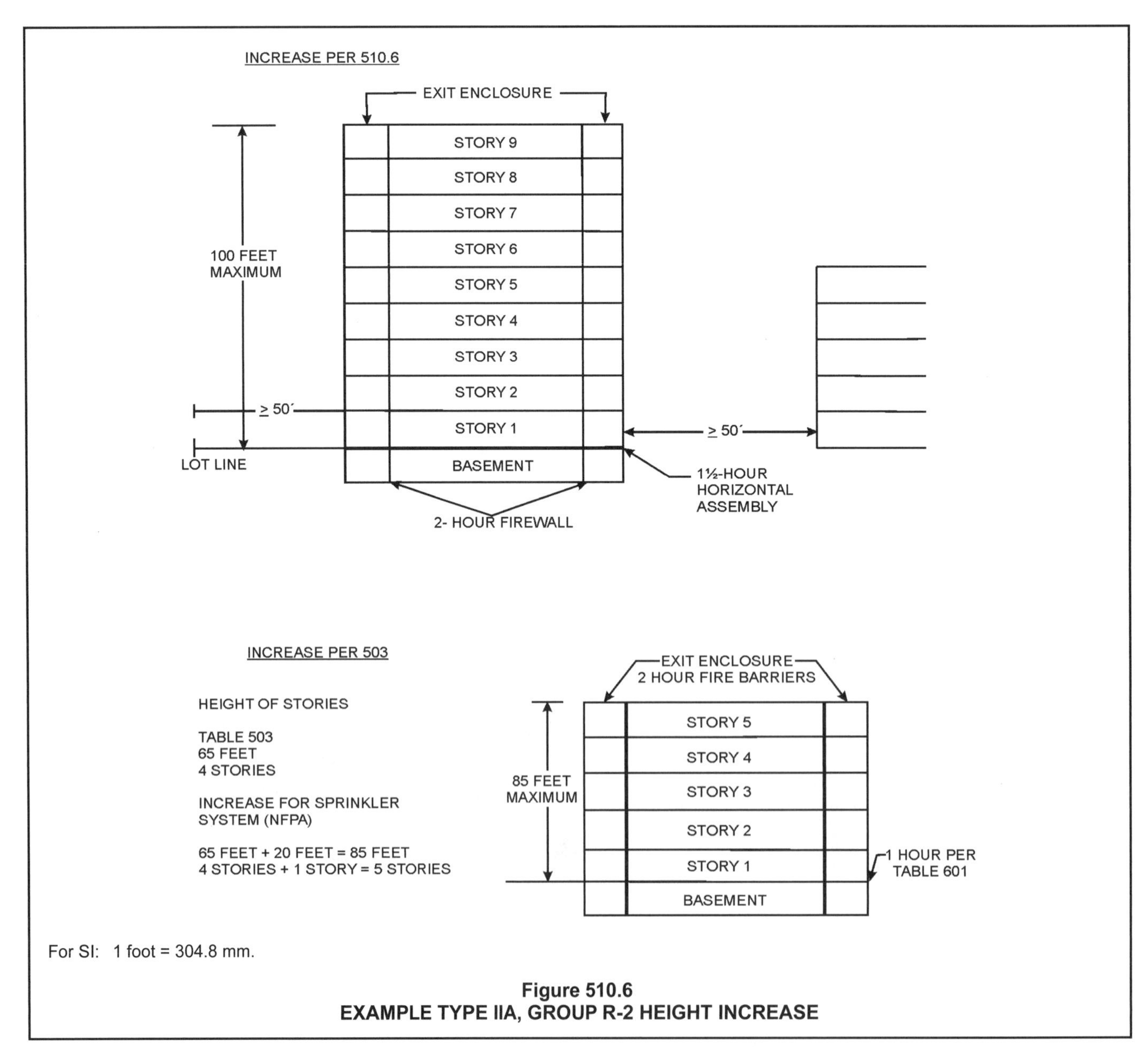

Figure 510.6
EXAMPLE TYPE IIA, GROUP R-2 HEIGHT INCREASE

with Section 716. *Means of egress* from the *open parking garage* shall comply with Section 406.5.

❖ This section contains additional conditions for the use of Section 510.7 as an alternative to the general mixed-use provisions in Section 508. It contains an additional five criteria:

1. The open parking structure and occupancy located above must be separated, both horizontally and vertically if necessary, by fire separation assemblies having a fire-resistance rating corresponding to that specified in Table 508.4.

   Example: An open parking structure, Group S-2, is located below an office (Group B); therefore, based on Table 508.4, all vertical and horizontal assemblies separating the two groups are required to have a minimum fire-resistance rating of 2 hours in a nonsprinklered building and 1 hour in a sprinklered building.

2. The upper and lower portions of the building may be of different types of construction (except as discussed in Item 5 below). The minimum type of construction for an open parking structure is Type IIB or IV, depending on the thresholds established in Table 406.5.4.

   Example: The open parking structure may be of Type IIB construction and the offices located above the open parking structure may be of Type IB construction.

3. Regardless of the construction types being used, all structural members, including main bracing in the open parking structure for the stability of the upper occupancy, must be rated in accordance with the most restrictive fire-resistance-rating requirement in accordance with Table 601, which would be that for Type IB construction.

   Example: Consider a building where the upper occupancy is of Type IB protected construction, and the open parking structure is of Type IIB unprotected construction. In accordance with Table 601, all load-bearing walls and structural frames, including columns and girders necessary to support the upper occupancy, must have at least a 2-hour fire-resistance rating, in accordance with the requirements of Type IB construction. This typically applies to the columns and bracing in the entire structure, beams supporting the floor separating the upper occupancy from the open parking structure and any transfer beams located in the open parking structure. The story of the open parking structure and any members supporting these stories, however, are permitted to have a zero fire-resistance rating in accordance with the requirements for Type IIB construction.

4. Means of egress facilities within and from the upper occupancy are to conform to Chapter 10. In addition, the egress facilities from the upper occupancy must be separated from the parking area by fire-resistance-rated wall assemblies of at least 2 hours, by fire barriers meeting the requirements of Section 707 or horizontal assemblies meeting Section 711. These egress

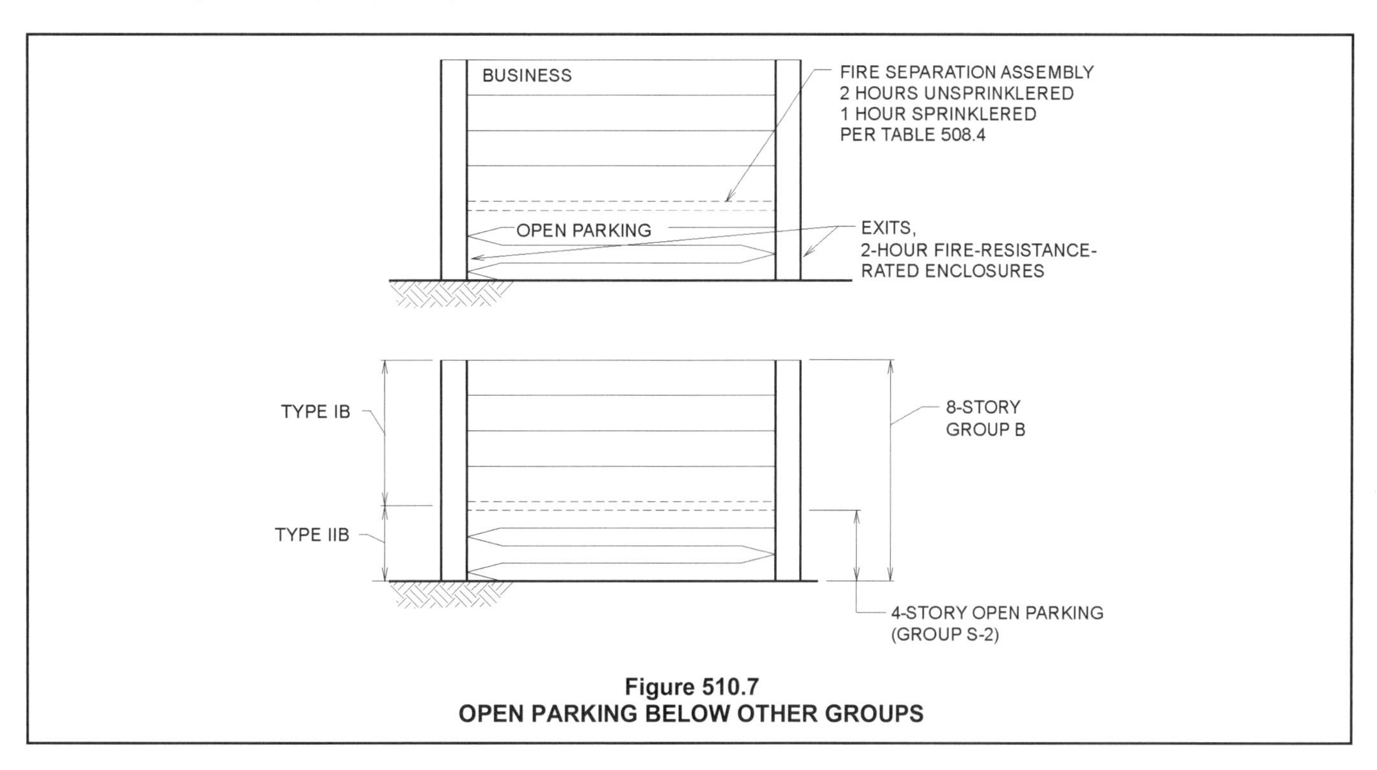

Figure 510.7
OPEN PARKING BELOW OTHER GROUPS

facilities are required to maintain the 2-hour protection for their full height and must be continuous to the level of exit discharge. The fire-resistance rating reduction of interior exit stairways for structures less than four stories in height (see Section 1023.2) is not applicable to an exit passing through the open parking structure in accordance with this section. Door openings to the interior exit stairway must have self-closing doors that comply with Section 716, with a fire protection rating of $1^1/_2$ hours in accordance with Table 716.5.

5. Means of egress facilities within and from the open parking structure are to conform to Section 406.5.7, which also references Chapter 10.

510.8 Group B or M buildings with Group S-2 open parking garage above. Group B or M occupancies located below a Group S-2 open parking garage of a lesser type of construction shall be considered as a separate and distinct building from the Group S-2 open parking garage for the purpose of determining the type of construction where all of the following conditions are met:

1. The buildings are separated with a *horizontal assembly* having a *fire-resistance rating* of not less than 2 hours.
2. The occupancies in the building below the *horizontal assembly* are limited to Groups B and M.
3. The occupancy above the *horizontal assembly* is limited to a Group S-2 *open parking garage*.
4. The building below the horizontal assembly is of Type IA construction.

   Exception: The building below the *horizontal assembly* shall be permitted to be of Type IB or II construction, but not less than the type of construction required for the Group S-2 *open parking garage* above, where the building below is not greater than one story in height above grade plane.
5. The height and area of the building below the *horizontal assembly* does not exceed the limits set forth in Section 503.
6. The height and area of the Group S-2 *open parking garage* does not exceed the limits set forth in Section 406.5. The height, in both feet and *stories*, of the Group S-2 *open parking garage* shall be measured from *grade plane* and shall include the building below the *horizontal assembly*.
7. *Exits* serving the Group S-2 *open parking garage* discharge directly to a street or *public way* and are separated from the building below the *horizontal assembly* by 2-hour *fire barriers* constructed in accordance with Section 707 or 2-hour *horizontal assemblies* constructed in accordance with Section 711, or both.

❖ This section addresses the inverse of the other circumstances in Section 510: the parking, which must be an open parking garage, is located above the other groups, in this case, Group B or M. This is a common type of construction in metropolitan areas. The parking garage is an open parking structure, with developers using the street level part of the building as an opportunity to provide retail space or other commercial space in a downtown application. The conditions under which this configuration of parking garage and Group B or M uses is similar to conditions for other circumstances is found in Section 510.

More specifically, there are seven criteria that must be met.

1. Similar to Item 1 of Section 510.2, a horizontal separation is required to essentially divide the buildings into separate buildings. Again it is like having a horizontal fire wall. The separation is only required to be 2 hours, which is less than the 3 hours required by Item 1 of Section 510.2.
2. The occupancies below the separation are limited to Groups B and M.
3. The occupancy above the horizontal assembly is limited to a Group S-2 open parking garage.
4. The building below is required to be of Type IA construction. The exception allows for Type IB or II construction if the building below is only one story; provided the construction type must always be at least that of the Group S-2 above.
5. The Group B and M occupancies must comply with the height (measured in feet and stories) and area requirements of Section 503. As written this would allow increases in area and height as appropriate.
6. The height (measured in tiers) and area of the Group S-2 occupancy shall not exceed that of Section 406.5. In addition, the overall height (measured feet and stories) is further restricted by the fact that the occupancies below the horizontal assembly must be included.
7. Lastly, similar to Section 510.7.1, the exits from the Group S-2 must be protected within 2-hour fire barriers or horizontal assemblies and extend directly to a street or public way.

See Commentary Figure 510.8 for an application of this section. See also Section 510.9 for where there are multiple buildings above the horizontal separation.

510.9 Multiple buildings above a horizontal assembly. Where two or more buildings are provided above the *horizontal assembly* separating a Group S-2 parking garage or building below from the buildings above in accordance with the special provisions in Section 510.2, 510.3 or 510.8, the buildings above the *horizontal assembly* shall be regarded as separate and distinct buildings from each other and shall comply with all other provisions of this code as applicable to each separate and distinct building.

❖ A very common practice when applying alternatives such as those found in Sections 510.2 and 510.3 is to have multiple buildings above the horizontal assemblies. Specifically, the concept in Section 510.2 is often called a "pedestal" building. In other words, the building below and including the horizontal assembly creates a "pedestal" on top of which many buildings, which are considered separate from one another, can be located. This section is simply clarifying that the multiple buildings located on top of the pedestal are considered separate from one another. This can be realized by there being multiple and distinct structures separated by yards or courts between different structures, or can be the result of the use of fire walls dividing a single structure into multiple buildings.

Similar to distinct buildings sitting on the ground, multiple buildings above the horizontal assemblies can be of different construction types from the pedestal building and of different construction types from each other (see Commentary Figure 510.9). This allowance is also applicable where Section 510.8 is utilized.

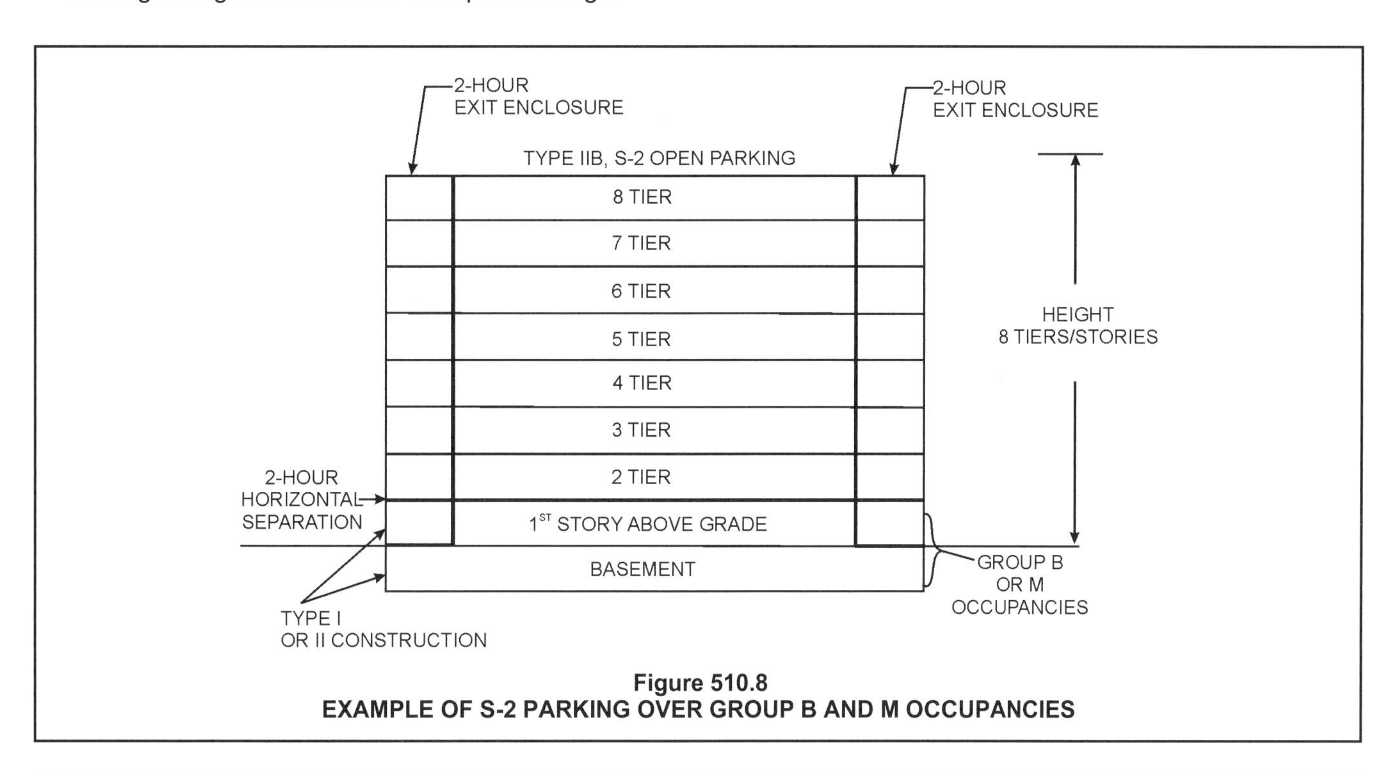

Figure 510.8
EXAMPLE OF S-2 PARKING OVER GROUP B AND M OCCUPANCIES

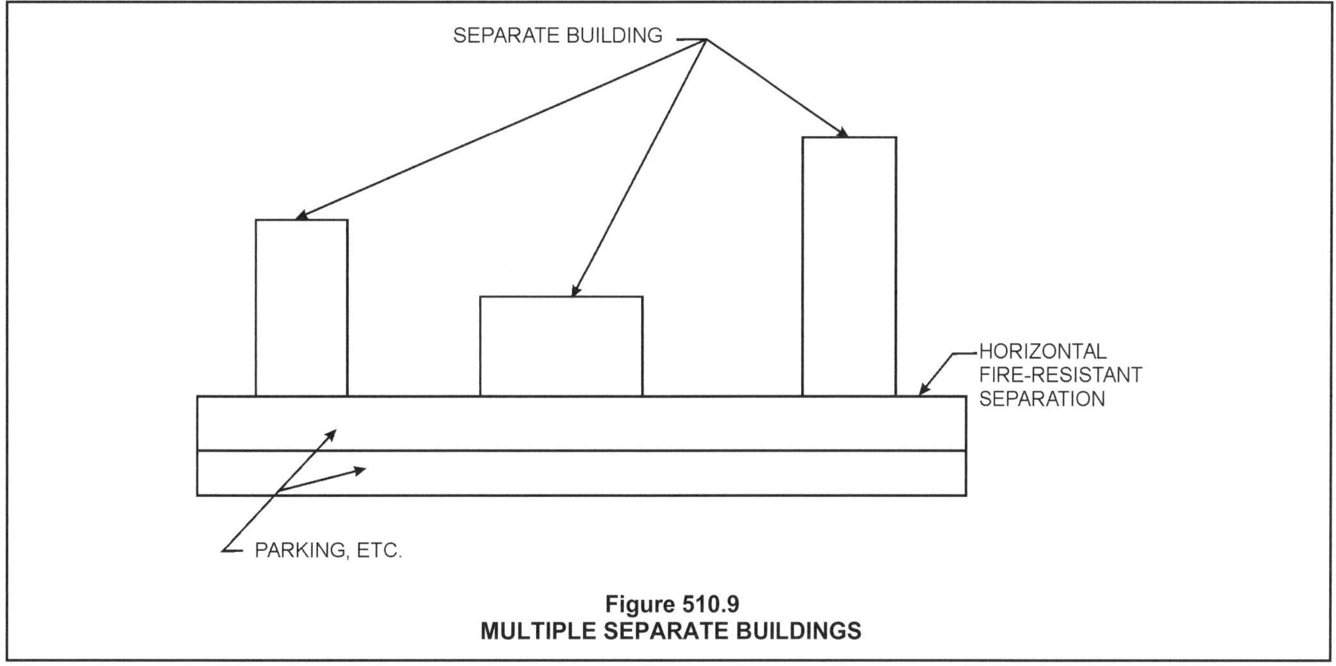

Figure 510.9
MULTIPLE SEPARATE BUILDINGS

Chapter 6: Types of Construction

General Comments

Chapter 6 contains the requirements to classify buildings into one of five types of construction. Tables 601 and 602 provide the minimum hourly fire-resistance ratings for the structural elements based on the type of construction of the building and fire separation distance. Section 602 describes each construction type in detail. Section 603 describes the permitted use of combustible materials in buildings of noncombustible construction.

Correct classification of a building by its type of construction is essential. Many code requirements applicable to a building, such as allowable height and area (see Chapter 5), are dependent on its type of construction. If a building is placed in an incorrect construction classification (for example, one that is overly restrictive), its owner may be penalized by increased construction costs. On the other hand, when a building is incorrectly classified in an overly lenient type of construction, it will not be constructed in a manner that takes into account the relative risks associated with its size or function. The provisions of this chapter, coupled with Chapters 3 and 5 and Tables 601 and 602, establish the basis for the "equivalent risk theory" on which the entire code is based.

Purpose

The purpose of classifying buildings or structures by their type of construction is to account for the response or participation that a building's structure will have in a fire condition originating within the building as a result of its occupancy or fuel load.

The code requires every building to be classified as one of five possible types of construction: Type I, II, III, IV or V. Each type of construction denotes the kinds of materials that are permitted to be used [i.e., noncombustible steel, concrete, masonry, combustible (wood, plastic) or heavy timber (HT)], and the minimum fire-resistance ratings that are associated with the structural elements in a building having that classification (i.e., 0, 1, $1^1/_2$, 2 or 3 hours). Type I and II construction consists of building elements that are noncombustible. Buildings of Type III construction have noncombustible exterior walls and combustible or noncombustible interior elements. Type IV construction identifies those structures with noncombustible exterior walls and heavy-timber interior elements. Type V buildings are permitted to have building elements that are either combustible or noncombustible, or a combination of both. Types I, II, III and V construction are further subdivided into two categories (IA and IB, IIA and IIB, IIIA and IIIB, VA and VB) that identify differences in the degree of fire resistance required.

SECTION 601 GENERAL

601.1 Scope. The provisions of this chapter shall control the classification of buildings as to type of construction.

❖ This section requires that all buildings be assigned a type of construction classification as indicated in the "General Comments" above.

TABLE 601. See page 6-2.

❖ Table 601 has three components: the top rows list the five general types of construction and their subclassifications; the left column lists the various building elements that are regulated by the table; and each cell in the table contains the minimum required fire-resistance rating, in hours, for the various elements based on the required type of construction of the building. "Building element" is defined in Section 202 as a fundamental component of building construction which needs to be constructed of the materials and have the fire-resistance rating specified in this chapter in order for a building to be classified in a type of construction. Notes a through f apply as specifically referenced in the table.

Types I, II, III and V construction are further subdivided into two categories (A and B). Type A and B construction are not defined in the code. The designations simply refer to the hourly fire-resistance rating required for the structural elements. A Type A designation will have a higher fire-resistance rating for the structural element than a Type B designation. In other than Type I construction, Types A and B are often referred to as protected and unprotected construction, respectively. Please note this terminology does not refer to whether the building is protected by an installed automatic sprinkler system.

The following describes the items in the left column of the table, titled "Building Element."

Row 1: *Primary structural frame.* This category includes the structural (load-bearing) components of the building frame. Definitions of "Primary structural frame" and "Secondary members" are found in Section 202. Any structural item that provides direct connections to columns and bracing members that are designed to carry a gravity load is considered part of the structural frame. To delay vertical (i.e., gravity) load-carrying collapse of a building due to fire exposure for a theoretical amount of time, the components that make up the primary structural frame are required to maintain a minimum degree of fire resistance. The components defined as part of the primary structural frame, with the exception of Type IV construction, must also comply with Section 704 [see the commentary to Section 602.4 for more information about Type IV (Heavy timber) construction]. Secondary members (e.g., floor or roof panels without a connection to the column) are not considered part of the structural frame (see Rows 5 and 6 of the table).

Row 2: *Bearing walls—exterior and interior.* Exterior bearing walls are the outermost walls that enclose the structure and support any structural load other than their own weight. Their required fire-resistance rating is established by the higher of two fire-resistance ratings. The first component of determining the fire-resistance rating is based on the type of construction of the building (Table 601). The second component of determining the fire-resistance rating is based on the exterior walls' fire separation distance (Table 602). Whichever of the two tables requires the higher fire-resistance rating will dictate the minimum required fire-resistance rating of the exterior wall.

In addition to Tables 601 and 602, exterior walls must comply with Section 705 and Chapter 14. In addition, Section 706.5.1 has fire-resistance-rating requirements for exterior walls on each side of the intersection of fire wall.

There are also several requirements related to exterior walls mentioned in Chapter 10. Section 1009.7 has fire-resistance-rating requirements for exterior walls adjacent to exterior areas for assisted rescue. Section 1023.7 has specific fire-resistance-rating requirements for exterior walls adjacent to an interior exit stairway. Section 1027.6 has fire-resistance-rating requirements for exterior walls adjacent to exterior exit stairways. Section 1028.4.2 has fire-resistance-rating requirements for exterior walls adjacent to an egress court.

Additionally, this category includes the structural (load-bearing) interior walls of a building. To delay vertical load-carrying collapse of a building due to fire exposure for a predetermined amount of time, the structural partitions are required to maintain a minimum degree of fire resistance. Primary structural frame elements supporting such walls must comply with Table 601, as well as have at least the same degree of fire resistance as the supported wall. See the commentary to Section 602.1 regarding opening and penetration protection of bearing walls.

Row 3: *Nonbearing walls and partitions—exterior.* This category includes all exterior walls that only support their own weight. The minimum required fire-resistance rating, unlike for load-bearing walls, is based solely on the exterior wall's fire separation distance (Table 602). Where nonbearing exterior walls occur in buildings of

TABLE 601
FIRE-RESISTANCE RATING REQUIREMENTS FOR BUILDING ELEMENTS (HOURS)

BUILDING ELEMENT	TYPE I		TYPE II		TYPE III		TYPE IV	TYPE V	
	A	B	A	B	A	B	HT	A	B
Primary structural frame[f] (see Section 202)	3[a]	2[a]	1	0	1	0	HT	1	0
Bearing walls Exterior[e, f] Interior	 3 3[a]	 2 2[a]	 1 1	 0 0	 2 1	 2 0	 2 1/HT	 1 1	 0 0
Nonbearing walls and partitions Exterior	See Table 602								
Nonbearing walls and partitions Interior[d]	0	0	0	0	0	0	See Section 602.4.6	0	0
Floor construction and associated secondary members (see Section 202)	2	2	1	0	1	0	HT	1	0
Roof construction and associated secondary members (see Section 202)	$1^{1}/_{2}$[b]	1[b,c]	1[b,c]	0[c]	1[b,c]	0	HT	1[b,c]	0

For SI: 1 foot = 304.8 mm.

a. Roof supports: Fire-resistance ratings of primary structural frame and bearing walls are permitted to be reduced by 1 hour where supporting a roof only.

b. Except in Group F-1, H, M and S-1 occupancies, fire protection of structural members shall not be required, including protection of roof framing and decking where every part of the roof construction is 20 feet or more above any floor immediately below. Fire-retardant-treated wood members shall be allowed to be used for such unprotected members.

c. In all occupancies, heavy timber shall be allowed where a 1-hour or less fire-resistance rating is required.

d. Not less than the fire-resistance rating required by other sections of this code.

e. Not less than the fire-resistance rating based on fire separation distance (see Table 602).

f. Not less than the fire-resistance rating as referenced in Section 704.10.

Type I or II construction, the walls can be constructed of fire-retardant-treated wood (FRTW) if no rating is required (see Section 603.1).

Row 4: *Nonbearing walls and partitions—interior.* This category includes all interior nonload-bearing walls and partitions (for example, the common wall separating two offices in the same suite). These walls need only comply with all of the material requirements associated with their type of construction classification and are not required to have a fire-resistance rating. Where they occur in buildings of Type I or II construction and are to be constructed of wood, the wood must be fire-retardant treated as indicated in Section 603.1. Nonload-bearing interior walls may be required to be fire-resistance rated when they also serve another purpose. Examples include interior walls also serving to separate mixed occupancies, dwelling units, sleeping units and incidental uses, as well as corridor walls required to have a fire-resistance rating.

Row 5: *Floor construction and associated secondary members.* Floor construction provides a natural fire compartment in a building by means of a horizontal barrier that retards the vertical passage of fire from floor to floor. In order to accomplish this, floor assemblies, including the beams and secondary structural members supporting the floor, must comply with Table 601 and Section 711. Ceilings are included if they are part of the tested assembly. A definition of "Secondary member" is provided in Section 202. Secondary members are not considered part of the primary structural frame.

Row 6: *Roof construction and associated secondary members.* Proper roof construction is necessary to prevent collapse from fire as well as potential impingement on adjacent buildings. Roof construction must comply with Table 601 and Section 711 (see the definition of "Secondary member" in Section 202). Additionally, when a portion of the roof construction is less than 20 feet (6096 mm) above the floor immediately below, the entire structural member must be protected to meet the minimum fire-resistance-rating requirements. In Commentary Figure 601(2), the roof construction is assumed to be one structural member. While a majority of the roof construction is greater than 20 feet (6096 mm) above the floor below, a portion of the roof construction is only 15 feet (4572 mm) above the mezzanine. As such, the entire length of the structural member, or in this case, the entire roof, must be rated for not less than 1 hour. The fire-resistance rating is not permitted to terminate at the portion where the roof is 20 feet (6096 mm) above the floor below unless the structural member ends at that point. The fire-resistance rating of a column must also be continuous for the full height of the column and not reduced or eliminated at a height of 20 feet (6096 mm) and above (see Note b). It should be noted that those members of the roof construction that are considered as primary structural frame members are not regulated by Row 5, but rather by Row 1.

Note a permits the fire-resistance ratings of primary structural frame and interior load-bearing walls in buildings of Type IA and IB construction to be reduced by 1 hour if the members are supporting only the roof.

Note b applies to the construction of the roof and related secondary members in all types of construction. It allows these elements to be of unprotected construction when all parts of the roof construction are more than 20 feet (6096 mm)

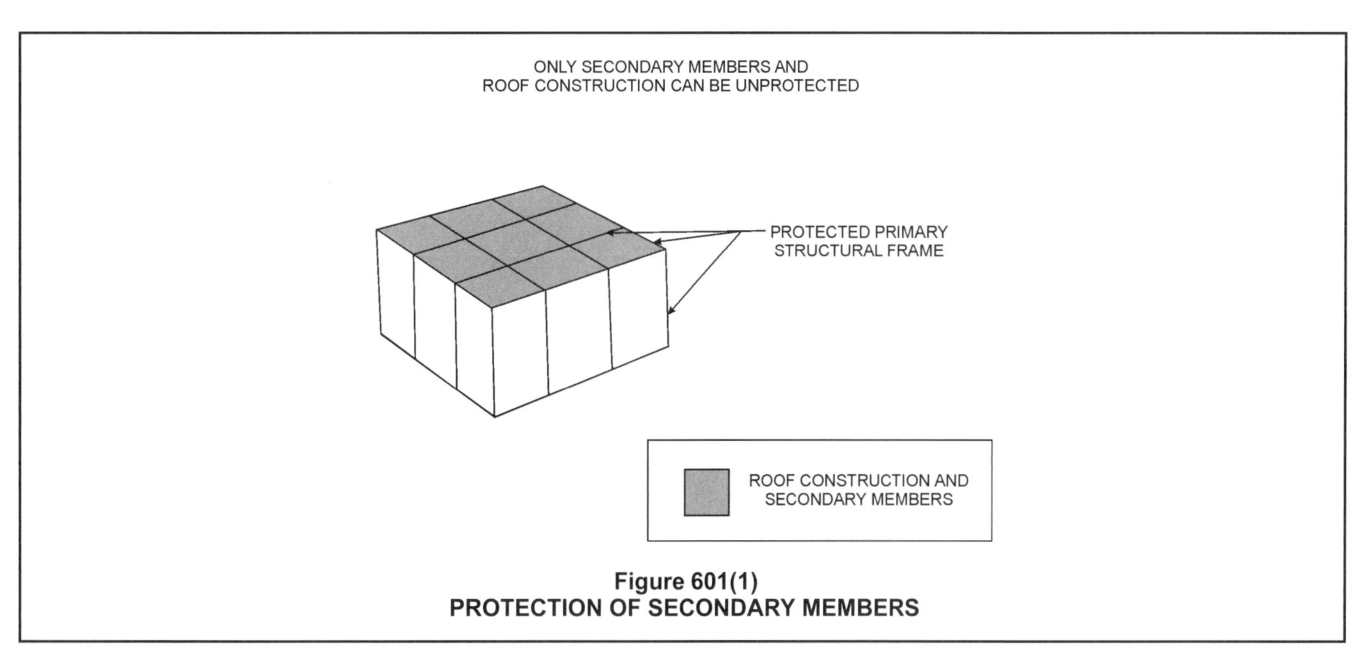

Figure 601(1)
PROTECTION OF SECONDARY MEMBERS

above any floor below. This allowance only applies to the secondary members of the roof structure and not to primary structural frame located within the roof or at the roof level [see Commentary Figure 601(1)]. This alternative is applicable for all occupancy classifications except Groups F-1, H, M and S-1. Commentary Figure 601(2) shows an example where a mezzanine reduces the clearance to the roof to less than 20 feet (6096 mm) for a portion of the total roof. Designs similar to Commentary Figure 601(2) do not comply with Note b, and elimination of fire resistance is not allowed for any portion of the roof.

In buildings of Type I and II construction, fire-retardant-treated wood (FRTW) may be utilized for unprotected roof members. Please note that FRTW is not required in the roof of buildings of Type IIIA or VA construction since combustible materials are already permitted by Sections 602.3 and 602.5, respectively.

Note c permits heavy timber (HT) construction to be utilized in the roof construction as an alternative to having a fire-resistance rating of 1 hour or less. Note that HT cannot be used in Type IA construction since the roof is required to have a rating greater than 1 hour. The intent of the note is to allow the substitution of HT for 1-hour construction in roof construction; it is not intended to say that HT is 1-hour-rated construction.

Note d is applicable only to interior nonload-bearing walls. While nonbearing interior walls are not required to have a rating in accordance with Table 601, other sections of the code may require a rating (e.g., corridor walls, dwelling unit separation, sleeping unit separation). If other sections of the code require a rating, this would override the requirements of Table 601 stating no rating is required. In addition, interior nonbearing walls in buildings of Type IV (Heavy Timber, HT) construction shall be of solid wood construction formed by not less than two layers of 1-inch (25 mm) matched boards or laminated construction 4 inches (102 mm) thick, or of 1-hour-rated construction (see Section 602.4.6).

Note e is a reminder that exterior load-bearing walls must also comply with the fire-resistance rating listed in Table 602 based on the fire separation distance. Exterior load-bearing walls must satisfy the higher of the fire-resistance ratings required by these two tables.

Note f is a reminder of the provisions of Section 704.10 that load-bearing structural members located within the exterior walls or "outside" of the structure need to comply with the highest of three fire-resistance ratings: the requirement for the primary structural frame; the requirement for exterior bearing walls; or the requirement from Table 602 based on the fire separation distance of the exterior wall. In Type III and IV construction, the rating for an exterior load-bearing wall is higher than that required for a primary structural frame alone.

TABLE 602. See page 6-5.

❖ Table 602 establishes the minimum fire-resistance ratings for both load-bearing and nonload-bearing exterior walls based on fire separation distance as defined in Section 202. The required ratings are based on the fuel load, probable fire intensity of the various occupancy classifications and the physical separation between the exterior wall and the line used to determine fire separation distance as established in Section 202 [see Commentary Figure 602(1)]. In using the table, the occupancy classifica-

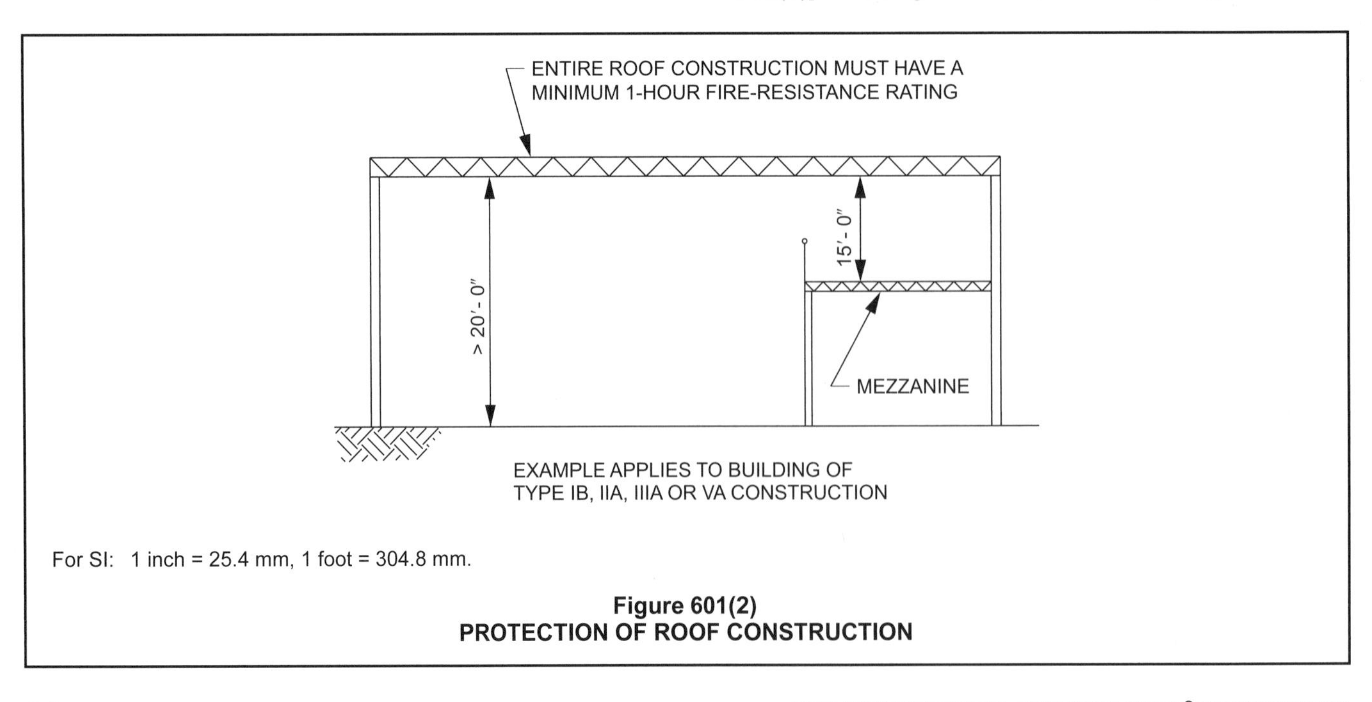

For SI: 1 inch = 25.4 mm, 1 foot = 304.8 mm.

Figure 601(2)
PROTECTION OF ROOF CONSTRUCTION

tion of the building and the fire separation distance of the walls must be determined. Once determined, the required fire-resistance rating is obtained by referring to the appropriate row and column corresponding to these parameters.

Please note that where there is more than one building on a lot, an imaginary lot line must be assumed between the buildings in accordance with Section 705.3. Commentary Figure 602(1) shows an assumed line halfway between the two buildings. The imaginary lot line can be at any location between the structures. Wherever the line is established, fire separation distance for each wall must be measured from the line, and wall and opening protection based on the distance to the assumed location.

Note a indicates that the fire-resistance-rating

TABLE 602
FIRE-RESISTANCE RATING REQUIREMENTS FOR EXTERIOR WALLS BASED ON FIRE SEPARATION DISTANCE[a, d, g]

FIRE SEPARATION DISTANCE = X (feet)	TYPE OF CONSTRUCTION	OCCUPANCY GROUP H[e]	OCCUPANCY GROUP F-1, M, S-1[f]	OCCUPANCY GROUP A, B, E, F-2, I, R, S-2, U[h]
$X < 5^b$	All	3	2	1
$5 \le X < 10$	IA Others	3 2	2 1	1 1
$10 \le X < 30$	IA, IB IIB, VB Others	2 1 1	1 0 1	1[c] 0 1[c]
$X \ge 30$	All	0	0	0

For SI: 1 foot = 304.8 mm.

a. Load-bearing exterior walls shall also comply with the fire-resistance rating requirements of Table 601.

b. See Section 706.1.1 for party walls.

c. Open parking garages complying with Section 406 shall not be required to have a fire-resistance rating.

d. The fire-resistance rating of an exterior wall is determined based upon the fire separation distance of the exterior wall and the story in which the wall is located.

e. For special requirements for Group H occupancies, see Section 415.6.

f. For special requirements for Group S aircraft hangars, see Section 412.4.1.

g. Where Table 705.8 permits nonbearing exterior walls with unlimited area of unprotected openings, the required fire-resistance rating for the exterior walls is 0 hours.

h. For a building containing only a Group U occupancy private garage or carport, the exterior wall shall not be required to have a fire-resistance rating where the fire separation distance is 5 feet or more.

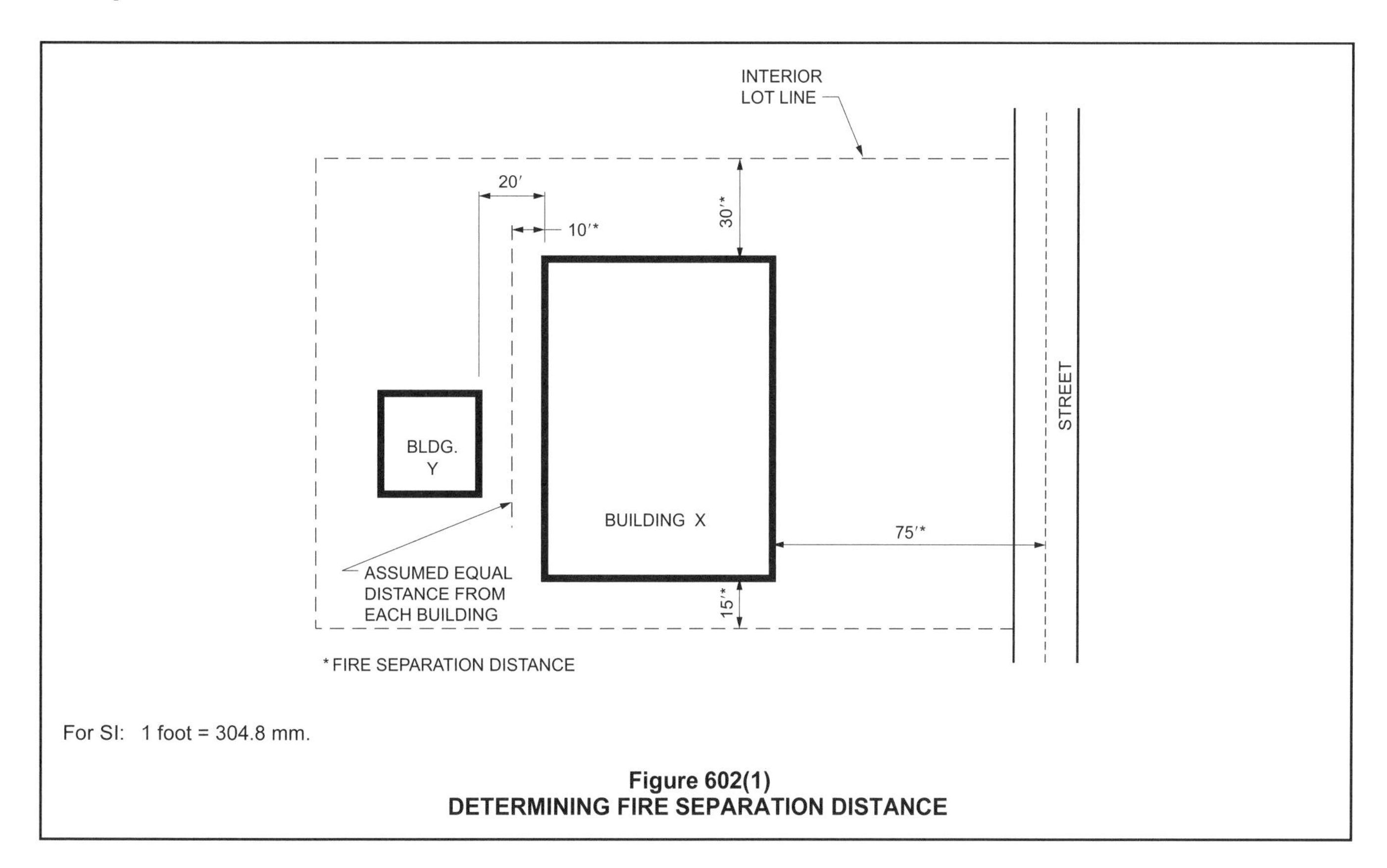

Figure 602(1)
DETERMINING FIRE SEPARATION DISTANCE

requirements of Table 601, which are based on construction type, also apply. Exterior load-bearing walls must conform to the higher of the fire-resistance ratings specified in Tables 601 and 602. Exterior non-load-bearing walls need only comply with Table 602.

Note b requires walls that are located on lot lines between adjacent buildings (i.e., zero fire separation distance) to be constructed as fire walls and be rated in accordance with Table 706.4.

Note c permits the exterior walls of an open parking garage to have no fire-resistance rating where the fire separation distance is at least 10 feet (3048 mm). Table 705.8 allows the openings to be unlimited for open parking garages at 10 feet (3048 mm) of fire separation distance. If the amount of openings is no longer limited, then the entire wall could be removed.

Note d clarifies that the required minimum fire-resistance rating of each exterior wall in each story of a building must be determined separately. Should a multistory building be configured such that the fire separation distance of each story is different, the required fire-resistance ratings associated with each of the exterior walls in each of those stories is established separately [see Commentary Figure 602(2)].

Note e provides a reference for Group H occupancies to Section 415.6 where there are specific standards for fire separation distances and wall and opening protection that supersede those of Table 602.

Note f provides a reference to Section 412.4.1 regulating aircraft hangars where there are specific standards for fire separation distances that supersede Table 602.

Note g allows for an elimination of the required wall protection for those exterior walls permitted to have unlimited openings in accordance with Table 705.8. For buildings protected by an automatic sprinkler system, Table 705.8 allows unlimited openings in walls with a fire separation distance of 20 feet (6096 mm) or more. This note is limited to the fire-resistance rating requirements of Table 602, and does not eliminate requirements established in Table 601 for load-bearing walls. Wall and opening protection for Group H-1, H-2 and H-3 occupancies is established by Section 415.6.

Note h addresses Group U occupancy garages and carports. Such structures are usually small and are either stand-alone or are attached to smaller apartment buildings or commercial buildings. This note would not apply to a Group U structure not used for the parking and storage of motor vehicles. Other Group U occupancies would have a minimum 1-hour fire-resistance rating for a fire separation distance of less than 10 feet (3048 mm).

SECTION 602
CONSTRUCTION CLASSIFICATION

602.1 General. Buildings and structures erected or to be erected, altered or extended in height or area shall be classified in one of the five construction types defined in Sections 602.2 through 602.5. The building elements shall have a *fire-resistance rating* not less than that specified in Table 601 and exterior walls shall have a *fire-resistance rating* not less than that specified in Table 602. Where required to have a *fire-resistance rating* by Table 601, building elements shall comply with the applicable provisions of Section 703.2. The protection of openings, ducts and air transfer openings in building elements shall not be required unless required by other provisions of this code.

❖ This section requires that each building or structure be put into one of five possible construction classifications: Type I, II, III, IV or V. Structural members are required to have a fire-resistance rating in accordance with Table 601. Additionally, the exterior walls of the structure must satisfy the requirements in Table 602, which bases the fire-resistance rating on the fire separation distance.

The use of multiple construction classifications in a single building is very limited and can only be done when specifically called out in the code. An example

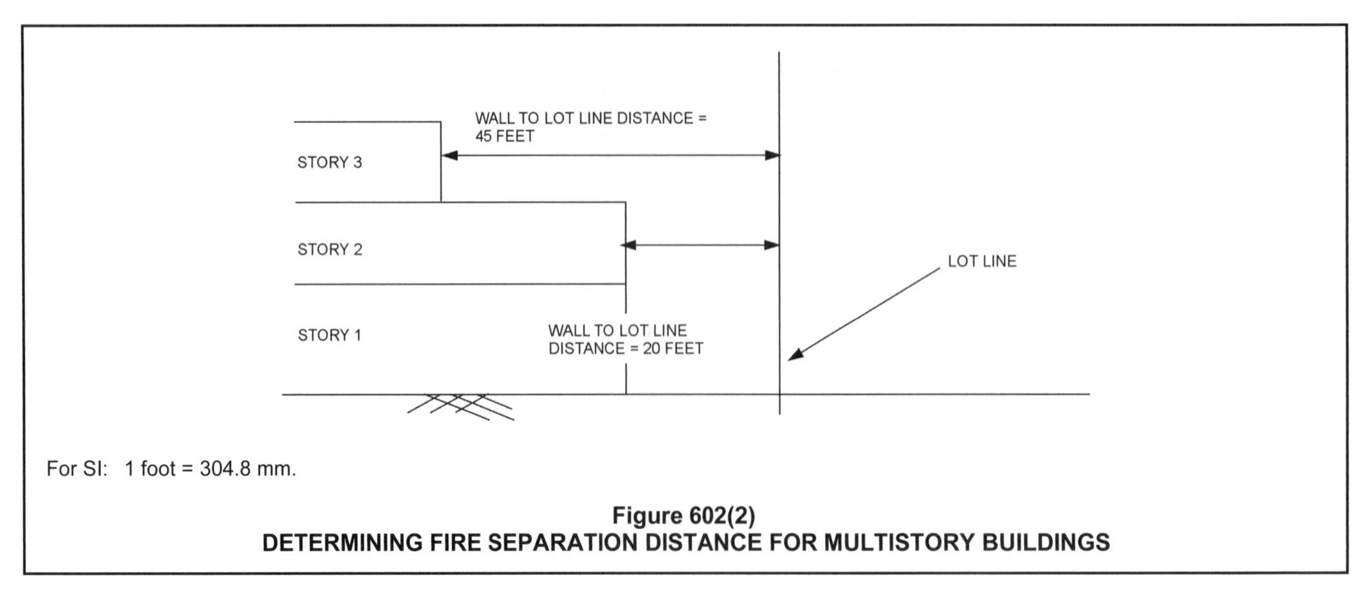

Figure 602(2)
DETERMINING FIRE SEPARATION DISTANCE FOR MULTISTORY BUILDINGS

of combining types of construction is an office building of Type IIA construction located above an open parking structure of Type IIB construction, as described in Sections 510.7 and 510.7.1. Several other special provisions found in Section 510 also permit a building of multiple construction classifications.

A more common example is where a single structure is divided into two parts by using a fire wall, resulting in two separate buildings or structures—each of which may be of a different type of construction. Where a structure contains more than one building (for example, separation by a fire wall), each building is to be individually assigned a type of construction.

Also, a building may have elements that comply with the requirements of more than one type of construction, in which case the building as a whole must be assigned the less restrictive type of construction. The designer may have intended, however, to comply with a higher type of construction, in which case those elements not in compliance with the intended type of construction are to be brought into compliance. The selection of a construction type is the prerogative of the permit applicant. The applicant should have the designer indicate on the submitted construction documents which construction type has been selected. This will expedite the compliance review by the plans examiner.

Section 602.1 applies to both new construction and additions. The provisions in Section 503 on general height and area limitations, Chapter 7 on fire and smoke protection features and construction, and other applicable portions of the code depend on the requirements of this section. Where Table 601 requires a building element to meet a fire-resistance rating, that rating is determined based on the criteria in Section 703. Building elements required to have a fire-resistance rating by Table 601 do not necessarily need opening protectives or dampers for penetrations through these elements. Opening protectives or dampers are only required for building elements listed in Table 601 that are also a specific type of wall or horizontal assembly that is required by other criteria in the code to have protected openings.

Where Table 601 requires a building element to meet a fire-resistance rating, that rating is determined based on the criteria in Section 703. Although a building element is required to have fire resistance by Table 601 this does not mean that the required openings in these building elements or duct penetrations through these elements have to be protected. However, a building element listed in Table 601 may also be a specific type of wall or horizontal assembly that is required by other criteria in the code to have protected openings. For example, an interior bearing wall inside a Type IIA building is required to have a fire-resistance rating of not less than 1 hour. However, the openings in that wall need not be protected unless that bearing wall is also serving another purpose. This wall could also be part of the walls establishing a control area on the second story. In accordance with Section 414.2.2, such a wall would also have to be a 1-hour-rated fire barrier, and in accordance with Table 716.5, openings in this 1-hour-rated fire barrier would need to have 45-minute-rated opening protection. It is essential to check the provisions of Chapter 7 to determine where building elements are required, or not required, to have protected openings, transfer openings, ducts, joints and other penetrations.

602.1.1 Minimum requirements. A building or portion thereof shall not be required to conform to the details of a type of construction higher than that type which meets the minimum requirements based on occupancy even though certain features of such a building actually conform to a higher type of construction.

❖ These requirements permit design flexibility by allowing various building materials and components to be used. A building must, as a minimum, meet all of the requirements of a given type of construction to be classified as such, even though portions of that building meet the criteria of a higher construction type (e.g., greater fire-resistance ratings). This is consistent with the concept that the code is a minimum requirement. For example, a building classified as Type III construction is not prohibited from having construction that is superior, but it could not be reclassified into a higher type of construction unless it met all of the requirements for that construction type. In a normal situation, the design professional has identified the construction classification on the drawings. When this assignment has not been made, the building official is placed in a position of verifying the designer's intent and selecting the least-restrictive type that will meet all of the code requirements.

602.2 Types I and II. Types I and II construction are those types of construction in which the building elements listed in Table 601 are of noncombustible materials, except as permitted in Section 603 and elsewhere in this code.

❖ Buildings of Types I and II construction are required to be constructed of noncombustible materials (see Section 703.5) and, therefore, are frequently referred to as "noncombustible construction." All Type I and II structural members have a fire-resistance rating as required by Table 601 and 602. A typical example of a building of Type IA, IB or IIA construction would be a high-rise structure or a very large low-rise structure. These buildings are permitted to be relatively large in height and area due to the fire resistance afforded the structure's components. The structural members of a building of Type IIB construction do not have the same fire resistance as structural members in a building of Type IA, IB or IIA construction. As such, the height and area requirements are not as large for Type IIB construction (see Commentary Figure 602.2 for an example of Type I or II construction).

Types I and II construction are divided into four subclassifications: Types IA, IB, IIA and IIB. The only difference among the four subclassifications is the degree of fire-resistance rating required for similar elements and assemblies. For example, the required rating for structural frame members in Type IA construction is 3 hours, for Type IB it is 2 hours, for Type IIA it is 1 hour, and for Type IIB no rating is required (0 hours). Often, the fire-resistance ratings required by Tables 601 and 602 for structural elements are achieved by "fireproofing" structural members. Fireproofing is typically the process of creating a fire-resistance-rated assembly that incorporates the structural member by encapsulating it, either by boxing it in or by spraying on a coating to achieve the required fire-resistance rating. It should be noted that when a protective covering is used to provide the fire-resistance rating, it must be of noncombustible material, except as indicated in Section 603.1, Item 21.

Fire-retardant-treated wood (FRTW), although combustible, is permitted in limited uses in buildings of Type I and II construction (see Section 603 and Table 601, Note b). While FRTW is permitted in certain applications in buildings of Type I and II construction, it is not assumed to be fire-resistance rated, and generally does not afford any higher fire-resistance rating than untreated wood material.

Other combustible items (as specified in Section 603.1) are also permitted in buildings of Type I or II construction.

602.3 Type III. Type III construction is that type of construction in which the exterior walls are of noncombustible materials and the interior building elements are of any material permitted by this code. *Fire-retardant-treated wood* framing complying with Section 2303.2 shall be permitted within *exterior wall* assemblies of a 2-hour rating or less.

❖ Buildings of Type III construction are typically constructed with both combustible and noncombustible materials. The exterior walls are required to be noncombustible with load-bearing exterior walls required to have a minimum 2-hour fire-resistance rating. Exterior nonloadbearing walls are not required by Table 601 to have a fire-resistance rating, but must comply with the provisions of Table 602. The elements within the perimeter established by the exterior walls (i.e., floors, roofs and walls) are permitted to be of combustible materials. An example of a typical building of Type III construction is a structure having its exterior walls constructed of concrete, masonry or other approved noncombustible materials, but with a wood frame floor, interior wall and roof construction (see Commentary Figure 602.3). The structural members of a building of Type IIIB construction are not

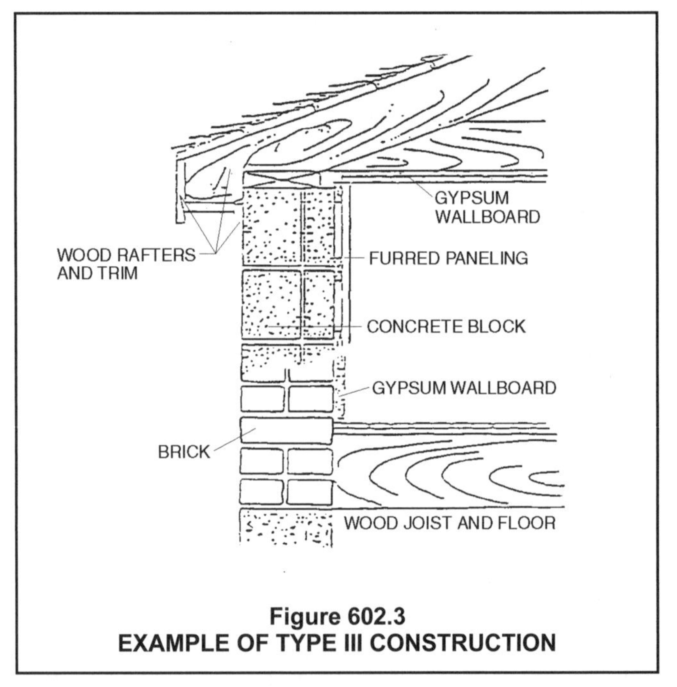

Figure 602.3
EXAMPLE OF TYPE III CONSTRUCTION

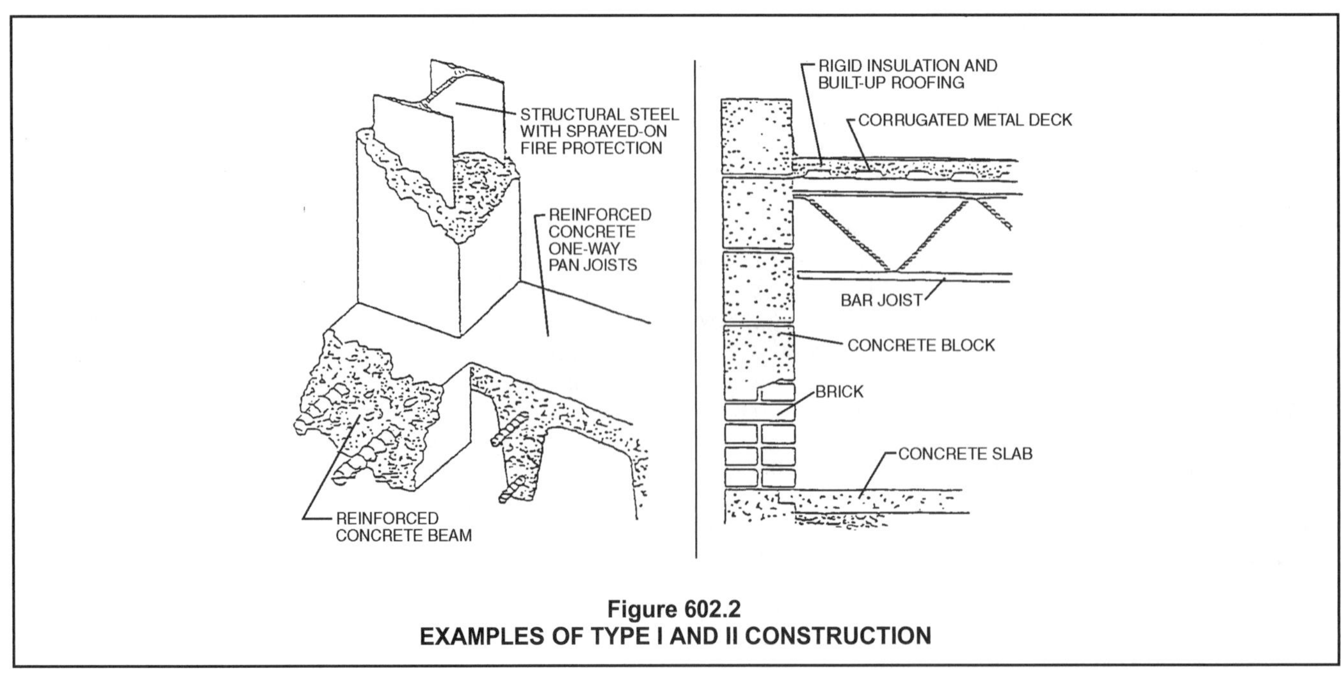

Figure 602.2
EXAMPLES OF TYPE I AND II CONSTRUCTION

required to have a fire-resistance rating, with the exception of the exterior load-bearing walls.

Although fire-retardant-treated wood (FRTW) does not meet the specifications of the code as a noncombustible material, it is permitted as a substitute for noncombustible materials for framing within exterior wall assemblies of Type III construction. The exterior surfaces of the walls must be of noncombustible materials. While the exterior walls are permitted to be either nonload-bearing or load-bearing, to apply the allowance for FRTW the required fire-resistance rating of the exterior wall must be no greater than 2 hours. FRTW is required to comply with the provisions in Section 2303.2.

602.4 Type IV. Type IV construction (Heavy Timber, HT) is that type of construction in which the exterior walls are of noncombustible materials and the interior building elements are of solid or laminated wood without concealed spaces. The details of Type IV construction shall comply with the provisions of this section and Section 2304.11. Exterior walls complying with Section 602.4.1 or 602.4.2 shall be permitted. Minimum solid sawn nominal dimensions are required for structures built using Type IV construction (HT). For glued-laminated members and structural composite lumber (SCL) members, the equivalent net finished width and depths corresponding to the minimum nominal width and depths of solid sawn lumber are required as specified in Table 602.4. *Cross-laminated timber* (CLT) dimensions used in this section are actual dimensions.

❖ This section provides the general regulations for Type IV (Heavy Timber, HT) construction. HT construction requires the exterior walls to be constructed of noncombustible materials. The interior elements are required to be constructed of solid or laminated wood without any concealed spaces. All of the combustible structural elements are permitted to be unprotected because of the massive element sizes and the requirement that there not be any concealed spaces, such as soffits, plenums or suspended ceilings. Because HT elements are unprotected, they do not have a fire-resistance rating. Therefore, compliance with Sections 703 and 704 is irrelevant and not required.

Sections 602.4.1 through 602.4.9 provide specific requirements for the connection of structural members and minimum dimensions. An examination of Tables 504.3. 504.4 and 506.2 indicates that the allowable height and area for Type IV construction is greater than that permitted for buildings of Type IIB construction. This distinction is based on testing that demonstrated that HT structural members perform better structurally under fire conditions than comparable unprotected steel structural members because of charring, which insulates the wood mass.

Historically, heavy timber was simply large pieces of sawn lumber from the large trees of North American forests. Over time, as such trees became less common, the industry has created three types of engineered wood products that can be used as "heavy timber": glued-laminated; structural composite lumber (SCL); and cross-laminated timber (CLT).

TABLE 602.4. See below.

❖ Solid sawn wood members, glued-laminated timbers, structural composite lumber and cross-laminated timber are manufactured using different methods and procedures and, therefore, do not have the same dimensions. However, they all have the same inherent fire-resistance capability. The dimensions noted in Sections 602.4.1 through 602.4.9 refer to the nominal dimensions of solid sawn lumber. These dimensions do not directly correlate to the dimensions of glued-laminated timbers, structural composite lumber or cross-laminated timber. Table 602.4 provides a simple procedure to determine the dimensions that are required for glued-laminated timbers and structural composite lumber when designing to meet the requirements of Type IV construction. Cross-laminated timber is not included in the table because the actual dimensions of CLT are to be used. The table provides minimum dimensions for glued-laminated wood members and for structural composite lumber for each set of dimensions specified in the specific provisions of Section 602.4. To use the table, compare the required minimum dimensions for sawn timber found in the two left-hand columns with the dimensions found in the two middle columns for glued-laminated wood members or the two right-hand columns for structural composite lumber. For example, where the code requires a minimum sawn timber of 8 inches by 8 inches (203 mm by 203 mm): for a glued-laminated wood member to be used, it would need to be a minimum of $6^3/_4$ inches wide by $8^1/_4$

TABLE 602.4
WOOD MEMBER SIZE EQUIVALENCIES

MINIMUM NOMINAL SOLID SAWN SIZE		MINIMUM GLUED-LAMINATED NET SIZE		MINIMUM STRUCTURAL COMPOSITE LUMBER NET SIZE	
Width, inch	Depth, inch	Width, inch	Depth, inch	Width, inch	Depth, inch
8	8	$6^3/_4$	$8^1/_4$	7	$7^1/_2$
6	10	5	$10^1/_2$	$5^1/_4$	$9^1/_2$
6	8	5	$8^1/_4$	$5^1/_4$	$7^1/_2$
6	6	5	6	$5^1/_4$	$5^1/_2$
4	6	3	$6^7/_8$	$3^1/_2$	$5^1/_2$

For SI: 1 inch = 25.4 mm.

inches deep (171 mm by 210 mm); and for structural composite lumber to be used, it would need to be a minimum of 7 inches wide by 7$^{1}/_{2}$ inches deep (178 mm by 190 mm).

602.4.1 Fire-retardant-treated wood in exterior walls. *Fire-retardant-treated wood* framing complying with Section 2303.2 shall be permitted within exterior wall assemblies with a 2-hour rating or less.

❖ As with Type III construction, fire-retardant-treated wood (FRTW) is permitted as a substitute for noncombustible materials within exterior wall assemblies of Type IV construction. Except as noted in Section 602.4.9, exterior structural members used externally must be noncombustible. While the exterior walls are permitted to be either nonload-bearing or load-bearing, to apply the allowance for FRTW the required fire-resistance rating of the exterior wall must be no greater than 2 hours. FRTW is required to comply with the provisions of Section 2303.2.

602.4.2 Cross-laminated timber in exterior walls. *Cross-laminated timber* complying with Section 2303.1.4 shall be permitted within exterior wall assemblies with a 2-hour rating or less, provided the exterior surface of the cross-laminated timber is protected by one the following:

1. *Fire-retardant-treated wood* sheathing complying with Section 2303.2 and not less than $^{15}/_{32}$ inch (12 mm) thick;
2. *Gypsum board* not less than $^{1}/_{2}$ inch (12.7 mm) thick; or
3. A noncombustible material.

❖ Similar to FRTW, cross-laminated timber can be used within the exterior walls of a Type IV building provided the CLT is protected by one of the three methods listed. In other words, the CLT can be part of the exterior wall provided it is not exposed but is covered.

602.4.3 Columns. Wood columns shall be sawn or glued laminated and shall be not less than 8 inches (203 mm), nominal, in any dimension where supporting floor loads and not less than 6 inches (152 mm) nominal in width and not less than 8 inches (203 mm) nominal in depth where supporting roof and ceiling loads only. Columns shall be continuous or superimposed and connected in an *approved* manner.

❖ Minimum construction requirements and dimensions for timber columns are provided in this section. Columns are required to be a minimum of 8 inches (203 mm) nominal in any dimension if they support floor loads, or a minimum of 6 by 8 inches (152 by 203 mm) nominal if they support no more than a roof and ceiling. Timber columns are required to be continuous or superimposed, positioned on or over each other, through floors for the entire height of the building. The design engineer or architect must provide details of all column connections. As with all structural members, each column must also be adequately

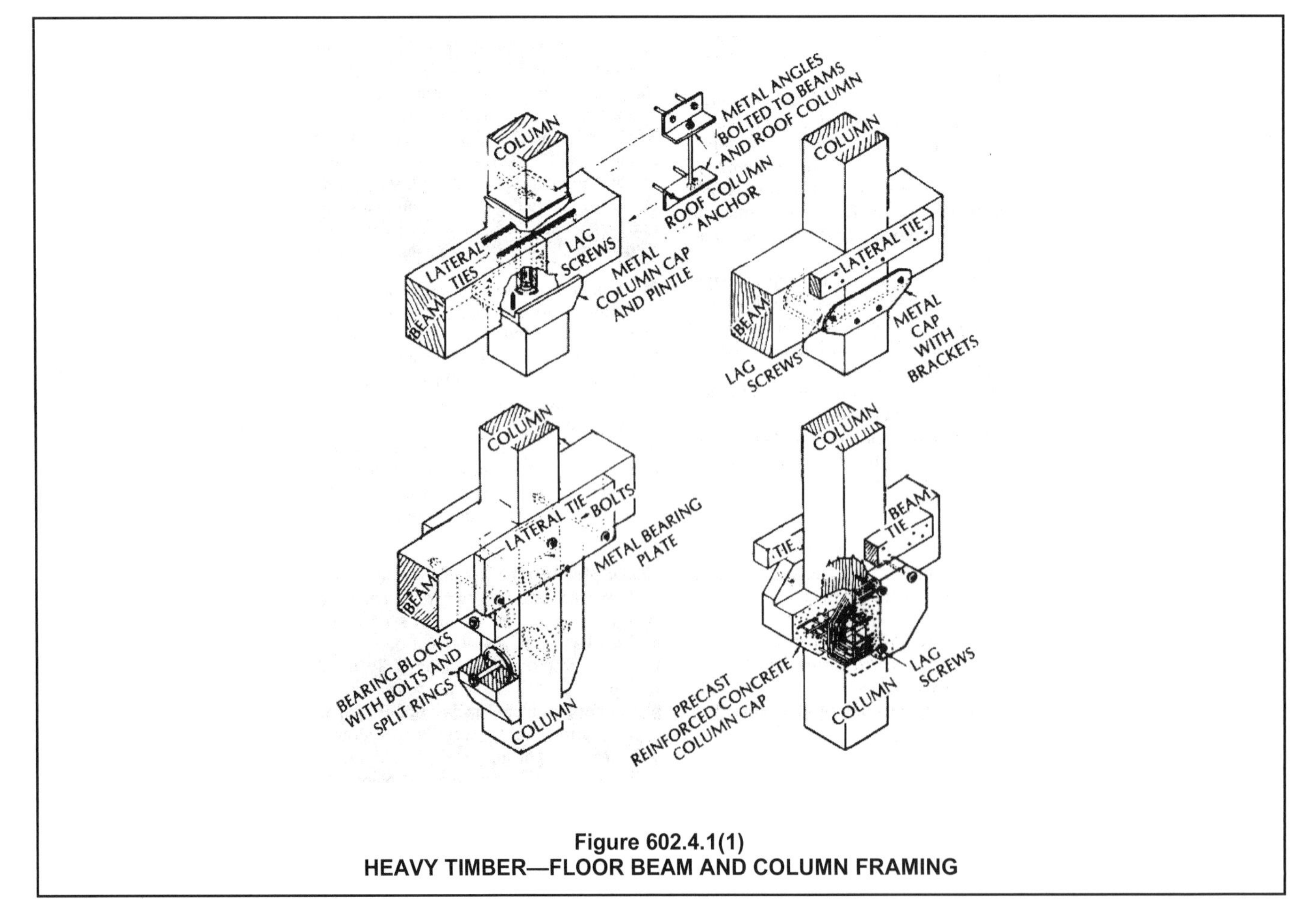

Figure 602.4.1(1)
HEAVY TIMBER—FLOOR BEAM AND COLUMN FRAMING

fastened to other structural members in order to withstand the loads that will be placed upon the column. Some typical examples include reinforced concrete or metal caps, steel or iron column caps and timber splice plates [see Commentary Figures 602.4.1(1) and 602.4.1(2)]. CLT is not included in this section because CLT is not used as a column or framing material.

602.4.4 Floor framing. Wood beams and girders shall be of sawn or glued-laminated timber and shall be not less than 6 inches (152 mm) nominal in width and not less than 10 inches (254 mm) nominal in depth. Framed sawn or glued-laminated timber arches, which spring from the floor line and support floor loads, shall be not less than 8 inches (203 mm) nominal in any dimension. Framed timber trusses supporting floor loads shall have members of not less than 8 inches (203 mm) nominal in any dimension.

❖ Minimum construction requirements and dimensions for floor framing are provided in this section. Girders are the principal horizontal structural members that support columns or beams. Beams are the structural members that support a floor or roof. Both girders and beams are required to be a minimum 6 inches (152 mm) wide and 10 inches (254 mm) deep. Both framed timber trusses supporting floor loads and framed sawn or glued-laminated timber arches that spring from the floor line and support floor loads are required to be at least 8 inches (203 mm) in any dimension. CLT is not included in this section because CLT is not used as a framing material.

602.4.5 Roof framing. Wood-frame or glued-laminated arches for roof construction, which spring from the floor line or from grade and do not support floor loads, shall have members not less than 6 inches (152 mm) nominal in width and have not less than 8 inches (203 mm) nominal in depth for the lower half of the height and not less than 6 inches (152 mm) nominal in depth for the upper half. Framed or glued-laminated arches for roof construction that spring from the top of walls or wall abutments, framed timber trusses and other roof framing, which do not support floor loads, shall have members not less than 4 inches (102 mm) nominal in width and not less than 6 inches (152 mm) nominal in depth. Spaced members shall be permitted to be composed of two or more pieces not less than 3 inches (76 mm) nominal in thickness where blocked solidly throughout their intervening spaces or where spaces are tightly closed by a continuous wood cover plate of not less than 2 inches (51 mm) nominal in thickness secured to the underside of the members. Splice plates shall be not less than 3 inches (76 mm) nominal in thickness. Where protected by *approved* automatic sprinklers under the roof deck, framing members shall be not less than 3 inches (76 mm) nominal in width.

❖ Minimum construction requirements and dimensions for arches and other types of roof framing are pro-

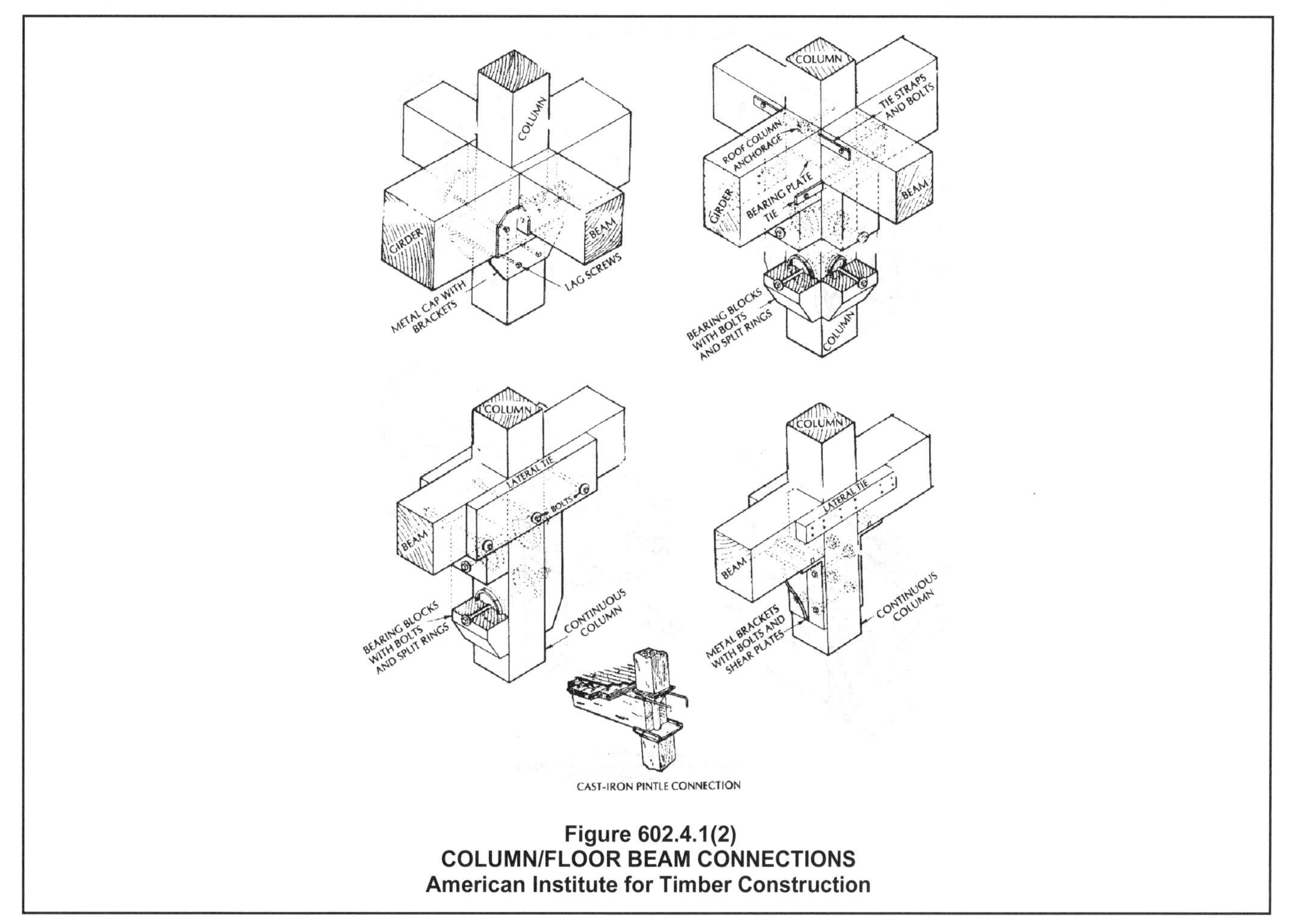

Figure 602.4.1(2)
COLUMN/FLOOR BEAM CONNECTIONS
American Institute for Timber Construction

vided in this section. Other types of roof framing included in this section are heavy timber trusses with spaced members. When the members of a heavy timber truss are split and placed on either side of a main member, such as a web connecting a chord, each component of the web must be 3 inches (76 mm) or more in nominal thickness. The space between the two web members must be protected with a 2-inch-thick (51 mm) cover plate [see Commentary Figure 602.4.3(1)], or solidly filled with blocking [see Commentary Figure 602.4.3(2)]. The size of the roof framing members is dependent on the configuration used and is regulated by this section. CLT is not included in this section because CLT is not used as a framing material.

If a building of Type IV construction is equipped with approved automatic sprinklers under the roof deck, the minimum required size of the roof framing members is reduced to 3 inches (76 mm). Roof framing members of a smaller size will have a lower resistance to fire than the 6-inch by 8-inch (152 mm by 203 mm) or 4-inch by 6-inch (102 mm by 152 mm) members required by this section where a sprinkler system is not present. However, the trade-off allowing smaller roof framing members when the building is equipped with an automatic sprinkler system is consistent with the concept of maintaining "equivalent risk" for the building.

602.4.6 Floors. Floors shall be without concealed spaces. Wood floors shall be constructed in accordance with Section 602.4.6.1 or 602.4.6.2.

602.4.6.1 Sawn or glued-laminated plank floors. Sawn or glued-laminated plank floors shall be one of the following:

1. Sawn or glued-laminated planks, splined or tongue-and-groove, of not less than 3 inches (76 mm) nominal in thickness covered with 1-inch (25 mm) nominal dimension tongue-and-groove flooring, laid crosswise

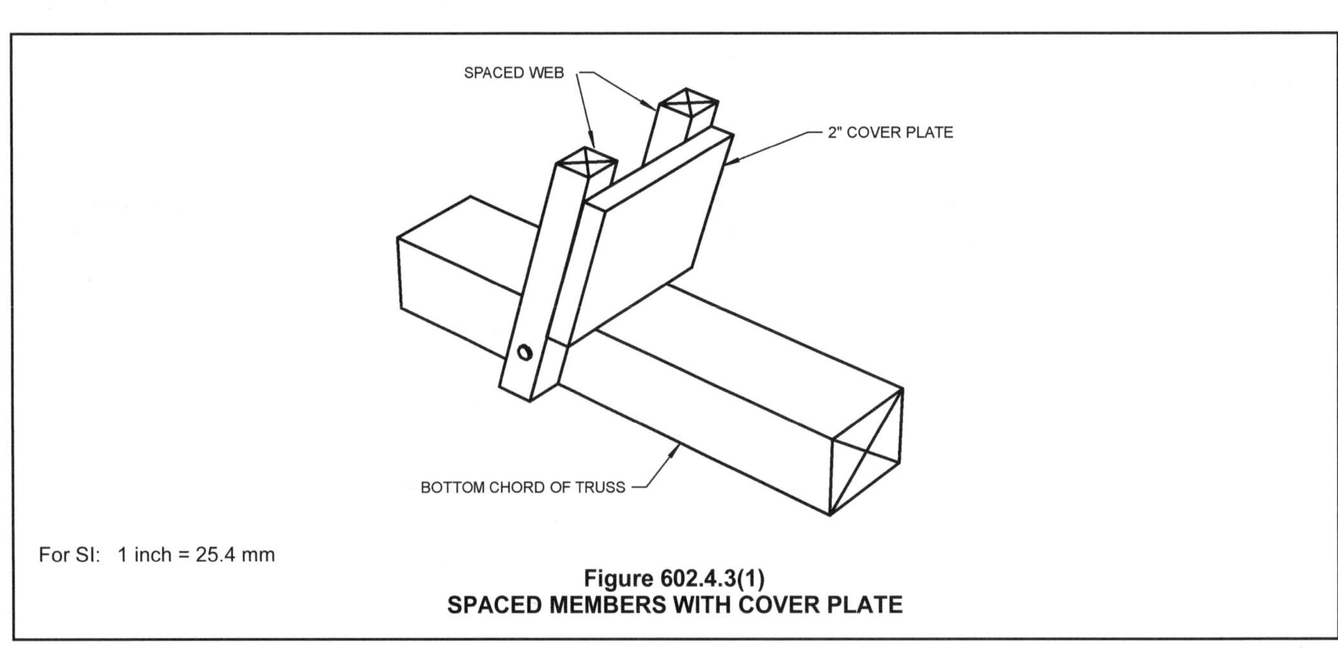

Figure 602.4.3(1)
SPACED MEMBERS WITH COVER PLATE

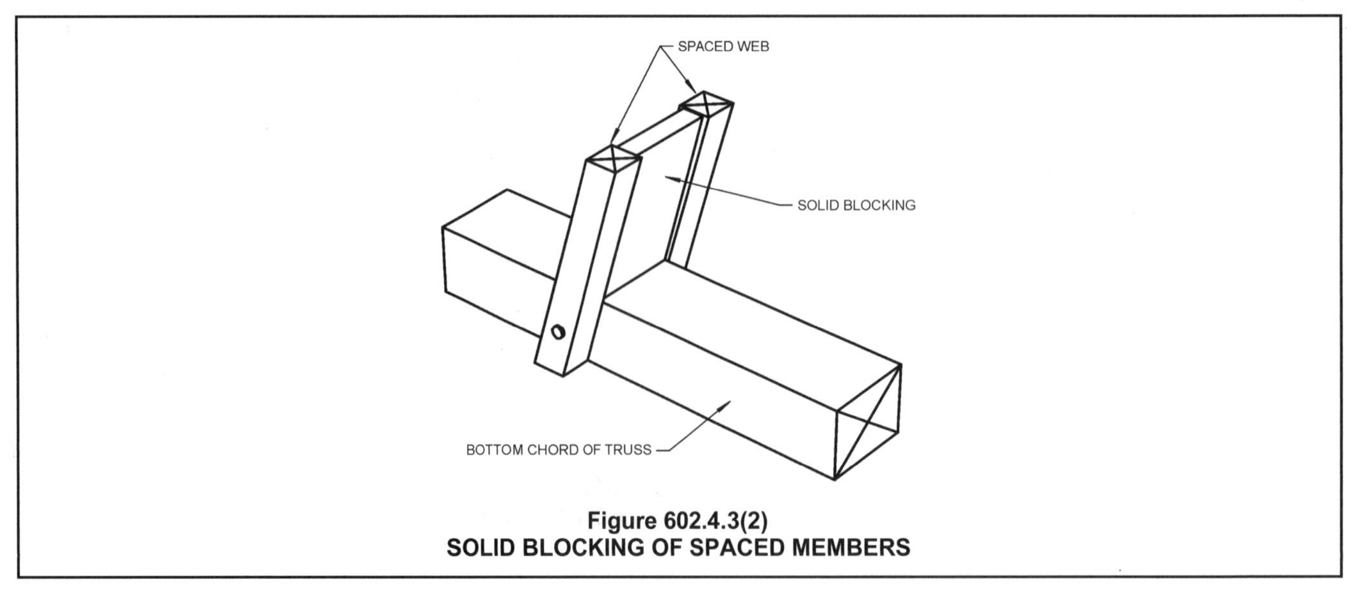

Figure 602.4.3(2)
SOLID BLOCKING OF SPACED MEMBERS

or diagonally, $^{15}/_{32}$-inch (12 mm) wood structural panel or $^{1}/_{2}$-inch (12.7 mm) particleboard.

2. Planks not less than 4 inches (102 mm) nominal in width set on edge close together and well spiked and covered with 1-inch (25 mm) nominal dimension flooring or $^{15}/_{32}$-inch (12 mm) wood structural panel or $^{1}/_{2}$-inch (12.7 mm) particleboard.

The lumber shall be laid so that no continuous line of joints will occur except at points of support. Floors shall not extend closer than $^{1}/_{2}$ inch (12.7 mm) to walls. Such $^{1}/_{2}$-inch (12.7 mm) space shall be covered by a molding fastened to the wall and so arranged that it will not obstruct the swelling or shrinkage movements of the floor. Corbelling of masonry walls under the floor shall be permitted to be used in place of molding.

❖ Heavy timber (HT) flooring is required to consist of minimum 3-inch-thick (76 mm) sawn or glued-laminated planks, splined floors or tongue-and-groove floors with an overlayment of 1-inch (25 mm) tongue-and-groove flooring, laid crosswise or diagonally. HT flooring may also consist of $^{1}/_{2}$-inch (12.7 mm) particleboard or planks at least 4 inches (102 mm) in width set on edge and secured together, with an appropriate overlayment, such as 1-inch (25 mm) hardwood flooring or a $^{15}/_{32}$-inch (12.7 mm) wood structural panel. Flooring in Type IV construction is not permitted to have concealed spaces because an undetected fire can spread quickly in combustible concealed floor spaces [see Commentary Figure 602.4.4(1)]. Because of the support afforded by adjacent members, continuous joints must only occur over supports.

Wood flooring must be fastened to supports that are perpendicular to the planking. Fastening must not be made to beams or girders that are parallel to the

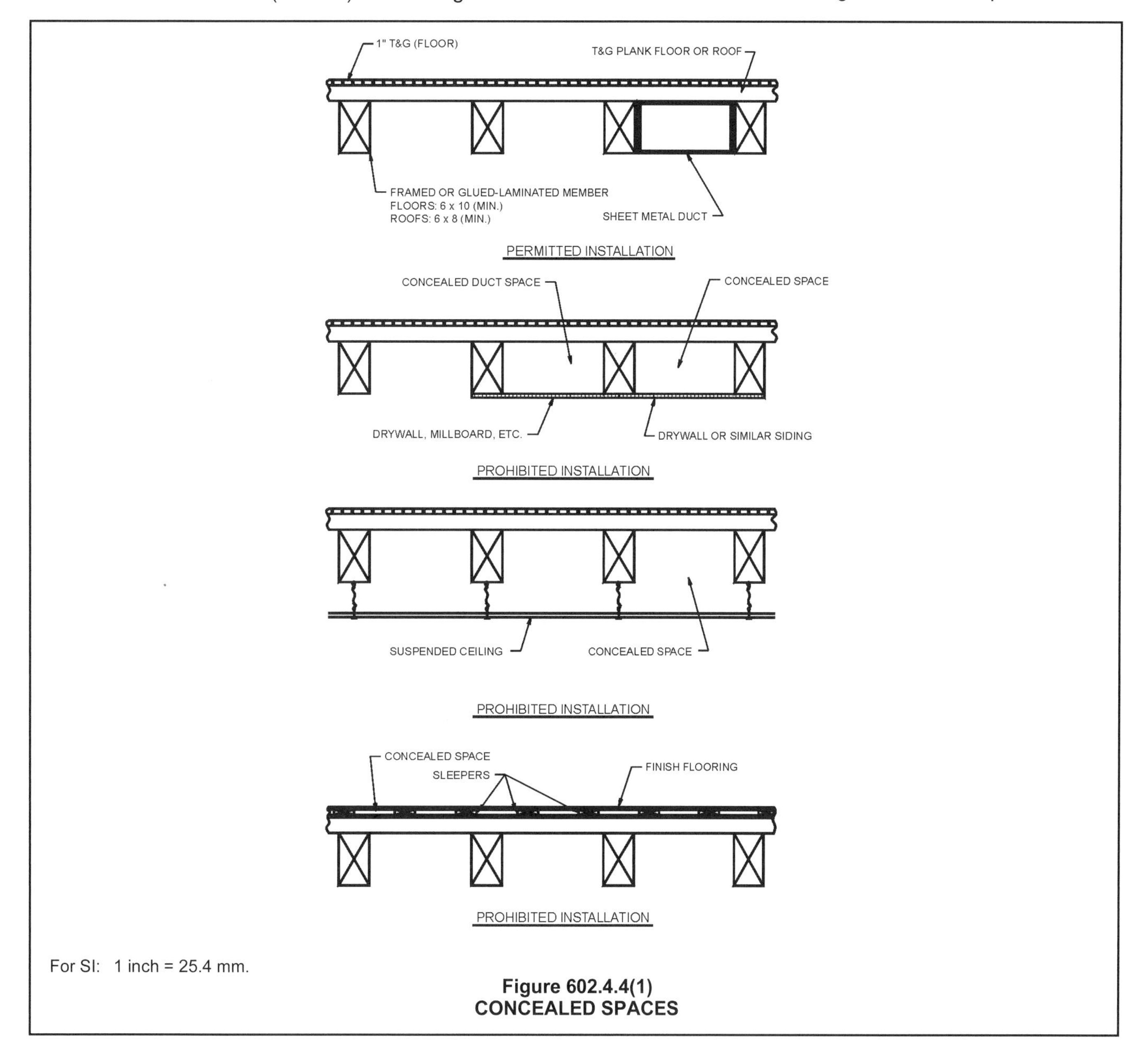

For SI: 1 inch = 25.4 mm.

Figure 602.4.4(1)
CONCEALED SPACES

planks [see Commentary Figure 602.4.4(2)]. This precaution is intended to prevent separation of the planks because of differential movement of the beam relative to the girders and possible expansion/contraction due to differing moisture or humidity levels. This section requires a minimum $^1/_2$-inch (12.7 mm) clearance between the wood flooring and exterior walls. This will prevent damage to the walls if the flooring expands or moves due to factors such as use and natural forces. It should be emphasized that the integrity of the floor assembly must be maintained to provide the equivalent of a 1-hour fire-resistance rating. In addition, the $^1/_2$-inch (12.7 mm) gap must be protected by a molding connected to the wall so that any possible contracting or expanding of the floor is not impeded. If masonry walls are utilized, corbeling of the masonry may be used as an alternate to the molding requirements.

602.4.6.2 Cross-laminated timber floors. *Cross-laminated timber* shall be not less than 4 inches (102 mm) in thickness. *Cross-laminated timber* shall be continuous from support to support and mechanically fastened to one another. *Cross-laminated timber* shall be permitted to be connected to walls without a shrinkage gap providing swelling or shrinking is considered in the design. Corbelling of masonry walls under the floor shall be permitted to be used.

❖ Cross-laminated timber is a large, thick panel composed of crosswise layers of dimension lumber bound with a structural adhesive. When used in floors, CLT is unlike traditional plank decking because the CLT panel doesn't have joints to protect. Therefore, there is no requirement for sheathing. Sheathing can be used on CLT, but is not necessary for compliance with this section.

602.4.7 Roofs. Roofs shall be without concealed spaces and wood roof decks shall be sawn or glued laminated, splined or tongue-and-groove plank, not less than 2 inches (51 mm) nominal in thickness; $1^1/_8$-inch-thick (32 mm) wood structural panel (exterior glue); planks not less than 3 inches (76 mm) nominal in width, set on edge close together and laid as required for floors; or of cross-laminated timber. Other types of decking shall be permitted to be used if providing equivalent fire resistance and structural properties.

Cross-laminated timber roofs shall be not less than 3 inches (76 mm) nominal in thickness and shall be continuous from support to support and mechanically fastened to one another.

❖ Minimum construction requirements and dimensions for roof decks are provided in this section. As required for floors, roofs are not permitted to have concealed spaces [see Commentary Figure 602.4.4(1)]. If the materials used in roof construction are different from those described in this section, the roof must have a minimum 1-hour fire-resistance rating and be of the same structural properties.

602.4.8 Partitions and walls. Partitions and walls shall comply with Section 602.4.8.1 or 602.4.8.2.

602.4.8.1 Interior walls and partitions. Interior walls and partitions shall be of solid wood construction formed by not less than two layers of 1-inch (25 mm) matched boards or laminated construction 4 inches (102 mm) thick, or of 1-hour fire-resistance-rated construction.

❖ Minimum construction requirements and dimensions for partitions in Type IV construction are provided in this section. Partitions must either be formed by not less than two layers of 1-inch (25 mm) matched boards or laminated construction 4 inches (102 mm)

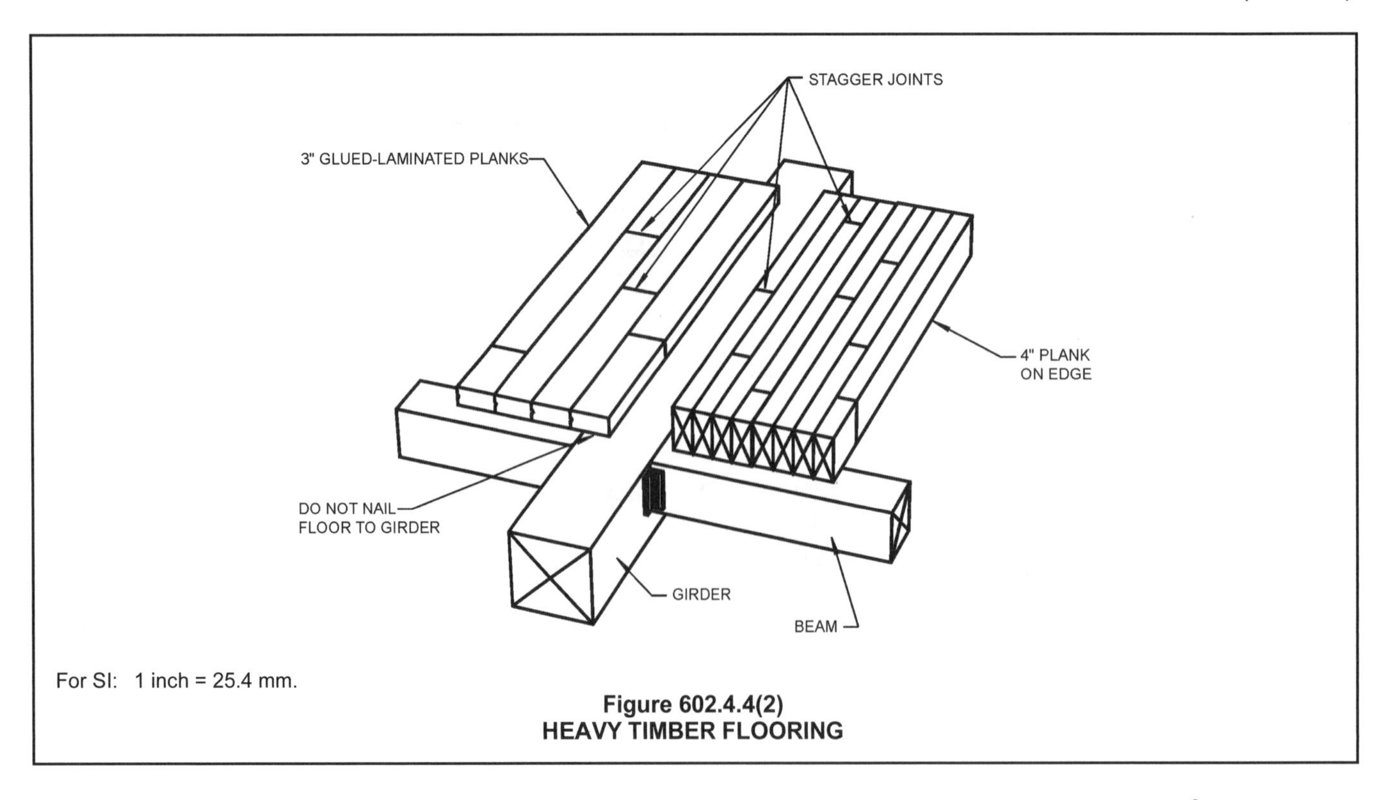

Figure 602.4.4(2)
HEAVY TIMBER FLOORING

thick if they are constructed of solid wood. Partitions are permitted to be constructed of materials other than solid wood if they have at least a 1-hour fire-resistance rating. An example of the use of alternative materials is when a fire-resistance rating for a corridor wall is required. It is common practice to utilize a 1-hour fire-resistance-rated stud and gypsum wallboard assembly between the exposed columns to form the walls of the corridor.

602.4.8.2 Exterior walls. Exterior walls shall be of one of the following:

1. Noncombustible materials.
2. Not less than 6 inches (152 mm) in thickness and constructed of one of the following:
   - 2.1. *Fire-retardant-treated wood* in accordance with Section 2303.2 and complying with Section 602.4.1.
   - 2.2. *Cross-laminated timber* complying with Section 602.4.2.

❖ Exterior walls of Type IV buildings must be of noncombustible materials, or can include either fire-retardant-treated wood in accordance with Section 602.4.1 or cross-laminated timber complying with Section 602.4.2. In addition, wood columns and arches that are of heavy timber sizes can be exposed in the exterior wall in accordance with Section 602.4.9.

602.4.9 Exterior structural members. Where a horizontal separation of 20 feet (6096 mm) or more is provided, wood columns and arches conforming to heavy timber sizes shall be permitted to be used externally.

❖ Heavy timber columns and arches that conform to minimum dimensional requirements may be used on the exterior if a fire separation distance of at least 20 feet (6096 mm) is maintained, although the exterior wall itself must be of noncombustible construction. If a fire separation distance of at least 20 feet (6096 mm) is maintained, the risk of exposure of the wood members to fire from an adjacent building is reduced, and the HT columns and arches are permitted to be exposed to the exterior.

If a building of Type IV construction has a fire separation distance of less than 20 feet (6096 mm), the wood columns and arches are to be located on the interior side of the exterior wall. The noncombustible construction of the exterior wall will provide some degree of protection to the interior timber members. Therefore, placing the wood structural members inside an exterior wall is preferable to placing them within 20 feet (6096 mm) of a lot line or adjacent building with no exposure protection.

602.5 Type V. Type V construction is that type of construction in which the structural elements, *exterior walls* and interior walls are of any materials permitted by this code.

❖ Type V construction allows the use of all types of materials, both noncombustible and combustible, but Type V buildings are most commonly constructed of dimensional lumber (see Commentary Figure 602.5 for an example of Type V construction). It is divided into two subclassifications: Types VA and VB. An example of a typical building of Type VA construction is a wood frame building in which the interior and exterior load-bearing walls, floors, roofs and all structural members are protected to provide a minimum 1-hour fire-resistance rating. Type V construction is required to comply with Table 601 and Chapter 23.

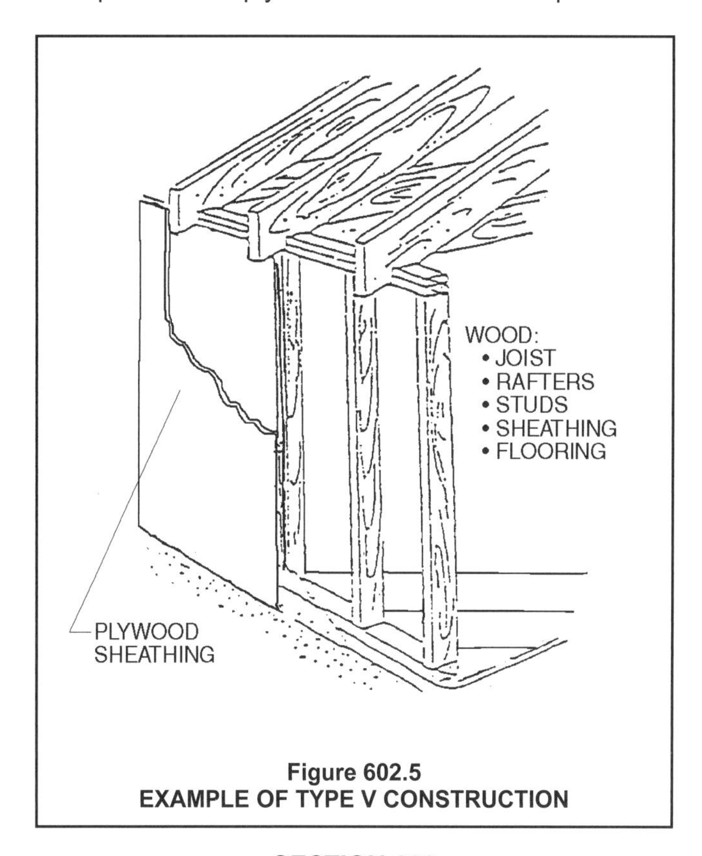

Figure 602.5
EXAMPLE OF TYPE V CONSTRUCTION

SECTION 603
COMBUSTIBLE MATERIAL IN TYPES I AND II CONSTRUCTION

603.1 Allowable materials. Combustible materials shall be permitted in buildings of Type I or II construction in the following applications and in accordance with Sections 603.1.1 through 603.1.3:

1. *Fire-retardant-treated wood* shall be permitted in:
   - 1.1. Nonbearing partitions where the required *fire-resistance rating* is 2 hours or less.
   - 1.2. Nonbearing *exterior walls* where fire-resistance-rated construction is not required.
   - 1.3. Roof construction, including girders, trusses, framing and decking.

     Exception: In buildings of Type IA construction exceeding two *stories above grade plane*, *fire-retardant-treated wood* is not permitted in roof construction where the vertical distance from the upper floor to the roof is less than 20 feet (6096 mm).

2. Thermal and acoustical insulation, other than foam plastics, having a *flame spread index* of not more than 25.

   Exceptions:

   1. Insulation placed between two layers of noncombustible materials without an intervening airspace shall be allowed to have a *flame spread index* of not more than 100.
   2. Insulation installed between a finished floor and solid decking without intervening airspace shall be allowed to have a *flame spread index* of not more than 200.

3. Foam plastics in accordance with Chapter 26.
4. Roof coverings that have an A, B or C classification.
5. *Interior floor finish* and floor covering materials installed in accordance with Section 804.
6. Millwork such as doors, door frames, window sashes and frames.
7. *Interior wall and ceiling finishes* installed in accordance with Sections 801 and 803.
8. *Trim* installed in accordance with Section 806.
9. Where not installed greater than 15 feet (4572 mm) above grade, show windows, nailing or furring strips and wooden bulkheads below show windows, including their frames, aprons and show cases.
10. Finish flooring installed in accordance with Section 805.
11. Partitions dividing portions of stores, offices or similar places occupied by one tenant only and that do not establish a *corridor* serving an *occupant load* of 30 or more shall be permitted to be constructed of *fire-retardant-treated wood*, 1-hour fire-resistance-rated construction or of wood panels or similar light construction up to 6 feet (1829 mm) in height.
12. Stages and platforms constructed in accordance with Sections 410.3 and 410.4, respectively.
13. Combustible *exterior wall coverings*, balconies and similar projections and bay or oriel windows in accordance with Chapter 14.
14. Blocking such as for handrails, millwork, cabinets and window and door frames.
15. Light-transmitting plastics as permitted by Chapter 26.
16. Mastics and caulking materials applied to provide flexible seals between components of *exterior wall* construction.
17. Exterior plastic veneer installed in accordance with Section 2605.2.
18. Nailing or furring strips as permitted by Section 803.11.
19. Heavy timber as permitted by Note c to Table 601 and Sections 602.4.7 and 1406.3.
20. Aggregates, component materials and admixtures as permitted by Section 703.2.2.
21. Sprayed fire-resistant materials and intumescent and mastic fire-resistant coatings, determined on the basis of *fire resistance* tests in accordance with Section 703.2 and installed in accordance with Sections 1705.14 and 1705.15, respectively.
22. Materials used to protect penetrations in fire-resistance-rated assemblies in accordance with Section 714.
23. Materials used to protect joints in fire-resistance-rated assemblies in accordance with Section 715.
24. Materials allowed in the concealed spaces of buildings of Types I and II construction in accordance with Section 718.5.
25. Materials exposed within plenums complying with Section 602 of the *International Mechanical Code*.
26. Wall construction of freezers and coolers of less than 1,000 square feet (92.9 m^2), in size, lined on both sides with noncombustible materials and the building is protected throughout with an *automatic sprinkler system* in accordance with Section 903.3.1.1.

❖ Section 603.1 provides a listing of circumstances where combustible materials can be used in buildings of Type I and II construction, which are otherwise required to be constructed of noncombustible materials. The list of 26 items frequently references the code user to other sections of the code where specific allowances are found. Most of these listed items, while of combustible materials, are either of minor quantities in the overall building or are of materials with fire-resistive properties.

Treated wood: Fire-retardant-treated wood (FRTW) does not meet the criteria in the code for a noncombustible material. It is, however, permitted as an alternative to noncombustible materials in specific locations in Type I and II construction (see Items 1, 11 and 13). For example, the use of FRTW in walls of Type I and II construction has been limited to nonload-bearing partitions with a fire-resistance rating of no greater than 2 hours and nonload-bearing exterior walls without a fire-resistance rating. Additionally, roofs in buildings of Type I and II construction are permitted to be constructed of FRTW. The exception to Item 1.3 does not permit FRTW in the roofs of buildings of Type IA construction over two stories in height if the distance from the uppermost floor to the roof is less than 20 feet (6096 mm). If the distance is 20 feet (6096 mm) or greater, then FRTW is acceptable in the roof. FRTW is permitted in the roof of any Type IA building two stories or less in height regardless of the distance from the floor to the roof. Similarly, FRTW is permitted in the roof of any Type IB or Type II building (no height restrictions) regardless of the distance from the floor to the roof.

Untreated wood: Numerous items in the list of Section 603.1 permit the use of untreated wood in Type I and II construction. Blocking or nailers used to support fixtures, railings, cabinets or interior and exterior finishes are permitted within walls and partitions required to be of noncombustible construction in accordance with Item 14. Item 18 permits combustible nailers and blocking as stipulated in accordance with Section 803.11. Section 803.11.1 indicates that "furring strips not exceeding 1.75 inches (44 mm)" are permitted to be used in concrete or masonry construction for securing trim and finishes. Although locating these combustible elements within noncombustible frame partitions is not specifically identified in this section, the presence of combustible nailers within noncombustible construction types, other than concrete and masonry, represents an equivalent circumstance. Therefore, it is the intent of the code to permit the use of combustible nailers and blocking within Type I and II construction.

According to Item 11, partitions within a small store or office can include untreated wood in their construction (or FTRW), provided the partitions do not exceed 6 feet (1829 mm) in height. The limitations to 30 occupants and to not forming a corridor restrict this exception from introducing extensive amounts of combustible materials. Item 9 permits the use of untreated wood for show windows and related construction up to a height 15 feet (4572 mm) above grade. This permits a broader range of materials for storefronts without jeopardizing the overall integrity of these buildings.

Item 19 references three different code sections where heavy timber construction can occur within and on the outside of noncombustible buildings.

Item 26 would allow wood construction in freezers and coolers provided the walls are lined with noncombustible materials. This would allow untreated wood as framing elements as well as trim and frames of these installations.

Plastics: Items 2, 3, 15 and 17 specifically address allowed plastics within or on the surfaces of Type I and II buildings. These items include allowances for thermal and acoustical insulation having a flame spread index of no greater than 25 when tested in accordance with ASTM E84 or UL 723; foam and light-transmitting plastics complying with Chapter 26; and exterior plastic veneers complying with Section 2605.2. Plastic materials are often included in installations allowed for interior finishes and within the construction of the building, in accordance with Item 7. Item 26 allows freezers and coolers which will often have foam plastic insulation within the walls.

Roof construction and roof coverings: While combustible roofs must be of FRTW if used in a Type I or II building, roof coverings, blocking, nailers and furring strips are also permitted to be combustible without the use of FRTW (see Items 4, 14 and 18).

"Roof covering" is defined as the membrane covering the roof that provides weather resistance, fire resistance and appearance. As long as a noncombustible roof deck (or FRTW) is provided as the structural element, foam plastic insulation, wood structural panels, nailing/furring strips and roof coverings may be applied.

Exterior wall coverings and projections: Item 13 provides a reference to Chapter 14 where provisions address combustible wall coverings and use of combustible materials in projections such as balconies and eaves. Combustible wall coverings need to be applied on top of a noncombustible wall meeting the required rating of Table 601. The coverings cannot be a substitute for the exterior face of a tested assembly. Plastic veneers (Item 17) are allowed in accordance with Section 2605.2.

Concealed spaces and construction: Many combustible materials are used, and allowed to be used within the elements of a building. Often these are used to help protect a fire-resistance-rated assembly where it is penetrated by pipes, ducts or conduit. Many of these materials are tested for the specific purposes and installations such as protection of joints or are used to actually resist, or react to, fire (see Items 16, 20, 21, 22, 23, 24 and 25).

Finish and interior materials: Finish materials for floors, walls and ceilings need to comply with various provisions of Chapter 8. Allowed materials do include combustible materials, but typically they are limited in flame spread index and smoke generation. Items 5, 7 and 8 address interior finishes. Stages and platforms can be combustible when in compliance with Section 410 (see Item 12). And, finally, doors and windows are not among the elements required to be noncombustible in accordance with Item 6.

603.1.1 Ducts. The use of nonmetallic ducts shall be permitted where installed in accordance with the limitations of the *International Mechanical Code.*

❖ Ducts are not addressed by construction-type requirements and their use is not controlled by construction-type provisions. The *International Mechanical Code*® (IMC®) provides requirements for nonmetallic ducts and addresses issues including flammability and flame spread. This section clarifies that this chapter is not intended to override those requirements.

603.1.2 Piping. The use of combustible piping materials shall be permitted where installed in accordance with the limitations of the *International Mechanical Code* and the *International Plumbing Code.*

❖ Piping is not addressed by construction-type requirements and the use of piping is not controlled by construction-type provisions. The *International Plumbing Code*® (IPC®) provides requirements for combustible piping materials, such as plastic, that address the issues of flammability and flame spread. This section clarifies that this chapter is not intended to override those requirements.

603.1.3 Electrical. The use of electrical wiring methods with combustible insulation, tubing, raceways and related components shall be permitted where installed in accordance with the limitations of this code.

❖ Electrical wiring and equipment are not addressed by construction-type requirements and their use is not controlled by construction-type provisions. NFPA 70, *National Electrical Code*, provides requirements for combustible wiring materials that address the issue of flammability and flame spread. This section clarifies that this chapter is not intended to override those requirements.

Bibliography

The following resource materials were used in the preparation of the commentary for this chapter of the code.

ASTM E84-13, *Test Methods for Surface Burning Characteristics of Building Materials.* West Conshohocken, PA: ASTM International, 2013.

DOC PS 20-05, *American Softwood Lumber Standard.* Washington, DC: United States Department of Commerce, 2005.

Fire Protection through Modern Building Codes. Washington, DC: American Iron and Steel Institute, 1981.

NFPA 13-13, *Standard for the Installation of Sprinkler Systems.* Quincy, MA: National Fire Protection Association, 2013.

NFPA 70-14, *National Electrical Code.* Quincy, MA: National Fire Protection Association, 2014.

Technical Report No. 1, *Comparative Fire Test on Wood and Steel Joists.* Washington, DC: National Forest Products Association, 1961.

Timber Construction Manual, 2nd edition. American Institute of Timber Construction. New York: John Wiley and Sons Inc., 1974.

Wood Construction Data No. 5, *Heavy Timber Construction.* Washington, DC: American Forest and Paper Association, 2004.

Chapter 7: Fire and Smoke Protection Features

General Comments

Chapter 7 provides detailed requirements for fire-resistance-rated construction, including structural members, walls, partitions and horizontal assemblies. Other portions of the code tell us when certain fire-resistance-rated elements are required. This chapter specifies how these elements are constructed, how openings in walls and partitions are protected, and how penetrations of such elements are protected.

Fire-resistance-rated construction is one form of fire protection in building design. It is often referred to as "passive protection." Fire-resistance-rated building elements provide resistance to the advance of fire, as opposed to active fire protection systems, such as automatic sprinkler systems, which actively attempt to suppress a fire.

At the core of fire-resistance-rated construction is test standard ASTM E119. This test standard specifies how to test different building elements for fire resistance, and defines what the level of fire-resistance performance is, based upon this test.

Restriction of fire growth, or passive protection, has been a key part of building codes since their inception. Other sections of the code require elements to have a fire-resistance rating or a degree of resistance to flame spread because of the relative hazard associated with the type of construction, the occupancy of the building or the function of the element.

The construction features dealt with in this chapter that are required to have some degree of fire resistance include structural members (see Section 704), exterior walls (Section 705), fire walls (Section 706), fire barriers (Section 707), fire partitions (Section 708), smoke barriers (Section 709), smoke partitions (Section 710), horizontal assemblies (Section 711), vertical openings (Section 712) and shaft enclosures (Section 713). Additionally, methods of protecting penetrations, joints, doors, windows, ducts and air transfer openings in fire-resistant elements are covered in Sections 714, 715, 716 and 717. This chapter also covers draftstopping and fireblocking of concealed spaces, in Section 718. The flame spread and smoke development of insulation are regulated by Section 720. As an alternative to tested assemblies, some prescriptive and calculated methods of determining fire-resistant assemblies are given in Sections 721 and 722.

SECTION 701 GENERAL

701.1 Scope. The provisions of this chapter shall govern the materials, systems and assemblies used for structural *fire resistance* and fire-resistance-rated construction separation of adjacent spaces to safeguard against the spread of fire and smoke within a building and the spread of fire to or from buildings.

❖ The provisions of Chapter 7 apply to the materials, assemblies and systems used to protect against the passage of fire and smoke. Each of the walls and partitions outlined herein provides various degrees of protection. The required fire-resistance rating varies with the potential fire hazard associated with the type of construction, occupancy, height and area of the building and degree of protection for different elements of the means of egress. The potential fire hazard associated with various occupancies is reflected in Tables 508.4, 509, 706.4, 707.3.10 and 1020.1. Chapter 7 provides the details and the extent of the protection (horizontal and vertical continuity); however, the actual fire-resistance-rated construction is mandated by provisions in Chapters 4, 5, 6, 7 and 10.

Whenever the code mandates the use of materials or assemblies that are noncombustible or have a degree of fire or smoke resistance, the performance of the material or assembly is required to be evaluated in accordance with the provisions of this chapter and the referenced standards.

The important criteria for the terms "combustible" and "noncombustible" used in this chapter and elsewhere throughout the code are contained in Section 703.5.

701.2 Multiple use fire assemblies. Fire assemblies that serve multiple purposes in a building shall comply with all of the requirements that are applicable for each of the individual fire assemblies.

❖ Some building elements are required by two or more code provisions to be fire-resistance rated. The same wall can serve as a load-bearing wall in Type IIA construction, an incidental use separation wall and a corridor wall. The structural load-bearing wall must meet Section 704.1, the incidental use separation wall is a fire barrier and must meet Section 707, while the corridor wall is a fire partition and must meet Section 708. Another example would be an elevator lobby with an adjacent enclosed exit stair open to a corridor system that is required to be fire-resistance rated. In this example the elevator shaft enclosure door opening and the exit stairway door opening would need to not only comply with the requirements for openings in a fire barrier, but additionally must meet the smoke and draft control requirements for an opening in a fire-resistance-rated corridor.

SECTION 702 DEFINITIONS

702.1 Definitions. The following terms are defined in Chapter 2:

❖ This is a list of terms applicable to Chapter 7. Those definitions and the commentary for them are in Chapter 2.

ANNULAR SPACE.

BUILDING ELEMENT.

CEILING RADIATION DAMPER.

COMBINATION FIRE/SMOKE DAMPER.

CORRIDOR DAMPER.

DAMPER.

DRAFTSTOP

F RATING.

FIRE BARRIER.

FIRE DAMPER.

FIRE DOOR.

FIRE DOOR ASSEMBLY.

FIRE PARTITION.

FIRE PROTECTION RATING.

FIRE-RATED GLAZING.

FIRE RESISTANCE.

FIRE-RESISTANCE RATING.

FIRE-RESISTANT JOINT SYSTEM.

FIRE SEPARATION DISTANCE.

FIRE WALL.

FIRE WINDOW ASSEMBLY.

FIREBLOCKING.

FLOOR FIRE DOOR ASSEMBLY.

HORIZONTAL ASSEMBLY.

JOINT.

L RATING.

MEMBRANE PENETRATION.

MEMBRANE-PENETRATION FIRESTOP.

MEMBRANE-PENETRATION FIRESTOP SYSTEM.

MINERAL FIBER.

MINERAL WOOL.

PENETRATION FIRESTOP.

SELF-CLOSING.

SHAFT.

SHAFT ENCLOSURE.

SMOKE BARRIER.

SMOKE COMPARTMENT.

SMOKE DAMPER.

SPLICE.

T RATING.

THROUGH PENETRATION.

THROUGH-PENETRATION FIRESTOP SYSTEM.

SECTION 703 FIRE-RESISTANCE RATINGS AND FIRE TESTS

703.1 Scope. Materials prescribed herein for *fire resistance* shall conform to the requirements of this chapter.

❖ Standard fire test methods for determining fire-resistance ratings and combustibility of materials are covered in this section. Section 703.2 addresses the acceptable testing method to be used in determining the fire-resistance ratings of building assemblies and structural elements, while Section 703.3 provides alternative methods. Section 703.4 clarifies that sprinklers cannot be utilized when testing passive fire-resistance ratings. Section 703.5 defines methods to determine if a material is combustible or noncombustible. Section 703.6 references the testing and labeling requirements for fire-resistance-rated glazing. The signage requirements for various fire-resistance-rated walls and partitions are in Section 703.7.

Fire-resistance ratings of structural members are among the factors taken into consideration when determining the allowable height and area of buildings. By using materials and construction methods that result in the required fire-resistance ratings (see commentary, Table 503), the inherent fire hazards associated with different uses are, in part, addressed. Other sections of the code also contain provisions for required fire-resistance ratings in specific applications based on the need to separate uses, processes, fire areas or building areas.

703.2 Fire-resistance ratings. The *fire-resistance rating* of building elements, components or assemblies shall be determined in accordance with the test procedures set forth in ASTM E119 or UL 263 or in accordance with Section 703.3. The *fire-resistance rating* of penetrations and fire-resistant joint systems shall be determined in accordance Sections 714 and 715, respectively.

❖ The fire-resistance properties of materials and assemblies must be measured and specified in accordance with a common test standard. For this reason, the fire-resistance ratings of building assemblies and structural elements are required to be determined in accordance with ASTM E119 or UL 263. The ASTM E119 and UL 263 test methods evaluate the ability of an assembly to contain a fire, to retain its structural integrity or both in terms of endurance time during the test conditions. The test standard also contains conditions for measuring heat transfer through membrane elements protecting framing or surfaces. The fire exposure is based on the standard time-temperature curve (see Commentary Figure 703.2). In addition, the data to indicate compliance with the requirements of Sections 714 and 715 for penetrations and joints must be submitted to verify that the required fire-resistance ratings are maintained. These sections give prescriptive and performance compliance criteria for penetrations and openings within or through fire-resistance-rated assemblies. The details of the penetrations and openings within the tested assembly must meet these minimum requirements.

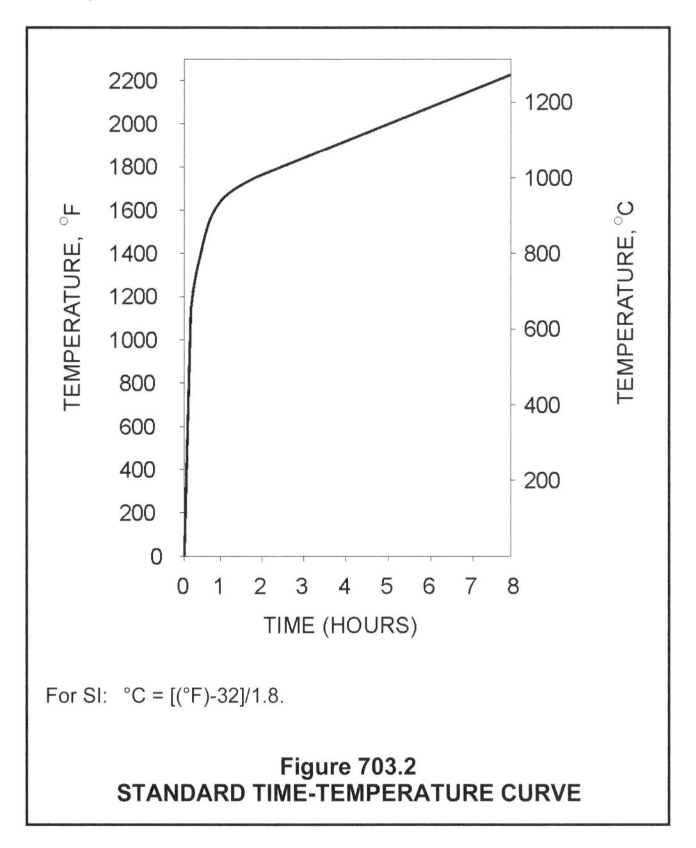

Figure 703.2
STANDARD TIME-TEMPERATURE CURVE

703.2.1 Nonsymmetrical wall construction. Interior walls and partitions of nonsymmetrical construction shall be tested with both faces exposed to the furnace, and the assigned *fire-resistance rating* shall be the shortest duration obtained from the two tests conducted in compliance with ASTM E119 or UL 263. Where evidence is furnished to show that the wall was tested with the least fire-resistant side exposed to the furnace, subject to acceptance of the *building official*, the wall need not be subjected to tests from the opposite side (see Section 705.5 for *exterior walls*).

❖ In applying Sections 703.2 and 703.2.1, it is the intent of the code that the required fire-resistance rating of interior walls be based on a fire exposure from either side. For example, it is appropriate to expect the required fire performance of an exit access corridor wall from the corridor side, as well as from the room side. If the corridor becomes involved in a fire, the corridor wall must restrict the spread of fire into adjacent spaces. However, Section 704.5 does allow exterior walls with a fire separation distance of more than 5 feet (1524 mm) to be fire-resistance rated from the interior only.

Section 703.2.1 permits testing from only one side of nonsymmetrical assemblies, provided it can be demonstrated that the tested side results in the lower value (i.e., the nontested side, if tested, would result in a higher value). For example, a 2-inch by 4-inch (51 mm by 102 mm) wood stud wall with two layers of $^{5}/_{8}$-inch (15.9 mm) Type X gypsum board on one side and only one layer on the opposite side is an illustration of a "nonsymmetrical" assembly. This assembly would be permitted to be tested only from the side with the single layer of gypsum board. This is because testing from the side with the two layers would result in a higher rating.

703.2.2 Combustible components. Combustible aggregates are permitted in gypsum and Portland cement concrete mixtures for fire-resistance-rated construction. Any component material or admixture is permitted in assemblies if the resulting tested assembly meets the fire-resistance test requirements of this code.

❖ The combustibility of materials is regulated by the code when it has been determined that the material or assembly could significantly contribute to a fire hazard. However, the limited use of combustible materials, such as combustible aggregates, is permitted as long as the desired fire-resistance rating can still be obtained. The use of a combustible aggregate in the material or assembly will not significantly contribute to fire growth or severity.

703.2.3 Restrained classification. Fire-resistance-rated assemblies tested under ASTM E119 or UL 263 shall not be considered to be restrained unless evidence satisfactory to the *building official* is furnished by the *registered design professional* showing that the construction qualifies for a restrained classification in accordance with ASTM E119 or UL 263. Restrained construction shall be identified on the *construction documents*.

❖ The effect on the fire-resistance rating of a structural element that is part of an assembly is addressed, to an extent, through the distinction in ASTM E119 or UL 263 between "restrained" and "unrestrained" conditions. The standard defines this distinction as follows: "A restrained condition in fire tests, as used in this test method, is one in which expansion of the supports of a load-carrying element resulting from the effects of the fire is resisted by forces external to the element. An unrestrained condition is one in which the load-carrying element is free to expand and rotate at its supports." In evaluating fire-resistance ratings of structural elements, the conditions of support must be consistent with the tested condition as described in the test report.

A restrained classification yields higher fire-resistance ratings than unrestrained; therefore, the code takes the conservative approach, and thus the lesser rating, by assuming the in-place conditions to be unrestrained unless structural documentation is provided that supports a restrained condition.

703.2.4 Supplemental features. Where materials, systems or devices that have not been tested as part of a fire-resistance-rated assembly are incorporated into the building element, component or assembly, sufficient data shall be made available to the *building official* to show that the required *fire-resistance rating* is not reduced.

❖ The extent to which the test specimen is representative of in-place conditions is determined by the test standard and what is actually tested. For example, the test standard does not provide for the effect on fire endurance of conventional openings in the assembly, such as electrical outlets, plumbing pipes, etc., unless specifically included in the construction assembly tested. The test reports will, however, typically indicate an allowance for openings for steel electrical outlet boxes that do not exceed 16 square inches (0.0103 m^2) in area, provided that the area of such openings does not exceed 100 square inches (0.0645 m^2) for any 100 square feet (9.3 m^2) of wall area. Since, in actuality, assemblies are often penetrated by cables, conduits, ducts, pipes, vents and similar materials, such penetrations that are not part of the tested assembly must be adequately protected and sealed for the assembly to perform as required in resisting the spread of fire. This section, therefore, also mandates that adequate documentation be submitted in order for the building official to verify that the fire-resistance rating of the assembly is not compromised by the penetration. The documentation submitted as evidence that the required fire-resistance ratings are not reduced can be in the form of specific product, assembly or system listings (see the definition of "Listed" in Section 202), engineering judgments provided by accredited testing laboratories, product manufacturers that are based on similar tested conditions or, less frequently, data reports for specific fire tests conducted to evaluate a specific penetration method. However, care must be taken to determine that the materials and methods of the proposed penetration or joint protection correspond to the details of the actual penetration or joint protection assembly that was tested. In the case of engineering judgments submitted for through-penetration firestops or fire-resistive joint systems, the publication *Recommended IFC Guidelines for Evaluating Firestop Engineering Judgments (EJ's)* can be used as a tool in evaluating the submitted documentation. It is available as a free download from the International Firestop Council (www.firestop.org).

ASTM E119 and UL 263 state that they are not intended to determine the following:

- Contribution of the tested assembly to the fire hazard;
- Measurement of the degree of control or limitation of the passage of smoke or products of combustion;
- Simulation of the fire behavior at joints between building elements, such as floor-to-wall or wall-to-wall connections; or
- Measurement of the flame spread over the surface of the tested element.

Documentation that a component or assembly has been tested to ASTM E119 or UL 263 is typically in the form of a test report issued by an approved, qualified testing agency. The test report contains certain information, including the time periods of fire resistance (e.g., 1 hour, $1^1/_2$ hours, 2 hours, etc.) and other significant details of the material or assembly. In lieu of an actual copy of a test report, the building official may accept the use of a design number found in a compendium of ASTM E119 or UL 263 test reports such as, but not limited to, the *Fire Resistance Design Manual* published by the Gypsum Association, the *Fire Resistance Directory* published by UL LLC, or certification listings published by Warnock Hersey. Each publication includes a detailed description of the assembly, its hourly fire-resistance rating and other pertinent details, such as specifications of materials and alternative assembly details.

These publications detail the design of columns, beams, walls, partitions and floor/ceiling and roof/ceiling assemblies that have been tested or analyzed in accordance with ASTM E119 or UL 263. The designs are classified in accordance with their use and fire-resistance rating. Often, wall, partition, floor/ceiling and roof/ceiling assemblies are further classified with sound transmission class (STC) ratings. Many floor/ceiling assemblies also include impact insulation class (IIC) ratings. The designs indicate whether an assembly was tested under load-bearing or nonload-bearing conditions, and what structural height limitations there are for nonload-bearing assemblies. Many of the publications have introductory material that provides details on the uses and limitations of the tested assemblies, such as the amount of penetrations permitted in an assembly for electrical outlet boxes.

703.2.5 Exterior bearing walls. In determining the *fire-resistance rating* of exterior bearing walls, compliance with the ASTM E119 or UL 263 criteria for unexposed surface temperature rise and ignition of cotton waste due to passage of flame or gases is required only for a period of time corresponding to the required *fire-resistance rating* of an exterior nonbearing wall with the same *fire separation distance*, and in a building of the same group. Where the *fire-resistance* *rating* determined in accordance with this exception exceeds the *fire-resistance rating* determined in accordance with ASTM E119 or UL 263, the fire exposure time period, water pressure and application duration criteria for the hose stream test of ASTM E119 or UL 263 shall be based on the *fire-resistance rating* determined in accordance with this section.

❖ This section applies to exterior load-bearing walls that are required to have a fire-resistance rating for structural integrity and not as a fire barrier (e.g., due to fire separation distance in accordance with Table 705.8). The exception is necessary because ASTM E119 and UL 263 do not distinguish between structural and fire-containment features of a wall. For example, assume that a concrete double-tee wall assembly was tested in accordance with ASTM E119 or UL 263. The flanges of the double-tees were $1^1/_2$ inches thick (38 mm) and the load applied to the wall was 9,600 pounds per lineal foot (14 150 kg/m). Because the temperature of the unexposed surface rose an average of 250°F (121°C) in 21 minutes, the fire-resistance rating assigned to the assembly was 21 minutes. The fire exposure was continued after the unexposed surface temperature criterion was exceeded and the wall assembly continued to support the applied load for 2 hours, at which time the test was terminated. Thus, the wall should be acceptable where structural performance for 2 hours is required and prevention of fire spread from building to building is not under consideration because of the fire separation distance. The exception, therefore, allows exterior load-bearing walls to be tested after the first failure criterion of ASTM E119 or UL 263 is reached, if the failure is other than a failure to sustain the applied load, and allows a structural fire-resistance rating to be determined. It should be noted that the hose stream test procedure is to be based on the fire exposure time period related to the structural fire-resistance rating.

703.3 Methods for determining fire resistance. The application of any of the methods listed in this section shall be based on the fire exposure and acceptance criteria specified in ASTM E119 or UL 263. The required *fire resistance* of a building element, component or assembly shall be permitted to be established by any of the following methods or procedures:

1. Fire-resistance designs documented in approved sources.
2. Prescriptive designs of fire-resistance-rated building elements, components or assemblies as prescribed in Section 721.
3. Calculations in accordance with Section 722.
4. Engineering analysis based on a comparison of building element, component or assemblies designs having *fire-resistance ratings* as determined by the test procedures set forth in ASTM E119 or UL 263.
5. Alternative protection methods as allowed by Section 104.11.
6. Fire-resistance designs certified by an approved agency.

❖ When based on the fire exposure and acceptance criteria of ASTM E119 or UL 263, an analytical method can be used in lieu of an actual ASTM E119 or UL 263 fire test to establish the fire-resistance rating of a single component or assembly.

1. The commentary to Section 703.2 also identifies additional resource materials relative to test reports.

2 & 3. Sections 721 and 722 provide further methods of determining fire-resistance ratings (see the commentary to these sections).

For example, methods for determining fire-resistance ratings of concrete or masonry assemblies have been standardized in ACI 216.1/TMS 0216.1 (see commentary, Section 722.1), as well as in the following:

- PCI MNL 124, Precast/Prestressed Concrete Institute.
- *Reinforced Concrete Fire Resistance*, Concrete Reinforcing Steel Institute.
- CMIFC SR267.01B, Concrete and Masonry Industry Fire Safety Committee.
- *A Compilation of Fire Tests on Concrete Masonry Assemblies*, National Concrete Masonry Association.
- NCMA TEK 6A, National Concrete Masonry Association.
- NCMA 35D, National Concrete Masonry Association.

4. An approved engineering analysis allows the fire protection properties of a component or assembly to be determined using information generated from previous fire tests through empirical calculations. There are several analytical methods available to designers for many types of construction materials. An engineering analysis is available for calculating fire-resistance ratings of heavy timber construction. This analysis uses an empirical mathematical method that generates a design for the minimum dimensions of a wood beam or column. These dimensions account for the changes that occur in the member due to charring of the wood after any given period of fire exposure. American Iron and Steel Institute (AISI) *Designing Fire Protection for Steel Columns* presents the factors that influence the fire-resistance ratings of steel columns in frequently used sizes and shapes. The publication also contains a discussion of the fire protection materials most often used with steel columns and provides accepted methods for calculating the fire-resistance rating of steel

columns predicated on the ASTM E119 or UL 263 fire exposure standard. Similarly, AISI's *Designing Fire Protection for Steel Beams* presents the factors that influence the fire-resistance ratings of steel beams in frequently used sizes. As typical test reports specify minimum beam sizes only, three different ways of evaluating beam substitutions are provided. Since the ASTM E119 or UL 263 test furnace is incapable of testing assemblies that incorporate large structural members, AISI's *Designing Fire Protection for Steel Trusses* proposes concepts for determining fire protection designs by applying existing test data. The application of these methods, as well as information on the fire resistance of archaic construction, is summarized in the International Code Council® (ICC®) publication, *2001 Guidelines for Determining Fire Resistance of Building Elements*. It is important to emphasize that all analytical methods of calculating fire-resistance ratings must be substantiated as being based on the fire exposure and acceptance criteria of ASTM E119 or UL 263. Upon submission of such documentation to the building official, the analytical methods mentioned herein, or any other such analytical methods, can be approved as the basis for showing compliance with the fire-resistance ratings required by the code.

5. As indicated in Section 104.11, alternative protection may be provided if approved by the building official. An example of an alternative protection method is the water-filled steel column system. The system consists of hollow, liquid-filled columns that are interconnected with pipe loops to allow the water to circulate freely. Engineering studies confirmed by tests have shown that critical temperatures are not reached in the steel columns as long as they remain filled with liquid.
6. This item allows the determination of fire-resistance design by qualified agencies. The term "certified by an approved agency" provides confidence that the design conforms to the test standard based on scientific data.

703.4 Automatic sprinklers. Under the prescriptive fire-resistance requirements of this code, the *fire-resistance rating* of a building element, component or assembly shall be established without the use of *automatic sprinklers* or any other fire suppression system being incorporated as part of the assembly tested in accordance with the fire exposure, procedures and acceptance criteria specified in ASTM E119 or UL 263. However, this section shall not prohibit or limit the duties and powers of the *building official* allowed by Sections 104.10 and 104.11.

❖ The fire-resistive requirements for structural members and walls and partitions separating areas in a building are passive fire-resistive measures and, when required, are required in addition to any other sprinkler requirements in the code. Required fire-resistance ratings must be established by tests not utilizing sprinklers. This section does not prohibit the building official from accepted alternatives in accordance with Sections 104.10 and 104.11.

703.5 Noncombustibility tests. The tests indicated in Sections 703.5.1 and 703.5.2 shall serve as criteria for acceptance of building materials as set forth in Sections 602.2, 602.3 and 602.4 in Type I, II, III and IV construction. The term "noncombustible" does not apply to the flame spread characteristics of *interior finish* or *trim* materials. A material shall not be classified as a noncombustible building construction material if it is subject to an increase in combustibility or flame spread beyond the limitations herein established through the effects of age, moisture or other atmospheric conditions.

❖ Restrictions on the use of combustible materials are primarily found in Chapter 6. However, certain sections of Chapter 7 (such as use of combustibles in Section 703.2.2 and fire walls in Section 706.3) contain restrictions on the use of combustible materials and separate provisions, depending on whether the material is combustible or noncombustible. Sections 703.5.1 and 703.5.2 contain the appropriate test criteria by which a material is to be evaluated to ascertain whether it is combustible or noncombustible.

Materials that are considered noncombustible must be capable of maintaining the required performance characteristics (noncombustibility and flame spread ratings) regardless of age, moisture or other atmospheric conditions. If exposure to atmospheric conditions results in an increase in combustibility or flame spread rating beyond the limitations specified, the material is considered combustible.

The criteria established by this section are primarily based on the National Board of Fire Underwriters' (NBFU) *National Building Code*, 1955 edition, which permitted a noncombustible material to contain a limited amount of combustible material, provided that it did not contribute to fire propagation. The requirement for noncombustibility is considered to apply to certain materials used for walls, roofs and other structural elements, but not to surface finish materials. The determination of whether a material is noncombustible from the standpoint of minimum required clearances to heating appliances, flues or other sources of high temperature is based on the ASTM E136 definition of "Noncombustibility," not the prescriptive definition in Section 703.5.2. Note that the *International*

Mechanical Code® (IMC®) contains a definition of "Noncombustible materials," which does not contain provisions for the use of composite materials (defined in Section 703.5.2) as acceptable noncombustible materials.

703.5.1 Elementary materials. Materials required to be noncombustible shall be tested in accordance with ASTM E136.

❖ Materials intended to be classified as noncombustible are to be tested in accordance with ASTM E136. In accordance with Section 8 of ASTM E136, such materials are acceptable as noncombustible when at least three of four specimens tested conform to all of the following criteria:

1. The recorded temperature of the surface and interior thermocouples shall not at any time during the test rise more than 54°F (30°C) above the furnace temperature at the beginning of the test.
2. There shall not be flaming from the specimen after the first 30 seconds.
3. If the weight loss of the specimen during testing exceeds 50 percent, the recorded temperature of the surface and interior thermocouples shall not at any time during the test rise above the furnace air temperature at the beginning of the test, and there shall not be flaming of the specimen.

The use of the test standard is limited to elementary materials and excludes laminated and coated materials because of the uncertainties associated with more complex materials and with products that cannot be tested in a realistic configuration. The test standard is also limited to solid materials and does not measure the self-heating tendencies of large masses of materials, such as resin-impregnated mineral fiber insulation.

The defined furnace temperature in the standard, 1,382°F (750°C), is representative of temperatures that are known to exist during building fires, although temperatures between 1,800°F (982°C) and 2,200°F (1204°C) are frequently achieved during intense fires. For most building materials, however, complete burning of the combustible fraction will occur as readily at 1,382°F (750°C) as at higher temperatures. The criterion requiring four specimens to be tested recognizes the variable nature of the measurements and the fact that there are difficulties in observing the presence and duration of flaming.

The need to measure and limit the duration of flaming and the rise in temperature recognizes that a brief period of flaming and a small amount of heating are not considered serious limitations on the use of building materials. Test results have shown that such criteria limit the combustible portion of noncombustible materials to a maximum of 3 percent. The 50-percent weight-loss limitation precludes the possibility that combustion of low-density materials will occur so rapidly that the recorded temperature rise and the measured flaming duration will be less than the prescribed limitations. The 50-percent limitation is considered appropriate for materials that contain appreciable quantities of combined water or gaseous components.

703.5.2 Composite materials. Materials having a structural base of noncombustible material as determined in accordance with Section 703.5.1 with a surfacing not more than 0.125 inch (3.18 mm) thick that has a *flame spread index* not greater than 50 when tested in accordance with ASTM E84 or UL 723 shall be acceptable as noncombustible materials.

❖ In recognition that an essentially noncombustible material with a thin combustible coating will not contribute appreciably to an ambient fire, this section provides criteria by which a composite material may be determined acceptable as a noncombustible material. The structural base of the composite material must meet the criteria for elementary materials (see Section 703.5.1). The material may have a surface not more than $^{1}/_{8}$ inch (3.2 mm) thick applied to the noncombustible base. This surface is required to have a flame spread rating no greater than 50 when tested in accordance with ASTM E84 or UL 723. For a discussion of both the ASTM E84 and UL 723 test methods, refer to the commentary to Section 803.1.

In accordance with this section, material such as gypsum board—which consists of a noncombustible base and a combustible (paper) surface—is considered to be a composite material. As noted in the commentary to Section 703.5, a composite material is considered a noncombustible material in the context of the code. The IMC, however, intentionally includes only limited applications of the criteria of this section in the definition of a noncombustible material. Therefore, composite materials are not considered acceptable noncombustible materials from the standpoint of clearances to heating appliances, flues or other sources of high temperature because of their potentially poor response to radiant heat exposure.

703.6 Fire-resistance-rated glazing. Fire-resistance-rated glazing, when tested in accordance with ASTM E119 or UL 263 and complying with the requirements of Section 707, shall be permitted. Fire-resistance-rated glazing shall bear a *label* marked in accordance with Table 716.3 issued by an agency and shall be permanently identified on the glazing.

❖ Fire-resistance-rated glazing is one of two designated types of fire-rated glazing—fire-protection-rated glazing being the other. Fire-rated glazing is allowed in three separate locations, depending on which tests the glazing has passed. The label is required to provide the location and the fire-resistance rating in minutes. There are door labels, window labels and wall labels. Section 703.6 concerns only the fire-resistance-rated glazing. Table 716.3 indicates that glazing meeting ASTM E119 or UL 263 shall bear a "W" label and is considered a wall; therefore, it is not subject to the limitations on doors and windows. This section indicates that fire-resistance-rated glazing with the "W" designation on the label is good for fire walls and fire barriers. It follows that it would be good for shaft walls and stair enclosures as well. It could also be used as a fire partition, smoke partition or smoke barrier.

Fire-resistance-rated glazing is also mentioned in Table 716.5 for doors and windows.

703.7 Marking and identification. Where there is an accessible concealed floor, floor-ceiling or *attic* space, *fire walls*, *fire barriers*, *fire partitions*, *smoke barriers* and smoke partitions or any other wall required to have protected openings or penetrations shall be effectively and permanently identified with signs or stenciling in the concealed space. Such identification shall:

1. Be located within 15 feet (4572 mm) of the end of each wall and at intervals not exceeding 30 feet (9144 mm) measured horizontally along the wall or partition.
2. Include lettering not less than 3 inches (76 mm) in height with a minimum $^3/_8$-inch (9.5 mm) stroke in a contrasting color incorporating the suggested wording, "FIRE AND/OR SMOKE BARRIER—PROTECT ALL OPENINGS," or other wording.

❖ This code requirement addresses the need for installed fire-resistance-rated assemblies to maintain their fire resistance over the life of the building. This identification will allow tradespeople, craftsmen, installers, maintenance workers or inspectors to know that the wall is a fire-resistance-rated wall and openings or penetrations of it must be protected. This identification is only required where there is an accessible concealed space associated with the wall construction.

SECTION 704
FIRE-RESISTANCE RATING OF STRUCTURAL MEMBERS

704.1 Requirements. The *fire-resistance ratings* of structural members and assemblies shall comply with this section and the requirements for the type of construction as specified in Table 601. The *fire-resistance ratings* shall be not less than the ratings required for the fire-resistance-rated assemblies supported by the structural members.

Exception: *Fire barriers*, *fire partitions*, *smoke barriers* and *horizontal assemblies* as provided in Sections 707.5, 708.4, 709.4 and 711.2, respectively.

❖ This section contains provisions that apply to structural members that are required to have a fire-resistance rating. The required rating of structural members is usually based on the type of construction and Table 601. Other sections of the code, such as the continuity provisions of Sections 707.5, 709.4, 710.4 and 711.2.2, may require structural members to have a fire-resistance rating because they support fire-resistance-rated construction. The minimum required fire-resistance rating for the structural member is, therefore, the more restrictive of the above two criteria.

There are instances in which the code does not require the supporting structural members to have a fire-resistance rating. An example includes the supporting construction for tenant and sleeping unit separations, exit access corridors and smoke partitions in Type IIB, IIIB, and VB construction (see Sections 708.4 and 709.4).

Protection of structural members differs for various members. Required column fire-resistance protection must be afforded in accordance with Section 704.2. Certain primary structural frames (see definition in Section 202) must be protected by individual encasement in accordance with Section 704.3. Secondary structural members (see definition in Section 202) may utilize membrane protection in accordance with Section 704.4. Special requirements apply to fire-resistance protection of trusses (see Section 704.5).

Section 704 addresses issues such as protection of attachments to structural members (see Section 704.6), embedment of service facilities (see Section 704.8), impact protection (see Section 704.9), exterior structural members (see Section 704.10), bottom flange protection (see Section 704.11), seismic isolation systems (see Section 704.12) and sprayed fire-resistant materials (SFRM) (see Section 704.13).

704.2 Column protection. Where columns are required to have protection to achieve a *fire-resistance rating*, the entire column shall be provided individual encasement protection by protecting it on all sides for the full column height, including connections to other structural members, with materials having the required *fire-resistance rating*. Where the column extends through a ceiling, the encasement protection shall be continuous from the top of the foundation or floor/ceiling assembly below through the ceiling space to the top of the column.

❖ Columns required to be fire-resistance rated must be encased on all sides. This is true even if the column is located on the exterior of the structure. The ceiling membrane or ceiling protection of a floor/ceiling or roof/ceiling fire-resistance-rated assembly is prohibited from being considered as the protection for columns required to have a fire-resistance rating. Therefore, the materials that encase or provide individual protection must continue through the concealed space above a ceiling, even if the ceiling membrane is part of a fire-resistance-rated assembly (see Commentary Figure 704.2). Connections of structural members to columns must also be individually protected (see Section 704.6). Columns that provide inherent fire resistance, without encasement, such as heavy timber, are considered as not requiring protection and do not need to comply with this section.

704.3 Protection of the primary structural frame other than columns. Members of the primary structural frame other than columns that are required to have protection to achieve a *fire-resistance rating* and support more than two floors or one floor and roof, or support a load-bearing wall or a nonload-bearing wall more than two stories high, shall be provided individual encasement protection by protecting them on all sides for the full length, including connections to other structural members, with materials having the required *fire-resistance rating*.

> **Exception:** Individual encasement protection on all sides shall be permitted on all exposed sides provided the extent of protection is in accordance with the required *fire-resistance rating*, as determined in Section 703.

❖ Not all primary structural frames need to comply with this section. Only those primary structural frame members that are required to have a fire-resistance rating and support more than two floors, more than one floor and a roof, a bearing wall of any height or a nonload-bearing wall more than two stories high, are required to be protected by individual encasement. Individual encasement, though not defined, is attained by using tested design assemblies that provide gypsum board applied directly to the member or to studs that are directly attached to all sides of the member, mastic and intumescent coatings, and sprayed fire-resistant materials.

Individual encasement is required for primary structural frame members receiving tributary loads from multiple levels in order to reduce the risk associated with catastrophic failure. The risk represented by structural collapse during a fire is significantly increased in multistory buildings, since the occupants require additional time of egress and a single structural element that supports multiple elements requires more conservative protection methods.

Though the text states that the encasement must be on all sides of the structural member, the exception will allow tested assemblies that have the encasement on only the exposed sides of a member. Commentary Figure 704.3 is an example of individual encasement protection.

Note that some structural members such as heavy wood beams may not require any protection (encasement or membrane) to attain a fire rating (see calculated fire resistance, Section 722.1).

704.4 Protection of secondary members. Secondary members that are required to have protection to achieve a *fire-resistance rating* shall be protected by individual encasement protection.

❖ Where secondary structural members are required to be fire-resistance-rated, the rating, in most cases, is required to be achieved by construction that encloses the individual member. See Commentary Figures 704.2 and 704.3 for examples of individual encasement. Situations where the fire rating can be achieved by a membrane are contained in Sections 704.4.1 and 704.4.2.

Remember that the fire-resistance requirements are established based upon a fire test in accordance with ASTM E119 or UL 263. Although the fire temperature in the furnace is the same, the test protocols in both standards have different temperature rise limits for the elements and different temperature couple

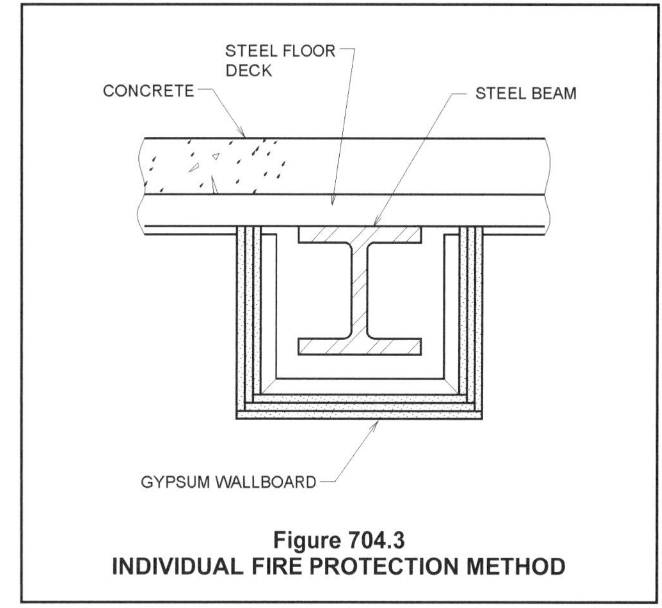

Figure 704.3
INDIVIDUAL FIRE PROTECTION METHOD

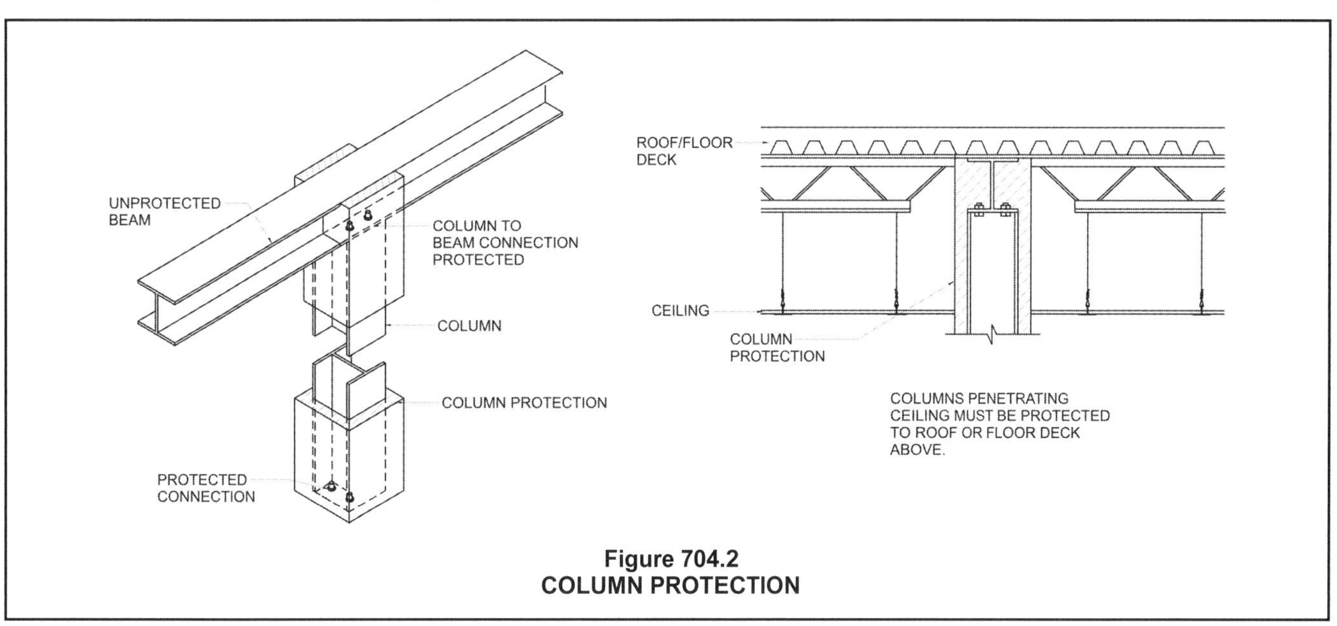

Figure 704.2
COLUMN PROTECTION

locations between a structural member test and a wall or horizontal assembly test. Therefore, although a wall may obtain a 1-hour fire-resistance rating, it does not mean that a column placed within that wall would also obtain a 1-hour fire-resistance rating. Also, a floor/ceiling assembly rated for 2 hours cannot be assumed to protect a beam that was not in the tested assembly. If the beam was in the tested assembly, the fire-resistance rating of the floor/ceiling may be different than the beam. It is important that the structural members located within the assemblies were also in the tested floor/ceiling and roof/ceiling assemblies.

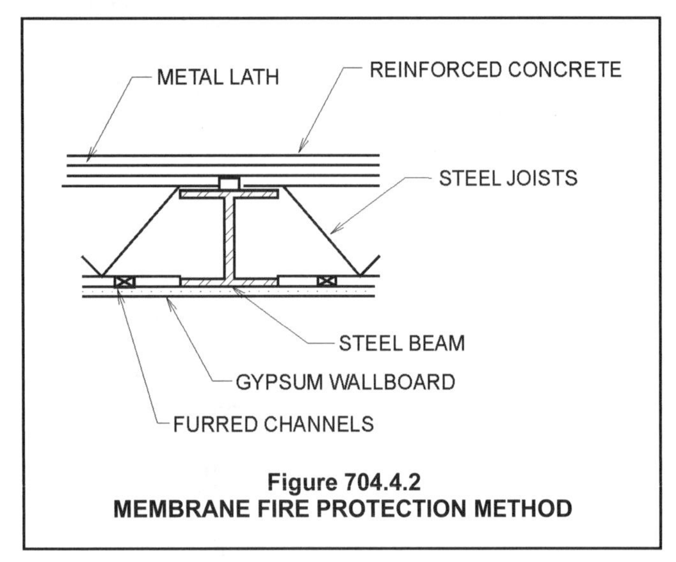

Figure 704.4.2
MEMBRANE FIRE PROTECTION METHOD

704.4.1 Light-frame construction. Studs and boundary elements that are integral elements in *load-bearing walls* of light-frame construction shall be permitted to have required *fire-resistance ratings* provided by the membrane protection provided for the *load-bearing wall.*

❖ Studs in light-frame construction are considered as standard parts of the wall assembly rather than as individual members acting as a typical column. Note that other vertical members that act as columns and are secondary members require individual encasement.

704.4.2 Horizontal assemblies. *Horizontal assemblies* are permitted to be protected with a membrane or ceiling where the membrane or ceiling provides the required *fire-resistance rating* and is installed in accordance with Section 711.

❖ The fire-resistance rating of secondary members within a horizontal assembly can be achieved by either a ceiling membrane or by individual encasement. Typically, protection by a ceiling membrane is the most economical, though it depends on the details of the specific condition. Commentary Figure 704.4.2 is an example of membrane protection.

704.5 Truss protection. The required thickness and construction of fire-resistance-rated assemblies enclosing trusses shall be based on the results of full-scale tests or combinations of tests on truss components or on *approved* calculations based on such tests that satisfactorily demonstrate that the assembly has the required *fire resistance*.

❖ Trusses can be of an almost infinite number of configurations and sizes. By nature, they are typically designed and constructed of different structural components. This leads to the impracticality of such trusses being tested. The code acknowledges this by allowing full-scale tests, tests on the individual components or some form of calculations that demonstrate that the structural elements that form the truss are provided with the requisite protection to attain the fire-resistance rating.

704.6 Attachments to structural members. The edges of lugs, brackets, rivets and bolt heads attached to structural members shall be permitted to extend to within 1 inch (25 mm) of the surface of the fire protection.

❖ This section is intended to limit the transfer of heat to the structural element from the connection elements by requiring the connection element to be protected with at least 1 inch (25 mm) of some form of fire-resistance protection. No attaching elements are allowed to the structural element that would allow heat to transfer through the protection and to the structural element. At least 1 inch (25 mm) of cover is required on all the attaching elements.

704.7 Reinforcing. Thickness of protection for concrete or masonry reinforcement shall be measured to the outside of the reinforcement except that stirrups and spiral reinforcement ties are permitted to project not more than 0.5-inch (12.7 mm) into the protection.

❖ Since load-bearing assemblies must sustain the superimposed load while being subjected to standard time-temperature fire conditions, concrete, masonry and reinforcing steel must be able to provide the required strength at elevated temperatures. This involves providing adequate cover over the steel reinforcement so that the stress induced in the reinforcement is less than its yield stress, which is commonly referred to as "yield strength." Tests of steel reinforcing bars show that at a temperature of approximately 1,100°F (593°C), the yield strength of steel is reduced to approximately 50 percent of that at ambient temperature conditions. Similar tests of prestressing tendons show that at approximately 800°F (427°C) the tensile strength is approximately 50 percent of that at ambient conditions. Cover requirements, therefore, are typically based on limiting the reinforcing and prestressing steel temperatures to 1,100 and 800°F (593 and 427°C), respectively. This section allows for stirrups and spiral column ties to project into the required cover by a minimal distance, acknowledging the practicality of construction tolerances while providing adequate cover to the main reinforcing steel, which is located inside of the stirrups and ties. This also acknowledges that the stirrups are not continuous for the length of the member and are spaced at intermittent intervals. Therefore, they only reduce the actual cover to the main reinforcing steel at a percentage of its length. The $^1/_2$-inch (12.7 mm) dimension reflects a No. 4 reinforcing bar.

704.8 Embedments and enclosures. Pipes, wires, conduits, ducts or other service facilities shall not be embedded in the required fire protective covering of a structural member that is required to be individually encased.

❖ Piping, wires, conduits, ducts and other service facilities are not to be embedded in the fire protective covering of a structural member that is required to be individually encased in accordance with Section 704.3. The fire protection performance of encasement materials is critical to achieve the required fire-resistance rating, and would be impaired if the continuity of the encasement is interrupted. This does not, however, prevent the installation if these items are installed in hollow spaces that may be created by the encasement method. When items are installed in such hollow spaces, penetrations of the protective assembly must be properly protected so that the fire-resistance rating is maintained.

704.9 Impact protection. Where the fire protective covering of a structural member is subject to impact damage from moving vehicles, the handling of merchandise or other activity, the fire protective covering shall be protected by corner guards or by a substantial jacket of metal or other noncombustible material to a height adequate to provide full protection, but not less than 5 feet (1524 mm) from the finished floor.

Exception: Corner protection is not required on concrete columns in open or enclosed parking garages.

❖ To aid in the reliability of the fire protective covering, impact protection must be provided if the covering is subject to impact from moving vehicles, handling of merchandise or other activities. The impact protection is to be provided to at least 5 feet (1524 mm) above the finished floor or as needed to prevent impact damage.

The type of protection required is not specified within the code, but it should be determined based on the type of fire protection used and also the expected impact that the protection may be expected to encounter. For example, if the fire-resistance protection of a column is provided by either spray-on fireproofing or gypsum board, that would be more likely to be damaged than a concrete-encased column. Additionally, if the column was in a grocery store and the only expected item to impact the column was a shopping cart, then it may be determined acceptable to simply cover the gypsum board with a piece of thin sheet metal. On the other hand, if the column is located in a warehouse or parking garage where a vehicle such as a forklift or automobile could strike it, then the protection may require the installation of steel plates or bollards.

704.10 Exterior structural members. Load-bearing structural members located within the *exterior walls* or on the outside of a building or structure shall be provided with the highest *fire-resistance rating* as determined in accordance with the following:

1. As required by Table 601 for the type of building element based on the type of construction of the building;
2. As required by Table 601 for exterior bearing walls based on the type of construction; and
3. As required by Table 602 for *exterior walls* based on the *fire separation distance*.

❖ Exterior load-bearing structural members, such as columns or girders, must have the same fire-resistance rating required for exterior load-bearing walls. As such, the required fire-resistance rating is the higher rating of that found in Table 601 for type of construction for structural elements or bearing walls or as required in Table 602 based on the fire separation distance.

704.11 Bottom flange protection. Fire protection is not required at the bottom flange of lintels, shelf angles and plates, spanning not more than 6 feet 4 inches (1931 mm) whether part of the primary structural frame or not, and from the bottom flange of lintels, shelf angles and plates not part of the structural frame, regardless of span.

❖ Structural frame elements located over an opening in a wall required to be fire-resistance rated, particularly a masonry wall with an exposed steel lintel or angle over the wall opening, are covered by this section. The bottom flanges of such lintels or angles if less than 6 feet 4 inches (1930 mm) in length are not required to be protected even if part of the primary structural frame. Regardless of the span, the bottom flange of such lintels or angles is not required to be protected if the member is not part of the primary structural frame.

704.12 Seismic isolation systems. *Fire-resistance ratings* for the isolation system shall meet the *fire-resistance rating* required for the columns, walls or other structural elements in which the isolation system is installed in accordance with Table 601. Isolation systems required to have a *fire-resistance rating* shall be protected with *approved* materials or construction assemblies designed to provide the same degree of *fire resistance* as the structural element in which the system is installed when tested in accordance with ASTM E119 or UL 263 (see Section 703.2).

Such isolation system protection applied to isolator units shall be capable of retarding the transfer of heat to the isolator unit in such a manner that the required gravity load-carrying capacity of the isolator unit will not be impaired after exposure to the standard time-temperature curve fire test prescribed in ASTM E119 or UL 263 for a duration not less than that required for the *fire-resistance rating* of the structure element in which the system is installed.

Such isolation system protection applied to isolator units shall be suitably designed and securely installed so as not to dislodge, loosen, sustain damage or otherwise impair its ability to accommodate the seismic movements for which the isolator unit is designed and to maintain its integrity for the purpose of providing the required fire-resistance protection.

❖ This section states that the fire-resistance requirements for seismic isolation systems are the same as those for other structural elements, as required in Table 601. Additionally, the fire-resistance method used must be able to withstand the anticipated movement of the system without the fire-resistance application methods and systems affecting its functionality during a seismic event. Elements of the system can either be individually protected or contained within a tested assembly. Another key element of this section is that, when tested under exposure to the standard time-temperature curve test, the structural elements are still able to carry the gravity load as designed.

704.13 Sprayed fire-resistant materials (SFRM). Sprayed fire-resistant materials (SFRM) shall comply with Sections 704.13.1 through 704.13.5.

❖ The intent of this code section is to increase the in-place durability of spray-applied fire-resistant materials (SFRM). The National Institute of Standards and Technology's (NIST) investigation on the World Trade Center tragedy documented that the proximate cause of the actual collapse was the action of a building contents fire on light steel members in the absence of spray-applied fire-resistive material, which had been dislodged. Events far less dramatic than an airplane attack have been known to dislodge SFRM. Events as simple as an elevator movement, building sway or maintenance activities can dislodge SFRM if it is not adhered properly. These code requirements as well as the special testing requirements in Section 1705.13 are necessary to increase the in-place capability of SFRM assemblies. Section 403.2.4 in the high-rise buildings section gives the bond strength requirements for SFRM.

704.13.1 Fire-resistance rating. The application of SFRM shall be consistent with the *fire-resistance rating* and the listing, including, but not limited to, minimum thickness and dry density of the applied SFRM, method of application, substrate surface conditions and the use of bonding adhesives, sealants, reinforcing or other materials.

❖ In order for the fire-resistance rating that is actually realized by sprayed fire-resistant materials to equal or exceed the required fire-resistance rating, it is necessary to install the materials in accordance with their listing. In this section, minimum thicknesses, proper conditions of the substrate, method of application, and correct bonding materials are all items that are necessary to ensure the proper performance of these materials.

704.13.2 Manufacturer's installation instructions. The application of SFRM shall be in accordance with the manufacturer's installation instructions. The instructions shall include, but are not limited to, substrate temperatures and surface conditions and SFRM handling, storage, mixing, conveyance, method of application, curing and ventilation.

❖ Sections 704.13.1 and 704.13.2 require that the application of SFRM be in accordance with all terms and conditions of the listing and the manufacturer's instruction.

704.13.3 Substrate condition. The SFRM shall be applied to a substrate in compliance with Sections 704.13.3.1 through 704.13.3.2.

❖ The in-place adhesion of SFRM can be reduced by a factor of 10 when applied over certain primers compared to the adhesion obtained by the rated material applied on bare, clean steel. This section gives the requirements for the surface condition of the substrate receiving the SFRM.

704.13.3.1 Surface conditions. Substrates to receive SFRM shall be free of dirt, oil, grease, release agents, loose scale and any other condition that prevents adhesion. The substrates shall be free of primers, paints and encapsulants other than those fire tested and *listed* by a nationally recognized testing agency. Primed, painted or encapsulated steel shall be allowed, provided that testing has demonstrated that required adhesion is maintained.

❖ This section gives the minimum temperature for the air and the material itself, which must be maintained during application and for 24 hours after, unless the manufacturer's instructions allow otherwise.

704.13.3.2 Primers, paints and encapsulants. Where the SFRM is to be applied over primers, paints or encapsulants other than those specified in the listing, the material shall be field tested in accordance with ASTM E736. Where testing of the SFRM with primers, paints or encapsulants demonstrates that required adhesion is maintained, SFRM shall be permitted to be applied to primed, painted or encapsulated wide flange steel shapes in accordance with the following conditions:

1. The beam flange width does not exceed 12 inches (305 mm); or
2. The column flange width does not exceed 16 inches (400 mm); or
3. The beam or column web depth does not exceed 16 inches (400 mm).
4. The average and minimum bond strength values shall be determined based on a minimum of five bond tests conducted in accordance with ASTM E736. Bond tests conducted in accordance with ASTM E736 shall indicate an average bond strength of not less than 80 percent and an individual bond strength of not less than 50 percent, when compared to the bond strength of the SFRM as applied to clean uncoated $^{1}/_{8}$-inch-thick (3.2 mm) steel plate.

❖ When there is some sort of material, such as paint or primer, on the steel structural members where it is desired to install SFRM, it is still possible to apply the SFRM to the steel without removing the material. However, this section specifies that a field test in accordance with ASTM E736 be done on the SFRM so applied. This ensures that a bond strength is achieved and that the material will remain in place. Note that there are limitations on the sizes of structural members that can utilize this field test (Conditions 1-3).

704.13.4 Temperature. A minimum ambient and substrate temperature of 40°F (4.44°C) shall be maintained during and for not fewer than 24 hours after the application of the SFRM, unless the manufacturer's instructions allow otherwise.

❖ This section gives the minimum temperature for the air and the material itself, which must be maintained during application and for 24 hours after, unless the manufacturer's instructions allow otherwise.

704.13.5 Finished condition. The finished condition of SFRM applied to structural members or assemblies shall not, upon complete drying or curing, exhibit cracks, voids, spalls, delamination or any exposure of the substrate. Surface irregularities of SFRM shall be deemed acceptable.

❖ This section and the special inspections required by Section 1704.12 establish the acceptance criteria for the finished condition of the SFRM.

SECTION 705 EXTERIOR WALLS

705.1 General. *Exterior walls* shall comply with this section.

❖ In order to accomplish these functions, this section regulates the fire protection capability of the exterior wall, including limitations on openings in the walls. The requirements contained herein and in Table 602 are based on the fire separation distance, the occupancy that represents a relative fuel loading and the percentage of openings in the wall. Section 705.8.5 also regulates the location of openings in order to prevent ignition of materials in adjacent stories.

All other applicable provisions of the code concerning exterior walls still apply. The provisions of this section refer to the fire-resistance requirements contained in Tables 601 and 602 (see Section 705.5). Table 601 establishes minimum fire-resistance ratings for load-bearing exterior walls based on the type of construction that addresses the structural integrity of the building under fire conditions. Table 602 establishes ratings based on fire separation distance in an effort to address conflagration concerns.

705.2 Projections. Cornices, eave overhangs, exterior balconies and similar projections extending beyond the exterior wall shall conform to the requirements of this section and Section 1406. Exterior egress balconies and exterior exit stairways and ramps shall comply with Sections 1021 and 1027, respectively. Projections shall not extend any closer to the line used to determine the fire separation distance than shown in Table 705.2.

❖ Horizontal projections, similar to walls and openings, are a potential source of the spread of fire from building to building. Some projections from the building exterior wall may even shelter combustible materials. As such, limits on the allowable projections are necessary. Examples of horizontal projections are roof overhangs, exterior balconies, canopies and covered porches. Balcony projections must also meet the requirements of Section 1406, while exterior egress balconies must meet Section 1021.

The first column in Table 705.2 is Fire Separation Distance (FSD) and is measured from the lot line or assumed lot line to the exterior wall—not to the edge of the overhang. The second column is distance from the line used to determine the FSD to the vertical edge of the overhang. For examples of the application of Table 705.2, see Commentary Figure 705.2.

The exception addresses buildings on the same lot that together qualify as one aggregate building under the limitations of Chapter 5. Projections on the exterior walls of these buildings that face each other are not limited in length.

TABLE 705.2
MINIMUM DISTANCE OF PROJECTION

FIRE SEPARATION DISTANCE (FSD)	MINIMUM DISTANCE FROM LINE USED TO DETERMINE FSD
0 feet to 2 feet	Projections not permitted
Greater than 2 feet to 3 feet	24 inches
Greater than 3 feet to less than 30 feet	24 inches plus 8 inches for every foot of FSD beyond 3 feet or fraction thereof
30 feet or greater	20 feet

For SI: 1 foot = 304.8 mm; 1 inch = 25.4 mm.

Exception: Buildings on the same lot and considered as portions of one building in accordance with Section 705.3 are not required to comply with this section for projections between the buildings.

705.2.1 Type I and II construction. Projections from walls of Type I or II construction shall be of noncombustible materials or combustible materials as allowed by Sections 1406.3 and 1406.4.

❖ The code requires projections from walls of Type I and II materials to be noncombustible. An exception for combustible materials is for balconies (Section 1406.3) and bay windows (Section 1406.4). Buildings of Type I through IV construction require the exterior walls to be noncombustible. However, the code places a higher restriction on projections on the two highest types of construction, while Section 705.2.2 allows combustible projections on buildings of Type III and IV construction. This approach mirrors the general requirements that all elements of Type I and II buildings be noncombustible, while those of Type III and IV are permitted to be constructed of combustible interior materials.

705.2.2 Type III, IV or V construction. Projections from walls of Type III, IV or V construction shall be of any *approved* material.

❖ See also Section 1406. When combustible materials are used, the provisions of Section 705.2.3 must be followed.

705.2.3 Combustible projections. Combustible projections extending to within 5 feet (1524 mm) of the line used to determine the *fire separation distance* shall be of not less than 1-hour *fire-resistance-rated* construction, Type IV construction, *fire-retardant-treated wood* or as required by Section 1406.3.

Exception: Type VB construction shall be allowed for combustible projections in Group R-3 and U occupancies with a fire separation distance greater than or equal to 5 feet (1524 mm).

❖ This section states the requirements for protecting combustible projections and does not state when combustible projections are allowed. That is covered in Sections 705.2.1, 705.2.2 and 1406.3. The location where protection is required is simply where the combustible projection extends to within 5 feet (1524 mm) of the line (either the property line or the assumed line) used to determine the fire separation distance (FSD) (see Section 202). Every projection, if allowed by Section 705.2.2 or 1406.3 to be of combustible construction, will have to be protected according to this section if the projection is within 5 feet (1524 mm) of the line used to determine the FSD.

A frequently asked question is: should the entire projection be protected or just the portion of the projection within the 5 feet (1524 mm)? From a practical standpoint the entire projection must be protected.

The four types of protection for the combustible projections enumerated in this section are:

- The projection must be 1-hour fire-resistance-rated construction;
- The projection must be of heavy timber size combustible members;
- The projection must be constructed of fire-retardant-treated wood; or
- The projection must be as required by Section 1406.3.

The first three are self-explanatory. Section 1406.3 has construction requirements for balconies and similar projections regardless of their location on the property. The only other protection method mentioned in Section 1406.3 is sprinklers. Balconies and similar projections in Type III, IV and V construction may be combustible if sprinklers are extended into the balcony area.

There is an exception to combustible projection

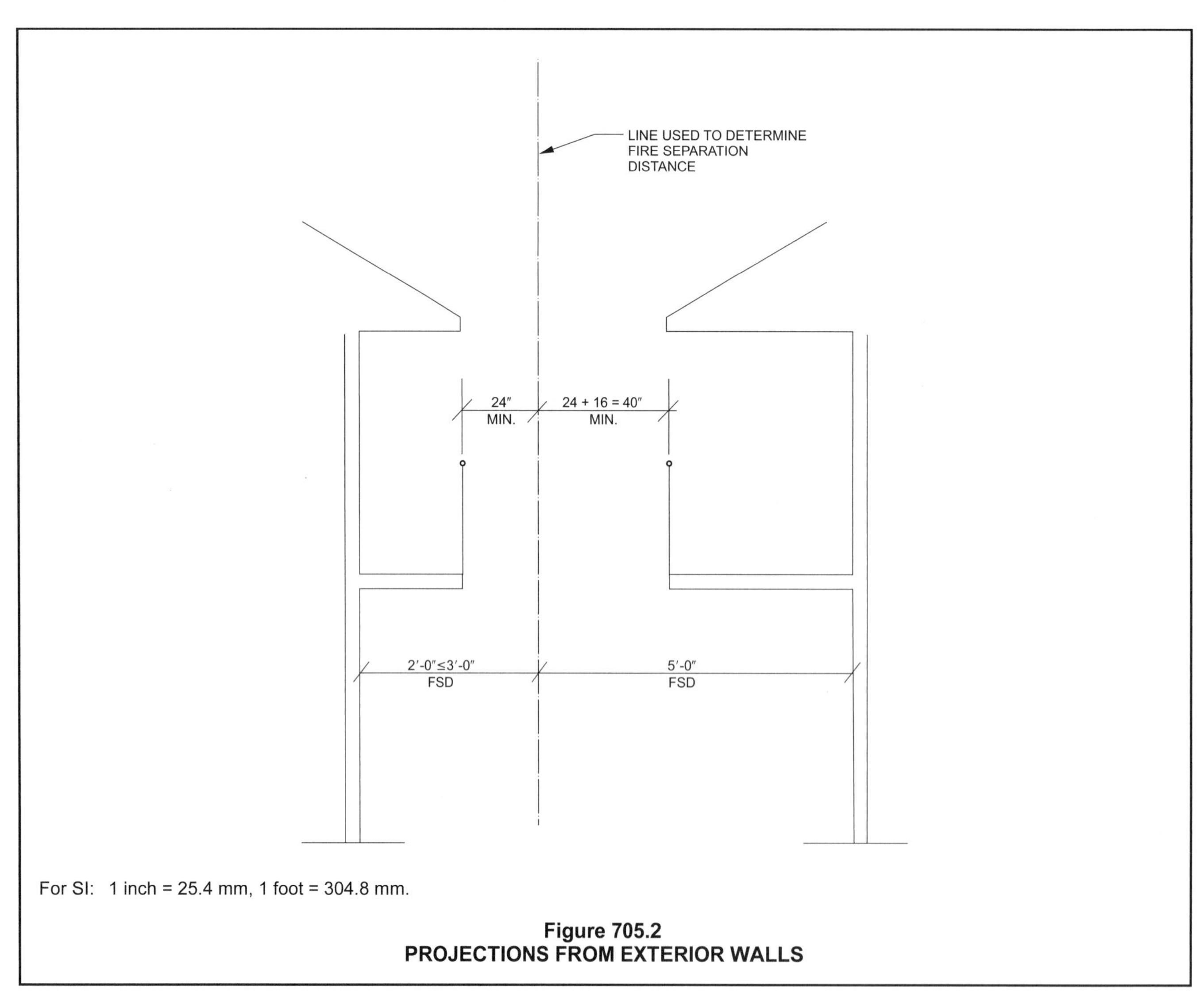

Figure 705.2
PROJECTIONS FROM EXTERIOR WALLS

protection for Type VB construction in Group R-3 and Group U occupancies with an FSD equal or greater than 5 feet (1524 mm).

705.3 Buildings on the same lot. For the purposes of determining the required wall and opening protection, projections and roof-covering requirements, buildings on the same lot shall be assumed to have an imaginary line between them.

Where a new building is to be erected on the same lot as an existing building, the location of the assumed imaginary line with relation to the existing building shall be such that the *exterior wall* and opening protection of the existing building meet the criteria as set forth in Sections 705.5 and 705.8.

Exceptions:

1. Two or more buildings on the same lot shall be either regulated as separate buildings or shall be considered as portions of one building if the aggregate area of such buildings is within the limits specified in Chapter 5 for a single building. Where the buildings contain different occupancy groups or are of different types of construction, the area shall be that allowed for the most restrictive occupancy or construction.
2. Where an S-2 parking garage of Construction Type I or IIA is erected on the same lot as a Group R-2 building, and there is no *fire separation distance* between these buildings, then the adjoining *exterior walls* between the buildings are permitted to have occupant use openings in accordance with Section 706.8. However, opening protectives in such openings shall only be required in the exterior wall of the S-2 parking garage, not in the exterior wall openings in the R-2 building, and these opening protectives in the exterior wall of the S-2 parking garage shall be not less than $1^1/_2$-hour *fire protection rating*.

❖ This section addresses buildings on the same lot and requires that an imaginary lot line be established between buildings in order to determine exterior wall fire ratings and opening protectives (see the definition of "Fire separation distance"). This section takes the approach of limiting the conflagration hazard between buildings on the same property.

Exception #1 permits two buildings on the same lot to be exempt from Sections 705.5 and 705.8 when considered as one building in accordance with Section 503.1.2. The provisions of Section 705.8.5 would still apply, since a need exists to restrict fire spread between stories within a building. Although not specifically identified in the exception, Section 705.11 would not apply, since it is dependent on the exterior wall being required to have a fire-resistance rating in accordance with Section 705.5, which relates to measurement of FSD. The last sentence of the exception reminds the user that normal code requirements would still be applicable once the two buildings are considered as one. Therefore, if two types of construction are involved, then the lowest type of construction would be assumed for the entire building because there is no fire wall between them. In applying this last sentence it is probably easiest to determine what the requirement is if the user imagines what the code requirement would be if the buildings were pushed together. In such a case, it would be easier to see that a fire wall would be needed between portions of a building that are of two different construction types or the entire building would have to be viewed as the lowest possible type of construction. Regarding occupancies and associated allowable areas, they would be separated per Section 508.4 or considered as nonseparated occupancies per Section 508.3 and, therefore, use the most restrictive allowable area.

In order to help make sense of the last sentence of the exception, the idea of imagining the two buildings being pushed together, as mentioned above, is helpful. Then realize that simply pulling the building apart should not increase or lessen the code requirements for that "single" building.

Exception #2 takes into account a practical design issue.

The great majority of multifamily projects are being built with parking garages beside apartment buildings. Access from the parking garage into the apartment unit's floor is provided at each garage floor onto the apartment's floor for convenience as well as for safety for the apartment dwellers. Many designs have one or more of the exterior walls of the parking garage and the apartment building at a 0-foot fire separation distance. The requirements within Table 705.8 would prohibit any openings in these exterior walls between the parking garage and the apartment building. Based on the protection afforded by the sprinkler system in the R-2 use and the inherent fire safety of the parking garage, this exception would allow these openings. See Commentary Figure 705.3 for an illustration of this condition.

705.4 Materials. *Exterior walls* shall be of materials permitted by the building type of construction.

❖ The material (combustible or noncombustible) requirements for exterior walls are found in Sections 602.1 through 602.5. Only Type V construction allows exterior walls to be combustible construction. Other types of construction require exterior walls, or at least the framing members, to be noncombustible. Type I and II construction allows limited use of fire-retardant wood for exterior nonload-bearing walls. All types of construction allow insulation, exterior wall coverings and interior finish to be combustible within limits. See Sections 603, 720 and 1405.5 for materials allowed within and on framed exterior walls.

705.5 Fire-resistance ratings. *Exterior walls* shall be fire-resistance rated in accordance with Tables 601 and 602 and this section. The required *fire-resistance rating* of *exterior walls* with a *fire separation distance* of greater than 10 feet (3048 mm) shall be rated for exposure to fire from the inside. The required *fire-resistance rating* of *exterior walls* with a *fire separation distance* of less than or equal to 10 feet (3048 mm) shall be rated for exposure to fire from both sides.

❖ Table 601 states the requirements for the fire-resistance ratings for load-bearing exterior walls. All load-bearing exterior walls except those in Type IIB and VB construction require some degree of fire-resistance rating. Table 602 states the requirements for the fire-resistance ratings for both load-bearing and nonload-bearing exterior walls. Whereas Table 602 is based on the fire separation distance, the type of construction and the occupancy group, Table 601 requirements are only a function of the type of construction. The table resulting in the highest fire-resistance rating shall be used.

Fire exposure from the outside of the building is thought to be from other buildings either on the same property or another property. Therefore, with increased FSD, the spread of fire from building to building is lessened. For years the code required the fire-resistance rating to be from both sides of the exterior wall when the FSD was 5 feet (1524 mm) or less. During the development of the IBC, it was decided that exterior walls that had no required fire-resistance rating according to Tables 601 and 602 could have unlimited unprotected openings (see Section 705.8.1, Exception 2). This would allow an exterior wall in a Type IIB or VB building of nearly any occupancy group to have unlimited openings and a nonfire-resistance-rated wall at an FSD of 10 feet (3048 mm). You could have another building 15-plus feet (4572 mm) from the first building with an exterior wall without fire-resistance rating from the exterior exposure because that building would have an FSD of 5 feet (1524 mm). The consensus now is that an FSD of 10 feet (3048 mm) is more practical for allowing the fire-resistance ratings of the exterior wall to be based on interior surface exposure only. Now the fire-resistance rating of the exterior wall is from both sides when the FSD is 10 feet (3048 mm) or less.

705.6 Structural stability. *Exterior walls* shall extend to the height required by Section 705.11. Interior structural elements that brace the exterior wall but that are not located within the plane of the exterior wall shall have the minimum *fire-resistance rating* required in Table 601 for that structural element. Structural elements that brace the exterior wall but are located outside of the exterior wall or within the plane of the exterior wall shall have the minimum *fire-resistance rating* required in Tables 601 and 602 for the exterior wall.

❖ Structural stability of fire-resistance-rated construction is an important concern. Section 705.6 requires elements providing bracing support to be fire-resistance rated for the same duration of time as the exterior wall. In light-frame platform construction, this will require that the band joist or beam supporting the floor and the wall above to also be of fire-resistant construction. Although the floor construction itself may not be required to be of fire-resistance-rated construction in Type IIB and VB construction, some effort must be made to ensure that the floor joists, at least at the exterior wall, are of fire-resistant-rated construction. Although the floor framing acts as a lateral support for the exterior wall, this section does not require that the entire floor system be of fire-resistance-rated construction. To state otherwise would prohibit Type IIB and VB buildings with an FSD of less than 10 feet (3048 mm). Only the structural element within the floor system that supports the vertical load of the wall must be of fire-resistance-rated construction.

For exterior walls, this section requires fire-resis-

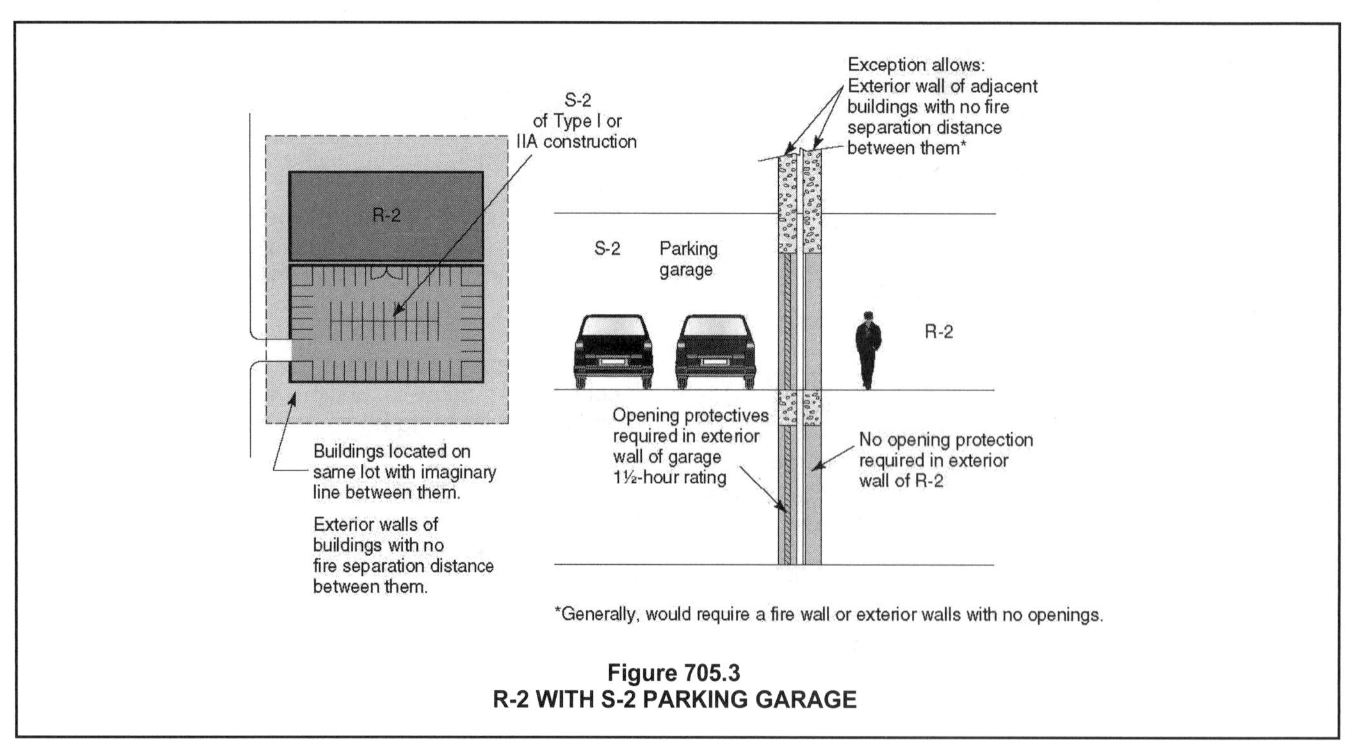

Figure 705.3
R-2 WITH S-2 PARKING GARAGE

tance-rated construction to extend to the roof construction or to the top of the parapet if a parapet is required (see Section 705.11). This begs the question—in conventional light-frame platform construction, is the floor system supported by the lower exterior wall and supporting the exterior wall above part of the exterior wall? And, if so, how and to what limits do you go to provide a fire-resistance rating? This is a valid concern in Type IIB and VB construction with an FSD of less than 10 feet (3048 mm) because the exterior wall is required to have a fire-resistance rating while the floor system is not. Both the continuity and the structural integrity issue are illustrated in Commentary Figure 705.6.

When parapet walls are not required, the exterior wall for fire-resistant rating purposes stops at the roof/ceiling construction.

Interior structural elements which brace an exterior wall do not require a fire-resistance rating equal to the wall it braces. However, the structural element must have a fire-resistance rating as required by Table 601 for that structural element.

This section does not require that the wall remain in place when the structure collapses. That language is only used for fire wall structural integrity.

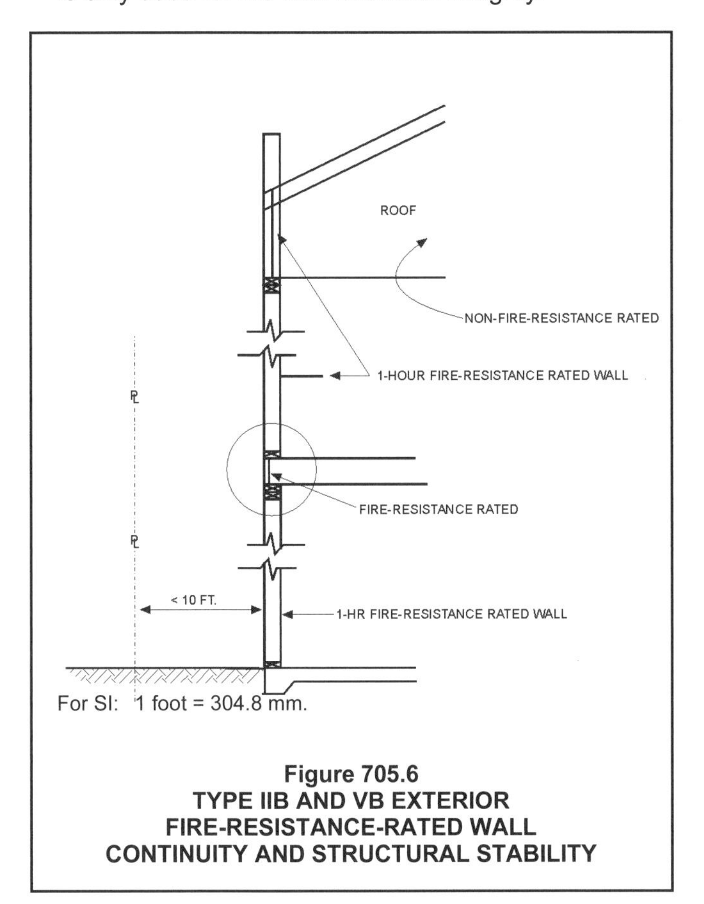

Figure 705.6
TYPE IIB AND VB EXTERIOR FIRE-RESISTANCE-RATED WALL CONTINUITY AND STRUCTURAL STABILITY

705.7 Unexposed surface temperature. Where protected openings are not limited by Section 705.8, the limitation on the rise of temperature on the unexposed surface of *exterior walls* as required by ASTM E119 or UL 263 shall not apply. Where protected openings are limited by Section 705.8, the limitation on the rise of temperature on the unexposed surface of *exterior walls* as required by ASTM E119 or UL 263 shall not apply provided that a correction is made for radiation from the unexposed *exterior wall* surface in accordance with the following formula:

$$A_e = A + (A_f \times F_{eo}) \quad \textbf{(Equation 7-1)}$$

where:

A_e = Equivalent area of protected openings.

A = Actual area of protected openings.

A_f = Area of *exterior wall* surface in the *story* under consideration exclusive of openings, on which the temperature limitations of ASTM E119 or UL 263 for walls are exceeded.

F_{eo} = An "equivalent opening factor" derived from Figure 705.7 based on the average temperature of the unexposed wall surface and the *fire-resistance rating* of the wall.

❖ The purpose of this provision is to allow the use of an exterior fire-resistance-rated wall assembly that does not meet all of the conditions of acceptance of ASTM E119 or UL 263. The only condition that may not apply is the limitation on the rise of temperature on the unexposed surface of the wall. To pass the standard, the rise of temperature cannot exceed 250°F (121°C) above the initial temperature. All listed fire-resistance-rated exterior walls and partitions meet all of the conditions of acceptance. If a particular required fire-resistance-rated exterior wall was not listed by an approved testing agency for the required duration of time, then this section could be applicable.

For exterior walls with FSDs of 20 feet (6096 mm) or greater, the limitation on the rise of temperature on the unexposed surface of the exterior wall does not apply. This is practical because at 20 feet (6096 mm) an unlimited area of protected openings can be installed; protected openings are treated as windows or doors and the standard fire test for windows and doors does not have a limit on the temperature rise. To accept a nonlisted exterior wall assembly, the failed test data would have to be provided showing that the assembly did indeed pass all of the conditions of acceptance except rise of temperature on the unexposed surface.

For required exterior walls with FSDs of less than 20 feet (6096 mm), the limitation on the rise of temperature on the unexposed surface of the exterior wall may not apply. To accept a nonlisted exterior wall assembly, the failed test data would have to be provided showing that the assembly did indeed pass all of the conditions of acceptance except rise of temperature on the unexposed surface. The test data must also state the average temperature of the unexposed surface at the time duration for which the exterior wall is required to be rated. We must make a correction for the radiation from the unexposed exterior wall surface. We convert the wall surface which has too much temperature rise to an equivalent amount of protected openings. Then we check for compliance with the opening provisions of Section 705.8.1 as always. See Example 705.7:

As shown in Commentary Figure 705.7, Example 705.7:

The exterior bearing wall of a Type VA nonsprinklered building with a Group B occupancy is 16 feet (4877 mm) from an interior lot line. The designer chose to use an exterior wall assembly which failed the ASTM E119 fire test because of the rise of temperature of the unexposed surface. The designer supplies the failed test data which clearly shows that the wall met all conditions of acceptance for a 1-hour fire-resistance rating except the rise of temperature on the unexposed side which was an average of 1200°F (649°C) at 1-hour fire duration. Since the temperature rise exceeds the allowed, we must treat the wall surface as though it had an equivalent area of protected openings.

Solution:

- From Table 602: the exterior wall must be fire-resistance rated for 1 hour.
- From Table 705.8: the wall may have 75-percent protected wall openings or 25-percent unprotected wall openings or a combination of the two in accordance with Section 705.8.4.
- From drawing of exterior wall elevation: total area of wall = 700 sq. ft. (65 m^2).

  total area of protected openings = 0
  total area of unprotected openings = 120 sq. ft. (11 m^2).
  total area of exterior wall surface minus the openings = 580 sq. ft. (54 m^2).
- First we compute the equivalent area of protected openings using Equation 7-1:
- We can calculate from the drawings the actual area of protected openings which is *A*. *A* = 0
- We can calculate *Af* from the drawings. It is the total area of the wall surface minus the openings. 700-120 = 540 sq. ft.
- We can obtain F_{eo} from Figure 705.7. Using the given average temperature rise on the unexposed side which was 1200°F and the curve on the figure for 1 hour and reading to the left legend we find that F_{eo} for this assembly is 0.34.
- Equation 7-1 yields: A_e = 0 + (540 x 0.34) or A_e = 197 sq. ft.

Having solved for the equivalent area of protected openings we can now proceed as normal by checking compliance with Section 705.8.4 since we have a mix

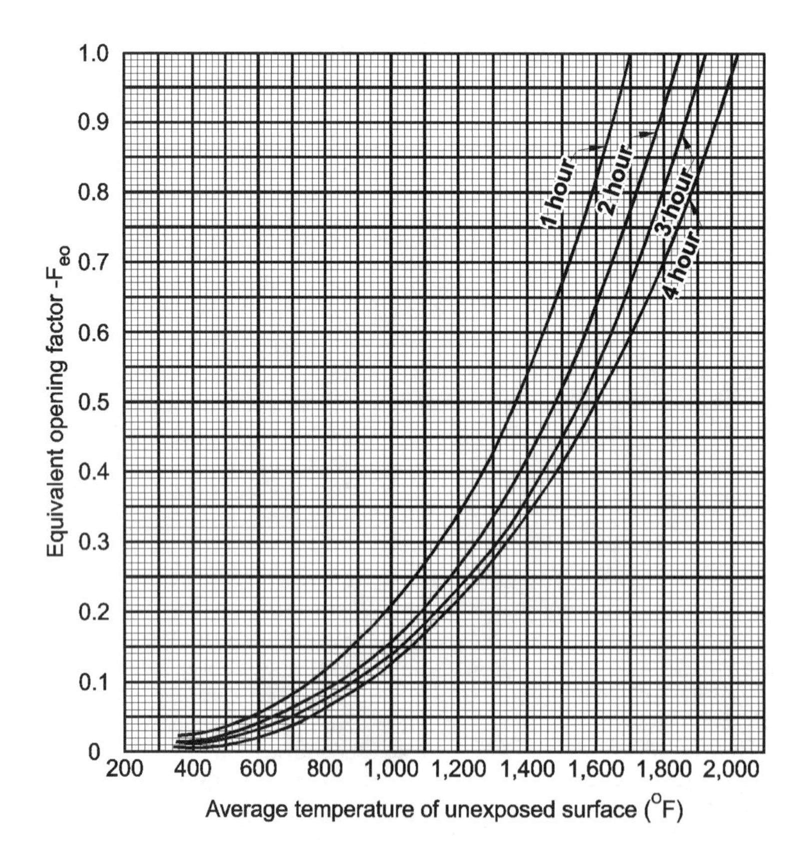

For SI: °C = [(°F) - 32] / 1.8.

FIGURE 705.7
EQUIVALENT OPENING FACTOR

of protected and unprotected openings in the fire-resistance-rated exterior wall.

- From Table 705.8: seventy-five percent of the total wall area may be protected openings or 0.75 x 700 = 525 sq. ft.
- From Table 705.8: twenty-five percent of the total wall area may be unprotected openings or 0.25 x 700 = 175 sq. ft.
- From Equation 7-2: $(A_p / a_p) + (A_u / a_u) \leq 1$ or we have too many openings in the wall.
- From Equation 7-2 and using A_e computed above for A_p we have $(197/525) + (120/175) \leq 1$ or 0.375 + 0.685 = 1.06 which is not less than or equal to 1. The wall does not meet the code. Too many openings.

705.8 Openings. Openings in *exterior walls* shall comply with Sections 705.8.1 through 705.8.6.

❖ The requirements of this section limit the allowable area of openings in exterior walls and are applicable to buildings with or without walls with FSDs less than 30 feet (9144 mm). The limitations on openings in exterior walls is a function of FSD (see definition this chapter) and the degree of protection provided for the opening. The degree of protection of the openings is either unprotected in a nonsprinklered building (UP, NS), unprotected in a sprinklered building (UP, S) or protected openings (P). Protected openings are openings with fire doors, fire shutters or fire window assemblies that comply with Sections 716.5 and 716.6. Sprinklered buildings are buildings with an NFPA 13 sprinkler system. Buildings with only an NFPA 13 R system shall be considered nonsprinklered for the purpose of opening limitations in exterior walls. The percentage of openings allowed is in Table 705.8, subject to the exceptions in Section 705.8.1.

TABLE 705.8. See page 7-20.

❖ Table 705.8, on page 7-20, provides allowable areas for exterior wall openings for structures based on the FSD of the exterior wall and the degree of protection afforded to the opening, such as fire shutters and automatic sprinkler systems. The more protection provided, the larger the area of openings can get.

705.8.1 Allowable area of openings. The maximum area of unprotected and protected openings permitted in an *exterior wall* in any *story* of a building shall not exceed the percentages specified in Table 705.8.

Exceptions:

1. In other than Group H occupancies, unlimited unprotected openings are permitted in the first *story* above grade plane either:
   1.1. Where the wall faces a street and has a *fire separation distance* of more than 15 feet (4572 mm); or
   1.2. Where the wall faces an unoccupied space. The unoccupied space shall be on the same lot or dedicated for public use, shall be not less than 30 feet (9144 mm) in width and shall have access from a street by a posted fire lane in accordance with the *International Fire Code*.
2. Buildings whose exterior bearing walls, exterior nonbearing walls and exterior primary structural frame are not required to be fire-resistance rated shall be permitted to have unlimited unprotected openings.

❖ The allowable area is given as a percentage of the total wall area at any story. Each story must comply independently.

Exception 1.1 only applies to the exterior walls on the first story above grade. For all occupancies except Group H, the first-story exterior wall that faces a street and that is at least 15 feet (4572 mm) from the wall to the centerline of the street shall be allowed unlimited unprotected openings.

Exception 1.2 also applies to the exterior walls on only the first story above grade. For all occupancies except Group H, the first-story exterior wall that faces a minimum 30-foot-wide (9144 mm) unoccupied space either on the same lot or dedicated for public use with access from a street by a fire lane, shall be allowed unlimited unprotected openings.

Since the first story above grade is generally read-

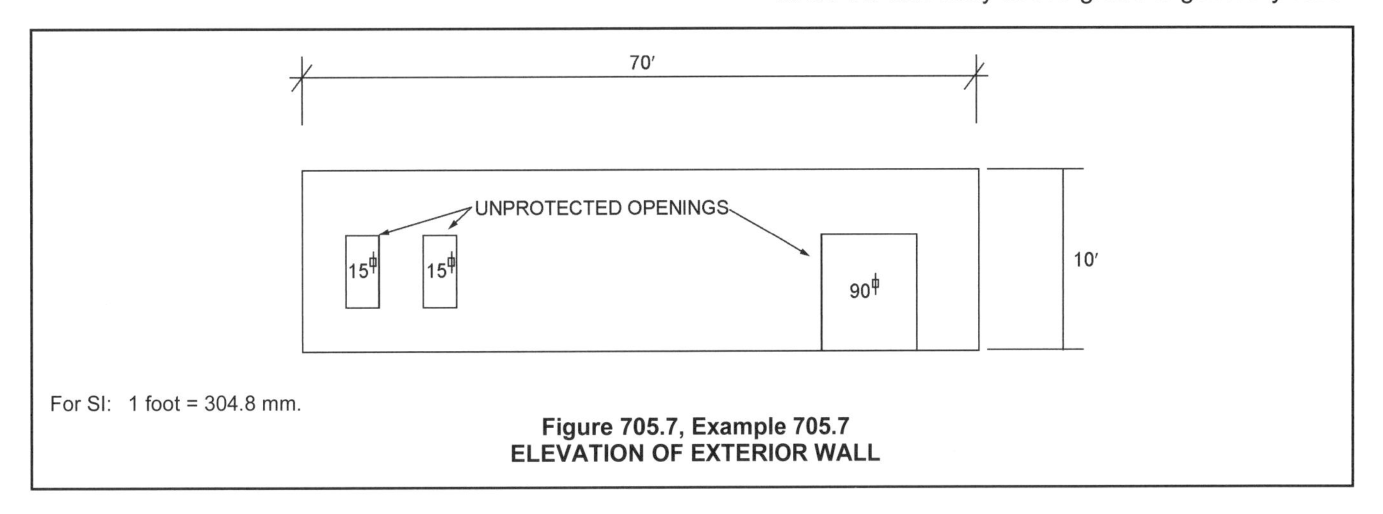

Figure 705.7, Example 705.7
ELEVATION OF EXTERIOR WALL

ily available for fire department access, unprotected openings are not limited in the above exception. See Commentary Figure 705.8.1 for examples of first-story openings.

Exception 2 is for openings in all exterior walls in all stories where Tables 601 and 602 do not require any exterior wall (bearing or nonbearing) or primary structural member to be fire-resistance rated. Therefore, this exception is only applicable to Type IIB and VB construction with an FSD of 10 feet (3048 mm) or greater. This allows unlimited unprotected openings in Type IIB and VB construction for all exterior walls facing an FSD of 10 feet (3048 mm) or more.

705.8.2 Protected openings. Where openings are required to be protected, *fire doors* and fire shutters shall comply with Section 716.5 and *fire window assemblies* shall comply with Section 716.6.

Exception: Opening protectives are not required where the building is equipped throughout with an *automatic*

TABLE 705.8
MAXIMUM AREA OF EXTERIOR WALL OPENINGS BASED ON FIRE SEPARATION DISTANCE AND DEGREE OF OPENING PROTECTION

FIRE SEPARATION DISTANCE (feet)	DEGREE OF OPENING PROTECTION	ALLOWABLE AREA[a]
0 to less than 3[b, c, k]	Unprotected, Nonsprinklered (UP, NS)	Not Permitted[k]
	Unprotected, Sprinklered (UP, S)[i]	Not Permitted[k]
	Protected (P)	Not Permitted[k]
3 to less than 5[d, e]	Unprotected, Nonsprinklered (UP, NS)	Not Permitted
	Unprotected, Sprinklered (UP, S)[i]	15%
	Protected (P)	15%
5 to less than 10[e, f, j]	Unprotected, Nonsprinklered (UP, NS)	10%[h]
	Unprotected, Sprinklered (UP, S)[i]	25%
	Protected (P)	25%
10 to less than 15[e, f, g, j]	Unprotected, Nonsprinklered (UP, NS)	15%[h]
	Unprotected, Sprinklered (UP, S)[i]	45%
	Protected (P)	45%
15 to less than 20[f, g, j]	Unprotected, Nonsprinklered (UP, NS)	25%
	Unprotected, Sprinklered (UP, S)[i]	75%
	Protected (P)	75%
20 to less than 25[f, g, j]	Unprotected, Nonsprinklered (UP, NS)	45%
	Unprotected, Sprinklered (UP, S)[i]	No Limit
	Protected (P)	No Limit
25 to less than 30[f, g, j]	Unprotected, Nonsprinklered (UP, NS)	70%
	Unprotected, Sprinklered (UP, S)[i]	No Limit
	Protected (P)	No Limit
30 or greater	Unprotected, Nonsprinklered (UP, NS)	No Limit
	Unprotected, Sprinklered (UP, S)[i]	No Limit
	Protected (P)	No Limit

For SI: 1 foot = 304.8 mm.

UP, NS = Unprotected openings in buildings not equipped throughout with an automatic sprinkler system in accordance with Section 903.3.1.1.

UP, S = Unprotected openings in buildings equipped throughout with an automatic sprinkler system in accordance with Section 903.3.1.1.

P = Openings protected with an opening protective assembly in accordance with Section 705.8.2.

a. Values indicated are the percentage of the area of the exterior wall, per story.

b. For the requirements for fire walls of buildings with differing heights, see Section 706.6.1.

c. For openings in a fire wall for buildings on the same lot, see Section 706.8.

d. The maximum percentage of unprotected and protected openings shall be 25 percent for Group R-3 occupancies.

e. Unprotected openings shall not be permitted for openings with a fire separation distance of less than 15 feet for Group H-2 and H-3 occupancies.

f. The area of unprotected and protected openings shall not be limited for Group R-3 occupancies, with a fire separation distance of 5 feet or greater.

g. The area of openings in an open parking structure with a fire separation distance of 10 feet or greater shall not be limited.

h. Includes buildings accessory to Group R-3.

i. Not applicable to Group H-1, H-2 and H-3 occupancies.

j. The area of openings in a building containing only a Group U occupancy private garage or carport with a fire separation distance of 5 feet or greater shall not be limited.

k. For openings between S-2 parking garage and Group R-2 building, see Section 705.3, Exception 2.

sprinkler system in accordance with Section 903.3.1.1 and the exterior openings are protected by a water curtain using automatic sprinklers *approved* for that use.

❖ To be considered protected, the opening must have a listed and labeled fire door, fire shutter or fire window meeting the requirements of Sections 716.5 and 716.6. The required fire protection rating of doors in protected openings of exterior walls shall be in accordance with Table 716.5. Protected window assemblies shall have the fire protection rating as indicated in Table 716.6.

The exception to Section 705.8.2 is irrelevant for most occupancies. Table 705.8 allows the same percentage of opening area for protected openings and unprotected openings in sprinklered buildings and Table 705.8 does not have the water curtain requirement as the exception does. In a building sprinklered throughout with a full NFPA 13 system, the same percentage of openings is allowed, whether they are protected or unprotected.

The exception to Section 705.8.2 would be relevant for Group H occupancies since Note i to Table 705.8 makes the unprotected sprinklered category not applicable to Group H-1, H-2, and H-3 occupancies.

705.8.3 Unprotected openings. Where unprotected openings are permitted, windows and doors shall be constructed of any *approved* materials. Glazing shall conform to the requirements of Chapters 24 and 26.

❖ Unprotected openings can be windows and doors of any approved material. Glazing must meet the structural requirements and the safety requirements as applicable in Chapters 24 and 26. All openings in wind-borne debris regions must be protected by impact-resistant material in accordance with Section 1609.1.2. For purposes of exterior wall fire-resistance requirements and opening limitations, an unprotected opening can have nothing at all in it as a partial enclosed building. Sheds without exterior walls must also meet Section 705.8 and Table 705.8.

705.8.4 Mixed openings. Where both unprotected and protected openings are located in the *exterior wall* in any *story* of a building, the total area of openings shall be determined in accordance with the following:

$$(A_p/a_p) + (A_u/a_u) \leq 1 \quad \textbf{(Equation 7-2)}$$

where:

A_p = Actual area of protected openings, or the equivalent area of protected openings, A_e (see Section 705.7).

a_p = Allowable area of protected openings.

A_u = Actual area of unprotected openings.

a_u = Allowable area of unprotected openings.

❖ The opening limitations in Table 705.8 vary according to the degrees of opening protection. The table establishes the acceptable percentage for openings if all openings were of one degree of protection. For instance, at 12 feet (3658 mm) if all openings are unprotected in a nonsprinklered building you can have only 15 percent of the wall area in unprotected openings. In that same exterior wall if all openings were protected, you could have 45 percent of the wall area in protected openings. If any exterior wall area in any story has both protected and unprotected openings, then the actual areas of both types of openings must meet Equation 7-2.

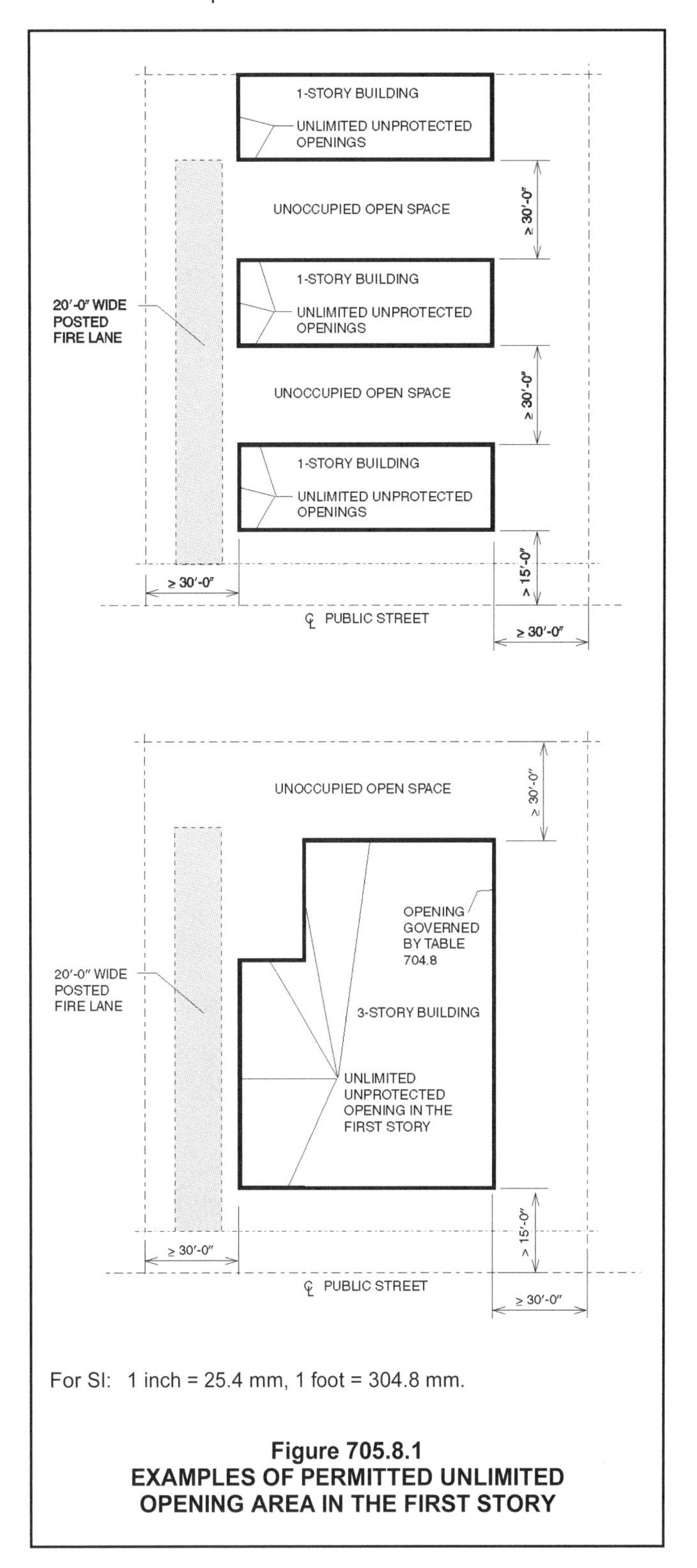

For SI: 1 inch = 25.4 mm, 1 foot = 304.8 mm.

Figure 705.8.1
EXAMPLES OF PERMITTED UNLIMITED OPENING AREA IN THE FIRST STORY

Example 705.8.4:

In a Type IIA building, the exterior wall is required to be of 1-hour fire-resistance-rated construction by Table 602. The wall has an FSD of 12 feet (3658 mm). The building is nonsprinklered. The wall in the first story has a total area of 1000 sq. ft. (93 m^2). The designer has 150 sq. ft. (14 m^2) of protected openings in the wall and 100 sq. ft. (9.3 m^2) of unprotected openings. Does this comply with the code, particularly Section 705.8.4?

Solution:

- Use Equation 7-2
- From the building in the example the protected openings are 150 sq. ft. (14 m^2). This value is A_p.
- From Table 705.8 the allowable area (if only protected openings are used) is 45 percent. Forty-five percent of 1000 sq. ft. (93 m^2) is 450 sq. ft. (42 m^2) (1000 x 0.45 = 450). This value is a_p in the equation.
- From the building in the example the unprotected openings are 100 sq. ft. (9.3 m^2) This value is A_u in the equation.
- From Table 705.8 the allowable area (if only unprotected openings are used) is 15 percent. Fifteen percent of 1000 sq. ft. (93 m^2) is 150 sq. ft. (14 m^2) (1000 x 0.15 = 150). This value is a_u in the equation.
- Equation 7-2 $(A_p/a_p) + (A_u/a_u) \leq 1$ or $(150/450) + (100/150) \leq 1$ or $0.333 + 0.666 = 1$

This amount of openings exactly complies with the code. Had the sum of the ratios exceeded 1 the opening mixture would not be allowed.

705.8.5 Vertical separation of openings. Openings in *exterior walls* in adjacent *stories* shall be separated vertically to protect against fire spread on the exterior of the buildings where the openings are within 5 feet (1524 mm) of each other horizontally and the opening in the lower *story* is not a protected opening with a *fire protection rating* of not less than $^3/_4$ hour. Such openings shall be separated vertically not less than 3 feet (914 mm) by spandrel girders, *exterior walls* or other similar assemblies that have a *fire-resistance rating* of not less than 1 hour, rated for exposure to fire from both sides, or by flame barriers that extend horizontally not less than 30 inches (762 mm) beyond the *exterior wall*. Flame barriers shall have a *fire-resistance rating* of not less than 1 hour. The unexposed surface temperature limitations specified in ASTM E119 or UL 263 shall not apply to the flame barriers or vertical separation unless otherwise required by the provisions of this code.

Exceptions:

1. This section shall not apply to buildings that are three *stories* or less above *grade plane*.
2. This section shall not apply to buildings equipped throughout with an *automatic sprinkler system* in accordance with Section 903.3.1.1 or 903.3.1.2.
3. Open parking garages.

❖ Where unprotected openings occur in adjacent stories, a fire that breaks out of an opening in a lower story can spread vertically to upper stories of the building. A fire at the Hilton Hotel in Las Vegas provides a good illustration of this problem. In that incident a fire broke out in an elevator lobby at a lower level and then quickly spread upward to a number of other stories simply due to the flame plume from the lower level impinging on the windows above and causing the combustible materials in those upper levels to ignite. A vertical panel or horizontal flame barrier is necessary to minimize the possibility of vertical flame spread. The vertical panel is to be at least 3 feet (914 mm) in height and have a fire-resistance rating of at least 1 hour [see Commentary Figure 705.8.5(1)]. Note that the fire rating is required from both sides of the assembly regardless of fire separation distance. Full-scale fire tests have shown that the vertical panel should extend at least 30 inches (762 mm) above the finished floor level. This is to prevent the transfer of radiant heat to furnishings and other combustibles located in the lower portion of the room in the story above the assumed fire exposure. Horizontal flame barriers are required to extend at least 30 inches (762 mm) beyond the face of the exterior wall and have a fire-resistance rating of at least 1 hour [see Commentary Figure 705.8.5(2)].

This section requires such protection when the opening on the lower level is not fire-protection rated for $^3/_4$ hour or more. The use of glazing is one example of why this section is so specific. Wired glass, for

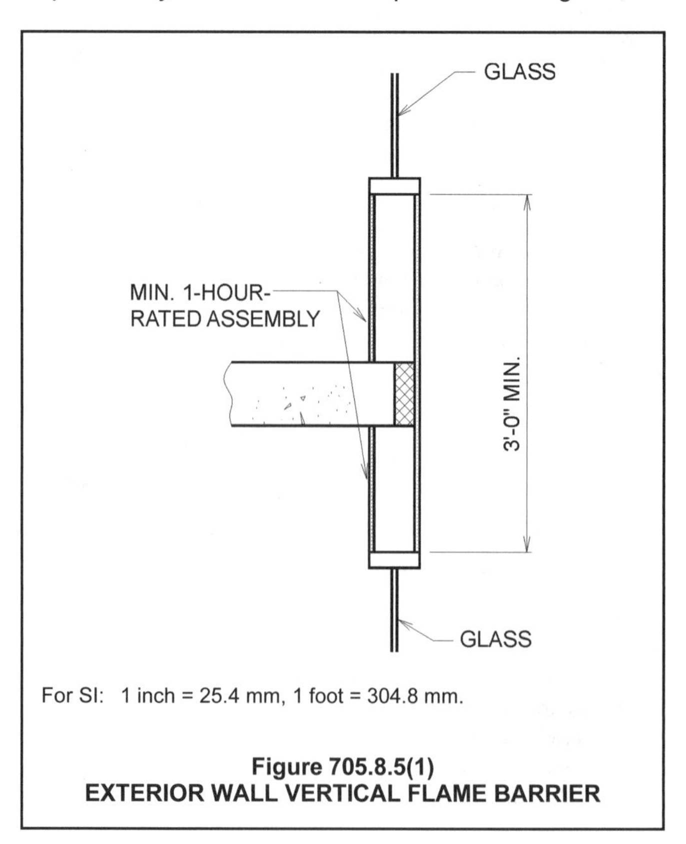

Figure 705.8.5(1)
EXTERIOR WALL VERTICAL FLAME BARRIER

example, has been shown to retain its integrity with limited fire exposure when exposed to a fire condition for a certain period of time. For this reason, wired glass has historically been recognized as a suitable exterior opening protective within size and mounting methods defined by its listing.

Wired glass permits a significant percentage, typically recognized as high as 50 percent, of radiant heat energy to pass through. Since the initial concern is heat radiated from the flame plume, wired glass does little to prevent ignition of combustibles in the room on the upper floor. This section permits the elimination of the unexposed surface temperature limitations prescribed by ASTM E119 for flame barriers and vertical shields. Fire spread between floors is primarily the result of radiant heat, not conductive heat transfer. Full-scale tests have shown that only a small degree of fire resistance is required for adequate protection since radiant energy is the primary concern and glazing in the unprotected openings above the story of fire origin may break within 1 minute.

Protection is not required for buildings three stories or less in height. This exception is consistent with Table 503, which permits buildings of three stories in height to be of Type IIB, IIIB and VB construction when these buildings do not require a fire-resistance rating for the floor construction.

The exception for buildings protected with an automatic sprinkler system is consistent with the code's approach to limiting fire size and, thus, limiting spread of fire beyond the area or room of origin. The exception is based on the effectiveness and reliability of automatic sprinkler systems in decreasing fire size. The reference to Section 903.3.1.1 or 903.3.1.2 reinforces that the system must be a code-compliant automatic sprinkler system per NFPA 13 as modified by Section 903.3.1.1 or in accordance with NFPA 13R as modified by Section 903.3.1.2. Exception 3 acknowledges the practicality of constructing open parking structures and the need to have a significant amount of openings for natural ventilation.

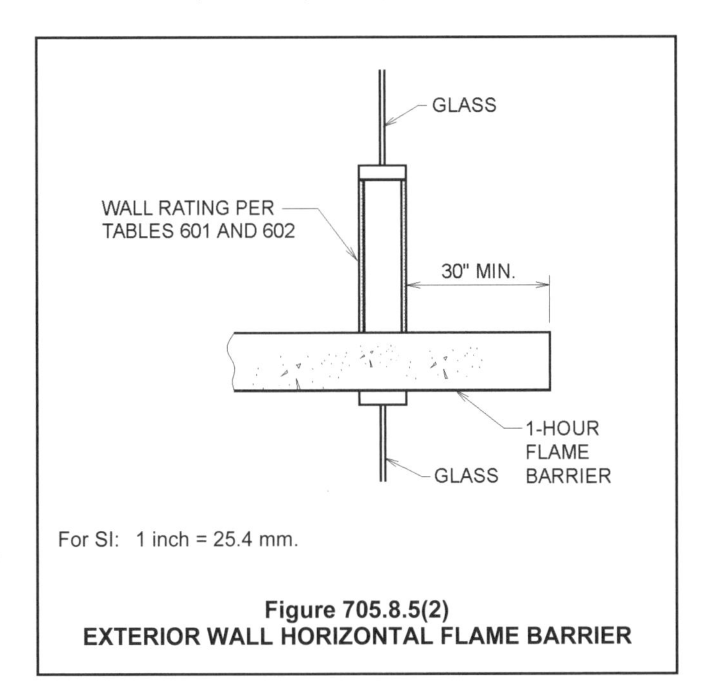

Figure 705.8.5(2)
EXTERIOR WALL HORIZONTAL FLAME BARRIER

705.8.6 Vertical exposure. For buildings on the same lot, opening protectives having a *fire protection rating* of not less than $^{3}/_{4}$ hour shall be provided in every opening that is less than 15 feet (4572 mm) vertically above the roof of an adjacent building or structure based on assuming an imaginary line between them. The opening protectives are required where the *fire separation distance* between the imaginary line and the adjacent building or structure is less than 15 feet (4572 mm).

Exceptions:

1. Opening protectives are not required where the roof assembly of the adjacent building or structure has a *fire-resistance rating* of not less than 1 hour for a minimum distance of 10 feet (3048 mm) from the *exterior wall* facing the imaginary line and the entire length and span of the supporting elements for the fire-resistance-rated roof assembly has a *fire-resistance rating* of not less than 1 hour.
2. Buildings on the same lot and considered as portions of one building in accordance with Section 705.3 are not required to comply with Section 705.8.6.

❖ A fire in a building that is adjacent to a taller building can be the source of fire exposure to openings in the taller building. Although the height of a fire plume is dependent on several factors, this section requires $^{3}/_{4}$-hour or greater opening protectives in the wall where the openings are within 15 feet vertically (4572 mm) above the roof of a building that is within a horizontal FSD of 15 feet (4572 mm) (see Commentary Figure 705.8.6). Full-scale tests have indicated that exterior flame plumes may extend higher than 16 feet (4877 mm) above a window using a fuel load of 8 pounds per square foot (psf) (39 kg/m^2). Since this provision is based on a fire exposure from the roof of the lower building, it does not apply where the roof construction is 1-hour fire-resistance rated, which reduces the potential for fire exposure.

705.9 Joints. Joints made in or between *exterior walls* required by this section to have a *fire-resistance rating* shall comply with Section 715.

Exception: Joints in *exterior walls* that are permitted to have unprotected openings.

❖ Joints, such as expansion or seismic, are another form of opening in exterior walls, and therefore, must be considered with regard to maintaining the fire-resistance ratings of the exterior walls. This section requires that all joints located in exterior walls required to be fire-resistance rated are to be protected by a joint system that has a fire-resistance rating and complies with the requirements of Section 715 (see commentary for Section 715). The exception to joint protection here and in Section 715 is for exterior walls that are permitted by Table 705.8 to have unprotected openings. Therefore, this section requires fire-resistant joint systems to protect joints in exterior walls of nonsprinklered buildings with an FSD of 5 feet (1524 mm) or less; in sprinklered buildings with an FSD of 3 feet (914 mm) or less; and in Group H-2 or H-3 buildings with an FSD of 15 feet (4572 mm) or less (see note e to Table 705.8). This implies that the building may have joints in fire-resistance-rated walls with an FSD of less than 3 feet (914 mm). In this context, joints are not treated like openings, which are not permitted in walls with an FSD of less than 3 feet (914 mm).

705.9.1 Voids. The void created at the intersection of a floor/ceiling assembly and an exterior curtain wall assembly shall be protected in accordance with Section 715.4.

❖ See the commentary to Section 715.4.

705.10 Ducts and air transfer openings. Penetrations by air ducts and air transfer openings in fire-resistance-rated *exterior walls* required to have protected openings shall comply with Section 717.

Exception: Foundation vents installed in accordance with this code are permitted.

❖ This section requires air outlets in exterior walls, including air intake and air exhaust outlets, to be protected by fire dampers where Table 705.8 requires protected openings. That is a wall with an FSD from 3 feet to 5 feet (914 mm to 1524 mm) for nonsprinklered buildings. For sprinklered buildings, protected openings are never required, so air outlets can be located in an exterior wall with an FSD of 3 feet (914 mm) or more. There can be no openings in a wall with an FSD of less than 3 feet (914 mm). Fire damper assembly standards are in accordance with Section 717. Exhaust outlets that cannot have fire dampers (such as commercial kitchen hood exhausts) cannot be located in exterior walls in locations where openings are required to be protected in accordance with Table 705.8. This is also stated in Section 506.3.12.2 of the IMC.

705.11 Parapets. Parapets shall be provided on *exterior walls* of buildings.

Exceptions: A parapet need not be provided on an *exterior wall* where any of the following conditions exist:

1. The wall is not required to be fire-resistance rated in accordance with Table 602 because of *fire separation distance*.
2. The building has an area of not more than 1,000 square feet (93 m^2) on any floor.
3. Walls that terminate at roofs of not less than 2-hour fire-resistance-rated construction or where the roof, including the deck or slab and supporting construc-

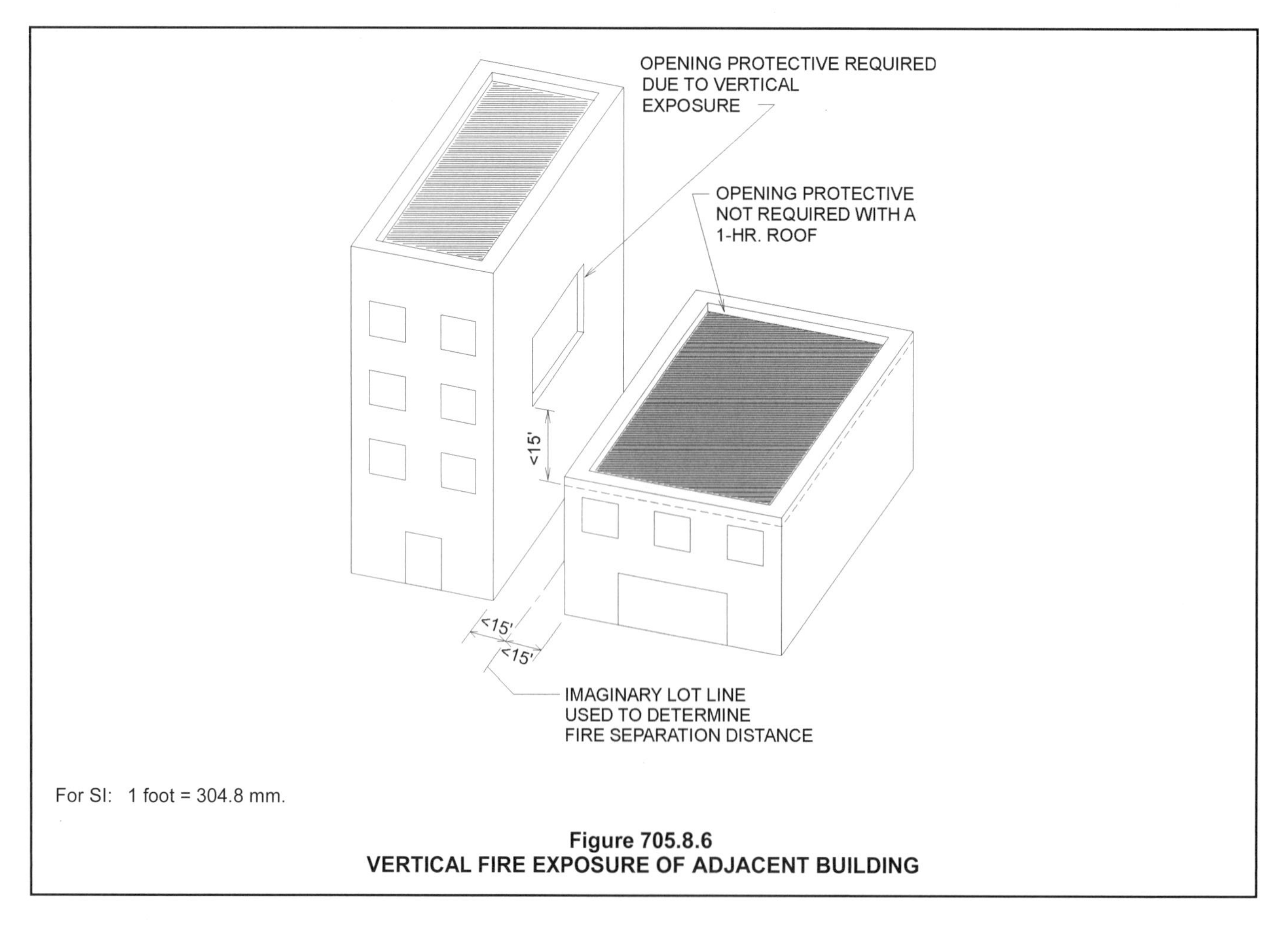

Figure 705.8.6
VERTICAL FIRE EXPOSURE OF ADJACENT BUILDING

tion, is constructed entirely of noncombustible materials.

4. One-hour fire-resistance-rated *exterior walls* that terminate at the underside of the roof sheathing, deck or slab, provided:
    4.1. Where the roof/ceiling framing elements are parallel to the walls, such framing and elements supporting such framing shall not be of less than 1-hour fire-resistance-rated construction for a width of 4 feet (1220 mm) for Groups R and U and 10 feet (3048 mm) for other occupancies, measured from the interior side of the wall.
    4.2. Where roof/ceiling framing elements are not parallel to the wall, the entire span of such framing and elements supporting such framing shall not be of less than 1-hour fire-resistance-rated construction.
    4.3. Openings in the roof shall not be located within 5 feet (1524 mm) of the 1-hour fire-resistance-rated *exterior wall* for Groups R and U and 10 feet (3048 mm) for other occupancies, measured from the interior side of the wall.
    4.4. The entire building shall be provided with not less than a Class B roof covering.
5. In Groups R-2 and R-3 where the entire building is provided with a Class C roof covering, the *exterior wall* shall be permitted to terminate at the underside of the roof sheathing or deck in Type III, IV and V construction, provided one or both of the following criteria is met:
    5.1. The roof sheathing or deck is constructed of *approved* noncombustible materials or of *fire-retardant-treated wood* for a distance of 4 feet (1220 mm).
    5.2. The roof is protected with 0.625-inch (16 mm) Type X gypsum board directly beneath the underside of the roof sheathing or deck, supported by not less than nominal 2-inch (51 mm) ledgers attached to the sides of the roof framing members for a minimum distance of 4 feet (1220 mm).
6. Where the wall is permitted to have not less than 25 percent of the *exterior wall* areas containing unprotected openings based on *fire separation distance* as determined in accordance with Section 705.8.

❖ Parapets are required to restrict spread of fire from building to building and are therefore a function of the FSD. Parapets are not required to extend above the roofs on all buildings. There are six exceptions. Exceptions 1 and 6 apply to all buildings.

- Exception 1 exempts exterior walls with a 30-foot (9144 mm) or greater FSD for all types of construction except Types IIB and VB. Exterior walls of Types IIB and VB (except for H occupancy) with greater than 10 feet (3048 mm) of FSD are exempt from parapets.
- Exception 6 exempts all sprinklered buildings from parapet walls where the exterior walls have an FSD of 5 feet (1524 mm) or greater. Exception 6 exempts all nonsprinklered buildings from parapets with an FSD of 15 feet (4572 mm) or greater. Exception 6 is the greater exemption for all but Type IIB and VB nonsprinklered buildings, when Exception 1 is better. These two exceptions yield the following:
- Parapet walls are not required on sprinklered buildings with an FSD of 5 feet (1524 mm) or greater.
- Parapet walls are not required for nonsprinklered Type IIB and VB construction with an FSD of 10 feet (3048 mm) or more.
- Parapet walls are not required for nonsprinklered buildings of all types of construction other than Types IIB and VB with an FSD of 15 feet (4572 mm) or more.
- Exception 2 exempts buildings with 1000 sq. ft. (93 m^2) or less on any floor.
- Exceptions 3, 4 and 5. While Exceptions 1, 2 and 6 completely exempt the exterior walls from parapets, Exceptions 3, 4 and 5 deal with special credit for the roof construction.

Exceptions 3, 4 and 5 require the exterior walls to extend to the roof deck or underside of roof sheathing and require special attention to the roof construction to eliminate the need for the parapet extension, thus they are not subject to any FSD requirements and can have an FSD of zero. Exceptions 1, 2 and 6 simply exempt the walls from extending above the roof without any special roof construction and do not require that the walls must extend to the roof deck.

Exception 1 applies to bearing and nonload-bearing walls and would also exempt bearing walls from parapets if the bearing wall was not required to have a fire-resistance rating by Table 602, even if said bearing wall was required to have a fire-resistance rating by Table 601.

705.11.1 Parapet construction. Parapets shall have the same *fire-resistance rating* as that required for the supporting wall, and on any side adjacent to a roof surface, shall have noncombustible faces for the uppermost 18 inches (457 mm), including counterflashing and coping materials. The height of the parapet shall be not less than 30 inches (762 mm) above the point where the roof surface and the wall intersect. Where the roof slopes toward a parapet at a slope greater than two units vertical in 12 units horizontal (16.7-percent slope), the parapet shall extend to the same height as any portion of the roof within a *fire separation distance* where protection of wall openings is required, but in no case shall the height be less than 30 inches (762 mm).

❖ Parapet wall construction shall be of combustible or noncombustible material depending on the exterior wall requirements of the type of construction and shall be of fire-resistance-rated construction as required for the exterior wall. The interior wall covering facing the roof, including the flashing, shall be noncombustible to a height of 18 inches (457 mm) above the roof. The required height of the parapet shall be 30 inches (762 mm) above the roof surface unless the roof slopes upward away from the wall on a pitch of 2 in 12 or greater. In some cases, the last part of this section requires a higher parapet, depending on the FSD. When the slope of the roof is over 2 in 12, the parapet shall extend to a height equal to the height of the roof at the point determined as follows:

- In a nonsprinklered building, Table 705.8 requires protected openings for any FSD between 3 and 5 feet (914 mm and 1524 mm). Therefore, the height of the roof at the 5-foot (1524 mm) FSD is the height to which the parapet wall extension must extend, but must always extend at least 30 inches (762 mm) above the roof at the exterior wall. For a nonsprinklered building, the height of the roof at the plane where protected openings are required according to Table 705.8 is the height at 5 feet (1524 mm) from the common lot line or assumed property line. See Commentary Figure 705.11.1(1).
- For a sprinklered building, Table 705.8 never requires protected openings. Therefore, for a sprinklered building the parapet need only extend 30 inches (762 mm) above the roof, regardless of the slope of the roof or the FSD. See Commentary Figure 705.11.1(2).

SECTION 706 FIRE WALLS

706.1 General. Each portion of a building separated by one or more *fire walls* that comply with the provisions of this section shall be considered a separate building. The extent and location of such *fire walls* shall provide a complete separation. Where a *fire wall* separates occupancies that are required to be separated by a *fire barrier* wall, the most restrictive requirements of each separation shall apply.

❖ Fire walls serve to create separate buildings for purposes of allowable area and type of construction requirements (see the definition of "Area, building" in

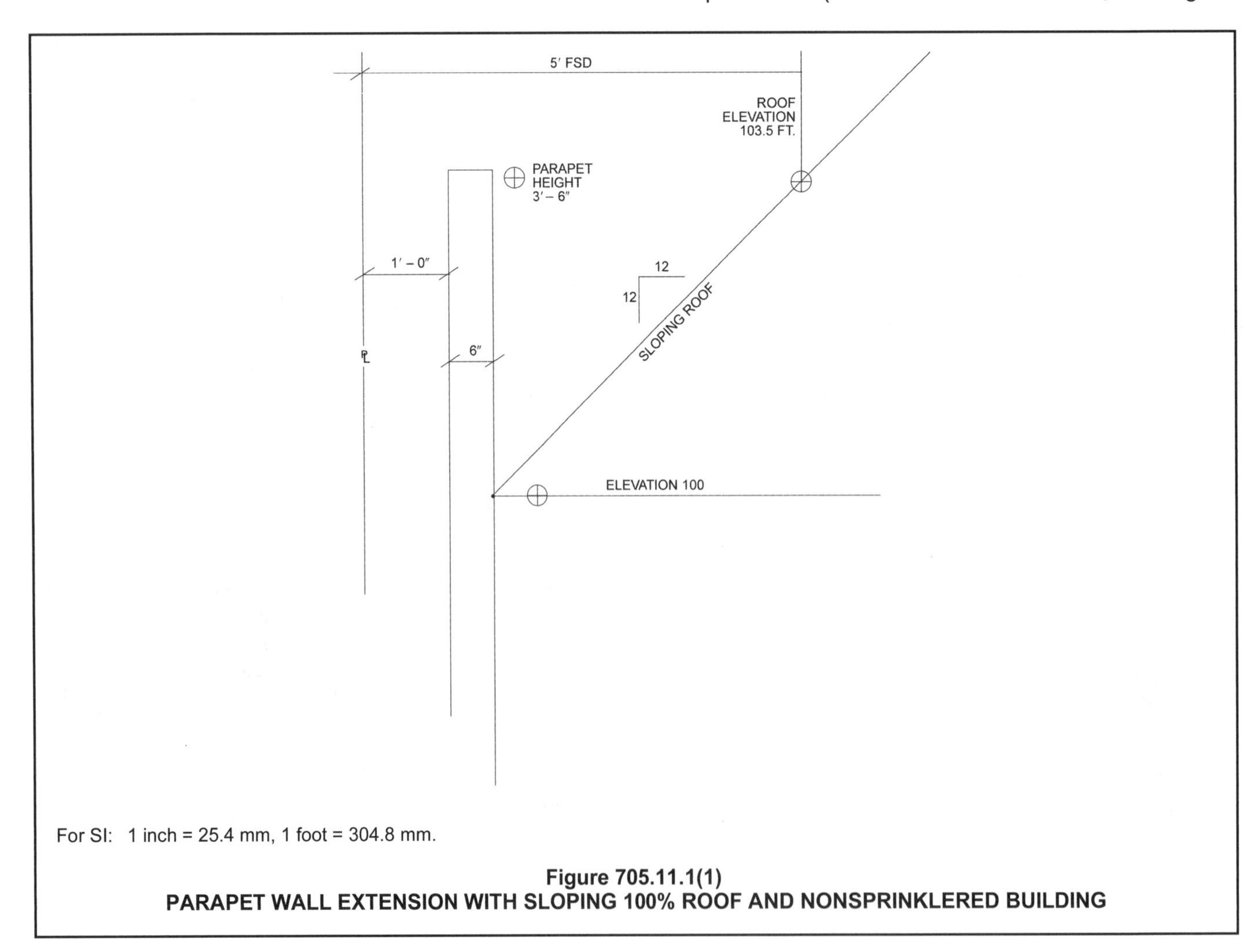

For SI: 1 inch = 25.4 mm, 1 foot = 304.8 mm.

Figure 705.11.1(1)
PARAPET WALL EXTENSION WITH SLOPING 100% ROOF AND NONSPRINKLERED BUILDING

Section 202). All code provisions that apply to buildings would be applied individually to the "building" on each side of the fire wall. Because of the special allowances for allowable areas, a fire wall must provide a higher level of fire safety, continuity and structural integrity than other types of fire-resistance-rated walls. When the designer utilizes a fire wall, he or she must design the fire wall to meet this section.

It is not intended that fire walls can be provided in the horizontal plane to create separate buildings. Fire walls cannot be utilized to increase the allowable building height. However, offsetting two vertical sections of fire wall is permissible as long as the required fire-resistance rating and structural stability are maintained.

706.1.1 Party walls. Any wall located on a *lot line* between adjacent buildings, which is used or adapted for joint service between the two buildings, shall be constructed as a *fire wall* in accordance with Section 706. Party walls shall be constructed without openings and shall create separate buildings.

Exception: Openings in a party wall separating an *anchor building* and a mall shall be in accordance with Section 402.4.2.2.1.

❖ A party wall is a fire wall on an interior lot line, adapted for joint use by both buildings. It is distinguished from other fire walls in that it is on the property line and serves to separate buildings usually owned by two separate parties. When two separate structures are built up to the property line, the designer has the option of using two separate exterior walls with zero FSD or a party wall. Since there is a real property line involved, the prohibition for openings between the two buildings is important and even utilities cannot penetrate the party wall. Unlike fire walls in other buildings, which can have openings in them, party walls cannot have any openings in them (see Section 706.8 for opening limits on other fire walls).

The exception to allow openings is important since many anchor stores are actually owned by the major department store while the mall is owned by a separate entity. The fact that there is a real property line at the separation walls between an anchor and a mall means that technically there is a party wall, but openings are normally present and a necessary function of the mall and anchor.

706.2 Structural stability. *Fire walls* shall be designed and constructed to allow collapse of the structure on either side without collapse of the wall under fire conditions. *Fire walls* designed and constructed in accordance with NFPA 221 shall be deemed to comply with this section.

❖ Since the collapse of one building from fire conditions should not cause the collapse of an adjacent building, a fire wall is required to be capable of withstanding the collapse of the construction on either side of the wall under fire conditions. An exception to the provision is found in Exception 5 to Section 706.6. In this situation, the issue of structural stability would still apply if the fire was in the buildings located above the horizontal separation required in Section 509.2, Item

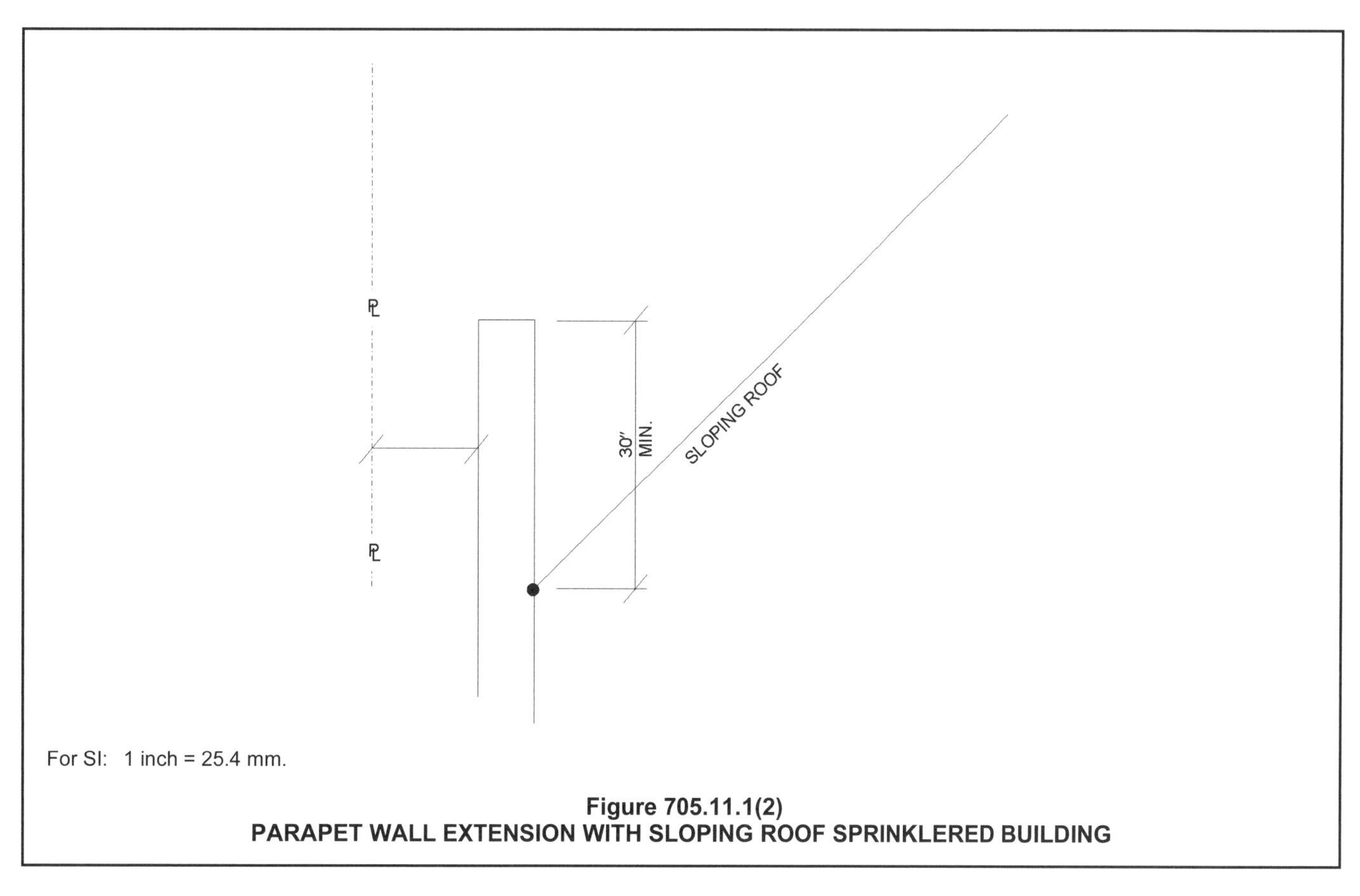

Figure 705.11.1(2)
PARAPET WALL EXTENSION WITH SLOPING ROOF SPRINKLERED BUILDING

1. However, it would clearly be impossible for the buildings above the separation or, for that matter, the fire wall to be able to withstand the collapse of the building beneath it. Since the building below is of Type IA construction and the separation has a 3-hour fire-resistance rating, this should not be a problem. See 706.6, Exception 5, regarding fire walls in this situation.

There are various methods of designing and constructing fire walls for structural stability during a fire. Among the systems used are cantilevered or freestanding walls, laterally supported and tied walls, and double wall construction. For masonry walls, the NCMA TEK Bulletin 5-8B contains helpful information. For an example with a masonry fire wall, see Commentary Figure 706.2(1). The Gypsum Association's *Fire Resistance Design Manual* contains construction details for area separation walls (party wall/fire walls) which have been accepted as fire walls. For an example of a gypsum board fire wall, see Commentary Figure 706.2(2).

The collapse of structural members at some distance from the wall can generally be ignored, but the structural members that are supported by the wall or

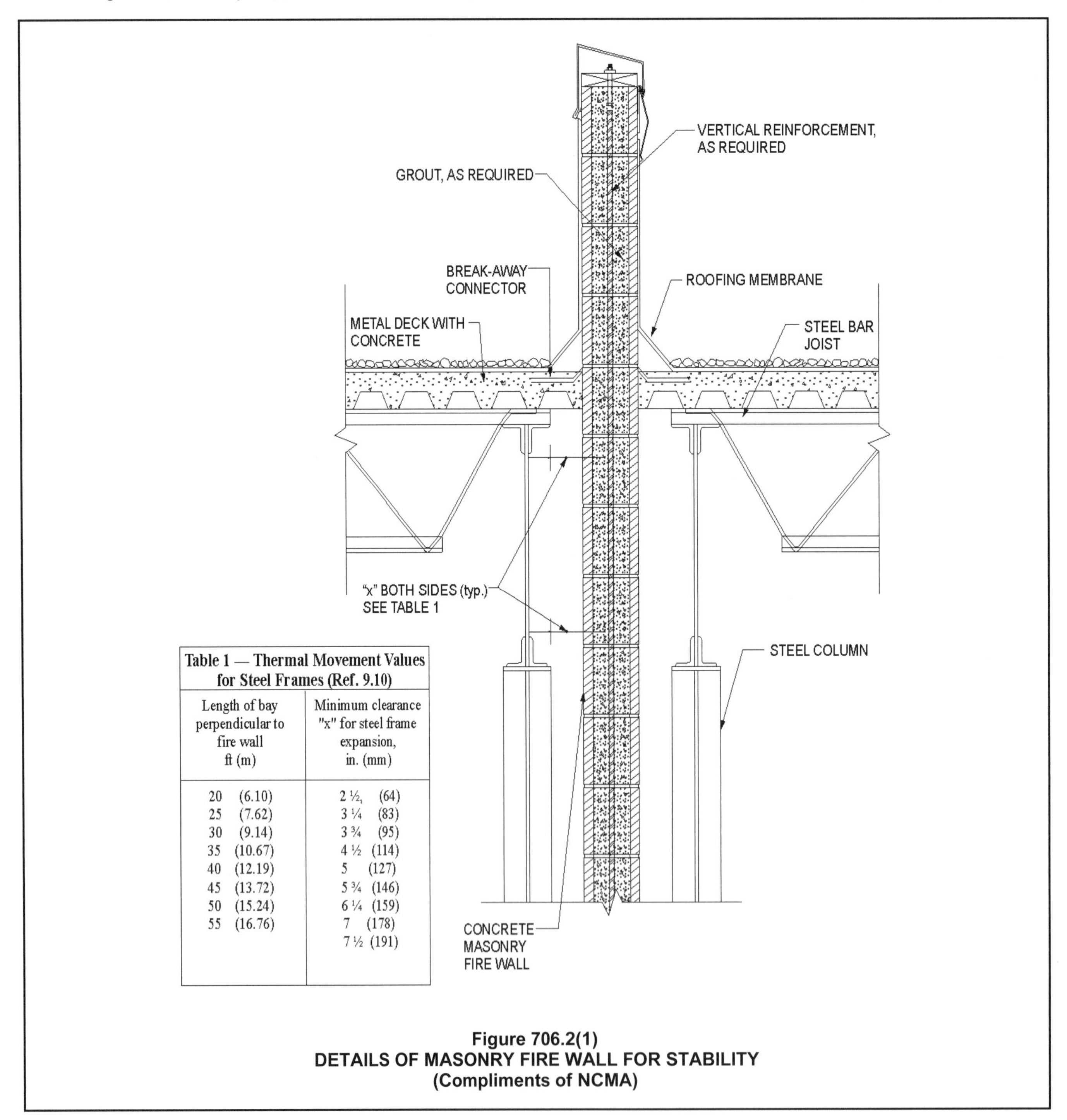

Table 1 — Thermal Movement Values for Steel Frames (Ref. 9.10)

Length of bay perpendicular to fire wall ft (m)	Minimum clearance "x" for steel frame expansion, in. (mm)
20 (6.10)	2 ½, (64)
25 (7.62)	3 ¼ (83)
30 (9.14)	3 ¾ (95)
35 (10.67)	4 ½ (114)
40 (12.19)	5 (127)
45 (13.72)	5 ¾ (146)
50 (15.24)	6 ¼ (159)
55 (16.76)	7 (178)
	7 ½ (191)

Figure 706.2(1)
DETAILS OF MASONRY FIRE WALL FOR STABILITY
(Compliments of NCMA)

that are attached directly to the wall for lateral support do require special attention.

Double fire walls shall be assumed to have the required structural stability if they are constructed in accordance with NFPA 221. NFPA 221 covers more than double fire walls, but the IBC only references it for double fire walls.

706.3 Materials. *Fire walls* shall be of any *approved* noncombustible materials.

Exception: Buildings of Type V construction.

❖ This section requires that fire walls be constructed of noncombustible materials unless both buildings are of Type V (combustible) construction. This is consistent with the provisions of Section 602, which require exterior walls of buildings of Type I, II, III and IV to be built of noncombustible construction, while buildings of Type V construction are permitted to be built of combustible exterior walls.

706.4 Fire-resistance rating. *Fire walls* shall have a *fire-resistance rating* of not less than that required by Table 706.4.

❖ The required fire-resistance rating must comply with the more restrictive occupancy of Table 706.4. For example, if two buildings of Type II construction with occupancies in Groups S-1 and B are built of any construction type and separated by a fire wall, the fire wall is required to have a fire-resistance rating of 3 hours (in accordance with Table 706.4 based on Group S-1). The provisions of Section 706.1 should also be reviewed when determining the required fire-resistance rating. If the occupancies involved require a separation, or are not permitted in a mixed occu-

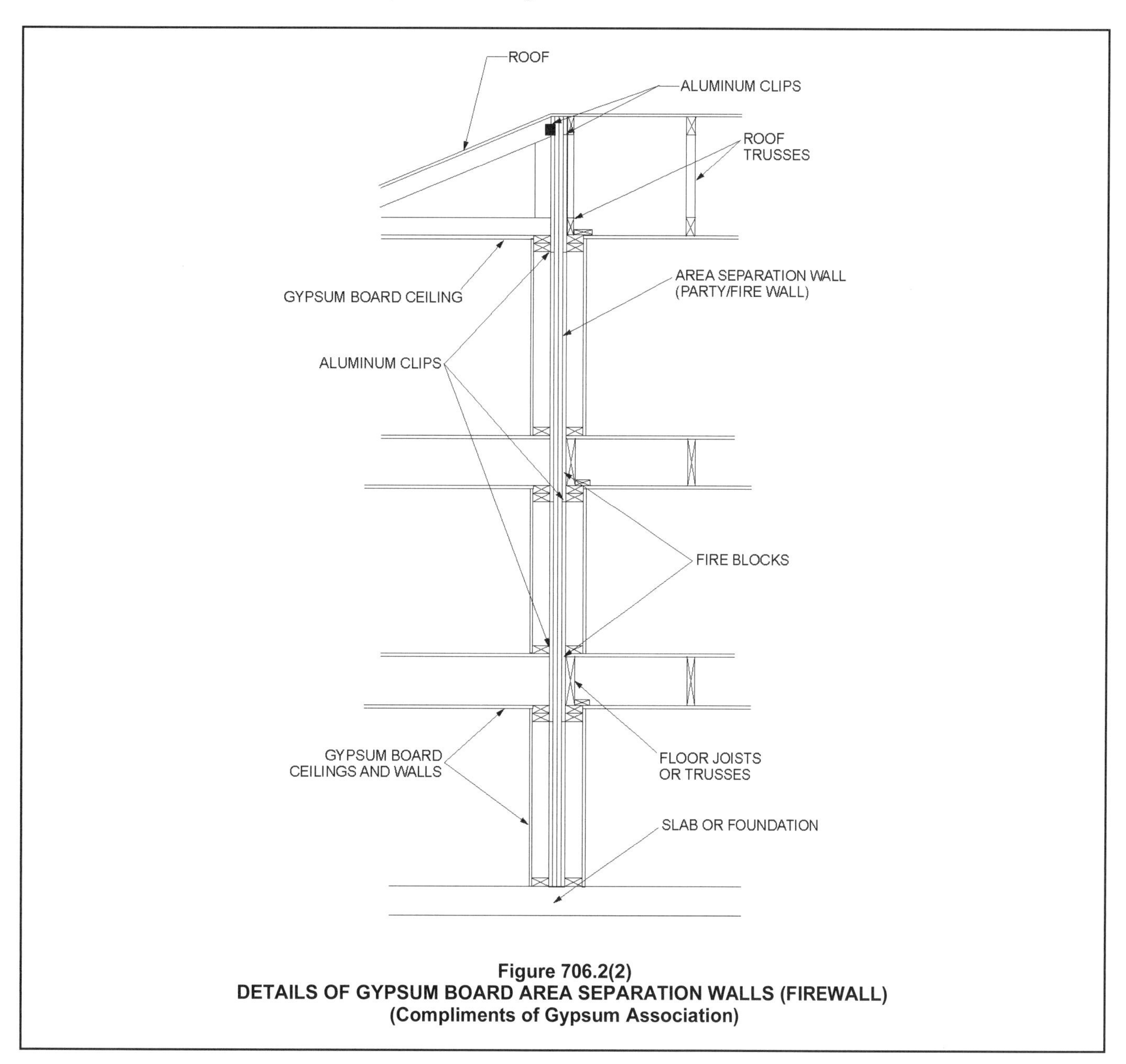

Figure 706.2(2)
DETAILS OF GYPSUM BOARD AREA SEPARATION WALLS (FIREWALL)
(Compliments of Gypsum Association)

pancy building, then the most restrictive requirement between that and the fire wall provisions will apply.

TABLE 706.4
FIRE WALL FIRE-RESISTANCE RATINGS

GROUP	FIRE-RESISTANCE RATING (hours)
A, B, E, H-4, I, R-1, R-2, U	3[a]
F-1, H-3[b], H-5, M, S-1	3
H-1, H-2	4[b]
F-2, S-2, R-3, R-4	2

a. In Type II or V construction, walls shall be permitted to have a 2-hour *fire-resistance rating*.

b. For Group H-1, H-2 or H-3 buildings, also see Sections 415.7 and 415.8.

❖ The fire-resistance ratings represent a relationship between the fuel load and an exposure to fire severity that is equivalent to the standard time-temperature curve. The relationship is based on a number of full-scale fire tests by the U.S. Department of Commerce's National Bureau of Standards (NBS) conducted in the 1920s to determine how actual building fires compared with the temperatures represented by the standard time-temperature curve (see commentary, Section 703.2).

An analysis of tests documented by S.H. Ingberg indicates that the weight per square foot (psf) of ordinary combustibles, such as wood and paper, with a heat of combustion of 7,000 to 8,000 British thermal units (Btu) per pound (16 282 to 18 608 kJ/kg), is related to hourly fire severity as described in Commentary Figure 706.4(1). By comparing Table 706.4 to Commentary Figure 706.4(2), it shows that the table is not based solely on the average fuel load of the occupancy. Not only are other factors, such as occupant density and evacuation capability, incorporated into the required fire-resistance rating, but it must also be recognized that fuel loads in a building are not evenly distributed. For example, the average fuel load for an occupancy in Group B may be 5 to 10 psf (24 to 49 kg/m²). The fuel load in file and storage rooms, however, may be 10 to 20 psf (49 to 98 kg/m²). While other sections of the code may address these areas by mandating an automatic fire suppression system with a partition capable of resisting the passage of smoke or a fire-resistance-rated enclosure of the incidental use area, the potential for higher fuel loads is also factored into determining the required fire-resistance rating.

The minimum fire-resistance ratings required in Table 706.4 generally correlate with the fire-resistance ratings required by Table 508.3.3. However, Table 706.4 contains provisions that acknowledge the need for more conservative ratings for fire walls due to their unique role in creating separate buildings.

Note a allows a reduction in the fire-resistance rating of the fire walls in buildings of Type II or V construction. For instance, the fire-resistance rating of a fire wall separating a Type II building from a Type V building can be only 2 hours. If the buildings on both sides of the fire wall are Type II construction, the same reduction is allowed. If the buildings on both sides of the fire wall are Type V construction, the same reduction is allowed. If the building on either side of the fire wall is of another type of construction than Type II or V, the greater fire-resistance rating is required.

Due to the unique nature of the hazards presented by occupancies of Groups H-1, H-2 and H-3, Table 706.4 includes Note b, which references Sections 415.7 and 415.8 for specific requirements relative to the construction of such buildings (see commentary, Sections 415.7 and 415.8).

Average fuel load psf	kg/m²	Equivalent fire endurance (hours)
5	24.4	1/2
7 1/2	36.6	3/4
10	48.8	1
15	73.2	1 1/2
20	97.6	2
30	146.5	3
40	195.3	4 1/2
50	244.1	6
60	292.9	7 1/2

For SI: 1 pound per square foot = 4.882 kg/m².

Figure 706.4(1)
RELATIONSHIP BETWEEN FUEL LOAD AND FIRE ENDURANCE

Occupancy	Combustibles in occupancy (psf)	Fire severity (hours)
Assembly	5 to 10	1/2 to 1
Business	5 to 10	1/2 to 1
Educational	5 to 10	1/2 to 1
Factory—Industrial		
Low hazard	0 to 10	0 to 1
Moderate hazard	10 to 25	1 to 3
Hazardous	Variable	Variable
Institutional	5 to 10	1/2 to 1
Mercantile	10 to 20	1 to 2
Residential	5 to 10	1/2 to 1
Storage		
Low hazard	0 to 10	0 to 1
Moderate hazard	10 to 30	1 to 1

For SI: 1 pound per square foot = 4.882 kg/m².

Figure 706.4(2)
OCCUPANCIES—FUEL LOAD-FIRE SEVERITY

706.5 Horizontal continuity. *Fire walls* shall be continuous from *exterior wall* to *exterior wall* and shall extend not less than 18 inches (457 mm) beyond the exterior surface of *exterior walls*.

Exceptions:

1. *Fire walls* shall be permitted to terminate at the interior surface of combustible exterior sheathing or siding provided the *exterior wall* has a *fire-resistance rating* of not less than 1 hour for a horizontal distance of not less than 4 feet (1220 mm) on both sides of the *fire wall*. Openings within such *exterior walls* shall be protected by opening protectives having a *fire protection rating* of not less than $^3/_4$ hour.
2. *Fire walls* shall be permitted to terminate at the interior surface of noncombustible exterior sheathing, exterior siding or other noncombustible exterior finishes provided the sheathing, siding or other exterior noncombustible finish extends a horizontal distance of not less than 4 feet (1220 mm) on both sides of the *fire wall*.
3. *Fire walls* shall be permitted to terminate at the interior surface of noncombustible exterior sheathing where the building on each side of the *fire wall* is protected by an *automatic sprinkler system* installed in accordance with Section 903.3.1.1 or 903.3.1.2.

❖ Historically, the codes have addressed the hazards of fire exposure at the fire wall from only a vertical perspective; namely, at the roof (see Section 706.6). Section 706.5 addresses a similar fire hazard concern from the horizontal perspective; namely, at the intersection of the fire wall and the exterior wall.

The 18-inch (457 mm) extension is intended to abate the potential for fire to travel from one building to the other around the fire wall. The 18-inch (457 mm) extension is required to extend the full height of the fire wall. The three exceptions acknowledge the effect certain types of construction will have on fire breaching the exterior wall and exposing the adjacent building. Specifically, fire-resistance-rated construction (see Exception 1), noncombustible finish materials (see Exception 2) and noncombustible sheathing materials coupled with sprinkler protection (see Exception 3) provide the necessary barrier to limit fire spread across the exterior surface. The difference between Exceptions 2 and 3 is that Exception 2 would not permit a combustible exterior finish to be placed over the noncombustible exterior wall construction, while Exception 3 would allow a combustible exterior finish, provided the building is sprinklered with either an NFPA 13 or 13R system.

706.5.1 Exterior walls. Where the *fire wall* intersects *exterior walls*, the *fire-resistance rating* and opening protection of the *exterior walls* shall comply with one of the following:

1. The *exterior walls* on both sides of the *fire wall* shall have a 1-hour *fire-resistance rating* with $^3/_4$-hour protection where opening protection is required by Section 705.8. The *fire-resistance rating* of the *exterior wall* shall extend not less than 4 feet (1220 mm) on each side of the intersection of the *fire wall* to *exterior wall*. *Exterior wall* intersections at *fire walls* that form an angle equal to or greater than 180 degrees (3.14 rad) do not need *exterior wall* protection.
2. Buildings or spaces on both sides of the intersecting *fire wall* shall assume to have an imaginary *lot line* at the *fire wall* and extending beyond the exterior of the *fire wall*. The location of the assumed line in relation to the *exterior walls* and the *fire wall* shall be such that the *exterior wall* and opening protection meet the requirements set forth in Sections 705.5 and 705.8. Such protection is not required for *exterior walls* terminating at *fire walls* that form an angle equal to or greater than 180 degrees (3.14 rad).

❖ This section deals only with fire walls that terminate at an exterior wall that does not continue in the same horizontal plane but the exterior walls on each side of the fire wall form an angle of less than 180 degrees (3.14 rad). For example, the required fire wall may end at an inside corner of an L-shaped building. Should this occur, not only does Section 706.5 apply, but this section has extra requirements to mitigate the exterior wall fire exposure from one building to another. This section states two methods of complying. See Commentary Figures 706.5.1(1), 706.5.1(2) and 706.5.1(3).

706.5.2 Horizontal projecting elements. *Fire walls* shall extend to the outer edge of horizontal projecting elements such as balconies, roof overhangs, canopies, marquees and similar projections that are within 4 feet (1220 mm) of the *fire wall*.

Exceptions:

1. Horizontal projecting elements without concealed spaces, provided the *exterior wall* behind and below the projecting element has not less than 1-hour fire-resistance-rated construction for a distance not less than the depth of the projecting element on both sides of the *fire wall*. Openings within such *exterior walls* shall be protected by opening protectives having a *fire protection rating* of not less than $^3/_4$ hour.
2. Noncombustible horizontal projecting elements with concealed spaces, provided a minimum 1-hour fire-resistance-rated wall extends through the concealed space. The projecting element shall be separated from the building by not less than 1-hour fire-resistance-rated construction for a distance on each side of the *fire wall* equal to the depth of the projecting element. The wall is not required to extend under the projecting element where the building *exterior wall* is not less than 1-hour fire-resistance rated for a distance on each side of the *fire wall* equal to the depth of the projecting element. Openings within such *exterior walls* shall be protected by opening protectives having a *fire protection rating* of not less than $^3/_4$ hour.

3. For combustible horizontal projecting elements with concealed spaces, the *fire wall* need only extend through the concealed space to the outer edges of the projecting elements. The *exterior wall* behind and below the projecting element shall be of not less than 1-hour fire-resistance-rated construction for a distance not less than the depth of the projecting elements on both sides of the *fire wall*. Openings within such *exterior walls* shall be protected by opening protectives having a fire-protection rating of not less than $^3/_4$ hour.

❖ Fire walls are typically used to reduce the building area for the purpose of code compliance. However, many structures divided by fire walls are still detailed as a contiguous structure with projecting elements located on the exterior wall that extend across the fire wall. These projecting elements represent a potential conduit for fire to be transferred from one side of the building to the other side of the building. Even if the projecting elements are terminated at a distance in close proximity to the fire wall, the potential for fire spread to the adjacent projecting element still exists. In this section, 4 feet (1219 mm) (measured from the fire wall) is considered the appropriate threshold.

This section requires the fire wall to extend to the outer edge of the projecting element when such an element is located within 4 feet (1219 mm) of the fire wall. For example, in a contiguous residential structure with interior lot lines and fire walls or party walls between each dwelling unit, and where balconies are located within 4 feet (1219 mm) of the party or fire wall, then the fire/party wall must be extended to the outer edge of the balcony or to a point in line with the outer edge of the balcony.

Exception 1 acknowledges the reduction in hazard due to the lack of a concealed space crossing the fire

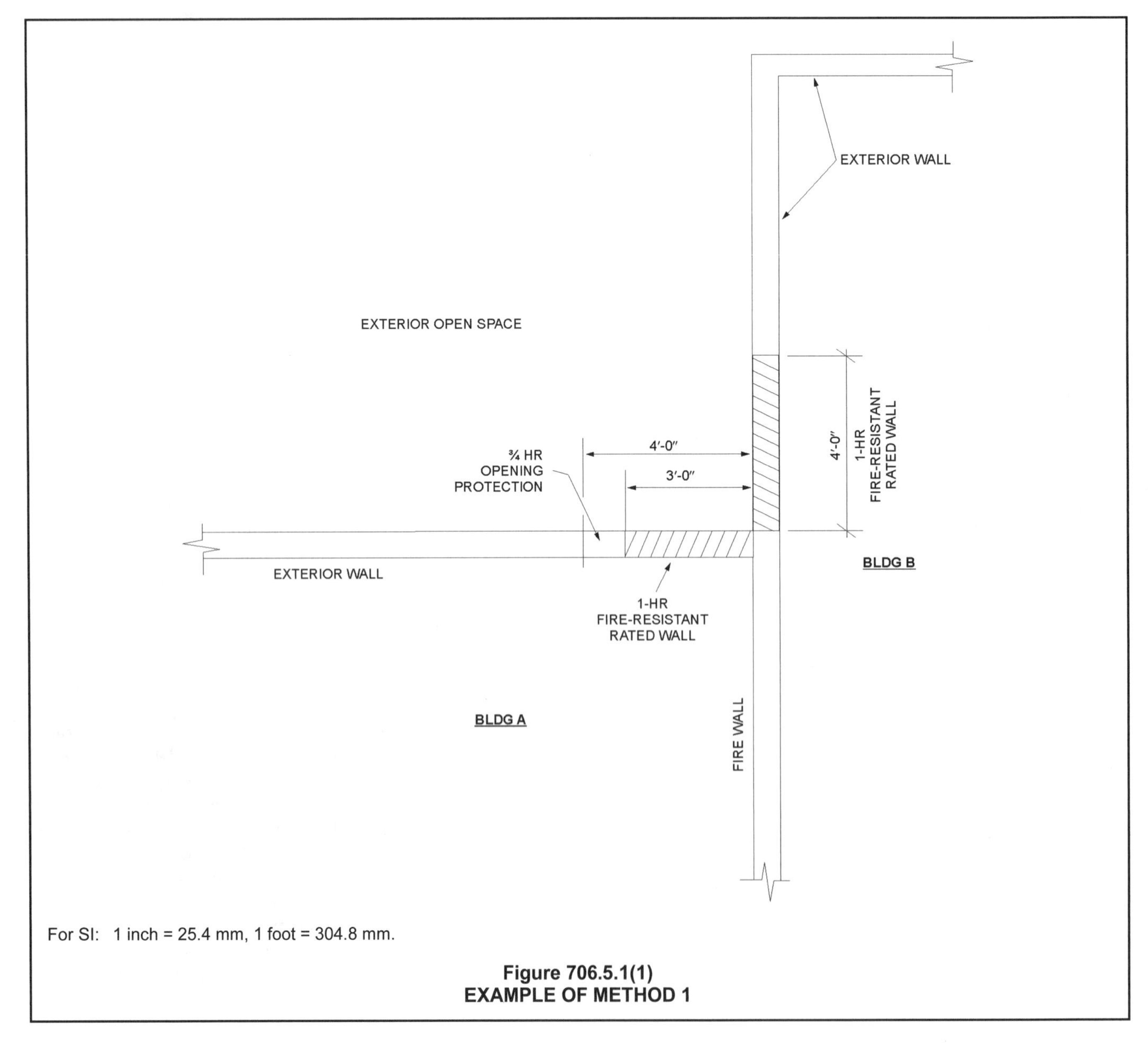

Figure 706.5.1(1)
EXAMPLE OF METHOD 1

wall. However, since the projecting element provides a connection between the buildings, there is still a potential for fire to transfer across the wall, thus the need for a rated exterior wall separation between the inside of the building and the projection. Exceptions 2 and 3 address projecting elements with concealed spaces of noncombustible and combustible construction, respectively.

706.6 Vertical continuity. *Fire walls* shall extend from the foundation to a termination point not less than 30 inches (762 mm) above both adjacent roofs.

Exceptions:

1. Stepped buildings in accordance with Section 706.6.1.
2. Two-hour fire-resistance-rated walls shall be permitted to terminate at the underside of the roof sheathing, deck or slab, provided:
   - 2.1. The lower roof assembly within 4 feet (1220 mm) of the wall has not less than a 1-hour *fire-resistance rating* and the entire length and span of supporting elements for the rated roof assembly has a *fire-resistance rating* of not less than 1 hour.
   - 2.2. Openings in the roof shall not be located within 4 feet (1220 mm) of the *fire wall.*
   - 2.3. Each building shall be provided with not less than a Class B roof covering.

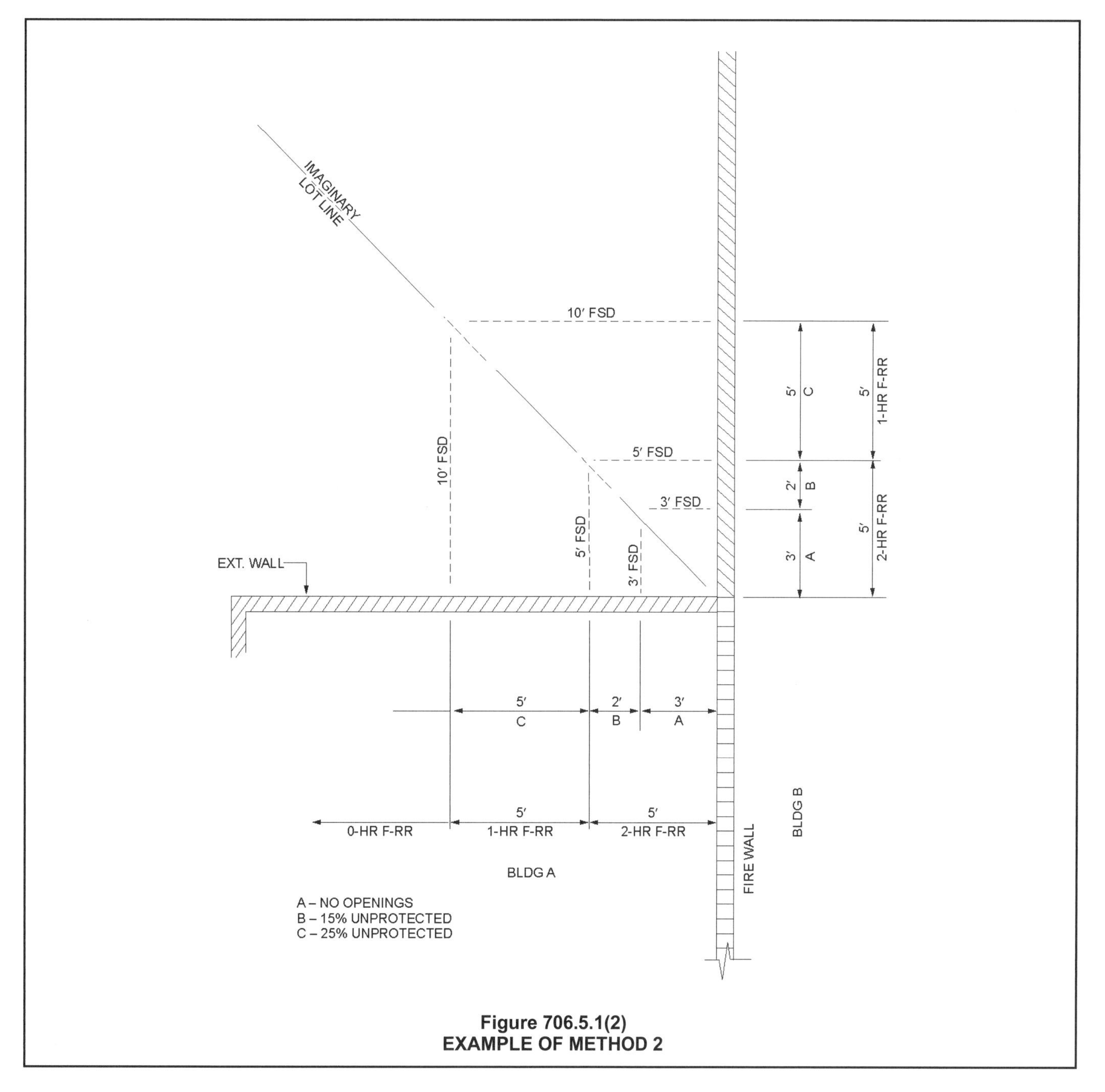

Figure 706.5.1(2)
EXAMPLE OF METHOD 2

3. Walls shall be permitted to terminate at the underside of noncombustible roof sheathing, deck or slabs where both buildings are provided with not less than a Class B roof covering. Openings in the roof shall not be located within 4 feet (1220 mm) of the *fire wall.*

4. In buildings of Type III, IV and V construction, walls shall be permitted to terminate at the underside of combustible roof sheathing or decks, provided:

   4.1. There are no openings in the roof within 4 feet (1220 mm) of the *fire wall,*

   4.2. The roof is covered with a minimum Class B roof covering, and

   4.3. The roof sheathing or deck is constructed of *fire-retardant-treated wood* for a distance of 4 feet (1220 mm) on both sides of the wall or the roof is protected with $^{5}/_{8}$-inch (15.9

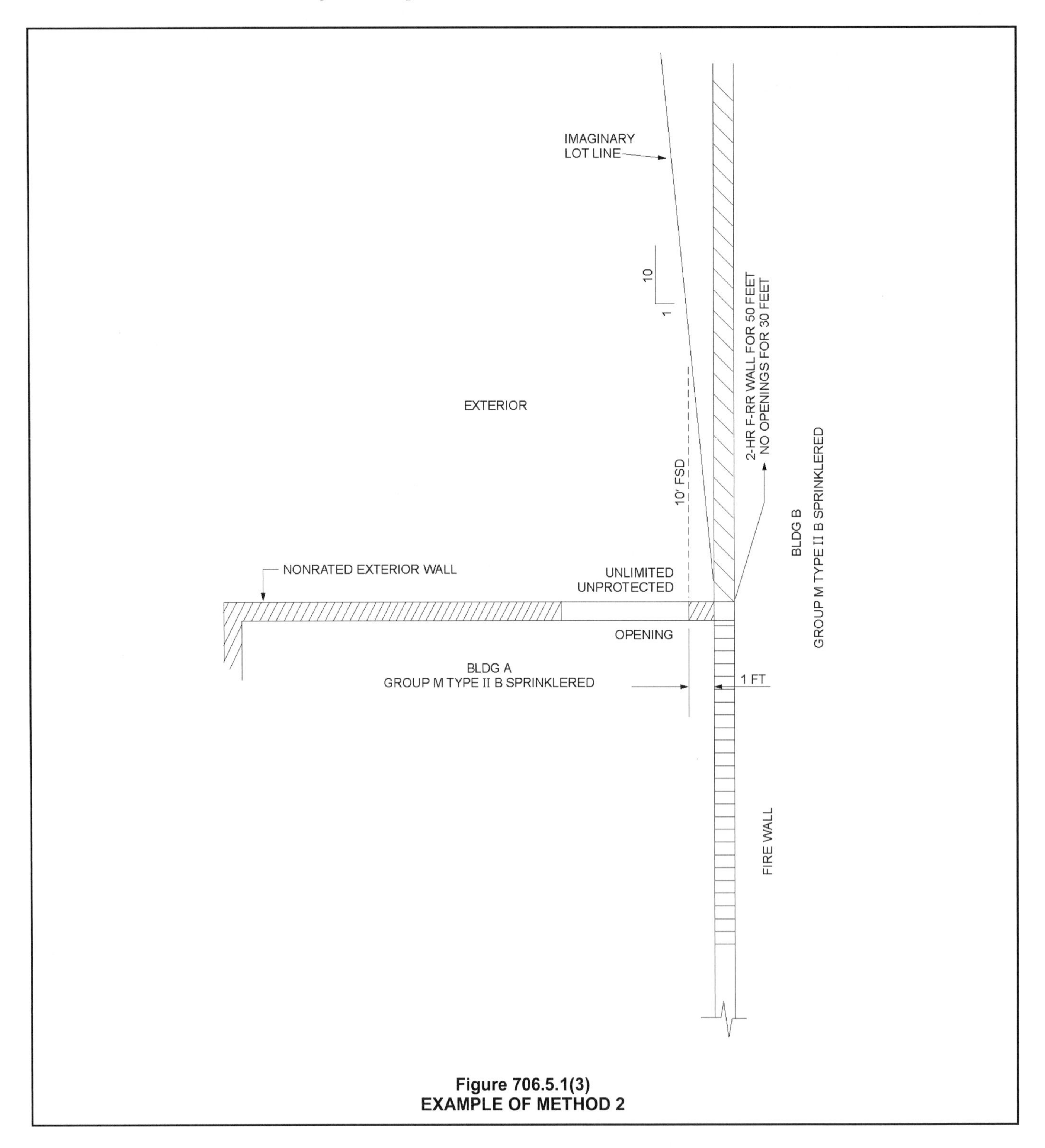

Figure 706.5.1(3)
EXAMPLE OF METHOD 2

mm) Type X gypsum board directly beneath the underside of the roof sheathing or deck, supported by not less than 2-inch (51 mm) nominal ledgers attached to the sides of the roof framing members for a distance of not less than 4 feet (1220 mm) on both sides of the *fire wall*.

5. In buildings designed in accordance with Section 510.2, *fire walls* located above the 3-hour *horizontal assembly* required by Section 510.2, Item 1 shall be permitted to extend from the top of this *horizontal assembly*.

6. Buildings with sloped roofs in accordance with Section 706.6.2.

❖ One of the primary purposes of a fire wall is to provide protection from an exposure fire. A fire wall must be continuous from the foundation to or through the roof. The fire wall is required to extend to 30 inches (762 mm) above the roof surface (see Commentary Figure 706.6 for a diagram of a typical fire wall). Where other provisions are met that will restrict the spread of fire, this 30-inch (762 mm) extension is not required (see Exceptions 1 through 5). Exception 1 stipulates that fire walls separating two buildings that have different roof levels must comply with the provisions of Section 706.6.1.

Exception 2 to this section's requirement for a parapet applies to buildings where: a. The maximum required fire-resistance rating of the fire wall is 2 hours; b. The roof assembly located within 4 feet (1219 mm) of the fire wall has a minimum fire-resistance rating of 1 hour (including all supporting elements and structural members); c. No roof openings are located within 4 feet (1219 mm) of the fire wall; and d. The buildings on either side of the fire wall have a minimum Class B roof covering. The provisions of Exception 2 are not applicable to buildings with combustible roof construction where a fire wall is required to have a fire-resistance rating of 3 hours or greater. Buildings where the fire wall is required to have a 3-hour or greater fire-resistance rating must comply with the general provisions of Section 706.6, or with Exception 3 or 4 in order to qualify for the parapet extension exception.

Exception 3 states that, since the intent of the parapet is to prevent fire spread over the roof, it is not required if the entire roof is of noncombustible construction; the fire wall is continuous to the underside of the roof deck, roof slab or roof sheathing; the roof-covering material has a minimum Class B classification in accordance with the provisions of Section 1505.3 (see commentary, Section 1505.3); and there are no roof openings located within 4 feet (1219 mm) on either side of the fire wall.

The intent of the requirement for a Class B roof covering is due to the fact that roof coverings are not considered part of the roof, and therefore, are not subject to the noncombustibility requirements of this section. This section requires that the roof covering

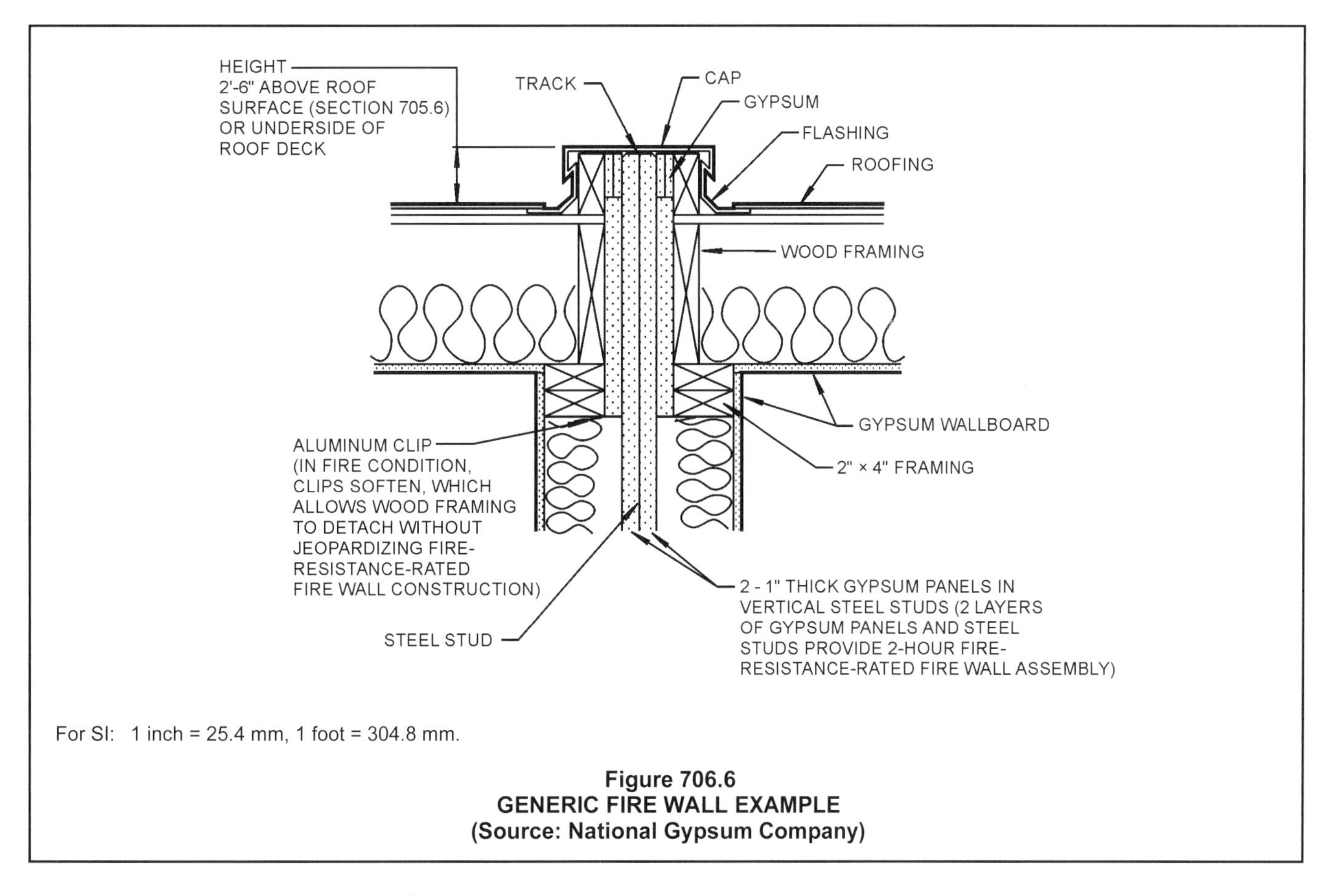

Figure 706.6
GENERIC FIRE WALL EXAMPLE
(Source: National Gypsum Company)

be effective against moderate fire test exposure to reduce the potential for fire spread across the roof. The restriction on the location of roof openings is intended to minimize the potential of the spread of fire from one building to an adjacent building via roof openings, such as skylights, penetrations or roof windows. This fire spread could occur from something such as a fan that breaches the roof opening in the involved building and exposes roof openings in the adjacent building to direct fire or to burning brands, which could land on the roof, burn through the roof opening and ignite combustibles in the interior of the adjacent building. The 4-foot (1219 mm) threshold provides a reasonable approximation of a fire plume, beyond which the exposure to fire of the adjacent building is reduced. Exception 4 to the requirement in Section 706.6 for a parapet is the condition in buildings of Type III, IV and V construction where the sheathing or deck is constructed of noncombustible materials or approved fire-retardant-treated wood for a distance of 4 feet (1219 mm) on both sides of the wall (see the commentary to Section 2303.2 for additional information on fire-retardant-treated wood) or where the prescribed application of $^5/_8$-inch (15.9 mm) Type X gypsum board is applied to the underside of the deck for a distance of 4 feet (1219 mm) on both sides of the wall. The roof coverings on both buildings must have a minimum Class B rating and no roof openings are to be located within 4 feet (1219 mm) of the fire wall (see commentary, Section 508.2). The intent is the same: to resist the passage of flame beyond the fire wall.

706.6.1 Stepped buildings. Where a *fire wall* serves as an *exterior wall* for a building and separates buildings having different roof levels, such wall shall terminate at a point not less than 30 inches (762 mm) above the lower roof level, provided the *exterior wall* for a height of 15 feet (4572 mm) above the lower roof is not less than 1-hour fire-resistance-rated construction from both sides with openings protected by fire assemblies having a *fire protection rating* of not less than $^3/_4$ hour.

Exception: Where the *fire wall* terminates at the underside of the roof sheathing, deck or slab of the lower roof, provided:

1. The lower roof assembly within 10 feet (3048 mm) of the wall has not less than a 1-hour *fire-resistance rating* and the entire length and span of supporting elements for the rated roof assembly has a *fire-resistance rating* of not less than 1 hour.
2. Openings in the lower roof shall not be located within 10 feet (3048 mm) of the *fire wall*.

❖ The basic provisions of Section 706.6 require that a parapet extend a minimum of 30 inches (762 mm) above the roof surfaces on both sides of the fire wall, which would then require fire walls separating buildings with different roof heights to extend 30 inches (762 mm) above the highest roof surface. The provisions of Section 706.6.1 address situations where a fire wall separates adjacent buildings with a difference in roof height. These provisions acknowledge that fire exposure to the exterior wall of an adjacent building from the roof of a lower building represents, to a certain extent, a reduced hazard.

This section retains the fire wall extension above the lower roof surface and places a 15-foot (4572 mm) limit on rated wall construction and opening protectives, while the exception allows the fire wall extension to be eliminated as long as a fire barrier, in the form of 1-hour-rated roof construction, is provided that extends a minimum of 10 feet (3048 mm) from the fire wall. The option provided by this section permits openings in an exterior wall that extend above the fire wall, where the fire wall extends 30 inches (762 mm) above the lower roof surface, and the exterior wall extending above the fire wall is constructed as a 1-hour fire-resistance-rated wall rated for exposure from both sides for a distance of 15 feet (4572 mm) above the lower roof surface. Where the fire wall and exterior wall comply with these requirements, openings are permitted in the exterior wall extension, provided that all openings located within 15 feet (4572 mm) of the lower roof surface are protected with $^3/_4$-hour-rated opening protectives. Openings located more than 15 feet (4572 mm) above the lower roof surface are not required to have opening protectives [see Commentary Figure 706.6.1(1)]. The provisions of this section are similar to those contained in Section 704.10 for vertical exposure of exterior openings. They require that when the difference in height between the roof surfaces of the adjacent buildings is less than 15 feet (4572 mm), the exterior wall extension is required to extend to the underside of the upper roof deck [see Commentary Figure 706.6.1(2)].The exception permits openings in the roof that extend above the fire wall, where: a. The fire wall terminates at the bottom of the roof deck of the lower roof; b. The lower roof assembly has a minimum 1-hour fire-resistance rating for a minimum distance of 10 feet (3048 mm) from the fire wall; and c. There are no openings located in the lower roof within 10 feet (3048 mm) of the fire wall. Openings located in the exterior wall above the lower roof surface are not required to have opening protectives [see Commentary Figure 706.6.1(3)]. It must be noted that all structural elements, including beams, columns and bearing walls that provide support for the fire-resistance-rated roof assembly, must also have a 1-hour fire-resistance rating for their entire length or span in order to maintain the effectiveness of the rated roof assembly. The provisions of this section are similar to those contained in Section 1023.7 for the protection of the exterior walls of exit stairways. The provisions for stepped buildings only apply where a fire wall is required in between the buildings and do not apply to party walls on real property lines.

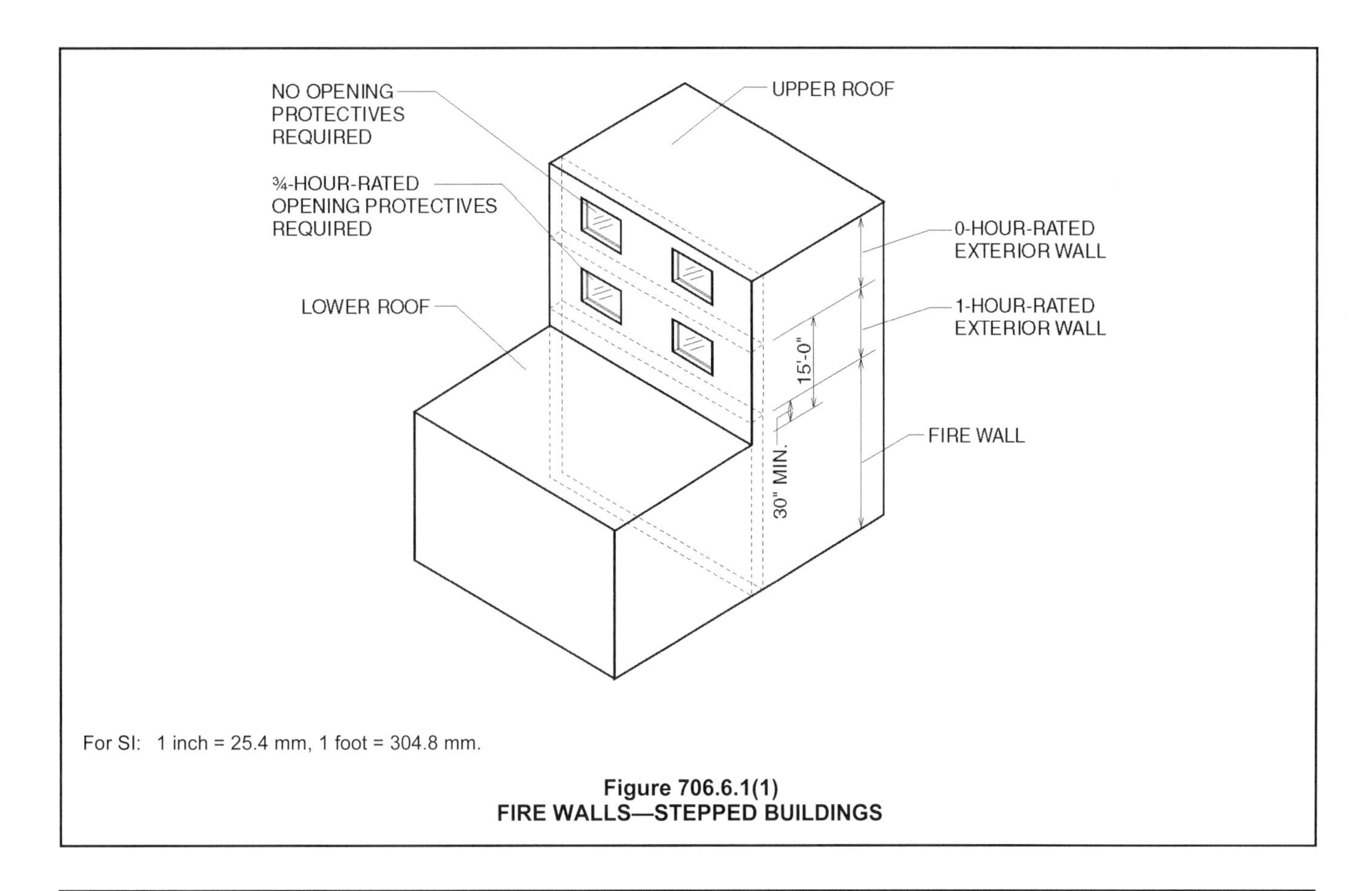

For SI: 1 inch = 25.4 mm, 1 foot = 304.8 mm.

Figure 706.6.1(1)
FIRE WALLS—STEPPED BUILDINGS

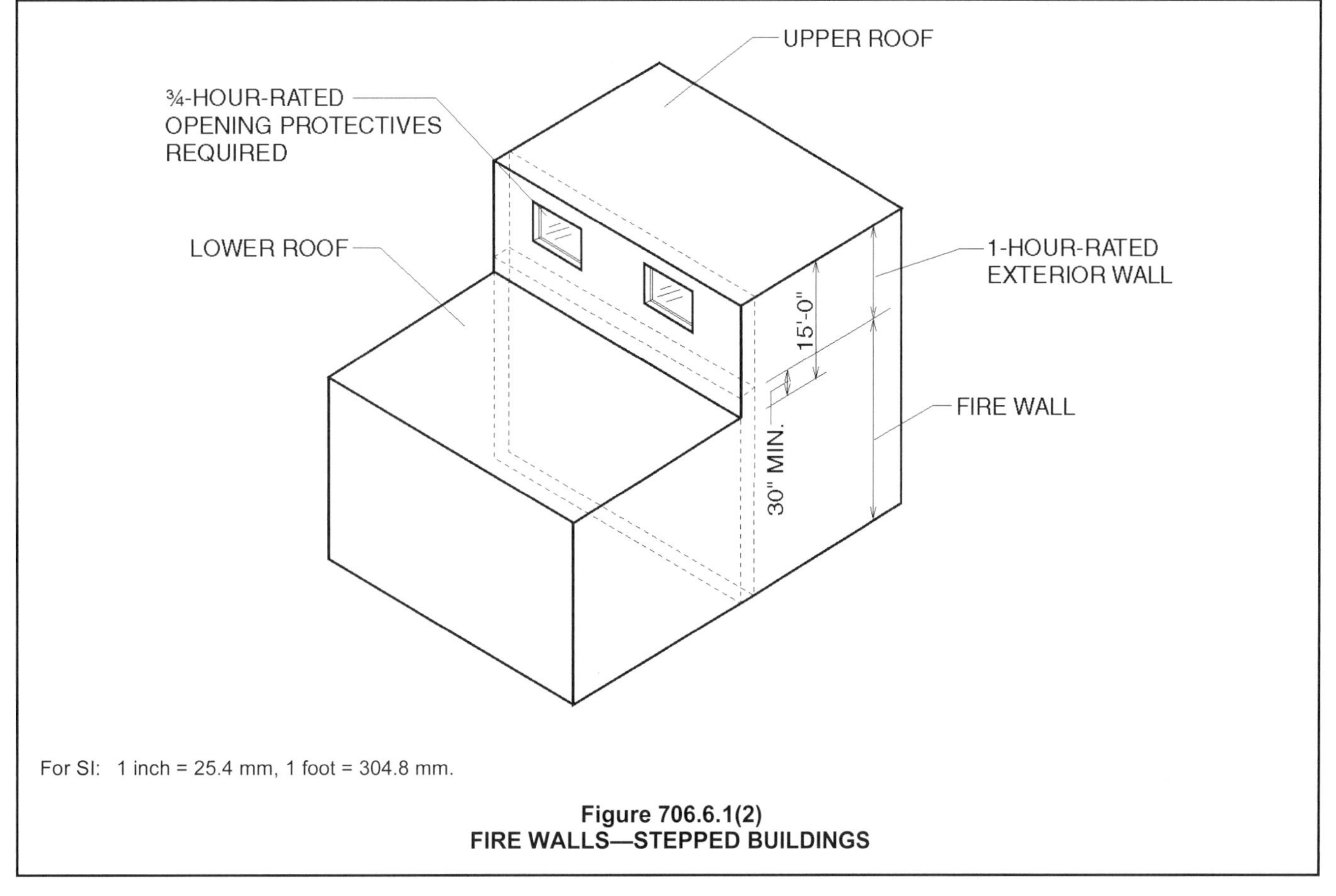

For SI: 1 inch = 25.4 mm, 1 foot = 304.8 mm.

Figure 706.6.1(2)
FIRE WALLS—STEPPED BUILDINGS

706.6.2 Buildings with sloped roofs. Where a *fire wall* serves as an interior wall for a building, and the roof on one side or both sides of the fire wall slopes toward the fire wall at a slope greater than two units vertical in 12 units horizontal (2:12), the *fire wall* shall extend to a height equal to the height of the roof located 4 feet (1219 mm) from the *fire wall* plus 30 inches (762 mm). In no case shall the extension of the fire wall be less than 30 inches (762 mm).

❖ This requirement is similar to that of stepped buildings. This section only applies if the roof slopes upward away from the fire wall at a slope greater than two units vertical in 12 units horizontal (2:12). The sloping roof presents a greater hazard to the building from fire exposure on either side than a roof at the same elevation on each side. The hazard would increase as the slope of the roof increases. Commentary Figure 706.6.2 illustrates this requirement. This section requires the extension above the roof regardless of the material of the roof construction on either side. For roofs sloping two units vertical in 12 units horizontal (2:12), fire wall extensions need only comply with Section 706.6. Section 706.6 may require a 30- inch (762 mm) height extension or may allow the fire wall to terminate at the roof deck under the conditions of one of the exceptions in Section 706.6.

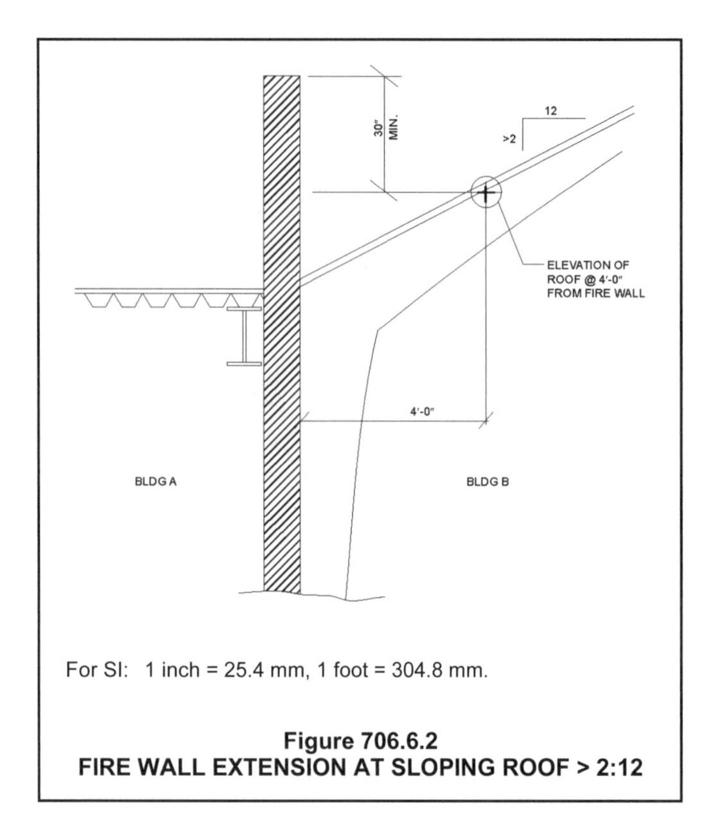

For SI: 1 inch = 25.4 mm, 1 foot = 304.8 mm.

Figure 706.6.2
FIRE WALL EXTENSION AT SLOPING ROOF > 2:12

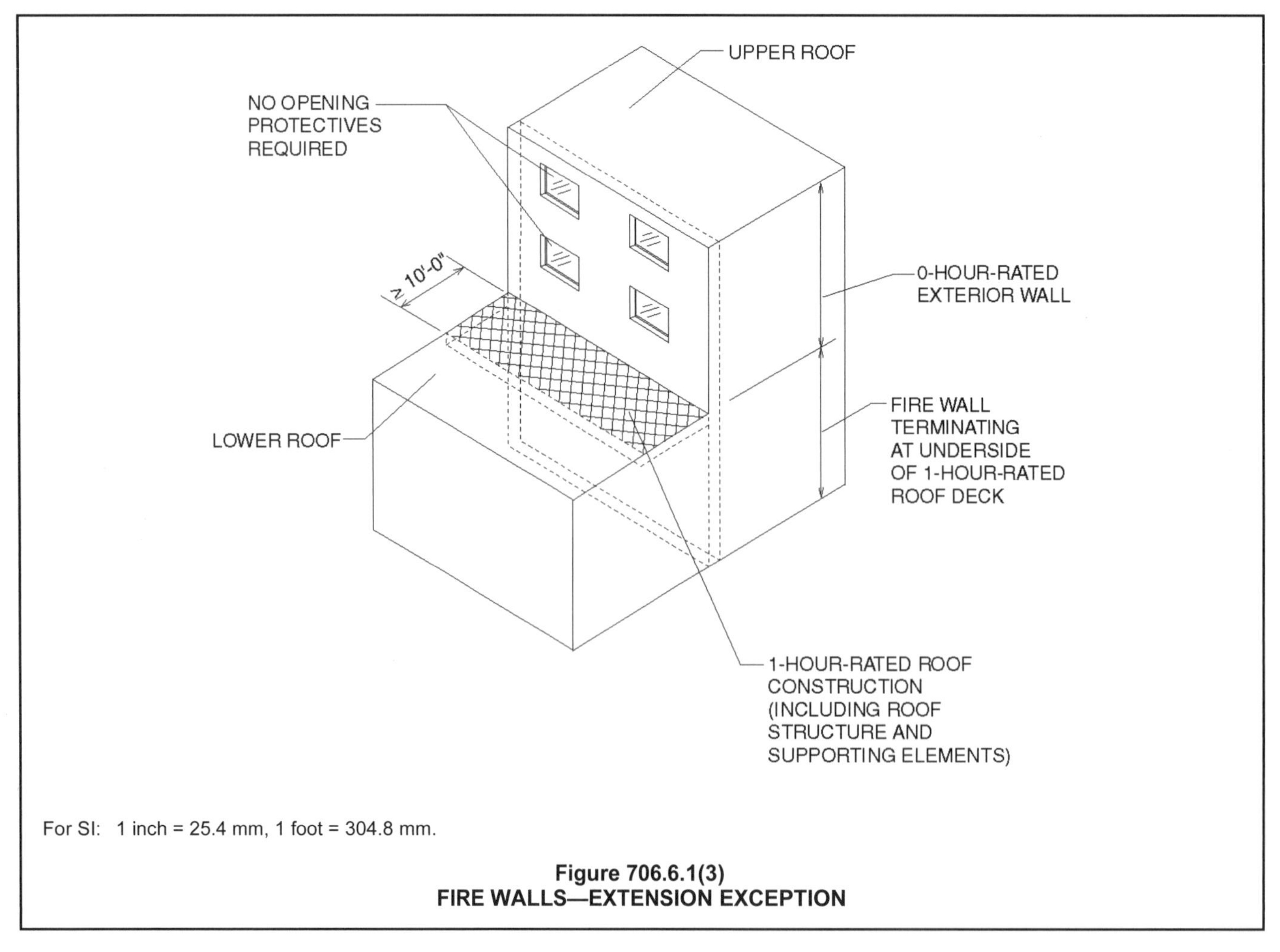

For SI: 1 inch = 25.4 mm, 1 foot = 304.8 mm.

Figure 706.6.1(3)
FIRE WALLS—EXTENSION EXCEPTION

706.7 Combustible framing in fire walls. Adjacent combustible members entering into a concrete or masonry *fire wall* from opposite sides shall not have less than a 4-inch (102 mm) distance between embedded ends. Where combustible members frame into hollow walls or walls of hollow units, hollow spaces shall be solidly filled for the full thickness of the wall and for a distance not less than 4 inches (102 mm) above, below and between the structural members, with noncombustible materials *approved* for fireblocking.

❖ In order to retain the fire-resistance capability of the wall where combustible members will frame into it, hollow walls or walls of hollow units must be solidly filled for the thickness of the wall and for a distance of not less than 4 inches (102 mm) above, below and between the structural members. Consistent with the construction of the walls, the fireblocking materials are to be noncombustible and approved for fireblocking in accordance with Section 717.2. If combustible members enter both sides of a fire wall, there must be at least 4 inches (102 mm) of masonry between the embedded ends.

In order to maintain structural integrity where wood beams or joists are supported in and on masonry fire walls, the joists and beams should be shaped so that they can rotate out of the pocket without exerting undue force on the wall.

706.8 Openings. Each opening through a *fire wall* shall be protected in accordance with Section 716.5 and shall not exceed 156 square feet (15 m^2). The aggregate width of openings at any floor level shall not exceed 25 percent of the length of the wall.

Exceptions:

1. Openings are not permitted in party walls constructed in accordance with Section 706.1.1.
2. Openings shall not be limited to 156 square feet (15 m^2) where both buildings are equipped throughout with an *automatic sprinkler system* installed in accordance with Section 903.3.1.1.

❖ In order to maintain the integrity of the fire wall, the maximum area and percent of openings in the wall are restricted. When provided, the openings must be properly protected so that the fire-resistance rating of the wall is maintained. This section prescribes the maximum area and the percent of openings that may be permitted in a fire wall at any one floor level. The provisions must be used in concert with Section 706.1.1, which limits openings for party walls.

Fire wall openings have restrictive limitations in their size and total area because of the critical function that a fire wall serves. To maintain the required fire performance of the fire wall, each opening through a fire wall is restricted in area to 156 square feet (11 m^2) and the aggregate width of all openings at any one floor level may not constitute more than 25 percent of the length of the wall. The 156-square-foot (11 m^2) limitation provides a reasonable size through which industrial machinery may pass and corresponds with the maximum area limitations of many tested fire doors.

Recognizing the effectiveness of automatic sprinklers, the 156-square-foot (11 m^2) opening limitation does not apply where the buildings on both sides of the fire wall are fully sprinklered (see Exception 2); however, the aggregate width of all openings in a fire wall at any one floor level is still limited to 25 percent of the length of the wall.

706.9 Penetrations. Penetrations of *fire walls* shall comply with Section 714.

❖ In order to maintain the integrity of the required fire-resistance rating, penetrations through the fire wall must be properly protected. Acceptable protection methods for various penetrations of fire walls are identified in Sections 714.2 and 714.3.

706.10 Joints. Joints made in or between *fire walls* shall comply with Section 715.

❖ Joints, such as expansion or seismic, are another form of openings in fire walls and, therefore, must be considered with regard to maintaining the fire-resistance ratings of fire walls. This section requires all joints that are located in fire walls to be protected by a joint system with a fire-resistance rating and to comply with the requirements of Section 715.

706.11 Ducts and air transfer openings. Ducts and air transfer openings shall not penetrate *fire walls*.

Exception: Penetrations by ducts and air transfer openings of *fire walls* that are not on a *lot line* shall be allowed provided the penetrations comply with Section 717. The size and aggregate width of all openings shall not exceed the limitations of Section 706.8.

❖ The general provisions of this section mirror those of Section 706.1.1 for party walls. The exception permits duct and transfer openings for fire walls not located on a lot line, provided the maximum aggregate area provisions of Section 706.8 are met and that the openings are protected in accordance with Section 717.

SECTION 707
FIRE BARRIERS

707.1 General. *Fire barriers* installed as required elsewhere in this code or the *International Fire Code* shall comply with this section.

❖ The provisions of this section apply to assemblies that are required to have a fire-resistance rating and are required to be constructed as "fire barriers." As addressed in Section 707.3, fire barriers are used for separating exits, incidental use areas, shafts, hazardous materials control areas and fire areas. Fire barriers provide a higher degree of protection than fire partitions (see Section 708), but lack the inherent structural integrity of fire walls. Fire barriers limit the number of openings. Fire barrier wall assemblies must be continuous from the top of a fire-resistance-rated floor/ceiling assembly to the bottom of the floor or roof slab/deck above. Unlike fire partitions, addressed in Section 708, there are no circumstances under which a fire barrier wall is permitted to terminate at a ceiling.

Fire barriers are used for a variety of purposes, such as mixed occupancies, shafts, exit access stairways and exit and floor opening enclosures. Fire barriers also include interior walls that serve to subdivide a space by separating one fire area from an adjacent fire area and for separating incidental use areas (see Section 509.4.1). Fire-resistance-rated assemblies used to separate exit access corridors in many applications as well as tenant, dwelling unit and guestroom separations are fire partitions (see Section 708). This section provides minimum requirements for the fire-resistance rating, continuity, combustibility and protection of openings and penetrations in order to help maintain the reliability of the fire separation assembly. As with any fire-resistance-rated assembly, consideration must be given to the openings and penetrations that are provided within the assembly. The intent is to maintain the fire-resistance rating of the assembly. These sections recognize that fire spread beyond a fire-resistance-rated compartment is often attributed to the protection given to any opening or penetration in the fire barrier, or the lack thereof.

Since the fire barrier is intended to provide a reliable subdivision of areas, the construction that structurally supports the assembly is required to provide at least the same hourly fire-resistance rating as the fire barrier. This is applicable regardless of the type of construction of the building. Structural stability is regulated by Section 707.5.

707.2 Materials. *Fire barriers* shall be of materials permitted by the building type of construction.

❖ The types of materials used in fire barriers are to be consistent with Sections 602 through 602.5 for the type of construction classification of the building. The fire-resistance ratings of fire barriers used to separate mixed occupancies are determined in accordance with Section 508.4 (see commentary, Section 508.4). Fire barriers are permitted to be of combustible materials in Type III, IV and V construction and are required to be of noncombustible materials in Type I and II construction.

707.3 Fire-resistance rating. The *fire-resistance rating* of *fire barriers* shall comply with this section.

❖ The elements that are identified in this section must be constructed as required for fire barriers and must be fire-resistance rated as required by the code sections referenced in this section.

707.3.1 Shaft enclosures. The *fire-resistance rating* of the *fire barrier* separating building areas from a shaft shall comply with Section 713.4.

❖ See Section 713.4 for the fire-resistance rating for shaft enclosures.

707.3.2 Interior exit stairway and ramp construction. The *fire-resistance rating* of the *fire barrier* separating building areas from an *interior exit stairway* or *ramp* shall comply with Section 1023.1.

❖ See Section 1023.1 for the fire-resistance rating of the fire barrier separating building areas from interior exit stairways and ramps (vertical exit enclosures).

707.3.3 Enclosures for exit access stairways. The *fire-resistance rating* of the *fire barrier* separating building areas from an *exit access stairway* or *ramp* shall comply with Section 713.4.

❖ See Sections 1023 for the enclosure requirements for exit access stairways and ramps. Unlike exit stairway enclosures, there are numerous exceptions to the enclosure requirements in Section 1023.

707.3.4 Exit passageway. The *fire-resistance rating* of the *fire barrier* separating building areas from an *exit* passageway shall comply with Section 1024.3.

❖ See Section 1024.3 for the fire-resistance rating for fire barriers forming exit passageways. Exit passageway enclosures may be fire barriers complying with this section or horizontal assemblies complying with Section 711.

707.3.5 Horizontal exit. The *fire-resistance rating* of the separation between building areas connected by a horizontal *exit* shall comply with Section 1026.1.

❖ See Section 1026.1 for the fire-resistance rating of the fire barriers forming horizontal exits. Horizontal exits may be formed by fire walls meeting Section 706, fire barriers complying with this section or horizontal assemblies complying with Sections 1026.2 and 711.

707.3.6 Atriums. The *fire-resistance rating* of the *fire barrier* separating atriums shall comply with Section 404.6.

❖ See Section 404.6 for the fire-resistance rating of the fire barrier walls separating the atrium area from other building use areas. Horizontal assemblies meeting Section 711 can also separate the atrium area.

707.3.7 Incidental uses. The *fire barrier* separating incidental uses from other spaces in the building shall have a *fire-resistance rating* of not less than that indicated in Table 509.

❖ Table 509 states the fire-resistance requirements for fire barrier walls separating incidental use areas. For some incidental uses there is a sprinkler option in lieu of the fire-resistance-rated fire barrier wall. With the sprinkler option, the wall is not required to be fire-resistance rated but must be capable of resisting the passage of smoke. Those nonrated wall assemblies are not required to be constructed as fire barriers.

707.3.8 Control areas. *Fire barriers* separating *control areas* shall have a *fire-resistance rating* of not less than that required in Section 414.2.4.

❖ Control areas refer to areas within a building where limited quantities of hazardous materials are stored or used. Section 414.2.4 references Table 414.2.4 for the design and number of control areas. The required fire-resistance rating for the fire barriers separating the control areas from the rest of the area is also found within the table.

707.3.9 Separated occupancies. Where the provisions of Section 508.4 are applicable, the *fire barrier* separating mixed occupancies shall have a *fire-resistance rating* of not less than that indicated in Table 508.4 based on the occupancies being separated.

❖ Where separation of mixed occupancies is the chosen option for mixed uses in a building, the required fire separation is either a fire barrier meeting the requirements of this section or a horizontal fire separation assembly meeting the requirements of Section 711 (see Section 508.4). The required fire-resistance rating for the fire barrier or horizontal assembly is found in Table 508.4, but in no case shall the fire-resistance rating be less than the highest value in Table 707.3.10 for the occupancies being separated.

707.3.10 Fire areas. The *fire barriers* or *horizontal assemblies*, or both, separating a single occupancy into different *fire areas* shall have a *fire-resistance rating* of not less than that indicated in Table 707.3.10. The *fire barriers* or *horizontal assemblies*, or both, separating *fire areas* of mixed occupancies shall have a *fire-resistance rating* of not less than the highest value indicated in Table 707.3.10 for the occupancies under consideration.

❖ One of the alternatives available in addressing fire protection systems in many buildings is to divide the building into separate fire areas (see the definition of "Fire areas" in Chapter 2). Since many of the fire suppression system thresholds (see Section 903.2) are based upon the fire area, separation of a single occupancy into small fire areas can be an acceptable method for avoiding the use of sprinklers. This is a classic type of design decision: sprinklers versus compartmentation. If the separation is provided in accordance with Table 707.3.10, each fire area may be evaluated separately for purposes of determining the applicable provisions of the code.

Areas separated with fire barriers are not considered separate buildings; they are considered separate fire areas. Two areas must be separated by a fire wall or exterior walls to be considered separate buildings. Two areas separated with fire barriers are still considered as part of a single building. This distinction is critical in determining compliance with allowable height and area and other code provisions.

TABLE 707.3.10
FIRE-RESISTANCE RATING REQUIREMENTS FOR FIRE BARRIER ASSEMBLIES OR HORIZONTAL ASSEMBLIES BETWEEN FIRE AREAS

OCCUPANCY GROUP	FIRE-RESISTANCE RATING (hours)
H-1, H-2	4
F-1, H-3, S-1	3
A, B, E, F-2, H-4, H-5, I, M, R, S-2	2
U	1

❖ Table 707.3.10 provides the minimum required fire-resistance ratings of the fire barrier wall or horizontal assembly separating two fire areas of the same occupancy group.

707.4 Exterior walls. Where exterior walls serve as a part of a required fire-resistance-rated shaft or stairway or ramp enclosure, or separation, such walls shall comply with the requirements of Section 705 for exterior walls and the fire-resistance-rated enclosure or separation requirements shall not apply.

Exception: Exterior walls required to be fire-resistance rated in accordance with Section 1021 for exterior egress balconies, Section 1023.7 for interior exit stairways and ramps and Section 1027.6 for exterior exit stairways and ramp.

❖ If an area is required to be enclosed by fire barriers and an exterior wall constitutes part of the enclosure, the exterior wall is only required to comply with the fire-resistance-rating requirements in Section 705, unless the exterior wall is protecting part of an exterior egress balcony, an exit enclosure or an exterior stairway or ramp (see commentary, Sections 1021, 1023.7 and 1027.6). The intent of the fire barrier requirements is to subdivide or enclose areas to protect them from a fire in the building. In general, the exterior wall only needs a fire-resistance rating if required for structural stability (see Table 601) or because of exterior exposure potential (see Table 602 and Section 705.5). The exception, however, points to three sections of the code where the exterior wall must be rated for reasons other than the FSD.

707.5 Continuity. *Fire barriers* shall extend from the top of the foundation or floor/ceiling assembly below to the underside of the floor or roof sheathing, slab or deck above and shall be securely attached thereto. Such *fire barriers* shall be continuous through concealed space, such as the space above a suspended ceiling. Joints and voids at intersections shall comply with Sections 707.8 and 707.9

Exceptions:

1. Shaft enclosures shall be permitted to terminate at a top enclosure complying with Section 713.12.
2. *Interior exit stairway* and *ramp* enclosures required by Section 1023 and *exit access stairway* and *ramp* enclosures required by Section 1019 shall be permit-

ted to terminate at a top enclosure complying with Section 713.12.

❖ To minimize the potential for fire spread from one area to another over a fire barrier wall, such fire barrier assemblies must be continuous from the fire-resistance-rated floor/ceiling assembly below to the underside of the floor slab or roof deck above (see Commentary Figure 707.5). To maintain the efficiency of the fire barrier, it must be continuous through all concealed spaces (such as a space above a suspended ceiling), be constructed tight and securely attached to the underside of the floor slab or roof deck. The joint at the intersection of fire barrier walls and other fire barriers, the underside of fire-resistance-rated floors or roof decks, slabs, decks and exterior walls must be an approved fire-resistance joint system meeting ASTM E119 or UL 263.

Exception 1 is an exception to the continuity requirement for shaft enclosure walls. Exception 2 allows exit stair enclosures and exit access stairway enclosures to terminate at a top enclosure complying with Section 713.12.

707.5.1 Supporting construction. The supporting construction for a *fire barrier* shall be protected to afford the required *fire-resistance rating* of the *fire barrier* supported. Hollow vertical spaces within a *fire barrier* shall be fireblocked in accordance with Section 718.2 at every floor level.

Exceptions:

1. The maximum required *fire-resistance rating* for assemblies supporting *fire barriers* separating tank storage as provided for in Section 415.9.1.2 shall be 2 hours, but not less than required by Table 601 for the building construction type.
2. Supporting construction for 1-hour *fire barriers* required by Table 509 in buildings of Type IIB, IIIB and VB construction is not required to be *fire-resistance rated* unless required by other sections of this code.

❖ In general, fire barriers must be supported by construction having an equivalent fire-resistance rating. If the supporting structure is a primary structural frame (see the definition in Chapter 2) and supports a fire barrier wall more than two stories in height, the fire-resistance rating for the supporting structure must be protected by the individual encasement method in Section 704.3. If the supporting members are a secondary structural member, then the supporting structure can be protected by a membrane as in Section 711 for horizontal assemblies. The intent of this requirement is to prevent the effectiveness of the assembly from being circumvented by a fire that threatens the supporting elements. The requirement for the supporting construction to be fire-resistance rated applies to buildings of all types of construction, even to buildings of Type IIB, IIIB and VB construction for all fire barrier walls except those separating incidental use areas.

Exception 1 is not an exception at all but a requirement that supporting structures for fire barriers separating flammable or combustible tank storage be 2-hour fire-resistance rated.

Exception 2 allows only incidental use area separation walls in Type IIB, IIIB and VB construction to be supported on nonfire-resistance-rated construction if no other code section requires the supporting elements to be fire-resistance rated.

Fire barrier walls will usually be built on top of a floor and will terminate at the floor above. Should a fire barrier wall, as in the case of some shaft walls, be constructed through a floor, any hollow space within that wall could provide a passage for fire or smoke and, therefore, must be fireblocked as specified in the last sentence of Section 707.5.1. Any hollow vertical

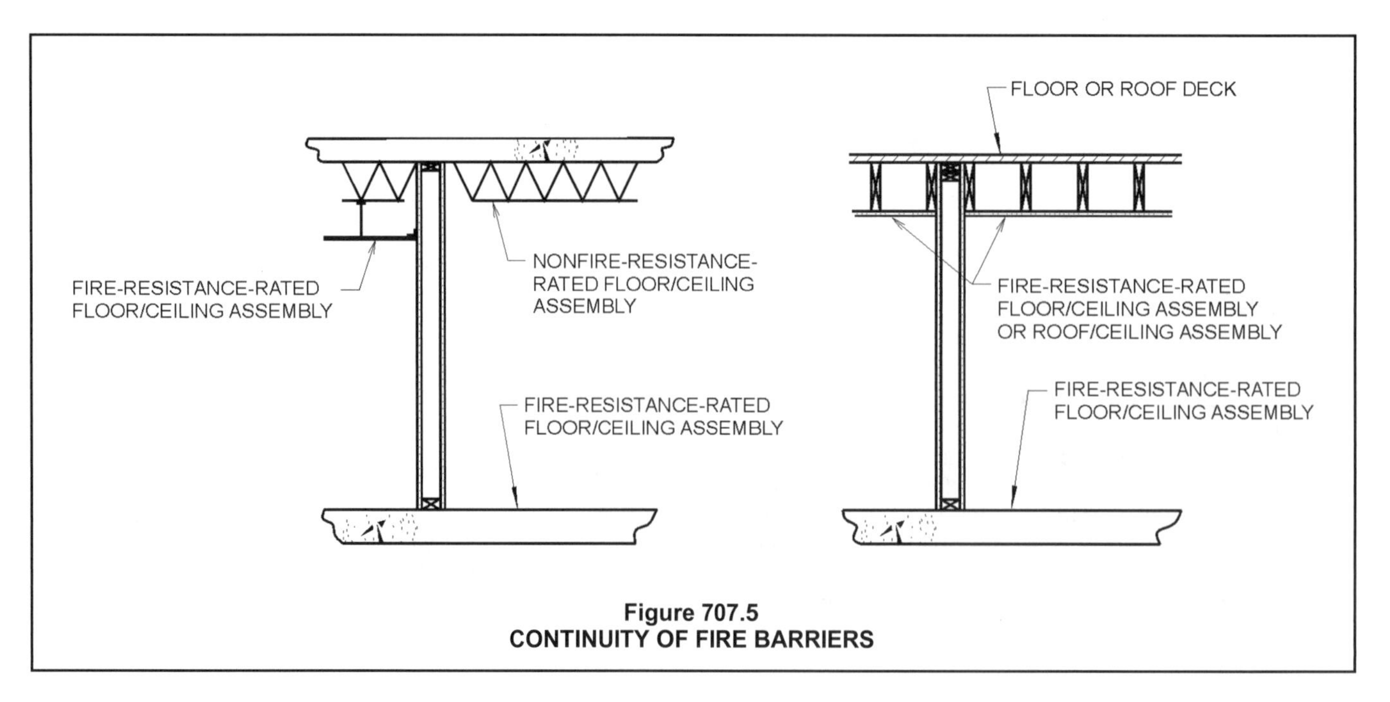

Figure 707.5
CONTINUITY OF FIRE BARRIERS

spaces within fire barrier walls must be fireblocked in accordance with Section 718.2

707.6 Openings. Openings in a *fire barrier* shall be protected in accordance with Section 716. Openings shall be limited to a maximum aggregate width of 25 percent of the length of the wall, and the maximum area of any single opening shall not exceed 156 square feet (15 m2). Openings in enclosures for *exit access stairways* and *ramps*, *interior exit stairways* and *ramps* and *exit passageways* shall also comply with Sections 1019, 1023.4 and 1024.5, respectively.

Exceptions:

1. Openings shall not be limited to 156 square feet (15 m^2) where adjoining floor areas are equipped throughout with an automatic sprinkler system in accordance with Section 903.3.1.1.
2. Openings shall not be limited to 156 square feet (15 m^2) or an aggregate width of 25 percent of the length of the wall where the opening protective is a fire door serving enclosures for exit access stairways and ramps, and interior exit stairways and ramps.
3. Openings shall not be limited to 156 square feet (15 m^2) or an aggregate width of 25 percent of the length of the wall where the opening protective has been tested in accordance with ASTM E119 or UL 263 and has a minimum *fire-resistance rating* not less than the *fire-resistance rating* of the wall.
4. Fire window assemblies permitted in atrium separation walls shall not be limited to a maximum aggregate width of 25 percent of the length of the wall.
5. Openings shall not be limited to 156 square feet (15 m^2) or an aggregate width of 25 percent of the length of the wall where the opening protective is a fire door assembly in a *fire barrier* separating an enclosure for *exit access* stairways and ramps, and interior exit stairways and ramps from an exit passageway in accordance with Section 1023.3.1.

❖ To function as intended, buildings must provide access into different building areas. This access requires openings within fire barriers. Section 707.6 defines what openings are permitted and how they have to be protected to maintain the integrity of the fire barrier. To maintain the viability of the fire barrier, the aggregate width of openings is restricted to a maximum of 25 percent of the length of the wall. This limitation is based on the fact that the criteria for opening protectives do not include limitations on unexposed surface temperature or radiant heat transfer. Consistent with typical listing limitations, a single opening protective is limited to a maximum of 156 square feet (15 m^2). It should be noted, however, that certain opening protectives, such as fire windows, are often limited to much smaller areas per opening.

Traditionally, the limiting size on openings in fire-resistance-rated walls has been based on the maximum sizes identified in the fire door listings. The previous size limitation was 120 square feet (11 m^2) due to the limitations of such listings. In 2006, the code increased the maximum permitted size of an opening protective to 156 square feet (15 m^2) in a fire barrier, based upon the current listing limitations of steel fire doors. The maximum length of any single opening is limited to 13 feet, 6 inches (4114 mm) for the same reason when used during the testing process.

The reference to Sections 1019, 1023.4 and 1024.5 specifies that in exit enclosures and exit passageways, only openings for the purpose of exiting from normally occupied spaces are permitted. Spaces that are not normally occupied, such as janitor closets or mechanical and electrical rooms, are not permitted to open directly into these protected exit systems since fire in those areas may produce large volumes of heat and smoke that could readily enter the exit enclosure and delay or prevent egress. In order to maintain the required fire-resistance rating of the assembly, opening protectives must have a fire protection rating in accordance with Section 716. The reference to Section 716 is intended to identify the required fire protection rating for the opening protective, as indicated in Table 716.5, installation requirements and the applicable test standards.

Openings in fire barriers are not limited to 156 square feet (15 m^2) when all fire areas separated by the assembly are equipped throughout with automatic sprinkler protection (see Exception 1). This exception is similar to the one made for fire walls (see Section 706.8), based on the historical fire record of automatic sprinkler systems. Although the openings in the fire barrier are not limited in size under this exception, they are still required to be protected by opening protectives that meet the requirements of Section 716.

Exception 2 acknowledges the practicality of the 25-percent limitation for walls enclosing interior exit stairways and ramps, and exit access stairways and ramps. Most exit enclosures are of such limited size that the placement of the fire door in the wall of the enclosure often exceeds 25 percent of the wall.

Exception 3 addresses new opening protective products that have been tested to the more rigorous provisions of ASTM E119 or UL 263 rather than, or in addition to, the opening protective standard NFPA 252 or UL 10C. Since the opening protective has been tested to the same standard as the wall itself, it is then logical to allow such an opening protective without restrictions (see commentary, Section 703.6).

Exception 4 acknowledges the inclusive requirements for openings into atriums addressed in Section 404 that include glazing material allowances not found elsewhere in Chapter 7.

Exception 5 is similar to Exception 2 in that the wall area between the exit enclosure and the exit passageway is usually so very small that the door opening will nearly always exceed the 25-percent limitation.

707.7 Penetrations. Penetrations of *fire barriers* shall comply with Section 714.

❖ In order to maintain the integrity of the fire barrier, penetrations into and through the fire-resistance-rated wall must be properly protected. Acceptable methods for various penetrations of fire barriers are identified in Section 714.3. The provisions of Section 707.7.1 must be used when the penetration is into an exit enclosure or exit passageway.

707.7.1 Prohibited penetrations. Penetrations into enclosures for *exit access stairways* and *ramps*, *interior exit stairways* and *ramps*, and *exit passageways* shall be allowed only where permitted by Sections 1019, 1023.5 and 1024.6, respectively.

❖ This section reminds the code user that although the penetration firestop systems do provide protection for the penetration, the code prohibits most penetrations through exit enclosures. Only penetrations of items such as sprinkler piping, necessary ductwork for stair pressurization, and electrical conduit that serve the exit enclosure are allowed. There can never be a penetration through a fire barrier that separates an adjacent exit enclosure.

707.8 Joints. Joints made in or between *fire barriers*, and joints made at the intersection of *fire barriers* with underside of a fire-resistance-rated floor or roof sheathing, slab or deck above, and the exterior vertical wall intersection shall comply with Section 715.

❖ This section contains requirements for joints or linear openings created between building assemblies, which are sometimes referred to as construction, expansion or seismic joints. These joints are most often created where the structural design of a building necessitates a separation between building components in order to accommodate anticipated structural displacements caused by thermal expansion and contraction, seismic activity, wind or other loads. Commentary Figure 715.1 illustrates some of the most common locations of these joints.

These linear openings create a "weak link" in fire-resistance-rated assemblies, which can compromise the integrity of the tested assembly by allowing an avenue for the passage of fire and the products of combustion through the assembly. In order to maintain the efficacy of the fire-resistance-rated assembly, these openings must be protected by a joint system with a fire-resistance rating equal to the adjacent assembly. It is not the intent of this section to regulate joints installed in assemblies that are provided to control shrinkage cracking, such as a saw-cut control joint in concrete (see Section 715).

707.9 Voids at intersections. The voids created at the intersection of a *fire barrier* and a nonfire-resistance-rated roof assembly or a nonfire-resistance-rated exterior wall assembly shall be filled. An approved material or system shall be used to fill the void, and shall be securely installed in or on the intersection for its entire length so as not to dislodge, loosen or otherwise impair its ability to accommodate expected building movements and to retard the passage of fire and hot gases.

❖ This new section is meant to give some prescriptive requirements for the joint formed by the intersection of a fire barrier wall and a nonfire-resistance-rated roof or wall assembly. Fire barrier walls are required to extend vertically to the roof deck. Fire barriers may also need to extend horizontally to an exterior wall. The method of sealing the resulting joint is prescribed in this section. Unlike a joint formed by a fire barrier wall and a fire-resistance-rated roof, which requires a joint meeting the test criteria of ASTM E1966 or UL 2079, this joint or void must only be filled with a material approved by the building official. It is not the intent of this section to require any performance-based tested assembly.

707.10 Ducts and air transfer openings. Penetrations in a *fire barrier* by ducts and air transfer openings shall comply with Section 717.

❖ Section 717 details the protection of ducts and air transfer openings at the point where they penetrate a fire-resistance-rated assembly. Section 717.5 indicates which situations will require the installation of a damper. As stated in Section 717.1.2, if a duct does not require a damper, the penetration of that duct through a fire-resistance-rated assembly must be protected as a through penetration in accordance with Section 714.

SECTION 708
FIRE PARTITIONS

708.1 General. The following wall assemblies shall comply with this section.

1. Separation walls as required by Section 420.2 for Groups I-1, R-1, R-2 and R-3.
2. Walls separating tenant spaces in *covered and open mall buildings* as required by Section 402.4.2.1.
3. Corridor walls as required by Section 1020.1.
4. Elevator lobby separation as required by Section 3006.2.
5. Egress balconies as required by Section 1019.2

❖ "Fire partitions," as defined in Section 202, are wall assemblies that enclose an exit access corridor, separate dwelling units, separate sleeping units, separate tenants in covered and open mall buildings and separate elevator lobbies. There are some exceptions to the requirement that corridor walls and elevator lobby walls be fire partitions. Those exceptions are found in Sections 3006.2 and 1020.1.

Corridor walls not required to be fire-resistance rated by Table 1020.1 are not required to meet this section. Elevator lobby walls not required to be fire-resistance rated by the exceptions to Section 3006.2 are not required to meet this section.

This section contains fire-resistance rating requirements, continuity requirements, opening requirements, penetration requirements, joint requirements and duct and air transfer opening requirements for fire partitions.

For horizontal assemblies separating dwelling and sleeping units, see Section 711.

708.2 Materials. The walls shall be of materials permitted by the building type of construction.

❖ The types of materials used in fire partitions are to be consistent with Sections 602.2 through 602.5 for the type of construction classification of the building. Fire partitions are permitted to be of combustible construction in Type III, IV and V construction and are required to be of noncombustible materials in Type I and II construction, except as allowed by Exception 1 to Section 603.1.

708.3 Fire-resistance rating. Fire partitions shall have a *fire-resistance rating* of not less than 1 hour.

Exceptions:

1. Corridor walls permitted to have a $^1/_2$-hour *fire-resistance rating* by Table 1020.1.
2. *Dwelling unit* and *sleeping unit* separations in buildings of Type IIB, IIIB and VB construction shall have *fire-resistance ratings* of not less than $^1/_2$ hour in buildings equipped throughout with an *automatic sprinkler system* in accordance with Section 903.3.1.1.

❖ This requirement is for fire partitions to be 1-hour fire-resistance rated. There are two exceptions.

Exception 1 states that corridor walls not required by Table 1020.1 to be fire-resistance rated need not comply with this section at all, since according to Section 1020.1 they are not fire partitions. Table 1020.1 also contains a reduction of the 1-hour fire-resistance rating down to 30 minutes for sprinklered Group R occupancies.

A reduction in the fire-resistance rating for dwelling and sleeping unit separation is allowed for Type IIB, IIIB and VB construction which is sprinklered with an NFPA 13 sprinkler system. This is not allowed with an NFPA 13R system according to Exception 2, which reduces the rating from 1 hour to only 30 minutes. There is a similar reduction for the horizontal assemblies separating dwelling and sleeping units in Section 711.2.4.3.

708.4 Continuity. Fire partitions shall extend from the top of the foundation or floor/ceiling assembly below to the underside of the floor or roof sheathing, slab or deck above or to the fire-resistance-rated floor/ceiling or roof/ceiling assembly above, and shall be securely attached thereto. In combustible construction where the *fire partitions* are not required to be continuous to the sheathing, deck or slab, the space between the ceiling and the sheathing, deck or slab above shall be fireblocked or draftstopped in accordance with Sections 718.2 and 718.3 at the partition line. The supporting construction shall be protected to afford the required *fire-resistance rating* of the wall supported, except for walls separating tenant spaces in *covered and open mall buildings*, walls separating *dwelling units*, walls separating *sleeping units* and *corridor* walls, in buildings of Type IIB, IIIB and VB construction.

Exceptions:

1. The wall need not be extended into the crawl space below where the floor above the crawl space has a minimum 1-hour *fire-resistance rating*.
2. Where the room-side fire-resistance-rated membrane of the *corridor* is carried through to the underside of the floor or roof sheathing, deck or slab of a fire-resistance-rated floor or roof above, the ceiling of the *corridor* shall be permitted to be protected by the use of ceiling materials as required for a 1-hour fire-resistance-rated floor or roof system.
3. Where the *corridor* ceiling is constructed as required for the *corridor* walls, the walls shall be permitted to terminate at the upper membrane of such ceiling assembly.
4. The fire partitions separating tenant spaces in a *covered or open mall building*, complying with Section 402.4.2.1, are not required to extend beyond the underside of a ceiling that is not part of a fire-resistance-rated assembly. A wall is not required in *attic* or ceiling spaces above tenant separation walls.
5. Attic fireblocking or draftstopping is not required at the partition line in Group R-2 buildings that do not exceed four *stories above grade plane*, provided the *attic* space is subdivided by draftstopping into areas not exceeding 3,000 square feet (279 m^2) or above every two *dwelling units*, whichever is smaller.
6. Fireblocking or draftstopping is not required at the partition line in buildings equipped with an *automatic sprinkler system* installed throughout in accordance with Section 903.3.1.1 or 903.3.1.2, provided that automatic sprinklers are installed in combustible floor/ceiling and roof/ceiling spaces.

❖ To minimize the potential for fire spread from the exposed side of the fire partition to the unexposed side, such partitions must be continuous from the floor assembly to the underside of a fire-resistance-rated floor/ceiling or roof/ceiling assembly. In the absence of a rated floor/ceiling or roof/ceiling assembly, the fire partition is to be continuous to the floor slab or roof deck above [see Commentary Figures 708.4(1) and 708.4(2)(A)]. All hollow vertical spaces in a combustible fire partition must be fireblocked at the ceiling and floor or roof levels in accordance with Section 718.2. If the fire-resistance-rated floor/ceiling or roof/ceiling assembly are of combustible construction, then draftstopping is required in the space above and in line with the partition. The draftstopping must meet Section 718.3. Fire partitions are not always required to be continuous through concealed spaces, whereas fire barriers are to be continuous (see Section 707.5). This is the primary difference between fire barriers and fire partitions.

Fire partitions serving as exit access corridor walls and tenant and guestroom separation walls in buildings of Type IIB, IIIB and VB construction are not required to be supported by structural elements having the same fire-resistance rating. The primary purpose of rated corridor walls is to prevent fire or smoke spread from a room to the corridor in order to maintain the protection of the means of egress. Secondly, the rated corridor wall also prevents fire or smoke spread from a corridor to adjacent rooms should the fire

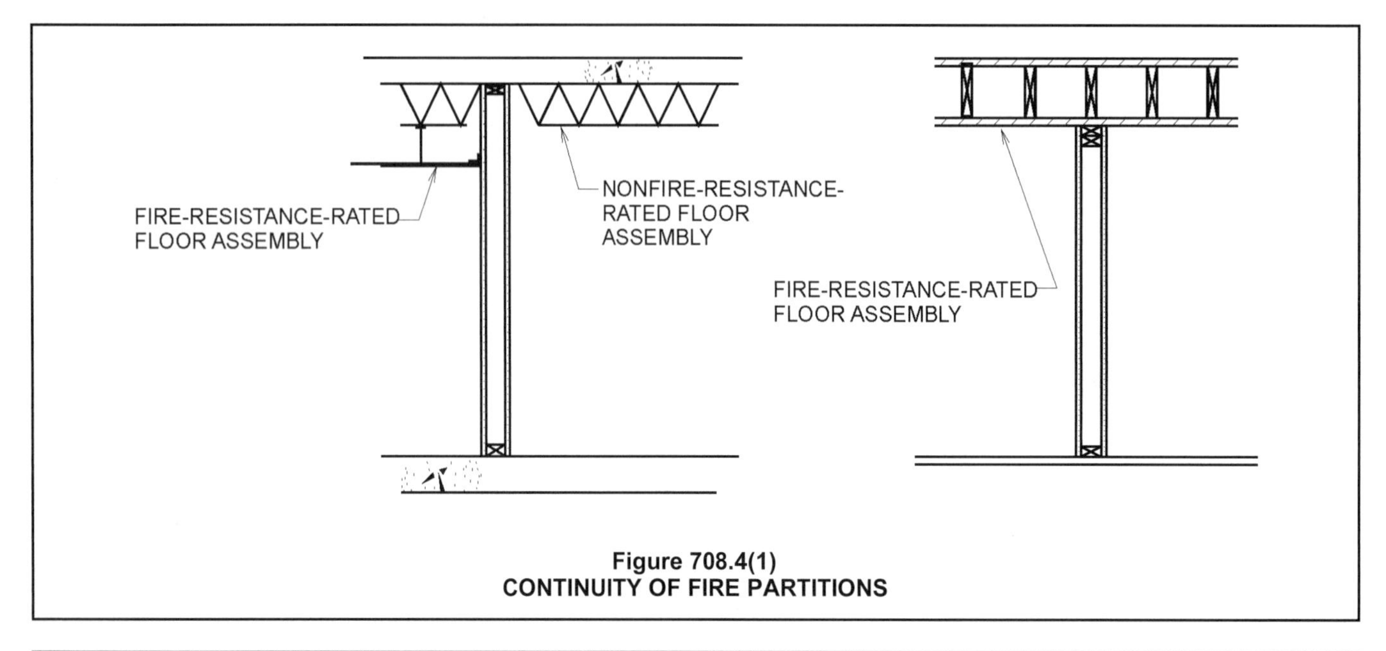

Figure 708.4(1)
CONTINUITY OF FIRE PARTITIONS

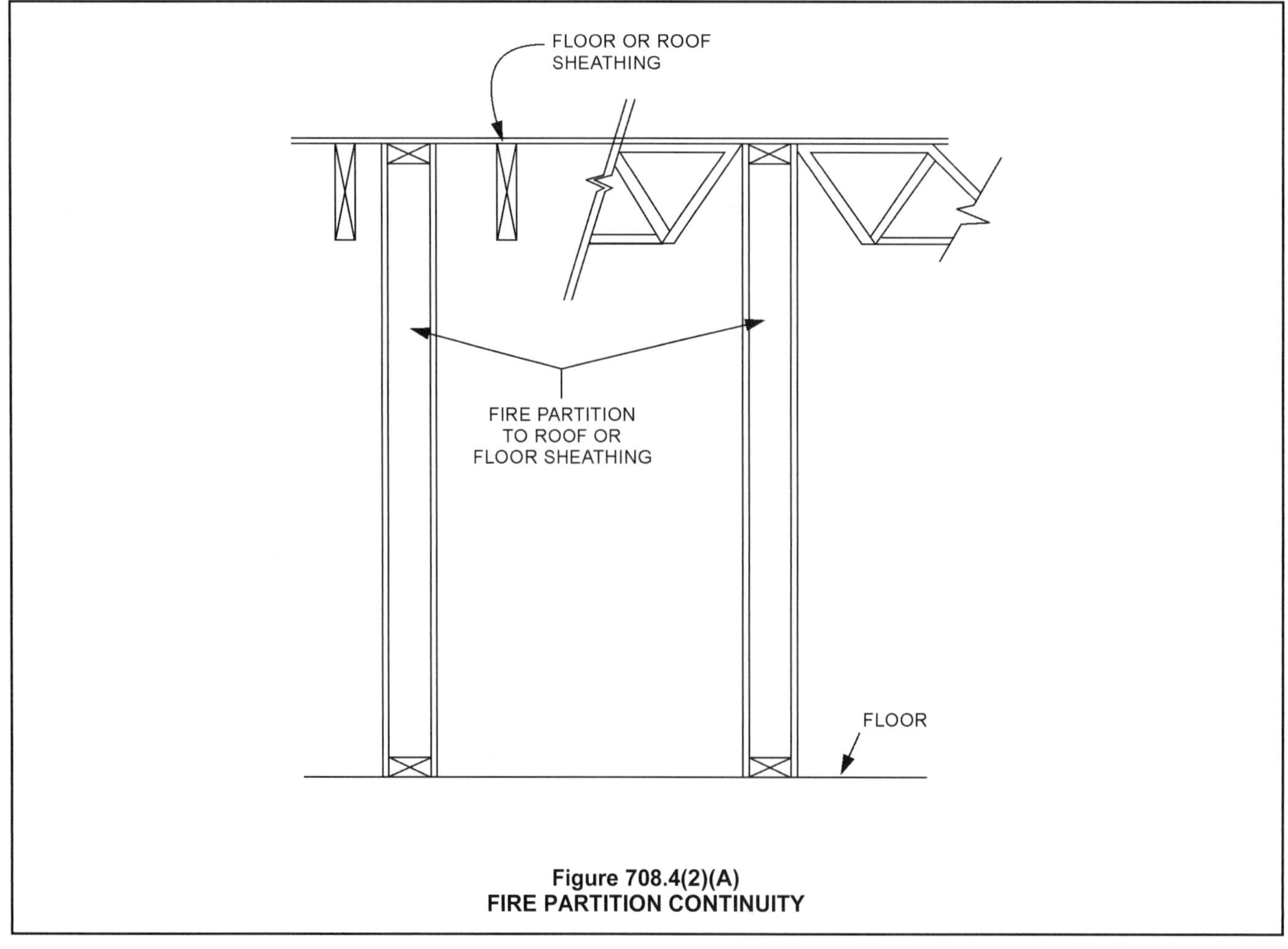

Figure 708.4(2)(A)
FIRE PARTITION CONTINUITY

involve the corridor. Therefore, the fire-resistance rating of the floor construction supporting the wall is not critical to the performance of the corridor wall.

The construction supporting tenant and guestroom separation walls is exempted from the requirements for a fire-resistance rating since the compartmentalization provided by these walls is similar to that provided by corridor walls. Additionally, if the supporting construction were required to be fire-resistance rated, this section would effectively result in all buildings containing guestrooms (occupancies in Groups R-1, R-2 and I-1) being built of protected construction only. The remaining construction types are not included, since the supporting structural elements (floors and columns) are required to have at least a 1-hour rating (per Table 601), which is the maximum fire-resistance rating required for corridor walls. Similarly, tenant separations in covered mall buildings of Type IIB construction are not required to be supported by construction having an equivalent fire-resistance rating.

While supporting members in Types IIB, IIIB and VB construction are not required to be protected by the same fire resistance as the separation walls between a dwelling and sleeping unit, the horizontal floor assembly between dwelling and sleeping units is required to be fire-resistance-rated construction by Section 711.2.4.3.

Exception 1 modifies the continuity requirement for the various tenant, dwelling unit or sleeping unit separations where a crawl space is found below. As discussed above, because corridors and elevator lobby separations are intended for protection from hazards on that floor level, they would not require a fire partition within the crawl space.

Although this exception provides a blanket exclusion for extending the partition into the crawl space, it may be prudent to consider each building on an individual basis. If the crawl space is arranged or provided so that each dwelling unit has access to it or it may be intended for storage or equipment, then it may be wise to consider the crawl space as a part of the dwelling unit; therefore, it may be reasonable to apply the exception to circumstances where the crawl space is not considered to be a part of the dwelling unit.

Although the exception requires that the floor above the crawl space have a 1-hour fire-resistance rating, it would still be permissible to use the provisions of Section 711.2.6 and eliminate the ceiling membrane if the crawl space is considered an unusable space.

Exception 2 allows for only one side (the room side) of a fire-resistance-rated wall assembly to be continuous to the rated assembly above. The membrane on the corridor side of the wall is carried up to the underside of the corridor ceilings, provided the corridor ceiling is constructed of materials that have been tested as part of a 1-hour assembly [see Commentary Figure 708.4(2)(B)].

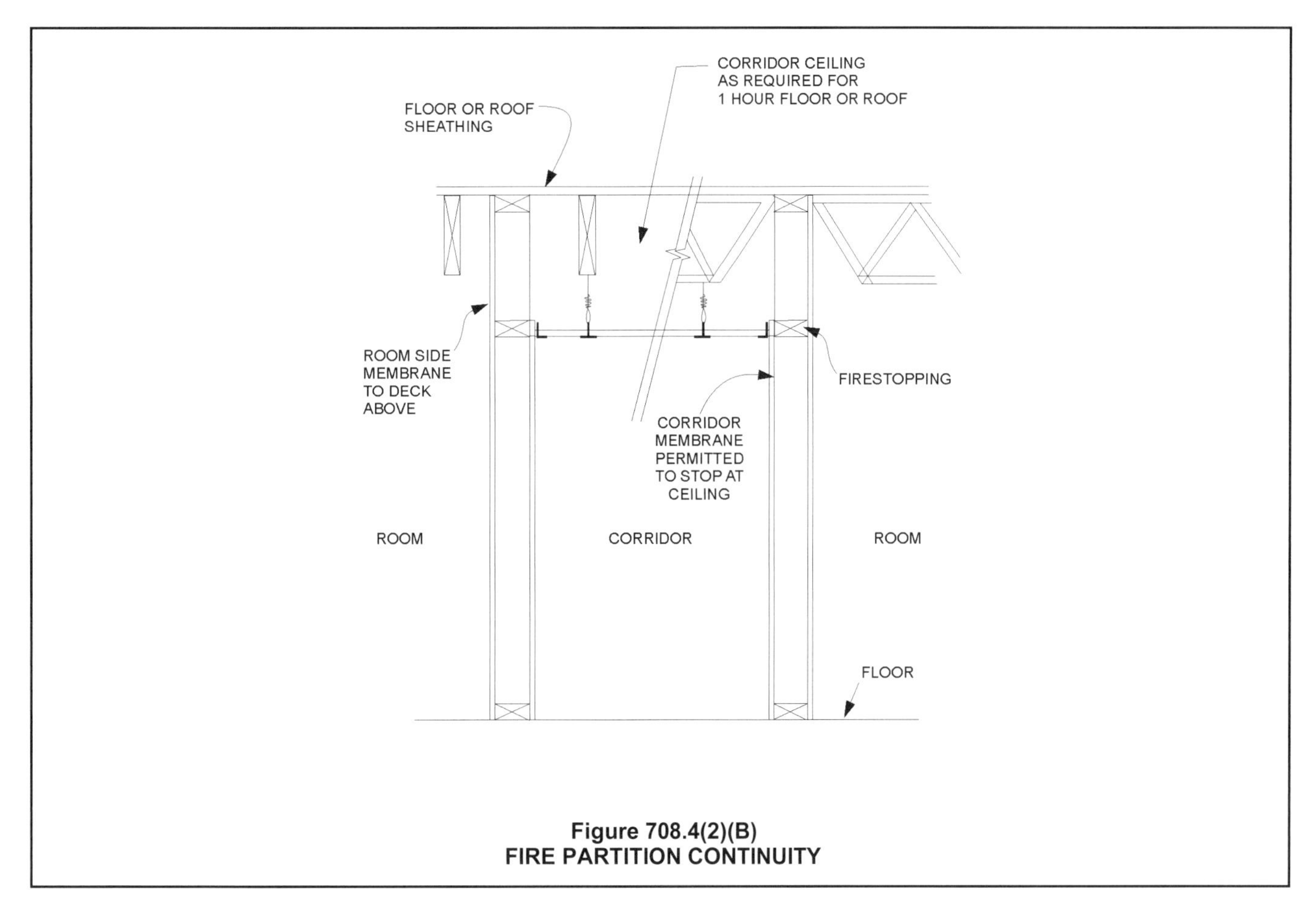

Figure 708.4(2)(B)
FIRE PARTITION CONTINUITY

Exception 3 allows for a corridor to be constructed in a "tunnel" fashion—with rated walls and a rated top. In these cases, the walls are not required to extend to the underside of a rated assembly. This is a unique provision in the code and the fact that it permits a wall assembly to be turned into a horizontal position should not be applied to other sections of the code or other rated assemblies. In this one situation, the code will permit an assembly that has been tested as a wall to be turned and used horizontally. Because of the differences in the fire test and the fact that a horizontal test is generally a more severe condition, most assemblies will not obtain the same fire-resistance rating when installed in a horizontal position. In this case, because one of the primary purposes of a corridor is to stop the spread of smoke versus necessarily stopping a fire, this level of protection and construction has been accepted [see Commentary Figure 708.4(2)(C)].

In order to limit the spread of smoke, tenant separation walls in a mall are required to be fire partitions (see Sections 702.4.2.1 and 708.1) having a fire-resistance rating of a least 1 hour and extending from the floor to the underside of the ceiling. Exception 4 acknowledges that extending tenant separations to the floor slab or roof deck above is not always practical for mall buildings because of the design and operation of the heating, ventilating and air-conditioning (HVAC) system. The effectiveness of the automatic sprinkler system as well as the location of the fuel loading are also reasons for not requiring tenant separations to extend above the ceiling, including attic spaces.

Within combustible construction in the situations where the fire partition is stopped at the ceiling line instead of being continuous up to the roof deck, Exception 5 permits draftstopping or fireblocking above every other dwelling unit or at a maximum spacing of 3,000 square feet (279 m^2). This design requirement allows for proper ventilation of the attic space above dwelling units. In typical multiple-family construction, with dwelling units along the front and back of the building, trusses run front to rear with soffit vents at each end. If draftstopping or fireblocking were required at each dwelling unit, it would block cross ventilation, eliminating the use of ridge vents. Soffit and ridge venting allows natural air circulation that, in turn, lowers the roof sheathing temperature in the winter, relieving many of the problems associated with ice dams. This exception is the same as permitted for draftstopping in Section 718.4.2, Exception 3.

Exception 6 recognizes the added protection afforded a building that is equipped throughout with an

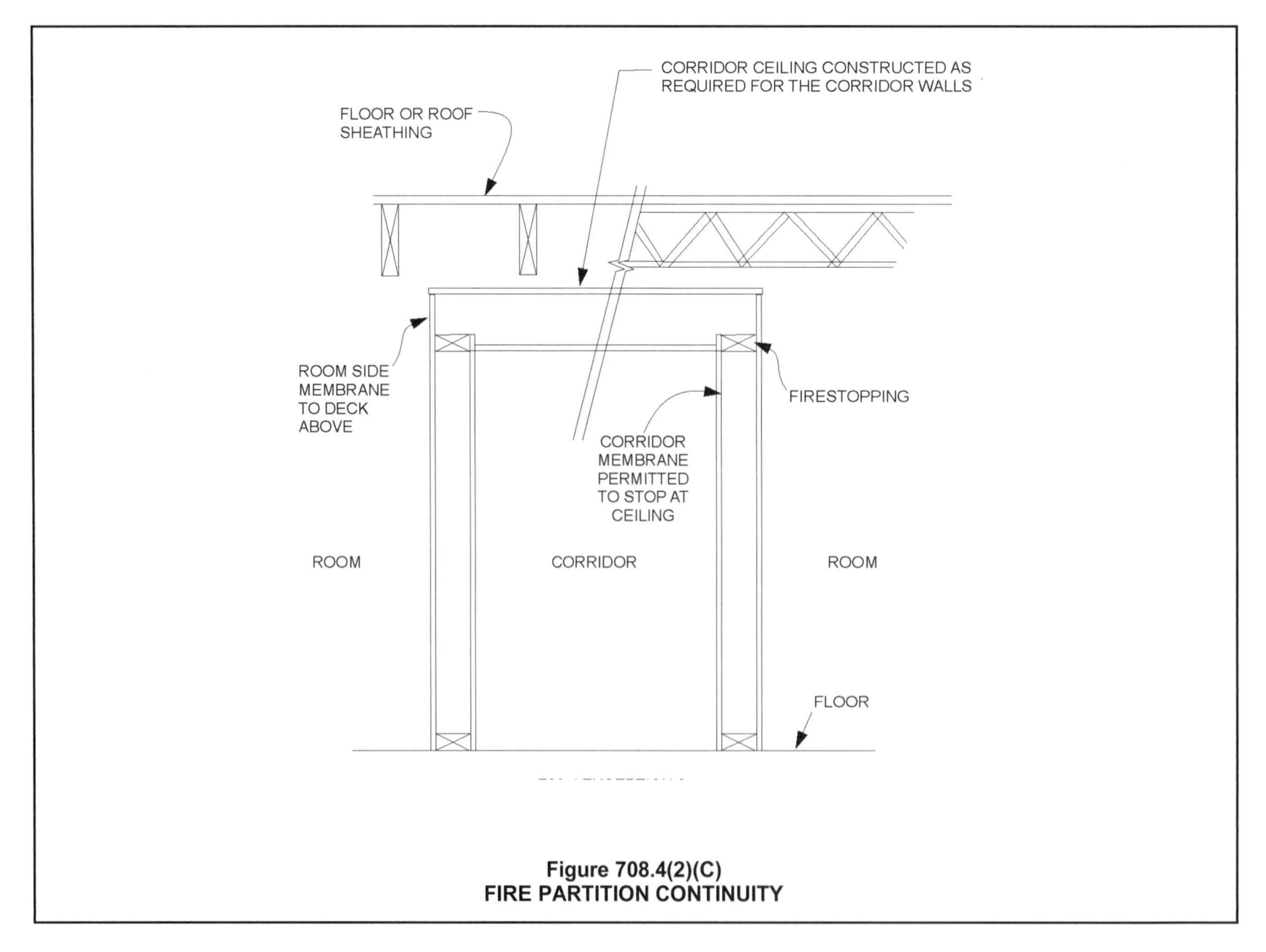

Figure 708.4(2)(C)
FIRE PARTITION CONTINUITY

automatic sprinkler system in accordance with Section 903.3.1.1 or 903.3.1.2 and NFPA 13 and 13R, respectively, and permits those fire partitions used for dwelling unit and guestroom separations to terminate at the underside of the ceiling membrane without the need for fireblocking or draftstopping above the partition. For buildings equipped with a sprinkler system that conforms to the NFPA 13 standard, the attic and other concealed areas are required to be sprinklered, thus providing protection that offsets the fact that the dwelling unit separations, fireblocking or draftstopping do not extend to the deck above. In buildings equipped throughout with an NFPA 13R system, the attic and other concealed areas may not be sprinklered. NFPA 13R allows the omission of sprinkler protection in combustible attics and concealed spaces, provided the space is not used for living purposes or storage. When using this exception, sprinkler protection, however, is necessary and required since the protection is considered available to control fires in the incipient stage and keep unoccupied concealed spaces and attic areas from becoming involved. For these buildings, the exception requires sprinkler protection and the fire partitions do not need to extend to the deck above.

Although Exception 6 provides a blanket exclusion for fireblocking or draftstopping within floor/ceiling and roof/ceiling spaces, this may need to be reviewed on a case-by-case basis. The intent of this section is to address concealed spaces that are not used for any purpose. When an attic space is provided, it would be necessary to decide if the attic is simply a concealed space or if it is an occupied portion of the dwelling unit. For example, if each dwelling unit had access to the attic above its unit by means of a pull-down ladder and an area of the attic was provided with flooring so that storage could be placed there, it would be prudent to consider that attic as being a part of the dwelling unit and, therefore, requiring a fire partition, instead of draftstopping, to be provided between the adjacent attics.

708.5 Exterior walls. Where *exterior walls* serve as a part of a required fire-resistance-rated separation, such walls shall comply with the requirements of Section 705 for *exterior walls*, and the fire-resistance-rated separation requirements shall not apply.

Exception: Exterior walls required to be fire-resistance rated in accordance with Section 1021.2 for exterior egress balconies, Section 1023.7 for interior exit stairways and ramps and Section 1027.6 for exterior exit stairways and ramps.

❖ This section pertains to exterior walls that are a part of an enclosure, such as corridors on the exterior face of a building or elevator lobbies that have exterior walls. Such exterior walls do not need to be fire-resistance-rated construction unless required by Table 601 or 602. The exception noted in this section is identical to the exceptions for Sections 707.4 and 713.6 for fire barriers and shaft enclosures. See the commentary for both of those sections.

708.6 Openings. Openings in a *fire partition* shall be protected in accordance with Section 716.

❖ Section 716.4 includes the requirements for fire doors in fire partitions. Generally, $^1/_3$-hour doors are required for corridors and $^3/_4$-hour doors for tenant dwelling and sleeping unit separations in accordance with Table 716.4. Where $^1/_2$-hour fire-resistance-rated assemblies are permitted for partitions, the doors are required to be $^1/_3$-hour doors. It is important to note that Section 716.4.3 requires smoke and draft control doors for corridors. Section 716.5 includes the requirements for windows located in a corridor.

708.7 Penetrations. Penetrations of *fire partitions* shall comply with Section 714.

❖ This section simply states that penetrations through fire partitions must conform to the requirements of Sections 714.2 and 714.3. This is the same as the penetration requirements for fire barriers.

708.8 Joints. Joints made in or between *fire partitions* shall comply with Section 715.

❖ Joints, such as expansion or seismic, are another form of openings in fire partitions and, therefore, must be considered with regard to maintaining the fire-resistance ratings of these walls. This section requires all joints that are located in fire partitions to be protected by a joint system that has a fire-resistance rating and complies with the requirements of Section 715 (see commentary, Section 715).

708.9 Ducts and air transfer openings. Penetrations in a *fire partition* by ducts and air transfer openings shall comply with Section 717.

❖ Section 717 details the protection of ducts and air transfer openings at the point where they penetrate a fire-resistance-rated assembly. Sections 717.5 and 717.5.4 indicate which situations will require the installation of a damper. As stated in Section 717.1.2, if a duct does not require a damper, the penetration of that duct through a fire-resistance-rated assembly must be protected as a through penetration in accordance with Section 714.

SECTION 709
SMOKE BARRIERS

709.1 General. Vertical and horizontal *smoke barriers* shall comply with this section.

❖ Smoke barriers divide areas of a building into separate smoke compartments both vertically and horizontally. A smoke barrier is designed to resist fire and smoke spread so that occupants can be evacuated or relocated to adjacent smoke compartments. This concept has proven effective in Group I-2 and I-3 occupancies. Sections 407.5 and 408.6 identify where smoke barriers are required. Also, while not cross-referenced in this section, smoke barriers may be utilized in other applications, including as part of smoke control systems (see Section 909.5), accessible means of egress (see Section 1009.6.4), accessible areas of refuge (see Section 1007.6), compartmentation of underground buildings (see Section 405.4.2) and elevator lobbies in underground buildings (see Section 405.4.3).

All of the elements in the smoke barrier that can potentially allow smoke travel through the smoke barrier are required to have a quantified resistance to leakage. This includes doors, joints, through penetrations and dampers. The maximum leakage limits are as established in the individual code sections referenced for each element.

709.2 Materials. *Smoke barriers* shall be of materials permitted by the building type of construction.

❖ The types of materials used in smoke barriers are to be consistent with Sections 602.2 through 602.5 for the type of construction classification of the building. Smoke barriers are permitted to be of combustible construction in Type III, IV and V construction and are required to be of noncombustible materials in Type I and II construction, except as permitted in the exceptions to Section 603.1.

709.3 Fire-resistance rating. A 1-hour *fire-resistance rating* is required for *smoke barriers*.

Exception: *Smoke barriers* constructed of minimum 0.10-inch-thick (2.5 mm) steel in Group I-3 buildings.

❖ Smoke barriers are intended to create an area of refuge; therefore, they are to be capable of resisting the passage of smoke (see Section 709.4). Smoke barriers are also required to have a fire-resistance rating of at least 1 hour. The smoke barrier is not intended or expected to be exposed to fire for extended periods and is, therefore, not required to have a fire-resistance rating exceeding 1 hour. The occupancies in which smoke barriers are required are also generally required to be sprinklered (see Section 903.2). The exception of this section allows smoke barriers in occupancies in Group I-3 to be constructed of nominal 0.10-inch (2.5 mm) steel plate. This exception to the 1-hour fire-resistance-rated assembly recognizes the security needs of such facilities and at the same time provides the requisite smoke barrier performance.

709.4 Continuity. *Smoke barriers* shall form an effective membrane continuous from the top of the foundation or floor/ceiling assembly below to the underside of the floor or roof sheathing, deck or slab above, including continuity through concealed spaces, such as those found above suspended ceilings, and interstitial structural and mechanical spaces. The supporting construction shall be protected to afford the required *fire-resistance rating* of the wall or floor supported in buildings of other than Type IIB, IIIB or VB construction. *Smoke barrier* walls used to separate smoke compartments shall comply with Section 709.4.1. *Smoke-barrier* walls used to enclose areas of refuge in accordance with Section 1009.6.4 or to enclose elevator lobbies in accordance with Section 405.4.3, 3007.6.2, or 3008.6.2 shall comply with Section 709.4.2.

Exception: *Smoke-barrier* walls are not required in interstitial spaces where such spaces are designed and constructed with ceilings or *exterior walls* that provide resistance to the passage of fire and smoke equivalent to that provided by the *smoke-barrier* walls.

❖ Smoke barriers are to be continuous from the top of the foundation or floor/ceiling assembly below to the underside of the floor or roof sheathing, deck or slab. Horizontal continuity requirements vary depending on what the smoke barrier is separating or enclosing (see sections 709.4.1 and 709.4.2). The provisions require the barrier to be continuous through all concealed and interstitial spaces, including suspended ceilings and the space between the ceiling and the floor or roof sheathing, deck or slab above; and the space between an interior wall finish and the exterior wall sheathing, such as in a wood or metal frame wall. Smoke barriers are not required to extend through interstitial spaces if the space is designed and constructed such that fire and smoke will not spread from one smoke compartment to another. Therefore, the construction assembly forming the bottom of the interstitial space must provide the required fire-resistance rating and be capable of resisting the passage of smoke from the spaces below.

As mentioned in the commentary for the general provisions of Section 709.1, the air-leakage performance of the smoke barrier itself is not typically regulated. This is because of the general assumption that a barrier which provides a fire-resistance rating will be capable of limiting the spread of smoke through it. When a smoke barrier is being used with a smoke control system, the barrier does have limitations on the air leakage (see Section 909.5). Therefore, under those circumstances and when using Exception 1 in Section 709.4, it may be necessary to provide a "hard ceiling" instead of a lay-in suspended ceiling in order to prevent the spread of smoke.

Since the primary performance of smoke barriers is to achieve protection on the fire floor, the supporting construction is not required to provide the same degree of fire resistance for buildings of Type IIB, IIIB and VB construction.

As with fire partitions serving as exit access corridor walls (see Section 709.4), these three construction types are identified because the floor construction is not otherwise required to have a fire-resistance rating, and it is not considered essential to only require fire-resistance-rated floor construction because the floor is supporting a smoke barrier.

When designing for occupancies in Groups I-2 and I-3, it is often desirable to incorporate a horizontal exit and a smoke barrier. The code does not prevent such a combination, but in such cases, the wall construction must meet the provisions of this section as well as Section 707 for fire barrier assemblies and Section

1026 for horizontal exits. The most restrictive provisions of each section would apply. For example, the building would be required to have a fire barrier with a fire-resistance rating of at least 2 hours (see Section 1026.2) and duct penetrations would need to be protected with a combination fire and smoke damper (see Sections 717.5.2 and 717.5.2.1) or with separate fire and smoke dampers. For requirements specific to smoke barriers used to separate smoke compartments and to enclose areas of refuge and elevator lobbies, see Sections 709.4.1 and 709.4.2, respectively.

Furthermore, the supporting construction would then be required to provide at least the same fire-resistance rating as the wall in all types of construction (see Section 707.5.1).

The exception recognizes that a ceiling membrane or exterior wall membrane may afford the same or greater resistance to the passage of fire and smoke that is provided by the smoke barrier. In these cases, continuity through the interstitial space is not required.

709.4.1 Smoke-barrier walls separating smoke compartments. *Smoke-barrier* walls used to separate smoke compartments shall form an effective membrane continuous from outside wall to outside wall.

❖ To completely separate smoke compartments within a building, the horizontal continuity of the smoke barrier walls must extend from exterior wall to exterior wall.

709.4.2 Smoke-barrier walls enclosing areas of refuge or elevator lobbies. *Smoke-barrier* walls used to enclose areas of refuge in accordance with Section 1009.6.4, or to enclose elevator lobbies in accordance with Section 405.4.3, 3007.6.2, or 3008.6.2, shall form an effective membrane enclosure that terminates at a *fire barrier* wall having a level of *fire protection rating* not less than 1 hour, another *smoke barrier* wall or an outside wall. A smoke and draft control door assembly as specified in Section 716.5.3.1 shall not be required at each elevator hoistway door opening or at each exit doorway between an area of refuge and the exit enclosure.

❖ Recognizing the differences between smoke compartments, areas of refuge and elevator lobby enclosures, this section provides alternative horizontal continuity compliance options. Since these areas can be located anywhere in a building, the enclosure can terminate at assemblies other than the exterior wall, such as a fire barrier or another smoke barrier.

709.5 Openings. Openings in a *smoke barrier* shall be protected in accordance with Section 716.

Exceptions:

1. In Group I-1 Condition 2, Group I-2 and *ambulatory care facilities*, where a pair of opposite-swinging doors are installed across a corridor in accordance with Section 709.5.1, the doors shall not be required to be protected in accordance with Section 716. The doors shall be close fitting within operational tolerances, and shall not have a center mullion or undercuts in excess of $^3/_4$ inch (19.1 mm), louvers or grilles. The doors shall have head and jamb stops, and astragals or rabbets at meeting edges. Where permitted by the door manufacturer's listing, positive-latching devices are not required.
2. In Group I-1 Condition 2, Group I-2 and *ambulatory care facilities*, horizontal sliding doors installed in accordance with Section 1010.1.4.3 and protected in accordance with Section 716.

❖ Section 709.5 requires openings in a smoke barrier to be protected in accordance with Section 716, Opening Protectives. Opening protectives include doors/door assemblies, glazed assemblies, fire shutters, chute intake and discharge doors, and other devices or assemblies. Within 716, Section 716.5.3 requires doors in corridors and smoke barriers to have a fire protection rating of 20 minutes. The requirements in 716.5.3 should be considered along with other requirements in Section 716 for doors required to have a fire protection rating. Such doors must also comply with UL 1784 in accordance with Section 716.5,3.1, which requires these doors to have a maximum leakage rating of 3 cfm per square foot [0.02 $m^3/(s \cdot m^2)$] or less as tested at a pressure of 0.10 inch of water (0.02 kPa).

The first exception to 709.5, along with the requirements of Section 709.5.1, specifically applies to Group I-1 Condition 2, Group I-2 and ambulatory care facilities. In order to maintain the integrity of the smoke barrier, opposite-swinging doors such as double-egress cross-corridor doors are required to be constructed and installed with some of the attributes of opening protectives, but would not be required to be listed. Since the primary purpose of a smoke barrier is to resist smoke spread, the doors are to be close fitting within operational tolerances, are not to have undercuts, louvers or grilles, and are to have stops at the head and jambs. The only openings permitted in the door assembly are those clearances that are necessary for the proper operation of the door. The clearance that occurs at the meeting edges of the door is to be protected with rabbets or astragals (see Commentary Figure 709.5).

The second exception to 709.5 applies in Group I-1 Condition 2, Group I-2 and ambulatory care facilities where special-purpose horizontal sliding, accordion, or folding doors are installed in accordance with Section 1010.1.4.3 and protected in accordance with Section 716. The requirements in Section 1010.1.4.3 include specific requirements for these special-purpose doors.

709.5.1 Group I-2 and ambulatory care facilities. In Group I-2 and *ambulatory care facilities*, where doors are installed across a corridor, the doors shall be automatic-closing by smoke detection in accordance with Section 716.5.9.3 and shall have a vision panel with fire-protection-rated glazing materials in fire-protection-rated frames, the area of which shall not exceed that tested.

❖ Section 709.5.1 requires all doors installed across a corridor in Group I-2 and ambulatory care facilities to be automatic closing upon smoke detection per the requirements in Section 716.5.9.3. In this manner, the doors will close completely, thus preventing the spread of smoke. Cross-corridor doors in Group I-2 and ambulatory care facilities are also required to have a vision panel consisting of fire-protection-rated glazing in a fire-protection-rated frame. The requirements for fire-protection-rated glazing are in Section 716.6. The size of the vision panel is limited to the maximum size tested in accordance with Section 716.6. Note: vision panels in doors (i.e., glazing in doors) are also required to meet the safety requirements for glazing in hazardous locations, per the requirements of Section 2406. The intent of requiring vision panels is to reduce the likelihood of an injury caused by opening the door into a person standing at or approaching the door from the opposite side. Vision panels also enable an individual to be alerted to a fire or smoke condition on the other side before opening the door.

709.6 Penetrations. Penetrations of *smoke barriers* shall comply with Section 714.

❖ The provisions for penetrations of smoke barriers are found in Sections 714.2, 714.3 and 714.5. To prevent through penetrations from becoming a source of significant smoke spread across the smoke barrier, the firestop system provided for each through penetration must have not only the appropriate F rating, but must also have a leakage rating (L rating) of 5 cfm per square foot [2.78 $m^3/(s \cdot m^2)$] or less as tested at a pressure of 0.30 inch of water (0.07 kPa) (see commentary, Section 713.5).

709.7 Joints. Joints made in or between *smoke barriers* shall comply with Section 715.

❖ Joints, such as expansion or seismic, are additional forms of openings in smoke barriers and, therefore, must be considered with regard to maintaining the fire-resistance ratings of these walls. This section requires all joints that are located in smoke barriers to be protected by a joint system that has a fire-resistance rating and complies with the requirements of Section 715 (see commentary, Section 715). To prevent joints from becoming a source of significant smoke spread across the smoke barrier, fire-resistant joint systems used in a smoke barrier must have a leakage rating (L rating) of 5 cfm per linear foot (0.00775 $m^3/s \cdot m$) or less (see commentary, Section 715.6).

709.8 Ducts and air transfer openings. Penetrations in a *smoke barrier* by ducts and air transfer openings shall comply with Section 717.

❖ To prevent ducts and air transfer openings from becoming a source of significant smoke spread across the smoke barrier, a listed smoke damper must be provided for duct or air transfer openings penetrating the smoke barrier (see Section 717.5.5). The smoke damper leakage rating, as tested per UL 555S; shall not be less than Class II (see Section 717.3.2.1).

Section 717.5.5 indicates which situations will require the installation of a damper. For smoke barriers, there is one exception to that requirement. If all the duct outlets are limited to one of the smoke compartments and the ducts are constructed of steel, then no smoke damper is required. As stated in Section 717.1.2, if a duct does not require a damper, the penetration of that duct through a fire-resistance-rated assembly must be protected as a through penetration in accordance with Section 714.

SECTION 710
SMOKE PARTITIONS

710.1 General. Smoke partitions installed as required elsewhere in the code shall comply with this section.

❖ Unlike a 1-hour fire-resistance-rated smoke barrier, smoke partitions are nonrated walls that serve to resist the spread of fire and the unmitigated move-

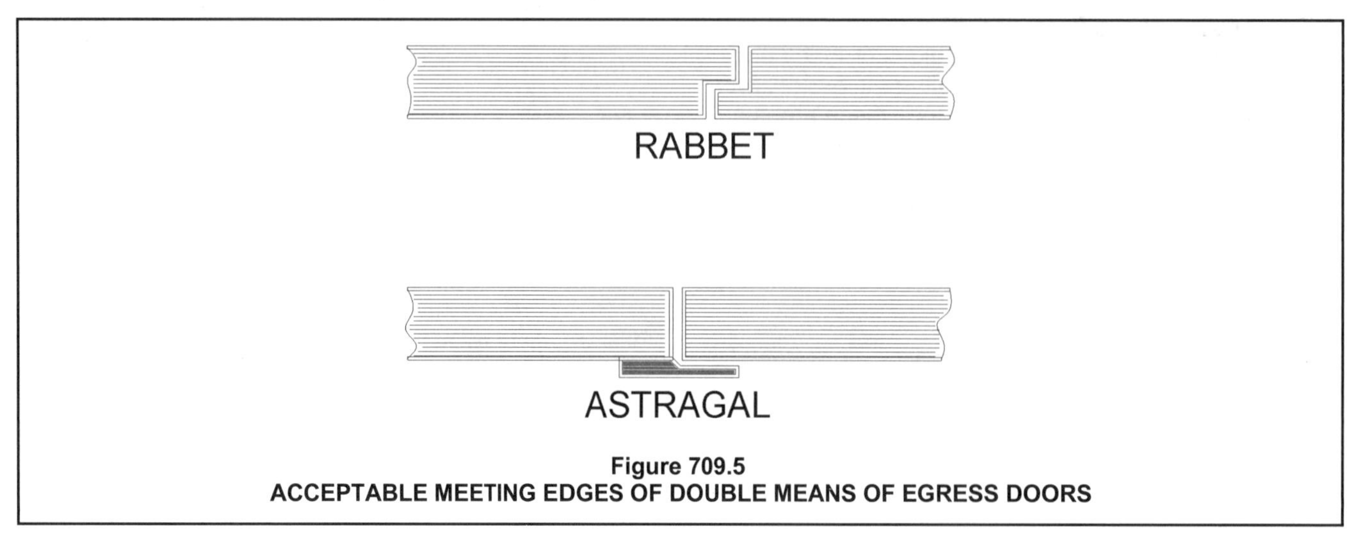

Figure 709.5
ACCEPTABLE MEETING EDGES OF DOUBLE MEANS OF EGRESS DOORS

ment of smoke for an unspecified period of time. A smoke partition is intended to provide less protection than a smoke barrier and is not required to be continuous through the concealed spaces and through the ceiling. The construction of a smoke partition is described in this section; however, the level of performance or a method of testing them is not provided. Some elements of this section (see Sections 710.5 and 710.8) will provide a performance level and test method to verify the protection provided by doors or dampers. In addition, some specific features of a smoke partition may be triggered by other sections of the code.

At this point, the application of smoke partitions is still fairly limited. The smoke partition requirements apply to Section 407.3, which permits corridor walls in Group I-2 to have no fire-resistance rating, but it does require that they be constructed as smoke partitions. A smoke partition can also be used under the provisions of Section 3006.2 to form the elevator lobby in a sprinklered building.

It is important to realize that, like all other elements found in Chapter 7, a smoke partition is not required unless the specific text of the code states that it is required. For example, the two sections listed in the previous paragraph do require the use of a smoke partition. On the other hand, Section 509.4.2, when dealing with incidental uses, will permit the elimination of a fire barrier and only require the wall constructed around the incidental use area to be "capable of resisting the passage of smoke." In this circumstance, the wall is not required to comply with the smoke partition requirements of Section 710 and is only required to comply with the details listed in Section 509.4.2.

710.2 Materials. The walls shall be of materials permitted by the building type of construction.

❖ As with most wall or partition assemblies, except for fire walls, there are no additional requirements or restrictions on the materials used, except that they meet the requirements for the type of construction.

710.3 Fire-resistance rating. Unless required elsewhere in the code, smoke partitions are not required to have a *fire-resistance rating*.

❖ The primary purpose of smoke partitions is to prevent the ready and quick passage of smoke into corridors in Group I-2 or for elevator lobby protection in a sprinklered building. The automatic sprinkler system installed in these instances eliminates the need for a fire-resistance-rated assembly; however, the issue of smoke propagation must be addressed with the use of a smoke partition. Because a performance level or a method of testing the partition is not established, any rated or nonrated assembly would be permitted, provided it is approved by the building official.

710.4 Continuity. Smoke partitions shall extend from the top of the foundation or floor below to the underside of the floor or roof sheathing, deck or slab above or to the underside of the ceiling above where the ceiling membrane is constructed to limit the transfer of smoke.

❖ The continuity provisions for smoke partitions are similar to those for fire partitions (see Section 708.4), except that the issue is the spread of smoke, not fire. Therefore, the allowance for termination at the underside of the ceiling membrane (as opposed to the underside of the floor or roof deck above) relates to the ability of the ceiling membrane to limit the spread of smoke. Typical "lay in" ceiling tiles, for instance, would probably not serve this function; however, a drywall ceiling that is taped and finished would be an example of construction that could resist the passage of smoke.

710.5 Openings. Openings in smoke partitions shall comply with Sections 710.5.1 and 710.5.2.

❖ Limiting smoke movement in buildings with smoke partitions includes opening protectives that resist the passage of smoke.

710.5.1 Windows. Windows in smoke partitions shall be sealed to resist the free passage of smoke or be automatic-closing upon detection of smoke.

❖ Windows in smoke partitions do not require a fire-resistance rating. There are no limits on the amount of glazing. They must be fixed glazing or automatic closing upon the activation of smoke detectors.

710.5.2 Doors. Doors in smoke partitions shall comply with Sections 710.5.2.1 through 710.5.2.3.

❖ Doors in smoke partitions do not require a fire-resistance rating and may not need self-closing devices.

710.5.2.1 Louvers. Doors in smoke partitions shall not include louvers.

❖ Louvers in doors allow smoke to move from one area to the remainder of the building. This would defeat the sole purpose of a smoke partition.

710.5.2.2 Smoke and draft control doors. Where required elsewhere in the code, doors in smoke partitions shall meet the requirements for a smoke and draft control door assembly tested in accordance with UL 1784. The air leakage rate of the door assembly shall not exceed 3.0 cubic feet per minute per square foot [$0.015424\ m^3/(s \cdot m^2)$] of door opening at 0.10 inch (24.9 Pa) of water for both the ambient temperature test and the elevated temperature exposure test. Installation of smoke doors shall be in accordance with NFPA 105.

❖ Only doors in smoke partitions that are required elsewhere in the code to be smoke and draft control doors must meet this section.

710.5.2.2.1 Smoke and draft control door labeling. Smoke and draft control doors complying only with UL 1784 shall be permitted to show the letter "S" on the manufacturer's labeling.

❖ This section clarifies that the use of the "S" letter mark is intended only to indicate compliance with UL 1784, and allows use of the marking on smoke partition doors that comply with that test standard.

710.5.2.3 Self- or automatic-closing doors. Where required elsewhere in the code, doors in smoke partitions shall be self- or automatic-closing by smoke detection in accordance with Section 716.5.9.3.

❖ The requirement for doors to be self-closing or automatic-closing does not automatically apply to all walls that must be built as smoke partitions. That requirement must be mandated elsewhere in the code for this provision to apply. In the case of corridor walls for Group I-2 occupancies, there is no requirement for self-closing or automatic-closing doors so as to accommodate the normal operational requirements of that occupancy. In that situation, it is anticipated that doors would be manually closed by personnel in the event of a fire emergency based upon a defend-in-place strategy. Having doors arranged for automatic closing or self-closing, however, is not prohibited as it would potentially increase the reliability of smoke partitions. There are exceptions to Section 3006.2 that allow elevator lobby smoke partition enclosures in lieu of fire partitions. In this case, closers on the doors to the lobby would be required as a part of the conditions for the exception.

710.6 Penetrations. The space around penetrating items shall be filled with an *approved* material to limit the free passage of smoke.

❖ There is no prescriptive requirement for a material used to seal penetrations in smoke partitions. The intent is to limit the transfer of smoke through the partition. See the commentary to Section 710.7 for more information.

710.7 Joints. Joints shall be filled with an *approved* material to limit the free passage of smoke.

❖ There is no requirement for a material used to seal penetrations or joints to be applied to both sides of the wall, as would usually be the case when the penetrations and joints are in a fire-resistance-rated wall. The intent of this section will have been met as long as the space around penetrating items and in joints is filled in a way that would prevent a continuous channel from one side of the wall to the other. In selecting a material to seal these gaps, consideration should be given to the fact that smoke may have a temperature that is above the normal ambient temperature. The code does not mandate any specific temperature resistance. However, as one possible reference point, UL uses a gas temperature of 325°F (180°C) above ambient to represent "warm smoke." With a common ambient temperature of 75°F (24°C), the "warm smoke" temperature would therefore be 400°F (204°C). Sealing materials that would not maintain their integrity to at least this temperature would be inferior choices for the purpose of limiting the free passage of smoke. Porous materials (e.g., fibrous insulation) that would not provide any significant resistance to the passage of smoke through them would also be inferior choices for sealing of joints and through penetrations for this purpose. Again, because the level of performance is not clearly established, the word "approved" and the building official's decision will determine what materials are required or accepted.

710.8 Ducts and air transfer openings. The space around a duct penetrating a smoke partition shall be filled with an *approved* material to limit the free passage of smoke. Air transfer openings in smoke partitions shall be provided with a *smoke damper* complying with Section 717.3.2.2.

> **Exception:** Where the installation of a *smoke damper* will interfere with the operation of a required smoke control system in accordance with Section 909, *approved* alternative protection shall be utilized.

❖ Smoke dampers are required to be provided to maintain the integrity of the smoke partition as a means to prevent the spread of smoke when an "air transfer opening" exists. If a ducted system is used, then the code specifies that the annular space around the duct must be protected. Because the code does not define what an "air transfer opening" is, the provision may not be consistently enforced. The general intent is that a ducted system does not require a damper, but an opening such as an air transfer grille or louvered opening would be dampered. A ducted system must be protected in accordance with Section 710.6. The exception relating to the required operation of a smoke control system is consistent with the exceptions given elsewhere regarding air transfer openings and ducts with fire dampers or smoke dampers (see Section 717). An effective smoke control system will protect the building against the spread of smoke and it is important to ensure such installation of a smoke damper will not lessen the effectiveness of a smoke control system.

SECTION 711
HORIZONTAL ASSEMBLIES

711.1 General. *Horizontal assemblies* shall comply with Section 711.2. Nonfire-resistance-rated floor and roof assemblies shall comply with Section 711.3.

❖ Horizontal assemblies by definition are fire-resistance-rated assemblies. A horizontal assembly may be either a floor or roof assembly and may rely on the ceiling membrane to achieve its fire-resistance-rating. The ceiling assembly is often an integral part of a fire-resistance-rated floor/ceiling or roof/ceiling assembly; therefore, the integrity of the ceiling assembly must be maintained in order to reduce the potential for premature failure of the floor or roof of a building. Section 711.2 contains requirements for horizontal assemblies.

Nonfire-resistance-rated floor/ceiling and roof/ceiling assemblies are not considered horizontal assemblies. These assemblies have requirements less stringent than horizontal assemblies. Section 711.3 contains requirements for nonfire-resistance-rated floor/ceiling and roof/ceiling assemblies.

711.2 Horizontal assemblies. *Horizontal assemblies* shall comply with Sections 711.2.1 through 711.2.6.

❖ This section is simply charging language for the subsections. See the commentary to Sections 711.2.1 through 711.2.6

711.2.1 Materials. Assemblies shall be of materials permitted by the building type of construction.

❖ The types of materials used in the construction of horizontal assemblies are to be consistent with Sections 602.2 through 602.5 for the type of construction classification of the building. Horizontal assemblies are permitted to be of combustible construction in Type III, IV and V construction and are required to be of noncombustible materials in Type I and II construction, except as permitted in the exceptions to Section 603.1 or in Notes c and d from Table 601.

711.2.2 Continuity. Assemblies shall be continuous without vertical openings, except as permitted by this section and Section 712.

❖ All floors, roofs and ceilings of horizontal assemblies are to be continuous without openings or penetrations, except as permitted by this section. The continuity of the assembly is critical to its ability to limit fire and smoke spread. Penetrations or openings of the assembly are permitted in accordance with Section 712 provided that the fire-resistance rating, is maintained. The fire-resistance rating required by Table 601 for roof construction is intended to minimize the threat of premature structural failure of the roof construction under fire conditions. The provisions for rated roof construction in Table 601 do not apply to openings in the roof. Code users should also review the fireblocking and draftstopping requirements that are found in Section 718. Fireblocking and draftstopping requirements apply to combustible concealed locations, are separate issues from fire-resistance ratings, and may impose additional requirements for the assembly.

711.2.3 Supporting construction. The supporting construction shall be protected to afford the required *fire-resistance rating* of the *horizontal assembly* supported.

Exception: In buildings of Type IIB, IIIB or VB construction, the construction supporting the *horizontal assembly* is not required to be *fire-resistance rated* at the following:

1. *Horizontal assemblies* at the separations of incidental uses as specified by Table 509 provided the required *fire-resistance rating* does not exceed 1 hour.
2. *Horizontal assemblies* at the separations of *dwelling units* and *sleeping units* as required by Section 420.3.
3. *Horizontal assemblies* at *smoke barriers* constructed in accordance with Section 709.

❖ The base requirement is that construction supporting horizontal assemblies be fire-resistance rated at least equal to the rating of the horizontal assembly. The exception deals with three specific applications of horizontal assemblies where it is unnecessary to provide a fire-resistance rating of the supporting construction of horizontal assemblies in buildings of Type IIB, IIIB or VB construction since Table 601 does not require the horizontal assembly or the supporting structural members to have a fire-resistance rating in these types of construction. This exception exempts the supporting construction of horizontal assemblies in the same manner that the code currently exempts the supporting construction of fire barriers, fire partitions and smoke barriers, but only in those circumstances where the horizontal assembly is a component of the same fire containment assembly as the fire barrier, fire partition or smoke barrier.

711.2.4 Fire-resistance rating. The *fire-resistance rating* of *horizontal assemblies* shall comply with Sections 711.2.4.1 through 711.2.4.6 but shall be not less than that required by the building type of construction.

❖ Floor and roof assemblies could have a fire-resistance rating for several reasons:

- Required for the type of construction in Table 601.
- Required for separation of mixed uses in Section 508.4.
- Separate fire areas of like or different occupancies in order to reduce the size of fire areas to below threshold values for sprinkler requirements in Section 903.
- Separation of dwelling units in the same building.
- Separation of sleeping units in Group R-1 hotels, R-2 and I-1.

711.2.4.1 Separating mixed occupancies. Where the *horizontal assembly* separates mixed occupancies, the assembly shall have a *fire-resistance rating* of not less than that required by Section 508.4 based on the occupancies being separated.

❖ Horizontal assemblies may be required to separate mixed occupancies. As an example, consider a nonsprinklered two-story building of Type IIA construction and a Group B occupancy, with the basement having a storage occupancy in Group S-2. Assume the designer has chosen the separated occupancies option for compliance with the mixed occupancy condition. In this case, even though Table 601 requires a 1-hour fire-resistance-rated floor throughout the building, the assembly separating the basement from the first floor of the building is also regulated by Section 508.4 for mixed occupancies. Ultimately, Section 508.4.4 references Table 508.4, which requires that the floor/ceiling construction separating Group S-2 from Group B provides a 2-hour fire-resistance-rated separation.

711.2.4.2 Separating fire areas. Where the *horizontal assembly* separates a single occupancy into different fire areas, the assembly shall have a *fire-resistance rating* of not less than that required by Section 707.3.10.

❖ A designer may choose to separate areas of the building into fire areas for reasons such as to incorporate horizontal exits or to create areas within the thresholds for nonsprinklered applications. Fire areas can separate the same or different groups. For these applications, the fire-resistance rating is required to comply with Section 707.3.10.

711.2.4.3 Dwelling units and sleeping units. *Horizontal assemblies* serving as dwelling or sleeping unit separations in accordance with Section 420.3 shall be not less than 1-hour *fire-resistance-rated* construction.

Exception: *Horizontal assemblies* separating *dwelling units* and *sleeping units* shall be not less than 1/2-hour fire-resistance-rated construction in a building of Type IIB, IIIB and VB construction, where the building is equipped throughout with an *automatic sprinkler system* in accordance with Section 903.3.1.1.

❖ The requirement for 1-hour floor assemblies separating dwelling and sleeping units mirrors that found in Section 708 for fire partitions. There are other sections of the code that require a fire-resistance rating for the floor or roof assembly, or where other provisions are dependent on the fire-resistance rating or construction of the assembly (examples of such sections include Sections 704.1, 705.6, 707.5.1, 708.4 and 709.4).

The exception for dwelling and sleeping unit separations in unprotected types of construction is based on the protection provided by a sprinkler system installed in accordance with Section 903.3.1.1 (NFPA 13). This presumably will reduce the potential fire exposure of the unit separation to that which makes a minimum $^{1}/_{2}$-hour fire-resistance rating adequate. The reason that only unprotected types of construction are permitted to take advantage of this trade-off is because Table 601 would not otherwise require the floor construction of such structures to be fire-resistance rated, whereas all other types of construction require the floor construction to provide at least a 1-hour fire-resistance rating.

711.2.4.4 Separating smoke compartments. Where the *horizontal assembly* is required to be a *smoke barrier*, the assembly shall comply with Section 709.

❖ Smoke compartments are required in Group I-2 and I-3 to create areas where the occupants can be relocated during an emergency without leaving the building. Horizontal assemblies that separate smoke compartments are required to comply with Section 709.

711.2.4.5 Separating incidental uses. Where the *horizontal assembly* separates incidental uses from the remainder of the building, the assembly shall have a *fire-resistance rating* of not less than that required by Section 509.

❖ Incidental uses are specific rooms or areas that require separation or separation and sprinkler fire protection. Horizontal assemblies that separate smoke compartments are required to comply with Section 509.

711.2.4.6 Other separations. Where a *horizontal assembly* is required by other sections of this code, the assembly shall have a *fire-resistance rating* of not less than that required by that section.

❖ Horizontal assemblies may be required elsewhere in the code, such as Chapter 6 as it relates to the building's type of construction. For these other separation requirements, the fire-resistance rating is as required by that section.

Chapter 6 also contains provisions governing materials that may be used in the construction of such assemblies based on the type of construction of the building and whether combustible or noncombustible materials are used.

711.2.5 Ceiling panels. Where the weight of lay-in ceiling panels, used as part of fire-resistance-rated floor/ceiling or roof/ceiling assemblies, is not adequate to resist an upward force of 1 pound per square foot (48 Pa), wire or other *approved* devices shall be installed above the panels to prevent vertical displacement under such upward force.

❖ Where a ceiling membrane constitutes part of a fire-resistance-rated floor/ceiling or roof/ceiling assembly, the ability of the ceiling membrane to not be displaced by the upward pressure of a fire condition is necessary to maintain the viability of the fire-resistance rating. Commentary Figure 711.2.5 shows a floor/ceiling assembly that uses hold-down clips to keep the ceiling panels in place. When the weight of the ceiling panel is less than 1 pound per square foot (psf) (48 Pa), wire or other approved means must be provided to prevent uplift of the ceiling panels. Manufacturers' literature can be used to determine the weight of the panel.

711.2.6 Unusable space. In 1-hour fire-resistance-rated floor/ceiling assemblies, the ceiling membrane is not required to be installed over unusable crawl spaces. In 1-hour fire-resistance-rated roof assemblies, the floor membrane is not required to be installed where unusable *attic* space occurs above.

❖ The section is a special exception for specific assemblies to allow the deletion of floor or roof decking or the ceiling membrane from a required fire-resistance-rated floor/ceiling or roof/ceiling assembly used in certain applications. In an attic application, the floor sheathing may be deleted from a fire-resistance-rated assembly as long as the joist and ceiling remain identical to the tested assembly and the attic space is not usable space where potential combustible materials or ignition sources may be located [see Commentary Figure 711.2.6(1)]. Over a crawl space, the ceiling membrane may be deleted subject to the same conditions [see Commentary Figure 711.2.6(2)]. In determining whether a space is "unusable," the building official must verify that combustible materials other than construction elements will not be located therein and ignition sources will be minimal. As such, pipes, conduit and ducts may be permitted in an unusable space.

711.3 Nonfire-resistance-rated floor and roof assemblies. Nonfire-resistance-rated floor, floor/ceiling, roof and roof/ceiling assemblies shall comply with Sections 711.3.1 and 711.3.2.

❖ Nonfire-resistance rated floor and roof assemblies, although not fire rated, do have minimum construction and continuity requirements as described in this section.

711.3.1 Materials. Assemblies shall be of materials permitted by the building type of construction.

❖ At a minimum, the materials used to construct these floor and roof assemblies need to be consistent with the materials required for the type of construction required based on the criteria in Chapters 5 and 6. For example, if a noncombustible type of construction is required, then the elements of the floor and roof assembly also need to be noncombustible unless otherwise allowed by the code, such as the allowances of Section 603.

711.3.2 Continuity. Assemblies shall be continuous without vertical openings, except as permitted by Section 712.

❖ Although these assemblies are not required to be fire rated, there is still a benefit to their being continuous. This benefit is a resistance to smoke migration. Openings are not prohibited and can be found in Section 712.

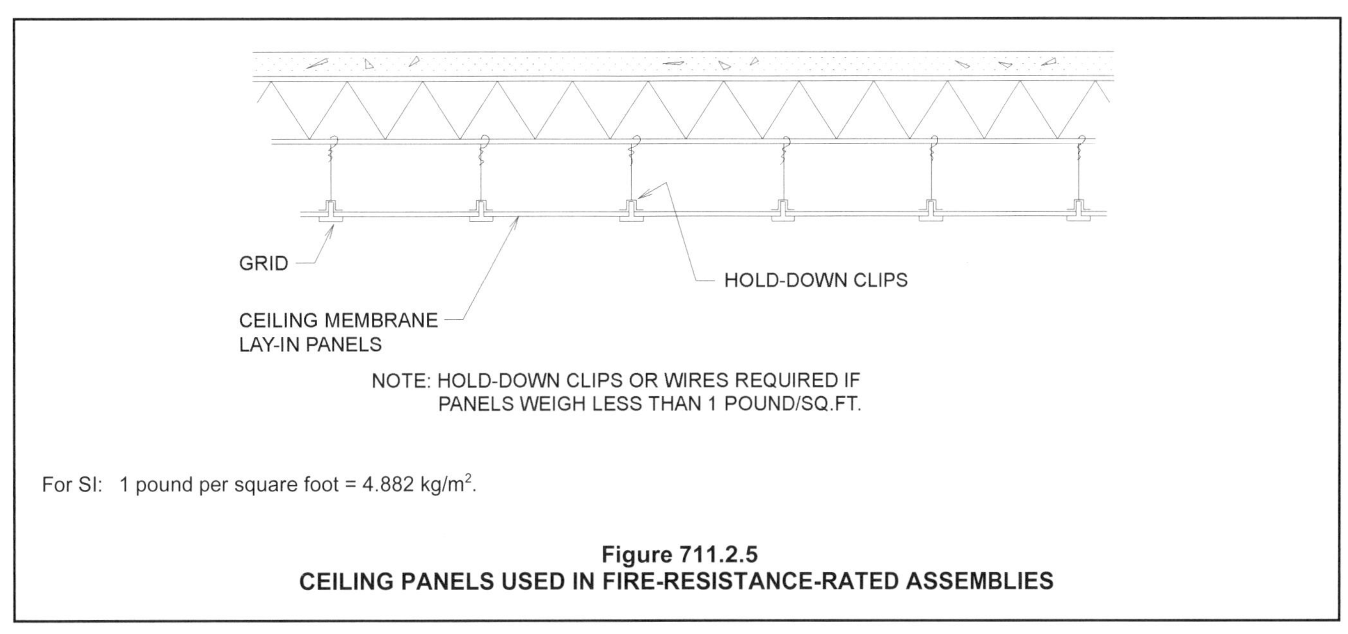

For SI: 1 pound per square foot = 4.882 kg/m^2.

Figure 711.2.5
CEILING PANELS USED IN FIRE-RESISTANCE-RATED ASSEMBLIES

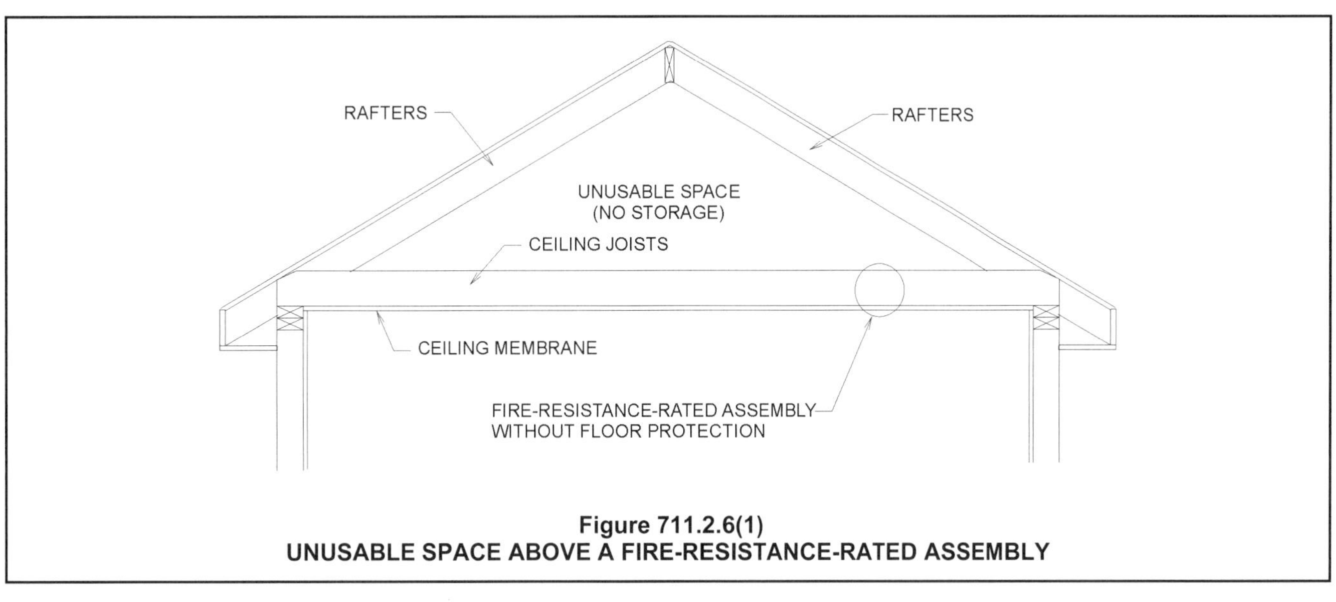

Figure 711.2.6(1)
UNUSABLE SPACE ABOVE A FIRE-RESISTANCE-RATED ASSEMBLY

SECTION 712
VERTICAL OPENINGS

712.1 General. Each vertical opening shall comply in accordance with one of the protection methods in Sections 712.1.1 through 712.1.16.

❖ Section 712 lists the applications when the code allows vertical openings between stories or levels within a building. The conditions that are necessary to allow the vertical openings are enumerated in this section or the appropriate sections are referenced. Please note that these are prescriptive options and are intended to be applied individually.

712.1.1 Shaft enclosures. Vertical openings contained entirely within a shaft enclosure complying with Section 713 shall be permitted.

❖ The first condition through which vertical openings between stories is allowed is fully enclosing the opening in shaft wall construction complying with Section 713. These types of openings contain ducts, vents, piping, chutes or elevator hoistways. Stairways are also vertical openings between floors and Section 713 references Sections 1009 and 1022 for stairway enclosures and the exceptions to their enclosures.

712.1.2 Individual dwelling unit. Unconcealed vertical openings totally within an individual residential dwelling unit and connecting four stories or less shall be permitted.

❖ The keyword in this section is "unconcealed." Vertical openings in view of the occupants, such as open exit access stairs, are allowed. The unconcealed floor opening connecting no more than four stories within a dwelling unit is permitted without a shaft enclosure. This is generally consistent with the permitted omission of stairway enclosures in such uses in accordance with Item 2 to Section 1019.3.

712.1.3 Escalator openings. Where a building is equipped throughout with an *automatic sprinkler system* in accordance with Section 903.3.1.1, vertical openings for escalators shall be permitted where protected in accordance with Section 712.1.3.1 or 712.1.3.2.

❖ Vertical openings containing escalators in buildings that are fully sprinklered in accordance with NFPA 13 are allowed under either of the two protection methods stated in Sections 712.1.3.1 and 712.1.3.2.

712.1.3.1 Opening size. Protection by a draft curtain and closely spaced sprinklers in accordance with NFPA 13 shall be permitted where the area of the vertical opening between stories does not exceed twice the horizontal projected area of the escalator. In other than Groups B and M, this application is limited to openings that do not connect more than four stories.

❖ Under this method a draft curtain is required to protect the vertical opening. The details of the draft curtain and sprinkler spacing are in NFPA 13. The size of the vertical opening cannot exceed twice the horizontal projection of the escalator (see Commentary Figure 712.1.3.1). This method allows such floor openings to occur for the full height of the building, regardless of the number of stories for Groups B and M. For all other occupancies, there is a four-story limitation (three floors permitted to contain openings). This is similar to Item 4 for exit access stairways in Section 1019.3.

712.1.3.2 Automatic shutters. Protection of the vertical opening by approved shutters at every penetrated floor shall be permitted in accordance with this section. The shutters shall be of noncombustible construction and have a *fire-resistance rating* of not less than 1.5 hours. The shutter shall be so constructed as to close immediately upon the actuation of a smoke detector installed in accordance with Section 907.3.1

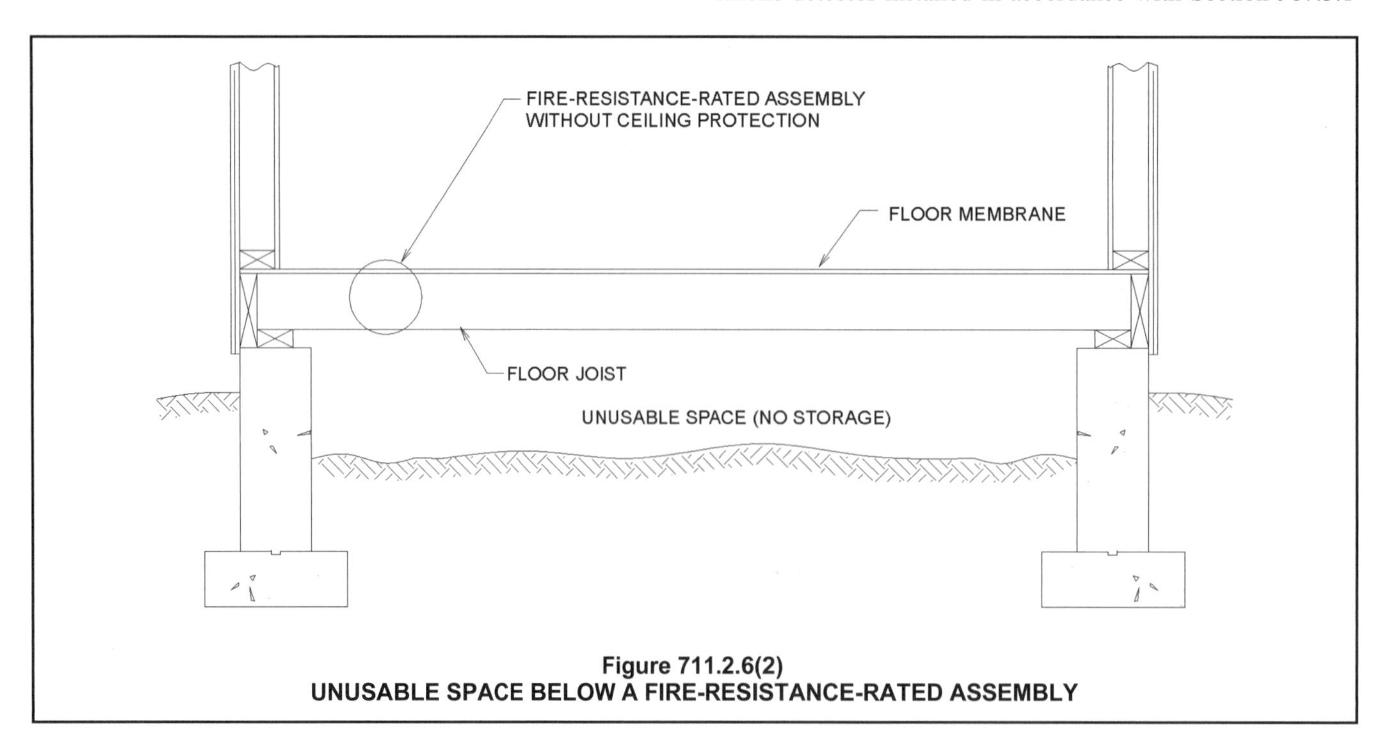

Figure 711.2.6(2)
UNUSABLE SPACE BELOW A FIRE-RESISTANCE-RATED ASSEMBLY

and shall completely shut off the well opening. Escalators shall cease operation when the shutter begins to close. The shutter shall operate at a speed of not more than 30 feet per minute (152.4 mm/s) and shall be equipped with a sensitive leading edge to arrest its progress where in contact with any obstacle, and to continue its progress on release there from.

❖ This method of protecting vertical openings containing escalators requires automatic shutters at each floor opening and is not limited as to height or occupancy [see Commentary Figure 712.1.3.2].

712.1.4 Penetrations. Penetrations, concealed and unconcealed, shall be permitted where protected in accordance with Section 714.

❖ Vertical openings are formed at the penetrations of horizontal assemblies by pipe, tube, conduit, wire, cable and other utilities. The annular spaces around the penetrating item must be protected in accordance with Section 714.4 or the penetrating items must be in shafts meeting Section 712.1.1.

712.1.5 Joints. Joints shall be permitted where complying with Section 712.1.5.1 or 712.1.5.2, as applicable.

❖ Joints are allowed in floor and roof assemblies where protected in accordance with this Section.

712.1.5.1 Joints in or between horizontal assemblies. Joints made in or between *horizontal assemblies* shall comply with Section 715. The void created at the intersection of a floor/ceiling assembly and an exterior curtain wall assembly shall be permitted where protected in accordance with Section 715.4.

❖ Joints, such as expansion or seismic, are additional forms of openings in horizontal assemblies and, therefore, must be considered with regard to maintaining the fire-resistance ratings of these floors or roofs. This section requires all joints that are located in rated horizontal assemblies to be protected by a joint system that has a fire-resistance rating and complies with the requirements of Section 715 (see commentary, Section 715). The provisions of this section

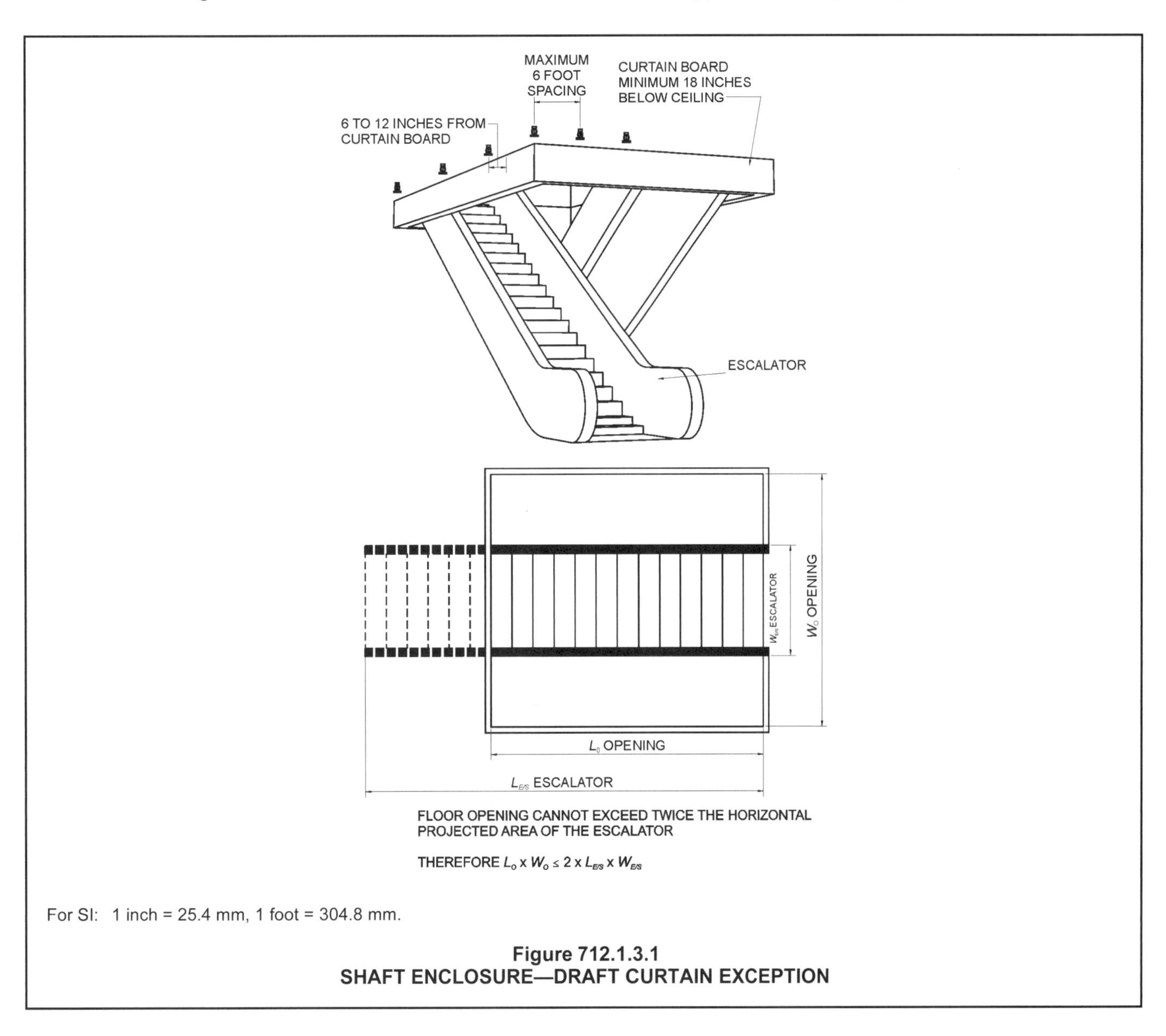

Figure 712.1.3.1
SHAFT ENCLOSURE—DRAFT CURTAIN EXCEPTION

also include the protection of joints/voids that may occur at the intersection of the horizontal assembly and an exterior wall.

712.1.5.2 Joints in or between nonfire-resistance-rated floor assemblies. Joints in or between floor assemblies without a required *fire-resistance rating* shall be permitted where they comply with one of the following:

1. The joint shall be concealed within the cavity of a wall.
2. The joint shall be located above a ceiling.
3. The joint shall be sealed, treated or covered with an *approved* material or system to resist the free passage of flame and the products of combustion.

Exception: Joints meeting one of the exceptions listed in Section 715.1.

❖ This section addresses joints in nonrated floor assemblies. The spread of fire from story to story is a concern for all buildings whether or not the floors are required to be fire-resistance rated or not. This is why shaft enclosures are required for buildings even with no requirement for fire-resistance-rated floors. Although applicable to fire-resistant joint systems, the exceptions in Section 715.1 are also applicable to joints in nonfire-resistance-rated floor assemblies.

712.1.6 Ducts and air transfer openings. Penetrations by ducts and air transfer openings shall be protected in accordance with Section 717. Grease ducts shall be protected in accordance with the *International Mechanical Code*.

❖ Ducts and air transfer openings for HVAC supply, return and exhaust through horizontal assemblies can be a source for the spread of smoke and fire from story to story. This section references Section 717.6, which states that the first option is to enclose the vertical ducts in shaft walls, but gives several other options as well (see commentary, Section 717.6).

712.1.7 Atriums. In other than Group H occupancies, atriums complying with Section 404 shall be permitted.

❖ "Atriums" are defined as an opening connecting two or more stories. Vertical openings connecting two or more stories may be allowed under the provisions and conditions in Section 404. All vertical openings connecting two or more stories are not required to meet Section 404. For openings connecting only two stories, the provisions of Section 712.1.9 may also be used to allow vertical openings. Vertical openings consisting of unenclosed exit access stairs may also be allowed when meeting specific portions of Section 1019.

712.1.8 Masonry chimney. Approved vertical openings for masonry chimneys shall be permitted where the annular space is fireblocked at each floor level in accordance with Section 718.2.5.

❖ A masonry chimney must meet Section 2113 in order to be approved. According to Section 718.2.5, the annular space surrounding the masonry chimney must be fireblocked with a material specifically tested in the form and manner intended for use to demonstrate its ability to remain in place and resist the free passage of flame and products of combustion. Furthermore, Section 2113.20 requires the annular space to be fireblocked with a noncombustible material at the ceiling and floor. This section does not

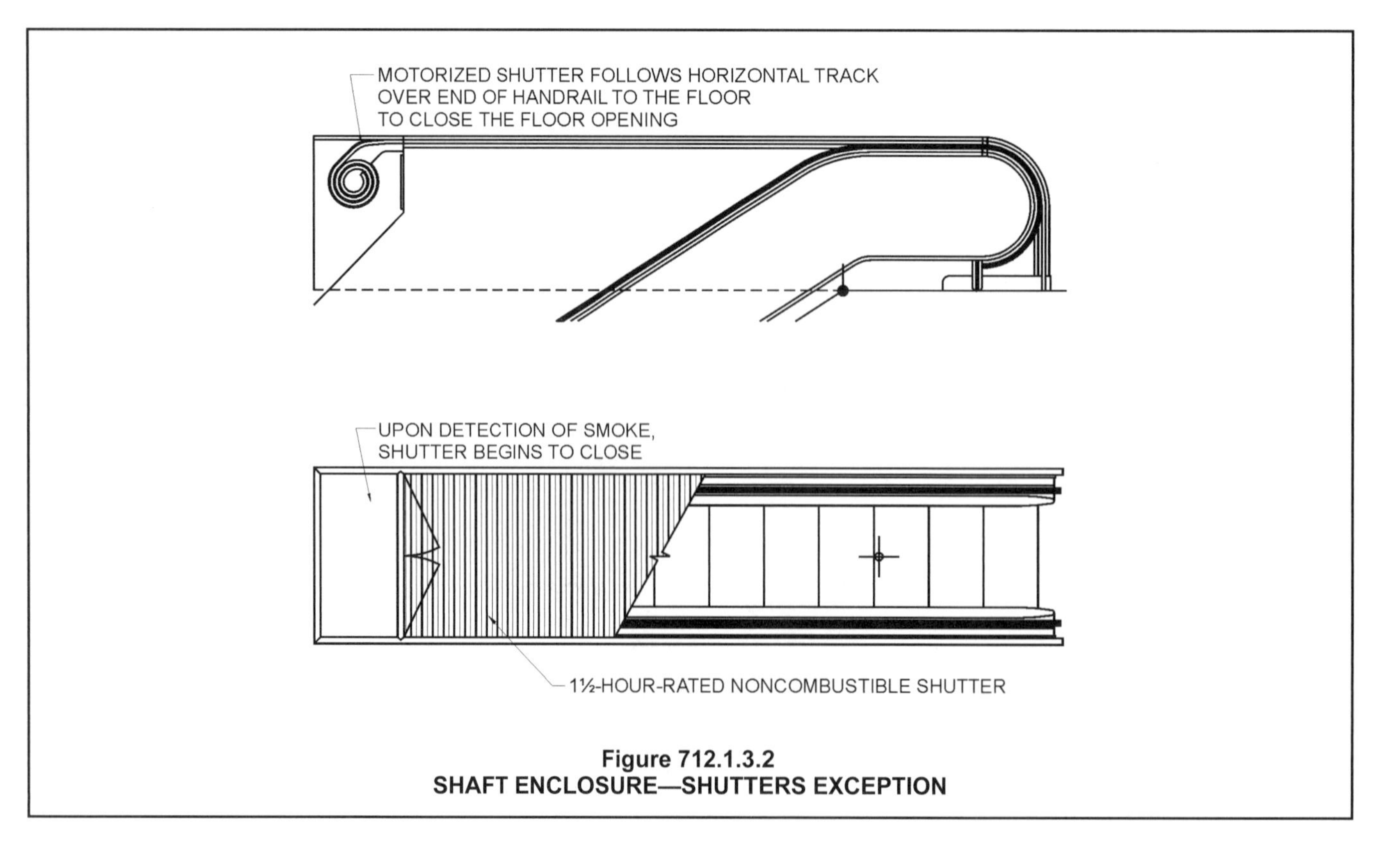

Figure 712.1.3.2
SHAFT ENCLOSURE—SHUTTERS EXCEPTION

apply to factory-built fireplaces and chimneys. Factory-built chimneys require shaft enclosures when penetrating floors.

712.1.9 Two-story openings. In other than Groups I-2 and I-3, a vertical opening that is not used as one of the applications listed in this section shall be permitted if the opening complies with all of the items below:

1. Does not connect more than two stories.
2. Does not penetrate a horizontal assembly that separates fire areas or smoke barriers that separate smoke compartments.
3. Is not concealed within the construction of a wall or a floor/ceiling assembly.
4. Is not open to a corridor in Group I and R occupancies.
5. Is not open to a corridor on nonsprinklered floors.
6. Is separated from floor openings and air transfer openings serving other floors by construction conforming to required shaft enclosures.

❖ The floor-opening allowance described in this section is not applicable to Groups I-2 and I-3. A floor opening connecting two stories that is not covered by one of the other provisions for vertical openings in Section 712 may use this provision. This provision addresses the issue of floor openings that are not a part of the required means of egress, such as unconcealed openings providing the same purpose as an atrium between two stories. Unlike the atrium provisions in Section 404, this section does not require sprinklers or smoke control. Such vertical openings cannot connect to exit access corridors in Group I or R because occupants can be sleeping and the integrity of the corridor system is especially important under such conditions. For a sprinklered Group B occupancy, Commentary Figure 712.1.9 illustrates the vertical opening connecting to an exit access corridor.

Under this provision, such two-story vertical openings cannot be connected to any other floor opening that connects to an additional floor level. Such vertical openings must be separated from vertical openings serving other stories by construction that complies with this section for shaft enclosures. This requirement limits the use of this provision so that an opening between two stories does not openly communicate with another opening to an additional story and, therefore, interconnect three or more different levels.

712.1.10 Parking garages. Vertical openings in parking garages for automobile ramps, elevators and duct systems shall comply with Section 712.1.10.1, 712.1.10.2 or 712.1.10.3, as applicable.

❖ Vertical openings between stories or tiers of parking in parking garages are permitted. It is impractical to consider not allowing those vertical openings created by the automobile ramps. While this section only applies to the vehicle ramps, Sections 712.1.15 and 712.1.10.3 apply to vertical openings for duct systems and elevator hoistways serving only the parking garage. Exit access stairways in enclosed parking garages may need enclosures if they do not meet one of the specific provisions in Section 1019.

712.1.10.1 Automobile ramps. Vertical openings for automobile ramps in open and enclosed parking garages shall be permitted where constructed in accordance with Sections 406.5 and 406.6, respectively.

❖ The special code requirements in Sections 406.5 and 406.6 recognize the unique features of parking garages. Adherence to those sections will mitigate the concerns about vertical openings.

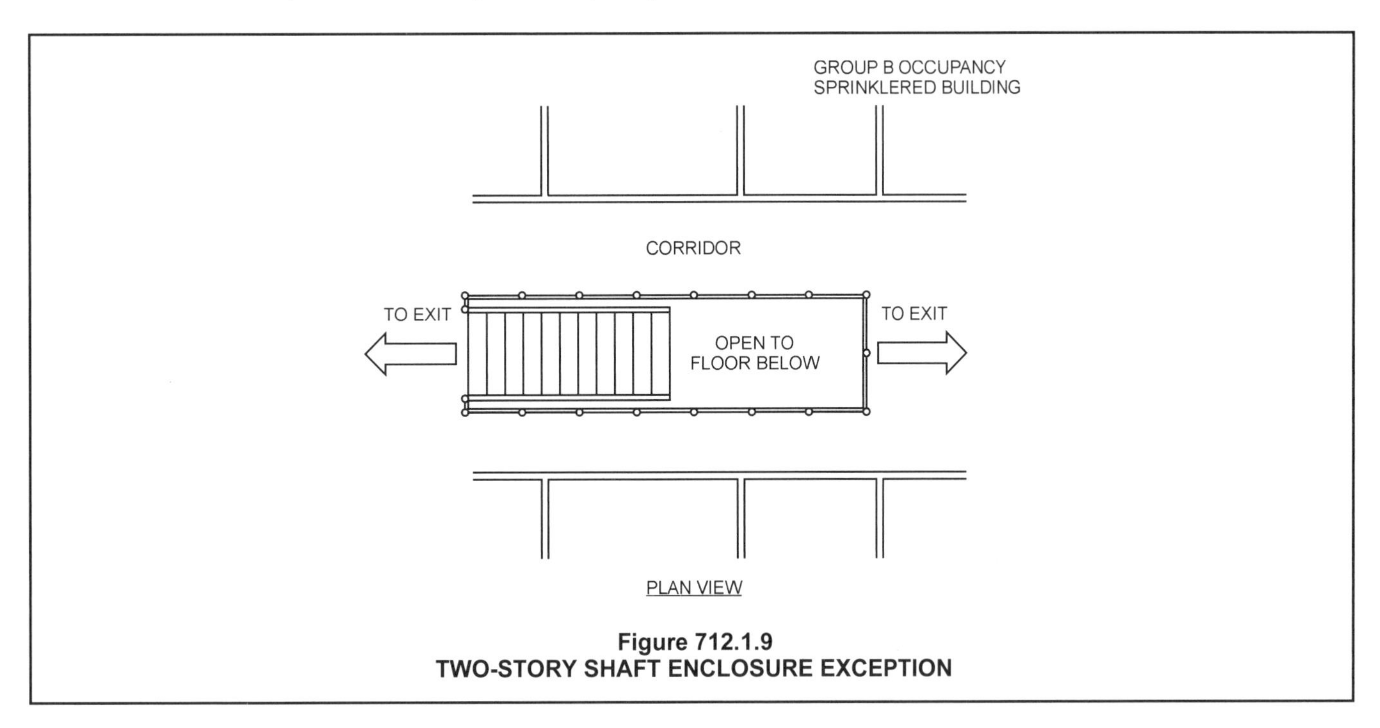

Figure 712.1.9
TWO-STORY SHAFT ENCLOSURE EXCEPTION

712.1.10.2 Elevators. Vertical openings for elevator hoistways in open or enclosed parking garages that serve only the parking garage, and complying with Sections 406.5 and 406.6, respectively, shall be permitted.

❖ Vertical openings that contain elevator hoistways that only serve a parking garage (open or enclosed) need not be enclosed in fire-resistance-rated shaft construction. The tiered levels of parking are already open; therefore, to provide fire-resistance-rated construction for the hoistway enclosures serves no useful purpose. Please note that if the elevator hoistway serves floors that are not part of the parking garage, then the fire-resistance-rated shafts would be required throughout the entire length of the elevator hoistway.

712.1.10.3 Duct systems. Vertical openings for mechanical exhaust or supply duct systems in open or enclosed parking garages complying with Sections 406.5 and 406.6, respectively, shall be permitted to be unenclosed where such duct system is contained within and serves only the parking garage.

❖ This section is similar to Section 712.1.10.2. Fire-resistance-rated shafts are not required for vertical duct work that is contained within and serves only the parking structure.

712.1.11 Mezzanine. Vertical openings between a mezzanine complying with Section 505 and the floor below shall be permitted.

❖ Section 712.1.11 recognizes that vertical openings between a mezzanine and the floor below are actually an advantage. The limited-size mezzanine is considered a part of the story below in accordance with Section 505.1. The openness between the mezzanine level and the floor below allows smoke or other hazards to be readily observable to occupants of both levels.

712.1.12 Exit access stairways and ramps. Vertical openings containing *exit access stairways* or *ramps* in accordance with Section 1019 shall be permitted.

❖ There are 10 separate conditions in Section 1019 that permit unenclosed exit access stairs or ramps. The vertical openings created by the unenclosed stairs that meet Section 1019 are allowed.

712.1.13 Openings. Vertical openings for floor fire doors and access doors shall be permitted where protected by Section 712.1.13.1 or 712.1.13.2.

❖ Vertical openings can be provided with fire doors in the plane of the floor. The door can be automatic closing. The floor fire door requirement is detailed in Section 712.1.13.1. These floor fire doors can be used to close a floor opening because they have a fire-resistance rating equal to that of the floor and, therefore, meet requirements that are similar to those of the floor itself. It should be noted that floor fire door assemblies must be installed in fire-resistance-rated floors and nonfire-resistance-rated floors.

712.1.13.1 Horizontal fire door assemblies. Horizontal *fire door* assemblies used to protect openings in fire-resistance-rated *horizontal assemblies* shall be tested in accordance with NFPA 288, and shall achieve a *fire-resistance rating* not less than the assembly being penetrated. Horizontal *fire door* assemblies shall be labeled by an *approved agency*. The *label* shall be permanently affixed and shall specify the manufacturer, the test standard and the *fire-resistance rating*.

❖ Horizontal fire doors are specifically addressed as an opening protective in horizontal floor assemblies. The code requires that these doors have a fire-resistance rating instead of just a fire protection rating. By requiring a fire-resistance rating, it will restore the opening in the horizontal assembly to the same level of protection that was originally established during the fire test by limiting the temperature transmission to the unexposed side of the assembly.

712.1.13.2 Access doors. Access doors shall be permitted in ceilings of fire-resistance-rated floor/ceiling and roof/ceiling assemblies, provided such doors are tested in accordance with ASTM E119 or UL 263 as horizontal assemblies and labeled by an approved agency for such purpose.

❖ Access doors are often necessary in order to service mechanical and plumbing systems above the ceiling. This section states that if such doors are used where the ceiling provides part of the protection, they must be tested in accordance with ASTM E119 or UL 263 as a horizontal assembly. This makes it clear that the standard fire test for doors (NFPA 80 or 257) is not acceptable. This ensures that the thermal transmission through the access door and its effect on the assembly is considered. The provisions of this section are not applicable if the ceiling membrane does not provide any portion of the fire-resistive protection. Therefore, in a nonrated ceiling, this access door requirement would not apply.

712.1.14 Group I-3. In Group I-3 occupancies, vertical openings shall be permitted in accordance with Section 408.5.

❖ Vertical openings within a Group I-3 housing unit are allowed under the conditions of Section 408.5.1. This is to allow tiered cells with observation by supervisory personnel.

712.1.15 Skylights. Skylights and other penetrations through a *fire-resistance-rated* roof deck or slab are permitted to be unprotected, provided that the structural integrity of the *fire-resistance-rated* roof assembly is maintained. Unprotected skylights shall not be permitted in roof assemblies required to be *fire-resistance rated* in accordance with Section 705.8.6. The supporting construction shall be protected to afford the required *fire-resistance rating* of the *horizontal assembly* supported.

❖ Fire-resistance-rated roof construction is not intended to create a barrier in order to contain the fire within the building, except for Exception 1 of Section 705.8.6 and the exception to Section 706.6.1. Non-fire-resistance-rated penetrations and skylight and roof window assemblies are, therefore, permitted to be installed in fire-resistance-rated roof assemblies, provided that the structural integrity of the roof assembly is not reduced and the provisions of Section 705.8.6 for protection of vertical exposure do not apply (see commentary, Section 705.8.6). The issue of structural integrity refers to the effect the collapse of a skylight assembly, under fire conditions, would have on the roof structure. Section 713.12, regarding the extension of a shaft to the roof level, would also permit the same unprotected openings to the exterior.

712.1.16 Openings otherwise permitted. Vertical openings shall be permitted where allowed by other sections of this code.

❖ With the 2012 code, an effort was made to include or reference all the conditions that would allow vertical openings in one section. Previous editions did not try to include vertical openings in one section; therefore, any other requirement is covered with this reference.

SECTION 713 SHAFT ENCLOSURES

713.1 General. The provisions of this section shall apply to shafts required to protect openings and penetrations through floor/ceiling and roof/ceiling assemblies. *Interior exit stairways* and *ramps* shall be enclosed in accordance with Section 1023.

❖ This section applies to vertical shafts enclosing vertical openings, such as vertical exhaust ducts; gas flues; metal chimneys; vertical supply ducts; return and outdoor air ducts; elevator hoistways, linen chutes and trash chutes. All openings or penetrations in floor/ceiling or roof/ceiling assemblies are required to be protected with a vertical shaft enclosure, except for the permitted uses in Section 712.

The purpose of shafts is to confine a fire to the floor of origin and to prevent the fire or the products of the fire (smoke, heat and hot gases) from spreading to other levels. For interior exit stairways and interior exit ramps, the requirements in Section 1023 are applicable.

713.2 Construction. Shaft enclosures shall be constructed as *fire barriers* in accordance with Section 707 or horizontal assemblies in accordance with Section 711, or both.

❖ Shafts are required to be enclosed in fire-resistance-rated fire barriers (Section 707) or a combination of fire barriers and horizontal assemblies (Section 712) where shafts are offset between stories.

713.3 Materials. The shaft enclosure shall be of materials permitted by the building type of construction.

❖ Material used for the construction of shaft walls must comply with the material requirements for partitions based on the building construction type in Section 602.

713.4 Fire-resistance rating. Shaft enclosures shall have a *fire-resistance rating* of not less than 2 hours where connecting four *stories* or more, and not less than 1 hour where connecting less than four *stories*. The number of *stories* connected by the shaft enclosure shall include any basements but not any *mezzanines*. Shaft enclosures shall have a *fire-resistance rating* not less than the floor assembly penetrated, but need not exceed 2 hours. Shaft enclosures shall meet the requirements of Section 703.2.1.

❖ The required fire-resistance rating for a shaft enclosure is related to the number of stories being connected. Please note that the shaft enclosure is required to be fire-resistance rated even though the floor may not be required to be fire-resistance rated, such as in a Type IIB, IIIB or VB construction. The fire-resistance rating of shaft enclosures must be 2 hours in buildings four or more stories in height. The fire-resistance rating of shaft enclosures that connect less than four stories is required to be only 1 hour. However, in the case of Type IA or IB construction, the shaft enclosure for a building height of three stories or less would have to be 2 hours. That is because the floors are 2-hour fire-resistance rated in those types of construction. The fire-resistance integrity of the floor construction must be maintained. The shaft wall construction shall never be allowed to be less than the fire-resistance rating of the floors being penetrated, but need not exceed 2 hours (see Commentary Figure 713.4). Another case where a two- or three-story building will require a 2-hour shaft enclosure is a mixed occupancy building using the separated option with a floor separating the occupancies rated for 2 hours.

There is one exception to the 2-hour fire-resistance rating for shaft enclosures in buildings meeting the high-rise provisions of Section 403. Section 403.2.1.2 allows a reduction to 1 hour in high-rise buildings less than 420 feet (128 m) in height for shaft enclosures other than exits and elevator hoistways where special sprinklers are installed in the shaft.

The reference to Section 703.2.1 is a 2009 revision to this section. The intent is that fire-resistance-rated shaft enclosure walls must be rated from fire exposure from both sides. That is, they must be symmetrical assemblies or assume the rating of the least-rated side. Chase walls are not to be confused with shaft walls.

713.5 Continuity. Shaft enclosures shall be constructed as *fire barriers* in accordance with Section 707 or *horizontal assemblies* constructed in accordance with Section 711, or both, and shall have continuity in accordance with Section 707.5 for *fire barriers* or Section 711.2.2 for *horizontal assemblies,* as applicable.

❖ This section references the fire barrier and horizontal assembly provisions for details on the continuity requirements. For fire barriers, the continuity section is Section 707.5, which also contains the supporting structure provisions. For horizontal assemblies, Section 711.2.2 describes the continuity provisions. See the commentary for both of those sections. Nothing in this section prohibits offsets in shaft enclosures, provided that the enclosures are continuous to the roof.

713.6 Exterior walls. Where *exterior walls* serve as a part of a required shaft enclosure, such walls shall comply with the requirements of Section 705 for *exterior walls* and the fire-resistance-rated enclosure requirements shall not apply.

Exception: Exterior walls required to be fire-resistance rated in accordance with Section 1021.2 for exterior egress balconies, Section 1023.7 for interior *exit* stairways and ramps and Section 1027.6 for exterior *exit* stairways and ramps.

❖ If an exterior wall constitutes part of a shaft enclosure, the exterior wall is only required to comply with the fire-resistance-rating requirements in Section 705 for exterior walls. The intent of the fire barrier requirements is to subdivide or enclose areas to protect them from a fire within the building. The exterior wall need only have a fire-resistance rating if required for structural stability (see Table 601) or because of an exterior exposure potential based on FSD. The exception to Section 713.6 recognizes that an exterior wall of a shaft can also be an egress balcony wall or a wall separating an exterior exit stair. Exterior balcony, exit enclosure and ramp and stair requirements are referenced by the exception as areas required to be fire-resistance rated for reasons other than structural or FSD. Section 701.2 also requires a wall serving multiple purposes to meet all the code requirements for each purpose.

713.7 Openings. Openings in a shaft enclosure shall be protected in accordance with Section 716 as required for *fire barriers*. Doors shall be self- or automatic-closing by smoke detection in accordance with Section 716.5.9.3.

❖ The integrity of the shaft enclosures must be maintained with approved opening protectives (see Section 716). An example of a protected opening is illustrated in Commentary Figure 713.7. Doors provided into shafts shall be self-closing or, if automatic closing, must be smoke activated in accordance with Section 716.5.9.3. Table 716.5 provides the fire pro-

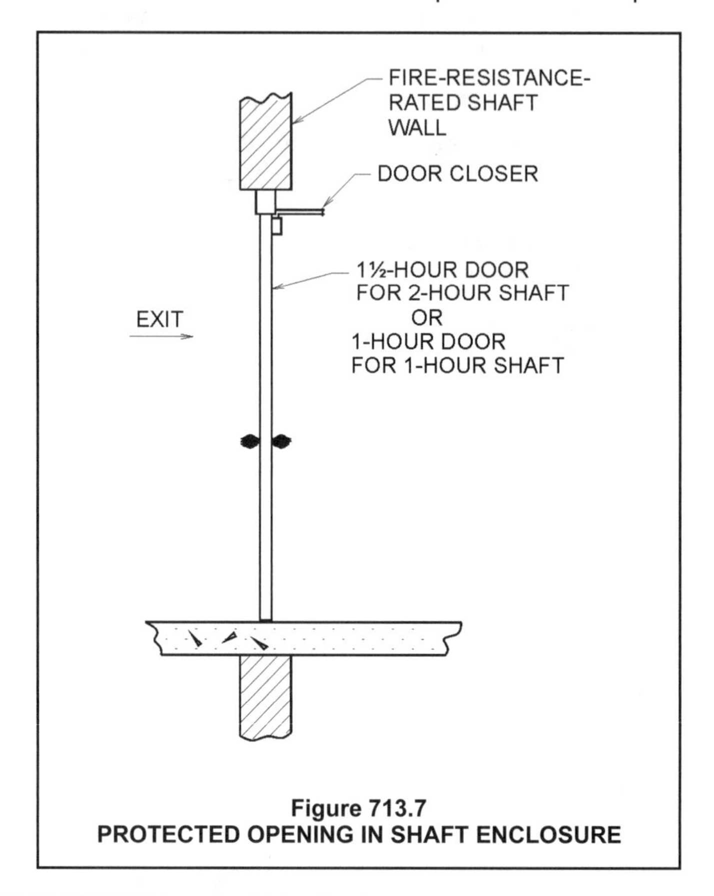

Figure 713.7
PROTECTED OPENING IN SHAFT ENCLOSURE

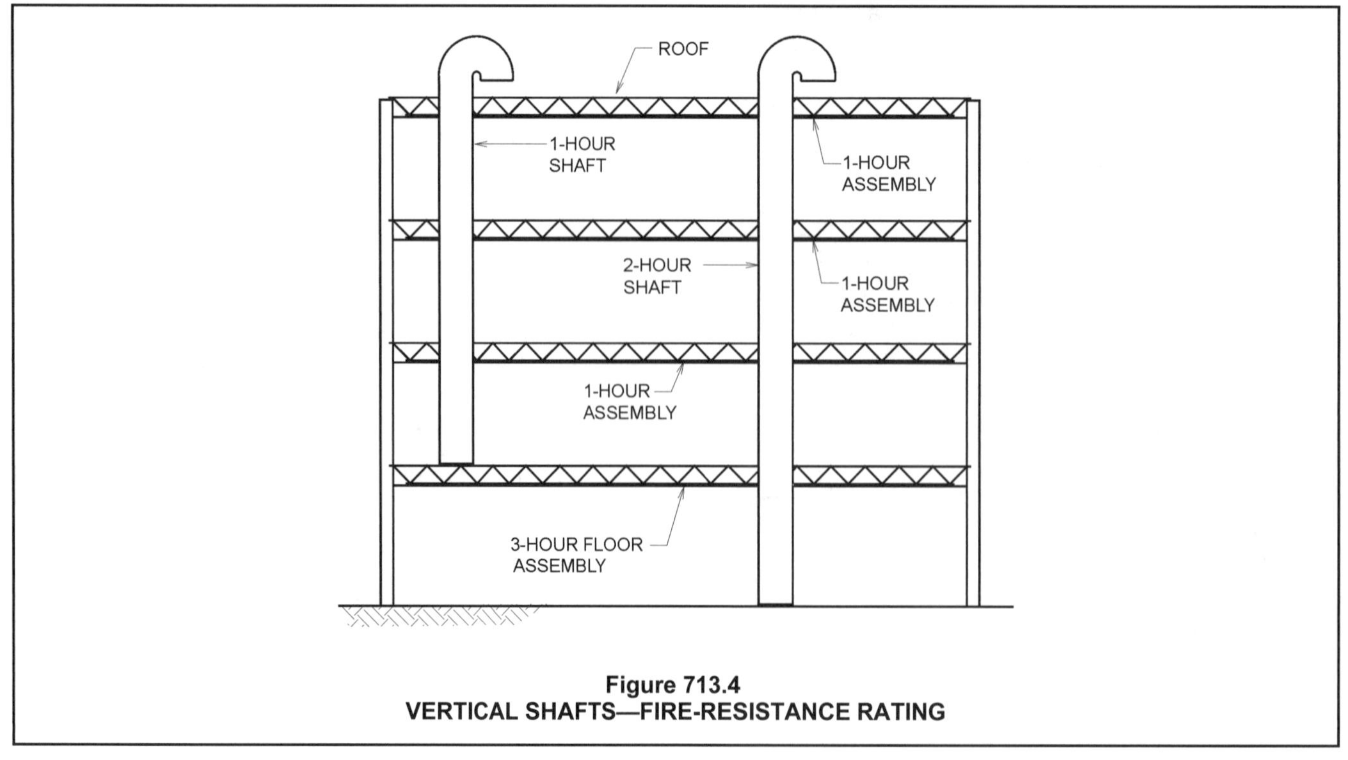

Figure 713.4
VERTICAL SHAFTS—FIRE-RESISTANCE RATING

tection rating required. There is no requirement for the doors to meet the smoke and draft control assembly testing, unless the shaft door also serves as an opening into a corridor or through a smoke barrier.

713.7.1 Prohibited openings. Openings other than those necessary for the purpose of the shaft shall not be permitted in shaft enclosures.

❖ In fire barrier walls forming shaft enclosures, openings are limited to only those necessary for the shaft to serve its intended purpose.

713.8 Penetrations. Penetrations in a shaft enclosure shall be protected in accordance with Section 714 as required for *fire barriers*. Structural elements, such as beams or joists, where protected in accordance with Section 714 shall be permitted to penetrate a shaft enclosure.

❖ Penetrations for items such as cables, cable trays, conduits, ducts and piping are required to be tested in accordance with ASTM E814 or UL 1479 with some minor exceptions (see Section 714.3.1). The change in the 2012 code permits structural members to penetrate shaft walls if the structural members have been tested in the assembly and meet ASTM E814 and UL 1479. This change is only for shafts in Section 713 and does not address stair shafts in Sections 1009 and 1022. Neither does this change affect penetrations of fire barriers in general.

713.8.1 Prohibited penetrations. Penetrations other than those necessary for the purpose of the shaft shall not be permitted in shaft enclosures.

❖ The provisions of Section 713.8.1 mirror those of Section 713.7.1 for openings. Because a shaft provides a potential method for a fire or the products of combustion to spread throughout a building, especially when driven due to the stack effect, the code restricts penetrations so that the effectiveness of the shaft is maintained. For instance, horizontal plumbing piping cannot penetrate the walls of a shaft serving vertical air ducts.

713.9 Joints. Joints in a shaft enclosure shall comply with Section 715.

❖ See the commentary to Section 715.

713.10 Duct and air transfer openings. Penetrations of a shaft enclosure by ducts and air transfer openings shall comply with Section 717.

❖ These provisions reference the applicable protection features of Section 717 for duct openings. Since a duct is also a penetration, the limitations of Section 713.8.1 are applicable. For exit enclosures, it is important to note that duct openings are limited to those allowed in Section 1023.4. As stated in Section 717.1.2, if a duct is exempted from the need for a damper, the penetration of that duct through a fire-resistance-rated assembly must be protected as a through penetration in accordance with Section 714. Section 717.5.3 requires shaft enclosure penetrations by ducts and air transfer openings to be protected by both fire dampers and smoke dampers. For the exceptions, see the commentary to Section 717.5.3

713.11 Enclosure at the bottom. Shafts that do not extend to the bottom of the building or structure shall comply with one of the following:

1. They shall be enclosed at the lowest level with construction of the same *fire-resistance rating* as the lowest floor through which the shaft passes, but not less than the rating required for the shaft enclosure.
2. They shall terminate in a room having a use related to the purpose of the shaft. The room shall be separated from the remainder of the building by *fire barriers* constructed in accordance with Section 707 or *horizontal assemblies* constructed in accordance with Section 711, or both. The *fire-resistance rating* and opening protectives shall be not less than the protection required for the shaft enclosure.
3. They shall be protected by *approved fire dampers* installed in accordance with their listing at the lowest floor level within the shaft enclosure.

Exceptions:

1. The fire-resistance-rated room separation is not required, provided there are no openings in or penetrations of the shaft enclosure to the interior of the building except at the bottom. The bottom of the shaft shall be closed off around the penetrating items with materials permitted by Section 718.3.1 for draftstopping, or the room shall be provided with an *approved automatic sprinkler system*.
2. A shaft enclosure containing a waste or linen chute shall not be used for any other purpose and shall discharge in a room protected in accordance with Section 713.13.4.
3. The fire-resistance-rated room separation and the protection at the bottom of the shaft are not required provided there are no combustibles in the shaft and there are no openings or other penetrations through the shaft enclosure to the interior of the building.

❖ In order to limit the spread of fire from story to story through vertical shafts, an enclosure is required throughout the length of the shaft, including the bottom, if the shaft does not extend to the bottom floor of the building. This section gives three separate methods of providing protection for the bottom of shafts that are not at the bottom of the building.

Method 1 is shown in Commentary Figure 713.11(1). Please note that the bottom of the shaft is supported by a horizontal assembly meeting Section 711. However, if the horizontal structural members are primary structural frames and not secondary structural members, and the structural floor member supports shaft walls three stories or more in height, then individual encasement is required (see Section 704.3).

Method 2 is illustrated in Commentary Figure 713.11(2). This figure shows a 1-hour fire-resistance-rated shaft, but this method is applicable to 2-hour fire-resistance-rated shafts as well, although the room separation would be 2 hours. This method recognizes that the purpose of some shafts cannot be accomplished if the bottom must be enclosed as in Method 1. Therefore, a room is permitted at the bottom of the shaft, but the function of the room must be related to the purpose of the shaft. Two common examples of this situation are either mechanical rooms that have vents from the room to the roof or refuse chutes that terminate at a room at the bottom for collection. Exceptions 1 and 2 to this section relate to this method.

Method 3 is similar to Method 1, but allows a fire damper to be provided at the bottom of the enclosure. Please note that this method predates the requirement for both smoke and fire dampers in ducts penetrating shaft walls (Section 717.5.3). The code user must comply with both code sections. Therefore, the code user must provide both fire and smoke dampers with this method unless the smoke damper is exempted by Section 717.5.3.

Exception 1 is illustrated by both Commentary Figures 713.11(3) and 713.11(4). This is an exception to Method 2 and eliminates the need for the room to be

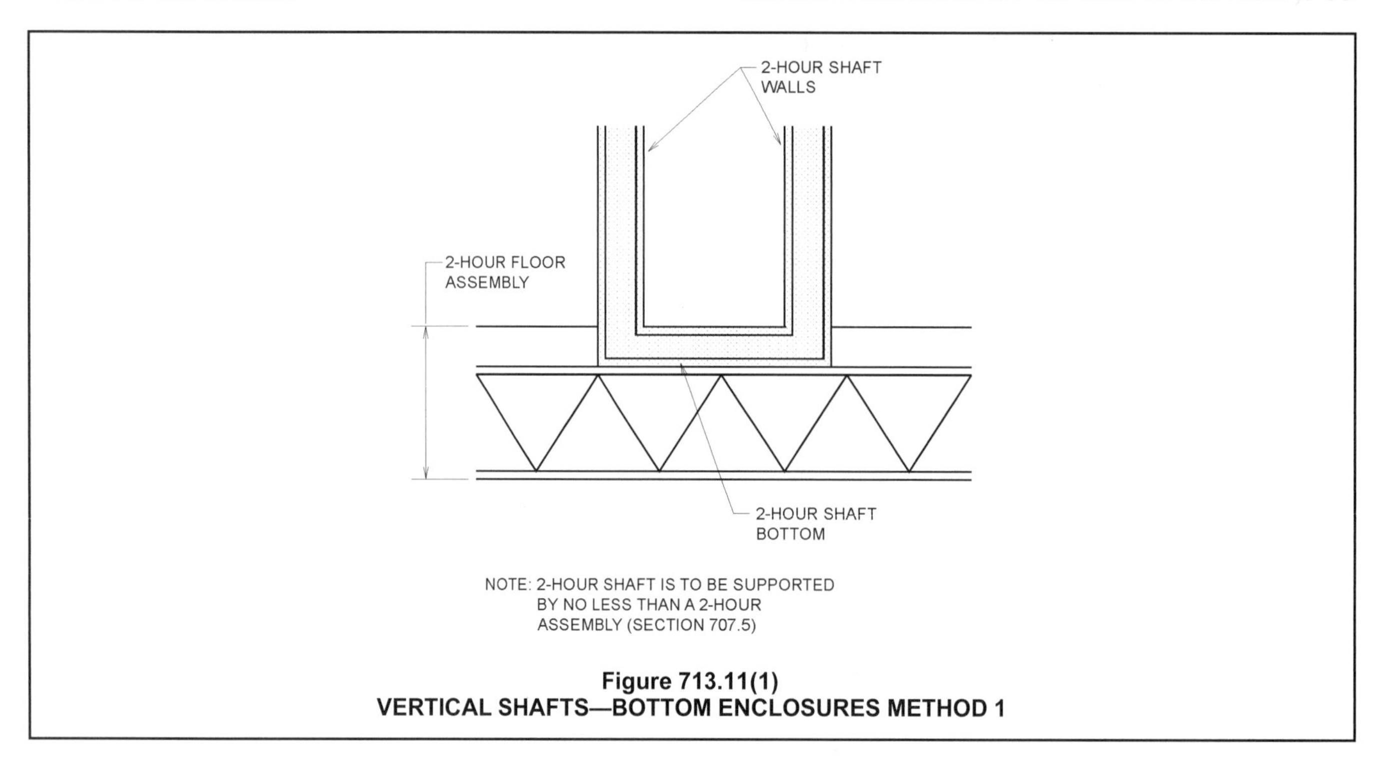

Figure 713.11(1)
VERTICAL SHAFTS—BOTTOM ENCLOSURES METHOD 1

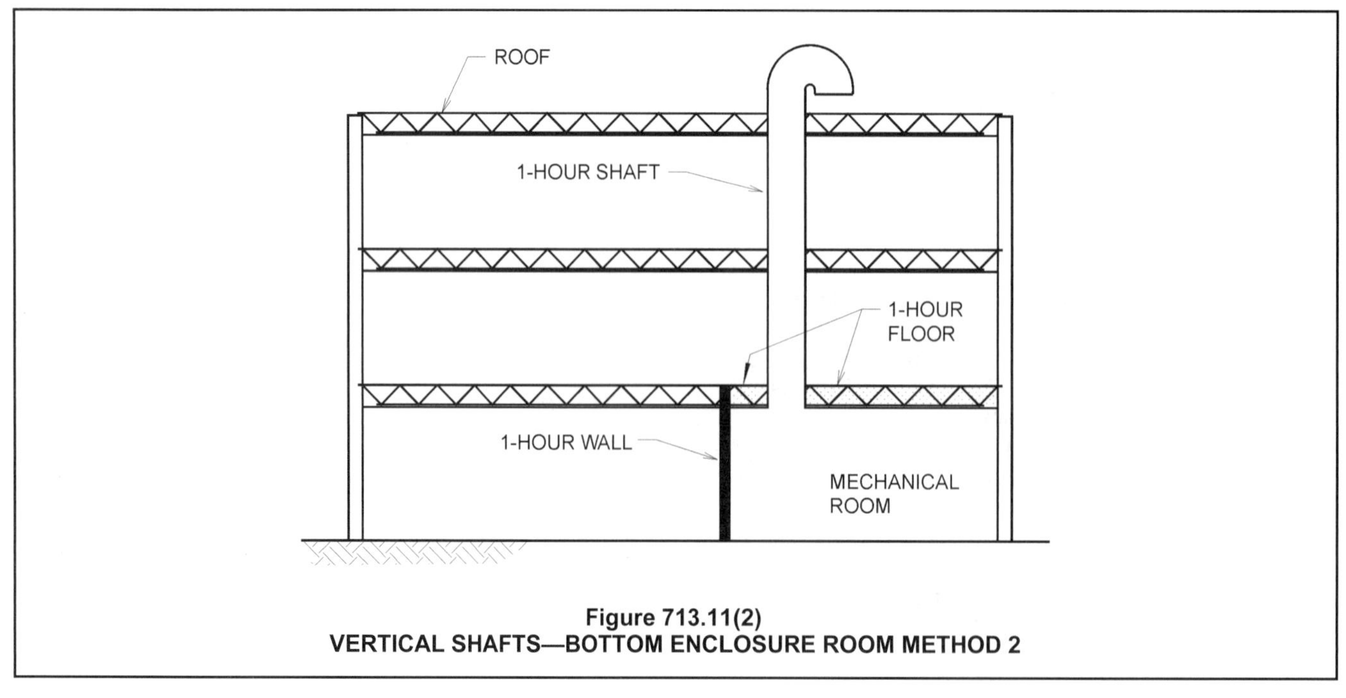

Figure 713.11(2)
VERTICAL SHAFTS—BOTTOM ENCLOSURE ROOM METHOD 2

separated. With this exception there cannot be any openings or penetrations into the shaft enclosures anywhere except at the bottom and the top. A gas-fired appliance flue and a commercial kitchen exhaust duct are good examples.

Exception 2 is specific to waste or linen chute termination rooms and requires compliance with Section 713.13.4.

Exception 3 is illustrated by Commentary Figure 713.11(5). This is an exception to the entire section and to all three methods. While the figure shows a very small light well shaft in a three-story building, this exception is not limited to three stories and the size of the shaft is not limited. The area and volume of the shaft can be considered as part of the story in which the bottom of the shaft terminates. The conditions of the exception are based on the fact that the shaft contents will not contribute to the fuel load and that the shaft serves and connects to only one story—the one in which it terminates. This is similar to an atrium with no openings except in the first story. Unlike an atrium, which is exempted from shaft enclosures, this actually is a shaft with shaft wall construction on all stories except the first.

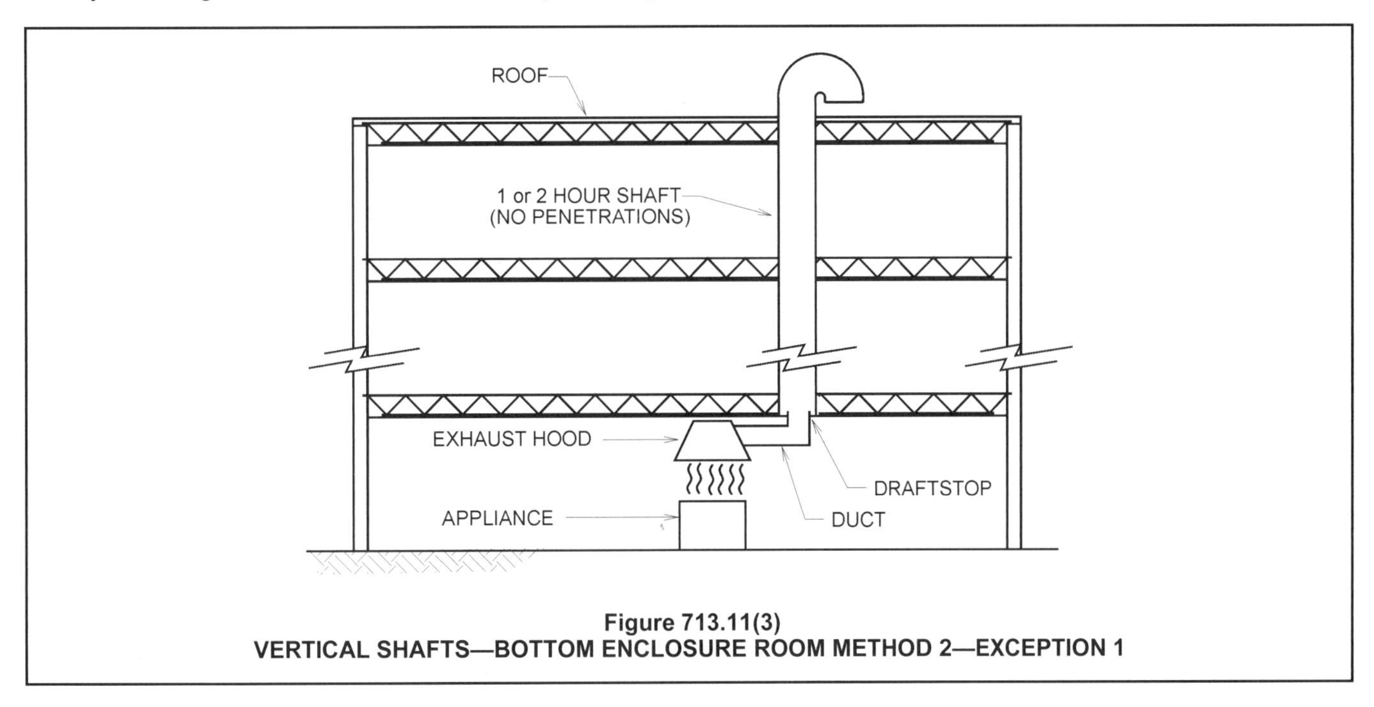

Figure 713.11(3)
VERTICAL SHAFTS—BOTTOM ENCLOSURE ROOM METHOD 2—EXCEPTION 1

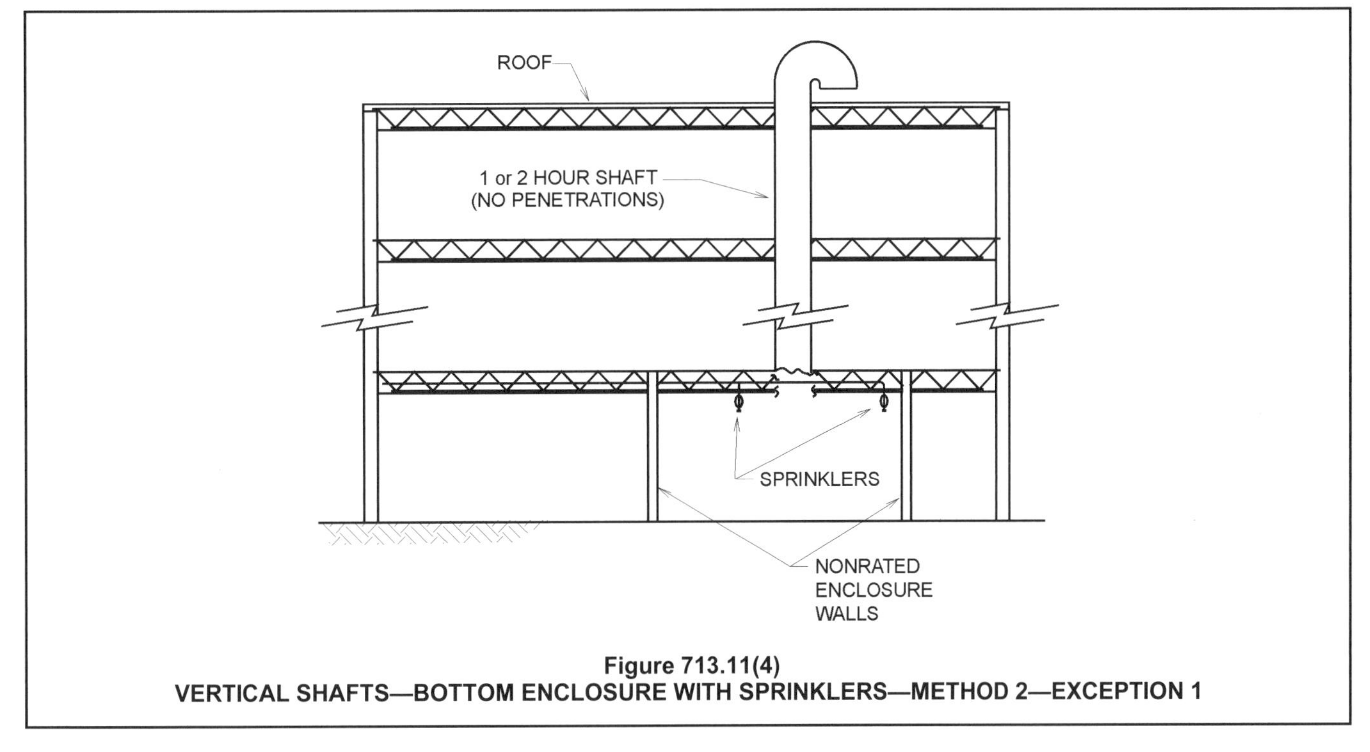

Figure 713.11(4)
VERTICAL SHAFTS—BOTTOM ENCLOSURE WITH SPRINKLERS—METHOD 2—EXCEPTION 1

713.12 Enclosure at top. A shaft enclosure that does not extend to the underside of the roof sheathing, deck or slab of the building shall be enclosed at the top with construction of the same *fire-resistance rating* as the topmost floor penetrated by the shaft, but not less than the *fire-resistance rating* required for the shaft enclosure.

❖ Proper shaft enclosures must include all sides and the top unless the top of the shaft is also the roof of the building. Because the purpose of the shaft is to limit the spread of fire within the building, if the top of the shaft does not extend to or through the underside of the roof sheathing, deck or slab, then the code requires a fire-resistance-rated horizontal assembly at the top. The fire-resistance rating for the top of the shaft must not be less than the required fire resistance of the shaft enclosure or the fire-resistance rating of the highest floor penetrated (see Commentary Figure 713.12). The required rating for the top of the shaft that extends to the sheathing or roof deck must be consistent with the requirements of Table 601 and Section 711 for roof construction. The top of the shaft

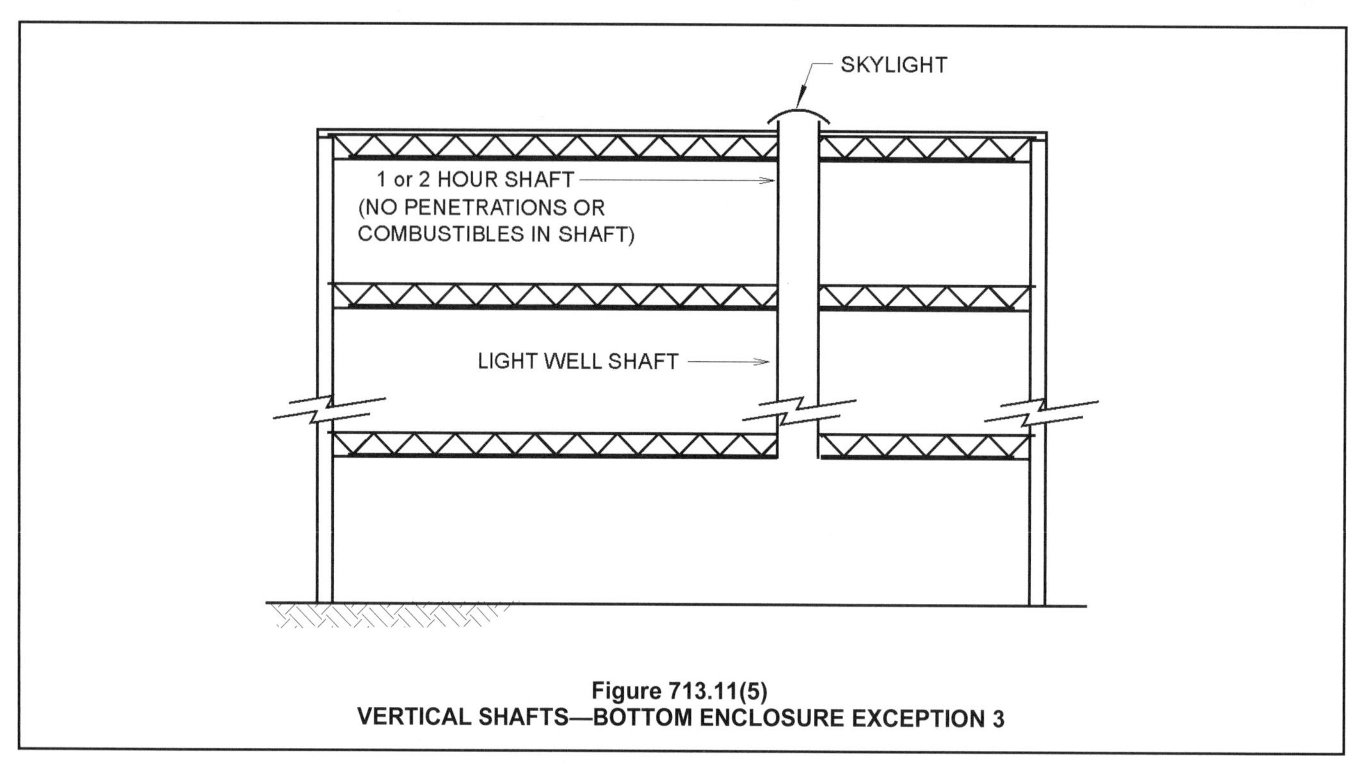

Figure 713.11(5)
VERTICAL SHAFTS—BOTTOM ENCLOSURE EXCEPTION 3

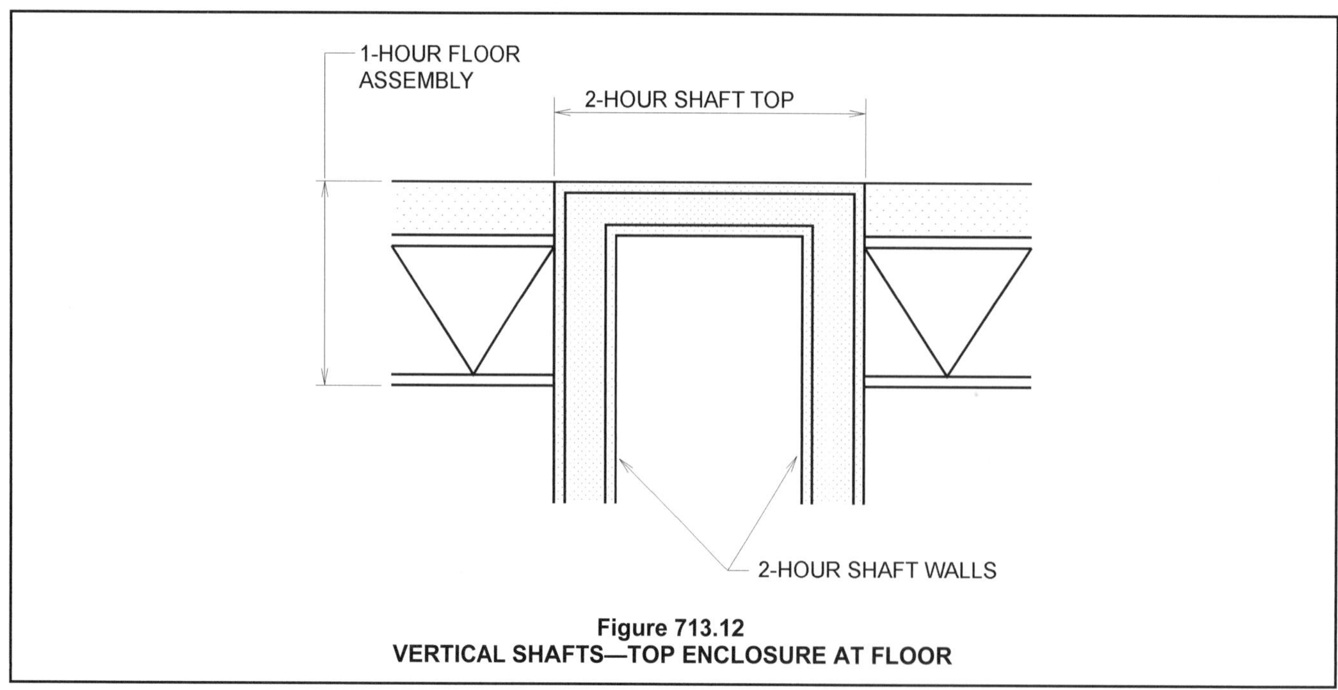

Figure 713.12
VERTICAL SHAFTS—TOP ENCLOSURE AT FLOOR

must be constructed using a horizontal assembly (see Section 711) with the proper fire-resistance rating. It is not permissible to simply take a fire barrier, such as the assembly used for the shaft wall, and turn it horizontally.

The code is silent on whether a fire damper is appropriate for use to enclose the top of a shaft. Fire dampers are expressly permitted to be used at the bottom of a fire-resistance-rated shaft in Section 713.11. The damper listing should be reviewed before any damper is installed at this location.

713.13 Waste and linen chutes and incinerator rooms. Waste and linen chutes shall comply with the provisions of NFPA 82, Chapter 5 and shall meet the requirements of Sections 713.13.1 through 713.13.6. Incinerator rooms shall meet the provisions of Sections 713.13.4 through 713.13.5.

Exception: Chutes serving and contained within a single dwelling unit.

❖ Waste- and linen-handling systems represent locations that provide rapid fire spread potential. The emphasis on the vertical chute systems in this section is because such systems interconnect multiple stories and can contain combustible material. Additionally, these systems all too often receive an ignition source, such as discarded chemicals and materials capable of producing spontaneous combustion, smoking materials that have not been extinguished and hot ashes, embers and charcoal. Such systems are also subject to acts of vandalism. The construction, fire suppression and termination requirements of waste- and linen-handling systems are interrelated in other code sections. The exception recognizes the small size, limited use and occupant control of refuse and laundry chutes commonly installed within dwelling units.

713.13.1 Waste and linen. A shaft enclosure containing a recycling, or waste or linen chute shall not be used for any other purpose and shall be enclosed in accordance with Section 713.4. Openings into the shaft, from access rooms and discharge rooms, shall be protected in accordance with this section and Section 716. Openings into chutes shall not be located in *corridors*. Doors into chutes shall be self-closing. Discharge doors shall be self- or automatic-closing upon the actuation of a smoke detector in accordance with Section 716.5.9.3, except that heat-activated closing devices shall be permitted between the shaft and the discharge room.

❖ This section requires that vertical systems or chutes be enclosed in a fire-resistance-rated shaft in accordance with Section 713.4. As with all shafts, the fire-resistance integrity of the shaft must be protected at all penetrations of the shaft enclosure. All access openings and disharge points must be treated as shaft enclosure penetrations requiring opening protectives. Waste and linen access openings are prohibited from being located in corridors or exits. While a corridor is a very convenient location for access to these chutes, in order to access the chute from a corridor, an intervening room (see Section 713.13.3) with an additional fire-rated opening protective provides redundancy that is justified by fire experience.

713.13.2 Materials. A shaft enclosure containing a waste, recycling, or linen chute shall be constructed of materials as permitted by the building type of construction.

❖ The types of materials used in shaft enclosures are to be consistent with Sections 602 through 602.5 for the type of construction of the building. Fire barriers or horizontal assemblies that form shaft walls are permitted to be of combustible construction in Type III, IV and V construction and are required to be of noncombustible materials in Type I and II construction.

713.13.3 Chute access rooms. Access openings for waste or linen chutes shall be located in rooms or compartments enclosed by not less than 1-hour *fire barriers* constructed in accordance with Section 707 or *horizontal assemblies* constructed in accordance with Section 711, or both. Openings into the access rooms shall be protected by opening protectives having a *fire protection rating* of not less than $^3/_4$ hour. Doors shall be self- or automatic-closing upon the detection of smoke in accordance with Section 716.5.9.3.

❖ The chutes themselves must have a rated fire door with closers in accordance with Section 713.13.1; therefore, the fire door in the shaft wall must be rated for 60 minutes or 90 minutes depending on the height of the shaft enclosure. In addition to this protection, such openings into the chute shaft must be located in a room or compartment. This room or compartment must be enclosed by fire barrier walls (Section 707) and horizontal assemblies (Section 711) with fire-resistance ratings of 1 hour. The door openings into the access room shall be fire-resistance rated for 45 minutes (see Commentary Figure 713.13.3). This requirement is necessary to provide an added measure of protection should the access door to the shaft fail to close due to waste being stuck in the door and because these rooms are a catch-all for combustible material.

713.13.4 Chute discharge room. Waste or linen chutes shall discharge into an enclosed room separated by *fire barriers* with a *fire-resistance rating* not less than the required fire rating of the shaft enclosure and constructed in accordance with Section 707 or *horizontal assemblies* constructed in accordance with Section 711, or both. Openings into the discharge room from the remainder of the building shall be protected by opening protectives having a *fire protection rating* equal to the protection required for the shaft enclosure. Doors shall be self- or automatic-closing upon the detection of smoke in accordance with Section 716.5.9.3. Waste chutes shall not terminate in an incinerator room. Waste and linen rooms that are not provided with chutes need only comply with Table 509.

❖ The chute discharge rooms, like the access rooms, are a repository of combustible material, and thus must be separated from all other parts of the building by fire barriers (Section 707) and horizontal assemblies (Section 711). The fire-resistance ratings are the same as for access rooms. See also Exception 2 of Section 713.11, which allows the shaft to discharge into such a room. The incidental use separation requirements for waste and linen separation rooms in Table 509 are only applicable to collection rooms not connected to chutes. The discharge room must never be connected to an incinerator room.

713.13.5 Incinerator room. Incinerator rooms shall comply with Table 509.

❖ This section refers to the incidental use separation table. Table 509 requires a 2-hour fire-resistance enclosure for such rooms and sprinkler protection.

713.13.6 Automatic sprinkler system. An *approved automatic sprinkler system* shall be installed in accordance with Section 903.2.11.2.

❖ In addition to meeting the requirements for shaft enclosures, laundry and refuse chutes must be provided with automatic sprinkler coverage. Most buildings with chutes will need to be fully sprinklered by other sections of the code. The chutes must have additional heads at the top, in the termination room and in the chute at every other floor level. The need for the sprinkler protection addresses the potential that a fire can occur in the chute or in the container below the chute.

713.14 Elevator, dumbwaiter and other hoistways. Elevator, dumbwaiter and other hoistway enclosures shall be constructed in accordance with Section 713 and Chapter 30.

❖ The hoistway enclosure is the fixed structure consisting of vertical walls or partitions that isolates the enclosure from all other building areas or from an adjacent enclosure in which the hoistway doors and door assemblies are installed. With the exception of observation elevators (usually in atriums), the hoistway is normally enclosed with fire barriers (see Section 707.3.1). Elevator hoistways in all parking structures that serve only the parking structure are exempt from enclosures by Section 712.1.15. In addition, shaft enclosures are not required for elevators located in an atrium since the hoistways are not concealed and there is no penetration of floor assemblies. Elevator hoistways are enclosed to ensure that flame, smoke and hot gases from a fire do not have an avenue of travel from one floor to another through a concealed space (see the discussion of stack effect in the commentary to Section 708.14.1). Enclosures are also provided to restrict contact with moving equipment and to prevent people from falling.

SECTION 714 PENETRATIONS

714.1 Scope. The provisions of this section shall govern the materials and methods of construction used to protect *through penetrations* and *membrane penetrations* of *horizontal assemblies* and fire-resistance-rated wall assemblies.

❖ This section addresses the specific requirements for maintaining the integrity of fire-resistance-rated assemblies at penetrations. The provisions of this section apply to penetrations of fire-resistance-rated walls (see Section 714.3), fire-resistance-rated horizontal assemblies (see Section 714.4.1) and nonfire-resistance-rated assemblies (see Section 714.5). Penetrations of fire-resistance-rated assemblies range from combustible pipe and tubing to noncombustible wiring with combustible covering to noncombustible items such as pipe, tube, conduit and ductwork.

Each type of penetration requires a specific method of protection, which is based on the type of fire-resistance-rated assembly that is penetrated and the type of penetrating item. To determine the type of

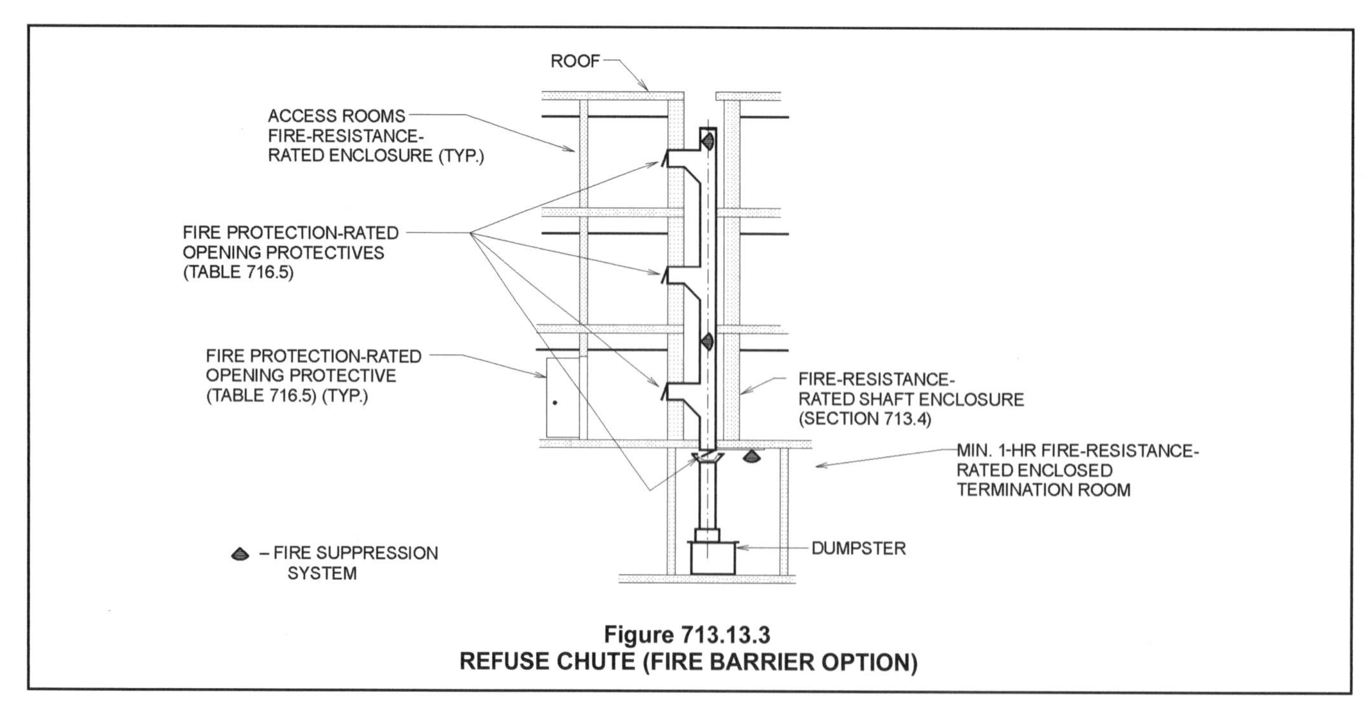

Figure 713.13.3
REFUSE CHUTE (FIRE BARRIER OPTION)

penetration protection required, the first step is to identify whether the penetrated assembly is a fire-resistance-rated wall, a fire-resistance-rated horizontal assembly or a nonfire-resistance-rated horizontal assembly. Next, identify the type of penetrating item in the applicable section and determine the applicable method of penetration protection necessary.

714.1.1 Ducts and air transfer openings. Penetrations of fire-resistance-rated walls by ducts that are not protected with *dampers* shall comply with Sections 714.2 through 714.3.3. Penetrations of *horizontal assemblies* not protected with a shaft as permitted by Section 717.6, and not required to be protected with fire *dampers* by other sections of this code, shall comply with Sections 714.4 through 714.5.2. Ducts and air transfer openings that are protected with *dampers* shall comply with Section 717.

❖ Duct penetrations are typically protected with fire and smoke dampers in accordance with Section 717.5. However, if the code does not require them and the designer does not provide such dampers, as it may interfere with the system design, Section 714.1.1 allows for such removal, provided the penetration is protected as part of a tested assembly (see Section 714.3.1.1) or is protected with some type of through-penetration firestop system (see Section 714.3.1.2).

This section coordinates with Section 717.1.2, which states the appropriate section to go to in order to determine the proper protection for ducts when they penetrate a fire-resistance-rated wall assembly. Where ducts or air transfer openings penetrate a fire-resistance-rated wall assembly, fire dampers can be installed to maintain the rating of the assembly as required by Section 717. However, in addition to maintaining the fire-resistant integrity of the floor, wall or roof/ceiling assembly, the damper also serves an additional function as an active component of the fire management design. Unlike through-penetration firestop systems, fire dampers can restrict the spread of fire through the air duct system within a building to areas remote from the fire or into a building from the outside. In addition, the air duct system can be used for the purpose of emergency smoke control, either manually or automatically. Although fire dampers and through-penetration firestop systems both are intended to maintain the rating of the assembly at the point where an opening occurs for the ventilation system, they are not optional or equivalent alternatives for one another.

The placement (or the elimination) of fire dampers within the ventilation system is based on additional factors other than the point at which the duct penetrates a fire-resistance-rated assembly.

Other sections of the code may permit fire dampers to be eliminated due to the installation of an automatic sprinkler system or the design may not include a fire damper to interface with the ventilation system and fire management design at specific locations. This section addresses the protection of the fire rating of the assembly when such conditions occur and provides that the duct penetration is protected as part of a tested assembly (see Section 714.3.1.1) or is protected with some type of through-penetration firestop system (see Section 714.3.1.1), tested and listed for use on ducts without fire dampers.

714.2 Installation details. Where sleeves are used, they shall be securely fastened to the assembly penetrated. The space between the item contained in the sleeve and the sleeve itself and any space between the sleeve and the assembly penetrated shall be protected in accordance with this section. Insulation and coverings on or in the penetrating item shall not penetrate the assembly unless the specific material used has been tested as part of the assembly in accordance with this section.

❖ Sleeves are typically installed when the opening penetrates an assembly with interior voids, such as a steel stud-framed wall. The sleeve must be securely fastened to the penetrated assembly to prevent the sleeve from becoming dislodged and adversely affecting the performance of the annular space protection. The space between the sleeve and the penetrating item, as well as between the sleeve and the rated assembly, must be protected in accordance with Section 714.3.1. Without attention to these conditions, a sleeved penetration may not perform as required by the code. A sleeve may also exist when a segment of plastic or metal pipe or tube is cast into a concrete floor, thus creating the needed opening for an anticipated penetration. In this case, the need for secure fastening to the underlying assembly is essentially guaranteed, and there would be no space needing to be sealed between the sleeve and the assembly being penetrated.

Section 714.3.1 requires that combustible penetrations be protected with assemblies that have been tested in accordance with ASTM E119 or UL 263 (see Section 714.3.1.1), or ASTM E814 or UL 1479 (see Section 714.3.1.2). If a sleeve is used in conjunction with the penetrating items, then the annular space within and around the sleeve must also be protected with materials that meet the conditions of the test standards. The purpose of this section is to provide the building official with the necessary text to enforce the conditions of the penetration protection standards.

If combustible insulation material or coverings on the penetrating item could provide a path for the line of fire to travel through the assembly, the method of protection must have also been fire tested using that specific type and thickness of insulation. This section clarifies that either the insulation or covering must be part of the tested assembly, whether the assembly was tested under ASTM E119 or UL 263 or is a through-penetration firestop system tested in accordance with ASTM E814 or UL 1479, or else the insulation or covering is not allowed to pass through the penetration. In the latter case, the insulation or covering should be removed from the penetrated item at the point of penetration through the fire-resistance-rated assembly, assuming that the removal would not cause other problems and that the removal would be compliant with other applicable codes [e.g., IMC or *International Plumbing Code®* (IPC®)].

714.3 Fire-resistance-rated walls. Penetrations into or through *fire walls*, *fire barriers*, *smoke barrier* walls and *fire partitions* shall comply with Sections 714.3.1 through 714.3.3. Penetrations in *smoke barrier* walls shall also comply with Section 714.4.4.

❖ In order to maintain the integrity of the fire-resistance-rated wall assembly (fire walls, fire barriers, smoke barriers and fire partitions), penetrations into and through the rated assembly must be properly protected. Acceptable protection methods for various penetrations of wall assemblies are identified in Sections 714.3.1 through 714.3.3. Additionally, penetrations of smoke barriers are required to meet the test requirements for air leakage given in Section 714.4.4.

714.3.1 Through penetrations. Through penetrations of fire-resistance-rated walls shall comply with Section 714.3.1.1 or 714.3.1.2.

Exception: Where the penetrating items are steel, ferrous or copper pipes, tubes or conduits, the *annular space* between the penetrating item and the fire-resistance-rated wall is permitted to be protected by either of the following measures:

1. In concrete or masonry walls where the penetrating item is a maximum 6-inch (152 mm) nominal diameter and the area of the opening through the wall does not exceed 144 square inches (0.0929 m^2), concrete, grout or mortar is permitted where installed the full thickness of the wall or the thickness required to maintain the *fire-resistance rating*.
2. The material used to fill the *annular space* shall prevent the passage of flame and hot gases sufficient to ignite cotton waste when subjected to ASTM E119 or UL 263 time-temperature fire conditions under a minimum positive pressure differential of 0.01 inch (2.49 Pa) of water at the location of the penetration for the time period equivalent to the *fire-resistance rating* of the construction penetrated.

❖ Combustible cables, wires, pipes, tubes and conduits that penetrate fire-resistance-rated walls are required to be properly protected [see Commentary Figures 714.3.1(1) and 714.3.1(2)]. These penetrations are required to be tested in accordance with ASTM E119 or UL 263 (see Section 714.3.1.1) or ASTM E814 or UL 1479 (see Section 714.3.1.2).

Because combustible materials have a greater propensity to spread fire through a penetration, the requirements for combustible penetrating items are considerably more restrictive than for noncombustible penetrating items, to which the exception can apply. Penetration of fire-resistance-rated walls by cables, wires, pipes, tubes, conduits and vents represents a weak link in the continuity of the required fire-resistance rating. These penetrations are to be protected with materials that will maintain the integrity of the wall for the duration of the required fire-resistance rating. In lieu of a system tested in accordance with ASTM E119 or UL 263 (see Section 714.3.1.1), or ASTM E814 or UL 1479 (see Section 713.3.1.2), penetrations with steel, ferrous or copper pipes, tubes or conduits may be protected by filling the annular space. Annular space protection is permitted for the protection of penetrations of wall assemblies by only the mentioned noncombustible items. "Annular space" is defined in Section 202 as the perimeter space between the penetrating item and the rated assembly. If sleeves are used, the annular space includes the space between the penetrating item and the sleeve, as well as the space between the sleeve and the rated assembly [see Commentary Figure 714.3.1(3)]. Item 1 of the exception permits the use of

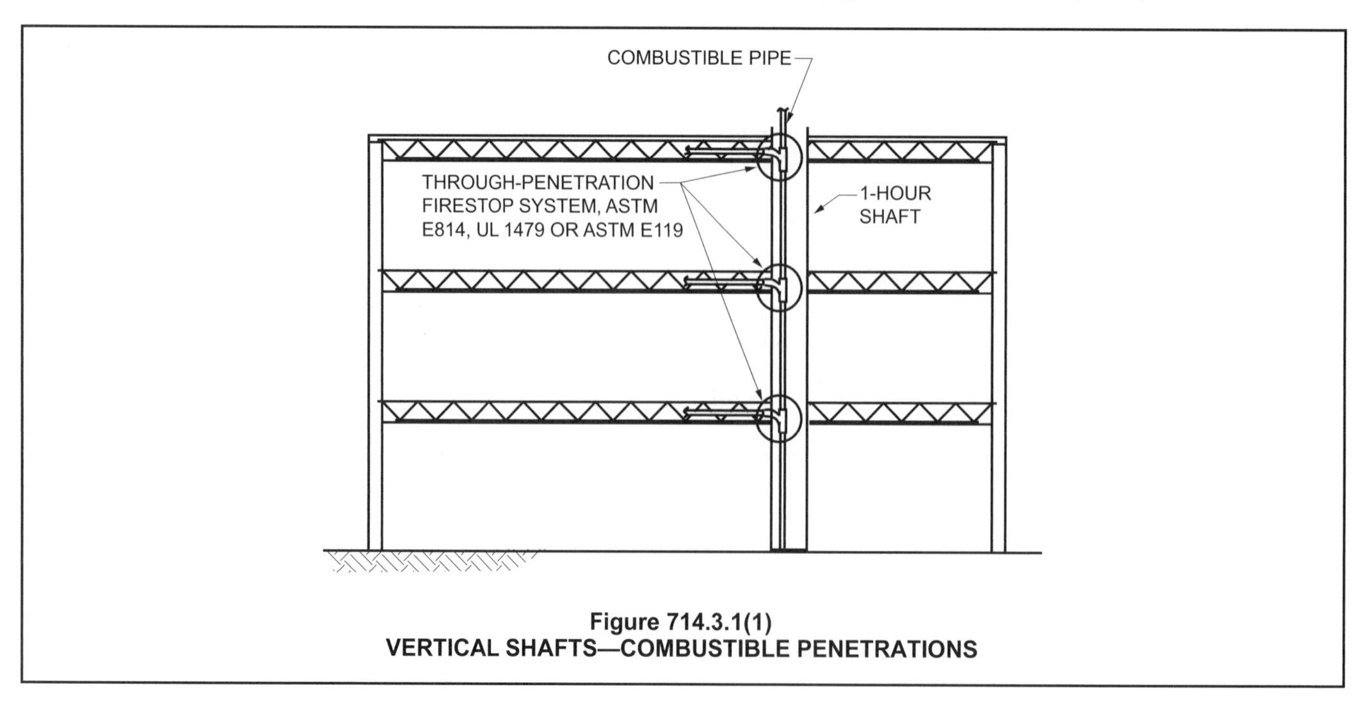

Figure 714.3.1(1)
VERTICAL SHAFTS—COMBUSTIBLE PENETRATIONS

concrete, grout or mortar to provide annular space protection for certain penetrations of concrete and masonry wall assemblies. The concrete, grout or mortar must be provided for the full thickness of the wall, unless evidence can be provided that demonstrates that the required fire-resistance rating can be achieved with a lesser depth.

Concrete, grout and mortar have traditionally been

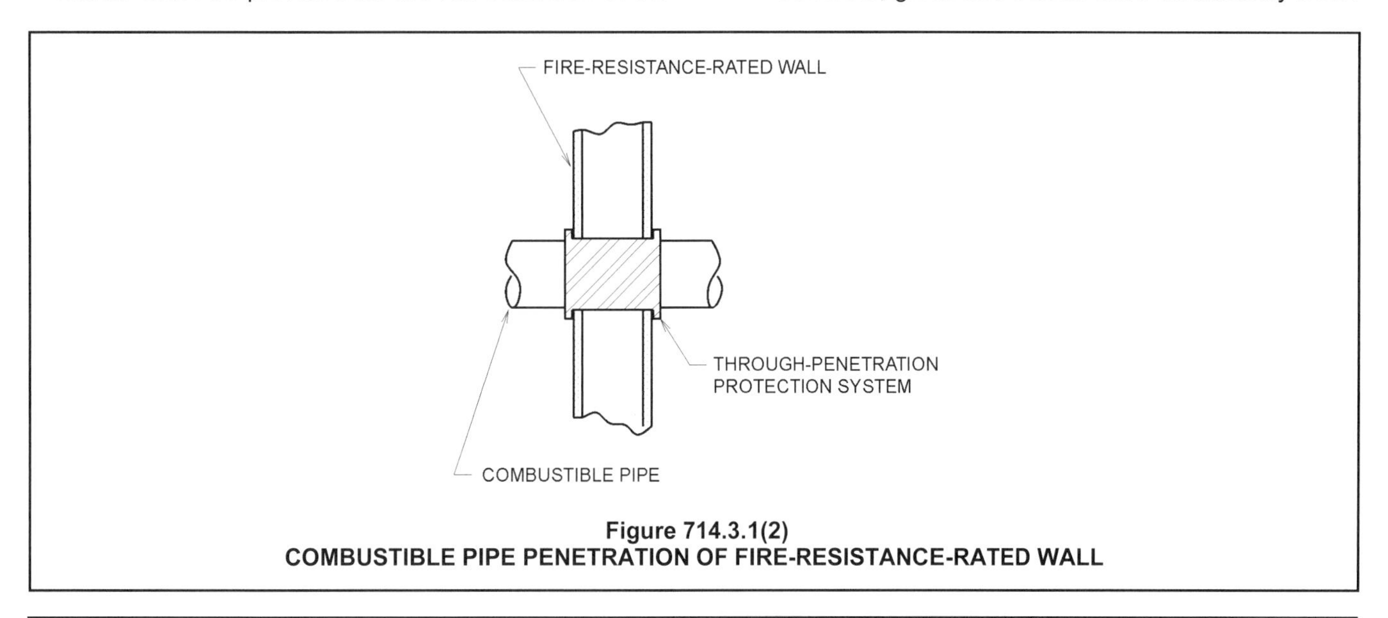

Figure 714.3.1(2)
COMBUSTIBLE PIPE PENETRATION OF FIRE-RESISTANCE-RATED WALL

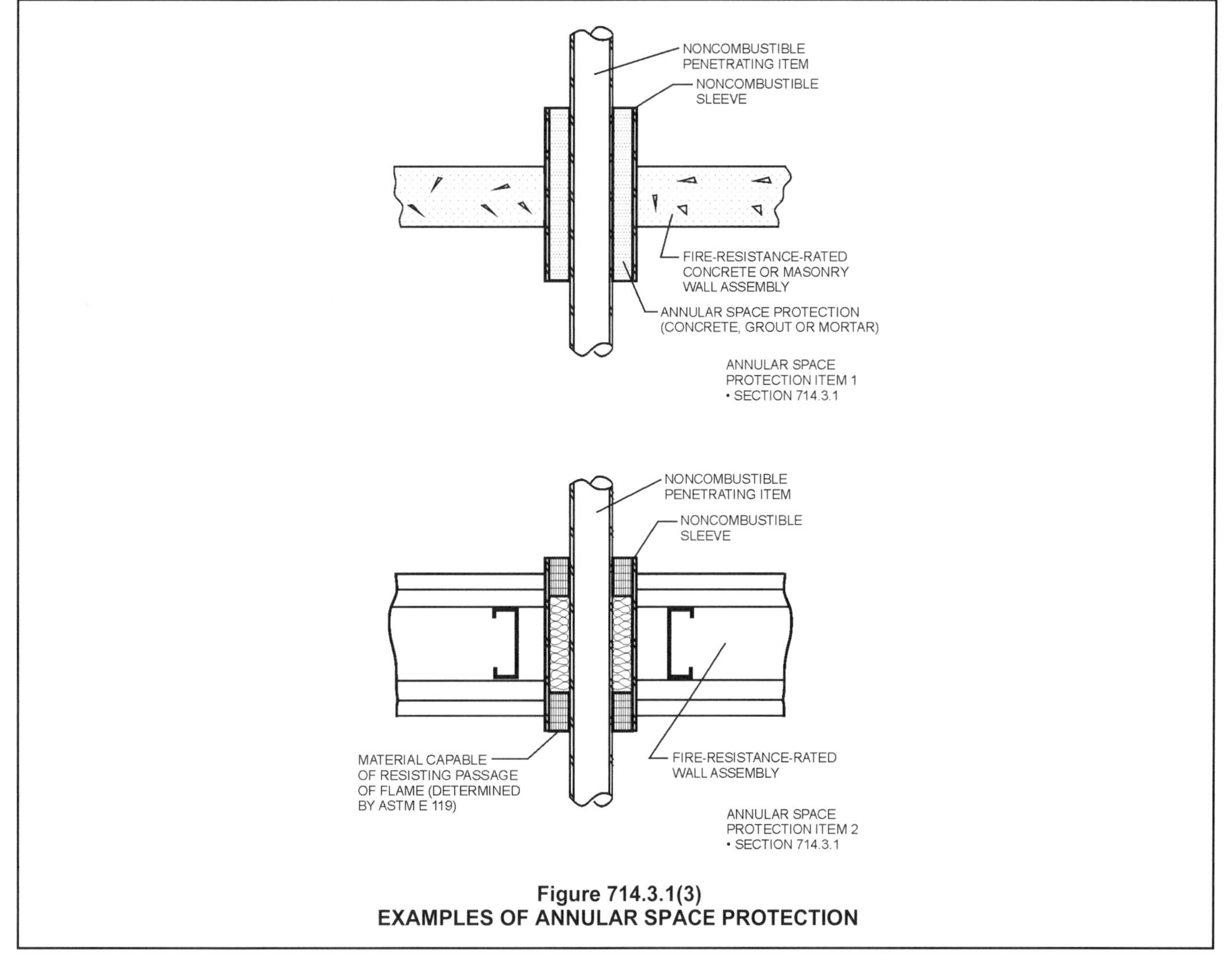

Figure 714.3.1(3)
EXAMPLES OF ANNULAR SPACE PROTECTION

used as protection for the annular space in penetrations of concrete and masonry walls. Experience has shown this form of protection to be viable.

Under Item 2 of the exception, because the penetrating items are limited and the annular space protection is not based on specific performance testing of each penetration arrangement and detail, the performance of the protection is dependent on the annular space protection material alone. As such, for walls other than concrete or masonry, Item 2 requires that the material be prequalified as to its ability to prevent the passage of flame and hot gases sufficient to ignite cotton waste when subjected to the time-temperature criteria of ASTM E119 or UL 263. This is consistent with the criteria required for through-penetration protection systems (ASTM E814 or UL 1479) in which the T rating is not required. Since it is very likely that the penetration in an actual fire will be exposed to a positive pressure, this section specifies that the test fire exposure include a positive pressure of 0.01 inch (0.25 mm) of water column as a further means to verify the performance of this protection method as required by the code.

714.3.1.1 Fire-resistance-rated assemblies. Penetrations shall be installed as tested in an *approved* fire-resistance-rated assembly.

❖ This section requires, as an option, that the tested assembly in accordance with ASTM E119 or UL 263 include the penetrations along with the proposed type of protection. In other words, the entire assembly is tested. As an option, this section is not used frequently due to the limitations placed on the tested assembly relative to its application. Penetration protection is most often provided in accordance with the exception to Section 714.3.1 and the provisions of Section 714.3.1.2.

714.3.1.2 Through-penetration firestop system. *Through penetrations* shall be protected by an *approved* penetration firestop system installed as tested in accordance with ASTM E814 or UL 1479, with a minimum positive pressure differential of 0.01 inch (2.49 Pa) of water and shall have an F rating of not less than the required *fire-resistance rating* of the wall penetrated.

❖ In order to maintain effective compartmentation of a building to restrict the spread of fire, all penetrations through fire-resistance-rated assemblies must be protected. In recent history, there have been several examples of buildings where extensive fire damage and loss of life was attributed, at least in part, to lack of or improper installation of penetration protection. In the absence of a through-penetration firestop system to protect penetrations of fire-resistance-rated assemblies, the potential exists for fire to spread beyond the initial area of fire origin. One report on the source of origin of fires in buildings indicated that 23 percent of all building fires originate from electrical systems. This fact, coupled with the number of penetrations through walls and floors created by electrical distribution piping, helps to underscore the necessity for the protection of all penetrations through rated assemblies. It must be noted that penetrations are not limited to electrical systems only, but that openings to accommodate plumbing and mechanical systems also contribute to the number of penetrations through any given assembly.

A through-penetration firestop system consists of specific materials or an assembly of materials that are designed to restrict the passage of fire and hot gases for a prescribed period of time through openings made in fire-resistance-rated walls or horizontal assemblies. In certain instances, the through-penetration firestop system is also required to limit the transfer of heat from the fire side to the unexposed side. In order to determine the effectiveness of a through-penetration firestop system in restricting the passage of fire and the transfer of heat, firestop systems are required to be subjected to fire testing. ASTM E814 and UL 1479 are the test methods developed specifically for the evaluation of a firestop system's ability to resist the passage of flame and hot gases, withstand thermal stresses and restrict transfer of heat through the penetrated assembly.

The basic provisions of ASTM E814 and UL 1479 require that a test assembly consisting of a specific wall or floor construction, containing through penetrations of various types and sizes, be constructed. The through-penetration firestop system to be tested is installed in accordance with the manufacturer's instructions around the penetrations. The test assembly is then exposed to a test fire that corresponds to the time-temperature curve established by ASTM E119 or UL 263 (see the commentary to Section 703.2 for more discussion on ASTM E119 or UL 263). After the fire test exposure, the through-penetration firestop system is subjected to a hose stream, which evaluates the ability of the through-penetration firestop system to resist the effects of erosion and thermal shock. After completion of the ASTM E814 or UL 1479 procedure, two ratings for the test subject are established, which indicate how the test specimen withstands exposure to the test fire for a specified period of time.

Ratings for the through-penetration firestop system are generated based on the results of the testing, and are reported as an F (flame) rating and a T (temperature) rating. The F rating indicates the period of time, in hours, the tested through-penetration firestop system remained in place without allowing the passage of fire during exposure or water during the hose stream. The T rating indicates the time, in hours, it took for the temperature, as recorded by thermocouples placed at specified locations on the unexposed side of the test assembly, to reach 325°F (162°C) above ambient. It must be noted that in order to obtain a T rating, the system must first obtain an F rating. F ratings are required for all through-penetration firestop systems and must be equal to the fire-resistance rating of the assembly penetrated. A T rating is not required for wall penetrations, but a minimum of 1 hour is required where the through-penetration firestop system is installed in a floor assembly where the penetrating item is a pipe, tube or conduit that is in direct contact with a combustible material. This requirement is intended to minimize the potential for ignition of the combustible material on the unexposed side of the assembly due to elevated temperatures transmitted via the pipe, tube or conduit.

An ASTM International (ASTM) standard practice is available that can be used in cases where a standardized methodology is desired for the inspection of installed through-penetration firestop systems. Such a standardized procedure can be particularly useful in cases where a third-party inspection firm is used. ASTM E2174, *Standard Practice for On-site Inspection of Installed Firestops*, includes items such as preconstruction meetings to review submittals, details, variances, building mock-ups for destructive testing and scheduling witnessed inspection of a specified percentage of installations or conducting destructive testing on a specified percentage of installations. This standard practice also identifies what actions should be taken based on the percentage of noncompliant installations identified, as well as providing a standardized reporting form.

Two of the most common types of materials used in through-penetration firestop systems are intumescent and endothermic materials. Intumescent materials expand to approximately eight to 10 times their original volume when exposed to temperatures exceeding 250°F (121°C). The expansion of the material fills the voids or openings within the penetration to resist the passage of flame, while the outer layer of the expanded intumescent material forms an insulating charred layer that assists in limiting the transfer of heat. The expansion properties of intumescent materials allow them to seal openings left by combustible penetrating items that burn away during a fire, but do not retard heat as well as endothermic materials. Intumescent materials are typically used with combustible penetrating items or where a higher T rating is not required.

Endothermic materials provide protection through chemically bound water released in the form of steam when exposed to temperatures exceeding 600°F (316°C). This released water provides a cooling of the penetration and retards heat transfer through the penetration. Endothermic materials tend to be superlatively resistant to heat transfer and have higher T ratings, but do not expand to fill voids left by combustible penetrating items that burn away during a fire. Endothermic materials are, therefore, typically used with noncombustible penetrating items and where a higher T rating is required.

714.3.2 Membrane penetrations. Membrane penetrations shall comply with Section 714.3.1. Where walls or partitions are required to have a *fire-resistance rating*, recessed fixtures shall be installed such that the required fire resistance will not be reduced.

Exceptions:

1. Membrane penetrations of maximum 2-hour fire-resistance-rated walls and partitions by steel electrical boxes that do not exceed 16 square inches (0.0 103 m^2) in area, provided the aggregate area of the openings through the membrane does not exceed 100 square inches (0.0645 m^2) in any 100 square feet (9.29 m^2) of wall area. The *annular space* between the wall membrane and the box shall not exceed $^1/_8$ inch (3.2 mm). Such boxes on opposite sides of the wall or partition shall be separated by one of the following:
   1.1. By a horizontal distance of not less than 24 inches (610 mm) where the wall or partition is constructed with individual noncommunicating stud cavities;
   1.2. By a horizontal distance of not less than the depth of the wall cavity where the wall cavity is filled with cellulose loose-fill, rockwool or slag mineral wool insulation;
   1.3. By solid fireblocking in accordance with Section 718.2.1;
   1.4. By protecting both outlet boxes with *listed* putty pads; or
   1.5. By other *listed* materials and methods.
2. Membrane penetrations by *listed* electrical boxes of any material, provided such boxes have been tested for use in fire-resistance-rated assemblies and are installed in accordance with the instructions included in the listing. The *annular space* between the wall membrane and the box shall not exceed $^1/_8$ inch (3.2 mm) unless *listed* otherwise. Such boxes on opposite sides of the wall or partition shall be separated by one of the following:
   2.1. By the horizontal distance specified in the listing of the electrical boxes;
   2.2. By solid fireblocking in accordance with Section 718.2.1;
   2.3. By protecting both boxes with *listed* putty pads; or
   2.4. By other *listed* materials and methods.
3. Membrane penetrations by electrical boxes of any size or type, that have been *listed* as part of a wall opening protective material system for use in fire-resistance-rated assemblies and are installed in accordance with the instructions included in the listing.
4. Membrane penetrations by boxes other than electrical boxes, provided such penetrating items and the *annular space* between the wall membrane and the box, are protected by an *approved mem-*

brane penetration firestop system installed as tested in accordance with ASTM E814 or UL 1479, with a minimum positive pressure differential of 0.01 inch (2.49 Pa) of water, and shall have an F and T rating of not less than the required *fire-resistance rating* of the wall penetrated and be installed in accordance with their listing.

5. The *annular space* created by the penetration of an automatic sprinkler, provided it is covered by a metal escutcheon plate.

6. Membrane penetrations of maximum 2-hour *fire resistance-rated* walls and partitions by steel electrical boxes that exceed 16 square inches (0.0 103 m^2) in area, or steel electrical boxes of any size having an aggregate area through the membrane exceeding 100 square inches (0.0645 m^2) in any 100 square feet (9.29 m^2) of wall area, provided such penetrating items are protected by listed putty pads or other listed materials and methods, and installed in accordance with the listing.

❖ "Membrane penetrations" are defined in Section 202 as "a breach in one side of a floor/ceiling, roof/ceiling or wall assembly to accommodate an item installed into or passing through the breach." Therefore they are different from a through penetration in that they do not pass through the entire assembly. Membrane penetrations are treated similarly to through penetrations. Specifically, noncombustible conduits, pipes and tubes that penetrate only one membrane of a fire-resistance-rated wall assembly are required to have the annular space protected in accordance with Section 714.3.1 or comply with the exception of that section. Combustible penetrations through only one membrane of a wall assembly would require either a tested assembly (see Section 714.3.1.1) or a through-penetration firestop system that complies with Section 714.3.1.2 for the penetrated membrane.

Penetrations, such as electrical outlet boxes, can affect the fire-resistance rating of an assembly. The criteria in Exception 1 [see Commentary Figure 714.3.2(1)] limits the size of steel electrical outlet boxes [16 square inches (.0103 m^2)] and the amount allowed in a 100-square-foot (9.3 m^2) area [100 square inches (0.0645 m^2)]. Although not directly stated in the code, based on the supporting text in one of the legacy codes, when membrane penetrations of electrical outlet boxes were first regulated in 1979, it would appear that the original intent was to permit 100 square inches (0.0645 m^2) of openings on each side of a 100-square-foot (9.3 m^2) wall section. These criteria are consistent with the criteria determined from fire tests that have generally shown that, within these limitations, these penetrations will not adversely affect the fire-resistance rating of the wall. The design information section of the *UL Fire Resistance Directory* states that the opening in a wallboard facing is to be cut such that the clearance between the box and the wallboard does not exceed $^1/_8$ inch (3.2 mm).

Electrical outlet boxes of steel or materials other

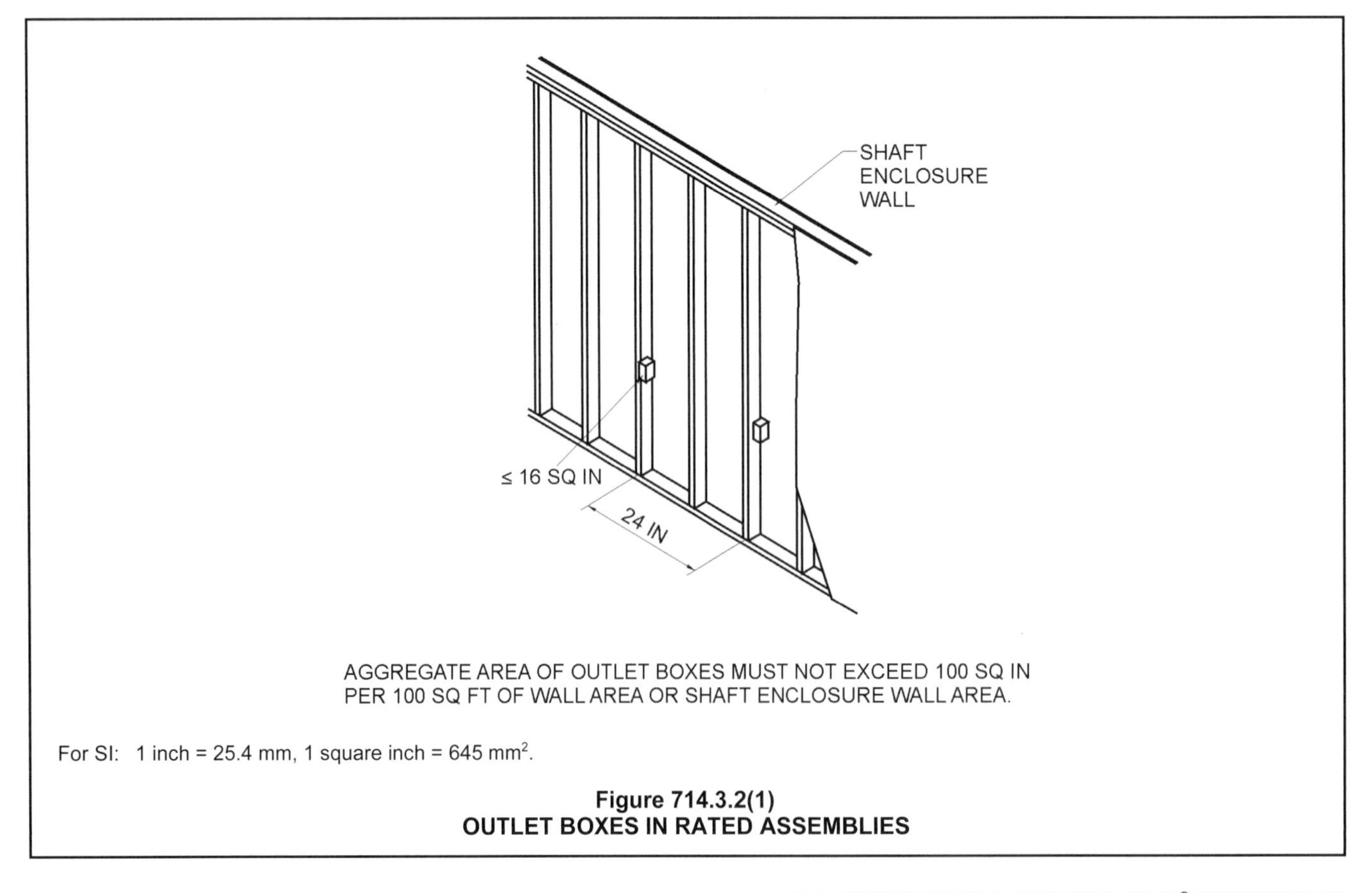

For SI: 1 inch = 25.4 mm, 1 square inch = 645 mm^2.

Figure 714.3.2(1)
OUTLET BOXES IN RATED ASSEMBLIES

than steel (see Exception 2) can be utilized in the fire-resistance-rated wall if they have been specifically tested as part of a fire-resistance-rated assembly. Steel outlet boxes that exceed the size limitations of Exception 1 would also require the same prequalification by testing. Exceptions 1 and 2 also allow electrical boxes on opposite sides of the wall if they meet one of several criteria, such as separation by a horizontal distance of 24 inches (610 mm).

Exception 3 relates to electrical boxes listed as part of an assembly by a testing agency. Certification and listing agencies have published listings covering proprietary compositions that are used to maintain the hourly ratings of fire-resistant walls and partitions incorporating flush-mounted devices, such as outlet boxes, electrical cabinets and mechanical cabinets penetrating membranes of fire-resistance-rated assemblies. The individual systems indicate the specific applications and the method of installation for which the materials have been evaluated. The basic standards used to investigate these products are UL 263 and ASTM E119. For example, UL classifies these materials and systems as "wall opening protective materials." This category includes classifications for both generic steel electrical boxes as well as specific types and models of outlet and switch boxes composed of other materials, all listed for specific usage in fire-resistance-rated wall assemblies. The UL listings for wall opening protective materials indicate that, depending upon the testing conducted for the individual listing, their use can allow for any combination of: 1. Reducing the spacing between boxes contained on opposite sides of the wall; 2. Increasing the size of the boxes; 3. Increasing the density of the boxes; and 4. Allowing the use of boxes on each side of staggered stud walls. Because these systems are tested for the specific end-use applications, the individual and aggregate restrictions on maximum sizes and quantities [i.e., 16 square inches (10 322 mm^2) for an individual box, and the aggregate maximum of 100 square inches (64 516 mm^2) per 100 square feet (9.3 m^2)] are not required for these systems to maintain the fire-resistance ratings of the assemblies penetrated.

Exception 4 creates a direct parallel between the requirements for electrical outlet boxes and other membrane penetration boxes, such as dryer exhaust boxes, washing machine hose connection boxes, fire or police alarm boxes, manual fire alarm boxes, switch boxes, valve boxes, special purpose boxes, electrical panels and hose cabinets. The protection systems are to be tested for use in fire-resistance-rated assemblies and installed in accordance with the instructions included in the listings. Because these utility boxes can exceed 100 square inches (64 516 mm^2) aggregate area, however, both an F and T rating are required in order to be directly equivalent to the fire-resistance rating of the assemblies penetrated. Given that these are membrane penetrations, there is a greater likelihood that someone could unknowingly place or store combustible materials, potentially even furniture and bedding, directly in contact with the unpenetrated membrane on the opposite side of the wall. This could significantly increase threat of a fire spread. The information provided for each classification would include the model numbers for the products, a description of the rated assemblies, the spacing limitations for the boxes and the installation details.

Exception 5 provides an alternative to the annular space protection provisions of Section 714.3.1 for fire sprinklers that penetrate a single membrane, provided that the annular space around the fire sprinkler head or fitting is completely covered by an escutcheon plate of noncombustible material [see Commentary Figure 714.3.2(2)]. The nature of the hazard posed by single membrane penetrations of sprinkler piping is limited due to the size of the opening, the potential amount of openings and the presence of a sprinkler system. The installation of a noncombustible escutcheon provides protection against free passage of fire through the annular space as well as allowing for the movement of sprinkler piping without breaking during a seismic event. These provisions correlate with the requirements of the National Earthquake Hazard Reduction Program (NEHRP) provisions recommended by the Building Seismic Safety Council (BSSC) for installation of sprinkler systems to resist seismic forces. Wall-mounted sprinkler heads installed back to back in fire-resistance-rated walls would be treated as through penetrations and not membrane penetrations.

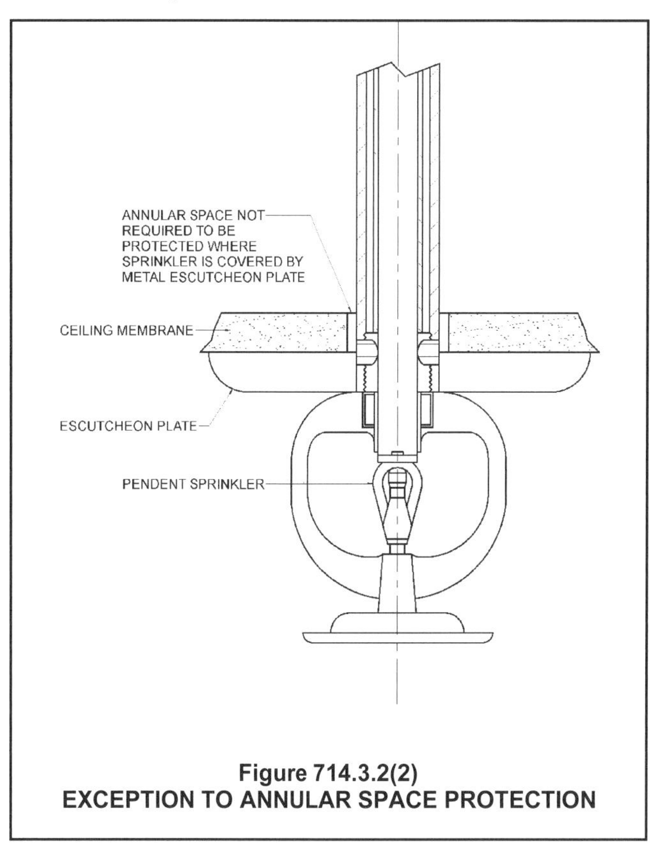

Figure 714.3.2(2)
EXCEPTION TO ANNULAR SPACE PROTECTION

Exception 6 reflects a common practice. This exception permits an additional allowance for steel electrical boxes exceeding 16 square inches (0.0 103 m^2) in size, and exceeding an aggregate area through the membrane of 100 square inches (0.0645 m^2) in any 100 square feet (9.29 m^2) of wall area based on testing and listing of the protection methods. The protection methods include putty pads or other such listed materials installed consistent with the listing.

714.3.3 Dissimilar materials. Noncombustible penetrating items shall not connect to combustible items beyond the point of firestopping unless it can be demonstrated that the fire-resistance integrity of the wall is maintained.

❖ This section limits the common practice of using a short metal nipple to penetrate a rated assembly, firestopping for the metal penetration (which is substantially less expensive than firestopping for plastic) and then connecting to plastic pipe or conduit on the room side of the wall. Arguably, there is a distance at which such connection is safe. However, this distance is variable and cannot be specified in the body of the code, hence the requirement for demonstration of fire-resistance integrity. An identical provision is found in Section 714.4.3 regarding penetrations of horizontal assemblies.

714.4 Horizontal assemblies. Penetrations of a *fire-resistance-rated* floor, floor/ceiling assembly or the ceiling membrane of a roof/ceiling assembly not required to be enclosed in a shaft by Section 712.1 shall be protected in accordance with Sections 714.4.1 through 714.4.4.

❖ Penetrations of horizontal assemblies are required to be protected by a shaft enclosure unless otherwise exempted by Section 712.1. The acceptable methods by which a floor or roof assembly can be interrupted by a penetration are identified in Sections 714.4.1 through 714.4.4.

As noted in the commentary to Section 711, roof construction that is required to have a fire-resistance rating is intended to minimize the threat of premature structural failure of the roof construction under fire conditions and not to create a barrier to contain fire within the building. Section 7112.1.15 permits unprotected penetrations in fire-resistance-rated roof assemblies, provided that the structural integrity of the roof assembly is not reduced.

714.4.1 Through penetrations. Through penetrations of *horizontal assemblies* shall comply with Section 714.4.1.1 or 714.4.1.2.

Exceptions:

1. Penetrations by steel, ferrous or copper conduits, pipes, tubes or vents or concrete or masonry items through a single fire-resistance-rated floor assembly where the *annular space* is protected with materials that prevent the passage of flame and hot gases sufficient to ignite cotton waste when subjected to ASTM E119 or UL 263 time-temperature fire conditions under a minimum positive pressure differential of 0.01 inch (2.49 Pa) of water at the location of the penetration for the time period equivalent to the *fire-resistance rating* of the construction penetrated. Penetrating items with a maximum 6-inch (152 mm) nominal diameter shall not be limited to the penetration of a single fire-resistance-rated floor assembly, provided the aggregate area of the openings through the assembly does not exceed 144 square inches (92 900 mm^2) in any 100 square feet (9.3 m^2) of floor area.
2. Penetrations in a single concrete floor by steel, ferrous or copper conduits, pipes, tubes or vents with a maximum 6-inch (152 mm) nominal diameter, provided the concrete, grout or mortar is installed the full thickness of the floor or the thickness required to maintain the *fire-resistance rating*. The penetrating items shall not be limited to the penetration of a single concrete floor, provided the area of the opening through each floor does not exceed 144 square inches (92 900 mm^2).
3. Penetrations by *listed* electrical boxes of any material, provided such boxes have been tested for use in fire-resistance-rated assemblies and installed in accordance with the instructions included in the listing.

❖ The code addresses the penetration of fire-resistance-rated horizontal separations in much the same way as penetrations through vertical assemblies. Specifically, the penetrations are required to either be part of a total assembly or be protected with a through-penetration firestop system (see commentary, Section 714.3.1).

Exception 1 is similar to Item 2 of the exception to Section 714.3.1. This exception allows the noncombustible penetrating item to connect two stories (penetrate one floor), regardless of the size of the penetrating item [see Commentary Figure 714.4.1(1)]. Where the size of the individual penetration [6-inch diameter (152 mm)] and the aggregate area [144 square inches (0.095 m^2)] is limited in any 100 square feet (9.3 m^2), the code allows such penetrations of noncombustible construction to be through an unlimited number of floors.

Exception 2 allows a 6-inch-diameter (152 mm) penetration through a concrete floor similar to Item 1 in the exception to Section 714.3.1. Where the area of the penetration is limited to 144 square inches (0.095 mm^2), such a penetration is permitted to be through an unlimited number of floors [see Commentary Figure 714.4.1(2) for methods of annular space protection for Exceptions 1 and 2].

Exception 3 mirrors Exception 2 to Section 714.3.2 for membrane penetrations, but is applicable to through penetrations of a horizontal assembly, provided the outlet box has been tested.

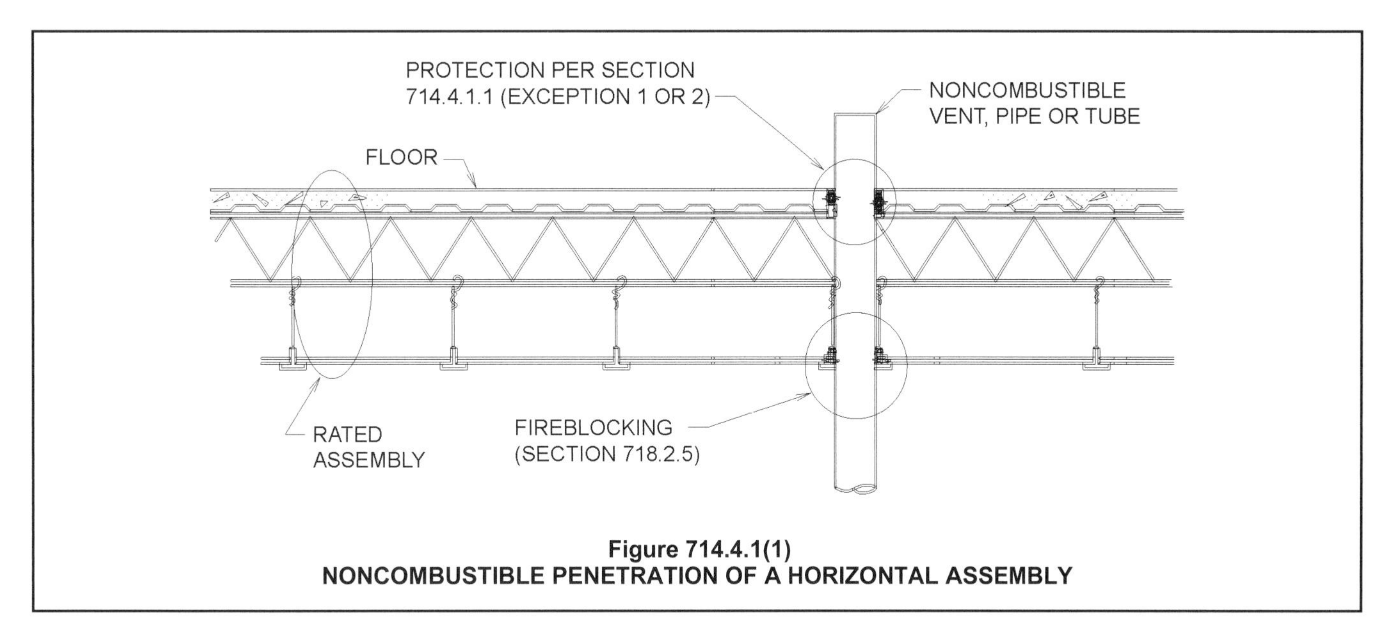

Figure 714.4.1(1)
NONCOMBUSTIBLE PENETRATION OF A HORIZONTAL ASSEMBLY

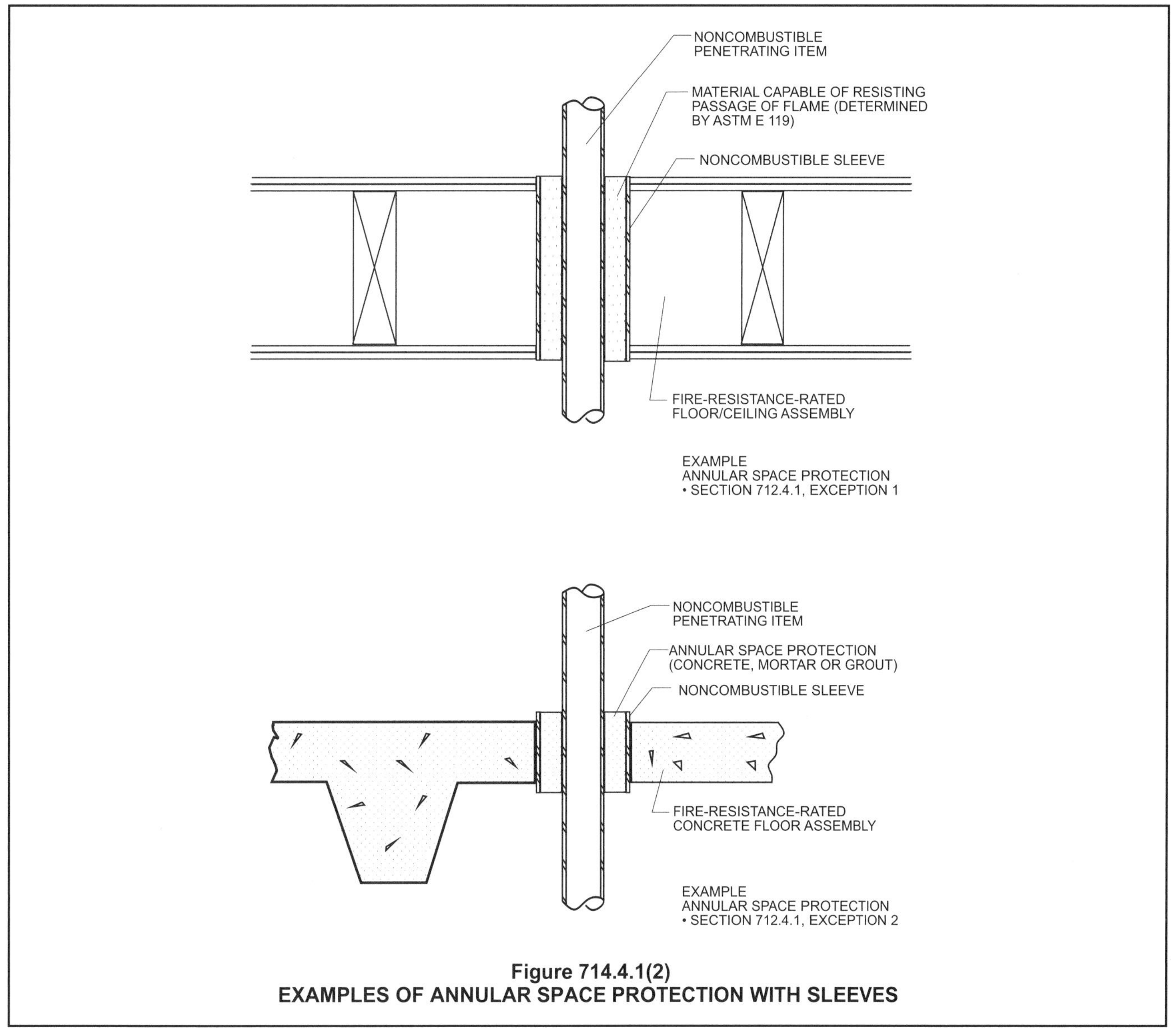

Figure 714.4.1(2)
EXAMPLES OF ANNULAR SPACE PROTECTION WITH SLEEVES

714.4.1.1 Installation. *Through penetrations* shall be installed as tested in the *approved* fire-resistance-rated assembly.

❖ This section addresses membrane penetrations that were tested as a part of the normal fire test of the assembly (see commentary, Section 714.3.1.1).

714.4.1.2 Through-penetration firestop system. *Through penetrations* shall be protected by an *approved through-penetration firestop system* installed and tested in accordance with ASTM E814 or UL 1479, with a minimum positive pressure differential of 0.01 inch of water (2.49 Pa). The system shall have an F rating/T rating of not less than 1 hour but not less than the required rating of the floor penetrated.

Exceptions:

1. Floor penetrations contained and located within the cavity of a wall above the floor or below the floor do not require a T rating.
2. Floor penetrations by floor drains, tub drains or shower drains contained and located within the concealed space of a horizontal assembly do not require a T rating.
3. Floor penetrations of maximum 4-inch (102 mm) nominal diameter penetrating directly into metal-enclosed electrical power switchgear do not require a T rating.

❖ This section differs slightly from Section 714.3.1.2 for wall penetrations in that a T rating is required for penetrations of horizontal assemblies, but is not required for penetrations of wall assemblies (see commentary, Section 714.3.1.2).

When permitted as an alternative to a shaft enclosure, cables, cable trays, conduits, tubes and pipes may penetrate a floor assembly when an approved through-penetration firestop system is used. An approved through-penetration firestop system is one that has been tested in accordance with ASTM E814 or UL 1479. The test method determines the performance of the protection system with respect to exposure to a standard time-temperature fire and hose stream test. The performance of the protection system is dependent on the specific assembly of materials tested, including the number, type and size of penetrations and the types of floors or walls in which it is installed. It should also be noted that tests have been conducted at various pressure differentials; however, the current criterion used is 0.01 inch (2.49 Pa) of water gauge, and only tests with such minimum pressure throughout the test period are to be accepted. In evaluating test reports, the building official must determine that the tested assembly is truly representative of the manner in which the system is to be installed. It should be noted that the ASTM E814 and UL 1479 test establishes two ratings: the F rating, which identifies the ability of the material to resist the passage of flame, and the T rating, which identifies the thermal transmission characteristics of the material or assembly. Exception 1 is intended to set aside the T-rating requirement where the penetrating item is located in a cavity of a wall and is separated from contact with adjacent materials in the occupiable floor area.

Exception 2 allows floor penetrations for plumbing drains located within a concealed horizontal assembly without a T rating. Meeting the T-rating requirement can be very challenging for penetrants that are metallic and therefore conduct heat. A through-penetration firestop system for such penetrants would not typically meet the T-rating requirement simply by sealing the penetrant within the floor. Additional insulation or covering of the through penetrant is typically required above and below the floor to prevent heat transmission via the penetrant. Listings for through-penetration firestop systems will indicate the T rating obtained for any specific system. Commentary Figure 714.4.1.2 shows an example of typical firestop details that would be needed for a cable bundle to meet the T rating requirement of Section 714.4.1.2. For other types of metallic penetrants, listed systems with suitable T ratings would typically have insulation wrapped around the penetrant for some distance above and below the floor.

Exception 3 permits an additional exception for metallic EMT or conduit penetrating a horizontal assembly that directly enters a metal-enclosed power switchgear assembly. The *National Electrical Code* (NEC) defines a metal-enclosed power switchgear as a switchgear assembly completely enclosed on all sides and top with sheet metal (except for ventilating openings and inspection windows), and containing primary power circuit switching, interrupting devices or both, with buses and connections. The assembly may include control and auxiliary devices. Access to the interior of the enclosure is provided by doors, removable covers or both. These devices consist of a substantial metal structure and a sheet metal enclosure. The NEC further requires that, where installed over a combustible floor, suitable protection to the floor and clearances for cable conductors entering these enclosures must be provided. The unobstructed space opposite terminals or raceways or cables entering a switchgear or control assembly must be adequate for the type of conductor and method of termination. Because the penetrating item (EMT) goes through the floor and enters directly into these robust enclosures, it is reasonable to provide an exemption to the T-rating requirements of the code in these conditions.

714.4.2 Membrane penetrations. Penetrations of membranes that are part of a *horizontal assembly* shall comply with Section 714.4.1.1 or 714.4.1.2. Where floor/ceiling assemblies are required to have a *fire-resistance rating*, recessed fixtures shall be installed such that the required *fire resistance* will not be reduced.

Exceptions:

1. *Membrane penetrations* by steel, ferrous or copper conduits, pipes, tubes or vents, or concrete or masonry items where the *annular space* is protected

either in accordance with Section 714.4.1 or to prevent the free passage of flame and the products of combustion. The aggregate area of the openings through the membrane shall not exceed 100 square inches (64 500 mm^2) in any 100 square feet (9.3 m^2) of ceiling area in assemblies tested without penetrations.

2. Ceiling *membrane penetrations* of maximum 2-hour *horizontal assemblies* by steel electrical boxes that do not exceed 16 square inches (10 323 mm^2) in area, provided the aggregate area of such penetrations does not exceed 100 square inches (44 500 mm^2) in any 100 square feet (9.29 m^2) of ceiling area, and the *annular space* between the ceiling membrane and the box does not exceed $^1/_8$ inch (3.2 mm).

3. *Membrane penetrations* by electrical boxes of any size or type, that have been *listed* as part of an opening protective material system for use in *horizontal assemblies* and are installed in accordance with the instructions included in the listing.

4. *Membrane penetrations* by *listed* electrical boxes of any material, provided such boxes have been tested for use in fire-resistance-rated assemblies and are installed in accordance with the instructions included in the listing. The *annular space* between the ceiling membrane and the box shall not exceed $^1/_8$ inch (3.2 mm) unless *listed* otherwise.

5. The *annular space* created by the penetration of a fire sprinkler, provided it is covered by a metal escutcheon plate.

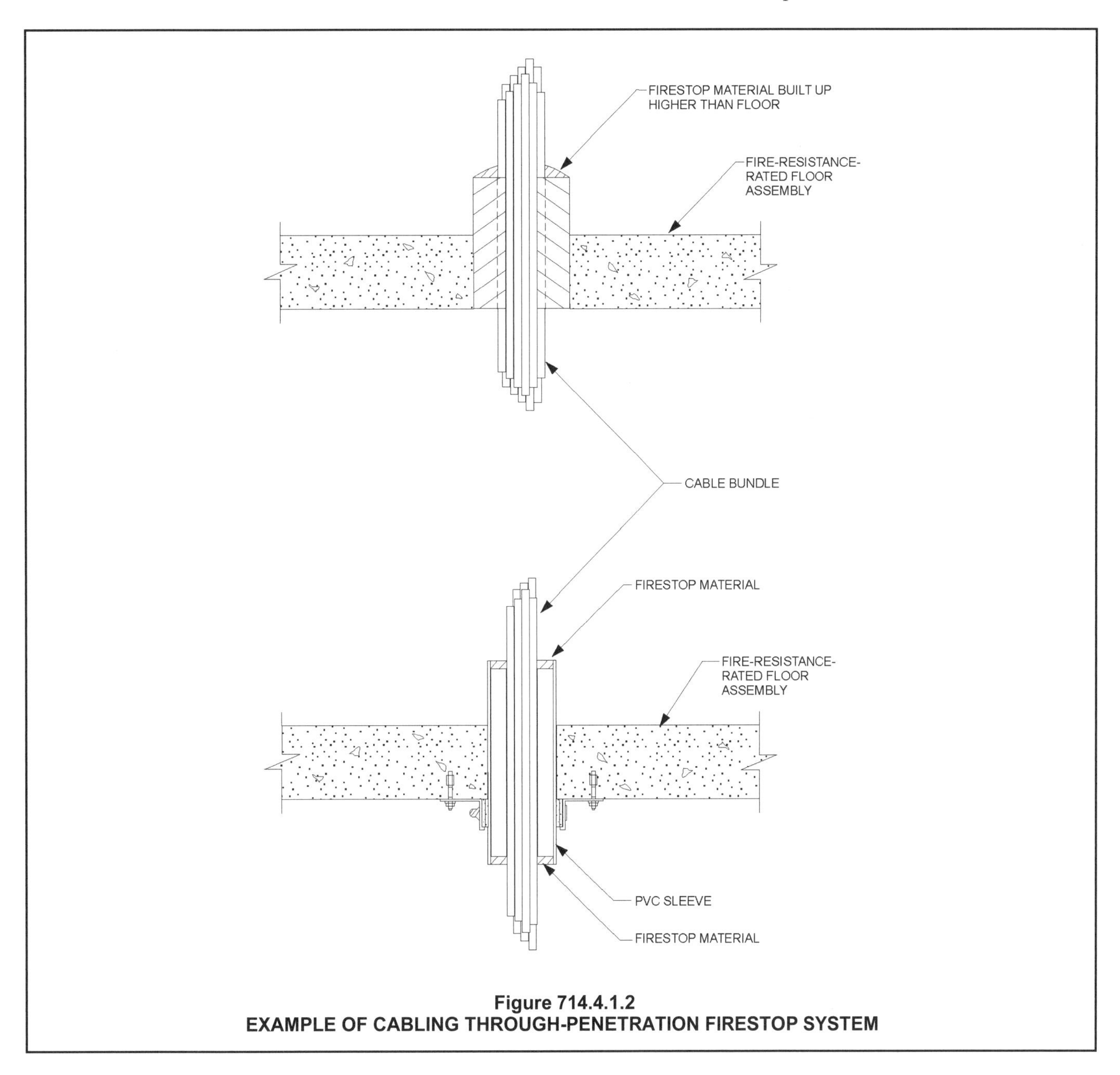

Figure 714.4.1.2
EXAMPLE OF CABLING THROUGH-PENETRATION FIRESTOP SYSTEM

6. Noncombustible items that are cast into concrete building elements and that do not penetrate both top and bottom surfaces of the element.

7. The ceiling membrane of 1- and 2-hour *fire-resistance-rated horizontal assemblies* is permitted to be interrupted with the double wood top plate of a wall assembly that is sheathed with Type X gypsum wallboard, provided that all penetrating items through the double top plates are protected in accordance with Section 714.4.1.1 or 714.4.1.2 and the ceiling membrane is tight to the top plates.

❖ Penetrations of fire-resistance-rated floor/ceiling and roof/ceiling assemblies that have a ceiling membrane as a component of the tested assembly, such as steel or wood joist assemblies using acoustical tiles or gypsum wallboard, are limited to those penetrations that are listed in the tested assembly (see Section 714.4.1.1) or those protected in accordance with ASTM E814 or UL 1479 [see Section 714.4.1.2 and Commentary Figure 714.4.2(1)]. Penetrations of a ceiling membrane by items that are not part of a tested assembly create a point for possible fire penetration into the space above the ceiling membrane, which in turn could jeopardize the fire-resistance rating of the assembly. The exceptions contain protection requirements for specific types of penetrations that are not required to be part of the tested assembly.

Exception 1 allows for a maximum aggregate amount of noncombustible pipe, tube, vents and conduit penetrations of a floor or ceiling that is part of a fire-resistance-rated assembly to be 100 square inches (0.0645 m^2) per 100 square feet (9.3 m^2) of ceiling area. The annular space of the penetration can be protected by either annular space protection (see Exception 1 of Section 714.4.1) or a through-penetration firestop system tested to ASTM E814 or UL 1479 [see Section 714.4.1.2 and Commentary Figure 714.4.2(2)].

Annular space protection is not an option for combustible items because it does not have the ability to protect the resulting void created when the combustible penetrating item burns away.

Exception 2 allows penetrations of electrical outlet boxes that can affect the fire-resistance rating of an assembly. The criteria of this section limit the size of noncombustible electrical outlet boxes that penetrate a ceiling membrane to 16 square inches and an aggregate area that shall not exceed 100 square inches (0.0645 m^2) in 100 square feet (9.3 m^2) of ceiling area. All of the openings at the penetrations of the outlets are required to be protected as mentioned above. The area limitations are consistent with the criteria determined from fire tests, which have shown that within these limitations, these penetrations will not adversely affect the fire-resistance rating of the floor/ceiling assembly.

The exception permits electrical outlet boxes and fittings, which have been evaluated and found suitable for this purpose, to be located in the floor or ceiling of the fire-resistance-rated assembly. Reference sources for fire-resistance-rated assemblies, such as the *UL Fire Resistance Directory*, state that certain tested floor assemblies will permit the installation of electrical and communication connection inserts that appear in the UL listing category "Outlet Boxes and Fittings Classified for Fire Resistance" without reducing the fire-resistance rating of the assembly. However, classified outlet boxes and fittings installed in floor/ceiling assemblies that do not specify their use will jeopardize the rating unless compensating protection is provided.

Exception 3 is related to membrane penetrations of fire-resistance-rated assemblies by electrical boxes of any size or type that have been listed for their particular use in a horizontal assembly.

Certification and listing agencies have published listings covering proprietary compositions that are used to maintain the hourly ratings of fire-resistant

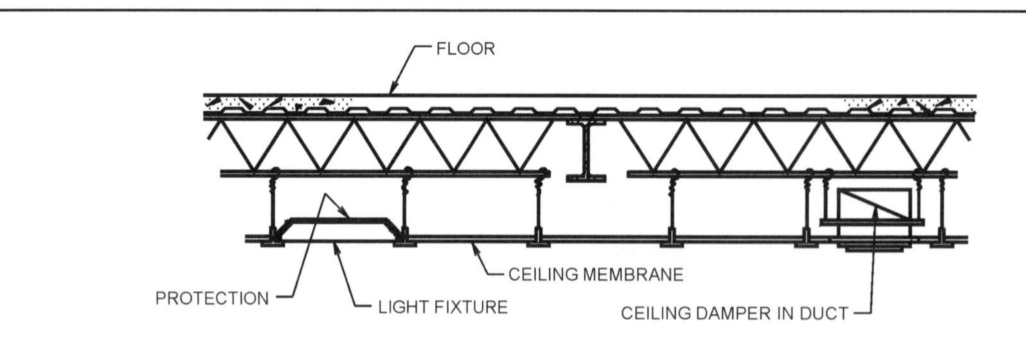

For SI: 1 square inch = 645 mm^2, 1 square foot = 0.0929 m^2.

Figure 714.4.2(1)
PROTECTION OF PENETRATIONS THROUGH CEILING MEMBRANE OF A HORIZONTAL ASSEMBLY

assemblies incorporating flush-mounted devices, such as outlet boxes, electrical cabinets and mechanical cabinets penetrating membranes of fire-resistance-rated assemblies. The individual systems indicate the specific applications and method of installation for which the materials have been evaluated. The basic standards used to investigate these products are ANSI/UL 263 and ASTM E119.

For example, UL classifies nonmetallic outlet boxes for installation in floors and ceilings in accordance with the provisions of NFPA 70, *National Electrical Code*® (NEC®). These systems are required to provide a degree of fire resistance when installed in specific floor assemblies. Listing information includes the model numbers for the products, a description of the rated assemblies in which they can be used, the spacing limitations for the boxes and the installation details.

Exception 4 is similar to Exception 3 and probably could be combined. This exception addresses the annular space between the box and the ceiling membrane and limits it to $^1/_8$ inch (3.2 mm) unless listed otherwise.

Exception 5 covers fire sprinkler heads in fire-resistant ceiling membranes and allows them without any testing where they are installed with metal escutcheon plates. This is identical to the membrane wall penetration for sprinkler heads in Exception 5 to Section 714.3.2 [see Commentary Figure 714.3.2(2)].

Exception 6 allows any noncombustible penetration that is cast into a concrete floor slab, provided that it penetrates only one surface (top or bottom). It may penetrate the top surface or the bottom surface but not both.

Exception 7 is a very important addition to the code, because it allows a common construction practice. For instance, a 1- or 2-hour fire-resistance-rated floor/ceiling or roof/ceiling assembly is tested and rated with a continuous $^5/_8$-inch (15.8 mm) gypsum wallboard ceiling. In conventional light platform framing, the bearing walls and nonbearing walls are framed with floor joists typically resting directly on studs. The ceiling membrane is then attached, which creates a membrane penetration where the top wall plates intersect the floor joists. This was previously a matter not provided for in the codes, requiring special analysis to be submitted to the building official for specific approval. This exception allows this typical framing detail, but gives special requirements. The wall must have double top plates (two-2Xs) and the wall must be sheathed with Type X gypsum wallboard.

714.4.3 Dissimilar materials. Noncombustible penetrating items shall not connect to combustible materials beyond the point of firestopping unless it can be demonstrated that the fire-resistance integrity of the *horizontal assembly* is maintained.

❖ This section mirrors that of Section 714.3.3 for direct penetrations through walls (see commentary, Section 714.3.3).

714.4.4 Penetrations in smoke barriers. Penetrations in *smoke barriers* shall be protected by an approved *through-penetration firestop system* installed and tested in accordance with the requirements of UL 1479 for air leakage. The *L rating* of the system measured at 0.30 inch (7.47 Pa) of water in both the ambient temperature and elevated temperature tests shall not exceed:

1. 5.0 cfm per square foot ($0.025\ m^3/s \cdot m^2$) of penetration opening for each *through-penetration firestop system*; or
2. A total cumulative leakage of 50 cfm ($0.024\ m^3/s$) for any 100 square feet ($9.3\ m^2$) of wall area, or floor area.

❖ Smoke barriers are intended to create compartments within a building where, under fire conditions, they provide protection from both fire and smoke. As such compartments are critical in providing the necessary

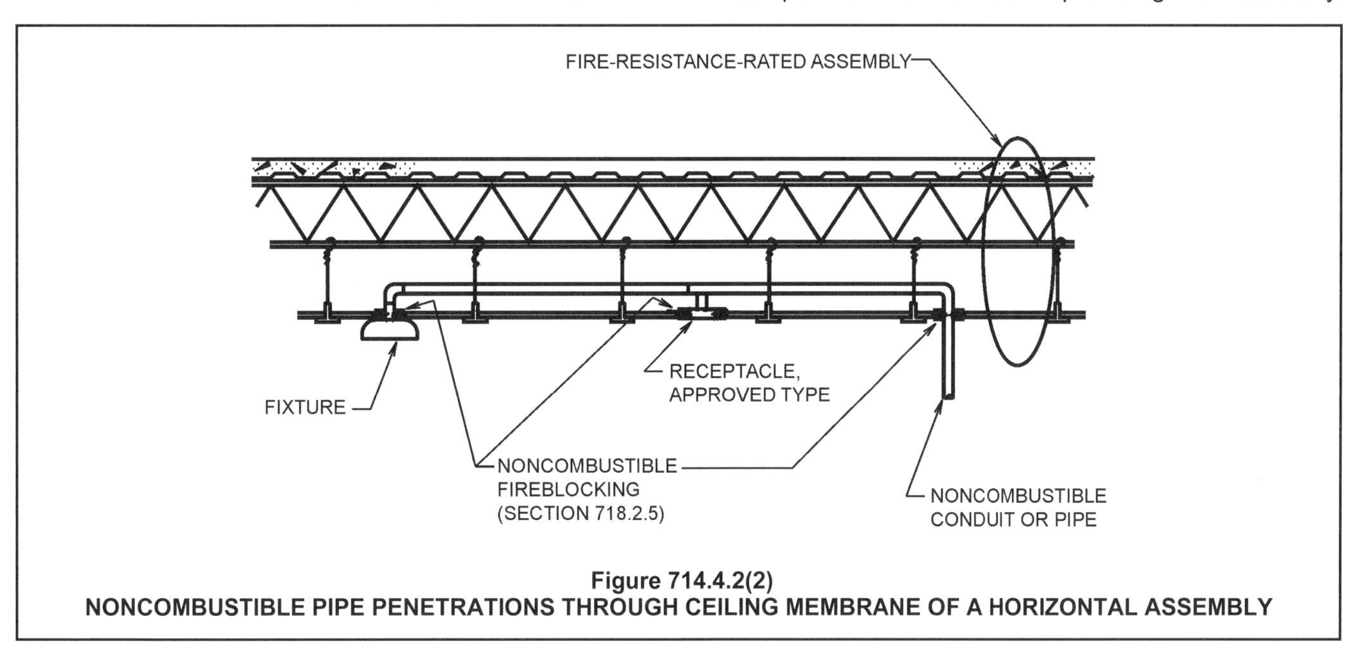

Figure 714.4.2(2)
NONCOMBUSTIBLE PIPE PENETRATIONS THROUGH CEILING MEMBRANE OF A HORIZONTAL ASSEMBLY

level of fire and smoke protection, enforceable language is deemed necessary to help ensure that the intention of the code is met. Previous language in the code included terms such as "limit," "restrict" and "resist" to indicate the degree of smoke protection required, creating inconsistent application of the provisions.

The L rating provides a quantitative indication of the through-penetration system's ability to resist the passage of smoke. Although the test is performed using air, the flow properties of air and of smoke are sufficiently close for engineering purposes as to provide a reasonable quantification of smoke leakage. The L-rating test is an optional test within the UL 1479 test standard; therefore, not all through-penetration firestop systems that are tested and listed will have this information. In the 2005 UL *Fire Resistance Directory (Vol. 2)*, there are approximately 550 through-penetration firestop systems that are listed that also have an L rating of 5 or less. This should provide a sufficient selection to allow for code compliance in all situations.

Smoke barriers, in general, have limited areas where smoke will leak through, such as at doors, dampers, penetrations and joints. Fire door assemblies and smoke dampers have previously had allowable leakage rates established by the code and referenced standards. Since the 2006 edition of the code added this text for penetrations and joint systems (see Section 715.6), all of the typical areas where leakage in smoke barriers can occur have been addressed.

714.5 Nonfire-resistance-rated assemblies. Penetrations of nonfire-resistance-rated floor or floor/ceiling assemblies or the ceiling membrane of a nonfire-resistance-rated roof/ceiling assembly shall meet the requirements of Section 713 or shall comply with Section 714.5.1 or 714.5.2.

❖ This section limits the penetrations of nonfire-resistance-rated floor assemblies to prevent the migration of smoke through a building, despite being of an unprotected type of construction. Penetrating items such as pipes are required to be in shafts or one of the permitted protection methods outlined in Sections 714.5.1 and 714.5.2.

Where penetrations of the ceiling membrane of a nonfire-resistance-rated roof/ceiling assembly occur, the penetrations are required to be protected by fireblocking in accordance with the requirements of Section 718.2.5 to restrict the spread of fire through the concealed roof space above the ceiling membrane (see commentary, Section 718.2.5).

714.5.1 Noncombustible penetrating items. Noncombustible penetrating items that connect not more than five *stories* are permitted, provided that the *annular space* is filled to resist the free passage of flame and the products of combustion with an *approved* noncombustible material or with a fill, void or cavity material that is tested and classified for use in *through-penetration firestop systems*.

❖ Noncombustible penetrations connecting five stories or less (penetrating four floors or less) are permitted when the annular space of the penetrating item is filled with an approved noncombustible material. As required in Section 718.2.5, this material shall be tested in the form and manner intended for use to demonstrate that it will stay in place and resist the passage of flame and products of combustion. It can also be filled with a material that is tested as a through-penetration firestop system (see Commentary Figure 714.5.1).

714.5.2 Penetrating items. Penetrating items that connect not more than two *stories* are permitted, provided that the *annular space* is filled with an *approved* material to resist the free passage of flame and the products of combustion.

❖ This section is not limited as to the combustibility of the penetrating item. This section allows both combustible and noncombustible penetrations of a single nonfire-resistance-rated floor (thus connecting two stories), provided the annular space is fireblocked in accordance with Section 718.2.5. This section does not require a noncombustible firestop to protect the annular space, while Section 714.5.1 does require a noncombustible firestop material.

SECTION 715
FIRE-RESISTANT JOINT SYSTEMS

715.1 General. Joints installed in or between fire-resistance-rated walls, floor or floor/ceiling assemblies and roofs or roof/ceiling assemblies shall be protected by an approved *fire-resistant joint system* designed to resist the passage of fire for a time period not less than the required *fire-resistance rating* of the wall, floor or roof in or between which the system is installed. *Fire-resistant joint systems* shall be tested in accordance with Section 715.3.

Exception: *Fire-resistant joint systems* shall not be required for joints in all of the following locations:

1. Floors within a single *dwelling unit*.
2. Floors where the joint is protected by a shaft enclosure in accordance with Section 713.
3. Floors within atriums where the space adjacent to the atrium is included in the volume of the atrium for smoke control purposes.
4. Floors within malls.
5. Floors and ramps within open and enclosed parking garages or structures constructed in accordance with Sections 406.5 and 406.6, respectively.
6. Mezzanine floors.
7. Walls that are permitted to have unprotected openings.

8. Roofs where openings are permitted.
9. Control joints not exceeding a maximum width of 0.625 inch (15.9 mm) and tested in accordance with ASTM E119 or UL 263.

❖ This section regulates joints or linear openings created between building assemblies, which are sometimes referred to as head-of-wall, expansion or seismic joints. These joints are most often created where the structural design of a building necessitates a separation between building components in order to accommodate anticipated structural displacements caused by thermal expansion and contraction, seismic activity, wind or other loads. Commentary Figure 715.1 illustrates some of the most common locations of these joints.

Seismic joints in multistory buildings are intended to allow differential lateral displacement of separate portions of a building during a seismic event. Expansion joints permit separate portions of the structural frame to contract and expand due to temperature change or wind sway without adversely affecting the building functions or structural integrity.

Joints can also occur in fire barriers at the intersection of the top of a wall and the underside of the floor or roof above (head of wall); within a wall or floor at any specific point to accommodate the structural design; or at the edge of the floor at the intersection of the floor and exterior wall. All of these linear openings create a “weak link” in fire-resistance-rated assemblies that can compromise the integrity of the tested assembly by allowing an avenue for the passage of fire and the products of combustion through the assembly. In order to maintain the efficacy of the fire-resistance-rated assembly, these openings must be protected by a fire-resistance-rated joint system with a rating equal to the assembly in the same plane. Where two assemblies intersect, the fire rating of the joint must be the same as the fire rating of the assembly (or assemblies) of the same plane as the assembly where the joint occurs. It is not the intent of this section to regulate joints installed in assemblies that are provided to control shrinkage cracking, such as a saw-cut control joint in concrete. This section con-

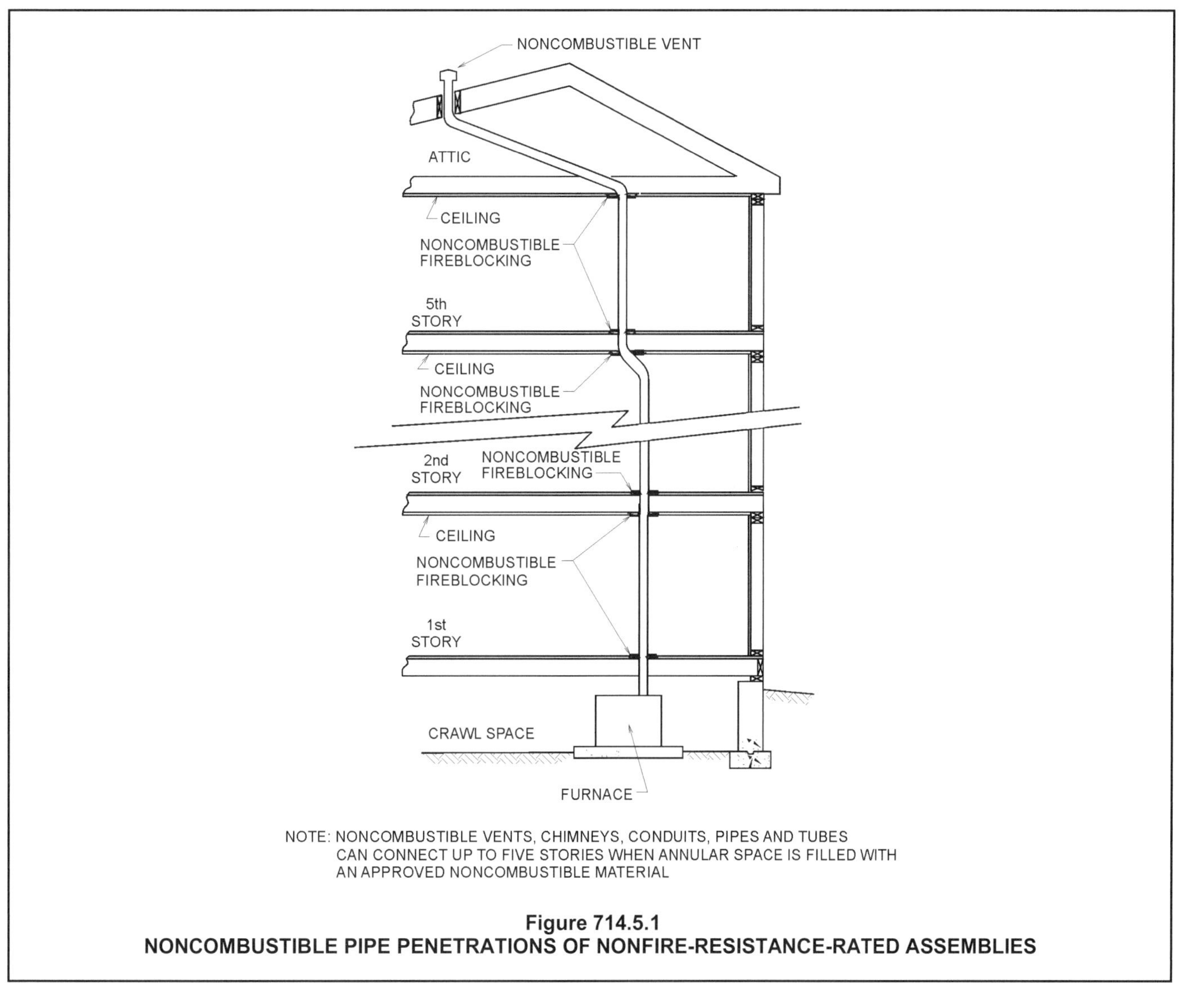

Figure 714.5.1
NONCOMBUSTIBLE PIPE PENETRATIONS OF NONFIRE-RESISTANCE-RATED ASSEMBLIES

tains nine locations where a fire-resistant joint system is not required to be installed to protect the joint. These generally are locations where a separation or protected opening is not required or where the joint occurs within an area that is bounded by other means of protection.

Exception 1 states that a fire-resistant joint is not required for joints contained within a single dwelling unit. This exception is similar to Section 712.1.2. Unlike Section 712.1.2, the joints can be in concealed spaces.

Exception 2 exempts fire-resistant joints completely enclosed in a shaft that complies with Section 713.

Exception 3 exempts the requirement for fire-resistant joints that are located in floors in an atrium that complies with Section 404. This exception is similar to Section 712.1.7.

Exception 4 exempts joints located in malls that comply with Section 402. Although this appears to be a general exception that would affect all floors within a covered or open mall building, it only applies to the mall. The mall is the common pedestrian area.

Exception 5 exempts joints located in open parking structures that comply with Section 406.5, and is similar to Section 712.1.9.

Exception 6 exempts joints in mezzanines that comply with Section 505, and is similar to Section 712.1.11. As a type of opening/penetration, a joint represents a similar hazard to exterior wall fire exposures as other openings.

Exception 7 acknowledges that if an exterior wall is permitted to have unprotected openings (see Table 705.8), then the joint does not require protection as well. A joint is a much smaller opening than an unprotected opening.

Exception 8 exempts joints located in roof decks that are not required to have protected openings, and is similar to that contained in Section 712.1.15, which allows skylights and other penetrations in rated roof decks, provided the structural integrity is not affected.

Exception 9 exempts control joints that have been tested as part of the overall assembly in accordance with ASTM E119 and UL 263. This allowance is in recognition of fire test data that has existed since the 1960s regarding control joints and perimeter relief joints in fire-rated systems, the performance of these devices and systems in the field and a rational approach regarding minor movement [defined here as a maximum of $^5/_8$ inch (15.9 mm)] where no appreciable damage or fatigue is incurred by the joint system that would have a profound impact on the joint's fire performance. Since all buildings move slightly, it is not prudent to have to cycle and test joint systems that only move a few millimeters. Although their movement is limited, control joint systems must still pass the rigors of ASTM E119 or UL 263 in accordance with this exception.

715.1.1 Curtain wall assembly. The void created at the intersection of a floor/ceiling assembly and an exterior curtain wall assembly shall be protected in accordance with Section 715.4.

❖ The void at the intersection of floor/ceiling assemblies and exterior curtain walls is a joint. To provide continuity for the horizontal assembly and block the verti-

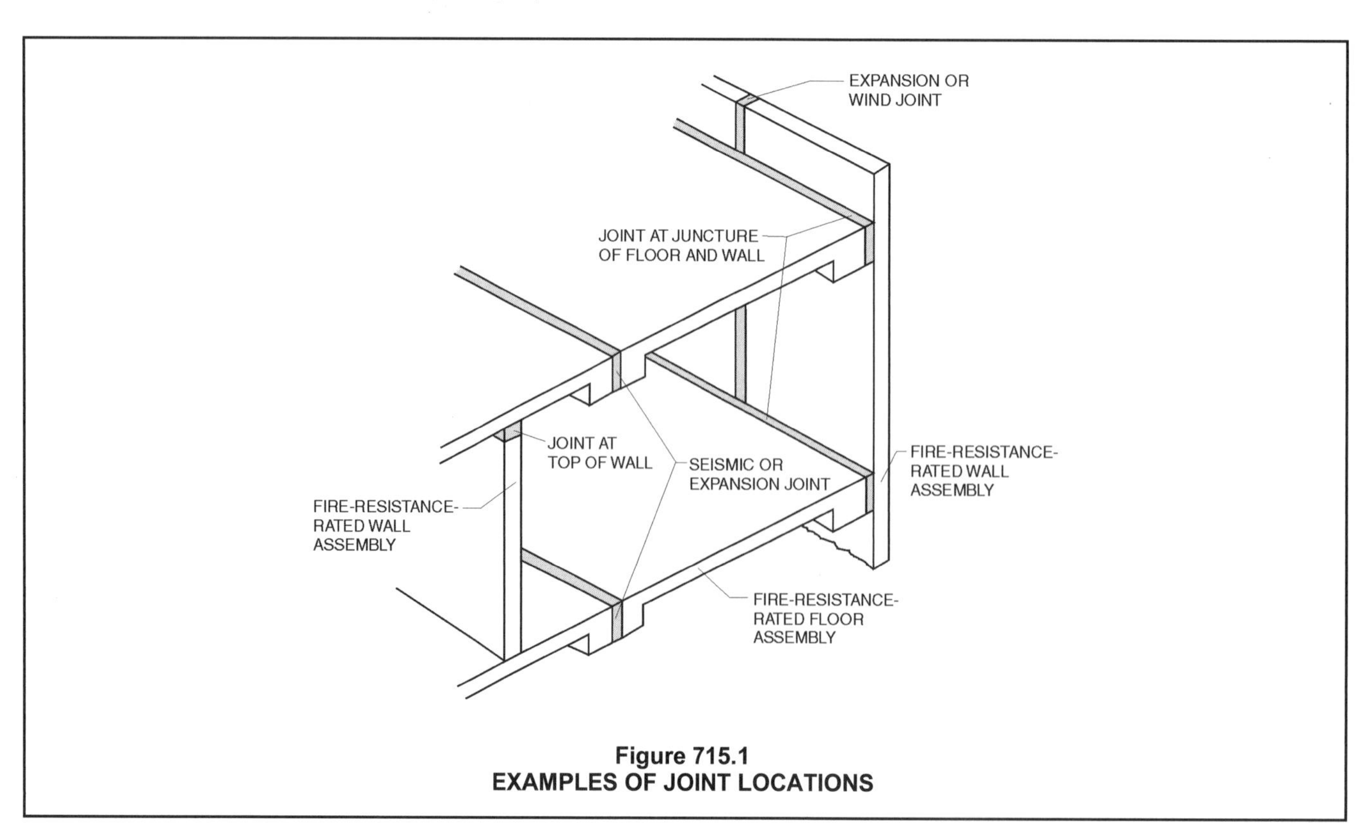

Figure 715.1
EXAMPLES OF JOINT LOCATIONS

cal opening, fire-resistance joint material is required. This section references Section 715.4 for the performance standard that must be met.

715.2 Installation. A *fire-resistant joint system* shall be securely installed in accordance with the listing criteria in or on the joint for its entire length so as not to dislodge, loosen or otherwise impair its ability to accommodate expected building movements and to resist the passage of fire and hot gases.

❖ This section requires that joint systems be installed for their full length or height, due to the fact that the openings to be protected by these types of joints are most often continuous. The required listing will have installation criteria.

715.3 Fire test criteria. *Fire-resistant joint systems* shall be tested in accordance with the requirements of either ASTM E1966 or UL 2079. Nonsymmetrical wall joint systems shall be tested with both faces exposed to the furnace, and the assigned *fire-resistance rating* shall be the shortest duration obtained from the two tests. Where evidence is furnished to show that the wall was tested with the least fire-resistant side exposed to the furnace, subject to acceptance of the *building official*, the wall need not be subjected to tests from the opposite side.

Exception: For *exterior walls* with a horizontal *fire separation distance* greater than 5 feet (1524 mm), the joint system shall be required to be tested for interior fire exposure only.

❖ In order to determine the anticipated protection provided by a given assembly, the hourly fire-resistance rating must be determined by testing the joint system in accordance with UL 2079 or ASTM E1966. This standard includes specific criteria for test specimen preparation, placement, configuration, size and testing conditions of joint systems. Also included in the test protocol is the minimum positive pressure differential to be used for the test, which is intended to assist in evaluating whether the joint will remain in place during a fire condition. A joint splice must be tested, since the presence and orientation of a splice in a joint system can affect the fire performance of the joint. These splices may occur where the length of the joint to be protected exceeds the length of a prefabricated joint or where a cold joint occurs in a field-installed system. The maximum joint width must be tested when in a fully expanded or extended condition in order to evaluate the fire-resistance rating. Joints that are designed to transfer structural building loads are required to have a superimposed load during the test, which is consistent with the requirements for the testing of load-bearing fire-resistance-rated assemblies. Finally, the test addresses requirements for joints that are intended to accommodate building movement, such as expansion, seismic and wind sway joints, including preconditioning cycling, which is intended to allow for evaluation of the joint's ability to withstand cyclical movement over its anticipated life.

Test data have indicated that the orientation of nonsymmetrical joints can have an effect on the performance of the joint. As a result, in accordance with this section, all joint systems must be tested for fire exposure from both sides so that the required protection will be provided regardless of which side of the joint is exposed to fire. The exception for joints in exterior walls correlates with Section 705.5 for the fire exposure rating of exterior walls.

Though not required by the code, an ASTM standard practice is available that can be used in cases where a standardized methodology is desired for the inspection of installed fire- resistant joint systems and perimeter fire barrier systems. Such a standardized procedure can be particularly useful in cases where a third-party inspection firm is used. ASTM E2393, *Standard Practice for On-Site Inspection of Installed Fire Resistive Joint Systems and Perimeter Fire Barriers*, includes items such as preconstruction meetings to review submittals, building mock-ups for destructive testing and scheduling on-site inspections of installed joint systems. This standard practice also identifies what actions should be taken based on the percentage of noncompliant installations identified, as well as providing a standardized reporting form.

715.4 Exterior curtain wall/floor intersection. Where fire resistance-rated floor or floor/ceiling assemblies are required, voids created at the intersection of the exterior curtain wall assemblies and such floor assemblies shall be sealed with an *approved* system to prevent the interior spread of fire. Such systems shall be securely installed and tested in accordance with ASTM E2307 to provide an *F rating* for a time period not less than the *fire-resistance rating* of the floor assembly. Height and fire-resistance requirements for curtain wall spandrels shall comply with Section 705.8.5.

Exception: Voids created at the intersection of the exterior curtain wall assemblies and such floor assemblies where the vision glass extends to the finished floor level shall be permitted to be sealed with an approved material to prevent the interior spread of fire. Such material shall be securely installed and capable of preventing the passage of flame and hot gases sufficient to ignite cotton waste where subjected to ASTM E119 time-temperature fire conditions under a minimum positive pressure differential of 0.01 inch (0.254 mm) of water column (2.5 Pa) for the time period not less than the *fire-resistance rating* of the floor assembly.

❖ The void created between a floor and a curtain wall requires sealing to prevent the spread of flames and products of combustion between adjacent stories. Most fire-rated horizontal assemblies are required to extend from outside wall to outside wall, that is from the inside faces of the exterior walls. The void in question is the one between the intersection of the edge of the fire-resistance-rated horizontal assembly and the interior face of the exterior wall or curtain wall. The void does not include the exterior wall cavity.

A fire test standard was specifically developed to

evaluate the interface between a fire-resistance-rated horizontal assembly and an exterior curtain wall. This particular standard is ASTM E2307, *Standard Test Method for Determining Fire Resistance of Perimeter Fire Barrier Systems Using Intermediate-Scale, Multistory Test Apparatus*. The "perimeter fire barrier system" is the assembly of materials that prevents the passage of flame and hot gases at the void space between the interior surface of the wall assembly and the adjacent edge of the floor. For the purposes of the ASTM E2307 test standard, the interior face is at the interior surface of the wall's framework. The width of the joint, which has maximum allowable dimensions specified in the perimeter fire barrier system listings, is therefore the distance between the edge of the framing nearest the floor and the adjacent floor edge. The void space or cavity between framing members is not considered joint space. Commentary Figure 715.4 shows the typical components that are specified in listings of tested perimeter fire barrier systems.

Tested and listed perimeter fire barrier systems do not include the interior finished wall (e.g., "knee wall") details. This makes the systems applicable to any and all finished wall configurations. The existence of the interior wall surface, even if made of fire-resistant materials such as fire-resistance-rated gypsum board, does not eliminate the need to have an appropriately tested material or system to protect the curtain wall from interior fire spread at the perimeter gap, unless that interior wall detail has been specifically tested and shown to meet the requirements of this section of the code.

Although not required by the code, an ASTM standard practice is available that can be used in cases where a standardized methodology is desired for the inspection of installed perimeter fire barrier systems. Such a standardized procedure can be particularly useful in cases where a third-party inspection firm is used. ASTM E2393, *Standard Practice for On-Site Inspection of Installed Fire Resistive Joint Systems*

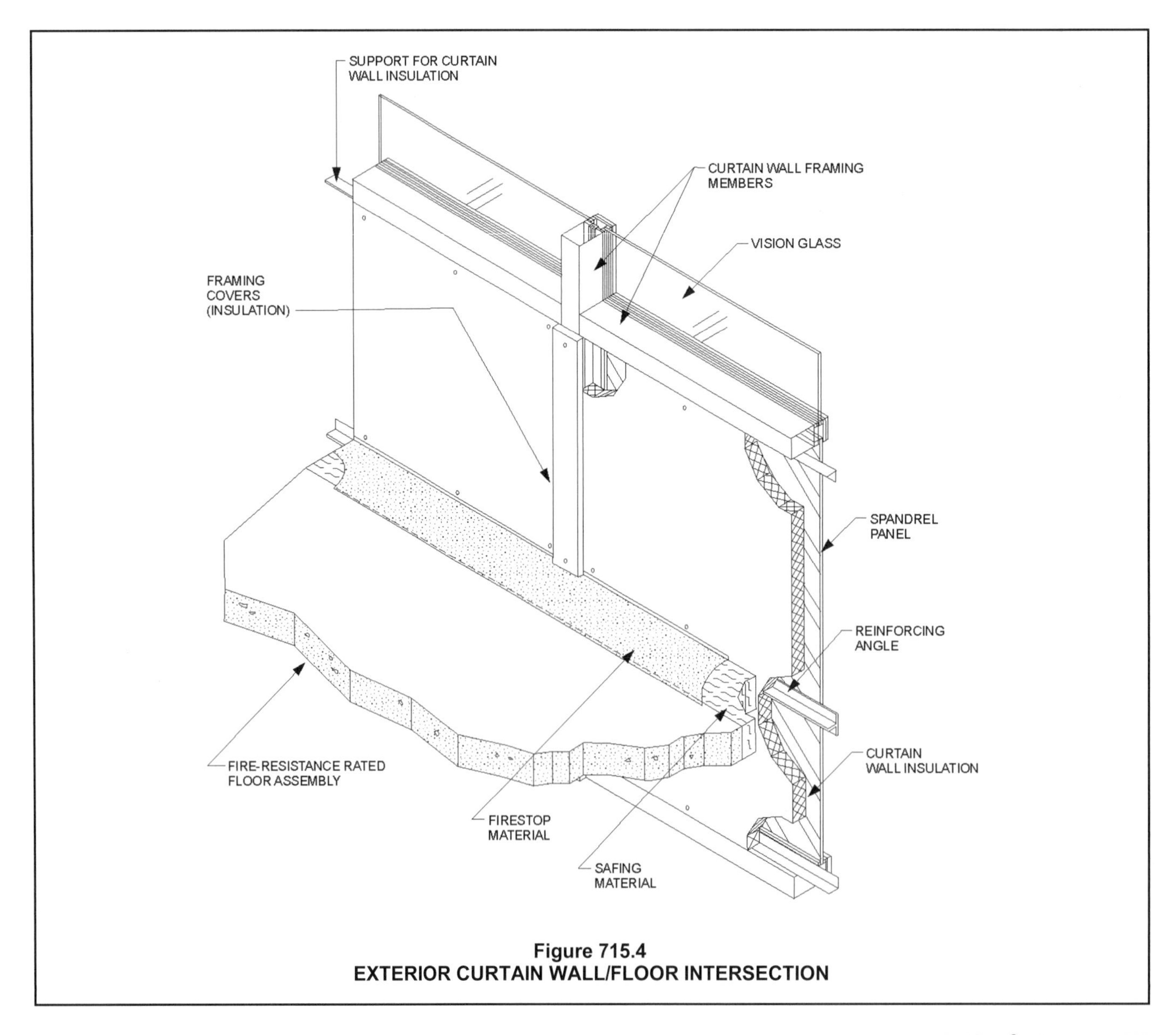

Figure 715.4
EXTERIOR CURTAIN WALL/FLOOR INTERSECTION

and Perimeter Fire Barriers, includes items such as preconstruction meetings to review submittals, details, variances, building mock-ups for destructive testing and scheduling a witnessed inspection of a specified percentage of installations or conducting destructive testing on a specified percentage of installations. This standard practice also identifies what actions should be taken based on the percentage of noncompliant installations identified, as well as providing a standardized reporting form.

The exception, which allows the option of using materials or systems that have been tested by subjecting them to an ASTM E119 or UL 263 time-temperature fire exposure under positive pressure, is the protection of the perimeter gap that existed in previous editions of the code before the 2009 edition. These systems and materials tested in the years before the formal issuance of ASTM E2307 can continue to be used to comply with Section 715.4. Those assemblies that were tested using the ASTM E119 or UL 263 time-temperature fire exposure in a multistory apparatus would be expected to be in substantial compliance with ASTM E2307 and are expected to provide an equivalent level of protection.

The option of using materials or systems that have been tested by subjecting them to an ASTM E119 or UL 263 time-temperature fire exposure under positive pressure is the requirement for protection of the perimeter gap that existed in previous editions of the code. These systems and materials tested in the years before the formal issuance of ASTM E2307 can continue to be used to comply with Section 714.4. Those assemblies that were tested using the ASTM E119 or UL 263 time-temperature fire exposure in a multistory apparatus would be expected to be in substantial compliance with ASTM E2307, and are expected to provide an equivalent level of protection.

The exterior wall construction must comply with Section 705.8.5. Section 705.8.5 requires a 3-foot-high (914 mm) 1-hour fire-resistance-rated spandrel wall in unsprinklered buildings four stories or greater in height.

715.4.1 Exterior curtain wall/nonfire-resistance-rated floor assembly intersections. Voids created at the intersection of exterior curtain wall assemblies and nonfire-resistance-rated floor or floor/ceiling assemblies shall be sealed with an *approved* material or system to retard the interior spread of fire and hot gases between *stories*.

❖ Where the joint between walls involves a nonfire-resistance-rated floor and an exterior curtain wall, there is no reason to try to maintain a fire-resistance rating with a rated joint system. However, spread of smoke is a concern and, therefore, the code calls for a tight joint to protect rapid spread of smoke from a floor of fire origin to other floors of the building. The exterior wall construction must also comply with Section 705.8.5 even though the floors are not fire-resistance rated.

715.4.2 Exterior curtain wall/vertical fire barrier intersections. Voids created at the intersection of nonfire-resistance-rated exterior curtain wall assemblies and *fire barriers* shall be filled. An approved material or system shall be used to fill the void and shall be securely installed in or on the intersection for its entire length so as not to dislodge, loosen or otherwise impair its ability to accommodate expected building movements and to retard the passage of fire and hot gases.

❖ This section specifically addresses the intersection of nonfire-resistance-rated exterior curtain wall and rated fire barriers. The proposed language provides clear performance requirements that can be applied and enforced in these conditions.

715.5 Spandrel wall. Height and fire-resistance requirements for curtain wall spandrels shall comply with Section 705.8.5. Where Section 705.8.5 does not require a fire-resistance-rated spandrel wall, the requirements of Section 715.4 shall still apply to the intersection between the spandrel wall and the floor.

❖ This provision serves as a cross reference to the vertical separation of openings in exterior wall requirements found in Section 705.8.5. This section still requires that where a floor intersects with a nonrated spandrel wall, the void space must be protected by an approved joint system.

715.6 Fire-resistant joint systems in smoke barriers. *Fire-resistant joint systems* in *smoke barriers*, and joints at the intersection of a horizontal *smoke barrier* and an exterior curtain wall, shall be tested in accordance with the requirements of UL 2079 for air leakage. The *L rating* of the joint system shall not exceed 5 cfm per linear foot (0.00775 m^3/s m) of joint at 0.30 inch (7.47 Pa) of water for both the ambient temperature and elevated temperature tests.

❖ The leakage, or L rating, provides a quantitative indication of a fire-resistant joint system's ability to resist the passage of smoke. Although the test is performed using air, the flow properties of air and smoke are sufficiently close for engineering purposes as to provide a reasonable quantification of smoke leakage. The L-rating test is an optional test within the UL 2079 test standard; therefore, not all fire-resistant joint systems that are tested and listed will have this information. In the 2005 UL *Fire Resistance Directory (Vol. 2)*, there are 335 fire-resistant joint systems that are listed that also have an L rating of 5 or less. This should provide a sufficient selection to allow for code compliance in all situations.

SECTION 716
OPENING PROTECTIVES

716.1 General. Opening protectives required by other sections of this code shall comply with the provisions of this section.

❖ This section regulates two types of opening protectives: fire doors (Section 716.5) and fire protection-rated glazing (Section 716.6). As covered in Section 716.2, fire-resistance-rated glazing is not addressed by the balance of the requirements in Section 716. Fire doors are a type of opening protective and are installed in openings in fire-resistance-rated assemblies, including fire walls, fire barriers, fire partitions and exterior walls where the openings are required to be protected. Where a load-bearing exterior wall requires a fire-resistance rating in accordance with Table 601, it is for structural integrity purposes. The openings are not required to be protected unless an opening protective is required by another section of the code, including Section 705.8.

In addition to fire doors, another form of an opening protective is fire protection-rated glazing (commonly referred to as "fire windows"). Fire windows refer to the entire assembly, which may consist of a frame and the glazing material and mounting components. A glass-block assembly is also considered a window assembly.

The fire protection ratings in this section are based on the acceptance criteria of NFPA 252, 257 and UL 10A, 14B or 14C. The fire protection rating acceptance criteria are not the same as that required for a fire-resistance rating for building structural elements. Fire-resistance ratings of building construction are determined by ASTM E119 or UL 263. The fire protection rating required for an opening protective is generally less than the required fire resistance of the wall (see Tables 716.5 and 716.6). This is based upon the ability of a wall to have material or a fuel package directly against the assembly while fire doors and windows are assumed to have the fuel package remote from the surface of the assembly. Sections 716.5 and 716.6 (and subsections) reference the appropriate test standards and require that all opening protectives be labeled (see Section 716.5.7.1 for doors, Section 716.5.8.3 for glazing in doors and Section 716.6.8 for glazing in windows), with the exception being oversized doors that require a certificate of inspection by an approved agency (see Section 716.5.7.2).

716.2 Fire-resistance-rated glazing. *Fire-resistance-rated* glazing tested as part of a *fire-resistance-rated* wall or floor/ceiling assembly in accordance with ASTM E119 or UL 263 and labeled in accordance with Section 703.6 shall not otherwise be required to comply with this section where used as part of a wall or floor/ceiling assembly. *Fire-resistance-rated* glazing shall be permitted in fire door and *fire window assemblies* where tested and installed in accordance with their listings and where in compliance with the requirements of this section.

❖ There are glazing materials used as and tested as a wall assembly under ASTM E119 or UL 263 that are not covered by Section 716. Because these materials can meet the same fire-resistance requirements of the wall, they are not required to be regulated as an opening that would require the lower level of protection that fire protection-rated glazing provides (see commentary, Section 703.6). The second sentence indicates that when fire-resistance-rated glazing is used as part of a fire door or fire window assembly, there are provisions in Section 716 that apply to its use.

716.3 Marking fire-rated glazing assemblies. *Fire-rated glazing* assemblies shall be marked in accordance with Tables 716.3, 716.5 and 716.6.

❖ With two types of fire-rated glazing, it is very important to understand the labeling system and the meaning of each mark. The type of wall, the type of opening and the size of the opening determine the minimum required type of fire-rated glazing. Table 716.3 defines the markings, while Tables 716.5 and 716.6 give the requirements for fire-rated glazing by type of wall and location in the wall. There are three different requirements depending on whether the glass is in a door, in a sidelite or transom door frame, or in a fire window. Commentary Figures 716.3(1) and 716.3(2) show the required fire-rated glazing markings by type of wall and location.

716.3.1 Fire-rated glazing identification. For *fire-rated glazing*, the *label* shall bear the identification required in Tables 716.3 and 716.5. "D" indicates that the glazing is permitted to be used in *fire door* assemblies and that the glazing meets the fire protection requirements of NFPA 252. "H" shall indicate that the glazing meets the hose stream requirements of NFPA 252. "T" shall indicate that the glazing meets the temperature requirements of Section 716.5.5.1. The placeholder "XXX" represents the fire-rating period, in minutes.

❖ The provisions of this section apply to fire-rated glazing used in fire door assemblies. By following the provisions of this section, the glazing utilized in fire door assemblies can be easily identified for verification of

TABLE 716.3
MARKING FIRE-RATED GLAZING ASSEMBLIES

FIRE TEST STANDARD	MARKING	DEFINITION OF MARKING
ASTM E119 or UL 263	W	Meets wall assembly criteria.
NFPA 257 or UL 9	OH	Meets fire window assembly criteria including the hose stream test.
NFPA 252 or UL 10B or UL 10C	D H T	Meets fire door assembly criteria. Meets fire door assembly hose stream test. Meets 450°F temperature rise criteria for 30 minutes
	XXX	The time in minutes of the fire resistance or fire protection rating of the glazing assembly.

For SI: °C = [(°F) - 32]/1.8.

its appropriate application. The "D" designation indicates the glazing can be used in a fire-door assembly, with the remaining identifiers providing specific information as to the glazing's capability to meet the hose stream test and temperature limits. See Section 716.5.8 for further discussion on the use of this glazing material.

716.3.2 Fire-protection-rated glazing identification. For *fire-protection-rated* glazing, the *label* shall bear the following identification required in Tables 716.3 and 716.6: "OH – XXX." "OH" indicates that the glazing meets both the fire protection and the hose-stream requirements of NFPA 257 or UL 9 and is permitted to be used in fire window openings. The placeholder "XXX" represents the fire-rating period, in minutes.

❖ The provisions of this section apply to fire-protection-rated glazing assemblies. Section 716.3.2 indicates that Tables 716.3 and 716.6 are the appropriate tables to be used for fire-protection-rated glazing and to provide details of the required label and standards for performance. The "OH" designation indicates the glazing meets both the fire protection and hose stream requirements for use in fire windows. See Section 716.5.8 for further discussion on the use of this glazing material.

716.3.3 Fire-rated glazing that exceeds the code requirements. *Fire-rated glazing* assemblies marked as complying with hose stream requirements (H) shall be permitted in applications that do not require compliance with hose stream requirements. *Fire-rated glazing* assemblies marked as complying with temperature rise requirements (T) shall be permitted in applications that do not require compliance with temperature rise requirements. *Fire-rated glazing* assemblies marked with ratings (XXX) that exceed the ratings required by this code shall be permitted.

❖ Fire-protection-rated glazing with an "H" marking has been tested and has passed the hose stream requirements of NFPA 252. Fire protection-rated glazing with a "T" marking has been tested and has passed the hose stream requirements of NFPA 252. For example, a door view panel less than 100 square inches (64 516 mm^2) in area in a 2-hour exit enclosure requires only a glazing mark of [D-H-90], but a glazing panel marked with [D-H-T-90] could also be used since it exceeds the requirements for the particular location. If Table 716.5 requires glazing with a marking [D-H-T-60], then a panel with a marking of [D-H-T-90] could be used.

716.4 Alternative methods for determining fire protection ratings. The application of any of the alternative methods *listed* in this section shall be based on the fire exposure and acceptance criteria specified in NFPA 252, NFPA 257 or UL 9. The required *fire resistance* of an opening protective shall be permitted to be established by any of the following methods or procedures:

1. Designs documented in *approved* sources.
2. Calculations performed in an *approved* manner.
3. Engineering analysis based on a comparison of opening protective designs having *fire protection ratings* as determined by the test procedures set forth in NFPA 252, NFPA 257 or UL 9.
4. Alternative protection methods as allowed by Section 104.11.

❖ Section 716.4 contains language similar to Section 703.3 regarding fire-resistance ratings. The intent of this section is to recognize approved calculations, an engineering analysis or alternative protection methods as permitted in Section 104.11 as an acceptable alternative means to determine a fire protection rating.

716.5 Fire door and shutter assemblies. Approved *fire door* and fire shutter assemblies shall be constructed of any material or assembly of component materials that conforms to the test requirements of Section 716.5.1, 716.5.2 or 716.5.3 and the *fire protection rating* indicated in Table 716.5. *Fire door*

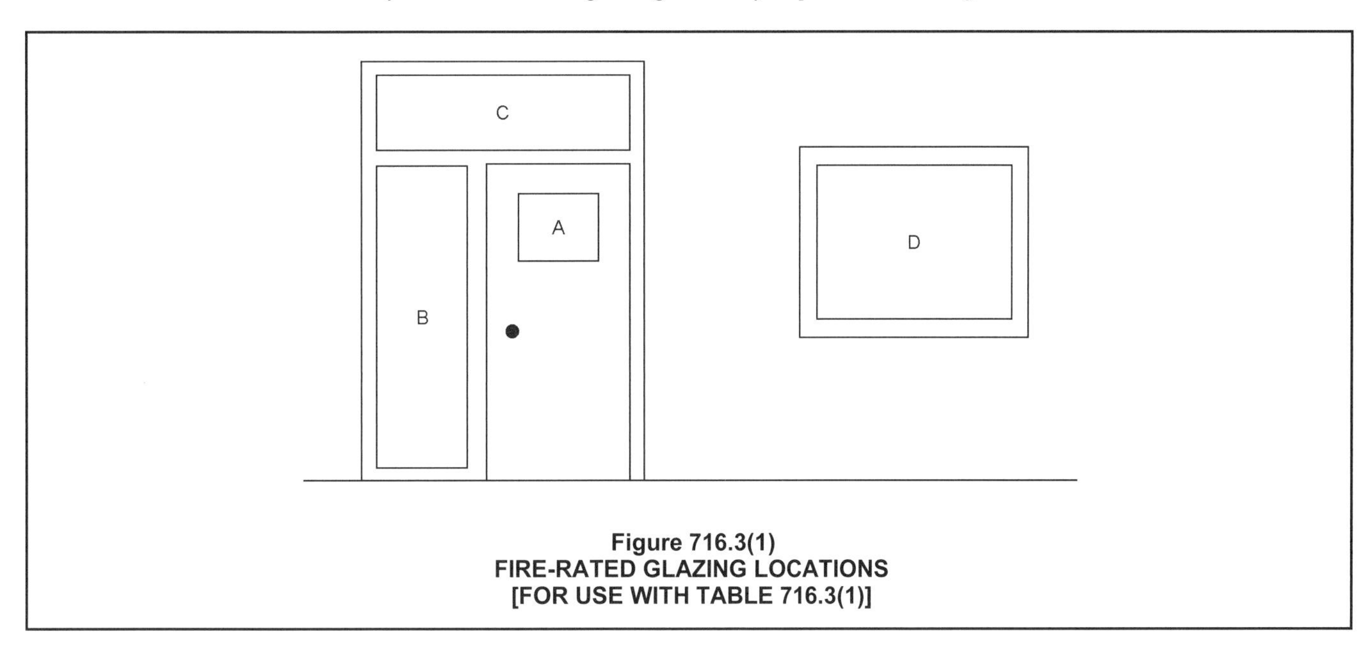

Figure 716.3(1)
FIRE-RATED GLAZING LOCATIONS
[FOR USE WITH TABLE 716.3(1)]

frames with transom lights, sidelights or both shall be permitted in accordance with Section 716.5.6. *Fire door* assemblies and shutters shall be installed in accordance with the provisions of this section and NFPA 80.

Exceptions:

1. Labeled protective assemblies that conform to the requirements of this section or UL 10A, UL 14B and UL 14C for tin-clad *fire door* assemblies.
2. Floor *fire door* assemblies in accordance with Section 712.1.13.1.

❖ When the code refers to a fire door assembly, the intent is that the term "fire door" (see the definition in Section 202) applies to the door itself, while the term "fire door assembly" includes the fire door, the frame, related hardware and accessories, unless otherwise noted (see the definition of "Fire door assembly" in Section 202). Therefore, it is important to realize that a fire door by itself is not generally accepted. An assembly that includes the door is needed. A door by itself cannot provide the protection. It needs to be supported by the hinges, held into a specific frame and generally have latching and closing hardware. By requiring an assembly tested to the specific standards, all of the components needed for protection will be provided. For example, although there is not a direct mention in the section relative to mandating latching hardware, the referenced installation standard, NFPA 80, addresses the requirements relative to positive latch devices.

Note that reference to Section 716.5.6 is made for fire door assemblies that include transoms and a sidelite. Fire door assemblies with sidelites or tran-

TYPE OF ASSEMBLY	REQUIRED WALL ASSEMBLY RATING (HRS)	DOOR VIEW PANEL A 100 IN^2 OR LESS	DOOR VIEW PANEL A > 100 IN^2	SIDELITE OR TRANSOM PANELS B & C	WINDOW ASSEMBLY D
Fire wall and fire barrier wall	4-hr	Not Permitted	Not Permitted	W-240[b]	W-240[b]
	3-hr	Not Permitted	Not Permitted	W-180[b]	W-180[b]
	2-hr	[D-H-90]	[D-H-W-90][c]	W-120[b]	W-120[b]
	$1^1/_2$-hr	[D-H-90]	[D-H-W-90][c]	W-90[b]	W-90[b]
Shaft, exit enclosures and exit passageway walls	2-hr	[D-H-90]	[D-H-W-90][c] or [D-H-T-W-90][c]	W-120[b, e]	W-120[b, e]
Fire barriers for enclosures of shafts, exit access stairways, exit access ramps, interior exit stairways and exit passageway walls	1-hr	[D-H-60]	[D-H-T-60] or [D-H-T-W-60][c]	W-60[b, e]	W-60[b, e]
Other fire barriers	1-hr	[D-H-NT-45]	[D-H-NT-45]	[D-H-NT-45]	W-60[b]
Fire barriers – incidental use areas, mixed occupancy separations	1-hr	[D-H-NT-45]	[D-H-NT-45]	[D-H-NT-45]	[OH-45] or W-60[b]
Fire partitions: Corridor walls	1-hr	[D-20]	[D-20]	[D-H-OH-45]	[OH-45] or W-60[b]
	0.5-hr	[D-20]	[D-20]	[D-H-OH-20]	[OH-20] or W-30[b]
Other fire partitions	1-hr	[D-H-45]	[D-H-45]	[D-H-45]	[OH-45] or W-60[b]
	0.5-hr	[D-H-20]	[D-H-20]	[D-H-20]	[OH-20] or W-30[b]
Smoke barriers	1-hr	[D-20]	[D-20]	[D-H-OH-45]	[OH-45] or W-60[b]
Exterior walls[d]	3-hr	[D-H-90]	[D-H-W-90][c]	W-180[b]	[OH-90] or W-180[b]
	2-hr	[D-H-90]	[D-H-W-90][c]	W-120[b]	[OH-90] or W-120[b]
	1-hr	[D-H-45]	[D-H-45]	[D-H-45]	[OH-45] or W-60[b]

For SI: 1 square inch = 645.2 mm^2.

a. Markings in brackets [ ] are fire-protection-rated glazing. Fire-protection-rated glazing in doors, sidelights and transoms meets NFPA 252. Fire protection-rated glazing in windows meets NFPA 257 or UL 9.

b. Fire-resistance-rated glazing meeting ASTM E119 or UL 263.

c. Glazing meeting both the standard for fire-protection-rated glazing in fire doors (NFPA 252) and the standard fire-resistance-rated glazing (ASTM E119 or UL 263).

d. Glazing requirements where protected openings are required by Table 705.8.

e. Only openings necessary for egress are allowed in exit access stairways and interior exit stairs.

Figure 716.3(2)
MARKINGS OF FIRE-RATED GLAZING[a]

soms must be fire-protection-rated glazing if the opening protective is 3/4 hour or less and fire-resistive rated if the open protective is 1 hour or more.

Labels that indicate compliance with UL 10A, 14B or 14C are acceptable (see Exception 1). Exception 2 exempts floor fire door assemblies from the requirements of this section, since Section 712.1.13.1 requires floor fire doors to comply with a different test standard (NFPA 288) and their rating must not be less than the rating of the horizontal assembly being penetrated (see the commentary to Section 712.1.13.1 for a discussion of floor fire doors).

TABLE 716.5. See below.

❖ This table lists the minimum fire protection ratings for fire doors relative to the nature and fire-resistance rating of the wall. Once the purpose and fire-resistance rating of the wall are identified, the minimum

TABLE 716.5
OPENING FIRE PROTECTION ASSEMBLIES, RATINGS AND MARKINGS

TYPE OF ASSEMBLY	REQUIRED WALL ASSEMBLY RATING (hours)	MINIMUM FIRE DOOR AND FIRE SHUTTER ASSEMBLY RATING (hours)	DOOR VISION PANEL SIZE[b]	FIRE-RATED GLAZING MARKING DOOR VISION PANEL[d]	MINIMUM SIDELIGHT/ TRANSOM ASSEMBLY RATING (hours)		FIRE-RATED GLAZING MARKING SIDELIGHT/TRANSOM PANEL	
					Fire protection	Fire resistance	Fire protection	Fire resistance
Fire walls and fire barriers having a required fire-resistance rating greater than 1 hour	4	3	See Note b	D-H-W-240	Not Permitted	4	Not Permitted	W-240
	3	3[a]	See Note b	D-H-W-180	Not Permitted	3	Not Permitted	W-180
	2	1 1/2	100 sq. in.	≤100 sq. in. = D-H-90 >100 sq. in.= D-H-W-90	Not Permitted	2	Not Permitted	W-120
	1 1/2	1 1/2	100 sq. in.	≤100 sq. in. = D-H-90 >100 sq. in.= D-H-W-90	Not Permitted	1 1/2	Not Permitted	W-90
Enclosures for shafts, interior exit stairways and interior exit ramps.	2	1 1/2	100 sq. in.	≤100 sq. in. = D-H-90 > 100 sq. in.= D-H-T-W-90	Not Permitted	2	Not Permitted	W-120
Horizontal exits in fire walls[e]	4	3	100 sq. in.	≤100 sq. in. = D-H-180 > 100 sq. in.= D-H-W-240	Not Permitted	4	Not Permitted	W-240
	3	3[a]	100 sq. in.	≤100 sq. in. = D-H-180 > 100 sq. in.= D-H-W-180	Not Permitted	3	Not Permitted	W-180
Fire barriers having a required fire-resistance rating of 1 hour: Enclosures for shafts, exit access stairways, exit access ramps, interior exit stairways and interior exit ramps; and exit passageway walls	1	1	100 sq. in.[c]	≤100 sq. in. = D-H-60 >100 sq. in.= D-H-T-W-60	Not Permitted	1	Not Permitted	W-60
					Fire protection			
Other fire barriers	1	3/4	Maximum size tested	D-H	3/4		D-H	
Fire partitions: Corridor walls	1	1/3[b]	Maximum size tested	D-20	3/4[b]		D-H-OH-45	
	0.5	1/3[b]	Maximum size tested	D-20	1/3		D-H-OH-20	
Other fire partitions	1	3/4	Maximum size tested	D-H-45	3/4		D-H-45	
	0.5	1/3	Maximum size tested	D-H-20	1/3		D-H-20	

(continued)

TABLE 716.5—continued
OPENING FIRE PROTECTION ASSEMBLIES, RATINGS AND MARKINGS

TYPE OF ASSEMBLY	REQUIRED WALL ASSEMBLY RATING (hours)	MINIMUM FIRE DOOR AND FIRE SHUTTER ASSEMBLY RATING (hours)	DOOR VISION PANEL SIZE[b]	FIRE-RATED GLAZING MARKING DOOR VISION PANEL[d]	MINIMUM SIDELIGHT/ TRANSOM ASSEMBLY RATING (hours)		FIRE-RATED GLAZING MARKING SIDELIGHT/TRANSOM PANEL	
					Fire protection	Fire resistance	Fire protection	Fire resistance
Exterior walls	3	$1^1/_2$	100 sq. in.[b]	≤100 sq. in. = D-H-90 >100 sq. in = D-H-W-90	Not Permitted	3	Not Permitted	W-180
	2	$1^1/_2$	100 sq. in.[b]	≤100 sq. in. = D-H-90 >100 sq. in.= D-H-W-90	Not Permitted	2	Not Permitted	W-120
					Fire protection			
	1	$^3/_4$	Maximum size tested	D-H-45	$^3/_4$		D-H-45	
Smoke barriers					Fire protection			
	1	$^1/_3$	Maximum size tested	D-20	$^3/_4$		D-H-OH-45	

For SI: 1 square inch = 645.2 mm^2.

a. Two doors, each with a fire protection rating of $1^1/_2$ hours, installed on opposite sides of the same opening in a fire wall, shall be deemed equivalent in fire protection rating to one 3-hour fire door.

b. Fire-resistance-rated glazing tested to ASTM E119 in accordance with Section 716.2 shall be permitted, in the maximum size tested.

c. Except where the building is equipped throughout with an automatic sprinkler and the fire-rated glazing meets the criteria established in Section 716.5.5.

d. Under the column heading "Fire-rated glazing marking door vision panel," W refers to the *fire-resistance rating* of the glazing, not the frame.

e. See Section 716.5.8.1.2.1.

required fire protection rating of the door can be determined using the table. Typically, the minimum permitted fire door fire protection rating is less than the required fire-resistance rating of the wall.

The allowable size and required marking of glazed door vision panels are shown in the fourth and fifth columns. The last column indicates the minimum glazing marking for the sidelite and transom. The glazing markings are explained in Section 716.5.8.3.1 and Table 716.3.

In addition to the fire protection ratings in Table 716.5, other sections of the code prescribe additional performance criteria for fire door assemblies. For example, Section 716.5.5 requires doors in exit enclosures to have a maximum temperature increase on the unexposed surface of 450°F (232°C) after 30 minutes. Although temperature rise is not a condition of acceptance, test standards require that unexposed surface temperatures be measured and documented in the test report. Another requirement not in the table is related to corridor doors. Required fire-rated corridor doors must also meet the requirements for smoke and draft control in Section 716.5.3.1.

The exterior wall opening protectives in Table 716.5 are not required where Section 705.8 does not require protected openings.

716.5.1 Side-hinged or pivoted swinging doors. *Fire door* assemblies with side-hinged and pivoted swinging doors shall be tested in accordance with NFPA 252 or UL 10C. After 5 minutes into the NFPA 252 test, the neutral pressure level in the furnace shall be established at 40 inches (1016 mm) or less above the sill.

❖ Fire door assemblies that contain side-hinged or pivoted swinging doors are to be tested in accordance with the referenced test standards.

When first introduced in one of the early editions of the code, the requirement for the door to be tested with a pressure differential (positive pressure at the top and a negative pressure at the bottom) and a specific neutral pressure point was a major revision in the way that doors were tested. The intent of this provision is to better approximate the conditions found in an actual building fire where smoke, heat and hot gasses rise to the top of the room and increase the pressure in the upper portion of the room. Test method UL 10C contains the requirements for positive pressure testing.

716.5.2 Other types of assemblies. *Fire door* assemblies with other types of doors, including swinging elevator doors, horizontal sliding fire door assemblies, and fire shutter assemblies, bottom and side-hinged chute intake doors, and top-hinged chute discharge doors, shall be tested in accordance with NFPA 252 or UL 10B. The pressure in the furnace shall be maintained as nearly equal to the atmospheric pressure as possible. Once established, the pressure shall be maintained during the entire test period.

❖ This section also references NFPA 252 or UL10B for fire door assemblies containing other than the typical side-hinged-type doors. NFPA 252 is a general test standard that is not specific to any type of door. UL 10B was developed for negative pressure testing of fire door and fire shutter assemblies and is viewed as a viable test method, provided that the pressure is maintained at or nearly at atmospheric pressure. It is important to distinguish that the pressure testing that is required for side-hinged or pivoted swinging doors by Section 716.5.1 is not applicable to this section and other door types.

716.5.3 Door assemblies in corridors and smoke barriers. *Fire door* assemblies required to have a minimum *fire protection rating* of 20 minutes where located in *corridor* walls or *smoke barrier* walls having a *fire-resistance rating* in accordance with Table 716.5 shall be tested in accordance with NFPA 252 or UL 10C without the hose stream test.

Exceptions:

1. Viewports that require a hole not larger than 1 inch (25 mm) in diameter through the door, have not less than a 0.25-inch-thick (6.4 mm) glass disc and the holder is of metal that will not melt out where subject to temperatures of 1,700°F (927°C).
2. *Corridor* door assemblies in occupancies of Group I-2 shall be in accordance with Section 407.3.1.
3. Unprotected openings shall be permitted for *corridors* in multitheater complexes where each motion picture auditorium has not fewer than one-half of its required *exit* or *exit access doorways* opening directly to the exterior or into an *exit* passageway.
4. Horizontal sliding doors in *smoke barriers* that comply with Sections 408.6 and 408.8.4 in occupancies in Group I-3.

❖ This section covers doors in fire partitions specifically for corridors and doors in smoke barriers. NFPA 252 and UL 10C standards require a hose stream test on all fire door assemblies. The hose stream test is intended to provide a measurement of structural integrity by evaluating the ability of the assembly to withstand impact. It has come to be accepted that a hose stream is not justified for 20-minute doors.

Exception 1 accepts certain viewports that are not required to be tested as a part of a door assembly. The use of viewports is fairly common in corridors of a Group R-1 occupancy or other similar locations. Because of the limited size and materials, the overall performance of the door will not be greatly affected by including the viewport.

Exception 2 serves as a cross reference to the requirements for a Group I-2 occupancy, which does not require the door to have a fire protection rating.

Exception 3 has little application now due to the requirements of Sections 1020.1 and 903.2.1.1, Item 4. This exception was originally intended to allow the removal of latching hardware on the corridor door and, therefore, reduce the level of noise that was made when it was opened during a movie. Since the code now requires all multitheater complexes to be sprinklered and corridors in sprinklered Group A occupancies do not require a fire-resistive rating, the other provisions of the code would eliminate the need for the door to have a fire-protection rating and the latch that was considered as the problem is then generally not required.

Exception 4 serves as a cross reference to Section 408 for Group I-3 occupancies that contain requirements specific to smoke-tight door assemblies for specific locations within resident housing areas.

716.5.3.1 Smoke and draft control. *Fire door* assemblies shall meet the requirements for a smoke and draft control door assembly tested in accordance with UL 1784. The air leakage rate of the door assembly shall not exceed 3.0 cubic feet per minute per square foot ($0.01524\ m^3/s \cdot m^2$) of door opening at 0.10 inch (24.9 Pa) of water for both the ambient temperature and elevated temperature tests. Louvers shall be prohibited. Installation of smoke doors shall be in accordance with NFPA 105.

❖ This section is only applicable to fire doors in corridors or smoke barrier walls. When a corridor wall is required to be fire protection rated, a degree of smoke control is needed to provide a tenable space for egress. Fire doors in corridors and smoke barriers are also required to meet the criteria of UL 1784. This standard measures the movement of smoke through a door assembly. The installation of the doors is to be done according to NFPA 105 requirements. This standard, entitled *Standard for the Installation of Smoke Door Assemblies*, is a companion to the previously referenced NFPA 80. The criteria for air leakage are also provided in this section.

716.5.3.2 Glazing in door assemblies. In a 20-minute *fire door assembly*, the glazing material in the door itself shall have a minimum fire-protection-rated glazing of 20 minutes and shall be exempt from the hose stream test. Glazing material in any other part of the door assembly, including transom lights and sidelights, shall be tested in accordance with NFPA 257 or UL 9, including the hose stream test, in accordance with Section 716.6.

❖ Glazing in 20-minute-rated doors must also have a 20-minute fire protection rating and is not required to pass the hose stream test. Sidelites and transoms, which are not part of the fire door but instead are part of the fire-door assembly (see definitions in Section 202), are required to pass the hose stream test as detailed in NFPA 257 or UL 9.

716.5.4 Door assemblies in other fire partitions. *Fire door* assemblies required to have a minimum fire protection rating of 20 minutes where located in other *fire partitions* having a fire-resistance rating of 0.5 hour in accordance with Table 716.5 shall be tested in accordance with NFPA 252, UL 10B or UL 10C with the hose stream test.

❖ Section 708.1 lists the locations requiring fire partitions. Fire partitions, other than corridor walls, would be dwelling unit separation walls, sleeping unit separation walls, tenant space walls in malls and elevator lobby separation walls. Unlike the corridor doors in Section 716.5.3, these doors must be tested with the hose stream.

716.5.5 Doors in interior exit stairways and ramps and exit passageways. *Fire door* assemblies in interior exit stairways and ramps and exit passageways shall have a maximum transmitted temperature rise of not more than 450°F (250°C) above ambient at the end of 30 minutes of standard fire test exposure.

Exception: The maximum transmitted temperature rise is not required in buildings equipped throughout with an *automatic sprinkler system* installed in accordance with Section 903.3.1.1 or 903.3.1.2.

❖ This section requires that door assemblies utilized in interior exit stairways, ramp enclosures and exit passageways comply with the requirements of Section 715 and be tested in accordance with either NFPA 252 or UL 10C. This section adds a requirement that door construction is limited to a temperature rise of 450°F (250°C) on the unexposed side during the first 30 minutes of the standard fire test (NFPA 252). The labels for these doors are required to indicate that the temperature on the unexposed side is 450°F (250°C) or less above ambient temperature. The temperature rise rating is not required on doors in sprinklered buildings.

It should be noted that the temperature rise criterion is not otherwise a limitation with respect to a door receiving a fire protection rating. Therefore, simply specifying a $1^1/_2$-hour fire protection rating on the door does not ensure the additional requirements of Section 715.4.4 are met. The basis for limiting the temperature rise of the unexposed surface of exit doors to 450°F (250°C) is that a higher allowable temperature would provide enough radiant heat to discourage or even prevent building occupants from closely approaching or passing by the door assembly during a fire emergency. The intent is to protect the occupants in the vertical exit enclosure and exit passageway from excessive radiant heat. The exception provides relief from providing doors meeting temperature rise ratings when buildings are equipped throughout with either an NFPA 13 or 13R automatic sprinkler system.

716.5.5.1 Glazing in doors. Fire-protection-rated glazing in excess of 100 square inches (0.065 m^2) is not permitted. Fire-resistance-rated glazing in excess of 100 square inches (0.065 m^2) shall be permitted in door *fire doors*. Listed *fire-resistance-rated* glazing in a *fire door* shall have a maximum transmitted temperature rise in accordance with Section 716.5.5 when the *fire door* is tested in accordance with NFPA 252, UL 10B or UL 10C.

❖ This section only addresses glazing in doors in stairway enclosures and exit passage enclosures. Glazing 100 square inches (0.065 m^2) or less may be fire-protection-rated glazing. Glazing over 100 square inches (0.065 m^2) must be fire-resistance-rated glazing. The larger glazing panels in fire doors shall have a temperature rise rating of 450°F (250°C). This increased glazing is to be tested as a component of the door. The test methods to be utilized are listed in Section 716.5 and include NFPA 252, UL 10B and UL10C. The 100-square-inch (0.065 m^2) threshold is consistent with the glazing size limitations contained in Section 716.5.8.1.2.1. In general, the glazing is required to be tested first, and, if it meets the ASTM E119 acceptance criteria, it is listed as a fire-resistance-rated glazing. That "listed fire-resistance-rated glazing" is then installed in a fire door and tested in accordance with NFPA 252, including tests for the maximum transmitted temperature rise requirements of Section 716.5.5.

716.5.6 Fire door frames with transom lights and sidelights. Door frames with transom lights, sidelights or both, shall be permitted where a $^3/_4$-hour *fire protection rating* or less is required in accordance with Table 716.5. *Fire door* frames with transom lights, sidelights, or both, installed with fire-resistance-rated glazing tested as an assembly in accordance with ASTM E119 or UL 263 shall be permitted where a fire protection rating exceeding $^3/_4$ hour is required in accordance with Table 716.5.

❖ The purpose of this section is to address the use of fire-resistance-rated glazing in fire door frames with transoms or sidelights where the fire protection rating exceeds $^3/_4$ hour. For instances where the required rating is $^3/_4$ hour or less, fire protection-rated glazing meeting the test requirements of NFPA 252 shall apply. Fire-resistance-rated glazing is required where the required rating exceeds $^3/_4$ hour. The assembly must be tested in accordance with ASTM E119 or UL 263. Testing the assemblies to these criteria exposes the glazing to the appropriate temperature rise to substantiate the higher fire protection rating.

716.5.7 Labeled protective assemblies. *Fire door* assemblies shall be labeled by an *approved agency*. The *labels* shall comply with NFPA 80, and shall be permanently affixed to the door or frame.

❖ The requirement that fire doors be labeled leads to the provisions of Section 1703.5 being applied. A label indicates that the door assembly has not only passed the required fire test but that a follow-up inspection was also performed during production (see Section 1703.5.2). The building official should, therefore, verify that fire door assemblies are properly labeled, and not rely solely on a test report. To provide specific guidance, compliance with NFPA 80 is required.

716.5.7.1 Fire door labeling requirements. *Fire doors* shall be labeled showing the name of the manufacturer or other identification readily traceable back to the manufacturer, the name or trademark of the third-party inspection agency, the *fire protection rating* and, where required for *fire doors* in interior exit stairways and ramps and exit passageways by Section 716.5.5, the maximum transmitted temperature end point. Smoke and draft control doors complying with UL 1784 shall be labeled as such and shall comply with Section 716.5.7.3. Labels shall be approved and permanently affixed.

The label shall be applied at the factory or location where fabrication and assembly are performed.

❖ Labels on fire doors apply to the door only. The building official should verify that the remaining portions of the assembly (door frame, hardware and accessories) are also labeled for use with a labeled fire door. In addition to the appropriate fire protection rating, the labels for interior exit stairways and ramps and exit passageway doors are required to indicate the temperature rise on the unexposed surface after 30 minutes (see commentary, Section 716.5.5). A label is also to serve as evidence of both a required fire protection rating and third-party inspection—not solely as a manufacturer's identification (see commentary, Section 1703.5). To ensure appropriate labeling, it is required to occur at the actual location where fabrication and assembly occur.

716.5.7.1.1 Light kits, louvers and components. Listed light kits and louvers and their required preparations shall be considered as part of the labeled door where such installations are done under the listing program of the third-party agency. *Fire doors* and door assemblies shall be permitted to consist of components, including glazing, vision light kits and hardware that are listed or classified and labeled for such use by different third-party agencies.

❖ The code allows the use of listed light kits, listed louvers and their required preparation material to be considered as part of a labeled door where such light kits and louvers are installed under the conditions of their listing. Where such components are listed for use in such doors, they may be listed, or classified, and labeled by different third-party agencies. Louvers would be required to be listed with shutters and could not be used where prohibited elsewhere in the code.

716.5.7.2 Oversized doors. Oversized *fire doors* shall bear an oversized *fire door label* by an *approved agency* or shall be provided with a certificate of inspection furnished by an *approved* testing agency. Where a certificate of inspection is furnished by an *approved* testing agency, the certificate shall state that the door conforms to the requirements of design, materials and construction, but has not been subjected to the fire test.

❖ Recognizing that doors may exceed the required tested size, the code indicates that oversized doors that have a certificate of inspection from an approved testing agency are acceptable. For example, rolling steel-type fire doors are normally listed for 120 square feet (11 m^2) with no dimension in excess of 12 feet (3658 mm). Rolling steel-type doors that exceed these dimensions are permitted as long as they have certification indicating that they are constructed of materials of the same grade, thickness, shape, etc., as labeled doors. This is normally indicated with a certificate that reads "Oversized Fire Door Certificate."

The code does not specify what is considered as an "oversized door." This information would be dependent on the test standard and the dimensions listed within it. Note that the provisions of Sections 706.8 and 707.6 allow openings up to 156 square feet (14 m^2) in area.

716.5.7.3 Smoke and draft control door labeling requirements. Smoke and draft control doors complying with UL 1784 shall be labeled in accordance with Section 716.5.7.1 and shall show the letter "S" on the fire-rating *label* of the door. This marking shall indicate that the door and frame assembly are in compliance where *listed* or labeled gasketing is installed.

❖ When doors are required to be approved for smoke and draft control, they require a special label containing the letter "S." Without this label it is very difficult to assess in the field that the door meets UL 1784 and that listed or labeled gasketing is also required to be installed with the door and frame.

716.5.7.4 Fire door frame labeling requirements. *Fire door* frames shall be labeled showing the names of the manufacturer and the third-party inspection agency.

❖ It is required that the fire door frame be labeled with names of the manufacturer and the third-party testing agency to assist in the assessment of the door in the field. The actual rating is not required to be included.

716.5.7.5 Fire door operator labeling requirements. *Fire door* operators for horizontal sliding doors shall be labeled and listed for use with the assembly.

❖ Section 716.5 requires fire door assemblies to be installed in accordance with NFPA 80. NFPA 80 requires fire door operators to be listed for use with the door. Therefore, the requirements of this section related to operators of horizontal sliding fire doors are consistent with the requirements in NFPA 80 that, by reference, are required.

716.5.8 Glazing material. *Fire-rated glazing* and *fire-resistance-rated* glazing conforming to the opening protection requirements in Section 716.5 shall be permitted in *fire door* assemblies.

❖ This section of the code addresses fire-resistance-rated glazing and fire protection-rated glazing material in door assemblies installed in fire walls and fire barriers. It applies to glazing in doors, sidelights and transoms, but not to glazing in fire windows, which is covered in Section 716.6. Wired glass is no longer specifically mentioned and, if used, must meet the same test standards as other glazing material (see Table C716.3).

716.5.8.1 Size limitations. *Fire-resistance-rated* glazing shall comply with the size limitations in Section 716.5.8.1.1. Fire-protection-rated glazing shall comply with the size limitations of NFPA 80, and as provided in Section 716.5.8.1.2.

❖ Fire-resistance-rated glazing is allowed in more applications than fire-protection-rated glazing. Therefore the more stringent fire testing methods of Section 716.5.8.1.1 are required. NFPA 80 permits fire-protection-rated glazing in fire doors with a fire protection rating up to 3 hours. For $^1/_3$-, $^1/_2$-, $^3/_4$-, 1- and $1^1/_2$-hour doors, the allowable area is a function of the glazed area tested. One-hour and $1^1/_2$- hour doors requiring a temperature rise rating are limited to 100 square inches (0.065 m^2) or less. While NFPA 80 allows a view panel 10 square inches (0.065 m^2) or less in 3-hour fire doors, Table 716.5 does not. The design must meet both NFPA 90 and Table 716.5.

716.5.8.1.1 Fire-resistance-rated glazing in door assemblies in fire walls and fire barriers rated greater than 1 hour. Fire-resistance-rated glazing tested to ASTM E119 or UL 263 and NFPA 252, UL 10B or UL 10C shall be permitted in *fire door assemblies* located in *fire walls* and in *fire barriers* in accordance with Table 716.5 to the maximum size tested and in accordance with their listings.

❖ Fire-resistance-rated glazing, not fire protection glazing, shall be permitted in fire walls and fire barrier walls to the maximum size tested and as allowed in Table 716.5. The glazing, if used in a fire door view panel, must be tested to two standards—both ASTM E119 or UL 263 and NFPA 252 or UL 10.

716.5.8.1.2 Fire-protection-rated glazing in door assemblies in fire walls and fire barriers rated greater than 1 hour. Fire-protection-rated glazing shall be prohibited in *fire walls* and *fire barriers* except as provided in Sections 716.5.8.1.2.1 and 716.5.8.1.2.2.

❖ Fire-protection-rated glazing, not fire-resistance-rated glazing, shall not be allowed in fire walls or fire barriers greater than 1 hour except for 100-square-inch (0.065 m^2) or less view panels in fire doors in horizontal exits and fire barrier walls. This view panel allowance recognizes the benefit of visual observance in an emergency situation.

716.5.8.1.2.1 Horizontal exits. Fire-protection-rated glazing shall be permitted as vision panels in *self-closing* swinging *fire door* assemblies serving as horizontal exits in *fire walls* where limited to 100 square inches (0.065 m^2) with no dimension exceeding 10 inches (0.3 mm).

❖ Only self-closing swinging fire doors and not all doors in horizontal exits are allowed the limited size view panel.

716.5.8.1.2.2 Fire barriers. Fire-protection-rated glazing shall be permitted in *fire doors* having a $1^1/_2$-hour *fire protection rating* intended for installation in *fire barriers*, where limited to 100 square inches (0.065 m^2).

❖ Only fire barrier doors required to be $1^1/_2$-hour fire rated are allowed to have the limited area view panel. Some fire barriers are required to be 3- and 4-hour rated, so doors with the 3-hour fire rating would not be allowed the view panels.

716.5.8.2 Elevator, stairway and ramp protectives. Approved fire-protection-rated glazing used in *fire door* assemblies in elevator, stairway and ramp enclosures shall be so located as to furnish clear vision of the passageway or approach to the elevator, stairway or ramp.

❖ The purpose of vision panels in exit doors and elevator doors is to permit observation by occupants who may be on the opposite side of the door before opening it. The limitations on the size of panels in Section 716.5.8.1 still apply.

716.5.8.3 Labeling. *Fire-rated glazing* shall bear a *label* or other identification showing the name of the manufacturer, the test standard and information required in Table 716.3 that shall be issued by an *approved agency* and shall be permanently identified on the glazing.

❖ This provision of the code maintains proper labeling for each piece of fire protection-rated glazing. The label must specify the name or fully identifying logo of the manufacturer, the test standard that was used to evaluate the glass and the rating established by the test.

716.5.8.4 Safety glazing. *Fire-protection-rated* glazing and *fire-resistance-rated* glazing installed in *fire door* assemblies shall comply with the safety glazing requirements of Chapter 24 where applicable.

❖ This section provides a cross reference to the safety glazing requirements of Chapter 24, specifically Section 2406. Not only do glass or glazed panels need to be fire-resistance rated, but the glass also must pass the test requirements of CPSC 16 CFR, Part 1201 if it is installed in locations subject to human impact loads. Locations where glazing presents a hazard to occupants upon impact are identified in Section 2406.4. This requirement reduces the hazards of someone falling into a fire-resistance-rated glass panel.

716.5.9 Door closing. *Fire doors* shall be latching and self- or automatic-closing in accordance with this section.

Exceptions:

1. *Fire doors* located in common walls separating *sleeping units* in Group R-1 shall be permitted without automatic- or *self-closing* devices.
2. The elevator car doors and the associated hoistway enclosure doors at the floor level designated for recall in accordance with Section 3003.2 shall be permitted to remain open during Phase I emergency recall operation.

❖ In order for fire doors to be effective, they must be in the closed position. Therefore, the preferred arrangement is to install self-closing doors. Recognizing that operational practices often require doors to be open for an extended period of time, automatic-closing doors are permitted as long as this opening will not pose a threat to occupant safety and the doors will self-latch. Automatic-closing devices enable the opening to be protected during a fire condition. The basic requirement for closing devices and specific requirements for automatic-closing and self-closing devices are given in NFPA 80 (see Section 716.5). The requirements for latching provides added confidence that the door will remain closed by being latched.

Exception 1 discusses doors in the separation walls of side-by-side sleeping units in Group R-1 occupancies. This allows two or more sleeping units to be opened to each other so that a suite can be established if desired. If the adjacent rooms are rented by separate people, the doors will generally be closed. If the rooms are rented together, this permits a larger room to be created simply by opening the door. Because most rooms only open to one adjacent room, most configurations would still result in a solid fire partition without any door openings between this suite of rooms and any other sleeping units.

Exception 2 coordinates with the requirements related to elevator recall that are found in Chapter 30. Since the hoistway doors generally provide the shaft opening protection, this exception is needed for this situation.

716.5.9.1 Latch required. Unless otherwise specifically permitted, single *fire doors* and both leaves of pairs of side-hinged swinging *fire doors* shall be provided with an active latch bolt that will secure the door when it is closed.

❖ This section merely reinforces the acceptance criteria of fire test standards for fire doors. A door that does not latch would generally be an ineffective barrier against the spread of a fire and would be unable to withstand the pressures of a fire in the adjacent space. Such hardware must be listed for use on fire doors.

716.5.9.1.1 Chute intake door latching. Chute intake doors shall be positive latching, remaining latched and closed in the event of latch spring failure during a fire emergency.

❖ Latching devices for chute intake doors must provide a positive latch and, in the event of a latch spring failure, shall remain closed. The proponent of this code change stated that simple technology exists that can accomplish this fail-safe provision.

716.5.9.2 Automatic-closing fire door assemblies. Automatic-closing *fire door* assemblies shall be *self-closing* in accordance with NFPA 80.

❖ NFPA 80 requires doors to be automatic closing with a closing device and a separate hold-release device or hold-open mechanism that closes upon activation of an automatic fire detector acceptable to the building official. This provision simply ensures that when the door is released it will move to the closed position.

716.5.9.3 Smoke-activated doors. Automatic-closing doors installed in the following locations shall be automatic-closing by the actuation of smoke detectors installed in accordance with Section 907.3 or by loss of power to the smoke detector or hold-open device. Doors that are automatic-closing by smoke detection shall not have more than a 10-second delay before the door starts to close after the smoke detector is actuated:

1. Doors installed across a *corridor*.
2. Doors installed in the enclosures of *exit access stairways* and *ramps* in accordance with Sections 1019 and 1023, respectively.
3. Doors that protect openings in *exits* or *corridors* required to be of fire-resistance-rated construction.
4. Doors that protect openings in walls that are capable of resisting the passage of smoke in accordance with Section 509.4.
5. Doors installed in *smoke barriers* in accordance with Section 709.5.
6. Doors installed in *fire partitions* in accordance with Section 708.6.
7. Doors installed in a *fire wall* in accordance with Section 706.8.
8. Doors installed in shaft enclosures in accordance with Section 713.7.
9. Doors installed in waste and linen chutes, discharge openings and access and discharge rooms in accordance with Section 713.13. Loading doors installed in waste and linen chutes shall meet the requirements of Sections 716.5.9 and 716.5.9.1.1.
10. Doors installed in the walls for compartmentation of underground buildings in accordance with Section 405.4.2.
11. Doors installed in the elevator lobby walls of underground buildings in accordance with Section 405.4.3.
12. Doors installed in smoke partitions in accordance with Section 710.5.2.3.

❖ Since the integrity of the means of egress or general occupant safety can be compromised by smoke, doors at certain locations are required to be automatic closing upon the detection of smoke. The automatic closer is also to activate upon loss of power to the smoke detector and to the hold-open device. NFPA 72, as referenced in Chapter 9, contains criteria relative to the number and location of detectors necessary for door release service. NFPA 80 also provides criteria for closing devices required for a fire door.

716.5.9.4 Doors in pedestrian ways. Vertical sliding or vertical rolling steel *fire doors* in openings through which pedestrians travel shall be heat activated or activated by smoke detectors with alarm verification.

❖ Where vertical sliding or rolling doors are provided as an opening protective through which pedestrian travel is intended, automatic-closing actuation is not permitted via a smoke detector unless alarm verification is provided. Sudden closing of these doors and the potential of false actuation of smoke detectors create a hazard to occupants moving through the opening. Although pedestrian travel is intended in openings regulated by this section, a common oversight is that the vertical rolling or sliding door is not permitted to be considered an egress component, as stated in Section 1008.1.2.

716.5.10 Swinging fire shutters. Where fire shutters of the swinging type are installed in exterior openings, not less than one row in every three vertical rows shall be arranged to be readily opened from the outside, and shall be identified by distinguishing marks or letters not less than 6 inches (152 mm) high.

❖ The fire department needs access to openings for ventilation purposes and entry. Therefore, if an extensive amount of swinging shutters is used, at least one in every three vertical rows must be readily openable from the outside. The operable shutter must also be easily recognized for use by the fire department. The identification is on the exterior.

716.5.11 Rolling fire shutters. Where fire shutters of the rolling type are installed, such shutters shall include *approved* automatic-closing devices.

❖ Rolling fire shutters require a detecting device to initiate the closing sequence. The actuation shall be by smoke detectors in locations required by Section 716.5.9.3.

716.6 Fire-protection-rated glazing. Glazing in *fire window assemblies* shall be fire protection rated in accordance with this section and Table 716.6. Glazing in *fire door* assemblies shall comply with Section 716.5.8. Fire-protection-rated glazing in fire window assemblies shall be tested in accordance with and shall meet the acceptance criteria of NFPA 257 or UL 9. Fire-protection-rated glazing shall comply with NFPA 80. Openings in nonfire-resistance-rated *exterior wall* assemblies that require protection in accordance with Section 705.3, 705.8, 705.8.5 or 705.8.6 shall have a fire protection rating of not less than $^3/_4$ hour. Fire-protection-rated glazing in 0.5-hour fire-resistance-rated partitions is permitted to have an 0.33-hour fire protection rating.

❖ This section covers glazing in fire windows, not fire doors. The appropriate test standards referenced for fire windows (fire-protection-rated glazing) are NFPA 257 or UL 9, which are pass/fail tests with a specified time of exposure. It is important to realize that this section only applies to materials that are tested in accordance with the two identified standards and that are intended as an opening protective in a wall that requires openings with a fire protection rating. The distinction between fire-protection-rated glazing and a fire-resistance-rated glazing must be understood. Glazing that has a fire-resistance rating is not required to comply with Section 716 if it is tested and installed in accordance with its listing (see Section 716.2). It is required in many openings where fire-protection-rated glazing is not allowed to be installed. Fire windows with fire-protection-rated glazing are only allowed in incidental use and occupancy separation fire barrier walls rated at 1 hour or less, fire partitions, smoke barriers and fire-resistance-rated exterior walls. In addition, Sections 1023.4 and 1024.5 would not allow fire windows in exit enclosures and exit passageway enclosures. The basic premise of the referenced standard is that it evaluates the effectiveness of windows when used as opening protectives to remain in place during the specified time of exposure. In addition to being subjected to a predetermined fire condition, the assembly is also subjected to a hose stream impact test. The test procedure does not measure or evaluate heat transmission or radiation through the assembly. Table 716.6 gives the fire ratings and the marking requirements for fire windows.

TABLE 716.6. See below.

❖ Table 716.6 prescribes the minimum fire protection ratings for fire windows, relative to the type and fire-resistance rating of the wall. Once the purpose and fire-resistance rating of the wall are identified, the

TABLE 716.6
FIRE WINDOW ASSEMBLY FIRE PROTECTION RATINGS

TYPE OF WALL ASSEMBLY	REQUIRED WALL ASSEMBLY RATING (hours)	MINIMUM FIRE WINDOW ASSEMBLY RATING (hours)	FIRE-RATED GLAZING MARKING
Interior walls			
Fire walls	All	NP[a]	W-XXX[b]
Fire barriers	>1 1	NP[a] NP[a]	W-XXX[b] W-XXX[b]
Incidental use areas (Section 707.3.7), Mixed occupancy separations (Section 707.3.9)	1	$^3/_4$	OH-45 or W-60
Fire partitions	1 0.5	$^3/_4$ $^1/_3$	OH-45 or W-60 OH-20 or W-30
Smoke barriers	1	$^3/_4$	OH-45 or W-60
Exterior walls	>1 1 0.5	$1^1/_2$ $^3/_4$ $^1/_3$	OH-90 or W-XXX[b] OH-45 or W-60 OH-20 or W-30
Party wall	All	NP	Not Applicable

NP = Not Permitted.

a. Not permitted except fire-resistance-rated glazing assemblies tested to ASTM E119 or UL 263, as specified in Section 716.2.

b. XXX = The fire rating duration period in minutes, which shall be equal to the *fire-resistance rating* required for the wall assembly.

minimum required fire protection rating of windows and the required marking are determined by using the table. The requirements in this table are not to be confused with the ratings for fire doors provided in Table 716.5 since the requirements do differ. For a comparison of the provisions of the two tables see Table C716.3.

716.6.1 Testing under positive pressure. NFPA 257 or UL 9 shall evaluate fire-protection-rated glazing under positive pressure. Within the first 10 minutes of a test, the pressure in the furnace shall be adjusted so not less than two-thirds of the test specimen is above the neutral pressure plane, and the neutral pressure plane shall be maintained at that height for the balance of the test.

❖ Under fire conditions there is likely to be positive pressure acting on the window. Therefore, this section requires that tested window assemblies be subjected to positive pressure conditions during the test. This requirement is consistent with standards used for testing window assemblies in the United States and other countries.

716.6.2 Nonsymmetrical glazing systems. Nonsymmetrical fire-protection-rated glazing systems in *fire partitions*, *fire barriers* or in *exterior walls* with a *fire separation distance* of 5 feet (1524 mm) or less pursuant to Section 705 shall be tested with both faces exposed to the furnace, and the assigned *fire protection rating* shall be the shortest duration obtained from the two tests conducted in compliance with NFPA 257 or UL 9.

❖ This section requires that glazing systems that by design are not the same on either side be tested on both sides. The rating resulting from these tests will be determined from the lower performing side. These requirements are for fire-type barriers (versus smoke) within buildings and when used on an exterior wall with a small separation distance [less than 5 feet (1524 mm)]. The concern here is the same as for nonsymmetrical wall assemblies in Section 703.2.1.

716.6.3 Safety glazing. *Fire-protection-rated* glazing and *fire-resistance-rated* glazing installed in *fire window assemblies* shall comply with the safety glazing requirements of Chapter 24 where applicable.

❖ This section provides a cross reference to the safety glazing requirements of Chapter 24, specifically Section 2406. Not only do glass or glazed panels need to be fire-resistance rated, but the glass also must pass the test requirements of CPSC 16 CFR, Part 1201 if it is installed in locations subject to human impact loads. Locations where glazing presents a hazard to occupants upon impact are identified in Section 2406.4. This requirement reduces the hazards of someone being injured by falling into a fire-resistance-rated glass panel.

716.6.4 Glass and glazing. Glazing in *fire window assemblies* shall be fire-protection-rated glazing installed in accordance with and complying with the size limitations set forth in NFPA 80.

❖ Nothing in this section prohibits fire-resistance-rated glazing from use in fire window assemblies installed in accordance with their listing. Fire protection-rated glazing shall be installed in accordance with NFPA 80.

716.6.5 Installation. Fire-protection-rated glazing shall be in the fixed position or be automatic-closing and shall be installed in *approved* frames.

❖ In order to adequately protect the opening, protection must be in place during a fire event. This is achieved by either having the glazing in a permanently fixed position or having the glazing protective actuated by a detection device.

716.6.6 Window mullions. Metal mullions that exceed a nominal height of 12 feet (3658 mm) shall be protected with materials to afford the same *fire-resistance rating* as required for the wall construction in which the protective is located.

❖ Fire windows are normally tested in panels not exceeding 1,296 square inches (0.85 m^2) and 54 inches (1372 mm) in width or length, respectively. The code does, however, permit panels to be installed adjacent to one another. This section requires that mullions higher than 12 feet (3658 mm) meet the criteria for fire-resistance-rated glazing (ASTM E119 or UL 263), not fire-protection-rated glazing (NFPA 257).

716.6.7 Interior fire window assemblies. Fire-protection-rated glazing used in *fire window assemblies* located in *fire partitions* and *fire barriers* shall be limited to use in assemblies with a maximum *fire-resistance rating* of 1 hour in accordance with this section.

❖ Since fire-protection-rated glazing has a maximum $^3/_4$-hour fire-resistance rating, fire windows using fire-protection-rated glazing may only be used in fire-partitions (1-hour rated in accordance with Section 708.3) and fire barriers having a fire-resistance rating of 1 hour. These limitations are indicated in Table 716.6.

716.6.7.1 Where $^3/_4$-hour fire protection window assemblies permitted. Fire-protection-rated glazing requiring 45-minute opening protection in accordance with Table 716.6 shall be limited to *fire partitions* designed in accordance with Section 708 and *fire barriers* utilized in the applications set forth in Sections 707.3.6 and 707.3.8 where the *fire-resistance rating* does not exceed 1 hour. Fire-resistance-rated glazing assemblies tested in accordance with ASTM E119 or UL 263 shall not be subject to the limitations of this section.

❖ Not all 1-hour fire-resistance-rated walls are allowed to have fire windows with fire-protection-rated glazing. All fire partitions and certain fire barrier walls may have fire protection-rated glazing. Only 1-hour fire-resistance-rated fire barriers in incidental use separations (Section 707.3.7) and mixed occupancy separations (Section 707.3.9) may have fire protection-rated glazing in fire windows. Shaft walls, exit access stairway enclosures, interior exit stairways, exit passageway enclosures and control area separation walls are considered fire barrier walls, though when rated for 1 hour, are not allowed to have fire windows with fire-protection-rated glazing. Of course, nothing in this section prohibits fire-resistance-rated glazing in fire windows in some of those applications.

716.6.7.2 Area limitations. The total area of the glazing in fire-protection-rated window assemblies shall not exceed 25 percent of the area of a common wall with any room.

❖ The area limitations for interior fire windows vary from those found in Section 707.6 for fire barriers. Openings of all types in fire barriers are limited to 25 percent of the length of the wall while this section bases the limitation on fire windows to the area of the wall. Therefore, when dealing with a fire barrier, both the 25-percent length and area would be applicable. When dealing with a fire partition, only the area limitation would apply.

The phrase "common wall" will limit the size to the actual visible wall that is shared and can be seen in the two rooms. The area of the wall that continues through the interstitial space above a ceiling would not be included when determining the wall size. In addition, if the ceiling height differs on the two sides of the wall, the lowest ceiling height would be used to determine what is considered the "common wall."

716.6.7.3 Where $^1/_3$-hour fire-protection window assemblies permitted. Fire-protection-rated glazing shall be permitted in window assemblies tested to NFPA 257 or UL 9 in *smoke barriers* and *fire partitions* requiring $^1/_3$-hour opening protection in accordance with Table 716.6.

❖ The previous sections dealt with fire partitions and fire barriers, whereas this section addresses smoke barriers and fire partitions requiring only 20-minute opening protectives. Fire windows with fire-protection-rated glazing are permitted in such partitions with the ratings and markings shown in Table 716.6.

716.6.8 Labeling requirements. Fire-protection-rated glazing shall bear a label or other identification showing the name of the manufacturer, the test standard and information required in Section 716.3.2 and Table 716.6 that shall be issued by an approved agency and permanently identified on the glazing.

❖ This section requires fire-protection-rated glazing (see Section 716.6) to be labeled by an approved testing agency. This requirement enables the code user to verify that a fire window has been tested to the correct standard and that the test was properly performed by experienced people with the correct equipment. Table 716.6 details the necessary information required on the fire window label. Each glazed panel is to be permanently identified. These descriptive requirements are used to verify that fire window assemblies have been correctly tested. The label on each and every fire window assembly must contain the manufacturer's name, the test standard used and the fire protection rating. The label must be applied at the manufacturer's plant by the third-party inspection agency. As a reminder, fire-resistance-rated glazing is not regulated by this section (see Section 716.2). Labeling requirements for fire-resistance-rated glazing can be found in Section 703.6.

SECTION 717
DUCTS AND AIR TRANSFER OPENINGS

717.1 General. The provisions of this section shall govern the protection of duct penetrations and air transfer openings in assemblies required to be protected and duct penetrations in nonfire-resistance-rated floor assemblies.

❖ Fire dampers, smoke dampers and combination fire/smoke dampers protect openings created by duct penetrations and transfer openings in assemblies required to be fire-resistance rated. Ceiling radiation dampers protect duct penetrations, which only penetrate the ceiling membrane of a fire-resistance-rated assembly. Penetrations with ducts through nonfire-resistance-rated floors are also regulated by this section. This section includes installation, testing, rating, and actuation requirements in Sections 717.2, 717.3.1, 717.3.2 and 717.3.3, respectively. Section 717.3.1 4 indicates that dampers are to bear the label of an approved agency. When dampers are provided, they must be properly maintained and, therefore, must be accessible (see Section 717.4). Section 717.5 indicates the conditions in which dampers are required at penetrations through vertical assemblies. Duct penetrations and transfer openings through horizontal assemblies, including requirements for ceiling dampers, are regulated in Section 717.6. The provisions of Section 717 are duplicated in Section 607 of the IMC. All the requirements for protection of duct penetrations and transfer openings are found in this section. This section is directly referenced by sections that address vertical and horizontal fire-resistance-rated assemblies, as follows:

Exterior walls	Section 705.10
Fire walls	Section 706.11
Fire barriers	Section 707.10
Fire partitions	Section 708.9
Smoke barriers	Section 709.8
Smoke partitions	Section 710.8
Vertical openings	Section 712.1.6
Shafts	Section 713.10
Penetrations	Section 714.1.1

717.1.1 Ducts and air transfer openings. Ducts transitioning horizontally between shafts shall not require a shaft enclosure provided that the duct penetration into each associated shaft is protected with *dampers* complying with this section.

❖ This section addresses a typical condition where the design requires HVAC ductwork to jog horizontally then continue vertically within the building. This is allowed without a continuous (horizontal) shaft enclosure as long as the penetrations through the vertical shafts are protected with the appropriate dampers. See Commentary Figure 717.1.1.

717.1.2 Ducts that penetrate fire-resistance-rated assemblies without dampers. Ducts that penetrate fire-resistance-rated assemblies and are not required by this section to have *dampers* shall comply with the requirements of Sections 714.2 through 714.3.3. Ducts that penetrate *horizontal assemblies* not required to be contained within a shaft and not required by this section to have *dampers* shall comply with the requirements of Sections 714.4 through 714.5.2.

❖ This section identifies how duct penetrations are to be properly protected, either in accordance with Section 717 using fire or fire/smoke dampers where required or, if dampers are not required, in accordance with Section 714 with materials that have been tested to either ASTM E814 or ASTM E119 or UL 263 to maintain the fire-resistance rating of the fire assembly. This section is coordinated with Section 714.3 (fire-resistance-rated walls) and Section 714.4 (horizontal assemblies). If a duct does not require a damper, the penetration of that duct through a fire-resistance-rated assembly must be protected as a through penetration in accordance with Section 714.

If a duct does require a fire damper, smoke damper or combination fire/smoke damper, the through-penetration protection requirements of Section 714 will not apply. The reason is that these dampers are fire tested as a complete assembly installed within a fire-resistance-rated wall assembly or horizontal assembly. The test and resulting listing will include not only the damper but also all required mounting hardware for the duct at the point where it penetrates the wall or floor. As long as the complete installation details in the listing are complied with, there is no need for any additional protection (e.g., per Section 714) to ensure that the fire-resistance rating of the wall assembly or horizontal assembly is maintained. However, it is possible to pass the fire test and obtain a listing for a fire damper or smoke damper without completely preventing smoke migration outside the duct, such as through the gap between the outside surface of the duct and the inside surface of the hole made to accommodate the duct. This can be quite apparent in some cases, as daylight might be visible from one side of the wall or floor assembly to the other side. This is because the tested dampers generally require a minimum gap to be provided between the damper and the wall or floor it is installed in. Although not required by the code, this gap is sometimes closed with a sealant in order to provide more complete smoke resistance for the wall or floor assembly. It is imperative that an intumescent firestop sealant not be used on the outside of the duct to seal this gap. The expansion pressure of the intumescent sealant can buckle the duct, thus hindering or preventing the

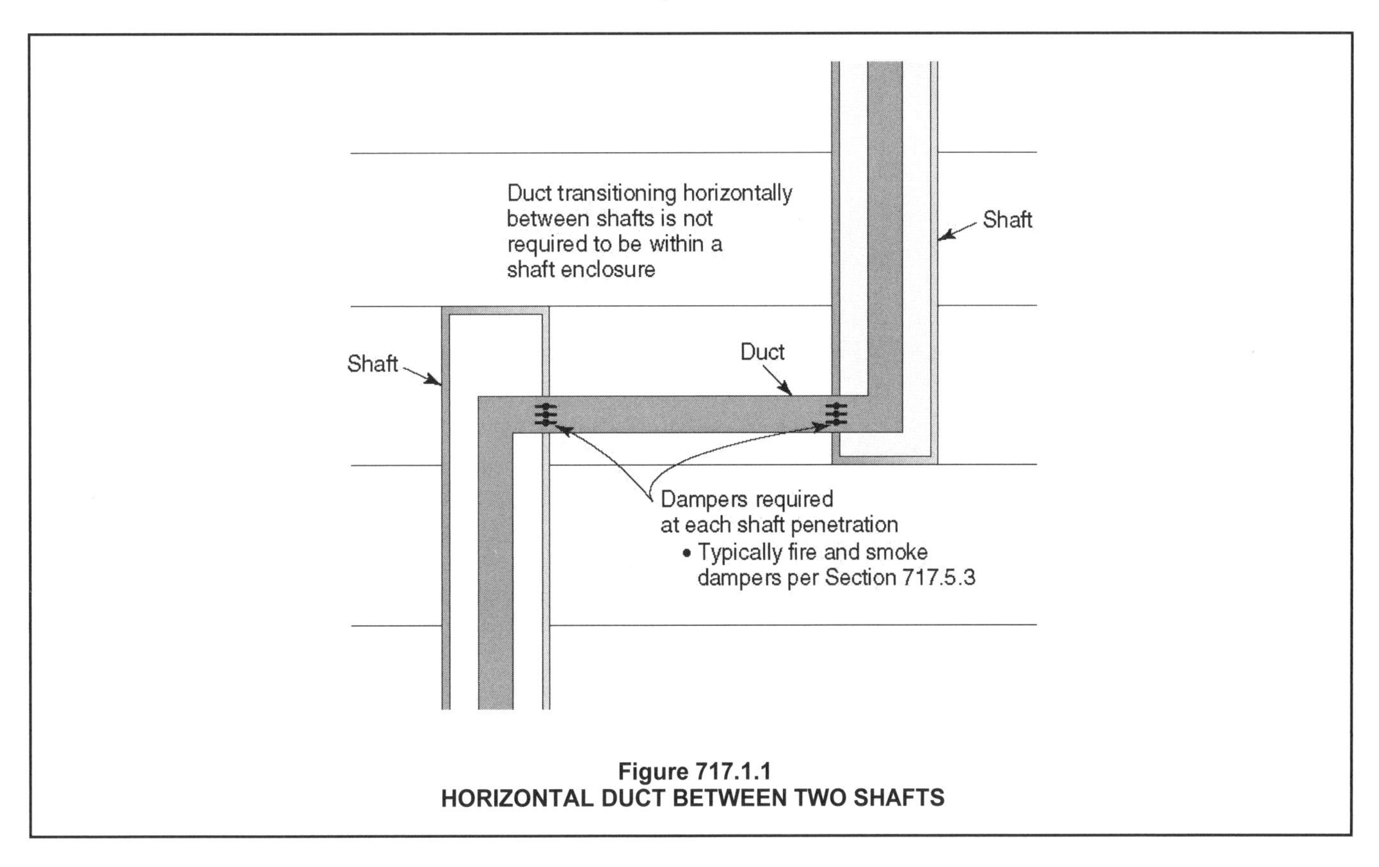

Figure 717.1.1
HORIZONTAL DUCT BETWEEN TWO SHAFTS

proper operation of the damper under fire conditions.

The purpose of this section is to point the user to the annular space protection requirements for ducts that penetrate nonfire-resistance-rated assemblies, when applicable. The referenced section, Section 717.6.3, provides options for protecting these types of penetrations, two of which contain requirements for annular space protection.

717.1.2.1 Ducts that penetrate nonfire-resistance-rated assemblies. The space around a duct penetrating a nonfire-resistance-rated floor assembly shall comply with Section 717.6.3.

❖ The purpose of this section is to point the user to the annular space protection requirements for ducts that penetrate nonfire-resistance-rated assemblies, when applicable. The referenced section, Section 717.6.3, provides options for protecting these types of penetrations, two of which contain requirements for annular space protection.

717.2 Installation. *Fire dampers, smoke dampers, combination fire/smoke dampers* and *ceiling radiation dampers* located within air distribution and smoke control systems shall be installed in accordance with the requirements of this section, the manufacturer's instructions and the *dampers'* listing.

❖ This section performs two regulating functions. First, it requires that dampers be installed in accordance with the manufacturer's installation instructions and listing. Such instructions will result in an installation that not only protects the penetration when the damper is actuated, but also create a protected opening should the duct collapse. Second, this section introduces Sections 717.2.1 and 717.2.2, which address the relationship of dampers with smoke control systems and hazardous exhaust ducts.

717.2.1 Smoke control system. Where the installation of a *fire damper* will interfere with the operation of a required smoke control system in accordance with Section 909, *approved* alternative protection shall be utilized. Where mechanical systems including ducts and *dampers* utilized for normal building ventilation serve as part of the smoke control system, the expected performance of these systems in smoke control mode shall be addressed in the rational analysis required by Section 909.4.

❖ Fire dampers are not permitted where they will obstruct or inhibit the proper operation of a required smoke control system. In a smoke control system, fire dampers are replaced with an alternative means of protection subject to the approval of the building official. Alternative protection may be in the form of a fire-resistance-rated shaft enclosure, steel subducts extending at least 22 inches (559 mm) vertically in an exhaust shaft having continuous upward airflow or a fire-resistance-rated horizontal duct enclosure. If a smoke control system is installed that is not required by the code, it is still subject to the same fire damper requirements as a required smoke control system; however, such nonmandated systems would be permitted to employ fire dampers instead of the mandatory alternative protection prescribed in this section for required smoke control systems. A required smoke control system is an important life-safety system and the code-mandated provisions here combine with Section 909 design requirements to ensure operability. This section does more than simply allow fire dampers to be omitted in required smoke control systems; it mandates the omission of fire dampers where such system operation could be jeopardized. Lastly, this section allows the normal building ventilation system (HVAC) to be employed to serve as the smoke control system, as long as the performance is substantiated.

717.2.2 Hazardous exhaust ducts. *Fire dampers* for hazardous exhaust duct systems shall comply with the *International Mechanical Code*.

❖ This section refers the user to the IMC. In particular, Section 510 of the IMC deals with hazardous exhaust systems. Hazardous exhaust can be emissions including flammable vapors, gases, fumes, mists or dusts, and airborne materials posing a health hazard. These types of duct systems are not allowed by Section 510.6.1 of the IMC to have fire or smoke dampers, but Section 510.6.3 of the IMC requires fire-resistance-rated duct enclosure construction from the point of penetrations of a fire-resistance-rated wall to the exterior. Section 510.4 of the IMC requires independent systems.

717.3 Damper testing, ratings and actuation. *Damper* testing, ratings and actuation shall be in accordance with Sections 717.3.1 through 717.3.3.

❖ The purpose of this section is to introduce the requirements for the testing, rating and actuation of fire dampers, smoke dampers, combination fire/smoke dampers and ceiling radiation dampers.

717.3.1 Damper testing. Dampers shall be listed and labeled in accordance with the standards in this section.

1. *Fire dampers* shall comply with the requirements of UL 555. Only *fire dampers* and *ceiling radiation dampers* labeled for use in dynamic systems shall be installed in heating, ventilation and air-conditioning systems designed to operate with fans on during a fire.
2. Smoke dampers shall comply with the requirements of UL 555S.
3. Combination fire/smoke dampers shall comply with the requirements of both UL 555 and UL 555S.
4. Ceiling radiation dampers shall comply with the requirements of UL 555C or shall be tested as part of a fire-resistance-rated floor/ceiling or roof/ceiling assembly in accordance with ASTM E119 or UL 263.
5. *Corridor dampers* shall comply with requirements of both UL 555 and UL 555S. *Corridor dampers* shall demonstrate acceptable closure performance when subjected to 150 feet per minute (0.76 mps) velocity across

the face of the damper during the UL 555 fire exposure test.

❖ The purpose of this section is to provide the physical testing requirements for fire dampers, smoke dampers, combination fire/smoke dampers and ceiling radiation dampers. A fire damper is a device designed to close automatically upon detection of heat, to interrupt migratory airflow and to restrict the passage of flame through duct penetrations of rated assemblies. Fire dampers are required to be tested in accordance with UL 555. The criteria for acceptance of fire dampers require that they close and latch automatically, limit movement (warping) to prescribed limitations and remain in the opening for the fire exposure period for which they are rated. In addition to being tested, fire dampers must be labeled and, in accordance with Section 1703.5, require a follow-up service.

Fire dampers and ceiling radiation dampers are classified by UL 555 for use in static and dynamic airflow conditions. Fire dampers and ceiling radiation dampers installed in air-distribution systems that remain in operation after smoke or heat from a fire is detected (a dynamic airflow condition) must be labeled for such use. Static fire dampers may not operate properly under dynamic conditions. Therefore, fire dampers used in systems designed with dynamic airflow must be tested and labeled for closure under the anticipated airflow and pressure conditions.

The manufacturer's installation instructions must be followed. Fire dampers are to be installed in the assembly such that an opening created by a duct collapse is properly protected. Dampers are to be installed in accordance with their tested application, either vertically or horizontally. Although the test criteria of UL 555 are not functions of a horizontal versus vertical application, there are specific tests required for spring-operated dampers that are not required for gravity operated types. Therefore, the type of damper used requires the building official to determine if it is acceptable for the given application.

Fire dampers are required to have the minimum fire protection rating established in Table 717.3.2.1. A smoke damper is a device installed in a duct, typically in smoke control systems or to protect smoke barrier penetrations, which operates to seal the duct against smoke leakage. Fire dampers tested to UL 555 are not rated for smoke leakage. Fire dampers are activated at a predetermined temperature using a fusible link or other heat-responsive device. Smoke dampers, on the other hand, are tested for leakage in accordance with the requirements of UL 555S and are activated by a smoke detector in the duct or area to be served.

Ceiling radiation dampers, formerly known as ceiling dampers, are designed to limit the radiative heat transfer through an air inlet/outlet opening in the ceiling membrane of a fire-resistance-rated floor/ceiling or roof/ceiling assembly. Ceiling radiation dampers are evaluated either in accordance with UL 555C or as part of the floor/ceiling or roof/ceiling assembly tested to ASTM E119 or UL 263. Ceiling radiation dampers evaluated in accordance with UL 555C are investigated for use in lieu of hinged-door-type dampers commonly specified in the listing of fire-resistance-rated floor/ceiling or roof/ceiling assemblies. UL 555C also provides criteria for the construction of ceiling dampers. Ceiling radiation dampers investigated as part of the fire-resistance-rated floor/ceiling or roof/ceiling assembly are specified in the listing of such assemblies. The description of the tested assembly will include a description of the ceiling damper and its installation. A fire damper is a device designed to close automatically upon detection of heat, to interrupt migratory airflow and to restrict the passage of flame through duct penetrations of rated assemblies. Since ceiling and fire dampers have different design criteria, a fire damper must not be used in place of a ceiling damper.

Corridor dampers are addressed in Item 5. This item describes the testing standards that corridor dampers must meet. Corridor dampers are also required to demonstrate acceptable closure performance when subjected to 150 fpm velocity across the face of the damper during fire exposure testing.

717.3.2 Damper rating. *Damper* ratings shall be in accordance with Sections 717.3.2.1 through 717.3.2.4.

❖ The purpose of this section is to provide direction to the requirements for the ratings of fire dampers (Section 717.3.2.1), smoke dampers (Section 717.3.2.2) and combination fire/smoke dampers (Section 717.3.2.3).

717.3.2.1 Fire damper ratings. *Fire dampers* shall have the minimum *fire protection rating* specified in Table 717.3.2.1 for the type of penetration.

❖ This section primarily serves the purpose of requiring fire damper ratings to be determined in accordance with Table 717.3.2.1. Table 717.3.2.1 sets the minimum damper ratings based upon the type of assemblies being penetrated.

TABLE 717.3.2.1
FIRE DAMPER RATING

TYPE OF PENETRATION	MINIMUM DAMPER RATING (hours)
Less than 3-hour fire-resistance-rated assemblies	1.5
3-hour or greater fire-resistance-rated assemblies	3

❖ The table summarizes the required hourly ratings for fire dampers based on the fire-resistance-rated assembly that is being penetrated by the air distribution system. The left-hand column lists the associated hourly ratings of either wall or horizontal assemblies. The code user enters the line item in the left-hand column based on the applicable fire-resistance rating of the element penetrated and reads across to the right column for the appropriate fire damper rating. These fire damper ratings are established based on tests conducted in accordance with UL 555 and represent the ratings necessary to maintain the integrity of the rated wall or horizontal assembly.

717.3.2.2 Smoke damper ratings. *Smoke damper* leakage ratings shall be Class I or II. Elevated temperature ratings shall be not less than 250°F (121°C).

❖ In accordance with UL 555S, smoke dampers are marked with a temperature rating starting at 250°F (121°C) and rising in increments of 100°F (56°C). Leakage ratings range from Classes I to III in accordance with the standard and are further identified for tested pressure ranging from 4 inches (102 mm) to 12 inches (305 mm) of water. Class I and II leakage ratings meet the code requirement for the damper to be not less than Class II.

717.3.2.3 Combination fire/smoke damper ratings. *Combination fire/smoke dampers* shall have the minimum *fire protection rating* specified for *fire dampers* in Table 717.3.2.1 for the type of penetration and shall have a minimum *smoke damper* rating as specified in Section 717.3.2.2.

❖ This section primarily serves the purpose of requiring the ratings of combination fire/smoke dampers to comply with the requirements for both fire and smoke dampers, including the fire protection rating specified in Table 717.3.2.1 and the leakage rating specified in Section 717.3.2.2.

717.3.2.4 Corridor damper ratings. *Corridor dampers* shall have the following minimum ratings:

1. One hour *fire-resistance rating*.
2. Class I or II leakage rating as specified in Section 717.3.2.2.

❖ This section describes the ratings that corridor dampers must meet. Corridor dampers are listed for both a fire-resistance rating of 1 hour, and a Class I or II leakage rating as defined by the Standard UL 555S. Leakage ratings of corridor dampers are determined at an elevated temperature of 250°F (121°C) or 350°F (177°C). Corridor dampers are only intended to be used to protect duct and air transfer openings in corridor ceilings, where the ceilings are constructed as required for the corridor walls (as permitted in Section 708.4, Exception 3).

717.3.3 Damper actuation. *Damper* actuation shall be in accordance with Sections 717.3.3.1 through 717.3.3.5 as applicable.

❖ The purpose of this section is to provide direction to the requirements for the actuation of fire dampers (Section 717.3.3.1), smoke dampers (Section 717.3.3.2), combination fire/smoke dampers (Section 717.3.3.3) and ceiling radiation dampers (Section 717.3.3.4).

717.3.3.1 Fire damper actuation device. The *fire damper* actuation device shall meet one of the following requirements:

1. The operating temperature shall be approximately 50°F (10°C) above the normal temperature within the duct system, but not less than 160°F (71°C).
2. The operating temperature shall be not more than 350°F (177°C) where located in a smoke control system complying with Section 909.

❖ The purpose of this section is to prescribe the actuation (operating temperature) requirements for fire dampers. The thresholds for actuation are based on UL 555. For static systems, UL 555 identifies 160°F (71°C) as the minimum and 215°F (102°C) as the maximum temperature (see Item 1). For dynamic systems, the standard includes a minimum value of 160°F (71°C) and a maximum value of 350°F (177°C) (see Item 2).

717.3.3.2 Smoke damper actuation. The *smoke damper* shall close upon actuation of a *listed* smoke detector or detectors installed in accordance with Section 907.3 and one of the following methods, as applicable:

1. Where a *smoke damper* is installed within a duct, a smoke detector shall be installed inside the duct or outside the duct with sampling tubes protruding into the duct. The detector or tubes within the duct shall be within 5 feet (1524 mm) of the *damper*. Air outlets and inlets shall not be located between the detector or tubes and the *damper*. The detector shall be *listed* for the air velocity, temperature and humidity anticipated at the point where it is installed. Other than in mechanical smoke control systems, *dampers* shall be closed upon fan shutdown where local smoke detectors require a minimum velocity to operate.
2. Where a *smoke damper* is installed above *smoke barrier* doors in a *smoke barrier*, a spot-type detector shall be installed on either side of the *smoke barrier* door opening. The detector shall be listed for releasing service if used for direct interface with the damper.
3. Where a *smoke damper* is installed within an air transfer opening in a wall, a spot-type detector shall be installed within 5 feet (1524 mm) horizontally of the *damper*. The detector shall be listed for releasing service if used for direct interface with the damper.
4. Where a *smoke damper* is installed in a *corridor* wall or ceiling, the *damper* shall be permitted to be controlled by a smoke detection system installed in the *corridor*.
5. Where a smoke detection system is installed in all areas served by the duct in which the damper will be located, the *smoke dampers* shall be permitted to be controlled by the smoke detection system.

❖ This section provides specific requirements to ensure that an installed detector will detect smoke (see Item 1). Additionally, for passive systems (not part of a smoke control system), the location of detectors is critical to their performance. Smoke dampers are installed in not only duct penetrations, but also transfer openings. A method for properly controlling dampers in such arrangements is provided in Item 3. A smoke detector provided throughout a corridor or an entire area detects smoke more effectively than a duct-type detector. Such detection is necessary for controlling dampers (Items 4 and 5). Smoke dampers may also be operated remotely, especially when used as part of a smoke control system. Combination fire/smoke dampers can be activated by either heat or smoke sensors. Once the primary heat-sensing device has been activated, a remote control system can be used to open the damper to permit its use as a smoke damper in the smoke removal system. Then, if the temperature rises to the damper's maximum degradation test temperature, the secondary heat sensor closes the damper again.

717.3.3.3 Combination fire/smoke damper actuation. *Combination fire/smoke damper* actuation shall be in accordance with Sections 717.3.3.1 and 717.3.3.2. *Combination fire/smoke dampers* installed in smoke control system shaft penetrations shall not be activated by local area smoke detection unless it is secondary to the smoke management system controls.

❖ This section primarily serves the purpose of requiring the actuation of combination fire/smoke dampers to comply with the requirements for both fire and smoke dampers, including the requirements in Sections 717.3.3.1 and 717.3.3.2. Further, when these dampers are used as part of a smoke control system, this section prohibits activation of these dampers by local area smoke detectors because this could render the smoke control system inoperable.

717.3.3.4 Ceiling radiation damper actuation. The operating temperature of a *ceiling radiation damper* actuation device shall be 50°F (27.8°C) above the normal temperature within the duct system, but not less than 160°F (71°C).

❖ The purpose of this section is to prescribe the actuation (operating temperature) requirements for ceiling radiation dampers. The thresholds for actuation are similar to those required for fire dampers used in static systems

717.3.3.5 Corridor damper actuation. *Corridor damper* actuation shall be in accordance with Sections 717.3.3.1 and 717.3.3.2.

❖ This section addresses the actuation criteria for corridor dampers using existing criteria for both fire dampers and smoke dampers.

717.4 Access and identification. Fire and smoke *dampers* shall be provided with an *approved* means of access that is large enough to *permit* inspection and maintenance of the *damper* and its operating parts. The access shall not affect the integrity of fire-resistance-rated assemblies. The access openings shall not reduce the *fire-resistance rating* of the assembly. Access points shall be permanently identified on the exterior by a *label* having letters not less than $^1/_2$ inch (12.7 mm) in height reading: FIRE/SMOKE DAMPER, SMOKE DAMPER or FIRE DAMPER. Access doors in ducts shall be tight fitting and suitable for the required duct construction.

❖ Fire and smoke dampers and combination dampers must be properly maintained so that they will operate as intended. The need to maintain dampers, as well as reset them and replace fusible links after operation, requires that they be accessible as a part of the initial design in the code. The access doors may need to be fire doors in accordance with Section 716, depending on the location of the door, in order to maintain a fire-resistance rating. Access doors in the duct itself do not need a fire protection rating.

717.5 Where required. *Fire, dampers, smoke dampers, combination fire/smoke dampers, ceiling radiation dampers* and *corridor dampers* shall be provided at the locations prescribed in Sections 717.5.1 through 717.5.7 and 717.6. Where an assembly is required to have both *fire dampers* and *smoke dampers, combination fire/smoke dampers* or a *fire damper* and a *smoke damper* shall be provided.

❖ In order to maintain the fire-resistance rating of fire and smoke vertical walls, barriers and partitions, code-required dampers must be provided for all duct penetrations. Fire dampers are to be installed in the assembly such that an opening created by a duct collapse is properly protected. The manufacturer's installation instructions, which should indicate how the damper is to be installed, must be followed so that the operation and functioning of the installed assembly will be consistent with that required by the code.

Commentary Figure 717.5 shows the difference between where a fire damper is required and where one is not. In this example, the designer has chosen to separate mixed occupancies in accordance with the option of Section 508.4.4 The occupancy separation wall (fire barrier) that is required to be fire-resistance rated is also required to have the duct penetration protected by a fire damper, while the nonrated wall requires no such protection. Occasionally, the visible gap between the outside of the dampered duct and the inside surface of the hole in a wall or floor that accommodates the duct is closed with a sealant. Although not required by the code, this is occasionally done to provide more complete smoke resistance for the wall or floor assembly. The installation of the damper assembly in accordance with the listing and manufacturer's instructions will provide the required fire resistance, even without sealing any apparent gaps between the duct and the hole. It is imperative that an intumescent firestop sealant not be used on the outside of the duct to seal this gap. The expansion pressure of the intumescent sealant can buckle the duct, thus hindering or preventing the proper operation of the damper under fire conditions.

The code requires that a fire damper not be used as a smoke damper unless the location of the installation is appropriate for the dual purpose. The most likely location for this to occur would be where the air distribution system also serves as a smoke control system and a duct penetrates a fire-resistance-rated assembly. Section 717.2.1 requires alternative protection to be provided where fire dampers will interfere with the operation of a smoke control system. A combination fire/smoke damper, such as that described above, may be approved for use as alternative protection.

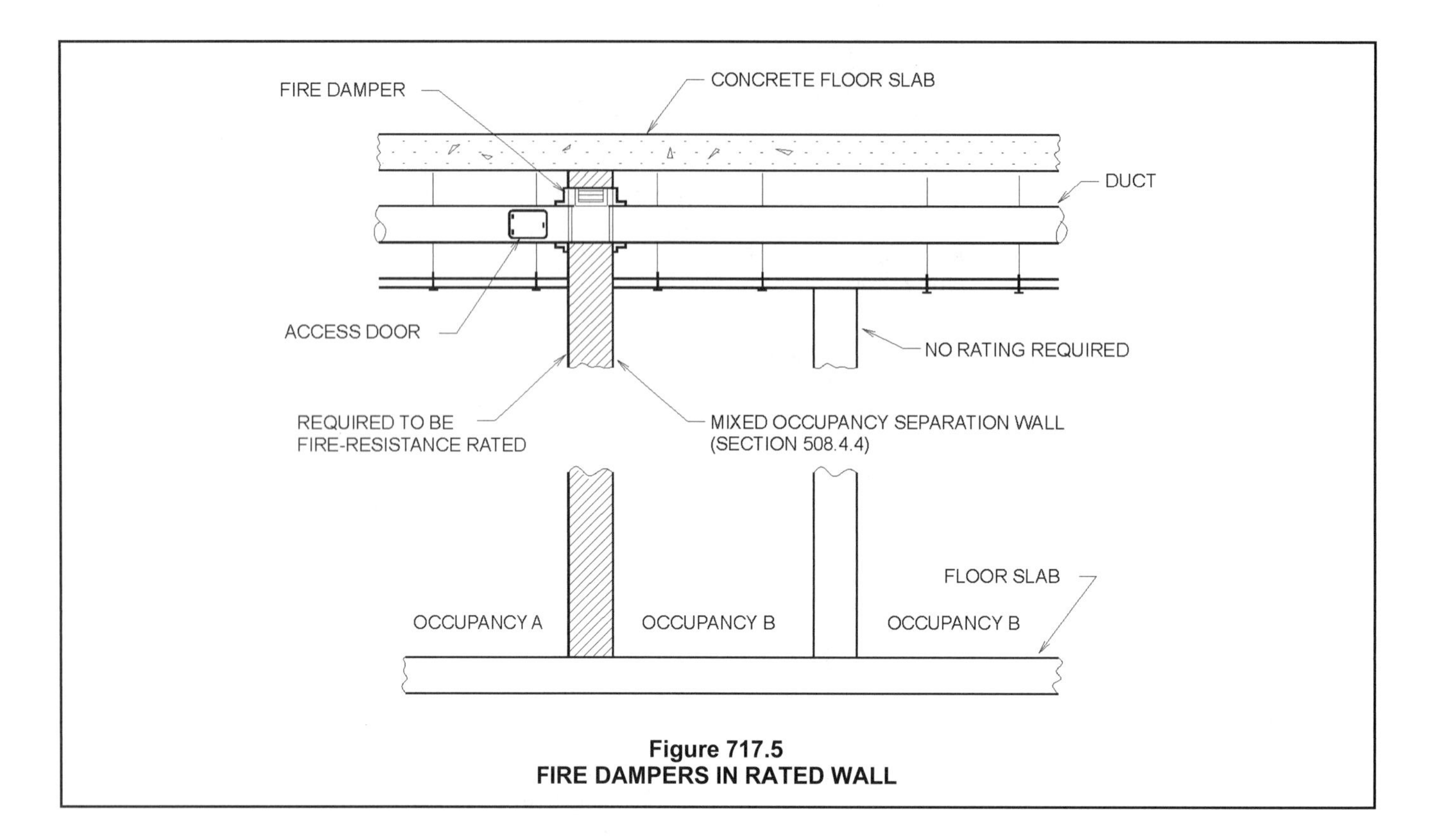

Figure 717.5
FIRE DAMPERS IN RATED WALL

717.5.1 Fire walls. Ducts and air transfer openings permitted in *fire walls* in accordance with Section 706.11 shall be protected with *listed fire dampers* installed in accordance with their listing.

❖ Fire walls can create separate buildings within a structure. These multiple buildings may be on a single lot or, in the case of zero lot line construction, the fire wall may be located on the lot line. In the case where the fire wall separates different buildings on different lots, the code does not permit openings (see Sections 706.1.1 and 706.11). In such instances, a duct penetration is not permitted. Where the fire wall is not located on a lot line, the wall is permitted to be penetrated by a duct or transfer opening, provided the opening is protected with a listed fire damper. The fire damper in 3-hour and greater fire-resistance-rated walls must be rated for 3 hours.

717.5.1.1 Horizontal exits. A *listed smoke damper* designed to resist the passage of smoke shall be provided at each point a duct or air transfer opening penetrates a *fire wall* that serves as a horizontal *exit*.

❖ Recognizing that fire walls can jointly serve as horizontal exits, this section requires ducts and air transfer opening penetrations of such walls to be provided with smoke dampers. The purpose of this section is to recognize that horizontal exits are required to resist the passage of smoke as well as fire.

717.5.2 Fire barriers. Ducts and air transfer openings of *fire barriers* shall be protected with *approved fire dampers* installed in accordance with their listing. Ducts and air transfer openings shall not penetrate enclosures for *interior exit stairways* and *ramps* and *exit passageways,* except as permitted by Sections 1023.5 and 1024.6, respectively.

Exception: *Fire dampers* are not required at penetrations of *fire barriers* where any of the following apply:

1. Penetrations are tested in accordance with ASTM E119 or UL 263 as part of the fire-resistance-rated assembly.
2. Ducts are used as part of an *approved* smoke control system in accordance with Section 909 and where the use of a *fire damper* would interfere with the operation of a smoke control system.
3. Such walls are penetrated by ducted HVAC systems, have a required *fire-resistance rating* of 1 hour or less, are in areas of other than Group H and are in buildings equipped throughout with an *automatic sprinkler system* in accordance with Section 903.3.1.1 or 903.3.1.2. For the purposes of this exception, a ducted HVAC system shall be a duct system for conveying supply, return or exhaust air as part of the structure's HVAC system. Such a duct system shall be constructed of sheet steel not less than No. 26 gage thickness and shall be continuous from the air-handling appliance or equipment to the air outlet and inlet terminals.

❖ Fire barriers require fire dampers at duct and transfer openings. However, ducts and transfer openings are not allowed to penetrate stairway and exit passageway enclosures even with fire dampers. Of course, independent stair pressurization ductwork may penetrate the stairway enclosure in accordance with Sections 1024.5 and 1025.6. Penetration of an exit has the potential to allow smoke or flame to be introduced into the enclosure from other parts of the building, effectively blocking the exit path.

Exception 1 states that fire dampers are not required for duct penetrations of fire barriers when the assembly has been tested in accordance with ASTM E119 or UL 263 without fire dampers and the required fire-resistance rating is obtained. In this case, the penetration assembly has been demonstrated as preserving the fire-resistance rating of the wall assembly.

Exception 2 reinforces the provisions of Section 717.2.1 (see commentary). A fire damper may interfere with the operation of the smoke control system; however, some form of alternative protection must be provided due to the duct penetrating the fire barrier.

Exception 3 states that, in fully sprinklered buildings containing occupancies other than Group H, fire dampers may be omitted in duct penetrations of walls having a fire-resistance rating of 1 hour or less. Such duct systems must be 26-gage steel continuous from the air handler to the air outlet with no flex duct connectors.

717.5.2.1 Horizontal exits. A *listed smoke damper* designed to resist the passage of smoke shall be provided at each point a duct or air transfer opening penetrates a *fire barrier* that serves as a horizontal *exit*.

❖ Walls separating buildings into areas of refuge in accordance with Section 1025 are specific types of fire barrier walls. The purpose of this section is to recognize that horizontal exits are required to resist the passage of smoke as well as fire. This section, therefore, requires that smoke dampers be installed at ducts or transfer openings that penetrate a fire barrier wall used as a horizontal exit.

717.5.3 Shaft enclosures. Shaft enclosures that are permitted to be penetrated by ducts and air transfer openings shall be protected with *approved* fire and smoke *dampers* installed in accordance with their listing.

Exceptions:

1. *Fire dampers* are not required at penetrations of shafts where any of the following criteria are met:
   - 1.1. Steel exhaust subducts are extended not less than 22 inches (559 mm) vertically in exhaust shafts, provided there is a continuous airflow upward to the outside.
   - 1.2. Penetrations are tested in accordance with ASTM E119 or UL 263 as part of the fire-resistance-rated assembly.
   - 1.3. Ducts are used as part of an *approved* smoke control system designed and installed in accordance with Section 909 and where the *fire damper* will interfere with the operation of the smoke control system.
   - 1.4. The penetrations are in parking garage exhaust or supply shafts that are separated from other building shafts by not less than 2-hour fire-resistance-rated construction.
2. In Group B and R occupancies equipped throughout with an *automatic sprinkler system* in accordance with Section 903.3.1.1, *smoke dampers* are not required at penetrations of shafts where all of the following criteria are met:
   - 2.1. Kitchen, clothes dryer, bathroom and toilet room exhaust openings are installed with steel exhaust subducts, having a minimum wall thickness of 0.0187-inch (0.4712 mm) (No. 26 gage).
   - 2.2. The subducts extend not less than 22 inches (559 mm) vertically.
   - 2.3. An exhaust fan is installed at the upper terminus of the shaft that is powered continuously in accordance with the provisions of Section 909.11, so as to maintain a continuous upward airflow to the outside.
3. *Smoke dampers* are not required at penetration of exhaust or supply shafts in parking garages that are separated from other building shafts by not less than 2-hour fire-resistance-rated construction.
4. *Smoke dampers* are not required at penetrations of shafts where ducts are used as part of an *approved* mechanical smoke control system designed in accordance with Section 909 and where the *smoke damper* will interfere with the operation of the smoke control system.
5. *Fire dampers* and *combination fire/smoke dampers* are not required in kitchen and clothes dryer exhaust systems where installed in accordance with the *International Mechanical Code*.

❖ This section requires both fire and smoke dampers at duct and air transfer openings in the shaft wall. The fire damper is required due to the penetration of a fire-resistance-rated wall. The smoke damper is required in order to limit the migration of smoke to other parts of the building via the shaft and the chimney effect (see commentary, Section 713.1). The exceptions allow the omission of fire dampers and smoke dampers under certain conditions. Exception 1 addresses fire dampers. Exceptions 2, 3 and 4 address smoke dampers. Exception 4 addresses the interface with smoke control systems as designed in Section 909, and coordinates with Section 717.2.1. Exception 5 addresses both fire and fire/smoke dampers, but not smoke dampers. When applying the exceptions, remember both fire and smoke dampers are required for shaft wall penetrations by ducts. Some exceptions only address one and not the other. Exceptions can be combined.

Exceptions 1.1 thru 1.4 only exempt fire dampers.

Exception 1.1 recognizes that the presence of a vertical subduct in a shaft will also offer some degree of fire resistance should the duct outside of the shaft collapse. Steel exhaust ducts that consist of a 22-inch

(559 mm) vertical upturn in the shaft need not be protected with a fire damper if there is a continuous upward airflow to the outside. The continuous airflow will create a negative pressure in the shaft as compared to adjacent spaces, thereby minimizing the spread of hot gases from the shaft [see Commentary Figure 717.5.3.1(1)].

Exceptions 1.2 and 1.3 are identical to Exceptions 1 and 2 to Section 717.5.2 (see commentary, Section 717.5.2).

Exception 1.4 states that fire dampers may also be omitted in exhaust and supply shafts that serve a garage and are separated from all other shafts in the building by a 2-hour fire separation assembly [see Commentary Figure 717.5.3.1(2)]. Requiring fire dampers in garage exhaust and supply shafts would not significantly prevent the spread of smoke and fire within the garage, since the vehicle ramp from floor to floor is a much greater conduit of smoke and fire in a garage.

Exception 2 is specific to Group B and R occupancies and exempts smoke dampers. Exception 2 has many more conditions than Exception 1.1, but together they can be used to exempt both fire and smoke dampers.

Exception 3 addresses parking garage shafts for exhaust and supply ducts and exempts smoke dampers, whereas Exception 1.4 exempts fire dampers.

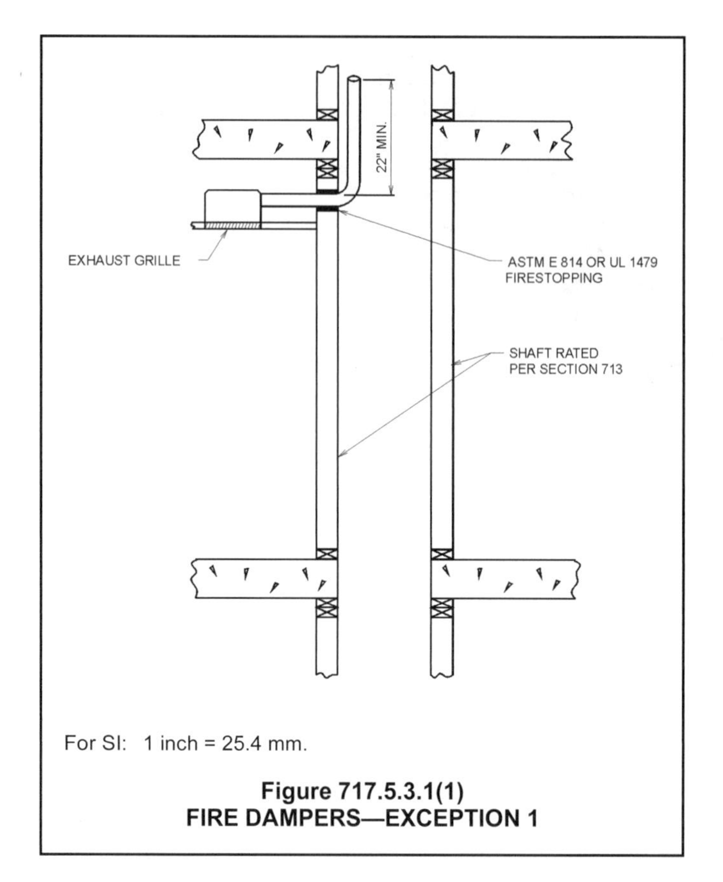

Figure 717.5.3.1(1)
FIRE DAMPERS—EXCEPTION 1

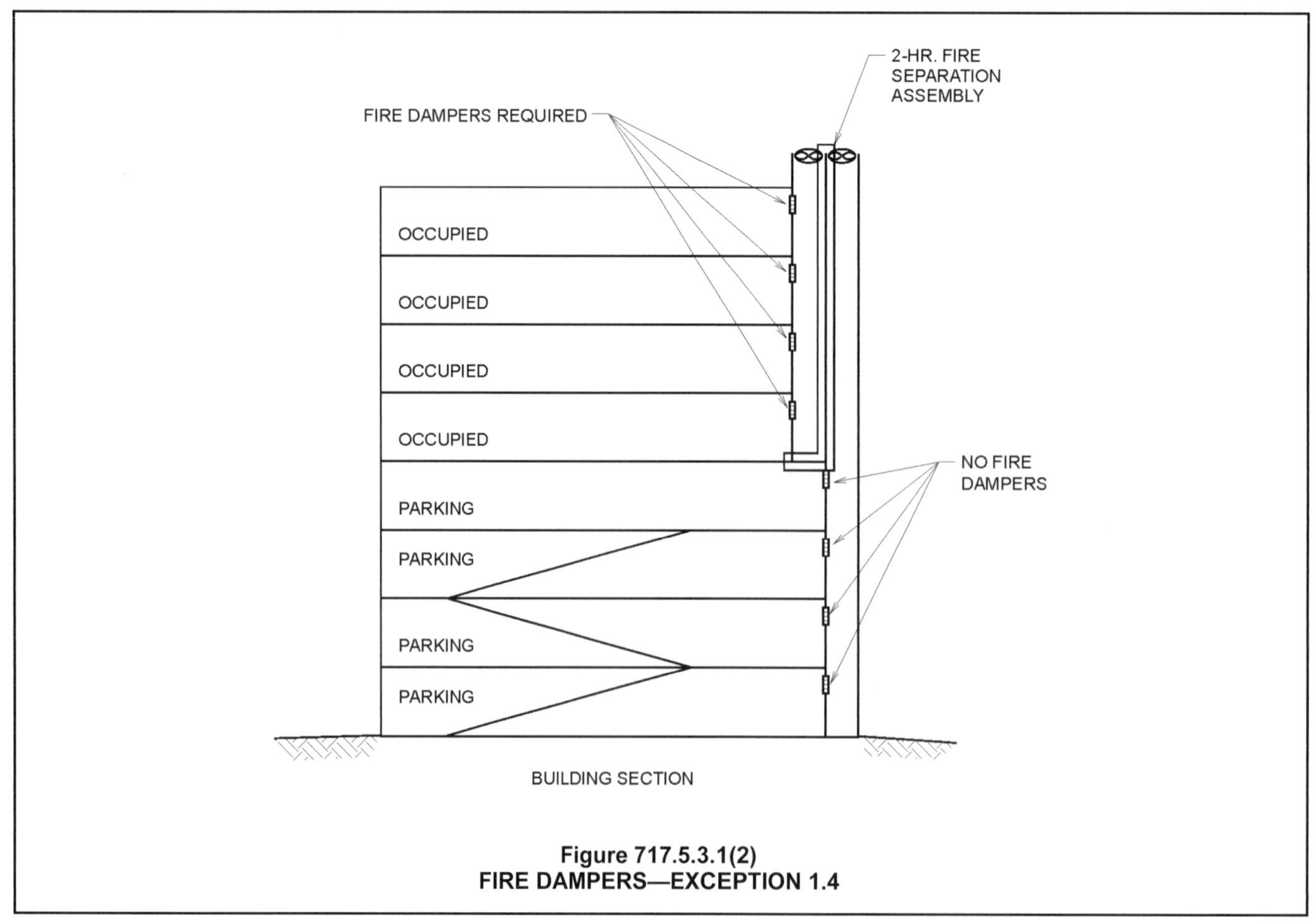

Figure 717.5.3.1(2)
FIRE DAMPERS—EXCEPTION 1.4

The conditions for the exception are the same.

Exception 4 is the counterpart to Exception 1.3. One exempts the fire damper and one exempts the smoke damper.

Exception 5 exempts fire dampers and combination fire/smoke dampers from kitchen and clothes dryer exhaust systems that are installed in accordance with the IMC. This is to recognize the hazards associated with obstructing the airflow for these installations. The IMC requires a shaft for multifloor kitchen and clothes dryer exhaust systems and requires 22-inch (559 mm) sub-ducts and special construction in accordance with Section 504.8 of the IMC. The fact that fire dampers and combination fire/smoke dampers are not required does not exempt the requirement for smoke dampers. However, Exception 2 of this section can be used.

717.5.4 Fire partitions. Ducts and air transfer openings that penetrate *fire partitions* shall be protected with *listed fire dampers* installed in accordance with their listing.

Exceptions: In occupancies other than Group H, *fire dampers* are not required where any of the following apply:

1. Corridor walls in buildings equipped throughout with an *automatic sprinkler system* in accordance with Section 903.3.1.1 or 903.3.1.2 and the duct is protected as a *through penetration* in accordance with Section 714.
2. Tenant partitions in *covered and open mall buildings* where the walls are not required by provisions elsewhere in the code to extend to the underside of the floor or roof sheathing, slab or deck above.
3. The duct system is constructed of *approved* materials in accordance with the *International Mechanical Code* and the duct penetrating the wall complies with all of the following requirements:
   - 3.1. The duct shall not exceed 100 square inches (0.06 m^2).
   - 3.2. The duct shall be constructed of steel not less than 0.0217 inch (0.55 mm) in thickness.
   - 3.3. The duct shall not have openings that communicate the *corridor* with adjacent spaces or rooms.
   - 3.4. The duct shall be installed above a ceiling.
   - 3.5. The duct shall not terminate at a wall register in the fire-resistance-rated wall.
   - 3.6. A minimum 12-inch-long (305 mm) by 0.060-inch-thick (1.52 mm) steel sleeve shall be centered in each duct opening. The sleeve shall be secured to both sides of the wall and all four sides of the sleeve with minimum $1^1/_2$-inch by $1^1/_2$-inch by 0.060-inch (38 mm by 38 mm by 1.52 mm) steel retaining angles. The retaining angles shall be secured to the sleeve and the wall with No. 10 (M5) screws. The *annular space* between the steel sleeve and the wall opening shall be filled with *mineral wool* batting on all sides.
4. Such walls are penetrated by ducted HVAC systems, have a required *fire-resistance rating* of 1 hour or less, and are in buildings equipped throughout with an automatic sprinkler system in accordance with Section 903.3.1.1 or 903.3.1.2. For the purposes of this exception, a ducted HVAC system shall be a duct system for conveying supply, return or exhaust air as part of the structure's HVAC system. Such a duct system shall be constructed of sheet steel not less than No. 26 gage thickness and shall be continuous from the air-handling appliance or equipment to the air outlet and inlet terminals.

❖ Fire dampers are required when a duct or air transfer opening penetrates a fire partition, and must be listed for the fire rating of the partition and installed in accordance with the listing. Please note that corridor walls, where required to be separated by Section 1020.1, must have duct penetrations or transfer openings protected with smoke dampers unless exempted by Section 717.5.4.1. This section includes four exceptions to the fire damper requirement.

Exception 1 addresses fire-resistance-rated corridor walls. In fully sprinklered buildings, corridor wall penetrations by ducts or air transfer openings need not have fire dampers but they still are required to have smoke dampers by Section 717.5.4.1 unless further exempted there. Remember that duct penetrations that are exempted from fire dampers are required by Section 717.1.2 to have through-penetration firestops [see Commentary Figure 717.5.4(1)].

Exception 2 coordinates with Exception 4 to Section 708.4 and addresses standard construction designs for mall buildings as they relate to open ceiling spaces. The exception removes an implied requirement for dampers where logic would dictate none is required. This provision recognizes that walls separating tenants in a covered mall are required to be rated but are allowed to stop at unrated ceilings and storefronts, providing no true separation. Therefore, it does not make sense to require a damper through the wall. If the tenant separation wall is required to be continuous to a roof deck or rated floor assembly for another purpose, the fire dampers would obviously be required.

Exception 3 states that fire dampers are not required where a steel duct penetrates a wall but does not have openings serving adjacent rooms or spaces [see Commentary Figure 717.5.4(2)] This exception has several criteria, one of which is to utilize a steel sleeve with mineral wool batting on all sides. Steel ducts have been shown to have the ability to remain in place during severe fire exposures. With no openings into the corridor to allow for smoke and heat transfer in the protected corridor enclosure, a fire damper is not necessary to attain the required fire performance of the penetration.

Exception 4 is similar to Exception 1. Exception 4 exempts all ducted penetrations of all fire partitions and not just corridor wall penetrations. Exception 4, in addition to the sprinkler provision, requires all steel ducts and does not allow flexible duct connectors. For corridors, Exception 1 is easier. Both exceptions would require through-penetration protection in accordance with Section 717.1.2. Exception 4 is identical to Exception 3 for fire barrier duct penetrations in Section 717.5.2.

717.5.4.1 Corridors. Duct and air transfer openings that penetrate *corridors* shall be protected with dampers as follows:

1. A *corridor damper* shall be provided where corridor ceilings, constructed as required for the corridor walls as permitted in Section 708.4, Exception 3, are penetrated.
2. A *ceiling radiation damper* shall be provided where the ceiling membrane of a *fire-resistance-rated* floor-ceil-

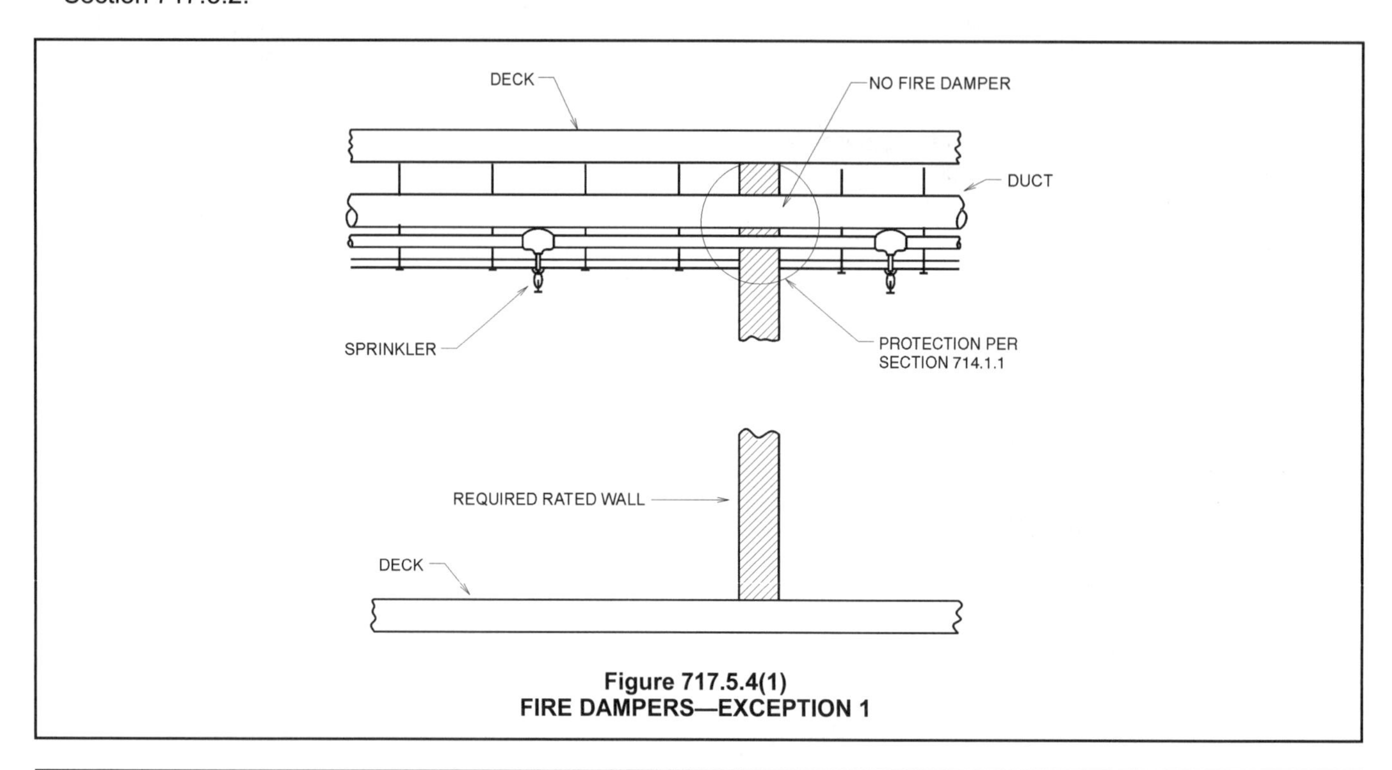

Figure 717.5.4(1)
FIRE DAMPERS—EXCEPTION 1

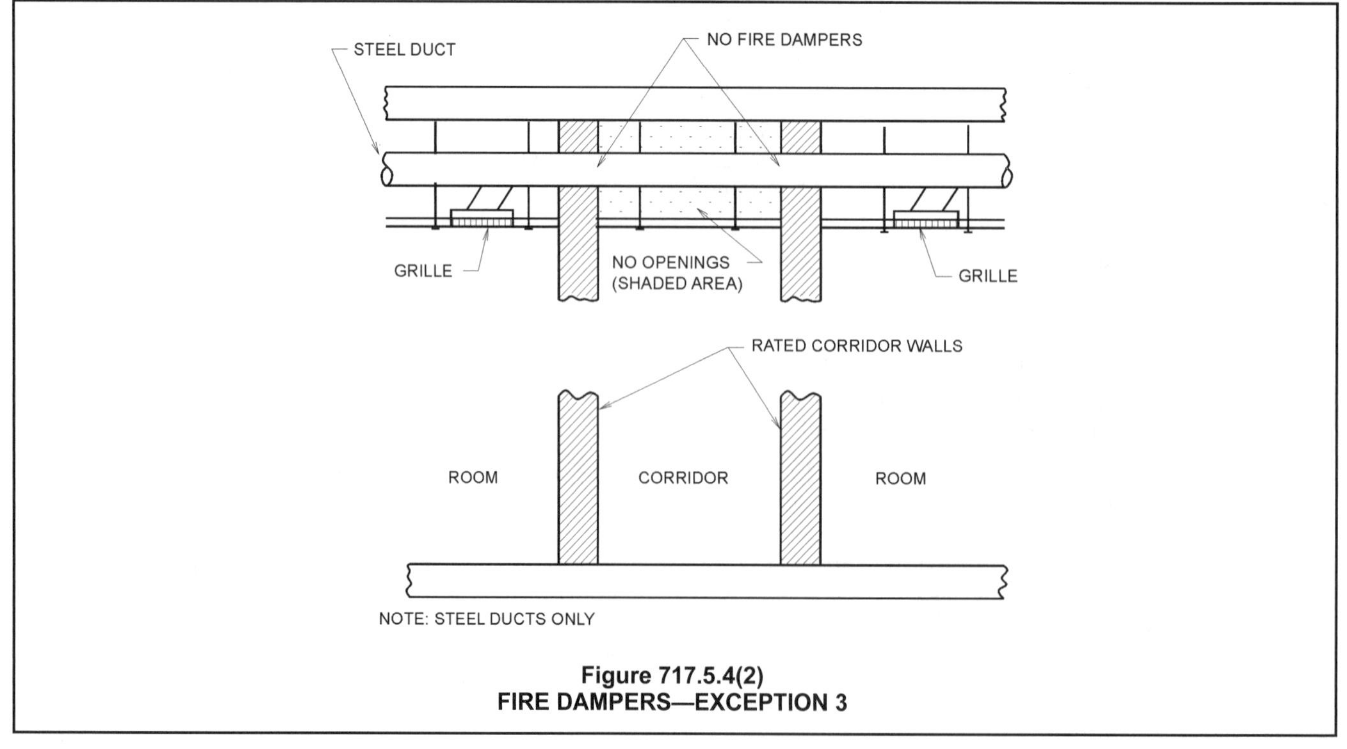

Figure 717.5.4(2)
FIRE DAMPERS—EXCEPTION 3

ing or roof-ceiling assembly, constructed as permitted in Section 708.4, Exception 2, is penetrated.

3. A listed smoke damper designed to resist the passage of smoke shall be provided at each point a duct or air transfer opening penetrates a corridor enclosure required to have smoke and draft control doors in accordance with Section 716.5.3.

Exceptions:

1. *Smoke dampers* are not required where the building is equipped throughout with an *approved* smoke control system in accordance with Section 909, and *smoke dampers* are not necessary for the operation and control of the system.
2. *Smoke dampers* are not required in *corridor* penetrations where the duct is constructed of steel not less than 0.019 inch (0.48 mm) in thickness and there are no openings serving the *corridor*.

❖ Protecting the corridors from smoke during a fire incident is important for egress from the buildings. Consistent with the door opening protection requirements of Section 716.5.3, penetration of ducts and air transfer openings require minimum protection. Therefore, all fire-rated corridor walls penetrated by ducts or air transfer openings must have listed dampers that are suited to three different conditions as follows:

1. Where the corridor ceiling is constructed to provide a fire rating equivalent to that of the corridor walls, then corridor dampers are required.
2. Where the corridor ceiling is protected by a ceiling membrane in accordance with Item 2 of Section 708.4, then ceiling radiation dampers are required.
3. For conditions other than 1 and 2 above, smoke dampers are required.

Exception 1 cites the concern of the damper interfering with the smoke control system (see commentary, Section 717.2.1).

Exception 2 exempts the duct penetration from smoke dampers when a steel duct penetrates the corridor walls and there are no duct outlets or openings serving the corridor.

717.5.5 Smoke barriers. A *listed smoke damper* designed to resist the passage of smoke shall be provided at each point a duct or air transfer opening penetrates a *smoke barrier*. *Smoke dampers* and *smoke damper* actuation methods shall comply with Section 717.3.3.2.

Exceptions:

1. *Smoke dampers* are not required where the openings in ducts are limited to a single *smoke compartment* and the ducts are constructed of steel.
2. *Smoke dampers* are not required in *smoke barriers* required by Section 407.5 for Group I-2, Condition 2—where the HVAC system is fully ducted in accordance with Section 603 of the *International Mechanical Code* and where buildings are equipped throughout with an *automatic sprinkler system* in accordance with Section 903.3.1.1 and equipped with quick-response sprinklers in accordance with Section 903.3.2.

❖ Smoke barriers are required for both Group I-2 (see Section 407.5) and I-3 (see Section 408.6) occupancies. See the commentary to Section 709.1 for other applications of smoke barriers. The concept of requiring smoke barriers is to limit the migration of smoke. In fact, Section 709.4 requires that smoke barriers form an effective continuous membrane from outside wall to outside wall and from the floor slab to the floor or roof deck above. This would include continuity through concealed spaces, such as those found above suspended ceilings and interstitial structural and mechanical spaces.

To prevent smoke migration across the smoke barrier where duct penetrations occur, an approved damper designed to resist the passage of smoke is to be provided at each point where a duct penetrates a smoke barrier. The damper closes upon detection of smoke by an approved smoke detector within the duct. If the duct penetration is above a smoke barrier door assembly, smoke detectors for adjacent doors may also be used to operate the damper and the duct detector may be eliminated. A reference to Section 717.3.3.2 is provided for activation mechanisms.

The designer must evaluate the specific conditions of the installation to determine which class of damper should be installed. If a fire damper is required (i.e., the smoke barrier is also a mixed occupancy separation or corridor wall), a combination fire/smoke damper or separate fire and smoke dampers may be installed.

Exception 1 eliminates the need for smoke dampers, provided the duct openings are limited to one smoke compartment and the ducts are of a construction that inherently resists the passage of smoke for a duration of time (steel).

Exception 2 eliminates the need for smoke dampers in care facilities and Group I-2 hospital occupancies under specific conditions. These facilities are fully protected with electronically supervised and tested and maintained quick-response automatic sprinkler systems. The meaningful time of the fire protection of the building occurs in the first 30 minutes of the fire incident, when decisions are made by fire professionals and the safety staff of the hospital in terms of status of the patients. Quick-response sprinklers are more often noted as the most important feature of the overall building fire protection system, and are demonstrated to be more effective in containing the spread of fire than dampering of the duct system.

717.5.6 Exterior walls. Ducts and air transfer openings in fire-resistance-rated *exterior walls* required to have protected openings in accordance with Section 705.10 shall be protected with *listed fire dampers* installed in accordance with their listing.

❖ The purpose of this section is to recognize that under certain conditions exterior walls are required to be fire-resistance rated. Depending on the fire separation distance, not all fire-resistance-rated exterior walls require protected openings. If protected openings are required by Section 705.10, then fire dampers are required.

717.5.7 Smoke partitions. A *listed smoke damper* designed to resist the passage of smoke shall be provided at each point that an air transfer opening penetrates a smoke partition. *Smoke dampers* and *smoke damper* actuation methods shall comply with Section 717.3.3.2.

Exception: Where the installation of a *smoke damper* will interfere with the operation of a required smoke control system in accordance with Section 909, *approved* alternative protection shall be utilized.

❖ The purpose of this section is to recognize that penetrations of smoke partitions by air transfer openings need to resist the transfer of smoke and, therefore, are required to be provided with a smoke damper. This is also consistent with the requirements in Section 710.8. This section also requires the actuation of the damper to comply with Section 717.3.3.2. The exception cites the concern of the damper interfering with the smoke control system (see commentary, Section 717.2.1).

717.6 Horizontal assemblies. Penetrations by ducts and air transfer openings of a floor, floor/ceiling assembly or the ceiling membrane of a roof/ceiling assembly shall be protected by a shaft enclosure that complies with Section 713 or shall comply with Sections 717.6.1 through 717.6.3.

❖ In general, all floor openings that connect two or more stories are to be enclosed in shafts that are constructed in accordance with Section 713. Section 712.1.6 references Section 717.6 for duct openings (penetration). Sections 717.6.1 through 717.6.3 identify conditions where a shaft is not required and either a fire damper (see Section 717.6.1) or ceiling radiation damper (see Section 717.6.2.1) is required. There is an exception within Section 717.6.1 that allows the elimination of the fire damper; however, ceiling radiation dampers would still be required in some cases (see Section 717.6.2.1).

717.6.1 Through penetrations. In occupancies other than Groups I-2 and I-3, a duct constructed of *approved* materials in accordance with the *International Mechanical Code* that penetrates a fire-resistance-rated floor/ceiling assembly that connects not more than two *stories* is permitted without shaft enclosure protection, provided a *listed fire damper* is installed at the floor line or the duct is protected in accordance with Section 714.4. For air transfer openings, see Section 712.1.9.

Exception: A duct is permitted to penetrate three floors or less without a *fire damper* at each floor, provided such duct meets all of the following requirements:

1. The duct shall be contained and located within the cavity of a wall and shall be constructed of steel having a minimum wall thickness of 0.0187 inches (0.4712 mm) (No. 26 gage).
2. The duct shall open into only one *dwelling or sleeping unit* and the duct system shall be continuous from the unit to the exterior of the building.
3. The duct shall not exceed 4-inch (102 mm) nominal diameter and the total area of such ducts shall not exceed 100 square inches (0.065 m^2) in any 100 square feet (9.3 m^2) of floor area.
4. The *annular space* around the duct is protected with materials that prevent the passage of flame and hot gases sufficient to ignite cotton waste where subjected to ASTM E119 or UL 263 time-temperature conditions under a minimum positive pressure differential of 0.01 inch (2.49 Pa) of water at the location of the penetration for the time period equivalent to the *fire-resistance rating* of the construction penetrated.
5. Grille openings located in a ceiling of a fire-resistance-rated floor/ceiling or roof/ceiling assembly shall be protected with a *listed ceiling radiation damper* installed in accordance with Section 717.6.2.1.

❖ Penetrations of a fire-resistance-rated horizontal assembly by air ducts are permitted where no more than two stories are connected (i.e., the duct penetrates a single floor) and a fire damper is provided at the floor line. The fire damper must be tested in accordance with UL 555 [see Commentary Figure 717.6.1(1)]. A roof assembly penetrated by a duct that is open to the atmosphere is not required to have a fire damper installed at the penetrations because it is considered acceptable for the fire to vent to the outside atmosphere and heat or fire extension will do no harm [see Commentary Figure 717.6.1(2)]. However, a ceiling radiation damper would still be needed to protect a roof/ceiling assembly per Section 717.6.2, Item 2. As noted, there is an exception to this section that allows the elimination of fire dampers at each floor when the ducts extend through a maximum of three floors. There are several criteria that must be followed to allow this, including:

- Minimum duct thickness (increases time for heat to penetrate and provides more durability) and enclosure within wall cavity.
- Limited to a single sleeping unit or dwelling unit (to reduce the likelihood that one dwelling unit or sleeping unit will be able to endanger another).

- Size of the duct opening (smaller opening means smaller amounts of fire effluents).
- Annular protection (reduces the chance of duct leakage from penetrated walls/barriers).
- Ceiling radiation dampers at grilles in fire-resistant ceilings (maintains ceiling fire resistance).
- All criteria center on limiting the impact of hot gases being present and spreading from one area to another. As noted, Section 712.1.8 addresses air transfer openings.

717.6.2 Membrane penetrations. Ducts and air transfer openings constructed of *approved* materials in accordance with the *International Mechanical Code* that penetrate the ceiling membrane of a fire-resistance-rated floor/ceiling or roof/ceiling assembly shall be protected with one of the following:

1. A shaft enclosure in accordance with Section 713.
2. A *listed ceiling radiation damper* installed at the ceiling line where a duct penetrates the ceiling of a fire-resistance-rated floor/ceiling or roof/ceiling assembly.
3. A *listed ceiling radiation damper* installed at the ceiling line where a diffuser with no duct attached penetrates the ceiling of a fire-resistance-rated floor/ceiling or roof/ceiling assembly.

❖ Unless the duct system is proteced with a shaft enclosure in accordance with Sections 713 and 714.4, a ceiling radiation damper, as described in Section 717.6.2.1, is to be installed at the line where

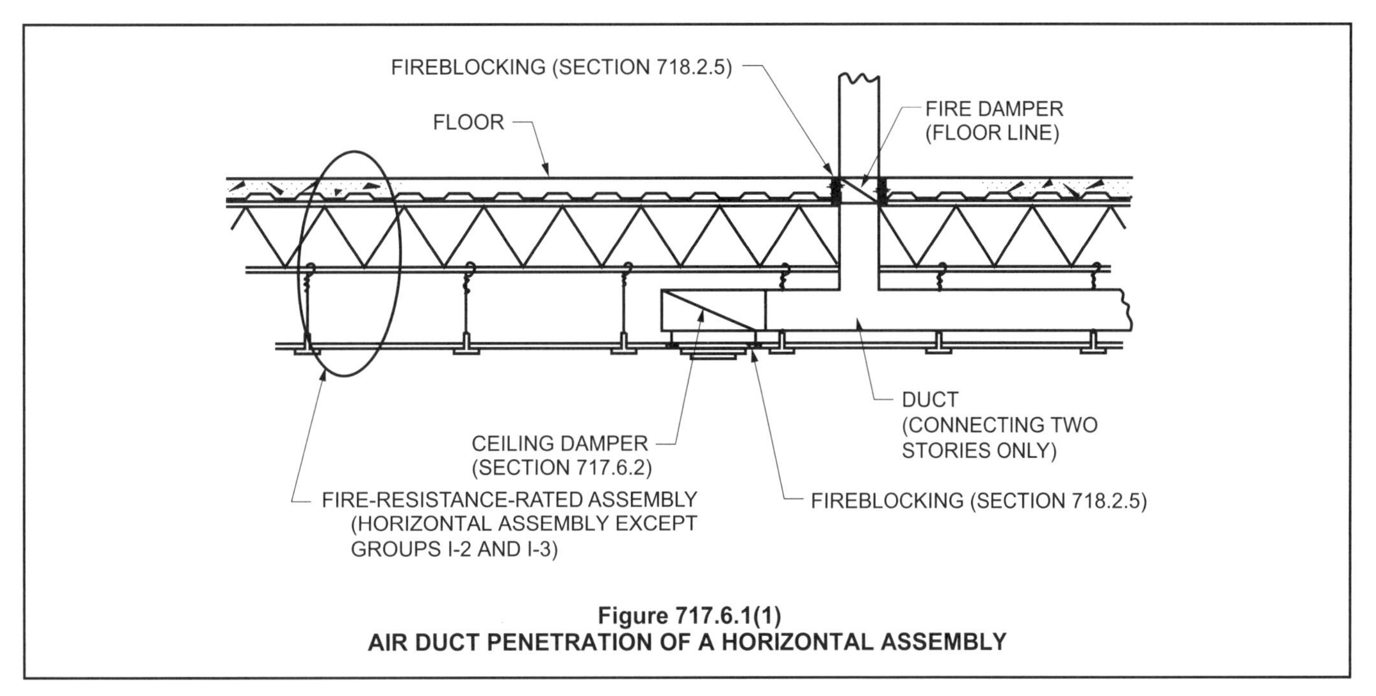

Figure 717.6.1(1)
AIR DUCT PENETRATION OF A HORIZONTAL ASSEMBLY

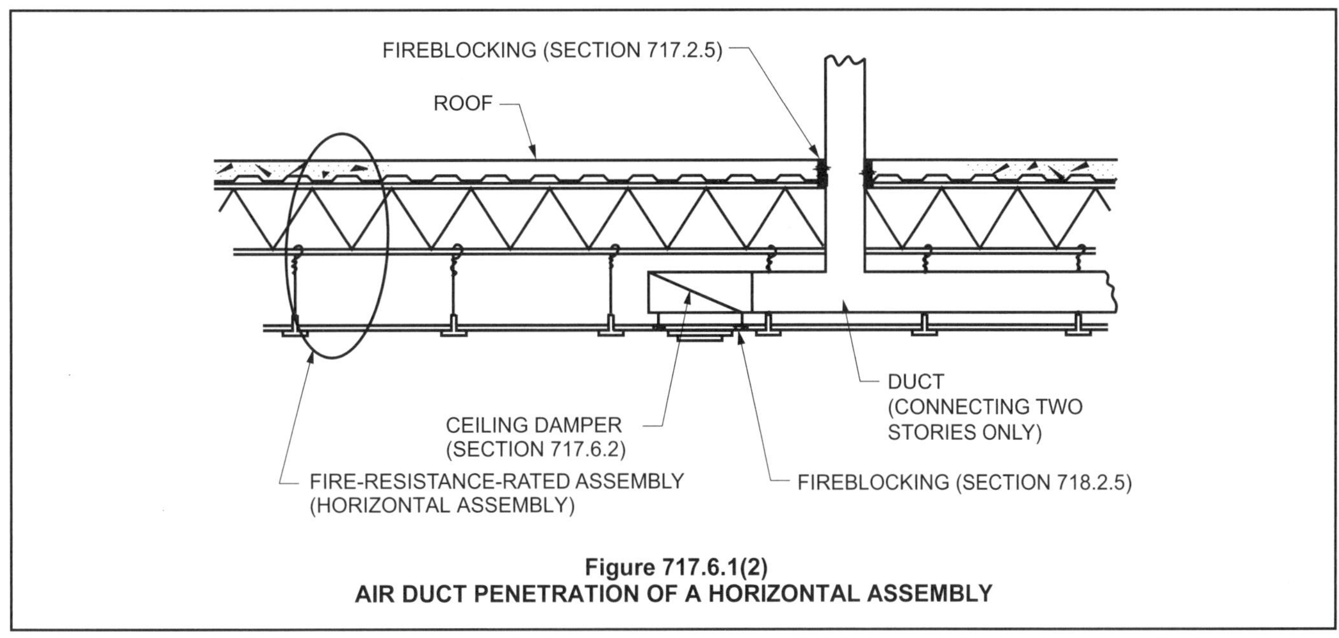

Figure 717.6.1(2)
AIR DUCT PENETRATION OF A HORIZONTAL ASSEMBLY

the ceiling is both penetrated by a noncombustible air duct and is an integral component of the fire-resistance rating.

The requirement for a ceiling radiation damper applies to ductwork that either penetrates a ceiling membrane from above and continues downward, or where a diffuser with no ductwork attached below penetrates a ceiling membrane. Ceiling diffusers without ducts attached above or below are also covered by this section. Section 717.6.1 contains the requirements for fire dampers where the duct penetrates through the entire assembly [see Commentary Figures 717.6.1(1) and 717.6.1(2) for ceiling radiation damper and fire damper locations].

Ceiling radiation dampers are designed to limit the radiative heat transfer through an air inlet or outlet opening in the ceiling membrane of a fire-resistance-rated floor/ceiling or roof/ceiling assembly.

717.6.2.1 Ceiling radiation dampers. *Ceiling radiation dampers* shall be tested in accordance with Section 717.3.1. *Ceiling radiation dampers* shall be installed in accordance with the details *listed* in the fire-resistance-rated assembly and the manufacturer's instructions and the listing. *Ceiling radiation dampers* are not required where one of the following applies:

1. Tests in accordance with ASTM E119 or UL 263 have shown that *ceiling radiation dampers* are not necessary in order to maintain the *fire-resistance rating* of the assembly.
2. Where exhaust duct penetrations are protected in accordance with Section 714.4.2, are located within the cavity of a wall and do not pass through another *dwelling unit* or tenant space.
3. Where duct and air transfer openings are protected with a duct outlet protection system tested as part of a *fire-resistance-rated* assembly in accordance with ASTM E119 or UL 263.

❖ Ceiling radiation dampers, formerly known as ceiling dampers, are evaluated either in accordance with UL 555C or as part of the floor/ceiling or roof/ceiling assembly tested to ASTM E119 or UL 263. Ceiling radiation dampers evaluated in accordance with UL 555C are investigated for use in lieu of hinged-door-type dampers commonly specified in the listing of fire-resistance-rated floor/ceiling or roof/ceiling assemblies. UL 555C also provides criteria for the construction of ceiling radiation dampers. If the fire-resistance-rated floor/ceiling or roof/ceiling assembly does not incorporate a hinged-door-type damper, a ceiling radiation damper may not be utilized in the assembly. Ceiling radiation dampers investigated as part of the fire-resistance-rated floor/ceiling or roof/ceiling assembly are specified in the listing of such assemblies. The description of the tested assembly will include a description of the ceiling radiation damper and its installation. Both types of ceiling radiation dampers shall be installed in accordance with the details listed in the fire-resistance-rated assembly and the manufacturer's installation instructions and listing.

The design information section of the *UL Fire Resistance Directory* also includes two duct outlet protection systems. "Duct Outlet Protection System A" may be used where specified in the fire-resistance-rated floor/ceiling or roof/ceiling assembly. "Duct Outlet Protection System B" may be used in lieu of hinged-door-type dampers commonly specified in the listing of fire-resistance-rated floor/ceiling or roof/ceiling assemblies. These methods meet ASTM E119 or UL 363.

One exception to the ceiling radiation dampers is for ceiling membrane penetrations by ducts or ceiling diffusers that have been tested in accordance with ASTM E119 or UL 263 as part of the overall fire-resistance-rated assembly.

Another exception is for exhaust ducts that penetrate a ceiling and: 1. Are protected in accordance with Section 714.4.2; 2. Are located within the cavity of a wall above; and 3. Do not pass through another dwelling unit or tenant space.

The last exception permits the use of duct protection methods other than ceiling radiation dampers for protecting ducts and air transfer openings through the ceiling membrane of a fire-resistance-rated floor/ceiling or roof/ceiling assembly. The acceptance of the alternative duct protection system is based on testing conducted in accordance with ASTM E119 or UL 263.

717.6.3 Nonfire-resistance-rated floor assemblies. Duct systems constructed of *approved* materials in accordance with the *International Mechanical Code* that penetrate nonfire-resistance-rated floor assemblies shall be protected by any of the following methods:

1. A shaft enclosure in accordance with Section 713.
2. The duct connects not more than two *stories*, and the *annular space* around the penetrating duct is protected with an *approved* noncombustible material that resists the free passage of flame and the products of combustion.
3. In floor assemblies composed of noncombustible materials, a shaft shall not be required where the duct connects not more than three stories, the annular space around the penetrating duct is protected with an approved noncombustible material that resists the free passage of flame and the products of combustion and a *fire damper* is installed at each floor line.

    Exception: *Fire dampers* are not required in ducts within individual residential *dwelling units*.

❖ Floor assemblies that are not rated still require isolation in some form based on the requirements of Section 717.6.3 when duct systems penetrate the assemblies. Three conditions or requirements are listed.

First, the ducts can be part of a shaft enclosure in accordance with Section 713. The shaft construction required in Section 713.4 would be at least 1 hour.

Second, duct systems that connect not more than two stories (penetrate one floor) are permitted, provided the annular space around the duct is sealed with a noncombustible material that will not allow free passage of flame or smoke. The two-story arrangement does not require a rated damper [see Commentary Figure 717.6.3(1)].

Third, when air ducts connect three levels or less and thus penetrate two floors or less, fire dampers must be provided at each floor line. A $1^1/_2$-hour-rated fire damper (see Table 717.3.2.1) is required even though the floor is not required to have a fire-resistance rating [see Commentary Figure 717.6.3(2)]. Fire dampers, unlike door opening protectives with multiple ratings, are only rated for $1^1/_2$ or 3 hours. Note that this method is limited to floor assemblies of noncombustible construction.

717.7 Flexible ducts and air connectors. Flexible ducts and air connectors shall not pass through any fire-resistance-rated assembly. Flexible air connectors shall not pass through any wall, floor or ceiling.

❖ Flexible air ducts are prohibited from penetrating fire-resistance-rated assemblies. Flexible air connectors are not permitted to pass through any wall, floor or ceiling (fire-resistance rated or not). An inadequate seal at the assembly penetration could allow smoke or flame to penetrate the assembly. Flexible air ducts and connectors can be constructed of both combustible and noncombustible components; therefore, the duct's resistance to the passage of fire could be less than the resistance of the penetrated assembly. All construction assemblies, whether fire-resistance rated or not, provide some inherent resistance to the spread of fire.

SECTION 718
CONCEALED SPACES

718.1 General. *Fireblocking* and draftstopping shall be installed in combustible concealed locations in accordance with this section. *Fireblocking* shall comply with Section 718.2. Draftstopping in floor/ceiling spaces and *attic* spaces shall comply with Sections 718.3 and 718.4, respectively. The permitted use of combustible materials in concealed spaces of buildings of Type I or II construction shall be limited to the applications indicated in Section 718.5.

❖ The key words in this section are "combustible concealed spaces." This section does not apply to noncombustible construction. During a fire, flame, smoke and gases will spread via the paths of least resistance. Certain assemblies create void spaces, which will not only affect the spread of fire but, since the voids are concealed, will also make access for suppression more difficult.

The code has established two means by which fire spread within void spaces can be controlled: fireblocking and draftstopping. Fireblocking involves the use of building materials to prevent the movement of flame and gases to other areas through concealed spaces. Draftstopping involves the use of building

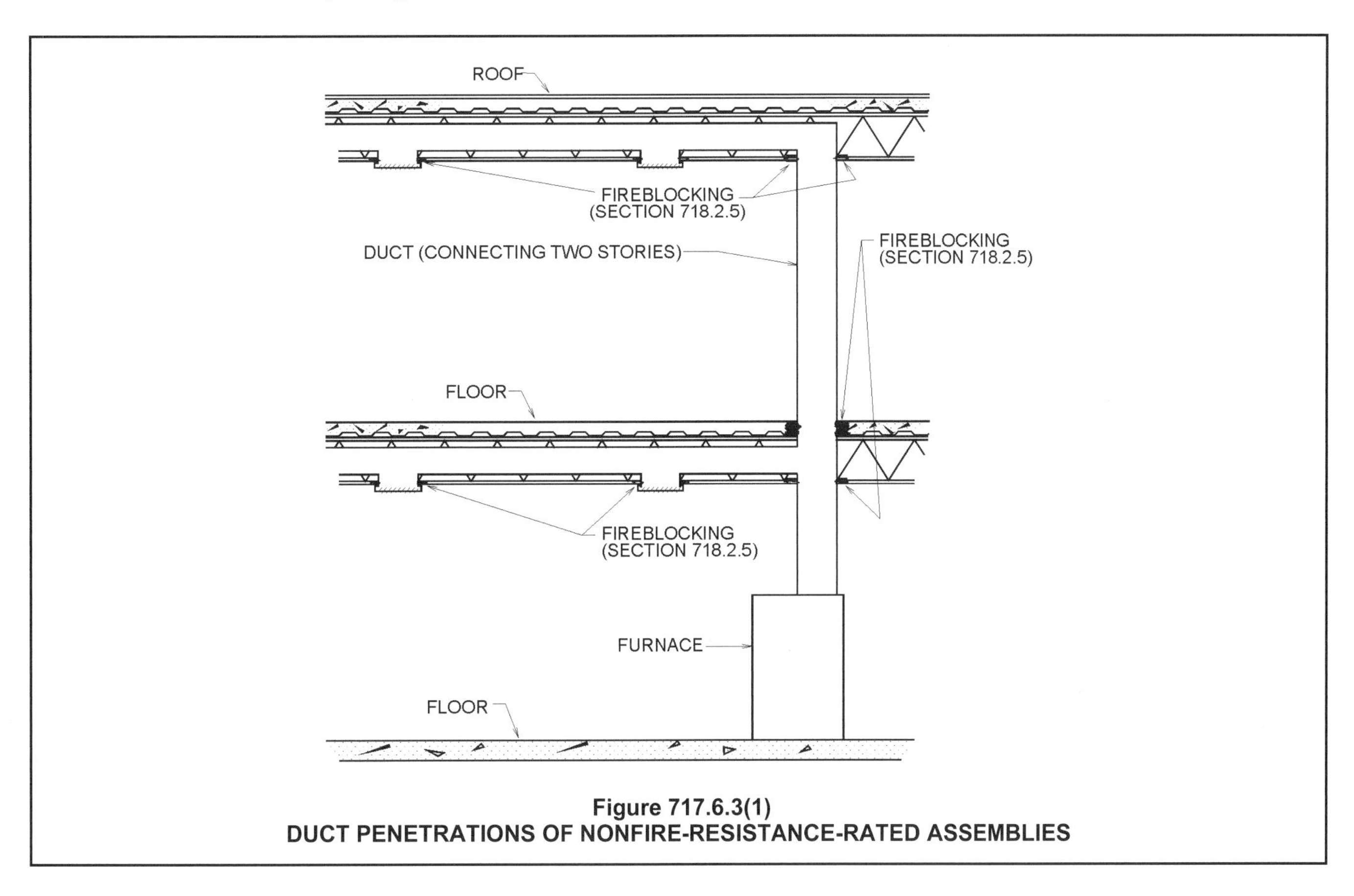

Figure 717.6.3(1)
DUCT PENETRATIONS OF NONFIRE-RESISTANCE-RATED ASSEMBLIES

materials to prevent the movement of air, smoke, gases and flames to other areas through large concealed spaces. For example, the protection in an attic space is draftstopping, while the protection in the cavity of a wall assembly is fireblocking.

Section 718.2.1 identifies the materials that are acceptable for use as fireblocks and Section 718.3.1 specifies the acceptable materials for draftstops. Sections 718.2, 718.3 and 718.4 address where fireblocking and draftstopping are required.

Fireblocks and draftstops are to be installed in combustible concealed spaces to prevent the spread of flame, smoke and gases. If a fire condition exists in a concealed space, the fireblocks and draftstops will help contain the fire until it can be suppressed. This section also includes provisions for permitted combustibles in concealed spaces in buildings of Type I and II construction (see Section 718.5 and Section 603.1, Exceptions 1 through 25).

718.2 Fireblocking. In combustible construction, *fireblocking* shall be installed to cut off concealed draft openings (both vertical and horizontal) and shall form an effective barrier between floors, between a top *story* and a roof or *attic* space. *Fireblocking* shall be installed in the locations specified in Sections 718.2.2 through 718.2.7.

❖ This section merely states the goals associated with fireblocking. The intent of fireblocking is to reduce the ability of fire, smoke and gases from moving to different parts of the building through combustible concealed spaces. Since there are no test standards for fireblocking as with firestopping materials, the requirements are prescriptive.

718.2.1 Fireblocking materials. *Fireblocking* shall consist of the following materials:

1. Two-inch (51 mm) nominal lumber.
2. Two thicknesses of 1-inch (25 mm) nominal lumber with broken lap joints.
3. One thickness of 0.719-inch (18.3 mm) wood structural panels with joints backed by 0.719-inch (18.3 mm) wood structural panels.
4. One thickness of 0.75-inch (19.1 mm) particleboard with joints backed by 0.75-inch (19 mm) particleboard.
5. One-half-inch (12.7 mm) gypsum board.
6. One-fourth-inch (6.4 mm) cement-based millboard.

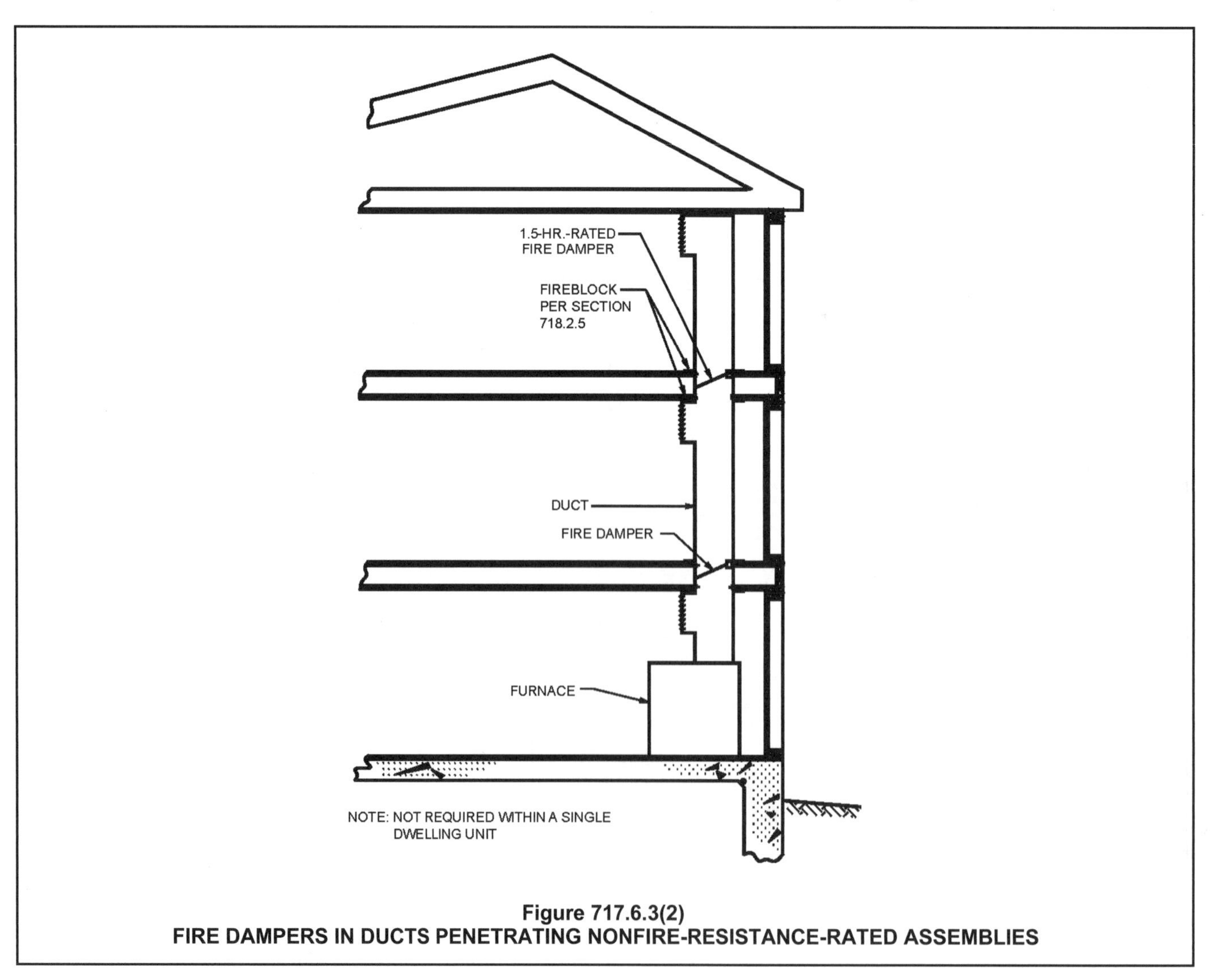

Figure 717.6.3(2)
FIRE DAMPERS IN DUCTS PENETRATING NONFIRE-RESISTANCE-RATED ASSEMBLIES

7. Batts or blankets of *mineral wool, mineral fiber* or other *approved* materials installed in such a manner as to be securely retained in place.
8. Cellulose insulation installed as tested for the specific application.

❖ The items on this approved list of fireblocking materials have not been tested for fire resistance, but are simply deemed as acceptable materials for slowing the spread of flame and products of combustion. Their effectiveness depends on their correct installation, something which must be inspected at the time of construction.

Various insulation materials have been evaluated for use as fireblocking and have been shown to perform well if installed in accordance with their specifications. For insulating material to function well as fireblocking, it must completely fill the area where fireblocking is required so that no unblocked passages exist.

Certain insulating materials have been evaluated for use as fireblocks. The evaluation reports that specify the configuration and attachment for the insulating materials in the construction assemblies are based on tests performed on the assemblies. Although the code does not specify a test standard for fireblocking, these materials have been tested using the ASTM E119 or UL 263 standard and their performance during the test is compared to the performance of other conventional fireblocking material. Nonrigid materials such as mineral wool, mineral fiber and fiber glass batts or blankets have somewhat limited applications. See Sections 718.2.1.1 and 718.2.1.2 for the limits on these materials. A new material has joined the list as possible fireblocking material—cellulose insulation. The code requires cellulose insulation to be tested for the specific application as fireblocking. Just because a space is filled with cellulose insulation does not guarantee that fireblocking will be accomplished.

718.2.1.1 Batts or blankets of mineral wool or mineral fiber. Batts or blankets of *mineral wool* or *mineral fiber* or other *approved* nonrigid materials shall be permitted for compliance with the 10-foot (3048 mm) horizontal *fireblocking* in walls constructed using parallel rows of studs or staggered studs.

❖ The use of mineral wool or mineral fiber or other approved nonrigid material is allowed to fulfill the requirement for fireblocking in the horizontal cavities formed by parallel rows of studs or staggered studs. Some type of fireblocking is required at 10-foot (3048 mm) intervals measured horizontally by Section 718.2.2. The code allows these types of batts or blankets, but the installation details must be approved and inspected in the field to verify that all concealed spaces are effectively fireblocked at the required locations. Simply because a space is filled with insulation does not guarantee that fireblocking will be accomplished.

718.2.1.2 Unfaced fiberglass. Unfaced fiberglass batt insulation used as *fireblocking* shall fill the entire cross section of the wall cavity to a minimum height of 16 inches (406 mm) measured vertically. Where piping, conduit or similar obstructions are encountered, the insulation shall be packed tightly around the obstruction.

❖ Unfaced fiberglass is a batt insulation that can be used as fireblocking when installed as described in this section. For fiberglass batt insulation to be effective it must fill the wall cavity at the location being fireblocked, not necessarily the full height of the wall. It must be a minimum of 16 inches (406 mm) vertically since it is assumed that this application is to prevent vertical spread of fire.

718.2.1.3 Loose-fill insulation material. Loose-fill insulation material, insulating foam sealants and caulk materials shall not be used as a fireblock unless specifically tested in the form and manner intended for use to demonstrate its ability to remain in place and to retard the spread of fire and hot gases.

❖ Loose-fill insulation materials are generally not suitable for use as fireblocking because, by their very nature, they are not intended to span an open gap and serve the purpose intended for fireblocking. However, the code does allow for the possibility that some types of materials with some cohesiveness might work; therefore, the code would require testing. This should be full-scale testing using a fire generated in accordance with ASTM E119 or UL 263.

718.2.1.4 Fireblocking integrity. The integrity of fireblocks shall be maintained.

❖ The continued integrity of the fireblocking over time is necessary to ensure that it is in place at the time a fire occurs. Some of these materials over time, with repair work or remodeling work being done, will be removed, have holes put in the material or be subject to other activity that will make the material ineffective as a fireblock.

718.2.1.5 Double stud walls. Batts or blankets of mineral or glass fiber or other *approved* nonrigid materials shall be allowed as *fireblocking* in walls constructed using parallel rows of studs or staggered studs.

❖ The fireblocking at horizontal intervals of 10 feet (3048 mm) or less in the cavities of double stud or staggered stud walls may be fiberglass, mineral fiber or other approved nonrigid materials (see commentary, Section 718.2.1.1).

718.2.2 Concealed wall spaces. *Fireblocking* shall be provided in concealed spaces of stud walls and partitions, including furred spaces, and parallel rows of studs or staggered studs, as follows:

1. Vertically at the ceiling and floor levels.
2. Horizontally at intervals not exceeding 10 feet (3048 mm).

❖ The intent of this section is to prevent the spread of fires within walls. Losses in this area have been seen in existing structures of balloon construction, but the requirements apply to platform framing as well. This section requires both vertical and horizontal blocking. Vertical blocking simply requires fireblocking in vertical wall cavities at floors and ceilings. Therefore, a tall warehouse-type building may not need any vertical fireblocking. Horizontal blocking, however, would be required every 10 feet (3048 mm). Section 718.2.1 provides the various methods for fireblocking. One method addressed is the use of mineral or glass fiber insulation, which is only allowed for horizontal fireblocking (see commentary, Section 717.2.1.1).

718.2.3 Connections between horizontal and vertical spaces. *Fireblocking* shall be provided at interconnections between concealed vertical stud wall or partition spaces and concealed horizontal spaces created by an assembly of floor joists or trusses, and between concealed vertical and horizontal spaces such as occur at soffits, drop ceilings, cove ceilings and similar locations.

❖ To prevent fire and smoke from spreading between vertical and horizontal spaces, fireblocks are required [see Commentary Figures 718.2.3(1 through 3)]. If not provided, a concealed fire could spread throughout the floor level because of the interconnection of concealed combustible wall and ceiling spaces.

In platform framing, where gypsum wallboard is continued to the top plate of the wall, typically either the gypsum board or the top plate will serve to cut off the connection between the vertical and horizontal construction. Therefore, with many typical details, no additional fireblocking is needed at this location.

718.2.4 Stairways. *Fireblocking* shall be provided in concealed spaces between *stair* stringers at the top and bottom of the run. Enclosed spaces under *stairways* shall comply with Section 1011.7.3.

❖ Fireblocks are required at the top and bottom of concealed spaces between stair stringers (see Commentary Figure 718.2.4). Similar to fireblocks between horizontal and vertical spaces, fireblocks at the top and bottom of a run will provide a barrier between the floor below the stair and the floor that the stairway serves. Additionally, a reference is made to Section 1011.7.3, which requires the interior of the space under stairways to contain 1-hour fire-resistant-rated construction or the stairway to be fire resistant (whichever is greater).

718.2.5 Ceiling and floor openings. Where required by Section 712.1.8, Exception 1 of Section 714.4.1.2 or Section 714.5, *fireblocking* of the *annular space* around vents, pipes, ducts, chimneys and fireplaces at ceilings and floor levels shall be installed with a material specifically tested in the form and manner intended for use to demonstrate its ability to remain in place and resist the free passage of flame and the products of combustion.

❖ Sections 712.1.8 and 714.4 allow penetrations of ceilings and floors to be fireblocked [see Figures 714.4.1(1), 714.4.2(2) and 714.5.1]. It should be noted that depending on the type of penetration and the number of floors penetrated (or stories connected), a shaft may be required (see Sections 712 and 713).

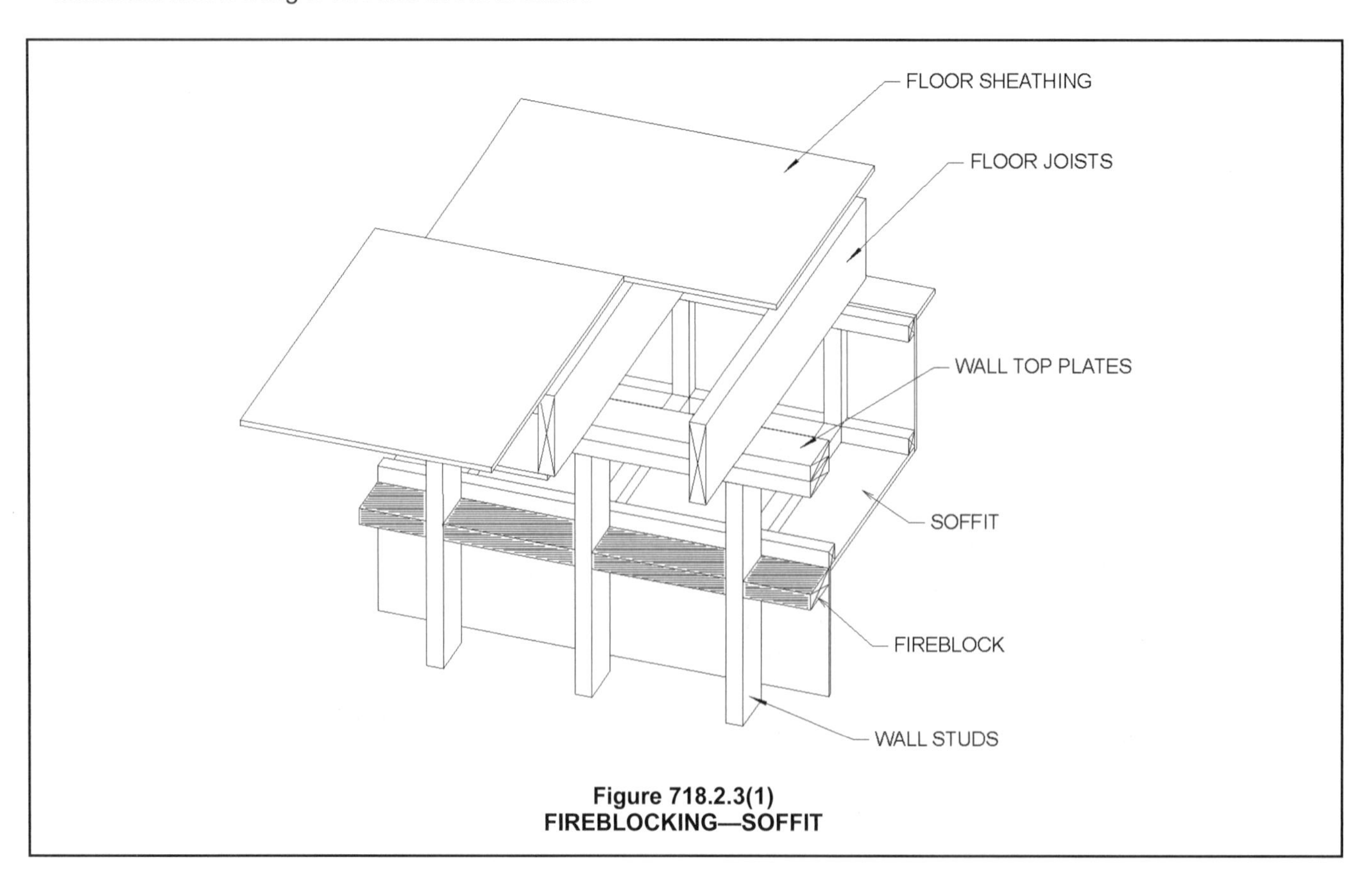

Figure 718.2.3(1)
FIREBLOCKING—SOFFIT

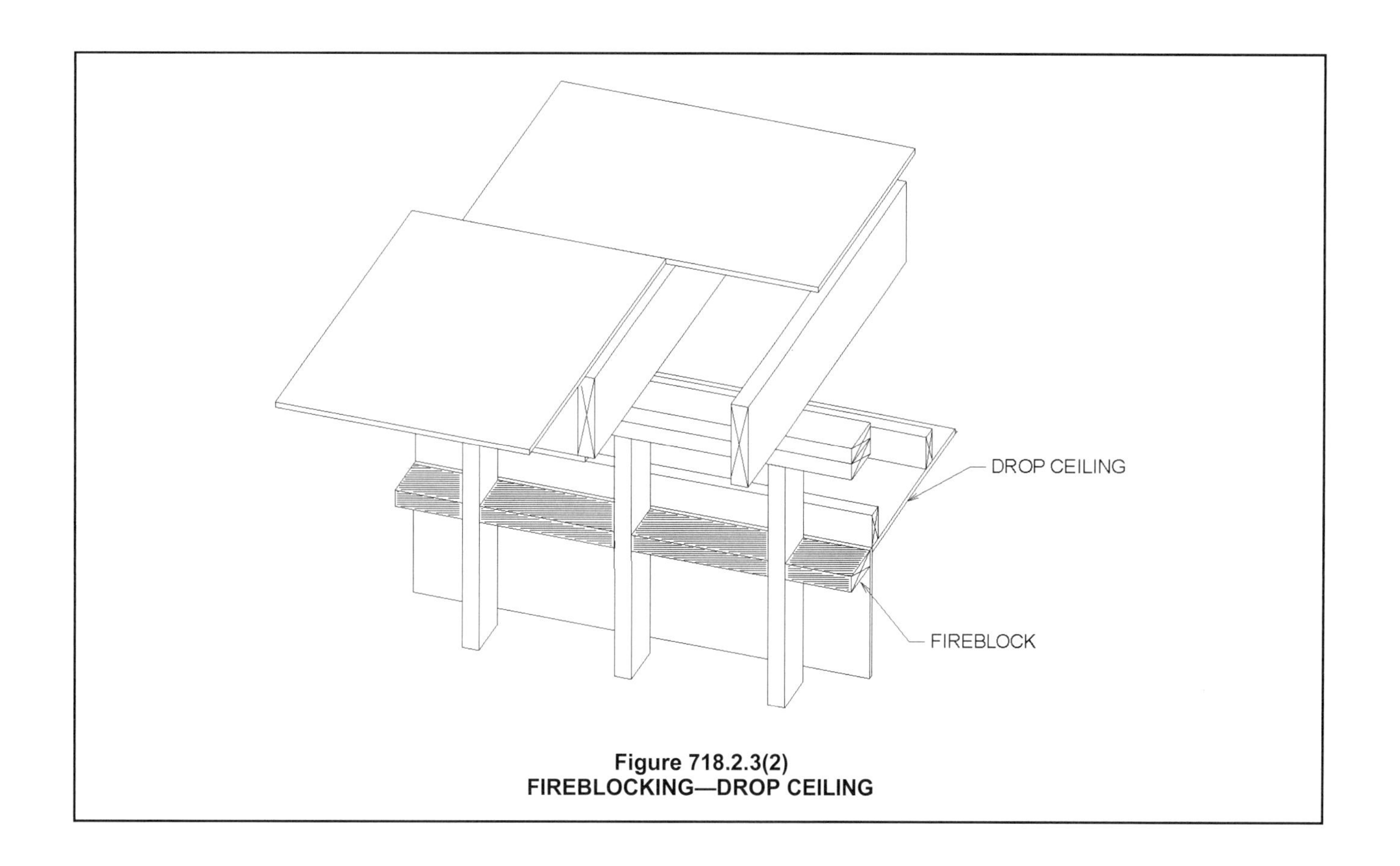

Figure 718.2.3(2)
FIREBLOCKING—DROP CEILING

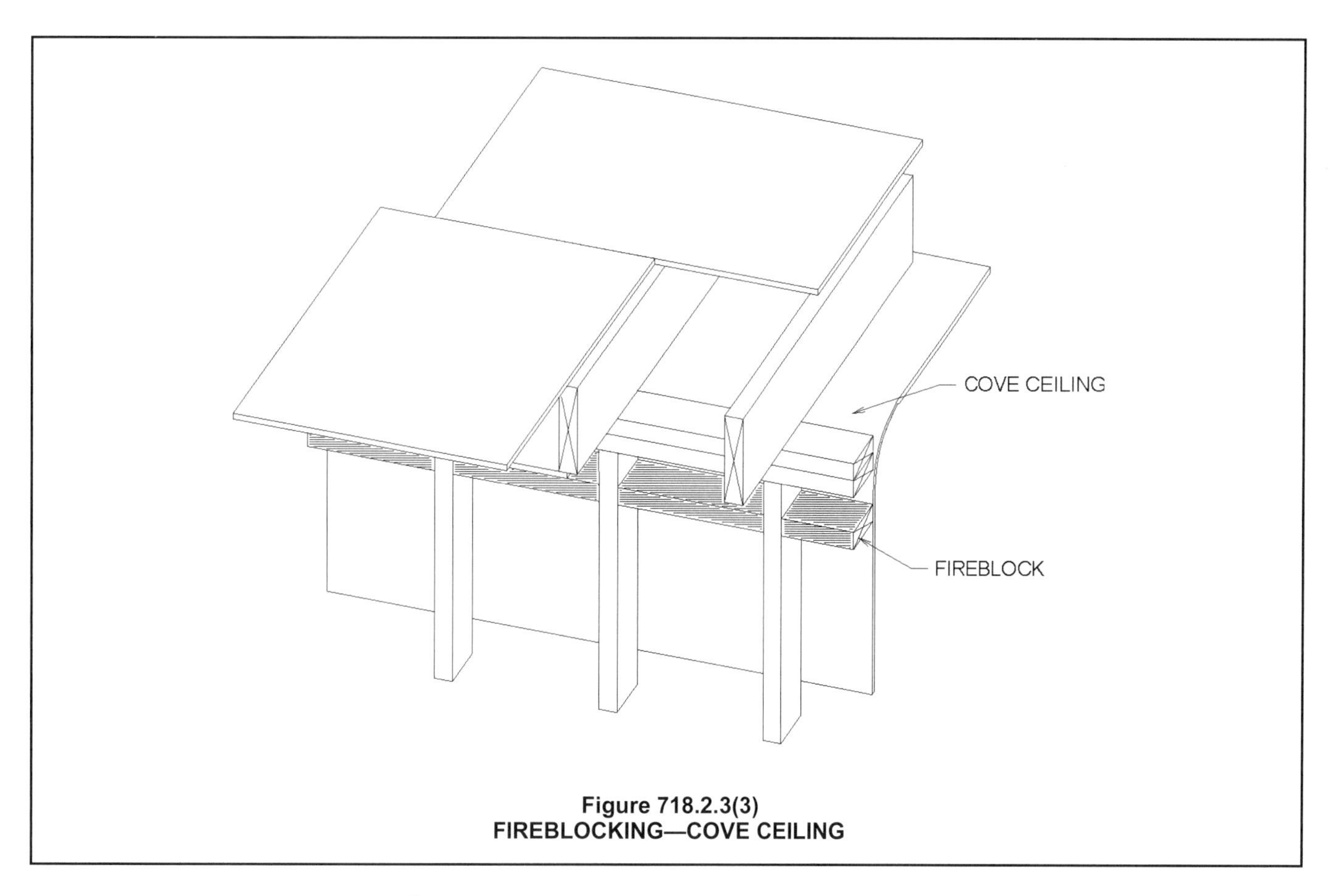

Figure 718.2.3(3)
FIREBLOCKING—COVE CEILING

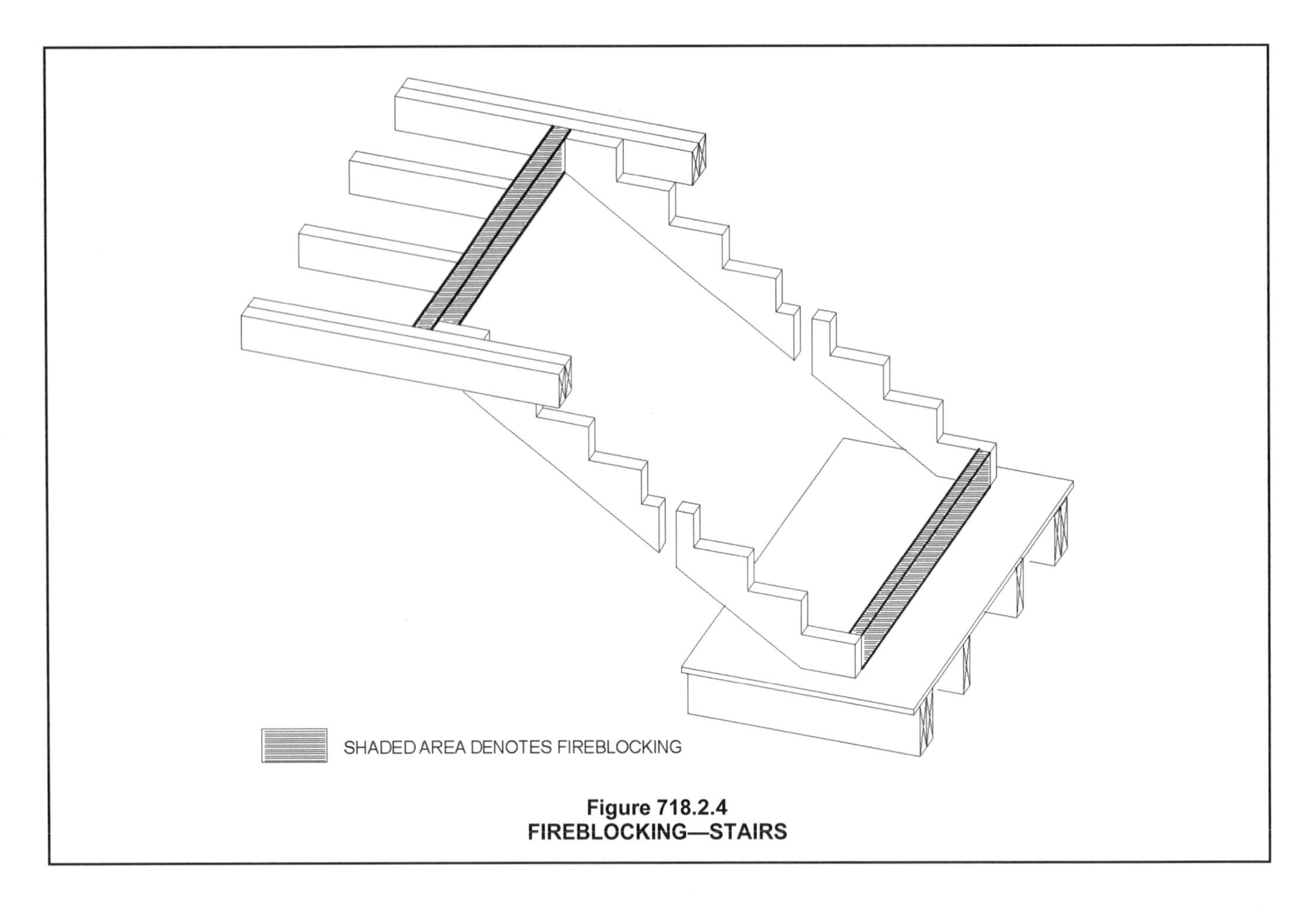

Figure 718.2.4
FIREBLOCKING—STAIRS

718.2.5.1 Factory-built chimneys and fireplaces. Factory-built chimneys and fireplaces shall be fireblocked in accordance with UL 103 and UL 127.

❖ UL 103 and 127 specify that only metal fireblocks and insulation shields are to be used with factory-built chimneys and fireplaces. The IMC contains a reference similar to that made in the UL standards. When fire-resistance-rated floor/ceiling assemblies are penetrated, the higher degree of protection afforded by tested through-penetration protection systems is required.

718.2.6 Exterior wall coverings. *Fireblocking* shall be installed within concealed spaces of exterior wall coverings and other exterior architectural elements where permitted to be of combustible construction as specified in Section 1406 or where erected with combustible frames. *Fireblocking* shall be installed at maximum intervals of 20 feet (6096 mm) in either dimension so that there will be no concealed space exceeding 100 square feet (9.3 m^2) between fireblocking. Where wood furring strips are used, they shall be of approved wood of natural decay resistance or preservative-treated wood. If noncontinuous, such elements shall have closed ends, with not less than 4 inches (102 mm) of separation between sections.

Exceptions:

1. *Fireblocking* of cornices is not required in single-family *dwellings*. *Fireblocking* of cornices of a two-family *dwelling* is required only at the line of *dwelling unit* separation.
2. *Fireblocking* shall not be required where the exterior wall covering is installed on noncombustible framing and the face of the exterior wall covering exposed to the concealed space is covered by one of the following materials:
   2.1. Aluminum having a minimum thickness of 0.019 inch (0.5 mm).
   2.2. Corrosion-resistant steel having a base metal thickness not less than 0.016 inch (0.4 mm) at any point.
   2.3. Other *approved* noncombustible materials.
3. *Fireblocking* shall not be required where the exterior wall covering has been tested in accordance with, and complies with the acceptance criteria of, NFPA 285. The exterior wall covering shall be installed as tested in accordance with NFPA 285.

❖ Combustible exterior wall finish and exterior architectural elements are required to be fireblocked at a maximum of 20-foot (6096 mm) intervals and a 100-square-foot (9.3 m^2) maximum area to prevent the spread of fire or smoke through concealed spaces. Adjacent sections of exterior trim need not be considered as contributing to the 20-foot (6096 mm) limitation if there is a minimum 4-inch (102 mm) separation between sections of exterior trim and the ends are closed. Trim on the building exterior presents the same concern as interior combustible concealed wall spaces and, therefore, requires fireblocking. To increase the long-term durability of wood used on the exterior, a certain level of decay resistance is required.

Exception 1 exempts single-family dwellings from the fireblocking requirement in the eaves or cornices.

Exception 2 exempts aluminum or steel siding installed over noncombustible wall framing.

Exception 3 exempts exterior wall coverings that have been tested and comply with NFPA 285, provided the coverings are installed in accordance with the listing. This exception is compatible to the metal composite material (MCM) wall coverings in Section 1407. Section 1407.10.4 also requires the NFPA 285 test.

718.2.7 Concealed sleeper spaces. Where wood sleepers are used for laying wood flooring on masonry or concrete fire-resistance-rated floors, the space between the floor slab and the underside of the wood flooring shall be filled with an *approved* material to resist the free passage of flame and products of combustion or fireblocked in such a manner that there will be no open spaces under the flooring that will exceed 100 square feet (9.3 m^2) in area and such space shall be filled solidly under permanent partitions so that there is no communication under the flooring between adjoining rooms.

Exceptions:

1. *Fireblocking* is not required for slab-on-grade floors in gymnasiums.
2. *Fireblocking* is required only at the juncture of each alternate lane and at the ends of each lane in a bowling facility.

❖ Concealed spaces created by floor sleepers must be fireblocked into areas no greater than 100 square feet (9.3 m^2) or such spaces must be filled with materials capable of resisting flame, smoke or gases (see Commentary Figure 718.2.7).

Since floor sleepers may cover large areas, this fireblock or material will restrict a fire from spreading within the concealed space to other areas of the floor. This section is commonly applied when a raised, wood-finished floor surface is installed in Type I or II construction.

718.3 Draftstopping in floors. In combustible construction, draftstopping shall be installed to subdivide floor/ceiling assemblies in the locations prescribed in Sections 718.3.2 through 718.3.3.

❖ Draftstopping subdivides combustible concealed horizontal spaces, unlike firestopping which subdivides vertical spaces. Draftstopping is required in combustible concealed floor construction in the locations required by Sections 718.3.2 and 718.3.3.

718.3.1 Draftstopping materials. Draftstopping materials shall be not less than $^1/_2$-inch (12.7 mm) gypsum board, $^3/_8$-inch (9.5 mm) wood structural panel, $^3/_8$-inch (9.5 mm) particleboard, 1-inch (25-mm) nominal lumber, cement fiberboard, batts or blankets of *mineral wool* or glass fiber, or other *approved* materials adequately supported. The integrity of *draftstops* shall be maintained.

❖ Similar to fireblocks in small concealed spaces, draftstops also act as a barrier to smoke and gases. Because of the large areas that require this barrier, the code permits certain types of sheathing material such as $^1/_2$-inch (12.7 mm) gypsum board, $^3/_8$-inch (9.5 mm) plywood and particleboard, 1-inch nominal lumber, cement fiberboard, mineral wool and glass fiber batts or blankets, or other approved materials. The draftstop material must be properly supported and capable of remaining in place when subjected to initial fire exposure. It is reasonable to assume that materials that are acceptable for fireblocking purposes (see Section 718.2.1) would perform as well for draftstopping situations, so many items are duplicated here. Although the code provides an extensive laundry list of accepted materials, it is important to note that the code allows the use of any other

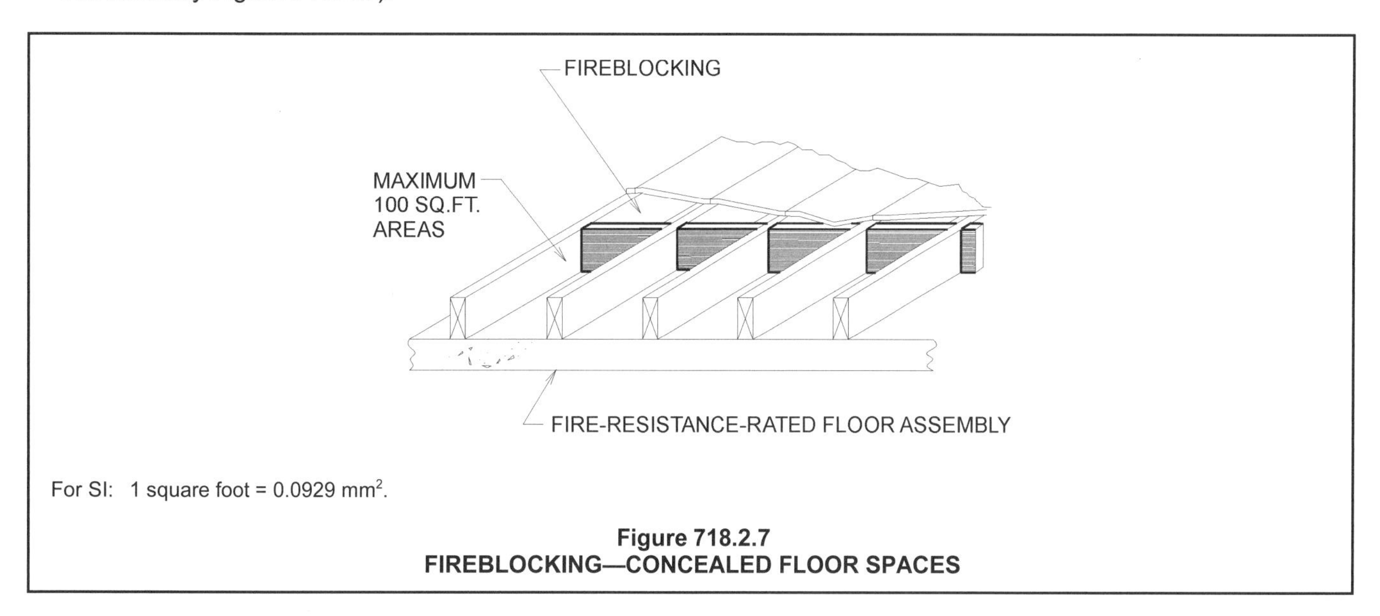

For SI: 1 square foot = 0.0929 mm^2.

Figure 718.2.7
FIREBLOCKING—CONCEALED FLOOR SPACES

approved material when it is adequately supported. Glass and mineral fiber batts are mentioned as an approved draftstopping material. Since the nonrigid glass or mineral fiber batts are not for insulation, they should be compressed to ensure that the void is completely filled. Batts used for insulation cannot be compressed.

718.3.2 Groups R-1, R-2, R-3 and R-4. Draftstopping shall be provided in floor/ceiling spaces in Group R-1 buildings, in Group R-2 buildings with three or more *dwelling units*, in Group R-3 buildings with two *dwelling units* and in Group R-4 buildings. Draftstopping shall be located above and in line with the *dwelling unit* and *sleeping unit* separations.

Exceptions:

1. Draftstopping is not required in buildings equipped throughout with an *automatic sprinkler system* in accordance with Section 903.3.1.1.
2. Draftstopping is not required in buildings equipped throughout with an *automatic sprinkler system* in accordance with Section 903.3.1.2, provided that automatic sprinklers are installed in the combustible concealed spaces where the draftstopping is being omitted.

❖ To maintain the integrity of dwelling or sleeping unit separation walls in buildings of Groups R-1, R-2, R-3 and R-4, draftstopping is to be provided when the dwelling, or sleeping-unit separation wall is not continuous to the floor sheathing above. The draftstopping must be installed directly above the dwelling or sleeping unit separation wall (see Commentary Figure 718.3.2). The dwelling unit or sleeping unit separation wall (see Section 708.4), plus the draftstopping above, are considered a barrier to the spread of fire and smoke. As such, the draftstop offers some level of protection to the occupants of one dwelling or sleeping unit from a fire occurring in another dwelling or sleeping unit.

The exception indicates that the draftstopping need not be provided if sprinklers are installed above and below the ceiling, since sprinkler activation will control the spread of fire. The sprinkler system must be installed in accordance with Section 903.3.1.1 or 903.3.1.2 and NFPA 13 or 13R. The NFPA 13 sprinkler system (see Section 903.3.1.1) will generally require the sprinklers to be installed within combustible floor spaces unless the space meets the exceptions found within the standard. The NFPA 13R sprinkler system (see Section 903.3.1.2) would generally not require the sprinkler system within the concealed floor space. However, when using the provisions of Exception 2, the NFPA 13R system must be extended into the floor space despite the normal exclusion within the standard. See Section 102.4, which requires this specific code requirement to apply beyond the requirements of the referenced standards.

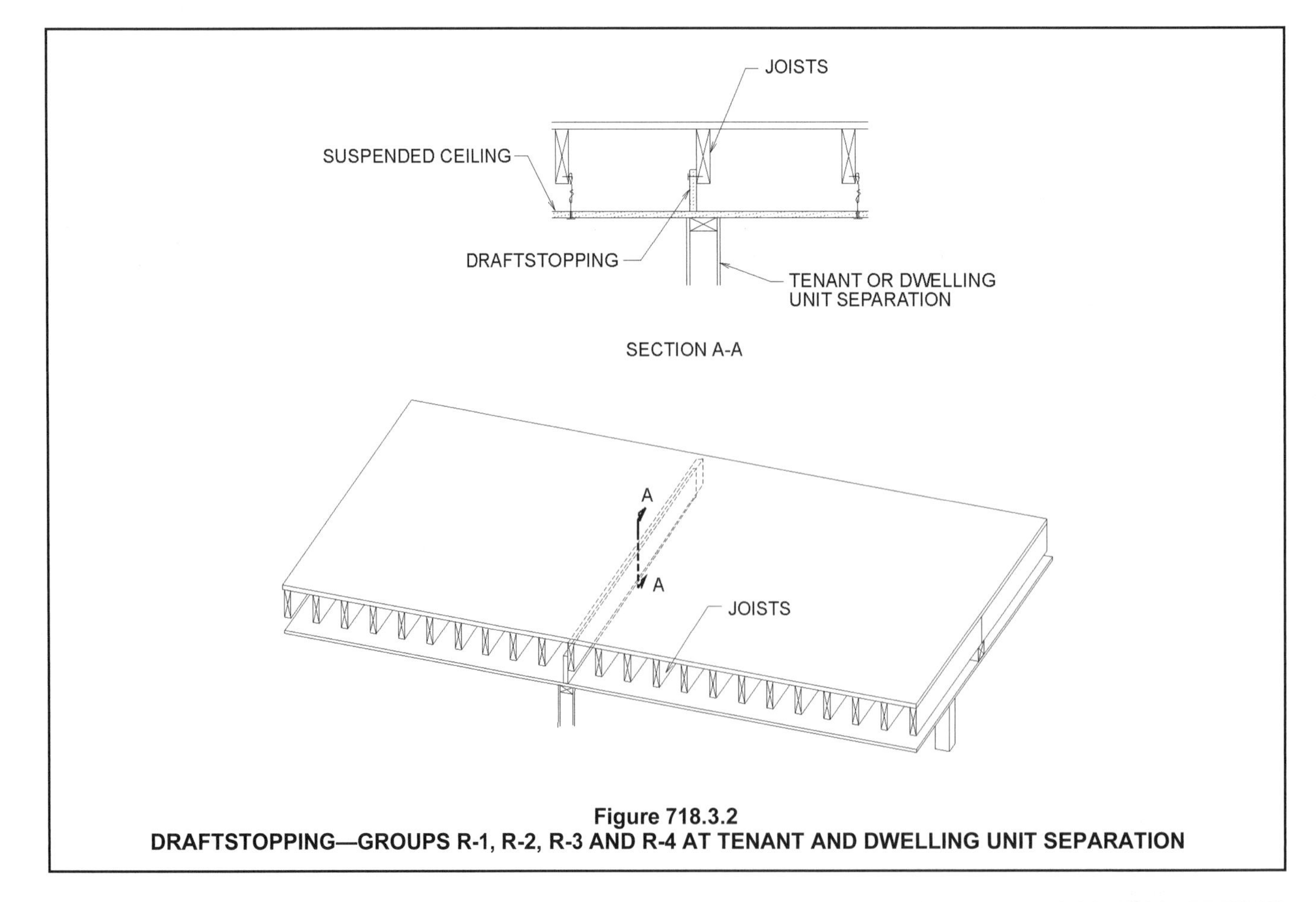

Figure 718.3.2
DRAFTSTOPPING—GROUPS R-1, R-2, R-3 AND R-4 AT TENANT AND DWELLING UNIT SEPARATION

718.3.3 Other groups. In other groups, draftstopping shall be installed so that horizontal floor areas do not exceed 1,000 square feet (93 m^2).

> **Exception:** Draftstopping is not required in buildings equipped throughout with an *automatic sprinkler system* in accordance with Section 903.3.1.1.

❖ Unless the buildings are sprinklered throughout with an NFPA 13 system, draftstopping is to be provided in all groups except Groups R-1, R-2, R-3 and R-4 such that the concealed combustible floor/ceiling space is subdivided into horizontal spaces not exceeding 1,000 square feet (93 m^2) (see Commentary Figure 718.3.3).

718.4 Draftstopping in attics. In combustible construction, draftstopping shall be installed to subdivide *attic* spaces and concealed roof spaces in the locations prescribed in Sections 718.4.2 and 718.4.3. Ventilation of concealed roof spaces shall be maintained in accordance with Section 1203.2.

❖ Fires that spread to attics that are not properly draftstopped often cause considerable damage. For this reason, draftstopping is required in attic and concealed roof spaces in accordance with Sections 718.4.2 and 718.4.3. The text also reminds the code user that, although the space must be compartmented, it still must be ventilated in accordance with the provisions of Chapter 12 in order to eliminate moisture or condensation or to cool the attic.

718.4.1 Draftstopping materials. Materials utilized for draftstopping of *attic* spaces shall comply with Section 718.3.1.

❖ See the commentary to Section 718.3.1

718.4.1.1 Openings. Openings in the partitions shall be protected by *self-closing* doors with automatic latches constructed as required for the partitions.

❖ Section 1209.2 requires attic access. The placement of draftstopping in the attic may interfere with the ability of the fire department to gain access to all attic spaces. This section requires that if draftstopping is provided with access openings to adjacent draftstopped portions of the attic, the openings must be constructed of draftstopping materials and be equipped with self-closing mechanisms in order to ensure that the draftstop will perform its intended function.

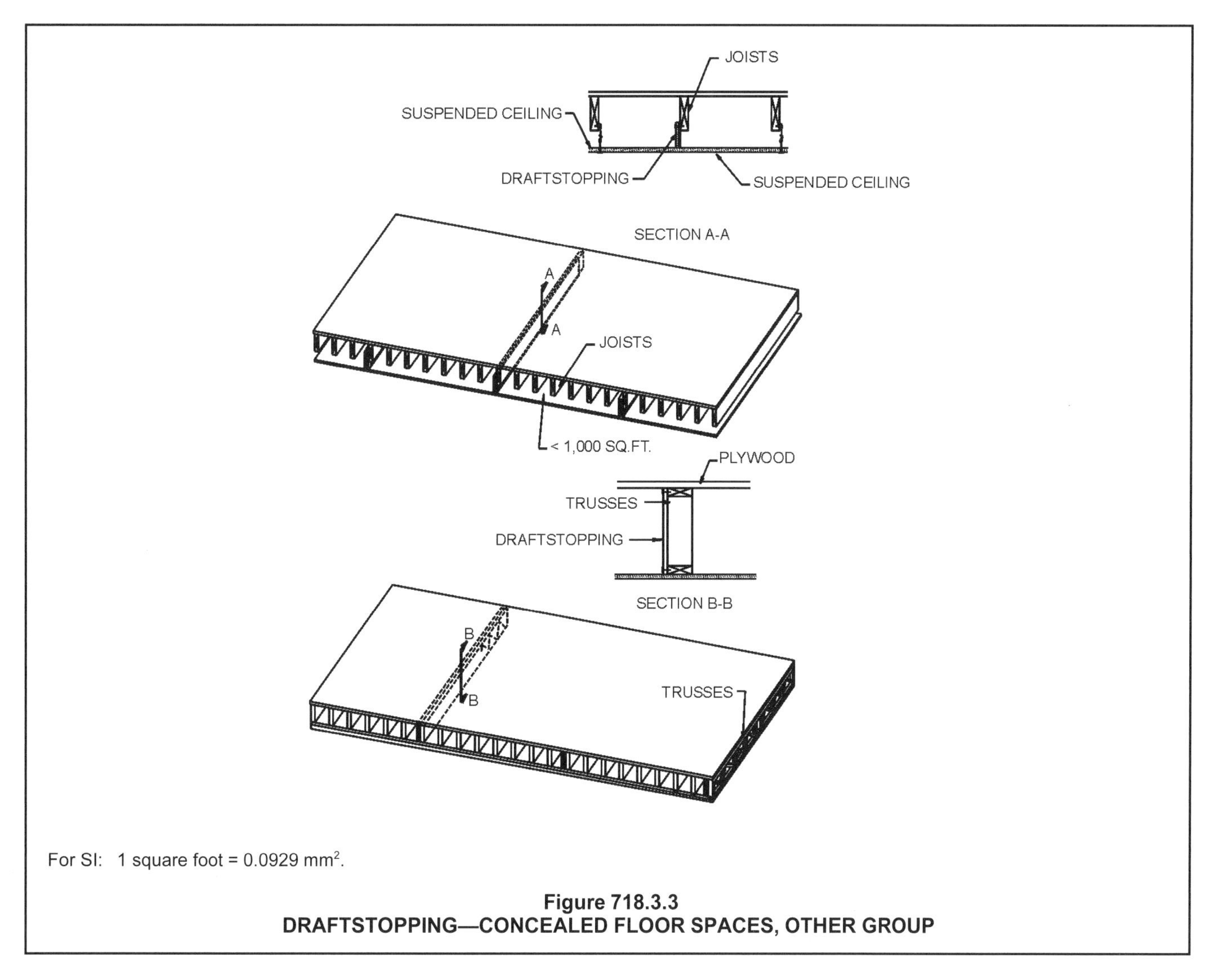

For SI: 1 square foot = 0.0929 mm^2.

Figure 718.3.3
DRAFTSTOPPING—CONCEALED FLOOR SPACES, OTHER GROUP

718.4.2 Groups R-1 and R-2. Draftstopping shall be provided in *attics*, mansards, overhangs or other concealed roof spaces of Group R-2 buildings with three or more *dwelling units* and in all Group R-1 buildings. Draftstopping shall be installed above, and in line with, *sleeping unit* and *dwelling unit* separation walls that do not extend to the underside of the roof sheathing above.

Exceptions:

1. Where *corridor* walls provide a *sleeping unit* or *dwelling unit* separation, draftstopping shall only be required above one of the *corridor* walls.
2. Draftstopping is not required in buildings equipped throughout with an *automatic sprinkler system* in accordance with Section 903.3.1.1.
3. In occupancies in Group R-2 that do not exceed four *stories above grade plane*, the *attic* space shall be subdivided by *draftstops* into areas not exceeding 3,000 square feet (279 m^2) or above every two *dwelling units*, whichever is smaller.
4. Draftstopping is not required in buildings equipped throughout with an *automatic sprinkler system* in accordance with Section 903.3.1.2, provided that automatic sprinklers are installed in the combustible concealed space where the draftstopping is being omitted.

❖ To maintain the integrity of dwelling- and sleeping-unit separation walls (see commentary, Section 718.3.2), draftstopping is to be provided above dwelling- or sleeping-unit separation walls that do not extend to the roof sheathing [see Commentary Figures 718.4.2(1) and 718.4.2(2)]. This provision is limited to Group R-1 and R-2 occupancies. Groups R-3 and R-4 are regulated by Section 718.4.3.

Exception 1 clarifies that for corridor walls that serve as dwelling- or sleeping-unit separations, draftstopping is only required above one of the two corridor walls. It should be noted that this option only applies to roof/attic spaces and not to floors (see Sections 718.3.2 and 708.4).

Exceptions 2 and 4 omit the requirement for draftstopping attics that are sprinklered in accordance with Section 903.3.1.1 (see Exception 2) or 903.3.1.2 (see Exception 4). This exception would require the attic to be sprinklered even though it may not be required by the standard. NFPA 13R typically would not require an attic to be sprinklered. This is consistent with the exceptions to Section 718.3.2 for draftstopping of floor spaces (see commentary, Section 718.3.2).

The draftstopping requirements can be reduced in Group R-2 occupancies of four stories or less by Exception 3. This exception allows the area between draftstopping to be increased to 3,000 square feet (279 m^2) or above every two dwelling units, whichever is smaller. This provides the opportunity for a simpler draftstopping installation when the draftstops are not required to be in line with the dwelling or sleeping unit

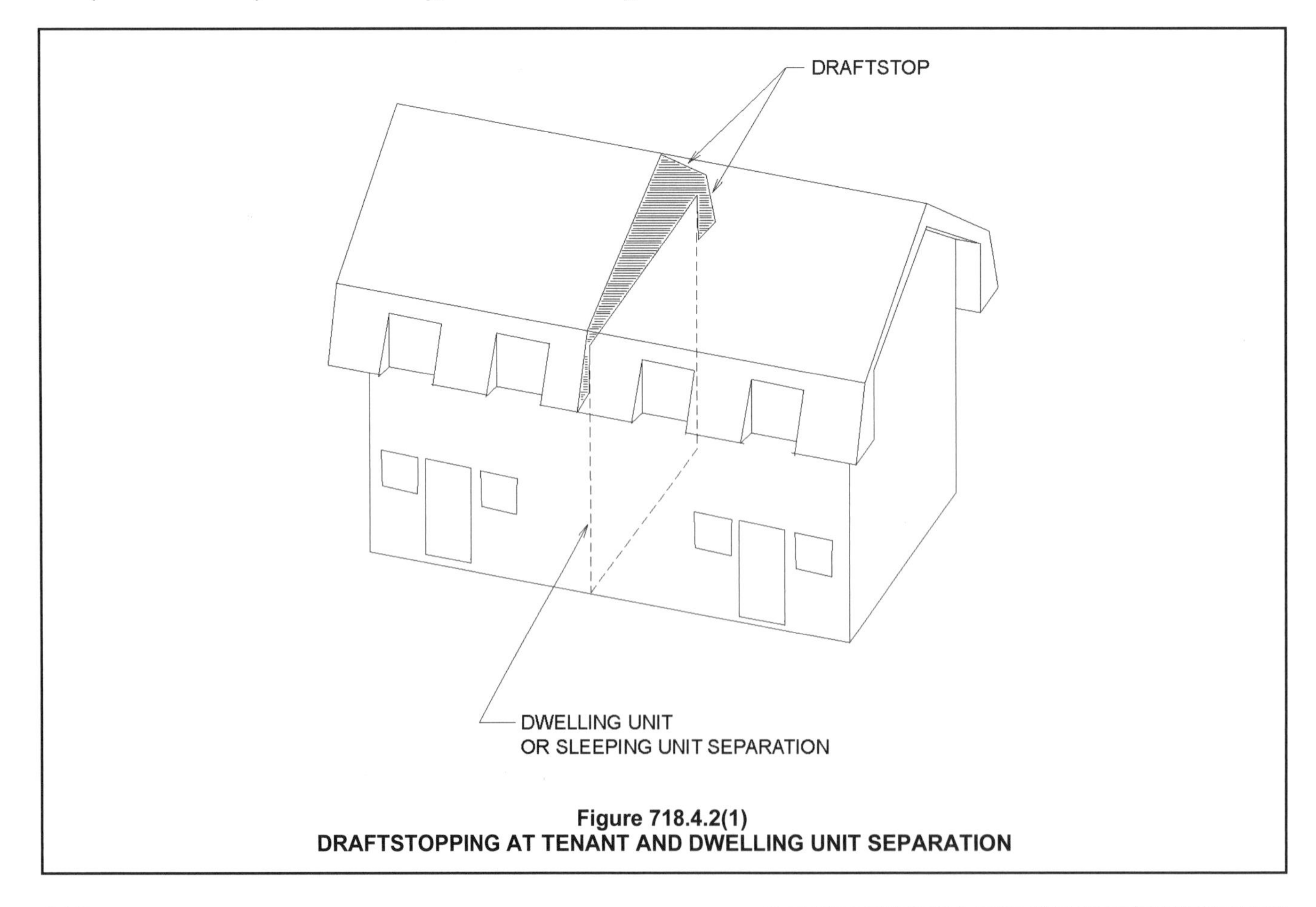

Figure 718.4.2(1)
DRAFTSTOPPING AT TENANT AND DWELLING UNIT SEPARATION

separation walls. Group R-2 buildings are required by Section 903.2.8 to be sprinklered; sprinklers reduce the need for attic draftstopping. The four-story restriction will match with the limitations of Section 903.3.1.2 for NFPA 13R sprinkler systems. Therefore, by providing this level of draftstopping, buildings that use the NFPA 13R systems will not need to comply with Exception 4, which is discussed above. If Section 903.3.1.1 (an NFPA 13 sprinkler system) is used, then sprinklers will generally be required within the attic space and Exception 2 (discussed above) would be applicable.

718.4.3 Other groups. Draftstopping shall be installed in *attics* and concealed roof spaces, such that any horizontal area does not exceed 3,000 square feet (279 m^2).

Exception: Draftstopping is not required in buildings equipped throughout with an *automatic sprinkler system* in accordance with Section 903.3.1.1.

❖ Since Group R-3, R-4 and nonresidential buildings are not always subdivided into small tenant spaces, concealed roof spaces and attics are to be subdivided into areas not exceeding 3,000 square feet (279 m^2). The exception indicates that draftstopping is not required when the spaces above and below the ceiling are provided coverage by an automatic sprinkler system, in accordance with NFPA 13.

718.5 Combustible materials in concealed spaces in Type I or II construction. Combustible materials shall not be permitted in concealed spaces of buildings of Type I or II construction.

Exceptions:

1. Combustible materials in accordance with Section 603.
2. Combustible materials exposed within plenums complying with Section 602 of the *International Mechanical Code*.
3. Class A *interior finish* materials classified in accordance with Section 803.
4. Combustible piping within partitions or shaft enclosures installed in accordance with the provisions of this code.
5. Combustible piping within concealed ceiling spaces installed in accordance with the *International Mechanical Code* and the *International Plumbing Code*.
6. Combustible insulation and covering on pipe and tubing, installed in concealed spaces other than plenums, complying with Section 720.7.

❖ This section is necessary because Section 718 would generally exclude Type I and II buildings from the concealed space draftstopping and fireblocking provisions. The use of combustibles in nonstructural applications is limited to the listed exceptions to prevent fire spread within building elements. The severity of a

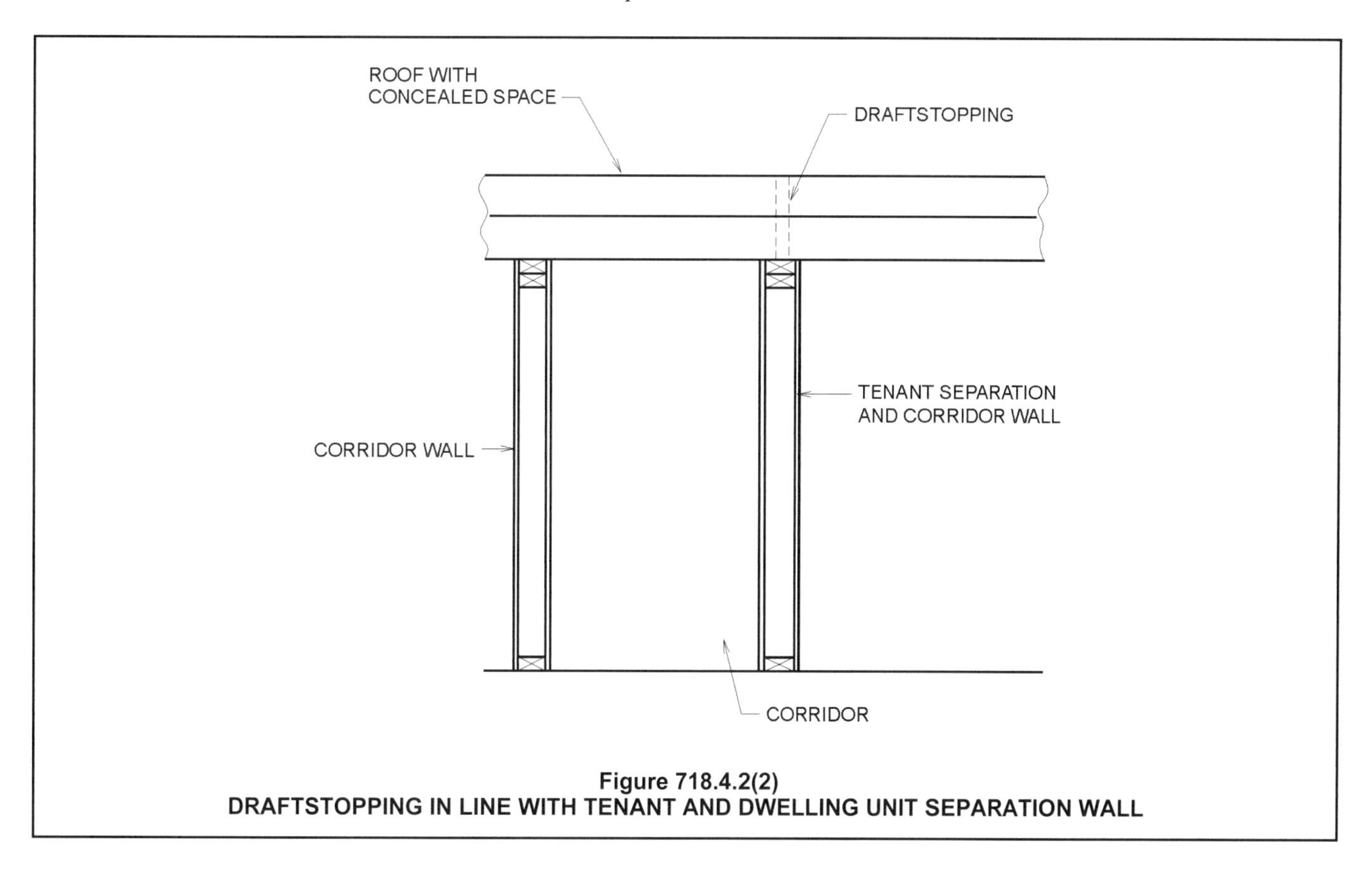

Figure 718.4.2(2)
DRAFTSTOPPING IN LINE WITH TENANT AND DWELLING UNIT SEPARATION WALL

potential problem increases when combustibles are located within concealed spaces that are inaccessible to manual fire fighting. This section regulates the use of combustibles within concealed spaces of noncombustible buildings. This section is referenced from Exception 24 of Section 603.1.

SECTION 719 FIRE-RESISTANCE REQUIREMENTS FOR PLASTER

719.1 Thickness of plaster. The minimum thickness of gypsum plaster or Portland cement plaster used in a fire-resistance-rated system shall be determined by the prescribed fire tests. The plaster thickness shall be measured from the face of the lath where applied to gypsum lath or metal lath.

❖ The fire resistance of plaster is dependent on material variations, such as the aggregate mixture (sand content) as well as bonding to the plaster base. The use of wire or wire fabric reinforces the plaster mixes applied to plaster bases and, therefore, ensures increased fire resistance. As such, the makeup of the plastered element is required when evaluating the fire-resistance rating of plaster. Because the type of plaster used affects the fire-resistance rating, it is essential that the tests conducted utilize a plaster mix similar to that which is to be used in actual construction. Depending on the type of plaster, additional tests may not be required. If the plaster is a type that is judged to be equivalent to a tested type (see Section 719.2), additional testing is not required.

Similar to the requirements of Section 714.2.5, it is the thickness of the protective covering that is important. Therefore, the plaster is to be measured from the face of the lath, which reinforces the plaster.

Code users should also be aware that Sections 721 and 722 contain a number of items related to the use of plaster in fire-resistance ratings. Section 720 contains a number of prescriptive assemblies that can be used to obtain a fire-resistance rating. Section 721 contains methods of calculating fire-resistant assemblies and the contribution of plaster to other assemblies. For more information, see Sections 722.2.1.4, 722.3.2 and 722.4.1. Although many people are not familiar with these provisions, they are permitted under Section 703.3 and provide another option for establishing rated assemblies.

719.2 Plaster equivalents. For fire-resistance purposes, $^1/_2$ inch (12.7 mm) of unsanded gypsum plaster shall be deemed equivalent to $^3/_4$ inch (19.1 mm) of one-to-three gypsum sand plaster or 1 inch (25 mm) of Portland cement sand plaster.

❖ This section indicates that prescribed thicknesses of unsanded gypsum plaster, one-to-three sanded gypsum plaster and Portland cement sand plaster are considered equivalent in terms of their fire-resistance rating characteristics. Once tests are conducted to determine the fire resistance of an assembly using one of these types, equivalency to the other prescribed types can be determined without subsequent tests.

719.3 Noncombustible furring. In buildings of Type I and II construction, plaster shall be applied directly on concrete or masonry or on *approved* noncombustible plastering base and furring.

❖ Buildings of Type I and II construction are required to be predominantly of noncombustible construction. This section requires the plastering base and furring to be noncombustible. The intent of this provision is to not permit a combustible cavity behind the finished plaster, which may be an area for a concealed fire buildup.

719.4 Double reinforcement. Plaster protection more than 1 inch (25 mm) in thickness shall be reinforced with an additional layer of *approved* lath embedded not less than $^3/_4$ inch (19.1 mm) from the outer surface and fixed securely in place.

Exception: Solid plaster partitions or where otherwise determined by fire tests.

❖ The additional reinforcement is intended to decrease the likelihood of the plaster loosening when subjected to high temperatures. The plaster relies on the lath to provide a finish that is structurally sound. As the thickness of the plaster increases, additional reinforcement is required to hold the plaster in place.

719.5 Plaster alternatives for concrete. In reinforced concrete construction, gypsum plaster or Portland cement plaster is permitted to be substituted for $^1/_2$ inch (12.7 mm) of the required poured concrete protection, except that a minimum thickness of $^3/_8$ inch (9.5 mm) of poured concrete shall be provided in reinforced concrete floors and 1 inch (25 mm) in reinforced concrete columns in addition to the plaster finish. The concrete base shall be prepared in accordance with Section 2510.7.

❖ Concrete elements, such as reinforced slabs and columns, rely in part on the thickness of the element to provide a fire-resistance rating. In order to reduce the concrete thickness, this section permits up to $^1/_2$ inch (12.7 mm) of plaster to be substituted for $^1/_2$ inch (12.7 mm) of concrete without affecting the fire-resistance rating of the element. While plaster can be substituted for concrete for fire-resistance purposes, a minimum amount of concrete is still required to be provided for structural integrity.

SECTION 720 THERMAL- AND SOUND-INSULATING MATERIALS

720.1 General. Insulating materials, including facings such as vapor retarders and *vapor-permeable membranes*, similar coverings and all layers of single and multilayer reflective foil insulations, shall comply with the requirements of this section. Where a flame spread index or a smoke-developed index is specified in this section, such index shall be determined in accordance with ASTM E84 or UL 723. Any material that is subject to an increase in flame spread index or smoke-developed index beyond the limits herein established through the effects of age, moisture or other atmospheric conditions shall not be permitted.

Exceptions:

1. Fiberboard insulation shall comply with Chapter 23.
2. Foam plastic insulation shall comply with Chapter 26.
3. Duct and pipe insulation and duct and pipe coverings and linings in plenums shall comply with the *International Mechanical Code*.
4. All layers of single and multilayer reflective plastic core insulation shall comply with Section 2613.

❖ This section addresses the potential fire or flame spread hazards of insulating materials installed in building spaces. Other provisions of the code or companion codes regulate insulating materials used for duct and plenum insulation [see the IMC for energy conservation purposes and the *International Energy Conservation Code*® (IECC®)] and foam plastic insulation (see Section 2603). Insulating materials can affect fire development and fire spread and are, therefore, regulated accordingly.

This section addresses the various insulating materials that may be installed in building spaces, including insulating batts, blankets, fills (including vapor barriers and vapor-permeable membranes) or other coverings. Fiberboard insulation is regulated by Chapter 23. Cellulose loose-fill insulation is regulated by Section 720.6. Foam plastic insulation is regulated by Chapter 26. The applicable test method is ASTM E84 or UL 723 (see commentary, Section 803.1).

The requirements of Section 703.2 regarding the potential that adding items into a fire-resistant assembly may reduce the rating of the tested assembly must be considered. The addition of insulation has the potential to either add to or reduce the fire-resistance rating of an assembly. The reduction of ratings is more of a problem when insulation is added into a horizontal assembly.

720.2 Concealed installation. Insulating materials, where concealed as installed in buildings of any type of construction, shall have a flame spread index of not more than 25 and a smoke-developed index of not more than 450.

Exception: Cellulosic fiber loose-fill insulation complying with the requirements of Section 720.6 shall not be required to meet a flame spread index requirement but shall be required to meet a smoke-developed index of not more than 450 when tested in accordance with CAN/ULC S102.2.

❖ Concealed insulation no longer acts as an interior finish unless the covering material is removed or the fire is in the concealed building space. To limit the contribution of the insulation to a fire condition, the material must have a flame spread index of no more than 25 and a smoke-developed index of no more than 450. This is illustrated in Commentary Figure 720.2. Cellulosic fiber loose-fill insulation is regulated by the U.S. Consumer Product Safety Commission (CPSC) in accordance with Section 720.6. These materials are exempt from the flame spread index requirements but still need to be tested for smoke-development in accordance with CAN/ULC S102.2. Spray-applied cellulose is regulated by Section 720.3.1 (see commentary, Section 720.4).

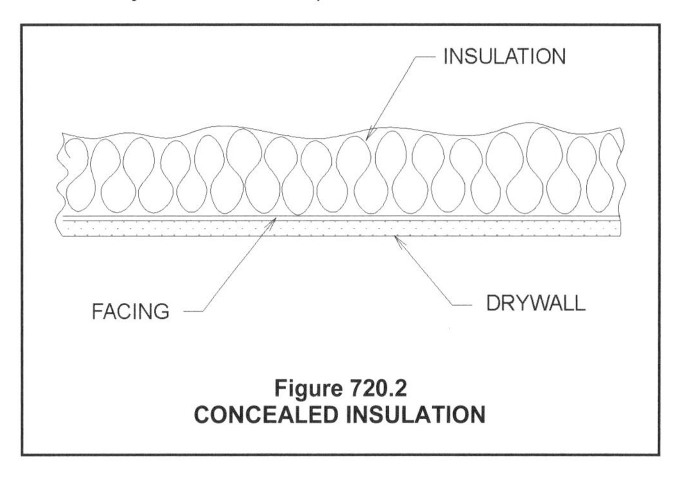

Figure 720.2
CONCEALED INSULATION

720.2.1 Facings. Where such materials are installed in concealed spaces in buildings of Type III, IV or V construction, the flame spread and smoke-developed limitations do not apply to facings, coverings, and layers of reflective foil insulation that are installed behind and in substantial contact with the unexposed surface of the ceiling, wall or floor finish.

Exception: All layers of single and multilayer reflective plastic core insulation shall comply with Section 2613.

❖ In buildings of Type III, IV and V construction, the flame spread indices do not apply to facing material, provided it is installed behind (and in substantial contact with) the unexposed surface of the ceiling, floor or wall finish. For example, when paper-backed insulation is placed directly on top of a ceiling, the paper facing is not required to meet the flame spread index; however, if the same material is applied to the underside of a roof deck and the paper facing is exposed to the attic space, the paper facing must meet the flame spread criteria. The potential for flame spread is greatly diminished when the facings are installed in direct contact with the finish material because of the lack of airspace to support a fire if it were to be exposed to a source of ignition.

The purpose of the vapor retarder being placed on the warm-in-winter side of the building element is to prevent moisture vapor in the warm interior from reaching the dew-point temperature on a cold surface in the roof, wall or floor system. The generally accepted rating of a vapor retarder is 1 perm [57 mg/(s • m^2 • Pa)]. It is assumed that the permeance of an adequate vapor barrier will not exceed 1 perm [57 mg/(s • m^2 • Pa)] (see commentary, Section 1203.2).

720.3 Exposed installation. Insulating materials, where exposed as installed in buildings of any type of construction, shall have a flame spread index of not more than 25 and a smoke-developed index of not more than 450.

Exception: Cellulosic fiber loose-fill insulation complying with the requirements of Section 720.6 shall not be required to meet a flame spread index requirement but shall be required to meet a smoke-developed index of not more than 450 when tested in accordance with CAN/ULC S102.2.

❖ Exposed insulating materials represent the same fire exposure hazard as any other exposed material, such as interior finish. The flame spread index for all applications of insulation materials is limited to 25, compared to 200 for interior finishes installed in rooms or spaces of certain occupancies (see commentary, Chapter 8). As with all interior finishes, the smoke-developed index is limited to 450. Cellulosic fiber loose-fill insulation is regulated by the CPSC in accordance with Section 720.6. These materials are exempt from the flame spread index requirements but still need to be tested for smoke development in accordance with CAN/ULC S102.2.Spray-applied cellulosic fiber is regulated by Section 720.3.1 (see commentary, Section 720.4).

It is important to remember that insulation materials that are open and exposed within an attic or crawl space are still considered as being "exposed" and, therefore, regulated by this section. The general requirements of this section do not differ from the provisions of Section 720.1. The main difference is that the facing materials within concealed installations may be exempt from the flame spread and smoke-developed rating (see Section 720.2.1) while the exposed facings would be limited as stated in this section.

720.3.1 Attic floors. Exposed insulation materials installed on *attic* floors shall have a critical radiant flux of not less than 0.12 watt per square centimeter when tested in accordance with ASTM E970.

❖ ASTM E970 is a test method that was developed by the insulation industry to evaluate the fire hazard of exposed attic insulation and is referenced in the material standards for insulation. Cellulosic fiber loose-fill insulation is required to comply with CPSC 16 CFR, Part 1209 (see Section 720.6), which requires testing by this method. Spray-applied cellulose insulation that is not subject to the CPSC standard is also subject to the ASTM E970 testing by virtue of this section.

720.4 Loose-fill insulation. Loose-fill insulation materials that cannot be mounted in the ASTM E84 or UL 723 apparatus without a screen or artificial supports shall comply with the flame spread and smoke-developed limits of Sections 720.2 and 720.3 when tested in accordance with CAN/ULC S102.2.

Exception: Cellulosic fiber loose-fill insulation shall not be required to meet a flame spread index requirement when tested in accordance with CAN/ULC S102.2, provided such insulation has a smoke-developed index of not more than 450 and complies with the requirements of Section 720.6.

❖ The exception serves to make a distinction between cellulose insulation, which is spray applied using a water-mist applicator, and cellulose or other loose-fill insulation, which is poured or blown in place. Spray-applied cellulose insulation can be exposed on vertical and horizontal ceiling-type surfaces so it is treated as any other insulating material in regard to testing. Cellulosic fiber loose-fill insulation, which is poured or blown in, is regulated by CPSC requirements (see Section 720.6) and is exempt from the test procedure described in this section. These materials are exempt from the flame spread index requirements but still need to be tested for smoke development in accordance with CAN/ULC S102.2.

720.5 Roof insulation. The use of combustible roof insulation not complying with Sections 720.2 and 720.3 shall be permitted in any type of construction provided that insulation is covered with *approved* roof coverings directly applied thereto.

❖ Foam plastic roof insulation is required to comply with Chapter 26. This section allows combustible insulation other than foam plastic insulation to be incorporated in the roof assembly without declassifying the type of construction. This provision coordinates with Section 603.1, Item 4.

720.6 Cellulosic fiber loose-fill insulation and self-supported spray-applied cellulosic insulation. Cellulosic fiber loose-fill insulation and self-supported spray-applied cellulosic insulation shall comply with CPSC 16 CFR Parts 1209 and 1404. Each package of such insulating material shall be clearly labeled in accordance with CPSC 16 CFR Parts 1209 and 1404.

❖ Cellulosic fiber loose-fill insulation is federally regulated by the CPSC. CPSC 16 CFR, Parts 1209 and 1404, contain various requirements that regulate the product to avoid excessive flammability or significant fire hazards. The smoke-developed index for cellulosic fiber loose-fill insulation must be determined by the ASTM E84 test and must be 450 or less. The intent of this section is that the procedure for the smoke-developed index should be done by an ASTM E84 test and not by the test procedures specified in Section 720.4.

720.7 Insulation and covering on pipe and tubing. Insulation and covering on pipe and tubing shall have a flame spread index of not more than 25 and a smoke-developed index of not more than 450.

Exception: Insulation and covering on pipe and tubing installed in plenums shall comply with the *International Mechanical Code.*

❖ This section maintains the general provision for insulation by requiring a maximum flame spread of 25 and a smoke-developed index of 450, as stated in

Sections 720.2 and 720.3; however, if exposed in a plenum, the IMC limits the smoke-developed index to 50.

SECTION 721
PRESCRIPTIVE FIRE RESISTANCE

721.1 General. The provisions of this section contain prescriptive details of fire-resistance-rated building elements, components or assemblies. The materials of construction listed in Tables 721.1(1), 721.1(2), and 721.1(3) shall be assumed to have the *fire-resistance ratings* prescribed therein. Where materials that change the capacity for heat dissipation are incorporated into a fire-resistance-rated assembly, fire test results or other substantiating data shall be made available to the *building official* to show that the required fire-resistance-rating time period is not reduced.

❖ Section 703.3 permits the fire-resistance ratings to be determined in a number of ways, including those found in Section 721 (see Section 703.3, Item 2). In this section, there are many prescriptive details for fire-resistance-rated construction, particularly those materials and assemblies listed in Table 721.1(1) for structural parts, Table 721.1(2) for walls and partitions and Table 721.1(3) for floor and roof systems. For the most part, the listed items have been tested in accordance with the fire-resistance ratings indicated. In addition, a similar footnote to all of the tables allows the acceptance of generic assemblies that are listed in GA 600, *Fire-Resistance Design Manual*. It is important to review all of the applicable footnotes when using a material or assembly from one of the tables.

As stated above, the fire-resistance ratings for the walls and partitions outlined in Table 721.1(2) are based on actual tests. For reinforced concrete walls, it is important to note the type of aggregate used. The difference in aggregates is quite significant for a 4-hour fire-resistance-rated wall, as it amounts to a difference in thickness of almost 2 inches (51 mm). For hollow-unit masonry walls, the thickness required for a particular fire-endurance rating is the equivalent thickness as defined in Section 722.3.1 for concrete masonry and Section 722.4.1.1 for clay masonry.

Table 721.1(3) provides fire-resistance ratings for floor/ceiling and roof/ceiling assemblies. Note n, which exempts unusable space from the flooring and ceiling requirements, is especially important. This exemption is consistent with Section 711.2.6.

Often, materials such as insulation are added to fire-resistance-rated assemblies. The code requires substantiating fire test data to show that when the materials are added, they do not reduce the required fire-endurance time period. As an example, adding insulation to a floor/ceiling assembly may change its capacity to dissipate heat and, particularly for noncombustible assemblies, the fire-resistance rating may be changed.

Although the primary intent of the provision is to cover those cases where thermal insulation is added, the language is intentionally broad so that it applies to any material that might be added to the assembly.

721.1.1 Thickness of protective coverings. The thickness of fire-resistant materials required for protection of structural members shall be not less than set forth in Table 721.1(1), except as modified in this section. The figures shown shall be the net thickness of the protecting materials and shall not include any hollow space in back of the protection.

❖ In accordance with this section, the required thickness of insulating material used to provide fire resistance to a structural member cannot be less than the dimension established by Table 721.1(1), except for permitted modifications. An example of the minimum thickness of concrete required for a structural steel column is shown in Commentary Figure 721.1.1(1). Commentary Figure 721.1.1(2) illustrates the minimum concrete thickness requirements for protecting reinforcing steel in concrete columns, beams, girders and trusses. Refer to Section 704 for additional provisions regarding structural members.

721.1.2 Unit masonry protection. Where required, metal ties shall be embedded in bed joints of unit masonry for pro-

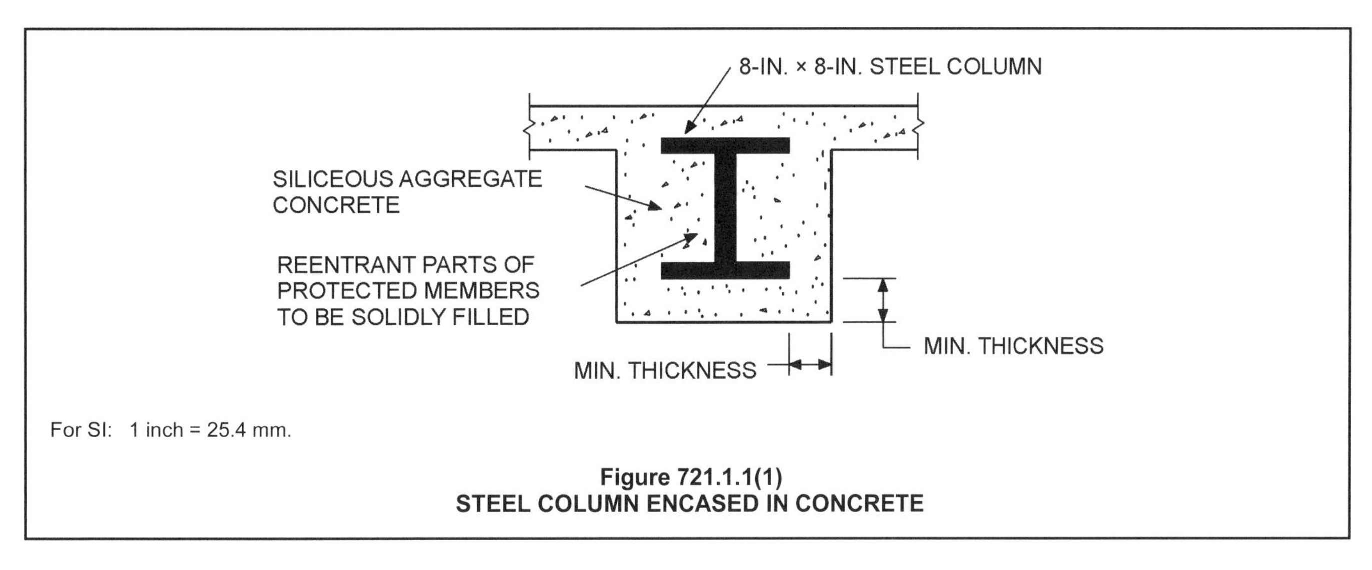

Figure 721.1.1(1)
STEEL COLUMN ENCASED IN CONCRETE

tection of steel columns. Such ties shall be as set forth in Table 721.1(1) or be equivalent thereto.

❖ Items 1-3.1 through 1-3.4 of Table 721.1(1) require horizontal joint reinforcement. This section stipulates that if ties are required to connect the masonry to the steel column, such ties are to be located in the bed joints, which is where the horizontal reinforcement is to be located.

721.1.3 Reinforcement for cast-in-place concrete column protection. Cast-in-place concrete protection for steel columns shall be reinforced at the edges of such members with wire ties of not less than 0.18 inch (4.6 mm) in diameter wound spirally around the columns on a pitch of not more than 8 inches (203 mm) or by equivalent reinforcement.

❖ In order for a concrete-encased steel column to maintain a fire-resistance rating, it is imperative that the concrete remain in place. Similar to a reinforced concrete column with spiral reinforcement (see Section 721.2.4.3), the ties required by this section are necessary for the concrete to remain in place around the column.

721.1.4 Plaster application. The finish coat is not required for plaster protective coatings where those coatings comply with the design mix and thickness requirements of Tables 721.1(1), 721.1(2) and 721.1(3).

❖ The thickness of plaster for prescriptive assemblies, such as Item 1-4.1 in Table 721.1(1), Item 12-1.1 in Table 721.1(2) and Item 11-1.1 in Table 721.1(3), does not include the finish coat. Only the scratch coat and brown coat are needed to achieve the indicated fire-resistance ratings in the tables.

721.1.5 Bonded prestressed concrete tendons. For members having a single tendon or more than one tendon installed with equal concrete cover measured from the nearest surface, the cover shall be not less than that set forth in Table 721.1(1). For members having multiple tendons installed with variable concrete cover, the average tendon cover shall be not less than that set forth in Table 721.1(1), provided:

1. The clearance from each tendon to the nearest exposed surface is used to determine the average cover.
2. In no case can the clear cover for individual tendons be less than one-half of that set forth in Table 721.1(1). A minimum cover of $^3/_4$ inch (19.1 mm) for slabs and 1 inch (25 mm) for beams is required for any aggregate concrete.
3. For the purpose of establishing a *fire-resistance rating*, tendons having a clear covering less than that set forth in Table 721.1(1) shall not contribute more than 50 percent of the required ultimate moment capacity for members less than 350 square inches (0.226 m^2) in cross-sectional area and 65 percent for larger members. For structural design purposes, however, tendons having a reduced cover are assumed to be fully effective.

❖ As the ultimate-moment capacity of the member is critical to its behavior under fire conditions, the code requires the reduction for those tendons having cover to be less than that specified by the code. Behavior at service loads, however, is less affected by the heat of a fire; therefore, the code permits those tendons with reduced cover to be fully effective.

TABLE 721.1(1). See page 7-135.

❖ An example of the minimum thickness of concrete required for a structural steel column is shown in Commentary Figure 721.1.1(1). As shown, this figure depicts Item 1-1.5 of Table 721.1(1). Commentary Figure 721.1.1(2) depicts a reinforced concrete column based on Item 5-1.1 (see commentary, Section 721.1).

TABLE 721.1(2). See page 7-139.

❖ An example of the minimum thickness of concrete required for a wall is shown in Commentary Figure 721.1(2)(a). As shown, this figure depicts Item 4-1.1 of Table 721.1(2). Commentary Figure 721.1(2)(b) depicts a noncombustible stud exterior wall based on Item 15-1.4 (see commentary, Section 721.1).

TABLE 721.1(3). See page 7-147.

❖ An example of a 1-hour-rated wood floor or roof truss is shown in Commentary Figure 721.1(3). As shown, this figure depicts Item 21-1.1 (see commentary, Section 721.1).

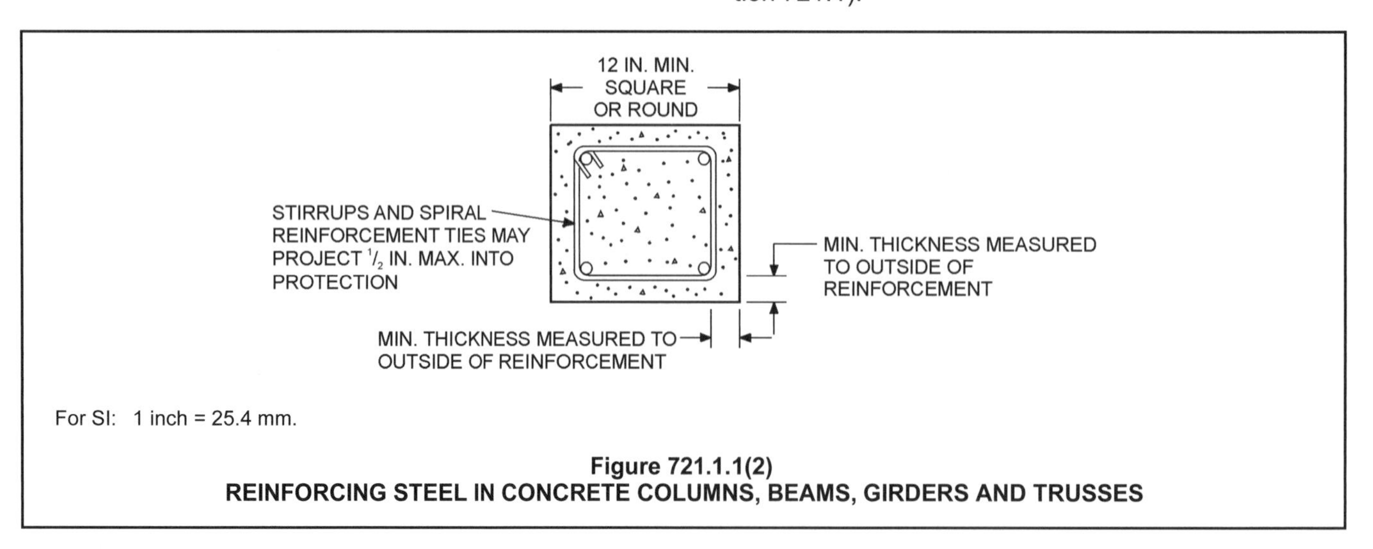

Figure 721.1.1(2)
REINFORCING STEEL IN CONCRETE COLUMNS, BEAMS, GIRDERS AND TRUSSES

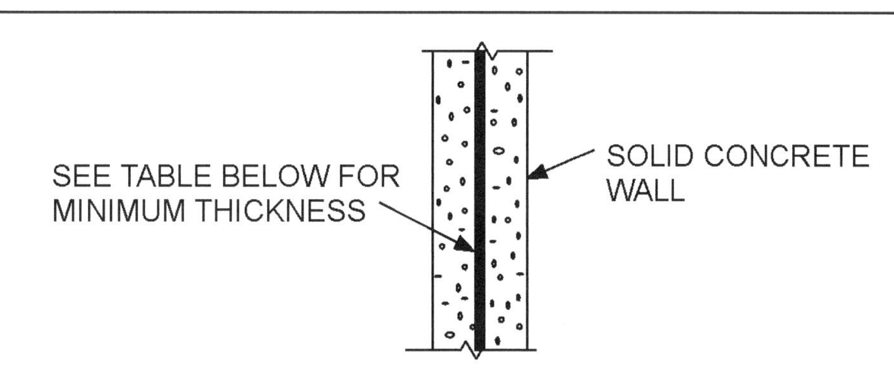

MATERIAL	ITEM NUMBER	CONSTRUCTION	MINIMUM FINISHED THICKNESS FACE-TO-FACE (inches)			
			4 hour	3 hour	2 hour	1 hour
Solid concrete	4-1.1	Siliceous aggregate concrete	7.0	6.2	5.0	3.5
		Carbondate aggregate concrete	6.6	5.7	4.6	3.2
		Sand-lightweight concrete	5.4	4.6	3.8	2.7
		Lightweight concrete	5.1	4.4	3.6	2.5

For SI: 1 inch = 25.4 mm.

Figure 721.1(2)(a)
CONCRETE WALL

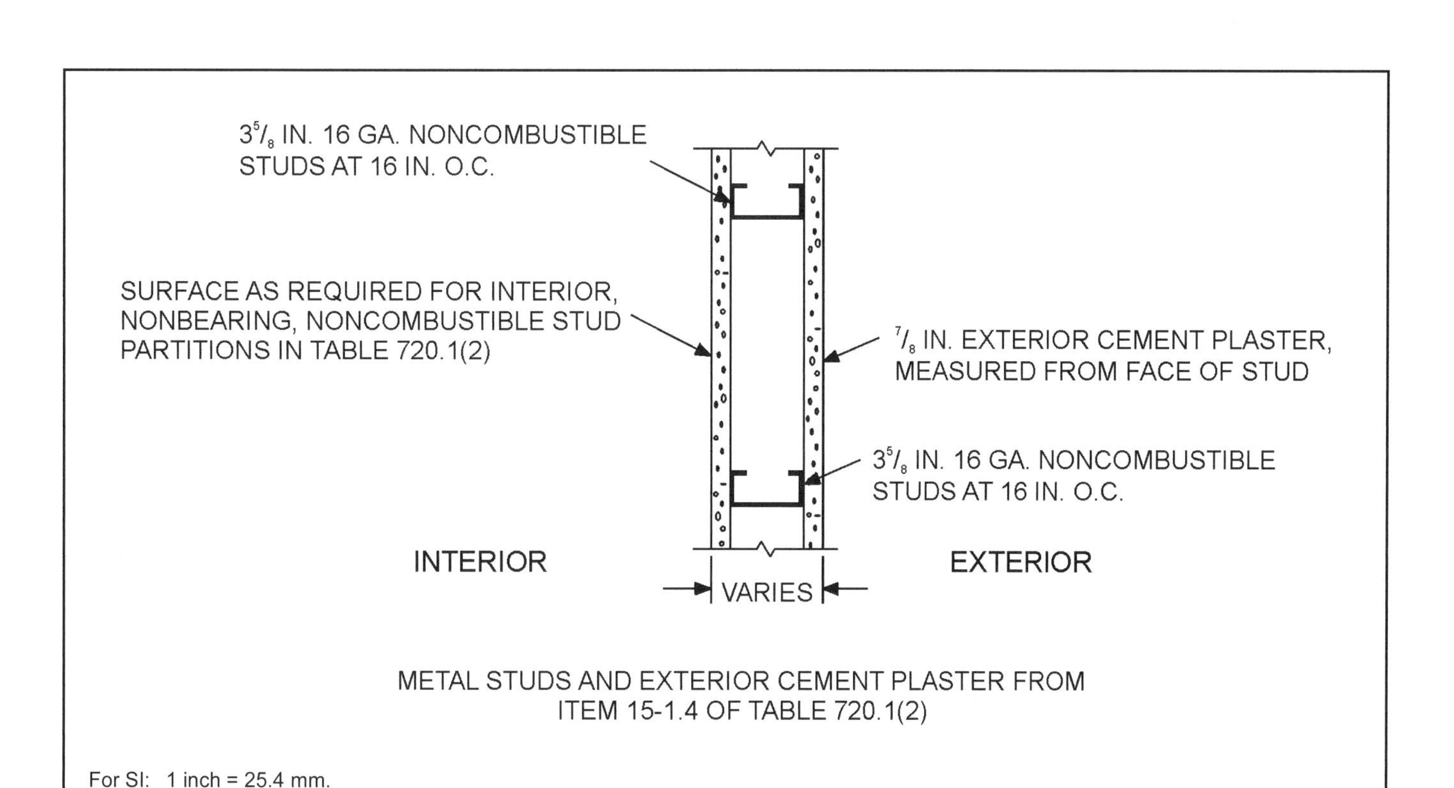

For SI: 1 inch = 25.4 mm.

Figure 721.1(2)(b)
METAL STUD WALL WITH PLASTER

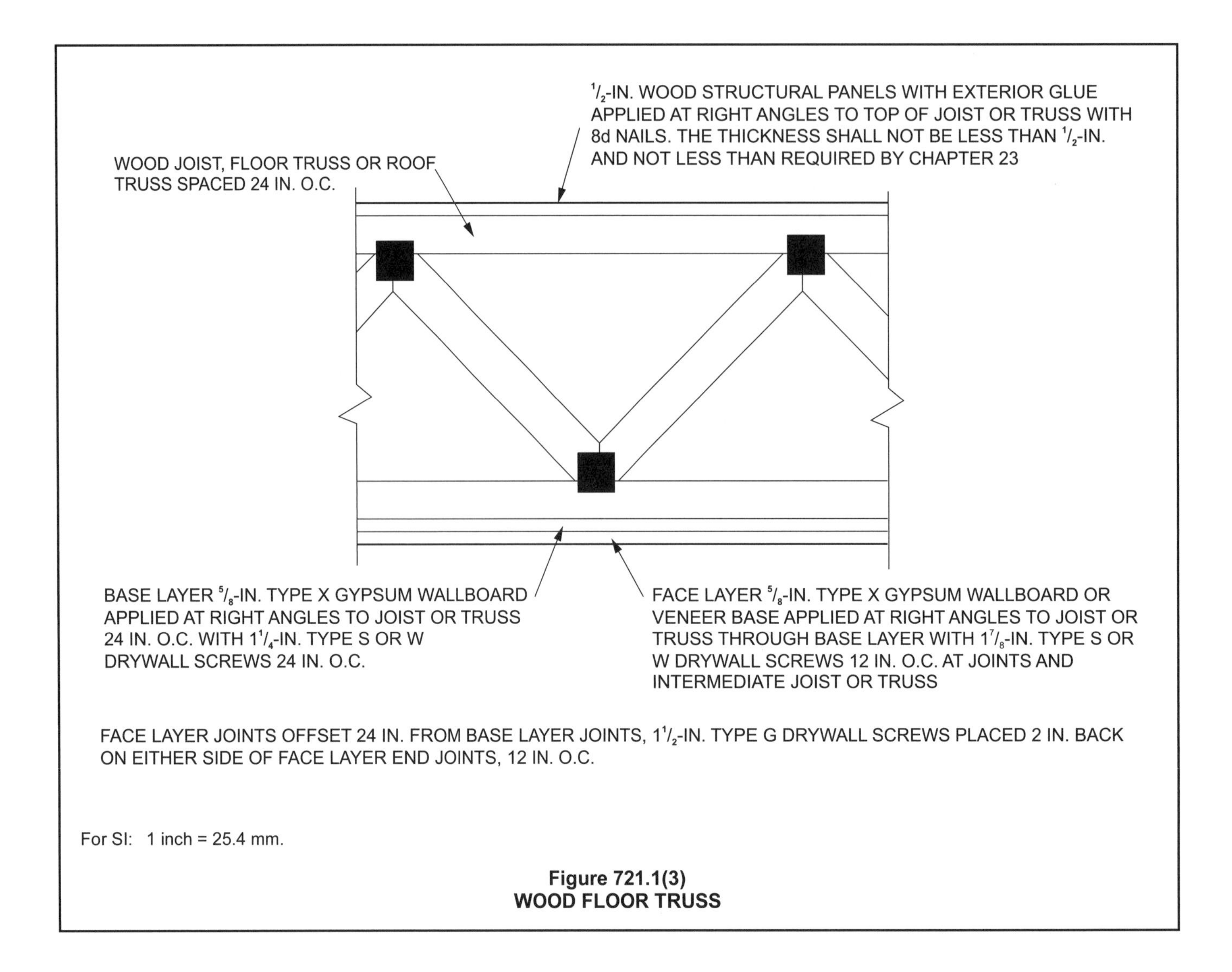

For SI: 1 inch = 25.4 mm.

Figure 721.1(3)
WOOD FLOOR TRUSS

TABLE 721.1(1)
MINIMUM PROTECTION OF STRUCTURAL PARTS BASED ON TIME PERIODS FOR VARIOUS NONCOMBUSTIBLE INSULATING MATERIALS[m]

STRUCTURAL PARTS TO BE PROTECTED	ITEM NUMBER	INSULATING MATERIAL USED	MINIMUM THICKNESS OF INSULATING MATERIAL FOR THE FOLLOWING FIRE-RESISTANCE PERIODS (inches)			
			4 hours	3 hours	2 hours	1 hour
1. Steel columns and all of primary trusses (continued)	1-1.1	Carbonate, lightweight and sand-lightweight aggregate concrete, members 6″ × 6″ or greater (not including sandstone, granite and siliceous gravel).[a]	$2^1/_2$	2	$1^1/_2$	1
	1-1.2	Carbonate, lightweight and sand-lightweight aggregate concrete, members 8″ × 8″ or greater (not including sandstone, granite and siliceous gravel).[a]	2	$1^1/_2$	1	1
	1-1.3	Carbonate, lightweight and sand-lightweight aggregate concrete, members 12″ × 12″ or greater (not including sandstone, granite and siliceous gravel).[a]	$1^1/_2$	1	1	1
	1-1.4	Siliceous aggregate concrete and concrete excluded in Item 1-1.1, members 6″ × 6″ or greater.[a]	3	2	$1^1/_2$	1
	1-1.5	Siliceous aggregate concrete and concrete excluded in Item 1-1.1, members 8″ × 8″ or greater.[a]	$2^1/_2$	2	1	1
	1-1.6	Siliceous aggregate concrete and concrete excluded in Item 1-1.1, members 12″ × 12″ or greater.[a]	2	1	1	1
	1-2.1	Clay or shale brick with brick and mortar fill.[a]	$3^3/_4$	—	—	$2^1/_4$
	1-3.1	4″ hollow clay tile in two 2″ layers; $^1/_2$″ mortar between tile and column; $^3/_8$″ metal mesh 0.046″ wire diameter in horizontal joints; tile fill.[a]	4	—	—	—
	1-3.2	2″ hollow clay tile; $^3/_4$″ mortar between tile and column; $^3/_8$″ metal mesh 0.046″ wire diameter in horizontal joints; limestone concrete fill[a]; plastered with $^3/_4$″ gypsum plaster.	3	—	—	—
	1-3.3	2″ hollow clay tile with outside wire ties 0.08″ diameter at each course of tile or $^3/_8$″ metal mesh 0.046″ diameter wire in horizontal joints; limestone or trap-rock concrete fill[a] extending 1″ outside column on all sides.	—	—	3	—
	1-3.4	2″ hollow clay tile with outside wire ties 0.08″ diameter at each course of tile with or without concrete fill; $^3/_4$″ mortar between tile and column.	—	—	—	2
	1-4.1	Cement plaster over metal lath wire tied to $^3/_4$″ cold-rolled vertical channels with 0.049″ (No. 18 B.W. gage) wire ties spaced 3″ to 6″ on center. Plaster mixed 1:2 $^1/_2$ by volume, cement to sand.	—	—	$2^1/_2$[b]	$^7/_8$
	1-5.1	Vermiculite concrete, 1:4 mix by volume over paperbacked wire fabric lath wrapped directly around column with additional 2″ × 2″ 0.065″ /0.065″ (No. 16/16 B.W. gage) wire fabric placed $^3/_4$″ from outer concrete surface. Wire fabric tied with 0.049″ (No. 18 B.W. gage) wire spaced 6″ on center for inner layer and 2″ on center for outer layer.	2	—	—	—
	1-6.1	Perlite or vermiculite gypsum plaster over metal lath wrapped around column and furred $1^1/_4$″ from column flanges. Sheets lapped at ends and tied at 6″ intervals with 0.049″ (No. 18 B.W. gage) tie wire. Plaster pushed through to flanges.	$1^1/_2$	1	—	—
	1-6.2	Perlite or vermiculite gypsum plaster over self-furring metal lath wrapped directly around column, lapped 1" and tied at 6″ intervals with 0.049″ (No. 18 B.W. gage) wire.	$1^3/_4$	$1^3/_8$	1	—
	1-6.3	Perlite or vermiculite gypsum plaster on metal lath applied to $^3/_4$″ cold-rolled channels spaced 24″ apart vertically and wrapped flatwise around column.	$1^1/_2$	—	—	—
	1-6.4	Perlite or vermiculite gypsum plaster over two layers of $^1/_2$″ plain full-length gypsum lath applied tight to column flanges. Lath wrapped with 1″ hexagonal mesh of No. 20 gage wire and tied with doubled 0.035" diameter (No. 18 B.W. gage) wire ties spaced 23″ on center. For three-coat work, the plaster mix for the second coat shall not exceed 100 pounds of gypsum to $2^1/_2$ cubic feet of aggregate for the 3-hour system.	$2^1/_2$	2	—	—

(continued)

TABLE 721.1(1)—continued
MINIMUM PROTECTION OF STRUCTURAL PARTS BASED ON TIME PERIODS FOR VARIOUS NONCOMBUSTIBLE INSULATING MATERIALS[m]

STRUCTURAL PARTS TO BE PROTECTED	ITEM NUMBER	INSULATING MATERIAL USED	MINIMUM THICKNESS OF INSULATING MATERIAL FOR THE FOLLOWING FIRE-RESISTANCE PERIODS (inches)			
			4 hours	3 hours	2 hours	1 hour
1. Steel columns and all of primary trusses	1-6.5	Perlite or vermiculite gypsum plaster over one layer of $^1/_2''$ plain full-length gypsum lath applied tight to column flanges. Lath tied with doubled 0.049″ (No. 18 B.W. gage) wire ties spaced 23″ on center and scratch coat wrapped with 1″ hexagonal mesh 0.035″ (No. 20 B.W. gage) wire fabric. For three-coat work, the plaster mix for the second coat shall not exceed 100 pounds of gypsum to $2^1/_2$ cubic feet of aggregate.	—	2	—	—
	1-7.1	Multiple layers of $^1/_2''$ gypsum wallboard[c] adhesively[d] secured to column flanges and successive layers. Wallboard applied without horizontal joints. Corner edges of each layer staggered. Wallboard layer below outer layer secured to column with doubled 0.049″ (No. 18 B.W. gage) steel wire ties spaced 15″ on center. Exposed corners taped and treated.	—	—	2	1
	1-7.2	Three layers of $^5/_8''$ Type X gypsum wallboard.[c] First and second layer held in place by $^1/_8''$ diameter by $1^3/_8''$ long ring shank nails with $^5/_{16}''$ diameter heads spaced 24″ on center at corners. Middle layer also secured with metal straps at mid-height and 18″ from each end, and by metal corner bead at each corner held by the metal straps. Third layer attached to corner bead with 1″ long gypsum wallboard screws spaced 12″ on center.	—	—	$1^7/_8$	—
	1-7.3	Three layers of $^5/_8''$ Type X gypsum wallboard,[c] each layer screw attached to $1^5/_8''$ steel studs 0.018" thick (No. 25 carbon sheet steel gage) at each corner of column. Middle layer also secured with 0.049″ (No. 18 B.W. gage) double-strand steel wire ties, 24″ on center. Screws are No. 6 by 1″ spaced 24″ on center for inner layer, No. 6 by $1^5/_8''$ spaced 12″ on center for middle layer and No. 8 by $2^1/_4''$ spaced 12″ on center for outer layer.	—	$1^7/_8$	—	—
	1-8.1	Wood-fibered gypsum plaster mixed 1:1 by weight gypsum-to-sand aggregate applied over metal lath. Lath lapped 1″ and tied 6″ on center at all end, edges and spacers with 0.049″ (No. 18 B.W. gage) steel tie wires. Lath applied over $^1/_2''$ spacers made of $^3/_4''$ furring channel with 2″ legs bent around each corner. Spacers located 1″ from top and bottom of member and a maximum of 40″ on center and wire tied with a single strand of 0.049″ (No. 18 B.W. gage) steel tie wires. Corner bead tied to the lath at 6″ on center along each corner to provide plaster thickness.	—	—	$1^5/_8$	—
	1-9.1	Minimum W8x35 wide flange steel column (w/d $\geq$ 0.75) with each web cavity filled even with the flange tip with normal weight carbonate or siliceous aggregate concrete (3,000 psi minimum compressive strength with 145 pcf $\pm$ 3 pcf unit weight). Reinforce the concrete in each web cavity with a minimum No. 4 deformed reinforcing bar installed vertically and centered in the cavity, and secured to the column web with a minimum No. 2 horizontal deformed reinforcing bar welded to the web every 18″ on center vertically. As an alternative to the No. 4 rebar, $^3/_4''$ diameter by 3″ long headed studs, spaced at 12″ on center vertically, shall be welded on each side of the web midway between the column flanges.	—	—	—	See Note n
2. Webs or flanges of steel beams and girders (continued)	2-1.1	Carbonate, lightweight and sand-lightweight aggregate concrete (not including sandstone, granite and siliceous gravel) with 3″ or finer metal mesh placed 1″ from the finished surface anchored to the top flange and providing not less than 0.025 square inch of steel area per foot in each direction.	2	$1^1/_2$	1	1
	2-1.2	Siliceous aggregate concrete and concrete excluded in Item 2-1.1 with 3″ or finer metal mesh placed 1″ from the finished surface anchored to the top flange and providing not less than 0.025 square inch of steel area per foot in each direction.	$2^1/_2$	2	$1^1/_2$	1
	2-2.1	Cement plaster on metal lath attached to $^3/_4''$ cold-rolled channels with 0.04″ (No. 18 B.W. gage) wire ties spaced 3″ to 6″ on center. Plaster mixed 1:2 $^1/_2$ by volume, cement to sand.	—	—	$2^1/_2$[b]	$^7/_8$

(continued)

TABLE 721.1(1)—continued
MINIMUM PROTECTION OF STRUCTURAL PARTS BASED ON TIME PERIODS FOR VARIOUS NONCOMBUSTIBLE INSULATING MATERIALS[m]

STRUCTURAL PARTS TO BE PROTECTED	ITEM NUMBER	INSULATING MATERIAL USED	MINIMUM THICKNESS OF INSULATING MATERIAL FOR THE FOLLOWING FIRE-RESISTANCE PERIODS (inches)			
			4 hours	3 hours	2 hours	1 hour
2. Webs or flanges of steel beams and girders	2-3.1	Vermiculite gypsum plaster on a metal lath cage, wire tied to 0.165″ diameter (No. 8 B.W. gage) steel wire hangers wrapped around beam and spaced 16″ on center. Metal lath ties spaced approximately 5″ on center at cage sides and bottom.	—	$^{7}/_{8}$	—	—
	2-4.1	Two layers of $^{5}/_{8}$″ Type X gypsum wallboard[c] are attached to U-shaped brackets spaced 24″ on center. 0.018″ thick (No. 25 carbon sheet steel gage) $1^{5}/_{8}$″ deep by 1″ galvanized steel runner channels are first installed parallel to and on each side of the top beam flange to provide a $^{1}/_{2}$″ clearance to the flange. The channel runners are attached to steel deck or concrete floor construction with approved fasteners spaced 12″ on center. U-shaped brackets are formed from members identical to the channel runners. At the bent portion of the U-shaped bracket, the flanges of the channel are cut out so that $1^{5}/_{8}$″ deep corner channels can be inserted without attachment parallel to each side of the lower flange. As an alternative, 0.021″ thick (No. 24 carbon sheet steel gage) 1″ × 2″ runner and corner angles shall be used in lieu of channels, and the web cutouts in the U-shaped brackets shall not be required. Each angle is attached to the bracket with $^{1}/_{2}$″-long No. 8 self-drilling screws. The vertical legs of the U-shaped bracket are attached to the runners with one $^{1}/_{2}$″ long No. 8 self-drilling screw. The completed steel framing provides a $2^{1}/_{8}$″ and $1^{1}/_{2}$″ space between the inner layer of wallboard and the sides and bottom of the steel beam, respectively. The inner layer of wallboard is attached to the top runners and bottom corner channels or corner angles with $1^{1}/_{4}$″-long No. 6 self-drilling screws spaced 16″ on center. The outer layer of wallboard is applied with $1^{3}/_{4}$″-long No. 6 self-drilling screws spaced 8″ on center. The bottom corners are reinforced with metal corner beads.	—	—	$1^{1}/_{4}$	—
	2-4.2	Three layers of $^{5}/_{8}$″ Type X gypsum wallboard[c] attached to a steel suspension system as described immediately above utilizing the 0.018″ thick (No. 25 carbon sheet steel gage) 1″ × 2″ lower corner angles. The framing is located so that a $2^{1}/_{8}$″ and 2″ space is provided between the inner layer of wallboard and the sides and bottom of the beam, respectively. The first two layers of wallboard are attached as described immediately above. A layer of 0.035″ thick (No. 20 B.W. gage) 1″ hexagonal galvanized wire mesh is applied under the soffit of the middle layer and up the sides approximately 2″. The mesh is held in position with the No. 6 $1^{5}/_{8}$″-long screws installed in the vertical leg of the bottom corner angles. The outer layer of wallboard is attached with No. 6 $2^{1}/_{4}$″-long screws spaced 8″ on center. One screw is also installed at the mid-depth of the bracket in each layer. Bottom corners are finished as described above.	—	$1^{7}/_{8}$	—	—
3. Bonded pre-tensioned reinforcement in prestressed concrete[e]	3-1.1	Carbonate, lightweight, sand-lightweight and siliceous[f] aggregate concrete Beams or girders	4[g]	3[g]	$2^{1}/_{2}$	$1^{1}/_{2}$
		Solid [h]		2	$1^{1}/_{2}$	1

(continued)

TABLE 721.1(1) —continued
MINIMUM PROTECTION OF STRUCTURAL PARTS BASED ON TIME PERIODS FOR VARIOUS NONCOMBUSTIBLE INSULATING MATERIALS[m]

STRUCTURAL PARTS TO BE PROTECTED	ITEM NUMBER	INSULATING MATERIAL USED	MINIMUM THICKNESS OF INSULATING MATERIAL FOR THE FOLLOWING FIRE-RESISTANCE PERIODS (inches)			
			4 hours	3 hours	2 hours	1 hour
4. Bonded or unbonded post-tensioned tendons in prestressed concrete[e, i]	4-1.1	Carbonate, lightweight, sand-lightweight and siliceous[f] aggregate concrete				
		Unrestrained members:				
		Solid slabs[h]	—	2	1 1/2	—
		Beams and girders[j]				
		8″ wide		4 1/2	2 1/2	1 3/4
		greater than 12″ wide	3	2 1/2	2	1 1/2
	4-1.2	Carbonate, lightweight, sand-lightweight and siliceous aggregate				
		Restrained members:[k]				
		Solid slabs[h]	1 1/4	1	3/4	—
		Beams and girders[j]				
		8″ wide	2 1/2	2	1 3/4	—
		greater than 12″ wide	2	1 3/4	1 1/2	—
5. Reinforcing steel in reinforced concrete columns, beams girders and trusses	5-1.1	Carbonate, lightweight and sand-lightweight aggregate concrete, members 12″ or larger, square or round. (Size limit does not apply to beams and girders monolithic with floors.)	1 1/2	1 1/2	1 1/2	1 1/2
		Siliceous aggregate concrete, members 12″ or larger, square or round. (Size limit does not apply to beams and girders monolithic with floors.)	2	1 1/2	1 1/2	1 1/2
6. Reinforcing steel in reinforced concrete joists[l]	6-1.1	Carbonate, lightweight and sand-lightweight aggregate concrete	1 1/4	1 1/4	1	3/4
	6-1.2	Siliceous aggregate concrete	1 3/4	1 1/2	1	3/4
7. Reinforcing and tie rods in floor and roof slabs[l]	7-1.1	Carbonate, lightweight and sand-lightweight aggregate concrete	1	1	3/4	3/4
	7-1.2	Siliceous aggregate concrete	1 1/4	1	1	3/4

For SI: 1 inch = 25.4 mm, 1 square inch = 645.2 mm^2, 1 cubic foot = 0.0283 m^3, 1 pound per cubic foot = 16.02 kg/m^3.

a. Reentrant parts of protected members to be filled solidly.

b. Two layers of equal thickness with a 3/4-inch airspace between.

c. For all of the construction with gypsum wallboard described in Table 721.1(1), gypsum base for veneer plaster of the same size, thickness and core type shall be permitted to be substituted for gypsum wallboard, provided attachment is identical to that specified for the wallboard and the joints on the face layer are reinforced, and the entire surface is covered with not less than 1/16-inch gypsum veneer plaster.

d. An approved adhesive qualified under ASTM E119 or UL 263.

e. Where lightweight or sand-lightweight concrete having an oven-dry weight of 110 pounds per cubic foot or less is used, the tabulated minimum cover shall be permitted to be reduced 25 percent, except that in no case shall the cover be less than 3/4 inch in slabs or 1 1/2 inches in beams or girders.

f. For solid slabs of siliceous aggregate concrete, increase tendon cover 20 percent.

g. Adequate provisions against spalling shall be provided by U-shaped or hooped stirrups spaced not to exceed the depth of the member with a clear cover of 1 inch.

h. Prestressed slabs shall have a thickness not less than that required in Table 721.1(3) for the respective fire-resistance time period.

i. Fire coverage and end anchorages shall be as follows: Cover to the prestressing steel at the anchor shall be 1/2 inch greater than that required away from the anchor. Minimum cover to steel-bearing plate shall be 1 inch in beams and 3/4 inch in slabs.

j. For beam widths between 8 inches and 12 inches, cover thickness shall be permitted to be determined by interpolation.

k. Interior spans of continuous slabs, beams and girders shall be permitted to be considered restrained.

l. For use with concrete slabs having a comparable fire endurance where members are framed into the structure in such a manner as to provide equivalent performance to that of monolithic concrete construction.

m. Generic *fire-resistance ratings* (those not designated as PROPRIETARY* in the listing) in GA 600 shall be accepted as if herein listed.

n. No additional insulating material is required on the exposed outside face of the column flange to achieve a 1-hour *fire-resistance rating*.

TABLE 721.1(2)
RATED FIRE-RESISTANCE PERIODS FOR VARIOUS WALLS AND PARTITIONS [a, o, p]

MATERIAL	ITEM NUMBER	CONSTRUCTION	MINIMUM FINISHED THICKNESS FACE-TO-FACE[b] (inches)			
			4 hours	3 hours	2 hours	1 hour
1. Brick of clay or shale	1-1.1	Solid brick of clay or shale[c].	6	4.9	3.8	2.7
	1-1.2	Hollow brick, not filled.	5.0	4.3	3.4	2.3
	1-1.3	Hollow brick unit wall, grout or filled with perlite vermiculite or expanded shale aggregate.	6.6	5.5	4.4	3.0
	1-2.1	4″ nominal thick units not less than 75 percent solid backed with a hat-shaped metal furring channel $^3/_4$″ thick formed from 0.021″ sheet metal attached to the brick wall on 24″ centers with approved fasteners, and $^1/_2$″ Type X gypsum wallboard attached to the metal furring strips with 1″-long Type S screws spaced 8″ on center.	—	—	5[d]	—
2. Combination of clay brick and load-bearing hollow clay tile	2-1.1	4″ solid brick and 4″ tile (not less than 40 percent solid).	—	8	—	—
	2-1.2	4″ solid brick and 8″ tile (not less than 40 percent solid).	12	—	—	—
3. Concrete masonry units	3-1.1[f, g]	Expanded slag or pumice.	4.7	4.0	3.2	2.1
	3-1.2[f, g]	Expanded clay, shale or slate.	5.1	4.4	3.6	2.6
	3-1.3[f]	Limestone, cinders or air-cooled slag.	5.9	5.0	4.0	2.7
	3-1.4[f, g]	Calcareous or siliceous gravel.	6.2	5.3	4.2	2.8
4. Solid concrete[h, i]	4-1.1	Siliceous aggregate concrete.	7.0	6.2	5.0	3.5
		Carbonate aggregate concrete.	6.6	5.7	4.6	3.2
		Sand-lightweight concrete.	5.4	4.6	3.8	2.7
		Lightweight concrete.	5.1	4.4	3.6	2.5
5. Glazed or unglazed facing tile, nonload-bearing	5-1.1	One 2″ unit cored 15 percent maximum and one 4″ unit cored 25 percent maximum with $^3/_4$″ mortar-filled collar joint. Unit positions reversed in alternate courses.	—	6$^3/_8$	—	—
	5-1.2	One 2″ unit cored 15 percent maximum and one 4″ unit cored 40 percent maximum with $^3/_4$″ mortar-filled collar joint. Unit positions side with $^3/_4$″ gypsum plaster. Two wythes tied together every fourth course with No. 22 gage corrugated metal ties.	—	6$^3/_4$	—	—
	5-1.3	One unit with three cells in wall thickness, cored 29 percent maximum.	—	—	6	—
	5-1.4	One 2″ unit cored 22 percent maximum and one 4″ unit cored 41 percent maximum with $^1/_4$″ mortar-filled collar joint. Two wythes tied together every third course with 0.030″ (No. 22 galvanized sheet steel gage) corrugated metal ties.	—	—	6	—
	5-1.5	One 4″ unit cored 25 percent maximum with $^3/_4$″ gypsum plaster on one side.	—	—	4$^3/_4$	—
	5-1.6	One 4″ unit with two cells in wall thickness, cored 22 percent maximum.	—	—	—	4
	5-1.7	One 4″ unit cored 30 percent maximum with $^3/_4$″ vermiculite gypsum plaster on one side.	—	—	4$^1/_2$	—
	5-1.8	One 4″ unit cored 39 percent maximum with $^3/_4$″ gypsum plaster on one side.	—	—	—	4$^1/_2$

(continued)

TABLE 721.1(2)—continued
RATED FIRE-RESISTANCE PERIODS FOR VARIOUS WALLS AND PARTITIONS [a, o, p]

MATERIAL	ITEM NUMBER	CONSTRUCTION	MINIMUM FINISHED THICKNESS FACE-TO-FACE[b] (inches)			
			4 hours	3 hours	2 hours	1 hour
6. Solid gypsum plaster	6-1.1	3/4″ by 0.055″ (No. 16 carbon sheet steel gage) vertical cold-rolled channels, 16″ on center with 2.6-pound flat metal lath applied to one face and tied with 0.049″ (No. 18 B.W. gage) wire at 6″ spacing. Gypsum plaster each side mixed 1:2 by weight, gypsum to sand aggregate.	—	—	—	2[d]
	6-1.2	3/4″ by 0.05″ (No. 16 carbon sheet steel gage) cold-rolled channels 16″ on center with metal lath applied to one face and tied with 0.049″ (No. 18 B.W. gage) wire at 6″ spacing. Perlite or vermiculite gypsum plaster each side. For three-coat work, the plaster mix for the second coat shall not exceed 100 pounds of gypsum to 2 1/2 cubic feet of aggregate for the 1-hour system.	—	—	2 1/2[d]	2[d]
	6-1.3	3/4″ by 0.055″ (No. 16 carbon sheet steel gage) vertical cold-rolled channels, 16″ on center with 3/8″ gypsum lath applied to one face and attached with sheet metal clips. Gypsum plaster each side mixed 1:2 by weight, gypsum to sand aggregate.	—	—	—	2[d]
	6-2.1	Studless with 1/2″ full-length plain gypsum lath and gypsum plaster each side. Plaster mixed 1:1 for scratch coat and 1:2 for brown coat, by weight, gypsum to sand aggregate.	—	—	—	2[d]
	6-2.2	Studless with 1/2″ full-length plain gypsum lath and perlite or vermiculite gypsum plaster each side.	—	—	2 1/2[d]	2[d]
	6-2.3	Studless partition with 3/8″ rib metal lath installed vertically adjacent edges tied 6″ on center with No. 18 gage wire ties, gypsum plaster each side mixed 1:2 by weight, gypsum to sand aggregate.	—	—	—	2[d]
7. Solid perlite and Portland cement	7-1.1	Perlite mixed in the ratio of 3 cubic feet to 100 pounds of Portland cement and machine applied to stud side of 1 1/2″ mesh by 0.058-inch (No. 17 B.W. gage) paper-backed woven wire fabric lath wire-tied to 4″-deep steel trussed wire[j] studs 16″ on center. Wire ties of 0.049″ (No. 18 B.W. gage) galvanized steel wire 6″ on center vertically.	—	—	3 1/8[d]	—
8. Solid neat wood fibered gypsum plaster	8-1.1	3/4″ by 0.055-inch (No. 16 carbon sheet steel gage) cold-rolled channels, 12″ on center with 2.5-pound flat metal lath applied to one face and tied with 0.049″ (No. 18 B.W. gage) wire at 6″ spacing. Neat gypsum plaster applied each side.	—	—	2[d]	—
9. Solid wallboard partition	9-1.1	One full-length layer 1/2″ Type X gypsum wallboard[e] laminated to each side of 1″ full-length V-edge gypsum coreboard with approved laminating compound. Vertical joints of face layer and coreboard staggered not less than 3″.	—	—	2[d]	—
10. Hollow (studless) gypsum wallboard partition	10-1.1	One full-length layer of 5/8″ Type X gypsum wallboard[e] attached to both sides of wood or metal top and bottom runners laminated to each side of 1″× 6″ full-length gypsum coreboard ribs spaced 2″ on center with approved laminating compound. Ribs centered at vertical joints of face plies and joints staggered 24″ in opposing faces. Ribs may be recessed 6″ from the top and bottom.	—	—	—	2 1/4[d]
	10-1.2	1″ regular gypsum V-edge full-length backing board attached to both sides of wood or metal top and bottom runners with nails or 1 5/8″ drywall screws at 24″ on center. Minimum width of rumors 1 5/8″. Face layer of 1/2″ regular full-length gypsum wallboard laminated to outer faces of backing board with approved laminating compound.	—	—	4 5/8[d]	—

(continued)

TABLE 721.1(2) —continued
RATED FIRE-RESISTANCE PERIODS FOR VARIOUS WALLS AND PARTITIONS [a, o, p]

MATERIAL	ITEM NUMBER	CONSTRUCTION	MINIMUM FINISHED THICKNESS FACE-TO-FACE[b] (inches)			
			4 hours	3 hours	2 hours	1 hour
11. Noncombustible studs-interior partition with plaster each side	11-1.1	$3^1/_4''$ × 0.044″ (No. 18 carbon sheet steel gage) steel studs spaced 24″ on center. $^5/_8''$ gypsum plaster on metal lath each side mixed 1:2 by weight, gypsum to sand aggregate.	—	—	—	$4^3/_4$[d]
	11-1.2	$3^3/_8''$ × 0.055″ (No. 16 carbon sheet steel gage) approved nailable[k] studs spaced 24″ on center. $^5/_8''$ neat gypsum wood-fibered plaster each side over $^3/_8''$ rib metal lath nailed to studs with 6d common nails, 8″ on center. Nails driven $1^1/_4''$ and bent over.	—	—	$5^5/_8$	—
	11-1.3	4″ × 0.044″ (No. 18 carbon sheet steel gage) channel-shaped steel studs at 16″ on center. On each side approved resilient clips pressed onto stud flange at 16″ vertical spacing, $^1/_4''$ pencil rods snapped into or wire tied onto outer loop of clips, metal lath wire-tied to pencil rods at 6″ intervals, 1″ perlite gypsum plaster, each side.	—	$7^5/_8$[d]	—	—
	11-1.4	$2^1/_2''$ × 0.044″ (No. 18 carbon sheet steel gage) steel studs spaced 16″ on center. Wood fibered gypsum plaster mixed 1:1 by weight gypsum to sand aggregate applied on $^3/_4$-pound metal lath wire tied to studs, each side. $^3/_4''$ plaster applied over each face, including finish coat.	—	—	$4^1/_4$[d]	—
12. Wood studs-interior partition with plaster each side	12-1.1[l, m]	2″ × 4″ wood studs 16″ on center with $^5/_8''$ gypsum plaster on metal lath. Lath attached by 4d common nails bent over or No. 14 gage by $1^1/_4''$ by $^3/_4''$ crown width staples spaced 6″ on center. Plaster mixed $1{:}1^1/_2$ for scratch coat and 1:3 for brown coat, by weight, gypsum to sand aggregate.	—	—	—	$5^1/_8$
	12-1.2[l]	2″ × 4″ wood studs 16″ on center with metal lath and $^7/_8''$ neat wood-fibered gypsum plaster each side. Lath attached by 6d common nails, 7″ on center. Nails driven $1^1/_4''$ and bent over.	—	—	$5^1/_2$[d]	—
	12-1.3[l]	2″ × 4″ wood studs 16″ on center with $^3/_8''$ perforated or plain gypsum lath and $^1/_2''$ gypsum plaster each side. Lath nailed with $1^1/_8''$ by No. 13 gage by $^{19}/_{64}''$ head plasterboard blued nails, 4″ on center. Plaster mixed 1:2 by weight, gypsum to sand aggregate.	—	—	—	$5^1/_4$
	12-1.4[l]	2″ × 4″ wood studs 16″ on center with $^3/_8''$ Type X gypsum lath and $^1/_2''$ gypsum plaster each side. Lath nailed with $1^1/_8''$ by No. 13 gage by $^{19}/_{64}''$ head plasterboard blued nails, 5″ on center. Plaster mixed 1:2 by weight, gypsum to sand aggregate.	—	—	—	$5^1/_4$
13. Noncombustible studs-interior partition with gypsum wallboard each side	13-1.1	0.018″ (No. 25 carbon sheet steel gage) channel-shaped studs 24″ on center with one full-length layer of $^5/_8''$ Type X gypsum wallboard[e] applied vertically attached with 1″-long No. 6 drywall screws to each stud. Screws are 8″ on center around the perimeter and 12″ on center on the intermediate stud. Where applied horizontally, the Type X gypsum wallboard shall be attached to $3^5/_8''$ studs and the horizontal joints shall be staggered with those on the opposite side. Screws for the horizontal application shall be 8″ on center at vertical edges and 12″ on center at intermediate studs.	—	—	—	$2^7/_8$[d]
	13-1.2	0.018″ (No. 25 carbon sheet steel gage) channel-shaped studs 25″ on center with two full-length layers of $^1/_2''$ Type X gypsum wallboard[e] applied vertically each side. First layer attached with 1″-long, No. 6 drywall screws, 8″ on center around the perimeter and 12″ on center on the intermediate stud. Second layer applied with vertical joints offset one stud space from first layer using $1^5/_8''$ long, No. 6 drywall screws spaced 9″ on center along vertical joints, 12″ on center at intermediate studs and 24″ on center along top and bottom runners.	—	—	$3^5/_8$[d]	—
	13-1.3	0.055″ (No. 16 carbon sheet steel gage) approved nailable metal studs[e] 24″ on center with full-length $^5/_8''$ Type X gypsum wallboard[e] applied vertically and nailed 7″ on center with 6d cement-coated common nails. Approved metal fastener grips used with nails at vertical butt joints along studs.	—	—	—	$4^7/_8$

(continued)

TABLE 721.1(2)—continued
RATED FIRE-RESISTANCE PERIODS FOR VARIOUS WALLS AND PARTITIONS [a, o, p]

MATERIAL	ITEM NUMBER	CONSTRUCTION	MINIMUM FINISHED THICKNESS FACE-TO-FACE[b] (inches)			
			4 hours	3 hours	2 hours	1 hour
14. Wood studs-interior partition with gypsum wallboard each side	14-1.1[h, m]	2″ × 4″ wood studs 16″ on center with two layers of $^3/_8$″ regular gypsum wallboard[e] each side, 4d cooler[n] or wallboard[n] nails at 8″ on center first layer, 5d cooler[n] or wallboard[n] nails at 8″ on center second layer with laminating compound between layers, joints staggered. First layer applied full length vertically, second layer applied horizontally or vertically.	—	—	—	5
	14-1.2[l, m]	2″ × 4″ wood studs 16″ on center with two layers $^1/_2$″ regular gypsum wallboard[e] applied vertically or horizontally each side[k], joints staggered. Nail base layer with 5d cooler[n] or wallboard[n] nails at 8″ on center face layer with 8d cooler[n] or wallboard[n] nails at 8″ on center.	—	—	—	$5^1/_2$
	14-1.3[l, m]	2″ × 4″ wood studs 24″ on center with $^5/_8$″ Type X gypsum wallboard[e] applied vertically or horizontally nailed with 6d cooler[n] or wallboard[n] nails at 7″ on center with end joints on nailing members. Stagger joints each side.	—	—	—	$4^3/_4$
	14-1.4[l]	2″ × 4″ fire-retardant-treated wood studs spaced 24″ on center with one layer of $^5/_8$″ Type X gypsum wallboard[e] applied with face paper grain (long dimension) parallel to studs. Wallboard attached with 6d cooler[n] or wallboard[n] nails at 7″ on center.	—	—	—	$4^3/_4$[d]
	14-1.5[l, m]	2″ × 4″ wood studs 16″ on center with two layers $^5/_8$″ Type X gypsum wallboard[e] each side. Base layers applied vertically and nailed with 6d cooler[n] or wallboard[n] nails at 9″ on center. Face layer applied vertically or horizontally and nailed with 8d cooler[n] or wallboard[n] nails at 7″ on center. For nail-adhesive application, base layers are nailed 6″ on center. Face layers applied with coating of approved wallboard adhesive and nailed 12″ on center.	—	—	6	—
	14-1.6[l]	2″ × 3″ fire-retardant-treated wood studs spaced 24″ on center with one layer of $^5/_8$″ Type X gypsum wallboard[e] applied with face paper grain (long dimension) at right angles to studs. Wallboard attached with 6d cement-coated box nails spaced 7″ on center.	—	—	—	$3^5/_8$[d]
15. Exterior or interior walls (continued)	15-1.1[l, m]	Exterior surface with $^3/_4$″ drop siding over $^1/_2$″ gypsum sheathing on 2″ × 4″ wood studs at 16″ on center, interior surface treatment as required for 1-hour-rated exterior or interior 2″ × 4″ wood stud partitions. Gypsum sheathing nailed with $1^3/_4$″ by No. 11 gage by $^7/_{16}$″ head galvanized nails at 8″ on center. Siding nailed with 7d galvanized smooth box nails.	—	—	—	Varies
	15-1.2[l, m]	2″ × 4″ wood studs 16″ on center with metal lath and $^3/_4$″ cement plaster on each side. Lath attached with 6d common nails 7″ on center driven to 1″ minimum penetration and bent over. Plaster mix 1:4 for scratch coat and 1:5 for brown coat, by volume, cement to sand.	—	—	—	$5^3/_8$
	15-1.3[l, m]	2″ × 4″ wood studs 16″ on center with $^7/_8$″ cement plaster (measured from the face of studs) on the exterior surface with interior surface treatment as required for interior wood stud partitions in this table. Plaster mix 1:4 for scratch coat and 1:5 for brown coat, by volume, cement to sand.	—	—	—	Varies
	15-1.4	$3^5/_8$″ No. 16 gage noncombustible studs 16″ on center with $^7/_8$″ cement plaster (measured from the face of the studs) on the exterior surface with interior surface treatment as required for interior, nonbearing, noncombustible stud partitions in this table. Plaster mix 1:4 for scratch coat and 1:5 for brown coat, by volume, cement to sand.	—	—	—	Varies[d]

(continued)

TABLE 721.1(2)—continued
RATED FIRE-RESISTANCE PERIODS FOR VARIOUS WALLS AND PARTITIONS [a, o, p]

MATERIAL	ITEM NUMBER	CONSTRUCTION	MINIMUM FINISHED THICKNESS FACE-TO-FACE[b] (inches)			
			4 hours	3 hours	2 hours	1 hour
15. Exterior or interior walls (continued)	15-1.5[m]	$2^1/_4$″ × $3^3/_4$″ clay face brick with cored holes over $^1/_2$″ gypsum sheathing on exterior surface of 2″ × 4″ wood studs at 16″ on center and two layers $^5/_8$″ Type X gypsum wallboard[e] on interior surface. Sheathing placed horizontally or vertically with vertical joints over studs nailed 6″ on center with $1^3/_4$″ × No. 11 gage by $^7/_{16}$″ head galvanized nails. Inner layer of wallboard placed horizontally or vertically and nailed 8″ on center with 6d cooler[n] or wallboard[n] nails. Outer layer of wallboard placed horizontally or vertically and nailed 8″ on center with 8d cooler[n] or wallboard[n] nails. Joints staggered with vertical joints over studs. Outer layer joints taped and finished with compound. Nail heads covered with joint compound. 0.035 inch (No. 20 galvanized sheet gage) corrugated galvanized steel wall ties $^3/_4$″ by $6^5/_8$″ attached to each stud with two 8d cooler[n] or wallboard[n] nails every sixth course of bricks.	—	—	10	—
	15-1.6[l, m]	2″ × 6″ fire-retardant-treated wood studs 16″ on center. Interior face has two layers of $^5/_8$″ Type X gypsum with the base layer placed vertically and attached with 6d box nails 12″ on center. The face layer is placed horizontally and attached with 8d box nails 8″ on center at joints and 12″ on center elsewhere. The exterior face has a base layer of $^5/_8$″ Type X gypsum sheathing placed vertically with 6d box nails 8″ on center at joints and 12″ on center elsewhere. An approved building paper is next applied, followed by self-furred exterior lath attached with $2^1/_2$″, No. 12 gage galvanized roofing nails with a $^3/_8$″ diameter head and spaced 6″ on center along each stud. Cement plaster consisting of a $^1/_2$″ brown coat is then applied. The scratch coat is mixed in the proportion of 1:3 by weight, cement to sand with 10 pounds of hydrated lime and 3 pounds of approved additives or admixtures per sack of cement. The brown coat is mixed in the proportion of 1:4 by weight, cement to sand with the same amounts of hydrated lime and approved additives or admixtures used in the scratch coat.	—	—	$8^1/_4$	—
	15-1.7[l, m]	2″ × 6″ wood studs 16″ on center. The exterior face has a layer of $^5/_8$″ Type X gypsum sheathing placed vertically with 6d box nails 8″ on center at joints and 12″ on center elsewhere. An approved building paper is next applied, followed by 1″ by No. 18 gage self-furred exterior lath attached with 8d by $2^1/_2$″ long galvanized roofing nails spaced 6″ on center along each stud. Cement plaster consisting of a $^1/_2$″ scratch coat, a bonding agent and a $^1/_2$″ brown coat and a finish coat is then applied. The scratch coat is mixed in the proportion of 1:3 by weight, cement to sand with 10 pounds of hydrated lime and 3 pounds of approved additives or admixtures per sack of cement. The brown coat is mixed in the proportion of 1:4 by weight, cement to sand with the same amounts of hydrated lime and approved additives or admixtures used in the scratch coat. The interior is covered with $^3/_8$″ gypsum lath with 1″ hexagonal mesh of 0.035 inch (No. 20 B.W. gage) woven wire lath furred out $^5/_{16}$″ and 1″ perlite or vermiculite gypsum plaster. Lath nailed with $1^1/_8$″ by No. 13 gage by $^{19}/_{64}$″ head plasterboard glued nails spaced 5″ on center. Mesh attached by $1^3/_4$″ by No. 12 gage by $^3/_8$″ head nails with $^3/_8$″ furrings, spaced 8″ on center. The plaster mix shall not exceed 100 pounds of gypsum to $2^1/_2$ cubic feet of aggregate.	—	—	$8^3/_8$	—

(continued)

TABLE 721.1(2)—continued
RATED FIRE-RESISTANCE PERIODS FOR VARIOUS WALLS AND PARTITIONS [a, o, p]

MATERIAL	ITEM NUMBER	CONSTRUCTION	MINIMUM FINISHED THICKNESS FACE-TO-FACE[b] (inches)			
			4 hours	3 hours	2 hours	1 hour
15. Exterior or interior walls (continued)	15-1.8[l, m]	2″ × 6″ wood studs 16″ on center. The exterior face has a layer of $^{5}/_{8}$″ Type X gypsum sheathing placed vertically with 6d box nails 8″ on center at joints and 12″ on center elsewhere. An approved building paper is next applied, followed by $1^{1}/_{2}$″ by No. 17 gage self-furred exterior lath attached with 8d by $2^{1}/_{2}$″ long galvanized roofing nails spaced 6″ on center along each stud. Cement plaster consisting of a $^{1}/_{2}$″ scratch coat, and a $^{1}/_{2}$″ brown coat is then applied. The plaster may be placed by machine. The scratch coat is mixed in the proportion of 1:4 by weight, plastic cement to sand. The brown coat is mixed in the proportion of 1:5 by weight, plastic cement to sand. The interior is covered with $^{3}/_{8}$″ gypsum lath with 1″ hexagonal mesh of No. 20 gage woven wire lath furred out $^{5}/_{16}$″ and 1″ perlite or vermiculite gypsum plaster. Lath nailed with $1^{1}/_{8}$″ by No. 13 gage by $^{19}/_{64}$″ head plasterboard glued nails spaced 5″ on center. Mesh attached by $1^{3}/_{4}$″ by No. 12 gage by $^{3}/_{8}$″ head nails with $^{3}/_{8}$″ furrings, spaced 8″ on center. The plaster mix shall not exceed 100 pounds of gypsum to $2^{1}/_{2}$ cubic feet of aggregate.	—	—	$8^{3}/_{8}$	—
	15-1.9	4″ No. 18 gage, nonload-bearing metal studs, 16″ on center, with 1″ Portland cement lime plaster (measured from the back side of the $^{3}/_{4}$-pound expanded metal lath) on the exterior surface. Interior surface to be covered with 1″ of gypsum plaster on $^{3}/_{4}$-pound expanded metal lath proportioned by weight-1:2 for scratch coat, 1:3 for brown, gypsum to sand. Lath on one side of the partition fastened to $^{1}/_{4}$″ diameter pencil rods supported by No. 20 gage metal clips, located 16″ on center vertically, on each stud. 3″ thick mineral fiber insulating batts friction fitted between the studs.	—	—	$6^{1}/_{2}$[d]	—
	15-1.10	Steel studs 0.060″ thick, 4″ deep or 6″ at 16″ or 24″ centers, with $^{1}/_{2}$″ Glass Fiber Reinforced Concrete (GFRC) on the exterior surface. GFRC is attached with flex anchors at 24″ on center, with 5″ leg welded to studs with two $^{1}/_{2}$″-long flare-bevel welds, and 4″ foot attached to the GFRC skin with $^{5}/_{8}$″ thick GFRC bonding pads that extend $2^{1}/_{2}$″ beyond the flex anchor foot on both sides. Interior surface to have two layers of $^{1}/_{2}$″ Type X gypsum wallboard.[e] The first layer of wallboard to be attached with 1″-long Type S buglehead screws spaced 24″ on center and the second layer is attached with $1^{5}/_{8}$″-long Type S screws spaced at 12″ on center. Cavity is to be filled with 5″ of 4 pcf (nominal) mineral fiber batts. GFRC has $1^{1}/_{2}$″ returns packed with mineral fiber and caulked on the exterior.	—	—	$6^{1}/_{2}$	—
	15-1.11	Steel studs 0.060″ thick, 4″ deep or 6″ at 16″ or 24″ centers, respectively, with $^{1}/_{2}$″ Glass Fiber Reinforced Concrete (GFRC) on the exterior surface. GFRC is attached with flex anchors at 24″ on center, with 5″ leg welded to studs with two $^{1}/_{2}$″-long flare-bevel welds, and 4″ foot attached to the GFRC skin with $^{5}/_{8}$″ -thick GFRC bonding pads that extend $2^{1}/_{2}$″ beyond the flex anchor foot on both sides. Interior surface to have one layer of $^{5}/_{8}$″ Type X gypsum wallboard[e], attached with $1^{1}/_{4}$″-long Type S buglehead screws spaced 12″ on center. Cavity is to be filled with 5″ of 4 pcf (nominal) mineral fiber batts. GFRC has $1^{1}/_{2}$″ returns packed with mineral fiber and caulked on the exterior.	—	—	—	$6^{1}/_{8}$
	15-1.12[q]	2″ × 6″ wood studs at 16″ with double top plates, single bottom plate; interior and exterior sides covered with $^{5}/_{8}$″ Type X gypsum wallboard, 4′ wide, applied horizontally or vertically with vertical joints over studs, and fastened with $2^{1}/_{4}$″ Type S drywall screws, spaced 12″ on center. Cavity to be filled with $5^{1}/_{2}$″ mineral wool insulation.	—	—	—	$6^{3}/_{4}$

(continued)

TABLE 721.1(2)—continued
RATED FIRE-RESISTANCE PERIODS FOR VARIOUS WALLS AND PARTITIONS [a, o, p]

MATERIAL	ITEM NUMBER	CONSTRUCTION	MINIMUM FINISHED THICKNESS FACE-TO-FACE[b] (inches)			
			4 hours	3 hours	2 hours	1 hour
15. Exterior or interior walls (continued)	15-1.13[q]	2″ × 6″ wood studs at 16″ with double top plates, single bottom plate; interior and exterior sides covered with $^5/_8$″ Type X gypsum wallboard, 4′ wide, applied vertically with all joints over framing or blocking and fastened with $2^1/_4$″ Type S drywall screws, spaced 12″ on center. R-19 mineral fiber insulation installed in stud cavity.	—	—	—	$6^3/_4$
	15-1.14[q]	2″ × 6″ wood studs at 16″ with double top plates, single bottom plate; interior and exterior sides covered with $^5/_8$″ Type X gypsum wallboard, 4′ wide, applied horizontally or vertically with vertical joints over studs, and fastened with $2^1/_4$″ Type S drywall screws, spaced 7″ on center.	—	—	—	$6^3/_4$
	15-1.15[q]	2″ × 4″ wood studs at 16″ with double top plates, single bottom plate; interior and exterior sides covered with $^5/_8$″ Type X gypsum wallboard and sheathing, respectively, 4′ wide, applied horizontally or vertically with vertical joints over studs, and fastened with $2^1/_4$″ Type S drywall screws, spaced 12″ on center. Cavity to be filled with $3^1/_2$″ mineral wool insulation.	—	—	—	$4^3/_4$
	15-1.16[q]	2″ x 6″ wood studs at 24″ centers with double top plates, single bottom plate; interior and exterior side covered with two layers of $^5/_8$″ Type X gypsum wallboard, 4′ wide, applied horizontally with vertical joints over studs. Base layer fastened with $2^1/_4$″ Type S drywall screws, spaced 24″ on center and face layer fastened with Type S drywall screws, spaced 8″ on center, wallboard joints covered with paper tape and joint compound, fastener heads covered with joint compound. Cavity to be filled with $5^1/_2$″ mineral wool insulation.	—	—	8	—
	15-2.1[d]	$3^5/_8$″ No. 16 gage steel studs at 24″ on center or 2″ × 4″ wood studs at 24″ on center. Metal lath attached to the exterior side of studs with minimum 1″ long No. 6 drywall screws at 6″ on center and covered with minimum $^3/_4$″ thick Portland cement plaster. Thin veneer brick units of clay or shale complying with ASTM C1088, Grade TBS or better, installed in running bond in accordance with Section 1405.10. Combined total thickness of the Portland cement plaster, mortar and thin veneer brick units shall be not less than $1^3/_4$″. Interior side covered with one layer of $^5/_8$″ thick Type X gypsum wallboard attached to studs with 1″ long No. 6 drywall screws at 12″ on center.	—	—	—	6
	15-2.2[d]	$3^5/_8$″ No. 16 gage steel studs at 24″ on center or 2″ × 4″ wood studs at 24″ on center. Metal lath attached to the exterior side of studs with minimum 1″ long No. 6 drywall screws at 6″ on center and covered with minimum $^3/_4$″ thick Portland cement plaster. Thin veneer brick units of clay or shale complying with ASTM C1088, Grade TBS or better, installed in running bond in accordance with Section 1405.10. Combined total thickness of the Portland cement plaster, mortar and thin veneer brick units shall be not less than 2″. Interior side covered with two layers of $^5/_8$″ thick Type X gypsum wallboard. Bottom layer attached to studs with 1″ long No. 6 drywall screws at 24″ on center. Top layer attached to studs with $1^5/_8$″ long No. 6 drywall screws at 12″ on center.	—	—	$6^7/_8$	—
	15-2.3[d]	$3^5/_8$″ No. 16 gage steel studs at 16″ on center or 2″× 4″ wood studs at 16″ on center. Where metal lath is used, attach to the exterior side of studs with minimum 1″ long No. 6 drywall screws at 6″ on center. Brick units of clay or shale not less than $2^5/_8$″ thick complying with ASTM C216 installed in accordance with Section 1405.6 with a minimum 1″ airspace. Interior side covered with one layer of $^5/_8$″ thick Type X gypsum wallboard attached to studs with 1″ long No. 6 drywall screws at 12″ on center.	—	—	—	$7^7/_8$

(continued)

TABLE 721.1(2)—continued
RATED FIRE-RESISTANCE PERIODS FOR VARIOUS WALLS AND PARTITIONS [a, o, p]

MATERIAL	ITEM NUMBER	CONSTRUCTION	MINIMUM FINISHED THICKNESS FACE-TO-FACE[b] (inches)			
			4 hours	3 hours	2 hours	1 hour
15. Exterior or interior walls	15-2.4[d]	$3^5/_8$″ No. 16 gage steel studs at 16″ on center or 2″ × 4″ wood studs at 16″ on center. Where metal lath is used, attach to the exterior side of studs with minimum 1″ long No. 6 drywall screws at 6″ on center. Brick units of clay or shale not less than $2^5/_8$″ thick complying with ASTM C216 installed in accordance with Section 1405.6 with a minimum 1″ airspace. Interior side covered with two layers of $^5/_8$″ thick Type X gypsum wallboard. Bottom layer attached to studs with 1″ long No. 6 drywall screws at 24″ on center. Top layer attached to studs with $1^5/_8$″ long No. 6 drywall screws at 12″ on center.	—	—	$8^1/_2$	—
16. Exterior walls rated for fire resistance from the inside only in accordance with Section 705.5.	16-1.1[q]	2″ × 4″ wood studs at 16″ centers with double top plates, single bottom plate; interior side covered with $^5/_8$″ Type X gypsum wallboard, 4″ wide, applied horizontally unblocked, and fastened with $2^1/_4$″ Type S drywall screws, spaced 12″ on center, wallboard joints covered with paper tape and joint compound, fastener heads covered with joint compound. Exterior covered with $^3/_8$″ wood structural panels, applied vertically, horizontal joints blocked and fastened with 6d common nails (bright) — 12″ on center in the field, and 6″ on center panel edges. Cavity to be filled with $3^1/_2$″ mineral wool insulation. Rating established for exposure from interior side only.	—	—	—	$4^1/_2$
	16-1.2[q]	2″ × 6″ wood studs at 16″ centers with double top plates, single bottom plate; interior side covered with $^5/_8$″ Type X gypsum wallboard, 4″ wide, applied horizontally or vertically with vertical joints over studs and fastened with $2^1/_4$″ Type S drywall screws, spaced 12″ on center, wallboard joints covered with paper tape and joint compound, fastener heads covered with joint compound, exterior side covered with $^7/_{16}$″ wood structural panels fastened with 6d common nails (bright) spaced 12″ on center in the field and 6″ on center along the panel edges. Cavity to be filled with $5^1/_2$″ mineral wool insulation. Rating established from the gypsum-covered side only.	—	—	—	$6^9/_{16}$
	16-1.3[q]	2″ × 6″ wood studs at 16″ centers with double top plates, single bottom plates; interior side covered with $^5/_8$″ Type X gypsum wallboard, 4″ wide, applied vertically with all joints over framing or blocking and fastened with $2^1/_4$″ Type S drywall screws spaced 7″ on center. Joints to be covered with tape and joint compound. Exterior covered with $^3/_8$″ wood structural panels, applied vertically with edges over framing or blocking and fastened with 6d common nails (bright) at 12″ on center in the field and 6″ on center on panel edges. R-19 mineral fiber insulation installed in stud cavity. Rating established from the gypsum-covered side only.	—	—	—	$6^1/_2$

For SI: 1 inch = 25.4 mm, 1 square inch = 645.2 mm^2, 1 cubic foot = 0.0283 m^3.

a. Staples with equivalent holding power and penetration shall be permitted to be used as alternate fasteners to nails for attachment to wood framing.

b. Thickness shown for brick and clay tile is nominal thicknesses unless plastered, in which case thicknesses are net. Thickness shown for concrete masonry and clay masonry is equivalent thickness defined in Section 722.3.1 for concrete masonry and Section 722.4.1.1 for clay masonry. Where all cells are solid grouted or filled with silicone-treated perlite loose-fill insulation; vermiculite loose-fill insulation; or expanded clay, shale or slate lightweight aggregate, the equivalent thickness shall be the thickness of the block or brick using specified dimensions as defined in Chapter 21. Equivalent thickness shall include the thickness of applied plaster and lath or gypsum wallboard, where specified.

c. For units in which the net cross-sectional area of cored brick in any plane parallel to the surface containing the cores is not less than 75 percent of the gross cross-sectional area measured in the same plane.

d. Shall be used for nonbearing purposes only.

e. For all of the construction with gypsum wallboard described in this table, gypsum base for veneer plaster of the same size, thickness and core type shall be permitted to be substituted for gypsum wallboard, provided attachment is identical to that specified for the wallboard, and the joints on the face layer are reinforced and the entire surface is covered with not less than $^1/_{16}$-inch gypsum veneer plaster.

f. The fire-resistance time period for concrete masonry units meeting the equivalent thicknesses required for a 2-hour *fire-resistance rating* in Item 3, and having a thickness of not less than $7^5/_8$ inches is 4 hours where cores that are not grouted are filled with silicone-treated perlite loose-fill insulation; vermiculite loose-fill insulation; or expanded clay, shale or slate lightweight aggregate, sand or slag having a maximum particle size of $^3/_8$ inch.

g. The *fire-resistance rating* of concrete masonry units composed of a combination of aggregate types or where plaster is applied directly to the concrete masonry shall be determined in accordance with ACI 216.1/TMS 0216. Lightweight aggregates shall have a maximum combined density of 65 pounds per cubic foot.

(continued)

TABLE 721.1(2)—continued
RATED FIRE-RESISTANCE PERIODS FOR VARIOUS WALLS AND PARTITIONS [a, o,]

h. See Note b. The equivalent thickness shall be permitted to include the thickness of cement plaster or 1.5 times the thickness of gypsum plaster applied in accordance with the requirements of Chapter 25.
i. Concrete walls shall be reinforced with horizontal and vertical temperature reinforcement as required by Chapter 19.
j. Studs are welded truss wire studs with 0.18 inch (No. 7 B.W. gage) flange wire and 0.18 inch (No. 7 B.W. gage) truss wires.
k. Nailable metal studs consist of two channel studs spot welded back to back with a crimped web forming a nailing groove.
l. Wood structural panels shall be permitted to be installed between the fire protection and the wood studs on either the interior or exterior side of the wood frame assemblies in this table, provided the length of the fasteners used to attach the fire protection is increased by an amount not less than the thickness of the wood structural panel.
m. For studs with a slenderness ratio, l_e/d, greater than 33, the design stress shall be reduced to 78 percent of allowable F'_c. For studs with a slenderness ratio, l_e/d, not exceeding 33, the design stress shall be reduced to 78 percent of the adjusted stress F'_c calculated for studs having a slenderness ratio l_e/d of 33.
n. For properties of cooler or wallboard nails, see ASTM C514, ASTM C547 or ASTM F1667.
o. Generic *fire-resistance ratings* (those not designated as PROPRIETARY* in the listing) in the GA 600 shall be accepted as if herein listed.
p. NCMA TEK 5-8A shall be permitted for the design of fire walls.
q. The design stress of studs shall be equal to a maximum of 100 percent of the allowable F'_c calculated in accordance with Section 2306.

TABLE 721.1(3)
MINIMUM PROTECTION FOR FLOOR AND ROOF SYSTEMS[a, q]

FLOOR OR ROOF CONSTRUCTION	ITEM NUMBER	CEILING CONSTRUCTION	THICKNESS OF FLOOR OR ROOF SLAB (inches)				MINIMUM THICKNESS OF CEILING (inches)			
			4 hours	3 hours	2 hours	1 hour	4 hours	3 hours	2 hours	1 hour
1. Siliceous aggregate concrete	1-1.1	Slab (no ceiling required). Minimum cover over nonprestressed reinforcement shall be not less than $^3/_4''$ [b].	7.0	6.2	5.0	3.5	—	—	—	—
2. Carbonate aggregate concrete	2-1.1		6.6	5.7	4.6	3.2	—	—	—	—
3. Sand-lightweight concrete	3-1.1		5.4	4.6	3.8	2.7	—	—	—	—
4. Lightweight concrete	4-1.1		5.1	4.4	3.6	2.5	—	—	—	—
5. Reinforced concrete	5-1.1	Slab with suspended ceiling of vermiculite gypsum plaster over metal lath attached to $^3/_4''$ cold-rolled channels spaced 12″ on center. Ceiling located 6″ minimum below joists.	3	2	—	—	1	$^3/_4$	—	—
	5-2.1	$^3/_8''$ Type X gypsum wallboard[c] attached to 0.018 inch (No. 25 carbon sheet steel gage) by $^7/_8''$ deep by $2^5/_8''$ hat-shaped galvanized steel channels with 1″-long No. 6 screws. The channels are spaced 24″ on center, span 35″ and are supported along their length at 35″ intervals by 0.033″ (No. 21 galvanized sheet gage) galvanized steel flat strap hangers having formed edges that engage the lips of the channel. The strap hangers are attached to the side of the concrete joists with $^5/_{32}''$ by $1^1/_4''$ long power-driven fasteners. The wallboard is installed with the long dimension perpendicular to the channels. End joints occur on channels and supplementary channels are installed parallel to the main channels, 12″ each side, at end joint occurrences. The finished ceiling is located approximately 12″ below the soffit of the floor slab.	—	—	$2^1/_2$	—	—	—	$^5/_8$	—

(continued)

TABLE 721.1(3)—continued
MINIMUM PROTECTION FOR FLOOR AND ROOF SYSTEMS[a, q]

FLOOR OR ROOF CONSTRUCTION	ITEM NUMBER	CEILING CONSTRUCTION	THICKNESS OF FLOOR OR ROOF SLAB (inches)				MINIMUM THICKNESS OF CEILING (inches)			
			4 hours	3 hours	2 hours	1 hour	4 hours	3 hours	2 hours	1 hour
6. Steel joists constructed with a poured reinforced concrete slab on metal lath forms or steel form units[d, e]	6-1.1	Gypsum plaster on metal lath attached to the bottom cord with single No. 16 gage or doubled No. 18 gage wire ties spaced 6″ on center. Plaster mixed 1:2 for scratch coat, 1:3 for brown coat, by weight, gypsum-to-sand aggregate for 2-hour system. For 3-hour system plaster is neat.	—	—	$2^1/_2$	$2^1/_4$	—	—	$^3/_4$	$^5/_8$
	6-2.1	Vermiculite gypsum plaster on metal lath attached to the bottom chord with single No.16 gage or doubled 0.049-inch (No. 18 B.W. gage) wire ties 6″ on center.	—	2	—	—	—	$^5/_8$	—	—
	6-3.1	Cement plaster over metal lath attached to the bottom chord of joists with single No. 16 gage or doubled 0.049″ (No. 18 B.W. gage) wire ties spaced 6″ on center. Plaster mixed 1:2 for scratch coat, 1:3 for brown coat for 1-hour system and 1:1 for scratch coat, 1:1 $^1/_2$ for brown coat for 2-hour system, by weight, cement to sand.	—	—	—	2	—	—	—	$^5/_8$[f]
	6-4.1	Ceiling of $^5/_8$″ Type X wallboard[c] attached to $^7/_8$″ deep by $2^5/_8$″ by 0.021 inch (No. 25 carbon sheet steel gage) hat-shaped furring channels 12″ on center with 1″ long No. 6 wallboard screws at 8″ on center. Channels wire tied to bottom chord of joists with doubled 0.049 inch (No. 18 B.W. gage) wire or suspended below joists on wire hangers.[g]	—	—	$2^1/_2$	—	—	—	$^5/_8$	—
	6-5.1	Wood-fibered gypsum plaster mixed 1:1 by weight gypsum to sand aggregate applied over metal lath. Lath tied 6″ on center to $^3/_4$″ channels spaced $13^1/_2$″ on center. Channels secured to joists at each intersection with two strands of 0.049 inch (No. 18 B.W. gage) galvanized wire.	—	—	$2^1/_2$	—	—	—	$^3/_4$	—
7. Reinforced concrete slabs and joists with hollow clay tile fillers laid end to end in rows $2^1/_2$″ or more apart; reinforcement placed between rows and concrete cast around and over tile.	7-1.1	$^5/_8$″ gypsum plaster on bottom of floor or roof construction.	—	—	8[h]	—	—	—	$^5/_8$	—
	7-1.2	None	—	—	—	$5^1/_2$[i]	—	—	—	—
8. Steel joists constructed with a reinforced concrete slab on top poured on a $^1/_2$″ deep steel deck.[e]	8-1.1	Vermiculite gypsum plaster on metal lath attached to $^3/_4$″ cold-rolled channels with 0.049″ (No. 18 B.W. gage) wire ties spaced 6″ on center.	$2^1/_2$[j]	—	—	—	$^3/_4$	—	—	—

(continued)

TABLE 721.1(3)—continued
MINIMUM PROTECTION FOR FLOOR AND ROOF SYSTEMS[a, q]

FLOOR OR ROOF CONSTRUCTION	ITEM NUMBER	CEILING CONSTRUCTION	THICKNESS OF FLOOR OR ROOF SLAB (inches)				MINIMUM THICKNESS OF CEILING (inches)			
			4 hours	3 hours	2 hours	1 hour	4 hours	3 hours	2 hours	1 hour
9. 3″ deep cellular steel deck with concrete slab on top. Slab thickness measured to top.	9-1.1	Suspended ceiling of vermiculite gypsum plaster base coat and vermiculite acoustical plaster on metal lath attached at 6″ intervals to $^3/_4$″ cold-rolled channels spaced 12″ on center and secured to $1^1/_2$″ cold-rolled channels spaced 36″ on center with 0.065″ (No. 16 B.W. gage) wire. $1^1/_2$″ channels supported by No. 8 gage wire hangers at 36″ on center. Beams within envelope and with a $2^1/_2$″ airspace between beam soffit and lath have a 4-hour rating.	$2^1/_2$	—	—	—	$1^1/_8$[k]	—	—	—
10. $1^1/_2$″-deep steel roof deck on steel framing. Insulation board, 30 pcf density, composed of wood fibers with cement binders of thickness shown bonded to deck with unified asphalt adhesive. Covered with a Class A or B roof covering.	10-1.1	Ceiling of gypsum plaster on metal lath. Lath attached to $^3/_4$″ furring channels with 0.049″ (No. 18 B.W. gage) wire ties spaced 6″ on center. $^3/_4$″ channel saddle tied to 2″ channels with doubled 0.065″ (No. 16 B.W. gage) wire ties. 2″ channels spaced 36″ on center suspended 2″ below steel framing and saddle-tied with 0.165″ (No. 8 B.W. gage) wire. Plaster mixed 1:2 by weight, gypsum-to-sand aggregate.	—	—	$1^7/_8$	1	—	—	$^3/_4$[l]	$^3/_4$[l]
11. $1^1/_2$″-deep steel roof deck on steel-framing wood fiber insulation board, 17.5 pcf density on top applied over a 15-lb asphalt-saturated felt. Class A or B roof covering.	11-1.1	Ceiling of gypsum plaster on metal lath. Lath attached to $^3/_4$″ furring channels with 0.049″ (No. 18 B.W. gage) wire ties spaced 6″ on center. $^3/_4$″ channels saddle tied to 2″ channels with doubled 0.065″ (No. 16 B.W. gage) wire ties. 2″ channels spaced 36″ on center suspended 2″ below steel framing and saddle tied with 0.165″ (No. 8 B.W. gage) wire. Plaster mixed 1:2 for scratch coat and 1:3 for brown coat, by weight, gypsum-to-sand aggregate for 1-hour system. For 2-hour system, plaster mix is 1:2 by weight, gypsum-to-sand aggregate.	—	—	$1^1/_2$	1	—	—	$^7/_8$[g]	$^3/_4$[l]

(continued)

TABLE 721.1(3) —continued
MINIMUM PROTECTION FOR FLOOR AND ROOF SYSTEMS[a, q]

FLOOR OR ROOF CONSTRUCTION	ITEM NUMBER	CEILING CONSTRUCTION	THICKNESS OF FLOOR OR ROOF SLAB (inches)				MINIMUM THICKNESS OF CEILING (inches)			
			4 hours	3 hours	2 hours	1 hour	4 hours	3 hours	2 hours	1 hour
12. $1^1/_2$″ deep steel roof deck on steel-framing insulation of rigid board consisting of expanded perlite and fibers impregnated with integral asphalt waterproofing; density 9 to 12 pcf secured to metal roof deck by $^1/_2$″ wide ribbons of waterproof, cold-process liquid adhesive spaced 6″ apart. Steel joist or light steel construction with metal roof deck, insulation, and Class A or B built-up roof covering.[e]	12-1.1	Gypsum-vermiculite plaster on metal lath wire tied at 6″ intervals to $^3/_4$″ furring channels spaced 12″ on center and wire tied to 2″ runner channels spaced 32″ on center. Runners wire tied to bottom chord of steel joists.	—	—	1	—	—	—	$^7/_8$	—
13. Double wood floor over wood joists spaced 16″ on center.[m,n]	13-1.1	Gypsum plaster over $^3/_8$″ Type X gypsum lath. Lath initially applied with not less than four $1^1/_8$″ by No. 13 gage by $^{19}/_{64}$″ head plasterboard blued nails per bearing. Continuous stripping over lath along all joist lines. Stripping consists of 3″ wide strips of metal lath attached by $1^1/_2$″ by No. 11 gage by $^1/_2$″ head roofing nails spaced 6″ on center. Alternate stripping consists of 3″ wide 0.049″ diameter wire stripping weighing 1 pound per square yard and attached by No.16 gage by $1^1/_2$″ by $^3/_4$″ crown width staples, spaced 4″ on center. Where alternate stripping is used, the lath nailing shall consist of two nails at each end and one nail at each intermediate bearing. Plaster mixed 1:2 by weight, gypsum-to-sand aggregate.	—	—	—	—	—	—	—	$^7/_8$
	13-1.2	Cement or gypsum plaster on metal lath. Lath fastened with $1^1/_2$″ by No. 11 gage by $^7/_{16}$″ head barbed shank roofing nails spaced 5″ on center. Plaster mixed 1:2 for scratch coat and 1:3 for brown coat, by weight, cement to sand aggregate.	—	—	—	—	—	—	—	$^5/_8$
	13-1.3	Perlite or vermiculite gypsum plaster on metal lath secured to joists with $1^1/_2$″ by No. 11 gage by $^7/_{16}$″ head barbed shank roofing nails spaced 5″ on center.	—	—	—	—	—	—	—	$^5/_8$
	13-1.4	$^1/_2$″ Type X gypsum wallboard[c] nailed to joists with 5d cooler[o] or wallboard[o] nails at 6″ on center. End joints of wallboard centered on joists.	—	—	—	—	—	—	—	$^1/_2$

(continued)

TABLE 721.1(3) —continued
MINIMUM PROTECTION FOR FLOOR AND ROOF SYSTEMS[a, q]

FLOOR OR ROOF CONSTRUCTION	ITEM NUMBER	CEILING CONSTRUCTION	THICKNESS OF FLOOR OR ROOF SLAB (inches)				MINIMUM THICKNESS OF CEILING (inches)			
			4 hours	3 hours	2 hours	1 hour	4 hours	3 hours	2 hours	1 hour
14. Plywood stressed skin panels consisting of $^5/_8$″ -thick interior C-D (exterior glue) top stressed skin on 2″ × 6″ nominal (minimum) stringers. Adjacent panel edges joined with 8d common wire nails spaced 6″ on center. Stringers spaced 12″ maximum on center.	14-1.1	$^1/_2$″ -thick wood fiberboard weighing 15 to 18 pounds per cubic foot installed with long dimension parallel to stringers or $^3/_8$″ C-D (exterior glue) plywood glued and/or nailed to stringers. Nailing to be with 5d cooler[o] or wallboard[o] nails at 12″ on center. Second layer of $^1/_2$″ Type X gypsum wallboard[c] applied with long dimension perpendicular to joists and attached with 8d cooler[o] or wallboard[o] nails at 6″ on center at end joints and 8″ on center elsewhere. Wallboard joints staggered with respect to fiberboard joints.	—	—	—	—	—	—	—	1
15. Vermiculite concrete slab proportioned 1:4 (Portland cement to vermiculite aggregate) on a $1^1/_2$″ -deep steel deck supported on individually protected steel framing. Maximum span of deck 6′-10″ where deck is less than 0.019 inch (No. 26 carbon steel sheet gage) or greater. Slab reinforced with 4″ × 8″ 0.109/0.083″ (No. $^{12}/_{14}$ B.W. gage) welded wire mesh.	15-1.1	None	—	—	—	3[j]	—	—	—	—
16. Perlite concrete slab proportioned 1:6 (Portland cement to perlite aggregate) on a $1^1/_4$″ -deep steel deck supported on individually protected steel framing. Slab reinforced with 4″ × 8″ 0.109/0.083″ (No. $^{12}/_{14}$ B.W. gage) welded wire mesh.	16-1.1	None	—	—	—	$3^1/_2$[j]	—	—	—	—

(continued)

TABLE 721.1(3)—continued
MINIMUM PROTECTION FOR FLOOR AND ROOF SYSTEMS[a, q]

FLOOR OR ROOF CONSTRUCTION	ITEM NUMBER	CEILING CONSTRUCTION	THICKNESS OF FLOOR OR ROOF SLAB (inches)				MINIMUM THICKNESS OF CEILING (inches)			
			4 hours	3 hours	2 hours	1 hour	4 hours	3 hours	2 hours	1 hour
17. Perlite concrete slab proportioned 1:6 (Portland cement to perlite aggregate) on a $^{9}/_{16}''$-deep steel deck supported by steel joists 4′ on center. Class A or B roof covering on top.	17-1.1	Perlite gypsum plaster on metal lath wire tied to $^{3}/_{4}''$ furring channels attached with 0.065″ (No. 16 B.W. gage) wire ties to lower chord of joists.	—	2[p]	2[p]	—	—	$^{7}/_{8}$	$^{3}/_{4}$	—
18. Perlite concrete slab proportioned 1:6 (Portland cement to perlite aggregate) on $1^{1}/_{4}''$-deep steel deck supported on individually protected steel framing. Maximum span of deck 6′-10″ where deck is less than 0.019″ (No. 26 carbon sheet steel gage) and 8′-0″ where deck is 0.019″ (No. 26 carbon sheet steel gage) or greater. Slab reinforced with 0.042″ (No. 19 B.W. gage) hexagonal wire mesh. Class A or B roof covering on top.	18-1.1	None	—	$2^{1}/_{4}$[p]	$2^{1}/_{4}$[p]	—	—	—	—	—
19. Floor and beam construction consisting of 3″-deep cellular steel floor unit mounted on steel members with 1:4 (proportion of Portland cement to perlite aggregate) perlite-concrete floor slab on top.	19-1.1	Suspended envelope ceiling of perlite gypsum plaster on metal lath attached to $^{3}/_{4}''$ cold-rolled channels, secured to $1^{1}/_{2}''$ cold-rolled channels spaced 42″ on center supported by 0.203 inch (No. 6 B.W. gage) wire 36″ on center. Beams in envelope with 3″ minimum airspace between beam soffit and lath have a 4-hour rating.	2[p]	—	—	—	1[l]	—	—	—

(continued)

TABLE 721.1(3)—continued
MINIMUM PROTECTION FOR FLOOR AND ROOF SYSTEMS[a, q]

FLOOR OR ROOF CONSTRUCTION	ITEM NUMBER	CEILING CONSTRUCTION	THICKNESS OF FLOOR OR ROOF SLAB (inches)				MINIMUM THICKNESS OF CEILING (inches)			
			4 hours	3 hours	2 hours	1 hour	4 hours	3 hours	2 hours	1 hour
20. Perlite concrete proportioned 1:6 (Portland cement to perlite aggregate) poured to $^1/_8$″ thickness above top of corrugations of $1^5/_{16}$″ -deep galvanized steel deck maximum span 8′-0″ for 0.024″ (No. 24 galvanized sheet gage) or 6′ 0″ for 0.019″ (No. 26 galvanized sheet gage) with deck supported by individually protected steel framing. Approved polystyrene foam plastic insulation board having a flame spread not exceeding 75 (1″ to 4″ thickness) with vent holes that approximate 3 percent of the board surface area placed on top of perlite slurry. A 2′ by 4′ insulation board contains six $2^3/_4$″ diameter holes. Board covered with $2^1/_4$″ minimum perlite concrete slab. Slab reinforced with mesh consisting of 0.042″ (No. 19 B.W. gage) galvanized steel wire twisted together to form 2″ hexagons with straight 0.065″ (No. 16 B.W. gage) galvanized steel wire woven into mesh and spaced 3″. Alternate slab reinforcement shall be permitted to consist of 4″ × 8″, 0.109/0.238″ (No. 12/4 B.W. gage), or 2″ × 2″, 0.083/0.083″ (No. 14/14 B.W. gage) welded wire fabric. Class A or B roof covering on top.	20-1.1	None	—	—	Varies	—	—	—	—	—
21. Wood joists, wood I-joists, floor trusses and flat or pitched roof trusses spaced a maximum 24″ o.c. with $^1/_2$″ wood structural panels with exterior glue applied at right angles to top of joist or top chord of trusses with 8d nails. The wood structural panel thickness shall be not less than nominal $^1/_2$″ nor less than required by Chapter 23.	21-1.1	Base layer $^5/_8$″ Type X gypsum wallboard applied at right angles to joist or truss 24″ o.c. with $1^1/_4$″ Type S or Type W drywall screws 24″ o.c. Face layer $^5/_8$″ Type X gypsum wallboard or veneer base applied at right angles to joist or truss through base layer with $1^7/_8$″ Type S or Type W drywall screws 12″ o.c. at joints and intermediate joist or truss. Face layer Type G drywall screws placed 2″ back on either side of face layer end joints, 12″ o.c.	—	—	—	Varies	—	—	—	$1^1/_4$

(continued)

TABLE 721.1(3)—continued
MINIMUM PROTECTION FOR FLOOR AND ROOF SYSTEMS[a, q]

FLOOR OR ROOF CONSTRUCTION	ITEM NUMBER	CEILING CONSTRUCTION	THICKNESS OF FLOOR OR ROOF SLAB (inches)				MINIMUM THICKNESS OF CEILING (inches)			
			4 hours	3 hours	2 hours	1 hour	4 hours	3 hours	2 hours	1 hour
22. Steel joists, floor trusses and flat or pitched roof trusses spaced a maximum 24″ o.c. with $^1/_2$″ wood structural panels with exterior glue applied at right angles to top of joist or top chord of trusses with No. 8 screws. The wood structural panel thickness shall be not less than nominal $^1/_2$″ nor less than required by Chapter 23.	22-1.1	Base layer $^5/_8$″ Type X gypsum board applied at right angles to steel framing 24″ on center with 1″ Type S drywall screws spaced 24″ on center. Face layer $^5/_8$″ Type X gypsum board applied at right angles to steel framing attached through base layer with $1^5/_8$″ Type S drywall screws 12″ on center at end joints and intermediate joints and $1^1/_2$″ Type G drywall screws 12 inches on center placed 2″ back on either side of face layer end joints. Joints of the face layer are offset 24″ from the joints of the base layer.	—	—	—	Varies	—	—	—	$1^1/_4$
23. Wood I-joist (minimum joist depth $9^1/_4$″ with a minimum flange depth of $1^5/_{16}$″ and a minimum flange cross-sectional area of 2.25 square inches) at 24″ o.c. spacing with a minimum 1 × 4 ($^3/_4$″ × 3.5″ actual) ledger strip applied parallel to and covering the bottom of the bottom flange of each member, tacked in place. 2″ mineral wool insulation, 3.5 pcf (nominal) installed adjacent to the bottom flange of the I-joist and supported by the 1 × 4 ledger strip.	23-1.1	$^1/_2$″ deep single leg resilient channel 16″ on center (channels doubled at wallboard end joints), placed perpendicular to the furring strip and joist and attached to each joist by $1^7/_8$″ Type S drywall screws. $^5/_8$″ Type C gypsum wallboard applied perpendicular to the channel with end joints staggered not less than 4′ and fastened with $1^1/_8$″ Type S drywall screws spaced 7″ on center. Wallboard joints to be taped and covered with joint compound.	—	—	—	Varies	—	—	—	$^5/_8$
24. Wood I-joist (minimum I-joist depth $9^1/_4$″ with a minimum flange depth of $1^1/_2$″ and a minimum flange cross-sectional area of 5.25 square inches; minimum web thickness of $^3/_8$″) @ 24″ o.c., $1^1/_2$″ mineral wool insulation (2.5 pcf-nominal) resting on hat-shaped furring channels.	24-1.1	Minimum 0.026″ thick hat-shaped channel 16″ o.c. (channels doubled at wallboard end joints), placed perpendicular to the joist and attached to each joist by $1^1/_4$″ Type S drywall screws. $^5/_8$″ Type C gypsum wallboard applied perpendicular to the channel with end joints staggered and fastened with $1^1/_8$″ Type S drywall screws spaced 12″ o.c. in the field and 8″ o.c. at the wallboard ends. Wallboard joints to be taped and covered with joint compound.	—	—	—	Varies	—	—	—	$^5/_8$
25. Wood I-joist (minimum I-joist depth $9^1/_4$″ with a minimum flange depth of $1^1/_2$″ and a minimum flange cross-sectional area of 5.25 square inches; minimum web thickness of $^7/_{16}$″) @ 24″ o.c., $1^1/_2$″ mineral wool insulation (2.5 pcf-nominal) resting on resilient channels.	25-1.1	Minimum 0.019″ thick resilient channel 16″ o.c. (channels doubled at wallboard end joints), placed perpendicular to the joist and attached to each joist by $1^5/_8$″ Type S drywall screws. $^5/_8$″ Type C gypsum wallboard applied perpendicular to the channel with end joints staggered and fastened with 1″ Type S drywall screws spaced 12″ o.c. in the field and 8″ o.c. at the wallboard ends. Wallboard joints to be taped and covered with joint compound.	—	—	—	Varies	—	—	—	$^5/_8$

(continued)

TABLE 721.1(3)—continued
MINIMUM PROTECTION FOR FLOOR AND ROOF SYSTEMS[a, q]

FLOOR OR ROOF CONSTRUCTION	ITEM NUMBER	CEILING CONSTRUCTION	THICKNESS OF FLOOR OR ROOF SLAB (inches)				MINIMUM THICKNESS OF CEILING (inches)			
			4 hours	3 hours	2 hours	1 hour	4 hours	3 hours	2 hours	1 hour
26. Wood I-joist (minimum I-joist depth $9^1/_4''$ with a minimum flange thickness of $1^1/_2''$ and a minimum flange cross-sectional area of 2.25 square inches; minimum web thickness of $^3/_8''$) @ 24″ o.c.	26-1.1	Two layers of $^1/_2''$ Type X gypsum wallboard applied with the long dimension perpendicular to the I-joists with end joints staggered. The base layer is fastened with $1^5/_8''$ Type S drywall screws spaced 12″ o.c. and the face layer is fastened with 2″ Type S drywall screws spaced 12″ o.c. in the field and 8″ o.c. on the edges. Face layer end joints shall not occur on the same I-joist as base layer end joints and edge joints shall be offset 24″ from base layer joints. Face layer to also be attached to base layer with $1^1/_2''$ Type G drywall screws spaced 8″ o.c. placed 6″ from face layer end joints. Face layer wallboard joints to be taped and covered with joint compound.	—	—	—	Varies	—	—	—	1
27. Wood I-joist (minimum I-joist depth $9^1/_2''$ with a minimum flange depth of $1^5/_{16}''$ and a minimum flange cross-sectional area of 1.95 square inches; minimum web thickness of $^3/_8''$) @ 24″ o.c.	27-1.1	Minimum 0.019″ thick resilient channel 16″ o.c. (channels doubled at wallboard end joints), placed perpendicular to the joist and attached to each joist by $1^1/_4''$ Type S drywall screws. Two layers of $^1/_2''$ Type X gypsum wallboard applied with the long dimension perpendicular to the I-joists with end joints staggered. The base layer is fastened with $1^1/_4''$ Type S drywall screws spaced 12″ o.c. and the face layer is fastened with $1^5/_8''$ Type S drywall screws spaced 12″ o.c. Face layer end joints shall not occur on the same I-joist as base layer end joints and edge joints shall be offset 24″ from base layer joints. Face layer to also be attached to base layer with $1^1/_2''$ Type G drywall screws spaced 8″ o.c. placed 6″ from face layer end joints. Face layer wallboard joints to be taped and covered with joint compound.	—	—	—	Varies	—	—	—	1

(continued)

TABLE 721.1(3)—continued
MINIMUM PROTECTION FOR FLOOR AND ROOF SYSTEMS[a, q]

FLOOR OR ROOF CONSTRUCTION	ITEM NUMBER	CEILING CONSTRUCTION	THICKNESS OF FLOOR OR ROOF SLAB (inches)				MINIMUM THICKNESS OF CEILING (inches)			
			4 hours	3 hours	2 hours	1 hour	4 hours	3 hours	2 hours	1 hour
28. Wood I-joist (minimum I-joist depth $9^1/_4''$ with a minimum flange depth of $1^1/_2''$ and a minimum flange cross-sectional area of 2.25 square inches; minimum web thickness of $^3/_8''$) @ 24″ o.c. Unfaced fiberglass insulation or mineral wool insulation is installed between the I-joists supported on the upper surface of the flange by stay wires spaced 12″ o.c.	28-1.1	Base layer of $^5/_8''$ Type C gypsum wallboard attached directly to I-joists with $1^5/_8''$ Type S drywall screws spaced 12″ o.c. with ends staggered. Minimum 0.0179″ thick hat-shaped $^7/_8$-inch furring channel 16″ o.c. (channels doubled at wallboard end joints), placed perpendicular to the joist and attached to each joist by $1^5/_8''$ Type S drywall screws after the base layer of gypsum wallboard has been applied. The middle and face layers of $^5/_8''$ Type C gypsum wallboard applied perpendicular to the channel with end joints staggered. The middle layer is fastened with 1″ Type S drywall screws spaced 12″ o.c. The face layer is applied parallel to the middle layer but with the edge joints offset 24″ from those of the middle layer and fastened with $1^5/_8''$ Type S drywall screws 8″ o.c. The joints shall be taped and covered with joint compound.	—	—	—	Varies	—	—	$2^3/_4$	—
29. Channel-shaped 18 gage steel joists (minimum depth 8″) spaced a maximum 24″ o.c. supporting tongue-and-groove wood structural panels (nominal minimum $^3/_4''$ thick) applied perpendicular to framing members. Structural panels attached with $1^5/_8''$ Type S-12 screws spaced 12″ o.c.	29-1.1	Base layer $^5/_8''$ Type X gypsum board applied perpendicular to bottom of framing members with $1^1/_8''$ Type S-12 screws spaced 12″ o.c. Second layer $^5/_8''$ Type X gypsum board attached perpendicular to framing members with $1^5/_8''$ Type S-12 screws spaced 12″ o.c. Second layer joints offset 24″ from base layer. Third layer $^5/_8''$ Type X gypsum board attached perpendicular to framing members with $2^3/_8''$ Type S-12 screws spaced 12″ o.c. Third layer joints offset 12″ from second layer joints. Hat-shaped $^7/_8$-inch rigid furring channels applied at right angles to framing members over third layer with two $2^3/_8''$ Type S-12 screws at each framing member. Face layer $^5/_8''$ Type X gypsum board applied at right angles to furring channels with $1^1/_8''$ Type S screws spaced 12″ o.c.	—	—	Varies	—	—	—	$3^3/_8$	—

(continued)

TABLE 721.1(3)—continued
MINIMUM PROTECTION FOR FLOOR AND ROOF SYSTEMS[a, q]

FLOOR OR ROOF CONSTRUCTION	ITEM NUMBER	CEILING CONSTRUCTION	THICKNESS OF FLOOR OR ROOF SLAB (inches)				MINIMUM THICKNESS OF CEILING (inches)			
			4 hours	3 hours	2 hours	1 hour	4 hours	3 hours	2 hours	1 hour
30. Wood I-joist (minimum I-joist depth $9^1/_2''$ with a minimum flange depth of $1^1/_2''$ and a minimum flange cross-sectional area of 2.25 square inches; minimum web thickness of $^3/_8''$) @ 24″ o.c. Fiberglass insulation placed between I-joists supported by the resilient channels.	30-1.1	Minimum 0.019″ thick resilient channel 16″ o.c. (channels doubled at wallboard end joints), placed perpendicular to the joists and attached to each joist by $1^1/_4''$ Type S drywall screws. Two layers of $^1/_2''$ Type X gypsum wallboard applied with the long dimension perpendicular to the I-joists with end joints staggered. The base layer is fastened with $1^1/_4''$ Type S drywall screws spaced 12″ o.c. and the face layer is fastened with $1^5/_8''$ Type S drywall screws spaced 12″ o.c. Face layer end joints shall not occur on the same I-joist as base layer end joints and edge joints shall be offset 24″ from base layer joints. Face layer to be attached to base layer with $1^1/_2''$ Type G drywall screws spaced 8″ o.c. placed 6″ from face layer end joints. Face layer wallboard joints to be taped and covered with joint compound.	—	—	—	Varies	—	—	—	1

For SI: 1 inch = 25.4 mm, 1 foot = 304.8 mm, 1 pound = 0.454 kg, 1 cubic foot = 0.0283 m^3,
1 pound per square inch = 6.895 kPa, 1 pound per linear foot = 1.4882 kg/m.

a. Staples with equivalent holding power and penetration shall be permitted to be used as alternate fasteners to nails for attachment to wood framing.

b. Where the slab is in an unrestrained condition, minimum reinforcement cover shall be not less than $1^5/_8$ inches for 4 hours (siliceous aggregate only); $1^1/_4$ inches for 4 and 3 hours; 1 inch for 2 hours (siliceous aggregate only); and $^3/_4$ inch for all other restrained and unrestrained conditions.

c. For all of the construction with gypsum wallboard described in this table, gypsum base for veneer plaster of the same size, thickness and core type shall be permitted to be substituted for gypsum wallboard, provided attachment is identical to that specified for the wallboard, and the joints on the face layer are reinforced and the entire surface is covered with not less than $^1/_{16}$-inch gypsum veneer plaster.

d. Slab thickness over steel joists measured at the joists for metal lath form and at the top of the form for steel form units.

e. (a) The maximum allowable stress level for H-Series joists shall not exceed 22,000 psi.
(b) The allowable stress for K-Series joists shall not exceed 26,000 psi, the nominal depth of such joist shall be not less than 10 inches and the nominal joist weight shall be not less than 5 pounds per linear foot.

f. Cement plaster with 15 pounds of hydrated lime and 3 pounds of approved additives or admixtures per bag of cement.

g. Gypsum wallboard ceilings attached to steel framing shall be permitted to be suspended with $1^1/_2$-inch cold-formed carrying channels spaced 48 inches on center, that are suspended with No. 8 SWG galvanized wire hangers spaced 48 inches on center. Cross-furring channels are tied to the carrying channels with No. 18 SWG galvanized wire hangers spaced 48 inches on center. Cross-furring channels are tied to the carrying channels with No. 18 SWG galvanized wire (double strand) and spaced as required for direct attachment to the framing. This alternative is applicable to those steel framing assemblies recognized under Note q.

h. Six-inch hollow clay tile with 2-inch concrete slab above.

i. Four-inch hollow clay tile with $1^1/_2$-inch concrete slab above.

j. Thickness measured to bottom of steel form units.

k. Five-eighths inch of vermiculite gypsum plaster plus $^1/_2$ inch of approved vermiculite acoustical plastic.

l. Furring channels spaced 12 inches on center.

m. Double wood floor shall be permitted to be either of the following:
(a) Subfloor of 1-inch nominal boarding, a layer of asbestos paper weighing not less than 14 pounds per 100 square feet and a layer of 1-inch nominal tongue-and-groove finished flooring; or
(b) Subfloor of 1-inch nominal tongue-and-groove boarding or $^{15}/_{32}$-inch wood structural panels with exterior glue and a layer of 1-inch nominal tongue-and-groove finished flooring or $^{19}/_{32}$-inch wood structural panel finish flooring or a layer of Type I Grade M-1 particleboard not less than $^5/_8$-inch thick.

n. The ceiling shall be permitted to be omitted over unusable space, and flooring shall be permitted to be omitted where unusable space occurs above.

o. For properties of cooler or wallboard nails, see ASTM C514, ASTM C547 or ASTM F1667.

p. Thickness measured on top of steel deck unit.

q. Generic *fire-resistance ratings* (those not designated as PROPRIETARY* in the listing) in the GA 600 shall be accepted as if herein listed.

SECTION 722 CALCULATED FIRE RESISTANCE

❖ Due to the prescriptive nature of Section 722, this section contains explanatory material, examples and reference materials related to calculating fire-resistance ratings of materials and assemblies. As such, commentary is provided where necessary as the basis for a provision or to illustrate the application of the requirement.

This section provides a number of useful alternatives for establishing compliance with the code. The provisions in this section can be used to calculate the fire-resistance rating for an assembly (concrete, concrete masonry, clay masonry, steel and wood). It also contains provisions that would permit modifications or changes to a tested assembly. For example, the steel assemblies (see Section 722.5) will allow the selection of different sized or shaped members depending on the weight-to-heated-perimeter ratio (see Sections 722.5.2.1.2 and 722.5.2.2). In addition, the amount of spray-applied fire-resistant materials can be modified from a tested assembly to either increase or decrease the coverage dependent on the member and the spray-applied material.

The section also addresses items that are much less exotic and may be helpful in a number of situations. Section 722.6 provides methods for calculating the fire resistance of wood assemblies, including light-framed walls and floors or exposed heavy-timber members. Section 722.4.1 can be used for masonry walls (including existing conditions) where additional protection such as filled cores or an additional wythe such as a brick veneer is added.

While this section is not often used by many people, those who have used it will find that it is often helpful in solving difficult situations related to the protection of certain assemblies. Therefore, taking the time to become familiar with this section can be very beneficial. The provisions of this section include methods for calculating fire resistance as follows:

Concrete Assemblies:
- Walls—722.2.1
- Floor and roof slabs—722.2.2
- Beams—722.2.3
- Columns—722.2.3

Concrete Masonry:
- Walls—722.3.2
- Lintels—722.3.4
- Columns—722.3.5

Clay Brick and Tile Masonry:
- Walls—722.4.1
- Lintels—722.4.3
- Columns—722.4.4

Steel Assemblies:
- Columns using:
  - Gypsum wallboard—722.5.1.2
  - Spray-applied materials—722.5.1.3
  - Concrete or masonry—722.5.1.4
- Beams using:
  - Spray-applied materials—722.5.2.2
- Trusses using:
  - Spray-applied materials—722.5.2.3

Wood Assemblies:
- Walls, floors and roofs—722.6.2

722.1 General. The provisions of this section contain procedures by which the *fire resistance* of specific materials or combinations of materials is established by calculations. These procedures apply only to the information contained in this section and shall not be otherwise used. The calculated *fire resistance* of concrete, concrete masonry and clay masonry assemblies shall be permitted in accordance with ACI 216.1/TMS 0216. The calculated *fire resistance* of steel assemblies shall be permitted in accordance with Chapter 5 of ASCE 29. The calculated *fire resistance* of exposed wood members and wood decking shall be permitted in accordance with Chapter 16 of ANSI/AF&PA *National Design Specification for Wood Construction (NDS)*.

❖ Section 703.3 permits fire-resistance ratings to be determined in a number of ways, including those found in Section 722 (see Section 703.3, Item 3). The provisions of this section contain procedures by which the fire resistance of specific materials or combinations of materials is established by calculations. These procedures apply only to the information contained in this section and are not to be otherwise used. The calculated fire resistance of concrete, concrete masonry and clay masonry assemblies is also permitted in accordance with ACI 216.1/TMS 0216.1 while steel assemblies can use ASCE 29 (see commentary, Section 703.3).

722.1.1 Definitions. The following terms are defined in Chapter 2:

❖ This section lists terms that are specifically associated with the subject matter of this chapter. It is important to emphasize that these terms are not exclusively related to this chapter but may or may not also be applicable where the term is used elsewhere in the code.

Definitions of terms can help in the understanding and application of the code requirements. The purpose for including a list within this chapter is to provide more convenient access to terms which may have a specific or limited application within this chapter. For the complete definition and associated commentary, refer back to Chapter 2. Terms that are italized provide a visual identification throughout the code that a definition exists for that term. The use and application of all defined terms are set forth in Section 201.

CERAMIC FIBER BLANKET.

CONCRETE, CARBONATE AGGREGATE.

CONCRETE, CELLULAR.

CONCRETE, LIGHTWEIGHT AGGREGATE.

CONCRETE, PERLITE.

CONCRETE, SAND-LIGHTWEIGHT.

CONCRETE, SILICEOUS AGGREGATE.

CONCRETE, VERMICULITE.

GLASS FIBERBOARD.

MINERAL BOARD.

722.2 Concrete assemblies. The provisions of this section contain procedures by which the *fire-resistance ratings* of concrete assemblies are established by calculations.

❖ See the commentary to Section 722.

722.2.1 Concrete walls. Cast-in-place and precast concrete walls shall comply with Section 722.2.1.1. Multiwythe concrete walls shall comply with Section 722.2.1.2. Joints between precast panels shall comply with Section 722.2.1.3. Concrete walls with gypsum wallboard or plaster finish shall comply with Section 722.2.1.4.

❖ See the commentary to Sections 702.2.1.1, 702.2.1.2, 702.2.1.3 and 702.2.1.4.

722.2.1.1 Cast-in-place or precast walls. The minimum equivalent thicknesses of cast-in-place or precast concrete walls for *fire-resistance ratings* of 1 hour to 4 hours are shown in Table 722.2.1.1. For solid walls with flat vertical surfaces, the equivalent thickness is the same as the actual thickness. The values in Table 722.2.1.1 apply to plain, reinforced or prestressed concrete walls.

❖ Although there have been few fire tests of concrete walls (other than concrete masonry), there have been many fire tests of concrete slabs tested as floors or roofs. Fire tests of floors or roofs are considered to be more severe than those of walls because of the loading conditions (tension cracks associated with bending moment). In addition, most ASTM E119 or UL 263 fire tests of floor assemblies have been conducted while the assembly was supported within restraining frames. As concrete assemblies are heated and expand, the expansion is resisted by the restraining frame. These restraining forces are usually much greater than the superimposed loads supported by load-bearing walls. Thus, floor or roof assemblies are subjected to both vertical superimposed loads and horizontal restraining loads during fire tests. By contrast, load-bearing walls are subjected only to superimposed loads.

The fire endurance of masonry or concrete walls is nearly always governed by the ASTM E119 or UL 263 criteria for temperature rise of the unexposed surface (i.e., the "heat transmission" end point). For flat concrete slabs or panels, the heat transmission fire endurance depends primarily on the aggregate type and thickness and is essentially the same for floors as it is for walls.

722.2.1.1.1 Hollow-core precast wall panels. For hollow-core precast concrete wall panels in which the cores are of constant cross section throughout the length, calculation of the equivalent thickness by dividing the net cross-sectional area (the gross cross section minus the area of the cores) of the panel by its width shall be permitted

❖ The method for determining equivalent thickness of masonry units was developed because the cores in masonry units taper. The method is, of course, also applicable to hollow-core precast concrete panels. However, because the cores in hollow-core panels do not taper, the equivalent thickness can be calculated by dividing the net cross-sectional area of the panel by its width.

TABLE 722.2.1.1
MINIMUM EQUIVALENT THICKNESS OF CAST-IN-PLACE OR PRECAST CONCRETE WALLS, LOAD-BEARING OR NONLOAD-BEARING

CONCRETE TYPE	MINIMUM SLAB THICKNESS (inches) FOR FIRE-RESISTANCE RATING OF				
	1 hour	$1^{1}/_{2}$ hours	2 hours	3 hours	4 hours
Siliceous	3.5	4.3	5.0	6.2	7.0
Carbonate	3.2	4.0	4.6	5.7	6.6
Sand-lightweight	2.7	3.3	3.8	4.6	5.4
Lightweight	2.5	3.1	3.6	4.4	5.1

For SI: 1 inch = 25.4 mm.

❖ The data in Table 722.2.1.1 were derived from Portland Cement Association (PCA) Bulletin 223, "Fire Endurances of Concrete Slabs as Influenced by Thickness, Aggregate Type, and Moisture," and PCA Publication T-140, *Fire Resistance of Reinforced Concrete Floors*.

722.2.1.1.2 Core spaces filled. Where all of the core spaces of hollow-core wall panels are filled with loose-fill material, such as expanded shale, clay or slag, or vermiculite or perlite, the *fire-resistance rating* of the wall is the same as that of a solid wall of the same concrete type and of the same overall thickness.

❖ The PCA report "Tests of Fire Resistance and Strength of Walls of Concrete Masonry Units" shows that filling cores of concrete masonry units with loose lightweight aggregates increases the fire endurance to a duration significantly longer than that of solid masonry units of the same total thickness. It is reasonable to assume that the same relationship exists for walls made of hollow-core panels.

722.2.1.1.3 Tapered cross sections. The thickness of panels with tapered cross sections shall be that determined at a distance $2t$ or 6 inches (152 mm), whichever is less, from the point of minimum thickness, where t is the minimum thickness.

❖ Some precast concrete wall panel sections (e.g., certain single-tee units) have tapered surfaces so the thickness varies, as shown in Code Figure 722.2.2.1.2. In fire tests, it has been customary to monitor the unexposed surface temperature at the location shown in the figure.

722.2.1.1.4 Ribbed or undulating surfaces. The equivalent thickness of panels with ribbed or undulating surfaces shall be determined by one of the following expressions:

For $s \geq 4t$, the thickness to be used shall be t

For $s \leq 2t$, the thickness to be used shall be t_e

For $4t > s > 2t$, the thickness to be used shall be

$$t + \left(\frac{4t}{s} - 1\right)(t_e - t) \quad \textbf{(Equation 7-3)}$$

where:

s = Spacing of ribs or undulations.

t = Minimum thickness.

t_e = Equivalent thickness of the panel calculated as the net cross-sectional area of the panel divided by the width, in which the maximum thickness used in the calculation shall not exceed $2t$.

❖ The portion of a ribbed panel that can be used in calculating equivalent thickness, t_e, is shown in Code Figure 722.2.2.1.3. Note that the procedure outlined gives no credit for stems of double tees or of similar ribbed panels and clearly indicates that the minimum thickness must be used for such sections.

722.2.1.2 Multiwythe walls. For walls that consist of two wythes of different types of concrete, the *fire-resistance ratings* shall be permitted to be determined from Figure 722.2.1.2.

❖ The graphs in the Code Figure 722.2.1.2 were derived from a report entitled "Fire Endurance of Two-Course Floors and Roofs" in the *ACI Journal.*

722.2.1.2.1 Two or more wythes. The *fire-resistance rating* for wall panels consisting of two or more wythes shall be permitted to be determined by the formula:

$$R = (R_1^{0.59} + R_2^{0.59} + ... + R_n^{0.59})^{1.7} \quad \textbf{(Equation 7-4)}$$

where:

R = The fire endurance of the assembly, minutes.

R_1, R_2, and R_n = The fire endurances of the individual wythes, minutes. Values of $R_n^{0.59}$ for use in Equation 7-4 are given in Table 722.2.1.2(1). Calculated *fire-resistance ratings* are shown in Table 722.2.1.2(2).

❖ Equation 7-4 was developed by the U.S. National Bureau of Standards (NBS) in the early 1940s and first appeared in Appendix B of BMS 92. Verification of the accuracy of the equation is given in "Fire Endurance of Two-Course Floors and Roofs" in the *ACI Journal.*

This section and the associated tables allow both the determination of the fire endurance of a particular assembly [see Table 722.2.1.2(1)] and what thickness is required to achieve a certain fire endurance [see Table 722.2.1.2(2)].

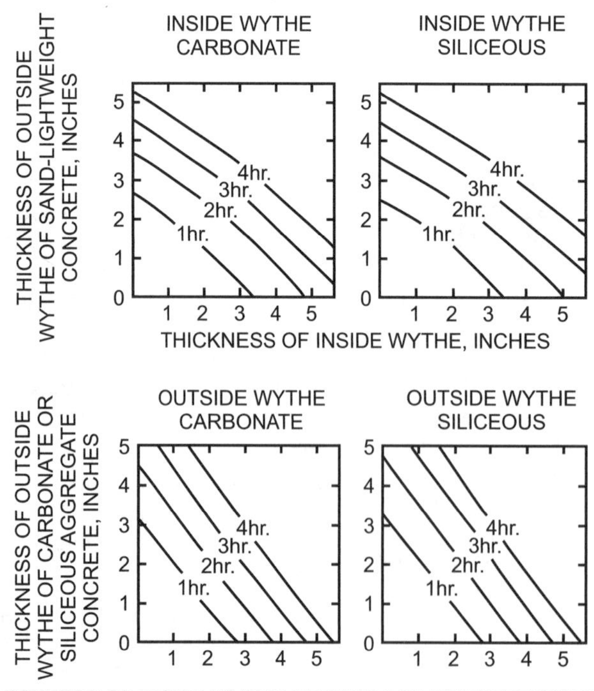

For SI: 1 inch = 25.4 mm.

FIGURE 722.2.1.2
FIRE-RESISTANCE RATINGS OF TWO-WYTHE CONCRETE WALLS

722.2.1.2.2 Foam plastic insulation. The *fire-resistance ratings* of precast concrete wall panels consisting of a layer of foam plastic insulation sandwiched between two wythes of concrete shall be permitted to be determined by use of Equation 7-4. Foam plastic insulation with a total thickness of less than 1 inch (25 mm) shall be disregarded. The R_n value for thickness of foam plastic insulation of 1 inch (25 mm) or greater, for use in the calculation, is 5 minutes; therefore $R_n^{0.59} = 2.5$.

❖ This value was for foam plastic insulation 1 inch (25 mm) thick and greater, determined from a fire test conducted on a panel that consisted of a 2-inch (51 mm) base slab of carbonate aggregate concrete, a 1-inch (25 mm) layer of cellular polystyrene insulation and a 2-inch (51 mm) face slab of carbonate aggregate concrete. The resulting fire endurance was 2 hours. From Equation 7-4, the contribution of the 1-inch (25 mm) layer of foam polystyrene was calculated to be 5 minutes.

Presumably, a comparable *R*-value for a 1-inch (25 mm) layer of foam polyurethane would be somewhat greater than for a 1-inch (25 mm) layer of foam polystyrene, but test values are not available. The above value for polystyrene is conservative for foam polyurethane.

Equation 7-4 can be rewritten as:

$$R^{0.59} = (R^{10.59} + R^{20.59} + \ldots R_n^{0.59})$$

This form of the equation is useful in making a quick determination of whether an assembly qualifies for a particular classification when used in conjunction with Tables 722.2.1.2(1) and 722.2.1.2(2).

EXAMPLE:

GIVEN: A sandwich wall panel consists of two $2^1/_2$-inch (63 mm) wythes of normal-weight concrete with a 2-inch (51 mm) layer of foam polystyrene between them.

FIND: Does the panel qualify for a 3-hour fire-resistance rating?

SOLUTION: From Table 722.2.1.2(1) in the code, the value of $R_n^{0.59}$ for a carbonate aggregate concrete is higher than for siliceous aggregate concrete, but because the type of concrete was not given, the value for siliceous should be used. From Section 722.2.1.2.2, the value of $R_n^{0.59}$ for the 2-inch (51 mm) layer of foam polystyrene is 2.5.

$R^{0.59} = 8.1 + 2.5 + 8.1 = 18.7$

From Table 722.2.1.2(2), a 3-hour rating is required to have an $R^{0.59}$ value of 21.41.

$18.7 < 21.41$, thus the wall does not qualify for a 3-hour rating.

722.2.1.3 Joints between precast wall panels. Joints between precast concrete wall panels that are not insulated as required by this section shall be considered as openings in walls. Uninsulated joints shall be included in determining the percentage of openings permitted by Table 705.8. Where openings are not permitted or are required by this code to be protected, the provisions of this section shall be used to determine the amount of joint insulation required. Insulated joints shall not be considered openings for purposes of determining

TABLE 722.2.1.2(1)
VALUES OF $R_n^{0.59}$ FOR USE IN EQUATION 7-4

TYPE OF MATERIAL	THICKNESS OF MATERIAL (inches)											
	$1^1/_2$	2	$2^1/_2$	3	$3^1/_2$	4	$4^1/_2$	5	$5^1/_2$	6	$6^1/_2$	7
Siliceous aggregate concrete	5.3	6.5	8.1	9.5	11.3	13.0	14.9	16.9	18.8	20.7	22.8	25.1
Carbonate aggregate concrete	5.5	7.1	8.9	10.4	12.0	14.0	16.2	18.1	20.3	21.9	24.7	27.2[c]
Sand-lightweight concrete	6.5	8.2	10.5	12.8	15.5	18.1	20.7	23.3	26.0[c]	Note c	Note c	Note c
Lightweight concrete	6.6	8.8	11.2	13.7	16.5	19.1	21.9	24.7	27.8[c]	Note c	Note c	Note c
Insulating concrete[a]	9.3	13.3	16.6	18.3	23.1	26.5[c]	Note c	Note c	Note c	Note c	Note c	Note c
Airspace[b]	—	—	—	—	—	—	—	—	—	—	—	—

For SI: 1 inch = 25.4 mm, 1 pound per cubic foot = 16.02 kg/m³.

a. Dry unit weight of 35 pcf or less and consisting of cellular, perlite or vermiculite concrete.

b. The $R_n^{0.59}$ value for one $^1/_2$" to $3^1/_2$" airspace is 3.3. The $R_n^{0.59}$ value for two $^1/_2$" to $3^1/_2$" airspaces is 6.7.

c. The *fire-resistance rating* for this thickness exceeds 4 hours.

TABLE 722.2.1.2(2)
FIRE-RESISTANCE RATINGS BASED ON $R^{0.59}$

R^a, MINUTES	$R^{0.59}$
60	11.20
120	16.85
180	21.41
240	25.37

compliance with the allowable percentage of openings in Table 705.8.

❖ See the commentary to Section 705.8.

722.2.1.3.1 Ceramic fiber joint protection. Figure 722.2.1.3.1 shows thicknesses of *ceramic fiber blankets* to be used to insulate joints between precast concrete wall panels for various panel thicknesses and for joint widths of $^3/_8$ inch (9.5 mm) and 1 inch (25 mm) for *fire-resistance ratings* of 1 hour to 4 hours. For joint widths between $^3/_8$ inch (9.5 mm) and 1 inch (25 mm), the thickness of *ceramic fiber blanket* is allowed to be determined by direct interpolation. Other tested and labeled materials are acceptable in place of *ceramic fiber blankets*.

❖ Code Figure 722.2.1.3.1 was derived from data in the report entitled "Fire Tests of Joints Between Precast Concrete Wall Units: Effect of Various Joint Treatments" in the *PCI Journal*.

EXAMPLE:

FIND: Determine the thickness of a ceramic fiber blanket needed for a 2-hour fire-resistance rating for joints between 5-inch-thick (127 mm) precast concrete wall panels made of siliceous aggregate concrete if the maximum joint width is $^7/_8$-inch (22.2 mm).

SOLUTION: Code Figure 722.2.1.3.1 indicates a minimum 0.7-inch (18 mm) thickness of ceramic fiber blanket for 5-inch (127 mm) panels for a 2-hour rating of a $^3/_4$-inch (19.5 mm) wide joint and 2.1 inches (53 mm) for a 1-inch (25 mm) wide joint. By interpolation, the thickness is computed as follows:

$t = 2.1 - (2.1 - 0.7)(1 - {}^7/_8) = 1.93$ inches

Therefore, the required thickness for a 2-hour rating is 1.93 inches (49 mm).

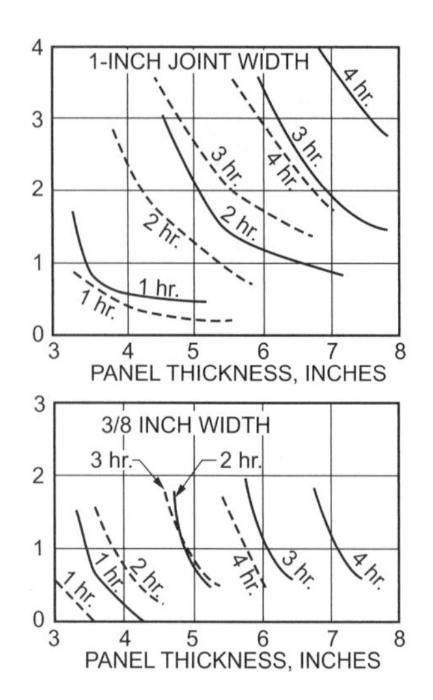

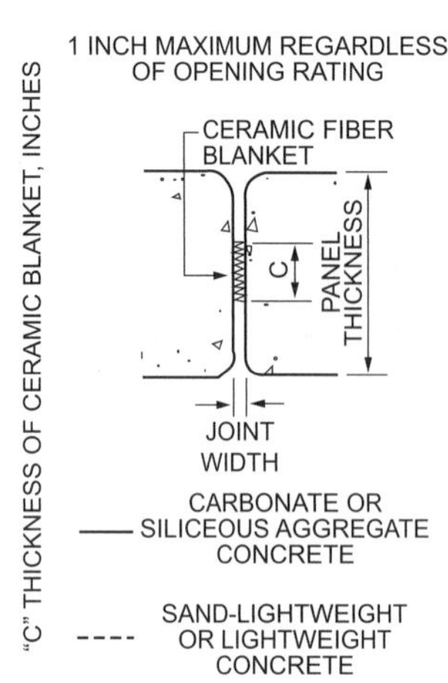

For SI: 1 inch = 25.4 mm.

FIGURE 722.2.1.3.1
CERAMIC FIBER JOINT PROTECTION

722.2.1.4 Walls with gypsum wallboard or plaster finishes. The *fire-resistance rating* of cast-in-place or precast concrete walls with finishes of gypsum wallboard or plaster applied to one or both sides shall be permitted to be calculated in accordance with the provisions of this section.

❖ The information contained in this section is based on fire endurance tests on unit masonry walls with gypsum wallboard and *The Supplement to the National Building Code of Canada 1980* (NRCC 17724) by the National Research Council of Canada.

For instance, a wall with a rating of 1 hour that has a 1-inch (25.4 mm) layer of portland cement-sand plaster applied on metal lath would add 30 minutes to the rating. Therefore, the fire-resistance rating would now be $1^1/_2$ hours.

722.2.1.4.1 Nonfire-exposed side. Where the finish of gypsum wallboard or plaster is applied to the side of the wall not exposed to fire, the contribution of the finish to the total *fire-resistance rating* shall be determined as follows: The thickness of the finish shall first be corrected by multiplying the actual thickness of the finish by the applicable factor determined from Table 722.2.1.4(1) based on the type of aggregate in the concrete. The corrected thickness of finish shall then be added to the actual or equivalent thickness of concrete and *fire-resistance rating* of the concrete and finish determined from Tables 722.2.1.1 and 722.2.1.2(1) and Figure 722.2.1.2.

❖ The fire resistance of concrete walls is generally determined by temperature rise on the unexposed surface (i.e., the heat transmission end point) (see commentary, Section 722.2.1.1). The time required to reach the heat transmission end point (fire-resistance rating) is primarily dependent upon the thickness of the concrete and the type of aggregate used to make the concrete. When additional finishes are applied to the unexposed side of the wall, the time required to reach the heat transmission end point is delayed and the fire-resistance rating of the wall is thus increased. The increase in the rating contributed by the finish can be determined by considering the finish as adding to the thickness of concrete. However, since the finish material and concrete may have different insulating properties, the actual thickness of the finish may need to be corrected to be compatible with the type of aggregate used in the concrete. The correction is made by multiplying the actual finish thickness by the factor determined from Table 722.2.1.4(1), and then adding the corrected thickness to the thickness of the concrete. This equivalent thickness is used to determine the fire-resistance rating from Table 722.2.1.1, Table 722.2.1.2(1) or Code Figure 722.21.2.

TABLE 722.2.1.4(1). See page 7-163.

❖ See the commentary to Section 722.2.1.4.1.

722.2.1.4.2 Fire-exposed side. Where gypsum wallboard or plaster is applied to the fire-exposed side of the wall, the contribution of the finish to the total *fire-resistance rating* shall be determined as follows: The time assigned to the finish as established by Table 722.2.1.4(2) shall be added to the *fire-*

resistance rating determined from Tables 722.2.1.1 and 722.2.1.2(1) and Figure 722.2.1.2 for the concrete alone, or to the rating determined in Section 722.2.1.4.1 for the concrete and finish on the nonfire-exposed side.

❖ Where finishes are added to the fire-exposed side of a concrete wall, their contribution to the total fire-resistance rating is based primarily upon the ability of the finish to remain in place, thus affording protection to the concrete wall. Table 722.2.1.4(2) lists the times that have been assigned to finishes on the fire-exposed side of the wall. These time-assigned values are based upon actual fire tests. The time-assigned values are added to the fire-resistance rating of the wall alone or to the rating determined for the wall and any finish on the unexposed side.

TABLE 722.2.1.4(2). See below.

❖ See the commentary to Section 722.2.1.4.2.

722.2.1.4.3 Nonsymmetrical assemblies. For a wall having no finish on one side or different types or thicknesses of finish on each side, the calculation procedures of Sections 722.2.1.4.1 and 722.2.1.4.2 shall be performed twice, assum-

TABLE 722.2.1.4(1)
MULTIPLYING FACTOR FOR FINISHES ON NONFIRE-EXPOSED SIDE OF WALL

TYPE OF FINISH APPLIED TO CONCRETE OR CONCRETE MASONRY WALL	TYPE OF AGGREGATE USED IN CONCRETE OR CONCRETE MASONRY			
	Concrete: siliceous or carbonate Concrete Masonry: siliceous or carbonate; solid clay brick	Concrete: sand-lightweight Concrete Masonry: clay tile; hollow clay brick; concrete masonry units of expanded shale and < 20% sand	Concrete: lightweight Concrete Masonry: concrete masonry units of expanded shale, expanded clay, expanded slag, or pumice < 20% sand	Concrete Masonry: concrete masonry units of expanded slag, expanded clay, or pumice
Portland cement-sand plaster	1.00	0.75[a]	0.75[a]	0.50[a]
Gypsum-sand plaster	1.25	1.00	1.00	1.00
Gypsum-vermiculite or perlite plaster	1.75	1.50	1.25	1.25
Gypsum wallboard	3.00	2.25	2.25	2.25

For SI: 1 inch = 25.4 mm.

a. For Portland cement-sand plaster $^5/_8$ inch or less in thickness and applied directly to the concrete or concrete masonry on the nonfire-exposed side of the wall, the multiplying factor shall be 1.00.

TABLE 722.2.1.4(2)
TIME ASSIGNED TO FINISH MATERIALS ON FIRE-EXPOSED SIDE OF WALL

FINISH DESCRIPTION	TIME (minutes)
Gypsum wallboard	
$^3/_8$ inch	10
$^1/_2$ inch	15
$^5/_8$ inch	20
2 layers of $^3/_8$ inch	25
1 layer of $^3/_8$ inch, 1 layer of $^1/_2$ inch	35
2 layers of $^1/_2$ inch	40
Type X gypsum wallboard	
$^1/_2$ inch	25
$^5/_8$ inch	40
Portland cement-sand plaster applied directly to concrete masonry	See Note a
Portland cement-sand plaster on metal lath	
$^3/_4$ inch	20
$^7/_8$ inch	25
1 inch	30
Gypsum sand plaster on $^3/_8$-inch gypsum lath	
$^1/_2$ inch	35
$^5/_8$ inch	40
$^3/_4$ inch	50
Gypsum sand plaster on metal lath	
$^3/_4$ inch	50
$^7/_8$ inch	60
1 inch	80

For SI: 1 inch = 25.4 mm.

a. The actual thickness of Portland cement-sand plaster, provided it is $^5/_8$ inch or less in thickness, shall be permitted to be included in determining the equivalent thickness of the masonry for use in Table 722.3.2.

ing either side of the wall to be the fire-exposed side. The *fire-resistance rating* of the wall shall not exceed the lower of the two values.

Exception: For an *exterior wall* with a *fire separation distance* greater than 5 feet (1524 mm) the fire shall be assumed to occur on the interior side only.

❖ Except for exterior walls having more than 5 feet (1524 mm) of horizontal separation, Section 705.5 requires that walls be rated for exposure to fire from both sides. Therefore, two calculations must be performed, on the assumption that each side of the wall is the fire-exposed side. Two calculations are not necessary for exterior walls with more than 5 feet (1524 mm) of horizontal separation or for other walls that are symmetrical (i.e., walls having the same type and thickness of finish on each side). The calculated fire-resistance rating of the wall is the lower of the two ratings determined, assuming that each side is the fire-exposed side.

722.2.1.4.4 Minimum concrete fire-resistance rating. Where finishes applied to one or both sides of a concrete wall contribute to the *fire-resistance rating*, the concrete alone shall provide not less than one-half of the total required *fire-resistance rating*. Additionally, the contribution to the *fire resistance* of the finish on the nonfire-exposed side of a *load-bearing wall* shall not exceed one-half the contribution of the concrete alone.

❖ Where gypsum wallboard or plaster finishes are applied to a concrete wall, the calculated fire-resistance rating for the concrete alone should not be less than one-half the required fire-resistance rating. This limitation is necessary to ensure that the concrete wall is of sufficient thickness to withstand fire exposure.

EXAMPLE:

GIVEN: An exterior bearing wall of a building of Type IB construction with 4 feet (1216 mm) of horizontal separation is required to have a 2-hour fire-resistance rating. The wall will be cast in place with siliceous aggregate concrete. The interior will be finished with a $^1/_2$-inch (12.7 mm) thickness of gypsum wallboard applied to steel furring members.

FIND: What is the minimum thickness of concrete required?

SOLUTION:

First calculation: Assume the interior to be the fire-exposed side.

1. From Table 722.2.1.4(2), the $^1/_2$-inch (12.7 mm) gypsum wallboard has a time-a.ssigned value of 15 minutes; therefore, the fire-resistance rating that must be developed by the concrete must not be less than $1^3/_4$ hours (2 hours - 15 minutes).
2. Since Table 722.2.1.1 does not include a minimum thickness requirement corresponding to $1^3/_4$ hours, direct interpolation between the values for $1^1/_2$ and 2 hours is acceptable. The interpolation results in a required thickness of 4.65 inches (118 mm) of concrete.

Second calculation: Assume the exterior to be the fire-exposed side.

1. From Table 722.2.1.4(1), the multiplying factor for gypsum wallboard and siliceous aggregate concrete is 1.25; therefore, the corrected thickness for $^1/_2$ inch (12.7 mm) of gypsum wallboard is 0.63 inch (16.02 mm) [1.25 inches (32 mm) × $^1/_2$ inch (12.7 mm)].
2. Table 722.2.1.1 requires 5 inches (127 mm) of siliceous aggregate concrete for a 2-hour fire-resistance rating. Therefore, the actual thickness of concrete required is 4.37 inches (111 mm) [5 inches (127 mm)–0.63 inches (16 mm)].
3. Since the thickness of concrete required when assuming the interior side to be the fire-exposed side is greater, the minimum concrete thickness required to achieve a 2-hour fire-resistance rating is 4.65 inches (118 mm).
4. Section 722.2.1.4.4 requires that the concrete alone provides no less than one-half the total required rating; thus, the concrete must provide at least a 1-hour rating. From Table 722.2.1.1, it can be seen that only 3.5 inches (89 mm) of siliceous aggregate concrete is required for 1 hour, whereas 4.65 inches (118 mm) will be provided.

❖ For a similar example problem, see the commentary to Section 722.3.2.4.

722.2.1.4.5 Concrete finishes. Finishes on concrete walls that are assumed to contribute to the total *fire-resistance rating* of the wall shall comply with the installation requirements of Section 722.3.2.5.

❖ See the commentary to Section 722.3.2.5.

722.2.2 Concrete floor and roof slabs. Reinforced and prestressed floors and roofs shall comply with Section 722.2.2.1. Multicourse floors and roofs shall comply with Sections 722.2.2.2 and 722.2.2.3, respectively.

❖ The fire test criteria for temperature rise of the unexposed surface and the ability to resist superimposed loads (heat transmission and structural criteria, respectively) must both be considered in determining the fire resistance of floors and roofs. Section 722.2.2 deals with heat transmission and Section 722.2.3 deals with structural criteria.

722.2.2.1 Reinforced and prestressed floors and roofs. The minimum thicknesses of reinforced and prestressed concrete floor or roof slabs for *fire-resistance ratings* of 1 hour to 4 hours are shown in Table 722.2.2.1.

> **Exception:** Minimum thickness shall not be required for floors and ramps within open and enclosed parking garages constructed in accordance with Sections 406.5 and 406.6, respectively.

❖ The criterion limiting the average temperature rise to 250°F (121°C) and the maximum rise at one point to 325°F (163°C) is often referred to as the "heat transmission end point." For solid concrete slabs, the heat transmission end point is primarily a function of slab thickness and aggregate type. Other factors that affect heat transmission to a lesser degree are moisture content of the concrete, aggregate size, mortar content and air content. Factors that have very little effect on heat transmission are cement content and strength; type; amount and location of reinforcement, provided these items are within the normal range of usage. The values in Table 722.2.2.1 apply to concrete slabs reinforced with bars or welded wire fabric, as well as to prestressed slabs.

The exception recognizes that Section 712.1.10.1 permits floor openings for automobile ramps in open and enclosed parking garages without shaft enclosures. Further, Exception 5 of Section 715.1 does not require fire-resistant joint systems for floors and ramps within open and enclosed parking garages or structures. Referenced standard ACI 216.1-07, *Standard Method for Determining Fire Resistance of Concrete and Masonry Construction Assemblies,* states that the purpose of the minimum thickness requirements (Section 722.2.2.1) is for "barrier fire resistance." It can be concluded from Sections 712.1.10.1 and 715.1 that there is no intent of creating a fire barrier between floors and ramps in open and enclosed parking garages. Therefore, there is no logic in requiring a minimum thickness for floors and ramps of open and enclosed parking garages.

TABLE 722.2.2.1
MINIMUM SLAB THICKNESS (inches)

CONCRETE TYPE	FIRE-RESISTANCE RATING (hours)				
	1	$1^1/_2$	2	3	4
Siliceous	3.5	4.3	5	6.2	7
Carbonate	3.2	4	4.6	5.7	6.6
Sand-lightweight	2.7	3.3	3.8	4.6	5.4
Lightweight	2.5	3.1	3.6	4.4	5.1

For SI: 1 inch = 25.4 mm.

❖ See the commentary to Section 722.2.2.1.

722.2.2.1.1 Hollow-core prestressed slabs. For hollow-core prestressed concrete slabs in which the cores are of constant cross section throughout the length, the equivalent thickness shall be permitted to be obtained by dividing the net cross-sectional area of the slab including grout in the joints, by its width.

❖ The method for determining equivalent thickness of masonry units was developed because the cores in masonry units taper. The method is, of course, also applicable to hollow-core precast concrete panels. However, because the cores in hollow-core panels do not taper, the equivalent thickness can be calculated by dividing the net cross-sectional area of the panel by its width.

722.2.2.1.2 Slabs with sloping soffits. The thickness of slabs with sloping soffits (see Figure 722.2.2.1.2) shall be determined at a distance $2t$ or 6 inches (152 mm), whichever is less, from the point of minimum thickness, where t is the minimum thickness.

❖ Some precast concrete wall panel sections (e.g., certain single-tee units) have tapered surfaces, so the thickness varies, as shown in Code Figure 722.2.2.1.2. In fire tests, it is customary to monitor the unexposed surface temperature at the location shown in the figure.

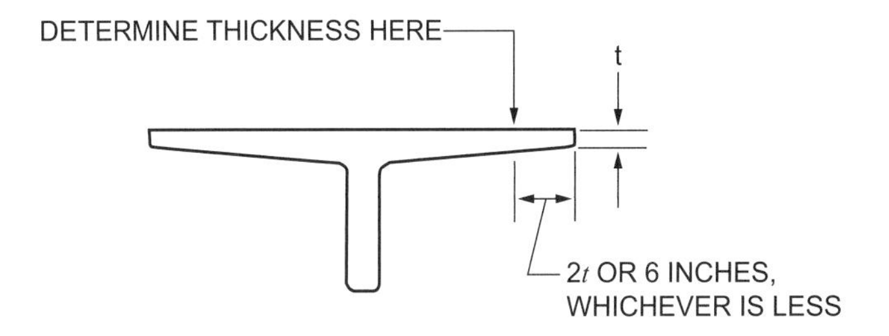

For SI: 1 inch = 25.4 mm.

FIGURE 722.2.2.1.2
DETERMINATION OF SLAB THICKNESS FOR SLOPING SOFFITS

722.2.2.1.3 Slabs with ribbed soffits. The thickness of slabs with ribbed or undulating soffits (see Figure 722.2.2.1.3) shall be determined by one of the following expressions, whichever is applicable:

For $s > 4t$, the thickness to be used shall be t

For $s \leq 2t$, the thickness to be used shall be t_e

For $4t > s > 2t$, the thickness to be used shall be

$$t + \left(\frac{4t}{s} - 1\right)(t_e - t) \qquad \textbf{(Equation 7-5)}$$

where:

s = Spacing of ribs or undulations.

t = Minimum thickness.

t_e = Equivalent thickness of the slab calculated as the net area of the slab divided by the width, in which the maximum thickness used in the calculation shall not exceed $2t$.

❖ The portion of a ribbed slab that can be used in calculating equivalent thickness, t_e, is shown in code Figure 722.2.2.1.3. Note that the procedure does not give credit for joists in one- or two-way joisted floors or in double tees. For such sections, the minimum thickness of the deck slab must be used.

EXAMPLE:

FIND: Determine the fire-resistance rating of the floor section shown in Code Figure 722.2.2.1.3 if the units were made of siliceous aggregate concrete.

SOLUTION:

s = 12 inches.

t = 4 inches.

Therefore:

$4t > s > 2t$ (16 inches > 12 inches > 8 inches)

$$t_e = \frac{(4'')(12'') + (5'')(1.6'') + \left(\frac{1}{2}''\right)(1'')(1.6'')}{12''}$$

t_e = 4.8 inches < $2t$.

Therefore the thickness to be used, t_s:

$$t_s = 4'' + \left(\frac{(4)(4'')}{12} - 1\right)(4.8'' - 4'')$$

t_s = 4.27 inches.

❖ From Table 722.2.2.1, using interpolation, the fire-resistance rating of the floor section shown is 1.48 hours or 1 hour, 29 minutes.

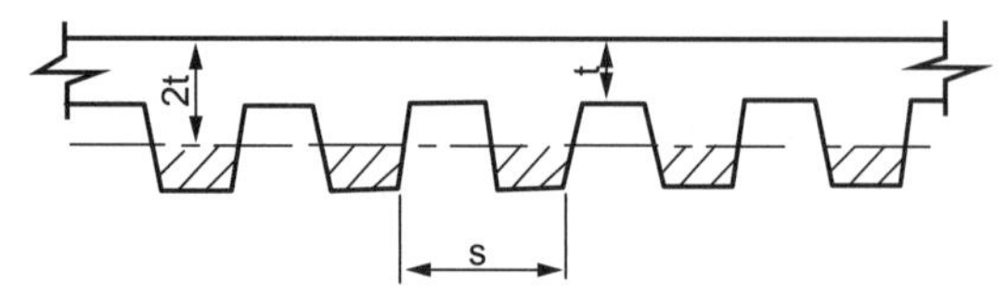

NEGLECT SHADED AREA IN CALCULATION OF EQUIVALENT THICKNESS

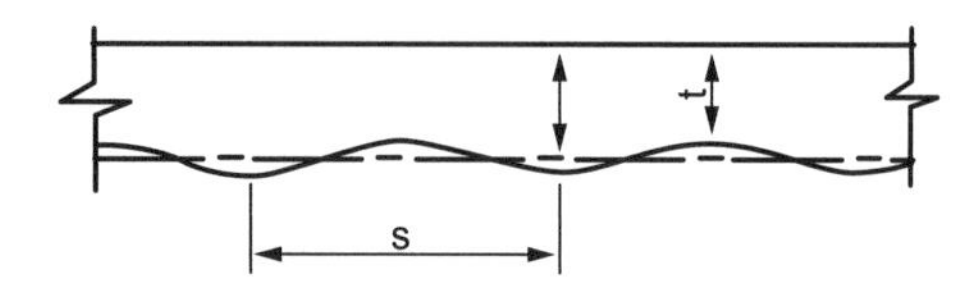

For SI: 1 inch = 25.4 mm.

FIGURE 722.2.2.1.3
SLABS WITH RIBBED OR UNDULATING SOFFITS

722.2.2.2 Multicourse floors. The *fire-resistance ratings* of floors that consist of a base slab of concrete with a topping (overlay) of a different type of concrete shall comply with Figure 722.2.2.2.

❖ The information in this section is based on the report entitled "Fire Endurance of Two-course Floors and Roofs" in the *ACI Journal*.

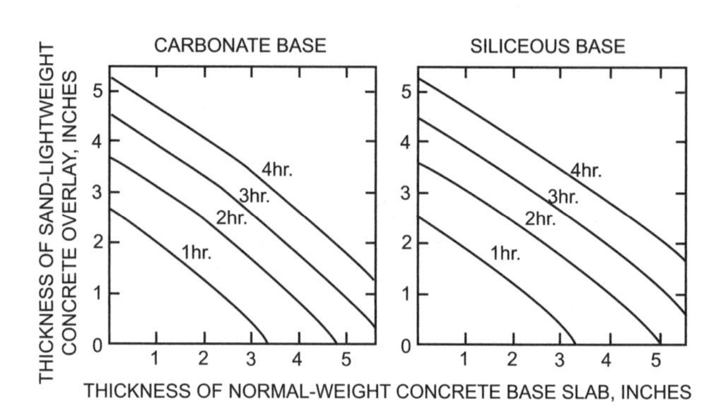

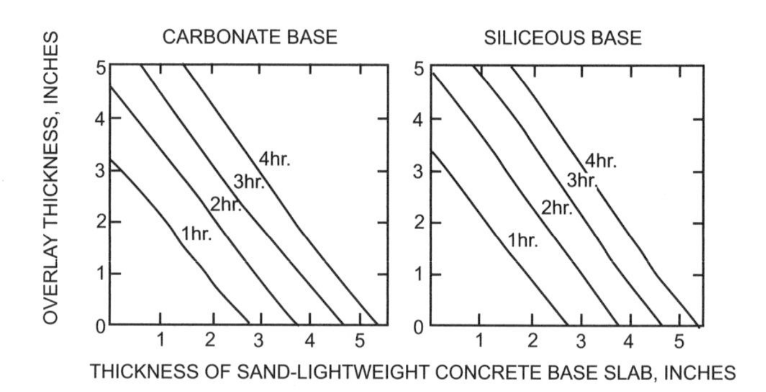

For SI: 1 inch = 25.4 mm.

FIGURE 722.2.2.2
FIRE-RESISTANCE RATINGS FOR TWO-COURSE CONCRETE FLOORS

722.2.2.3 Multicourse roofs. The *fire-resistance ratings* of roofs that consist of a base slab of concrete with a topping (overlay) of an insulating concrete or with an insulating board and built-up roofing shall comply with Figures 722.2.2.3(1) and 722.2.2.3(2).

❖ The information in this section is based on the report entitled "Fire Endurance of Two-course Floors and Roofs" in the *ACI Journal*.

722.2.2.3.1 Heat transfer. For the transfer of heat, three-ply built-up roofing contributes 10 minutes to the *fire-resistance rating*. The *fire-resistance rating* for concrete assemblies such as those shown in Figure 722.2.2.3(1) shall be increased by 10 minutes. This increase is not applicable to those shown in Figure 722.2.2.3(2).

❖ See the commentary to Section 722.2.2.3.

722.2.2.4 Joints in precast slabs. Joints between adjacent precast concrete slabs need not be considered in calculating the slab thickness provided that a concrete topping not less than 1 inch (25 mm) thick is used. Where no concrete topping is used, joints must be grouted to a depth of not less than one-third the slab thickness at the joint, but not less than 1 inch (25 mm), or the joints must be made fire resistant by other *approved* methods.

❖ Based on data developed by UL, where a concrete topping is not used over precast concrete floors, joints must be grouted. If a concrete topping at least 1 inch thick (25 mm) is used, the joints need not be grouted.

722.2.3 Concrete cover over reinforcement. The minimum thickness of concrete cover over reinforcement in concrete slabs, reinforced beams and prestressed beams shall comply with this section.

❖ See the commentary to Sections 722.2.3.1, 722.2.3.2 and 722.2.3.3.

722.2.3.1 Slab cover. The minimum thickness of concrete cover to the positive moment reinforcement shall comply with Table 722.2.3(1) for reinforced concrete and Table 722.2.3(2) for prestressed concrete. These tables are applicable for solid or hollow-core one-way or two-way slabs with flat undersurfaces. These tables are applicable to slabs that are either cast in place or precast. For precast prestressed concrete not covered elsewhere, the procedures contained in PCI MNL 124 shall be acceptable.

❖ The temperature of the tensile reinforcement depends on the thickness of cover and aggregate type. For unrestrained slabs, the values shown in Tables 722.2.3(1) and 722.2.3(2) are the cover thickness needed to keep the tensile reinforcement below 1,100°F (593°C) and 800°F (427°C), respectively. For restrained slabs, the temperature of the tensile reinforcement is not as critical, and thus, a minimum cover of $^{3}/_{4}$ inch (19.5 mm) is specified in PCA Bulletin 223, "Fire Endurance of Concrete Slabs as Influenced by Thickness, Aggregate Type and Moisture."

TABLE 722.2.3(1). See page 7-168.

❖ See the commentary to Section 722.2.3.1.

TABLE 722.2.3(2). See page 7-168.

❖ See the commentary to Section 722.2.3.1.

TABLE 722.2.3(3). See page 7-168.

❖ See the commentary to Section 722.2.3.2.

TABLE 722.2.3(4). See page 7-168.

❖ See the commentary to Section 722.2.3.2.

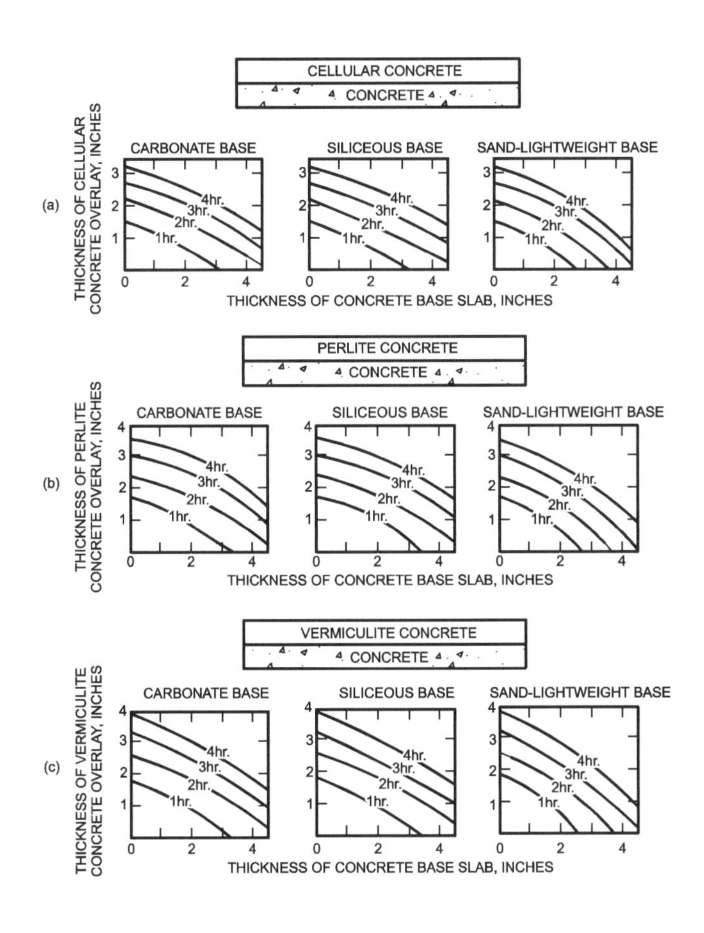

For SI: 1 inch = 25.4 mm.

**FIGURE 722.2.2.3(1)
FIRE-RESISTANCE RATINGS FOR
CONCRETE ROOF ASSEMBLIES**

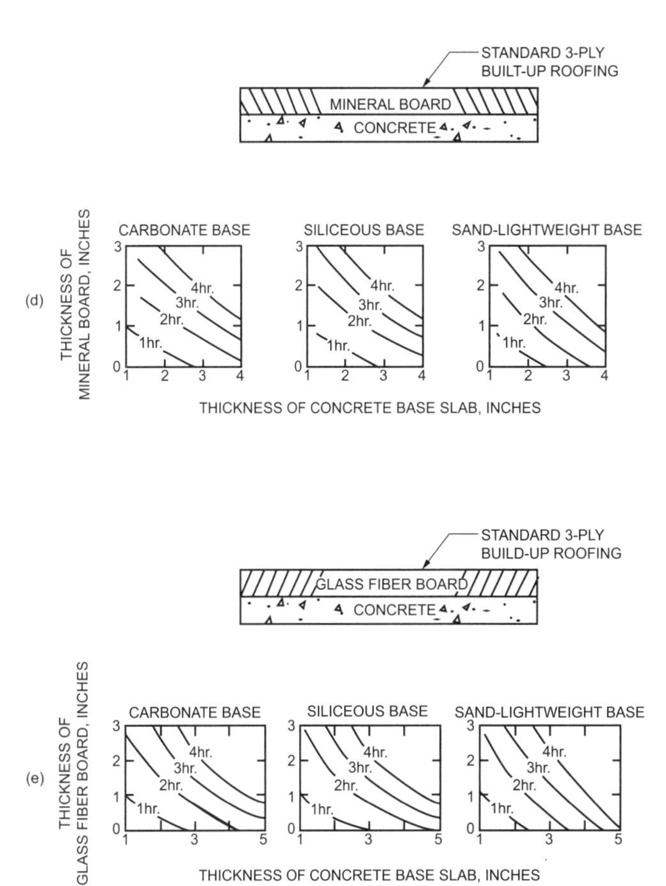

For SI: 1 inch = 25.4 mm.

**FIGURE 722.2.2.3(2)
FIRE-RESISTANCE RATINGS
FOR CONCRETE ROOF ASSEMBLIES**

TABLE 722.2.3(1)
COVER THICKNESS FOR REINFORCED CONCRETE FLOOR OR ROOF SLABS (inches)

CONCRETE AGGREGATE TYPE	FIRE-RESISTANCE RATING (hours)									
	Restrained					Unrestrained				
	1	1 1/2	2	3	4	1	1 1/2	2	3	4
Siliceous	3/4	3/4	3/4	3/4	3/4	3/4	3/4	1	1 1/4	1 5/8
Carbonate	3/4	3/4	3/4	3/4	3/4	3/4	3/4	3/4	1 1/4	1 1/4
Sand-lightweight or lightweight	3/4	3/4	3/4	3/4	3/4	3/4	3/4	3/4	1 1/4	1 1/4

For SI: 1 inch = 25.4 mm.

TABLE 722.2.3(2)
COVER THICKNESS FOR PRESTRESSED CONCRETE FLOOR OR ROOF SLABS (inches)

CONCRETE AGGREGATE TYPE	FIRE-RESISTANCE RATING (hours)									
	Restrained					Unrestrained				
	1	1 1/2	2	3	4	1	1 1/2	2	3	4
Siliceous	3/4	3/4	3/4	3/4	3/4	1 1/8	1 1/2	1 3/4	2 3/8	2 3/4
Carbonate	3/4	3/4	3/4	3/4	3/4	1	1 3/8	1 5/8	2 1/8	2 1/4
Sand-lightweight or lightweight	3/4	3/4	3/4	3/4	3/4	1	1 3/8	1 1/2	2	2 1/4

For SI: 1 inch = 25.4 mm.

TABLE 722.2.3(3)
MINIMUM COVER FOR MAIN REINFORCING BARS OF REINFORCED CONCRETE BEAMS[c]
(APPLICABLE TO ALL TYPES OF STRUCTURAL CONCRETE)

RESTRAINED OR UNRESTRAINED[a]	BEAM WIDTH[b] (inches)	FIRE-RESISTANCE RATING (hours)				
		1	1 1/2	2	3	4
Restrained	5	3/4	3/4	3/4	1[a]	1 1/4[a]
	7	3/4	3/4	3/4	3/4	3/4
	≥ 10	3/4	3/4	3/4	3/4	3/4
Unrestrained	5	3/4	1	1 1/4	—	—
	7	3/4	3/4	3/4	1 3/4	3
	≥ 10	3/4	3/4	3/4	1	1 3/4

For SI: 1 inch = 25.4 mm, 1 foot = 304.8 mm.

a. Tabulated values for restrained assemblies apply to beams spaced more than 4 feet on center. For restrained beams spaced 4 feet or less on center, minimum cover of 3/4 inch is adequate for ratings of 4 hours or less.

b. For beam widths between the tabulated values, the minimum cover thickness can be determined by direct interpolation.

c. The cover for an individual reinforcing bar is the minimum thickness of concrete between the surface of the bar and the fire-exposed surface of the beam. For beams in which several bars are used, the cover for corner bars used in the calculation shall be reduced to one-half of the actual value. The cover for an individual bar must be not less than one-half of the value given in Table 722.2.3(3) nor less than 3/4 inch.

TABLE 722.2.3(4)
MINIMUM COVER FOR PRESTRESSED CONCRETE BEAMS 8 INCHES OR GREATER IN WIDTH[b]

RESTRAINED OR UNRESTRAINED[a]	CONCRETE AGGREGATE TYPE	BEAM WIDTH (inches)	FIRE-RESISTANCE RATING (hours)				
			1	1 1/2	2	3	4
Restrained	Carbonate or siliceous	8	1 1/2	1 1/2	1 1/2	1 3/4[a]	2 1/2[a]
	Carbonate or siliceous	≥ 12	1 1/2	1 1/2	1 1/2	1 1/2	1 7/8[a]
	Sand lightweight	8	1 1/2	1 1/2	1 1/2	1 1/2	2[a]
	Sand lightweight	≥ 12	1 1/2	1 1/2	1 1/2	1 1/2	1 5/8[a]
Unrestrained	Carbonate or siliceous	8	1 1/2	1 3/4	2 1/2	5[c]	—
	Carbonate or siliceous	≥ 12	1 1/2	1 1/2	1 7/8[a]	2 1/2	3
	Sand lightweight	8	1 1/2	1 1/2	2	3 1/4	—
	Sand lightweight	≥ 12	1 1/2	1 1/2	1 5/8	2	2 1/2

For SI: 1 inch = 25.4 mm, 1 foot = 304.8 mm.

a. Tabulated values for restrained assemblies apply to beams spaced more than 4 feet on center. For restrained beams spaced 4 feet or less on center, minimum cover of 3/4 inch is adequate for 4-hour ratings or less.

b. For beam widths between 8 inches and 12 inches, minimum cover thickness can be determined by direct interpolation.

c. Not practical for 8-inch-wide beam but shown for purposes of interpolation.

TABLE 722.2.3(5). See below.

❖ See the commentary to Section 722.2.3.2.

722.2.3.2 Reinforced beam cover. The minimum thickness of concrete cover to the positive moment reinforcement (bottom steel) for reinforced concrete beams is shown in Table 722.2.3(3) for *fire-resistance ratings* of 1 hour to 4 hours.

❖ For reinforced concrete beams, the critical steel temperature is 1,100°F (593°C), so the effect of aggregate type is minimal. For prestressed concrete, the comparable temperature is 800°F (427°C), so aggregate type must be considered. The data in Tables 722.2.3(3) and 722.2.3(4) were derived from the fire tests of a series of beam specimens that ranged in width between 2 and 24 inches (51 and 607 mm). Other variables in the series included aggregate type and amount of reinforcement. Tests were conducted at the PCA. The results of the fire tests of beams conducted at UL were also analyzed. Charts showing the resulting data are shown in Figure A4 of PCI Manual 124, *Design for Fire Resistance of Precast Prestressed Concrete*.

722.2.3.3 Prestressed beam cover. The minimum thickness of concrete cover to the positive moment prestressing tendons (bottom steel) for restrained and unrestrained prestressed concrete beams and stemmed units shall comply with the values shown in Tables 722.2.3(4) and 722.2.3(5) for *fire-resistance ratings* of 1 hour to 4 hours. Values in Table 722.2.3(4) apply to beams 8 inches (203 mm) or greater in width. Values in Table 722.2.3(5) apply to beams or stems of any width, provided the cross-section area is not less than 40 square inches (25 806 mm^2). In case of differences between the values determined from Table 722.2.3(4) or 722.2.3(5), it is permitted to use the smaller value. The concrete cover shall be calculated in accordance with Section 722.2.3.3.1. The minimum concrete cover for nonprestressed reinforcement in prestressed concrete beams shall comply with Section 722.2.3.2.

❖ See the commentary to Section 721.2.3.2.

722.2.3.3.1 Calculating concrete cover. The concrete cover for an individual tendon is the minimum thickness of concrete between the surface of the tendon and the fire-exposed surface of the beam, except that for ungrouted ducts, the assumed cover thickness is the minimum thickness of concrete between the surface of the duct and the fire-exposed surface of the beam. For beams in which two or more tendons are used, the cover is assumed to be the average of the minimum cover of the individual tendons. For corner tendons (tendons equal distance from the bottom and side), the minimum cover used in the calculation shall be one-half the actual value. For stemmed members with two or more prestressing tendons located along the vertical centerline of the stem, the average cover shall be the distance from the bottom of the member to the centroid of the tendons. The actual cover for any individual tendon shall be not less than one-half the smaller value shown in Tables 722.2.3(4) and 722.2.3(5), or 1 inch (25 mm), whichever is greater.

722.2.4 Concrete columns. Concrete columns shall comply with this section.

❖ Most building code provisions for reinforced concrete columns are based on two reports: "Fire Tests of Building Columns" by the Associated Factory Mutual Insurance Companies and "Fire Resistance of Concrete Columns" by the NBS. Sizes of the tested columns were 12 inches (305 mm), 16 inches (406 mm) and 18 inches (457 mm). Nearly all of the columns withstood 4-hour fire tests that were essentially conducted in accordance with ASTM E119 or UL 263. Most of the tests were stopped after 4 hours, but some were continued to 8 hours. The shortest duration was 3 hours. At the time of the tests few, if any, concrete columns were smaller than 12 inches (305 mm), but smaller columns have been in use for many years since then.

Fire tests conducted in Europe on smaller columns indicate that fire endurance is greater for larger columns.

TABLE 722.2.3(5)
MINIMUM COVER FOR PRESTRESSED CONCRETE BEAMS OF ALL WIDTHS

RESTRAINED OR UNRESTRAINED[a]	CONCRETE AGGREGATE TYPE	BEAM AREA[b] A (square inches)	FIRE-RESISTANCE RATING (hours)				
			1	$1^1/_2$	2	3	4
Restrained	All	$40 \le A \le 150$	$1^1/_2$	$1^1/_2$	2	$2^1/_2$	—
	Carbonate or siliceous	$150 < A \le 300$	$1^1/_2$	$1^1/_2$	$1^1/_2$	$1^3/_4$	$2^1/_2$
		$300 < A$	$1^1/_2$	$1^1/_2$	$1^1/_2$	$1^1/_2$	2
	Sand lightweight	$150 < A$	$1^1/_2$	$1^1/_2$	$1^1/_2$	$1^1/_2$	2
Unrestrained	All	$40 \le A \le 150$	2	$2^1/_2$	—	—	—
	Carbonate or siliceous	$150 < A \le 300$	$1^1/_2$	$1^3/_4$	$2^1/_2$	—	—
		$300 < A$	$1^1/_2$	$1^1/_2$	2	3[c]	4[c]
	Sand lightweight	$150 < A$	$1^1/_2$	$1^1/_2$	2	3[c]	4[c]

For SI: 1 inch = 25.4 mm, 1 foot = 304.8 mm.

a. Tabulated values for restrained assemblies apply to beams spaced more than 4 feet on center. For restrained beams spaced 4 feet or less on center, minimum cover of $^3/_4$ inch is adequate for 4-hour ratings or less.

b. The cross-sectional area of a stem is permitted to include a portion of the area in the flange, provided the width of the flange used in the calculation does not exceed three times the average width of the stem.

c. U-shaped or hooped stirrups spaced not to exceed the depth of the member and having a minimum cover of 1 inch shall be provided.

722.2.4.1 Minimum size. The minimum overall dimensions of reinforced concrete columns for *fire-resistance ratings* of 1 hour to 4 hours for exposure to fire on all sides shall comply with this section.

❖ Column requirements are based on provisions in ACI 216.1-07/TMS 0216.1-07, *Code Requirements for Determining Fire Resistance of Concrete and Masonry Construction Assemblies*, the successor to ACI 216.1-97/TMS 0216.1-97.

722.2.4.1.1 Concrete strength less than or equal to 12,000 psi. For columns made with concrete having a specified compressive strength, f'_c, of less than or equal to 12,000 psi (82.7 MPa), the minimum dimension shall comply with Table 722.2.4.

❖ For concrete with compressive strength less than 12,000 psi (82.7 MPa), the minimum dimensions necessary to achieve certain fire-resistance ratings are given by Table 722.2.4. Concrete compressive strengths used in building construction are normally in the range of 4,000 to 6,000 psi (27.6 to 41.4 MPa) so this section and Table 722.2.4 are operative most of the time in determining fire-resistance ratings for concrete.

TABLE 722.2.4
MINIMUM DIMENSION OF CONCRETE COLUMNS (inches)

TYPES OF CONCRETE	FIRE-RESISTANCE RATING (hours)				
	1	$1^1/_2$	2[a]	3[a]	4[b]
Siliceous	8	9	10	12	14
Carbonate	8	9	10	11	12
Sand-lightweight	8	$8^1/_2$	9	$10^1/_2$	12

For SI: 1 inch = 25 mm.

a. The minimum dimension is permitted to be reduced to 8 inches for rectangular columns with two parallel sides not less than 36 inches in length.

b. The minimum dimension is permitted to be reduced to 10 inches for rectangular columns with two parallel sides not less than 36 inches in length.

❖ Table 722.2.4 reflects the information cited in the reports listed in the commentary to Section 722.2.4 and in the report "Investigations on Building Fires, Part VI: Fire Resistance of Reinforced Concrete Columns" in the National Building Studies Research Paper No. 18. This table is for columns that may be exposed to fire on all sides. Notes a and b to Table 722.2.4 permit a smaller minimum dimension for columns than required in the table, provided the longer dimension of the column is at least 36 inches (914 mm).

722.2.4.1.2 Concrete strength greater than 12,000 psi. For columns made with concrete having a specified compressive strength, f'_c, greater than 12,000 psi (82.7 MPa), for *fire-resistance ratings* of 1 hour to 4 hours the minimum dimension shall be 24 inches (610 mm).

❖ Concrete with a compressive strength greater than 12,000 psi (82.7 MPa) is very rarely used in building construction. The concrete is very stiff, with a very low water-to-cement ratio, and therefore is very difficult to work with. Because it contains very small air pockets and voids, the material conducts heat relatively quickly when compared to more common concrete with compressive strengths in the 4,000 to 6,000 psi (27.6 to 41.4 MPa) range. Therefore, the minimum thickness necessary to achieve a fire-resistance rating of 1 to 4 hours is considerably higher.

722.2.4.2 Minimum cover for R/C columns. The minimum thickness of concrete cover to the main longitudinal reinforcement in columns, regardless of the type of aggregate used in the concrete and the specified compressive strength of concrete, f'_c, shall be not less than 1 inch (25 mm) times the number of hours of required *fire resistance* or 2 inches (51 mm), whichever is less.

❖ The minimum concrete cover refers to the main longitudinal reinforcement, not ties or spirals.

722.2.4.3 Tie and spiral reinforcement. For concrete columns made with concrete having a specified compressive strength, f'_c, greater than 12,000 psi (82.7 MPa), tie and spiral reinforcement shall comply with the following:

1. The free ends of rectangular ties shall terminate with a 135-degree (2.4 rad) standard tie hook.
2. The free ends of circular ties shall terminate with a 90-degree (1.6 rad) standard tie hook.
3. The free ends of spirals, including at lap splices, shall terminate with a 90-degree (1.6 rad) standard tie hook.

The hook extension at the free end of ties and spirals shall be the larger of six bar diameters and the extension required by Section 7.1.3 of ACI 318. Hooks shall project into the core of the column.

❖ The intent of the provisions in this section is to prevent ties or spirals from disengaging from the longitudinal reinforcement should the concrete cover over the ties or spirals be lost during a fire. The section expands on the provisions found in ACI 216.1/TMS 0216.1 by addressing spiral reinforcement, which is typically used for lateral reinforcement in round columns.

722.2.4.4 Columns built into walls. The minimum dimensions of Table 722.2.4 do not apply to a reinforced concrete column that is built into a concrete or masonry wall provided all of the following are met:

1. The *fire-resistance rating* for the wall is equal to or greater than the required rating of the column;
2. The main longitudinal reinforcing in the column has cover not less than that required by Section 722.2.4.2; and
3. Openings in the wall are protected in accordance with Table 716.5.

Where openings in the wall are not protected as required by Section 716.5, the minimum dimension of columns required to have a *fire-resistance rating* of 3 hours or less shall be 8 inches (203 mm), and 10 inches (254 mm) for columns required to have a *fire-resistance rating* of 4 hours, regardless of the type of aggregate used in the concrete.

❖ Table 722.2.4 does not apply to columns built into concrete or masonry walls that meet all three of the conditions in this section.

722.2.4.5 Precast cover units for steel columns. See Section 722.5.1.4.

❖ See the commentary to Section 722.5.1.4.

722.3 Concrete masonry. The provisions of this section contain procedures by which the *fire-resistance ratings* of concrete masonry are established by calculations.

❖ See the commentary to Section 722.

722.3.1 Equivalent thickness. The equivalent thickness of concrete masonry construction shall be determined in accordance with the provisions of this section.

❖ See the commentary to Section 722.

722.3.1.1 Concrete masonry unit plus finishes. The equivalent thickness of concrete masonry assemblies, T_{ea}, shall be computed as the sum of the equivalent thickness of the concrete masonry unit, T_e, as determined by Section 722.3.1.2, 722.3.1.3 or 722.3.1.4, plus the equivalent thickness of finishes, T_{ef}, determined in accordance with Section 722.3.2:

$$T_{ea} = T_e + T_{ef} \qquad \textbf{(Equation 7-6)}$$

❖ See the commentary to Section 722.

722.3.1.2 Ungrouted or partially grouted construction. T_e shall be the value obtained for the concrete masonry unit determined in accordance with ASTM C140.

❖ See the commentary to Section 722.

722.3.1.3 Solid grouted construction. The equivalent thickness, T_e, of solid grouted concrete masonry units is the actual thickness of the unit.

❖ See the commentary to Section 722.

722.3.1.4 Airspaces and cells filled with loose-fill material. The equivalent thickness of completely filled hollow concrete masonry is the actual thickness of the unit where loose-fill materials are: sand, pea gravel, crushed stone, or slag that meet ASTM C33 requirements; pumice, scoria, expanded shale, expanded clay, expanded slate, expanded slag, expanded fly ash, or cinders that comply with ASTM C331; or perlite or vermiculite meeting the requirements of ASTM C549 and ASTM C516, respectively.

❖ See the commentary to Section 722.

722.3.2 Concrete masonry walls. The *fire-resistance rating* of walls and partitions constructed of concrete masonry units shall be determined from Table 722.3.2. The rating shall be based on the equivalent thickness of the masonry and type of aggregate used.

❖ It has been accepted practice to determine the fire-resistance rating of concrete masonry walls based on the type of aggregate used to manufacture them and the equivalent thickness of solid material in the wall. Equivalent thicknesses shown in Table 722.3.2 have been developed and refined through actual fire testing.

To determine the minimum equivalent thickness, Section 722.3.1.2 references ASTM C140.

EXAMPLE:

FIND: Determine the equivalent thickness of a standard 8-inch by 8-inch by 16-inch (203 mm by 203 mm by 406 mm) concrete masonry unit. The unit is normal weight with sand and gravel aggregate.

SOLUTION: The equivalent thickness is actually the average thickness of the solid material in the unit. The equivalent thickness is determined by the following:

$T_e = 1728\ A/(L \times H)$

where:

T_e = Equivalent thickness, in.

A = Net volume of unit, ft^3.

L = Length of unit, in.

H = Height of unit, in.

Net volume, A, is determined by the following equation:

$A = C/D$

where:

C = Dry weight of unit, pounds.

D = Density of unit, pounds per cubic foot.

From data furnished by the manufacturer, the dry unit weight is 44 pounds (20 kg) and the density is 135 pounds per cubic foot (pcf) (2115 kg/m^2).

$$T_e = \frac{(1,728 \text{ in.}^3/\text{ft.}^3)(0.326 \text{ ft.}^3)}{(15.625 \text{ in.})(7.625 \text{ in.})}$$

T_e = 4.73 inches (120 mm).

The fire-resistance rating for this type of unit with siliceous gravel aggregate can now be determined from Table 722.3.2. In this case, the fire-resistance rating will fall between 2.25 and 2.5 hours.

TABLE 722.3.2. See page 7-172.

❖ See the commentary to Section 722.3.2.

722.3.2.1 Finish on nonfire-exposed side. Where plaster or gypsum wallboard is applied to the side of the wall not exposed to fire, the contribution of the finish to the total *fire-resistance rating* shall be determined as follows: The thickness of gypsum wallboard or plaster shall be corrected by multiplying the actual thickness of the finish by applicable factor determined from Table 722.2.1.4(1). This corrected thickness of finish shall be added to the equivalent thickness of masonry and the *fire-resistance rating* of the masonry and finish determined from Table 722.3.2.

❖ The information contained in this section is based on *Fire Endurance Tests on Unit Masonry Walls with Gypsum Wallboard* and *The Supplement to the*

TABLE 722.3.2
MINIMUM EQUIVALENT THICKNESS (inches) OF BEARING OR NONBEARING CONCRETE MASONRY WALLS[a,b,c,d]

TYPE OF AGGREGATE	FIRE-RESISTANCE RATING (hours)														
	1/2	3/4	1	1 1/4	1 1/2	1 3/4	2	2 1/4	2 1/2	2 3/4	3	3 1/4	3 1/2	3 3/4	4
Pumice or expanded slag	1.5	1.9	2.1	2.5	2.7	3.0	3.2	3.4	3.6	3.8	4.0	4.2	4.4	4.5	4.7
Expanded shale, clay or slate	1.8	2.2	2.6	2.9	3.3	3.4	3.6	3.8	4.0	4.2	4.4	4.6	4.8	4.9	5.1
Limestone, cinders or unexpanded slag	1.9	2.3	2.7	3.1	3.4	3.7	4.0	4.3	4.5	4.8	5.0	5.2	5.5	5.7	5.9
Calcareous or siliceous gravel	2.0	2.4	2.8	3.2	3.6	3.9	4.2	4.5	4.8	5.0	5.3	5.5	5.8	6.0	6.2

For SI: 1 inch = 25.4 mm.

a. Values between those shown in the table can be determined by direct interpolation.

b. Where combustible members are framed into the wall, the thickness of solid material between the end of each member and the opposite face of the wall, or between members set in from opposite sides, shall be not less than 93 percent of the thickness shown in the table.

c. Requirements of ASTM C55, ASTM C73, ASTM C90 or ASTM C744 shall apply.

d. Minimum required equivalent thickness corresponding to the hourly *fire-resistance rating* for units with a combination of aggregate shall be determined by linear interpolation based on the percent by volume of each aggregate used in manufacture.

National Building Code of Canada 1980 by the National Research Council of Canada.

The fire resistance of concrete masonry walls is generally determined by temperature rise on the unexposed surface (i.e., the "heat transmission" end point). The time required to reach the heat transmission end point (fire-resistance rating) is primarily dependent upon the equivalent thickness of the masonry and the type of aggregate used to make the concrete. When additional finishes are applied to the unexposed side of the wall, the time required to reach the heat transmission end point is delayed; thus, the fire-resistance rating of the wall is increased. The increase in the rating contributed by the finish can be determined by considering the finish as an addition to the equivalent thickness of masonry. However, since the finish material and masonry may have different insulating properties, the actual thickness of the finish must be multiplied by a correction factor to be compatible with the type of aggregate used in the concrete. The correction is made by multiplying the actual finish thickness by the factor determined from Table 722.2.1.4(1), then adding the modified thickness to the equivalent thickness of masonry. This combined equivalent thickness is used to determine the fire-resistance rating from Table 722.3.2.

722.3.2.2 Finish on fire-exposed side. Where plaster or gypsum wallboard is applied to the fire-exposed side of the wall, the contribution of the finish to the total *fire-resistance rating* shall be determined as follows: The time assigned to the finish as established by Table 722.2.1.4(2) shall be added to the *fire-resistance rating* determined in Section 722.3.2 for the masonry alone, or in Section 722.3.2.1 for the masonry and finish on the nonfire-exposed side.

❖ Where finishes are added to the fire-exposed side of a concrete masonry wall, their contribution to the total fire-resistance rating is based primarily upon their ability to remain in place, thus affording protection to the masonry wall. Table 722.2.1.4(2) lists the times that have been assigned to finishes on the fire-exposed side of the wall. These time-assigned values are based upon actual fire tests. The time-assigned values are added to the fire-resistance rating of the wall alone or to the rating determined for the wall and any finish on the unexposed side.

722.3.2.3 Nonsymmetrical assemblies. For a wall having no finish on one side or having different types or thicknesses of finish on each side, the calculation procedures of this section shall be performed twice, assuming either side of the wall to be the fire-exposed side. The *fire-resistance rating* of the wall shall not exceed the lower of the two values calculated.

Exception: For *exterior walls* with a *fire separation distance* greater than 5 feet (1524 mm), the fire shall be assumed to occur on the interior side only.

❖ Except for exterior walls having more than 5 feet (1524 mm) of FSD, Section 704.5 requires that walls be rated for exposure to fire from both sides. Therefore, two calculations must be performed, on the assumption that each side is the fire-exposed side. Two calculations are not necessary for exterior walls with an FSD of more than 5 feet (1524 mm) or for other walls that are symmetrical (i.e., walls having the same type and thickness of finish on each side). The calculated fire-resistance rating must not exceed the lower of the two ratings determined, assuming that each side is the fire-exposed side.

722.3.2.4 Minimum concrete masonry fire-resistance rating. Where the finish applied to a concrete masonry wall contributes to its *fire-resistance rating*, the masonry alone shall provide not less than one-half the total required *fire-resistance rating*.

❖ Where gypsum wallboard or plaster finishes are applied to a concrete masonry wall, the calculated fire-resistance rating for the concrete alone should be not less than one-half the required fire-resistance rating. This limitation is necessary to ensure that the masonry wall is of sufficient thickness to withstand fire exposure.

EXAMPLE:

GIVEN: A wall required to have a 4-hour fire-resistance rating will be constructed with concrete masonry units of expanded shale aggregate. The wall will be finished on each side with a layer of $^1/_2$-inch (12.7 mm) gypsum wallboard.

FIND: What is the minimum equivalent thickness of concrete masonry required?

SOLUTION: Since the wall has the same type and thickness of finish on each side, only one calculation is required.

1. The $^1/_2$-inch (12.7 mm) gypsum wallboard on the fire-exposed side has a time-assigned value of 15 minutes in accordance with Table 722.2.1.4(2).
2. Therefore, the fire resistance required to be provided by the masonry and gypsum wallboard on the unexposed side is 3 hours and 45 minutes (4 hours - 15 minutes).
3. From Table 722.2.1.4(1), the corrected thickness of gypsum wallboard on the unexposed side is $^1/_2$ inch (12.7 mm) (1.00 × $^1/_2$ inch).
4. From Table 722.3.2, the minimum equivalent thickness of masonry, including the corrected thickness of gypsum wallboard, required for a rating of 3 hours and 45 minutes is 4.9 inches (124 mm).
5. Therefore, the equivalent thickness of masonry required is 4.4 inches (112 mm) [4.9 inches (124 mm) - $^1/_2$ inch (12.7 mm)].
6. From Table 722.3.2, it can be determined that 4.4 inches (112 mm) of expanded shale aggregate concrete masonry will provide a fire resistance of 3 hours. Therefore, the requirement that the masonry alone provide at least one-half of the total required rating is satisfied.

For a similar example problem, see the commentary to Section 722.2.1.4.4.

722.3.2.5 Attachment of finishes. Installation of finishes shall be as follows:

1. Gypsum wallboard and gypsum lath applied to concrete masonry or concrete walls shall be secured to wood or steel furring members spaced not more than 16 inches (406 mm) on center (o.c.).
2. Gypsum wallboard shall be installed with the long dimension parallel to the furring members and shall have all joints finished.
3. Other aspects of the installation of finishes shall comply with the applicable provisions of Chapters 7 and 25.

❖ This section provides prescriptive installation requirements for gypsum wallboard when installed on concrete masonry or concrete walls. Specification for gypsum wallboard materials and accessories are contained in Chapter 25.

722.3.3 Multiwythe masonry walls. The *fire-resistance rating* of wall assemblies constructed of multiple wythes of masonry materials shall be permitted to be based on the *fire-resistance rating* period of each wythe and the continuous airspace between each wythe in accordance with the following formula:

$$R_A = (R_1^{0.59} + R_2^{0.59} + ... + R_n^{0.59} + A_1 + A_2 + ... + A_n)^{1.7}$$

(Equation 7-7)

where:

R_A = *Fire-resistance rating* of the assembly (hours).

$R_1, R_2, ..., R_n$ = *Fire-resistance rating* of wythes for 1, 2, n (hours), respectively.

$A_1, A_2, ..., A_n$ = 0.30, factor for each continuous airspace for 1, 2, ...n, respectively, having a depth of $^1/_2$ inch (12.7 mm) or more between wythes.

❖ This calculation method was first published in Report BMS 92 of the NBS.

722.3.4 Concrete masonry lintels. *Fire-resistance ratings* for concrete masonry lintels shall be determined based upon the nominal thickness of the lintel and the minimum thickness of concrete masonry or concrete, or any combination thereof, covering the main reinforcing bars, as determined in accordance with Table 722.3.4, or by *approved* alternate methods.

❖ See the commentary to Section 722.

TABLE 722.3.4
MINIMUM COVER OF LONGITUDINAL REINFORCEMENT IN FIRE-RESISTANCE-RATED REINFORCED CONCRETE MASONRY LINTELS (inches)

NOMINAL WIDTH OF LINTEL (inches)	FIRE-RESISTANCE RATING (hours)			
	1	2	3	4
6	$1^1/_2$	2	—	—
8	$1^1/_2$	$1^1/_2$	$1^3/_4$	3
10 or greater	$1^1/_2$	$1^1/_2$	$1^1/_2$	$1^3/_4$

For SI: 1 inch = 25.4 mm.

❖ See the commentary to Section 722.

722.3.5 Concrete masonry columns. The *fire-resistance rating* of concrete masonry columns shall be determined based upon the least plan dimension of the column in accordance with Table 722.3.5 or by *approved* alternate methods.

❖ See the commentary to Section 722.

TABLE 722.3.5
MINIMUM DIMENSION OF CONCRETE MASONRY COLUMNS (inches)

FIRE-RESISTANCE RATING (hours)			
1	2	3	4
8 inches	10 inches	12 inches	14 inches

For SI: 1 inch = 25.4 mm.

❖ See the commentary to Section 722.

722.4 Clay brick and tile masonry. The provisions of this section contain procedures by which the *fire-resistance ratings* of clay brick and tile masonry are established by calculations.

❖ See the commentary to Section 722.

722.4.1 Masonry walls. The *fire-resistance rating* of masonry walls shall be based upon the equivalent thickness as calculated in accordance with this section. The calculation shall take into account finishes applied to the wall and airspaces between wythes in multiwythe construction.

❖ See the commentary to Section 722.

722.4.1.1 Equivalent thickness. The *fire-resistance ratings* of walls or partitions constructed of solid or hollow clay masonry units shall be determined from Table 722.4.1(1) or 722.4.1(2). The equivalent thickness of the clay masonry unit shall be determined by Equation 7-8 where using Table 722.4.1(1). The *fire-resistance rating* determined from Table 722.4.1(1) shall be permitted to be used in the calculated *fire-resistance rating* procedure in Section 722.4.2.

$$T_e = V_n/LH \qquad \textbf{(Equation 7-8)}$$

where:

T_e = The equivalent thickness of the clay masonry unit (inches).

V_n = The net volume of the clay masonry unit ($inch^3$).

L = The specified length of the clay masonry unit (inches).

H = The specified height of the clay masonry unit (inches).

❖ It has been accepted practice to determine the fire-resistance ratings of clay masonry walls by the ASTM E119 or UL 263 test method. In fact, Table 722.4.1(1) is based on tests performed by various laboratories in conformance with ASTM E119 or UL 263. Most notable are the NBS, Ohio State University Experiment Station and the University of California at Berkeley. New methods, however, have been developed to evaluate the fire resistance of clay masonry by the equivalent thickness method, similar to concrete masonry. This approach was used in conjunction with extensive testing previously performed on clay masonry to develop Table 722.4.1(1).

TABLE 722.4.1(1). See below.

❖ See the commentary to Section 722.4.1.1.

TABLE 722.4.1(2). See page 7-175.

❖ See the commentary to Section 722.4.1.1.

722.4.1.1.1 Hollow clay units. The equivalent thickness, T_e, shall be the value obtained for hollow clay units as determined in accordance with Equation 7-8. The net volume, V_n, of the units shall be determined using the gross volume and percentage of void area determined in accordance with ASTM C67.

❖ See the commentary to Section 722.4.1.1.

722.4.1.1.2 Solid grouted clay units. The equivalent thickness of solid grouted clay masonry units shall be taken as the actual thickness of the units.

❖ See the commentary to Section 722.4.1.1.

722.4.1.1.3 Units with filled cores. The equivalent thickness of the hollow clay masonry units is the actual thickness of the unit where completely filled with loose-fill materials of: sand, pea gravel, crushed stone, or slag that meet ASTM C33 requirements; pumice, scoria, expanded shale, expanded clay, expanded slate, expanded slag, expanded fly ash, or cinders in compliance with ASTM C331; or perlite or vermiculite meeting the requirements of ASTM C549 and ASTM C516, respectively.

❖ See the commentary to Section 722.4.1.1.

722.4.1.2 Plaster finishes. Where plaster is applied to the wall, the total *fire-resistance rating* shall be determined by the formula:

$$R = (R_n^{0.59} + pl)^{1.7} \qquad \textbf{(Equation 7-9)}$$

where:

R = The *fire-resistance rating* of the assembly (hours).

R_n = The *fire-resistance rating* of the individual wall (hours).

pl = Coefficient for thickness of plaster.

TABLE 722.4.1(1)
FIRE-RESISTANCE PERIODS OF CLAY MASONRY WALLS

MATERIAL TYPE	MINIMUM REQUIRED EQUIVALENT THICKNESS FOR FIRE RESISTANCE[a, b, c] (inches)			
	1 hour	2 hours	3 hours	4 hours
Solid brick of clay or shale[d]	2.7	3.8	4.9	6.0
Hollow brick or tile of clay or shale, unfilled	2.3	3.4	4.3	5.0
Hollow brick or tile of clay or shale, grouted or filled with materials specified in Section 722.4.1.1.3	3.0	4.4	5.5	6.6

For SI: 1 inch = 25.4 mm.

a. Equivalent thickness as determined from Section 722.4.1.1.

b. Calculated fire resistance between the hourly increments listed shall be determined by linear interpolation.

c. Where combustible members are framed in the wall, the thickness of solid material between the end of each member and the opposite face of the wall, or between members set in from opposite sides, shall be not less than 93 percent of the thickness shown.

d. For units in which the net cross-sectional area of cored brick in any plane parallel to the surface containing the cores is not less than 75 percent of the gross cross-sectional area measured in the same plane.

Values for $R_n^{0.59}$ for use in Equation 7-9 are given in Table 722.4.1(3). Coefficients for thickness of plaster shall be selected from Table 722.4.1(4) based on the actual thickness of plaster applied to the wall or partition and whether one or two sides of the wall are plastered.

❖ The variables for use in Equation 7-9 for determining the fire-resistance rating of plastered clay masonry walls are based on BMS 92, *Fire Resistance Classifications of Building Construction,* of the NBS. These values were derived from fire test results. The average thickness of plaster applied in the series of tests ranged from $^1/_2$ inch (12.7 mm) to $^3/_4$ inch (19.1 mm). The thickness for which the coefficients of plaster in Table 722.4.1(4) are given are those most likely to be applied to a clay masonry wall. Ratings for other thicknesses provided can be obtained by substituting the appropriate coefficients in the formula.

A test of four hollow concrete unit walls shows the effect of one coat of plaster on the fire-exposed side to be about the same as for one coat of plaster on the unexposed side. Tests have not been made with plaster on the unexposed side of only clay hollow walls. However, the ratings resulting from Equation 7-9 for clay or tile masonry walls for plaster on one side are believed to have a sufficient margin of safety to be applicable to either the exposed or unexposed side of the wall.

EXAMPLE:

GIVEN: A 4-inch (102 mm) solid brick wall is required to have a 3-hour fire-resistance rating with clay masonry units.

FIND: The thickness of one side of plaster required to attain a 3-hour rating.

SOLUTION:

1. From Table 722.4.1(1), a 2-hour solid brick wall is 3.8 inches (99 mm) of equivalent thickness and a 3-hour wall is 4.9 inches (124 mm). Through interpolation (Note a), a 4-inch (102 mm) wall is approximately a 131-minute (2.18 hours) fire rating.
2. From Table 722.4.1(4), plaster has a coefficient of 0.3 for $^1/_2$-inch thick (12.7 mm) plaster, 0.37 for $^5/_8$-inch thick (15.9 mm) plaster or 0.45 for $^3/_4$-inch (19.1 mm) thick plaster.
3. From Equation 7-9:

   $R = (R_n.59 + pl)1.7$

   where:

   $R = 3$

   $R_n = 2.18$

   $3 = [(2.18)^{0.59} + pl]^{1.7}$

   $\sqrt[1.7]{3} = \sqrt[1.7]{[(1.58 + pl)^{1.7}]}$

   $1.91 = 1.584 + pl$

   $pl = .326.$
4. Therefore, one coat of $^5/_8$-inch-thick (15.9 mm) sanded gypsum plaster on a 4-inch (102 mm) solid brick masonry wall would result in a 3-hour fire-resistance rating.

TABLE 722.4.1(3)
VALUES OF $R_n^{0.59}$

R_n 0.59	R (hours)
1	1.0
2	1.50
3	1.91
4	2.27

❖ See the commentary to Section 722.4.1.2.

TABLE 722.4.1(2)
FIRE-RESISTANCE RATINGS FOR BEARING STEEL FRAME BRICK VENEER WALLS OR PARTITIONS

WALL OR PARTITION ASSEMBLY	PLASTER SIDE EXPOSED (hours)	BRICK FACED SIDE EXPOSED (hours)
Outside facing of steel studs: $^1/_2$″ wood fiberboard sheathing next to studs, $^3/_4$″ airspace formed with $^3/_4$″ × $1^5/_8$″ wood strips placed over the fiberboard and secured to the studs; metal or wire lath nailed to such strips, $3^3/_4$″ brick veneer held in place by filling $^3/_4$″ airspace between the brick and lath with mortar. Inside facing of studs: $^3/_4$″ unsanded gypsum plaster on metal or wire lath attached to $^5/_{16}$″ wood strips secured to edges of the studs.	1.5	4
Outside facing of steel studs: 1″ insulation board sheathing attached to studs, 1″ airspace, and $3^3/_4$″ brick veneer attached to steel frame with metal ties every 5th course. Inside facing of studs: $^7/_8$″ sanded gypsum plaster (1:2 mix) applied on metal or wire lath attached directly to the studs.	1.5	4
Same as above except use $^7/_8$″ vermiculite-gypsum plaster or 1″ sanded gypsum plaster (1:2 mix) applied to metal or wire.	2	4
Outside facing of steel studs: $^1/_2$″ gypsum sheathing board, attached to studs, and $3^3/_4$″ brick veneer attached to steel frame with metal ties every 5th course. Inside facing of studs: $^1/_2$″ sanded gypsum plaster (1:2 mix) applied to $^1/_2$″ perforated gypsum lath securely attached to studs and having strips of metal lath 3 inches wide applied to all horizontal joints of gypsum lath.	2	4

For SI: 1 inch = 25.4 mm.

TABLE 722.4.1(4)
COEFFICIENTS FOR PLASTER, pl [a]

THICKNESS OF PLASTER (inch)	ONE SIDE	TWO SIDES
$^1/_2$	0.3	0.6
$^5/_8$	0.37	0.75
$^3/_4$	0.45	0.90

For SI: 1 inch = 25.4 mm.
a. Values listed in the table are for 1:3 sanded gypsum plaster.

❖ See the commentary to Section 722.4.1.2.

TABLE 722.4.1(5)
REINFORCED MASONRY LINTELS

NOMINAL LINTEL WIDTH (inches)	MINIMUM LONGITUDINAL REINFORCEMENT COVER FOR FIRE RESISTANCE (inches)			
	1 hour	2 hours	3 hours	4 hours
6	$1^1/_2$	2	NP	NP
8	$1^1/_2$	$1^1/_2$	$1^3/_4$	3
10 or more	$1^1/_2$	$1^1/_2$	$1^1/_2$	$1^3/_4$

For SI: 1 inch = 25.4 mm.
NP = Not permitted.

❖ See the commentary to Section 722.4.1.2.

TABLE 722.4.1(6)
REINFORCED CLAY MASONRY COLUMNS

COLUMN SIZE	FIRE-RESISTANCE RATING (hours)			
	1	2	3	4
Minimum column dimension (inches)	8	10	12	14

For SI:1 inch = 25.4 mm.

❖ See the commentary to Section 722.4.1.2.

722.4.1.3 Multiwythe walls with airspace. Where a continuous airspace separates multiple wythes of the wall or partition, the total *fire-resistance rating* shall be determined by the formula:

$$R = (R_1^{0.59} + R_2^{0.59} + ...+R_n^{0.59} + as)^{1.7} \quad \textbf{(Equation 7-10)}$$

where:

R = The *fire-resistance rating* of the assembly (hours).

R_1, R_2 and R_n = The *fire-resistance rating* of the individual wythes (hours).

as = Coefficient for continuous airspace.

Values for $R_n^{0.59}$ for use in Equation 7-10 are given in Table 722.4.1(3). The coefficient for each continuous airspace of $^1/_2$ inch to $3^1/_2$ inches (12.7 to 89 mm) separating two individual wythes shall be 0.3.

❖ Tests have shown that a continuous airspace separating two wythes of masonry can also increase the fire-resistance rating of a clay masonry wall. According to BMS 92 of the NBS, the coefficient for continuous airspace (as) between $^1/_2$ inch (12.7 mm) and $3^1/_2$ inches (89 mm) is 0.3. Equation 7-10 provides the formula for calculating the fire-resistance ratings involving an airspace if the fire-resistance rating of the wythes separated by the airspace is known. The fire-resistance ratings for various wythes can be obtained from Tables 722.4.1(1) and 722.4.1(2). The procedure for using Equation 7-10 is the same as Equation 7-9, except that a coefficient for airspace (as) is substituted for the coefficient of plaster (pl).

722.4.1.4 Nonsymmetrical assemblies. For a wall having no finish on one side or having different types or thicknesses of finish on each side, the calculation procedures of this section shall be performed twice, assuming either side to be the fire-exposed side of the wall. The *fire resistance* of the wall shall not exceed the lower of the two values determined.

Exception: For *exterior walls* with a *fire separation distance* greater than 5 feet (1524 mm), the fire shall be assumed to occur on the interior side only.

❖ Except for exterior walls having an FSD of more than 5 feet (1524 mm), Section 704.5 requires that walls be rated for exposure to fire from both sides. Therefore, two calculation procedures must be performed, on the assumption that each side is the fire-exposed side. Two calculations are not necessary for exterior walls with an FSD of more than 5 feet (1524 mm) or for other walls that are symmetrical (i.e., having the same type and thickness of finish on each side). The calculated fire-resistance rating must not exceed the lower of the two values determined, assuming that each side is the fire-exposed side.

722.4.2 Multiwythe walls. The *fire-resistance rating* for walls or partitions consisting of two or more dissimilar wythes shall be permitted to be determined by the formula:

$$R = (R_1^{0.59} + R_2^{0.59} + ...+R_n^{0.59})^{1.7} \quad \textbf{(Equation 7-11)}$$

where:

R = The *fire-resistance rating* of the assembly (hours).

R_1, R_2 and R_n = The *fire-resistance rating* of the individual wythes (hours).

Values for $R_n^{0.59}$ for use in Equation 7-11 are given in Table 722.4.1(3).

❖ Typically, the fire-resistance-rating period for clay masonry walls is determined by the temperature rise on the unexposed side of the wall. This criterion is what Equation 7-11 is based on. According to the general theory of heat transmission, if walls of the same material are exposed to a heat source that maintains a constant surface temperature on the exposed side and the unexposed side is protected against heat loss, the time in which a given temperature will be attained on the unexposed side will vary as the square of the wall thickness does.

In the standard ASTM E119 or UL 263 test, which involves specified conditions of temperature measurement and a fire that increases the temperature at the exposed surface of the wall as the test proceeds, the time required to attain a given temperature rise on the unexposed side will differ when the temperature on the exposed side remains constant at the initial exposure temperature for any period. A degree of correlation between test results and calculations can be obtained by assuming that the variation is a lower power of n than the second power. The fire resistance of the wall can then be expressed by the formula:

$R = (CV)^n$

where:

R = Fire-resistance period;

C = Coefficient depending on the material, design of the wall and the units of measurement of R and V;

V = Volume of solid material per unit area of wall surface; and

n = Exponent depending on the rate of increase of temperature at the exposed face of the wall.

For walls of a given material and design, it was found that an increase of 50 percent in volume of solid material per unit area of wall surface resulted in a 100-percent increase in the fire-resistance period. This correlates to a value of 1.7 for n. A value of n less than 2 is expected, since an increasing temperature on the exposed surface would tend to shorten the fire-resistance rating of walls that would qualify for a higher rating.

The fire-resistance rating of a wall may be expressed in terms of the fire-resistance rating of the conjoined wythes or laminae of the wall as follows:

$R_1 = (C_1V_1)^n$

$R_2 = (C_2V_2)^n$

$R_n = (C_nV_n)^n$

where R_1, R_2 and R_n are the fire resistances of each conjoined wythe.

The fire-resistance period of the composite wall will be:

$$R = \left(\sum_{i=1}^{n} C_iV_i\right)^{n_0}$$

$$R = (C_1V_1 + C_2V_2 + \ldots C_nV_n)^{n_0}$$

Substituting for C_1V_1, C_2V_2 and C_nV_n from the above relations yields:

$$R = (R_1^{1/n_0} + R_2^{1/n_0} + R_n^{1/n_0})^{n_0}$$

Substituting 1.7 for n_0 and 0.59 for $1/n_0$, the general formula becomes:

$R = (R_1^{0.59} + R_2^{0.59} + \ldots R_n^{0.59})^{1.7}$

Equation 7-11 was developed by the NBS in the early 1940s and first appeared in Appendix B of BMS 92. It is noted that the fire-resistance rating has been expressed in terms of the fire-resistance rating of the component laminae of the wall, which need not be of the same material and design.

722.4.2.1 Multiwythe walls of different material. For walls that consist of two or more wythes of different materials (concrete or concrete masonry units) in combination with clay masonry units, the *fire-resistance rating* of the different materials shall be permitted to be determined from Table 722.2.1.1 for concrete; Table 722.3.2 for concrete masonry units or Table 722.4.1(1) or 722.4.1(2) for clay and tile masonry units.

❖ A multiwythe wall (i.e., a wall consisting of two or more dissimilar materials) has a greater fire-resistance rating than a simple summation of the fire-endurance rating of the various layers. Equation 7-11 permits a calculated fire-resistance rating if the fire-resistance endurance rating is known for each dissimilar material. This section lists the applicable tables from Section 722 that can be used in conjunction with clay masonry walls to determine the total fire-resistance rating of a multiwythe wall composed of a combination of concrete, concrete masonry or clay masonry units.

722.4.3 Reinforced clay masonry lintels. *Fire-resistance ratings* for clay masonry lintels shall be determined based on the nominal width of the lintel and the minimum covering for the longitudinal reinforcement in accordance with Table 722.4.1(5).

❖ See the commentary to Section 722.

722.4.4 Reinforced clay masonry columns. The *fire-resistance ratings* shall be determined based on the last plan dimension of the column in accordance with Table 722.4.1(6). The minimum cover for longitudinal reinforcement shall be 2 inches (51 mm).

❖ See the commentary to Section 722.

722.5 Steel assemblies. The provisions of this section contain procedures by which the *fire-resistance ratings* of steel assemblies are established by calculations.

❖ See the commentary to Section 722.

722.5.1 Structural steel columns. The *fire-resistance ratings* of structural steel columns shall be based on the size of the element and the type of protection provided in accordance with this section.

❖ See the commentary to Section 722.

722.5.1.1 General. These procedures establish a basis for determining the *fire resistance* of column assemblies as a function of the thickness of fire-resistant material and, the weight, W, and heated perimeter, D, of structural steel columns. As used in these sections, W is the average weight of a structural steel column in pounds per linear foot. The heated perimeter, D, is the inside perimeter of the fire-resistant material in inches as illustrated in Figure 722.5.1(1).

❖ See the commentary to Section 722.

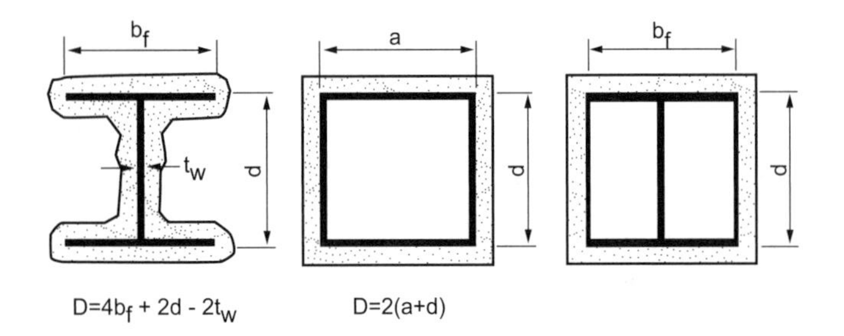

FIGURE 722.5.1(1)
DETERMINATION OF THE HEATED PERIMETER OF STRUCTURAL STEEL COLUMNS

722.5.1.1.1 Nonload-bearing protection. The application of these procedures shall be limited to column assemblies in which the fire-resistant material is not designed to carry any of the load acting on the column.

❖ See the commentary to Section 722.

722.5.1.1.2 Embedments. In the absence of substantiating fire-endurance test results, ducts, conduit, piping, and similar mechanical, electrical, and plumbing installations shall not be embedded in any required fire-resistant materials.

❖ See the commentary to Section 722.

722.5.1.1.3 Weight-to-perimeter ratio. Table 722.5.1(1) contains weight-to-heated-perimeter ratios (*W/D*) for both contour and box fire-resistant profiles, for the wide flange shapes most often used as columns. For different fire-resistant protection profiles or column cross sections, the weight-to-heated-perimeter ratios (*W/D*) shall be determined in accordance with the definitions given in this section.

❖ See the commentary to Section 722.

722.5.1.2 Gypsum wallboard protection. The *fire resistance* of structural steel columns with weight-to-heated-perimeter ratios (*W/D*) less than or equal to 3.65 and that are protected with Type X gypsum wallboard shall be permitted to be determined from the following expression:

$$R = 130\left[\frac{h(W'/D)^{0.75}}{2}\right] \quad \textbf{(Equation 7-12)}$$

where:

R = Fire resistance (minutes).

h = Total thickness of gypsum wallboard (inches).

D = Heated perimeter of the structural steel column (inches).

W' = Total weight of the structural steel column and gypsum wallboard protection (pounds per linear foot).

W' = $W + 50hD/144$.

722.5.1.2.1 Attachment. The gypsum board or gypsum panel products shall be supported as illustrated in either Figure 722.5.1(2) for *fire-resistance ratings* of 4 hours or less, or Figure 722.5.1(3) for *fire-resistance ratings* of 3 hours or less.

❖ **EXAMPLE**:

GIVEN: A *W*12 × 136 column protected by 1/2-inch (12.7 mm) Type X gypsum wallboard as shown in Commentary Figure 722.5.1.2.1.

FIND: The fire-resistance rating of the column.

SOLUTION:

1. The heated perimeter D = 2(12.4 inches + 13.4 inches) = 51.6 inches (1311 mm).
2. W' = 136 + (50)(0.5)(51.6)/144 = 144.96.
3. W' = 144.96 / 51.62 = 2.81.
4. R = 130[(0.5)(2.81)/2]$^{0.75}$ = 99.7 minutes (1.66 hours).

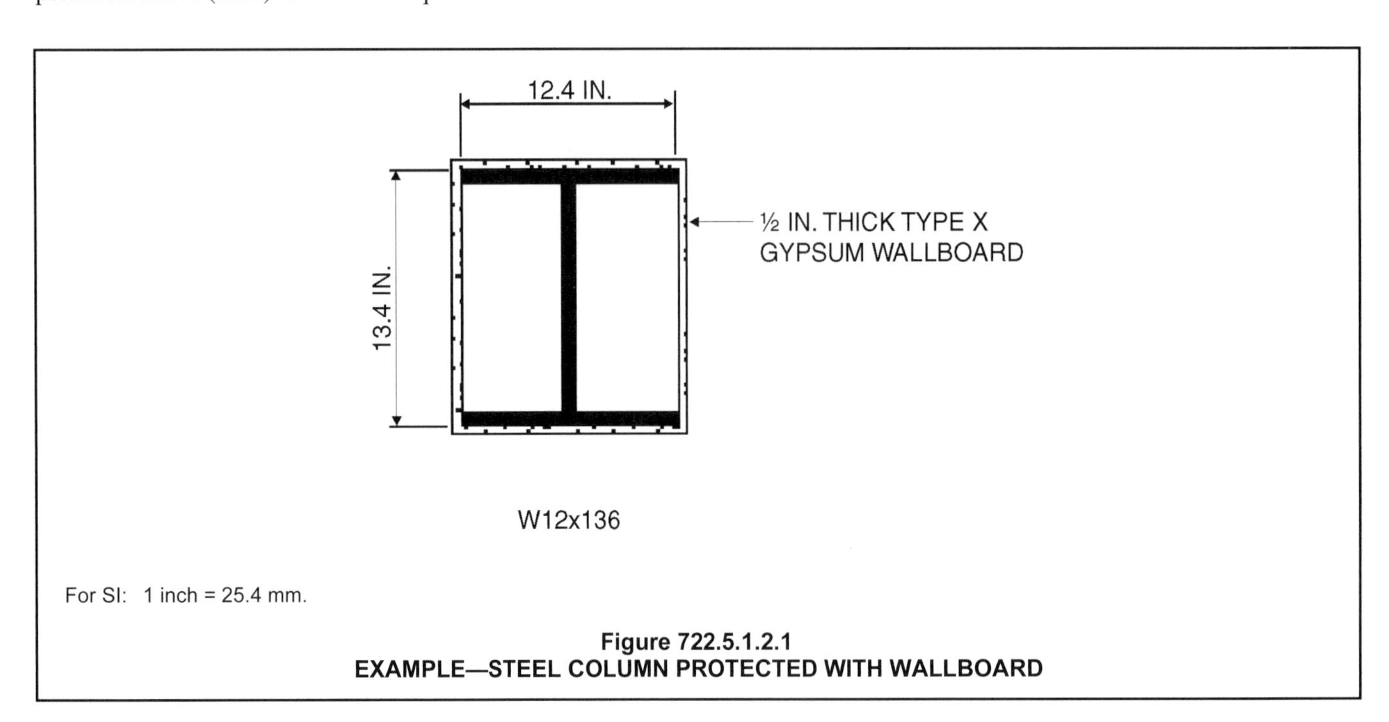

For SI: 1 inch = 25.4 mm.

Figure 722.5.1.2.1
EXAMPLE—STEEL COLUMN PROTECTED WITH WALLBOARD

5. Alternatively, from Table 722.5.1(1), the *W/D* ratio for a *W*12 × 136 is tabulated as 2.63. From Code Figure 722.5.1(4) using a *W/D* ratio of 2.63 and $^1/_2$-inch (12.7 mm) gypsum wallboard, an *R*-value of approximately 1.6 hours is obtained.

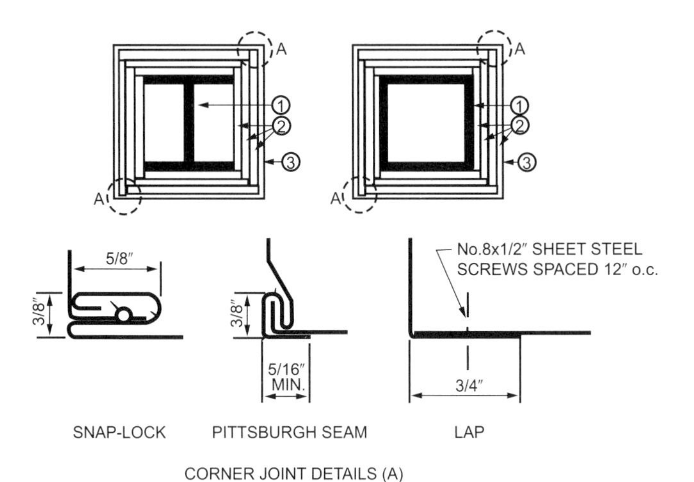

FIGURE 722.5.1(2)
GYPSUM-PROTECTED STRUCTURAL STEEL COLUMNS WITH SHEET STEEL COLUMN COVERS

For SI: 1 inch = 25.4 mm, 1 foot = 305 mm.

1. Structural steel column, either wide flange or tubular shapes.
2. Type X gypsum board or gypsum panel products in accordance with ASTM C1177, C1178, C1278, C1396 or C1658. The total thickness of gypsum board or gypsum panel products calculated as h in Section 722.5.1.2 shall be applied vertically to an individual column using one of the following methods:
    1. As a single layer with no horizontal joints.
    2. As multiple layers with no horizontal joints permitted in any layer.
    3. As multiple layers with horizontal joints staggered not less than 12 inches vertically between layers and not less than 8 feet vertically in any single layer. The total required thickness of gypsum board or gypsum panel products shall be determined on the basis of the specified fire-resistance rating and the weight-to-heated-perimeter ratio (W/D) of the column. For fire-resistance ratings of 2 hours or less, one of the required layers of gypsum board or gypsum panel product may be applied to the exterior of the sheet steel column covers with 1-inch long Type S screws spaced 1 inch from the wallboard edge and 8 inches on center. For such installations, 0.0149-inch minimum thickness galvanized steel corner beads with $1^1/_2$-inch legs shall be attached to the wallboard with Type S screws spaced 12 inches on center.
3. For *fire-resistance ratings* of 3 hours or less, the column covers shall be fabricated from 0.0239-inch minimum thickness galvanized or stainless steel. For 4-hour *fire-resistance ratings*, the column covers shall be fabricated from 0.0239-inch minimum thickness stainless steel. The column covers shall be erected with the Snap Lock or Pittsburgh joint details.

   For *fire-resistance ratings* of 2 hours or less, column covers fabricated from 0.0269-inch minimum thickness galvanized or stainless steel shall be permitted to be erected with lap joints. The lap joints shall be permitted to be located anywhere around the perimeter of the column cover. The lap joints shall be secured with $^1/_2$-inch-long No. 8 sheet metal screws spaced 12 inches on center.

   The column covers shall be provided with a minimum expansion clearance of $^1/_8$ inch per linear foot between the ends of the cover and any restraining construction.

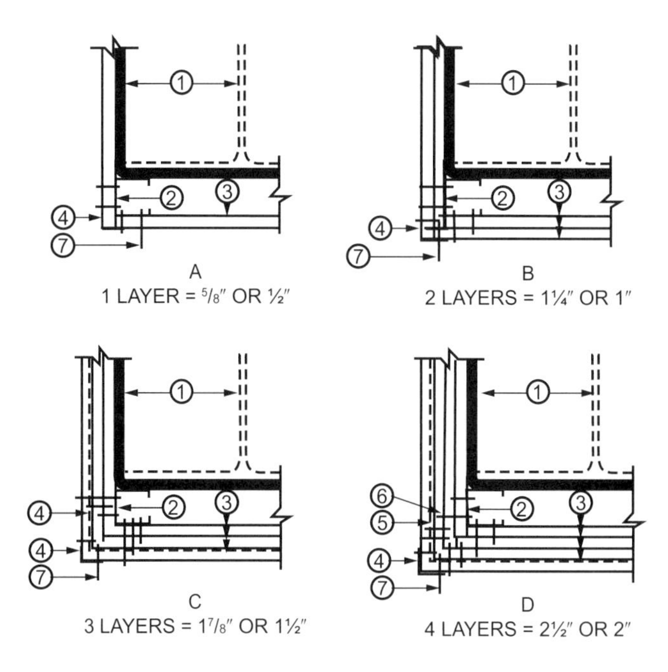

FIGURE 722.5.1(3)
GYPSUM-PROTECTED STRUCTURAL STEEL COLUMNS WITH STEEL STUD/SCREW ATTACHMENT SYSTEM

For SI: 1 inch = 25.4 mm, 1 foot = -305 mm.

1. Structural steel column, either wide flange or tubular shapes.
2. $1^5/_8$-inch deep studs fabricated from 0.0179-inch minimum thickness galvanized steel with $1^5/_{16}$ or $1^7/_{16}$-inch legs. The length of the steel studs shall be $^1/_2$ inch less than the height of the assembly.
3. Type X gypsum board or gypsum panel products in accordance with ASTM C177, C1178, C1278, C1396 or C1658. The total thickness of gypsum board or gypsum panel products calculated as *h* in Section 722.5.1.2 shall be applied vertically to an individual column using one of the following methods:
    1. As a single layer with no horizontal joints.
    2. As multiple layers with no horizontal joints permitted in any layer.
    3. As multiple layers with horizontal joints staggered not less than 12 inches vertically between layers and not less than 8 feet vertically in any single layer. The total required thickness of gypsum board or gypsum panel products shall be determined on the basis of the specified *fire-resistance rating* and the weight-to-heated-perimeter ratio (W/D) of the column.
4. Galvanized 0.0149-inch minimum thickness steel corner beads with $1^1/_2$-inch legs attached to the gypsum board or gypsum panel products with 1-inch-long Type S screws spaced 12 inches on center.
5. No. 18 SWG steel tie wires spaced 24 inches on center.
6. Sheet metal angles with 2-inch legs fabricated from 0.0221-inch minimum thickness galvanized steel.
7. Type S screws, 1 inch long, shall be used for attaching the first layer of gypsum board or gypsum panel product to the steel studs and the third layer to the sheet metal angles at 24 inches on center. Type S screws $1^3/_4$-inch long shall be used for attaching the second layer of gypsum board or gypsum panel product to the steel studs and the fourth layer to the sheet metal angles at 12 inches on center. Type S screws $2^1/_4$ inches long shall be used for attaching the third layer of gypsum board or gypsum panel product to the steel studs at 12 inches on center.

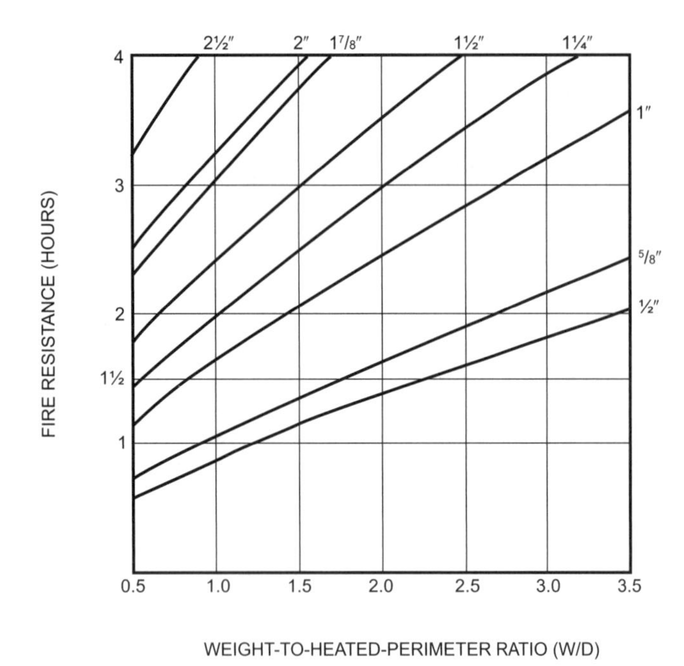

For SI: 1 inch = 25.4 mm, 1 pound per linear foot/inch = 0.059 kg/m/mm.

FIGURE 722.5.1(4)
FIRE RESISTANCE OF STRUCTURAL STEEL COLUMNS PROTECTED WITH VARIOUS THICKNESSES OF TYPE X GYPSUM WALLBOARD

a. The W/D ratios for typical wide flange columns are listed in Table 722.5.1(1). For other column shapes, the W/D ratios shall be determined in accordance with Section 722.5.1.1.

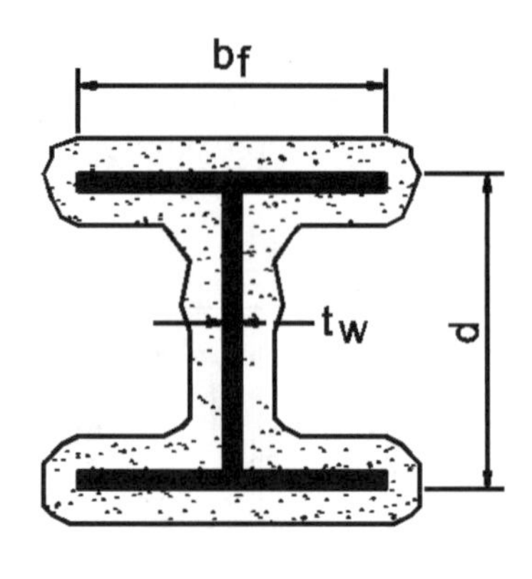

FIGURE 722.5.1(5)
WIDE FLANGE STRUCTURAL STEEL COLUMNS WITH SPRAYED FIRE-RESISTANT MATERIALS

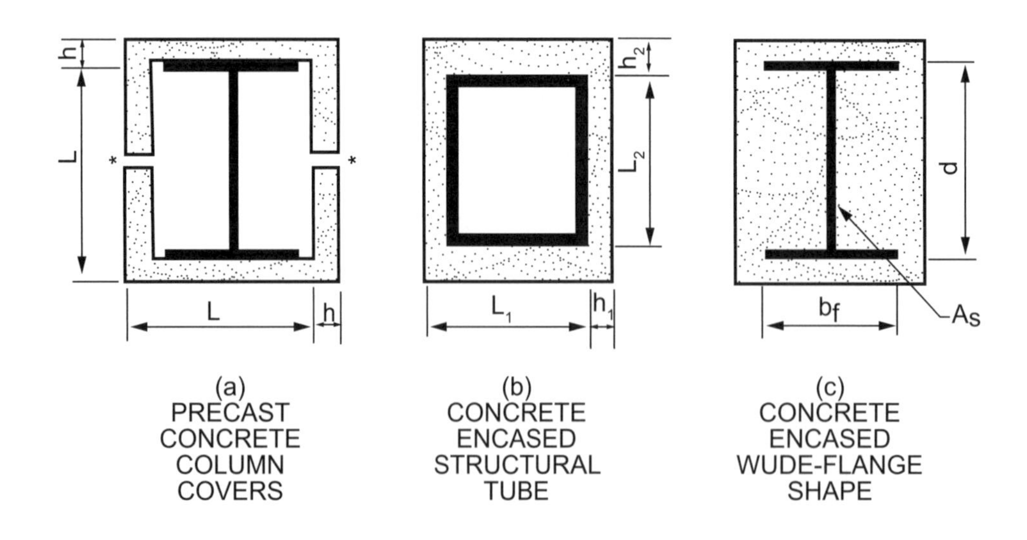

FIGURE 722.5.1(6)
CONCRETE PROTECTED STRUCTURAL STEEL COLUMNS[a,b]

a. When the inside perimeter of the concrete protection is not square, L shall be taken as the average of L_1 and L_2. When the thickness of concrete cover is not constant, h shall be taken as the average of h_1 and h_2.

b. Joints shall be protected with a minimum 1 inch thickness of ceramic fiber blanket but in no case less than one-half the thickness of the column cover (see Section 722.2.1.3).

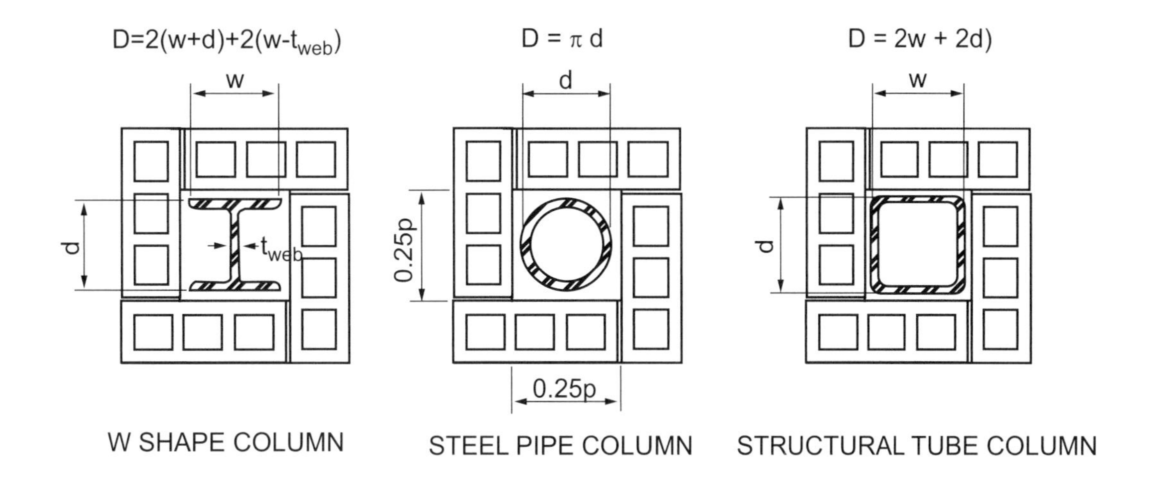

FIGURE 722.5.1(7)
CONCRETE OR CLAY MASONRY PROTECTED STRUCTURAL STEEL COLUMNS

For SI: 1 inch = 25.4 mm.
d = Depth of a wide flange column, outside diameter of pipe column, or outside dimension of structural tubing column (inches).
t_{web} = Thickness of web of wide flange column (inches).
w = Width of flange of wide flange column (inches).

TABLE 722.5.1(1)
W/D RATIOS FOR STEEL COLUMNS

STRUCTURAL SHAPE	CONTOUR PROFILE	BOX PROFILE	STRUCTURAL SHAPE	CONTOUR PROFILE	BOX PROFILE
W14 × 233	2.55	3.65	W10 × 112	1.81	2.57
× 211	2.32	3.35	× 100	1.64	2.33
× 193	2.14	3.09	× 88	1.45	2.08
× 176	1.96	2.85	× 77	1.28	1.85
× 159	1.78	2.60	× 68	1.15	1.66
× 145	1.64	2.39	× 60	1.01	1.48
× 132	1.56	2.25	× 54	0.922	1.34
× 120	1.42	2.06	× 49	0.84	1.23
× 109	1.29	1.88	× 45	0.888	1.24
× 99	1.18	1.72	× 39	0.78	1.09
× 90	1.08	1.58	× 33	0.661	0.93
× 82	1.23	1.68			
× 74	1.12	1.53	W8 × 67	1.37	1.94
× 68	1.04	1.41	× 58	1.20	1.71
× 61	0.928	1.28	× 48	1.00	1.44
× 53	0.915	1.21	× 40	0.849	1.23
× 48	0.835	1.10	× 35	0.749	1.08
× 43	0.752	0.99	× 31	0.665	0.97
			× 28	0.688	0.96
W12 × 190	2.50	3.51	× 24	0.591	0.83
× 170	2.26	3.20	× 21	0.577	0.77
× 152	2.04	2.90	× 18	0.499	0.67
× 136	1.86	2.63			
× 120	1.65	2.36	W6 ×25	0.696	1.00
× 106	1.47	2.11	× 20	0.563	0.82
× 96	1.34	1.93	× 16	0.584	0.78
× 87	1.22	1.76	× 15	0.431	0.63
× 79	1.11	1.61	× 12	0.448	0.60
× 72	1.02	1.48	× 9	0.338	0.46
× 65	0.925	1.35			
× 58	0.925	1.31	W5 ×19	0.644	0.93
× 53	0.855	1.20	× 16	0.55	0.80
× 50	0.909	1.23			
× 45	0.829	1.12	W4 ×13	0.556	0.79
× 40	0.734	1.00			

For SI: 1 pound per linear foot per inch = 0.059 kg/m/mm.

TABLE 722.5.1(2)
PROPERTIES OF CONCRETE

PROPERTY	NORMAL-WEIGHT CONCRETE	STRUCTURAL LIGHTWEIGHT CONCRETE
Thermal conductivity (k_c)	0.95 Btu/hr · ft · °F	0.35 Btu/hr · ft · °F
Specific heat (c_c)	0.20 Btu/lb °F	0.20 Btu/lb °F
Density (P_c)	145 lb/ft^3	110 lb/ft^3
Equilibrium (free) moisture content (m) by volume	4%	5%

For SI: 1 inch = 25.4 mm, 1 foot = 304.8 mm, 1 lb/ft^3 = 16.0185 kg/m^3, Btu/hr • ft • °F = 1.731 W/(m • K).

TABLE 722.5.1(3)
THERMAL CONDUCTIVITY OF CONCRETE OR CLAY MASONRY UNITS

DENSITY (d_m) OF UNITS (lb/ft^3)	THERMAL CONDUCTIVITY (*K*) OF UNITS (Btu/hr · ft · °F)
Concrete Masonry Units	
80	0.207
85	0.228
90	0.252
95	0.278
100	0.308
105	0.340
110	0.376
115	0.416
120	0.459
125	0.508
130	0.561
135	0.620
140	0.685
145	0.758
150	0.837
Clay Masonry Units	
120	1.25
130	2.25

For SI: 1 pound per cubic foot = 16.0185 kg/m^3, Btu/hr • ft • °F = 1.731 W/(m • K).

TABLE 722.5.1(4)
WEIGHT-TO-HEATED-PERIMETER RATIOS (*W/D*) FOR TYPICAL WIDE FLANGE BEAM AND GIRDER SHAPES

STRUCTURAL SHAPE	CONTOUR PROFILE	BOX PROFILE	STRUCTURAL SHAPE	CONTOUR PROFILE	BOX PROFILE
W36 x 300	2.50	3.33	W24 x 68	0.942	1.21
x 280	2.35	3.12	x 62	0.934	1.14
x 260	2.18	2.92	x 55	0.828	1.02
x 245	2.08	2.76			
x 230	1.95	2.61	W21 x 147	1.87	2.60
x 210	1.96	2.45	x 132	1.68	2.35
x 194	1.81	2.28	x 122	1.57	2.19
x 182	1.72	2.15	x 111	1.43	2.01
x 170	1.60	2.01	x 101	1.30	1.84
x 160	1.51	1.90	x 93	1.40	1.80
x 150	1.43	1.79	x 83	1.26	1.62
x 135	1.29	1.63	x 73	1.11	1.44
			x 68	1.04	1.35

(continued)

TABLE 722.5.1(4)—continued
WEIGHT-TO-HEATED-PERIMETER RATIOS (*W/D*) FOR TYPICAL WIDE FLANGE BEAM AND GIRDER SHAPES

STRUCTURAL SHAPE	CONTOUR PROFILE	BOX PROFILE	STRUCTURAL SHAPE	CONTOUR PROFILE	BOX PROFILE
W33 x 241	2.13	2.86	W21 x 62	0.952	1.23
x 221	1.97	2.64	x 57	0.952	1.17
x 201	1.79	2.42	x 50	0.838	1.04
x 152	1.53	1.94	x 44	0.746	0.92
x 141	1.43	1.80			
x 130	1.32	1.67	W18 x 119	1.72	2.42
x 118	1.21	1.53	x 106	1.55	2.18
			x 97	1.42	2.01
W30 x 211	2.01	2.74	x 86	1.27	1.80
x 191	1.85	2.50	x 76	1.13	1.60
x 173	1.66	2.28	x 71	1.22	1.59
x 132	1.47	1.85	x 65	1.13	1.47
x 124	1.39	1.75	x 60	1.04	1.36
x 116	1.30	1.65	x 55	0.963	1.26
x 108	1.21	1.54	x 50	0.88	1.15
x 99	1.12	1.42	x 46	0.878	1.09
			x 40	0.768	0.96
W27 x 178	1.87	2.55	x 35	0.672	0.85
x 161	1.70	2.33			
x 146	1.55	2.12	W16 x 100	1.59	2.25
x 114	1.39	1.76	x 89	1.43	2.03
x 102	1.24	1.59	x 77	1.25	1.78
x 94	1.15	1.47	x 67	1.09	1.56
x 84	1.03	1.33	x 57	1.09	1.43
			x 50	0.962	1.26
			x 45	0.870	1.15
W24 x 162	1.88	2.57	x 40	0.780	1.03
x 146	1.70	2.34	x 36	0.702	0.93
x 131	1.54	2.12	x 31	0.661	0.83
x 117	1.38	1.91	x 26	0.558	0.70
x 104	1.24	1.71			
x 94	1.28	1.63	W14 x 132	1.89	3.00
x 84	1.15	1.47	x 120	1.71	2.75
x 76	1.05	1.34	x 109	1.57	2.52
W14 x 99	1.43	2.31	W10 x 30	0.806	1.12
x 90	1.31	2.11	x 26	0.708	0.98
x 82	1.45	2.12	x 22	0.606	0.84
x 74	1.32	1.93	x 19	0.607	0.78
x 68	1.22	1.78	x 17	0.543	0.70
x 61	1.10	1.61	x 15	0.484	0.63
x 53	1.06	1.48	x 12	0.392	0.51
x 48	0.970	1.35			

(continued)

TABLE 722.5.1(4)—continued
WEIGHT-TO-HEATED-PERIMETER RATIOS (*W/D*) FOR TYPICAL WIDE FLANGE BEAM AND GIRDER SHAPES

STRUCTURAL SHAPE	CONTOUR PROFILE	BOX PROFILE	STRUCTURAL SHAPE	CONTOUR PROFILE	BOX PROFILE
W14 x 43	0.874	1.22	W8 x 67	1.65	2.55
x 38	0.809	1.09	x 58	1.44	2.26
x 34	0.725	0.98	x 48	1.21	1.91
x 30	0.644	0.87	x 40	1.03	1.63
x 26	0.628	0.79	x 35	0.907	1.44
x 22	0.534	0.68	x 31	0.803	1.29
			x 28	0.819	1.24
W12 x 87	1.47	2.34	x 24	0.704	1.07
x 79	1.34	2.14	x 21	0.675	0.96
x 72	1.23	1.97	x 18	0.583	0.84
x 65	1.11	1.79	x 15	0.551	0.74
x 58	1.10	1.69	x 13	0.483	0.65
x 53	1.02	1.55	x 10	0.375	0.51
x 50	1.06	1.54			
x 45	0.974	1.40	W6 x 25	0.839	1.33
x 40	0.860	1.25	x 20	0.678	1.09
x 35	0.810	1.11	x 16	0.684	0.96
x 30	0.699	0.96	x 15	0.521	0.83
x 26	0.612	0.84	x 12	0.526	0.75
x 22	0.623	0.77	x 9	0.398	0.57
x 19	0.540	0.67			
x 16	0.457	0.57	W5 x 19	0.776	1.24
x 14	0.405	0.50	x 16	0.664	1.07
W10 x 112	2.17	3.38	W4 x 13	0.670	1.05
x 100	1.97	3.07			
x 88	1.74	2.75			
x 77	1.54	2.45			
x 68	1.38	2.20			
x 60	1.22	1.97			
x 54	1.11	1.79			
x 49	1.01	1.64			
x 45	1.06	1.59			
x 39	0.94	1.40			
x 33	0.77	1.20			

For SI: 1 pound per linear foot per inch = 0.059 kg/m/mm.

TABLE 722.5.1(5)
FIRE RESISTANCE OF CONCRETE MASONRY PROTECTED STEEL COLUMNS

COLUMN SIZE	CONCRETE MASONRY DENSITY POUNDS PER CUBIC FOOT	MINIMUM REQUIRED EQUIVALENT THICKNESS FOR FIRE-RESISTANCE RATING OF CONCRETE MASONRY PROTECTION ASSEMBLY, T_e (inches)				COLUMN SIZE	CONCRETE MASONRY DENSITY POUNDS PER CUBIC FOOT	MINIMUM REQUIRED EQUIVALENT THICKNESS FOR FIRE-RESISTANCE RATING OF CONCRETE MASONRY PROTECTION ASSEMBLY, T_e (inches)			
		1 hour	2 hours	3 hours	4 hours			1 hour	2 hours	3 hours	4 hours
W14 × 82	80	0.74	1.61	2.36	3.04	W10 × 68	80	0.72	1.58	2.33	3.01
	100	0.89	1.85	2.67	3.40		100	0.87	1.83	2.65	3.38
	110	0.96	1.97	2.81	3.57		110	0.94	1.95	2.79	3.55
	120	1.03	2.08	2.95	3.73		120	1.01	2.06	2.94	3.72
W14 × 68	80	0.83	1.70	2.45	3.13	W10 × 54	80	0.88	1.76	2.53	3.21
	100	0.99	1.95	2.76	3.49		100	1.04	2.01	2.83	3.57
	110	1.06	2.06	2.91	3.66		110	1.11	2.12	2.98	3.73
	120	1.14	2.18	3.05	3.82		120	1.19	2.24	3.12	3.90
W14 × 53	80	0.91	1.81	2.58	3.27	W10 × 45	80	0.92	1.83	2.60	3.30
	100	1.07	2.05	2.88	3.62		100	1.08	2.07	2.90	3.64
	110	1.15	2.17	3.02	3.78		110	1.16	2.18	3.04	3.80
	120	1.22	2.28	3.16	3.94		120	1.23	2.29	3.18	3.96
W14 × 43	80	1.01	1.93	2.71	3.41	W10 × 33	80	1.06	2.00	2.79	3.49
	100	1.17	2.17	3.00	3.74		100	1.22	2.23	3.07	3.81
	110	1.25	2.28	3.14	3.90		110	1.30	2.34	3.20	3.96
	120	1.32	2.38	3.27	4.05		120	1.37	2.44	3.33	4.12
W12 × 72	80	0.81	1.66	2.41	3.09	W8 × 40	80	0.94	1.85	2.63	3.33
	100	0.91	1.88	2.70	3.43		100	1.10	2.10	2.93	3.67
	110	0.99	1.99	2.84	3.60		110	1.18	2.21	3.07	3.83
	120	1.06	2.10	2.98	3.76		120	1.25	2.32	3.20	3.99
W12 × 58	80	0.88	1.76	2.52	3.21	W8 × 31	80	1.06	2.00	2.78	3.49
	100	1.04	2.01	2.83	3.56		100	1.22	2.23	3.07	3.81
	110	1.11	2.12	2.97	3.73		110	1.29	2.33	3.20	3.97
	120	1.19	2.23	3.11	3.89		120	1.36	2.44	3.33	4.12
W12 × 50	80	0.91	1.81	2.58	3.27	W8 × 24	80	1.14	2.09	2.89	3.59
	100	1.07	2.05	2.88	3.62		100	1.29	2.31	3.16	3.90
	110	1.15	2.17	3.02	3.78		110	1.36	2.42	3.28	4.05
	120	1.22	2.28	3.16	3.94		120	1.43	2.52	3.41	4.20
W12 × 40	80	1.01	1.94	2.72	3.41	W8 × 18	80	1.22	2.20	3.01	3.72
	100	1.17	2.17	3.01	3.75		100	1.36	2.40	3.25	4.01
	110	1.25	2.28	3.14	3.90		110	1.42	2.50	3.37	4.14
	120	1.32	2.39	3.27	4.06		120	1.48	2.59	3.49	4.28
4 × 4 × $^1/_2$ wall thickness	80	0.93	1.90	2.71	3.43	4 double extra strong 0.674 wall thickness	80	0.80	1.75	2.56	3.28
	100	1.08	2.13	2.99	3.76		100	0.95	1.99	2.85	3.62
	110	1.16	2.24	3.13	3.91		110	1.02	2.10	2.99	3.78
	120	1.22	2.34	3.26	4.06		120	1.09	2.20	3.12	3.93
4 × 4 × $^3/_8$ wall thickness	80	1.05	2.03	2.84	3.57	4 extra strong 0.337 wall thickness	80	1.12	2.11	2.93	3.65
	100	1.20	2.25	3.11	3.88		100	1.26	2.32	3.19	3.95
	110	1.27	2.35	3.24	4.02		110	1.33	2.42	3.31	4.09
	120	1.34	2.45	3.37	4.17		120	1.40	2.52	3.43	4.23

(continued)

TABLE 722.5.1(5)—continued
FIRE RESISTANCE OF CONCRETE MASONRY PROTECTED STEEL COLUMNS

COLUMN SIZE	CONCRETE MASONRY DENSITY POUNDS PER CUBIC FOOT	MINIMUM REQUIRED EQUIVALENT THICKNESS FOR FIRE-RESISTANCE RATING OF CONCRETE MASONRY PROTECTION ASSEMBLY, T_e (inches)				COLUMN SIZE	CONCRETE MASONRY DENSITY POUNDS PER CUBIC FOOT	MINIMUM REQUIRED EQUIVALENT THICKNESS FOR FIRE-RESISTANCE RATING OF CONCRETE MASONRY PROTECTION ASSEMBLY, T_e (inches)			
		1 hour	2 hours	3 hours	4 hours			1 hour	2 hours	3 hours	4 hours
4 × 4 × $^1/_4$ wall thickness	80	1.21	2.20	3.01	3.73	4 standard 0.237 wall thickness	80	1.26	2.25	3.07	3.79
	100	1.35	2.40	3.26	4.02		100	1.40	2.45	3.31	4.07
	110	1.41	2.50	3.38	4.16		110	1.46	2.55	3.43	4.21
	120	1.48	2.59	3.50	4.30		120	1.53	2.64	3.54	4.34
6 × 6 × $^1/_2$ wall thickness	80	0.82	1.75	2.54	3.25	5 double extra strong 0.750 wall thickness	80	0.70	1.61	2.40	3.12
	100	0.98	1.99	2.84	3.59		100	0.85	1.86	2.71	3.47
	110	1.05	2.10	2.98	3.75		110	0.91	1.97	2.85	3.63
	120	1.12	2.21	3.11	3.91		120	0.98	2.02	2.99	3.79
6 × 6 × $^3/_8$ wall thickness	80	0.96	1.91	2.71	3.42	5 extra strong 0.375 wall thickness	80	1.04	2.01	2.83	3.54
	100	1.12	2.14	3.00	3.75		100	1.19	2.23	3.09	3.85
	110	1.19	2.25	3.13	3.90		110	1.26	2.34	3.22	4.00
	120	1.26	2.35	3.26	4.05		120	1.32	2.44	3.34	4.14
6 × 6 × $^1/_4$ wall thickness	80	1.14	2.11	2.92	3.63	5 standard 0.258 wall thickness	80	1.20	2.19	3.00	3.72
	100	1.29	2.32	3.18	3.93		100	1.34	2.39	3.25	4.00
	110	1.36	2.43	3.30	4.08		110	1.41	2.49	3.37	4.14
	120	1.42	2.52	3.43	4.22		120	1.47	2.58	3.49	4.28
8 × 8 × $^1/_2$ wall thickness	80	0.77	1.66	2.44	3.13	6 double extra strong 0.864 wall thickness	80	0.59	1.46	2.23	2.92
	100	0.92	1.91	2.75	3.49		100	0.73	1.71	2.54	3.29
	110	1.00	2.02	2.89	3.66		110	0.80	1.82	2.69	3.47
	120	1.07	2.14	3.03	3.82		120	0.86	1.93	2.83	3.63
8 × 8 × $^3/_8$ wall thickness	80	0.91	1.84	2.63	3.33	6 extra strong 0.432 wall thickness	80	0.94	1.90	2.70	3.42
	100	1.07	2.08	2.92	3.67		100	1.10	2.13	2.98	3.74
	110	1.14	2.19	3.06	3.83		110	1.17	2.23	3.11	3.89
	120	1.21	2.29	3.19	3.98		120	1.24	2.34	3.24	4.04
8 × 8 × $^1/_4$ wall thickness	80	1.10	2.06	2.86	3.57	6 standard 0.280 wall thickness	80	1.14	2.12	2.93	3.64
	100	1.25	2.28	3.13	3.87		100	1.29	2.33	3.19	3.94
	110	1.32	2.38	3.25	4.02		110	1.36	2.43	3.31	4.08
	120	1.39	2.48	3.38	4.17		120	1.42	2.53	3.43	4.22

For SI: 1 inch = 25.4 mm, 1 pound per cubic feet = 16.02 kg/m^3.

Note: Tabulated values assume 1-inch air gap between masonry and steel section.

TABLE 722.5.1(6)
FIRE RESISTANCE OF CLAY MASONRY PROTECTED STEEL COLUMNS

COLUMN SIZE	CLAY MASONRY DENSITY, POUNDS PER CUBIC FOOT	MINIMUM REQUIRED EQUIVALENT THICKNESS FOR FIRE-RESISTANCE RATING OF CLAY MASONRY PROTECTION ASSEMBLY, T_e (inches)			
		1 hour	2 hours	3 hours	4 hours
W14 × 82	120	1.23	2.42	3.41	4.29
	130	1.40	2.70	3.78	4.74
W14 × 68	120	1.34	2.54	3.54	4.43
	130	1.51	2.82	3.91	4.87
W14 × 53	120	1.43	2.65	3.65	4.54
	130	1.61	2.93	4.02	4.98
W14 × 43	120	1.54	2.76	3.77	4.66
	130	1.72	3.04	4.13	5.09
W12 × 72	120	1.32	2.52	3.51	4.40
	130	1.50	2.80	3.88	4.84
W12 × 58	120	1.40	2.61	3.61	4.50
	130	1.57	2.89	3.98	4.94
W12 × 50	120	1.43	2.65	3.66	4.55
	130	1.61	2.93	4.02	4.99
W12 × 40	120	1.54	2.77	3.78	4.67
	130	1.72	3.05	4.14	5.10

COLUMN SIZE	CLAY MASONRY DENSITY, POUNDS PER CUBIC FOOT	MINIMUM REQUIRED EQUIVALENT THICKNESS FOR FIRE-RESISTANCE RATING OF CLAY MASONRY PROTECTION ASSEMBLY, T_e (inches)			
		1 hour	2 hours	3 hours	4 hours
W10 × 68	120	1.27	2.46	3.26	4.35
	130	1.44	2.75	3.83	4.80
W10 × 54	120	1.40	2.61	3.62	4.51
	130	1.58	2.89	3.98	4.95
W10 × 45	120	1.44	2.66	3.67	4.57
	130	1.62	2.95	4.04	5.01
W10 × 33	120	1.59	2.82	3.84	4.73
	130	1.77	3.10	4.20	5.13
W8 × 40	120	1.47	2.70	3.71	4.61
	130	1.65	2.98	4.08	5.04
W8 × 31	120	1.59	2.82	3.84	4.73
	130	1.77	3.10	4.20	5.17
W8 × 24	120	1.66	2.90	3.92	4.82
	130	1.84	3.18	4.28	5.25
W8 × 18	120	1.75	3.00	4.01	4.91
	130	1.93	3.27	4.37	5.34

STEEL TUBING

NOMINAL TUBE SIZE (inches)	CLAY MASONRY DENSITY, POUNDS PER CUBIC FOOT	MINIMUM REQUIRED EQUIVALENT THICKNESS FOR FIRE-RESISTANCE RATING OF CLAY MASONRY PROTECTION ASSEMBLY, T_e (inches)			
		1 hour	2 hours	3 hours	4 hours
4 × 4 × $^1/_2$ wall thickness	120	1.44	2.72	3.76	4.68
	130	1.62	3.00	4.12	5.11
4 × 4 × $^3/_8$ wall thickness	120	1.56	2.84	3.88	4.78
	130	1.74	3.12	4.23	5.21
4 × 4 × $^1/_4$ wall thickness	120	1.72	2.99	4.02	4.92
	130	1.89	3.26	4.37	5.34
6 × 6 × $^1/_2$ wall thickness	120	1.33	2.58	3.62	4.52
	130	1.50	2.86	3.98	4.96
6 × 6 × $^3/_8$ wall thickness	120	1.48	2.74	3.76	4.67
	130	1.65	3.01	4.13	5.10
6 × 6 × $^1/_4$ wall thickness	120	1.66	2.91	3.94	4.84
	130	1.83	3.19	4.30	5.27
8 × 8 × $^1/_2$ wall thickness	120	1.27	2.50	3.52	4.42
	130	1.44	2.78	3.89	4.86
8 × 8 × $^3/_8$ wall thickness	120	1.43	2.67	3.69	4.59
	130	1.60	2.95	4.05	5.02
8 × 8 × $^1/_4$ wall thickness	120	1.62	2.87	3.89	4.78
	130	1.79	3.14	4.24	5.21

STEEL PIPE

NOMINAL PIPE SIZE (inches)	CLAY MASONRY DENSITY, POUNDS PER CUBIC FOOT	MINIMUM REQUIRED EQUIVALENT THICKNESS FOR FIRE-RESISTANCE RATING OF CLAY MASONRY PROTECTION ASSEMBLY, T_e (inches)			
		1 hour	2 hours	3 hours	4 hours
4 double extra strong 0.674 wall thickness	120	1.26	2.55	3.60	4.52
	130	1.42	2.82	3.96	4.95
4 extra strong 0.337 wall thickness	120	1.60	2.89	3.92	4.83
	130	1.77	3.16	4.28	5.25
4 standard 0.237 wall thickness	120	1.74	3.02	4.05	4.95
	130	1.92	3.29	4.40	5.37
5 double extra strong 0.750 wall thickness	120	1.17	2.44	3.48	4.40
	130	1.33	2.72	3.84	4.83
5 extra strong 0.375 wall thickness	120	1.55	2.82	3.85	4.76
	130	1.72	3.09	4.21	5.18
5 standard 0.258 wall thickness	120	1.71	2.97	4.00	4.90
	130	1.88	3.24	4.35	5.32
6 double extra strong 0.864 wall thickness	120	1.04	2.28	3.32	4.23
	130	1.19	2.60	3.68	4.67
6 extra strong 0.432 wall thickness	120	1.45	2.71	3.75	4.65
	130	1.62	2.99	4.10	5.08
6 standard 0.280 wall thickness	120	1.65	2.91	3.94	4.84
	130	1.82	3.19	4.30	5.27

For SI: 1 inch = 25.4 mm, 1 pound per cubic foot = 16.02 kg/m^3.

TABLE 722.5.1(7)
MINIMUM COVER (inch) FOR STEEL COLUMNS ENCASED IN NORMAL-WEIGHT CONCRETE[a] [FIGURE 722.5.1(6)(c)]

<table>
<tr><th rowspan="2">STRUCTURAL SHAPE</th><th colspan="5">FIRE-RESISTANCE RATING (hours)</th></tr>
<tr><th>1</th><th>1 1/2</th><th>2</th><th>3</th><th>4</th></tr>
<tr><td>W14 × 233</td><td rowspan="7">1</td><td rowspan="5">1</td><td rowspan="3">1</td><td rowspan="2">1 1/2</td><td>2</td></tr>
<tr><td>× 176</td><td rowspan="3">2 1/2</td></tr>
<tr><td>× 132</td><td rowspan="3">2</td></tr>
<tr><td>× 90</td><td rowspan="4">1 1/2</td></tr>
<tr><td>× 61</td><td rowspan="3">3</td></tr>
<tr><td>× 48</td><td rowspan="2">1 1/2</td><td rowspan="2">2 1/2</td></tr>
<tr><td>× 43</td></tr>
<tr><td>W12 × 152</td><td rowspan="5">1</td><td rowspan="2">1</td><td rowspan="2">1</td><td rowspan="3">2</td><td rowspan="2">2 1/2</td></tr>
<tr><td>× 96</td></tr>
<tr><td>× 65</td><td rowspan="3">1 1/2</td><td rowspan="3">1 1/2</td><td rowspan="3">3</td></tr>
<tr><td>× 50</td><td rowspan="2">2 1/2</td></tr>
<tr><td>× 40</td></tr>
<tr><td>W10 × 88</td><td>1</td><td rowspan="5">1 1/2</td><td rowspan="4">1 1/2</td><td>2</td><td rowspan="3">3</td></tr>
<tr><td>× 49</td><td rowspan="4">1</td><td rowspan="4">2 1/2</td></tr>
<tr><td>× 45</td></tr>
<tr><td>× 39</td><td rowspan="2">3 1/2</td></tr>
<tr><td>× 33</td><td>2</td></tr>
<tr><td>W8 × 67</td><td rowspan="6">1</td><td rowspan="2">1</td><td rowspan="3">1 1/2</td><td rowspan="3">2 1/2</td><td rowspan="2">3</td></tr>
<tr><td>× 58</td></tr>
<tr><td>× 48</td><td rowspan="4">1 1/2</td><td rowspan="3">3 1/2</td></tr>
<tr><td>× 31</td><td rowspan="3">2</td><td rowspan="3">3</td></tr>
<tr><td>× 21</td></tr>
<tr><td>× 18</td><td>4</td></tr>
<tr><td>W6 × 25</td><td rowspan="3">1</td><td>1 1/2</td><td>2</td><td rowspan="2">3</td><td>3 1/2</td></tr>
<tr><td>× 20</td><td rowspan="4">2</td><td rowspan="4">2 1/2</td><td rowspan="4">4</td></tr>
<tr><td>× 16</td><td rowspan="3">3 1/2</td></tr>
<tr><td>× 15</td><td rowspan="2">1 1/2</td></tr>
<tr><td>× 9</td></tr>
</table>

For SI: 1 inch = 25.4 mm.

a. The tabulated thicknesses are based upon the assumed properties of normal-weight concrete given in Table 722.5.1(2).

TABLE 722.5.1(8)
MINIMUM COVER (inch) FOR STEEL COLUMNS ENCASED IN STRUCTURAL LIGHTWEIGHT CONCRETE[a] [FIGURE 722.5.1(6)(c)]

<table>
<tr><th rowspan="2">STRUCTURAL SHAPE</th><th colspan="5">FIRE-RESISTANCE RATING (HOURS)</th></tr>
<tr><th>1</th><th>1 1/2</th><th>2</th><th>3</th><th>4</th></tr>
<tr><td>W14 × 233</td><td rowspan="5">1</td><td rowspan="5">1</td><td rowspan="4">1</td><td>1</td><td rowspan="2">1 1/2</td></tr>
<tr><td>× 193</td><td rowspan="3">1 1/2</td></tr>
<tr><td>× 74</td><td>2</td></tr>
<tr><td>× 61</td><td rowspan="2">2 1/2</td></tr>
<tr><td>× 43</td><td>1 1/2</td><td>2</td></tr>
<tr><td>W12 × 65</td><td rowspan="3">1</td><td rowspan="3">1</td><td rowspan="2">1</td><td>1 1/2</td><td>2</td></tr>
<tr><td>× 53</td><td rowspan="2">2</td><td rowspan="2">2 1/2</td></tr>
<tr><td>× 40</td><td>1 1/2</td></tr>
<tr><td>W10 × 112</td><td rowspan="4">1</td><td rowspan="4">1</td><td rowspan="3">1</td><td rowspan="2">1 1/2</td><td rowspan="2">2</td></tr>
<tr><td>× 88</td></tr>
<tr><td>× 60</td><td rowspan="2">2</td><td rowspan="2">2 1/2</td></tr>
<tr><td>× 33</td><td>1 1/2</td></tr>
<tr><td>W8 × 35</td><td rowspan="4">1</td><td rowspan="3">1</td><td rowspan="4">1 1/2</td><td rowspan="2">2</td><td>2 1/2</td></tr>
<tr><td>× 28</td><td rowspan="3">3</td></tr>
<tr><td>× 24</td><td rowspan="2">2 1/2</td></tr>
<tr><td>× 18</td><td>1 1/2</td></tr>
</table>

For SI: 1 inch = 25.4 mm.

a. The tabulated thicknesses are based upon the assumed properties of structural lightweight concrete given in Table 722.5.1(2).

TABLE 722.5.1(9)
MINIMUM COVER (inch) FOR STEEL COLUMNS IN NORMAL-WEIGHT PRECAST COVERS[a] [FIGURE 722.5.1(6)(a)]

<table>
<tr><th rowspan="2">STRUCTURAL SHAPE</th><th colspan="5">FIRE-RESISTANCE RATING (hours)</th></tr>
<tr><th>1</th><th>1½</th><th>2</th><th>3</th><th>4</th></tr>
<tr><td>W14 × 233</td><td rowspan="8">1½</td><td rowspan="4">1½</td><td rowspan="2">1½</td><td rowspan="3">2½</td><td>3</td></tr>
<tr><td>× 211</td><td rowspan="4">3½</td></tr>
<tr><td>× 176</td><td rowspan="2">2</td></tr>
<tr><td>× 145</td><td rowspan="3">3</td></tr>
<tr><td>× 109</td><td rowspan="4">2</td><td rowspan="4">2½</td></tr>
<tr><td>× 99</td><td rowspan="2">4</td></tr>
<tr><td>× 61</td><td rowspan="2">3½</td></tr>
<tr><td>× 43</td><td>4½</td></tr>
<tr><td>W12 × 190</td><td rowspan="7">1½</td><td rowspan="4">1½</td><td>1½</td><td rowspan="2">2½</td><td rowspan="2">3½</td></tr>
<tr><td>× 152</td><td rowspan="3">2</td></tr>
<tr><td>× 120</td><td rowspan="2">3</td><td rowspan="3">4</td></tr>
<tr><td>× 96</td></tr>
<tr><td>× 87</td><td rowspan="3">2</td><td rowspan="3">2½</td><td rowspan="3">3½</td></tr>
<tr><td>× 58</td><td rowspan="2">4½</td></tr>
<tr><td>× 40</td></tr>
<tr><td>W10 × 112</td><td rowspan="5">1½</td><td rowspan="2">1½</td><td rowspan="2">2</td><td rowspan="3">3</td><td>3½</td></tr>
<tr><td>× 88</td><td rowspan="3">4</td></tr>
<tr><td>× 77</td><td rowspan="3">2</td><td rowspan="3">2½</td></tr>
<tr><td>× 54</td><td rowspan="2">3½</td></tr>
<tr><td>× 33</td><td>4½</td></tr>
<tr><td>W8 × 67</td><td rowspan="6">1½</td><td>1½</td><td>2</td><td>3</td><td rowspan="3">4</td></tr>
<tr><td>× 58</td><td rowspan="3">2</td><td rowspan="3">2½</td><td rowspan="4">3½</td></tr>
<tr><td>× 48</td></tr>
<tr><td>× 28</td><td rowspan="3">4½</td></tr>
<tr><td>× 21</td><td rowspan="2">2½</td><td rowspan="2">3</td></tr>
<tr><td>× 18</td><td>4</td></tr>
<tr><td>W6 × 25</td><td rowspan="3">1½</td><td>2</td><td>2½</td><td rowspan="2">3½</td><td rowspan="4">4½</td></tr>
<tr><td>× 20</td><td rowspan="4">2½</td><td rowspan="4">3</td></tr>
<tr><td>× 16</td><td rowspan="3">4</td></tr>
<tr><td>× 12</td><td rowspan="2">2</td></tr>
<tr><td>× 9</td><td>5</td></tr>
</table>

For SI: 1 inch = 25.4 mm.

a. The tabulated thicknesses are based upon the assumed properties of normal-weight concrete given in Table 722.5.1(2).

TABLE 722.5.1(10)
MINIMUM COVER (inch) FOR STEEL COLUMNS IN STRUCTURAL LIGHTWEIGHT PRECAST COVERS[a] [FIGURE 722.5.1(6)(a)]

<table>
<tr><th rowspan="2">STRUCTURAL SHAPE</th><th colspan="5">FIRE-RESISTANCE RATING (hours)</th></tr>
<tr><th>1</th><th>1½</th><th>2</th><th>3</th><th>4</th></tr>
<tr><td>W14 × 233</td><td rowspan="8">1½</td><td rowspan="8">1½</td><td rowspan="5">1½</td><td rowspan="2">2</td><td>2½</td></tr>
<tr><td>× 176</td><td rowspan="5">3</td></tr>
<tr><td>× 145</td><td rowspan="5">2½</td></tr>
<tr><td>× 132</td></tr>
<tr><td>× 109</td></tr>
<tr><td>× 99</td><td rowspan="3">2</td></tr>
<tr><td>× 68</td><td rowspan="2">3½</td></tr>
<tr><td>× 43</td><td>3</td></tr>
<tr><td>W12 × 190</td><td rowspan="8">1½</td><td rowspan="8">1½</td><td rowspan="6">1½</td><td rowspan="3">2</td><td rowspan="2">2½</td></tr>
<tr><td>× 152</td></tr>
<tr><td>× 136</td><td rowspan="2">3</td></tr>
<tr><td>× 106</td><td rowspan="4">2½</td></tr>
<tr><td>× 96</td><td rowspan="4">3½</td></tr>
<tr><td>× 87</td></tr>
<tr><td>× 65</td><td rowspan="2">2</td></tr>
<tr><td>× 40</td><td>3</td></tr>
<tr><td>W10 × 112</td><td rowspan="7">1½</td><td rowspan="6">1½</td><td rowspan="3">1½</td><td>2</td><td rowspan="3">3</td></tr>
<tr><td>× 100</td><td rowspan="4">2½</td></tr>
<tr><td>× 88</td></tr>
<tr><td>× 77</td><td rowspan="4">2</td><td rowspan="4">3½</td></tr>
<tr><td>× 60</td></tr>
<tr><td>× 39</td><td rowspan="2">3</td></tr>
<tr><td>× 33</td><td>2</td></tr>
<tr><td>W8 × 67</td><td rowspan="5">1½</td><td rowspan="3">1½</td><td>1½</td><td>2½</td><td>3</td></tr>
<tr><td>× 48</td><td rowspan="3">2</td><td rowspan="4">3</td><td rowspan="3">3½</td></tr>
<tr><td>× 35</td></tr>
<tr><td>× 28</td><td rowspan="2">2</td></tr>
<tr><td>× 18</td><td>2½</td><td>4</td></tr>
<tr><td>W6 × 25</td><td rowspan="3">1½</td><td rowspan="3">2</td><td>2</td><td rowspan="2">3</td><td>3½</td></tr>
<tr><td>× 15</td><td rowspan="2">2½</td><td rowspan="2">4</td></tr>
<tr><td>× 9</td><td>3½</td></tr>
</table>

For SI: 1 inch = 25.4 mm.

a. The tabulated thicknesses are based upon the assumed properties of structural lightweight concrete given in Table 722.5.1(2).

722.5.1.2.2 Gypsum wallboard equivalent to concrete. The determination of the *fire resistance* of structural steel columns from Figure 722.5.1(4) is permitted for various thicknesses of gypsum wallboard as a function of the weight-to-heated-perimeter ratio (*W/D*) of the column. For structural steel columns with weight-to-heated-perimeter ratios (*W/D*) greater than 3.65, the thickness of gypsum wallboard required for specified *fire-resistance ratings* shall be the same as the thickness determined for a *W*14 × 233 wide flange shape.

❖ See the commentary to Section 722.

722.5.1.3 Sprayed fire-resistant materials. The *fire resistance* of wide-flange structural steel columns protected with sprayed fire-resistant materials, as illustrated in Figure 722.5.1(5), shall be permitted to be determined from the following expression:

$$R = [C_1(W/D) + C_2)h \qquad \textbf{(Equation 7-13)}$$

where:

R = Fire resistance (minutes).

h = Thickness of sprayed fire-resistant material (inches).

D = Heated perimeter of the structural steel column (inches).

C_1 and C_2 = Material-dependent constants.

W = Weight of structural steel columns (pounds per linear foot).

The *fire resistance* of structural steel columns protected with intumescent or mastic fire-resistant coatings shall be determined on the basis of fire-resistance tests in accordance with Section 703.2.

❖ This section sets forth procedures for determining the fire resistance of structural steel columns protected with spray-applied cementitious or mineral fiber fire protection materials. These procedures do not apply to intumescent or mastic fire-resistant coatings, which must be tested in accordance with the requirements found in Section 703.2. These procedures are based upon an empirical equation, which includes two material-dependent constants. As a result, in order to apply this equation, the values of these two constants must be determined for specific fire protection materials. The purpose of this section is to provide guidance for the determination of these constants so that the resulting equation will be reasonably accurate, and yet slightly conservative, over the range of column shapes for the test data that is available.

Two different techniques are available for determining the two constants. The first requires a knowledge of thermal conductivity and specific heat of the fire protection material at elevated temperatures. Data of this nature is both difficult and expensive to obtain with any reasonable degree of accuracy. As a result, this technique will probably not be widely used, and accordingly, it will not be described in this section.

The second technique involves the use of Equation 7-13 as a means for interpolating between ASTM E119 or UL 263 fire endurance test results on different structural steel columns. Since this technique will undoubtedly be the most widely used, it is described in detail in this section. It is, however, important to recognize that a wide variety of both large- and small-scale tests can be used to accurately determine the required constants. Inherent in Equation 7-13 is the fact that the ratio of fire endurance time to the thickness of fire protection material (*R/h*) varies linearly as a function of the increasing *W/D* ratio of the protected steel column. These concepts are graphically illustrated in the example that follows. It has been found that this assumption is reasonably accurate for lightweight [density less than 50 pcf (801 kg/m^3)] spray-applied materials.

If ASTM E119 or UL 263 fire endurance test results are available for a specific fire protection material on two different structural steel column shapes, the constants C_1 and C_2 can be determined directly. The resulting equation can then be used to determine the thickness of fire protection material required for any specified fire endurance rating when applied to structural steel columns with *W/D* ratios between the largest and smallest column for the actual test results that are available.

To determine the constants, at least four ASTM E119 or UL 263 fire endurance tests or a combination of two ASTM E119 or UL 263 fire endurance tests and six small-scale tests [3-foot-long (914 mm) specimens] are conducted. If the results of the small-scale tests are used, at least two of the test assemblies are necessary to duplicate the ASTM E119 or UL 263 test assemblies for the purpose of establishing correlation. Regardless of the combination of small- and large-scale tests selected, at least two tests are conducted on the largest and two on the smallest columns to establish the limits of applicability to the resulting equation. The constants C_1 and C_2 are to be determined on the basis of the lowest ratios of fire endurance time to fire protection thickness (*R/h*) for these columns.

In addition, the test data need to be evaluated with respect to the assumption that the ratio of fire endurance to fire protection thickness (*R/h*) is reasonably constant for a given column shape [(*W/D*) ratio]. The tests conducted on columns of the same shape are designed so that the resulting fire endurance times are approximately $1^1/_2$ and $3^1/_2$ hours. In evaluating the *R/h* ratios resulting from tests on the same column shape, differences in the range of 10 percent are typical. Differences greater than 20 percent may, however, suggest that Equation 7-13 is not applicable to the specific fire protection material under consideration, and further examination of the test data is warranted.

EXAMPLE:

GIVEN: The fire endurance test data given in Commentary Figure 722.5.1.3.

FIND: The thickness of a spray-applied cementitious material required for a 3-hour rating on a W12 × 136.

SOLUTION:

1. Determine C_1 and C_2. From the end points of the graph:

   R/h = 75 for a W/D ratio of 0.6 and

   R/h = 200 for a W/D ratio of 2.5.

   From this, two equations with two unknowns are created:

   $75 = C_1 (0.6) + C_2$.

   $200 = C_1 (2.5) + C_2$.

   Solving these results in: C_1 = 65.79; C_2 = 35.53 and Equation 7-13, with h in the denominator, becomes:

   $R/h = 65.79 (W/D) + 35.53$.

2. From Table 722.5.1(1), a W12 × 136 has a W/D ratio of 1.82.

3. The required fire resistance is 180 minutes. Calculating for h yields:

   $180 = [(65.79)(1.82) + 35.53]h$

   h = 1.16 inches (29 mm).

Therefore, the required thickness of the spray-applied material is 1.16 inches (29 mm).

722.5.1.3.1 Material-dependent constants. The material-dependent constants, C_1 and C_2, shall be determined for specific fire-resistant materials on the basis of standard fire endurance tests in accordance with Section 703.2. Unless evidence is submitted to the *building official* substantiating a broader application, this expression shall be limited to determining the *fire resistance* of structural steel columns with weight-to-heated-perimeter ratios (W/D) between the largest and smallest columns for which standard fire-resistance test results are available.

❖ See the commentary to Section 722.

722.5.1.3.2 Identification. Sprayed fire-resistant materials shall be identified by density and thickness required for a given *fire-resistance rating*.

❖ See the commentary to Section 722.

722.5.1.4 Concrete-protected columns. The *fire resistance* of structural steel columns protected with concrete, as illustrated in Figure 722.5.1(6)(a) and (b), shall be permitted to be determined from the following expression:

$$R = R_o(1 + 0.03_m) \qquad \textbf{(Equation 7-14)}$$

where:

$$R_o = 10\,(W/D)^{0.7} + 17\,(h^{1.6}/k_c^{\ 0.2}) \times [1 + 26\,\{H/p_c c_c h\,(L + h)\}^{0.8}]$$

As used in these expressions:

R = Fire endurance at equilibrium moisture conditions (minutes).

R_o = Fire endurance at zero moisture content (minutes).

m = Equilibrium moisture content of the concrete by volume (percent).

W = Average weight of the structural steel column (pounds per linear foot).

D = Heated perimeter of the structural steel column (inches).

h = Thickness of the concrete cover (inches).

k_c = Ambient temperature thermal conductivity of the concrete (Btu/hr ft °F).

H = Ambient temperature thermal capacity of the steel column = 0.11W (Btu/ ft °F).

p_c = Concrete density (pounds per cubic foot).

c_c = Ambient temperature specific heat of concrete (Btu/lb °F).

L = Interior dimension of one side of a square concrete box protection (inches).

❖ **EXAMPLE:**

GIVEN: A W8 × 28 steel column encased in lightweight concrete [density = 110 pcf (1762 kg/m^3)] shown in Code Figure 722.5.1(4), with all reentrant spaces filled with concrete cover 1.25 inches (32 mm)

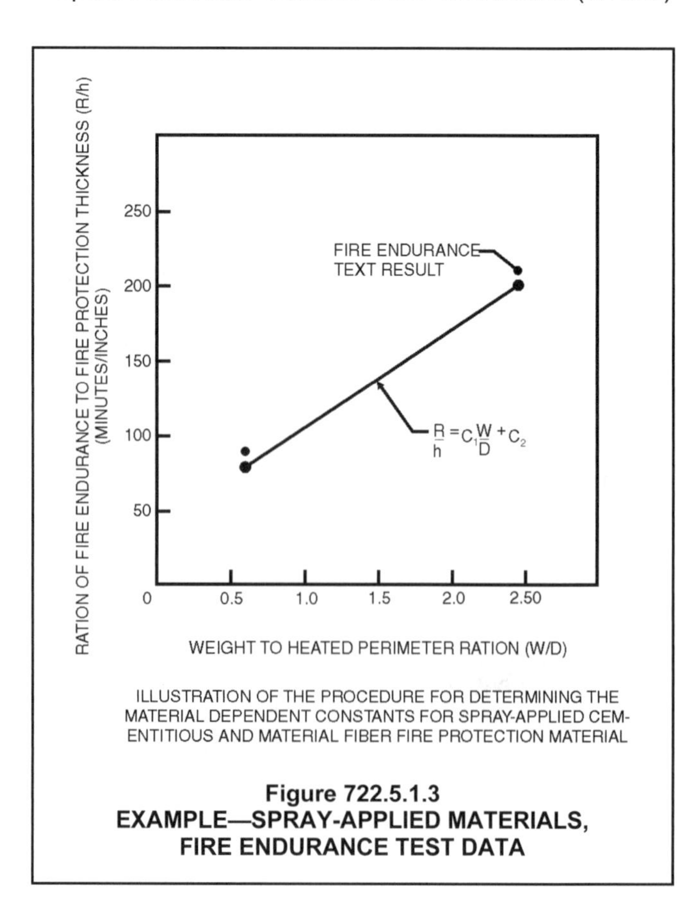

Figure 722.5.1.3
EXAMPLE—SPRAY-APPLIED MATERIALS, FIRE ENDURANCE TEST DATA

and a moisture content of 5 percent. The web thickness is 0.285[2].

FIND: The fire-resistance rating.

SOLUTION:

D = 4(6.535) + 2(8.06) - 2(0.285) = 41.69 inches (1059 mm).

W/D = 28 lb/ft/41.69² = 0.67 lb/ft-in.

h = 1.25 inches (32 mm).

k_c = 0.35 Btu/hr × °F.

c_c = 0.20 Btu/lb × °F.

r_c = 110 pcf (1762 kg/m³).

L = (6.535 + 8.06)/2 = 7.30 inches (185 mm).

A_s = 8.25 square inches (.005 m²).

H = 11(28) + [(110)(0.20)/144][(6.535)(8.06) - 8.25] = 9.87

$$R_0 = 10(0.67)^{0.7} + 17\left(\frac{(1.25)^{1.6}}{(0.35)^{0.2}}\right)$$

$$\left(1 + 26\left[\frac{9.87}{(110)(0.2)(1.25)(7.3 + 1.25)}\right]^{0.8}\right)$$

R_o = 99 minutes.

Therefore,

R = 99[1 + 0.33(5)] = 114 minutes.

For comparison purposes, the minimum cover requirement for a *W*8 × 28 steel column from Table 722.5.1(8) is 1 inch (25 mm) for a 1¹/₂-hour rating. The column does not quite meet the required fire rating for a 2-hour column since 1.5 inches (38 mm) is required. Therefore, the fire resistance of the column is 1¹/₂ hours.

722.5.1.4.1 Reentrant space filled. For wide-flange structural steel columns completely encased in concrete with all reentrant spaces filled [Figure 722.5.1(6)(c)], the thermal capacity of the concrete within the reentrant spaces shall be permitted to be added to the thermal capacity of the steel column, as follows:

$$H = 0.11\ W + (p_c c_c/144)\ (b_f d - A_s)$$

(Equation 7-15)

where:

b_f = Flange width of the structural steel column (inches).

d = Depth of the structural steel column (inches).

A_s = Cross-sectional area of the steel column (square inches).

❖ See the commentary to Section 722.

722.5.1.4.2 Concrete properties unknown. If specific data on the properties of concrete are not available, the values given in Table 722.5.1(2) are permitted.

❖ See the commentary to Section 722.

722.5.1.4.3 Minimum concrete cover. For structural steel column encased in concrete with all reentrant spaces filled, Figure 722.5.1(6)(c) and Tables 722.5.1(7) and 722.5.1(8) indicate the thickness of concrete cover required for various *fire-resistance ratings* for typical wide-flange sections. The thicknesses of concrete indicated in these tables apply to structural steel columns larger than those listed.

❖ See the commentary to Section 722.

722.5.1.4.4 Minimum precast concrete cover. For structural steel columns protected with precast concrete column covers as shown in Figure 722.5.1(6)(a), Tables 722.5.1(9) and 722.5.1(10) indicate the thickness of the column covers required for various *fire-resistance ratings* for typical wide-flange shapes. The thicknesses of concrete given in these tables apply to structural steel columns larger than those listed.

❖ See the commentary to Section 722.

722.5.1.4.5 Masonry protection. The *fire resistance* of structural steel columns protected with concrete masonry units or clay masonry units as illustrated in Figure 722.5.1(7) shall be permitted to be determined from the following expression:

$$R = 0.17\ (W/D)^{0.7} + [0.285\ (T_e^{1.6}/K^{0.2})]\ [1.0 + 42.7\ \{(A_s/d_m\ T_e)/(0.25p + T_e)\}^{0.8}]$$

(Equation 7-16)

where:

R = *Fire-resistance rating* of column assembly (hours).

W = Average weight of structural steel column (pounds per foot).

D = Heated perimeter of structural steel column (inches) [see Figure 722.5.1(7)].

T_e = Equivalent thickness of concrete or clay masonry unit (inches) (see Table 722.3.2 Note a or Section 722.4.1).

K = Thermal conductivity of concrete or clay masonry unit (Btu/hr · ft · °F) [see Table 722.5.1(3)].

A_s = Cross-sectional area of structural steel column (square inches).

d_m = Density of the concrete or clay masonry unit (pounds per cubic foot).

p = Inner perimeter of concrete or clay masonry protection (inches) [see Figure 722.5.1(7)].

❖ See the commentary to Section 722.

722.5.1.4.6 Equivalent concrete masonry thickness. For structural steel columns protected with concrete masonry, Table 722.5.1(5) gives the equivalent thickness of concrete masonry required for various *fire-resistance ratings* for typical column shapes. For structural steel columns protected with clay masonry, Table 722.5.1(6) gives the equivalent thickness of concrete masonry required for various *fire-resistance ratings* for typical column shapes.

❖ See the commentary to Section 722.

722.5.2 Structural steel beams and girders. The *fire-resistance ratings* of structural steel beams and girders shall be based upon the size of the element and the type of protection provided in accordance with this section.

❖ See the commentary to Section 722.

722.5.2.1 Determination of *fire resistance*. These procedures establish a basis for determining resistance of structural steel beams and girders that differ in size from that specified in *approved* fire-resistance-rated assemblies as a function of the thickness of fire-resistant material and the weight (W) and heated perimeter (D) of the beam or girder. As used in these sections, W is the average weight of a *structural steel element* in pounds per linear foot (plf). The heated perimeter, D, is the inside perimeter of the fire-resistant material in inches as illustrated in Figure 722.5.2.

❖ See the commentary to Section 722.

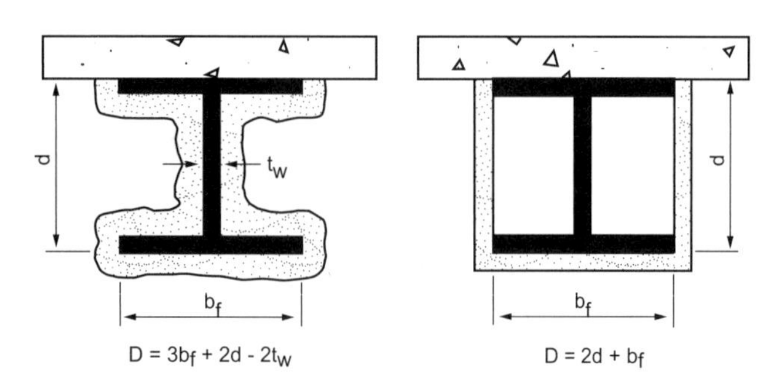

FIGURE 722.5.2
DETERMINATION OF THE HEATED PERIMETER OF STRUCTURAL STEEL BEAMS AND GIRDERS

722.5.2.1.1 Weight-to-heated perimeter. The weight-to-heated-perimeter ratios (W/D), for both contour and box fire-resistant protection profiles, for the wide flange shapes most often used as beams or girders are given in Table 722.5.1(4). For different shapes, the weight-to-heated-perimeter ratios (W/D) shall be determined in accordance with the definitions given in this section.

❖ See the commentary to Section 722.

722.5.2.1.2 Beam and girder substitutions. Except as provided for in Section 722.5.2.2, structural steel beams in *approved* fire-resistance-rated assemblies shall be considered the minimum permissible size. Other beam or girder shapes shall be permitted to be substituted provided that the weight-to-heated-perimeter ratio (W/D) of the substitute beam is equal to or greater than that of the beam specified in the *approved* assembly.

❖ This section defines a general rule for the substitution of different steel beam and girder shapes in all fire-resistant assemblies. In the past, the substitution of larger beams has been permitted based upon the thickness of web and flange elements. Extensive research at UL has proven that the heat transfer to a protected steel beam or girder is a direct function of the W/D ratio. As a result, beam substitutions should be based upon W/D ratios, as defined in this section. The significance of the thickness of web and flange elements is inherently included in the determination of W/D ratios. Code Figure 722.5.2 illustrates the procedure for determining heated perimeters (D), and Table 722.5.1(4) provides W/D ratios for the most commonly used wide flange beam and girder shapes.

722.5.2.2 Sprayed fire-resistant materials. The provisions in this section apply to structural steel beams and girders protected with sprayed fire-resistant materials. Larger or smaller beam and girder shapes shall be permitted to be substituted for beams specified in *approved* unrestrained or restrained fire-resistance-rated assemblies, provided that the thickness of the fire-resistant material is adjusted in accordance with the following expression:

$$h_2 = h_1\ [(W_1 / D_1) + 0.60] / [(W_2 / D_2) + 0.60]$$

(Equation 7-17)

where:

h = Thickness of sprayed fire-resistant material in inches.

W = Weight of the structural steel beam or girder in pounds per linear foot.

D = Heated perimeter of the structural steel beam in inches.

Subscript 1 refers to the beam and fire-resistant material thickness in the *approved* assembly.

Subscript 2 refers to the substitute beam or girder and the required thickness of fire-resistant material.

The *fire resistance* of structural steel beams and girders protected with intumescent or mastic fire-resistant coatings shall be determined on the basis of fire-resistance tests in accordance with Section 703.2.

❖ This section defines an equation for adjusting the thickness of spray-applied cementitious and mineral fiber materials as a function of W/D ratios. This equation was developed by UL, and appropriate limitations have been included. The minimum W/D ratio of 0.37 in Section 722.5.2.2.1 will prevent the use of this equation for determining the fire resistance of very small shapes that have not been tested. The $^3/_8$-inch (9.5 mm) minimum thickness of protection is a practical limit based upon the most commonly used spray-applied fire protection materials.

EXAMPLE:

GIVEN: Determine the thickness of spray-applied fire protection required to provide a 2-hour fire-resistance rating for a W12 × 16 beam to be substituted for a W8 × 15 beam requiring 1.44 inches (37 mm) of protection for the same rating.

SOLUTION:

From Table 722.5.1(4):

W_1/D_1 = 0.54 for W8 × 15

W_2/D_2 = 0.45 for W12 × 16

h_1 = 1.44 inches.

$$h_2 = \left[\frac{0.54 + 0.60}{0.45 + 0.60}\right] + 1.44$$

= 1.56 inches (39 mm).

It is noted within this section that intumescent or mastic fire-resistant coatings are required to be tested in accordance with the requirements of Section 703.2.

722.5.2.2.1 Minimum thickness. The use of Equation 7-17 is subject to the following conditions:

1. The weight-to-heated-perimeter ratio for the substitute beam or girder (W_2/D_2) shall be not less than 0.37.
2. The thickness of fire protection materials calculated for the substitute beam or girder (T_1) shall be not less than $^3/_8$ inch (9.5 mm).
3. The unrestrained or restrained beam rating shall be not less than 1 hour.
4. Where used to adjust the material thickness for a restrained beam, the use of this procedure is limited to structural steel sections classified as compact in accordance with AISC 360.

❖ See the commentary to Section 722.5.2.2.

722.5.2.3 Structural steel trusses. The *fire resistance* of structural steel trusses protected with fire-resistant materials sprayed to each of the individual truss elements shall be permitted to be determined in accordance with this section. The thickness of the fire-resistant material shall be determined in accordance with Section 722.5.1.3. The weight-to-heated-perimeter ratio (*W/D*) of truss elements that can be simultaneously exposed to fire on all sides shall be determined on the same basis as columns, as specified in Section 722.5.1.1. The weight-to-heated-perimeter ratio (*W/D*) of truss elements that directly support floor or roof assembly shall be determined on the same basis as beams and girders, as specified in Section 722.5.2.1.

The *fire resistance* of structural steel trusses protected with intumescent or mastic fire-resistant coatings shall be determined on the basis of fire-resistance tests in accordance with Section 703.2.

❖ This section describes the application of spray-applied fire protection to structural steel trusses when each truss element is individually protected with spray-applied materials. The thickness of protection is determined using the column equation specified in Section 722.5.1.3. For trusses, the column equation is more technically correct than the beam equation, since it requires greater thicknesses of protection than the beam equation. Most truss elements can be exposed to fire on all four sides simultaneously. As a result, the heated perimeter of truss elements should be determined in the same manner as for columns. However, an exception is included for top chord elements that directly support floor or roof construction. The heated perimeter of such elements should be determined in the same manner as for beams and girders.

This section notes that intumescent and mastic fire-resistant materials must be tested in accordance with the requirements in Section 703.2.

722.6 Wood assemblies. The provisions of this section contain procedures by which the *fire-resistance ratings* of wood assemblies are established by calculations.

❖ The information contained in this section is based on Technical Paper No. 222 of the Division of Building Research Council of Canada entitled "Fire Endurance of Light Framed and Miscellaneous Assemblies" and NBS Report BMS 92. The fire-resistance rating is equal to the sum of the time assigned to the membranes [see Table 722.6.2(1)], the time assigned to the framing members [see Table 722.6.2(2)] and the time assigned for additional contribution by other protective measures, such as insulation [see Table 722.6.2(5)]. The membrane on the unexposed side is not included in the calculations. It is assumed that once the structural members fail, the entire assembly fails. The time assigned to an individual membrane is the individual contribution to the overall fire-resistance rating of the complete assembly. The assigned time is not to be confused with the fire-resistance rating of the membranes. Fire-resistance rating takes into account the rise in temperature on the unexposed side of the membrane. Times that have been assigned to membranes on the fire-exposed side of the assembly are based on their ability to remain in place during fire tests.

The fire-resistance rating of wood-frame assemblies is equal to the sum of the time assigned to the various components (membranes) on the fire-exposed side and the structural members. Interior walls and partitions, exterior walls within 5 feet (1524 mm) of a lot line or assumed lot line and some shaft walls must be designed for fire resistance by assuming that either side of the wall is exposed to fire.

Mineral fiber insulation provides additional protection to wood studs by shielding the studs from exposure to the furnace, thus delaying the time of collapse. The use of reinforcement in the membrane exposed to fire also adds to the fire resistance by extending the time to failure. Special care must be taken to ensure that all insulation materials used in conjunction with this method satisfy the weight criteria of Table 722.6.2(5).

The following are examples of how to use this section:

EXAMPLE 1:

GIVEN: The exterior wall assembly shown in Commentary Figure 722.6 having a layer of $^{15}/_{32}$-inch (12 mm) wood structural panel bonded with exterior glue covered with a layer of $^5/_8$-inch (15.8 mm) gypsum wallboard and attached to studs spaced at 16 inches (406 mm) on center (o.c.).

FIND: Does the wall assembly qualify as a 1-hour fire-resistant wall assembly?

SOLUTION: From Tables 722.6.2(1) and 722.6.2(2):

Wood structural panel	= 10 minutes.
Gypsum wallboard	= 30 minutes.
Wood studs	= 20 minutes.
Total	= 60 minutes.

Therefore, the wall does qualify as a 1-hour wall. If the wall is an interior wall, both sides would be required to be fire protected with at least 40 minutes of membrane coverings (60 minutes - 20 minutes for wood frame).

It should be noted that Section 722.6.2.3 requires the exterior side to be protected in accordance with Table 722.6.2(3) or any membrane that is assigned a time of at least 15 minutes as listed in Table 722.6.2(1).

It should also be noted that if the wall cavities between the studs are filled with mineral fiber batts weighing no less than that specified in Table 722.6.2(5), the $^{15}/_{32}$-inch (12 mm) plywood membrane layer could be eliminated because the insulation adds 15 minutes of fire resistance, as indicated in Section 722.6.2.5 and Table 722.6.2(5). Thus, adding the contribution times for the $^{5}/_{8}$-inch (15.9 mm) gypsum board, the wood framing and the insulation (30 minutes + 20 minutes + 15 minutes), the resultant rating for the wall would be 65 minutes and meet a 1-hour fire-resistance rating.

EXAMPLE 2:

GIVEN: A floor/ceiling assembly using wood joists spaced at 16 inches (406 mm) o.c., protected on the bottom side (ceiling side) with two layers of $^{1}/_{2}$-inch (12.7 mm) Type X gypsum wallboard and protected on the upper side (floor side) with a $^{15}/_{32}$-inch (12 mm) plywood subfloor, a $^{3}/_{8}$-inch (15.9 mm) panel-type underlayment and carpet.

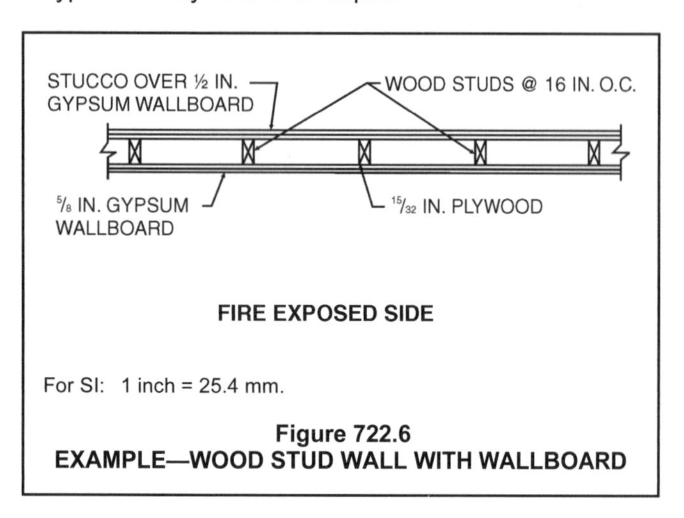

For SI: 1 inch = 25.4 mm.

Figure 722.6
EXAMPLE—WOOD STUD WALL WITH WALLBOARD

FIND: Does the floor/ceiling assembly meet the requirements of a 1-hour fire-resistance-rated assembly?

SOLUTION: Referring to Sections 722.6.2.1 and 722.6.2.4, Table 722.6.2(1) indicates that the time contribution for each layer of $^{1}/_{2}$-inch (12.7 mm) Type X gypsum wallboard is 25 minutes. The time of contribution for wood joists 16 inches (406 mm) o.c. is listed in Table 722.6.2(2) as 10 minutes. By adding the two layers of gypsum board (2 × 25 minutes) to the wood frame (10 minutes), a fire-resistance rating of 60 minutes, or 1 hour, can be obtained. It should be noted that Section 722.6.2.4 requires the upper membrane to be specified as in Table 722.6.2(4) or any membrane that has a time of contribution of at least 15 minutes as listed in Table 722.6.2(1).

If the above example had been a roof/ceiling assembly, the upper membrane would have been treated the same. If the proposed assembly is a ceiling with an attic above, Section 722.6.2.4 notes the exception to Section 711.2.6, which allows the elimination of the upper membrane.

TABLE 722.6.2(1)
TIME ASSIGNED TO WALLBOARD MEMBRANES[a, b, c, d]

DESCRIPTION OF FINISH	TIME[e] (minutes)
$^{3}/_{8}$-inch wood structural panel bonded with exterior glue	5
$^{15}/_{32}$-inch wood structural panel bonded with exterior glue	10
$^{19}/_{32}$-inch wood structural panel bonded with exterior glue	15
$^{3}/_{8}$-inch gypsum wallboard	10
$^{1}/_{2}$-inch gypsum wallboard	15
$^{5}/_{8}$-inch gypsum wallboard	30
$^{1}/_{2}$-inch Type X gypsum wallboard	25
$^{5}/_{8}$-inch Type X gypsum wallboard	40
Double $^{3}/_{8}$-inch gypsum wallboard	25
$^{1}/_{2}$-inch + $^{3}/_{8}$-inch gypsum wallboard	35
Double $^{1}/_{2}$-inch gypsum wallboard	40

For SI: 1 inch = 25.4 mm.

a. These values apply only where membranes are installed on framing members that are spaced 16 inches o.c. or less.

b. Gypsum wallboard installed over framing or furring shall be installed so that all edges are supported, except $^{5}/_{8}$-inch Type X gypsum wallboard shall be permitted to be installed horizontally with the horizontal joints staggered 24 inches each side and unsupported but finished.

c. On wood frame floor/ceiling or roof/ceiling assemblies, gypsum board shall be installed with the long dimension perpendicular to framing members and shall have all joints finished.

d. The membrane on the unexposed side shall not be included in determining the fire resistance of the assembly. Where dissimilar membranes are used on a wall assembly, the calculation shall be made from the least fire-resistant (weaker) side.

e. The time assigned is not a finished rating.

The fastening requirements for assemblies developed by Section 722.6 should be in accordance with Chapter 23 as stated in Section 722.6.2.6.

722.6.1 General. This section contains procedures for calculating the *fire-resistance ratings* of walls, floor/ceiling and roof/ceiling assemblies based in part on the standard method of testing referenced in Section 703.2.

❖ See the commentary to Section 722.6.

722.6.1.1 Maximum fire-resistance rating. *Fire-resistance ratings* calculated for assemblies using the methods in Section 722.6 shall be limited to a maximum of 1 hour.

❖ See the commentary to Section 722.6.

722.6.1.2 Dissimilar membranes. Where dissimilar membranes are used on a wall assembly that requires consideration of fire exposure from both sides, the calculation shall be made from the least fire-resistant (weaker) side.

❖ All wood wall assemblies that require fire resistance from both sides, by which the fire-resistance rating is calculated, must have a membrane on each side. However, where dissimilar membranes are utilized, the calculation shall be made from the least fire-resistant side.

722.6.2 Walls, floors and roofs. These procedures apply to both load-bearing and nonload-bearing assemblies.

❖ See the commentary to Section 722.6.

TABLE 722.6.2(2)
TIME ASSIGNED FOR CONTRIBUTION OF WOOD FRAME [a, b, c]

DESCRIPTION	TIME ASSIGNED TO FRAME (minutes)
Wood studs 16 inches o.c.	20
Wood floor and roof joists 16 inches o.c.	10

For SI: 1 inch = 25.4 mm.

a. This table does not apply to studs or joists spaced more than 16 inches o.c.

b. All studs shall be nominal 2 × 4 and all joists shall have a nominal thickness of not less than 2 inches.

c. Allowable spans for joists shall be determined in accordance with Sections 2308.4.2.1, 2308.7.1 and 2308.7.2.

TABLE 722.6.2(3)
MEMBRANE[a] ON EXTERIOR FACE OF WOOD STUD WALLS

SHEATHING	PAPER	EXTERIOR FINISH
$^5/_8$-inch T & G lumber $^5/_{16}$-inch exterior glue wood structural panel $^1/_2$-inch gypsum wallboard $^5/_8$-inch gypsum wallboard $^1/_2$-inch fiberboard	Sheathing paper	Lumber siding Wood shingles and shakes $^1/_4$-inch fiber-cement lap, panel or shingle siding $^1/_4$-inch wood structural panels-exterior type $^1/_4$-inch hardboard Metal siding Stucco on metal lath Masonry veneer Vinyl siding
None	—	$^3/_8$-inch exterior-grade wood structural panels

For SI: 1 inch = 25.4 mm.

a. Any combination of sheathing, paper and exterior finish is permitted.

TABLE 722.6.2(4)
FLOORING OR ROOFING OVER WOOD FRAMING[a]

ASSEMBLY	STRUCTURAL MEMBERS	SUBFLOOR OR ROOF DECK	FINISHED FLOORING OR ROOFING
Floor	Wood	$^{15}/_{32}$-inch wood structural panels or $^{11}/_{16}$-inch T & G softwood	Hardwood or softwood flooring on building paper resilient flooring, parquet floor felted-synthetic fiber floor coverings, carpeting, or ceramic tile on $^1/_4$-inch-thick fiber-cement underlayment or $^3/_8$-inch-thick panel-type underlay Ceramic tile on $1^1/_4$-inch mortar bed
Roof	Wood	$^{15}/_{32}$-inch wood structural panels or $^{11}/_{16}$-inch T & G softwood	Finished roofing material with or without insulation

For SI: 1 inch = 25.4 mm.

a. This table applies only to wood joist construction. It is not applicable to wood truss construction.

722.6.2.1 Fire-resistance rating of wood frame assemblies. The *fire-resistance rating* of a wood frame assembly is equal to the sum of the time assigned to the membrane on the fire-exposed side, the time assigned to the framing members and the time assigned for additional contribution by other protective measures such as insulation. The membrane on the unexposed side shall not be included in determining the *fire resistance* of the assembly.

❖ See the commentary to Section 722.6.

722.6.2.2 Time assigned to membranes. Table 722.6.2(1) indicates the time assigned to membranes on the fire-exposed side.

❖ See the commentary to Section 722.6.

722.6.2.3 Exterior walls. For an exterior wall with a *fire separation distance* greater than 10 feet (3048 mm), the wall is assigned a rating dependent on the interior membrane and the framing as described in Tables 722.6.2(1) and 722.6.2(2). The membrane on the outside of the nonfire-exposed side of exterior walls with a *fire separation distance* greater than 10 feet (3048 mm) shall consist of sheathing, sheathing paper and siding as described in Table 722.6.2(3).

❖ See the commentary to Section 722.6.

722.6.2.4 Floors and roofs. In the case of a floor or roof, the standard test provides only for testing for fire exposure from below. Except as noted in Section 703.3, Item 5, floor or roof assemblies of wood framing shall have an upper membrane consisting of a subfloor and finished floor conforming to Table 722.6.2(4) or any other membrane that has a contribution to *fire resistance* of not less than 15 minutes in Table 722.6.2(1).

❖ See the commentary to Section 722.6.

722.6.2.5 Additional protection. Table 722.6.2(5) indicates the time increments to be added to the *fire resistance* where glass fiber, rockwool, slag mineral wool or cellulose insulation is incorporated in the assembly.

❖ See the commentary to Section 722.6.

722.6.2.6 Fastening. Fastening of wood frame assemblies and the fastening of membranes to the wood framing members shall be done in accordance with Chapter 23.

❖ See the commentary to Section 722.6.

Bibliography

The following resource materials were used in the preparation of the commentary for this chapter of the code.

Allen L.W., M. Galbreath and W.W. Stanzak. *Fire Endurance Tests on Unit Masonry Walls with Gypsum Wallboard (NRCC 13901).* Division of Building Research, National Research Council of Canada, 1980.

Analytical Methods of Determining Fire Endurance of Concrete and Masonry Members — Model Code Approved Procedures (SR267.01B). Skokie, IL: Concrete and Masonry Industry Fire Safety Committee, 1987.

ASTM E119-2012A, *Test Methods for Fire Tests of Building Construction and Materials.* West Conshohocken, PA: ASTM International, 2012.

ASTM E136-12, *Test Method for Behavior of Materials in a Vertical Tube Furnace at 750°C.* West Conshohocken, PA: ASTM International, 2012.

ASTM E814-13, *Test Method for Fire Tests of Through-Penetration Fire Stops.* West Conshohocken, PA: ASTM International, 2013.

TABLE 722.6.2(4)
FLOORING OR ROOFING OVER WOOD FRAMING[a]

ASSEMBLY	STRUCTURAL MEMBERS	SUBFLOOR OR ROOF DECK	FINISHED FLOORING OR ROOFING
Floor	Wood	$^{15}/_{32}$-inch wood structural panels or $^{11}/_{16}$-inch T & G softwood	Hardwood or softwood flooring on building paper resilient flooring, parquet floor felted-synthetic fiber floor coverings, carpeting, or ceramic tile on $^{1}/_{4}$-inch-thick fiber-cement underlayment or $^{3}/_{8}$-inch-thick panel-type underlay Ceramic tile on $1^{1}/_{4}$-inch mortar bed
Roof	Wood	$^{15}/_{32}$-inch wood structural panels or $^{11}/_{16}$-inch T & G softwood	Finished roofing material with or without insulation

For SI: 1 inch = 25.4 mm.

a. This table applies only to wood joist construction. It is not applicable to wood truss construction.

TABLE 722.6.2(5)
TIME ASSIGNED FOR ADDITIONAL PROTECTION

DESCRIPTION OF ADDITIONAL PROTECTION	FIRE RESISTANCE (minutes)
Add to the *fire-resistance rating* of wood stud walls if the spaces between the studs are completely filled with glass fiber mineral wool batts weighing not less than 2 pounds per cubic foot (0.6 pound per square foot of wall surface) or rockwool or slag material wool batts weighing not less than 3.3 pounds per cubic foot (1 pound per square foot of wall surface), or cellulose insulation having a nominal density not less than 2.6 pounds per cubic foot.	15

For SI: 1 pound/cubic foot = 16.0185 kg/m^3.

BMS 92, *Fire Resistance Classifications, Building Materials and Structures*. National Bureau of Standards, 1942.

CPSC 16 CFR, Part 1209-(2002), *Interim Safety Standard for Cellulose Insulation*. Washington, DC: Consumer Product Safety Commission, 2002.

CPSC 16 CFR, Part 1404-(2002), *Cellulose Insulation*. Washington, DC: Consumer Product Safety Commission, 2002.

Designing Fire Protection for Steel Beams. Washington, DC: American Iron and Steel Institute, 1984.

Designing Fire Protection for Steel Columns. Washington, DC: American Iron and Steel Institute, 1980.

Designing Fire Protection for Steel Trusses. Washington, DC: American Iron and Steel Institute, 1981.

"Fire Endurance of Concrete Slabs as Influenced by Thickness, Aggregate Type and Moisture." *PCA Bulletin 223*. Skokie, IL: Portland Cement Association.

Fire Resistance Classifications of Building Construction. National Bureau of Standards, 1942.

Fire Resistance Directory. Northbrook, IL: Underwriters Laboratories Inc., 1996.

Fire Tests of Building Columns. Associated Factory Mutual Insurance Companies, 1921.

"Fire Tests of Joints Between Precast Concrete Wall Units: Effect of Various Joint Treatments." *PCI Journal*. September-October, 1975.

GA 600-12, *Fire Resistance Design Manual*. Washington, DC: Gypsum Association, 2012.

Galbreath, Murdock. "Fire Resistance of Light Framed and Miscellaneous Assemblies." Technical Paper No. 222 of the Division of Building Research, National Research Council of Canada.

Hull and Inburg. "Fire Resistance of Concrete Columns." Technological Papers of the Bureau of Standards, No. 271, February 24, 1925.

"Investigations on Building Fires, Part VI: Fire Resistance of Reinforced Concrete Columns," *National Building Studies Research Paper No. 18*. London, England: Her Majesty's Stationary Office, 1953.

NCMA TEK 6A-91, *Fire Resistance Rating of Concrete Masonry Assembly*. Herndon, VA: National Concrete Masonry Association, 1991.

NFPA 72-13, *National Fire Alarm Code*. Quincy, MA: National Fire Protection Association, 2013.

NFPA 80-13, *Fire Doors and Windows*. Quincy, MA: National Fire Protection Association, 2013.

NFPA 252-12, *Methods of Fire Tests of Door Assemblies*. Quincy, MA: National Fire Protection Association, 2012.

NRCC 17724, *Supplement to the National Building Code of Canada 1980*. Associate Committee on the National Building Code. National Research Council of Canada, 1980.

PCA Publication T-140, *Fire Resistance of Referenced Concrete Floors*. Skokie, IL: Portland Cement Association.

PCI MNL 124-11, *Design for Fire Resistance of Precast Prestressed Concrete*. Chicago: Precast/Prestressed Concrete Institute, 2011.

Reinforced Concrete Fire Resistance. Schaumburg, IL: Concrete Reinforcing Steel Institute, 1980.

UL 103-2010, *Chimneys, Factory Built, Residential Type and Building Heating Appliance—with Revisions through February 1996*. Northbrook, IL: Underwriters Laboratories Inc., 2010.

UL 127-2011, *Factory-Built Fireplaces*. Northbrook, IL: Underwriters Laboratories Inc., 2011.

UL 263-11, *Standard for Fire Tests of Building Construction and Materials*. Northbrook, IL: Underwriters Laboratories Inc., 2011.

UL 555-2006, *Fire Dampers*. Northbrook, IL: Underwriters Laboratories Inc., 2006.

UL 555C-2006, *Ceiling Dampers*. Northbrook, IL: Underwriters Laboratories Inc., 2006.

UL 555S-99, *Leakage Rated Dampers for Use in Smoke Control Systems*. Northbrook, IL: Underwriters Laboratories Inc., 1999.

Chapter 8: Interior Finishes

General Comments

Past fire experience has shown that interior finish and decorative materials are key elements in the development and spread of fire. In some cases, as with Boston's devastating Cocoanut Grove fire in 1942, interior decorations were the first materials ignited. In many other cases, the interior finish materials became involved in the early stages of fire and contributed to its early growth and spread.

The provisions of Chapter 8 require materials used as interior finishes and decorations to have a flame spread index or meet certain flame propagation criteria based on the relative fire hazard associated with the occupancy.

The performance of the material is evaluated based on test standards.The design professional or permit applicant is responsible for determining and providing documentation on the flame spread potential and ability to propagate flames of a material in order to support the permit application.The building official then evaluates the information supplied to ascertain that compliance with the applicable code provisions has been achieved. It is also critical that a field inspection verifies that the material is installed in accordance with the approved documents and the code.

Purpose

This chapter contains the performance requirements for controlling fire growth within buildings by restricting interior finish materials and decorative materials.

SECTION 801 GENERAL

801.1 Scope. The provisions of this chapter shall govern the use of materials used as *interior finishes*, *trim* and *decorative materials*.

❖ This chapter contains requirements for materials used as interior finish, trim or decorative materials. These materials must conform to the flame spread index limitations, heat release and flashover limitations or flame propagation limitations established by this chapter. This chapter is similar to Chapter 8 of the *International Fire Code*® (IFC®) in intent, but is focused only on new construction and only on interior finish, trim and a limited amount of decorative materials. The IFC addresses both new and existing buildings and also some key building contents.

801.2 Interior wall and ceiling finish. The provisions of Section 803 shall limit the allowable fire performance and smoke development of *interior wall and ceiling finish* materials based on occupancy classification.

❖ The provisions of this chapter require materials used as interior finishes to have limited flame spread and smoke-developed indexes (or a limited heat and smoke release), based on the relative fire hazard and occupant vulnerability associated with the occupancy classification. Section 803 provides the necessary performance of wall and ceiling finishes required in test standards based upon the occupancy classification. The primary test standard is ASTM E84 or UL 723).

801.3 Interior floor finish. The provisions of Section 804 shall limit the allowable fire performance of *interior floor finish* materials based on occupancy classification.

❖ Just as for wall and ceiling finish materials, Chapter 8 provides limitations on flame propagation for materials used as floor finish. Section 804 provides the test requirements and criteria for floor finish materials.

[F] 801.4 Decorative materials and trim. *Decorative materials* and *trim* shall be restricted by combustibility, fire performance or flame propagation performance criteria in accordance with Section 806.

❖ This section is specific to decorative materials and trim, and their specific hazards versus the interior finish within a space. Section 806 addresses the tests for these materials, such as NFPA 701, which provides an indication of the potential for flame propagation of materials such as draperies. Additionally, materials, such as foam plastic used as interior trim, are limited in their combustibility through size, density and coverage limits (see commentary, Section 806).

801.5 Applicability. For buildings in flood hazard areas as established in Section 1612.3, *interior finishes*, *trim* and *decorative materials* below the elevation required by Section 1612 shall be flood-damage-resistant materials.

❖ All building materials located below the elevation required by Section 1612 (which refers to ASCE 24 for specific elevation requirements), including interior finishes, trim and decorative materials, must be resistant to flood damage. Many elevated buildings located in flood hazard areas have enclosed lower areas below the elevated floor. "Flood-damage-resistant materials" are defined as "any construction material capable of withstanding direct and prolonged contact with floodwaters without sustaining any damage that requires more than cosmetic repair." In ASCE 24, the term "prolonged contact" means at least 72 hours, and the term "cosmetic repair" means cleaning the affected surfaces with commercially available cleaners and repainting as necessary. For further guidance, refer to FEMA TB #2, *Flood Damage-Resistant Material Requirements for Buildings Located in Special Flood Hazard Areas.*

801.6 Application. Combustible materials shall be permitted to be used as finish for walls, ceilings, floors and other interior surfaces of buildings.

❖ This section simply points out that this chapter allows combustible materials as interior finish. This would include the finish of walls that are required to be "noncombustible" due to the building's type of construction or other code requirements.This is supported by the provisions of Section 603.1, which allows limited combustible materials in Type I and II construction. Item 5 of Section 603.1 lists interior floor finish and interior finish, trim and millwork, such as doors, door frames, window sashes and frames, as allowable combustible materials. The majority of materials that are used as finish materials are combustible in nature. However, all materials in regulated environments are required to exhibit maximum flame spread and smoke-developed indexes or an equivalent performance on heat and smoke release in an appropriate room-corner fire test. Section 803 clarifies the limits on such materials by limiting the flame spread index and smoke-developed index of such materials in certain portions of the building based upon the occupancy.

801.7 Windows. Show windows in the exterior walls of the first *story* above grade plane shall be permitted to be of wood or of unprotected metal framing.

❖ First-story windows used as display spaces are exempt from finish requirements. These areas are subject to frequent redecoration; however, the enforcement of finish requirements would be impractical. The exemption is limited to the first story above grade plane, which is considered to provide ready access for the fire department from the exterior for effective manual fire suppression efforts.

801.8 Foam plastics. Foam plastics shall not be used as *interior finish* except as provided in Section 803.4. Foam plastics shall not be used as interior *trim* except as provided in Section 806.5 or 2604.2. This section shall apply both to exposed foam plastics and to foam plastics used in conjunction with a textile or vinyl facing or cover.

❖ This section establishes that foam plastic materials are not permitted to be used as interior finish materials unless the foam plastic complies with Section 2603.4, 2603.10 or 2604. Sections 2603.4 and 2603.4.1 generally require that foam plastic materials be covered by an approved thermal barrier, such as 0.5-inch (12.7 mm) gypsum wallboard. When covered in this manner, the foam plastic is no longer the interior finish material. Section 2603.10 provides a means by which foam plastics can be installed without a thermal barrier. A thermal barrier is not required when foam plastic insulation has met the following two criteria: 1. It has been tested in accordance with FM 4880, UL 1040 or UL 1715 or with NFPA 286 (with the pass/fail criteria of Section 803.1.2.1); and 2. The foam plastic meets the flame spread requirements of Section 803 (see commentary, Section 2603.10).

The reference to Section 2604 is for the use of foam plastic as trim.

Sections 806.5 and 2604.2 do allow the use of a limited amount of foam plastic as interior trim when certain criteria are met. The IFC contains the same allowance for interior trim (see commentary, Section 806.5).

It should be noted that the IFC further regulates some building contents with regard to foam plastic. More specifically, Section 807.4.2.1 of the IFC places maximum heat release rate criteria on foam plastic used for decorative purposes, stage scenery or exhibit booths when tested in accordance with UL 1975. This code typically does not regulate building contents.

SECTION 802
DEFINITIONS

802.1 Definitions. The following terms are defined in Chapter 2:

❖ Definitions facilitate the understanding of code provisions and minimize potential confusion. To that end, this section lists definitions of terms associated with interior finishes. Note that these definitions are found in Chapter 2. The use and application of defined terms, as well as undefined terms, are set forth in Section 201.

EXPANDED VINYL WALL COVERING.

FLAME SPREAD.

FLAME SPREAD INDEX.

INTERIOR FINISH.

INTERIOR FLOOR FINISH.

INTERIOR FLOOR-WALL BASE.

INTERIOR WALL AND CEILING FINISH.

SITE-FABRICATED STRETCH SYSTEM.

SMOKE-DEVELOPED INDEX.

TRIM.

SECTION 803
WALL AND CEILING FINISHES

803.1 General. *Interior wall and ceiling finish* materials shall be classified for fire performance and smoke development in accordance with Section 803.1.1 or 803.1.2, except as shown in Sections 803.2 through 803.13. Materials tested in accordance with Section 803.1.2 shall not be required to be tested in accordance with Section 803.1.1.

❖ Wall and ceiling interior finish and trim materials are required to have limits on flame spread and smoke-developed indexes as prescribed in Section 803.11 based upon occupancy. ASTM E84 or UL 723 is one of the tests available to demonstrate compliance with the requirements of Section 803. This test method has been around since 1944 [see Commentary Figure 803.1(1)]. This is the primary method used, but NFPA 286 is an alternate optional test that is discussed in the commentary for Section 803.1.2. ASTME 84 is intended to determine the relative burning behavior of materials on exposed surfaces, such as walls and ceilings, by visually observing the flame spread along the test specimen [see Commentary Figure 803.1(2)]. Flame spread and smoke-developed indexes are then reported. The test method is not appropriate for materials that are not capable of supporting themselves or of being supported in the test tunnel. There may also be concerns with materials that drip, melt or delaminate and with very thin materials. A distinction is made, therefore, for textiles, expanded vinyl wall or ceiling coverings and foam plastic insulation materials (see Sections 803.6, 803.7 and 806.5).

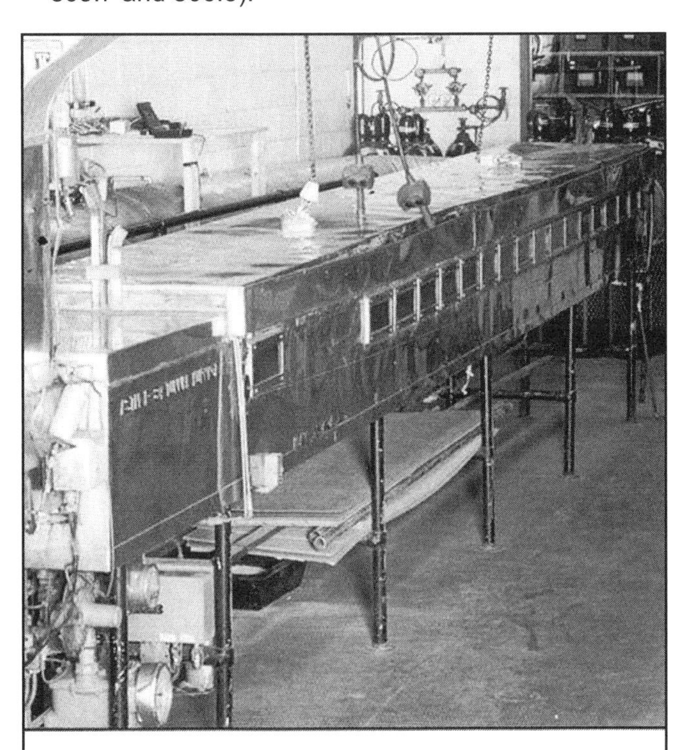

Figure 803.1(1)
ASTM E84 TUNNEL TEST

ASTM E84 establishes a flame spread index based on the area under a curve when the actual flame spread distance is plotted as a function of time. The code has divided the acceptable range of flame spread indexes (0-200) into three classes: Class A (0-25), Class B (26-75) and Class C (76-200). For all three classes the code has established a common acceptable range of smoke-developed index: 0-450. An indication of relative fire performance is as follows: an inorganic reinforced-cement board has a flame spread index of zero, while select grade red oak wood flooring has a flame spread index of 100. Not to preclude more detailed information resulting from an ASTME 84 test report, Table 803.1(3) identifies the typical flame spread properties of certain building materials.

803.1.1 Interior wall and ceiling finish materials. Interior wall and ceiling finish materials shall be classified in accordance with ASTM E84 or UL 723. Such *interior finish* materials shall be grouped in the following classes in accordance with their flame spread and *smoke-developed indexes*.

Class A: = Flame spread index 0-25; smoke-developed index 0-450.

Class B: = Flame spread index 26-75; smoke-developed index 0-450.

Class C: = Flame spread index 76-200; smoke-developed index 0-450.

Exception: Materials tested in accordance with Section 803.1.2.

❖ Wall and ceiling interior finish and trim materials are required to have limits on flame spread and smoke-developed indexes as prescribed in Section 803.11, based upon occupancy. ASTM E84 (or UL 723) is one of the tests available to demonstrate compliance with the requirements of Section 803. This test method has been around since 1944 [see Commentary Figure 803.1(1)]. This is the primary method used, but, as noted earlier, NFPA 286 is an alternative test, which is discussed in the commentary for Section 803.1.2. ASTME 84 is intended to determine the relative burning behavior of materials on exposed

Figure 803.1(2)
FLAME IN TUNNEL TEST

surfaces, such as ceilings and walls, by visually observing the flame spread along the test specimen [see Commentary Figure 803.1(2)]. Flame spread and smoke-developed indexes are then reported. The test method is not appropriate for materials that are not capable of supporting themselves or of being supported in the test tunnel. There may also be concerns with materials that drip, melt or delaminate and that are very thin. A distinction is made, therefore, for textile wall or ceiling coverings, expanded vinyl wall or ceiling coverings and foam plastic insulation materials (see Sections 803.5, 803.6, 803.7 and 803.8).

ASTME 84 establishes a flame spread index based on the area under a curve when the actual flame spread distance is plotted as a function of time. The code has divided the acceptable range of flame spread indexes (0-200) into three classes: Class A (0-25), Class B (26-75) and Class C (76-200). For all three classes the code has established a common acceptable range of smoke-developed index of 0-450. An indication of relative fire performance is as follows: an inorganic reinforced-cement board has a flame spread and smoke-developed index of zero, while select grade red oak wood flooring has a flame spread and smoke-developed index of 100. Not to preclude more detailed information resulting from an ASTME 84 test report, Figure 803.1(3) identifies the typical flame spread properties of certain building materials.

803.1.2 Room corner test for interior wall or ceiling finish materials. *Interior wall or ceiling finish* materials shall be permitted to be tested in accordance with NFPA 286. Interior wall or ceiling finish materials tested in accordance with NFPA 286 shall comply with Section 803.1.2.1.

❖ The alternative test method for determining compliance for interior wall and ceiling finish and trim, other than textiles, is found in test standard NFPA 286. This test is known as a "room-corner" fire test and is similar to that referenced for textile wall coverings in Section 803.1.3.1 (NFPA 265) [see Commentary Figures 803.1.2(1) and 803.1.2(2)]. In this test, materials are mounted covering three walls of the compartment (excluding the wall containing the door) and the ceiling. In the case where testing is only for ceiling finish properties, the sample only needs to be mounted on the ceiling. Then a fire source consisting of a gas burner is placed in one corner, flush against both walls (furthest from the doorway) of the compartment with the following exposure conditions:

- 40 kilowatts (kW) for 5 minutes, then
- 160 kW for 10 minutes.

The test then measures heat release and smoke release through the collection of the fire effluents and measurement of oxygen concentrations in the exhaust duct. Heat release is calculated by the oxygen consumption principle, which has shown that heat release is a function of the decrease in oxygen concentration in the fire effluents. Thus, exhaust duct measurements include temperatures, pressures and smoke values for use in the calculations. Temperatures and heat fluxes are also measured in the room. This generally provides a more realistic understanding of the fire hazard associated with the materials.

The NFPA 286 test method does not contain pass/fail criteria. However, the code provides such criteria in Section 803.1.2.1.

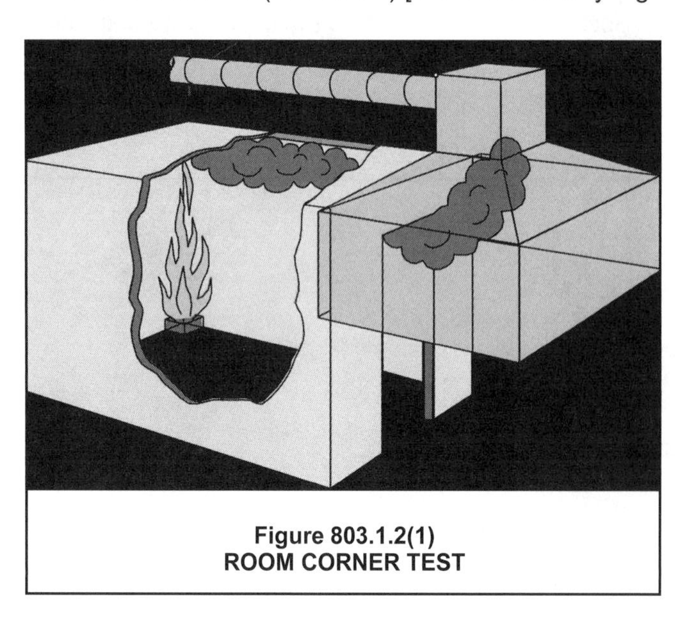

Figure 803.1.2(1)
ROOM CORNER TEST

Figure 803.1.2(2)
FLAME IN ROOM CORNER TEST

803.1.2.1 Acceptance criteria for NFPA 286. The interior finish shall comply with the following:

1. During the 40 kW exposure, flames shall not spread to the ceiling.
2. The flame shall not spread to the outer extremity of the sample on any wall or ceiling.
3. Flashover, as defined in NFPA 286, shall not occur.
4. The peak heat release rate throughout the test shall not exceed 800 kW.
5. The total smoke released throughout the test shall not exceed 1,000 m^2.

❖ The alternative test method for determining compliance for interior wall and ceiling finish and trim, other than textiles, is found in test standard NFPA 286. This test is known as a "room corner" fire test and is similar to that referenced for textile wall coverings in Section 803.1.3.1 (NFPA 265) [see Commentary Figures 803.1.2(1) and 803.1.2(2)]. In this test, materials are mounted covering three walls of the compartment (excluding the wall containing the door) and the ceiling. In the case where testing is only for ceiling finish properties, the sample only needs to be mounted on the ceiling. Then a fire source consisting of a gas burner is placed in one corner, flush against both walls (furthest from the doorway) of the compartment with the following exposure conditions:

- 40 kilowatts (kW) for 5 minutes; then
- 160 kW for 10 minutes.

The test then measures heat release and smoke release through the collection of the fire effluents and measurement of oxygen concentrations in the exhaust duct. Heat release is calculated by the oxygen consumption principle, which has shown that heat release is a function of the decrease in oxygen concentration in the fire effluents. Thus, exhaust duct measurements include temperatures, pressures and smoke values for use in the calculations. Temperatures and heat fluxes are also measured in the room. This generally provides a more realistic understanding of the fire hazard associated with the materials.

The NFPA 286 test method does not contain pass/fail criteria. However, the code provides such criteria in Section 803.1.2.1.

803.1.3 Room corner test for textile wall coverings and expanded vinyl wall coverings. Textile wall coverings and expanded vinyl wall coverings shall meet the criteria of Section 803.1.3.1 when tested in the manner intended for use in accordance with the Method B protocol of NFPA 265 using the product-mounting system, including adhesive.

❖ This particular section allows textile materials and expanded vinyl wall coverings when tested in accordance with Method B protocol of NFPA 265. NFPA 265 is known as a full-scale "room corner" fire test. Past research conducted with this kind of configuration has shown that flame spread indexes produced by ASTM E84 may not reliably predict the fire behavior of textile wall coverings. A more reliable test procedure was developed at the University of California and involved the use of a room corner fire test with a gas diffusion burner(s).The research findings are described in a report from the University of California Fire Research Laboratory titled, "Room Fire Experiments of Textile Wall Coverings." The NFPA 265 test method is only slightly different from NFPA 286. NFPA 286 is more severe in three ways:

1. The gas diffusion burner is used to expose the material on the wall in the room. The fire test starts with a heat release rate exposure of 40 kW for the first 5 minutes in both tests, but is then followed by 150 kW for 10 minutes in NFPA 265 and by 160 kW for 10 minutes in NFPA 286.
2. The gas burner is placed 2 inches (51 mm) away from each of the walls in NFPA 265, whereas, it is placed flush against both walls in NFPA 286.
3. The test sample is mounted on the walls only in accordance with NFPA 265, while it is mounted both on the walls and ceiling in accordance with NFPA 286.

A key result of the difference in intensity and location of the gas burner is that the burner flame does not reach the ceiling during the 150 kW exposure (while it does reach the ceiling during the 160 kW exposure in NFPA 286). Therefore, NFPA 265 is not considered to be suitable for testing ceiling coverings.

Based on the use of test Method B from NFPA 265, the sample must meet the criteria of Section 803.1.3.1. This test helps to determine the contribution of textile wall coverings to overall fire growth and spread in a compartment fire. This test also exposes the textile to an ignition source while mounted on the wall. It should be noted that the IFC allows the use of test Method A or B in NFPA 265. The testing requirements in the IFC apply specifically to existing buildings. In the Method A test protocol, 2-foot-wide (610 mm) strips of the material are mounted on the two walls closest to the corner with the burner, whereas, in the Method B test protocol, the sample is mounted completely covering three walls (except for the wall containing the door). Therefore, Method B test protocol is more severe. As noted, these textiles are exposed to a prescribed heat release rate of 40 kW for 5 minutes, which is then followed by a heat release rate of 150 kW for 10 minutes.

803.1.3.1 Acceptance criteria for NFPA 265. The interior finish shall comply with the following:

1. During the 40 kW exposure, flames shall not spread to the ceiling.
2. The flame shall not spread to the outer extremities of the samples on the 8-foot by 12-foot (203 by 305 mm) walls.
3. Flashover, as defined in NFPA 265, shall not occur.

4. The total smoke released throughout the test shall not exceed 1,000 m^2.

❖ To pass the Method B test protocol of NFPA 265, there are several criteria, including avoiding flashover, as defined in the standard. See the commentary to Section 803.1.3, which discusses limitations on the flame spread of the sample in both the 40 kW and 160 kW exposure and limits the total smoke released to 1000 m^2, just as in NFPA 286.

803.1.4 Acceptance criteria for textile and expanded vinyl wall or ceiling coverings tested to ASTM E84 or UL 723. Textile wall and ceiling coverings and expanded vinyl wall and ceiling coverings shall have a Class A flame spread index in accordance with ASTM E84 or UL 723 and be protected by an *automatic sprinkler system* installed in accordance with Section 903.3.1.1 or 903.3.1.2. Test specimen preparation and mounting shall be in accordance with ASTM E2404.

❖ Textile wall coverings and expanded vinyl wall coverings dramatically increase the fuel load in a room. As such, additional protection is required as a safety factor for rooms finished with these types of material. ASTM E84 alone is not considered a sufficient test of the flame spread properties of textile wall coverings because they are too thin when tested, and thus sprinklers would be required in addition to a Class A flame spread rating as a safety factor.

803.2 Thickness exemption. Materials having a thickness less than 0.036 inch (0.9 mm) applied directly to the surface of walls or ceilings shall not be required to be tested.

❖ Thin materials used as interior finish, such as wallpaper, are exempt from the testing requirements of this section. These materials represent a very low fuel load and minimal risk of significant flame spread.

803.3 Heavy timber exemption. Exposed portions of building elements complying with the requirements for buildings of Type IV construction in Section 602.4 shall not be subject to *interior finish* requirements.

❖ This section is simply intended to clarify the code regarding heavy timber, which is commonly exposed to the interior without any interior finish. The concerns over the contribution of heavy timber building elements to flame spread in a structure are dealt with in the general requirements of Chapter 6 and other provisions of the code that allow heavy timber.

803.4 Foam plastics. Foam plastics shall not be used as *interior finish* except as provided in Section 2603.9. This section shall apply both to exposed foam plastics and to foam plastics used in conjunction with a textile or vinyl facing or cover.

❖ This section is simply intended to serve as a reminder that the provisions of Chapter 26 apply when the material is foam plastic. The special approval provisions of Section 2603.10 would be required in order to allow a foam plastic to be used in an exposed interior application. The special approval involves testing in accordance with NFPA 286, which is also used for acceptance of other materials in this section.

803.5 Textile wall coverings. Where used as interior wall finish materials, textile wall coverings, including materials having woven or nonwoven, napped, tufted, looped or similar surface and carpet and similar textile materials, shall be tested in the manner intended for use, using the product mounting system, including adhesive, and shall comply with the requirements of Section 803.1.2, 803.1.3 or 803.1.4.

❖ This section recognizes that the methods for attaching interior finishes could contribute to the flame spread characteristics of the interior finish material. For textiles and expanded vinyl wall coverings, the variety of adhesives and mounting methods are such that the entire system needs to be tested in the Steiner tunnel tests or the room corner test specified in Section 803.1.3 or 803.1.4.

803.6 Textile ceiling coverings. Where used as interior ceiling finish materials, textile ceiling coverings, including materials having woven or nonwoven, napped, tufted, looped or similar surface and carpet and similar textile materials, shall be tested in the manner intended for use, using the product mounting system, including adhesive, and shall comply with the requirements of Section 803.1.2 or 803.1.4.

❖ This section recognizes that the methods for attaching interior finishes could contribute to the flame spread characteristics of the interior finish material. For textile ceiling coverings, the variety of adhesives and mounting methods are such that the entire system needs to be tested in the Steiner tunnel tests or the room corner test specified in Section 803.1.3 or 803.1.4.

803.7 Expanded vinyl wall coverings. Where used as interior wall finish materials, expanded vinyl wall coverings shall be tested in the manner intended for use, using the product mounting system, including adhesive, and shall comply with the requirements of Section 803.1.2, 803.1.3 or 803.1.4.

❖ This section recognizes that the methods for attaching interior finishes could contribute to the flame spread characteristics of the interior finish material. For expanded vinyl wall coverings, the variety of adhesives and mounting methods are such that the entire system needs to be tested in the Steiner tunnel tests or the room corner test specified in Section 803.1.3 or 803.1.4.

803.8 Expanded vinyl ceiling coverings. Where used as interior ceiling finish materials, expanded vinyl ceiling coverings shall be tested in the manner intended for use, using the product mounting system, including adhesive, and shall comply with the requirements of Section 803.1.2 or 803.1.4.

❖ This section recognizes that the methods for attaching interior finishes could contribute to the flame spread characteristics of the interior finish material. For expanded vinyl ceiling coverings, the variety of adhesives and mounting methods are such that the entire system needs to be tested in the Steiner tunnel tests or the room corner test specified in Section 803.1.3 or 803.1.4.

803.9 High-density polyethylene (HDPE) and polypropylene (PP). Where high-density polyethylene or polypropylene is used as an interior finish it shall comply with Section 803.1.2.

❖ High-density polyethylene (HDPE) and polypropylene (PP) are thermoplastic materials that, when burned, give off considerable energy and produce flammable liquid fires. Recent full-scale room corner tests using NFPA 286 have demonstrated a significant hazard with some of these (HDPE) materials. Extensive flammable liquid fires occurred during the tests. The Steiner tunnel test, however, is not a suitable measure of the performance of this material because the calculation of flame spread index does not take into account the unique hazards known for this material. Therefore, these materials are required to be tested in a full-scale room corner test in accordance with Section 803.1.2.

803.10 Site-fabricated stretch systems. Where used as interior wall or interior ceiling finish materials, site-fabricated stretch systems containing all three components described in the definition in Chapter 2 shall be tested in the manner intended for use, and shall comply with the requirements of Section 803.1.1 or 803.1.2. If the materials are tested in accordance with ASTM E84 or UL 723, specimen preparation and mounting shall be in accordance with ASTM E2573.

❖ The ASTM E05 committee on fire standards has issued a standard practice, ASTM E2573, for specimen preparation and mounting of site-fabricated stretch systems. Until now, there was no correct mandatory way to test these systems. These systems are now being used extensively because they can stretch to cover decorative walls and ceilings with unusual looks and shapes. The systems consist of three parts: a fabric (or vinyl), a frame and an infill core material. The testing has often been done on each component separately, which is inappropriate and unsafe, instead of testing the composite system.

803.11 Interior finish requirements based on group. *Interior wall and ceiling finish* shall have a flame spread index not greater than that specified in Table 803.11 for the group and location designated. *Interior wall and ceiling finish* materials tested in accordance with NFPA 286 and meeting the acceptance criteria of Section 803.1.2.1, shall be permitted to be used where a Class A classification in accordance with ASTM E84 or UL 723 is required.

❖ The requirements for flame spread indexes for interior finish materials applied to walls and ceilings are contained in Table 803.11 (see Sections 804 and 805 for floor finish requirements). The referenced test for determining flame spread indexes is ASTM E84 or UL 723, which establishes a relative measurement of flame spread across the surface of the material. The classifications used in Table 803.11 are defined in Section 803.1. See the commentary to Section 803.1 for additional information on the uses and limitations of the test procedure. This section also allows the use of NFPA 286 as noted in the exception to Section 803.1 and Section 803.2 (without the exception). Section 803.5, in particular, states that materials that pass NFPA 286 in accordance with the pass/fail criteria in Section 803.2.1 can be used as an alternative to ASTM E84. Passing NFPA 286 means that the material would be considered equivalent to Class A. In other words, it can be used as a replacement for all categories, as Class A is the most restrictive (see commentary, Section 803.1.2 for more detail on the test).

When evaluating test reports, the method of application and substrate material are considered. For example, adhesives that soften at moderate temperatures will allow wall or ceiling finishes to drop or peel. This not only increases the susceptibility of the material to ignition but also exposes the substrate material. The substrate material can also affect the interior finish rating of the surface material. It is not uncommon for a thin, combustible finish applied to a noncombustible material to obtain a low flame spread index, while the same material applied to a combustible substrate or applied without a substrate could exhibit a significantly high tendency to spread flame.

TABLE 803.11. See page 8-8.

❖ This table prescribes the minimum requirements for interior finishes applied to walls and ceilings. Class A material has the most restrictive flame spread index and Class C has the least restrictive flame spread index. Therefore, the use of a Class A material in an area that requires a minimum Class B material is always allowed. Likewise, when the table requires Class C materials, Classes A and B can also be used.

The requirements are based on the use of the space. To determine the applicable criteria, first determine whether the space is an exit stairway, ramp or passageway; a corridor, exit access stairway or ramp; or a room or enclosed space. For definitions of these various spaces, see Chapter 2. For the minimum requirements for these spaces, see Chapter 10.

Interior finishes in spaces that are not separated from a corridor (for example, waiting areas in business or health care facilities) must comply with the requirements for a corridor space. As shown in the table, the code places more emphasis on the allowable flame spread index for exits than for enclosed rooms because of the critical nature and relative importance that is placed on maintaining the integrity of exits to evacuate the building.

Numerous notes amend the basic requirements of Table 803.11. Notes a through l apply only where they are specifically referenced in the table.

TABLE 803.11
INTERIOR WALL AND CEILING FINISH REQUIREMENTS BY OCCUPANCY[k]

GROUP	SPRINKLERED[l]			NONSPRINKLERED		
	Interior exit stairways and ramps and exit passageways[a, b]	Corridors and enclosure for exit access stairways and ramps	Rooms and enclosed spaces[c]	Interior exit stairways and ramps and exit passageways[a, b]	Corridors and enclosure for exit access stairways and ramps	Rooms and enclosed spaces[c]
A-1 & A-2	B	B	C	A	A[d]	B[e]
A-3[f], A-4, A-5	B	B	C	A	A[d]	C
B, E, M, R-1	B	C	C	A	B	C
R-4	B	C	C	A	B	B
F	C	C	C	B	C	C
H	B	B	C[g]	A	A	B
I-1	B	C	C	A	B	B
I-2	B	B	B[h, i]	A	A	B
I-3	A	A[j]	C	A	A	B
I-4	B	B	B[h, i]	A	A	B
R-2	C	C	C	B	B	C
R-3	C	C	C	C	C	C
S	C	C	C	B	B	C
U	No restrictions			No restrictions		

For SI: 1 inch = 25.4 mm, 1 square foot = $0.0929m^2$.

a. Class C interior finish materials shall be permitted for wainscotting or paneling of not more than 1,000 square feet of applied surface area in the grade lobby where applied directly to a noncombustible base or over furring strips applied to a noncombustible base and fireblocked as required by Section 803.13.1.

b. In other than Group I-3 occupancies in buildings less than three stories above grade plane, Class B interior finish for nonsprinklered buildings and Class C interior finish for sprinklered buildings shall be permitted in interior exit stairways and ramps.

c. Requirements for rooms and enclosed spaces shall be based upon spaces enclosed by partitions. Where a fire-resistance rating is required for structural elements, the enclosing partitions shall extend from the floor to the ceiling. Partitions that do not comply with this shall be considered enclosing spaces and the rooms or spaces on both sides shall be considered one. In determining the applicable requirements for rooms and enclosed spaces, the specific occupancy thereof shall be the governing factor regardless of the group classification of the building or structure.

d. Lobby areas in Group A-1, A-2 and A-3 occupancies shall not be less than Class B materials.

e. Class C interior finish materials shall be permitted in places of assembly with an occupant load of 300 persons or less.

f. For places of religious worship, wood used for ornamental purposes, trusses, paneling or chancel furnishing shall be permitted.

g. Class B material is required where the building exceeds two stories.

h. Class C interior finish materials shall be permitted in administrative spaces.

i. Class C interior finish materials shall be permitted in rooms with a capacity of four persons or less.

j. Class B materials shall be permitted as wainscotting extending not more than 48 inches above the finished floor in corridors and exit access stairways and ramps.

k. Finish materials as provided for in other sections of this code.

l. Applies when protected by an automatic sprinkler system installed in accordance with Section 903.3.1.1 or 903.3.1.2.

803.12 Stability. *Interior finish* materials regulated by this chapter shall be applied or otherwise fastened in such a manner that such materials will not readily become detached where subjected to room temperatures of 200°F (93°C) for not less than 30 minutes.

❖ Interior finishes are not to become detached under exposure to elevated temperatures [200°F (93°C)] for 30 minutes.There is not a standard test method yet developed to evaluate this requirement. Other sections of the code, however, offer some additional guidance. For example, the performance of the method of attachment of finish materials during a fire-resistance test will usually be an adequate indication of performance. Section 2506 contains requirements for gypsum wallboard that is not part of a fire-resistance-rated assembly. Interior finishes that become detached may add to the spread of fire, as well as create a hazard for fire fighters.

803.13 Application of interior finish materials to fire-resistance-rated or noncombustible building elements. Where *interior finish* materials are applied on walls, ceilings or structural elements required to have a *fire-resistance rating* or to be of noncombustible construction, these finish materials shall comply with the provisions of this section.

❖ Where an assembly is required to have a fire-resistance rating or be of noncombustible materials, interior finish materials are required to be applied in accordance with Sections 803.4.1 through 803.4.4. These methods are to prevent other types of fire spread onto the exposed surface through the use of fireblocking and similar methods.

803.13.1 Direct attachment and furred construction. Where walls and ceilings are required by any provision in this code to be of fire-resistance-rated or noncombustible construction, the *interior finish* material shall be applied directly against such construction or to furring strips not exceeding $1^3/_4$ inches (44 mm), applied directly against such surfaces.

❖ Interior finish materials are required to be applied directly to the exposed surface of structural elements or to the furring attached to such surfaces. By applying the finish directly to the surface, the potential for the back side to contribute to a fire is diminished and, as a result, the effects on the fire-resistance rating and the creation of combustible cavities in noncombustible walls are minimized.

803.13.1.1 Furred construction. If the interior finish material is applied to furring strips, the intervening spaces between such furring strips shall comply with one of the following:

1. Be filled with material that is inorganic or noncombustible;
2. Be filled with material that meets the requirements of a Class A material in accordance with Section 803.1.1 or 803.1.2; or
3. Be fireblocked at a maximum of 8 feet (2438 mm) in every direction in accordance with Section 718.

❖ Interior finishes are allowed to be applied to furring strips under specific limitations. All concealed spaces created by furring are to be fireblocked at a maximum of 8-foot (2438 mm) intervals in any direction. Likewise, fireblocks in furred construction also diminish the likelihood of fire involvement between the finish and the interior wall surface. The use of wood nailing strips is intended to be limited to that which is necessary for the installation of finish and trim materials. As such, the amount of combustible material is restricted so that the performance of the element to which it is attached will not be adversely affected. It should be noted that the type of construction classification of the building is not intended to regulate the combustibility of the interior finish materials. The fire performance characteristics of the interior finish materials are regulated by Section 803.1.

803.13.2 Set-out construction. Where walls and ceilings are required to be of fire-resistance-rated or noncombustible construction and walls are set out or ceilings are dropped distances greater than specified in Section 803.13.1, Class A finish materials, in accordance with Section 803.1.1 or 803.1.2, shall be used.

Exceptions:

1. Where *interior finish* materials are protected on both sides by an *automatic sprinkler system* in accordance with Section 903.3.1.1 or 903.3.1.2.
2. Where *interior finish* materials are attached to noncombustible backing or furring strips installed as specified in Section 803.13.1.1.

❖ Where walls and ceilings are required to be of fire-resistance-rated or noncombustible construction and walls are set out or ceilings are dropped distances greater than specified in Section 803.13.1, finish materials with the most restrictive flame spread index (Class A) shall be used. Exceptions 1 and 2 describe alternatives to providing the Class A finish material. The first is to provide the protection afforded by an automatic sprinkler system. Note that this protection is required on each side of the finish material. The second is to secure the finish to noncombustible backing or furring strips in accordance with Section 803.13.1.1 (see Commentary Figure 803.13.2).

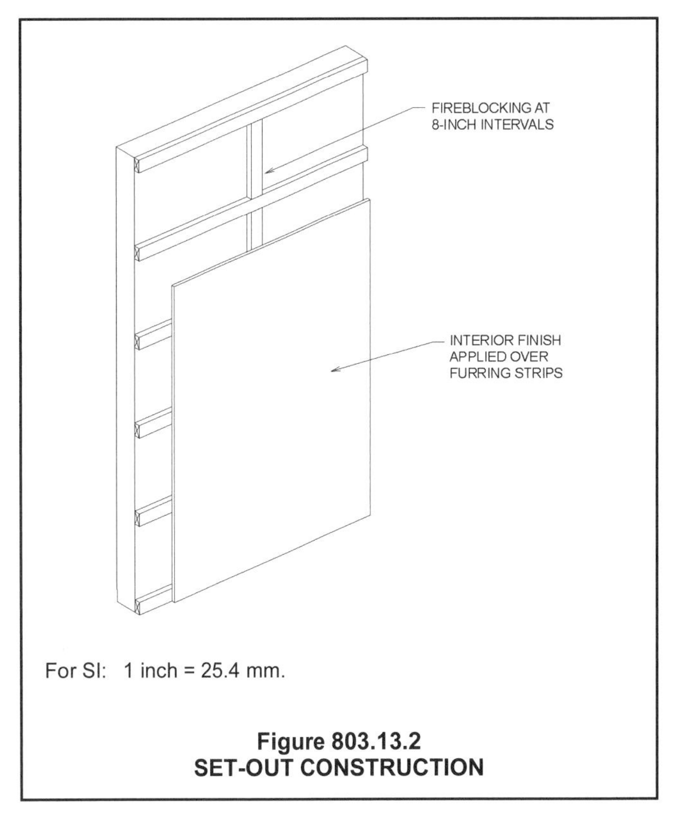

For SI: 1 inch = 25.4 mm.

Figure 803.13.2
SET-OUT CONSTRUCTION

803.13.2.1 Hangers and assembly members. The hangers and assembly members of such dropped ceilings that are below the horizontal fire-resistance-rated floor or roof assemblies shall be of noncombustible materials. The construction of each set-out wall and horizontal fire-resistance-rated floor or roof assembly shall be of fire-resistance-rated construction as required elsewhere in this code.

Exception: In Type III and V construction, *fire-retardant-treated wood* shall be permitted for use as hangers and assembly members of dropped ceilings.

❖ The hangers and assembly members of dropped ceilings that are below the main ceiling line are required to be noncombustible materials to ensure that the ceiling will remain in place for some time if a fire occurs. The exception allows fire-retardant-treated wood (FRTW) in Type III and V construction. This is consistent with other provisions in the code dealing with options to noncombustible materials used in Type III and V construction.

803.13.3 Heavy timber construction. Wall and ceiling finishes of all classes as permitted in this chapter that are installed directly against the wood decking or planking of Type IV construction or to wood furring strips applied directly to the wood decking or planking shall be fireblocked as specified in Section 803.13.1.1.

❖ Attachment directly to heavy timber decking or planking eliminates concealed spaces (see Section 803.3). Furring strips of any depth are allowed when concealed spaces are treated according to Section 803.13.1.1. Again, the intent is to reduce the likelihood of fire spread between these spaces.

803.13.4 Materials. An interior wall or ceiling finish material that is not more than $^1/_4$ inch (6.4 mm) thick shall be applied directly onto the wall, ceiling or structural element without the use of furring strips and shall not be suspended away from the building element to which that finish material it is applied.

Exceptions:

1. Noncombustible interior finish materials.
2. Materials that meet the requirements of Class A materials in accordance with Section 803.1.1 or 803.1.2 where the qualifying tests were made with the material furred out from the noncombustible backing shall be permitted to be used with furring strips.
3. Materials that meet the requirements of Class A materials in accordance with Section 803.1.1 or 803.1.2 where the qualifying tests were made with the material suspended away from the noncombustible backing shall be permitted to be used suspended away from the building element.

❖ Combustible materials that have a maximum thickness of $^1/_4$ inch (6.4 mm) are required to be applied directly against a noncombustible backing and not to an added substrate or furred away from the backing, which may significantly increase the spread of fire (see Commentary Figure 803.13.4). Exception 1 allows noncombustible interior finish materials to be furred away from the noncombustible backing. Exceptions 2 and 3 allow materials that, in tests, perform similarly to directly attached materials when furred or suspended from a noncombustible backing, respectively.

SECTION 804
INTERIOR FLOOR FINISH

804.1 General. *Interior floor finish* and floor covering materials shall comply with Sections 804.2 through 804.4.2.

Exception: Floor finishes and coverings of a traditional type, such as wood, vinyl, linoleum or terrazzo, and resilient floor covering materials that are not comprised of fibers.

❖ This section regulates the design and installation of floor finish and floor covering materials. Traditional floor coverings such as wood, vinyl, terrazzo and other resilient floor covering material must be exempt from this section. Smooth-surface floor coverings generally contribute minimally to a fire. The focus is more upon textile floor coverings, such as carpets.

804.2 Classification. *Interior floor finish* and floor covering materials required by Section 804.4.2 to be of Class I or II materials shall be classified in accordance with NFPA 253. The classification referred to herein corresponds to the classi-

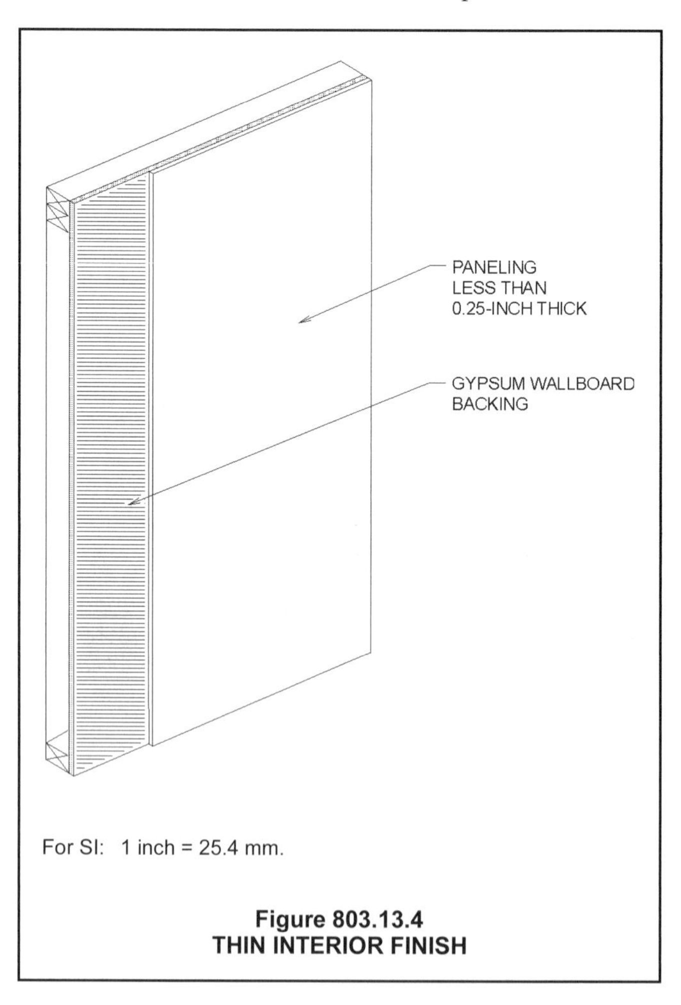

Figure 803.13.4
THIN INTERIOR FINISH

fications determined by NFPA 253 as follows: Class I, 0.45 watts/cm^2 or greater; Class II, 0.22 watts/cm^2 or greater.

❖ The use of a classification system eliminates the need to state the actual critical radiant flux value for a product to meet the identification requirements of Section 804. Over the years, a classification system has been found to be much easier for the industry to follow and still provides the building official with the information required to verify compliance. The test required to measure the combustibility of floor coverings is NFPA 253. This standard is a radiant floor panel test, which basically simulates materials subjected to heat from a fire above.The primary concern with flooring is related to the spread of a fire that has already been ignited within a space or room to a different room. Commentary Figure 804.2 shows the test apparatus. The critical heat flux indicates the threshold value above which flame spread occurs in the testing environment.

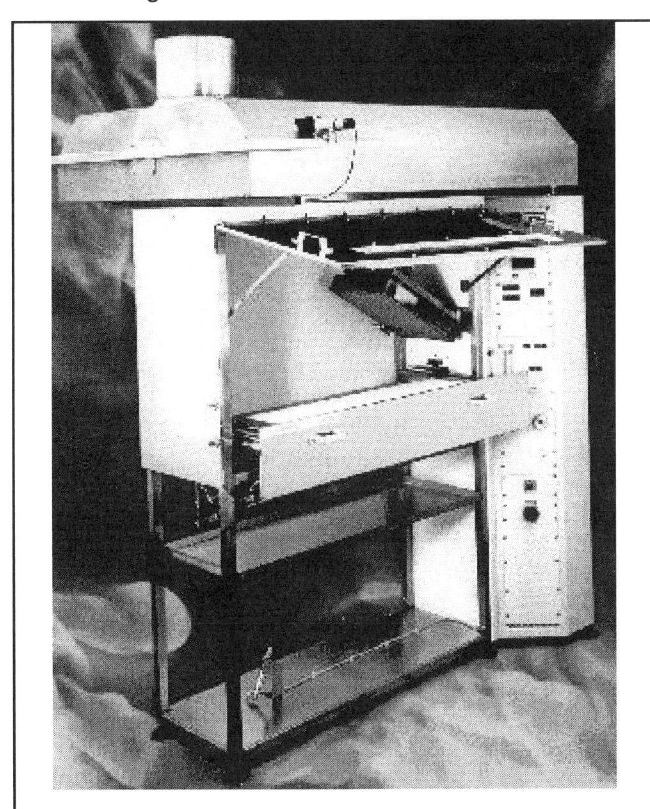

Figure 804.2
RADIANT FLOOR PANEL TEST (NFPA 253)

804.3 Testing and identification. *Interior floor finish* and floor covering materials shall be tested by an agency in accordance with NFPA 253 and identified by a hang tag or other suitable method so as to identify the manufacturer or supplier and style, and shall indicate the *interior floor finish* or floor covering classification in accordance with Section 804.2. Carpet-type floor coverings shall be tested as proposed for use, including underlayment. Test reports confirming the information provided in the manufacturer's product identification shall be furnished to the building official upon request.

❖ The only method to ascertain that a floor meets the criteria of this section is to request a copy of the test report for the specific material being installed. Therefore, it is critical that the carpeting be properly identified in order to verify that acceptable materials are being provided in the appropriate locations. The identification is to be provided on the material itself since a manufacturer's designation is required.

804.4 Interior floor finish requirements. Interior floor covering materials shall comply with Sections 804.4.1 and 804.4.2 and interior floor finish materials shall comply with Section 804.4.2.

❖ Regardless of occupancy, all interior floor covering materials need to comply with one of two test requirements as indicated in Section 804.4.1, both of which deal with ignition performance of the floor covering. The finish flooring requirements for means of egress components, such as enclosed exit stairways and ramps, exit passageways, corridors and spaces open to the corridor, are regarded as more critical than other building floor surfaces. As such, these finishes and coverings are also required to comply with the minimum critical radiant flux provisions in Section 804.4.2.

804.4.1 Test requirement. In all occupancies, interior floor covering materials shall comply with the requirements of the DOC FF-1 "pill test" (CPSC 16 CFR Part 1630) or with ASTM D2859.

❖ This section contains two test methods for interior floor covering materials: DOC FF-1 and ASTM D2859. All carpeting greater than 24 square feet (2.2 m^2) in area sold in the United States is required by federal law to pass DOC FF-1 as a minimum. DOC FF-1, also referred to as the "Methenamine Pill Test," was developed as a means of preventing the distribution of highly flammable soft-floor coverings within the United States. ASTM D2859 has been determined to be equivalent to DOC FF-1 and can be used as an alternative test method. Both test methods essentially evaluate the performance of the floor covering when subject to a cigarette-type ignition by using a small methenamine tablet.

804.4.2 Minimum critical radiant flux. In all occupancies, interior floor finish and floor covering materials in enclosures for stairways and ramps, exit passageways, corridors and rooms or spaces not separated from corridors by partitions extending from the floor to the underside of the ceiling shall withstand a minimum critical radiant flux. The minimum critical radiant flux shall be not less than Class I in Groups I-1, I-2 and I-3 and not less than Class II in Groups A, B, E, H, I-4, M, R-1, R-2 and S.

Exception: Where a building is equipped throughout with an automatic sprinkler system in accordance with Section 903.3.1.1 or 903.3.1.2, Class II materials are permitted in any area where Class I materials are required, and materials complying with DOC FF-1 "pill test" (CPSC 16 CFR Part 1630) or with ASTM D2859 are permitted in any area where Class II materials are required.

❖ This section prescribes the minimum requirements for interior floor finish materials used in enclosed exit stairways and ramps, passageways, corridors and rooms or spaces open to the corridor. The criteria are based on the occupancy classification and the relationship of the space to the egress system. Similar to Table 803.11, the occupancy classification designation is meant to apply to the actual occupancy of the space and not necessarily the overall building classification.

Classifications I and II, as used in this section, are defined in Section 804.2 and are based on the results of the NFPA 253 test procedure.

Recognizing the ability of automatic sprinkler systems to control a fire and the minimal contribution of interior floor finishes to the early stages of fire growth, the exception allows the required interior floor finish ratings to be reduced when an automatic sprinkler system is provided throughout the building. The reference to Section 903.3.1.1 or 903.3.1.2 clarifies that the system is to be installed in accordance with NFPA 13 or 13R. In cases where Class II materials are required and an automatic sprinkler system is provided, the minimum requirement is that the material meet either the DOC FF-1 or the ASTM D2859 test criteria. For a discussion of these two methods, see the commentary to Section 804.4.1.

SECTION 805 COMBUSTIBLE MATERIALS IN TYPES I AND II CONSTRUCTION

805.1 Application. Combustible materials installed on or embedded in floors of buildings of Type I or II construction shall comply with Sections 805.1.1 through 805.1.3.

Exception: Stages and platforms constructed in accordance with Sections 410.3 and 410.4, respectively.

❖ This section provides several methods of allowing combustible materials in the floors of Type I or II construction. Generally, the focus is on fire spread in concealed spaces. If not properly addressed, combustible floor surfaces may eventually contribute to a fire within a concealed space. Sections 410.3 and 410.4 permit wood flooring in stages.

805.1.1 Subfloor construction. Floor sleepers, bucks and nailing blocks shall not be constructed of combustible materials, unless the space between the fire-resistance-rated floor assembly and the flooring is either solidly filled with noncombustible materials or fireblocked in accordance with Section 718, and provided that such open spaces shall not extend under or through permanent partitions or walls.

❖ Sleepers, bucks and nailing blocks are permitted to be of combustible materials only where the void space is filled with a noncombustible material or is fireblocked in accordance with Section 718. Section 718 permits a maximum concealed space area of 100 square feet (9.3 m^2). In either case, the open spaces cannot extend under or through permanent partitions or walls. The purpose of the fill or fireblocking is to reduce the impact of a fire in a concealed combustible space in the floor. Likewise, fire spread around partitions or walls through the concealed space in the floor is also intended to be prevented.

805.1.2 Wood finish flooring. Wood finish flooring is permitted to be attached directly to the embedded or fireblocked wood sleepers and shall be permitted where cemented directly to the top surface of fire-resistance-rated floor assemblies or directly to a wood subfloor attached to sleepers as provided for in Section 805.1.1.

❖ Wood floor finish materials may be applied directly to the wood sleepers, provided that the space is protected in accordance with Section 805.1.1. Wood floor finish materials may also be applied directly to the top surface of a fire-resistance-rated assembly or directly to a wood subfloor that is attached to sleepers that are either fireblocked or provide noncombustible fill between the sleepers in accordance with Section 805.1.1.

805.1.3 Insulating boards. Combustible insulating boards not more than $^1/_2$ inch (12.7 mm) thick and covered with finish flooring are permitted where attached directly to a noncombustible floor assembly or to wood subflooring attached to sleepers as provided for in Section 805.1.1.

❖ The addition of combustible insulating boards with a maximum thickness of $^1/_2$ inch (12.7 mm) will not significantly increase the fuel load when applied as prescribed by this section or attached to sleepers as provided for in Section 805.1.1.

In all occupancies, interior floor finish in exit enclosures, exit passageways, exit access corridors and rooms or spaces not separated from exit access corridors by full-height partitions extending from the floor to the underside of the ceiling must withstand a minimum critical radiant flux as specified in Section 804.4.2. The focus surrounding the critical radiant flux is combustibility and potential for flame spread of the exposed materials. Section 805 focuses upon fire spread through concealed spaces created by floors.

SECTION 806 DECORATIVE MATERIALS AND TRIM

[F] 806.1 General. Combustible decorative materials, other than decorative vegetation, shall comply with Sections 806.2 through 806.8.

❖ The requirements in this section apply to decorative materials other than decorative vegetation. The bulk of the requirements in this section are applicable to all groups. Section 806.3 is applicable to all groups except Group I-3. Section 806.6 is applicable to

Group A occupancies. Decorative materials must be noncombustible or meet the flame propagation performance criteria of Section 806.4 and NFPA 701.

[F] 806.2 Noncombustible materials. The permissible amount of noncombustible materials shall not be limited.

❖ Where decorative materials are classified as noncombustible, it is presumed that they will contribute little, if at all, to the growth and spread of fire. Therefore, the quantity of noncombustible decorative materials is not limited.

[F] 806.3 Combustible decorative materials. In other than Group I-3, curtains, draperies, fabric hangings and similar combustible decorative materials suspended from walls or ceilings shall comply with Section 806.4 and shall not exceed 10 percent of the specific wall or ceiling area to which such materials are attached.

Fixed or movable walls and partitions, paneling, wall pads and crash pads applied structurally or for decoration, acoustical correction, surface insulation or other purposes shall be considered *interior finish* shall comply with Section 803 and shall not be considered *decorative materials* or furnishings.

Exceptions:

1. In auditoriums in Group A, the permissible amount of curtains, draperies, fabric hangings and similar combustible decorative materials suspended from walls or ceilings shall not exceed 75 percent of the aggregate wall area where the building is equipped throughout with an *approved automatic sprinkler system* in accordance with Section 903.3.1.1, and where the material is installed in accordance with Section 803.13 of this code.
2. In Group R-2 dormitories, within sleeping units and dwelling units, the permissible amount of curtains, draperies, fabric hangings and similar decorative materials suspended from walls or ceiling shall not exceed 50 percent of the aggregate wall areas where the building is equipped throughout with an *approved automatic sprinkler system* installed in accordance with Section 903.3.1.
3. In Group B and M occupancies, the amount of combustible fabric partitions suspended from the ceiling and not supported by the floor shall comply with Section 806.4 and shall not be limited.

❖ In any occupancy classification where a movable wall, partition, paneling, wall pads or crash pads cover larger areas, they need to be dealt with as interior finishes instead of as decorative materials. Such areas would not be considered decorative materials where the area covered is greater than 10 percent of the wall or ceiling area. This is clarified in the paragraph addressing Exception 3. Meeting the flame propagation performance criteria of NFPA 701 does not mean that materials will not burn, only that they are going to spread flame relatively slowly. These materials are, therefore, limited to a maximum of 10 percent of the total wall and ceiling area of the space under consideration. Unlike incidental trim, decorative materials are not necessarily distributed evenly throughout the room. Additionally, consideration of the long-term maintenance of the materials, including possible periodic retreatment, should be taken into account.

There are two exceptions to the 10-percent limitation in Section 806.3. Exception 1 is for Group A auditoriums and would allow 75-percent coverage of the walls and ceilings (instead of the limit of 10 percent) if the space is sprinklered in accordance with NFPA 13 and the material is applied in accordance with Section 803.4. Exception 2 correlates with the general requirements in Section 806.1 and further emphasizes that an unlimited amount of fabric ceiling partitions are allowed as long as they meet the flame propagation performance of NFPA 701.

Exception 3 of this section also clarifies that, in Group B and M occupancies, fabric partitions suspended from the ceiling but not physically contacting the floor should be treated as decorative materials similar to curtains or draperies and comply with the flame propagation performance heat release criteria of Section 806.4.

[F] 806.4 Acceptance criteria and reports. Where required to exhibit improved fire performance, curtains, draperies, fabric hangings and similar combustible decorative materials suspended from walls or ceilings shall be tested by an *approved agency* and meet the flame propagation performance criteria of Test 1 or 2, as appropriate, of NFPA 701, or exhibit a maximum heat release rate of 100 kW when tested in accordance with NFPA 289, using the 20 kW ignition source. Reports of test results shall be prepared in accordance with the test method used and furnished to the *building official* upon request.

❖ One of the standard test methods to be used to evaluate the ability of a material to propagate flame is NFPA 701, which contains two test methods. Test 1 is a less severe test than Test 2 and uses smaller test specimens. Test 1 is intended for lighter-weight materials and single-layer fabrics: the limit is a linear density of 700 g/m^2 (21 oz/yd^2). Test 2 is more severe than Test 1 and uses larger test specimens. It is intended for higher-density materials and multilayered fabrics. It also applies to vinyl-coated fabric blackout linings and lined draperies using any density, because they have been shown to give misleading results using Test 1. NFPA 701 sets out the types of materials, including fabrics, that should be tested using each method.

Essentially, NFPA 701 provides a mechanism to distinguish between materials that allow flames to spread quickly and those that do not when using a small fire exposure.

Materials tested only to NFPA 701 are not permitted for use as interior finish materials, but instead are generally used as shades, swags, curtains and other similar materials. These tests are used to determine whether materials propagate flame beyond the area exposed to the ignition source. They are not intended to indicate whether the material tested will resist the propagation of flame under fire exposures more extreme than the test conditions.

The other standard test method that can be used is NFPA 289. This test method quantifies the contribution of materials to heat and smoke release when subjected to different ignition sources. This test method also determines the potential for growth of a fire and the fire spread of a given material when exposed to an ignition source in a controlled environment.

[F] 806.5 Foam plastic. Foam plastic used as *trim* in any occupancy shall comply with Section 2604.2.

❖ This section establishes that some dense foam plastic materials may be used as interior trim if the thickness, width and area of coverage are specifically limited as required in Section 2604.2.

Section 2604.2 provides the following maximum values: $^{1}/_{2}$-inch (12.7 mm) thickness, 8-inch (203 mm) width and 10 percent of the aggregate wall and ceiling area. Section 2604.2 also establishes a minimum density for foam permitted for use as trim: 20 pounds per cubic foot (320 kg/m^3). Minimum instead of maximum density is specified because the denser the foam plastic material, the less likely it is that it will generate misleading ASTM E84 test results due to having insufficient material for fire testing of the foam [see Commentary Figures 806.7(1) and 806.7(2)].

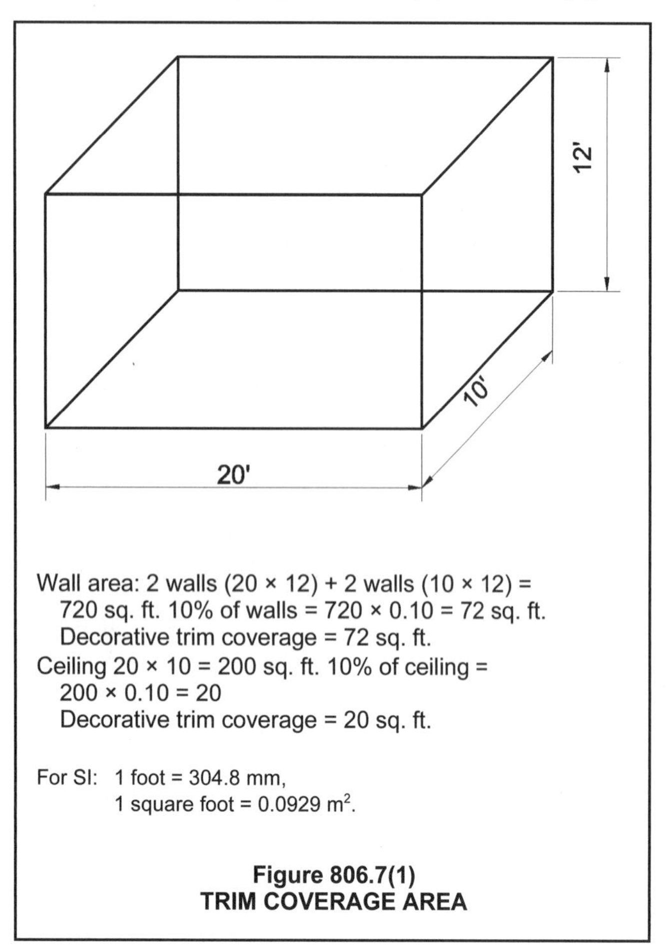

Figure 806.7(1)
TRIM COVERAGE AREA

This section specifically calls out a numerical flame spread index limitation of 75 for foam plastic used as trim, which is a Class B flame spread index. No thermal barrier is required for the use of foam plastic as trim and the smoke-developed index is not limited. It should be noted that Section 803.7.3 of the IFC, which regulates the use of foam plastic trim for existing construction, has essentially identical requirements to those found in this section. Neither this section nor Section 2603.4.1.11 requires a thermal barrier for interior trim.

[F] 806.6 Pyroxylin plastic. Imitation leather or other material consisting of or coated with a pyroxylin or similarly hazardous base shall not be used in Group A occupancies.

❖ Use of pyroxylin plastics, also known as cellulose nitrate plastics, as imitation leather (or coated to imitation leather) is strictly prohibited in Group A occupancies because of the normally high occupant loads of such occupancies. This type of material is very hazardous because it has the potential to develop explosive atmospheres with high heat emission and it will begin decomposition at temperatures starting at 300°F (149°C). In actual fact, cellulose nitrate use has generally declined considerably because it tends to become somewhat unstable and is easily ignitable. Specifically, use of cellulose nitrate for motion picture film was discontinued in 1951.

[F] 806.7 Interior trim. Material, other than foam plastic used as interior *trim*, shall have a minimum Class C flame spread and smoke-developed index when tested in accor-

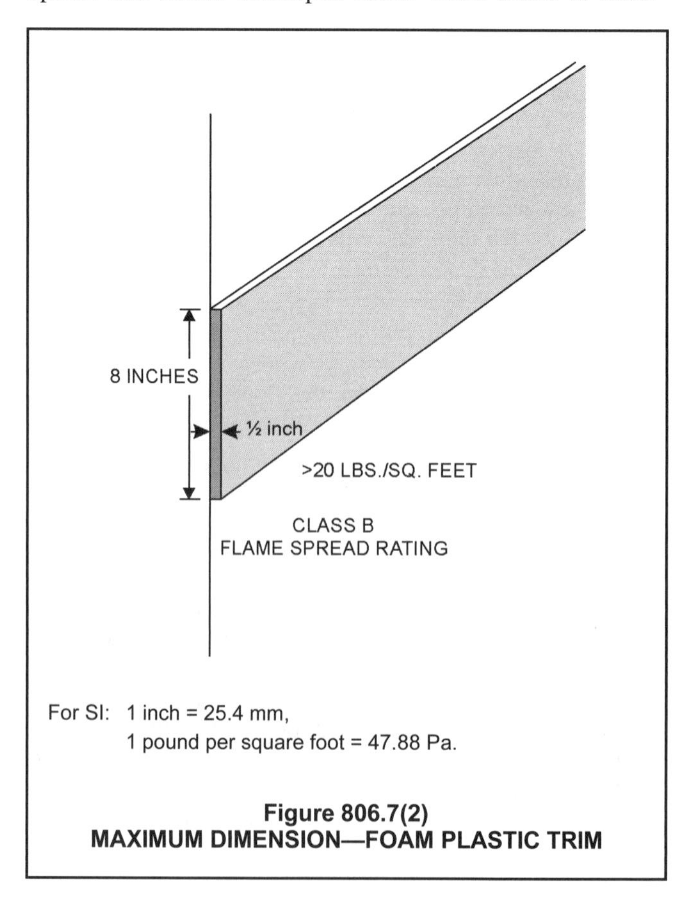

Figure 806.7(2)
MAXIMUM DIMENSION—FOAM PLASTIC TRIM

dance with ASTM E84 or UL 723, as described in Section 803.1.1. Combustible *trim*, excluding handrails and guardrails, shall not exceed 10 percent of the specific wall or ceiling area in which it is attached.

❖ In occupancies of any group, unless otherwise noted in the code, the minimum classification of all trim must be Class C. Additionally, combustible trim may not exceed 10 percent of the area of the aggregate wall or ceiling area. The 10-percent calculation does not need to include handrails and guardrails. Although a Class C rating may be lower than the rating required for a particular building or facility, this quantity of combustible material will not significantly increase the fuel load and flame spread hazards [see Commentary Figures 806.7(1) and 806.7(2)].

[F] 806.8 Interior floor-wall base. *Interior floor-wall base* that is 6 inches (152 mm) or less in height shall be tested in accordance with Section 804.2 and shall be not less than Class II. Where a Class I floor finish is required, the floor-wall base shall be Class I.

> **Exception:** Interior *trim* materials that comply with Section 806.7.

❖ Section 806.8 addresses the issue of testing and regulation of interior floor-wall base trim materials. In many cases, the floor covering material is just seamlessly turned up or used at the intersection of the floor and the wall and thus it becomes the floor-wall base trim.

In the 2006 code, these materials could have been considered interior trim in accordance with Sections 804.1 and 806.7 and would have been required to be tested in accordance with ASTM E84 or UL 723, even though the floor covering may have been required to be tested in accordance with NFPA 253. The testing apparatus and specimen mounting requirements of ASTM E84 make it difficult to obtain reliable test results for these smaller floor-wall base profiles. Because of their location, at the floor line, floor-wall base materials are not likely to be involved in a fire until the floor covering is also involved, usually at room flashover. Thus, it is reasonable that floor-wall base materials meet the same criteria as floor coverings. The section specifies that floor-wall base materials 6 inches (152 mm) or less in height be tested in accordance with NFPA 253, and the proposal provides requirements for this application.

SECTION 807
INSULATION

807.1 Insulation. Thermal and acoustical insulation shall comply with Section 720.

❖ This section simply references Section 720 for the requirements for insulation materials and clarifies that these materials can be used when they meet the requirements of that section.

SECTION 808
ACOUSTICAL CEILING SYSTEMS

808.1 Acoustical ceiling systems. The quality, design, fabrication and erection of metal suspension systems for acoustical tile and lay-in panel ceilings in buildings or structures shall conform to generally accepted engineering practice, the provisions of this chapter and other applicable requirements of this code.

❖ This section regulates the design and installation of metal suspension systems for acoustical tile and acoustical lay-in panel ceilings. Suspended ceiling systems basically consist of a grid of metal channels or T-bars suspended by steel hangers (wires, rods or flats) from the structure above. Light fixtures, air diffusers and acoustical panels may be placed between the metal framing members.

808.1.1 Materials and installation. Acoustical materials complying with the *interior finish* requirements of Section 803 shall be installed in accordance with the manufacturer's recommendations and applicable provisions for applying *interior finish*.

❖ Acoustical materials used in metal suspension systems are regulated for the purpose of limiting flame spread and smoke development in the occupancies shown in Table 803.11.

Metal suspension system components for acoustical ceilings that might be subjected to the severe environmental conditions of high humidity (fog) or salt spray may have coatings ranked according to their ability to protect the components from deterioration. The installation of metal suspension systems in any exterior application must be considered as an installation in a severe environment (see ASTM C635).

808.1.1.1 Suspended acoustical ceilings. Suspended acoustical ceiling systems shall be installed in accordance with the provisions of ASTM C635 and ASTM C636.

❖ ASTM C635 covers metal ceiling suspension systems used primarily to support acoustical tile or acoustical lay-in panels. Suspension systems have three structural classifications, as described below:

- Light-duty systems—Used primarily for residential and light commercial structures where ceiling loads other than acoustical tile or lay-in panels are not anticipated.
- Intermediate-duty systems—Used primarily for ordinary commercial structures where some ceiling loads, due to light fixtures and air diffusers, are anticipated.
- Heavy-duty systems—Used primarily for commercial structures in which the quantities and weights of ceiling fixtures (lights, air diffusers, etc.) are greater than those for an ordinary commercial structure.

ASTM C636 covers the installation of suspension systems for acoustical tile and lay-in panels.

While the practices described in this standard have equal application to fire-resistance-rated suspension systems, additional requirements may be imposed to obtain the fire endurance classification of particular floor/ceiling or roof/ceiling assemblies (see Section 808.1.1.2).

808.1.1.2 Fire-resistance-rated construction. Acoustical ceiling systems that are part of fire-resistance-rated construction shall be installed in the same manner used in the assembly tested and shall comply with the provisions of Chapter 7.

❖ A suspended acoustical ceiling system by itself does not possess a fire-resistance rating. Acoustical ceiling systems are part of an assembly that has a specified fire-resistance rating. Acoustical lay-in panels must remain in place for the tested assembly to be effective.

Bibliography

The following resource materials were used in the preparation of the commentary for this chapter of the code.

ASTM C635-13, *Specification for the Manufacture, Performance, and Testing of Metal Suspension Systems for Acoustical Tile and Lay-In Panel Ceilings*. West Conshohocken, PA: ASTM International, 2013.

ASTM D2859-06(2011), *Standard Test Method for Ignition Characteristics of Finished Textile Floor Covering Materials*. West Conshohocken, PA: ASTM International, 2011.

ASTM E84-13A, *Test Method for Surface-burning Characteristics of Building Materials*. West Conshohocken, PA: ASTM International, 2013.

DOC FF-1 (CPSC 16 CFR, Part 1630-07), *Standard for the Surface Flammability of Carpets and Rugs*. Bethesda, MD: Consumer Product Safety Commission, 2007.

Fisher, F.L., B. McCracken and R.B. Williamson. "Room Fire Experiments of Textile Wall Coverings, Final Report of All Materials Tested Between March 1985 and January 1986." *Service to Industry Report*. ES-7853, Report 86-2, p.142. Berkley, CA: California University, March 1986.

NFPA 253-11, *Test for Critical Radiant Flux of Floor Covering Systems Using a Radiant Heat Energy Source*. Quincy, MA: National Fire Protection Association, 2011.

NFPA 265-11, *Standard Fire Tests for Evaluating Room Fire Growth Contribution of Textile Wall Coverings*. Quincy, MA: National Fire Protection Association, 2011.

NFPA 286-11, *Standard Methods of Fire Tests for Evaluating Contribution of Wall and Ceiling Interior Finish to Room Fire Growth*. Quincy, MA: National Fire Protection Association, 2011.

NFPA 701-2010, *Standard Methods of Fire Tests for Flame Propagation of Textiles and Films*. Quincy, MA: National Fire Protection Association, 2010.

UL 723-2008, *Standard for Test for Surface Burning Characteristics of Building Materials—with Revisions through May 2005*. Northbrook, IL: Underwriters Laboratories Inc., 2008.

Chapter 9: Fire Protection Systems

General Comments

The provisions required by Chapter 9 are just one aspect of the overall fire protection system of a building or structure. All fire protection requirements contained in the code must be considered as a package or overall system. Noncompliance with any part of the overall system may cause other parts of the system to fail. Failure to install the systems in accordance with code provisions may result in an increased loss of life and property due to a reduction in the level of protection provided.

It is important that every effort be made to verify the proper design and installation of a given fire protection system, especially those that result in construction alternatives and other code tradeoffs.

The requirements found in Chapter 9 can be considered active fire safety provisions. They are provisions directed at containing and abating the fire once it has erupted. This chapter is almost a direct copy of Chapter 9 in the *International Fire Code*® (IFC®). The IFC, however, contains specific provisions that are only applicable to existing buildings. The IFC also contains periodic testing criteria that are not duplicated in this chapter. Proper testing, inspection and maintenance of the various systems, however, are critical to establish the reliability of the system. Additionally, Chapter 9 references and adopts numerous National Fire Protection Association (NFPA) standards, including the acceptance testing criteria within the standard. The referenced standards will also contain more specific design and installation criteria than are found in this chapter. As noted in Section 102.4, where differences occur between code provisions and the referenced standards, the code provisions apply.

Purpose

Fire protection systems may serve one or more purposes in providing adequate protection from fire and hazardous material exposure. The purpose of Chapter 9 is to prescribe the minimum requirements for an active system or systems of fire protection to perform the following functions: to detect a fire; to alert the occupants or fire department of a fire emergency; to control smoke and to control or extinguish the fire. Generally, the requirements are based on the occupancy and height and area of the building, as these are the factors that most affect fire-fighting capabilities and the relative hazard of a specific space or area.

SECTION 901 GENERAL

901.1 Scope. The provisions of this chapter shall specify where *fire protection systems* are required and shall apply to the design, installation and operation of *fire protection systems*.

❖ Chapter 9 contains requirements for fire protection systems that may be provided in a building. These include automatic suppression systems; standpipe systems; fire alarm and detection systems; smoke control systems; smoke and heat vents and portable fire extinguishers and emergency alarm systems. Besides indicating the conditions under which respective systems are required, this chapter contains the design, installation, maintenance and operational criteria for fire protection systems. While the chapter requires proper maintenance for the reliability of the systems, the actual maintenance provisions (periodic testing, inspections and maintenance) are found in the IFC.

This chapter also addresses the requirements for fire command centers, fire department connections, fire pumps and emergency radio systems. These features all directly relate to the proper function of fire protection systems.

901.2 Fire protection systems. *Fire protection systems* shall be installed, repaired, operated and maintained in accordance with this code and the *International Fire Code*.

Any *fire protection system* for which an exception or reduction to the provisions of this code has been granted shall be considered to be a required system.

Exception: Any *fire protection system* or portion thereof not required by this code shall be permitted to be installed for partial or complete protection provided that such system meets the requirements of this code.

❖ The fire protection system is an integral component of the protection features of the building and is required to be properly installed, repaired, operated and maintained in accordance with the code and the IFC. Improperly installed or maintained systems can negate any anticipated protection and, in fact, create a hazard in itself.

While the code may not require a protection system for a specific building or portion thereof, due to its occupancy, the fire protection system would still be considered a required system if some other code tradeoff, exception or reduction was taken based on the installation of the fire protection system. See Commentary Figure 904.2.1. For example, a typical small office building may not require an automatic sprinkler system solely because of its Group B occupancy classification; however, if an exit access corridor fire-resistance-rating reduction is taken in accordance with Table 1020.1 for buildings equipped throughout with an NFPA 13 sprinkler system, that sprinkler system would be considered a required system.

The exception acknowledges that a building owner or designer may elect to install a fire protection system that is not required by the code. Even though such a system is not required, it must comply with the applicable provisions of this chapter and the IFC. This requirement is predicated on the concept that any fire protection system not installed in accordance with the code is intrinsically lacking because it could give a false impression of properly installed protection.

For example, if a building owner chooses to provide sprinkler protection in a certain area and such protection is not required by the code, the system must be installed in accordance with the applicable NFPA sprinkler standard (13, 13R or 13D) and other applicable requirements of the code, such as water supply and supervision. The extent of the protection thus provided would not be regulated.

901.3 Modifications. Persons shall not remove or modify any *fire protection system* installed or maintained under the provisions of this code or the *International Fire Code* without approval by the *building official*.

❖ This section emphasizes the principle that systems installed and maintained in compliance with the codes and standards in effect at the time they were placed in service must remain in an operative condition at all times. Protection must not be diminished in any existing building except for the purpose of conducting tests, maintenance or repairs. The length of service interruptions should be kept to a minimum. The building official must be notified of any service interruptions and should carefully evaluate the continued operation or occupancy of buildings and structures where protection is interrupted.

901.4 Threads. Threads provided for fire department connections to sprinkler systems, standpipes, yard hydrants or any other fire hose connection shall be compatible with the connections used by the local fire department.

❖ Incompatible fire service threads have been a major problem throughout the history of modern fire suppression. Efforts to standardize fire service threads in the United States began as early as 1898. Following the Great Baltimore Fire of 1904, NFPA adopted a national standard for fire hose thread with diameters $2^1/_2$ inches (64 mm) and larger; however, to this day many jurisdictions continue to use their own (historic) thread standards. A number of jurisdictions have pioneered the use of non-threaded, quarter-turn or "Storz" fire hose couplings in place of the national and local traditional thread standards.

901.5 Acceptance tests. *Fire protection systems* shall be tested in accordance with the requirements of this code and the *International Fire Code*. When required, the tests shall be conducted in the presence of the *building official*. Tests required by this code, the *International Fire Code* and the standards listed in this code shall be conducted at the expense of the owner or the owner's authorized agent. It shall be unlawful to occupy portions of a structure until the required *fire protection systems* within that portion of the structure have been tested and *approved*.

❖ All fire protection systems are to be subjected to an acceptance test to determine that the system will operate in the manner required by the code. Acceptance tests are usually part of the final inspection procedures required by Section 110.3.10. Specific acceptance test procedures are provided in other sections as well as the referenced standards. In most instances, the acceptance test procedures require 100-percent operation of the testable system components to determine that they are functioning as required. Often, the design professional may require additional testing to verify that the system operates as designed, which may be beyond the code requirements.

The inclusion of the requirement for acceptance tests in the code is not intended to assign responsibility for witnessing the tests. The responsibility to witness the acceptance test is an administrative issue that each municipality must address. Because the acceptance test is critical during design and construction and is a requirement of occupancy, the requirement is located in the code. The section also clarifies that it is the owner or owner's authorized agent's responsibility to conduct the test and the role of the building official to witness—not conduct—the test. Typically, the owner will assign the responsibility to the installing contractor.

Access to all flow test connections, valves and points of fluid discharge required by other sections is to be provided at locations acceptable to the building official. Consideration should be given to the location of sprinkler system inspector test connections and the main drain valves, as well as to the valves necessary to test standpipe and fire pump installations.

Consideration should also be given to the discharge of water from the various test connections for aqueous systems that should be arranged to discharge to the outside or to a drain with sufficient capacity to handle the anticipated water flow. If the test connection is piped to the outside, then protection should be furnished to prevent unnecessary damage to the property by providing splash blocks or similar protection. When the building drainage system is used, the indirect waste provisions of Chapter 7 of the *International Plumbing Code*® (IPC®) are to apply.

Partial occupancy of any structure should not be permitted unless all fire protection systems for the occupied areas have been tested and approved. Even so, the code assumes that full protection for all areas will be provided as expeditiously as possible. The installation of many fire protection systems and the associated code tradeoffs permitted for a given occupancy assume complete building protection, not just in the occupied areas. Final approval of all partial occupancy conditions is subject to the building official.

901.6 Supervisory service. Where required, *fire protection systems* shall be monitored by an approved supervising station in accordance with NFPA 72.

❖ All required fire protection systems are to be supervised as a means of determining at any time that the system is operational. Acceptable methods of supervision are provided in NFPA 72.

901.6.1 Automatic sprinkler systems. *Automatic sprinkler systems* shall be monitored by an *approved* supervising station.

Exceptions:

1. A supervising station is not required for *automatic sprinkler systems* protecting one- and two-family dwellings.
2. Limited area systems serving fewer than 20 sprinklers.

❖ This section highlights that automatic sprinkler systems must be monitored by an approved supervising station. All water supply control valves and waterflow switches are required to be electrically supervised as indicated in Section 903.4. A central station, remote supervising station or proprietary supervising station is an approved service recognized in NFPA 72 (see commentary, Section 903.4.1).

Automatic sprinkler systems in one- and two-family dwellings are typically designed in accordance with NFPA 13D, which does not require electrical supervision (see Exception 1). Limited area sprinkler systems are generally supervised by their connection to the domestic water service (see Exception 2). As such, limited area sprinkler systems with fewer than 20 sprinkler heads are not required to have local alarms or a fire department connection.

901.6.2 Fire alarm systems. Fire alarm systems required by the provisions of Section 907.2 of this code and Sections 907.2 and 907.9 of the *International Fire Code* shall be monitored by an *approved* supervising station in accordance with Section 907.6.6.

Exceptions:

1. Single- and multiple-station smoke alarms required by Section 907.2.11.
2. Smoke detectors in Group I-3 occupancies.
3. Supervisory service is not required for *automatic sprinkler systems* in one- and two-family dwellings.

❖ Fire alarm systems are required to automatically transmit an alarm and any trouble signals to the fire department through one of the approved methods in NFPA 72. This includes systems in new buildings where required by Section 907.2 of both the code and the IFC and in existing buildings where required by Section 907.3 of the IFC.

Exception 1 exempts single- and multiple-station smoke alarms from being supervised due to the potential for unwanted false alarms. Similarly, due to the concern over unwanted alarms, smoke detectors in Group I-3 occupancies need only sound an approved alarm signal, which automatically notifies staff (see Section 907.2.6.3.1). Smoke detectors in Group I-3 occupancies are typically subject to misuse and abuse. Frequent unwanted alarms would negate the effectiveness of the system (see Exception 2).

Exception 3 clarifies that sprinkler systems in one- and two-family dwellings are not part of a dedicated fire alarm system and are typically designed in accordance with NFPA 13D, which does not require electrical supervision.

901.6.3 Group H. Supervision and monitoring of emergency alarm, detection and automatic fire-extinguishing systems in Group H occupancies shall be in accordance with the *International Fire Code*.

❖ Given the varied nature and quantity of hazardous materials that could be present, all emergency alarm, detection and automatic fire-extinguishing systems in Group H occupancies are required to be both supervised and monitored by an approved supervising station. Section 5004.10 of the IFC correlates with this section and further allows the building official to approve an on-site monitoring system. Many companies that routinely deal with hazardous materials employ their own fire brigades and have emergency response procedures in place.

901.7 Fire areas. Where buildings, or portions thereof, are divided into *fire areas* so as not to exceed the limits established for requiring a *fire protection system* in accordance with this chapter, such *fire areas* shall be separated by *fire barriers* constructed in accordance with Section 707 or *horizontal assemblies* constructed in accordance with Section 711, or both, having a *fire-resistance rating* of not less than that determined in accordance with Section 707.3.10.

❖ This section provides specific guidance on how a building needs to be divided into fire areas in order to avoid requiring a fire protection system to be installed. A single occupancy group would require fire barriers or horizontal assemblies in order to create multiple fire areas (see Table 707.3.10), each having an area below the threshold for fire protection system installation.

[F] 901.8 Pump and riser room size. Where provided, fire pump rooms and *automatic sprinkler system* riser rooms shall be designed with adequate space for all equipment necessary for the installation, as defined by the manufacturer, with sufficient working room around the stationary equipment. Clearances around equipment to elements of permanent construction, including other installed equipment and appliances, shall be sufficient to allow inspection, service, repair or replacement without removing such elements of permanent construction or disabling the function of a required fire-resistance-rated assembly. Fire pump and *automatic sprinkler system* riser rooms shall be provided with a door(s) and unobstructed passageway large enough to allow removal of the largest piece of equipment.

❖ Section 901.8 establishes that, where a pump or rise room is necessary, the rooms housing fire protection system risers or fire pumps and their components have adequate space to facilitate their maintenance. This section does not require the construction of a room to house fire protection systems; however, if a room is provided, this section requires that it be adequately sized to allow for maintenance.

Instead of prescribing arbitrary dimensions, this provision bases the room area on clearances specified by the equipment manufacturers to ensure adequate space is available for its installation or removal. The design must provide enough area so that walls, finish materials and doors are not required to be removed during maintenance activities. The provision also prescribes that the size of the door serving a riser or pump room is of a size to accommodate the removal of the largest piece of equipment.

Given that the design of fire protection systems generally commences during the period that building construction drawings and specifications are being reviewed by the jurisdiction, it will be especially important for the building's designer to establish dialogue with the fire protection system contractor early in the design process to ensure that the room and at least one door can accommodate the largest equipment and provide the space needed for maintenance.

SECTION 902 DEFINITIONS

902.1 Definitions. The following terms are defined in Chapter 2:

❖ Definitions of terms can help in the understanding and application of the code requirements. This section directs the code user to Chapter 2 for the proper application of the indicated terms used in this chapter. Terms may be defined in Chapter 2 of another International Code as indicated in Section 201.3 or the dictionary meaning may be all that is needed (see commentary, Sections 201 through 201.4).

[F] ALARM NOTIFICATION APPLIANCE.

[F] ALARM SIGNAL.

[F] ALARM VERIFICATION FEATURE.

[F] ANNUNCIATOR.

[F] AUDIBLE ALARM NOTIFICATION APPLIANCE.

[F] AUTOMATIC.

[F] AUTOMATIC FIRE-EXTINGUISHING SYSTEM.

[F] AUTOMATIC SMOKE DETECTION SYSTEM.

[F] AUTOMATIC SPRINKLER SYSTEM.

[F] AUTOMATIC WATER MIST SYSTEM.

[F] AVERAGE AMBIENT SOUND LEVEL.

[F] CARBON DIOXIDE EXTINGUISHING SYSTEMS.

[F] CEILING LIMIT.

[F] CLEAN AGENT.

[F] COMMERCIAL MOTOR VEHICLE.

[F] CONSTANTLY ATTENDED LOCATION.

[F] DELUGE SYSTEM.

[F] DETECTOR, HEAT.

[F] DRY-CHEMICAL EXTINGUISHING AGENT.

[F] ELECTRICAL CIRCUIT PROTECTIVE SYSTEM.

[F] ELEVATOR GROUP.

[F] EMERGENCY ALARM SYSTEM.

[F] EMERGENCY VOICE/ALARM COMMUNICATIONS.

[F] FIRE ALARM BOX, MANUAL.

[F] FIRE ALARM CONTROL UNIT.

[F] FIRE ALARM SIGNAL.

[F] FIRE ALARM SYSTEM.

FIRE AREA.

[F] FIRE COMMAND CENTER.

[F] FIRE DETECTOR, AUTOMATIC.

[F] FIRE PROTECTION SYSTEM.

[F] FIRE SAFETY FUNCTIONS.

[F] FOAM-EXTINGUISHING SYSTEM.

[F] HALOGENATED EXTINGUISHING SYSTEM.

[F] INITIATING DEVICE.

[F] MANUAL FIRE ALARM BOX.

[F] MULTIPLE-STATION ALARM DEVICE.

[F] MULTIPLE-STATION SMOKE ALARM.

[F] NOTIFICATION ZONE.

[F] NUISANCE ALARM.

PRIVATE GARAGE.

[F] RECORD DRAWINGS.

[F] SINGLE-STATION SMOKE ALARM.

[F] SMOKE ALARM.

[F] SMOKE DETECTOR.

[F] SMOKEPROOF ENCLOSURE.

[F] STANDPIPE SYSTEM, CLASSES OF.

Class I system.

Class II system.

Class III system.

[F] STANDPIPE, TYPES OF.

Automatic dry.

Automatic wet.

Manual dry.

Manual wet.

Semiautomatic dry.

[F] SUPERVISING STATION.

[F] SUPERVISORY SERVICE.

[F] SUPERVISORY SIGNAL.

[F] SUPERVISORY SIGNAL-INITIATING DEVICE.

[F] TIRES, BULK STORAGE OF.

[F] TROUBLE SIGNAL.

[F] VISIBLE ALARM NOTIFICATION APPLIANCE.

[F] WET CHEMICAL EXTINGUISHING SYSTEM.

[F] WIRELESS PROTECTION SYSTEM.

[F] ZONE.

[F] ZONE, NOTIFICATION.

❖ Certain requirements in the code are based on code provisions in the IFC, *International Mechanical Code®* (IMC®) and IPC. A review of definitions included in those codes will aid in the understanding of many of the requirements contained in the code.

SECTION 903
AUTOMATIC SPRINKLER SYSTEMS

[F] 903.1 General. *Automatic sprinkler systems* shall comply with this section.

❖ This section identifies the conditions requiring an automatic sprinkler system for all occupancies. The need for an automatic sprinkler system may depend on not only the occupancy but also the occupant load, fuel load, height and area of the building as well as fire-fighting capabilities. Section 903.2 addresses all occupancy conditions requiring an automatic sprinkler system. Section 903.3 contains the installation requirements for all sprinkler systems in addition to the requirements of NFPA 13, NFPA 13R and NFPA 13D. The supervision and alarm requirements for sprinkler systems are contained in Section 903.4, whereas Section 903.5 refers to the IFC, which references testing and maintenance requirements for sprinkler systems found in Section 901 and NFPA 25.

Unless specifically allowed by the code, residential sprinkler systems installed in accordance with NFPA 13R or NFPA 13D are not recognized for reductions or exceptions permitted by other sections of the code. NFPA 13 systems provide the level of protection associated with adequate fire suppression for all occupancies. NFPA 13R and NFPA 13D systems are intended more to provide adequate time for egress but not necessarily for complete suppression of the fire. Commentary Figure 903.2 lists examples of where the various sprinkler thresholds differ in application.

The area values contained in this section are intended to apply to fire areas, which are comprised of all floor areas bounded by fire barriers, fire walls or exterior walls. The minimum required fire-resistance rating of fire barrier assemblies that define a fire area is specified in Table 707.3.10. Because the areas are defined as fire areas, fire barriers, horizontal assemblies, fire walls or exterior walls are the only acceptable means of subdividing a building into smaller areas instead of installing an automatic sprinkler system. Where fire barriers and exterior walls define multiple fire areas within a single building, a fire wall defines separate buildings within one structure. Also note that some of the threshold limitations result in a requirement to install an automatic sprinkler system throughout the building while others may require only specific fire areas to be sprinklered.

Another important point is that one fire area may include floor areas in more than one story of a building (see the commentary to the definition of "Fire area" in Section 202).

The application of mixed occupancies and fire areas must be carefully researched. Often the required separation between occupancies for the purposes of applying the separated mixed-use option in Section 508.4 will result in a separation that is less than what is required to define the boundaries of a fire area. It is possible to have two different occupancies within a given fire area, treated as separated uses but with code requirements applicable to both, since they are not separated by the rating required for fire areas.

[F] 903.1.1 Alternative protection. Alternative *automatic fire-extinguishing systems* complying with Section 904 shall be permitted instead of automatic sprinkler protection where recognized by the applicable standard and *approved* by the fire code official.

❖ This section permits the use of an alternative automatic fire-extinguishing system when approved by the fire code official as a means of compliance with the occupancy requirements of Section 903. Although the use of an alternative extinguishing system allowed by Section 904, such as a carbon dioxide system or clean-agent system, would satisfy the requirements of Section 903.2, it would not be considered an acceptable alternative for the purposes of exceptions, reductions or other code alternatives that would be applicable if an automatic sprinkler system were installed.

[F] 903.2 Where required. Approved *automatic sprinkler systems* in new buildings and structures shall be provided in the locations described in Sections 903.2.1 through 903.2.12.

Exception: Spaces or areas in telecommunications buildings used exclusively for telecommunications equipment, associated electrical power distribution equipment, batteries and standby engines, provided those spaces or areas are equipped throughout with an *automatic smoke detection* *system* in accordance with Section 907.2 and are separated from the remainder of the building by not less than 1-hour *fire barriers* constructed in accordance with Section 707 or not less than 2-hour *horizontal assemblies* constructed in accordance with Section 711, or both.

❖ Sections 903.2.1 through 903.2.12 identify the conditions requiring an automatic sprinkler system (see Commentary Figure 903.2). The type of sprinkler system must be one that is permitted for the specific occupancy condition. An NFPA 13R sprinkler system, for example, may not be installed to satisfy the sprinkler threshold requirements for a mercantile occupancy (see Section 903.2.7). As indicated in Section 903.3.1.2, the use of an NFPA 13R sprinkler system is limited to Group R occupancies not exceeding four stories in height.

There is one exception for those spaces or areas used exclusively for telecommunications equipment. The telecommunications industry has continually stressed the need for the continuity of telephone service, and the ability to maintain this service is of prime importance. This service is a vital link between the community and the various life safety services, including fire, police and emergency medical services. The integrity of this communications service can be jeopardized not only by fire, but also by water, from whatever the source.

It must be emphasized that the exception applies only to those spaces or areas that are used exclusively for telecommunications equipment. Historically, those spaces have a low incidence of fire events. Fires in telecommunications equipment are difficult to start and, if started, grow slowly, thus permitting early detection. Such fires are typically of the smoldering type, do not spread beyond the immediate area and generally self-extinguish.

Note, however, that the exception requires a fire-resistance-rated separation from other portions of the building and, most importantly, that the building cannot qualify for any code tradeoffs for fully sprinklered buildings.

[F] 903.2.1 Group A. An *automatic sprinkler system* shall be provided throughout buildings and portions thereof used as Group A occupancies as provided in this section. For Group A-1, A-2, A-3 and A-4 occupancies, the *automatic sprinkler system* shall be provided throughout the story where the *fire area* containing the Group A-1, A-2, A-3 or A-4 occupancy is located, and throughout all stories from the Group A occupancy to, and including, the *levels of exit discharge* serving the Group A occupancy. For Group A-5 occupancies, the *automatic sprinkler system* shall be provided in the spaces indicated in Section 903.2.1.5.

❖ Group A occupancies are characterized by a significant number of people who are not familiar with their surroundings. The requirement for a suppression system reflects the additional time needed for egress. The protection is also intended to extend to the occupants of the assembly group from unobserved fires in other building areas located between the story containing the assembly occupancy and the levels of exit discharge serving such occupancies. The only exception to the coverage is for Group A-5 occupancies that are open to the atmosphere. Such occupancies require only certain aspects to be sprinklered, such as certain concession stands and other enclosed use areas (see commentary, Section 903.2.1.5).

The requirement for sprinklers is based on the location and function of the space. It is not dependent on whether or not the area is provided with exterior walls. IFC Committee Interpretation No. 25-05 to this section discusses this issue and states, in part, that "where no surrounding exterior walls are provided along the perimeter of the building, the building area is used to identify and determine applicable fire area." Outdoor areas such as pavilions and patios may have no walls but will have an occupant load and other factors that identify the assembly occupancy as such. If any of the thresholds are reached requiring sprinkler protection, then sprinkler protection must be provided whether there are exterior walls or not.

Occupancy	Threshold	Exception
All occupancies	Buildings with floor level ≥ 55 feet above fire department vehicle access and occupant load ≥ 30.	Open parking structures. F-2
Assembly (A-1, A-3, A-4)	Fire area > 12,000 sq. ft. or fire area occupant load > 300 or fire area above/below level of exit discharge. Multitheater complex (A-1 only)	None
Assembly (A-2)	Fire area > 5,000 sq. ft. or fire area occupant load > 100 or fire area above/below level of exit discharge.	None
Assembly (A-5)	Accessory areas > 1,000 sq. ft.	None
Ambulatory care facility (B)	≥ 4 care recipients incapable of self-preservation or any care recipients incapable of self-preservation above or below level of exit discharge.	None
Educational (E)	Fire area > 12,000 sq. ft. or below level of exit discharge.	Each classroom has exterior door at grade.
Factory (F-1) Mercantile (M) Storage (S-1)	Fire area > 12,000 sq. ft. or building > three stories or combined fire area > 24,000 sq. ft. Woodworking > 2,500 sq. ft. (F-1 only). Manufacture > 2,500 sq. ft. (F-1), display and sale > 5,000 sq. ft. (M), storage > 2,500 sq. ft. (S-1) of upholstered furniture or mattresses. Bulk storage of tires > 20,000 cu. ft. (S-1 only).	None
High hazard (H-1, H-2, H-3, H-4, H-5)	Sprinklers required.	None
Institutional (I-1, I-2, I-3, I-4)	Sprinklers required.	Day Care at level of exit discharge and each classroom has exterior exit door.
Residential (R)	Sprinklers required.	None
Repair garage (S-1)	Fire area > 12,000 sq. ft. or ≥ two stories (including basement) with fire area > 10,000 sq. ft. or repair garage servicing vehicles in basement or servicing commercial motor vehicles in fire area > 5,000 sq. ft.	None
Parking garage (S-1)	Commercial motor vehicles parking area > 5,000 sq. ft.	None
Parking garage (S-2)	Fire area > 12,000 sq. ft. or fire area > 5,000 sq. ft. for storage of commercial motor vehicles; or beneath other groups. (enclosed parking)	Not if beneath Group R-3
Covered and open malls (402.5)	Sprinklers required.	Attached open parking structures.
High-rises (403.3)	Sprinklers required.	Open garages; certain telecommunications buildings
Unlimited area buildings (507)	A-3, A-4, B, F, M, S: one story. B, F, M, S: two story.	One story F-2 or S-2.

Note: Thresholds located in Section 903.2 unless noted. See also Table 903.2.11.6 for additional required suppression systems.

For SI: 1 foot = 304.8 mm, 1 square foot = 0.0929 m^2.

Figure 903.2
SUMMARY OF OCCUPANCY-RELATED AUTOMATIC SPRINKLER THRESHOLDS

[F] 903.2.1.1 Group A-1. An *automatic sprinkler system* shall be provided for *fire areas* containing Group A-1 occupancies and intervening floors of the building where one of the following conditions exists:

1. The *fire area* exceeds 12,000 square feet (1115 m^2).
2. The *fire area* has an *occupant load* of 300 or more.
3. The *fire area* is located on a floor other than a *level of exit discharge* serving such occupancies.
4. The *fire area* contains a multitheater complex.

❖ Group A-1 occupancies are identified as assembly occupancies with fixed seating, such as theaters. In addition to the high occupant load associated with these types of facilities, egress is further complicated by the possibility of low lighting levels customary during performances. The fuel load in these buildings is usually of a type and quantity that would support fairly rapid fire development and sustained duration.

Theaters with stages pose a greater hazard. Sections 410.7 and 410.8 require stages to be equipped with an automatic sprinkler system and standpipe system, respectively. The proscenium opening must also be protected. These features compensate for the additional hazards associated with stages in Group A-1 occupancies.

This section lists four conditions that require installing a suppression system in a Group A-1 occupancy including the entire story where the A-1 occupancy is located and all intervening floors. Condition 1 requires that, if any one fire area of Group A-1 exceeds 12,000 square feet (1115 m^2), the automatic sprinkler system is to be installed throughout the entire story where a Group A-1 occupancy is located, regardless of whether the building is divided into more than one fire area. However, if all fire areas are less than 12,000 square feet (1115 m^2) (and less than the other thresholds), then sprinklers would not be required. Compartmentalization into multiple fire areas in compliance with Chapter 7 is deemed an adequate alternative to sprinkler protection.

Condition 2 establishes the minimum number of occupants for which an automatic sprinkler system is considered necessary. The determination of the actual occupant load must be based on Section 1004.

Condition 3 accounts for occupant egress delay when traversing a stairway requiring a sprinkler system, regardless of the size of occupant load. In such cases alternative emergency escape elements such as windows may not be available, making the suppression needs all the greater. It is not necessary for the occupant load to exceed 300 on a level other than the level of exit discharge serving such occupancy. Any number of Group A-1 occupants on the alternative level would be cause to apply the requirement for sprinklers. The text does not make reference to "story" but uses the term "floor," which could include mezzanines and basements.

Condition 4 states that a sprinkler system is required for multiplex theater complexes to account for the delay associated with the notification of adjacent compartmentalized spaces where the occupants may not be immediately aware of an emergency.

[F] 903.2.1.2 Group A-2. An *automatic sprinkler system* shall be provided for *fire areas* containing Group A-2 occupancies and intervening floors of the building where one of the following conditions exists:

1. The *fire area* exceeds 5,000 square feet (464.5 m^2).
2. The *fire area* has an *occupant load* of 100 or more.
3. The *fire area* is located on a floor other than a *level of exit discharge* serving such occupancies.

❖ Group A-2 assembly occupancies are intended for food or drink consumption, such as banquet halls, nightclubs and restaurants. Occupancies in Group A-2 involve life safety factors such as a high occupant density, flexible fuel loading, movable furnishings and limited lighting; therefore, they must be protected with an automatic sprinkler system under any of the listed conditions.

In the case of an assembly use, the purpose of the automatic sprinkler system is to provide life safety from fire as well as preserving property. By requiring fire suppression in areas through which the occupants may egress, including the level of exit discharge serving such occupancies, the possibility of unobserved fire development affecting occupant egress is minimized.

The 5,000-square-foot (464 m^2) threshold for the automatic sprinkler system reflects the higher degree of life safety hazard associated with Group A-2 occupancies. As alluded to earlier, Group A-2 occupancies could have low lighting levels, loud music, late hours of operation, dense seating with ill-defined aisles and alcoholic beverage service. These factors in combination could delay fire recognition, confuse occupant response and increase egress time.

Although the calculated occupant load for a 5,000 square-foot (465 m^2) space at 15 square feet (1.4 m^2) per occupant would be over 100, the occupant load threshold in Condition 2 is meant to reflect the concern for safety in these higher density occupancies. Although the major reason for establishing the occupant threshold at 100 was because of several recent nightclub incidents, the requirement is not limited to nightclubs or banquet facilities but to all Group A-2 occupancies. Any restaurant with an occupant load greater than 100 would require sprinkler protection as well. This includes fast food facilities with no low lighting or alcohol sales. The similar intent of Condition 3 is addressed in the commentary to Section 903.2.1.1.

Note that as with Group A-1 occupancies, when sprinklers are required they are required on the story where the Group A-2 occupancy is located and on all intervening floors leading to the levels of exit discharge.

[F] 903.2.1.3 Group A-3. An *automatic sprinkler system* shall be provided for *fire areas* containing Group A-3 occupancies and intervening floors of the building where one of the following conditions exists:

1. The *fire area* exceeds 12,000 square feet (1115 m^2).
2. The *fire area* has an *occupant load* of 300 or more.
3. The *fire area* is located on a floor other than a *level of exit discharge* serving such occupancies.

❖ Group A-3 occupancies are assembly occupancies intended for worship, recreation or amusement and other assembly uses not classified elsewhere in Group A, such as churches, museums and libraries. While Group A-3 occupancies could potentially have a high occupant load, they normally do not have the same potential combination of life safety hazards associated with Group A-2 occupancies. As with most assembly occupancies, however, most of the occupants are typically not completely familiar with their surroundings. When any of the three listed conditions are applicable, an automatic sprinkler system is required throughout the fire area containing the Group A-3 occupancy, including the entire story where the Group A-3 occupancy is located and throughout all floors between the Group A occupancy and exit discharge that serves that occupancy (see commentary, Sections 903.2.1 and 903.2.1.1).

[F] 903.2.1.4 Group A-4. An *automatic sprinkler system* shall be provided for *fire areas* containing Group A-4 occupancies and intervening floors of the building where one of the following conditions exists:

1. The *fire area* exceeds 12,000 square feet (1115 m^2).
2. The *fire area* has an *occupant load* of 300 or more.
3. The *fire area* is located on a floor other than a *level of exit discharge* serving such occupancies.

❖ Group A-4 occupancies are assembly uses intended for viewing of indoor sporting events and activities such as arenas, skating rinks and swimming pools. The occupant load density may be high depending on the extent and style of seating, such as bleachers or fixed seats, and the potential for standing-room viewing.

When any of the three listed conditions are applicable, an automatic sprinkler system is required throughout the fire area containing the Group A-4 occupancy, including the entire story where the Group A-4 occupancy is located, and in all floors between the Group A occupancy and exit discharge (see commentary, Sections 903.2.1 and 903.2.1.1).

[F] 903.2.1.5 Group A-5. An *automatic sprinkler system* shall be provided for Group A-5 occupancies in the following areas: concession stands, retail areas, press boxes and other accessory use areas in excess of 1,000 square feet (93 m^2).

❖ Group A-5 occupancies are assembly uses intended for viewing of outdoor activities. This occupancy classification could include amusement park structures, grandstands and open stadiums. A sprinkler system is not required in the open area of Group A-5 occupancies because the buildings would not accumulate smoke and hot gases. A fire in open areas would also be obvious to all spectators.

Enclosed areas such as retail areas, press boxes and concession stands require sprinklers if they are in excess of 1,000 square feet (93 m^2). The 1,000-square-foot (93 m^2) accessory use area is not intended to be an aggregate condition but rather per space. Thus, a press box that is 2,500 square feet (232 m^2) in area would need to be subdivided into areas less than 1,000 square feet (93 m^2) each in order to be below the threshold for sprinklers. There is no specific requirement for the separation of these spaces. It is assumed, however, that the separation would be a solid barrier of some type but without a required fire-resistance rating.

The provision is meant to mirror that in Section 1029.6.2.3, which exempts press boxes and storage facilities less than 1,000 square feet (93 m^2) in area from sprinkler requirements in smoke-protected assembly seating areas.

[F] 903.2.1.6 Assembly occupancies on roofs. Where an occupied roof has an assembly occupancy with an *occupant load* exceeding 100 for Group A-2 and 300 for other Group A occupancies, all floors between the occupied roof and the *level of exit discharge* shall be equipped with an *automatic sprinkler system* in accordance with Section 903.3.1.1 or 903.3.1.2.

Exception: Open parking garages of Type I or Type II construction.

❖ Frequently, roof tops are being used and occupied as assembly occupancies. Building owners will provide an open air roof-top bar or lounge, or other use similar to a Group A-2 occupancy on the roof of a building. A roof does not meet the definition of a fire area. As such, protection of the occupants can be less than what would otherwise be required were the occupancy located on a floor rather than on the roof. In addition, even if a fire occurs within the building itself, it puts these occupants at risk. The provisions requiring sprinklers are based on the type of assembly occupancy located on the roof. The roof itself is not required to be sprinklered. The reference to Section 903.3.1.2 is added, since this use can occur on the roof of multi-family housing facilities.

The exception for open parking garages is consistent with the existing code requirement exception for open parking garages under Section 903.2.11.3 for "Buildings 55 feet or more in height." It is becoming more common in the urban renewal areas throughout the U.S. that jurisdictions are asking developers to provide additional recreational and green spaces for their citizens to enjoy within their own communities. Because of the limited space available, it is not uncommon for such recreational and green spaces to be provided on the roofs of open parking garages. Based on the existing wording of this section, these recreational and green spaces greater than 700 square feet (based on 7 square feet net per occupant) or 1,500 square feet (based on 15 square feet net per occupant) would now require the open parking garage to be sprinklered. In other words, an open recreational or green space on the roof of an open parking structure that is more than approximately 26 to 39 feet square would require the garage to be sprinklered with a dry pipe sprinkler system that is initially a major cost to the project as well as a major monthly and yearly maintenance expense. Such an expense would most likely have an adverse affect on developers doing major city urban renewal projects from agreeing to provide such amenities for the local jurisdiction.

Additionally, there are considerable data supporting the exception's elimination of automatic sprinkler systems in open parking garages. Two sample reports that evaluated fire behavior in parking garages are:

1. 2006 NFPA Fire Data Report, "Structure and Vehicle Fires in General Vehicle Parking Garages."
2. 2008 Parking Consultants Council Fire Safety Committee Report, "Parking Structure Fire Facts."

These reports provided the following conclusions:

- There was an average of only 660 fires per year in all types of parking garages in the U.S. This represented a mere 0.006 percent of all fires annually.
- There were no fire fatalities in open parking garages of Type I or II construction. On average, there were only 2 injuries per year.
- There was no structural damage in 98.7 percent of the fires in parking garages.
- Vehicle fires in parking garages typically do not spread from vehicle to vehicle. Fire spread from vehicle to vehicle occurred in only 7 percent of the incidents.

Automatic sprinkler systems are required in occupancies other than open parking garages to protect the assembly occupancy above the fire and to protect the means of egress. Based on the inherent fire safety provided by open parking garages of Type I or II construction, an automatic sprinkler system is not required when an assembly use is located on the roof.

903.2.1.7 Multiple fire areas. An *automatic sprinkler system* shall be provided where multiple fire areas of Group A-1, A-2, A-3 or A-4 occupancies share exit or exit access components and the combined *occupant load* of theses fire areas is 300 or more.

❖ There are two conditions required to trigger the requirements of this section. The first is that exit or exit access components of Group A-1, A-2, A-3 or A-4 fire areas are shared. The second is that the combined occupant load of the fire areas that share these components exceeds 300 persons.

This section addresses the issue of multiple small assembly occupancies placed in a single-story building and not triggering a sprinkler system requirement because of the installation of a rated corridor and separation walls. The code now requires that sprinkler systems be added when the convergence of more than 300 persons share an exit. This is consistent with the intent of automatic sprinkler systems being required for life safety and to maintain tenable exiting in a fire event. A fire event that is near an exit is the same whether there are 300 occupants in one room or three rooms with 100 occupants each sharing an exit. This is also consistent with the requirement in the "multi-theater complex" for Group A-1, which is a requirement for anytime two or more theaters are in the same tenancy and does not consider occupant load as a trigger.

This will still allow those single-story buildings with multiple tenancies that have separate exits and utilize the fire area separation concept, such as buildings with multiple restaurants with separate entrances and strip-style mall buildings.

[F] 903.2.2 Ambulatory care facilities. An *automatic sprinkler system* shall be installed throughout the entire floor containing an *ambulatory care facility* where either of the following conditions exist at any time:

1. Four or more care recipients are incapable of self-preservation, whether rendered incapable by staff or staff has accepted responsibility for care recipients already incapable.
2. One or more care recipients that are incapable of self-preservation are located at other than the level of exit discharge serving such a facility.

In buildings where ambulatory care is provided on levels other than the *level of exit discharge*, an *automatic sprinkler system* shall be installed throughout the entire floor where such care is provided as well as all floors below, and all floors between the level of ambulatory care and the nearest *level of exit discharge*, including the *level of exit discharge*.

❖ Ambulatory care facilities are Group B occupancies, which have an enhanced set of requirements that account for the fact that patients may be incapable of self-preservation and require rescue by other occupants or fire personnel. There are several aspects to the enhanced features, including smoke compartments, sprinklers and fire alarms. More specifically, the requirements for sprinklers are based on the presence of four or more care recipients at any given time that are incapable of self-preservation or any number of care recipients that are incapable of self-preservation located on a floor other than the level of exit discharge that serves the ambulatory care facility. The sprinkler requirement is limited to the floor area that contains the Group B ambulatory care facility and any floors between the ambulatory care facility and level of exit discharge (see commentary, Section 422).

[F] 903.2.3 Group E. An *automatic sprinkler system* shall be provided for Group E occupancies as follows:

1. Throughout all Group E *fire areas* greater than 12,000 square feet (1115 m^2) in area.
2. Throughout every portion of educational buildings below the lowest *level of exit discharge* serving that portion of the building.

   Exception: An *automatic sprinkler system* is not required in any area below the lowest *level of exit discharge* serving that area where every classroom throughout the building has not fewer than one exterior *exit* door at ground level.

❖ Group E occupancies are limited to educational purposes through the 12th grade and day care centers serving children older than 2$^1/_2$ years of age. The 12,000-square-foot (1115 m^2) fire area threshold for the sprinkler system was established to allow smaller schools and day care centers to be nonsprinklered to minimize the economic impact on these facilities. The 12,000-square-foot (1115 m^2) threshold is similar to that used for several other occupancies, such as Group M occupancies.

Sprinklers would also be required in portions of the building located below the level of exit discharge serving that occupancy. However, there is an exception that would allow the omission of the automatic sprinkler system for the Group E fire area if there is a direct exit to the exterior from each classroom at ground level. The occupants must be able to go from the classroom directly to the outside without passing through intervening corridors, passageways or interior exit stairways.

[F] 903.2.4 Group F-1. An *automatic sprinkler system* shall be provided throughout all buildings containing a Group F-1 occupancy where one of the following conditions exists:

1. A Group F-1 *fire area* exceeds 12,000 square feet (1115 m^2).
2. A Group F-1 *fire area* is located more than three stories above *grade plane*.
3. The combined area of all Group F-1 *fire areas* on all floors, including any mezzanines, exceeds 24,000 square feet (2230 m^2).
4. A Group F-1 occupancy used for the manufacture of upholstered furniture or mattresses exceeds 2,500 square feet (232 m^2).

❖ Group F-1 occupancies must meet several different conditions as to when the fire area or occupancy must be sprinklered. The first three conditions are related to the difficulty of manually suppressing a fire involving a large area. Therefore, occupancies of Group F-1 must be protected throughout with an automatic sprinkler system if the fire area is in excess of 12,000 square feet (1115 m^2); if the total of all fire areas is in excess of 24,000 square feet (2230 m^2); or if the fire area is located more than three stories above grade plane. This is one of the few locations in the code where the total floor area of the building is aggregated for application of a code requirement. The stipulated conditions for when an automatic sprinkler system is required also apply to Group M (see Section 903.2.7) and S-1 (see Section 903.2.9) occupancies. Condition 4 for sprinklering a Group F-1 occupancy relates to the requirement for Group F-1 occupancies in excess of 2,500 square feet (232 m^2) that are used for the manufacture of upholstered furniture or mattresses. Note that this requirement is based simply on the square footage of the Group F-1 occupancy and is not related to fire areas. Upholstered furniture has the potential for rapid-growing and high-heat-release fires. This hazard is increased substantially when there are numerous upholstered furniture or mattresses being manufactured. Such fires put the occupants and emergency responders at risk. This requirement exists regardless of whether the upholstered furniture has passed any fire-retardant tests. See the commentary for Section 903.2.7 for more discussion on the subject of upholstered furniture. See the commentary to Section 903.2.9 for discussion of the formal interpretation and applicability to the code and the IFC.

The following examples illustrate how the criteria of this section are intended to be applied:

- If a building contains a single fire area of Group F-1 and the fire area is 13,000 square feet (1208 m^2), an automatic sprinkler system is required throughout the entire building; however, if this fire area is separated into two fire areas and neither is in excess of 12,000 square feet (1115 m^2), an automatic fire sprinkler system is not required. To be considered separate fire areas, the areas must be separated by fire barriers or horizontal assemblies having a fire-resistance rating as required in Table 707.3.10.
- If a 30,000-square-foot (2787 m^2) Group F-1 building was equally divided into separate fire areas of 10,000 square feet (929 m^2) each, an automatic sprinkler system would still be required throughout the entire building. Because the aggregate area of all fire areas exceeds 24,000 square feet (2230 m^2), additional compartmentation will not eliminate the need for an automatic sprinkler system. However, the use of a fire wall to separate the structure into two buildings would reduce the aggregate area of each building to less than 24,000 square feet (2230 m^2) and each fire area to less than 12,000 square feet (1115 m^2), which would offset the need for an automatic sprinkler system.

[F] 903.2.4.1 Woodworking operations. An *automatic sprinkler system* shall be provided throughout all Group F-1 occupancy *fire areas* that contain woodworking operations in excess of 2,500 square feet (232 m^2) in area that generate finely divided combustible waste or use finely divided combustible materials.

❖ Because of the potential amount of combustible dust that could be generated during woodworking operations, an automatic sprinkler system is required throughout a fire area when it contains a woodworking operation that exceeds 2,500 square feet (232 m^2) in area. Facilities where woodworking operations take place, such as cabinet making, are considered Group F-1 occupancies. The intent of the phrase "finely divided combustible waste" is to describe particle concentrations that are in the explosive range (see Chapter 22 of the IFC for discussion of dust-producing operations).

The extent of sprinkler coverage is only intended to be for the Group F-1 occupancy involved in the woodworking activity. If the fire area is larger than 2,500 square feet (232 m^2) but the woodworking area is 2,500 square feet (232 m^2) or less, sprinklers are not required. It is not the intent to require the installation of sprinklers throughout the building but rather in the fire area where the hazard may be present.

[F] 903.2.5 Group H. *Automatic sprinkler systems* shall be provided in high-hazard occupancies as required in Sections 903.2.5.1 through 903.2.5.3.

❖ Group H occupancies are those intended for the manufacturing, processing or storage of hazardous materials that constitute a physical or health hazard. To be considered a Group H occupancy, the amount of hazardous materials is assumed to be in excess of the maximum allowable quantities permitted by Tables 307.1(1) and 307.1(2).

[F] 903.2.5.1 General. An *automatic sprinkler system* shall be installed in Group H occupancies.

❖ This section requires an automatic sprinkler system in all Group H occupancies. Even though in some instances the hazard associated with the occupancy may be one that is not a fire hazard, an automatic sprinkler system is still required to minimize the potential for fire spreading to the high-hazard use; that is, the sprinklers protect the high-hazard area from fire outside the area. This section does not prohibit the use of an alternative automatic fire-extinguishing system in accordance with Section 904. When a water-based system is not compatible with the hazardous materials involved and thus creates a dangerous condition, an alternative fire-extinguishing system should be used. For example, combustible metals, such as magnesium and titanium, have a serious record of involvement with fire and are typically not compatible with water (see commentary, Chapter 59 of the IFC).

Where control areas are used to regulate the quantity of hazardous material within a building, the building is not considered a Group H occupancy. Unless a building would be required by some other code provision to be protected with sprinklers, control areas can be used to control the allowable quantities of hazardous materials in a building so as to not warrant a Group H classification and its mandatory sprinkler requirements.

[F] 903.2.5.2 Group H-5 occupancies. An *automatic sprinkler system* shall be installed throughout buildings containing Group H-5 occupancies. The design of the sprinkler system shall be not less than that required by this code for the occupancy hazard classifications in accordance with Table 903.2.5.2.

Where the design area of the sprinkler system consists of a *corridor* protected by one row of sprinklers, the maximum number of sprinklers required to be calculated is 13.

❖ Group H-5 occupancies are structures that are typically used as semiconductor fabrication facilities and comparable research laboratory facilities that use hazardous production materials (HPM). Many of the materials used in semiconductor fabrication present unique hazards. Many of the materials are toxic, while some are corrosive, water reactive or pyrophoric. Fire protection for these facilities is aimed at preventing incidents from escalating and producing secondary threats beyond a fire, such as the release of corrosive or toxic materials. Because of the nature of Group H-5 facilities, the overall amount of hazardous materials can far exceed the maximum allowable quantities given in Tables 307.1(1) and 307.1(2). Although the amount of HPM material is restricted in fabrication areas, the quantities of HPM in storage rooms normally will be in excess of those allowed by the tables. Additional requirements for Group H-5 facilities are located in Chapter 27 of the IFC and Section 415.11 of the code.

This section also specifies the sprinkler design criteria, based on NFPA 13, for various areas in a Group H-5 occupancy (see commentary, Table 903.2.5.2). When the corridor design area sprinkler option is used, a maximum of 13 sprinklers must be calculated. This exceeds the requirements of NFPA 13 for typical egress corridors, which require a maximum of either five or seven calculated sprinklers, depending on the extent of protected openings in the corridor. The increased number of calculated corridor sprinklers is based on the additional hazard associated with the movement of hazardous materials in corridors of Group H-5 facilities.

[F] TABLE 903.2.5.2
GROUP H-5 SPRINKLER DESIGN CRITERIA

LOCATION	OCCUPANCY HAZARD CLASSIFICATION
Fabrication areas	Ordinary Hazard Group 2
Service corridors	Ordinary Hazard Group 2
Storage rooms without dispensing	Ordinary Hazard Group 2
Storage rooms with dispensing	Extra Hazard Group 2
Corridors	Ordinary Hazard Group 2

❖ Table 903.2.5.2 designates the appropriate occupancy hazard classification for the various areas within a Group H-5 facility. The listed occupancy hazard classifications correspond to specific sprinkler system design criteria in NFPA 13. Ordinary Hazard Group 2 occupancies, for example, require a minimum design density of 0.20 gpm/ft^2 (8.1 L/min/m^2) with a minimum design area of 1,500 square feet

(139 m^2). An Extra Hazard Group 2 occupancy, in turn, requires a minimum design density of 0.40 gpm/ft^2 (16.3 L/min/m^2) with a minimum operating area of 2,500 square feet (232 m^2). The increased overall sprinkler demand for Extra Hazard Group 2 occupancies is based on the potential use and handling of substantial amounts of hazardous materials, such as flammable or combustible liquids.

[F] 903.2.5.3 Pyroxylin plastics. An *automatic sprinkler system* shall be provided in buildings, or portions thereof, where cellulose nitrate film or pyroxylin plastics are manufactured, stored or handled in quantities exceeding 100 pounds (45 kg).

❖ Cellulose nitrate (pyroxylin) plastics pose unusual and substantial fire risks. Pyroxylin plastics are the most dangerous and unstable of all plastic compounds. The chemically bound oxygen in their structure permits them to burn vigorously in the absence of atmospheric oxygen. Although these compounds produce approximately the same amount of energy as paper when they burn, pyroxylin plastics burn at a rate as much as 15 times greater than comparable common combustibles. When burning, these materials release highly flammable and toxic combustion byproducts. Consequently, cellulose nitrate fires are very difficult to control. Although this section specifies a sprinkler threshold quantity of 100 pounds, the need for additional fire protection should be considered for pyroxylin plastics in any amount.

Although the code includes cellulose nitrate "film" in its requirements, cellulose nitrate motion picture film has not been used in the United States since the 1950s. All motion picture film produced since that time is what is typically called "safety film." Consequently, the only application for this section relative to motion picture film is where it may be used in laboratories or storage vaults that are dedicated to film restoration and archives. The protection of these facilities is addressed in Sections 306.2 and 6504.2, both in the IFC.

[F] 903.2.6 Group I. An *automatic sprinkler system* shall be provided throughout buildings with a Group I *fire area*.

Exceptions:

1. An *automatic sprinkler system* installed in accordance with Section 903.3.1.2 shall be permitted in Group I-1 Condition 1 facilities.
2. An *automatic sprinkler system* is not required where Group I-4 day care facilities are at the *level of exit discharge* and where every room where care is provided has not fewer than one exterior exit door.
3. In buildings where Group I-4 day care is provided on levels other than the *level of exit discharge*, an *automatic sprinkler system* in accordance with Section 903.3.1.1 shall be installed on the entire floor where care is provided, all floors between the level of care and the level of *exit discharge*, and all floors below the *level of exit discharge* other than areas classified as an open parking garage.

❖ The Group I occupancy is divided into four individual occupancy classifications based on the degree of detention, supervision and physical mobility of the occupants. The evacuation difficulties associated with the building occupants creates the need to incorporate a defend-in-place philosophy of fire protection in occupancies of Group I. For this reason, all such occupancies are to be protected with an automatic sprinkler system. Note that this section is applicable to the entire building that contains a Group I occupancy.

Of particular note, this section encompasses all Group I-3 occupancies where more than five persons are detained (see Section 308.5). There has been considerable controversy concerning the use of automatic sprinklers in detention and correctional occupancies. Special design considerations can be taken into account to alleviate the perceived problems with sprinklers in sleeping units. Sprinklers that reduce the likelihood of vandalism as well as the potential to hang oneself are commercially available. Knowledgeable designers can incorporate certain design features to increase reliability and decrease the likelihood of damage to the system.

Group I-4 occupancies would include either adult-only care facilities or occupancies that provide personal care for more than five children $2^1/_2$ years of age or less on a less-than-24-hour basis. Because the degree of assistance and the time needed for egress cannot be gauged, an automatic sprinkler system is required.

There are three exceptions to this section. Exception 1 permits Group I-1 Condition 1 occupancies to be protected throughout with an NFPA 13R system instead of an NFPA 13 system. This is the lower risk condition for Group I-1 occupancies. Group I-1 Condition 2 occupancies would be required to use an NFPA 13 system.

Exception 2 exempts sprinkler systems completely if the day care center is at the level of exit discharge and every room has at least one exterior exit door. Note that day cares to which this section applies are considered by Section 308.6.1 to be Group E occupancies. An automatic sprinkler system would not be required unless dictated by the requirements in Section 903.2.2 (see the commentary for Section 308.6.1).

Exception 3 is also related to day cares that are still classified as Group I-4 by nature of the location in the building. In that case, an NFPA 13 system would be required on the floor where the center is located and all floors between and including the level of exit discharge. This is less stringent than the main requirement in Section 903.2.6 that requires the entire building to be sprinklered. As defined in Section 202, a Group I-4 child care facility located at the level of exit discharge and accommodating no more than 100 children, with each child care room having an exit directly to the exterior, would be classified as a Group E occupancy.

[F] 903.2.7 Group M. An *automatic sprinkler system* shall be provided throughout buildings containing a Group M occupancy where one of the following conditions exists:

1. A Group M *fire area* exceeds 12,000 square feet (1115 m^2).
2. A Group M *fire area* is located more than three stories above *grade plane*.
3. The combined area of all Group M *fire areas* on all floors, including any mezzanines, exceeds 24,000 square feet (2230 m^2).
4. A Group M occupancy used for the display and sale of upholstered furniture or mattresses exceeds 5,000 square feet (464 m^2).

❖ The sprinkler threshold requirements for Group M occupancies are identical to those of Group F-1 and S-1 occupancies (see commentary, Section 903.2.4). The one exception is that Group M occupancies are provided with an increased area for display of upholstered furniture and mattresses of 5,000 square feet (464 m^2) versus 2,500 square feet (232 m^2) required for Group F-1 and S-1 occupancies. As noted in the commentary for Group F-1 occupancies, upholstered furniture and mattresses have the potential for rapidly growing and high-heat-release fires. This hazard is increased substantially when there are numerous upholstered furniture items or mattresses on display. Such fires put the occupants and emergency responders at risk. This requirement exists regardless of whether the upholstered furniture has passed any fire-retardant tests.

The code does not specifically address what constitutes upholstered furniture, but by simple dictionary definition, upholstered furniture has seats covered with padding, springs, webbing and fabric or leather covers. The code does not make any distinction between levels of padding and upholstery provided on furniture, which was intentional. The proponent's reason statement for code change F135-07/08 stated, in part, "the American Home Furnishings Alliance (AHFA) and the National Home Furnishings Association (NHFA) have examined proposals for exempting vendors of certain constructions of furniture and concluded that such exemptions would be impractical for local code officials to enforce. This is the case because the internal construction of furniture cannot be established reliably without deconstructing it."

Note that, as with Group F-1 occupancies, the criteria is written such that any Group M occupancy, not the fire area, over 5,000 square feet (464 m^2) used for the display and sale of upholstered furniture and mattresses shall be sprinklered throughout. This is regardless of the quantity of upholstered furniture and mattresses actually available for purchase. The reason these requirements were placed into the code and the IFC was based on a large fire in Charleston, South Carolina that killed nine fire fighters. The facility was a combination furniture showroom and associated storage area. The building did not provide an automatic sprinkler system. See the commentary to Section 903.2.9 for discussion of a formal interpretation dealing with Group S-1 occupancies and applicability to the code and the IFC.

Automatic sprinkler systems for mercantile occupancies are typically designed for an Ordinary Hazard Group 2 classification in accordance with NFPA 13. If high-piled storage (see Section 903.2.7.1) is anticipated, additional levels of fire protection may be required. Also, some merchandise in mercantile occupancies, such as aerosols, rubber tires, paints and certain plastic commodities, even at limited storage heights, are considered beyond the standard Class I through IV commodity classification assumed for mercantile occupancies in NFPA 13 and may warrant additional fire protection.

[F] 903.2.7.1 High-piled storage. An *automatic sprinkler system* shall be provided in accordance with the *International Fire Code* in all buildings of Group M where storage of merchandise is in high-piled or rack storage arrays.

❖ Regardless of the size of the Group M fire area, an automatic sprinkler system may be required in a high-piled storage area. High-piled storage includes piled, palletized, bin box, shelf or rack storage of Class I through IV commodities to a height greater than 12 feet (3658 mm) and certain high-hazard commodities greater than 6 feet (1829 mm). Chapter 23 of the IFC provides a package of requirements that may include sprinkler protection depending on the size of the high-piled storage area. The design standard for the sprinkler protection of high-piled storage is NFPA 13, which addresses the many different types and configurations of high-piled storage.

[F] 903.2.8 Group R. An *automatic sprinkler system* installed in accordance with Section 903.3 shall be provided throughout all buildings with a Group R *fire area*.

❖ This section requires sprinklers in any building that contains a Group R fire area. This includes uses such as hotels, apartment buildings, group homes and dormitories. There are no minimum criteria and no exceptions.

It should be noted that buildings constructed under the *International Residential Code*® (IRC®) are not included in Group R and would not, therefore, be subject to these particular requirements. The 2009 IRC required sprinklers in all new townhouses and, beginning January 1, 2011, in all new one- and two-family dwellings. The IRC is a stand-alone code for the construction of detached one- and two-family dwellings and multiple single-family dwellings (townhouses) no more than three stories in height with a separate means of egress and addresses the requirements for sprinklers in a different way. That is, all of the provisions for new construction that affect those buildings are to be covered exclusively by the IRC and are not to be covered by another *International Code*. Buildings that do not fall within the scope of the IRC would be classified in Group R and be subject to these provisions. This is stated clearly in IFC Committee Interpretation No. 29-03.

With respect to life safety, the need for a sprinkler system is dependent on the occupants' proximity to the fire and the ability to respond to a fire emergency. Group R occupancies could contain occupants who may require assistance to evacuate, such as infants, those with a disability or who may simply be asleep. While the presence of a sprinkler system cannot always protect occupants in residential buildings who are aware of the ignition and either do not respond or respond inappropriately, it can prevent fatalities outside of the area of fire origin regardless of the occupants' response. Section 903.3.2 requires quick-response or residential sprinklers in all Group R occupancies. Full-scale fire tests have demonstrated the ability of quick-response and residential sprinklers to maintain tenability from flaming fires in the room of fire origin.

Where a different occupancy is located in a building with a residential occupancy, the provisions of this section still apply and the entire building is required to be provided with an automatic sprinkler system regardless of the type of mixed-use condition considered. This is consistent with the mixed-use provisions in Chapter 5. The type of sprinkler system permitted in the different types of Group R occupancies is further clarified in Sections 903.2.8.1 through 903.2.8.4.

[F] 903.2.8.1 Group R-3. An *automatic sprinkler system* installed in accordance with Section 903.3.1.3 shall be permitted in Group R-3 occupancies.

❖ Group R-3 occupancies are essentially one- and two-family dwellings that fall outside the scope of the IRC; thus an NFPA 13D system is appropriate. It should be noted there is no restriction on the use of NFPA 13 or NFPA 13R systems.

[F] 903.2.8.2 Group R-4 Condition 1. An *automatic sprinkler system* installed in accordance with Section 903.3.1.3 shall be permitted in Group R-4 Condition 1 occupancies.

❖ Group R-4 Condition 1 is the lesser of the risk categories for Group R-4 occupancies. The occupants are more capable of evacuating without assistance. Therefore they are treated no differently than a Group R-3 occupancy.

[F] 903.2.8.3 Group R-4 Condition 2. An *automatic sprinkler system* installed in accordance with Section 903.3.1.2 shall be permitted in Group R-4 Condition 2 occupancies. Attics shall be protected in accordance with Section 903.2.8.3.1 or 903.2.8.3.2.

❖ In Group R-4 Condition 2 occupancies, the occupants need more assistance evacuating a building; therefore, a more robust sprinkler system is required. An NFPA 13R system is required. It should be noted that there are some concerns with NFPA 13R systems not adequately addressing attic spaces as typically NFPA 13R systems focus primarily on the main habitable portion of the building. Specific compliance conditions are provided in Sections 903.2.8.3.1 and 908.2.8.3.2.

[F] 903.2.8.3.1 Attics used for living purposes, storage or fuel-fired equipment. Attics used for living purposes, storage or fuel-fired equipment shall be protected throughout with an *automatic sprinkler system* installed in accordance with Section 903.3.1.2.

❖ This section clarifies that if the attic is used for living purposes or if fuel-fired equipment or storage is in these areas, full coverage in accordance with NFPA 13R is required.

[F] 903.2.8.3.2 Attics not used for living purposes, storage or fuel-fired equipment. Attics not used for living purposes, storage or fuel-fired equipment shall be protected in accordance with one of the following:

1. Attics protected throughout by a heat detector system arranged to activate the building fire alarm system in accordance with Section 907.2.10.
2. Attics constructed of noncombustible materials.
3. Attics constructed of fire-retardant-treated wood framing complying with Section 2303.2.
4. The *automatic sprinkler system* shall be extended to provide protection throughout the attic space.

❖ In attics where people are not expected and where storage or fuel-fired equipment is not located, some protection is required on top of what NFPA 13R would require. Four different options of protection are provided. The first is simply to provide more warning time to the occupants if a fire should occur in the attic via a heat detector that activates the fire alarm system. The second is simply to reduce the risk of fire by requiring noncombustible construction materials. The third, similar to the second, is reducing the fire hazard by using fire-retardant-treated wood. This will slow the growth of a fire should one occur or prevent the start of a fire. The final option is simply to provide sprinkler protection to the attic. However if a sprinkler system is provided in the attic, issues such as freezing temperatures need to be addressed.

[F] 903.2.8.4 Care facilities. An *automatic sprinkler system* installed in accordance with Section 903.3.1.3 shall be permitted in care facilities with five or fewer individuals in a single-family dwelling.

❖ This section is similar to Sections 903.2.8.1 and 903.2.8.2 and allows the use of an NFPA 13D system in place of an NFPA 13 or 13R system. In this case, it is specific to smaller care facilities with five or fewer residents. Again, while not technically a single-family dwelling, they are very similar in nature based on the type and actual use of the building.

[F] 903.2.9 Group S-1. An *automatic sprinkler system* shall be provided throughout all buildings containing a Group S-1 occupancy where one of the following conditions exists:

1. A Group S-1 *fire area* exceeds 12,000 square feet (1115 m^2).
2. A Group S-1 *fire area* is located more than three stories above *grade plane*.
3. The combined area of all Group S-1 *fire areas* on all floors, including any mezzanines, exceeds 24,000 square feet (2230 m^2).
4. A Group S-1 *fire area* used for the storage of commercial motor vehicles where the *fire area* exceeds 5,000 square feet (464 m^2).
5. A Group S-1 occupancy used for the storage of upholstered furniture or mattresses exceeds 2,500 square feet (232 m^2).

❖ An automatic sprinkler system must be provided throughout all buildings containing a Group S-1 occupancy fire area where the fire area exceeds 12,000 square feet (1115 m^2); is more than three stories above grade plane; combined, on all floors including mezzanines, exceeds 24,000 square feet (2230 m^2); or is used for the storage of commercial motor vehicles and exceeds 5,000 square feet (464 m^2). See the commentary to the definition of "Commercial motor vehicle" in Chapter 2.

The first three sprinkler threshold requirements for Group S-1 occupancies are identical to those of Groups F-1 and M (see commentary, Sections 903.2.4 and 903.2.7). Group S-1 occupancies, such as warehouses and self-storage buildings, are assumed to be used for the storage of combustible materials. While high-piled storage does not change the Group S-1 occupancy classification, sprinkler protection, if required, may have to comply with the additional requirements of Chapter 32 of the IFC. High-piled stock or rack storage in any occupancy must comply with the code and the IFC. The fifth sprinkler threshold is the same as for Group F-1 except that, in this case, upholstered furniture and mattresses are being stored and not manufactured. Group M has a similar threshold, but is required for larger occupancies containing such items with an area of 5,000 square feet (464 m^2) versus what is required for Groups S-1 and F-1 occupancies of 2,500 square feet (232 m^2). See the commentary for Group M and Group F-1 definitions for more discussion on this issue. Again, it is important to note that the threshold is based upon the square footage of the occupancy and not upon the size of the fire area. A formal interpretation (IFC Interpretation 20-14) has been issued on this section. The formal interpretation addresses self storage warehouses specifically and whether such a facility between 2500 and 12000 square feet would require an automatic sprinkler system. This is based upon the fact that upholstered furniture may be stored in such units. The response provided noted that a sprinkler system would be required based on the fact the requirements are focused on the square footage of the occupancy and are not based on fire area or the amount of upholstered furniture or mattresses present.

[F] 903.2.9.1 Repair garages. An *automatic sprinkler system* shall be provided throughout all buildings used as repair garages in accordance with Section 406, as shown:

1. Buildings having two or more *stories above grade plane*, including basements, with a *fire area* containing a repair garage exceeding 10,000 square feet (929 m^2).
2. Buildings not more than one *story above grade plane*, with a *fire area* containing a repair garage exceeding 12,000 square feet (1115 m^2).
3. Buildings with repair garages servicing vehicles parked in basements.
4. A Group S-1 *fire area* used for the repair of commercial motor vehicles where the *fire area* exceeds 5,000 square feet (464 m^2).

❖ Automatic sprinklers may be required in repair garages, depending on the quantity of combustibles present, their location and floor area. In addition, any Group S-1 fire area intended for the repair of commercial motor vehicles that exceeds 5,000 square feet (464 m^2) would require sprinklers. This is the same criteria as Group S-1 occupancies and Group S-2 enclosed parking garages storing commercial motor vehicles. Repair garages may contain significant quantities of flammable liquids and other combustible materials. These occupancies are typically considered Ordinary Hazard Group 2 occupancies as defined in NFPA 13. Portions of repair garages used for parts cleaning using flammable or combustible liquids may require automatic sprinkler protection. If quantities of hazardous materials exceed the limitations in Section 307 for maximum allowable quantities per control area, the repair garage would be reclassified as a Group H occupancy. Note that the term "commercial motor vehicles" is specially defined in Chapter 2.

[F] 903.2.9.2 Bulk storage of tires. Buildings and structures where the area for the storage of tires exceeds 20,000 cubic feet (566 m^3) shall be equipped throughout with an *automatic sprinkler system* in accordance with Section 903.3.1.1.

❖ This section specifies when an automatic sprinkler system is required for the bulk storage of tires based on the volume of the storage area as opposed to a specific number of tires. Even in fully sprinklered buildings, tire fires pose significant problems to fire departments. Tire fires produce thick smoke and are difficult to extinguish by sprinklers alone. NFPA 13 contains specific fire protection requirements for the storage of rubber tires.

Whether the volume of tires is divided into different fire areas or not is irrelevant to the application of this section. If the total for all areas where tires are stored is great enough that the resultant storage volume exceeds 20,000 cubic feet (566 m^3), the building must be sprinklered throughout. See the commentary to the Section 202 definition of "Tires, bulk storage of" for further information.

[F] 903.2.10 Group S-2 enclosed parking garages. An *automatic sprinkler system* shall be provided throughout buildings classified as enclosed parking garages in accordance with Section 406.6 where either of the following conditions exists:

1. Where the *fire area* of the enclosed parking garage exceeds 12,000 square feet (1115 m^2).
2. Where the enclosed parking garage is located beneath other groups.

   Exception: Enclosed parking garages located beneath Group R-3 occupancies.

❖ Fire records have shown that fires in parking structures typically fully involve only a single automobile with minor damage to adjacent vehicles. An enclosed parking garage, however, does not allow the dissipation of smoke and hot gases as readily as an open parking structure, which is also considered a Group S-2 occupancy. If the enclosed parking garage has a fire area greater than 12,000 square feet (1115 m^2) or is located beneath another occupancy group, the enclosed parking garage must be protected with an automatic sprinkler system. This requirement that the enclosed parking garage located beneath other occupancy groups is required to be sprinklered is based on the potential for a fire to develop undetected, which would endanger the occupants of the other occupancy. The 12,000-square-foot (1115 m^2) threshold is similar to other occupancies such as Groups M and S-1.

It should be noted that while open parking garages are considered a Group S-2 occupancy, they are not required by the provisions of this section to be equipped with an automatic sprinkler system.

The exception exempts enclosed garages in buildings where the garages are located below a Group R-3 occupancy. The exception is essentially moot since the code requires all buildings with a Group R occupancy to be sprinklered throughout. Because the entire building with the residential occupancy is required to be sprinklered according to Section 903.2.8, the garage would be sprinklered as well. It should be noted that if the Group R-3 occupancy was protected with an NFPA 13D system, the enclosed parking garage would not require sprinklers.

[F] 903.2.10.1 Commercial parking garages. An *automatic sprinkler system* shall be provided throughout buildings used for storage of commercial motor vehicles where the *fire area* exceeds 5,000 square feet (464 m^2).

❖ Because of the larger-sized vehicles involved in commercial parking structures, such as those housing commercial motor vehicles as defined in Section 202, a more stringent sprinkler threshold is required. Bus garages may also be located adjacent to passenger terminals (Group A-3) that have a substantial occupant load. Commercial parking requires only a single vehicle in order to be classified as commercial parking.

The criterion for sprinkler protection is based on the size of the fire area and not the size of the commercial parking. If the commercial parking involves only 1,000 square feet (93 m^2) but the fire area exceeds 5,000 square feet (464 m^2), sprinkler protection is required.

[F] 903.2.11 Specific building areas and hazards. In all occupancies other than Group U, an *automatic sprinkler system* shall be installed for building design or hazards in the locations set forth in Sections 903.2.11.1 through 903.2.11.6.

❖ Sections 903.2.11.1 through 903.2.11.2 specify certain conditions under which an automatic sprinkler system is required, even in otherwise nonsprinklered buildings. As indicated, the listed conditions in the noted sections are applicable to all occupancies except Group U. Most structures that qualify as Group U do not typically have the type of conditions stipulated in Sections 903.2.11.1 through 903.2.11.1.3.

[F] 903.2.11.1 Stories without openings. An *automatic sprinkler system* shall be installed throughout all *stories*, including basements, of all buildings where the floor area exceeds 1,500 square feet (139.4 m^2) and where there is not provided not fewer than one of the following types of *exterior wall* openings:

1. Openings below grade that lead directly to ground level by an exterior *stairway* complying with Section 1009 or an outside ramp complying with Section 1010. Openings shall be located in each 50 linear feet (15 240 mm), or fraction thereof, of *exterior wall* in the *story* on at least one side. The required openings shall be distributed such that the lineal distance between adjacent openings does not exceed 50 feet (15 240 mm).
2. Openings entirely above the adjoining ground level totaling not less than 20 square feet (1.86 m^2) in each 50 linear feet (15 240 mm), or fraction thereof, of *exterior wall* in the story on at least one side. The required openings shall be distributed such that the lineal distance between adjacent openings does not exceed 50 feet (15 240 mm). The height of the bottom of the clear opening shall not exceed 44 inches (1118 mm) measured from the floor.

❖ Because of both the lack of openings in exterior walls for access by the fire department for fire fighting and rescue and the problems associated with venting the products of combustion during fire suppression operations, all stories, including any basements of buildings that do not have adequate openings as defined in this section, must be equipped with an automatic sprinkler system. This section applies to stories without sufficient exterior openings where the floor area exceeds 1,500 square feet (139 m^2) and where the building is not otherwise required to be fully sprinklered. The requirement for an automatic sprinkler system in this section applies only to the affected area and does not mandate sprinkler protection throughout the entire building.

Stories without openings, as defined in this section, are stories that do not have at least 20 square feet (1.9 m^2) of opening leading directly to ground level in each 50 lineal feet (15 240 mm) or fraction thereof on at least one side. Since exterior doors will provide openings of 20 square feet (1.9 m^2), or slightly less in some occupancies, exterior stairways and ramps in each 50 lineal feet (15 240 mm) are considered acceptable.

This section specifically states that the required openings be distributed such that the lineal distance between adjacent openings does not exceed 50 feet (15 240 mm). If the openings in the exterior wall are located without regard to the location of the adjacent openings, it is possible that segments of the exterior wall will not have the required access to the interior of the building for fire-fighting purposes. Any arrangement of required stairways, ramps or openings that results in a portion of the wall 50 feet (15 240 mm) or more in length with no openings to the exterior does not meet the intent of the code that access be provided in each 50 lineal feet (15 240 mm) (see Commentary Figure 903.2.11.1).

There is a further restriction on openings that are entirely above grade. More specifically, to support fire-fighting operations the openings need to be accessible and usable. Therefore, Item 2 specifies that the maximum sill height be 44 inches (1118 mm) above the floor. This height is consistent with the height provided for emergency escape and rescue windows in Section 1030.3.

One application of this section has been addressed in the 2009 edition of the *International Code Interpretations* book and deals with automotive service shops that have below-grade service areas where employees perform oil changes and other minor maintenance services. The below-grade areas are typically open to the grade-level service bays via openings providing access to the underside of the vehicles without requiring the vehicle to be lifted into the air. Inasmuch as the below-grade space has no openings directly to the exterior, the question was asked if it would be regulated as a windowless story and thus be required to be equipped with an automatic fire suppression system in accordance with Section 903.2.11.1.

The answer to that question is no. Because of the openness between the adjacent service levels, the below-grade area would be more appropriately regulated similar to a mezzanine rather than a story. A mezzanine is not regulated as a separate story but rather as part of the same story that it serves. Therefore, if the below-grade service level is in compliance with the applicable provisions of Section 505, the windowless story provisions of Section 903.2.11.1 would be evaluated based on the exterior wall openings of the main level and not the service mezzanine below. The direct interconnections between the two adjacent floor levels by multiple service openings provide access to the lower service area for fire-fighting and rescue operations. As such, it would not be regulated as a windowless story.

The requirement to sprinkler the basement is independent of mixed-use conditions. Whether the basement is separated or nonseparated is irrelevant to the need for sprinkler protection, nor does the requirement to provide sprinklers in the basement imply that sprinklers must be provided elsewhere. This requirement is applicable to the basement or any story without openings irrespective of other code provisions.

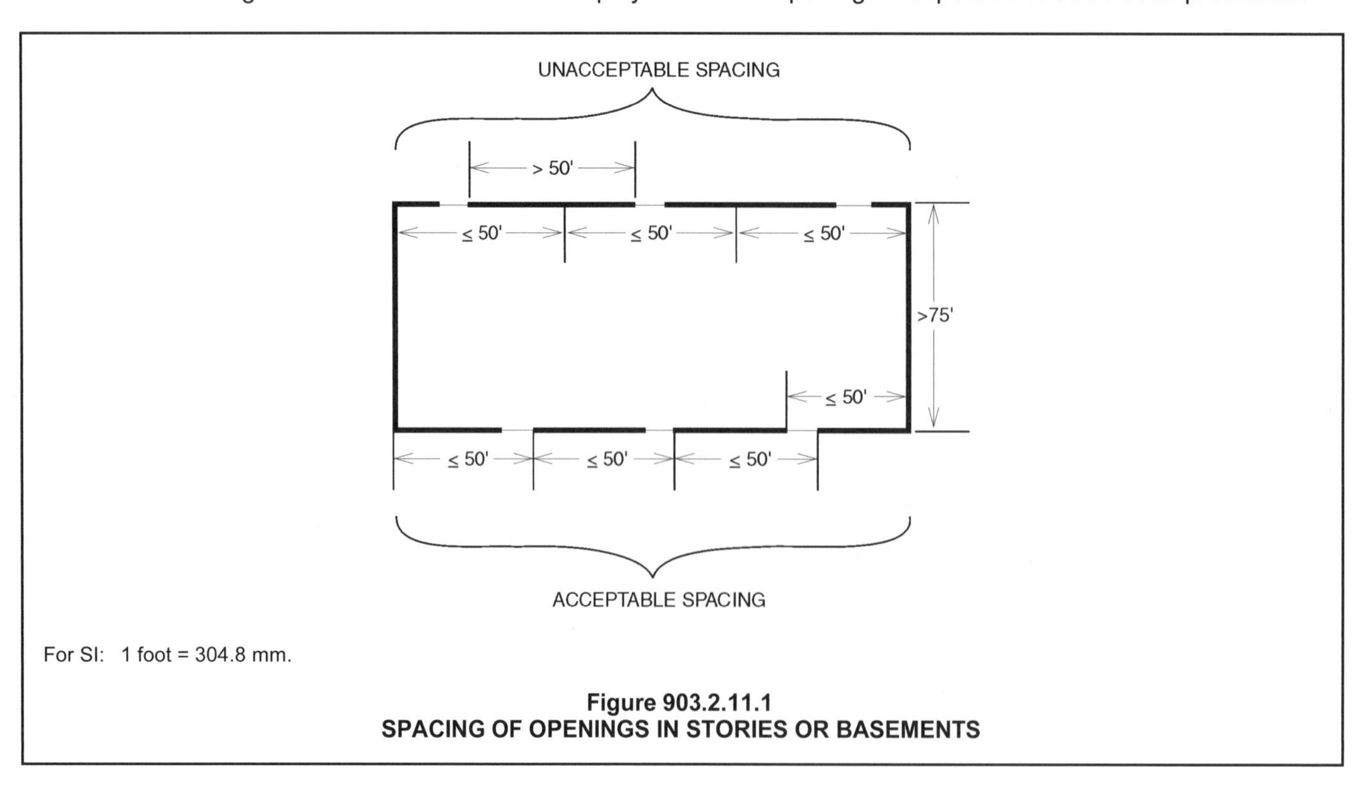

Figure 903.2.11.1
SPACING OF OPENINGS IN STORIES OR BASEMENTS

Also, these provisions are not based on the size of a fire area but rather on the size of the basement. Thus, subdividing the basement into multiple fire areas would have no effect on the requirement. However, one benefit of the multiple fire areas could be that each fire area could have a separate limited area sprinkler system with less than 20 sprinklers.

[F] 903.2.11.1.1 Opening dimensions and access. Openings shall have a minimum dimension of not less than 30 inches (762 mm). Such openings shall be accessible to the fire department from the exterior and shall not be obstructed in a manner such that fire fighting or rescue cannot be accomplished from the exterior.

❖ To qualify, an opening must not be less than 30 inches (762 mm) in least dimension and must be accessible to the fire department from the exterior. The minimum opening dimension gives fire department personnel access to the interior of the story or basement for fire-fighting and rescue operations and provides openings that are large enough to vent the products of combustion.

[F] 903.2.11.1.2 Openings on one side only. Where openings in a *story* are provided on only one side and the opposite wall of such *story* is more than 75 feet (22 860 mm) from such openings, the *story* shall be equipped throughout with an *approved automatic sprinkler system*, or openings as specified above shall be provided on not fewer than two sides of the *story*.

❖ If openings are provided on only one side, an automatic sprinkler system would still be required if the opposite wall of the story is more than 75 feet (22 860 mm) from existing openings. An alternative to providing the automatic sprinkler system would be to design openings on at least two sides of the exterior of the building. As long as the story being considered is not a basement, the openings on two sides can be greater than 75 feet (22 860 mm) from any portion of the floor. In basements, if any portion is more than 75 feet (22 860 mm) from the openings, the entire basement must be equipped with an automatic sprinkler system, as indicated in Section 903.2.11.1.3. Providing openings on more than one wall allows cross ventilation to vent the products of combustion [see Commentary Figures 903.2.11.1.2(1–4)].

[F] 903.2.11.1.3 Basements. Where any portion of a *basement* is located more than 75 feet (22 860 mm) from openings required by Section 903.2.11.1, or where walls, partitions or other obstructions are installed that restrict the application of water from hose streams, the *basement* shall be equipped throughout with an *approved automatic sprinkler system*.

❖ The 75-foot (22 860 mm) distance is intended to be measured in the line of travel—not in a straight line perpendicular to the wall. Where obstructions, such as walls or other partitions, are present in a basement, the walls and partitions enclosing any room or space must have openings that provide an equivalent degree of fire department access to that provided by the openings prescribed in Section 903.2.11.1 for exterior walls. When obstructions such as walls or partitions are installed in the basement, the ability to apply hose streams through these openings and

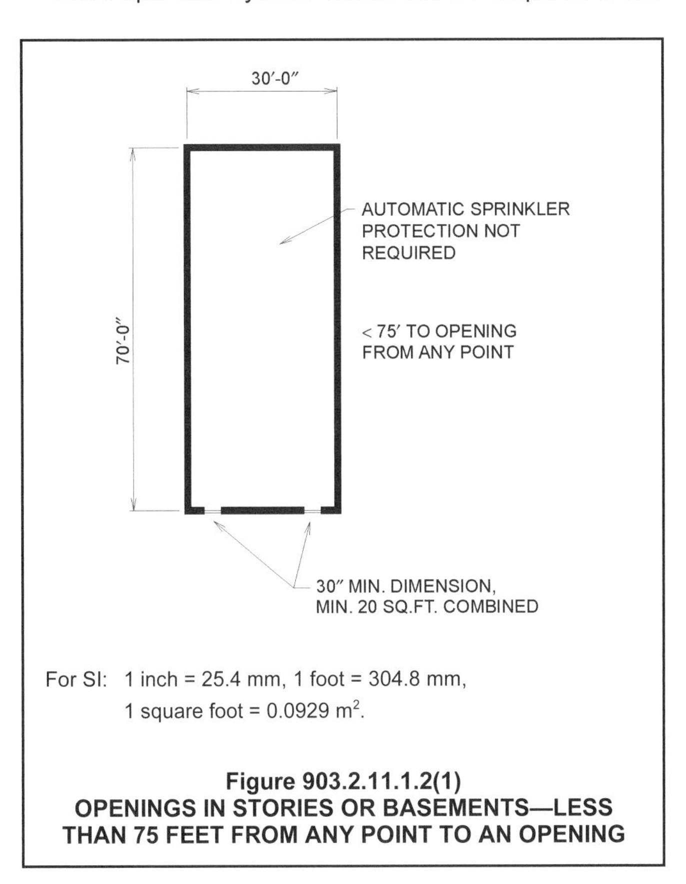

For SI: 1 inch = 25.4 mm, 1 foot = 304.8 mm, 1 square foot = 0.0929 m^2.

Figure 903.2.11.1.2(1)
OPENINGS IN STORIES OR BASEMENTS—LESS THAN 75 FEET FROM ANY POINT TO AN OPENING

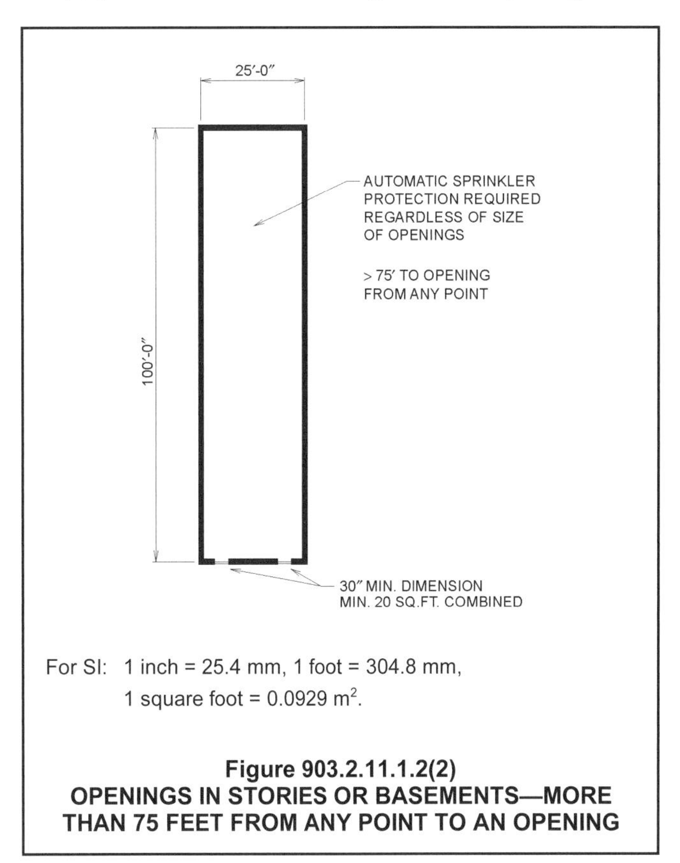

For SI: 1 inch = 25.4 mm, 1 foot = 304.8 mm, 1 square foot = 0.0929 m^2.

Figure 903.2.11.1.2(2)
OPENINGS IN STORIES OR BASEMENTS—MORE THAN 75 FEET FROM ANY POINT TO AN OPENING

reach the basement area is reduced or eliminated. The configuration and clear-opening requirements become useless when an interior wall or other obstruction is placed inside the basement. In that case, it is reasonable to require automatic fire sprinklers to provide adequate protection in the basement. If an equivalent degree of fire department access to all portions of the floor area is not provided, the basement would require an automatic sprinkler system.

[F] 903.2.11.2 Rubbish and linen chutes. An *automatic sprinkler system* shall be installed at the top of rubbish and linen chutes and in their terminal rooms. Chutes shall have additional sprinkler heads installed at alternate floors and at the lowest intake. Where a rubbish chute extends through a building more than one floor below the lowest intake, the extension shall have sprinklers installed that are recessed from the drop area of the chute and protected from freezing in accordance with Section 903.3.1.1. Such sprinklers shall be installed at alternate floors, beginning with the second level below the last intake and ending with the floor above the discharge. Chute sprinklers shall be accessible for servicing.

❖ Gravity rubbish (waste) and linen chutes can pose a significant hazard to building occupants if they are not properly installed and protected. Generally, these systems are installed in high-occupancy buildings where the occupants will be sleeping or are incapable of self-rescue, such as in Group I, R-1 and R-2 occupancies. For occupant convenience, openings to the chutes are commonly provided in areas accessible to the public and, in older buildings, the chute opening may be located in an exit access corridor. In comparison to other building shafts, gravity rubbish and linen chutes always contain fuel. As bags of waste debris or linen fall through the chute, they can deposit fluids such as waste cooking oil, which adheres to the shaft surface. This waste material and other debris provide fuel that can support and accelerate vertical fire spread. The greatest accumulation of fuel will be in the termination room; however, a significant amount of fuel that covers the interior surface area of the chute will be found in the sections of chutes closest to the collection or termination room. Therefore, it is important that the automatic sprinklers be properly placed and protected so they are available in the event of a fire in the termination room and to protect waste compaction equipment where such equipment is installed.

Installation of gravity chutes for rubbish or linen requires compliance with the code, the IFC and Chapter 6 of NFPA 82. Under the code, permanent rubbish and linen chutes are constructed inside of a fire-resistance-rated shaft assembly with a minimum 1-hour fire-resistance rating in buildings less than four stories in height; in buildings four or more stories in height, the fire-resistance rating is increased to 2

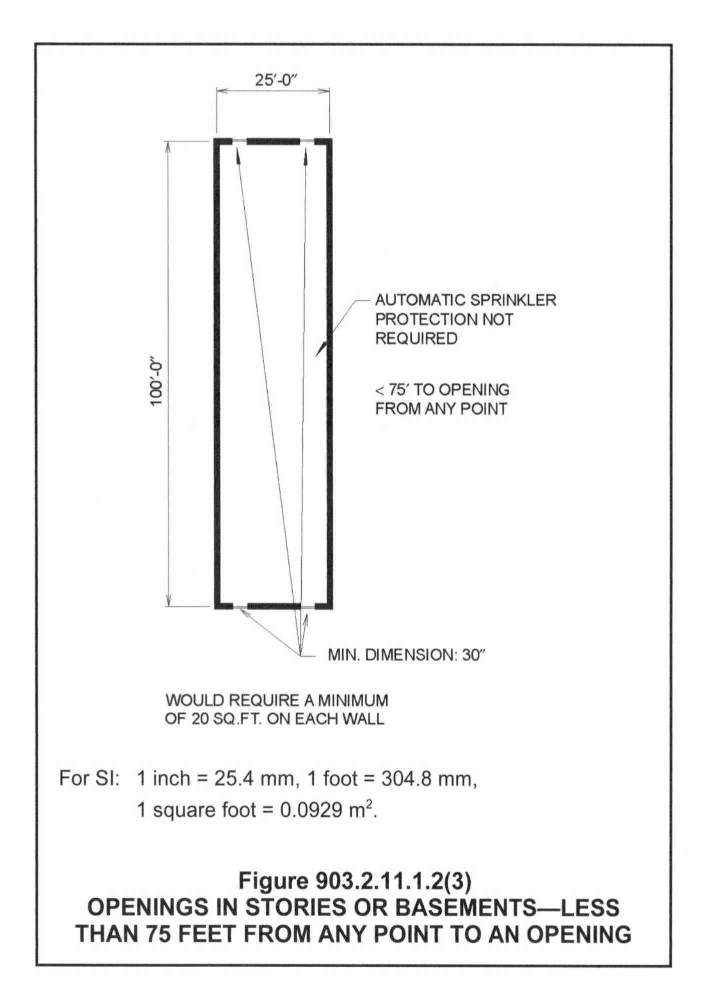

For SI: 1 inch = 25.4 mm, 1 foot = 304.8 mm, 1 square foot = 0.0929 m^2.

Figure 903.2.11.1.2(3)
OPENINGS IN STORIES OR BASEMENTS—LESS THAN 75 FEET FROM ANY POINT TO AN OPENING

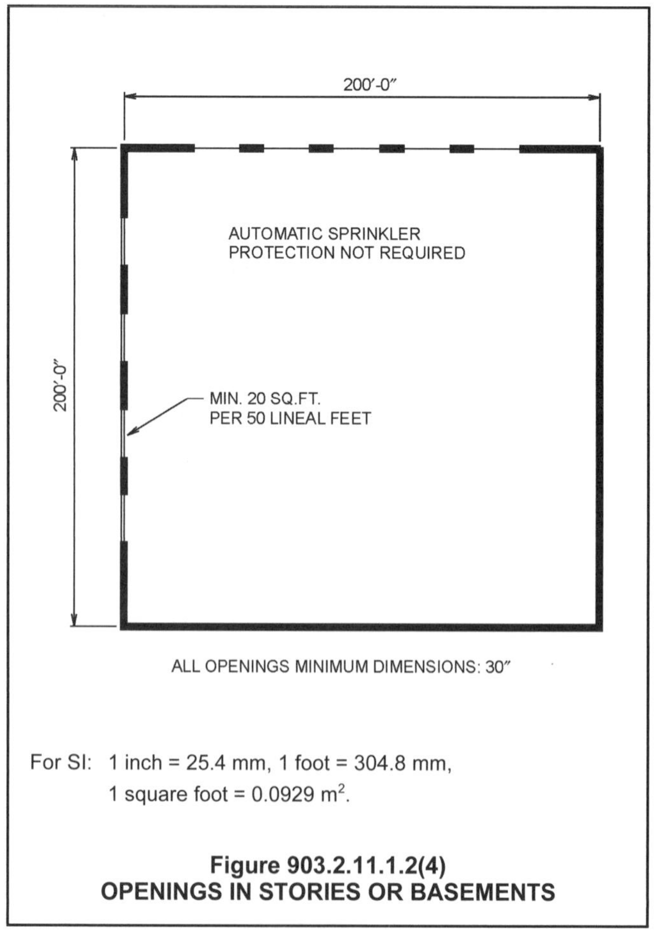

For SI: 1 inch = 25.4 mm, 1 foot = 304.8 mm, 1 square foot = 0.0929 m^2.

Figure 903.2.11.1.2(4)
OPENINGS IN STORIES OR BASEMENTS

hours by Section 713.4. The design of the shaft system and its openings must also comply with the requirements in Sections 713.11 and 713.13, which require the termination room receiving the discharged material to be separated from the building by a fire-resistance rating equivalent to that of the shaft that it serves.

Section 713.13.6 requires the installation of an automatic sprinkler system in rubbish and linen chutes to comply with the requirements of Section 903.2.11.2. Section 903.2.11.2 correlates with the requirements in Chapter 22 of NFPA 13. Chapter 22 of NFPA 13 contains the special occupancy requirements for all buildings, including gravity waste and linen chutes. The provisions align the code and IFC requirements with those in NFPA 82 and NFPA 13.

A critical term in this section is "extension." The word was selected to address chutes installed in buildings of pedestal construction or other designs in which the fire-resistant construction shaft and chute pass through a less hazardous occupancy, such as a Group S-2 parking garage, or other floors that do not have access to the shaft. In these areas, chute openings are generally not provided. As a result, this section now contains a specific provision that may impose a requirement for sprinklers in the portion of the chute that serves as an extension beyond the last intake and the termination room or discharge area.

Because objects will be falling through the chute, the code requires the chute sprinklers to be recessed and protected from impact. Sprinklers are not required at every story housing a chute. The code requires automatic sprinklers at the top of the chute and at its termination. In addition, sprinkler heads are required at alternate floors within the chute, with a head being installed at the floor level with the lowest intake point into the chute. Previously, these additional sprinkler heads were only required where the shaft extended through three or more floors. These revisions, plus the previously discussed requirements for extensions, may result in additional sprinkler heads within some shafts as compared to the previous requirements.

Sprinklers in chutes that are in locations subject to freezing require freeze protection in accordance with the requirements of Section 903.3.1.1 and, therefore, the NFPA 13 standard. This can be accomplished using a dry-pendant sprinkler or constructing a dry-pipe sprinkler system.

[F] 903.2.11.3 Buildings 55 feet or more in height. An *automatic sprinkler system* shall be installed throughout buildings that have one or more stories with an *occupant load* of 30 or more located 55 feet (16 764 mm) or more above the lowest level of fire department vehicle access, measured to the finished floor.

Exceptions:

1. Open parking structures.
2. Occupancies in Group F-2.

❖ Because of the difficulties associated with manual suppression of a fire in mid-rise buildings in excess of 55 feet (16 764 mm) above the lowest level of fire department vehicle access, an automatic sprinkler system is required throughout the building regardless of occupancy. Buildings that qualify for a sprinkler system under this section are not necessarily high-rise buildings as defined in Section 202 and are focused also on those with occupants located on the upper floors. These provisions apply only to buildings with occupied floors having an occupant load of 30 or more located on stories 55 feet or greater from fire department vehicle access. The 55 feet is measured to the finished floor (see Commentary Figure 903.2.11.3).

The listed exceptions are occupancies that, based on height only, do not require an automatic sprinkler system. Open parking structures are also exempt from the high-rise provisions of Section 403. Although an automatic sprinkler system is not required in open parking structures, a sprinkler system may still be needed, depending on the building construction type and the area and number of parking tiers (see Table 406.3.5).

[F] 903.2.11.4 Ducts conveying hazardous exhausts. Where required by the *International Mechanical Code*, automatic sprinklers shall be provided in ducts conveying hazardous exhaust or flammable or combustible materials.

Exception: Ducts where the largest cross-sectional diameter of the duct is less than 10 inches (254 mm).

❖ Section 510 of the IMC addresses the requirements for hazardous exhaust systems. To protect against the spread of fire within a hazardous exhaust system and to prevent a duct fire from involving the building, an automatic sprinkler system must be installed to protect the exhaust duct system. Where materials conveyed in the ducts are not compatible with water, alternative extinguishing agents should be used. The fire suppression requirement is intended to apply to exhaust systems having an actual fire hazard. An automatic sprinkler system in the duct would be of little value for an exhaust system that conveys only nonflammable or noncombustible materials, fumes, vapors or gases.

The exception recognizes the reduced hazard associated with smaller ducts and the impracticality of installing sprinkler protection. Another exception in the IMC indicates that laboratory hoods that meet specific provisions of the IMC are not required to be suppressed. Because the IMC is more specific in this regard, it should be consulted for the proper application of the exception.

[F] 903.2.11.5 Commercial cooking operations. An *automatic sprinkler system* shall be installed in commercial kitchen exhaust hood and duct systems where an *automatic sprinkler system* is used to comply with Section 904.

❖ An automatic suppression system is required for commercial kitchen exhaust hood and duct systems where required by Section 609 of the IFC or by the IMC to have a Type I hood. Type I hoods are required for commercial cooking equipment that produces grease-laden vapors or smoke. Section 904.12 recognizes that alternative extinguishing systems other than an automatic sprinkler system may be used. Where an automatic sprinkler system is used for commercial cooking operations, it must comply with the requirements identified in Section 904.11.4.

[F] 903.2.11.6 Other required suppression systems. In addition to the requirements of Section 903.2, the provisions indicated in Table 903.2.11.6 require the installation of a fire suppression system for certain buildings and areas.

❖ In addition to Section 903.2, requirements for automatic fire suppression systems are also found elsewhere in the code as indicated in Table 903.2.11.6.

TABLE 903.2.11.6. See next column.

❖ Table 903.2.11.6 identifies other sections of the code that require an automatic fire suppression system based on the specific occupancy or use because of the unique hazards of such use or occupancy. The table does not identify the various sections of the code that contain design alternatives based on the use of an automatic fire suppression system, typically an automatic sprinkler system.

[F] TABLE 903.2.11.6
ADDITIONAL REQUIRED SUPPRESSION SYSTEMS

SECTION	SUBJECT
402.5, 402.6.2	Covered and open mall buildings
403.3	High-rise buildings
404.3	Atriums
405.3	Underground structures
407.6	Group I-2
410.7	Stages
411.4	Special amusement buildings
412.3.6	Airport traffic control towers
412.4.6, 412.4.6.1, 412.6.5	Aircraft hangars
415.11.11	Group H-5 HPM exhaust ducts
416.5	Flammable finishes
417.4	Drying rooms
419.5	*Live/work units*
424.3	Children's play structures
507	Unlimited area buildings
509.4	Incidental uses
1029.6.2.3	Smoke-protected assembly seating
IFC	Sprinkler system requirements as set forth in Section 903.2.11.6 of the *International Fire Code*

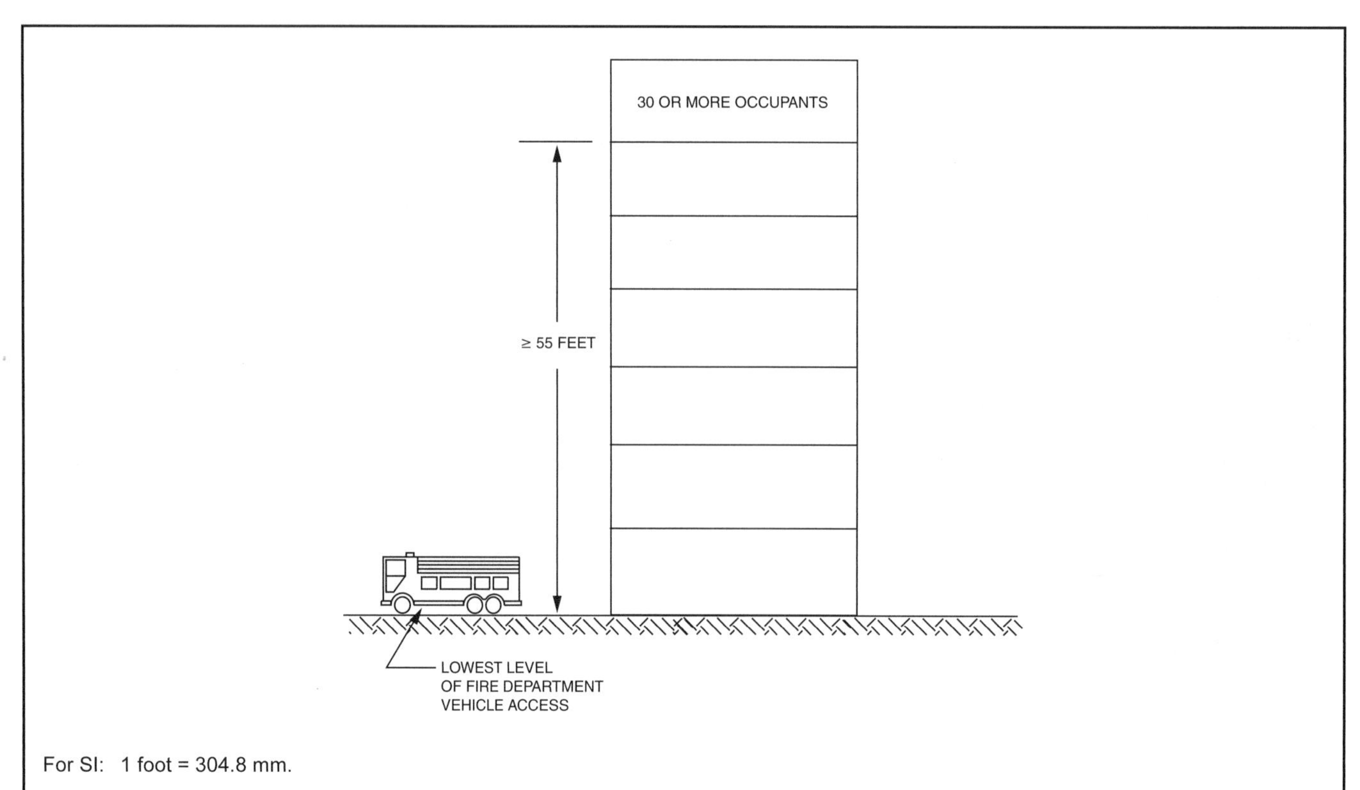

For SI: 1 foot = 304.8 mm.

Figure 903.2.11.3
SPRINKLER REQUIREMENTS: OCCUPANTS 55 FEET OR GREATER FROM FIRE DEPARTMENT VEHICLE ACCESS

[F] 903.2.12 During construction. *Automatic sprinkler systems* required during construction, *alteration* and demolition operations shall be provided in accordance with Chapter 33 of the *International Fire Code*.

❖ Chapter 33 of the code and Chapter 14 of the IFC address fire safety requirements during construction, alteration or demolition work. Working sprinkler systems should remain operative at all times unless it is absolutely necessary to shut down the system because of the proposed work. All sprinkler system impairments should be rectified as quickly as possible unless specific prior approval has been obtained from the fire code official. Buildings with a required sprinkler system should not be occupied unless the sprinkler system has been installed and tested consistent with Section 901.5. If the system must be placed out of service, the requirements of Section 901.7 of the IFC are necessary to address the temporary impairment to the fire protection system.

[F] 903.3 Installation requirements. *Automatic sprinkler systems* shall be designed and installed in accordance with Sections 903.3.1 through 903.3.8.

❖ Specific design, installation and testing criteria are given for automatic sprinkler systems in the sections and subsections that follow, as well as an indication of the applicability of a nationally recognized standard in the area. The information required to complete a thorough review of an automatic sprinkler system is listed in Commentary Figure 903.3.

[F] 903.3.1 Standards. Sprinkler systems shall be designed and installed in accordance with Section 903.3.1.1 unless otherwise permitted by Sections 903.3.1.2 and 903.3.1.3 and other chapters of this code, as applicable.

❖ Automatic sprinkler systems are to be installed to comply with the code and NFPA 13, 13R or 13D. As provided for in Section 102.4, where differences occur between the code and NFPA 13, 13R or 13D, the code applies. The fire code official also has the authority to approve the type of sprinkler system to be installed. See Commentary Figure 903.3.1 for typical design parameters for each type of sprinkler system.

This section also provides a pointer to other sections of the code that might provide more specific or detailed sprinkler requirements such as those found in Chapter 4 of the code.

[F] 903.3.1.1 NFPA 13 sprinkler systems. Where the provisions of this code require that a building or portion thereof be equipped throughout with an *automatic sprinkler system* in accordance with this section, sprinklers shall be installed throughout in accordance with NFPA 13 except as provided in Sections 903.3.1.1.1 and 903.3.1.1.2.

❖ NFPA 13 contains the minimum requirements for the design and installation of automatic water sprinkler systems and exposure protection sprinkler systems. The requirements contained in the standard include the character and adequacy of the water supply and the selection of sprinklers, piping, valves and all of the materials and accessories. The standard does not include requirements for installation of private fire service mains and their appurtenances; installation of fire pumps or construction and installation of gravity and pressure tanks and towers.

NFPA 13 defines seven classifications or types of water sprinkler systems: wet pipe [see Commentary Figure 903.3.1.1], dry pipe; preaction or deluge; combined dry pipe and preaction; antifreeze systems; sprinkler systems that are designed for a special purpose and outside sprinklers for exposure protection. While numerous variables must be considered in selecting the proper type of sprinkler system, the wet-pipe sprinkler system is recognized as the most effective and efficient. The wet-pipe system is also the most reliable type of sprinkler system, because water under pressure is available at the sprinkler. Therefore, wet-pipe sprinkler systems are recommended wherever possible.

The extent of coverage and distribution of sprinklers is based on the NFPA 13 standard. Numerous conditions exist in the standard where sprinklers are specifically required and also where they may or may not be located. Once it is determined that the sprinkler system is to be in accordance with NFPA 13, that standard must be reviewed for installation details. For example, exterior spaces such as combustible canopies are required to be equipped with sprinklers according to Section 8.15.7 of NFPA 13 where the canopy extends for a distance of 4 feet (1219 mm) or more. A 3-foot (914 mm) combustible canopy would not require sprinklers nor would a 6-foot (1829 mm) canopy constructed of noncombustible materials, provided there is no combustible storage under the canopy.

Because installation is required to be in accordance with NFPA 13, if the standard allows for the omission of sprinklers in any location, then the building is still considered as sprinklered throughout. For example, Section 8.15.8.1.1 of NFPA 13 allows sprinklers to be omitted from bathrooms in dwelling units in motels and hotels. If sprinklers are not provided in the bathrooms because of the conditions stipulated in NFPA 13, the building would still be considered as sprinklered throughout in accordance with the code, NFPA 13 and the IFC.

Exceptions for the use of NFPA 13R and 13D systems are addressed throughout the code when exceptions based on the use of sprinklers are provided. More specifically, if the use of these other standards is appropriate it will be noted within the exception. For a building to be considered "equipped throughout" with an NFPA 13 sprinkler system, complete protection must be provided in accordance with the referenced standard, subject to the exempt locations indicated in Section 903.3.1.1.1. See Commentary Figure 904.2.1 for examples of requirements modified through the use of sprinkler systems.

1. Information required on shop drawings includes:
— Name of owner and occupant
— Location, including street address
— Point of compass
— Graphic indication of scale
— Ceiling construction
— Full-height cross section
— Location of fire walls
— Location of partitions
— Occupancy of each area or room
— Location and size of blind spaces and closets
— Any questionable small enclosures in which no sprinklers are to be installed
— Size of city main in street, pressure and whether dead end or circulation and, if dead end, direction and distance to nearest circulating main, city main test results
— Other source of water supply, with pressure or elevation
— Make, type and orifice size of sprinkler
— Temperature rating and location of high-temperature sprinklers
— Limitations on extended coverage sprinklers or other special sprinkler types
— Number of sprinklers on each riser and on each system by floors and total area by each system on each floor
— Make, type, model and size of alarm or dry pipe valve
— Make, type, model and size of preaction or deluge valve
— Type and location of alarm bells
— Backflow prevention method and details
— Total number of sprinklers on each dry pipe system or preaction deluge system
— Approximate capacity in gallons or each dry pipe system
— Setting for pressure-reducing valves
— Pipe size, type, and schedule of wall thickness
— Cutting lengths of pipe (or center-to-center dimensions)
— Type of fittings, riser nipples and size, and all welds and bends
— Type and location of hangers, inserts and sleeves
— Calculations of loads and details for sway bracing
— All control valves, checks, drain pipes, flushing, and test pipes
— Size and location of standpipe risers and hose outlets
— Small hand-hose equipment
— Underground pipe size, length, location, weight, material, point of connection to city main; the type of valves, meters and valve pits; and the depth that top of the pipe is laid below grade
— Size and location of hydrants along with hose-houses
— Size and location of fire department connections
— When the equipment is to be installed as an addition to an old group of sprinklers without additional feed from the yard system, enough of the old system shall be indicated on the plans to show the total number of sprinklers to be supplied and to make all connections clear
— Information to be provided on the hydraulic nameplate
— Name, address and phone number of contractor and sprinkler designer
— Hydraulic reference points shall be shown by a number and/or letter designation and shall correspond with comparable reference points shown on the hydraulic calculation sheets
— System design criteria showing the minimum rate of water application (density), the design area of water application and the water required for hose streams both inside and outside
— Actual calculated requirements showing the total quantity of water and the pressure required at a common reference point for each system
— Elevation data showing elevations of sprinklers, junction points and supply or reference points
— Protected wall openings if room design method is used

2. Information required on calculations includes:
— Location
— Name of owner and occupant
— Building identification
— Description of hazard
— Name and address of contractor and designer
— Name of approving agency

3. System design requirements include:
— Design area of water application
— Minimum rate of water application (density)
— Area of sprinkler coverage
— Hazard or commodity classification
— Building height
— Storage height
— Storage method
— Total water requirements, as calculated, including allowance for hose demand water supply information and allowance for in-rack sprinklers
— Location and elevation static and residual test gauge with relation to the riser reference point
— Size and location of hydrants used for flow test data
— Flow location
— Static pressure, psi
— Residual pressure, psi
— Flow, gpm
— Date
— Time
— Test conducted by whom
— Sketch to accompany gridded system calculations to indicate flow quantities and directions for lines with sprinklers operated in the remote area

4. Additional information necessary for complete review includes:
— Sprinkler description and discharge constant (K-value)
— Hydraulic reference points
— Flow, gpm
— Pipe diameter (actual internal diameter)
— Pipe length
— Equivalent pipe length for fittings and components
— Friction loss in psi per foot of pipe
— Total friction loss between reference points
— Elevation difference between reference points
— Required pressure in psi at each reference point
— Velocity pressures and normal pressure if included in calculations
— Notes to indicate starting points, reference to other sheets or clarification of data
— Information on antifreeze solution (type and quantity)
— Water treatment system information including reason for treatment and program details

5. Included with the submittal must be a graph sheet showing water supply curves and system requirements including:
— Hose demand plotted on semilogarithmic graph paper so as to present a graphic summary of the complete hydraulic calculations
— Sprinkler system demand including in-rack sprinklers (if applicable)

Figure 903.3
SAMPLE SPRINKLER SYSTEM DRAWING AND DATA SUBMITTALS

[F] 903.3.1.1.1 Exempt locations. Automatic sprinklers shall not be required in the following rooms or areas where such rooms or areas are protected with an *approved* automatic fire detection system in accordance with Section 907.2 that will respond to visible or invisible particles of combustion. Sprinklers shall not be omitted from a room merely because it is damp, of fire-resistance-rated construction or contains electrical equipment.

1. A room where the application of water, or flame and water, constitutes a serious life or fire hazard.
2. A room or space where sprinklers are considered undesirable because of the nature of the contents, where *approved* by the fire code official.
3. Generator and transformer rooms separated from the remainder of the building by walls and floor/ceiling or roof/ceiling assemblies having a *fire-resistance rating* of not less than 2 hours.
4. Rooms or areas that are of noncombustible construction with wholly noncombustible contents.
5. Fire service access elevator machine rooms and machinery spaces.
6. Machine rooms, machinery spaces, control rooms and control spaces associated with occupant evacuation elevators designed in accordance with Section 3008.

❖ This section allows the omission of sprinkler protection in certain locations if an approved automatic fire detec-

	NFPA 13	NFPA 13R	NFPA 13D
Extent of protection	Equip throughout (Section 903.3.1.1)	Occupied spaces (Section 903.3.1.2)	Occupied spaces (Section 903.3.1.3)
Scope	All occupancies	Low-rise residential	One- and two-family dwellings
Sprinkler design	Density/area concept	4-head design	2-head design
Sprinklers	All types	Residential only	Residential only
Duration	30 minutes (minimum)	30 minutes	10 minutes
Advantages	Property and life protection	Life safety/tenability	Life safety/tenability

Figure 903.3.1
NFPA 13, NFPA 13R AND NFPA 13D SYSTEMS

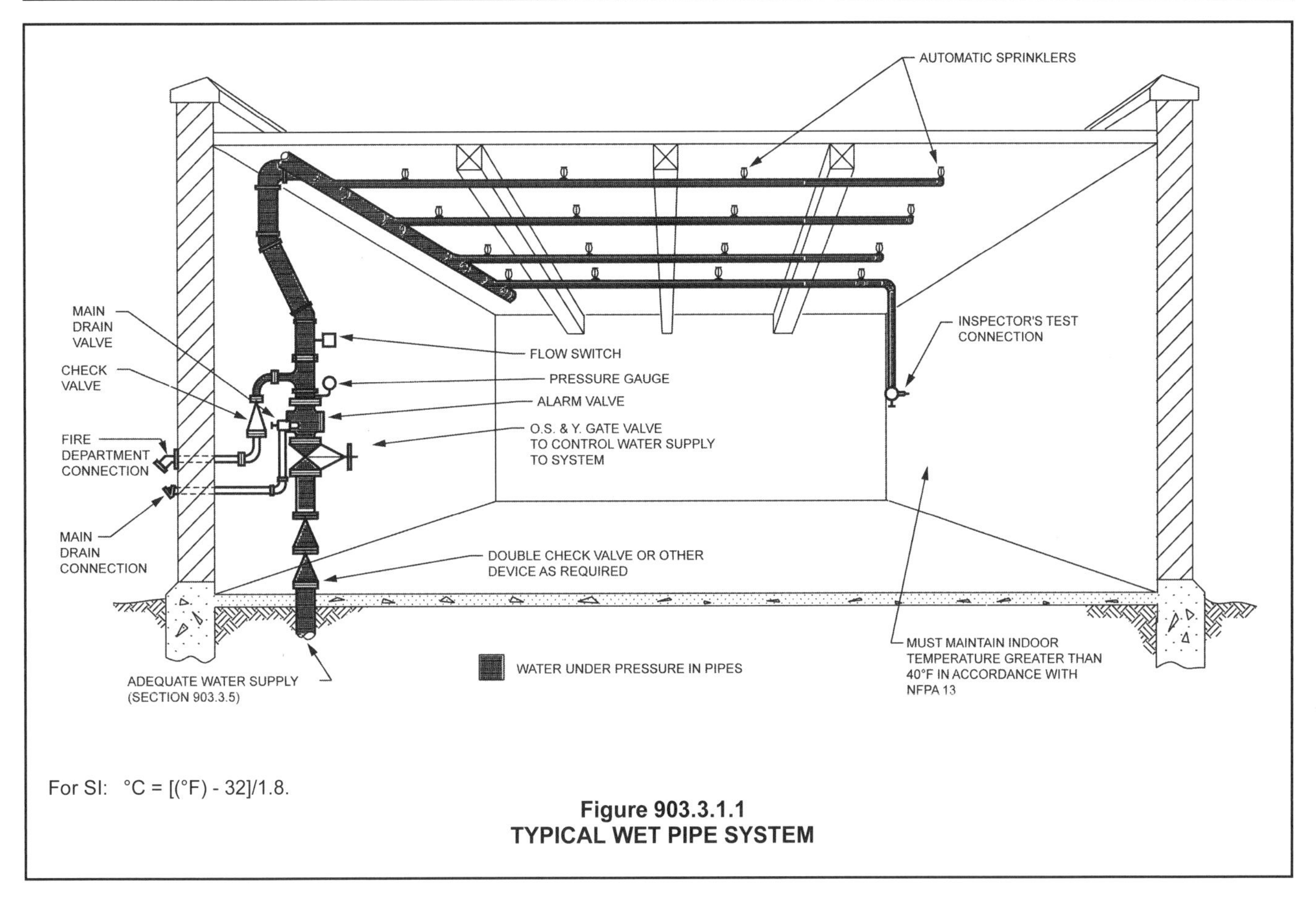

For SI: °C = [(°F) - 32]/1.8.

Figure 903.3.1.1
TYPICAL WET PIPE SYSTEM

tion system is installed. Buildings in compliance with one of the six listed conditions would still be considered fully sprinklered throughout in compliance with the code and NFPA 13 and thus are eligible for all applicable code alternatives, exceptions or reductions. Elimination of the sprinkler system in a sensitive area is subject to the approval of the fire code official.

Condition 1 addresses restrictions where the application of water could create a hazardous condition. For example, sprinkler protection is to be avoided where it is not compatible with certain stored materials (i.e., some water-reactive hazardous materials such as calcium carbide). Combustible metals, such as magnesium and aluminum, may burn so intensely that the use of water to attempt fire control will only intensify the reaction.

It is not the intent of Condition 2 to omit sprinklers solely because of a potential for water damage. A desire to not sprinkler a certain area (such as a computer room or operating room) does not fall within the limitations of the exception unless there is something unique about the space that would result in water being incompatible. A computer room can be adequately protected using an automatic sprinkler system or an alternative gaseous suppression agent system or a combination of these systems. The intent of Condition 2 is to consider whether or not the contents would react adversely to the application of water. It is important to note that the fire code official must approve the use of this item. Note also that with respect to computer rooms, NFPA 75 (*Protection of Information Technology Equipment*) (not a referenced standard) recognizes automatic sprinklers as the primary fire protection system for computer rooms.

Condition 3 recognizes the low fuel load and low occupancy hazards associated with generator and transformer rooms and, therefore, allows the omission of sprinkler protection if the rooms are separated from adjacent areas by 2-hour fire-resistance-rated construction. This condition assumes the room is not used for any combustible storage. This condition is similar to Section 8.15.11.3 of NFPA 13, which exempts electrical equipment rooms from sprinkler protection, provided the room is dedicated to the use of dry-type electrical equipment, is constructed as a 2-hour fire-resistance-rated enclosure, and is not used for combustible storage.

Condition 4 requires the construction of the room or area, as well as the contents, to be noncombustible. An example would be an area in an unprotected steel-frame building (Type IIB construction) used for steel or concrete block storage. Neither involves any significant combustible packaging or sources of ignition, and few combustibles are present (see Commentary Figure 903.3.1).

Condition 5 addresses the concern for elevator machine rooms and machinery spaces associated with fire service access elevators as required for buildings with occupied floors greater than 120 feet (36.58 m) from the lowest level of fire department access by Sections 403.6.1 and 3007. These elevators need to work during fire situations and their operation cannot be threatened by the activation of a sprinkler in a machine room or space that may affect the operation of the elevator. Fire service access elevators are required to be continuously monitored at the fire command center in accordance with Section 3007.7.

Condition 6, similar to Condition 5, exempts sprinklers from the machine rooms, machinery spaces, control rooms and control spaces for occupant evacuation elevators. Like fire service access elevators, these elevators need to work during fire situations and their operation cannot be threatened by the activation of a sprinkler in machine rooms, machinery spaces, control rooms and control spaces. Such elevators are required to be monitored at the fire command center in accordance with Section 3008.7.

[F] 903.3.1.1.2 Bathrooms. In Group R occupancies, other than Group R-4 occupancies, sprinklers shall not be required in bathrooms that do not exceed 55 square feet (5 m^2) in area and are located within individual *dwelling units* or *sleeping units*, provided that walls and ceilings, including the walls and ceilings behind a shower enclosure or tub, are of noncombustible or limited-combustible materials with a 15-minute thermal barrier rating.

❖ This provision was added to the code to reinstate an exception of NFPA 13 that had been in existence since 1976 but was deleted for all but dwelling units in motels and hotels (NFPA 8.15.8.1.1) from the 2013 edition of NFPA 13. Although reinstating the small bathroom exception will have a limited impact on new construction because many modern bathrooms exceed the 55-square-foot area limit to accommodate wheelchair access, the more important consequence will be removing an unnecessary cost increase for building owners who choose to retrofit existing properties with small bathrooms that were built before it was common to provide wheelchair access. Codes and standards should not erect any unnecessary barriers to retrofitting sprinklers into existing properties, such as existing high-rise buildings.

In the 1976 edition of the Life Safety Code (NFPA 101), to encourage cost-effective fire protection systems for apartment buildings, Section 11-3.8.3.4.1 provided an exception to permit bathrooms that did not exceed 55 square feet within individual dwelling units to omit sprinklers when the apartment building was sprinklered in accordance with NFPA 13. The basis of the 55-square-foot area is that this area accommodates a "typical" small bathroom that contains a standard tub, a toilet and a sink and nothing more. This exception was later duplicated from NFPA 101 into the 1991 edition of NFPA 13 with the understanding that the next edition of NFPA 101 (1994) could delete the exception since NFPA 13 would have it covered. NFPA 101-1994 then, as planned, deleted the exception.

The history of apartment unit bathroom fires is statistically minimal. According to the NFPA Home

Structure Fire Report, January 2009, Table 9B, "Reported Apartment Structure Fires by Area of Origin 2003-2006 Annual Averages," out of 113,000 fires/year, only 1,600 (1 percent) are in bathrooms. Given that there is more than 35 years of experience with the bathroom sprinkler exception being in place (since it was put into NFPA 101 in 1976), it would certainly be reasonable to expect anecdotal or statistical experience to indicate the existence of a problem if there were one.

[F] 903.3.1.2 NFPA 13R sprinkler systems. *Automatic sprinkler systems* in Group R occupancies up to and including four stories in height in buildings not exceeding 60 feet (18 288 mm) in height above grade plane shall be permitted to be installed throughout in accordance with NFPA 13R.

The number of stories of Group R occupancies constructed in accordance with Sections 510.2 and 510.4 shall be measured from the horizontal assembly creating separate buildings.

❖ NFPA 13R contains design and installation requirements for a sprinkler system to aid in the detection and control of fires in low-rise (four stories or less) residential occupancies.

Sprinkler systems designed in accordance with NFPA 13R are intended to prevent flashover (total involvement) in the room of fire origin and to improve the chance for occupants to escape or be evacuated. The design criteria in NFPA 13R are similar to those in NFPA 13 except that sprinklers may be omitted from areas in which fatal fires in residential occupancies do not typically originate (bathrooms, closets, attics, porches, garages and concealed spaces).

A common question is whether a mixed occupancy building which contains a Group R occupancy could still use NFPA 13R for the design. If one of the mixed-use occupancies would require a sprinkler system throughout the building in accordance with NFPA 13, then a 13R system would not be allowed. If, however, the only reason a sprinkler system is being installed is because there is a Group R fire area within the building, then an NFPA 13R system would be an appropriate design choice. The areas that are not classified as Group R would require protection in accordance with NFPA 13.

It must be noted that although the building would be considered sprinklered throughout in accordance with NFPA 13R, not all of the code sprinkler alternatives could be applied. Any alternative that requires the installation of an NFPA 13 system would not be applicable if a portion of the building utilizes an NFPA 13R system.

The code provisions that allow for an increase in building height according to Section 504.3 do not compound this section. NFPA 13R is applicable to buildings that are up to four stories and 60 feet (18 288 mm) in height above grade plane. If the design of a residential building intends to take advantage of the sprinkler height increase so that the building is five stories or more, the sprinkler system must be an NFPA 13 system. Because this section limits the height to four stories, that is the maximum height for a building that can utilize an NFPA 13R system. This is consistent with the scoping provisions in the NFPA 13R standard.

The limitation of four stories in height is to be measured with respect to the established grade plane, which is consistent with IFC Interpretation No. 43-03. As such, a basement would not be considered a story above grade for purposes of determining the applicability of this section.

The second paragraph recognizes the application of the requirements of Sections 510.2 and 510.4, which are essentially exceptions to the height and area requirements. This exception is based on providing a horizontal fire separation similar to the concept of a fire wall to create separate buildings. This establishes that the height in stories can be measured from the horizontal assembly instead of from grade plane. The height, in feet, would still be limited to being measured from grade plane. Such buildings are often referred to as "pedestal buildings."

[F] 903.3.1.2.1 Balconies and decks. Sprinkler protection shall be provided for exterior balconies, decks and ground floor patios of *dwelling units* and *sleeping units* where the building is of Type V construction, provided there is a roof or deck above. Sidewall sprinklers that are used to protect such areas shall be permitted to be located such that their deflectors are within 1 inch (25 mm) to 6 inches (152 mm) below the structural members and a maximum distance of 14 inches (356 mm) below the deck of the exterior balconies and decks that are constructed of open wood joist construction.

❖ Balconies, decks and patios in buildings of Type V construction and used for Group R occupancies are required to have sprinkler protection when there is a roof or deck above. This is in addition to the requirements of NFPA 13R, which primarily addresses the life safety of occupants and not property protection. The intent is to address hazards such as grilling and similar activities. Since NFPA 13R does not require such coverage, there is potential that a fire on a balcony could grow much too large for the system within the building to handle. The concern is that a potential exterior balcony fire could spread to unprotected floor/ceiling assemblies and attic spaces and result in major property damage. Section 308.1.4 of the IFC specifically addresses restrictions on open flame cooking devices used on combustible balconies. Note that sprinklers are not intended to be provided in closets found on such balconies.

Regardless of whether the exterior walking surface is attached to the building and called a balcony or is a freestanding structure such as a deck or patio the concern for fire ignition in the area adjacent to the exterior wall is the same. Sidewall sprinklers should be selected based on the area of coverage and climate. If the potential for freezing exists, a dry sidewall sprinkler should be used. Where the overhanging deck or balcony is extensive, an extended coverage sprinkler should be selected.

[F] 903.3.1.2.2 Open-ended corridors. Sprinkler protection shall be provided in *open-ended corridors* and associated *exterior stairways* and *ramps* as specified in Section 1027.6, Exception 3.

❖ This section is simply emphasizing the fact that, since there is no separation from the exterior exit stairways, sprinklers would be required when using an NFPA 13R system. Section 1027.6 Exception 3 allows the separation between the open-ended corridor and exterior exit stairway to be omitted but only where several conditions are met. The primary condition is that the corridor be sprinklered. A definition of "open-ended corridor" is provided in Chapter 2. See Commentary Figure 903.3.1.2.2 for an example of an open-ended corridor.

Figure 903.3.1.2.2
OPEN-ENDED CORRIDOR

[F] 903.3.1.3 NFPA 13D sprinkler systems. *Automatic sprinkler systems* installed in one- and two-family *dwellings*; Group R-3, Group R-4 Condition 1 and *townhouses* shall be permitted to be installed throughout in accordance with NFPA 13D.

❖ NFPA 13D contains design and installation requirements for a sprinkler system to aid in the detection and control of fires in one- and two-family dwellings, mobile homes and townhouses. This section also specifically allows the use of an NFPA 13D system for occupancies classified as Group R-3 and Group R-4 Condition 1. Group R-3 occupancies are one- and two-family dwellings that fall outside the scope of the IRC and also those buildings housing small congregate living facilities and boarding houses. These facilities operate very similarly to a single-family home, and the level of the ability of the occupants is such that little assistance is needed for self-evacuation. This is consistent with the NFPA 13D requirements and is also consistent with FHA court cases based on nondiscrimination for group homes.

Similar to NFPA 13R, sprinkler systems designed in accordance with NFPA 13D are intended to prevent flashover (total involvement) in the room of fire origin and to improve the chance for occupants to escape or be evacuated. Although the allowable omission of sprinklers in certain areas of the dwelling unit in NFPA 13D is similar to that in NFPA 13R, the water supply requirements are less restrictive. NFPA 13D uses a two-head sprinkler design with a 10-minute duration requirement, while NFPA 13R uses a four-head sprinkler design with a 30-minute duration requirement. The decreased water supply requirement emphasizes the main intent of NFPA 13D to control the fire and maintain tenability during evacuation of the residence.

Since the fire code official has the authority to approve the type of sprinkler system, this Section may be used to prevent the use of a specific type of sprinkler system that may be inappropriate for a particular type of occupancy.

[F] 903.3.2 Quick-response and residential sprinklers. Where *automatic sprinkler systems* are required by this code, quick-response or residential automatic sprinklers shall be installed in all of the following areas in accordance with Section 903.3.1 and their listings:

1. Throughout all spaces within a smoke compartment containing care recipient *sleeping units* in Group I-2 in accordance with this code.
2. Throughout all spaces within a smoke compartment containing treatment rooms in ambulatory care facilities.
3. *Dwelling units* and *sleeping units* in Group I-1 and R occupancies.
4. Light-hazard occupancies as defined in NFPA 13.

❖ This section requires the use of either listed quick-response or residential automatic sprinklers, depending on the type of sprinkler system required to achieve faster and more effective suppression in certain areas. Residential sprinklers are required in all types of residential buildings that would permit the use of an NFPA 13R or 13D sprinkler system.

Quick-response and residential sprinklers are similar in nature. They use a lighter material for the operating mechanism, thus reducing the heat lag in the element. The faster the heat can be absorbed, the sooner the sprinkler will begin to discharge water. Quick-response sprinklers have shown that they operate up to 25 percent faster than traditional sprinklers and create conditions in the room of origin that significantly increase the tenability of the environment. In tests performed by Factory Mutual (FM) for the Federal Emergency Management Agency (FEMA), the gas temperature in the room of origin was 550°F (288°C) with quick-response sprinklers, while it was 1,470°F (799°C) for conventional sprinklers at the time of sprinkler activation. More importantly, while the carbon monoxide (CO) level was 1,860 ppm for conventional sprinklers, the CO level when tested with quick-response sprinklers was only around 350 ppm. Comparatively, the National Institute of Occupational Safety and Health (NIOSH) considers the IDLH (immediately dangerous to life and health) level of CO to be 1,200 ppm. Thus, quick-response sprinklers have been shown to add significantly to the life safety effects of standard sprinkler systems.

Condition 1 requires the use of approved quick-response or residential sprinklers in smoke compartments containing care recipient sleeping units in Group I-2 occupancies. Even though properly operating standard sprinklers are effective, the extent of fire growth and smoke production that can occur before sprinkler activation creates the need for early warning to enable faster response by care providers and initiation of egress that is critical in occupancies containing persons incapable of self-preservation. The faster response time associated with quick-response or residential sprinklers increases the probability that the sprinklers will actuate before the care recipient's life would be threatened by a fire in his or her room.

Condition 2 requires the use of approved quick-response or residential sprinklers in smoke compartments containing treatment rooms in ambulatory care facilities. The justification is the same as that for Condition 1. When there is a potential for care recipients to be incapable of self-preservation, the use of residential sprinklers or quick-response sprinklers is critical.

Because of the kind of occupants sleeping in Group R and I-1 occupancies, as indicated in Condition 3, a faster responding type of sprinkler is desirable. Similar to the first condition, because occupants will be sleeping, the use of quick-response sprinklers creates additional safety by reducing sprinkler response time, thereby increasing the time available for egress and allowing for the time necessary for occupants to wake up and recognize the emergency event.

Condition 4 recognizes light-hazard occupancies in accordance with NFPA 13. These could include restaurants, schools, office buildings, places of religious worship and similar occupancies where the fire load and potential heat release of combustible contents are low.

[F] 903.3.3 Obstructed locations. Automatic sprinklers shall be installed with due regard to obstructions that will delay activation or obstruct the water distribution pattern. Automatic sprinklers shall be installed in or under covered kiosks, displays, booths, concession stands, or equipment that exceeds 4 feet (1219 mm) in width. Not less than a 3-foot (914 mm) clearance shall be maintained between automatic sprinklers and the top of piles of combustible fibers.

Exception: Kitchen equipment under exhaust hoods protected with a fire-extinguishing system in accordance with Section 904.

❖ To provide adequate sprinkler coverage, sprinkler protection must be extended under any obstruction that exceeds 4 feet (1219 mm) in width. Large air ducts are another common obstruction where sprinklers are routinely extended beneath the duct. The 3-foot (914 mm) storage clearance requirement for combustible fibers is caused by their potential high heat release. Most storage conditions require only a minimum 18-inch (457 mm) storage clearance to combustibles, depending on the type of sprinklers used and their actual storage conditions.

The exception recognizes that an alternative extinguishing system is permitted for commercial cooking systems in place of sprinkler protection for exhaust hoods that may be more than 4 feet (1219 mm) wide.

The application of this section is more critical to the ongoing use of the space. The obstruction conditions, therefore, should have already been addressed during plan review and installation inspection. This section gives the fire official and building owner adequate information to avoid the most typical obstruction-related issues in terms of proper sprinkler coverage.

[F] 903.3.4 Actuation. *Automatic sprinkler systems* shall be automatically actuated unless specifically provided for in this code.

❖ The intent of this section is to eliminate the need for occupant intervention during a fire. As such, it is assumed that it will not be necessary for a person to manually open a valve or perform some other physical activity in order to allow the sprinkler system to activate.

Wet-pipe and dry-pipe sprinkler systems, for example, are essentially fail-safe systems in the sense that, if the system is in proper operating condition, it will operate once a sprinkler fuses. Dry systems have an inherent time lag for water to reach the sprinkler; therefore, the response is not as fast as for a wet-pipe system. Other types of sprinkler systems, such as preaction and deluge, rely on the actuation of a detection system to operate the sprinkler valve.

[F] 903.3.5 Water supplies. Water supplies for *automatic sprinkler systems* shall comply with this section and the standards referenced in Section 903.3.1. The potable water supply shall be protected against backflow in accordance with the requirements of this section and the *International Plumbing Code*. For connections to public waterworks systems, the water supply test used for design of fire protection systems shall be adjusted to account for seasonal and daily pressure fluctuations based on information from the water supply authority and as approved by the fire code official.

❖ To be effective, all sprinkler systems must have an adequate supply of water. The criteria for an acceptable water supply are contained in the standards referenced in Section 903.3.1. For example, NFPA 13 contains criteria for different types of water supplies as well as the methods to determine the pressure, flow capabilities and capacity necessary to get the intended performance from a sprinkler system. An acceptable water supply could consist of a reliable municipal supply, a gravity tank or a fire pump with a pressure tank or a combination of these.

This section also establishes the requirements for protecting the potable water system against a nonpotable source, such as stagnant water retained within the sprinkler piping. As stated in Section 608.16.4 of the IPC, an approved double check valve device or reduced pressure principle backflow preventer is required.

The other issue addressed by this section is fluctuations in water pressure. Information on pressure fluctuation is necessary to not only ensure that the minimum required pressure will be available for the automatic sprinkler system, but also to ensure that high pressures do not exceed the pressure limitations of the sprinkler system. If the water pressure on a sprinkler system exceeds 100 psi, changes in the hanging methods are required. Also, if a fire pump is provided, it might be possible to exceed 175 psi, which is typically considered the maximum working pressure for a sprinkler system. These are just additional reasons why it is critical to account for pressure fluctuations in the water supply. Obviously, fire flows can be affected by this, as well as other water-based fire protection systems, such as standpipes, which require a minimum 100 psi at the roof of high-rise buildings and such may not be available due to pressure fluctuations. With regards to gathering of data, this requirement is simply intended to make sure that pressure fluctuations are addressed in accordance with the water supply authority to the extent that such information is available. Further authority is provided to the fire code official to accept other documentation.

[F] 903.3.5.1 Domestic services. Where the domestic service provides the water supply for the *automatic sprinkler system*, the supply shall be in accordance with this section.

❖ This section establishes the scope of domestic services for residential combination services. Essentially, compliance with Section 903.3.5 is required.

[F] 903.3.5.2 Residential combination services. A single combination water supply shall be allowed provided that the domestic demand is added to the sprinkler demand as required by NFPA 13R.

❖ NFPA 13R permits a common supply main to a building to serve both the sprinkler system and domestic services if the domestic demand is added to the sprinkler demand. NFPA 13R systems do not provide the same level of property protection as NFPA 13 systems.

[F] 903.3.6 Hose threads. Fire hose threads and fittings used in connection with *automatic sprinkler systems* shall be as prescribed by the fire code official.

❖ The threads on connections and fittings that the fire department will use to connect a hose must be compatible with the fire department threads.

Design documents must specify the type of thread to be used in order to be compatible with the local fire department equipment after consultation and coordination with the fire code official. The criteria typically apply to fire department connections for sprinkler and standpipe systems, standpipe hose connections, yard hydrants and wall hydrants.

The majority of fire departments in the United States use the American National Fire Hose Connection Screw Thread also commonly known as NST and NS. NFPA 1963 gives the screw thread dimensions and the thread size of threaded connections, with nominal sizes ranging from $^{3}/_{4}$ inch (19 mm) to 6 inches (152 mm) for the NS thread. Although efforts to standardize fire hose threads began after the Boston conflagration in 1872, there are still many different screw threads, some of which give the appearance of compatibility with the NH thread. While NFPA 1963 may be used as a guide, the code does not require that any particular standard be used. Rather, it is important that the fire code official be consulted for the appropriate thread selection. The intent is that the threads match those of the local department identically so that adapters are not required within the fire department's own district.

[F] 903.3.7 Fire department connections. Fire department connections for *automatic sprinkler systems* shall be installed in accordance with Section 912.

❖ Section 912, to which this section points, provides a comprehensive set of requirements for FDCs reducing the opportunity for any of its requirements to be overlooked. See the commentary for Section 912.

[F] 903.3.8 Limited area sprinkler systems. Limited area sprinkler systems shall be in accordance with the standards listed in Section 903.3.1 except as provided in Sections 903.3.8.1 through 903.3.8.5.

❖ The use of limited area sprinkler systems is restricted to cases in which the code requires a limited number of sprinklers to protect a specific hazard or area and not a complete automatic sprinkler system. For example, limited area sprinkler systems may be used to protect areas including, but not limited to, stages; storage and workshop areas; painting rooms; trash rooms and chutes; furnace rooms; kitchens and hazardous exhaust systems and incidental uses as regulated in Section 509.

903.3.8.1 Number of sprinklers. Limited area sprinkler systems shall not exceed six sprinklers in any single *fire area*.

❖ In the 2015 edition of the code, the number of sprinklers allowed on a limited area sprinkler system in a fire area has been reduced to only six. In previous editions, up to 19 sprinklers were allowed. This reduced number will limit the type of applications for such systems. In the past, 19 sprinklers may have been able to address an entire building, depending on the system design, more easily allowing the use of the domestic water supply.

903.3.8.2 Occupancy hazard classification. Only areas classified by NFPA 13 as Light Hazard or Ordinary Hazard Group 1 shall be permitted to be protected by limited area sprinkler systems.

❖ The use of limited area systems is restricted to only the two lowest hazard occupancy classifications in accordance with NFPA 13. Such systems are fairly limited and can only contain six sprinklers per fire area. Because of this, the types of hazards protected should be limited.

903.3.8.3 Piping arrangement. Where a limited area sprinkler system is installed in a building with an automatic wet standpipe system, sprinklers shall be supplied by the standpipe system. Where a limited area sprinkler system is installed in a building without an automatic wet standpipe system, water shall be permitted to be supplied by the plumbing system provided that the plumbing system is capable of simultaneously supplying domestic and sprinkler demands.

❖ Two options are provided for how water to the limited area sprinkler system is to be supplied. If the building contains automatic wet standpipes, then the system must be supplied by the standpipe system. If there is no standpipe system available, connection to the domestic water supply is permitted. The water supply must be analyzed to determine if it is sufficient to supply simultaneously both domestic usage and the sprinkler system demand. This will be mean looking at peak water use throughout the day. See Section 903.3.5 and the *International Plumbing Code*.

903.3.8.4 Supervision. Control valves shall not be installed between the water supply and sprinklers unless the valves are of an *approved* indicating type that are supervised or secured in the open position.

❖ No shutoff valves are permitted in the sprinkler system piping unless the valves are specifically approved and are either supervised or secured in the open position. These restrictions increase the likelihood that the sprinkler system will be operational should a fire occur. Valve supervision or securing a valve in the open position are considered equally reliable by this section.

903.3.8.5 Calculations. Hydraulic calculations in accordance with NFPA 13 shall be provided to demonstrate that the available water flow and pressure are adequate to supply all sprinklers installed in any single *fire area* with discharge densities corresponding to the hazard classification.

❖ Hydraulic calculations are required to be in accordance with NFPA 13 to demonstrate that the water system is adequate to supply the sprinkler demand in any particular fire area.

[F] 903.4 Sprinkler system supervision and alarms. Valves controlling the water supply for *automatic sprinkler systems*, pumps, tanks, water levels and temperatures, critical air pressures and waterflow switches on all sprinkler systems shall be electrically supervised by a *listed* fire alarm control unit.

Exceptions:

1. *Automatic sprinkler systems* protecting one- and two-family *dwellings*.
2. Limited area sprinkler systems in accordance with Section 903.3.8.
3. *Automatic sprinkler systems* installed in accordance with NFPA 13R where a common supply main is used to supply both domestic water and the *automatic sprinkler system*, and a separate shutoff valve for the *automatic sprinkler system* is not provided.
4. Jockey pump control valves that are sealed or locked in the open position.
5. Control valves to commercial kitchen hoods, paint spray booths or dip tanks that are sealed or locked in the open position.
6. Valves controlling the fuel supply to fire pump engines that are sealed or locked in the open position.
7. Trim valves to pressure switches in dry, preaction and deluge sprinkler systems that are sealed or locked in the open position.

❖ The reliability data on automatic sprinkler systems clearly indicate that a closed valve is the leading cause of sprinkler system failure. There are also a number of other critical elements that contribute to successful sprinkler system operation, including, but not limited to, pumps, water tanks and air pressure maintenance devices; therefore, this section requires that the various critical elements that contribute to an available water supply and to the function of the sprinkler system be electrically supervised.

Automatic sprinkler systems in one- and two-family dwellings are typically designed to comply with NFPA 13D, which does not require electrical supervision (see Exception 1).

Limited area sprinkler systems are generally supervised by their connection to the domestic water service although the use of a supervised indicating valve is permitted. Compliance with Section 903.3.8 means that the alarm provisions of this section are not applicable to limited area systems. Consequently, limited area sprinkler systems do not require local alarms or supervision. Again, electrical supervision is required only if a control valve is installed between the riser control valve and the sprinkler system piping.

Similar to limited area sprinkler systems, electrical supervision is not required for NFPA 13R residential combination services when a shutoff valve is not installed (see Exception 3). NFPA 13R sprinkler systems are supervised in that the only way to shut off the sprinkler system is to also shut off the domestic water supply.

The valves discussed in Exceptions 4 through 7 can be sealed or locked in the open position because they do not control the sprinkler system water supply.

[F] 903.4.1 Monitoring. Alarm, supervisory and trouble signals shall be distinctly different and shall be automatically transmitted to an *approved* supervising station or, where *approved* by the fire code official, shall sound an audible signal at a *constantly attended location*.

Exceptions:

1. Underground key or hub valves in roadway boxes provided by the municipality or public utility are not required to be monitored.
2. Backflow prevention device test valves located in limited area sprinkler system supply piping shall be locked in the open position. In occupancies required to be equipped with a fire alarm system, the backflow preventer valves shall be electrically supervised by a tamper switch installed in accordance with NFPA 72 and separately annunciated.

❖ Automatic sprinkler systems must be supervised as a means of determining that the system is operational. A valve supervisory switch operating as a normally open or normally closed switch is usually used. NFPA 72 does not permit valve supervisory switches to be connected to the same zone circuit as the waterflow switch unless it is specifically arranged to actuate a signal that is distinctive from the circuit trouble condition signal.

Required sprinkler systems are to be monitored by an approved supervising service to comply with NFPA 72. Types of supervising stations recognized in NFPA 72 include central stations, remote supervising stations or proprietary supervising stations.

A central station is an independent off-site facility operated and maintained by personnel whose primary business is to furnish, maintain, record and supervise a signaling system. A proprietary system is similar to a central station system; however, a proprietary system is typically an on-site facility monitoring a number of buildings on the same site for the same owner. A remote station system has an alarm signal that is transmitted to a remote location acceptable to the authority having jurisdiction and that is attended 24 hours a day. The receiving equipment is usually located at a fire station, police station, regional emergency communications center or telephone answering service. An alternative use to the three previous supervising methods is an audible signal that can be transmitted to a constantly attended location approved by the fire code official.

Exception 1 recognizes that underground key or hub valves in roadway boxes are not normally supervised or required to be supervised by this section or NFPA 13.

Exception 2 acknowledges that local water utilities and environmental authorities in many instances require, by local ordinances, that backflow prevention devices be installed in limited-area sprinkler system piping. To make the testing and maintenance of backflow prevention devices easier, test valves are installed on each side of the device. These valves are typically indicating-type valves and can function as shutoff valves for the sprinkler system, and therefore require some level of supervision.

Because these infrequently used valves may be the only feature of protection requiring supervision in occupancies not otherwise required to be equipped with a fire alarm system, Exception 2 permits these valves to be locked in the open position; however, if the occupancy is protected by a fire alarm system, these valves must be equipped with approved valve supervisory devices connected to the fire alarm control panel on a separate (supervisory) zone so that the supervisory signal is transmitted to the designated receiving station. Installation and testing of backflow preventers in sprinkler systems are regulated in Sections 312.10 (testing) and 608.16.4 (devices) of the IPC.

[F] 903.4.2 Alarms. An approved audible device, located on the exterior of the building in an approved location, shall be connected to each *automatic sprinkler system*. Such sprinkler waterflow alarm devices shall be activated by water flow equivalent to the flow of a single sprinkler of the smallest orifice size installed in the system. Where a fire alarm system is installed, actuation of the *automatic sprinkler system* shall actuate the building fire alarm system.

❖ The audible alarm, sometimes referred to as the "outside ringer" or "water-motor gong," sounds when the sprinkler system has activated. The alarm device may be electrically operated or it may be a true water-motor gong operated by a paddle-wheel-type attachment to the sprinkler system riser that responds to the flow of water in the piping. Though no longer the alarm device of choice, water-motor gongs do have the advantage of not being subject to power failures within or outside the protected building (see Sections 6.9 and 8.17 of NFPA 13 for further information on these devices). The alarm must be installed on the exterior of the building in a location approved by the fire code official. This location is often in close proximity to the fire department connection (FDC), serving a collateral function of helping the responding fire apparatus engineer more promptly locate the FDC.

The alarm is not intended to be an evacuation alarm. The requirement is also not intended to be an indirect requirement for a fire alarm system. Unless a fire alarm system is required by some other code provision, only the exterior alarm device is required. However, when a fire alarm system is installed, the sprinkler system must also be interconnected with the fire alarm system so that when the sprinkler system actuates, it sounds the evacuation alarms required for the fire alarm system.

The primary purpose of the exterior alarm is to notify people outside the building that the sprinkler system is in operation. Originally, it was to act as a supplemental alert so that passersby could notify the fire department of the condition. However, because the code now requires electronic supervision of sprinkler systems, that function is mostly moot. The exterior notification now primarily serves the function of alerting the arriving fire department of which building or sprinkler system is in operation before staging firefighting activities for the building.

[F] 903.4.3 Floor control valves. *Approved* supervised indicating control valves shall be provided at the point of connection to the riser on each floor in high-rise buildings.

❖ In high-rise buildings, sprinkler control valves with supervisory initiating devices must be installed at the point of connection to the riser on each floor. Sprinkler control valves on each floor are intended to permit servicing activated systems without impairing the water supply to large portions of the building.

[F] 903.5 Testing and maintenance. Sprinkler systems shall be tested and maintained in accordance with the *International Fire Code*.

❖ Section 901 contains requirements for the testing and maintenance of sprinkler systems. Acceptance tests are necessary to verify that the system performs as intended by design and by the code. Periodic testing and maintenance are essential to verify that the level of protection designed into the building will be operational whenever a fire occurs. Water-based extinguishing systems must be tested and maintained as required by NFPA 25.

SECTION 904 ALTERNATIVE AUTOMATIC FIRE-EXTINGUISHING SYSTEMS

[F] 904.1 General. Automatic fire-extinguishing systems, other than *automatic sprinkler systems*, shall be designed, installed, inspected, tested and maintained in accordance with the provisions of this section and the applicable referenced standards.

❖ Section 904 covers alternative fire-extinguishing systems that use extinguishing agents other than water. Alternative automatic fire-extinguishing systems include wet-chemical, dry-chemical, foam, carbon dioxide, halon and clean-agent suppression systems. In addition to the provisions of Section 904, the indicated referenced standards include specific installation, maintenance and testing requirements for all systems.

[F] 904.2 Where permitted. Automatic fire-extinguishing systems installed as an alternative to the required *automatic sprinkler systems* of Section 903 shall be *approved* by the fire code official.

❖ One of the main considerations in selecting an extinguishing agent should be the compatibility of the agent with the hazard. The fire code official is responsible for approving an alternative extinguishing agent. The approval should be based on the compatibility of the agent with the hazard and the potential effectiveness of the agent to suppress a fire involving the hazards present.

[F] 904.2.1 Restriction on using automatic sprinkler system exceptions or reductions. Automatic fire-extinguishing systems shall not be considered alternatives for the purposes of exceptions or reductions allowed for *automatic sprinkler systems* or by other requirements of this code.

❖ Although Section 904.2 allows the use of alternative fire-extinguishing systems with specific approval, this section prohibits the use of such systems to allow reductions or exceptions allowed for automatic sprinkler systems throughout the code. Therefore the building will not be considered as equipped throughout with an automatic sprinkler system when using systems such as automatic water mist or other alternative systems. See Commentary Figure 904.2.1.

[F] 904.2.2 Commercial hood and duct systems. Each required commercial kitchen exhaust hood and duct system required by Section 609 of the *International Fire Code* or Chapter 5 of the *International Mechanical Code* to have a Type I hood shall be protected with an approved automatic fire-extinguishing system installed in accordance with this code.

❖ This section requires an effective suppression system to combat fire on the cooking surfaces of grease-producing appliances and within the hood and exhaust system of a commercial kitchen installation. Type I hoods, including the duct system, must be protected with an approved automatic fire-extinguishing system because they are used for handling grease-laden vapors or smoke, whereas Type II hoods handle fumes, steam, heat and odors. Type I hoods are typically required for commercial food heat-processing equipment, such as deep fryers, griddles, charbroilers, broilers and open burner stoves and ranges. For additional guidance on the requirements for Type I and II hoods, see the commentary to Section 507 of the IMC.

[F] 904.3 Installation. Automatic fire-extinguishing systems shall be installed in accordance with this section.

❖ The installation of automatic fire-extinguishing systems must comply with the requirements of Sections 904.3.1 through 904.3.5 in addition to the installation criteria contained in the referenced standard for the proposed type of alternative extinguishing system.

CODE SECTION[a]	MODIFICATION	NFPA 13	NPFA 13R	NFPA 13D
Increases				
504.3 and Table 504.3	Height increase in feet	yes	yes	no
504.4 and Table 504.4	Height increase in number of stories	yes	yes	no
506.2 and Table 506.2	Area increase	yes	yes	no
1005.3.1, 1005.3.2[b]	Egress width	yes	yes	no
Table 1017.2[b]	Travel distance	yes	yes	no
Rating Reductions				
Table 508.4	Separated occupancies	yes	no	no
708.3	Fire partitions (dwelling units, sleeping units)	yes	no	no
Table1020.1[b]	Corridor walls	yes	yes	no
Miscellaneous				
Tables 307.1(1), 307.1(2)	Hazardous material increase	yes	no	no
404.2	Atriums	yes	no	no
507.4, 507.5	Unlimited area buildings	yes	no	no
Table 705.8	Allowable area of openings	yes	no	no
705.8.5	Vertical separation of openings	yes	yes	no
718.3.2	Residential floor/ceiling draftstopping	yes	yes[c]	no
718.3.3	Nonresidential draftstopping	yes	no	no
718.4.2	Groups R-1, R-2 attic draftstopping	yes	yes[c]	no
718.4.3	Other group draftstopping	yes	no	no
Table 803.9	Interior finish	yes	yes	no
804.4.2	Floor finish	yes	yes	no
907.2.1 – 907.2.10[b]	Manual fire alarm system	yes (A, B, E, F, M)	yes (R-1, R-2)	no
1009.2.1[b]	Accessible egress	yes	yes	no
1028.1[b]	Exit discharge	yes	yes	no
1406.3	Balconies	yes	yes[c, d]	yes[c]

a. Section numbers refer to sections in the code.
b. Section numbers in Chapters 9 and 10 apply to both the IFC and the code.
c. Sprinkler protection must be extended to the affected areas.
d. For additional balcony requirements, see Section 903.3.1.2.1.

Figure 904.2.1
SELECTED EXAMPLES OF REQUIREMENTS MODIFIED THROUGH USE OF AUTOMATIC SPRINKLER SYSTEMS

[F] 904.3.1 Electrical wiring. Electrical wiring shall be in accordance with NFPA 70.

❖ NFPA 70 regulates the design and installation of electrical systems and equipment. All electrical work must also be in compliance with any specific electrical classifications and conditions contained in the referenced standards for each type of system.

Chapter 27 of the code and Section 605 of the IFC contain provisions that also reference NFPA 70. Those sections also contain additional information that must be applied when addressing electrical issues.

[F] 904.3.2 Actuation. Automatic fire-extinguishing systems shall be automatically actuated and provided with a manual means of actuation in accordance with Section 904.12.1. Where more than one hazard could be simultaneously involved in fire due to their proximity, all hazards shall be protected by a single system designed to protect all hazards that could become involved.

Exception: Multiple systems shall be permitted to be installed if they are designed to operate simultaneously.

❖ Section 904.3.2 requires alternative fire-extinguishing systems to be designed for automatic activation. Activation commonly occurs when a heat, fire or smoke detection system operates. In Type I commercial kitchen hoods, Section 904.12 requires a manual and automatic means of activating the fire-extinguishing system. Designing a fire-extinguishing system to only operate upon manual actuation is prohibited by the code and many of the NFPA fire protection system standards.

The requirements for fire-extinguishing system actuation correlate with the requirements of NFPA 17 and NFPA 17A. The requirement prescribes that when a hazard is protected by two or more fire-extinguishing systems, all of the systems must be designed to operate simultaneously. The reason for the revision is that a typical alternative automatic fire-extinguishing system has a limited amount of fire-extinguishing agent. The amount of agent that is available is based on the area or volume of the hazard and the fire behavior of the fuel. Because the amount of agent is limited, the simultaneous operation of all the fire-extinguishing systems ensures that enough agent is applied to extinguish the fire and prevent its spread from the area of origin.

It is fairly common for a single hazard to be protected by two or more alternative automatic fire-extinguishing systems. For example, protection of a spray booth used for the application of flammable finishes using dry chemical commonly requires two or three alternative automatic fire-extinguishing systems. The reason is that many dry-chemical and all wet-chemical systems are preengineered systems. Utilizing listed nozzles, preengineered systems are designed and constructed based on the manufacturer's installation requirements. Because these systems are assembled using listed nozzles and extinguishing agents, one system may not be able to protect the spraying space and exhaust plenum. As a result, two or more systems may be required as a provision of an extinguishing system's listing to protect certain hazards.

Another example is commercial kitchen cooking operations. Consider a flat grill broiler and a deep fat fryer located beneath the same Type I hood. It is quite common for each of these commercial cooking appliances to be protected by separate automatic fire-extinguishing systems. Based on the revision to Section 904.3.2, both extinguishing systems must simultaneously operate in the event a fire involves either of the example appliances (see commentary, Section 904.12.1).

[F] 904.3.3 System interlocking. Automatic equipment interlocks with fuel shutoffs, ventilation controls, door closers, window shutters, conveyor openings, smoke and heat vents and other features necessary for proper operation of the fire-extinguishing system shall be provided as required by the design and installation standard utilized for the hazard.

❖ Shutting off fuel supplies will eliminate potential ignition sources in the protected area. Automatic door and window closers and dampers for forced-air ventilation systems are intended to maintain the desired concentration level of the extinguishing agent in the protected area. See the commentary for Section 904.12.2 for information on system interconnections in commercial cooking fire-extinguishing systems.

[F] 904.3.4 Alarms and warning signs. Where alarms are required to indicate the operation of automatic fire-extinguishing systems, distinctive audible and visible alarms and warning signs shall be provided to warn of pending agent discharge. Where exposure to automatic-extinguishing agents poses a hazard to persons and a delay is required to ensure the evacuation of occupants before agent discharge, a separate warning signal shall be provided to alert occupants once agent discharge has begun. Audible signals shall be in accordance with Section 907.5.2.

❖ Safeguards are necessary to prevent injury or death to personnel in areas where the atmosphere will be made hazardous by oxygen depletion due to agent discharge in a confined space. The "where alarms are required" phrase is referring to requirements that will be found in the referenced installation standards indicated in Sections 904.5 through 904.12, as applicable. Predischarge alarms that will operate on fire detection system activation must be installed within and at entrances to the affected areas.

Where required by the appropriate installation standard, an extinguishing agent discharge delay feature shall also be provided to allow evacuation of personnel prior to agent discharge. Warning and instructional signs are also to be posted, preferably at the entrances to and within the protected area. See Section 4.5.6.1 of NFPA 12 for additional information on carbon dioxide system alarms, Section 4.3.5 of NFPA 12A for additional information on Halon system alarms and Section 4.3.5 of NFPA 2001 for additional information on clean agent system alarms.

[F] 904.3.5 Monitoring. Where a building fire alarm system is installed, automatic fire-extinguishing systems shall be monitored by the building fire alarm system in accordance with NFPA 72.

❖ Automatic fire-extinguishing systems need not be electrically supervised unless the building is equipped with a fire alarm system. This section recognizes the fact that a fire alarm system is not required in all buildings. However, because most alternative fire-extinguishing systems require the space to be evacuated before the system is discharged, they are equipped with evacuation alarms. Interconnection of the fire-extinguishing system evacuation alarm with the building evacuation alarm results in an increased level of hazard notification for the occupants in addition to the electrical supervision of the fire-extinguishing system.

[F] 904.4 Inspection and testing. Automatic fire-extinguishing systems shall be inspected and tested in accordance with the provisions of this section prior to acceptance.

❖ The completed installation must be tested and inspected to determine that the system has been installed in compliance with the code and will function as required. Full-scale acceptance tests must be conducted as required by the applicable referenced standard.

[F] 904.4.1 Inspection. Prior to conducting final acceptance tests, all of the following items shall be inspected:

1. Hazard specification for consistency with design hazard.
2. Type, location and spacing of automatic- and manual-initiating devices.
3. Size, placement and position of nozzles or discharge orifices.
4. Location and identification of audible and visible alarm devices.
5. Identification of devices with proper designations.
6. Operating instructions.

❖ This section identifies those items that need to be verified or visually inspected prior to the final acceptance tests. All equipment should be listed, approved and installed in accordance with the manufacturer's recommendations.

[F] 904.4.2 Alarm testing. Notification appliances, connections to fire alarm systems and connections to *approved* supervising stations shall be tested in accordance with this section and Section 907 to verify proper operation.

❖ Components of fire-extinguishing systems related to alarm devices and their supervision must be tested before the system is approved. Alarm devices must be tested to satisfy the requirements of NFPA 72.

[F] 904.4.2.1 Audible and visible signals. The audibility and visibility of notification appliances signaling agent discharge or system operation, where required, shall be verified.

❖ This section requires verification upon completion of the system installation of the audibility and visibility of notification appliances in the area affected by the extinguishing agent discharge of the alternative automatic fire-extinguishing system.

[F] 904.4.3 Monitor testing. Connections to protected premises and supervising station fire alarm systems shall be tested to verify proper identification and retransmission of alarms from automatic fire-extinguishing systems.

❖ Where monitoring of fire-extinguishing systems is required, such as by Section 904.3.5, all connections related to the supervision of the system must be tested to verify they are in proper working order.

[F] 904.5 Wet-chemical systems. Wet-chemical extinguishing systems shall be installed, maintained, periodically inspected and tested in accordance with NFPA 17A and their listing. Records of inspections and testing shall be maintained.

❖ NFPA 17A contains minimum requirements for the design, installation, operation, testing and maintenance of wet-chemical preengineered extinguishing systems. Equipment that is typically protected with wet-chemical extinguishing systems includes restaurant, commercial and institutional hoods; plenums; ducts and associated cooking equipment. Strict compliance with the manufacturer's installation instructions is vital for a viable installation.

Wet-chemical solutions used in extinguishing systems are relatively harmless and there is usually no lasting significant effect on a person's skin, respiratory system or clothing. These solutions may produce a mild, temporary irritation but the symptoms will usually disappear when contact is eliminated.

This section also specifically requires that records be maintained of inspections and testing to increase the effectiveness of such systems. Without records of such inspections and testing it is more difficult to determine if the systems will be effective when activated and when future inspection and testing is necessary.

[F] 904.6 Dry-chemical systems. Dry-chemical extinguishing systems shall be installed, maintained, periodically inspected and tested in accordance with NFPA 17 and their listing. Records of inspections and testing shall be maintained.

❖ NFPA 17 contains the minimum requirements for the design, installation, testing, inspection, approval, operation and maintenance of dry-chemical extinguishing systems.

The fire code official has the authority to approve

the type of dry-chemical extinguishing system to be used. NFPA 17 identifies three types of dry-chemical extinguishing systems: total flooding, local application and hand hose-line systems. Only total flooding and local application systems are considered automatic extinguishing systems.

The types of hazards and equipment that can be protected with dry-chemical extinguishing systems include: flammable and combustible liquids and combustible gases; combustible solids, which melt when involved in a fire; electrical hazards, such as transformers or oil circuit breakers; textile operations subject to flash surface fires; ordinary combustibles such as wood, paper or cloth and restaurant and commercial hoods, ducts and associated cooking appliance hazards, such as deep fat fryers and some plastics, depending on the type of material and configuration.

Total flooding dry-chemical extinguishing systems are used only where there is a permanent enclosure surrounding the hazard that is adequate to enable the required concentration to be built up. The total area of unclosable openings must not exceed 15 percent of the total area of the sides, top and bottom of the enclosure. Consideration must be given to eliminating the probable sources of re-ignition within the enclosure because the extinguishing action of dry-chemical systems is transient.

Local application of dry-chemical extinguishing systems is to be used for extinguishing fires where the hazard is not enclosed or where the enclosure does not conform to the requirements for total flooding systems. Local application systems have successfully protected hazards involving flammable or combustible liquids, gases and shallow solids, such as paint deposits.

NFPA 17 also discusses pre-engineered dry-chemical systems consisting of components designed to be installed in accordance with pretested limitations as tested and labeled by a testing agency. Pre-engineered systems must be installed within the limitations that have been established by the testing agency and may include total flooding, local application or a combination of both types of systems.

The type of dry chemical used in the extinguishing system is a function of the hazard to be protected. The type of dry chemical used in a system should not be changed unless it has been proven changeable by a testing laboratory, is recommended by the manufacturer of the equipment and is acceptable to the fire code official for the hazard being protected. Additional guidance on the use of various dry-chemical agents can be found in NFPA 17.

This section also specifically requires that records be maintained of inspections and testing to increase the effectiveness of such systems. Without records of such inspections and testing it is more difficult to determine if the systems will be effective when activated and when future inspection and testing is necessary.

[F] 904.7 Foam systems. Foam-extinguishing systems shall be installed, maintained, periodically inspected and tested in accordance with NFPA 11 and NFPA 16 and their listing. Records of inspections and testing shall be maintained.

❖ NFPA 11 covers the characteristics of foam-producing materials used for fire protection and the requirements for design, installation, operation, testing and maintenance of equipment and systems, including those used in combination with other fire-extinguishing agents. The minimum requirements are covered for flammable and combustible liquid hazards in local areas within buildings, storage tanks and indoor and outdoor processing areas.

Low-expansion foam is defined as an aggregation of air-filled bubbles resulting from the mechanical expansion of a foam solution by air with a foam-to-solution volume ratio of less than 20:1. It is most often used to protect flammable and combustible liquid hazards. Also, low-expansion foam may be used for heat radiation protection. Combined-agent systems involve the application of low-expansion foam to a hazard simultaneously or sequentially with dry-chemical powder.

NFPA 11 gives minimum requirements for the installation, design, operation, testing and maintenance of medium- and high-expansion foam systems. Medium-expansion foam is defined as an aggregation of air-filled bubbles resulting from the mechanical expansion of a foam solution by air or other gases with a foam-to-solution volume ratio of 20:1 to 200:1. High-expansion foam has a foam-to-solution volume ratio of 200:1 to approximately 1,000:1.

Medium-expansion foam may be used on solid fuel and liquid fuel fires where some degree of in-depth coverage is necessary (for example, for the total flooding of small, enclosed or partially enclosed volumes, such as engine test cells, transformer rooms, etc.). High-expansion foam is most suitable for filling volumes in which fires exit at various levels. For example, high-expansion foam can be used effectively against high-rack storage fires in enclosures such as in underground passages, where it may be dangerous to send personnel to control fires involving liquefied natural gas (LNG) and liquefied petroleum gas (LP-gas), and to provide vapor dispersion control for LNG and ammonia spills. High-expansion foam is particularly suited for indoor fires in confined spaces, since it is highly susceptible to wind and lack-of-confinement effects.

NFPA 16 contains the minimum requirements for open-head deluge-type foam-water sprinkler systems and foam-water spray systems. The systems are especially applicable to the protection of most flammable liquid hazards and have been used successfully to protect aircraft hangars and truck loading racks.

This section also specifically requires that records be maintained of inspections and testing to increase the effectiveness of such systems. Without records of such inspections and testing it is more difficult to determine if the systems will be effective when activated and when future inspection and testing is necessary.

[F] 904.8 Carbon dioxide systems. Carbon dioxide extinguishing systems shall be installed, maintained, periodically inspected and tested in accordance with NFPA 12 and their listing. Records of inspections and testing shall be maintained.

❖ NFPA 12 provides minimum requirements for the design, installation, testing, inspection, approval, operation and maintenance of carbon dioxide extinguishing systems.

Carbon dioxide extinguishing systems are useful in extinguishing fires in specific hazards or equipment in occupancies where an inert electrically nonconductive medium is essential or desirable and where cleanup of other extinguishing agents, such as dry-chemical residue, presents a problem. Carbon dioxide systems have satisfactorily protected the following: flammable liquids; electrical hazards, such as transformers, oil switches, rotating equipment and electronic equipment; engines using gasoline and other flammable liquid fuels; ordinary combustibles, such as paper, wood and textiles and hazardous solids.

The fire code official has the authority to approve the type of carbon dioxide system to be installed. NFPA 12 defines four types of carbon dioxide systems: total flooding, local application, hand hose lines and standpipe and mobile supply systems. Only total flooding and local application systems are automatic suppression systems.

Total-flooding systems may be used where there is a permanent enclosure around the hazard that is adequate to allow the required concentration to be built up and maintained for the required period of time, which varies for different hazards. Examples of hazards that have been successfully protected by total flooding systems include rooms, vaults, enclosed machines, ducts, ovens and containers and their contents.

Local application systems may be used for extinguishing surface fires in flammable liquids, gases and shallow solids where the hazard is not enclosed or the enclosure does not conform to the requirements for a total-flooding system. Examples of hazards that have been successfully protected by local application systems include dip tanks, quench tanks, spray booths, oil-filled electric transformers and vapor vents.

This section also specifically requires that records be maintained of inspections and testing to increase the effectiveness of such systems. Without records of such inspections and testing it is more difficult to determine if the systems will be effective when activated and when future inspection and testing is necessary.

[F] 904.9 Halon systems. Halogenated extinguishing systems shall be installed, maintained, periodically inspected and tested in accordance with NFPA 12A and their listing. Records of inspections and testing shall be maintained.

❖ NFPA 12A contains minimum requirements for the design, installation, testing, inspection, approval, operation and maintenance of Halon 1301 extinguishing systems. Halon 1301 fire-extinguishing systems are useful in specific hazards, equipment or occupancies where an electrically nonconductive medium is essential or desirable and where cleanup of other extinguishing agents presents a problem.

Halon 1301 systems have satisfactorily protected gaseous and liquid flammable materials; electrical hazards, such as transformers, oil switches and rotating equipment; engines using gasoline and other flammable fuels; ordinary combustibles, such as paper, wood and textiles and hazardous solids. Halon 1301 systems have also satisfactorily protected electronic computers, data processing equipment and control rooms.

The fire code official has the authority to approve the type of halogenated extinguishing system to be installed. NFPA 12A defines two types of halogenated extinguishing systems: total flooding and local application. Total-flooding systems may be used where there is a fixed enclosure around the hazard that is adequate to enable the required halon concentration to be built up and maintained for the required period of time to enable the effective extinguishing of the fire. Total-flooding systems may provide fire protection for rooms, vaults, enclosed machines, ovens, containers, storage tanks and bins.

Local application systems are used where there is not a fixed enclosure around the hazard or where the fixed enclosure around the hazard is not adequate to enable an extinguishing concentration to be built up and maintained in the space. Hazards that may be successfully protected by local application systems include dip tanks, quench tanks, spray booths, oil-filled electric transformers and vapor vents.

Two other considerations in selecting the proper extinguishing system are ambient temperature and the personnel hazards associated with the agent. The ambient temperature of the enclosure for a total-flooding system must be above 70°F (21°C) for halon 1301 systems. Special consideration must also be given to the use of halon systems when the temperatures are in excess of 900°F (482°C) because halon will readily decompose at such temperatures and the products of decomposition can be extremely irritating if inhaled, even in small amounts.

Halon 1301 total-flooding systems must not be used in concentrations greater than 10 percent in normally occupied areas. Where personnel cannot vacate the area within 1 minute, Halon 1301 total-flooding systems must not be used in normally occupied areas with concentrations greater than 7 percent. Halon 1301 total-flooding systems may be used with concentrations of up to 15 percent if the area is not normally occupied and the area can be evacuated within 30 seconds.

The use of halogenated extinguishing systems has become a concern with respect to the potential environmental effects of halon. Halongenated fire-extinguishing agents have been identified as a source of emissions, resulting in the depletion of the stratospheric ozone layer and, in accordance with the Montreal Protocol, the ceasing of its production in January 1994. Therefore, the supply of halon is limited and new supplies of halogenated extinguishing agents will not be available in the future. Existing supplies of halon can, however, continue to be used in existing, undischarged systems or to recharge discharged systems. This newfound need for halon supplies has given rise to new industries geared to the ranking, recycling and reclamation of existing halon supplies. Alternative "clean agent" extinguishing agents have been developed to replace halogenated agents (see Section 904.10).

This section also specifically requires that records be maintained of inspections and testing to increase the effectiveness of such systems. Without records of such inspections and testing it is more difficult to determine if the systems will be effective when activated and when future inspection and testing is necessary.

[F] 904.10 Clean-agent systems. Clean-agent fire-extinguishing systems shall be installed, maintained, periodically inspected and tested in accordance with NFPA 2001 and their listing. Records of inspections and testing shall be maintained.

❖ NFPA 2001 contains minimum requirements for the design, installation, testing, inspection and operation of clean-agent fire-extinguishing systems. A clean agent is an electrically nonconducting suppression agent that is volatile or gaseous at discharge and does not leave a residue on evaporation. Clean-agent fire-extinguishing systems are installed in locations that are enclosed and have openings in the protected area that can be sealed on activation of the alarm to provide effective clean-agent concentrations. A clean-agent fire-extinguishing system should not be installed in locations that cannot be sealed unless testing has shown that adequate concentrations can be developed and maintained.

The two categories of clean agents are halocarbon compounds and inert gas agents. Halocarbon compounds include bromine, carbon, chlorine, fluorine, hydrogen and iodine. Halocarbon compounds suppress fire by a combination of breaking the chemical chain reaction of the fire, reducing the oxygen supporting the fire and reducing the ambient temperature of the fire origin to reduce the propagation of the fire. Inert gas agents contain primary components consisting of helium, neon, argon or a combination of these. Inert gases work by reducing the oxygen concentration around the fire origin to a level that does not support combustion.

Clean-agent fire-extinguishing systems were developed in response to the demise of halon as an acceptable fire-extinguishing agent because of its harmful effect on the environment. Although the original hope for a halon substitute was that these new clean agents could be directly and proportionally substituted for halon agents in existing systems (drop in replacements), research has shown that clean agents are less efficient in extinguishing fires than are the halons they were intended to replace and require approximately 60 percent more agent by weight and volume in storage to do the same job. Additionally, the physical and chemical characteristics of clean agents differ sufficiently from halon to require different nozzles in addition to the need for larger storage vessels. Existing piping systems should be salvaged for use with clean agents only if they are carefully evaluated and determined to be hydraulically compatible with the flow characteristics of the new agent.

This section also relies on strict adherence to the system manufacturer's design and installation instructions for code compliance. As with many of the alternative fire suppression systems covered in this chapter, clean-agent systems are, for the most part, subjected by their manufacturers to a testing and listing program conducted by an approved testing agency. In such testing and listing programs, the clean agent is listed for use with specific equipment and equipment is listed for use with specific clean agents. The resultant listings include reference to the manufacturer's installation manuals, thereby giving the fire code official another valuable resource for reviewing and approving clean-agent systems.

Although clean agents have found a limited market for local application uses, such as a replacement for Halon 1211 in portable fire extinguishers, their primary application is in total-flooding systems and they are available in both engineered and preengineered configurations.

Engineered clean-agent systems are specifically designed for protection of a particular hazard, whereas preengineered systems are designed to operate within predetermined limitations up to the noted maximums, thus allowing broader applicability to a variety of hazard applications.

Total flooding systems are used where there is a fixed enclosure around the hazard that is adequate to enable the required clean-agent concentration to build up and be maintained within the space long enough to extinguish the fire. Such applications can include vaults, ovens, containers, tanks, computer rooms, paint lockers or enclosed machinery. In selecting the clean agent to be used in a given application, careful consideration must be given to whether the protected area is a normally occupied space, because different agents have different levels of concentration at which they may be a health hazard to occupants of the area.

The fire code official has the authority to approve the type of clean-agent system to be installed and should become familiar with the unique characteristics and hazards of clean-agent extinguishing systems using all available resources on the subject.

This section also specifically requires that records be maintained of inspections and testing to increase the effectiveness of such systems. Without records of such inspections and testing it is more difficult to determine if the systems will be effective when activated and when future inspection and testing is necessary.

[F] 904.11 Automatic water mist systems. *Automatic water mist systems* shall be permitted in applications that are consistent with the applicable listing or approvals and shall comply with Sections 904.11.1 through 904.11.3.

❖ This section provides the ability to use automatic water mist systems in specific applications. These installations are required to be consistent with the listings or approvals to which such systems have been tested. See Commentary Figure 904.11 for a picture showing the discharge of a water mist nozzle.

Figure 904.11
WATER MIST NOZZLE DISCHARGE

[F] 904.11.1 Design and installation requirements. *Automatic water mist systems* shall be designed and installed in accordance with Sections 904.11.1.1 through 904.11.1.4.

❖ The subsections to follow provide the various installation and design requirements related to water mist systems. This relates to the systems themselves and providing a reliable water supply.

[F] 904.11.1.1 General. *Automatic water mist systems* shall be designed and installed in accordance with NFPA 750 and the manufacturer's instructions.

❖ This section simply provides reference to the subsections with the detailed requirements related to the design and installation of water mist systems.

[F] 904.11.1.2 Actuation. *Automatic water mist systems* shall be automatically actuated.

❖ If a water mist system is used, it is required to be automatically activated.

[F] 904.11.1.3 Water supply protection. Connections to a potable water supply shall be protected against backflow in accordance with the *International Plumbing Code*.

❖ The water supply quality is more critical with automatic water mist systems than automatic sprinkler systems. Backflow prevention is required as it would be for water supply for automatic sprinkler systems.

[F] 904.11.1.4 Secondary water supply. Where a secondary water supply is required for an *automatic sprinkler system*, an *automatic water mist system* shall be provided with an *approved* secondary water supply.

❖ Since these systems work to protect buildings and spaces in a fashion similar to a sprinkler system and possibly as an alternative to automatic sprinkler systems, the water supply must be appropriate. If an automatic sprinkler system is required to have a secondary water supply in accordance with Section 403.3.3, then the water mist system must also provide an approved secondary water supply.

[F] 904.11.2 Water mist system supervision and alarms. Supervision and alarms shall be provided as required for *automatic sprinkler systems* in accordance with Section 903.4.

❖ This section plays a similar role to an automatic sprinkler system with regard to supervision and alarms. It is critical to make sure valves are open. In addition, an alarm to notify of an activation must be provided as it is for a sprinkler system. Note that this is not intended to be an occupant notification system. See the commentary to Section 903.4.

[F] 904.11.2.1 Monitoring. Monitoring shall be provided as required for *automatic sprinkler systems* in accordance with Section 903.4.1.

❖ See the commentary to Sections 904.11.2 and 903.4.1.

[F] 904.11.2.2 Alarms. Alarms shall be provided as required for *automatic sprinkler systems* in accordance with Section 903.4.2.

❖ Again this is not intended to be occupant notification but to simply alert someone that the system has activated. See the commentary to Sections 904.11.2 and 903.4.2.

[F] 904.11.2.3 Floor control valves. Floor control valves shall be provided as required for *automatic sprinkler systems* in accordance with Section 903.4.3.

❖ See the commentary to Section 903.4.3.

[F] 904.11.3 Testing and maintenance. *Automatic water mist systems* shall be tested and maintained in accordance with the *International Fire Code*.

❖ Testing and maintenance is critical to the long-term viability of automatic water mist systems. This section references the IFC for such provisions since that code focuses not simply on initial installation but on maintenance.

[F] 904.12 Commercial cooking systems. The automatic fire-extinguishing system for commercial cooking systems shall be of a type recognized for protection of commercial cooking equipment and exhaust systems of the type and arrangement protected. Preengineered automatic dry- and wet-chemical extinguishing systems shall be tested in accordance with UL 300 and *listed* and *labeled* for the intended application. Other types of automatic fire-extinguishing systems shall be *listed* and *labeled* for specific use as protection for commercial cooking operations. The system shall be installed in accordance with this code, its listing and the manufacturer's installation instructions. Automatic fire-extinguishing systems of the following types shall be installed in accordance with the referenced standard indicated, as follows:

1. Carbon dioxide extinguishing systems, NFPA 12.
2. *Automatic sprinkler systems*, NFPA 13.
3. Foam-water sprinkler system or foam-water spray systems, NFPA 16.
4. Dry-chemical extinguishing systems, NFPA 17.
5. Wet-chemical extinguishing systems, NFPA 17A.

Exception: Factory-built commercial cooking recirculating systems that are tested in accordance with UL 710B and *listed*, *labeled* and installed in accordance with Section 304.1 of the *International Mechanical Code*.

❖ The history of commercial kitchen exhaust systems shows that the mixture of flammable grease and effluents carried by such systems and the potential for the cooking equipment to act as an ignition source contribute to a higher level of hazard for kitchen exhaust systems than is normally found in many other exhaust systems. Furthermore, fire in a grease exhaust duct can produce temperatures of 2,000°F (1093°C) or more and heat radiating from the duct can ignite nearby combustibles. As a result, the code requires exhaust systems serving grease-producing equipment to include fire suppression to protect the cooking surfaces, hood, filters and exhaust duct to confine a fire to the hood and duct system, thus reducing the likelihood of it spreading to the structure.

In addition to the general requirements of this section, five industry standards are referenced for the installation of fire-extinguishing systems protecting commercial food heat-processing equipment and kitchen exhaust systems. Design professionals should specify and design fire-extinguishing systems to comply with these referenced standards. Only the installation of fire-extinguishing systems is regulated by these references. Where preengineered automatic dry- and wet-chemical extinguishing systems are installed, they must be listed and labeled for the specific cooking operation and tested in accordance with UL 300. Design and construction requirements for the specific types of fire-extinguishing systems are found in the respective sections of the referenced standards.

Regulatory requirements for the approval and installation of fire-extinguishing systems are the same as the approval required for all mechanical equipment and appliances. This section, therefore, requires extinguishing systems to be listed and labeled by an approved agency and installed in accordance with their listing and the manufacturer's installation instructions.

The exception allows factory-built commercial cooking recirculating systems to be installed if they have been tested and listed in accordance with UL 710B. It is important that they be installed in accordance with the manufacturer's installation instructions so that the listing requirements are met. An improper installation could result in hazardous vapors being discharged back into the kitchen.

Commercial cooking recirculating systems consist of an electric cooking appliance and an integral or matched packaged hood assembly. The hood assembly consists of a fan, collection hood, grease filter, fire damper, fire-extinguishing system and air filter, such as an electrostatic precipitator. These systems are tested for fire safety and emissions. The grease vapor (condensible particulate matter) in the effluent at the system discharge is not allowed to exceed a concentration of 5.0 mg/m^3. Recirculating systems are not used with fuel-fired appliances because the filtering systems do not remove combustion products. Kitchens require ventilation in accordance with Chapter 4 of the IMC.

Although the provisions in Section 904.12 address many of the specifics for commercial kitchens, additional information regarding commercial cooking suppression systems is located in Sections 904.2 and 904.3. This information is supplemental to that and should be considered together in developing the design for commercial cooking suppression systems.

[F] 904.12.1 Manual system operation. A manual actuation device shall be located at or near a *means of egress* from the cooking area not less than 10 feet (3048 mm) and not more than 20 feet (6096 mm) from the kitchen exhaust system. The manual actuation device shall be installed not more than 48 inches (1200 mm) or less than 42 inches (1067 mm) above the floor and shall clearly identify the hazard protected. The manual actuation shall require a maximum force of 40 pounds (178 N) and a maximum movement of 14 inches (356 mm) to actuate the fire suppression system.

Exception: *Automatic sprinkler systems* shall not be required to be equipped with manual actuation means.

❖ The manual device, usually a pull station, mechanically activates the suppression system. The typical system uses a mechanical circuit of cables under tension to hold the system in the armed (cocked) mode. Melting of a fusible link or actuation of a manual pull station causes the cable to lose tension, which, in turn, starts the discharge of the suppression agent. The manual actuation device must be readily and easily usable by the building occupants; therefore, the device must not require excessive force or range of movement to cause actuation.

In order to allow the actuation device to be used most effectively, the specified mounting height is intended to be consistent with the NFPA 17A standards and be handicapped accessible. This includes the requirement to identify the actuation device with the hazard protected. Where multiple kitchen appliances are provided, properly identifying which device relates to which appliance is very important. Required signage should be readily visible in the hazard area and capable of conveying information quickly and concisely.

Manual actuation is not required for automatic sprinkler systems because the typical system design will employ closed heads and wet system piping. A manual actuation valve would serve no purpose because sprinkler heads are already supplied with pressurized water and will discharge water only when the individual fusible elements open the heads.

[F] 904.12.2 System interconnection. The actuation of the fire suppression system shall automatically shut down the fuel or electrical power supply to the cooking equipment. The fuel and electrical supply reset shall be manual.

❖ The actuation of any fire suppression system must automatically shut off all sources of fuel or power to all cooking equipment located beneath the exhaust hood protected by the suppression system. This requirement is intended to shut off all heat sources that could reignite or intensify a fire. Shutting off a fuel and power supply to cooking appliances will eliminate an ignition source and allow the cooking surfaces to cool down. This shutdown is accomplished with mechanical or electrical interconnections between the suppression system and a shutoff valve or switch located on the fuel or electrical supply.

Common fuel shutoff valves include mechanical-type gas valves and electrical solenoid-type gas valves. Contactor-type switches or shunt-trip circuit breakers can be used for electrically heated appliances. The fuel or electric source must not be automatically restored after the suppression system has been actuated.

Chemical-type fire-extinguishing systems discharge for only a limited time and can discharge only once before recharge and reset; therefore, precautions must be taken to prevent a fire from reigniting. After a fire is detected and the initial suppressant discharge begins, the fuel and power supply will be locked out, thereby preventing the operation of the appliances until all systems are again ready for operation. Fuel and power supply shutoff must be manually restored by resetting a mechanical linkage or holding (latching)-type circuit.

[F] 904.12.3 Carbon dioxide systems. Where carbon dioxide systems are used, there shall be a nozzle at the top of the ventilating duct. Additional nozzles that are symmetrically arranged to give uniform distribution shall be installed within vertical ducts exceeding 20 feet (6096 mm) and horizontal ducts exceeding 50 feet (15 240 mm). *Dampers* shall be installed at either the top or the bottom of the duct and shall be arranged to operate automatically upon activation of the fire-extinguishing system. Where the *damper* is installed at the top of the duct, the top nozzle shall be immediately below the *damper*. Automatic carbon dioxide fire-extinguishing systems shall be sufficiently sized to protect against all hazards venting through a common duct simultaneously.

❖ This section states specific design requirements for nozzle locations, dampers and ducts for carbon dioxide extinguishing systems that may be used to protect commercial cooking systems. These requirements are intended to supersede similar, more general provisions in NFPA 12. Because carbon dioxide (CO_2) is a gaseous suppressant, dampers are required in the ductwork to define the atmosphere where the fire event would be. A specific concentration of CO_2 is necessary and dampers are required to define and contain the suppressant. The discharge cools exposed surfaces in addition to depriving the fire of oxygen. Although not mentioned specifically in this section, the applicable provisions of NFPA 12 should also be applied because the system is a CO_2 system as referenced in Section 904.8.

[F] 904.12.3.1 Ventilation system. Commercial-type cooking equipment protected by an automatic carbon dioxide-extinguishing system shall be arranged to shut off the ventilation system upon activation.

❖ Shutting down the ventilation system upon activation of the CO_2 extinguishing system maintains the desired concentration of carbon dioxide to suppress the fire. Leakage of gas from the protected area should be kept to a minimum. Where leakage is anticipated, additional quantities of carbon dioxide must be provided to compensate for any losses.

[F] 904.12.4 Special provisions for automatic sprinkler systems. *Automatic sprinkler systems* protecting commercial-type cooking equipment shall be supplied from a separate, readily accessible, indicating-type control valve that is identified.

❖ This section requires a separate control valve in the water line to the sprinklers protecting the cooking and ventilating system. The additional valve allows the flexibility to shut off the system for repairs or for cleanups after sprinkler discharge without taking the entire system out of service.

[F] 904.12.4.1 Listed sprinklers. Sprinklers used for the protection of fryers shall be tested in accordance with UL 199E, *listed* for that application and installed in accordance with their listing.

❖ Sprinklers specifically listed for such use must be used when protecting deep-fat fryers. These specially listed sprinklers use finer water droplets than standard spray sprinklers. The water spray lowers the temperature below a point where the fire can sustain itself and reduces the possibility of expanding the fire. UL 199E addresses these special sprinklers and includes performance tests for deep-fat fryer extinguishment and also deep-fat fryer cooking temperature splash. The selection of inappropriate sprinklers for deep-fat fryer protection can increase the hazards during water application rather than suppressing the fire.

[F] 904.13 Domestic cooking systems in Group I-2 Condition 1. In Group I-2 Condition 1, occupancies where cooking facilities are installed in accordance with Section 407.2.6 of this code, the domestic cooking hood provided over the cooktop or range shall be equipped with an automatic fire-extinguishing system of a type recognized for protection of domestic cooking equipment. Preengineered automatic extinguishing systems shall be tested in accordance with UL 300A and listed and labeled for the intended application. The system shall be installed in accordance with this code, its listing and the manufacturer's instructions.

❖ As nursing homes move away from institutional models, it is critical to have a functioning kitchen that can serve as the "hearth of the home." Instead of a large, centralized, institutional kitchen where all meals are prepared and delivered to a central dining room or the resident's room, the new "household model" nursing home uses de-centralized kitchens and small dining areas to create the feeling and focus of home. For persons with dementia, it is particularly important to have spaces that look familiar, like the kitchen in their former home, to increase their understanding and ability to function at their highest level. This section addresses the fire protection system needs for these kitchens. Note that these occupancies already contain quick-response sprinklers. This section provides the requirements for the hood fire-extinguishing system needed for the cooking hood located over the cook top or range. That system is required to be in compliance with UL 300A.

[F] 904.13.1 Manual system operation and interconnection. Manual actuation and system interconnection for the hood suppression system shall be installed in accordance with Sections 904.12.1 and 904.12.2, respectively.

❖ To activate such systems manually, the requirements of Sections 904.12.1 and 904.12.2 would apply. This provides for the location of the manual shutdown and also directs what equipment and power needs to automatically shut down. See the commentary to Sections 904.12.1 and 904.12.2.

[F] 904.13.2 Portable fire extinguishers for domestic cooking equipment in Group I-2 Condition 1. A portable fire extinguisher complying with Section 906 shall be installed within a 30-foot (9144 mm) distance of travel from domestic cooking appliances.

❖ This requirement for a portable fire extinguisher relates to the added safety of allowing a residential-type kitchen in nursing homes. This provides another line of defense and the location is focused on the distance from the cooking appliances.

SECTION 905
STANDPIPE SYSTEMS

[F] 905.1 General. Standpipe systems shall be provided in new buildings and structures in accordance with Sections 905.2 through 905.10. In buildings used for high-piled combustible storage, fire protection shall be in accordance with the *International Fire Code*.

❖ Standpipe systems are required in buildings to provide a quick, convenient water source for fire department use where hose lines would otherwise be impractical, such as in high-rise buildings. Standpipe systems can also be used prior to deployment of hose lines from fire department apparatus. The requirements for stand- pipes are based on practical requirements of typical fire-fighting operations and the nationally recognized standard NFPA 14.

The threads on connections to which the fire department may connect a hose must be compatible with the fire department hose threads (see commentary, Section 903.3.6). Chapter 23 of the IFC requires a Class I standpipe system in exit passageways of buildings used for high-piled storage. Note that if a building containing high-piled storage does not contain an exit passageway, then standpipes would not be required. High-piled storage involves the solid-piled, bin box, palletized or rack storage of Class I through IV commodities over 12 feet (3658 mm) high. High-hazard commodities stored higher than 6 feet (1829 mm) are also considered high piled.

[F] 905.2 Installation standard. Standpipe systems shall be installed in accordance with this section and NFPA 14. Fire department connections for standpipe systems shall be in accordance with Section 912.

❖ This section requires the installation of standpipe systems to comply with the applicable provisions of NFPA 14 in addition to Section 905. NFPA 14 contains the minimum requirements for the installation of standpipe and hose systems for buildings and structures. The standard addresses additional requirements not addressed in the code, such as pressure limitations, minimum flow rates, piping specifications, hose connection details, valves, fittings, hangers and the testing and inspection of standpipes. The periodic inspection, testing and maintenance of standpipe systems must comply with NFPA 25.

Section 905 and NFPA 14 recognize three classes of standpipe systems: Class I, II or III. The type of system required depends on building height, building area, type of occupancy and the extent of automatic sprinkler protection. Section 905 also recognizes five types of standpipe systems: automatic dry, automatic wet, manual dry, manual wet and semiautomatic dry. The use of each type of system is limited to the building conditions and locations identified in Section 905.3. The classes and types of standpipe systems are defined in Section 202.

Section 912, to which this section refers, provides a comprehensive set of requirements for FDCs, reducing the opportunity for any of its requirements to be overlooked. See the commentary to Section 912.

[F] 905.3 Required installations. Standpipe systems shall be installed where required by Sections 905.3.1 through 905.3.8. Standpipe systems are allowed to be combined with *automatic sprinkler systems.*

Exception: Standpipe systems are not required in Group R-3 occupancies.

❖ Standpipe systems are installed in buildings based on the occupancy, fire department accessibility and special conditions that may require manual fire suppression exceeding the capacity of a fire extinguisher. Standpipe systems are most commonly required for buildings that exceed the height threshold requirement in Section 905.3.1 or due to features of a specific occupancy or the building such as covered and open mall buildings, stages and underground buildings.

This section also states that a standpipe system does not have to be separate from an installed sprinkler system. It is common practice in multistory buildings for the standpipe system risers to also serve as risers for the automatic sprinkler systems.

In these instances, precautions need to be taken so that the operation of one system will not interfere with the operation of the other system. Therefore, control valves for the sprinkler system must be installed where the sprinklers are connected to the standpipe riser at each floor level. This allows the standpipe system to remain operational, even if the sprinkler system is shut off at the floor control valve.

The exception recognizes that standpipe systems in Group R-3 occupancies would be of minimal value to the fire department and would send the wrong message to the occupants of a dwelling unit. In the case of multiple single-family dwellings, each dwelling unit has a separate entrance and is separated from the other units by 1-hour fire partitions. These conditions permit ready access to fires and also provide for a degree of fire containment through compartmentation, which is not always present in other occupancies.

[F] 905.3.1 Height. Class III standpipe systems shall be installed throughout buildings where the floor level of the highest *story* is located more than 30 feet (9144 mm) above the lowest level of fire department vehicle access, or where the floor level of the lowest *story* is located more than 30 feet (9144 mm) below the highest level of fire department vehicle access.

Exceptions:

1. Class I standpipes are allowed in buildings equipped throughout with an *automatic sprinkler system* in accordance with Section 903.3.1.1 or 903.3.1.2.
2. Class I manual standpipes are allowed in *open parking garages* where the highest floor is located not more than 150 feet (45 720 mm) above the lowest level of fire department vehicle access.
3. Class I manual dry standpipes are allowed in *open parking garages* that are subject to freezing temperatures, provided that the hose connections are located as required for Class II standpipes in accordance with Section 905.5.
4. Class I standpipes are allowed in basements equipped throughout with an *automatic sprinkler system.*
5. In determining the lowest level of fire department vehicle access, it shall not be required to consider either of the following:
   5.1. Recessed loading docks for four vehicles or less.
   5.2. Conditions where topography makes access from the fire department vehicle to the building impractical or impossible.

❖ Given the available manpower on the fire department vehicle, standard fire-fighting operations and standard hose sizes, a 30-foot (9144 mm) vertical distance is generally considered the maximum height to which a typical fire department engine company can practically and readily extend its hose lines. Thus, the maximum vertical travel (height) threshold is based on the time it would take a typical fire department engine (pumper) company to manually suppress a fire. The standpipe connection reduces the time needed for the fire department to extend hose lines up or down stairways to advance and apply water to the fire. For this use, a minimum Class III standpipe system is required.

With respect to the height of the building, the threshold is measured from the level at which the fire department can gain access to the building directly from its vehicle and begin vertical movement. Floor levels above grade are measured from the lowest level of fire department vehicle access to the highest floor level above [see Commentary Figure 905.3.1(1)]. If a building contains floor levels below the level of fire department vehicle access, the measurement is made from the highest level of fire department vehicle access to the lowest floor level. In cases where a building has more than one level of fire department vehicle access, the most restrictive measurement is used because it is not known at which level the fire department will access the building. In other words, the vertical distance is to be measured from the more restrictive level of fire department vehicle access to the level of the highest (or lowest, if below) floor [see Commentary Figure 905.3.1(2)].

The threshold based on the height of the building is independent of the occupancy of the building, the area of the building or the presence of an automatic sprinkler system. This is based on the universal need to be able to provide a water supply for fire suppression in any building and on the limitations of the physical effort necessary to extend hose lines vertically.

Before discussing the exceptions it is important to understand the differences between the different classes and operational characteristics of standpipes. More detailed information is included in Section 202 for the definitions of the different classes and types of standpipes.

Standpipes can be dry or wet, manual, automatic or semiautomatic. Automatic systems can be either

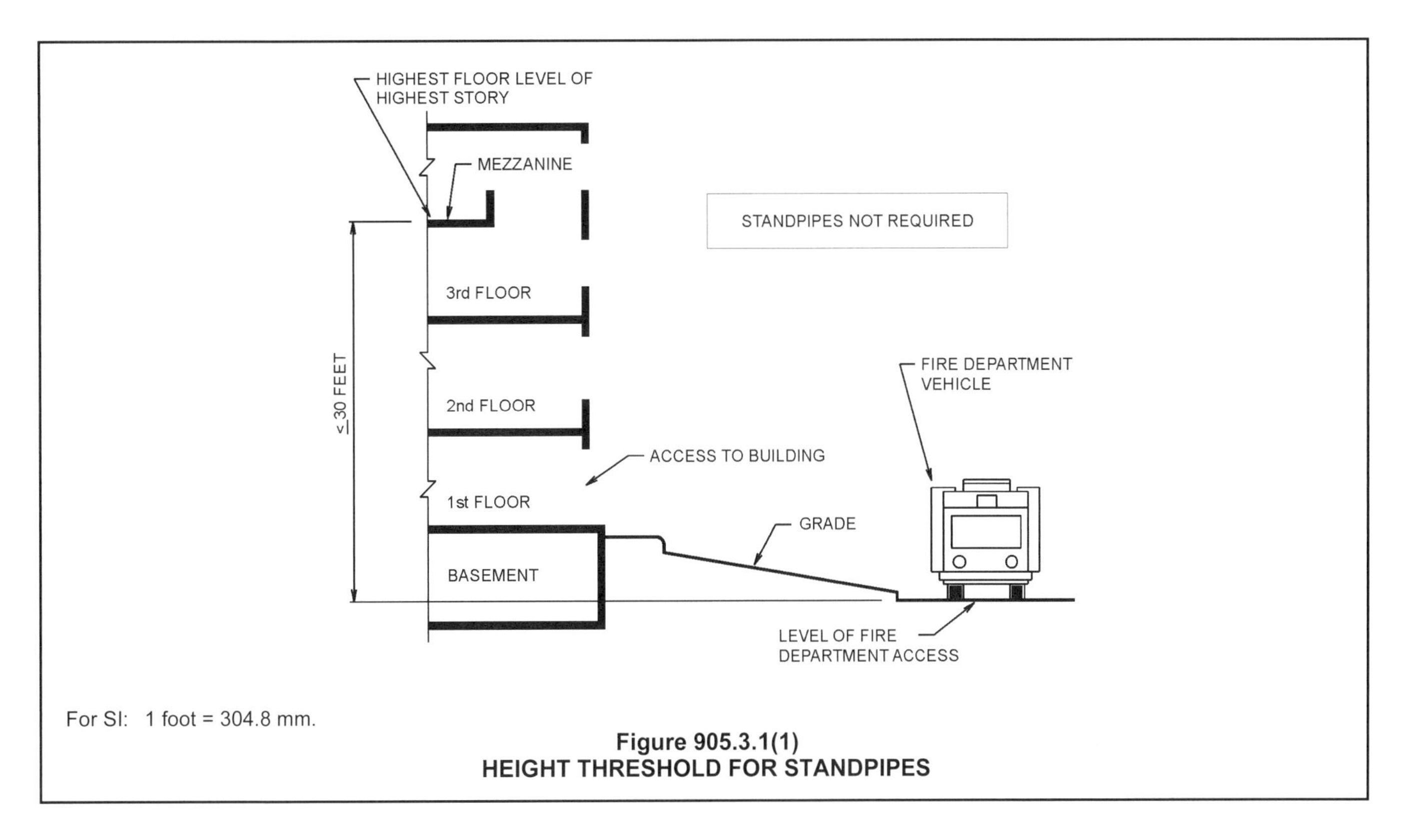

For SI: 1 foot = 304.8 mm.

Figure 905.3.1(1)
HEIGHT THRESHOLD FOR STANDPIPES

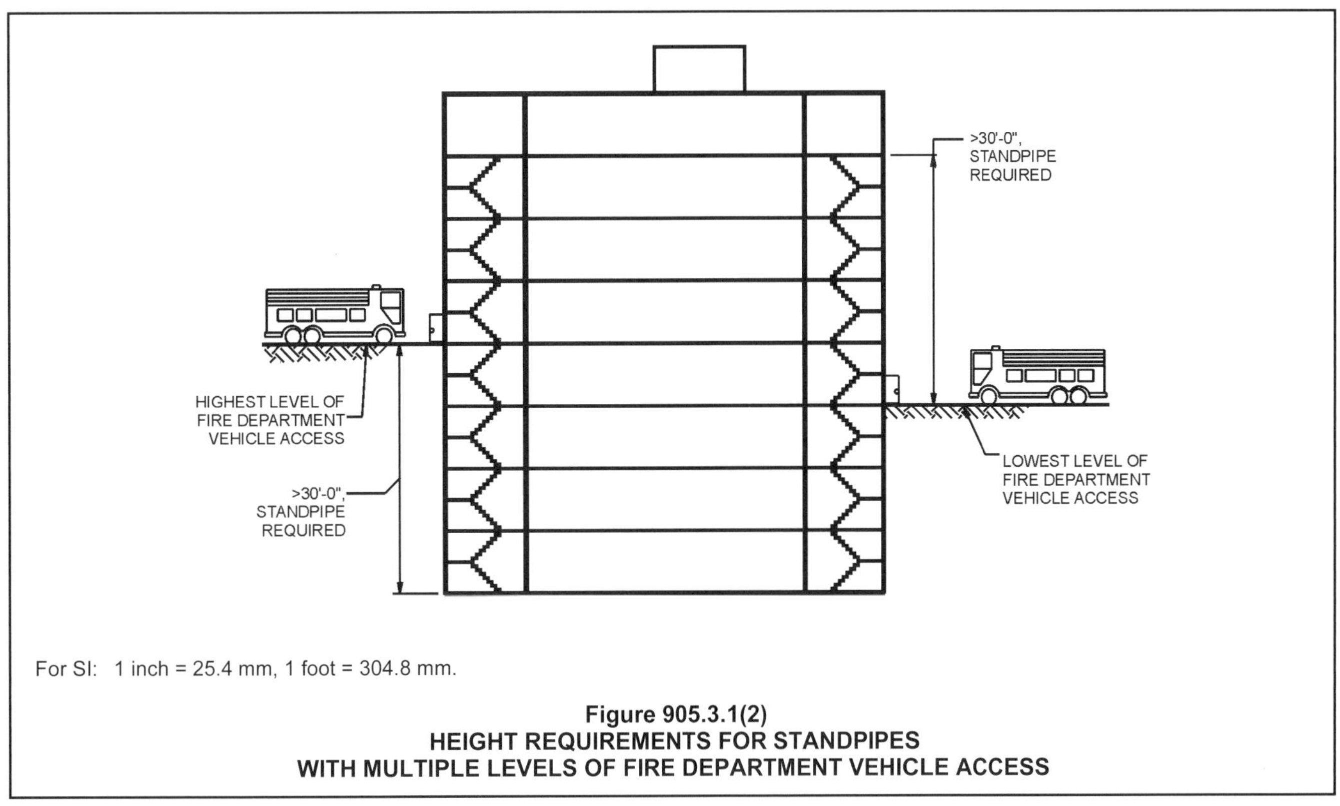

For SI: 1 inch = 25.4 mm, 1 foot = 304.8 mm.

Figure 905.3.1(2)
HEIGHT REQUIREMENTS FOR STANDPIPES WITH MULTIPLE LEVELS OF FIRE DEPARTMENT VEHICLE ACCESS

wet or dry. Manual systems can be either wet or dry. A semiautomatic system is always in association with a dry system.

The code is written such that it could be assumed the default is an automatic wet system. This is, however, not the case. The requirement is left to the design standard, NFPA 14. Section 5.4.1.1 of NFPA 14 indicates that Class I standpipes can be manual if the building is not a high rise. Section 5.4.1.4 of the standard indicates that a Class I standpipe must be wet except where the pipe is subject to freezing. Thus, where a Class I standpipe is installed, possibly as a part of Exception 1, the system can be manual wet if the building is not a high rise. This is consistent with IFC Committee Interpretation No. 33-03. As long as the building is not a high rise, it can be provided with a Class I standpipe system that is manual wet.

Class II and III standpipes are required to be automatic-wet or semiautomatic wet except where the piping is subject to freezing according to Section 5.4.2 of NFPA 14. They cannot be manual. Only Class I standpipes can be manual and only be used under the conditions noted in this code. Note that other sections of the code may specify whether the system must be automatic or not. If the requirement is not noted elsewhere in the code, then the decision to use an automatic or manual system is left to the designer.

Exception 1 recognizes that with a fully operational automatic sprinkler system, the time that the fire department has to extend hoses within the building is substantially increased and that the amount of effort required is greatly reduced. Consequently a single Class I connection can be provided. The second, $1^1/_2$-inch (38 mm) connection is allowed to be omitted. NFPA 14 also has a similar provision but is more restrictive as it only eliminates the hose station and additionally requires a $2^1/_2$ inch by $1^1/_2$ inch (65 mm by 40 mm) reducer and a cap attached with a chain (see NFPA 14, Section 7.3.4.1). In accordance with Section 102.4, the code would take precedence and the reducer and cap would not be required.

Exception 2 identifies one of the issues relative to open parking garages. This exception allows for the garage, when not more than 150 feet (45 720 mm) in height above the lowest level of fire department access, to have a wet standpipe but without additional operating pressure until the fire department connects and begins pumping into the system. This makes sense since normal operations typically do not begin until after the fire department is on the scene and has made its initial assessments. This is generally considered to be the maximum safe height for pumpers to overcome the hydrostatic head presented by 150 feet (45 720 mm) of water. Careful considerations should be made since not all fire departments have equipment capable of this type of pumping capacity.

Exception 3 is similar to the prior exception but with the added provision that the standpipe can be dry if subject to freezing, regardless of height. Because the standpipe will be without water and be dependent on the fire department to provide both water and pressure, standpipe outlets must be spaced more frequently, as noted in Section 905.5 for Class II standpipes, so that fire fighters can connect and begin operations quicker. The exception does not require Class II outlets; only that the spacing be consistent with the requirement for Class II.

Exception 4 is similar to Exception 1 but only addresses sprinklers in the basement. Thus it is possible to use this exception if only the basement is protected by automatic sprinklers. However, Class I connections can only be provided in the basements—not on the upper floors. The exception cannot be used for stories above grade unless the entire building is sprinklered and, therefore, compliant with Exception 1.

Exception 5 provides additional information about what must be considered when determining building height with respect to the level of fire department vehicle access. The first item is a practical one that excludes loading docks of a limited size. The second item notes that although it may be possible to have a fire department vehicle arrive adjacent to the building at a low level, it may not be possible for the fire department to access the building from that level. An example of this condition would be where a road surface is located below a building built on a bluff. Although the fire department vehicles can approach from the lower road, fire department personnel cannot access the building from that lower level. Thus, the standpipe requirement would not be based on the road below the bluff.

[F] 905.3.2 Group A. Class I automatic wet standpipes shall be provided in nonsprinklered Group A buildings having an *occupant load* exceeding 1,000 persons.

Exceptions:

1. Open-air-seating spaces without enclosed spaces.
2. Class I automatic dry and semiautomatic dry standpipes or manual wet standpipes are allowed in buildings that are not high-rise buildings.

❖ The main concern in assembly occupancies with a high occupant load is evacuation. Many occupants may not be familiar with either their surroundings or the egress arrangement in the building. This section also assumes the building is not sprinklered; therefore, control and suppression of the fire is left to the fire department.

Exception 1 exempts open-air seating without enclosed spaces, such as grandstands and bleachers. In such occupancies, a buildup of smoke and hot gases is not possible because these structures are open to the atmosphere.

Exception 2 states that in lieu of a Class I automatic wet standpipe, automatic dry and semiautomatic dry Class I standpipes are permitted in buildings that are not considered to be a high rise.

[F] 905.3.3 Covered and open mall buildings. Covered mall and open mall buildings shall be equipped throughout with a standpipe system where required by Section 905.3.1. Mall buildings not required to be equipped with a standpipe system by Section 905.3.1 shall be equipped with Class I hose connections connected to the *automatic sprinkler system* sized to deliver water at 250 gallons per minute (946.4 L/min) at the most hydraulically remote hose connection while concurrently supplying the automatic sprinkler system demand. The standpipe system shall be designed to not exceed a 50 pounds per square inch (psi) (345 kPa) residual pressure loss with a flow of 250 gallons per minute (946.4 L/min) from the fire department connection to the hydraulically most remote hose connection. Hose connections shall be provided at each of the following locations:

1. Within the mall at the entrance to each *exit* passageway or *corridor*.
2. At each floor-level landing within *interior exit stairways* opening directly on the mall.
3. At exterior public entrances to the mall of a covered mall building.
4. At public entrances at the perimeter line of an open mall building.
5. At other locations as necessary so that the distance to reach all portions of a tenant space does not exceed 200 feet (60 960 mm) from a hose connection.

❖ Covered and open mall buildings are only required to have a standpipe system if Section 905.3.1 requires such features. If standpipes are not required because of building height, Class I hose connections that are connected to the automatic sprinkler system are still required. Also to ensure that both the sprinkler system and hose connections will function at an acceptable level, the system must be sized for both the sprinkler demand and the hose connection demand. This section specifies a minimum flow rate and a maximum pressure loss to the most remote hose connection so that the fire department can gain full use of the hose connection during a fire. Hose connections are required when a standpipe system is not at key locations, such as entrances to exit passageways and at entrances to the covered or open mall. Note that these locations are essentially the same locations required for Class I hose connections in Section 905.4, except that this section also requires that the distance of all portions of tenant spaces does not exceed 200 feet (60 960 mm) from a hose connection.

[F] 905.3.4 Stages. Stages greater than 1,000 square feet in area (93 m^2) shall be equipped with a Class III wet standpipe system with $1^1/_2$-inch and $2^1/_2$-inch (38 mm and 64 mm) hose connections on each side of the stage.

> **Exception:** Where the building or area is equipped throughout with an *automatic sprinkler system*, a $1^1/_2$-inch (38 mm) hose connection shall be installed in accordance with NFPA 13 or in accordance with NFPA 14 for Class II or III standpipes.

❖ Because of the potentially large fuel load and three-dimensional aspect of the fire hazard associated with stages greater than 1,000 square feet (93 m^2) in area, Class III standpipes are required on each side of these large stages. The standpipes must be equipped with a $1^1/_2$-inch (38 mm) hose connection and a $2^1/_2$-inch (64 mm) hose connection. The $1^1/_2$-inch (38 m^2) connection is for the hose requirement in Section 905.3.4.1. The $2^1/_2$-inch (64 mm) connection is to provide greater flexibility for the fire department in its fire-fighting operations.

Stages, as used in this section, are those stages defined in Section 410.2, which include overhead hanging curtains, drops, scenery or stage effects other than lighting and sound. These were traditionally referred to as "legitimate stages." It is not an appropriate application of this section to require standpipes for elevated areas in banquet rooms, or theatrical platforms where the higher fuel loads associated with a legitimate stage do not exist.

The exception recognizes the benefit of the building or area being sprinklered. If so, then only a single $1^1/_2$-inch (38 mm) connection is required. This hose connection is intended to be used by the fire department and apply less water from the hose because of the suppression activity of the sprinkler system. Hose threads must be compatible with those of the fire department as required in Section 903.3.6.

In a fully sprinklered building it is acceptable to supply the hose connections through the same standpipe as the sprinklers. This is reflected in the reference to both NFPA 13, which acknowledges this concept, and NFPA 14, which contains similar provisions. If the provisions of NFPA 14 are used, although the standpipe must be wet and Class II in its installation, the design of the water supply and interconnection of systems can be in accordance with the requirements for Class II as well as for Class III standpipes.

[F] 905.3.4.1 Hose and cabinet. The $1^1/_2$-inch (38 mm) hose connections shall be equipped with sufficient lengths of $1^1/_2$-inch (38 mm) hose to provide fire protection for the stage area. Hose connections shall be equipped with an *approved* adjustable fog nozzle and be mounted in a cabinet or on a rack.

❖ The $1^1/_2$-inch (38 mm) standpipe hose installed for stages greater than 1,000 square feet (93 m^2) in area is intended for use by stage personnel who have been trained to use it. The length of hose provided is a function of the size and configuration of the stage. This includes by definition the entire performance area and adjacent backstage and support areas not fire separated from the performance area. The effective reach of the fire stream from the fog nozzle is a function of the available water supply, and in particular, the pressure. Fog nozzles typically require 100 pounds per square inch (psi) (690 kPa) for optimum performance.

[F] 905.3.5 Underground buildings. Underground buildings shall be equipped throughout with a Class I automatic wet or manual wet standpipe system.

❖ Underground buildings pose unique hazards to life safety because of their isolation and inaccessibility. Additional fire protection and fire-fighting measures for the fire department are required to compensate for the lack of exterior access for fire suppression and rescue operations (see Section 405).

[F] 905.3.6 Helistops and heliports. Buildings with a rooftop *helistop* or *heliport* shall be equipped with a Class I or III standpipe system extended to the roof level on which the *helistop* or *heliport* is located in accordance with Section 2007.5 of the *International Fire Code*.

❖ Buildings containing rooftop helistops or heliports are required to be equipped with a Class I or III standpipe. A heliport is a distinct hazard that will involve flammable fuels. In the event of an emergency, rapid deployment of hand hose lines will be necessary to attack a resulting fire, effectuate rescue and to protect exposures and the remainder of the building.

The requirement results in a standpipe system throughout the building, not just a connection at the roof level. This is critical in fire-fighting operations because many times the connection below the rooftop level may be needed just to gain access onto the roof. If the only connection is on the roof, it is of no use if the fire fighters cannot get to it.

Additionally, a heliport includes fueling operations. It is entirely possible for a spill to not only affect the rooftop, but also floors below as the liquid fuel spreads. The standpipe system will again be utilized in these situations.

Section 2007.5 of the IFC requires a $2^1/_2$-inch (64 mm) standpipe outlet to be within 150 feet (45 675 mm) of all portions of the heliport or helistop area and be either Class I or III.

[F] 905.3.7 Marinas and boatyards. Standpipes in marinas and boatyards shall comply with Chapter 36 of the *International Fire Code*.

❖ Section 3604.2 of the IFC contains the specifics as to when standpipes are required at marinas. Marinas and boatyards have unique challenges for fire fighting. Although there is water readily available, it is not easily or effectively capable of being applied to a fire at such a facility. A fire in such facilities can spread from structure to structure and from vessel to vessel with no effective way to attack and control it. Section 3604.2 of the IFC references NFPA 303 for the standpipe requirements and additionally requires that no point on the marina pier or float system exceed 150 feet (45 720 mm) from a standpipe hose connection (see commentary, Section 3604.2 of the IFC).

[F] 905.3.8 Rooftop gardens and landscaped roofs. Buildings or structures that have rooftop gardens or landscaped roofs and that are equipped with a standpipe system shall have the standpipe system extended to the roof level on which the rooftop garden or landscaped roof is located.

❖ This section requires that if the building is equipped with a standpipe system, whether or not such systems are required, it must be extended to a roof containing a garden or that which is landscaped. These requirements relate to the requirements in Section 317 of the IFC that address the increased fuel load being added to roofs.

[F] 905.4 Location of Class I standpipe hose connections. Class I standpipe hose connections shall be provided in all of the following locations:

1. In every required *interior exit stairway*, a hose connection shall be provided for each story above and below grade. Hose connections shall be located at an intermediate landing between stories, unless otherwise *approved* by the fire code official.
2. On each side of the wall adjacent to the *exit* opening of a *horizontal exit*.

   Exception: Where floor areas adjacent to a *horizontal exit* are reachable from an *interior exit stairway* hose connection by a 30-foot (9144 mm) hose stream from a nozzle attached to 100 feet (30 480 mm) of hose, a hose connection shall not be required at the *horizontal exit*.
3. In every *exit* passageway, at the entrance from the *exit* passageway to other areas of a building.

   Exception: Where floor areas adjacent to an *exit* passageway are reachable from an *interior exit stairway* hose connection by a 30-foot (9144 mm) hose stream from a nozzle attached to 100 feet (30 480 mm) of hose, a hose connection shall not be required at the entrance from the *exit* passageway to other areas of the building.
4. In covered mall buildings, adjacent to each exterior public entrance to the mall and adjacent to each entrance from an exit passageway or exit corridor to the mall. In open mall buildings, adjacent to each public entrance to the mall at the perimeter line and adjacent to each entrance from an exit passageway or exit corridor to the mall.
5. Where the roof has a slope less than four units vertical in 12 units horizontal (33.3-percent slope), a hose connection shall be located to serve the roof or at the highest landing of an *interior exit stairway* with access to the roof provided in accordance with Section 1011.12.
6. Where the most remote portion of a nonsprinklered floor or *story* is more than 150 feet (45 720 mm) from a hose connection or the most remote portion of a sprinklered floor or *story* is more than 200 feet (60 960 mm) from a hose connection, the fire code official is authorized to require that additional hose connections be provided in *approved* locations.

❖ Hose connections are required for the fire department to make use of the standpipe system. Since the fire department will typically access the building using the stairways, and most fire departments do not permit entry to the fire floor without an operating hose line, a hose connection must be installed for each floor level of each required enclosed stairway.

Item 1 also specifies that the hose connections are to be located at intermediate landings between stories. This reduces congestion at the stairway door and may reduce the hose lay distance. The hose connections, however, are still permitted at each story instead of at the intermediate landing if this arrangement is approved by the fire code official.

Because horizontal exits are also primary entrances to the fire floor, Item 2 states that hose connections must also be provided at each horizontal exit. The construction of the fire separation assembly used as the horizontal exit will protect the fire fighters while they are connecting to the standpipe system. The hose connections are to be located on each side of the hori-zontal exit to enable fire fighters to be in a protected area, regardless of the location of the fire. The exception acknowledges that there may already be a hose connection in close proximity to the horizontal exit if there is a stairway adjacent to the horizontal exit. The intent is to allow fewer standpipe outlets if the area can be adequately covered by the standpipes in stairways since those are the standpipes typically used by the fire department.

Item 3 states that an exit passageway in a building required to have a standpipe system is typically used as an extension of a required exit stairway. This allows use of the exit passageway for fire-fighting staging operations in the same way as an exit stair. The exception acknowledges that there may already be a hose connection in close proximity to the exit passageway. If there is a stairway containing a hose connection in close proximity to the exit passageway that can meet the 30-foot (9144 mm) hose stream from a nozzle attached to 100 feet (30 480 mm) of hose, then an additional standpipe is not required. The intent is to allow fewer standpipe outlets if the area can be adequately covered by the standpipes in stairways since those are the standpipes typically used by the fire department.

In covered and open mall buildings, Item 4 requires hose connections at each entrance to an exit passageway or exit corridor. In addition, covered mall buildings would be required to have connections at each exterior public entrance. Open malls would require connections at the public entrance perimeter line. These locations allow fire personnel to have a support line as soon as they enter the building.

Item 5 is consistent with NFPA 14 regarding the installation of Class I standpipe hose connections on the roofs of buildings. This requirement requires only one standpipe to extend to the roof level or highest landing of the stairway serving the roof. This coordinates with Section 1011.12, which only requires one stairway to extend to the roof.

Hose connections in each interior exit stairway result in hose connections being located based on the travel distances permitted in Table 1017.2, which recognizes that most fire departments carry standpipe hose packs with 150 feet (45 720 mm) of hose or possibly with 100 feet (30 480 mm) of hose and an additional 50-foot (15 240 mm) section that could be easily connected.

With the typical travel distance permitted in nonsprinklered buildings of 200 feet (60 960 mm), reasonable coverage is provided when the effective reach of a fire stream is considered. Depending on the arrangement of the floor, however, all areas may not be effectively protected. Although this situation could easily be corrected by locating additional hose connections on the floor, such connections may rarely be used because of the difficulty in identifying their location during a fire and the fact that most fire departments require an operational hose line before they enter the fire floor. Because longer travel distances are allowed in sprinklered buildings, the problem is increased, but the need for prompt manual suppression is reduced by the presence of the sprinkler system. Item 6 gives the fire code official the authority to require additional hose connections if needed.

[F] 905.4.1 Protection. Risers and laterals of Class I standpipe systems not located within an *interior exit stairway* shall be protected by a degree of *fire resistance* equal to that required for vertical enclosures in the building in which they are located.

> **Exception:** In buildings equipped throughout with an *approved automatic sprinkler system*, laterals that are not located within an *interior exit stairway* are not required to be enclosed within fire-resistance-rated construction.

❖ To minimize the potential for damage to the standpipe systems from a fire, the risers and laterals (i.e., the horizontal segments of standpipe system piping) must be located in an enclosure having the same fire-resistance rating as required for a vertical or shaft enclosure within the building. The required fire-resistance rating for the enclosure can be determined as detailed in Section 713.4.

The exception states that the enclosure for laterals is not required if the building is equipped throughout with an approved automatic sprinkler system. The potential for damage to the standpipe system is minimized by the protection provided by the sprinkler system. The automatic sprinkler system may be either an NFPA 13 or 13R system, depending on what was permitted for the building occupancy.

If the interior exit stairway is not required to have a rated enclosure, such as in an open parking garage, the laterals are similarly not required to be in an enclosure. The protection afforded the vertical riser in the stairway must be the same as that afforded the laterals. If the stairway is not required by other sections of the code to be located in a rated enclosure then the laterals are not required to be in rated protection either.

[F] 905.4.2 Interconnection. In buildings where more than one standpipe is provided, the standpipes shall be interconnected in accordance with NFPA 14.

❖ In cases where there are multiple Class I standpipe risers, the risers must be supplied from and interconnected to a common supply line. The required fire department connection must serve all of the sprinklers or standpipes in the building.

[F] 905.5 Location of Class II standpipe hose connections. Class II standpipe hose connections shall be accessible and located so that all portions of the building are within 30 feet (9144 mm) of a nozzle attached to 100 feet (30 480 mm) of hose.

❖ Sections 905.5.1 through 905.5.3 specify the requirements for Class II standpipe hose connections. Class II standpipe systems are primarily intended for use by the building occupants.

This section for Class II standpipes does not specifically require the hose station and uses the term "hose connection" with a location based upon 100 feet (30 480 mm) of hose. However, the definition of Class II and III standpipes and Section 7.3.3.1 of NFPA 14 specifically require hose stations. Section 905.2 specifically references NFPA 14.

Although NFPA 14 requires a hose station, the decision as to whether a hose station is required may be one that is affected by the policies and procedures of the local fire department. It should be remembered that Class II hose connections and hose stations are intended for occupant use and not necessarily for fire department use. The fire department typically uses the Class I connection that is compatible with $2^1/_2$ inch (64 mm) hose.

[F] 905.5.1 Groups A-1 and A-2. In Group A-1 and A-2 occupancies having *occupant loads* exceeding 1,000 persons, hose connections shall be located on each side of any stage, on each side of the rear of the auditorium, on each side of the balcony and on each tier of dressing rooms.

❖ Because of the high occupant load density in Group A-1 and A-2 occupancies, providing additional means for controlling fires in their initial stage is important to enable prompt evacuation of the building. This section is independent of the Class I standpipe requirement for stages based on square footage as indicated in Section 905.3.4.

[F] 905.5.2 Protection. Fire-resistance-rated protection of risers and laterals of Class II standpipe systems is not required.

❖ Class II standpipe systems are normally not located in exit stairways; standpipe hose connections are located near the protected area to allow quick access. Therefore, it is likely that neither the risers nor the laterals would be located in any enclosure.

[F] 905.5.3 Class II system 1-inch hose. A minimum 1-inch (25 mm) hose shall be allowed to be used for hose stations in light-hazard occupancies where investigated and *listed* for this service and where *approved* by the fire code official.

❖ This section permits the use of 1-inch (25 mm) listed noncollapsible hose as an alternative to $1^1/_2$-inch (38 mm) hose, subject to the approval of the fire code official. This alternative is limited to light-hazard occupancies, such as office buildings and certain assembly occupancies that tend to have lower fuel loads, since a smaller hose can discharge less water.

[F] 905.6 Location of Class III standpipe hose connections. Class III standpipe systems shall have hose connections located as required for Class I standpipes in Section 905.4 and shall have Class II hose connections as required in Section 905.5.

❖ Class III standpipe systems that have both a $2^1/_2$-inch (64 mm) hose connection and a $1^1/_2$-inch (38 mm) hose connection must comply with the applicable requirements of Sections 905.4, 905.5 and 905.6. Thus, it is necessary to review and comply with all applicable provisions.

[F] 905.6.1 Protection. Risers and laterals of Class III standpipe systems shall be protected as required for Class I systems in accordance with Section 905.4.1.

❖ Because Class III standpipe systems are intended for use by fire-suppression personnel, they must be located in construction that has a fire-resistance rating equivalent to that of the vertical or shaft enclosure requirements of the building (see commentary, Section 905.4.1).

[F] 905.6.2 Interconnection. In buildings where more than one Class III standpipe is provided, the standpipes shall be interconnected in accordance with NFPA 14.

❖ As indicated in Section 905.4.2 for Class I standpipe systems, multiple standpipe risers must be interconnected with a common supply line. An indicating valve is typically installed at the base of each riser so that individual risers can be taken out of service without affecting the water supply or the operation of other standpipe risers.

[F] 905.7 Cabinets. Cabinets containing fire-fighting equipment such as standpipes, fire hoses, fire extinguishers or fire department valves shall not be blocked from use or obscured from view.

❖ This section does not require that cabinets be provided to contain fire protection equipment. However, if they are provided, cabinets must be readily visible and accessible at all times. Sections 905.7.1 and 905.7.2 contain additional criteria for the construction and identification of the cabinets. Where cabinets are located in fire-resistance-rated assemblies, the integrity of the assembly must be maintained. Cabinet design for hose connections, control valves or other devices that require manual operation should be such that there is sufficient clearance between the cabinet body and the device to allow grasping of the device (quite likely with a gloved hand) and prompt operation of it.

[F] 905.7.1 Cabinet equipment identification. Cabinets shall be identified in an *approved* manner by a permanently attached sign with letters not less than 2 inches (51 mm) high in a color that contrasts with the background color, indicating the equipment contained therein.

Exceptions:

1. Doors not large enough to accommodate a written sign shall be marked with a permanently attached pictogram of the equipment contained therein.
2. Doors that have either an *approved* visual identification clear glass panel or a complete glass door panel are not required to be marked.

❖ This section specifies the minimum criteria to make the signs readily visible. Different color combinations may be approved by the fire code official if the color contrast between the letters and the background is vivid enough to make the sign visible at an approved distance. The exceptions address alternatives to letter signage if the cabinet is still conspicuously identified or the contents are readily visible.

[F] 905.7.2 Locking cabinet doors. Cabinets shall be unlocked.

Exceptions:

1. Visual identification panels of glass or other *approved* transparent frangible material that is easily broken and allows access.
2. *Approved* locking arrangements.
3. Group I-3.

❖ Ready access to all fire-fighting equipment in the cabinet is essential. The exceptions, however, recognize the need to lock cabinets for security reasons and to prevent theft or vandalism (see commentary, Section 906.8).

[F] 905.8 Dry standpipes. Dry standpipes shall not be installed.

Exception: Where subject to freezing and in accordance with NFPA 14.

❖ Wet standpipe systems are preferred because they tend to be the most reliable type of standpipe system; therefore, dry standpipes are prohibited unless subject to freezing. For example, Class I manual standpipe systems, which do not have a permanent water supply, are permitted in open parking structures. This recognizes that open parking structures are not heated and that most fires are limited to the vehicle of origin. The use of any dry standpipe system instead of a wet standpipe should take into consideration the added response time and its effect on the occupancy characteristics of the building.

[F] 905.9 Valve supervision. Valves controlling water supplies shall be supervised in the open position so that a change in the normal position of the valve will generate a supervisory signal at the supervising station required by Section 903.4. Where a fire alarm system is provided, a signal shall be transmitted to the control unit.

Exceptions:

1. Valves to underground key or hub valves in roadway boxes provided by the municipality or public utility do not require supervision.
2. Valves locked in the normal position and inspected as provided in this code in buildings not equipped with a fire alarm system.

❖ As with sprinkler systems, water control valves for standpipe systems must be electrically supervised as a means of determining that the system is operational (see commentary, Section 903.4).

Exception 1 recognizes that underground key or hub valves in roadway boxes are not normally supervised or need to be supervised whether the building contains a standpipe system or an automatic sprinkler system.

Exception 2 does not require the control valves for the standpipes to be electrically monitored if they are locked in the normal position and a fire alarm system is not installed in the building. When a fire alarm system is installed, the control valves for the standpipes must be electrically monitored and tied into the supervision required for the fire alarm system.

[F] 905.10 During construction. Standpipe systems required during construction and demolition operations shall be provided in accordance with Section 3311.

❖ As stated in Section 3311, at least one standpipe is required during construction of buildings four stories or more in height or during demolition of standpipe-equipped buildings. Standpipe systems must be accessible and operable during construction and demolition operations to assist in any potential fire (see commentary, Sections 3311.1 and 3311.2 of the code, and Sections 3313.1 and 3313.2 of the IFC).

SECTION 906 PORTABLE FIRE EXTINGUISHERS

[F] 906.1 Where required. Portable fire extinguishers shall be installed in all of the following locations:

1. In Group A, B, E, F, H, I, M, R-1, R-2, R-4 and S occupancies.

   Exception: In Group R-2 occupancies, portable fire extinguishers shall be required only in locations specified in Items 2 through 6 where each *dwelling unit* is provided with a portable fire extinguisher having a minimum rating of 1-A:10-B:C.

2. Within 30 feet (9144 mm) of commercial cooking equipment.
3. In areas where flammable or combustible liquids are stored, used or dispensed.
4. On each floor of structures under construction, except Group R-3 occupancies, in accordance with Section 3315.1 of the *International Fire Code*.

5. Where required by the *International Fire Code* sections indicated in Table 906.1.

6. Special-hazard areas, including but not limited to laboratories, computer rooms and generator rooms, where required by the fire code official.

❖ Portable fire extinguishers are required in certain instances to give the occupants the means to suppress a fire in its incipient stage. The capability for manual fire suppression can contribute to the protection of the occupants, especially if there are evacuation difficulties associated with the occupancy or the specific hazard in the area. To be effective, personnel must be properly trained in the use of portable fire extinguishers.

Because of the high-hazard nature of building contents, portable fire extinguishers are required in occupancies in Group H.

Portable fire extinguishers are required in occupancies in Groups A, B, E, F, I, M, R-1, R-2, R-4 and S because of the need to control the fire in its early stages and because evacuation can be slowed by the density of the occupant load, the capability of the occupants to evacuate or the overall fuel load in the building. Because the code typically focuses upon new buildings, this section is applicable to new buildings.

Portable fire extinguishers are required in areas containing special hazards, such as commercial cooking equipment, and specific hazardous operations, as indicated in Table 906.1. Because of the potentially extreme fire hazard associated with such areas or occupancy conditions, prompt extinguishment of the fire is critical.

Portable fire extinguishers are required in all buildings under construction, except in occupancies in Group R-3. The extinguishers are intended for use by construction personnel to suppress a fire in its incipient stages.

Portable fire extinguishers are also required in laboratories, computer rooms and other work spaces in which fire hazards may exist based on the use of the space. Many of these will be addressed by the required occupancy group criteria or by the specific hazard provisions of Table 906.1. Laboratories, for example, may not be considered Group H, but still use limited amounts of hazardous materials that would make manual means of fire extinguishment desirable.

The exception to Item 1 permits smaller PFEs in dwelling units of Group R-2 occupancies instead of larger PFEs in the common areas. Under the exception, the installation of 1-A:10-B:C PFEs within individual dwelling units allows apartment owners to eliminate their installation in common areas such as corridors, laundry rooms and swimming pool areas. PFEs in these areas are susceptible to vandalism or theft. Another issue is that larger PFEs are more difficult for the infirm and elderly to safely deploy and operate.

For the period of 2003 through 2007, NFPA reported that approximately 38,000 fires occurred annually in apartment buildings. Sixty percent of these fires occurred inside of dwelling units versus 14 percent that occurred in common areas covered by Items 3 and 6 of Section 906.1. It is more logical to place PFEs inside of dwelling units versus common areas because it locates the extinguisher in an area where statistically most fires occur. If the occupant cannot control the fire using the PFE, he or she can escape and allow the automatic sprinkler system to operate and control the fire. This exception improves the safety of Group R-2 residents because it does not require them to leave a dwelling involved in a fire, find a PFE and then return to the fire-involved dwelling unit to attempt incipient fire attack.

Including this requirement in the code alerts designers and building officials that the extinguishers are required. This will allow designers to plan for thicker walls where recessed cabinets may be used or to design locations where the extinguishers will not project into or obstruct the egress or circulation path.

TABLE 906.1. See page 9-53.

❖ Table 906.1 lists those sections of the IFC that represent specific occupancy conditions requiring portable fire extinguishers for incipient fire control. Wherever the code requires a fire extinguisher because of one of the listed occupancy conditions, it may identify the required rating of the extinguisher that is compatible with the hazard involved in addition to referencing Section 906.

[F] 906.2 General requirements. Portable fire extinguishers shall be selected and installed in accordance with this section and NFPA 10.

Exceptions:

1. The distance of travel to reach an extinguisher shall not apply to the spectator seating portions of Group A-5 occupancies.

2. In Group I-3, portable fire extinguishers shall be permitted to be located at staff locations.

❖ NFPA 10 contains minimum requirements for the selection, installation and maintenance of portable fire extinguishers. Portable fire extinguishers are investigated and rated in conformance to NFPA 10 and listed under a variety of standards. Portable fire extinguishers must be labeled and rated for use on fires of the type, severity and hazard class protected.

This section references NFPA 10 for the selection and installation of portable fire extinguishers and states that the code is focused on the initial construction versus long-term maintenance of a building. The maintenance of portable fire extinguishers is dealt with in the IFC.

Exception 1 recognizes the openness to the atmosphere associated with Group A-5 occupancies. A fire in open areas is more obvious to all spectators. Group A-5 occupancies also do not accumulate smoke and hot gases because they are not enclosed spaces. These reasons, in addition to the large and expansive layout within seating areas, make it reasonable and practical not to apply the distance of travel to a PFE criteria in Group A-5. Revised distance of travel allowances would need to be approved by the code official. Group A-5 occupancies also tend to be more subject to the corrosive conditions of an outdoor environment, and may include freeze/thaw cycles that can be detrimental to fire extinguishers.

Exception 2 recognizes that portable fire extinguishers located throughout the facility are at times tampered with, removed or used for weapons by inmates in a detention or correctional setting. This exception would protect the extinguishers from damage or removal by inmates while still making them available to staff and employees for use in an emergency situation.

[F] 906.3 Size and distribution. The size and distribution of portable fire extinguishers shall be in accordance with Sections 906.3.1 through 906.3.4.

❖ Proper selection and distribution of portable fire extinguishers are essential to having adequate protection for the building structure and the occupancy conditions within. This section introduces the sections that provide those requirements. Determination of the desired type of portable fire extinguisher depends on the character of the fire anticipated, building occupancy, specific hazards and ambient temperature conditions [see commentary, Tables 906.3(1) and 906.3(2)].

[F] 906.3.1 Class A fire hazards. The minimum sizes and distribution of portable fire extinguishers for occupancies that involve primarily Class A fire hazards shall comply with Table 906.3(1).

❖ Class A fires generally involve materials considered to be "ordinary combustibles," such as wood, cloth, paper, rubber and most plastics [see commentary, Table 906.3(1)].

[F]TABLE 906.3(1). See page 9-54.

❖ Table 906.3(1), which parallels Table 6.2.1.1 of NFPA 10, establishes the minimum number and rating of fire extinguishers for Class A fires in any particular occupancy. The occupancy classifications are further defined in NFPA 10. The maximum area that a single fire extinguisher can protect is determined based on the rating of the fire extinguisher. The distance of travel limitation of 75 feet (22 860 mm) is intended to be the actual walking distance along a normal path of travel to the extinguisher. For this reason, it is necessary to select fire extinguishers that comply with both the distribution criteria and distance of travel limitation for a specific occupancy classification.

[F] TABLE 906.1
ADDITIONAL REQUIRED PORTABLE FIRE EXTINGUISHERS IN THE INTERNATIONAL FIRE CODE

IFC SECTION	SUBJECT
303.5	Asphalt kettles
307.5	Open burning
308.1.3	Open flames—torches
309.4	Powered industrial trucks
2005.2	Aircraft towing vehicles
2005.3	Aircraft welding apparatus
2005.4	Aircraft fuel-servicing tank vehicles
2005.5	Aircraft hydrant fuel-servicing vehicles
2005.6	Aircraft fuel-dispensing stations
2007.7	Heliports and helistops
2108.4	Dry cleaning plants
2305.5	Motor fuel-dispensing facilities
2310.6.4	Marine motor fuel-dispensing facilities
2311.6	Repair garages
2404.4.1	Spray-finishing operations
2405.4.2	Dip-tank operations
2406.4.2	Powder-coating areas
2804.3	Lumberyards/woodworking facilities
2808.8	Recycling facilities
2809.5	Exterior lumber storage
2903.5	Organic-coating areas
3006.3	Industrial ovens
3104.12	Tents and membrane structures
3206.10	High-piled storage
3315.1	Buildings under construction or demolition
3317.3	Roofing operations
3408.2	Tire rebuilding/storage
3504.2.6	Welding and other hot work
3604.4	Marinas
3703.6	Combustible fibers
5703.2.1	Flammable and combustible liquids, general
5704.3.3.1	Indoor storage of flammable and combustible liquids
5704.3.7.5.2	Liquid storage rooms for flammable and combustible liquids
5705.4.9	Solvent distillation units
5706.2.7	Farms and construction sites—flammable and combustible liquids storage
5706.4.10.1	Bulk plants and terminals for flammable and combustible liquids
5706.5.4.5	Commercial, industrial, governmental or manufacturing establishments—fuel dispensing
5706.6.4	Tank vehicles for flammable and combustible liquids
5906.5.7	Flammable solids
6108.2	LP-gas

[F] 906.3.2 Class B fire hazards. Portable fire extinguishers for occupancies involving flammable or combustible liquids with depths less than or equal to 0.25-inch (6.4 mm) shall be selected and placed in accordance with Table 906.3(2).

Portable fire extinguishers for occupancies involving flammable or combustible liquids with a depth of greater than 0.25-inch (6.4 mm) shall be selected and placed in accordance with NFPA 10.

❖ Class B fires involve flammable and combustible liquids, oil-based paints, alcohols, solvents, flammable gases and similar materials. Selection of these extinguishers is made based on the depth of the liquid that could become involved in a fire. If the depth is $^1/_4$-inch (6.35 mm) or less, selection is made using Table 906.3(2). Class B extinguishers for greater liquid depth, characterized in NFPA 10 as "appreciable depth," must be selected and installed in accordance with Section 6.3.2 of NFPA 10 [see commentary, Table 906.3(2)].

TABLE 906.3(2). See next column.

❖ Fires involving flammable or combustible liquids present a severe hazard challenge regardless of occupancy. Table 906.3(2), which parallels Table 6.3.1.1 of NFPA 10, prescribes the minimum portable fire extinguisher requirements where flammable or combustible liquids are limited in depth [0.25 inch (6 mm) or less]. As can be seen in the table, the size of the extinguisher is directly related to the distance of travel to the extinguisher for each given occupancy classification. These fire extinguisher provisions are independent of whether other fixed automatic fire-extinguishing systems are installed. For occupancy conditions involving flammable or combustible liquids in potential depths greater than 0.25 inch (6 mm), the selection and spacing criteria of NFPA 10 must be used in addition to any applicable requirements in Chapter 57 of the IFC and NFPA 30.

[F] 906.3.3 Class C fire hazards. Portable fire extinguishers for Class C fire hazards shall be selected and placed on the basis of the anticipated Class A or B hazard.

❖ Class C fires involve energized electrical equipment where the electrical nonconductivity of the extinguishing agent is critical. The need for this class of extinguisher is simply based on the presence of the hazard in an occupancy and no numerical rating is required.

[F] 906.3.4 Class D fire hazards. Portable fire extinguishers for occupancies involving combustible metals shall be selected and placed in accordance with NFPA 10.

❖ Class D fires involve flammable solids, the bulk of which are combustible metals including, but not limited to, magnesium, potassium, sodium and titanium. Most Class D extinguishers will have a special low-velocity nozzle or discharge wand to gently apply the agent in large volumes to avoid disrupting any finely divided burning materials. Extinguishing agents are also available in bulk and can be applied with a scoop or shovel. While Class D extinguishers are often referred to as "dry chemical" fire extinguishers, they are more properly called "dry powder" fire extinguishers because their mechanism of extinguishment is by a smothering action rather than by chemical reaction with the combustion process.

There are several Class D fire-extinguisher agents available—some will handle multiple types of metal fires, others will not. Sodium carbonate-based extinguishers are used to control sodium, potassium, and sodium-potassium alloy fires but have limited use on other metals. This material smothers and forms a crust. Sodium chloride-based extinguishers contain sodium chloride salt and a thermoplastic additive. The plastic melts to form an oxygen-excluding crust over the metal, and the salt dissipates heat. This powder is useful on most alkali metals, including magnesium, titanium, aluminum, sodium, potassium and zirconium. Graphite-based extinguishers contain dry graphite powder that smothers burning metals.

[F] TABLE 906.3(1)
FIRE EXTINGUISHERS FOR CLASS A FIRE HAZARDS

	LIGHT (Low) HAZARD OCCUPANCY	ORDINARY (Moderate) HAZARD OCCUPANCY	EXTRA (High) HAZARD OCCUPANCY
Minimum rated single extinguisher	2-A[c]	2-A	4-A[a]
Maximum floor area per unit of A	3,000 square feet	1,500 square feet	1,000 square feet
Maximum floor area for extinguisher[b]	11,250 square feet	11,250 square feet	11,250 square feet
Maximum distance of travel to extinguisher	75 feet	75 feet	75 feet

For SI: 1 foot = 304.8 mm, 1 square foot = 0.0929m^2, 1 gallon = 3.785 L.

a. Two $2^1/_2$-gallon water-type extinguishers shall be deemed the equivalent of one 4-A rated extinguisher.

b. Annex E.3.3 of NFPA 10 provides more details concerning application of the maximum floor area criteria.

c. Two water-type extinguishers each with a 1-A rating shall be deemed the equivalent of one 2-A rated extinguisher for Light (Low) Hazard Occupancies.

[F] TABLE 906.3(2)
FIRE EXTINGUISHERS FOR FLAMMABLE OR COMBUSTIBLE LIQUIDS WITH DEPTHS LESS THAN OR EQUAL TO 0.25 INCH

TYPE OF HAZARD	BASIC MINIMUM EXTINGUISHER RATING	MAXIMUM DISTANCE OF TRAVEL TO EXTINGUISHERS (feet)
Light (Low)	5-B 10-B	30 50
Ordinary (Moderate)	10-B 20-B	30 50
Extra (High)	40-B 80-B	30 50

For SI: 1 inch = 25.4 mm, 1 foot = 304.8 mm.

Note: For requirements on water-soluble flammable liquids and alternative sizing criteria, see Section 5.5 of NFPA 10.

Unlike sodium chloride powder extinguishers, the graphite powder fire extinguishers can be used on very hot burning metal fires such as lithium, but the powder will not stick to and extinguish flowing or vertical lithium fires. The graphite powder acts as a heat sink as well as smothering the metal fire. See the commentary to Section 5906.5.7 of the IFC for a discussion of extinguishing flammable solid fires.

[F] 906.4 Cooking grease fires. Fire extinguishers provided for the protection of cooking grease fires shall be of an *approved* type compatible with the automatic fire-extinguishing system agent and in accordance with Section 904.12.5 of the *International Fire Code*.

❖ The combination of high-efficiency cooking appliances and hotter burning cooking media creates a potentially severe fire hazard. Although commercial cooking systems must have an approved exhaust hood and be protected by an approved automatic fire-extinguishing system, a manual means of extinguishment is desirable to attack a fire in its incipient stage.

As indicated in Section 904.12.5, a Class K-rated portable fire extinguisher must be located within 30 feet (9144 mm) of travel distance of commercial-type cooking equipment. Class K-rated extinguishers have been specifically tested on commercial cooking appliances using vegetable or animal oils or fats. These portable fire extinguishers are usually of sodium bicarbonate or potassium bicarbonate dry-chemical type.

[F] 906.5 Conspicuous location. Portable fire extinguishers shall be located in conspicuous locations where they will be readily accessible and immediately available for use. These locations shall be along normal paths of travel, unless the fire code official determines that the hazard posed indicates the need for placement away from normal paths of travel.

❖ Fire extinguishers must be located in readily accessible locations along normal egress paths. This increases the occupants, familiarity with the location of the fire extinguishers. When considering location, the most frequent occupants should be considered. These are the occupants who would become most familiar with the fire-extinguisher placement. For most buildings, it is the employees who are most familiar with their surroundings; therefore, a good understanding of employee operations is important for proper extinguisher placement.

[F] 906.6 Unobstructed and unobscured. Portable fire extinguishers shall not be obstructed or obscured from view. In rooms or areas in which visual obstruction cannot be completely avoided, means shall be provided to indicate the locations of extinguishers.

❖ Portable fire extinguishers must be located where they are readily visible at all times. If visual obstruction cannot be avoided, the location of the extinguishers must be marked by an approved means of identification. This could include additional signage, lights, arrows or other means approved by the fire code official. Unobstructed does not necessarily mean visible from all angles within the space. Often, columns or furnishings may obscure the extinguisher from one direction or another. These are not by themselves obstructions. The intent is that the extinguisher is not hidden but rather can be readily found. If the extinguisher is placed in the wall behind a door, it is clearly obstructed since it cannot be easily viewed. An extinguisher on a wall that is visible from most of the space would be considered unobstructed.

[F] 906.7 Hangers and brackets. Hand-held portable fire extinguishers, not housed in cabinets, shall be installed on the hangers or brackets supplied. Hangers or brackets shall be securely anchored to the mounting surface in accordance with the manufacturer's installation instructions.

❖ Portable fire extinguishers not housed in cabinets are usually mounted on walls or columns using securely fastened hangers. Brackets must be used where the fire extinguishers need to be protected from impact or other potential physical damage.

[F] 906.8 Cabinets. Cabinets used to house portable fire extinguishers shall not be locked.

Exceptions:

1. Where portable fire extinguishers subject to malicious use or damage are provided with a means of ready access.
2. In Group I-3 occupancies and in mental health areas in Group I-2 occupancies, access to portable fire extinguishers shall be permitted to be locked or to be located in staff locations provided the staff has keys.

❖ Cabinets housing fire extinguishers must not be locked in order to provide quick access in an emergency. Exception 1, however, allows the cabinets to be locked in occupancies where vandalism, theft or other malicious behavior is possible. Exception 2 also permits cabinets housing fire extinguishers to be locked or to be located in staff locations in Group I-3 occupancies and mental health areas in Group I-2 occupancies. Occupants in Group I-3 areas of jails, prisons or similar restrained occupancies should not have access to fire extinguishers because they could possibly be used as a weapon or be subject to vandalism. Staff adequately trained in the use of fire extinguishers are assumed to have ready access to the keys for the cabinets at all times.

[F] 906.9 Extinguisher installation. The installation of portable fire extinguishers shall be in accordance with Sections 906.9.1 through 906.9.3.

❖ This section introduces the installation criteria for portable fire extinguishers based on the weight of the unit.

[F] 906.9.1 Extinguishers weighing 40 pounds or less. Portable fire extinguishers having a gross weight not exceeding 40 pounds (18 kg) shall be installed so that their tops are not more than 5 feet (1524 mm) above the floor.

❖ Because of the varying height and physical strength levels of persons who might be called on to operate a portable fire extinguisher, the mounting height of the extinguisher must be commensurate with its weight so that it may be easily retrieved by anyone from its mounting location and placed into use.

[F] 906.9.2 Extinguishers weighing more than 40 pounds. Hand-held portable fire extinguishers having a gross weight exceeding 40 pounds (18 kg) shall be installed so that their tops are not more than 3.5 feet (1067 mm) above the floor.

❖ See the commentary to Section 906.9.1.

[F] 906.9.3 Floor clearance. The clearance between the floor and the bottom of installed hand-held portable fire extinguishers shall be not less than 4 inches (102 mm).

❖ The clearance between the floor and the bottom of installed hand-held extinguishers must not be less than 4 inches (102 mm) to facilitate cleaning beneath the unit and reduce the likelihood of the extinguisher becoming dislodged during cleaning operations (floor mopping, sweeping, etc.).

[F] 906.10 Wheeled units. Wheeled fire extinguishers shall be conspicuously located in a designated location.

❖ Wheeled fire extinguishers consist of a large-capacity (up to several hundred pounds of agent) fire extinguisher assembly (either stored-pressure or pressure-transfer type) equipped with a carriage and wheels and discharge hose. They are constructed so that one able-bodied person could move the unit to the fire area and begin extinguishment unassisted. Wheeled fire extinguishers are capable of delivering greater flow rates and stream range for various extinguishing agents than hand-held portable fire extinguishers. Wheeled fire extinguishers are generally more effective in high-hazard areas and, as with any extinguisher, must be readily available and stored in an approved location. The wheeled fire extinguisher should be located a safe distance from the hazard area so that it will not become involved in the fire or access to it compromised by a fire. These units are typically found at airport fueling ramps, refineries, bulk plants and similar locations where high-challenge fires may be encountered. The extinguishing agents available in wheeled units include carbon dioxide, dry chemical, dry powder and foam.

SECTION 907
FIRE ALARM AND DETECTION SYSTEMS

[F] 907.1 General. This section covers the application, installation, performance and maintenance of fire alarm systems and their components.

❖ Fire alarm systems in new buildings, which typically include manual fire alarm systems and automatic smoke detection systems, must be installed in accordance with Section 907 and NFPA 72. Fire alarm systems in existing buildings are regulated by Chapter 11 of the IFC.

Manual fire alarm systems are installed in buildings to limit fire casualties and property losses. Fire alarm systems do this by promptly notifying the occupants of the building of an emergency, which increases the time available for evacuation. Similarly, when fire alarm systems are supervised, the fire department will be promptly notified and its response time relative to the onset of the fire will be reduced.

Automatic smoke detection systems are required under certain conditions to increase the likelihood that a fire is detected and occupants are given an early warning. The detection system is a system of devices and associated hardware that activates the alarm system. The automatic detecting devices are to be smoke detectors, unless a condition exists that calls for the use of a different type of detector.

[F] 907.1.1 Construction documents. *Construction documents* for fire alarm systems shall be of sufficient clarity to indicate the location, nature and extent of the work proposed and show in detail that it will conform to the provisions of this code, the *International Fire Code* and relevant laws, ordinances, rules and regulations, as determined by the fire code official.

❖ Construction documents for fire alarm systems must be submitted for review to determine compliance with the code, the IFC and NFPA 72. All of the information required by this section may not be available during the design stage and initial permit process. Later submission of more detailed shop drawings may be required in accordance with Section 907.1.2. These provisions are intended to reflect the minimum scope of information needed to determine code compliance. When the work can be briefly described on the application form, the fire code official may utilize judgment in determining the need for more detailed documents.

[F] 907.1.2 Fire alarm shop drawings. Shop drawings for fire alarm systems shall be submitted for review and approval prior to system installation, and shall include, but not be limited to, all of the following where applicable to the system being installed:

1. A floor plan that indicates the use of all rooms.
2. Locations of alarm-initiating devices.
3. Locations of alarm notification appliances, including candela ratings for visible alarm notification appliances.
4. Design minimum audibility level for occupant notification.
5. Location of fire alarm control unit, transponders and notification power supplies.
6. Annunciators.
7. Power connection.
8. Battery calculations.
9. Conductor type and sizes.
10. Voltage drop calculations.
11. Manufacturers' data sheets indicating model numbers and listing information for equipment, devices and materials.

12. Details of ceiling height and construction.
13. The interface of fire safety control functions.
14. Classification of the supervising station.

❖ Since the fire protection contractor(s) may not have been selected at the time a permit is issued for construction of a building, detailed shop drawings for fire alarm systems may not be available. Because they provide the information necessary to determine code compliance, as specified in this section, they must be submitted and approved by the fire code official before the contractor can begin installing the system.

[F] 907.1.3 Equipment. Systems and components shall be *listed* and *approved* for the purpose for which they are installed.

❖ The components of the fire alarm system must be approved for use in the planned system. NFPA 72 requires all devices, combinations of devices, appliances and equipment to be labeled for their proposed use. The testing agency will test the components for use in various types of systems and stipulate the use of the component on the label. Evidence of listing and labeling of the system components must be submitted with the shop drawings. In some instances, the entire system may be labeled.

At least one major testing agency, Underwriters Laboratories, Inc. (UL), has a program in which alarm installation and service companies are issued a certificate and become listed by the agency as being qualified to design, install and maintain local, auxiliary, remote station or proprietary fire alarm systems. The listed companies may then issue a certificate showing that the system is in compliance with Section 907. Terms of the company certification by UL include the company being responsible for keeping accurate system documentation, including as-built record drawings, acceptance test records and complete maintenance records on a given system. The company is also responsible for the required periodic inspection and testing of the system under contract with the owner. A similar program has been available for many years for central station alarm service, whereas the UL program is relatively new to the industry. Even though this company and system listing program is not required by the code or NFPA 72, it can be a valuable tool for the fire code official in determining compliance with the referenced standard.

Another issue that must be considered is the compatibility of the system components as required by NFPA 72. The labeling of system components discussed above should include any compatibility restrictions for components. Compatibility is primarily an issue of the ability of smoke detectors and fire alarm control panels (FACPs) to function properly when interconnected and affects the two-wire type of smoke detectors, which obtain their operating power over the same pair of wires used to transmit signals to the FACP (the control unit initiating device circuits). Laboratories will test for component compatibility either by actual testing or by reviewing the circuit parameters of both the detector and the FACP. Generally, if both the two-wire detector and the FACP are of the same brand, there should not be a compatibility problem. Nevertheless, the fire code official must be satisfied that the components are listed as being compatible. Failure to comply with the compatibility requirements of NFPA 72 can lead to system malfunction or failure at a critical time.

[F] 907.2 Where required—new buildings and structures. An *approved* fire alarm system installed in accordance with the provisions of this code and NFPA 72 shall be provided in new buildings and structures in accordance with Sections 907.2.1 through 907.2.23 and provide occupant notification in accordance with Section 907.5, unless other requirements are provided by another section of this code.

Not fewer than one manual fire alarm box shall be provided in an *approved* location to initiate a fire alarm signal for fire alarm systems employing automatic fire detectors or waterflow detection devices. Where other sections of this code allow elimination of fire alarm boxes due to sprinklers, a single fire alarm box shall be installed.

Exceptions:

1. The manual fire alarm box is not required for fire alarm systems dedicated to elevator recall control and supervisory service.
2. The manual fire alarm box is not required for Group R-2 occupancies unless required by the fire code official to provide a means for fire watch personnel to initiate an alarm during a sprinkler system impairment event. Where provided, the manual fire alarm box shall not be located in an area that is accessible to the public.

❖ This section specifies the occupancies or conditions in new buildings or structures that require some form of fire alarm system which is either a manual fire alarm system (manual fire alarm boxes) or an automatic smoke detection system. These systems must, upon activation, provide occupant notification throughout the area protected by the system unless other alternative provisions are allowed by this section.

Manual fire alarm systems must be installed in certain occupancies depending on the number of occupants, capabilities of the occupants and height of the building. An automatic smoke detection system must be installed in those occupancies and conditions where the need to detect the fire is essential to evacuation or protection of the occupants. The requirements for automatic smoke detection are generally based on the evacuation needs of the occupants and whether the occupancy includes sleeping accommodations.

Fire alarm systems must be installed in accordance with the code and NFPA 72. NFPA 72 identifies the minimum performance, location, mounting, testing and maintenance requirements for fire alarm systems. Smoke detectors must be used, except when ambient conditions would prohibit their use. In that case other detection methods may be used. The manufacturer's literature will identify the limitations on the use of smoke detectors, including environmental conditions such as humidity, temperature and airflow.

Only certain occupancies are required to have either a manual fire alarm or automatic fire detection system installed (see Commentary Figure 907.2). The need for either system is generally determined by the number of occupants, the height of the building or the ability of the occupants for self-preservation.

Note that generally the fire alarm requirements are based on occupancy and not on fire area. Commentary Figure 907.2 contains the conditions that require when either system must be installed in a building. The extent that an alarm system must be installed in a building once it has been determined that such a system is required is based on a several factors. One, if it is the only occupancy in the building, then it would be required throughout the building. Two, if the building is a mixed occupancy, it can either be separated or nonseparated. If the occupancy is separated in accordance with Section 508.4, then the alarm system is only required within that separated portion of the building. If the building is considered a nonseparated, mixed occupancy building, then Section 508.3.1 states that the code apply to each portion of the building based on the occupancy classification of that space and that the most restrictive provisions of Chapter 9 shall apply to the building or portion thereof in which the nonseparated occupancies are located. Therefore, if you had a Group A occupancy in a nonseparated mixed occupancy (containing other occupancies such as Groups B and M) where the Group A occupancy exceeds an occupant load of 300, then the entire nonseparated mixed occupancy would require the alarm system. Note that Section 508.3.1 focuses on each space to determine occupancy and requirements. Once the occupant load is determined, then any requirements, such as fire alarms, would be required throughout.

The code does not address whether or not a nonseparated mixed occupancy has a completely independent means of egress such as in a strip mall. Additionally, in a building containing primarily Group A occupancies, the code does not clearly address whether such occupancies within a building should be looked at as an aggregate or individual space. This issue has been clarified in the 2012 edition of the code through the use of the fire area concept for Group A occupancies, but only for the basic manual fire alarm requirements in Section 907.2.1. The emergency voice/alarm communication requirements in Section 907.2.1.1 still simply provide a criteria of 1,000 or more occupants in a Group A occupancy. A building with multiple Group A occupancies without using the concept of separation of egress paths would need to be reviewed in aggregate. Fire area separation could not be used to provide separation of occupancies in this case.

Commentary Figure 907.2 contains the threshold requirements for when a manual fire alarm system or an automatic fire detection system is required based on the occupancy group. It is important to remember that although the requirement for manual pull stations may not apply (e.g., sprinklered buildings), alarm and occupant notification may still be required. Sections 907.2.11 through 907.2.23 contain additional requirements for fire alarm systems depending on special occupancy conditions such as atriums, high-rise buildings or covered mall buildings.

The single manual fire alarm box required by this section is needed to provide a means of manually activating a fire alarm system that only contains automatic devices such as sprinkler waterflow switches or smoke detectors. Its primary use is for alarm system maintenance technicians to be able to manually activate the fire alarm system in the event of a fire during the time the system or portions of the system is down for maintenance. Note that this requirement is not subject to any of the exceptions in Sections 907.2.1 through 907.2.23 that might waive the need for manual fire alarm boxes in certain buildings.

Exception 1 recognizes the specialized nature of fire alarm systems installed only for emergency elevator control and supervision.

Exception 2 waives the single manual fire alarm box but gives the fire code official authority to require it in sprinklered buildings for use by fire watch personnel or sprinkler maintenance personnel to be able to manually activate the fire alarm system in the event of a fire during the time the sprinkler system is down for maintenance.

[F] 907.2.1 Group A. A manual fire alarm system that activates the occupant notification system in accordance with Section 907.5 shall be installed in Group A occupancies where the occupant load due to the assembly occupancy is 300 or more. Group A occupancies not separated from one another in accordance with Section 707.3.10 shall be considered as a single occupancy for the purposes of applying this section. Portions of Group E occupancies occupied for assembly purposes shall be provided with a fire alarm system as required for the Group E occupancy.

> **Exception:** Manual fire alarm boxes are not required where the building is equipped throughout with an *automatic sprinkler system* installed in accordance with Section 903.3.1.1 and the occupant notification appliances will activate throughout the notification zones upon sprinkler water flow.

❖ Group A occupancies are typically occupied by a significant number of people who are not completely familiar with their surroundings. The provisions of this section address three separate situations regarding the application of the alarm requirements for Group A occupancies. The three situations addressed by the provisions are: (1) where an assembly occupancy and another occupancy are involved; (2) where multiple assembly areas exist in a building; and (3) where the assembly use occurs in and is a part of a Group E occupancy.

In situations where an assembly area and another occupancy are involved, the code specifies that it is the occupant load "due to the assembly occupancy" that would need to be 300 or more before the manual alarm system is required. For example, if the building is constructed with an assembly occupancy, such as a restaurant, with an occupant load of 250 and an adjacent office area with an occupant load of 100, the assembly space would not require an alarm system because the occupant load "due to the assembly occupancy" is less than 300. This is really simply a clarification of the way the provisions were intended to be applied. This would be the intended way to apply the provision whether the building was constructed using the accessory-, separated- or nonseparated-occupancy requirements of Chapter 5.

MANUAL FIRE ALARM SYSTEM	
Occupancy Group(s)	**Threshold**
Assembly (A-1, A-2, A-3, A-4, A-5)	All with an occupant load of > 300 (907.2.1)
Business (B)	Total Group B occupant load of > 500; or, > 100 above/below level of exit discharge; or, in Group B fire areas containing an ambulatory care facility. (907.2.2)
Educational (E)	> 50 occupants (several exceptions for manual fire alarm box placement) (907.2.3)
Factory (F-1, F-2)	> 2 stories with occupant load of > 500 above/below lowest level of exit discharge (exception for sprinklers) (907.2.4)
High hazard (H)	Group H-5 and in occupancies for manufacture of organic coatings. (907.2.5)
Institutional (I-1, I-2, I-3, I-4)	All (exceptions for I-1 and I-2 manual fire alarm box placement and private mode signaling) (907.2.6)
Mercantile (M)	Total Group M occupant load of > 500; or, occupant load of >100 above/below level of exit discharge (907.2.7)
Hotels (R-1)	All (exceptions for < 2 stories with sleeping units having exit directly to exterior; sprinklers) (907.2.8.1)
Multifamily (R-2)	If units > 3 stories above lowest level of exit discharge; or, > 1 story below highest level of exit discharge; or, > 16 units (exceptions for < 2 stories with sleeping units having exit directly to exterior; sprinklers); open-ended corridors/no corridor (907.2.9.1)
Residential care/assisted living (R-4)	All (exceptions for sprinklers, manual fire alarm boxes at staff locations, direct exit to exterior < 2 stories) (907.2.10.1)
AUTOMATIC SMOKE DETECTION SYSTEM	
Business (B) Ambulatory care facilities	Facility, plus public use areas outside of it including public corridors and elevator lobbies (exception for sprinklers) (907.2.2.1)
High hazard (H)	Highly toxic gases, organic peroxides, oxidizers (907.2.5)
Institutional (I-1, I-2, I-3)	All, in specific areas by occupancy (907.2.6.1, 907.2.6.2, 907.2.6.3.3)
Hotels (R-1)	All, in interior corridors (exception for buildings without interior corridors and with sleeping units having exit directly to exterior) (907.2.8.2)
Residential care/assisted living (R-4)	All, in corridors, waiting areas open to corridors, nonsleeping area habitable spaces and kitchens (exceptions for sprinklers and sleeping units having exit directly to exterior) (907.2.10.2)
College and university buildings (R-2 dormitories)	Common spaces; laundry, mechanical and storage rooms; interior corridors (exception for no interior corridors and each room has direct exit access or exit).

Figure 907.2
SUMMARY OF MANUAL FIRE ALARM AND AUTOMATIC SMOKE DETECTION SYSTEM THRESHOLDS

In buildings that contain multiple assembly areas, the second portion of the code text requires that the aggregate occupant load of the assembly areas is used unless the spaces are separated as required for fire areas in Section 707.3.10. Consider two examples to address this portion of the requirements. In a multi-theater complex, the auditoriums are generally not separated from each other by the 2-hour fire-resistance rating that Table 707.3.10 would require; therefore, the aggregate occupant load of all of the assembly spaces would be combined to determine if the occupant load was 300 or more. If it was, then the manual alarm would be required in all of the assembly spaces. As another example, consider a strip mall shopping center with a restaurant at one end of the building with an occupant load of 200 and a different restaurant with an occupant load of 150 at the other end of the building. Even though these are two completely separate establishments and have an amount of retail occupancy between them, the occupant load of the assembly areas does exceed 300. Therefore, unless a 2-hour fire-resistance rated separation complying with Section 707.3.10 is provided somewhere between the two restaurants to separate them into different fire areas, a manual fire alarm would be required in the Group A occupancies. If a complying separation is provided at some point in the building, then each assembly space can be reviewed independently and would not require the installation of the alarm system. Be aware that the separation of assembly spaces or the need to aggregate the occupant loads from them could occur not only on the same floor within a building, but also to assembly uses located on different stories.

The exception allows the omission of manual fire alarm boxes in buildings equipped throughout with an automatic sprinkler system if activation of the sprinkler system will activate the building evacuation alarms associated with the manual fire alarm system.

This section also permits assembly type areas in Group E occupancies to comply with Section 907.2.3 instead of the requirements of this section. A typical high school, for example, contains many areas used for assembly purposes such as a gymnasium, cafeteria, auditorium or library; however, they all exist to serve as an educational facility as their main function. The exception does not eliminate the fire alarm system and occupant notification system, but rather permits them to be initiated automatically by the sprinkler waterflow switch(es) instead of by the manual fire alarm boxes. It also reduces the possibility of mischievous or malicious false alarms being turned on by manual fire alarm boxes in venues where large numbers of people congregate.

[F] 907.2.1.1 System initiation in Group A occupancies with an occupant load of 1,000 or more. Activation of the fire alarm in Group A occupancies with an *occupant load* of 1,000 or more shall initiate a signal using an emergency voice/alarm communications system in accordance with Section 907.5.2.2.

> **Exception:** Where *approved*, the prerecorded announcement is allowed to be manually deactivated for a period of time, not to exceed 3 minutes, for the sole purpose of allowing a live voice announcement from an *approved, constantly attended location*.

❖ In order to afford authorized personnel the ability to selectively evacuate or manage occupant relocation in large assembly venues, this section requires the fire alarm system to operate through an emergency voice/alarm communications system. The exception allows the automatic alarm signals to be overridden for live voice instructions if the live voice instructions do not exceed 3 minutes. The location from which the live voice announcement originates must be constantly attended and approved by the fire code official (see also commentary, Section 907.5.2.2). In terms of the applicability of this section, it is not as specific as Section 907.2.1. More specifically, the concept of fire areas does not apply. Credit is not given to reduce the occupant load through the use of the fire area concept (see commentary, Section 907.2).

[F] 907.2.1.2 Emergency voice/alarm communication captions. Stadiums, arenas and grandstands required to caption audible public announcements shall be in accordance with Section 907.5.2.2.4.

❖ A 2008 U.S. Federal Court case prompted a change to the code and the IFC. The court ruled that persons with hearing impairments who attend events at stadiums, grandstands and arenas require a means of equivalent communications in lieu of the public address system. Providing occupant notification in these structures is challenging because of the building area and the number and diversity of occupants. Provisions were added in the code to require captioned messages in these buildings and grandstands when public address (PA) systems are prescribed by the accessibility requirements.

Section 1108.2.7.3 sets forth requirements for audible PA systems in stadiums, arenas and grandstands. It requires that equivalent text information be provided to the audience and that the delivery time for these messages be the same as those broadcast from the PA system. The requirements apply to prerecorded and real-time messages. Section 1108.2.7.3 also requires captioning of messages in stadiums, arenas and grandstands that have more than 15,000 fixed seats.

Because messages being broadcast can include instructions to building or site occupants explaining the actions they need to take in the event of an emergency, the requirements of NFPA 72 are applicable for captioning systems. Such a system falls within the scope of Chapter 24 of NFPA 72 entitled "Emergency Communication Systems." NFPA 72 defines an emergency communications system (ECS) as a system designed for life safety that indicates the existence of an emergency and communicates the appropriate response and action. The ECS is required to be classified as either a one- or two-way path system. It can be within a building or over a wide area or can be targeted to a particular group of recipients. The messages that will be broadcast are based on an emergency response plan developed during a risk analysis by the project stakeholders and is approved by the fire code official.

The NFPA 72 ECS requirements are based on in-building or wide-area occupant notification. Wide-area systems could include the entire area of a jurisdiction. For a stadium, arena or grandstand captioning, NFPA 72 defines these as mass notification systems (MNS). In the context of the NFPA 72 requirements, this particular code change requires a one-way MNS where instructions are broadcasted by personnel authorized to distribute messages. This could include fire fighters during a fire event; also the system could be used by law enforcement officers during a domestic terrorism incident or a weather event like a tornado warning.

The design of the compliant MNS Chapter 24 of NFPA 72 is not prescriptive—an MNS is a performance-based design. Accordingly, fire code officials should require their design to be sealed by a registered professional engineer. Section 24.7 of the standard requires the preparation of a risk analysis based on the nature and anticipated risks of the facility. The risk analysis is part of the design brief, which will serve as the basis of the system design and is a required design document. NFPA 72 requires the following elements included in the risk analysis:

- The number of persons within the building, area, space, campus or region;
- The character of the occupancy, such as unique hazards and the rate at which the hazard can develop;
- The anticipated threats, including natural, technological and intentional events;
- The reliability and performance of the MNS;
- Security of the MNS and its components;
- How the building or staff implement the risk analysis, including the use of the MNS and
- How emergency services, such as the fire service and law enforcement agencies can employ the MNS.

In a stadium or arena, the captioning system is required to be a component of the emergency voice/communications alarm system (EV/ACS). Such a system is required by Section 907.2.1.1 in Group A occupancies with an occupant load of 1,000 or more.

The requirement in Section 907.5.2.2.4 specifies the captioning system would be connected to the EV/ACS. The fire alarm control unit will require a listed interface unit capable of displaying text messages. Textual visible appliances are allowed by NFPA 72 when used in conjunction with audible, visual or both types of notification appliances. In the public mode, textual visible appliances are located to ensure readability by the building occupants. Such a system can display messages using televisions or light emitting diode (LED) marquee signs.

The design concept of MNS is relatively new in the design community. Captioning systems might utilize components that are not listed for fire alarm service so the design will be required to comply with Chapter 24 of NFPA 72 for textual visible notification appliances. Emergency textual messages take precedence over any nonemergency text messages. Under NFPA 72 these devices require a primary and secondary power supply. If the devices are not monitored for integrity or loss of communications by an autonomous control or a fire alarm control unit, the appliance must clearly display its status. The size of characters displayed must comply with the requirements in NFPA 72. The NFPA 72 size, character and font requirements are based on the location of the display in relation to the height and distance from the persons viewing it.

[F] 907.2.2 Group B. A manual fire alarm system shall be installed in Group B occupancies where one of the following conditions exists:

1. The combined Group B *occupant load* of all floors is 500 or more.
2. The Group B *occupant load* is more than 100 persons above or below the lowest *level of exit discharge*.
3. The *fire area* contains an ambulatory care facility.

Exception: Manual fire alarm boxes are not required where the building is equipped throughout with an *automatic sprinkler system* installed in accordance with Section 903.3.1.1 and the occupant notification appliances will activate throughout the notification zones upon sprinkler water flow.

❖ Group B occupancies generally involve individuals or groups of people in separate office areas. As a result, the occupants are not necessarily aware of what is going on in other parts of the building. Group B buildings with large occupant loads, even in single-story buildings, or where a substantial number of occupants are above or below the level of exit discharge, increase the difficulty of alerting the occupants of a fire. This is especially true in nonsprinklered buildings with given occupant load thresholds. Group B occupancies include a specific use called ambulatory care facilities which present a higher level of life hazard than the typical Group B occupancy. The fact that the care recipients of such facilities may be rendered incapable of self-preservation for limited periods of time makes the need for a fire alarm system critical. Section 907.2.2 requires a manual alarm system any time a fire area contains a ambulatory care facility. See the commentary to Section 202, definition of "Ambulatory care facility" and Section 907.2.2.1.

The exception does not eliminate the fire alarm

system, but rather permits it to be initiated automatically by the sprinkler waterflow switch(es) instead of by the manual fire alarm boxes.

[F] 907.2.2.1 Ambulatory care facilities. *Fire areas* containing ambulatory care facilities shall be provided with an electronically supervised automatic smoke detection system installed within the ambulatory care facility and in public use areas outside of tenant spaces, including public *corridors* and elevator lobbies.

Exception: Buildings equipped throughout with an *automatic sprinkler system* in accordance with Section 903.3.1.1, provided the occupant notification appliances will activate throughout the notification zones upon sprinkler waterflow.

❖ Years ago, few surgical procedures were performed outside of a hospital. Today, complex outpatient surgeries conducted outside of a hospital are commonplace. They are performed in facilities often called "day surgery centers" or "ambulatory surgical centers" because patients are able to walk in and walk out the same day. Procedures render care recipients temporarily incapable of self-preservation by application of nerve blocks, sedation or anesthesia; however, they do typically recover quickly.

The code identifies health care Group I occupancies as including a 24-hour stay. Without a 24-hour stay, these surgery centers were classified as Group B which allowed the care providers to render an unlimited number of people incapable of self-preservation with no more protection than a business office. Since these types of facilities contain distinctly different hazards to life safety than other Group B occupancies, they are now required to have a higher level of life safety and fire protection as evidenced by the requirements of this section as well as Section 903.2.2 and the construction provisions of the code.

This section more specifically states that any time a fire area contains an ambulatory care facility, the fire area should be provided with a supervised smoke detection system in the ambulatory care facility and in public use areas outside of tenant spaces. Therefore, in a medical office building, for example, the ambulatory care facility contained within would have a full coverage system. The other offices in the building would not require smoke detection in the individual tenant spaces, but instead in the public areas, such as lobby or lounge areas.

The exception does not eliminate the fire alarm system, but rather permits it to be initiated automatically by the sprinkler waterflow switch(es) instead of by the smoke detection system.

[F] 907.2.3 Group E. A manual fire alarm system that initiates the occupant notification signal utilizing an emergency voice/alarm communication system meeting the requirements of Section 907.5.2.2 and installed in accordance with Section 907.6 shall be installed in Group E occupancies. When *automatic sprinkler systems* or smoke detectors are installed, such systems or detectors shall be connected to the building fire alarm system.

Exceptions:

1. A manual fire alarm system is not required in Group E occupancies with an *occupant load* of 50 or less.
2. Emergency voice/alarm communication systems meeting the requirements of Section 907.5.2.2 and installed in accordance with Section 907.6 shall not be required in Group E occupancies with occupant loads of 100 or less, provided that activation of the manual fire alarm system initiates an *approved* occupant notification signal in accordance with Section 907.5.
3. Manual fire alarm boxes are not required in Group E occupancies where all of the following apply:
   - 3.1. Interior *corridors* are protected by smoke detectors.
   - 3.2. Auditoriums, cafeterias, gymnasiums and similar areas are protected by *heat detectors* or other *approved* detection devices.
   - 3.3. Shops and laboratories involving dusts or vapors are protected by *heat detectors* or other *approved* detection devices.
4. Manual fire alarm boxes shall not be required in Group E occupancies where all of the following apply:
   - 4.1. The building is equipped throughout with an *approved automatic sprinkler system* installed in accordance with Section 903.3.1.1.
   - 4.2. The emergency voice/alarm communication system will activate on sprinkler waterflow.
   - 4.3. Manual activation is provided from a normally occupied location.

❖ Section 404.2.3 of the IFC addresses the development and implementation of lockdown plans. These requirements were developed to ensure that the level of life safety inside of the building is not reduced or compromised during a lockdown. In order for a building to safely function in a lockdown condition, the code requires a means of communication between the established central location and each secured area. Section 404.3.3.1 of the IFC does not prescribe the means of communication, which could include the use of text messages to cell phones/mobile devices, e-mail messages or the use of preestablished audio or visual signals. The provisions in Section 404.2.3 of the IFC are not specific to Group E occupancies—they are applicable to all occupancies that develop and implement lockdown plans.

Because of concerns of school campus safety serving kindergarten through 12th grade students, specific requirements were put into the 2012 edition of the code and the IFC for enhanced communication between the school administrators, teachers and students when a lockdown plan is activated in Group E occupancies. As a result, emergency voice/alarm communication systems (EV/ACS) are prescribed in Group E occupancies. Previously, the code would have permitted the manual fire alarm system to use audible and visible alarm notification appliances and did not require the added capabilities that an EV/ACS provides.

Section 907.2.3 sets forth the requirements for automatic fire alarm and detection system requirements in Group E occupancies and prescribes the installation of an EV/ACS as opposed to a traditional horn/strobe occupant notification system.

Exception 1 exempts Group E occupancies from requiring a fire alarm system when the occupant load is less than 50. This would exempt small day care centers that serve children older than $2^1/_2$ years of age or a small Sunday school classroom at a place of religious worship.

Exception 2 provides relief for smaller schools. If the occupant load is 100 or less, notification is not required to be via an emergency voice/alarm communication system. A school with 100 occupants only has a couple classrooms of children. Communication is simplified and an emergency voice/alarm communication system is considered to be excessive.

Exception 3 exempts manual fire alarm boxes in interior corridors, laboratories, auditoriums, cafeterias, gymnasiums and similar spaces based on the installation of heat/smoke detectors. This is not an exception from the EV/ACS but simply an exemption of locations requiring manual fire alarm boxes. The applicability of Exception 2 is independent of whether an automatic sprinkler system is installed. If an automatic smoke detection system is installed, it must be connected to the building fire alarm system.

Exception 4 allows the omission of the manual fire alarm boxes in Group E occupancies equipped throughout with an automatic sprinkler system if the actuation of the sprinkler system will activate the EV/ ACS . See Section 903.2.3 for sprinkler requirements in Group E buildings.

[F] 907.2.4 Group F. A manual fire alarm system that activates the occupant notification system in accordance with Section 907.5 shall be installed in Group F occupancies where both of the following conditions exist:

1. The Group F occupancy is two or more *stories* in height.
2. The Group F occupancy has a combined *occupant load* of 500 or more above or below the lowest *level of exit discharge*.

Exception: Manual fire alarm boxes are not required where the building is equipped throughout with an *automatic sprinkler system* installed in accordance with Section 903.3.1.1 and the occupant notification appliances will activate throughout the notification zones upon sprinkler water flow.

❖ This section is intended to apply to large multistory manufacturing facilities. For this reason, a manual fire alarm system would be required only if the building were at least two stories in height and had 500 or more occupants above or below the level of exit discharge. An unlimited area, two-story Group F occupancy complying with Section 507.5 of the code would be indicative of an occupancy requiring a manual fire alarm system.

Buildings in compliance with Section 507.5 of the code, and large manufacturing facilities in general, however, must be fully sprinklered and would thus be eligible for the exception. The exception does not eliminate the fire alarm system but rather permits it to be initiated automatically by the sprinkler system waterflow switch(es) instead of by the manual fire alarm boxes.

[F] 907.2.5 Group H. A manual fire alarm system that activates the occupant notification system in accordance with Section 907.5 shall be installed in Group H-5 occupancies and in occupancies used for the manufacture of organic coatings. An automatic smoke detection system shall be installed for highly toxic gases, organic peroxides and oxidizers in accordance with Chapters 60, 62 and 63, respectively, of the *International Fire Code*.

❖ Because of the nature and potential quantity of hazardous materials in Group H-5 occupancies, a manual means of activating an occupant notification system is essential for the safety of the occupants. In accordance with Section 415.10.8, the activation of the alarm system must initiate a local alarm and transmit a signal to the emergency control station. The manual fire alarm system requirement for the building is in addition to the emergency alarm requirements in Section 415.11.3.5 (see also Section 908.2).

Occupancies involved in the manufacture of organic coatings present special hazardous conditions because of the unstable character of the materials, such as nitrocellulose. Good housekeeping and control of ignition sources is critical. Section 418 of the code and Chapter 29 of the IFC contain additional requirements for organic coating manufacturing processes.

This section also requires an automatic smoke detection system in certain occupancy conditions involving either highly toxic gases or organic peroxides and oxidizers. The need for the automatic smoke detection system may depend on the class of materials and additional levels of fire protection provided. This requirement also assumes the quantity of materials is in excess of the maximum allowable quantities shown in Tables 307.1(1) and 307.1(2).

[F] 907.2.6 Group I. A manual fire alarm system that activates the occupant notification system in accordance with Section 907.5 shall be installed in Group I occupancies. An automatic smoke detection system that activates the occupant notification system in accordance with Section 907.5 shall be provided in accordance with Sections 907.2.6.1, 907.2.6.2 and 907.2.6.3.3.

Exceptions:

1. Manual fire alarm boxes in sleeping units of Group I-1 and I-2 occupancies shall not be required at *exits* if located at all care providers' control stations or other constantly attended staff locations, provided such stations are visible and continuously accessible and that the distances of travel required in Section 907.4.2.1 are not exceeded.
2. Occupant notification systems are not required to be activated where private mode signaling installed in accordance with NFPA 72 is *approved* by the fire code official and staff evacuation responsibilities are included in the fire safety and evacuation plan required by Section 404 of the *International Fire Code*.

❖ Because the protection and possible evacuation of the occupants in Group I occupancies are most often dependent on the response by care providers, occupancies in Group I must be protected with a manual fire alarm system and in certain instances, as described in Sections 907.2.6.1, 907.2.6.2 and 907.2.6.3, an automatic smoke detection system. In Group I-1, smoke alarms are also required in accordance with Section 907.2.6.1.1.

It is not the intent of this section to require a smoke detection system throughout all Group I occupancies. Smoke detectors are only generally required in the corridors and in waiting rooms that are open to corridors, unless noted otherwise. IFC Committee Interpretation No. 36-03 makes it clear that the Group I provisions only require a manual fire alarm system with smoke detectors in selected areas. To reduce the potential for unwanted alarms, manual fire alarm boxes may be located at the care providers' control stations or another constantly attended location.

Exception 1 reduces the likelihood of accidental or malicious false alarm system activations by manual means by allowing the pull stations to be located in a more controlled area. It assumes the approved location is always accessible by care providers and within a distance of travel of 200 feet (60 960 mm).

Exception 2 allows the common practice in Group I occupancies of only notifying the care providers instead of all building occupants in the event of a fire, subject to the approval of the fire code official. In order to have confidence that the actions taken will be appropriate, the code also requires that the responsibilities of the staff be documented in the fire safety and evacuation plan for the facility. These requirements are found in Section 404 of the IFC. This will increase the likelihood for any staff training to be linked to the allowance of private mode signaling.

[F] 907.2.6.1 Group I-1. In Group I-1 occupancies, an automatic smoke detection system shall be installed in *corridors*, waiting areas open to *corridors* and *habitable spaces* other than *sleeping units* and kitchens. The system shall be activated in accordance with Section 907.5.

Exceptions:

1. For Group I-1 Condition 1 occupancies, smoke detection in *habitable spaces* is not required where the facility is equipped throughout with an *automatic sprinkler system* installed in accordance with Section 903.3.1.1.
2. Smoke detection is not required for exterior balconies.

❖ Occupancies in Group I-1 tend to be compartmentalized into small rooms so that a fire in one area of the building would not easily be noticed by occupants in another part of the building. Therefore, smoke detection is required in areas other than sleeping units and kitchens. Sleeping units are required by Section 907.2.6.1.1 to be equipped with single- and multiple-station smoke alarms in accordance with Section 907.2.11.

Since Group I-1 occupancies may not be supervised by care providers and to reduce the likelihood that a fire within a waiting area open to the corridor or the corridor itself could develop beyond the incipient stage, thereby jeopardizing the building egress, these areas must be equipped with automatic smoke detection.

Exception 1 allows smoke detectors to be eliminated from habitable spaces of Group I-1 Condition 1 occupancies if the building is equipped throughout with an NFPA 13 automatic sprinkler system. The sprinkler system should control any fire and perform occupant notification through actuation of the waterflow switch and subsequent activation of the building alarm notification appliances. A sprinkler system is required for all Group I occupancies in accordance with Section 903.2.6. It should be noted that Group I-1 Condition 1 is the lower risk Group I-1 occupancy where residents are able to evacuate without assistance. See the commentary to Section 308.3.

Exception 2 allows for omitting smoke detectors from exterior balconies for environmental reasons and does not require the installation of an alternative type of detector. The exterior balconies are assumed to be sufficiently open to the atmosphere to readily allow the dissipation of smoke and hot gases.

[F] 907.2.6.1.1 Smoke alarms. Single- and multiple-station smoke alarms shall be installed in accordance with Section 907.2.11.

❖ As with dwelling units or sleeping units in any occupancy, this section requires that single- and multiple-station smoke alarms be installed in accordance with Section 907.2.11. Section 907.2.11.2 deals specifically with the requirements for Group I-1.

[F] 907.2.6.2 Group I-2. An automatic smoke detection system shall be installed in *corridors* in Group I-2 Condition 1 facilities and spaces permitted to be open to the *corridors* by Section 407.2. The system shall be activated in accordance with Section 907.4. Group I-2 Condition 2 occupancies shall be equipped with an automatic smoke detection system as required in Section 407.

Exceptions:

1. Corridor smoke detection is not required in smoke compartments that contain sleeping units where such units are provided with smoke detectors that comply with UL 268. Such detectors shall provide a visual display on the corridor side of each sleeping unit and shall provide an audible and visual alarm at the care providers' station attending each unit.
2. Corridor smoke detection is not required in smoke compartments that contain sleeping units where sleeping unit doors are equipped with automatic door-closing devices with integral smoke detectors on the unit sides installed in accordance with their listing, provided that the integral detectors perform the required alerting function.

❖ Automatic smoke detection is required in areas permitted to be open to corridors in occupancies classified as Group I-2 Condition 2 and corridors in Group I-2 Condition 1 occupancies (e.g., nursing homes, long-term care facilities and detoxification facilities). In recognition of quick-response sprinkler technology and the fact that the sprinkler system is electronically supervised, and because the doors to care recipient's sleeping units are continuously supervised by care providers when in the open position, smoke detectors are not required for adequate fire safety in care recipient sleeping units.

In Group I-2 Condition 1 occupancies (nursing homes, long-term care facilities and detoxification facilities), some redundancy is appropriate because such facilities typically have less control over furnishings and personal items, thereby resulting in a less predictable and usually higher fire hazard load than Group I-2 Condition 2 occupancies (hospitals). Also, there is generally less care provider's supervision in these facilities than in other health care facilities and thus less control over care recipient smoking and other fire causes. Therefore, to provide additional protection against fires spreading from the room of origin, smoke detection is required in corridors of nursing homes, long-term care facilities and detoxification facilities.

Smoke detection is not required in corridors of Group I-2 Condition 2 occupancies except where otherwise specifically required in the code. Similarly, because areas open to the corridor very often are the room of fire origin, and such areas are no longer required by the code to be under visual supervision by care providers, some redundancy to protection by the sprinkler system is requested. Accordingly, all areas open to corridors must be protected by an automatic smoke detection system. This requirement provides an additional level of protection against sprinkler system failures or lapses in care provider supervision.

These requirements are not applicable to Group I-2 Condition 2 (hospitals). The scope of this section clearly indicates that its provisions are only applicable to detoxification facilities and nursing homes (Group I-2 Condition 1). Hospitals are noted as being subject to the provisions in Section 407.2. IFC Committee Interpretation No. 37-03 addresses this issue. Section 407.2 notes that smoke detection is only required for spaces open to corridors, such as waiting areas and mental health treatment areas where patients are not capable of self-preservation (see commentary, Section 407.2).

There are two exceptions to the requirement for an automatic fire detection system in corridors of nursing homes, long-term care facilities and detoxification facilities. Both exceptions provide an alternative method for redundant protection in care recipient sleeping units. For this reason, they provide either a backup to the notification of a fire or containment of fire in the room of origin.

Exception 1 requires smoke detectors in sleeping units that activate both a visual display on the corridor side of the care recipient sleeping unit and a visual and audible alarm at the care provider's station serving the room. Detectors complying with UL 268 are intended for open area protection and for connection to a normal power supply or as part of a fire alarm system. This exception, however, is specifically designed not to require the detectors to activate the building fire alarm system where approved care recipient sleeping unit smoke detectors are installed and where visual and audible alarms are provided. This is in response to the concern over unwanted alarms. The required alarm signals will not necessarily indicate to care providers that a fire emergency exists because the care provider call system may typically be used to identify numerous conditions within the room.

Exception 2 addresses the situation where smoke detectors are incorporated within automatic door-closing devices. The units are acceptable as long as the required alarm functions are still provided. Such units are usually listed as combination door closer and hold-open devices.

[F] 907.2.6.3 Group I-3 occupancies. Group I-3 occupancies shall be equipped with a manual fire alarm system and automatic smoke detection system installed for alerting staff.

❖ Because of the evacuation difficulties associated with Group I-3 occupancies and the dependence on adequate staff response, a manual fire alarm system and an automatic smoke detection system are required subject to the special occupancy conditions in Sections 907.2.6.3.1 through 907.2.6.3.3. This section recognizes that the evacuation of Group I-3 occupancies depends on an effective staff response. The requirements in Chapter 4 and specifically Section

403.8.3 of the IFC contains the requirements for an emergency plan, including employee training, staff availability, the need for occupants to notify staff and the need for the proper keys for unlocking doors for staff in Group I-3 occupancies.

[F] 907.2.6.3.1 System initiation. Actuation of an automatic fire-extinguishing system, *automatic sprinkler system*, a manual fire alarm box or a fire detector shall initiate an approved fire alarm signal that automatically notifies staff.

❖ This section specifies the systems that, upon activation, must initiate the required alarm signal immediately and automatically to the staff so that staff will respond in a timely manner.

[F] 907.2.6.3.2 Manual fire alarm boxes. Manual fire alarm boxes are not required to be located in accordance with Section 907.4.2 where the fire alarm boxes are provided at staff-attended locations having direct supervision over areas where manual fire alarm boxes have been omitted.

❖ Because of the potential for intentional false alarms and the resulting disruption to the facility, manual fire alarm boxes in Group I-3 occupancies may be either locked or made inaccessible to the occupants.

[F] 907.2.6.3.2.1 Manual fire alarm boxes in detainee areas. Manual fire alarm boxes are allowed to be locked in areas occupied by detainees, provided that staff members are present within the subject area and have keys readily available to operate the manual fire alarm boxes.

❖ The locking of manual fire alarm boxes is permitted only in areas where staff members are present and keys are readily available to them to unlock the boxes, or where the alarm boxes are located in a manned staff location that has direct supervision of the Group I-3 area.

[F] 907.2.6.3.3 Automatic smoke detection system. An automatic smoke detection system shall be installed throughout resident housing areas, including *sleeping units* and contiguous day rooms, group activity spaces and other common spaces normally accessible to residents.

Exceptions:

1. Other *approved* smoke detection arrangements providing equivalent protection, including, but not limited to, placing detectors in exhaust ducts from cells or behind protective guards *listed* for the purpose, are allowed when necessary to prevent damage or tampering.
2. *Sleeping units* in Use Conditions 2 and 3 as described in Section 308.
3. Smoke detectors are not required in *sleeping units* with four or fewer occupants in smoke compartments that are equipped throughout with an *automatic sprinkler system* installed in accordance with Section 903.3.1.1.

❖ Evacuation of Group I-3 facilities is impractical because of the need to maintain security. An automatic smoke detection system is therefore required to provide early warning of a fire.

As indicated in Exception 1, the installation of automatic smoke detectors must take into account the need to protect the detector from vandalism by residents. As a result, detectors may have to be located in return air ducts or be protected by a substantial physical barrier.

Since occupants in Use Condition 2 or 3 are not locked in their sleeping units, Exception 2 reduces the need for smoke detection.

Exception 3 allows smoke detectors to be omitted in sleeping units housing no more than four occupants on the basis that in a building that is protected throughout with an approved automatic sprinkler system, the system will provide both detection and suppression functions. Group I facilities are assumed to be fully sprinklered throughout in accordance with NFPA 13 as required by Section 903.2.6. The limitation of four occupants reduces the potential fuel load (mattresses, clothes, etc.) and the likelihood of involvement over an extended area.

[F] 907.2.7 Group M. A manual fire alarm system that activates the occupant notification system in accordance with Section 907.5 shall be installed in Group M occupancies where one of the following conditions exists:

1. The combined Group M *occupant load* of all floors is 500 or more persons.
2. The Group M *occupant load* is more than 100 persons above or below the lowest *level of exit discharge*.

Exceptions:

1. A manual fire alarm system is not required in *covered or open mall buildings* complying with Section 402.
2. Manual fire alarm boxes are not required where the building is equipped throughout with an *automatic sprinkler system* installed in accordance with Section 903.3.1.1 and the occupant notification appliances will automatically activate throughout the notification zones upon sprinkler water flow.

❖ Group M occupancies have the potential for large numbers of occupants who may not be familiar with their surroundings. The installation of a fire alarm system increases the ability to alert the occupants of a fire. Note that the occupant thresholds must be considered independently. If the total occupant load is 500 or more persons, a manual fire alarm system is required. If more than 100 persons are above or below the level of exit discharge, a manual fire alarm system is required.

This section also specifies that the manual fire alarm boxes must, upon activation, provide occupant notification throughout the Group M occupancy.

The extent of fire alarm application is based on the area in which the Group M occupancy is located. If the building is considered as a separated mixed occupancy, then the fire alarm system is only required in the individual occupancy in which the occupant load exceeds the threshold quantity. The rest of the building would not require a fire alarm system. This approach is noted in Section 508.4.1, which states that each separated space must comply with the code based upon the occupancy classification of that portion of the building. If the Group M occupancy was part of a non separated mixed use building then the alarm system would be required in the entire building in accordance with Section 508.3.1. The determination as to when such a system is required would be based solely upon the Group M occupant load.

Exception 1 recognizes the increased level of fixed automatic protection inherently required in covered mall buildings including an automatic sprinkler system, and, possibly, a smoke control system. Covered mall buildings are also required to contain an emergency voice communication system (see Section 907.2.20).

Exception 2 does not eliminate the fire alarm system, but rather allows it to be initiated automatically by sprinkler system waterflow switch(es) instead of by manual fire alarm boxes. Buildings with a fire area containing a Group M occupancy in excess of 12,000 square feet (1115 m^2) must be equipped with an automatic sprinkler system complying with Section 903.2.7.

[F] 907.2.7.1 Occupant notification. During times that the building is occupied, the initiation of a signal from a manual fire alarm box or from a waterflow switch shall not be required to activate the alarm notification appliances when an alarm signal is activated at a *constantly attended location* from which evacuation instructions shall be initiated over an emergency voice/alarm communication system installed in accordance with Section 907.5.2.2.

❖ Occupants in a mercantile occupancy may assume the alarm is a false alarm or act inappropriately and thus delay evacuation of the building. To prevent such a dangerous situation, the manual fire alarm system may be part of an EV/ACS. The signal is to be sent to a constantly attended location on site from which evacuation instructions can be given.

It should be noted that, although the alarm notification alternative allows for the manual use of an EV/ACS, the alternative does not remove the requirement for audible and visual notification devices.

[F] 907.2.8 Group R-1. Fire alarm systems and smoke alarms shall be installed in Group R-1 occupancies as required in Sections 907.2.8.1 through 907.2.8.3.

❖ Because residents of Group R-1 occupancies may be asleep and are usually transients who are unfamiliar with the building, and because such buildings contain numerous small rooms so that the occupants may not notice a fire in another part of the building, occupancies in Group R-1 must have a manual fire alarm system and an automatic smoke detection system installed throughout. Requirements for single- or multiple-station smoke alarms in sleeping units are contained in Section 907.2.11.1.

[F] 907.2.8.1 Manual fire alarm system. A manual fire alarm system that activates the occupant notification system in accordance with Section 907.5 shall be installed in Group R- 1 occupancies.

Exceptions:

1. A manual fire alarm system is not required in buildings not more than two *stories* in height where all individual *sleeping units* and contiguous *attic* and crawl spaces to those units are separated from each other and public or common areas by not less than 1-hour *fire partitions* and each individual *sleeping unit* has an *exit* directly to a *public way*, *egress court* or *yard*.
2. Manual fire alarm boxes are not required throughout the building where all of the following conditions are met:
   - 2.1. The building is equipped throughout with an *automatic sprinkler system* installed in accordance with Section 903.3.1.1 or 903.3.1.2.
   - 2.2. The notification appliances will activate upon sprinkler water flow.
   - 2.3. Not fewer than one manual fire alarm box is installed at an *approved* location.

❖ This section is specific to manual fire alarm systems and requires such systems in all Group R-1 occupancies, with two exceptions.

Exception 1 eliminates the requirement for a manual fire alarm system if the sleeping units have an exit discharging directly to a public way, exit court or yard. Even though the building may be two stories in height, the sleeping units on each floor must have access directly to an approved exit at grade level. The use of an exterior exit access balcony with exterior stairs serving the second floor does not constitute an exit directly at grade. The minimum 1-hour fire-resistance rating required for adequate separation of the sleeping units must be maintained.

Exception 2 does not omit the fire alarm system but rather permits it to be initiated automatically by sprinkler system waterflow switch(es) in lieu of manual fire alarm boxes. The sprinkler system must activate the occupant notification system and at least one manual fire alarm box shall be installed at an approved location. See the commentary to Section 907.2 for a discussion of the single manual fire alarm box.

The exceptions do not affect the independent provision in Section 907.2.11 for single- or multiple-station smoke alarms.

[F] 907.2.8.2 Automatic smoke detection system. An automatic smoke detection system that activates the occupant notification system in accordance with Section 907.5 shall be installed throughout all interior *corridors* serving *sleeping units*.

Exception: An automatic smoke detection system is not required in buildings that do not have interior *corridors* serving *sleeping units* and where each *sleeping unit* has a *means of egress* door opening directly to an *exit* or to an exterior *exit access* that leads directly to an *exit*.

❖ This section requires an automatic smoke detection system within interior corridors. Such systems make use of smoke detectors for alarm initiation in accordance with Section 907.2, with one exception.

The exception provides that automatic fire detectors are not required in motels and hotels that do not have interior corridors and in which sleeping units have a door opening directly to an exterior exit access that leads directly to the exits. The intent of the exception is that the exit access from the sleeping unit door be exterior and not require reentering the building prior to entering the exit. Since the exit access is outside, the need for detectors other than the smoke alarms required by Section 907.2.8.3 in sleeping units is greatly reduced.

[F] 907.2.8.3 Smoke alarms. Single- and multiple-station smoke alarms shall be installed in accordance with Section 907.2.11.

❖ The actual requirements for single- and multiple-station smoke alarms are located in Section 907.2.11. That section requires that the single- and multiple-station smoke alarms within sleeping units be connected to the emergency electrical system. Automatic activation of the fire alarm system is avoided to reduce unnecessary alarms within such buildings.

[F] 907.2.9 Group R-2. Fire alarm systems and smoke alarms shall be installed in Group R-2 occupancies as required in Sections 907.2.9.1 through 907.2.9.3.

❖ This section introduces the fire alarm system and smoke alarm requirements for Group R-2 occupancies. This includes Group R-2 occupancies in general and also Group R-2 college and university buildings.

[F] 907.2.9.1 Manual fire alarm system. A manual fire alarm system that activates the occupant notification system in accordance with Section 907.5 shall be installed in Group R-2 occupancies where any of the following conditions apply:

1. Any *dwelling unit* or *sleeping unit* is located three or more *stories* above the lowest *level of exit discharge*.
2. Any *dwelling unit* or *sleeping unit* is located more than one *story* below the highest *level of exit discharge* of *exits* serving the *dwelling unit* or *sleeping unit*.
3. The building contains more than 16 *dwelling units* or *sleeping units*.

Exceptions:

1. A fire alarm system is not required in buildings not more than two *stories* in height where all *dwelling units* or *sleeping units* and contiguous *attic* and crawl spaces are separated from each other and public or common areas by not less than 1-hour *fire partitions* and each *dwelling unit* or *sleeping unit* has an *exit* directly to a *public way*, *egress court* or *yard*.
2. Manual fire alarm boxes are not required where the building is equipped throughout with an *automatic sprinkler system* installed in accordance with Section 903.3.1.1 or 903.3.1.2 and the occupant notification appliances will automatically activate throughout the notification zones upon a sprinkler water flow.
3. A fire alarm system is not required in buildings that do not have interior *corridors* serving *dwelling units* and are protected by an *approved automatic sprinkler system* installed in accordance with Section 903.3.1.1 or 903.3.1.2, provided that *dwelling units* either have a *means of egress* door opening directly to an exterior *exit access* that leads directly to the *exits* or are served by open-ended *corridors* designed in accordance with Section 1027.6, Exception 3.

❖ The occupants of Group R-2 occupancies are not considered to be as transient as those of Group R-1, which increases the probability that residents can more readily notify each other of a fire. Therefore, Group R-1 occupancies must have a manual fire alarm system with audible and visual notification appliances subject to the exceptions in Section 907.2.8.1, whereas Group R-2 occupancies are required to have only a manual fire alarm system as stipulated in one of the three listed conditions. The threshold conditions are meant to be applied independently of each other.

Exception 1 eliminates the requirement for a manual fire alarm system if the sleeping units have an exit discharging directly to a public way, exit court or yard. Even though the building may be two stories in height, the sleeping units on each floor must have access directly to an approved exit at grade level. The use of an exterior exit access balcony with exterior stairs serving the second floor does not constitute an exit directly at grade. The minimum 1-hour fire-resistance rating required for adequate separation of the sleeping units must be maintained.

Exception 2 does not omit the fire alarm system but rather permits it to be initiated automatically by sprinkler system waterflow switch(es) in lieu of manual fire alarm boxes. The sprinkler system must activate the occupant notification system. This exception does not affect the independent provisions of Section 907.2.11.

Exception 3 allows the omission of a fire alarm system in fully sprinklered buildings (NFPA 13 or 13R) with no interior corridors and that exit directly to an exterior exit access or have open-ended corridors. The important thing to note is that the sprinkler system is not required to activate alarm notification appliances since a fire alarm system would not be required. Only the sprinkler alarms required by Section 903.4 would be required.

[F] 907.2.9.2 Smoke alarms. Single- and multiple-station smoke alarms shall be installed in accordance with Section 907.2.11.

❖ The actual requirements for single- and multiple-station smoke alarms are located in Section 907.2.11. That section requires that the single- and multiple-station smoke alarms within sleeping units be connected to the emergency electrical system. Automatic activation of the fire alarm system is avoided to reduce unnecessary alarms within such buildings.

[F] 907.2.9.3 Group R-2 college and university buildings. An automatic smoke detection system that activates the occupant notification system in accordance with Section 907.5 shall be installed in Group R-2 occupancies operated by a college or university for student or staff housing in all of the following locations:

1. Common spaces outside of *dwelling units* and *sleeping units*.
2. Laundry rooms, mechanical equipment rooms and storage rooms.
3. All interior corridors serving *sleeping units* or *dwelling units*.

   Exception: An automatic smoke detection system is not required in buildings that do not have interior *corridors* serving *sleeping units* or *dwelling units* and where each *sleeping unit* or *dwelling unit* either has a *means of egress* door opening directly to an exterior *exit access* that leads directly to an *exit* or a *means of egress* door opening directly to an *exit*.

Required smoke alarms in *dwelling units* and *sleeping units* in Group R-2 occupancies operated by a college or university for student or staff housing shall be interconnected with the fire alarm system in accordance with NFPA 72.

❖ This section requires an automatic smoke detection system in Group R-2 occupancies that are operated by a college or university to provide student or staff housing. It also requires the smoke alarms in individual units to be interconnected with the fire alarm system. This interconnection is only for the purpose of making occupants within each unit aware of the fire alarm activation in the building. The intent was not to activate the building fire alarm system by smoke alarms in each unit. This is more restrictive than a Group R-2 occupancy in general as typically the requirements are limited to a manual fire alarm system and smoke alarms in the individual sleeping or dwelling units.

The smoke detection system is focused on common areas, such as interior corridors, lounge areas, laundry, and areas such as mechanical rooms, which could be the source of a fire, especially in these specific types of Group R-2 occupancies.

In a study completed by the New York State Governor's Task Force on Campus Fire Safety it was cited that 43 percent of fires in college dormitories are located in dorm rooms or kitchens, leaving the other 57 percent to be located in areas that would not require smoke detection under the current code. The study also showed that there were approximately 300 fires on college campus over a three-year period while only 160 were reported to the fire department. The Center for Campus Fire Safety states that 99 deaths have been "reported" in fires in student housing since 2000.

An NFPA study on student housing showed 3,300 structural fires in dormitories, fraternities, sororities and barracks between 2002 and 2005. Since 1980, there has been an increase of 3 percent in reported fires in dormitory-type occupancies, while there has been a 52-percent decrease in overall reported structural fires.

The requirements in this section are very similar to the recommendations of the study done in New York State. It is important to note that the recommendations for that study were specifically aimed at the properties of the colleges and universities so it was not the study's intent to cover off-campus housing in this particular regard. There were recommendations for off-campus housing, such as sororities and fraternities to have annual inspections. This particular distinction was not addressed in detail during the code development process initially; however, since this is somewhat of a continuation of requirements that were added into Chapter 4 of the 2006 edition of the IFC for emergency preparedness and planning, and those requirements were intended to deal with buildings that were college or university property, it seemed reasonable to interpret that this requirement is also limited to the buildings that are college or university property and does not apply to Group R-2 occupancies that are not college or university property. This has since been clarified through the code development process and the section now specifically notes that the Group R-2 occupancies are specifically operated by a college or university.

The exception allows for the elimination of the smoke detection system in the specific situation where there are no interior corridors and the occupants essentially exit directly to the outside. The lack of interior corridors and exterior exits reduces the amount of smoke one unit will expose to another. Note that smoke alarms within the units are still required by Section 907.2.11.2. Note also that it is entirely possible that some areas that are required to have smoke detection may have ambient conditions that warrant a different type of alarm-initiating device. Section 907.4.3 addresses this concern.

[F] 907.2.10 Group R-4. Fire alarm systems and smoke alarms shall be installed in Group R-4 occupancies as required in Sections 907.2.10.1 through 907.2.10.3.

❖ This section, based on the Group R-2 requirements for manual fire alarm systems and Group I-1 requirements for automatic smoke detection systems, contains manual fire alarm and automatic smoke detection system requirements for new Group R-4 occupancies. Reviewing the occupancy categories in

Chapter 3 of the code, a Group R-4 could be considered either a small Group I-1 or a Group R-2 with occupants that have special needs or limitations. A further review found that both Group I-1 and R-2 occupancies had fire alarm requirements for new buildings, but Group R-4 did not, even though the IFC required a fire alarm system retroactively in existing Group R-4 occupancies (see Section 1103.7.7 of the IFC).

[F] 907.2.10.1 Manual fire alarm system. A manual fire alarm system that activates the occupant notification system in accordance with Section 907.5 shall be installed in Group R-4 occupancies.

Exceptions:

1. A manual fire alarm system is not required in buildings not more than two *stories* in height where all individual *sleeping units* and contiguous *attic* and crawl spaces to those units are separated from each other and public or common areas by not less than 1-hour *fire partitions* and each individual *sleeping unit* has an *exit* directly to a *public way*, *egress court* or *yard*.
2. Manual fire alarm boxes are not required throughout the building where all of the following conditions are met:
   - 2.1. The building is equipped throughout with an *automatic sprinkler system* installed in accordance with Section 903.3.1.1 or 903.3.1.2.
   - 2.2. The notification appliances will activate upon sprinkler water flow.
   - 2.3. Not fewer than one manual fire alarm box is installed at an *approved* location.
3. Manual fire alarm boxes in resident or patient sleeping areas shall not be required at *exits* where located at all nurses' control stations or other constantly attended staff locations, provided such stations are visible and continuously accessible and that the distances of travel required in Section 907.4.2.1 are not exceeded.

❖ This section is specific to manual fire alarm systems and requires such systems in all Group R-4 occupancies, with three exceptions.

Exception 1 eliminates the requirement for a manual fire alarm system if the sleeping units have an exit discharging directly to a public way, exit court or yard. Even though the building may be two stories in height, the sleeping units on each floor must have access directly to an approved exit at grade level. The use of an exterior exit access balcony with exterior stairs serving the second floor does not constitute an exit directly at grade. The minimum 1-hour fire-resistance rating required for adequate separation of the sleeping units must be maintained.

Exception 2 does not omit the fire alarm system but rather permits it to be initiated automatically by sprinkler system waterflow switch(es) in lieu of manual fire alarm boxes. The sprinkler system must activate the occupant notification system and at least one manual fire alarm box shall be installed at an approved location. See the commentary to Section 907.2 for a discussion of the single manual fire alarm box.

Exception 3 reduces the likelihood of accidental or malicious false alarm system activations by manual means by allowing the pull stations to be located in a more controlled area. It assumes the approved location is always accessible by staff and within a distance of travel of 200 feet (60 960 mm).

[F] 907.2.10.2 Automatic smoke detection system. An automatic smoke detection system that activates the occupant notification system in accordance with Section 907.5 shall be installed in *corridors*, waiting areas open to *corridors* and *habitable spaces* other than *sleeping units* and kitchens.

Exceptions:

1. Smoke detection in *habitable spaces* is not required where the facility is equipped throughout with an *automatic sprinkler system* installed in accordance with Section 903.3.1.1.
2. An automatic smoke detection system is not required in buildings that do not have interior *corridors* serving *sleeping units* and where each *sleeping unit* has a *means of egress* door opening directly to an *exit* or to an exterior *exit access* that leads directly to an *exit*.

❖ Occupancies in Group R-4 can be compartmentalized into small rooms so that a fire in one area of the building would not easily be noticed by occupants in another part of the building. Therefore, smoke detection is required in areas other than sleeping units and kitchens. Sleeping units are required by Section 907.2.10.3 to be equipped with single- and multiple-station smoke alarms in accordance with Section 907.2.11.

Since Group R-4 occupancies may not be supervised by staff and to reduce the likelihood that a fire within a waiting area open to the corridor or the corridor itself could develop beyond the incipient stage, thereby jeopardizing the building egress, these areas must be equipped with automatic smoke detection.

Exception 1 allows smoke detectors to be eliminated from habitable spaces if the building is equipped throughout with an NFPA 13 automatic sprinkler system. The sprinkler system should control any fire and perform occupant notification through actuation of the waterflow switch and subsequent activation of the building alarm notification appliances. A sprinkler system is required for all Group R occupancies in accordance with Section 903.2.8.

The exception provides that automatic fire detectors are not required in buildings that do not have interior corridors and in which sleeping units have a door opening to an exterior exit access that leads directly to the exits. The intent of the exception is that the exit access from the sleeping unit door be exterior and not require reentering the building prior to entering the exit. Since the exit access is outside, the need for detectors other than the smoke alarms required by Section 907.2.10.3 in sleeping units is greatly reduced.

[F] 907.2.10.3 Smoke alarms. Single- and multiple-station smoke alarms shall be installed in accordance with Section 907.2.11.

❖ The actual requirements for single- and multiple-station smoke alarms are located in Section 907.2.11. That section requires that the single- and multiple-station smoke alarms within sleeping units be connected to the emergency electrical system. Automatic activation of the fire alarm system is avoided to reduce unnecessary alarms within such buildings.

[F] 907.2.11 Single- and multiple-station smoke alarms. *Listed* single- and multiple-station smoke alarms complying with UL 217 shall be installed in accordance with Sections 907.2.11.1 through 907.2.11.6 and NFPA 72.

❖ Single- and multiple-station smoke alarms have evolved as one of the most important fire safety features in residential and similar occupancies having sleeping occupants. The value of early fire warning in these occupancies has been repeatedly demonstrated in fires involving both successful and unsuccessful smoke alarm performance.

For successful smoke alarm operation and performance, single- and multiple-station smoke alarms must be listed in accordance with UL 217 and installed to comply with the code and Chapter 11 of NFPA 72, which contains the minimum requirements for the selection, installation, operation and maintenance of fire warning equipment for use in family living units. These devices are called “smoke alarms” rather than “smoke detectors” because they are independent of a fire alarm system and include an integral alarm notification device.

[F] 907.2.11.1 Group R-1. Single- or multiple-station smoke alarms shall be installed in all of the following locations in Group R-1:

1. In sleeping areas.
2. In every room in the path of the *means of egress* from the sleeping area to the door leading from the *sleeping unit*.
3. In each *story* within the *sleeping unit*, including basements. For *sleeping units* with split levels and without an intervening door between the adjacent levels, a smoke alarm installed on the upper level shall suffice for the adjacent lower level provided that the lower level is less than one full *story* below the upper level.

❖ Because the occupants of a sleeping unit or suite may be asleep and unaware of a fire developing in the room or in the egress path, single- or multiple-station smoke alarms must be provided in the sleeping unit and in any intervening room between the sleeping unit and the exit access door from the room. If the sleeping unit or suite involves more than one level, a smoke alarm must also be installed on every level. See the commentary to Section 202, definition of “Sleeping unit.”

Smoke alarms are required in split-level arrangements, except those that meet the conditions described in Item 3. In accordance with Section 907.2.11.5, all smoke alarms within a sleeping unit or suite must be interconnected so that actuation of one alarm will actuate all smoke alarms within the sleeping unit or suite.

[F] 907.2.11.2 Groups R-2, R-3, R-4 and I-1. Single- or multiple-station smoke alarms shall be installed and maintained in Groups R-2, R-3, R-4 and I-1 regardless of *occupant load* at all of the following locations:

1. On the ceiling or wall outside of each separate sleeping area in the immediate vicinity of bedrooms.
2. In each room used for sleeping purposes.
3. In each *story* within a *dwelling unit*, including basements but not including crawl spaces and uninhabitable *attics*. In *dwellings* or *dwelling units* with split levels and without an intervening door between the adjacent levels, a smoke alarm installed on the upper level shall suffice for the adjacent lower level provided that the lower level is less than one full *story* below the upper level.

❖ Because the occupants of a dwelling unit may be asleep and unaware of a fire developing in the room or in an area within the dwelling unit that will affect their ability to escape, single- or multiple-station smoke alarms must be installed in every bedroom, in the vicinity of all bedrooms (e.g., hallways leading to the bedrooms) and on each story of the dwelling unit (see Commentary Figure 907.2.11.2 and the commentary to Section 202 definition of “Dwelling unit”).

If a sprinkler system was installed throughout the building in accordance with NFPA 13, 13R or 13D, if applicable, smoke alarms would still be required in the bedrooms even if residential sprinklers were used.

Smoke alarms are required in split-level arrangements. As required by Section 907.2.11.5, all smoke alarms within a dwelling unit must be interconnected so that actuation of one alarm will actuate the alarms in all detectors within the dwelling unit.

These provisions do not apply to one- and two-family dwellings and multiple single-family dwellings (townhouses) not more than three stories in height with a separate means of egress that are regulated by the IRC. The IRC is intended to be a stand-alone document but if the residential units do not fall within the scope of the IRC or for other reasons are intended to be subject to this code, then the requirements of this section would apply. IFC Committee Interpretation No. 42-03 addresses this condition and contains additional explanatory information about the IRC and its relationship to the other *International Codes*.

Although the occupants of a sleeping unit in a Group I-1 occupancy may be asleep, they are still considered capable of self-preservation. Regardless, smoke alarms are required in sleeping units. The exception allows single- or multiple-station smoke alarms to be eliminated in the room if an automatic fire detection system that includes in-room system smoke detectors is installed as required by Section 907.2.6.

[F] 907.2.11.3 Installation near cooking appliances. Smoke alarms shall not be installed in the following locations unless this would prevent placement of a smoke alarm in a location required by Section 907.2.11.1 or 907.2.11.2:

1. Ionization smoke alarms shall not be installed less than 20 feet (6096 mm) horizontally from a permanently installed cooking appliance.
2. Ionization smoke alarms with an alarm-silencing switch shall not be installed less than 10 feet (3048 mm) horizontally from a permanently installed cooking appliance.
3. Photoelectric smoke alarms shall not be installed less than 6 feet (1829 mm) horizontally from a permanently installed cooking appliance.

❖ This requirement is intended to reduce nuisance alarms attributed to locating smoke alarms in close proximity to cooking appliances and bathrooms in which steam is produced. These provisions are based on the findings in the Task Group Report - Minimum Performance Requirements for Smoke Alarm Detection Technology - February 22, 2008, and are consistent with similar requirements included in Section 29.8.3.4 of NFPA 72.

[F] 907.2.11.4 Installation near bathrooms. Smoke alarms shall be installed not less than 3 feet (914 mm) horizontally from the door or opening of a bathroom that contains a bathtub or shower unless this would prevent placement of a smoke alarm required by Section 907.2.11.1 or 907.2.11.2.

❖ See the commentary to Section 907.2.11.3. Sections 907.2.11.3 and 907.2.11.4 are provided to reduce nuisance alarms.

[F] 907.2.11.5 Interconnection. Where more than one smoke alarm is required to be installed within an individual *dwelling unit* or *sleeping unit* in Group R or I-1 occupancies, the smoke alarms shall be interconnected in such a manner that the activation of one alarm will activate all of the alarms in the individual unit. Physical interconnection of smoke alarms shall not be required where listed wireless alarms are installed and all alarms sound upon activation of one alarm. The alarm shall be clearly audible in all bedrooms over background noise levels with all intervening doors closed.

❖ The installation of smoke alarms in areas remote from the sleeping area will be of minimal value if the alarm is not heard by the occupants. Interconnection of multiple smoke alarms within an individual dwelling unit or sleeping unit is required in order to alert a sleeping occupant of a remote fire within the unit before the combustion products reach the smoke alarm in the sleeping area and thus provide additional time for evacuation.

The term "interconnection" refers to either hard-wired systems or listed wireless systems. UL has listed smoke detectors that use this technology. It is presumed that on safely evacuating the unit or room of fire origin, an occupant will notify other occupants by actuating the manual fire alarm system or using other available means. Section 907.7.1 addresses the testing of the smoke alarms to demonstrate that interconnection of such smoke alarms is properly functioning.

Similar requirements can now be found in the IRC, IFC and *International Existing Building Code*®

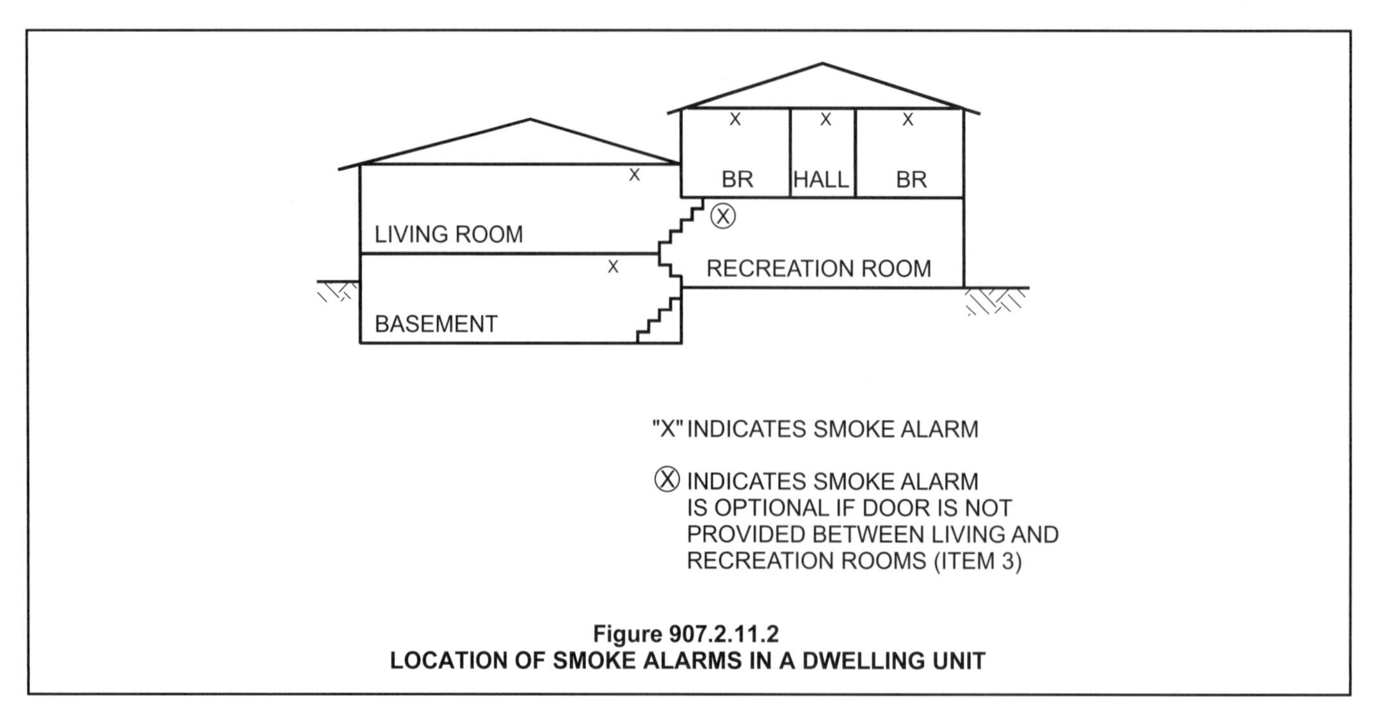

Figure 907.2.11.2
LOCATION OF SMOKE ALARMS IN A DWELLING UNIT

(IEBC®) for both new and existing buildings to allow interconnection with wireless technology of smoke alarms. All wirelessly interconnected smoke alarms are listed to UL 217 and are classified by NFPA 72 as low-power systems.

[F] 907.2.11.6 Power source. In new construction, required smoke alarms shall receive their primary power from the building wiring where such wiring is served from a commercial source and shall be equipped with a battery backup. Smoke alarms with integral strobes that are not equipped with battery backup shall be connected to an emergency electrical system in accordance with Section 2702. Smoke alarms shall emit a signal when the batteries are low. Wiring shall be permanent and without a disconnecting switch other than as required for overcurrent protection.

Exception: Smoke alarms are not required to be equipped with battery backup where they are connected to an emergency electrical system that complies with Section 2702.

❖ Smoke alarms are required to use AC as a primary power source and battery power as a secondary source to improve their reliability. For example, during a power outage, the probability of fire is increased because of the use of candles or lanterns for temporary light. Required backup battery power is intended to provide continued functioning of the smoke alarms. Smoke alarms are commonly designed to emit a recurring signal when batteries are low and need to be replaced.

Certain occupancies may already have an emergency electrical system in the building to monitor other building system conditions. The emergency electrical system provides a level of reliability equivalent to battery backup.

[F] 907.2.11.7 Smoke detection system. Smoke detectors listed in accordance with UL 268 and provided as part of the building *fire alarm system* shall be an acceptable alternative to single- and multiple-station *smoke alarms* and shall comply with the following:

1. The *fire alarm system* shall comply with all applicable requirements in Section 907.
2. Activation of a smoke detector in a *dwelling unit* or *sleeping unit* shall initiate alarm notification in the *dwelling unit* or *sleeping unit* in accordance with Section 907.5.2.
3. Activation of a smoke detector in a *dwelling unit* or *sleeping unit* shall not activate alarm notification appliances outside of the *dwelling unit* or *sleeping unit*, provided that a supervisory signal is generated and monitored in accordance with Section 907.6.6.

❖ This section specifically allows the use of an automatic smoke detection system as an alternative to smoke alarms. In the past, when this concept was proposed, it was only allowed through an alternative method and materials approach (see Section 104.11 of this code), even though, in concept, it provided the same level of protection. Such systems provide the same safety features necessary for occupants but are simply part of a fire alarm system. Note that if a detector activates within a sleeping or dwelling unit, the occupant notification system is not intended to activate. This is consistent with the operation of smoke alarms. Item 3 specifically requires the notification to be only to occupants of the sleeping unit or dwelling unit.

[F] 907.2.12 Special amusement buildings. An automatic smoke detection system shall be provided in *special amusement buildings* in accordance with Sections 907.2.12.1 through 907.2.12.3.

❖ Special amusement buildings are buildings in which the means of egress is not readily apparent, is intentionally confounded or is not readily available. Special amusement buildings must also comply with the provisions of Section 411.

The approved automatic smoke detection system is required to provide early warning of a fire. The detection system is required regardless of the presence of staff in the building. The exception recognizes that the ambient conditions in some special amusement buildings may preclude the use of automatic smoke detectors. In those instances, an alternative detection device must be used for early detection of a fire.

[F] 907.2.12.1 Alarm. Activation of any single smoke detector, the *automatic sprinkler system* or any other automatic fire detection device shall immediately activate an audible and visible alarm at the building at a constantly attended location from which emergency action can be initiated, including the capability of manual initiation of requirements in Section 907.2.12.2.

❖ On activation of either a smoke detector or other automatic fire detection device or the automatic sprinkler system, an alarm must activate both an audible and visible alarm at a constantly attended location. The staff at the location is expected to be capable of then providing the required egress illumination, stopping the conflicting or confusing sounds and distractions and activating the exit marking required by Section 907.2.12.2. The staff is also expected to be capable of preventing additional people from entering the building.

[F] 907.2.12.2 System response. The activation of two or more smoke detectors, a single smoke detector equipped with an alarm verification feature, the *automatic sprinkler system* or other *approved* fire detection device shall automatically do all of the following:

1. Cause illumination of the *means of egress* with light of not less than 1 footcandle (11 lux) at the walking surface level.
2. Stop any conflicting or confusing sounds and visual distractions.
3. Activate an *approved* directional *exit* marking that will become apparent in an emergency.
4. Activate a prerecorded message, audible throughout the *special amusement building*, instructing patrons to pro-

ceed to the nearest *exit*. Alarm signals used in conjunction with the prerecorded message shall produce a sound that is distinctive from other sounds used during normal operation.

❖ Once a fire has been detected, measures must be taken to stop the confusion or distractions. Additionally, the egress path must be illuminated and marked. These measures must occur automatically on detection of the fire or sprinkler waterflow. A prerecorded message that can be heard throughout the building instructing the occupants to proceed to the nearest exit must be automatically activated. The message and alarm signals should be designed to prevent panic. The prerecorded message capability is in addition to the EV/ACS requirement of Section 907.2.12.3. The wiring of all devices must comply with NFPA 72.

[F] 907.2.12.3 Emergency voice/alarm communication system. An emergency voice/alarm communication system, which is also allowed to serve as a public address system, shall be installed in accordance with Section 907.5.2.2 and be audible throughout the entire *special amusement building*.

❖ Because of the problem associated with evacuating special amusement buildings, an emergency voice/alarm communication system is required (see also Section 907.5.2.2). This section allows the system to also serve as a public address (PA) system to have the capability to alert the occupants of a fire and give them evacuation instructions. The system must be designed so that once the voice alarm is activated, the typical public address function is superseded by the voice alarm. Because a manual override must be provided, it is possible that the same microphone used for the public address can be used for the override. However, a separate action would be necessary so that the override function can be used once the voice alarm is active.

[F] 907.2.13 High-rise buildings. High-rise buildings shall be provided with an automatic smoke detection system in accordance with Section 907.2.13.1, a fire department communication system in accordance with Section 907.2.13.2 and an emergency voice/alarm communication system in accordance with Section 907.5.2.2.

Exceptions:

1. Airport traffic control towers in accordance with Sections 412 and 907.2.22.
2. *Open parking garages* in accordance with Section 406.5.
3. Buildings with an occupancy in Group A-5 in accordance with Section 303.1.
4. Low-hazard special occupancies in accordance with Section 503.1.1.
5. Buildings with an occupancy in Group H-1, H-2 or H-3 in accordance with Section 415.
6. In Group I-1 and I-2 occupancies, the alarm shall sound at a *constantly attended location* and occupant notification shall be broadcast by the emergency voice/alarm communication system.

❖ High-rise buildings require additional fire protection systems because of the difficulties with smoke movement, egress time and fire department access. As a result, this section requires both an automatic fire alarm system and an emergency voice/alarm communication system (see commentary, Section 907.5.2.2). Exceptions 1 through 5 are the same as those in Section 403.1 regarding the applicability of the high-rise provisions.

Exception 1 addresses airport traffic control towers and is based on the limited fuel load and the limited number of persons occupying the tower.

Open parking garages and places of outdoor assembly (Group A-5) are exempted by Exceptions 2 and 3, respectively, because of the free ventilation to the outside that exists in such structures.

In Exception 4, low-hazard special industrial occupancies may be exempted when approved by the fire code official. Such buildings should be evaluated based on the occupant load and the hazards of the occupancy and its contents to determine whether the protection features required by Section 403 of the code are necessary.

Buildings with occupancies in Groups H-1, H-2 and H-3 are excluded from the requirements of this section by Exception 5 because the fire hazard characteristics of these occupancies have not yet been considered in high-rise buildings.

Exception 6 recognizes the supervised environment typical of institutional uses and the reliance placed on staff to act appropriately in an emergency. As is the case for most voice alarms, the key is in being able to deliver specific information to the people who can affect a safe egress—whether it be the public, employees, or both.

[F] 907.2.13.1 Automatic smoke detection. Automatic smoke detection in high-rise buildings shall be in accordance with Sections 907.2.13.1.1 and 907.2.13.1.2.

❖ This section simply introduces the fire alarm and detection system requirements for high-rise buildings.

[F] 907.2.13.1.1 Area smoke detection. Area smoke detectors shall be provided in accordance with this section. Smoke detectors shall be connected to an automatic fire alarm system. The activation of any detector required by this section shall activate the emergency voice/alarm communication system in accordance with Section 907.5.2.2. In addition to smoke detectors required by Sections 907.2.1 through 907.2.10, smoke detectors shall be located as follows:

1. In each mechanical equipment, electrical, transformer, telephone equipment or similar room that is not provided with sprinkler protection.
2. In each elevator machine room, machinery space, control room and control space and in elevator lobbies.

❖ Automatic smoke detectors are required in all high-rise buildings in certain locations so that a fire will be detected in its early stages of development. The detectors must be connected to the automatic fire alarm system and be capable of initiating operation of the EV/ACS.

This section divides the automatic smoke detection requirement into two categories. Smoke detectors must be installed in rooms that are not typically occupied. This includes rooms used for: mechanical equipment, electrical equipment, transformer equipment and telephone equipment where such rooms do not have automatic sprinkler protection. In most cases, these rooms will have sprinkler protection by virtue of being in a high-rise building and will therefore not require smoke detectors. However, in elevator machine rooms, machinery spaces, control rooms, control spaces and elevator lobbies, smoke detectors are required regardless of sprinkler protection.

Note that smoke detection and smoke alarms may be required based on occupancy-related requirements elsewhere in Section 907.2.

[M] 907.2.13.1.2 Duct smoke detection. Duct smoke detectors complying with Section 907.3.1 shall be located as follows:

1. In the main return air and exhaust air plenum of each air-conditioning system having a capacity greater than 2,000 cubic feet per minute (cfm) (0.94 m^3/s). Such detectors shall be located in a serviceable area downstream of the last duct inlet.
2. At each connection to a vertical duct or riser serving two or more stories from a return air duct or plenum of an air-conditioning system. In Group R-1 and R-2 occupancies, a smoke detector is allowed to be used in each return air riser carrying not more than 5,000 cfm (2.4 m^3/s) and serving not more than 10 air-inlet openings.

❖ Smoke detectors must be installed in the main return air and exhaust air plenum of each air-conditioning system having a design capacity exceeding 2,000 cubic feet per minute (cfm) (0.94 m^3/s). Systems with design capacities equal to or less than 2,000 cfm (0.94 m^3/s) are exempt from this requirement because their small size limits their capacity for spreading smoke to parts of the building not already involved with fire.

The area that could be served by a 2,000-cfm (0.94 m^3) system (approximately 5 tons of cooling capacity) is comparatively small; therefore, the distribution of smoke in a system of that size would be minimal. Smoke detectors must be located so that they monitor the total airflow within the system. If a single detector is unable to sample the total airflow at all times, then multiple detectors are required. The smoke detectors must be made accessible for maintenance and inspection. Many failures and false alarms are caused by a lack of maintenance and cleaning of the smoke detectors.

Consistent with Section 606.2.3 of the IMC, return air risers serving two or more stories must have smoke detectors installed at each story. Item 2 allows the use of a single listed smoke detector in each return air riser in a Group R-1 or R-2 occupancy if the capacity of each riser does not exceed 5,000 cfm (2.4 m^3/s) and does not serve more than 10 air-inlet openings. This alternative recognizes that it is not as necessary in buildings dedicated to residential occupancies only to monitor the return air from each story prior to intermixing the return air in the common riser.

[F] 907.2.13.2 Fire department communication system. Where a wired communication system is *approved* in lieu of an emergency responder radio coverage system in accordance with Section 510 of the *International Fire Code*, the wired fire department communication system shall be designed and installed in accordance with NFPA 72 and shall operate between a fire command center complying with Section 911, elevators, elevator lobbies, emergency and standby power rooms, fire pump rooms, *areas of refuge* and inside *interior exit stairways*. The fire department communication device shall be provided at each floor level within the *interior exit stairway*.

❖ High-rise buildings have posed a challenge to the traditional communication systems used by the fire service for fire-to-ground communications to assist fire ground officers in communicating with the fire fighters working in various areas of the building. Where testing of the emergency responder radio coverage system required by Section 403.4.5 of the code shows that the signal strengths are not satisfactory, Section 510.1, Exception 1, of the IFC allows for the alternative of installation of a wired communication system designed in accordance with this section. The system must be capable of operating between the fire command center and every elevator, elevator lobby, emergency/standby power room, fire pump room, area of refuge and interior exit stairway. Note that this section does not offer specific criteria as to what constitutes an acceptable wired communication system or its components. It could be a component of an emergency voice/alarm communication system that complies with Section 907.5.2.2 or a building's telephone system. In any event, when applying Section 510.1, Exception 1, of the IFC and this section, the concurrent approval of the fire and building code officials is required.

[F] 907.2.14 Atriums connecting more than two stories. A fire alarm system shall be installed in occupancies with an atrium that connects more than two *stories*, with smoke detection installed in locations required by a rational analysis in Section 909.4 and in accordance with the system operation requirements in Section 909.17. The system shall be activated in accordance with Section 907.5. Such occupancies in Group A, E or M shall be provided with an emergency voice/alarm communication system complying with the requirements of Section 907.5.2.2.

❖ Buildings containing an atrium that connects more than two stories are to be equipped with a fire alarm system that can be used to notify building occupants to begin evacuating in case of a fire. The other critical part of such fire alarm systems is to activate the smoke control system. The system is to be activated by smoke detection designed and installed in accordance with the rational analysis as required in Section 909.4. More specifically smoke control systems are engineered systems that are activated by carefully placed and zoned smoke detection. If improperly designed and installed, the system may not be effective. For instance, wrongly placed or inappropriate smoke detection technology may not activate fast enough and the smoke control system would be overwhelmed. This section goes on to state that the alarm system must be initiated in accordance with Section 907.5, which requires that in buildings containing an atrium, the alarm system is to be initiated by the sprinkler system and any automatic or manual fire alarm-initiating devices found in the atrium as well as elsewhere in the building. It does not intend to require certain features to be installed within the atrium but rather is simply requiring that any such features present initiate the occupant notification system. It would not necessarily be appropriate to also initiate the smoke control system on activation of the alarm system within a building containing an atrium (see Section 909.12.3). The alarm system needs to be carefully zoned in such buildings to avoid an inappropriate activation of the smoke control system from a space not associated with the atrium.

Groups A, E and M must have an emergency voice/alarm communication system that complies with Section 907.5.2.2 because of the number of persons to be evacuated and the lack of familiarity with the location of exits that is typical of occupants in Groups A and M.

[F] 907.2.15 High-piled combustible storage areas. An automatic smoke detection system shall be installed throughout high-piled combustible storage areas where required by Section 3206.5 of the *International Fire Code*.

❖ Section 3206.5 of the IFC requires an automatic fire detection system in high-piled combustible storage areas depending on the commodity class, the size of the high-piled storage area and the presence of an automatic sprinkler system. High-piled storage is the storage of Class I through IV commodities in piles, bin boxes, on pallets or in racks more than 12 feet (3658 mm) high or for high-hazard commodities stored higher than 6 feet (1829 mm). Chapter 32 of the IFC and NFPA 13 contain additional requirements for all high-piled storage conditions.

[F] 907.2.16 Aerosol storage uses. Aerosol storage rooms and general-purpose warehouses containing aerosols shall be provided with an *approved* manual fire alarm system where required by the *International Fire Code*.

❖ Chapter 32 of the IFC and NFPA 30B contain additional guidance on the storage of and fire protection requirements for aerosol products. The requirements for storing the various levels of aerosol products are dependent on the level of sprinkler protection, the type of storage and the quantity of aerosol products. Although aerosol product fires generally involve property loss as opposed to loss of life, installation of a manual fire alarm system could aid in the prompt evacuation of the occupants. Fires involving aerosol products can spread rapidly through a building that is not properly protected and controlled.

[F] 907.2.17 Lumber, wood structural panel and veneer mills. Lumber, wood structural panel and veneer mills shall be provided with a manual fire alarm system.

❖ Any facility using mechanical methods to process wood into finished products produces debris and the potential for combustible dust. Such facilities include mills that produce solid wood lumber and wood veneers as well as those that manufacture structural wood panels such as waferboard, oriented strandboard, composite wood panels or plywood. Good housekeeping and control of ignition sources are therefore essential. To aid in the quick evacuation of occupants in an emergency, Section 2804.2.1 of the IFC requires a manual fire alarm system in lumber, wood structural panel and veneer mills that contain product dryers because of their potential as a source of ignition. A manual fire alarm system is not required, however, if the dryers and all other potential sources of ignition are protected by a supervised automatic sprinkler system.

[F] 907.2.18 Underground buildings with smoke control systems. Where a smoke control system is installed in an underground building in accordance with this code, automatic smoke detectors shall be provided in accordance with Section 907.2.18.1.

❖ As indicated in Section 405.5.2, each compartment of an underground building must have a smoke control/exhaust system that can be activated both automatically and manually. Floor levels more than 60 feet (18 288 mm) below the lowest level of exit discharge must be compartmented. Compartmentation is a key element in the egress and fire access plan for floor areas in an underground building. The smoke control system must not only facilitate egress during a fire, but also improve fire department access to the fire source by maintaining visibility that is otherwise impossible given the inability of the fire service to manually ventilate the underground portion of the building (see commentary, Section 405.4.1).

[F] 907.2.18.1 Smoke detectors. Not fewer than one smoke detector *listed* for the intended purpose shall be installed in all of the following areas:

1. Mechanical equipment, electrical, transformer, telephone equipment, elevator machine or similar rooms.
2. Elevator lobbies.
3. The main return and exhaust air plenum of each air-conditioning system serving more than one *story* and located in a serviceable area downstream of the last duct inlet.

4. Each connection to a vertical duct or riser serving two or more floors from return air ducts or plenums of heating, ventilating and air-conditioning systems, except that in Group R occupancies, a *listed* smoke detector is allowed to be used in each return air riser carrying not more than 5,000 cfm (2.4 m^3/s) and serving not more than 10 air-inlet openings.

❖ Automatic smoke detectors are required in certain locations in all underground buildings so that a fire will be detected in its early stages of development. Underground buildings are similar to high-rise buildings in that they present an unusual hazard by being virtually inaccessible to exterior fire department suppression and rescue operations with the increased potential to trap occupants inside the structure. For this reason, the smoke detector location requirements for underground buildings are similar to those in Section 907.2.13.1 for high-rise buildings (see commentary, Section 907.2.13.1).

The requirement for a smoke detector in the main return and exhaust air plenum of an air-conditioning system in an underground building, however, differs from that of a high-rise building in that it is not a function of capacity [2,000 cfm (0.94 m^3/s)] but rather a function of whether the system serves more than one floor level. There is more concern over the threat of smoke movement from floor to floor because the products of combustion cannot be vented directly to the atmosphere.

[F] 907.2.18.2 Alarm required. Activation of the smoke control system shall activate an audible alarm at a *constantly attended location*.

❖ The audible alarm is required to notify qualified personnel immediately that the smoke control system has been activated and to put emergency procedures into action quickly.

[F] 907.2.19 Deep underground buildings. Where the lowest level of a structure is more than 60 feet (18 288 mm) below the finished floor of the lowest *level of exit discharge*, the structure shall be equipped throughout with a manual fire alarm system, including an emergency voice/alarm communication system installed in accordance with Section 907.5.2.2.

❖ The ability to communicate and offer warning of a fire can increase the time available for egress from the building. Underground structures located more than 60 feet (18 288 mm) below the level of exit discharge must therefore have a manual fire alarm system. An emergency voice/alarm communication system is also required as part of this system (see commentary, Section 907.5.2.2).

[F] 907.2.20 Covered and open mall buildings. Where the total floor area exceeds 50,000 square feet (4645 m^2) within either a covered mall building or within the perimeter line of an open mall building, an emergency voice/alarm communication system shall be provided. Emergency voice/alarm communication systems serving a mall, required or otherwise, shall be accessible to the fire department. The system shall be provided in accordance with Section 907.5.2.2.

❖ Because of the potentially large number of occupants and their unfamiliarity with their surroundings, an EV/ACS, accessible by the fire department, is required to aid in evacuation of covered mall buildings exceeding 50,000 square feet (4645 m^2) in total floor area or an open mall exceeding 50,000 square feet (4645 m^2) measured within the perimeter lines of the open mall. Anchor stores are not included as part of the covered or open mall building (see commentary, Section 202) definition of "Covered mall building".

[F] 907.2.21 Residential aircraft hangars. Not fewer than one single-station smoke alarm shall be installed within a residential aircraft hangar as defined in Chapter 2 and shall be interconnected into the residential smoke alarm or other sounding device to provide an alarm that will be audible in all sleeping areas of the *dwelling*.

❖ Residential aircraft hangars are assumed to be on the same property as a one- or two-family dwelling. Section 412.5 contains additional requirements for the construction of residential aircraft hangars. The hangar could be located immediately adjacent to the dwelling unit if it is separated by 1-hour fire-resistance-rated construction. Because of the potentially close proximity of the aircraft and its flammability and fuel source, at least one smoke alarm is required in the hangar that is interconnected to the residential smoke alarms. It should be noted, however, that the requirement for a smoke alarm is also applicable to residential aircraft hangars that are detached from the dwelling unit. Because a minimum separation distance is not specified, a fire in the hangar could still present a serious fire hazard to the dwelling unit.

[F] 907.2.22 Airport traffic control towers. An automatic smoke detection system that activates the occupant notification system in accordance with Section 907.5 shall be provided in airport control towers in accordance with Sections 907.2.22.1 and 907.2.22.2.

Exception: Audible appliances shall not be installed within the control tower cab.

❖ Airport traffic control towers must be designed to comply with Section 412.3. These structures are unique in that they can be built to excessive heights and are often permitted to have one exit stairway. Section 412.3 requires that airport traffic control towers with an occupied floor more than 35 feet above fire department vehicle access be equipped throughout with an automatic sprinkler system. The requirements for detection systems and associated occupant notification are addressed based on whether the airport traffic control tower is equipped throughout with an automatic sprinkler system and whether multiple exits are provided.

The exception recognizes the sensitive nature of the operations that take place in the cab located at the top of the tower and prohibits the installation of audible alarm notification devices there. Notification of occupants within the cab is to be by visual notification appliances only.

[F] 907.2.22.1 Airport traffic control towers with multiple exits and automatic sprinklers. Airport traffic control towers with multiple *exits* and equipped throughout with an *automatic sprinkler system* in accordance with Section 903.3.1.1 shall be provided with smoke detectors in all of the following locations:

1. Airport traffic control cab.
2. Electrical and mechanical equipment rooms.
3. Airport terminal radar and electronics rooms.
4. Outside each opening into *interior exit stairways*.
5. Along the single *means of egress* permitted from observation levels.
6. Outside each opening into the single *means of egress* permitted from observation levels.

❖ This section addresses airport traffic control towers that are both equipped throughout with an automatic sprinkler system and provide multiple exits. The requirements are less restrictive than Section 907.2.22.2 because two important safety aspects are provided.

The first three items address occupiable or equipment-related rooms where the fires are more likely to start. The last two items address the paths of egress. Item 4 requires one detector outside each entrance to the interior stairway. Item 5 is specific to providing detection along the entire means of egress path from the observation levels.

[F] 907.2.22.2 Other airport traffic control towers. Airport traffic control towers with a single *exit* or where sprinklers are not installed throughout shall be provided with smoke detectors in all of the following locations:

1. Airport traffic control cab.
2. Electrical and mechanical equipment rooms.
3. Airport terminal radar and electronics rooms.
4. Office spaces incidental to the tower operation.
5. Lounges for employees, including sanitary facilities.
6. *Means of egress*.
7. Accessible utility shafts.

❖ This section addresses airport traffic control towers that have only a single exit or where an automatic sprinkler system is not provided. Essentially all of the items addressed are the areas permitted in an airport traffic control tower in accordance with Section 412.3. Items 1 through 3 are the same as Section 907.2.22.1. These are the critical occupiable spaces and equipment spaces where fires have the greatest effect on the operation of airport traffic control towers. Since this is a single exit tower and possibly not equipped throughout with an automatic sprinkler system, other occupiable areas such as offices and lounges for employees must also provide smoke detection. Since there are limited exits or no systems able to control the fire, early warning of a fire becomes more critical. Item 6 requires smoke detection along the means of egress path. The intent is to address the exit access path leading to the interior exit stairway. Finally, Item 7 addresses the potential for fires in any utility shaft that may be accessible to building occupants.

[F] 907.2.23 Battery rooms. An automatic smoke detection system shall be installed in areas containing stationary storage battery systems with a liquid capacity of more than 50 gallons (189 L).

❖ Stationary lead-acid battery systems are commonly used for standby power, emergency power or uninterrupted power supplies. The release of hydrogen gas during battery system operation is usually minimal. Adequate ventilation will disperse the small amounts of liberated hydrogen. Because standby power and emergency power systems control many important building emergency systems and functions, a supervised automatic smoke detection system is required for early warning notification of a hazardous condition. Section 608 of the IFC contains additional requirements, including the need for safety venting; room enclosure requirements; spill control and neutralization provisions; ventilation criteria; signage and seismic protection. Section 509 also requires that such rooms in certain occupancies be separated by 1-hour construction.

[F] 907.3 Fire safety functions. Automatic fire detectors utilized for the purpose of performing fire safety functions shall be connected to the building's fire alarm control unit where a fire alarm system is required by Section 907.2. Detectors shall, upon actuation, perform the intended function and activate the alarm notification appliances or activate a visible and audible supervisory signal at a *constantly attended location*. In buildings not equipped with a fire alarm system, the automatic fire detector shall be powered by normal electrical service and, upon actuation, perform the intended function. The detectors shall be located in accordance with NFPA 72.

❖ When the code requires installation of automatic fire detectors to perform a specific function, such as elevator recall or smokeproof enclosure ventilation, or when detectors are installed to comply with a permitted alternative, such as door-closing devices, these detectors must be connected to the building's automatic fire alarm system if the building is required by the code to have such a system.

In addition to performing its intended function (for example, closing a door), if a detector is activated, it must also activate either the building alarm devices (if one is present) or a supervisory signal at a constantly attended location. This requirement recognizes that these detectors and the devices they control are part of the building fire protection system and are expected to perform as designed. If they are connected to a fire alarm system, they will have the supervision necessary for operational reliability. If they are not connected to and supervised by a fire alarm system, they still must be supervised through the constantly attended location.

An exception is provided for fire safety function detectors in buildings not required to have a fire alarm system. The fire safety function detectors must be powered by the building electrical system and be located as required by NFPA 72. Without this exception, these detectors could not be expected to perform as intended because there would be no power supply.

Note that in the IFC, this section is entitled "Where required in existing buildings and structures." but since the IBC does not apply retroactively, such requirements do not appear here. That is why all sections of Section 907 following Section 907.2.23 do not exactly correlate with the content of Section 907 of the IFC.

[F] 907.3.1 Duct smoke detectors. Smoke detectors installed in ducts shall be *listed* for the air velocity, temperature and humidity present in the duct. Duct smoke detectors shall be connected to the building's fire alarm control unit when a fire alarm system is required by Section 907.2. Activation of a duct smoke detector shall initiate a visible and audible supervisory signal at a *constantly attended location* and shall perform the intended fire safety function in accordance with this code and the *International Mechanical Code*. In facilities that are required to be monitored by a supervising station, duct smoke detectors shall report only as a supervisory signal and not as a fire alarm. They shall not be used as a substitute for required open area detection.

Exceptions:

1. The supervisory signal at a *constantly attended location* is not required where duct smoke detectors activate the building's alarm notification appliances.
2. In occupancies not required to be equipped with a fire alarm system, actuation of a smoke detector shall activate a visible and an audible signal in an *approved* location. Smoke detector trouble conditions shall activate a visible or audible signal in an *approved* location and shall be identified as air duct detector trouble.

❖ It is not the intent of this section to send a signal to the fire department or to activate the alarm notification devices within a building. Instead, this section requires that a supervisory signal be sent to a constantly attended location. Smoke detectors must be connected to a fire alarm system where such systems are installed. Connection to the fire alarm system will activate a visible and audible supervisory signal at a constantly attended location, which will alert building supervisory personnel that a smoke alarm has activated and will also provide electronic supervision of the duct detectors, thereby indicating any problems that may develop in the detector system circuitry or power supply.

Exception 1 allows activation of the building alarm notification appliances in place of a supervisory signal. Causing the occupant notification system to sound would alert the occupants of the building that an alarm condition exists within the air distribution system, thereby performing the same function as a supervisory signal sent to a constantly attended location.

Exception 2 recognizes that not all buildings are required to have a fire alarm system. A visible and audible signal must be activated at an approved location that will alert building supervisory personnel to take action. Additionally, the duct smoke detectors must be electronically supervised to indicate trouble (system fault) in the detector system circuitry or power supply. A trouble condition must activate a distinct visible or audible signal at a location that will alert the responsible personnel.

[F] 907.3.2 Delayed egress locks. Where delayed egress locks are installed on *means of egress* doors in accordance with Section 1010.1.9.7, an automatic smoke or heat detection system shall be installed as required by that section.

❖ This section alerts the code user to additional requirements in Section 1010.1.9.7 that tie the operation of egress doors into the activation of an automatic fire detection system. A smoke or heat detection system is required to unlock delayed egress locks upon activation. The heat detection system can be the sprinkler system. For example, a similar requirement is found in Section 1010.1.4.3 that requires horizontal sliding doors used as a component of the means of egress, where required to be rated, to be self-closing or automatic-closing upon smoke detection. Also, electrically locked egress doors in occupancies as required by Section 1010.1.9.8 must be capable of being automatically unlocked by activation of an automatic fire detection system, if one is installed.

[F] 907.3.3 Elevator emergency operation. Automatic fire detectors installed for elevator emergency operation shall be installed in accordance with the provisions of ASME A17.1 and NFPA 72.

❖ This section provides correlation with Section 607.1 of the IFC by making it clear that automatic fire detection devices used to initiate Phase I emergency recall of elevators are to be installed in accordance with both ASME A17.1 and NFPA 72.

[F] 907.3.4 Wiring. The wiring to the auxiliary devices and equipment used to accomplish the fire safety functions shall be monitored for integrity in accordance with NFPA 72.

❖ In order to provide a reasonable level of integrity and reliability to the installation of automatic fire detection devices and related equipment installed to perform various fire safety functions in accordance with Section 907.3, this section requires that all wiring interconnecting such devices and equipment be monitored for integrity in accordance with NFPA 72.

[F] 907.4 Initiating devices. Where manual or automatic alarm initiation is required as part of a fire alarm system, the initiating devices shall be installed in accordance with Sections 907.4.1 through 907.4.3.1.

❖ This section introduces Sections 907.4.1 through 907.4.3, which contain requirements for the various types of manual or automatic fire alarm-initiating devices.

[F] 907.4.1 Protection of fire alarm control unit. In areas that are not continuously occupied, a single smoke detector shall be provided at the location of each fire alarm control unit, notification appliance circuit power extenders, and supervising station transmitting equipment.

> **Exception:** Where ambient conditions prohibit installation of a smoke detector, a *heat detector* shall be permitted.

❖ This section requires a smoke detector at the fire alarm control unit. This is consistent with Section 10.4.4 of NFPA 72. This smoke detector will activate the fire alarm control unit and allow it to either notify occupants or transmit a signal to a remote monitoring location before the fire impairs the fire alarm control unit. The exception parallels Section 907.4.3 by allowing a heat detector to be installed in lieu of a smoke detector in areas where the ambient environment is hostile to smoke detectors and could lead to unwanted alarm activations. This exception is also allowed by NFPA 72.

[F] 907.4.2 Manual fire alarm boxes. Where a manual fire alarm system is required by another section of this code, it shall be activated by fire alarm boxes installed in accordance with Sections 907.4.2.1 through 907.4.2.6.

❖ This section specifies the requirements for manual fire alarm boxes that are part of a required manual fire alarm system.

[F] 907.4.2.1 Location. Manual fire alarm boxes shall be located not more than 5 feet (1524 mm) from the entrance to each *exit*. In buildings not protected by an *automatic sprinkler system* in accordance with Section 903.3.1.1 or 903.3.1.2, additional manual fire alarm boxes shall be located so that the *exit access* travel distance to the nearest box does not exceed 200 feet (60 960 mm).

❖ Manual fire alarm boxes must be located in the path of egress and be readily accessible to the occupants. They must be located within 5 feet (1524 mm) of the entrance to each exit on every story of the building. This would include the need to locate manual fire alarm boxes near each horizontal exit, as well as entrances to stairs and exit doors to the exterior.

Manual fire alarm boxes are located near exits so that an adequate number of devices are available in the path of egress to transmit an alarm in a timely manner. These locations also encourage the actuation of a manual fire alarm box on the fire floor prior to entering the stair, resulting in the alarm being received from the actual fire floor and not another floor along the path of egress.

The location also presumes that individuals will be evacuating the area where the fire originated. When evacuation of the fire area is unlikely, consideration could be given to putting manual fire alarm boxes in more convenient places. Examples of such instances would be officer stations in Group I-3 occupancies and care provider's stations in Group I-2 occupancies.

The 200-foot (60 960 mm) exit access travel distance limitation is consistent with the exit access travel distance permitted for most nonsprinklered occupancies. If the 200-foot (60 960 mm) travel distance to a manual fire alarm box is exceeded in a nonsprinklered building, additional manual fire alarm boxes would be required.

[F] 907.4.2.2 Height. The height of the manual fire alarm boxes shall be not less than 42 inches (1067 mm) and not more than 48 inches (1372 mm) measured vertically, from the floor level to the activating handle or lever of the box.

❖ Manual fire alarm boxes must be reachable by the occupants of the building. They must also be mounted high enough to reduce the likelihood of damage or false alarms from something accidentally striking the device. Therefore, manual fire alarm boxes must be mounted a minimum of 42 inches (1067 mm) and a maximum of 48 inches (1372 mm) above the floor level. The 48-inch (1372 mm) measurement corresponds to the maximum unobstructed side-reach height by a person in a wheelchair.

[F] 907.4.2.3 Color. Manual fire alarm boxes shall be red in color.

❖ Manual fire alarm boxes are to be painted or manufactured in a distinctive and traditional red color to provide a visual cue to help building occupants identify the device.

[F] 907.4.2.4 Signs. Where fire alarm systems are not monitored by a supervising station, an *approved* permanent sign shall be installed adjacent to each manual fire alarm box that reads: WHEN ALARM SOUNDS CALL FIRE DEPARTMENT.

> **Exception:** Where the manufacturer has permanently provided this information on the manual fire alarm box.

❖ This section has limited application because, as indicated in Section 907.6.6, fire alarm systems generally must be monitored by an approved supervising station. When a system is not monitored, such as possibly a fire alarm system that is not required by code, adequate signage must be displayed to tell occupants what response actions must be taken. Most building occupants assume that when an alarm device is activated, the fire department will automatically be notified as well. The sign must be conspicuously located next to the manual fire alarm box unless it is mounted on the manual fire alarm box itself by the manufacturer.

[F] 907.4.2.5 Protective covers. The fire code official is authorized to require the installation of *listed* manual fire alarm box protective covers to prevent malicious false alarms or to provide the manual fire alarm box with protection from physical damage. The protective cover shall be transparent or red in color with a transparent face to permit visibility of the manual fire alarm box. Each cover shall include proper operating instructions. A protective cover that emits a local alarm signal shall not be installed unless *approved*. Protective covers shall not project more than that permitted by Section 1003.3.3.

❖ Although manual fire alarm boxes must be readily available to all occupants in buildings required to have a manual fire alarm system, this section allows the use of protective covers if they are approved by the fire code official. Protective covers are commonly used to reduce either the potential for intentional false alarms or vandalism. They also provide protection in locations where the manual fire alarm boxes may be exposed to physical damage, such as in gymnasiums, indoor tennis courts and the like.

[F] 907.4.2.6 Unobstructed and unobscured. Manual fire alarm boxes shall be accessible, unobstructed, unobscured and visible at all times.

❖ This section addresses the concern that manual fire alarm boxes be kept clear and unobstructed. It is recommended that a minimum of 3 feet (914 mm) be kept clear but more may be needed. NFPA 72 addresses the need for manual fire alarm boxes to be unobstructed in Section 17.14.8.2 and states that manual fire alarm boxes be conspicuous, unobstructed and accessible. This requirement will assist during the design, construction, inspection and future maintenance of manual fire alarm boxes when they are located where they will be provided with enough space to access and will not be obstructed.

[F] 907.4.3 Automatic smoke detection. Where an automatic smoke detection system is required it shall utilize smoke detectors unless ambient conditions prohibit such an installation. In spaces where smoke detectors cannot be utilized due to ambient conditions, *approved* automatic *heat detectors* shall be permitted.

❖ Smoke detectors must be used, except when ambient conditions would prohibit their use. This section would allow a heat detector to be installed in lieu of a smoke detector in areas where the ambient environment is hostile to smoke detectors and could lead to unwanted alarm activations. The smoke detector manufacturer's literature will identify the limitations on the use of smoke detectors, including environmental conditions such as humidity, temperature and airflow.

[F] 907.4.3.1 Automatic sprinkler system. For conditions other than specific fire safety functions noted in Section 907.3, in areas where ambient conditions prohibit the installation of smoke detectors, an *automatic sprinkler system* installed in such areas in accordance with Section 903.3.1.1 or 903.3.1.2 and that is connected to the fire alarm system shall be *approved* as automatic heat detection.

❖ This section states that automatic heat detection is not required when buildings are fully sprinklered in accordance with NFPA 13 or 13R. The presence of a sprinkler system exempts areas where a heat detector can be installed in place of a smoke detector, such as in storage or furnace rooms. The sprinkler head in this case essentially acts as a heat detection device. Note that this provision does not apply to the fire safety functions indicated in Section 907.3.

[F] 907.5 Occupant notification systems. A fire alarm system shall annunciate at the fire alarm control unit and shall initiate occupant notification upon activation, in accordance with Sections 907.5.1 through 907.5.2.3.3. Where a fire alarm system is required by another section of this code, it shall be activated by:

1. Automatic fire detectors.
2. *Automatic sprinkler system* waterflow devices.
3. Manual fire alarm boxes.
4. Automatic fire-extinguishing systems.

Exception: Where notification systems are allowed elsewhere in Section 907 to annunciate at a *constantly attended location*.

❖ This section makes it clear that fire alarm system activation begins first by activating the fire alarm control unit, then by notifying the occupants of an alarm condition and then goes on to introduce all of the components of an occupant notification system contained in Sections 907.5.1 through 907.5.2.3.3.

It also lists the system components that are to act as alarm initiation devices. The exception is a recognition that there are places in the code where an alternative to occupant notification is an alarm notification at a constantly attended location. The exception is intended to clarify the code so that there is no question as to whether this general provision for alarm activation is superseded by the other sections addressing the alarm notification at a constantly attended location.

[F] 907.5.1 Presignal feature. A presignal feature shall not be installed unless *approved* by the fire code official and the fire department. Where a presignal feature is provided, a signal shall be annunciated at a *constantly attended location approved* by the fire department so that occupant notification can be activated in the event of fire or other emergency.

❖ A presignal feature on a fire alarm system allows the occupant notification devices to activate in selected, constantly attended locations only and from which human intervention is required to activate a general occupant notification signal. Alternatively, this feature can be programmed to delay the general alarm notification for more than 1 minute before it will automatically be activated by the control panel. In either presignal scenario, remote transmission of the alarm signal to the fire department is immediate. See NFPA 72 for additional information on the presignal feature.

Improper use of the presignal feature has been a contributing factor in several multiple-death fire incidents. In most instances, the staff failed to activate the general alarm quickly and the occupants of the building were unaware of the fire. Therefore, the use of a presignal feature is discouraged by the code. A presignal feature may be used only if it is approved by the fire code official and the fire department.

[F] 907.5.2 Alarm notification appliances. Alarm notification appliances shall be provided and shall be *listed* for their purpose.

❖ The code requires that fire alarm systems be equipped with approved alarm notification appliances so that in an emergency, the fire alarm system will notify the occupants of the need for evacuation or implementation of the fire emergency plan. Alarm notification devices required by the code are of two general types: visible and audible. Except for voice/alarm signaling systems, once the system has been activated, all visible and audible alarms are required to activate. Voice/alarm signaling systems are special signaling systems that are activated selectively in response to specific emergency conditions.

[F] 907.5.2.1 Audible alarms. Audible alarm notification appliances shall be provided and emit a distinctive sound that is not to be used for any purpose other than that of a fire alarm.

Exceptions:

1. Audible alarm notification appliances are not required in critical care areas of Group I-2 Condition 2 occupancies that are in compliance with Section 907.2.6, Exception 2.
2. A visible alarm notification appliance installed in a nurses' control station or other continuously attended staff location in a Group I-2 Condition 2 suite shall be an acceptable alternative to the installation of audible alarm notification appliances throughout the suite in Group I-2 Condition 2 occupancies that are in compliance with Section 907.2.6, Exception 2.
3. Where provided, audible notification appliances located in each occupant evacuation elevator lobby in accordance with Section 3008.9.1 shall be connected to a separate notification zone for manual paging only.

❖ To attract the attention of building occupants, audible alarms must be distinctive, using a sound that is unique to the fire alarm system and used for no other purpose than alerting occupants to a fire emergency. Other emergencies, such as tornados, etc., must be signaled by another sound different from the fire signal.

Exception 1 recognizes that the occupants in critical care areas of Group I-2 occupancies are usually incapacitated. The audible alarms may have the effect of unnecessarily disrupting the care recipients who are most likely not capable of self-preservation. Likewise, audible alarms in operating theaters of hospitals could be hazardous because an alarm activation could startle a surgeon during a delicate procedure. Critical care areas are also assumed to be adequately staffed at all times. Section 907.2.6 Exception 2 allows the use of private mode signaling in accordance with NFPA 72 and also requires that staff evacuation responsibilities be included in the fire safety and evacuation plan. See the commentary to Section 907.2.6. In private mode, as permitted by Section 907.2.6, there is still a requirement for an audible alarm notification from appliances, though at a much lower decibel level meant to alert staff of the alarm activation. Allowing the audible alarm to be eliminated from critical care areas (operating rooms) in exchange for a visual notification device is also not appropriate since the visual signal device also creates a distraction in critical care areas that may not be able to immediately stop a patient procedure. The emergency action plan would include provisions for alerting of critical area staff and the actions to be taken.

Exception 2 allows hospital care suites to eliminate audible alarms where visible alarm notification appliances are located at a continuously attended staff location or nurses control station. In a suite arrangement the "control area" is the centrally manned location for staff monitoring patients in the separate rooms. An alarm indicator at this location will alert staff for response in a more effective and efficient manner. Similar to exception 1 audible alarms can unnecessarily disrupt the care recipients. In addition simply providing visible alarm notification throughout the suite is not appropriate. The patient's evacuation depends on staff since they are not capable of self-preservation. As with Exception 1, compliance with Exception 2 to Section 907.2.6 would be required in order to take advantage of this exception.

Exception 3 is intended to address the concern that automatic emergency voice/alarm messages do not interfere with operation of the two-way communication associated with the occupant evacuation elevators. Live voice messages would be appropriate in the lobbies.

[F] 907.5.2.1.1 Average sound pressure. The audible alarm notification appliances shall provide a sound pressure level of 15 decibels (dBA) above the average ambient sound level or 5 dBA above the maximum sound level having a duration of not less than 60 seconds, whichever is greater, in every occupiable space within the building.

❖ To attract the attention of building occupants, this section requires that the distinctive audible alarms must be capable of being heard above the ambient noise level in a space. The indicated levels are considered the minimum pressure differential that will be perceivable by most people. It prescribes that the sound pressure level (SPL) for notification appliances shall be a minimum of 15 decibels measured in the A-scale (dBA) above the ambient SPL or 5 dBA above the maximum SPL in every space that can be occupied in a building. These SPLs are based on a minimum 1-minute measurement period. SPLs for Group R and I-1 occupancies, mechanical rooms and other occupancies are no longer stipulated as they had been in previous editions of the code.

The values mandated in Section 907.5.2.1.1 in previous editions of the code were not consistent with the notification appliance SPL requirements in NFPA 72. NFPA 72 requirements for the audible notification appliances are based on if the devices emit alert or evacuation tones, voice messages or audible notifications for exit markings. The provisions in Section 907.5.2.1.1 would apply to all notification appliances designed to operate in either public- or private-mode. In sleeping areas, the minimum SPL is no longer specified in Section 907.5.2.1.1; however, for smoke alarms, Section 907.2.11 and NFPA 72 require a minimum 75 dBA SPL at the pillow.

Also note that the 2010 American with Disabilities Act *Standard for Accessible Design* has an exception for medical care facilities following industry practice that will allow a dependence upon staff. The activation of either audible or visible alarms could be detrimental to the care recipients in locations like operating rooms and intensive or critical care units.

[F] 907.5.2.1.2 Maximum sound pressure. The maximum sound pressure level for audible alarm notification appliances shall be 110 dBA at the minimum hearing distance from the audible appliance. Where the average ambient noise is greater than 95 dBA, visible alarm notification appliances shall be provided in accordance with NFPA 72 and audible alarm notification appliances shall not be required.

❖ In no case may the sound pressure level exceed 110 dBA at the minimum hearing distance from the audible appliance. This is consistent with Americans with Disabilities Act (ADA) requirements. Sound pressures above that level can cause pain or even permanent hearing loss. In such cases, audible alarms are not required to be installed but visual alarms would be necessary to compensate for the lack of audibility.

It should also be noted that in certain work areas, Occupational Safety and Health Administration (OSHA) requires employees to wear hearing protection, possibly preventing them from hearing an audible alarm. Additionally, the noise factor in these areas is high enough that an audible alarm may not be discernible. In these areas, as well as in others, the primary method of indicating a fire can be by a visible signal. Employees must be capable of identifying such a signal as indicating a fire.

[F] 907.5.2.2 Emergency voice/alarm communication systems. Emergency voice/alarm communication systems required by this code shall be designed and installed in accordance with NFPA 72. The operation of any automatic fire detector, sprinkler waterflow device or manual fire alarm box shall automatically sound an alert tone followed by voice instructions giving *approved* information and directions for a general or staged evacuation in accordance with the building's fire safety and evacuation plans required by Section 404 of the *International Fire Code*. In high-rise buildings, the system shall operate on at least the alarming floor, the floor above and the floor below. Speakers shall be provided throughout the building by paging zones. At a minimum, paging zones shall be provided as follows:

1. Elevator groups.
2. *Interior exit stairways.*
3. Each floor.
4. *Areas of refuge* as defined in Chapter 2.

Exception: In Group I-1 and I-2 occupancies, the alarm shall sound in a constantly attended area and a general occupant notification shall be broadcast over the overhead page.

❖ The primary purpose of an emergency voice/alarm communication system is to provide dedicated manual and automatic facilities for the origination, control and transmission of information and instructions pertaining to a fire alarm emergency to the occupants of a building. This section identifies that notification speakers are required throughout the building with a minimum of one speaker in each paging zone when an emergency voice/alarm communication system is required. The system may sound a general alarm or be a selective system in which only certain areas of the building receive the alarm indication for staged evacuation. See Chapter 4 of the IFC for evacuation plan requirements. The intent is to provide the capability to send out selective messages to individual areas; however, it does not prohibit the same message to be sent to all areas. In high-rise buildings, a minimum area of notification must include the alarming floor and one floor above and one floor below it.

This section also identifies the minimum paging zone arrangement. This does not preclude further zone divisions for logical staged evacuation in accordance with an approved evacuation plan.

This section also indicates that the emergency voice/alarm system is to be initiated as all other fire alarm systems are initiated. The functional operation of the system begins with an alert tone (usually 3 to 10 seconds in duration) followed by the evacuation signal (message). It is important to remember that the voice alarm system is not an "audible alarm." It has its own specific criteria for installation and approval according to NFPA 72. Consequently the sound pressure requirements for audible alarms do not apply to voice alarm systems. For voice alarm systems, the intent is communication and an understanding of what is being said, not volume.

The exception is similar to the one to Section 907.5.2.1 and recognizes the supervised environment typical of institutional uses and the reliance placed on staff to act appropriately in an emergency. As is the case for most voice alarms, the key is in being able to deliver specific information to the people who can affect a safe egress—whether this is the public, employees, or both.

[F] 907.5.2.2.1 Manual override. A manual override for emergency voice communication shall be provided on a selective and all-call basis for all paging zones.

❖ The intent of this section is to provide the ability to transmit live voice instructions over any previously initiated signals or prerecorded messages for all zones. This would include the ability to override the voice message at once throughout the building or to be able to select individual paging zones for the message override.

[F] 907.5.2.2.2 Live voice messages. The emergency voice/alarm communication system shall have the capability to broadcast live voice messages by paging zones on a selective and all-call basis.

❖ This would include the ability to provide the live voice message at once throughout the building or to be able to select individual paging zones to receive the message. Speakers used for background music must not be used unless specifically listed for fire alarm system use. NFPA 72 has additional requirements for the placement, location and audibility of speakers used as part of an emergency voice/alarm communication system.

[F] 907.5.2.2.3 Alternate uses. The emergency voice/alarm communication system shall be allowed to be used for other announcements, provided the manual fire alarm use takes precedence over any other use.

❖ In certain circumstances which should be approved by the fire code officials, the emergency voice/alarm communication system could be used to convey information other than fire alarm-related items. This could include severe weather warnings that might require evacuation or relocation, lockdown instructions (see commentary, Section 404.2.3 of the IFC) and similar approved messages. In the event of such usage, the system must respond immediately to manual fire alarm box activations.

[F] 907.5.2.2.4 Emergency voice/alarm communication captions. Where stadiums, arenas and grandstands are required to caption audible public announcements in accordance with Section 1108.2.7.3, the emergency/voice alarm communication system shall be captioned. Prerecorded or live emergency captions shall be from an *approved* location constantly attended by personnel trained to respond to an emergency.

❖ This provision links the EV/ACS with the requirements for captioning in Section 1108.2.7.3. Section 1108.2.7.3 requires that stadiums, arenas and grandstands have 15,000 fixed seats to provide captioning for audible announcements (see commentary, Section 907.2.1.2).

[F] 907.5.2.2.5 Emergency power. Emergency voice/alarm communications systems shall be provided with emergency power in accordance with Section 2702. The system shall be capable of powering the required load for a duration of not less than 24 hours, as required in NFPA 72.

❖ Because the emergency voice/alarm communication system is a critical aid in evacuating the building, the system must be connected to an approved emergency power source complying with Section 2702. The section also clarifies that the duration of the load for EV/ACS is a minimum of 24 hours.

[F] 907.5.2.3 Visible alarms. Visible alarm notification appliances shall be provided in accordance with Sections 907.5.2.3.1 through 907.5.2.3.3.

Exceptions:

1. Visible alarm notification appliances are not required in *alterations*, except where an existing fire alarm system is upgraded or replaced, or a new fire alarm system is installed.
2. Visible alarm notification appliances shall not be required in *exits* as defined in Chapter 2.
3. Visible alarm notification appliances shall not be required in elevator cars.
4. Visual alarm notification appliances are not required in critical care areas of Group I-2 Condition 2 occupancies that are in compliance with Section 907.2.6, Exception 2.

❖ This section contains alarm system requirements for occupants who are hearing impaired. Visible alarm notification appliances are to be installed in conjunction with the audible devices and located and oriented so that they will display alarm signals throughout a space. It is not the intent of the code to offer visible alarm signals as an option to audible alarm signals. Both are required. However, the code acknowledges conditions when audible alarms may be of little or no value, such as when the ambient sound level exceeds 105 dBA. In such cases, Section 907.5.2.1, similar to NFPA 72, allows for visible alarm notification appliances in the area.

Exception 1 states that visible alarm devices are not required in previously approved existing fire alarm systems or as part of minor alterations to existing fire alarm systems. Extensive modifications to an existing fire alarm system such as an upgrade or replacement would require the installation of visible alarm devices even if the previous existing system neither had them nor required them. The main reason is a combination of simple economics and practical application. Many existing systems that do not have visible signal devices do not have the wiring capability to include such devices. To make the necessary changes to the existing system, a total replacement of the existing system may need to take place. In many cases this is cost prohibitive. Thus, if the alteration is small, the system can be left as is, without the visual devices. The second consideration is scope. If the alteration involves only a limited area, it could be confusing to have part of the area equipped with visual devices and part without. This is not good practice, as the alarm could be confusing. If an entire floor is being altered, then it becomes subject to consideration for an upgrade to an alarm system with visual devices. If only an office is being remodeled, then the implication is that the upgrade to visual devices may not be warranted. This determination will be subjective in many cases and should be applied based on the life safety benefit and financial expenses involved and whether adequate audible devices are present for full coverage.

In Exception 2, visible alarm devices are not required in exit elements because of the potential distraction during evacuation. Exits, as defined in Chapter 2, could include interior exit stairways or exit passageways but not exit access corridors. In tall buildings, exiting may be phased based on alarm zone. If the alarm floor and adjacent floors are notified of the emergency but the remainder of the building is not, then a visual device in the stairway would be confusing to those people who may not be coming from the alarm floor.

Previously, some jurisdictions were requiring visible alarm notification appliances to be installed in elevator cars since there was no exception in the code or NFPA 72 to allow omission of this type of notification appliance in elevator cars. Exception 3 eliminates any confusion regarding the need to install visible notification appliances in elevator cars. The rationale for not installing visible notification appliances in elevator cars is the same as for interior exit stairways; high light intensity from these notification appliances may cause confusion and disorientation. Also, elevator passengers are "captive" in that they cannot respond to such devices until the elevator arrives at its destination or is recalled by the Phase I emergency recall feature, which could lead to passenger panic.

Exception 4 was added to eliminate visual alarm notification appliances in critical care areas of Group I-2 Condition 2 occupancies (hospitals) because of the hazards they may pose to the occupants. Such occupancies already have staff procedures in place during fires that more than compensate for visual alarm notification appliances. In addition in these areas of the hospital the patients are not typically ambulatory. This allowance includes specific requirements for detailing of the staff evacuation responsibilities based on the requirements in Section 907.2.6. See also the commentary to Exception 1 of Section 907.5.2.1.

[F] 907.5.2.3.1 Public use areas and common use areas. Visible alarm notification appliances shall be provided in *public use areas* and *common use areas*.

Exception: Where employee work areas have audible alarm coverage, the notification appliance circuits serving the employee work areas shall be initially designed with not less than 20-percent spare capacity to account for the potential of adding visible notification appliances in the future to accommodate hearing-impaired employee(s).

❖ Visible alarm notification appliances must provide coverage in all areas open to the public (use areas) as well as all shared or common use areas (e.g., corridors, public restrooms, shared offices, classrooms, medical exam rooms, etc.). Areas where visible alarm notification appliances are not required include private offices, mechanical rooms or similar spaces. The intent with this section is to replicate the provisions included in the Americans with Disabilities Act *Accessibility Guidelines for Buildings and Facilities* (ADAAG).

The exception allows employee work areas to provide only for spare capacity on notification circuits to allow for those with hearing impairments to be accommodated as necessary. This spare capacity is intended to eliminate the potential for overloading notification circuits when a hearing-impaired person is hired and needs to be accommodated, but reduces the initial construction cost as such alarm notification appliances may not be necessary in every situation.

[F] 907.5.2.3.2 Groups I-1 and R-1. Group I-1 and R-1 *dwelling units* or *sleeping units* in accordance with Table 907.5.2.3.2 shall be provided with a visible alarm notification appliance, activated by both the in-room smoke alarm and the building fire alarm system.

❖ Fire alarm systems in Group I-1 and R-1 sleeping accommodations must be equipped with visible alarms to the extent stated in Table 907.5.2.3.2. The visible alarm notification devices in these rooms are to be activated by both the required in-room smoke alarm and the building fire alarm system. All visible alarm notification appliances in a building, however, need not be activated by individual room smoke alarms. It is not a requirement that the accessible sleeping units be provided with visible alarm notification appliances even though some elderly patients or residents may be both mobility and hearing impaired.

[F] TABLE 907.5.2.3.2
VISIBLE ALARMS

NUMBER OF SLEEP UNITS	SLEEPING ACCOMMODATIONS WITH VISIBLE ALARMS
6 to 25	2
26 to 50	4
51 to 75	7
76 to 100	9
101 to 150	12
151 to 200	14
201 to 300	17
301 to 400	20
401 to 500	22
501 to 1,000	5% of total
1,001 and over	50 plus 3 for each 100 over 1,000

❖ This table specifies the minimum number of sleeping units that are to be equipped with visible and audible alarms. The numbers are based on the total number of sleeping accommodations in the facility. The requirements in this table are intended to be consistent with the ADAAG.

[F] 907.5.2.3.3 Group R-2. In Group R-2 occupancies required by Section 907 to have a fire alarm system, all *dwelling units* and *sleeping units* shall be provided with the capability to support visible alarm notification appliances in accordance with Chapter 10 of ICC A117.1. Such capability shall be permitted to include the potential for future interconnection of the building fire alarm system with the unit smoke alarms, replacement of audible appliances with combination audible/visible appliances, or future extension of the existing wiring from the unit smoke alarm locations to required locations for visible appliances.

❖ Group R-2 occupancies with a fire alarm system are required to have the capability to support visual alarm notification appliances in accordance with Chapter 10 of ICC A117.1. This requirement has been in the IBC and the language added in the 2012 edition is intended to provide more specific guidance as to what is meant by "capability." Note that this requirement includes all dwelling and sleeping units, not just those classified as either Type A or B. Sections 1006.2 through 1006.4.4 of ICC A117.1 address smoke and fire alarm requirements as they pertain to accessible communication features. More specifically, Section 1006.2 states that when unit smoke detection is provided, it shall provide audible notification in compliance with NFPA 72. Section 1006.3 is focused on buildings where fire alarm systems are provided. If a fire alarm system is provided in the building, ICC A117.1 requires that the wiring be extended to a point within the unit in the vicinity of the smoke detection system. Based on the type of unit and the strategy used by the designer, this location may vary. Section 1006.4 addresses the visible alarm requirements specifically and has various issues it addresses, as follows:

1. Complies with Section 702 of ICC A117.1, which focuses on the requirements of NFPA 72, and that such notification devices be hardwired.
2. Addresses the fact that all visible notification devices be activated within the unit either when the smoke alarms in the unit activate or when that portion of the building fire alarm system in that portion of the building activate.
3. Allows the same visible notification for the smoke alarms in the unit and the building fire alarm system.
4. Prohibits the use of the visible notification for anything other than the operation of the smoke alarms in the unit or the building fire alarm system.

In terms of the specific capability requirements, this section has been clarified to provide direction as to what may be meant by bringing the wiring to the unit. There has been confusion in the past and it has been interpreted that all units are required to be prewired for visible appliances, which was not the intent of ICC A117.1. More specifically, now the requirements provide essentially three options for future capability, as follows:

- Potential for future interconnection of the building fire alarm system with the unit smoke alarms.
- Replacement of audible appliance with combination audible/visible appliances.
- Extension of wiring from the unit smoke alarm locations to required locations of visible appliances.

It is important to remember that the location of visible notification devices, if installed, are driven by the requirements of NFPA 72 and may vary the approach taken, based upon the configuration of the space.

[F] 907.6 Installation and monitoring. A fire alarm system shall be installed and monitored in accordance with Sections 907.6.1 through 907.6.6.2 and NFPA 72.

❖ This section specifies the requirements for fire alarm system installation and monitoring and also references the installation requirements of NFPA 72.

[F] 907.6.1 Wiring. Wiring shall comply with the requirements of NFPA 70 and NFPA 72. Wireless protection systems utilizing radio-frequency transmitting devices shall comply with the special requirements for supervision of low-power wireless systems in NFPA 72.

❖ Wiring for fire alarm systems must be installed so that it is secure and will function reliably in an emergency. The code requires that the wiring for fire alarm systems meet the requirements of NFPA 70 and NFPA 72. This requirement is in addition to the general requirements for electrical installations set forth in Chapter 27 of the IBC. For reliability, systems that use radio-frequency transmitting devices for signal transmission are required to have supervised transmitting and receiving equipment that conforms to the special requirements contained in NFPA 72. This requirement is in addition to the general requirements for supervision in Section 907.6.6.

[F] 907.6.2 Power supply. The primary and secondary power supply for the fire alarm system shall be provided in accordance with NFPA 72.

> **Exception:** Back-up power for single-station and multiple-station smoke alarms as required in Section 907.2.11.6.

❖ The operation of fire alarm systems is essential to life safety in buildings and must be reliable in the event the normal power supply fails. For proper operation of fire alarm systems, this section requires that the primary and secondary power supplies comply with NFPA 72. This is in addition to the general requirements for electrical installations in Chapter 27. NFPA 72 offers three alternatives for secondary supply: a

24-hour storage battery; storage batteries with a 4-hour capacity; and a generator or multiple generators.

NFPA 72 requires that the primary and secondary power supplies for remotely located control equipment essential to the system operation must conform to the requirements for primary and secondary power supplies for the main system. Also, NFPA 72 contains requirements for monitoring the integrity of primary power supplies and requires a backup power supply.

[F] 907.6.3 Initiating device identification. The fire alarm system shall identify the specific initiating device address, location, device type, floor level where applicable and status including indication of normal, alarm, trouble and supervisory status, as appropriate.

Exceptions:

1. Fire alarm systems in single-story buildings less than 22,500 square feet (2090 m2) in area.
2. Fire alarm systems that only include manual fire alarm boxes, waterflow initiating devices and not more than 10 additional alarm-initiating devices.
3. Special initiating devices that do not support individual device identification.
4. Fire alarm systems or devices that are replacing existing equipment.

❖ Current technology makes identification of alarm-initiating devices much easier. This section takes advantage of this technology to improve the ability of emergency responders to rapidly identify the location and status of initiating devices at the time of an emergency. It will also help identify problematic alarm-initiating devices and thus reduce nuisance alarms. It also eliminates the requirements for providing zone indication of system status. This is considered particularly important in high-rise buildings, where the number of initiating devices and the geometry of the building warrant a need for point monitoring of individual devices, which is not currently accommodated by single floor zones.

This section allows the building official the flexibility to not require individual detection device identification in smaller buildings where the source of alarm and trouble signals can be more easily determined.

The 22,500-square-foot limitation noted in Exception 1 relates to the size of a typical fire alarm zone (see Section 907.6.4) and represents a small building. Exception 2 addresses manual fire alarm systems in which the location of initiation may not be an indicator of where the fire actually is or an automatic sprinkler waterflow alarm-initiating device which could be annunciating an entire building, along with a very limited number of other alarm initiating devices Exception 3 recognizes that some devices will not work with this requirement. Finally, Exception 4 provides flexibility to existing system replacement. Replacement should be encouraged and requiring identification of alarm-initiating device locations is considered onerous.

[F] 907.6.3.1 Annunciation. The initiating device status shall be annunciated at an *approved* on-site location.

❖ This section specifically notes that the alarm-initiating device status of trouble versus alarm needs to be provided in an approved location to enable rapid identification of problematic devices or alarm conditions by first responders.

[F] 907.6.4 Zones. Each floor shall be zoned separately and a zone shall not exceed 22,500 square feet (2090 m^2). The length of any zone shall not exceed 300 feet (91 440 mm) in any direction.

Exception: *Automatic sprinkler system* zones shall not exceed the area permitted by NFPA 13.

❖ Since the fire alarm system also aids emergency personnel in locating the fire, the system must be zoned to shorten response time to the fire location. Zoning is also critical if the fire alarm system initiates certain other fire protection systems or control features, such as smoke control systems.

At a minimum, each floor of a building must constitute one zone of the system. If the floor area exceeds 22,500 square feet (2090 m^2), additional zones per floor are required. The maximum length of a zone is 300 feet (91 440 mm).

The exception states that NFPA 13 defines the maximum areas to be protected by one sprinkler system and that the sprinkler system need not be designed to meet the 22,500-square-foot (2090 m^2) area limitations for a fire alarm system zone. For example, NFPA 13 permits a sprinkler system riser in a light-hazard occupancy to protect an area of 52,000 square feet (4831 m^2) per floor. In accordance with the exception, a single waterflow switch, and consequently a single fire alarm system zone, would be acceptable. If other alarm-initiating devices are present on the floor, they would need to be zoned separately to meet the 22,500-square-foot (2098 m^2) limitation.

It is not intended that this section apply to sprinkler systems. This section only applies where a fire alarm system is required in accordance with Section 907. Unless the building is categorized as a high rise and must comply with Section 907.6.4.2, the code does not mandate the zoning of sprinkler systems per floor. With today's fully addressable fire alarm systems, each detector effectively becomes its own zone. The intent with zoning is to identify and limit the search area for fire alarm systems. Addressable devices will indicate the precise location of the alarm condition, thereby eliminating the need for the zoning contemplated by this section when approved by the fire code official in accordance with Section 104.11.

[F] 907.6.4.1 Zoning indicator panel. A zoning indicator panel and the associated controls shall be provided in an *approved* location. The visual zone indication shall lock in until the system is reset and shall not be canceled by the operation of an audible-alarm silencing switch.

❖ The zoning indicator panel, which can be the fire alarm control unit or a separate fire alarm annunciator panel (FAAP), must be installed in a location approved by the fire code official. One of the key considerations in determining panel placement is whether or not the panel is located to permit ready access by emergency responders. Once an alarm-initiating device within a zone has been activated, the annunciation of the zone must lock in until the system is reset.

[F] 907.6.4.2 High-rise buildings. In high-rise buildings, a separate zone by floor shall be provided for each of the following types of alarm-initiating devices where provided:

1. Smoke detectors.
2. Sprinkler waterflow devices.
3. Manual fire alarm boxes.
4. Other *approved* types of automatic fire detection devices or suppression systems.

❖ High-rise buildings must have a separate zone by floor for each indicated type of alarm-initiating device. Although this feature may be desirable in all buildings, the incremental cost difference is substantially higher in low-rise buildings in which basic fire alarm systems are installed. State-of-the-art fire alarm systems installed in high-rise buildings are addressable and by their nature automatically provide this minimum zoning.

[F] 907.6.5 Access. Access shall be provided to each fire alarm device and notification appliance for periodic inspection, maintenance and testing.

❖ Automatic fire detectors, especially smoke detectors, require periodic cleaning to reduce the likelihood of malfunction. Section 907.8 and NFPA 72 require inspection and testing at regular intervals. Access to perform the required inspections, necessary maintenance and testing is a particularly important consideration for those detectors that are installed within a concealed space, such as an air duct.

[F] 907.6.6 Monitoring. Fire alarm systems required by this chapter or by the *International Fire Code* shall be monitored by an *approved* supervising station in accordance with NFPA 72.

Exception: Monitoring by a supervising station is not required for:

1. Single- and multiple-station smoke alarms required by Section 907.2.11.
2. Smoke detectors in Group I-3 occupancies.
3. *Automatic sprinkler systems* in one- and two-family dwellings.

❖ Fire alarm systems required by Section 907 are required to be electrically supervised by one of the methods prescribed in NFPA 72.

Exception 1 exempts single- and multiple-station smoke alarms from being supervised because of the potential for unwanted false alarms.

Exception 2 recognizes a similar problem in Group I-3 occupancies. Accordingly, because of the concern over unwanted alarms, smoke detectors in Group I-3 occupancies need only sound an approved alarm signal that automatically notifies staff (see Section 907.2.6.3.1). Smoke detectors in such occupancies are typically subject to misuse and abuse, and frequent unwanted alarms would negate the effectiveness of the system.

Exception 3 clarifies that sprinkler systems in one- and two-family dwellings are not part of a dedicated fire alarm system and are typically designed in accordance with NFPA 13D, which does not require electrical supervision.

[F] 907.6.6.1 Automatic telephone-dialing devices. Automatic telephone-dialing devices used to transmit an emergency alarm shall not be connected to any fire department telephone number unless *approved* by the fire chief.

❖ On initiation of an alarm, supervisory or trouble signal, an automatic telephone-dialing device takes control of the telephone line for the reliability of transmission of all signals. The device, however, must not be connected to the fire department telephone number unless specifically approved by the fire department because that could disrupt any potential emergency (911) calls. NFPA 72 contains additional guidance on such devices, including digital alarm-communicator systems.

[F] 907.6.6.2 Termination of monitoring service. Termination of fire alarm monitoring services shall be in accordance with Section 901.9 of the *International Fire Code*.

❖ This is simply a cross link to a requirement in the IFC related to the termination of monitoring service. Section 901.9 of the IFC requires the monitoring service itself to notify the fire code official of the service being terminated. Although the ultimate responsibility rests with the building owner, he or she is not cited in this section since if they discontinued the service, they would likely not understand the implications, and if they did, would have no incentive to contact the fire code official.

[F] 907.7 Acceptance tests and completion. Upon completion of the installation, the fire alarm system and all fire alarm components shall be tested in accordance with NFPA 72.

❖ A complete performance test of the fire alarm system must be conducted to determine that the system is operating as required by the code. The acceptance test must include a test of each circuit, alarm-initiating device, alarm notification appliance and any supplementary functions, such as activation of closers and dampers. The operation of the primary and secondary (emergency) power supplies must also be tested, as well as the supervisory function of the control panel. Section 901.5 assigns responsibility for conducting the acceptance tests to the owner or the owner's representative.

NFPA 72 contains specific acceptance test procedures. Additional guidance on periodic testing and inspection can also be obtained from Section 907.8 and NFPA 72.

[F] 907.7.1 Single- and multiple-station alarm devices. When the installation of the alarm devices is complete, each device and interconnecting wiring for multiple-station alarm devices shall be tested in accordance with the smoke alarm provisions of NFPA 72.

❖ To determine that smoke alarms have been properly installed and are ready to function as intended, they must be actuated during an acceptance test. The test also confirms that interconnected detectors will operate simultaneously as required. The responsibility for conducting the acceptance tests rests with the owner or the owner representative as stated in Section 901.5.

[F] 907.7.2 Record of completion. A record of completion in accordance with NFPA 72 verifying that the system has been installed and tested in accordance with the *approved* plans and specifications shall be provided.

❖ In accordance with NFPA 72, this section requires a written statement from the installing contractor that the fire alarm system has been tested and installed in compliance with the approved plans and the manufacturer's specifications. Any deviations from the approved plans or the applicable provisions of NFPA 72 are to be noted in the record of completion.

[F] 907.7.3 Instructions. Operating, testing and maintenance instructions and record drawings ("as-builts") and equipment specifications shall be provided at an *approved* location.

❖ To permit adequate testing, maintenance and troubleshooting of the installed fire alarm system, an owner's manual with complete installation instructions must be kept on site or in another approved location. The instructions include a description of the system, operating procedures and testing and maintenance requirements.

[F] 907.8 Inspection, testing and maintenance. The maintenance and testing schedules and procedures for fire alarm and fire detection systems shall be in accordance with Section 907.8 of the *International Fire Code*.

❖ Fire alarms and smoke detection systems are to be inspected, tested and maintained in accordance with Sections 907.8.1 through 907.8.5 of the IFC and the applicable requirements of NFPA 72. It is the building owner's responsibility to keep these systems operable at all times.

SECTION 908
EMERGENCY ALARM SYSTEMS

[F] 908.1 Group H occupancies. Emergency alarms for the detection and notification of an emergency condition in Group H occupancies shall be provided in accordance with Section 415.5.

❖ Emergency alarm systems provide indication and warning of emergency situations involving hazardous materials. An emergency alarm system is required in all Group H occupancies, as indicated in Section 415.5 as well as Group H-5 HPM facilities as indicated in Sections 415.11.3.5 and 908.2. The Group H occupancy classification assumes the storage or use of hazardous materials exceeds the maximum allowable quantities specified in Tables 307.1(1) and 307.1(2).

An emergency alarm system should include an emergency alarm-initiating device outside each interior door of hazardous material storage areas, a local alarm device and adequate supervision.

Even though ozone gas-generator rooms (see Section 908.4), repair garages (see Section 908.5) and refrigeration systems (see Section 908.6) are not typically classified as Group H occupancies, the potential hazards associated with these occupancy conditions are great enough to require additional means of early warning detection.

[F] 908.2 Group H-5 occupancy. Emergency alarms for notification of an emergency condition in an HPM facility shall be provided as required in Section 415.11.3.5. A continuous gas detection system shall be provided for HPM gases in accordance with Section 415.11.7.

❖ In addition to hazardous material storage areas as regulated by Section 415.5.1 of the IFC, Section 415.11.3.5 also requires emergency alarms for service corridors, exit access corridors and interior exit stairways because of the potential transport of hazardous materials through these areas. Section 415.11.7 requires a continuous gas detection system for early detection of leaks in areas where HPM gas is used. Gas detection systems are required to initiate a local alarm and transmit a signal to the emergency control station on detection (see commentary, Sections 415.11.3.5 and 415.11.7.

[F] 908.3 Highly toxic and toxic materials. A gas detection system shall be provided to detect the presence of *highly toxic* or *toxic* gas at or below the permissible exposure limit (PEL) or ceiling limit of the gas for which detection is provided. The system shall be capable of monitoring the discharge from the treatment system at or below one-half the immediately dangerous to life and health (IDLH) limit.

Exception: A gas detection system is not required for *toxic* gases when the physiological warning threshold level for the gas is at a level below the accepted PEL for the gas.

❖ A gas detection system in the room or area utilized for indoor storage or the use of highly toxic or toxic gases provides early notification of a leak that is occurring before the escaping gas reaches hazardous exposure concentration levels. The exception recognizes that certain toxic compressed gases do not pose a severe exposure hazard. Those toxic gases whose properties under standard conditions are still below the 8-hour weighted average concentration for the permitted exposure limit (PEL) are exempt from the requirement for a gas detection system.

This section also specifies the discharge requirements for treatment system performance to establish a maximum allowable concentration of highly toxic or toxic gases at the point of discharge to the atmosphere. The concentration level of one-half the immediately dangerous to life and health (IDLH) limit represents a minimum acceptable level of dilution at the point of discharge where the location of discharge is away from the general public. Where the treatment system processes more than one type of compressed gas, the maximum allowable concentration must be based on the release rate, quantity and IDLH for the gas that poses the worst-case release scenario.

[F] 908.3.1 Alarms. The gas detection system shall initiate a local alarm and transmit a signal to a constantly attended control station when a short-term hazard condition is detected. The alarm shall be both visible and audible and shall provide warning both inside and outside the area where gas is detected. The audible alarm shall be distinct from all other alarms.

Exception: Signal transmission to a constantly attended control station is not required when not more than one cylinder of *highly toxic* or *toxic* gas is stored.

❖ The required local alarm is intended to alert occupants to a hazardous condition in the vicinity of the inside storage room or area. The alarm is not intended to be an evacuation alarm; however, it is required to be monitored to hasten emergency personnel response. The exception allows the omission of supervision for the gas detection system where a single cylinder of highly toxic or toxic gas is stored. It should be noted that this section is only intended to apply when the maximum allowable quantity of a specific gas per control area is exceeded. A single cylinder would not require a gas detection system if it contained less than the maximum allowable quantities indicated in Tables 307.1(1) and 307.1(2).

[F] 908.3.2 Shutoff of gas supply. The gas detection system shall automatically close the shutoff valve at the source on gas supply piping and tubing related to the system being monitored for whichever gas is detected.

Exception: Automatic shutdown is not required for reactors utilized for the production of *highly toxic* or *toxic* compressed gases where such reactors are:

1. Operated at pressures less than 15 pounds per square inch gauge (psig) (103.4 kPa).
2. Constantly attended.
3. Provided with readily accessible emergency shutoff valves.

❖ Actuation of the gas detection system is required to close automatically all valves controlling highly toxic or toxic gases. This degree of protection is deemed necessary to provide a life safety measure in areas where highly toxic and toxic gases are being utilized for filling, dispensing or other operations.

In most situations, dispensing and use operations are assumed to be constantly attended because of the potential hazard and need to monitor operations involving highly toxic and toxic compressed gases.

The exception recognizes the decreased level of hazard for constantly attended operations involving reactors in manufacturing processes with low operating pressures. With low operating pressures, the rate at which the volume of the affected space would be filled by the released gas is minimized. Readily accessible manual shutoff valves would thus be permitted in lieu of an automatic-closing shutoff valve.

[F] 908.3.3 Valve closure. The automatic closure of shutoff valves shall be in accordance with the following:

1. When the gas-detection sampling point initiating the gas detection system alarm is within a gas cabinet or exhausted enclosure, the shutoff valve in the gas cabinet or exhausted enclosure for the specific gas detected shall automatically close.
2. Where the gas-detection sampling point initiating the gas detection system alarm is within a gas room and compressed gas containers are not in gas cabinets or exhausted enclosures, the shutoff valves on all gas lines for the specific gas detected shall automatically close.
3. Where the gas-detection sampling point initiating the gas detection system alarm is within a piping distribution manifold enclosure, the shutoff valve for the compressed container of specific gas detected supplying the manifold shall automatically close.

Exception: When the gas-detection sampling point initiating the gas detection system alarm is at a use location or within a gas valve enclosure of a branch line downstream of a piping distribution manifold, the shutoff valve in the gas valve enclosure for the branch line located in the piping distribution manifold enclosure shall automatically close.

❖ This section specifies the conditions where the shutoff valve for the gas supply is required to automatically close. The requirement depends on the location of the gas detection sampling point, which would initiate the actuation of the gas detection system, and the use of gas cabinets, exhaust enclosures or a piping distribution manifold enclosure.

[F] 908.4 Ozone gas-generator rooms. Ozone gas-generator rooms shall be equipped with a continuous gas detection system that will shut off the generator and sound a local alarm when concentrations above the PEL occur.

❖ To monitor the potential buildup of dangerous levels of ozone, a gas detection system is required to, on actuation, shut off the generator and sound a local alarm. Ozone gas generators are commonly used in water treatment applications. The ozone gas-generator room should not be a normally occupied area or be used for the storage of combustibles or other hazardous materials. Section 6005 of the IFC contains additional requirements for ozone gas generators.

[F] 908.5 Repair garages. A flammable-gas detection system shall be provided in repair garages for vehicles fueled by nonodorized gases in accordance with Section 406.8.5.

❖ As indicated in Section 406.8.5, an approved flammable-gas detection system is required for garages used for repair of vehicles fueled by nonodorized gases, such as hydrogen and nonodorized LNG. To prevent a hazardous potential buildup of flammable gas caused by normal leakage and use conditions, the flammable-gas detection system is required to activate when the level of flammable gas exceeds 25 percent of the lower explosive limit (LEL) (see commentary, Section 406.8.5).

[F] 908.6 Refrigerant detector. Machinery rooms shall contain a refrigerant detector with an audible and visual alarm. The detector, or a sampling tube that draws air to the detector, shall be located in an area where refrigerant from a leak will concentrate. The alarm shall be actuated at a value not greater than the corresponding TLV-TWA values for the refrigerant classification shown in the *International Mechanical Code* for the refrigerant classification. Detectors and alarms shall be placed in *approved* locations. The detector shall transmit a signal to an *approved* location.

❖ A refrigerant-specific detector is required for the purpose of leak detection, early warning and actuation of emergency exhaust systems. Depending on the density of the refrigerants, leakage may collect near the floor, near the ceiling or disperse equally throughout. Refrigerant detector locations must be carefully considered. Most refrigerants are heavier than air; thus, floor depressions and pits are natural areas for accumulation. The code is silent on the location of sensors due to the endless variety of equipment room designs. The key to properly locating a detector in the machinery room is to remember that occupant safety is the primary objective, and the danger is in breathing refrigerant. Placing the sensor below the common breathing height of 5 feet (1525 mm) provides an additional safety margin because all commonly used halocarbon refrigerants are three to five times heavier than air. When undistributed by airflow, such escaping refrigerant will fall to the floor, seeking the lowest levels, and fill the room from the bottom up. Since pits, stairwells or trenches are likely to fill with refrigerant first, detectors should be also be placed in any of these areas that may be occupied. The alarm actuation threshold is dictated by the TLV-TWA values for the refrigerant classification in accordance with Table 1103.1 of the IMC.

Manufacturer's instructions for detectors provide installation guidance for the location and required number of detectors for any given room size.

The exception recognizes that ammonia refrigerant is "self-alarming" because of its strong odor and that ammonia machinery rooms are required to be continuously ventilated in accordance with the IMC.

Most general machinery rooms are unoccupied for long periods of time. Because of this, a refrigeration leak may go undetected, allowing a buildup of refrigerant that can pose a threat to the building occupants and the maintenance personnel who will be required to enter the machinery room. Also, the refrigerants may or may not be detectable by the sense of smell, depending upon the chemical nature and concentration in air of the refrigerant. This can be especially critical when a toxic refrigerant is used in the refrigeration system.

There are three levels of refrigerant exposure that are defined by the American Conference of Government Industrial Hygienists (ACGIH). Level 1 is the allowable exposure limit (AEL), which is the level at which a person can be exposed for 8 hours per day for 40 hours per week without having an adverse effect on health. At this level, a person exposed to the refrigerant should not suffer any adverse health effects. Level 2 is the short-term exposure limit (STEL), which is defined as three times the AEL. At this level, a person should not be exposed for more than 30 minutes at a time. Persons working in a machinery room having this concentration of refrigerant should be equipped with respiratory protection. Level 3 is the emergency exposure limit (EEL). At this level, persons should not be in the room without a self-contained breathing apparatus.

Early detection of leaking refrigerant is dependent upon the location of the refrigerant detectors. If improperly located, a refrigerant leak could go undetected for an undesirable period of time, thus allowing a significant amount of refrigerant to escape. It is therefore advantageous to locate the detectors in positions that will provide early detection and minimize the amount of refrigerant leakage.

Factors to be considered when choosing locations for detectors are the airflow patterns of the room, the particular refrigerant density and the fact that the primary hazard to the occupants is through inhalation. Specifically, halocarbon refrigerants are oxygen displacers; that is, they are heavier than air and will occupy the lowest areas of the room first. This suggests that the location of detectors should be below the breathing zone height. Additionally, detectors should be located so as to prevent the normal ventilation system from interfering with detection. Placing detectors between the refrigeration system and exhaust fan inlets should help to ensure that the presence of refrigerant will be detected. Depending on the size of the machinery room and the number and type of refrigeration systems, more than one detector may be necessary. Manufacturer's installation instructions for the refrigeration detection system should be followed when choosing the location and number of sensors for a particular machinery room application.

[F] 908.7 Carbon dioxide (CO_2) systems. Emergency alarm systems in accordance with Section 5307.5.2 of the *International Fire Code* shall be provided where required for compliance with Section 5307.5 of the *International Fire Code*.

❖ This is simply a cross reference to requirements in the IFC related to the danger of CO_2 asphyxiation. The requirement is intended to address fatal CO_2 poisoning incidents in restaurants where CO_2 leaked from large storage tanks for beverage mixing and displaced oxygen in these areas. Section 5307.5 of the IFC addresses CO_2 systems with capacities of 100 pounds or greater. The hazard is required to be addressed for indoor areas or areas where CO_2 can collect. The options include ventilation in accordance with IFC Section 5307.5.1 or an emergency alarm system in accordance with IFC Section 5307.5.2.

SECTION 909 SMOKE CONTROL SYSTEMS

[F] 909.1 Scope and purpose. This section applies to mechanical or passive smoke control systems where they are required by other provisions of this code. The purpose of this section is to establish minimum requirements for the design, installation and acceptance testing of smoke control systems that are intended to provide a tenable environment for the evacuation or relocation of occupants. These provisions are not intended for the preservation of contents, the timely restoration of operations or for assistance in fire suppression or overhaul activities. Smoke control systems regulated by this section serve a different purpose than the smoke- and heat-venting provisions found in Section 910. Mechanical smoke control systems shall not be considered exhaust systems under Chapter 5 of the *International Mechanical Code*.

❖ This section is clarifying the intent of smoke control provisions, which is to provide a tenable environment to occupants during evacuation and relocation and not to protect the contents, enable timely restoration of operations or facilitate fire suppression and overhaul activities. There are provisions for high-rise buildings in Section 403.4.7 that are focused upon the removal of smoke for post fire and overhaul operations, which is very different than the smoke control provisions in Section 909. Another element addressed in this section is that smoke control systems serve a different purpose than smoke and heat vents (see Section 910). This eliminates any confusion that smoke and heat vents can be used as a substitution for smoke control. Additionally, a clarification is provided to note that smoke control systems are not considered an exhaust system in accordance with Chapter 5 of the IMC. This is because such systems are unique in their operation and are not necessarily designed to exhaust smoke but are focused on tenability for occupants during egress. It should be noted that the smoke control provisions are duplicated in Chapter 5 of the IMC.

It is important to note that these provisions only apply when smoke control is required by other sections of the code. The code requires smoke management within atrium spaces (see Section 404.5) and underground buildings (see Section 405.5). High-rise facilities require smokeproof exit enclosures in accordance with Sections 909.20 and 1019.1.8 (see Section 403.5.4). Also, covered mall buildings that contain atriums that connect more than two stories require smoke control (see Section 402.7.2).

Section 909 focuses primarily on mechanical smoke control systems, but there are many instances within the code where smoke is required to be managed in a passive way through the use of concepts such as smoke compartments. Smoke compartments are formed through the use of smoke barriers in accordance with Section 709. Smoke barriers can be used simply as a passive smoke management system or can be a design component of a mechanical smoke control system in accordance with Section 909. Some examples of occupancies requiring passive systems include hospitals, nursing homes and similar facilities (Group I-2 occupancies) and detention facilities (Group I-3 occupancies) (see Sections 407.5 and 408.6).

In some cases, mechanical smoke control in accordance with Section 909 is allowed as an option for compliance. More specifically if a Group I-3 contains windowless areas of the facility, natural or mechanical smoke management is required (see Section 408.9).

In the last several years, smoke control provisions have become more complex. The reason is related to the fact that smoke is a complex problem, while a generic solution of six air changes has repeatedly and scientifically been shown to be inadequate. Six air changes per hour does not take into account factors such as buoyancy; expansion of gases; wind; the geometry of the space and of communicating spaces; the dynamics of the fire, including heat release rate; the production and distribution of smoke and the interaction of the building systems.

Smoke control systems can be either passive or active. Active systems are sometimes referred to as mechanical. Passive smoke control systems take advantage of smoke barriers surrounding the zone in which the fire event occurs or high bay areas that act as reservoirs to control the movement of smoke to other areas of the building. Active systems utilize pressure differences to contain smoke within the event zone or exhaust flow rates sufficient to slow the descent of the upper-level smoke accumulation to some predetermined position above necessary exit paths through the event zone. On rare occasions, there is also a possibility of controlling the movement of smoke horizontally by opposed airflow, but this method requires a specific architectural geometry to function properly that does not create an even greater hazard.

Essentially, there are three methods of mechanical or active smoke control that can be used separately or in combination within a design: pressurization, exhaust and, in rare and very special circumstances, opposed airflow.

Of course, all of these active approaches can be used in combination with the passive method.

Typically, the mechanical pressurization method is used in high-rise buildings when pressurizing stairways and for zoned smoke control. Pressurization is not practical in large open spaces such as atriums or malls, since it is difficult to develop the required pressure differences due to the large volume of the space.

The exhaust method is typically used in large open spaces such as atriums and malls. As noted, the pressurization method would not be practical within large spaces. The opposed airflow method, which basically uses a velocity of air horizontally to slow the movement of smoke, is typically applied in combination with either a pressurization method or exhaust method within hallways or openings into atriums and malls.

The application of each of these methods will be dependent on the specifics of the building design. Smoke control within a building is fundamentally an architecturally driven problem. Different architectural geometries first dictate the need or lack thereof for smoke control, and then define the bounds of available solutions to the problem.

[F] 909.2 General design requirements. Buildings, structures or parts thereof required by this code to have a smoke control system or systems shall have such systems designed in accordance with the applicable requirements of Section 909 and the generally accepted and well-established principles of engineering relevant to the design. The *construction documents* shall include sufficient information and detail to adequately describe the elements of the design necessary for the proper implementation of the smoke control systems. These documents shall be accompanied by sufficient information and analysis to demonstrate compliance with these provisions.

❖ This section simply states that when smoke control systems are required by the code, the design is required to be in accordance with the provisions of this section. As noted in the commentary to Section 909.1, there are instances within the code that have smoke management systems that are purely passive in nature and do not reference Section 909.

This section stresses that designs in accordance with this section need to follow "generally accepted and well-established principles of engineering relevant to the design," essentially requiring a certain level of qualifications in the applicable areas of engineering to prepare such designs. The primary engineering disciplines tend to be fire engineering and mechanical engineering. It should be noted that each state in the U.S. typically requires minimum qualifications to undertake engineering design. Two important resources when designing smoke control systems are the International Code Council's (ICC) *A Guide to Smoke Control in the 2006 IBC®* and American Society of Heating, Refrigerating and Air-Conditioning Engineers' (ASHRAE) *Design of Smoke Management Systems*. Additionally, Section 909.8 requires the use of NFPA 92 for the design of smoke control systems using the exhaust method. This standard has many relevant aspects beyond the design that are beneficial. In particular, Annex B provides resources in terms of determination of fire size for design. ICC's *A Guide to Smoke Control in the 2006 IBC®* also provides guidance on design fires.

A key element covered in this section is the need for detailed and clear construction documents so that the system is installed correctly. In most complex designs, the key to success is appropriate communication to the contractors as to what needs to be installed. The more complex a design becomes, the more likely there is to be construction errors. Most smoke control systems are complex, which is why special inspections in accordance with Section 909.3 and Chapter 17 are critical for smoke control systems. Additionally, in order for the design to be accepted, analyses and justifications need to be provided in enough detail to evaluate for compliance. Adequate documentation is critical to the commissioning, inspection, testing and maintenance of smoke control systems and significantly contributes to the overall reliability and effectiveness of such systems.

[F] 909.3 Special inspection and test requirements. In addition to the ordinary inspection and test requirements that buildings, structures and parts thereof are required to undergo, smoke control systems subject to the provisions of Section 909 shall undergo *special inspections* and tests sufficient to verify the proper commissioning of the smoke control design in its final installed condition. The design submission accompanying the *construction documents* shall clearly detail procedures and methods to be used and the items subject to such inspections and tests. Such commissioning shall be in accordance with generally accepted engineering practice and, where possible, based on published standards for the particular testing involved. The special inspections and tests required by this section shall be conducted under the same terms in Section 1704.

❖ Because of the complexity and uniqueness of each design, special inspection and testing must be conducted. The designer needs to provide specific recommendations for special inspection and testing within his or her documentation. In fact, the code specifies in Chapter 17 that special inspection agencies for smoke control have expertise in fire protection engineering, mechanical engineering and certification as air balancers. Since the designs are unique to each building, there probably will not be a generic approach available to inspect and test such systems. The designer can and should, however, use any available published standards or guides when developing the special inspection and testing requirements for that particular design. ICC's *A Guide to Smoke Control in the 2006 IBC®* provides some background on such inspections, Also, ASHRAE Guideline 5 is a good starting place, but only as a general outline. In addition, NFPA 92A and NFPA 92B also have extensive testing, documentation and maintenance requirements that may be a good resource. NFPA 92B is referenced in Section 909.8 for the design of smoke control systems using the exhaust method. Each system will require a unique commissioning plan that can be developed only after careful and thoughtful examination of the final design and all of its components and interrelationships. Generally, these provisions may be included in design standards or engineering guides.

[F] 909.4 Analysis. A rational analysis supporting the types of smoke control systems to be employed, their methods of operation, the systems supporting them and the methods of construction to be utilized shall accompany the submitted *construction documents* and shall include, but not be limited to, the items indicated in Sections 909.4.1 through 909.4.7.

❖ This section indicates that simply determining airflow, exhaust rates and pressures to maintain tenable conditions is not adequate. There are many factors that could alter the effectiveness of a smoke control system, including stack effect, temperature effect of fire, wind effect, heating, ventilating and air-conditioning (HVAC) system interaction and climate, as well as the placement, quantity of inlets/outlets and velocity of supply and exhaust air. These factors are addressed in the sections that follow. Additionally, the duration of operation of any smoke control system is mandated at a minimum of 20 minutes or 1.5 times the egress time, whichever is less. The code cannot reasonably anticipate every conceivable building arrangement and condition the building may be subject to over its life and must depend on such factors being addressed through a rational analysis.

[F] 909.4.1 Stack effect. The system shall be designed such that the maximum probable normal or reverse stack effect will not adversely interfere with the system's capabilities. In determining the maximum probable stack effect, altitude, elevation, weather history and interior temperatures shall be used.

❖ Stack effect is the tendency for air to rise within a heated building when the temperature is colder on the exterior of the building. Reverse stack effect is the tendency for air to flow downward within a building when the interior is cooler than the exterior of the building. This air movement can affect the intended operation of a smoke control system. If stack effect is great enough, it may overcome the pressures determined during the design analyses and allow smoke to enter areas outside the zone of origin (see Commentary Figure 909.4.1).

[F] 909.4.2 Temperature effect of fire. Buoyancy and expansion caused by the design fire in accordance with Section 909.9 shall be analyzed. The system shall be designed such that these effects do not adversely interfere with the system's capabilities.

❖ This section requires that the design account for the effect temperature may have on the success of the system. When air or any gases are heated they will expand. This expansion makes the gases lighter and therefore more buoyant. The buoyancy of hot gases is important when the design is to exhaust such gases from a location in or close to the ceiling; therefore, if sprinklers are part of the design, as required by Section 909, the gases may be significantly cooler than an unsprinklered fire, making it more difficult to remove the smoke and alter the plume dynamics. The fact that air expands when heated needs to be accounted for in the design.

When using the pressurization method, the expansion of hot gases needs to be accounted for, since it will take a larger volume of air to create the neces-

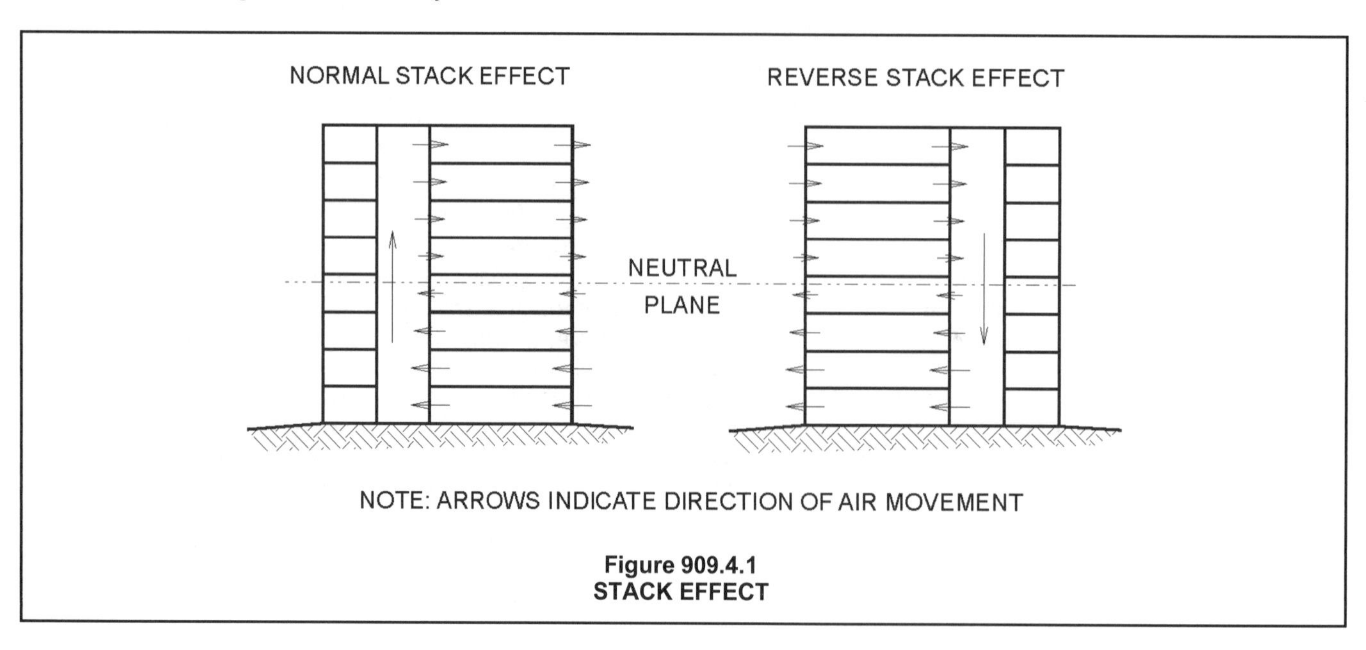

Figure 909.4.1
STACK EFFECT

sary pressure differences to maintain the area of fire origin in negative pressure. The expansion of the gases has the effect of pushing the hot gases out of the area of fire origin. Since sprinklers will tend to cool the gases, the effect of expansion is lower. The pressure differences required in Section 909.6.1 are specifically based on a sprinklered building. If the building is nonsprinklered, higher pressure differences may be required. The minimum pressure difference for certain unsprinklered ceiling height buildings is as follows:

Ceiling height (feet)	Minimum pressure difference (inch water gage)
9	.10
15	.14
21	.18

This is a very complex issue that needs to be part of the design analysis. It needs to address the type and reaction of the fire protection systems, ceiling heights and the size of the design fire.

[F] 909.4.3 Wind effect. The design shall consider the adverse effects of wind. Such consideration shall be consistent with the wind-loading provisions of Chapter 16.

❖ The effect of wind on a smoke control system within a building is very complex. It is generally known that wind exerts a load upon a building. The loads are looked at as windward (positive pressure) and leeward (negative pressure). The velocity of winds will vary based on the terrain and the height above grade; therefore, the height of the building and surrounding obstructions will have an effect on these velocities. These pressures alter the operation of fans, especially propeller fans, thus altering the pressure differences and airflow direction in the building. There is not an easy solution to dealing with these effects. In fact, little research has been done in this area.

It should be noted that in larger buildings a wind study is normally undertaken for the structural design. The data from those studies can be used in the analysis of the effects on the pressures and airflow within the building with regard to the performance of the smoke control system.

[F] 909.4.4 HVAC systems. The design shall consider the effects of the heating, ventilating and air-conditioning (HVAC) systems on both smoke and fire transport. The analysis shall include all permutations of systems status. The design shall consider the effects of the fire on the HVAC systems.

❖ If not properly configured to shut down or be included as part of the design, the HVAC system can alter the smoke control design. More specifically, if dampers are not provided between smoke zones within the HVAC system ducts, smoke could be transported from one zone to another. Additionally, if the HVAC system places more supply air than assumed for the smoke control system design, the velocity of the air may adversely affect the fire plume or a positive pressure may be created. Generally, an analysis of the smoke control design and the HVAC system in all potential modes should occur and be noted within the design documentation as well as incorporated into inspection, testing and maintenance procedures. This is critical as these systems need to be maintained and tested to help ensure that they operate and shut down systems as required.

[F] 909.4.5 Climate. The design shall consider the effects of low temperatures on systems, property and occupants. Air inlets and exhausts shall be located so as to prevent snow or ice blockage.

❖ This section is focused on properly protecting equipment from weather conditions that may affect the reliability of the design. For instance, extremely cold or hot air may damage critical equipment within the system when pulled directly from the outside. Some listings of duct smoke detectors are for specific temperature ranges; therefore, placing such detectors within areas exposed to extreme temperatures may void the listing. Also, the equipment and air inlets and outlets should be designed and located so as to not collect snow and ice that could block air from entering or exiting the building.

[F] 909.4.6 Duration of operation. All portions of active or engineered smoke control systems shall be capable of continued operation after detection of the fire event for a period of not less than either 20 minutes or 1.5 times the calculated egress time, whichever is greater.

❖ The intent of the smoke control provisions is to provide a tenable environment for occupants to either evacuate or relocate to a safe place. Evacuation and relocation activities include notifying occupants, possible investigation time for occupants, decision time and the actual travel time. In order to achieve this goal, the code has established 20 minutes or 1.5 times the calculated egress time, whichever is greater, as a minimum time for evacuation or relocation. Basically this allows a designer to undertake an analysis to more closely determine the necessary time required. The code provides a safety factor of 1.5 times the egress time to account for uncertainty related to human behavior. It is stressed that the 20-minute duration as well as the calculated egress time, whichever approach is chosen, begins after the detection of the fire event and notification to the building occupants to evacuate has occurred, since occupants need to be alerted before evacuation can occur. The calculation of evacuation time needs to include delays with notification and the start of evacuation (i.e., pre-movement time, etc.). It is stressed that the code states 20 minutes or 1.5 times the egress time, whichever is greater (i.e., 20 minutes is a minimum). Egress of occupants can be addressed through hand calculations or through the use of computerized egress models. Some of the more advanced models can address a variety of factors, including the building layout, different sizes of people, different movement speeds and different egress paths available. With these types of programs, the actual time can be even more precisely calculated. Of course it is cautioned that in many cases these models provide the optimal time for egress. The safety factor of 1.5 within the code is intended to address many of these uncertainties.

Note that this section applies to all types of smoke control designed in accordance with Section 909. Also, most smoke control systems will typically have the ability to run for much longer than 20 minutes as they are on standby power and may be able to continue to achieve the tenability goals.

System response as required in Section 909.17 needs to be accounted for when determining the ability of the smoke control system to keep the smoke layer interface at the appropriate level (see commentary, Section 909.17).

909.4.7 Smoke control system interaction. The design shall consider the interaction effects of the operation of multiple smoke control systems for all design scenarios.

❖ The focus of this section is related to the interaction of multiple mechanical smoke control systems by asking for a specific analysis of the interaction of such systems similar to that required for the interaction of HVAC systems. Where hoistway pressurization is chosen as an option for compliance with the hoistway opening protection requirements, the potential exists for a pressurized stair system to also be present. These two systems need to be able to operate at the same time without a negative impact on either system. It is also possible that an atrium with a smoke control system is located in a building containing a stair pressurization system.

[F] 909.5 Smoke barrier construction. *Smoke barriers* required for passive smoke control and a smoke control system using the pressurization method shall comply with Section 709, and shall be constructed and sealed to limit leakage areas exclusive of protected openings. The maximum allowable leakage area shall be the aggregate area calculated using the following leakage area ratios:

1. Walls $A/A_w = 0.00100$
2. Interior *exit stairways* and *ramps* and *exit passageways*: $A/A_w = 0.00035$
3. Enclosed *exit access stairways* and *ramps* and all other shafts: $A/A_w = 0.00150$
4. Floors and roofs: $A/A_F = 0.00050$

where:

A = Total leakage area, square feet (m²).

A_F = Unit floor or roof area of barrier, square feet (m²).

A_w = Unit wall area of barrier, square feet (m²).

The leakage area ratios shown do not include openings due to gaps around doors and operable windows. The total leakage area of the *smoke barrier* shall be determined in accordance with Section 909.5.1 and tested in accordance with Section 909.5.2.

❖ Part of the strategy of smoke control systems, particularly smoke control systems using the pressurization method (often termed zoned smoke control) is the use of smoke barriers to divide a building into separate smoke zones (or compartments). This strategy is used in both passive and mechanical systems. It should be noted that not all walls, ceilings or floors would be considered smoke barriers. Only walls that designate separate smoke zones within a building need to be constructed as smoke barriers. This section is simply providing requirements for walls, floors and ceilings that are used as smoke barriers. It should be noted that it is possible that a smoke control system utilizing the exhaust method may not need to utilize a smoke barrier to divide the building into separate smoke zones; therefore, the evaluation of barrier construction and leakage area may not be necessary and, as noted, is primarily focused on designs using a passive approach or the pressurization method.

In order for smoke to not travel from one smoke zone to another, specific construction requirements are necessary in accordance with the code. It should be noted that openings such as doors and windows are dealt with separately within Section 909.5.3 from openings such as cracks or penetrations.

[F] 909.5.1 Total leakage area. Total leakage area of the barrier is the product of the *smoke barrier* gross area multiplied by the allowable leakage area ratio, plus the area of other openings such as gaps around doors and operable windows.

❖ It is impossible for walls and floors to be constructed that are completely free from openings that may allow the migration of smoke; therefore, leakage needs to be compensated for within the design by calculating the leakage area of walls, ceilings and floors. The factors provided in Section 909.5, which originate from ASHRAE's provisions on leaky buildings, are used to calculate the total leakage area. The total leakage area is then used in the design process to determine the proper amount of air to create the required pressure differences across these surfaces that form smoke zones.

Additionally, Section 909.5 provides ratios to determine the maximum allowable leakage in walls, interior exit stairways, shafts, floors and roofs. These leakage areas are critical in determining whether the proper pressure differences are provided when utilizing the pressurization method of smoke control. Pressure differences will decrease as the openings get larger.

[F] 909.5.2 Testing of leakage area. Compliance with the maximum total leakage area shall be determined by achieving the minimum air pressure difference across the barrier with the system in the smoke control mode for mechanical smoke control systems utilizing the pressurization method. Compliance with the maximum total leakage area of passive smoke control systems shall be verified through methods such as door fan testing or other methods, as *approved* by the fire code official.

❖ These leakage criteria need to be evaluated through testing. For the case of a pressurization system, pressure differences need to be verified when the system is in smoke control mode. In the case of passive smoke control systems, pressure testing through tests such as the door fan method is necessary.

[F] 909.5.3 Opening protection. Openings in *smoke barriers* shall be protected by automatic-closing devices actuated by the required controls for the mechanical smoke control system. Door openings shall be protected by *fire door assemblies* complying with Section 716.5.3.

Exceptions:

1. Passive smoke control systems with automatic-closing devices actuated by spot-type smoke detectors *listed* for releasing service installed in accordance with Section 907.3.
2. Fixed openings between smoke zones that are protected utilizing the airflow method.
3. In Group I-1 Condition 2, Group I-2 and ambulatory care facilities, where a pair of opposite-swinging doors are installed across a corridor in accordance with Section 909.5.3.1, the doors shall not be required to be protected in accordance with Section 716. The doors shall be close-fitting within operational tolerances and shall not have a center mullion or undercuts in excess of $^3/_4$ inch (19.1 mm), louvers or grilles. The doors shall have head and jamb stops and astragals or rabbets at meeting edges and, where permitted by the door manufacturer's listing, positive-latching devices are not required.
4. In Group I-2 and ambulatory care facilities, where such doors are special-purpose horizontal sliding, accordion or folding door assemblies installed in accordance with Section 1010.1.4.3 and are automatic closing by smoke detection in accordance with Section 716.5.9.3.
5. Group I-3.
6. Openings between smoke zones with clear ceiling heights of 14 feet (4267 mm) or greater and bank-down capacity of greater than 20 minutes as determined by the design fire size.

❖ Similar to concerns of smoke leakage between smoke zones, openings may compromise the necessary pressure differences between smoke zones. Openings in smoke barriers, such as doors and windows, must be either constantly or automatically closed when the smoke control system is operating. This section requires that doors be automatically closed through the activation of an automatic-closing device linked to the smoke control system. Essentially, when the smoke control system is activated, all openings are automatically closed. This most likely would mean that the mechanism that activates the smoke control system would also automatically close all openings. The smoke control system will be activated by a specifically zoned smoke detection or sprinkler system as required by Sections 909.12.3 and 909.12.4.

In terms of actual opening protection, Section 909.5.2 is simply referring the user to Section 716.5.3 for specific construction requirements for doors located in smoke barriers. Note that smoke barriers are different from fire barriers, since the intended measure of performance is different. One is focused on fire spread from the perspective of heat, the other from the perspective of smoke passage. Smoke barriers do require a 1-hour fire-resistance rating.

There are several exceptions to this particular section. Exception 1 is specifically for passive systems. Passive systems, as noted, are systems in which there is no use of mechanical systems. Instead, the system operates primarily upon the configuration of barriers and layout of the building to provide smoke control. Passive systems can use spot-type detectors to close doors that constitute portions of a smoke barrier. Essentially, this means a full fire alarm system would not be required. Instead, single station detectors would be allowed to close the doors. Such doors would need to fail in the closed position if power is lost. The specifics as to approved devices would be found in NFPA 72.

Exception 2 is based on the fact that some systems take advantage of the opposed airflow method such that smoke is prevented from migrating past the doors. Therefore, since the design already accounts for potential smoke migration at these openings through the use of air movement, it is unnecessary to require the barrier to be closed.

Exception 3 is specifically related to the unique requirements for Group I-1 Condition 2, I-2 occupancies and ambulatory care occupancies. Essentially, a very specific alternative, which meets the functional needs of these occupancy types, is provided. Opposite-swinging doors are allowed without meeting the specific requirements of Section 716. Note that Group I-2 and ambulatory care facilities utilize the concept of smoke compartments, which is a form of a passive smoke control system. The requirements of this section are focused on openings in smoke barriers that need to close upon activation of a pressurization system.

Exception 4 provides a specific allowance for horizontal sliding doors in Group I-2 Condition 2 occupancies and ambulatory care occupancies. These doors are commonly used in such occupancies and have very specific installation requirements in Section 1010.1.4.3. These doors are an alternative to pivoted or side-hinged, swinging-type doors. This exception requires compliance with Section 716.5.9.3 for automatic closing upon detection of smoke.

Exception 5 allows an exemption from the automatic-closing requirements for all Group I-3 occupancies. This is because facilities that have occupants under restraint or with specific security restrictions have unique requirements in accordance with Section 408 of the code. These requirements accomplish the intent of providing reliable barriers between each smoke zone since, for the most part, such facilities will have a majority of doors closed and in a locked position because of the nature of the facility. The staff very closely controls these types of facilities.

Exception 6 relates to the behavior of smoke. The assumption is that smoke rises because of the buoyancy of hot gases, and if the ceiling is sufficiently high, the smoke layer will be contained for a longer period of time before it begins to move into the next smoke zone. Therefore, it is not as critical that the doors automatically close. This allowance is dependent on the specific design fire for a building. See Section 909.9 for more information on design fire determination. Different size design fires create different amounts of smoke that, depending on the layout of the building, may migrate in different ways throughout the building. This section mandates that smoke cannot begin to migrate into the next smoke zone for at least 20 minutes. This is consistent with the 20-minute minimum duration of operation of smoke control systems required in Section 909.4.6. It should be noted that a minimum of 14-foot (4267 mm) ceilings are required to take advantage of this exception. This exception would require an engineering analysis.

909.5.3.1 Group I-1 Condition 2; Group I-2 and ambulatory care facilities. In Group I-1 Condition 2, Group I-2 and *ambulatory care facilities*, where doors are installed across a *corridor*, the doors shall be automatic closing by smoke detection in accordance with Section 716.5.9.3 and shall have a vision panel with fire protection-rated glazing materials in fire protection-rated frames, the area of which shall not exceed that tested.

❖ Part of the alternative allowed for horizontal sliding doors in Exception 3 of Section 909.5.3 for Group I-2 Condition 2, Group I-1 Condition 2 and ambulatory care facilities is the requirement that vision panels be provided. These vision panels need to be approved fire-protection-rated glazing and be within frames of a size that does not exceed the frame size that was used for testing of the glazing.

[F] 909.5.3.2 Ducts and air transfer openings. Ducts and air transfer openings are required to be protected with a minimum Class II, 250°F (121°C) *smoke damper* complying with Section 717.

❖ Another factor that adds to the reliability of smoke barriers is the protection of ducts and air transfer openings within smoke barriers. Left open, these openings may allow the transfer of smoke between smoke zones. These ducts and air transfer openings most often are part of the HVAC system. Damper operation and the reaction with the smoke control system will be evaluated during acceptance testing. It should be noted that there are duct systems used within a smoke control design that are controlled by the smoke control system and should not automatically close upon detection of smoke via a smoke damper.

It should be noted that a smoke damper works differently than a fire damper. Fire dampers react to heat via a fusible link, while smoke dampers activate upon the detection of smoke. The smoke dampers used should be rated as Class II, 250°F (121°C). The class of the smoke damper refers to its level of performance relative to leakage. The temperature rating is related to its ability to withstand the heat of smoke resulting from a fire. It should be noted that although smoke barriers are only required to utilize smoke dampers, there may be many instances where a fire damper is also required. For instance, the smoke barrier may also be used as a fire barrier. Also, Section 717.5.3 would require penetration of shafts to contain both a smoke and fire damper. Therefore, in some cases both a smoke damper and fire damper would be required. There are listings specific to combination smoke and fire dampers. Note that the exceptions to Section 717.5.3 recognize that smoke and fire dampers may interfere with a smoke control design.

More specific requirements about dampers can be found in Chapter 7 of this code and Chapter 6 of the IMC.

[F] 909.6 Pressurization method. The primary mechanical means of controlling smoke shall be by pressure differences across smoke barriers. Maintenance of a tenable environment is not required in the smoke control zone of fire origin.

❖ There are several methods or strategies that may be used to control smoke movement. One of these methods is pressurization, wherein the system primarily utilizes pressure differences across smoke barriers to control the movement of smoke. Basically, if the area of fire origin maintains a negative pressure, then the smoke will be contained to that smoke zone. A typical approach used to obtain a negative pressure is to exhaust the fire floor. This is a fairly common practice in high-rise buildings. Interior exit stairways also utilize the concept of pressurization by keeping the interior exit stairways under positive pressure. The pressurization method in large open spaces, such as malls and atria, is impractical since it would take a large quantity of supply air to create the necessary pressure differences. It should be noted that pressurization is mandated as the primary method for mechanical smoke control design, but this is related to the primary methods historically used for smoke control in high-rise buildings. Currently high-rise buildings do not require smoke control. Airflow and exhaust methods are only allowed when appropriate. The exhaust method is the most commonly applied method based on the use of the atrium provisions in Section 404.5.

The pressurization method does not require that tenable conditions be maintained in the smoke zone where the fire originates. Maintaining this area tenable would be impossible, because pressures from the surrounding smoke zones would be placing a negative pressure within the zone of origin to keep the smoke from migrating.

Pressurization is used often with interior exit stairways. This method provides a positive pressure within the interior exit stairways to resist the passage of smoke. Stair pressurization is one method of compliance for stairways in high-rise or underground buildings where the floor surface is located more than 75 feet (22 860 mm) above the lowest level of fire department vehicle access or more than 30 feet (9144 mm) below the floor surface of the lowest level of exit discharge. It should be noted that there are two methods found in the code that address smoke movement—smokeproof enclosures or pressurized stairs. A smokeproof enclosure requires a certain fire-resistance rating along with access through a ventilated vestibule or an exterior balcony. The vestibule can be ventilated in two ways: using natural ventilation or mechanical ventilation as outlined in Sections 909.20.3 and 909.20.4. The pressurization method requires a sprinklered building and a minimum pressure difference of 0.15 inch (37 Pa) of water and a maximum of 0.35 inch (87 Pa) of water. These pressure differences are to be available with all doors closed under maximum stack pressures (see Sections 909.20 and 1023.11 for more details).

As noted, the pressurization method utilizes pressure differences across smoke barriers to achieve control of smoke. Sections 909.6.1 and 909.6.2 provide the criteria for smoke control design in terms of minimum and maximum pressure differences.

In summary, the pressurization method is used in two ways. The first is through the use of smoke zones where the zone of origin is exhausted, creating a negative pressure. The second is stair pressurization that creates a positive pressure within the stair to avoid the penetration of smoke. Note that the code allows the use of a smokeproof enclosure instead of pressurization.

[F] 909.6.1 Minimum pressure difference. The minimum pressure difference across a *smoke barrier* shall be 0.05-inch water gage (0.0124 kPa) in fully sprinklered buildings.

In buildings permitted to be other than fully sprinklered, the smoke control system shall be designed to achieve pressure differences not less than two times the maximum calculated pressure difference produced by the design fire.

❖ The minimum pressure difference is established as 0.05-inch water gage (12 Pa) in fully sprinklered buildings. This particular criterion is related to the pressures needed to overcome buoyancy and the pressures generated by the fire, which include expansion. This particular criterion is based on a sprinklered building. The pressure difference would need to be higher in a building that is not sprinklered. Additionally, the pressure difference needs to be provided based on the possible stack and wind effects present.

[F] 909.6.2 Maximum pressure difference. The maximum air pressure difference across a *smoke barrier* shall be determined by required door-opening or closing forces. The actual force required to open *exit* doors when the system is in the smoke control mode shall be in accordance with Section 1010.1.3. Opening and closing forces for other doors shall be determined by standard engineering methods for the resolution of forces and reactions. The calculated force to set a side-hinged, swinging door in motion shall be determined by:

$$F = F_{dc} + K(WA\Delta P)/2(W-d) \qquad \textbf{(Equation 9-1)}$$

where:

A = Door area, square feet (m^2).

d = Distance from door handle to latch edge of door, feet (m).

F = Total door opening force, pounds (N).

F_{dc} = Force required to overcome closing device, pounds (N).

K = Coefficient 5.2 (1.0).

W = Door width, feet (m).

ΔP = Design pressure difference, inches of water (Pa).

❖ The maximum pressure difference is based primarily on the force needed to open and close doors. The code establishes maximum opening forces for doors. This maximum opening force cannot be exceeded, taking into account the pressure differences across a doorway in a pressurized environment. Essentially, based on the opening force requirements of Section 1010.1.3, the maximum pressure difference can be calculated in accordance with Equation 9-1. In accordance with Chapter 10, the maximum opening force of a door has three components, including:

Door latch release:
Maximum of 15 pounds (67 N)

Set door in motion:
Maximum of 30 pounds (134 N)

Swing to full open position:
Maximum of 15 pounds (67 N)

Equation 9-1 is used to calculate the total force to set the door into motion when in the smoke control mode; therefore, the limiting criteria would be 30 pounds (134 N). It should be noted that although the accessibility requirements related to door opening force are more restrictive in Section 404.2.8 of ICC A117.1, fire doors do not require compliance with these requirements.

[F] 909.6.3 Pressurized stairways and elevator hoistways. Where stairways or elevator hoistways are pressurized, such pressurization systems shall comply with Section 909 as smoke control systems, in addition to the requirements of Sections 909.20 of this code and 909.21 of the *International Fire Code*.

❖ The purpose of this section is to clarify that pressurized stairways and pressurized hoistways are smoke control systems and must be addressed in the same way that a pressurization system in accordance with

Section 909.6 addresses such systems. This would require compliance with various sections but in particular the requirements for a rational analysis in accordance with Section 909.4.

[F] 909.7 Airflow design method. Where *approved* by the fire code official, smoke migration through openings fixed in a permanently open position, which are located between smoke control zones by the use of the airflow method, shall be permitted. The design airflow shall be in accordance with this section. Airflow shall be directed to limit smoke migration from the fire zone. The geometry of openings shall be considered to prevent flow reversal from turbulent effects. Smoke control systems using the airflow method shall be designed in accordance with NFPA 92.

❖ This method is only allowed when approved by the building official. As the title states, this method utilizes airflow to avoid the migration of smoke across smoke barriers. This has been referred to as opposed airflow. Specifically, this method is suited for the protection of smoke migration through doors and related openings fixed in a permanently open position. This method consists of providing a particular velocity of air based on the temperature of the smoke and the height of the opening. The temperature of the smoke will depend on the design fire that is established for the particular building. The higher the temperature of the smoke and the larger the opening, the higher the velocity necessary to maintain the smoke from migrating into the smoke zone. It should be noted that the airflow method seldom works for large openings, since the velocity to oppose the smoke becomes too high. This method tends to work better for smaller openings, such as pass-through windows. Reference is made to NFPA 92 to determine the minimum velocity required to limit smoke spread.

[F] 909.7.1 Prohibited conditions. This method shall not be employed where either the quantity of air or the velocity of the airflow will adversely affect other portions of the smoke control system, unduly intensify the fire, disrupt plume dynamics or interfere with exiting. In no case shall airflow toward the fire exceed 200 feet per minute (1.02 m/s). Where the calculated airflow exceeds this limit, the airflow method shall not be used.

❖ The airflow method has a limitation on maximum velocity. This limitation is because air may distort the flame and cause additional entrainment and turbulence; therefore, having a high velocity of air entering the zone of fire origin has the potential of increasing the amount of smoke produced. The velocity may also interact with other portions of the smoke control design. For instance, the pressure differences in other areas of the building may be altered, which may exceed the limitations of Sections 909.6.1 and 909.6.2. This section requires that when a velocity of over 200 feet per minute (1.02 m/sec) is calculated, the airflow method is not allowed. The solution may result in requiring a barrier such as a wall or door.

If the airflow design method is chosen to protect areas communicating with an atrium, the air added to the smoke layer needs to be accounted for in the exhaust rate.

[F] 909.8 Exhaust method. Where *approved* by the fire code official, mechanical smoke control for large enclosed volumes, such as in atriums or malls, shall be permitted to utilize the exhaust method. Smoke control systems using the exhaust method shall be designed in accordance with NFPA 92.

❖ This method is only allowed when approved by the building official. The primary application of the exhaust method is in large spaces, such as atriums and malls and is the most widely used method in the IBC. The strategy of this method is to keep the smoke layer at a certain level within the space. This is primarily accomplished through exhausting smoke. The amount of exhaust depends on the design fire [see Commentary Figure 909.8(1)]. Essentially, fires produce different amounts and properties of smoke based on the material being burned, and size and placement of the fire; therefore, NFPA 92 is referenced for the design of such systems. NFPA 92 presents several ways to address the control of smoke, which includes the use of the following tools:

- Scale Modeling (Small-scale testing)—Utilizes the concept of scaling to allow small-scale tests to be conducted to understand smoke movement within a space.
  - Benefits—More realistic understanding of smoke movement in spaces with unusual configurations or projections than algebraic calculations.
  - Disadvantages—Expensive and the application of results is limited to the uniqueness of the space being analyzed.
- Algebraic (Calculations—Empirically derived (based on testing) modeling in its simplest form.
  - Benefits—Simple, cost-effective analysis.
  - Disadvantages—Limited applicability because of the range of values they were derived from. Only appropriate with certain types of design fires, typically over conservative outputs that increase equipment needs, equipment costs and can impact aesthetics and architectural design.
- Computer Modeling [Computational Fluid Dynamics (CFD) or zone models]—Combination of theory and empirical values to determine the smoke movement and fire-induced conditions within a space and effectiveness of the smoke control system.
  - Benefits—More realistic understanding of smoke movement in spaces with unusual configurations or projections and less expensive than scale modeling. Helps significantly in designing smoke control sys-

tems tailored to spaces and achieving cost-effective designs, and can help limit the impact to architectural design.

- Disadvantages—Computing time and cost can be longer than algebraic calculations but benefits typically outweigh this disadvantage. Early planning is important and can limit these adverse impacts.

In terms of computer modeling, as noted, there are essentially two methods that include zone models and CFD models. Zone models are based on the unifying assumption that in any room or space where the effects of the fire are present, there are distinct layers (hot upper layer, cool lower layer). In real life such distinct layers do not exist. Some examples of zone models used in such applications include Consolidated Model of Fire Growth and Smoke Transport (C-FAST) and Available Safe Egress Time (ASET). See Section 3-7 of the *SFPE Handbook of Fire Protection Engineering* for further information. CFD models take this much further and actually divide the space into thousands or millions of interconnected "cells" or "fields." The model then evaluates the fire dynamics and heat and mass in each individual cell and how it interacts with those adjacent to it. The use of such models becomes more accurate with more numerous and smaller cells but the computing power and expertise required is much higher than for zone models. As noted the use of either types of models can be advantageous but such use must be undertaken by someone qualified. Proper review and verification of the input and output is critical. The most popular model in the area of CFD with regard to fire is the Fire Dynamics Simulator (FDS) developed by NIST. Other models such as Fluent are sometimes used (Fluent Inc.).

Depending on the space being evaluated, some design strategies may provide a better approach than others. Past editions of the code smoke control provisions for the exhaust method mandated the use of the algebraic methods with a steady fire. This also mandated that a mechanical system be used, whereas NFPA 92 allows an overall review of smoke layer movement and whether the design goals, which in this case are mandated by the code, can be met. Therefore, if it can be shown that the smoke layer interface can be held at 6 feet (1829 mm) as mandated in Section 909.8.1 for the design operation time required by Section 909.4.6 without mechanical ventilation, then the space would comply with Section 909. NFPA 92 presents several design approaches. This allows more flexibility in design than that found in previous editions of the code.

NFPA 92 as a standard does not set the minimum smoke layer interface height or duration for system operation. Such criteria are found within Sections 909.8.1 and 909.4.6, respectively. See the commentary for those sections.

If the algebraic approach is used, consideration of three types of fire plumes may be required to determine which one is the most demanding in terms of smoke removal needs based on the space being assessed. They include:

Axisymmetric plumes—Smoke rises unimpeded by walls, balconies or similar projections [see Commentary Figure 909.8(2)].

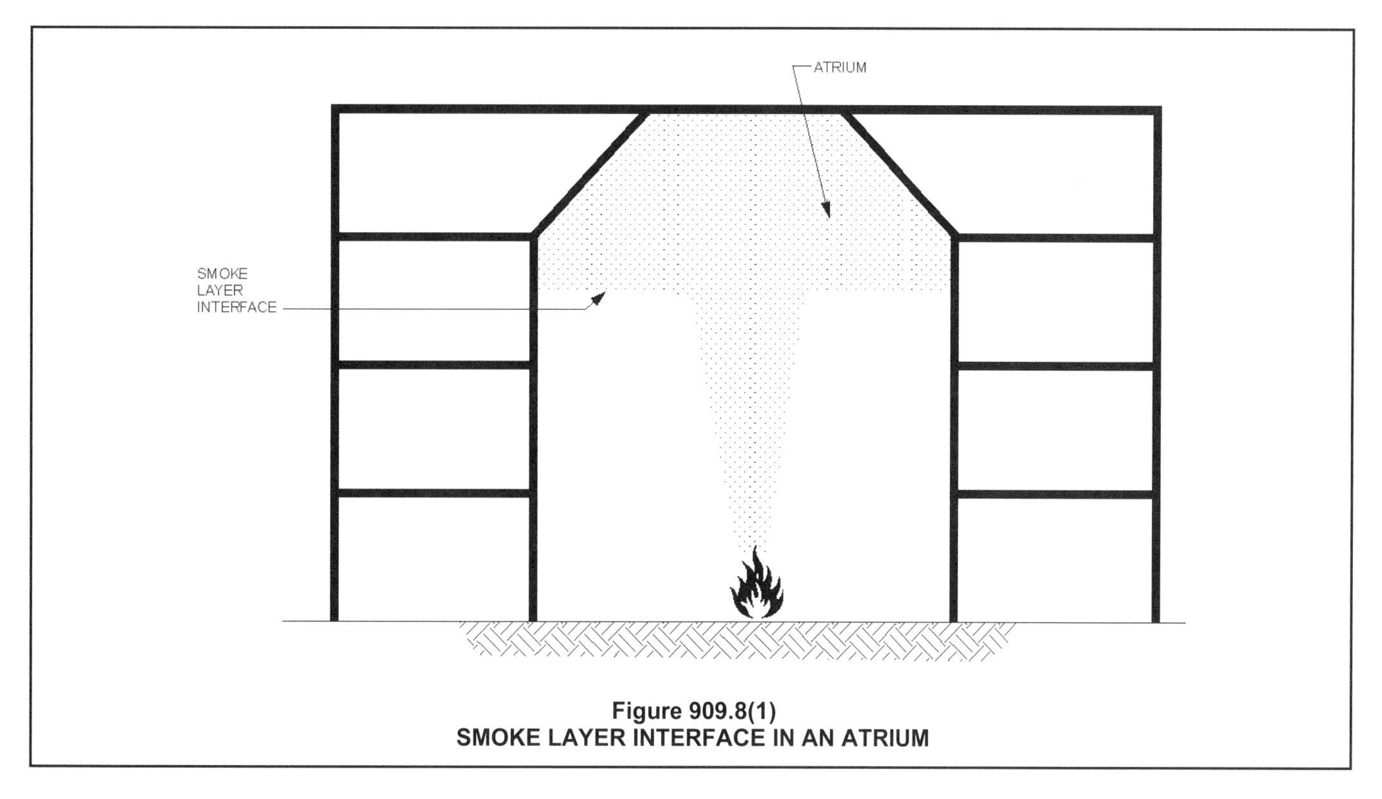

Figure 909.8(1)
SMOKE LAYER INTERFACE IN AN ATRIUM

Balcony spill plumes—Smoke flows under and around edges of a horizontal projection [see Commentary Figure 909.8(3)].

Window plumes—Smoke flows through an opening into a large-volume space [see Commentary Figure 909.8(4)].

It should be noted that prior to the reference to NFPA 92 in the code, the balcony spill and window plume calculations had been eliminated from the smoke control requirements of the code because of concerns with the applicability of those calculations. The major difference is that NFPA 92 does not mandate the use of such equations as did previous editions of the IBC. The use of such equations will depend on the design fires agreed upon for the particular design and whether an algebraic approach is chosen. These equations are used to determine a mass flow rate of smoke to ultimately determine the required exhaust volume for that space. If the potential for a balcony or window spill plumes is known to exist within the space, then appropriate measures need to be taken to address these, as they typically result in more onerous exhaust and supply requirements. Part of the reason for the initial deletion of these equations was the fact that such scenarios are not as likely or their impact is significantly reduced in sprinklered buildings. There is also some concern with the applicability of the balcony spill plume equation in a variety of applications. These potential fire scenarios and resulting plumes may further the need to undertake a CFD analysis to address such hazards more appropriately and effectively.

Another key aspect that NFPA 92 included within

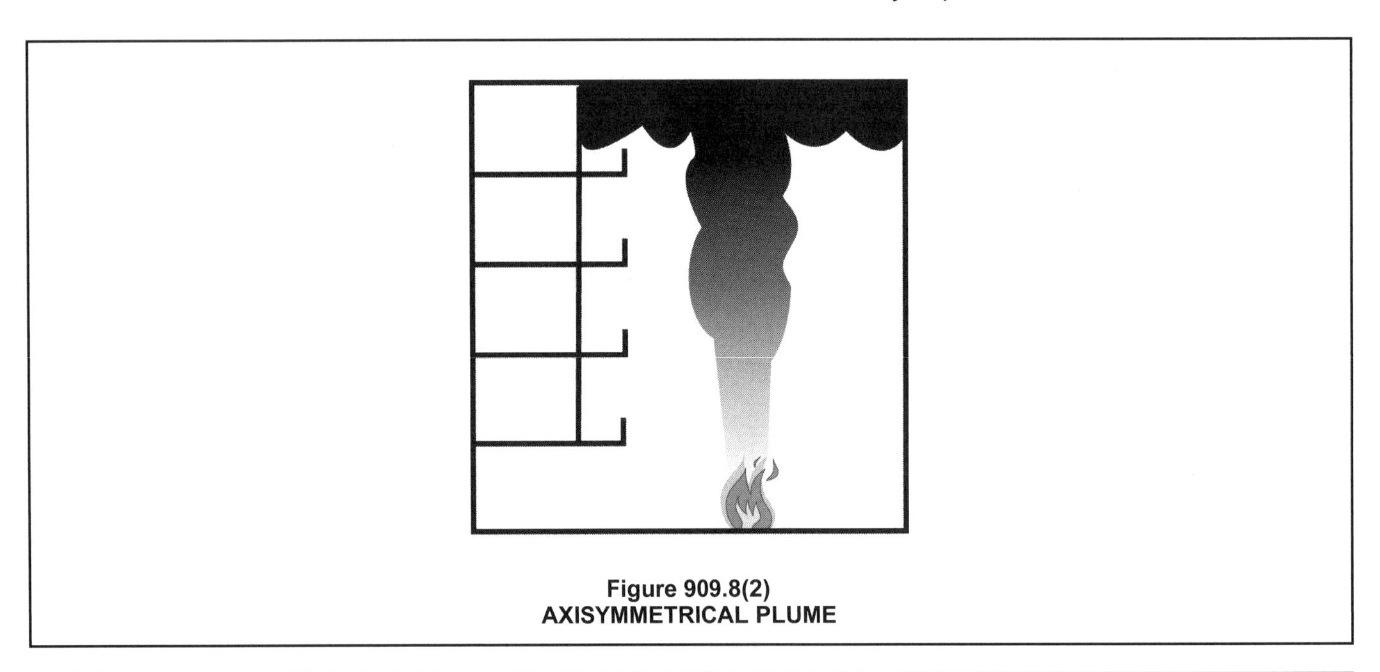

Figure 909.8(2)
AXISYMMETRICAL PLUME

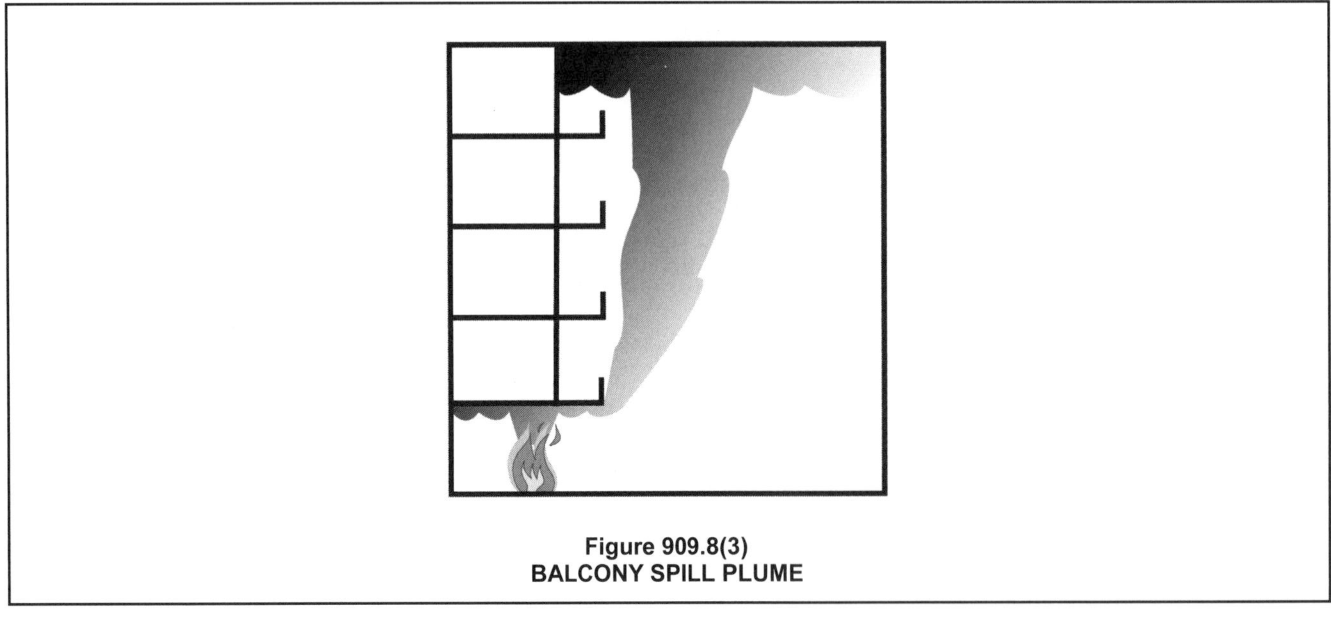

Figure 909.8(3)
BALCONY SPILL PLUME

the algebraic methods is equations to determine that a minimum number of exhaust inlets are available to prevent plugholing. Plugholing occurs when air from below the smoke layer is pulled through the smoke layer into the smoke exhaust inlets. As such, if plugholing occurs, some of the fan capacity is used to exhaust air rather than smoke and thus can affect the ability to maintain the smoke layer at or above the design height. Scale modeling and computer fire modeling would demonstrate these potential problems during the testing and analysis, respectively [see Commentary Figure 909.8(5)].

It should be noted that this section specifically references NFPA 92 for the design of smoke control using the exhaust method. Therefore the requirements in NFPA 92 related to testing, documentation and maintenance would not be applicable though they may be a good resource. Equipment and controls would be part of the design; therefore, related provisions of NFPA 92 would apply. Generally, the code addresses equipment and controls in a similar fashion.

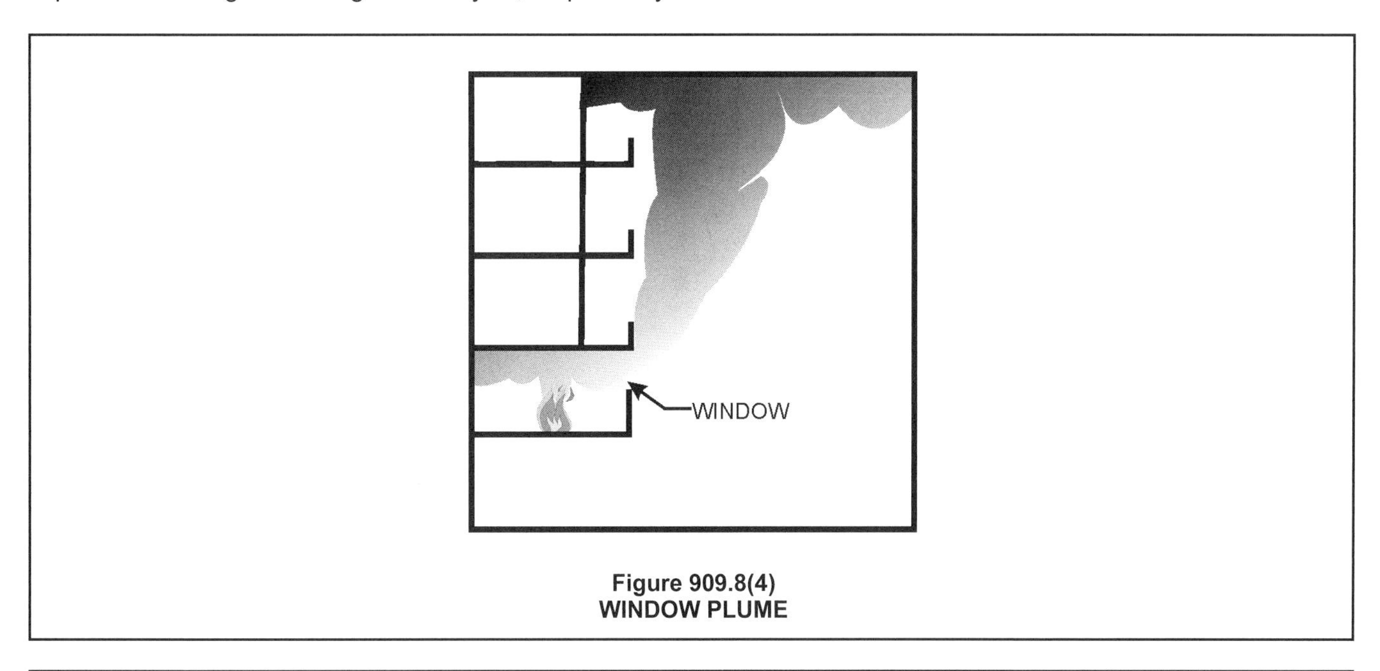

Figure 909.8(4)
WINDOW PLUME

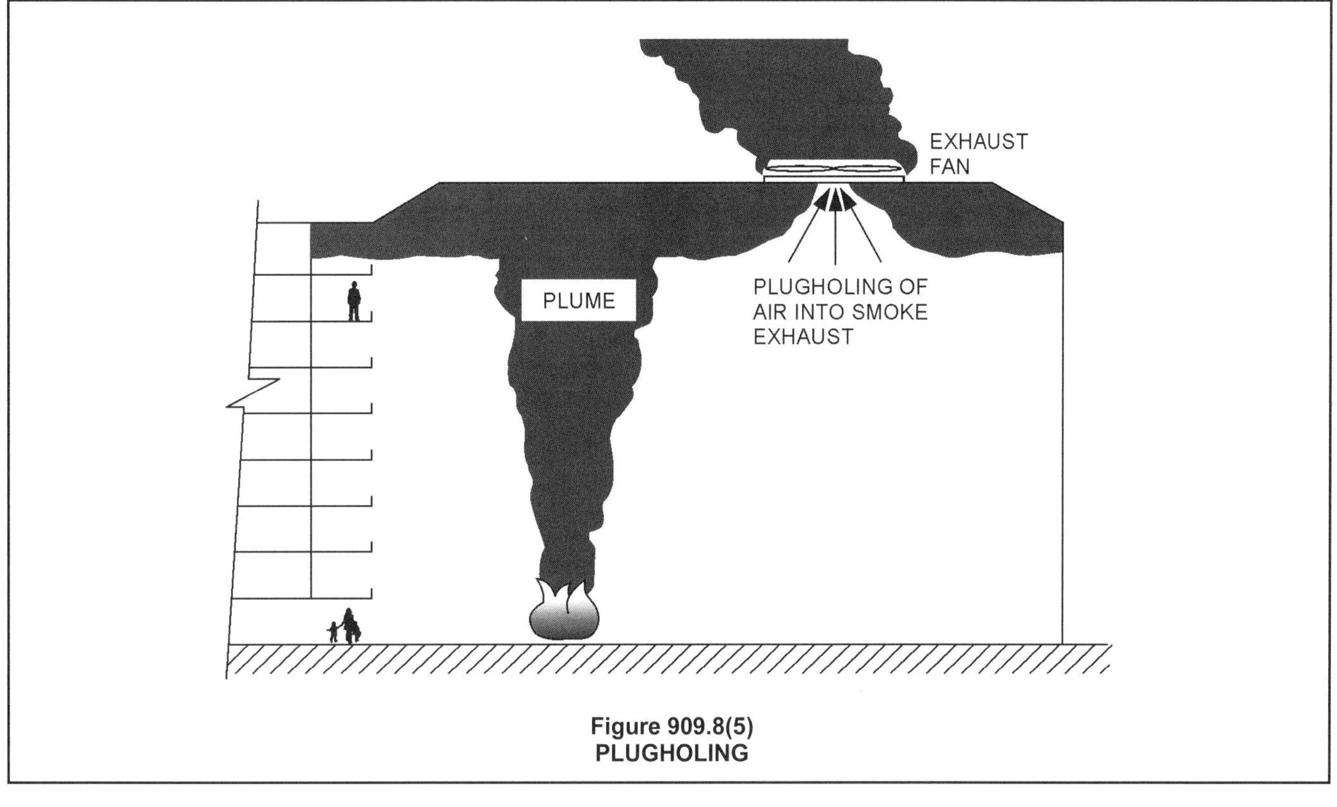

Figure 909.8(5)
PLUGHOLING

[F] 909.8.1 Smoke layer. The height of the lowest horizontal surface of the smoke layer interface shall be maintained not less than 6 feet (1829 mm) above a walking surface that forms a portion of a required egress system within the smoke zone.

❖ The design criteria to be used when applying NFPA 92 is to maintain the smoke layer interface at least 6 feet (1829 mm) above any walking surface that is considered part of the required egress within the particular smoke zone, such as an atrium, for 20 minutes or 1.5 times the calculated egress time (see Section 909.4.6). Chapter 10 considers the majority of occupiable space as part of the means of egress system. Also keep in mind that the criteria of 6 feet (1829 mm) does not apply just to the main floor surface of the mall or atrium but to any level where occupants may be exposed (for example, balconies) [see Commentary Figure 909.8.1(1)].

The code uses the terminology "lowest horizontal surface of the accumulating smoke layer interface." NFPA 92 has several definitions related to smoke layer, which include the following:

Smoke layer. The accumulated thickness of smoke below a physical barrier.

Smoke layer interface. The theoretical boundary between a smoke layer and the smoke-free air. (Note: This boundary is at the beginning of the transition zone.)

First indication of smoke. The boundary between the transition zone and the smoke-free air.

Transition zone. The layer between the smoke layer interface and the first indication of smoke in which the smoke layer temperature decreases to ambient. The transition zone may be several feet thick (large open space) or may barely exist (small area with intense fire).

See also Commentary Figure 909.8.1(2).

NFPA 92 provides algebraic equations to determine the first indication of smoke but is limited to very specific conditions such as a uniform cross section, specific aspect ratios, steady or unsteady fires and no smoke exhaust operating. When using algebraic equations for smoke layer interface looking at different types of plumes, the smoke layer interface terminology is used and the user enters the desired smoke layer interface height. Zone models use simplifying assumptions so the layers are distinct from one another. In contrast, when CFD or scale modeling is used, the data must be analyzed to verify that the smoke layer interface is located at or above 6 feet (1829 mm) during the event. This is not a simple analysis, as CFD and scale modeling provide more detail on actual smoke behavior; therefore, the location of the smoke layer interface may not be initially clear without some level of analysis. Again it depends on the depth of the transition layer. This may require reviewing tenability within the transition zone. Tenability limits need to be agreed upon by the stakeholders involved. Using CFD or scale modeling would likely need to occur through the alternative methods and materials section (see Section 104.11) because of the need to review tenability limits. It

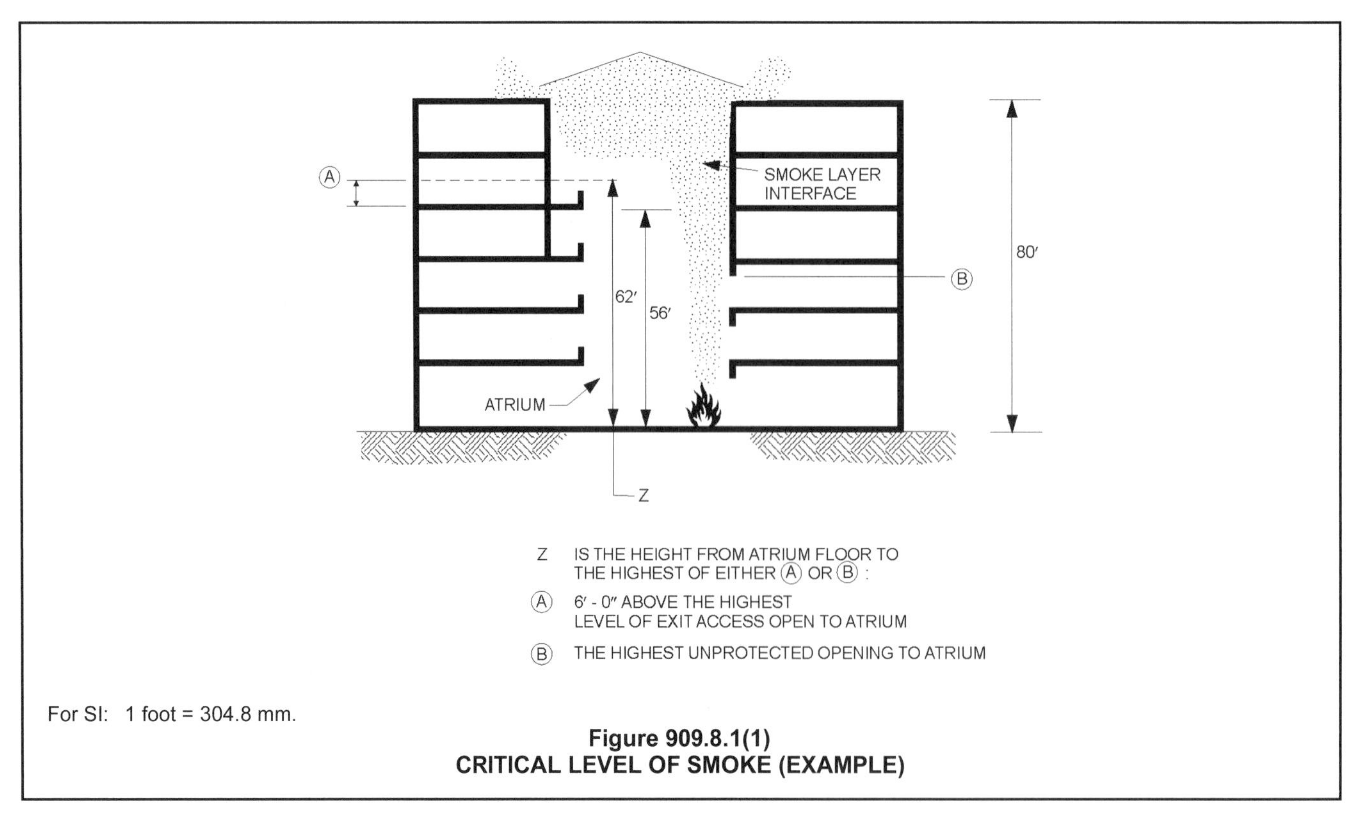

Figure 909.8.1(1)
CRITICAL LEVEL OF SMOKE (EXAMPLE)

should be noted that NFPA 92 Annex A suggests that there are methods to determine where the smoke layer interface and first indication of smoke are located when undertaking CFD and scale modeling using a limited number of point measurements.

Also, Section 909.8.1 specifies a minimum distance for the smoke layer interface from any walking surface whereas Section 4.5.4.2 of NFPA 92 has provisions that simply allow the analysis to demonstrate tenability regardless of where the layer height is located above the floor. Defining tenability can be more difficult as there is not a standard definition. Any design using that approach would need to be addressed through Section 104.11.

Note that the response time of the system components (detection, activation, ramp up time, shutting down HVAC, opening/closing doors and dampers, etc.) needs to be accounted for when analyzing the location of the smoke layer interface in relation to the duration of operations stated in Section 909.4.6 (see commentary, Section 909.17).

[F] 909.9 Design fire. The design fire shall be based on a rational analysis performed by the *registered design professional* and *approved* by the fire code official. The design fire shall be based on the analysis in accordance with Section 909.4 and this section.

❖ The design fire is the most critical element in the smoke control system design. The fire is what produces the smoke to be controlled by the system; thus, the size of the fire directly impacts the quantity of smoke being produced. This section ensures that the design fire be determined through a rational analysis by a registered design professional with knowledge in this area. Such professionals should have experience in the area of fire dynamics, fire engineering and general building design, including mechanical systems. When determining the design fire, the designer should work with various stakeholders to determine the types of hazards and combustible materials (fire scenarios) on a permanent as well as temporary basis (i.e., Christmas/holiday decorative materials or scenery, temporary art exhibits) that may be present throughout the use of the building once occupied. Those hazards then need to be translated to potential design fires to be used when determining the smoke layer interface height for the duration as determined by Section 909.4.6. See the commentary for Section 909.9.3 for potential sources when determining design fires.

This section also does not mandate the type of fire (i.e., steady versus unsteady). A steady fire assumes a constant heat release rate over a period of time, where unsteady fires do not. An unsteady fire includes the growth and decay phases of the fire, as well as the peak heat release rate. An unsteady fire will hit a peak heat release rate when burning in the open, like an axisymmetric fire. An unsteady fire is a more realistic view of how fires actually burn. It should be noted that fires can be a combination of unsteady and steady fires when sufficient fuel is available. In other words, the fire initially grows (unsteady) then reaches a steady state and burns for sometime at a particular heat release rate before decay occurs.

Design fire information should therefore typically include growth rate, peak heat release rate, duration and decay as well as information related to fire locations and products of combustion yield (CO, smoke, etc.) that are produced by the various design fires that are deemed credible for the space.

To provide an order of magnitude of fire sizes obtained from various combustibles, the following data from fire tests are provided. The following heat release rates, found in Section 3, Chapter 3-1 of the

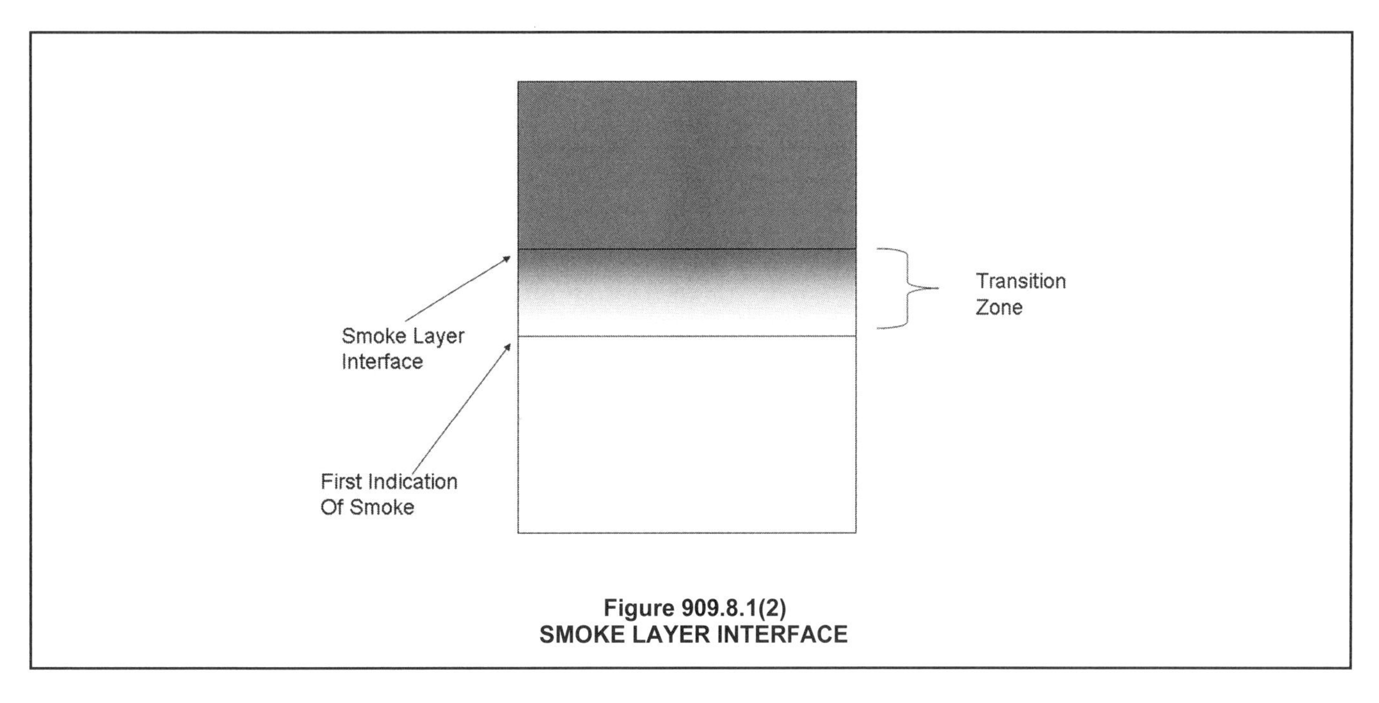

Figure 909.8.1(2)
SMOKE LAYER INTERFACE

4th edition of the *SFPE Handbook of Fire Protection Engineering*, are peak heat release rates:

- Plastic trash bags/paper trash:
  Approximately 120-350 kw
- Latex foam pillow:
  Approximately 120 kw
- Dry Douglas fir Christmas tree:
  Approximately 3000 kw at 20% moisture
- Polyurethane foam (PVC Ticking material)
  Mattress:
  Approximately 2630 kw
- Plywood wardrobe:
  Approximately 2900-6400 kw

[F] 909.9.1 Factors considered. The engineering analysis shall include the characteristics of the fuel, fuel load, effects included by the fire and whether the fire is likely to be steady or unsteady.

❖ This section simply provides more detail on the factors that should be taken under consideration when determining the design fire size. To determine the appropriate fire size, an engineering analysis is necessary that takes into account the following elements: fuel (potential burning rates), fuel load (how much), effects included by the fire (smoke particulate size and density), steady or unsteady (burn steadily or simply peak and dissipate) and likelihood of sprinkler activation (based on height and distance from the fire).

[F] 909.9.2 Design fire fuel. Determination of the design fire shall include consideration of the type of fuel, fuel spacing and configuration.

❖ The design fire size may also be affected by surrounding combustibles, which may have the effect of increasing the fire size. More specifically, there is concern that if sufficient separation is not maintained between combustibles, then a larger design fire is likely. The code does not provide extensive detail on this as such determination is left to the rational analysis undertaken by the design professional. NFPA 92 provides one method in which you determine the critical separation distance, *R*. This is based on fire size and the critical radiant heat flux for nonpiloted ignition. Nonpiloted ignition means the radiated heat from the fire without direct flame contact will ignite adjacent combustibles.

[F] 909.9.3 Heat-release assumptions. The analysis shall make use of best available data from *approved* sources and shall not be based on excessively stringent limitations of combustible material.

❖ This section stresses that data obtained for use in a rational analysis need to come from relevant and appropriate sources. Data can be obtained from groups such as NIST or from Annex B of NFPA 92. Data from fire tests are available and are a good resource for such analysis. As noted earlier, such data are not prevalent (see also Chapter 8, Analysis of Design Fires, from *A Guide to Smoke Control in the 2006 IBC®* and Section 3, Chapter 3-1, of the *SFPE Handbook of Fire Protection Engineering*).

[F] 909.9.4 Sprinkler effectiveness assumptions. A documented engineering analysis shall be provided for conditions that assume fire growth is halted at the time of sprinkler activation.

❖ This section raises a few questions regarding activation of sprinklers and their impact on the fire both in terms of their ability to "control" as well as "extinguish" a fire. The first is concerning an assumption that sprinklers will immediately control the fire as soon as they are activated (i.e., control results in limiting further growth and maintaining the heat release rate at approximately the same fire size as when the sprinklers activated). This assumption may be true in some cases, but for high ceilings, the sprinkler may not activate or may be ineffective. Sprinklers may be ineffective in high spaces, since by the time they are activated the fire is too large to control. Essentially, the fire plume may push away and evaporate the water before it actually reaches the seat of the fire. In addition, the fire may be shielded from sprinkler spray so that insufficient quantities of water reach the fuel. These are common problems with high-piled storage as well as other fires, including retail, and has been shown in actual tests. Also, if the fire becomes too large before the sprinklers are activated, the available water supply and pressure for the system may be compromised. Additionally, based on the layout of the room and the movement of the fire effluents, the wrong sprinklers could be activated, which leads to a larger fire size and depletion of the available water supply and pressure.

Another issue is whether the sprinklers "control" or "extinguish" the fire. Typical sprinklers are assumed only to control fires as opposed to extinguishing them. Sprinklers may be able to extinguish the fire but it should not automatically be assumed. A fire that is controlled will achieve steady state and maintain a certain fire size, which is very different from a fire that is actually extinguished.

Based on these concerns, each scenario needs to be looked at individually to determine whether sprinklers would be effective in halting the growth or extinguishing the fire. More specifically, the evaluation should include droplet size, density and area of coverage and should also be based on actual test results.

[F] 909.10 Equipment. Equipment including, but not limited to, fans, ducts, automatic *dampers* and balance *dampers*, shall be suitable for its intended use, suitable for the probable exposure temperatures that the rational analysis indicates and as *approved* by the fire code official.

❖ Section 909.10 and subsequent sections are primarily related to the reliability of the system components to provide a smoke control system that works according to the design. One of the largest concerns when using smoke control provisions is the overall reliability of the system. Such systems have many different components, such as smoke and fire dampers; fans; ducts and controls associated with such components. The more components a system has, the less reliable it becomes. In fact, one approach in providing a higher level of reliability is utilizing the normal building systems such as the HVAC to provide the smoke control system. Basically, systems used every day are more likely to be working appropriately, since they are essentially being tested daily; however, there are many components that are specific to the smoke control system, such as exhaust fans in an atrium or the smoke control panel.

Also, there is not a generic prescriptive set of requirements as to how all smoke control system elements should operate, since each design may be fairly unique. The specifics on operation of such a system need to be included within the design and construction documents. Most components used in smoke control systems are elements used in many other applications such as HVAC systems; therefore, the basic mechanisms of a fan used in a smoke control system may not be different, although they may be applied differently.

[F] 909.10.1 Exhaust fans. Components of exhaust fans shall be rated and certified by the manufacturer for the probable temperature rise to which the components will be exposed. This temperature rise shall be computed by:

$$T_s = (Q_c/mc) + (T_a) \quad \textbf{(Equation 9-2)}$$

where:

c = Specific heat of smoke at smoke layer temperature, Btu/lb°F (kJ/kg · K).

m = Exhaust rate, pounds per second (kg/s).

Q_c = Convective heat output of fire, Btu/s (kW).

T_a = Ambient temperature, °F (K).

T_s = Smoke temperature, °F (K).

Exception: Reduced T_s as calculated based on the assurance of adequate dilution air.

❖ Fans used for smoke control systems must be able to tolerate the possible elevated temperatures to which they will be exposed. Again, like many other factors this depends on the specifics of the design fire. Essentially, Equation 9-2 requires the calculation of the potential temperature rise. The exhaust fans must be specifically rated and certified by the manufacturer to be able to handle these rises in temperature. There is an exception that allows reduction of the temperature if it can be shown that adequate temperature reduction will occur. In many cases if the exhaust fans are near the ceiling, the smoke will be much cooler than the value resulting from Equation 9-2 since the smoke may cool considerably by the time it reaches the ceiling. Also, sprinkler activation will assist in cooling the smoke further.

[F] 909.10.2 Ducts. Duct materials and joints shall be capable of withstanding the probable temperatures and pressures to which they are exposed as determined in accordance with Section 909.10.1. Ducts shall be constructed and supported in accordance with the *International Mechanical Code*. Ducts shall be leak tested to 1.5 times the maximum design pressure in accordance with nationally accepted practices. Measured leakage shall not exceed 5 percent of design flow. Results of such testing shall be a part of the documentation procedure. Ducts shall be supported directly from fire-resistance-rated structural elements of the building by substantial, noncombustible supports.

Exception: Flexible connections, for the purpose of vibration isolation, complying with the *International Mechanical Code* and that are constructed of *approved* fire-resistance-rated materials.

❖ The next essential component of a smoke control system is the integrity of the ducts to transport supply and exhaust air. The integrity of ducts is also important for an HVAC system, but is more critical in this case since it is not simply a comfort issue but one of life safety. The key concern with ducts in smoke control systems is that they can withstand elevated temperatures and that there will be minimal leakage. The concern with leakage is the potential of leaking smoke into another smoke zone or not providing the proper amount of supply air to support the system.

More specifically, all ducts need to be leak tested to 1.5 times the maximum static design pressure. The leakage resulting should be no more than 5 percent of the design flow. For example, a duct that has a design flow of 300 cubic feet per minute (cfm) (0.141 m^3/s) would be allowed 15 cfm (0.007 m^3/s) of leakage when exposed to a pressure equal to 1.5 times the design pressure for that duct. The tests should be in accordance with nationally accepted practices. This criterion will often limit ductwork for smoke control systems to lined systems, since the amount of leakage in such systems is much less.

As part of the concern for possible exposure to fire and fire products, the ducts are required to be supported by way of substantial noncombustible supports connected to the fire-resistance-rated structural elements of the building. As noted, the system needs to able to run for 20 minutes starting from the detection of the fire. The exception to this section is really more of an acknowledgement that flexible connections for vibration isolation are acceptable when constructed of approved fire-resistance-rated materials. More specifically, it is often necessary to use such connections for connecting the duct to the fan. These connections cannot necessarily meet the requirements of the main section, but are a minimal part of the ductwork and as long as they perform adequately with regard to fire resistance they are permitted. Note that the term "approved" is used to determine the required fire resistance, therefore, flexibility is provided. The code does not specifically address this determination but perhaps a relationship to the duration or operation and these flexible connections could be made to determine the necessary performance.

[F] 909.10.3 Equipment, inlets and outlets. Equipment shall be located so as to not expose uninvolved portions of the building to an additional fire hazard. Outside air inlets shall be located so as to minimize the potential for introducing smoke or flame into the building. Exhaust outlets shall be so located as to minimize reintroduction of smoke into the building and to limit exposure of the building or adjacent buildings to an additional fire hazard.

❖ The intent of this section is to minimize the likelihood of smoke being reintroduced into the building due to poorly placed outdoor air inlets and exhaust air outlets; therefore, placing one right next to another on the exterior of the building would be inappropriate. In addition, wind and other adverse conditions should be considered when choosing locations for these inlets and outlets. Particular attention should be paid to introducing exhausted smoke into another smoke zone. Also, smoke should be exhausted in a direction that will not introduce it into surrounding buildings or facilities. Within the building itself, the supply air and exhaust outlets should also be strategically located. The exhaust inlets and supply air should be evenly distributed to reduce the likelihood of a high velocity of air that may disrupt the fire plume and also push smoke back into occupied areas. See the commentary to Section 909.8 for discussion on avoiding plugholing.

[F] 909.10.4 Automatic dampers. Automatic *dampers*, regardless of the purpose for which they are installed within the smoke control system, shall be *listed* and conform to the requirements of *approved*, recognized standards.

❖ This section addresses the reliability of any dampers used within a smoke control system. This particular provision requires that the dampers be listed and conform to the appropriate recognized standards. More specifically, Section 717 contains more detailed information on the specific requirements for smoke and fire dampers. Smoke and fire dampers should be listed in accordance with UL 555S and 555, respectively. Also, remember that each smoke control design is unique and the sequence and methods used to activate the dampers may vary from design to design. This information needs to be addressed in the construction documents.

Another factor to take into account, with regard to timing of the system, is the fact that some dampers react more quickly than others, simply due to the particular smoke damper characteristics. Additionally, during the commissioning of the system, the damper is going to be exposed to many repetitions. These repetitions need to be accounted for in the overall reliability of the system.

[F] 909.10.5 Fans. In addition to other requirements, belt-driven fans shall have 1.5 times the number of belts required for the design duty, with the minimum number of belts being two. Fans shall be selected for stable performance based on normal temperature and, where applicable, elevated temperature. Calculations and manufacturer's fan curves shall be part of the documentation procedures. Fans shall be supported and restrained by noncombustible devices in accordance with the requirements of Chapter 16.

Motors driving fans shall not be operated beyond their nameplate horsepower (kilowatts), as determined from measurement of actual current draw, and shall have a minimum service factor of 1.15.

❖ Part of the overall reliability requires that fans used to provide supply air and exhaust capacity will be functioning when necessary; therefore, a safety factor of 1.5 is placed on the required belts for fans. All fans used as part of a smoke control system must provide 1.5 times the number of required belts with a minimum of two belts for all fans.

This section also points out that the fan chosen should fit the specific application. It should be able to withstand the temperature rise as calculated in Section 909.10.1 and generally be able to handle typical exposure conditions, such as location and wind. For instance, propeller fans are highly sensitive to the effects of wind. When located on the windward side of a building, wall-mounted, nonhooded propeller fans are not able to compensate for wind effects. Additionally, even hooded propeller fans located on the leeward side of the building may not adequately compensate for the decrease in pressure caused by wind effects. In general, when designing a system, it should be remembered that field conditions might vary from the calculations; therefore, flexibility should be built into the design that would account for things such as variations in wind conditions.

Finally, this section stresses that fan motors not be operated beyond their rated horsepower.

[F] 909.11 Standby power. Smoke control systems shall be provided with standby power in accordance with Section 2702.

❖ This section references Chapter 27 for the specifics as to what is required for standby power. As with any life safety system, a level of redundancy with regard to power supply is required to enable the functioning of the system during a fire. The primary source is the building's normal power system. The secondary power system is by means of standby power. One of the key elements is that standby power systems are intended to operate within 60 seconds of loss of primary power. It should be noted that the primary difference between standby power and emergency power is that emergency power must operate within 10 seconds of loss of primary power versus 60 seconds.

909.11.1 Equipment room. The standby power source and its transfer switches shall be in a room separate from the normal power transformers and switch gears and ventilated directly to and from the exterior. The room shall be enclosed with not less than 1-hour *fire barriers* constructed in accordance with Section 707 or *horizontal assemblies* constructed in accordance with Section 711, or both.

❖ This section requires isolation from normal building power systems via a 1-hour fire barrier, 1-hour horizontal assembly, or both, depending on the location within the building. This increases the reliability and reduces the likelihood that a single event could remove both power supplies. The intent of the ventilation is focused on the proper function of the standby power source in terms of engine-driven generators having appropriate cooling air and combustion air. The requirement that it be from the outside is related to the protection of such ventilation from the effects of fire.

[F] 909.11.2 Power sources and power surges. Elements of the smoke control system relying on volatile memories or the like shall be supplied with uninterruptable power sources of sufficient duration to span 15-minute primary power interruption. Elements of the smoke control system susceptible to power surges shall be suitably protected by conditioners, suppressors or other *approved* means.

❖ Smoke control systems have many components, sometimes highly sensitive electronics, that are adversely affected by any interruption in or sudden surges of power. Therefore, Section 909.11.1 requires that any components of a smoke control system, such as volatile memories, be supplied with an uninterruptible power system for the first 15 minutes of loss of primary power. Volatile memory components will lose memory upon any loss of power, no matter how short the time period. Once the 15 minutes elapses, these elements can be transitioned to the already operating standby power supply.

With regard to components sensitive to power surges, they need to be provided with surge protection in the form of conditioners, suppressors or other approved means.

[F] 909.12 Detection and control systems. Fire detection systems providing control input or output signals to mechanical smoke control systems or elements thereof shall comply with the requirements of Section 907. Such systems shall be equipped with a control unit complying with UL 864 and *listed* as smoke control equipment.

❖ This section is focused on proper monitoring of the fire detection systems that activate the smoke control system through compliance with Section 907 and UL 864. This requires a specific listing of the fire alarm control unit as smoke control equipment. UL 864 has a subcategory (UUKL) specific to fire alarm control panels for smoke control system applications.

909.12.1 Verification. Control systems for mechanical smoke control systems shall include provisions for verification. Verification shall include positive confirmation of actuation, testing, manual override and the presence of power downstream of all disconnects. A preprogrammed weekly test sequence shall report abnormal conditions audibly, visually and by printed report. The preprogrammed weekly test shall operate all devices, equipment and components used for smoke control.

Exception: Where verification of individual components tested through the preprogrammed weekly testing sequence will interfere with, and produce unwanted effects to, normal building operation, such individual components are permitted to be bypassed from the preprogrammed weekly testing, where *approved* by the building official and in accordance with both of the following:

1. Where the operation of components is bypassed from the preprogrammed weekly test, presence of power downstream of all disconnects shall be verified weekly by a listed control unit.
2. Testing of all components bypassed from the preprogrammed weekly test shall be in accordance with Section 909.20.6 of the *International Fire Code*.

❖ This section addresses the function of the mechanical elements of the smoke control system once the system is activated. In particular, there is a focus on verification of activities. Verification would include the following two aspects:

1. The system is able to verify actuations, testing, manual overrides and the presence of power downstream. This would require information reported back to the smoke control panel, which can be accomplished via the weekly test sequence or through full electronic monitoring of the system.
2. Conduct a preprogrammed weekly test that simulates an actual (smoke) event to test the components of the system. These components would include elements such as smoke dampers, fans and doors. Abnormal conditions need to be reported in three ways:
   a. Audibly;
   b. Visually; and
   c. Printed report.

It should be noted that electrical monitoring of the control components is not required (supervision). Such supervision verifies integrity of the conductors from a fire alarm control unit to the control system input. The weekly test is considered sufficient verification of system performance and is often termed end-to-end verification. In other words, the control system input provides the expected results. Verification can be accomplished through any sensor that is calibrated to distinguish between the difference between proper operation and a fault condition. For fans, proper operation means that the fan is moving air within the intent of its design. Fault conditions include power failure, broken fan belts, adverse wind effects, a locked rotor condition and/or filters or large ducts that are blocked causing significantly reduced airflow. In addition to differential pressure transmitters and sail switches, this can be accomplished by state-of-the-art current sensors. More discussion on verification for elements such as ducts and fire doors is discussed in Chapter 9 of *A Guide to Smoke Control in the 2006 IBC®*.

Also, the fact that a smoke control system is non-dedicated (integrated with an HVAC system) does not mean that it is automatically being tested on a daily basis. It is cautioned that simply depending on occupant discomfort, for example, is sometimes an insufficient indicator of a fully functioning smoke control system. There may be various modes in which the HVAC system could operate that may not exercise the smoke control features and the sequence in which the system should operate. An example is an air-conditioning system operating only in full recirculating mode versus exhaust mode. This failure will likely not affect occupants and will not exercise the exhaust function. Plus, doors, which may be part of the smoke barrier, may not need to be closed in normal building operations but would need to be closed during smoke control system operation. This is why this section does not necessarily differentiate between dedicated and nondedicated smoke control systems and requires the system components to be tested.

It is important to note that this weekly test sequence is not an actual smoke event and is only intended to activate the system to ensure that the components are working correctly.

The exception addresses the impracticality of requiring a weekly test for many buildings. For many systems, the weekly test requires the introduction of untreated air into the smoke zone. This can be impractical in areas with cold or hot climates and for buildings that require close control of temperature and humidity, such as art museums and similar facilities. The introduction of the untreated air can also result in wasting energy to reheat, re-cool, humidify, or dehumidify the smoke control zone.

The intent of the exception is to provide means to verify that the required systems will be available when needed. The code requires control units to comply with UL 864, thus all components of the control system will be supervised. The exception includes requirements for verification of the power downstream of all disconnects, such as power breakers, power disconnects, automatic transfer switches, motor starters, and motor controls. This will provide reasonable assurance that power will be available for all smoke control components, such as fans, dampers, doors, and windows. The exception also adds the semi-annual requirement for a complete system test by reference to the IFC Section 909.20.6. This allows the building owner to schedule complete system testing on days that will reduce the impact to the building and energy needs. The combination of additional supervision and additional testing provides a reasonable alternative to weekly testing.

[F] 909.12.2 Wiring. In addition to meeting requirements of NFPA 70, all wiring, regardless of voltage, shall be fully enclosed within continuous raceways.

❖ Wiring is required to be placed within continuous raceways, which provides an additional level of reliability for the system. The definition of the term "raceway" in NFPA 70 lists several acceptable types of complying raceway that can be used; however, manufactured cable assemblies such as metal-clad cable (Type MC) or armored cable (Type AC) are not included.

[F] 909.12.3 Activation. Smoke control systems shall be activated in accordance with this section.

❖ The activation of a smoke control system is dependent on when such a system is required. Mechanical smoke control systems, which could include pressurization, airflow or exhaust methods, require an automatic activation mechanism. When using a passive system which depends on compartmentation, spot-type detectors are acceptable for the release of door closers and similar openings. With more complex mechanical systems, such activation needs to go beyond single-station detectors and be part of an automatic coordinated system.

[F] 909.12.3.1 Pressurization, airflow or exhaust method. Mechanical smoke control systems using the pressurization, airflow or exhaust method shall have completely automatic control.

❖ Automatic activation of such systems is especially critical as tenability is much more difficult to achieve if a delay occurred waiting for manual activation of the system. See Sections 909.6 for the pressurization method, 909.7 for the airflow design method and 909.8 for the exhaust method.

[F] 909.12.3.2 Passive method. Passive smoke control systems actuated by *approved* spot-type detectors *listed* for releasing service shall be permitted.

❖ This section recognizes that a passive system does not address smoke containment through mechanical means; therefore, it does not need to be "automatically activated" except in cases where smoke barriers have openings. These openings would be required to have smoke detectors to close openings where required by the design. Although spot-type detectors are technically automatic, they are not part of a more coordinated system of activation as needed for mechanical smoke control systems. Such detectors are simply standalone devices that fail in the fail-safe position. In other words, if the power were lost, a door on a magnetic hold would simply close.

[F] 909.12.4 Automatic control. Where completely automatic control is required or used, the automatic-control sequences shall be initiated from an appropriately zoned *automatic sprinkler system* complying with Section 903.3.1.1, manual controls that are readily accessible to the fire department and any smoke detectors required by engineering analysis.

❖ When automatic activation is required, it must be accomplished by a properly zoned automatic sprinkler system and, if the engineering analysis requires them, smoke detectors. Manual control for the fire department needs to be provided. An important point with this particular requirement is that smoke control systems are engineered systems and a prescribed smoke detection system may not fit the needs of the specific design. Other types of detectors, such as beam detectors (within an atrium), may be used and could be more useful and more practical from a maintenance standpoint. Also, it may not be practical or appropriate for the building's fire alarm system to activate such systems, as it may alter the effectiveness of the system by pulling smoke through the building versus removing or containing the smoke. For example, a building with an atrium may have several floors below the space. If a fire occurs in one of the floors not associated with the atrium, the atrium smoke control system could possibly pull smoke throughout the building if the detection is zoned incorrectly.

[F] 909.13 Control air tubing. Control air tubing shall be of sufficient size to meet the required response times. Tubing shall be flushed clean and dry prior to final connections and shall be adequately supported and protected from damage. Tubing passing through concrete or masonry shall be sleeved and protected from abrasion and electrolytic action.

❖ Control tubing is a method that uses pneumatics to operate components such as the opening and closing of dampers. Because of the sophistication of electronic systems today, control tubing is becoming less common.

These particular requirements provide the criteria for properly designing and installing control tubing. Essentially, it is up to the design professional to determine the size requirements and to properly design appropriate supports. This information needs to be detailed within the construction documents. Additionally, because of the effect of moisture and other contaminants on control tubing, it must be flushed clean then dried before installation.

[F] 909.13.1 Materials. Control-air tubing shall be hard-drawn copper, Type L, ACR in accordance with ASTM B42, ASTM B43, ASTM B68, ASTM B88, ASTM B251 and ASTM B280. Fittings shall be wrought copper or brass, solder type in accordance with ASME B16.18 or ASME B16.22. Changes in direction shall be made with appropriate tool bends. Brass compression-type fittings shall be used at final connection to devices; other joints shall be brazed using a BCuP-5 brazing alloy with solidus above 1,100°F (593°C) and liquids below 1,500°F (816°C). Brazing flux shall be used on copper-to-brass joints only.

Exception: Nonmetallic tubing used within control panels and at the final connection to devices provided all of the following conditions are met:

1. Tubing shall comply with the requirements of Section 602.2.1.3 of the *International Mechanical Code*.
2. Tubing and connected devices shall be completely enclosed within a galvanized or paint-grade steel enclosure having a minimum thickness of 0.0296 inch (0.7534 mm) (No. 22 gage). Entry to the enclosure shall be by copper tubing with a protective grommet of neoprene or Teflon or by suitable brass compression to male barbed adapter.
3. Tubing shall be identified by appropriately documented coding.
4. Tubing shall be neatly tied and supported within the enclosure.Tubing bridging cabinets and doors or moveable devices shall be of sufficient length to avoid tension and excessive stress. Tubing shall be protected against abrasion. Tubing serving devices on doors shall be fastened along hinges.

❖ This section addresses the materials allowed for control air tubing along with approved methods of connection. All of this information needs to be documented, as it will be subject to review by the special inspector.

[F] 909.13.2 Isolation from other functions. Control tubing serving other than smoke control functions shall be isolated by automatic isolation valves or shall be an independent system.

❖ This section requires separation of control tubing used for other functions through the use of isolation valves or a completely separate system. This is caused by the difference in requirements for control tubing used in a smoke control system versus other building systems. The isolation of the control air tubing for a smoke control system needs to be specifically noted on the construction documents.

[F] 909.13.3 Testing. Control air tubing shall be tested at three times the operating pressure for not less than 30 minutes without any noticeable loss in gauge pressure prior to final connection to devices.

❖ As part of the acceptance testing of the smoke control system, the control air tubing will be pressure tested three times the operating pressure for 30 minutes or more. The performance criterion as to whether the control tubing is considered a failure is when there is any noticeable loss in gauge pressure prior to final connection of devices during the 30-minute duration test.

[F] 909.14 Marking and identification. The detection and control systems shall be clearly marked at all junctions, accesses and terminations.

❖ This section requires that all portions of the fire detection system that activate the smoke control system be marked and identified appropriately. This includes all applicable fire alarm-initiating devices, the respective junction boxes, all data-gathering panels and fire alarm control panels. Additionally, all components of the smoke control system, which are not considered a fire detection system, are required to be properly identified and marked. This would include all applicable junction boxes, control tubing, temperature control modules, relays, damper sensors, automatic door sensors and air movement sensors.

[F] 909.15 Control diagrams. Identical control diagrams showing all devices in the system and identifying their location and function shall be maintained current and kept on file with the fire code official, the fire department and in the fire command center in a format and manner *approved* by the fire chief.

❖ The purpose of control diagrams is to provide consistent information on the system in several key locations, including the building department, the fire department and the fire command center. If a fire command center is not required or provided, the diagrams need to be located such that they can be readily accessed during an emergency. Some possible locations may be the security office, the building manager's office or, if possible, within the smoke control panel. This information is intended to assist in the use and operation of the smoke control system. The format of the control diagram is as approved by the fire chief. This is necessary since the fire department is the agency that will be using such a system during a fire and when the system is tested in the future. The more clearly the information is communicated, the more effective the smoke control system will be.

It should be noted that the fire department may want all smoke control systems within a jurisdiction to follow a particular protocol for control diagrams. Generally, the control diagrams should indicate the required reaction of the system in all scenarios. The status or position of every fan and damper in every scenario must be clearly identified.

[F] 909.16 Fire fighter's smoke control panel. A fire fighter's smoke control panel for fire department emergency response purposes only shall be provided and shall include manual control or override of automatic control for mechanical smoke control systems. The panel shall be located in a fire command center complying with Section 911 in high-rise buildings or buildings with smoke-protected assembly seating. In all other buildings, the fire fighter's smoke control panel shall be installed in an *approved* location adjacent to the fire alarm control panel. The fire fighter's smoke control panel shall comply with Sections 909.16.1 through 909.16.3.

❖ One of the elements that makes a smoke control system effective is that its activity is successfully communicated to the fire department and the fire department is able to manually operate the system. The following sections provide requirements for a control panel specifically for smoke control systems. This panel is required to be located within a fire command center when it is located in a high-rise building or there is smoke-protected seating. Section 403.4.6 would require a fire command center for high-rise buildings. Smoke-protected seating does not require a fire command center in Chapter 10 but this provision would ensure that one exists and contains the smoke control panel. Facilities with smoke-protected seating tend to be larger facilities that, at the very least, would already have a central security center if not a fire command center as required by the jurisdiction. All other locations would only need to provide the panel in an approved location as long as it is located with the fire alarm panel. The specific location will depend on the needs of the fire department in that jurisdiction. The reason not all fire-fighter smoke control panels need to be located in a fire command center is that many smoke control systems are located in a building containing an atrium that may only be three stories in height. A 200-square-foot (19 m^2) fire command center would be excessive for such buildings. There are two components that include the requirements for the display and for the controls. This control panel will provide an ability to override any other controls whether manual or automatic within the building as they relate to the smoke control system.

Note that the publication *A Guide to Smoke Control in the 2006 IBC®* goes into more detail about the fire fighter smoke control panel requirements.

[F] 909.16.1 Smoke control systems. Fans within the building shall be shown on the fire fighter's control panel. A clear indication of the direction of airflow and the relationship of components shall be displayed. Status indicators shall be provided for all smoke control equipment, annunciated by fan and zone, and by pilot-lamp-type indicators as follows:

1. Fans, *dampers* and other operating equipment in their normal status—WHITE.
2. Fans, *dampers* and other operating equipment in their off or closed status—RED.
3. Fans, *dampers* and other operating equipment in their on or open status—GREEN.
4. Fans, *dampers* and other operating equipment in a fault status—YELLOW/AMBER.

❖ This section denotes what should be displayed on the control panel. The display is required to include all fans, an indication of the direction of airflow and the relationship of the components. Also, status lights are required, and this section sets out specific standardized colors to indicate normal status, closed status, open status and fault status. A standardized approach increases the likelihood that the fire department will be able to quickly become familiar with a system. Since the fire department has the ability to override the automatic functions of the system, this information is critical.

[F] 909.16.2 Smoke control panel. The fire fighter's control panel shall provide control capability over the complete smoke control system equipment within the building as follows:

1. ON-AUTO-OFF control over each individual piece of operating smoke control equipment that can also be controlled from other sources within the building. This includes *stairway* pressurization fans; smoke exhaust fans; supply, return and exhaust fans; elevator shaft fans and other operating equipment used or intended for smoke control purposes.
2. OPEN-AUTO-CLOSE control over individual *dampers* relating to smoke control and that are also controlled from other sources within the building.
3. ON-OFF or OPEN-CLOSE control over smoke control and other critical equipment associated with a fire or smoke emergency and that can only be controlled from the fire fighter's control panel.

Exceptions:

1. Complex systems, where *approved*, where the controls and indicators are combined to control and indicate all elements of a single smoke zone as a unit.
2. Complex systems, where *approved*, where the control is accomplished by computer interface using *approved*, plain English commands.

❖ This section sets the requirements as to which controls need to be provided for the fire department on the control panel.

There are two aspects to the controls. The controls will include on-auto-off and open-auto-close settings or will be strictly on-off or open-close. If the system or component can be set on automatic (auto), it can be controlled from other locations beyond the fire command center. This would include an automatic smoke detection system or by manual activation. If a control only contains on-off or open-close settings, the only way the system component can be controlled is in the fire command center.

It should be noted that components such as fans are usually associated with on-off-type controls, whereas components such as dampers are associated with open-close-type controls.

[F] 909.16.3 Control action and priorities. The fire-fighter's control panel actions shall be as follows:

1. ON-OFF and OPEN-CLOSE control actions shall have the highest priority of any control point within the building. Once issued from the fire fighter's control panel, automatic or manual control from any other control point within the building shall not contradict the control action. Where automatic means are provided to interrupt normal, nonemergency equipment operation or produce a specific result to safeguard the building or equipment including, but not limited to, duct freezestats, duct smoke detectors, high-temperature cutouts, temperature-actuated linkage and similar devices, such means shall be capable of being overridden by the fire fighter's control panel. The last control action as indicated by each fire fighter's control panel switch position shall prevail. Control actions shall not require the smoke control system to assume more than one configuration at any one time.

   Exception: Power disconnects required by NFPA 70.

2. Only the AUTO position of each three-position fire-fighter's control panel switch shall allow automatic or manual control action from other control points within the building. The AUTO position shall be the NORMAL, nonemergency, building control position. Where a fire fighter's control panel is in the AUTO position, the actual status of the device (on, off, open, closed) shall continue to be indicated by the status indicator described in Section 909.16.1. Where directed by an automatic signal to assume an emergency condition, the NORMAL position shall become the emergency condition for that device or group of devices within the zone. Control actions shall not require the smoke control system to assume more than one configuration at any one time.

❖ This section clarifies that when a component of the system is placed in an on-off or open-close configuration, no other control point in the building, whether automatic or manual, can override the action established in the fire command center. If a system component is configured in auto mode, it can be controlled from locations within the building beyond the fire command center. Some controls are specifically designed to only allow an action from the fire command center.

[F] 909.17 System response time. Smoke-control system activation shall be initiated immediately after receipt of an appropriate automatic or manual activation command. Smoke control systems shall activate individual components (such as *dampers* and fans) in the sequence necessary to prevent physical damage to the fans, *dampers*, ducts and other equipment. For purposes of smoke control, the fire fighter's control panel response time shall be the same for automatic or manual smoke control action initiated from any other building control point. The total response time, including that necessary for detection, shutdown of operating equipment and smoke control system startup, shall allow for full operational mode to be achieved before the conditions in the space exceed the design smoke condition. The system response time for each component and their sequential relationships shall be detailed in the required rational analysis and verification of their installed condition reported in the required final report.

❖ This particular section provides the criteria as to when the smoke control system is required to begin operation. Whether or not the activation is manual or automatic, these criteria clarify that the system be initiated immediately. Also, it requires that components activate in a sequence that will not potentially damage the fans, dampers, ducts and other equipment. Unrealistic timing of the system has the potential of creating an unsuccessful system. Delays in the system can be seen in slow dampers, fans that ramp up or down, systems that poll slowly and intentional built-in delays. These factors can add significantly to the reaction time of the system and may hamper achieving the design goals.

The key element is that the system be fully operational before the smoke conditions exceed the design parameters. The design should include these possible delays when analyzing the smoke layer interface location. The sequence of events needs to be justified within the design analysis and described clearly in the construction documents.

[F] 909.18 Acceptance testing. Devices, equipment, components and sequences shall be individually tested. These tests, in addition to those required by other provisions of this code, shall consist of determination of function, sequence and, where applicable, capacity of their installed condition.

❖ In order to achieve a certain level of performance, the smoke control system needs to be thoroughly tested. Section 909.18 requires that all devices, equipment components and sequences be individually tested.

[F] 909.18.1 Detection devices. Smoke or fire detectors that are a part of a smoke control system shall be tested in accordance with Chapter 9 in their installed condition. Where applicable, this testing shall include verification of airflow in both minimum and maximum conditions.

❖ Detection devices are required to be tested in accordance with the fire protection requirements found in Chapter 9. Also, since such detectors may be subject to higher air velocities than typical detectors, their operation needs to be verified in the minimum and maximum anticipated airflow conditions.

[F] 909.18.2 Ducts. Ducts that are part of a smoke control system shall be traversed using generally accepted practices to determine actual air quantities.

❖ This section requires ducts that are part of the smoke control system to be tested to show that the proper amount of air is flowing. It should be noted that Section 909.10.2 requires that the ducts be leak tested to 1.5 times the maximum design pressure. Such leakage is not allowed to exceed 5 percent of the design flow.

[F] 909.18.3 Dampers. *Dampers* shall be tested for function in their installed condition.

❖ This section notes that all dampers need to be inspected to meet the function for which they are installed. For instance, a damper that is to be open when the system is in smoke control mode should be verified to be open when testing the system. Also, a damper may have a specific timing associated with its operation that would need to be verified through testing.

[F] 909.18.4 Inlets and outlets. Inlets and outlets shall be read using generally accepted practices to determine air quantities.

❖ Similar to ducts, the appropriate amount of air that is entering or exiting the inlets and outlets, respectively, must be checked.

[F] 909.18.5 Fans. Fans shall be examined for correct rotation. Measurements of voltage, amperage, revolutions per minute (rpm) and belt tension shall be made.

❖ This section requires the testing of fans for the following: correct rotation, voltage, amperage, revolutions per minute and belt tension. These features are key to having the system run as designed.

A common problem with fans is that they are often installed in the reverse direction. Also, to verify the reliability of the fans, elements such as the appropriate voltage and belt tension need to be tested.

[F] 909.18.6 Smoke barriers. Measurements using inclined manometers or other *approved* calibrated measuring devices shall be made of the pressure differences across *smoke barriers*. Such measurements shall be conducted for each possible smoke control condition.

❖ As discussed in Section 909.5.2, the testing of pressure differences across smoke barriers needs to be measured in the smoke control mode. As noted in Section 909.18.6, such testing is to be performed for every possible smoke control condition, and the measurements will be taken using an inclined manometer or other approved methods. Electronic devices are also available. Qualified individuals must calibrate these types of devices. Additionally, before using an alternative method of testing, the building official needs to approve such a method.

[F] 909.18.7 Controls. Each smoke zone equipped with an automatic-initiation device shall be put into operation by the actuation of one such device. Each additional device within the zone shall be verified to cause the same sequence without requiring the operation of fan motors in order to prevent damage. Control sequences shall be verified throughout the system, including verification of override from the fire-fighter's control panel and simulation of standby power conditions.

❖ This section requires the overall testing of the system. More specifically, each zone needs to individually initiate the smoke control system by the activation of an automatic initiation device. Once that has occurred, all other devices within each zone need to be verified that they will activate the system, but to avoid damage, the fans do not need to be activated.

In addition to determining that all of the appropriate devices initiate the system, it must also be verified that all of the controls on the fire-fighter control panel initiate the appropriate aspects of the smoke control system, including the override capability.

Finally, the initiation and availability of the standby power system need to be verified.

[F] 909.18.8 Testing for smoke control. Smoke control systems shall be tested by a special inspector in accordance with Section 1705.18.

❖ Smoke control systems require testing by a special inspector since they are unique and complex life safety systems. Section 1705.18 provides the same requirements for testing as presented in Sections 909.18.8.1 and 909.18.8.2.

[F] 909.18.8.1 Scope of testing. Testing shall be conducted in accordance with the following:

1. During erection of ductwork and prior to concealment for the purposes of leakage testing and recording of device location.
2. Prior to occupancy and after sufficient completion for the purposes of pressure-difference testing, flow measurements, and detection and control verification.

❖ Special inspections need to occur at two different stages during construction to facilitate the necessary inspections. The first round of testing occurs before concealment of the ductwork or fire protection elements. The special inspector needs to verify the leakage in accordance with Section 909.10.2. Additionally, the location of all fire protection devices needs to be verified and documented at this time.

The second round of testing occurs just prior to occupancy. The testing includes the verification of pressure differences across smoke barriers, as required in Sections 909.5.2 and 909.18.6, the verification of appropriate volumes of airflow as noted in the design, and finally the verification of the appropriate operation of the detection and control mechanisms as required in Sections 909.18.1 and 909.18.7. These tests need to occur just prior to occupancy, since the test results will more clearly represent actual conditions. This also makes a strong design and quality assurance during construction critical, as it is very costly and difficult in most cases to make changes at this stage. Note that the test does not actually place smoke into the space demonstrate or the smoke layer interface location. Instead, the testing is focused on all of the elements of the design such as airflow and duct closure as prescribed by the specific design.

[F] 909.18.8.2 Qualifications. *Approved* agencies for smoke control testing shall have expertise in fire protection engineering, mechanical engineering and certification as air balancers.

❖ As noted in Section 909.3, special inspections are required for smoke control systems. This means a certain level of qualification that would include the need for expertise in fire protection engineering, mechanical engineering and certification as air balancers.

[F] 909.18.8.3 Reports. A complete report of testing shall be prepared by the *approved* agency. The report shall include identification of all devices by manufacturer, nameplate data, design values, measured values and identification tag or *mark*. The report shall be reviewed by the responsible *registered design professional* and, when satisfied that the design intent has been achieved, the responsible *registered design professional* shall sign, seal and date the report.

❖ Once the testing by the special inspector is complete, documentation of the activity is required. This documentation is to be prepared in the form of a report that identifies all devices by manufacturer, nameplate data, design values, measured values and identification or mark.

[F] 909.18.8.3.1 Report filing. A copy of the final report shall be filed with the fire code official and an identical copy shall be maintained in an *approved* location at the building.

❖ The report needs to be reviewed, approved and then signed, sealed and dated. This report is to be provided to the building official and a copy is also to remain in the building in an approved location. When a fire command center is required, this is the best location for such documents. Otherwise, a location such as the security office or building manager's office might be appropriate.

[F] 909.18.9 Identification and documentation. Charts, drawings and other documents identifying and locating each component of the smoke control system, and describing its proper function and maintenance requirements, shall be maintained on file at the building as an attachment to the report required by Section 909.18.8.3. Devices shall have an *approved* identifying tag or *mark* on them consistent with the other required documentation and shall be dated indicating the last time they were successfully tested and by whom.

❖ Additional documentation that needs to be maintained includes charts, drawings and other related documentation that assists in the identification of each aspect of the smoke control system. This documentation is where information, such as the last time a device or component was successfully tested and by whom, is recorded. This will serve as the main documentation for the system. Again, the fire command center, if required, is the most appropriate location for such information (see commentary, Section 909.18.8.3.1).

[F] 909.19 System acceptance. Buildings, or portions thereof, required by this code to comply with this section shall not be issued a certificate of occupancy until such time that the fire code official determines that the provisions of this section have been fully complied with and that the fire department has received satisfactory instruction on the operation, both automatic and manual, of the system and a written maintenance program complying with the requirements of Section 909.20.1 of the *International Fire Code* has been submitted and approved by the fire code official.

> **Exception:** In buildings of phased construction, a temporary certificate of occupancy, as *approved* by the fire code official, shall be allowed provided that those portions of the building to be occupied meet the requirements of this section and that the remainder does not pose a significant

hazard to the safety of the proposed occupants or adjacent buildings.

❖ This section stipulates that the certificate of occupancy cannot be issued unless the smoke control system has been accepted. It is essential that the system be inspected and approved since it is a life safety system. There is an exception for buildings that are constructed in phases where a temporary certificate of occupancy is allowed. For example, a building where the portion requiring smoke control is not yet occupied so egress concerns through that space are not relevant. This space needs to be separated by smoke barriers (different smoke zone). The IFC also requires a maintenance program for smoke control systems since the long-term success of such systems depends heavily on proper maintenance in addition to rigorous acceptance testing. The IBC simply provides a reference to that section of the IFC.

909.20 Smokeproof enclosures. Where required by Section 1023.11, a smokeproof enclosure shall be constructed in accordance with this section. A smokeproof enclosure shall consist of an *interior exit stairway* or *ramp* that is enclosed in accordance with the applicable provisions of Section 1023 and an open exterior balcony or ventilated vestibule meeting the requirements of this section. Where access to the roof is required by the *International Fire Code*, such access shall be from the smokeproof enclosure where a smokeproof enclosure is required.

❖ In a building that serves stories where the floor surface is located more than 75 feet (22 860 mm) above the level of fire department access or more than 30 feet (9144 mm) below the level of exit discharge stairways, either a smokeproof enclosure or a pressurized stairway is required.

A smokeproof enclosure essentially takes advantage of fire-resistance-rated construction surrounding the interior exit stairway shaft and a buffer zone created by an outside balcony or ventilated vestibule to avoid the accumulation of smoke within the interior exit stairway. The premise is that if access to the smokeproof enclosure provides a sufficient buffer to the effects of smoke via ventilation or simply being located outside, the smoke will not migrate into the interior exit stairway.

The vestibule as noted is required to be ventilated. There are two alternatives provided in Section 909.20. They include natural ventilation, covered in Section 909.20.3, and mechanical ventilation, covered in Section 909.20.4. It should be noted that a smokeproof enclosure is not considered a pressurized stairway. The method of pressurizing stairways is recognized as an alternative to smokeproof enclosures in Section 909.20.5, but it is not required.

909.20.1 Access. Access to the *stairway* or *ramp* shall be by way of a vestibule or an open exterior balcony. The minimum dimension of the vestibule shall be not less than the required width of the *corridor* leading to the vestibule but shall not have a width of less than 44 inches (1118 mm) and shall not have a length of less than 72 inches (1829 mm) in the direction of egress travel.

❖ As noted, access to the stairway in a smokeproof enclosure is via an exterior balcony or a ventilated vestibule. This section provides the minimum dimensions for access to the smokeproof enclosure through a vestibule and then travel space to the entrance into the stairway. If a vestibule is chosen, the minimum width is established as the larger of 44 inches (1118 mm) or the width of the corridor. Additionally, when entering a vestibule there needs to be at least 72 inches (1829 mm) in the direction of travel (see Commentary Figure 909.20.1). It should be noted that the

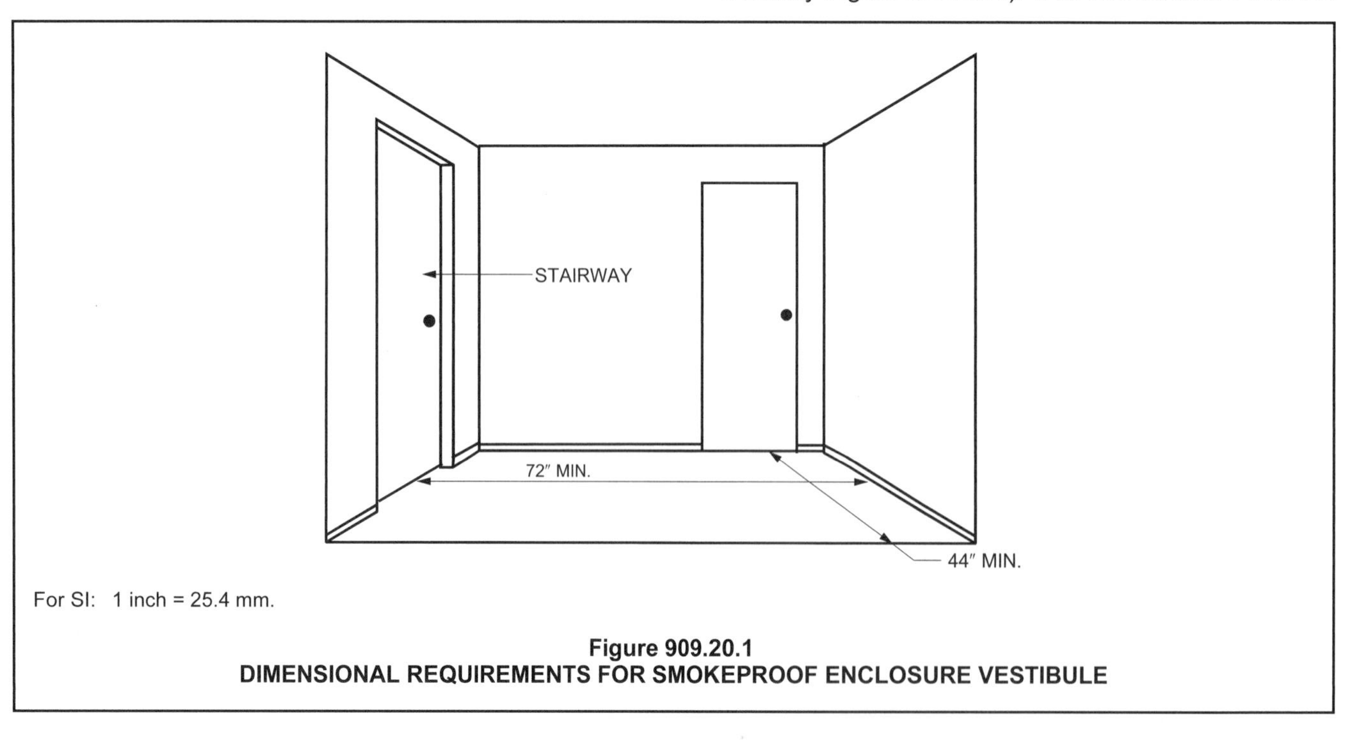

Figure 909.20.1
DIMENSIONAL REQUIREMENTS FOR SMOKEPROOF ENCLOSURE VESTIBULE

pressurized stairway alternative would not require a vestibule.

909.20.2 Construction. The smokeproof enclosure shall be separated from the remainder of the building by not less than 2-hour *fire barriers* constructed in accordance with Section 707 or *horizontal assemblies* constructed in accordance with Section 711, or both. Openings are not permitted other than the required *means of egress* doors. The vestibule shall be separated from the *stairway* or *ramp* by not less than 2-hour *fire barriers* constructed in accordance with Section 707 or *horizontal assemblies* constructed in accordance with Section 711, or both. The open exterior balcony shall be constructed in accordance with the *fire-resistance rating* requirements for floor assemblies.

❖ This section sets the basic construction requirements for the stairway and the associated vestibule or exterior balcony. Essentially, a 2-hour fire-resistance-rated fire barrier or horizontal assembly is required between the enclosure and the rest of the building.

Additionally, vestibules need to be separated from the stairway by a 2-hour fire barrier or horizontal assembly. The exterior balcony, if used, would only be required to comply with the fire-rating requirements for floor construction. The opening protection for doors into the stairway from the vestibule are only required to be fire protection rated for 20 minutes (see Sections 909.20.3.2 and 909.20.4.1).

909.20.2.1 Door closers. Doors in a smokeproof enclosure shall be self- or automatic closing by actuation of a smoke detector in accordance with Section 716.5.9.3 and shall be installed at the floor-side entrance to the smokeproof enclosure. The actuation of the smoke detector on any door shall activate the closing devices on all doors in the smokeproof enclosure at all levels. Smoke detectors shall be installed in accordance with Section 907.3.

❖ In order to maintain the separation between the smokeproof enclosure and the rest of the building, the doors need to be self-closing or automatic closing through the actuation of a smoke detector. Also, when a smoke detector is activated on any floor, doors at all levels are required to close for that particular smokeproof enclosure.

909.20.3 Natural ventilation alternative. The provisions of Sections 909.20.3.1 through 909.20.3.3 shall apply to ventilation of smokeproof enclosures by natural means.

❖ Smoke and hot gases generated by fire may enter smokeproof enclosures through the doorways. This section provides for the diffusion of smoke and gases to the outside of the building by means of unenclosed openings to the outside.

909.20.3.1 Balcony doors. Where access to the *stairway* or *ramp* is by way of an open exterior balcony, the door assembly into the enclosure shall be a *fire door assembly* in accordance with Section 716.5.

❖ Where an open exterior balcony is used for entry into the stairway, the doorways to the stairway are required to have a $1^1/_2$-hour fire protection rating as required for 2-hour fire-resistance-rated wall assemblies in accordance with the requirements of Section 716.5. If a vestibule is used, the rating of the door from the vestibule into the stairway would only be required to be fire protection rated for 20 minutes. The door into the vestibule would require a fire protection rating of $1^1/_2$ hours.

909.20.3.2 Vestibule doors. Where access to the *stairway* or *ramp* is by way of a vestibule, the door assembly into the vestibule shall be a *fire door assembly* complying with Section 716.5. The door assembly from the vestibule to the *stairway* shall have not less than a 20-minute *fire protection rating* complying with Section 716.5.

❖ When a smokeproof enclosure is of the type that employs a vestibule as the means of access to an enclosed exit stairway, the entry into the vestibule from the exit access must consist of a fire door assembly having a $1^1/_2$-hour fire protection rating. This requirement is commensurate with the 2-hour fire-resistance- rated smokeproof enclosure construction requirement and the associated fire protection ratings specified in Table 716.5. Essentially, the fire door assemblies and the enclosure walls serve as the primary exit enclosure protection from fires occurring in adjacent spaces.

The entry from a vestibule into the stairway requires a 20-minute fire-protection-rated door assembly. These doorways are not normally subject to direct fire exposure and, therefore, only a nominal fire protection rating is required as a standard of construction that would minimize air leakage and thus reduce the possibilities of smoke and hot gases penetrating the stairway in amounts that could become a threat to life safety.

Door assemblies must comply with the requirements of Section 715.4 (see Commentary Figure 909.20.3.2).

909.20.3.3 Vestibule ventilation. Each vestibule shall have a minimum net area of 16 square feet (1.5 m^2) of opening in a wall facing an outer *court*, *yard* or *public way* that is not less than 20 feet (6096 mm) in width.

❖ Natural ventilation is allowed for vestibules that have a minimum net opening of 16 square feet (1.49 m^2) and open into an outer court, yard or public way that is at least 20 feet (6096 mm) in width. Basically, this option is intended to take advantage of the fact that smoke may disperse into an open area before entering the stair enclosure.

909.20.4 Mechanical ventilation alternative. The provisions of Sections 909.20.4.1 through 909.20.4.4 shall apply to ventilation of smokeproof enclosures by mechanical means.

❖ When an exterior balcony is not possible or when the clear open area cannot be achieved, then the mechanical ventilation alternative is necessary.

909.20.4.1 Vestibule doors. The door assembly from the building into the vestibule shall be a *fire door assembly* complying with Section 716.5.3. The door assembly from the vestibule to the *stairway* or *ramp* shall not have less than a 20-

minute *fire protection rating* and shall meet the requirements for a smoke door assembly in accordance with Section 716.5.3. The door shall be installed in accordance with NFPA 105.

❖ The entry into the vestibule from the exit access must consist of a fire door assembly having a $1^1/_2$-hour fire protection rating. This requirement is commensurate with the 2-hour fire-resistance-rated smokeproof enclosure construction requirement and the associated fire protection ratings specified in Table 716.5. Essentially, the fire door assemblies and the enclosure walls serve as the primary interior exit stairway protection from fires occurring in adjacent spaces.

The entry from a vestibule into the stairway requires a 20-minute fire-protection-rated door assembly that also meets the requirements for a smoke door assembly. Vestibule doors into stairways for the natural ventilation option (see Section 909.20.3.2) would not require the smoke door assembly rating in accordance with Section 716.5.3. The natural ventilation option requires an opening to the outside that is intended to compensate for this difference (see Commentary Figure 909.20.4.1).

909.20.4.2 Vestibule ventilation. The vestibule shall be supplied with not less than one air change per minute and the exhaust shall be not less than 150 percent of supply. Supply air shall enter and exhaust air shall discharge from the vestibule through separate, tightly constructed ducts used only for that purpose. Supply air shall enter the vestibule within 6 inches (152 mm) of the floor level. The top of the exhaust register shall be located at the top of the smoke trap but not more than 6 inches (152 mm) down from the top of the trap, and shall be entirely within the smoke trap area. Doors in the open position shall not obstruct duct openings. Duct openings with controlling *dampers* are permitted where necessary to meet the design requirements, but *dampers* are not otherwise required.

❖ This section provides the basic criteria for the mechanical ventilation of the vestibule during a fire situation. Such ventilation is required to be activated via a smoke detector in accordance with Section 909.20.6. The basic requirements are for one air change per minute within the vestibule. Also the exhaust is required to be 150 percent of the supply. The exhaust is required to exceed the supply so that air will not be pushed into the stairway due to a positive pressure being created within the vestibule. It should be noted that there is not a minimum pressure difference requirement. Instead, the code simply prescribes a particular air change rate (exhaust).

This section also prescribes criteria as to where supply air and exhaust air are to enter and exit, respectively, and requires that those paths are clear (see Commentary Figure 909.20.4.2).

It should be noted that this section does not specify when and how such systems should activate. Activation is addressed by Section 909.20.6.

909.20.4.2.1 Engineered ventilation system. Where a specially engineered system is used, the system shall exhaust a quantity of air equal to not less than 90 air changes per hour from any vestibule in the emergency operation mode and shall be sized to handle three vestibules simultaneously. Smoke detectors shall be located at the floor-side entrance to each vestibule and shall activate the system for the affected vestibule. Smoke detectors shall be installed in accordance with Section 907.3.

❖ If the method prescribed in Section 909.20.4.2 for mechanical ventilation of vestibules is not chosen,

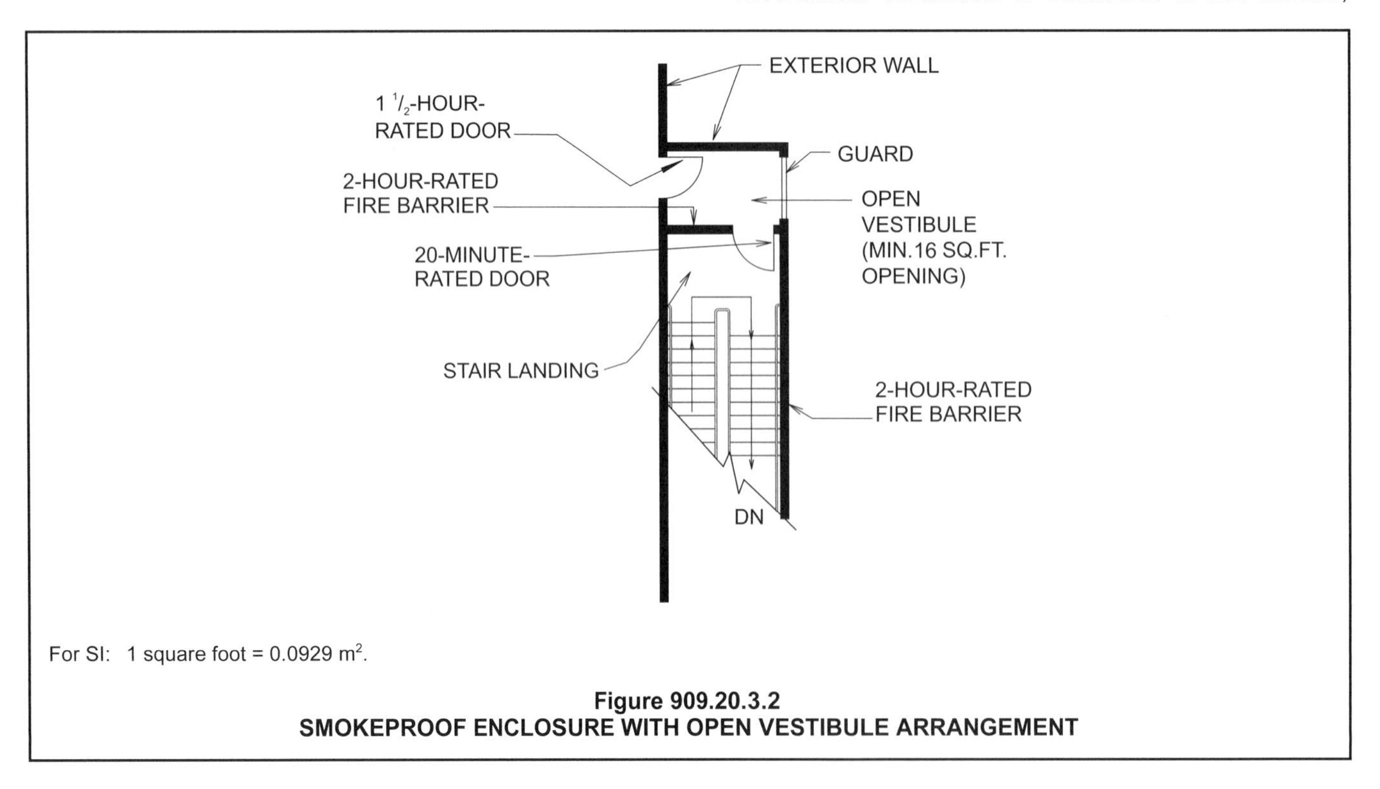

Figure 909.20.3.2
SMOKEPROOF ENCLOSURE WITH OPEN VESTIBULE ARRANGEMENT

then minimum criteria are provided for engineered systems. Essentially, the designer can create a design that meets the flexibility needs of the building and at the same time meets basic criteria. Once designed and installed, the vestibules must be ventilated at 90 air changes per hour, which is 30 more air changes per hour than the prescriptive solution would require. The exhaust rate is not mandated. It is simply left to the designer to achieve the 90 air changes per hour within the space (vestibule). Additionally, this

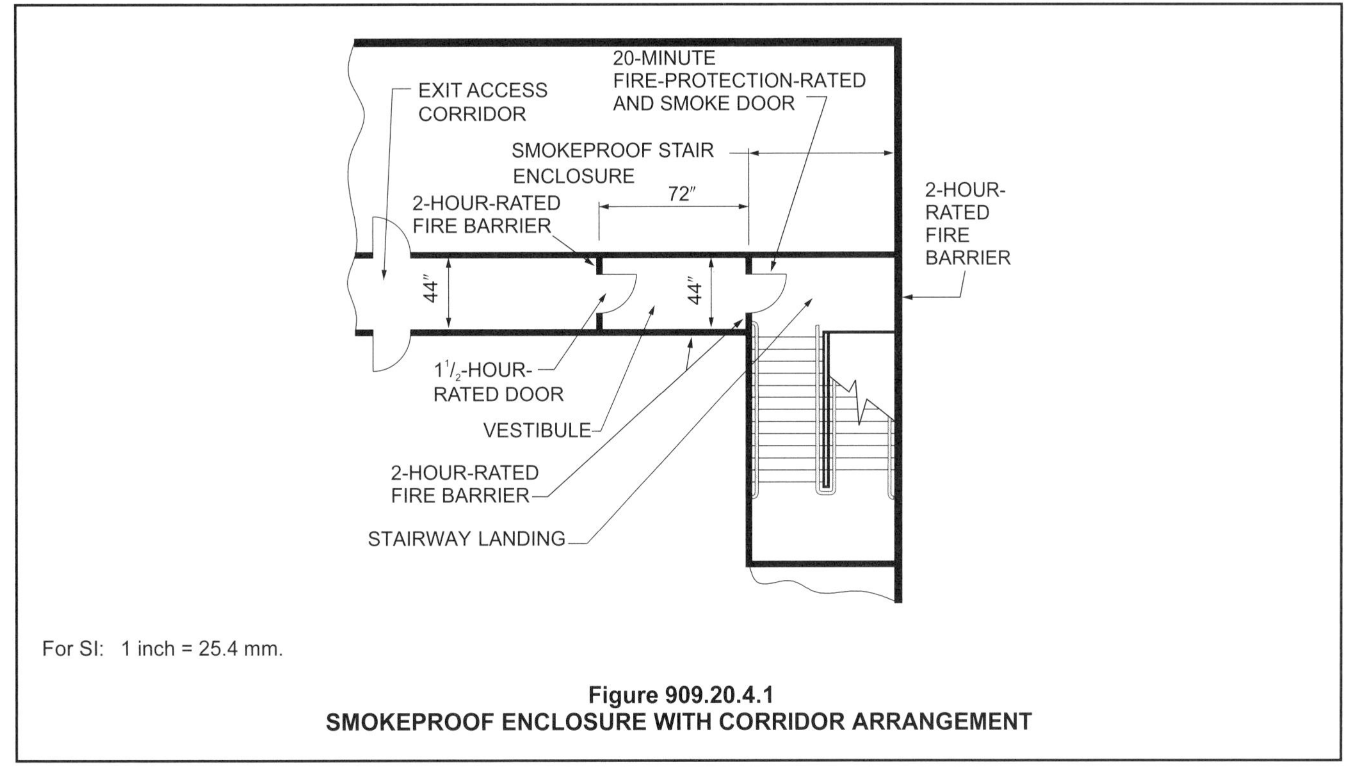

For SI: 1 inch = 25.4 mm.

Figure 909.20.4.1
SMOKEPROOF ENCLOSURE WITH CORRIDOR ARRANGEMENT

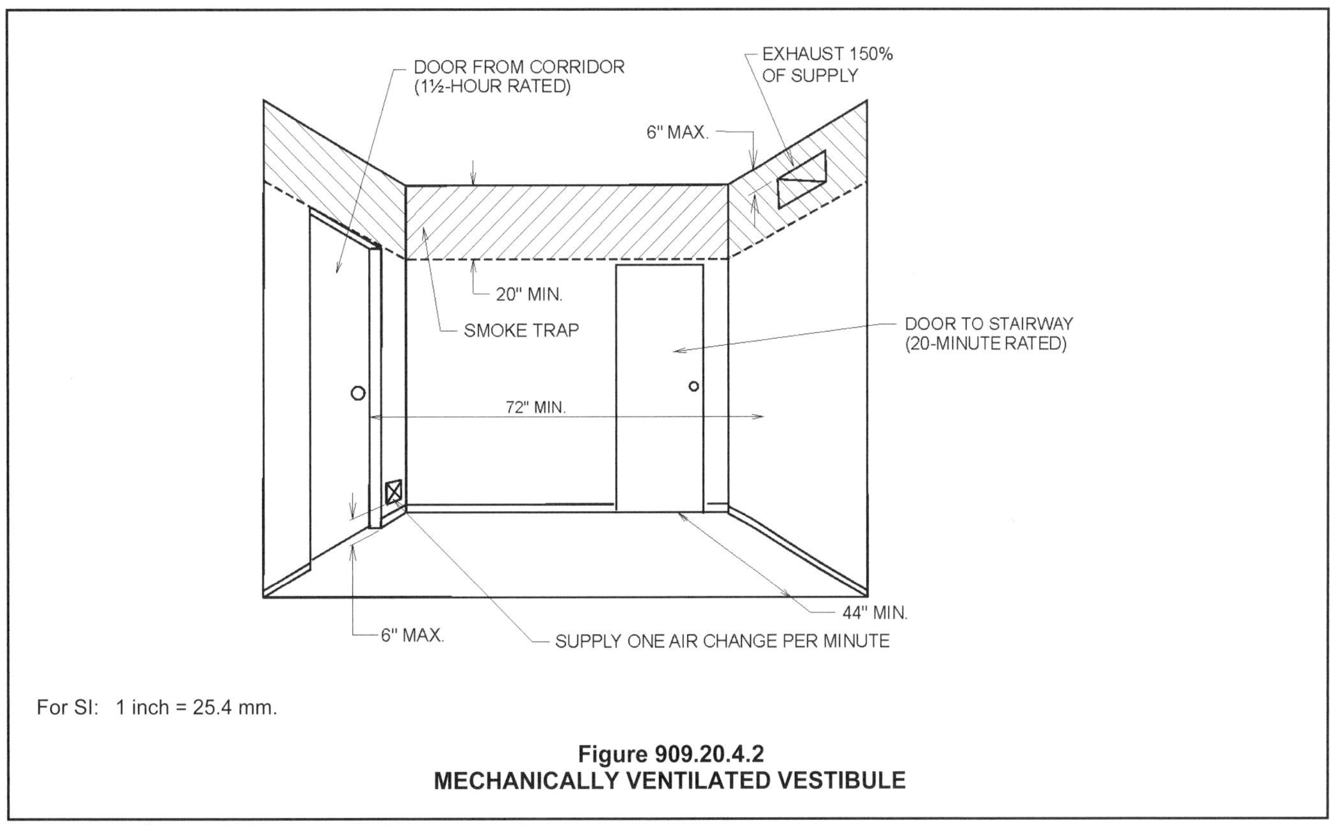

For SI: 1 inch = 25.4 mm.

Figure 909.20.4.2
MECHANICALLY VENTILATED VESTIBULE

system needs to be sized to handle at least three vestibules simultaneously. Finally, the engineered systems are required to operate in accordance with the initiation of a smoke detector located on the floor-side entrance of the affected vestibule; therefore, a smoke detector would be required on each floor at the entrance to each vestibule. The system would only need to be activated on the floors where smoke is detected.

909.20.4.3 Smoke trap. The vestibule ceiling shall be not less than 20 inches (508 mm) higher than the door opening into the vestibule to serve as a smoke and heat trap and to provide an upward-moving air column. The height shall not be decreased unless *approved* and justified by design and test.

❖ This section requires a 20-inch (508 mm) difference in height between the door openings into the vestibule and the ceiling. This is intended to allow an additional safeguard to assist in the containment of smoke inside the vestibule and outside the stair shaft. Essentially, as the title states, it provides a smoke trap. This arrangement also has the effect of keeping the air movement upward.

909.20.4.4 Stairway or ramp shaft air movement system. The *stairway* or *ramp* shaft shall be provided with a dampered relief opening and supplied with sufficient air to maintain a minimum positive pressure of 0.10 inch of water (25 Pa) in the shaft relative to the vestibule with all doors closed.

❖ In order to prevent smoke from migrating into the stairway from the vestibule, a positive pressure of 0.10 inch (25 Pa) of water in the stair shaft relative to the vestibule is required. This pressure difference is to be available when all doors are closed. This pressure difference is lower than that required by the stairway pressurization alternative. This would not be considered a pressurized stairway. This pressure difference would need to be tested to obtain approval once constructed.

909.20.5 Stairway and ramp pressurization alternative. Where the building is equipped throughout with an *automatic sprinkler system* in accordance with Section 903.3.1.1, the vestibule is not required, provided each interior *exit stairway* or *ramp* is pressurized to not less than 0.10 inch of water (25 Pa) and not more than 0.35 inches of water (87 Pa) in the shaft relative to the building measured with all *interior exit stairway* and *ramp* doors closed under maximum anticipated conditions of stack effect and wind effect.

❖ This method is allowed only when the building is equipped throughout with an automatic sprinkler system. This is partially related to the fact that these pressure differences were developed based on a sprinklered fire. It should be noted that smokeproof enclosures are not required to be in buildings equipped throughout with an automatic sprinkler system, but the areas where smokeproof enclosures are required are often sprinklered buildings (i.e., high-rise buildings). This alternative would not require vestibules or an exterior exit balcony. The criteria for smoke control design are provided in terms of minimum and maximum pressure differences of 0.10 inch (37 Pa) of water and 0.35 inch (87 Pa) of water, respectively, between the shaft and the building. This pressure difference is to be achieved when all doors are closed and maximum conditions of wind and stack effect have been taken into account. It should be noted that additional limitations may be placed on the maximum pressure differences for pressurized stairways due to the lower opening forces required in order to comply with Section 1010.1.3. If the maximum pressure difference of 35 would exceed the requirements of Section 1010.1.3, the maximum pressure difference would need to be lowered. Also note that Section 404.2.8 of ICC A117.1 would not require opening forces to be lowered for accessibility purposes if the door is a fire door. Finally, as with all other smoke control systems addressed in Section 909, such systems need to be designed through a rational analysis, tested and documented as such. See also commentary for Section 909.6.3.

909.20.6 Ventilating equipment. The activation of ventilating equipment required by the alternatives in Sections 909.20.4 and 909.20.5 shall be by smoke detectors installed at each floor level at an *approved* location at the entrance to the smokeproof enclosure. When the closing device for the *stairway* and *ramp* shaft and vestibule doors is activated by smoke detection or power failure, the mechanical equipment shall activate and operate at the required performance levels. Smoke detectors shall be installed in accordance with Section 907.3.

❖ This section clarifies that the activation mechanism for both mechanical means of smoke management for interior exit stairways in Sections 909.20.4 and 909.20.5 should be via a smoke detector located at each level outside the door leading into the vestibule and stairway, respectively. For systems that use automatic-closing devices on the doors, whether for vestibules in smokeproof enclosures or for pressurized stairways, if the door closes, the system must activate. This includes normal activation of the smoke detector or in the event of a power failure. Essentially, if there is a power failure it will operate in a fail-safe manner even if there is not a fire.

909.20.6.1 Ventilation systems. Smokeproof enclosure ventilation systems shall be independent of other building ventilation systems. The equipment, control wiring, power wiring and ductwork shall comply with one of the following:

1. Equipment, control wiring, power wiring and ductwork shall be located exterior to the building and directly connected to the smokeproof enclosure or connected to the smokeproof enclosure by ductwork enclosed by not less than 2-hour *fire barriers* constructed in accordance with Section 707 or *horizontal assemblies* constructed in accordance with Section 711, or both.

2. Equipment, control wiring, power wiring and ductwork shall be located within the smokeproof enclosure with intake or exhaust directly from and to the outside or through ductwork enclosed by not less than 2-hour *fire*

barriers constructed in accordance with Section 707 or *horizontal assemblies* constructed in accordance with Section 711, or both.

3. Equipment, control wiring, power wiring and ductwork shall be located within the building if separated from the remainder of the building, including other mechanical equipment, by not less than 2-hour *fire barriers* constructed in accordance with Section 707 or *horizontal assemblies* constructed in accordance with Section 711, or both.

Exceptions:

1. Control wiring and power wiring utilizing a 2-hour rated cable or cable system.
2. Where encased with not less than 2 inches (51 mm) of concrete.
3. Control wiring and power wiring protected by a listed electrical circuit protective system with a fire-resistance rating of not less than 2 hours.

❖ Smokeproof enclosures and pressurized stair shaft ventilation systems must be independent of other building ventilating systems. This section provides three options for the location of the ductwork associated with smokeproof enclosures and stair pressurization mechanical equipment. The three options include the following:

- Located on the exterior of the building and directly connected to the smokeproof enclosure;
- Within the smokeproof enclosure; or
- Within the building but separated from the remainder of the building by 2-hour fire-barriers or horizontal barriers, or both.

A question that often arises with the inclusion of the terms "control wiring" and "power wiring" is whether this addresses the smoke detection and fire alarm equipment associated with these ventilation systems. The original intent of the addition of these terms was focused primarily on the fans and related duct work and not necessarily on the fire alarm system components related to the activation of such systems. This is not specifically discussed within the code itself however. Code change FS172-07/08 placed this language in the code.

There are three exceptions to this section. The first is allowing the use of 2-hour-rated cable. This provides the protection required in terms of rating but provides more flexibility in the placement of such wiring.

Exception 2 allows a simple exception that provides a sufficient but not necessarily listed fire-resistance-rated approach by using concrete to encase the equipment or wiring.

Exception 3 is intended to add the option of using conventional cables with a protective material applied to them. These materials are called "electrical circuit protective systems" and are recognized by NFPA 70 *National Electrical Code* for protection of fire pump control wiring, emergency system circuit wiring, and critical operations power system circuit wiring.

909.20.6.2 Standby power. Mechanical vestibule and *stairway* and *ramp* shaft ventilation systems and automatic fire detection systems shall be provided with standby power in accordance with Section 2702.

❖ This section requires standby power for mechanical systems for pressurized stair shafts, mechanically ventilated vestibules, stair shafts for smokeproof enclosures and any automatic fire detection systems related to the activation of such systems.

909.20.6.3 Acceptance and testing. Before the mechanical equipment is *approved*, the system shall be tested in the presence of the *building official* to confirm that the system is operating in compliance with these requirements.

❖ This section requires verification of successful testing of the system to the building official. The requirements of Sections 909.18 and 909.19 should be addressed when testing smokeproof enclosures or pressurized stairs. It should be noted that the IFC does not contain the provisions for smokeproof enclosures and pressurized stairways, but it may be in the best interest to also notify the fire department of such testing and acceptance.

909.21 Elevator hoistway pressurization alternative. Where elevator hoistway pressurization is provided in lieu of required enclosed elevator lobbies, the pressurization system shall comply with Sections 909.21.1 through 909.21.11.

❖ This section sets out the requirements for hoistway pressurization when used as a method to protect hoistway openings as required by Section 3006. The minimum and maximum pressures are similar to those for pressurized stairs in Section 909.20.5. The provisions are located in Section 909 to group all of the smoke control techniques and requirements into the same section. This will promote more consistency in code application when addressing smoke control.

909.21.1 Pressurization requirements. Elevator hoistways shall be pressurized to maintain a minimum positive pressure of 0.10 inch of water (25 Pa) and a maximum positive pressure of 0.25 inch of water (67 Pa) with respect to adjacent occupied space on all floors. This pressure shall be measured at the midpoint of each hoistway door, with all elevator cars at the floor of recall and all hoistway doors on the floor of recall open and all other hoistway doors closed. The pressure differentials shall be measured between the hoistway and the adjacent elevator landing. The opening and closing of hoistway doors at each level must be demonstrated during this test. The supply air intake shall be from an outside, uncontaminated source located a minimum distance of 20 feet (6096 mm) from any air exhaust system or outlet.

Exceptions:

1. On floors containing only Group R occupancies, the pressure differential is permitted to be measured between the hoistway and a *dwelling unit* or *sleeping unit.*

2. Where an elevator opens into a lobby enclosed in accordance with Section 3007.6 or 3008.6, the pressure differential is permitted to be measured between the hoistway and the space immediately outside the door(s) from the floor to the enclosed lobby.
3. The pressure differential is permitted to be measured relative to the outdoor atmosphere on floors other than the following:
    3.1. The fire floor.
    3.2. The two floors immediately below the fire floor.
    3.3. The floor immediately above the fire floor.
4. The minimum positive pressure of 0.10 inch of water (25 Pa) and a maximum positive pressure of 0.25 inch of water (67 Pa) with respect to occupied floors are not required at the floor of recall with the doors open.

❖ This section states the minimum and the maximum positive pressure that must be achieved by the smoke control mechanical pressurization system. The minimum positive pressure is 0.10 inch of water column (0.02 kPa), the same as required for stairway pressurization in Section 909.20.5. The maximum pressure is 0.25 inch of water column (0.06 kPa) which is a little less than the maximum allowed for stairway pressurization. The minimum pressure is to ensure that the stack effect is overcome and the maximum pressure is an upper limit to ensure that the doors will operate properly. This section requires a test when the system is complete. The pressures are measured at the midpoint of each hoistway door with all elevator cars at the recall floor and all the hoistway doors open on that level. This simulates the Phase I recall requirements in Section 3003.2. Hoistway doors are then tested on each level to ensure proper operation.

The supply air intake for the pressurization system must be located at least 20 feet (6096 mm) away from any source of contamination to ensure that the hoistway remains tenable through the fire event or well into it before the elevators can no longer be used. Also, if smoke is drawn into the supply air, the system will only spread smoke and not prevent its spread.

There are four exceptions to the requirements of Section 909.21.1. The first two provide exceptions as to where the pressure differential can be measured. Exception 3 allows the pressure differences only to be measured at the fire floor and several designated floors. Finally, Exception 4 allows the pressure differential measurement to be omitted for the floor of recall.

The first three exceptions originated from the City of Seattle, Washington, which has had a long history of requiring pressurized hoistways in high-rise buildings to prevent smoke migration. In 2005, the City of Seattle Department of Planning & Development (DPD) convened a committee which included representatives from industry, the Seattle Fire Department, and DPD, to decide whether to recommend changes to the high-rise smoke migration control requirements in place at that time. The committee also consulted with Dr. John Klote, who suggested the approach that Seattle eventually adopted with some small modifications. These requirements are an adaptation of the Seattle approach.

During the 2009/2010 code change cycle, a proposal was made to delete the hoistway pressurization requirements in the IBC without substitution (FS51-09/10), based on a study conducted by Drs. Miller and Beasley. This study showed that requiring the pressure differential of 0.10 inch of water column to be maintained at the recall floor with the elevator doors in the open position resulted in over pressurization of all the other floors—meaning the current standards in the code cannot be met. Based on further modeling by Dr. Miller, the proponent for FS51-09/10 submitted a public comment introducing Seattle's requirements into the IBC. The reason statement for the public comment stated Dr. Miller "concluded that the 'Seattle approach' does indeed meet all the prescriptive requirements of the IBC 2009." The proposal and its public comment were ultimately withdrawn by the proponent in anticipation of further review of the overall elevator lobby provisions.

The intent of the code is to keep smoke out of the hoistway, so the pressure should be measured between the elevator hoistway and the elevator landing/lobby. However, Exception 1 allows the pressure to be measured between the hoistway and sleeping or dwelling units in residential buildings, since they are highly compartmented. In addition, the fire source is most likely to be in the dwelling or sleeping unit, and providing positive pressure in the corridor/hallway outside the units (via leakage through the elevator hoistway doors) will help reduce the smoke migrating from the affected unit. Exception 2, which is specific to elevator lobbies associated with fire service access elevators (FSAE) and occupant evacuation elevators, allows the pressure to be measured between the hoistway and the space on the outside of the smoke barrier that forms the lobby. It should be noted that hoistway pressurization is not a design alternative to enclosed lobbies for FSAEs and occupant evacuation elevators. Enclosed elevator lobbies are always required for these types of elevators. This exception would only apply if such a system was provided.

Exception 3 allows the 0.10-inch-water-column pressure differential between the hoistway and the floor be met only on the four most critical floors—the floor of fire origin, the two floors immediately below, and one floor immediately above. For all other stories, the pressure differential is allowed to be measured between the hoistway and the outside of the building. The purpose of this requirement is to maintain a slightly positive pressure in the building relative to atmospheric, so as to lower the neutral pressure plane in the building, which then reduces the driving force of stack effect. This exception is intended to be permitted to be used in conjunction with Exceptions 1 and 2. The engineers who design this system begin by modeling one floor as the "notionalized" fire floor, and designing the system (fans, dampers, etc.) accordingly. Each floor is subsequently modeled as the notionalized fire floor, and the system is checked to make sure the maximum and minimum pressure differentials are met. (Note that actual models may not have to be run for each floor, if it is clear the worst case has been covered.) Ultimately, the system will need to be designed so it will correctly configure itself for a fire originating on any floor in the building.

The fourth exception omits the need to measure the pressure differential at the floor of elevator recall where the door is typically open. Section 909.21.1 requires the pressure difference, required for the pressurization alternative, to be measured at the midpoint of each hoistway door, with all elevator cars at the floor of recall and all hoistway doors on the floor of recall open and all other hoistway doors closed.

Elevator hositway pressurization is intended to minimize smoke movement into an elevator shaft when a lobby is not provided. Meeting the required pressure difference on the recall floor with the hoistway doors open is not necessary, because the recall floor is protected by smoke detectors that will not allow the hoistway doors to open if smoke is present.

The pressurization method is based on using pressure differences produced by fans to minimize the spread of smoke across a barrier. A barrier will not exist on the recall floor when the hoistway doors are open and smoke detectors used for elevator recall prevent the doors from opening when smoke is present.

909.21.1.1 Use of ventilation systems. Ventilation systems, other than hoistway supply air systems, are permitted to be used to exhaust air from adjacent spaces on the fire floor, two floors immediately below and one floor immediately above the fire floor to the building's exterior where necessary to maintain positive pressure relationships as required in Section 909.21.1 during operation of the elevator shaft pressurization system.

❖ This section allows the use of the general building HVAC system to exhaust air to create/maintain the required pressure differential. It is to be noted that the requirements of the rest of Section 909.21, in particular, Section 909.21.10 regarding protection of equipment, would still apply to these components.

909.21.2 Rational analysis. A rational analysis complying with Section 909.4 shall be submitted with the *construction documents*.

❖ Section 909.4 recognizes that there are many factors involved in a smoke control system, including stack effect due to height, temperature effect of fire, wind effect, interaction of the HVAC system, the weather and the egress time, all of which must be evaluated. The report must be submitted with the permit documents. Most importantly, the duration of operation of the smoke control system is a function of 1.5 times the egress time or 20 minutes, whichever is less. More discussion on the duration of operation is found in the commentary for Section 909.4.6.

909.21.3 Ducts for system. Any duct system that is part of the pressurization system shall be protected with the same *fire-resistance rating* as required for the elevator shaft enclosure.

❖ All ductwork necessary for hoistway pressurization must be protected from the effects of fire by enclosure in fire-resistance-rated construction equivalent to that required for the elevator hoistway shaft enclosure.

909.21.4 Fan system. The fan system provided for the pressurization system shall be as required by Sections 909.21.4.1 through 909.21.4.4.

❖ This section details the requirements for the mechanical system used for pressurization of the hoistway enclosure.

909.21.4.1 Fire resistance. Where located within the building, the fan system that provides the pressurization shall be protected with the same *fire-resistance rating* required for the elevator shaft enclosure.

❖ The only way to ensure that the mechanical pressurization system can operate during a fire is to locate it in a safe place. If located within the building it must be in an enclosed room protected with the same fire-resistance-rated construction required for the hoistway enclosure.

909.21.4.2 Smoke detection. The fan system shall be equipped with a smoke detector that will automatically shut down the fan system when smoke is detected within the system.

❖ The airflow must be free of smoke or it will only increase the likelihood of smoke spreading throughout the building. The smoke detector required by this section should be located on the intake side of the blower fan.

909.21.4.3 Separate systems. A separate fan system shall be used for each elevator hoistway.

❖ This section requires that each hoistway enclosure have its own mechanical system. This provides a more redundant system and helps to increase the likelihood that fans will be operational during a fire.

909.21.4.4 Fan capacity. The supply fan shall be either adjustable with a capacity of not less than 1,000 cfm (0.4719 m^3/s) per door, or that specified by a *registered design professional* to meet the requirements of a designed pressurization system.

❖ The fan capacity should be as specified by the registered design professional to meet the operational ranges of pressure at each door or be adjustable with a capacity of at least 1,000 cfm (0.4719 m^3/s) per hoistway door. In either case, it is subject to field testing and adjustments to meet the pressure ranges.

909.21.5 Standby power. The pressurization system shall be provided with standby power in accordance with Section 2702.

❖ The elevator hoistway pressurization system is an emergency system and must have provisions for standby power like other emergency systems. Section 2702 states the requirements that standby power systems must meet. It is consistent with other smoke control systems required by Section 909 in that such systems have standby power as they are life-safety systems.

909.21.6 Activation of pressurization system. The elevator pressurization system shall be activated upon activation of either the building fire alarm system or the elevator lobby smoke detectors. Where both a building fire alarm system and elevator lobby smoke detectors are present, each shall be independently capable of activating the pressurization system.

❖ This section requires that the pressurization system will be activated when the general building fire alarm system or an elevator lobby smoke detector is activated. All buildings using this pressurization option will more than likely be required to have both. High-rise buildings require elevator lobby smoke detectors, but other buildings may not. Section 909.12 requires smoke detectors to activate the pressurization system if the design requires it to operate to remove the smoke.

909.21.7 Testing. Testing for performance shall be required in accordance with Section 909.18.8. System acceptance shall be in accordance with Section 909.19.

❖ Testing will be required to evaluate the performance of the completed system (see commentary, Sections 909.18 and 909.19).

909.21.8 Marking and identification. Detection and control systems shall be marked in accordance with Section 909.14.

❖ See the commentary to Section 909.14.

909.21.9 Control diagrams. Control diagrams shall be provided in accordance with Section 909.15.

❖ See the commentary to Section 909.15.

909.21.10 Control panel. A control panel complying with Section 909.16 shall be provided.

❖ See the commentary to Section 909.16.

909.21.11 System response time. Hoistway pressurization systems shall comply with the requirements for smoke control system response time in Section 909.17.

❖ See the commentary to Section 909.17.

SECTION 910
SMOKE AND HEAT REMOVAL

[F] 910.1 General. Where required by this code, smoke and heat vents or mechanical smoke removal systems shall conform to the requirements of this section.

❖ This section essentially requires either smoke and heat vents or a mechanical smoke removal system where required by Section 910.2. It should be noted that where high-piled combustible storage is involved, Chapter 32 of the IFC also applies.

The purpose of smoke and heat vents or smoke removal systems has historically been related to the needs of fire fighters. More specifically, smoke and heat vents or smoke removal systems, when activated, have the potential effect of lifting the height of the smoke layer and providing more tenable conditions to undertake fire-fighting activities. Other potential benefits include a reduction in property damage and the creation of more tenable conditions for occupants.

These provisions are based on research on the interaction of sprinklers, roof vents and draft curtains funded by the National Fire Protection Research Foundation (NFPRF) and conducted at Underwriters Laboratories (UL) in 1997/1998. This research is summarized in a document referred to as National Institute of Science and Technology Interagency Report (NISTIR) 6196-1 dated September 1998. The current provisions were also based on the following:

- Provisions for the use of roof vents in sprinklered buildings included in the 2010 and 2013 edition of NFPA 13, including the substantiation statement for the NFPA 13 roof vent provisions.
- The capability of standard spray sprinklers to both control and extinguish a fire within 30 minutes of sprinkler operation without supplemental fire department activity has been documented.
- Recommendations contained in National Institute for Occupational Safety and Health (NIOSH) 2005-132, *Preventing Injuries and Deaths of Fire Fighters Due to Truss Systems*, and NIOSH 2010-153, *Preventing Deaths and Injuries of Fire Fighters using Risk Management Principles at Structure Fires*.
- Recommendations contained in the Initial Report of the Federal Emergency Management Agency (FEMA)/National Fallen Firefighter Foundation (NFFF®) Firefighter Life Safety Summit held on April 14, 2004, in Tampa, Florida.

The primary purpose of smoke and heat removal from the perspective of the building code requirement is to assist fire-fighting operations after control of the fire has been achieved by the automatic sprinkler system. Automatic smoke and heat vents and automatic sprinkler systems were developed independently of one another and their interaction has been a concern for many years. Even today, there is no accepted method of analyzing their interaction and, therefore, the installation standards for each (NFPA 204 and NFPA 13, respectively) give cautions to the designers of buildings having both systems. Note that NFPA 204 is not referenced in Section 910.

Manually activated mechanical smoke removal systems can perform the same function as roof vents. Mechanical smoke removal systems as required in Section 910 provide fire-rated, grade-level enclosures for the control of the mechanical smoke removal system. This provides greater control of the system for the fire incident commander and reduces the need to place fire fighters on roofs or in other hazardous situations to operate smoke and heat venting systems. This methodology is consistent with the latest recommendations from NIOSH and NFFF for fire fighter safety, risk management and recommended fire-fighting tactics.

[F] 910.2 Where required. Smoke and heat vents or a mechanical smoke removal system shall be installed as required by Sections 910.2.1 and 910.2.2.

Exceptions:

1. Frozen food warehouses used solely for storage of Class I and II commodities where protected by an *approved automatic sprinkler system*.
2. Smoke and heat removal shall not be required in areas of buildings equipped with early suppression fast-response (ESFR) sprinklers.
3. Smoke and heat removal shall not be required in areas of buildings equipped with control mode special application sprinklers with a response time index of 50 $(m \cdot s)^{1/2}$ or less that are listed to control a fire in stored commodities with 12 or fewer sprinklers.

❖ Sections 910.2.1 and 910.2.2 provide the locations where such smoke removal or smoke and heat vents would be required. There are three overall exceptions to the application of Section 910.

Exception 1 recognizes the "building-within-a-building" nature of typical frozen food warehouses. As such, smoke from a fire within a freezer would be contained within the freezer, thus negating the usefulness of smoke and heat vents at the roof level.

Exception 2 recognizes the negative effect that smoke and heat vents can have on the operation of early suppression fast-response (ESFR) sprinklers. Those negative effects include diverting heat away from the sprinklers, which could delay their activation or result in the activation of more sprinklers in areas away from the source of the fire, which may overwhelm the system.

Exception 3 recognizes a new category of automatic sprinkler that shares the key characteristics of ESFR sprinklers, i.e., thermal elements that have a response time index (RTI) of 50 or less and are listed to protect a design area that involves 12 or fewer sprinklers. These control mode special application (CMSA) sprinklers, while not called ESFR, still require similar precautions to ESFR sprinklers with respect to not introducing unknowns, such as smoke and heat removal, that were not present in the full-scale fire tests that determined their listing parameters. Such unknowns can lead to sprinkler "skipping" and exceeding the 12 sprinkler design area, which was the exact concern that led to the ESFR-related provisions that are currently in this chapter and Chapter 32. Note that CMSA sprinklers must have both an RTI of 50 or less and be listed to control or suppress a fire with 12 or fewer sprinklers to qualify for this exception. Any sprinkler listed as "quick response" will satisfy the "50 RTI or less" criterion based on the definition of "quick response" in NFPA 13 Section 3.6.4.7. The number of operating sprinklers will be indicated in the listing criterion for each sprinkler.

910.2.1 Group F-1 or S-1. Smoke and heat vents installed in accordance with Section 910.3 or a mechanical smoke removal system installed in accordance with Section 910.4 shall be installed in buildings and portions thereof used as a Group F-1 or S-1 occupancy having more than 50,000 square feet (4645 m^2) of undivided area. In occupied portions of a building equipped throughout with an *automatic sprinkler system* in accordance with Section 903.3.1.1 where the upper surface of the story is not a roof assembly, a mechanical smoke removal system in accordance with Section 910.4 shall be installed.

Exception: Group S-1 aircraft repair hangars.

❖ Large-area buildings with moderate to heavy fire loads present special challenges to the fire department in disposing of the smoke generated in a fire. In order to provide the fire department with the ability to rapidly and efficiently dispose of smoke in large-area Group F-1 and S-1 buildings exceeding 50,000 square feet (4645 m^2) in undivided area without the exposure of personnel to the dangers associated with cutting ventilation holes in the roof, smoke and heat vents or a smoke removal system must be provided.

The code is not clear on what is meant by the term "undivided area." However the intent is to provide the ability to manage the smoke in large spaces. Draft curtains or potentially any physical separation (regardless of rating) would provide such division. Draft curtains are typically constructed of sheet metal, lath and plaster, gypsum board or other materials that resist the passage of smoke. Typically draft curtains are at least 6 feet deep (1829 mm) from the ceiling. To keep with the concern for managing smoke, the joints and connections should be smoke tight.

Based on the intent of "undivided area," a fire barrier, smoke barrier, fire partition or smoke partition would be more than what is required and would therefore be an acceptable method of dividing the area.

This requirement is independent of the requirements related to high-piled storage in Section 910.2.2. High-piled combustible storage is not occupancy specific.

This section also addresses multistory buildings where the Group F-1 or S-1 occupancy is not the uppermost story and therefore would not have a roof in which to place smoke and heat vents. This section would require that a smoke removal system be installed. This issue was not clearly addressed in previous editions of the IBC.

[F] 910.2.2 High-piled combustible storage. Smoke and heat removal required by Table 3206.2 of the *International Fire Code* for buildings and portions thereof containing high-piled combustible storage shall be installed in accordance with Section 910.3 in unsprinklered buildings. In buildings and portions thereof containing high-piled combustible storage equipped throughout with an *automatic sprinkler system* in accordance with Section 903.3.1.1, a smoke and heat removal system shall be installed in accordance with Section 910.3 or 910.4. In occupied portions of a building equipped throughout with an *automatic sprinkler system* in accordance with Section 903.3.1.1, where the upper surface of the story is not a roof assembly, a mechanical smoke removal system in accordance with Section 910.4 shall be installed.

❖ This section requires smoke and heat removal as it is required in Chapter 32 of the IFC for high-piled combustible storage. Specifically Table 3206.2 sets out when smoke and heat removal is required. If Table 3206.2 does not require such protection, compliance with Section 910 is not necessary. The requirement in IFC Table 3206.2 is based primarily on the size of the high-piled combustible storage area and whether or not the area is equipped with an automatic sprinkler system.

There are several requirements provided within this section. The first addresses nonsprinklered high-piled storage areas. Such areas are required to use smoke and heat vents as mechanical smoke removal systems are designed for use in sprinklered buildings. The rationale for this provision is that a mechanical smoke removal system capable of handling temperatures between 1,000°F and 2,000°F cannot be practically provided at a reasonable cost.

Where high-piled storage areas are equipped with an automatic sprinkler system, smoke and heat vents or a smoke removal system are required to comply with this section. However, if the high-piled storage area is located in a multistory building where the storage area is not located on the uppermost story, this section requires that a smoke removal system be used. This was not clearly addressed in past editions of the IBC.

[F] 910.3 Smoke and heat vents. The design and installation of smoke and heat vents shall be in accordance with Sections 910.3.1 through 910.3.3.

❖ This section simply sets out the subsections that must be addressed to comply with the requirements for the installation of smoke and heat vents.

[F] 910.3.1 Listing and labeling. Smoke and heat vents shall be *listed* and labeled to indicate compliance with UL 793 or FM 4430.

❖ This section specifically requires that all smoke and heat vents be both listed and labeled in accordance with UL 793 or FM 4430. This provides consistency and a level of quality when smoke and heat vents are required. The standard addresses smoke and heat vents that automatically operate during fires via nonelectrical means. Automatic vents listed and labeled to this standard can be operated both automatically and manually. There are two main mechanisms for activation that include a heat-responsive device or on the action of a plastic cover shrinking and falling out of place because of fire exposure.

[F] 910.3.2 Smoke and heat vent locations. Smoke and heat vents shall be located 20 feet (6096 mm) or more from adjacent *lot lines* and *fire walls* and 10 feet (3048 mm) or more from *fire barriers*. Vents shall be uniformly located within the roof in the areas of the building where the vents are required to be installed by Section 910.2 with consideration given to roof pitch, sprinkler location and structural members.

❖ This section has two functions, the first being to focus on hazards to adjacent buildings and the second being proper function of smoke and heat vents through proper placement.

In terms of adjacent properties, this section requires a minimum distance to lot lines and fire walls and then a minimum distance to fire barriers. The first set of distances focuses on separate buildings and exposures, whereas the distance to fire barriers is less restrictive since it focuses on different uses and occupancies within the same building (see Commentary Figure 910.3.2).

To enhance vent performance within the area containing the smoke and heat vents, such vents need to be uniformly spaced. Consideration of issues such as sprinkler location and roof pitch are also essential to proper vent location.

910.3.3 Smoke and heat vents area. The required aggregate area of smoke and heat vents shall be calculated as follows:

For buildings equipped throughout with an *automatic sprinkler system* in accordance with Section 903.3.1.1:

$A_{VR} = V/9000$ **(Equation 9-3)**

where:

A_{VR} = The required aggregate vent area (ft^2).

V = Volume (ft^3) of the area that requires smoke removal.

For unsprinklered buildings:

$A_{VR} = A_{FA}/50$ **(Equation 9-4)**

where:

A_{VR} = The required aggregate vent area (ft^2).

A_{FA} = The area of the floor in the area that requires smoke removal.

❖ This section provides the design criteria to determine the area of smoke and heat vents required. The requirements are based on whether the area requiring smoke and heat vents is equipped with an automatic sprinkler system.

The design of roof vents in buildings protected by an automatic sprinkler system require that the area of roof vents provide equivalent venting to that required for the mechanical smoke removal system (two air changes per hour) based on an assumption that each square foot of vent area will provide 300 cubic feet per minute (cfm) of ventilation. The reason for this requirement is that the roof vents should at least provide venting equivalent to the minimum venting provided by the mechanical smoke removal system. A factor of 300 cfm of venting per square foot of vent area was included in the 2012 edition of the code, although the use of this conversion factor is questionable at best. The actual ventilation provided by each square foot of vent area will depend on the temperature differential between ambient conditions and the smoke layer under the roof deck or the pressure achieved if positive pressure ventilation is utilized. If the prescribed value is not practical for a given building design, designers have the option of demonstrating other values that provide the same performance under Section 104.11 of the code, which allows alternative methods and designs.

The design of roof vents in buildings not protected by a sprinkler system requires that the ratio of the area of the vents to the floor area be a minimum of 1:50. The rationale is that the case where roof vents will be provided without sprinkler protection will be rare (e.g., buildings that contain high-piled storage with an area between 2,500 and 12,000 square feet). Given that this situation will be rare, a complex analysis to determine the required area of roof vents was believed to be unnecessary. The ratio of vent area to floor area of 1:50 is conservative based on the requirements that were included in the 2012 code and the IFC.

[F] 910.4 Mechanical smoke removal systems. Mechanical smoke removal systems shall be designed and installed in accordance with Sections 910.4.1 through 910.4.7.

❖ Mechanical smoke removal systems are considered to be equivalent to smoke and heat vents in terms of code compliance. In multistory buildings, those areas requiring smoke and heat removal will require the use of a mechanical smoke removal system because of its location on stories other than the uppermost story. This section provides the various design requirements for such systems.

910.4.1 Automatic sprinklers required. The building shall be equipped throughout with an *approved automatic sprinkler system* in accordance with Section 903.3.1.1.

❖ The rationale for this provision is that a mechanical smoke removal system capable of handling temperatures between 1,000°F and 2,000°F cannot be practi-

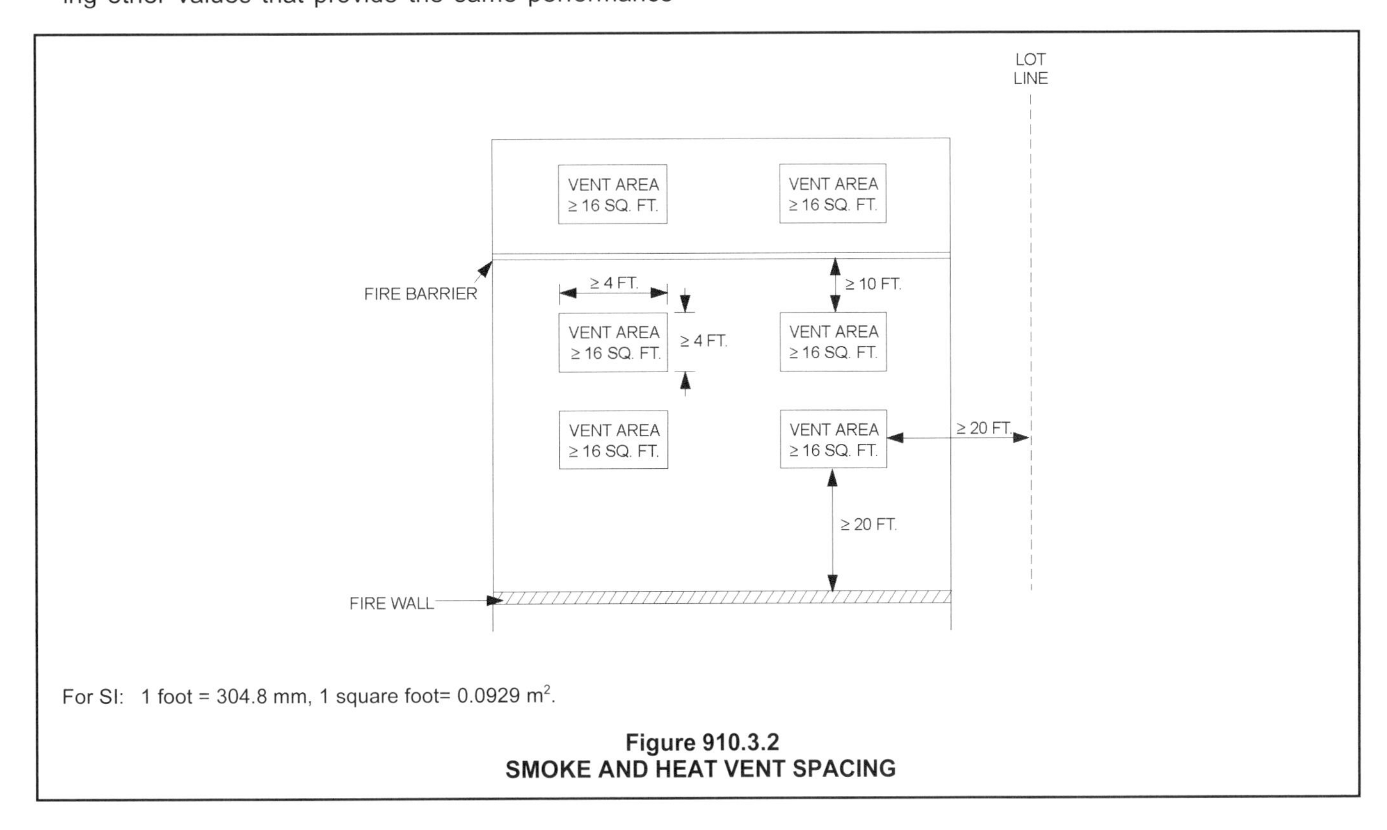

For SI: 1 foot = 304.8 mm, 1 square foot= 0.0929 m^2.

Figure 910.3.2
SMOKE AND HEAT VENT SPACING

cally provided at a reasonable cost. Therefore, in order to allow the use of a mechanical smoke removal system, the building is required to be equipped throughout with an automatic sprinkler system.

910.4.2 Exhaust fan construction. Exhaust fans that are part of a mechanical smoke removal system shall be rated for operation at 221°F (105°C). Exhaust fan motors shall be located outside of the exhaust fan air stream.

❖ This section requires exhaust fan motors to be located out of the exhaust stream to protect the mechanical equipment from excessive heat. Provisions for the mechanical smoke removal system permit the system to be designed to handle air at ambient temperature provided that the fan motors are located outside the air stream. The basis for this provision is the thermocouple temperature data for the large-scale fire tests conducted at UL in 1997/1998, specifically Tests P-1 and P-4. (In Tests P-1 and P-4, no vents opened so the ceiling temperatures recorded would be unaffected by the activation of vents. See Pages 40 and 52 of the NISTIR 6196-1 report dated September 1998 for the thermocouple temperature data recorded as a function of time.)

The exposing temperatures and time periods were reviewed and not considered to pose a threat to the building structure, fans or power wiring.

The sprinkler activation times and ceiling temperature data for the five large-scale fire tests summarized in NISTIR 6196-1 indicate that the exposure of mechanical exhaust fans and ducts located at the ceiling to high temperatures will be relatively short. Since it is anticipated that the exhaust system will only be activated after the arrival of fire fighters at the scene (estimated to be 7 minutes or longer after ignition), ceiling temperatures should be reduced sufficiently to allow fans rated for only ambient temperatures to be used for the exhaust system.

910.4.3 System design criteria. The mechanical smoke removal system shall be sized to exhaust the building at a minimum rate of two air changes per hour based upon the volume of the building or portion thereof without contents. The capacity of each exhaust fan shall not exceed 30,000 cubic feet per minute (14.2 m^3/sec).

❖ Sections 910.4.3 and 910.4.3.1 specify the design requirement for the minimum number of air changes, maximum fan capacity, and requirements for the provision of makeup air.

These provisions require that the mechanical smoke removal system be sized to provide a minimum exhaust rate of two air changes per hour based on the enclosed volume of the building space to be exhausted, without any deductions for the space occupied by storage or equipment. An exhaust rate of two air changes per hour is based on an analysis assuming a conservative approach using a Factory Mutual Research Corporation (FMRC) Standard Plastic Commodity (polystyrene cups in compartmented cartons). This commodity is recognized to represent a severe fire hazard of high-density plastics.

In a calculation based on this commodity, a maximum of 68,960 cfm of smoke was generated by the design fire. Based on an empty building volume of 2.659 million cubic feet, the exhaust rate required to achieve two air changes per hour is 88,633 cfm. Because no single fan can exceed 30,000 cfm, this building required five fans, each exhausting 25,570 cfm for a total of 127,850 cfm. This exceeds the minimum two air changes per hour by more than 40 percent. Even at the minimum required rate of two air changes per hour, the calculation results show that the mechanical smoke removal system will be capable of removing the smoke from the building faster than it will be generated, ultimately removing smoke from the building.

910.4.3.1 Makeup air. Makeup air openings shall be provided within 6 feet (1829 mm) of the floor level. Operation of makeup air openings shall be manual or automatic. The minimum gross area of makeup air inlets shall be 8 square feet per 1,000 cubic feet per minute (0.74 m^2 per 0.4719 m^3/s) of smoke exhaust.

❖ In order for a mechanical smoke removal system to work properly, makeup air at the proper location and volume needs to be provided. Generally, makeup air inlets need to be located much lower than where the smoke is exhausted to get the proper movement of air. In this case the required location is within 6 feet of the floor.

The derivation of the gross vent area is based on NFPA 92-2012. Specifically, Annex Section A-4.4.4.1.4 states that the maximum air velocity through the makeup air inlet is 1 m/sec or 200 ft/min. This is the same limitation found in Section 909.7.1, which is for the airflow method of smoke control. The area requirement is then derived as follows:

- Effective Vent Area = (1,000 ft^3/min)/(200 ft/min) = 5 ft^2 per 1,000 cfm
- Assume an orifice coefficient of 0.6
- Gross Vent Area = 5 ft^2/(0.6) = 8.33 ft^2 per 1,000 cfm, which is rounded down because of the conservative nature of the requirement

The reason for this limitation is to prevent significant deflection of the plume, which will cause more air entrainment into the plume and more smoke production. This is the same limitation found in Section 909.7.1, which is for the airflow method of smoke control. This criterion is conservative as the requirement above assumes an active fire and the design philosophy for this requirement is to provide post-fire smoke exhaust.

910.4.4 Activation. The mechanical smoke removal system shall be activated by manual controls only.

❖ This section requires that mechanical systems are to be activated manually so that the fire department is in control of the system. In some situations, automatic operation could cause a fire to grow or spread, opening an excessive number of sprinklers. Automatic operation of the mechanical smoke removal system could also be detrimental to the operation of the sprinkler system in a manner similar to draft curtains. The effect of the automatic mechanical smoke removal system on sprinkler operation would depend on when the system was activated. The sooner the system is automatically activated, the greater the detrimental effect. The fire department will retain the option to shut down the exhaust system.

910.4.5 Manual control location. Manual controls shall be located so as to be accessible to the fire service from an exterior door of the building and protected against interior fire exposure by not less than 1-hour *fire barriers* constructed in accordance with Section 707 or *horizontal assemblies* constructed in accordance with Section 711, or both.

❖ This section establishes the required placement, access and protection of the manual controls to ensure that fire fighters will have quick and protected access to them.

[F] 910.4.6 Control wiring. Wiring for operation and control of mechanical smoke removal systems shall be connected ahead of the main disconnect in accordance with Section 701.12E of NFPA 70 and be protected against interior fire exposure to temperatures in excess of 1,000°F (538°C) for a period of not less than 15 minutes.

❖ Unless the mechanical smoke removal system also functions as a component of a smoke control system, standby power is not specifically required (see commentary, Sections 909.11 and 2702). In order to provide an enhanced level of operational reliability, this section requires that the power supply to smoke exhaust fans must be provided from a circuit connected on the supply side (i.e., ahead) of the building's main electrical service disconnecting means. Note that this is one of the sources of standby power recognized by NFPA 70, Section 701.12(E). Such a circuit connected "ahead of the main" must still have its own approved over current protection.

The provisions for the design of a mechanical smoke removal system indicate that wiring providing power to exhaust fans located in the interior of the building is to be protected by materials that will provide a 15-minute finish rating protection. The ceiling temperature data collected in the five large-scale fire tests summarized in NISTIR 6196-1 (cited above) show that temperatures at the ceiling will be far less than the exposure temperatures defined by the ASTM E119 time-temperature curve and that the ceiling temperatures will rapidly decrease once sprinklers activate. The ceiling temperature data included in NISTIR 6196-1 indicate that providing a 15-minute finish rating protection for the interior electrical power supply is more than adequate to prevent damage to the power supply wiring for the exhaust system.

[F] 910.4.7 Controls. Where building air-handling and mechanical smoke removal systems are combined or where independent building air-handling systems are provided, fans shall automatically shut down in accordance with the *International Mechanical Code*. The manual controls provided for the smoke removal system shall have the capability to override the automatic shutdown of fans that are part of the smoke removal system.

❖ This section requires that if a mechanical smoke removal system is integrated with a standard HVAC system, then the system must shut down upon detection of smoke as required by the IMC. This relates to the requirement in Section 910.4.4 that mechanical smoke removal systems shall be manually operated only.

The concern is that HVAC systems should not work against the intended operation of the smoke exhaust system. In some cases the system may be a combination system where shutdown is not necessary or appropriate. It really depends on how the smoke exhaust system has been designed.

910.5 Maintenance. Smoke and heat vents and mechanical smoke removal systems shall be maintained in accordance with the *International Fire Code*.

❖ As with any system or equipment, the long-term operation is critical. This section is simply referring to the IFC for the maintenance provisions. Section 910.5 of that code goes into more extensive detail as to what needs to be tested, frequency of testing and necessary records that are required to be kept of such testing.

SECTION 911
FIRE COMMAND CENTER

[F] 911.1 General. Where required by other sections of this code and in buildings classified as high-rise buildings by this code, a fire command center for fire department operations shall be provided and shall comply with Sections 911.1.1 through 911.1.6.

❖ Fireground operations usually involve establishing an incident command post where the incident command officer can observe what is happening; control arriving personnel and equipment and direct the resources and fire-fighting operations effectively. Because of the difficulties in controlling a fire in a high-rise building, a protected, readily accessible, separate room for this purpose within the building must be established to assist the incident command officer (see commentary, Section 202 for the definition of "Fire command center").

A fire command center is also required in buildings containing smoke-protected assembly seating to house the fire-fighter's smoke control panel. Facilities with smoke-protected seating tend to be larger facilities that, at the very least, would already have a central security center which could also function as a fire command center where approved by the jurisdiction (see commentary, Section 909.16).

[F] 911.1.1 Location and access. The location and accessibility of the fire command center shall be *approved* by the fire chief.

❖ Because of its importance to fire suppression and rescue operations, the fire command center must be provided at a location that is acceptable to the fire department, usually near the front of the building near the main entrance, so that the first arriving command officer can access it quickly and undertake operations. Since fireground operations are based on local operational procedures, it is only reasonable that the fire chief of the jurisdiction have approval authority over the location of and access to the fire command center.

[F] 911.1.2 Separation. The fire command center shall be separated from the remainder of the building by not less than a 1-hour *fire barrier* constructed in accordance with Section 707 or *horizontal assembly* constructed in accordance with Section 711, or both.

❖ Again, because of its importance to fire suppression and rescue operations, the fire command center must be separated from the remainder of the building by 1-hour fire barriers and horizontal assemblies, including opening protectives, to protect the room, its contents and the occupants from an incident in adjacent areas of the building and to limit noise and distractions during command operations within the room.

[F] 911.1.3 Size. The room shall be not less than 200 square feet (19 m^2) with a minimum dimension of 10 feet (3048 mm).

❖ This section is intended to provide a minimum size and configuration of the fire command center that allows sufficient space for the necessary command personnel to effectively perform the required tasks associated with a fire command center without interfering with each other. Fire command centers need to be designed to accommodate several emergency response commanders wearing full protective equipment and also provide space to review building emergency plans during incidents, colocate decision makers within the Incident Command System (ICS) and interpret fire protection system and building system information generated by the features required by Section 911.1.5. Given the multiple uses of the fire command center, a room any smaller would compromise the effectiveness of incident management.

[F] 911.1.4 Layout approval. A layout of the fire command center and all features required by this section to be contained therein shall be submitted for approval prior to installation.

❖ The flow of critical tactical information into, within and out of a fire command center is, by its very nature, both high in volume and intense in nature and has a direct bearing on the safety of building occupants and the emergency response forces at work at an incident. For that reason, the layout and arrangement of the fire command center must comport with the operational procedures of the local fire department to optimize the receipt, processing and dissemination of operational information and orders. Accordingly, the fire code official must review and approve the arrangement of the fire command center prior to the installation of any of the controls and features required by Section 911.1.5. Consistent with Section 911.1.1, given the operational importance of the fire command center, the fire code official should work closely with the jurisdiction's fire chief to make sure that all operational needs are identified and met during the design stages.

[F] 911.1.5 Storage. Storage unrelated to operation of the fire command center shall be prohibited.

❖ Fire command centers are unique rooms in unique buildings and are strictly reserved for emergency management operations. As such, they must be neat and orderly at all times so as not to obstruct or limit access to all of the system controls that they contain. This section supports that need by prohibiting the storage within a fire command center of anything not directly related to the function of fire command.

[F] 911.1.6 Required features. The fire command center shall comply with NFPA 72 and shall contain all of the following features:

1. The emergency voice/alarm communication system control unit.
2. The fire department communications system.
3. Fire detection and alarm system annunciator.
4. Annunciator unit visually indicating the location of the elevators and whether they are operational.
5. Status indicators and controls for air distribution systems.
6. The fire fighter's control panel required by Section 909.16 for smoke control systems installed in the building.
7. Controls for unlocking *interior exit stairway* doors simultaneously.
8. Sprinkler valve and waterflow detector display panels.
9. Emergency and standby power status indicators.
10. A telephone for fire department use with controlled access to the public telephone system.
11. Fire pump status indicators.
12. Schematic building plans indicating the typical floor plan and detailing the building core, *means of egress*, fire protection systems, fire fighter air replenishment system, fire-fighting equipment and fire department access and the location of *fire walls, fire barriers, fire partitions, smoke barriers* and smoke partitions.
13. An *approved* Building Information Card that contains, but is not limited to, the following information:
    13.1. General building information that includes: property name, address, the number of floors in the building above and below grade, use and occupancy classification (for mixed uses,

identify the different types of occupancies on each floor), and the estimated building population during the day, night and weekend.

13.2. Building emergency contact information that includes: a list of the building's emergency contacts including but not limited to building manager and building engineer and their respective work phone number, cell phone number, e-mail address.

13.3. Building construction information that includes: the type of building construction including but not limited to floors, walls, columns, and roof assembly.

13.4. *Exit access* and *exit stairway* information that includes: number of *exit access* and *exit stairways* in the building, each *exit access* and *exit stairway* designation and floors served, location where each *exit access* and *exit stairway* discharges, *interior exit stairways* that are pressurized, *exit* stairways provided with emergency lighting, each *exit stairway* that allows reentry, *exit stairways* providing roof access; elevator information that includes: number of elevator banks, elevator bank designation, elevator car numbers and respective floors that they serve; location of elevator machine rooms, control rooms and control spaces; location of sky lobby, location of freight elevator banks.

13.5. Building services and system information that includes: location of mechanical rooms, location of building management system, location and capacity of all fuel oil tanks, location of emergency generator, location of natural gas service.

13.6. Fire protection system information that includes: location of standpipes, location of fire pump room, location of fire department connections, floors protected by automatic sprinklers, location of different types of *automatic sprinkler systems* installed including, but not limited to, dry, wet and pre-action.

13.7 Hazardous material information that includes: location of hazardous material, quantity of hazardous material.

14. Work table.

15. Generator supervision devices, manual start and transfer features.

16. Public address system, where specifically required by other sections of this code.

17. Elevator fire recall switch in accordance with ASME A17.1.

18. Elevator emergency or standby power selector switch(es), where emergency or standby power is provided.

❖ The fire command center must contain all equipment necessary to enable the incident commander to monitor or control fire protection and other building service systems as listed in this section (also see commentary, Section 909.16). This room houses fire protection, smoke control and building system controls, as well as a work space for emergency responders. The room also contains schematic plans and a work table so that responders have the basic layout and geometry of the building and can identify locations of utility controls, standby or emergency power systems, and where hazardous materials are stored or used. Providing concise information in a uniform format is essential to fire fighters and emergency responders and improves their ability to utilize building systems to their advantage. This was confirmed during the National Institute of Science and Technology (NIST) investigations of the World Trade Center attacks on September 11, 2001. The Final Report on the collapse of the World Trade Center contained 30-key recommendations compiled by NIST designed to address the building vulnerabilities learned in that tragedy. Three of those 30 recommendations embrace increasing situational awareness and emergency communications of first responders in large-scale emergencies. As a result of that investigation, this section includes an Item 13 that prescribes requirements for the Building Information Card (BIC).

The BIC is divided into multiple information areas and is intended to be formatted as a single form to provide a quick, concise source of information about the building. The code does not prescribe any particular format or layout for the BIC and does not have any limits on the level of information required to satisfy the requirements. It should be recognized that the intent of the BIC is to provide an easily understood and consistent tool to emergency responders who are taking control of systems in high-rise and smoke-protected assembly buildings. Jurisdictions should develop a policy to ensure that BICs are prepared in standard, consistent format to avoid confusing the responders, and yet provide the minimum information required so they can correctly and efficiently utilize all of the building features. The number and types of features required by this section can create a large volume of data, thus reinforcing the need for an approved layout as required by Section 911.1.4.

SECTION 912
FIRE DEPARTMENT CONNECTIONS

[F] 912.1 Installation. Fire department connections shall be installed in accordance with the NFPA standard applicable to the system design and shall comply with Sections 912.2 through 912.6.

❖ An FDC is required as part of a water-based suppression system as the auxiliary water supply. These connections give the fire department the capability of supplying the necessary water to the automatic sprinkler or standpipe system at a sufficient pressure. The FDC also serves as an alternative source of water should a valve in the primary water supply be closed. A fire department connection does not, however, constitute an automatic water source. See Commentary Figure 903.3.1.1 for a typical FDC arrangement on a wet pipe sprinkler system.

The requirements for the FDC depend on the type of sprinkler system installed and whether a standpipe system is installed. NFPA 13 and 13R, for example, include design considerations for FDCs that are an auxiliary water supply source for automatic sprinkler systems; NFPA 14 is the design standard to use for FDCs serving standpipe systems. Threads for FDCs to sprinkler systems, standpipes, yard hydrants or any other fire hose connection must be approved (NFPA 1963 may be utilized as part of the approval or as otherwise approved) and be compatible with the connections used by the local fire department (see commentary, Sections 903.3.6 and 905.1).

[F] 912.2 Location. With respect to hydrants, driveways, buildings and landscaping, fire department connections shall be so located that fire apparatus and hose connected to supply the system will not obstruct access to the buildings for other fire apparatus. The location of fire department connections shall be *approved* by the fire chief.

❖ This section specifies that the FDC must be located so that vehicles and hose lines will not interfere with access to the building for the use of other fire department apparatus. The location of potential connected hose lines to the FDC and hydrants must be preplanned with the fire department. Many fire departments have a policy restricting the distance that a FDC may be from a fire hydrant. Some also have policies that indicate the maximum distance from the nearest point of fire department vehicle access (often, the curb). Since fireground operations are based on local operational procedures, it is only reasonable that the fire chief of the jurisdiction have approval authority over the location of and access to the FDC.

Landscaping can also be a hindrance to fire department operations. Even where the FDC is visible, the extensive use of landscaping may make access difficult. Landscaping also changes over time. What may not have been an obstruction when it was planted can sometimes grow into an obstruction over time.

[F] 912.2.1 Visible location. Fire department connections shall be located on the street side of buildings, fully visible and recognizable from the street or nearest point of fire department vehicle access or as otherwise *approved* by the fire chief.

❖ FDCs must be readily visible and easily accessed. A local policy constituting what is readily visible and accessible needs to be established. While the intent is clearly understandable, its application can vary widely. A precise policy is the best way to avoid ambiguous directives that result in inconsistent and arbitrary enforcement. Usually, the policy will address issues such as location on the outside of the building and proximity to fire hydrants.

Landscaping is often used to hide the FDCs from the public. This can greatly hamper the efforts of the fire department in staging operations and supplying water to the fire protection systems. Landscaping must be designed so that it does not obstruct the visibility of the FDC. Since fireground operations are based on local operational procedures, it is only reasonable that the fire chief of the jurisdiction have final approval authority over the visibility of and access to the FDC.

[F] 912.2.2 Existing buildings. On existing buildings, wherever the fire department connection is not visible to approaching fire apparatus, the fire department connection shall be indicated by an *approved* sign mounted on the street front or on the side of the building. Such sign shall have the letters "FDC" not less than 6 inches (152 mm) high and words in letters not less than 2 inches (51 mm) high or an arrow to indicate the location. Such signs shall be subject to the approval of the fire code official.

❖ The section acknowledges that FDCs on existing buildings may not always be readily visible from the street or nearest point of fire department vehicle access. In those instances, the location of the connection must be clearly marked with signage. The FDC may be located on the side of the building or in an alley, not visible to arriving fire-fighting forces. A sign is necessary so that those driving the arriving apparatus know where to maneuver the vehicle to get close to the FDC.

[F] 912.3 Fire hose threads. Fire hose threads used in connection with standpipe systems shall be *approved* and shall be compatible with fire department hose threads.

❖ There are several sections in the code that contain requirements for fire department connections. This section simply correlates with those requirements by further clarifying that fire hose threads for standpipe systems must be approved.

[F] 912.4 Access. Immediate access to fire department connections shall be maintained at all times and without obstruction by fences, bushes, trees, walls or any other fixed or moveable object. Access to fire department connections shall be *approved* by the fire chief.

Exception: Fences, where provided with an access gate equipped with a sign complying with the legend requirements of this section and a means of emergency operation. The gate and the means of emergency operation shall be *approved* by the fire chief and maintained operational at all times.

❖ The FDC must be readily accessible to fire fighters and allow fire-fighting personnel an adequate area to maneuver a hose for the connection. Landscaping design must not block a clear view of the FDC from arriving fire department vehicles. Depending on the type of landscaping materials, an active maintenance program may be necessary to maintain ready access over time. This section also recognizes that the obstructing objects regulated here can be either fixed or moveable (such as outdoor furnishings, shopping cart queue areas, etc.). Note that no specific dimension is given as was the case in previous editions of the code. This performance language avoids previous misinterpretations that the code intended to allow obstructions to FDC access as long as they were kept 3 feet (914 mm) away. Since fireground operations are based on local operational procedures, it is only reasonable that the fire chief of the jurisdiction have final approval authority over the access to the FDC.

The exception recognizes the practical fact that sometimes, security or other considerations make installation of a fence around a building necessary as long as the fence meets the stated criteria. The sign requirement intends to provide a visual location cue to approaching fire apparatus where the height of the fence may obscure the visibility of the FDC.

[F] 912.4.1 Locking fire department connection caps. The fire code official is authorized to require locking caps on fire department connections for water-based *fire protection systems* where the responding fire department carries appropriate key wrenches for removal.

❖ This section allows for the FDC caps to be equipped with locks as long as the fire departments that respond to that building or facility have the appropriate key wrenches. This avoids vandalism and affords a more functional FDC when needed. Locking caps, even more so than regular FDC caps, need proper maintenance so that they can be removed when required. Any time that an additional mechanical function is added to something that is exposed to the elements, it must be done with the understanding that the corrosive nature of the elements can place the FDC out of commission if the cap cannot be removed (see Commentary Figure 912.4.1).

[F] 912.4.2 Clear space around connections. A working space of not less than 36 inches (762 mm) in width, 36 inches (914 mm) in depth and 78 inches (1981 mm) in height shall be provided and maintained in front of and to the sides of wall-mounted fire department connections and around the circumference of free-standing fire department connections, except as otherwise required or *approved* by the fire chief.

❖ Care must be taken so that fences, utility poles, barricades and other obstructions do not prevent access to and use of FDCs. A clear space of 3 feet (914 mm) must be maintained in front of and to either side of wall-mounted FDCs and around free-standing FDCs to allow easy hose connections to the fitting and efficient use of spanner wrenches and other tools needed by the apparatus engineer.

Though not specifically mentioned in this section, it is also important that FDCs be installed with the hose connections well above adjoining grade to accommodate the free turning of a spanner wrench when connecting hoses to the FDC.

[F] 912.4.3 Physical protection. Where fire department connections are subject to impact by a motor vehicle, vehicle impact protection shall be provided in accordance with Section 312 of the *International Fire Code.*

❖ Section 312 of the IFC requires vehicle impact protection by placing steel posts filled with concrete around the FDC. Section 312 of the IFC gives the specifications for the posts.

[F] 912.5 Signs. A metal sign with raised letters not less than 1 inch (25 mm) in size shall be mounted on all fire department connections serving automatic sprinklers, standpipes or fire pump connections. Such signs shall read: AUTOMATIC SPRINKLERS or STANDPIPES or TEST CONNECTION or a combination thereof as applicable. Where the fire department connection does not serve the entire building, a sign shall be provided indicating the portions of the building served.

❖ The purpose of the sign is to provide the responding fire fighters with the correct information on which portions of a building are served by the fire department connection. They identify the type of system or zone served by a given FDC. Many buildings include multiple sets of fire department connections which are not interconnected, such as separate connections for the

Figure 912.4.1
LOCKING FDC CAPS
(Photo courtesy of Knox Company)

building sprinkler system and the dry standpipe system in open parking structures. Some buildings may have only a partial sprinkler system, such as rehabilitated buildings where a sprinkler system is only installed on certain floors or a building that only has basement sprinklers in accordance with Section 903.2.11.

Signs may also distinguish FDCs from fire pump test headers. Usually, FDCs may be distinguished from fire pump test headers by the types of couplings provided. FDCs are customarily equipped with female couplings, while fire pump test headers usually have separately valved male couplings. Furthermore, fire pump test headers are equipped with one $2^1/_2$-inch (64 mm) outlet for each 250 gallon per minute (gpm) (16 L/s) of rated capacity.

Raised letters are required so that any repainting or fading of the colors on the sign will not affect its ability to be read. Each letter must be at least 1 inch (25 mm) in height so that the wording is clear. Often the wording may be abbreviated such that "AUTOMATIC SPRINKLERS" reads as "AUTO. SPKR." Existing signs may use language slightly different than that noted in the code. As long as the information is adequately communicated, there should be no reason to require new signage to replace existing ones (see Commentary Figure 912.5).

Figure 912.5
FIRE DEPARTMENT CONNECTION WITH SIGN

[P] 912.6 Backflow protection. The potable water supply to automatic sprinkler and standpipe systems shall be protected against backflow as required by the *International Plumbing Code*.

❖ Section 608.16.4 of the IPC requires all connections to automatic sprinkler systems and standpipe systems to be equipped with a means to protect the potable water supply. The means of backflow protection can be either a double check-valve assembly or a reduced-pressure-principle backflow preventer. This, in general, assumes a FDC is required. For example, a limited-area sprinkler system off the domestic supply does not necessarily require a FDC and would not require backflow protection.

SECTION 913
FIRE PUMPS

[F] 913.1 General. Where provided, fire pumps shall be installed in accordance with this section and NFPA 20.

❖ This section contains specific installation requirements for fire pumps supplying water to fire protection systems. Inspection, testing and maintenance requirements comply with NFPA 20 unless noted otherwise. Applicable maintenance standards are also identified.

Fire pumps are installed in sprinkler and standpipe systems to pressurize the water supply for the minimum required sprinkler and standpipe operation. They are considered a design feature or component of the system. Fire pumps can improve only the pressure of the incoming water supply, not the volume of water available.

When the volume from a water supply is not adequate to supply sprinkler or standpipe demand, water tanks for private fire protection, improvements in the size and capacity of fire mains or water distribution systems or all of these for the installation of a fire pump are needed.

When fire pumps are required to meet the pressure requirements of sprinkler and standpipe systems, they must be installed and tested in accordance with NFPA 20.

[F] 913.2 Protection against interruption of service. The fire pump, driver and controller shall be protected in accordance with NFPA 20 against possible interruption of service through damage caused by explosion, fire, flood, earthquake, rodents, insects, windstorm, freezing, vandalism and other adverse conditions.

❖ This section lists hazards that must be taken into account when determining the extent of protection required for the fire pump and its auxiliary equipment. A pump room in a building that is protected against the listed hazards in compliance with the code would be considered in compliance. Because fire pumps are also typically located in separate detached structures, geographical and security issues must also be considered.

913.2.1 Protection of fire pump rooms. Fire pumps shall be located in rooms that are separated from all other areas of the building by 2-hour *fire barriers* constructed in accordance with Section 707 or 2-hour *horizontal assemblies* constructed in accordance with Section 711, or both.

Exceptions:

1. In other than high-rise buildings, separation by 1-hour *fire barriers* constructed in accordance with

Section 707 or 1-hour *horizontal assemblies* constructed in accordance with Section 711, or both, shall be permitted in buildings equipped throughout with an *automatic sprinkler system* in accordance with Section 903.3.1.1 or 903.3.1.2.

2. Separation is not required for fire pumps physically separated in accordance with NFPA 20.

❖ This section correlates the NFPA 20 requirements for separation of fire pumps with the IFC and specifies the required degree of fire resistance as a requirement in the code so that designers and building officials are made aware of it. It requires fire pumps to be separated using either fire-resistive barriers or horizontal assemblies when the pump is located inside of a building or by using spatial separation (physical distance) when the fire pump is located outside of the building it serves.

The requirements are based on the criteria found in NFPA 20. Table 508.2.5, when used as a means for dealing with mixed uses in a building, and this section require fire pump units located inside of high-rise buildings to have a minimum 2-hour separation from the remainder of the building. The 2-hour separation requirement is consistent with the requirements in NFPA 20. Separation must be in the form of fire barriers used in conjunction with a horizontal assembly, as applicable, with both being of 2-hour fire-resistance-rated construction.

In all other buildings, consistent with similar exceptions elsewhere in the code, Exception 1 and Table 508.2.5, if applicable, recognize the protection afforded by complete automatic sprinkler protection in accordance with NFPA 13 or 13R as being equivalent to 1 hour of fire resistance.

A fire pump unit also may be located outside of a building. This is a common practice because it limits the exposure of a fire pump unit from a fire inside of the building it is serving. When a fire pump unit is located outdoors, Exception 2 recognizes spatial separation of at least 50 feet (15 240 mm) in accordance with Section 5.12 of NFPA 20 as an acceptable alternative to fire-resistance-rated protection.

[F] 913.2.2 Circuits supplying fire pumps. Cables used for survivability of circuits supplying fire pumps shall be *listed* in accordance with UL 2196. Electrical circuit protective systems shall be installed in accordance with their listing requirements.

❖ This section is provided to better protect cables used for survivability of circuits associated with fire pumps by referencing the appropriate standard. UL 2196 is the ANSI-approved standard for tests of fire-resistive cables. NFPA 20 includes selective survivability requirements to ensure integrity of certain critical circuits. NFPA 70 does not specify the applicable standard within its mandatory provisions but recognizes electrical circuit protective systems as an alternative to listed cables. An electrical circuit protective system is a field assembly of components that must be installed according to the listing requirements and manufacturer's instructions in order to maintain the listing for the system. There are more than two dozen electrical circuit protective systems listed in the UL Fire Resistance Directory.

[F] 913.3 Temperature of pump room. Suitable means shall be provided for maintaining the temperature of a pump room or pump house, where required, above 40°F (5°C).

❖ As previously noted for sprinkler systems, standpipe systems and other water-based fire protection systems, pump rooms or pump houses must be maintained at a temperature of 40°F (4°C) or above to prevent the system from freezing. This is consistent with Section 5.12.2.1 of NFPA 20.

[F] 913.3.1 Engine manufacturer's recommendation. Temperature of the pump room, pump house or area where engines are installed shall never be less than the minimum recommended by the engine manufacturer. The engine manufacturer's recommendations for oil heaters shall be followed.

❖ The engine manufacturer's recommendation must be compiled with where the recommended minimum temperature is higher than the minimum established in Section 913.3. Maintaining the desired engine temperature enhances the startability of the engine. Maintaining water heaters and oil heaters as required for diesel engines, for example, will improve the starting capabilities of the fire pump and reduce engine wear and the drain on batteries.

[F] 913.4 Valve supervision. Where provided, the fire pump suction, discharge and bypass valves, and isolation valves on the backflow prevention device or assembly shall be supervised open by one of the following methods:

1. Central-station, proprietary or remote-station signaling service.
2. Local signaling service that will cause the sounding of an audible signal at a *constantly attended location.*
3. Locking valves open.
4. Sealing of valves and *approved* weekly recorded inspection where valves are located within fenced enclosures under the control of the owner.

❖ As was the case with sprinkler systems, water control valves that are a part of the fire pump installation must be supervised in the open position so that the system is operational when needed and also to reduce the chance of a system failure (see commentary, Section 903.4). In most cases the required water-based extinguishing system, of which the fire pump is an integral component, will be electrically supervised. Locking or sealing valves open as the only means of supervision may not be permitted, depending on the type of valve. Section 903.4, for example, specifically exempts jockey pump control valves from being electrically supervised if they are sealed or locked in the open position.

[F] 913.4.1 Test outlet valve supervision. Fire pump test outlet valves shall be supervised in the closed position.

❖ Fire pump test outlet valves are for performance testing of the fire pump and do not control the available water supply to either a sprinkler system or a standpipe system. These valves are normally in a closed position and are supervised accordingly.

[F] 913.5 Acceptance test. Acceptance testing shall be done in accordance with the requirements of NFPA 20.

❖ Chapter 14 of NFPA 20 details the procedure for conducting a fire pump acceptance test. This test is run to determine that the installation matches the sprinkler or standpipe system design criteria, the approved shop drawings and the pump manufacturer's performance specifications. The test is to be conducted in the presence of the building official in accordance with Section 901.5 by the installing contractor and representatives of the pump manufacturer and the controller manufacturer. Where the pump engine and/ or transfer switch are separately supplied components, their manufacturer representatives must also be present.

SECTION 914
EMERGENCY RESPONDER SAFETY FEATURES

[F] 914.1 Shaftway markings. Vertical shafts shall be identified as required by Sections 914.1.1 and 914.1.2.

❖ This section was developed to prevent fire fighters from falling through shafts when entering buildings from ladders placed on the exterior of the building or from passing through an interior opening that leads to a shaft.

[F] 914.1.1 Exterior access to shaftways. Outside openings accessible to the fire department and that open directly on a hoistway or shaftway communicating between two or more floors in a building shall be plainly marked with the word "SHAFTWAY" in red letters not less than 6 inches (152 mm) high on a white background. Such warning signs shall be placed so as to be readily discernible from the outside of the building.

❖ All exterior wall openings that are accessible to fire fighters by way of ladders and aerial equipment and open directly into shafts or hoistways communicating between two or more floors must be clearly marked (see Commentary Figure 914.1.1). The markings serve to warn emergency responders that the opening is unsafe for laddering to gain access to upper floors and could result in a fall to the bottom of the shaft if such a mistake were made.

[F] 914.1.2 Interior access to shaftways. Door or window openings to a hoistway or shaftway from the interior of the building shall be plainly marked with the word "SHAFTWAY" in red letters not less than 6 inches (152 mm) high on a white background. Such warning signs shall be placed so as to be readily discernible.

Exception: Markings shall not be required on shaftway openings that are readily discernible as openings onto a shaftway by the construction or arrangement.

❖ Openings into shaftways from the interior of the building pose a threat to fire fighters when visibility is poor. Interior shaft openings must be marked so that they are plainly visible from the interior of the building. If fire fighters can readily identify an opening into a shaft by the way the opening is constructed, the shaft opening need not be marked, keeping in mind that the fire fighter may be feeling his or her way in heavy smoke or darkness.

[F] 914.2 Equipment room identification. Fire protection equipment shall be identified in an *approved* manner. Rooms containing controls for air-conditioning systems, sprinkler risers and valves or other fire detection, suppression or control elements shall be identified for the use of the fire department. *Approved* signs required to identify fire protection equipment and equipment location shall be constructed of durable materials, permanently installed and readily visible.

❖ In an emergency, it is vitally important that the fire department and other emergency responders be able to quickly locate and access critical controls for fire protection systems. Obstructed or poorly marked equipment can cause delays in fire-fighting operations while fire fighters locate other hose stations and stretch additional hose. Valves and other controls are often located in rooms or other enclosures and their location must be identified with written or pictographic signs, which must be clearly visible and legible. Signs using the NFPA 170 symbols for fire protection equipment can provide standardized markings throughout a jurisdiction. White reflective symbols on a red reflective background are effective. For exterior signs, heavy-gage, sign-grade aluminum is recom-

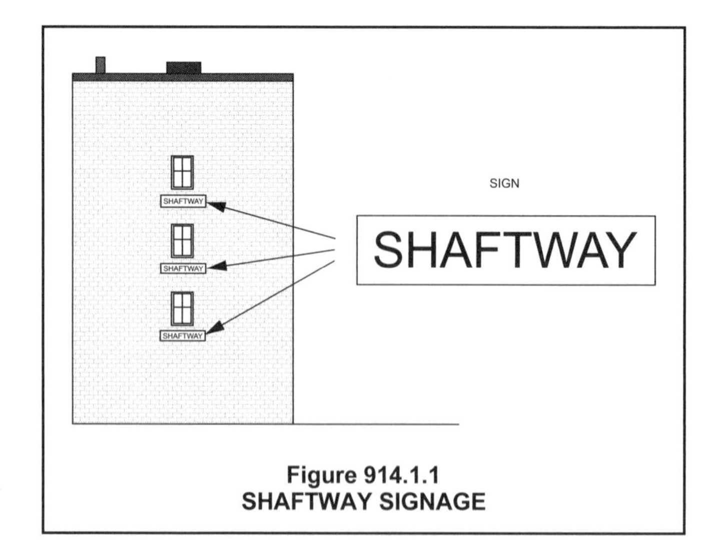

Figure 914.1.1
SHAFTWAY SIGNAGE

mended. Interior signs may be constructed of plastic, light-gage aluminum or other approved, durable, water-resistant material. As a general rule, fire protection piping, cabinets, enclosures, wiring, equipment and accessories are red or are identified by red or red/white markings. The manner of identification is subject to the approval of the building official.

SECTION 915 CARBON MONOXIDE DETECTION

[F] 915.1 General. Carbon monoxide detection shall be installed in new buildings in accordance with Sections 915.1.1 through 915.6. Carbon monoxide detection shall be installed in existing buildings in accordance with Chapter 11 of the *International Fire Code.*

❖ These provisions were added to the code and the IFC to be consistent with the requirements for carbon monoxide (CO) detectors in all new construction of one- and two-family dwellings that had been added to the IRC in the 2009 edition. Another reason for its approval was technical data in a 1998 article published by the *Journal of the American Medical Association* that stated that approximately 2,100 deaths occur annually as a result of CO poisoning. That annual number is based on the findings of a paper prepared by the U.S. Department of Health Centers for Disease Control (CDC). That paper documented epidemiological research by two CDC physicians who examined 56,133 death certificates over a 10-year period. Excluding suicides, homicides, structure fires and deaths resulting from CO poisoning in motor vehicles, the death rate steadily decreased for the sample period, from a value of 1,513 people in 1979 to 878 in 1988. The highest death rates occurred in winter and among males, African Americans, the elderly and residents in northern states.

CO is a colorless, tasteless, odorless gas that interrupts the attachment of oxygen molecules to hemoglobin in blood cells and can cause headaches, confusion and dizziness. At higher concentrations CO can cause loss of consciousness and eventual death. Exposures above 100 parts/million are dangerous to human health. It is not a toxic or highly toxic gas as defined in Chapter 2 but is classified as a flammable gas.

These provisions detail what occupancies require CO detection and where that detection is specifically to be located within the building. Also flexibility on the use of single- and multiple-station alarms versus CO detection systems is provided.

Retroactive requirements are found in Chapter 11 of the IFC. These requirements are essentially the same except that Group E classrooms are not addressed and the CO alarms can be battery powered.

[F] 915.1.1 Where required. Carbon monoxide detection shall be provided in Group I-1, I-2, I-4 and R occupancies and in classrooms in Group E occupancies in the locations specified in Section 915.2 where any of the conditions in Sections 915.1.2 through 915.1.6 exist.

❖ CO detection is provided to protect occupants of dwelling units and sleeping units within Group I-1, I-4, and R occupancies and classrooms in Group E occupancies. These are locations where occupants are likely to be sleeping or that young children may be at risk. Sections 915.1.2 through 915.1.5 address the different scenarios that warrant CO detection and are more specific than past editions of the code and the IFC.

[F] 915.1.2 Fuel-burning appliances and fuel-burning fireplaces. Carbon monoxide detection shall be provided in *dwelling units*, *sleeping units* and classrooms that contain a fuel-burning appliance or a fuel-burning fireplace.

❖ This section clarifies that CO detection is required where the dwelling or sleeping unit or the classroom actually contains a fuel-burning appliance or fuel-burning fireplace. Section 915.1.3 discusses dwelling and sleeping units and classrooms that are served by fuel-burning, forced-air furnaces.

[F] 915.1.3 Forced-air furnaces. Carbon monoxide detection shall be provided in *dwelling units*, *sleeping units* and classrooms served by a fuel-burning, forced-air furnace.

> **Exception:** Carbon monoxide detection shall not be required in *dwelling units*, *sleeping units* and classrooms if carbon monoxide detection is provided in the first room or area served by each main duct leaving the furnace, and the carbon monoxide alarm signals are automatically transmitted to an approved location.

❖ This section addresses forced-air furnaces that physically move air. This potential source of CO is more indirect than having the fuel-burning appliance or fuel-burning fireplace within the space itself, but still creates a potential hazard.

The exception addresses the fact that if detection is provided closer to the source, then further detection within dwelling units, sleeping units and Group E classrooms is not necessary provided that a CO alarm signal is transmitted to an approved location from which emergency actions can be initiated and occupants in the portions of the building farther from the initial detection can be notified. The CO will continue to be generated until action is taken.

[F] 915.1.4 Fuel-burning appliances outside of dwelling units, sleeping units and classrooms. Carbon monoxide detection shall be provided in *dwelling units*, *sleeping units* and classrooms located in buildings that contain fuel-burning appliances or fuel-burning fireplaces.

Exceptions:

1. Carbon monoxide detection shall not be required in *dwelling units*, *sleeping units* and classrooms where there are no communicating openings between the fuel-burning appliance or fuel-burning fireplace and the *dwelling unit*, *sleeping unit* or classroom.
2. Carbon monoxide detection shall not be required in *dwelling units*, *sleeping units* and classrooms where

carbon monoxide detection is provided in one of the following locations:

2.1. In an approved location between the fuel-burning appliance or fuel-burning fireplace and the *dwelling unit*, *sleeping unit* or classroom.

2.2. On the ceiling of the room containing the fuel-burning appliance or fuel-burning fireplace.

❖ This section focuses on the presence of fuel-burning equipment in buildings containing dwelling units, sleeping units and classrooms but with the fuel-burning equipment located outside of those areas. CO detection is required in every dwelling unit, sleeping unit and classroom, even if that equipment does not serve those spaces. A good example of this is a multistory hotel that has all electric HVAC in the sleeping units, but perhaps a fireplace in the lobby, forced-air heating in the common area, and a boiler in an equipment room.

There are two exceptions provided that recognize, based on several factors, that CO is a very low risk to the spaces required to be protected by Section 915.

Exception 1 addresses buildings where there are no communicating openings between the location of the fuel-burning appliance and dwelling units, sleeping units and classrooms. The intent is that if the appliance is not serving the space and has no communication with the space, CO should not enter those spaces. This covers situations where CO emanating from the fuel-burning appliance has no direct path to a dwelling unit or sleeping unit, such as a water heater in an equipment room that only has access from the exterior of the building, and no openings through which the CO can get to dwelling units or sleeping units. An interior door, between the equipment room and a dwelling unit, even if it is self-closing, would not allow this exception to be used.

Exception 2 is similar to the exception to Section 915.1.3. This exception requires the installation of one or more CO alarms in approved locations between fuel-burning appliances and the nearest dwelling unit, sleeping unit or classroom or on the ceiling of the room in which a fuel-burning appliance is located. CO alarms are only required where there are communicating openings including ducts, concealed spaces, interior hallways, stairs and spaces between the fuel-burning appliance or fuel-burning fireplace and the dwelling unit or sleeping unit where air can flow from the appliance to the dwelling unit or sleeping unit.

[F] 915.1.5 Private garages. Carbon monoxide detection shall be provided in *dwelling units*, *sleeping units* and classrooms in buildings with attached private garages.

Exceptions:

1. Carbon monoxide detection shall not be required where there are no communicating openings between the private garage and the *dwelling unit*, *sleeping unit* or classroom.
2. Carbon monoxide detection shall not be required in *dwelling units*, *sleeping units* and classrooms located more than one story above or below a private garage.
3. Carbon monoxide detection shall not be required where the private garage connects to the building through an open-ended corridor.
4. Where carbon monoxide detection is provided in an approved location between openings to a private garage and *dwelling units*, *sleeping units* or classrooms, carbon monoxide detection shall not be required in the *dwelling units*, *sleeping units* or classrooms.

❖ This section addresses attached private garages. Private garages, a term defined in Chapter 2, are often sources of CO generation. When attached to a building containing dwelling units, sleeping units or classrooms in Group E, CO detection is required.

There are several exceptions to the requirement for CO detection. These are similar to those found in Sections 915.1.3 and 915.1.4.

Exception 1 is similar to Exception 1 in Section 915.1.4. If there are no communicating openings, then CO detection is not required. The likelihood of CO entering the dwelling units, sleeping units or classrooms is greatly reduced.

Exception 2 addresses buildings where the dwelling units, sleeping units and classrooms are two stories above or below the location of a fuel-burning appliance. The CO would need to travel through an entire story before reaching the areas of concern.

Exception 3 does not require CO alarms to be provided when the private garage is attached to the building by an open-ended corridor (a term defined in Chapter 2 and the IFC, and commonly called a breeze way). This allows the CO concentration to dissipate before becoming a hazard to building occupants.

Exception 4 is similar to the exception to Section 915.1.3 and recognizes that the detection can be located closer to the source of the hazard rather than requiring detection in every dwelling unit, sleeping unit or classroom.

[F] 915.1.6 Exempt garages. For determining compliance with Section 915.1.5, an *open parking garage* complying with Section 406.5 or an enclosed parking garage complying with Section 406.6 shall not be considered a private garage.

❖ This section is only provided to distinguish open parking garages and enclosed parking garages from private garages. Open parking garages and private garages are defined terms in Section 202 of the code.

[F] 915.2 Locations. Where required by Section 915.1.1, carbon monoxide detection shall be installed in the locations specified in Sections 915.2.1 through 915.2.3.

❖ Sections 915.2.1 through 915.2.3 provide more detail on where detection is specifically required to be installed. This is a clarification from previous editions. NFPA 720 does not provide detailed enough information as to where detection should be located, so specific direction is provided in the code. In some cases the locations differ from that provided in NFPA 720.

[F] 915.2.1 Dwelling units. Carbon monoxide detection shall be installed in dwelling units outside of each separate sleeping area in the immediate vicinity of the bedrooms. Where a fuel-burning appliance is located within a bedroom or its attached bathroom, carbon monoxide detection shall be installed within the bedroom.

❖ The language is similar to that required for smoke alarms, but the detection is only required outside of each sleeping area. If the fuel-burning appliance is within the bedroom or the attached bathroom, detection is required within the bedroom itself. Commentary Figure 915.2.1 shows the layout of carbon monoxide alarms versus smoke alarms. Note that if they are combined (smoke and CO alarms), the smoke alarm placement would be more restrictive.

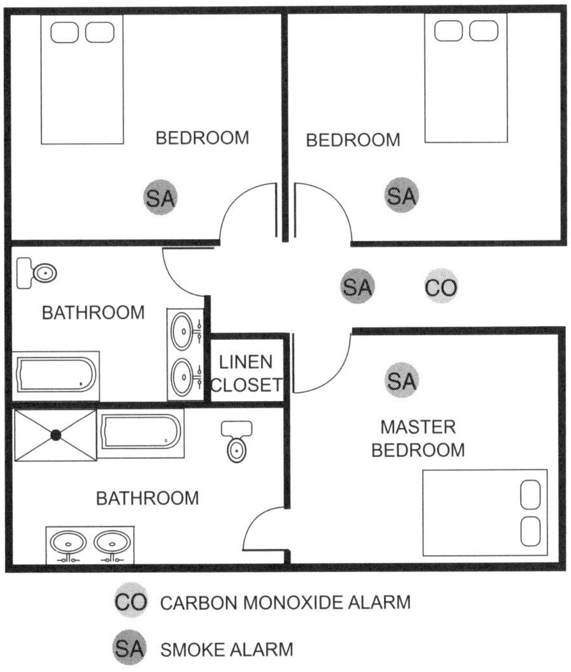

Figure 915.2.1
CARBON MONOXIDE ALARM PLACEMENT VERSUS SMOKE DETECTION PLACEMENT

[F] 915.2.2 Sleeping units. Carbon monoxide detection shall be installed in *sleeping units*.

Exception: Carbon monoxide detection shall be allowed to be installed outside of each separate sleeping area in the immediate vicinity of the *sleeping unit* where the *sleeping unit* or its attached bathroom does not contain a fuel-burning appliance and is not served by a forced air furnace.

❖ The intent of this section is the same as Section 915.2.1 except that it is applicable to sleeping units. Sleeping units can be as simple as a hotel room or a suite-type layout. This allows the detection to be either in the sleeping area itself, as with a simple hotel room, or outside the sleeping area as is the case for dwelling units.

[F] 915.2.3 Group E occupancies. Carbon monoxide detection shall be installed in classrooms in Group E occupancies. Carbon monoxide alarm signals shall be automatically transmitted to an on-site location that is staffed by school personnel.

Exception: Carbon monoxide alarm signals shall not be required to be automatically transmitted to an on-site location that is staffed by school personnel in Group E occupancies with an occupant load of 30 or less.

❖ This section clarifies that the detection is to be located within the classroom and that a signal from the CO detection be automatically transmitted to a location on-site where school personnel can react to the situation, such as the school office. The exception provides relief to small schools where it is very easy to communicate a hazardous situation throughout the building.

[F] 915.3 Detection equipment. Carbon monoxide detection required by Sections 915.1 through 915.2.3 shall be provided by carbon monoxide alarms complying with Section 915.4 or carbon monoxide detection systems complying with Section 915.5.

❖ This section provides the option of providing the CO detection through single- or multiple-station CO alarms or through a CO detection system. This is a similar concept to smoke alarms and smoke detection systems. Each type of CO detector is listed to a different UL standard depending on whether it is to be used as a stand-alone or interconnected detector (UL 2034) or a detector that is part of a CO detection system (UL 2075). Section 915.4 addresses CO alarms and Section 915.5 addresses CO detection systems.

[F] 915.4 Carbon monoxide alarms. Carbon monoxide alarms shall comply with Sections 915.4.1 through 915.4.3.

❖ If the option of using CO alarms is chosen, compliance is required with Sections 915.4.1 through 915.4.3. CO alarms can be stand-alone or interconnected (single- or multiple-station). Section 915 does not address whether such alarms are required to be interconnected.

CO alarms are designed to initiate an audible alarm when the level of CO is below that which can cause a loss of the ability to react to the dangers of CO exposure. UL specifies that CO alarms activate at a level where the CO concentration over a given period of time can achieve 10-percent carboxyhemoglobin (COHb) in the body. Ten-percent COHb will not cause physiological injury, but is a level at which increases in the CO concentration will begin to affect the human body.

[F] 915.4.1 Power source. Carbon monoxide alarms shall receive their primary power from the building wiring where such wiring is served from a commercial source, and when primary power is interrupted, shall receive power from a battery. Wiring shall be permanent and without a disconnecting switch other than that required for overcurrent protection.

Exception: Where installed in buildings without commercial power, battery-powered carbon monoxide alarms shall be an acceptable alternative.

❖ This section is very similar to that required for smoke alarms (i.e., Section 907.2.11.6). The power supply must be provided by the building with battery backup. As with smoke alarms, if commercial power is not available, battery power is acceptable.

[F] 915.4.2 Listings. Carbon monoxide alarms shall be listed in accordance with UL 2034.

❖ Single- or multiple-station CO alarms are required to be listed in accordance with UL 2034. This standard is specific to CO alarms that provide both detection and notification.

[F] 915.4.3 Combination alarms. Combination carbon monoxide/smoke alarms shall be an acceptable alternative to carbon monoxide alarms. Combination carbon monoxide/smoke alarms shall be listed in accordance with UL 2034 and UL 217.

❖ Since smoke alarms are required in many occupancies, often a single combination alarm is desired. In fact, in some jurisdictions the CO alarms are required to be combined with smoke alarms. To meet this requirement, the combination CO and smoke alarm must be listed to both UL 2034 and UL 217. See Commentary Figure 915.4.3 for an example of a combination CO and smoke alarm.

[F] 915.5 Carbon monoxide detection systems. Carbon monoxide detection systems shall be an acceptable alternative to carbon monoxide alarms and shall comply with Sections 915.5.1 through 915.5.3.

❖ If a CO detection system is preferred over CO alarms, compliance with Sections 915.5.1 through 915.5.3 is required.

[F] 915.5.1 General. Carbon monoxide detection systems shall comply with NFPA 720. Carbon monoxide detectors shall be listed in accordance with UL 2075.

❖ There are two standards applicable to CO detection systems. The systems are required to comply with NFPA 720 and the detectors themselves are required to comply with UL 2075.

[F] 915.5.2 Locations. Carbon monoxide detectors shall be installed in the locations specified in Section 915.2. These locations supersede the locations specified in NFPA 720.

❖ This section is provided to clarify that the locations required for CO detectors in Section 915.2 differ from NFPA 720 and that those locations will supersede the standard. This simply reaffirms the provisions of Section 102.4.1 regarding conflicts between the code and a referenced standard.

[F] 915.5.3 Combination detectors. Combination carbon monoxide/smoke detectors installed in carbon monoxide detection systems shall be an acceptable alternative to carbon monoxide detectors, provided they are listed in accordance with UL 2075 and UL 268.

❖ Similar to CO alarms, there is often a desire or a requirement to combine smoke detection and CO detection systems. If the detectors themselves are combined, they need to be listed in accordance with both UL 2075 and UL 268. NFPA 720 requires the CO alarm to be capable of transmitting a distinct audible signal that is different from the smoke alarm signal.

[F] 915.6 Maintenance. Carbon monoxide alarms and carbon monoxide detection systems shall be maintained in accordance with the *International Fire Code*.

❖ This section is simply a reference to the IFC maintenance requirements that require compliance with NFPA 720. In addition, when detectors become inoperable or start producing end-of-life signals, they are required to be replaced.

Figure 915.4.3
COMBINATION CARBON MONOXIDE AND SMOKE ALARM

SECTION 916
EMERGENCY RESPONDER RADIO COVERAGE

[F] 916.1 General. Emergency responder radio coverage shall be provided in all new buildings in accordance with Section 510 of the *International Fire Code*.

❖ The provisions of Section 915 are new to the 2009 edition and are concerned with the reliability of portable radios used by emergency responders inside of buildings. This is in keeping with the philosophy inherent in the *International Codes* that, when a facility grows too large or complex for effective fire response, fire protection features must be provided within the building. See the commentary to Section 510 of the IFC (and its companion Appendix J) for complete information on this topic.

Bibliography

The following resource materials were used in the preparation of the commentary for this chapter of the code.

Americans With Disabilities Act (ADA), *Standard for Accessible Design* 2010. Washington, DC: US Department of Justice, 2010.

ASHRAE Guideline 5-1994 (RA 2001), *Commissioning Smoke Management Systems*. Atlanta, GA: American Society of Heating, Refrigerating and Air-Conditioning Engineers, 2001.

Cobb, Nathaniel, and Ruth A. Etzel. "Unintentional Carbon Monoxide–Related Deaths in the United States, 1979 through 1988." *Journal of the American Medical Association*, August 7 1991, Vol. 266, No. 5, pp. 659-663.

Fire Protection Handbook, 19th edition. Quincy, MA: National Fire Protection Association, 2003.

International Code Interpretations. Washington, DC: International Code Council, 2009.

Grosshandler, William. *The Use of Portable Fire Extinguishers in Nightclubs*. Gaithersburg, MD: National Institute of Science and Technology, July 2008, pg. 5.

Klote, J., and D. Evans. *A Guide to Smoke Control in the 2006 IBC*. Washington, DC: International Code Council, 2007.

Klote, J., and J. Milke. *Principles of Smoke Management*. Atlanta, GA: American Society of Heating, Refrigerating and Air-Conditioning Engineers, 2002.

NFPA 75-13, Standard for the Fire Protection of Information Technology Equipment. Quincy, MA: National Fire Protection Association, 2013.

NIST Special Publication 1066: Residential Kitchen Fire Suppression Research Needs, Madrzykowski, Hamins & Mehta, Feb. 2007.

NIST IR 6196-1 report dated September 1998.

Occupational Safety and Health (NIOSH) 2005-132, *Preventing Injuries and Deaths of Fire Fighters Due to Truss Systems*, and NIOSH 2010-153, *Preventing Deaths and Injuries of Fire Fighters using Risk Management Principles at Structure Fires*.

Recommendations contained in the Initial Report of the Federal Emergency Management Agency (FEMA)/ National Fallen Firefighter Foundation (NFFF®) Firefighter Life Safety Summit held on April 14, 2004, in Tampa, Florida.

SFPE *Engineering Guide to Performance-based Fire Protection Analysis and Design of Buildings*. Quincy, MA: National Fire Protection Association, 2004.

The SFPE Handbook of Fire Protection Engineering, 4th edition. Quincy, MA: National Fire Protection Association, 2008.

Chapter 10: Means of Egress

General Comments

The general criteria set forth in Chapter 10 regulating the design of the means of egress are established as the primary method for protection of people in buildings. Chapter 10 provides the minimum requirements for means of egress in all buildings and structures. Both prescriptive and performance language is utilized in this chapter to provide a basic approach in determining a safe exiting system for all occupancies. It addresses all portions of the egress system and includes design requirements as well as provisions regulating individual components. The requirements detail the size, arrangement, number and protection of means of egress components. Functional and operational characteristics are also specified for the components, which will permit their safe use without special knowledge or effort.

A zonal approach to egress provides a general basis for the chapter's format. A means of egress system has three parts: exit access, exit and exit discharge.

Section 1001 includes the administrative provisions. Section 1002 shows a list of defined terms that are primarily associated with Chapter 10. For commentary on these definitions, see Chapter 2.

Sections 1003 through 1015 include general provisions that apply to all three components of an egress system. This includes general means of egress requirements; occupant loads; means of egress sizing; the number of exits and exit access doorways and their configuration; illumination for the means of egress; specific requirements for accessible means of egress; doors, gates and turnstiles; provisions for stairways and ramps along with their associated handrail and guard requirements; and exit signage (see commentary, Section 1003).

The exit access requirements are in Sections 1016 through 1021. This includes general exit access requirements; exit access travel distance; aisles (for other than assembly spaces); specific requirements for stairways and ramps that serve as part of the exit access; corridors; and egress balconies (see commentary, Section 1016).

The exit requirements are in Sections 1022 through 1027. Exit information includes provisions for exits; interior and exterior stairways and ramps that serve as an exit element; exit passageways; luminous egress path markings that are required on the exit stairways in high-rise buildings (see Section 403); and horizontal exits (see commentary, Section 1022).

The exit discharge requirements are in Section 1028.

Section 1029 includes those means of egress requirements that are unique to spaces used for assembly purposes.

Emergency escape and rescue opening requirements are in Section 1030.

Chapter 10 requirements are replicated in Chapter 10 of the *International Fire Code*® (IFC®). The IFC has one additional section at the end of the chapter dealing with maintenance of the means of egress (see commentary, Section 1001.3). For means of egress requirements in existing buildings, refer to the *International Existing Building Code*® (IEBC®) or Chapter 11 of the IFC.

The evolution of means of egress requirements has been influenced by lessons learned from real fire incidents. While contemporary fires may reinforce some of these lessons, one must view each incident as an opportunity to assess critically the safety and reasonability of current regulations.

Cooperation among the developers of model codes and standards has resulted in agreement on many basic terms and concepts. The text of the code, including this chapter, is consistent with these national uniformity efforts.

National uniformity in an area such as means of egress has many benefits for the building official and other code users. At the top of the list are the lessons to be learned from experiences throughout the nation and the world, which can be reported in commonly used terminology and conditions that we can all relate to and clearly understand.

Purpose

A primary purpose of codes in general and building codes in particular is to safeguard life in the presence of a fire. Integral to this purpose is the path of egress travel for occupants to escape and avoid a fire. Means of egress can be considered the lifeline of a building. The principles on which means of egress are based and that form the fundamental criteria for requirements are to provide a system that:

1. Will give occupants alternative paths of travel to a place of safety to avoid fire.
2. Will shelter occupants from fire and the products of combustion.
3. Will accommodate all occupants of a structure.
4. Is clear, unobstructed, well marked and illuminated and in which all components are under con-

trol of the user without requiring any tools, keys or special knowledge or effort.

History is marked with the severe loss of life from fire. Early, as well as contemporary, multiple-fatality fires can be traced to a compromise of one or more of the above principles.

Life safety from fire is a matter of successfully evacuating or relocating the occupants of a building to a place of safety. As a result, life safety is a function of time: time for detection, time for notification and time for safe egress. The fire growth rate over a period of time is also a critical factor in addressing life safety. Other sections of the code, such as protection of vertical openings (see Chapter 7), interior finish (see Chapter 8), fire suppression and detection systems (see Chapter 9) and numerous others, also have an impact on life safety. This chapter addresses the issues related to the means available to relocate or evacuate building occupants.

SECTION 1001 ADMINISTRATION

1001.1 General. Buildings or portions thereof shall be provided with a *means of egress* system as required by this chapter. The provisions of this chapter shall control the design, construction and arrangement of *means of egress* components required to provide an *approved means of egress* from structures and portions thereof.

❖ The minimum requirements for emergency evacuation, or means of egress, are to be incorporated in all new structures as specified in this chapter. The system shall include exit access, exit and exit discharge and address the needs for all occupants of the facility. Such application would be effective on the date the code is adopted and placed into effect.

1001.2 Minimum requirements. It shall be unlawful to alter a building or structure in a manner that will reduce the number of *exits* or the minimum width or required capacity of the *means of egress* to less than required by this code.

❖ A fundamental concept in life safety design is that the means of egress system is to be constantly available throughout the life of a building. Any change in the building or its contents, by physical reconstruction, alteration or a change of occupancy, is cause to review the resulting egress system. At a minimum, a building's means of egress is to be continued as initially approved. If a building or portion thereof has a change of occupancy, the complete egress system is to be evaluated and approved for compliance with the current code requirements for new occupancies (see the IEBC and the IFC Section 1031 and Chapter 11).

The means of egress in an existing building that experiences a change of occupancy, such as from Group S-2 (storage) to A-3 (assembly), would require reevaluation for code compliance based on the new occupancy. Similarly, the means of egress in an existing Group A-3 occupancy in which additional seating is to be provided, thereby increasing the occupant load, would require reevaluation for code compliance based on the increased occupant load.

The temptation is to temporarily remove egress components or other fire protection features from service during an alteration, repair to or temporary occupancy of a building. During such times, a building is frequently more vulnerable to fire and the rapid spread of products of combustion. Either the occupants should not occupy those spaces where the means of egress has been compromised by the construction or compensating fire safety features should be considered that will provide equivalent safety for the occupants. Occupants in adjacent areas may also require access to the egress facilities in the area under construction.

[F] 1001.3 Maintenance. *Means of egress* shall be maintained in accordance with the *International Fire Code*.

❖ This section provides a cross reference to the code requirements that address the maintenance of the means of egress in an existing building. The means of egress must be maintained so that occupants are not prevented from exiting the building quickly in case of an emergency.

Sections 1003 through 1029 in the code are repeated in the IFC. These sections are maintained by the IBC Means Of Egress Committee so that there will be consistency between the two documents. Note the [BE] in front of the main section headings in the IFC. Note the [F] in front of this section. This means that this section is maintained by the International Fire Code Development Committee. Additionally, the IFC includes Section 1031, which applies to maintenance of the means of egress. For means of egress in existing buildings, refer to Chapter 11 of the IFC and the IEBC.

[F] 1001.4 Fire safety and evacuation plans. Fire safety and evacuation plans shall be provided for all occupancies and buildings where required by the *International Fire Code*. Such fire safety and evacuation plans shall comply with the applicable provisions of Sections 401.2 and 404 of the *International Fire Code*.

❖ Preplanning how to handle emergencies and practicing those plans are important parts of managing emergencies in order to minimize the hazard to occupants. The reference to Section 401.2 of the IFC clarifies that fire and safety evacuation plans must be approved by the fire official. Section 404 of the IFC lists the facilities for which fire and safety evacuation plans are required. Fire evacuation plans include information on: emergency egress routes; phased evacuation options; operation of critical equipment; assisted rescue; verification of evacuation; notification to occupants; and responders and responsible personnel. Fire safety plans include information on: reporting a fire; life-safety strategy; notifying, relocating or evacuating occupants; assisted rescue; site plans indicating occupancy assembly point, fire hydrants and fire department vehicle access; floor plans indicating exits, evacuation routes (general and accessible), fire alarm boxes, portable fire extinguishers, fire hose stations, fire alarms and controls; major fire hazards in the building; sprinkler systems; alarm and detection equipment; and responsible personnel. The IFC also addresses locations, such as hospitals or nursing homes, where it may be more appropriate to defend in place.

Some facilities, such as schools and courthouses, have developed lockdown plans for other emergencies. Rather than evacuating, the appropriate response may be to defend in place. Typically, there are different levels, depending on the emergency. For example a school may have: 1. Police emergency action in town; 2. Police emergency action in the immediate area; and 3. Police emergency in the building. Lockdown plans include information on: initiation of the plan; persons accountable; recall; communication; and practice drills.

These plans must be reevaluated when necessary and, at a minimum, on an annual basis. Copies must be provided to building tenants and the fire department. As with all plans, practice drills are an important part of full communication of the plans to everyone involved.

SECTION 1002 DEFINITIONS

1002.1 Definitions. The following terms are defined in Chapter 2:

ACCESSIBLE MEANS OF EGRESS.

AISLE.

AISLE ACCESSWAY.

ALTERNATING TREAD DEVICE.

AREA OF REFUGE.

BLEACHERS.

BREAKOUT.

COMMON PATH OF EGRESS TRAVEL.

CORRIDOR.

DOOR, BALANCED.

EGRESS COURT.

EMERGENCY ESCAPE AND RESCUE OPENING.

EXIT.

EXIT ACCESS.

EXIT ACCESS DOORWAY.

EXIT ACCESS RAMP.

EXIT ACCESS STAIRWAY.

EXIT DISCHARGE.

EXIT DISCHARGE, LEVEL OF.

EXIT, HORIZONTAL.

EXIT PASSAGEWAY.

EXTERIOR EXIT RAMP.

EXTERIOR EXIT STAIRWAY.

FIRE EXIT HARDWARE.

FIXED SEATING.

FLIGHT.

FLOOR AREA, GROSS.

FLOOR AREA, NET.

FOLDING AND TELESCOPIC SEATING.

GRANDSTAND.

GUARD.

HANDRAIL.

INTERIOR EXIT RAMP.

INTERIOR EXIT STAIRWAY.

LOW ENERGY POWER-OPERATED DOOR.

MEANS OF EGRESS.

MERCHANDISE PAD.

NOSING.

OCCUPANT LOAD.

OPEN-ENDED CORRIDOR.

PANIC HARDWARE.

PHOTOLUMINESCENT.

POWER-ASSISTED DOOR.

POWER-OPERATED DOOR.

PUBLIC WAY.

RAMP.

SCISSOR STAIRWAY.

SELF-LUMINOUS.

SMOKE-PROTECTED ASSEMBLY SEATING.

STAIR.

STAIRWAY.

STAIRWAY, SPIRAL.

WINDER.

❖ This section lists terms that are specifically associated with the subject matter of this chapter. It is important to emphasize that these terms are not exclusively related to this chapter, but may or may not also be applicable where the term is used elsewhere in the code.

Definitions of terms can help in the understanding and application of the code requirements. The purpose for including a list within this chapter is to provide more convenient access to terms that may have a specific or limited application within this chapter.

For the complete definition and associated commentary, refer back to Chapter 2. Terms that are italicized provide a visual identification throughout the code that a definition exists for that term. The use and application of all defined terms are set forth in Section 201.

SECTION 1003
GENERAL MEANS OF EGRESS

1003.1 Applicability. The general requirements specified in Sections 1003 through 1015 shall apply to all three elements of the *means of egress* system, in addition to those specific requirements for the *exit access*, the *exit* and the *exit discharge* detailed elsewhere in this chapter.

❖ The requirements in the chapter address the three parts of a means of egress system: the exit access, the exit and the exit discharge. This section specifies that the requirements of Sections 1003 through 1015 apply to the components of all three parts of the system. For example, the stair tread and riser dimensions in Section 1009 apply to interior exit access stairways, such as those leading from a mezzanine, and also apply to enclosed exit stairways per Section 1022, exterior exit stairways per Section 1026 and steps in the exit discharge per Section 1027.

The following sections are applicable for all parts of the means of egress:

- Section 1003 deals with the path for means of egress to remain free of obstructions and tripping hazards.
- Section 1004 provides criteria for determining occupant loads for a space. These numbers are used for determining means of egress, as a threshold for some suppression requirements and to determine the required plumbing fixture count.
- Section 1005 deals with the required size (i.e., width) of the path of travel for emergency evacuation. It is important not to create a "bottleneck" that could increase the amount of time necessary for occupants to exit the buildings.
- Section 1006 deals with the number of ways out of a space or off a floor, either by exit elements or exit access elements.
- Section 1007 provides placement and remoteness requirements for the exit and exit access elements prescribed in Section 1006.
- Section 1008 deals with illumination for the path of travel for the means of egress. Both general lighting and emergency backup lighting are addressed.
- Section 1009 – Chapter 11 indicates how to get people with mobility impairments into a building. Section 1009 explains the options to allow people with mobility impairments to self-evacuate or how to arrange for assisted rescue. The accessible means of egress is an important part of the fire and safety evacuation plans (see Section 1001.4).
- Section 1010 includes requirements for doors, gates and turnstiles that are part of the path of travel from any occupied spaces. For example, doors that lead to a walk-in closet must comply with this section, but doors for reach-in closets are exempted.
- Section 1011 provides information on all types of stairways: interior and exterior and from one riser to stairways with multiple flights and landings. Stepped aisles for areas within assembly seating are specifically addressed in Section 1029. For protection of the stairways between stories, see Sections 1019, 1023 and 1027.
- Section 1012 deals with ramps. Ramped aisles serving assembly seating areas are specifically addressed in Section 1029. The ramp provisions are coordinated with ICC A117.1 and the 2010 Standard for Accessible Design [formerly the Americans with Disabilities Act Accessibility Guidelines (ADAAG), now referred to as the 2010 ADA Standard]. For protection of the ramp between stories, see Sections 1019, 1023 and 1027.
- Section 1013 describes where exit signs are required and what criteria they need to meet to be readily visible.
- Section 1014 describes handrail requirements for stairways and ramps. Handrails are important for guidance and to arrest a possible fall.
- Section 1015 provides criteria for the vertical portions of barriers that serve to protect people from possible falls at dropoffs greater than 30 inches (762 mm).

1003.2 Ceiling height. The *means of egress* shall have a ceiling height of not less than 7 feet 6 inches (2286 mm).

Exceptions:

1. Sloped ceilings in accordance with Section 1208.2.
2. Ceilings of *dwelling units* and *sleeping units* within residential occupancies in accordance with Section 1208.2.
3. Allowable projections in accordance with Section 1003.3.
4. *Stair* headroom in accordance with Section 1011.3.
5. Door height in accordance with Section 1010.1.1.
6. *Ramp* headroom in accordance with Section 1012.5.2.
7. The clear height of floor levels in vehicular and pedestrian traffic areas of public and private parking garages in accordance with Section 406.4.1.
8. Areas above and below *mezzanine* floors in accordance with Section 505.2.

❖ Generally, the specified ceiling height is the minimum allowed in any part of the egress path. The exceptions are intended to address conditions where the code allows the ceiling height to be lower than specified in this section.

This section is consistent with the minimum ceiling height for other areas as specified in Section 1208. The exceptions are pointers to the lower headroom areas permitted in the code. For example, the headroom above and below a mezzanine is 7 feet (2134 mm) minimum.

1003.3 Protruding objects. Protruding objects on *circulation paths* shall comply with the requirements of Sections 1003.3.1 through 1003.3.4.

❖ This section begins the provisions that apply to protruding objects and helps to improve awareness of these safety and accessibility-related provisions. The intent of the phrase "on circulation paths" is intended to allow for judgment determining where people walk versus all floor surfaces. For example, a drinking fountain in an alcove is over a floor, but it is not over a circulation path; therefore, it typically it would not be considered a protruding object.

1003.3.1 Headroom. Protruding objects are permitted to extend below the minimum ceiling height required by Section 1003.2 where a minimum headroom of 80 inches (2032 mm) is provided over any walking surface, including walks, *corridors*, *aisles* and passageways. Not more than 50 percent of the ceiling area of a *means of egress* shall be reduced in height by protruding objects.

Exception: Door closers and stops shall not reduce headroom to less than 78 inches (1981 mm).

A barrier shall be provided where the vertical clearance is less than 80 inches (2032 mm) high. The leading edge of such a barrier shall be located 27 inches (686 mm) maximum above the floor.

❖ This provision is applicable to all routes that make up components of the means of egress. Specifically, the limitations in this section and those in Sections 1003.3.2 and 1003.3.3 provide a reasonable level of safety for people with vision impairments as well as during emergency events when vision may be obscured by smoke or low lighting.

Minimum dimensions for headroom clearance are specified in this section. The minimum headroom clearance over all walking surfaces or circulation paths is required to be maintained at 80 inches (2032 mm). This minimum headroom clearance is consistent with the requirements in Section 1011.3 for stairs and Section 1012.5.2 for ramps. Allowance must be made for door closers and stops, since their design and function necessitates placement within the door opening. The minimum headroom clearance for door closers and stops is allowed to be 78 inches (1981 mm) [see Commentary Figure 1003.3.1(1)]. The 2-inch (51 mm) projection into the doorway height is reasonable since these devices are normally mounted away from the center of the door opening, thus minimizing the potential for contact with a person moving through the opening. This is consistent with the exception to Section 1010.1.1.1.

The limitation on overhangs is of primary importance to individuals with visual impairments. When vertical clearance along a walking surface is less than 80 inches (2032 mm), such as underneath the stairway on the ground floor, some sort of barrier that is detectable by a person using a cane must be provided. This can be a full-height wall, a rail at or below 27 inches (686 mm), a planter, fixed seating, etc. A low curb is not effective as a barrier. A person with visual impairments might mistake it for a stair tread, step up onto it and strike their head. A rail at handrail height would not be detectable by a person using a cane, and he or she could possibly walk into the rail before detecting it. Also, when making decisions on the choice of type of barrier, keep in mind that persons of shorter stature and children have a detectable range that may be below 27 inches (686 mm) [see Commentary Figure 1003.3.1(2)].

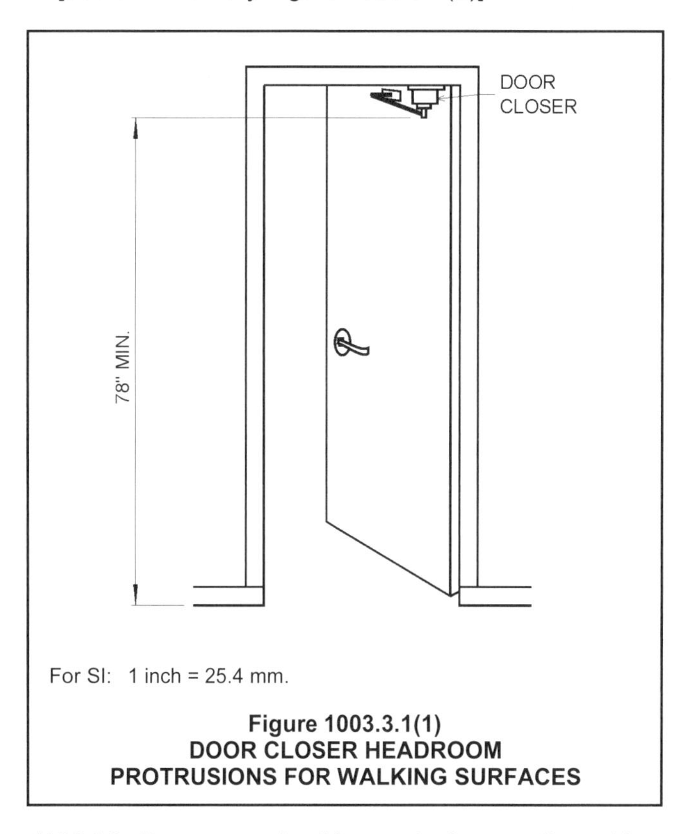

Figure 1003.3.1(1)
DOOR CLOSER HEADROOM PROTRUSIONS FOR WALKING SURFACES

1003.3.2 Post-mounted objects. A free-standing object mounted on a post or pylon shall not overhang that post or pylon more than 4 inches (102 mm) where the lowest point of the leading edge is more than 27 inches (686 mm) and less than 80 inches (2032 mm) above the walking surface. Where a sign or other obstruction is mounted between posts or pylons and the clear distance between the posts or pylons is greater than 12 inches (305 mm), the lowest edge of such sign or obstruction shall be 27 inches (686 mm) maximum or 80 inches (2032 mm) minimum above the finished floor or ground.

Exception: These requirements shall not apply to sloping portions of *handrails* between the top and bottom riser of *stairs* and above the *ramp* run.

❖ Post-mounted objects, such as signs or some types of drinking fountains or phone boxes, are not permitted to overhang more than 4 inches (102 mm) past the post where the bottom edge is located higher than 27 inches (686 mm) above the walking surface [see Commentary Figure 1003.3.2(1)]. Since the minimum required height of doorways, stairways and ramps in the means of egress is 80 inches (2032 mm), protruding objects located higher than 80 inches (2032 mm) above the walking surface are not regulated. Protrusions that are located lower than 27 inches (686 mm) above the walking surface are also permitted since they are more readily detected by a person using a long cane, provided that the minimum required width of the egress element is maintained. This is consistent with the post-mounted objects requirements in Section 307.3 of ICC A117.1, *Accessible and Usable Buildings and Facilities*. The intent is to reduce the potential for accidental impact for a person who is visually impaired.

When signs are provided on multiple posts, the posts must be located closer than 12 inches (305 mm) apart, or the bottom edge of the sign must be lower than 27 inches (686 mm) so it is within detectable cane range or above 80 inches (2032 mm) so that it is above headroom clearances [see Commentary Figure 1003.3.2(2)].

The exception is intended for handrails that are located along the run of a stairway flight or ramp run. The extensions at the top and bottom of stairways and ramps must meet the requirements for protruding objects where people walk perpendicular to the stair or ramp.

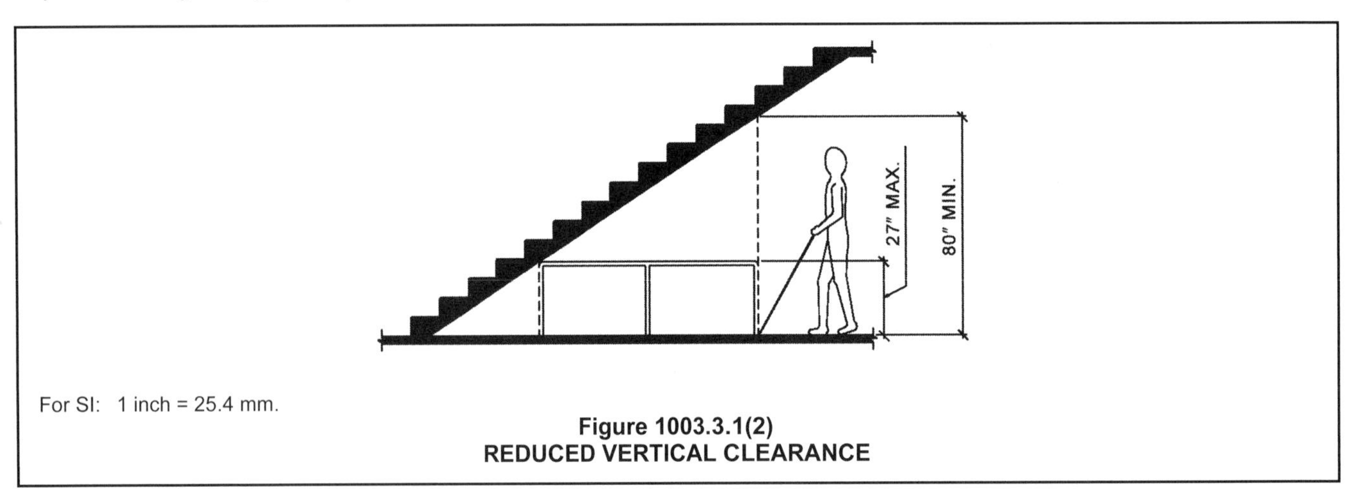

For SI: 1 inch = 25.4 mm.

Figure 1003.3.1(2)
REDUCED VERTICAL CLEARANCE

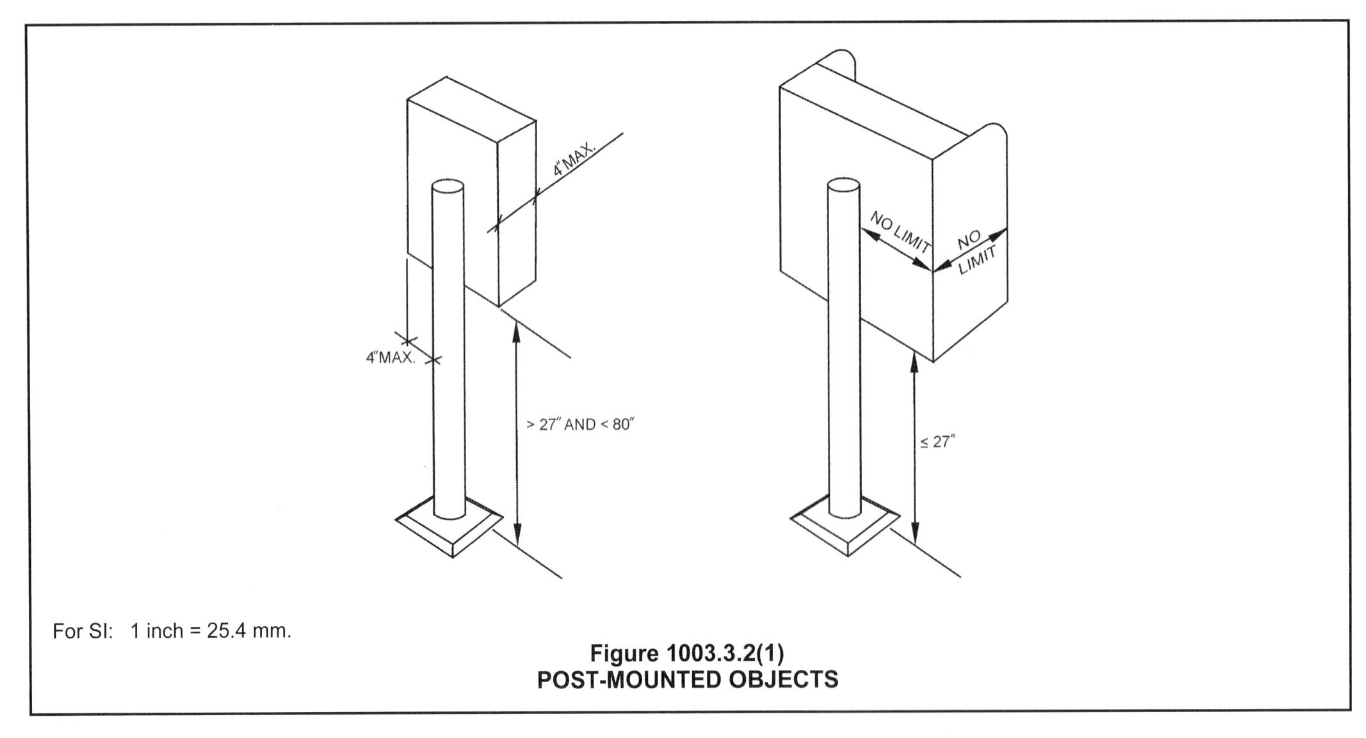

For SI: 1 inch = 25.4 mm.

Figure 1003.3.2(1)
POST-MOUNTED OBJECTS

1003.3.3 Horizontal projections. Objects with leading edges more than 27 inches (685 mm) and not more than 80 inches (2030 mm) above the floor shall not project horizontally more than 4 inches (102 mm) into the *circulation path*.

Exception: *Handrails* are permitted to protrude $4^1/_2$ inches (114 mm) from the wall.

❖ Protruding objects could slow the egress flow through a corridor or passageway and injure someone hurriedly passing by or someone with a visual impairment. Persons with a visual impairment who use a long cane for guidance must have sufficient warning of a protruding object. Where protrusions are located higher than 27 inches (686 mm) above the walking surface, the cane will most likely not encounter the protrusion before the person collides with the object.

Additionally, people with poor visual acuity or poor depth perception may have difficulty identifying protruding objects higher than 27 inches (686 mm). Therefore, objects such as lights, signs and door hardware, located between 27 inches (686 mm) and 80 inches (2032 mm) above the walking surface, are not permitted to extend more than 4 inches (102 mm) from each wall (see Commentary Figure 1003.3.3). The requirement for protrusions into the door clear width in Section 1010.1.1.1 is different because it deals with allowances for panic hardware on a door. It is not the intent of this section to prohibit columns,

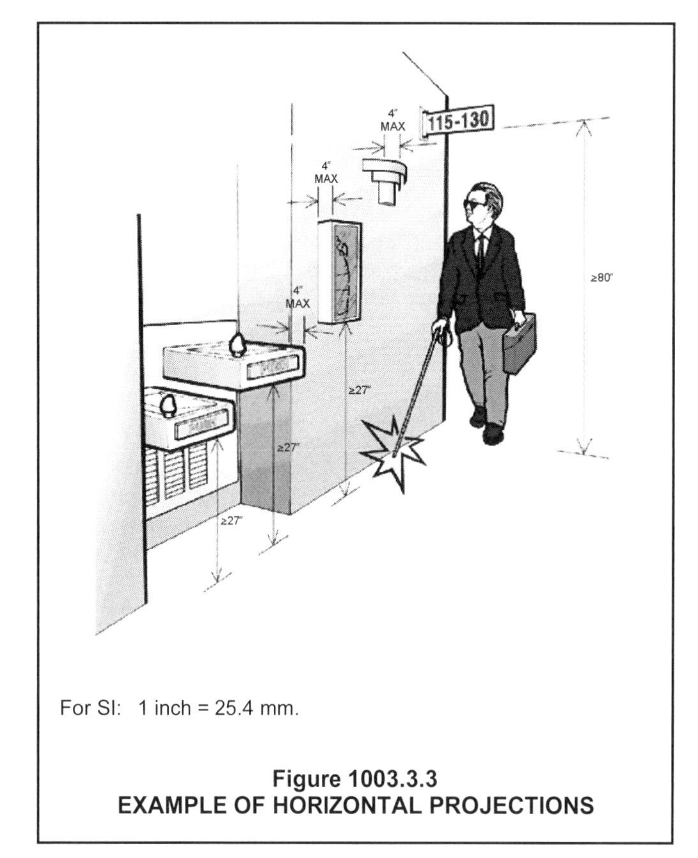

For SI: 1 inch = 25.4 mm.

Figure 1003.3.3
EXAMPLE OF HORIZONTAL PROJECTIONS

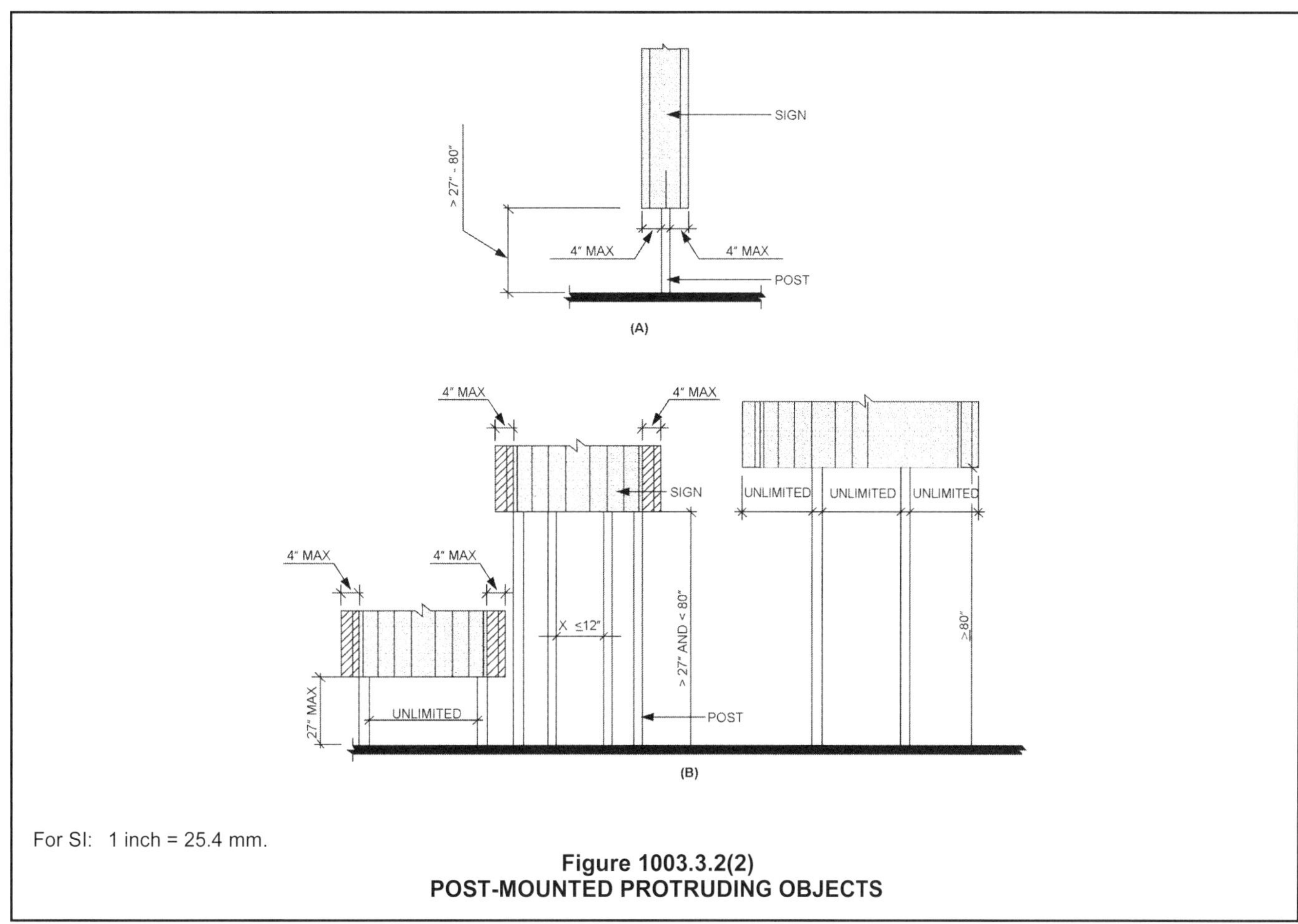

For SI: 1 inch = 25.4 mm.

Figure 1003.3.2(2)
POST-MOUNTED PROTRUDING OBJECTS

pilasters or wing walls to project into a corridor as long as adequate egress width is maintained. These types of structural elements are detectable by persons using a long cane.

The exception is an allowance for handrails when they are provided along a wall, such as in some hospitals or nursing homes. The $4^1/_2$-inch (114 mm) measurement is intended to be consistent with projections by handrails into the required width of stairways and ramps in Section 1014.8. There are additional requirements when talking about the required width (see Section 1005.2).

1003.3.4 Clear width. Protruding objects shall not reduce the minimum clear width of *accessible routes*.

❖ The intent of this section is to limit the projections into an accessible route so that a minimum clear width of 36 inches (914 mm) is maintained along the route. ICC A117.1 is referenced by Chapter 11 for technical requirements for accessibility. ICC A117.1, Section 403.5, allows the accessible route to be reduced in width to 32 inches (914 mm) for segments not to exceed 24 inches (635 mm) in length and spaced a minimum of 48 inches (1219 mm) apart. This allows for movement through a doorway or through a gap in planters or counters.

1003.4 Floor surface. Walking surfaces of the *means of egress* shall have a slip-resistant surface and be securely attached.

❖ As the pace of exit travel becomes hurried during emergency situations, the probability of slipping on smooth or slick floor surfaces increases. To minimize the hazard, all floor surfaces in the means of egress are required to be slip resistant. The use of hard floor materials with highly polished, glazed, glossy or finely finished surfaces should be avoided.

Field testing and uniform enforcement of the concept of slip resistance are not practical. One method used to establish slip resistance is that the static coefficient of friction between leather [Type 1 (Vegetable Tanned) of Federal Specification KK-L-165C] and the floor surface is greater than 0.5. Laboratory test procedures, such as ASTM D2047, can determine the static coefficient of resistance. Bulletin No. 4, "Surfaces," issued by the U.S. Architectural and Transportation Barriers Compliance Board (ATBCB or Access Board) contains further information regarding slip resistance.

1003.5 Elevation change. Where changes in elevation of less than 12 inches (305 mm) exist in the *means of egress*, sloped surfaces shall be used. Where the slope is greater than one unit vertical in 20 units horizontal (5-percent slope), *ramps* complying with Section 1012 shall be used. Where the difference in elevation is 6 inches (152 mm) or less, the *ramp* shall be equipped with either *handrails* or floor finish materials that contrast with adjacent floor finish materials.

Exceptions:

1. A single step with a maximum riser height of 7 inches (178 mm) is permitted for buildings with occupancies in Groups F, H, R-2, R-3, S and U at exterior doors not required to be *accessible* by Chapter 11.
2. A *stair* with a single riser or with two risers and a tread is permitted at locations not required to be *accessible* by Chapter 11 where the risers and treads comply with Section 1011.5, the minimum depth of the tread is 13 inches (330 mm) and not less than one *handrail* complying with Section 1014 is provided within 30 inches (762 mm) of the centerline of the normal path of egress travel on the *stair*.
3. A step is permitted in *aisles* serving seating that has a difference in elevation less than 12 inches (305 mm) at locations not required to be *accessible* by Chapter 11, provided that the risers and treads comply with Section 1029.13 and the *aisle* is provided with a *handrail* complying with Section 1029.15.

Throughout a story in a Group I-2 occupancy, any change in elevation in portions of the *means of egress* that serve nonambulatory persons shall be by means of a *ramp* or sloped walkway.

❖ Minor changes in elevation, such as a single step that is located in any portion of the means of egress (i.e., exit access, exit or exit discharge), may not be readily apparent during normal use or emergency egress and are considered to present a potential tripping hazard. Where the elevation change is less than 12 inches (305 mm), a ramp or sloped surface is specified to make the transition from higher to lower levels. This is intended to reduce accidental falls associated with the tripping hazard of an unseen step. Ramps must be constructed in accordance with Section 1012.1. Ramp provisions do not require handrails for ramps with a rise of 6 inches or less. However, the presence of the ramp must be readily apparent from the directions from which it is approached. Handrails are one method of identifying the change in elevation. In lieu of handrails, the surface of the ramp must be finished with materials that visually contrast with the surrounding floor surfaces. The walking surface of the ramp should contrast both visually and physically.

None of the exceptions are permitted along an accessible route required for either entry or egress from a space or building (see Section 1009 and Chapter 11).

Exception 1 allows up to a 7-inch (178 mm) step at exterior doors to avoid blocking the outward swing of the door by a buildup of snow or ice in locations that are not used by the public on a regular basis (see Commentary Figure 1003.5). This exception is coordinated with Exception 2 of Section 1010.1.5, and is only applicable in occupancies that have relatively low occupant densities, such as factory and industrial structures. This exception is not applicable to exterior doors that are required to serve as an accessible entrance or that are part of a required accessible route. If this exception is utilized at a Group R-2 or R-3 occupancy, the designer may want to consider the issues of potential tripping hazards if this is a common entrance for a large number of occupants.

Exception 2 allows the transition from higher to lower elevations to be accomplished through the construction of stairs with one or two risers. The pitch of the stairway, however, must be shallower than that required for typical stairways (see Section 1011.5.2). Since the total elevation change is limited to 12 inches (305 mm), each riser must be approximately 6 inches (152 mm) in height. The elevation change must be readily apparent from the directions from which it is approached. At least one handrail is required. It must be constructed in accordance with Section 1014 and located so as to provide a graspable surface from the normal walking path.

Exception 3 is basically a cross reference to the assembly provisions for stepped aisles in Section 1029.

None of the exceptions are permitted in a Group I-2 occupancy (e.g., nursing home, hospital) in areas where nonambulatory persons may need access. The mobility impairments of these individuals require additional consideration.

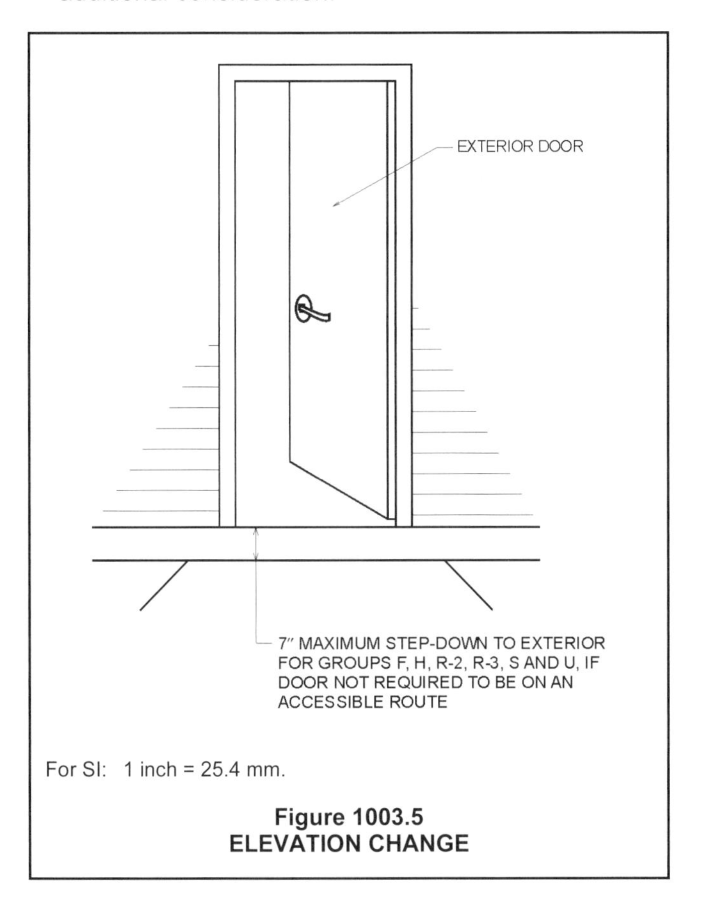

Figure 1003.5
ELEVATION CHANGE

1003.6 Means of egress continuity. The path of egress travel along a *means of egress* shall not be interrupted by a building element other than a *means of egress* component as specified in this chapter. Obstructions shall not be placed in the minimum width or required capacity of a *means of egress* component except projections permitted by this chapter. The minimum width or required capacity of a *means of egress* system shall not be diminished along the path of egress travel.

❖ This section requires that the entire means of egress path be clear of obstructions that could reduce the egress path to below the minimum width at any point. The egress path is also not allowed to be reduced in width such that the design occupant load (required capacity) would not be served. Note, however, that the egress path could be reduced in width in situations where it is wider than required by the code based on the occupant load. For example, if the required width of a corridor was 52 inches (1321 mm) based on the number of occupants using the corridor and the corridor provided was 96 inches (2438 mm) in width, the corridor would be allowed to be reduced to the minimum required width of 52 inches (1321 mm) since that width would still serve the number of occupants required by the code. In the context of this section, a "means of egress component" would most likely be a door or doorway.

1003.7 Elevators, escalators and moving walks. Elevators, escalators and moving walks shall not be used as a component of a required *means of egress* from any other part of the building.

Exception: Elevators used as an accessible *means of egress* in accordance with Section 1009.4.

❖ Generally, the code does not allow elevators, escalators and moving sidewalks to be used as a required means of egress. The concern is that, because of possible power outages, escalators and moving sidewalks may not provide a safe and reliable means of egress that is available for use at all times.

Elevators are not typically used for unassisted evacuation during fire emergencies. However, in taller buildings, fire fighters use the elevators for both staging to fight the fire and assisted evacuation. They can verify that the shaft is not full of smoke, that the elevators will remain operational and, since they know the fire location, at which floors the elevator can be safely accessed. In accordance with the exception, elevators are allowed to be part of an accessible means of egress (i.e., assisted evacuation), provided they comply with the requirements of Section 1009.4. Where elevators are required to serve as part of the accessible means of egress is addressed in Section 1009.2.1. There are new provisions for fire service elevators and occupant evacuation elevators for high rises in Sections 403, 3007 and 3008. These specific provisions will provide a level of safety that would meet the intent of the means of egress provisions in Chapter 10.

SECTION 1004
OCCUPANT LOAD

1004.1 Design occupant load. In determining *means of egress* requirements, the number of occupants for whom *means of egress* facilities are provided shall be determined in accordance with this section.

❖ The design occupant load is the number of people intended to occupy a building or portion thereof at any one time; essentially the number for which the means of egress is to be designed. It is the largest number derived by the application of Sections 1004.1 through 1004.6. Occupant density is limited to ensure a reasonable amount of freedom of movement (see Section 1004.2). The design occupant load is also utilized to determine the required plumbing fixture count (see commentary, Chapter 29) and other building requirements, such as automatic sprinkler systems and fire alarm and detection systems (see Chapter 9).

The intent of this section is to indicate the procedure by which design occupant loads are determined. This is particularly important because accurate determination of design occupant load is fundamental to the proper design of any means of egress system.

1004.1.1 Cumulative occupant loads. Where the path of egress travel includes intervening rooms, areas or spaces, cumulative *occupant loads* shall be determined in accordance with this section.

❖ When occupants from an accessory area move through another area to exit, the combined number of occupants must be utilized to determine the capacity that the egress components must be designed to accommodate. It is not the intent of this section to "double count" occupants. For example, the means of egress from a lobby must be sized for the cumulative occupant load of the adjacent office spaces if the occupants must travel through the lobby to reach an exit. Likewise, if an adjacent room has an egress route independent of the lobby, the occupant load of that room would not be combined with the occupant loads of the other rooms that pass through the lobby. If a portion of the adjacent room's occupant load is to travel through the lobby, only that portion would be combined with the lobby occupant load for determining lobby egress (see Commentary Figure 1004.1.1). This is particularly important in determining the number of ways out of a space or off a story and the required capacity of those elements.

1004.1.1.1 Intervening spaces or accessory areas. Where occupants egress from one or more rooms, areas or spaces through others, the design *occupant load* shall be the combined *occupant load* of interconnected accessory or intervening spaces. Design of egress path capacity shall be based on the cumulative portion of *occupant loads* of all rooms, areas or spaces to that point along the path of egress travel.

❖ An example of intervening spaces could be small tenant spaces within a large mercantile. It is common for banks or coffee shops to be located within large grocery stores. Another example would be a dentist's office where people in the staff and exam room areas would egress through the reception area.

1004.1.1.2 Adjacent levels for mezzanines. That portion of the *occupant load* of a *mezzanine* with required egress through a room, area or space on an adjacent level shall be added to the *occupant load* of that room, area or space.

❖ The egress requirements for mezzanines that use exit access stairways to move to the ground level are handled similarly to those spaces with accessory areas addressed in Section 1004.1.1.1, versus the requirements for exiting from multiple stories in Sections 1004.1.1.3, 1005.4.1 and 1006. That is, that portion of the mezzanine occupant load that travels

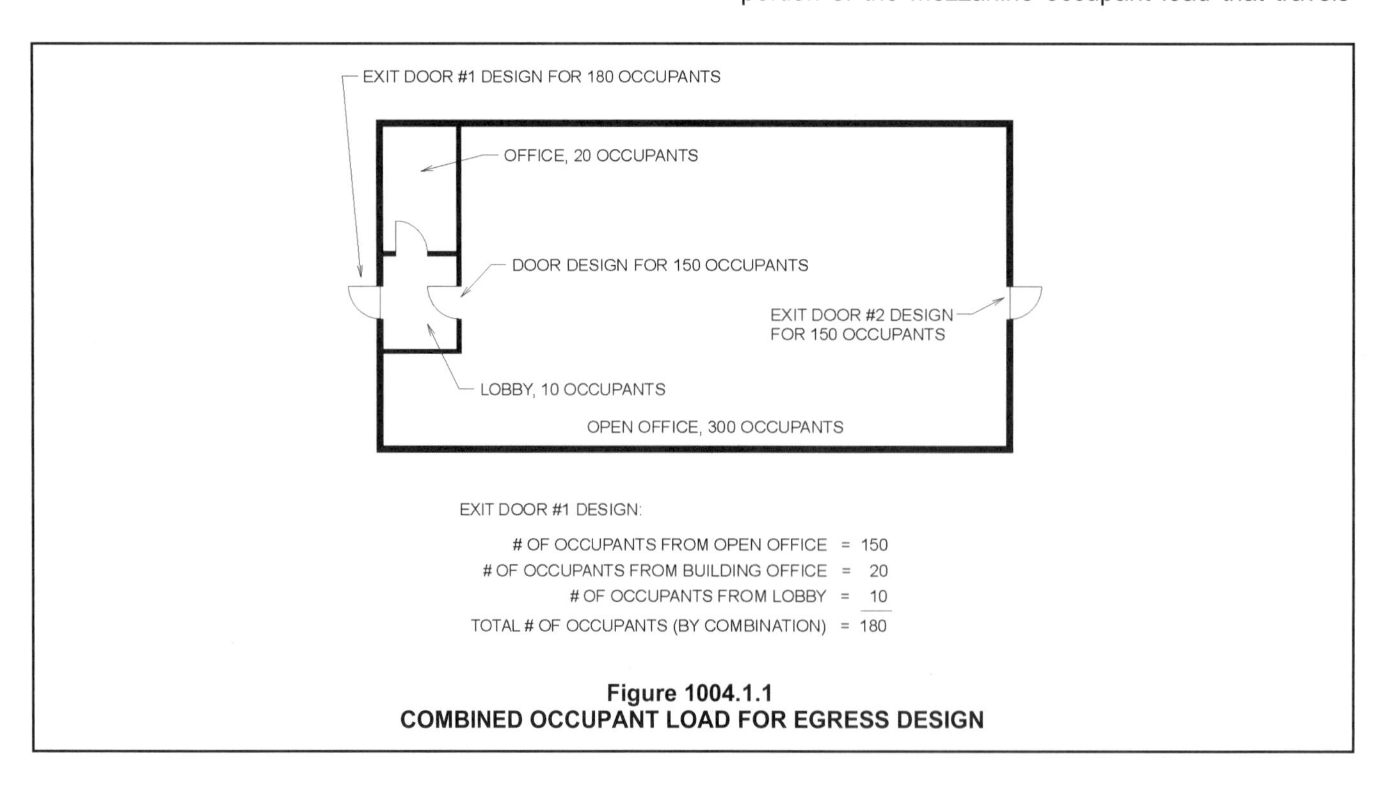

Figure 1004.1.1
COMBINED OCCUPANT LOAD FOR EGRESS DESIGN

through the space below to get to the exit is to be added to the occupant load of the space on the floor below. The sizing and number of the egress components must reflect this combined occupant load. Section 505.2.3 contains additional criteria for the means of egress from mezzanines or a portion of it is enclosed.

1004.1.1.3 Adjacent stories. Other than for the egress components designed for convergence in accordance with Section 1005.6, the *occupant load* from separate stories shall not be added.

❖ Second-floor egress requirements that use exit access stairways to move to the ground level are coordinated with the requirements for exiting from multiple stories found in Sections 1005.3.1, 1005.4.1 and 1006. That is, the portion of the second floor occupant load that travels through the floor below to the exit is not to be added to the occupant load of the space on the floor below as you would for a mezzanine. The sizing and number of the egress components do not have to reflect this combined occupant load. The exception to the rule is where there would be egress convergence of two stories at a level between the two stories (see Section 1005.6).

1004.1.2 Areas without fixed seating. The number of occupants shall be computed at the rate of one occupant per unit of area as prescribed in Table 1004.1.2. For areas without *fixed seating*, the occupant load shall be not less than that number determined by dividing the floor area under consideration by the *occupant load* factor assigned to the function of the space as set forth in Table 1004.1.2. Where an intended function is not listed in Table 1004.1.2, the *building official* shall establish a function based on a listed function that most nearly resembles the intended function.

Exception: Where *approved* by the *building official*, the actual number of occupants for whom each occupied space, floor or building is designed, although less than those determined by calculation, shall be permitted to be used in the determination of the design *occupant load*.

❖ The numbers for floor area per occupant load factor in Table 1004.1.2 reflect common and traditional occupant density based on empirical data for the density of similar spaces. The number determined using the occupant load factors in Table 1004.1.2 generally establishes the minimum occupant load for which the egress facilities of the rooms, spaces and building must be designed. The design occupant load is also utilized for other code requirements, such as determining the required plumbing fixture count (see commentary, Chapter 29) and other building requirements, including automatic sprinkler systems and alarm and detection systems (see Chapter 9).

It is difficult to predict the many conditions by which a space within a building will be occupied over time. An assembly banquet room in a hotel, for example, could be arranged with rows of chairs to host a business seminar one day and with mixed tables and chairs to host a dinner reception the next day. In some instances, the room will be arranged with no tables and very few chairs to accommodate primarily standing occupants. In such a situation, the egress facilities must safely accommodate the maximum number of persons permitted to occupy the space. When determining the occupant load of this type of occupancy, the various arrangements (e.g., tables and chairs, chairs only, standing space) should be recognized. The worst-case scenario should be utilized to determine the requirements for the means of egress elements. This is consistent with the requirements for multiple-use spaces addressed in Section 302.1.

While some of the values in the table utilize the net floor area, most utilize the gross floor area. See the commentary to Table 1004.1.2 and the definitions for "Floor area, gross" and "Floor area, net" in Chapter 2 for additional discussion and examples.

The occupant load determined in accordance with this section is typically the minimum occupant load on which means of egress requirements are to be based. Some occupancies may not typically contain an occupant load totally consistent with the occupant load density factors of Table 1004.1.2. The exception is intended to address the limited circumstances where the actual occupant load is less than the calculated occupant load. Previously, designing for a reduced occupant load was permitted only through the variance process. With this exception, the building official can make a determination if a design that would use the actual occupant load was permissible. The building official may want to create specific conditions for approval. For example, the building official could choose to permit the actual occupant load to be utilized to determine the plumbing fixture count, but not the means of egress or sprinkler design; the determination could be that the reduced occupant load may be utilized in a specific area, such as in the storage warehouse, but not in the factory or office areas. Another point to consider would be the potential of the space being utilized for different purposes at different times, or the potential of a future change of tenancy without knowledge of the building department. Any special considerations for such unique uses must be documented and justified. Additionally, the owner must be aware that such special considerations will impact the future use of the building with respect to the means of egress and other protection features.

TABLE 1004.1.2. See page 10-12.

❖ Table 1004.1.2 establishes minimum occupant densities based on the function or actual use of the space (not group classification).The table presents the maximum floor area allowance per occupant (i.e., occupant load factor) based on studies and counts of the number of occupants in typical buildings. The use of this table, then, results in the minimum occupant load for which rooms, spaces and the building must be designed. While an assumed normal occupancy may be viewed as somewhat less than that determined by the use of the table factors, such a normal occupant load is not necessarily an appropriate design criterion. The greatest hazard to the occupants occurs when an unusually large crowd is present. The code does not limit the occupant load density of an area, except as provided for in Section 1004.2, but once the occupant load is established, the means of egress must be designed for at least that capacity. If it is intended that the occupant load will exceed that calculated in accordance with the table, then the occupant load is to be based on the estimated actual number of people, but not to exceed the maximum allowance in accordance with Section 1004.2. Therefore, the occupant load of the office or business areas in a storage warehouse or nightclub is to be determined using the occupant load factor most appropriate to that space—one person for each 100 square feet (9 m^2) of gross floor area.

The use of net and gross floor areas as defined in Chapter 2 is intended to provide a refinement in the occupant load determination. The gross floor area technique applied to a building only allows the deduction of the plan area of the exterior walls, vent shafts and interior courts from the plan area of the building.

The net floor area permits the exclusion of certain spaces that would be included in the gross floor area. The net floor area is intended to apply to the actual occupied floor areas. The area used for permanent building components, such as shafts, fixed equipment, thicknesses of walls, corridors, stairways, toilet rooms, mechanical rooms and closets, is not included in net floor area. For example, consider a restaurant dining area with dimensions measured from the inside of the enclosing walls of 80 feet by 60 feet (24 384 mm by 18 288 mm) (see Commentary Figure 1004.1.2). Within the restaurant area is a 6-inch (152 mm) privacy wall running the length of the room [80 feet by 0.5 feet = 40 square feet (3.7 m^2)], a fireplace [40 square feet (3.7 m^2)] and a cloak room [60 square feet (5.6 m^2)]. Each of these areas is deducted from the restaurant area, resulting in a net floor area of 4,660 square feet (433 m^2). Since the restaurant intends to have unconcentrated seating that involves loose tables and chairs, the resulting occupant load is 311 persons (4,660 divided by 15). As the definition of "Floor area, net" indicates, certain spaces are to be excluded from the gross floor area to derive the net floor area. The key point in this definition is that the net floor area is to include the actual occupied area and does not include spaces uncharacteristic of that occupancy.

In determining the occupant load of a building with mixed groups, each floor area of a single occupancy must be separately analyzed, such as required by Section 1004.6. The occupant load of the business portion of an office/warehouse building is determined at a rate of one person for each 100 square feet (9 m^2) of office space, whereas the occupant load of the warehouse portion is determined at the rate of one person for each 300 square feet (28 m^2). There may even be different uses within the same room. For example, a restaurant dining room would have seat-

TABLE 1004.1.2
MAXIMUM FLOOR AREA ALLOWANCES PER OCCUPANT

FUNCTION OF SPACE	OCCUPANT LOAD FACTOR[a]
Accessory storage areas, mechanical equipment room	300 gross
Agricultural building	300 gross
Aircraft hangars	500 gross
Airport terminal	
Baggage claim	20 gross
Baggage handling	300 gross
Concourse	100 gross
Waiting areas	15 gross
Assembly	
Gaming floors (keno, slots, etc.)	11 gross
Exhibit Gallery and Museum	30 net
Assembly with fixed seats	See Section 1004.4
Assembly without fixed seats	
Concentrated (chairs only-not fixed)	7 net
Standing space	5 net
Unconcentrated (tables and chairs)	15 net
Bowling centers, allow 5 persons for each lane including 15 feet of runway, and for additional areas	7 net
Business areas	100 gross
Courtrooms—other than fixed seating areas	40 net
Day care	35 net
Dormitories	50 gross
Educational	
Classroom area	20 net
Shops and other vocational room areas	50 net
Exercise rooms	50 gross
H-5 Fabrication and manufacturing areas	200 gross
Industrial areas	
Inpatient treatment areas	240 gross
Outpatient areas	100 gross
Sleeping areas	120 gross
Kitchens, commercial	200 gross
Library	
Reading rooms	50 net
Stack area	100 gross
Mall buildings—covered and open	See Section 402.8.2
Mercantile	
Areas on other floors	60 gross
Basement and grade floor areas	30 gross
Storage, stock, shipping areas	300 gross
Parking garages	200 gross
Residential	200 gross
Skating rinks, swimming pools	
Rink and pool	50 gross
Decks	15 gross
Stages and platforms	15 net
Warehouses	500 gross

For SI: 1 square foot = 0.0929 m^2.

ing but may also have a waiting area with standing room, a take-out window with a queue line or employee areas behind a bar or reception desk.

If a specific type of facility is not found in the table, the occupancy it most closely resembles should be utilized. For example, a training room in a business office may utilize the 20-square-feet (1.86 m^2) net established for educational classroom areas, or a dance or karate studio may use the occupant load for rinks and pools for the studio areas.

Table 1004.1.2 presents a method of determining the absolute base minimum occupant load of a space that the means of egress is to accommodate.

The table occupant loads are based on the stereotypical configuration of spaces. For example, the dorm requirements were written based on dormitories with sleeping rooms with two to four students, a gang bathroom and a meeting/study lounge on each floor. Dormitory buildings that operate like army barracks may have a heavier occupant load, while facilities with groups of rooms with private bathrooms, living and even kitchenette areas may have a lower occupant load. Industrial facilities are based on typical fabricating plants. Warehouses are based on consistent in and out movement of product by employees. Factories with largely mechanized operations or warehouses that contain long-term storage are other examples where discussion with the building official and the application of the exception in Section 1004.1.2 might be considered.

In addition to the table, Section 402 contains the basis for calculating the occupant load of a covered mall building; however, Table 1004.1.2 should be used for determining the occupant load of each anchor store.

1004.2 Increased occupant load. The *occupant load* permitted in any building, or portion thereof, is permitted to be increased from that number established for the occupancies in Table 1004.1.2, provided that all other requirements of the code are met based on such modified number and the *occupant load* does not exceed one occupant per 7 square feet (0.65 m^2) of occupiable floor space. Where required by the *building official*, an *approved aisle*, seating or fixed equipment diagram substantiating any increase in *occupant load*

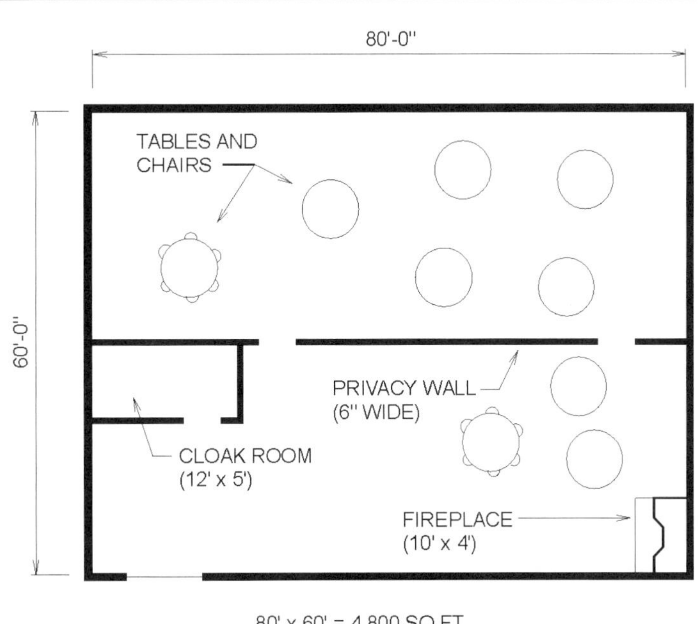

80' x 60' = 4,800 SQ.FT.

PRIVACY WALL:	40 SQ.FT.
FIREPLACE:	40 SQ.FT.
CLOAK ROOM:	60 SQ.FT.
TOTAL:	140 SQ.FT.

(TOTAL AREA WITHIN WALLS) - (EXCLUDED ITEMS) = (NET FLOOR AREA)
4,800 SQ.FT. - 140 SQ.FT. = 4,660 SQ.FT.

(NET FLOOR AREA)/(TABLE 1004.1.2 VALUE) = (OCCUPANT LOAD)
4,660 SQ.FT./15 SQ.FT. PER OCCUPANT = 311 OCCUPANTS

For SI: 1 inch = 25.4 mm, 1 foot = 304.8 mm,
1 square foot = 0.0929 m^2.

Figure 1004.1.2
TYPICAL NET FLOOR AREA OCCUPANT LOAD CALCULATION

shall be submitted. Where required by the *building official*, such diagram shall be posted.

❖ An increased occupant load is permitted above that developed by using Table 1004.1.2, for example, by utilizing the actual occupant load. However, if the occupant load exceeds that which is determined in accordance with Section 1004.1.2, the building official has the authority to require aisle, seating and equipment diagrams to confirm that all occupants have access to an exit, the exits provide sufficient capacity for all occupants and compliance with this section is attained.

The maximum area of 7 square feet (0.65 m^2) per occupant should allow for sufficient occupant movement in actual fire situations. This is not a conflict with the standing space provisions of 5 square feet (0.46 m^2) net in accordance with Table 1004.1.2. Standing space is typically limited to a portion of a larger area, such as the area immediately in front of the bar or the waiting area in a restaurant, while the rest of the dining area would use 15 square feet (1.4 m^2) net per occupant.

1004.3 Posting of occupant load. Every room or space that is an assembly occupancy shall have the *occupant load* of the room or space posted in a conspicuous place, near the main *exit* or *exit access doorway* from the room or space. Posted signs shall be of an approved legible permanent design and shall be maintained by the owner or the owner's authorized agent.

❖ Each room or space used for an assembly occupancy is required to display the approved occupant load. The placard must be posted in a visible location (near the main entrance) (see Commentary Figure 1004.3 for an example of an occupant load limit sign).

The posting is required to provide a means by which to determine that the maximum approved occupant load is not exceeded. This permanent and readily visible sign provides a constant reminder to building personnel and is a reference for building officials during periodic inspections. The posted occupant load could also be an indication that the room was designed for a layout of just tables and chairs, not a layout of chairs only (see Section 302.1).

While the composition and organization of information in the sign are not specified, information must be recorded in a permanent manner. This means that a sign with changeable numbers would not be acceptable.

Figure 1004.3
EXAMPLE OF OCCUPANT LOAD LIMIT SIGN

1004.4 Fixed seating. For areas having *fixed seats* and *aisles*, the *occupant load* shall be determined by the number of *fixed seats* installed therein. The *occupant load* for areas in which *fixed seating* is not installed, such as waiting spaces, shall be determined in accordance with Section 1004.1.2 and added to the number of *fixed seats*.

The *occupant load* of *wheelchair spaces* and the associated companion seat shall be based on one occupant for each *wheelchair space* and one occupant for the associated companion seat provided in accordance with Section 1108.2.3.

For areas having *fixed seating* without dividing arms, the *occupant load* shall be not less than the number of seats based on one person for each 18 inches (457 mm) of seating length.

The *occupant load* of seating booths shall be based on one person for each 24 inches (610 mm) of booth seat length measured at the backrest of the seating booth.

❖ The occupant load in an area with fixed seats is readily determined. In spaces with a combination of fixed and loose seating, the occupant load is determined by a combination of the occupant density number from Table 1004.1.2 and a count of the fixed seats.

For bleachers, booths and other seating facilities without dividing arms, the occupant load is simply based on the number of people that can be accommodated in the length of the seat. Measured at the hips, an average person occupies about 18 inches (457 mm) on a bench. In a booth, additional space is necessary for "elbow room" while eating. In a circular or curved booth or bench, the measurement should be taken just a few inches from the back of the seat, which is where a person's hips would be located (see Commentary Figure 1004.4).

Some assembly spaces may have areas for standing or waiting. For example, some large sports stadiums have "standing room only" areas used for sell-out games. The Globe Theater in England has standing room in an area at the front of the theater. This section is not intended to assign an occupant load to the typical circulation aisles in an assembly space. Occupant load for wheelchair spaces should be based on the number of wheelchairs and companion seats that the space was designed for. As specified in Section 1004.6, if the wheelchair spaces may also be utilized for standing space or removable seating, the occupant load must be determined by the worst-case scenario.

1004.5 Outdoor areas. *Yards*, patios, *courts* and similar outdoor areas accessible to and usable by the building occupants shall be provided with *means of egress* as required by this chapter. The *occupant load* of such outdoor areas shall be assigned by the *building official* in accordance with the anticipated use. Where outdoor areas are to be used by persons in addition to the occupants of the building, and the path of egress travel from the outdoor areas passes through the building, *means of egress* requirements for the building shall be based on the sum of the *occupant loads* of the building plus the outdoor areas.

Exceptions:

1. Outdoor areas used exclusively for service of the building need only have one *means of egress*.
2. Both outdoor areas associated with Group R-3 and individual dwelling units of Group R-2.

❖ This section addresses the means of egress of outdoor areas such as yards, patios and courts. The primary concern is for those outdoor areas, used for functions that may include occupants other than the building occupants or the building occupants alone, where egress from the outdoor area is back through the building to reach the exit discharge. An example is an interior court of an office building where assembly functions are held during normal business hours for persons other than the building occupants. When court occupants must egress from the interior court back through the building, the building's egress system is to be designed for the building occupants, plus the assembly occupants from the interior court. Another example would be an outdoor dining area that exited back through the restaurant.

The occupant load is to be assigned by the building official based on use. It is suggested that the design occupant load be determined in accordance with Section 1004.1.2.

Exception 1 describes conditions where the occupant load is very limited, such as areas where an interior courtyard had strictly plants or mechanical equipment. If the courtyard was open for building occupants, other than maintenance personnel, to use the space, the space must be designed with the occupant loads in Table 1004.1.2. Balconies or patios associated with individual dwelling units, in Exception 2, would typically be used by the occupants of the unit. Means of egress can be back through the building in accordance with Section 1016.2.

1004.6 Multiple occupancies. Where a building contains two or more occupancies, the *means of egress* requirements shall apply to each portion of the building based on the occupancy of that space. Where two or more occupancies utilize portions of the same *means of egress* system, those egress components shall meet the more stringent requirements of all occupancies that are served.

❖ Since the means of egress systems are designed for the specific occupancy of a space, the provisions of this chapter are to be applied based on the actual

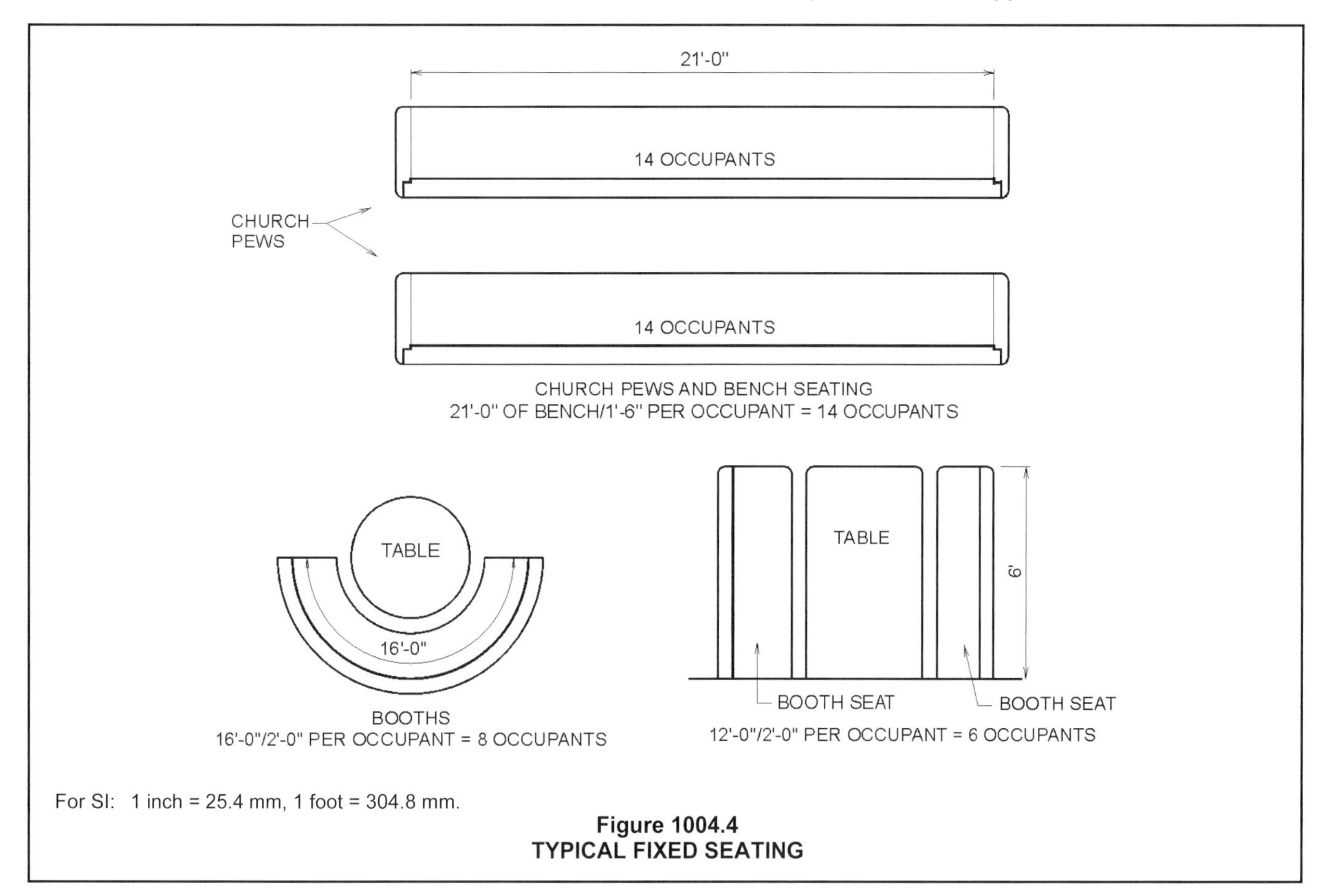

Figure 1004.4
TYPICAL FIXED SEATING

occupancy conditions of the space served.

For example, a hospital is classified as Group I-2 and normally includes the associated administrative or business functions found in the same building. Chapter 3 would permit the entire building to be constructed to the more restrictive provisions for Group I-2; however, each area of the building need only have the means of egress designed in accordance with the actual occupancy conditions, such as Groups I-2 and B. If the corridor serves only the occupants in the business use (i.e., administrative staff), and is not intended to serve as a required means of egress for patients, the corridor need only be 36 or 44 inches (914 or 1118 mm) in width, depending on the occupant load.

Where the corridor is used by both Group I-2 and B occupancies, it must meet the most stringent requirement. For example, if a corridor in the business area is also used for the movement of beds (i.e., exit access from a patient care area), it would need to be a minimum of 96 inches (2438 mm) in clear width.

SECTION 1005 MEANS OF EGRESS SIZING

1005.1 General. All portions of the *means of egress* system shall be sized in accordance with this section.

Exception: *Aisles* and *aisle accessways* in rooms or spaces used for assembly purposes complying with Section 1029.

❖ This section is a charging paragraph for sizing for the means of egress system in a tenant space, floor or building. The exception is based on the understanding that means of egress paths within assembly areas have unique criteria based on the high occupant load and possibility of stepped or sloped aisles.

1005.2 Minimum width based on component. The minimum width, in inches (mm), of any *means of egress* components shall be not less than that specified for such component, elsewhere in this code.

❖ The code requires the utilization of two methods to determine the minimum width of egress components. While this section provides a methodology for determining required widths based on the design occupant load calculated in accordance with Section 1004.1, other sections provide minimum widths of various components. The actual width that is provided is to be the larger of the two widths.

1005.3 Required capacity based on occupant load. The required capacity, in inches (mm), of the *means of egress* for any room, area, space or story shall be not less than that determined in accordance with Sections 1005.3.1 and 1005.3.2.

❖ For this section, the sum of the capacities of the means of egress components that serve each space must equal or exceed the occupant load of that space. For example, the combined width of all of the exit stairways from a floor needs to be considered to determine if the stairways have adequate capacity for everyone to evacuate the building. All elements must meet the minimum width requirements specified in other sections (e.g., Section 1010.1.1 for doors; Sections 1009.3 and 1011.2 for stairs).

This section establishes the necessary width of each egress component on a "per-occupant" basis. Means of egress components are separated between "stairs" and "other;" "other" being doors, doorways, corridors, ramps, aisles, etc.

The traditional unit of measurement of egress capacity was based on a "unit exit width" that was to simulate the body ellipse with a basic dimensional width of 22 inches (559 mm)—approximately the shoulder width of an average adult male. This unit exit width was combined with assumed egress movement (such as single file or staggered file) to result in an egress capacity per unit exit width for various occupancies. This assumption simplifies the dynamic egress process since contemporary studies have indicated that people do not egress in such precise and predictable movements. As traditionally used in the codes, the method of determining capacity per unit of clear width implies a higher level of accuracy than can realistically be achieved. The resulting factors preserve the features of the past practices that can be documented, while providing a more straightforward method of determining egress capacity.

1005.3.1 Stairways. The capacity, in inches, of *means of egress stairways* shall be calculated by multiplying the *occupant load* served by such *stairways* by a means of egress capacity factor of 0.3 inch (7.6 mm) per occupant. Where *stairways* serve more than one story, only the occupant load of each story considered individually shall be used in calculating the required capacity of the *stairways* serving that story.

Exceptions:

1. For other than Group H and I-2 occupancies, the capacity, in inches, of *means of egress stairways* shall be calculated by multiplying the *occupant load* served by such *stairways* by a means of egress capacity factor of 0.2 inch (5.1 mm) per occupant in buildings equipped throughout with an automatic sprinkler system installed in accordance with Section 903.3.1.1 or 903.3.1.2 and an *emergency voice/alarm communication* system in accordance with Section 907.5.2.2.
2. Facilities with *smoke-protected assembly seating* shall be permitted to use the capacity factors in Table 1029.6.2 indicated for stepped aisles for *exit access* or *exit stairways* where the entire path for *means of egress* from the seating to the *exit discharge* is provided with a smoke control system complying with Section 909.
3. Facilities with outdoor *smoke-protected assembly seating* shall be permitted to the capacity factors in Section 1029.6.3 indicated for stepped aisles for *exit access* or *exit stairways* where the entire path for

means of egress from the seating to the *exit discharge* is open to the outdoors.

❖ The capacity factor for stairways is larger than "other egress components" because of the slowdown of travel to negotiate the steps. When the required occupant capacity of an egress component is determined, multiplication by the appropriate factor results in the required clear width of the component in inches, based on capacity. Similarly, if the clear width of a component is known, division by the appropriate factor results in the permitted capacity of that component.

Per Exception 1, other than in Group H or I-2, if the building is sprinklered and has an emergency voice/alarm communication system, the capacity factor for stairways is permitted to be reduced to 0.2 inches (5.1 mm) per occupant.

Smoke-protected seating is permitted to use a lower capacity number to determine the width of the egress components within the seating bowl. When designing the stepped aisles within the seating bowl, the provisions for Section 1029 are applicable. When designing components outside of the seating bowl, Section 1005.3 is applicable. Per Exceptions 2 and 3, if both the seating bowl and the entire route out of the building is smoke protected, the entire route can use the lower capacity numbers specified in Section 1029. For example, if an outdoor stadium has an enclosed concourse that spectators use to enter and exit the stadium seating, the smoke-protected capacity numbers can be used to design the stepped aisles in the stadium, but the higher numbers for stairways must be used to size the stairways in the concourse. Only if the concourse is also open to the outside air can the smoke-protected seating capacity numbers be used for the entire means of egress route. See the definition of "Smoke-protected assembly seating." See also Sections 1029.6.2 and 1029.6.3.

The following illustrate typical calculations for stairways from a nonsprinklered, two-story, two-exit office building:

1. Determine the minimum required stairway width with a second-floor occupant load of 350:
   - 350 occupants divided by 0.3 inch = 105 inches (2667 mm) minimum;
   - 105 inches divided by two stairways is $52^1/_2$ inches (1334 mm) minimum per stairway; or
   - Section 1009.1 prescribes that the width of an interior stairway cannot be less than 44 inches (1118 mm).

The capacity criteria are more restrictive and, therefore, the minimum required width for each stairway is $52^1/_2$ inches (1334 mm).

2. Determine the minimum required stairway width with a second-floor occupant load of 90:
   - 90 occupants divided by 0.3 inches (7.62 mm) = 27 inches (686 mm) minimum;
   - 27 inches (686 mm) divided by two stairways is $13^1/_2$ inches (343 mm); or
   - Section 1011.1 prescribes that the width of an interior stairway cannot be less than 44 inches (1118 mm). Note that the stair width reduction in Section 1011.2, Exception 1, is applicable only when the entire occupant load of a story is less than 50.

The minimum clear width requirements are more restrictive and, therefore, the minimum required width for each stairway is 44 inches (1118 mm).

The maximum capacity of a 44-inch (1118 mm) stairway is 44 inches divided by 0.3 inches (7.62 mm) per occupant = 146 occupants. Therefore, a floor level with two exit stairways could have 292 occupants before the capacity would control the stairway egress width.

Using the exception for sprinklered buildings, a 44-inch (1118 mm) stairway divided by 0.2 inches (5.08 mm) per occupant = 220 occupants. Therefore, a floor level with two exit stairways could have 440 occupants before the capacity would control the stairway egress width.

Keep in mind that accessible means of egress stairways in nonsprinklered buildings require a minimum clear width of 48 inches (1219 mm) between handrails.

1005.3.2 Other egress components. The capacity, in inches, of *means of egress* components other than *stairways* shall be calculated by multiplying the *occupant load* served by such component by a means of egress capacity factor of 0.2 inch (5.1 mm) per occupant.

Exceptions:

1. For other than Group H and I-2 occupancies, the capacity, in inches, of *means of egress* components other than *stairways* shall be calculated by multiplying the *occupant load* served by such component by a means of egress capacity factor of 0.15 inch (3.8 mm) per occupant in buildings equipped throughout with an automatic sprinkler system installed in accordance with Section 903.3.1.1 or 903.3.1.2 and an *emergency voice/alarm communication* system in accordance with Section 907.5.2.2.
2. Facilities with *smoke-protected assembly seating* shall be permitted to use the capacity factors in Table 1029.6.2 indicated for level or ramped *aisles* for *means of egress* components other than *stairways* where the entire path for *means of egress* from the seating to the *exit discharge* is provided with a smoke control system complying with Section 909.

3. Facilities with outdoor *smoke-protected assembly seating* shall be permitted to the capacity factors in Section 1029.6.3 indicated for level or ramped *aisles* for *means of egress* components other than *stairways* where the entire path for *means of egress* from the seating to the *exit discharge* is open to the outdoors.

❖ The capacity factor for "other egress components" (e.g., doors, gates, corridors, aisles, ramps) is less than stairways because of the slowdown of travel to negotiate the steps. When the required occupant capacity of an egress component is determined, multiplication by the appropriate factor results in the required clear width of the component in inches, based on capacity. Similarly, if the clear width of a component is known, division by the appropriate factor results in the permitted capacity of that component.

Per Exception 1, other than in Group H or I-2, if the building is sprinklered and has an emergency voice/alarm communication system, the capacity factor for doors, corridors, aisles, etc., is permitted to be reduced to 0.15 inches (3.8 mm) per occupant.

Smoke-protected seating is permitted to use a lower capacity number to determine the width of the egress components. When designing the ramped or level aisles and aisle accessways within the seating bowl, the provisions for Section 1029 are applicable. When designing components outside of the seating bowl, Section 1005.3 is applicable. Per Exceptions 2 and 3, if both the seating bowl and the entire route out of the building is smoke protected, the entire route can use the lower capacity numbers. For example, if an outdoor stadium has an enclosed concourse that spectators use to enter and exit the stadium seating, the smoke-protected capacity numbers can be used to design the ramped or level aisles in the stadium, but the higher numbers for other egress components must be used to size the corridors or ramps in the concourse. Only if the concourse is also open to the outside air can the smoke-protected seating capacity numbers be used for the entire means of egress route. See the definition of "Smoke-protected assembly seating." See also Sections 1029.6.2 and 1029.6.3.

For example, two exit access doorways from a room with an occupant load of 300 would each have a required capacity of not less than 150. Based on the minimum required clear door width [32-inch (813 mm) clear width per door divided by 0.2 inch (5.08 mm) per occupant = 160 occupants], two 32-inch (813 mm) clear width doors would meet both the minimum clear width (Section 1010.1.1) and the capacity requirements. Two exits from a space with an occupant load of 450 would each have a required capacity of not less than 225, necessitating more doors or larger door leaves.

Using the exception, the door capacity would increase [32-inch (813 mm) clear width per door divided by 0.15 inch (3.08 mm) per occupant = 213 occupants].

1005.4 Continuity. The minimum width or required capacity of the *means of egress* required from any story of a building shall not be reduced along the path of egress travel until arrival at the public way.

❖ The requirement that both the minimum widths and required capacity from any floor are to be provided all the way along the exit to the termination, typically down the stairway to the exterior exit door at the level of exit discharge, results in an egress width that is adequate for the exit discharge.

The sum total capacity of the exits that serve a floor is not to be less than the occupant load of the floor as determined by Section 1004.1. If an exit, such as a stairway, also serves a second floor and the required capacity of the exit serving the occupants of the second floor is greater than the first floor, the greater capacity would govern the egress components that the occupants of the floors share. For example, if an exit stairway serves two floors, with occupant loads of 300 on the lower floor and 500 on the upper floor, assuming that two stairways serve each floor, the two stairways would be designed for a capacity of 250 people each, using the upper-floor occupant load of 500 as the basis of determination. Note that the doors to the stairways on the lower floor would be designed for a capacity of 150 and the doors to the stairways on the upper floor would be designed for a capacity of 250. Reversing these two floors would result in the portion of the stairways that serves the upper floor to be designed for a capacity of 150 and the stairways that serve the lower floor to be designed for 250. Requiring the egress component to be designed for the largest tributary occupant load accommodates the worst-case situation.

Also note that the capacity of the exits is based on the occupant load of one floor. The occupant loads are not combined with other floors for the exit design. It is assumed that the peak demand or flow of occupants from more than one floor level at a common point in the means of egress will not occur simultaneously, except as provided for in Sections 1005.6 (Egress convergence) and 1004.1.1.2 (Adjacent levels for mezzanines).

1005.5 Distribution of minimum width and required capacity. Where more than one *exit*, or access to more than one *exit*, is required, the *means of egress* shall be configured such that the loss of any one *exit*, or access to one *exit*, shall not reduce the available capacity or width to less than 50 percent of the required capacity or width.

❖ It is critical that the distribution of both egress capacity and minimum width are examined. Where multiple means of egress are required, the loss of any one path cannot reduce the available capacity or width to less than 50 percent. The 50-percent minimum of the required egress capacity and width results in a fairly uniform distribution of egress paths. This requirement does not, however, require that the capacities be equally distributed when *more* than two means of egress are provided. An egress design with a dramatic imbalance of egress component capacities relative to occupant load distribution should be reviewed closely to avoid a needless delay in egressing a story or area. The balancing of the means of egress components, in accordance with the distribution of the occupant load, is reasonable and, in some cases, necessary for facilities having mixed occupancies with dramatically different occupant loads.

1005.6 Egress convergence. Where the *means of egress* from stories above and below converge at an intermediate level, the capacity of the *means of egress* from the point of convergence shall be not less than the largest minimum width or the sum of the required capacities for the *stairways* or *ramps* serving the two adjacent stories, whichever is larger.

❖ Convergence of occupants can occur whenever the occupants of an upper floor travel down and occupants of a lower floor travel up and meet at a common, intermediate egress component on the route to the exit discharge. The intermediate component may or may not be another occupiable floor and, most often, is an exit door [see Commentary Figures 1005.6(1) and 1005.6(2)].

The entire premise of egress convergence is based on the assumption of simultaneous notification (i.e., all occupants of all floors begin moving toward the exits at the same time). As illustrated in Commentary Figure 1005.6(3), the occupants of the first floor will have exited the building by the time most of the occupants of the second floor have reached the exit discharge door. However, as illustrated in Commentary Figure 1005.6(1), the occupants of a basement will reach the discharge door simultaneously with the second-floor occupants, thereby creating a bottleneck and the need for sizing the affected component for a larger combined occupant load.

An egress convergence situation can also be created when an intermediate floor level is not present, as illustrated in Commentary Figure 1005.6(2). Again, under the assumption of simultaneous notification, occupants of both floors would reach the exit discharge door at approximately the same time, invoking the requirements for increased egress capacity.

1005.7 Encroachment. Encroachments into the required *means of egress* width shall be in accordance with the provisions of this section.

❖ This section addresses maximum encroachment into the required width along the path of travel for means of egress. Types of encroachment are door leaves, door hardware, handrails, trim and protruding objects. These requirements are referenced for aisles (Section 1018.1), corridors (Section 1020.3), exit passageways (Section 1024.2) and exit courts (Section 1028.4.1). Along stairways, handrail projections are permitted per Section 1014.8.

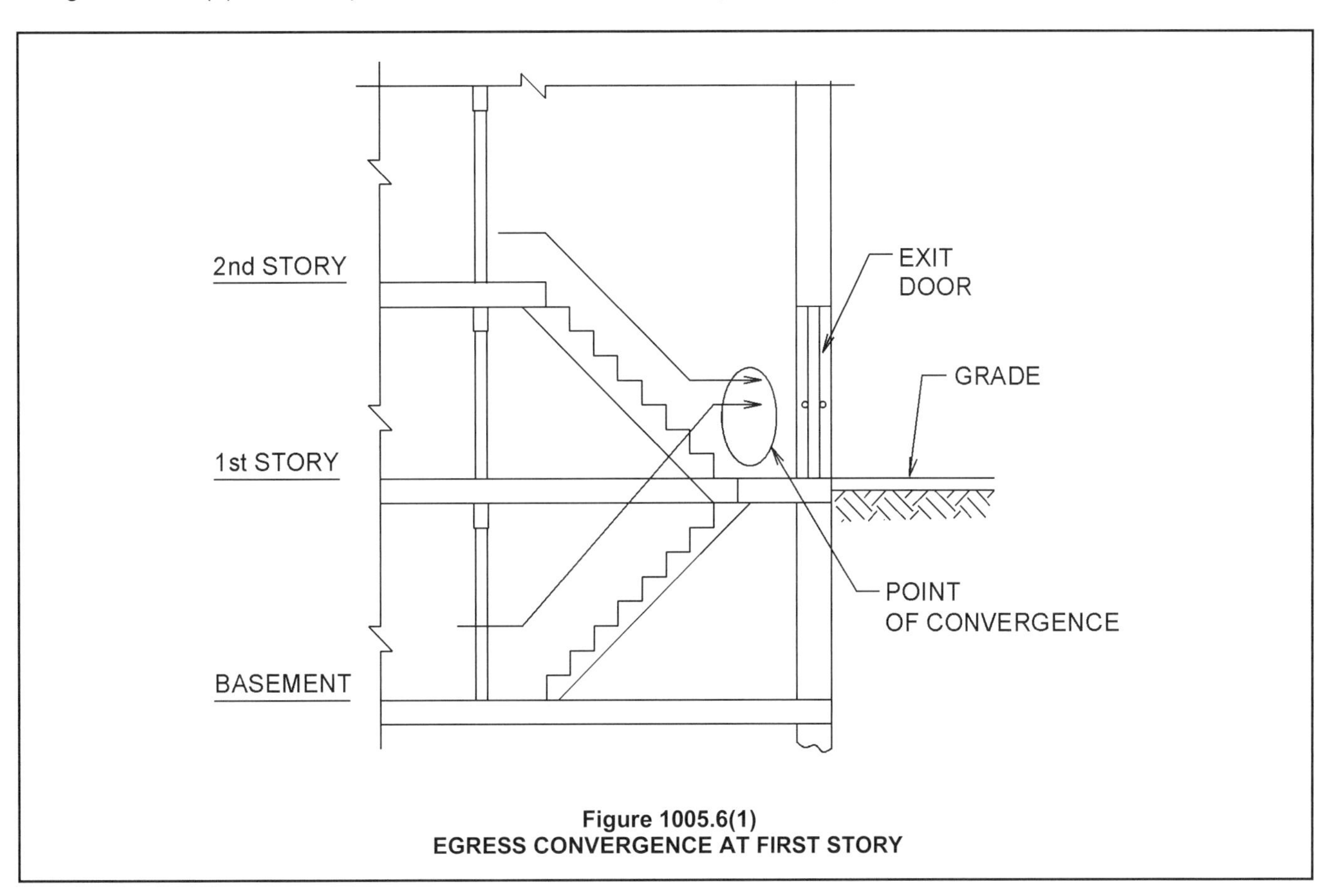

Figure 1005.6(1)
EGRESS CONVERGENCE AT FIRST STORY

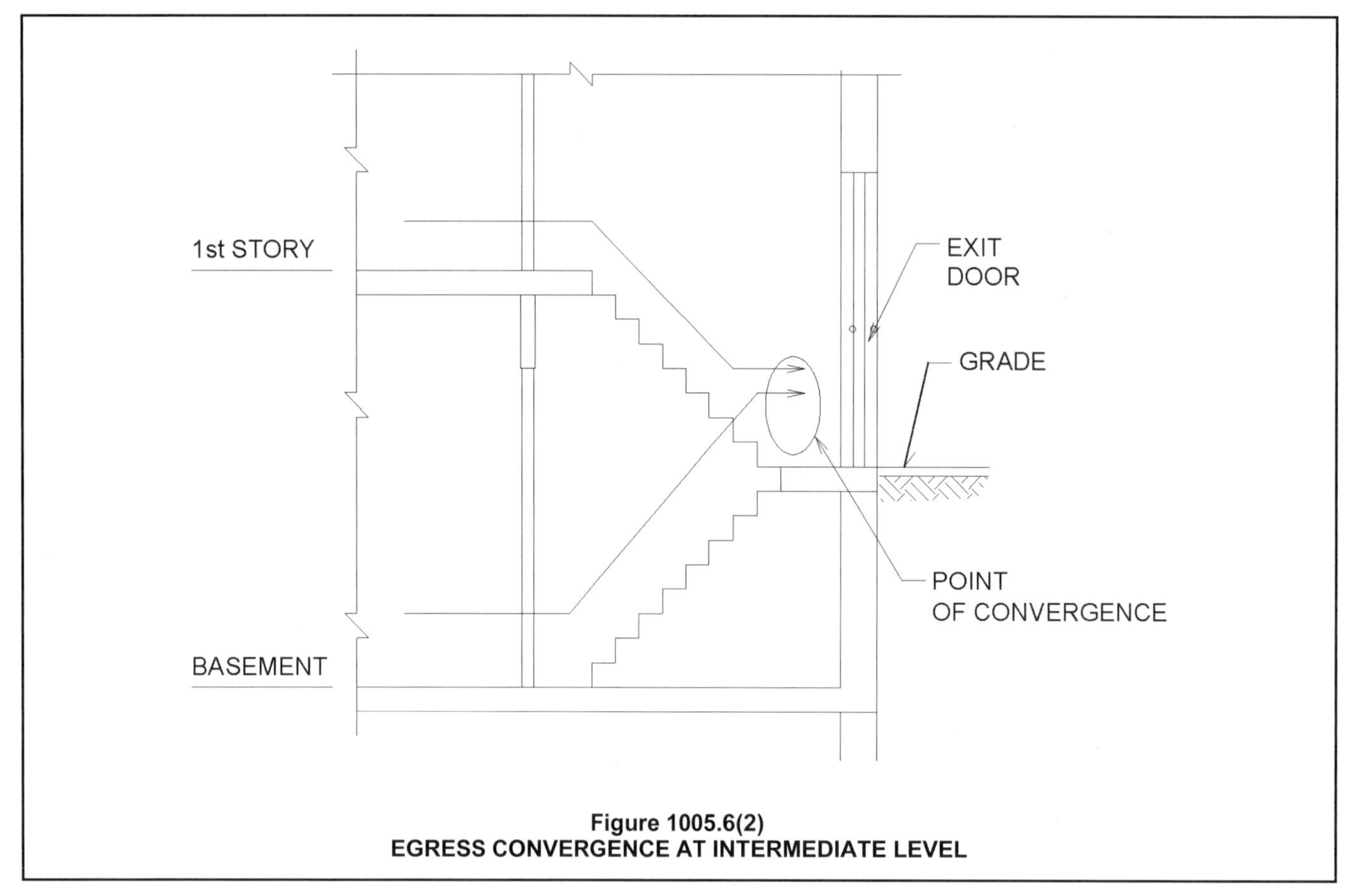

Figure 1005.6(2)
EGRESS CONVERGENCE AT INTERMEDIATE LEVEL

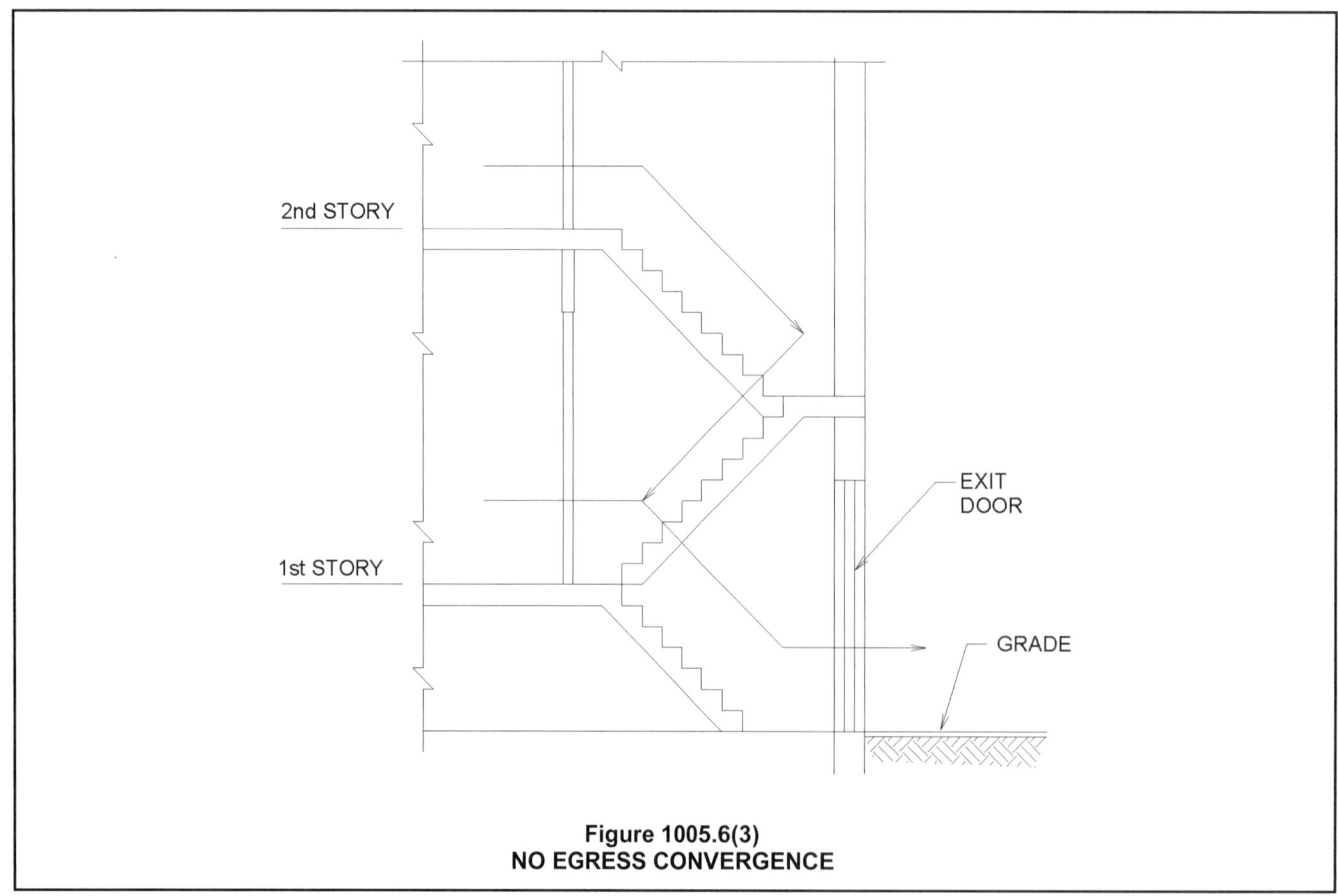

Figure 1005.6(3)
NO EGRESS CONVERGENCE

1005.7.1 Doors. Doors, when fully opened, shall not reduce the required width by more than 7 inches (178 mm). Doors in any position shall not reduce the required width by more than one-half.

Exceptions:

1. Surface-mounted latch release hardware shall be exempt from inclusion in the 7-inch maximum (178 mm) encroachment where both of the following conditions exist:
   1.1. The hardware is mounted to the side of the door facing away from the adjacent wall where the door is in the open position.
   1.2. The hardware is mounted not less than 34 inches (865 mm) nor more than 48 inches (1219 mm) above the finished floor.
2. The restrictions on door swing shall not apply to doors within individual *dwelling units* and *sleeping units* of Group R-2 occupancies and *dwelling units* of Group R-3 occupancies.

❖ Projections or restrictions in the required width can impede and restrict occupant travel, causing egress to occur less efficiently than expected. The swinging of a door, such as from a room into a corridor, and any handrails along the route are permitted projections.

Historically, this section has looked at doors on one wall at a time. Doors located across the hall from one another are not considered additive when considering protrusion limits. Doors would not typically be opened to the full extent at exactly the same moment, nor can they remain open at 90 degrees and totally blocking the hall because of the maximum limitation of 7 inches (178 mm) when fully open (typically approaching 180 degrees). Regarding door encroachment, there are two tests. The arc created by the door's outside edge cannot project into more than one-half of the required corridor width. When opened to its fullest extent, the door cannot project more than 7 inches (178 mm) into the required width, which is the dimension of the leaf thickness excluding the hardware as shown in Commentary Figure 1005.7.1. Door hardware encroachment is addressed separately in Exception 1. These projections are permitted because they are considered to be temporary and do not significantly impede the flow. Occupants will compensate for the projection by a reduction in the natural cushion they retain between themselves and a boundary, known as the edge effect.

Per Exception 2, the door swing restrictions do not apply within dwelling units since the occupant load is very low. Based on the intent of this section, other situations that could be approved by the official having jurisdiction would be situations where the opening door would not block the egress, such as the door at the end of a corridor, or the room was not typically occupied, such as a janitor's closet.

The provision in Exception 1 indicates that hardware facing the corridor when the door is fully open need not be considered when determining the allowable door encroachment into a corridor of 7 inches (178 mm) maximum. The allowance is applicable provided the hardware is mounted within a height range of 34 inches to 48 inches (865 to 1220 mm), which is consistent with the range for means of egress door hardware height as established in Section 1010.1.9.2. Where hardware extends across a door, such as panic hardware, the 4-inch (102 mm) projection in the door opening is addressed in Section 1010.1.1.1.

1005.7.2 Other projections. *Handrail* projections shall be in accordance with the provisions of Section 1014.8. Other nonstructural projections such as trim and similar decorative

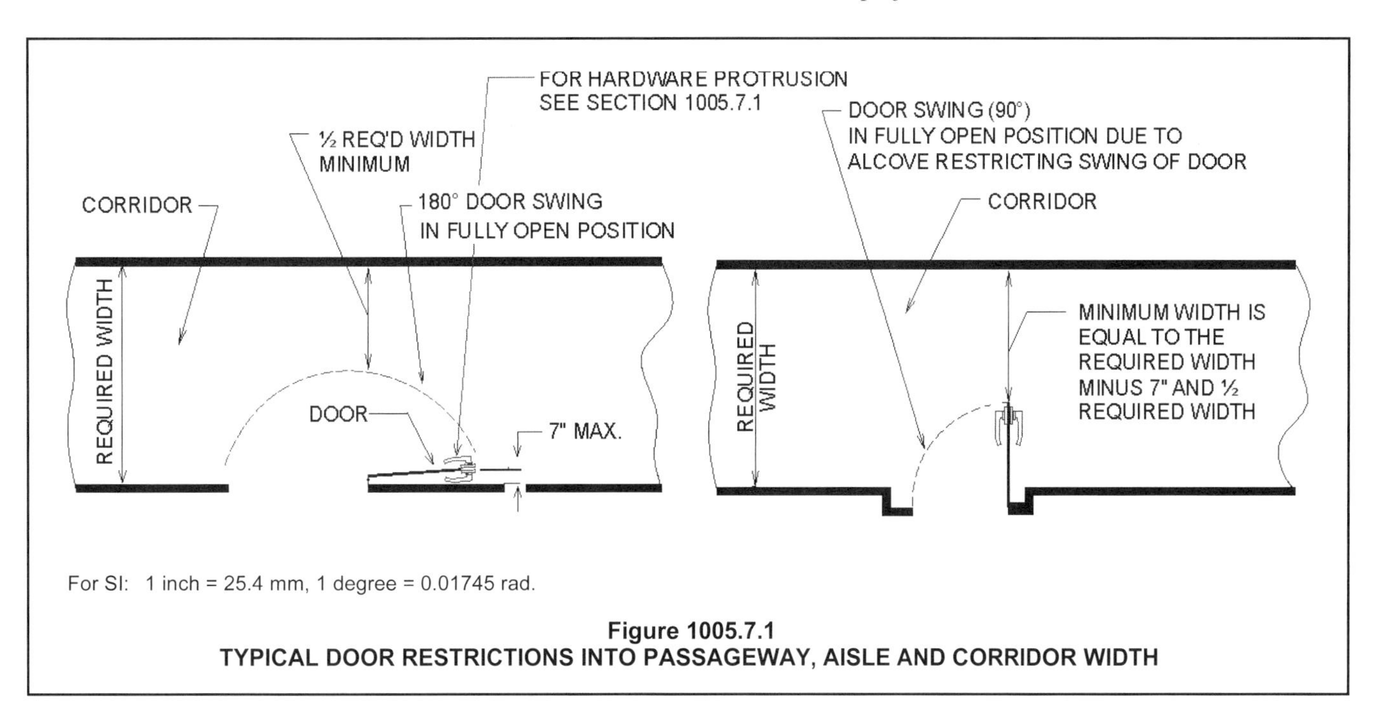

For SI: 1 inch = 25.4 mm, 1 degree = 0.01745 rad.

Figure 1005.7.1
TYPICAL DOOR RESTRICTIONS INTO PASSAGEWAY, AISLE AND CORRIDOR WIDTH

features shall be permitted to project into the required width not more than $1^1/_2$ inches (38 mm) on each side.

Exception: Projections are permitted in corridors within Group I-2 Condition 1 in accordance with Section 407.4.3.

❖ Handrails are not required along corridors, level aisles, exit passageways and exit corridors; however, if provided, Section 1014.8 would be applicable. Handrails are sometimes provided along the hallways in hospitals or nursing homes to aid the residents. Bumper guards along the walls are not handrails.

Items such as baseboards, chair rails, pilasters, etc., are limited to protruding over the *required* width of the corridor a maximum of $1^1/_2$ inches (38 mm); however, once again, Section 1003.3.3 would be applicable when the corridor was wider than required.

The exception is in recognition of a situation unique to nursing homes (Group I-2, Condition 1). Section 407.4.3 includes allowances for furniture in corridors to address the need of patients needing a place to sit to rest as well as the new style of design that emphasizes the residential aspects of the environment.

1005.7.3 Protruding objects. Protruding objects shall comply with the applicable requirements of Section 1003.3.

❖ This section is a reminder that protruding objects are applicable when looking at encroachments into a confined path of travel. The difference, however, is that door and other projections are applied to the required minimum width, while protruding object provisions apply to paths of travel even when wider than required.

SECTION 1006 NUMBER OF EXITS AND EXIT ACCESS DOORWAYS

1006.1 General. The number of *exits* or *exit access doorways* required within the *means of egress* system shall comply with the provisions of Section 1006.2 for spaces, including *mezzanines*, and Section 1006.3 for *stories*.

❖ The criteria in this section to determine the number of ways to leave rooms or spaces (including mezzanines) and stories are based on an empirical judgment of the associated risks.

1006.2 Egress from spaces. Rooms, areas or spaces, including *mezzanines*, within a *story* or *basement* shall be provided with the number of *exits* or access to *exits* in accordance with this section.

❖ This section dictates the minimum number of paths of travel an occupant is to have available to avoid a fire incident in the occupied room or space. While providing multiple egress doorways from every room is unrealistic, a point does exist where alternative egress paths must be provided based on the number of occupants at risk, the distance any one occupant must travel to reach a doorway and the relative hazards associated with the occupancy of the space. Generally, the number of egress doorways required from any room or space coincides with the occupant load threshold criteria set forth for the minimum number of exits required from a story (see Section 1006.3).

1006.2.1 Egress based on occupant load and common path of egress travel distance. Two *exits* or *exit access doorways* from any space shall be provided where the design *occupant load* or the *common path of egress travel* distance exceeds the values listed in Table 1006.2.1.

Exceptions:

1. In Group R-2 and R-3 occupancies, one *means of egress* is permitted within and from individual *dwelling units* with a maximum *occupant load* of 20 where the *dwelling unit* is equipped throughout with an *automatic sprinkler* system in accordance with Section 903.3.1.1 or 903.3.1.2 and the *common path of egress travel* does not exceed 125 feet (38 100 mm).
2. *Care suites* in Group I-2 occupancies complying with Section 407.4.

❖ This section dictates the minimum number of paths of travel an occupant is to have available to avoid a fire incident in the occupied room or space. While providing multiple egress doorways from every room is unrealistic, a point does exist where alternative egress paths must be provided based on the number of occupants at risk, the distance any one occupant must travel to reach a doorway and the relative hazards associated with the occupancy of the space. Generally, the number of egress doorways required from any room or space coincides with the occupant load threshold criteria set forth for determining the minimum number of exits required from a story (see Section 1006.3).

The limiting criteria in Table 1006.2.1 for rooms or spaces permitted to have a single exit access doorway are based on an empirical judgment of the associated risks.

If the occupants of a room are required to egress through another room, as permitted in Sections 1004.1.1.1 and 1016.2, the rooms are to be combined to determine if multiple doorways are required from the combined rooms. For example, if a suite of offices shares a common reception area, the entire suite with the reception area must meet both the occupant load and the travel distance criteria. The same logic would hold true for a space with a mezzanine (see Section 1004.1.1.2).

It should be noted that where two doorways are required, the remoteness requirement of Section 1007.1 is applicable.

The common path of travel is the distance measured from the most remote point in a space to the point in the exit path where the occupant has access to two required exits in separate directions. The distance limitations are applicable to all paths of travel that lead out of a space or building where two exits are required. An illustration of this distance is found in

Commentary Figure 1006.2.1. The illustration reflects two examples of a common path of travel where the occupants at points A and B are able to travel in only one direction before they reach a point at which they have a choice of two paths of travel to the required exits from the building. Note that from point A, the occupants have two available paths, but these merge to form a single path out of the space. This is also considered a common path of travel. The common path of travel is considered part of the overall travel distance limitations in Section 1017.2.

While a Group R-3 occupancy is typically a single-exit space, it is included in the table to address mixed-use buildings.

Exception 1 allows for individual dwelling units to be considered a space with one means of egress, provided they meet the same common path of travel requirements in Table 1006.2.1. If the building is sprinklered with an NFPA 13 or 13R system, the occupant load for the unit can be 20 (4,000-square-foot apartment/200 square feet per occupant = 20 occupants), whereas if the unit complies with the table, the occupant load is limited to 10 people per unit (2,000-square-foot apartment/200 square feet per occupant = 10 occupants).

Exception 2 allows for hospital patient rooms and care suites to egress in accordance with the specific criteria in Section 407.4, including the common path of egress travel provisions (Table 1006.2.1, Note d). Other areas in Group I-2 occupancies are addressed in Table 1006.2.1.

TABLE 1006.2.1. See page 10-24.

❖ The table represents an empirical judgment of the risks associated with a single means of egress from a room or space based on the occupant load in the room, the travel distance to the exit access door and the inherent risks associated with the occupancy (such as occupant mobility, occupant familiarity with the building, occupant response and the fire growth rate). The number 49 is for consistency with other occupant load thresholds, such as panic hardware (see Section 1010.1.10).

Since the occupants of Groups I and R may be sleeping and, therefore, not able to detect a fire in its early stages without staff supervision or room detectors, the number of occupants in a single egress room or space is limited to 10. See the exceptions to Section 1006.2.1 for Group R-2 and R-3 individual units and Group I-2 patient rooms and care suites.

Because of the potential for rapidly developing hazardous conditions, the single egress condition in Groups H-1, H-2 and H-3 is limited to a maximum of three persons. Because the materials contained in Groups H-4 and H-5 do not represent the same fire hazard potential as those found in Groups H-1, H-2 and H-3, the occupant load for spaces with one means of egress is increased.

Because of the reduced occupant density in Group S and the occupants' normal familiarity with the building, the single egress condition is permitted with an occupant load of 29.

In nonsprinklered business, storage and utility buildings, the length of the common path of egress travel is greater for single-exit spaces when the occupant load for that space is 30 or less. Business, factories and nontransient residential buildings get an increase in the common path of travel in recognition of the additional fire safety offered by a fully sprinklered building. While a Group R-4 is required to be sprinklered, the additional travel distance is not granted where they choose to use an NFPA 13D system as permitted in Section 903.3.1.3. Common path of egress travel does not apply to stories or buildings with one exit. The definition for "Common path" indicates the provisions are only applicable when access

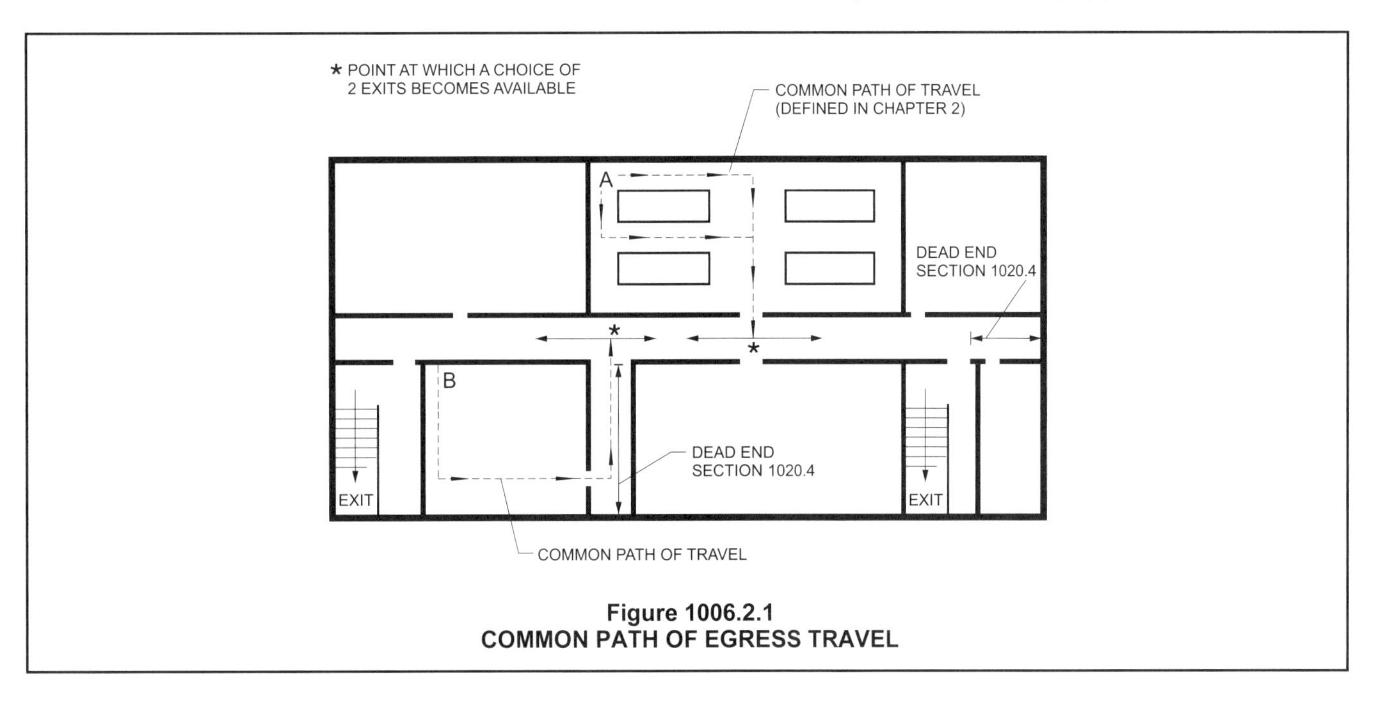

Figure 1006.2.1
COMMON PATH OF EGRESS TRAVEL

to two or more exits is required. See Section 1021.2 for travel limitations for single-exit one-, two- and three-story buildings and Section 1015.1 for spaces with one exit.

Note a indicates if the travel distance increase is based on an NFPA 13 or NFPA 13R sprinkler system being provided throughout the building, whichever is applicable to that occupancy. Where a Group R-4 occupancy can use an NFPA 13D system, the travel distance is indicated in the "without sprinkler system" column. The table does not currently provide information for a Group R-3 occupancy with an NFPA 13D system. Note b is a general reminder for special requirements for sprinklers in Group H.

The reference in Note c to Section 1029.8 is to allow for the unique common path of travel requirements in spaces with assembly seating, such as in a lecture room or sports facility.

Note d is a reference to the common path of travel provisions specific to Group I-2 care rooms and suites that are specifically addressed in Section 407.4.

As indicated in Note e, while Group R-3 dwellings are typically only required to have one exit (see 1006.2.1 Exception 1 and Section 1006.3.2, Exception 5), there can be a situation where a Group R-3 unit is included in a mixed-use building. In these situations, the travel distance limitations for common path of travel in the table are applicable.

Note f is an allowance for the common path of travel in an open parking garage to be 100 feet (30 480 mm) when the occupant load is greater than 30 and there is no sprinkler system provided. This is in recognition of the minimal possibility of smoke accumulation due to the openness requirements and the low fuel loads for open parking garages.

1006.2.1.1 Three or more exits or exit access doorways. Three *exits* or *exit access doorways* shall be provided from any space with an occupant load of 501 to 1,000. Four *exits* or *exit access doorways* shall be provided from any space with an occupant load greater than 1,000.

❖ Large facilities with high occupant loads are required to have more than two exits leading from each story. This is so that at least one exit will be available in case of a fire emergency and to increase the likelihood that a large number of occupants can be accommodated by the remaining exits when one exit is not available. Section 1005.5 specifies that the loss of one exit must not reduce the available exit capacity by more than 50 percent. This is reiterated in Sections 1029.2 and 1029.3 for spaces with assembly seating with more than 300 occupants. Exits should be separated in accordance with Section 1007.1.2. While an equal distribution of exit capacity among all the exits is not required, a proper design would con-

TABLE 1006.2.1
SPACES WITH ONE EXIT OR EXIT ACCESS DOORWAY

OCCUPANCY	MAXIMUM OCCUPANT LOAD OF SPACE	MAXIMUM COMMON PATH OF EGRESS TRAVEL DISTANCE (feet)		
		Without Sprinkler System (feet)		With Sprinkler System (feet)
		Occupant Load		
		OL ≤ 30	OL > 30	
A[c], E, M	49	75	75	75[a]
B	49	100	75	100[a]
F	49	75	75	100[a]
H-1, H-2, H-3	3	NP	NP	25[b]
H-4, H-5	10	NP	NP	75[b]
I-1, I-2[d], I-4	10	NP	NP	75[a]
I-3	10	NP	NP	100[a]
R-1	10	NP	NP	75[a]
R-2	10	NP	NP	125[a]
R-3[e]	10	NP	NP	125[a]
R-4[e]	10	75	75	125[a]
S[f]	29	100	75	100[a]
U	49	100	75	75[a]

For SI: 1 foot = 304.8 mm.

NP = Not Permitted.

a. Buildings equipped throughout with an *automatic sprinkler system* in accordance with Section 903.3.1.1 or 903.3.1.2. See Section 903 for occupancies where *automatic sprinkler systems* are permitted in accordance with Section 903.3.1.2.

b. Group H occupancies equipped throughout with an *automatic sprinkler system* in accordance with Section 903.2.5.

c. For a room or space used for assembly purposes having *fixed seating*, see Section 1029.8.

d. For the travel distance limitations in Group I-2, see Section 407.4.

e. The length of *common path of egress travel* distance in a Group R-3 occupancy located in a mixed occupancy building or within a Group R-3 or R-4 *congregate living facility*.

f. The length of *common path of egress travel* distance in a Group S-2 *open parking garage* shall be not more than 100 feet.

sider occupant load distribution as well as reasonable capacity distribution so as to avoid a severe dependence on one exit or bottlenecks in anticipated high-use areas.

1006.2.2 Egress based on use. The numbers of *exits* or access to *exits* shall be provided in the uses described in Sections 1006.2.2.1 through 1006.2.2.5.

❖ Five types of spaces, because of their levels of hazard, have egress requirements based on use rather than occupant load and travel distance.

1006.2.2.1 Boiler, incinerator and furnace rooms. Two *exit access doorways* are required in boiler, incinerator and furnace rooms where the area is over 500 square feet (46 m^2) and any fuel-fired equipment exceeds 400,000 British thermal units (Btu) (422 000 KJ) input capacity. Where two *exit access doorways* are required, one is permitted to be a fixed ladder or an *alternating tread device. Exit access doorways* shall be separated by a horizontal distance equal to one-half the length of the maximum overall diagonal dimension of the room.

❖ This section requires two exit access doorways for the specified mechanical equipment spaces because of the level of hazards in this type of space. A fixed ladder or an alternating tread device is permitted for service personnel to egress where two doorways are required. The remoteness of the exit access doorways specified in this section provides two paths of travel to exit the room so that if one doorway is not available, the alternate path can be used.

1006.2.2.2 Refrigeration machinery rooms. Machinery rooms larger than 1,000 square feet (93 m2) shall have not less than two *exits* or *exit access doorways*. Where two *exit access doorways* are required, one such doorway is permitted to be served by a fixed ladder or an *alternating tread device. Exit access doorways* shall be separated by a horizontal distance equal to one-half the maximum horizontal dimension of the room.

All portions of machinery rooms shall be within 150 feet (45 720 mm) of an *exit* or *exit access doorway*. An increase in *exit access* travel distance is permitted in accordance with Section 1017.1.

Doors shall swing in the direction of egress travel, regardless of the *occupant load* served. Doors shall be tight fitting and self-closing.

❖ The reasons for these requirements are the same as for Section 1006.2.2.1. Travel distance is to be limited in accordance with Section 1017.2. For example, the travel distance limit for a large refrigeration machinery room classified as Group F-1 that has a sprinkler system throughout the entire building in accordance with NFPA 13 would be 250 feet (76 200 mm) based on Table 1017.2. The 150-foot (45 720 mm) maximum distance to an exit or exit access doorway that is specified in this section would not apply in this example. The 150-foot (45 720 mm) travel distance is intended to be applied where a sprinkler system is not installed and to shorten the time that occupants would be exposed to the hazards within the machinery room.

1006.2.2.3 Refrigerated rooms or spaces. Rooms or spaces having a floor area larger than 1,000 square feet (93 m^2), containing a refrigerant evaporator and maintained at a temperature below 68°F (20°C), shall have access to not less than two *exits* or *exit access doorways*.

Exit access travel distance shall be determined as specified in Section 1017.1, but all portions of a refrigerated room or space shall be within 150 feet (45 720 mm) of an *exit* or *exit access doorway* where such rooms are not protected by an approved *automatic sprinkler system*. Egress is allowed through adjoining refrigerated rooms or spaces.

> **Exception:** Where using refrigerants in quantities limited to the amounts based on the volume set forth in the *International Mechanical Code*.

❖ Refrigeration rooms also have a higher hazard level. The exception is intended to apply if Chapter 11 of the *International Mechanical Code*® (IMC®) does not require a separate refrigeration machinery room due to the small amount of refrigerant used (see the commentary to Section 1104 of the IMC for further explanation of the machinery room requirements).

1006.2.2.4 Day care means of egress. Day care facilities, rooms or spaces where care is provided for more than 10 children that are $2^1/_2$ years of age or less, shall have access to not less than two *exits* or *exit access doorways*.

❖ Day care occupancies are limited to a maximum of 10 occupants for rooms or spaces with infants and toddlers before two exits are required from a room. This limit is in consideration of needing a quick means of egress for children who would need to be carried or led for evacuation (i.e., children under $2^1/_2$ years of age).

There is an exception for the Group I-4 classification in Section 308.6.1 that allows day care facilities with up to 100 children $2^1/_2$ years of age or less with care rooms having direct access to the exterior to be classified as Group E. An exterior door to the outside from infant and toddler rooms can serve as the second exit to meet this section. It is not the intent that a day care classified as Group E, with children $2^1/_2$ years of age or less, could use the means of egress requirements for Group E for these rooms.

1006.2.2.5 Vehicular ramps. Vehicular ramps shall not be considered as an *exit access ramp* unless pedestrian facilities are provided.

❖ A vehicle-only ramp may be considered as one of the required exit access ramps if pedestrian walkways are provided along the ramp. The low-slope ramps that are lined with parking spaces are not considered vehicle ramps. In open parking garages, according to Section 1019.3, Exception 6, the exit access stairways and ramps are not required to be enclosed since an open parking structure is designed to permit the ready ventilation of the products of combustion to the outside by exterior wall openings (see Section

406.5.2). Also, parking structures are characterized by open floor areas that allow the occupants to observe a fire condition and choose a travel path that would avoid the fire threat.

1006.3 Egress from stories or occupied roofs. The *means of egress* system serving any *story* or occupied roof shall be provided with the number of *exits* or access to *exits* based on the aggregate *occupant load* served in accordance with this section. The *path of egress travel* to an *exit* shall not pass through more than one adjacent *story*.

❖ Emergency evacuation from a multistory building will typically involve stairways or ramps as the vertical element for the means of egress route. The number of required ways off the story (via exit or exit access elements) is based on the occupant loads shown in Table 1006.3.1. These stairways and ramps must comply with the general provisions (Sections 1011 and 1012 respectively), and can be exit access (Section 1019) or exits (Section 1023 and 1027). When exit access stairways or ramps are part of that route, the measurement of the exit access travel distance will include travel from the most remote point on the floor, to and down the exit access stairway or ramp and from the bottom of the stairway or ramp to an enclosure for an exit stairway or ramp or exterior exit door (see Section 1017.3). Vertical travel is slower than horizontal travel, so the exit access stairway or ramp in a building with two or more exits should not be used for more than one story before an exit is reached.

1006.3.1 Egress based on occupant load. Each *story* and occupied roof shall have the minimum number of independent *exits*, or access to *exits*, as specified in Table 1006.3.1. A single *exit* or access to a single *exit* shall be permitted in accordance with Section 1006.3.2. The required number of *exits*, or *exit access stairways* or *ramps* providing access to *exits*, from any *story* or occupied roof shall be maintained until arrival at the *exit discharge* or a *public way*.

❖ This section starts out saying that every occupant on a story must have access to the required number of means of egress as specified:

- Per Table 1006.3.1—at least two means of egress per floor;
- Three means of egress for floors with greater than 500 occupants and four means of egress for floors with greater than 1,000 occupants;
- Per Sections 1006.3.2.1 and 1006.3.2.2—limited allowance for a single means of egress from a floor.

Saying "means of egress" instead of "exits" acknowledges that direct entrances to exits on each story are not always mandatory since access to exits (i.e., exit access stairways or ramps) on some stories is permissible within certain limitations. This allows some freedom of design when, for instance, required exit stairways may not be available without passing through other tenant spaces. The intent is to allow for a balance of security concerns providing adequate safety for emergency evacuation.

If a designer chooses to use horizontal exits instead of exit stairways or exit ramps, those specific provisions are addressed in Section 1026. For buildings built into a hillside, an exit door directly to the outside would also provide the same or better level of protection as an interior exit element. It is not the intent of this provision to prohibit horizontal exits or direct exterior exit doors as an option.

Once the number of means of egress is determined, those paths must remain available until occupants leave the building (see Commentary Figure 1006.3.1). The need for exits to be independent of each other cannot be overstated. Each occupant of each floor must be provided with the required number of exits without having to pass through one exit to gain access to another. Each exit is required to be independent of other exits to prohibit such areas from merging downstream and becoming, in effect, one exit.

TABLE 1006.3.1
MINIMUM NUMBER OF EXITS OR ACCESS TO EXITS PER STORY

OCCUPANT LOAD PER STORY	MINIMUM NUMBER OF EXITS OR ACCESS TO EXITS FROM STORY
1-500	2
501-1,000	3
More than 1,000	4

❖ Table 1006.3.1 specifies that the minimum number of exits available to each occupant of a floor is based on the total occupant load of that floor. This is so that at least one exit will be available in case of a fire emergency and to increase the likelihood that a larger number of occupants can be accommodated by the remaining exits when one exit is not available. While an equal distribution of exit capacity among all of the exits is not required, a proper design would not only balance capacity with the occupant load distribution, but also consider a reasoned distribution of capacity to avoid a severe dependence on one exit.

1006.3.2 Single exits. A single *exit* or access to a single *exit* shall be permitted from any *story* or occupied roof where one of the following conditions exists:

1. The *occupant load*, number of *dwelling units* and *exit access* travel distance do not exceed the values in Table 1006.3.2(1) or 1006.3.2(2).
2. Rooms, areas and spaces complying with Section 1006.2.1 with *exits* that discharge directly to the exterior at the *level of exit discharge*, are permitted to have one *exit* or access to a single *exit*.
3. Parking garages where vehicles are mechanically parked shall be permitted to have one *exit* or access to a single *exit*.
4. Group R-3 and R-4 occupancies shall be permitted to have one *exit* or access to a single *exit*.
5. Individual single-story or multistory *dwelling units* shall be permitted to have a single exit or access to a

single *exit* from the *dwelling unit* provided that both of the following criteria are met:

5.1. The *dwelling unit* complies with Section 1006.2.1 as a space with one *means of egress*.

5.2. Either the *exit* from the *dwelling unit* discharges directly to the exterior at the *level of exit discharge*, or the *exit access* outside the dwelling unit's entrance door provides access to not less than two approved independent *exits*.

❖ The base assumption is that all stories of a building shall have access to at least two separate ways out for emergencies.

Single-exit stories can have access to an exit from any floor, therefore, single-exit stories can use an open exit access stairway for as many stories as permitted by Sections 1006.3.2 and 1019.3.3 provided they meet the exit access of travel distance limitations for that use in Sections 1006.2.1 and 1006.3.2 and Tables 1006.2.1, 1006.3.2(1) and 1006.3.2(2).

A story can have a single exit if the design meets one of the five items listed.

Item 1 states what situations permit one exit by a reference to Tables 1006.3.2(1) and 1006.3.2(2). If a story can meet the provisions for occupant load, number of units and travel distance in Table 1006.3.2(1) or 1006.3.2(2), then that story can have one means of egress. See the commentary for Tables 1006.3.2(1) and 1006.3.2(2) for information on single-exit buildings.

Item 2 references Table 1006.2.1 for single-exit spaces. Table 1006.2.1 is intended to be applicable to rooms and spaces on a floor, but not to an entire floor level. One of the main concerns has been that vertical travel takes longer than horizontal travel in emergency exiting situations. However, if the single-exit space can exit directly to the exterior rather than egress into an interior corridor, a higher level of safety is provided. While the term "building" limits the area addressed to that bordered by exterior walls or fire walls, a common application of Item 2 is on a tenant-by-tenant basis. For example, a single-story strip mall may not meet the provisions for a building with one means of egress but each tenant area meets the provisions for a space with one means of egress in accordance with Section 1006.1. This tenant could exist as either a stand-alone single-exit building or as a single-exit tenant space that exits into an interior corridor. Is it not just as safe to permit this tenant to exist as part of a larger building with the door exiting directly to the exterior? See also the commentary to Tables 1006.3.2(1) and 1006.3.2(2).

While not specifically stated in this section, there is a situation where the single means of egress can be used from a multilevel tenant space. When the combined occupant load of the space and a mezzanine meets the occupant load in Table 1006.2.1 and the common path of travel distance measured from the most remote point down the exit access stairway and to the exterior exit doorway, the ground floor space and its mezzanine can be considered a space with one means of egress.

Item 3 allows for one exit from all stories in a parking garage where the cars are mechanically parked. This is in recognition of the extremely low occupant load in this unique type of building. The single exit

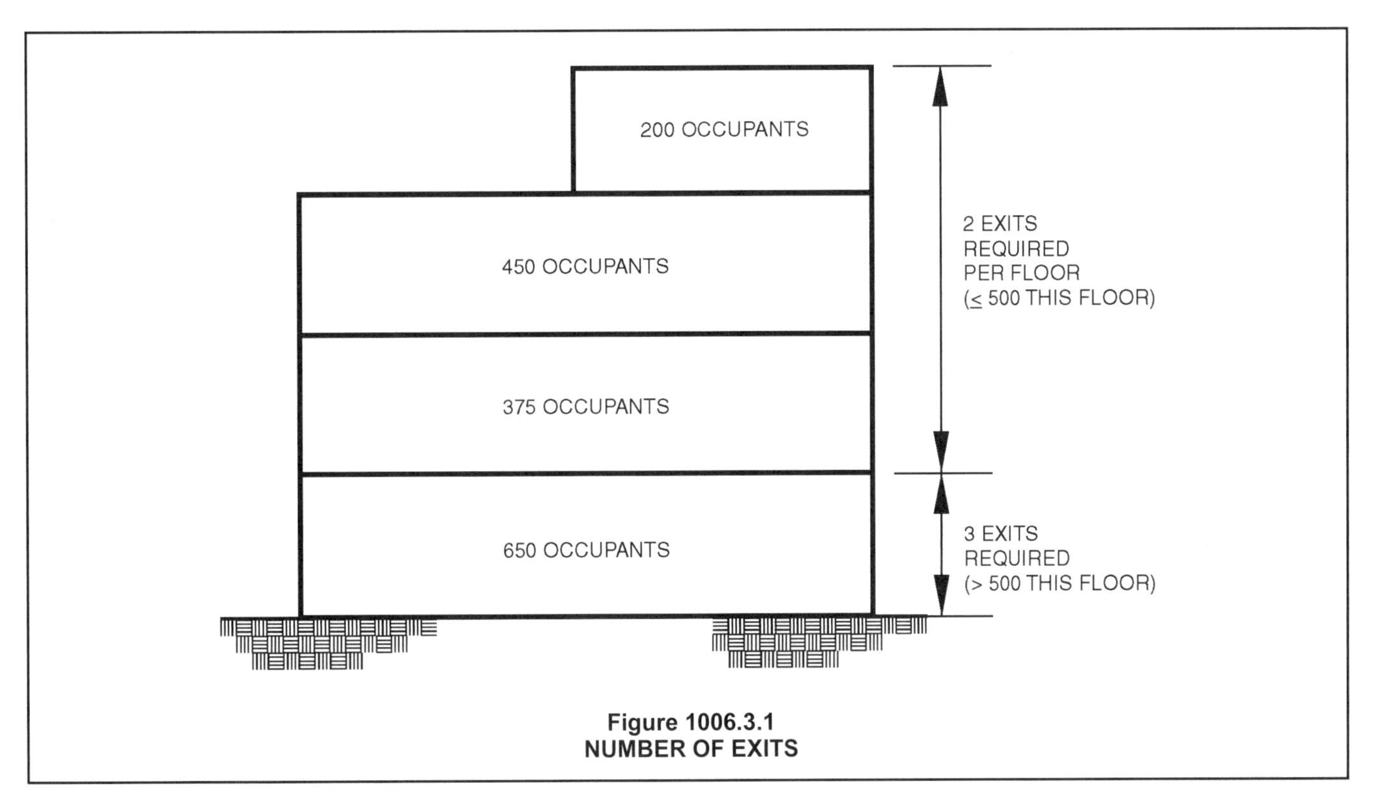

Figure 1006.3.1
NUMBER OF EXITS

would be for maintenance and service personnel who could be on the different levels. Exit access travel would still have to meet Table 1017.2. If this facility is an open parking garage, it could have one exit access stairway compliant with Sections 1017.2, Exception 1, and 1006.3.2, Exception 1.

Group R-3 is limited to no more than two dwelling units per building. Often these units are townhouse style with direct exits at grade or one unit on top of another. Residents in Group R-3 and R-4 congregate residences typically operate as a single-family home. The Fair Housing Act includes no discrimination based on familial status (i.e., family cannot be determined by blood or marriage). Many court cases have been filed under the Fair Housing Act (FHA) requiring that group homes be permitted to operate similar to a single-family home as a point of nondiscrimination. For additional information on the FHA, see the commentaries at the beginning of Chapter 11 and under Section 1107. In either configuration, per Item 4, each unit is required to have only one exit leading from the building. In a multistory unit, any interior stairway would be considered an exit access stairway (Section 1019.3, Exception 2) and would be part of the exit access travel distance in Table 1017.2.

Item 5 is based on decades of practice within an individual dwelling. Item 4 addresses single exits for a Group R-3 building; one or two apartments could be provided in a mixed-use building and still be considered Group R-3. This exception could be used within the unit. Multistory apartments may also be provided within Group R-2 high-rise residential buildings.

In Group R-2, R-3 and R-4 facilities with multistory dwelling units, the means of egress from a dwelling or sleeping unit is typically permitted to be from one level only. In a Group R-2 apartment- or townhouse-style building, if the building is sprinklered with an NFPA 13 or 13R system and the common path of travel from the most remote point on any level to the exit door from the unit itself is 125 feet (22 860 mm) maximum (see Section 1006.2.1), that unit may have only one means of egress. Section 1006.2.1, Exception 1, provides for an occupant load of 20, rather than the 10 occupants allowed in Table 1006.3(2). Therefore, this could allow apartments of up to 4,000 square feet (185.81 m^2). An exit access stairway would be permitted within the unit as part of the exit access travel distance (Section 1019.3, Exception 2). Once the occupants exit the unit itself, however, they must be outside at grade or the floor level must have access to two or more means of egress for all tenants, depending on the number required for the building as a whole [see Commentary Figures 1006.3.2(1) and 1006.3.2(3)]. The common path of egress within the unit would be part of the overall exit access travel distance of 250 feet (45 720 mm) required in Table 1017.2). It is not the intent to allow this exception for apartment-style dwelling units in conjunction with the allowances for a single-exit building. The emergency escape and rescue opening addressed in Section 1029 does not count toward the required number of exits.

TABLE 1006.3.2(1). See below.

❖ Per Note b, this table addresses when single exits can be provided from stories in Group R-2 occupancies having dwelling units only, such as apartments and condominiums. For Group R-2 occupancies with sleeping units, such as dormitories, sororities, fraternities, convents, monasteries or boarding houses, see Table 1006.3.2(2). The second row is to clarify that buildings containing Group R-2 single-exit buildings cannot be four stories or taller. In addition to meeting the number of dwelling units per floor and exit access travel distance, these buildings must be equipped throughout with an NFPA 13 or 13R sprinkler system and an emergency escape and rescue opening must be provided in every bedroom (Note a). The exit access travel distance would be measured from the most remote point in the unit to the exit from the floor. This is different from the travel distance in Section 1006.2.1, Exception 1 where the travel distance is measured to the door of the unit. However, in order to be able to use the single-exit building provisions, each unit must also meet the single-exit space requirements, so the occupant load of each apartment is limited. See Commentary Figures 1006.3.2(2) and 1006.3.2(3) for examples of the differences. Formal committee interpretation 21-14 states that this table allows for groups of four units on a story to have access to a single exit. These units would have to be separated in accordance with Section 420, but would not have to be separated by fire barriers or fire walls.

See the commentary to Table 1006.3.2(2) for additional discussion on single-exit building options.

This table is not intended to work in combination

TABLE 1006.3.2(1)
STORIES WITH ONE EXIT OR ACCESS TO ONE EXIT FOR R-2 OCCUPANCIES

STORY	OCCUPANCY	MAXIMUM NUMBER OF DWELLING UNITS	MAXIMUM COMMON PATH OF EGRESS TRAVEL DISTANCE
Basement, first, second or third story above grade plane	R-2[a, b]	4 dwelling units	125 feet
Fourth story above grade plane and higher	NP	NA	NA

For SI: 1 foot = 3048 mm.

NP = Not Permitted.

NA = Not Applicable.

a. Buildings classified as Group R-2 equipped throughout with an *automatic sprinkler system* in accordance with Section 903.3.1.1 or 903.3.1.2 and provided with *emergency escape and rescue openings* in accordance with Section 1030.

b. This table is used for R-2 occupancies consisting of *dwelling units*. For R-2 occupancies consisting of *sleeping units*, use Table 1006.3.2(2).

with Section 1006.3.2, Exception 5. That section requires a minimum of two exits from the floor.

TABLE 1006.3.2(2). See page 10-30.

❖ Buildings with one exit are permitted where the configuration and occupancy meet certain characteristics so as not to present an unacceptable fire risk to the occupants. Buildings that are relatively small in size have a shorter travel distance and fewer occupants; thus, having access to a single exit does not significantly compromise the safety of the occupants since they will also be alerted to and get away from the fire more quickly. It is important to note that the provisions in Section 1006.3.2 apply to individual stories. Multiple single-exit spaces or units may exist in the same building, including those cases where differing occupancies exist. Therefore, Tables 1006.3.2(1) and 1006.3.2(2) can address mixed occupancy buildings (see Section 1006.3.2.1).

Occupants of a story of limited size and configuration may have access to a single exit, provided that the building does not have more than one level below

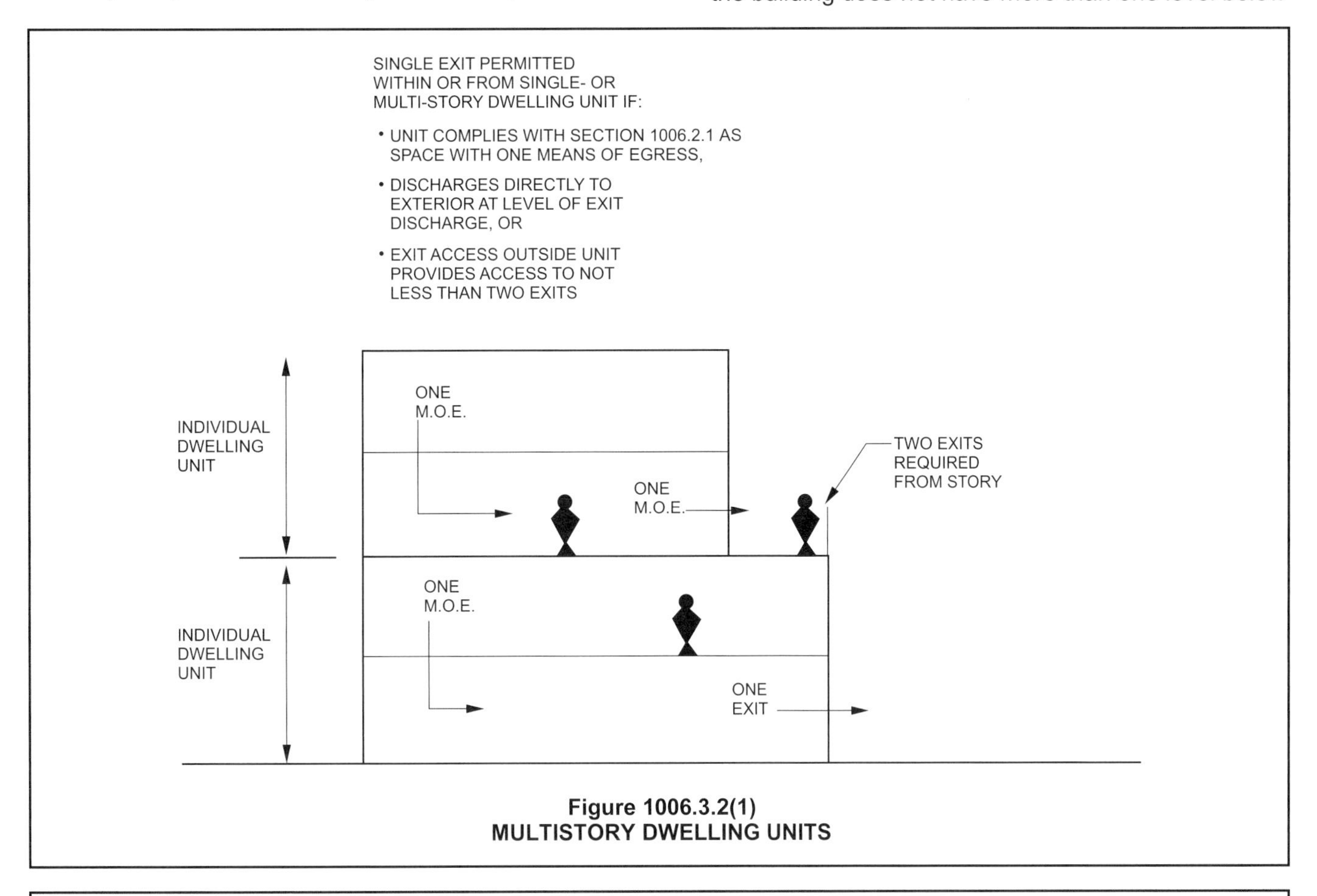

Figure 1006.3.2(1)
MULTISTORY DWELLING UNITS

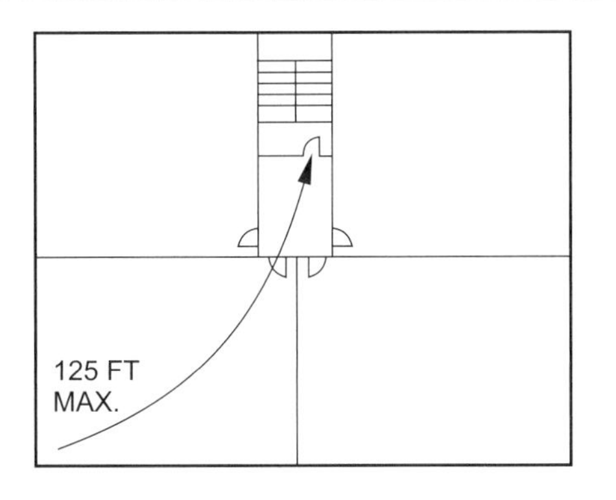

R-2 DWELLING UNITS

SECOND OR THIRD STORY
SINGLE EXIT PERMITTED

- MAXIMUM OF 4 UNITS AND 3 STORIES
- 125 FT. MAX. TRAVEL DISTANCE TO EXIT
- COMPLY WITH 1006.2.1, EXP 1.
  20 OCCUPANTS MAX.

For SI: 1 foot = 304.8 mm.

Figure 1006.3.2(2)
EXAMPLE OF SINGLE-EXIT APARTMENT BUILDING

the first story above the grade plane. The limitation on the number of levels above and below the first story is intended to limit the vertical travel an occupant must accomplish to reach the exit discharge in a single-exit building (see Section 1006.3.2.2).

Only one exit is required from a story where permitted by Table 1006.3.2(1) or 1006.3.2(2) regardless of the number of exits required from other stories in the building. For example, a Group B occupancy on the second floor of a two-story building is only required to have one exit from the story, provided its occupant load does not exceed 29 and the maximum exit access travel distance to the exit stairway does not exceed 75 feet (22 860 mm). The number of occupants and travel distances on the first floor do not affect the determination of the second story as a single-exit story. Other stories are also regulated independently as to number of exits. For mixed-occupancy floors or floors where tenant spaces have separate exits, see Section 1006.3.2.1.

Tables 1006.3.2(1) and 1006.3.2(2) list the characteristics a building must have to be of single-exit construction, including occupancy, maximum height of building above grade plane, maximum occupants or dwelling units per floor and exit access travel distance per floor. The occupant load of each floor is determined in accordance with the provisions of Section 1004.1. The exit access travel distance is measured along the natural and unobstructed path to the exit, as described in Section 1017.3. If the occupant load or common path of travel is exceeded, two exits are required from each floor in the building.

The enclosure required for the exit in a two- or three-story, single-exit building is identical to any other complying exit (e.g., interior stairs, exterior stairs, etc.). Similarly, the fire-resistance rating required for opening protectives is identical to that required by Section 716. Exit access stairways could be used in two-story buildings, with or without basements, where permitted by Sections 1019.3 and 1006.3.1.

Per Note c, Group R-2 sleeping units, such as congregate residences, fraternities, sororities, dormitories, convents or monasteries, use Table 1021.2(2).

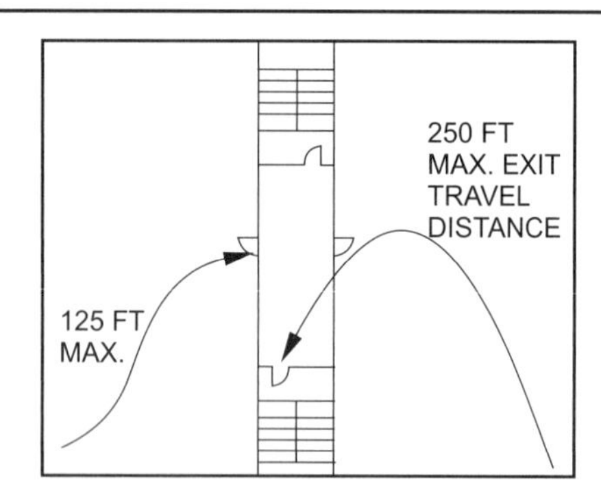

SINGLE EXIT WITHIN AND FROM UNIT

- 250 FT. MAX. EXIT TRAVEL DISTANCE
- NO LIMIT ON NUMBER OF UNITS
- 125 FT. MAX. COMMON PATH OF TRAVEL (TO APARTMENT DOOR)
- 20 OCCUPANT MAX. PER UNIT PER SECTION 1006.3.2, ITEM 5

For SI: 1 foot = 304.8 mm.

Figure 1006.3.2(3)
EXAMPLE OF SINGLE-EXIT APARTMENT WITHIN A TWO-EXIT BUILDING

TABLE 1006.3.2(2)
STORIES WITH ONE EXIT OR ACCESS TO ONE EXIT FOR OTHER OCCUPANCIES

STORY	OCCUPANCY	MAXIMUM OCCUPANT LOAD PER STORY	MAXIMUM COMMON PATH OF EGRESS TRAVEL DISTANCE (feet)
First story above or below grade plane	A, B[b], E F[b], M, U	49	75
	H-2, H-3	3	25
	H-4, H-5, I, R-1, R-2[a, c], R-4	10	75
	S[b, d]	29	75
Second story above grade plane	B, F, M, S[d]	29	75
Third story above grade plane and higher	NP	NA	NA

For SI: 1 foot = 304.8 mm.
NP = Not Permitted.
NA = Not Applicable.

a. Buildings classified as Group R-2 equipped throughout with an *automatic sprinkler system* in accordance with Section 903.3.1.1 or 903.3.1.2 and provided with *emergency escape and rescue openings* in accordance with Section 1030.

b. Group B, F and S occupancies in buildings equipped throughout with an *automatic sprinkler system* in accordance with Section 903.3.1.1 shall have a maximum *exit access* travel distance of 100 feet.

c. This table is used for R-2 occupancies consisting of *sleeping units*. For R-2 occupancies consisting of *dwelling units*, use Table 1006.3.2(1).

d. The length of *exit access* travel distance in a Group S-2 *open parking garage* shall be not more than 100 feet.

Table 1006.3.2(1) addresses dwelling units in Group R-2 occupancies. Effectively, this section allows for one-story congregate residences to have a single exit where there are 10 or fewer occupants on that first floor or basement. The travel distance from the most remote part of a unit to the exit must be less than 75 feet (22 860 mm). The maximum occupant load would limit the congregate residence to 2,000 square feet (185.81 m^2) maximum per floor. Per Note a, the congregate residence must be sprinklered throughout with an NFPA 13 or 13R system and all bedrooms must have an emergency escape and rescue opening (Section 1029.1). Alternatively, a congregate residence could use Section 1006.3.2, Item 4. Emergency escape windows would be required, but a single exit for upper floors would be allowed as well as an NFPA 13D sprinkler system where permitted by Section 903.2.8.

Table 1006.3.2(2) allows for two-story business, factory, mercantile and storage facilities to have a single-exit stairway or exit access stairway (see Section 1019.3) from the second floor if the space meets the maximums of 29 occupants and a 75-foot (22 860 mm) travel distance. The increased travel distance in Note b is not permitted for the second floor. However, there is an increase for travel distance permitted for Group S-2 open parking garages in Note d.

Table 1006.3.2(2) allows for a variety of single-story buildings, with or without basements, which meet the maximum occupant load and exit access travel distance specified. Exit access stairways from the basement would be a viable option where permitted by Section 1019.3.

Again, these tables are based on per-story criteria. These tables would also allow for mixed occupancies, such as four apartments over a restaurant or a business over a day care, as long as occupant loads and travel distances are met. Means of egress could be by a stairway through the floor below or a separate exterior stairway from the upper level. For additional information on a mixed-use floor, see Section 1006.3.2.1.

Formal code interpretation 21-14 states that this table allows for an exit configuration for multiple single-exit spaces for Groups B, M, F and S on the second floor where each group meets Table 1006.3.2(2).

1006.3.2.1 Mixed occupancies. Where one *exit*, or *exit access stairway* or *ramp* providing access to *exits* at other *stories*, is permitted to serve individual *stories*, mixed occupancies shall be permitted to be served by single *exits* provided each individual occupancy complies with the applicable requirements of Table 1006.3.2(1) or 1006.3.2(2) for that occupancy. Where applicable, cumulative *occupant loads* from adjacent occupancies shall be considered in accordance with the provisions of Section 1004.1. In each *story* of a mixed occupancy building, the maximum number of occupants served by a single *exit* shall be such that the sum of the ratios of the calculated number of occupants of the space divided by the allowable number of occupants indicated in Table 1006.3.2(2) for each occupancy does not exceed one. Where *dwelling units* are located on a story with other occupancies, the actual number of *dwelling units* divided by four plus the ratio from the other occupancy does not exceed one.

❖ Where multiple tenants or occupancies are located on a specific story, they are to be regulated by a "unity" formula if they want to use the same exit. If they have separate exits, they will be evaluated separately. This would allow for a second floor of a mixed-use occupancy to be evaluated similarly to Section 1016.2, which applies where multiple spaces combine. For example, the second story of a building houses two tenants, one business and one mercantile. Each tenant would be permitted a single, but separate, exit, provided each had an occupant load of less than 30 and a travel distance not exceeding 75 feet (22.860 m) (see Commentary Figure 1006.3.2.1, Option 2). However, if the tenants wanted to share

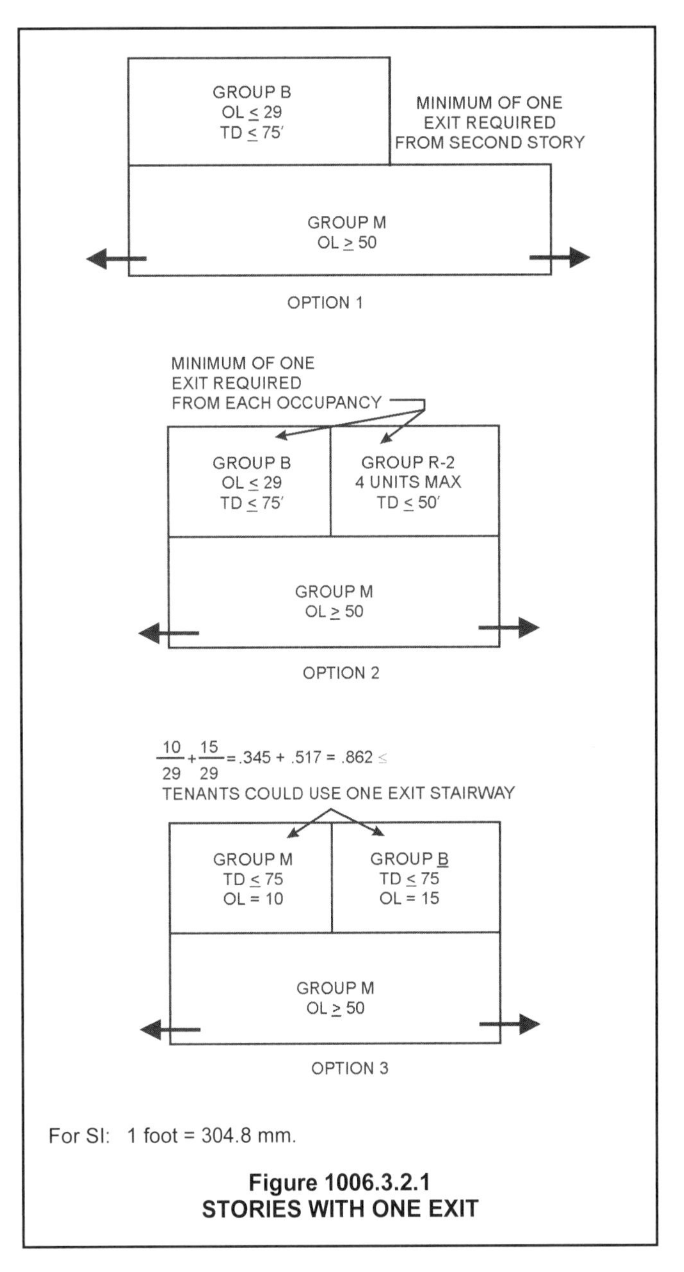

Figure 1006.3.2.1
STORIES WITH ONE EXIT

the same single exit, the combined occupant load would have to be less than 30 (see Commentary Figure 1006.3.2.1, Option 3).

Where the mixed-use stories includes dwelling units, since four dwelling units are allowed per floor in a single-exit building, for the unity formula use 0.25 for one unit, 0.50 for two units and 0.75 for three units.

SECTION 1007 EXIT AND EXIT ACCESS DOORWAY CONFIGURATION

1007.1 General. *Exits*, *exit access doorways*, and *exit access stairways* and *ramps* serving spaces, including individual building *stories*, shall be separated in accordance with the provisions of this section.

❖ The final exits, as well as the doors, stairways and ramps that get occupants to those exits (exit access elements), need to be unobstructed and obvious at all times for occupants to evacuate the building safely in an emergency situation. This is consistent with the requirements in Section 1010.1 that exit or exit access doors are not to be concealed by curtains, drapes, decorations or mirrors. Whether the doors from the space are exit access doors leading to a hallway or actual exit doors leading to an exit enclosure or directly to the outside, they must be located in accordance with this section.

The need for exits from a space or story to be independent of each other cannot be overstated. Each occupant of each floor must be provided with the required number of exits without having to pass through one exit to gain access to another. Each exit is required to be independent of other exits to prohibit such areas from merging downstream and becoming, in effect, one exit.

The requirement for exits to be continuous is consistent with the exit termination requirements in Section 1023.3 and the exit discharge termination requirements in Section 1024.4. The intent is to provide safety in all portions of the exit by requiring continuity of the fire protection characteristics of the enclosure for the exit stairway or ramp. Exit passageways (see Section 1024) are a continuation of an exit enclosure. This would include, but not be limited to, the fire-resistance rating of the exit enclosure walls and the opening protection rating of the doors.

Section 1028.1, Exceptions 1 and 2, allows for an alternative for direct access to the outside via an intervening lobby or vestibule. Horizontal exits (see Section 1026), while not providing direct access to the outside of the structure, do move occupants to another "building" by moving through a fire wall (see Section 1026 and Section 1028.1, Exception 3) or into a refuge area protected by fire barriers and horizontal assemblies. Horizontal exits are commonly used in hospitals and jails for a defend-in-place type of protection.

1007.1.1 Two exits or exit access doorways. Where two *exits*, *exit access doorways*, *exit access stairways* or *ramps*, or any combination thereof, are required from any portion of the *exit access*, they shall be placed a distance apart equal to not less than one-half of the length of the maximum overall diagonal dimension of the building or area to be served measured in a straight line between them. Interlocking or *scissor stairways* shall be counted as one *exit stairway*.

Exceptions:

1. Where interior *exit stairways* or *ramps* are interconnected by a 1-hour fire-resistance-rated corridor conforming to the requirements of Section 1020, the required exit separation shall be measured along the shortest direct line of travel within the corridor.
2. Where a building is equipped throughout with an *automatic sprinkler system* in accordance with Section 903.3.1.1 or 903.3.1.2, the separation distance shall be not less than one-third of the length of the maximum overall diagonal dimension of the area served.

❖ This section provides a method to determine, quantitatively, remoteness between exits and exit access doors based on the dimensional characteristics of the space served. This measure has been common practice for some years with significant success. Very simply, the method involves determining the maximum dimension between any two points in a floor or a room (e.g., a diagonal between opposite corners in a rectangular room or building or the diameter in a circular room or building). If two doors or exits are required from the room or building (see Sections 1006.2 and 1006.3), the straight-line distance between the center of the thresholds of the doors must be at least one-half of the maximum dimension [see Commentary Figure 1007.1.1(1)].

While technical proof is not available to substantiate this method of determining remoteness, it has been found to be realistic and practical for most building designs. Buildings with exits in a center core and occupied spaces around the perimeter are addressed in Exception 1.

If a scissor stairway is utilized, regardless of the separation of the two entrances, the scissor stair may only be counted as one exit. Two independent stairways within the same enclosure could result in both stairways being usable in an emergency where smoke penetrates the single enclosure (see the definition for "Scissor stairways" in Chapter 2). Interlocking stairways that occur over the same building footprint but within separate enclosures are not "scissor stairways" and can count as two independent exits. Due to concern about smoke migration, careful review of the construction details and verification that they meet all the provisions for fire barriers and horizontal assemblies must be made. Of special concern would be the provisions for continuity, penetrations and joints.

The entrance to the enclosures for exit stairways or

ramps shall meet the same arrangement as exit and exit access doorways. The need for exits to be independent of each other cannot be overstated.

In Exception 1, a method of permitting the distance between exits to be measured along a complying corridor connecting the enclosure for the exit stairway or ramp has served to mitigate the disruption to this design concept [see Commentary Figure 1007.1.1(2)].

As reflected in Exception 2, the protection provided by an automatic sprinkler system can reduce the threat of fire buildup so that the reduction in remoteness to one-third of the diagonal dimension is not unreasonable, based on the presumption that it provides the occupants with an acceptable level of safety from fire [see Commentary Figure 1007.1.1(3)]. The automatic sprinkler system must be installed throughout the building in accordance with NFPA 13 or 13R. This reduced separation (one-third diagonal) may also be used when applying the requirements of Exception 1.

In applying the provisions of this section, it is important to recognize any convergence of egress paths that may exist. Commentary Figure 1007.1.1(4) illustrates an assembly room with remotely located exit access doors, but the doors from the entire space are not considered remote in accordance with this section.

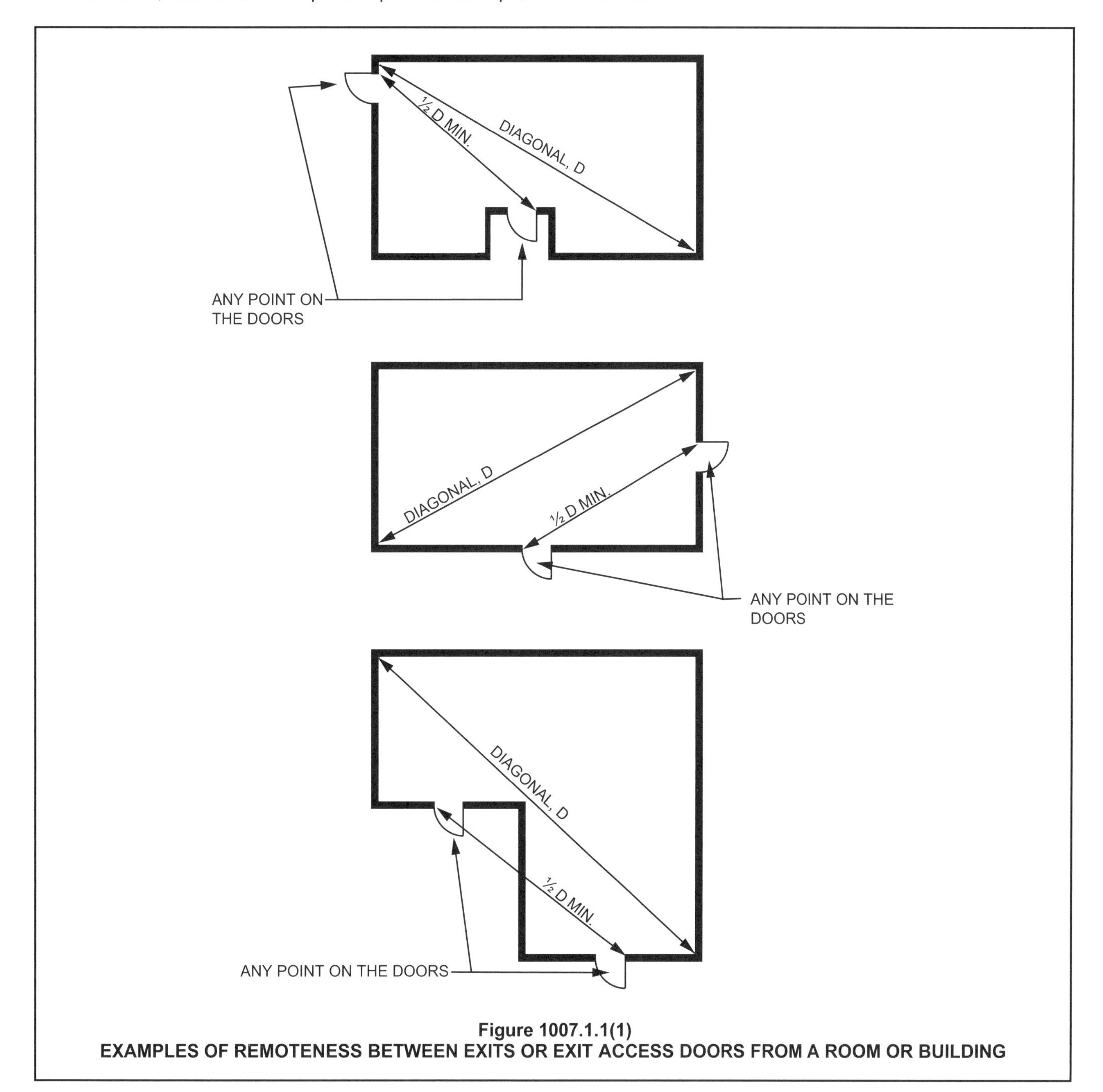

Figure 1007.1.1(1)
EXAMPLES OF REMOTENESS BETWEEN EXITS OR EXIT ACCESS DOORS FROM A ROOM OR BUILDING

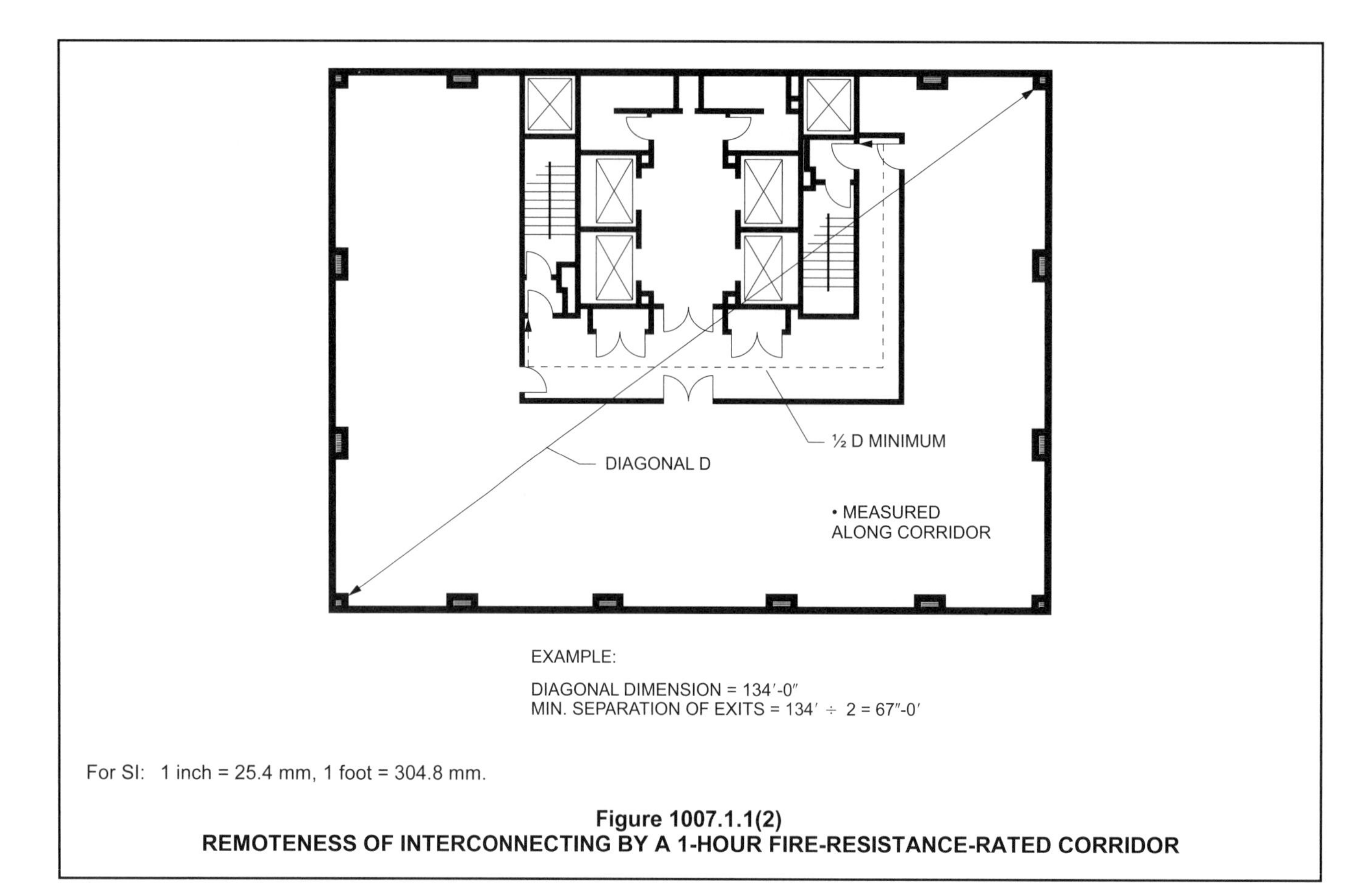

For SI: 1 inch = 25.4 mm, 1 foot = 304.8 mm.

Figure 1007.1.1(2)
REMOTENESS OF INTERCONNECTING BY A 1-HOUR FIRE-RESISTANCE-RATED CORRIDOR

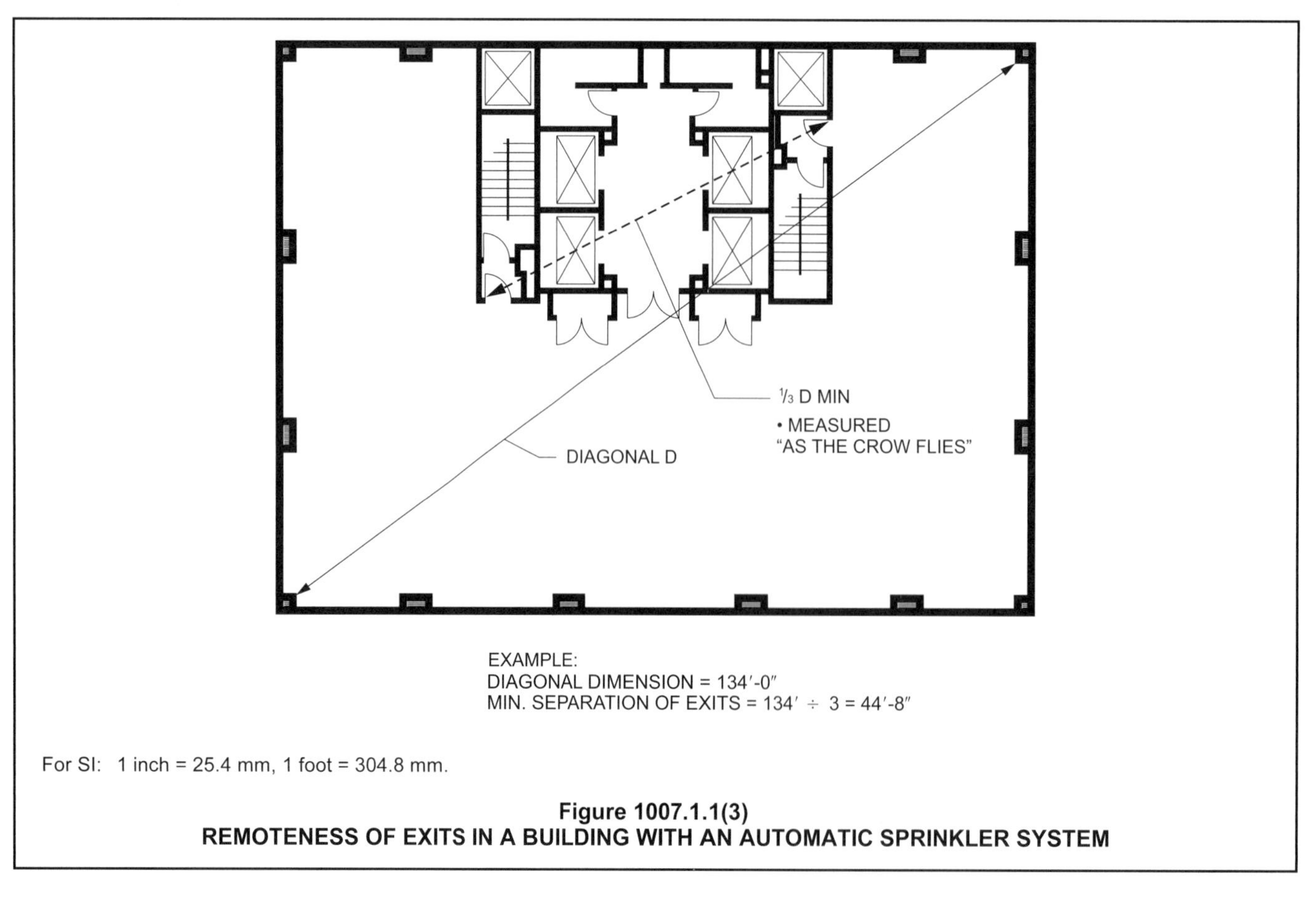

For SI: 1 inch = 25.4 mm, 1 foot = 304.8 mm.

Figure 1007.1.1(3)
REMOTENESS OF EXITS IN A BUILDING WITH AN AUTOMATIC SPRINKLER SYSTEM

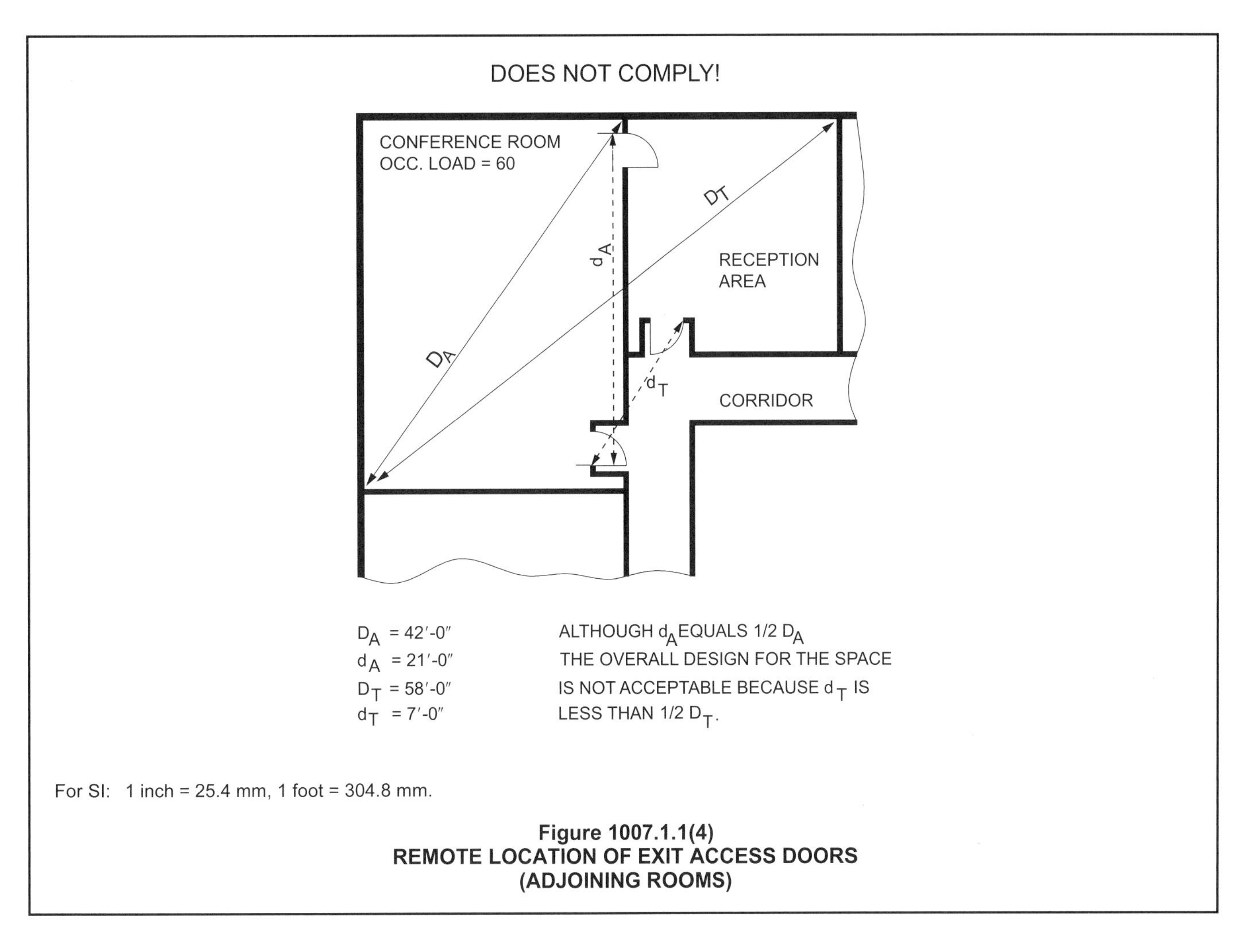

Figure 1007.1.1(4)
REMOTE LOCATION OF EXIT ACCESS DOORS (ADJOINING ROOMS)

1007.1.1.1 Measurement point. The separation distance required in Section 1007.1.1 shall be measured in accordance with the following:

1. The separation distance to *exit* or *exit access doorways* shall be measured to any point along the width of the doorway.
2. The separation distance to *exit access stairways* shall be measured to the closest riser.
3. The separation distance to *exit access ramps* shall be measured to the start of the ramp run.

❖ Where exit access stairways are permitted to be unenclosed, the remoteness measurement for doorways shall begin at the center of the top riser of the unenclosed stairways; this is consistent with the exit access travel distance measurement in Section 1017.3. When enclosure is provided, distances are measured to the door of the enclosure. Typically, travel distance is measured to the center of the door, ramp or stairway; however, there is an allowance for that point to be at any location along the door, stairway or ramp. The intent is to reduce subjectivity in the determination of exit/exit access configuration. The result is that a designer could literally measure to the far extreme edge of the two doorways leading out of a room to meet the separation requirements.

1007.1.2 Three or more exits or exit access doorways. Where access to three or more *exits* is required, not less than two *exit* or *exit access doorways* shall be arranged in accordance with the provisions of Section 1007.1.1. Additional required *exit* or *exit access doorways* shall be arranged a reasonable distance apart so that if one becomes blocked, the others will be available.

❖ When there are three or more required exits from a building or exit access doors from a room, they are to be analyzed identically to the method described in Section 1007.1.1. Two of the exits or exit access doors must meet the remoteness test. Any additional exits or doors can be located anywhere within the floor plan that meets the code requirements, including independence, accessibility, capacity and continuity.

There is no specific separation requirement provided in Section 1007.1.2 for the third exit, but the exits must be located so that one fire event will not block two exits; thus, two doors immediately adjacent would not be acceptable. The appropriate separation is subjective and would be partially dependent on the layout of the space. Using some other provisions in the code for guidance can help. Section 1005.5 states that multiple means of egress need to be sized such that the loss of any one will not reduce the capacity below 50 percent. This intent is repeated for an assembly space with more than 300 occupants. Sections 1029.2 and 1029.3 can require this space to have a main exit that accommodates one-half of the total occupant load and then require the balance of the exits (means of egress) to provide for the remaining one-half of the occupant load. The "main exit" is typically the result of a single "main entrance" for fee or ticket entry. If there is not a main exit, the exceptions to both Sections 1029.2 and 1029.3 require the means of egress to be "distributed around the perimeter of the building."

1007.1.3 Remoteness of exit access stairways or ramps. Where two *exit access stairways* or *ramps* provide the required *means of egress* to *exits* at another *story*, the required separation distance shall be maintained for all portions of such *exit access stairways* or *ramps*.

❖ The intent of these provisions are to prohibit open stairways and ramps that meet the separation distance at the first riser or start of the ramp run from converging toward one another such that the separation distance is reduced as the occupants follow the egress path to the lower level. Exit access stairways and ramps have to maintain the separation distance until the travel to the lower floor is complete (see Commentary Figure 1007.1.3).

1007.1.3.1 Three or more exit access stairways or ramps. Where more than two *exit access stairways* or *ramps* provide the required *means of egress*, not less than two shall be arranged in accordance with Section 1007.1.3.

❖ See the commentary to Section 1007.1.3, Remoteness of exit access stairways and ramps.

SECTION 1008 MEANS OF EGRESS ILLUMINATION

1008.1 Means of egress illumination. Illumination shall be provided in the *means of egress* in accordance with Section 1008.2. Under emergency power, means of egress illumination shall comply with Section 1008.3.

❖ This section is split into two distinct requirements: Section 1008.2 for egress lighting during typical lighting situations and Section 1008.3 for egress lighting for emergencies when the building has lost normal power.

1008.2 Illumination required. The *means of egress* serving a room or space shall be illuminated at all times that the room or space is occupied.

Exceptions:

1. Occupancies in Group U.
2. *Aisle accessways* in Group A.
3. *Dwelling units* and *sleeping units* in Groups R-1, R-2 and R-3.
4. *Sleeping units* of Group I occupancies.

❖ All portions of the means of egress serving a space must be illuminated by artificial lighting when that space is occupied, so that the paths of exit travel are always visible and available for evacuation of the occupants during emergencies. The intent is to allow portions of a building to have the lights turned off when that portion is unoccupied.

Three of the exceptions are for occupancies where the constant illumination of the means of egress would interfere with the use of space, such as sleeping areas or theater seating during a performance.

Bear in mind that means of egress lighting is not emergency lighting. For emergency lighting requirements, see Section 1008.3.

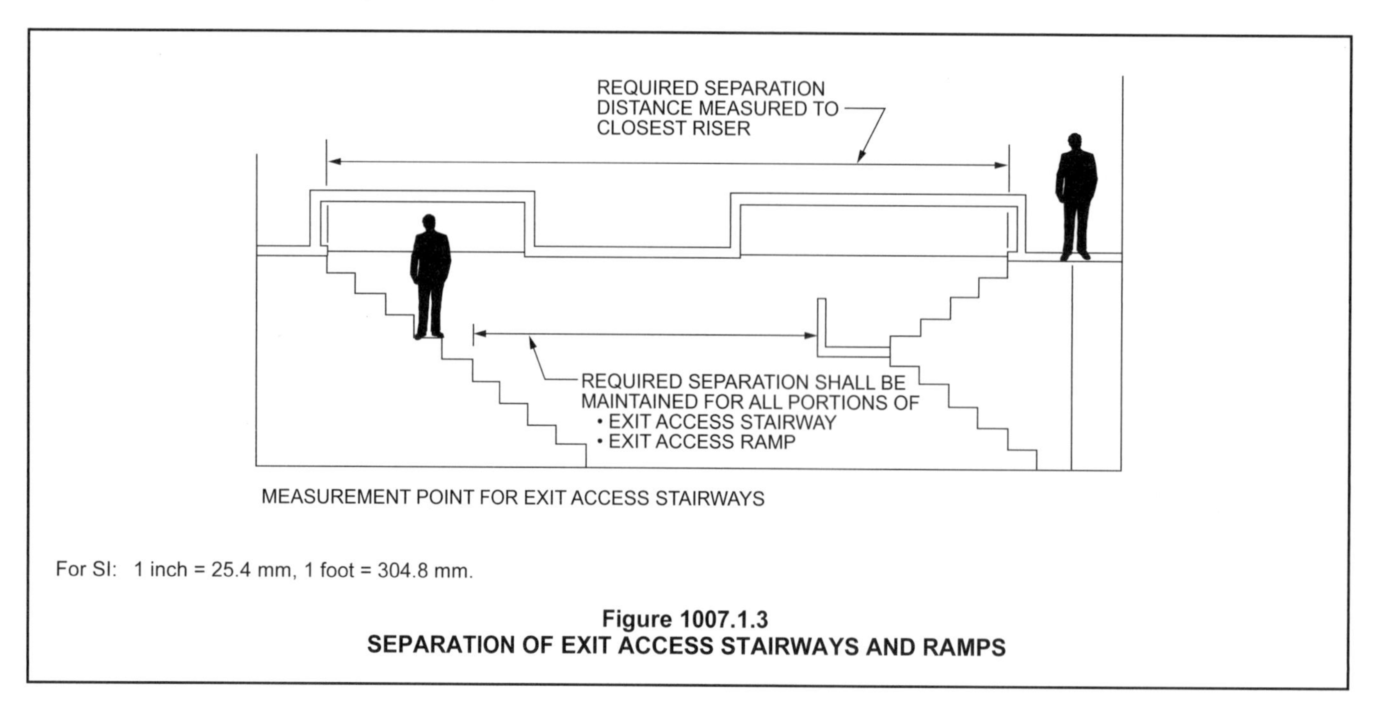

Figure 1007.1.3
SEPARATION OF EXIT ACCESS STAIRWAYS AND RAMPS

The high-rise provisions in Section 403 utilize the means of egress lighting and emergency lighting in this section. High-rise buildings also require a secondary backup system of luminous egress path markings within enclosed exit stairways (Section 1025). Section 1025.5 requires the enclosed exit stairways to be illuminated for 60 minutes prior to the daily occupancy and during the entire time the building is occupied so that the self-luminous stripes can charge. In a building that is always occupied, such as a hospital or hotel, this would require at least some of the lights in the stairway to be on 24 hours per day.

1008.2.1 Illumination level under normal power. The *means of egress* illumination level shall be not less than 1 footcandle (11 lux) at the walking surface.

Exception: For auditoriums, theaters, concert or opera halls and similar assembly occupancies, the illumination at the walking surface is permitted to be reduced during performances by one of the following methods provided that the required illumination is automatically restored upon activation of a premises' fire alarm system:

1. Externally illuminated walking surfaces shall be permitted to be illuminated to not less than 0.2 footcandle (2.15 lux).
2. Steps, landings and the sides of ramps shall be permitted to be marked with self-luminous materials in accordance with Sections 1025.2.1, 1025.2.2 and 1025.2.4 by systems listed in accordance with UL 1994.

❖ The intensity of lighting along the entire means of egress, including open plan spaces, aisles, corridors and passageways, exit stairways, exit doors and places of exit discharge at the walking surface or floor level must not be less than 1 footcandle (11 lux). One footcandle (11 lux) is approximately the same lighting level as found outdoors at twilight. It has been found that even this relatively low level of lighting renders enough visibility for occupants to evacuate a building safely.

It is important to note that this lighting level is measured at the floor in order to make the floor surface visible. Levels of illumination above the floor may be higher or lower, thus allowing lights along steps to be used rather than general area lighting.

The exception addresses assembly occupancies where low light level is needed for the function of the space. It is not the intent of the exception to require a fire alarm system but to require a connection to the egress lighting where a fire alarm system is provided. There are two options for that path lighting. Per Exception 1, the level of intensity of aisle lighting in such spaces may be reduced to 0.2 footcandle (2.15 lux), but only during the time of a performance. This intensity of illumination is sufficient to distinguish the aisles and stairs leading to the egress doors and not be a source of distraction during a performance. The option in Exception 2 allows for internally illuminated luminous path stripes (see Section 1025). Since the illumination in the space between shows and the duration of the shows may not be compatible with luminous path markings that charge from a light source, this option is currently not permitted.

1008.2.2 Exit discharge. In Group I-2 occupancies where two or more exits are required, on the exterior landings required by Section 1010.6.1, means of egress illumination levels for the exit discharge shall be provided such that failure of any single lighting unit shall not reduce the illumination level on that landing to less than 1 footcandle (11 lux).

❖ Section 1006.1, by requiring the means of egress to be illuminated, would include the exit discharge. Often, when moving to the outside of the building, the lighting from the parking lot or street lights will sufficiently light the exit discharge without additional lighting from the building itself. The transition moving from a lighted space to the outside may be an issue at the exit door, therefore, many facilities do provide a light over the exit door on the outside. For hospitals, the fixture provided must have at least two light bulbs, so that if one burns out, the other will still provide sufficient light for this transition.

This proposal was submitted by the Adhoc Health Care Committee, so it is limited to their scope of work. To reduce potential hazards at this important juncture, this could be considered best practice for other occupancies.

1008.3 Emergency power for illumination. The power supply for means of egress illumination shall normally be provided by the premises' electrical supply.

❖ The main routes for means of egress must be illuminated in times of emergency when the occupants must have a lighted path of exit travel in order to evacuate the building safely. The code is very specific in the description of the areas that are required to be illuminated by the emergency, not the standby power system (see Section 2702.2.11).

1008.3.1 General. In the event of power supply failure in rooms and spaces that require two or more means of egress, an emergency electrical system shall automatically illuminate all of the following areas:

1. *Aisles.*
2. *Corridors.*
3. *Exit access stairways* and *ramps.*

❖ Where two means of egress are required from a room or space, emergency lighting must be provided along aisle corridors, exit access stairways and ramps. It is not the intent to require the entire space to be lit as during normal use situations addressed in Section 1008.2. However, the main egress paths out of the two exit spaces must be illuminated. This would always include the main corridor in a story with two exits required, and could include paths off that main corridor. For example, an aisle in an open office plan must be illuminated, but not the path within individual offices that egress through the open area. Exit access stairways from a mezzanine or from a second floor (as permitted by Section 1019.3) and the route from the bottom of the stairway to the exit must also be illuminated.

1008.3.2 Buildings. In the event of power supply failure in buildings that require two or more *means of egress*, an emergency electrical system shall automatically illuminate all of the following areas:

1. *Interior exit access stairways* and *ramps*.
2. *Interior* and *exterior exit stairways* and *ramps*.
3. *Exit passageways*.
4. Vestibules and areas on the level of discharge used for *exit discharge* in accordance with Section 1028.1.
5. Exterior landings as required by Section 1010.1.6 for *exit doorways* that lead directly to the *exit discharge*.

❖ Where two or more exits are required from a story, essential portions of the interior egress system must be illuminated. This includes:

- Exit access stairways from a second floor (as permitted by Section 1019.3) as well as the route from the bottom of the stairway to the exit (see Section 1008.3.1).
- All exit stairways and ramps along their entire length.
- Exit passageways used in facilities with long travel distances, such as malls, or as an extension of the exit stairway enclosure.
- Means of egress systems that use exit balconies (see Section 1021) or open exterior exit stairways or ramps (see Section 1028).
- Interior exit discharge elements, such as lobbies and vestibules (see Section 1028.1), where stairways discharge into these elements instead of directly to the exterior of a building.
- Exterior portions of the exit discharge. Note that only the portion of the exterior discharge that is immediately adjacent to the building exit discharge door is required to have emergency illumination and not the entire exterior discharge path to the public way.

1008.3.3 Rooms and spaces. In the event of power supply failure, an emergency electrical system shall automatically illuminate all of the following areas:

1. Electrical equipment rooms.
2. Fire command centers.
3. Fire pump rooms.
4. Generator rooms.
5. Public restrooms with an area greater than 300 square feet (27.87 m^2).

❖ The intent of Items 1 through 4 is to have emergency lighting in areas significant for emergency responders or maintenance personnel who may be trying to locate and fix the loss of power issue for the building.

Item 5 requires an emergency light in large restrooms. Given the activity, people may need some illumination to quickly get themselves ready to be able to evacuate.

1008.3.4 Duration. The emergency power system shall provide power for a duration of not less than 90 minutes and shall consist of storage batteries, unit equipment or an on-site generator. The installation of the emergency power system shall be in accordance with Section 2702.

❖ So that there will be a continuing source of electrical energy for maintaining the illumination of the means of egress when there is a loss of the main power supply, the means of egress lighting system must be connected to an emergency electrical system that consists of storage batteries, unit equipment or an on-site generator. This emergency power-generating facility must be capable of supplying electricity for at least 90 minutes, thereby giving the occupants sufficient time to leave the premises. In most cases, where the loss of the main electrical supply is attributed to a malfunction in the distribution system of the electric power company, experience has shown that such power outages do not usually last as long as 90 minutes.

The IFC requirements for emergency power for emergency egress lighting and exit signage in existing buildings only require a 60-minute time duration. This is not a conflict, but rather recognition of the loss of battery storage capability over a length of time.

1008.3.5 Illumination level under emergency power. Emergency lighting facilities shall be arranged to provide initial illumination that is not less than an average of 1 footcandle (11 lux) and a minimum at any point of 0.1 footcandle (1 lux) measured along the path of egress at floor level. Illumination levels shall be permitted to decline to 0.6 footcandle (6 lux) average and a minimum at any point of 0.06 footcandle (0.6 lux) at the end of the emergency lighting time duration. A maximum-to-minimum illumination uniformity ratio of 40 to 1 shall not be exceeded. In Group I-2 occupancies, failure of any single lighting unit shall not reduce the illumination level to less than 0.2 footcandle (2.2 lux).

❖ This section provides the criteria of the illumination levels of the emergency lighting system. The initial average level for the main egress paths is the same as for the overall means of egress illumination in Section 1008.2.1. The reduction of illumination recognizes the performance characteristics over time of some types of power supplies, such as batteries. The minimum levels are sufficient for the occupants to egress from the building. In addition, the emergency lighting system is a secondary system that is spaced along the main egress routes. It will not provide the same level of general lighting over the route as what can be provided by the building lighting system.

The maximum illumination uniformity ratio of 40 means that the variation in the illumination levels is not to exceed that number. For example, a minimum of 0.06 footcandle (0.6 lux) would establish a maximum illumination of 2.4 footcandles (24 lux) in an adjacent area. This is to establish a variation limit such that the means of egress can be seen as a person walks from bright to darker areas along the egress path.

Group I-2 nursing homes and hospitals typically use a defend-in-place strategy for patients and residents who require assistance from a staff trained in the fire and safety evacuation plans. Given the possible critical nature of some of the patients and the need to move some people with life-sustaining equipment, there is an additional requirement that the lighting in the main corridors and along the exit path will always have some redundancy in the fixtures providing that illumination. This could be done with either two bulb fixtures, or locating emergency lighting fixtures close enough that if one burned out, light from fixtures on each side would overlap enough to not result in a dark spot.

SECTION 1009 ACCESSIBLE MEANS OF EGRESS

1009.1 Accessible means of egress required. Accessible *means of egress* shall comply with this section. *Accessible* spaces shall be provided with not less than one accessible *means of egress*. Where more than one *means of egress* are required by Section 1006.2 or 1006.3 from any *accessible* space, each *accessible* portion of the space shall be served by not less than two accessible *means of egress*.

Exceptions:

1. Accessible *means of egress* are not required to be provided in existing buildings.
2. One accessible *means of egress* is required from an *accessible mezzanine* level in accordance with Section 1009.3, 1009.4 or 1009.5.
3. In assembly areas with ramped *aisles* or stepped *aisles*, one accessible *means of egress* is permitted where the *common path of egress travel* is *accessible* and meets the requirements in Section 1029.8.

❖ The Access Board has revised and updated its accessibility guidelines for buildings and facilities covered by the Americans with Disabilities Act of 1990 (ADA) and the Architectural Barriers Act of 1968 (ABA). The final ADA/ABA Guidelines, published by the Access Board in July 2004, serve as the basis for the minimum standards when adopted by other federal agencies responsible for issuing enforceable standards. The plan is to eventually use this new document in place of the Uniform Federal Accessibility Standard (UFAS) and ADAAG. The U.S. Department of Justice officially adopted the new federal requirements on September 15, 2010. The new name is the *2010 Standard for Accessible Design,* otherwise known as the 2010 ADA Standard. The 2010 ADA Standard, Section 207/F207, references the 2000 edition of the IBC with 2001 Supplement, as well as the 2003 edition of the IBC, for accessible means of egress requirements. The International Code Council® (ICC®) is very proud to be recognized for its work regarding accessible means of egress in this manner. Refer to the Access Board website (www.access-board.gov) for more specific information and the current status of this adoption process. Based on the date of the publication and the adoption process, later editions of the IBC are not specifically referenced; however, none of the later editions reduced the accessible means of egress requirements found in the referenced IBC editions, so they should be able to be considered as providing equivalent or better accessibility.

The accessible means of egress locations may not be near the accessible route used for ingress into the building (see Sections 1104 and 1105). For example, a two-story building requires one accessible route to connect all accessible spaces within the building. The accessible route to the second level is typically by an elevator. During a fire emergency, persons with mobility impairments on the second level should move to the exit stairways for assisted rescue, not back the way they came in via the elevator. Signage at the elevator will direct occupants to an exit stairway.

This section establishes the minimum requirements for means of egress facilities serving all spaces that are required to be accessible to people with physical disabilities. Previously, attention had been focused on response to the civil-rights-based issue of providing adequate access for people with physical disabilities into and throughout buildings. Concerns about life safety and evacuation of people with mobility impairments were frequently cited as reasons for not embracing widespread building accessibility, in the best interests of the disabled community.

The provisions for accessible means of egress are predominantly, though not exclusively, intended to address the safety of persons with a mobility impairment. These requirements reflect the balanced philosophy that accessible means of egress are to be provided for occupants who have gained access into the building but are incapable of independently utilizing the typical means of egress facilities, such as the exit stairways. By making such provisions, the code now addresses means of egress for all building occupants, with and without physical disabilities.

Any space that is not required by the code to be accessible in accordance with Chapter 11 is not required to be provided with accessible means of egress. This may include an entire story, a portion of a story, a mezzanine or an individual room. For example, a mechanical penthouse is not required to be accessible in accordance with Section 1103.2.9; therefore, the mechanical penthouse is not required to have an accessible means of egress.

In new construction and additions, at least two accessible means of egress are required. For example, in buildings, stories or spaces required by Section 1006.3.1 to have three or more exits or exit access doors, a minimum of two accessible means of egress is required. The accessibility requirements are based on the required means of egress from both individual spaces and the building as a whole. Therefore, facilities with multiple large assembly rooms, such as banquet halls or multiplex theaters where the second exit from the space is often a door directly to the outside, may require additional accessible means of egress from the building because of space requirements.

While there are no dispersement requirements specific to accessible means of egress or travel distance limitations where there is no area of refuge requirement (see Sections 1009.3, 1009.4 and 1009.6), the code requires all exits to be distinct, separated and independent. The main intent is that a person with mobility impairments will always have options. If not all exits are accessible, possible entrapment should be a consideration in determining which exits are to be made accessible. In most buildings, the upper floors will already have an accessible route to all stairways. On the first floor, if the issue is accessible exit discharge, see Section 1009.7.

An accessible means of egress is required to provide a continuous path of travel to a public way. This principle is consistent with the general requirements for all means of egress, as reflected in Section 1003.1 and in the definition of "Means of egress" in Chapter 2. This section also emphasizes the intent that accessible means of egress must be available to a person with a mobility impairment, such as a person using a wheelchair, scooter or walker. Some mobility impairments do not allow for self-evacuation along a stairway; therefore, utilization of the exit and exit discharge may require assistance. The safety and fire evacuation plans (see Section 1001.4 of the code and Section 404 of the IFC) require planning for all occupants of a building. This assistance is typically with the fire department or other trained personnel, either along the exit stairways or in buildings five stories or taller, with the elevator system or a combination of both (see commentary, Section 1009.2.1). It is required that accessible routes, areas of refuge and exterior areas of rescue assistance are indicated on these plans. These plans must be approved by the local fire official and reviewed annually.

The exceptions address special situations where accessible means of egress requirements need special consideration. Note that these are exceptions for accessible means of egress; not exceptions for accessible entrance requirements (see Section 1105).

Exception 1 indicates that existing buildings undergoing alterations are not required to be provided with accessible means of egress as part of that alteration. In many cases, meeting the requirements for accessible means of egress, especially the 48-inch (1219 mm) clear stair width required in nonsprinklered buildings, would be considered technically infeasible. However, if an accessible means of egress was part of the original construction, it must be maintained in accordance with IEBC Section 410.2 or 705.1.

Exception 2 is a special consideration for mezzanines. The size of mezzanines is limited to a portion of the space below (see Section 505). Most are open to the space below; thus, with the same atmosphere and line of sight, fire recognition is quicker than in a two-story situation. If the elevator used for ingress has not gone into fire department recall, that system could be used for self-evacuation. There are three different scenarios:

1. If the mezzanine is exempted from accessibility, such as a mechanical mezzanine (see Section 1103.2.9), or small enough not to be required to be accessed by an accessible route (see Section 1104.4, Exception 1), no accessible means of egress is required.

2. If a mezzanine is required to be accessible (see Section 1104.4) and meets the provisions for spaces with one means of egress (see Section 1006.2), the exit access stairway must meet the provisions of Section 1009.3. Per Section 1009.3, Exception 1, this can be an open exit access stairway. Per Exception 4, in a nonsprinklered building, a designer would have the option of either an area of refuge or two-way communication at the elevator. Per Exception 5, in sprinklered buildings, the area of refuge would not be required.

3. If a mezzanine is required to be accessible (see Section 1104.4) and is required to have two means of egress, at least one of the means of egress stairways must meet the provisions of Section 1009.3. Per Section 1009.3, Exception 1, this can be an open exit access stairway. Per Exception 4, in a nonsprinklered building, a designer would have the option of either an area of refuge or two-way communication at the elevator. Per Exception 5, in sprinklered buildings, the area of refuge would not be required. Practically speaking, in a sprinklered building, both stairways from the mezzanine will meet Section 1009.3.

While under Scenarios 2 and 3 it is optional to have the elevator meet the requirements of Section 1009.4, the provisions for standby power at elevators are based on fire department assisted rescue. This is an expensive option that would likely never be used during a fire event. Where platform lifts can be used in new construction (see Section 1109.8), they are so limited that it is not likely that they will provide the accessible route to a mezzanine. If the platform lift serves as part of the route to the space, Section 1009.5 allows for platform lifts to serve as part of the accessible route for accessible means of egress when they have standby power.

Exception 3 is in consideration of the practical difficulties of providing accessible routes in assembly areas with sloped floors and stepped aisles. Rooms with more than 50 persons are required to have two means of egress; therefore, each accessible seating location is required to have access to two accessible means of egress. Depending on the slope of the seating arrangement, providing an accessible route to both distinct exits can be difficult to achieve, especially in small theaters. A maximum travel distance of 30 feet (9144 mm) for ambulatory persons moving from the last seat in dead-end aisles or from box-type seating arrangements to where they have access to a choice of means of egress routes has been established in Section 1029.8. In accordance with Exception 3, persons using wheelchair seating spaces have the same maximum 30-foot (9144 mm) travel distance from the accessible seating locations to a cross aisle or out of the room to an adjacent corridor or space where two choices for accessible means of egress are provided. Note that there are increases in travel distance for smoke-protected seating and small spaces, such as boxes, galleries or balconies. For additional information, see Section 1029.8.

1009.2 Continuity and components. Each required accessible *means of egress* shall be continuous to a *public way* and shall consist of one or more of the following components:

1. *Accessible routes* complying with Section 1104.
2. *Interior exit stairways* complying with Sections 1009.3 and 1023.
3. *Exit access stairways* complying with Sections 1009.3 and 1019.3 or 1019.4.
4. *Exterior exit stairways* complying with Sections 1009.3 and 1027 and serving levels other than the *level of exit discharge*.
5. Elevators complying with Section 1009.4.
6. Platform lifts complying with Section 1009.5.
7. *Horizontal exits* complying with Section 1026.
8. *Ramps* complying with Section 1012.
9. *Areas of refuge* complying with Section 1009.6.
10. Exterior areas for assisted rescue complying with Section 1009.7 serving exits at the *level of exit discharge*.

❖ This section identifies the various building features that can serve as elements of an accessible means of egress. Accessible routes are readily recognizable as to how they can provide accessible means of egress; however, some nontraditional principles have been established for the total concept of accessible means of egress. This is evident in that stairways and elevators are also identified as elements that can be part of an accessible means of egress. For example, elevators are generally not available for egress during a fire, while stairways are not independently usable by a person using a wheelchair. The concept of accessible means of egress includes the idea that evacuating people with a mobility impairment may require the assistance of others. In some situations, provisions are also included for creating an area of refuge or exterior areas of rescue assistance wherein people can safely await either further instructions or evacuation assistance. Larger refuge areas can also be established by utilizing horizontal exits. All of these elements can be arranged in the manner prescribed in this section to provide accessible means of egress.

It is important to note that the subjects of Item 10, exterior areas for assisted rescue, are intended to be used only at the level of exit discharge (see commentary in Chapter 2 for the defined term, "Level of exit discharge"). Exterior exit stairways that are from other floors are addressed by Item 4.

1009.2.1 Elevators required. In buildings where a required *accessible* floor is four or more *stories* above or below a *level of exit discharge*, not less than one required accessible *means of egress* shall be an elevator complying with Section 1009.4.

Exceptions:

1. In buildings equipped throughout with an *automatic sprinkler system* installed in accordance with Section 903.3.1.1 or 903.3.1.2, the elevator shall not be required on floors provided with a *horizontal exit* and located at or above the *levels of exit discharge*.
2. In buildings equipped throughout with an *automatic sprinkler system* installed in accordance with Section 903.3.1.1 or 903.3.1.2, the elevator shall not be required on floors provided with a *ramp* conforming to the provisions of Section 1012.

❖ Elevators are the most common and convenient means of providing access to the upper floors in multistory buildings. As such, elevators represent a prime candidate for accessible means of egress from such buildings, especially in light of the difficulties involved in carrying a person up or down a stairway for multiple levels. The primary consideration for elevators as an accessible means of egress is that the elevator will be available and protected during a fire event to allow for fire department assisted rescue. Typically it is not the intent that people use the elevator for self-evacuation due to the hazards associated with smoke in the elevator shaft or the elevator taking people to the floor with a direct fire hazard. There are some new technological advances for "fire service access elevators" and "occupant evacuation elevators" that are discussed in Sections 403, 3007 and 3008.

This section addresses where an elevator must serve as part of an accessible means of egress. See Section 1104 for when elevators are required for the accessible route into a building. By a reference to Section 1009.4, both an area of refuge and a standby source of power for the elevator are required. The standby power requirement establishes a higher degree of reliability that the elevator will be available and usable by reducing the likelihood of power loss caused by fire or other conditions of power failure.

The code defines "Exit Discharge, Level of" as the story at which an exit terminates. In buildings having four or more stories above or below the level of exit discharge, it is unreasonable to rely solely on exit stairways for all of the required accessible means of egress. This is the point at which complete reliance on assisted evacuation down the stairs will not be effective or adequate because of the limited availability of either experienced personnel who are trained to carry people safely (e.g., fire fighters) or the availability of special devices (i.e., self-braking stairway descent equipment or evacuation chairs). In this case, the code requires that at least one elevator, serving all floors of the building, is to serve as one of the required accessible means of egress. This should not represent a hardship, since elevators are typically provided in such buildings for the convenience of the occupants.

On a flat site, "buildings with four or more stories above the level of exit discharge" would typically be a five-story building. The level of exit discharge is the entire first story level (not merely the plane or level of the first floor); therefore, the fifth floor is the fourth story above the level of exit discharge. In a building with multiple basements, a story four stories below the level of exit discharge would be the fourth basement level. The actual vertical distance is the same from the fifth floor above grade down to discharge, as it is from the fourth floor below grade up to discharge. The verbiage is such that a building built on a sloped site can take into consideration that people may be exiting the building from different levels on different sides of the building (see Commentary Figure 1009.2.1).

Exception 1 establishes that accessible egress elevator service to floor levels at or above the level of exit discharge is not necessary under specified conditions. The conditions are that the building is equipped throughout with an automatic sprinkler system in accordance with NFPA 13 or NFPA 13R (see Section 903.3.1.1 or 903.3.1.2) and that floors not serviced by an accessible egress elevator are provided with a horizontal exit. The presence of an automatic sprinkler system significantly reduces the potential fire hazard and provides for increased evacuation time. The combination of automatic sprinklers and a horizontal exit provides adequate protection for the occupants despite their distance to the level of exit discharge. This exception does not apply to floor lev-

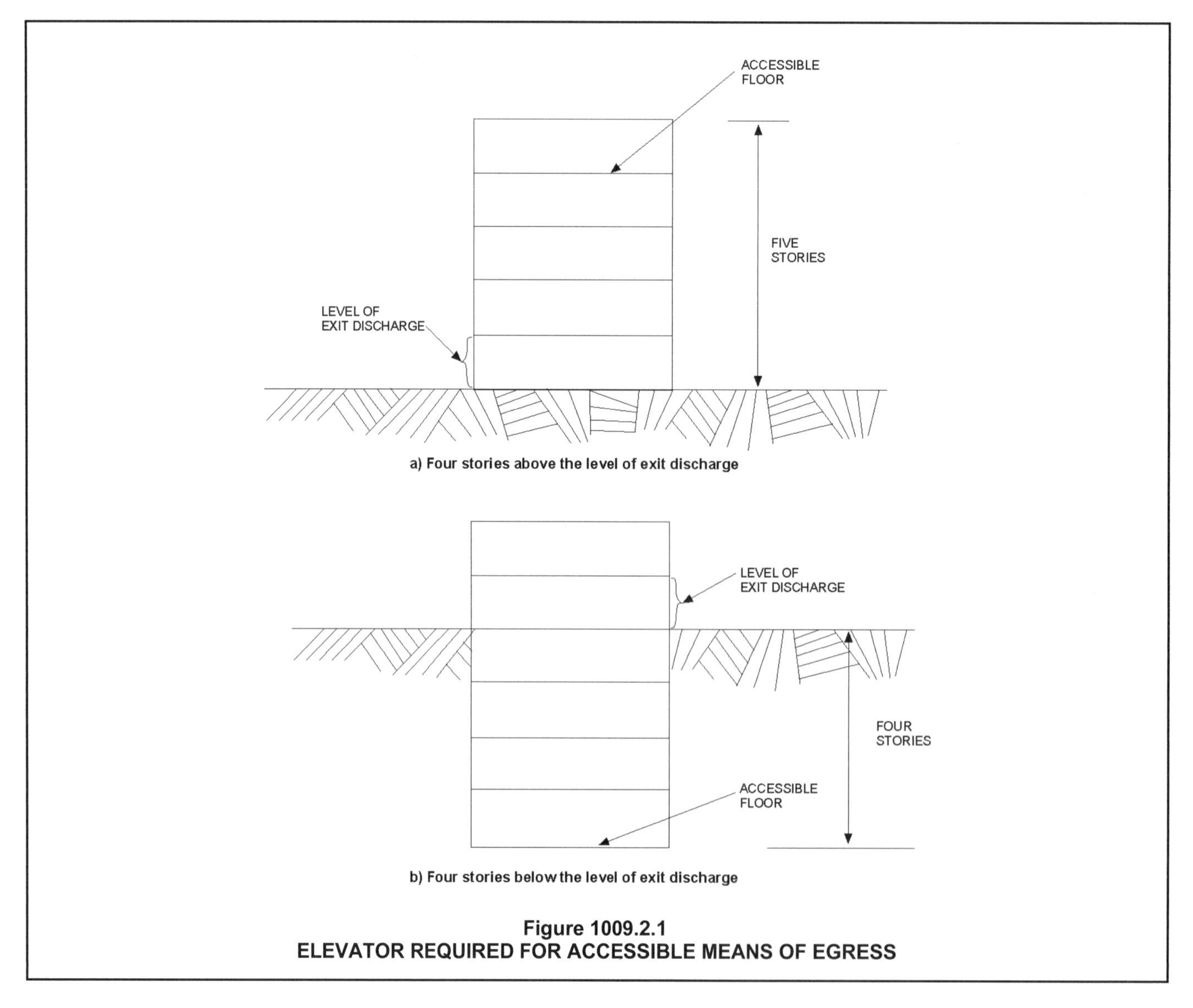

Figure 1009.2.1
ELEVATOR REQUIRED FOR ACCESSIBLE MEANS OF EGRESS

els below the level of exit discharge, since such levels are typically below grade and do not have the added advantage of exterior openings that are available for fire-fighting or rescue purposes. This option is most often utilized when a defend-in-place approach to occupant protection is utilized, such as in a hospital, nursing home or jail. Keep in mind that the horizontal exit (see Section 1026) creates large refuge areas that have separation requirements and capacity requirements that exceed area of refuge requirements.

Exception 2 specifies that a building sprinklered throughout in accordance with NFPA 13 or NFPA 13R (see Section 903.3.1.1 or 903.3.1.2), with ramp access to each level, such as in a sports stadium, is not required to also have an elevator for accessible means of egress. The reasoning behind this is that the issue of carrying people down stairways does not occur because the ramps may be utilized instead.

1009.3 Stairways. In order to be considered part of an accessible *means of egress*, a *stairway* between *stories* shall have a clear width of 48 inches (1219 mm) minimum between *handrails* and shall either incorporate an *area of refuge* within an enlarged floor-level landing or shall be accessed from an *area of refuge* complying with Section 1009.6. *Exit access stairways* that connect levels in the same *story* are not permitted as part of an accessible *means of egress*.

Exceptions:

1. *Exit access stairways* providing *means of egress* from *mezzanines* are permitted as part of an accessible *means of egress*.
2. The clear width of 48 inches (1219 mm) between *handrails* is not required in buildings equipped throughout with an *automatic sprinkler* system installed in accordance with Section 903.3.1.1 or 903.3.1.2.
3. The clear width of 48 inches (1219 mm) between *handrails* is not required for *stairways* accessed from a refuge area in conjunction with a *horizontal exit*.
4. *Areas of refuge* are not required at *exit access stairways* where two-way communication is provided at the elevator landing in accordance with Section 1009.8.
5. *Areas of refuge* are not required at *stairways* in buildings equipped throughout with an *automatic sprinkler system* installed in accordance with Section 903.3.1.1 or 903.3.1.2.
6. *Areas of refuge* are not required at *stairways* serving *open parking garages*.
7. *Areas of refuge* are not required for *smoke-protected assembly seating* areas complying with Section 1029.6.2.
8. *Areas of refuge* are not required at *stairways* in Group R-2 occupancies.
9. *Areas of refuge* are not required for *stairways* accessed from a refuge area in conjunction with a *horizontal exit*.

❖ This section addresses stairways between floor levels or to a mezzanine level (see Exception 1). The last sentence indicates that steps that connect raised or lowered areas on the same level are not permitted to be part of an accessible means of egress. People with mobility impairments cannot be asked to wait at the top of steps that may be anywhere in the building; they must be able to get to the stairways where the fire department will be coming into the building.

Stairways (exit or exit access) between floor levels, while not part of an accessible route, can serve as part of the accessible means of egress when they are used as part of an assisted evacuation route. The starting point for these requirements is that the stairways must be 48 inches (1219 mm) clear width between handrails; and either include or be accessed directly by a location where people can wait for assisted evacuation. This place to wait can be either an "area of refuge" (see Section 1009.6) or a "refuge area" created by a horizontal exit (see Exception 9 and Section 1026).

There are many mobility impairments that can limit or negate a person's ability to walk up and down the stairs. The taller the building, the higher the percentage of the population that will be affected. For example, an elderly person or a person with a broken foot may be able to get down a couple of flights, but not from an upper floor in a high-rise.

Note that this section is for exit stairways as addressed in Sections 1023 and 1027 and exit access stairways as addressed in Section 1019. Therefore, exit access stairways between stories can be considered part of an accessible means of egress, but exit access steps within the same level, such as steps in a corridor or room leading to an exit or exit access doorway, cannot. Stairways that lead from the level of exit discharge (see the commentary to Chapter 2 for the defined term, "Exit discharge, level of") to grade are considered part of the exit discharge. When the exit discharge is not accessible, the options are an interior area of refuge complying with Section 1009.6 or an exterior area of assisted rescue in accordance with Section 1009.7. Do not use the provisions in this section.

The dimension of 48 inches (1219 mm) clear width between handrails is sufficient to enable two or three persons to carry a person up or down to the level of exit discharge where access to a public way is afforded.

The enclosed exit stairway, in combination with an area of refuge, can provide for safety from fire in one of two ways. One approach is for the fire-resistance-rated stairway enclosure to afford the necessary safety. To accomplish this, the landing within the stairway enclosure must be able to contain the wheelchair. The concept is that the person in the wheelchair will remain on the stairway landing for a period of time awaiting further instructions or evacuation assistance; therefore, the stairway landing must be able to accommodate the wheelchair without obstructing the use of the stairway by other egressing occupants. An enlarged, story-level landing is required within the stairway enclosure and must be of sufficient size to accommodate the number of wheelchairs [see Section 1009.6 and Commentary Figure 1009.3(1)].

The other approach is to utilize an enclosed exit stairway that is accessed from an area of refuge complying with Section 1009.6. Under this approach, the stairway is made safe by virtue of its access being in an area that is separated and protected from the point of fire origin. An area of refuge can be created by constructing a vestibule adjacent and with direct access to the stair enclosure [see Section 1009.6 and Commentary Figure 1009.3(2)]. This is similar in theory to the approach of an enlarged landing within the stairway enclosure. Again, the general means of egress path must be available past the wheelchair spaces.

Exceptions may be combined. Exceptions 2 through 9 are either the area of refuge or the 48-inch (1219 mm) stairway width requirement. It is very important to note that an exception for the area of refuge in Section 1009.3 or 1009.4 is not an exception for the accessible means of egress. The accessible route must be available to the stairway or elevator so that people with mobility impairments and emergency responders can meet up as soon as possible.

The exceptions are applicable to all stairways between stories (i.e., exit access, interior exit and exterior exit). The same stairway that serves the ambulatory population for between floor levels can also serve as part of the accessible means of egress. Open exit access stairway requirements for mezzanines and in assembly seating arrangements are addressed in Exceptions 2 and 3 to Section 1009.1.

While the last sentence of Section 1009.3 says that exit access stairways between levels cannot serve as part of an accessible means of egress, Exception 1 allows for exit access stairways between a floor level and a mezzanine. Section 505.1 requires at least a 7-foot clearance below and above a mezzanine. Thus, while it is considered part of the floor below for other parts of the code, there are allowances for the height difference being similar to a story change in level and the difficulty of providing a ramp for that height. Therefore, the open exit access stairway from the mezzanine can serve as part of an accessible means of egress. In a sprinklered building, Exceptions 2 and

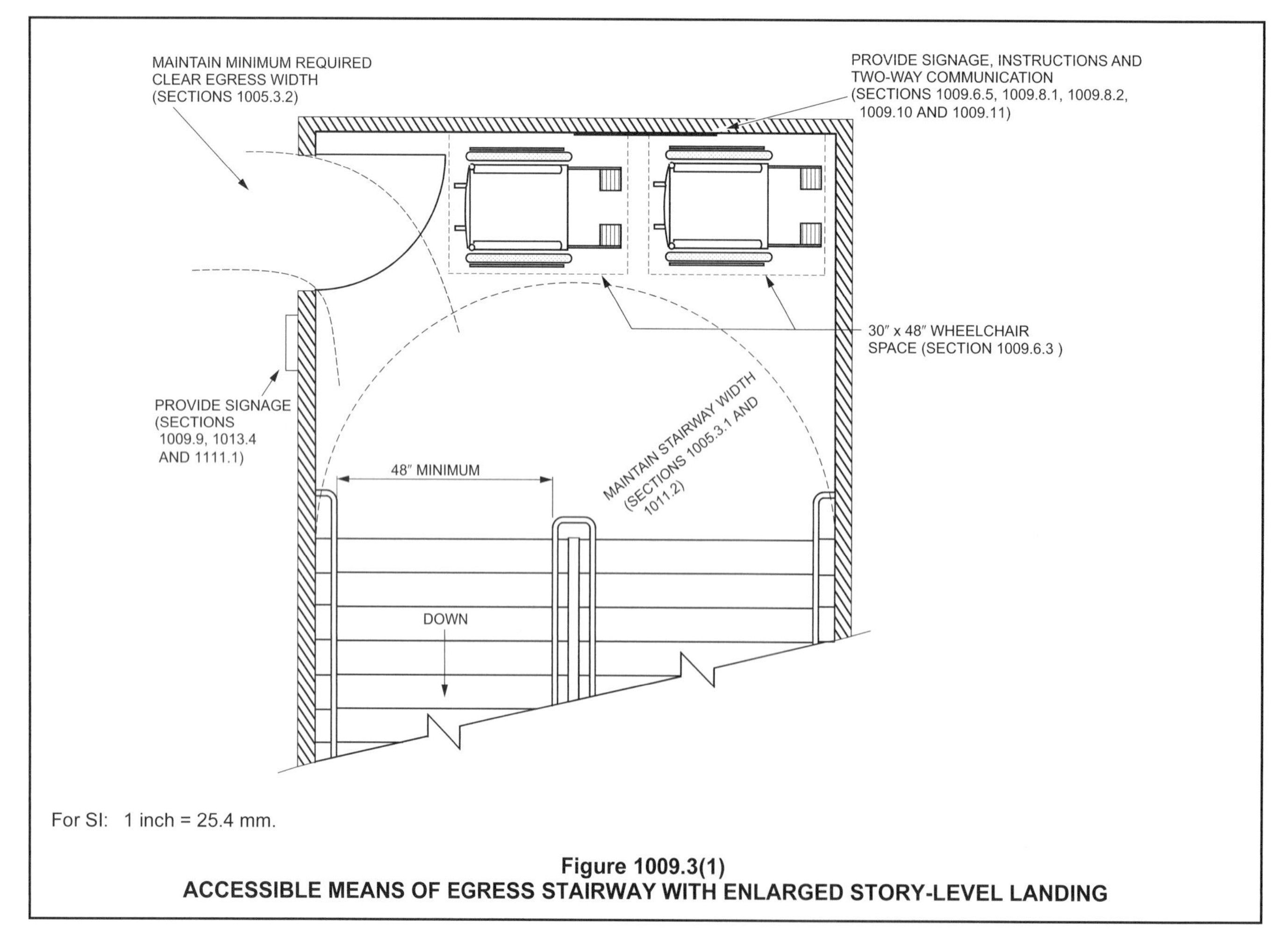

Figure 1009.3(1)
ACCESSIBLE MEANS OF EGRESS STAIRWAY WITH ENLARGED STORY-LEVEL LANDING

5 would allow for a narrower stairway and no area of refuge. In a nonsprinklered building, Exception 4 would allow for two-way communication at the elevator as an alternative to providing an area of refuge at the top of the open stairway.

Exceptions 2 and 5 are in recognition of the increased level of safety and evacuation time that are afforded in a sprinklered occupancy. The expectation is that a supervised system will reduce the threat of fire by reliably controlling and confining the fire to the immediate area of origin. There is also additional safety afforded by sprinkler system requirements for automatic notification when the system is activated. This has been substantiated by a study of accessible means of egress conducted for the General Services Administration (GSA). A report issued by the National Institute for Standards and Technology (NIST), NIST IR 4770, *Staging Areas for Persons with Mobility Limitations*, concluded that the operation of a properly designed sprinkler system eliminates the life threat to all building occupants, regardless of their individual physical abilities, and is a superior form of protection as compared to areas of refuge. It was deemed that the ability of a properly designed and operational automatic sprinkler system to control a fire at its point of origin and to limit production of toxic products to a level that is not life-threatening to all occupants of the building, including persons with disabilities, eliminates the need for areas of refuge.

Exceptions 2 and 3 deal with stairway width. Exceptions 4 through 9 are exceptions for the area of refuge. These are not exceptions for the accessible route to the exit, just the area of refuge at the exit.

Exception 2 allows the stairway width to go back to the base requirements in Section 1011.2 in buildings sprinklered in accordance with NFPA 13 or NFPA 13R for both unenclosed and enclosed exit and exit access stairways (see Sections 1011, 1022 and 1027). Exception 2 is often used in conjunction with Exception 5. With the sprinkler system in place, there is more opportunity for the fire department to bring in evacuation chairs or possibly bring people to the elevator for evacuation, thus the extra width for carrying someone down the stairway is not needed. This is safer for both emergency responders and the evacuees.

Exception 5 is for the area of refuge for all exit access stairways (see Section 1011) and exit stairways, interior and exterior (see Sections 1023 and 1027), where the building is sprinklered throughout with an NFPA 13 or 13R system. Again, this is not an exception for the accessible route to the exit, just the area of refuge at the exit.

Exception 3 allows the stairway width to go back to

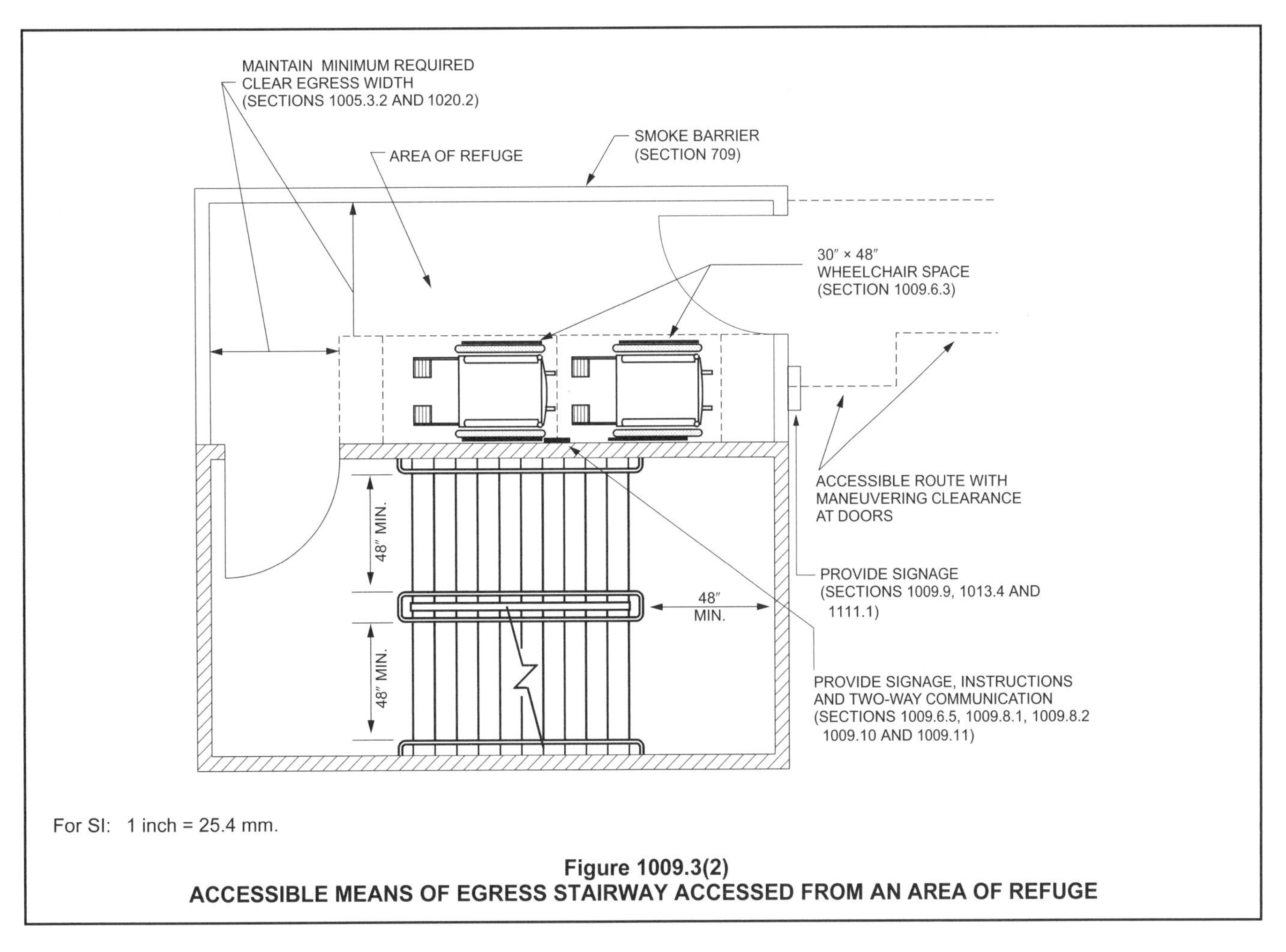

Figure 1009.3(2)
ACCESSIBLE MEANS OF EGRESS STAIRWAY ACCESSED FROM AN AREA OF REFUGE

the base requirements in Section 1011.2 when the stairway is within the refuge area created by a horizontal exit. This exception considers that the extra exiting time will permit the egress down the stairway to be more deliberate. Horizontal exits are often used in hospitals or jails when the defense scenario is defend in place rather than evacuation.

Exception 5 can be used in conjunction with Exception 9. In the case of a horizontal exit [see Commentary Figure 1009.3(3)], each floor area on either side of the exit is considered a refuge area (see commentary, Section 1026.1) by virtue of the construction and separation requirements for horizontal exits. The discharge area is always assumed to be the nonfire side and, therefore, is protected from fire. Therefore, per Exception 9, stairways within this refuge area are not required to have areas of refuge.

Exceptions 6 and 7 are for structures where the natural ventilation of the products of combustion will be afforded by the exterior openings or smoke protection required of such structures (see Sections 406.3, 909 and 1029.6.2). The most immediate hazard for occupants in a fire incident is exposure to smoke and fumes. Floor areas in open parking structures communicate sufficiently with the outdoors such that the need for protection from smoke is not necessary; therefore, open parking garages are exempted from the requirements for an area of refuge (see also the exception to Section 1009.6.4). Because of this level of natural ventilation, parking garage exit stairways are not required to be enclosed (see Section 1019.3, Exception 6). The logic for exterior sports facilities and smoke-protected seating is the same: if there is no accumulation of smoke, there is no need for areas of refuge, even when a sprinkler system is not included (see Section 1019.3, Exception 7).

Exception 8 is in recognition of the dwelling unit separation and fire-resistance-rated corridors in Group R-2 facilities (see Sections 420 and 1020). Effectively, each dwelling unit can serve as a protected area. Since the current text requires all Group R structures to be sprinklered (see Section 903.2.8), Exceptions 2 and 5 could also be utilized.

1009.4 Elevators. In order to be considered part of an accessible *means of egress*, an elevator shall comply with the emergency operation and signaling device requirements of Section 2.27 of ASME/CSA B44 A17.1. Standby power shall be provided in accordance with Chapter 27 and Section 3003. The elevator shall be accessed from an *area of refuge* complying with Section 1009.6.

Exceptions:

1. *Areas of refuge* are not required at the elevator in *open parking garages*.
2. *Areas of refuge* are not required in buildings and facilities equipped throughout with an *automatic sprinkler system* installed in accordance with Section 903.3.1.1 or 903.3.1.2.
3. *Areas of refuge* are not required at elevators not required to be located in a shaft in accordance with Section 712.
4. *Areas of refuge* are not required at elevators serving *smoke-protected assembly seating* areas complying with Section 1029.6.2.

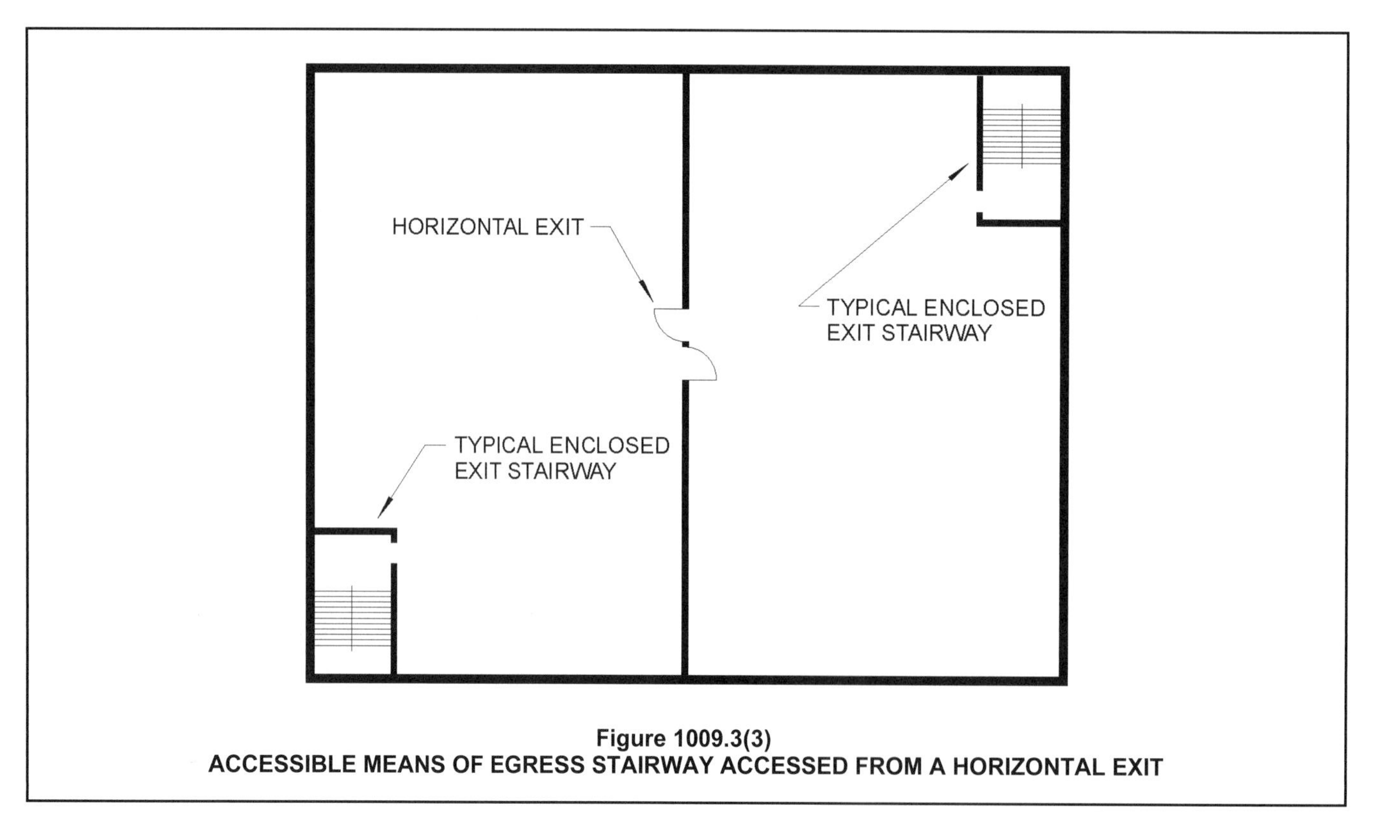

Figure 1009.3(3)
ACCESSIBLE MEANS OF EGRESS STAIRWAY ACCESSED FROM A HORIZONTAL EXIT

5. *Areas of refuge* are not required for elevators accessed from a refuge area in conjunction with a *horizontal exit.*

❖ Elevators are the most common and convenient means of providing access to upper and lower floors in multistory buildings. As such, elevators represent a prime candidate for accessible means of egress from such buildings, especially in light of the difficulties involved in carrying a person in a wheelchair up or down a stairway. The primary consideration for elevators as an accessible means of egress is that the elevator will be available and protected during a fire event. See Sections 403, 3007 and 3008 for new provisions in high-rise buildings for "fire service access elevators" and "occupant evacuation elevators."

This section addresses the use of an elevator as part of an accessible means of egress by requiring both a backup source of power for the elevator and access to the elevator from an area of refuge. For situations where elevators are required to be part of one of the accessible means of egress, see Section 1009.2.1. Note that an elevator lobby that is off a fire-resistance-rated corridor must also comply with Section 713.14 and Chapter 30. The backup power requirement establishes a higher degree of reliability that the elevator will be available and usable by reducing the likelihood of power loss caused by fire or other conditions. Requiring access from an area of refuge affords the same degree of fire safety as described for stairways (see commentary, Section 1009.3). Additionally, the reference to Chapter 27 and Section 3003 clarifies that the elevator will comply with the emergency operation features that relate to operating an elevator under fire conditions (see commentary, Sections 2702.2.2 and 3003). Elevators on an accessible route are also required to meet the accessibility provisions of ICC A117.1 (see commentary, Sections 1109.7 and 3001.3).

Exception 2 is for the area of refuge for all elevators where the building is sprinklered throughout with an NFPA 13 or 13R system. Again, this is not an exception for the accessible route to the exit, just the area of refuge at the elevator with standby power. Exception 2 is in recognition of the increased level of safety and evacuation time that is afforded in a sprinklered occupancy. The expectation is that a supervised system will reduce the threat of fire by reliably controlling and confining the fire to the immediate area of origin. There is also additional safety afforded by sprinkler system requirements for automatic notification when the system is activated. This has been substantiated by a study of accessible means of egress conducted for the GSA. NIST IR 4770 concluded that the operation of a properly designed sprinkler system eliminates the life threat to all building occupants, regardless of their individual physical abilities, and is a superior form of protection as compared to areas of refuge. It was deemed that the ability of a properly designed and operational automatic sprinkler system to control a fire at its point of origin and to limit production of toxic products to a level that is not life threatening to all occupants of the building, including persons with disabilities, eliminates the need for areas of refuge.

If a level in an open parking garage contains accessible parking spaces or is part of the route to and from those spaces, that level is required to have accessible means of egress. Exception 1, for open parking structures, is in recognition of the natural ventilation of the products of combustion that will be afforded by the exterior openings required of such structures (see Section 406.5.2). The most immediate hazard for occupants in a fire incident is exposure to smoke and fumes. Floor areas in open parking structures are sufficiently exposed to the outdoors; thus, the need for protection from smoke is not necessary. Therefore, open parking garages are exempt from the requirements for an area of refuge to access an elevator that is utilized as part of the accessible means of egress. The same idea holds true for smoke-protected seating areas, in accordance with Exception 4. The protection offered by the smoke control system allows for adequate evacuation time before there is danger from smoke and fume accumulation.

Exception 3 allows elevators not required to be enclosed by Section 712 to not have an area of refuge or be accessed by a horizontal exit. If there is no shaft enclosure around the elevator, construction of a smoke-tight compartment immediately in front of the elevator doors would be very difficult. While there are many items listed under Section 712, combined with the height requirements in Section 1009.2.1, typically this would be elevators in atriums or in open and enclosed parking garages. Again, the nature of the location adjacent to an atrium or with the open ramps in parking garages would minimize the chances of smoke accumulation at the elevators.

In the case of a horizontal exit [see Commentary Figure 1009.3(3)], each floor area on either side of the exit is considered a refuge area (see commentary, Section 1026.1) by virtue of the construction and separation requirements for horizontal exits. The discharge area is always assumed to be the nonfire side and, therefore, is protected from fire. Therefore, per Exception 5, any elevator within this refuge area is not required to have areas of refuge.

1009.5 Platform lifts. Platform lifts shall be permitted to serve as part of an accessible *means of egress* where allowed as part of a required *accessible route* in Section 1109.8 except for Item 10. Standby power for the platform lift shall be provided in accordance with Chapter 27.

❖ Previously, there have been concerns about whether a platform lift will be reliably available at all times. However, ASME A18.1, the standard for platform lifts, no longer requires key operation. It is important to note that platform lifts are not prohibited by the code. They simply cannot be counted as a required accessible means of egress in other than locations where they are allowed as part of the accessible route into a space (see commentary, Section 1109.8). When platform lifts are utilized as part of an accessible means of egress, they must come equipped with standby power. Per ASME A18.1, the standby power needs to be sufficient to run the platform lift for at least five round trips. Note that platform lifts cannot be used to meet accessible means of egress requirements for a situation that utilizes Section 1109.8, Item 10. Accessible means of egress must be provided at other locations.

Section 1109.8, Item 10, recognizes that existing site constraints may make installation of a ramp or elevator infeasible. An example would be the situation of dealing with existing public sidewalks, easements and public ways in downtown urban areas. This situation would be most common in hilly areas where the street and sidewalk follow grade and the building's floor is level, resulting in steps up or down at entrances. The concern for allowing this as part of the accessible means of egress is due to standby power requiring only five cycles. If a platform lift was utilized to provide entry for a large building, or for a building with a large number of occupants that may use mobility devices, such as a hospital, it was deemed that there might not be sufficient time or power for everyone to evacuate safely.

In existing buildings undergoing alterations, platform lifts are allowed as part of an accessible route into a building (see commentary, IEBC Sections 410.8.3 and 705.1.3) at any location as long as they are compliant with ASME A18.1. Note that accessible means of egress are not required in existing buildings undergoing an alteration or a change of occupancy (see Sections 1007.1 and IEBC Sections 410.6 and 705.1).

1009.6 Areas of refuge. Every required *area of refuge* shall be accessible from the space it serves by an accessible *means of egress*.

❖ Areas of refuge, when provided, are an important component of fire and safety evacuation plans for buildings. These areas must be included in the plans required by Section 1001.4 of the code and Section 404 of the IFC.

An area of refuge is of no value as part of an accessible means of egress if it is not accessible. The code states an obvious but essential requirement: the path that leads to an area of refuge must qualify as an accessible means of egress. This provision is required so that there will be an accessible route leading from every accessible space to each required area of refuge. See Commentary Figure 1009.6.

1009.6.1 Travel distance. The maximum travel distance from any *accessible* space to an *area of refuge* shall not exceed the *exit access* travel distance permitted for the occupancy in accordance with Section 1017.1.

❖ For consistency in principle with the general means of egress design concepts, the code also limits the travel distance to the area of refuge. The limitation is the same distance as specified in Section 1017.2 for maximum exit access travel distance. This equates the maximum travel distance required to reach an exit with the maximum distance required to reach an area of refuge. It should be noted that an area of refuge is not necessarily an exit in the classic sense. For example, when the area of refuge is an enlarged, story-level landing within an exit stairway, the area of refuge is within the exit and the maximum travel distance for both the conventional exit and the accessible area of refuge is measured to the same point (the entrance to the exit stairway). If the area of refuge is a vestibule immediately adjacent to an enclosed exit stairway, the maximum travel distance for the required accessible means of egress is measured to the entrance of the area of refuge and, for the travel distance to the exit, to the entrance of the exit stairway (see Commentary Figure 1009.6). In the case of accessible means of egress with an elevator, the maximum travel distance may end up being measured along two different paths, with the only consistency between the conventional means of egress and the accessible means of egress being the maximum travel distance (see Commentary Figure 1009.6). The travel distance within an area of refuge is not directly regulated, but will be limited by the general provisions for maximum exit access travel distance, which are always applicable.

In summary, the code takes a reasonably consistent approach for both conventional and accessible means of egress by limiting the distance one must travel to reach a safe area from which further egress to a public way is available.

1009.6.2 Stairway or elevator access. Every required *area of refuge* shall have direct access to a *stairway* complying with Sections 1009.3 and 1023 or an elevator complying with Section 1009.4.

❖ To ensure that there is continuity in an accessible means of egress, the code requires that every area of refuge have direct access to either an exit stairway (see Section 1009.3) or an elevator (see Section 1009.4). This, again, may be viewed as stating the obvious, but it is necessary so that the egress layout does not involve entering an area of refuge and then having to leave that protected area before gaining access to a stairway or elevator. Once an occupant reaches the safety of an area of refuge, that level of protection must be continuous until the vertical transportation element (the stairway or elevator) is reached.

If one chooses to comply with accessible means of egress requirements by providing an accessible elevator with an area of refuge in the form of an elevator lobby, the elevator shaft and the lobby are required to be constructed in accordance with Sections 713.14 and Chapter 30. The requirements provide additional assurance that the elevator will not be rendered unavailable because of smoke movement into the elevator shaft. If the elevator is in a refuge area that is formed by the use of a horizontal exit (i.e., fire walls or fire barriers in accordance with Section 1026.2) or smoke compartments formed by smoke barriers (see Sections 407.5 and 408.6), it is presumed that the refuge area is relatively free from smoke; therefore, the extra protection of Section 713.14 and Chapter 30 may not be needed.

1009.6.3 Size. Each *area of refuge* shall be sized to accommodate one *wheelchair space* of 30 inches by 48 inches (762 mm by 1219 mm) for each 200 occupants or portion thereof, based on the *occupant load* of the *area of refuge* and areas served by the *area of refuge*. Such *wheelchair spaces* shall not reduce the *means of egress* minimum width or required capacity. Access to any of the required *wheelchair spaces* in an *area of refuge* shall not be obstructed by more than one adjoining *wheelchair space*.

❖ The number of wheelchair spaces that are required to be provided in an area of refuge is intended to represent broadly the expected population of the average building. As one point of measurement, a 1977 survey conducted by the National Center for Health indicated that one in 333 civilian, noninstitutionalized persons uses a wheelchair. The 1990 ADA currently utilizes the criterion of one space for each 200 occupants, based on the space served by the area of refuge. Given the variations and difficulties involved in accurately predicting a representative ratio for application to all occupancies, it was concluded that a requirement for one space for each 200 occupants based on the area of refuge itself, plus the areas served by the area of refuge, represents a reasonable criterion. Very few buildings would ever require more than four wheelchair spaces on a floor, since nearly all buildings with an occupant load greater than 400 per floor would be sprinklered and using Exception 5 in Section 1009.3.

Arrangement of the required wheelchair spaces is critical so as not to interfere with the means of egress for ambulatory occupants (see Section 1011.6). Since the design concept is that wheelchair occupants will move to the area of refuge and await further instructions or evacuation assistance, the spaces must be located so as not to reduce the required means of egress width of the stairway, door, corridor or other egress path through the exterior area of rescue.

In order to provide for orderly maneuvering of wheelchairs, this section states that access to any of the required wheelchair spaces cannot be obstructed by more than one adjoining wheelchair space. For example, this precludes an arrangement that three or more wheelchairs could be stacked down a dead-end corridor. This also effectively limits the difficulty any given wheelchair occupant would have in reaching or leaving a given wheelchair space, as well as providing easier access to all wheelchair spaces by persons providing evacuation assistance.

1009.6.4 Separation. Each *area of refuge* shall be separated from the remainder of the story by a *smoke barrier* complying with Section 709 or a *horizontal* exit complying with Section 1026. Each *area of refuge* shall be designed to minimize the intrusion of smoke.

Exceptions:

1. *Areas of refuge* located within an enclosure for *interior exit stairways* complying with Section 1023.

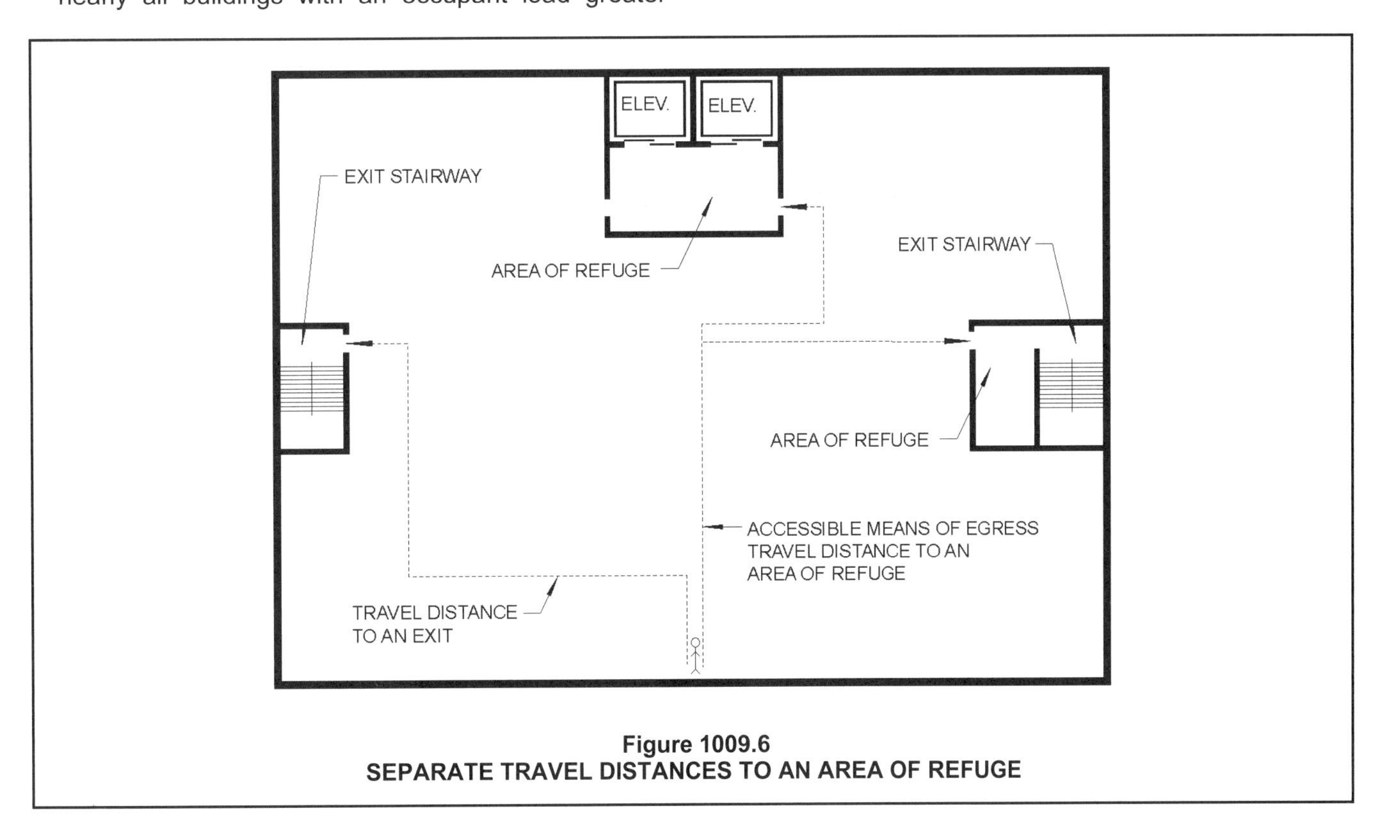

Figure 1009.6
SEPARATE TRAVEL DISTANCES TO AN AREA OF REFUGE

2. *Areas of refuge* in outdoor facilities where *exit access* is essentially open to the outside.

❖ The minimum standard for construction of an area of refuge is a smoke barrier, in accordance with Section 709. This establishes a minimum degree of performance by means of a 1-hour fire-resistance rating, including opening protectives and a minimum degree of performance against the intrusion of smoke into an enclosed area of refuge, as specified in Sections 709.4 and 709.5. By the nature of the connection to the stair enclosure or elevator shaft, the normal smoke barrier requirement for extension from exterior wall to exterior wall is replaced by connection to the shaft enclosure.

An alternative is to provide a refuge area created by a horizontal exit complying with Section 1026. Horizontal exits are formed by fire walls or fire barriers with a minimum fire-resistance rating of 2 hours. The horizontal exit separation must extend vertically through all levels of the building, unless floor assemblies have a fire-resistance rating of not less than 2 hours with no unprotected openings (see Section 1026.2). The other provisions for horizontal exits for additional egress elements, opening protection and capacity must also be complied with.

This section does not require an area of refuge within an exit stairway to be designed to prevent the intrusion of smoke. This was based on a study of areas of refuge conducted by the NIST for the GSA, which concluded that a story-level landing within a fire-resistance-rated exit stairway would provide a satisfactory staging area for evacuation assistance (see Exception 1).

Exception 2 is in recognition that, despite not being a sprinklered venue, where the entire facility is protected from smoke and fumes by the nature of being open to the outside a separation for areas of refuge is not required.

1009.6.5 Two-way communication. *Areas of refuge* shall be provided with a two-way communication system complying with Sections 1009.8.1 and 1009.8.2.

❖ If a building includes areas of refuge at the stairway or elevators, each area of refuge must include a two-way communication system. If the building uses one of the exceptions for areas of refuge, Section 1007.8 would still require a two-way communication system at the elevator. This way anyone needing assistance can communicate with a person at a constantly attended location to request evacuation assistance. This system is an important part of the fire and safety evacuation plans required by Section 1001.4 of the code and Section 404 of the IFC. See Sections 1009.8.1 and 1009.8.2 for specific requirements for this system.

1009.7 Exterior areas for assisted rescue. Exterior areas for assisted rescue shall be accessed by an *accessible route* from the area served.

Where the *exit discharge* does not include an *accessible route* from an *exit* located on the *level of exit discharge* to a *public way*, an exterior area of assisted rescue shall be provided on the exterior landing in accordance with Sections 1009.7.1 through 1009.7.4.

❖ Section 1009.2 requires the accessible means of egress to have an accessible route along the path for exit access, exit and exit discharge. Stairways that lead from the level of exit discharge (see the commentary to Chapter 2 for the defined term, "Exit discharge, level of") to grade are considered part of the exit discharge. When the exit discharge is not accessible, the options are an interior area of refuge complying with Section 1009.6 or an exterior area of assisted rescue in accordance with Section 1009.7. The provisions for stairways in Section 1009.3, including the exceptions for sprinklered buildings, are not an option to address nonaccessible exit discharge.

Exterior areas of assisted rescue are intended to be open-air locations for persons with physical disabilities to wait for assisted rescue. There must be an interior or exterior accessible route along the path of travel for access to this location. This allows a person unable to negotiate the exit discharge to get to a location where they can be quickly discovered by the fire department or other emergency responders.

In most situations, interior areas of refuge are not a positive alternative. Tenants tend to use such areas as convenient storage areas. Where persons with mobility impairments can wait for assisted rescue outside of the building, they are effectively protected from interior smoke and fumes—the deadliest of the fire hazards. Being immediately visible at an exit should also result in a shorter period of time before assisted rescue is achieved.

Exterior areas for assisted rescue, when provided, are an important component of the fire and safety evacuation plans for the buildings. These areas must be included in the plans required by Section 1001.4 of the code and Section 404 of the IFC.

The option under Section 1009.7 is commonly used only at the level of exit discharge for the second exit out the back of a building or tenant space. This will be either a single-story building, or the first level of a multistory building, where the secondary exit discharge is not accessible due to changes in elevation around the perimeter. If this is an exterior exit stairway (i.e., more than one story of vertical travel), the provisions in Section 1009.3 would be applicable.

Examples of this issue are:

Example 1:
A strip mall would have accessible entrances to each tenant in the front (Section 1105.6). Many have service entrances or loading bays across the back, so the second exit door leads to steps. A ramp installed for accessible exit discharge could: be a prohibitively large structure due to the elevation change; block access to the loading docks; get damaged by the maneuvering trucks; or, be impossible because of space restrictions in a narrow alley.

Example 2:
An office building has a second exit that leads to a concrete stoop and the exit discharge is sloped, is uneven or may be blocked by snow. This is just as impassible for a person using a wheelchair as a series of steps. Providing a sidewalk all the way to the front of the building may not be practical because of adjacent buildings or because someone could be travelling adjacent to a burning building.

Sections 1009.7.1 through 1009.7.4 provide criteria for a safe place to wait temporarily for assistance. These address size, separation/protection, openness and any steps leading from the exterior area for assisted rescue.

1009.7.1 Size. Each exterior area for assisted rescue shall be sized to accommodate *wheelchair spaces* in accordance with Section 1009.6.3.

❖ The exterior area for assisted rescue must have an enlarged landing area with space for at least one wheelchair for every 200 occupants that will be using that exit. The wheelchair spaces must be located so that they do not obstruct the general means of egress. If these spaces are confined by walls, guards or edges, they must also meet the alcove provisions in ICC A117.1 so that persons using wheelchairs can maneuver into the space [see Commentary Figure 1009.7.1].

1009.7.2 Separation. Exterior walls separating the exterior area of assisted rescue from the interior of the building shall have a minimum *fire-resistance rating* of 1 hour, rated for exposure to fire from the inside. The fire-resistance-rated exterior wall construction shall extend horizontally 10 feet (3048 mm) beyond the landing on either side of the landing or equivalent fire-resistance-rated construction is permitted to extend out perpendicular to the exterior wall 4 feet (1220 mm) minimum on the side of the landing. The *fire-resistance-rated* construction shall extend vertically from the ground to a point 10 feet (3048 mm) above the floor level of the area for assisted rescue or to the roof line, whichever is lower. Openings within such *fire-resistance-rated* exterior walls shall be protected in accordance with Section 716.

❖ The protection provided by an exterior area for assisted rescue would be equivalent to that required

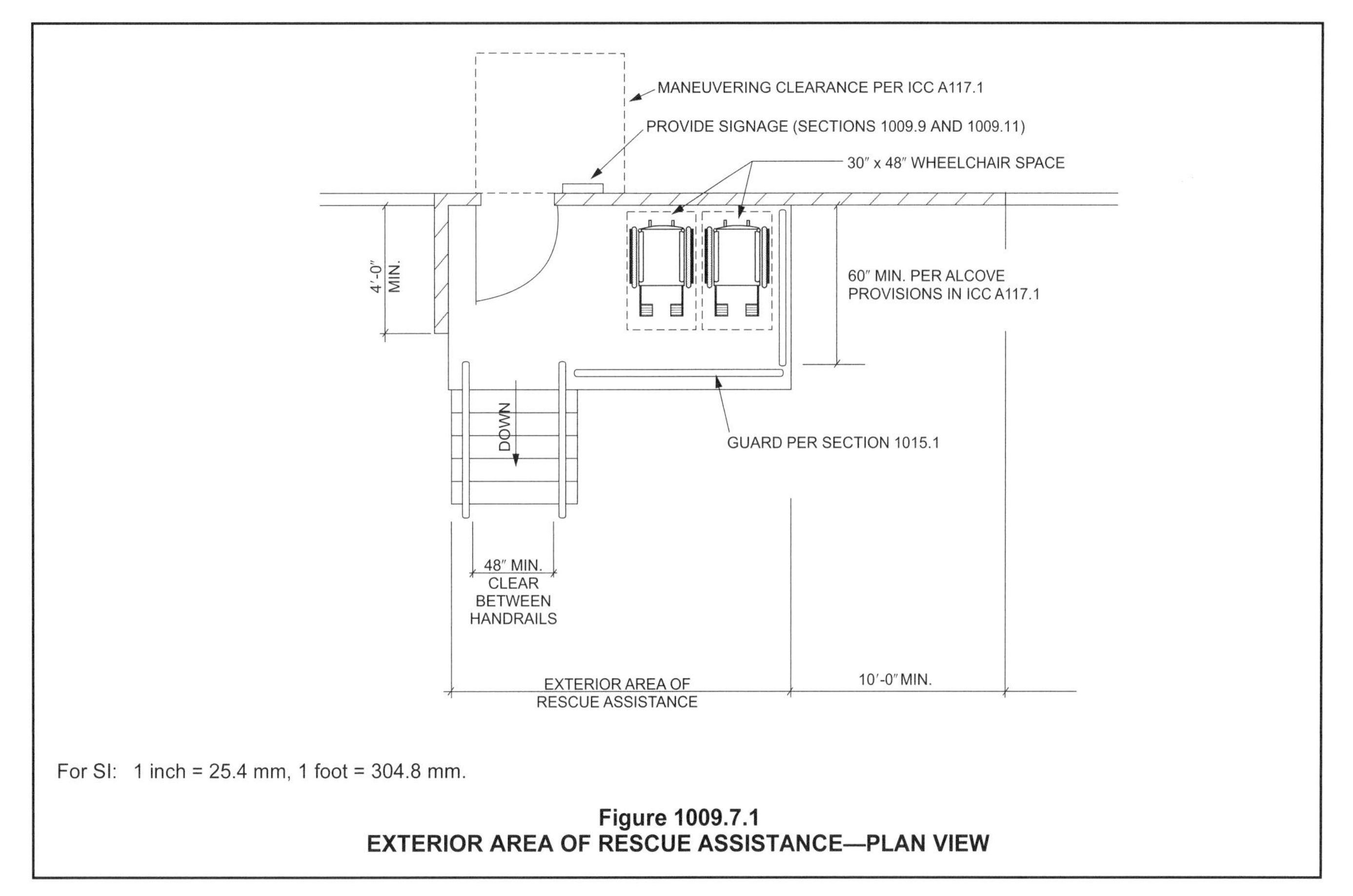

For SI: 1 inch = 25.4 mm, 1 foot = 304.8 mm.

Figure 1009.7.1
EXTERIOR AREA OF RESCUE ASSISTANCE—PLAN VIEW

for an interior area of refuge. Note that there is no exception for the exterior area of assisted rescue for buildings that contain sprinkler systems. The separation requirements are similar to exterior exit stairways, rated walls and protected openings for 10 feet (3048 mm) above, below and to the sides of the landing (see Sections 1023.7 and 1027.6). The exceptions for exterior exit stairway protection in Section 1027.6 would not be applicable where the area includes an exterior area for assisted rescue (see Commentary Figure 1009.7.2).

The current separation requirements address typical rear exit situations. Other locations may require alternative protection measures to "shield" an exterior area for assisted rescue. The principle of "wing" walls (as used at a fire wall extension) suggests an alternative to $^{3}/_{4}$-hour opening protectives at dock doors adjacent to an exterior area for assisted rescue.

Note that providing a rescue location 10 feet (3048 mm) away from an exterior wall does not serve as a viable alternative to a fire-resistance-rated exterior wall. Persons waiting for assistance must have a minimum level of shielding from a fire in the building.

A common situation is for the path of egress travel from the first floor to move through the bottom level of an enclosed exit stairway. If there are steps outside the exit door of the stairway, there is no accessible path for exit discharge. Where approved by the code official (see Section 104.11), the stairway enclosure could be considered equivalent to a protected exterior wall. It is assumed that the fire is in the building somewhere, not in the stairway. A person with a mobility impairment could be provided with a place to wait either inside the stairway enclosure, or outside the building with the stairway enclosure as the separation between them and the fire – as an alternative to Section 1009.7.2. The exterior wall of an interior exit stairway can be nonfire-resistance rated where permitted by Section 1023.7.

1009.7.3 Openness. The exterior area for assisted rescue shall be open to the outside air. The sides other than the separation walls shall be not less than 50 percent open, and the open area shall be distributed so as to minimize the accumulation of smoke or toxic gases.

❖ The openness criteria for exterior areas of assisted rescue are similar to the requirements for exterior balconies. The purpose is to ensure that a person at an exterior area of rescue assistance is not in danger from smoke and fumes. The criteria are to address the situation where the rescue area is open to outside air, but a combination of roof overhangs and perimeter walls or guards could still trap enough smoke that the safety of the occupants would be jeopardized.

1009.7.4 Stairways. *Stairways* that are part of the *means of egress* for the exterior area for assisted rescue shall provide a clear width of 48 inches (1220 mm) between *handrails.*

> **Exception:** The clear width of 48 inches (1220 mm) between *handrails* is not required at *stairways* serving buildings equipped throughout with an *automatic sprinkler system* installed in accordance with Section 903.3.1.1 or 903.3.1.2.

❖ Any steps that lead from an exterior area for assisted rescue to grade must have a clear width of 48 inches (1219 mm) between handrails. The additional width is to permit adequate room to assist a mobility-impaired person down the steps and to a safe location.

If the building is sprinklered, the exception allows for the stairway to utilize the minimum widths required in Section 1011.2. This is consistent with Section 1009.3, Exceptions 2 and 3.

1009.8 Two-way communication. A two-way communication system complying with Sections 1009.8.1 and 1009.8.2

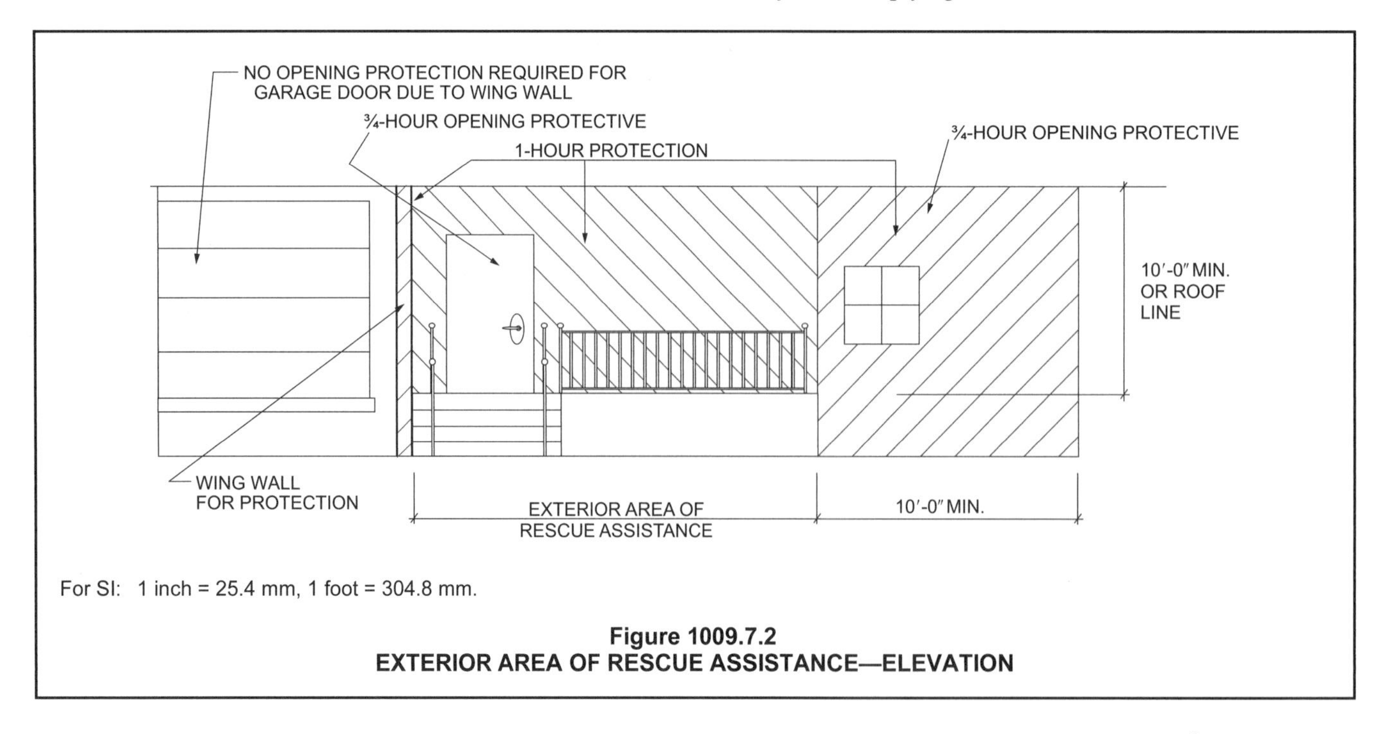

Figure 1009.7.2
EXTERIOR AREA OF RESCUE ASSISTANCE—ELEVATION

shall be provided at the landing serving each elevator or bank of elevators on each accessible floor that is one or more stories above or below the *level of exit discharge*.

Exceptions:

1. Two-way communication systems are not required at the landing serving each elevator or bank of elevators where the two-way communication system is provided within *areas of refuge* in accordance with Section 1009.6.5.
2. Two-way communication systems are not required on floors provided with *ramps* conforming to the provisions of Section 1012.
3. Two-way communication systems are not required at the landings serving only service elevators that are not designated as part of the accessible *means of egress* or serve as part of the required *accessible route* into a facility.
4. Two-way communication systems are not required at the landings serving only freight elevators.
5. Two-way communication systems are not required at the landing serving a private residence elevator.

❖ Unless provided in areas of refuge, in multistory buildings a two-way communication system must be located at the elevator landing of each accessible floor level other than the level of exit discharge. The system is intended to offer a means of communication to individuals with mobility impairment, either permanent or temporary, who need assistance during an emergency situation. Such a system can be useful not only in the event of a fire, but also in the case of a natural or technological disaster by providing emergency responders with the location of individuals who will require assistance in being evacuated from floor levels above or below the discharge level. The ability of emergency responders to locate persons needing assistance quickly is an important part of the fire and safety plan. The two-way communication system is a critical element in that plan.

Exception 1 exempts the requirement for locating the communication systems at the elevator landings where the building is provided with complying areas of refuge. Since areas of refuge are required by Section 1009.6.5 to be equipped with two-way communication systems, there is limited need to provide such additional systems at the elevator landings. However, where multistory buildings are not provided with areas of refuge, such as is the case with most sprinklered buildings, the installation of communications systems at the elevator landings is important to those individuals unable to negotiate egress stairways during an emergency. As a result, both sprinklered and nonsprinklered multistory buildings will be provided with the means for two-way communication at all accessible floor levels other than the level of exit discharge.

Exception 2 applies to floor levels that utilize ramps as vertical accessible means of egress elements. Where complying ramps are available for independent evacuation, such as occurs in a sports stadium, the two-way communication system is not required at the elevator landings.

Because persons at an exterior area of rescue assistance provided at ground level (i.e., level of exit discharge) are immediately visible and such locations are at high risk for vandalism, two-way communication systems are not required for exterior areas of assisted rescue.

If the option of horizontal exits is utilized, the code does not currently address whether a two-way communication system should be provided within a refuge area without an elevator. Since the horizontal exit is not typically recognizable by a person not familiar with the building plan, the most logical location for the two-way communication, if provided, would seem to be adjacent to the exit stairway that was located within the refuge area.

Exceptions 3, 4 and 5 address types of elevators where two way communications are not required. The two-way communication is intended for anyone to be able to communicate with emergency responders. If it is located in the lobby of the public elevator, a system at a back of house service elevator would be redundant and not easy for most occupants to find. A freight elevator cannot be part of an accessible route, so again, this is not the elevator that occupants would typically use. The ASME A17.1 limits the use of private residence elevators to within or serving individual dwelling units. If a person lives in a unit with an elevator, it is not unreasonable to expect them to address communication needs that may arise on their own, such as carrying a portable or cell phone with them.

1009.8.1 System requirements. Two-way communication systems shall provide communication between each required location and the *fire command center* or a central control point location *approved* by the fire department. Where the central control point is not a *constantly attended location*, a two-way communication system shall have a timed automatic telephone dial-out capability to a monitoring location or 9-1-1. The two-way communication system shall include both audible and visible signals.

❖ Use of an elevator, stair enclosure or other area of refuge as part of an accessible means of egress requires a person to wait for evacuation assistance or relevant instructions. The two-way communication system allows this person to inform emergency personnel of his or her location and to receive additional instructions or assistance as needed.

The arrangement and design of the two-way communication system is specified in Section 1009.8.1. In addition to the required locations specified in Section 1009.6.5 for areas of refuge or Section 1009.8 for elevator landings, a communication device is also required to be located in a high-rise building's fire command center or at a central control point whose location is approved by the fire department (see Sections 403.4.4 and 907.2.13.2). "Central control point" is not a defined term. However, given the intent and function of the two-way communication system, a central control point is a location where an individual answers the call for assistance and either provides or requests aid for a person who needs help. A suitable central control point is often not available in low-rise buildings or in a high-rise building where the central control point may not be manned on a 24-hour basis. In order that a caller may reach an appropriate emergency location, the fire department must approve the configuration of the system. A central control point could be the lobby of a building constantly staffed by a security officer, an alarm company, a public safety answering point such as a 9-1-1 center or a central supervising station in a Group I occupancy. There could be a combination solution—such as a system configured to automatically call 9-1-1 when the central control point within the building is not manned. The communication system provides visual signals for the hearing impaired and audible signals to assist the vision impaired.

1009.8.2 Directions. Directions for the use of the two-way communication system, instructions for summoning assistance via the two-way communication system and written identification of the location shall be posted adjacent to the two-way communication system. Signage shall comply with the ICC A117.1 requirements for visual characters.

❖ Guidance to the users of a two-way communication system is also specified. Operating instructions for the two-way communication system must be posted and the instructions are to include a means of identifying the physical location of the communication device. If a signal from a two-way communication system terminates to a public safety answering point, such as a fire department communication center, current 9-1-1 telephony technology only reports the address of the location of the emergency—it does not report a floor or area from the address reporting the emergency. The "identification of the location" posted adjacent to the communication system should ensure that most discrete location information can be provided to the central control point. This will aid emergency responders, especially in high-rise buildings or corporate campuses with multiple multistory structures. The signage is not required to be raised letters or braille, but is required to meet the style, size and contrast requirements for visual signage in A117.1.

1009.9 Signage. Signage indicating special accessibility provisions shall be provided as shown:

1. Each door providing access to an *area of refuge* from an adjacent floor area shall be identified by a sign stating: AREA OF REFUGE.
2. Each door providing access to an exterior area for assisted rescue shall be identified by a sign stating: EXTERIOR AREA FOR ASSISTED RESCUE.

Signage shall comply with the *ICC A117.1* requirements for visual characters and include the International Symbol of Accessibility. Where exit sign illumination is required by Section 1013.3, the signs shall be illuminated. Additionally, visual characters, raised character and braille signage complying with ICC A117.1 shall be located at each door to an *area of refuge* and exterior area for assisted rescue in accordance with Section 1013.4.

❖ Signage enables an occupant to become aware of an area of refuge and/or the exterior area for rescue assistance. The assistance areas must provide signage on or above the door stating either "AREA OF REFUGE" or "EXTERIOR AREA FOR ASSISTED RESCUE" and includes the International Symbol of Accessibility. The approach that the code takes for identification of the area of refuge is comparable to the general provisions for identification of exits, including the requirement for lighted signage. Raised letters and braille stating "EXIT" are also required adjacent to the door for the benefit of persons with a visual impairment.

The current text does not clearly indicate how to identify a refuge area formed by a horizontal exit. In hospitals and jails, where this option is typically utilized, the location of the horizontal exits must be part of the staff training for the fire safety and evacuation plans.

1009.10 Directional signage. Directional signage indicating the location of all other *means of egress* and which of those are accessible *means of egress* shall be provided at the following:

1. At *exits* serving a required *accessible* space but not providing an approved accessible *means of egress.*
2. At elevator landings.
3. Within *areas of refuge.*

❖ The additional signage required by this section is intended to advise persons of the locations of all means of egress and which of those also serve as accessible means of egress. Since not all of the exits will necessarily be accessible means of egress, it is appropriate to provide this information at exit stairways and, particularly, at all elevators, regardless of whether they are part of an accessible means of egress. Directional signage is not required to meet raised character or braille signage requirements. Depending on the facility, this could be as simple as a basic block plan of the main corridors and stairways in the building in relation to the elevator.

1009.11 Instructions. In *areas of refuge* and exterior areas for assisted rescue, instructions on the use of the area under emergency conditions shall be posted. Signage shall comply with the ICC A117.1 requirements for visual characters. The instructions shall include all of the following:

1. Persons able to use the *exit stairway* do so as soon as possible, unless they are assisting others.
2. Information on planned availability of assistance in the use of *stairs* or supervised operation of elevators and how to summon such assistance.

3. Directions for use of the two-way communication system where provided.

❖ The instructions provided at the exterior area of rescue assistance and the areas of refuge will differ. The required instructions on the proper use of the area of refuge and the communication system provide a greater likelihood that the communication system will accomplish its intended function and occupants will behave as expected. A two-way communication system will not be of much value if a person in that area does not know how to operate it. Also, since the area of refuge is required by Section 1009.9 to be identified as such, ambulatory occupants may mistakenly conclude that they should remain in that area. The instructions remind ambulatory occupants that they should continue to egress as soon as possible.

For an exterior area of assisted rescue at grade level, a two-way communication system is not required so this portion of the instructions is not needed. However, instructions for any ambulatory persons to move to the exit discharge are still required, as well as information on how assistance will be provided at this location.

Since each building's means of egress and fire and safety evacuation plans are unique, specific requirements for verbiage are not indicated, but will depend on the situation. The signage is not required to be raised letters or braille, but is required to meet the style, size and contrast requirements for visual signage in A117.1.

SECTION 1010 DOORS, GATES AND TURNSTILES

1010.1 Doors. *Means of egress* doors shall meet the requirements of this section. Doors serving a *means of egress* system shall meet the requirements of this section and Section 1022.2. Doors provided for egress purposes in numbers greater than required by this code shall meet the requirements of this section.

Means of egress doors shall be readily distinguishable from the adjacent construction and finishes such that the doors are easily recognizable as doors. Mirrors or similar reflecting materials shall not be used on *means of egress* doors. *Means of egress* doors shall not be concealed by curtains, drapes, decorations or similar materials.

❖ The general requirements for doors are in this section and the following subsections. The reference to Section 1022.2 is intended to emphasize that exterior exit doors must lead to a route that will allow a path to a public street or alley (see definition for "Public way"). A door that is intended to be used for egress purposes, even though that door may not be required by the code, is also required to meet the requirements of this section. An example may be an assembly occupancy where four doors would be required to meet the required capacity of the occupant load. But assume the designer elects to provide six doors for aesthetic reasons or occupant convenience. All six doors must comply with the requirements of this section.

Doors need to be easily recognizable for immediate use in an emergency condition. Thus, the code specifies that doors are not to be hidden in such a manner that a person would have trouble seeing where to egress.

1010.1.1 Size of doors. The required capacity of each door opening shall be sufficient for the *occupant load* thereof and shall provide a minimum clear width of 32 inches (813 mm). Clear openings of doorways with swinging doors shall be measured between the face of the door and the stop, with the door open 90 degrees (1.57 rad). Where this section requires a minimum clear width of 32 inches (813 mm) and a door opening includes two door leaves without a mullion, one leaf shall provide a clear opening width of 32 inches (813 mm). The maximum width of a swinging door leaf shall be 48 inches (1219 mm) nominal. *Means of egress* doors in a Group I-2 occupancy used for the movement of beds shall provide a clear width not less than $41^1/_2$ inches (1054 mm). The height of door openings shall be not less than 80 inches (2032 mm).

Exceptions:

1. The minimum and maximum width shall not apply to door openings that are not part of the required *means of egress* in Group R-2 and R-3 occupancies.
2. Door openings to resident *sleeping units* in Group I-3 occupancies shall have a clear width of not less than 28 inches (711 mm).
3. Door openings to storage closets less than 10 square feet (0.93 m^2) in area shall not be limited by the minimum width.
4. Width of door leaves in revolving doors that comply with Section 1010.1.4.1 shall not be limited.
5. Door openings within a *dwelling unit* or *sleeping unit* shall be not less than 78 inches (1981 mm) in height.
6. Exterior door openings in *dwelling units* and *sleeping units*, other than the required *exit* door, shall be not less than 76 inches (1930 mm) in height.
7. In other than Group R-1 occupancies, the minimum widths shall not apply to interior egress doors within a *dwelling unit* or *sleeping unit* that is not required to be an *Accessible unit*, *Type A unit* or *Type B unit*.
8. Door openings required to be *accessible* within *Type B units* shall have a minimum clear width of 31.75 inches (806 mm).
9. Doors to walk-in freezers and coolers less than 1,000 square feet (93 m^2) in area shall have a maximum width of 60 inches (1524 mm).
10. In Group R-1 *dwelling units* or *sleeping units* not required to be *Accessible units*, the minimum

width shall not apply to doors for showers or saunas.

❖ The size of a door opening determines its capacity as a component of egress and its ability to fulfill its function in normal use. A door opening must meet certain minimum criteria as to its width and height in order to be used safely and to provide accessibility to people with physical disabilities. Doorways that are not in the means of egress are not limited in size by this section. However, doors that are used for egress purposes, including additional doors over and above the number of means of egress required by the code, are required to meet the requirements of this section unless one of the exceptions applies.

The minimum clear width of an egress doorway for occupant capacity is based on the portion of the occupant load (see Section 1004.1) intended to utilize the doorway for egress purposes, multiplied by the egress width per occupant from Section 1005.1. The capacity of a 32-inch (813 mm) clear width door is 32/0.2 = 160 occupants. The 0.15-inch (3.81 mm) allowance for capacity is permitted in sprinklered buildings (32/0.15 = 213 occupants). The clear width of a swinging door opening is the horizontal dimension measured between the face of the door and the door stops when the door is in the 90-degree (1.57 rad) position [see Commentary Figure 1010.1.1(1)].

Using the face of the door as the measurement point is consistent with the provisions of ICC A117.1 and the ADAAG Review Advisory Committee. Further, this measurement is not intended to prohibit other projections into the required clear width, such as latching or panic hardware [see the commentary to Section 1010.1.1.1 and Commentary Figure 1010.1.1(2) for further discussion on the specific projections allowed in the required clear width]. For nonswinging means of egress doors, such as a sliding door, the clear width is to be measured from the face of the door jambs.

The minimum clear width in a doorway of 32 inches (813 mm) is to allow passage of a wheelchair as well as persons utilizing walking devices or other support apparatus. Similarly, because of the difficulties that a person with physical disabilities would have in opening a pair of doors simultaneously, the 32-inch (813 mm) minimum must be provided by a single door leaf.

Note that in some cases, with standard door construction and hardware, a 36-inch-wide (914 mm) door is the narrowest door that can be used while still providing the minimum clear width of 32 inches (813 mm). A standard 34-inch-wide (864 mm) door has less than a 32-inch (813 mm) clear opening depending on the thickness of the opposing doorstop, the door thickness and the type of hinge. The building designer must verify that the swinging door specified will in fact provide the required clear width.

A minimum clear width of $41^1/_2$ inches (1054 mm) is required for doors in any portion of Group I-2 where patients may need to be moved in beds. This is especially important for evacuating patients from the area in the event of a fire.

The maximum width for a means of egress door leaf in a swinging door is 48 inches (1219 mm) because larger doors are difficult to handle and are of sizes that typically are not fire tested. The maximum width only applies to swinging doors and not to horizontal sliding doors.

Minimum door heights are required to provide clear headroom for the users. A minimum height of 80 inches (2032 mm) has been empirically derived as sufficient for most users. Note that although the clear height of a doorway is not specified, typical door frame dimensions will render an opening very close to 80 inches (2032 mm) in clear height. The exception in Sections 1003.3.1 and 1010.1.1.1 allows for door closers and doorstops to be as low as 78 inches (1981 mm).

Exception 1 is very limited in scope and is primarily

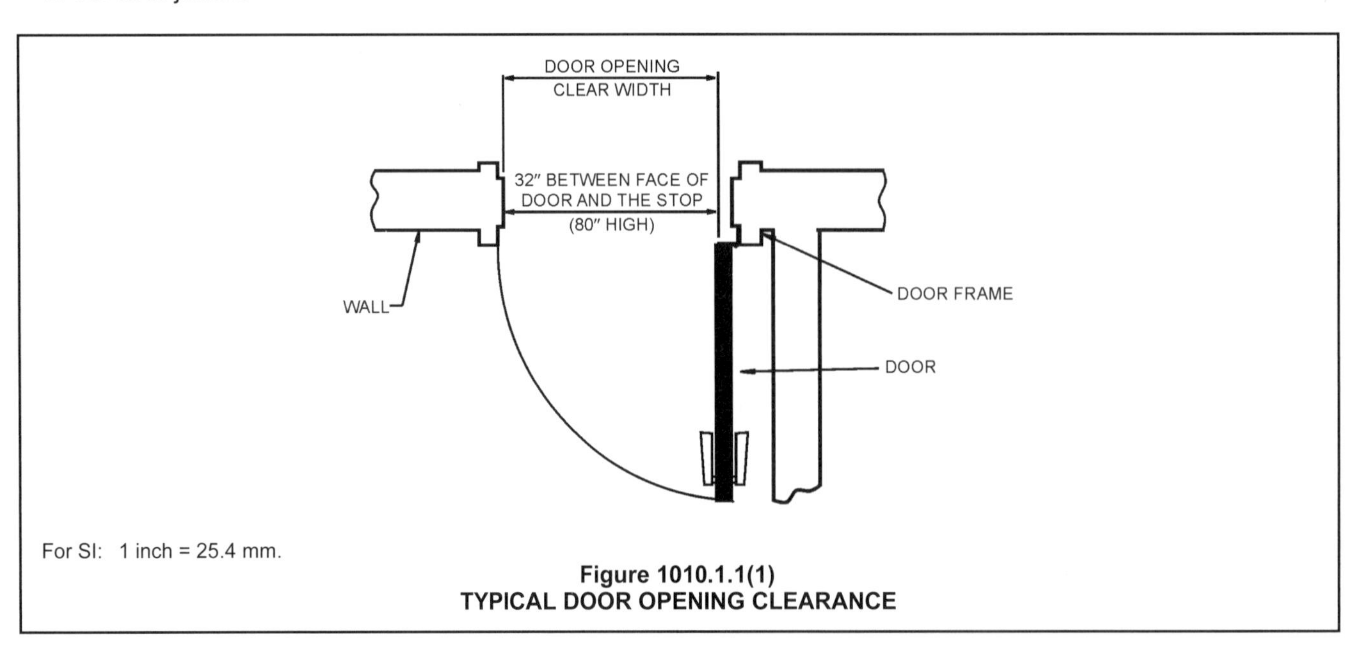

Figure 1010.1.1(1)
TYPICAL DOOR OPENING CLEARANCE

intended to permit decorative-type doors, e.g., café doors, in dwelling units. This exception addresses spaces that are provided with two or more doors when only one means of egress is required. These nonrequired doors are exempted from the minimum and maximum dimensions.

Exception 2 permits the continued use of doors to resident sleeping rooms (cells) in jails (Group I-3) to be a minimum of 28 inches (711 mm), according to current practices.

Exception 3 is for all occupancies. This exception permits doors to storage closets less than 10 square feet (0.9 m^2) in area to be less than 32 inches (813 mm). This provision is intended to include those closets that can be reached in an arm's length and thus do not require full passage into the closet to be functional.

Exception 4 permits the door leaves in a revolving door assembly to comply with Section 1010.1.4.1, which provides for adequate egress width when the revolving door is collapsed into a breakout position.

Exception 5 permits the doorway within a dwelling or sleeping unit to be a minimum of 78 inches (1981 mm) in clear height. This is deemed acceptable because of the familiarity persons in a dwelling or sleeping unit usually have with the egress system and the lack of adverse injury statistics relating to such doors. Note that this exception does not apply to exterior doors of a townhouse or the main entrance doors leading to the hallway in hotels or apartment buildings. However, exterior doors could use the limited exception for doorstops and closers in Sections 1003.3.1 and 1010.1.1.1.

Exception 6 permits exterior doorways to a dwelling or sleeping unit, except for the required exit door, to be a minimum of 76 inches (1930 mm) in clear height. Accordingly, the required exterior exit door to a dwelling or sleeping unit must be 80 inches (2032 mm) in height (exterior doors are not within the scope of Exception 5), but other exterior doors are allowed to be a height of only 76 inches (1930 mm). This provision allows for the continued use of 76-inch-high (1930 mm) sliding patio doors and swinging doors sized to replace such doors.

Exception 7 allows interior means of egress doors within dwelling or sleeping units to have a clear width less than 32 inches (813 mm). If the dwelling or sleeping unit is required to be an Accessible, Type A or Type B unit, this exception is not applicable. ICC A117.1 requires door openings within Accessible and Type A units to be 32 inches (813 mm) clear and doors within Type B units to be 31$^3/_4$ inches (806 mm) clear. This exception is not applicable to Group R-1. The requirement for all doorways within a Group R-1 unit to be sized to provide access to persons with physical disabilities is applicable to both entrance doors to the units and all doors to rooms in the unit (e.g., bathroom doors). Because of the social interaction and visitation that often occur in lodging facilities, a door opening sized for accessibility (e.g., wheelchairs, walkers, canes, crutches) is deemed necessary to allow people with disabilities to visit a friend's, colleague's or relative's unit. In addition, wider doors provide an additional benefit to all persons handling luggage and bulky items, or for the situation when an Accessible unit is not available. This requirement for Group R-1 occupancies is consistent with the 2010 ADA Standard.

Exception 10 has put in an allowance for sauna and steam room doors in hotels rooms that are not Accessible units. The doors to the bathroom would still have to provide a 32-inch clear width.

Exception 8 addresses the clear width of doors within a Type B dwelling or sleeping unit. The 31$^3/_4$-inch (806

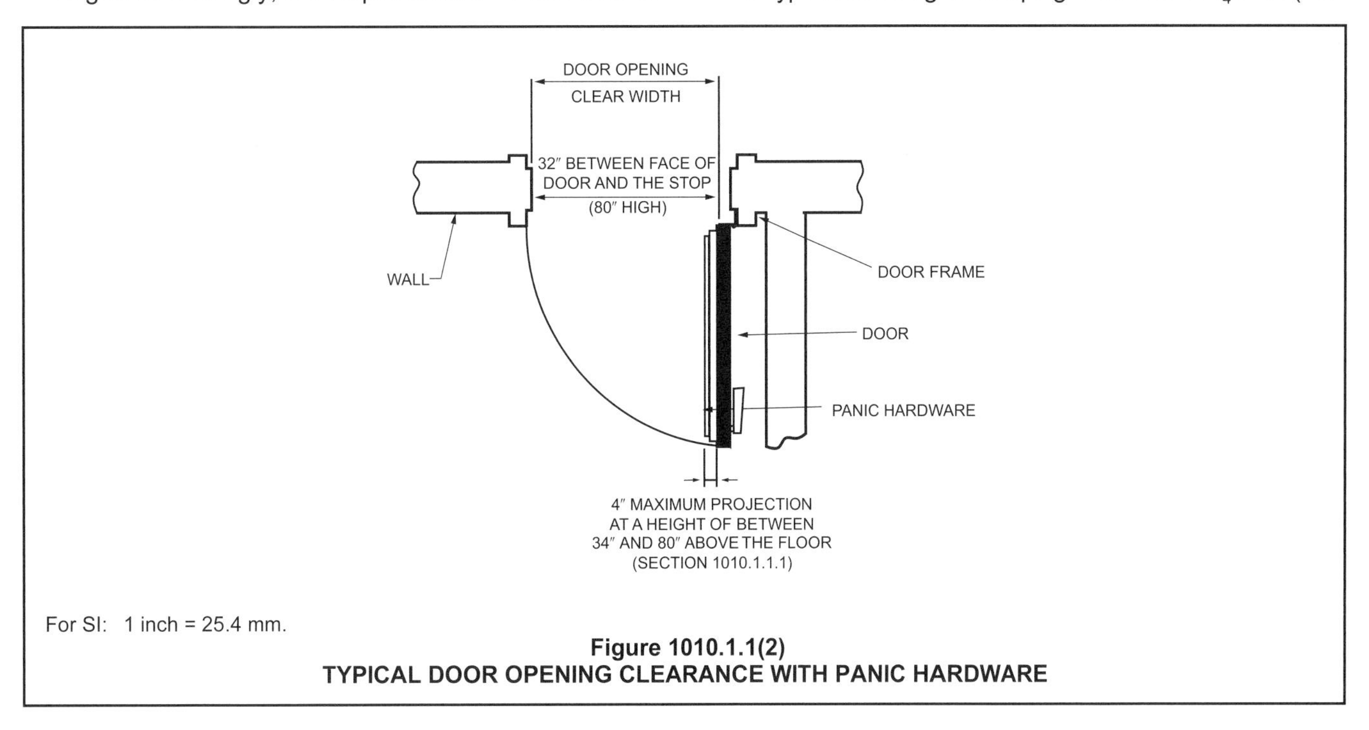

Figure 1010.1.1(2)
TYPICAL DOOR OPENING CLEARANCE WITH PANIC HARDWARE

mm) dimension effectively allows for 2-foot, 10-inch (864 mm) doors to be used inside the unit. Again, note that the exterior door to the garden-style apartments or the main door to the hallway from units in an apartment building are not covered by this exception. ICC A117.1 requires the exterior doors of Type B dwelling units to provide a 32-inch (813 mm) clear width. This is consistent with the correlative text in ICC A117.1 for Type B units. Refer to Chapter 11 for additional information related to Type B dwelling and sleeping units.

Exception 9 allows for doors on walk-in freezers and coolers that would allow for the use of small carts to move supplies in and out. Such doors would still have to meet the force requirements in Section 1010.1.3.

1010.1.1.1 Projections into clear width. There shall not be projections into the required clear width lower than 34 inches (864 mm) above the floor or ground. Projections into the clear opening width between 34 inches (864 mm) and 80 inches (2032 mm) above the floor or ground shall not exceed 4 inches (102 mm).

Exception: Door closers and door stops shall be permitted to be 78 inches (1980 mm) minimum above the floor.

❖ This section of the code provides specific allowances for projection into the required clear widths of means of egress doors. These allowances directly correspond with the method of measuring the required clear width of the door as specified in Section 1010.1.1. A reasonable range of projections for door hardware and trim has been established by these requirements. The use of the means of egress door by a wheelchair occupant will not be significantly impacted by small projections located in inconspicuous areas. The key to these allowances is their location. Projections are allowed at a height between 34 inches (864 mm) and 80 inches (2032 mm). Below the 34-inch (864 mm) height, the code does not permit any projections since they would decrease the available width for wheelchair operation. The full 32-inch (813 mm) width must be provided at this location. At 34 inches (864 mm) and higher, projections of up to and including 4 inches (102 mm) are permitted. The 4-inch (102 mm) projection is consistent with the allowances of Section 1003.3.3. This section permits door hardware, such as panic hardware, to extend into the clear width, yet maintain accessibility for persons with physical disabilities [see Commentary Figure 1010.1.1(2)].

Allowance must be made for door closers and stops, since their design and function necessitates placement within the door opening. The minimum headroom clearance for door closers and stops is allowed to be 78 inches (1981 mm) [see Commentary Figure 1003.3.1(1)]. The 2-inch (51 mm) projection into the doorway height is reasonable since these devices are normally mounted away from the center of the door opening, thus minimizing the potential for contact with a person moving through the opening. This is consistent with the exception in Section 1003.3.1. Other items that are mounted at the top of the door opening, such as an electromagnetic lock on a pair of doors, would still require an 80-inch (2032 mm) minimum headroom.

While this section deals with door hardware projection within the clear door opening width, door hardware projection into the required width of corridors, aisles, exit passageways and exit discharge is addressed in Section 1005.7.1.

1010.1.2 Door swing. Egress doors shall be of the pivoted or side-hinged swinging type.

Exceptions:

1. Private garages, office areas, factory and storage areas with an *occupant load* of 10 or less.
2. Group I-3 occupancies used as a place of detention.
3. Critical or intensive care patient rooms within suites of health care facilities.
4. Doors within or serving a single *dwelling unit* in Groups R-2 and R-3.
5. In other than Group H occupancies, revolving doors complying with Section 1010.1.4.1.
6. In other than Group H occupancies, special purpose horizontal sliding, accordion or folding door assemblies complying with Section 1010.1.4.3.
7. Power-operated doors in accordance with Section 1010.1.4.2.
8. Doors serving a bathroom within an individual *sleeping unit* in Group R-1.
9. In other than Group H occupancies, manually operated horizontal sliding doors are permitted in a *means of egress* from spaces with an *occupant load* of 10 or less.

❖ Generally, egress doors are required to be the side-swinging type. The swinging hardware can be either a hinge or a pivot type (see the definition for "Balanced door"). Side-swinging doors are familiar to all occupants in the method of operation. Door designs with pivots are permitted by this section since the door action itself has little difference between the side-hinged-type door.

The code has several conditions where it allows doors that are not side-hinged-swinging types.

Examples of the doors permitted in Exception 1 are overhead garage doors and horizontal sliding doors. Exception 1 allows doors other than the swinging type for the listed uses where the number of occupants is very low.

Exception 2 allows for the sliding-type doors that are commonly used in prisons and jails.

Exception 3 allows for sliding doors between nursing areas and patient rooms in critical care and intensive care suites. Patients are not typically moving around on their own in these areas, visitors are extremely limited, the glass doors allow a better view for nurse supervision and the sliding option allows for equipment locations unaffected by door swing. See also Sections 407, 1010.1.9.6 and 1010.1.9.8 for these types of areas.

Exception 4 allows for sliding-type doors or pocket-type doors within or serving individual units in a non-transient residential occupancy. Residents are typically familiar with the door operation. The use of sliding doors on the interior of dwelling units is permitted by the Fair Housing Accessibility Guidelines (FHAG) and by ICC A117.1 for Accessible, Type A and Type B units.

Exception 5 allows for revolving doors that meet the requirements of Section 1010.1.4.1. Revolving doors are not permitted for egress from high-hazard spaces.

Exception 6 allows for special-purpose horizontal sliding, accordion or folding doors that meet the requirements of Section 1010.1.4.3 to be used in the means of egress. The doors addressed by Section 101.4.3 are commonly in the normally-open position (hidden in their enclosure). In the event of fire or smoke, where these doors are installed in the means of egress, Section 1010.1.4.3 requires the doors to be power operated but also openable manually to the required minimum egress width. This exception is intended to allow wide span openings to be used in a means of egress.

Exception 7 allows for power-operated doors that meet the requirements of Section 1010.1.4.2. This is to enhance the movement of the general population as well as people with mobility impairments to areas of safety without obstructions, since the specified doors afford simple operation by persons for both typical and emergency operation.

Exception 8 allows for pocket doors between the bathrooms and living or sleeping space within hotel rooms. Since the bathroom is most commonly placed immediately inside the entrance to the room, a side-swinging door could be an obstruction for a person entering carrying suitcases. Familiarity with these types of doors and minimal occupant loads makes this situation acceptable.

Exception 9 partially overlaps the allowances in Exception 1 for horizontal sliding doors by matching the 10 or less occupant load, but extends the use to all other groups except for high hazard. For example, some emergency rooms or clinics use glazed horizontal sliding doors to divide patient care rooms providing for increased privacy and infection control while still allowing visual supervision. Another example would be that a pocket door may be used for access to a bathroom within a private office. The allowance for such a manually operated horizontal sliding door provides greater design flexibility and efficiency, while at the same time maintaining an acceptable level of safety.

1010.1.2.1 Direction of swing. Pivot or side-hinged swinging doors shall swing in the direction of egress travel where serving a room or area containing an occupant load of 50 or more persons or a Group H occupancy.

❖ A side-hinged door must swing in the direction of egress travel where the required occupant capacity of the room is 50 or more. As such, a room with two doors and an occupant load of 99 would require both doors to swing in the direction of egress travel, even though each door has a calculated occupant usage of less than 50. At this level of occupant load, the possibility exists that, in an emergency situation, a compact line of people could form at a closed door that swings in a direction opposite the egress flow. This could delay or eliminate the first person's ability to open the door inward with the rest of the queue behind the person.

In a Group H occupancy, the threat of rapid fire buildup, or worse, is such that any delay in egress caused by door swing may jeopardize the opportunity for all occupants to evacuate the premises. For this reason, all egress doors in Group H occupancies are to swing in the direction of egress.

1010.1.3 Door opening force. The force for pushing or pulling open interior swinging egress doors, other than fire doors, shall not exceed 5 pounds (22 N). These forces do not apply to the force required to retract latch bolts or disengage other devices that hold the door in a closed position. For other swinging doors, as well as sliding and folding doors, the door latch shall release when subjected to a 15-pound (67 N) force. The door shall be set in motion when subjected to a 30-pound (133 N) force. The door shall swing to a full-open position when subjected to a 15-pound (67 N) force.

❖ The ability of all potential users to be physically capable of opening an egress door is a function of the forces required to open the door. The 5-pound (22 N) maximum force for pushing and pulling interior swinging doors without closers that are part of the means of egress inside a building is based on that which has been deemed appropriate for people with a physical limitation due to size, age or disability. The operating force is permitted to be higher for all exterior doors, interior swinging doors that are not part of the means of egress, doors that are part of the means of egress but also serve as opening protectives in fire-resistance-rated walls (i.e., fire doors), sliding doors and folding doors. This recognizes that doors with closers, particularly fire doors, require greater operating forces in order to close fully in an emergency where combustion gases may be exerting pressure on the door assembly. Similarly, exterior doors are exempted because air pressure differentials and strong winds may prevent doors from being fully automatically closed.

The opening force is different than the force to retract bolts or operate other types of door hardware. A maximum force of 15 pounds (67 N) is required for operating the latching mechanism. Once unlatched, a maximum force of 30 pounds (133 N) is applied to the latch side of the leaf to start the door in motion by overcoming its stationary inertia. Once in motion, it must not take more than 15 pounds (67 N) of force to keep the door in motion until it reaches its full open position and the required clear width is available. To conform to this requirement on a continual basis, door closers must be adjusted periodically and door fits must also be checked and adjusted when necessary.

1010.1.3.1 Location of applied forces. Forces shall be applied to the latch side of the door.

❖ See the commentary for door opening forces in Section 1010.1.3.

1010.1.4 Special doors. Special doors and security grilles shall comply with the requirements of Sections 1010.1.4.1 through 1010.1.4.4.

❖ This section simply defines the scope of the code requirements for special doors such as revolving doors, power-operated swinging doors, power-operated horizontal sliding doors and security grilles

1010.1.4.1 Revolving doors. Revolving doors shall comply with the following:

1. Revolving doors shall comply with BHMA A156.27 and shall be installed in accordance with the manufacturer's instructions.
2. Each revolving door shall be capable of *breakout* in accordance with BHMA A156.27 and shall provide an aggregate width of not less than 36 inches (914 mm).
3. A revolving door shall not be located within 10 feet (3048 mm) of the foot or top of *stairways* or escalators. A dispersal area shall be provided between the *stairways* or escalators and the revolving doors.
4. The revolutions per minute (rpm) for a revolving door shall not exceed the maximum rpm as specified in BHMA A156.27. Manual revolving doors shall comply with Table 1010.1.4.1(1). Automatic or power-operated revolving doors shall comply with Table 1010.1.4.1(2).
5. An emergency stop switch shall be provided near each entry point of power or automatic operated revolving doors within 48 inches (1220 mm) of the door and between 24 inches (610 mm) and 48 inches (1220 mm) above the floor. The activation area of the emergency stop switch button shall be not less than 1 inch (25 mm) in diameter and shall be red.
6. Each revolving door shall have a side-hinged swinging door that complies with Section 1010.1 in the same wall and within 10 feet (3048 mm) of the revolving door.
7. Revolving doors shall not be part of an *accessible route* required by Section 1009 and Chapter 11.

❖ Revolving doors must comply with all seven provisions.

Item 1: BHMA A156.27 is the revolving door industry standard and includes numerous safety-related requirements for revolving doors. For example, BHMA A156.27 requires manually operated revolving doors to contain governors to limit the rotational speed of the door. For automatic, or power-operated, revolving doors, BHMA A156.27 includes requirements for numerous sensors and switches, and complex motor controls to safely operate the door.

Item 2: One of the causes contributing to the loss of lives in the 1942 Cocoanut Grove fire in Boston was that the revolving doors at the club's entrance could not collapse (*breakout*) for emergency egress and there was not an alternative means of egress adjacent to the revolving doors. Thus, in the panic of the fire, the door became jammed and the club's occupants were trapped.

As a result of this fire experience, all revolving doors, including those for air structures, now are required to be equipped with a breakout feature. A breakout operation is where all leaves collapse parallel to each other and to the direction of egress [see Commentary Figure 1010.1.4.1(1)]. A breakout operation creates two openings of approximately equal width. The sum of the widths is not to be less than 36 inches (914 mm) so that a stream of pedestrians may use each side of the opening. BHMA A156.27 includes explicit breakout requirements for the wide range of sizes and configurations of manual and automatic revolving doors.

Item 3: If a stairway or escalator delivers users to a landing in front of a revolving door at a greater rate than the capacity of the door, a compact line of people will develop. Lines of people formed on a stairway or escalator create an unsafe situation, since stairways and escalators are not intended to be used as standing space for persons who may be waiting to

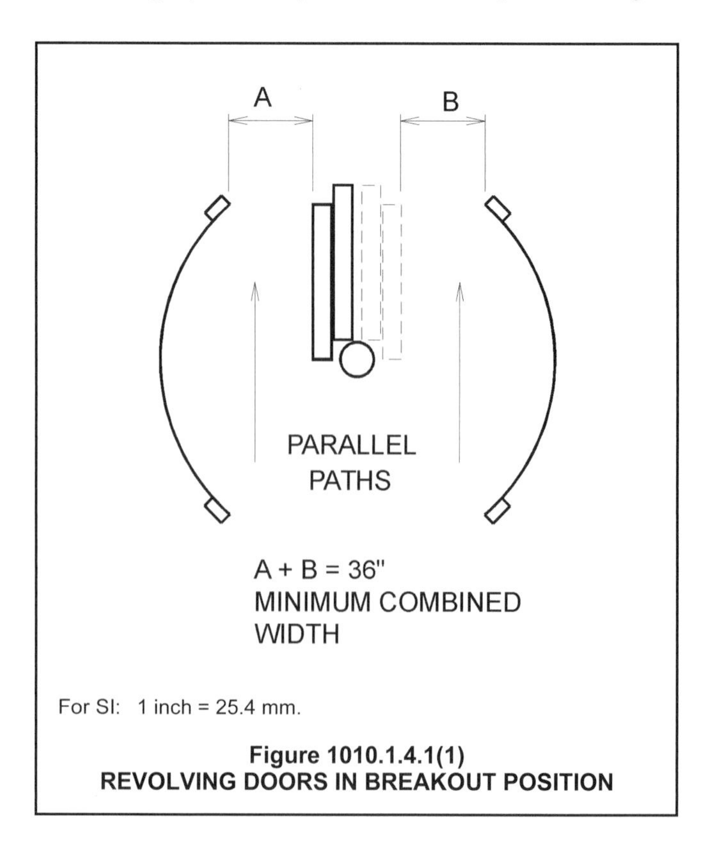

For SI: 1 inch = 25.4 mm.

Figure 1010.1.4.1(1)
REVOLVING DOORS IN BREAKOUT POSITION

use the revolving doors. Therefore, to avoid congestion at a revolving door that under normal operation has a maximum delivery capacity of users, a dispersal area is required between the stairways or escalators and the revolving doors to allow for the queuing of people as they enter the door. Accordingly, to create a dispersal area for users of a revolving door, the door is not to be placed closer than 10 feet (3048 mm) from the foot or top of a stairway or escalator.

Item 4: Door speeds also directly relate to the capacity of a revolving door, which is calculated by multiplying the number of leaves (wings) by the revolutions per minute (rpm). For example, if you have a four-leaf door (four-bay door) moving at 10 rpm, the door will allow 40 people to move in either direction in 1 minute. The larger revolving doors are designed to allow more than one person in each bay, and this should be taken into account when calculating the capacity of a revolving door.

Item 5: An emergency stop switch near each entry point of automatic or power-operated revolving doors provides a method to stop the door's operation.

Item 6: In case a revolving door malfunctions or becomes obstructed, the adjacent area is to be equipped with a conventional side-hinged door to provide users with an immediate alternative way to exit a building. The side-hinged door is intended to be used as a relief device for people lined up to use the revolving door or who desire to avoid it because of a physical disability or other reason. It also can be used when the revolving door is obstructed or out of service. The swinging door is to be immediately adjacent to the revolving door so that its availability is obvious [see Commentary Figure 1010.1.4.1(2)]. A single swinging door can be located between side-by-side revolving doors in order to comply with this provision.

Item 7: While some revolving doors may be considered part of a means of egress, they cannot be considered part of a required accessible route for either ingress or egress. This requirement is consistent with ICC A117.1, which also prohibits revolving gates and turnstiles along the only accessible route. The side-swinging door required by Item 6 can serve as the accessible entrance or exit required by Sections 1009 and 1105.

A route through a hinged or sliding door differs remarkably from that provided through a revolving door. For a revolving door, the route includes a turn into the doorway, an arcing path of travel as the door revolves, followed by a change of direction when leaving the door. Items that may cause difficulty for anyone with mobility impairments could involve the overall doorway diameter, the number of leaves and their relative angle, and the configuration of the return walls surrounding the revolving door. Additionally, the speed of the door movement if motorized, or the force required for movement if not motorized, would be a concern for anyone who needed to keep both hands on their device to move forward (e.g., walker or wheelchair).

Automatic revolving doors, if large enough, may be usable by many people who use wheelchairs. How-

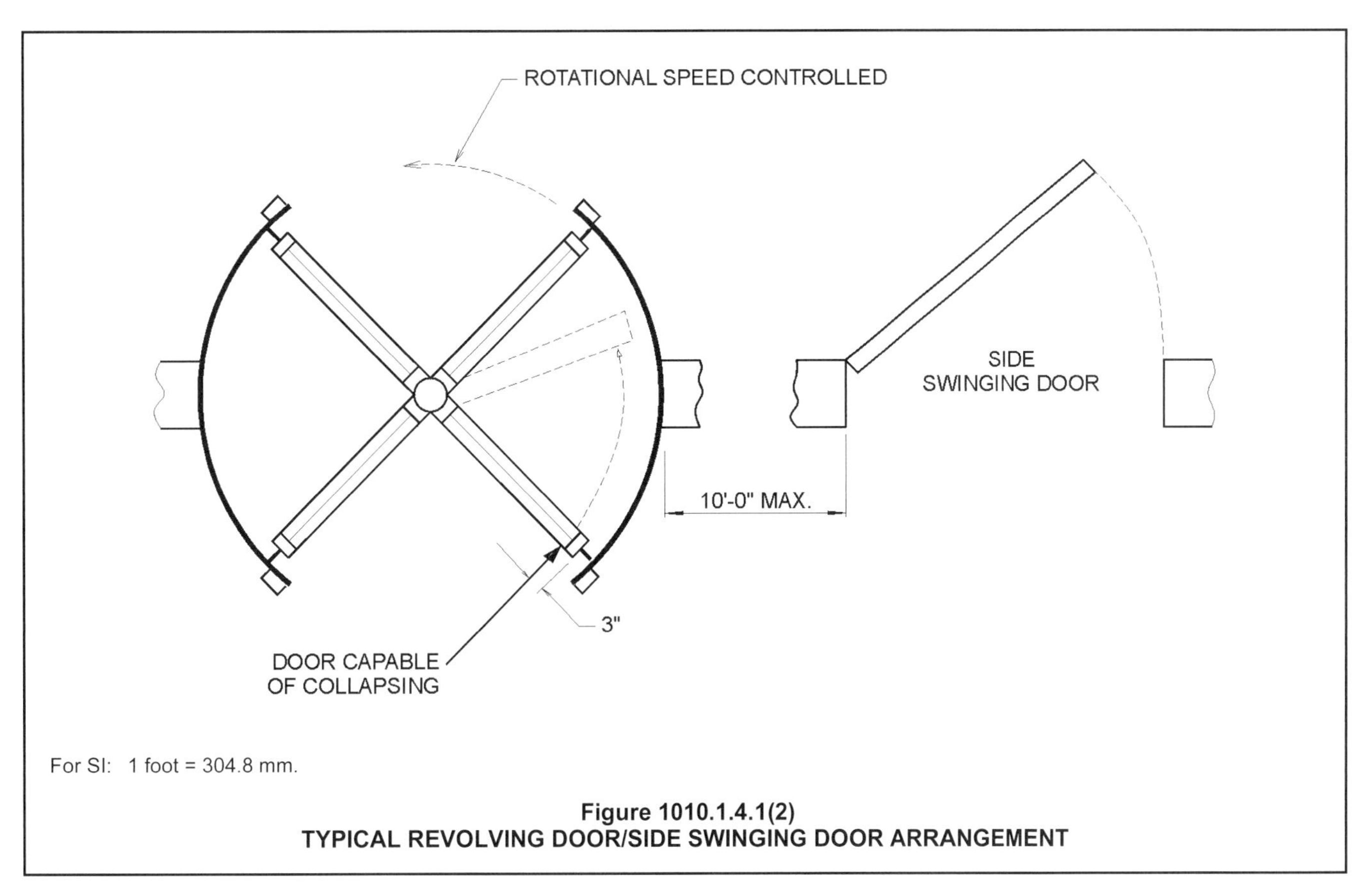

Figure 1010.1.4.1(2)
TYPICAL REVOLVING DOOR/SIDE SWINGING DOOR ARRANGEMENT

ever, the intent of this section is that these types of doors not be the only means of passage at an entrance or exit. An alternative door in full compliance with this section is considered necessary because some people with disabilities may be uncertain of the usability, or may not have enough strength or speed to use them. Although manufacturers have developed safety criteria, certain questions remain, such as the appropriate maximum and minimum speeds that would work for persons trying to maneuver a wheelchair through a revolving door.

Figure 1010.1.4.1(3)
EXAMPLE OF MANUAL REVOLVING DOOR (7-FOOT DIAMETER)

Figure 1010.1.4.1(4)
EXAMPLE OF AUTOMATIC REVOLVING DOOR (8-FOOT DIAMETER)

Revolving doors range from (smaller) manually operated revolving door systems to automatic (power operated) revolving doors of small (8-foot) to large (24-foot) diameter. Three configurations of revolving doors are illustrated in Commentary FIgures 1010.1.4(3) through 1010.1.4(5).

TABLE 1010.1.4.1(1)
MAXIMUM DOOR SPEED MANUAL REVOLVING DOORS

REVOLVING DOOR MAXIMUM NOMINAL DIAMETER (FT-IN)	MAXIMUM ALLOWABLE REVOLVING DOOR SPEED (RPM)
6-0	12
7-0	11
8-0	10
9-0	9
10-0	8

For SI: 1 inch = 25.4 mm, 1 foot = 304.8 mm.

❖ See the commentary for Section 1010.1.4 and Table 1010.1.4.1(2).

TABLE 1010.1.4.1(2)
MAXIMUM DOOR SPEED AUTOMATIC OR POWER-OPERATED REVOLVING DOORS

REVOLVING DOOR MAXIMUM NOMINAL DIAMETER (FT-IN)	MAXIMUM ALLOWABLE REVOLVING DOOR SPEED (RPM)
8-0	7.2
9-0	6.4
10-0	5.7
11-0	5.2
12-0	4.8
12-6	4.6
14-0	4.1
16-0	3.6
17-0	3.4
18-0	3.2
20-0	2.9
24-0	2.4

For SI: 1 inch = 25.4 mm, 1 foot = 304.8 mm.

❖ Door speeds also directly relate to the capacity of a revolving door, which is calculated for smaller revolving doors by multiplying the number of leaves (wings) by the revolutions per minute (rpm). For example, if you have an 8-foot diameter four-leaf door (four-bay door) moving at 10 rpm, the door will allow 40 people to move in either direction in 1 minute. Larger revolving doors may be designed to allow for more than one person in each bay as the door rotates, which should be taken into account when determining maximum egress capacity of the revolving door.

Figure 1010.1.4.1(5)
EXAMPLE OF AUTOMATIC REVOLVING DOOR (20-FOOT DIAMETER)

1010.1.4.1.1 Egress component. A revolving door used as a component of a *means of egress* shall comply with Section 1010.1.4.1 and the following three conditions:

1. Revolving doors shall not be given credit for more than 50 percent of the minimum width or required capacity.
2. Each revolving door shall be credited with a capacity based on not more than a 50-person *occupant load*.
3. Each revolving door shall provide for egress in accordance with BHMA A156.27 with a *breakout* force of not more than 130 pounds (578 N).

❖ A revolving door can be incorporated, to a very limited extent, in a means of egress. Compliance with these three additional conditions is required.

Condition 1 limits the exit capacity that revolving doors can provide in a building. This is so that 50 percent of the capacity has conventional egress components and is not dependent on mechanical devices or fail-safe mechanisms.

Condition 2 limits the capacity of any one revolving door for the same reasons as stated in Condition 1. Each revolving door is therefore limited to a 50-person capacity.

Condition 3 limits the breakout force to 130 pounds (578 N), as opposed to the 180-pound (792 N) value listed in Section 1010.1.4.1.2. Revolving doors used as means of egress are not permitted to have the breakout force exceed 130 pounds (578 N) under any circumstances.

1010.1.4.1.2 Other than egress component. A revolving door used as other than a component of a *means of egress* shall comply with Section 1010.1.4.1. The *breakout* force of a revolving door not used as a component of a *means of egress* shall not be more than 180 pounds (801 N).

Exception: A *breakout* force in excess of 180 pounds (801 N) is permitted if the collapsing force is reduced to not more than 130 pounds (578 N) when not less than one of the following conditions is satisfied:

1. There is a power failure or power is removed to the device holding the door wings in position.
2. There is an actuation of the *automatic sprinkler system* where such system is provided.
3. There is an actuation of a smoke detection system that is installed in accordance with Section 907 to provide coverage in areas within the building that are within 75 feet (22 860 mm) of the revolving doors.
4. There is an actuation of a manual control switch, in an approved location and clearly identified, that reduces the *breakout* force to not more than 130 pounds (578 N).

❖ This section addresses revolving doors that are not used to serve any portion of the occupant egress capacity. For example, where adjacent side-hinged doors have more than the required egress capacity, the revolving door would not be part of the required means of egress.

The maximum breakout force of 180 pounds (792 N), applied within 3 inches (76 mm) of the outer edge of a wing, is based on industry standards to accommodate normal use conditions and other forces that may act on the leaves, such as those caused by wind or air pressure. An exception for revolving doors that are not a component of a required means of egress allows the breakout force to exceed 180 pounds (792 N) in normal operating conditions provided that a force of not more than 130 pounds (578 N) is required whenever any one of the listed conditions is satisfied.

1010.1.4.2 Power-operated doors. Where *means of egress* doors are operated or assisted by power, the design shall be such that in the event of power failure, the door is capable of being opened manually to permit *means of egress* travel or closed where necessary to safeguard *means of egress*. The forces required to open these doors manually shall not exceed those specified in Section 1010.1.3, except that the force to set the door in motion shall not exceed 50 pounds (220 N). The door shall be capable of swinging open from any position to the full width of the opening in which such door is installed when a force is applied to the door on the side from which egress is made. Power-operated swinging doors, power-operated sliding doors and power-operated folding doors shall comply with BHMA A156.10. Power-assisted swinging doors and low-energy power-operated swinging doors shall comply with BHMA A156.19.

Exceptions:

1. Occupancies in Group I-3.

2. Horizontal sliding doors complying with Section 1010.1.4.3.

3. For a biparting door in the emergency breakout mode, a door leaf located within a multiple-leaf opening shall be exempt from the minimum 32-inch (813 mm) single-leaf requirement of Section 1010.1.1, provided a minimum 32-inch (813 mm) clear opening is provided when the two biparting leaves meeting in the center are broken out.

❖ For convenience purposes, power-operated doors are intended to facilitate the normal nonemergency flow of persons through a doorway. Where a power-operated or power-assisted door is also required to be an egress door, the door must conform to the requirements of this section. The essential characteristic is that the door is to be manually openable from any position to its full open position at any time, with or without a power failure or a failure of a door mechanism. Hence, both swinging and horizontal sliding doors, complying with this section, may be used, provided the door can be operated manually from any position as a swinging door and that the minimum required clear width for egress capacity is not less than 32 inches (813 mm). Note that the opening forces of Section 1010.1.3 are applicable, except that the 30-pound (133 N) force needed to set the door in motion is increased to 50 pounds (220 N) as an operational tolerance in the design of the power-operated door.

Definitions for the different types of power-operated doors were added to the 2015 IBC to help clarify which standard (BHMA A156.10 or BHMA A156.19) is applicable to which type of power-operated door.

Power-operated doors are required to comply with BHMA A156.10. These doors open automatically when approached by a person or upon an action by a person, close automatically and include provisions such as presence sensors to prevent entrapment. *Low-energy power-operated doors* are required to comply with BHMA A56.19. These doors are always swinging doors which open automatically upon an action by a person such as pressing a push plate or waving a hand in front of a sensor. Additionally, these doors close automatically and operate with decreased forces and decreased speeds (compared to *power-operated doors*). Least common are *power-assisted doors*, which are required to comply with BHMA A156.19. These doors are swinging doors which open by reduced pushing or pulling force on the door-operating hardware, close automatically after the pushing or pulling force is released and function with decreased forces.

In accordance with Exception 1, power-operated doors in detention and correctional occupancies (Group I-3) are not required to be manually operable by the occupants (inmates) for security reasons, but otherwise are required to conform to Section 408.

Exception 2 states that power-operated, special-purpose horizontal sliding accordion or folding doors that meet the requirements of Section 1010.1.4.3 are not required to meet the requirements of Section 1010.1.4.2. This is consistent with the option offered in Section 1010.1.2 by Exception 6.

Exception 3 allows an individual leaf of a four-panel biparting door to be less than 32 inches (813 mm) wide, provided 32 inches (813 mm) of clear space is available when the two center biparting leaves are broken out as part of the emergency breakaway feature.

1010.1.4.3 Special purpose horizontal sliding, accordion or folding doors. In other than Group H occupancies, special purpose horizontal sliding, accordion or folding door assemblies permitted to be a component of a *means of egress* in accordance with Exception 6 to Section 1010.1.2 shall comply with all of the following criteria:

1. The doors shall be power operated and shall be capable of being operated manually in the event of power failure.

2. The doors shall be openable by a simple method from both sides without special knowledge or effort.

3. The force required to operate the door shall not exceed 30 pounds (133 N) to set the door in motion and 15 pounds (67 N) to close the door or open it to the minimum required width.

4. The door shall be openable with a force not to exceed 15 pounds (67 N) when a force of 250 pounds (1100 N) is applied perpendicular to the door adjacent to the operating device.

5. The door assembly shall comply with the applicable *fire protection rating* and, where rated, shall be self-closing or automatic closing by smoke detection in accordance with Section 716.5.9.3, shall be installed in accordance with NFPA 80 and shall comply with Section 716.

6. The door assembly shall have an integrated standby power supply.

7. The door assembly power supply shall be electrically supervised.

8. The door shall open to the minimum required width within 10 seconds after activation of the operating device.

❖ Special purpose horizontal sliding, accordion or folding doors are permitted in the means of egress, in other than rooms or areas of Group H, under the conditions set forth in this section. Special purpose horizontal sliding, accordion or folding doors are not permitted to be used in Group H occupancies because of the potential for delaying or impeding egress from those areas and the additional risk to occupants in hazardous occupancies. Note that this section regulates doors that are part of the means of egress to meet a set of requirements different than Section 1010.1.4.2 (e.g., a power-operated horizontal sliding door that does not have "breakout" capabilities to allow the door panels to swing if power is lost). The doors addressed by Section 1010.1.4.3 are commonly in the normally open position (hidden in their enclosure).

All eight of the criteria listed in this section must be met for a special purpose horizontal sliding, accordion or folding door since there is a concern that it must be able to be easily opened to the minimum required width under all conditions.

Additionally, the door must be openable even if a force of 250 pounds (1100 N) is being applied perpendicular to it, as may occur if a group of people were pushing on it.

Since the doors are manually operable, they need not automatically open or close during a loss of power; however, a standby power supply must be provided. The primary power supply must be supervised so that an alarm is received at a constantly attended location (such as a security desk) on loss of the primary power. If the doors are also serving as opening protective (i.e., fire doors), they must be automatic closing or self-closing in accordance with Section 716.

Since the maximum swinging door leaf width limitations of Section 1008.1.1 do not apply, a maximum opening time of 10 seconds is permitted. It should be noted, however, that the door need not open fully within the 10 seconds; rather, it must open to the minimum required width. For example, if the door is protecting an opening that is 10 feet (3048 mm) wide, but the minimum required width of the opening is 32 inches (813 mm) (as determined by Section 1008.1.1), the door need only open 32 inches (813 mm) within the 10-second criterion. In fact, the door may have controls such that the automatic opening feature only opens the door to a width of 32 inches (813 mm). If additional width is required, it can be accomplished by manual means and, possibly, by an additional activation of the operating device.

1010.1.4.4 Security grilles. In Groups B, F, M and S, horizontal sliding or vertical security grilles are permitted at the main *exit* and shall be openable from the inside without the use of a key or special knowledge or effort during periods that the space is occupied. The grilles shall remain secured in the full-open position during the period of occupancy by the general public. Where two or more *means of egress* are required, not more than one-half of the *exits* or *exit access doorways* shall be equipped with horizontal sliding or vertical security grilles.

❖ This section really functions as an exception to several sections, including Sections 1010.1.2 (Door swing) and 1010.1.9.3 (Locks and latches), and permits the use of these security grilles under conditions that are similar to those found in Section 402 for covered mall buildings. These security grilles will be open when the space is occupied and will, therefore, not obstruct any egress path. Since the building may be partially used (e.g., team practice in a football stadium) when not fully occupied, no more than one-half of the exits from the building can be through security grilles.

1010.1.5 Floor elevation. There shall be a floor or landing on each side of a door. Such floor or landing shall be at the same elevation on each side of the door. Landings shall be level except for exterior landings, which are permitted to have a slope not to exceed 0.25 unit vertical in 12 units horizontal (2-percent slope).

Exceptions:

1. Doors serving individual *dwelling units* in Groups R-2 and R-3 where the following apply:
   1.1. A door is permitted to open at the top step of an interior *flight* of *stairs*, provided the door does not swing over the top step.
   1.2. Screen doors and storm doors are permitted to swing over *stairs* or landings.
2. Exterior doors as provided for in Section 1003.5, Exception 1, and Section 1022.2, which are not on an *accessible route*.
3. In Group R-3 occupancies not required to be *Accessible units*, *Type A units* or *Type B units*, the landing at an exterior doorway shall be not more than $7^3/_4$ inches (197 mm) below the top of the threshold, provided the door, other than an exterior storm or screen door, does not swing over the landing.
4. Variations in elevation due to differences in finish materials, but not more than $^1/_2$ inch (12.7 mm).
5. Exterior decks, patios or balconies that are part of *Type B dwelling units*, have impervious surfaces and that are not more than 4 inches (102 mm) below the finished floor level of the adjacent interior space of the dwelling unit.
6. Doors serving equipment spaces not required to be *accessible* in accordance with Section 1103.2.9 and serving an occupant load of five or less shall be permitted to have a landing on one side to be not more than 7 inches (178 mm) above or below the landing on the egress side of the door.

❖ Changes in floor surface elevation at a door, however small, often are slip or trip hazards. This is because persons passing through a door, including those who may have some mobility impairments, usually do not expect changes in floor surface elevation or are not able to recognize them because of the intervening door leaf. Under emergency conditions, a fall in a doorway could result not only in injury to the falling occupant but also interruption of orderly egress by other occupants. The exterior landing is allowed to slope to drain.

The size of this landing is set by Section 1010.1.6. In accordance with Exception 4, the floor surface elevation of the landing is to be at the same elevation plus or minus $^1/_2$ inch (12.7 mm) (see Commentary Figure 1010.1.5).

Note that some of the exceptions indicate which direction the door swings to allow the exceptions while others do not limit the door swing direction.

Exception 1, which applies to nontransient residential occupancies, recognizes that occupants are familiar with the stair and landing arrangements. In Exception 1.1, an interior stairway can start immediately at the door, provided the door leaf does not swing over the stairway. Exception 1.2 clarifies that a screen door or storm door would not be considered the main door prohibited from swinging over the stairway by Exception 1.1.

Exception 2 references two other locations. Section 1003.5, Exception 1, permits a 7-inch (178 mm) change in elevation at exterior doors in Groups F, H, R-2, R-3, S and U if they are not on an accessible route. The door could swing in either direction for this exception and may actually be required to swing out in accordance with Section 1010.1.2. A reference to Section 1022.2 does not address a change in elevation but does address exterior exit doors.

In accordance with Exception 3, for a residential unit, the step-down is limited to $7^3/_4$ inches (197 mm) measured from the top of the threshold rather than the interior floor surface. In addition, the exterior door cannot swing over the exterior landing. A screen door or storm door could swing over the exterior landing. This is consistent with the exception to Section 1010.1.7.

Exception 4 addresses a change in floor finish material (see Commentary Figure 1010.1.5).

In accordance with Exception 5, Type B dwelling or sleeping units are not required to have level landings on both sides of some of the exterior doors. Please note that this exception is not applicable for the primary entrance door to the unit (see Section 1105.1.7). Exterior doors that open out onto an exterior deck, patio or balcony are allowed a 4-inch (102 mm) step down. Type B units are established by Chapter 11 for residential occupancies containing four or more dwelling or sleeping units. In order to use this exception, the exterior decks, patios or balconies must be of solid and impervious construction, such as concrete or wood. A 4-inch (102 mm) step from inside the unit down to the exterior surfaces is allowed for weather purposes. This allowance is consistent with the provisions of ICC A117.1 and Fair Housing Accessibility Guidelines (FHAG).

While Exception 2 would allow for exterior doors of Group F or S to have a step down, it would not have this same allowance for small equipment spaces within a building. Sometimes these rooms have a step down to allow for spills within the space to not leak out under the door. This space is not required to be accessible (Section 1103.2.9) and the space is only accessed by maintenance and service personnel, so a step down would not be a barrier.

1010.1.6 Landings at doors. Landings shall have a width not less than the width of the *stairway* or the door, whichever is greater. Doors in the fully open position shall not reduce a required dimension by more than 7 inches (178 mm). Where a landing serves an *occupant load* of 50 or more, doors in any position shall not reduce the landing to less than one-half its required width. Landings shall have a length measured in the direction of travel of not less than 44 inches (1118 mm).

> **Exception:** Landing length in the direction of travel in Groups R-3 and U and within individual units of Group R-2 need not exceed 36 inches (914 mm).

❖ Door landings are at either side of the door. Landings can overlap floor surfaces within a room or corridor, overlap an exterior porch or balcony, or share the landings for stairways. The 7-inch (178 mm) encroachment and one-half required width limitations are consistent with Section 1005.7 for door encroachment. Section 1005.7 deals with egress width and capacity and is referenced from aisles (see Section 1018.1), corridors (see Section 1020.3), exit passageways (see Section 1024.2) and egress courts (see Section 1028.4.1).

This section also is intended to address landings at

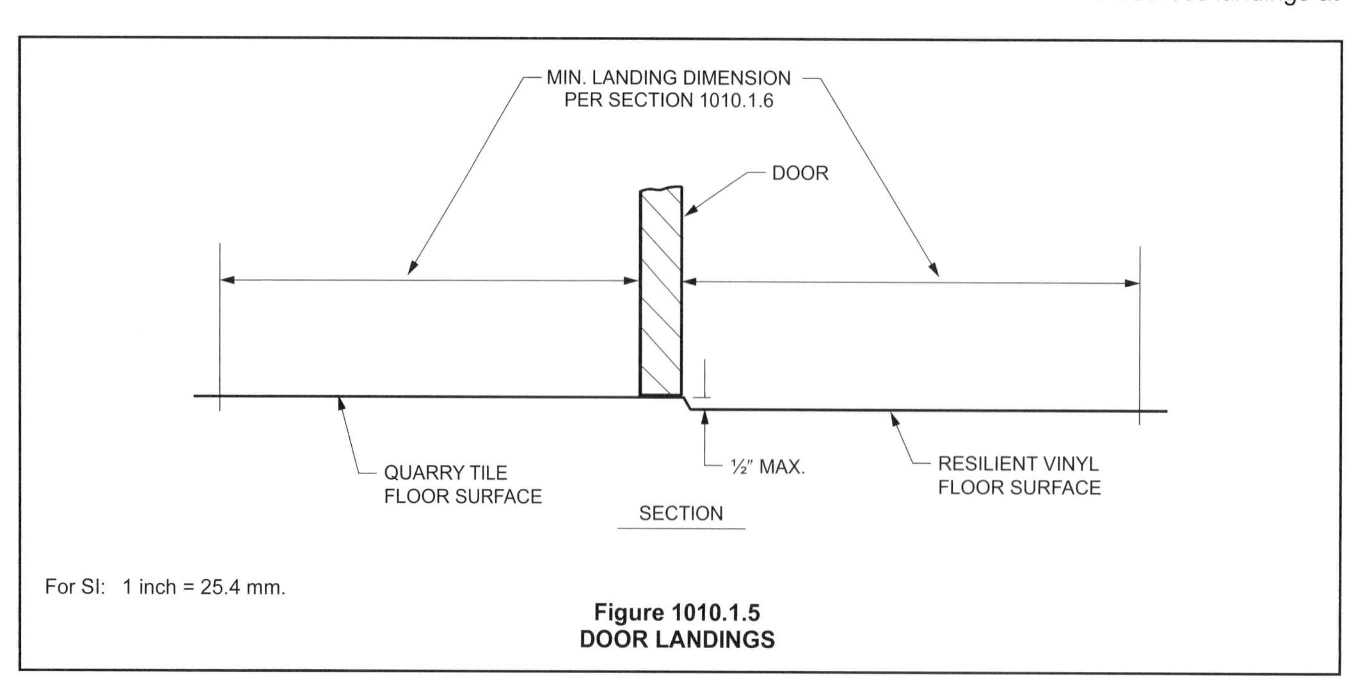

Figure 1010.1.5
DOOR LANDINGS

the entrance door to enclosed stairways (also see Section 1011.6 for stairway landings). The width of a landing at a door in a stairway is to be not less than the width of the stairway or the door, whichever is greater [see Commentary Figure 1011.6(4) for an example of these provisions].

No matter what size the door or stair landing is, door landings are to have the floor elevation requirements of Section 1010.1.5 extending at least 44 inches (1118 mm) in the direction of egress travel.

The reduction in landing length from 44 inches minimum to 36 inches minimum for certain residential occupancies allowed by the exception is in recognition of their low occupant load.

1010.1.7 Thresholds. Thresholds at doorways shall not exceed $^{3}/_{4}$ inch (19.1 mm) in height above the finished floor or landing for sliding doors serving *dwelling units* or $^{1}/_{2}$ inch (12.7 mm) above the finished floor or landing for other doors. Raised thresholds and floor level changes greater than $^{1}/_{4}$ inch (6.4 mm) at doorways shall be beveled with a slope not greater than one unit vertical in two units horizontal (50-percent slope).

Exceptions:

1. In occupancy Group R-2 or R-3, threshold heights for sliding and side-hinged exterior doors shall be permitted to be up to $7^{3}/_{4}$ inches (197 mm) in height if all of the following apply:
   1.1. The door is not part of the required *means of egress*.
   1.2. The door is not part of an *accessible route* as required by Chapter 11.
   1.3. The door is not part of an *Accessible unit, Type A unit* or *Type B unit*.
2. In *Type B units*, where Exception 5 to Section 1010.1.5 permits a 4-inch (102 mm) elevation change at the door, the threshold height on the exterior side of the door shall not exceed $4^{3}/_{4}$ inches (120 mm) in height above the exterior deck, patio or balcony for sliding doors or $4^{1}/_{2}$ inches (114 mm) above the exterior deck, patio or balcony for other doors.

❖ A threshold is a potential tripping hazard and a barrier to accessibility by people with mobility impairments. For these reasons, thresholds for all doorways, except exterior sliding doors serving dwelling units, are to be a maximum of $^{1}/_{2}$ inch (12.7 mm) high. Exterior sliding doors serving dwelling units, however, are permitted to be $^{3}/_{4}$ inch (19.1 mm) high because of practical design considerations, concern for deterioration of the doorway because of snow and ice buildup and lack of adequate drainage in severe climates. Raised threshold and floor level changes at doorways without beveled edge treatment [see Commentary Figure 1010.1.7(1)] are limited to $^{1}/_{4}$ inch (6.3 mm) high vertically.

Commentary Figures 1010.1.7(2), 1010.1.7(3) and 1010.1.7(4) illustrate configurations where the change in elevation is between $^{1}/_{4}$ inch (6.4 mm) and $^{1}/_{2}$ inch (12.7 mm). The slope of required beveled edges cannot exceed one unit vertical in two units horizontal (1:2 or 50-percent slope) but lesser slopes are fully compliant. This kind of threshold treatment provides for mini-

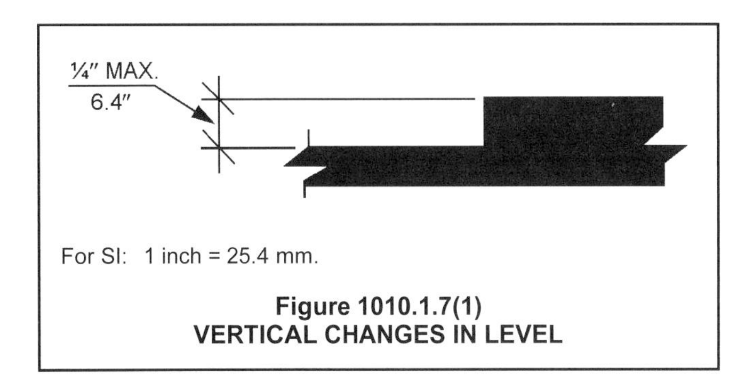

For SI: 1 inch = 25.4 mm.

Figure 1010.1.7(1)
VERTICAL CHANGES IN LEVEL

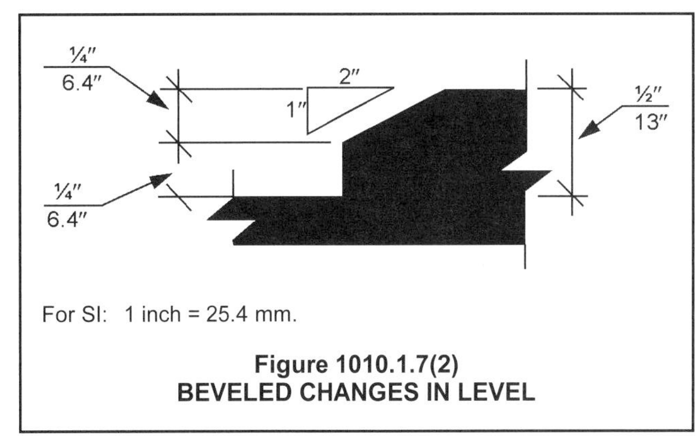

For SI: 1 inch = 25.4 mm.

Figure 1010.1.7(2)
BEVELED CHANGES IN LEVEL

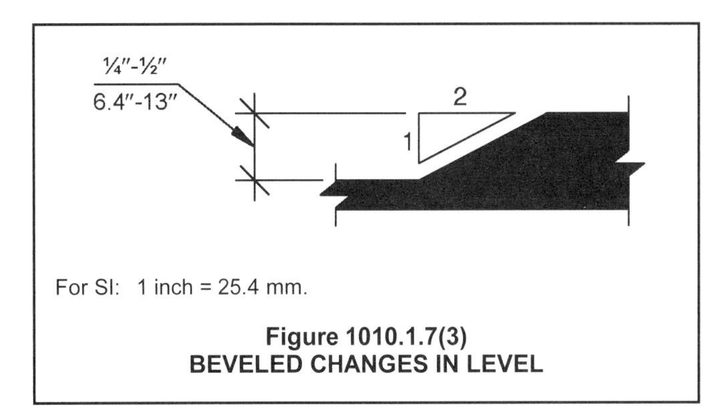

For SI: 1 inch = 25.4 mm.

Figure 1010.1.7(3)
BEVELED CHANGES IN LEVEL

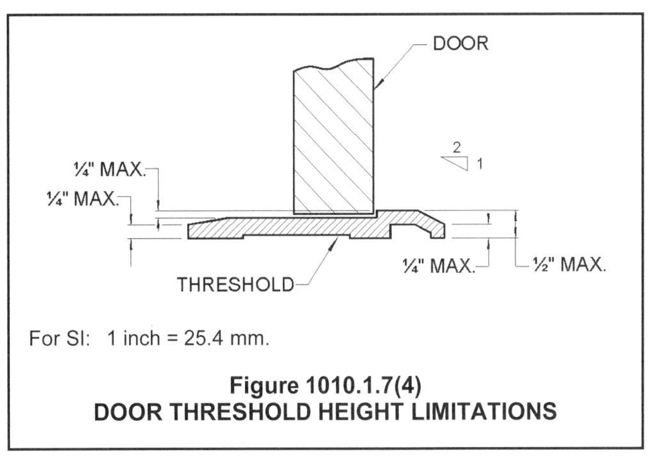

For SI: 1 inch = 25.4 mm.

Figure 1010.1.7(4)
DOOR THRESHOLD HEIGHT LIMITATIONS

mum obstructions for wheelchair users and limits the trip hazard for those with other mobility disabilities.

The first exception permits a $7^3/_4$-inch (197 mm) threshold at both side-swinging and sliding exterior doors in Group R-2 and R-3 buildings. However, this door cannot be part of the required means of egress out of a building; cannot be part of an accessible route required into a building or unit (see Sections 1104 and 1105) and cannot serve an Accessible, Type A or Type B unit. The result would be that this exception would be permitted at the "back door," stoop or patio in single-family homes (see Section 1103.2.3) or upper floor balconies within individual apartments in buildings without elevators (see Section 1107). This terminology is consistent with the $7^3/_4$-inch (197 mm) step down in Section 1010.1.5, Exception 3; however, this exception is to the threshold requirements, not the landing elevations.

The second exception is for doors where a 4-inch step down is permitted in Type B dwelling units between the interior finished floor surface and the exterior surface of exterior decks, patios and balconies (Section 1010.1.5 Exception 5). This section would permit the height of the threshold itself to exceed $^1/_2$ or $^3/_4$ inch in height, as long as the resultant profile from the interior floor to the exterior surface is maintained.

The threshold itself can be higher than $^1/_2$ or $^3/_4$ inch measured from the outside. The additional height, however, is contained within the 4-inch step down. The height of the threshold is limited to $^1/_2$ inch or $^3/_4$ inch above the interior floor and the total height must not be more than $4^1/_2$ or $4^3/_4$ inches above the exterior surface, depending upon the type of door. If the threshold is greater than $^1/_4$ inch above the interior floor, it is to be beveled. See Commentary Figure 1010.1.7(5).

1010.1.8 Door arrangement. Space between two doors in a series shall be 48 inches (1219 mm) minimum plus the width of a door swinging into the space. Doors in a series shall swing either in the same direction or away from the space between the doors.

Exceptions:

1. The minimum distance between horizontal sliding power-operated doors in a series shall be 48 inches (1219 mm).
2. Storm and screen doors serving individual *dwelling units* in Groups R-2 and R-3 need not be spaced 48 inches (1219 mm) from the other door.
3. Doors within individual *dwelling units* in Groups R-2 and R-3 other than within *Type A dwelling units*.

❖ Door arrangement is required to be such that an occupant's use of a means of egress doorway is not hampered by the operation of a preceding door located in the same line of travel so that the occupant flow can be smooth through the openings. Successive doors in a single egress path (i.e., in a series) can cause such interference. The 4-foot (1219 mm) clear distance between doors when the first door is open at 90 degrees allows an occupant, including a person using a wheelchair, to move past one door and its swing before beginning the operation of the next door [see Commentary Figure 1010.1.8(1)]. Note that where doors in a series are not arranged in a straight line, the intent of the code is to provide sufficient space to enable occupants to negotiate the second door without being encumbered by the first door's swing arc. To facilitate accessibility, the space between doors should provide sufficient clear space for a wheelchair [30 inches by 48 inches (762 mm by 1219 mm)] beyond the arc of the door swing [see Commentary Figure 1010.1.8(2)]. Additionally, the approach and access provisions of ICC A117.1 must be considered for all doors along an accessible route.

The exception is to permit horizontal sliding power-operated doors (see Section 1010.1.4.2) to be designed with a lesser distance between them in a series arrangement because they are customarily designed to open simultaneously or in sequence

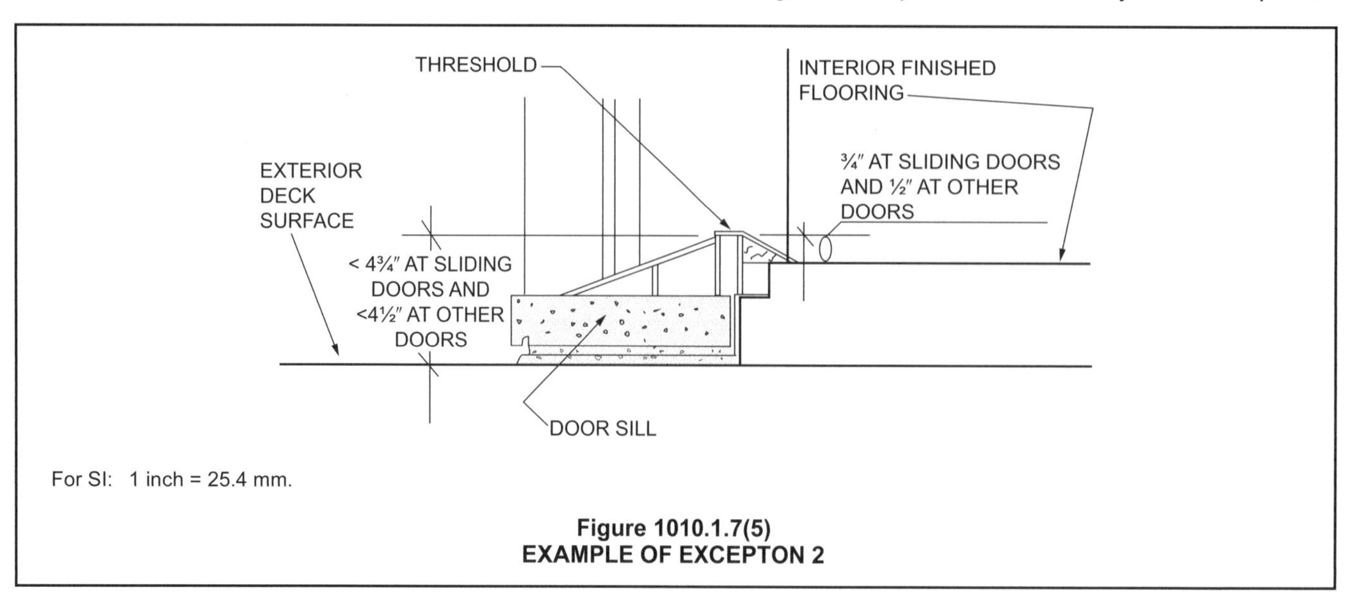

Figure 1010.1.7(5)
EXAMPLE OF EXCEPTON 2

such that movement through them is unhampered. Storm and screen doors on residential dwelling units need not be spaced at 48 inches (1219 mm) since it would be impractical, and they do not operate the same as doors in a series. Doors within dwelling units of Group R-2 or R-3 that are not Type A dwelling units (see Section 1107) are also permitted to have a lesser distance between doors, because the accessibility provisions do not apply. There are requirements in Chapter 10 of the ICC A117.1 for door arrangements within Accessible and Type A dwelling and sleeping units.

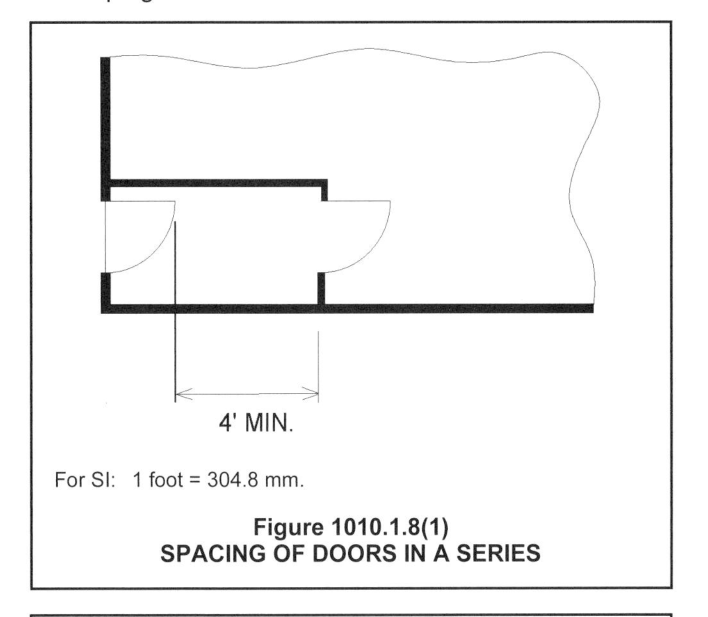

For SI: 1 foot = 304.8 mm.

Figure 1010.1.8(1)
SPACING OF DOORS IN A SERIES

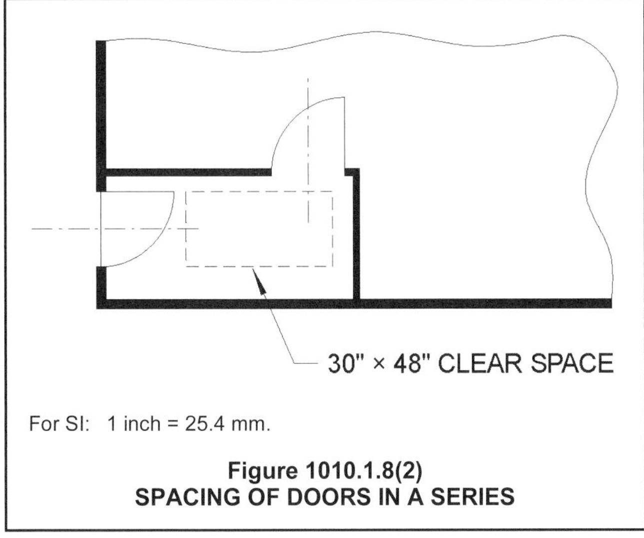

For SI: 1 inch = 25.4 mm.

Figure 1010.1.8(2)
SPACING OF DOORS IN A SERIES

1010.1.9 Door operations. Except as specifically permitted by this section, egress doors shall be readily openable from the egress side without the use of a key or special knowledge or effort.

❖ When installed for security purposes, locks and latches can intentionally prohibit the use of an egress door and thus interfere with or prevent the egress of occupants at the time of a fire. While the security of property is important, the life safety of occupants is essential. Where security and life safety objectives conflict, alternative measures, such as those permitted by each of the exceptions in Section 1010.1.9.3, may be applicable.

Egress doors are permitted to be locked, but must be capable of being unlocked and readily openable from the side from which egress is to be made. The outside of a door can be key locked as long as the inside—the side from which egress is to be made—can be unlocked without the use of tools, keys or special knowledge or effort. For example, an unlocking operation that is integral with an unlatching operation is acceptable.

Examples of special knowledge would be a combination lock or an unlocking device or deadbolt in an unknown, unexpected or hidden location. Special effort would dictate the need for unusual and unexpected physical ability to unlock or make the door fully available for egress.

Where a pair of egress door leaves is installed, with or without a center mullion, the general requirement is that each leaf must be provided with its own releasing or unlatching device so as to be readily openable. Door arrangements or devices that depend on the release of one door before the other can be opened are not to be used except as permitted by Section 1010.1.9.4.

1010.1.9.1 Hardware. Door handles, pulls, latches, locks and other operating devices on doors required to be *accessible* by Chapter 11 shall not require tight grasping, tight pinching or twisting of the wrist to operate.

❖ Any doors that are located along an accessible route for ingress or egress must have door hardware that is easy to operate by a person with limited mobility or dexterity. This would include all elements of the door hardware used in typical door operation, such as door levers, locks, security changes, etc. This requirement is also an advantage for persons with arthritis in their hands. Items such as small, full-twist thumb turns or smooth circular knobs are examples of hardware that is not acceptable. There are many types latching or locking devices can be operated without tight grasping, tight pinching, or twisting of the wrist what would comply with the requirements of this section.

Some people with disabilities are unable to grasp objects with their hands or twist their wrists. Such people are unable to operate, or have great difficulty operating, door hardware other than lever-operated mechanisms, push-type mechanisms and U-shaped door pulls. Door hardware that can be operated with a closed fist or a loose grip accommodates the greatest range of users. Hardware operated by simultaneous hand and finger movement requires greater dexterity and coordination and should be avoided for doors along an accessible route (see Commentary Figure 1010.1.9.1).

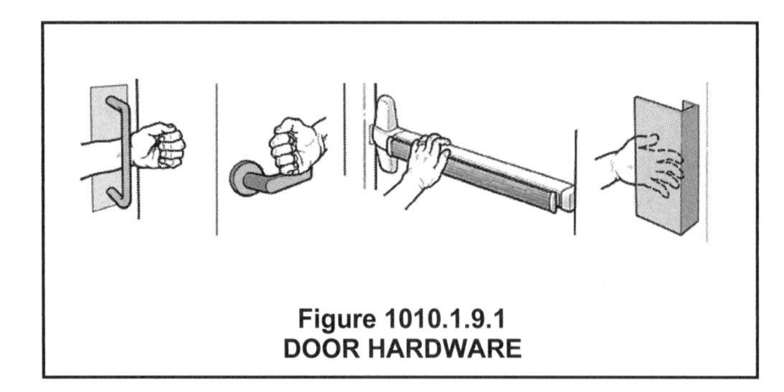

Figure 1010.1.9.1
DOOR HARDWARE

1010.1.9.2 Hardware height. Door handles, pulls, latches, locks and other operating devices shall be installed 34 inches (864 mm) minimum and 48 inches (1219 mm) maximum above the finished floor. Locks used only for security purposes and not used for normal operation are permitted at any height.

Exception: Access doors or gates in barrier walls and fences protecting pools, spas and hot tubs shall be permitted to have operable parts of the release of latch on self-latching devices at 54 inches (1370 mm) maximum above the finished floor or ground, provided the self-latching devices are not also self-locking devices operated by means of a key, electronic opener or integral combination lock.

❖ The requirements in this section place the door hardware at a level that is usable by most people, including a person using a wheelchair. The exception allows security locks to be placed at any height. An example would be an unframed glass door at the front door of a tenant space in a mall that has the lock near the floor level. The lock is only used when the store is not open for business. Such locks are not required for the normal operation of the door.

The exception permits a special allowance for security latches at pools, spas and hot tubs. The concern is that the 48-inch (1219 mm) maximum height would place the security latch within reach of children. The 54-inch (1372 mm) maximum height is intended to override the maximum 48-inch (1219 mm) reach range in ICC A117.1. This compromise addresses concerns for children's safety and still maintains accessibility to a reasonable level. Based on the last phrase in the exception, if the gate hardware also had a locking function and a key or other similar device would be needed to unlock the hardware to allow unlatching the latch, then the exception to allow the hardware to be 54-inches (1372 mm) above the floor is not applicable. A reference to this exception is found in Section 1109.13, Exception 7, for the accessibility requirements for operable parts. This is consistent with the 2010 Standard and ANSI/NSPI-8 1996, *Model Barrier Code for Residential Swimming Pools, Spas, and Hot Tubs.*

1010.1.9.3 Locks and latches. Locks and latches shall be permitted to prevent operation of doors where any of the following exist:

1. Places of detention or restraint.
2. In buildings in occupancy Group A having an *occupant load* of 300 or less, Groups B, F, M and S, and in *places of religious worship*, the main door or doors are permitted to be equipped with key-operated locking devices from the egress side provided:
   - 2.1. The locking device is readily distinguishable as locked.
   - 2.2. A readily visible durable sign is posted on the egress side on or adjacent to the door stating: THIS DOOR TO REMAIN UNLOCKED WHEN THIS SPACE IS OCCUPIED. The sign shall be in letters 1 inch (25 mm) high on a contrasting background.
   - 2.3. The use of the key-operated locking device is revokable by the *building official* for due cause.
3. Where egress doors are used in pairs, *approved* automatic flush bolts shall be permitted to be used, provided that the door leaf having the automatic flush bolts does not have a doorknob or surface-mounted hardware.
4. Doors from individual *dwelling* or *sleeping units* of Group R occupancies having an *occupant load* of 10 or less are permitted to be equipped with a night latch, dead bolt or security chain, provided such devices are openable from the inside without the use of a key or tool.
5. *Fire doors* after the minimum elevated temperature has disabled the unlatching mechanism in accordance with *listed fire door* test procedures.

❖ Where security and life safety objectives conflict, alternative measures, such as those permitted by each of the listed situations, may be applicable.

Item 1 is needed for jails and prisons or locations where someone must be kept inside for their own safety (i.e., dementia wards, psychiatric wards).

Item 2 permits a locking device, such as a double-cylinder dead bolt, on the main entrance door to a building or space. It must be immediately apparent that these doors are locked. For example, such locking devices may have an integral indicator that automatically reflects the "locked" or "unlocked" status of the device. In addition, a sign must be provided that clearly states that the door is to be unlocked when the building or space is occupied. The sign on or adjacent to the door not only reminds employees to unlock the door, but also advises the public that an unacceptable arrangement exists if one finds the door locked. Ideally, the individual who encounters the locked door will notify management and possibly the building official. Note that the use of the key-locking device is revocable by the building official. The locking arrangement is not permitted on any door other than the main exit and, therefore, the employees, security and cleaning crews will have access to other exits without requiring the use of a key. This allowance is not limited just to multiple-exit buildings but also to small buildings with one exit. This option is an alternative to the panic hardware required by Section 1010.1.10.

In Item 3, an automatic flush bolt device is one that is internal to the inactive leaf of a pair of doors. The device has a small "knuckle" that extends from the inactive leaf into an opening in the active leaf. When the active leaf is opened, the bolt is automatically retracted. When the active leaf is closed, the knuckle is pressed into the inactive leaf by the active leaf, extending the flush bolt(s), in the head or sill of the inactive leaf (see Commentary Figure 1010.1.9.3).

Automatic flush bolts on one leaf of a pair of egress doors are acceptable, provided the leaf with the automatic flush bolts is not equipped with a doorknob or other hardware that would imply to the user that the door leaf is unlatched independently of the companion leaf.

Item 4 addresses the need for security in residential dwelling and sleeping units such as hotel rooms, apartments, dormitory rooms or townhouses. The occupants are familiar with the operation of the indicated devices, which are intended to be relatively simple to operate without the use of a key or tool. Note that this item only applies to the door leading from individual dwelling or sleeping units in a building. This item would not be applicable for doors locked as part of a security system in a multiunit building.

Item 5 is in recognition of required test procedures (UL 10B or UL 10C) for listed fire doors, which include the disabling of the locking mechanism when a fire door is exposed to the elevated temperatures of a fire.

1010.1.9.4 Bolt locks. Manually operated flush bolts or surface bolts are not permitted.

Exceptions:

1. On doors not required for egress in individual *dwelling units* or *sleeping units*.
2. Where a pair of doors serves a storage or equipment room, manually operated edge- or surface-mounted bolts are permitted on the inactive leaf.
3. Where a pair of doors serves an *occupant load* of less than 50 persons in a Group B, F or S occupancy, manually operated edge- or surface-mounted bolts are permitted on the inactive leaf. The inactive leaf shall not contain doorknobs, panic bars or similar operating hardware.
4. Where a pair of doors serves a Group B, F or S occupancy, manually operated edge- or surface-mounted bolts are permitted on the inactive leaf provided such inactive leaf is not needed to meet egress capacity requirements and the building is equipped throughout with an *automatic sprinkler system* in accordance with Section 903.3.1.1. The inactive leaf shall not contain doorknobs, panic bars or similar operating hardware.
5. Where a pair of doors serves patient care rooms in Group I-2 occupancies, self-latching edge- or surface-mounted bolts are permitted on the inactive leaf provided that the inactive leaf is not needed to meet egress capacity requirements and the inactive leaf

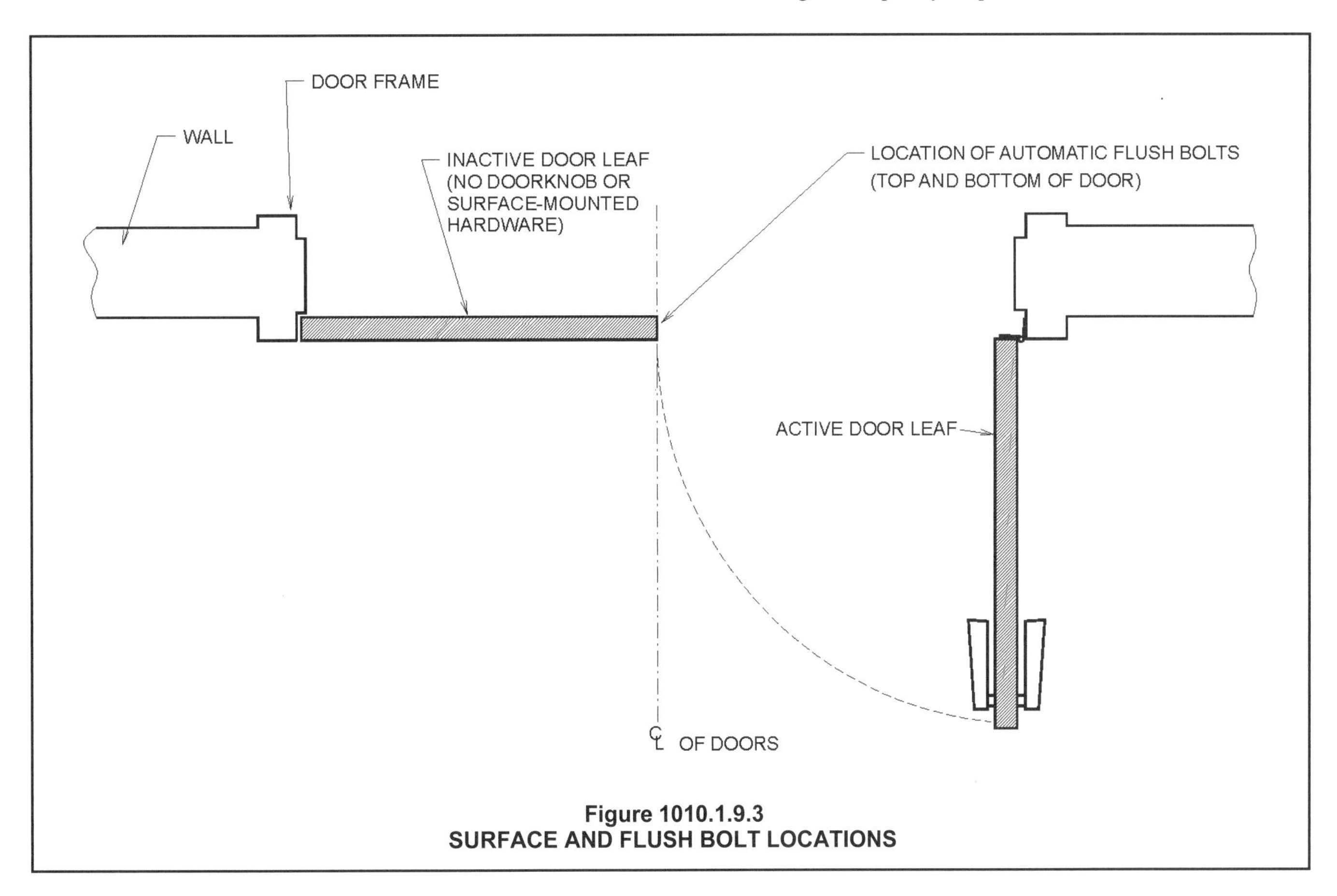

Figure 1010.1.9.3
SURFACE AND FLUSH BOLT LOCATIONS

shall not contain doorknobs, panic bars or similar operating hardware.

❖ This section is applicable to doors that are required to be for means of egress purposes or are identified as a means of egress, such as by an "Exit" sign or other device. Doors, as well as a second leaf in a doorway that is provided for a purpose other than means of egress, such as for convenience or building operations, should be arranged or identified so as not to be mistaken as a means of egress. The use of manually operated flush bolts or surface bolts on means of egress doors have traditionally been prohibited because of the inability of users to quickly identify and operate such devices under emergency conditions.

This section prohibits installation of manually operated flush and surface bolts except in limited situations. The exceptions allowing the use of such hardware are intended to expand the use of manually operated edge- or surface-mounted bolts under specified conditions while maintaining an appropriate degree of safety for the building occupants. Flush and surface bolts represent locking devices that may be difficult to operate because of their location and operation (see Commentary Figure 1010.1.9.4).

Where edge-mounted or surface-mounted manually operated bolts are installed on the inactive leaf of a pair of doors, per the exceptions below, the manually operated bolts must have no effect on the egress operation of the active leaf.

Exception 1 allows manual flush bolts and surface bolts at some doors within an individual dwelling or sleeping unit. Even then, such bolts may only be used on doors not required for egress (see Section 1010.1.9.3, Item 4, for security of doors from individual dwelling and sleeping units).

Exception 2 provides for edge-mounted or surface-mounted bolts on the inactive leaf of a pair of doors from storage or equipment areas. Double doors are often provided to allow for the easy removal or replacement of large pieces of equipment or bulk movement of goods.

Exceptions 3 and 4 offer two options for limited doors in Group B, F and S occupancies. Again, the wider door is sometimes needed for the movement of equipment. Automatic flush bolts and removable center posts can be easily damaged and difficult to maintain in areas of frequent door usage. Revisions to the requirements for door hardware on such pairs of doors will increase building functionality while maintaining a very high degree of occupant safety.

In Exception 3, the number of occupants within the space must be less than 50, the active leaf must meet means of egress requirements and the inactive leaf must not have any operating hardware so that it could be mistaken for an egress door.

In accordance with Exception 4, if the Group B, F or S building is sprinklered throughout with an NFPA 13 system, the room served by the double door can have any occupant load if the inactive leaf is not required for egress capacity and has no operating hardware.

Exception 5 is in recognition of the need to move equipment into some patient sleeping and treatment rooms in hospital and nursing home environments. Again, the inactive leaf must not be needed for means of egress or have any operating hardware. This is consistent with Section 407.1.1, which allows for staff to operate patient sleeping and treatment room doors during emergency events. The doors would still have to meet smoke barrier opening protective requirements. The clear width requirement of $41^1/_2$ inches (1054 mm) in Section 1010.1.1 would still have to be met with the active door leaf.

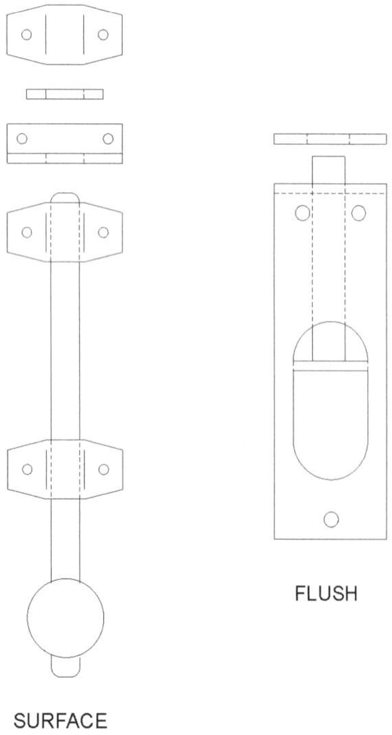

SURFACE

FLUSH

Figure 1010.1.9.4
TYPICAL MANUAL BOLT HARDWARE

1010.1.9.5 Unlatching. The unlatching of any door or leaf shall not require more than one operation.

Exceptions:

1. Places of detention or restraint.
2. Where manually operated bolt locks are permitted by Section 1010.1.9.4.
3. Doors with automatic flush bolts as permitted by Section 1010.1.9.3, Item 3.
4. Doors from individual *dwelling units* and *sleeping units* of Group R occupancies as permitted by Section 1010.1.9.3, Item 4.

❖ The code prohibits the use of locks or latching devices that require more than one operation on any door required or used for egress, which could be a safety hazard in an emergency situation. The exceptions address locations where multiple locks or latching devices which require more than one operation are acceptable. See the referenced sections for additional commentary.

1010.1.9.5.1 Closet and bathroom doors in Group R-4 occupancies. In Group R-4 occupancies, closet doors that latch in the closed position shall be openable from inside the closet, and bathroom doors that latch in the closed position shall be capable of being unlocked from the ingress side.

❖ The intent of this provision is to address possible entrapment concerns in group homes. If a closet door has a door latch, the closet door must be openable from both inside and outside. This will ensure that someone cannot get stuck inside a closet by accident. If a closet does not latch, no interior hardware is required. In case a resident needs assistance in a bathroom, the bathroom door must have a type of hardware that would allow the door to be unlocked from the outside by staff. This requirement is unique to Group R-4. There are not similar requirements for other Group R occupancies, Group I-1 assisted living facilities or Group I-2 nursing homes; however, some assisted living or nursing home facilities install such devices to increase patient safety. Section 2902.3.5 and the equivalent section in the *International Plumbing Code*® (IPC®) (Section 403.3.6) prohibits the doors leading from multistall bathrooms from being lockable from the inside.

1010.1.9.6 Controlled egress doors in Groups I-1 and I-2. Electric locking systems, including electro-mechanical locking systems and electromagnetic locking systems, shall be permitted to be locked in the means of egress in Group I-1 or I-2 occupancies where the clinical needs of persons receiving care require their containment. Controlled egress doors shall be permitted in such occupancies where the building is equipped throughout with an *automatic sprinkler system* in accordance with Section 903.3.1.1 or an *approved automatic smoke* or *heat detection system* installed in accordance with Section 907, provided that the doors are installed and operate in accordance with all of the following:

1. The door locks shall unlock on actuation of the *automatic sprinkler system* or *automatic fire detection system*.
2. The door locks shall unlock on loss of power controlling the lock or lock mechanism.
3. The door locking system shall be installed to have the capability of being unlocked by a switch located at the *fire command center*, a nursing station or other approved location. The switch shall directly break power to the lock.
4. A building occupant shall not be required to pass through more than one door equipped with a controlled egress locking system before entering an exit.
5. The procedures for unlocking the doors shall be described and approved as part of the emergency planning and preparedness required by Chapter 4 of the *International Fire Code*.
6. All clinical staff shall have the keys, codes or other means necessary to operate the locking systems.
7. Emergency lighting shall be provided at the door.
8. The door locking system units shall be listed in accordance with UL 294.

Exceptions:

1. Items 1 through 4 shall not apply to doors to areas occupied by persons who, because of clinical needs, require restraint or containment as part of the function of a psychiatric treatment area.
2. Items 1 through 4 shall not apply to doors to areas where a *listed* egress control system is utilized to reduce the risk of child abduction from nursery and obstetric areas of a Group I-2 hospital.

❖ The intent of these provisions is to address the special safety needs for wards, units, or areas in assisted living facilities, nursing homes and hospitals where egress may need to be controlled for the safety of the occupants or where specialized protective measures are needed for patients. "Controlled egress" means simply that the ability of occupants, such as patients or residents, to leave a space is controlled by others, such as staff. The areas where controlled egress may be permitted include psychiatric areas, dementia units, Alzheimer's units, maternity units, and newborn nurseries. Code officials may also permit these provisions in other areas such as emergency departments or pediatric areas where the safety and/or security of the occupants are of primary concern. In all situations, there must be a balance between maintaining a safe and secure environment and providing for emergency egress.

The requirements of this section apply to locking systems controlling egress. The functions of an ingress control locking system are not addressed in the codes and are unrelated as long as egress is provided as required or permitted by this section and other applicable provisions of the code.

In areas where additional security may be a concern, a door hardware system may include an unlocking device, such as a keypad, card reader, eye scanner, thumbprint scanner or other credential device. Where these credential unlocking systems are only on the ingress side of a building or space, there are no limitations in the codes. Where these credential unlocking systems are on the egress side of a space (perhaps to monitor personnel movement), egress must be available through one of the locking systems addressed in Section 1010.1.9.7, 1010.1.9.8 or 1010.1.9.9. Such a credential unlocking device could also be included in the controlled locking arrangements of Sections 408.4 and 1010.1.9.6 (Item 6).

IBC Section 907.6 requires a fire alarm system to be installed in accordance with the requirements of that section and NFPA 72. NFPA 72 includes specific requirements for unlocking doors in the direction of egress that are consistent with the requirements in this section for controlled egress locking systems. Additionally, NFPA 72 includes specific requirements and guidance for backup power where the backup power is used to keep these doors in a locked condition in the direction of egress.

Items 1 through 3 address when the controlled egress locks would be required to automatically unlock allowing unrestricted egress.

Item 4 requires that no occupant shall have to pass through more than one controlled egress lock before entering an exit.

Item 5 requires the procedures for unlocking the doors to be described, approved by the code official, and included in the emergency planning and preparedness plan required by Chapter 4 of the *International Fire Code*.

Item 6 requires clinical staff to have the means necessary to operate the controlled egress locking system, such as keys, codes, etc.

Item 7 requires emergency lighting at the controlled egress door to ensure visibility for egress during a possible power outage.

Item 8 requires the units of the controlled egress locking system to be listed to UL 294. The UL 294 Access Control Systems standard applies to construction, performance, and operation of systems which control passage through a door, and electrical, electronic, or mechanical units of these systems.

The first exception allows for the automatic unlocking requirements to be omitted for doors in a psychiatric treatment area in hospitals or nursing homes due to additional safety concerns for the residents.

The second exception allows for automatic unlocking requirements to be omitted where a listed egress control system is installed for the specific purposes of reducing child abductions from nursery and obstetric areas of hospitals.

1010.1.9.7 Delayed egress. Delayed egress locking systems shall be permitted to be installed on doors serving any occupancy except Group A, E and H in buildings that are equipped throughout with an *automatic sprinkler system* in accordance with Section 903.3.1.1 or an *approved automatic smoke* or *heat detection system* installed in accordance with Section 907. The locking system shall be installed and operated in accordance with all of the following:

1. The delay electronics of the delayed egress locking system shall deactivate upon actuation of the *automatic sprinkler system* or *automatic fire detection system*, allowing immediate, free egress.
2. The delay electronics of the delayed egress locking system shall deactivate upon loss of power controlling the lock or lock mechanism, allowing immediate free egress.
3. The delayed egress locking system shall have the capability of being deactivated at the *fire command center* and other *approved* locations.
4. An attempt to egress shall initiate an irreversible process that shall allow such egress in not more than 15 seconds when a physical effort to exit is applied to the egress side door hardware for not more than 3 seconds. Initiation of the irreversible process shall activate an audible signal in the vicinity of the door. Once the delay electronics have been deactivated, rearming the delay electronics shall be by manual means only.

   Exception: Where approved, a delay of not more than 30 seconds is permitted on a delayed egress door.

5. The egress path from any point shall not pass through more than one delayed egress locking system.

   Exception: In Group I-2 or I-3 occupancies, the egress path from any point in the building shall pass through not more than two delayed egress locking systems provided the combined delay does not exceed 30 seconds.

6. A sign shall be provided on the door and shall be located above and within 12 inches (305 mm) of the door exit hardware:
   - 6.1. For doors that swing in the direction of egress, the sign shall read: PUSH UNTIL ALARM SOUNDS. DOOR CAN BE OPENED IN 15 [30] SECONDS.
   - 6.2. For doors that swing in the opposite direction of egress, the sign shall read: PULL UNTIL ALARM SOUNDS. DOOR CAN BE OPENED IN 15 [30] SECONDS.
   - 6.3. The sign shall comply with the visual character requirements in ICC A117.1.

   Exception: Where approved, in Group I occupancies, the installation of a sign is not required where care recipients who because of clinical needs require restraint or containment as part of the function of the treatment area.

7. Emergency lighting shall be provided on the egress side of the door.
8. The delayed egress locking system units shall be listed in accordance with UL 294.

❖ The intent of these provisions is to address special needs where there are concerns about internal security. Delays in egress, as allowed in this section, are not considered detrimental to occupant evacuation. The intent of these provisions is also to address special needs where occupants may need to be protected from harm because of their own actions. For example, residents or patients in some Group I-1 or Group I-2 occupancies may present a danger to themselves of elopement from the building into traffic, weather and other environmental hazards.

Delayed egress locking systems are permitted for doors in a means of egress serving occupancies other than those in Groups A, E and H. Delayed egress locking systems are not permitted in assembly or educational occupancies because the resulting delay in egress is not acceptable given the greater number of occupants who may be unfamiliar with the space or of a young age. Also, the delay from Group

H would be unreasonable given the potential for rapid fire buildup in such areas.

This locking system is called "delayed egress" because of Item 4, which permits a fixed amount of time to pass prior to allowing egress.

The requirements of this section apply to locking systems which delay egress. The functions of an ingress control locking system are unrelated and are not addressed as long as egress is provided as required or permitted by this section and other applicable provisions of the code.

The requirements of this section apply to electromagnetic locking systems which open on hardware activation. The functions of an ingress control locking system are not addressed in the codes and are unrelated as long as egress is provided as required or permitted by this section and other applicable provisions of the code.

In areas where additional security may be a concern, a door hardware system may include an unlocking device, such as a keypad, card reader, eye scanner, thumbprint scanner or other credential device. Where these credential unlocking systems are only on the ingress side of a building or space, there are no limitations in the codes. Where these credential unlocking systems are on the egress side of a space (perhaps to monitor personnel movement), egress must be available through one of the locking systems addressed in Section 1010.1.9.7, 1010.1.9.8 or 1010.1.9.9. Such a credential unlocking device could also be included in the controlled locking arrangements of Sections 408.4 and 1010.1.9.6 (Item 6).

IBC Section 907.6 requires a fire alarm system to be installed in accordance with the requirements of that section and NFPA 72. NFPA 72 includes specific requirements for unlocking doors in the direction of egress that are consistent with the requirements in this section for delayed egress locking systems. Additionally, NFPA 72 includes specific requirements and guidance for backup power where the backup power is used to keep these doors in a locked condition in the direction of egress.

Because of the egress delay caused by the system on the egress door, the building must be provided throughout with compensating fire protection features to promptly warn occupants of a fire condition. All listed conditions must be met in order to permit use of a delayed egress locking system.

Item 1 interconnects the locking system with an automatic sprinkler system in accordance with NFPA 13 or, alternatively, an automatic fire detection system in accordance with Section 907, which is required to be installed throughout the building. Such systems are to provide occupants with an early warning of a fire event, and thus additional time for egress. Note that actuation of the automatic sprinkler system or actuation of the automatic fire detection system must eliminate the delay of the delayed egress locking system so the door immediately allows egress (without delay).

Item 2 specifies the delay is to be eliminated on loss of power. Since the operation of the delayed egress locking system is dependent on electrical power, in the event of electrical power loss to the lock or locking mechanism, the doors must immediately allow egress (without delay).

Item 3 specifies that the delay in the delayed egress locking system is capable of being eliminated by a signal sent from a fire command center or other approved locations. Personnel at the fire command center location would normally be the first alerted to an emergency event and would be expected to take appropriate action to allow immediate egress at all doors equipped with delayed egress locking systems. Other locations may be alternates to or back up the fire command center. Item 3 facilitates remotely eliminating the delay in the delayed egress locking system in the event of a fire or in the event of other nonfire emergencies.

Item 4 specifies the operational characteristics of the delayed egress locking system. The delay timer of the system and the audible signal (alarm) at the door may be configured one of two ways. In some occupancies, the delay timer is initiated and the alarm sounded immediately upon an attempt to open the door by pushing on the panic bar, or causing a slight movement of the door. In other occupancies, to prevent nuisance alarms from inadvertent bumps or accidental contact, the initiation of the delay timer and sounding of the alarm may be deferred by up to 3 seconds, requiring the occupant to attempt to operate the door hardware for up to, but not more than, 3 seconds.

Once the delay timer starts, the door is required to be openable from the egress side in not more than 15 seconds (or not more than 30 seconds where approved by the code official, per the exception). At the end of the delay, the door's locking system is required to allow the door to be opened by the occupant operating the egress side door hardware (i.e. pushing on the panic bar), allowing egress. The unlocking cycle is irreversible; once it is started, it does not stop. Once the door is openable from the egress side at the end of the delay, it remains openable allowing immediate egress until someone comes to the door and manually rearms the delay. The first user to the door may face a delay, but after that other users would be able to exit immediately. A method of automatically rearming the delayed egress locking system at the door from a remote location such as a central control station or security office is not permitted.

The exception to Item 4 permits the building official to allow the time delay prior to allowing egress to be increased beyond 15 seconds, but not in excess of 30 seconds. This exception is more often granted for the safety/security of occupants (e.g., to reduce or prevent elopement of patients) than for loss or theft prevention.

Item 5 limits the egress path to not more than one door with a delayed egress locking system. However, the exception to Item 5 allows not more than two doors with delayed egress locking systems in Groups I-2 and I-3 with the delay of the two systems to be not more than 30 seconds total.

Having multiple doors can help with preventing resident elopement and yet the overall delay does not exceed the previously accepted time period. An example of where the two-door arrangement may be helpful is a multistory facility where both the door from the story and the door from the building could be controlled.

Item 6 requires a sign to inform occupants how to operate the delayed egress locking system and when that door will become available for egress. An undesirable consequence of the door not unlocking immediately is if the user assumes it will never be available for egress and then proceeds to another exit door. In some occupancies, delayed egress locking systems may be utilized on doors not required to swing in the direction of egress travel. Options 6.1 and 6.2 allow for signs appropriate for the swing of the door. The required sign is typically supplied with the delayed egress locking system. The reference to ICC A117.1 visual requirements would not require raised letters or braille, but would require readable text, with good finish and contrast. The exception in Item 6 allows the sign to be omitted if the clinical needs of occupants require restraint or containment, as these patients may be capable of reading and following the sign's instructions then potentially putting themselves in harm's way. Based on the level of staff training within these facilities, the need to protect the patients by preventing elopement and the facts that these systems are required to be interconnected with the sprinkler or fire detection systems and unlock upon loss of power, it was determined that eliminating the sign in these facilities is reasonable.

Item 7 requires emergency lighting on the egress side of the door so that the user can read the sign required by Item 5.

Item 8 requires the units of the delayed egress locking system to be listed to UL 294. The UL 294 Access Control Systems standard applies to construction, performance and operation of systems which control passage through a door, and electrical, electronic or mechanical units of these systems.

1010.1.9.8 Sensor release of electrically locked egress doors. The electric locks on sensor released doors located in a *means of egress* in buildings with an occupancy in Group A, B, E, I-1, I-2, I-4, M, R-1 or R-2 and entrance doors to tenant spaces in occupancies in Group A, B, E, I-1, I-2, I-4, M, R-1 or R-2 are permitted where installed and operated in accordance with all of the following criteria:

1. The sensor shall be installed on the egress side, arranged to detect an occupant approaching the doors. The doors shall be arranged to unlock by a signal from or loss of power to the sensor.
2. Loss of power to the lock or locking system shall automatically unlock the doors.
3. The doors shall be arranged to unlock from a manual unlocking device located 40 inches to 48 inches (1016 mm to 1219 mm) vertically above the floor and within 5 feet (1524 mm) of the secured doors. Ready access shall be provided to the manual unlocking device and the device shall be clearly identified by a sign that reads "PUSH TO EXIT." When operated, the manual unlocking device shall result in direct interruption of power to the lock—independent of other electronics—and the doors shall remain unlocked for not less than 30 seconds.
4. Activation of the building *fire alarm system*, where provided, shall automatically unlock the doors, and the doors shall remain unlocked until the fire alarm system has been reset.
5. Activation of the building *automatic sprinkler system* or *fire detection system*, where provided, shall automatically unlock the doors. The doors shall remain unlocked until the *fire alarm system* has been reset.
6. The door locking system units shall be listed in accordance with UL 294.

❖ The intent of this section is to provide consistent requirements where an electronically locked door is unlocked by activating devices mounted somewhere other than on the door itself. The unlocking activation is designed to be from a passive action by the occupant (e.g., walking to the door triggering a sensor), but the system includes a required nearby manual unlocking device (such as a push button) as a secondary electrical lock release device.

This section permits doors in a means of egress and entrance doors to tenant spaces in occupancies of Groups A, B, E, I-1, I-2, I-4, M, R-1 and R-2 to be electronically secured (locked) to control ingress while maintaining the doors as a means of egress. Typically these systems are used in high-security areas where a record of who has entered and left a space is desired. Or this system may be used where there is a concern for elopement or child abduction (Group I-1, I-2 or I-4).

This section of the IBC does not limit the number of doors in a means of egress that may be equipped with sensor-released electrically locked doors.

IBC Section 907.6 requires that a fire alarm system, where provided, is installed in accordance with the requirements of that section and NFPA 72. NFPA 72 includes specific requirements for unlocking doors in the direction of egress, which are consistent with the requirements in this section for sensor release of electrically locked egress doors. Additionally, NFPA 72 includes specific requirements and guidance for backup power where the backup power is used to keep these doors in a locked condition.

The requirements of this section apply to locking systems with a sensor release. The functions of an ingress control locking system are not addressed in the codes and are unrelated as long as egress is provided as required or permitted by this section and other applicable provisions of the code.

In areas where additional security may be a concern, a door hardware system may include an unlocking device, such as a keypad, card reader, eye scanner, thumbprint scanner or other credential device. Where these credential unlocking systems are only on the ingress side of a building or space, there are no limitations in the codes. Where these credential unlocking systems are on the egress side of a space (perhaps to monitor personnel movement), egress must be available through one of the locking systems addressed in Section 1010.1.9.7, 1010.1.9.8 or 1010.1.9.9. Such a credential unlocking device could also be included in the controlled locking arrangements of Sections 408.4 and 1010.1.9.6 (Item 6).

In areas where additional security may be a concern, a door hardware system may include a remote unlocking device, such as a keypad, card reader, eye scanner or thumbprint scanner. Where these systems are only on the ingress side of a building or space, there are no limitations in the codes. Where these systems are on the egress side of a space, free egress must still be available through one of the electromagnetic systems addressed in Section 11010.1.9.7, 1010.1.9.8 or 1010.1.9.9. Such a remote unlocking device could also be included in the controlled locking arrangements for Sections 408.4 and 1010.9.6.

Items 1 through 6 provide operational criteria to ensure egress during normal and emergency situations. Occupancies in Groups F, S and H are not included here because of their increased hazard caused by fuel load and other potentially life-threatening factors.

Item 1 requires that such doors be provided with an occupant sensor on the egress side of the door. These sensors typically operate on an infrared, microwave or sonic principle, but other technologies may be available. This sensor is required to automatically release the electrical lock on approach of an occupant from the egress side or when there is a loss of power to the sensor. This provision is written as "performance-based," where any means of sensor design can be utilized to cause the doors to unlock, allowing immediate egress. This section does not indicate at what distance the sensor should be set to operate. The sensor may be set to detect an approaching occupant in time to unlock the electrical locks prior to the occupant reaching the door, to permit egress without the occupant realizing the doors were electrically locked. In other applications, the sensor may be set to require the occupant to be closer to the door prior to unlocking the electrical lock to allow egress.

Item 2 states that if there is a loss of power to the electrical lock or to the locking system, the electrical lock on the door must unlock. These doors are commonly secured with fail-safe devices which prioritize life safety over security (such as electromagnetic locks or fail-safe power bolts) so that the electrical locks on these doors will automatically unlock when power to the electrical locking device or locking system is interrupted. In some instances, the locking system controller may be powered from a different source than the electrical locking device itself. In these cases, a loss of power to the locking system controller (while power remains applied to the locking device) must also cause the electrical locking device on the egress door to automatically unlock.

Item 3 requires that there be a manual unlocking device (push button), within 5 feet of the door, mounted 40 to 48 inches above the floor, unobstructed and with a clearly identifiable sign that says "PUSH TO EXIT." When operated, the manual unlocking device is to directly interrupt, independent of other electronics, the power to the electrical lock and cause the doors to remain electrically unlocked for a minimum of 30 seconds. To achieve a minimum 30-second delay independent of other electronics, the push button unlocking device should be designed and installed to provide this delay. The 30-second minimum is to allow adequate time for an individual to operate the manual unlocking device and then to egress through the door.

Items 4 and 5 require the building fire alarm system, automatic fire detection system or sprinkler system, if provided, to be interfaced with the door's electric locking system to unlock automatically on activation. The electrical locks on the doors are to remain unlocked until the fire alarm system is reset, ensuring egress is not impeded by the electric locking system.

Item 6 requires the units of the locking system to be listed to UL 294. The UL 294 Access Control System standard applies to construction, performance, and operation of systems which control passage through a door, and electrical, electronic, or mechanical units of these systems.

To summarize, it is important to keep in mind that an egress door equipped with a sensor release system must always allow egress whether power is present or not. The egress door must be "fail-safe" and must assume this "fail-safe" condition when power is removed from any part of the senor release system. In other words, if the sensor release system loses power, the egress door must be capable of being opened. People must be kept from being involuntarily locked inside buildings.

1010.1.9.9 Electromagnetically locked egress doors. Doors in the *means of egress* in buildings with an occupancy in Group A, B, E, I-1, I-2, I-4, M, R-1 or R-2 and doors to tenant spaces in Group A, B, E, I-1, I-2, I-4, M, R-1 or R-2 shall be permitted to be locked with an electromagnetic locking system where equipped with hardware that incorporates a built-in switch and where installed and operated in accordance with all of the following:

1. The hardware that is affixed to the door leaf has an obvious method of operation that is readily operated under all lighting conditions.

2. The hardware is capable of being operated with one hand.

3. Operation of the hardware directly interrupts the power to the electromagnetic lock and unlocks the door immediately.

4. Loss of power to the locking system automatically unlocks the door.

5. Where *panic* or *fire exit hardware* is required by Section 1010.1.10, operation of the *panic* or *fire exit hardware* also releases the electromagnetic lock.

6. The locking system units shall be listed in accordance with UL 294.

❖ The intent of this section is to provide consistent requirements where an electromagnetic lock is released by door-mounted hardware such as a panic bar, lockset/latchset, or touch-sense bar, all of which would be equipped with an integral switch that, when actuated by the normal actions of opening the door, causes the electromagnetic lock to release allowing immediate egress. In the identified occupancy groups, doors in the means of egress are permitted to be locked with an electromagnetic locking system where equipped with door hardware that incorporates a built-in switch, provided all the specified conditions are met. Additionally, doors to tenant spaces of the identified occupancy groups may be equipped with electromagnetic locks, provided all of the specified conditions are met. The use of this type of locking system may provide for a greater degree of security preventing or controlling access or ingress than that offered by a door with mechanical locking devices alone. The allowance for electromagnetically locked egress doors is limited to low- and moderate-hazard occupancies where security may be a concern.

It may be important to note other “shall be permitted” locking arrangements in the code may also use electromagnetic locks as part of their system. Controlled egress locking systems (Section 1010.1.9.6), delayed egress locking systems (Section 1010.1.9.7), and sensor release of electrically locked door systems (Section 1010.1.9.8) all frequently use electromagnetic locks.

The requirements of this section apply to electromagnetic locking systems which open on hardware activation. The functions of an ingress control locking system are not addressed in the codes and are unrelated as long as egress is provided as required or permitted by this section and other applicable provisions of the code.

In areas where additional security may be a concern, a door hardware system may include an unlocking device, such as a keypad, card reader, eye scanner, thumbprint scanner or other credential device. Where these credential unlocking systems are only on the ingress side of a building or space, there are no limitations in the codes. Where these credential unlocking systems are on the egress side of a space (perhaps to monitor personnel movement), egress must be available through one of the locking systems addressed in Section 1010.1.9.7, 1010.1.9.8 or 1010.1.9.9. Such a credential unlocking device could also be included in the controlled locking arrangements of Sections 408.4 and 1010.1.9.6 (Item 6).

When the occupant prepares to egress through the door, the method of operating the door hardware must be obvious, even under poor lighting conditions. The operation shall be accomplished through the use of a single motion and meet the general requirement that the door be readily openable without the use of special knowledge or effort. The release of the electromagnetic lock on the door must occur immediately on the operation of the door hardware by interrupting the power supply to the electromagnetic lock. This requirement is the same regardless of the type of door hardware: panic hardware, fire exit hardware, a latchset/lockset, or a touch-sense bar. As an additional safeguard, the loss of power to the locking system is required to automatically release the electromagnetic lock on the door.

A properly designed and installed electromagnetic locking system complying with the requirements of this section of the IBC may not be obvious to the occupants, as the door unlatches/unlocks (electromagnetically and mechanically) allowing egress through the normal operation of the door hardware (panic or fire exit hardware, latchset/lockset, or touch-sense bar). Considering these performance requirements, this section of the IBC does not include a requirement that the electromagnetic lock be unlocked (released) on activation of the building fire alarm system. Also, this section of the IBC does not limit the number of doors in the means of egress which may be equipped with electromagnetic locking systems.

The units of the electromagnetic locking system are required to be listed in accordance with UL 294. The UL 294 Access Control Systems standard applies to construction, performance, and operation of systems which control passage through a door, and electrical, electronic, or mechanical units of these systems. Where these special provisions are utilized, the requirements of Section 1010.1.10 regarding panic hardware remain applicable. In Group A and E occupancies having occupant loads of 50 or more, the door hardware must also comply with the requirements for panic hardware.

1010.1.9.10 Locking arrangements in correctional facilities. In occupancies in Groups A-2, A-3, A-4, B, E, F, I-2, I-3, M and S within correctional and detention facilities, doors in *means of egress* serving rooms or spaces occupied by persons whose movements are controlled for security reasons shall be permitted to be locked where equipped with egress control devices that shall unlock manually and by not less than one of the following means:

1. Activation of an *automatic sprinkler system* installed in accordance with Section 903.3.1.1.

2. Activation of an *approved manual fire alarm box*.

3. A signal from a *constantly attended location*.

❖ Correctional facilities can include a variety of uses where detainees may be gathered for eating, recreational activities, education, technical training, job training, etc. Correctional facilities can also contain types of support services, such as a store, storage areas or hospital area. Security is still a concern within these areas. This provision will allow the correctional facility to maintain security on all areas. Most commonly the doors would be opened by staff from a central control point under Item 3, but Items 1 and 2 allow for other alternatives in lower security facilities. This provision is not intended to apply to these groups when located outside of a detention or correctional facility.

1010.1.9.11 Stairway doors. Interior *stairway means of egress* doors shall be openable from both sides without the use of a key or special knowledge or effort.

Exceptions:

1. *Stairway* discharge doors shall be openable from the egress side and shall only be locked from the opposite side.
2. This section shall not apply to doors arranged in accordance with Section 403.5.3.
3. In *stairways* serving not more than four stories, doors are permitted to be locked from the side opposite the egress side, provided they are openable from the egress side and capable of being unlocked simultaneously without unlatching upon a signal from the *fire command center*, if present, or a signal by emergency personnel from a single location inside the main entrance to the building.
4. *Stairway exit* doors shall be openable from the egress side and shall only be locked from the opposite side in Group B, F, M and S occupancies where the only interior access to the tenant space is from a single *exit stairway* where permitted in Section 1006.3.2.
5. *Stairway exit* doors shall be openable from the egress side and shall only be locked from the opposite side in Group R-2 occupancies where the only interior access to the *dwelling unit* is from a single *exit stairway* where permitted in Section 1006.3.2.

❖ Based on adverse fire experience where occupants have become trapped in smoke-filled stairway enclosures, stairway doors generally must be arranged to permit reentry into the building without the use of any tools, keys or special knowledge or effort. For security reasons, this restriction does not apply to the discharge door from the stairway enclosure to the outside or into an exit passageway (Exception 1). Section 403 for high-rise buildings permits locking doors from the stairway side, provided the doors are capable of being unlocked from a fire command station and there is a communication system within the stairway enclosure that allows contact with the fire command station (Exception 2).

Exception 3 addresses the need for security. The exception is limited to four-story buildings to provide a short travel distance to the stairway discharge door for the building occupants. In addition, to allow quick entrance for fire fighters and emergency responders, a means of simultaneously unlocking all of the doors by emergency personnel must be provided. This provision further requires that the stairway doors be unlocked without unlatching. Stairway doors will typically be fire door assemblies, and their continued latching is necessary to maintain the integrity of the fire-resistive separation for the exit enclosure. The remote unlocking signal shall be initiated from the fire command station, if provided, or a single point of signal initiation at an approved location inside the building's main entrance.

Exceptions 4 and 5 allow for stairways in single-exit buildings to have doorways that lead to multiple tenants and dwelling units. For security reasons, those doors can remain locked from the stairway side so no one can enter another tenant space or dwelling unit from the exit stairway.

1010.1.10 Panic and fire exit hardware. Doors serving a Group H occupancy and doors serving rooms or spaces with an *occupant load* of 50 or more in a Group A or E occupancy shall not be provided with a latch or lock other than *panic hardware* or *fire exit hardware*.

Exceptions:

1. A main *exit* of a Group A occupancy shall be permitted to be locking in accordance with Section 1010.1.9.3, Item 2.
2. Doors serving a Group A or E occupancy shall be permitted to be electromagnetically locked in accordance with Section 1010.1.9.9.

Electrical rooms with equipment rated 1,200 amperes or more and over 6 feet (1829 mm) wide, and that contain overcurrent devices, switching devices or control devices with *exit* or *exit access doors*, shall be equipped with *panic hardware* or *fire exit hardware*. The doors shall swing in the direction of egress travel.

❖ Doors that are part of a means of egress from the locations listed in this section shall not be provided with a latch or lock other than panic hardware or fire exit hardware unless one of the two exceptions is met. Fire exit hardware is essentially panic hardware with internal modifications for use on fire-rated doors. Also see the commentaries to the definitions for "Fire exit hardware" or "Panic hardware" and Sections 1010.1.10.1 and 1010.1.10.2.

For all Group H occupancies, regardless of the occupant load, if latching (or locking) hardware is installed, it must be panic hardware or fire exit hardware because of the physical hazards of these spaces.

For all doors that provide means of egress for rooms and spaces of assembly and educational

(Group A and E) occupancies with an occupant load of 50 or more, if latching (or locking) hardware is installed, it must be panic hardware or fire exit hardware. This would include large assembly spaces in mixed-use buildings. These uses are characterized by higher occupant load densities. Whereas doors from an assembly or educational room with an occupant load of less than 50 do not require panic hardware or fire exit hardware, a door that provides means of egress for two or more such rooms would require panic hardware or fire exit hardware when the combination of spaces has a total occupant load of 50 or more.

The first exception clarifies that the provisions for key-locking hardware at the main exit in Group A occupancies are permitted instead of panic hardware at those specific locations. (For the Group A exception, see the commentary to Section 1010.1.9.3, Item 2.)

The second exception resolves a potential conflict with Section 1010.1.9.9, and specifically Item 5, which allows an electromagnetic lock to be installed in addition to panic or fire exit hardware where panic or fire exit hardware would be required as long as the operation of the panic hardware or fire exit hardware releases the electromagnetic lock.

Certain electrical rooms are required to have panic hardware. Refer to IBC Chapter 27 and *NFPA 70: National Electrical Code*® (NEC) for specific requirements for when panic hardware or fire exit hardware is required. This requirement is applicable only where multiple conditions are present. The type of room regulated creates a potentially hazardous environment. In the event of an electrical accident, the more immediate egress provided by the panic hardware is desirable.

1010.1.10.1 Installation. Where *panic* or *fire exit hardware* is installed, it shall comply with the following:

1. *Panic hardware* shall be *listed* in accordance with UL 305.
2. *Fire exit hardware* shall be *listed* in accordance with UL 10C and UL 305.
3. The actuating portion of the releasing device shall extend not less than one-half of the door leaf width.
4. The maximum unlatching force shall not exceed 15 pounds (67 N).

❖ As its name implies, panic hardware is special unlatching and unlocking hardware that is intended to simplify the unlatching and unlocking operation to a single, no more than 15-pound (67 N), force applied in the direction of egress [see Commentary Figures 1010.1.10.1(1) and (2)]. In a panic situation with a rush of persons trying to utilize a door, devices such as doorknobs or thumb turns may cause sufficient delay so as to create a crush at the door and prevent or slow the opening operation.

The locational specifications for the activating panel or bar are based on ready availability and access to the unlatching device. Note that the section requires the width of the actuating portion of the panic hardware or fire exit hardware to measure at least one-half the width of the door leaf. For example, on a

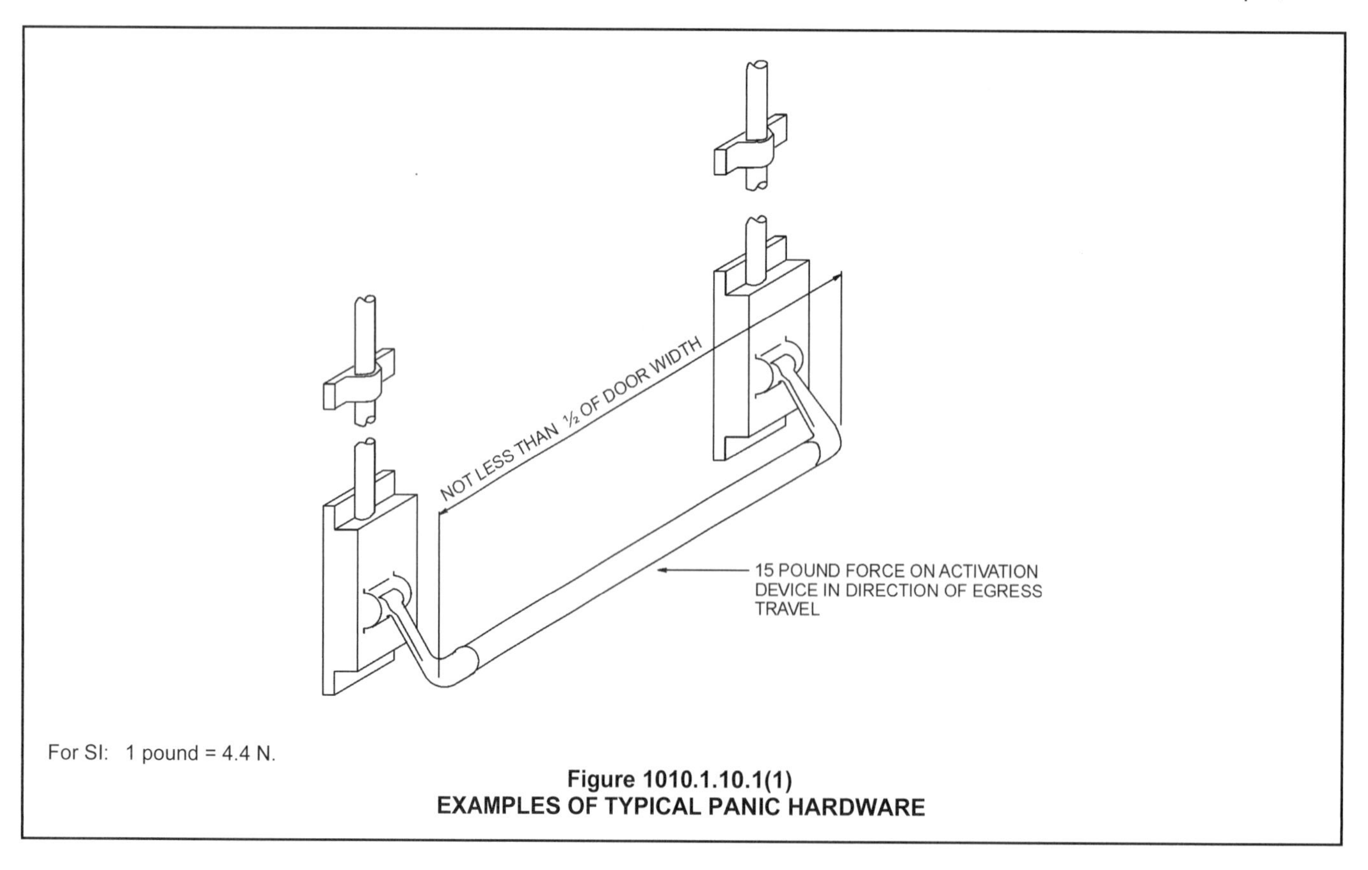

Figure 1010.1.10.1(1)
EXAMPLES OF TYPICAL PANIC HARDWARE

3-0 door (nominal 36 inches wide), the activating portion of the panic or fire hardware would measure at least 18 inches. Panic and fire exit hardware must be listed. UL 305, *Standard for Safety Panic Hardware*, includes construction and performance requirements dealing with endurance, emergency operation, elevated ambient exposure and low-temperature impact tests to ensure that the panic device operates properly (for panic hardware on a balanced door, see Section 1010.1.10.2). The activation device must be mounted between 34 inches and 48 inches (864 mm and 1219 mm) above the floor in accordance with Section 1010.1.9.2. Section 1010.1.1.1 allows the panic hardware to extend the full width of the door as long as it does not protrude more than 4 inches (102 mm) into the door's required minimum clear width.

Standard panic hardware or "listed panic hardware" is not approved for use on fire door assemblies. Panic hardware and fire exit hardware can be similar in appearance.

Where a fire door, such as to an exit stairway, is required to be equipped with panic hardware, the hardware must accomplish the dual objectives of panic hardware and continuity of the enclosure in which it is located—thus the reference to UL 10C, *Standard for Safety Positive Pressure Fire Tests of Door Assemblies*. In this case, fire exit hardware is to be provided that meets both objectives and requirements, since panic hardware is not tested for use on fire doors. There are standard test procedures designed to evaluate the performance of panic and fire exit hardware from the panic standpoint as well as from a fire protection standpoint. "Fire door assemblies" are defined in Section 716.5 as a combination of doors, frame, hardware and other accessories required to provide a specific degree of fire and smoke barrier protection to the opening in a fire wall, fire barrier, fire partition, smoke barrier or exterior wall required to have a fire-resistance rating.

Fire doors must close and positively latch in order to protect exit stairways, corridors and other areas of the building from the spread of smoke and fire. Additionally, fire doors are required to self-close and automatically latch after each use. Positive latching of fire doors is not related to the locking of the door and should never be confused with locking or security issues.

The requirement for positive latching means that dogging devices are not permitted on fire exit hardware. A dogging device is an option on the hardware that allows for the panic hardware to be locked in the fully depressed position. A dogging device mechanically defeats the latching feature of panic hardware preventing the door from positively latching when in the closed position. The dogging device is typically manually activated with a small wrench or tool and is

Figure 1010.1.10.1(2)
EXAMPLES OF TYPICAL PANIC HARDWARE

activated through a hole adjacent to the activation bar. Dogging capability is often provided on exterior doors that are intended as building entrances.

Fire exit hardware must be labeled. Typical locations are on either end of the hardware. Information on the label must include the words "listed" and "fire exit hardware" and indicate a control or serial number. The label on the fire door itself should indicate that it is a fire door suitable for use with fire exit hardware.

1010.1.10.2 Balanced doors. If *balanced doors* are used and *panic hardware* is required, the *panic hardware* shall be the push-pad type and the pad shall not extend more than one-half the width of the door measured from the latch side.

❖ The provisions for balanced doors ensure that the occupants push only on the latch side of the door since the hinge side of a balanced door pivots "against" the direction of egress (see the commentary for the definition of "Door, balanced" in Chapter 2).

1010.2 Gates. Gates serving the *means of egress* system shall comply with the requirements of this section. Gates used as a component in a *means of egress* shall conform to the applicable requirements for doors.

Exception: Horizontal sliding or swinging gates exceeding the 4-foot (1219 mm) maximum leaf width limitation are permitted in fences and walls surrounding a stadium.

❖ This section specifies that all requirements for doors also apply to gates, except that gates surrounding a stadium are allowed to exceed 4 feet (1219 mm) in width. Usually a large gate is required to adequately serve a stadium crowd for egress purposes.

1010.2.1 Stadiums. *Panic hardware* is not required on gates surrounding stadiums where such gates are under constant immediate supervision while the public is present, and where safe dispersal areas based on 3 square feet (0.28 m^2) per occupant are located between the fence and enclosed space. Such required safe dispersal areas shall not be located less than 50 feet (15 240 mm) from the enclosed space. See Section 1028.5 for *means of egress* from safe dispersal areas.

❖ Panic hardware is impractical for large gates that surround stadiums. Normally, these gates are opened and closed by the stadium's grounds crew, which is constantly in attendance during the use of such gates. The safe dispersal area requirement provides for the safety of the crowd if for some reason the gate is not open. The safe dispersal area is to be between the stadium enclosure and the surrounding fence and the area to be occupied is not to be closer than 50 feet (15 240 mm) to the stadium enclosure.

See the commentary for Section 1028.5 for access to a safe dispersal area when access to a public way is not available.

1010.3 Turnstiles. Turnstiles or similar devices that restrict travel to one direction shall not be placed so as to obstruct any required *means of egress*.

Exception: Each turnstile or similar device shall be credited with a capacity based on not more than a 50-person *occupant load* where all of the following provisions are met:

1. Each device shall turn free in the direction of egress travel when primary power is lost and on the manual release by an employee in the area.
2. Such devices are not given credit for more than 50 percent of the required egress capacity or width.
3. Each device is not more than 39 inches (991 mm) high.
4. Each device has not less than $16^1/_2$ inches (419 mm) clear width at and below a height of 39 inches (991 mm) and not less than 22 inches (559 mm) clear width at heights above 39 inches (991 mm).

Where located as part of an *accessible route*, turnstiles shall have not less than 36 inches (914 mm) clear at and below a height of 34 inches (864 mm), not less than 32 inches (813 mm) clear width between 34 inches (864 mm) and 80 inches (2032 mm) and shall consist of a mechanism other than a revolving device.

❖ This section provides for a limited use of turnstiles to serve as a means of egress component. The exception to this section limits each turnstile to a maximum egress capacity of 50 persons. The turnstile must comply with all four listed items to be considered as serving any part of the occupant load for means of egress. The turnstiles must rotate freely both when there is a loss of power and when they are manually released. Note that the 50-person limit applies to each individual turnstile. These provisions are similar to the revolving door provisions in Section 1010.1.4.1.

If turnstiles are located along an accessible route, the route for persons using mobility devices must be something other than a revolving device, such as a swinging gate. A common example would be the turnstiles for automatic ticket taking, such as at the entrance to a mass transit platform.

1010.3.1 High turnstile. Turnstiles more than 39 inches (991 mm) high shall meet the requirements for revolving doors.

❖ Where a turnstile is higher than 39 inches (991 mm), the restriction to egress is much like a revolving door. Thus, the egress limitations for revolving doors in Section 1010.1.4.1 apply to this type of turnstile. If a high turnstile does not meet the revolving door requirements for doors that are an egress component, it is not to be included as serving a portion of the means of egress. It would be necessary to provide doors in these areas for egress. High turnstiles may not be part of an accessible route for ingress or egress.

1010.3.2 Additional door. Where serving an *occupant load* greater than 300, each turnstile that is not portable shall have a side-hinged swinging door that conforms to Section 1010.1 within 50 feet (15 240 mm).

❖ This section addresses a common egress condition for sports arenas where a number of turnstiles are installed for ticket taking. Portable turnstiles are moved from the egress path for proper exiting capacity. Permanent turnstiles are not considered as providing any of the required egress capacity when serving an occupant load greater than 300, no matter how many turnstiles are installed. Doors are required to provide occupants with a path of egress other than through the turnstiles. The doors are to be located within 50 feet (15 240 mm) of the turnstiles.

SECTION 1011 STAIRWAYS

1011.1 General. *Stairways* serving occupied portions of a building shall comply with the requirements of Sections 1011.2 through 1011.13. *Alternating tread devices* shall comply with Section 1011.14. Ship's ladders shall comply with Section 1011.15. Ladders shall comply with Section 1011.16.

Exception: Within rooms or spaces used for assembly purposes, stepped aisles shall comply with Section 1029.

❖ It is important for stairway safety that all stairways meet the provisions in this section. This would include all elevation changes using stairways; everything from one riser to multiple flights and landings between stories (see the definitions for "Stair" and "Stairway" in Chapter 2). These provisions will be applicable for interior exit access stairways, interior exit stairways and exterior exit stairways, as well as any steps along the paths for exit access or exit discharge. It is intended that this section be applicable to required stairways as well as what can be called "convenience" stairways.

It is important to understand the terminology. Exit stairways are stairways that provide a protected path of travel between the exit access and the exit discharge. Interior exit stairways are required to be enclosed in accordance with Section 1022. Exterior exit stairways are protected by the exterior wall of the building and must comply with Section 1027. Exit access stairways are typically unenclosed interior stairways and comply with Section 1019 when they provide access between stories. Exit access travel distance stops at an exit stairway enclosure, but includes any travel down an exit access stairway. Stairways that are outside and provide a route from the level of exit discharge to grade are considered part of the exit discharge. See the commentary in Chapter 2 for the defined term, "Exit discharge, level of."

Sections 1011.2 through 1011.13 provide criteria for the typical stairway. Special provisions are provided for curved stairways (Section 1011.9), spiral stairways (Section 1011.10), and stairways to the roof for fire department access and for elevator equipment service (Sections 1011.12 and 1011.13).

Items that provide vertical access similar to stairways are also addressed in this section: alternating tread devices (Section 1011.14), ship's ladders (Section 1011.15) and ladders (Section 1011.16). These devices are only permitted to provide access to very limited spaces.

The exception indicates that stepped aisles (previously called aisle stairs) are addressed in Section 1029. Having this exception at the beginning of the stairway section negated the need for repeated exceptions throughout the stairway provisions. While both stairways and stepped aisles are a series of treads and risers, how occupants move on and off and configurations are very different. Occupants leave and join stepped aisles along the entire run, while occupants only enter the stairways at the top and bottom. Stepped aisles have center handrails with breaks to allow for access into the seating, while most stairways have handrails on both sides. Stepped aisles can be nonuniform in some locations to allow for parabolic seating bowls, while with stairways, uniform tread and riser configurations are required. Section 1029 should be used for stepped aisles between and immediately adjacent to seating or where the steps are a direct continuation of the stepped aisles and lead to a level cross aisle or floor. Section 1011 is used for stairways that lead from the balcony, concourse or cross aisle to a floor level above or below the seating areas (see Sections 1011.5.2, 1011.5.4 and 1011.6).

1011.2 Width and capacity. The required capacity of *stairways* shall be determined as specified in Section 1005.1, but the minimum width shall be not less than 44 inches (1118 mm). See Section 1009.3 for accessible *means of egress stairways*.

Exceptions:

1. *Stairways* serving an *occupant load* of less than 50 shall have a width of not less than 36 inches (914 mm).
2. *Spiral stairways* as provided for in Section 1011.10.
3. Where an incline platform lift or stairway chairlift is installed on *stairways* serving occupancies in Group R-3, or within *dwelling units* in occupancies in Group R-2, a clear passage width not less than 20 inches (508 mm) shall be provided. Where the seat and platform can be folded when not in use, the distance shall be measured from the folded position.

❖ To provide adequate space for occupants traveling in opposite directions and to permit the intended full egress capacity to be developed, minimum dimensions are dictated for means of egress stairways. A minimum width of 44 inches (1118 mm) is required for stairway construction to permit two columns of users to travel in the same or opposite directions. The reference to Section 1005.1 is for the determination of stairway width based on the occupant load it will serve (i.e, capacity). The larger of the two widths is to be used.

Exception 1 recognizes the relatively small occupant loads of less than 50 that permit a staggered file of users when traveling in the same direction. When traveling in opposite directions, one column of users must stop their ascent (or descent) to permit the opposite column to continue. Again, considering the relatively small occupant loads, any disruption of orderly flow will be infrequent. The use of this exception is limited to buildings where the entire occupant load of each upper story and/or basement is less than 50.

Exception 2 permits a spiral stairway to have a minimum width of 26 inches (660 mm) when it conforms to Section 1011.10, on the basis that the configuration of a spiral stairway will allow nothing other than single-file travel.

Exception 3 addresses the use of inclined platform lifts or stairway chairlifts for individual dwelling units. For clarification on the types of lifts, see the commentary to Section 1109.8. Both types of lifts may be installed to aid persons with mobility impairments in their homes. The code and ASME A18.1 allow for a reduction in the width of the stair to a minimum of 20 inches (508 mm) of clear passageway to be maintained on a stairway where a lift is located. If a portion of the lift, such as a platform or seat, can be folded, the minimum clear dimension is to be measured from the folded position. If the lift cannot be folded, then the 20 inches (508 mm) is measured from the fixed position. The track for these lifts typically extends 9 to 12 inches (229 to 305 mm) from the wall, making the 20-inch (508 mm) clear measurement actually 24 to 27 inches (610 to 686 mm) from the edge of the track.

The code does not have any specific provisions for where incline platform lifts are utilized along stairways in locations other than within dwelling units. Section 1109.8 limits the use of platform lifts in new construction to mainly areas with minimal occupant loads or where elevators and ramps are impracticable. IEBC Sections 410.8.3 and 705.1.3 allows for platform lifts anywhere in existing buildings in order to gain accessibility for persons with mobility impairments. When in the closed and off position, the platform lifts should not block the clear width required for the stairway, or use of the handrails. The industry is currently working on different options to address the concern that the lift may be in operation during an event that requires evacuation.

1011.3 Headroom. *Stairways* shall have a headroom clearance of not less than 80 inches (2032 mm) measured vertically from a line connecting the edge of the *nosings*. Such headroom shall be continuous above the *stairway* to the point where the line intersects the landing below, one tread depth beyond the bottom riser. The minimum clearance shall be maintained the full width of the *stairway* and landing.

Exceptions:

1. *Spiral stairways* complying with Section 1011.10 are permitted a 78-inch (1981 mm) headroom clearance.
2. In Group R-3 occupancies; within *dwelling units* in Group R-2 occupancies; and in Group U occupancies that are accessory to a Group R-3 occupancy or accessory to individual *dwelling units* in Group R-2 occupancies; where the *nosings* of treads at the side of a *flight* extend under the edge of a floor opening through which the *stair* passes, the floor opening shall be allowed to project horizontally into the required headroom not more than $4^3/_4$ inches (121 mm).

❖ This headroom requirement is necessary to avoid an obstruction to orderly flow and to provide visibility to the users so that the desired path of travel can be planned and negotiated. Height is a vertical measurement above every point along the stairway stepping and walking surfaces, with minimum height measured vertically from the tread nosing or from the surface of a landing or platform up to the ceiling [see Commentary Figure 1011.3(1)].

Sections 1003.2 and 1208.2 require a minimum ceiling height within a room of 7 feet, 6 inches (2307 mm). A bulkhead or doorway at the bottom of the stairway would be allowed to meet the minimum headroom height of 80 inches (2032 mm), as permitted in Section 1003.3.

Exception 1, allowing for a clear headroom of 6 feet, 6 inches (1981 mm) for spiral stairs, correlates with the provisions of Section 1011.10.

Exception 2 recognizes a common method of stairwell construction in which the stringer on the open side of a stair is supported by the same floor joists or wall that supports the edge of the opening through which the stairway passes to the floor above, thus resulting in the stairway being wider at the lower portion than at the top portion. In this case, headroom is not required for a distance of up to $4^3/_4$ inches (121 mm) measured horizontally from the edge of the opening above to the handrail or guard system, which limits the clear width on the lower open sides of the stairway. The $4^3/_4$ inches (121 mm) maximum is derived from the finished width of a typical 2 by 4 supporting wall and is not critical to obstructing orderly flow or visibility in the desired path of travel [see Commentary Figure 1011.3(2)].

1011.4 Walkline. The walkline across *winder* treads shall be concentric to the direction of travel through the turn and located 12 inches (305 mm) from the side where the *winders* are narrower. The 12-inch (305 mm) dimension shall be measured from the widest point of the clear *stair* width at the walking surface of the *winder*. Where *winders* are adjacent within the *flight*, the point of the widest clear *stair* width of the adjacent *winders* shall be used.

❖ This requirement is essential for smooth, consistent travel on stairs that turn with winder treads. It provides a standard location for the regulation of the uniform tread depth of winders. Because of the wide range of anthropometrics of stairway users, there is no one line that all persons will travel on stairs; however, the code recognizes a standard location of a walkline is essential to design and enforcement. Each footfall of the user through the turn can be associated with an arc to describe the path traveled. As a user ascends or descends the flight, the turning at each step should be consistent through the turn. The walkline is established concentric to, or having the same center (approximately parallel) as, the arc of travel of the user. The tread depth dimension at the walkline is one of two tread depths across the width of the stair at which winder tread depth is regulated, cited in Section 1011.5.2. The second is the minimum tread depth. Regulation at these two points controls the angularity of the turn and the configuration of the flight. In order to establish consistently shaped winders, tread depths must always be measured concentric to the arc of travel. The walkline is unique as the only line or path of travel where winder tread depth is controlled by the same minimum tread depth as rectangular treads. However, Exception 2 of Section 1011.5.4 recognizes that winder tread depth need not be compared to rectangular tread depths for dimensional uniformity in the same flight because the location of the walkline is chosen for the purpose of

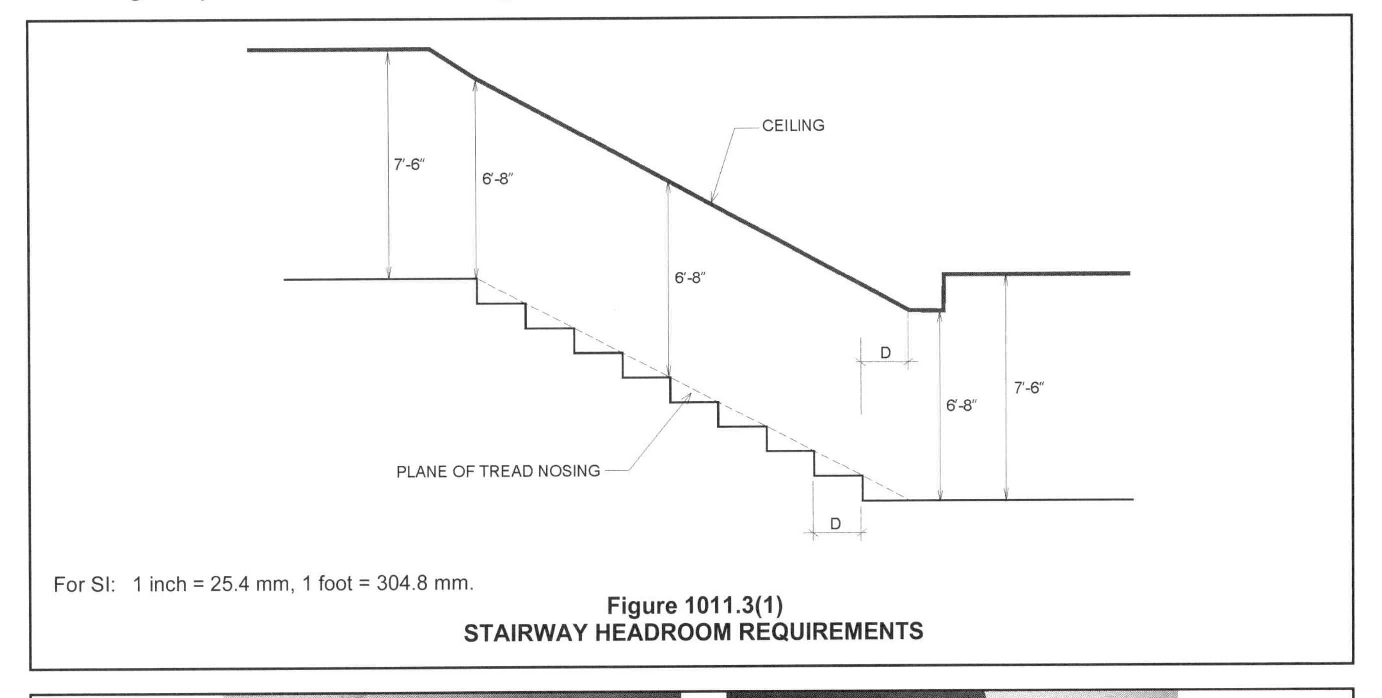

For SI: 1 inch = 25.4 mm, 1 foot = 304.8 mm.

Figure 1011.3(1)
STAIRWAY HEADROOM REQUIREMENTS

Figure 1011.3(2)
EXAMPLE OF SECTION 1011.3, EXCEPTION 2

providing a standard and cannot be specific to the variety of actual paths followed by all users. This specific line location is determined by measuring along each nosing edge 12 inches (305 mm) from the extreme of the clear width of the stair at the surface of the winder tread or the limit of where the foot might be placed in use of the stair. If adjacent winders are present, the point of the widest clear stair width at the surface of the tread in the group of adjacent consecutive winders is used to provide the reference from which the 12-inch (305 mm) dimension will be measured along each nosing. The tread depth may be determined by measuring between adjacent nosings at these determined intersections of the nosings with the walkline. It is important to note that the clear stair width is only that portion of the stair width that is clear for passage. Portions of the stair beyond the clear width are not consequential to use of the stair, consistent travel or location of the walkline.

1011.5 Stair treads and risers. *Stair* treads and risers shall comply with Sections 1011.5.1 through 1011.5.5.3.

❖ The provisions for treads and risers contribute to the efficient use of the stairway, facilitating smooth and consistent travel. This section provides dimensional ranges and tolerances for the component elements to allow the flexibility required to design and construct a stair or a flight of stairs that are elements of a stairway. The allowed proportion of maximum riser height and minimum tread depth provide for a maximum angle of ascent but there is no maximum tread depth to consider with the minimum riser height that would define a minimum angle for a stairway. Nor is the proportion of riser height to tread depth compared with the limitations of the length of the user's stride on stairways, which is significantly foreshortened from the user's stride on the level. For this reason, care should be taken when incorporating larger tread depths and controlling the point at which a tread might be wide enough to require more than one step to cross, which can vary significantly when considering ascent and descent movement patterns. Especially in areas where all segments of the public might use the stairs, those persons requiring two smaller sequential steps to cross the tread would progress at significantly different rates than those who might be able to stretch or jump and lead to dangerous complications, especially in egress. Of equal significance is the use of shorter risers without increasing tread depth resulting in a proportion that could cause overstepping. With these same limitations for proportion in mind, however, by controlling the minimum depth of rectangular treads and the minimum depth and angularity of winder treads, these components can control the configuration of the plan of a flight of stairs to provide for smooth and consistent travel.

Section 1011.5.1 provides for consistent identification of the surfaces that are to be measured. This is as critical to the users' experience throughout the built environment as it is to determining all the dimensions in this section as they must relate to each other and the ultimate design and construction of a safe stairway.

The remaining sections address the uniformity and essential attributes of the tread and riser. Of particular note is the nosing or leading edge of the tread. The nosing shape and projection affect the determination of the tread depth and the riser height and are regulated at all steps throughout the stairway, including the nosings at landings to provide for smooth and consistent travel. Furthermore, the line connecting the nosings is used to determine handrail height, guard height and headroom, causing consistent nosings to be a dependent integral of every element of the stairway design, construction and regulation.

1011.5.1 Dimension reference surfaces. For the purpose of this section, all dimensions are exclusive of carpets, rugs or runners.

❖ Carpets, rugs and runners, like furniture, are frequently changed by the occupants and are not regulated by the code. For this reason it is essential that the riser height and tread depth be regulated exclusive of these transitory surfaces to provide an enforceable standard. This practice minimizes the possible variation because of the removal of nonpermanent carpeting throughout the life of a structure and provides a standard enforcement methodology that will provide consistency across the build environment for all users. When owners or occupants add carpeting, rugs or runners, they need to add it to all tread and landing surfaces in the stairway. It is important that the tread and landing surfaces are consistent and comply with the code prior to the addition of carpet. This methodology of enforcement makes it unnecessary to reconstruct floor and stair elevations in the stairway when nonpermanent carpet surfaces are changed that do not require a building permit and eliminates the resulting variations in the built environment that will not comply with the tolerance in Section 1011.5.4 (see Commentary Figure 1011.5.1).

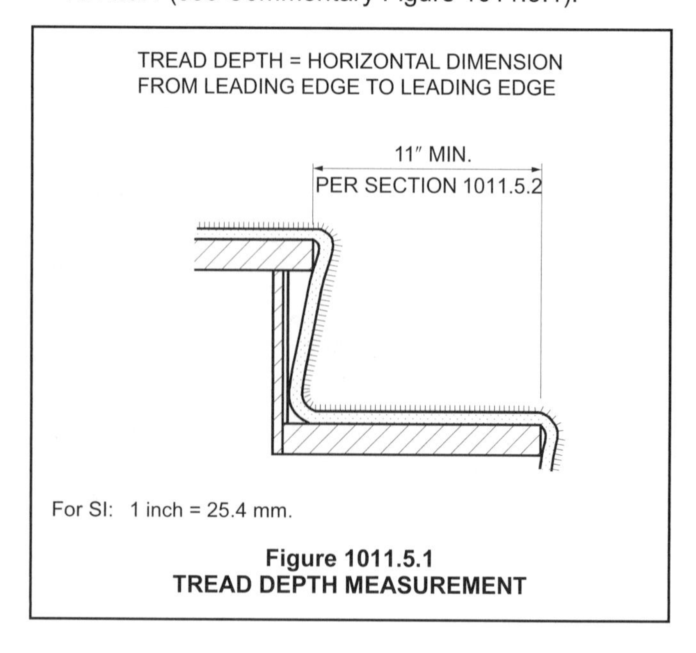

Figure 1011.5.1
TREAD DEPTH MEASUREMENT

1011.5.2 Riser height and tread depth. *Stair* riser heights shall be 7 inches (178 mm) maximum and 4 inches (102 mm) minimum. The riser height shall be measured vertically between the *nosings* of adjacent treads. Rectangular tread depths shall be 11 inches (279 mm) minimum measured horizontally between the vertical planes of the foremost projection of adjacent treads and at a right angle to the tread's *nosing*. *Winder* treads shall have a minimum tread depth of 11 inches (279 mm) between the vertical planes of the foremost projection of adjacent treads at the intersections with the walkline and a minimum tread depth of 10 inches (254 mm) within the clear width of the *stair*.

Exceptions:

1. *Spiral stairways* in accordance with Section 1011.10.
2. *Stairways* connecting stepped *aisles* to cross *aisles* or concourses shall be permitted to use the riser/tread dimension in Section 1029.13.2.
3. In Group R-3 occupancies; within *dwelling units* in Group R-2 occupancies; and in Group U occupancies that are accessory to a Group R-3 occupancy or accessory to individual *dwelling units* in Group R-2 occupancies; the maximum riser height shall be $7^3/_4$ inches (197 mm); the minimum tread depth shall be 10 inches (254 mm); the minimum *winder* tread depth at the walkline shall be 10 inches (254 mm); and the minimum *winder* tread depth shall be 6 inches (152 mm). A *nosing* projection not less than $^3/_4$ inch (19.1 mm) but not more than $1^1/_4$ inches (32 mm) shall be provided on *stairways* with solid risers where the tread depth is less than 11 inches (279 mm).
4. See Section 403.1 of the *International Existing Building Code* for the replacement of existing *stairways*.
5. In Group I-3 facilities, *stairways* providing access to guard towers, observation stations and control rooms, not more than 250 square feet (23 m^2) in area, shall be permitted to have a maximum riser height of 8 inches (203 mm) and a minimum tread depth of 9 inches (229 mm).

❖ The riser height—the vertical dimension from tread surface to tread surface or tread surface to landing surface—is typically limited to not more than 7 inches (178 mm) or less than 4 inches (102 mm). The minimum tread depth—the horizontal distance from the leading edge (nosing) of one tread to the leading edge (nosing) of the next adjacent tread or landing—is typically limited to not less than 11 inches (279 mm) [see Commentary Figure 1011.5.2]. The minimum tread depth of 11 inches (279 mm) is intended to accommodate the largest shoe size found in 95 percent of the adult population, allowing for an appropriate overhang of the foot beyond the tread nosing while descending a stairway. Tread depths under 11 inches (279 mm) could cause a larger overhang (depending on the size of the foot) and could force users with larger feet to increase the angle of their foot to the line of travel while descending a stairway. Based on the probability of adequate foot placement, the rate of misstep with various step sizes and consideration for the user's comfort and energy expenditure, it was agreed that the 11-inch (279 mm) minimum tread depth and maximum 7-inch (178 mm) riser height resulted in the reasonable proportion of riser height and tread depth for stairway construction. A minimum riser height of 4 inches (102 mm) is considered to allow the visual identification of the presence of the riser in ascent or descent.

The precise location of rectangular tread depth and riser measurements is to be perpendicular to the

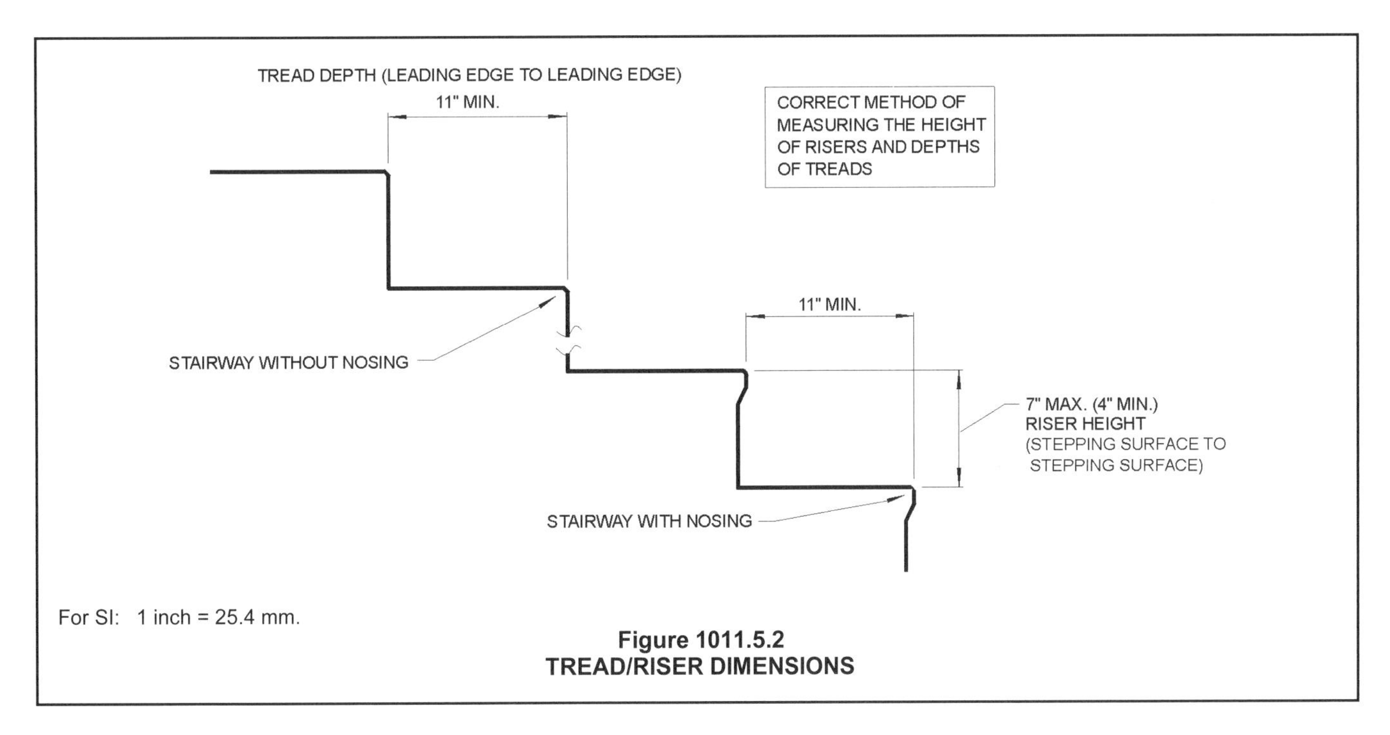

Figure 1011.5.2
TREAD/RISER DIMENSIONS

tread's nosing or leading edge. This is to duplicate the user's anticipated foot placement in traveling the stairway.

The size for a winder tread is also considered for proper foot placement along the walkline [see Commentary Figure 1011.9 and the commentary for Section 1011.4]. The dimensional requirements are consistent with the straight tread.

The exceptions apply only to the extent of the text of each exception. For example, the entire text of Section 1011.5.2 is set aside for spiral stairways conforming to Section 1011.10 (see Exception 3). However, Exception 3 allows a different maximum riser and minimum tread under limited conditions, but retains the minimum riser height and measurement method of Section 1011.5.2.

The requirements for dimensional uniformity are found in Sections 1011.5.4 and 1011.5.4.1.

Exception 1 is for spiral staircases, a unique type of stairway. Section 1011.5.2 is not applicable to this stair type, again because of construction issues and limited applications. For a discussion on spiral staircases, see Section 1011.10.

Exception 2 provides a practical exception where assembly facilities are designed for viewing. See Sections 1029.13 through 1029.13.2.4 for assembly stepped aisle-walking surfaces. This exception is limited to when stairways are a direct continuation of the path of travel from the level cross aisle to the stepped aisles. It is not permitted for other stairways within the assembly space.

Exception 3 allows revisions to the 7 inches/11 inches (178 mm/279 mm) riser/tread requirements for Group R-3 and any associated utility (such as barns, connected garages or detached garages) and within individual units of Group R-2 and their associated utility areas (such as attached garages). This change is allowed because of the low occupant load and the high degree of occupant familiarity with the stairways. When this exception is taken for stairways that have solid risers, each tread is required to have a nosing projection with a minimum dimension of $^3/_4$ inch (19.1 mm) and maximum dimension of $1^1/_4$ inches (32 mm) where the tread depth is less than 11 inches (279 mm). Nosing projections are created where the nosing of the tread above extends beyond the trailing edge of the tread below or when a solid riser is angled under the tread above and connected to the trailing edge of the tread below. Nosing projections are not required for residential stairs with open risers and 10-inch (254 mm) treads. A nosing projection provides a greater stepping surface for those ascending the stairway. For users descending the stairway, the nosing projection allows the toe of the foot to be placed further away from the riser above, providing the necessary clearance for the heel of the foot as it swings down in an arc to its position on the tread (see Commentary Figure 1011.5.3).

Exception 4 allows for the replacement of an existing stair. Where a change of occupancy would require compliance with current standards, this exception allows a stairway that may be steeper than that permitted, provided it does not constitute a hazard [see *International Existing Building Code*® (IEBC®) Section 403.1].

Exception 5 allows steeper stairs in spaces of not more than 250 square feet (23 m^2) in correctional facilities (Group I-3) with a maximum riser height of 8 inches (203 mm) and a minimum tread depth of 9 inches (229 mm) because of the minimal occupant load and the familiarity of the users with the stairway. Although not stated in this exception, utilizing a nosing projection to provide effective tread depth, as stated in Exception 3 for tread depths less than 11 inches (279 mm), is a good design practice.

1011.5.3 Winder treads. *Winder* treads are not permitted in *means of egress stairways* except within a *dwelling unit.*

Exceptions:

1. Curved *stairways* in accordance with Section 1011.9.
2. *Spiral stairways* in accordance with Section 1011.10.

❖ The intent of this section is to coordinate the general provisions for stairway tread and riser dimensions in Section 1011.5.2 with the provisions for winder treads permitted in curved and spiral stairways (see Sections 1011.9 and 1011.10). Winders are permitted in means of egress stairways within dwelling units where occupant loads are smaller and occupants have increased familiarity (see Commentary Figure 1011.5.3 and Section 1011.5.2, Exception 3). This is consistent with provisions in the *International Residential Code*® (IRC®).

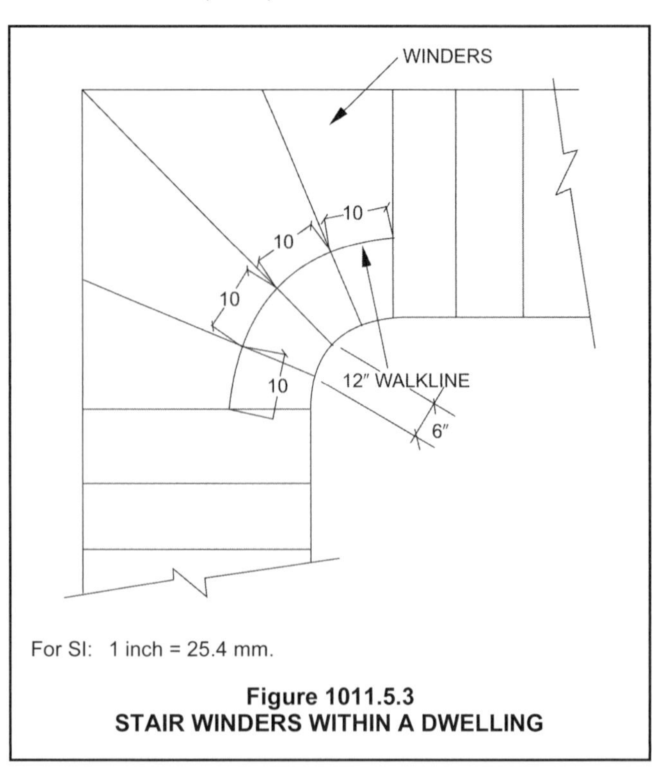

For SI: 1 inch = 25.4 mm.

Figure 1011.5.3
STAIR WINDERS WITHIN A DWELLING

Winders are used to change the direction of a flight by introducing a consistent incremental turn associated with each tread. The risk of injury in the use of stairways constructed with winders is considered to be greater than for stairways constructed as straight runs where users may be restricted by the presence of other users limiting visual clues or influencing the rate of travel. Additional user attention in the turn and the aid of the turn in arresting falls similar to turns at landings is also understood to negate this.

The employment of winders in stairway construction may necessitate the change of the user's gait in both ascent and descent where the tread depth of the winder is not equal to the tread depth of any rectangular treads in the same flight. For example, a person descending a straight flight of stairs will develop a particular gait conforming to the proportion of the riser height and tread depth that will be consistent throughout the flight. However, in a flight that includes winders and rectangular treads, the user must accommodate a change in the proportion of the riser height and tread depth as determined by the path of travel chosen. Visual clues are important to the users' instinctive responses to alter the path of travel, the length of the stride, or a combination of both that may result in nonconcentric movement. To ensure users of the visual clues necessary to alter their gait and limit the need to alter the path of travel in conditions of higher occupant loading, flights with winders must meet the specific safety provisions listed for curved or spiral stairways unless they are within a dwelling unit.

1011.5.4 Dimensional uniformity. *Stair* treads and risers shall be of uniform size and shape. The tolerance between the largest and smallest riser height or between the largest and smallest tread depth shall not exceed $^3/_8$ inch (9.5 mm) in any *flight* of *stairs*. The greatest *winder* tread depth at the walkline within any *flight* of *stairs* shall not exceed the smallest by more than $^3/_8$ inch (9.5 mm).

Exceptions:

1. *Stairways* connecting stepped *aisles* to cross *aisles* or concourses shall be permitted to comply with the dimensional nonuniformity in Section 1029.13.2.
2. Consistently shaped *winders*, complying with Section 1011.5, differing from rectangular treads in the same *flight* of *stairs*.
3. Nonuniform riser dimension complying with Section 1011.5.4.1.

❖ Dimensional uniformity in the design and construction of stairways contributes to safe stairway use. When ascending or descending a stair, users establish a gait based on the instinctive expectation or "feel" that each step taken will be at the same height and will land in approximately the same position on the tread as the previous steps in the pattern. A change in tread or riser dimensions in a stairway flight in excess of the allowed dimensional tolerance can break the rhythm and cause a misstep, stumbling or physical strain that may result in a fall or serious injury. Therefore, this section limits the dimensional variations to a tolerance of $^3/_8$ inch (9.5 mm) between the largest and smallest riser or tread dimension in a flight of stairs. A "flight" of stairs is defined as a run of stairs between landings.

For special conditions of construction and as a practical matter, this section allows some greater variations in stairway tread and riser dimensions than the general limitations specified above.

Exception 1 provides a practical exception where assembly facilities are designed for viewing. See Sections 1029.13 through 1029.13.2.4 for assembly stepped aisle walking surfaces. This exception is limited to when stairways are a direct continuation of the path of travel from the level cross aisle to the stepped aisles. It is not permitted for other stairways within the assembly space.

Exception 2 addresses winder treads, which must be consistent along the walkline (see Commentary Figure 1011.5.4) when compared to other winder treads in the same flight but are not required to meet the tolerance when compared to the uniform dimension of rectangular treads in the same flight.

Exception 3 is in recognition of the situation where a stairway moves down to a surface that slopes up or down perpendicular to the stairway. See the commentary to Section 1105.4.1.

1011.5.4.1 Nonuniform height risers. Where the bottom or top riser adjoins a sloping *public way*, walkway or driveway having an established grade and serving as a landing, the bottom or top riser is permitted to be reduced along the slope to less than 4 inches (102 mm) in height, with the variation in height of the bottom or top riser not to exceed one unit vertical in 12 units horizontal (8-percent slope) of *stair* width. The *nosings* or leading edges of treads at such nonuniform height risers shall have a distinctive marking stripe, different from any other *nosing* marking provided on the *stair flight*. The distinctive marking stripe shall be visible in descent of the *stair* and shall have a slip-resistant surface. Marking stripes shall have a width of not less than 1 inch (25 mm) but not more than 2 inches (51 mm).

❖ This section addresses the situation where the bottom riser of a flight of stairways meets a sloped landing, such as a public way, walk or driveway (see Commentary Figure 1011.5.4.1). Because the sidewalk landing is sloped perpendicular to the stairway run, stepping off the bottom tread on one side will result in a higher riser than stepping off the bottom tread on the other side. This is permitted provided the bottom riser is marked so that someone using the stairs will be aware of the hazard of a nonuniform riser.

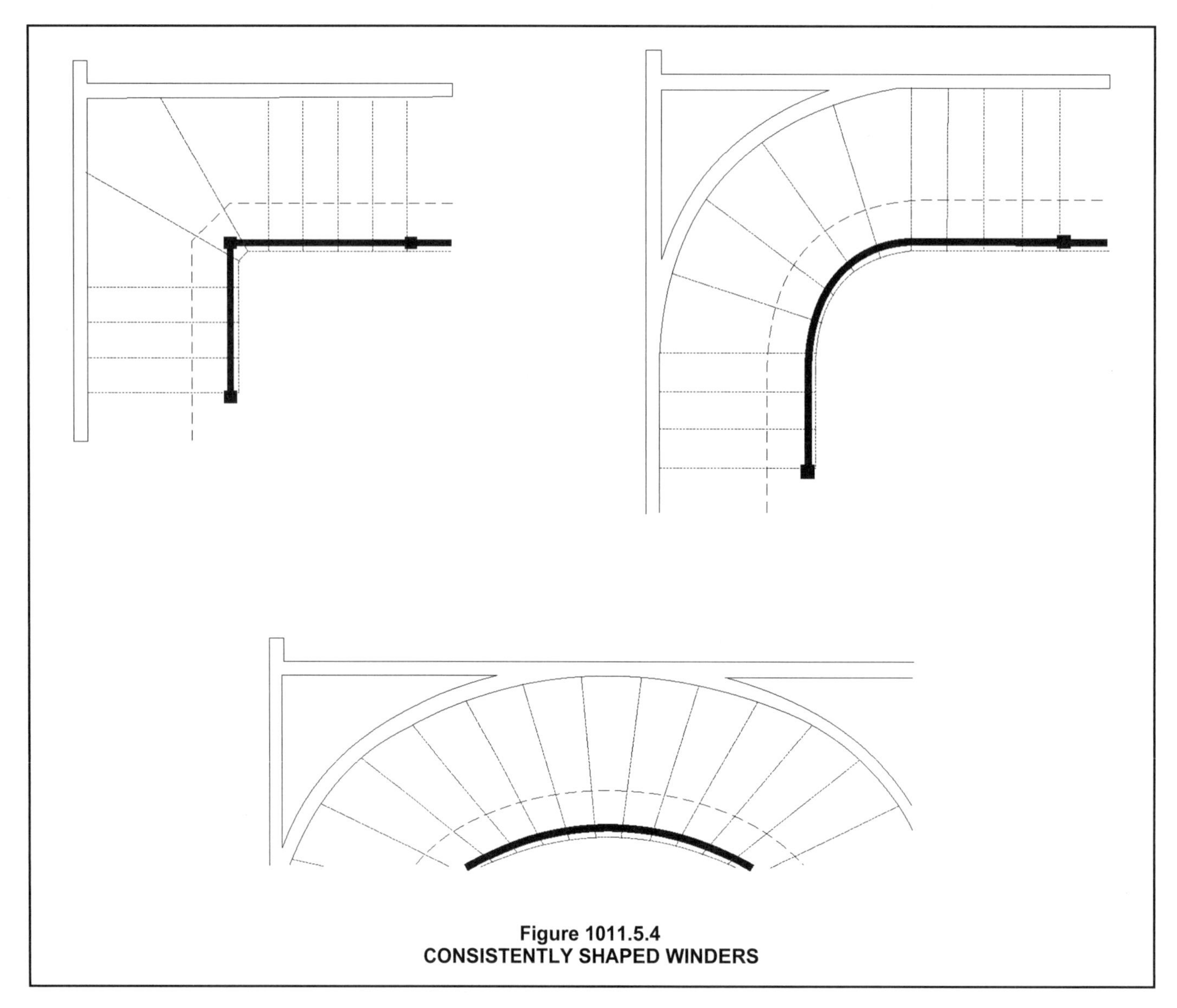

Figure 1011.5.4
CONSISTENTLY SHAPED WINDERS

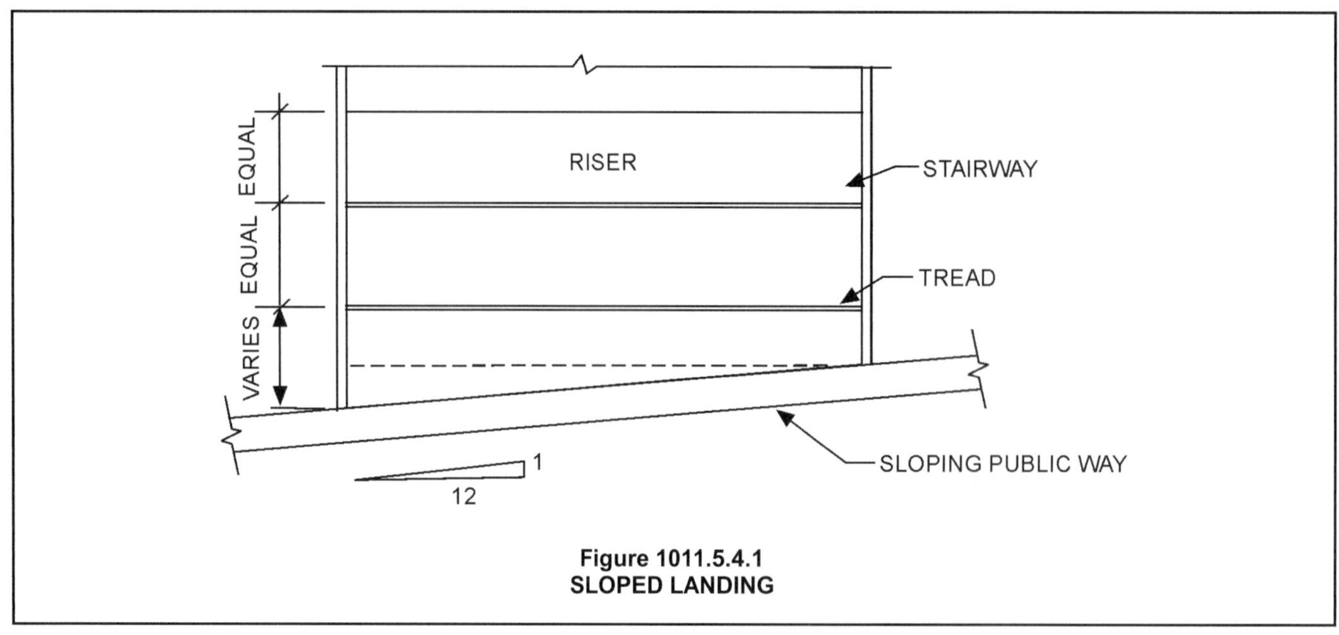

Figure 1011.5.4.1
SLOPED LANDING

1011.5.5 Nosing and riser profile. *Nosings* shall have a curvature or bevel of not less than $^1/_{16}$ inch (1.6 mm) but not more than $^9/_{16}$ inch (14.3 mm) from the foremost projection of the tread. Risers shall be solid and vertical or sloped under the tread above from the underside of the *nosing* above at an angle not more than 30 degrees (0.52 rad) from the vertical.

❖ The profiles of treads and risers contribute to stairway safety. The radius or bevel of the nosing eases the otherwise square edge of the tread and prevents irregular chipping and wear that can become a maintenance issue and seriously affect the safe use of the stair. The minimum curvature or bevel of $^1/_{16}$ inch (1.6 mm) eliminates a sharp square edge that will cause greater injury in falls and provides a certain contrast from the other surfaces of the stair for easier visual location of the start of the tread surface. The $^9/_{16}$-inch (14.3 mm) limit of beveling and maximum radius of curvature at the leading edge of the tread is intended to allow descending foot placement on a surface that does not pitch the foot forward or allow the ball of the foot to slide off the treads and ascending foot placement to slide onto the tread without catching on a square edge. This section also states that risers shall be solid; however, Section 1011.5.5.3 specifically states exceptions to this requirement that are to be applied. The sloping of risers allows the step profile to have a nosing projection without a lip that might cause a foot to catch when dragged up the face of the riser. Such designs are subject to the maximum nosing projection stated in Section 1011.5.5.1 that must be considered when choosing the angle to slope the riser.

1011.5.5.1 Nosing projection size. The leading edge (*nosings*) of treads shall project not more than $1^1/_4$ inches (32 mm) beyond the tread below.

❖ A nosing projection allows the descending foot to be placed further forward on the tread and the heel to then clear the nosing of the tread above as it swings down in an arc, landing on a tread that is effectively deeper than if no nosing projection is used. Nosing projections are so common in stair design that they are usually only noticed by users when they are absent since the lack of nose projection can affect one's gait. Treads with vertical risers are allowed with or without a nosing projection. A nosing projection may also be accommodated by slanting the riser under the tread above. The nosing projection is limited to $1^1/_4$ inch (32 mm) maximum. Treads designed with rounding or bevel on the underside would reduce the chance that a user's toe might catch while ascending the stairway (see Commentary Figure 1011.5.5.1).

1011.5.5.2 Nosing projection uniformity. *Nosing* projections of the leading edges shall be of uniform size, including the projections of the *nosing's* leading edge of the floor at the top of a *flight*.

❖ See the commentary to Section 1011.5.5.1.

1011.5.5.3 Solid risers. Risers shall be solid.

Exceptions:

1. Solid risers are not required for *stairways* that are not required to comply with Section 1009.3, provided that the opening between treads does not permit the passage of a sphere with a diameter of 4 inches (102 mm).
2. Solid risers are not required for occupancies in Group I-3 or in Group F, H and S occupancies other than areas accessible to the public. There are no restrictions on the size of the opening in the riser.
3. Solid risers are not required for *spiral stairways* constructed in accordance with Section 1011.10.

❖ The code does not address when a riser could contain openings and still be considered "solid." How-

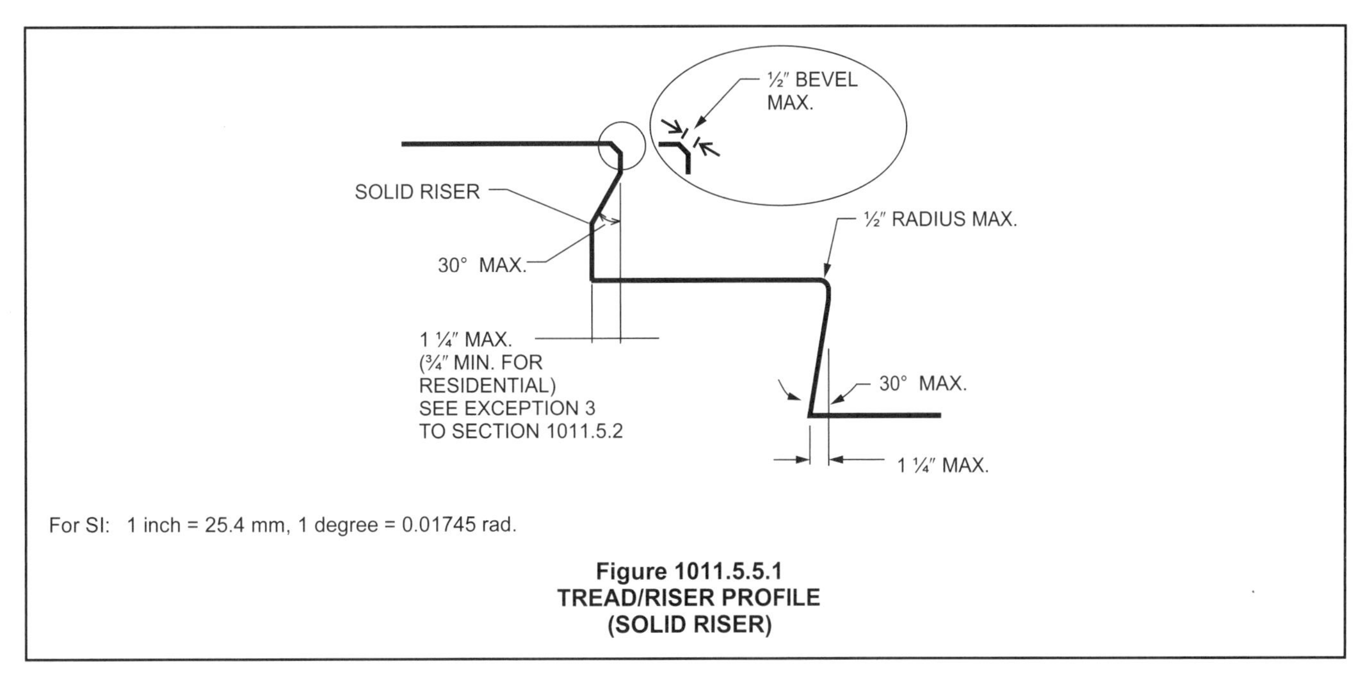

Figure 1011.5.5.1
TREAD/RISER PROFILE
(SOLID RISER)

ever, the intent is so that someone would not catch their toe as they moved up the stairway (see the commentary for nosing projections in Section 1011.5.5). It is not the intent to prohibit risers made of grills or other designs where a toe would not catch. Grill stairways are often used in exterior locations to allow for the passage of snow or rain and decrease the chance of accumulation and possible slips and falls on the stairways.

Exception 1 allows the use of open risers on all stairs that are not part of an accessible means of egress. Where the riser is allowed to be open, the opening is limited to be consistent with the requirements for guards (see Commentary Figure 1011.5.5.3). While not required, the second option shown in Commentary Figure 1011.5.5.3 would limit the possibility of a toe catch as someone moved up the stairways and would be a safer design. The code does not reference ICC A117.1 for stairways, because stairways are not part of an accessible route; however, the code and standard provide opening limitations in tread surfaces. Section 1011.7.1 does allow for treads to have a maximum opening that allows for a $^1/_2$-inch (12.7 mm) sphere.

Exception 2 recognizes that open risers are commonly used for stairs in occupancies such as detention facilities, storage, industrial and high-hazard areas for practical reasons. In detention facilities, open risers provide a greater degree of security and supervision because people cannot effectively conceal themselves behind the stair. Factories, high-hazard buildings and storage facilities have areas where workers may need the open risers to decrease the chance of spillage, water or snow accumulating on the stairs. See Section 1011.7.1 for permitted openings in the treads.

Exception 3 recognizes open risers as necessary for adequate foot placement in spiral stairways. The 4-inch (102 mm) opening limitations of Exception 1 are not applicable to spiral stairways.

1011.6 Stairway landings. There shall be a floor or landing at the top and bottom of each *stairway*. The width of landings shall be not less than the width of *stairways* served. Every landing shall have a minimum width measured perpendicular to the direction of travel equal to the width of the *stairway*. Where the *stairway* has a straight run the depth need not exceed 48 inches (1219 mm). Doors opening onto a landing shall not reduce the landing to less than one-half the required width. When fully open, the door shall not project more than 7 inches (178 mm) into a landing. Where *wheelchair spaces* are required on the *stairway* landing in accordance with Section 1009.6.3, the *wheelchair space* shall not be located in the required width of the landing and doors shall not swing over the *wheelchair spaces*.

> **Exception:** Where *stairways* connect stepped *aisles* to cross *aisles* or concourses, *stairway* landings are not required at the transition between *stairways* and stepped *aisles* constructed in accordance with Section 1029.

❖ A level portion of a stairway provides users with a place to rest in their ascent or descent, to enter a stairway and to adjust their gait before continuing. Landings also break up the run of a stairway, especially at a turn, to aid in the arrest of falls that may occur (see Section 1011.8).

The minimum size (width and depth) of all landings in a stairway is determined by the actual width of the stairway. If Section 1011.2 requires a stairway to have a width of at least 44 inches (1118 mm) and the stairway is constructed with that minimum width, then all landings serving that stairway must be at least 44 inches (1118 mm) wide and 44 inches (1118 mm) deep [see Commentary Figure 1011.6(1)]. If a stairway is constructed wider than required, landings must increase accordingly so as to not create a bottleneck situation in the egress travel. However, when a stairway is configured so that it has a straight run, the depth of the landing between flights in the direction of egress travel is not required to exceed 48 inches (1219 mm), even though the actual width of

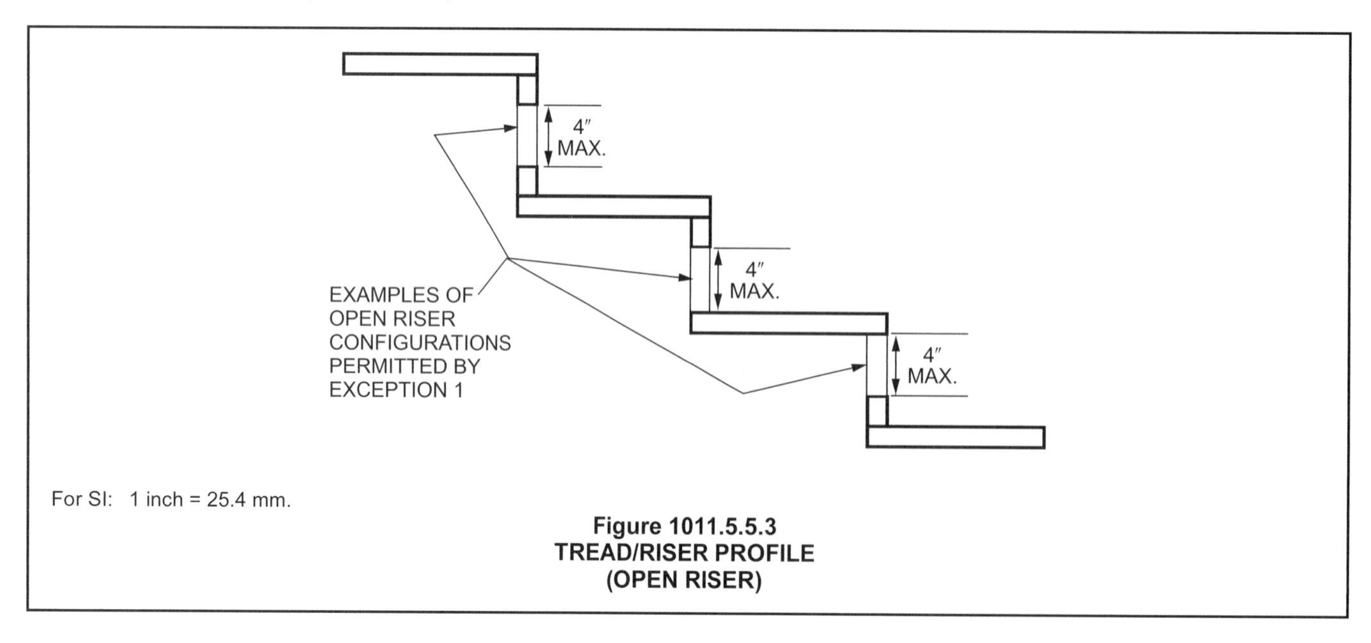

Figure 1011.5.5.3
TREAD/RISER PROFILE
(OPEN RISER)

the stair may exceed 48 inches (1219 mm) [see Commentary Figure 1011.6(2)].

It is not the intent of this section to require that a stairway landing be shaped as a square or rectangle. A landing turning the stairway 90 degrees (1.57 rad) or more with a curved or segmented outside periphery would be permitted, as long as the landing provides an area described by an arc with a radius equal to the actual stairway width [see Commentary Figure 1011.6(3)]. In this case, the space necessary for means of egress will be available.

The last portion of the requirement limits the extent to which doors that swing onto landings may interfere or encroach upon the required landing space. This limits the arc of the door swing on a landing, so that the effect on the means of egress is minimized [see Commentary Figure 1011.6(4)]. This is consistent with a door opening into an exit access corridor in Section 1005.7. For safety reasons and to ensure the means of egress is continually available for everyone, where an area of refuge/wheelchair space must be located on a landing, the wheelchair spaces must not be within the required landing width and the entrance door to the stair enclosure may not swing over the wheelchair spaces [see Commentary Figure 1009.3(1)].

Exception 1 provides a practical exception where assembly facilities are designed for viewing. See Sections 1029.13 through 1029.13.2.4 for assembly stepped aisle walking surfaces. This exception is limited to when stairways are a direct continuation of the path of travel from the level cross aisle to the stepped aisles. It is not permitted for other stairways within the assembly space.

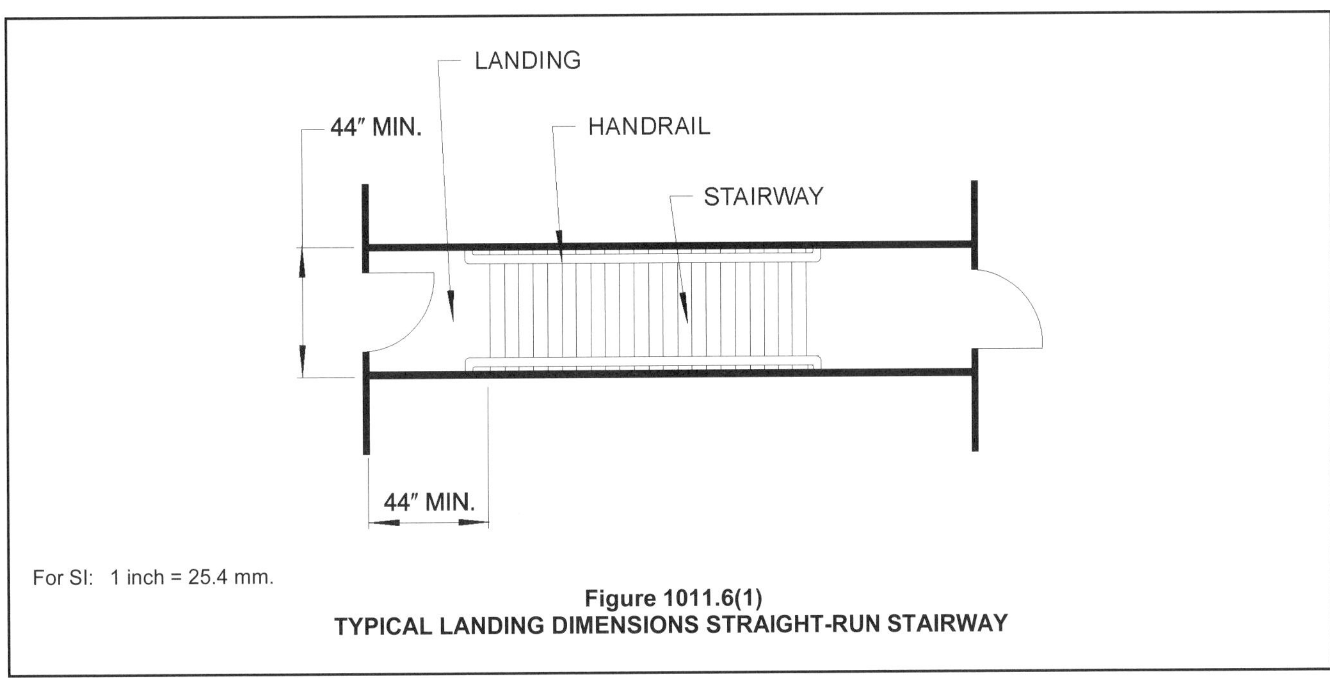

Figure 1011.6(1)
TYPICAL LANDING DIMENSIONS STRAIGHT-RUN STAIRWAY

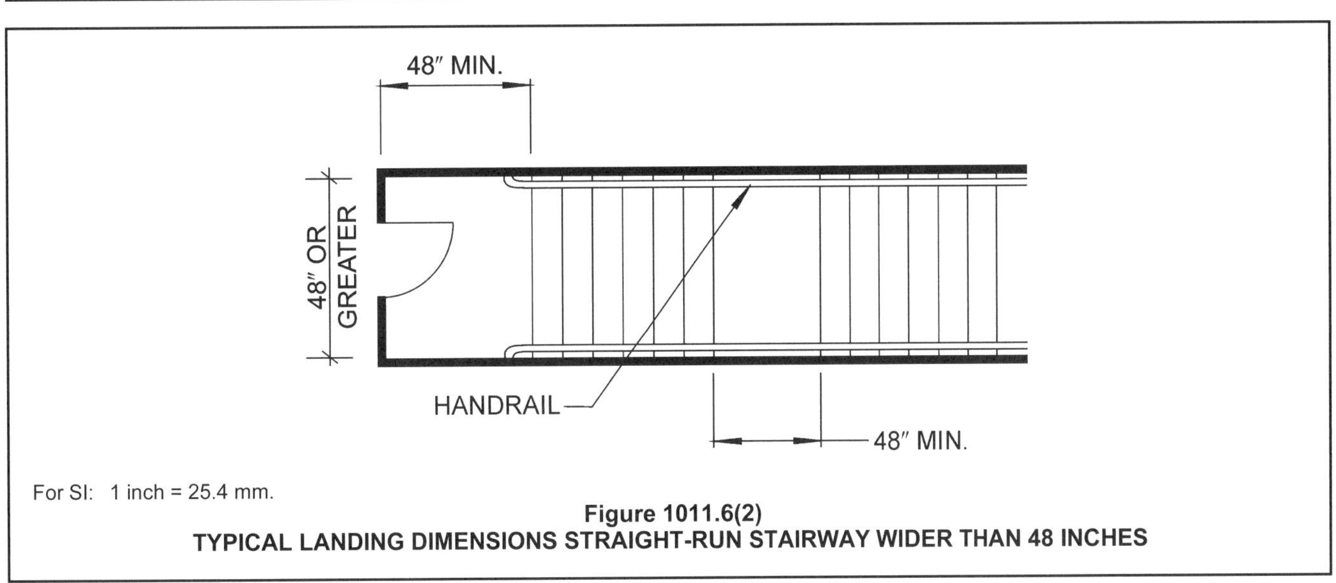

Figure 1011.6(2)
TYPICAL LANDING DIMENSIONS STRAIGHT-RUN STAIRWAY WIDER THAN 48 INCHES

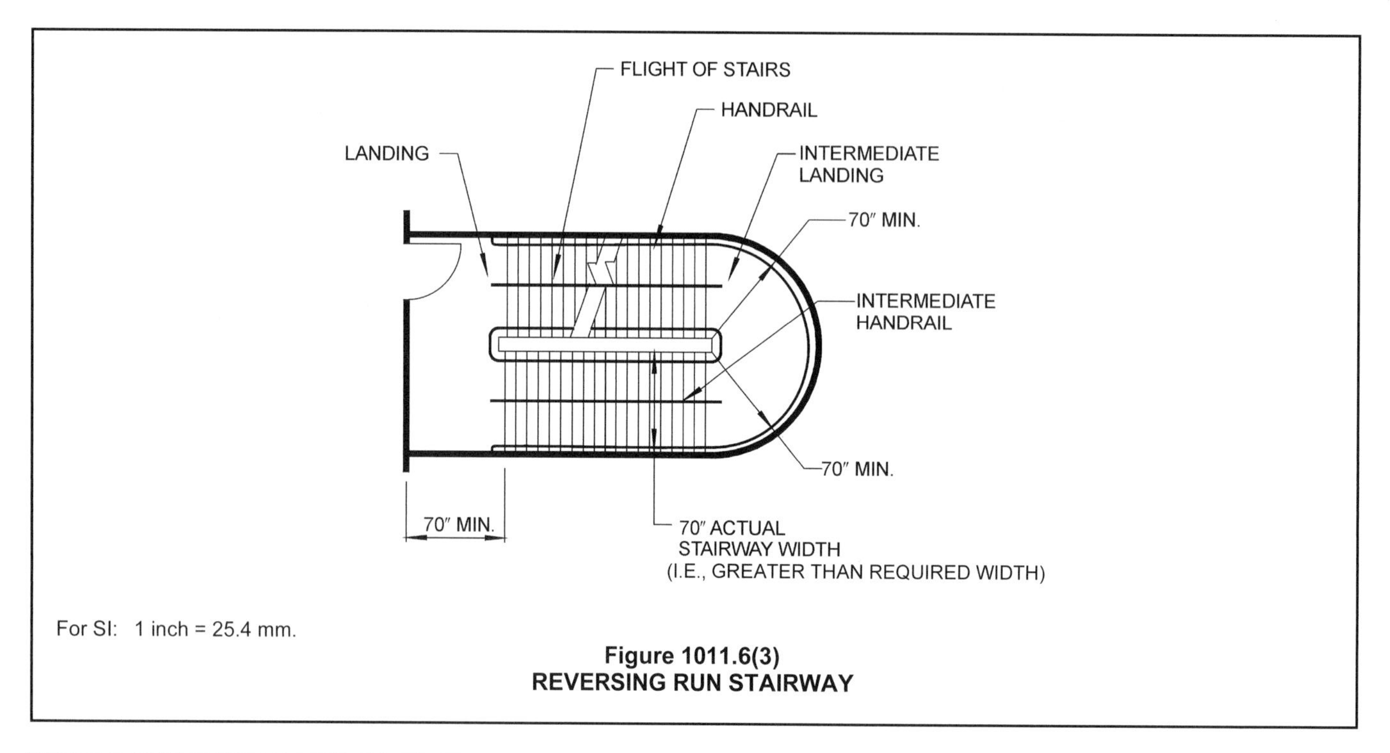

For SI: 1 inch = 25.4 mm.

Figure 1011.6(3)
REVERSING RUN STAIRWAY

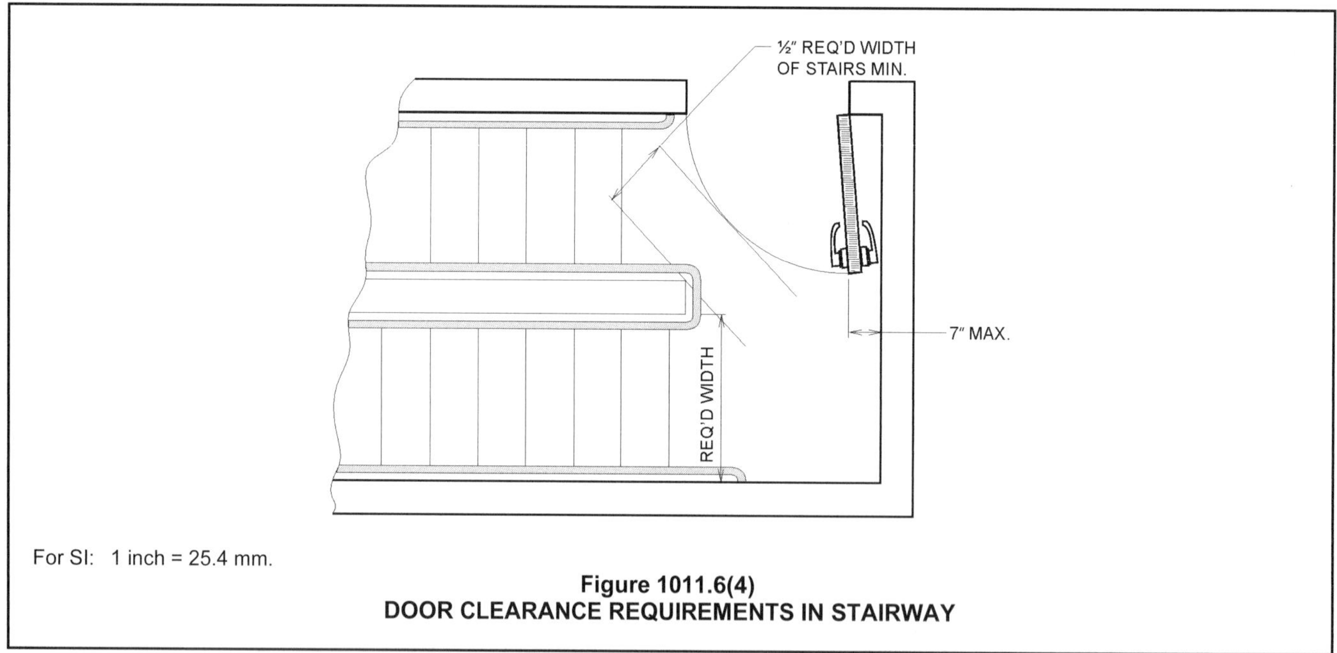

For SI: 1 inch = 25.4 mm.

Figure 1011.6(4)
DOOR CLEARANCE REQUIREMENTS IN STAIRWAY

1011.7 Stairway construction. *Stairways* shall be built of materials consistent with the types permitted for the type of construction of the building, except that wood *handrails* shall be permitted for all types of construction.

❖ In keeping with the different levels of fire protection provided by each of the five basic types of construction designated in Chapter 6, the materials used for stairway construction must meet the appropriate combustibility/noncombustibility requirements indicated in Section 602 for the particular type of construction of the building in which the stairway is located. This is required whether or not the stair is part of the required means of egress. Any structure supporting the stairway and the stairway enclosure must be fire-resistance rated consistent with the construction type; however, the stairway components inside the enclosure need only comply with the material limits for the type of construction.

If desired, wood handrails may be used on the basis that the fuel load contributed by this combustible component of stairway construction is insignificant and will not pose a fire hazard.

1011.7.1 Stairway walking surface. The walking surface of treads and landings of a *stairway* shall not be sloped steeper than one unit vertical in 48 units horizontal (2-percent slope) in any direction. *Stairway* treads and landings shall have a solid surface. Finish floor surfaces shall be securely attached.

Exceptions:

1. Openings in *stair* walking surfaces shall be a size that does not permit the passage of $^1/_2$-inch-diameter (12.7 mm) sphere. Elongated openings shall be placed so that the long dimension is perpendicular to the direction of travel.
2. In Group F, H and S occupancies, other than areas of parking structures accessible to the public, openings in treads and landings shall not be prohibited provided a sphere with a diameter of $1^1/_8$ inches (29 mm) cannot pass through the opening.

❖ It is the intent of this section that both landing and stair treads be solid and level with firmly attached surface materials; however, the 1:48 slope should be adequate to allow for drainage to limit the chance for an accumulation of water where someone might slip.

The exceptions permit the use of open grate-type material or slotted grill for stairway treads and landings in two different situations.

Exception 1 allows for a maximum $^1/_2$-inch (12.7 mm) opening on stairway treads in public areas and serving any use (see Commentary Figure 1011.7.1). This is very beneficial on exterior stairways where snow, ice or water may accumulate. The $^1/_2$-inch (12.7 mm) limitation is based on the size of a crutch or cane tip and is consistent with ICC A117.1 and federal accessibility requirements. The opening limitation is also small enough that most shoe heels will not get stuck. If a slotted grill pattern is used, the slots must run side to side on the stairway tread, not nosing to back.

Exception 2 is applicable in factory, industrial, storage and high-hazard occupancies. This provision is intended to apply primarily to stairs that provide access to areas not required to be accessible, such as pits, catwalks, tanks, equipment platforms, roofs or mezzanines. Walking surfaces with limited-size openings are typically used because open grate-type material is less susceptible to accumulation of dirt, debris or moisture, as well as being more resistant to corrosion. Most commercially available grate material is manufactured with a maximum nominal 1-inch (25 mm) opening; therefore, the limitation that the openings not allow the passage of a sphere of $1^1/_8$ inches (29 mm) diameter allows the use of most material as well as accounting for manufacturing tolerances.

The allowances for openings in risers is addressed in Section 1011.5.5.3.

Figure 1011.7.1
OPEN TREAD IN ACCORDANCE WITH EXCEPTION 1

1011.7.2 Outdoor conditions. Outdoor *stairways* and outdoor approaches to *stairways* shall be designed so that water will not accumulate on walking surfaces.

❖ Outdoor stairways and approaches to stairways are to be constructed with a slope that complies with Section 1011.7.1 or are required to be protected such that walking surfaces do not accumulate water. While not specifically stated, any interior locations, such as near a pool, should also have the stair designed to limit the accumulation of water in order to maintain slip resistance (see Section 1003.4).

Where exterior stairways are used in moderate or severe climates, there may also be a concern to protect the stairway from accumulations of snow and ice to provide a safe path of egress travel at all times. Maintenance of the means of egress in the IFC requires an unobstructed path to allow for full instant use in case of a fire or emergency (see Section 1031.3 of the IFC). Typical methods for protecting these egress elements include roof overhangs or canopies; heated slabs; grated treads and landings; or, when approved by the building official, a reliable snow removal maintenance program.

1011.7.3 Enclosures under interior stairways. The walls and soffits within enclosed usable spaces under enclosed and unenclosed stairways shall be protected by 1-hour fire-resistance-rated construction or the fire-resistance rating of the stairway enclosure, whichever is greater. Access to the enclosed space shall not be directly from within the stairway enclosure.

Exception: Spaces under *stairways* serving and contained within a single residential dwelling unit in Group R-2 or R-3 shall be permitted to be protected on the enclosed side with $^1/_2$-inch (12.7 mm) gypsum board.

❖ This section addresses the fire hazard of storage under an interior stairway, whether it is an exit access stairway or exit stairway. The stairway must be protected from a storage area under it, even if the stair is not required to be enclosed. The section also requires that the storage area not open into a stairway enclosure. This limits the potential of a fire that starts in the storage area from affecting the means of egress. The exception provides specific criteria for separation for storage areas under an interior stairway for the indicated residential occupancies.

1011.7.4 Enclosures under exterior stairways. There shall not be enclosed usable space under *exterior exit stairways* unless the space is completely enclosed in 1-hour fire-resistance-rated construction. The open space under *exterior stairways* shall not be used for any purpose.

❖ If the space under an exterior stairway is to be used, such as for storage, the area below the stairway must be separated from the stairway with walls and a ceiling with a fire-resistance rating of at least 1 hour. If the space under the exterior stairway is open, it must remain free and clear. A fire occurring in this space would jeopardize the use of the stairway for exiting during an emergency.

1011.8 Vertical rise. A flight of stairs shall not have a vertical rise greater than 12 feet (3658 mm) between floor levels or landings.

Exception: Spiral stairways used as a means of egress from technical production areas.

❖ Between landings and platforms, the vertical rise is to be measured from one landing walking surface to another (see Commentary Figure 1011.8). The limited height provides a reasonable interval for users with physical limitations to rest on a level surface and also serves to alleviate potential negative psychological effects of long and uninterrupted stairway flights.

The exception allows for spiral stairways that only serve technical production areas to eliminate intermediate landings regardless of height. These stairways typically have limited use and serve areas such as catwalks or lighting booths in stadiums and theaters. In addition, there is the technical difficulty of maintaining proper headroom in a spiral stairway with an intermediate landing.

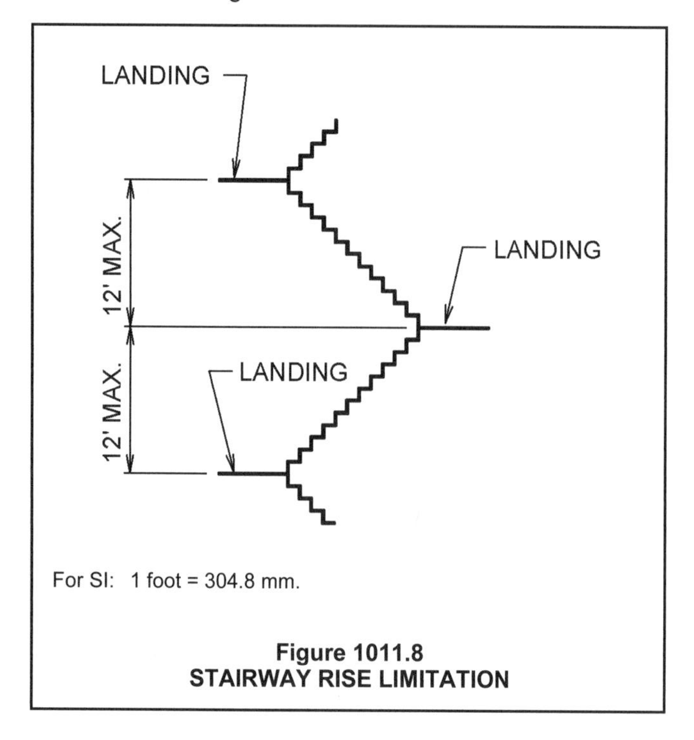

For SI: 1 foot = 304.8 mm.

Figure 1011.8
STAIRWAY RISE LIMITATION

1011.9 Curved stairways. Curved stairways with winder treads shall have treads and risers in accordance with Section 1011.5 and the smallest radius shall be not less than twice the minimum width or required capacity of the stairway.

Exception: The radius restriction shall not apply to curved stairways in Group R-3 and within individual dwelling units in Group R-2.

❖ Curved stairway construction consists of a series of winder treads that form a stairway configuration. Options are many, including circular, S-shaped, oval, elliptical, hourglass, etc. The commentary to Section 1011.5.3 regarding the possible event of nonconcentric movement on stairways with winders also applies to curved stairways. This type of stairway is allowed to be used as a component of a means of egress when tread and riser dimensions meet the requirements or exceptions of Section 1011.5. This section also requires that the shorter radius must be equal to or greater than twice the required width (see Section 1011.2) of the stairway to limit the degree of turning thereby expediting egress from higher occupancies (see Commentary Figure 1011.9).

The exception for residential units eliminates the minimum radius requirement where the occupants are familiar with the extent of the turning of the stair through the curve.

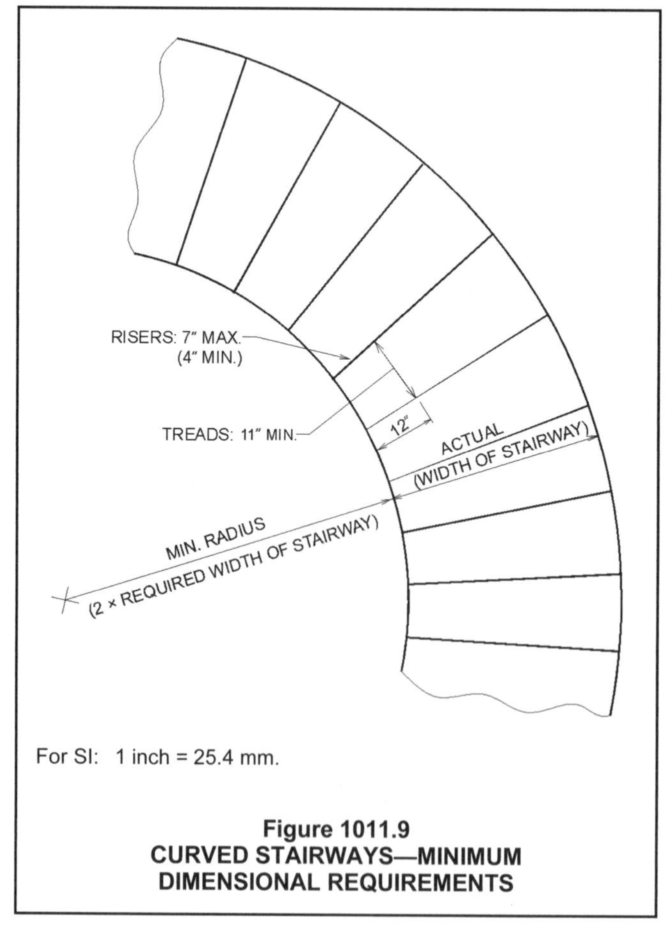

For SI: 1 inch = 25.4 mm.

Figure 1011.9
CURVED STAIRWAYS—MINIMUM DIMENSIONAL REQUIREMENTS

1011.10 Spiral stairways. *Spiral stairways* are permitted to be used as a component in the *means of egress* only within *dwelling units* or from a space not more than 250 square feet (23 m^2) in area and serving not more than five occupants, or from *technical production areas* in accordance with Section 410.6.

A *spiral stairway* shall have a $7^1/_2$-inch (191 mm) minimum clear tread depth at a point 12 inches (305 mm) from the narrow edge. The risers shall be sufficient to provide a headroom of 78 inches (1981 mm) minimum, but riser height shall not be more than $9^1/_2$ inches (241 mm). The minimum *stairway* clear width at and below the *handrail* shall be 26 inches (660 mm).

❖ Spiral stairways can be used within an individual dwelling unit, from small spaces in other occupancies and from technical production areas in spaces such as theaters. Spiral stairways are permitted to provide access between the levels within a live/work unit (see Section 419.3.2).

Spiral stairways are generally constructed with a fixed center pole that serves as either the primary or the only means of support from which pie-shaped treads radiate to form a winding stairway. The term "spiral" in the geometrical sense describes a curve that diminishes in radius and relates to the form of the stair as viewed in perspective from above or below; however, "spiral" does not describe the actual geometry of the stair. The unique turning of spiral stairs allows the center pole to act as the guard at the inside of the stair and the typically narrow width requires users to choose a walkline along the outer perimeter near the only required handrail. On spiral stairways of larger widths, where two users can pass and restrict access to the single handrail, the provision of a handrail at both sides should be considered.

The commentary to Section 1011.5.3 regarding the possible event of nonconcentric movement on stairways with winders also applies to spiral stairways. The nature of stairway construction is such that it does not serve well when used in emergencies that require immediate evacuation, nor does a spiral stairway configuration permit the handling of a large occupant load in an efficient and safe manner. Furthermore, it is impossible for fire service personnel to use a spiral stairway at the same time and in a direction opposite that being used by occupants to exit the premises, possibly causing a serious delay in firefighting operations. Therefore, this section allows only very limited use of spiral stairways when used as part of a required means of egress. Spiral stairways may be used in any occupancy as long as such stairways are not a component of a required means of egress.

Spiral stairways are required to have dimensional uniformity. The stairway must have a clear width of at least 26 inches (660 mm) at and below the handrail. The depth of the treads must not be less than $7^1/_2$ inches (191 mm) measured at a point that is 12 inches (305 mm) out from the narrow edge (see Commentary Figure 1011.10). Riser heights are required to be the same throughout the stairway, but are not to exceed $9^1/_2$ inches (241 mm). Minimum headroom of 6 feet, 6 inches (1981 mm) is required.

1011.11 Handrails. Stairways shall have handrails on each side and shall comply with Section 1014. Where glass is used to provide the handrail, the handrail shall comply with Section 2407.

Exceptions:

1. Stairways within dwelling units and spiral stairways are permitted to have a handrail on one side only.
2. Decks, patios and walkways that have a single change in elevation where the landing depth on each side of the change of elevation is greater than what is required for a landing do not require handrails.
3. In Group R-3 occupancies, a change in elevation consisting of a single riser at an entrance or egress door does not require handrails.
4. Changes in room elevations of three or fewer risers within dwelling units and sleeping units in Group R-2 and R-3 do not require handrails.

❖ Handrails are required along each side of a flight of stairs; however, handrails are not required along stairway landings. Handrail continuity and extensions that will overlap the landings are addressed in Sections 1014.4 and 1014.6.

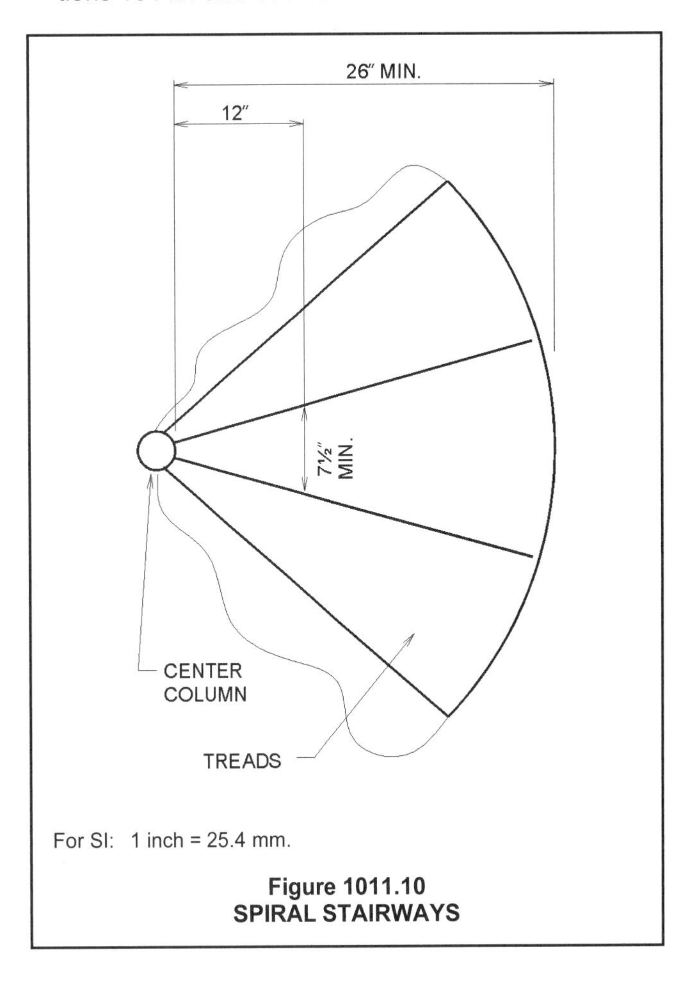

Figure 1011.10
SPIRAL STAIRWAYS

Handrails have four recognized functions in stairway use. First, they serve to guide persons in ascent and descent along the path of egress travel, especially important for those with low vision, and in cases of fire where vision might be obscured by smoke. Second, they provide a tool for the user to exert stabilizing forces longitudinally (along the length of the rail), vertically and, most importantly, transversely (perpendicular) to the rail as the body transfers weight from side to side with each leg swing of the unique gait used on stairs. Third, they provide for pulling when arms are used to augment legs in ascent of steeper angles or when such climbing strategies result in more efficient use of the strengths of the user. Fourth, they are a tool that can be utilized to help in the arrest of a fall. In these capacities handrails serve to aid in the use of the stairway and are required on both sides of stairways in compliance with Section 1014 to allow passing users unencumbered access to a handrail. Finally, when glass is the material used to provide the handrail, it must comply with Section 2407.

Note that if the handrail extension is at a location that could be considered a protruding object, the handrail must return to the post at a height of less than 27 inches (686 mm) above the floor. Handrails along the stair runs are not considered protruding objects.

The exceptions state conditions where handrails are only required on one side or are not needed at all. By the nature of their construction, spiral stairways can only have a single handrail (see Exception 1). In accordance with Exceptions 1 and 4, within dwelling units, all stairways can have a handrail on one side only, and stairs with three or fewer rises are not required to have any handrails. Since "Stair" is defined as one or more risers, Exceptions 2 and 3 are necessary. Exception 3 exempts the single step at the front or back door of a Group R-3 dwelling unit (i.e., townhouse). Decks, patios and walkways often move down with the grade. When there are single steps, either off a patio or deck to grade, or along the surface, a handrail is not required (Exception 2, see Commentary Figure 1011.11). Many of these exceptions dealing with residential units are consistent with the IRC.

For guard requirements at stairways, see Section 1011.13.

1011.12 Stairway to roof. In buildings four or more stories above *grade plane*, one *stairway* shall extend to the roof surface unless the roof has a slope steeper than four units vertical in 12 units horizontal (33-percent slope).

> **Exception:** Other than where required by Section 1011.12.1, in buildings without an occupied roof access to the roof from the top story shall be permitted to be by an *alternating tread device*, a ship's ladder or a permanent ladder.

❖ Because of safety considerations, roofs used for habitable purposes such as roof gardens, observation decks, sporting facilities (including jogging or walking tracks and tennis courts) or similar uses, must be provided with conventional stairways that will serve as required means of egress. Access by ladders or an alternating tread device for such uses is not permitted.

In buildings four or more stories high, roofs that are not used for habitable purposes must be provided with ready access by conventional stairways or by an alternating tread device (see Section 1011.14). If this stair is also to provide access to an elevator penthouse on the roof, see additional requirements in Section 1011.12.1. Two reasons for this are access for roof or rooftop equipment repair and fire department access during a fire event. Sloping roofs with a rise greater than 4 inches (102 mm) for every 12 inches (305 mm) in horizontal measurement (4:12) are exempt from the requirements of this section because of the steepness of the construction and the inherent dangers to life safety.

While it is not specifically required that roof access be through an exit stairway enclosure, since part of the intent is for fire department access to the roof, it is strongly advised. Section 1023.9 requires signage at the level of exit discharge indicating whether the stairway has roof access.

1011.12.1 Stairway to elevator equipment. Roofs and penthouses containing elevator equipment that must be accessed for maintenance are required to be accessed by a stairway.

❖ The requirement for a stair to the roof for maintaining elevator equipment correlates the code with ASME A17.1/CSA B44, *Safety Code for Elevators and Escalators.* This referenced standard (see Section 3001.2) has required stairs and a door to access elevator equipment since 1955. More specifically, Section 2.27.3.2.1 of ASME A17.1/CSA B44 states the following: "a stairway with a swinging door and platform at

Figure 1011.11
EXAMPLE OF SECTION 1011.11, EXCEPTION 2

the top level, conforming to 2.7.3.3 shall be provided from the top floor of the building to the roof level. Hatch covers as a means of access to the roofs shall not be permitted." Alternating tread devices or ladders are not permitted as an alternative to the stairway for access to the elevator penthouse. This provision is more specific; therefore, while not prohibiting using the same stairway for access to the roof and the elevator penthouse (see Sections 1011.12 and 1011.12.2), access to that elevator penthouse must be via a stairway with door access, not an alternating tread device and hatch.

1011.12.2 Roof access. Where a stairway is provided to a roof, access to the roof shall be provided through a penthouse complying with Section 1510.2.

Exception: In buildings without an occupied roof, access to the roof shall be permitted to be a roof hatch or trap door not less than 16 square feet (1.5 m^2) in area and having a minimum dimension of 2 feet (610 mm).

❖ The purpose of the penthouse or stairway bulkhead requirement in this section is to protect the walking surface of the stairway to the roof. The exception provides for situations when roof access is only needed for service or maintenance purposes, and where the access may be permitted by alternatives such as alternating tread devices, ship's ladders or ladders.

1011.13 Guards. Guards shall be provided along stairways and landings where required by Section 1015 and shall be constructed in accordance with Section 1015. Where the roof hatch opening providing the required access is located within 10 feet (3049 mm) of the roof edge, such roof access or roof edge shall be protected by guards installed in accordance with Section 1015.

❖ While guards are required at the edge of a normally occupied roof by Section 1015, there is also a safety concern for roof areas that need to be accessed by service personnel, inspectors and emergency responders. This requirement for guards provides a minimum measure of safety when the roof access is close to the roof edge. This is consistent with the requirements at mechanical equipment in Section 1015.6 and 1015.7.

1011.14 Alternating tread devices. *Alternating tread devices* are limited to an element of a *means of egress* in buildings of Groups F, H and S from a mezzanine not more than 250 square feet (23 m^2) in area and that serves not more than five occupants; in buildings of Group I-3 from a guard tower, observation station or control room not more than 250 square feet (23 m^2) in area and for access to unoccupied roofs. *Alternating tread devices* used as a means of egress shall not have a rise greater than 20 feet (6096 mm) between floor levels or landings.

❖ This type of device is constructed in such a way that each tread alternates with each adjacent tread so that the device consists of a system of right-footed and left-footed treads (see Commentary Figure 1011.14).

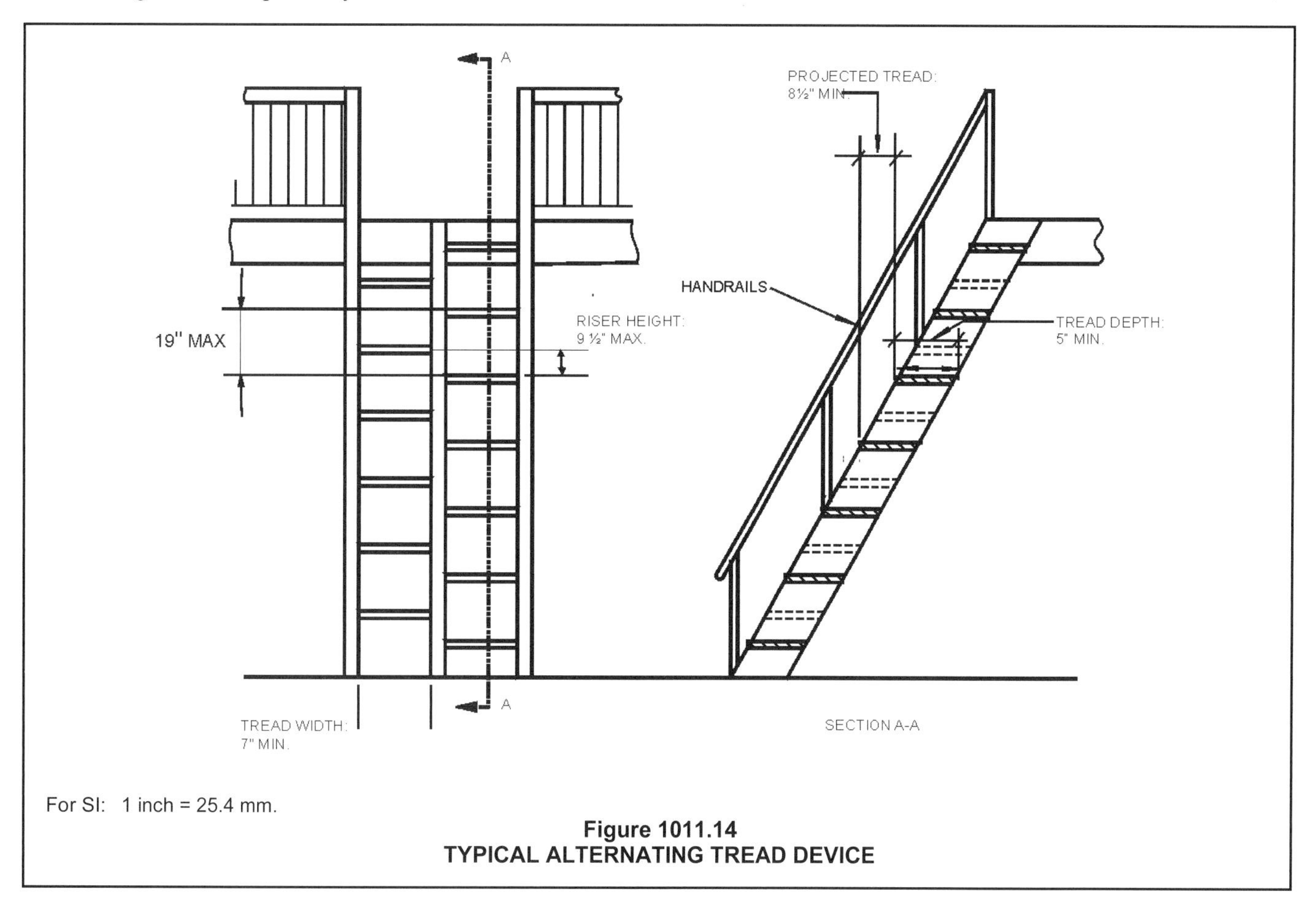

Figure 1011.14
TYPICAL ALTERNATING TREAD DEVICE

The use of center stringer construction, half-treads and an incline that is considerably steeper than allowed for ordinary stairway construction makes the alternating tread device unique. However, because of its structural feature, only single-file use of the device (between handrails) is possible, thus preventing the occupants from passing one another. The pace of occupant travel is set by the slowest user, a condition that could become critical in an emergency situation. Furthermore, it is impossible for fire service personnel to use an alternating tread device at the same time and in a direction opposite that being used by occupants to exit the premises, possibly causing a serious delay in fire-fighting operations. For these reasons, this section greatly restricts the use of alternating tread devices as a means of egress. Alternating tread devices may be used in any occupancy as long as such stairways are not a component of a required means of egress.

Alternating tread devices are considered a modest improvement to ladder construction and, therefore, can be used as an unoccupied roof access in accordance with the requirements of Section 1011.12.

Alternating tread devices are permitted 20 feet (6096 mm) between landings given their limited application and low occupant loads. In addition, it is recognized a vertical rise higher than the typical stairway is needed for these steeper devices which are used where space is often too restrictive for a regular stairway.

1011.14.1 Handrails of alternating tread devices. Handrails shall be provided on both sides of alternating tread devices and shall comply with Section 1014.

❖ For the safety of occupants, this section references the dimensional requirements for handrail locations to be used in conjunction with the special construction features of alternating tread devices provided in Section 1011.14. Because of the steepness of these devices, handrails on both sides are essential for safe functional use and additional clearances are required so that hand movement will not be encumbered by obstructions. Section 1015.3, Exception 5, permits the handrails along alternating tread devices to also serve as guards.

1011.14.2 Treads of alternating tread devices. *Alternating tread devices* shall have a minimum tread depth of 5 inches (127 mm), a minimum projected tread depth of $8^1/_2$ inches (216 mm), a minimum tread width of 7 inches (178 mm) and a maximum riser height of $9^1/_2$ inches (241 mm). The tread depth shall be measured horizontally between the vertical planes of the foremost projections of adjacent treads. The riser height shall be measured vertically between the leading edges of adjacent treads. The riser height and tread depth provided shall result in an angle of ascent from the horizontal of between 50 and 70 degrees (0.87 and 1.22 rad). The initial tread of the device shall begin at the same elevation as the platform, landing or floor surface.

Exception: *Alternating tread devices* used as an element of a *means of egress* in buildings from a mezzanine area not more than 250 square feet (23 m^2) in area that serves not more than five occupants shall have a minimum tread depth of 3 inches (76 mm) with a minimum projected tread depth of $10^1/_2$ inches (267 mm). The rise to the next alternating tread surface shall not exceed 8 inches (203 mm).

❖ Alternating tread stairways (see Section 1009.13) are required to have tread depths of at least 5 inches (127 mm). Tread projections are not to be less than $3^1/_2$ inches (89 mm) when measured from tread nosing to tread nosing [next adjacent tread to the left or right to provide a minimum projected tread depth of $8^1/_2$ inches (216 mm)] (see Commentary Figure 1011.14).

The risers are to be not more than $9^1/_2$ inches (241 mm) when measured from tread to alternating tread (next adjacent tread to the left or right). The rise between treads on the same side would be 19 inches (482 mm) maximum. Applying the limiting dimensions stated above results in a device with a very steep incline that is common to ladders; however, because the device may be walked facing down in descent, it is considered a type of stairway in the code.

Tread widths are required to be a minimum of 7 inches (178 mm) or more. With a center support, the total width will be more than 15 inches (381 mm). Although no maximum width of the tread is stated, the device must be of a width to provide for functional use of both handrails at the same time in ascent and descent. For this same reason, handrail heights for alternating tread devices are modified from those stairways in Section 1014.2.

Just using the dimensions could result in an alternating tread device with an angle greater than 75 degrees (1.3 rad). In any case, the overall angle of the device must be between 50 and 70 degrees (0.87 and 1.22 rad).

For alternating tread devices used as a means of egress from small-area mezzanines as prescribed in the exception, the treads must project at least $7^1/_2$ inches (191 mm) as compared to the $3^1/_2$ inches (89 mm) stated above; treads are to be at least 3 inches (76 mm) in depth [compared to 5 inches (127 mm)] and risers are not to exceed 8 inches (203 mm) in height [compared to $9^1/_2$ inches (341 mm)].

1011.15 Ship's ladders. Ship's ladders are permitted to be used in Group I-3 as a component of a *means of egress* to and from control rooms or elevated facility observation stations not more than 250 square feet (23 m^2) with not more than three occupants and for access to unoccupied roofs. The minimum clear width at and below the *handrails* shall be 20 inches (508 mm).

❖ Ship's ladders can be used in correctional facilities for access to small control rooms, observation stations and unoccupied roofs. Where approved by the code official, ship's ladders could be used for access to unoccupied roofs in other occupancies. Ship's ladders are of similar gradient or pitch to alternating tread devices; however, the treads span the full width like that of a ladder rather than being staggered to either side (see Commentary Figure 1011.15).

1011.15.1 Handrails of ship's ladders. *Handrails* shall be provided on both sides of ship's ladders.

❖ Handrails are needed on both sides to assist in ascent and descent and the absence of a maximum width. Section 1015.3, Exception 5, permits the handrails along ship's ladders to also serve as guards.

1011.15.2 Treads of ship's ladders. Ship's ladders shall have a minimum tread depth of 5 inches (127 mm). The tread shall be projected such that the total of the tread depth plus the *nosing* projection is not less than $8^1/_2$ inches (216 mm). The maximum riser height shall be $9^1/_2$ inches (241 mm).

❖ See Commentary Figure 1011.15 for an example of this configuration.

1011.16 Ladders. Permanent ladders shall not serve as a part of the *means of egress* from occupied spaces within a building. Permanent ladders shall be permitted to provide access to the following areas:

1. Spaces frequented only by personnel for maintenance, repair or monitoring of equipment.
2. Nonoccupiable spaces accessed only by catwalks, crawl spaces, freight elevators or very narrow passageways.
3. Raised areas used primarily for purposes of security, life safety or fire safety including, but not limited to, observation galleries, prison guard towers, fire towers or lifeguard stands.
4. Elevated levels in Group U not open to the general public.
5. Nonoccupied roofs that are not required to have *stairway* access in accordance with Section 1011.12.1.
6. Ladders shall be constructed in accordance with Section 306.5 of the *International Mechanical Code*.

❖ Permanent ladders are permitted as a means of ingress and egress to very limited spaces. Typically, these spaces are not considered occupied and, as such, are not required to have a means of egress. While the term "technical production areas" is not used in this section, Section 410.6.3.4, Item 6, allows for ladders to be used to access technical production areas. Item 6 of this section references the IMC for when ladders can be used to access mechanical equipment that is located in an elevated space or in a room. The details and construction requirements for a permanent ladder are also found in that section. This will help make sure that permanent ladders are safe and useable, while providing consistency for both the designer and the building official.

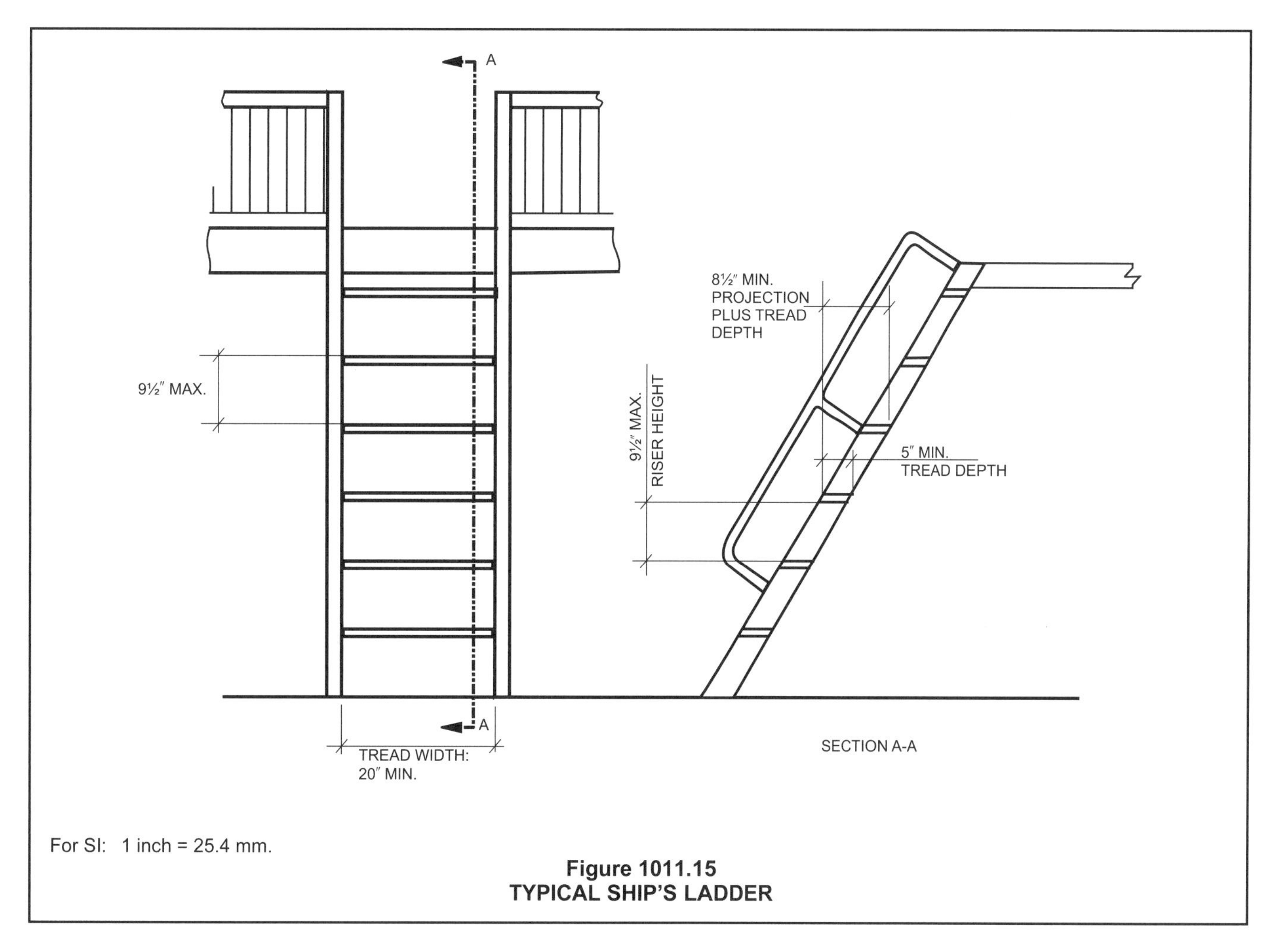

Figure 1011.15
TYPICAL SHIP'S LADDER

SECTION 1012
RAMPS

1012.1 Scope. The provisions of this section shall apply to ramps used as a component of a *means of egress*.

Exceptions:

1. Ramped *aisles* within assembly rooms or spaces shall comply with the provisions in Section 1029.
2. Curb ramps shall comply with ICC A117.1.
3. Vehicle ramps in parking garages for pedestrian *exit access* shall not be required to comply with Sections 1012.3 through 1012.10 where they are not an *accessible route* serving *accessible* parking spaces, other required *accessible* elements or part of an accessible *means of egress*.

❖ Ramps provide an alternative method of vertical means of access to or egress from a building. Ramps are required for access to building areas for persons who are mobility impaired (see Chapter 11) and for small changes in floor elevations that are a safety hazard in themselves (see Section 1003.5). All ramps intended for pedestrian usage, whether required or otherwise provided, must comply with the requirements of this section. The code considers any walking surface that has a slope steeper than one unit vertical in 20 units horizontal (5-percent slope) to be a ramp (see the definition for "Ramp" in Chapter 2).

As with stairways, it is important to understand the terminology. Exit ramps are ramps that provide a protected path of travel between the exit access and the exit discharge. Interior exit ramps are required to be enclosed in accordance with Section 1022. Exterior exit ramps are protected by the exterior wall of the building and must comply with Section 1027. Exit access ramps are typically unenclosed interior ramps and comply with Section 1019 when they provide access between stories. Exit access travel distance stops at an exit ramp enclosure, but includes any travel down an exit access ramp. Ramps that are outside and provide a route from the level of exit discharge to grade are considered part of the exit discharge. See the commentary in Chapter 2 for the defined term, "Exit discharge, level of."

Exception 1 indicates that ramped aisles are addressed in Section 1029. Having this exception at the beginning of the ramp section negated the need for repeated exceptions throughout the ramp provisions. While ramps and ramped aisles may look similar, configurations and how occupants move on and off those walking surfaces are very different. Occupants leave and join ramped aisles along the entire run, while occupants only enter the ramp at the top and bottom. Ramped aisles may have no or only one handrail in order to allow for access to the seats, while most ramps have handrails on both sides. Ramped aisles can have steeper slopes to allow for seating bowls to address line of sight. Section 1029 should be used for ramped aisles between and immediately adjacent to seating or where the ramps are a direct continuation of the ramped aisles and lead to a level cross aisle or floor. Section 1012 is used for ramps that lead from the balcony, concourse or cross aisle to a floor level above or below the seating areas.

Exception 2 references specific curb cut requirements found in Section 406 of ICC A117.1. It is important to realize there are different provisions for curb ramps and ramps. For example, a curb ramp can have a rise of any height and not require handrails. Ramps require handrails when the rise is more than 6 inches.

Exception 3 addresses parking garages. An accessible route is required to and from any accessible parking space, and all ramp provisions must be followed. However, ramps that provide access to and from nonaccessible spaces in the remainder of the parking garage need only comply with the provisions for slope and guard requirements. This permits nonaccessible portions of garages to be constructed as a continuous slope. Ramps that are strictly for vehicles, such as jump ramps, are not required to meet any of the ramp provisions.

1012.2 Slope. *Ramps* used as part of a *means of egress* shall have a running slope not steeper than one unit vertical in 12 units horizontal (8-percent slope). The slope of other pedestrian *ramps* shall not be steeper than one unit vertical in eight units horizontal (12.5-percent slope).

❖ Maximum slope is limited to facilitate the ease of ascent and to control the descent of persons with or without a mobility impairment. The maximum slope of a ramp in the direction of travel is limited to one unit vertical in 12 units horizontal (1:12) (see Commentary Figure 1012.2). Ramps in existing buildings may be permitted to have a steeper slope at small changes in elevation (see Sections 410.8.5 and 705.1.4 of the IEBC). An example of a ramp that is not part of a means of egress and, therefore, allowed to be a maximum slope of 1:8, is a loading dock or delivery ramp where the ramp is not part of any required exit discharge.

1012.3 Cross slope. The slope measured perpendicular to the direction of travel of a *ramp* shall not be steeper than one unit vertical in 48 units horizontal (2-percent slope).

❖ The limitation of one unit vertical in 48 units horizontal on the slope across the direction of travel is to prevent a severe cross slope that would pitch a user to one side (see Commentary Figure 1012.2).

1012.4 Vertical rise. The rise for any *ramp* run shall be 30 inches (762 mm) maximum.

Because pushing a wheelchair up a ramp requires a great deal of energy, landings must be situated so that a person can rest after each 30-inch (762 mm) elevation change (see Commentary Figure 1012.2).

1012.5 Minimum dimensions. The minimum dimensions of *means of egress ramps* shall comply with Sections 1012.5.1 through 1012.5.3.

❖ These minimum dimension requirements allow the ramp to function as a means of egress and an accessible route.

1012.5.1 Width and capacity. The minimum width and required capacity of a *means of egress ramp* shall be not less than that required for *corridors* by Section 1020.2. The clear width of a *ramp* between *handrails*, if provided, or other permissible projections shall be 36 inches (914 mm) minimum

❖ .The requirements for the width of a means of egress ramp is based on the width required for minimum width [typically 36 inches (941 mm)] and the capacity based on the occupant load to be served (see Section 1005.3.2). Note that the clear width of 36 inches (914 mm) is required between the handrails and any other obstructions (e.g., handrail supports, curbs) for proper clearance for a person in a wheelchair. This is different from stairways where handrails are permitted to project into the required width. The 36-inch (914 mm) minimum clear width between handrails is consistent with ICC A117.1 and the federal 2010 ADA Standard.

1012.5.2 Headroom. The minimum headroom in all parts of the *means of egress ramp* shall be not less than 80 inches (2032 mm).

❖ The requirement for headroom on any part of an egress ramp is identical to the requirement of a conventional (nonspiral) stairway (see Section 1011.3). General headroom heights along the means of egress are addressed in Section 1003.2.

1012.5.3 Restrictions. *Means of egress ramps* shall not reduce in width in the direction of egress travel. Projections into the required *ramp* and landing width are prohibited. Doors opening onto a landing shall not reduce the clear width to less than 42 inches (1067 mm).

❖ The purpose of not allowing ramps to reduce in width in the direction of egress travel is to prevent a restriction that would interfere with the flow of occupants out of a facility. This would include ramp landings in accordance with Section 1012.6.2. Handrails are the only exception in accordance with Sections 1012.5.1 and 1014.8.

Doors that open onto a ramp landing, including those at the top and bottom landings, must not reduce the clear width to less than 42 inches (1067 mm). This is a more restrictive provision than for cor-

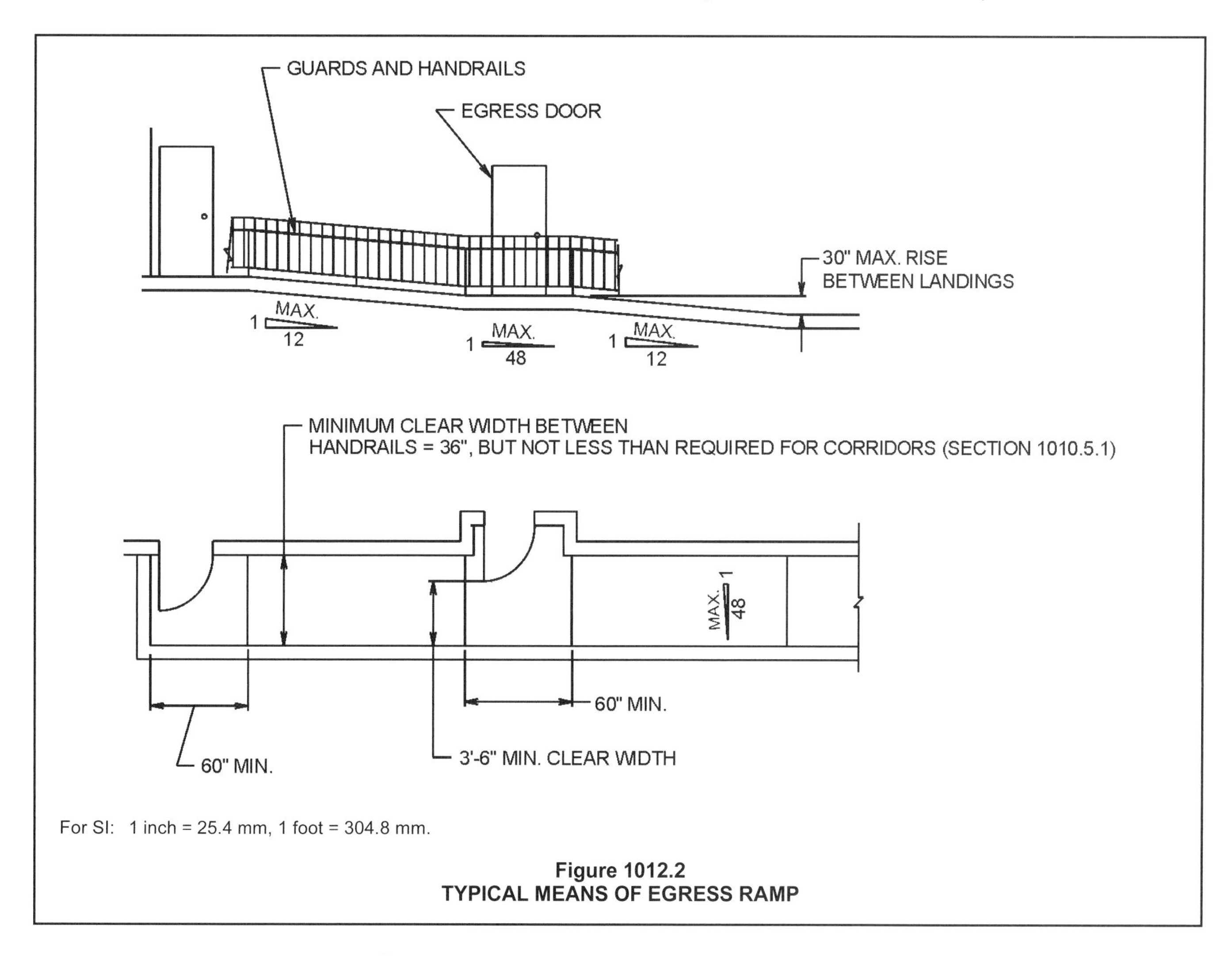

For SI: 1 inch = 25.4 mm, 1 foot = 304.8 mm.

Figure 1012.2
TYPICAL MEANS OF EGRESS RAMP

ridors that would permit the reduction to one-half the required width (see Section 1005.7). Since one of the purposes of a ramp is to accommodate persons with physical disabilities, it must provide the additional clear width for access by those confined to wheelchairs without the interference or potential blockage caused by the swing of a door (see Commentary Figures 1012.2 and 1012.5.3).

1012.6 Landings. *Ramps* shall have landings at the bottom and top of each *ramp*, points of turning, entrance, exits and at doors. Landings shall comply with Sections 1012.6.1 through 1012.6.5.

❖ Landings must be provided to allow users of a ramp to rest on a level floor surface and to adjust to the change in floor surface pitch.

Landings are required at the top and bottom of each ramp run (see Commentary Figure 1012.6). In addition, Section 1012.4 requires a landing every 30 inches (762 mm) of vertical rise of the ramp. The requirements for landings allow those occupants of the structure the ability to negotiate all changes in direction, and prepare themselves to either ascend or descend the ramp and to rest. If there is a door at the top or the bottom of the ramp, there are additional requirements in Section 1012.5.3 for door swing over the landing and Section 405 of ICC A117.1 for maneuvering space and turning space at the door.

1012.6.1 Slope. Landings shall have a slope not steeper than one unit vertical in 48 units horizontal (2-percent slope) in any direction. Changes in level are not permitted.

❖ Landings must be almost flat. This allows persons confined to a wheelchair to come to a complete stop without having to activate the brake or hold themselves stationary at the landing. The maximum slope or cross slope of the landing in any direction is 1:48 (see Commentary Figure 1012.2). This minimum slope is to allow for drainage to limit the accumulation of water on the landing surface.

1012.6.2 Width. The landing width shall be not less than the width of the widest *ramp* run adjoining the landing.

❖ The width of all landings must be consistently as wide as the widths of the ramp runs leading to them. Means of egress ramps cannot be reduced in width in the direction of egress travel. This is also applicable to the landings connecting the ramp runs (see Commentary Figure 1012.6).

1012.6.3 Length. The landing length shall be 60 inches (1525 mm) minimum.

Exceptions:

1. In Group R-2 and R-3 individual *dwelling* and *sleeping units* that are not required to be *Accessible units*, *Type A units* or *Type B units* in accordance with Sec-

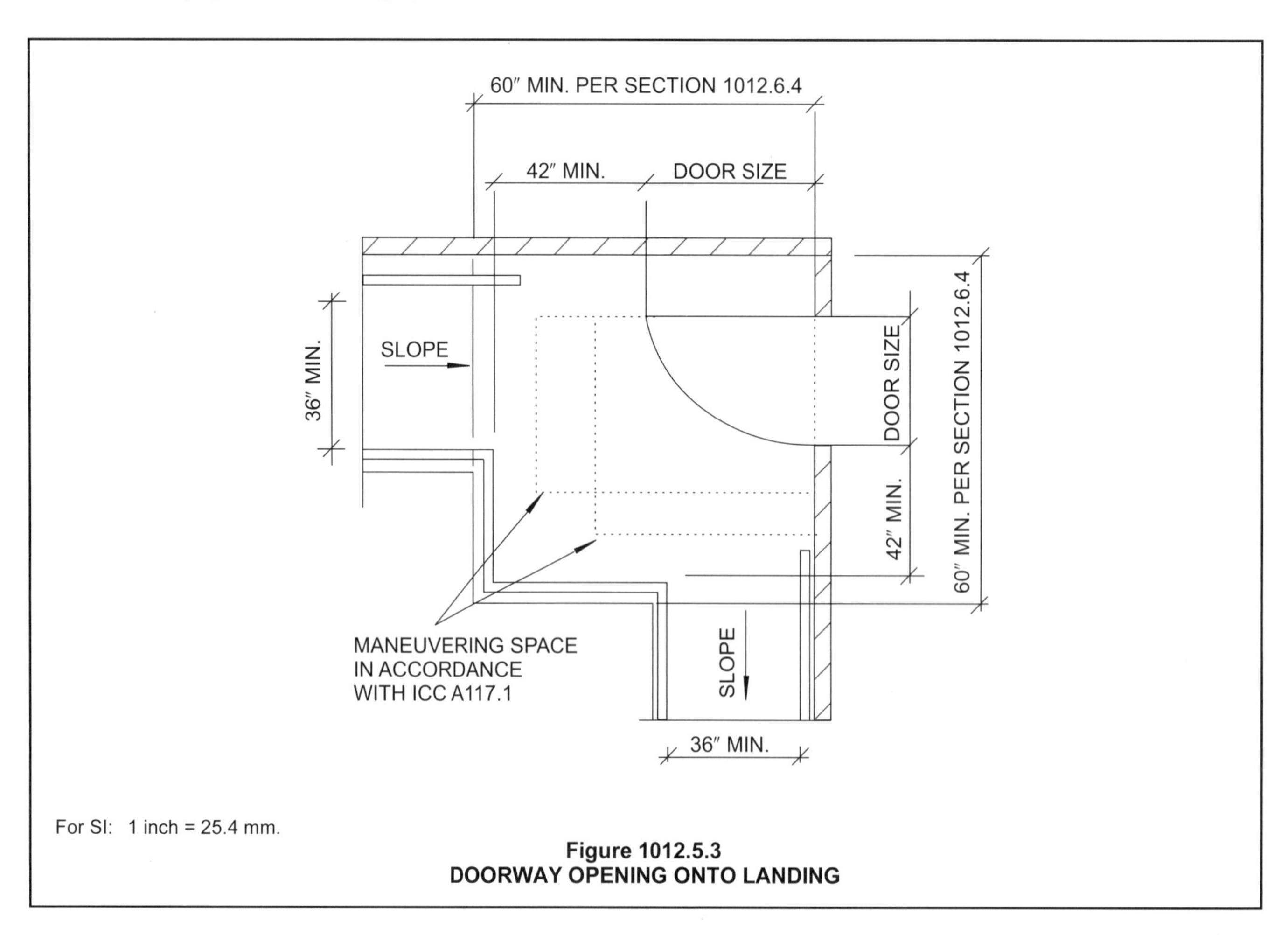

Figure 1012.5.3
DOORWAY OPENING ONTO LANDING

tion 1107, landings are permitted to be 36 inches (914 mm) minimum.

2. Where the *ramp* is not a part of an *accessible route*, the length of the landing shall not be required to be more than 48 inches (1220 mm) in the direction of travel.

❖ The landings for ramps must be at least 60 inches (1524 mm) long (see Commentary Figure 1012.6). This allows persons confined to wheelchairs a sufficient distance to stop and rest along with any persons who may be assisting them. This requirement is directly applicable to straight-run ramps that may require an intermediate landing at every 30 inches (762 mm) of vertical rise (see Commentary Figure 1012.2). If the landing is also to be used to negotiate a change in the ramp's direction, Section 1012.6.4 is applicable. If a door overlaps the landing, Section 1012.5.3 is applicable.

The exceptions provide for smaller landings in dwelling and sleeping units and other locations where the ramp is not part of an accessible route. Exception 1 is consistent with the IRC. Exception 2 would be applicable in areas such as service ramps and ramps serving assembly seating areas that do not contain any wheelchair spaces.

1012.6.4 Change in direction. Where changes in direction of travel occur at landings provided between *ramp* runs, the landing shall be 60 inches by 60 inches (1524 mm by 1524 mm) minimum.

Exception: In Group R-2 and R-3 individual *dwelling* or *sleeping units* that are not required to be *Accessible units*, *Type A units* or *Type B units* in accordance with Section 1107, landings are permitted to be 36 inches by 36 inches (914 mm by 914 mm) minimum.

❖ When a change in direction is made in the ramp at a landing, the landing must be a square of at least 60 inches (1524 mm). This allows the person confined to a wheelchair enough room to negotiate the turn with minimal effort. The length of the landing may need to exceed 60 inches (1524 mm) to match the widths of the two ramp runs. In any case, the landing would still need to be 60 inches (1524 mm) wide (see Commentary Figures 1012.5.3 and 1012.6). If a door overlaps the landing, Section 1012.5.3 is applicable. It is not the intent of this provision to prohibit curved ramps. As long as the cross slope meets the limitations in Section 1010.4, the curved ramp is permitted.

The exception provides for smaller landings in dwelling and sleeping units where the ramp is not part of an accessible route. This is consistent with requirements in the IRC.

1012.6.5 Doorways. Where doorways are located adjacent to a *ramp* landing, maneuvering clearances required by *ICC A117.1* are permitted to overlap the required landing area.

❖ This section specifies that the area required for maneuvering to open the door and the area of the landing are allowed to overlap. It is not necessary to provide the sum of the two area requirements (see Commentary Figure 1012.5.3). Requirements for maneuvering space and turning space at the top and bottom of ramps are found in Section 405 of ICC A117.1. ICC A117.1 requires a turning space at the top or bottom landing of a ramp where the door may be locked. This allows people to turn around to travel back along the ramp.

1012.7 Ramp construction. *Ramps* shall be built of materials consistent with the types permitted for the type of construction of the building, except that wood *handrails* shall be permitted for all types of construction.

❖ Material requirements for the type of construction as required by Section 602 for floors are also the material requirements for ramp construction.

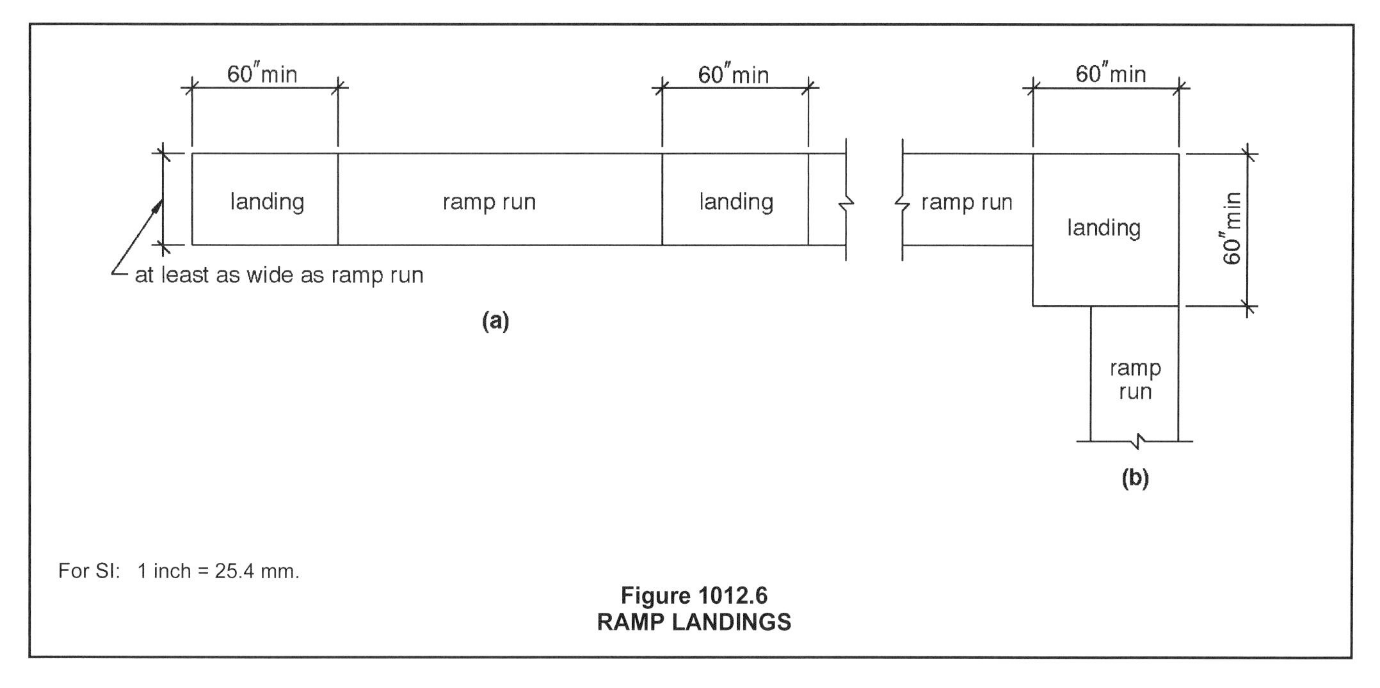

Figure 1012.6
RAMP LANDINGS

1012.7.1 Ramp surface. The surface of *ramps* shall be of slip-resistant materials that are securely attached.

❖ As the pace of exit travel becomes hurried during emergency situations, the probability of slipping on smooth or slick floor surfaces increases. To minimize the hazard, all floor surfaces in the means of egress are required to be slip resistant. The use of hard floor materials with highly polished, glazed, glossy or finely finished surfaces should be avoided. This is consistent with Section 1003.4.

Field testing and uniform enforcement of the concept of slip resistance is not practical. One method used to establish slip resistance is that the static coefficient of friction between leather [Type 1 (Vegetable Tanned) of Federal Specification KK-L-165C] and the floor surface is greater than 0.5. Laboratory test procedures such as ASTM D2047 can determine the static coefficient of resistance. Bulletin No. 4 entitled "Surfaces" issued by the U.S. Access Board (ATBCB) contains further information regarding slip resistance.

1012.7.2 Outdoor conditions. Outdoor *ramps* and outdoor approaches to *ramps* shall be designed so that water will not accumulate on walking surfaces.

❖ Outdoor ramps, landings and the approaches to the ramp must be sloped so that surfaces do not accumulate water so as to provide a safe path of egress travel at all times. While not specifically stated, any interior locations, such as near a pool, should also have the ramps designed to limit the accumulation of water in order to maintain slip resistance (see Sections 1003.4 and 1012.7.1).

Where exterior ramps are used in moderate or severe climates, there may also be a concern to protect the ramp from accumulations of snow and ice to provide a safe path of egress travel at all times, including inclement weather. Maintenance of the means of egress in the IFC requires an unobstructed path to allow for full instant use in case of a fire or emergency (see Section 1031.3 of the IFC). Typical methods for protecting these egress elements include roof overhangs or canopies, heated slabs and, when approved by the building official, a reliable snow removal maintenance program.

1012.8 Handrails. *Ramps* with a rise greater than 6 inches (152 mm) shall have *handrails* on both sides. *Handrails* shall comply with Section 1014.

❖ To aid in the use of a ramp, handrails are to be provided. Handrails are intended to provide the user with a graspable surface for guidance and support. All ramps with a vertical rise greater than 6 inches (152 mm) between landings are to be provided with handrails on both sides [see Commentary Figures 1012.8(1) and 1014.2]. General strength requirements for handrails are found in Section 1014 with a reference to Section 1607.8. Note that if the handrail extension is at a location that could be considered a protruding object, the handrail extension must return to the post at a height of less than 27 inches (686 mm) above the floor. Handrails along the ramp runs are not considered protruding objects.

Depending on the configuration of the ramp and the adjacent walking surface, ramps may require a combination of handrails, edge protection and guards. See Commentary Figures 1012.8(1), 1012.8(2), 1012.8(3) and 1012.8(4) for illustrations of some alternatives.

1012.9 Guards. *Guards* shall be provided where required by Section 1015 and shall be constructed in accordance with Section 1015.

❖ To protect the user from falls to surfaces below, guards are to be provided where the sides of a ramp or landing are more than 30 inches (762 mm) above the adjacent grade. Guards are to be constructed in accordance with Section 1015, including the mini-

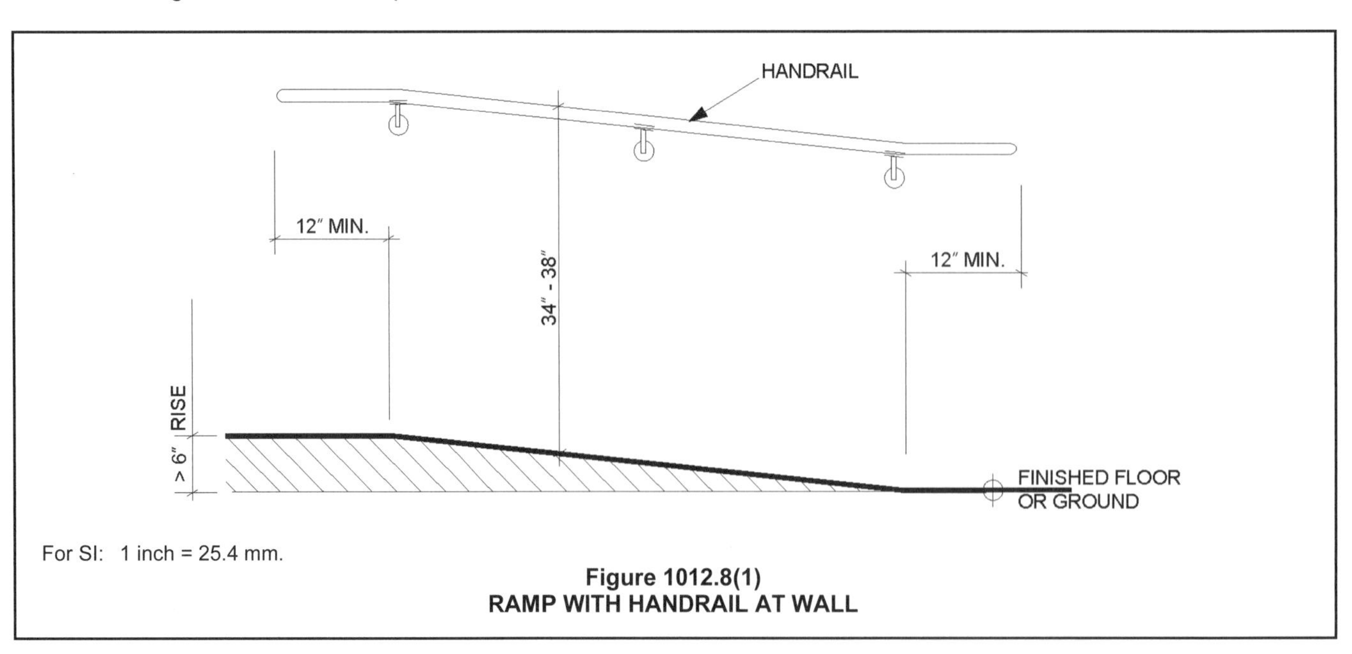

For SI: 1 inch = 25.4 mm.

Figure 1012.8(1)
RAMP WITH HANDRAIL AT WALL

mum height of 42 inches (1067 mm) [see Commentary Figure 1012.8(4)].

Depending on the configuration of the ramp and the adjacent walking surface, ramps may require a combination of handrails, edge protection and guards. See Commentary Figures 1012.8(1), 1012.8(2), 1012.8(3) and 1012.8(4) for illustrations of some alternatives.

1012.10 Edge protection. Edge protection complying with Section 1012.10.1 or 1012.10.2 shall be provided on each side of *ramp* runs and at each side of *ramp* landings.

Exceptions:

1. Edge protection is not required on *ramps* that are not required to have *handrails*, provided they have flared sides that comply with the *ICC A117.1* curb ramp provisions.
2. Edge protection is not required on the sides of *ramp* landings serving an adjoining *ramp* run or *stairway*.
3. Edge protection is not required on the sides of *ramp* landings having a vertical dropoff of not more than $^1/_2$ inch (12.7 mm) within 10 inches (254 mm) horizontally of the required landing area.

❖ This section of the code now addresses the comprehensive requirements for edge protection for all ramps. It must be noted that edge protection is not the same as the requirements for guards. The presence of a guard does not necessarily provide adequate edge protection and the presence of adequate edge protection does not satisfy the requirements for a guard. Edge protection is necessary to prevent the wheels of a wheelchair from leaving the ramp surface or becoming lodged between the edge of the ramp and any adjacent construction. For example, a ramp may be located relatively adjacent to the exterior wall of a building. However, between the ramp edge and the exterior wall, there is a strip of earth for landscape purposes. Without adequate edge protection, per-

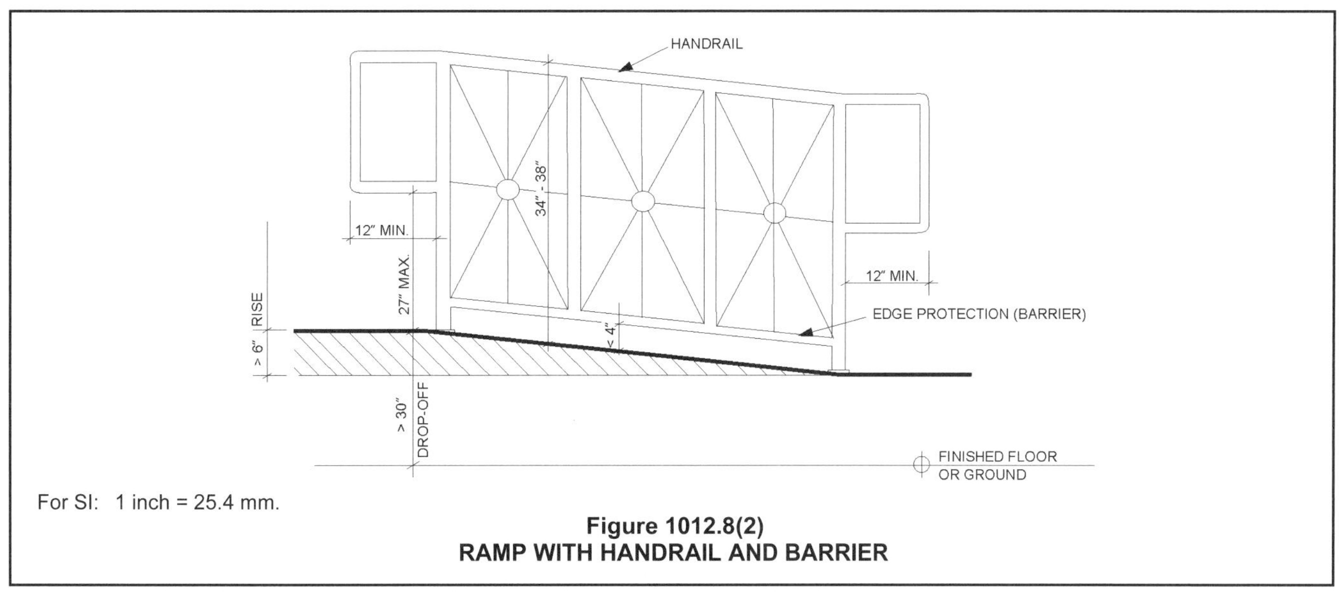

For SI: 1 inch = 25.4 mm.

Figure 1012.8(2)
RAMP WITH HANDRAIL AND BARRIER

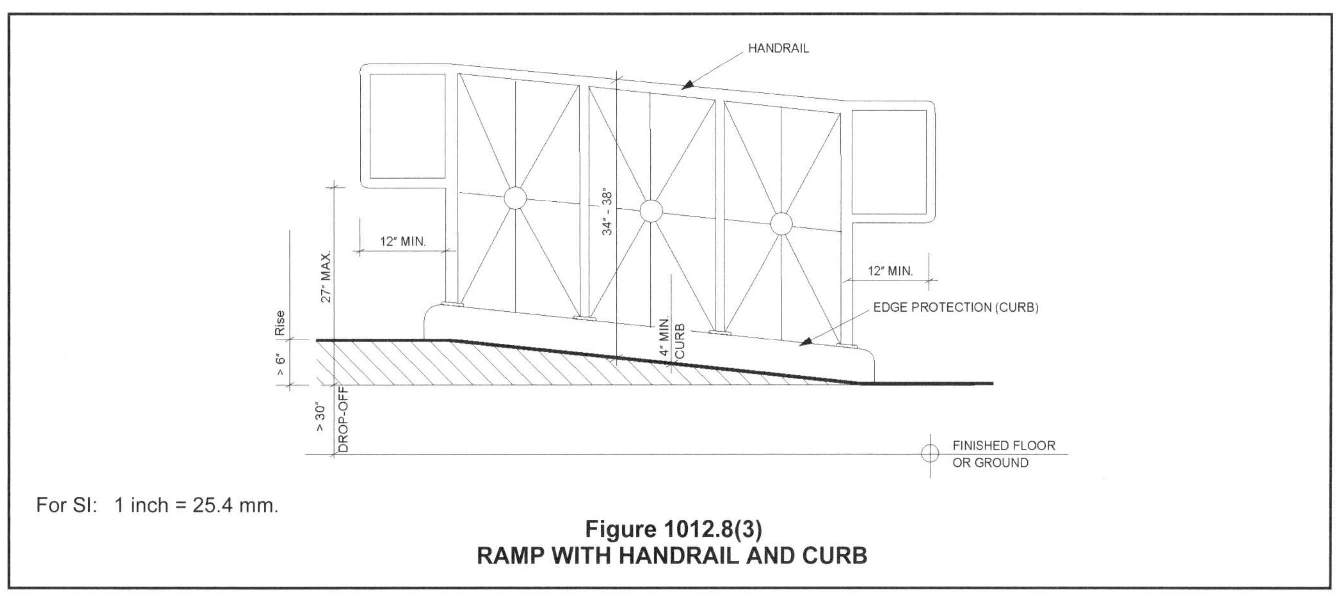

For SI: 1 inch = 25.4 mm.

Figure 1012.8(3)
RAMP WITH HANDRAIL AND CURB

sons confined to wheelchairs could possibly have their wheels run off the side of the ramp into the landscape, causing them to tip. These requirements are consistent with Section 405 of ICC A117.1 and those in the federal 2010 ADA Standard.

Exception 1 allows a ramp to have minimal edge protection as long as its vertical rise is 6 inches (152 mm) or less. The exception is predicated on the ramp not needing any handrails, which is established by the provisions of Section 1012.8. Such a ramp would only need flared sides or returned curbs. Edge protection without handrails or guards could be a tripping hazard for ambulatory persons. For specific details of these types of edge protection, the provisions of Section 406 of ICC A117.1 for curb ramps must be followed.

Exception 2 reiterates that edge protection is not literally required entirely around a ramp landing. Obviously, edge protection is not required along that portion of the landing that directly adjoins a ramp run; it is only required along the the edges of the landing with a dropoff (other than steps of ramp runs).

Exception 3 states that edge protection is not required for those sides of a ramp landing directly adjacent to the ground surface that gently slopes away from the edge of the landing. If the grade adjacent to the ramp landing slopes no more than 1/2:10 (which equates to 1:20) away from the landing, additional edge protection is not required. Such a gradual slope would not be detrimental to persons confined to wheelchairs as they negotiate the ramp landing. Note that this exception is limited to landings, not the ramp surface itself. The ramp must meet the edge protection in Section 1012.10.1 or 1012.10.2.

Depending on the configuration of the ramp and the adjacent walking surface, ramps may require a combination of handrails, edge protection and guards. See Commentary Figures 1012.8(1), 1012.8(2), 1012.8(3) and 1012.8(4) for illustrations of some alternatives.

1012.10.1 Curb, rail, wall or barrier. A curb, rail, wall or barrier shall be provided to serve as edge protection. A curb shall be not less than 4 inches (102 mm) in height. Barriers shall be constructed so that the barrier prevents the passage of a 4-inch-diameter (102 mm) sphere, where any portion of the sphere is within 4 inches (102 mm) of the floor or ground surface.

❖ Edge protection for ramps and ramp landings may be achieved with a built-up curb or other barrier, such as a rail, wall or guard. The barrier must be located near the surface of the ramp and landing such that a 4-inch-diameter (102 mm) sphere cannot pass through any openings. An example of an effective barrier would be the bottom rail of a guard system. If the bottom rail is located less than 4 inches (102 mm) above the ramp and landing surface, edge protection has been provided. If a curb option is used, the curb must be a minimum of 4 inches (102 mm) high. The curb or barrier prevents the wheel of a wheelchair from running off the edge of the surface and provides people with visual disabilities a toe stop at the edge of the walking surface (see Commentary Figure 1012.10.1).

1012.10.2 Extended floor or ground surface. The floor or ground surface of the *ramp* run or landing shall extend 12 inches (305 mm) minimum beyond the inside face of a *handrail* complying with Section 1014.

❖ An alternative to providing some type of barrier at the edge of the ramp (see Section 1012.10.1) is to make the ramp surface wider than the handrails provided at either side. The combination of the wider surface and the handrail barrier would assist in preventing a wheelchair or crutch tip from moving very far off the ramp during a temporary slip (see Commentary Figure 1012.10.1).

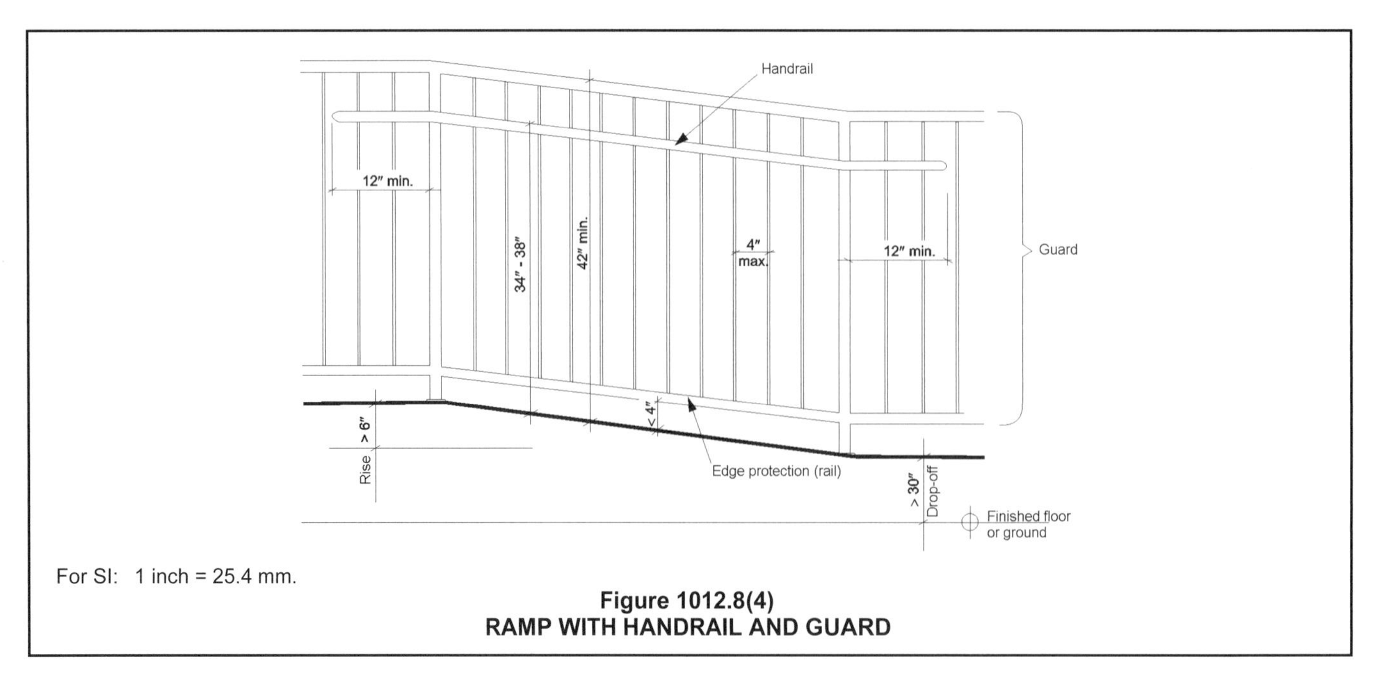

Figure 1012.8(4)
RAMP WITH HANDRAIL AND GUARD

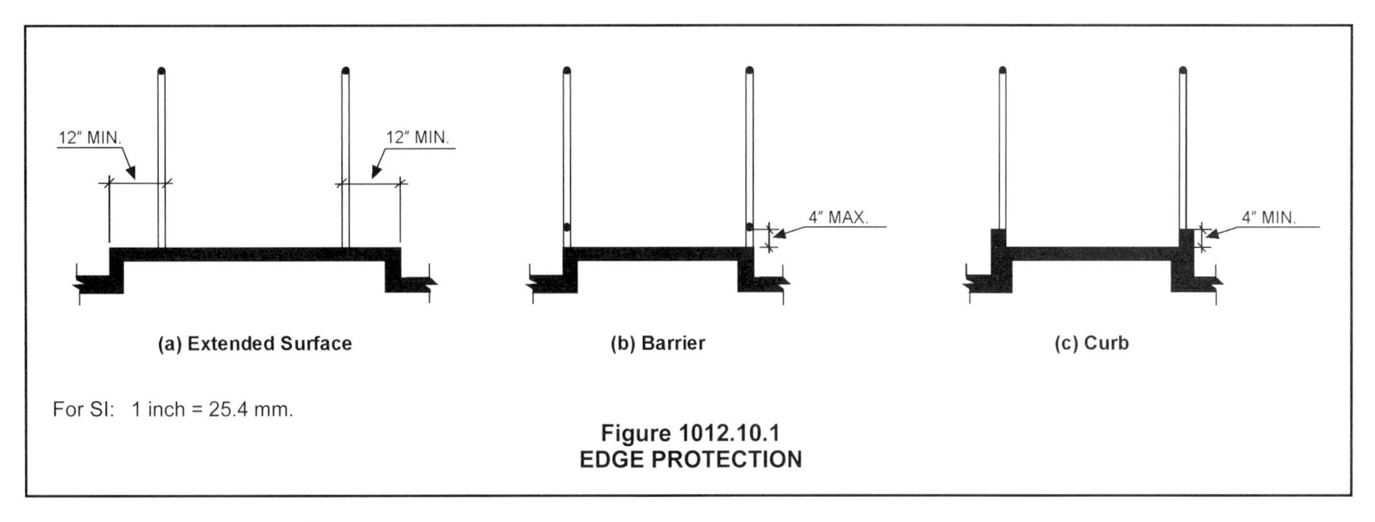

Figure 1012.10.1
EDGE PROTECTION

SECTION 1013
EXIT SIGNS

1013.1 Where required. *Exits* and *exit access* doors shall be marked by an *approved* exit sign readily visible from any direction of egress travel. The path of egress travel to *exits* and within *exits* shall be marked by readily visible exit signs to clearly indicate the direction of egress travel in cases where the *exit* or the path of egress travel is not immediately visible to the occupants. Intervening *means of egress* doors within *exits* shall be marked by exit signs. Exit sign placement shall be such that no point in an *exit access corridor* or *exit passageway* is more than 100 feet (30 480 mm) or the *listed* viewing distance for the sign, whichever is less, from the nearest visible exit sign.

Exceptions:

1. Exit signs are not required in rooms or areas that require only one *exit* or *exit access*.
2. Main exterior *exit* doors or gates that are obviously and clearly identifiable as *exits* need not have exit signs where *approved* by the *building official*.
3. Exit signs are not required in occupancies in Group U and individual *sleeping units* or *dwelling units* in Group R-1, R-2 or R-3.
4. Exit signs are not required in dayrooms, sleeping rooms or dormitories in occupancies in Group I-3.
5. In occupancies in Groups A-4 and A-5, exit signs are not required on the seating side of vomitories or openings into seating areas where exit signs are provided in the concourse that are readily apparent from the vomitories. Egress lighting is provided to identify each vomitory or opening within the seating area in an emergency.

❖ Where an occupancy has two or more required exits or exit accesses, the means of egress must be provided with illuminated signs that readily identify the location of, and indicate the path of travel to, the exits. The signs must be illuminated with letters reading "Exit." The illumination may be internal or external to the sign. The signs should be visible from all directions in the exit access route. In cases where the signs are not visible to the occupants because of turns in the corridor or for other reasons, additional illuminated signs must be provided indicating the direction of egress to an exit. Exit signs must be located so that, where required, the nearest one is within 100 feet (30 480 mm), of the sign's listed viewing distance. UL 924 permits exit signs to be listed with a viewing distance of less than 100 feet (30 480 mm) (see Section 1013.5). When a sign is listed for a viewing distance of less than 100 feet (30 480 mm) the label on the sign will indicate the appropriate viewing distance. If such a sign is used, the spacing of the signs should be based on the listed viewing distance.

Typically, once an occupant enters an exit enclosure, exit signs are no longer needed; however, in buildings with more complicated egress layouts, it is possible that the direction for egress travel within the exit is not immediately apparent. For example, exit passageways can be part of the path of exit travel at the level of exit discharge or transfer floors. Evacuees may hesitate or be confused when the vertical travel becomes horizontal travel, which may result in a delay in evacuation. In these situations, exit signs may be needed within the exit enclosure (see Commentary Figure 1013.1).

The exceptions identify conditions where exit signs are not necessary since they would not increase the safety of the egress path.

For Exceptions 1 and 3, the assumption is that the occupants are familiar enough with the space to know the way out and/or the exits are obvious. In addition, in most cases, the way out is the same as the way in.

In accordance with Exception 2, when the main exterior door through which occupants enter the building is obviously an exit, exit signs are not required. For example, a two-story Group B building has a main employee/customer entrance. The entrance consists of a storefront arrangement with glass doors and sidelights. The entrance is centrally located within the building. These main exterior exit doors can be quickly observed as being an exit and would not need to be marked with an exit sign.

In accordance with Exception 4, exit signs are not required in detainee living and sleeping room areas of Group I-3 buildings. In cases of emergency, occupants in Group I-3 are escorted by staff to the exits and to safety. exit sign materials can also be potential weapons when they are accessible to the detainees.

In the Group A-4 and A-5 occupancies described in Exception 5, the egress path is obvious and thus exit signs are not needed. Additionally, because of the configuration of the vomitories, the exit signs are not readily visible to the persons immediately adjacent to or above the vomitory.

1013.2 Floor-level exit signs in Group R-1. Where exit signs are required in Group R-1 occupancies by Section 1013.1, additional low-level exit signs shall be provided in all areas serving guest rooms in Group R-1 occupancies and shall comply with Section 1013.5.

The bottom of the sign shall be not less than 10 inches (254 mm) nor more than 12 inches (305 mm) above the floor level. The sign shall be flush mounted to the door or wall. Where mounted on the wall, the edge of the sign shall be within 4 inches (102 mm) of the door frame on the latch side.

❖ Because people may be sleeping and because most residents are transient (and thus unfamiliar with the space) in hotels (i.e., Group R-1), low-level exit signs must be provided for emergency exit routes from guestrooms. When smoke at the ceiling obscures the exit signs required by Section 1013.1, these signs will serve as a backup identification of the exit door. By the reference to Section 1013.5, these signs must be internally illuminated and listed and labeled in accordance with UL 924. It is not the intent of this section to require low level exit signs in the guestrooms. The low level exit signs are required only in areas where Section 1013.1 requires exit signs and only leading from the guestroom area, not throughout the entire hotel.

The exit signs must be mounted on the exit door itself or to the side of the exit door on the latch side. The height of the bottom of the exit sign will allow for exit signs not to conflict with accessible route requirements. ICC A117.1 requires doors on an accessible route to have a smooth surface for the bottom 10 inches (254 mm) so that someone could use the footplates on their wheelchair to assist in opening the door.

The requirements do not indicate what low-level signage is appropriate for exit signs that provide direction rather than at the exit. It is also not clear on how an exit sign on a door will be visible from down the corridor if the door is perpendicular to the direction of the hallway.

1013.3 Illumination. Exit signs shall be internally or externally illuminated.

Exception: Tactile signs required by Section 1013.4 need not be provided with illumination.

❖ This section simply provides the scope for illumination of regulated exit signs. Exit signs must be illuminated so that they are readily apparent in situations where the lights may be off or the building has lost power. Exit signs with raised letters and braille are specifically addressed in Section 1013.4.

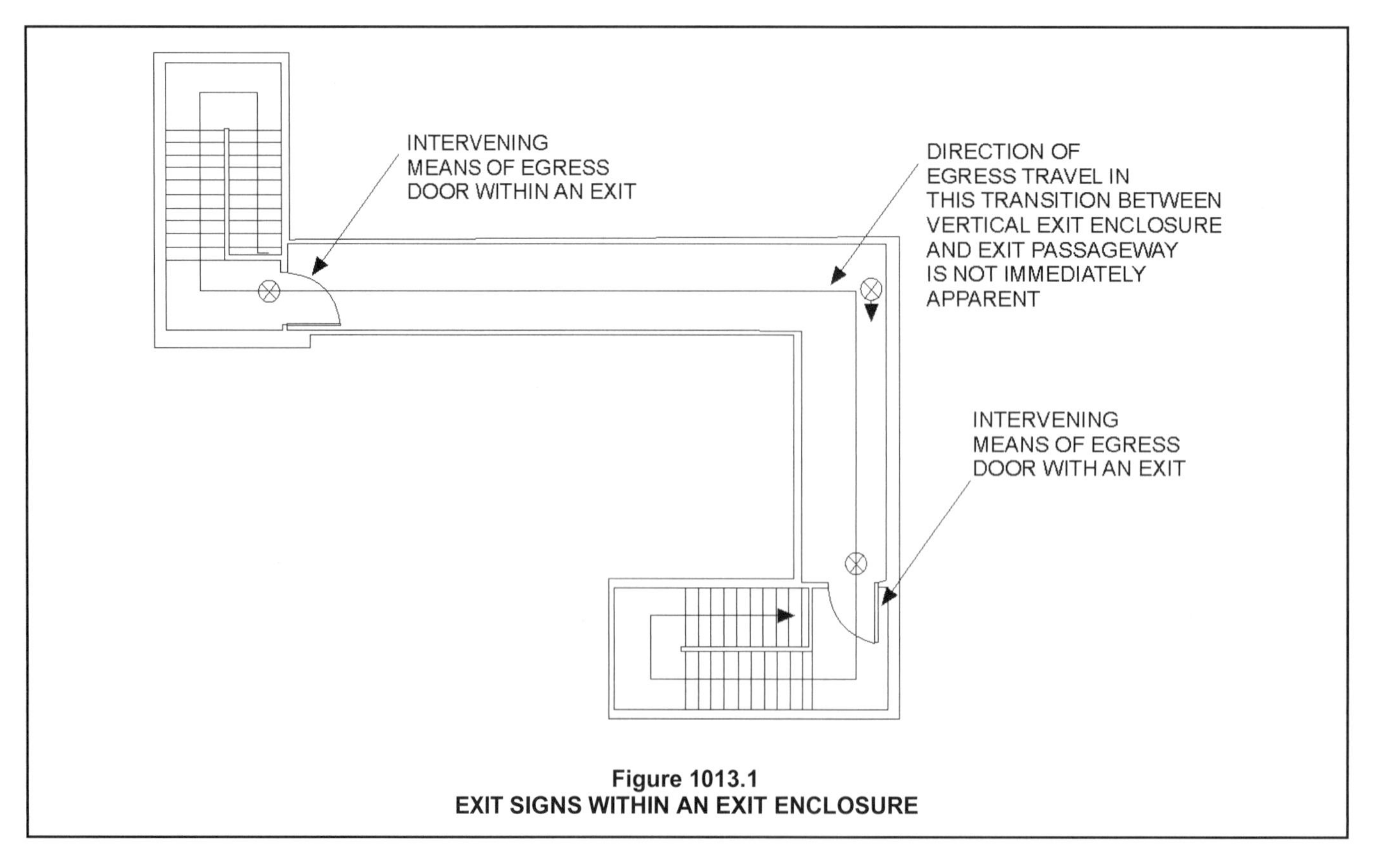

Figure 1013.1
EXIT SIGNS WITHIN AN EXIT ENCLOSURE

1013.4 Raised character and braille exit signs. A sign stating EXIT in visual characters, raised characters and braille and complying with *ICC A117.1* shall be provided adjacent to each door to an *area of refuge*, an exterior area for assisted rescue, an *exit stairway* or *ramp*, an *exit passageway* and the *exit discharge*.

❖ The purpose of this sign is to serve as way finding for a person with vision impairments. This is in addition to the exit signs required by Section 1013.1.

"Tactile" is defined as "used for the sense of touch;" therefore, signage that has either raised letters or braille is considered tactile. For exit signage, visual, raised letters and braille are required. ICC A117.1 includes requirements for the sign and the correct placement. Typically, the sign is placed at about 4 feet, 6 inches (1375 mm) above the floor and on the wall at the latch side of the door. While not required to be illuminated by these provisions, illumination would be advantageous for a person with partial sight. High contrast is important (see Commentary Figure 1013.4).

This signage is needed to indicate which doors are serving as exits for those persons with visual impairments. Signs are needed on the required exit doors in the building: including at doors leading to an interior or exterior exit stairway or ramp; doors leading to an exit passageway; within the exit enclosures leading to the outside or to an exit passageway; and exit doors that lead directly to the outside.

While an area of refuge may be located within an enclosure for an exit stairway, Section 1009.3 also allows the area of refuge to be located immediately outside of the enclosure for the exit stairway. In this situation, exit signage with visual, raised letters and braille would be required both at the door leading into the area of refuge and the door leading to the exit stairway. Exterior areas for assisted rescue are typically located immediately outside of an exit door (see Section 1009.7).

This is not intended to preclude the signage from including additional information as long as "Exit" is first. For example, labeling the door to the exit enclosure as "Exit Stairway" would indicate to the visually impaired person that once they moved through the door they would be dealing with vertical travel. This could be considered an additional safety feature. This section is also referenced in Section 1111.3. For additional way-finding signage inside the stairway enclosure, see Section 1023.9.

Figure 1013.4
EXAMPLE OF EXIT SIGNAGE WITH VISUAL AND RAISED CHARACTERS WITH BRAILLE

1013.5 Internally illuminated exit signs. Electrically powered, *self-luminous* and *photoluminescent* exit signs shall be *listed* and *labeled* in accordance with UL 924 and shall be installed in accordance with the manufacturer's instructions and Chapter 27. Exit signs shall be illuminated at all times.

❖ All exit signage must be listed and labeled as indicated in UL 924, *Standard for Safety of Emergency Lighting and Power Equipment*. Listed "Exit" signs are required by UL 924 to meet the same graphics, illumination and power sources defined in Sections 1013.6.1 through 1013.6.3 for externally illuminated signs. Internal illumination may be electrically powered or be of a self-luminous or photoluminescent product. Electrically powered would include LED, incandescent, fluorescent and electroluminescent types of signs. If a sign is photoluminescent, the "charging" source must be continually available (see the definitions in Chapter 2 for "Photoluminescent" and "Self-luminous"). "Exit" signs must be illuminated at all times, including when the building may not be fully occupied. If a fire occurs late at night, there may be cleaning crews or persons working overtime in the building who will need to be able to find the exits. The reference to Chapter 27 is so the signs will be equipped with a connection to an emergency power supply.

1013.6 Externally illuminated exit signs. Externally illuminated exit signs shall comply with Sections 1013.6.1 through 1013.6.3.

❖ Externally illuminated exit signage must meet the graphic, illumination and emergency power requirements in the referenced sections. The requirements are the same as for internally illuminated signage.

1013.6.1 Graphics. Every exit sign and directional exit sign shall have plainly legible letters not less than 6 inches (152 mm) high with the principal strokes of the letters not less than $^3/_4$ inch (19.1 mm) wide. The word "EXIT" shall have letters having a width not less than 2 inches (51 mm) wide, except the letter "I," and the minimum spacing between letters shall be not less than $^3/_8$ inch (9.5 mm). Signs larger than the minimum established in this section shall have letter widths, strokes and spacing in proportion to their height.

The word "EXIT" shall be in high contrast with the background and shall be clearly discernible when the means of exit sign illumination is or is not energized. If a chevron directional indicator is provided as part of the exit sign, the construction shall be such that the direction of the chevron directional indicator cannot be readily changed.

❖ Every exit sign and directional sign located in the exit access or exit route is required to have a color contrast vivid enough to make the signs readily visible, even when not illuminated. Letters must be at least 6 inches (152 mm) high and their stroke not less than $^3/_4$ inch (19.1 mm) wide (see Commentary Figure 1013.6.1). The sizing of the letters is predicated on the readability of the wording from a distance of 100 feet (30 480 mm).

While red letters are common for exit signs, sometimes green on black is used in auditorium areas with low-lighting levels, such as theaters, because that color combination tends not to distract the audience's attention. It is more important that the exit sign be readily visible with respect to the background.

Exit signs may be larger than the minimum size specified; however, the standardized proportion of the letters must be maintained. Externally illuminated signage that is smaller could use the requirements in UL 924 for guidance; however, sign spacing would need to be adjusted, and alternative approval would be through the building official having jurisdiction.

A "chevron directional indicator" is the same as a directional arrow. The language is intended to be consistent with UL 924.

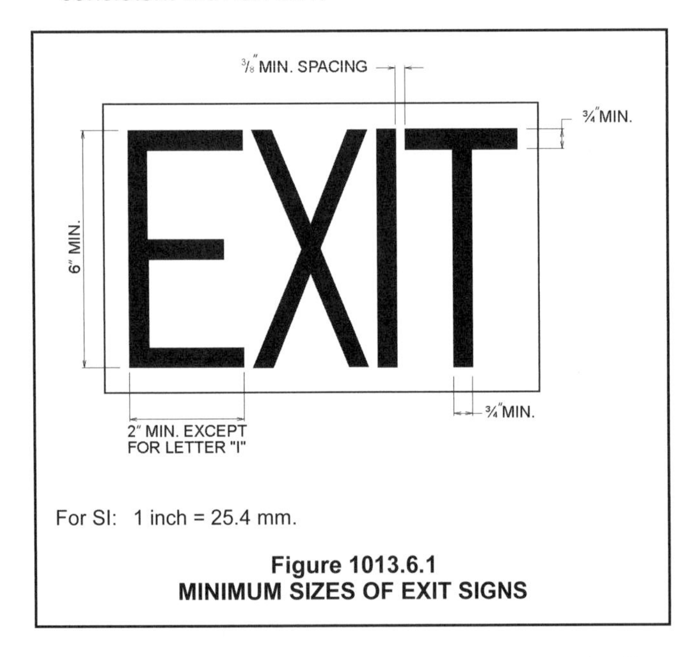

For SI: 1 inch = 25.4 mm.

Figure 1013.6.1
MINIMUM SIZES OF EXIT SIGNS

1013.6.2 Exit sign illumination. The face of an exit sign illuminated from an external source shall have an intensity of not less than 5 footcandles (54 lux).

❖ Every exit sign and directional sign must be continuously illuminated to provide a light intensity at the illuminated surface of at least 5 footcandles (54 lux). It is not a requirement that the exit signs be internally illuminated. An external illumination source with the power capabilities specified by Section 1013.6.3 is acceptable.

1013.6.3 Power source. Exit signs shall be illuminated at all times. To ensure continued illumination for a duration of not less than 90 minutes in case of primary power loss, the sign illumination means shall be connected to an emergency power system provided from storage batteries, unit equipment or an on-site generator. The installation of the emergency power system shall be in accordance with Chapter 27.

Exceptions:

1. *Approved* exit sign illumination means that provide continuous illumination independent of external power sources for a duration of not less than 90 minutes, in case of primary power loss, are not required to be connected to an emergency electrical system.
2. Group I-2 Condition 2 exit sign illumination shall not be provided by unit equipment battery only.

❖ Exit signs must be illuminated on a continuous basis so that when a fire emergency occurs, occupants will be able to identify the locations of the exits. The reliability of the power sources supplying the electrical energy required for maintaining the illumination of exit signs is important. When power interruptions occur, exit sign illumination must be obtained from an emergency power system. This does not imply that the sign must be internally illuminated. Whatever illumination system is used, whether internal or external, it must be connected to a system designed to pick up the power load required by the exit signs after loss of the normal power supply.

Per Exception 1, where self-luminous signs are used, connection to the emergency electrical supply system is not required. A trickle-charge battery to illuminate the exit sign is another option.

Exception 2 is really a more restrictive requirement for hospitals. Hospitals cannot only rely on equipment batteries for the illumination of their exit signs. Typically, hospitals connect illumination of their exit signage to the emergency on-site generator.

The IFC requirements for emergency power for emergency egress lighting and exit signage in existing buildings only requires a 60-minute time duration. This is not a conflict, but rather recognition of the loss of battery storage capability over a length of time.

SECTION 1014
HANDRAILS

1014.1 Where required. *Handrails* serving *stairways*, *ramps*, stepped *aisles* and ramped *aisles* shall be adequate in strength and attachment in accordance with Section 1607.8. *Handrails* required for *stairways* by Section 1011.11 shall comply with Sections 1014.2 through 1014.9. *Handrails* required for *ramps* by Section 1012.8 shall comply with Sections 1014.2 through 1014.8. *Handrails* for stepped *aisles* and ramped *aisles* required by Section 1029.15 shall comply with Sections 1014.2 through 1014.8.

❖ Handrails are required at stairways and ramps. In all situations, they must be designed in accordance with the structural requirements in Section 1607.8. There are, however, distinct differences in how the handrail requirements are applied in stairways, ramps, stepped aisles and ramped aisles. The specific section references allow for this consideration. Where and how many handrails for stairways and ramps are specified in Sections 1003.5, 1011.11 and 1012.8. Where handrails are required in assembly seating for stepped aisles and ramped aisles is specified in Section 1029.15.

Stairways and their handrails are not part of an accessible route and are not subject to the stairway technical requirements in ICC A117.1. Standards are only referenced to the extent specified by the code (see Section 102.4).

Handrails are also very distinct from guards, even though they are sometimes incorrectly called "guardrails." The "handrail" is the element that is grasped during vertical travel for guidance, stabilization, pulling and as an aid in arresting a possible fall. Guards are located near the side of an elevated walking surface to minimize the possibility of a fall to a lower level and are discussed in Section 1015. However, in residential applications, in some locations in assembly seating and along alternating tread devices and ship's ladders, the top rail of a guard may also serve as a handrail (see Section 1015.3, Exceptions 3, 4 and 5, and Section 1029.16).

1014.2 Height. *Handrail* height, measured above *stair* tread *nosings*, or finish surface of *ramp* slope, shall be uniform, not less than 34 inches (864 mm) and not more than 38 inches (965 mm). *Handrail* height of *alternating tread devices* and ship's ladders, measured above tread *nosings*, shall be uniform, not less than 30 inches (762 mm) and not more than 34 inches (864 mm).

Exceptions:

1. Where handrail fittings or bendings are used to provide continuous transition between *flights*, the fittings or bendings shall be permitted to exceed the maximum height.
2. In Group R-3 occupancies; within *dwelling units* in Group R-2 occupancies; and in Group U occupancies that are associated with a Group R-3 occupancy or associated with individual *dwelling units* in Group R-2 occupancies; where handrail fittings or bendings are used to provide continuous transition between *flights*, transition at *winder* treads, transition from *handrail* to *guard*, or where used at the start of a *flight*, the *handrail* height at the fittings or bendings shall be permitted to exceed the maximum height.
3. *Handrails* on top of a guard where permitted along stepped aisles and ramped aisles in accordance with Section 1029.15.

❖ It has been demonstrated that for safe use, the height of handrails must not be less than 34 inches (864 mm) nor more than 38 inches (965 mm) above the leading edge of stairway treads, landings or other walking surfaces (see Commentary Figure 1014.2). This requirement is applicable for all uses, including handrails within a dwelling unit.

The fundamental stairway requirements are not appropriate for alternating tread devices and ship's ladders since they differ significantly from other types of stairways. The permitted range for handrail height for alternating tread devices and ship's ladders allows for a lower height above the tread nosings. The minimum required height is reduced from 34 inches (864 mm) to 30 inches (762 mm), with a maximum permitted height of 34 inches (864 mm). The special features of an alternating tread device or ship's ladder result in differences of handrail use, such as different arm posture, the hand gripping the handrail near a higher part of the body and the use of handrails under the arms for stabilization when descending. Therefore, a lower handrail height is more appropriate.

Exceptions 1 and 2 allow for the use of common fittings and bendings to provide for continuous transition of the handrail at specified turns and pitch changes in the stairway where the fitting or bending might exceed the maximum handrail height. In some

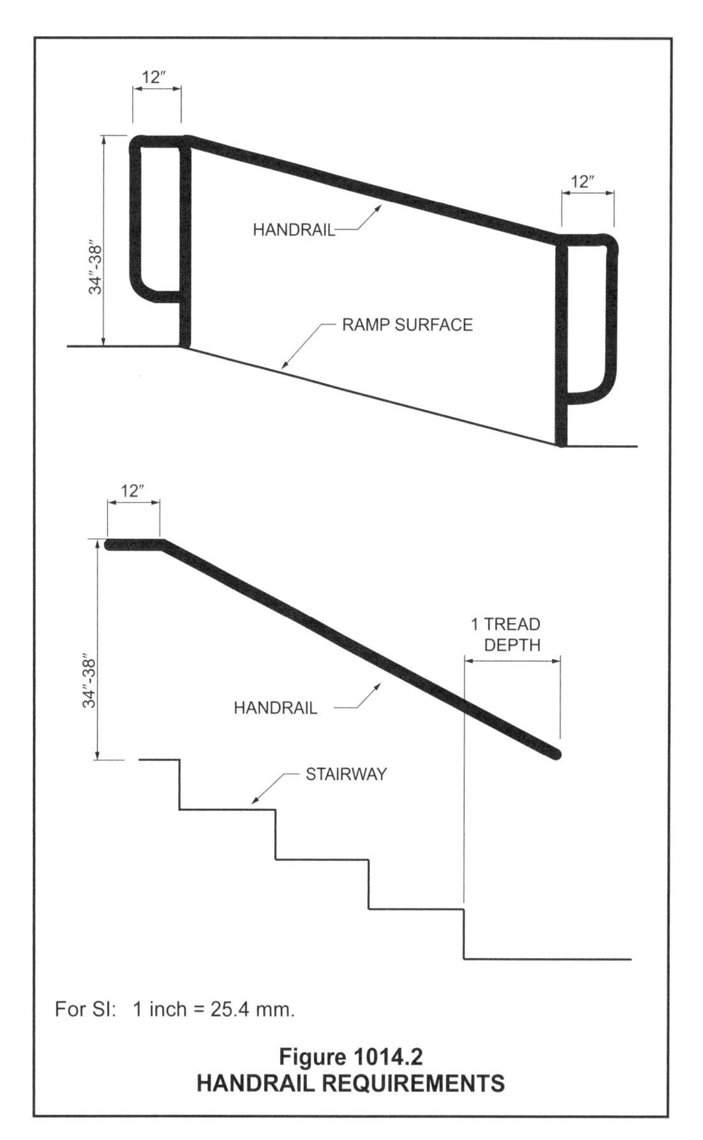

Figure 1014.2
HANDRAIL REQUIREMENTS

cases, there is no reference for the height when the fitting or handrail might extend beyond the nosings at the intersection of adjoining flights. Exception 1 applies to all stairways between flights. This typically occurs at the center handrail at the landings on a dogleg or switchback stairway configuration. Although the handrails are not required on the landing, the code requirement for handrail extensions or for handrail continuity often creates the need for some type of transition, especially at turns (see also Section 1014.6). This exception would allow for an easing or gooseneck riser over the landing to be used for a smooth transition of the handrail. The use of the new exceptions may permit a more gradual variation in the height even though it will allow for portions of the handrail to exceed the normal 38-inch (965 mm) maximum height. The belief is that handrail continuity is more important than staying within the height limitation.

Exception 2 applies to dwelling units with the intent to allow transition elements at common locations, such as the start of the flight, at winder treads or from handrail to guard at landings. Combined with the guard height reduction in Section 1015.3, Exception 1, the end result may be fewer transition pieces. This is consistent with provisions in the IRC.

Exception 3 is in recognition of the special allowances for the top rail of a guard to serve as a handrail in limited situations in an assembly seating area. See Section 1029.15 for specifics.

1014.3 Handrail graspability. Required *handrails* shall comply with Section 1014.3.1 or shall provide equivalent graspability.

> **Exception:** In Group R-3 occupancies; within *dwelling units* in Group R-2 occupancies; and in Group U occupancies that are accessory to a Group R-3 occupancy or accessory to individual *dwelling units* in Group R-2 occupancies; *handrails* shall be Type I in accordance with Section 1014.3.1, Type II in accordance with Section 1014.3.2 or shall provide equivalent graspability.

❖ The abilities to grasp a handrail firmly and slide a hand along the rail's gripping surface without meeting obstructions are important factors in the safe use of stairways and ramps. These properties are largely functions of the shape of the handrail. Handrails for stairways and ramps must meet the specifications of Section 1014.3.1 or be determined to have grasping properties and attributes equivalent to profiles allowed in Section 1014.3.1. Such determinations of equivalence are an allowed option made by local building officials based on the profile presented and the building official's evaluation of its properties. A complete evaluation will consider the four basic functions of a handrail: guidance, stabilization, pulling and aid in arresting a fall. The determination is best made by comparative use on stairs or ramps of properly mounted samples. Handrails that meet neither Type I nor Type II characteristics may be considered to have equivalent graspability. Complete evaluation will also consider the mounting of the handrail and understanding the interference of handrail mounts on the gripping surfaces of smaller profiles. For a discussion of this, see Section 1014.4.

The exception allows for an alternative, Type II handrails, within residential units and their associated structures. A handrail on common stairways within an apartment building or on the steps to the front door of a townhouse could not use this exception unless approved by the building official as being equivalent to Type I. The residential allowance is consistent with the IRC.

1014.3.1 Type I. *Handrails* with a circular cross section shall have an outside diameter of not less than $1^1/_4$ inches (32 mm) and not greater than 2 inches (51 mm). Where the *handrail* is not circular, it shall have a perimeter dimension of not less than 4 inches (102 mm) and not greater than $6^1/_4$ inches (160 mm) with a maximum cross-sectional dimension of $2^1/_4$ inches (57 mm) and minimum cross-sectional dimension of 1 inch (25 mm). Edges shall have a minimum radius of 0.01 inch (0.25 mm).

❖ Handrails have traditionally been regulated as either circular or noncircular rails. The noncircular rails have previously been limited to a maximum perimeter dimension of $6^1/_4$ inches (160 mm), with other limitations addressing minimum perimeter and minimum and maximum cross-sectional dimensions. These handrails shapes are now referred to as Type I handrails.

Type I handrails include circular cross sections with an outside diameter of at least $1^1/_4$ inches (32 mm) but not greater than 2 inches (51 mm). This limits the perimeter of the cross section such that the gripping surface incorporates the bottom of the rail. A handrail with either a very narrow or a large cross section is not graspable in a power grip by all able-bodied users and certainly not by those with hand-strength or flexibility deficiencies. A power grip typically accesses the bottom surface of the handrail, such as around a bar. A pinching grip is not as effective when there is a need to arrest a fall. An example of a pinching grip would be where a 2x4 stud on edge was used as a handrail. This would not comply with graspability concerns.

Noncircular Type I cross sections must meet the alternative noncircular criteria in this section, and the bottom of the rail must be considered part of the suitable gripping surface. Of note is that this criteria now includes a minimum cross section of 1 inch (25 mm) to provide for designs that reduce the interference of fingers and an opposing thumb that occurs when small objects are tightly grasped. Edges must be slightly rounded and not sharp. An example is shown in Commentary Figure 1014.3.1.

Ramp requirements in ICC A117.1 are referenced as part of the accessible route requirements in Chapter 11. Current handrail provisions are largely coordinated with the Type I handrail requirements in this section. Section 103 of ICC A117.1 would also permit alternative handrail shapes if they provided equivalent or better graspability.

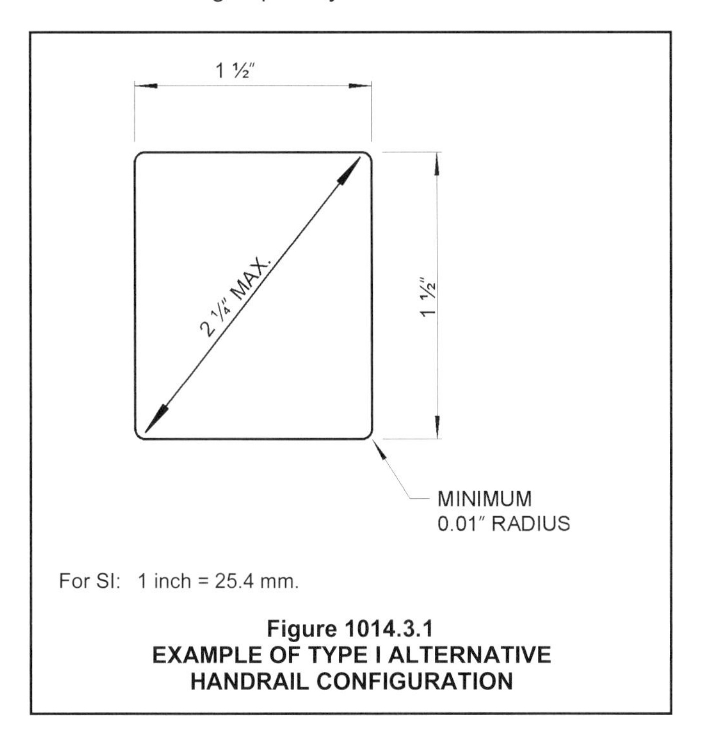

Figure 1014.3.1
EXAMPLE OF TYPE I ALTERNATIVE HANDRAIL CONFIGURATION

1014.3.2 Type II. *Handrails* with a perimeter greater than $6^1/_4$ inches (160 mm) shall provide a graspable finger recess area on both sides of the profile. The finger recess shall begin within a distance of $^3/_4$ inch (19 mm) measured vertically from the tallest portion of the profile and achieve a depth of not less than $^5/_{16}$ inch (8 mm) within $^7/_8$ inch (22 mm) below the widest portion of the profile. This required depth shall continue for not less than $^3/_8$ inch (10 mm) to a level that is not less than $1^3/_4$ inches (45 mm) below the tallest portion of the profile. The width of the *handrail* above the recess shall be not less than $1^1/_4$ inches (32 mm) to not greater than $2^3/_4$ inches (70 mm). Edges shall have a minimum radius of 0.01 inch (0.25 mm).

❖ Handrail profiles having a perimeter dimension greater than $6^1/_4$ inches (160 mm), identified as Type II handrails, are acceptable within dwelling units and their associated structures when complying with all of the specific dimensional requirements.

Research has shown that Type II handrails have graspability that is essentially equal to or greater than the graspability of handrails meeting the long-accepted and codified shape and size now defined as Type I.

The key features of the graspability of Type II handrails are graspable finger recesses on both sides of the handrail. These recesses allow users to firmly grip a properly proportioned grasping surface on the top of the handrail, ensuring that the user can tightly retain a power-span grip on the handrail for all forces that are associated with attempts to arrest a fall.

This class of handrails incorporates a grip surface with controlled recesses for the purchase of the fingers and opposing thumb. These handrail shapes allow the use of power-span grips that need not encompass the bottom surfaces of the rail, allowing the design of taller cross sections that can eliminate the interference of mountings and provide a completely uninterrupted gripping surface for the user. The limits of the position and depth of the required recesses represent the minimum standard. Optimizing the design within the parameters with larger recesses and complete finger clearance from mountings will enhance the performance of these profiles. Although this standard allows design flexibility, it is important to follow the specifications accurately to comply. Each drawing in Commentary Figure 1014.3.2 illustrates the requirements of each sentence for clarity.

1014.4 Continuity. Handrail gripping surfaces shall be continuous, without interruption by newel posts or other obstructions.

Exceptions:

1. *Handrails* within *dwelling units* are permitted to be interrupted by a newel post at a turn or landing.
2. Within a *dwelling unit*, the use of a volute, turnout, starting easing or starting newel is allowed over the lowest tread.
3. Handrail brackets or balusters attached to the bottom surface of the *handrail* that do not project horizontally beyond the sides of the *handrail* within $1^1/_2$ inches (38 mm) of the bottom of the *handrail* shall not be considered obstructions. For each $^1/_2$ inch (12.7 mm) of additional handrail perimeter dimension above 4 inches (102 mm), the vertical clearance dimension of $1^1/_2$ inches (38 mm) shall be permitted to be reduced by $^1/_8$ inch (3.2 mm).
4. Where *handrails* are provided along walking surfaces with slopes not steeper than 1:20, the bottoms of the handrail gripping surfaces shall be permitted to be obstructed along their entire length where they are integral to crash rails or bumper guards.
5. *Handrails* serving stepped *aisles* or ramped *aisles* are permitted to be discontinuous in accordance with Section 1029.15.1.

❖ Handrails must be usable for their entire length without requiring the users to release their grasp. Typically, when using the handrail while traveling the means of egress, an individual's arm is extended to lead the body with the hand forming a loose grip on the top and sides of the rail. If handrails are to be of service to users, they must be uninterrupted and continuous. Oversize newels, changes in the guard system, or excessive supports at the bottom of small perimeter handrails can cause interruption of the handrail, requiring the occupants to release their grip [see Commentary Figure 1014.4(1)]. Exception 1 allows the interruption of the handrail by a newel post at the intersection of two handrail sections at a turn within a flight or at a landing in dwelling units; however, this exception is not applicable to curved or spiral stairs. Exception 2 provides for familiar and historical handrail details, often combined with guards, that have been used for decades in dwelling units without substantiated detriment to the safety of the occupants. Exception 3 provides specifications for the attachment of brackets and balusters to the bottom of handrails to provide for minimum finger clearance such that they will not be considered obstructions. This method of handrail support limits interruptions of the grip surfaces at the bottom of smaller perimeter shapes that otherwise would deter or impede the user's ability to attain a stabilizing grip essential to safe stairway use. Larger handrail sizes permit shorter brackets since geometrically the finger

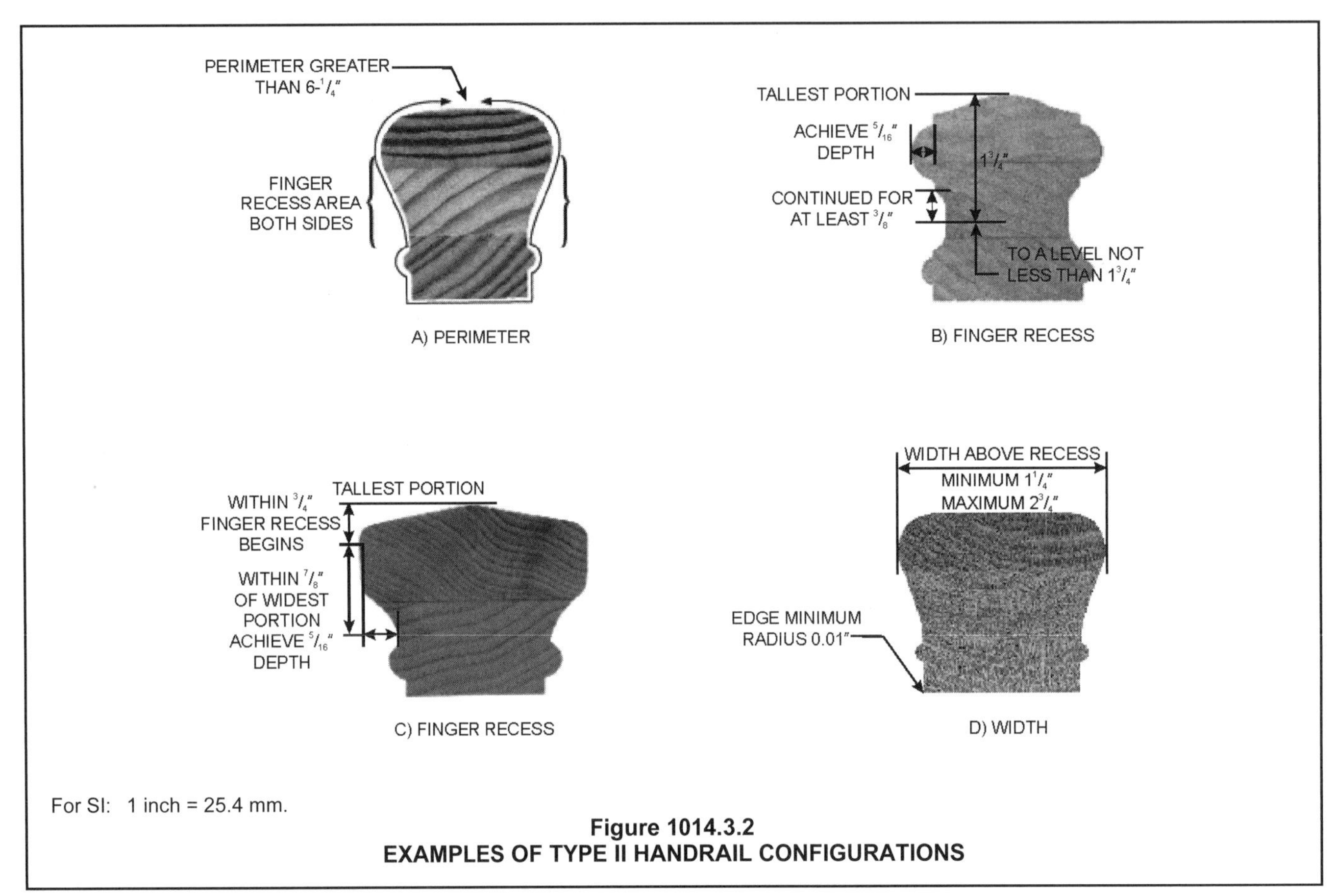

Figure 1014.3.2
EXAMPLES OF TYPE II HANDRAIL CONFIGURATIONS

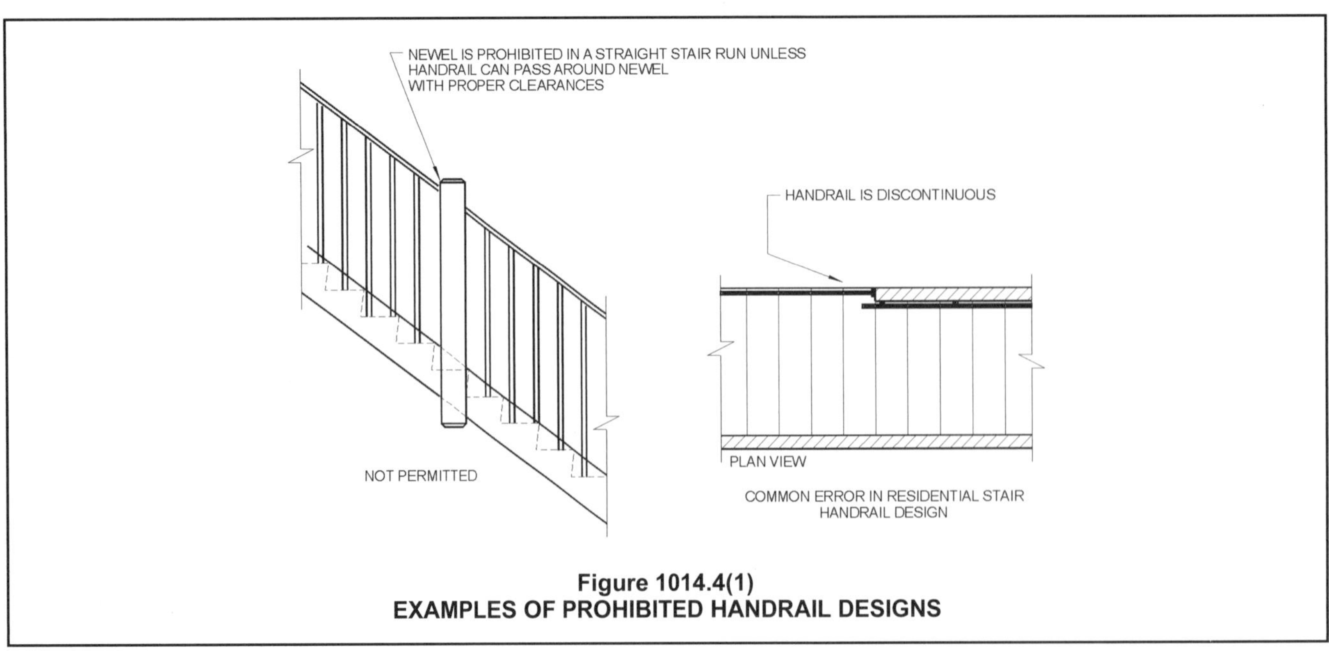

Figure 1014.4(1)
EXAMPLES OF PROHIBITED HANDRAIL DESIGNS

clearance is still maintained [see Commentary Figure 1014.4(2)]. For example, a Type II handrail may elevate the fingertips completely above the supports, allowing the rail to be mounted by other means that do not obstruct when providing the clearances specified. Exception 4 allows for products that serve dual purposes, such as the bumper guard/handrail found in hospitals and nursing homes along corridors, to have a continuous bottom support. Since these are only permitted on slopes that are less than what is defined as a ramp, the handrails are more for assistance in mobility or balance rather than to arrest a fall.

Exception 5 references back to provisions for stepped and ramped aisles in assembly seating. Discontinuous handrails are permitted to allow for access to seats. Such discontinuous handrails must still extend the full run of the aisle stairs or ramp. See Section 1029.15.1.

1014.5 Fittings. *Handrails* shall not rotate within their fittings.

❖ Fittings are those component pieces of a continuous handrail that are shaped or bent and attached to the longer sections of straight or curved handrail to provide for transition at changes in pitch, direction or to provide for termination of a continuous handrail. Fittings and handrails must be securely joined to ensure a stable handrail that does not allow any portion to rotate when grasped.

1014.6 Handrail extensions. *Handrails* shall return to a wall, *guard* or the walking surface or shall be continuous to the handrail of an adjacent *flight* of *stairs* or *ramp* run. Where *handrails* are not continuous between *flights*, the *handrails* shall extend horizontally not less than 12 inches (305 mm) beyond the top riser and continue to slope for the depth of one tread beyond the bottom riser. At *ramps* where *handrails* are not continuous between runs, the *handrails* shall extend horizontally above the landing 12 inches (305 mm) minimum beyond the top and bottom of *ramp* runs. The extensions of *handrails* shall be in the same direction of the *flights* of *stairs* at *stairways* and the *ramp* runs at *ramps*.

Exceptions:

1. *Handrails* within a *dwelling unit* that is not required to be *accessible* need extend only from the top riser to the bottom riser.
2. *Handrails* serving aisles in rooms or spaces used for assembly purposes are permitted to comply with the handrail extensions in accordance with Section 1029.15.
3. *Handrails* for *alternating tread devices* and ship's ladders are permitted to terminate at a location vertically above the top and bottom risers. *Handrails* for *alternating tread devices* are not required to be continuous between *flights* or to extend beyond the top or bottom risers.

❖ The purpose of return requirements at handrail ends is to prevent a person from catching an article of clothing or satchel straps, or from being injured by falling on the extended end of a handrail.

The length that a handrail extends beyond the top and bottom of a stairway, ramp or intermediate landing where handrails are not continuous to another stair flight or ramp run is an important factor for the safety of the users. An occupant must be able to grasp securely a handrail beyond the last riser of a stairway or the last sloped segment of a ramp. Handrail terminations that bend around a corner do not provide this stability; therefore, the handrail must extend in the direction of the stair flight or ramp run. The handrail extensions are not required where a user could keep his or her hand on the handrail, such as the continuous handrail at the landing of a switchback stairway or ramp (see Section 1014.2, Exception 1).

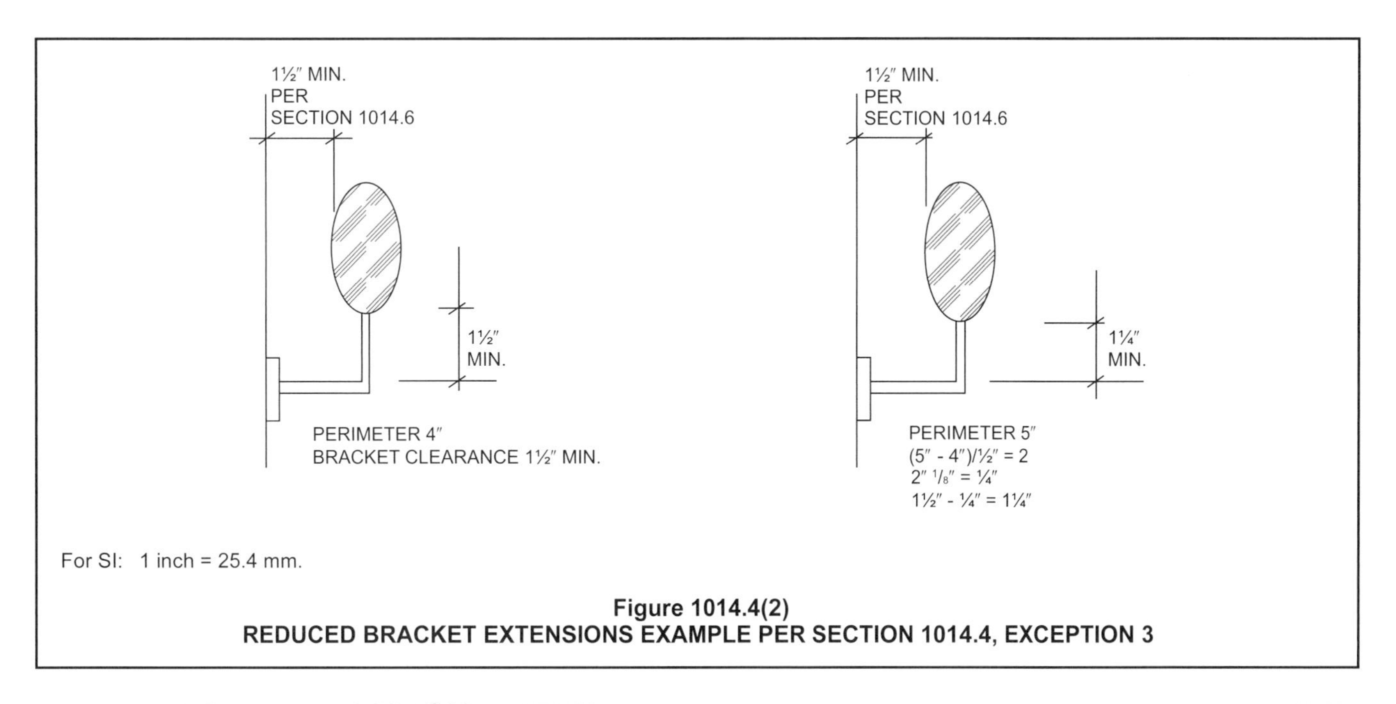

Figure 1014.4(2)
REDUCED BRACKET EXTENSIONS EXAMPLE PER SECTION 1014.4, EXCEPTION 3

For stairways, handrails must be extended 12 inches (305 mm) horizontally beyond the top riser and sloped a distance of one tread depth beyond the bottom riser. For ramps, handrails must be extended 12 inches (305 mm) horizontally beyond the last sloped ramp segment at both the top and bottom locations. These handrail extensions are not only required at the top and bottom on both sides of stairways and ramps, but also at other places where handrails are not continuous, such as landings and platforms. These requirements are intended to reflect the current provisions of ICC A117.1 (see Commentary Figure 1014.2) and the 2010 ADA Standard. Note that if the handrail extension is at a location that could be considered a protruding object, the handrail must return to the post at a height of less than 27 inches (686 mm) above the floor (see Sections 1003.3.2 and 1003.3.3).

In accordance with Exception 1, handrail extensions are not required where a dwelling unit is not required to meet any level of accessibility (i.e., Accessible unit, Type A unit or Type B unit). Handrail extensions are permitted to end at a newel post or turnaround.

Exception 2 provides for handrails along ramped or stepped aisles in assembly seating configurations, such as in sports facilities, theaters and lecture halls. It is necessary to limit handrail extensions in assembly aisles so that circulation in cross aisles that run perpendicular to the stepped or sloped aisles is not compromised.

Exception 3 allows for the unique construction considerations for alternating tread devices and ship's ladders. Again, usage of these devices is very limited. Since alternative tread devices and ship's ladders are typically utilized as a safer alternative to a vertical ladder, they are often located in tight spaces where traditional-type stairs cannot be used or are not required. With a much steeper angle than traditional stairs and differing usage, handrail extensions and continuity provisions are not practical.

1014.7 Clearance. Clear space between a handrail and a wall or other surface shall be not less than $1^1/_2$ inches (38 mm). A handrail and a wall or other surface adjacent to the *handrail* shall be free of any sharp or abrasive elements.

❖ A clear space is needed between a handrail and the wall or other surface to allow the user to slide his or her hand along the rail with fingers in the gripping position without contacting the wall surface, which could have an abrasive texture. In climates where persons may be expected to be wearing heavy gloves during the winter, an open design with greater clearance would be desirable at an exterior stairway, or a stairway directly inside the entrance to a building. [See Commentary Figures 1014.4(2) and 1014.8(2) for an illustration of handrail clearance.]

1014.8 Projections. On *ramps* and on ramped *aisles* that are part of an *accessible route*, the clear width between *handrails* shall be 36 inches (914 mm) minimum. Projections into the required width of *aisles*, *stairways* and *ramps* at each side shall not exceed $4^1/_2$ inches (114 mm) at or below the handrail height. Projections into the required width shall not be limited above the minimum headroom height required in Section 1011.3. Projections due to intermediate *handrails* shall not constitute a reduction in the egress width. Where a pair of intermediate *handrails* are provided within the *stairway* width without a walking surface between the pair of intermediate *handrails* and the distance between the pair of intermediate *handrails* is greater than 6 inches (152 mm), the available egress width shall be reduced by the distance between the closest edges of each such intermediate pair of *handrails* that is greater than 6 inches (152 mm).

❖ Handrails may not project more than $4^1/_2$ inches (114 mm) into the required width of a stairway, so that the clear width of the passage will not be seriously reduced [see Commentary Figure 1014.8(1)]. This is consistent with Section 1003.3.3. This projection may exist below the handrail height as well [see Commentary Figure 1014.8(2)]. The projection can be greater on stairways or ramps exceeding minimum width requirements.

The requirement for some stairways to have center handrails (Section 1014.9) could result in a single or double handrail down the center of the flight of stairs. Since these stairways are 60 inches (1525 mm) or wider, a center handrail would typically not be an obstruction and, therefore, would not reduce the capacity of the stairway. Schools and large assembly spaces commonly put in center handrails to aid in the flow up and down the stairs during peak usage times. If the center handrail is a double rail with a spacing of more than 6 inches (152 mm), the width of the stairway many need to be adjusted to compensate for the loss of available width.

1014.9 Intermediate handrails. *Stairways* shall have intermediate *handrails* located in such a manner that all portions of the *stairway* minimum width or required capacity are within 30 inches (762 mm) of a handrail. On monumental *stairs*, *handrails* shall be located along the most direct path of egress travel.

❖ In order to always be available to the user of the stairway, the maximum horizontal distance to a handrail from within the width required must not exceed 30 inches (762 mm). People tend to walk close to handrails, and if intermediate handrails are not provided for very wide stairways, the center portion of such stairways will normally receive less use. More importantly, in emergencies, the center portions of wide stairways with handrails would enhance egress travel rather than delay it by overcrowding at the sides with the handrails. This would especially be true under panic conditions where the use of wide interior stairways could become particularly hazardous.

The distance to the handrail applies to the "required" width of the stairway. If a stairway is greater than 60 inches (1524 mm) in width, but only 60 inches (1524 mm) total width is required based on occupant load (see Section 1005.3.1), intermediate handrails are not required. Adequate safety is provided since every user is within 30 inches (762 mm) of a handrail.

Monumental stairways are typically provided for architectural effect and may or may not be considered required egress stairs. The criteria for monumental egress stairways deal with the very wide stairway in relation to the required width. While handrails on both sides of the stairway may be sufficient to accommodate the required width, the handrails may not be near the stream of traffic or even apparent to the user. In this case, the handrails are to be placed in a location more reflective of the egress path (see Commentary Figure 1014.9).

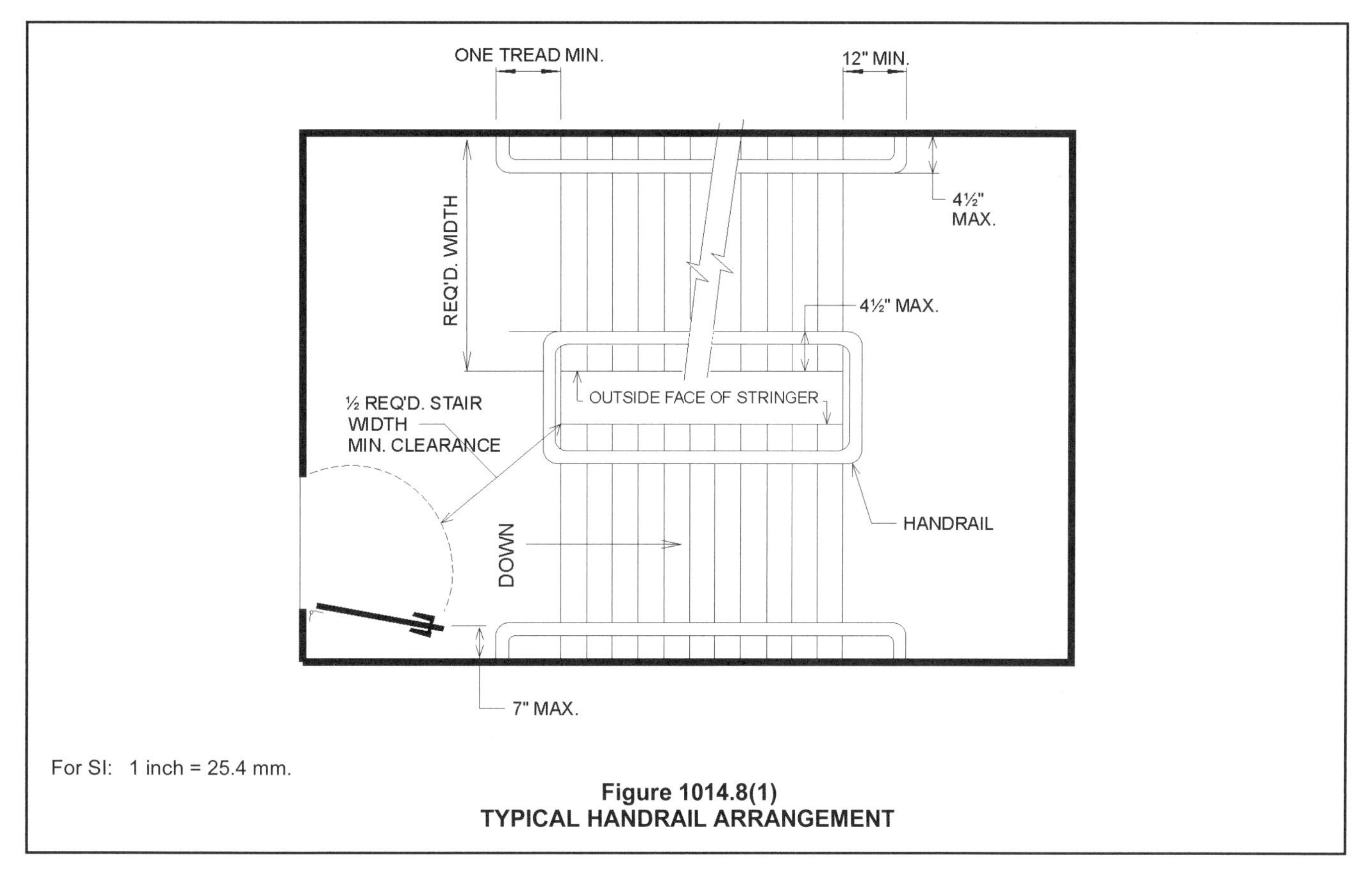

For SI: 1 inch = 25.4 mm.

Figure 1014.8(1)
TYPICAL HANDRAIL ARRANGEMENT

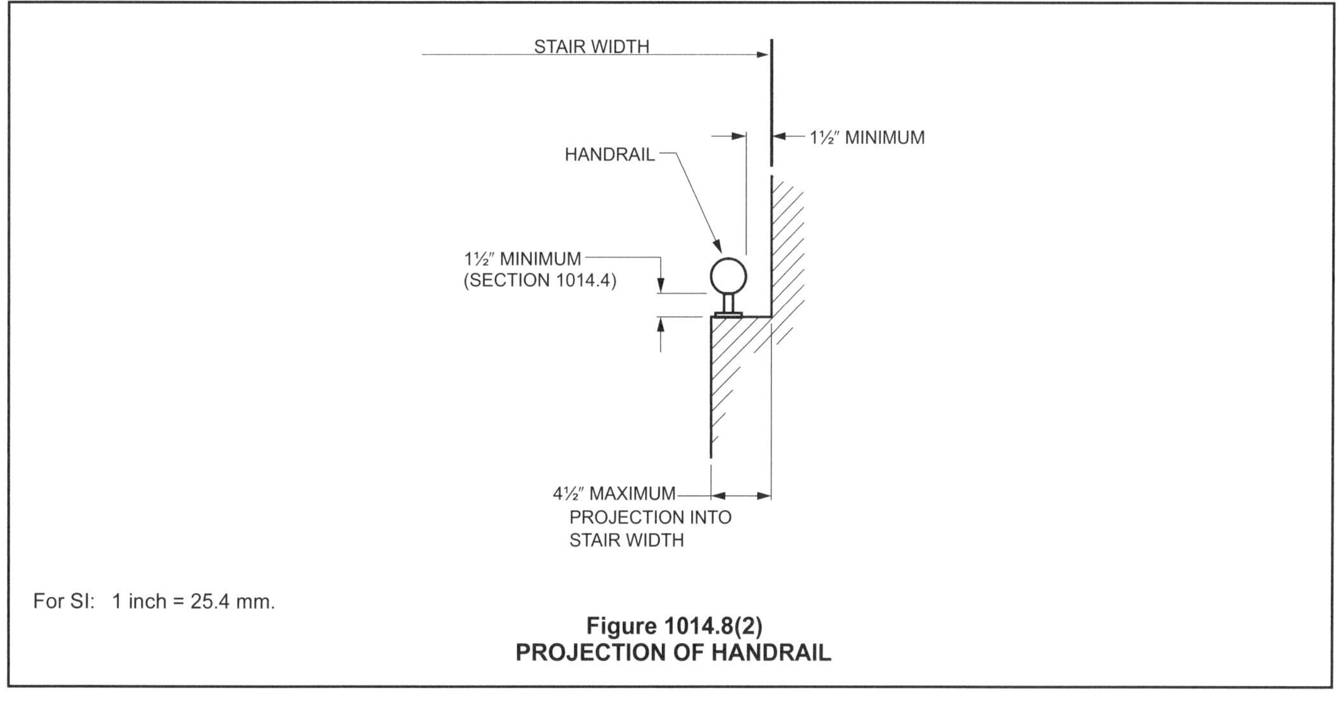

For SI: 1 inch = 25.4 mm.

Figure 1014.8(2)
PROJECTION OF HANDRAIL

SECTION 1015
GUARDS

1015.1 General. *Guards* shall comply with the provisions of Sections 1015.2 through 1015.7. Operable windows with sills located more than 72 inches (1829 mm) above finished grade or other surface below shall comply with Section 1015.8.

❖ Guards required along dropoffs must comply with the provisions for height, strength and opening limitations. Special provisions are provided for unique locations, such as screened-in porches, around mechanical equipment on platforms or roofs and at hatch openings.

Where there are operable windows on upper floors, there is a concern over the possibility of a child falling; thus, the placement of this requirement is in the guard section. Such windows must comply with Section 1015.8.

1015.2 Where required. *Guards* shall be located along open-sided walking surfaces, including *mezzanines*, *equipment platforms*, *aisles*, *stairs*, *ramps* and landings that are located more than 30 inches (762 mm) measured vertically to the floor or grade below at any point within 36 inches (914 mm) horizontally to the edge of the open side. *Guards* shall be adequate in strength and attachment in accordance with Section 1607.8.

Exception: *Guards* are not required for the following locations:

1. On the loading side of loading docks or piers.
2. On the audience side of *stages* and raised *platforms*, including *stairs* leading up to the *stage* and raised *platforms*.
3. On raised *stage* and *platform* floor areas, such as runways, *ramps* and side *stages* used for entertainment or presentations.
4. At vertical openings in the performance area of *stages* and *platforms*.
5. At elevated walking surfaces appurtenant to *stages* and *platforms* for access to and utilization of special lighting or equipment.
6. Along vehicle service pits not accessible to the public.
7. In assembly seating areas at cross aisles in accordance with Section 1029.16.2.

❖ Where one or more sides of a walking surface are open to the floor level or grade below, a guard system must be provided to minimize the possibility of occupants accidentally falling to the surface below. A guard is required only where the difference in elevation between the higher walking surface and the surface below is greater than 30 inches (762 mm). When the ground slopes away from the edge, the vertical distance from the walking surface to the grade or floor below must also be more than 30 inches (762 mm) on the lowest point within a 36-inch (914 mm) radius measured horizontally from the edge of the open-sided walking surface (see Commentary Figure 1015.2).

The loads for guard design are addressed in Section 1607 and are typically 50 plf (222 N) along the top with a 200-pound (90.72 kg) concentrated force. If glazing is used as part of a guard system, or windows are located adjacent to stairways or ramps, the guard must also comply with Section 1015.2.1 (see Sections 1015.2.1, 1607.8 and 2407).

Most of the exceptions identify situations when guards are not practical, such as along loading docks, stages and their approaches, and vehicle service pits. Exception 7 references assembly spaces where a lower guard is permitted or the alternative of seat backs where the dropoff is adjacent to a cross aisle.

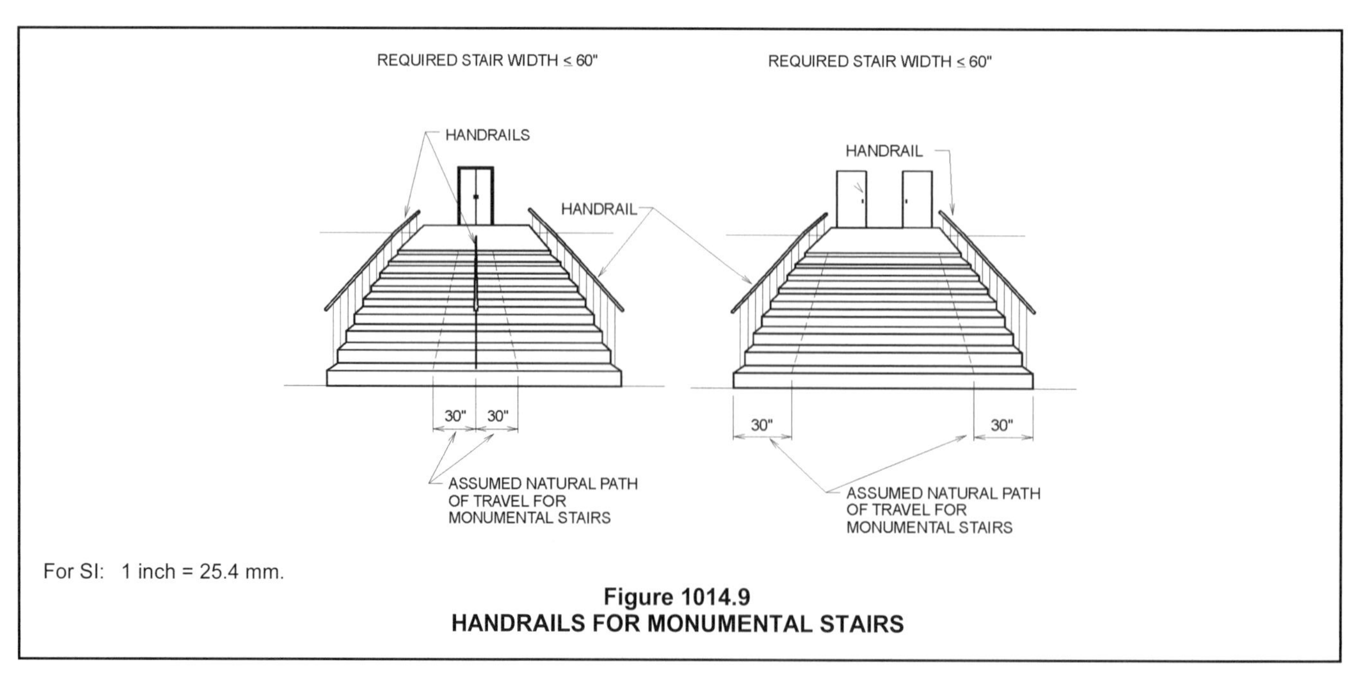

Figure 1014.9
HANDRAILS FOR MONUMENTAL STAIRS

1015.2.1 Glazing. Where glass is used to provide a *guard* or as a portion of the *guard* system, the *guard* shall comply with Section 2407. Where the glazing provided does not meet the strength and attachment requirements of Section 1607.8, complying *guards* shall be located along glazed sides of open-sided walking surfaces.

❖ Glazing in guards may be infill or structural. The loads for guard design in Section 1015.2, which references Section 1607, are typically 50 plf (222 N) along the top with a 200-pound (90.72 kg) concentrated force. Two different situations are addressed with glazing: where glazing is installed in a guard on the side of a stairway, ramp or landing; or when a stairway, ramp or landing is immediately adjacent to a window where the glazing has not been designed to resist the forces from a fall (see Sections 1607.8 and 2407 and Commentary Figure 1015.2.1).

1015.3 Height. Required *guards* shall be not less than 42 inches (1067 mm) high, measured vertically as follows:

1. From the adjacent walking surfaces.
2. On *stairways* and stepped *aisles*, from the line connecting the leading edges of the tread *nosings*.
3. On *ramps* and ramped *aisles*, from the *ramp* surface at the *guard*.

Exceptions:

1. For occupancies in Group R-3 not more than three stories above grade in height and within individual *dwelling units* in occupancies in Group R-2 not more than three stories above grade in height with separate *means of egress*, required *guards* shall be not less than 36 inches (914 mm) in height measured vertically above the adjacent walking surfaces or adjacent *fixed seating*.
2. For occupancies in Group R-3, and within individual *dwelling units* in occupancies in Group R-2, *guards*

Figure 1015.2.1
GUARD SYSTEM WITH GLAZING

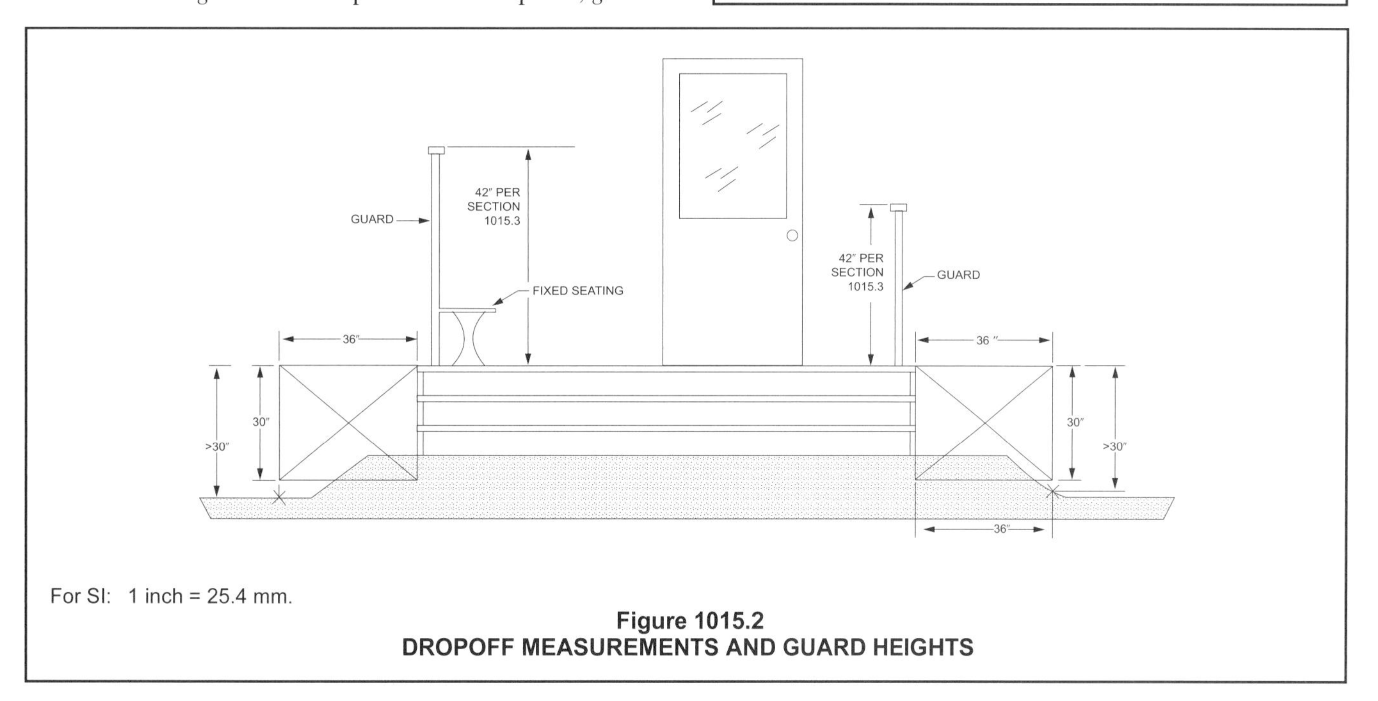

For SI: 1 inch = 25.4 mm.

Figure 1015.2
DROPOFF MEASUREMENTS AND GUARD HEIGHTS

on the open sides of *stairs* shall have a height not less than 34 inches (864 mm) measured vertically from a line connecting the leading edges of the treads.

3. For occupancies in Group R-3, and within individual *dwelling units* in occupancies in Group R-2, where the top of the *guard* also serves as a *handrail* on the open sides of *stairs*, the top of the *guard* shall be not less than 34 inches (864 mm) and not more than 38 inches (965 mm) measured vertically from a line connecting the leading edges of the treads.
4. The *guard* height in assembly seating areas shall comply with Section 1029.16 as applicable.
5. Along *alternating tread devices* and ship's ladders, *guards* where the top rail also serves as a *handrail* shall have height not less than 30 inches (762 mm) and not more than 34 inches (864 mm), measured vertically from the leading edge of the device tread *nosing*.

❖ Guards must not be less than 42 inches (1067 mm) in height as measured vertically from the top of the guard down to the sloped line connecting the leading edge of the tread along stairways or stepped aisles or to an adjacent walking surface for floors, ramps and ramped aisles [see Commentary Figures 1015.2, 1015.4(1) and (2) and 1012.8(4)]. Experience has shown that 42 inches (1067 mm) or more provides adequate height to minimize accidental falls in occupancies where crowding is more likely to occur. This puts the top of the guard above the center of gravity of the average adult. The height requirement is not intended to consider such items as planters or loose furniture next to the dropoff as walking surfaces. Because of safety concerns, the designer often chooses to install a barrier where there is a dropoff of less than 30 inches (762 mm). Decorative barriers may be utilized to support handrails or serve as part of the edge protection along a ramp. Where nonrequired guards/barriers are provided, provisions for guard height, openings and strength may be used for design, but are not required. It is common practice to follow the strength provisions, but not always the opening or height provisions for nonrequired guards.

Exception 1 measures the guard height differently. In Group R-2 and R-3 units that are three stories or less, where there is a fixed seat along the edge of deck or balcony, it is reasonable to consider the additional exposure for young children that may use the seat as standing or walking surfaces. In this situation, while the dropoff is measured from the floor, the guard height is measured from the top of the seat surface (see Commentary Figure 1015.3). Within dwelling units the incidence of the exposure to crowds and to egress from higher occupancy spaces is extremely limited, thus a minimum guard height of 36 inches (914 mm) is appropriate. Exception 1 is consistent with provisions in the 2012 IRC, however, the requirement to measure the guard height above a fixed seat has been deleted in the 2015 IRC.

Where the minimum guard height requirement is 42 inches (1067 mm), where both a guard and handrail are required, the handrail cannot be the top of the guard, but must be placed along the inside face of the guard unless specifically permitted in Exceptions 2 through 5.

Exceptions 2 and 3 are for nontransient residential occupancies and only address guard heights along the stairways, not at other dropoffs such as balconies or second-floor landings. For Group R-2, these guard height exceptions are only permitted for stairways within individual dwelling units, not common stairways within the building. The handrail provisions

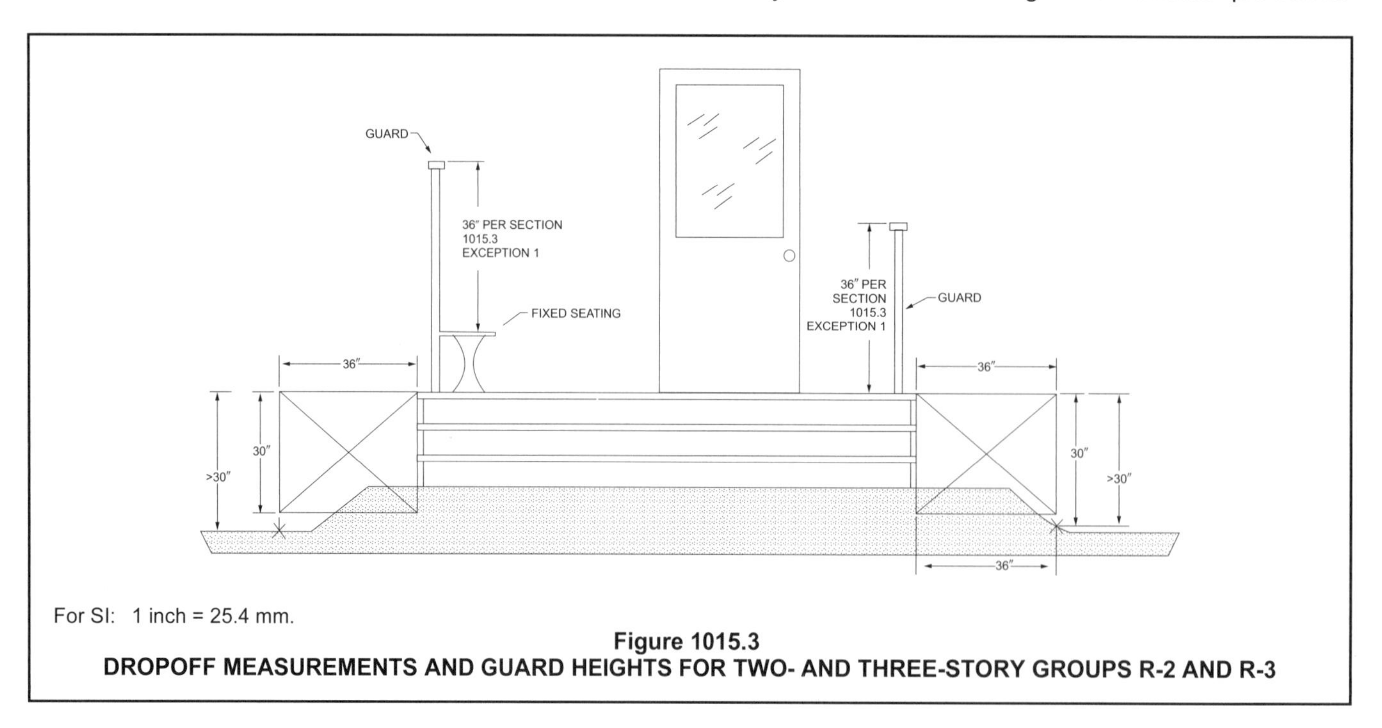

For SI: 1 inch = 25.4 mm.

Figure 1015.3
DROPOFF MEASUREMENTS AND GUARD HEIGHTS FOR TWO- AND THREE-STORY GROUPS R-2 AND R-3

allow some residential stairways to only have one handrail (see Section 1011.11). Exceptions 2 and 3 allow for a reduced guard height when the guard is also used as a handrail and when it just serves the purpose of a guard along a stairway. The reduced allowable guard height along stairways is consistent with current construction practice. Unless a dwelling could comply with Exception 1, the guard height along other dropoffs would have to be 42 inches (1067 mm) measured from the floor.

Exception 4 references the lower guards permitted at limited locations where a line of sight for assembly spaces is part of the consideration. There are also allowances in Section 1029.16 that would allow for the top of the guard to serve as a handrail.

Exception 5 permits a reduction in guard heights based on the limited used and unique design considerations for alternating tread devices and ship's ladders (see Sections 1011.14 and 1011.15).

1015.4 Opening limitations. Required *guards* shall not have openings that allow passage of a sphere 4 inches (102 mm) in diameter from the walking surface to the required *guard* height.

Exceptions:

1. From a height of 36 inches (914 mm) to 42 inches (1067 mm), *guards* shall not have openings that allow passage of a sphere $4^3/_8$ inches (111 mm) in diameter.
2. The triangular openings at the open sides of a *stair*, formed by the riser, tread and bottom rail shall not allow passage of a sphere 6 inches (152 mm) in diameter.
3. At elevated walking surfaces for access to and use of electrical, mechanical or plumbing systems or equipment, *guards* shall not have openings that allow passage of a sphere 21 inches (533 mm) in diameter.
4. In areas that are not open to the public within occupancies in Group I-3, F, H or S, and for *alternating tread devices* and ship's ladders, *guards* shall not have openings that allow passage of a sphere 21 inches (533 mm) in diameter.
5. In assembly seating areas, *guards* required at the end of aisles in accordance with Section 1029.16.4 shall not have openings that allow passage of a sphere 4 inches (102 mm) in diameter up to a height of 26 inches (660 mm). From a height of 26 inches (660 mm) to 42 inches (1067 mm) above the adjacent walking surfaces, *guards* shall not have openings that allow passage of a sphere 8 inches (203 mm) in diameter.
6. Within individual *dwelling units* and *sleeping units* in Group R-2 and R-3 occupancies, *guards* on the open sides of *stairs* shall not have openings that allow passage of a sphere $4^3/_8$ (111 mm) inches in diameter.

❖ The opening limitations in a guard are based on anthropometric research that indicates children in the 99th percentile who have developed to the point of being able to crawl will have chest depth and head size of at least $4^3/_4$ inches. Both the 4-inch (102 mm) and the $4^3/_8$-inch (111 mm) sphere rules are intended to provide an additional margin of safety. Note that the opening limitations are stated as preventing the passage of a sphere, requiring the openings to be smaller than the dimension stated.

Exception 1 allows the $4^3/_8$-inch (203 mm) opening limitation at heights where falling through the guard is not a risk for children of early development who cannot access the opening [i.e., above 36 inches (914 mm) in height] [see Commentary Figure 1015.4(1)].

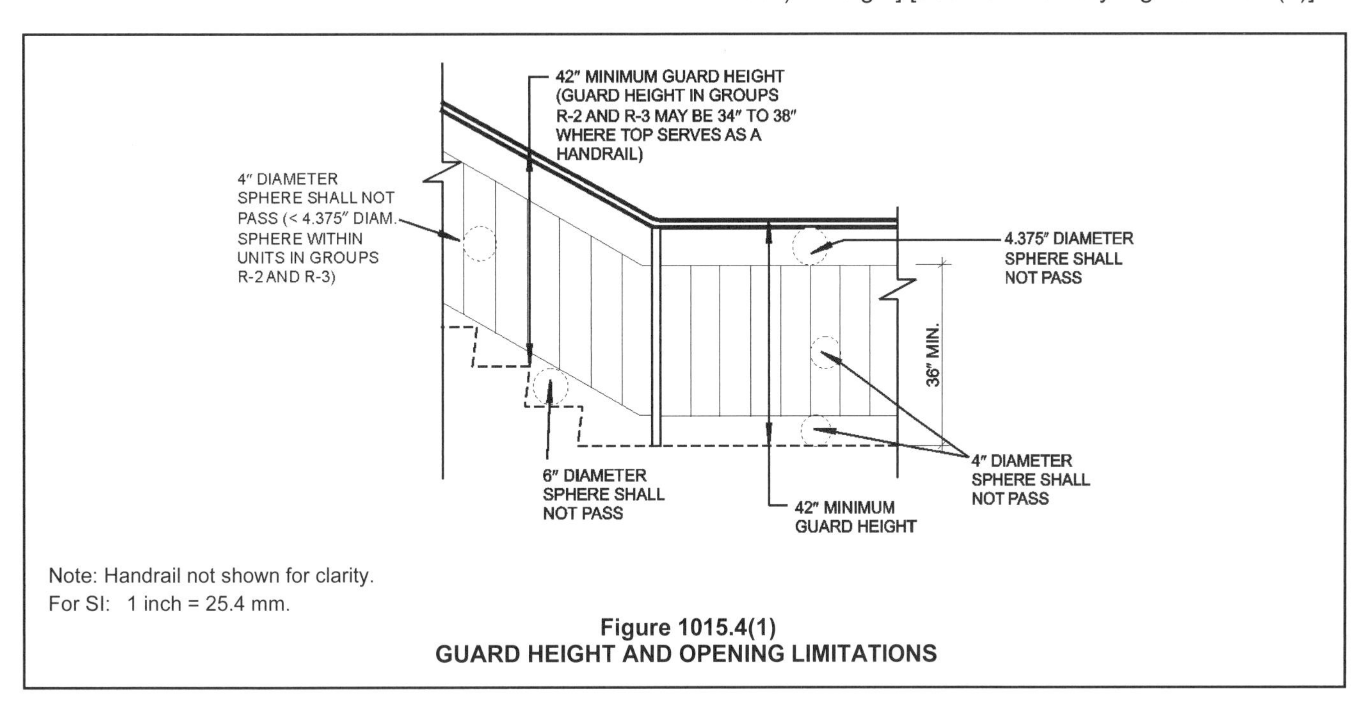

Figure 1015.4(1)
GUARD HEIGHT AND OPENING LIMITATIONS

Exception 2 allows a 6-inch (152 mm) opening limitation at openings formed by the riser, tread and bottom rail of guards at the open side of a stairway. The geometry of these openings is such that the entire body cannot pass through the triangular opening, and further limiting such openings is impractical to achieve with a sloped bottom member in the guard without intersecting the tread nosings [see Commentary Figure 1015.4(1)].

Exceptions 3 and 4 address areas where the presence of small children is unlikely and often prohibited. Guards along walkways leading to electrical, mechanical and plumbing systems or equipment and in occupancies in Groups I-3, F, H and S may be constructed in such a way that a sphere 21 inches (533 mm) in diameter will not pass through any of the openings [see Commentary Figure 1015.4(2)]. This requirement allows the use of one horizontal intermediate member with the standard top-of-guard height of 42 inches (1067 mm).

Exception 5, for the guard infill near the top of the aisle guard in assembly seating areas, is provided to reduce sightline problems in limited locations (see Section 1029.16.4).

Exception 6 recognizes a standard construction practice within residential units. In practicality this allows a stairway with the 7-inch riser height/11-inch tread depth (178 mm riser height/279 mm tread depth) step geometry to have two common $1^{1}/_{4}$-inch (32 mm) balusters per stair tread instead of three. Where the $7^{3}/_{4}$-inch riser height/10-inch tread depth (197 mm riser height/254 mm tread depth) step geometry (see Section 1011.5.2, Exception 3) is utilized, the two balusters would meet the 4-inch (102 mm) opening limitation and most profiled or turned balusters would meet the $4^{3}/_{8}$-inch (111 mm) sphere rule.

21″ SPHERE CANNOT PASS THROUGH

42″ MIN. GUARD HEIGHT

APPLICABLE TO GROUPS I-3, F, H AND S NOT OPEN TO THE PUBLIC, ACCESS TO EQUIPMENT, ALTERNATING TREAD DEVICES AND SHIP'S LADDERS

For SI: 1 inch = 25.4 mm.

Figure 1015.4(2)
GUARD OPENING LIMITATIONS, SECTION 1015.4, EXCEPTIONS 3 AND 4

The provisions of guards are to minimize accidental falls through or over a guard. Opening limitations do not prohibit the use of horizontal members or ornamentation infill as guard components. Research has shown that no practical design for guard infill, including solid panels, can prevent climbing; but good design practices can greatly reduce the opportunity for small children to "climb" the guard [see Commentary Figure 1015.4(3)]. In this example, the handrail stops a child from climbing the guard.

1015.5 Screen porches. Porches and decks that are enclosed with insect screening shall be provided with *guards* where the walking surface is located more than 30 inches (762 mm) above the floor or grade below.

❖ Insect screening located on the open sides of porches and decks does not provide an adequate barrier to reasonably protect an occupant from falling to the surface below. Guards are required on the open sides of porches and decks where the floor is located more than 30 inches (762 mm) above the surface below. The guards must comply with all of the provisions of Section 1015.

1015.6 Mechanical equipment, systems and devices. *Guards* shall be provided where various components that require service are located within 10 feet (3048 mm) of a roof edge or open side of a walking surface and such edge or open side is located more than 30 inches (762 mm) above the floor, roof or grade below. The *guard* shall extend not less than 30 inches (762 mm) beyond each end of such components. The *guard* shall be constructed so as to prevent the passage of a sphere 21 inches (533 mm) in diameter.

Exception: *Guards* are not required where permanent fall arrest/restraint anchorage connector devices that comply with ANSI/ASSE Z 359.1 are affixed for use during the

Figure 1015.4(3)
EXAMPLE OF GUARD WITH HORIZONTAL MEMBERS

entire roof covering lifetime. The devices shall be reevaluated for possible replacement when the entire roof covering is replaced. The devices shall be placed not more than 10 feet (3048 mm) on center along hip and ridge lines and placed not less than 10 feet (3048 mm) from the roof edge or open side of the walking surface.

❖ The purpose of this requirement is to protect workers from falls off of roofs or from open-sided walking surfaces when doing maintenance work on equipment. The guard opening is allowed to be up to 21 inches (533 mm) since children are not likely to be in such areas. Either the equipment should be located so that it is more than 10 feet from the roof edge, or a guard or raised parapet must be provided to prevent falls. The guard also has to extend at least 30 inches (762 mm) in all directions past the corners of the equipment (see Commentary Figure 1015.7).

The exception for the guards is an alternative of a personal fall arrest system with anchorage points, more commonly called tie down points for restraining harnesses. The code does not specify who has to provide the equipment that fits the maintenance crew and attaches to the system. If this option is chosen, when the roof covering is replaced, the system must also be reevaluated and possibly replaced. This system is more commonly used on sloped roofs while guards are more often used on flat roofs.

1015.7 Roof access. *Guards* shall be provided where the roof hatch opening is located within 10 feet (3048 mm) of a roof edge or open side of a walking surface and such edge or open side is located more than 30 inches (762 mm) above the floor, roof or grade below. The *guard* shall be constructed so as to prevent the passage of a sphere 21 inches (533 mm) in diameter.

Exception: *Guards* are not required where permanent fall arrest/restraint anchorage connector devices that comply with ANSI/ASSE Z 359.1 are affixed for use during the entire roof covering lifetime. The devices shall be reevaluated for possible replacement when the entire roof covering is replaced. The devices shall be placed not more than 10 feet (3048 mm) on center along hip and ridge lines and placed not less than 10 feet (3048 mm) from the roof edge or open side of the walking surface.

❖ The code already requires guards around equipment on the roof; this section is intended to provide the same level of safety at the hatch opening service personnel use to access the roof (see Commentary Figure 1015.7 and the commentary to Section 1015.6). While not specifically indicated for roof hatches, Section 1015.6 requires the guard to extend at least 30 inches (752 mm) past the corners of the equipment.

1015.8 Window openings. Windows in Group R-2 and R-3 buildings including *dwelling units*, where the top of the sill of an operable window opening is located less than 36 inches above the finished floor and more than 72 inches (1829 mm) above the finished grade or other surface below on the exterior of the building, shall comply with one of the following:

1. Operable windows where the top of the sill of the opening is located more than 75 feet (22 860 mm) above the finished grade or other surface below and that are provided with window fall prevention devices that comply with ASTM F2006.

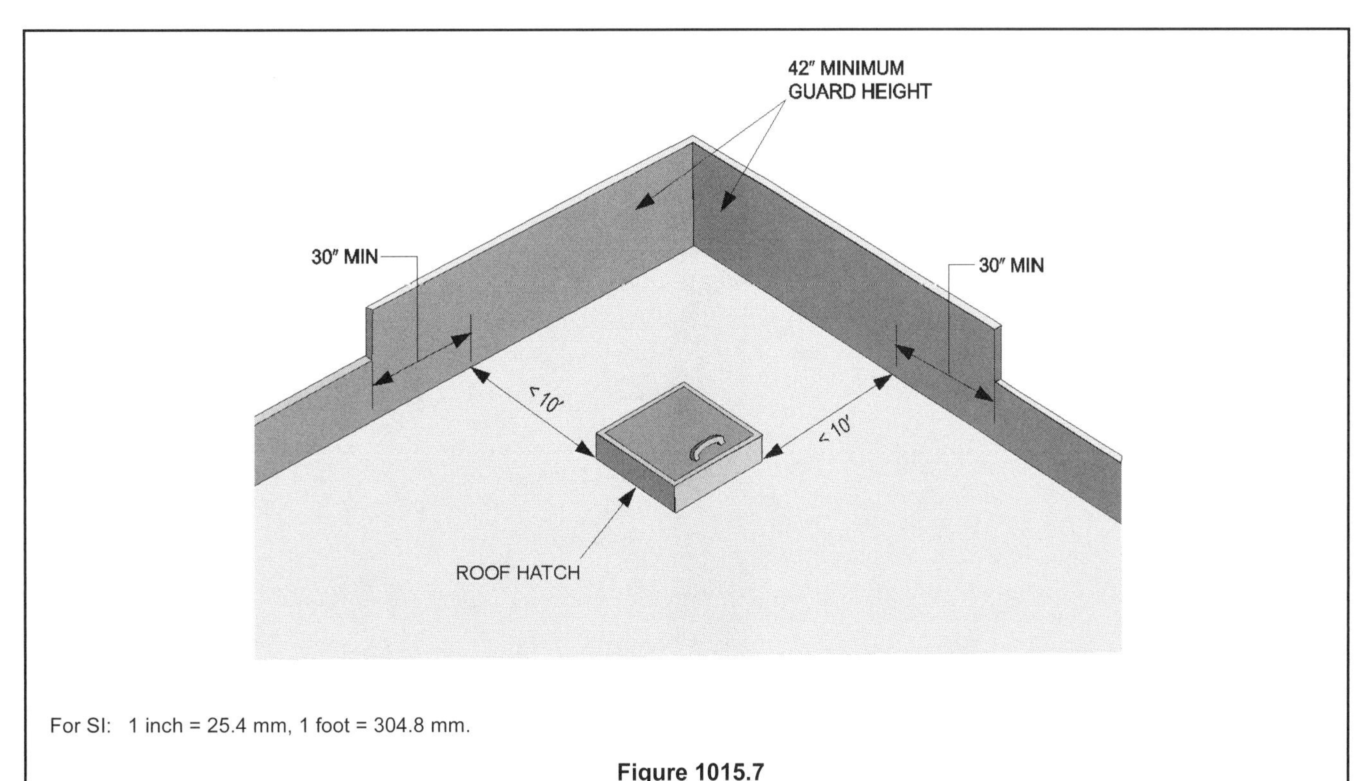

For SI: 1 inch = 25.4 mm, 1 foot = 304.8 mm.

Figure 1015.7
PROTECTION AT ROOF-HATCH OPENING

2. Operable windows where the openings will not allow a 4-inch-diameter (102 mm) sphere to pass through the opening when the window is in its largest opened position.

3. Operable windows where the openings are provided with window fall prevention devices that comply with ASTM F2090.

4. Operable windows that are provided with window opening control devices that comply with Section 1015.8.1.

❖ The window limitations specified here are intended for Group R-2 and R-3 units. These facilities have the highest potential for infants and toddlers being present for an extended period of time. The requirement is intended to provide a level of protection to children and to help limit the chances of them falling through window openings. In most cases, these provisions are not applicable to first-floor windows. Typically, the 72 inches (1829 mm) to finished grade would make these provisions applicable for windows starting at the second floor. For windows in bedrooms that may also be required to serve as emergency escape and rescue openings, see Section 1030.

There are basically five options offered:

One option is to locate the window so that any opening is at least 36 inches (915 mm) above the floor. By raising the lowest operable portion of a window to 36 inches (915 mm) or more, the sill height is above the center of gravity of smaller children. The National Ornamental & Miscellaneous Metals Association (NOMMA) commissioned a paper on child safety related to falls. The report indicates that the standing center of gravity of children aged 2 to 3.5 years is 24.1 inches (612 mm) [50th percentile is 22.2 inches (564 mm)] and of children aged 3.5 to 4.5 years is 25.2 inches (640 mm) [50th percentile is 23.6 inches (599 mm)]. The 36-inch (915 mm) sill height was chosen to reduce the ability of a child to climb onto the sill, enabling the fall through the opening. Windows that are also to serve as emergency escape windows must also comply with Section 10309.3 sill height requirements for 44 inches (1118 mm) maximum. Note that Section R612.2 of the IRC requires a minimum sill height of 24 inches (610 mm).

A second option (Item 1) would be for windows with an opening lower than 36 inches (915 mm) to limit the opening to 4 inches (102 mm) maximum. This opening size is consistent with the guard opening provisions (see Section 1015.4). Many awning or hopper-type casement windows have control arms that limit the opening width.

The third option offered (Item 2) allows for fixed fall-prevention devices to be installed in accordance with ASTM F2006. The limitation for window sills over 75 feet (22 860 mm) is for consistency with Section 1.3 of ASTM F2006, which states: "This safety specification applies only to devices intended to be applied to windows installed at height of more than 75 feet above the ground level in multiple family dwelling buildings. This safety specification is not intended to apply to windows below 75 feet (22 860 mm) because all windows below 75 feet (22 860 mm) that are operable could be used as a possible secondary means of escape." Since these devices will always prevent the opening of the window, they cannot be used for windows that are required to also serve as emergency escape openings.

ASTM F2090 includes window fall prevention devices (Item 3) and window opening control devices (Item 4) (see Section 1015.8.1). Window fall-prevention devices (such as a window guard) must be removable from the interior of the building so the window can be used for emergency escape. Window opening-control devices allow the window to be opened beyond 4 inches, so that a window can be used for emergency escape. This standard is specifically written for window openings within 75 feet (22 860 mm) of grade and specifically allows for windows to be used for emergency escape and rescue. Both the code and IRC reference ASTM F2090, *Specification for Window Fall Prevention Devices with Emergency Escape (Egress Release Mechanisms)*. This standard was updated in 2008 to address window opening control devices. Opening control devices allow for normal operation to result in a 4-inch (102 mm) maximum opening (Section 1015.8 and Section R612.4.1 of the IRC). This control device can be released from the inside to allow the window to be fully opened in order to comply with the emergency escape provisions in both the code (Section 1030.2) and IRC (Section R310.1.1).

Criteria have also been added to IEBC to address window opening controls in existing buildings.

1015.8.1 Window opening control devices. Window opening control devices shall comply with ASTM F2090. The window opening control device, after operation to release the control device allowing the window to fully open, shall not reduce the minimum net clear opening area of the window unit to less than the area required by Section 1030.2.

❖ See the commentary to Section 1015.8.

SECTION 1016
EXIT ACCESS

1016.1 General. The *exit access* shall comply with the applicable provisions of Sections 1003 through 1015. *Exit access* arrangement shall comply with Sections 1016 through 1021.

❖ Sections 1016 through 1021 include the design requirements for exit access and exit access components. The general requirements that also apply to the exit access are in Sections 1003 through 1015.

The following sections are included under exit access:

- Section 1016 deals with egress through intervening spaces, as well as travel and separation of the common path of travel to the exit.

- Section 1017 lists the total exit access travel distance from an occupied space to an exit. This distance includes the common path of travel addressed in Sections 1006.2.1 and 1006.3.2.
- Section 1018 takes a look at requirements for aisles and aisle accessways for occupancies other than assembly spaces. Aisle and aisle accessways in spaces used for assembly purposes are specifically addressed in Section 1029.
- Section 1019 provides criteria for when an open stairway between floors can serve as a required exit access stairway.
- Section 1020 deals with another type of confined part of the exit access path: corridors.
- Section 1021 addresses egress balconies, where the path of travel to the exterior exit stairway is partially open to the exterior.

1016.2 Egress through intervening spaces. Egress through intervening spaces shall comply with this section.

1. *Exit access* through an enclosed elevator lobby is permitted. Access to not less than one of the required *exits* shall be provided without travel through the enclosed elevator lobbies required by Section 3006. Where the path of exit access travel passes through an enclosed elevator lobby, the level of protection required for the enclosed elevator lobby is not required to be extended to the *exit* unless direct access to an *exit* is required by other sections of this code.
2. Egress from a room or space shall not pass through adjoining or intervening rooms or areas, except where such adjoining rooms or areas and the area served are accessory to one or the other, are not a Group H occupancy and provide a discernible path of egress travel to an *exit*.

   Exception: *Means of egress* are not prohibited through adjoining or intervening rooms or spaces in a Group H, S or F occupancy where the adjoining or intervening rooms or spaces are the same or a lesser hazard occupancy group.
3. An *exit access* shall not pass through a room that can be locked to prevent egress.
4. *Means of egress* from *dwelling units* or sleeping areas shall not lead through other sleeping areas, toilet rooms or bathrooms.
5. Egress shall not pass through kitchens, storage rooms, closets or spaces used for similar purposes.

   Exceptions:

   1. *Means of egress* are not prohibited through a kitchen area serving adjoining rooms constituting part of the same *dwelling unit* or *sleeping unit.*
   2. *Means of egress* are not prohibited through stockrooms in Group M occupancies where all of the following are met:

      2.1. The stock is of the same hazard classification as that found in the main retail area.

      2.2. Not more than 50 percent of the *exit access* is through the stockroom.

      2.3. The stockroom is not subject to locking from the egress side.

      2.4. There is a demarcated, minimum 44-inch-wide (1118 mm) *aisle* defined by full- or partial-height fixed walls or similar construction that will maintain the required width and lead directly from the retail area to the *exit* without obstructions.

❖ This section allows adjoining spaces to be considered a part of the room or space from which egress originates, provided that there are reasonable assurances that the continuous egress path will always be available. The code does not limit the number of intervening or adjoining rooms through which egress can be made, provided that all other code requirements (e.g., travel distance, number of doorways, etc.) are met. An exit access route, for example, may be laid out such that an occupant leaves a room or space, passes through an adjoining space, enters an exit access corridor, passes through another room and, finally, into an exit [see Commentary Figure 1016.2(2)], as long as all other code requirements are satisfied.

The intent of Item 1 is to correlate the provision for corridor continuity (Section 1020.6) and elevator lobbies. This is especially important since elevator lobbies for fire service access elevators require direction connection to a stairway. Therefore, occupants may need to egress through an elevator lobby to get to the exit stairway. In a two-exit building, only one exit can be through an elevator lobby, but at the same time, the elevator lobby will not be considered an intervening room [see Commentary Figure 1020.6(3)]. This also clarifies that the protection requirements for corridors would not be put on top of protection requirements for the elevator lobbies.

The intent of Item 2 is not that the accessory space be limited to the 10-percent area in Section 508.2.3, but that the spaces be interrelated so that doors between the spaces will not risk being blocked or locked. For example, a conference room and managers' offices could exit through the secretary's office to reach the exit access corridor; or several office spaces could exit through a common reception/lobby area. Requiring occupants to egress from an area and pass through an adjoining Group H that can be characterized by rapid fire buildup places them in an unreasonable risk situation [see Commentary Figure 1016.2(1)]; therefore, this illustrated egress path would be prohibited. As an exception to Item 2, in facilities that may contain a Group H area, buildings of Group H, S or F can exit through adjoining rooms or spaces that have the same or lesser hazard. For example, a person exiting from a Group H storage room (see Section 415) could egress either through a similar Group H storage area or through the factory to get to an exit, but the person in the factory could not egress through the Group H storage rooms to get to the outside.

As expressed in Item 3, a common code enforcement problem is a locked door in the egress path. Twenty-five workers perished in September 1991 when they were trapped inside the Imperial Food Processing Plant in Hamlet, North Carolina, in part because of locked exit doors. As long as the egress door is readily openable in the direction of egress travel without the use of keys, special knowledge or effort (see Section 1010.1.9.5), the occupants can move unimpeded away from a fire emergency. Relying on an egress path through an adjacent dwelling unit to be available at all times is not a reasonable expectation. Egress through an adjacent business tenant space can be unreasonable given the security and privacy measures the adjacent tenant may take to secure such a space. However, egress through a reception area that serves a suite of offices of the same tenant is clearly accessible and is permitted.

Item 4 addresses concerns along the path of

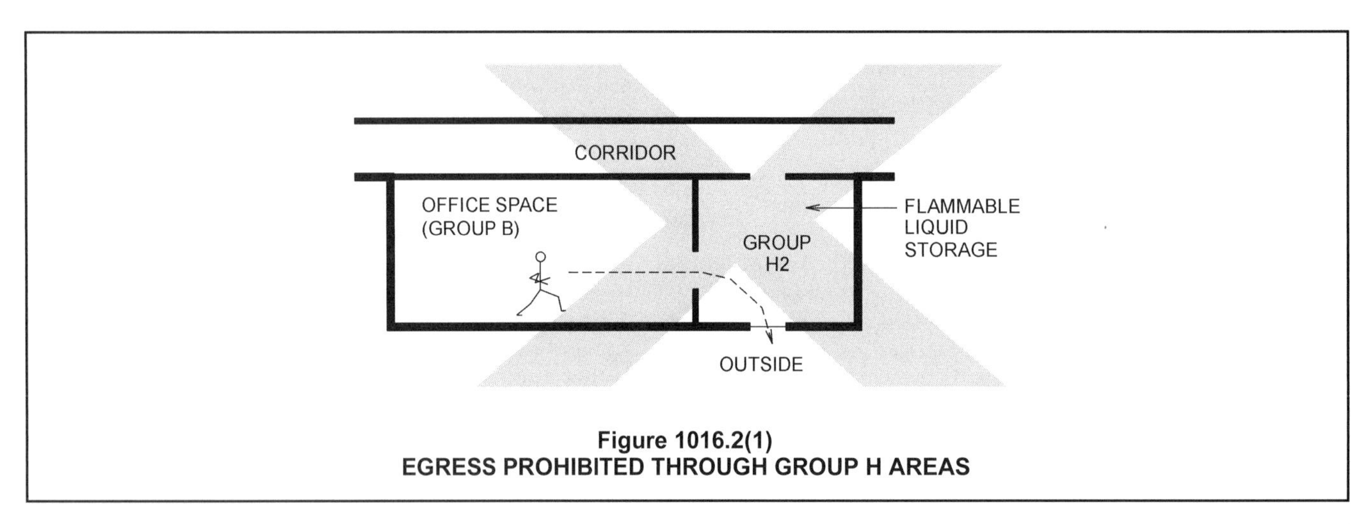

Figure 1016.2(1)
EGRESS PROHIBITED THROUGH GROUP H AREAS

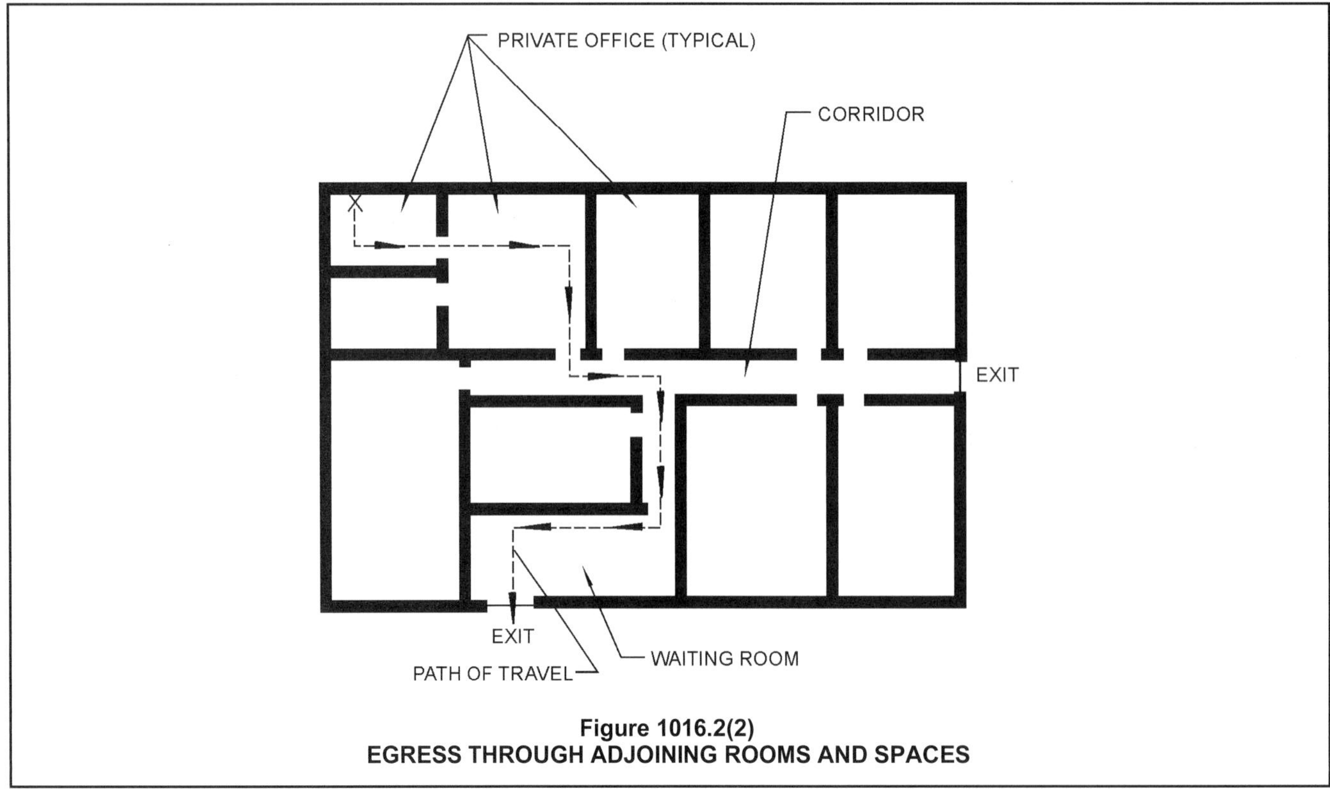

Figure 1016.2(2)
EGRESS THROUGH ADJOINING ROOMS AND SPACES

egress travel within individual dwelling or sleeping units. The concern once again is possible locking devices. Egress for one bedroom should not be through another bedroom or bathroom.

The concern in Item 5 is that kitchens, storage rooms and similar spaces may be subject to locking or blockage of the exit access path. This is not a general provision for all Group S occupancies; therefore, it is not the intent of this provision to address the situation of egress for offices through an associated warehouse space. Item 5, Exception 1, does not apply this same prohibition to areas within dwelling or sleeping units. However, for other spaces, a customer means of egress should not be through the working portions of a commercial kitchen in a restaurant or the stock storage area of a storage room in a mercantile occupancy. A dedicated path must be established through such space. The four items listed in Item 5, Exception 2, are intended to provide measurable criteria to increase the likelihood that the exit access path of travel would always be available and identifiable through the stock room of a store. It is not acceptable to just mark the path on the floor. Whatever defines the route must permanently establish the egress path in a manner to maintain the minimum required unobstructed width.

1016.2.1 Multiple tenants. Where more than one tenant occupies any one floor of a building or structure, each tenant space, *dwelling unit* and *sleeping unit* shall be provided with access to the required *exits* without passing through adjacent tenant spaces, *dwelling units* and *sleeping units*.

> **Exception:** The *means of egress* from a smaller tenant space shall not be prohibited from passing through a larger adjoining tenant space where such rooms or spaces of the smaller tenant occupy less than 10 percent of the area of the larger tenant space through which they pass; are the same or similar occupancy group; a discernible path of egress travel to an *exit* is provided; and the *means of egress* into the adjoining space is not subject to locking from the egress side. A required *means of egress* serving the larger tenant space shall not pass through the smaller tenant space or spaces.

❖ Where a floor is occupied by multiple tenants, each tenant must be provided with full and direct access to the required exits serving that floor without passing through another tenant space. Tenants typically lock the doors to their spaces for privacy and security. Should an egress door that is shared by both tenants be locked, occupants in one of the spaces could be trapped and unable to reach a secondary exit. Therefore, an egress layout where occupants from one tenant space travel through another tenant space to gain access to one of the required exits from that floor is prohibited.

This limitation is so that occupants from all tenant spaces will have unrestricted access to the required egress elements while maintaining the security and privacy of the individual tenants. This limitation is based on one of the fundamental principles of egress: to provide a means of egress where all components are capable of being used by the occupants without keys, tools, special knowledge or special effort (see Section 1010.1.9.5).

A common practice is to have a bank or small restaurant located within a large grocery store or department store. These can be separate tenants. In these situations, the small tenants are not open when the main store is closed. The intent of the exception is to allow those small tenants to egress through the large tenant. Since there may be times when the larger tenant is open and the smaller is closed (e.g., bank holidays), the larger tenant cannot exit through the smaller tenant.

SECTION 1017 EXIT ACCESS TRAVEL DISTANCE

1017.1 General. Travel distance within the *exit access* portion of the *means of egress* system shall be in accordance with this section.

❖ "Exit access" is defined as "that portion of a means of egress system that leads from any occupied portion of a building or structure to an exit" (see the commentary for the definition in Chapter 2). Exit access includes rooms, spaces, aisles and corridors that an occupant would travel along to get to an exit. This can also include stairways and ramps between levels, or between stories where permitted by Section 1019. Doors and doorways along this route are exit access doorways, but may sometimes be called "exit doors." True exits for this exit access travel can be: 1. An exterior exit door at grade; 2. The door to an enclosure for an interior exit stairway, ramp or exit passageway; 3. The exit door leading to an exterior exit door or ramp; or 4. A door leading through a horizontal exit. How exit access travel distance is measured is one of the key differences between interior exit access stairways/ramps (see Section 1017.3.1) and interior exit stairways/ramps (see Section 1017.3).

It is important to understand the relationship between the common path of travel limitations of Sections 1006.2.1 and 1006.3.2 and the exit access travel distance limitations of this section. Measurements start at the same location, i.e., the most remote location in any occupied space. Both are measured in the exit access portion of the means of egress system. The common path of travel is measured to the point where the occupant has two distinct paths of travel, which will lead to two distinct exits. Travel distance is measured all the way until the exit is reached. The common path of travel measurement can end within a space or at a corridor where a single means of egress space (Section 1006.2.1) has its door to a corridor that provides access to two exits.

1017.2 Limitations. *Exit access* travel distance shall not exceed the values given in Table 1017.2.

❖ The table includes the travel distance measurements for buildings with or without sprinkler systems. "Not Permitted" is in support of the Chapter 9 requirements for all Group I and H occupancies to be sprinklered. While the other occupancies may also be required to be sprinklered, the exit access travel distance is indicated for existing buildings.

TABLE 1017.2. See next column.

❖ This table reflects the maximum distance a person is allowed to travel from any point in a building floor area to the nearest exit along a natural and unobstructed path. While quantitative determinations or formulas are not available to substantiate the tabular distances, empirical factors are utilized to make relative judgments as to reasonable limitations. Such considerations include the nature and fitness of the occupants; the typical configuration within the space; the level of fire hazard with respect to the specific uses of the facilities, including fire spread and the potential intensity of a fire. The inclusion of an automatic sprinkler system throughout the building can serve to control, confine or possibly eliminate the fire threat to the occupants so an increased travel distance is permitted. Increased travel distances are permitted when an automatic sprinkler system is installed in accordance with NFPA 13 or 13R.

When measuring travel distance, it is important to consider the natural path of travel [see Commentary Figure 1017.3(1)]. In many cases, the actual layout of furnishings and equipment is not known or is not identified on the plans submitted with the permit application. In such instances, it may be necessary to measure travel distance using the legs of a triangle instead of the hypotenuse [see Commentary Figure 1017.3(2)]. Since most people tend to migrate to more open spaces while egressing, measurement of the natural path of travel typically excludes floor areas within 1 foot (305 mm) of walls, corners, columns and other permanent construction. Where the travel path includes passage through a doorway, the natural route is generally measured through the centerline of the door openings.

The common path of travel addressed in Sections 1006.2.1 and 1006.3.2 is part of the overall exit access travel distance, with both starting at the same point. Common path of travel stops when the occupant has a choice of at least two exits, and overall travel distance stops when an occupant gets to the closest exit. The references in Note a are to sections where specific requirements are addressed—this can be an increase or decrease in travel distance.

Note a is a reference to other travel distance limitations in the code. Notes b and c are simply references to the allowed types of sprinkler system—NFPA 13 or 13R. Some travel distance increases are based on the type of sprinkler system provided. The table does not currently address a building with an NFPA 13D sprinkler system.

Note d addresses the sprinkler requirements in Group H. Group H occupancies are only required to be sprinklered within Group H and not throughout the building; therefore, Note d distinguishes this requirement from Notes b and c. The travel distance is the same if the path is through Group H or leaves Group H and continues through a nonsprinklered occupancy (see Section 1018.2, Item 2.)

TABLE 1017.2
EXIT ACCESS TRAVEL DISTANCE[a]

OCCUPANCY	WITHOUT SPRINKLER SYSTEM (feet)	WITH SPRINKLER SYSTEM (feet)
A, E, F-1, M, R, S-1	200	250[b]
I-1	Not Permitted	250[b]
B	200	300[c]
F-2, S-2, U	300	400[c]
H-1	Not Permitted	75[d]
H-2	Not Permitted	100[d]
H-3	Not Permitted	150[d]
H-4	Not Permitted	175[d]
H-5	Not Permitted	200[c]
I-2, I-3, I-4	Not Permitted	200[c]

For SI: 1 foot = 304.8 mm.

a. See the following sections for modifications to *exit access* travel distance requirements:

Section 402.8: For the distance limitation in malls.

Section 404.9: For the distance limitation through an atrium space.

Section 407.4: For the distance limitation in Group I-2.

Sections 408.6.1 and 408.8.1: For the distance limitations in Group I-3.

Section 411.4: For the distance limitation in special amusement buildings.

Section 412.7: For the distance limitations in aircraft manufacturing facilities.

Section 1006.2.2.2: For the distance limitation in refrigeration machinery rooms.

Section 1006.2.2.3: For the distance limitation in refrigerated rooms and spaces.

Section 1006.3.2: For buildings with one exit.

Section 1017.2.2: For increased distance limitation in Groups F-1 and S-1.

Section 1029.7: For increased limitation in assembly seating.

Section 3103.4: For temporary structures.

Section 3104.9: For pedestrian walkways.

b. Buildings equipped throughout with an *automatic sprinkler system* in accordance with Section 903.3.1.1 or 903.3.1.2. See Section 903 for occupancies where *automatic sprinkler systems* are permitted in accordance with Section 903.3.1.2.

c. Buildings equipped throughout with an *automatic sprinkler system* in accordance with Section 903.3.1.1.

d. Group H occupancies equipped throughout with an *automatic sprinkler system* in accordance with Section 903.2.5.1.

1017.2.1 Exterior egress balcony increase. *Exit access* travel distances specified in Table 1017.2 shall be increased up to an additional 100 feet (30 480 mm) provided the last portion of the *exit access* leading to the *exit* occurs on an exterior egress balcony constructed in accordance with Section 1021. The length of such balcony shall be not less than the amount of the increase taken.

❖ This section allows an additional travel distance on exterior egress balconies since smoke disperses rapidly. Note that the length of the increase is not to be more than the length of the exterior balcony. For example, if the length of the balcony is 75 feet (22 860 mm), the additional travel distance is limited to 75 feet (22 860 mm). In order for the increase to apply, the exterior balcony must be located at the end of the path of egress travel and not in some other portion of the egress path.

1017.2.2 Group F-1 and S-1 increase. The maximum *exit access* travel distance shall be 400 feet (122 m) in Group F-1 or S-1 occupancies where all of the following conditions are met:

1. The portion of the building classified as Group F-1 or S-1 is limited to one story in height.
2. The minimum height from the finished floor to the bottom of the ceiling or roof slab or deck is 24 feet (7315 mm).
3. The building is equipped throughout with an *automatic sprinkler system* in accordance with Section 903.3.1.1.

❖ This section provides the criteria for an increased exit access travel distance of 400 feet in Group F-1 and S-1 occupancies when three criteria are met: the S-1/F-1 area is one story, has a ceiling height of at least 24 feet, and is sprinklered.

The travel distance increase is only applicable to portions of the building that are one story in height. This is not intended to preclude a building with a one-story storage warehouse or factory area and a two-story office or a mezzanine from also utilizing this section. The section is written so that the one-story limitation is only applicable to the area where the 400-foot travel distance is utilized.

The 24 feet of clearance is based on the "Fire Modeling Analysis Report" by Aon Fire Protection Engineering. The ceiling height is used to provide a volume for the smoke to accumulate during the fire and provide time for egress, much like the concept used for smoke-protected seating. Control mode sprinklers were utilized in the fire modeling to demonstrate the more conservative approach.

The building is required to sprinklered in accordance with NFPA 13 requirements. While not required, ESFR or specialty sprinklers would be more effective.

1017.3 Measurement. *Exit access* travel distance shall be measured from the most remote point within a story along the natural and unobstructed path of horizontal and vertical egress travel to the entrance to an *exit*.

Exception: In *open parking garages*, *exit access* travel distance is permitted to be measured to the closest riser of an *exit access stairway* or the closest slope of an *exit access ramp*.

❖ The length of travel, as measured from the most remote point within a structure to an exit, is limited to restrict the amount of time that the occupant is exposed to a potential fire condition [see Commentary Figure 1017.3(1)]. The route must be assumed to be the natural path of travel without obstruction. This commonly results in a rectilinear path similar to what can be experienced in most occupancies, such as a schoolroom or an office with rows of desks [see Commentary Figure 1017.3(2)]. The "arc" method, using an "as the crow flies" linear measurement, must be used with caution, as it seldom represents typical floor design and room layout and, in most cases, would not be the natural, unobstructed path.

The travel distance is measured from every occupiable point on a floor to the closest exit. While each occupant may be required to have access to a second or third exit, the travel distance limitation is only applicable to the distance to the nearest exit. In effect, this means that the distance an occupant must travel to the second or third exit is not regulated.

For exit access travel distances that include vertical elements, such as stairways or ramps, see Section 1017.3.1. For outdoor assembly seating, see Section 1029.7.

The exception provides for a travel distance terminating at the top of an open exit access stairway in an open parking structure (see Section 1019.3, Exception 6). This is appropriate in view of the low hazard and minimal possible smoke accumulation in these facilities. While Section 1006.3 does say that an open exit access stairway can only be used for one story, travel distance typically does include the exit access stairway. With the exception here for travel distance, the intent is to allow open parking garages to have open exit access stairways all the way down and out of the building. The number of means of egress requirements are met when the exit access travel distance requirement is met on each floor.

The distance of travel within an exit enclosure (e.g., enclosed interior exit stairway or ramp or exit passageway) and in the exit discharge portion of the means of egress are also not regulated. Section 1006.3.2 permits certain buildings to be provided with a single exit. In instances where there is a single exit, travel distances less than those permitted in Table 1017.2 apply [see Tables 1006.3.2(1) and 1006.3.2(2)].

1017.3.1 Exit access stairways and ramps. Travel distance on *exit access stairways* or *ramps* shall be included in the *exit access* travel distance measurement. The measurement along *stairways* shall be made on a plane parallel and tangent to the *stair* tread *nosings* in the center of the *stair* and landings. The measurement along *ramps* shall be made on the walking surface in the center of the *ramp* and landings.

❖ Travel distance is measured along the exit access path. Exit access travel distance may include travel on an interior stairway or ramp if it is not enclosed and constructed to meet the definition of an exit (see Section 1023). An example of exit access stairways would be unenclosed exit access stairways from a mezzanine level, open exit access stairways from a second floor (see Section 1019.3, Exception 1), or steps along the path of travel in a split floor-level situation. An example of an open exit access ramp would be a ramp between levels or ramps leading from the upper levels in an open parking garage (see Section 1019.3, Exception 7). For the last example, Section 1016.3, Exception 1, allows for the exit access travel distance to be measured to the top of the ramp or stairway rather than down the ramp or stairway and to the exit.

When Section 1019.3 permits an exit access stairway to be unenclosed, the travel distance would also include travel down the open stairway and to an enclosed exit stairway, a horizontal exit or an exit door to the outside. An example of this would be an open exit stairway within an individual dwelling unit (see Section 1019.3, Exception 2) or an open exit stairway from a press box (see Section 1019.3, Exception 8).

SECTION 1018
AISLES

1018.1 General. *Aisles* and *aisle accessways* serving as a portion of the *exit access* in the *means of egress* system shall comply with the requirements of this section. *Aisles* or *aisle accessways* shall be provided from all occupied portions of the *exit access* that contain seats, tables, furnishings, displays and similar fixtures or equipment. The minimum width or required capacity of *aisles* shall be unobstructed.

Exception: Encroachments complying with Section 1005.7.

❖ This section addresses aisles and aisle accessways, primarily in occupancies other than assembly seating areas. Current provisions address aisles in all uses, but only address aisle accessways for assembly spaces and mercantile.

"Aisle accessway" is defined in Chapter 2 as "that portion of exit access that leads to an aisle." The term "Aisle" is defined in Chapter 2 as "an exit access component that defines and provides a path of egress travel." Given the many possible configurations of fixtures and furniture, both permanent and

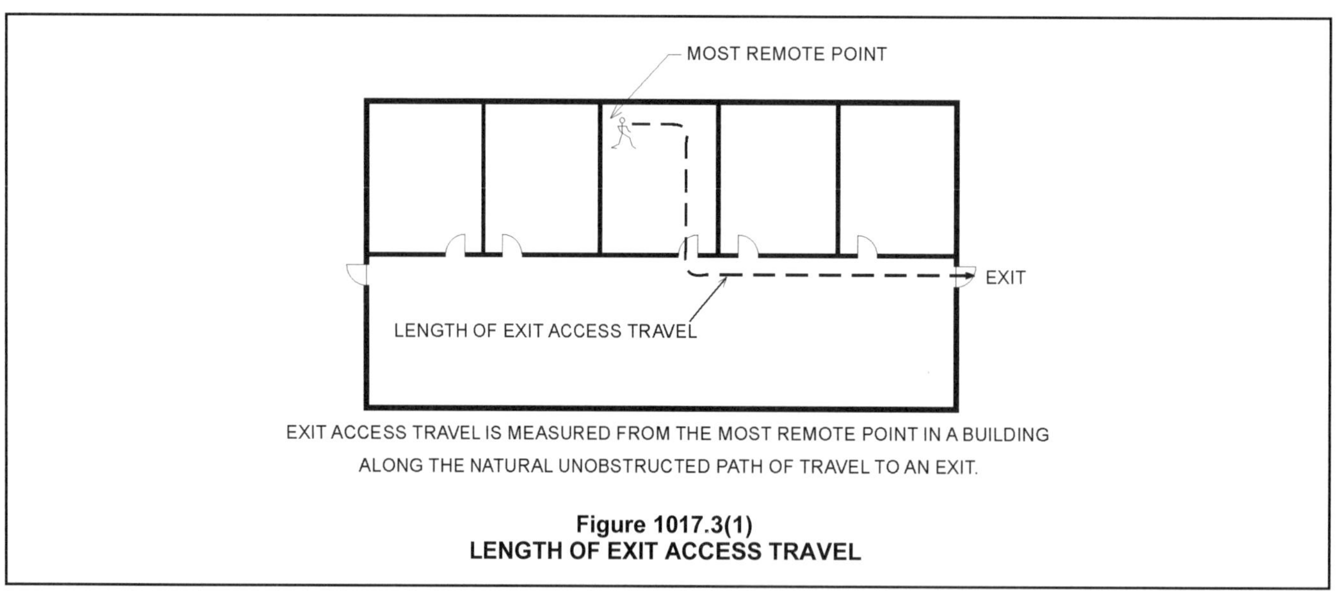

Figure 1017.3(1)
LENGTH OF EXIT ACCESS TRAVEL

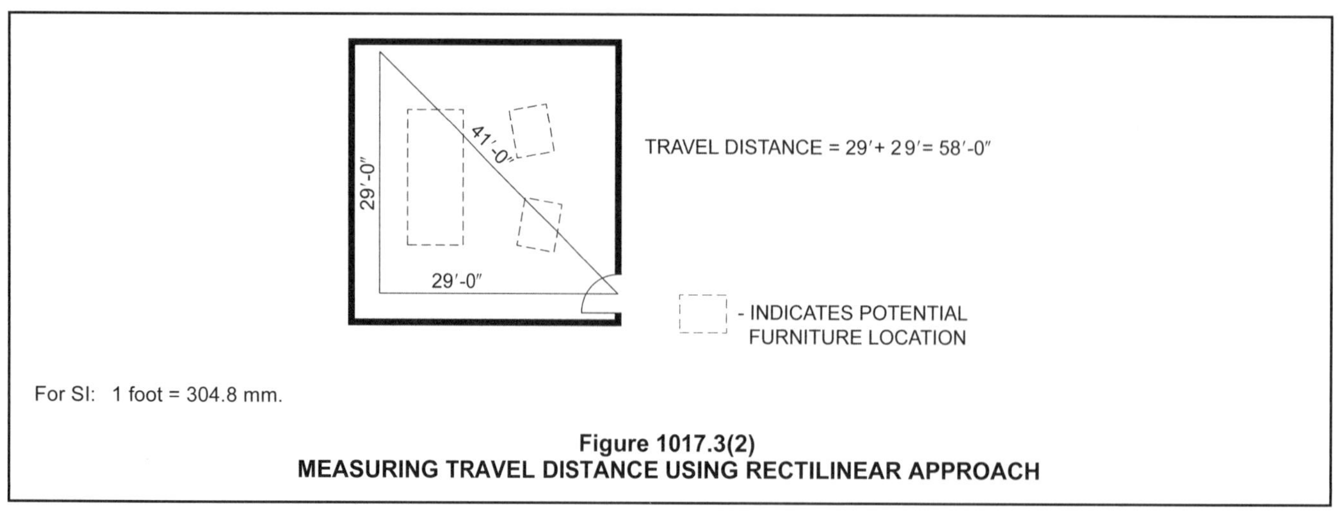

Figure 1017.3(2)
MEASURING TRAVEL DISTANCE USING RECTILINEAR APPROACH

movable, the determination of where aisle accessways stop and aisles begin is often subject to interpretation.

Typically, the aisle accessways lead to the aisles, which in turn lead to the exits. Since the aisle serves as a path for means of egress similar to a corridor, the requirements for doors obstructing the aisle are the same (see Section 1005.7).

A cross reference back to Section 1005.7 from the exceptions for width in aisles (see Section 1018.1), corridors (see Section 1020.3), exit passageways (see Section 1024.2) and exit courts (see Section 1028.4.1) reinforces the fact that encroachment limits are generally applicable for these types of confined routes.

1018.2 Aisles in assembly spaces. *Aisles and aisle accessways* serving a room or space used for assembly purposes shall comply with Section 1029.

❖ The provisions for aisles and aisle accessways in spaces with assembly seating, such as restaurants, theaters and sports arenas, are unique. See Section 1029 for criteria.

1018.3 Aisles in Groups B and M. In Group B and M occupancies, the minimum clear aisle width shall be determined by Section 1005.1 for the *occupant load* served, but shall be not less than that required for corridors by Section 1020.2.

> **Exception:** Nonpublic *aisles* serving less than 50 people and not required to be *accessible* by Chapter 11 need not exceed 28 inches (711 mm) in width.

❖ This requirement establishes aisle-width criteria for Group B and M occupancies based on the occupant load served by the aisle. While not providing as confined a path as corridors, the displays or equipment would limit the choice of paths, so the minimum width is the same as corridors (see Section 1020.2). The reference to Section 1005.1 would trigger a requirement for aisles wider than 44 inches (914 mm) when the anticipated occupant load that the aisle served was larger than 220 (44 inches/0.2 = 220 occupants for nonsprinklered buildings and 44 inches/0.15 = 293 occupants in sprinklered buildings). When an aisle allows for access to exits in two directions, the occupant load could be split, similar to corridors. The exception addresses aisles that may be found in an archival file room or stock storage racks.

For mercantile, if fixtures are permanent, such as in a typical grocery store or office cubicles in a business, the aisle provisions would be applicable throughout. In a situation where there were groups of displays separated by aisles, the area within the displays may be considered aisle accessways (see Section 1018.4).

1018.4 Aisle accessways in Group M. An *aisle accessway* shall be provided on not less than one side of each element within the *merchandise pad*. The minimum clear width for an *aisle accessway* not required to be *accessible* shall be 30 inches (762 mm). The required clear width of the *aisle accessway* shall be measured perpendicular to the elements and merchandise within the *merchandise pad*. The 30-inch (762 mm) minimum clear width shall be maintained to provide a path to an adjacent *aisle* or *aisle accessway*. The *common path of egress travel* shall not exceed 30 feet (9144 mm) from any point in the *merchandise pad*.

> **Exception:** For areas serving not more than 50 occupants, the *common path of egress travel* shall not exceed 75 feet (22 860 mm).

❖ The definition for "Merchandise pad" can be found in Chapter 2. The idea is that a merchandise pad contains movable displays and aisle accessways. A surrounding aisle or permanent walls or displays would define the extent of the merchandise pad. Large department stores will have numerous merchandise pads (see Commentary Figure 1018.3). In accordance with Section 105.2, Item 13, movable cases, counters and partitions not over 5 feet, 9 inches (1753 mm) in height do not require a building permit to move, add or alter. Every element within a merchandise pad must adjoin a minimum 30-inch-wide (762 mm) aisle accessway on at least one side. Travel within a merchandise pad is limited, with a maximum common path of travel of 30 feet (9144 mm). The common path of travel limitation is extended to 75 feet (22 m) in those areas serving a maximum occupant load of 50.

Figure 1018.3
AISLES AND AISLE ACCESSWAYS IN MERCANTILE

1018.5 Aisles in other than assembly spaces and Groups B and M. In other than rooms or spaces used for assembly purposes and Group B and M occupancies, the minimum clear *aisle* capacity shall be determined by Section 1005.1 for the occupant load served, but the width shall be not less than that required for corridors by Section 1020.2.

Exception: Nonpublic *aisles* serving less than 50 people and not required to be *accessible* by Chapter 11 need not exceed 28 inches (711 mm) in width.

❖ Aisles can occur in other occupancies when there is a confined path of travel to the exit access or exit door leading from a space.

While not providing as confined a path as corridors, the displays or equipment would limit the choice of paths, so the minimum width is the same as corridors (see Section 1020.2). The reference to Section 1005.1 would trigger a requirement for aisles wider than 44 inches (914 mm) when the anticipated occupant load that the aisle served was larger than 220 (44 inches/0.2 = 220 occupants for nonsprinklered buildings and 44 inches/0.15 = 293 occupants in sprinklered buildings). When an aisle allows for access to exits in two directions, the occupant load could be split, similar to corridors. The exception addresses aisles that may be found in an archival file room or stock storage racks.

SECTION 1019 EXIT ACCESS STAIRWAYS AND RAMPS

1019.1 General. *Exit access stairways* and *ramps* serving as an *exit access* component in a *means of egress* system shall comply with the requirements of this section. The number of stories connected by *exit access stairways* and *ramps* shall include *basements*, but not *mezzanines*.

❖ This is a general scoping section. Exit access stairways can be between levels or between stories. Section 1019.3 addresses egress between stories. It is important to clarify that basements are considered a story, but not mezzanines. While the exit access travel distance would be measured along the open stairway or ramp for a change in level, a mezzanine, a basement or a second floor, a mezzanine is a space that is considered part of the room to which it is open and is addressed accordingly. Exit stairways are addressed under Sections 1023 and 1027. All stairways and ramps are required to comply with Sections 1011 and 1012, respectively.

1019.2 All occupancies. *Exit access stairways* and *ramps* that serve floor levels within a single story are not required to be enclosed.

❖ Exit access stairways and ramps between levels on the same story and between a story and an associated mezzanine are always permitted to be open unless they are part of a fire-resistance-rated corridor (see Section 1020.6).

1019.3 Occupancies other than Groups I-2 and I-3. In other than Group I-2 and I-3 occupancies, floor openings containing *exit access stairways* or *ramps* that do not comply with one of the conditions listed in this section shall be enclosed with a shaft enclosure constructed in accordance with Section 713.

1. *Exit access stairways* and *ramps* that serve or atmospherically communicate between only two stories. Such interconnected stories shall not be open to other stories.
2. In Group R-1, R-2 or R-3 occupancies, *exit access stairways* and *ramps* connecting four stories or less serving and contained within an individual *dwelling unit* or *sleeping unit* or *live/work unit.*
3. *Exit access stairways* serving and contained within a Group R-3 congregate residence or a Group R-4 facility are not required to be enclosed.
4. *Exit access stairways* and *ramps* in buildings equipped throughout with an *automatic sprinkler system* in accordance with Section 903.3.1.1, where the area of the vertical opening between stories does not exceed twice the horizontal projected area of the *stairway* or *ramp* and the opening is protected by a draft curtain and closely spaced sprinklers in accordance with NFPA 13. In other than Group B and M occupancies, this provision is limited to openings that do not connect more than four stories.
5. *Exit access stairways* and *ramps* within an *atrium* complying with the provisions of Section 404.
6. *Exit access stairways* and *ramps* in *open parking garages* that serve only the parking garage.
7. *Exit access stairways* and *ramps* serving open-air seating complying with the *exit access* travel distance requirements of Section 1029.7.
8. *Exit access stairways* and *ramps* serving the balcony, gallery or press box and the main assembly floor in occupancies such as theaters, *places of religious worship*, auditoriums and sports facilities.

❖ This section includes conditions which permit unprotected floor openings for exit access stairways and ramps. Groups I-2 and I-3 use smoke compartments as part of their defend-in-place strategies; as such, they are addressed separately in Section 1019.4.

The allowances listed are for "interior exit access stairways and ramps" that are between stories. Exit access stairways and ramps between elevation changes on the same story or serving mezzanines do not serve different stories; therefore, the enclosure requirements in this section are not applicable. The general provisions for stairways in Section 1011 and ramps in Section 1012 would be applicable. For simplicity in the commentary to Sections 1019.3 and 1019.4, when interior exit access stairways are mentioned, let it be understood that the same rules apply to interior exit access ramps.

The primary difference between "interior exit access stairways" and "interior exit stairways" is how exit access travel distance is measured (see Section 1017.3). Where a designer chooses to use an exit access stairway, the exit access travel distance includes travel along the slope of the exit access stairway, similar to stairways leading from an open mezzanine or steps at a change in level on a floor.

The base requirement is that exit access stairways between floors are protected in a similar manner as floor openings (see Section 713). The enclosure is needed because a stairway penetrates the floor/ceiling assemblies between the levels, thus creating a vertical opening or shaft. In cases of fire, a vertical opening may act as a chimney, causing smoke, hot gases and light-burning products to flow upward (buoyant force). If an opening is unprotected, these products of combustion will be forced by positive pressure differentials to spread horizontally into the building spaces. There are exceptions for shaft protection around openings in Section 712.1 or exit access stairways as permitted in this section in Items 1 through 8 (see also Section 712.1.12).

Exit access stairways are not required to discharge directly to the exterior as are interior exit stairways (see Section 1023.3). Instead, exit access stairways are part of the route that leads to the exit (i.e., exterior exit door, horizontal exit or enclosed interior exit stairway). See the commentary to Sections 712 and 713 for a discussion of other differences in protection permitted for the enclosure of exit access stairways versus exit stairways.

While this section does not state that exit access stairways cannot be used for any purpose other than means of egress (see Section 1021.1), the provisions for no obstructions in the path of travel or reduction of the capacity for means of egress (Section 1003.6) are applicable. If one of the items eliminating enclosures for a stairway is utilized, the exit access stairway openings would still have to meet the exit and exit access door separation requirement specified in Section 1006 for the space or floor.

It is important to remember that while exit access stairways may be open for multiple floors, Section 1006.3 states “Access to exits at other levels shall be from an adjacent story.” However, single-exit buildings do allow for exit access from any story. Therefore, if there are provisions in both Sections 1006.3.3 and 1019.3, the exit access travel distance could be measured all the way down the open exit access stairway.

Item 1 allows an open exit access stairway when the opening is only between two floors. There cannot be any other unprotected openings that connect to other floors since this could create a staggered stack effect for the movement of smoke between multiple stories. In two-story buildings, this would allow for open stairways between the basement and ground level or between the first and second floors. Another example would be an open exit access stairway between the fifth and sixth floors of a building, provided there were no other unprotected openings between the fourth and fifth floor or the sixth and seventh floors. This is consistent with Section 712.1.9 for openings between two stories.

In Item 2, for residential occupancies, exit access stairways within a single-family home or townhouse are not required to be enclosed because of the small occupant load and resident familiarity with the space. Examples of “within an individual dwelling or sleeping unit” would be a two-story hotel suite or a multistory apartment unit. See Section 1006.3.3, Items 3 and 7, for the allowance for these to be single-exit stories. Live/work units call for the egress to be designed for the “function served,” and vertical opening are not required to be enclosed (Sections 419.3 and 419.4); therefore, it is appropriate to allow open exit access stairways to serve the dwelling portion of the live/work unit. Note that Section 712.1.2 limits any vertical opening within dwelling units, including stairways, to four stories or less.

Item 3 - A congregate residence would not be considered an individual dwelling unit so Item 2 would not be applicable; however, these small facilities operate similarly to a single-family home. Thus, they have an allowance for open exit access stairways for the same reasons as dwelling units in Item 2. See Section 1006.3.2, Item 4, which allows a single exit in Group R-3 and R-4 congregate residences. These congregate residences are limited to 16 occupants that are capable of self-preservation in Sections 308.3.3, 310.5 and 310.6. If a facility offers medical care, it needs to comply with Section 1019.4.

Item 4 discusses buildings sprinklered throughout with an NFPA 13 system. The size of the opening is limited to twice the size of the stairway footprint. The opening must be protected with a draft curtain and closely spaced sprinklers (see Commentary Figure 1019.3). This allows for such an opening to extend the entire height of the building; however, to serve as part of the means of egress, the exit access stairway must still be able to meet the exit access travel distance and provide the number of exits or access to exits from each story (see Section 1006.3). This item is similar to what is permitted for escalators (see Sections 712.1.3 and 712.1.3.1). The power-operated automatic shutter permitted in Section 712.1.3.2 would not be an option for exit access stairways since they must be available for exit access travel. For groups other than business and mercantile, there is the additional limitation of a maximum of four stories.

Item 5 allows for exit access stairways to travel down through an atrium. Atriums often penetrate more than two floors. Atriums must meet the provisions in Section 404 for sprinklers, fire alarms, smoke control, enclosure, limitations on flame spread and smoke development of interior finishes, and the reduction in the normal exit access travel distance. In no case can the exit access travel distance through the atrium, including travel along the stairway, be more than 200 feet (60 960 mm). If the stairway in the atrium serves as part of the required means of egress, each story must also meet the number of exits or access to exits required by Section 1006.

In Item 6, exit access stairways located in open parking structures are exempt from the enclosure requirements because of the ease of accessibility by the fire services, the natural ventilation of such structures, the low level of fire hazard, the small number of people using the structure at any one time and the excellent fire record of such structures. The exception to Section 1017.3 permits the exit access travel distance to be measured to the top step of the exit access stairway in open parking garages. In essence, open parking garages using this item can have open stairways for the full height of the building. By the travel distance being met at the top of the stairway, the means of egress requirements have been met.

In Item 7, stairways in outdoor facilities (i.e., Group A-5) in which the means of egress is essentially open to the outside need not be enclosed because of the ability to vent the fire to the outside. This item is coordinated with the requirements for open-air seating as regulated by Section 1028.7, which allows unlimited travel distance in noncombustible construction that has open-air seating and 400 feet in combustible construction.

Item 8 addresses the unique situation for large indoor assembly seating areas. Since press boxes, galleries and balconies all have the same atmosphere and fire recognition ability as the rest of the seating, the stairways serving these spaces are protected by the same system as the seating (e.g., open to the interior volume and smoke protected; open to the outside); therefore, it is logical to treat the access to these spaces the same as the seating bowl. Open exit access stairways can serve the press box where those stairways move the occupants directly from the press box into the seating bowl. See Sections 1029.5 and 1029.8 for the number of exit access stairways required.

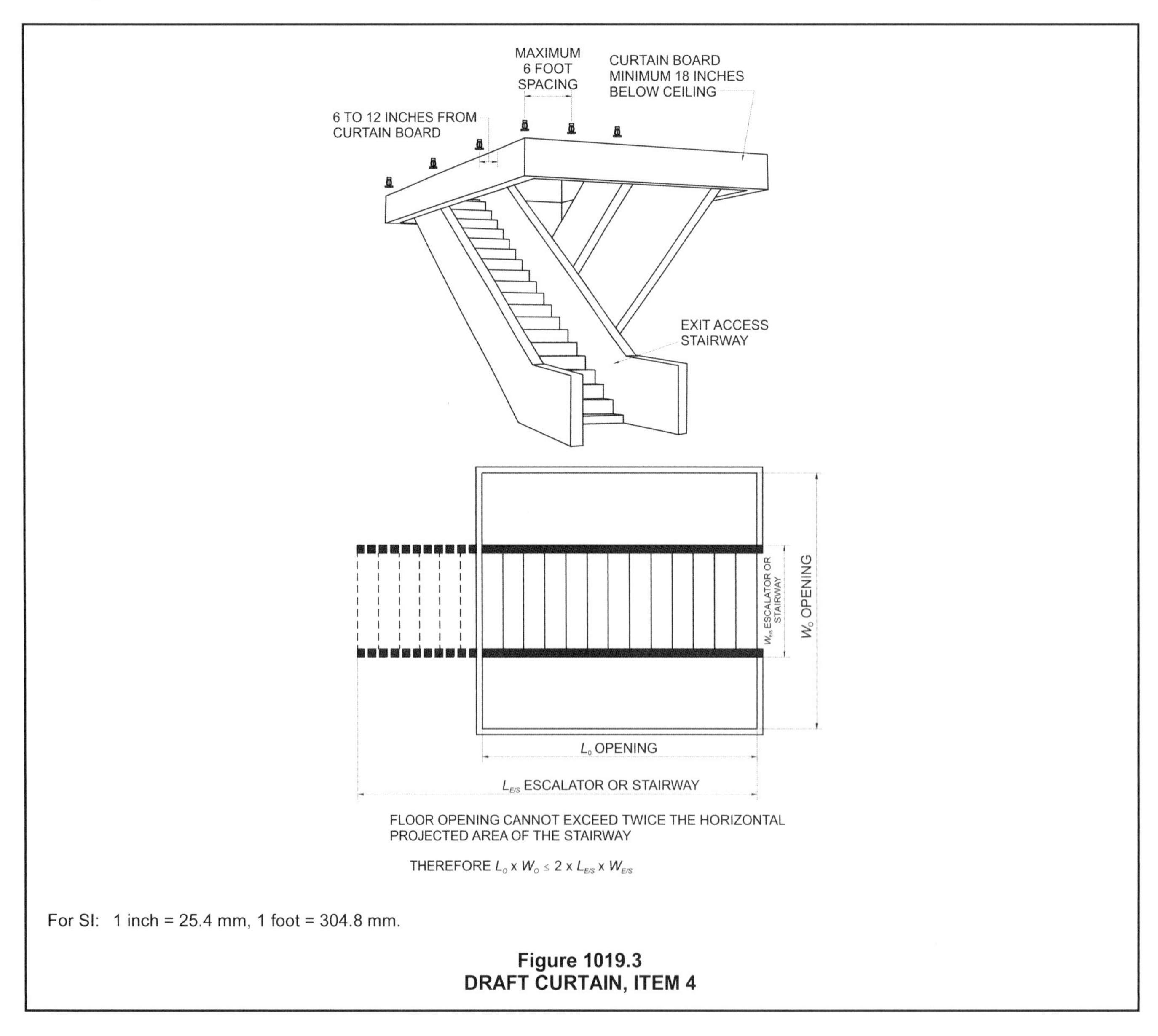

For SI: 1 inch = 25.4 mm, 1 foot = 304.8 mm.

Figure 1019.3
DRAFT CURTAIN, ITEM 4

1019.4 Group I-2 and I-3 occupancies. In Group I-2 and I-3 occupancies, floor openings between stories containing *exit access stairways* or *ramps* are required to be enclosed with a shaft enclosure constructed in accordance with Section 713.

Exception: In Group I-3 occupancies, *exit access stairways* or *ramps* constructed in accordance with Section 408 are not required to be enclosed.

❖ Groups I-2 and I-3 use smoke compartments as part of their defend-in-place strategies; as such, they are addressed separately in this section. Any openings in the floors must be protected the same as a shaft.

The exception allows for observation and security needs in detention facilities, and exit access stairways within a housing unit to not be required to be enclosed (see Section 408.5.1). Section 1023.2 also includes a limited exception for enclosure of exit stairways within jails per Section 408.3.8.

SECTION 1020 CORRIDORS

1020.1 Construction. *Corridors* shall be fire-resistance rated in accordance with Table 1020.1. The *corridor* walls required to be fire-resistance rated shall comply with Section 708 for *fire partitions*.

Exceptions:

1. A *fire-resistance rating* is not required for *corridors* in an occupancy in Group E where each room that is used for instruction has not less than one door opening directly to the exterior and rooms for assembly purposes have not less than one-half of the required *means of egress* doors opening directly to the exterior. Exterior doors specified in this exception are required to be at ground level.
2. A *fire-resistance rating* is not required for *corridors* contained within a *dwelling unit* or *sleeping unit* in an occupancy in Groups I-1 and R.
3. A *fire-resistance rating* is not required for *corridors* in *open parking garages*.
4. A *fire-resistance rating* is not required for *corridors* in an occupancy in Group B that is a space requiring only a single *means of egress* complying with Section 1006.2.
5. *Corridors* adjacent to the *exterior walls* of buildings shall be permitted to have unprotected openings on unrated *exterior walls* where unrated walls are permitted by Table 602 and unprotected openings are permitted by Table 705.8.

❖ It is not the intent of this section to require corridors. Once corridors are provided, so that occupants are limited to a confined path of travel, then these provisions apply.

The purpose of corridor enclosures is to provide fire protection to occupants as they travel the confined path, perhaps unaware of a fire buildup in an adjacent floor area. The base protection is a fire partition having a 1-hour fire-resistance rating (see Table 1020.1). The table allows a reduction or elimination of the fire-resistance rating depending on the occupant load and the presence of an NFPA 13 or 13R automatic sprinkler system throughout the building.

Section 708 addresses the continuity of fire partitions serving as corridor walls. In addition to allowing the fire partitions to terminate at the underside of a fire-resistance-rated floor/ceiling or roof/ceiling assembly, the supporting construction need not have the same fire-resistance rating in buildings of Type IIB, IIIB and VB construction as specified in Section 708. If such walls were required to be supported by fire-resistance-rated construction, the use of these construction types would be severely restricted when the corridors are required to have a fire-resistance rating. Section 407.3 requires that corridor walls in Group I-2 occupancies that are required to have a fire-resistance rating must be continuous to the underside of the floor or roof deck above or at a smoke-limiting ceiling membrane. Continuity is required because of the defend-in-place protection strategy utilized in such buildings. Requirements for corridor construction within Group I-3 occupancies are found in Section 408.8. Dwelling unit separation in Groups I-1, R-1, R-2 and R-3 is found in Sections 420.2 and 420.3. Ambulatory care facilities have special requirements in Section 422.2 when some of the patients can be incapable of self-preservation. For additional requirements for an elevator lobby that is adjacent to or part of a corridor, see the commentaries to Sections 713.14 and 3006.

Exception 1 indicates a fire-resistance rating is not required for corridors in Group E when any room adjacent to the corridor that is used for instruction or assembly purposes has a door directly to the outside. The need for a fire-resistance-rated corridor is eliminated because these rooms are provided with an alternative egress path because of the requirement for exterior exits. This option is typically utilized in nonsprinklered buildings, such as day care facilities, since a sprinkler system would also allow for unrated corridors in Group E (see Table 1020.1). Grade schools and high schools have security concerns that sometimes make the outdoor exit for every classroom not a preferred option.

In accordance with Exception 2, a fire-resistance rating for a corridor contained within a single dwelling unit (e.g., apartment, townhouse) or sleeping unit (e.g., hotel guestroom, assistive living suite) is not required for practical reasons. It is unreasonable to expect fire doors and the associated hardware and closing devices to be within dwellings and similar occupancies.

Given the relatively smoke-free environment of open parking structures, Exception 3 does not require rated corridors in these types of facilities.

If an office suite is small enough that only one means of egress is required from the suite, Exception 4 indicates that a rated corridor would not be required in that area. The main corridor that connected these suites to the exits would be evaluated in accordance with Table 1020.1.

Exception 5 addresses when the exterior wall of a building is also the wall of the corridor. The exterior wall is not required to be rated by the corridor provisions where there is a sufficient fire separation distance for the exterior wall to be able to have unprotected openings. This is similar to the exterior wall for an enclosed interior exit stairway. The fire is assumed to be inside the building, so that is where the protection is required.

TABLE 1020.1
CORRIDOR FIRE-RESISTANCE RATING

OCCUPANCY	OCCUPANT LOAD SERVED BY CORRIDOR	REQUIRED FIRE-RESISTANCE RATING (hours)	
		Without sprinkler system	With sprinkler system[c]
H-1, H-2, H-3	All	Not Permitted	1
H-4, H-5	Greater than 30	Not Permitted	1
A, B, E, F, M, S, U	Greater than 30	1	0
R	Greater than 10	Not Permitted	0.5
I-2[a], I-4	All	Not Permitted	0
I-1, I-3	All	Not Permitted	1[b]

a. For requirements for occupancies in Group I-2, see Sections 407.2 and 407.3.

b. For a reduction in the *fire-resistance rating* for occupancies in Group I-3, see Section 408.8.

c. Buildings equipped throughout with an *automatic sprinkler system* in accordance with Section 903.3.1.1 or 903.3.1.2 where allowed.

❖ The required fire-resistance ratings of corridors serving adjacent spaces are provided in Table 1020.1. The fire-resistance rating is based on the group classification (considering characteristics such as occupant mobility, density and familiarity with the building as well as the fire hazard associated with the classification), the total occupant load served by the corridor and the presence of an automatic sprinkler system.

Where the corridor serves a limited number of people (second column in Table 1020.1), the fire-resistance rating is eliminated because of the limited size of the facility and the likelihood that the occupants would become aware of a fire buildup in sufficient time to exit the structure safely. The total occupant load that the corridor serves is used to determine the requirement for a rated corridor enclosure. The number of occupants served is the total occupants that will move into the corridor to egress. Corridors serving a total occupant load equal to or less than that indicated in the second column of Table 1020.1 are not required to be enclosed with fire-resistance-rated construction. For example, a corridor serving an occupant load of 30 or less in an unsprinklered Group B occupancy is not required to be enclosed with fire-resistance-rated construction. This example is illustrated in Commentary Figure 1020.1.

The purpose of corridor enclosures is to provide fire protection to occupants as they travel the confined path, perhaps unaware of a fire buildup in an adjacent floor area. The base protection is a fire partition having a 1-hour fire-resistance rating. The table allows a reduction or elimination of the fire-resistance rating depending on the occupant load and the presence of an NFPA 13 or 13R automatic sprinkler system throughout the building.

A common mistake is assuming a building is sprinklered throughout and utilizing the corridor rating reductions, when in fact certain requirements in NFPA 13 would not consider the building sprinklered throughout. For example, a health club installs a sprinkler system, but chooses to eliminate the sprinklers over the swimming pool in accordance with the exception in Section 507.4. Any corridors within the building that serve greater than 30 occupants must be rated because the building would not be considered sprinklered throughout in accordance with NFPA 13 requirements.

Note that because of the hazardous nature of occupancies in Groups H-1, H-2 and H-3, fire-resistance-rated corridors are required under all conditions. Regardless of the presence of a fire sprinkler system, a 1-hour-rated corridor enclosure is required in high-hazard occupancies with detonation, deflagration,

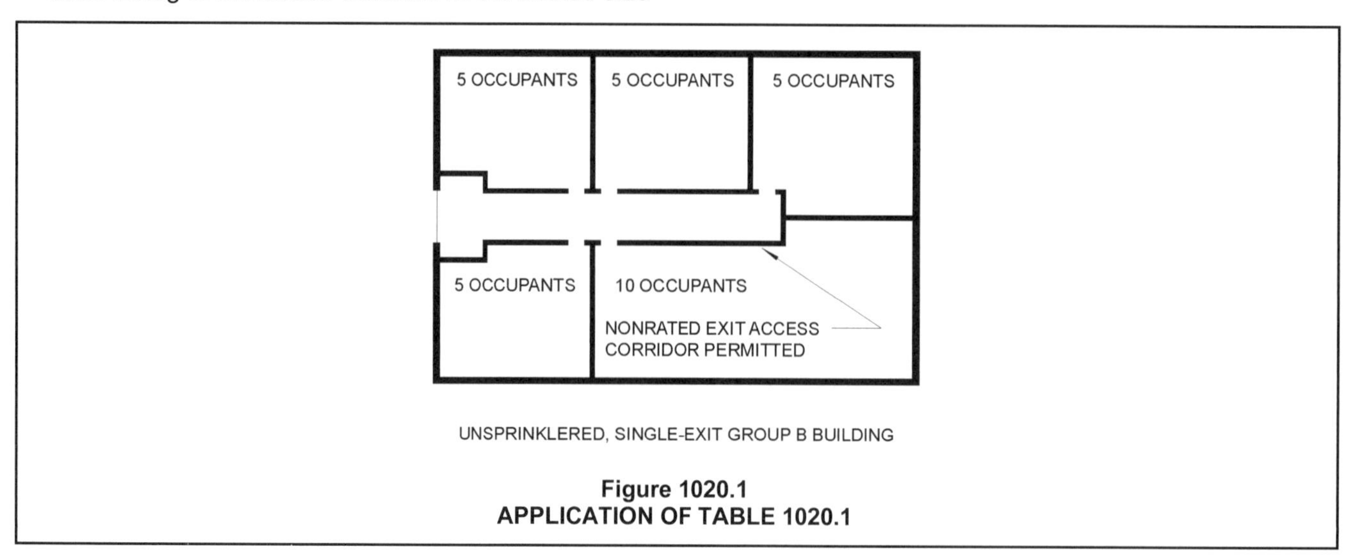

Figure 1020.1
APPLICATION OF TABLE 1020.1

accelerated burning, readily supported combustion or physical hazards. Group H-4 and H-5 occupancies that contain semiconductor fabrication materials or operations constituting a health hazard do not pose the same relative fire or explosion hazard as Group H-1, H-2 or H-3 materials. As such, in Group H-4 or H-5, where the corridor serves a total occupant load of 30 or less, a fire-resistance-rated enclosure is not required. The "not permitted" in the third column is in coordination with Section 903.2.5, which requires all Group H buildings to be fully sprinklered.

The code acknowledges that an automatic sprinkler system can serve to control or eliminate fire development that could threaten the exit access corridor. Most occupancies where sleeping rooms are not present (Groups A, B, E, F, M, S and U) are permitted to have nonfire-resistance-rated corridors if a sprinkler system is installed throughout the building in accordance with NFPA 13.

In residential facilities, the response time to a fire may be delayed because the residents may be sleeping. With this additional safety concern, the requirements for corridors are more restrictive than for nonresidential occupancies. If the corridor serves more than 10 occupants, it is required to be rated for 1 hour. If the building is sprinklered throughout with either an NFPA 13 or 13R system, then the rating of the corridor may be reduced to $^1/_2$ hour. Note the exception for fire-resistance-rated corridors within an individual dwelling or sleeping unit in Section 1020.1. Also note that the reduction in the rating of the corridor walls is not permitted when an NFPA 13D sprinkler system is provided.

While all Group I facilities are supervised environments, the level of supervision in Group I-2 and I-4 occupancies would permit assisted evacuation by staff in an emergency; therefore, corridors are not required to be rated. Corridors in Groups I-2 are also regulated by Section 407.3. Because of the lower staff/resident ratio in Group I-1 and the limitation on free egress in Group I-3, corridors must have a 1-hour fire-resistance rating (see Section 408.8 for a reduction in the corridors in Group I-3). The "not permitted" in the third column is in coordination with Section 903.2.6, which requires all Group I buildings to be fully sprinklered.

1020.2 Width and capacity. The required capacity of *corridors* shall be determined as specified in Section 1005.1, but the minimum width shall be not less than that specified in Table 1020.2.

> **Exception:** In Group I-2 occupancies, *corridors* are not required to have a clear width of 96 inches (2438 mm) in areas where there will not be stretcher or bed movement for access to care or as part of the defend-in-place strategy.

❖ The corridor widths specified in Table 1020.2 are long-established minimums originally derived from human dimensions, practical concerns, occupant loads and psychological considerations. Additional corridor capacity, when necessary for large crowds, is determined in accordance with Section 1005.

The number of occupants using a corridor for egress establishes the required capacity of a corridor, as well as for any specific portion of a multileg corridor system. Portions of a corridor system may differ in width for a variety of reasons not related to code minimums. The designer and building official are expected to verify that corridor widths and corridor fire-resistance ratings are in accordance with Sections 1005 and 1020, whichever is more restrictive.

The required occupant capacity of a corridor is based on the total occupant load of the rooms and spaces served by the corridor as determined by Section 1004. Where a corridor is served by two exits in opposite directions, the corridor capacity is split to determine the minimum required width of those exits (i.e., exit door, exit stairway) at each end of the corridor. The total occupant load served by a corridor is not split to establish the corridor fire-resistance rating (see Section 1020.1) or the required capacity.

The exception is because of an item in Table 1020.2 that requires a 96-inch-wide (2438 mm) corridor in Group I-2 facilities. In a hospital, while this is required in the area of patient sleeping rooms and patient care areas, there are a large number of areas that will not be for the movement of beds, either for patient care or for the hospital's defend-in-place strategies. For example, the hospital may have office or therapy areas where patients are brought in walking on their own or using wheelchairs. Many nursing homes do not move patients in beds at all. The intent of the exception is to clarify that the 96-inch (2438 mm) corridor width is not a minimum for all corridors throughout a hospital or nursing home, but only in certain areas.

TABLE 1020.2
MINIMUM CORRIDOR WIDTH

OCCUPANCY	MINIMUM WIDTH (inches)
Any facilities not listed below	44
Access to and utilization of mechanical, plumbing or electrical systems or equipment	24
With an occupant load of less than 50	36
Within a *dwelling unit*	36
In Group E with a *corridor* having an occupant load of 100 or more	72
In *corridors* and areas serving stretcher traffic in occupancies where patients receive outpatient medical care that causes the patient to be incapable of self-preservation	72
Group I-2 in areas where required for bed movement	96

For SI: 1 inch = 25.4 mm.

❖ The widths of passageways, aisles and corridors are functional elements of building construction that allow the occupants to circulate freely and comfortably throughout the floor area under nonemergency conditions. Under emergency situations, the egress paths must provide the needed width to accommodate the number of occupants that must utilize the corridor for egress.

When the occupant load of the space exceeds 49, the minimum width of the passageway, aisle or corridor serving that space is required to be at least 44 inches (1118 mm) to permit two unimpeded parallel columns of users to travel in opposite directions. When the total occupant load served by a corridor is 49 or less, a minimum width of 36 inches (914 mm) is permitted and the users are expected to encounter some intermittent travel interference from fellow users, but the lower occupant load makes those occasions infrequent and tolerable. The 36-inch (914 mm) minimum width is also required within a dwelling unit.

Passageways that lead to building equipment and systems must be at least 24 inches (610 mm) in width to provide a means to access and service the equipment when needed. Because of the frequency of the servicing intervals and the limited number of occupants in these normally unoccupied areas, a reduced width is warranted. This minimum width criteria applies to many common situations, such as stage lighting and special-effects catwalks; catwalks leading to heating and cooling equipment; as well as passageways providing access to boilers, furnaces, transformers, pumps, piping and other equipment.

Except for small buildings, Group E occupancies are required to have minimum 72-inch-wide (1829 mm) corridors where the corridors serve educational areas. This width is needed not only for proper functional use, but also because of the edge effect caused by student lockers and other boundary attractions and objects. Service and other corridors outside of educational areas, such as an administrative area, would be regulated consistent with their use. Note that Section 1020.3 would not allow the wall lockers to overlap the required corridor width.

In Group I-2 occupancies, where the corridor is utilized during a fire emergency for moving patients confined to beds, it is required to be at least 96 inches (2438 mm) in clear width. This width requirement is applicable to all areas where there are patient sleeping rooms, and may also be required in some of the treatment room areas where in-house patients will be brought in on beds or rolling stretchers. This minimum width allows two rolling beds or stretchers to pass in a corridor and permits the movement of a bed/stretcher into the corridor through a room door. In Group I-2 and ambulatory care center areas, where the movement of beds is not anticipated, such as administrative and some outpatient areas of a hospital or clinic, the corridor would not be required to be 96 inches (2438 mm) wide. The minimum width would be determined by one of the appropriate applicable criteria. For outpatient medical care, where the patient may be incapable of self-preservation, such as some outpatient surgery areas or dialysis treatment areas, the 72-inch-wide (1829 mm) corridor is required. This would include Group I surgical areas, areas such as MRI suites or dialysis centers, emergency rooms or Group B ambulatory care centers.

1020.3 Obstruction. The minimum width or required capacity of *corridors* shall be unobstructed.

Exception: Encroachments complying with Section 1005.7.

❖ It is important to maintain required corridor width so that the path of travel to an exit is continually available and unobstructed. Because corridors tend to be lined with user passage doors, there are allowances under Section 1005.7. In no case may a door block more than 50 percent of the required corridor width. In addition, when fully open, the doors must not protrude more than 7 inches (178 mm) into the required width. Where doors swing out into the corridor, options would be to move doors back into alcoves or to provide corridors wider than the required width. The alcoves would have to be deep enough to also meet the 7-inch (118 mm) maximum protrusion when doors are open, or at least 29 inches (737 mm) deep. For an example of the wider corridor: a standard door is 36 inches (914 mm) wide, and the typical minimum corridor width is 44 inches (1118 mm). By adding the door leaf and half the required corridor width (36+22=58), a designer could provide a corridor width of 58 inches (1473 mm) and not have any issues with encroachment of doors. This is consistent with the provisions in aisles, corridors, stairways, ramps, exit passageways and exit discharge courts.

A cross reference back to Section 1005.7 from the exceptions for width in (see Section 1019.1), corridors (see Section 1020.3), aisles exit passageways (see Section 1024.2) and exit courts (see Section 1028.4.1) reinforces the fact that the protrusion limits provision is generally applicable for these types of confined routes.

1020.4 Dead ends. Where more than one *exit* or *exit access doorway* is required, the *exit access* shall be arranged such that there are no dead ends in *corridors* more than 20 feet (6096 mm) in length.

Exceptions:

1. In occupancies in Group I-3 of Condition 2, 3 or 4, the dead end in a *corridor* shall not exceed 50 feet (15 240 mm).
2. In occupancies in Groups B, E, F, I-1, M, R-1, R-2, R-4, S and U, where the building is equipped throughout with an *automatic sprinkler system* in accordance with Section 903.3.1.1, the length of the dead-end *corridors* shall not exceed 50 feet (15 240 mm).
3. A dead-end *corridor* shall not be limited in length where the length of the dead-end *corridor* is less

than 2.5 times the least width of the dead-end *corridor*.

❖ The requirements of this section apply where a space is required to have more than one means of egress according to Section 1006.2.

Dead ends in corridors and passageways can seriously increase the time needed for an occupant, especially if unfamiliar with the space, to locate the exits. More importantly, dead ends will allow a single fire event to eliminate access to all of the exits by trapping the occupants in the dead-end area. A dead end exists whenever a user of the corridor or passageway has only one direction to travel to reach any building exit [see Commentary Figure 1020.4(1)]. While a preferred building layout would be one without dead ends, a maximum dead-end length of 20 feet (6096 mm) is permitted and is to be measured from the extreme point in the dead end to the point where occupants have a choice of two directions to separate exits. Having to go back only 20 feet (6096 mm) after coming to a dead end is not such a significant distance as to cause a serious delay in reaching an exit during an emergency situation.

A dead end results whether or not egress elements open into it. A dead end is a hazard for occupants who enter the area from adjacent spaces, travel past an exit into a dead end or enter a dead end with the mistaken assumption that an exit is directly accessible from the dead end.

Note that Section 402.8.6 deals with dead-end distances in a covered mall and assumes that, with a sufficiently wide mall in relation to its length, alternative paths of travel will be available in the mall itself to reach an exit (i.e., the common mall area is not to be construed as a corridor).

Under special conditions, exceptions to the 20-foot (6096 mm) dead-end limitation apply.

Exception 1 is permitted based on the considerations of the functional needs of Group I-3 Occupancy Condition 2, 3 or 4, the requirements for smoke compartmentalization in Section 408.6 and the requirement for automatic sprinkler protection of the facility in Section 903.2.6.

Exception 2 recognizes the fire protection benefits and performance history of automatic fire sprinkler systems. While the degree of hazard in Group B, E, F, M, S and U occupancies does not initially require an automatic fire suppression system, the length of a dead-end corridor or passageway is permitted to be extended to 50 feet (15 240 mm) where an automatic fire sprinkler system in accordance with NFPA 13 is provided throughout the building. This exception is also permitted in Group I-1, R-1, R-2 and R-4 occupancies, but only when they use an NFPA 13 system, not an NFPA 13R system. In addition, these provisions are consistent with those in the IEBC and IFC in the regulation of dead-end corridors in existing buildings undergoing alterations. Dead-end provisions are not applicable in single-exit spaces; therefore, dead-end provisions are not applicable in Group R-3 and R-4 occupancies.

Exception 3 addresses the condition presented by "cul-de-sac" elevator lobbies directly accessible from exit access corridors. In such an elevator lobby, lengths of 20 to 30 feet (6096 to 9144 mm) are common for three- or four-car elevator banks. Typically, the width of this elevator lobby is such that the possibility of confusion with a path of egress is minimized. Below the $2^1/_2$:1 ratio, the dead end becomes so wide that it is less likely to be perceived as a corridor leading to an exit. For example, based on the $2^1/_2$:1 ratio limitation, a 25-foot-long (7620 mm) dead end over 10 feet (3048 mm) in width would not be considered a dead-end corridor [see Commentary Figure 1020.4(2)]. For additional elevator lobby requirements, see the commentaries to Sections 713.14 and 3006.

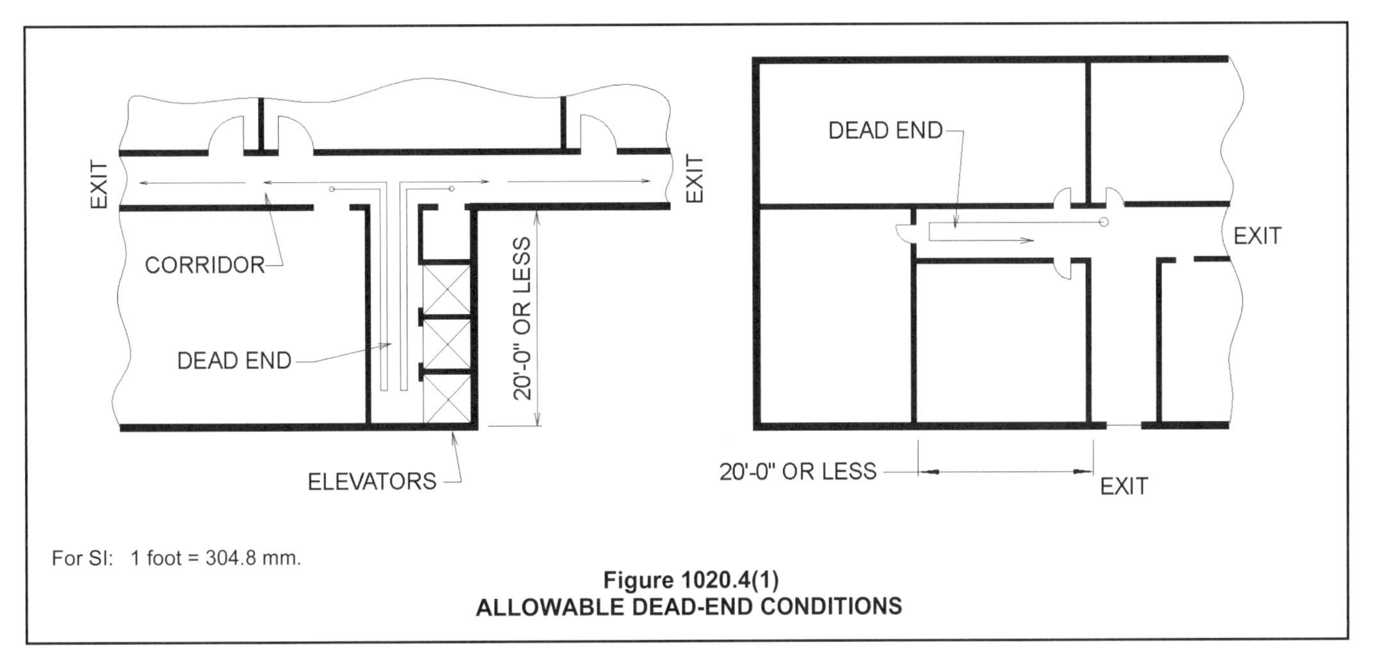

Figure 1020.4(1)
ALLOWABLE DEAD-END CONDITIONS

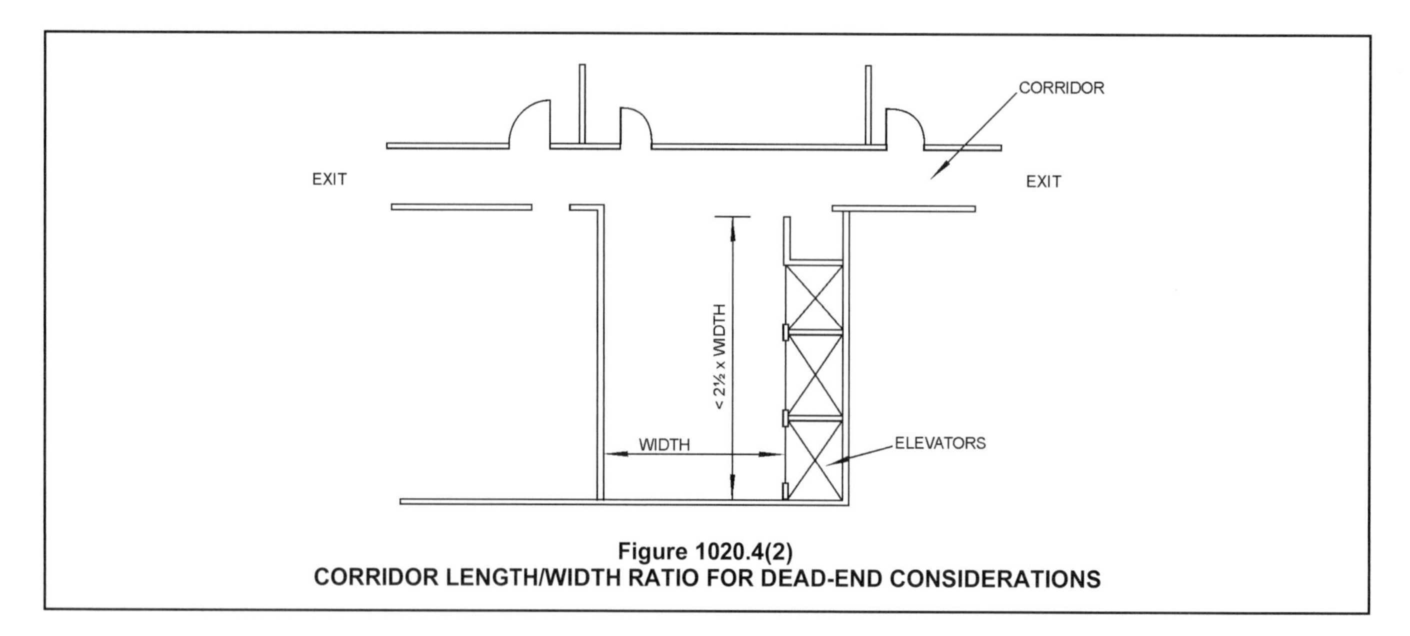

Figure 1020.4(2)
CORRIDOR LENGTH/WIDTH RATIO FOR DEAD-END CONSIDERATIONS

1020.5 Air movement in corridors. *Corridors* shall not serve as supply, return, exhaust, relief or ventilation air ducts.

Exceptions:

1. Use of a *corridor* as a source of makeup air for exhaust systems in rooms that open directly onto such *corridors*, including toilet rooms, bathrooms, dressing rooms, smoking lounges and janitor closets, shall be permitted, provided that each such *corridor* is directly supplied with outdoor air at a rate greater than the rate of makeup air taken from the *corridor*.
2. Where located within a *dwelling unit*, the use of *corridors* for conveying return air shall not be prohibited.
3. Where located within tenant spaces of 1,000 square feet (93 m^2) or less in area, utilization of *corridors* for conveying return air is permitted.
4. Incidental air movement from pressurized rooms within health care facilities, provided that the *corridor* is not the primary source of supply or return to the room.

❖ Two of the most critical elements of the means of egress are the required exit stairways and corridors. Exit stairways serve as protected areas in the building that provide occupants with safe passage to the level of exit discharge. Since required exits and corridors are critical elements in the means of egress, the potential spread of smoke and fire into these spaces must be minimized. The scope of this section is corridors. For requirements for the exits, see Section 1023.

The use of these corridors as part of the air distribution system could render those egress elements unusable. The intent is to have positive pressure in the corridors. Therefore, any air movement condition that could introduce smoke into these vital egress elements is prohibited. It is not the intent of this section to prohibit the air movement necessary for ventilation and space conditioning of corridors, but rather to prevent those spaces from serving as conduits for the distribution of air to, or the collection of air from, adjacent spaces. This restriction also extends to door transoms and door grilles that would allow the spread of smoke into a corridor. This limitation is not, however, intended to restrict slight pressure differences across corridor doors, such as a negative pressure differential maintained in kitchens to prevent odor migration into dining rooms. Note that air distribution via ducted systems located in or above corridors is acceptable since the corridor itself would not be functioning as a duct.

The four exceptions to this section identify conditions where a corridor can be utilized as part of the air distribution system. The exceptions apply only to exit access corridors, not to exit passageways.

Exception 1 addresses the common practice of using air from the corridor as makeup air for small exhaust fans in adjacent rooms. Where the corridor is supplied directly with outdoor air at a rate equal to or greater than the makeup air rate, negative pressure will not be created in the corridor with respect to the adjoining rooms and smoke would generally not be drawn into the corridor.

Regarding Exception 2, it is common practice to locate return air openings in the corridors of dwelling units and draw return air from adjoining spaces through the corridor. Such use of dwelling unit corridors for conveying return air is not considered to be a significant hazard and is permitted. Individual dwelling units are permitted to have unprotected openings between floors. Corridors within dwelling units that serve small occupant loads are short in length and are not required to be fire-resistance rated. For these reasons, the use of the corridor or the space above a corridor ceiling for conveying return air does not constitute an unacceptable hazard.

Exception 3 permits corridors located in small tenant spaces to be used for conveying return air based on the relatively low occupant load and the relatively short length of the corridor. These conditions do not pose a significant hazard. In the event of an emergency, the occupants of the space would tend to simply retrace their steps to the entrance.

Health care facilities require direct pressurization control of certain rooms to provide a clean and sterile environment for patients. For example, operating rooms and pharmacies are required to have positive air pressure in the room, resulting in a general air movement out of the room. This ensures that airborne contaminants do not infect a sterile procedure or supplies. Pressurization is achieved by supplying air at a greater or lesser rate than the return air. Exception 4 recognizes the need of infection control and clarifies that the corridor should not be the primary source of supply return. There should be supply and return air in the room.

1020.5.1 Corridor ceiling. Use of the space between the *corridor* ceiling and the floor or roof structure above as a return air plenum is permitted for one or more of the following conditions:

1. The *corridor* is not required to be of *fire-resistance-rated* construction.
2. The *corridor* is separated from the plenum by *fire-resistance-rated* construction.
3. The air-handling system serving the *corridor* is shut down upon activation of the air-handling unit *smoke detectors* required by the *International Mechanical Code*.
4. The air-handling system serving the *corridor* is shut down upon detection of sprinkler water flow where the building is equipped throughout with an *automatic sprinkler system*.
5. The space between the *corridor* ceiling and the floor or roof structure above the *corridor* is used as a component of an approved engineered smoke control system.

❖ This section identifies five different conditions where the space above the corridor ceiling is permitted to serve as a return air plenum. Since a return air plenum operates at a negative pressure with respect to the corridor, any smoke and gases within the plenum should be contained within that space. Conversely, a supply plenum operates at a positive pressure with respect to the corridor, thus increasing the likelihood that smoke and gases will infiltrate the corridor enclosure. Where any one of the five conditions is present, the use of the corridor ceiling space as a return air plenum is permitted. This is consistent with IMC Section 601.2.1.

Where the corridor is permitted to be constructed without a fire-resistance rating (see Section 1020.1), Item 1 permits the space above the ceiling to be utilized as a return air plenum without requiring it to be separated from the corridor with fire-resistance-rated construction.

Item 2 is only applicable to corridors that are required to be enclosed with fire-resistance-rated construction. Compliance with this item requires the plenum to be separated from the corridor by fire-resistance-rated construction equivalent to the rating of the corridor enclosure itself. Therefore, the ceiling membrane itself must provide the fire-resistance rating required of the corridor enclosure. Section 708.4, Exception 3, is an example of this method of construction.

Items 3 and 4 recognize that the hazard associated with smoke spread through a plenum is minimized if the air movement is stopped.

It is not uncommon for an above-ceiling plenum to be utilized as part of the smoke removal system. This practice is permitted by Item 5. Because of the way these systems are designed, the higher equipment ratings and the power supply provisions, this is considered acceptable.

1020.6 Corridor continuity. *Fire-resistance-rated corridors* shall be continuous from the point of entry to an *exit*, and shall not be interrupted by intervening rooms. Where the path of egress travel within a *fire-resistance-rated corridor* to the exit includes travel along unenclosed *exit access stairways* or *ramps*, the *fire-resistance rating* shall be continuous for the length of the *stairway* or *ramp* and for the length of the connecting *corridor* on the adjacent floor leading to the *exit*.

Exceptions:

1. Foyers, lobbies or reception rooms constructed as required for *corridors* shall not be construed as intervening rooms.
2. Enclosed elevator lobbies as permitted by Item 1 of Section 1016.2 shall not be construed as intervening rooms.

❖ This section requires that where fire protection is offered by a corridor, it is to be continuous from the point of entry into the corridor to an exit. This is to protect occupants from the accumulation of smoke or fire exposure and to allow for sufficient time to evacuate the building. Where a corridor is served by two or more exits, only one of the exits is required to be accessed directly from the corridor. Other exits may be accessed through intervening spaces in accordance with Section 1016.2, provided that there is an opening protective at the end of the corridor to separate the rated corridor from the intervening rooms. Thus, occupants will always have their protected path to an exit, and at the same time a reasonable degree of design freedom is allowed [see Commentary Figures 1020.6(1) and 1020.6(3)].

When a level is permitted to have unenclosed exit access stairways in accordance with Section 1019.3, the corridor protection is required to continue down the exit access stairway to an enclosure for an interior exit stairway or to an exit door leading to the outside. Since the exit access stairway effectively becomes part of the corridor, doors would not be required at the top and bottom of the open exit access stairway as they are when entering enclosures for exit stairways.

Exception 1 allows a foyer, lobby or reception room to be located on the path of egress from a corridor or as part of the fire-resistance-rated corridor, provided the room has the same fire-resistance-rated walls and doors as required for the corridor. The use of this provision should be viewed as limiting the types of uses that may occur within the protected corridor. Occupied spaces within the corridor should have very limited uses and hazards. Foyers and lobbies are included in this exception based on the low fire hazard of the contents in such rooms [see Commentary Figure 1020.6(2)].

Another consideration is corridor continuity at an elevator opening. When an elevator opens into a corridor that is required to be of fire-resistance-rated construction, the opening between the elevator shaft and the corridor must be protected to meet not only the shaft's fire protection rating but also the additional smoke and draft protection requirements necessary to limit the spread of smoke into the corridor. This

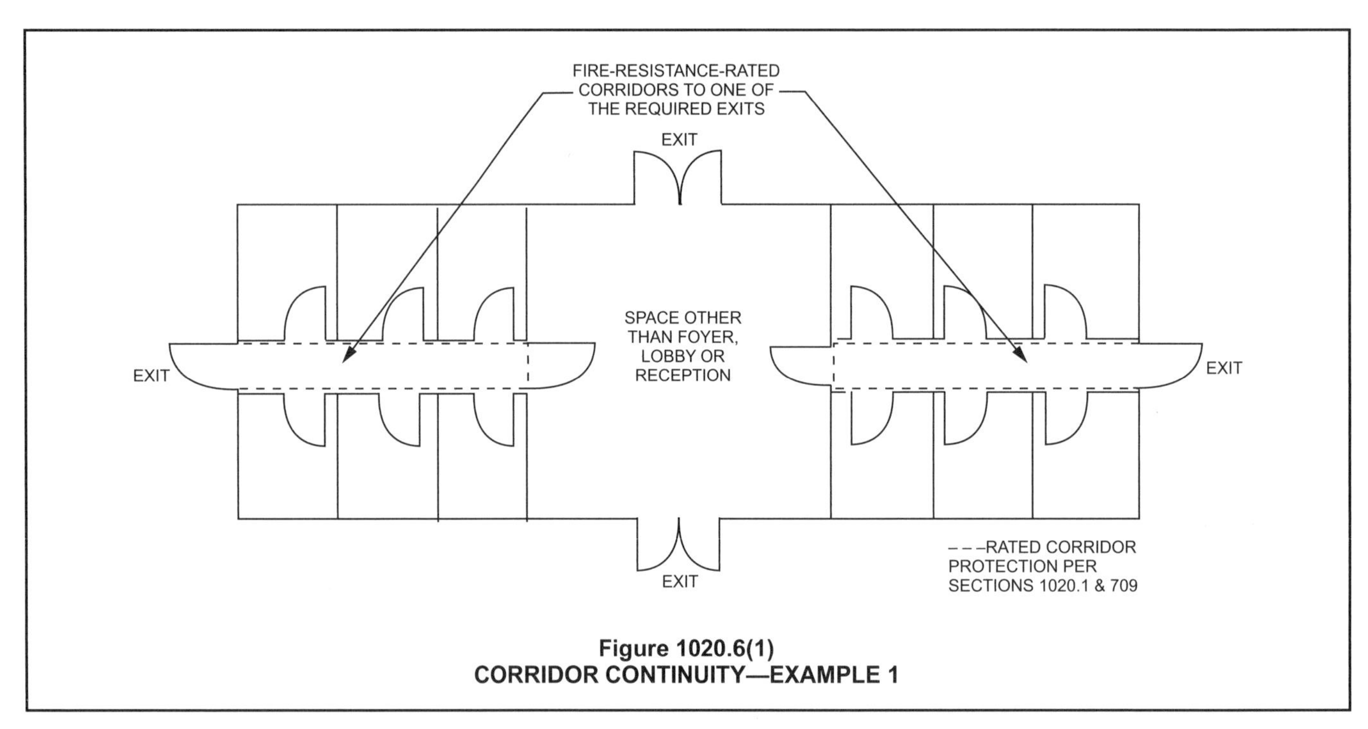

Figure 1020.6(1)
CORRIDOR CONTINUITY—EXAMPLE 1

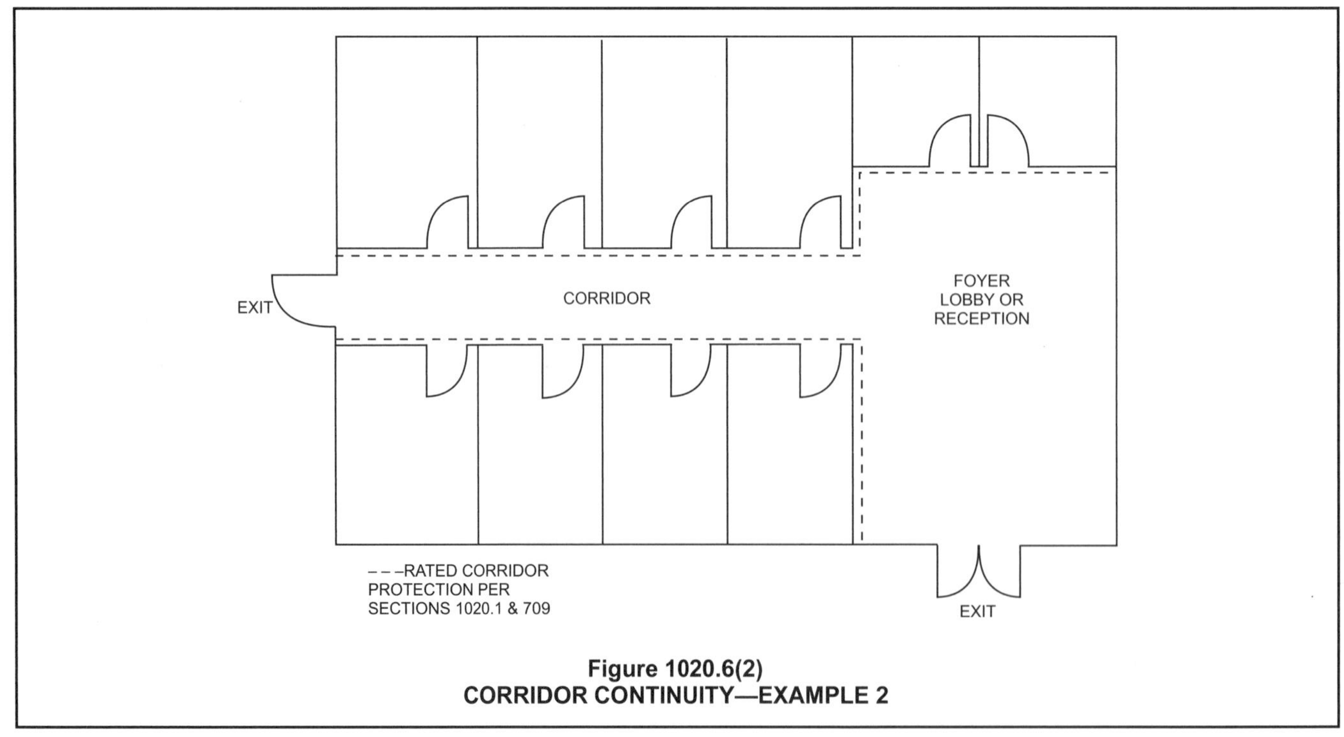

Figure 1020.6(2)
CORRIDOR CONTINUITY—EXAMPLE 2

additional smoke and draft control requirement is found in Section 716.5.3.1. Because elevator hoistway doors do not typically comply as smoke- and draft-control assemblies, they would not be able to open directly into a corridor that is required to have protected openings. The provisions in Sections 713.14 and 3006 waiving the requirements for an elevator lobby do not waive the corridor opening protection requirements. Therefore, to maintain the integrity of the corridor, the elevator hoistway shaft doors opening into such rated corridors will need to be separated from the corridor by one of the following methods of protection:

1. A lobby needs to be provided with the appropriate walls and doors [see Commentary Figure 1020.6(4) and Section 3006.3, Items 1 and 2] to separate the lobby from the corridor.
2. Additional doors must be provided at the hoistway [see Commentary Figure 1020.6(5) and

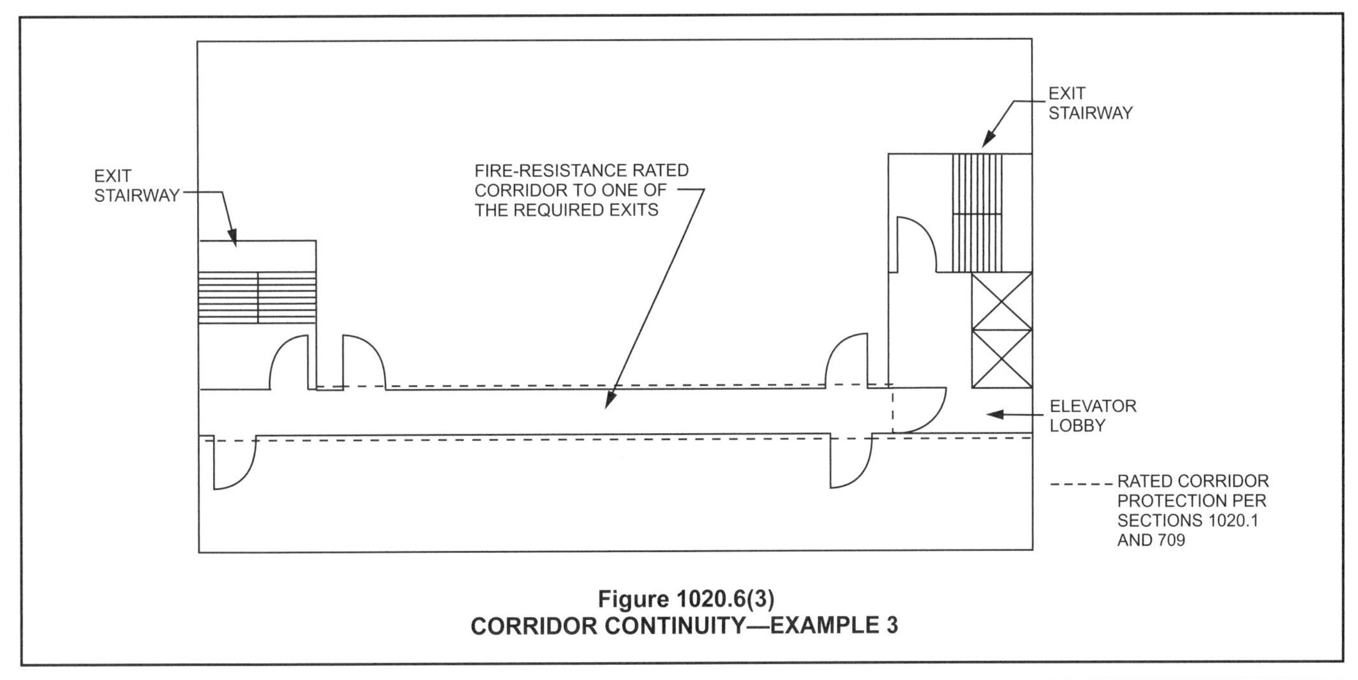

Figure 1020.6(3)
CORRIDOR CONTINUITY—EXAMPLE 3

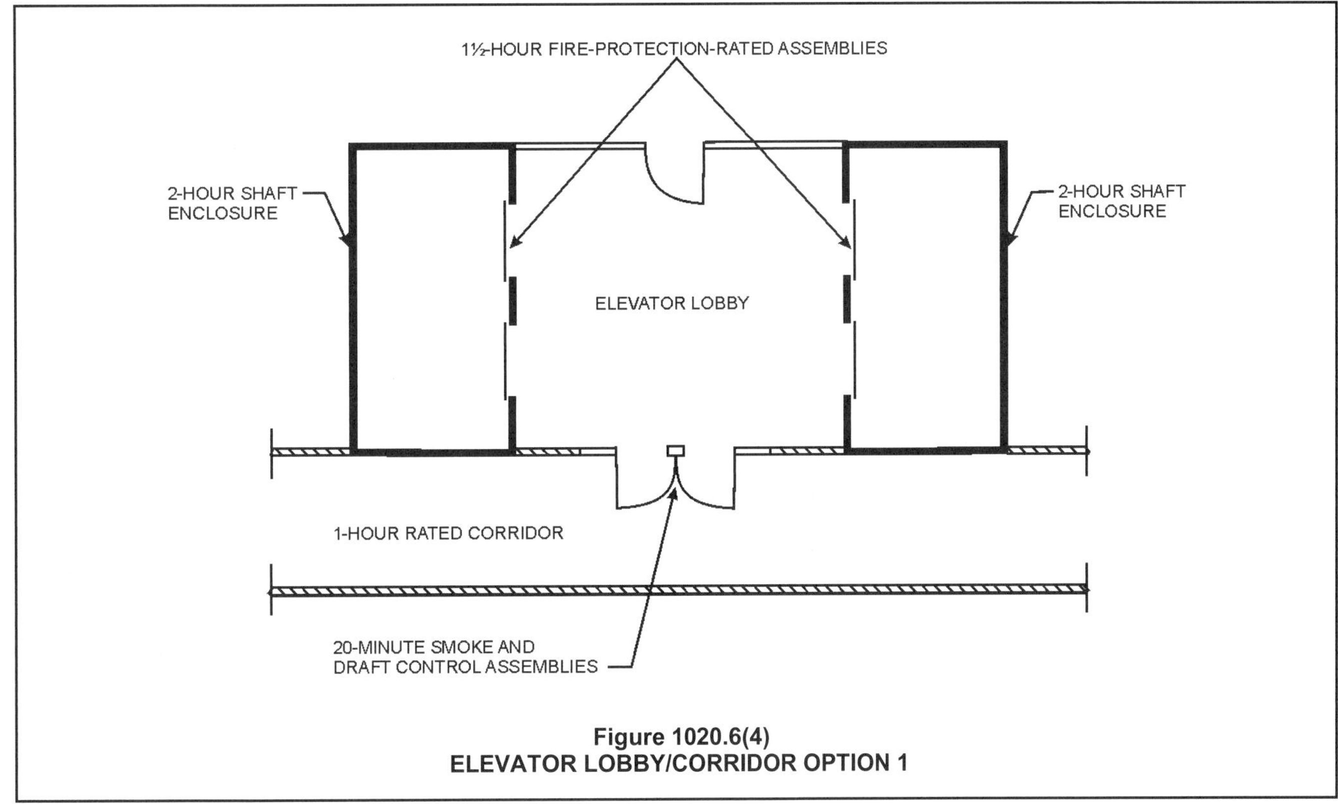

Figure 1020.6(4)
ELEVATOR LOBBY/CORRIDOR OPTION 1

Sections 3006.3 Item 3] that will protect the shaft openings the same as required for the corridor doors.

3. An elevator shaft door meeting both the smoke and draft protection requirements for corridor doors in Section 716.5.3.1 and the appropriate fire protection rating of Table 716.5 for the shaft must be provided.

4. The corridor must be separated from the lobby [see Commentary Figure 1020.6(6) and Section 3006.3, Items 1 and 2]. Per Exception 2 of Section 1020.6, Option 4 is permitted for corridor

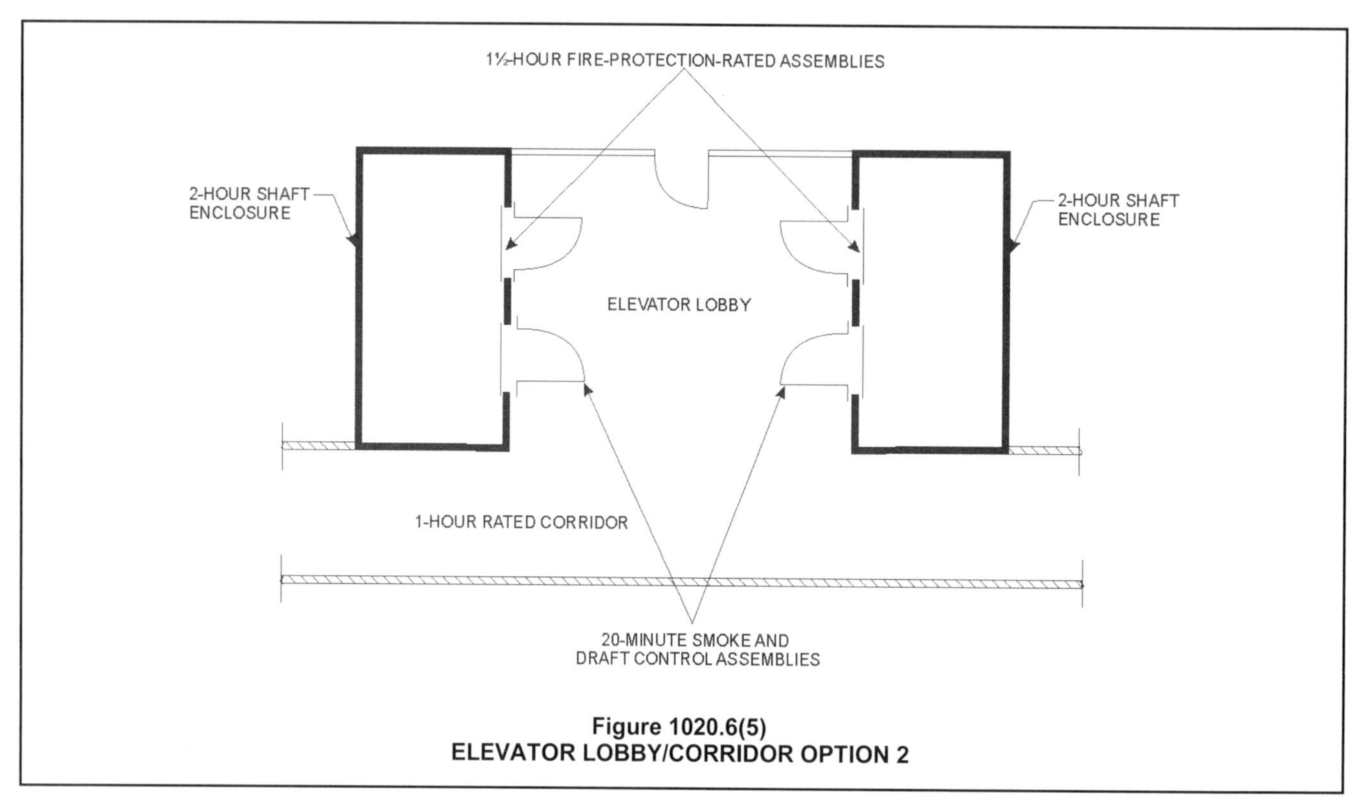

Figure 1020.6(5)
ELEVATOR LOBBY/CORRIDOR OPTION 2

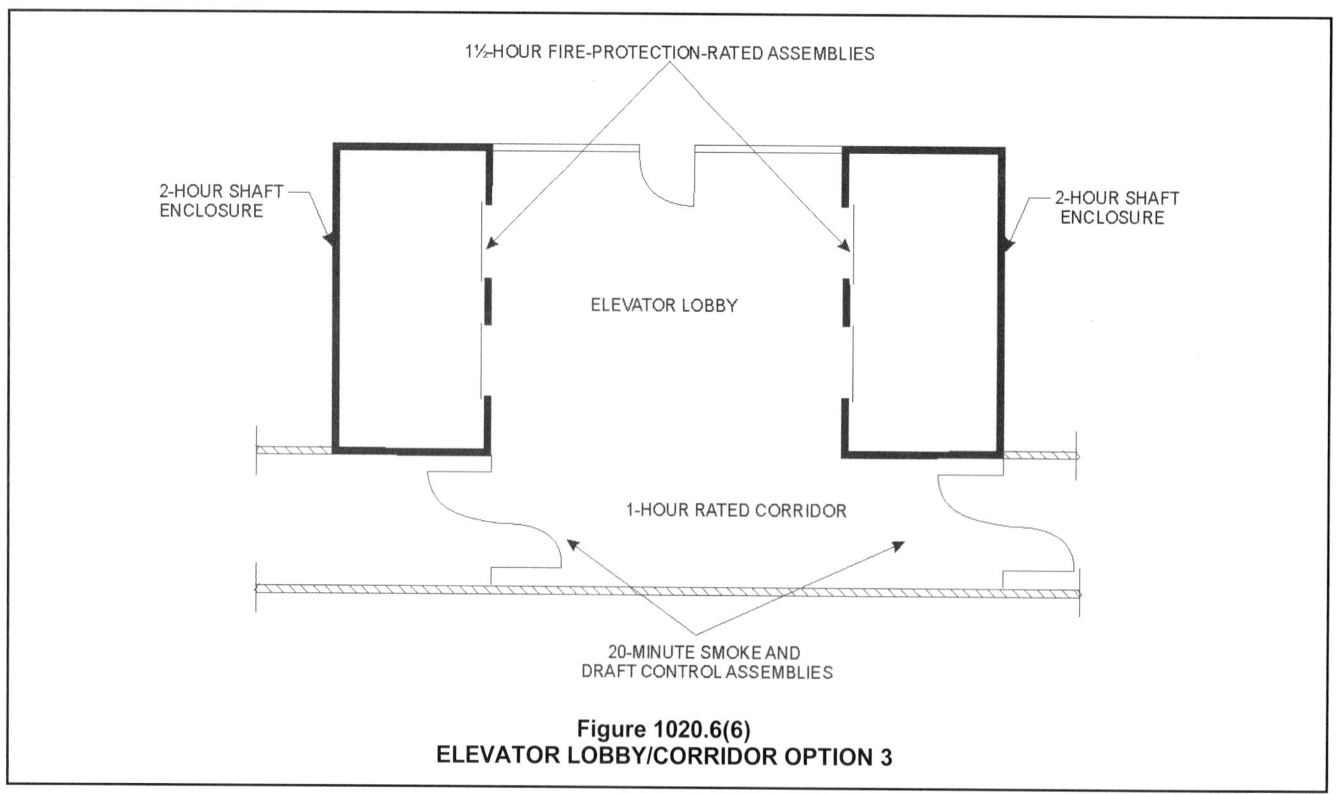

Figure 1020.6(6)
ELEVATOR LOBBY/CORRIDOR OPTION 3

continuity since the elevator lobby is also not considered an intervening room. Section 1016.2, Item 1, states that at least one end of a fire-resistance-rated corridor leads directly to an exit without going through the lobby. Alternatively, the second means of egress serving the corridor can be an elevator lobby with direct access to an exit [see Commentary Figure 1020.6(3)].

5. The elevator hoistway can be pressurized so that smoke will not move up the shaft (see Section 3006.3, Item 4).

While many elevator hoistway shaft doors are tested and labeled for the 1-hour or $1^1/_2$-hour fire-resistance rating (see Section 716.5), very few, if any of the doors typically sold in the U.S. will also meet the smoke and draft requirements (see Section 716.5.3.1) that would allow them to open directly into a fire-resistance-rated corridor. Because of this, Items 1, 2 and 4 above will be the general methods for protecting such openings.

For additional explanation of the requirements for elevator lobbies that are adjacent to rated corridors, see Section 3006. For requirements for exit enclosures, see Section 1023.

SECTION 1021
EGRESS BALCONIES

1021.1 General. Balconies used for egress purposes shall conform to the same requirements as *corridors* for minimum width, required capacity, headroom, dead ends and projections.

❖ This section regulates balconies that are used as an exit access element. Requirements are the same as exit access corridors, except for the enclosure.

Where exterior egress balconies are used in moderate or severe climates, there may also be a concern to protect the egress balcony from accumulations of snow and ice to provide a safe path of egress travel at all times, including winter. Maintenance of the means of egress in the IFC requires an unobstructed path to allow for full instant use in case of a fire or emergency. Typical methods for protecting these egress elements include roof overhangs or canopies, a heated slab and, when approved by the building official, a reliable snow removal maintenance program.

1021.2 Wall separation. Exterior egress balconies shall be separated from the interior of the building by walls and opening protectives as required for *corridors*.

Exception: Separation is not required where the exterior egress balcony is served by not less than two *stairways* and a dead-end travel condition does not require travel past an unprotected opening to reach a *stairway*.

❖ An exterior exit access balcony has a valuable attribute in that the products of combustion may be freely vented to the open air. In the event of a fire in an adjacent space, the products of combustion would not be expected to build up in the balcony area as would commonly occur in an interior corridor. However, there is still a concern for the egress of occupants who must use the balcony for exit access, and, consequently, may have to pass the room or space where the fire is located. Therefore, an exterior exit access balcony is required to be separated from interior spaces by fire partitions, as is required for interior corridors. The other provisions of Section 1020 relative to dead ends and opening protectives also apply.

If there are no dead-end conditions that require travel past an unprotected opening and the balcony is provided with at least two stairways, then the wall separating the balcony from the interior spaces need not have a fire-resistance rating (see Commentary Figure 1021.2). Such an arrangement reduces the probability that occupants will need to pass the area with the fire to gain access to an exit.

1021.3 Openness. The long side of an egress balcony shall be at least 50 percent open, and the open area above the *guards* shall be so distributed as to minimize the accumulation of smoke or toxic gases.

❖ This section provides an opening requirement that is intended to preclude the rapid buildup of smoke and toxic gases. A minimum of one side of the exterior

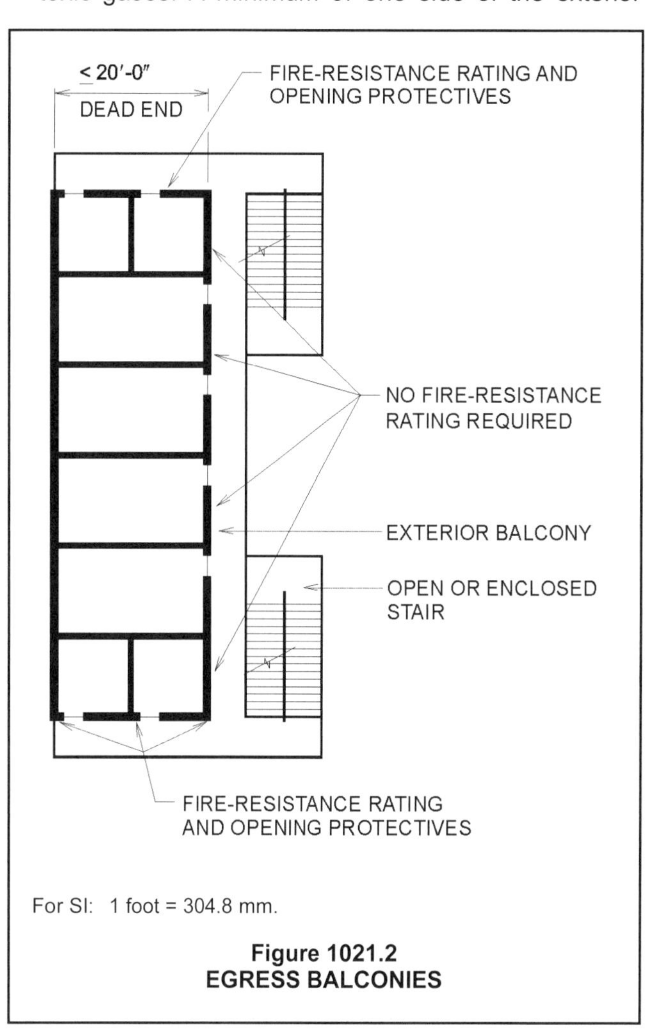

Figure 1021.2
EGRESS BALCONIES

balcony is required to have a minimum open exterior area of 50 percent of the side area of the balcony. The side openings are to be uniformly distributed along the length of the balcony.

1021.4 Location. Exterior egress balconies shall have a minimum *fire separation distance* of 10 feet (3048 mm) measured at right angles from the exterior edge of the egress balcony to the following:

1. Adjacent *lot lines*.
2. Other portions of the building.
3. Other buildings on the same lot unless the adjacent building *exterior walls* and openings are protected in accordance with Section 705 based on *fire separation distance*.

For the purposes of this section, other portions of the building shall be treated as separate buildings.

❖ The location requirements for exterior egress balconies given by this section protect the users of the egress balcony from the effects of a fire in another building on the same lot or an adjacent lot. The separation distance reduces the exposure to heat and smoke. If the egress balcony is closer than specified, then the adjacent buildings' exterior walls and openings are to be protected in accordance with Section 705 so that the users of the egress balcony are protected. The reason for the required distance to a lot line is to provide for a future building that could be built on the adjacent lot. While buildings on the same lot can be considered one building for height and area limitations (see Section 503.1.2), they must be separated by a minimum of 10 feet (3048 mm) if there is a path for exit discharge between them. The purpose of the last sentence is to clarify that an egress balcony needs a minimum 10-foot separation where a building wraps around on itself, such as a U-shaped building. It is not intended that the distance be measured to the imaginary lot line between buildings on the same lot.

Requirements are the same for exterior exit stairways and ramps. For an illustration of how exterior egress balconies and exterior exit stairways work together, see Commentary Figure 1027.5.

SECTION 1022 EXITS

1022.1 General. *Exits* shall comply with Sections 1022 through 1027 and the applicable requirements of Sections 1003 through 1015. An *exit* shall not be used for any purpose that interferes with its function as a *means of egress*. Once a given level of *exit* protection is achieved, such level of protection shall not be reduced until arrival at the *exit discharge*. *Exits* shall be continuous from the point of entry into the *exit* to the *exit discharge*.

❖ This group of sections is applicable to the "exit" portion of the three-part means of egress system. Sections 1003 through 1015 are also applicable to exits. The following sections are covered in this group:

- Section 1022 provides general requirements for exterior exit doorways.
- Section 1023 provides criteria for interior exit stairways and ramps.
- Section 1024 covers horizontal portions of the exit and exit passageways.
- Section 1025 is referenced by the high-rise provisions in Section 403. This section addresses a backup system for means of egress lighting (see Section 1008). "Glow-in-the-dark" stripes identify steps, handrails, obstructions and doorways within the enclosure for interior exit stairways.
- Section 1026 discusses the option of horizontal exits. This type of system is most commonly used for buildings such as hospitals and jails where building evacuation is not always the best option.
- Section 1027 provides criteria for stairways or ramps that are primarily open to the exterior, thus reducing the chance of accumulation of smoke or fumes.

The use of required exterior exit doors, exit stairways, exit passageways and horizontal exits for any purpose other than exiting is prohibited, because it might interfere with use as an exit. This is *not* intended to prohibit a door or stairway being used as part of normal circulation patterns, such as the exit doors also serving as entrances, or using the stairway to move between floors when there is not an emergency. However, these spaces must not include furniture, storage or work space. For example, the use of an exit stairway landing for storage, vending machines, copy machines, displays or any purpose other than for exiting is not permitted. Such a situation could not only lead to obstruction of the path of exit travel, thereby creating a hazard to life safety, but if the contents consist of combustible materials, then the use of the stairway as a means of egress could be jeopardized by fire or smoke in the exit enclosure.

It is recognized that standpipe risers are provided within the stair enclosure and that vertical electrical conduit may be necessary for power or lighting. However, such risers must be located so as not to interfere with the required clear width of the exit. For example, a standpipe riser located in the corner of a stairway will not reduce the required clear radius of the landing. This also applies when the stairway landing is used as an area of refuge. The spaces for wheelchairs must not obstruct the general path of egress travel [see Commentary Figures 1009.3(1) and 1009.3(2)]. Electrical conduit and mechanical equipment are permitted when necessary to serve the exit enclosure.

In existing buildings, sometimes the only viable option for providing access into the space is a platform lift. These platform lifts can be located within the enclosure for the exit (see IEBC Sections 410.8.3 and 705.1.3). Regulations in ASME A18.1 would limit the potential for any concerns for platform lifts being a fuel load in an exit stairway. However, the platform lift must be located so that it is not an obstruction to the exit pathway. Typically, this is reviewed when the platform lift is in the off and folded position.

Sections 1022 through 1027 apply to all exits but do not apply to elements of the means of egress that are not actually exits, such as exit access stairways, ramps, corridors and passageways or elements of the exit discharge. Once an exit is entered, that same level of protection must be available until the occupants leave the building at the level of exit discharge. For exit discharge options for the enclosure, other than a door leading directly to the outside, see the requirements for exit passageways in Section 1024 or the options permitting usage of a lobby or vestibule in the exceptions to Section 1028.1.

1022.2 Exterior exit doors. Buildings or structures used for human occupancy shall have not less than one exterior door that meets the requirements of Section 1010.1.1.

❖ The purpose of this section is to specify that at least one exterior exit door is required to meet the door size requirements in Section 1010.1.1. It is not the intent of this section to specify the number of exit doors required, which is addressed in Section 1006.

1022.2.1 Detailed requirements. Exterior *exit* doors shall comply with the applicable requirements of Section 1010.1.

❖ The purpose of this section is simply to provide a cross reference from the exit section to all of the detailed requirements for doors that are included in Section 1010.1 and all of its subsections. For example, the requirements for door operation on exterior exit doors are controlled by Section 1010.1.9.

1022.2.2 Arrangement. Exterior *exit* doors shall lead directly to the *exit discharge* or the *public way*.

❖ The exterior exit door is to be the entry point of the exit discharge or lead directly to the public way. When a person reaches the exterior exit door, he or she is directly outside, where smoke and toxic gases are not a health hazard. Additionally, this section will keep exterior doors at other locations, such as to an egress balcony, from being viewed as an exit element.

SECTION 1023 INTERIOR EXIT STAIRWAYS AND RAMPS

1023.1 General. *Interior exit stairways* and *ramps* serving as an *exit* component in a *means of egress* system shall comply with the requirements of this section. *Interior exit stairways* and *ramps* shall be enclosed and lead directly to the exterior of the building or shall be extended to the exterior of the building with an *exit passageway* conforming to the requirements of Section 1024, except as permitted in Section 1028.1. An *interior exit stairway* or *ramp* shall not be used for any purpose other than as a *means of egress* and a circulation path.

❖ The first sentence is a general reference to the rest of the section. Sections 1023.2 through 1023.7 deal with the construction of the walls and ceiling that enclose the exit stairway or exit ramp. Sections 1023.8 and 1023.9 deal with information required in the enclosure for safe exiting. Section 1023.10 provides criteria for stairways in high-rise or underground buildings where a smokeproof enclosure and pressurization of the stair tower are required (see Sections 403.5.4 and 405.7.2).

Most exit stairways or ramps have an exterior door at the level of exit discharge that leads directly to the outside. From this doorway, there must be a path for exit discharge that leads to the public way; however, there are other options. Exit passageways are considered an extension of the exit enclosure. The stairway enclosure discharges into the exit passageway, which in turn leads to the outside of the building. There are limited allowances for the exit enclosure to discharge through a lobby or vestibule (see Section 1028.1, Exceptions 1 and 2). A stairway enclosure could also discharge through a horizontal exit (see Section 1028.1, Exception 3). The termination requirements are the same as stated in Section 1023.3.

It is important that an exit stairway or ramp not be used for any purpose other than as a means of egress. For example, there is a tendency to use stairway landings for storage purposes. Such a situation obstructs the path of exit travel and if the stored contents consist of combustible materials, the use of the exit stairway as part of the path for a means of egress may be jeopardized, creating a hazard to life safety. It is not the intent of these provisions to prohibit an exit stairway from being used as part of the normal building circulation system. If the tenants or building owner have security concerns that would prompt them to wish to limit stairway access, consult Section 1010.1.9.11.

It is not the intent of this provision to exclude inclined platform lifts in the enclosure for the stairway; however, it is important that when not in operation they do not block access to the exit stairway or handrails. The referenced technical standard, ASME A18.1, basically requires noncombustible elements, so there is not a fire load issue associated with the lifts. Platform lifts are an important option for providing accessibility in a building. Section 1109.7 limits the use of platform lifts in new construction to mainly areas with minimal occupant loads or where elevators and ramps are impracticable. Platform lifts can be part of an accessible means of egress in limited situations (Section 1007.5). IEBC Sections 410.8.3 and 705.1.3 allow for platform lifts anywhere in existing buildings in order to gain accessibility for persons with mobility impairments. The industry is currently working on different options to address the concern that the lift may be in operation during an event that requires evacuation.

1023.2 Construction. Enclosures for *interior exit stairways* and *ramps* shall be constructed as *fire barriers* in accordance with Section 707 or *horizontal assemblies* constructed in accordance with Section 711, or both. *Interior exit stairway* and *ramp* enclosures shall have a *fire-resistance rating* of not less than 2 hours where connecting four stories or more and not less than 1 hour where connecting less than four stories. The number of stories connected by the *interior exit stairways* or *ramps* shall include any *basements*, but not any *mezzanines*. *Interior exit stairways* and *ramps* shall have a *fire-resistance rating* not less than the floor assembly penetrated, but need not exceed 2 hours.

Exceptions:

1. *Interior exit stairways* and *ramps* in Group I-3 occupancies in accordance with the provisions of Section 408.3.8.
2. *Interior exit stairways* within an *atrium* enclosed in accordance with Section 404.6.

❖ This section requires that all interior exit stairways or ramps are to be enclosed with rated walls (i.e., fire barriers) and floor/ceiling assemblies (i.e., horizontal assemblies). The fire-resistance rating required depends on the number of connected stories and the required fire-resistance rating of the penetrated floors. The minimum fire-resistance rating of an enclosure for an exit stairway or ramp is at least 1 hour. The fire-resistance rating of the enclosure must be increased to 2 hours if the stairway or ramp connects four or more stories or if it penetrates a floor system with a fire-resistance rating of 2 hours or more (see Table 602 for Type I construction). Note that the criteria are based on the number of stories connected by the stairway or ramp and not the height of the building. Therefore, a building that has three stories located entirely above the grade plane and a basement would require an enclosure with a 2-hour fire-resistance rating if the stairway or ramp connects all four stories. Where the floor construction penetrated by the enclosure has a fire-resistance rating, the enclosure must have the same minimum rating. For example, an enclosure that penetrates a 2-hour floor assembly must have a minimum fire-resistance rating of 2 hours, regardless of the number of stories the enclosure connects. The fire-resistance rating of an enclosure need never exceed 2 hours. If the floor assembly penetrated requires a minimum 3-hour fire-resistance rating, the enclosure rating is only required to be 2-hour fire-resistance rated. All linear voids at joints between fire-resistance-rated wall and floor/ceiling assemblies and where an enclosure would intersect with an exterior wall must be filled so that the integrity of the enclosure is maintained (see Section 715). The fire-resistance-rated requirements for enclosures for exit stairways and ramps are consistent with those for shaft enclosures and enclosures for exit access stairways and ramps.

The enclosure is needed because an exit stairway or ramp penetrates the floor/ceiling assemblies between the levels, thus creating a vertical opening or shaft. In cases of fire, a vertical opening may act as a chimney, causing smoke, hot gases and light-burning products to flow upward (buoyant force). If an opening is unprotected, these products of combustion will be forced by positive pressure differentials to spread horizontally into the building spaces. There are exceptions for shaft protection around stairways and ramps that are not part of a required means of egress in Section 712 or exit access stairways as permitted in Section 1019.3.

The enclosure of interior stairways or ramps with construction having a fire-resistance rating is intended to prevent the spread of fire from floor to floor. Another important purpose is to provide a safe path of travel for the building occupants and to serve as a protected means of access to the fire floor by fire department personnel. For this reason, Sections 1023.4 through 1023.6 limit the penetrations and openings permitted in the enclosure for an exit stairway or ramp.

For travel distance measurements at the exit stairways, see the commentary to Section 1017.3. While not specifically mentioned as an exception, Section 1027 for exterior exit stairways is considered to provide an equivalent level of protection to enclosed interior stairways.

Per Exception 1, because of security needs in detention facilities, one of the exit stairways is permitted to be glazed in a manner similar to atrium enclosures. Specific limitations and requirements are discussed in Section 408.3.8. Exit access stairways within housing units are addressed in Sections 408.5.1 and 1019.4.

Exception 2 is in consideration of an increased level of safety in an atrium. Section 404 for atriums requires the space to be enclosed by a 1-hour passive enclosure and also protected by various active systems including fire suppression and smoke control features. The natural configuration of an atrium affords building occupants immediate views of the entire egress to the bottom of the atrium.

1023.3 Termination. *Interior exit stairways* and *ramps* shall terminate at an *exit discharge* or a *public way*.

Exception: A combination of *interior exit stairways*, *interior exit ramps* and *exit passageways*, constructed in accordance with Sections 1023.2, 1023.3.1 and 1024, respectively, and forming a continuous protected enclosure, shall be permitted to extend an *interior exit stairway* or *ramp* to the *exit discharge* or a *public way*.

❖ The intent of this section is to provide safety in all portions of the exit by requiring continuity of the fire protection characteristics of the enclosure for the exit stairway. Exit passageways (see Section 1024) are a continuation of the enclosure for the exit stairway. This would include, but not be limited to, the fire-resistance rating of the exit enclosure walls and the opening protection rating of the doors. While an exit passageway is most commonly found on the level of exit discharge as a means to connect a stairway enclosure to the exterior, in buildings that step back in footprint as the rise, the stairways may not be totally vertical shafts, but move out as they move down to keep the required separation distance. Exit passageways can be at any level to connect the stair towers. There are special exit signage considerations for this particular issue in Section 1013.1 (see Commentary Figure 1013.1).

Section 1028.1 would allow for an alternative for direct access to the outside via an intervening lobby or vestibule.

Horizontal exits (see Section 1026), while not providing direct access to the outside of the structure, do move occupants to another "building" by moving through a fire wall (see Sections 1026 and 1028.1, Exception 3) into a refuge area protected by fire barriers and horizontal assemblies. Horizontal exits are commonly used in hospitals and jails for a defend-in-place type of protection.

1023.3.1 Extension. Where *interior exit stairways* and *ramps* are extended to an *exit discharge* or a *public way* by an *exit passageway*, the *interior exit stairway* and *ramp* shall be separated from the *exit passageway* by a *fire barrier* constructed in accordance with Section 707 or a *horizontal assembly* constructed in accordance with Section 711, or both. The *fire-resistance rating* shall be not less than that required for the *interior exit stairway* and *ramp*. A *fire door* assembly complying with Section 716.5 shall be installed in the *fire barrier* to provide a *means of egress* from the *interior exit stairway* and *ramp* to the *exit passageway*. Openings in the *fire barrier* other than the *fire door* assembly are prohibited. Penetrations of the *fire barrier* are prohibited.

Exceptions:

1. Penetrations of the *fire barrier* in accordance with Section 1023.5 shall be permitted.
2. Separation between an *interior exit stairway* or *ramp* and the *exit passageway* extension shall not be required where there are no openings into the *exit passageway* extension.

❖ Once a person enters the enclosure surrounding an exit stairway, that same level of protection should be provided to them until they can leave the building. When an enclosure for an exit stairway connects to an exit passageway, either at the ground level or at an intermediate transition floor, the exit passageway must provide the same level of protection as the enclosure for the exit stairway, including fire-resistance of the walls, floor, ceiling and supporting construction and protection of any openings. At the junction between the enclosure for the exit stairway and the exit passageway, there must be both a rated fire barrier and a fire door. This has the additional benefit of preventing any smoke that may migrate into the exit passageway from also moving up the exit stairway or ramp. Permitted penetrations for the exit passageway are the same as those limitations set for the enclosure for the exit stairway. See Sections 1013.1 and 1025 for egress markings and signage within these types of spaces.

For the situation when an exit stairway is constructed as a smokeproof enclosure or a pressurized stairway, see Section 1023.10.1.

Exception 1—The penetrations for the exit passageway are allowed to be the same as permitted for the stairway enclosure.

Exception 2—The purpose in having a door at this interface in the existing requirement is to prevent smoke from a possible open door or other penetration in the passageway from traveling up the exit enclosure. This is prevented if there are no openings or penetrations in the exit passageway. Then the exit passageway is horizontal offset of the exit enclosure and does not propose the same hazard. Egress can proceed faster if there are no intermediate doors contained at the enclosure transitions.

1023.4 Openings. *Interior exit stairway* and *ramp* opening protectives shall be in accordance with the requirements of Section 716.

Openings in *interior exit stairways* and *ramps* other than unprotected exterior openings shall be limited to those necessary for *exit access* to the enclosure from normally occupied spaces and for egress from the enclosure.

Elevators shall not open into *interior exit stairways* and *ramps*.

❖ In order for fire doors to be effective, they must be in the closed position; therefore, the preferred arrangement is to install self-closing doors. Recognizing that operational practices often require doors to be open for an extended period of time, automatic-closing doors are permitted as long as this opening will not pose a threat to occupant safety and the doors will be self-latching. Automatic-closing devices enable the opening to be protected during a fire condition. The basic requirement for closing devices and specific requirements for automatic-closing and self-closing devices are given in NFPA 80. Automatic-closing doors that are provided at protected openings in exits are also required to close on the actuation of smoke detectors or loss of power to the smoke detectors (see Section 716.5.9.3).

The only openings that are permitted in fire-resistance-rated enclosures for exit stairways or ramps are doors that lead from normally occupied spaces into the enclosure and doors leading out of the enclosure to the outside. This restriction on openings essentially prohibits the use of windows in an exit enclosure except for those exterior windows that are not exposed to any hazards. This requirement is not intended to prohibit windows or other openings in the exterior walls of the exit enclosure. The verbiage "unprotected exterior openings" includes windows or doors not required to be protected by either Section 705.8 or 1023.7. The only exception would be window assemblies that have been tested as wall assemblies in accordance with ASTM E119 or UL 263. The objective of this provision is to minimize the possibility of fire spreading into an enclosure and endangering the occupants or even preventing the use of the exit at a time when it is most needed. The limitation on openings applies regardless of the fire protection rating of the opening protective. The limitation on openings from normally occupied areas is intended to reduce the probability of a fire occurring in an unoccupied area, such as a storage closet, which has an opening into the stairway, thereby possibly resulting in fire spread into the stairway. Other spaces that are not normally occupied include, but are not limited to, toilet rooms, electrical/mechanical equipment rooms and janitorial closets. For connection between the vertical exit enclosure and an exit passageway, see Section 1023.3.1.

Elevators may not open into exit enclosures. The difficulty is to have elevator doors that can meet the opening protectives for a fire barrier, but still operate effectively as elevator doors. For additional information on elevator lobbies and doors, see the commentary for Sections 713.14, 1020.6 and 3006.

These opening limitations are very similar to those required for an exit passageway (see Section 1024.5).

1023.5 Penetrations. Penetrations into or through *interior exit stairways* and *ramps* are prohibited except for equipment and ductwork necessary for independent ventilation or pressurization, sprinkler piping, standpipes, electrical raceway for fire department communication systems and electrical raceway serving the *interior exit stairway* and *ramp* and terminating at a steel box not exceeding 16 square inches (0.010 m^2). Such penetrations shall be protected in accordance with Section 714. There shall not be penetrations or communication openings, whether protected or not, between adjacent *interior exit stairways* and *ramps*.

Exception: Membrane penetrations shall be permitted on the outside of the *interior exit stairway* and *ramp*. Such penetrations shall be protected in accordance with Section 714.3.2.

❖ This section specifically lists the items that are allowed to penetrate the walls and ceiling of the enclosure for the exit stairway. This is consistent for all types of enclosures for exits, including interior exit stairways or interior exit ramps and exit passageways (see Section 1024.6). In general, only portions of the building service systems that serve the enclosure are allowed to penetrate the enclosure. As indicated in the commentary to Section 1022.1, standpipe systems are commonly located in the exit stair enclosures. If two exit enclosures are adjacent to one another, there must be no penetrations between them, thereby limiting the chances of smoke being in both stairwells.

Section 714 addresses through penetrations and membrane penetrations for fire-resistance-rated wall and floor/ceiling assemblies. The intent is to maintain the integrity of the enclosure for the exit access stairway. This section and Section 1023.6 are meant to work together. Penetrations in exterior walls are addressed in Section 1023.7.

The exception allows for electrical boxes, "Exit" signs or fire alarm pull stations to be installed on the outside of the enclosure, provided that the boxes are installed so that the required fire-resistance rating is not reduced (see Section 714.3.2).

1023.6 Ventilation. Equipment and ductwork for *interior exit stairway* and *ramp* ventilation as permitted by Section 1023.5 shall comply with one of the following items:

1. Such equipment and ductwork shall be located exterior to the building and shall be directly connected to the *interior exit stairway* and *ramp* by ductwork enclosed in construction as required for shafts.
2. Where such equipment and ductwork is located within the *interior exit stairway* and *ramp*, the intake air shall be taken directly from the outdoors and the exhaust air shall be discharged directly to the outdoors, or such air shall be conveyed through ducts enclosed in construction as required for shafts.
3. Where located within the building, such equipment and ductwork shall be separated from the remainder of the building, including other mechanical equipment, with construction as required for shafts.

In each case, openings into the *fire-resistance-rated* construction shall be limited to those needed for maintenance and operation and shall be protected by opening protectives in accordance with Section 716 for shaft enclosures.

The *interior exit stairway* and *ramp* ventilation systems shall be independent of other building ventilation systems.

❖ The purpose of the requirements for ventilation system equipment and ductwork is to maintain the fire resistance of the enclosure for the exit stairway. The ventilation system serving the enclosure is to be independent of other building systems to prevent smoke in the enclosure from traveling to other areas of the building. This section and Section 1023.5 are meant to work together. Where ductwork penetrates the outside wall, if it is not required to be rated by Section 1023.7, then the duct does not require a fire and smoke damper. If the ductwork serving the shaft does penetrate a rated wall or rated floor/ceiling assembly, a fire and smoke damper would be required by Section 717.5.3. Openings required for access to the ventilation system for maintenance and operation shall be protected in accordance with Section 716.

1023.7 Interior exit stairway and ramp exterior walls. *Exterior walls* of the *interior exit stairway* or *ramp* shall comply with the requirements of Section 705 for *exterior walls*. Where nonrated walls or unprotected openings enclose the exterior of the *stairway* or *ramps* and the walls or openings are exposed by other parts of the building at an angle of less than 180 degrees (3.14 rad), the building *exterior walls* within 10 feet (3048 mm) horizontally of a nonrated wall or unprotected opening shall have a *fire-resistance rating* of not less than 1 hour. Openings within such *exterior* walls shall be protected by opening protectives having a *fire protection rating* of not less than $^3/_4$ hour. This construction shall extend vertically from the ground to a point 10 feet (3048 mm) above the topmost landing of the *stairway* or *ramp*, or to the roof line, whichever is lower.

❖ This section does not require exterior walls of an enclosure for an exit stairway or ramp to have the same fire-resistance rating as the interior walls. Table 602 and Section 705 establish when exterior walls are required to be rated and openings limited due to adjacent buildings or lot lines. This exposure is different from exterior load-bearing walls that are required to be rated because of the type of construction per Table 601. What is unique to exterior walls at exit stairways is the need to stop fires from burning through an exterior wall adjacent to the enclosure, which may then jeopardize the occupant's ability to continue to use that exit stairway. Essentially, there are two alternatives where an exposure hazard exists: 1. Provide protection to the stairway by having a fire-resistance rating on its exterior wall; or 2. Provide a fire-resistance rating to the walls adjacent to the stairway. The ratings apply for a distance of 10 feet (3048 mm) measured horizontally and vertically from the stairway enclosure where those walls are at an angle of less than 180 degrees (3.14 rad) from the exterior wall portion of the enclosure [see Condition 1 in Commentary Figure 1023.7(1)]. When the adjacent exterior wall is protected in lieu of the stairway enclosure wall, the protection is to extend from the ground to a level of 10 feet (3048 mm) above the highest landing of the stairway. However, the protection is not required to extend beyond the normal roof line of the building.

The 180-degree (3.14 rad) angle criterion is based on the scenario where the exterior wall of the stair enclosure is in the same plane and flush with the exterior wall of the building [see Conditions 2 and 3 in Commentary Figure 1023.7(1)]. In this scenario, heat or fire would need to travel 180 degrees (3.14 rad) around in order to impinge on the stair. Based on studies of existing buildings, this 180-degree (3.14 rad) spread of fire does not appear to be a problem. This criterion is only applicable when the angle between the walls is 180 degrees (3.14 rad) or less.

As the fire exposure on the exterior is different than can be expected on the interior, the fire-resistance rating of the exterior wall is not required to exceed 1 hour, regardless of whether it is the stairway enclosure wall or the adjacent exterior wall, unless the exterior wall is required by other sections of the code to have a higher fire-resistance rating (see Tables 601 and 602). The fire protection rating on any openings in the exterior wall of a stairway enclosure or adjacent exterior wall within 180 degrees (3.14 rad) is to be a minimum of $^3/_4$ hour [see Commentary Figure 1023.7(2)].

In a situation where the upper levels are smaller than lower levels, an interior stairway can end up having an exterior wall when it moves above the roof of the lower levels. In this situation, the question is the rating requirements for the exterior wall of the stairway over the roof. Therefore, the exterior wall of the stairway must meet the vertical opening provisions in Section 705.8.6.

1023.8 Discharge identification. An *interior exit stairway* and *ramp* shall not continue below its *level of exit discharge* unless an *approved* barrier is provided at the *level of exit discharge* to prevent persons from unintentionally continuing into levels below. Directional exit signs shall be provided as specified in Section 1013.

❖ So that building occupants using an exit stairway during an emergency situation will be prevented from going past the level of exit discharge, the run of the stairway is to be interrupted by a partition, door, gate or other approved means. These devices help the users of the stairway to recognize when they have reached the point that is the level of exit discharge. Exit signs, including raised letters and braille, are to be provided for occupant guidance at the door leading to the way out (i.e., directly to the exterior, or via an exit passageway, lobby or vestibule). Furthermore, signs are to be placed at each floor landing in all interior exit stairways connecting more than three floor levels, designating the level or story of the landings in accordance with Section 1023.9.

The code does not specify the type of material or construction of the barrier used to identify the level of exit discharge. The key issues to be considered in the selection and approval of the type of barrier to be used are: 1. Whether the barrier provides a visible and physical means of alerting occupants who are exiting under emergency conditions that they have reached the level of exit discharge; and 2. Whether the barrier is constructed of materials that are permitted by the construction type of the building. In an emergency situation, some occupants are likely to come in contact with the barrier during exiting before realizing that they are at the level of exit discharge. Therefore, the barrier should be constructed in a manner that is substantial enough to withstand the anticipated physical contact, such as pushing or shoving. It would be reasonable, as a minimum, to design the barrier to withstand the structural load requirements of Section 1607.5 for interior walls and partitions. The barrier could be opaque (such as gypsum wallboard and stud framing) or not (such as a wire grid-type material).

The use of signage only or relatively insubstantial barriers, such as ropes or chains strung across the opening, is typically not sufficient to prevent occupants from attempting to continue past the level of exit discharge during an emergency.

Commentary Figure 1023.8 is an example of one method of discharge identification.

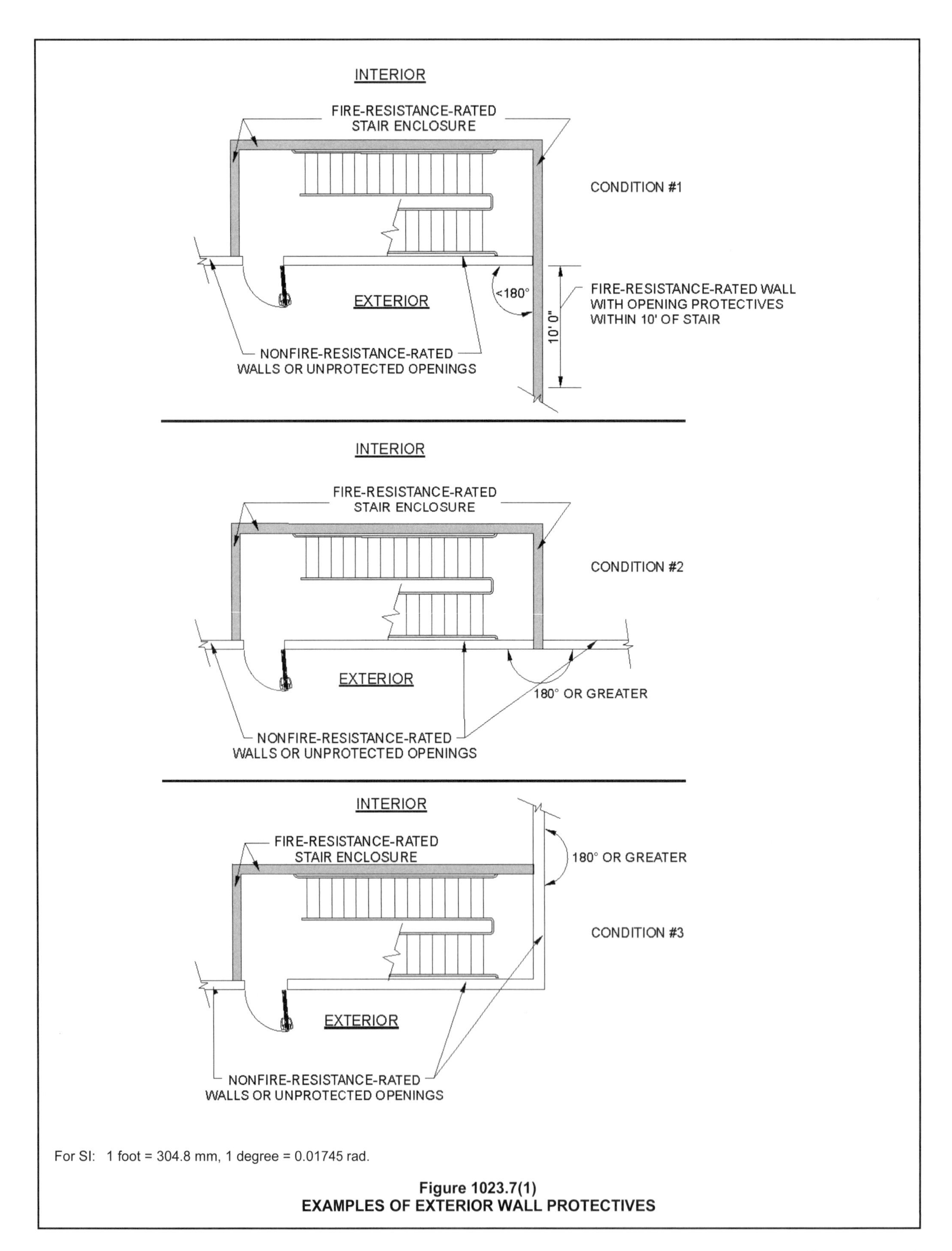

For SI: 1 foot = 304.8 mm, 1 degree = 0.01745 rad.

Figure 1023.7(1)
EXAMPLES OF EXTERIOR WALL PROTECTIVES

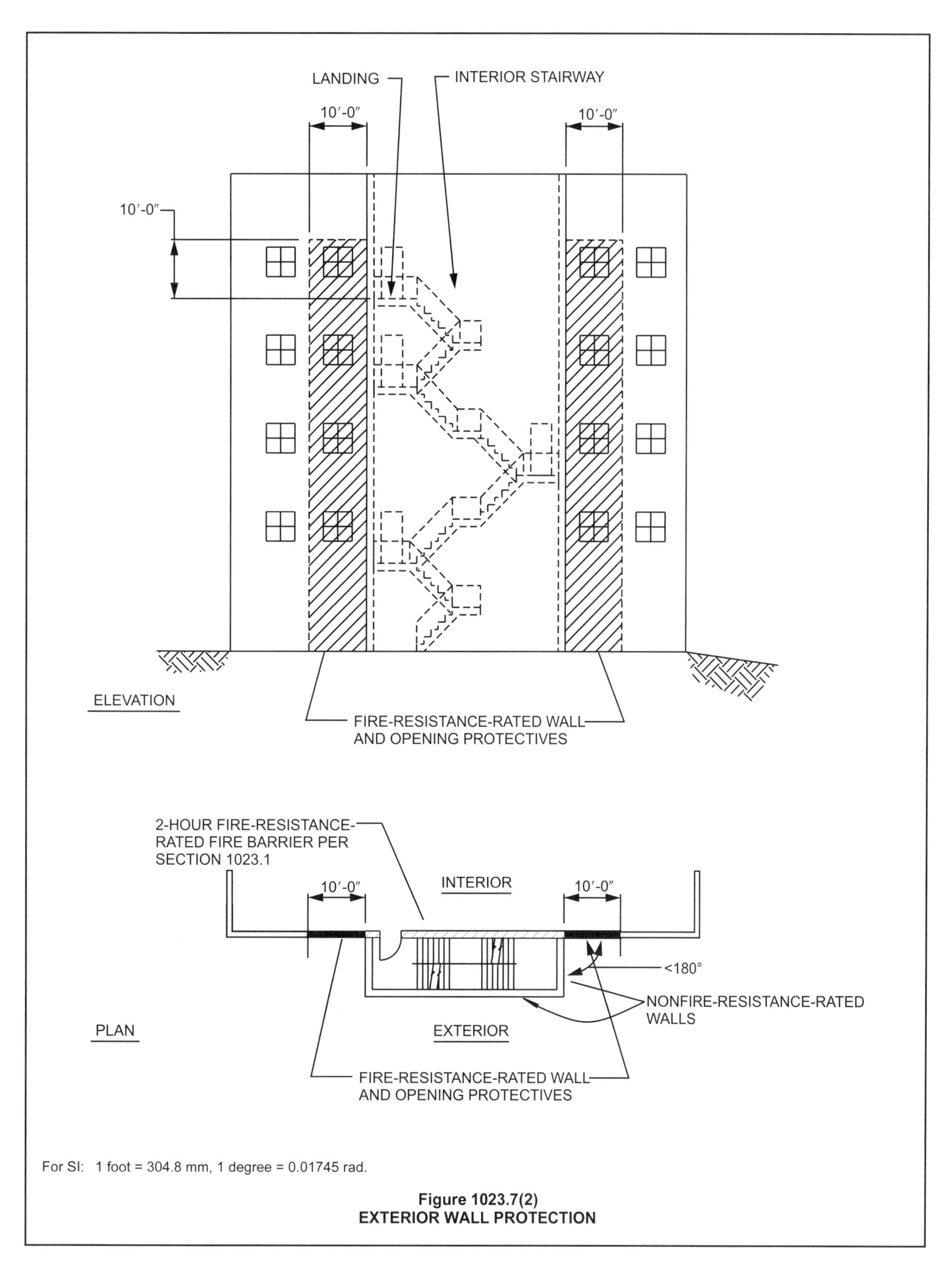

For SI: 1 foot = 304.8 mm, 1 degree = 0.01745 rad.

Figure 1023.7(2)
EXTERIOR WALL PROTECTION

1023.9 Stairway identification signs. A sign shall be provided at each floor landing in an *interior exit stairway* and *ramp* connecting more than three stories designating the floor level, the terminus of the top and bottom of the *interior exit stairway* and *ramp* and the identification of the *stairway* or *ramp*. The signage shall also state the story of, and the direction to, the *exit discharge* and the availability of roof access from the *interior exit stairway* and *ramp* for the fire department. The sign shall be located 5 feet (1524 mm) above the floor landing in a position that is readily visible when the doors are in the open and closed positions. In addition to the *stairway* identification sign, a floor-level sign in visual characters, raised characters and braille complying with *ICC A117.1* shall be located at each floor-level landing adjacent to the door leading from the *interior exit stairway* and *ramp* into the *corridor* to identify the floor level.

❖ This section discusses two distinct sign requirements that have totally different purposes.

Signs are to be placed at each floor landing in all exit stairways connecting more than three stories. The signs are to designate the level or story of the landings above or below the level of exit discharge. The purpose is to inform the occupants of their location with respect to the level of exit discharge as they use the stairway to leave the building. More importantly, it allows the fire service to locate and gain quick access to the fire floor. At each level, the direction to the exit discharge is required to be indicated. The identification of the level that is the exit discharge is also to be indicated at each level. The identification of the roof access availability is for the fire department (see Commentary Figure 1023.9.1). Roof access is required by Section 1011.13. For visibility, the signs are required to be located approximately 5 feet (1524 mm) above the floor surface and to be vis-

Figure 1023.9.1
STAIRWAY IDENTIFICATION SIGN

Figure 1023.8
EXAMPLE OF A STAIRWAY BARRIER AT THE LEVEL OF EXIT DISCHARGE

ible when the stairway door is open. The need to designate levels remaining to reach the level of exit discharge may mean that the numbering is other than the normal designation used by building management. For example, a designation of P1, P2, P3, etc., would not be acceptable for stairways in the basement parking garage, since in themselves they do not designate the floor level below the level of exit discharge.

To aid people with vision impairments, the floor designation must also be available in visual and raised letters and braille at each door. The intent is to let persons with vision impairments be able to find out what floor they are on without leaving the stairway enclosure. Visual and raised letters and braille signage indicating the door leading to the exterior is covered in Section 1013.4.

1023.9.1 Signage requirements. *Stairway* identification signs shall comply with all of the following requirements:

1. The signs shall be a minimum size of 18 inches (457 mm) by 12 inches (305 mm).
2. The letters designating the identification of the *interior exit stairway* and *ramp* shall be not less than $1^1/_2$ inches (38 mm) in height.
3. The number designating the floor level shall be not less than 5 inches (127 mm) in height and located in the center of the sign.
4. Other lettering and numbers shall be not less than 1 inch (25 mm) in height.
5. Characters and their background shall have a nonglare finish. Characters shall contrast with their background, with either light characters on a dark background or dark characters on a light background.
6. Where signs required by Section 1023.9 are installed in the *interior exit stairways* and *ramps* of buildings subject to Section 1025, the signs shall be made of the same materials as required by Section 1025.4.

❖ The requirements for stairway identification signage will provide for a consistent approach. The intent is to make signs visible and immediately recognizable to occupants and emergency responders using the stairway (see Commentary Figure 1023.9.1).

In addition, if the building is a high-rise and luminous egress path markings are required (see Section 1025.1), the stairway identification signage must also be self-luminous or photoluminescent. In order to also meet the contrast requirements in Item 5, typically the sign will have dark letters on a glow-in-the-dark background.

1023.10 Elevator lobby identification signs. At landings in *interior exit stairways* where two or more doors lead to the floor level, any door with direct access to an enclosed elevator lobby shall be identified by signage located on the door or directly adjacent to the door stating "Elevator Lobby." Signage shall be in accordance with Section 1023.9.1, Items 4, 5 and 6.

❖ This section is mainly related to fire service access elevators required in Section 403.6.1 for a building with an occupied floor more than 120 feet (3048 mm) above the street. The elevator lobby is required to have direct access to an exit stairway (Section 3007.6.1). In addition, that same exit enclosure has to have access to the floor without going back through the elevator lobby (see Section 3007.9.1). This leads to two doors at each floor landing. Elevator lobby signs identify the correct door though which fire fighters should access the floor so that lobby smoke protection is maintained.

Since typically the fire service access elevators will use the same lobby as the occupant evacuation elevators (when provided), this signage would effectively be required for the doors in the stairway enclosure that lead to occupant evacuation elevator lobbies as well.

1023.11 Smokeproof enclosures. Where required by Section 403.5.4 or 405.7.2, *interior exit stairways* and *ramps* shall be *smokeproof enclosures* in accordance with Section 909.20.

❖ While smokeproof enclosures and pressurized stairways for exiting can, at the designer's option, be used in buildings of any occupancy, height or area, this section specifically requires smokeproof enclosures or pressurized stairways to be provided when either of two conditions occur.

The first condition requires all exit stairways in high-rise buildings [i.e., buildings with floor levels higher than 75 feet (22 860 mm) above the level of exit discharge (see Section 403.5.4)] to be smokeproof enclosures or pressurized stairways. The reason for this provision is that in very tall buildings, often during fire emergencies, total and immediate evacuation of the occupants cannot be readily accomplished. In such situations, exit stairways become places of safety for the occupants and must be adequately protected with smokeproof enclosures or pressurization to provide a safe egress environment. In order to provide this safe environment, the enclosure must be constructed to resist the migration of smoke caused by the "stack effect." Stack effect occurs in tall enclosures such as chimneys, when a fluid such as smoke, which is less dense than the ambient air, is introduced into the enclosure. The smoke will rise because of the effects of buoyancy and will induce additional flow into the enclosure through openings of any size at the lower levels.

The second condition applies for underground buildings [i.e., when an occupiable floor level is located more than 30 feet (9144 mm) below the level of exit discharge (see Section 405.7.2)]. Stairways serving those levels are also required to be protected by smokeproof enclosures or pressurization because underground portions of a building present unique problems in providing not only for life safety but also access for fire-fighting purposes. The choice of a 30-foot (9144 mm) threshold for this requirement is intended to provide a reasonable limitation on vertical travel distance before the requirement applies.

Detailed system requirements for a smokeproof enclosure or pressurization are in Section 909.20. Note that Sections 403 and 405 each have exceptions for specific building types that are not to be regulated by those sections. Likewise, the requirements of this section do not apply to buildings identified in those exceptions.

1023.11.1 Termination and extension. A *smokeproof enclosure* shall terminate at an *exit discharge* or a *public way*. The *smokeproof enclosure* shall be permitted to be extended by an *exit passageway* in accordance with Section 1023.3. The *exit passageway* shall be without openings other than the *fire door assembly* required by Section 1023.3.1 and those necessary for egress from the *exit passageway*. The *exit passageway* shall be separated from the remainder of the building by 2-hour *fire barriers* constructed in accordance with Section 707 or *horizontal assemblies* constructed in accordance with Section 711, or both.

Exceptions:

1. Openings in the *exit passageway* serving a *smokeproof enclosure* are permitted where the *exit passageway* is protected and pressurized in the same manner as the *smokeproof enclosure*, and openings are protected as required for access from other floors.
2. The *fire barrier* separating the *smokeproof enclosure* from the *exit passageway* is not required, provided the *exit passageway* is protected and pressurized in the same manner as the *smokeproof enclosure*.
3. A *smokeproof enclosure* shall be permitted to egress through areas on the *level of exit discharge* or vestibules as permitted by Section 1028.

❖ The walls forming the smokeproof enclosure, which includes the stairway shaft and the vestibules, must be fire barriers having a fire-resistance rating of at least 2 hours. This level of fire endurance is specified because exit stairways in high-rise buildings serve as principal components of the egress system and as the source of fire service access to the fire floor. This supersedes any allowed reduction of enclosure rating, even if the stair from the level that is more than 30 feet (9144 mm) below exit discharge connects three stories or less. A pressurized stairway is a special case of a smokeproof enclosure; therefore, the requirements would be the same.

The first exception applies to openings in the exit passageway that are permitted, provided the exit passageway is protected and pressurized. If the exit stairway enclosure is connected to an exit passageway, Exception 2 allows the elimination of a door between, since this would interfere with the pressurization of the stairway and passageway as a combined exit system. In accordance with Exception 3, 50 percent of the stairways in smokeproof enclosures or pressurized stairways can use the exit discharge exceptions in Section 1028 to egress through a lobby or vestibule.

1023.11.2 Enclosure access. Access to the *stairway* or *ramp* within a *smokeproof enclosure* shall be by way of a vestibule or an open exterior balcony.

Exception: Access is not required by way of a vestibule or exterior balcony for *stairways* and *ramps* using the pressurization alternative complying with Section 909.20.5.

❖ See Commentary Figures 1023.11.2(1) and 1023.11.2(2) for illustrations of access to the smokeproof stairway or ramp by way of a vestibule or an exterior balcony. The purpose of this requirement is to keep the enclosure clear of smoke. Where a pressurized stairway is used, these elements are not necessary.

SECTION 1024
EXIT PASSAGEWAYS

1024.1 Exit passageways. *Exit passageways* serving as an exit component in a *means of egress* system shall comply with the requirements of this section. An *exit passageway* shall not be used for any purpose other than as a *means of egress* and a *circulation path*.

❖ This section provides acceptable methods of continuing the protected path of travel for building occupants. The building designer or owner is given different options for achieving this protected path of travel. See Commentary Figure 1024.1 for an illustration of an exit passageway arrangement. In the case of office buildings or similar structures, the exit stairways are often located at the central core or in line with the centrally located exit access corridors. Exit passageways may be used to connect the exit stair to the exterior exit door or to connect enclosures for exit stairways that are not vertically continuous. Such an arrangement provides great flexibility in the design use of the building. Without an exit passageway at the grade floor or the level of exit discharge, the occupants of the upper floors or basement levels would have to leave the safety of the exit stairway to travel to the exterior doors. Such a reduction of protection is not acceptable (see Section 1028 for exit discharge alternatives).

Exit passageways may also be used on their own in locations not connected with an enclosure for an exit stairway. The exit passageway is often used as a protected horizontal exit path, such as in a mall or unlimited area building. Sometimes on large floor plans, an exit passageway may be used to extend an exit into areas that would not otherwise be able to meet the travel distance requirements. Like exit stairways, there is no travel distance limitation within an exit passageway.

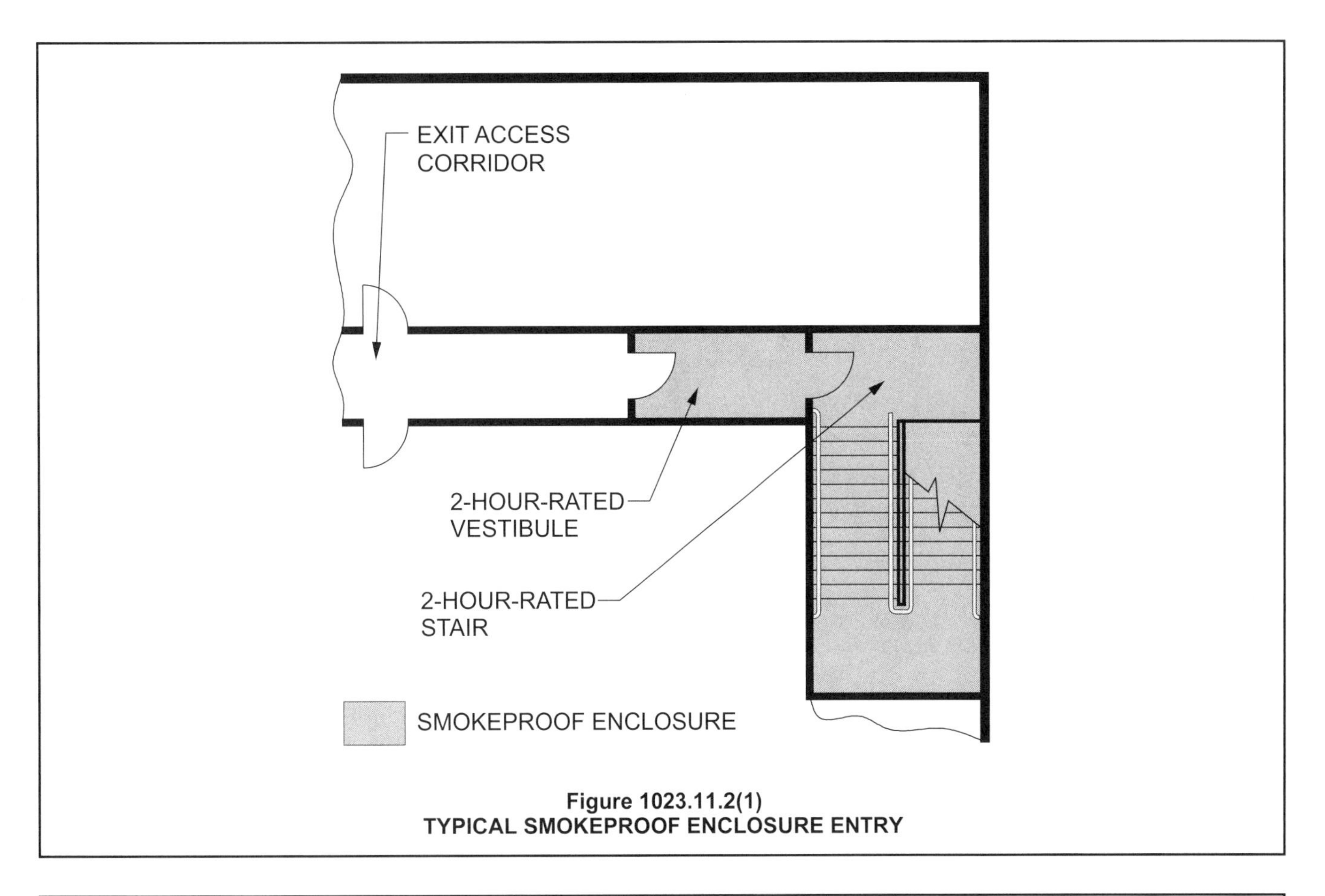

Figure 1023.11.2(1)
TYPICAL SMOKEPROOF ENCLOSURE ENTRY

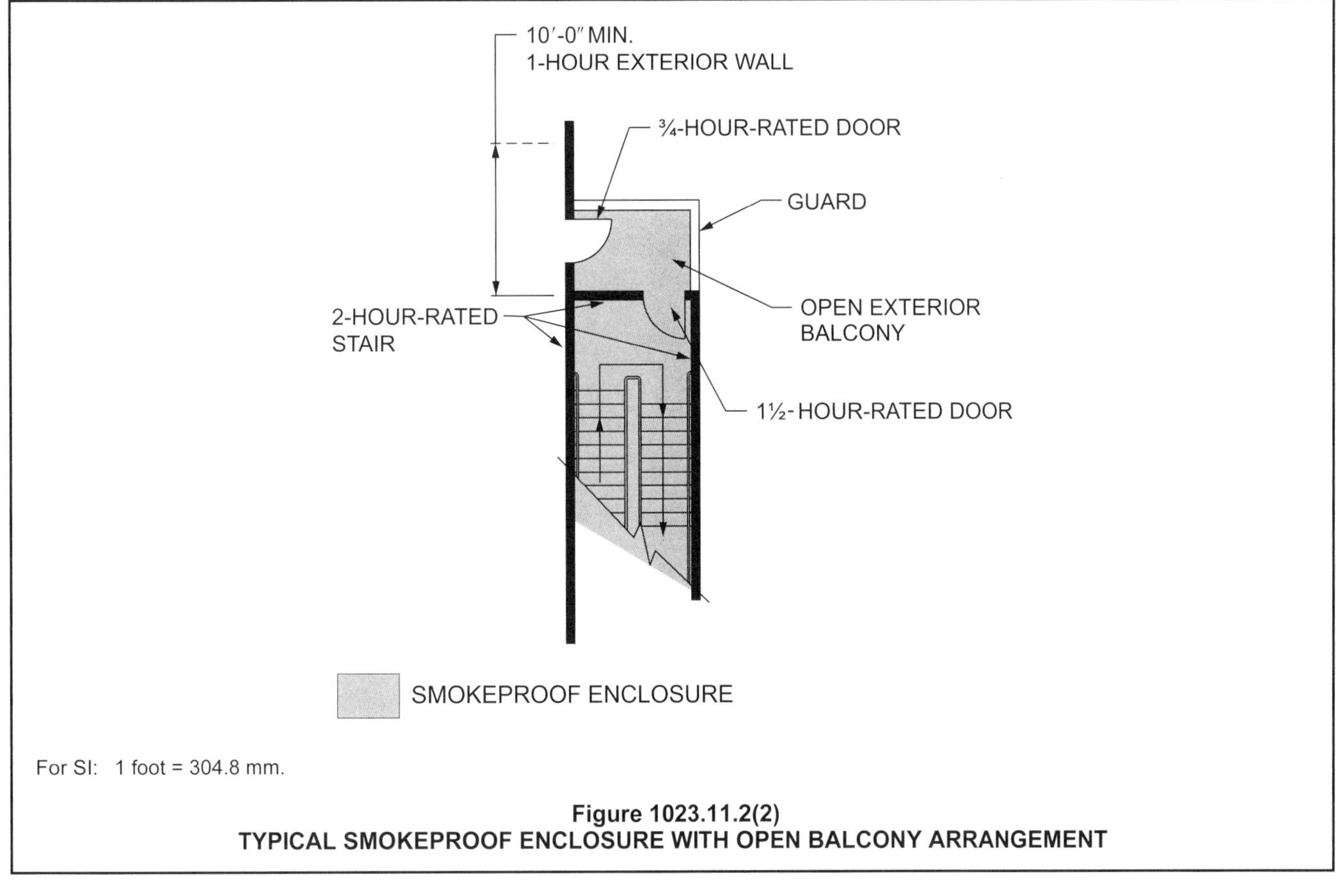

For SI: 1 foot = 304.8 mm.

Figure 1023.11.2(2)
TYPICAL SMOKEPROOF ENCLOSURE WITH OPEN BALCONY ARRANGEMENT

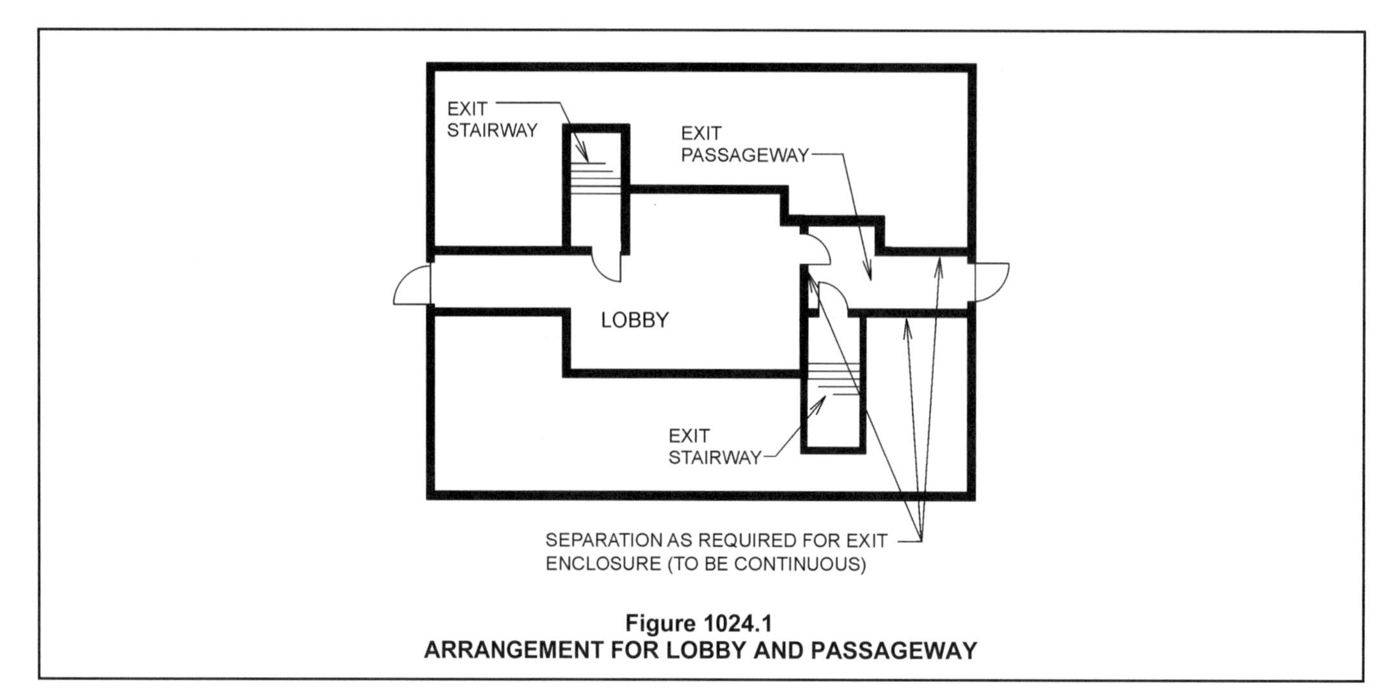

Figure 1024.1
ARRANGEMENT FOR LOBBY AND PASSAGEWAY

1024.2 Width. The required capacity of *exit passageways* shall be determined as specified in Section 1005.1 but the minimum width shall be not less than 44 inches (1118 mm), except that *exit passageways* serving an occupant load of less than 50 shall be not less than 36 inches (914 mm) in width. The minimum width or required capacity of *exit passageways* shall be unobstructed.

Exception: Encroachments complying with Section 1005.7.

❖ The width of an exit passageway is to be determined in accordance with Section 1005.1, based on the number of occupants served in the same manner as for corridors. The greater of the minimum required width or the width determined based on occupancy is to be used. In situations where the exit passageway also serves as an exit access corridor for the first floor, the corridor width must comply with the stricter requirement.

A cross reference back to Section 1005.7 from the exceptions for encroachments in the required width in aisles (see Section 1011.1), corridors (see Section 1020.3), exit passageways (see Section 1024.2) and exit courts (see Section 1028.4.1) reinforces the fact that the protrusion limits provision is generally applicable for these types of confined routes.

1024.3 Construction. *Exit passageway* enclosures shall have walls, floors and ceilings of not less than a 1-hour *fire-resistance rating*, and not less than that required for any connecting *interior exit stairway* or *ramp*. *Exit passageways* shall be constructed as *fire barriers* in accordance with Section 707 or *horizontal assemblies* constructed in accordance with Section 711, or both.

❖ The entire exit passageway enclosure is to be fire-resistance rated as specified. The floors and ceilings are required to be rated in addition to the walls. When used separately, a minimum 1-hour fire-resistance rating is required. Where extending an enclosure for an exit stairway, the rating must not be less than the enclosure for the exit stairway so that the degree of protection is kept at the same level. Remember that if the exit passageway extends over a lower level, such as a basement, all supporting construction is to have the same fire-resistance rating as the elements supported in accordance with the continuity requirements for fire barriers and horizontal assemblies (see Sections 707.5 and 711.2.3). The continuity requirements would also be a concern for the rated ceiling of the exit passageway. An alternative for the ceiling of the exit passageway could be a top-of-shaft enclosure (see Section 713.12).

1024.4 Termination. *Exit passageways* on the *level of exit discharge* shall terminate at an *exit discharge*. *Exit passageways* on other levels shall terminate at an *exit*.

❖ This is consistent with the exit continuity and enclosure requirements in Section 1023.3. Section 1023.3 has an exception that allows for combinations of interior exit stairways and ramps with the exit passageway. The intent of this section is to provide safety in all portions of the exit by requiring continuity of the fire protection characteristics of the enclosure for the exit stairway in combination with an exit passageway. This would include, but not be limited to, the fire-resistance rating of the enclosure walls for the exit stairways and the opening protection rating of the doors. When an exit passageway is supported by the structure, the supporting construction must be fire-resistance rated equal to the walls and floors of the passageway being supported.

Section 1028.1, Exceptions 1 and 2, allow for an alternative for direct access to the outside via an intervening lobby or vestibule.

Horizontal exits, while not providing direct access to the outside of the structure, do move occupants to another "building" by moving through a fire wall (see Sections 1026 and 1028.1, Exception 3) into a refuge area protected by fire barriers and horizontal assemblies. Horizontal exits are commonly used in hospitals and jails for a defend-in-place type of protection.

An exit passageway may be used to move between vertical enclosures for exit stairways. This may occur when a building is shaped like a wedding cake, so the stairway towers move outward as the building gets larger toward grade level. If this situation happens, it is very important to let occupants know how to proceed. See Section 1013.1 for exit signage requirements for within the exit passageways.

1024.5 Openings. *Exit passageway* opening protectives shall be in accordance with the requirements of Section 716.

Except as permitted in Section 402.8.7, openings in *exit passageways* other than unprotected exterior openings shall be limited to those necessary for *exit access* to the *exit passageway* from normally occupied spaces and for egress from the *exit passageway*.

Where an *interior exit stairway* or *ramp* is extended to an *exit discharge* or a *public way* by an *exit passageway*, the *exit passageway* shall comply with Section 1023.3.1.

Elevators shall not open into an *exit passageway*.

❖ In order for fire doors to be effective, they must be in the closed position; therefore, the preferred arrangement is to install self-closing doors. Recognizing that operational practices often require doors to be open for an extended period of time, automatic-closing doors are permitted as long as this opening will not pose a threat to occupant safety and the doors will be self-latching. Automatic-closing devices enable the opening to be protected during a fire condition. The basic requirement for closing devices and specific requirements for automatic-closing and self-closing devices are given in NFPA 80. Automatic-closing doors that protect openings into exits are also required to close on the actuation of smoke detectors or loss of power to the smoke detectors (see Section 716.5.9.3).

The requirements for exit passageways are very similar to those required for enclosures for interior exit stairways (see Section 1023.4). The only openings that are permitted in fire-resistance-rated exit passageways are doors that lead from normally occupied spaces, from the enclosure for the exit stairway and to the outside. This restriction on openings essentially prohibits the use of windows in an exit passageway except for those exterior windows that are not exposed to any hazards. This requirement is not intended to prohibit windows or other openings in the exterior walls of the exit enclosure. The verbiage "unprotected exterior openings" includes windows or doors not required to be protected by either Section 705.8 or 1023.7. The only exception would be window assemblies that have been tested as wall assemblies in accordance with ASTM E119 and UL 263. The objective of this provision is to minimize the possibility of fire spreading into an exit passageway and endangering the occupants or even preventing the use of the exit at a time when it is most needed. The limitation on openings applies regardless of the fire protection rating of the opening protective. The limitation on openings from normally occupied areas is intended to reduce the probability of a fire occurring in an unoccupied area, such as a storage closet, that has an opening into the exit passageway, thereby resulting in smoke spreading into the exit passageway. Other spaces that are not normally occupied include, but are not limited to, toilet rooms, electrical/mechanical equipment rooms and janitorial closets. Note that exit passageways prohibit elevators from opening directly into the passageway. There are some exceptions for these unoccupied spaces in exit passageways in covered malls (see Section 402.4.6).

The third paragraph addresses when the vertical enclosure for an exit stairway or ramp transitions to the exit passageway. While the exit passageway is an extension of the protection offered by the vertical exit, there must still be a door (i.e., opening protective) between the bottom of the stairway or ramp enclosure and the exit passageway. This is to prevent any smoke that may migrate into the exit passageway from also moving up the exit stairway or ramp. This door is not required if the exit stairway and exit passageway are protected by pressurization (see Section 1023.10.1).

Elevators may not open into exit passageways. The difficulty is to have elevator doors that can meet the opening protectives for a fire barrier, but still operate effectively. For additional information on elevator lobbies and doors, see the commentary to Sections 713.14 and 1020.6.

These opening limitations are very similar to those required for an exit enclosure (see Section 1023.3).

1024.6 Penetrations. Penetrations into or through an *exit passageway* are prohibited except for equipment and ductwork necessary for independent pressurization, sprinkler piping, standpipes, electrical raceway for fire department communication and electrical raceway serving the *exit passageway* and terminating at a steel box not exceeding 16 square inches (0.010 m2). Such penetrations shall be protected in accordance with Section 714. There shall not be penetrations or communicating openings, whether protected or not, between adjacent *exit passageways*.

Exception: Membrane penetrations shall be permitted on the outside of the *exit passageway*. Such penetrations shall be protected in accordance with Section 714.3.2.

❖ This section specifically lists the items that are allowed to penetrate an exit passageway. This is consistent for all types of exit enclosures, including enclosures for exit stairways or ramps (see Section 1023.5) and exit passageways. In general, only portions of the building service systems that serve the exit passageways are allowed to penetrate the exit passageway. Section 714 addresses through penetrations and membrane penetrations for fire-resistance-rated wall and floor/ceiling assemblies. The intent is to maintain the integrity of the enclosure for the exit access stairway. If the ductwork serving the exit passageway penetrates a rated wall or floor/ceiling assembly, a fire and smoke damper is required by Section 717.5.3. This requirement is not intended to prohibit windows in the exterior walls of exit passageways that are not required to be rated.

The exception allows for electrical boxes or fire alarm pull stations to be installed on the outside of the enclosure, provided that the boxes are installed so that the required fire-resistance rating is not reduced (see Section 714.3.2).

1024.7 Ventilation. Equipment and ductwork for *exit passageway* ventilation as permitted by Section 1024.6 shall comply with one of the following:

1. The equipment and ductwork shall be located exterior to the building and shall be directly connected to the *exit passageway* by ductwork enclosed in construction as required for shafts.
2. Where the equipment and ductwork is located within the *exit passageway*, the intake air shall be taken directly from the outdoors and the exhaust air shall be discharged directly to the outdoors, or the air shall be conveyed through ducts enclosed in construction as required for shafts.
3. Where located within the building, the equipment and ductwork shall be separated from the remainder of the building, including other mechanical equipment, with construction as required for shafts.

In each case, openings into the fire-resistance-rated construction shall be limited to those needed for maintenance and operation and shall be protected by opening protectives in accordance with Section 716 for shaft enclosures.

Exit passageway ventilation systems shall be independent of other building ventilation systems.

❖ As a continuation of an exit stairway enclosure, an exit passageway has the same considerations for protection. While an exit passageway may look like a regular hallway, it is important that it be maintained as free of smoke as the exit stairway. See the commentary to Section 1023.6.

SECTION 1025
LUMINOUS EGRESS PATH MARKINGS

1025.1 General. *Approved* luminous egress path markings delineating the exit path shall be provided in *high-rise buildings* of Group A, B, E, I, M, and R-1 occupancies in accordance with Sections 1025.1 through 1025.5.

Exception: Luminous egress path markings shall not be required on the *level of exit discharge* in lobbies that serve as part of the exit path in accordance with Section 1028.1, Exception 1.

❖ Improved safety for individuals negotiating stairs during egress of a high-rise building is provided by improving the visibility of stair treads and handrails under emergency conditions. A second source of emergency power for exit illumination, exit signs and stair shaft pressurization systems in smokeproof enclosures is currently mandated for high-rise buildings. In the event of an emergency that disconnects utility power, the emergency power source should engage, causing the stair shaft to be illuminated and kept smoke free by the pressurization system. Unfortunately, such systems can fail under demand conditions. The provisions of Section 1025 add an additional level of safety to the egress path by requiring the installation of photoluminescent or self-illuminating marking systems that do not require electrical power and its associated wiring and circuits. An additional means for ensuring that occupants can safely egress a building via exit stairways is now available even if the emergency power supply and system fails to operate. The groups indicated have a high anticipated occupant load or occupants may not be as familiar with the space. Note that the provisions only require these markings within the enclosure for the exit stairway, exit ramp and the exit passageway used for enclosure continuation. The markings are not required before reaching the exit (i.e., exit access) or after leaving the exit (i.e., exit discharge).

The exception indicates that if the exit stairway enclosure discharges through the lobby (see Section 1028.1, Exception 1), the egress markings would not be required outside the stairway enclosure for the portion from the stairway enclosure to the door leading to the outside. If the exit stairway discharges through an exit passageway, the exit path markings must continue to the door leading to the outside. The current text is silent for the options of vestibules and horizontal exits, but since Sections 1025.1 through 1025.5 only address within the enclosure for the exit, and not the exit discharge, exit path markings in the vestibule or after the horizontal exit are not required.

1025.2 Markings within exit components. Egress path markings shall be provided in *interior exit stairways*, *interior exit ramps* and *exit passageways*, in accordance with Sections 1025.2.1 through 1025.2.6.

❖ Luminous egress path markings are required inside the enclosure for exit stairways or ramps for all floors. If the stairway connects to an exit passageway as part of the travel down to the level of exit discharge or on the level of exit discharge, the path markings must also be continued in the exit passageway.

The subsections include marking the tread nosings, the surrounding edges of landings and any exit passageways, handrails and any protruding objects within the protected enclosure.

All exit path markings are required to be solid and continuous stripes. A key requirement for marking systems is that their design must be uniform. The placement and dimensions of markings must be consistent along the entire exit stairway. By specifying a standard marking dimension, the requirements will ensure that the marking is visible during dark conditions and provides consistent and standard application in the design and enforcement of exit path markings. Markings installed on stair steps, perimeter demarcation lines and handrails must have a minimum width of 1 inch (25 mm). For stair steps and perimeter demarcation lines, their maximum width cannot exceed 2 inches (51 mm). The provisions for stair steps, perimeter demarcation lines and handrails allow the width of the marking to be reduced to less than 1 inch (25 mm) when marking stripes are listed in accordance with UL 1994.

1025.2.1 Steps. A solid and continuous stripe shall be applied to the horizontal leading edge of each step and shall extend for the full length of the step. Outlining stripes shall have a minimum horizontal width of 1 inch (25 mm) and a maximum width of 2 inches (51 mm). The leading edge of the stripe shall be placed not more than $^1/_2$ inch (12.7 mm) from the leading edge of the step and the stripe shall not overlap the leading edge of the step by not more than $^1/_2$ inch (12.7 mm) down the vertical face of the step.

> **Exception:** The minimum width of 1 inch (25 mm) shall not apply to outlining stripes listed in accordance with UL 1994.

❖ Luminous stripes are required the full width of the stairway on all tread nosings and along the leading edge of stair landings. These demarcation lines serve to identify the transition from the stair steps to the landing, which is important to minimize the risk of a fall inside of a stairway enclosure that is not illuminated. In order to clearly identify the leading edge of the step, the front edge of the stripe must be within $^1/_2$ inch (13 mm) plus or minus of the leading edge of the tread (see Commentary Figure 1025.2.1).

The code does not specify any minimum slip-resistance requirements for luminous products installed on walking surfaces. However, Section 1003.4 requires all walking surfaces for means of egress to be slip resistant. Persons with vision impairments often rely on high-contrast elements to delineate changes in elevation such as that required in Sections 1011.5.4 and 1029.13.2.2. In medium light conditions, the luminous materials may be hard to discern. Luminous materials installed adjacent to dark contrasting materials may help with both situations.

The provisions for stair steps, perimeter demarcation lines and handrails allow the width of the marking to be reduced to less than 1 inch (25 mm) when marking stripes are listed in accordance with UL 1994.

1025.2.2 Landings. The leading edge of landings shall be marked with a stripe consistent with the dimensional requirements for steps.

❖ The edge of the landing at the top of the steps must be marked in the same manner as the tread nosing so that a person can tell where the steps start. See the commentary to Section 1025.2.1 and Commentary Figure 1025.2.1.

1025.2.3 Handrails. *Handrails* and handrail extensions shall be marked with a solid and continuous stripe having a minimum width of 1 inch (25 mm). The stripe shall be placed on the top surface of the *handrail* for the entire length of the *handrail*, including extensions and newel post caps. Where *handrails* or handrail extensions bend or turn corners, the stripe shall not have a gap of more than 4 inches (102 mm).

> **Exception:** The minimum width of 1 inch (25 mm) shall not apply to outlining stripes listed in accordance with UL 1994.

❖ The handrail must have a glow-in-the-dark stripe down the entire length. The 1-inch (25 mm) minimum width stripe must be on the top of the handrail so it can charge and maximize its visibility in the dark. Bends or turns may result in a break in the marking, which must not be more than 4 inches (102 mm) maximum (see Commentary Figure 1025.2.3).

The provisions for stair steps, perimeter demarcation lines and handrails allow the width of the marking to be reduced to less than 1 inch (25 mm) when marking stripes are listed in accordance with UL 1994.

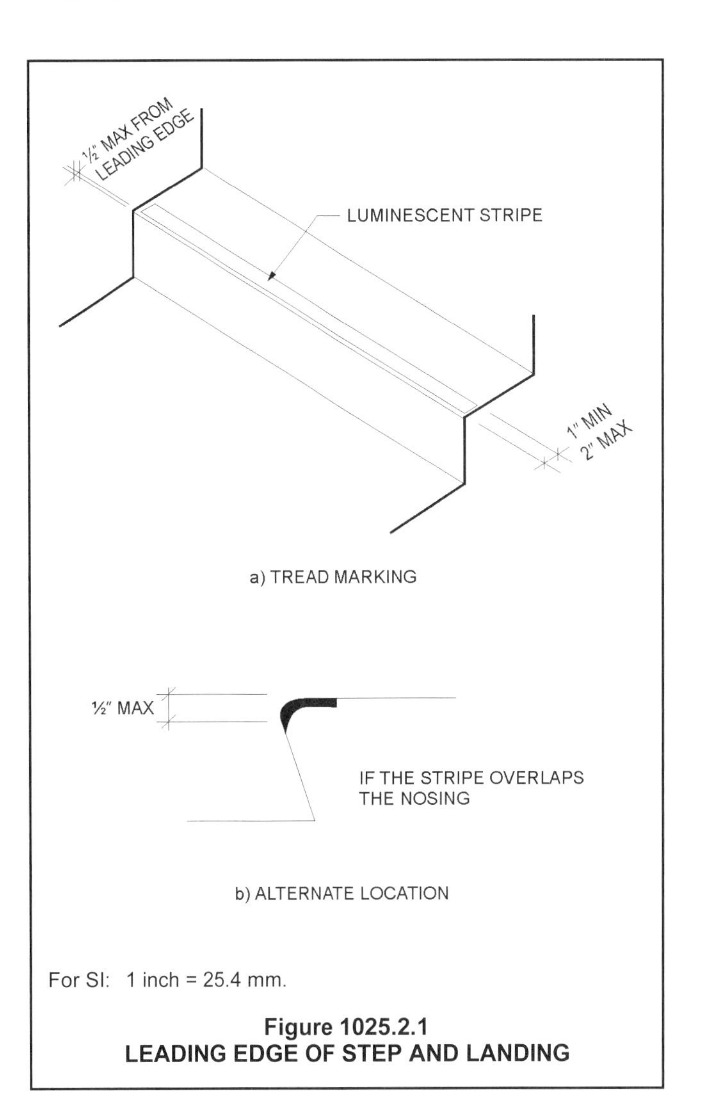

Figure 1025.2.1
LEADING EDGE OF STEP AND LANDING

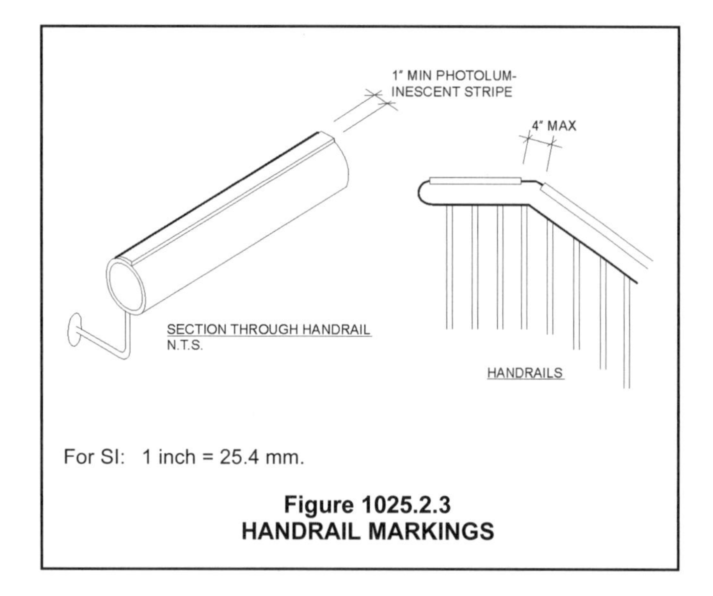

For SI: 1 inch = 25.4 mm.

Figure 1025.2.3
HANDRAIL MARKINGS

1025.2.4 Perimeter demarcation lines. Stair landings and other floor areas within *interior exit stairways*, *interior exit ramps* and *exit passageways*, with the exception of the sides of steps, shall be provided with solid and continuous demarcation lines on the floor or on the walls or a combination of both. The stripes shall be 1 to 2 inches (25 mm to 51 mm) wide with interruptions not exceeding 4 inches (102 mm).

> **Exception:** The minimum width of 1 inch (25 mm) shall not apply to outlining stripes listed in accordance with UL 1994.

❖ In addition to the leading edge of the landing, the landing must have a luminous stripe all the way around the edge. If the enclosure includes any type of exit passageway, that corridor must also have a perimeter stripe. The stripe can be on the floor or on the wall at baseboard height (see Sections 1025.2.4.1 and 1025.2.4.2).

The provisions for stair steps, perimeter demarcation lines and handrails allow the width of the marking to be reduced to less than 1 inch (25 mm) when marking stripes are listed in accordance with UL 1994.

1025.2.4.1 Floor-mounted demarcation lines. Perimeter demarcation lines shall be placed within 4 inches (102 mm) of the wall and shall extend to within 2 inches (51 mm) of the markings on the leading edge of landings. The demarcation lines shall continue across the floor in front of all doors.

> **Exception:** Demarcation lines shall not extend in front of *exit discharge* doors that lead out of an *exit* and through which occupants must travel to complete the exit path.

❖ Luminous stripes shall extend all the way around any stair landings. This section specifies the option of stripes on the floor (see Commentary Figure 1025.2.4.1). On a typical landing, the stripe will extend across the front of the door, indicating to someone moving down the stairway that they should continue (see Commentary Figure 1025.2.4.3). At the level of exit discharge the occupant has a different visual cue. The line should not extend in front of the door leading to the exterior (see Commentary Figure 1025.2.6). This should allow people to understand which door they need to move through to get to safety.

1025.2.4.2 Wall-mounted demarcation lines. Perimeter demarcation lines shall be placed on the wall with the bottom edge of the stripe not more than 4 inches (102 mm) above the finished floor. At the top or bottom of the *stairs*, demarcation lines shall drop vertically to the floor within 2 inches (51 mm) of the step or landing edge. Demarcation lines on walls shall transition vertically to the floor and then extend across the floor where a line on the floor is the only practical method of outlining the path. Where the wall line is broken by a door, demarcation lines on walls shall continue across the face of the door or transition to the floor and extend across the floor in front of such door.

> **Exception:** Demarcation lines shall not extend in front of *exit discharge* doors that lead out of an *exit* and through which occupants must travel to complete the exit path.

❖ Luminous stripes that surround the stairway landings can be on the floor or on the wall. This section specifies the option of stripes on the wall (see Commentary Figure 1025.2.4.2). See the commentary to Sections 1025.2.4 and 1025.2.4.1.

1025.2.4.3 Transition. Where a wall-mounted demarcation line transitions to a floor-mounted demarcation line, or vice versa, the wall-mounted demarcation line shall drop vertically to the floor to meet a complimentary extension of the

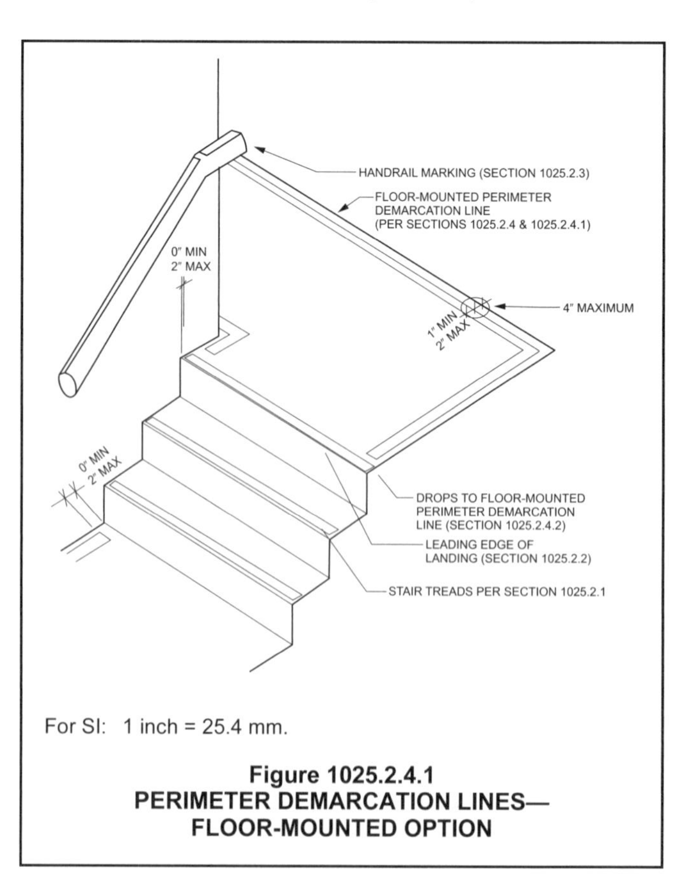

For SI: 1 inch = 25.4 mm.

Figure 1025.2.4.1
PERIMETER DEMARCATION LINES—FLOOR-MOUNTED OPTION

floor-mounted demarcation line, thus forming a continuous marking.

❖ While the luminous perimeter stripes on a landing can be on either the wall or the floor, when they transition from one to another, the lines should appear continuous. See Section 1025.2.4.2 for special requirements for when the wall perimeter marking transitions to the floor stripe indicating the leading edge of the landing.

When perimeter lines cross a door frame, the lines can transition or stay on the same plain (see Commentary Figure 1025.2.4.3).

1025.2.5 Obstacles. Obstacles at or below 6 feet 6 inches (1981 mm) in height and projecting more than 4 inches (102 mm) into the egress path shall be outlined with markings not less than 1 inch (25 mm) in width comprised of a pattern of alternating equal bands, of luminous material and black, with the alternating bands not more than 2 inches (51 mm) thick and angled at 45 degrees (0.79 rad). Obstacles shall include, but are not limited to, standpipes, hose cabinets, wall projections and restricted height areas. However, such markings shall not conceal any required information or indicators including but not limited to instructions to occupants for the use of standpipes.

❖ Any obstacles within the stairway must be marked with a dashed line of diagonal slashes (see Commentary Figure 1025.2.5). The markings of obstacles are consistent with the intent of protruding object provisions in Section 1003.3. However, there is a difference in the height of 6 feet, 6 inches (1981 mm) instead of 6 feet, 8 inches (2032 mm). Items permitted within the enclosures for exit stairways are limited by Sections 1023.2 and 1023.5.

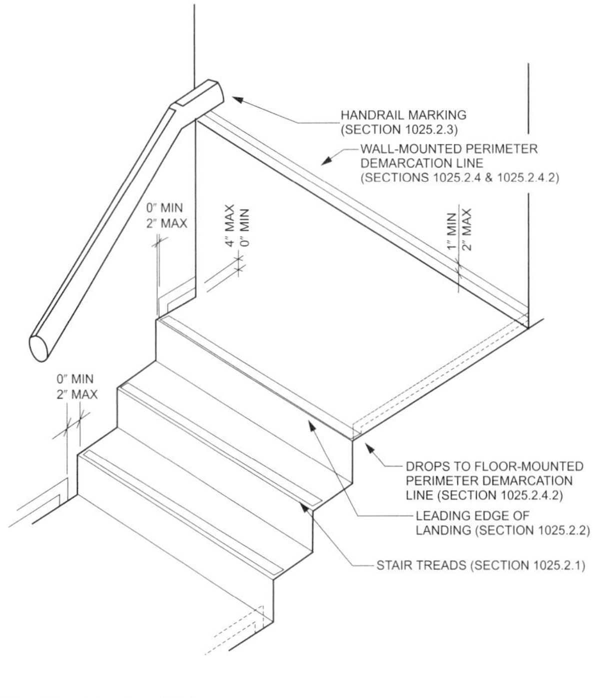

For SI: 1 inch = 25.4 mm.

Figure 1025.2.4.2
PERIMETER DEMARCATION LINES—WALL-MOUNTED OPTION

1025.2.6 Doors within the exit path. Doors through which occupants must pass in order to complete the exit path shall be provided with markings complying with Sections 1025.2.6.1 through 1025.2.6.3.

❖ Doors within an enclosure for an exit stairway can:

- Lead directly to the exterior.
- Pass through a horizontal exit (Section 1028.1, Exception 3).
- Lead to an exit passageway on the level of exit discharge, to a lobby or vestibule at the level of exit discharge (Section 1028.1, Exceptions 2 and 3).
- Lead to an exit passageway on a transition level.

Doors at these locations should be marked as indicated in the following three subsections. Combined with the landing markings not extending across the bottom of this door, this will provide several visual cues to indicate that this is the door to continue through to get out of the building (see Commentary Figure 1025.2.6).

The NIST egress study of the World Trade Center indicated that transition floors can cause delays in egress because people hesitate when it is unclear about which way they should continue. Effectively marking these doors in four different ways will decrease that hazard. Which doors to mark is consistent with the exit sign requirements within exit enclosures in Section 1013.1.

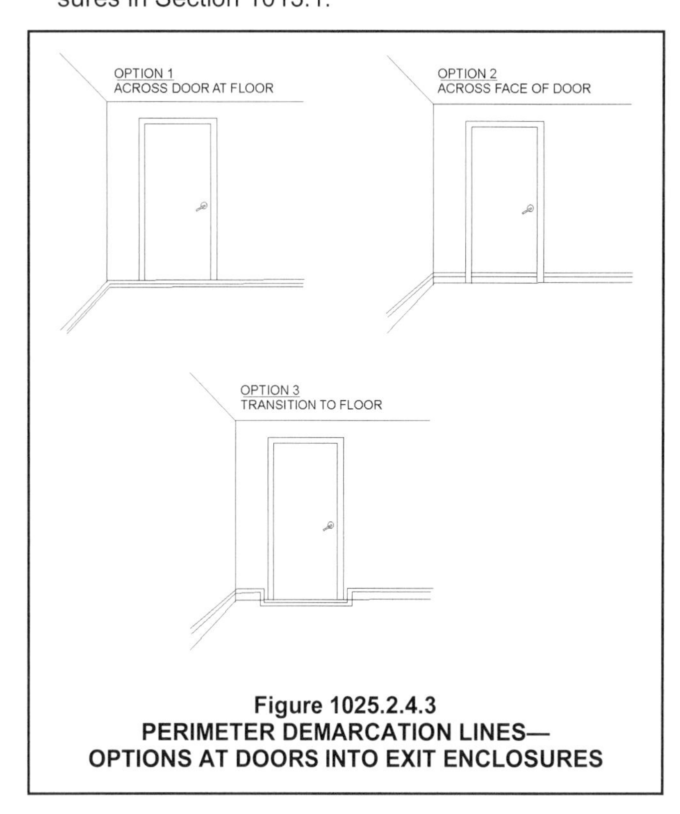

Figure 1025.2.4.3
PERIMETER DEMARCATION LINES—OPTIONS AT DOORS INTO EXIT ENCLOSURES

1025.2.6.1 Emergency exit symbol. The doors shall be identified by a low-location luminous emergency exit symbol complying with NFPA 170. The exit symbol shall be not less than 4 inches (102 mm) in height and shall be mounted on the door, centered horizontally, with the top of the symbol not higher than 18 inches (457 mm) above the finished floor.

❖ The door shall include a low luminous level exit symbol within 18 inches (457 mm) of the floor. For an example of the emergency exit symbol, see Commentary Figure 1025.2.6.1. The sign on the door can be just this symbol or it can also contain additional information such as a directional arrow or "EXIT."

1025.2.6.2 Door hardware markings. Door hardware shall be marked with not less than 16 square inches (406 mm^2) of luminous material. This marking shall be located behind, immediately adjacent to, or on the door handle or escutcheon. Where a panic bar is installed, such material shall not be less than 1 inch (25 mm) wide for the entire length of the actuating bar or touchpad.

❖ Door hardware locations must be clearly visible. If a panic bar is used, a luminous stripe with a minimum width of 1 inch (25 mm) should be provided down the entire length of the activation bar/paddle [see Commentary Figure 1025.2.6, Example 1]. If lever hardware is used, a donut, square or rectangle with a minimum area of 16 square inches (10 322 mm^2) should be provided behind the hardware [see Commentary Figure 1025.2.6, Example 2]. There is also the option of marking the door handle itself, but it would be difficult to get 16 square inches (10 322 mm^2) of visible surface area. Plus, over time, a finish on the hardware has a greater chance of wearing off with normal use. The language allows the designer the freedom to decide (with wear and hardware options) which configuration would give the best results.

1025.2.6.3 Door frame markings. The top and sides of the door frame shall be marked with a solid and continuous 1-inch- to 2-inch-wide (25 mm to 51 mm) stripe. Where the door molding does not provide sufficient flat surface on which to locate the stripe, the stripe shall be permitted to be located on the wall surrounding the frame.

❖ Doors must be marked along the sides and top with stripes similar to those provided on the stair nosing.

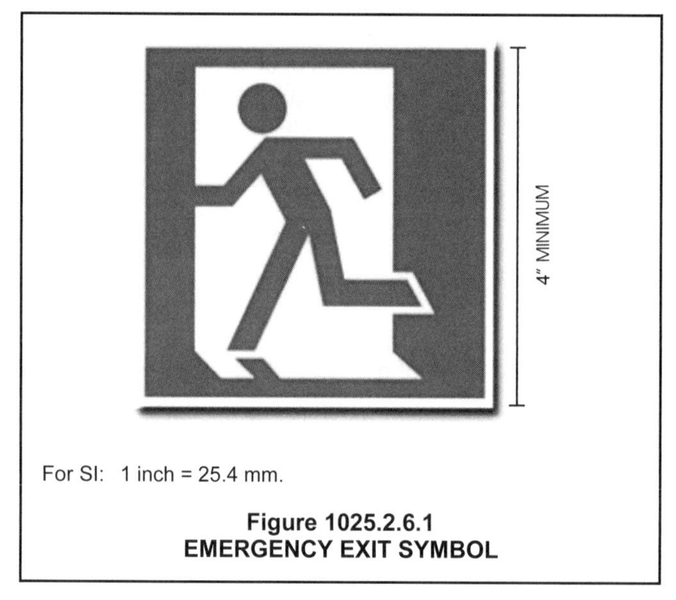

For SI: 1 inch = 25.4 mm.

Figure 1025.2.6.1
EMERGENCY EXIT SYMBOL

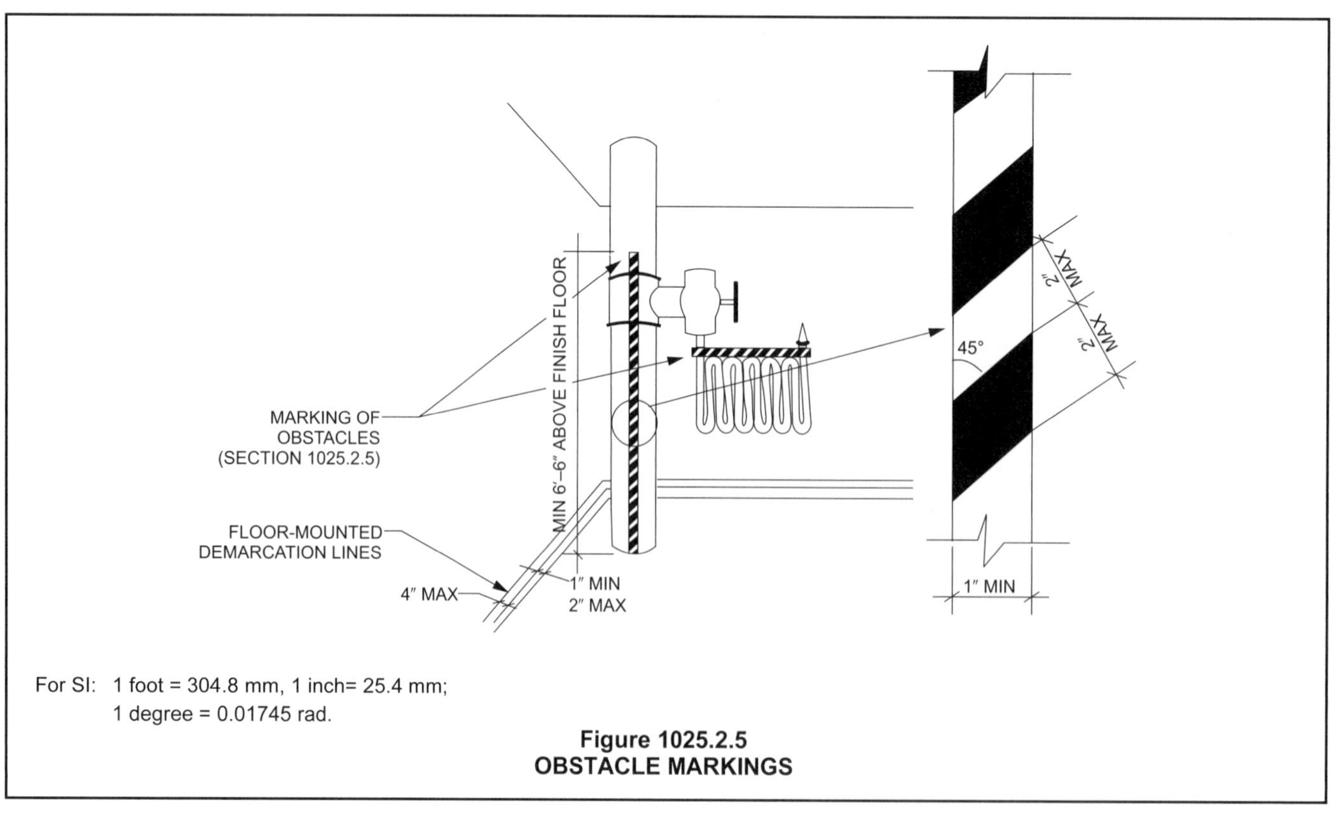

For SI: 1 foot = 304.8 mm, 1 inch= 25.4 mm;
1 degree = 0.01745 rad.

Figure 1025.2.5
OBSTACLE MARKINGS

Door frames come in a variety of shapes. If there is not space on the door for the marking stripe, the stripe can be around the perimeter of the door [see Commentary Figure 1025.2.6, Examples 1 and 2].

1025.3 Uniformity. Placement and dimensions of markings shall be consistent and uniform throughout the same enclosure.

❖ All exit path markings are required to be solid and have continuous stripes. A key requirement for marking systems is that their design must be uniform. The placement and dimensions of markings must be consistent for the path of travel along the exit stairway. By specifying a standard marking dimension, the requirements will ensure that the marking is visible during dark conditions and provides consistent and standard application in the design and enforcement of exit path markings. Markings installed on stair steps, perimeter demarcation lines and handrails must have a minimum width of 1 inch (25 mm). For

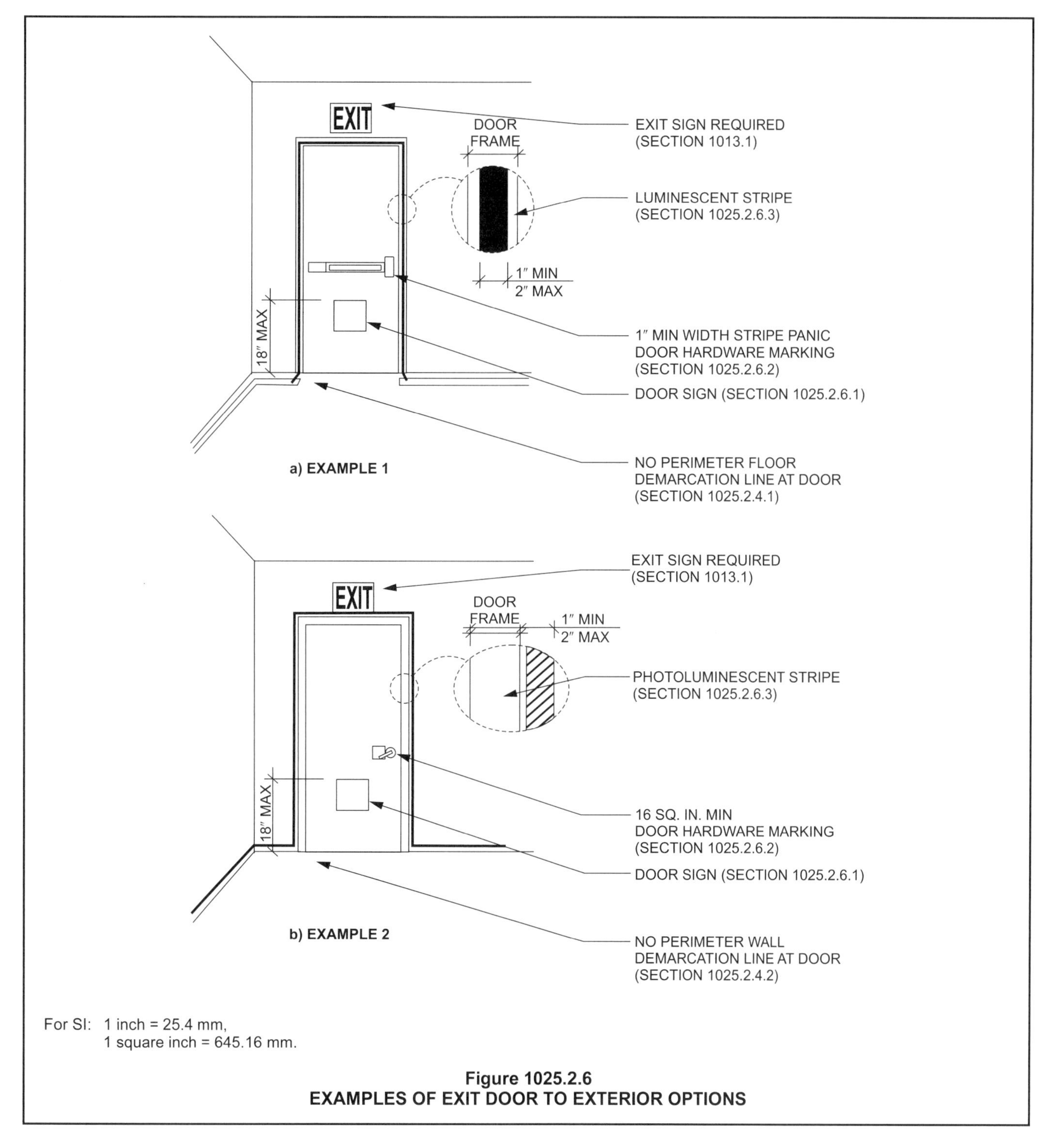

Figure 1025.2.6
EXAMPLES OF EXIT DOOR TO EXTERIOR OPTIONS

stair steps and perimeter demarcation lines, their maximum width cannot exceed 2 inches (51 mm). The provisions for stair steps, perimeter demarcation lines and handrails allow the width of the marking to be reduced to less than 1 inch (25 mm) when marking stripes are listed in accordance with UL 1994.

1025.4 Self-luminous and photoluminescent. Luminous egress path markings shall be permitted to be made of any material, including paint, provided that an electrical charge is not required to maintain the required luminance. Such materials shall include, but not be limited to, *self-luminous* materials and *photoluminescent* materials. Materials shall comply with either of the following standards:

1. UL 1994.
2. ASTM E2072, except that the charging source shall be 1 footcandle (11 lux) of fluorescent illumination for 60 minutes, and the minimum luminance shall be 30 milicandelas per square meter at 10 minutes and 5 milicandelas per square meter after 90 minutes.

❖ Products utilized to meet the requirements for luminous egress path markings in high-rise buildings (see Sections 411.7 and 1025) or exit signs (see Section 1013.5) may be photoluminescent or self-luminous (see definitions in Chapter 2). An example of photoluminescent material is paint or tape that is charged by exposure to light. When the lights are turned off, the product will "glow" in the dark. Self-luminous products do not need an outside light source to charge them like photoluminescent materials do.

A variety of materials can comply with the referenced standards for egress path markings (UL 1994, and ASTM E2072) and for signs (UL 924).

ASTM E2072 allows the use of paints and coatings, which can be useful because it avoids a potential tripping hazard, especially in locations where the surface substrate may not be even. The luminescence of the selected marking system must provide an illumination of 1 footcandle (11 lux) for 60 minutes, which is consistent with the requirement in Section 1008.2 for the illumination of walking surfaces. Section 1008.3 requires the emergency lighting system to have power for 90 minutes; however, because of normal battery considerations, the IFC only requires a 60-minute duration in existing buildings.

1025.5 Illumination. Where *photoluminescent* exit path markings are installed, they shall be provided with not less than 1 footcandle (11 lux) of illumination for not less than 60 minutes prior to periods when the building is occupied and continuously during occupancy.

❖ Analogous to rechargeable batteries, many photoluminescent egress path markings require exposure to light to perform properly. Thus, photoluminescent egress path markings must be exposed to a minimum 1 footcandle (11 lux) of light energy at the walking surface for at least 60 minutes prior to the building being occupied. The charging rate for photoluminescent egress path markings is based on the wattage of lamps used to provide egress path illumination. Therefore it is important to verify that the specified lamps have sufficient wattage to meet the specified time period. This requirement may be a concern for buildings developed with the IECC or trying for LEED certification.

Note that this requirement does not apply to self-luminous materials since these materials operate independently of the external power source. See the definitions for "Photoluminescent" and "Self-luminous" in Chapter 2.

SECTION 1026
HORIZONTAL EXITS

1026.1 Horizontal exits. *Horizontal exits* serving as an *exit* in a *means of egress* system shall comply with the requirements of this section. A *horizontal exit* shall not serve as the only *exit* from a portion of a building, and where two or more *exits* are required, not more than one-half of the total number of *exits* or total *exit* minimum width or required capacity shall be *horizontal exits*.

Exceptions:

1. *Horizontal exits* are permitted to comprise two-thirds of the required *exits* from any building or floor area for occupancies in Group I-2.
2. *Horizontal exits* are permitted to comprise 100 percent of the *exits* required for occupancies in Group I-3. Not less than 6 square feet (0.6 m^2) of accessible space per occupant shall be provided on each side of the *horizontal exit* for the total number of people in adjoining compartments.

❖ Horizontal exits can provide up to 50 percent of the exits from a given area of a building. The percentage is higher for Group I-2 and I-3 occupancies where the evacuation strategy is defend in place rather than direct egress (see Commentary Figure 1026.1 for a typical horizontal exit arrangement). However, a horizontal exit cannot serve as the only exit from a single exit space. Section 1026.4 allows for some areas to have all the exits from a space to be horizontal exits under specific conditions. A horizontal exit can be designed for either one-way or two-way operation, depending on the exiting needs of each side of the wall providing the horizontal exit.

A horizontal exit may be an element of a means of egress when in compliance with the requirements of this section. The actual horizontal exit is the protected door opening in a wall, or the open-air balcony or bridge that separates two areas of a building. A horizontal exit is often used in hospitals and in prisons where it is not feasible or desirable that all occupants exit the facility (see Chapter 2 for the definition of a "Horizontal exit").

Horizontal exits and their associated "refuge areas" are considered to provide the same or higher level of protection as an "area of refuge" for people who cannot use the egress system. Sections 1009.3 and 1009.4 allow for a horizontal exit or an area of refuge as alternatives. See these sections for exceptions for buildings with sprinkler systems and/or where the path of travel has protection from the accumulation of smoke (i.e., open parking garages, open air assembly seating, smoke-protected assembly seating).

1026.2 Separation. The separation between buildings or refuge areas connected by a *horizontal exit* shall be provided by a *fire wall* complying with Section 706; or by a *fire barrier* complying with Section 707 or a *horizontal assembly* complying with Section 711, or both. The minimum *fire-resistance rating* of the separation shall be 2 hours. Opening protectives in *horizontal exits* shall also comply with Section 716. Duct and air transfer openings in a *fire wall* or *fire barrier* that serves as a *horizontal exit* shall also comply with Section 717. The *horizontal exit* separation shall extend vertically through all levels of the building unless floor assemblies have a *fire-resistance rating* of not less than 2 hours with no unprotected openings.

Exception: A *fire-resistance rating* is not required at *horizontal exits* between a building area and an above-grade *pedestrian walkway* constructed in accordance with Section 3104, provided that the distance between connected buildings is more than 20 feet (6096 mm).

Horizontal exits constructed as *fire barriers* shall be continuous from *exterior wall* to *exterior wall* so as to divide completely the floor served by the *horizontal exit*.

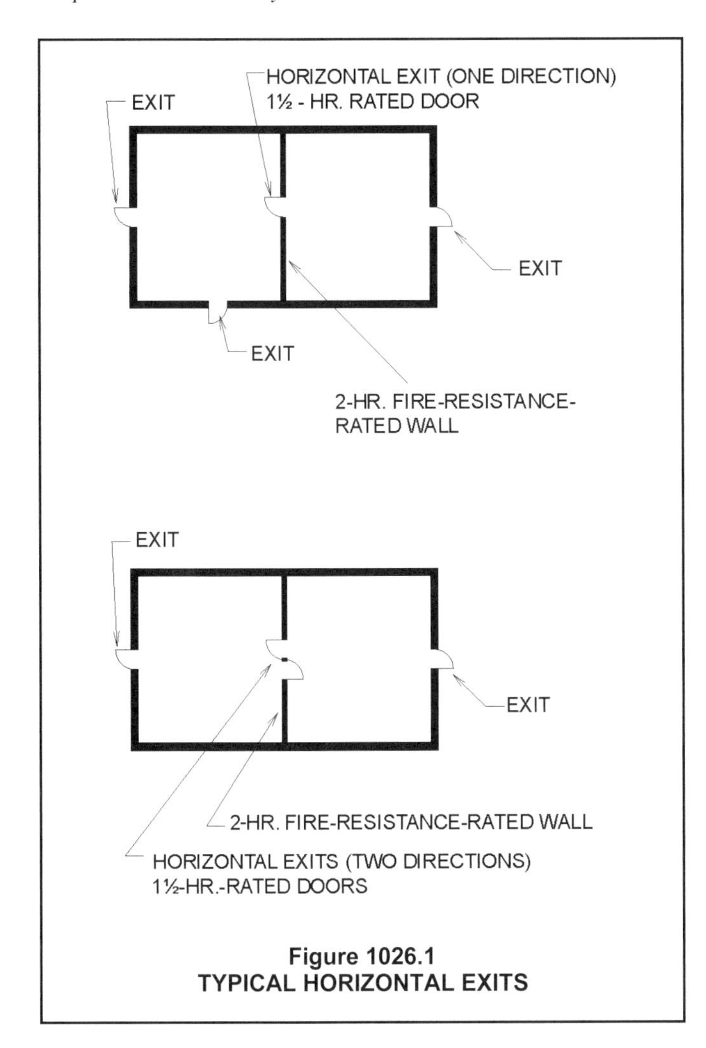

Figure 1026.1
TYPICAL HORIZONTAL EXITS

❖ The basic concept of a horizontal exit is that during a fire emergency, the occupants of a floor will transfer from one fire area to another. Separation between areas of a building can be accomplished by a fire wall (see Section 706), a fire barrier (see Section 707), horizontal assemblies (see Section 711) or a combination thereof, with a fire-resistance rating not less than 2 hours. Any fire shutters or fire doors must have an opening protective of not less than $1^1/_2$ hours (see Table 716.5). Ducts and air transfer openings must comply with Section 717.

In buildings of Groups I-2 and I-3, it may also be desirable (while not mandatory) for the horizontal exit to serve as a smoke barrier. In such cases, the wall containing the horizontal exit must also comply with the requirements for a smoke barrier (see Section 709).

In order to decrease the amount of smoke able to migrate around the edges of a horizontal exit, the horizontal exit must extend from at least the floor to the deck above (i.e., fire barrier), as well as across the floor level from one side of the building to another. Moving up from floor to floor, there are two choices. One option is that the horizontal exit can extend vertically through all levels of the building (i.e., fire wall or fire barriers). The second option is to utilize fire barriers that are not aligned vertically (i.e., a combination of fire barriers and horizontal assemblies), but then the floor must have a 2-hour fire-resistance rating and no unprotected openings are permitted between any two refuge areas. The supporting construction would also have to be a minimum of 2 hours.

The exception is permitting a pedestrian walkway or sky bridge to act as a horizontal exit when buildings are at least 20 feet (6096 mm) apart.

1026.3 Opening protectives. *Fire doors* in *horizontal exits* shall be self-closing or automatic-closing when activated by a *smoke detector* in accordance with Section 716.5.9.3. Doors, where located in a cross-corridor condition, shall be automatic-closing by activation of a *smoke detector* installed in accordance with Section 716.5.9.3.

❖ For the safety of occupants using a horizontal exit, it is important for the doors to be fire doors that are self-closing or automatic-closing by activation of a smoke detector. Smoke detectors that initiate automatic-closing should be located at both sides of the doors (see the commentary to Section 907.3 for an additional explanation of the installation requirements). Any openings in the fire barriers or fire walls used as horizontal exits must be protected in coordination with the rating of the wall. There is a reference to Section 716 for opening protectives.

1026.4 Refuge area. The refuge area of a *horizontal exit* shall be a space occupied by the same tenant or a public area and each such refuge area shall be adequate to accommodate the original *occupant load* of the refuge area plus the *occupant load* anticipated from the adjoining compartment. The anticipated *occupant load* from the adjoining compartment shall be based on the capacity of the *horizontal exit doors* entering the refuge area.

❖ The building area on the discharge side of a horizontal exit must serve as a refuge area for the occupants of both sides of the floor areas connected by the horizontal exit. Therefore, adequate space must be available on each side of the wall to hold the full occupant load of that side, plus the number of occupants from the other side that may be required to use the horizontal exit. Explaining the anticipated occupant load is easiest with an example. If one side of the horizontal exit contained an assembly space with 750 occupants, three exits would be required. If one of the exits through the horizontal exit was a double door, that double exit door would have a capacity of 320 (64 inches/0.2 = 320 occupants). Therefore, the refuge area must be sized for the occupant load in the space, plus the 320 people who might come through the horizontal exit. This is a higher number than would be anticipated if the occupant load of the assembly space was divided equally between the three exits, but it is also not the entire occupant load of the assembly space.

1026.4.1 Capacity. The capacity of the refuge area shall be computed based on a *net floor area* allowance of 3 square feet (0.2787 m^2) for each occupant to be accommodated therein.

Exceptions: The *net floor area* allowable per occupant shall be as follows for the indicated occupancies:

1. Six square feet (0.6 m^2) per occupant for occupancies in Group I-3.
2. Fifteen square feet (1.4 m^2) per occupant for ambulatory occupancies in Group I-2.
3. Thirty square feet (2.8 m^2) per occupant for nonambulatory occupancies in Group I-2.

❖ These refuge areas are meant to hold the occupants temporarily in a safe place until they can evacuate the premises in an orderly manner or, in the case of hospitals and like facilities, to hold bedridden patients and other nonambulatory occupants in a protected area until the fire emergency has ended. This is commonly referred to as a defend-in-place strategy. The size of the refuge area is based on the nature of the expected occupants. In the case of Group I-3, the area will be used to hold the occupants until deliberate egress can be accomplished with staff assistance or supervision. In other cases, it is assumed that the occupants simply wait in line to egress through the required exit facilities provided on the discharge side. Although similar language is used in describing the "area of refuge" for an accessible means of egress, Section 1009.6 specifies area requirements that are insufficient for use as a "refuge area" for a horizontal exit. Care must be taken when applying both principles to the same horizontal exit.

The 3-square-feet (0.28 m^2) per occupant requirement is based on the maximum permitted occupant density at which orderly movement to the exits is reasonable. The 30-square-feet (2.8 m^2) per hospital or nursing home patient requirement is based on the space necessary for a bed or litter. It should be noted that 30 square feet (2.8 m^2) is not based on the total occupant load, as would be determined in accordance with Section 1004.1, but rather on the number of nonambulatory patients. The 15-square-foot (1.4 m^2) requirement for occupancies in Group I-2 facilities is based on each ambulatory patient having a staff attendant.

1026.4.2 Number of exits. The refuge area into which a *horizontal exit* leads shall be provided with *exits* adequate to meet the occupant requirements of this chapter, but not including the added *occupant load* imposed by persons entering the refuge area through *horizontal exits* from other areas. Not less than one refuge area exit shall lead directly to the exterior or to an *interior exit stairway* or *ramp*.

Exception: The adjoining compartment shall not be required to have a *stairway* or door leading directly outside, provided the refuge area into which a *horizontal exit* leads has *stairways* or doors leading directly outside and are so arranged that egress shall not require the occupants to return through the compartment from which egress originates.

❖ In a single-tenant facility, any of the spaces that are constantly available (i.e., not lockable) can be used as places of refuge. However, in spaces housing more than one tenant, public refuge areas, such as corridors or passageways, must be provided and be accessible at all times. This requirement is necessary because if a horizontal exit connected two areas occupied by different tenants, the tenants could (for privacy and security purposes) render the necessary free access through the horizontal exit ineffective. When the horizontal exit discharges into a public or common space, such as a corridor leading to an exit, each tenant can obtain the desired security.

Note that the capacity of exits (such as an exit stairway) from a refuge area into which a horizontal exit leads is required to be sufficient for the design occupant load in the area, and does not include those who come into the space from other areas via the horizontal exit. This is because the adjacent refuge area is of sufficient safety to house occupants during a fire or until the egress system is available.

The door through the horizontal exit and the second exit must meet the separation requirements in Section 1007.1.1. Measurement of the travel distance stops at the doorway that serves as the horizontal exit. There are no requirements for travel distance from the horizontal exit to the exit (i.e., exterior exit doorway, exit stairway or exit ramp) on the other side;

however, the areas on each side need to be evaluated for all means of egress requirements individually.

When there is one horizontal exit and two fire compartments, at least one exit from each side of the horizontal exit must go directly to the outside of an exit stairway or ramp enclosure (see Commentary Figure 1026.1). The exception allows for a central building/fire area with access to two horizontal exits and no direct exterior exit door or exit stairway/ramp enclosure as long as the piece on each side has access to exterior exits or exit stairways/ramps (see Commentary Figure 1026.4.2).

SECTION 1027
EXTERIOR EXIT STAIRWAYS AND RAMPS

1027.1 Exterior exit stairways and ramps. *Exterior exit stairways* and *ramps* serving as an element of a required *means of egress* shall comply with this section.

❖ Stairways and ramps can be exit access, exit or exit discharge elements. Exterior exit access and exit discharge stairways and ramps typically involve a change of elevation of less than a story. Exterior exit access stairways or ramps between stories must comply with Section 1019.3 and, where permitted as part of the required means of egress, are limited by Section 1006.3. Exit stairways and ramps traverse a full story or more. Interior exit stairways and ramps must be enclosed in accordance with Section 1023. This section addresses exterior stairways and ramps that function as exit elements.

Exterior exit stairways and ramps are an important element of the means of egress system and must be designed and constructed so that they will serve as a safe path of travel. The general requirements in Section 1011 also apply to exterior stairways (for ramp provisions, see Section 1012).

Outdoor stadiums and open parking garages are examples of buildings that may appear to have exterior exit stairways, but actually have open exit access stairways (Sections 1017, 1019 and 1029).

1027.2 Use in a means of egress. *Exterior exit stairways* shall not be used as an element of a required *means of egress* for Group I-2 occupancies. For occupancies in other than Group I-2, *exterior exit stairways* and *ramps* shall be permitted as an element of a required *means of egress* for buildings not exceeding six stories above *grade plane* or that are not *high-rise buildings*.

❖ This section specifies the conditions where exterior exit ramps or stairways can be used as required exits. Exterior exit stairways are not permitted for Group I-2 since quick evacuation of nonambulatory patients from buildings using exterior stairways is impractical. Some patients may not be capable of self-preservation and, therefore, may require assistance from staff. The period of evacuation of nonambulatory patients could become lengthy, especially in bad weather conditions.

Exterior stairways or ramps are not allowed to be required exits in buildings that exceed six stories in height because of the hazard of using such a stairway or ramp in poor weather. Some persons may not be willing to use such a stair due to vertigo. When

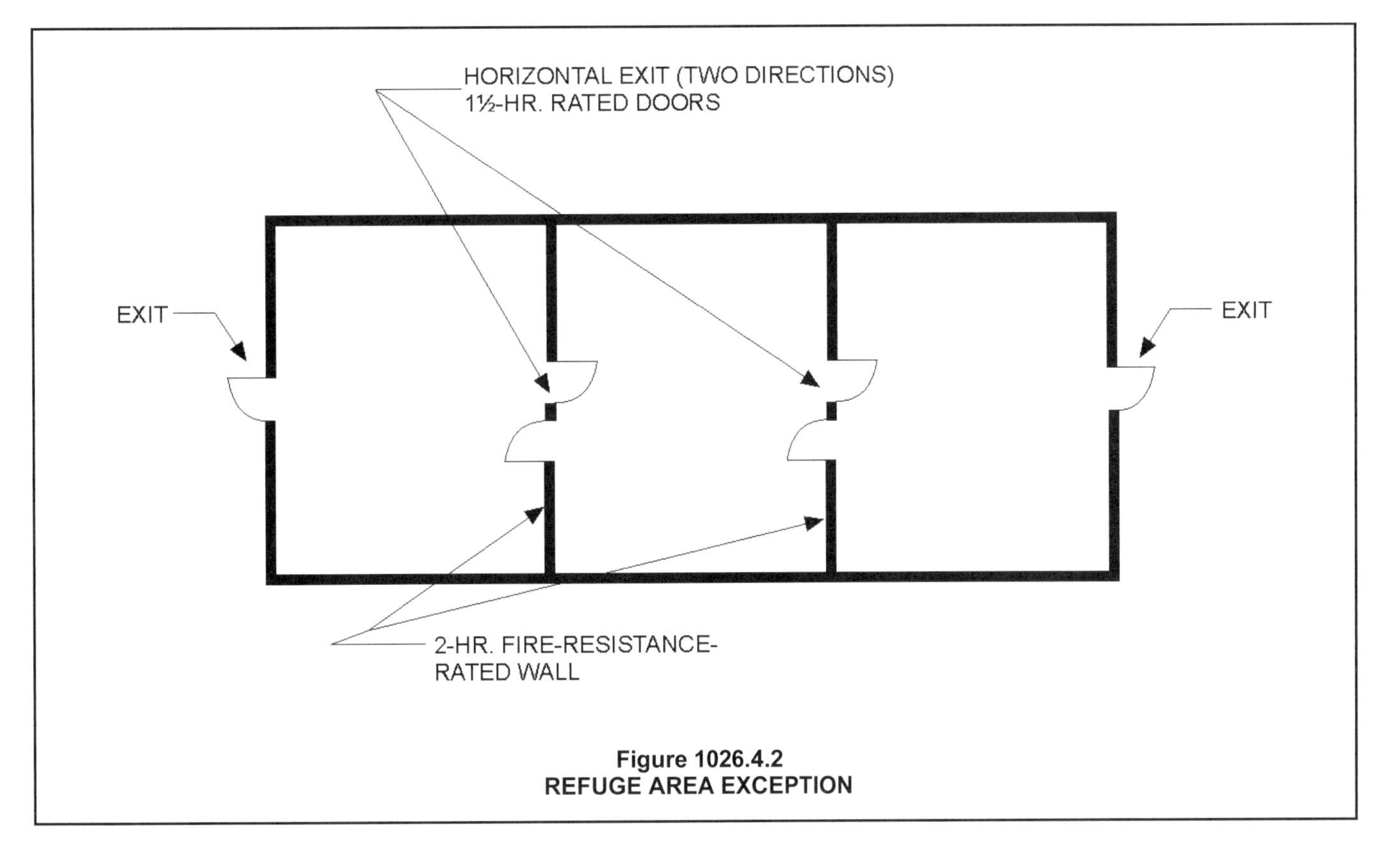

Figure 1026.4.2
REFUGE AREA EXCEPTION

confronted with a view from a great height, vertigo sufferers can become confused, disoriented and dizzy. They could injure themselves, become disoriented or refuse to move (freeze). In a fire situation, they could become an obstruction in the path of travel, possibly causing panic and injuries to other users of the exit.

1027.3 Open side. *Exterior exit stairways* and *ramps* serving as an element of a required *means of egress* shall be open on not less than one side, except for required structural columns, beams, *handrails* and *guards*. An open side shall have not less than 35 square feet (3.3 m^2) of aggregate open area adjacent to each floor level and the level of each intermediate landing. The required open area shall be located not less than 42 inches (1067 mm) above the adjacent floor or landing level.

❖ An important factor in exterior exit stairways or ramps is natural ventilation. Sufficient natural ventilation is necessary so that smoke will not be trapped above the stairway or ramp walking surfaces, thereby compromising safe egress.

The exterior exit stairway or ramp must have at least one of its sides directly facing an outer court, yard or public way. This will allow the products of combustion escaping from the interior of the building to quickly vent to the outdoor atmosphere and let the building occupants egress down the exterior exit stairway or ramp. Since exterior exit stairways or ramps are partially bounded by exterior walls, a minimum amount of exterior openness is specified by the code.

The openings on each and every floor level and landing must total 35 square feet (3.3 m^2) or greater. The openings for which credit is given must occur higher than 42 inches (1067 mm) above each floor and intermediate landing level. [The bottom edge of the opening is consistent with the height requirements for guards (see Section 1015.3)]. With a standard 8-foot (2438 mm) ceiling height minus the 42-inch-high (1067 mm) guard and a typical 8-foot-wide (2438 mm) opening, the result would be $4^1/_2$ feet x 8 feet = 36 square feet (3.34 m^2). Openings of this height and area readily dissipate the smoke buildup from the exterior exit stairway or ramp (see Commentary Figure 1027.3).

1027.4 Side yards. The open areas adjoining *exterior exit stairways* or *ramps* shall be either *yards*, *courts* or *public ways*; the remaining sides are permitted to be enclosed by the *exterior walls* of the building.

❖ This section simply specifies the type of areas that the exterior opening of the exterior exit stairway or ramp is to adjoin. These open spaces will enable the smoke to dissipate from the exterior exit stairway or ramp so it will be usable as a required exit. See Section 1027.3 for a discussion of the opening requirements. See Sections 1027.4 and 1206 for the minimum sizes of yards and courts.

1027.5 Location. *Exterior exit stairways* and *ramps* shall have a minimum fire separation distance of 10 feet (3048 mm) measured at right angles from the exterior edge of the *stairway* or *ramps*, including landings, to:

1. Adjacent *lot lines*.
2. Other portions of the building.
3. Other buildings on the same lot unless the adjacent building *exterior walls* and openings are protected in accordance with Section 705 based on *fire separation distance*.

For the purposes of this section, other portions of the building shall be treated as separate buildings.

❖ The location requirements of this section protect the users of the exterior exit stairway or ramp from the effects of a fire in another building on the same lot or an adjacent lot. The separation distance reduces the exposure to heat and smoke. If the exterior exit stairway or ramp is closer than specified, then adjacent buildings' exterior walls and openings are to be protected in accordance with Section 705 so that the users of the exterior exit stairway or ramp are protected. The reason for a minimum required distance to a lot line is to provide for a future building that could be built on an adjacent lot. While buildings on the same lot can be considered one building for height and area limitations (see Section 503.1.2), they must be separated by a minimum of 10 feet (3048 mm) if there is a path for exit discharge between them. The purpose of the last sentence is to clarify that an exterior exit stairway or ramp needs a minimum 10-foot separation where a building wraps around on itself, such as a U-shaped building. It is not intended that the distance be measured to an imaginary lot line between buildings on the same lot.

Requirements are the same for exterior egress balconies (Section 1021.4). For an illustration of exterior egress balconies and exterior exit stairways working together, see Commentary Figure 1027.5.

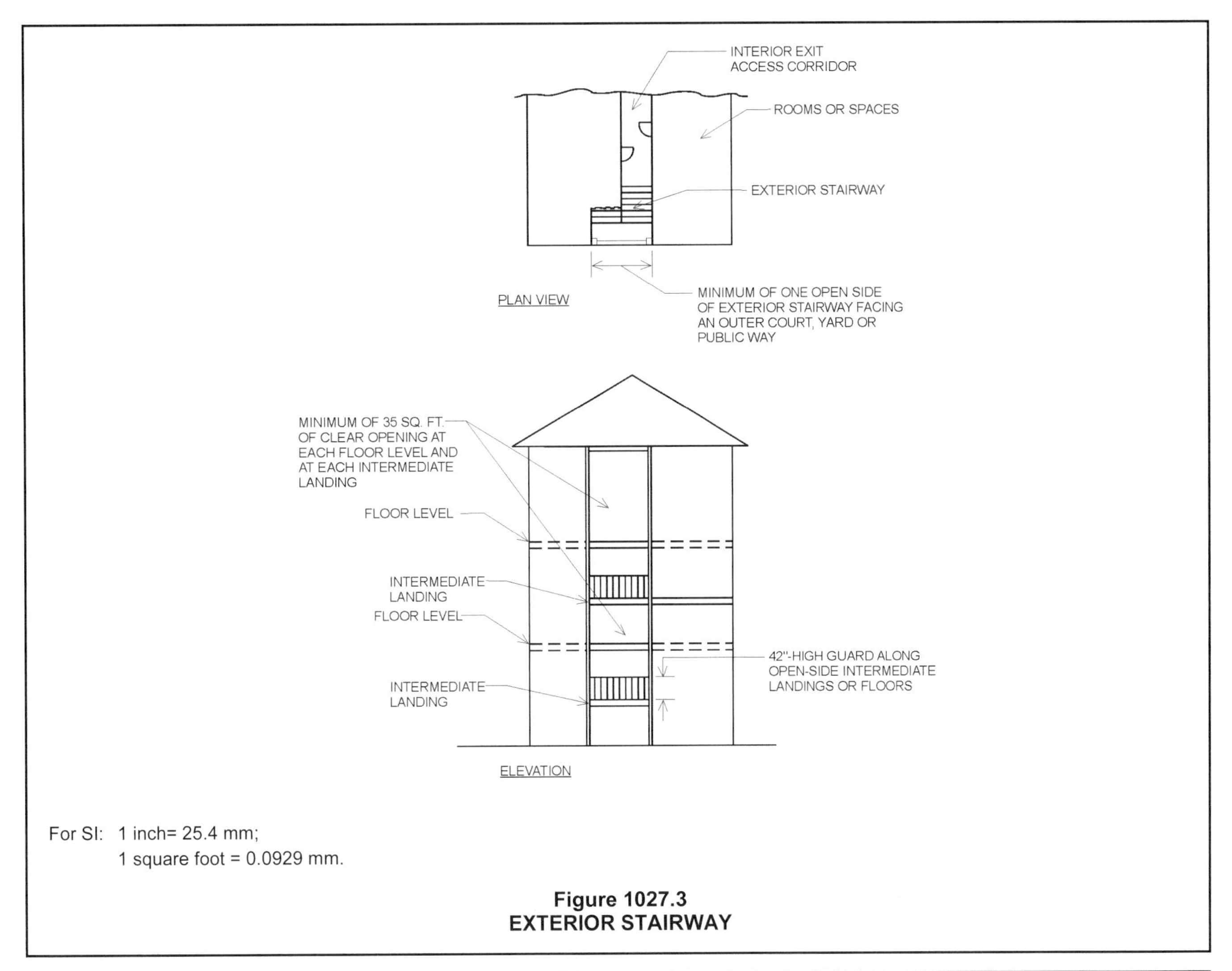

For SI: 1 inch= 25.4 mm;
1 square foot = 0.0929 mm.

Figure 1027.3
EXTERIOR STAIRWAY

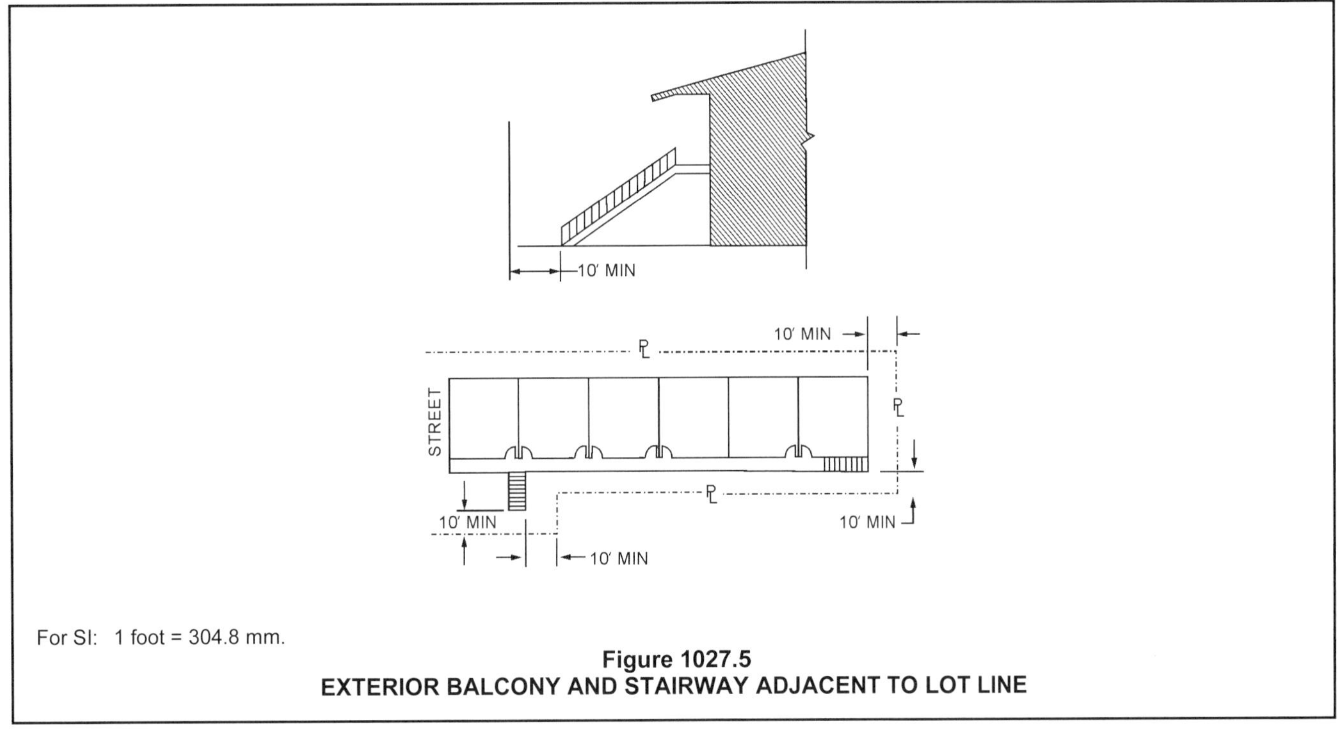

For SI: 1 foot = 304.8 mm.

Figure 1027.5
EXTERIOR BALCONY AND STAIRWAY ADJACENT TO LOT LINE

1027.6 Exterior exit stairway and ramp protection. *Exterior exit stairways* and *ramps* shall be separated from the interior of the building as required in Section 1023.2. Openings shall be limited to those necessary for egress from normally occupied spaces. Where a vertical plane projecting from the edge of an *exterior exit stairway* or *ramp* and landings is exposed by other parts of the building at an angle of less than 180 degrees (3.14 rad), the exterior wall shall be rated in accordance with Section 1023.7.

Exceptions:

1. Separation from the interior of the building is not required for occupancies, other than those in Group R-1 or R-2, in buildings that are not more than two stories above *grade plane* where a *level of exit discharge* serving such occupancies is the first story above *grade plane*.
2. Separation from the interior of the building is not required where the *exterior exit stairway* or *ramp* is served by an *exterior exit ramp* or balcony that connects two remote *exterior exit stairways* or other *approved exits* with a perimeter that is not less than 50 percent open. To be considered open, the opening shall be not less than 50 percent of the height of the enclosing wall, with the top of the openings not less than 7 feet (2134 mm) above the top of the balcony.
3. Separation from the open-ended *corridor* of the building is not required for *exterior exit stairways* or *ramps*, provided that Items 3.1 through 3.5 are met:
   - 3.1. The building, including open-ended *corridors*, and *stairways* and *ramps*, shall be equipped throughout with an *automatic sprinkler system* in accordance with Section 903.3.1.1 or 903.3.1.2.
   - 3.2. The open-ended *corridors* comply with Section 1020.
   - 3.3. The open-ended *corridors* are connected on each end to an *exterior exit stairway* or *ramp* complying with Section 1027.
   - 3.4. The *exterior walls* and openings adjacent to the *exterior exit stairway* or *ramp* comply with Section 1023.7.
   - 3.5. At any location in an open-ended *corridor* where a change of direction exceeding 45 degrees (0.79 rad) occurs, a clear opening of not less than 35 square feet (3.3 m^2) or an *exterior stairway* or *ramp* shall be provided. Where clear openings are provided, they shall be located so as to minimize the accumulation of smoke or toxic gases.

❖ Exterior exit stairways or ramps must be protected from interior fires that may project through windows or other openings adjacent to the exit stairway or ramp, possibly endangering the occupants using this means of egress to reach grade. The protection of an exterior exit stairway or ramp is to be obtained by separating the exterior exit from the interior of the building using exterior walls having a fire-resistance rating of at least 1 hour with opening protectives. Consistent with the protection required in Sections 1023.2 and 1023.7 for interior exit stairways, the fire-resistance rating must be provided for a distance of 10 feet (3048 mm) horizontally and vertically from the ramp or stairway edges, and from the ground to a level of 10 feet (3048 mm) above the highest landing.

All window and door openings falling inside the 10-foot (3048 mm) horizontal separation distance as well as all window and door openings 10 feet (3048 mm) above the topmost landing and below the stairway must be protected with minimum $^3/_4$-hour fire protection-rated opening protectives [see Commentary Figure 1027.6(1)]. The last sentence is similar to Section 1023.7 except that instead of measuring the angle between the building exterior walls and the unprotected walls at the exterior of the stairway or ramp, the measurement is between the building exterior walls and a vertical projection for the planes of the guard of the exterior stairway and ramp including landings.

Openings within the width of the stairway must only be from normally occupied spaces. This is consistent with the requirements for vertical exit enclosures (see Sections 1023.4 and 1023.5).

Exception 1 indicates that opening protectives are not required for occupancies (other than Groups R-1 and R-2) that are two stories or less above grade when the level of exit discharge is at the lower story. The reason for this exception is that in cases of fire in low buildings, the occupants are usually able to evacuate the premises before the fire can emerge through exterior wall openings and endanger the exit ramp or stairways. In hotels and apartments, however, the occupants' response to a fire emergency could be significantly reduced because they may be either unfamiliar with the surroundings or sleeping.

Exception 2 allows the opening protectives to be omitted when an exterior exit access balcony is served by two exits and when the exits are remote from each other. Remoteness is regulated by Section 1007. This exception is applicable to all groups. In such instances, it is unlikely that the users of the exterior stairway or ramp will become trapped by fire, since they have the option of using the balcony to gain access to either of the two available exits, and the products of combustion will be vented directly to the outside (see Section 1021.1 regarding exterior balconies). At least one-half of the total perimeter of the exterior balcony must be permanently open to the outside. The requirement for at least one-half the height of that level to be open allows for columns, solid guards and architectural or decorative elements, such as arches. With the top of the opening at least 7 feet (2134 mm) above the walking surface, products of combustion can vent and allow occupant passage below the smoke layer [see Commentary Figure 1027.6(2)].

An open-ended corridor is not defined in Chapter 2. Exception 3 deletes the requirement for a separation between the interior of the building and the exterior wall area immediately adjacent to the exit stairway where an open-ended corridor (breezeway) interfaces with an exterior stairway or ramp. In other words, a door is not required between the exterior exit stairway and the open-ended corridor. The separation is not needed because of the NFPA 13 or 13R sprinkler system is required in all areas of the building, including the open-ended corridor. The other characteristics of the open-ended corridor described in this exception are needed so that it is safe to be used in the event of a fire. The requirements for an exterior stairway or ramp at each end, and the additional openings or exterior stairways or ramps where the open-ended corridor has a change of direction of greater than 45 degrees (0.79 rad), are for adequate ventilation of the open-ended corridor. The corridor may be required to be rated per Table 1020.1. The reference back to Section 1023.7 results in any exterior walls that form an angle less than 180 degrees (3.14 rad) from the side of the exterior exit stairway to be protected for a distance of 10 feet (3048 mm) [see Commentary Figure 1027.6(3)]. Exit access travel distance on an open-ended corridor is measured to the first riser of an exterior stair or the beginning slope of an exterior ramp.

Similar separation language is used in describing the exterior wall requirements for an exterior area for assisted rescue that is provided at an exit located above grade level. Where there are steps, they are merely part of the exit discharge, not exterior exit stairways as addressed in this section.

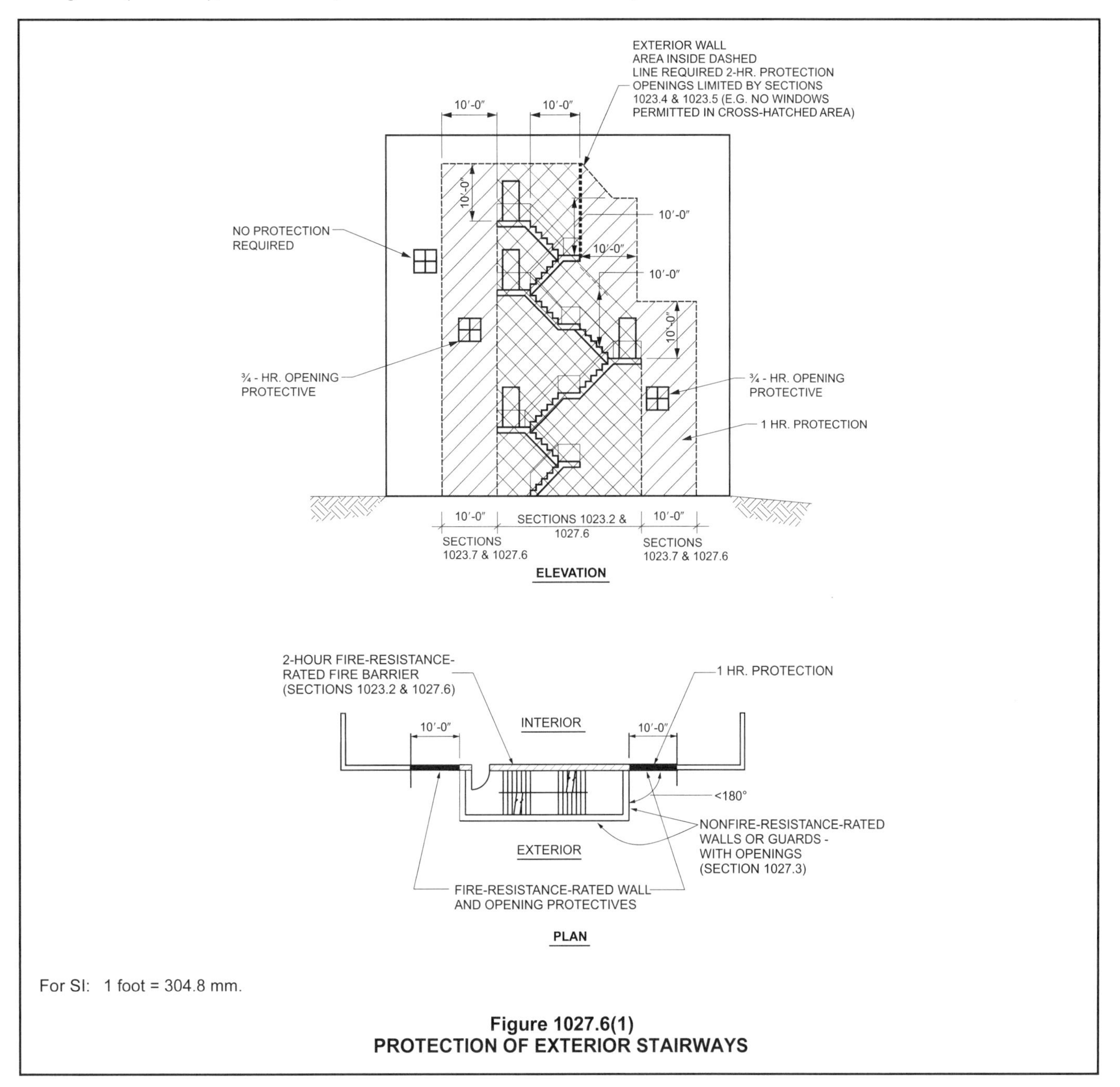

For SI: 1 foot = 304.8 mm.

Figure 1027.6(1)
PROTECTION OF EXTERIOR STAIRWAYS

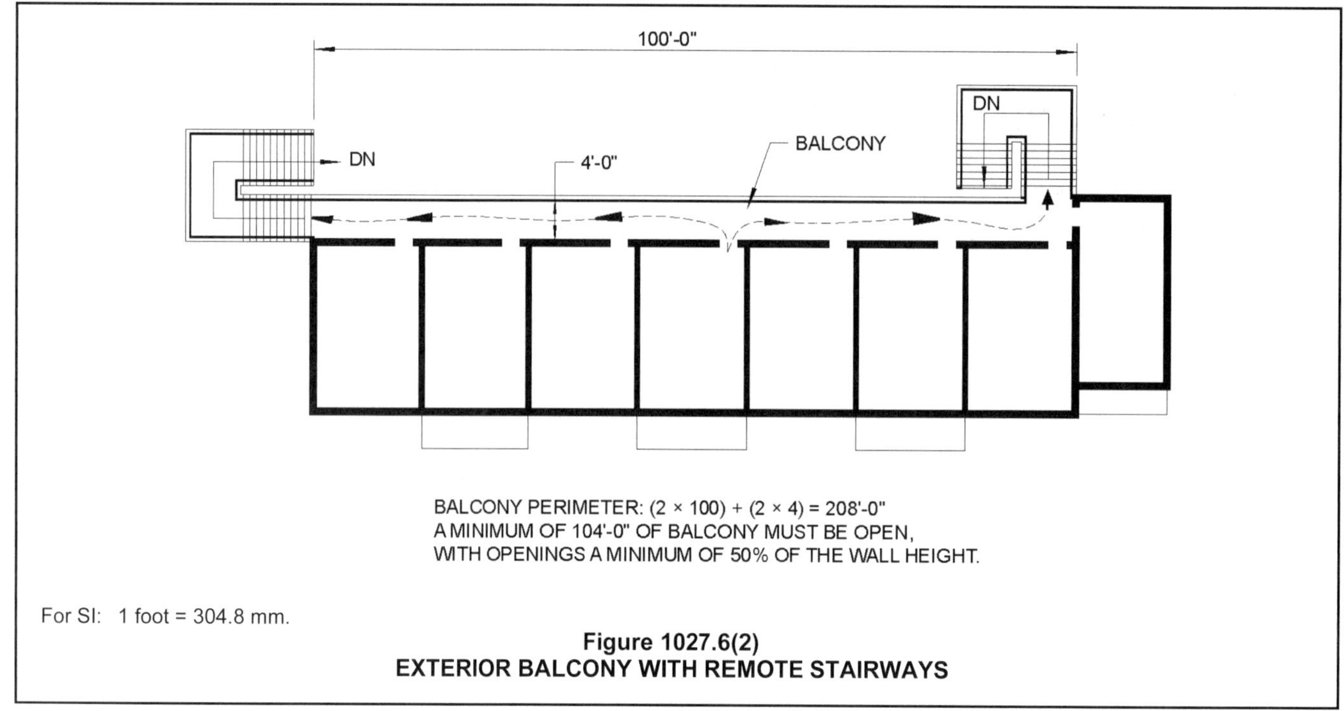

For SI: 1 foot = 304.8 mm.

Figure 1027.6(2)
EXTERIOR BALCONY WITH REMOTE STAIRWAYS

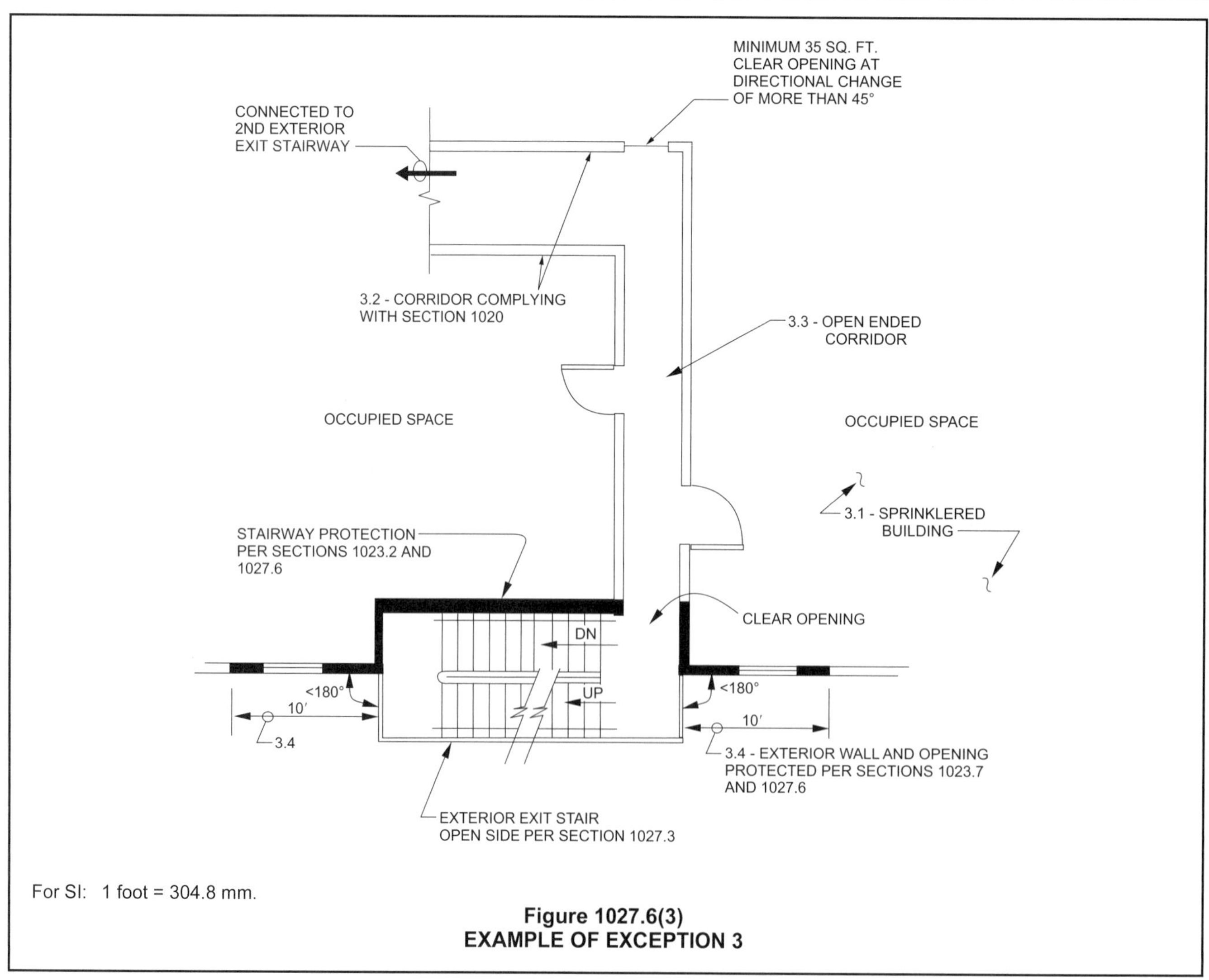

For SI: 1 foot = 304.8 mm.

Figure 1027.6(3)
EXAMPLE OF EXCEPTION 3

SECTION 1028
EXIT DISCHARGE

1028.1 General. *Exits* shall discharge directly to the exterior of the building. The *exit discharge* shall be at grade or shall provide a direct path of egress travel to grade. The *exit discharge* shall not reenter a building. The combined use of Exceptions 1 and 2 shall not exceed 50 percent of the number and minimum width or required capacity of the required exits.

Exceptions:

1. Not more than 50 percent of the number and minimum width or required capacity of *interior exit stairways* and *ramps* is permitted to egress through areas on the level of discharge provided all of the following conditions are met:
    1.1. Discharge of *interior exit stairways* and *ramps* shall be provided with a free and unobstructed path of travel to an exterior *exit* door and such *exit* is readily visible and identifiable from the point of termination of the enclosure.

    1.2. The entire area of the *level of exit discharge* is separated from areas below by construction conforming to the *fire-resistance rating* for the enclosure.

    1.3. The egress path from the *interior exit stairway* and *ramp* on the *level of exit discharge* is protected throughout by an *approved automatic sprinkler system*. Portions of the *level of exit discharge* with access to the egress path shall be either equipped throughout with an *automatic sprinkler system* installed in accordance with Section 903.3.1.1 or 903.3.1.2, or separated from the egress path in accordance with the requirements for the enclosure of *interior exit stairways* or *ramps*.

    1.4. Where a required *interior exit stairway* or *ramp* and an *exit access stairway* or *ramp* serve the same floor level and terminate at the same *level of exit discharge*, the termination of the *exit access stairway* or *ramp* and the *exit discharge* door of the *interior exit stairway* or *ramp* shall be separated by a distance of not less than 30 feet (9144 mm) or not less than one-fourth the length of the maximum overall diagonal dimension of the building, whichever is less. The distance shall be measured in a straight line between the *exit discharge* door from the *interior exit stairway* or *ramp* and the last tread of the *exit access stairway* or termination of slope of the *exit access ramp*.
2. Not more than 50 percent of the number and minimum width or required capacity of the *interior exit stairways* and *ramps* is permitted to egress through a vestibule provided all of the following conditions are met:
    2.1. The entire area of the vestibule is separated from areas below by construction conforming to the *fire-resistance rating* of the *interior exit stairway* or *ramp enclosure*.

    2.2. The depth from the exterior of the building is not greater than 10 feet (3048 mm) and the length is not greater than 30 feet (9144 mm).

    2.3. The area is separated from the remainder of the *level of exit discharge* by a *fire partition* constructed in accordance with Section 708.

    Exception: The maximum transmitted temperature rise is not required.

    2.4. The area is used only for *means of egress* and *exits* directly to the outside.
3. *Horizontal exits* complying with Section 1026 shall not be required to discharge directly to the exterior of the building.

❖ The exit discharge is the third piece of the means of egress system, which includes exit access, exit and exit discharge. The general provisions for means of egress in Sections 1003 through 1015 are applicable to the exit discharge. The basic provision is that exits must discharge directly to the outside of the building. The exit discharge is the path from the termination of the exit to the public way. When it is not practical to discharge directly to the outside, there are four alternatives: an exit passageway (see Section 1024), an exit discharge lobby (see Section 1028.1, Exception 1), an exit discharge vestibule (see Section 1028.1, Exception 2) or a horizontal exit (see Sections 1026 and 1028.1, Exception 3). While Exceptions 1 and 2 could be applicable for exit passageways and exit ramps, they are most often applied for exit stairways. The commentary for Section 1027 will be limited to interior exit stairways that are enclosed in accordance with Section 1023 (see Sections 1006 and 1019.3 for exit access stairway requirements). Up to 50 percent of the interior exit stairways in a building may use either Exception 1 or 2; therefore, neither exception is viable for a single-exit building. In a two- or three-exit building, either a lobby or a vestibule can be used for exit discharge for one of the exit stairways. In a four-exit building, two of the exit stairways can use either a lobby or a vestibule for exit discharge.

An interior exit discharge lobby is permitted to receive the discharge from an exit stairway in lieu of the stairway discharging directly to the exterior. A fire door must be provided at the point where the exit stairway discharges into the lobby. Without an opening protective between the stairway and a lobby, it would be possible for the stairway to be directly exposed to smoke movement from a fire in the lobby. The opening protective provides for full continuity of the vertical component of the exit arrangement. Additionally, in buildings where stair towers must be pressurized, pressurization would not be possible without a door at the lobby level.

An exit discharge lobby is the sole location recognized in the code where an exit element can be used for purposes other than pedestrian travel for means of egress. The lobby may contain furniture or decoration and nonoccupiable spaces may open directly into the lobby. The lobby, and all other areas on the same level that are not separated from the lobby by fire barriers consistent with the rating of the stair enclosure, must be sprinklered in accordance with an NFPA 13 or NFPA 13R system [see Commentary Figure 1028.1(1)]. If the entire level is sprinklered, no separation is required. In this case, the automatic sprinkler system is anticipated to control and (perhaps) eliminate the fire threat so as not to jeopardize the path of egress of the occupants. The lobby floor and any supporting construction must be rated the same as the stairway enclosure. If the lobby is slab on grade, this requirement is not applicable. This is consistent with the fundamental concept that an exit enclosure provides the necessary level of protection from adjacent areas. A path of travel through the lobby must be continually clear and available. The exit door leading out of the building must be visible and identifiable immediately when a person leaves the exit. This does not mean the exterior exit door must be directly in front of the door at the bottom of the stairway, but the intent is that it should be within the general range of vision. A person should not have to turn completely around or go around a corner to be able to see the way out.

Item 1.4 addresses when an exit access stairway and an exit stairway both discharge into the same lobby on the ground floor. For example, many hotels have meeting rooms on the level immediately above the lobby. This heavier occupant load, or circulation considerations, may result in an exit access stairway coming down from that second level into the same lobby that is being used for discharge from the upper floors for one of the required exit stairways. This limitation for a 30-foot (9144) separation is to prevent an exit access stairway and an exit stairway from termination too close together on the level of exit discharge. The intent is that one localized fire event in the lobby will not jeopardize the use of both means of egress components. The 30-foot or $^{1}/_{4}$ diagonal separation distances were based on the 30-foot or $^{1}/_{4}$ diagonal that is specified for separation of interior stairways in a high-rise building (see Section 403.5.1). This measurement is taken only at the bottom of the exit access stairway and the door to the

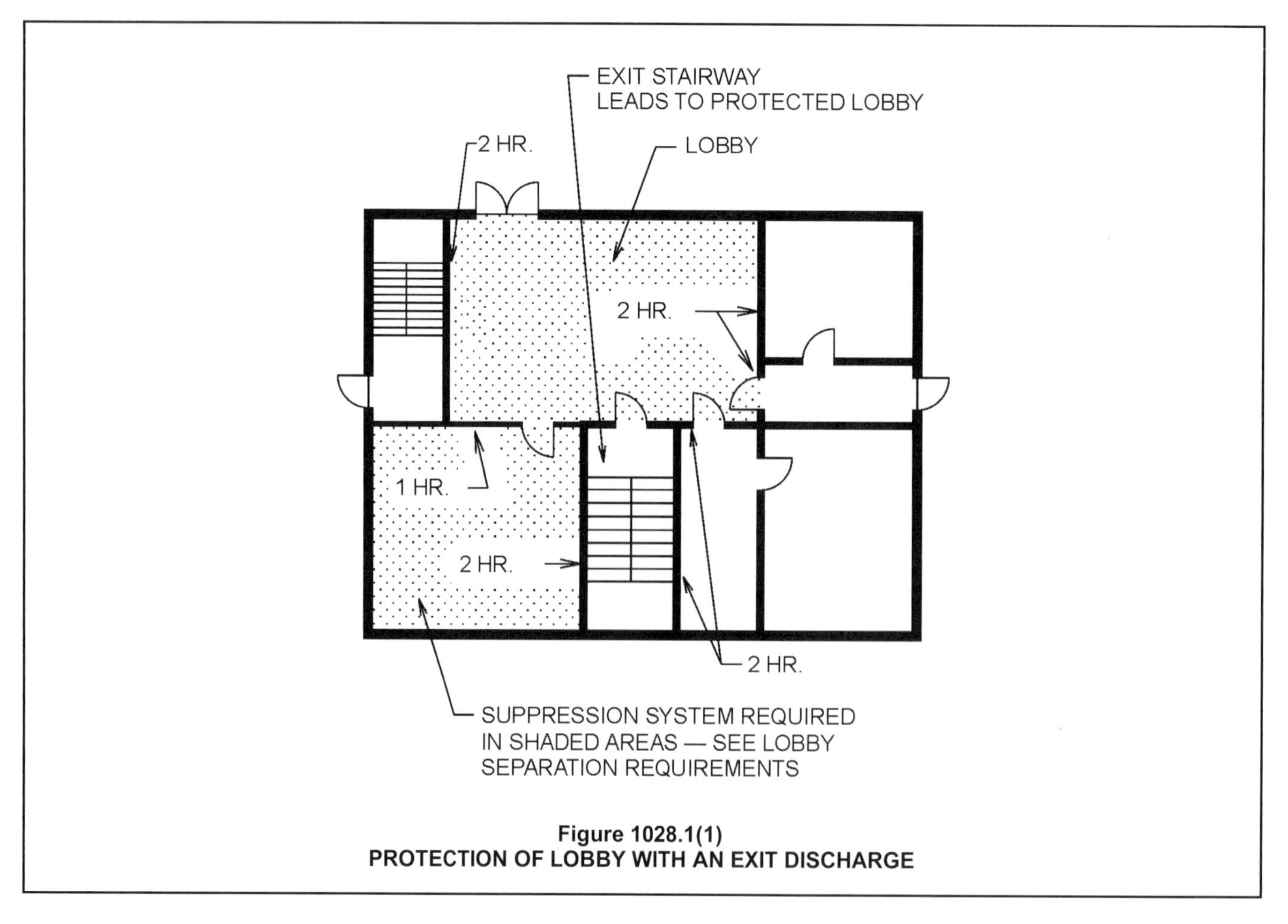

Figure 1028.1(1)
PROTECTION OF LOBBY WITH AN EXIT DISCHARGE

exit enclosure. It is not required to be maintained the entire length of the stairway as required for two exit access stairways in Section 1007.1.3. The enclosure of the exit stairway offers sufficient separation protection. The intent of the exception is to allow for glazing that meets the smoke partition requirements.

An exit is also allowed to discharge through a vestibule, provided it complies with the specified requirements of Exception 2. Vestibules utilizing this provision are not to be used for other purposes, such as access to closets, furniture/seating, drinking fountains, vending machines, etc. The vestibule floor and any supporting construction must be rated the same as the stairway enclosure. If the vestibule is slab on grade, this requirement is not applicable. The size of the vestibule is limited so that it cannot be used for other activities, and the travel distance from the exit stairway to the exterior exit doorway is limited [see Commentary Figures 1028.1(2) and 1028.1(3)]. The interior walls of the vestibule must be constructed as fire partitions with a fire-protection rating of at least 1 hour. Section 708 does not reference a specific test standard for fire partitions. As such, fire partitions must comply with ASTM E119 or UL 263 per Section 703.2 (see Section 703.2 for additional commentary on the tests). ASTM E119 and UL 263 have a transmission of heat criterion for nonbearing partitions to be considered as having passed the test. The exception exempts fire partitions for the separation of a vestibule from having to comply with the transmission of heat criterion. Glass panels set in a metal frame can be used for the vestibule as long as they meet the rest of the ASTM E119 criteria.

Exception 3 acknowledges that horizontal exits offer refuge areas that will have access to an exit on the other side; therefore, a stairway that exits through a horizontal exit is not required to exit to the exterior. Many hospitals and correctional facilities use horizontal exits to "defend in place" rather than require an immediate building evacuation. There are exit stairways or exits available from the refuge areas, so occupants can move to the outside if needed (see Section 1026 for additional information).

1028.2 Exit discharge width or capacity. The minimum width or required capacity of the *exit discharge* shall be not less than the minimum width or required capacity of the *exits* being served.

❖ This section specifies the exit discharge width based on minimum width and the number of occupants exiting (capacity). The exit discharge is required to be designed for the occupant from all of the exits it serves. If the exit discharge serves two exits, it is to be designed for the sum of the occupants served by both exits. Note that the capacity of the exit discharge is not required to match the total provided capacity of both exits, which is typically higher than the sum of the occupants served by both exits.

1028.3 Exit discharge components. *Exit discharge* components shall be sufficiently open to the exterior so as to minimize the accumulation of smoke and toxic gases.

❖ An exit discharge component could be a large open space where people could discharge in a number of different directions or it could be limited to a narrower path by landscaping or walls (i.e., egress court). In all cases, the space must be open enough to the outside that smoke and fumes will vent upward and away from people evacuating the building.

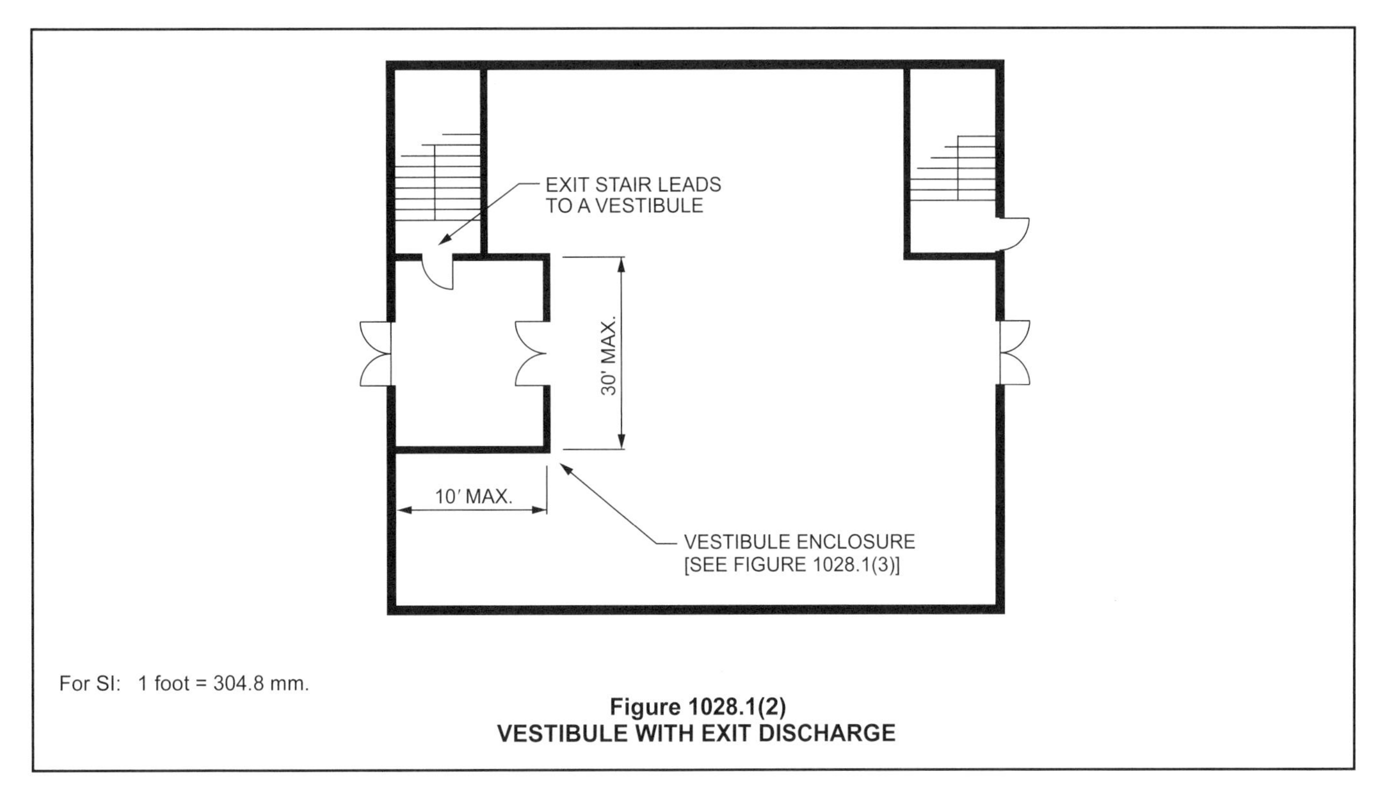

Figure 1028.1(2)
VESTIBULE WITH EXIT DISCHARGE

1028.4 Egress courts. *Egress courts* serving as a portion of the *exit discharge* in the *means of egress* system shall comply with the requirements of Sections 1028.4.1 and 1028.4.2.

❖ A portion of the exit discharge that is partially confined by exterior walls or other elements that confine the discharge path to a single narrow route is regulated as an egress court.

This section and the following subsections address the detailed requirements for egress courts. It is essential that exterior egress courts serving occupants from an exit to a public way be sufficiently open to prevent the accumulation of smoke and toxic gases in the event of a fire as well as wide enough to accommodate the number of occupants leaving in that direction.

See Section 1206 for additional minimum width and openness requirements when yards and courts are needed for natural light or ventilation.

1028.4.1 Width or capacity. The required capacity of *egress courts* shall be determined as specified in Section 1005.1, but the minimum width shall be not less than 44 inches (1118 mm), except as specified herein. *Egress courts* serving Group R-3 and U occupancies shall be not less than 36 inches (914 mm) in width. The required capacity and width of *egress courts* shall be unobstructed to a height of 7 feet (2134 mm).

Exception: Encroachments complying with Section 1005.7.

Where an *egress court* exceeds the minimum required width and the width of such *egress court* is then reduced along the path of exit travel, the reduction in width shall be gradual. The transition in width shall be affected by a guard not less than 36 inches (914 mm) in height and shall not create an angle of more than 30 degrees (0.52 rad) with respect to the axis of the *egress court* along the path of egress travel. The width of the *egress court* shall not be less than the required capacity.

❖ The width of an exterior court is to be determined in the same fashion as for an interior corridor. The width is not to be less than required to serve the number of occupants from the exit or exits and not less than the minimum specified in this section (see also Section 1206). A cross reference back to Section 1005.7 for obstructions in the width in aisles (see Section 1018.1), corridors (see Section 1020.3), exit passageways (see Section 1024.2) and exit courts (see Section 1028.4.1) reinforces the fact that the protrusion limits provision is generally applicable for these types of confined routes.

Many egress courts are significantly larger than required. Thus, the code allows such an egress court to decrease in width along the path of travel to the public way. The gradual transition requirement is so the flow of the occupants will be uniform without pockets of congestion. The transition requirements should be applied to egress courts where a reduction results in a width that is near the minimum based on the number of occupants served. It is this condition where the uniform flow of occupants is essential.

1028.4.2 Construction and openings. Where an *egress court* serving a building or portion thereof is less than 10 feet (3048 mm) in width, the *egress court* walls shall have not less than 1-hour *fire-resistance-rated* construction for a distance of 10 feet (3048 mm) above the floor of the *egress court*. Openings

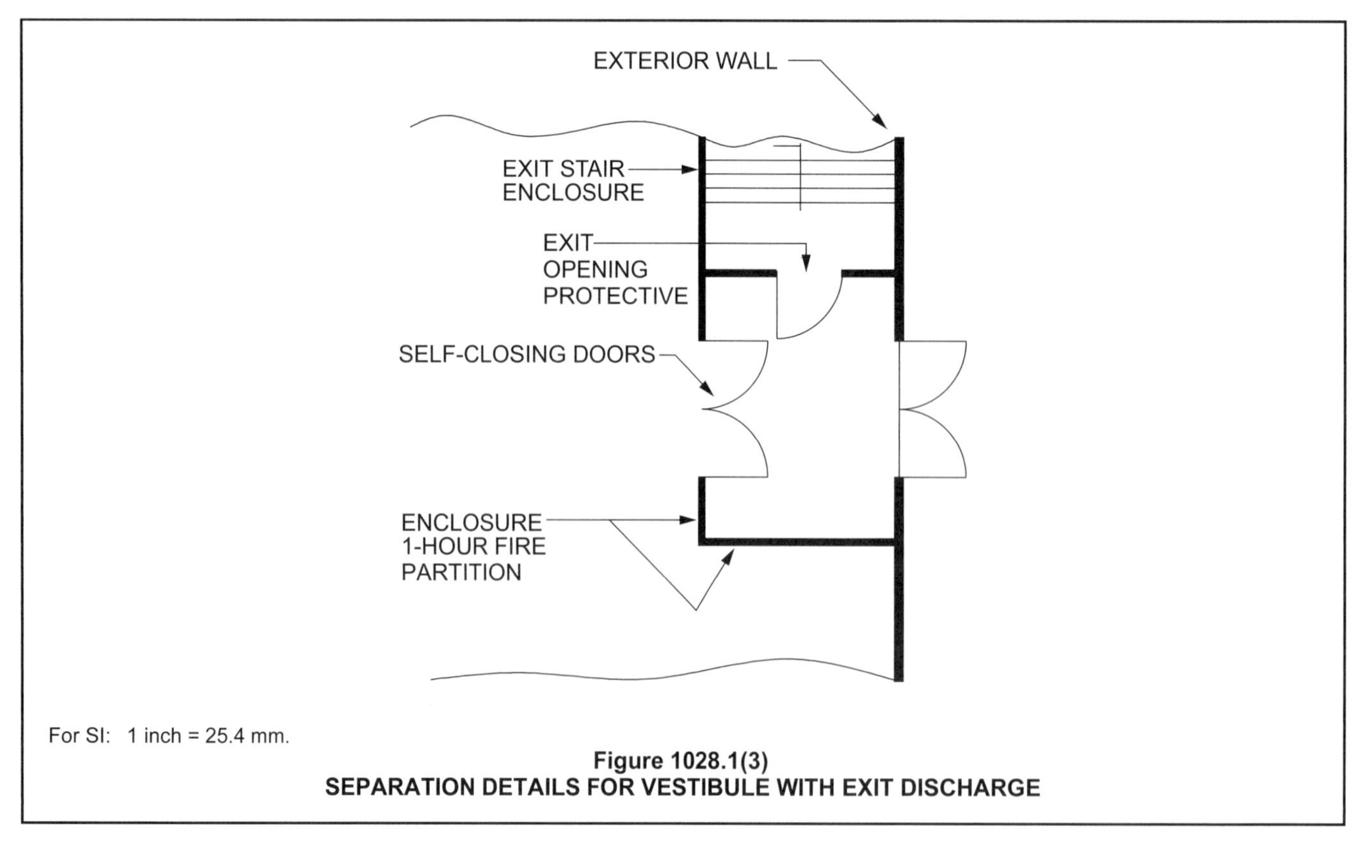

Figure 1028.1(3)
SEPARATION DETAILS FOR VESTIBULE WITH EXIT DISCHARGE

within such walls shall be protected by opening protectives having a fire protection rating of not less than $^3/_4$ hour.

Exceptions:

1. *Egress courts* serving an *occupant load* of less than 10.
2. *Egress courts* serving Group R-3.

❖ The purpose of this section is to protect the occupants served by the egress court from the building that they are exiting from. If occupants must walk closely by the exterior walls of the court, the walls are required to have the specified fire-resistance rating and the openings are required to be protected as specified. This requirement is only for the first 10 feet (3048 mm) above the level of the egress court since the exposure hazard from walls and openings above 10 feet (3048 mm) is reduced. The two exceptions provide for egress courts that serve a very low number of occupants and the specified residential occupancy where the protection requirement would be located.

1028.5 Access to a public way. The *exit discharge* shall provide a direct and unobstructed access to a *public way*.

Exception: Where access to a *public way* cannot be provided, a safe dispersal area shall be provided where all of the following are met:

1. The area shall be of a size to accommodate not less than 5 square feet (0.46 m2) for each person.
2. The area shall be located on the same lot not less than 50 feet (15 240 mm) away from the building requiring egress.
3. The area shall be permanently maintained and identified as a safe dispersal area.
4. The area shall be provided with a safe and unobstructed path of travel from the building.

❖ There are instances where the path of travel to the public way is not safe or not achievable because of site constraints or security concerns. The provisions in this section specify what would constitute a safe area to allow occupants of a building to assemble in an emergency. The requirement of 5 square feet (0.28 m^2) would allow adequate space for standing persons as well as some space for persons in wheelchairs or on stretchers. Everyone who is expected to wait in this dispersal area for fire department assistance must be a minimum of 50 feet (15 240 mm) away from the building. This refuge must always remain open and not be used for parking, storage or temporary structures. A safe dispersal area is commonly found at schools or jails. Stadiums are more specifically addressed in Section 1010.2.1. Walls or fences may surround the building and part of the site due to other safety concerns. These walls and fences could stop occupants from reaching the public way.

SECTION 1029 ASSEMBLY

1029.1 General. A room or space used for assembly purposes that contains seats, tables, displays, equipment or other material shall comply with this section.

❖ Any room that is used for assembly purposes, regardless of the occupancy of the rest of the building, must comply with this section. Spaces used for assembly seating may appear in buildings of other occupancy types; for example, a library in a school, or a meeting room in an office building. This includes spaces with less than 50 occupants in other occupancies. For evaluation of the occupant load and the means of egress in these spaces, these spaces are regulated based on their function, rather than their occupancy group.

Although most of the provisions in Section 1029 focus on fixed seating auditoriums or theaters, this section also addresses loose seats, tables, displays, equipment, etc. Rooms or spaces used for assembly purposes contain elements that would affect the path of travel for the means of egress. These spaces require special consideration because of the larger occupant loads and possible low lighting (e.g., nightclubs, theaters), which can possibly lead to slower fire recognition or crowd concerns.

Since this section is extensive, here is a basic breakdown:

- Sections 1029.1.1 and 1029.1.1.1 deal with bleachers and grandstands.
- Sections 1029.3 through 1029.5 deal with number and dispersement of exits.
- Section 1029.6 and subsections discuss aisle widths based on the required capacity.
- Sections 1029.7 and 1029.8 are for travel distances and aisle accessways.
- Section 1029.9 deals with where aisles are required, minimum widths, and layouts. This includes provisions for where stairways are a direct continuation of a stepped aisle (Sections 1029.9.7 through 1029.10.3).
- Section 1029.11 discusses types of materials and walking surfaces.
- Section 1029.12 covers the aisle accessways that lead to the main aisles as they move through tables (Section 1029.12.1) and seating in rows (Section 1029.12.2).
- Section 1029.13 deals with slope, landings and edge protection for ramped aisles (Sections 1029.13.1 through 1029.13.1.3) and treads and risers for stepped aisles (Sections 1029.13.2 through 1029.13.2.4).
- Section 1029.14 discusses where seating needs to be fastened to the floor in order for aisles and accessways to be maintained.

- Section 1029.15 includes handrail provisions for ramped and stepped aisles.
- Section 1029.16 states where guards are required.

1029.1.1 Bleachers. *Bleachers*, *grandstands* and *folding and telescopic seating*, that are not building elements, shall comply with ICC 300.

❖ On February 24, 1999, the Bleacher Safety Act of 1999 was introduced in the House of Representatives. The bill, which cites the ICC and the code, authorizes the U.S. Consumer Product Safety Commission (CPSC) to issue a standard for bleacher safety. This was in response to concerns relative to accidents on bleacher-type structures. As a result, the CPSC developed and revised the *Guidelines for Retrofitting Bleachers*. The ICC Board of Directors decided that a comprehensive standard dealing with all aspects of both new and existing bleachers was warranted and authorized the formation of the ICC Consensus Committee on Bleacher Safety. The committee is composed of 12 members, including the requisite balance of general, user interest and producer interest.

ICC 300 was completed in December 2001, and submitted to ANSI on January 1, 2002. ICC 300 was reissued with some revisions in 2007 and 2012. While the term "bleachers" is generic, the standard addresses all aspects of tiered seating associated with bleachers, grandstands, and folding and telescopic seating. These types of seating are supported on dedicated structural systems, which in turn may sit on the ground or on a building floor system. Single seats or bench seats bolted down to a stepped floor are not considered a bleacher or grandstand and should comply with Section 1029. See the definitions in Chapter 2 for "Building element," "Bleachers," "Grandstands" and "Folding and telescopic seating." While ICC 300 is consistent and also relies on Chapter 10 of the code for some provisions, the standard addresses items specific to these types of seating arrangements. For example, the minimum number of exits from a bleacher is addressed in ICC 300, Section 404.1; however, to determine the minimum number of exits from the room the bleacher is located in, Section 1006 is applicable. The bleacher standard references Chapter 11 of the code and ICC A117.1 for accessibility requirements.

The ICC 300 has minimum requirements for new, alterations, repair, operation and maintenance of bleacher systems. A bleacher or grandstand is defined as "Tiered seating supported on a dedicated structural system and two or more rows high and is not a building element." The intent of the terms "dedicated structural system" and "not a building element" in the definition is to recognize that bleacher systems sit on the floor or ground and have a support system separate from the building system. However, the bleacher could rely on the building system for lateral or gravity support. The intent of "two or more rows" is so that a tiered floor system with a bench or row of seats on each tier would not be considered a bleacher.

The criteria in ICC 300 include provisions for construction; means of egress within the bleacher system; inspection and maintenance for existing bleachers; and for when seating systems are relocated.

1029.1.1.1 Spaces under grandstands and bleachers. Where spaces under *grandstands* or *bleachers* are used for purposes other than ticket booths less than 100 square feet (9.29 m^2) and toilet rooms, such spaces shall be separated by *fire barriers* complying with Section 707 and *horizontal assemblies* complying with Section 711 with not less than 1-hour *fire-resistance-rated* construction.

❖ Sometimes spaces under grandstands are used for other purposes such as bathrooms, concession stands, storage, etc. If that space caught on fire, it could jeopardize the safe evacuation options for persons on the bleachers. For safety, the spaces below must be separated from the bleachers by fire-resistance-rated construction. This is typically the roof and back walls of the concession stand or storage room. If the space below is either a small ticket booth or bathrooms of any size, the potential fire load is low enough that these spaces are not required to be separated.

Note that Section 903.2.1.5 requires enclosed spaces with an area of over 1,000 square feet (93 m^2) and under an outdoor bleacher system to be sprinklered.

While the path for means of egress passing under the bleachers (i.e., vomitory) is not specifically addressed in this section, when a bleacher system is outside, the capacity factors for determining minimum egress width [Table 404.5(3) of ICC 300] are based on the assumption that the egress route is essentially open to the outside and, therefore, has a limited chance for the accumulation of smoke along that route. Two of the three legacy codes specifically exempted open means of egress routes under bleachers from separation requirements.

1029.2 Assembly main exit. A building, room or space used for assembly purposes that has an *occupant load* of greater than 300 and is provided with a main *exit*, that main *exit* shall be of sufficient capacity to accommodate not less than one-half of the *occupant load*, but such capacity shall be not less than the total required capacity of all *means of egress* leading to the *exit*. Where the building is classified as a Group A occupancy, the main *exit* shall front on not less than one street or an unoccupied space of not less than 10 feet (3048 mm) in width that adjoins a street or *public way*. In a building, room or space used for assembly purposes where there is not a well-defined main *exit* or where multiple main *exits* are provided, *exits* shall be permitted to be distributed around the perimeter of the building provided that the total capacity of egress is not less than 100 percent of the required capacity.

❖ Assembly buildings, as well as other buildings including spaces that function as assembly spaces (e.g., the band classroom in a school, the training room in an office, the cafeteria in a large factory), present an unusual life safety problem that includes frequent higher occupant densities and, therefore, larger occupant loads and the opportunity for irrational mass response to a perceived emergency (i.e., panic). For this reason, the code requires a specific arrangement of the exits. Studies have indicated that in any emergency, occupants will tend to egress via the same path of travel used to enter the room and building. Therefore, a main entrance to the building or space must also be designed as the main exit to accommodate this behavior, even if the required exit capacity might be more easily accommodated elsewhere. The main entrance (and exit) must be sized to accommodate at least 50 percent of the total occupant load of the structure and must front on a large, open space, such as a street or lobby, for rapid dispersal of the occupants outside the building or space. The remaining exits must also accommodate at least 50 percent of the total occupant load from each level (see Commentary Figure 1029.2). The total occupant load includes those within the theater seating area, the foyer and any other space (e.g., ticket booth, concession stand, offices, storage and the like). When the assembly space is within a mixed use building, the intent is that the main exits from the space comply with these provisions for one-half the capacity, but not necessarily that they lead directly to the outside. Egress requirements from the building would depend on how it was anticipated for the assembly space occupants to disperse. For example, an office building may have a large training/conference room where the path of exit access travel from the room goes out a main exit from the space and then disseminates into the general floor egress system. The room exit access doors may need to meet the 50-percent criterion, but once the occupants leave the room and enter the general floor egress system, exit capacity can be dispersed.

The required width of the means of egress in places of assembly is more often determined by the occupant load than in most other occupancies. In other occupancies, the minimum required widths and the travel distances will often determine the required widths and locations of the exits.

This section only requires the main exit to accommodate 50 percent of the occupant load when there is a single main entrance. Therefore, a large stadium or civic center, in which there are numerous entrances (and exits), need not comply with the main entrance criteria.

1029.3 Assembly other exits. In addition to having access to a main *exit*, each level in a building used for assembly purposes having an *occupant load* greater than 300 and provided with a main *exit*, shall be provided with additional *means of egress* that shall provide an egress capacity for not less than one-half of the total *occupant load* served by that level and shall comply with Section 1007.1. In a building used for assembly purposes where there is not a well-defined main *exit* or where multiple main *exits* are provided, *exits* for each level shall be permitted to be distributed around the perimeter

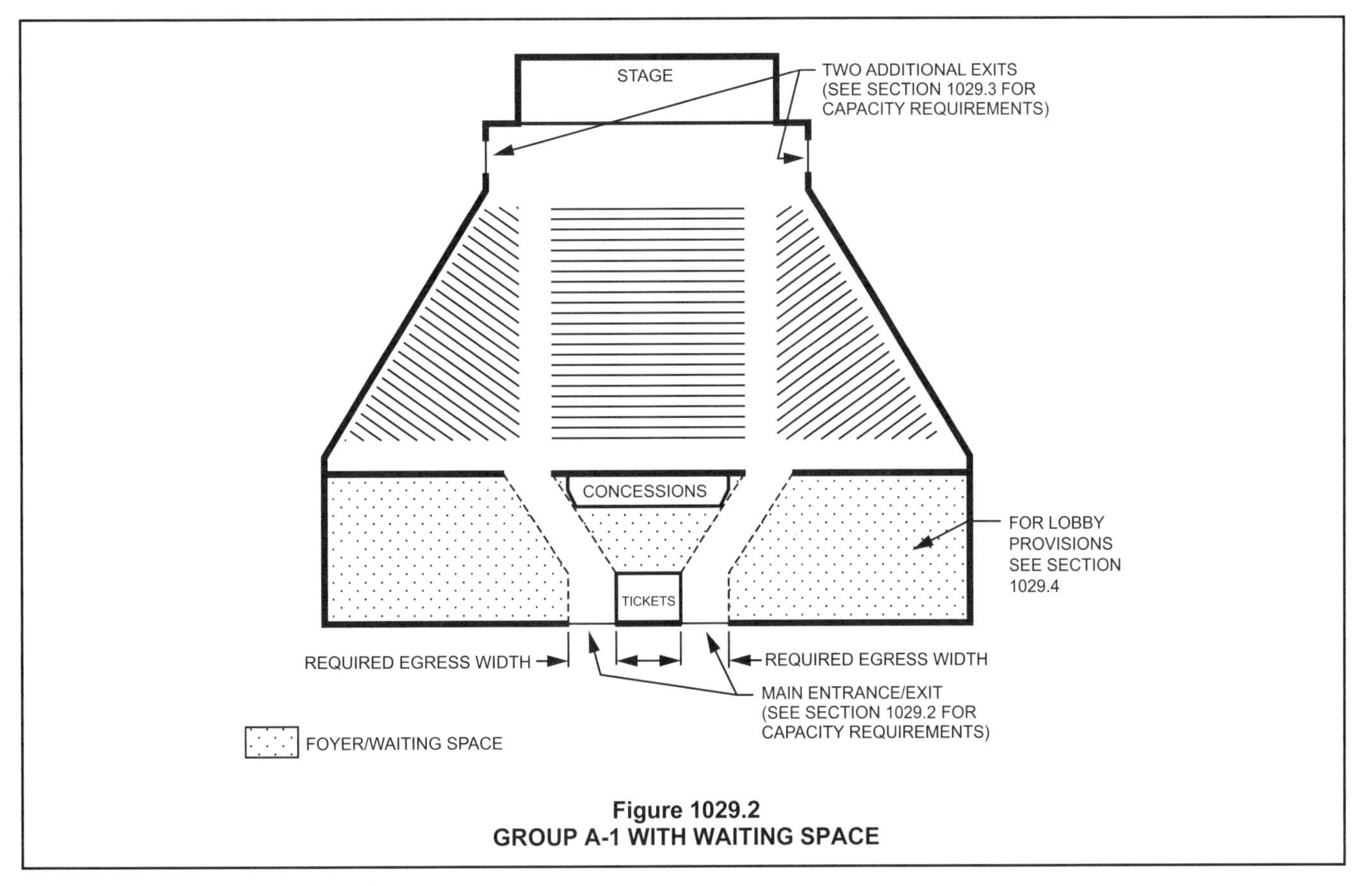

Figure 1029.2
GROUP A-1 WITH WAITING SPACE

of the building, provided that the total width of egress is not less than 100 percent of the required width.

❖ This section provides for the egress of one-half of the total occupant load by way of exits other than the main exit that is described in Section 1029.2. Assembly buildings, as well as other buildings including spaces that function as assembly spaces, that are provided with multiple entrances but no single main entrance do not provide a well-defined main exit; therefore, the total required exit width needs to be distributed around the perimeter of the space or building. Examples of these assemblies would be a school gymnasium or a large stadium or civic center in which there are numerous entrances (and exits).

1029.4 Foyers and lobbies. In Group A-1 occupancies, where persons are admitted to the building at times when seats are not available, such persons shall be allowed to wait in a lobby or similar space, provided such lobby or similar space shall not encroach upon the minimum width or required capacity of the *means of egress*. Such foyer, if not directly connected to a public street by all the main entrances or *exits*, shall have a straight and unobstructed *corridor* or path of travel to every such main entrance or *exit*.

❖ In theaters, people may arrive and wait for the next show while another group has yet to exit. This is extremely common in multiplex theater complexes. In every case, the main entrance (exit) and all other exits are to be constantly available for the entire building occupant load.

For example, because of the queuing of large crowds, particularly in theaters where a performance may be in progress and people must wait to attend the next one, standing space is often provided. For reasons of safety, such spaces cannot be located in or interfere with established paths of egress from the assembly areas. While a facility may choose to separate the route for means of egress using partitions or railings from the general lobby space to allow for easy traffic flow through the lobby to the street, it is not required to designate these areas (see Commentary Figure 1029.2).

1029.5 Interior balcony and gallery means of egress. For balconies, galleries or press boxes having a seating capacity of 50 or more located in a building, room or space used for assembly purposes, not less than two *means of egress* shall be provided, with one from each side of every balcony, gallery or press box.

❖ This section states the threshold where two means of egress are required based on the occupant load of the interior balcony, gallery or press box. Those two exits need to be dispersed. Section 1019.3, Exception 8, allows for both stairways to be unenclosed exit access stairways where they are effectively part of the main seating bowl. These requirements will ensure that at least one path of travel is always available and occupants face a minimum number of hazards.

For balconies, galleries or press boxes with 50 or fewer occupants, see Section 1029.8. When balconies, galleries or press boxes contain wheelchair spaces, the area must also meet the accessible means of egress requirements (Sections 1009.1, Exception 3, and 1029.8).

1029.6 Capacity of aisle for assembly. The required capacity of *aisles* shall be not less than that determined in accordance with Section 1029.6.1 where *smoke-protected assembly seating* is not provided and with Section 1029.6.2 or 1029.6.3 where *smoke-protected assembly seating* is provided.

❖ The means of egress width for spaces used for assembly is to be in accordance with this section and the referenced sections instead of the criteria specified in Section 1005 when dealing with the means of egress within the seating area. The width factors in Section 1029.6 and its subsections apply to those doorways, passageways, stepped aisles, ramped aisles and level aisles that are within the assembly seating areas.

The Board for the Coordination of Model Codes (BCMC) issued a report on means of egress dated June 10, 1985. The provisions in Section 1029 are based on this report. This report limits the application of these provisions to aisles and aisle accessways that provide exit access within the room or space with the assembly seating. This would include aisle accessways, level aisles, stepped aisles and ramped aisles. The primary concern for occupant safety would be that where the different provisions for the capacity requirements in Sections 1005 and 1029 were utilized, there would not be a bottleneck in the path of travel for means of egress. For example, a common configuration at football fields is to have the seating area raised several feet above grade. Where a step or series of steps leading to grade from a raised seating area are a continuation of the stepped aisle, the width of the stepped aisle between seats and the continuation without adjacent seats should be the same. In theaters, commonly the occupants leave the seating area for a concourse or lobby area that leads to exit stairways between floor levels. In this situation, the capacity requirement in Section 1005.3 would be applicable for the exit stairway. For the many situations between these two scenarios, the decision would be based on the configuration of the seating and exit stairways between levels.

For example, in a facility without smoke protection and an occupant load of 800, doorways would need to be calculated based on 0.20 inches (5.1 mm) per occupant (see Item 4 of Section 1029.6.1). An occupant load of 800 x 0.2 inches (4064 mm) = 160 inches (4064 mm) of egress width capacity, which translates into not less than five doors [assuming each with a minimum 32-inch (813 mm) clear opening]. This facility needs to have not less than three distinct means of egress (Section 1006.3). A main exit must accommodate not less than one-half of the occupant load (Section 1029.2). Therefore, the result would be not less than six doors: three located at the main exit and three others distributed to at least two other locations (Section 1029.3).

Different means of egress width criteria are also specified for assembly seating where smoke protection is provided versus areas where smoke protection is not provided. The egress width for smoke-protected seating is allowed to be less than for areas where smoke protection is not provided, since the smoke level is required to be maintained at least 6 feet (1829 mm) above the floor of the means of egress, according to Section 1029.6.2.1.

1029.6.1 Without smoke protection. The required capacity in inches (mm) of the *aisles* for assembly seating without smoke protection shall be not less than the *occupant load* served by the egress element in accordance with all of the following, as applicable:

1. Not less than 0.3 inch (7.6 mm) of *aisle* capacity for each occupant served shall be provided on stepped *aisles* having riser heights 7 inches (178 mm) or less and tread depths 11 inches (279 mm) or greater, measured horizontally between tread *nosings*.
2. Not less than 0.005 inch (0.127 mm) of additional *aisle* capacity for each occupant shall be provided for each 0.10 inch (2.5 mm) of riser height above 7 inches (178 mm).
3. Where egress requires stepped *aisle* descent, not less than 0.075 inch (1.9 mm) of additional *aisle* capacity for each occupant shall be provided on those portions of *aisle* capacity having no *handrail* within a horizontal distance of 30 inches (762 mm).
4. Ramped *aisles*, where slopes are steeper than one unit vertical in 12 units horizontal (8-percent slope), shall have not less than 0.22 inch (5.6 mm) of clear *aisle* capacity for each occupant served. Level or ramped *aisles*, where slopes are not steeper than one unit vertical in 12 units horizontal (8-percent slope), shall have not less than 0.20 inch (5.1 mm) of clear *aisle* capacity for each occupant served.

❖ This section prescribes the criteria needed to calculate the clear widths of aisles and aisle accessways in order to provide sufficient capacity to handle the occupant loads established by the "catchment areas" described in Section 1029.9.2. Clear width is to be measured to walls, edges of seating and tread edges.

The criteria for determining the required widths are based on analytical studies and field tests that used people to model egress situations [see Commentary Figures 1029.6.1(1) and 1029.6.1(2)].

Criterion 1 addresses the method for determining the required egress width for stepped aisles. This method corresponds with the requirements of Section 1005.3.1 for egress width per occupant of stairways in an unsprinklered building.

Criterion 2 addresses the method for determining the additional stepped aisle width required where the risers along that stepped aisle are greater than 7 inches (178 mm) high.

Criterion 3 addresses the method for determining the additional stepped aisle width where a handrail is not located within 30 inches (762 mm). With a center handrail, that would be stepped aisles wider than 60 inches (1524 mm). Side aisles have handrails along the wall. If the handrail is the top of the guard, then this requirement would apply to any side aisles greater than 30 inches (762 mm) wide. Because of the increased chance of falling and not being within reach of a handrail, the capacity per occupant is increased by 25 percent for the entire stepped aisle. For example, to calculate the capacity of a 72-inch-wide (1829 mm) stepped aisle, the answer would be 72 inches/0.375 inch = 192 occupants.

Criterion 4 addresses the method for determining the required widths for level or ramped and level means of egress. Where slopes are less than 1:12 (see definition for "Ramp"), the capacity requirements are also less. Level floors are quicker to negotiate than ramped surfaces for persons with limited mobility. Ramped aisle slopes are addressed in Section 1029.13.1.

These provisions are applicable within the seating area itself. When the occupants have left the seating bowl and have moved to stairways outside of the room, or have moved to cross aisles and then stairways or ramps leaving the balcony level, Section 1005.3 would be applicable. See Sections 1029.9.6.1

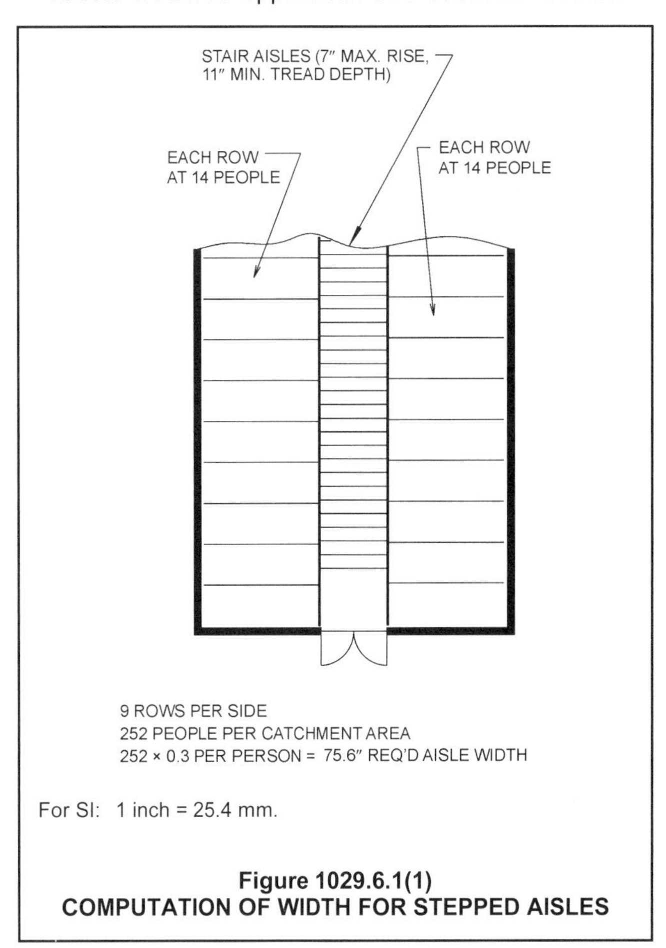

Figure 1029.6.1(1)
COMPUTATION OF WIDTH FOR STEPPED AISLES

through 1029.10.3 for stairways that are a direct continuation of stepped aisles.

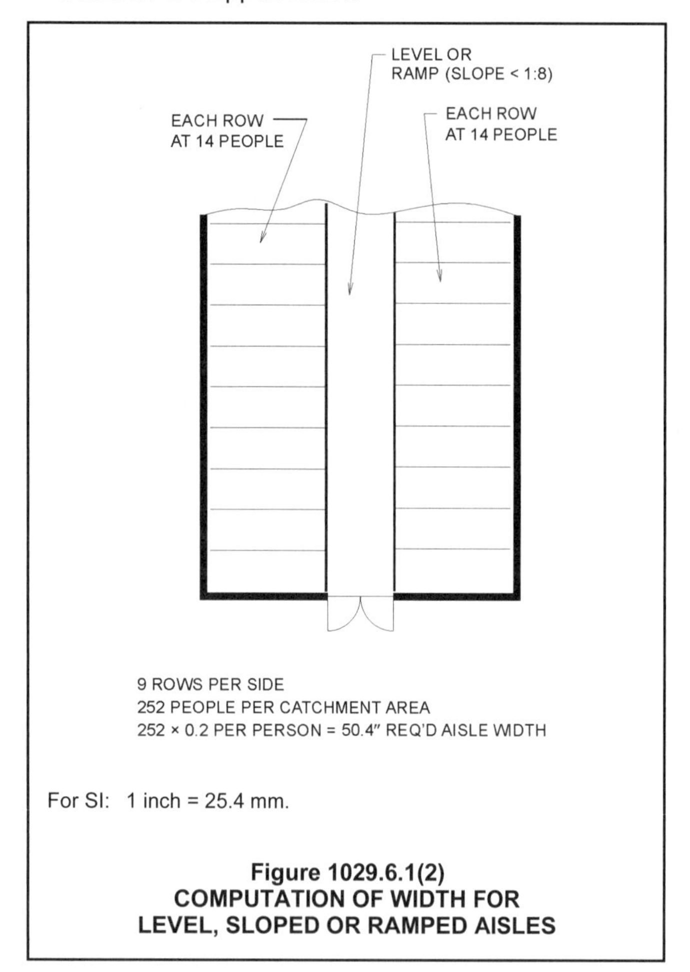

Figure 1029.6.1(2)
COMPUTATION OF WIDTH FOR LEVEL, SLOPED OR RAMPED AISLES

1029.6.2 Smoke-protected assembly seating. The required capacity in inches (mm) of the aisle for *smoke-protected assembly seating* shall be not less than the occupant load served by the egress element multiplied by the appropriate factor in Table 1029.6.2. The total number of seats specified shall be those within the space exposed to the same smoke-protected environment. Interpolation is permitted between the specific values shown. A life safety evaluation, complying with NFPA 101, shall be done for a facility utilizing the reduced width requirements of Table 1029.6.2 for *smoke-protected assembly seating*.

> **Exception:** For outdoor *smoke-protected assembly seating* with an *occupant load* not greater than 18,000, the required capacity in inches (mm) shall be determined using the factors in Section 1029.6.3.

❖ Special consideration is given to facilities with features that will prevent the means of egress from being blocked by smoke. Facilities to be considered smoke protected by Sections 1029.6.2.1 through 1029.6.2.3 are permitted increases in travel distance, egress capacity, and dead-end aisle and row lengths. Smoke control increases allowable egress time. Typically, model codes based on research by Dr. John Fruin and others recognize the need for occupants exposed to the fire environment to evacuate to a safe area within 90 seconds of notification and to reach the exterior or enclosed exit stairway within 5 minutes. With the increases permitted for smoke-protected facilities, these times are effectively doubled since the time available for safe egress also increases.

The exception is a pointer to the specific criteria for outdoor seating areas. For outdoor stadiums with 18,000 seats or greater, use Table 1029.6.2.

TABLE 1029.6.2. See page 10-187.

❖ This section requires the egress component to be of adequate size to accommodate the occupant load. For smoke-protected seating, the egress width per occupant is based on Table 1029.6.2. Typically, the larger the facility, the higher the ceiling; therefore, more space for smoke containment is associated with more time for egress. The egress width per occupant for nonsmoke-protected seating is to be based on Section 1029.6.1 and is similar to the provisions in Section 1005.3.

Where the entire means of egress is smoke protected, the concourse and any stairways and ramps can also use the same capacity numbers (see Section 1005.3.1, Exception 2, and Section 1005.3.2, Exception 2) to determine egress width. If the concourse surrounding the smoke-protected seating bowl is not also smoke protected, the requirements in Sections 1005.3.2 and 1005.3.1 would be applicable to determine widths required for capacity of the concourse, stairways and ramps providing means of egress from the seating bowl to the exterior of the building. See Sections 1029.9.6.1 through 1029.10.3 for stairways that are a direct continuation of stepped aisles.

1029.6.2.1 Smoke control. *Aisles* and *aisle accessways* serving a *smoke-protected assembly seating* area shall be provided with a smoke control system complying with Section 909 or natural ventilation designed to maintain the smoke level not less than 6 feet (1829 mm) above the floor of the *means of egress*.

❖ The means of egress (aisles and aisle accessways) within the assembly seating area are required to have some type of smoke control system that will prevent smoke buildup from encroaching on the egress path. This may be a mechanical smoke control system, designed in accordance with Section 909, or natural ventilation as in open-air stadiums.

In either type of system, the major consideration is that a smoke-free environment be maintained at least 6 feet (1829 mm) above the floor of the means of egress for a period of at least 20 minutes.

1029.6.2.2 Roof height. A *smoke-protected assembly seating* area with a roof shall have the lowest portion of the roof deck not less than 15 feet (4572 mm) above the highest *aisle* or *aisle accessway*.

> **Exception:** A roof canopy in an outdoor stadium shall be permitted to be less than 15 feet (4572 mm) above the

highest *aisle* or *aisle accessway* provided that there are no objects less than 80 inches (2032 mm) above the highest *aisle* or *aisle accessway*.

❖ One element of a smoke-protected assembly seating facility is that the lowest portion of the roof is required to be at least 15 feet (4572 mm) above the highest aisle or aisle accessway. The objective of this provision is to have a minimum 6-foot (1829 mm) smoke-free height to accommodate safe egress through the area. The additional 9 feet (2743 mm) of height is to provide a volume of space that will act to dissipate smoke. The measurement of the height is shown in Commentary Figures 1029.6.2.2(1) and 1029.6.2.2(2).

1029.6.2.3 Automatic sprinklers. Enclosed areas with walls and ceilings in buildings or structures containing *smoke-protected assembly seating* shall be protected with an *approved automatic sprinkler system* in accordance with Section 903.3.1.1.

Exceptions:

1. The floor area used for contests, performances or entertainment provided the roof construction is more than 50 feet (15 240 mm) above the floor level and the use is restricted to low-fire-hazard uses.
2. Press boxes and storage facilities less than 1,000 square feet (93 m^2) in area.
3. Outdoor seating facilities where seating and the *means of egress* in the seating area are essentially open to the outside.

❖ If there are areas in the smoke-protected assembly seating structure enclosed by walls and ceilings, the entire structure is to be provided with an automatic sprinkler designed to meet the requirements of NFPA 13.

Exception 1 indicates that the area over the playing field or performance area is not required to be sprinklered if the use of the floor area is restricted. If the facility is used for conventions, trade shows, displays or similar purposes, sprinklers are required throughout, since the occupancy would no longer be a low-fire-hazard use. A characteristic of a low-fire-hazard occupancy is that the fuel load caused by combustibles is approximately 2 pounds per square foot (9.8 kg/m^2) or less.

In order for the contest, performance or entertainment area to be nonsprinklered, the roof over that area must be at least 50 feet (15 240 mm) above the

TABLE 1029.6.2
CAPACITY FOR AISLES FOR SMOKE-PROTECTED ASSEMBLY

TOTAL NUMBER OF SEATS IN THE SMOKE-PROTECTED ASSEMBLY SEATING	INCHES OF CAPACITY PER SEAT SERVED			
	Stepped aisles with handrails within 30 inches	Stepped aisles without handrails within 30 inches	Level aisles or ramped aisles not steeper than 1 in 10 in slope	Ramped aisles steeper than 1 in 10 in slope
Equal to or less than 5,000	0.200	0.250	0.150	0.165
10,000	0.130	0.163	0.100	0.110
15,000	0.096	0.120	0.070	0.077
20,000	0.076	0.095	0.056	0.062
Equal to or greater than 25,000	0.060	0.075	0.044	0.048

For SI: 1 inch = 25.4 mm.

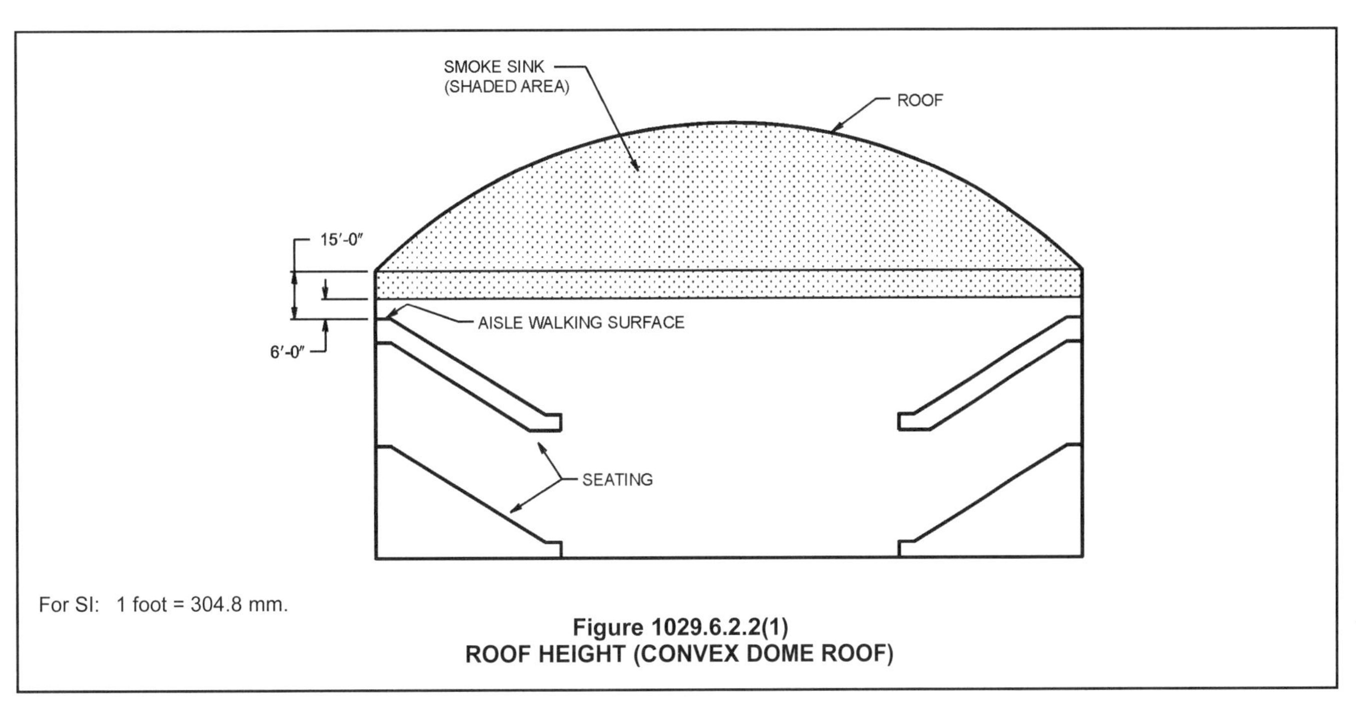

Figure 1029.6.2.2(1)
ROOF HEIGHT (CONVEX DOME ROOF)

floor in addition to the floor area meeting the-low fire-hazard criteria. The 50-foot (15 240 mm) criterion was selected because the response time for sprinklers at this height is extremely slow. It is estimated that the response time for standard sprinklers [50 feet (15 240 mm) above a floor with a fire having a heat release rate of 5 British thermal units (Btu) per square foot per second] exceeds 15 minutes. Therefore, it is not reasonable to install sprinklers at that height with little expectation of timely activation [see Commentary Figure 1029.6.2.3(1)]. Note that if this exception is utilized, the tradeoffs for a fully sprinklered building, such as increased height and area limitations or decreased corridor ratings, are no longer permitted.

Exception 2 indicates that automatic sprinklers are not required in small spaces in buildings. Sprinklers are required in press box and storage areas of outdoor facilities when the area exceeds 1,000 square feet (93 m^2). The primary reason for sprinklers in these areas is that both are anticipated to have a relatively large combustible load when compared to the main seating and participant areas. Additionally, in the case of storage areas, there is an increased potential for an undetected fire condition to occur [see Commentary Figure 1029.6.2.3(2)].

Exception 3 provides for outdoor seating facilities where smoke entrapment is not a safety concern.

1029.6.3 Outdoor smoke-protected assembly seating. The required capacity in inches (mm) of *aisles* shall be not less than the total *occupant load* served by the egress element multiplied by 0.08 (2.0 mm) where egress is by stepped *aisle* and multiplied by 0.06 (1.52 mm) where egress is by level *aisles* and ramped *aisles*.

Exception: The required capacity in inches (mm) of *aisles* shall be permitted to comply with Section 1029.6.2 for the number of seats in the outdoor *smoke-protected assembly seating* where Section 1029.6.2 permits less capacity.

❖ This section lists the coefficients determining the width of aisles required for outdoor smoke-protected assembly seating areas. Note that the coefficients are significantly less when compared to the values in Section 1029.6.1 for assembly areas without smoke protection. The coefficients are also less than those for smoke-protected assembly seating in Table 1029.6.2 except for very large assembly areas. The exception in this section would apply where the coefficients in Table 1029.6.2 are less than those in this section.

Low coefficients are a result of the very low hazards for outdoor smoke-protected assembly areas and increased egress time. Where the entire means of egress is essentially open to the exterior, the concourse and any stairways and ramps can also use the same capacity numbers (see Section 1005.3.1, Exception 2, and Section 1005.3.2, Exception 2) to determine egress width. If the concourse surrounding the outdoor seating bowl is enclosed, the requirements in Sections 1005.3.2 and 1005.3.1 would be applicable to determine widths required for capacity of the concourse, stairways and ramps providing means of egress from the seating bowl to the exterior of the building. See Sections 1029.9.6.1 through 1029.10.3 for stairways that are a direct continuation of stepped aisles.

Generally, an outdoor assembly area meets the smoke control requirements of Section 1029.6.2 by natural ventilation and does not require an automatic sprinkler system according to Section 1029.6.2.3, Exception 3.

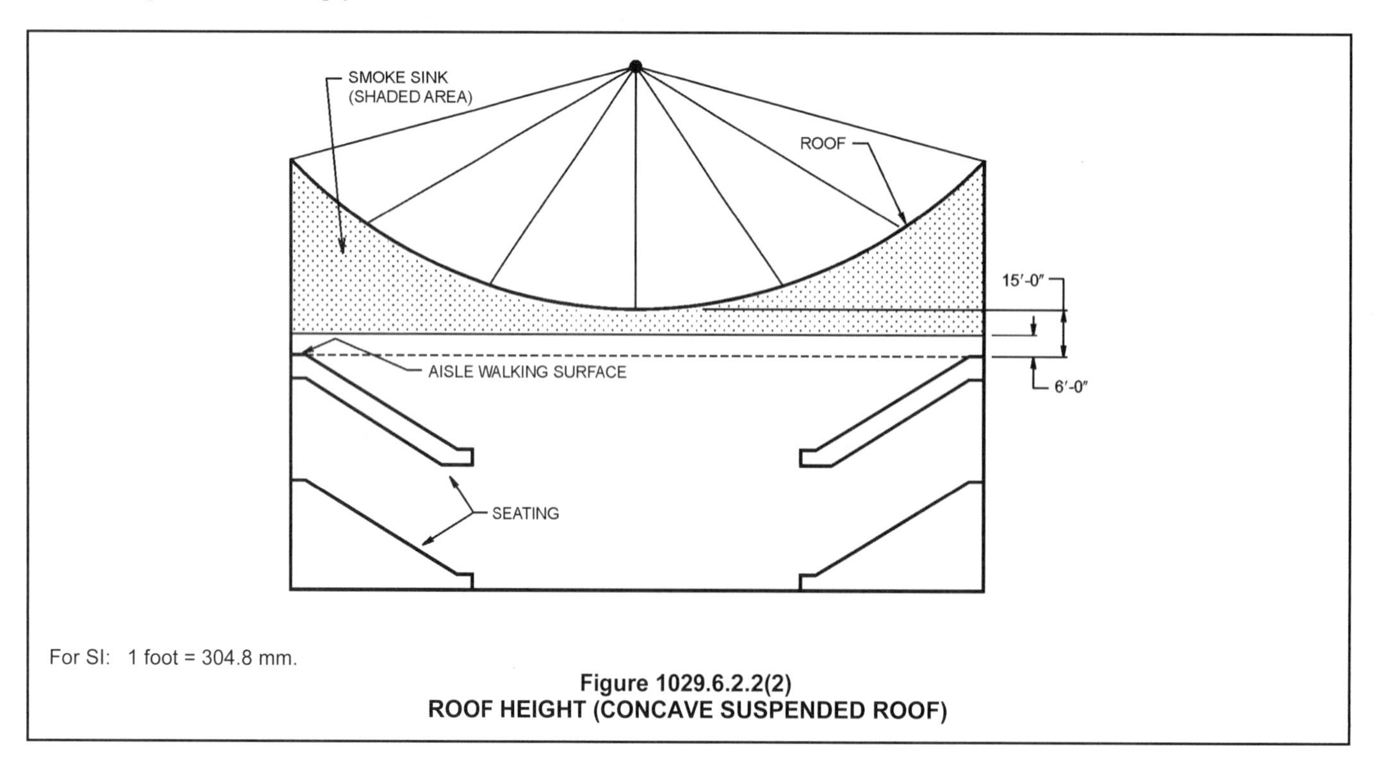

Figure 1029.6.2.2(2)
ROOF HEIGHT (CONCAVE SUSPENDED ROOF)

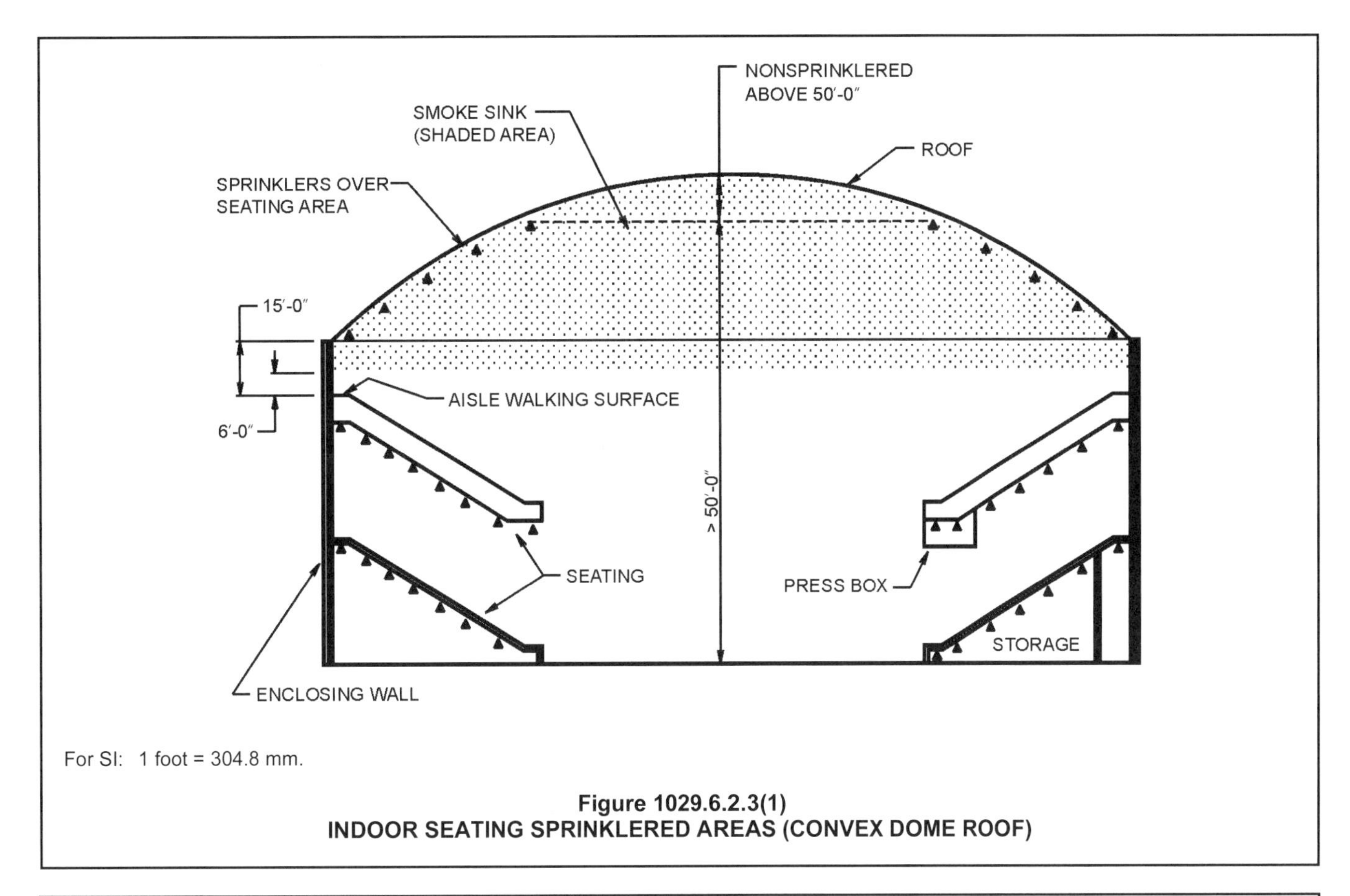

For SI: 1 foot = 304.8 mm.

Figure 1029.6.2.3(1)
INDOOR SEATING SPRINKLERED AREAS (CONVEX DOME ROOF)

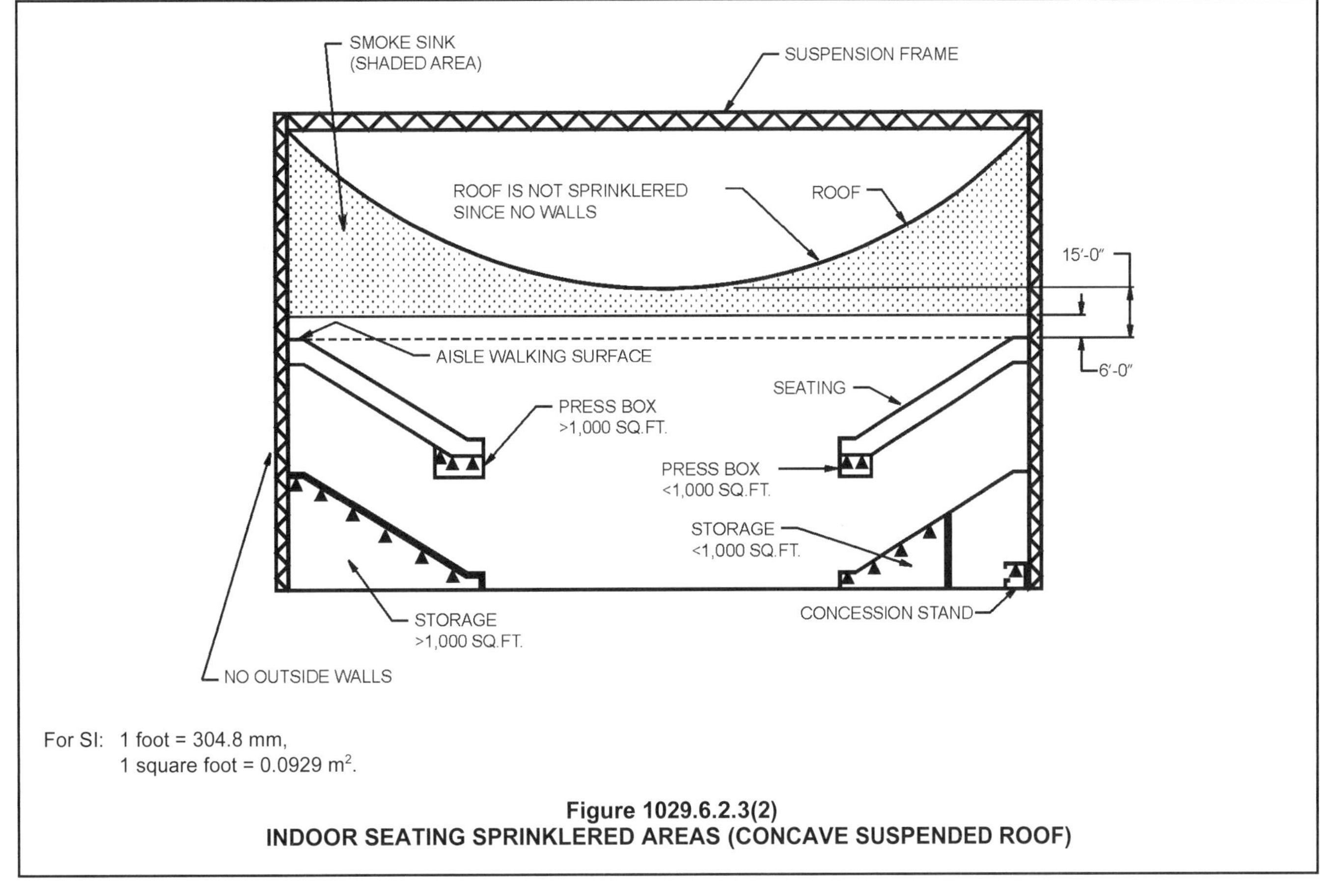

For SI: 1 foot = 304.8 mm,
1 square foot = 0.0929 m^2.

Figure 1029.6.2.3(2)
INDOOR SEATING SPRINKLERED AREAS (CONCAVE SUSPENDED ROOF)

1029.7 Travel distance. *Exits* and *aisles* shall be so located that the travel distance to an *exit* door shall be not greater than 200 feet (60 960 mm) measured along the line of travel in nonsprinklered buildings. Travel distance shall be not more than 250 feet (76 200 mm) in sprinklered buildings. Where *aisles* are provided for seating, the distance shall be measured along the *aisles* and *aisle accessways* without travel over or on the seats.

Exceptions:

1. *Smoke-protected assembly seating*: The travel distance from each seat to the nearest entrance to a vomitory or concourse shall not exceed 200 feet (60 960 mm). The travel distance from the entrance to the vomitory or concourse to a *stairway*, *ramp* or walk on the exterior of the building shall not exceed 200 feet (60 960 mm).
2. Open-air seating: The travel distance from each seat to the building exterior shall not exceed 400 feet (122 m). The travel distance shall not be limited in facilities of Type I or II construction.

❖ This section includes the travel distance limits for an assembly occupancy, which are the same as those in Table 1017.2 for Group A. The travel distance is to be measured along the same path the occupants would normally take to exit the facility.

Exception 1 provides an extended travel distance for smoke-protected assembly seating that meets the requirements of Sections 1029.6.2 through 1029.6.2.3.

Exception 2 applies to open-air seating areas where the smoke and fire hazard is very low. The Type I and II construction referred to in this exception is described in Section 602.

1029.8 Common path of egress travel. The *common path of egress travel* shall not exceed 30 feet (9144 mm) from any seat to a point where an occupant has a choice of two paths of egress travel to two *exits*.

Exceptions:

1. For areas serving less than 50 occupants, the *common path of egress travel* shall not exceed 75 feet (22 860 mm).
2. For *smoke-protected assembly seating*, the *common path of egress travel* shall not exceed 50 feet (15 240 mm).

❖ The maximum travel distance down a single aisle accessway between rows of seating to a location where a patron would have two choices for a way out of the space is 30 feet (9144 mm). In smoke-protected seating, the common path of travel can be up to 50 feet (15 240 mm).

If the room or space (e.g., box, gallery or balcony) has 50 or fewer occupants, the travel distance can be increased to 75 feet (22 860 mm). For example, this allows for a path of travel from a box seat, out of the box and to a main aisle or even a corridor located outside the assembly room itself.

When this section is referenced for accessible means of egress (see Section 1009.1, Exception 3), the utilization of Exception 1 would include the entire occupant load of the seating area, box, gallery or balcony, not just the number of wheelchair spaces and/ or companion seats. Wheelchair spaces that are integrated into the general seating would have the same common path of travel distance of 30 feet (9144 mm) before the person needing the accessible route could choose two different paths for accessible means of egress. This provides the same level of protection for the persons in the accessible seating as provided for others within the space.

Sections 1006.2.1, 1006.3.2 and 1029.8 must be considered when determining the common path of travel requirements for a building. Section 1006.2.1 regulates the means of egress from a room or space. Section 1029.8, however, regulates the distance to a decision point within a room or space used for assembly seating where a single access row of seating is provided to a location where a patron would have two choices for a way out of the space.

1029.8.1 Path through adjacent row. Where one of the two paths of travel is across the *aisle* through a row of seats to another *aisle*, there shall be not more than 24 seats between the two *aisles*, and the minimum clear width between rows for the row between the two *aisles* shall be 12 inches (305 mm) plus 0.6 inch (15.2 mm) for each additional seat above seven in the row between *aisles*.

Exception: For *smoke-protected assembly seating* there shall be not more than 40 seats between the two *aisles* and the minimum clear width shall be 12 inches (305 mm) plus 0.3 inch (7.6 mm) for each additional seat.

❖ In establishing the point where the occupants of a row served by a single access aisle have two distinct paths of travel, the code allows one of those paths to be through the rows of an adjacent seating area or section. This requirement increases the row widths for the single-access seating section and the adjacent dual-access seating section. This allows the occupants to either travel down the single access aisle or readily traverse the oversized row widths to gain access to a second means of egress (see Commentary Figure 1029.8.1). This exception allows a greater number of seats spaced with a minimum clearance of 12 inches (305 mm) for smoke-protected assembly seating that complies with Sections 1029.6.2 through 1029.6.2.3 or Section 1029.6.3. For the base width requirements for single- and dual-access rows, see the commentary to Sections 1029.12.2 through 1029.12.2.2.

1029.9 Assembly aisles are required. Every occupied portion of any building, room or space used for assembly purposes that contains seats, tables, displays, similar fixtures or equipment shall be provided with *aisles* leading to *exits* or *exit access doorways* in accordance with this section.

❖ This section requires that each assembly space have designated aisles. For aisle accessway requirements, see Section 1029.10. For aisles in other occupancies, see Section 1018.

1029.9.1 Minimum aisle width. The minimum clear width for *aisles* shall comply with one of the following:

1. Forty-eight inches (1219 mm) for stepped *aisles* having seating on each side.

   Exception: Thirty-six inches (914 mm) where the stepped *aisles* serve less than 50 seats.

2. Thirty-six inches (914 mm) for stepped *aisles* having seating on only one side.

   Exception: Twenty-three inches (584 mm) between a stepped *aisle handrail* and seating where a stepped *aisle* does not serve more than five rows on one side.

3. Twenty-three inches (584 mm) between a stepped *aisle handrail* or *guard* and seating where the stepped *aisle* is subdivided by a mid-aisle *handrail.*

4. Forty-two inches (1067 mm) for level or ramped *aisles* having seating on both sides.

   Exceptions:

   1. Thirty-six inches (914 mm) where the *aisle* serves less than 50 seats.
   2. Thirty inches (762 mm) where the *aisle* does not serve more than 14 seats.

5. Thirty-six inches (914 mm) for level or ramped *aisles* having seating on only one side.

   Exception: For other than ramped *aisles* that serve as part of an *accessible route*, 30 inches (762 mm) where the ramped *aisle* does not serve more than 14 seats.

❖ The clear widths of aisles established by the formulas given in Section 1029.6 must not be less than the minimum width requirements of this section. The development of minimum width requirements is based on the association of aisle capacity with the path of exit travel as influenced by the different features of aisle construction. The purpose is to create an aisle system that would provide an even flow of occupant egress. The minimum width of the aisles is also based on an anticipated movement of people in two directions.

Items 1, 2 and 3 deal with stepped aisles. Items 4 and 5 deal with ramped and level aisles. Items 1, 3 and 4 deal with aisles with seating on both sides. Items 2 and 5 deal with aisles with seating on one

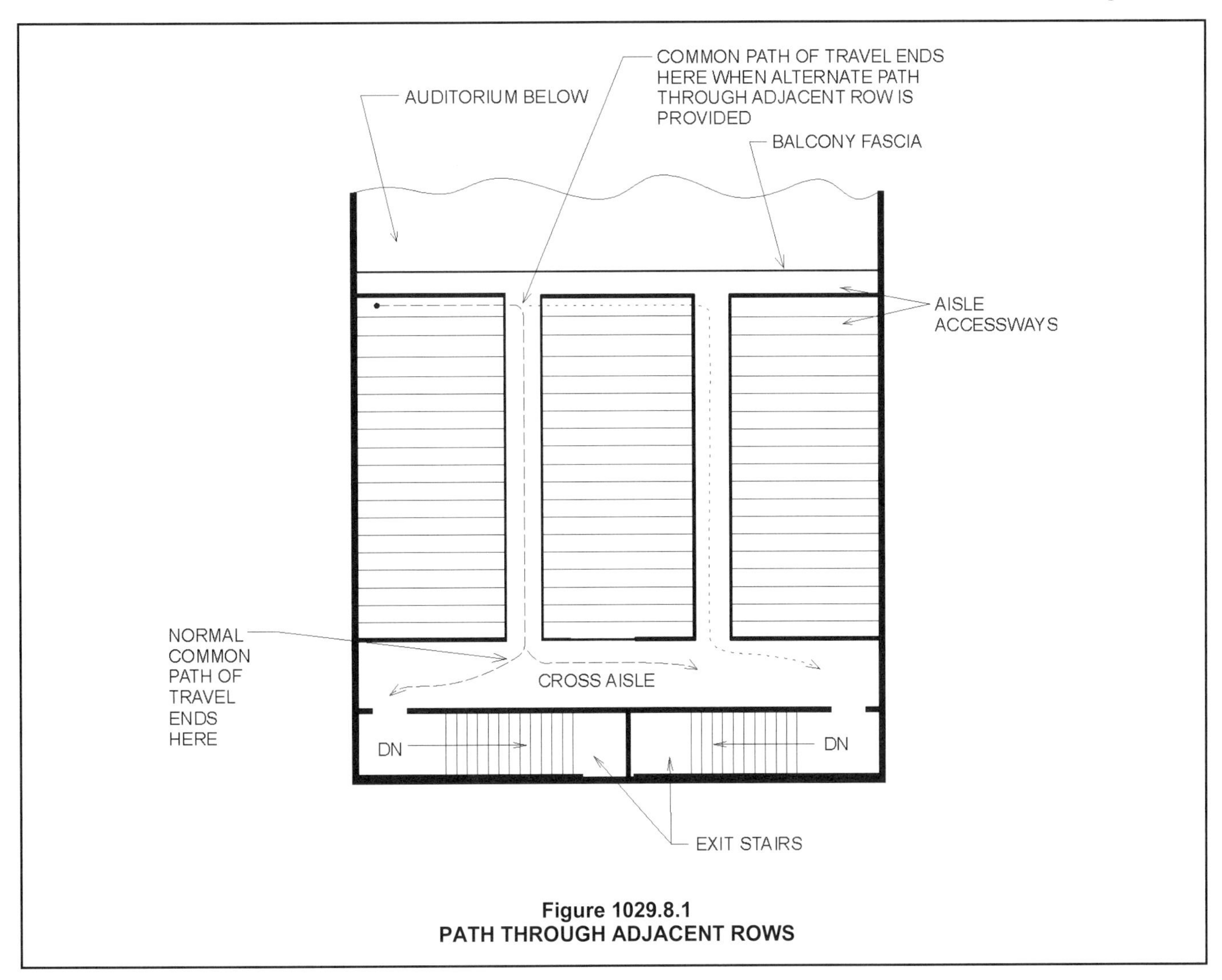

Figure 1029.8.1
PATH THROUGH ADJACENT ROWS

side. Note that each exception is only applicable to the item directly above. For example, per Item 4, where there is seating on both sides, the minimum ramped or level aisle width is 42 inches (1219 mm), except where that aisle serves less than 50 seats it can be 36 inches (914 mm) wide (Exception 1) and where that aisle serves less than 15 seats it can be 30 inches (762 mm) wide (Exception 2).

1029.9.2 Aisle catchment area. The *aisle* shall provide sufficient capacity for the number of persons accommodated by the catchment area served by the *aisle*. The catchment area served by an *aisle* is that portion of the total space served by that section of the *aisle*. In establishing catchment areas, the assumption shall be made that there is a balanced use of all *means of egress*, with the number of persons in proportion to egress capacity.

❖ The determination of required aisle and aisle accessway width is a function of the occupant load. In calculating the required widths, the assumption is that in a system or network of aisles and aisle accessways serving an occupied area, people will normally exit the area in a way that will distribute the occupant load throughout the system in proportion to the egress capacity of the aisles and aisle accessways. Each aisle and aisle accessway would take its tributary share (catchment area) of the total occupant load (see Commentary Figure 1029.9.2).

In addition to the provisions in this section, the requirement for the capacity of the main exit and other exits must also be considered (see Sections 1029.2 and 1029.3). While this section assumes an equal distribution, Section 1029.2 requires that where the facility has a main exit, the main exit and the access thereto must be capable of handling 50 percent of the occupant load.

1029.9.3 Converging aisles. Where *aisles* converge to form a single path of egress travel, the required capacity of that path shall be not less than the combined required capacity of the converging aisles.

❖ Where one or more aisles or aisle accessways meet to form a single path of egress travel, that path must be sized to handle the combined occupant capacity of the converging aisles and aisle accessways (see Commentary Figure 1029.9.3). The reason for this requirement is to maintain the natural pace of travel all the way through the aisle accessways or aisles to the exits and to minimize the queuing of occupants.

This section requires combining the required occupant capacity of converging aisles and aisle accessways, but not necessarily the required widths. For example, if two 48-inch (1219 mm) aisles converge, the result need not be a 96-inch (2438 mm) aisle unless the 48-inch (1219 mm) width of the aisles is required based on the requirements of Section 1029.6 for the actual occupant load served. However, if the 48-inch (1219 mm) width is not based on the occupant load but is required to comply with the minimum aisle width requirements of Section 1029.9.1, the resulting aisle width must be sized for the total occupant load served by the converging aisles, as determined by Section 1029.6, but not less than the minimum widths of Section 1029.9.1.

1029.9.4 Uniform width and capacity. Those portions of *aisles*, where egress is possible in either of two directions, shall be uniform in minimum width or required capacity.

❖ Aisles that connect or lead to opposite exits must, at a minimum, be of uniform width throughout their entire lengths to allow for exit travel in two directions without creating a traffic bottleneck (see Commentary Figure 1029.9.4). They may need to be wider based on Section 1029.9.2 or 1029.9.3.

1029.9.5 Dead end aisles. Each end of an *aisle* shall be continuous to a cross *aisle*, foyer, doorway, vomitory, concourse or *stairway* in accordance with Section 1029.9.7 having access to an *exit*.

Exceptions:

1. Dead-end *aisles* shall be not greater than 20 feet (6096 mm) in length.
2. Dead-end *aisles* longer than 16 rows are permitted where seats beyond the 16th row dead-end *aisle* are not more than 24 seats from another *aisle*, measured along a row of seats having a minimum clear width of 12 inches (305 mm) plus 0.6 inch (15.2 mm) for each additional seat above seven in the row where seats have backrests or beyond 10 where seats are without backrests in the row.
3. For *smoke-protected assembly seating*, the dead end *aisle* length of vertical *aisles* shall not exceed a distance of 21 rows.
4. For *smoke-protected assembly seating*, a longer dead-end *aisle* is permitted where seats beyond the 21-row dead-end *aisle* are not more than 40 seats from another *aisle*, measured along a row of seats having an *aisle* accessway with a minimum clear width of 12 inches (305 mm) plus 0.3 inch (7.6 mm) for each additional seat above seven in the row where seats have backrests or beyond 10 where seats are without backrests in the row.

❖ Both ends of a cross aisle must terminate at either an intersecting aisle, a foyer, a doorway or a vomitory (lane) that gives access to an exit(s). Each exception allows an aisle to have a dead end of limited length. Exceptions 1 and 2 address dead-end aisles in assembly spaces with or without smoke protection. Exceptions 3 and 4 address dead-end aisles only in smoke-protected assembly seating. In accordance with Exception 1, dead-end aisles (similar to corridors and passageways) that terminate at one end of a cross aisle or at a foyer, doorway or vomitory must not be more than 20 feet (6096 mm) in length. The intent of the row width requirements in the exceptions is to provide sufficient clear width between rows of seating to allow the occupants in times of emergency to pass quickly from a dead-end aisle to the aisle at the opposite end.

In Exception 2, the 0.6-inch (15 mm) increase beyond seven seats with backrests is consistent with the minimum width determined in accordance with Section 1029.12.2 for single access rows. The code recognizes that one dead-end aisle may not be usable, thus creating a single access row condition. There is a greater allowance for seating that looks similar to bleacher style seating. This is consistent with ICC 300 (see Section 1029.1.1).

Exceptions 3 and 4 allow longer dead-end aisles for smoke-protected assembly seating that complies with Sections 1029.6.2 through 1029.6.2.3 or Section 1029.6.3 (see Commentary Figure 1029.9.5). In Exception 4, there is a greater allowance for seating that looks similar to bleacher style seating. This is consistent with ICC 300 (see Section 1029.1.1).

The overall purpose of this section is to provide aisle/seating arrangements that would allow the occupants to seek safe and rapid passage to exits in case of fire or other emergency.

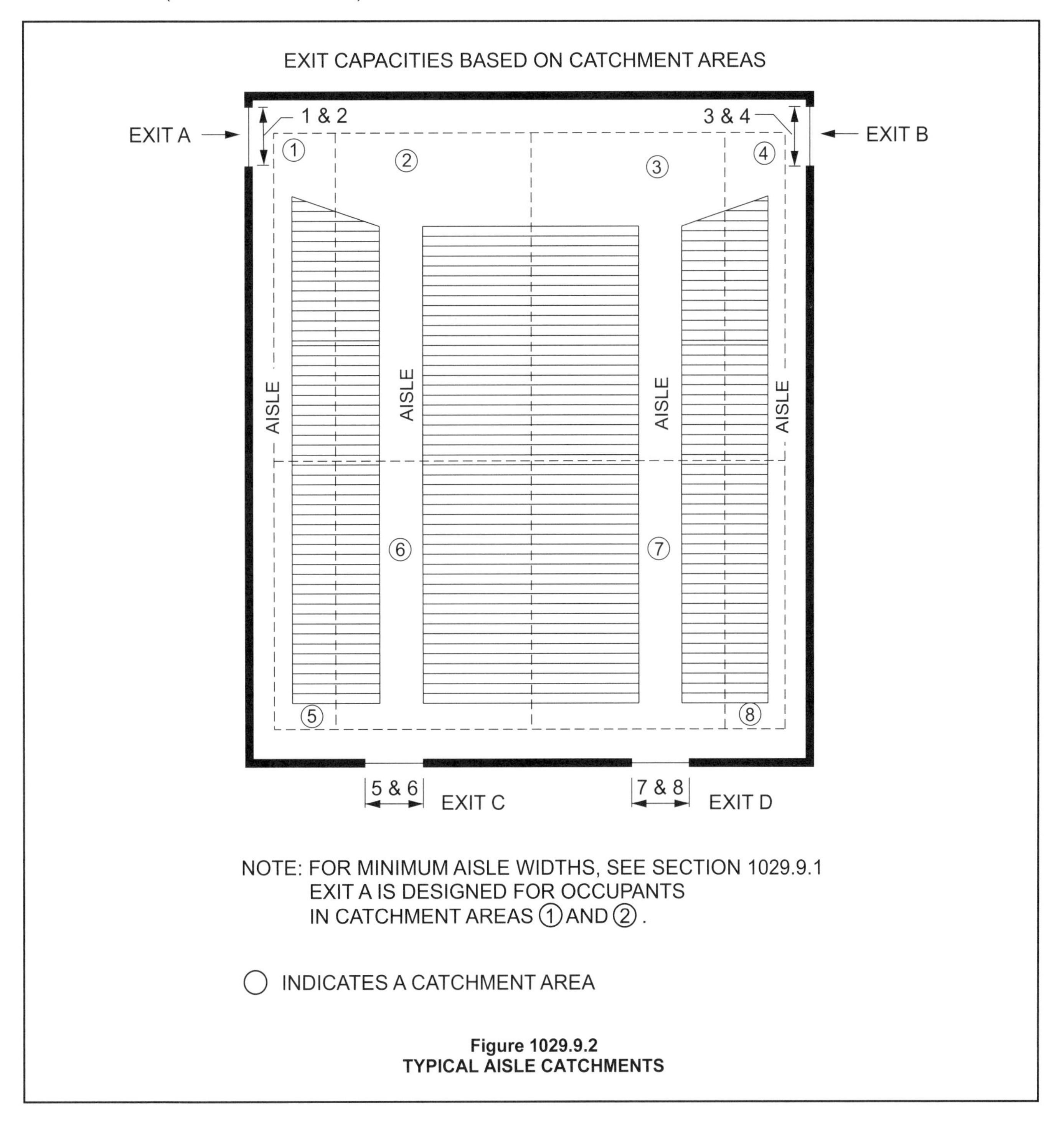

Figure 1029.9.2
TYPICAL AISLE CATCHMENTS

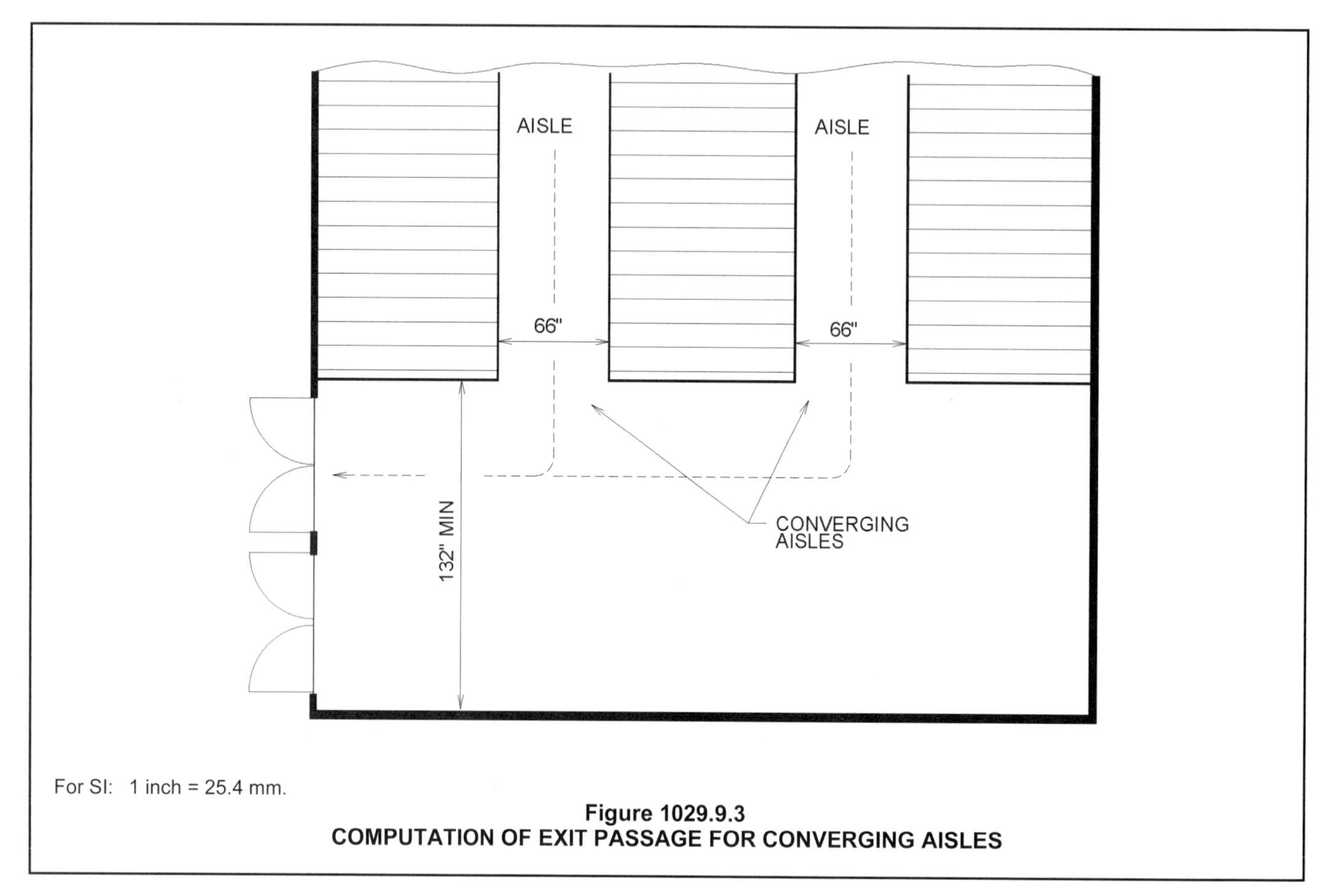

Figure 1029.9.3
COMPUTATION OF EXIT PASSAGE FOR CONVERGING AISLES

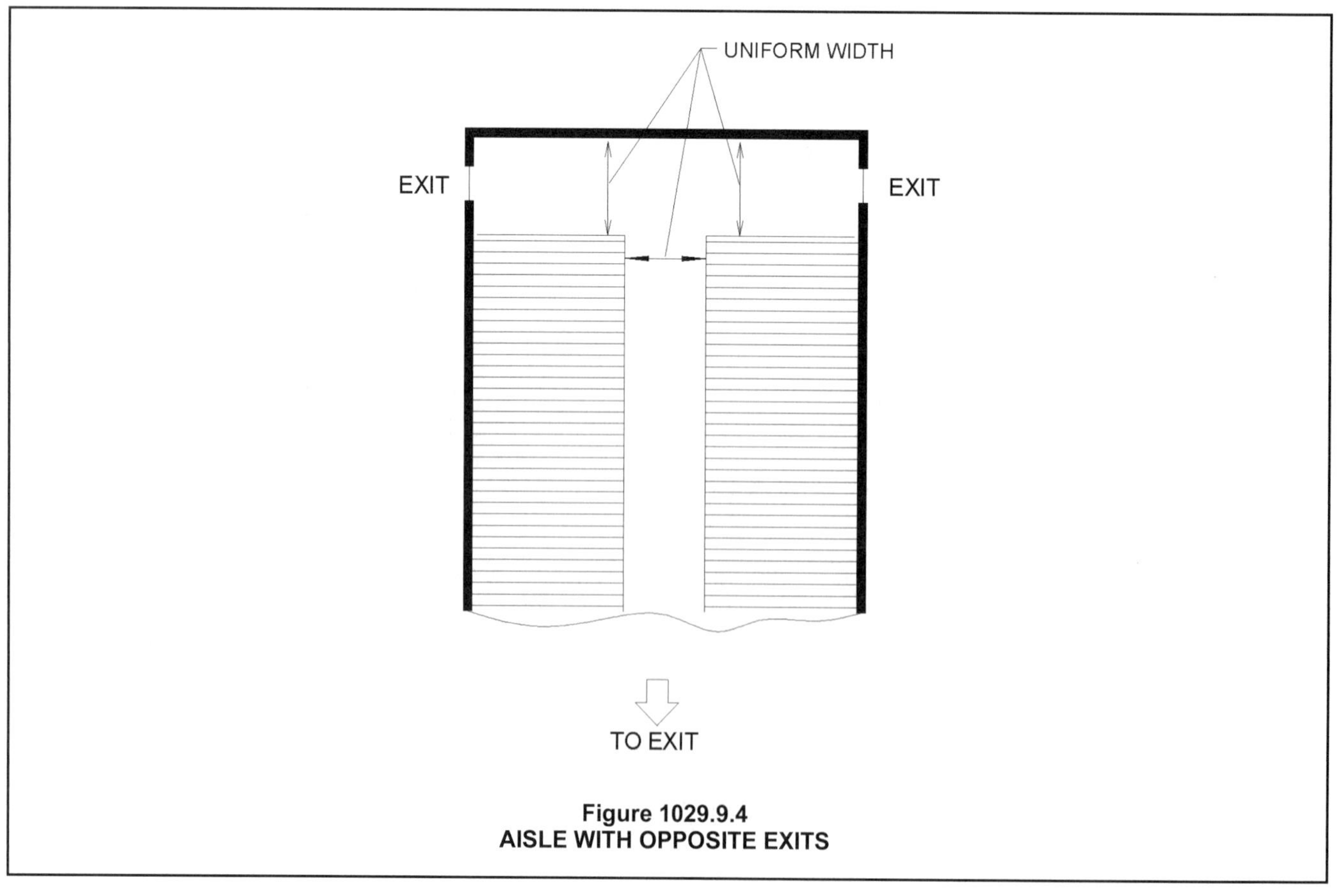

Figure 1029.9.4
AISLE WITH OPPOSITE EXITS

1029.9.6 Aisle measurement. The clear width for *aisles* shall be measured to walls, edges of seating and tread edges except for permitted projections.

Exception: The clear width of *aisles* adjacent to seating at tables shall be permitted to be measured in accordance with Section 1029.12.1.

❖ The clear width for stepped, ramped and level aisles is measured consistently with stairways and ramps to ensure a clear width for egress. The exception is dealing with aisles in dining areas and how to measure with loose tables and chairs.

1029.9.6.1 Assembly aisle obstructions. There shall not be obstructions in the minimum width or required capacity of *aisles*.

Exception: *Handrails* are permitted to project into the required width of stepped *aisles* and ramped *aisles* in accordance with Section 1014.8.

❖ Except for handrails, aisles are required to be clear of any obstructions so that the full width is available for egress purposes. Handrails are allowed to project into the required aisle width in the same manner as handrail projections in stairways.

1029.9.7 Stairways connecting to stepped aisles. A *stairway* that connects a stepped *aisle* to a cross *aisle* or concourse shall be permitted to comply with the assembly *aisle* walking surface requirements of Section 1029.13. Transitions between *stairways* and stepped *aisles* shall comply with Section 1029.10.

❖ Stairways that are a direct continuation of a stepped aisle, either at the top or bottom of the stepped aisle, and provide access to a cross aisle, shall be handled the same way as a stepped aisle for treads and risers (see Section 1029.13). For the transitions between the stepped aisle and the stairway, see Section 1029.10. For examples, see Commentary Figures 1029.9.7(1) and 1029.9.7(2).

1029.9.8 Stairways connecting to vomitories. A *stairway* that connects a vomitory to a cross aisle or concourse shall be permitted to comply with the assembly *aisle* walking surface requirements of Section 1029.13. Transitions between *stairways* and stepped *aisles* shall comply with Section 1029.10.

❖ A vomitory is an entrance that pierces the back of a seating bowl of a theater or stadium that allows for entrance into the seating area. Stairways that provide a direct connection between that vomitory and an aisle or concourse shall be handled the same way as a stepped aisle for treads and risers (see Section 1029.13). For the transitions between the stepped aisle and the stairway, see Section 1029.10.

1029.10 Transitions. Transitions between *stairways* and stepped *aisles* shall comply with either Section 1029.10.1 or 1029.10.2.

❖ Line of sight in assembly seating is an important issue that needs to be balanced with safe egress for people seated in the assembly spaces. In order to maintain line of site, sometime it is necessary to have a transition between stairways and stepped aisles that does not meet the standard tread, riser and landing provisions for stairway. This section replaces the standard landing requirements for stairways with a transition tread/landing.

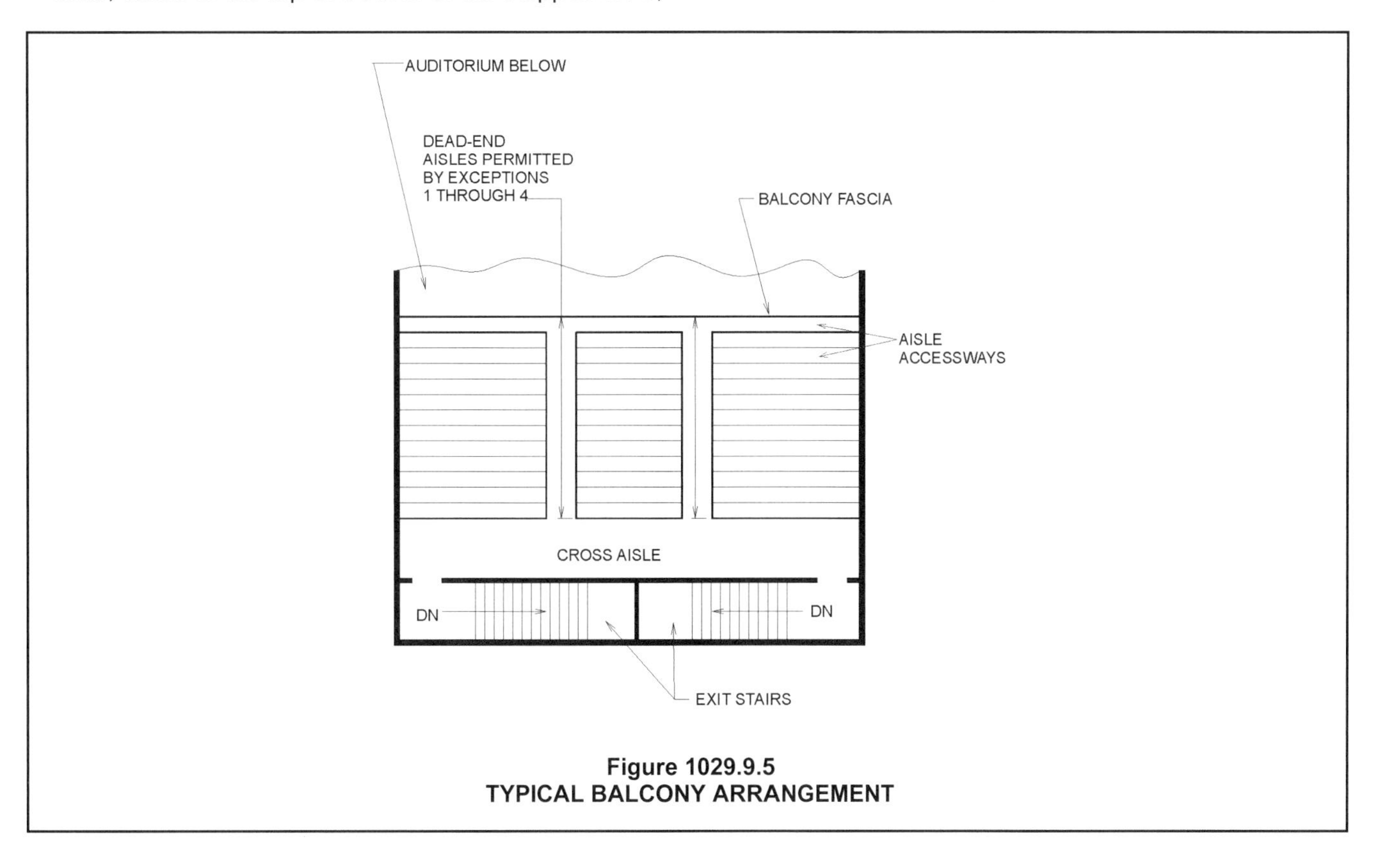

Figure 1029.9.5
TYPICAL BALCONY ARRANGEMENT

1029.10.1 Transitions and stairways that maintain stepped aisle riser and tread dimensions. Stepped *aisles*, transitions and *stairways* that maintain riser and tread dimensions shall comply with Section 1029.13 as one *exit access* component.

❖ Where the treads and risers on the stairway and the treads and risers on the stepped aisle are the same, the whole run can be considered as one long stepped aisle.

1029.10.2 Transitions to stairways that do not maintain stepped aisle riser and tread dimensions. Transitions to *stairways* from stepped *aisles* with riser and tread dimensions that differ from the *stairways* shall comply with Sections 1029.10.2.1 through 1029.10.3.

❖ Where the stepped aisle and directly connected stairways do not have the same tread and riser dimensions, there are options for the transition between the two sections. This transition is the tread(s) between the stepped aisle and the stairway.

1029.10.2.1 Stairways and stepped aisles in a straight run. Transitions where the *stairway* is a straight run from the stepped *aisle* shall have a minimum depth of 22 inches (559 mm) where the treads on the descending side of the transition have greater depth and 30 inches (762 mm) where the treads on the descending side of the transition have lesser depth.

❖ Typically, the treads on the stairway will have a smaller depth than the treads at the seating. Assuming that is the case, where the stairway is at the high end of the stepped aisle, a transition tread 22 inches (559 mm) in depth shall be provided between the stairway and the stepped aisle [see Commentary Figure 1029.10.2.1(1)]. Where the stairway is at the low end of the stepped aisle, a transition tread a minimum of 30 inches (762 mm) in depth shall be provided between the stepped aisle and the stairway [see Commentary Figure 1029.10.2.1(2)].

1029.10.2.2 Stairways and stepped aisles that change direction. Transitions where the *stairway* changes direction from the stepped *aisle* shall have a minimum depth of 11 inches (280 mm) or the stepped *aisle* tread depth, whichever is greater, between the stepped *aisle* and *stairway*.

❖ Where the stepped aisle takes a turn or takes a turn and becomes a stairway, the aisle width must be maintained (see Sections 1029.6 and 1029.9.1). The transition between the stepped aisle and the turn must continue the tread depth of the stepped aisle, or have a minimum depth of 11 inches (280 mm) to match the stairway treads, whichever is greater. There may be more than one transition between the stepped aisle and the turn [see Commentary Figures 1029.10.2.2(1) and 1029.10.2.2(2)].

1029.10.3 Transition marking. A distinctive marking stripe shall be provided at each *nosing* or leading edge adjacent to the transition. Such stripe shall be not less than 1 inch (25 mm), and not more than 2 inches (51 mm), wide. The edge marking stripe shall be distinctively different from the stepped *aisle* contrasting marking stripe.

❖ At these transitions there may be a change in riser height, even if the tread depth stayed the same. The stripe is to draw attention to the transition to reduce the chance of a trip and fall. This is different from the stripes required for the stepped aisles in Sections 1029.13.2.2.1 and 1029.13.2.3.

1029.11 Construction. *Aisles*, stepped *aisles* and ramped *aisles* shall be built of materials consistent with the types permitted for the type of construction of the building.

> **Exception:** Wood *handrails* shall be permitted for all types of construction.

❖ The construction materials permitted for stepped, ramped and level aisles are consistent with stairways and ramps.

Figure 1029.9.7(1)
EXAMPLE OF A CONTINUATION OF A STEPPED AISLE UP TO A CROSS AISLE

Figure 1029.9.7(2)
EXAMPLE OF A CONTINUATION OF A STEPPED AISLE DOWN TO A FLOOR

1029.11.1 Walking surface. The surface of *aisles*, stepped *aisles* and ramped *aisles* shall be of slip-resistant materials that are securely attached. The surface for stepped *aisles* shall comply with Section 1011.7.1.

❖ It is the intent of this section that for walking surfaces, stepped aisles will be addressed similar to stairways (Section 1011.7.1), level aisles the same as floors (Section 1003.4) and ramped aisles the same as ramps (Section 1012.7.1). See the commentary for these sections.

1029.11.2 Outdoor conditions. Outdoor *aisles*, stepped *aisles* and ramped *aisles* and outdoor approaches to *aisles*, stepped *aisles* and ramped *aisles* shall be designed so that water will not accumulate on the walking surface.

❖ Where stepped, ramped and level aisles are located in an outdoor situation, there is the same concern for the accumulation of ice and snow as there would be for exterior stairways or ramps. See Sections 1011.7.2 and 1012.7.2 regarding maximum slope to allow for drainage and when grated walking surfaces are permitted.

1029.12 Aisle accessways. *Aisle accessways* for seating at tables shall comply with Section 1029.12.1. *Aisle accessways* for seating in rows shall comply with Section 1029.12.2.

❖ Aisle accessways are paths for means of egress between rows of seats or between tables. The aisle accessway leads to aisles, which in turn lead toward exits from the space. This is the same idea as moving through a room full of furniture to a corridor to reach an exit. In both situations, the path for means of egress is confined. For aisle accessway requirements in other occupancies, see Section 1018.

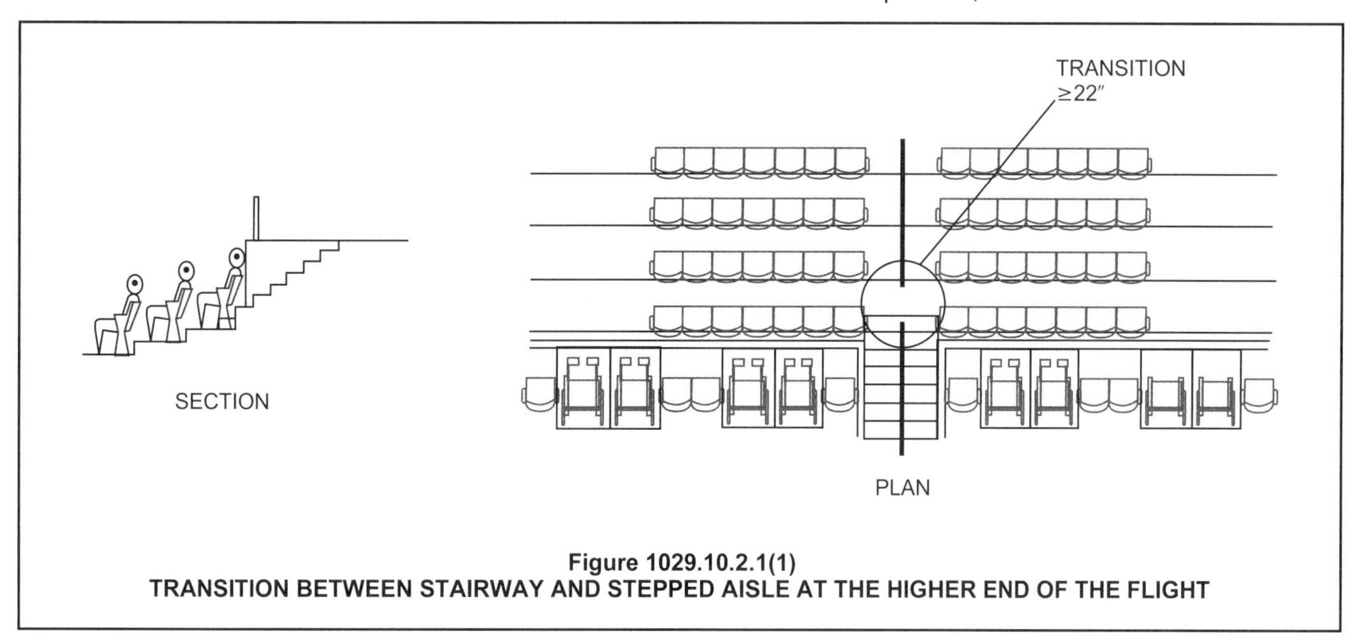

Figure 1029.10.2.1(1)
TRANSITION BETWEEN STAIRWAY AND STEPPED AISLE AT THE HIGHER END OF THE FLIGHT

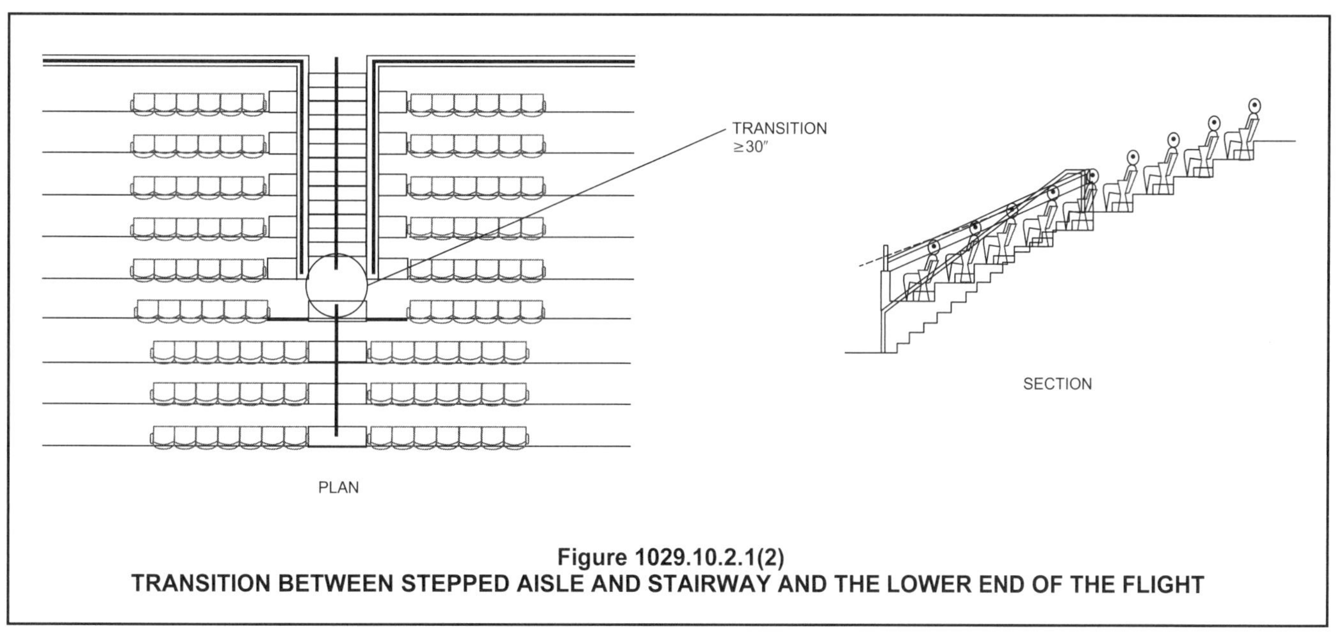

Figure 1029.10.2.1(2)
TRANSITION BETWEEN STEPPED AISLE AND STAIRWAY AND THE LOWER END OF THE FLIGHT

1029.12.1 Seating at tables. Where seating is located at a table or counter and is adjacent to an *aisle* or *aisle accessway*, the measurement of required clear width of the *aisle* or *aisle accessway* shall be made to a line 19 inches (483 mm) away from and parallel to the edge of the table or counter. The 19-inch (483 mm) distance shall be measured perpendicular to the side of the table or counter. In the case of other side boundaries for *aisles* or *aisle accessways*, the clear width shall be measured to walls, edges of seating and tread edges.

Exception: Where tables or counters are served by *fixed seats*, the width of the *aisle* or *aisle accessway* shall be measured from the back of the seat.

❖ Most seating at tables should be adjacent to aisles or aisle accessways. In measuring the width of an aisle or aisle accessway for movable seating, the measurement is taken at a distance of 19 inches (483 mm) perpendicular to the side of the table or counter. This 19-inch (483 mm) space from the edge of the table or counter to the line where the aisle or aisle accessway measurement begins is intended to represent the space occupied by a typical seated occupant. This dimension is also considered to be adequate to accommodate seats with armrests that are too high to fit under the table where fixed seats are used. The aisle width is permitted to be measured from the back of the seat based on the exception. As indicated in Commentary Figure 1029.12.1, where seating abuts an aisle or aisle accessway, 19 inches (483 mm) must be added to the required aisle or aisle accessway width for seating on only one side and 38 inches (965 mm) for seating on both sides. When

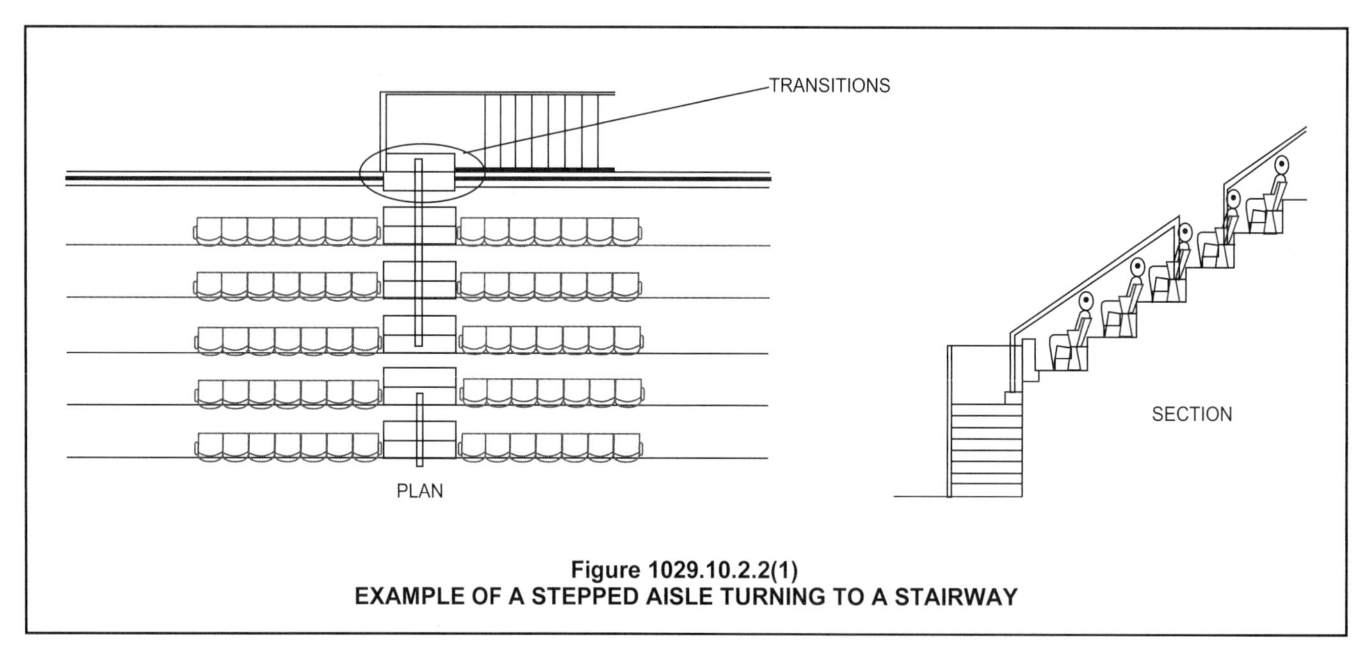

Figure 1029.10.2.2(1)
EXAMPLE OF A STEPPED AISLE TURNING TO A STAIRWAY

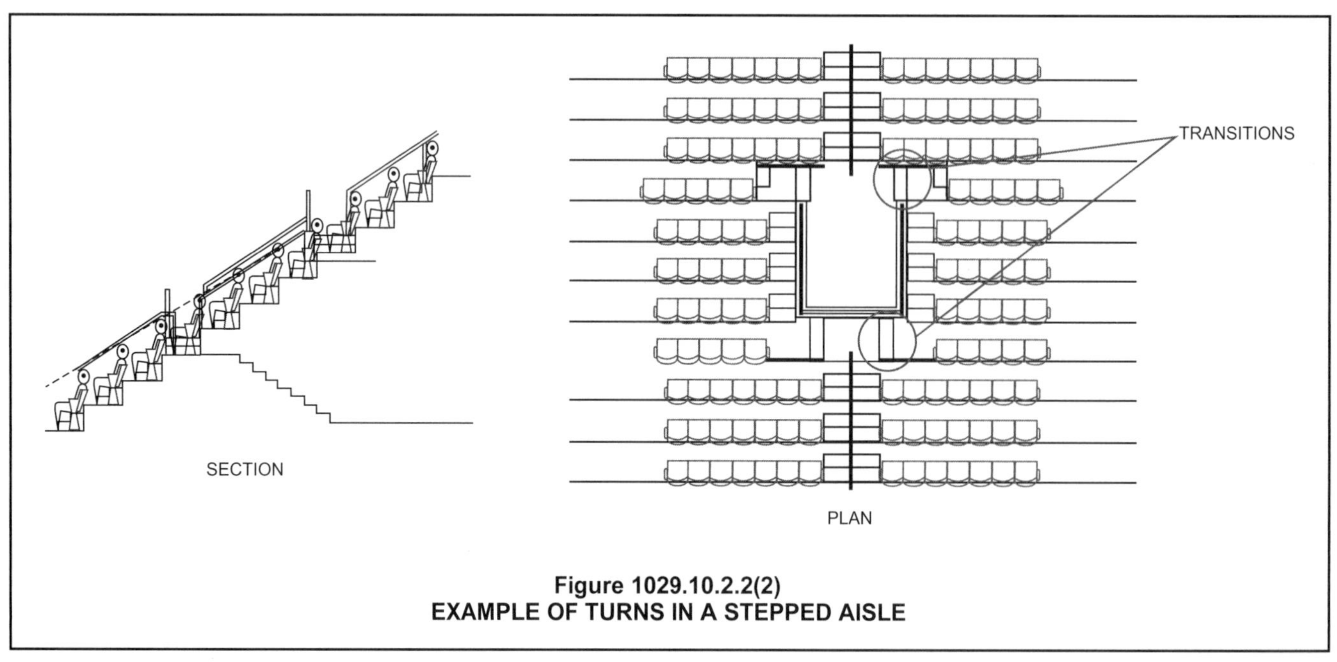

Figure 1029.10.2.2(2)
EXAMPLE OF TURNS IN A STEPPED AISLE

seating will not be adjacent to the aisles or aisle passageways, as is the case when tables are at an angle to the aisle or aisle accessway, the measurement may be taken to the edge of the seating, table, counter or tread. Sections 1029.12.1.1 and 1029.12.1.2 address width and travel along aisle accessways. For aisles between tables, see the general requirements for aisles in assembly spaces.

1029.12.1.1 Aisle accessway capacity and width for seating at tables. *Aisle accessways* serving arrangements of seating at tables or counters shall comply with the capacity requirements of Section 1005.1 but shall not have less than 12 inches (305 mm) of width plus $^{1}/_{2}$ inch (12.7 mm) of width for each additional 1 foot (305 mm), or fraction thereof, beyond 12 feet (3658 mm) of *aisle accessway* length measured from the center of the seat farthest from an *aisle*.

Exception: Portions of an *aisle accessway* having a length not exceeding 6 feet (1829 mm) and used by a total of not more than four persons.

❖ This section specifies two criteria for the determination of the required width of aisle accessways at tables: the requirements of Section 1005.1 for capacity based on the number of occupants and the option described in this section. The aisle accessway width is to be the wider of the two requirements. The aisle accessway width between tables is determined similarly to the aisle accessway between rows of seats that view an event.

The relationship of tables and seating sometimes results in a situation in which it is difficult to determine which chairs are served by which aisle accessway; therefore, the width of the aisle accessway is a function of the distance from the aisle. The same minimum 12 inches (305 mm) is used and is increased $^{1}/_{2}$ inch (12.7 mm) for each additional foot of travel beyond 12 feet (3658 mm).

Recognizing that the normal use of table and chair seating will require some clearance for access and service, the exception eliminates the minimum width criteria if the distance to the aisle [or an aisle accessway of at least 12 inches (305 mm)] is less than 6 feet (1829 mm) and the number of people served is not more than four. Therefore, the first 6 feet (1829 mm) are not required to meet any minimum width criteria. After the first 6 feet (1829 mm), the requirements for an aisle accessway will apply. The length of the aisle accessway is then restricted by Section 1029.12.1.2. When the maximum length of the aisle accessway is reached, an aisle, corridor or exit access door must be provided (see Commentary Figure 1029.12.1.1).

1029.12.1.2 Seating at table aisle accessway length. The length of travel along the *aisle accessway* shall not exceed 30 feet (9144 mm) from any seat to the point where a person has a choice of two or more paths of egress travel to separate *exits*.

❖ At some point in the exit access travel, it is necessary to reach an aisle complying with the minimum widths of Section 1029.9.1. Aisle accessway travel distance for seating at tables is not to exceed 30 feet (9144 mm), which may represent a dead-end condition (see Commentary Figure 1029.12.1.1).

More and more sports facilities are starting to add venues that include dining while watching the sporting event. At this time, the code does not include provisions for dining that occurs in smoke-protected areas.

1029.12.2 Clear width of aisle accessways serving seating in rows. Where seating rows have 14 or fewer seats, the minimum clear *aisle accessway* width shall be not less than 12 inches (305 mm) measured as the clear horizontal distance from the back of the row ahead and the nearest projection of the row behind. Where chairs have automatic or self-rising seats, the measurement shall be made with seats in the raised position. Where any chair in the row does not have an automatic or self-rising seat, the measurements shall be made with the seat in the down position. For seats with folding tab-

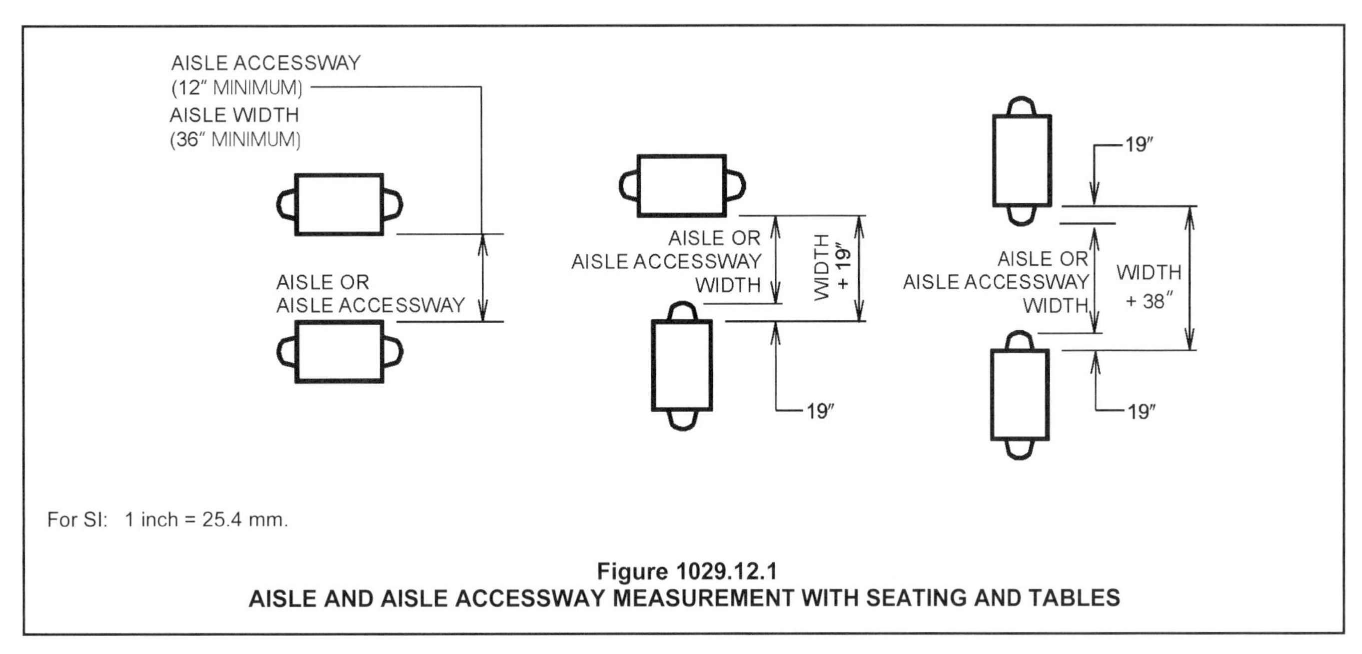

Figure 1029.12.1
AISLE AND AISLE ACCESSWAY MEASUREMENT WITH SEATING AND TABLES

let arms, row spacing shall be determined with the tablet arm in the used position.

> **Exception:** For seats with folding tablet arms, row spacing is permitted to be determined with the tablet arm in the stored position where the tablet arm when raised manually to vertical position in one motion automatically returns to the stored position by force of gravity.

❖ The requirements of this section are applicable to theater-type seating arrangements. This includes both "continental" and "traditional" seating arrangements. Theater-type seating is characterized by a number of seats arranged side by side and in rows. In this type of seating arrangement, the potential exists for a large number of occupants to be present in a confined environment where the ability of the occupants to move quickly is limited. In order to egress, people are required to move single file within a narrow row (i.e., aisle accessway) before reaching an aisle; both the aisle or aisle accessway limit movement toward an exit. To provide adequate passage between rows of seats, this section requires that the clear width between the back of a row to the nearest projection of the seating immediately behind must be at least 12 inches (305 mm) [see Commentary Figure 1029.12.2(1)]. Where chairs are manufactured with automatic or self-lifting seats, the minimum width requirement may be measured with the seats in a raised position. Commonly used in college lecture halls, seats with built-in tablet arms are provided so that students can take notes. For an example of a type of tablet arm that complies with the exception, see Commentary Figure 1029.12.2(2).

When tablet arm chairs are used, the required

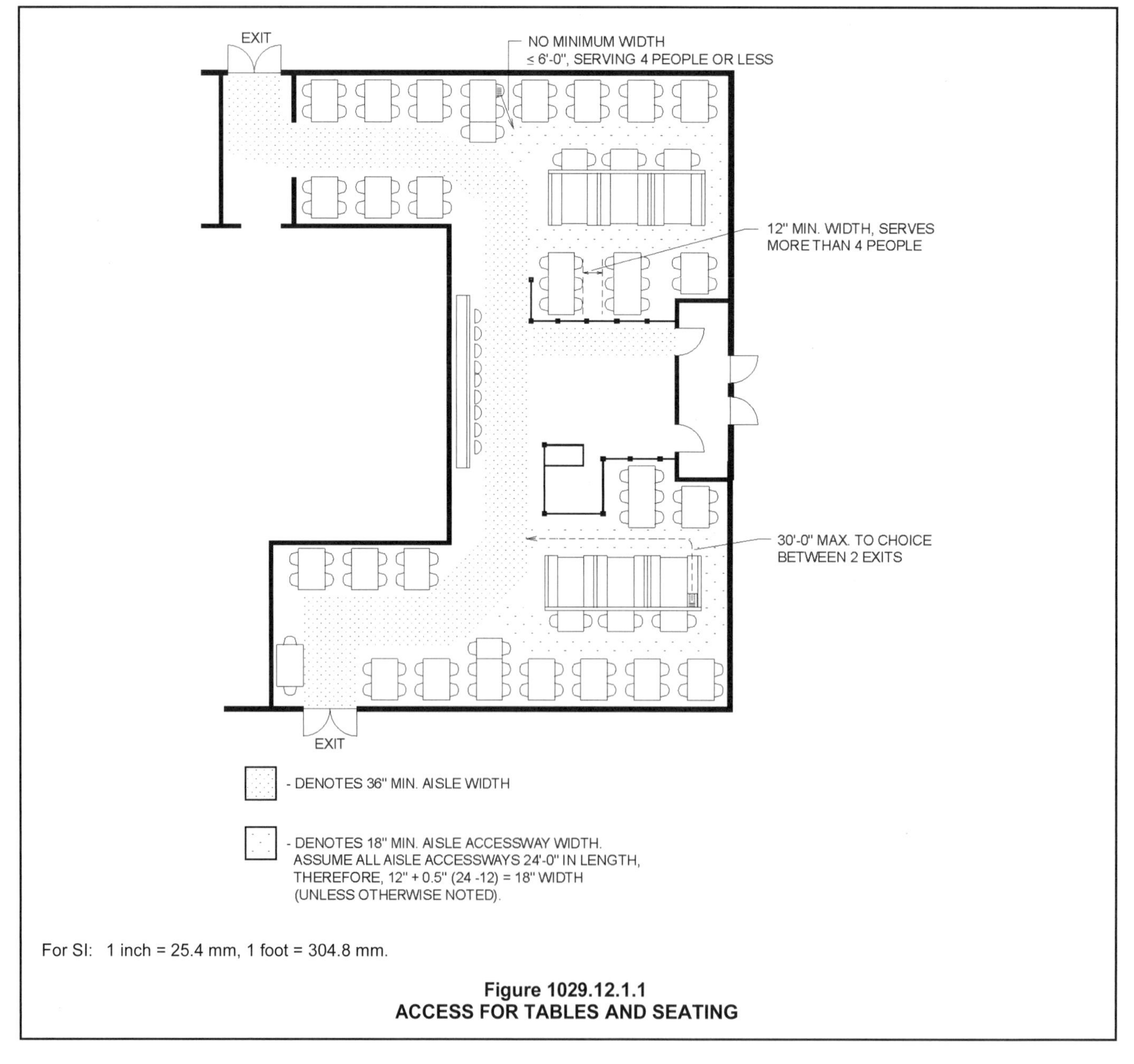

For SI: 1 inch = 25.4 mm, 1 foot = 304.8 mm.

Figure 1029.12.1.1
ACCESS FOR TABLES AND SEATING

width is to be determined with the tablet arm in its usable position. The exception allows for folding tablet arms that fall back into the stored position when a person rises out of the seat. With seats occupied and tablets raised, students egressing en masse down the row would at most encounter one tablet to move out of the way. With these types of arms, the aisle accessway can be measured for the seat or arm as indicated in Commentary Figure 1029.12.2(1).

With respect to self-rising seats, ASTM F851 provides one method of determining acceptability.

1029.12.2.1 Dual access. For rows of seating served by *aisles* or doorways at both ends, there shall be not more than 100 seats per row. The minimum clear width of 12 inches (305 mm) between rows shall be increased by 0.3 inch (7.6 mm) for every additional seat beyond 14 seats where seats have backrests or beyond 21 where seats are without backrests. The minimum clear width is not required to exceed 22 inches (559 mm).

Exception: For *smoke-protected assembly seating*, the row length limits for a 12-inch-wide (305 mm) *aisle accessway*, beyond which the *aisle accessway* minimum clear width shall be increased, are in Table 1029.12.2.1.

❖ Where rows of seating are served by aisles or doorways located at both ends of the path of row travel, the number of seats that may be used in a row may be up to, but not more than, 100 (continental seating) and the minimum required clear width aisle accessway of 12 inches (305 mm) between rows of seats must be increased by 0.3 inch (8 mm) for every additional seat with backrests beyond 14, but not more than a total of 22 inches (559 mm) (see Commentary Figure 1029.12.2.1). The increase for seating without backrests is to allow for bench seating similar to bleacher requirements in ICC 300 (see Section 1029.1.1). For example, in a row of 24 seats, the minimum clear width would compute to 15 inches (381 mm) [12 + (0.3 x 10)]. For a row of 34 seats, a clear width of 18 inches (457 mm) would be required. Increases in the clear width between rows of seats would occur up to a row of 46 seats. From 47 to 100 seats, a maximum clear width between rows of 22 inches (559 mm) would apply.

Since the row is to provide access to an aisle in both directions, the minimum width applies to the entire length of the row aisle accessway.

The exception allows more seats in a row with the minimum 12-inch (305 mm) seat spacing since safe egress time is extended for this condition.

When a second means of egress for occupants of a single access row is possible through an adjacent dual access row, see Section 1029.8.1.

TABLE 1029.12.2.1. See page 10-202.

❖ Table 1029.10.1 recognizes the increased egress time available in smoke-protected assembly seating areas. Therefore, the table permits greater lengths of rows that have the minimum 12 inches (305 mm) of clear width. When a row exceeds the lengths identified in the table, the row width is to be increased in accordance with Section 1029.12.2.1 [0.3 inch (8

For SI: 1 inch = 25.4 mm.

Figure 1029.12.2(2)
EXAMPLE OF FOLDING TABLET ARMS

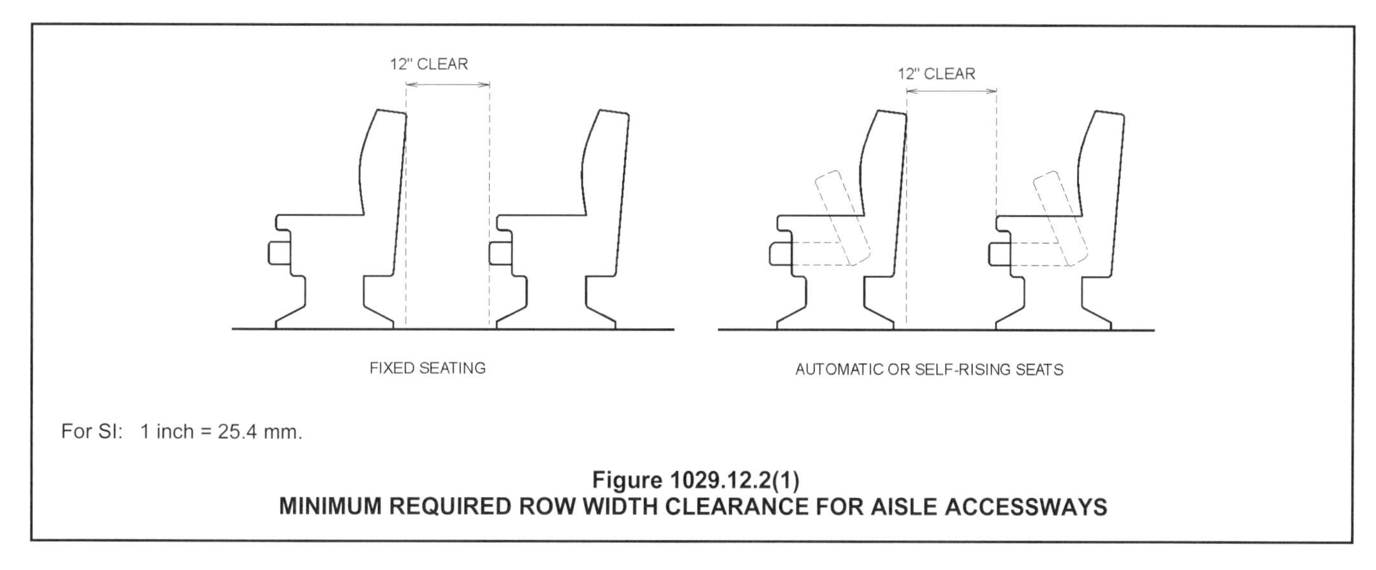

For SI: 1 inch = 25.4 mm.

Figure 1029.12.2(1)
MINIMUM REQUIRED ROW WIDTH CLEARANCE FOR AISLE ACCESSWAYS

mm) per additional seat] for dual-access rows and Section 1029.12.2.2 [0.6 inch (15 mm) per additional seat] for single-access rows. Column one in this table is based on the total number of seats contained within the assembly space, not the seats per level. The increase for seating without backrests is to allow for bench seating similar to bleacher requirements in ICC 300 (see Section 1029.1.1).

1029.12.2.2 Single access. For rows of seating served by an *aisle* or doorway at only one end of the row, the minimum clear width of 12 inches (305 mm) between rows shall be increased by 0.6 inch (15.2 mm) for every additional seat beyond seven seats where seats have backrests or beyond 10 where seats are without backrests. The minimum clear width is not required to exceed 22 inches (559 mm).

Exception: For *smoke-protected assembly seating*, the row length limits for a 12-inch-wide (305 mm) *aisle accessway*, beyond which the *aisle accessway* minimum clear width shall be increased, are in Table 1029.12.2.1.

❖ Where rows of seating are served by an aisle or doorway at only one end of a row, the minimum clear width of 12 inches (305 mm) between rows of seats must be increased by 0.6 inch (15 mm) for every additional seat beyond seven for seats with backrests, but not more than a total of 22 inches (559 mm) (see Commentary Figure 1029.12.2.2). The increase for seating without backrests is to allow for bench seating similar to bleacher requirements in ICC 300 (see Section 1029.1.1). While this section does not specify the maximum number of seats permitted in a row, the 30-foot (9144 mm) common path of travel limitation (see Section 1029.8) essentially restricts the single-access row to approximately 20 seats, based on an 18-inch (457 mm) width per seat. A row

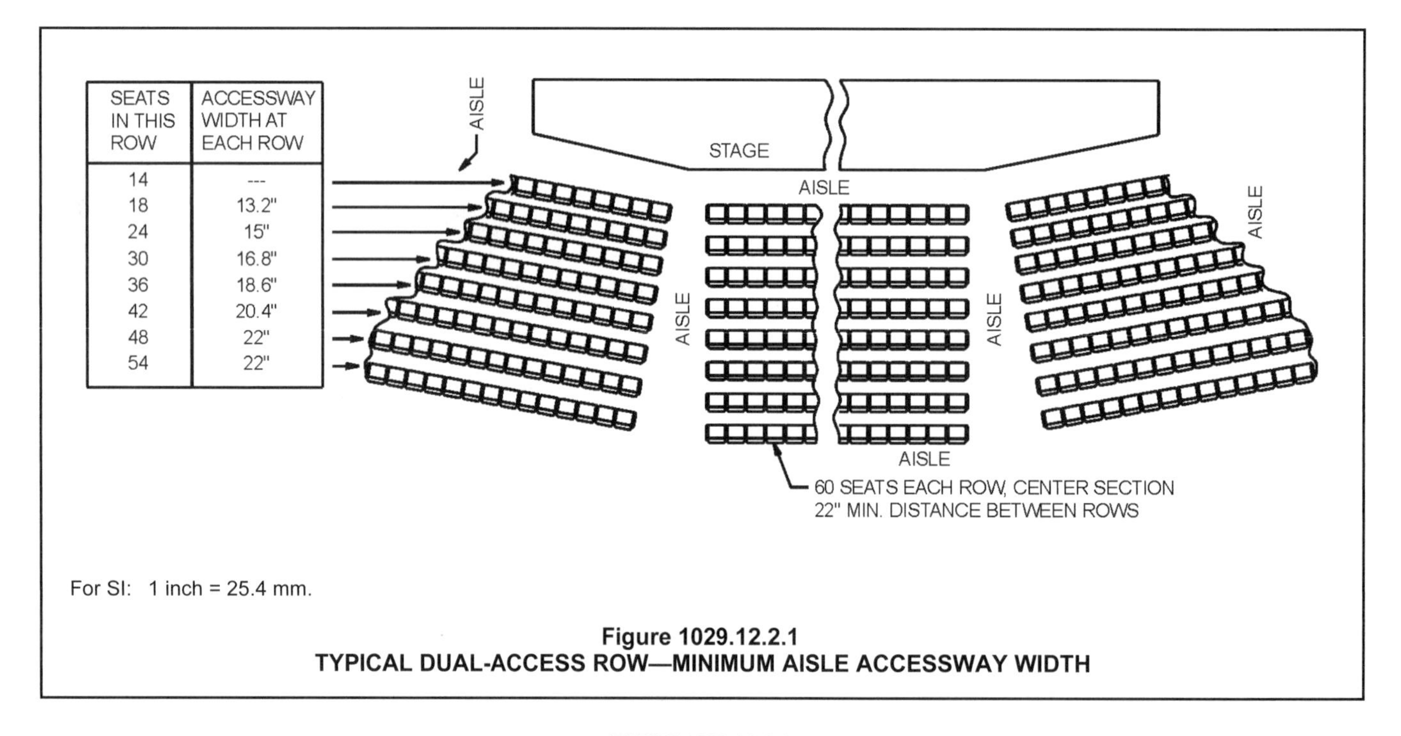

Figure 1029.12.2.1
TYPICAL DUAL-ACCESS ROW—MINIMUM AISLE ACCESSWAY WIDTH

TABLE 1029.12.2.1
SMOKE-PROTECTED ASSEMBLY AISLE ACCESSWAYS

TOTAL NUMBER OF SEATS IN THE SMOKE-PROTECTED ASSEMBLY SEATING	MAXIMUM NUMBER OF SEATS PER ROW PERMITTED TO HAVE A MINIMUM 12-INCH CLEAR WIDTH AISLE ACCESSWAY			
	Aisle or doorway at both ends of row		Aisle or doorway at one end of row only	
	Seats with backrests	Seats without backrests	Seats with backrests	Seats without backrests
Less than 4,000	14	21	7	10
4,000	15	22	7	10
7,000	16	23	8	11
10,000	17	24	8	11
13,000	18	25	9	12
16,000	19	26	9	12
19,000	20	27	10	13
22,000 and greater	21	28	11	14

For SI: 1 inch = 25.4 mm.

of 12 seats with backrests would compute to a required minimum width of 15 inches [12 + (0.5 x 5)]. Similarly, a row of 17 seats with backrests would require a clear width of 18 inches (457 mm) and so on. Since dual access is not provided, incremental increases would be permitted in the aisle accessway width as shown in Commentary Figure 1029.12.2.2. Incremental increases in the required width would occur up to the maximum number of seats, which is determined by the 30-foot (9144 mm) dead-end limitation.

The reason for increasing the row accessway widths incrementally with increases in the number of seats per row is to provide more efficient passage for the occupants who are using the aisle accessway. As a practical matter, where dual-access (see Section 1029.12.2.1) and single-access seating arrangements are used together, the largest computed clear width dimension would normally be applied by the designer to both arrangements so that the rows of seats will be in alignment. When a second means of egress for occupants of a single-access row is possible through an adjacent dual access row, see Section 1029.8.1.

1029.13 Assembly aisle walking surfaces. Ramped *aisles* shall comply with Sections 1029.13.1 through 1029.13.1.3. Stepped *aisles* shall comply with Sections 1029.13.2 through 1029.13.2.4.

❖ Assembly facilities such as theaters and auditoriums often require sloping or stepped floors to provide seated occupants with preferred sightlines for viewing presentations (for sightlines for wheelchair spaces, see Section 1108.2). Aisles must, therefore, be designed to accommodate the changing elevations of the floor in such a manner that the path of travel will allow occupants to leave the area at a rapid pace with minimal possibilities for stumbling or falling during times of emergency.

1029.13.1 Ramped aisles. *Aisles* that are sloped more than one unit vertical in 20 units horizontal (5-percent slope) shall be considered a ramped *aisle*. Ramped *aisles* that serve as part of an *accessible route* in accordance with Sections 1009 and 1108.2 shall have a maximum slope of one unit vertical in 12 units horizontal (8-percent slope). The slope of other ramped *aisles* shall not exceed one unit vertical in 8 units horizontal (12.5-percent slope).

❖ Similar to the definition for ramp in Chapter 2, aisles that slope 1:20 or less are considered sloped aisles, but are not ramps. Note that ramps that serve as part of an accessible route to and from accessible wheelchair spaces (Sections 1009 and 1108.2) must comply with the more restrictive requirements for ramps in Section 1012. This section requires that aisles with a gradient from 1:20 to 1:8 (12.5 percent slope) must meet the ramped aisle provisions in this section. Aisles with a gradient exceeding one unit vertical and eight units horizontal (12.5-percent slope) must consist of a series of treads and risers that comply with the requirements of stepped aisles in Section 1029.13.2.

1029.13.1.1 Cross slope. The slope measured perpendicular to the direction of travel of a ramped *aisle* shall not be steeper than one unit vertical in 48 units horizontal (2-percent slope).

❖ The limitation of one unit vertical in 48 units horizontal on the slope across the direction of travel is to prevent a severe cross slope that would pitch a user to one side (see Commentary Figure 1012.2).

1029.13.1.2 Landings. Ramped *aisles* shall have landings in accordance with Sections 1012.6 through 1012.6.5. Landings for ramped *aisles* shall be permitted to overlap required *aisles* or cross *aisles*.

❖ The reference for ramp landings to Sections 1012.6 through 1012.6.5 picks up the requirements for ramps to have a landing at the top and bottom of each run and when ramps change direction. This also includes requirements for landing slope, width and length. This does not pick up the requirement for ramps to have a landing at every 30 inches (762 mm) in rise as indicated in Section 1012.4. This is in con-

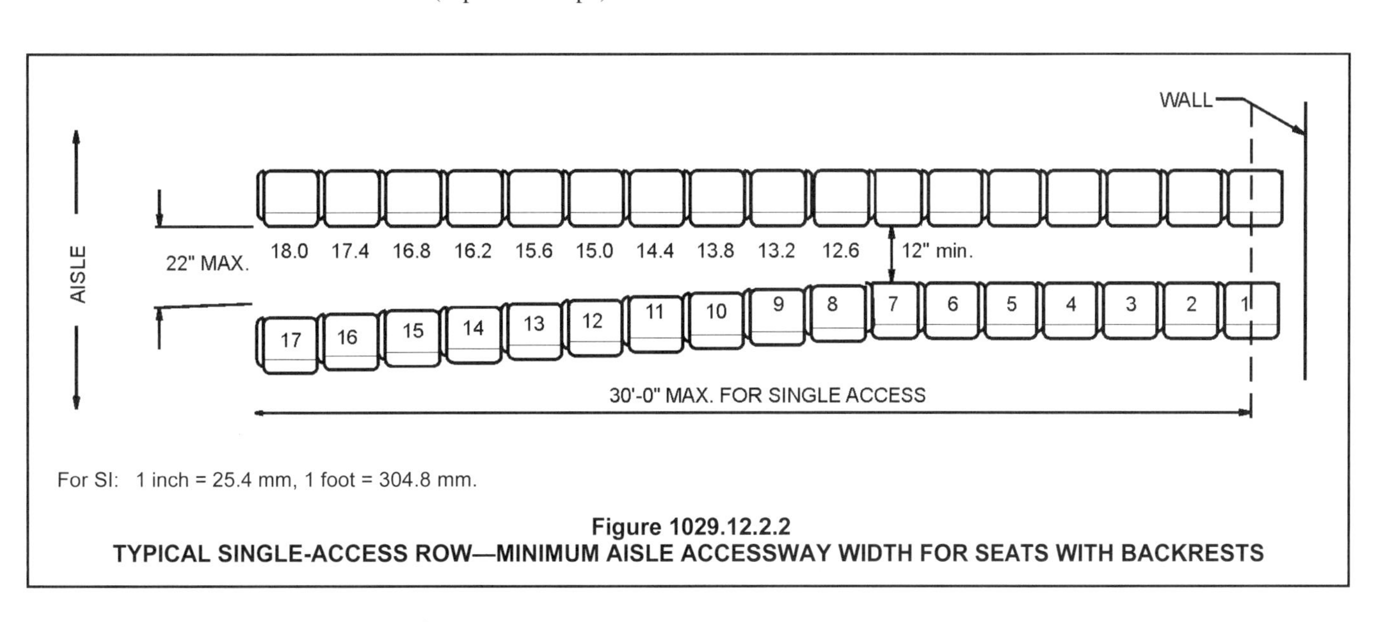

Figure 1029.12.2.2
TYPICAL SINGLE-ACCESS ROW—MINIMUM AISLE ACCESSWAY WIDTH FOR SEATS WITH BACKRESTS

sideration of the line of sight requirements for the different venues and the ability to provide a safe and smooth transition between the ramped aisles and the adjacent aisle accessways as patrons move into and out of their rows. A cross aisle or aisle can also serve as the ramp landing as long as it meets the 1:48 maximum slope provisions for landings.

1029.13.1.3 Edge protection. Ramped *aisles* shall have edge protection in accordance with Sections 1012.10 and 1012.10.1.

> **Exception:** In assembly spaces with *fixed seating*, edge protection is not required on the sides of ramped *aisles* where the ramped *aisles* provide access to the adjacent seating and *aisle accessways*.

❖ Where an aisle has a dropoff on either side, it needs edge protection. In a seating venue, this would typically be the ramped aisle at the perimeter of the seating. This edge protection can be a wall, a horizontal rail that prevents the passage of a 4-inch (102 mm) sphere between the ramp surface and the rail, or a minimum 4-inch-high (102 mm) curb (see Commentary Figure 1012.10.1). Edge protection is not required between the ramped aisle and adjacent seating, including wheelchair seating spaces, as this could be a tripping hazard for people coming in and out of the rows.

1029.13.2 Stepped aisles. *Aisles* with a slope exceeding one unit vertical in eight units horizontal (12.5-percent slope) shall consist of a series of risers and treads that extends across the full width of *aisles* and complies with Sections 1029.13.2.1 through 1029.13.2.4.

❖ What must be recognized here is that stepped aisles are part of the floor construction and are intended to provide horizontal egress. Tread and riser construction for this purpose should not be directly compared to the requirements for treads and risers in conventional stairways that serve as means of vertical egress. Sometimes, because of design considerations, the gradient of an aisle is required to change from a level floor to a ramp and then to steps. In cases where there is no uniformity in the path of travel, occupants tend to be considerably more cautious, particularly in the use of stepped aisles, than they would normally be in the use of conventional stairways.

This section requires aisles with a slope of greater than 1:8 to use steps. Aisles with slopes of greater than 1:20 to 1:8 must comply with provisions for ramped aisles in Section 1029.13.1.

1029.13.2.1 Treads. Tread depths shall be not less than 11 inches (279 mm) and shall have dimensional uniformity.

> **Exception:** The tolerance between adjacent treads shall not exceed $^{3}/_{16}$ inch (4.8 mm).

❖ Depths of treads are not to be less than 11 inches (279 mm) and uniform throughout each flight, except that a variance of not more than $^{3}/_{16}$ inch (4.8 mm) is permitted between adjacent treads to accommodate variations in construction. While the minimum tread depth provision is the same as the limiting dimension for treads in interior stairways (see Section 1011.5.2), it rarely applies in the construction of stepped aisles. A more common form of stepped aisle construction is to provide a tread depth equal to the back-to-back distance between rows of seats. This way the treads can be extended across the full length of the row and serve as a supporting platform for the seats. Other arrangements might require two treads between rows of seats, each tread equaling one-half of the depth of the seat row.

In theaters, for example, the back-to-back distance between rows of fixed seats usually ranges somewhere between 3 and 4 feet (914 and 1219 mm), depending on seat style and seat dimensions as well as the ease of passage between the rows (see Commentary Figure 1029.13.2.1). The selection of single-tread or two-tread construction between rows of seats depends on the gradient and suitable riser height (see Section 1029.13.2.2), as needed for sightlines.

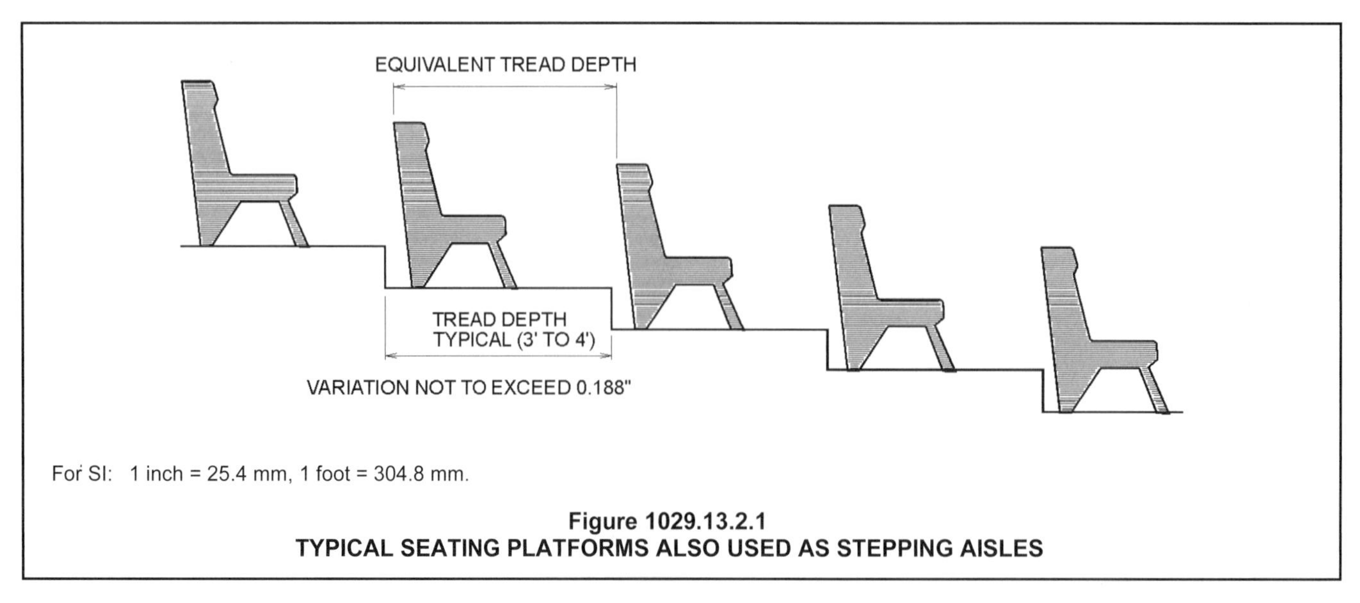

Figure 1029.13.2.1
TYPICAL SEATING PLATFORMS ALSO USED AS STEPPING AISLES

In comparing this section to Section 1029.13.2.2, it is significant to note the emphasis placed on the tread dimension. While not desirable, the code permits riser heights to deviate; however, tread dimensions must not vary beyond the $^3/_{16}$-inch (4.8 mm) tolerance.

1029.13.2.2 Risers. Where the gradient of stepped *aisles* is to be the same as the gradient of adjoining seating areas, the riser height shall be not less than 4 inches (102 mm) nor more than 8 inches (203 mm) and shall be uniform within each *flight*.

Exceptions:

1. Riser height nonuniformity shall be limited to the extent necessitated by changes in the gradient of the adjoining seating area to maintain adequate sightlines. Where nonuniformities exceed $^3/_{16}$ inch (4.8 mm) between adjacent risers, the exact location of such nonuniformities shall be indicated with a distinctive marking stripe on each tread at the *nosing* or leading edge adjacent to the nonuniform risers. Such stripe shall be not less than 1 inch (25 mm), and not more than 2 inches (51 mm), wide. The edge marking stripe shall be distinctively different from the contrasting marking stripe.
2. Riser heights not exceeding 9 inches (229 mm) shall be permitted where they are necessitated by the slope of the adjacent seating areas to maintain sightlines.

❖ In stepped aisles where the gradient of the aisle is the same as the gradient of the adjoining seating area, riser heights are not to be less than 4 inches (102 mm) nor more than 8 inches (203 mm) (see Commentary Figure 1029.13.2.2). For the safety of the occupants, risers should have uniform heights, where possible, throughout each flight. However, nonuniformity of riser heights is permitted in cases where changes to the gradient in the adjoining seating area are required because of sightlines and other seating layout considerations.

Where variations in height exceed $^3/_{16}$ inch (4.8 mm) between adjacent risers, a distinctive marking stripe between 1 inch (25 mm) and 2 inches (51 mm) wide is to be located on the nosings of each tread as a visual warning to the occupants to be cautious. Frequently, this is done with "runway" lights. Note that this stripe must be different from the tread contrast marking stripes required for transitions in Section 1029.10.3 and the tread stripes in Section 1029.13.2.2.1. All of these stripes must be visible in lighted conditions; therefore, they are not required to comply with the provisions for luminous tread markings in Section 1025.

While the riser height may vary to adjust to sight lines of the associated seating, there is a maximum change for adjacent treads so that there is consistency in the flight (see Section 1029.13.2.2.1).

In comparing this section with Section 1029.13.2.1, it is significant to note the emphasis placed on the tread dimension. While not desirable, the code permits riser heights to deviate; however, Section 1029.13.2.1 does not permit tread dimensions to vary beyond the $^3/_{16}$-inch (4.8 mm) tolerance.

1029.13.2.2.1 Construction tolerances. The tolerance between adjacent risers on a stepped *aisle* that were designed to be equal height shall not exceed 3/16 inch (4.8 mm). Where the stepped *aisle* is designed in accordance with Exception 1 of Section 1029.13.2.2, the stepped *aisle* shall be constructed so that each riser of unequal height, determined in the direction of descent, is not more than 3/8 inch (9.5 mm) in height different from adjacent risers where stepped *aisle* treads are less than 22 inches (560 mm) in depth and 3/4 inch (19.1 mm) in height different from adjacent risers where stepped *aisle* treads are 22 inches (560 mm) or greater in depth.

❖ Where risers in a stepped aisle are consistent, the construction tolerance between adjacent treads allows for a maximum difference of $^3/_{16}$ inch (4.8 mm). Stairways allow for $^3/_8$ inch (9.5 mm), but this is for the flight of stairs (see Section 1011.5.4). Where the

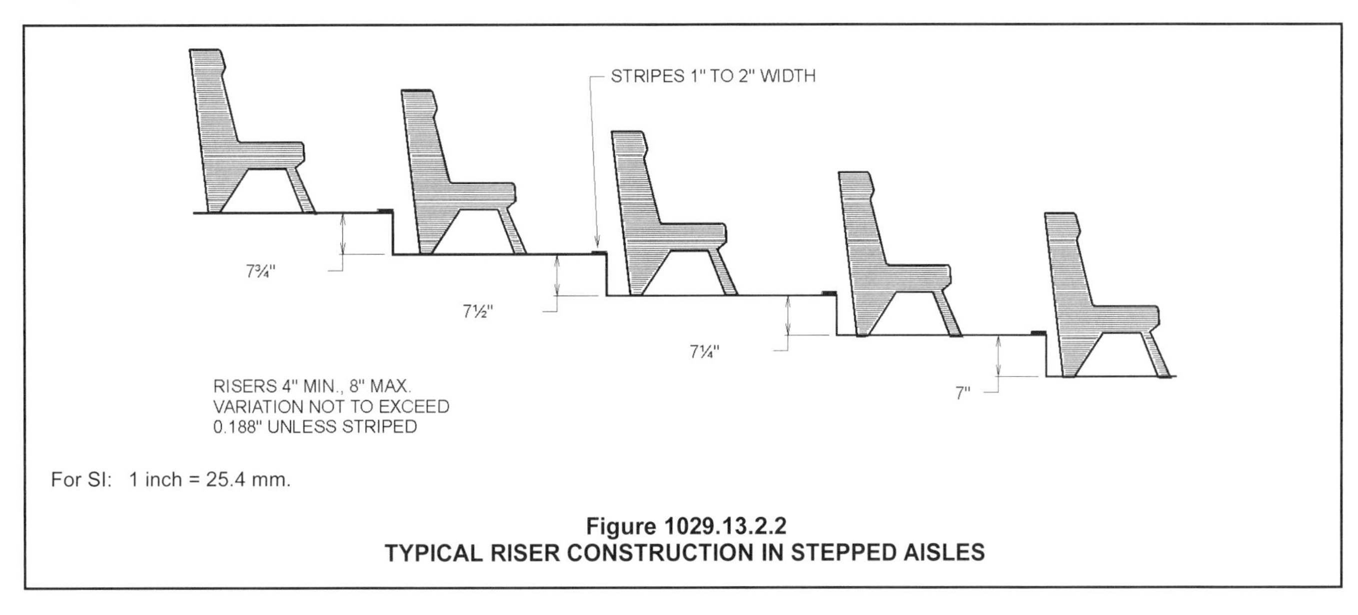

Figure 1029.13.2.2
TYPICAL RISER CONSTRUCTION IN STEPPED AISLES

seating arrangement allows for unequal heights, the difference between adjacent risers is greater for situations with deeper treads. With a tread of 22 inches (560 mm) or greater, most people will be taking more than one step on each tread.

1029.13.2.3 Tread contrasting marking stripe. A contrasting marking stripe shall be provided on each tread at the *nosing* or leading edge such that the location of each tread is readily apparent when viewed in descent. Such stripe shall be not less than 1 inch (25 mm), and not more than 2 inches (51 mm), wide.

Exception: The contrasting marking stripe is permitted to be omitted where tread surfaces are such that the location of each tread is readily apparent when viewed in descent.

❖ The exception provides for the omission of the contrasting marking stripe where the tread is readily apparent, such as when aisle stair treads are provided with a roughened metal nosing strip or where lighted nosings occur. In this situation, the user is aware of the treads without the marking stripe. This stripe must be different from the marking stripe required for nonuniform risers in Section 1029.13.2.2.1and the tread contrast marking stripes required for transitions in Section 1029.10.3

These stripes must be visible in lighted conditions; therefore, these stripes are not required to comply with the provisions for luminous tread markings in Section 1025.

1029.13.2.4 Nosing and profile. *Nosing* and riser profile shall comply with Sections 1011.5.5 through 1011.5.5.3.

❖ The profiles of treads and risers contribute to stepped aisle safety. The radius or bevel of the nosing eases the otherwise square edge of the tread and prevents irregular chipping and wear that can become a maintenance issue and seriously affect the safe use of the stepped aisles. Since safety considerations for stepped aisles and stairways are the same for nosing and riser profiles, the provisions in stairways are referenced for this requirement.

1029.14 Seat stability. In a building, room or space used for assembly purposes, the seats shall be securely fastened to the floor.

Exceptions:

1. In a building, room or space used for assembly purposes or portions thereof without ramped or tiered floors for seating and with 200 or fewer seats, the seats shall not be required to be fastened to the floor.
2. In a building, room or space used for assembly purposes or portions thereof without ramped or tiered floors for seating, the seats shall not be required to be fastened to the floor.
3. In a building, room or space used for assembly purposes or portions thereof without seating at tables and without ramped or tiered floors for seating and with greater than 200 seats, the seats shall be fastened together in groups of not less than three or the seats shall be securely fastened to the floor.
4. In a building, room or space used for assembly purposes where flexibility of the seating arrangement is an integral part of the design and function of the space and seating is on tiered levels, not more than 200 seats shall not be required to be fastened to the floor. Plans showing seating, tiers and *aisles* shall be submitted for approval.
5. Groups of seats within a building, room or space used for assembly purposes separated from other seating by railings, *guards*, partial height walls or similar barriers with level floors and having not more than 14 seats per group shall not be required to be fastened to the floor.
6. Seats intended for musicians or other performers and separated by railings, *guards*, partial height walls or similar barriers shall not be required to be fastened to the floor.

❖ The purpose of this section is to require that assembly seating be fastened to the floor where it would be a significant hazard if loose and subject to tipping over. The exceptions allow loose assembly seating for situations where the hazard is lower, such as floors where ramped or tiered seating is not used, where not more than 200 seats are used and for box seating arrangements and where a limited number of seats are within railings, guards or partial height walls.

1029.15 Handrails. Ramped *aisles* having a slope exceeding one unit vertical in 15 units horizontal (6.7-percent slope) and stepped *aisles* shall be provided with *handrails* in compliance with Section 1014 located either at one or both sides of the *aisle* or within the *aisle* width.

Exceptions:

1. *Handrails* are not required for ramped *aisles* with seating on both sides.
2. *Handrails* are not required where, at the side of the *aisle*, there is a *guard* with a top surface that complies with the graspability requirements of *handrails* in accordance with Section 1014.3.
3. *Handrail* extensions are not required at the top and bottom of stepped *aisles* and ramped *aisles* to permit crossovers within the *aisles*.

❖ For the safety of occupants, handrails must be provided in aisles where ramps exceed a gradient of one unit vertical in 15 units horizontal (6.67-percent slope) (see Commentary Figure 1029.15). All stepped aisles are required to have handrails. Handrails can be on one side, both sides or in the center of the aisle. Typically, handrails are in the center of the aisle when there is seating on both sides or on one side when the aisle is adjacent to the side walls of the room or a guard. While Sections 1011.11 and 1012.8 specify that handrails are required on both sides of stairways or ramps, the exceptions in Sections 1011.1 and 1012.1 and throughout Section 1014 allow for Section 1029.13 to be utilized for handrail location when dealing with stepped aisles, transitions and ramped aisles within assembly seating areas. This would include requiring handrails being within 30 inches (762 mm) (Section 1014.2) of the required stairway width. Safety for aisles with center handrails wider than 60 inches (1524 mm) and side aisles wider than 30 inches (762 mm) is specifically addressed with the capacity increases in Sections 1029.6.1 and 1029.6.2. What the reference to Section 1014 will require are the provisions for handrail height, graspability and fittings. Continuity, extensions, clearance and projections are more specifically addressed in Section 1029.9.1, this section and subsequent subsections.

Exception 1 omits the handrail requirements where ramped aisles are steeper, but the seats on both sides of the aisle effectively reduce the fall hazard.

Exception 2 allows handrails to be omitted where there is a guard at the side of the ramped or stepped aisle with a top rail that complies with the requirements for handrail graspability (see Section 1014.3). Note that the guard must meet the height and opening requirements specified in Section 1015 or 1029.16, as applicable.

While Section 1029.15.1 allows for discontinuous handrails, and Exception 3 (as well as Section 1014.6, Exception 2) exempts handrail extensions where they could block access to the seating, the handrail must extend the full run of the aisle stair. Stopping the handrail short of the bottom riser flight (except where permitted for mid-aisle handrails by Section 1029.15.3) would be considered a code violation. Handrails along the wall adjacent to stepped aisles or ramps should include the handrail extensions where feasible.

1029.15.1 Discontinuous handrails. Where there is seating on both sides of the *aisle*, the mid-aisle *handrails* shall be discontinuous with gaps or breaks at intervals not exceeding five rows to facilitate access to seating and to permit crossing from one side of the *aisle* to the other. These gaps or breaks shall have a clear width of not less than 22 inches (559 mm) and not greater than 36 inches (914 mm), measured horizontally, and the mid-aisle *handrail* shall have rounded terminations or bends.

❖ Where aisles have seating on both sides, handrails may be located at the sides of the aisles, but are typically located in the center of the aisle. (Handrails on both sides of a stepped aisle will typically either block access to the aisle accessways or not be able to meet the gap requirements. Transitions per Section 1029.10 can have handrails on both sides or mid-width.) The width of each section of the subdivided aisle between the handrail and the edge of seating is to be not less than 23 inches (584 mm) measured to the handrail centerline (see Section 1029.9.1, Item 3).

For reasons of life safety in fire situations and also as a practical matter in the efficient use of the facility, a handrail down the middle of an aisle should not be continuous along its entire length. Crossovers must be provided by means of gaps or breaks in the handrail installation. Such openings must not be less than 22 inches (559 mm) or more than 36 inches (914 mm) wide, and must be provided at intervals not exceeding the distance of five rows of seats (see Commentary Figure 1029.15.1). All handrail terminations should be designed to have rounded ends or bends to avoid possible injury to the occupants (see Commentary Figure 1029.15).

1029.15.2 Handrail termination. *Handrails* located on the side of stepped *aisles* shall return to a wall, *guard* or the walking surface or shall be continuous to the *handrail* of an adjacent stepped *aisle flight.*

❖ The purpose of return requirements at handrail ends is to prevent a person from catching an article of clothing or satchel straps or from being injured by falling on the extended end of a handrail. When a handrail is on a wall adjacent to a stepped aisle, it can return the same as a handrail on a stairway.

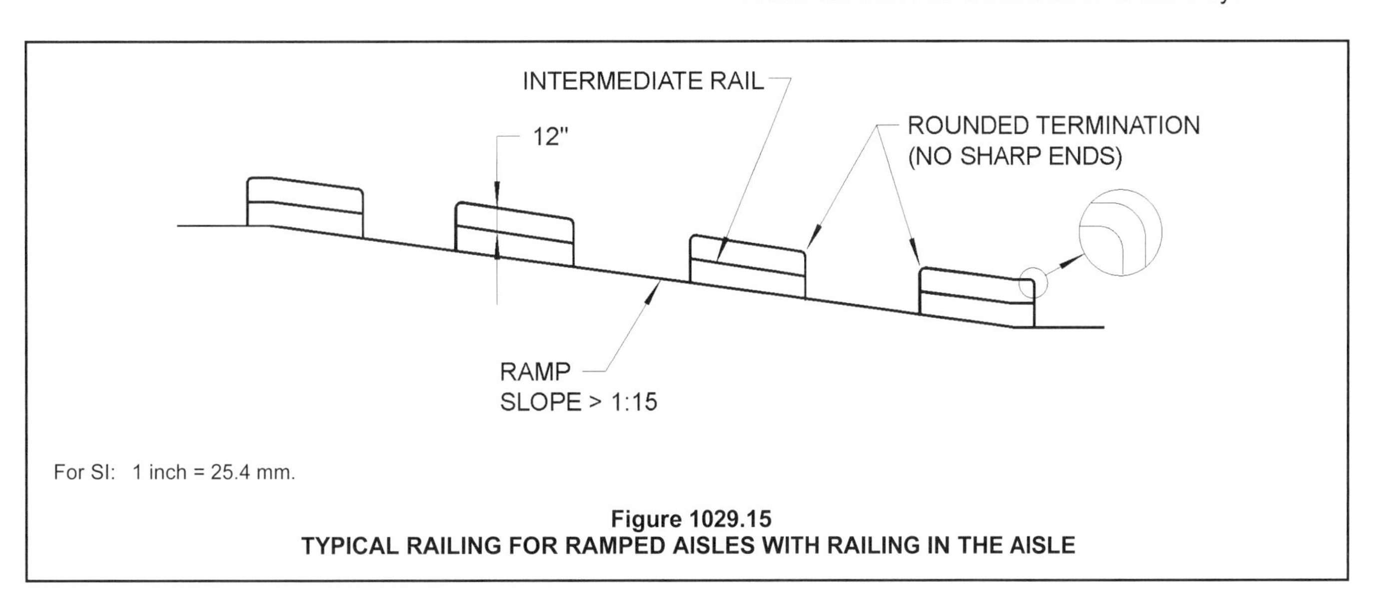

Figure 1029.15
TYPICAL RAILING FOR RAMPED AISLES WITH RAILING IN THE AISLE

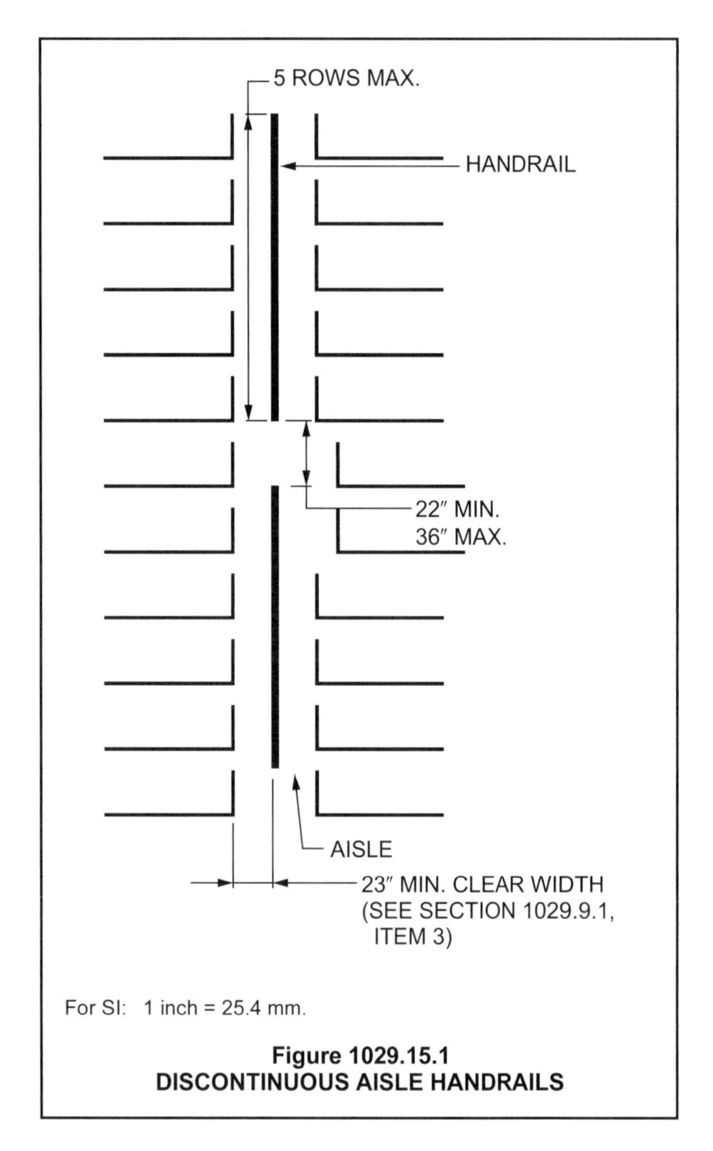

Figure 1029.15.1
DISCONTINUOUS AISLE HANDRAILS

1029.15.3 Mid-aisle termination. Mid-aisle *handrails* shall not extend beyond the lowest riser and shall terminate within 18 inches (381 mm), measured horizontally, from the lowest riser. *Handrail* extensions are not required.

> **Exception:** Mid-aisle *handrails* shall be permitted to extend beyond the lowest riser where the *handrail* extensions do not obstruct the width of the cross *aisle*.

❖ The intent is to clarify how far a mid-aisle handrail can stop from the bottom of a stepped aisle and still allow for a person to get past the front of the rail to enter the first row of seating. Since the purpose of a handrail is to arrest a fall or provide stability for someone moving up or down the stepped aisle, the handrails should also not start so far back that someone could not reach the handrail from the cross aisle, or have to let go too soon before they get to the cross aisle. The intent of the exception is to allow for a mid-aisle handrail to have a bottom extension if it will not block the cross aisle. However, if a designer wants to not have a handrail extension in order to provide access to the front row of raised seats, the handrail must extend to at least 18 inches (381 mm) from the nosing of the bottom riser.

1029.15.4 Rails. Where mid-aisle *handrails* are provided in stepped *aisles*, there shall be an additional rail located approximately 12 inches (305 mm) below the *handrail*. The rail shall be adequate in strength and attachment in accordance with Section 1607.8.1.2.

❖ Handrail installations down the middle of an aisle must be constructed with intermediate rails located 12 inches (305 mm) below and parallel to main handrails. The rail below the handrail is to stop people from going under the handrail or swinging on the handrail (see Commentary Figure 1029.15). The lower rail is not intended to meet all handrail provisions for graspability; however, it does have to meet the same strength criteria as a handrail.

1029.16 Assembly guards. *Guards* adjacent to seating in a building, room or space used for assembly purposes shall be provided where required by Section 1015 and shall be constructed in accordance with Section 1015 except where provided in accordance with Sections 1029.16.1 through 1029.16.4. At *bleachers*, *grandstands* and *folding and telescopic seating*, *guards* must be provided where required by ICC 300 and Section 1029.16.1.

❖ This section establishes the scope of the guard provisions within assembly seating. Depending on the event, the sightline constraints can be for a wide viewing angle. For example, to see the entire football field, a person may have to look right and left. Good design will establish a balance between safety and sight issues that will need to be evaluated on a case-by-case basis.

Some situations unique to assembly make it necessary to strike a balance between safety requirements and line of sight issues. Section 1029.16.1 addresses the guards around the outside edge of a seating area, such as a bleacher or tiered seating arrangement that is on each side of a high-school football field, or along the first- and third-base lines at a baseball field. The perimeter guards are required for assembly seating addressed in Section 1029 and seating covered by ICC 300. The purpose of Section 1029.16.2 is to provide for occupant safety with guards along elevated cross aisles, typically in aisles that occupants use to move side to side across the seating bowl. If there is only an aisle accessway between the seats and the guard, typically found at the front of a balcony or raised section, then Section 1029.16.3 is applicable. If the aisle is moving down through the seating toward a dropoff, then Section 1029.16.4 is applicable.

1029.16.1 Perimeter guards. Perimeter *guards* shall be provided where the footboards or walking surface of seating facilities are more than 30 inches (762 mm) above the floor or grade below. Where the seatboards are adjacent to the perimeter, *guard* height shall be 42 inches (1067 mm) high minimum, measured from the seatboard. Where the seats are self-rising, *guard* height shall be 42 inches (1067 mm) high minimum, measured from the floor surface. Where there is an *aisle* between the seating and the perimeter, the *guard* height shall be measured in accordance with Section 1015.2.

Exceptions:

1. *Guards* that impact sightlines shall be permitted to comply with Section 1029.16.3.
2. *Bleachers*, *grandstands* and *folding and telescopic seating* shall not be required to have perimeter *guards* where the seating is located adjacent to a wall and the space between the wall and the seating is less than 4 inches (102 mm).

❖ The intent of perimeter guards is to address the risk of falling from the side and back of a seating area. This question is applicable for bleacher, grandstands and folding and telescopic seating addressed in ICC 300 and other assembly seating arrangements.

ICC 300 requires guards with 4-inch openings where the floor surface has an adjacent 30-inch dropoff (ICC 300, Section 408). The dropoff is measured from the floor rather than the seatboard because the ICC 300 committee did not believe it was appropriate to require guards in a two- or three-row bleacher system. For fixed seats and benches, the height for perimeter guards is measured from the seatboard, to address when people stand on the seats. Where seats are self-rising, the guard height would be measured from the floor. Self-rising seats have backs and are very difficult to stand on.

Exception 1 is to allow for the limited situation where guards at the sides of the seating may affect the line of sight in wide venues.

Exception 2 will permit bleacher systems constructed inside the building to use the building walls as perimeter guards if the opening between the bleacher and the wall is less than the opening permitted for guards.

1029.16.2 Cross aisles. Cross *aisles* located more than 30 inches (762 mm) above the floor or grade below shall have *guards* in accordance with Section 1015.

Where an elevation change of 30 inches (762 mm) or less occurs between a cross *aisle* and the adjacent floor or grade below, *guards* not less than 26 inches (660 mm) above the *aisle* floor shall be provided.

Exception: Where the backs of seats on the front of the cross *aisle* project 24 inches (610 mm) or more above the adjacent floor of the *aisle*, a *guard* need not be provided.

❖ The purpose of this section is to provide for occupant safety with guards along elevated cross aisles. The minimum height of the guard is a function of the cross-aisle elevation above the adjacent floor or grade below [i.e., 42 inches (1067 mm) high with more than a 30-inch (762 mm) dropoff and 26 inches (660 mm) high with a 30-inch (762 mm) or less dropoff]. When the seatbacks adjacent to cross aisles are a minimum of 24 inches (610 mm) above the floor level of the cross aisle, they will serve as the guard (see Commentary Figure 1029.16.3 for an illustration of the requirements in this section).

1029.16.3 Sightline-constrained guard heights. Unless subject to the requirements of Section 1029.16.4, a fascia or railing system in accordance with the *guard* requirements of Section 1015 and having a minimum height of 26 inches (660 mm) shall be provided where the floor or footboard elevation is more than 30 inches (762 mm) above the floor or grade below and the fascia or railing would otherwise interfere with the sightlines of immediately adjacent seating.

❖ The purpose of this section is to provide for occupant safety with guards along elevated seating for areas other than cross aisles, such as the front of a raised area or balcony level. The seats would only have an access aisle between the seat and the dropoff.

This section specifies a height of 26 inches (660 mm) for guards along fascias. The guard height and the dropoff are measured from the floor or footboard. This is to provide a reasonable degree of safety while improving sightlines for persons seated immediately behind a fascia or balcony edge. The guard opening configuration must comply with Section 1015.3 (see Commentary Figure 1029.16.3 for an illustration of the requirements in this section).

1029.16.4 Guards at the end of aisles. A fascia or railing system complying with the *guard* requirements of Section 1015 shall be provided for the full width of the *aisle* where the foot of the *aisle* is more than 30 inches (762 mm) above the floor or grade below. The fascia or railing shall be a minimum of 36 inches (914 mm) high and shall provide a minimum 42 inches (1067 mm) measured diagonally between the top of the rail and the *nosing* of the nearest tread.

❖ This section applies only where the foot end of aisles (the lower end) is greater than 30 inches (762 mm) above the adjacent floor or grade below. This typically occurs where aisles move down through rows of seating to the front edge of a balcony or raised seating area. The guard must satisfy both of the specified height requirements to provide safety for persons at the end of the aisle in case someone trips moving down the stepped or ramped aisle. The 36-inch (914 mm) minimum height is measured from the floor vertically to the top of the guard. The minimum 42-inch (1067 mm) diagonal dimension from the nosing of the nearest stair tread to the top of the fascia or guard provides sufficient height to arrest a fall from the nearest aisle riser (see Commentary Figure 1029.16.3 for an illustration of the requirements in this section). The end result could be a guard with a vertical height of more than 36 inches (914 mm).

SECTION 1030 EMERGENCY ESCAPE AND RESCUE

1030.1 General. In addition to the *means of egress* required by this chapter, provisions shall be made for *emergency escape and rescue openings* in Group R-2 occupancies in accordance with Tables 1006.3.2(1) and 1006.3.2(2) and Group R-3 occupancies. *Basements* and sleeping rooms below the fourth story above *grade plane* shall have at least one exterior *emergency escape and rescue opening* in accordance with this section. Where *basements* contain one or more sleeping rooms, *emergency escape and rescue openings* shall be required in each sleeping room, but shall not be required in adjoining areas of the *basement*. Such openings shall open directly into a *public way* or to a *yard* or *court* that opens to a *public way*.

Exceptions:

1. *Basements* with a ceiling height of less than 80 inches (2032 mm) shall not be required to have *emergency escape and rescue openings*.
2. *Emergency escape and rescue openings* are not required from *basements* or sleeping rooms that have an *exit* door or *exit access* door that opens directly into a *public way* or to a *yard*, *court* or exterior exit balcony that opens to a *public way*.
3. *Basements* without *habitable spaces* and having not more than 200 square feet (18.6 m^2) in floor area shall not be required to have *emergency escape and rescue openings*.

❖ Emergency escape and rescue openings (EERO) are required in single-exit residential buildings where occupants may be sleeping during a potential fire buildup (Groups R-2, R-3 and R-4). Group R-2 apartment buildings permitted to have a single exit from a story are required to have EERO by Table 1006.3.2(1), Note a. Group R-2 congregate residences permitted to have a single exit are required to have an EERO by Table 1006.3.2(2), Note a. All Group R-4 congregate residences would also be required to have EERO because Group R-4 follows Group R-3 provisions unless specifically mentioned otherwise (Section 310.6).

All basements and each bedroom/sleeping room are to be provided with an exterior window or door that meets the minimum size requirements and is operable for emergency escape by methods that are obvious and clearly understood by all users. Sleeping rooms four stories or more above grade are not required to be so equipped, since fire service access at that height, as well as escape through such an opening, may not be practical or reliable. Since single-exit apartment buildings are limited to three stories and most Group R-3 and R-4 buildings are two or three stories, this limit is not applicable very often. Section 1018.3 limits Groups R-3 and R-4 to four stories where using a single open exit access stairway.

The provision for basements is in recognition that they typically have only a single means of egress without alternative routes through standard windows. Many times a basement is finished at a later time; therefore, as a safety precaution; at least one EERO is required in every basement. If bedrooms are provided within the basement, the location and number of EERO are determined by the bedrooms.

It is important to note that this window is only an element of escape and is not part of the means of egress required from the story unless it is a door conforming to normal egress requirements.

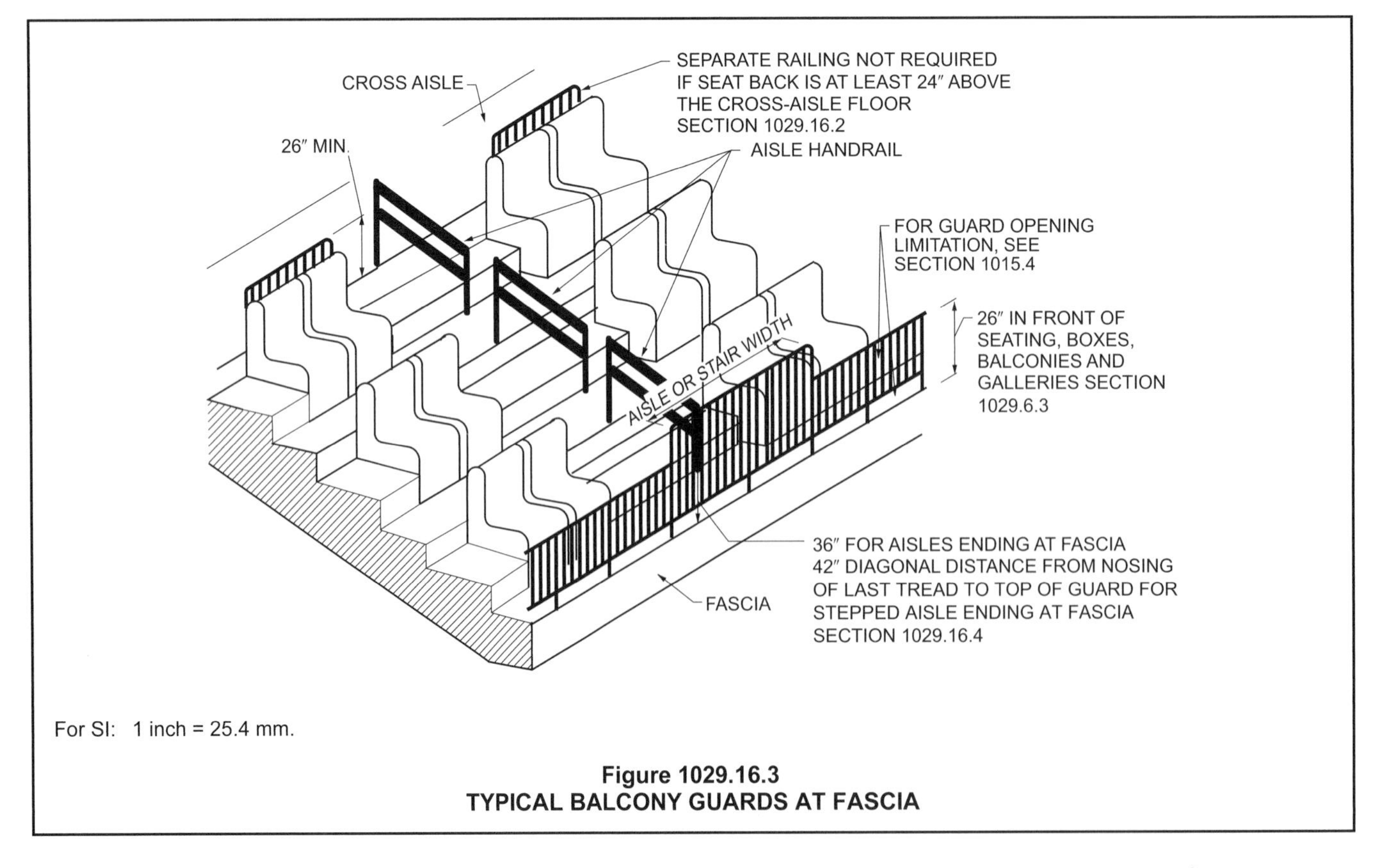

Figure 1029.16.3
TYPICAL BALCONY GUARDS AT FASCIA

Exceptions 1 and 3 are intended to exempt basements that would not be likely to be finished as living space, and thus have sleeping rooms in them.

The intent of Exception 2 is to permit sleeping rooms with a door that has direct access to an exterior-type environment, such as a street or exit balcony, to not have an EERO. The open atmosphere of the escape route would increase the likelihood that the means of egress would be available even with the delayed response time for sleeping residents. This would also exempt walk-out basements that did not include bedrooms.

1030.2 Minimum size. *Emergency escape and rescue openings* shall have a minimum net clear opening of 5.7 square feet (0.53 m^2).

Exception: The minimum net clear opening for grade-floor *emergency escape and rescue openings* shall be 5 square feet (0.46 m^2).

❖ The dimensional criteria of the openings are intended to permit fire service personnel (in full protective clothing with a breathing apparatus) to enter from a ladder, as well as permit occupants to escape. The net clear opening area and minimum dimensions are intended to provide a clear opening through which an occupant can pass to escape the building or a fire fighter can pass to enter the building for rescue or fire suppression activities. Since the emergency escape windows must be usable to all occupants, including children and guests, the required opening dimensions must be achieved by the normal operation of the window from the inside (e.g., sliding, swinging or lifting the sash). It is impractical to assume that all occupants can operate a window that requires a special sequence of operations to achieve the required opening size. While most occupants are familiar with the normal operation by which to open the window, children and guests are frequently unfamiliar with any special procedures that may be necessary to remove or tilt the sashes. The time spent in comprehending the special operation unnecessarily delays egress from the bedroom and could lead to panic and further confusion. Thus, windows that achieve the required opening dimensions only through operations such as the removal of sashes or mullions are not permitted. It should be noted that the minimum area cannot be achieved by using both the minimum height and minimum width specified in Section 1030.2.1 (see Commentary Figure 1030.2).

1030.2.1 Minimum dimensions. The minimum net clear opening height dimension shall be 24 inches (610 mm). The minimum net clear opening width dimension shall be 20 inches (508 mm). The net clear opening dimensions shall be the result of normal operation of the opening.

❖ Note that both the minimum dimensions in this section and the minimum area requirements in Section 1030.2 apply. Thus, a grade-floor window that is only 24 inches (610 mm) in height must be 30 inches (762 mm) wide to meet the 5-square-foot (0.46 m^2) area requirement of Section 1030.2 for grade-floor window (see Commentary Figure 1030.2).

1030.3 Maximum height from floor. *Emergency escape and rescue openings* shall have the bottom of the clear opening not greater than 44 inches (1118 mm) measured from the floor.

❖ This section limits the height of the bottom of the clear opening to 44 inches (1118 mm) or less such that it can be used effectively as an emergency escape (see Commentary Figure 1030.2).

Windows in Group R-2, R-3 and R-4 dwelling units may also have to meet window-opening limitations because of concerns about child falls (see Section 1015.8). Only some of the window opening-control devices will also work for emergency escape and rescue openings.

1030.4 Operational constraints. *Emergency escape and rescue openings* shall be operational from the inside of the room without the use of keys or tools. Bars, grilles, grates or similar devices are permitted to be placed over *emergency escape and rescue openings* provided the minimum net clear opening size complies with Section 1030.2 and such devices shall be releasable or removable from the inside without the use of a key, tool or force greater than that which is required for normal operation of the *emergency escape and rescue opening*. Where such bars, grilles, grates or similar devices are installed in existing buildings, *smoke alarms* shall be installed

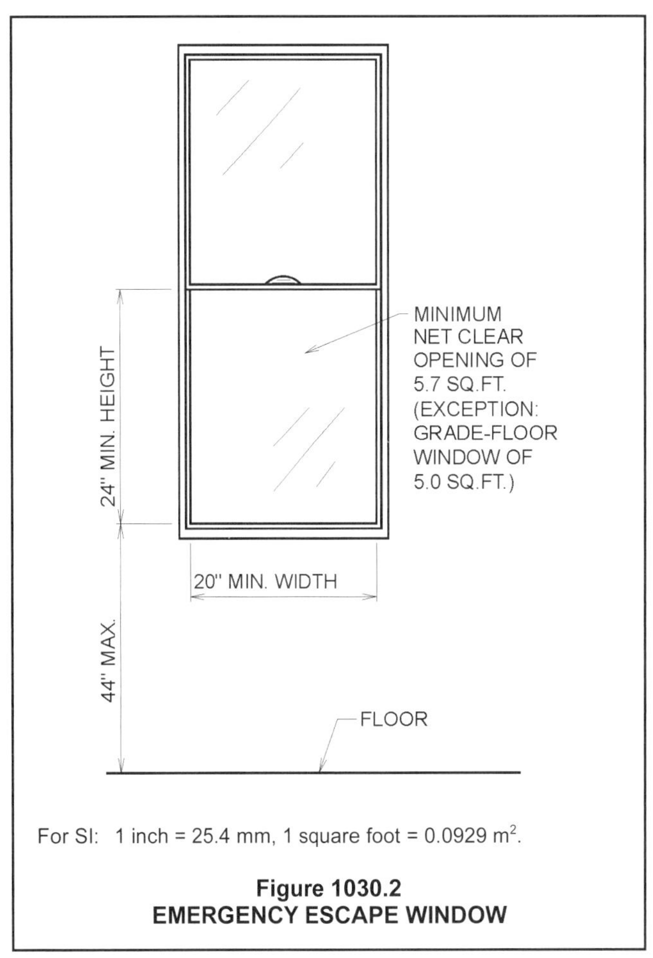

Figure 1030.2
EMERGENCY ESCAPE WINDOW

in accordance with Section 907.2.11 regardless of the valuation of the *alteration*.

❖ If security grilles, decorations or similar devices are installed on escape windows, such items must be readily removable to permit occupant escape without the use of any tools, keys or a force greater than that required for the normal operation of the window. This would include any window fall-prevention devices and window opening-control devices required by Section 1013.8.

Where bars, grilles, grates or other devices that prevent full operation of the window are placed over the emergency escape and rescue opening, it is important that they are easily removable. Thus, the requirements for ease of operation are the same as required for windows.

Windows in Group R-2, R-3 and R-4 dwelling units may also have to meet window opening limitations because of concerns about child falls (see Section 1015.8). Only some of the window opening-control devices will also work for emergency escape and rescue openings.

Where smoke alarms are not already provided and when items that could possibly slow the opening of emergency escape and rescue windows are installed in existing buildings, smoke alarms must also be installed. Smoke alarms are necessary to provide advance warning of a fire for safety purposes.

1030.5 Window wells. An *emergency escape and rescue opening* with a finished sill height below the adjacent ground level shall be provided with a window well in accordance with Sections 1030.5.1 and 1030.5.2.

❖ Emergency escape and rescue openings that are partially or completely below grade need to have window wells so that they can be used effectively (see Commentary Figure 1030.5).

1030.5.1 Minimum size. The minimum horizontal area of the window well shall be 9 square feet (0.84 m^2), with a minimum dimension of 36 inches (914 mm). The area of the window well shall allow the *emergency escape and rescue opening* to be fully opened.

❖ This section specifies the size of the window well that is needed for a rescue person in full protective clothing and breathing apparatus to use the rescue opening. The required 9 square feet (0.84 m^2) is the size of

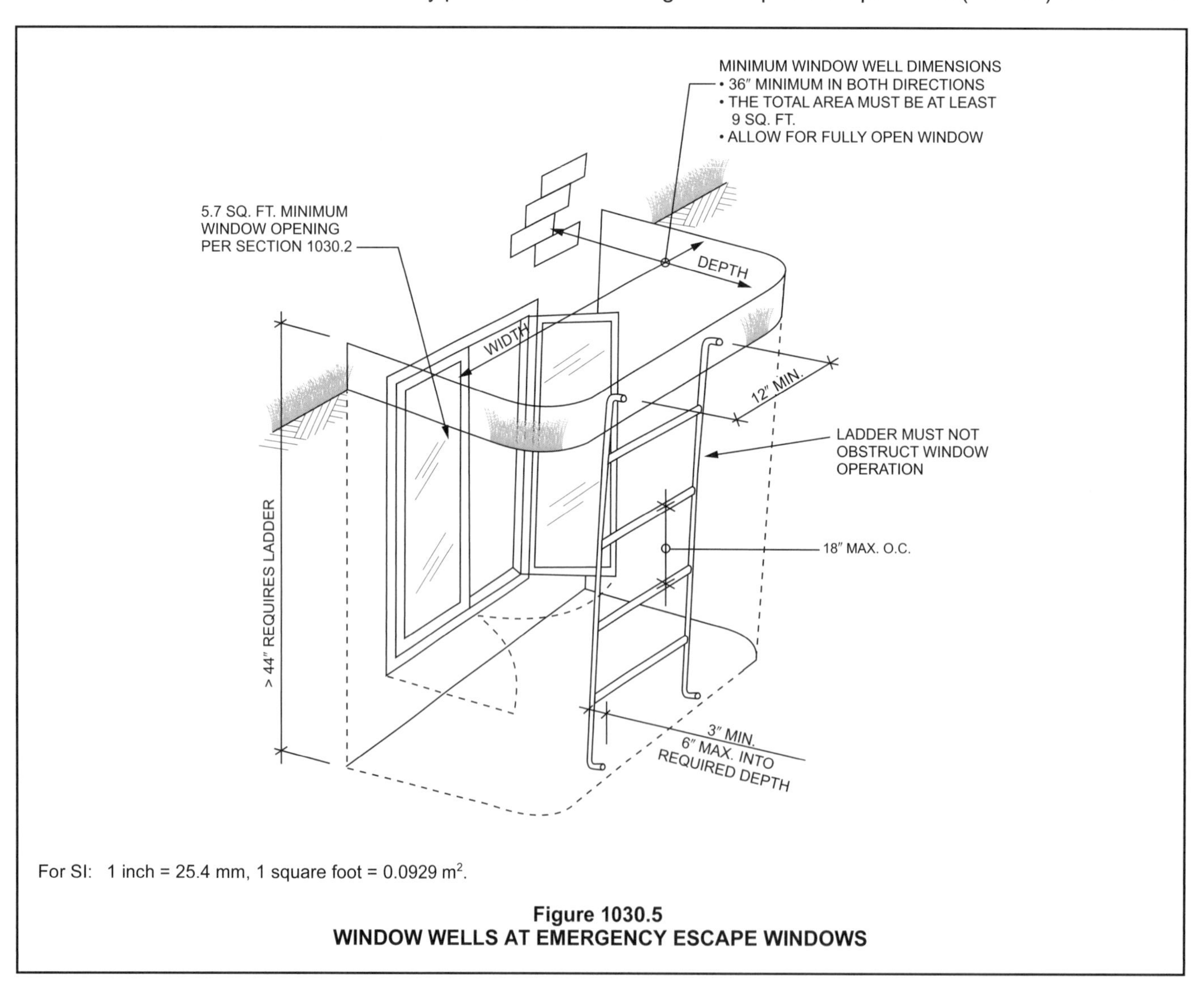

Figure 1030.5
WINDOW WELLS AT EMERGENCY ESCAPE WINDOWS

the window well. Thus, the window well must project away from the plane of the window at least 3 feet (914 mm), and the required dimension in the plane of the window along the wall is also 3 feet (914 mm) (see Commentary Figure 1030.5).

1030.5.2 Ladders or steps. Window wells with a vertical depth of more than 44 inches (1118 mm) shall be equipped with an *approved* permanently affixed ladder or steps. Ladders or rungs shall have an inside width of at least 12 inches (305 mm), shall project at least 3 inches (76 mm) from the wall and shall be spaced not more than 18 inches (457 mm) on center (o.c.) vertically for the full height of the window well. The ladder or steps shall not encroach into the required dimensions of the window well by more than 6 inches (152 mm). The ladder or steps shall not be obstructed by the *emergency escape and rescue opening*. Ladders or steps required by this section are exempt from the *stairway* requirements of Section 1011.

❖ This section specifies that a ladder or steps be provided for ease of getting into and out of window wells that are more than 44 inches (1118 mm) deep.

Usually, ladder rungs are embedded in the wall of the window well. The 44-inch (1118 mm) dimension is the depth of the window well, not the distance from the bottom of the window well to grade. Thus, if the floor of a window well is 40 inches (1016 mm) below grade, but the wall of the window well projects above grade by 6 inches (152 mm), steps or a ladder are required since the vertical depth is 46 inches (1168 mm).

It is important that the ladder not obstruct the operation of the emergency escape window (see Commentary Figure 1030.5).

Bibliography

The following resource materials were used in the preparation of the commentary for this chapter of the code.

24 CFR, *Fair Housing Accessibility Guidelines* (FHAG). Washington, DC: Department of Housing and Urban Development, 1991.

28 CFR Part 35 and 36 Final Rule, *2010 Standard for Accessible Design*. Washington, DC: U.S. Department of Justice, September 15, 2010.

36 CFR Parts 1190 and 1191 Final Rule, *The Americans with Disabilities Act (ADA) Accessibility Guidelines; Architectural Barriers Act (ABA) Accessibility Guidelines*. Washington, DC: Architectural and Transportation Barriers Compliance Board, July 23, 2004.

Appendix B to Part 36, *Analysis and Commentary on the 2010 ADA Standards for Accessible Design*. Washington, DC: Department of Justice, September 15, 2010.

Architectural and Transportation Barriers Compliance Board, 42 USC 3601-88, *Fair Housing Amendments Act (FHAA)*. Washington, DC: United States Code, 1988.

ASME A17.1-2010/CSA B44-10, *Safety Code for Elevators and Escalators*. New York: American Society of Mechanical Engineers, 2010.

ASME A18.1-2008, *Safety Standard for Platform Lifts and Stairway Chairlifts*. New York: American Society of Mechanical Engineers, 2008.

ASTM D2047-04, *Standard Test Method for Static Coefficient of Friction Polish-coated Flooring Surfaces as Measured by the James Machine*. West Conshohocken, PA: ASTM International, 2004.

ASTM E119-10b, *Standard Test Methods for Fire Tests of Building Construction and Materials*. West Conshohocken, PA: ASTM International, 2010.

ASTM E2072-10, *Standard Specification for Photoluminescent (Phosphorescent) Safety Markings*. West Conshohocken, PA: ASTM International, 2010.

ASTM F851-87 (Reapproved 2005), *Standard Test Method for Self-Rising Seat Mechanisms*. West Conshohocken, PA: ASTM International, 2005.

DOJ 28 CFR, Part 36 (Appendix A)-91, *ADA Accessibility Guidelines for Buildings and Facilities*. Washington, DC: U.S. Department of Justice, 1991.

DOJ 28 CFR, Part 36-91, *Americans with Disabilities Act* (ADA). Washington, DC: U.S. Department of Justice, 1991.

Final Report of the HUD Review of the Fair Housing Accessibility Requirements and the 2003 International Building Code (IBC). February 18, 2005 (Docket No FR-4943-N-02).

Final Report of the HUD Review of the Fair Housing Accessibility Requirements and the 2006 International Building Code (IBC). May 31, 2007 (Docket No FR-5136-N-01).

Fire Modeling Analysis Report, July 20, 2011 and *The Report to the California State Fire Marshal on Exit Access Travel Distance of 400 Feet by Task Group 400*, December 20, 2010 by Aon Fire Protection Engineering. The complete report can be found on the California State Fire Marshal's website at: http://osfm.fire.ca.gov/codedevelopment/pdf/2010interimcodeadoption/Part-9_ISOR_Attachment_A_rev20110720comp.pdf.

HUD 24 CFR, Part 100, *Federal Fair Housing Accessibility Guidelines*. Washington, DC: U.S. Department of Housing and Urban Development.

NIST IR 4770-92, *Report on Staging Areas for Persons with Mobility Limitations*. Washington, DC: National Institute of Standards and Technology, 1992.

"Review of Fall Safety of Children Between the Ages of 18 months and 4 Years in Relation to Guards and Climbing in the Built Environment," prepared for the National Ornamental & Miscellaneous Metals Association (NOMMA). Upper Marlboro, MD: NAHB Research Center, Inc., 2007.

UL 10C-09, *Positive Pressure Fire Tests of Door Assemblies*. Northbrook, IL: Underwriters Laboratories, Inc., 2009.

UL 305-07, *Panic Hardware*. Northbrook, IL: Underwriters Laboratories, Inc., 2007.

UL 924-06, *Standard for Safety Emergency Lighting and Power Equipment—with Revisions through 2009*. Northbrook, IL: Underwriters Laboratories Inc., 2009.

UL 1994-04, *Standard for Luminous Egress Path Marking Systems—with Revisions through 2010*. Northbrook, IL: Underwriters Laboratories, Inc., *2010*.

Chapter 11: Accessibility

General Comments

Chapter 11 contains provisions that set forth requirements for accessibility of buildings and their associated sites and facilities for people with physical disabilities. Existing building criteria are addressed in the *International Existing Building Code*® (IEBC®). Appendix E is included in the code to address accessibility for items in the 2010 ADA Standard that were not typically enforceable through the standard traditional building code enforcement approach system (e.g., beds, room signage).

In July 2004, the United States Architectural and Transportation Barriers Compliance Board (Access Board) published new design guidelines under the name *Americans with Disabilities Act and Architectural Barriers Act Accessibility Guidelines*, otherwise known as the "ADA/ABA Guidelines." On September 15, 2010, the U.S. Department of Justice (DOJ) published/adopted the 2004 ADA/ABA Guidelines, officially renaming them the *2010 ADA Standards for Accessible Design*. For purposes of the following discussion, the 2010 document will be referred to as the "2010 ADA Standard," while the original guidelines of 1991 are referred to as "ADAAG." For additional technical assistance and information on the 2010 ADA Standard, visit the Access Board website at www.access-board.gov.

The *International Residential Code*® (IRC®) references Chapter 11 for accessibility provisions. Therefore, this chapter may be applicable to housing covered under the IRC (see commentary, Sections 1107.6 and 1107.6.3). Structures referenced to Chapter 11 from the IRC would be considered Group R-3.

Section 1101 contains the broad scope statement of the chapter and identifies the baseline criteria for accessibility as being in compliance with this chapter and 2009 ICC A117.1. ICC A117.1, *Accessible and Useable Buildings and Facilities,* is the consensus national standard that sets forth the technical details, dimensions and construction specifications for accessibility.

Section 1102 contains a list of terms that are associated with accessibility. For the complete definitions and their commentary, see Chapter 2 of this code.

Section 1103 describes the applicability of the provisions of this chapter. Accessibility is broadly required in all buildings, structures, sites and facilities. Those specific circumstances in which accessibility is not required (or is limited) are set forth as exceptions.

Section 1104 contains the requirements for interior and exterior accessible routes. An accessible route is a key component of the built environment that provides a person with a disability access to spaces, elements, facilities and buildings. Note that ramps are addressed in Section 1010.

Section 1105 contains requirements for accessible entrances to buildings and structures. Note that requirements for accessible means of egress are addressed in Section 1009.

Section 1106 sets forth the requirements for accessible parking facilities and passenger loading zones.

Section 1107 contains various accessibility requirements that are unique to occupancies that contain dwelling units and sleeping units and are applicable in addition to other general requirements of this chapter. Specific provisions unique to Group I and R occupancies are included. The Type B dwelling- and sleeping-unit requirements in this section are coordinated with the requirements found in the *Fair Housing Accessibility Guidelines* (FHAG). Requirements for institutional and transient lodging are also consistent with the 2010 ADA Standard. When federal funding is provided for the project, there may be a higher level of accessibility required.

Section 1108 contains various accessibility requirements that are unique to specific occupancies, other than Groups I and R, and are applicable in addition to all other general requirements of Chapter 11. Specific provisions unique to assembly seating, performance areas, dining areas, self-storage and judicial facilities are included.

Section 1109 contains various requirements that are applicable to features and facilities that are not occupancy related, including requirements for toilet and bathing facilities; sinks; kitchens and kitchenettes; drinking fountains; sauna/steam rooms; elevators; platform lifts; storage facilities; detectable warnings; seating at tables, counters and work surfaces; service facilities; controls and operating mechanisms; fuel dispensing systems; and gaming machines and gaming tables.

Section 1110 provides scoping for a variety of different types of recreational facilities. Basically all recreational facilities are required to have an accessible route to them. Section 1110.4 includes additional requirements specific to certain types or recreational facilities (e.g., pools, miniature golf courses) or exceptions (e.g., raised boxing rings, water slides and diving boards).

Section 1111 sets forth requirements for signage identifying certain required accessible elements or where communication must be offered by alternative means.

History—Building Codes and Federal Laws

Access to buildings and structures for people with physical disabilities has been a subject that building codes have regulated since the early 1970s. The codes have consistently relied on a consensus national standard, CABO/ANSI A117.1, as the technical basis for accessibility. The title of CABO/ANSI A117.1 is now ICC A117.1 to reflect that the International Code Council® (ICC®) is the secretariat for this standard. Accessibility is not a new subject to the construction regulatory community. There has been a great deal of emphasis and awareness placed on the subject of accessibility through the passage of two federal laws. The Americans with Disabilities Act (ADA) and the Fair Housing Amendment Act (FHA) are two examples of federal regulations that affect building construction as it relates to accessibility.

The ADA, signed into law in July 1990, is a very broad civil rights law designed to protect persons with disabilities. There are five sections of the ADA that address different aspects of civil rights for people with disabilities. Title I deals with employment and generally prohibits discrimination against disabled people in employment. Title II deals with public services, including access to all services and facilities receiving federal funding, and access to public transportation, including buses, rail lines and passenger transit facilities. Title III requires access to a large category of buildings and structures in a manner that is equal or comparable to that available to the general public. Title IV deals with accessible telecommunication facilities and Title V further defines miscellaneous issues dealing with ADA compliance.

Note that the FHA is a companion law to the ADA. The FHA deals with residential and institutional living arrangements that, for the most part, are not covered under the ADA. However, there is some overlap. See the commentary to Section 1107 for additional information. Titles II and III of the ADA overlap numerous requirements customarily regulated by building codes. When the ADA was originally written, Title II facilities had the option to follow either the FED-STD-795-88, *Uniform Federal Accessibility Standards* (UFAS) or ADAAG for compliance with accessibility requirements. Title II facilities include those receiving federal funding for the building or for programs offered in the building, i.e., colleges, schools, park districts, local and state governments, hospitals, etc. Implementation of Title III of the ADA was based on the requirements of ADAAG. Both UFAS and ADAAG set forth scoping and technical requirements for building design and construction. The federal government is interested in bringing more uniformity to federal accessibility standards and is continuously looking at ways in which harmonization of all federal and private-sector standards can be accomplished. In many ways, UFAS and ADAAG are similar to a building code and in fact cover many of the same matters that are dealt with in such codes. With this in mind, the Access Board developed new and additional requirements within the past decade. The new requirements are organized in a building code format. The ICC and Access Board both worked toward harmonization between the federal and building code requirements. The 2010 ADA Standard replaced UFAS (1984) and ADAAG (1991) as the standard referenced for compliance with the ADA and another federal law, the Architectural Barriers Act (ABA, 1968). From September 15, 2010, to March 15, 2012, compliance with either the 1991 ADAAG or the 2010 ADA Standard was an option. Beginning March 15, 2012, compliance with the 2010 ADA Standard is required for all new construction and as well as for alterations to existing facilities.

It is important to emphasize, however, that the ADA (a law) is written as civil rights legislation and not as a building code. The enforcement mechanism for civil rights laws is significantly different than that for traditional building regulation. The ADA does not preempt the adoption or enforcement of accessibility-related codes by state and local governments. All buildings constructed within a jurisdiction must comply with locally adopted building codes and any applicable state codes, as well as the 2010 ADA Standard. However, as a federal law, state and local governments have neither the authority nor the responsibility to enforce the ADA. Enforcement of the ADA is the responsibility of the U.S. DOJ. As previously described, the ADA covers a large range of issues dealing with disabilities. Under Titles II and III, the scoping and technical requirements of the 2010 ADA Standard deal directly with accessibility of buildings. Typically, enforcement of these regulations takes place after the construction process. Any differences between the regulations and proposed design must be ultimately reconciled by the building owner or designer. Building code compliance is typically evaluated during the plan review and construction process. This is a less costly time to make changes to a facility that are necessary for compliance. Again, the building official is expected to enforce the code adopted by the state or local jurisdiction, and any applicable state laws. The building official is not responsible for interpretation or enforcement of the ADA unless the state or local jurisdiction has referenced ADAAG or the 2010 ADA Standard in a state's accessibility laws. No attempt should be made to represent that the 2010 ADA Standard is being interpreted or enforced by the building official or the jurisdiction when the 2010 ADA Standard has not been specifically adopted by the jurisdiction.

The concept of two completely separate and independent, broad-based sets of regulations that affect building design and construction may sound onerous, but the situation may not be quite as bad as it appears. The reason is that the code has a high level of consistency with the 2010 ADA Standard. Since the early 1990s, code change activity has incorporated the recommendations of the Council of American Building Officials' (CABO) Board for the Coordination of the Model Codes (BCMC) [succeeded in 1995 by the ICC Board for the Development of the Model Codes (BDMC)]. The BDMC undertook the effort to review comprehensively all facets of the accessibility issue. This effort included a comparison of ADAAG with the provisions of the model codes in use at that time. The BDMC's recommendations were ultimately reflected in codes. This work has continued to evolve as the industry learns and new technologies become readily available. There are still differences between the code and the 2010 ADA Standard, but they are minimal and will become less in future code editions. The ICC Code Technology Committee (CTC) currently has a study group that is reviewing the IBC, IEBC and the 2010 ADA Standard, and proposing code changes to address any areas in the IBC or IEBC that may fall below the 2010 ADA Standard requirements. In addition, interested parties have proposed changes to clarify requirements as well as address new technologies. An example of clarification would be the specific requirements for 4-inch (102 mm) minimum-height curbs when used as edge protection for a ramp (see Section 1012.10.1). An example of new technology would be the requirements for variable message signs (see Section 1111.4). There are also types of facilities that the IBC covers, that are exempted under the ADA. This includes church facilities and private clubs. The IBC sees these buildings as public buildings and requires full compliance, including accessibility. See the ADA for specific requirements for these exceptions.

Efforts for harmonization with the federal accessibility requirements are ongoing. Representatives from interested accessibility groups, the Department of Housing and Urban Development (HUD) and the Access Board have been attending and participating in the code change process for the code and ICC A117.1. Additionally, the ICC has participated in the public comment process on the development of federal regulations. The ICC has worked toward, and will continue to strive for, accessibility regulations that reflect the highest possible degree of consistency with federal regulations and, more importantly, reasonable and appropriate provisions to meet the needs of people with disabilities.

The FHA is federal legislation promulgated by HUD that extends fair housing protection against discrimination on the basis of family status or persons with disabilities. The standard for the FHA is the Fair Housing Accessibility Guidelines (FHAG), which sets forth accessibility scoping and technical requirements for a broad category of residential construction. The scope of ADAAG generally excludes occupancies that are covered by FHAG; however, the FHA is being interpreted to overlap the scope of the coverage of the ADA in some respects (e.g., dorms, nursing homes). Residential buildings that are covered by FHAG must also comply with the locally adopted building code. The same circumstances with respect to differences in the requirements and responsibility for enforcement described for ADA also exist with the FHA.

Again, as a federal law, building officials have neither the responsibility nor the authority to enforce the requirements of FHAG. Inquiries regarding those laws should be referred to the DOJ or HUD. Until such time that their laws are coordinated with and consider traditional code enforcement mechanisms and methods, building officials are advised not to attempt to interpret or enforce these laws. Permit applicants should be advised that the work they propose has not been reviewed for compliance with FHAG.

Due to concerns about residential construction complying with all applicable codes and laws, the National Association of Home Builders (NAHB), along with others, approached the model code groups about incorporating the FHAG requirements into the model code requirements. Since the FHA is a civil rights law rather than a building code, careful study was needed to interpret them into enforceable language that could be utilized in model codes. Through the efforts of the BDMC, recommendations to the model codes were proposed in 1994. The "adaptable" dwelling unit requirements were carried forward as a Type A unit, with the FHA requirements reflected in the Type B units.

In 2000, HUD reviewed the 2000 edition of the IBC and the referenced accessibility standard ICC/ANSI A117.1-1998 for compliance with FHAG. Based on HUD's report, a series of modifications were proposed as part of the 2000 code change cycle. The proposed modifications were accepted by the voting members and were incorporated into the 2001 Supplement to the IBC and eventually into all subsequent editions of the code. As a result of the earlier code changes, HUD eventually issued press releases stating that the 2000 code with the 2001 Supplement and 2003 code (with the ICC/ANSI A117.1-1998 accessibility standard) and the 2006 IBC (with the ICC/ANSI A117.1-2003 accessibility standard) could be considered "safe harbor" for anyone wanting to comply with FHAG scoping and technical requirements. The original safe harbor status that was bestowed on the 2000 IBC and its accessory documents was identified in the preamble of HUD's *Fair Housing Act Design Manual*.

With the 2003 and the 2006 editions of the code being safe harbor documents, architects and developers could design and be reviewed for compliance with the code through the typical building code review process. At the same time, they would also be in compliance with the building requirements for the FHAG. At the time this commentary was written, HUD had not completed its review of the 2009, 2012 or 2015 editions of the IBC code or the 2009 ICC A117.1 for safe-harbor status. Since the code is developed through a code change/public comment process, there have been changes to items in Chapter 11. This is typically a good way for new concerns, concepts or procedures to be incorporated into the requirements. It is the ICC's opinion that none of these changes have jeopardized the safe-harbor status between FHA and the code requirements in this edition.

Philosophy

The fundamental philosophy of the code on the subject of accessibility is that everything is required to be accessible. This is reflected in the basic applicability requirement (see Section 1103.1). The code's scoping requirements then address the conditions under which accessibility is not required in terms of exceptions to this general mandate. In the early 1990s, building codes tended to describe where accessibility was required in each occupancy, and any circumstance not specifically identified was excluded. The more recent codes represent a fundamental change in approach. Now one must think of accessibility in terms of "if it is not specifically exempted, it must be accessible."

Another important concept is that of "mainstreaming." There are many accessibility issues that not only benefit people with disabilities, but also provide a tangible benefit to people who do not consider themselves disabled, especially as the population ages. This type of requirement can be set forth in the code as generally applicable without necessarily identifying it specifically as an accessibility-related issue. Such a requirement would then be considered as having been mainstreamed. For example, the limitation on objects protruding into corridors, aisles and passageways (see Section 1003.3) is intended to aid people with a vision impairment by reducing the potential for unintended contact that may cause injury. Clearly, this has an additional benefit to people without a vision impairment. A protruding object can be encountered by a sighted person who is simply not paying close attention, or is exiting a building during an event where smoke may obscure vision. The concept of mainstreaming is responsive to the desire of people with disabilities to not be singled out and categorized separately from the remainder of society. It is therefore important to recognize that, while the provisions of Chapter 11 are specifically grouped and identified as accessibility related, they do not represent all of the issues that must be taken into consideration when evaluating accessibility in buildings.

There are many items that have some basis in accessibility requirements, but actually result in more "user-friendly" buildings for all of us. Physical disabilities can be permanent or temporary; can affect all age groups; can involve all levels of abilities; and can range from persons with minor visual, hearing or mobility impairments to persons who are blind, deaf or confined to a wheelchair. Everyone will benefit from accessible features in buildings, either directly or indirectly. With some foresight in designing our built environment, access into and throughout buildings can be better for everyone throughout their lifetimes.

Purpose

The purpose of this chapter is to set forth requirements for accessibility applicable to those elements of the built environment that are included within the scope of the code. There are two categories or types of requirements that must be addressed in order to accomplish accessibility: scoping and technical. Scoping requirements describe what and where accessibility is required, or how many accessible features or elements must be provided. For example, a requirement that at least one of the first 25 sleeping units in a hotel must be accessible is referred to as a scoping requirement because it describes what and how many accessible features are required. Another example of scoping is the exception for detached one- and two-family dwellings from accessibility requirements because it defines what is not required to be accessible. A technical requirement is intended to refer to a statement indicating how accessibility is to be accomplished. For example, a requirement indicating that in order for a parking space to be accessible, it must have a width of 96 inches (2438 mm) with an adjacent 60-inch (1524 mm) access aisle is considered a technical requirement. The provisions of Chapter 11 are scoping requirements in that they all address what and where accessibility is required or how many accessible elements are required. ICC A117.1 indicates how to make something accessible and is the consensus national standard that is referenced for establishing technical requirements. In addition to the provisions of Chapter 11, there are accessibility-related technical requirements elsewhere in the code. These are subject-matter-specific provisions that have been mainstreamed into the chapter or section of the code that deals with those subjects. As such, the code and its references regulate accessibility comprehensively by setting forth the scoping and technical requirements that establish the minimum level of accessibility required in the built environment.

SECTION 1101 GENERAL

1101.1 Scope. The provisions of this chapter shall control the design and construction of facilities for accessibility for individuals with disabilities.

❖ This section establishes the scope of the chapter as providing for design and construction of facilities for accessibility for persons with a disability. The scope is broadly inclusive of all aspects of construction that affect the ability of people with disabilities to approach, enter and utilize a facility. The term "facility," as defined in Chapter 2, includes not only buildings and structures, but also the site on which they are located. Features of a site, such as parking areas and paths of travel from a public way to a structure, affect accessibility and are, therefore, within the scope of Chapter 11. Chapter 11, in conjunction with mainstreamed provisions throughout the code, sets forth scoping requirements.

1101.2 Design. Buildings and facilities shall be designed and constructed to be *accessible* in accordance with this code and ICC A117.1.

❖ This section establishes the primary and fundamental relationship of ICC A117.1 to the code. The code text is intended to "scope" or provide thresholds for application of required accessibility features. The referenced standard contains technical provisions indicating how compliance with the code is achieved. In short, Chapter 11 specifies what, when and how many accessible features are required; the referenced standard indicates how to make that feature accessible. Compliance with both the code and the standard is required.

In accordance with Section 102.4, standards are utilized only to the extent that they are referenced. Note that ICC A117.1 includes technical criteria for several items that are not actually scoped in the code. Such items are not required to comply with ICC A117.1 criteria unless specifically scoped by the authority have jurisdiction. Some examples of items in ICC A117.1 that are not referenced by the code are:

- Section 403.6 – Corridor handrails
- Section 504 – Stairways
- Section 704 – Telephones (scoped in IBC Appendix E)
- Section 707 – Automatic Teller Machines (scoped in IBC Appendix E)

SECTION 1102 DEFINITIONS

1102.1 Definitions. The following terms are defined in Chapter 2:

ACCESSIBLE.

ACCESSIBLE ROUTE.

ACCESSIBLE UNIT.

AREA OF SPORT ACTIVITY.

CIRCULATION PATH.

COMMON USE.

DETECTABLE WARNING.

EMPLOYEE WORK AREA.

FACILITY.

INTENDED TO BE OCCUPIED AS A RESIDENCE.

MULTILEVEL ASSEMBLY SEATING.

MULTISTORY UNIT.

PUBLIC ENTRANCE.

PUBLIC-USE AREAS.

RESTRICTED ENTRANCE.

SELF-SERVICE STORAGE FACILITY.

SERVICE ENTRANCE.

SITE.

TYPE A UNIT.

TYPE B UNIT.

WHEELCHAIR SPACE.

❖ This section lists terms that are specifically associated with the subject matter of this chapter. It is important to emphasize that these terms are not exclusively related to this chapter but may or may not also be applicable where the term is used elsewhere in the code. For example, the defined term "Accessible" is applicable to where a space or element is required to be designed to address the needs of persons with mobility, hearing or vision impairments. However, when other chapters provide criteria for equipment or construction elements to be accessible for maintenance or inspections (e.g., Section 110.1), the intent is for an inspector or maintenance person to have physical and visual access to that element—not wheelchair access.

Definitions of terms can help in the understanding and application of the code requirements. The purpose for including a list within this chapter is to provide more convenient access to terms that may have a specific or limited application within this chapter. For the complete definition and associated commentary, refer back to Chapter 2. Terms that are italicized provide a visual identification throughout the code that a definition exists for that term. The use and application of all defined terms are set forth in Section 201.

SECTION 1103 SCOPING REQUIREMENTS

1103.1 Where required. *Sites*, buildings, *structures*, *facilities*, elements and spaces, temporary or permanent, shall be *accessible* to individuals with disabilities.

❖ This section establishes the broad principle that all buildings, structures and their associated sites and facilities are required to be accessible to persons with disabilities. Such persons would include anyone who utilizes a space, including occupants, employees, students, spectators, participants and visitors. The approach taken by the code on the subject of accessibility is to require all construction to be accessible and then allow for the level of inaccessibility that is reasonable and logical. In codes created before the early 1990s, the approach was to list the conditions and occupancies to which the accessibility requirements applied. This is no longer practical, since the exceptions are far fewer than the circumstances to which accessibility applies. The 14 exceptions to this section, Sections 1103.2.1 to 1103.2.14 reflect the extent to which accessibility in construction is either exempted or reduced in scope.

1103.2 General exceptions. *Sites*, buildings, *structures*, *facilities*, elements and spaces shall be exempt from this chapter to the extent specified in this section.

❖ Accessibility is generally applicable to all building sites and all spaces and elements within constructed facilities except as specifically exempted in the subsections to Section 1103.2.

Note that many of the exceptions say "are not required to comply with this chapter." This means that not only are the areas themselves exempted, but also that an accessible route is not required to those elements.

1103.2.1 Specific requirements. *Accessibility* is not required in buildings and *facilities*, or portions thereof, to the extent permitted by Sections 1104 through 1111.

❖ This section provides a reference to the various sections in Chapter 11 that identify when the intended number of accessible elements in various occupancies is less than 100 percent. The number of accessible fixtures and elements required by the referenced sections are deemed to provide adequate accessibility for those circumstances. For example, Section 1106 does not require all parking spaces in a parking facility to be accessible, Section 1107.6.1 does not require all sleeping units in Group R-1 to be fully accessible and Section 1109.12.1 does not require all dressing rooms to be accessible.

1103.2.2 Employee work areas. Spaces and elements within *employee work areas* shall only be required to comply with Sections 907.5.2.3.1, 1009 and 1104.3.1 and shall be designed and constructed so that individuals with disabilities can approach, enter and exit the work area. Work areas, or portions of work areas, other than raised courtroom stations in accordance with Section 1108.4.1.4, that are less than 300 square feet ($30 m^2$) in area and located 7 inches (178 mm) or more above or below the ground or finished floor where the change in elevation is essential to the function of the space shall be exempt from all requirements.

❖ This section states that elements within individual work areas are not required to be fully accessible, but must be provided with accessible means of egress (see Section 1009) and accessible circulation paths (see Section 1104.3.1). In addition, these spaces must be provided with visible alarms (see Section 907.5.2.3). The assumption is that the employment nondiscrimination requirements of the ADA will provide for "reasonable accommodations" to the disability of the employee and the area in which they work. In other words, employers will modify individual work areas for the specific requirements of the individual utilizing the space. An accessible route will be required to each work area. An example of this is an individual work station in a laboratory. Installing sinks and built-in counters at accessible levels (see commentary, Sections 1109.3 and 1109.11) could make the station impractical for use by a person who is standing. Ergonomic researchers used to recommend a standing counter at a minimum of 36 inches (914 mm), but are now advising even higher counters to reduce fatigue. When a station is required to be adapted for an individual, it would be revised based on the individual's needs and abilities. An accessible route to each work station in the laboratory would be required so that access to and from that station would be available. Note that the 36-inch (914 mm) clear width for the accessible route is the same as the minimum required width of an exit access aisle. For exceptions to the route to work areas, see Section 1104.3.1.

There is an additional exception for work areas that are required to be raised or lowered 7 inches (178 mm) or more above the floor and have an area of less than 300 square feet ($30 m^2$). The key word is "required" to be raised, not just that some decided to make that work station raised. Examples of raised work areas required to be raised would include an area around a metal stamping machine, a safety manager's observation station on a production line or the pulpit area in a church. Raised courtroom areas are specifically addressed in Section 1108.4.1.4. An example of a work area required to be lowered might be a pit under a car in a car-repair station.

An employee's work area can be something as simple as an office or cubicle, or it can extend over a much wider area. For example, an office space might include copy rooms, file rooms, mail rooms, etc., or a nurse or doctor in a clinic would have part of their work area in each exam room as well as at a central station area. In addition, not all employee-only spaces are considered work areas. Areas used by employees not for work (i.e., break rooms, bathrooms, corridors) would not fall under this exception (see the definition for "Employee work area").

1103.2.3 Detached dwellings. Detached one- and two-family *dwellings*, their accessory structures and their associated *sites* and *facilities* are not required to comply with this chapter.

❖ This section exempts detached one- and two-family dwellings from accessibility requirements. The key word here is "detached." For example, a structure containing four or more dwelling units, even if the dwelling units are separated by fire walls, would still be required to be accessible as indicated in Section 1107.

Although one- and two-family detached dwellings are typically regulated by the IRC, this exception in the code is still necessary. Single-family dwellings or duplexes that are four stories or higher would be designed and constructed under Group R-3 requirements of the code. The IRC references Chapter 11 of the code for accessibility requirements in Section R320. Those multiple-family structures with four or more dwelling units that qualify as townhouses or congregate residences with four or more sleeping units under the IRC are required to comply with the requirements for Group R-3 in Section 1107.6.3.

1103.2.4 Utility buildings. Group U occupancies are not required to comply with this chapter other than the following:

1. In agricultural buildings, access is required to paved work areas and areas open to the general public.
2. Private garages or carports that contain required *accessible* parking.

❖ This section exempts Group U from accessibility requirements except as indicated, on the basis that such structures are a low priority when considering the need for accessibility. Areas in utility buildings that are required to be accessible would be paved work areas or areas open to the general public, such as if a farmer included a farmstand within his or her barn or buildings in which accessible parking spaces are located.

1103.2.5 Construction sites. Structures, *sites* and equipment directly associated with the actual processes of construction including, but not limited to, scaffolding, bridging, materials hoists, materials storage or construction trailers are not required to comply with this chapter.

❖ This section exempts structures directly associated with the construction process because the need for accessibility on a continuous or regular basis in those circumstances is unlikely to arise. Portable toilets that are for the construction workers would also be exempted as part of this exception. Note that all structures that may be involved during a construction project are not exempt—only those specifically involved in the actual process of construction. For example, if mobile units are brought into house classrooms during a school addition, these classrooms must be accessible. A construction trailer that also serves as a sales office would also be required to be accessible.

1103.2.6 Raised areas. Raised areas used primarily for purposes of security, life safety or fire safety including, but not limited to, observation galleries, prison guard towers, fire towers or lifeguard stands are not required to comply with this chapter.

❖ If there is a reason to elevate an area for concerns about security or safety, these areas are not required to be accessible (see Commentary Figure 1103.2.6).

Figure 1103.2.6
EXAMPLE OF THE EXCEPTION FOR RAISED AREA

1103.2.7 Limited access spaces. Spaces accessed only by ladders, catwalks, crawl spaces, freight elevators or very narrow passageways are not required to comply with this chapter.

❖ Limited access spaces are considered areas where work could not reasonably be performed by a person using a wheelchair because of the physical requirements of the job activity. These areas are not required to be accessible.

1103.2.8 Areas in places of religious worship. Raised or lowered areas, or portions of areas, in *places of religious worship* that are less than 300 square feet (30 m^2) in area and located 7 inches (178 mm) or more above or below the finished floor and used primarily for the performance of religious ceremonies are not required to comply with this chapter.

❖ While the ADA does not cover churches and other places of worship, under the IBC such facilities must comply with the codes including accessibility provisions. This exception is similar to the allowance for employee work areas in Section 1103.2.2, however, in a religious facility, a raised area may be utilized by anyone participating in a service, not just employees of the church.

There are some areas of a religious facility that are

required to be raised to allow for a line of site (e.g., a pulpit), or raised as part of the tradition or symbology in the religious facility (e.g., minibar in a mosque, high altars, torah arks in a synagogue) (see Commentary Figure 1103.2.8). There are also areas in a religious facility where everyone will be assisted (e.g., a full-immersion baptistery). Where the area is used primarily for the performance of religious ceremonies, that raised or lowered area is not required to be accessible. Discussions during the code change process indicated that many jurisdictions were allowing some exceptions for religious facilities, but it varied greatly by jurisdiction and involved the religious facilities going through the variance process. The new allowance will provide guidance and create consistency.

It is not the intent of this exception to exempt the entire sanctuary area from being accessible. Allowing participation by everyone in the church, mosque or synagogue should be considered.

Figure 1103.2.8
EXAMPLE OF A RAISED AREA IN A PLACE OF RELIGIOUS WORSHIP

1103.2.9 Equipment spaces. Spaces frequented only by service personnel for maintenance, repair or occasional monitoring of equipment are not required to comply with this chapter.

❖ Spaces that only contain heating, ventilation and air conditioning (HVAC), electrical, elevator, communication and similar types of equipment are considered areas where work could not reasonably be performed by a person utilizing a wheelchair because of the physical requirements of the job activity. Such spaces include, but are not limited to, elevator pits; elevator penthouses; mechanical, electrical or communications equipment rooms; piping or equipment catwalks; water or sewage treatment pump rooms and stations; electric substations and transformer vaults; and highway and tunnel utility facilities. These areas are not required to be accessible.

1103.2.10 Highway tollbooths. Highway tollbooths where the access is provided only by bridges above the vehicular traffic or underground tunnels are not required to comply with this chapter.

❖ This section addresses facilities such as toll booths, which are raised to facilitate access to higher vehicles as well as to provide protection from being struck by vehicles. Access to these spaces are often by tunnel or bridge. Similar facilities could also be addressed by Sections 1103.2.2 and 1104.3.1.

1103.2.11 Residential Group R-1. Buildings of Group R-1 containing not more than five *sleeping units* for rent or hire that are also occupied as the residence of the proprietor are not required to comply with this chapter.

❖ This section exempts small bed-and-breakfast or transient boarding house facilities that are also the home of the owner. These types of facilities can be constructed using the IRC (see Section 101.2). The IRC also exempts these same facilities from accessibility in Section 320.

1103.2.12 Day care facilities. Where a day care facility is part of a *dwelling unit*, only the portion of the structure utilized for the day care facility is required to comply with this chapter.

❖ When an adult day care or child day care facility is part of a person's home, the accessibility requirements are only applicable to the portion of the home that constitutes the day care facility, not the home itself. For other home-based businesses, see the discussion on live/work units in Sections 419 and 1107.6.2.1.

1103.2.13 Detention and correctional facilities. In detention and correctional facilities, *common use* areas that are used only by inmates or detainees and security personnel, and that do not serve holding cells or housing cells required to be *Accessible units*, are not required to comply with this chapter.

❖ Section 1107.5.5 addresses when sleeping units, special holding or housing cells and medical care units in detention and correctional facilities are required to be accessible. If the purpose of any common or shared space is to serve only the associated cells that are not required to be accessible (e.g., shared bathrooms or living space serving a specific group of cells), then those common or shared spaces are not required to be accessible.

1103.2.14 Walk-in coolers and freezers. Walk-in coolers and freezers intended for employee use only are not required to comply with this chapter.

❖ Walk-in coolers and freezers usually have features that make accessibility difficult. For thermal efficiency they may have raised floors, special door seals, unconventional door-operating hardware, tight internal storage, wheeled racking systems, etc. For these and other reasons they are not required to have an accessible entry or to be accessible within. Walk-in coolers and freezers may also be considered exempted under Sections 1103.2.2 and 1104.3.1. Coolers that are accessed from the front by customers would need to comply with operable parts and display criteria on the public side (see Sections 1109.9.2 and 1109.13).

SECTION 1104 ACCESSIBLE ROUTE

1104.1 Site arrival points. At least one *accessible route* within the *site* shall be provided from public transportation stops, *accessible* parking, *accessible* passenger loading zones, and public streets or sidewalks to the *accessible* building entrance served.

Exception: Other than in buildings or *facilities* containing or serving *Type B units*, an *accessible route* shall not be required between *site* arrival points and the building or *facility* entrance if the only means of access between them is a vehicular way not providing for pedestrian access.

❖ The intent of this section is to require an accessible route from the point at which one enters the site to any buildings or facilities that are required to be accessible on that site. It is presumed that people with disabilities are capable of gaining access to the site from such locations as accessible parking, public transportation stops, loading zones, public streets or sidewalks. Due to possible site limiting factors, at least one from each arrival point, but not every route available, is required to be an accessible route. When not every route is accessible, good signage, while not required at site arrival points, would aid persons with mobility impairments in finding the routes without backtracking.

The exception addresses vehicular routes that provide the only route between an arrival point and an accessible entrance. An accessible route for pedestrian access is not required except in buildings or structures having or serving Type B units. For example, if there is a bus stop at the front of an industrial complex, but the only route to the building entrance is via a long driveway, an accessible pedestrian route to that entrance from the bus stop is not required. For special considerations in residential developments, see the commentaries to Sections 1107.4, 1107.7.4, 1107.7.5, and 1110.4.1.

1104.2 Within a site. At least one *accessible route* shall connect *accessible* buildings, *accessible* facilities, *accessible* elements and *accessible* spaces that are on the same *site*.

Exceptions:

1. An *accessible route* is not required between *accessible* buildings, *accessible* facilities, *accessible* elements and *accessible* spaces that have, as the only means of access between them, a vehicular way not providing for pedestrian access.
2. An *accessible route* to recreational facilities shall only be required to the extent specified in Section 1110.

❖ Developments may include several buildings on the same site. The intent of this section is to require an accessible route to all facilities offered on a site. Often sites are designed such that the only way to reach a building or facility is by automobile. If there are multiple, separated parking areas serving one or more buildings on a site, an accessible route is required between the parking facilities and the buildings they serve. If there is an exterior feature, such as a swimming pool, located on a site containing multiple buildings, an accessible route is required from each building to the swimming pool. For residential facilities, also see Section 1107.3.

Exception 1 clarifies that an accessible route is not required where no pedestrian access (i.e., sidewalk) is otherwise intended or provided to a particular building or feature on the site. For example, on a multi-building site with a road between and no sidewalks, it is not the intent to require a sidewalk be built strictly for accessible route requirements. If it is expected that everyone will drive between buildings, then accessible parking at each building is sufficient.

Per Exception 2, for special considerations for the route to recreational facilities, see the commentary to Section 1110.

1104.3 Connected spaces. When a building or portion of a building is required to be *accessible*, at least one *accessible route* shall be provided to each portion of the building, to *accessible* building entrances connecting *accessible* pedestrian walkways and to the *public way*.

Exceptions:

1. *Stories* and *mezzanines* exempted by Section 1104.4.
2. In a building, room or space used for assembly purposes with *fixed seating*, an *accessible route* shall not be required to serve levels where *wheelchair spaces* are not provided.
3. Vertical access to elevated employee work stations within a courtroom complying with Section 1108.4.1.4.
4. An *accessible route* to recreational facilities shall only be required to the extent specified in Section 1110.

❖ This section requires that there be at least one route from an accessible entrance to all required accessible features within a building. If an area is addressed with a specific exception, then an accessible route is not required. For example, mechanical penthouses are exempt under Section 1103.2.9; therefore, an accessible route would not be required to this area.

Once someone gets into a space, they must also

be able to evacuate in an emergency situation. This may or may not be via the same route. For accessible means of egress requirements, see Section 1009.

Exception 1 correlates with Section 1104.4 so that it is clear that the requirements for connecting spaces within a building are not intended to override the exception for an accessible route between stories and mezzanines.

Exception 2 reemphasizes that levels in assembly seating that do not contain wheelchair seating locations are not required to be accessed by an accessible route. However, the types of services available in the facility must be considered when determining what services must be available to accessible seats. A route is required to services from accessible seats in accordance with Section 1108.2.1.

Exception 3 is a pointer to the more specific criteria for courtroom work stations in Section 1108.4.1.4.

Exception 4 deals with recreational facilities that may be within a building. For example, the same accessible route requirements apply for an ice rink, regardless of whether it was located inside or outside. See Section 1110 for information specific to recreational facilities.

1104.3.1 Employee work areas. *Common use circulation paths* within *employee work areas* shall be *accessible routes.*

Exceptions:

1. *Common use circulation paths*, located within *employee work areas* that are less than 1,000 square feet (93 m^2) in size and defined by permanently installed partitions, counters, casework or furnishings, shall not be required to be *accessible routes*.
2. *Common use circulation paths*, located within *employee work areas*, that are an integral component of equipment, shall not be required to be *accessible routes*.
3. *Common use circulation paths*, located within exterior *employee work areas* that are fully exposed to the weather, shall not be required to be *accessible routes*.

❖ This requirement for common use circulation paths within employee work areas is consistent with the exception in Section 1103.2.2. An accessible route is required to each employee work area. When employees share work areas, an accessible route must be available throughout that area. Note that the accessible route minimum width of 36 inches (914 mm) clear is consistent with the minimum means of egress pathways.

Exception 1 addresses the accessible route within small employee work areas. Shared work areas that are less than 1,000 square feet (93 m^2) and confined by walls, partitions, permanently installed equipment, cabinets and counters are not required to have an accessible route through that particular area. The square footage limitation does not always pertain to a room, but also to part of a room. The intent was to allow such areas as the last two work stations down an aisle, two or three work stations in a small office, the employee side of a beverage bar, portions of commercial kitchens, etc. An accessible route is required to these areas, but not necessarily throughout the area. There can be multiple exceptions within the same area, such as in a commercial kitchen. Again, modifications would be performed at a later date based on employee needs.

Exception 2 permits nonaccessible areas around and through pieces of equipment. An example would be the shared work areas around a piece of assembly equipment in a factory where one or more persons are required to monitor and operate the machine and incoming or outgoing product.

Exception 3 is an exception for outdoor work areas. This would be applicable to landscapers, sewer workers, gravel pit crews, etc.

1104.3.2 Press boxes. Press boxes in a building, room or space used for assembly purposes shall be on an *accessible route*.

Exceptions:

1. An *accessible route* shall not be required to press boxes in *bleachers* that have a single point of entry from the *bleachers*, provided that the aggregate area of all press boxes for each playing field is not more than 500 square feet (46 m^2).
2. An *accessible route* shall not be required to free-standing press boxes that are more than 12 feet (3660 mm) above grade provided that the aggregate area of all press boxes for each playing field is not more than 500 square feet (46 m^2).

❖ Press boxes are required to be served by an accessible route. If the occupant load is five or less, this could be provided by a platform lift (see Section 1109.8, Item 3).

Exception 1 is mainly applicable to press boxes located at the back of the bleacher seating in an outdoor sports facility. The "single point of entry" refers to access to the press box directly from the bleacher seating (see Commentary Figure 1104.3.2).

Exception 2 is mainly applicable to the free-standing press box. If that press box is either less than 12 feet (3677 mm) above the ground or more than 500 square feet (46 m^2) in area, it must be served by an accessible route. Historically, the 12-foot (3677 mm) height limit was chosen because that was the maximum rise for a platform lift in previous editions of ASME A18.1.

For both exceptions, if one press box is provided, the total area of the press box must be less than 500 square feet (46 m^2). If more than one press box overlooks the same playing field, the aggregate area of both press boxes must be less than 500 square feet (46 m^2). The aggregate area is not intended to be applicable to press boxes that happen to be on the same multiple-facility site. For example, if a high school has a football field with two press boxes, the area of those two press boxes must be added together. However, if the same high school also has a press box for the baseball field, the area of the press box in the baseball field would not be included with the area of the press boxes in the football field.

Figure 1104.3.2
EXAMPLE OF PRESS BOX ACCESSED FROM BLEACHER SEATING

1104.4 Multistory buildings and facilities. At least one *accessible route* shall connect each *accessible story* and *mezzanine* in multilevel buildings and *facilities*.

Exceptions:

1. An *accessible route* is not required to *stories* and *mezzanines* that have an aggregate area of not more than 3,000 square feet (278.7 m^2) and are located above and below *accessible* levels. This exception shall not apply to:
   1.1. Multiple tenant facilities of Group M occupancies containing five or more tenant spaces used for the sales or rental of goods and where at least one such tenant space is located on a floor level above or below the *accessible* levels;
   1.2. *Stories* or *mezzanines* containing offices of health care providers (Group B or I);
   1.3. Passenger transportation facilities and airports (Group A-3 or B); or
   1.4. Government buildings.
2. *Stories* or *mezzanines* that do not contain *accessible* elements or other spaces as determined by Section 1107 or 1108 are not required to be served by an *accessible route* from an *accessible* level.
3. In air traffic control towers, an *accessible route* is not required to serve the cab and the floor immediately below the cab.
4. Where a two-story building or facility has one *story* or *mezzanine* with an *occupant load* of five or fewer persons that does not contain *public use* space, that *story* or *mezzanine* shall not be required to be connected by an *accessible route* to the *story* above or below.

❖ At least one accessible route is required between stories or mezzanines in a facility. Other changes in level within a story are addressed in Section 1104.3. This requirement does not mandate an elevator. The accessible route between levels can be via ramps, platform lifts (where permitted), limited use/limited access (LULA) elevators, passenger elevators, etc. The intent of the exceptions is to allow limited areas to be inaccessible without restricting access to services available to the general public.

Exception 1 addresses conditions under which it may not be practical or economical to provide an accessible route in multilevel buildings. The primary economic consideration is that, in the vast majority of circumstances, the means of providing an accessible route to floor levels above or below the entrance level of the building will be by an elevator. This exception applies to levels above and below the entry level that have an aggregate area of 3,000 square feet (279 m^2) or less. The 3,000 square feet does not include the area of the accessible level. For example, if a building had a second floor that is 2,000 square feet (186 m^2) and a basement that is 2,000 square feet (186 m^2), since the aggregate area of the basement and second floor is 4,000 square feet (372 m^2), at least one of the two floor areas would be required to be connected to the entrance level by an accessible route. There are certain facilities that, despite being relatively small in size, are not exempt from the requirement for an accessible route. Because of the critical nature of the services provided, offices of health care providers, passenger transportation facilities and airports are not included in this exception. Multitenant mercantile occupancies (i.e., five or more tenants in the facility), which include facilities such as shopping malls, are not included in the exception because individual retail stores have limited types of commodities (i.e., shoes, eyeglasses, sporting goods, etc.), and the exception would mean that people with disabilities would not have access to the same range of goods that are available to the general public. In these cases, access must be provided such that people with disabilities may patronize all the establishments within the facility. The 3,000-square-foot (279 m^2) exception can be applied to a mercantile facility with fewer than five tenants, or where all tenants have space on an accessible ground level.

A mezzanine is required by Section 505.2 to have a clear height below of at least 7 feet (2134 mm). Therefore, Exception 1 cannot be applied to platforms or raised or lowered floor areas.

Government buildings, such as state or city office buildings, police stations, fire stations or schools, cannot use the 3,000-square-foot exception. This is intended to be consistent with what the ADA would refer to as a Title II building.

Exception 2 is a direct reference to the specific items addressed in Sections 1107 and 1108. It is not a general exception for Groups A, I, R and S. The purpose of Exception 2 is to address areas within specific occupancies where a single element would be repeated multiple times on a level. Where Sections 1107 and 1108 would allow construction of a floor level with no accessible features, this section is not intended to trigger an accessible route to that otherwise inaccessible level. For example, an accessible route would not be required to upper levels in a hotel with all public spaces and all Accessible sleeping units (see Section 1197.6.1.1) located on the ground floor. Another example: Section 1108.2 requires wheelchair spaces on different levels, and access to all services. ICC A117.1 requires dispersion of the wheelchair space locations. Once those criteria are met, the tiers that do not contain wheelchair space locations or the associated services are not required to be served by an accessible route. This exception does not have a square footage limitation.

Exception 3 is specific to the unique situations found in air traffic control towers. Because air traffic control towers are required to have a 360-degree line of sight for air traffic safety, an elevator or the elevator penthouse cannot be located so that it could block the line of sight and create a safety issue.

Exception 4 permits small nonpublic second floors or mezzanines to not be served by an accessible route. An example would be the second floor in a doctor's office that is used only for storage. The occupant load table in Section 1004.1.1 would limit this storage area to 1,500 square feet (139 m^2) [i.e., 300 square feet (28 m^2) per occupant x 5 occupants maximum = 1,500 square feet (28 m^2)]. If the second floor also contained a mechanical room, that area would be exempt under Section 1103.2.9. However, if the doctor chose later to expand exam rooms into this second floor, he or she would have to provide an accessible route. Another example would be a second level in an airport that included only operational offices. In this case, the area would be limited by the occupant load table in Section 1004.1.1 to 500 square feet (46 m^2) [i.e., 100 square feet (9 m^2) per occupant x 5 occupants maximum = 500 square feet (46 m^2)]. While the exception is specifically stated as applicable only to second-story buildings or mezzanines, it could be interpreted to apply to a one-story building with a basement level.

1104.5 Location. *Accessible routes* shall coincide with or be located in the same area as a general *circulation path*. Where the *circulation path* is interior, the *accessible route* shall also be interior. Where only one *accessible route* is provided, the *accessible route* shall not pass through kitchens, storage rooms, restrooms, closets or similar spaces.

Exceptions:

1. *Accessible routes* from parking garages contained within and serving *Type B units* are not required to be interior.
2. A single *accessible route* is permitted to pass through a kitchen or storage room in an *Accessible unit*, *Type A unit* or *Type B unit*.

❖ One of the objectives of accessibility requirements is to normalize, to the extent possible, the facilities provided for disabled persons and those comparable facilities that exist for people without disabilities. In addition, the intent of this section is to avoid the circumstance where an interior path between facilities is provided but the only accessible route between those same facilities is an exterior path.

When only one accessible route is provided, this section puts limits on the types of spaces through which that accessible route may pass so that it is readily available. Spaces, such as storage rooms, restrooms, closets and kitchens, can be subject to being locked or their availability may be restricted to authorized personnel; therefore, they would not serve as reliable accessible routes for all building occupants.

Exception 1 addresses individual dwelling units that include their own garages and residential facilities with shared garage levels. This exception is limited to accessible parking for Type B units. It is intended to allow a person to exit the garage and enter through the front door, instead of requiring access from the garage into the unit directly. Section 1106.2 requires that when parking is provided within a building, accessible parking must also be located within the building. If this exception is utilized in accordance with Sections 1104.1 and 1107.4, an accessible route must be provided from the accessible parking spaces to the front door.

Exception 2 allows a single accessible route to pass through a kitchen or storage room within a unit that is required to be an Accessible, Type A or Type B dwelling or sleeping unit. Given the space limitations and typical arrangements of dwelling units, it is not unreasonable to allow access, for example, through a kitchen, to reach an eating area or an exterior patio as the only means of access.

1104.6 Security barriers. Security barriers including, but not limited to, security bollards and security check points shall not obstruct a required *accessible route* or accessible *means of egress*.

Exception: Where security barriers incorporate elements that cannot comply with these requirements, such as certain metal detectors, fluoroscopes or other similar devices, the *accessible route* shall be permitted to be provided adjacent to security screening devices. The *accessible route* shall permit persons with disabilities passing around security barriers to maintain visual contact with their personal items to the same extent provided others passing through the security barrier.

❖ This requirement provides guidance for when a security feature is required along an accessible route. An example would be the security checkpoints in airports. The intent is that the route for a person with mobility impairments should move through security checkpoints as close to the typical route as possible. It is recognized that some people using wheelchairs or with braces could not move through standard security barriers, such as metal detectors.

SECTION 1105 ACCESSIBLE ENTRANCES

1105.1 Public entrances. In addition to *accessible* entrances required by Sections 1105.1.1 through 1105.1.7, at least 60 percent of all *public entrances* shall be *accessible*.

Exceptions:

1. An *accessible* entrance is not required to areas not required to be *accessible*.
2. Loading and *service entrances* that are not the only entrance to a tenant space.

❖ A facility is not accessible if the entrances into it are inaccessible. This section establishes reasonable criteria for providing accessible entrances. A facility is not required to have all of its entrances accessible in order to provide reasonable accommodation to disabled persons. If a facility has multiple public entrances, it is not considered unreasonable, as a minimum, to require at least 60 percent of the entrances to be accessible. In addition to the 60-percent overall requirements, entrances that have a specific function or provide access to only certain portions of the facility must be addressed (see commentary, Sections 1105.1.1 through 1105.1.7). Sections 1111.1 and 1111.2 require signage at all entrances if all entrances are not accessible. "Public entrance," "Restricted entrance" and "Service entrance" are defined in Chapter 2.

Doors that are only for means of egress are not considered when determining the number of entrances required to be accessible. However, these doors may need to be on an accessible route for accessible means of egress requirements (see Section 1009). Depending on the arrangement, sometimes the entrance requirements are more restrictive, and sometimes the means of egress requirements are more restrictive. For example, take a small tenant space with two exterior doors—the front door is the public entrance and the back door is the service entrance. Section 1105.1 along with Exception 2 would require only the front door to be on an accessible route for ingress into the building. However, when a space has two means of egress required, Section 1009.1 would require both doors to be on an accessible route for means of egress purposes.

Exception 1 is self-evident in that if a facility or portion of a facility is not required to be accessible, then the entrances to such facilities or spaces are not required to be accessible. An example would be an exterior entrance to the sprinkler room (see Section 1103.2.9).

Exception 2 exempts loading and service entrances from the accessibility requirement on the basis that such entrances are unlikely to be used on a regular basis and may be raised to allow truck unloading. Loading and service entrances are required to be accessible if they are the only means of access into a facility or tenant space. This is consistent with Section 1105.1.5.

The intent of the reference to Sections 1105.1.1 through 1105.1.7 is to provide reasonable and convenient availability of an accessible entrance from the accessible facilities provided on the site. All entrances that serve distinct arrival points on the site are required to be accessible. For example, a public entrance from a public parking structure must be accessible (see Section 1105.1.1), as well as the employees-only entrance adjacent to separate employee parking (see Section 1105.1.3). This should deter the formerly common practice of designating an entrance as "the handicapped entrance" (which is inappropriate) without regard to its location relative to such arrival points as accessible parking facilities, transportation facilities, passenger loading zones, taxi stands, public streets or sidewalks and tunnels or elevated walkways. In conjunction with Section 1104.5, this section is also intended to provide reasonable and convenient availability of an accessible entrance to the building's accessible vertical movement elements, such as the elevators. This would allow a person with disabilities to move readily throughout all levels of the building.

1105.1.1 Parking garage entrances. Where provided, direct access for pedestrians from parking structures to buildings or facility entrances shall be *accessible*.

❖ Occasionally, a parking garage is attached to an office building or mall. An accessible route must be provided from the accessible parking spaces in that parking garage to an accessible entrance into the office or mall portion of the building.

1105.1.2 Entrances from tunnels or elevated walkways. Where direct access is provided for pedestrians from a pedestrian tunnel or elevated walkway to a building or facility, at least one entrance to the building or facility from each tunnel or walkway shall be *accessible*.

❖ Elevated walkways are sometimes provided between adjacent buildings for easy access or protection from the weather for people who must commonly move between certain buildings. Tunnels may serve the same purpose. Washington, DC, has tunnels between many of its congressional buildings. Chicago has a "pedway" system connecting the commuter train stations with the basement level of many buildings in the downtown area. Minneapolis deals with the extreme winter weather conditions in that city by having a system of elevated walkways that connect many of the downtown buildings (see Commentary Figure 1105.1.2). When these types of walkways or tunnels are provided, at least one of the building entrances off of these walkways or tunnels must be an accessible entrance.

Figure 1105.1.2
EXAMPLE OF ELEVATED WALKWAY ENTRANCE

1105.1.3 Restricted entrances. Where *restricted entrances* are provided to a building or facility, at least one *restricted entrance* to the building or facility shall be *accessible*.

❖ Where access into a specific entrance is controlled (see the definition for "Restricted entrance" in Chapter 2), at least one of that type of entrance must be accessible. An entrance that is locked is not necessarily a restricted entrance. For example, an apartment building may have multiple entrances for tenants where they enter with their key; however, there may be only one entrance for visitors if they need to access a tenant intercom system to gain entrance. The one visitor entrance is considered to be the restricted entrance and is required to be accessible.

1105.1.4 Entrances for inmates or detainees. Where entrances used only by inmates or detainees and security personnel are provided at judicial facilities, detention facilities or correctional facilities, at least one such entrance shall be *accessible*.

❖ Where access to a specific entrance is limited for security reasons to inmates or detainees and the staff supervising them, that entrance must be accessible. This type of situation could occur at courthouses, jails and police stations (see also Sections 1103.2.13, 1107.5.5 and 1108.4.2).

1105.1.5 Service entrances. If a *service entrance* is the only entrance to a building or a tenant space in a facility, that entrance shall be *accessible*.

❖ Loading and service entrances are required to be accessible if they are the only means of access into a facility or tenant space. Typically, loading and service entrances are exempted from the accessibility requirement on the basis that such entrances are unlikely to be used on a regular basis and may be raised to allow for truck unloading (see the definition for "Service entrance" in Chapter 2). This is consistent with Section 1105.1, Exception 2.

1105.1.6 Tenant spaces. At least one *accessible* entrance shall be provided to each tenant in a facility.

Exception: An *accessible* entrance is not required to self-service storage facilities that are not required to be *accessible*.

❖ Each tenant space must have at least one accessible entrance. For multiple-tenant buildings with shared entrances, the 60-percent entrance requirement in Section 1105.1 is also applicable. Per the exception, in a self-storage facility, the specific criteria in Section 1108.3 would allow for some of the self-storage facilities to not have an accessible entrance.

1105.1.7 Dwelling units and sleeping units. At least one *accessible* entrance shall be provided to each *dwelling unit* and *sleeping unit* in a facility.

Exception: An *accessible* entrance is not required to *dwelling units* and *sleeping units* that are not required to be *Accessible units*, *Type A units* or *Type B units*.

❖ Dwelling and sleeping units that are Accessible, Type A or Type B units must have at least one accessible entrance. If a dwelling unit or sleeping unit does not have accessibility requirements, then an accessible entrance is not required to that unit. For multiunit buildings with shared entrances, the 60-percent entrance requirement in Section 1105.1 for the building as a whole would also be applicable.

SECTION 1106
PARKING AND PASSENGER LOADING FACILITIES

1106.1 Required. Where parking is provided, *accessible* parking spaces shall be provided in compliance with Table 1106.1, except as required by Sections 1106.2 through 1106.4. Where more than one parking facility is provided on a *site*, the number of parking spaces required to be *accessible* shall be calculated separately for each parking facility.

Exception: This section does not apply to parking spaces used exclusively for buses, trucks, other delivery vehicles, law enforcement vehicles or vehicular impound and motor pools where lots accessed by the public are provided with an *accessible* passenger loading zone.

❖ Accessible parking facilities are an important component of building and site accessibility. This section addresses the number of parking spaces required. The location of such parking spaces in relation to accessible building entrances is addressed in Section 1106.6. Parking facility requirements are primarily intended to facilitate accessibility for people with mobility impairments, not just persons who use wheelchairs.

This section is not intended to require that parking facilities be provided. Rather, when parking is provided, the minimum number of accessible parking spaces specified herein must be provided. ICC A117.1 contains detailed requirements on the size and configuration of accessible parking spaces, including the required access aisle adjacent to the space, as well as the dimensional requirements for van-accessible parking spaces required by Section 1106.5.

Multiple-tenant facilities, such as shopping malls, or large facilities, such as hospitals or assembly occupancies, may have multiple parking lots or parking garages. If multiple parking lots, or a combination of parking garages and parking lots are provided for a facility, the number of spaces in each separate parking facility is utilized to determine the number of accessible spaces required. Accessible car and van parking would then be provided in each parking facility accordingly, except as otherwise permitted by Section 1106.6. What constitutes a multiple facility is somewhat subjective. Putting a line of trees or landscaping along the side of the parking spaces does not always constitute a separate facility. From a practical viewpoint, the separate facilities would serve specific groups of people (e.g., employee-only parking), provide different types of parking (e.g., surface lots, garage parking, covered parking) or serve distinct areas of the building (e.g., upper and lower levels as you move around a large mall).

In certain types of lots, it is not logical to require accessible parking spaces. The exception helps to clarify these situations. Parking lots that contain only parking for buses, trucks and delivery vehicles would not need accessible parking spaces. Lots where the parked cars are limited to vehicles used by law enforcement, impound lots or motor pools are also not required to provide accessible spaces. An accessible passenger dropoff is required to be provided on the site, so that when a person with mobility impairments comes to that site, he or she has a place to exit the arrival vehicle safely and enter a vehicle coming from the lot.

TABLE 1106.1. See next column.

❖ The required number of accessible parking spaces is based on the accessible parking requirements of the 2010 ADA Standard. It does not reflect the demographic statistics on wheelchair usage that were used to scope other requirements in Chapter 11, because the majority of disabled parking permit and license plate holders in most states are ambulatory, mobility-impaired persons. The required ratios are intended to be responsive to the anticipated demand for all facilities, such that accessible parking spaces will be reasonably available on demand. Section 1111.1 states that signage is not required on the one required accessible parking space when the total number of parking spaces provided is four or less. This could be burdensome for the building tenant in that the accessible parking space, which is restricted for use only by authorized vehicles, could constitute anywhere from 25 to 100 percent of the available parking. This may unduly restrict the availability of parking for all other vehicles and patrons of the facility. While not reserved by signage, the space must still be sized in accordance with a van-accessible space.

1106.2 Groups I-1, R-1, R-2, R-3 and R-4. *Accessible* parking spaces shall be provided in Group I-1, R-1, R-2, R-3 and R-4 occupancies in accordance with Items 1 through 4 as applicable.

1. In Group R-2, R-3 and R-4 occupancies that are required to have *Accessible*, *Type A* or *Type B dwelling units* or *sleeping units*, at least 2 percent, but not less than one, of each type of parking space provided shall be *accessible*.
2. In Group I-1 and R-1 occupancies, *accessible* parking shall be provided in accordance with Table 1106.1.
3. Where at least one parking space is provided for each *dwelling unit* or *sleeping unit*, at least one *accessible* parking space shall be provided for each *Accessible* and *Type A unit*.
4. Where parking is provided within or beneath a building, *accessible* parking spaces shall also be provided within or beneath the building.

❖ This section provides a separate criterion for the required number of accessible parking spaces for occupancies in Groups I-1, R-1, R-2, R-3 and R-4 that include Accessible, Type A or Type B units.

The 2-percent requirement in Item 1 for R-2, R-3 and R-4 is based on HUD's FHAG. Section 1107.7 identifies buildings where Type A and Type B units may not be required. For example, a townhouse development may not have any Type A or Type B dwelling units required, therefore, no accessible parking spaces are required. Designers should keep in mind that asking for accessible parking spaces is a common accommodation asked for by residents in both townhouse and apartment developments. While not required, it would be good design practice to exceed code and at least have space on the parking lot to add accessible parking when requested.

TABLE 1106.1
ACCESSIBLE PARKING SPACES

TOTAL PARKING SPACES PROVIDED IN PARKING FACILITIES	REQUIRED MINIMUM NUMBER OF ACCESSIBLE SPACES
1 to 25	1
26 to 50	2
51 to 75	3
76 to 100	4
101 to 150	5
151 to 200	6
201 to 300	7
301 to 400	8
401 to 500	9
501 to 1,000	2% of total
1,001 and over	20, plus one for each 100, or fraction thereof, over 1,000

Per Item 2, assisted living facilities (Group I-1) and hotels and motels (Group R-1) should use Table 1106.1 to determine the number of accessible parking spaces required.

Due to the higher anticipated need, per Item 3, when a residential parking lot provides one or more spaces for each dwelling or sleeping unit, there should be accessible parking spaces for each Accessible or Type A unit in the facility, in addition to the 2 percent required by Items 1 or 2. For example, a 100-unit hotel has 100 parking spaces for the guests. Four Accessible guestrooms are required. Table 1106.1 would require four accessible spaces. Item 3 would require an additional four accessible spaces. Therefore the hotel will have to provide eight accessible parking spaces, two sized for a van.

Per Item 4, where parking is provided within or beneath a building, accessible parking spaces also are to be provided within or beneath the building. If a combination of surface and covered parking is provided, accessible parking may be provided in both locations. This is intended to establish consistency in the type and location of parking spaces available to all people. If parking is provided in individual private parking garages, 2 percent of the parking garages would have to contain accessible parking spaces (see the exception to Section 1106.5).

In a development, typically parking for dwelling units is considered on a site basis rather than a building by building basis. Accessible parking should be dispersed throughout the development so as to provide the best access possible. It is not the intent to require accessible parking spaces at the entrance to every building, or within every strip of parking garages. For example, it would not be logical to ask for a surface space and a garage space for each building in developments with multiple four-unit buildings. See Section 1106.5 for a discussion of the distribution of van-accessible spaces.

1106.3 Hospital outpatient facilities. At least 10 percent, but not less than one, of care recipient and visitor parking spaces provided to serve hospital outpatient facilities shall be *accessible*.

❖ Certain facilities can be expected to have a higher demand for accessible parking spaces than that reflected in Table 1106.1. Medical outpatient facilities are one such type of facility. This section requires that at least 10 percent of the parking provided for patients and visitors of these types of facilities be accessible. Parking provided for the employees of these facilities should comply with Table 1106.1. If a hospital contains multiple clinics or types of facilities, the amount of parking dedicated to the outpatient facility is interpretive. Options for determining the number of accessible spaces could be based on the area percentage of the outpatient facilities to other facilities, or the anticipated number of patients and visitors to these facilities in relation to other facilities.

1106.4 Rehabilitation facilities and outpatient physical therapy facilities. At least 20 percent, but not less than one, of the portion of care recipient and visitor parking spaces serving rehabilitation facilities specializing in treating conditions that affect mobility and outpatient physical therapy facilities shall be *accessible*.

❖ Medical facilities that specialize in treatment or other services for people with mobility impairments can be expected to have a higher demand for accessible parking spaces. In these cases, at least 20 percent of the parking spaces provided for visitors and care recipients are required to be accessible. Parking provided for the employees of these facilities should comply with Table 1106.1. If a hospital contains multiple types of facilities, the amount of parking dedicated to the rehabilitation and physical therapy facility is interpretive. Options for determining the number of accessible spaces could be based on the area percentage of the rehabilitation and physical therapy facilities to other facilities or the anticipated number of care recipients and visitors to these facilities in relation to other facilities.

1106.5 Van spaces. For every six or fraction of six *accessible* parking spaces, at least one shall be a van-accessible parking space.

Exception: In Group R-2 and R-3 occupancies, van-accessible spaces located within private garages shall be permitted to have vehicular routes, entrances, parking spaces and access aisles with a minimum vertical clearance of 7 feet (2134 mm).

❖ Vans that are specially equipped to accommodate a person using a wheelchair require wider parking spaces than the typical accessible parking space. This is because vans are usually equipped with a mechanized loading/unloading platform that extends outward from the passenger side of the vehicle. ICC A117.1 contains the dimensional requirements for van-accessible spaces. In order to accommodate the growing usage of these specially equipped vans, this section requires that one in every six accessible parking spaces, or a fraction thereof, be of a size to accommodate these vans. The requirement for van-accessible spaces is not intended to require a greater number of accessible spaces than required by Table 1106.1. For example, if Table 1106.1 requires six accessible parking spaces, one of those six must be van accessible. If seven accessible spaces are required by Table 1106.1, two of those seven are required to be van accessible. On the other hand, if only one accessible parking space is required by Table 1106.1, that space would be a van-accessible space.

The exception modifies the height criteria for van-accessible spaces in ICC A117.1 for private garages serving dwelling units. (See Section 406.3 for the criteria for what can be considered a private garage.) This is a reduction from the 98-inch (2489 mm) height requirement in ICC A117.1. Since many vans are now converted by lowering the floor rather than raising the roof, the 82-inch (2083 mm) height should accommodate most private passenger vans with lifts.

Note that the determination of the number of van-accessible spaces is not tied to separate facilities as the number of spaces is in Sections 1106.1 and 1106.2. However, distribution of van-accessible spaces throughout the accessible parking space in large facilities is good design practice. Van-accessible spaces can be used by cars with accessible placards. It is just common courtesy to leave the van-accessible space for vans when there are other accessible spaces available. However, the van space is not reserved for vans only; in small parking lots the van space may be the only accessible parking provided.

1106.6 Location. *Accessible* parking spaces shall be located on the shortest *accessible route* of travel from adjacent parking to an *accessible* building entrance. In parking facilities that do not serve a particular building, *accessible* parking spaces shall be located on the shortest route to an *accessible* pedestrian entrance to the parking facility. Where buildings have multiple *accessible* entrances with adjacent parking, *accessible* parking spaces shall be dispersed and located near the *accessible* entrances.

Exceptions:

1. In multilevel parking structures, van-accessible parking spaces are permitted on one level.
2. *Accessible* parking spaces shall be permitted to be located in different parking facilities if substantially equivalent or greater accessibility is provided in terms of distance from an *accessible* entrance or entrances, parking fee and user convenience.

❖ As previously stated, the majority of disabled parking permits and license plate holders in most states are ambulatory, mobility-impaired persons. Travel distance, as well as severe weather conditions encountered when traversing from the parking lot to the building entrance, are more difficult for persons with mobility impairments to deal with than the general population. The intent of this section is to locate the parking so that the people utilizing the accessible parking spaces have to travel a minimum distance to an accessible entrance (see Commentary Figure 1106.6). This requirement is stated in performance terms and requires a degree of subjective judgment on the part of both the designer and the building official in determining the appropriate location for accessible spaces that meets the intent of this section. If a facility has multiple accessible entrances, accessible parking spaces are to be dispersed consistent with the location of these entrances.

Exception 1 is intended to acknowledge a practical difficulty associated with multilevel parking structures. In many cases, a multilevel parking structure will serve accessible building entrances on more than one or all of its parking levels. While more common now for multiperson vans rather than private vehicles, vans equipped to transport persons using wheelchairs may be modified by raising the roof of the vehicle in order to provide greater interior headroom. Consequently, such vans require greater vertical clearances. Typical parking structure design may not easily accommodate the necessary vertical clearance for accessible vans due to their low floor-to-ceiling heights. It would be impractical and economically unjustified to require parking structures to be designed solely for the purpose of enabling van-accessible spaces to be located on upper levels. Accordingly, the exception allows the required van-accessible parking spaces to be located on only one level of a multilevel parking structure. The route to and from the space, as well as at the van space and associated access aisle, must meet the minimum clear height of 98 inches (2489 mm) as specified in Section 502.6 of ICC A117.1. This will usually be the entry level of the parking facility. While directional signage is not required, best practice would be to indicate the routes in and out for someone who has a vehicle where the headroom height may be an issue.

Exception 2 addresses sites where multiple parking facilities are provided to serve a single destination or facility. Since one or more of the parking lots or parking garages may be more attractive to users for various reasons, it is acceptable to locate the required accessible parking based on perceived user convenience and preferences, including distance, parking fees and amenities. For example, if a hospital has a remote parking garage, and a surface lot close an accessible entrance, it can be argued that the shorter route provides a higher level of access. Therefore, all or a portion of the accessible spaces could be located on the surface lot.

Figure 1106.6
EXAMPLE OF ACCESSIBLE PARKING SPACES

1106.7 Passenger loading zones. Passenger loading zones shall be *accessible*.

❖ This section does not require passenger loading zones to be provided; however, where provided, they must be accessible in accordance with Section 503 of ICC A117.1. For locations where passenger loading zones are required, see Sections 1106.7.1 through 1106.7.4.

1106.7.1 Continuous loading zones. Where passenger loading zones are provided, one passenger loading zone in every continuous 100 linear feet (30.4 m) maximum of loading zone space shall be *accessible*.

❖ The intent of this section is to address the continuous loading zones found at most larger airports as well as some other types of transportation facilities, such as commuter train "kiss-n-ride" drop-off areas. At least one 20-foot (6096 mm) section for every 100 feet (30 480 mm) of passenger loading zone must meet the accessibility provisions in Section 503 of ICC A117.1.

1106.7.2 Medical facilities. A passenger loading zone shall be provided at an *accessible* entrance to licensed medical and long-term care facilities where people receive physical or medical treatment or care and where the period of stay exceeds 24 hours.

❖ The requirement for accessible passenger loading zones is applicable to both medical and long-term care facilities. The most common examples are hospitals and nursing homes; however, this could also be applicable to some rehabilitation and assisted living facilities (typically Group I-2). Medical and long-term care facilities typically have a higher percentage of people with mobility impairments living in the facility or visiting on a regular basis. Because of their mobility impairments, some people will be driven by a friend or family member, and it is desirable and convenient to first drop them off at the entrance and then park, rather than first parking and then requiring travel from the parking lot to the facility's entrance. As such, this section requires that a passenger loading zone be provided at no less than one accessible entrance to the facility.

1106.7.3 Valet parking. A passenger loading zone shall be provided at valet parking services.

❖ Valet parking is typically the public access point to a facility. When valet parking is provided, the passenger loading zone must be accessible so that it may be utilized by persons requiring wheelchairs. This would include the access aisle and an accessible route from the access aisle to the front entrance. Note that providing valet parking services does not eliminate the need for accessible parking spaces where parking is provided. Some vehicles that are modified for persons with disabilities may not be drivable by a valet attendant. In addition, valet parking may only be provided during certain periods of time, such as during dinner hours at a restaurant.

1106.7.4 Mechanical access parking garages. Mechanical access parking garages shall provide at least one passenger loading zone at vehicle drop-off and vehicle pick-up areas.

❖ Mechanical parking garages allow a much higher density of vehicles in a given footprint because of the elimination of ramps and drive lanes. With respect to the customer interface, mechanical garages are similar to valet parking and, therefore, require similar scoping for accessibility.

SECTION 1107 DWELLING UNITS AND SLEEPING UNITS

❖ The development of this section was part of the coordination effort with the FHA and the FHAG. For further explanation, see the general comments for this chapter.

One of the phrases developed as part of this coordination was the defined phrase "Intended to be occupied as a residence." This defines a type of dwelling or sleeping unit that is a person's home or place of abode. The intent of this language is to clarify that, consistent with the FHA, all dwelling units or sleeping units that can or will be used as a place of abode—even for a short period of stay—must meet the requirements for Type B dwellings. There is a presumption in the IBC that if a building is nontransient (i.e., occupants stay more than 30 days), it is covered by the requirements in this section. In addition, in coordination with the FHA, occupancies that allow stays of fewer than 30 days may be required to meet Type B design and construction requirements.

Accordingly, this commentary will provide guidance for determining when short-term occupancies that might not typically be viewed as housing are covered by this section. Such short-term Group R occupancies may include residential hotels and motels; corporate housing; seasonal vacation units; timeshares; boarding houses; dormitories and migrant farmworker housing. Also included are some occupancies in Group I, such as nursing homes, assisted living facilities, hospices and homeless shelters.

The key factor in determining whether any of these occupancies are subject to Type B requirements is whether the occupant will have the right to return to the property and whether he or she would have anywhere else to which to return. If it is intended that an occupant will have a right to return to the property and will not have anywhere else to return, the unit is "intended to be occupied as a residence" and must meet the Type B requirements regardless of the length of stay. Thus, for example, homeless shelters where occupants have a right to return nightly must comply with Type B unit criteria even if the occupants stay only a few nights. Additionally, nursing homes in which a resident moves after vacating his or her primary residence are subject to Type B requirements.

If the occupants have a right to return to the property but also have another place to which to return, the unit may still be subject to Type B criteria. Additional factors must be considered to determine whether the property is a short-term dwelling or sleeping unit that must meet Type B criteria, or a transient property that is not required to comply with Type B criteria. These factors must be considered by owners, builders, developers, architects and other designers to determine whether a building must be designed and constructed in accordance with Type B unit requirements.

For additional discussion on the elements related to hotels, motels, corporate housing and seasonal vacation units and timeshares, see the commentary to Section 1107.6.1 (Group R-1). With respect to most other occupancies, the following factors are also relevant to determine whether the units are subject to Type B requirements:

1. Whether the property is to be marketed as short-term housing;
2. Whether the terms and length of occupancy will be established through a written agreement;
3. How payment will be calculated (e.g., on a daily, weekly, monthly or yearly basis); and
4. What types of amenities and services are offered with the occupancy.

For example, an assisted living facility that provides sleeping units and custodial services to its occupants, and bills them on a monthly basis, is subject to Type B design and construction requirements. In addition, housing for migrant farm workers that is provided in conjunction with the worker's employment (whether or not rent is paid) and contains amenities for cooking and sleeping would be subject to Type B criteria.

1107.1 General. In addition to the other requirements of this chapter, occupancies having *dwelling units* or *sleeping units* shall be provided with *accessible* features in accordance with this section.

❖ There are two basic types of facilities that this section covers: dwelling units and sleeping units. A "Dwelling unit" is defined in Chapter 2 as a single unit that contains permanent provisions for "living, sleeping, eating, cooking and sanitation." A "Sleeping unit" is defined in Chapter 2 as a room in which people sleep that can include some of the provisions found in a dwelling unit but not all. Occupancy of dwelling units or sleeping units can be transient or nontransient. Dwelling units are typically apartments, condominiums, detached homes or townhouses. Dwelling units can be located in hotels that offer cabins, suites or rooms with kitchen facilities. Bedrooms within dwelling units are not considered sleeping units.

A sleeping unit could be a typical hotel guestroom; a bedroom in a congregate residence, such as a dorm, sorority house, fraternity house, convent, monastery or boarding house; a nursing home room; or a jail cell.

1107.2 Design. *Dwelling units* and *sleeping units* that are required to be *Accessible units*, *Type A units* and *Type B units* shall comply with the applicable portions of Chapter 10 of ICC A117.1. Units required to be *Type A units* are permitted to be designed and constructed as *Accessible units*. Units required to be *Type B units* are permitted to be designed and constructed as *Accessible units* or as *Type A units*.

❖ There are three levels of accessibility that can be required in a dwelling unit or sleeping unit: Accessible units, Type A units and Type B units. All three types are defined in Chapter 2.

An Accessible unit is constructed for full accessibility in accordance with Section 1002 of ICC A117.1. For example, grab bars are in place in the bathrooms, a clear floor space is provided for front approach at the kitchen sink and bathroom lavatories, 32-inch (813 mm) clear width doors with maneuvering clearances and lever hardware are provided, etc. None of the elements in the unit are constructed for adaptability. The requirements for an Accessible unit provide a higher level of accessibility than either a Type A unit or a Type B unit.

A Type A unit has some elements that are constructed accessible [e.g., 32-inch (813 mm) clear width doors with maneuvering clearances and lever hardware] and some elements designed to be added or altered when needed (e.g., grab bars can be easily added in bathrooms since blocking in the walls is in place). Type A units follow the technical criteria in Section 1003 of ICC A117.1. This type of unit is less accessible than an Accessible unit and more accessible than a Type B unit.

The scoping or technical requirements for Type B units are consistent with the requirements for units required by the FHAG (see the definition for "Type B units" in Chapter 2). A Type B unit is constructed to a lower level of accessibility than either an Accessible unit or a Type A unit. While a person who uses a wheelchair could maneuver in a Type B unit, the technical requirements are geared more toward persons with lesser mobility impairments. Type B units follow the technical requirements in Section 1004 of ICC A117.1. Areas of a Type B unit are allowed to be totally nonaccessible (e.g., sunken living room, extra bedrooms on a mezzanine level). Side approach is permitted to sinks in the kitchen and lavatories in the bathroom rather than planning for a front approach. Some elements are constructed with a minimal level of accessibility [e.g., doors within the unit have a $31^3/_4$-inch (806 mm) clear width but do not require maneuvering clearances], while some elements are designed to be altered when needed (e.g., blocking in the walls of the bathroom for future installation of grab bars).

This section also takes into consideration the fact that Accessible unit requirements provide a higher level of accessibility than Type A requirements, and Type A requirements provide a higher level of accessibility than Type B requirements. Units are permitted to be constructed to a higher level of accessibility than required. In other words, Accessible units are constructed wheelchair ready. Type A units are constructed wheelchair friendly and Type B units are constructed wheelchair usable.

The technical criteria for each type of unit in ICC A117.1 is organized in the same order and section number for each element, making comparisons between types easier. For example, looking at the door provisions in Sections 1002.5, 1003.5 and 1004.5 of ICC A117.1 would clarify that the maneuvering clearances are required at all doors that are part of the accessible route through an Accessible unit or a Type A unit, but for the Type B unit, the maneuvering clearance is only required at the front door to the unit, not within the unit. Another example would be the requirements for operable parts: Sections 1002.9 and 1003.9 address requirements for plumbing fixture and appliance controls, while in accordance with Section 1004.9, Type B units do not require plumbing fixture and appliance controls to meet operable parts requirements.

1107.3 Accessible spaces. Rooms and spaces available to the general public or available for use by residents and serving *Accessible units*, *Type A units* or *Type B units* shall be *accessible*. *Accessible* spaces shall include toilet and bathing rooms, kitchen, living and dining areas and any exterior spaces, including patios, terraces and balconies.

Exceptions:

1. *Stories* and *mezzanines* exempted by Section 1107.4.
2. Recreational facilities in accordance with Section 1110.2.
3. Exterior decks, patios or balconies that are part of *Type B units* and have impervious surfaces, and that are not more than 4 inches (102 mm) below the finished floor level of the adjacent interior space of the unit.

❖ Spaces available for use by residents or the general public that are associated with Accessible units, Type A units or Type B units are required to be accessible. This would include spaces within the unit and outside the unit in the same building, as well as facilities that may be outside the building somewhere else on the site. The intent is that a person with a disability could take full advantage of the amenities associated with his or her dwelling or sleeping unit. While it is fairly clear which elements within a unit must be accessible, the associated facilities outside the unit vary depending on the type of building and facilities provided. Public use spaces within a building would include lobby areas while spaces for residents could include shared storage areas or refuse rooms. Shared spaces in nursing homes or assisted living facilities might be dining rooms, therapy rooms and recreational rooms. In dormitories, residents may share study rooms, laundry rooms or mail rooms and facilities located remotely, such as cafeterias. Examples of shared spaces in congregate residences could include community kitchens, living rooms and shared bathrooms. Examples in apartment complexes could include grouped mailboxes, rental offices, exercise rooms and exterior spaces, such as community buildings and swimming pools.

Exception 1 is a correlation with Section 1107.4, so that it is clear that the requirements for connecting spaces within a building are not intended to override the exception for an accessible route between stories and mezzanines.

Exception 2 allows for limited access to exterior recreational facilities when multiples of the same type of facility are provided. An example would be multiple tennis courts on the same site for use by residents in a complex (see Section 1110 for additional information). The net result is that a person with a disability, either a resident or guest, can access all Accessible, Type A and Type B units, as well as any associated facilities and public areas on the site.

Exception 3 is intended for application to Type B units with a private deck, patio or balcony within an individual unit, or a shared deck, patio or balcony in a "residents only" area of a congregate residence. Decks, patios or balconies located in areas such as the community building or swimming pool should be constructed accessible. When the outside walking surface is impervious to water, such as concrete, there may be a 4-inch (102 mm) step-down between the inside finished floor surface and the outside floor surface. The step-down is permitted to address the concern of water infiltration under the doors at such locations. With a maximum 4-inch (102 mm) difference in elevation, adaptation for accessibility is achievable with raised sleepers and decking. Normal door threshold criteria apply.

1107.4 Accessible route. At least one *accessible route* shall connect *accessible* building or facility entrances with the primary entrance of each *Accessible unit*, *Type A unit* and *Type B unit* within the building or facility and with those exterior and interior spaces and facilities that serve the units.

Exceptions:

1. If due to circumstances outside the control of the owner, either the slope of the finished ground level between *accessible* facilities and buildings exceeds one unit vertical in 12 units horizontal (1:12), or where physical barriers or legal restrictions prevent the installation of an *accessible route*, a vehicular route with parking that complies with Section 1106 at each *public* or *common use* facility or building is permitted in place of the *accessible route*.
2. In Group I-3 facilities, an *accessible route* is not required to connect *stories* or *mezzanines* where *Accessible units*, all *common use* areas serving *Accessible units* and all *public use* areas are on an *accessible* route.
3. In Group R-2 facilities with *Type A units* complying with Section 1107.6.2.2.1, an *accessible route* is not required to connect *stories* or *mezzanines* where *Type A units*, all *common use* areas serving *Type A*

units and all public use areas are on an *accessible route*.

4. In other than Group R-2 dormitory housing provided by places of education, in Group R-2 facilities with *Accessible units* complying with Section 1107.6.2.3.1, an *accessible route* is not required to connect *stories* or *mezzanines* where *Accessible units*, all *common use* areas serving *Accessible units* and all *public use* areas are on an *accessible* route.

5. In Group R-1, an *accessible route* is not required to connect *stories* or *mezzanines* within individual units, provided the *accessible* level meets the provisions for *Accessible units* and sleeping accommodations for two persons minimum and a toilet facility are provided on that level.

6. In congregate residences in Groups R-3 and R-4, an *accessible route* is not required to connect *stories* or *mezzanines* where *Accessible units* or *Type B units*, all *common use* areas serving *Accessible units* and *Type B units* and all *public use* areas serving *Accessible units* and *Type B units* are on an *accessible route*.

7. An *accessible route* between *stories* is not required where *Type B units* are exempted by Section 1107.7.

❖ Section 1107.3 indicates which spaces associated with accessible dwelling and sleeping units (i.e., Accessible, Type A and Type B units) are required to be accessible. Section 1105 indicates which building entrances are required to be accessible. The intent of this section is twofold: that there will be at least one accessible route that connects required accessible building entrances with the entrance of all accessible dwelling and sleeping units, and that there also must be accessible routes connecting accessible unit entrances with all interior and exterior common and public use spaces and facilities that serve the accessible dwelling or sleeping unit. Some additional facilities that serve units that are not really considered spaces would be accessible parking (see Section 1106.1), passenger loading zones (see Section 1106.7) and public transportation stops located either on site or adjacent to the perimeter. In determining the site arrival points that serve an accessible unit, the designer should make a reasonable determination of how pedestrians and vehicles would arrive at the site.

HUD's review of the 2003 and 2006 editions of the code found that the codes meet or exceed the seven design and construction requirements of the FHA and are "safe harbor" documents for complying with the FHAG. To read the final report of the HUD review, go to http://www.hud.gov/offices/fheo/disabilities/modelcodes/. HUD stated, "Having a more recent edition of the IBC recognized by HUD as a safe harbor will ensure that covered dwelling units will be built with the accessible features required by the Act." For additional information, see the commentary at the beginning of this chapter. At this writing, the 2009, 2012 and 2015 IBC and the 2009 ICC A117.1 are being reviewed by HUD to be considered as "safe harbor" documents. There is no set timeline for HUD to complete their review of these documents.

(*The following commentary concerning the accessible route serving dwelling units and sleeping units has been provided by HUD.*)

The five examples serve to illustrate a variety of conditions that may be encountered on nonlevel building sites where Exception 1 is applicable. These examples pertain to the accessible route between facilities on a site. In addition to the on-site accessible routes, the accessible route(s) between the site arrival point(s) and facilities on the site must be considered. Note that the exception in Section 1104.1 does not apply to facilities containing or serving Type B units.

The intent of this section is to ensure that there will be at least one accessible route that connects all accessible building and facility entrances with the entrance of all Accessible, Type A and Type B units. To qualify as an accessible route, that route must serve pedestrians (e.g., sidewalk or other walkway). People with disabilities who need the features of an Accessible, Type A or Type B dwelling or sleeping unit cannot use them if accessible routes are not provided from the entrances to buildings or facilities to the primary entrance of their dwelling or sleeping unit. There also must be accessible routes connecting accessible building or facility entrances with all interior and exterior spaces and facilities that serve such dwelling or sleeping units. For example, if a development has a recreational facility, such as a community center, persons with disabilities who need the features of an Accessible, Type A or Type B unit need an accessible route from their dwelling unit to that community center.

Exception 1 is intended to provide consistency with the federal FHA, which recognizes that, in very rare circumstances, an accessible pedestrian route between an accessible entrance to an accessible unit or an accessible entrance to a building containing accessible units and an exterior public use or common use facility may be impractical because of factors outside the control of the owner. Section 1107.4 requires an accessible pedestrian route between covered dwelling units and public use or common use areas and facilities that are required to be accessible, except in rare circumstances that are outside the control of the owner where extreme terrain or impractical site characteristics result in a finished grade exceeding 8.33 percent or physical barriers or legal restrictions prevent the installation of an accessible pedestrian route. In these cases, Exception 1 allows access to be provided by means of a vehicular route leading from the accessible parking serving the accessible unit to the accessible parking serving the public use or common use facility. Accessible parking complying with Section 1106 must be provided in each parking area. If a building containing accessible units also contains accessible features that are required by other code provisions or federal, state or local laws, then Exception 1 may not apply at all.

It is important to understand that compliance with the accessible design and construction requirements of the FHA is a legal obligation applicable to all architects, engineers, builders, developers and others involved in the design and construction of housing. HUD's regulations implementing the FHA make it clear that the burden of showing the applicability of exceptions is the responsibility of those individuals and entities involved in the design and construction of such housing. In order to ensure compliance with FHA, architects, engineers, developers, builders and others who use the code must make accessibility a priority at the planning and design phase of Group I and R developments, including the siting of housing and public use or common use areas. To do this, at the initial stage of site planning and design for all sites, before considering whether Exception 1 applies, persons and entities involved in the design of covered residential occupancies must determine whether and how the exceptions of Sections 1107.7.4 and 1107.7.5 apply.

After careful site planning and design has been completed, the following factors may then be considered to determine whether it is outside the control of the owner to provide an accessible pedestrian route between a building/Type B dwelling unit entrance and a given public use or common use facility. Each such route must be analyzed individually. Exception 1 will only apply when at least one of the following factors is present:

1. Legal restrictions outside the control of the owner. These include setback requirements, tree-save ordinances, easements, environmental restrictions and other limitations that prevent installation of an accessible pedestrian route without violating the law.
2. Physical barriers outside the control of the owner. These include physical characteristics of the site, which are outside the control of the owner that prevent the installation of an accessible pedestrian route.
3. On sites that qualify for the exceptions of Sections 1107.7.4 and 1107.7.5, the presence of extreme terrain or other unusual site characteristics (e.g., flood plain, wetlands) outside the control of the owner that would require substantial additional grading to achieve a slope that will allow for an accessible pedestrian route.

In considering whether the additional grading is substantial enough to qualify for Exception 1, the extent to which the builder has elected to grade the site for other purposes unassociated with accessibility must be considered. If grading for those other purposes is extensive, then substantial additional grading would be necessary to provide the required accessible pedestrian route. If grading for other purposes is not extensive, and substantial additional grading is necessary to provide an accessible pedestrian route, then reliance on Exception 1 would be appropriate. Note that when determining whether the additional grading is substantial, one may not consider the grading that the builder must perform to provide accessible pedestrian routes from site arrival points to the accessible entrances of accessible dwelling or sleeping units. If none of the factors above are present, Exception 1 does not apply. If one or more of these factors is present, then the next step in determining whether Exception 1 applies (i.e., the vehicular route is the only feasible option) is to consider alternative locations and designs for buildings, facilities and accessible pedestrian routes connecting each accessible building and accessible dwelling unit entrance and each public use or common use area required to be accessible to ensure that there is no other way to provide the required accessible pedestrian routes. It is important to recognize that if a road sloping 8.33 percent or less can be provided, then an accessible pedestrian route would also be feasible and must be provided.

Following are some examples to illustrate the proper application of Exception 1:

Example 1: An undisturbed site has slopes of 8.33 percent or less between required accessible entrances to Type B dwelling units and public use or common use areas and there are no legal restrictions or other unique characteristics preventing the construction of accessible routes. For aesthetic reasons, the developer would like to create some hills or decorative berms on the site. Because there are no extreme site conditions (severe terrain or unusual site characteristics such as flood plains), and no legal barriers that prevent installation of an accessible pedestrian route between the buildings/Type B dwelling units and any planned public use or common use facilities, the developer will still be obligated to provide accessible pedestrian routes. Exception 1 to Section 1107.4 is inapplicable in this circumstance.

Example 2: A developer plans to construct several buildings with Type B units clustered in a level area of a site that has some slopes of 10 percent. A swimming pool and tennis court will be added on the two opposing sides of the site. The builder plans grading that will result in a finished grade exceeding a slope of 8.33 percent along the route between the Type B units and the swimming pool and tennis court. There are no physical barriers or legal restrictions outside the control of the owner or builder that prevent him or her from reducing the existing grade to provide an accessible pedestrian route between the Type B units and the pool and tennis court. Therefore, the builder's building plan would not be approved under the code because it is within the owner's control to have the final grading fall below 8.33 percent and meet the slope and other requirements for an accessible pedestrian route. Accessible pedestrian routes between the Type B units, pool and tennis court must be provided.

Example 3: A multiple-family housing complex is built on two sections of a large piece of property, which is divided by a wide stream running through protected wetlands. Both sections of the property are at the same relative elevation and have dwelling units with accessible routes from site arrival points; however, a combination clubhouse and swimming pool is located on one section of the property. Access to each section is provided by an existing public road outside the boundary of the site, which includes a bridge over the stream. Environmental restrictions prevent construction of any type of paved surface between the two sections within the boundary of the site. If environmental restrictions do not prevent the construction of an accessible pedestrian route, such as a boardwalk, through the wetlands connecting the two sections, then the accessible pedestrian route must be provided even if a road cannot be provided. If construction of any type of pedestrian route is prohibited, then a vehicular route that utilizes the public road and bridge is permitted with parking complying with Section 1106 located at the clubhouse/swimming pool, even though the vehicular route relies on a public road instead of a road through the development.

Example 4: A narrow and deep site has a level section in the front taking up most of the site and another level section at the back that is located up a steep incline. The developer chooses to place all of the buildings/Type B dwelling units on the front section to provide for accessible routes from site arrival points to building entrances. After considering all options for siting buildings and facilities in different locations, including the priority of accessibility, the most feasible location for a planned swimming pool is at the top of the higher section to the rear of the property. Because of the narrowness of the site and the relative elevation of the upper level at the rear of the property, it is impracticable to construct an accessible pedestrian route to the pool; however, a road that slopes more than 8.33 percent can be provided. Under these circumstances, Exception 1 is applicable and access to the swimming pool on the upper level of the site may be provided by means of a vehicular route with parking complying with Section 1106 provided at the Type B dwelling units and the pool.

Example 5: A developer plans to build a multiple-family housing complex with nonelevator buildings on a site with hilly terrain. All of these buildings will have some Type B dwelling units. The developer plans to locate a tennis court on the site. There are gentle slopes exceeding 8.33 percent with existing trees between the entrances to the Type B units and the tennis court. There is also a tree-save ordinance in place. If the builder can grade the site to allow for an accessible pedestrian route to the tennis court without disturbing the trees in violation of the tree-save ordinance, then an accessible pedestrian route between the Type B units and the planned location of the tennis court must be provided. If, however, the grading necessary to reduce the slope of the site near the trees to provide an accessible route would cause tree loss or damage in violation of the ordinance, then the developer cannot grade without violating the tree-save ordinance. The developer must then consider whether the tennis court can be relocated so it is served by an accessible pedestrian route, and if so, the tennis court must be relocated. If the tennis court cannot be relocated so it can be served by an accessible pedestrian route, then the developer may provide a vehicular route from the Type B dwelling units to the tennis court with parking complying with Section 1106 at the Type B dwelling units and the tennis court. Note, however, that if the developer can provide an accessible pedestrian route from some of the buildings without violating the ordinance, the developer must do so, even if it is necessary to provide a vehicular route from other buildings. Additionally, if the grading and construction of the proposed vehicular route can be limited to 8.33 percent by design and would not violate the tree-save ordinance, it is likely that an additional accessible walkway adjacent to the vehicular route would also fall under the scope of work that would not violate the tree-save ordinance and, therefore, must be provided, eliminating the use of Exception 1.

(*This completes the guidance on this topic provided by HUD for inclusion in this commentary.*)

Exceptions 2 through 7 are intended to clarify when an elevator is required between floors. Exceptions 2, 3, 4 and 6 provide exceptions for the accessible route to upper levels where the units required to be accessible, all common use areas associated with those units and all public use areas are on an accessible level.

Exception 2 is coordination with jail (Group I-3) provisions in Sections 1103.2.13 and 1107.5.5. Due to safety and security concerns, Accessible units are required in jails, and must be dispersed by type, but not by location. This would result in areas of a jail having only cells (i.e., sleeping units) that are exempted from accessibility. An accessible route is not required to those cells or the common areas associated with those cells. The public and administrative side of the jail cannot use this exception.

Exception 3 deals with large apartment buildings or complexes (i.e., more than dwelling units), or convents and monasteries where Type A units are required. If the common use and public use spaces associated with the Type A units are on an accessible level, Section 1107.4 would not require an elevator between stories. Section 1107.7.1 would possibly allow for no elevator in the building for the upper floors.

Exception 4 is a similar idea to Exception 3, only dealing with Group R-2 larger congregate residences (i.e., more than 16 residents, per Section 310.4), such as sororities, fraternities and boarding houses where Accessible units are required. If the common use and public use spaces associated with the Accessible units are on an accessible level, Section 1107.4 would not require an elevator between stories. Section 1107.7.1 would possibly allow for no elevator in the building for the upper floors. When applying Exception 4 it is important to recognize that dormitories at colleges or boarding schools are excluded from this exception. Based upon the DOJ adoption of the 2010 ADA Standard, these facilities are required to provide elevators by federal law. The

IBC is consistent with that intent, but it does not differentiate between public and private schools. Many colleges now provide "living communities," or living arrangements for students with shared studies or interests, on different levels within dormitories. This accessible route requirement is intended to allow for full participation in all activities, as well as to visit or study with students on other levels in the dormitory.

Exception 5 is applicable within individual units in a hotel where the unit itself is multistory or has a mezzanine. If a sleeping area and toilet are located on the accessible level of the unit, there may be other amenities or bedrooms on the nonaccessible level. This exception was added to coordinate with the 2010 ADA Standard Section 206.2.3, Exception 5.

Exception 6 is consistent with Exceptions 3 and 4; this time dealing with congregate residences with 16 or fewer residents (see Sections 310.5 and 310.6). In a Group R-3, at least one Type B sleeping unit would have to be provided on the accessible level. In a Group R-4, at least one Accessible sleeping unit would have to be provided on the accessible level. If the common use and public use spaces associated with the those particular units are on an accessible level, Section 1107.4 would not require an elevator between stories. Section 1107.7.1 would possibly allow for no elevator in the building for the upper floors.

Exception 7 is a correlation with Section 1107.7, so that it is clear that the requirements for connecting spaces within a building are not intended to override the exception for an accessible route between stories and mezzanines.

1107.5 Group I. *Accessible units* and *Type B units* shall be provided in Group I occupancies in accordance with Sections 1107.5.1 through 1107.5.5.

❖ This section introduces five subsections that set forth the threshold for accessibility in institutional occupancies.

Note that among the various categories of institutional facilities, only Accessible units and Type B units are scoped. There are no requirements for Type A units in institutional occupancies. The criteria for Accessible units and Type B units are specified in Chapter 10 of ICC A117.1.

1107.5.1 Group I-1. *Accessible units* and *Type B units* shall be provided in Group I-1 occupancies in accordance with Sections 1107.5.1.1 and 1107.5.1.2.

❖ Group I-1 is comparable in many respects to Group R-2 in that it is a residential setting. However, the difference is that Group I-1 is a supervised environment where custodial care is provided due to the reduced physical or mental abilities of the occupants (see Section 308.3).

1107.5.1.1 Accessible units. In Group I-1 Condition 1, at least 4 percent, but not less than one, of the *dwelling units* and *sleeping units* shall be *Accessible units*. In Group I-1 Condition 2, at least 10 percent, but not less than one, of the *dwelling units* and *sleeping units* shall be *Accessible units*.

❖ The threshold for accessibility in apartments (Group R-2), in accordance with Section 1107.6.2.1.1, is that 2 percent of the units are required to be Type A units. It is anticipated that Group I-1 will experience a greater demand for accessible facilities because of the nature of the occupants and the frequency of occupant turnover. As such, this section requires that 4 percent of the sleeping units and their associated facilities (e.g., bathing and toilet facilities) must meet the criteria for Accessible units in Group I-1 Condition 1, and 10 percent of the sleeping units and their associated facilities in Group I-1 Condition 2. To understand the care level and resident capability thresholds demarcating a Condition 1 or 2 facility, see the commentary to Section 308.3. The higher level of accessible units is based on the anticipated need. Group I-1 Condition 2 is not the same level of care as a nursing home. The main difference is custodial versus medical care provided for residents of the facility. These Group I-1 types of facilities are not scoped in the 2010 ADA Standard.

1107.5.1.2 Type B units. In structures with four or more *dwelling units* or *sleeping units intended to be occupied as a residence*, every *dwelling unit* and *sleeping unit intended to be occupied as a residence* shall be a *Type B unit*.

Exception: The number of *Type B units* is permitted to be reduced in accordance with Section 1107.7.

❖ As discussed in the general comments to this section, most Group I-1 assisted living facilities are intended to serve as a person's place of residence. Since a Group I-1 starts at an occupant load of more than 16 residents, the Type B criteria for any units that are not Accessible units are almost always required. The exception is a general reference to Section 1107.7, which addresses situations where it is logical to back off on the requirements for Type B units within a structure.

1107.5.2 Group I-2 nursing homes. *Accessible units* and *Type B units* shall be provided in nursing homes of Group I-2 occupancies in accordance with Sections 1107.5.2.1 and 1107.5.2.2.

❖ The requirements for accessibility in Group I-2 are based on the anticipated frequency of usage by disabled persons. There are different thresholds established for different types of health care occupancies, based on the nature of their activity (see Section 308.4). A nursing home is anticipated to have a high percentage of residents who use wheelchairs or some type of mobility aid.

1107.5.2.1 Accessible units. At least 50 percent but not less than one of each type of the *dwelling units* and *sleeping units* shall be *Accessible units*.

❖ Nursing homes, which generally provide long-term care, are more likely to have patients with physical disabilities. As such, 50 percent of the patient sleeping rooms and their associated facilities (e.g., bathing and toilet facilities) must meet the criteria for Accessible units. Housings for the elderly should not automatically be classified as a "nursing home." Some residential care/assisted living facilities provide quarters for residents that more closely represent a residential setting (i.e., sleeping, eating, cooking and sanitation within the individual living accommodations) and the residents are more ambulatory than in a nursing home. Based on the anticipated frequency of residents needing wheelchairs, these facilities might be more appropriately classified within the 4- or 10-percent requirement for Accessible units (see Sections 1107.5.1.1 and 1107.5.3.1).

1107.5.2.2 Type B units. In structures with four or more *dwelling units* or *sleeping units intended to be occupied as a residence*, every *dwelling unit* and *sleeping unit intended to be occupied as a residence* shall be a *Type B unit.*

Exception: The number of *Type B units* is permitted to be reduced in accordance with Section 1107.7.

❖ As discussed in the general comments to this section, most Group I-2 nursing homes are intended to serve as a person's place of residence. The Type B criteria for any units that are not Accessible units are almost always required. The exception is a general reference to Section 1107.7, which addresses situations where it is logical to back off on the requirements for Type B units within a structure.

1107.5.3 Group I-2 hospitals. *Accessible units* and *Type B units* shall be provided in general-purpose hospitals, psychiatric facilities and detoxification facilities of Group I-2 occupancies in accordance with Sections 1107.5.3.1 and 1107.5.3.2.

❖ The requirements for accessibility in Group I-2 are based on the anticipated number of disabled occupants. There are different thresholds established for different types of health care occupancies based on the nature of their services and level of care provided. The anticipated need in hospitals is partially based on patients needing varying levels of assistance due to temporary disabilities resulting from illness or surgery.

1107.5.3.1 Accessible units. At least 10 percent, but not less than one, of the *dwelling units* and *sleeping units* shall be *Accessible units*.

Exception: Entry doors to *Accessible dwelling units* or *sleeping units* shall not be required to provide the maneuvering clearance beyond the latch side of the door.

❖ While the anticipated percentage of patients in general-purpose hospitals is typically higher than 10 percent, there is also an anticipation that many of the patients will be routinely assisted by staff. Section 1109.2, Exception 5, exempts the bathrooms associated with critical care or intensive care.

The exception recognizes the practical implications of door maneuvering clearances on the room side at in-swinging doors in smaller-sized patient sleeping rooms in hospitals. The maneuvering clearance is still required on the outside of the Accessible patient sleeping room door. Since the maneuvering clearance in ICC A117.1 is based on the size of the door, there is no credit provided for the wider $41^1/_2$-inch (1054 mm) clear width door (required in Section 1010.1.1 in Group I-2 for movement of beds) versus the standard 32-inch (813 mm) clear width door. Without this exception, a patient room would likely require extra width for maneuvering clearance at the door when a toilet room is located on the same wall as the door. Due to operational procedures and observation needs, the door position is normally open. Therefore, the need for door maneuvering clearance on the interior side is reduced. At hospitals, the patient room door is usually closed only when privacy is needed during procedures.

Maneuvering clearances are part of an accessible route, and accessible routes connect accessible elements; therefore, maneuvering clearances at patient sleeping rooms that are not Accessible or Type B units are not required since an accessible route is not required. See Section 1107.5.3.2 regarding Type B units in hospitals.

1107.5.3.2 Type B units. In structures with four or more *dwelling units* or *sleeping units intended to be occupied as a residence*, every *dwelling unit* and *sleeping unit intended to be occupied as a residence* shall be a *Type B unit.*

Exception: The number of *Type B units* is permitted to be reduced in accordance with Section 1107.7.

❖ As discussed in the general comments to this section, most Group I-2 hospitals are not intended to serve as a person's place of residence; therefore, Type B units are not required in most hospitals. However, whether a Group I-2 psychiatric or detoxification facility is a person's place of abode is subject for interpretation. Please see the general discussion to this section for additional information. The exception is a general reference to Section 1107.7, which addresses situations where it is logical to back off on the requirements for Type B units within a structure.

1107.5.4 Group I-2 rehabilitation facilities. In hospitals and rehabilitation facilities of Group I-2 occupancies that specialize in treating conditions that affect mobility, or units within either that specialize in treating conditions that affect mobility, 100 percent of the *dwelling units* and *sleeping units* shall be *Accessible units*.

❖ The requirements for accessibility in Group I-2 are based on the anticipated frequency of usage by disabled persons. There are different thresholds established for different types of health care occupancies based on the nature of their activity. In rehabilitation facilities that specialize in treating conditions that affect mobility, the anticipated need is very high compared to other Group I-2 facilities. In such cases, it is realistic to presume that the majority, if not all, of the patients will have some degree of mobility impairment and will, thus, require accessible features. Accordingly, all patient rooms and their associated bathing and toilet facilities are required to be Accessible units.

1107.5.5 Group I-3. *Accessible units* shall be provided in Group I-3 occupancies in accordance with Sections 1107.5.5.1 through 1107.5.5.3.

❖ The requirement for accessibility in Group I-3 facilities was taken from the 2010 ADA Standard (see Section 308.5 for Group I-3). Section 1107.5.5.1 contains requirements for general housing cells. Sections 1107.5.5.2 and 1107.5.5.3 are additional requirements for special types of cells or holding areas. While jail cells may be considered a detainee's residence, since the FHA does not cover detention and correctional facilities, Type B units are not required in Group I-3 occupancies. For visiting areas in jails, see Section 1109.12.2.

1107.5.5.1 Group I-3 sleeping units. In Group I-3 occupancies, at least 3 percent of the total number of *sleeping units* in the facility, but not less than one unit in each classification level, shall be *Accessible units*.

❖ The requirement for sleeping units or cells in Group I-3 is for 3 percent of the general housing cells and their associated facilities to be Accessible units. The current trend in jail design is to group housing cells into "pods." Since the administration of the correctional facility will determine where detainees will reside, there are no requirements for distribution of the Accessible units within the same basic security level or type (e.g., woman, men, juvenile). Most jails are all one security level and only house male or female inmates. A correctional facility may decide to locate all its Accessible cells in one pod due to suicide prevention and other safety concerns. Sections 1103.2.13 and 1107.4 have exceptions for spaces shared by detainees for areas within the jail where Accessible units are not provided.

1107.5.5.2 Special holding cells and special housing cells or rooms. In addition to the *Accessible units* required by Section 1107.5.5.1, where special holding cells or special housing cells or rooms are provided, at least one serving each purpose shall be an *Accessible unit*. Cells or rooms subject to this requirement include, but are not limited to, those used for purposes of orientation, protective custody, administrative or disciplinary detention or segregation, detoxification and medical isolation.

Exception: Cells or rooms specially designed without protrusions and that are used solely for purposes of suicide prevention shall not be required to include grab bars.

❖ If special types of holding or housing cells are provided within a facility (e.g., orientation, protective custody, disciplinary detention, segregation, detoxification, medical isolation), at least one of each type must be an Accessible unit. A cell that is designed specifically for a suicide watch is required to meet Accessible unit criteria, but if a toilet is provided, grab bars are not required.

1107.5.5.3 Medical care facilities. Patient *sleeping units* or cells required to be *Accessible units* in medical care facilities shall be provided in addition to any medical isolation cells required to comply with Section 1107.5.5.2.

❖ When medical care facilities are provided within a detention or correctional facility, 10 percent of the units are required to be Accessible units. The intent is to be consistent with the hospital requirements found in Section 1107.5.3.1.

1107.6 Group R. *Accessible units*, *Type A units* and *Type B units* shall be provided in Group R occupancies in accordance with Sections 1107.6.1 through 1107.6.4.

❖ This section introduces four subsections that address accessibility requirements for occupancies in Group R. Criteria for Accessible units, Type A units and Type B units are specified in Chapter 10 of ICC A117.1.

The IRC references this chapter for accessibility provisions. Therefore, this chapter may be applicable to housing covered under the IRC (see commentary, Section 1107.6.3). Structures referenced to Chapter 11 from the IRC would be considered Group R-3 for purposes of accessibility only.

1107.6.1 Group R-1. *Accessible units* and *Type B units* shall be provided in Group R-1 occupancies in accordance with Sections 1107.6.1.1 and 1107.6.1.2.

❖ The terms "dwelling unit" and "sleeping unit" are utilized instead of "guestrooms" so that the amenities in the units themselves are reflected. An exception applicable to small bed-and-breakfast style hotels is provided in Section 1103.2.11.

Hotels, motels, boarding houses and other short-term housing types are required to provide Accessible units. Where such facilities are also intended to be occupied as a residence, Type B unit criteria are applicable. For additional discussion, please see the general comments at the beginning of this chapter.

The following factors should be considered where persons can stay more or less than 30 days, including hotels, motels, corporate housing and seasonal vacation units, in determining the applicability of Type B unit criteria:

1. Amenities included inside the units, including kitchen facilities;
2. Whether the property is to be marketed to the public as short-term housing;
3. Whether the terms and length of the occupancy will be established through a lease or other written agreement; and
4. How payment is calculated (e.g., on a daily, weekly, monthly or yearly basis).

If the amenities and operation of the units are closer to those of apartments than of hotels, they are subject to Type B requirements. For example, if a hotel is marketed as short-term housing, payment is made monthly, and if the units contain kitchens, the hotel would be subject to Type B unit criteria. For additional information see the commentary at the beginning of this chapter and Section 1107.

1107.6.1.1 Accessible units. *Accessible dwelling units* and *sleeping units* shall be provided in accordance with Table 1107.6.1.1. Where buildings contain more than 50 *dwelling units* or *sleeping units*, the number of *Accessible units* shall be determined per building. Where buildings contain 50 or fewer *dwelling units* or *sleeping units*, all *dwelling units* and *sleeping units* on a *site* shall be considered to determine the total number of *Accessible units*. *Accessible units* shall be dispersed among the various classes of units.

❖ The required number of Accessible dwelling or sleeping units in Group R-1 is indicated in Table 1107.6.1.1. The table requires Accessible units to include a variety of bathing options. Where roll-in-type showers are provided, ICC A117.1 requires seats in roll-in showers. These showers can then serve the dual purpose of roll-in and transfer showers.

For sites where the hotel rooms are provided in multiple buildings, all units on the site should be considered to determine the number of Accessible units if each building has 50 or fewer units. For example, if a hotel consists of several small buildings on a site, the same number of Accessible units would be required as if the hotel was constructed as a single structure. If the hotel has buildings with more than 50 units, those buildings must be considered separately to determine the number of Accessible units. See Commentary Figure 1107.6.1.1.

Accessible units must be dispersed among the different types of units. This is not an automatic requirement to disperse the Accessible units to different floor levels (see Section 1107.4). When different classes of dwelling or sleeping units are available, special amenities must also be made available in the Accessible units, including suites or larger rooms, kitchenettes, executive levels, etc. For example, if sleeping unit options include single rooms, suites or kitchenettes, and two Accessible units are required, the Accessible units must be provided in two of the three options. If no other types of amenities are provided on other levels, the hotel could provide the Accessible units on the main level. Some hotels provide all Accessible units on the main level due to concerns for disabled guests' ease of escape from the building during an emergency, such as a fire. Keep in mind that services available to guests (e.g., pools, exercise rooms, laundry rooms) must also be on an accessible route for guests in the Accessible units.

Note that, in accordance with Section 1010.1.1, all doors within a hotel room are required to have a clear width of 32 inches, even in the nonaccessible rooms. The U.S. Access Board and DOJ had received numerous complaints that persons who requested Accessible hotel rooms were not provided with those rooms when they arrived at the hotel. With a 32-inch clear width on all doors, a person using a mobility device should at least be able to get into the bathroom in the nonaccessible units.

TABLE 1107.6.1.1. See page 11-28.

❖ The number of Accessible dwelling and sleeping units required is coordinated with the 2010 ADA Standard requirements. The ADAAG Review Federal Advisory Committee reviewed all information and data provided by the industry in support of a lower

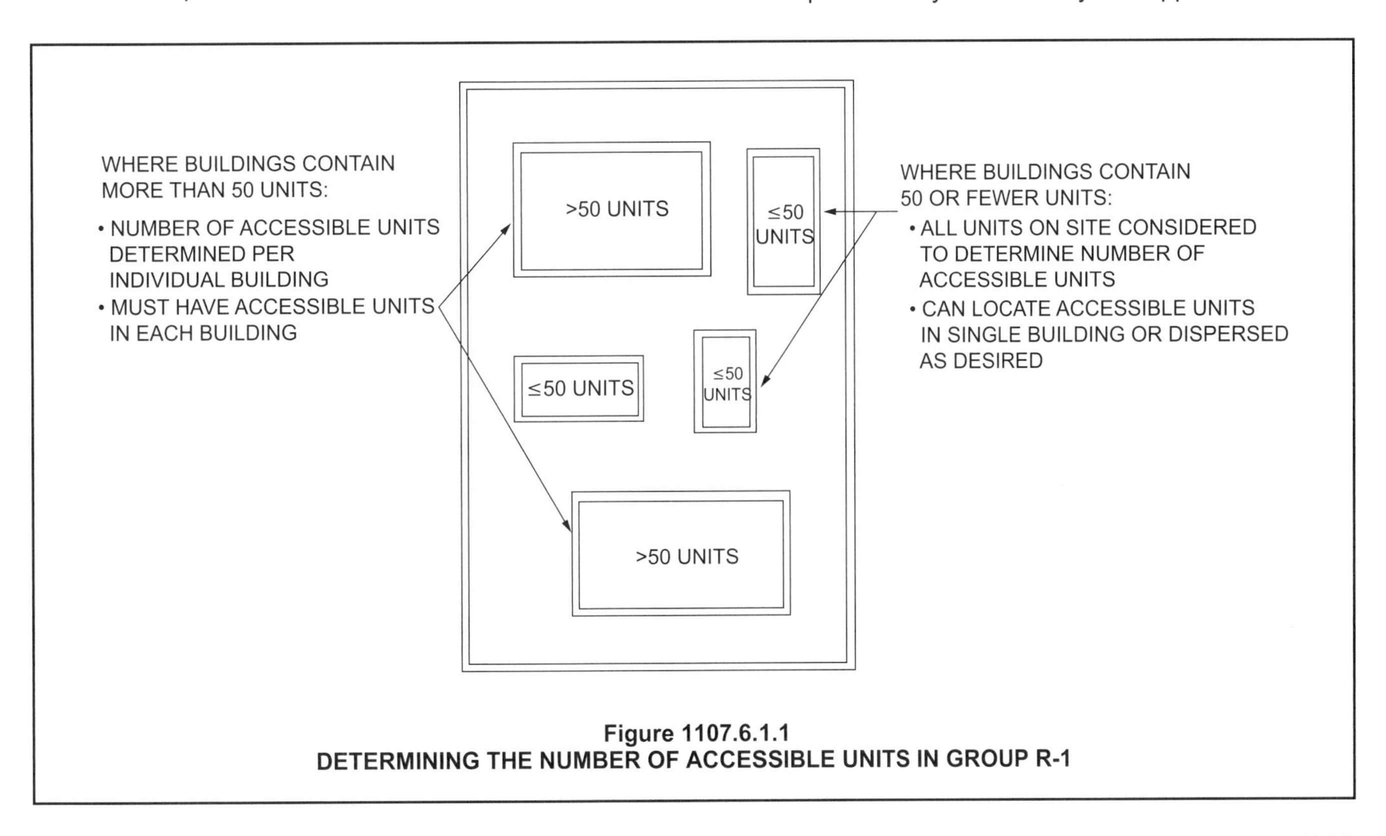

Figure 1107.6.1.1
DETERMINING THE NUMBER OF ACCESSIBLE UNITS IN GROUP R-1

number of Accessible units being required. The committee determined that accessible features are useful to persons with disabilities beyond those who use wheelchairs, and that good design would lessen the "institutional" impression that may make Accessible units less attractive to nondisabled guests. To accommodate wheelchair users in relatively large facilities, this table requires a minimum number of units to be equipped with roll-in-type showers. The intent of the second column in this table is that different options (e.g., roll-in shower, transfer shower or bathtubs) are available in the Accessible units. It is not the intent of this column to prohibit roll-in showers with seats when there are only one or two Accessible units in the hotel.

1107.6.1.2 Type B units. In structures with four or more *dwelling units* or *sleeping units intended to be occupied as a residence*, every *dwelling unit* and *sleeping unit intended to be occupied as a residence* shall be a *Type B unit.*

Exception: The number of *Type B units* is permitted to be reduced in accordance with Section 1107.7.

❖ Most hotels and motels are not intended to be occupied as a residence; therefore, the criteria for Type B units would not be applicable. However, in certain situations, extended-stay hotels or similar corporate housing arrangements are required to meet Type B unit criteria. For additional information on how to determine if this section is applicable, see the comments at the beginning of this chapter and Sections 1107 and 1107.6.1. The Accessible sleeping units required by Section 1107.6.1.1 exceed Type B unit requirements. The exception is a general reference to Section 1107.7, which addresses situations where it is logical to back off on the requirements for Type B units within a structure.

1107.6.2 Group R-2. *Accessible units*, *Type A units* and *Type B units* shall be provided in Group R-2 occupancies in accordance with Sections 1107.6.2.1 through 1107.6.2.3.

❖ The terms "dwelling unit" and "sleeping unit" are utilized instead of "apartment," "dorm room," "sleeping accommodations," "suites," etc., so that the amenities in the units themselves are reflected and for consistent application within all Group R-2 facilities. Since these types of facilities are intended to be occupied as a residence, Type B unit criteria would be applicable. For a complete discussion, please see the comments to this chapter and this section.

Timeshare properties require a different analysis. Timeshare owners have an interest in the property, yet they typically stay less than 30 days. For a timeshare property to be subject to Type B requirements, the owners must have an ownership interest in the property itself, rather than in just coming to the area for a vacation without ties to a particular property. The following additional factors must be considered to determine whether timeshare units must meet Type B design and construction requirements:

1. Whether traditional rights of ownership are to be unrestricted (e.g., whether the timeshare owner has the right to occupy, alter or exercise control over a particular unit over a period of time);
2. The nature of the ownership interest conveyed (e.g., fee simple); and
3. The extent to which operations resemble those of a hotel, motel or inn (e.g., reservation, central registration, meals, laundry service).

If an owner's rights regarding the units are subject to few restrictions and the operation of the units is closer to that of condominiums/apartments than hotels, the units are subject to Type B requirements.

TABLE 1107.6.1.1
ACCESSIBLE DWELLING UNITS AND SLEEPING UNITS

TOTAL NUMBER OF UNITS PROVIDED	MINIMUM REQUIRED NUMBER OF ACCESSIBLE UNITS WITHOUT ROLL-IN SHOWERS	MINIMUM REQUIRED NUMBER OF ACCESSIBLE UNITS WITH ROLL-IN SHOWERS	TOTAL NUMBER OF REQUIRED ACCESSIBLE UNITS
1 to 25	1	0	1
26 to 50	2	0	2
51 to 75	3	1	4
76 to 100	4	1	5
101 to 150	5	2	7
151 to 200	6	2	8
201 to 300	7	3	10
301 to 400	8	4	12
401 to 500	9	4	13
501 to 1,000	2% of total	1% of total	3% of total
Over 1,000	20, plus 1 for each 100, or fraction thereof, over 1,000	10 plus 1 for each 100, or fraction thereof, over 1,000	30 plus 2 for each 100, or fraction thereof, over 1,000

1107.6.2.1 Live/work units. In *live/work units* constructed in accordance with Section 419, the nonresidential portion is required to be *accessible*. In a structure where there are four or more *live/work units intended to be occupied as a residence*, the residential portion of the *live/work unit* shall be a *Type B unit*.

Exception: The number of *Type B units* is permitted to be reduced in accordance with Section 1107.7.

❖ Live/work units are dwelling units in which a significant portion of the space includes a nonresidential use operated by the tenant/owner. Although the entire unit is classified as a Group R-2 occupancy for construction purposes, for accessibility purposes it is viewed more as a mixed-use condition. The residential portion of the unit is regulated differently for accessibility purposes than the nonresidential portion.

When live/work units are built in a strip, similar to townhouses, the exceptions for Type B units set forth in Section 1107.7 would also exempt such units where applicable. For example, a two-story dwelling unit above the work unit in a nonelevator building would not be subject to Type B requirements because of the exemption in Section 1107.7.2. The code does not clarify if a two-story live/work unit with the business on the entire first floor and the residence on the entire second floor would be considered a multistory dwelling unit for purposes of the exception in Section 1107.7.2. Where a portion of the dwelling unit was on the first floor and all of the second floor, the unit would be exempted as a multistory unit without elevator service. If the dwelling unit was the entire second and third floors, the same exception would be applicable. Just as a reminder, Type B units are applied per structure, not per building, so legal property lines may not be relevant when evaluating accessibility requirements (see Section 1107).

In the nonresidential portion of the unit, full accessibility would be required based on the intended use. For example, if the nonresidential area of the unit is utilized for hair care services, all elements related to the service activity must be accessible. This would include site parking where provided, site and building accessible routes, the public entrance, and applicable patron services. In essence, the work portion of the live/work unit would be regulated in the same manner as a stand-alone commercial occupancy.

1107.6.2.2 Apartment houses, monasteries and convents. *Type A units* and *Type B units* shall be provided in apartment houses, monasteries and convents in accordance with Sections 1107.6.2.2.1 and 1107.6.2.2.2.

❖ Unless they receive some type of federal funding, apartments and condominiums are not covered by the 2010 ADA Standard. Note that the code deals with apartments and condominiums in the same manner when dealing with accessibility requirements.

Because convents and monasteries are associated with religious organizations, they also are not covered by ADA requirements. Since all of these facilities may serve as a person's residence, they are covered by FHAG. A convent or monastery cannot be sued under the FHA for selectively limiting its residents. However, a convent or monastery must comply with the FHAG construction requirements for the residential units that it provides. The scope for Type B units is consistent with FHAG requirements.

1107.6.2.2.1 Type A units. In Group R-2 occupancies containing more than 20 *dwelling units* or *sleeping units*, at least 2 percent but not less than one of the units shall be a *Type A unit*. All Group R-2 units on a *site* shall be considered to determine the total number of units and the required number of *Type A units*. *Type A units* shall be dispersed among the various classes of units. Bedrooms in monasteries and convents shall be counted as *sleeping units* for the purpose of determining the number of units. Where the *sleeping units* are grouped into suites, only one *sleeping unit* in each suite shall count towards the number of required *Type A units*.

Exceptions:

1. The number of *Type A units* is permitted to be reduced in accordance with Section 1107.7.
2. *Existing structures* on a *site* shall not contribute to the total number of units on a *site*.

❖ To be able to better accommodate a person who uses a wheelchair, 2 percent of apartments and condominiums are required to provide Type A units within the development. Congregate residences, such as convents and monasteries, must also provide Type A units. (Other types of congregate living arrangements are addressed in Section 1107.6.2.2.) The housing industry has been concerned that dwelling units equipped with the full range of features to accomplish accessibility will not be marketable to people without disabilities. The Type A requirements are intended to establish a middle ground that will satisfy both needs. Allowances made during the construction process for future conversion for accessibility, such as reinforcement in bathroom walls for future installation of grab bars, will allow a unit to be altered later at a considerably lesser cost.

Type A units are required when the site contains more than 20 dwelling units or sleeping units. So that there is a consistent number of Type A units in multiple-building sites based on the size of a development as a whole, all the buildings are added together to determine the number required. For example, a 300-unit building would require six Type A units, and a development with 75 buildings with four units per building would also require six Type A units. Exception 2 is in recognition that a development may be built in stages and does not exempt existing buildings that are being altered from complying with the requirement for Type A units. Only the units that are being constructed as part of that development phase are considered in determining the number of Type A units. For specific provisions see the IEBC Sections 410.8.6, 705.1.8 and 1105.3.

There has been some discussion that congregate living facilities, such as convents and monasteries, should be considered dwelling units instead of a group of sleeping units with common-use living areas. As a single-family dwelling unit, they would not have to provide any level of accessibility (see Section 1103.2.3). This is not the correct interpretation for providing accessibility features. The last sentence is intended to make clear that for the purpose of determining the number of Type A units, the bedrooms in the congregate residence must be counted separately. See the commentaries to Sections 1107.3 and 1107.4 regarding what shared elements must be accessible.

The Type A units must be dispersed among the classes of units provided. For example, if one-, two- and three-bedroom units are available with the development and two Type A units are required, it is the designer's choice as to which two of the options to provide as Type A units. The designer, however, cannot choose to only provide the one-bedroom option with both Type A units. This is not intended to require Type A units to be provided in different buildings in a multiple-building site. Many times in multiple-building developments, there are shared facilities such as clubhouses or pools. The designer may choose to locate the Type A units in the building closest to those amenities for ease of access for the residents. This is acceptable as long as the dispersion-by-type requirement is met.

In limited circumstances, townhouse-style units can be Group R-2 facilities (see Section 310.4). If multistory units are provided for the Type A units, ICC A117.1 would require an accessible route between levels. Within the dwelling, this can be provided by a passenger elevator, a LULA, a private residence elevator (see Section 1109.7) or a platform lift (see Section 1109.8, Item 4). A possible alternative would be to provide some single-story units within the development to meet Type A unit requirements. If the amenities and size are the same, they could be considered the same type as a multilevel unit. The multistory exception in Section 1107.7.2 is not applicable to Type A units.

Exception 1 allows the number of Type A units to be reduced in accordance with Section 1107.7.5 when the building's first-floor elevation is required to be raised due to flood-plain regulations. This is the only exception that allows for a reduction in the number of Type A units required.

1107.6.2.2.2 Type B units. Where there are four or more *dwelling units* or *sleeping units intended to be occupied as a residence* in a single structure, every *dwelling unit* and *sleeping unit intended to be occupied as a residence* shall be a *Type B unit.*

Exception: The number of *Type B units* is permitted to be reduced in accordance with Section 1107.7.

❖ When four or more dwelling or sleeping units are provided in a single structure, those units must meet Type B criteria (note the use of the term "structure" instead of "building"). The criteria are applicable if four dwelling units are built together, regardless of fire walls [see Commentary Figures 1107.6.2.2.2(1) and 1107.6.2.2.2(2)]. The exception is a general reference to Section 1107.7, which addresses situations where it is logical to back off on the requirements for Type B units within a structure.

Type A units provided in accordance with Section 1107.6.2.1.1 exceed Type B unit requirements.

1107.6.2.3 Group R-2 other than live/work units, apartment houses, monasteries and convents. In Group R-2 occupancies, other than *live/work units*, apartment houses, monasteries and convents falling within the scope of Sections 1107.6.2.1 and 1107.6.2.2, *Accessible units* and *Type B units* shall be provided in accordance with Sections 1107.6.2.3.1 and 1107.6.2.3.2. Bedrooms within congregate living facilities shall be counted as *sleeping units* for the purpose of determining the number of units. Where the *sleeping units* are grouped into suites, only one *sleeping unit* in each suite shall be permitted to count towards the number of required *Accessible units.*

❖ This would include congregate housing types, such as dormitories, boarding houses, fraternities and sororities. Nontransient hotels and motels would typically be grouped with these requirements since they may also be operating as transient and nontransient. A building

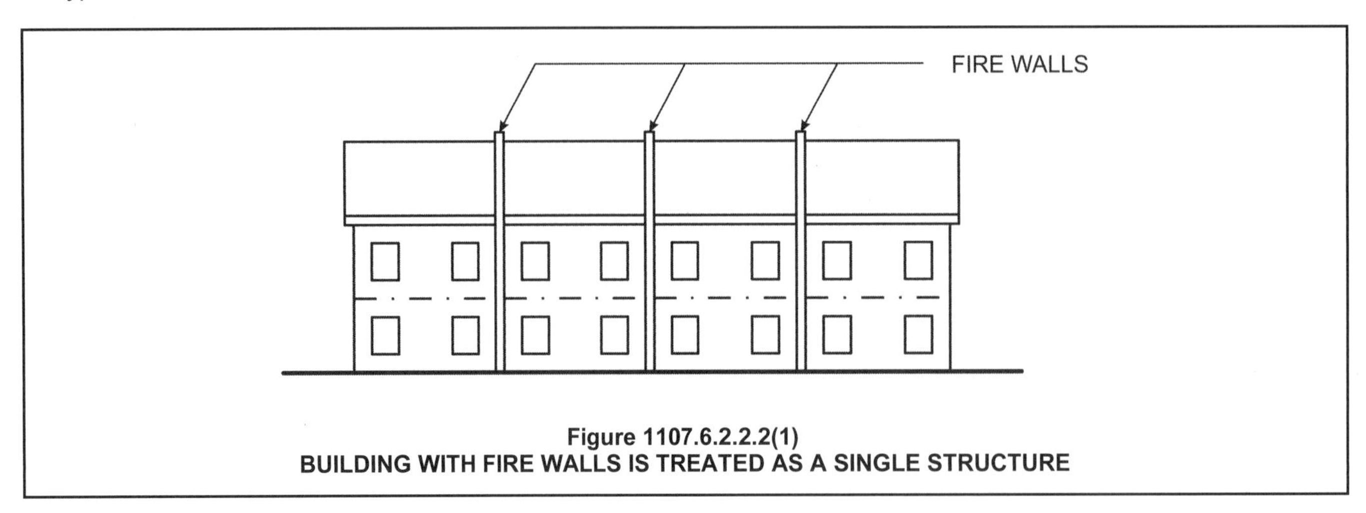

Figure 1107.6.2.2.2(1)
BUILDING WITH FIRE WALLS IS TREATED AS A SINGLE STRUCTURE

used for two purposes must be designed for the most restrictive provisions (see Section 302.1). A time-share should comply with Section 1107.6.2.1 or 1107.6.2.2, based on whichever it most closely represents (see also commentary, Section 1107.6.2).

Dormitories are typically located in universities. Universities are required to make all programs accessible. Therefore, dormitories are covered by ADA requirements. Boarding houses, fraternity houses and sorority houses have similar types of living arrangements. It seems logical that the same anticipated need should result in the same level of required Accessible units. Since all these facilities typically serve as a person's residence, they are covered by FHAG. The scope for Accessible units is consistent with ADA requirements for dormitories. The scope for Type B units is consistent with FHAG requirements.

There has been some discussion that congregate living facilities, such as sororities and fraternities, should be considered dwelling units instead of a group of sleeping units with common-use living areas. As a single-family dwelling unit, they would not have to provide any level of accessibility (see Section 1103.2.3). This is not the correct interpretation for providing accessibility features. The new style of dormitory living is groups of bedrooms that share a common living space, sometimes with a kitchen or kitchenette. If these are counted as dwelling units instead of sleeping units, the number of Accessible rooms available may be less, or multiple bedrooms in the same suite would be interpreted to all be Accessible units. Administratively, when housing students, the universities still treat this style of units as the old-style dorm rooms down a long hallway. The last sentence is intended to make clear that for the purpose of determining the number of Accessible units, the bedrooms in any congregate residence must be counted separately. In a suite configuration, only one of the sleeping units can be counted as part of the required number of Accessible units. See the commentary to Sections 1107.3 and 1107.4 regarding what shared elements must be accessible.

1107.6.2.3.1 Accessible units. *Accessible dwelling units* and *sleeping units* shall be provided in accordance with Table 1107.6.1.1.

❖ The number of Accessible units required in these Group R-2 facilities is the same as that required in hotels and motels. All associated and shared areas (e.g., bathrooms, kitchens, study rooms) must also be accessible in accordance with Section 1002 of ICC A117.1. A certain number of the Accessible units must have associated bathrooms equipped with a roll-in shower. See the commentary for Section 1107.6.2.3 regarding how to count the number of units to determine the number of Accessible units required.

1107.6.2.3.2 Type B units. Where there are four or more *dwelling units* or *sleeping units intended to be occupied as a residence* in a single structure, every *dwelling unit* and every *sleeping unit intended to be occupied as a residence* shall be a *Type B unit*.

Exception: The number of *Type B units* is permitted to be reduced in accordance with Section 1107.7.

❖ The FHA specifically considers dormitories as places where students live, even if it is just for the school year. Therefore, when four or more sleeping units are provided in a single structure, those units must meet Type B criteria. The Accessible sleeping units required by Section 1107.6.2.2.1 exceed Type B unit requirements. The exception is a general reference to Section 1107.7, which addresses situations where it is logical to back off on the requirements for Type B units within a structure. See the commentary to Section 1107.4 regarding elevator requirements in dormitories.

1107.6.3 Group R-3. In Group R-3 occupancies where there are four or more *dwelling units* or *sleeping units intended to be occupied as a residence* in a single structure, every *dwell-*

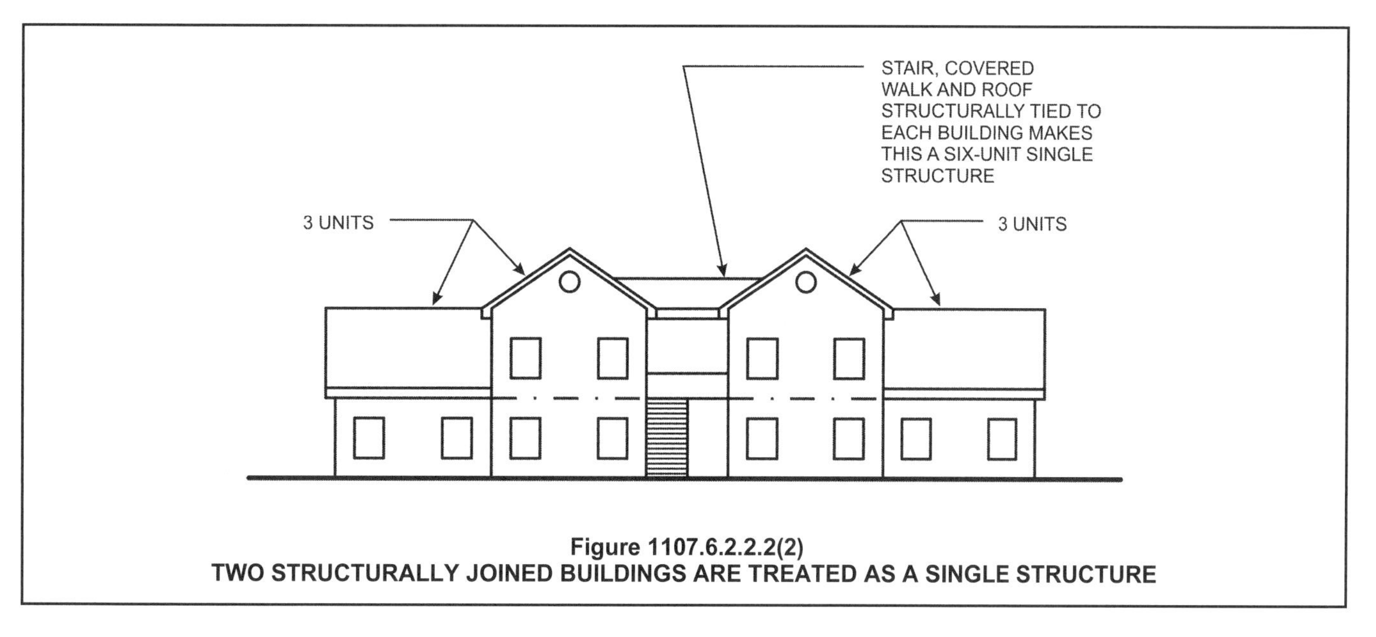

Figure 1107.6.2.2.2(2)
TWO STRUCTURALLY JOINED BUILDINGS ARE TREATED AS A SINGLE STRUCTURE

ing unit and *sleeping unit intended to be occupied as a residence* shall be a *Type B unit*. Bedrooms within congregate living facilities shall be counted as *sleeping units* for the purpose of determining the number of units.

Exception: The number of *Type B units* is permitted to be reduced in accordance with Section 1107.7.

❖ When four or more dwelling or sleeping units are provided in a single structure, those units must meet Type B criteria (note the use of the term "structure" instead of "building"). Since Group R-3 structures can be a series of one or two dwellings per building separated by fire walls (see Section 310.5), the provisions for Type B units would apply to groups of R-3 units. The criteria are applicable if four dwelling units are built together, regardless of fire walls. The exception is a general reference to Section 1107.7, which addresses situations where it is logical to back off on the requirements for Type B units within a structure.

Transient congregate residences with 10 or fewer occupants and nontransient congregate residences with 16 or fewer occupants are permitted to comply with Group R-3 construction requirements (see Section 310.5). Small bed-and-breakfast-type hotels are exempted from accessibility under Section 1103.2.11 (and IRC Section 320). There has been some discussion that congregate living facilities, such as group homes and boarding houses, should be considered dwelling units instead of a group of sleeping units with common-use living areas. As a single-family dwelling unit, they would not have to provide any level of accessibility (see Section 1103.2.3). This is not the correct interpretation for providing accessibility features. The last sentence is intended to make clear that for the purpose of determining the number of Accessible units, the bedrooms in the congregate residence must be counted separately. See the commentaries to Sections 1107.3 and 1107.4 regarding what shared elements must be accessible.

Section 101.2, Exception 1 states that detached one- and two-family dwellings and townhouses that are both three stories or less and have an independent means of egress must comply with the IRC. Section 1103.2.3 exempts detached single-family homes and duplexes from accessibility requirements, but not townhouses. In the IRC, townhouses are further defined as extending from foundation to roof and open on at least two sides. Therefore, the typical side-by-side townhouse is constructed using the IRC. Any configuration of townhouses that does not meet all four criteria has to be constructed under the IBC as Group R-2 or possibly Group R-3. If the structure is divided into one or two dwelling units per building with fire walls, the structure is a Group R-3 (see commentaries, Sections 310.4 and 310.5). Group I-1 and I-2 institutional-type facilities with five or fewer residents have the option of complying with the IBC or the IRC, since they often operate similarly to a single-family home (see Sections 308.3.4, 308.4.2 and 310.5.1). Day care facilities within single-family homes are specifically addressed under Section 1103.2.12. Live/work units are specifically addressed under Section 1107.6.2.1.

Section 320.1 of the IRC refers any structure with four or more sleeping units or dwelling units back to the IBC as Group R-3 buildings for accessibility requirements. Since these types of facilities typically serve as a person's permanent residence, the units must meet Type B criteria. The exception is a general reference to Section 1107.7, which addresses situations where it is logical to back off on the requirements for Type B units within a structure.

1107.6.4 Group R-4. *Accessible units* and *Type B units* shall be provided in Group R-4 occupancies in accordance with Sections 1107.6.4.1 and 1107.6.4.2. Bedrooms in Group R-4 facilities shall be counted as *sleeping units* for the purpose of determining the number of units.

❖ Group R-4 facilities are limited to between six and 16 residents. A Group R-4 is basically a small Group I-1 facility (see Sections 308.3.3 and 310.6). The Group R-4 occupancy was originally developed in response to some lawsuits filed under FHA concerning homes for mentally disabled adults. Under the past legacy codes, a structure where persons live in a supervised environment is a Group I occupancy. In neighborhoods zoned residential only, a variance is required if a group of mentally disabled adults wants to move into a single-family-style dwelling. Additionally, Group I facilities are typically required to be sprinklered throughout with an NFPA 13 system. This type of sprinkler system is not what is typically used in a single-family home. The 16-resident criterion is based on two things: 1. In the last census, 98 percent of the households in the United States that identified themselves as single family had 16 or fewer residents; and 2. In facilities where residents are capable of self-preservation, the number 16 also happens to be the limit of residents permitted in a building where an NFPA 13D sprinkler system can be utilized. Establishing these types of facilities as part of Group R eliminates potential conflict with the zoning issue. The limit on the number of residents and allowing the alternative sprinkler system addresses the discrimination in housing concerns and provides a reasonable level of sprinkler protection for the residents.

There has been some discussion that congregate living facilities, such as group homes, should be considered dwelling units instead of a group of sleeping units with common-use living areas. As a single-family dwelling unit, they would not have to provide any level of accessibility (see Section 1103.2.3). This is not the correct interpretation for providing accessibility features. The last sentence is intended to make clear that for the purpose of determining the number of Accessible units, the bedrooms in the congregate residence must be counted separately. See the commentaries to Sections 1107.3 and 1107.4 regarding what shared elements must be accessible.

1107.6.4.1 Accessible units. In Group R-4 Condition 1, at least one of the *sleeping units* shall be an *Accessible unit*.

❖ The requirement for one sleeping unit and its associated facilities (e.g., bathrooms) to meet Accessible unit criteria is consistent with Group I-1 Condition 1 requirements, and for two units, it is consistent with Group I-2 Condition 2 requirements. All common rooms are required to be fully accessible (see Section 1107.3). Section 1002 of ICC A117.1 provides the technical criteria for Accessible sleeping units.

1107.6.4.2 Type B units. In structures with four or more *sleeping units intended to be occupied as a residence*, every *sleeping unit intended to be occupied as a residence* shall be a *Type B unit*.

Exception: The number of *Type B units* is permitted to be reduced in accordance with Section 1107.7.

❖ Since these types of facilities typically serve as a person's permanent residence, the sleeping unit must meet Type B criteria. The Accessible sleeping unit required by Section 1107.6.4.1 exceeds Type B unit requirements. The exception is a general reference to Section 1107.7, which addresses situations where it is logical to back off on the requirements for Type B units within a structure. Since Group R-4 is a congregate residence, this section would be applicable when four or more sleeping units are provided.

1107.7 General exceptions. Where specifically permitted by Section 1107.5 or 1107.6, the required number of *Type A units* and *Type B units* is permitted to be reduced in accordance with Sections 1107.7.1 through 1107.7.5.

❖ Sections 1107.5 and 1107.6 establish when Accessible, Type A or Type B units are expected to be provided. Section 1107.7 covers the general exceptions where it is reasonable and logical to not provide accessibility for some of the units. This would be consistent with the general provisions of Section 1103.2.1. Code users should start out with the assumption that everything, in this case Group I and R dwelling and sleeping units, is required to be accessible, and then back off from that level of accessibility when specific exceptions are indicated. Note that there are no exceptions for Accessible units. While Type A units are mentioned in Section 1107.7.1, the only exception for Type A units is found in Section 1107.7.5.

Sections 1107.7.1 through 1107.7.4 primarily deal with nonelevator buildings. If elevators are provided, except as addressed in Section 1107.7.3, all units in the building are required to meet Type B criteria or better. Providing access to the upper floors in these cases would require either an elevator or a ramp system (or multiple elevators or ramps), both of which would be unreasonable. These exceptions are intended to be consistent with the scope of FHAG. It should be noted that these are only exceptions to the requirements for Type B dwelling units.

In accordance with Section 1107.2, Accessible and Type A units provide a higher level of accessibility than Type B units. Therefore, if Accessible or Type A units are provided on the accessible levels, this would exceed Type B unit requirements. For clarity in the commentary for the exceptions, only Type B units will be mentioned.

1107.7.1 Structures without elevator service. Where no elevator service is provided in a structure, only the *dwelling units* and *sleeping units* that are located on stories indicated in Sections 1107.7.1.1 and 1107.7.1.2 are required to be *Type A units* and *Type B units*, respectively. The number of *Type A units* shall be determined in accordance with Section 1107.6.2.2.1.

❖ Only the units located on the floor levels defined in Sections 1107.7.1.1 and 1107.7.1.2 are required to contain Type A or Type B units when no elevator service is provided in the building. Floor levels that do not meet the criteria are not required to have an accessible route to that level, nor do the dwelling units or sleeping units on those levels have to meet Type A or Type B unit criteria. The building must still have the minimum number of Type A units required on the first floor.

1107.7.1.1 One story with Type B units required. At least one *story* containing *dwelling units* or *sleeping units intended to be occupied as a residence* shall be provided with an *accessible* entrance from the exterior of the structure and all units *intended to be occupied as a residence* on that *story* shall be *Type B units*.

❖ This section basically states that Type B units must be provided on at least one level of a building that is not equipped with elevator service. For example, on a flat site, a two-story structure with Type B dwelling units or sleeping units on the first floor would not require an accessible route to, or Type B units on, the second floor (see Commentary Figure 1107.7.1.1).

1107.7.1.2 Additional stories with Type B units. On all other stories that have a building entrance in proximity to arrival points intended to serve units on that *story*, as indicated in Items 1 and 2, all *dwelling units* and *sleeping units intended to be occupied as a residence* served by that entrance on that *story* shall be *Type B units*.

1. Where the slopes of the undisturbed *site* measured between the planned entrance and all vehicular or pedestrian arrival points within 50 feet (15 240 mm) of the planned entrance are 10 percent or less, and
2. Where the slopes of the planned finished grade measured between the entrance and all vehicular or pedestrian arrival points within 50 feet (15 240 mm) of the planned entrance are 10 percent or less.

Where no such arrival points are within 50 feet (15 240 mm) of the entrance, the closest arrival point shall be used unless that arrival point serves the *story* required by Section 1107.7.1.1.

❖ This section addresses the idea that a building could have two levels that would be provided with accessible routes. This could be a structure built into a hill or a building with the first level a few feet down and the second level a few feet up from grade level [see Commentary Figure 1107.7.1.2(1)].

The basic test for multiple accessible levels is to check for the slope between the building entrance and any arrival points within a 50-foot (15 240 mm) arc. If no arrival points (e.g., sidewalk, parking) are within that 50-foot (15 240 mm) arc, the closest arrival point should be used as a reference. If the slope between these two points is 10 percent or less both before and after grading of the site, then an accessible route is required to that level [see Commentary Figures 1107.7.1.2(2) and 1107.7.1.2(3)].

In the case of sidewalks, the closest point to the entrance will be where a public sidewalk entering the site intersects with the sidewalk to the entrance. In the case of resident parking areas, the closest point to the planned entrance will be measured from the entry point to the parking area that is located closest to the planned entrance. The measurement for elevation should be taken from the center of the entrance door to the top of the pavement at the arrival point.

1107.7.2 Multistory units. A *multistory dwelling unit* or *sleeping unit* that is not provided with elevator service is not required to be a *Type B unit*. Where a *multistory unit* is provided with external elevator service to only one floor, the floor provided with elevator service shall be the primary entry to the unit, shall comply with the requirements for a

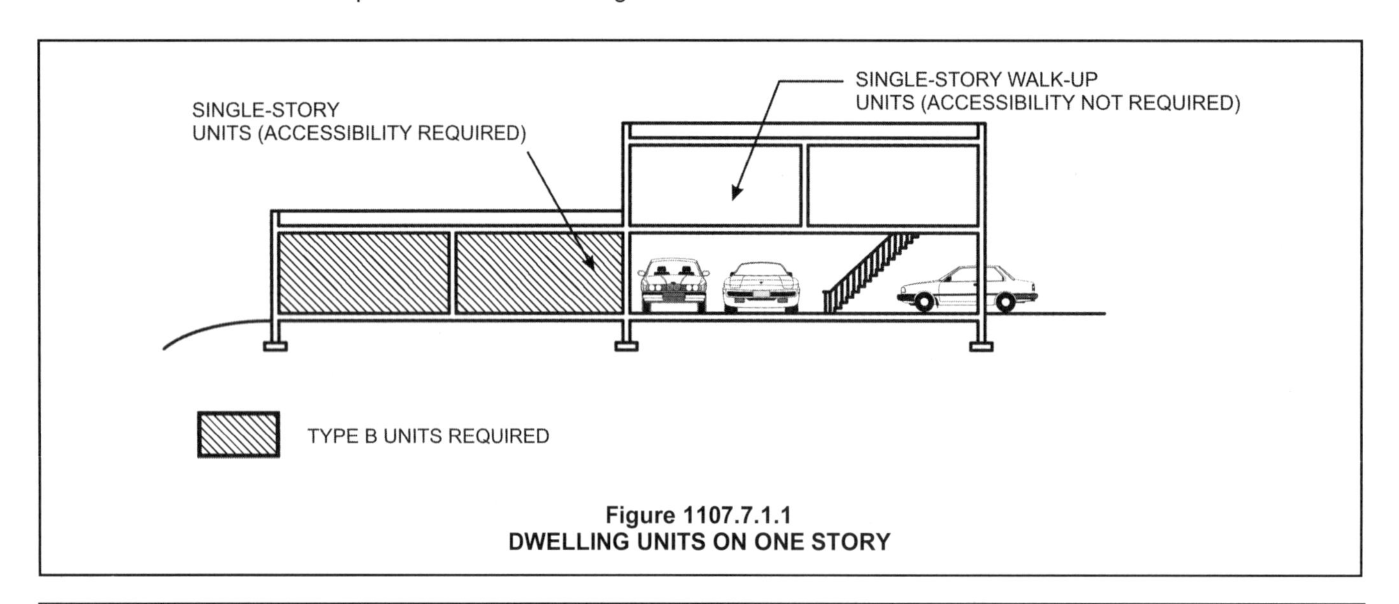

Figure 1107.7.1.1
DWELLING UNITS ON ONE STORY

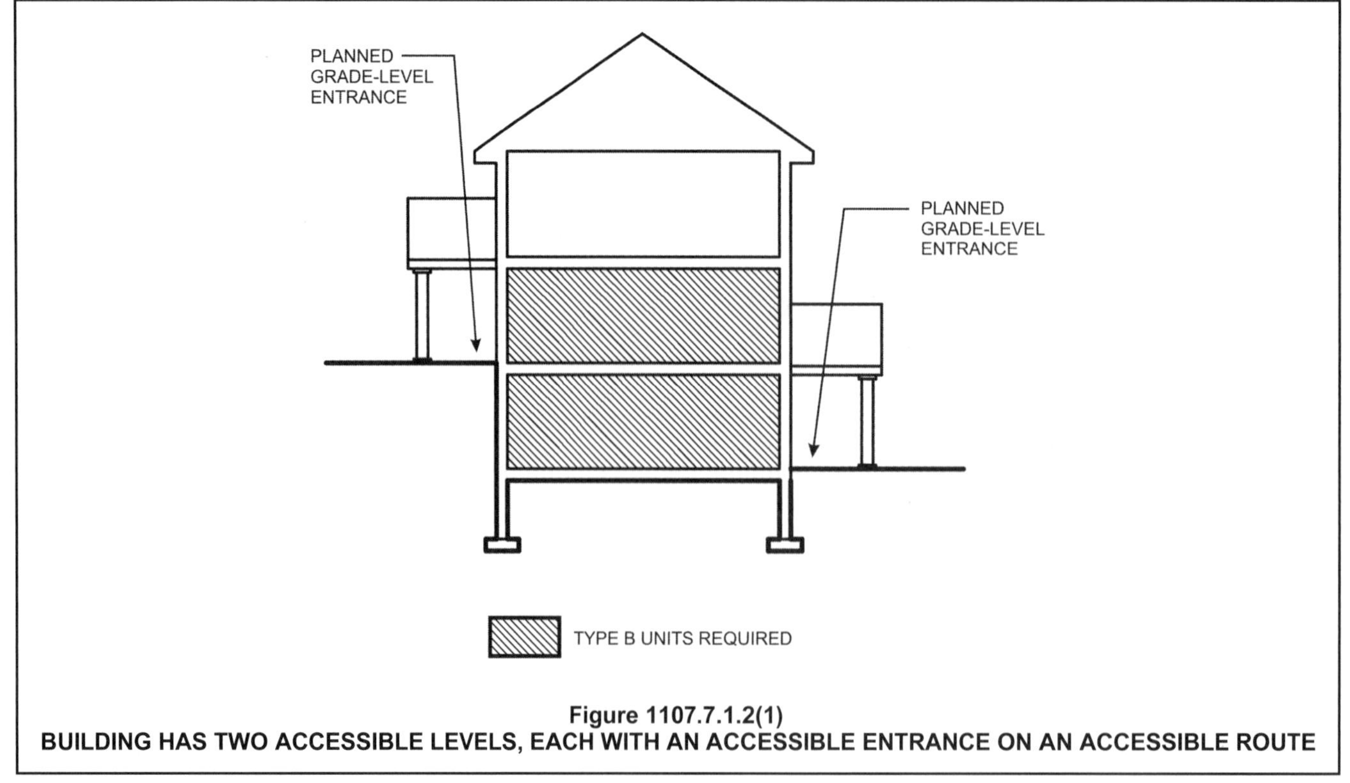

Figure 1107.7.1.2(1)
BUILDING HAS TWO ACCESSIBLE LEVELS, EACH WITH AN ACCESSIBLE ENTRANCE ON AN ACCESSIBLE ROUTE

Type B unit and, where provided within the unit, a living area, a kitchen and a toilet facility shall be provided on that floor.

❖ Section 1107.7.2 addresses "multistory units" (see the definition in Chapter 2). For example, the typical townhouse scenario, where there is a series of multistory units adjacent to each other in a single structure and no elevators are provided, would not require Type B units. If a townhouse has a private residence elevator, it would have to meet Type B criteria. In a multistory dwelling or sleeping unit that only has elevator access at one floor level, that level must be accessible and contain a living area kitchen and toilet facility [see Commentary Figures 1107.7.2(1) and 1107.7.2(2)].

1107.7.3 Elevator service to the lowest story with units. Where elevator service in the building provides an *accessible route* only to the lowest *story* containing *dwelling units* or *sleeping units intended to be occupied as a residence*, only the units on that *story* that are *intended to be occupied as a residence* are required to be *Type B units*.

❖ Section 1107.7.3 exempts dwelling units located on upper floors as long as the dwelling units on the first floor containing dwelling units are at least Type B units. For example, a three-story structure has a business on the first floor with apartments on the second and third levels. An accessible route is required to the second level, and all dwelling units on that level are required to be Type B units; however, an accessible route and Type B units are not required to the third level. (Note: If an elevator is utilized to provide access to only the lowest level containing dwelling or sleeping units, this building would not be considered an "elevator" building) [see Commentary Figures 1107.7.3(1) and 1107.7.3(2)].

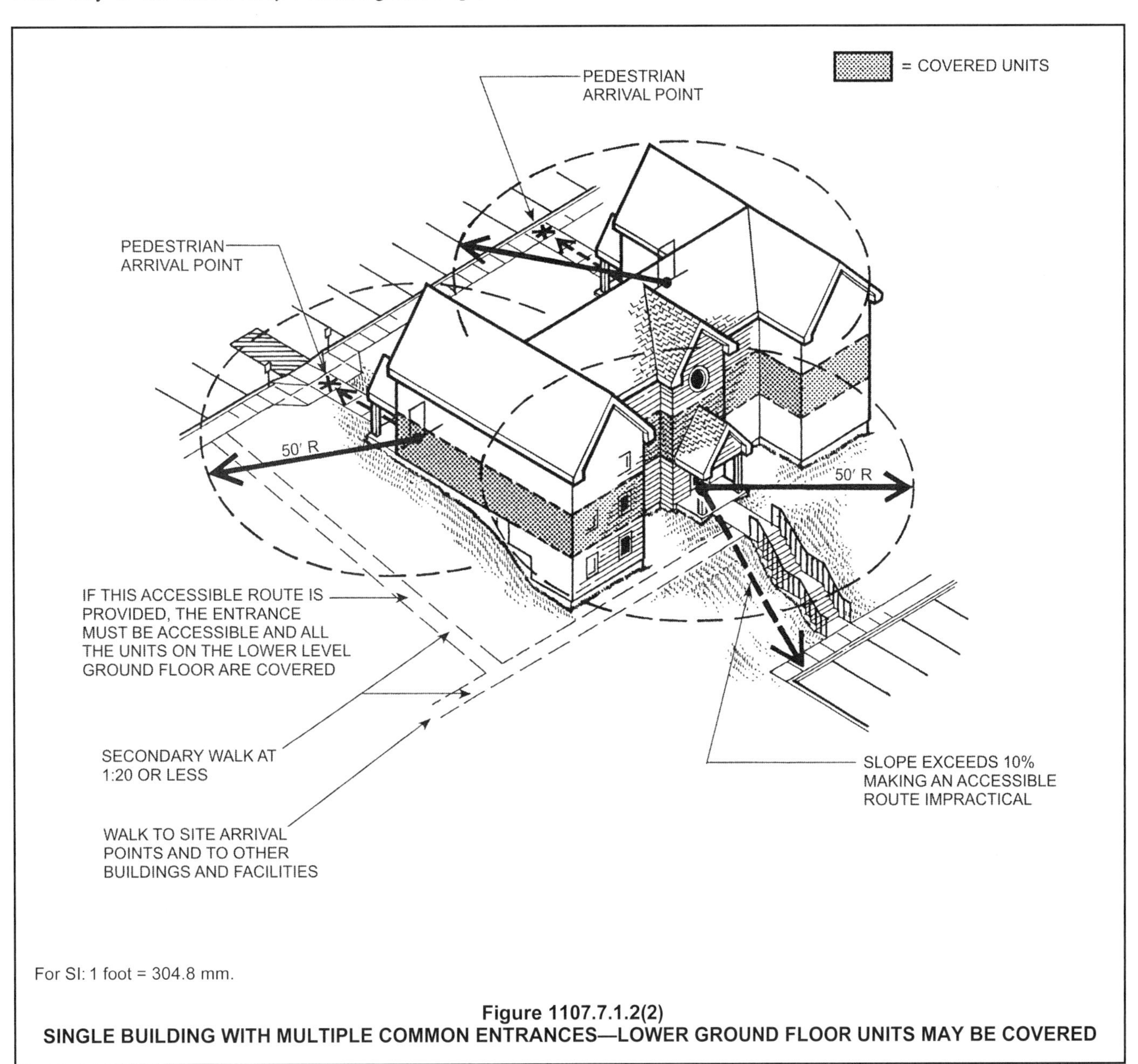

Figure 1107.7.1.2(2)
SINGLE BUILDING WITH MULTIPLE COMMON ENTRANCES—LOWER GROUND FLOOR UNITS MAY BE COVERED

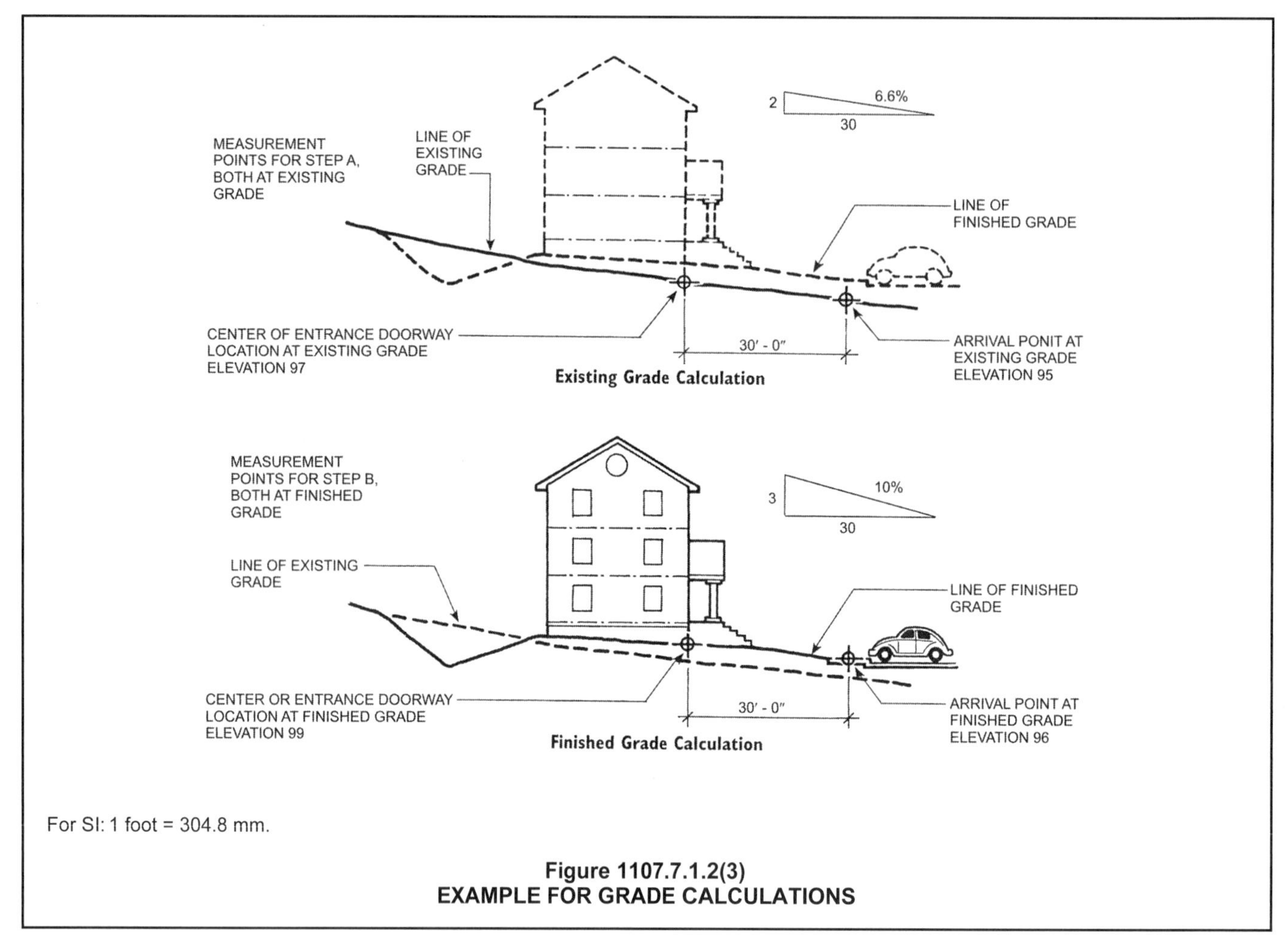

For SI: 1 foot = 304.8 mm.

Figure 1107.7.1.2(3)
EXAMPLE FOR GRADE CALCULATIONS

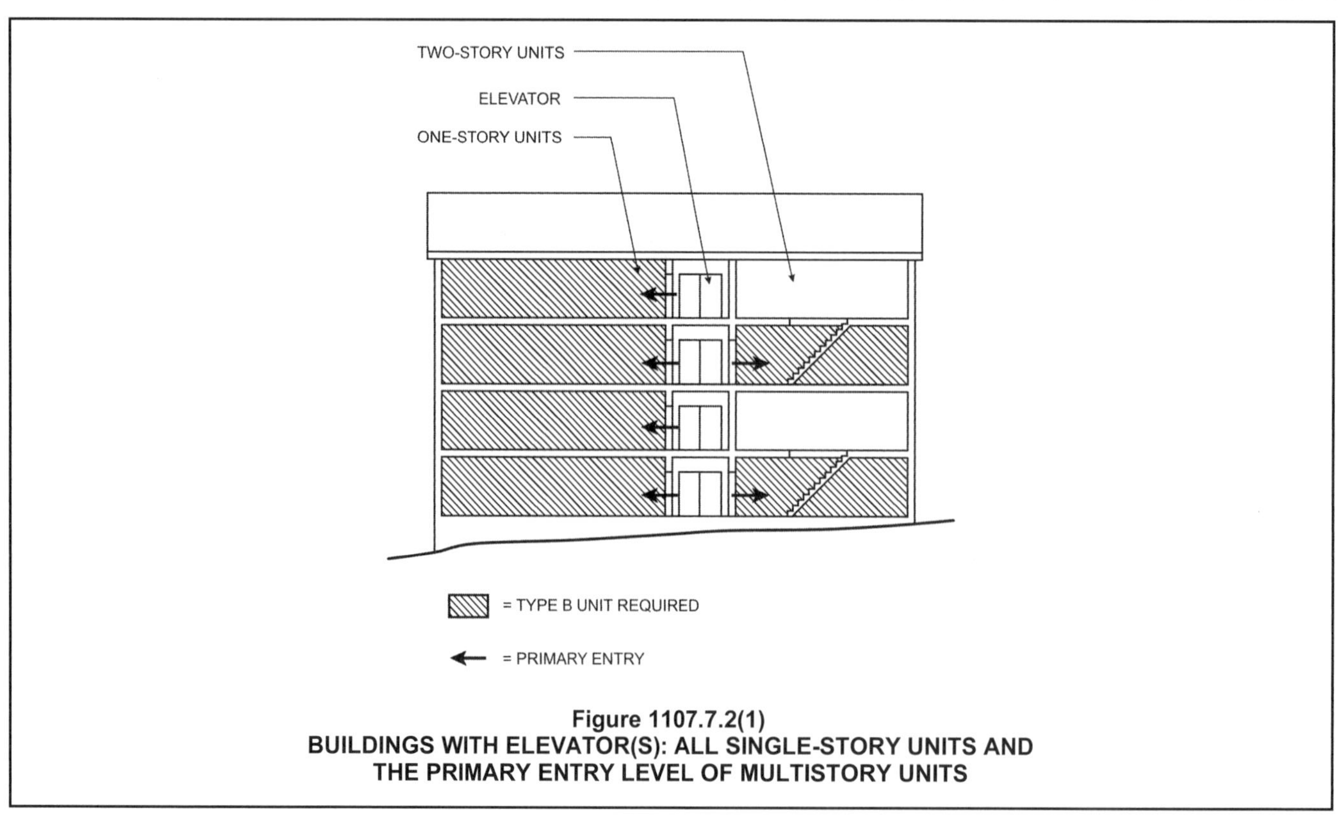

Figure 1107.7.2(1)
BUILDINGS WITH ELEVATOR(S): ALL SINGLE-STORY UNITS AND THE PRIMARY ENTRY LEVEL OF MULTISTORY UNITS

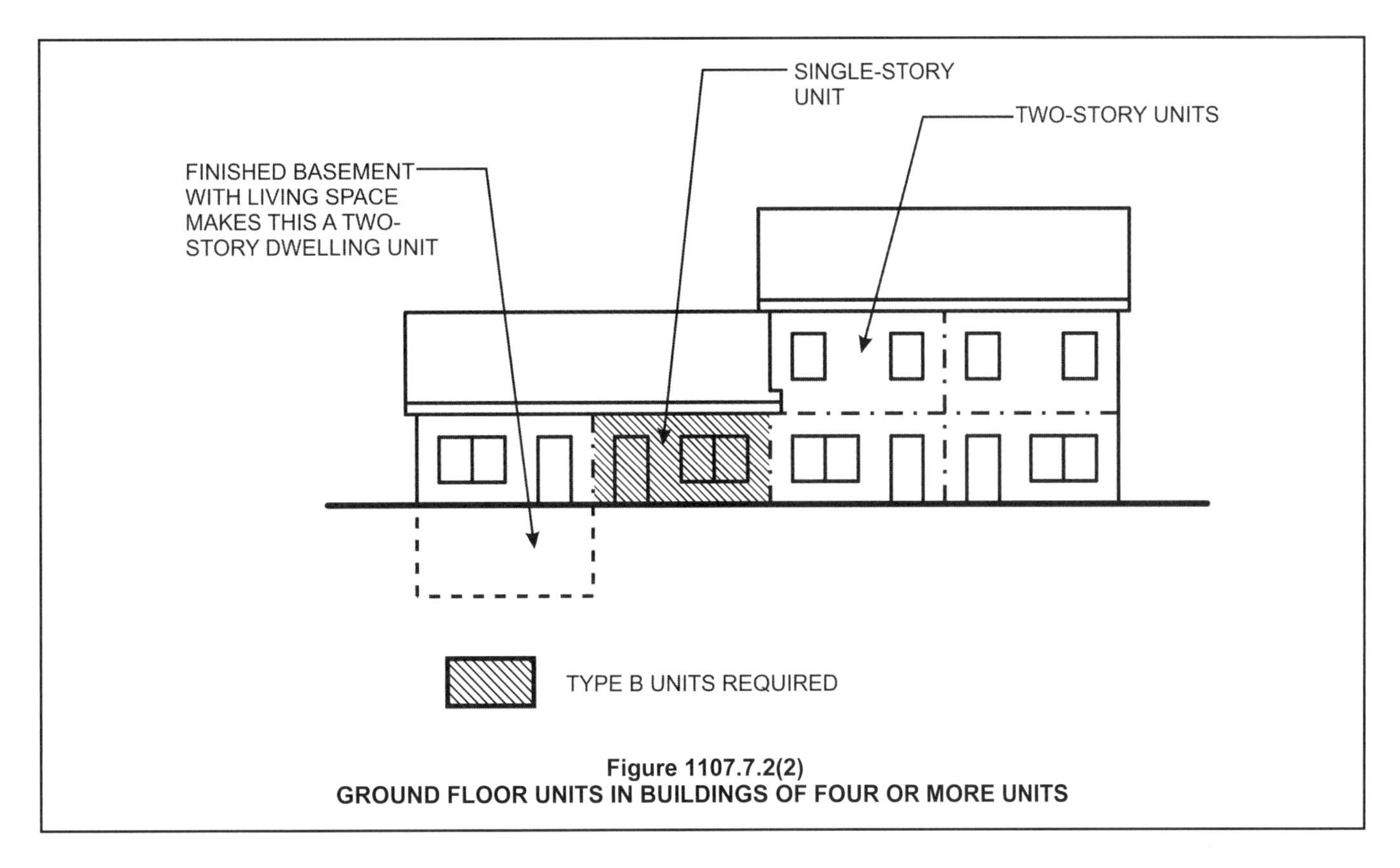

Figure 1107.7.2(2)
GROUND FLOOR UNITS IN BUILDINGS OF FOUR OR MORE UNITS

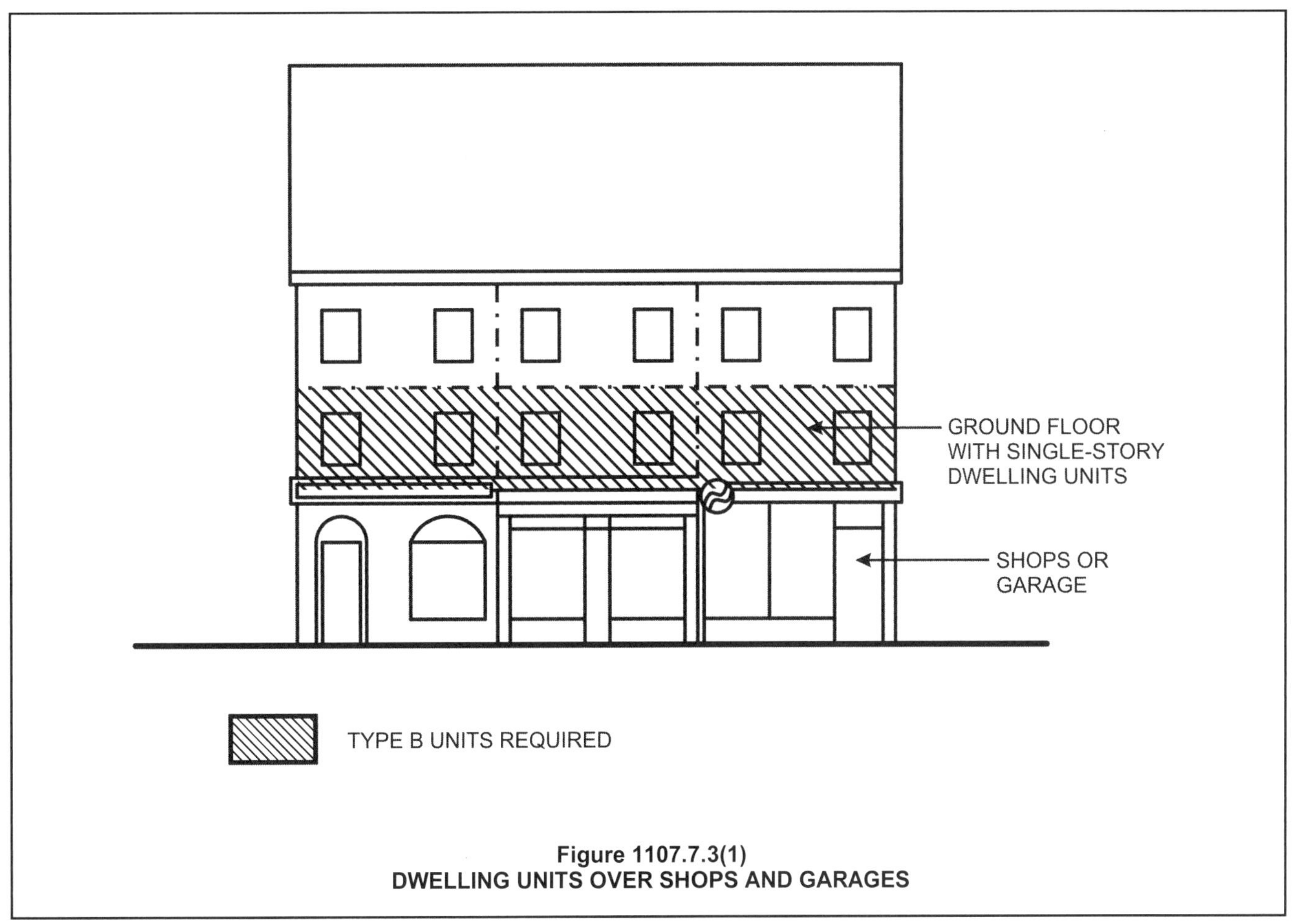

Figure 1107.7.3(1)
DWELLING UNITS OVER SHOPS AND GARAGES

1107.7.4 Site impracticality. On a *site* with multiple nonelevator buildings, the number of units required by Section 1107.7.1 to be *Type B units* is permitted to be reduced to a percentage that is equal to the percentage of the entire *site* having grades, prior to development, that are less than 10 percent, provided that all of the following conditions are met:

1. Not less than 20 percent of the units required by Section 1107.7.1 on the *site* are *Type B units*;
2. Units required by Section 1107.7.1, where the slope between the building entrance serving the units on that *story* and a pedestrian or vehicular arrival point is no greater than 8.33 percent, are *Type B units*;
3. Units required by Section 1107.7.1, where an elevated walkway is planned between a building entrance serving the units on that *story* and a pedestrian or vehicular arrival point and the slope between them is 10 percent or less, are *Type B units*; and
4. Units served by an elevator in accordance with Section 1107.7.3 are *Type B units*.

❖ Section 1107.7.4 addresses multiple buildings on a sloping site. For example, if an apartment complex was built on a steep or hilly site, the number of Type B dwelling units required on the ground floor could be reduced because of the difficulty of providing accessible routes. A minimum of 20 percent of the total ground floor units on the site must be Type B units, regardless of site complications (see Commentary Figure 1107.7.4).

1107.7.5 Design flood elevation. The required number of *Type A units* and *Type B units* shall not apply to a *site* where the required elevation of the lowest floor or the lowest horizontal structural building members of nonelevator buildings are at or above the *design flood elevation* resulting in:

1. A difference in elevation between the minimum required floor elevation at the primary entrances and vehicular and pedestrian arrival points within 50 feet (15 240 mm) exceeding 30 inches (762 mm), and
2. A slope exceeding 10 percent between the minimum required floor elevation at the primary entrances and vehicular and pedestrian arrival points within 50 feet (15 240 mm).

Where no such arrival points are within 50 feet (15 240 mm) of the primary entrances, the closest arrival points shall be used.

❖ Residential structures in flood hazard areas must be elevated (see Section 1612). If, based on the required floor elevation, the criteria in either Item 1 or 2 are met, it is considered that an accessible route to the units is not feasible. This section applies to both Type A and Type B dwelling units [see Commentary Figures 1107.7.5(1) and 1107.7.5(2)].

Flood-resistant requirements address, in part, concerns about public safety. The code provides for reductions in the level of accessibility in residential facilities without elevators that are elevated to comply with Section 1612. Note that the exception is based on the required design flood elevation. If the first-floor elevation is higher than the design flood elevation by choice, the measurement would still be from the design flood elevation rather than the actual floor elevation.

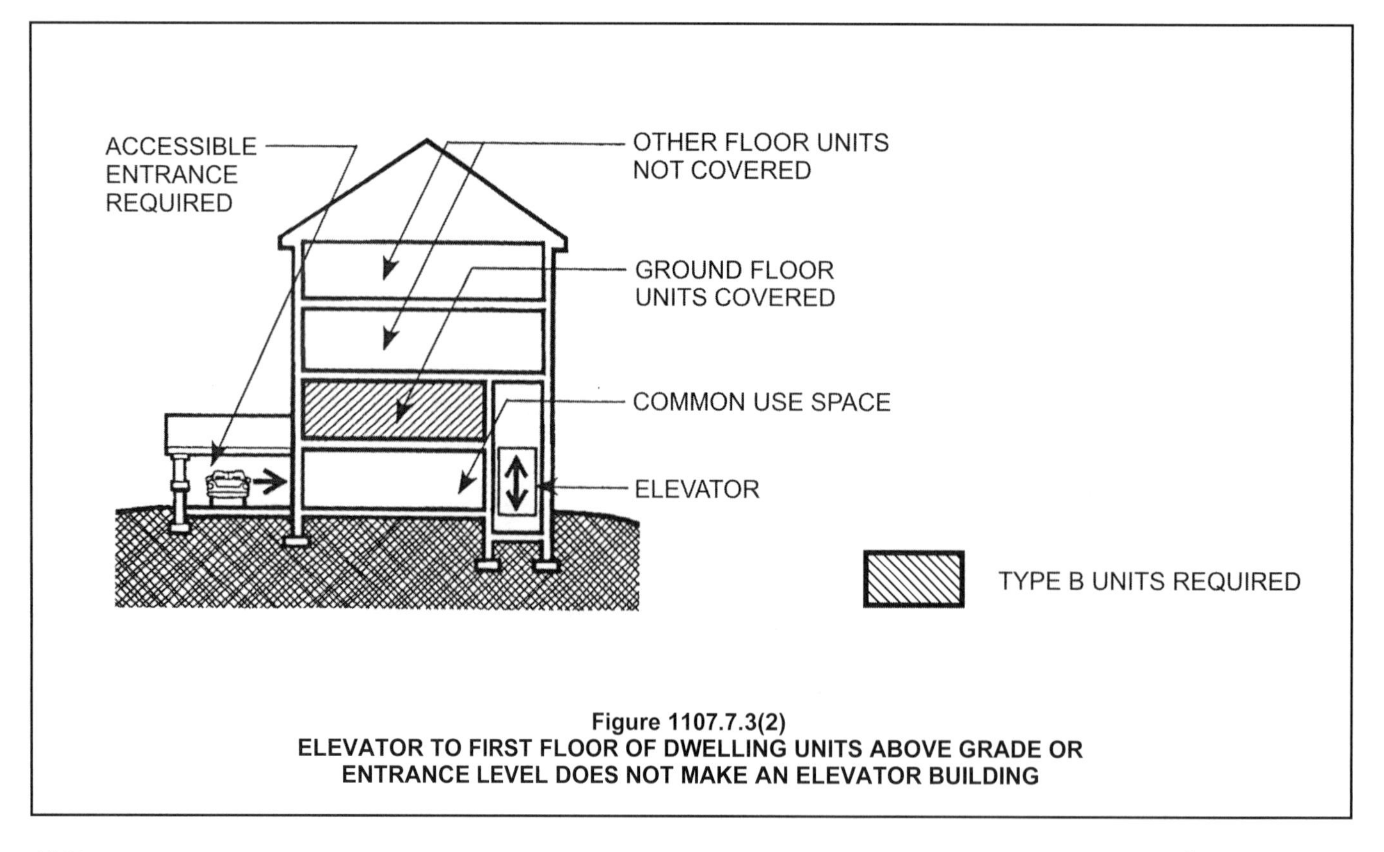

Figure 1107.7.3(2)
ELEVATOR TO FIRST FLOOR OF DWELLING UNITS ABOVE GRADE OR ENTRANCE LEVEL DOES NOT MAKE AN ELEVATOR BUILDING

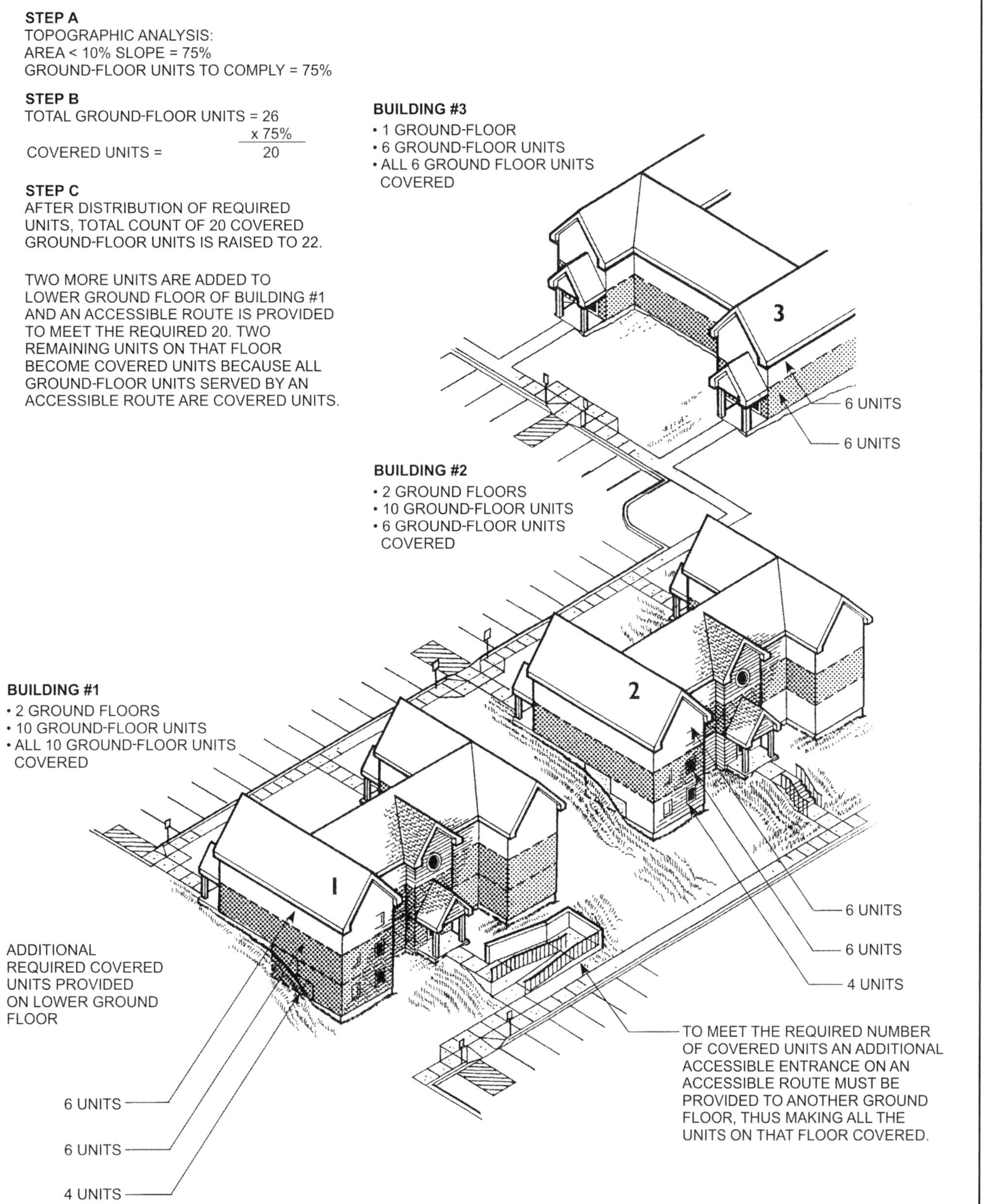

STEP A
TOPOGRAPHIC ANALYSIS:
AREA < 10% SLOPE = 75%
GROUND-FLOOR UNITS TO COMPLY = 75%

STEP B
TOTAL GROUND-FLOOR UNITS = 26
x 75%
COVERED UNITS = 20

STEP C
AFTER DISTRIBUTION OF REQUIRED UNITS, TOTAL COUNT OF 20 COVERED GROUND-FLOOR UNITS IS RAISED TO 22.

TWO MORE UNITS ARE ADDED TO LOWER GROUND FLOOR OF BUILDING #1 AND AN ACCESSIBLE ROUTE IS PROVIDED TO MEET THE REQUIRED 20. TWO REMAINING UNITS ON THAT FLOOR BECOME COVERED UNITS BECAUSE ALL GROUND-FLOOR UNITS SERVED BY AN ACCESSIBLE ROUTE ARE COVERED UNITS.

BUILDING #3
- 1 GROUND-FLOOR
- 6 GROUND-FLOOR UNITS
- ALL 6 GROUND FLOOR UNITS COVERED

3

6 UNITS

6 UNITS

BUILDING #2
- 2 GROUND FLOORS
- 10 GROUND-FLOOR UNITS
- 6 GROUND-FLOOR UNITS COVERED

2

BUILDING #1
- 2 GROUND FLOORS
- 10 GROUND-FLOOR UNITS
- ALL 10 GROUND-FLOOR UNITS COVERED

1

6 UNITS

6 UNITS

4 UNITS

ADDITIONAL REQUIRED COVERED UNITS PROVIDED ON LOWER GROUND FLOOR

TO MEET THE REQUIRED NUMBER OF COVERED UNITS AN ADDITIONAL ACCESSIBLE ENTRANCE ON AN ACCESSIBLE ROUTE MUST BE PROVIDED TO ANOTHER GROUND FLOOR, THUS MAKING ALL THE UNITS ON THAT FLOOR COVERED.

6 UNITS

6 UNITS

4 UNITS

Figure 1107.7.4
SITE ANALYSIS TEST: THE NUMBER OF COVERED UNITS

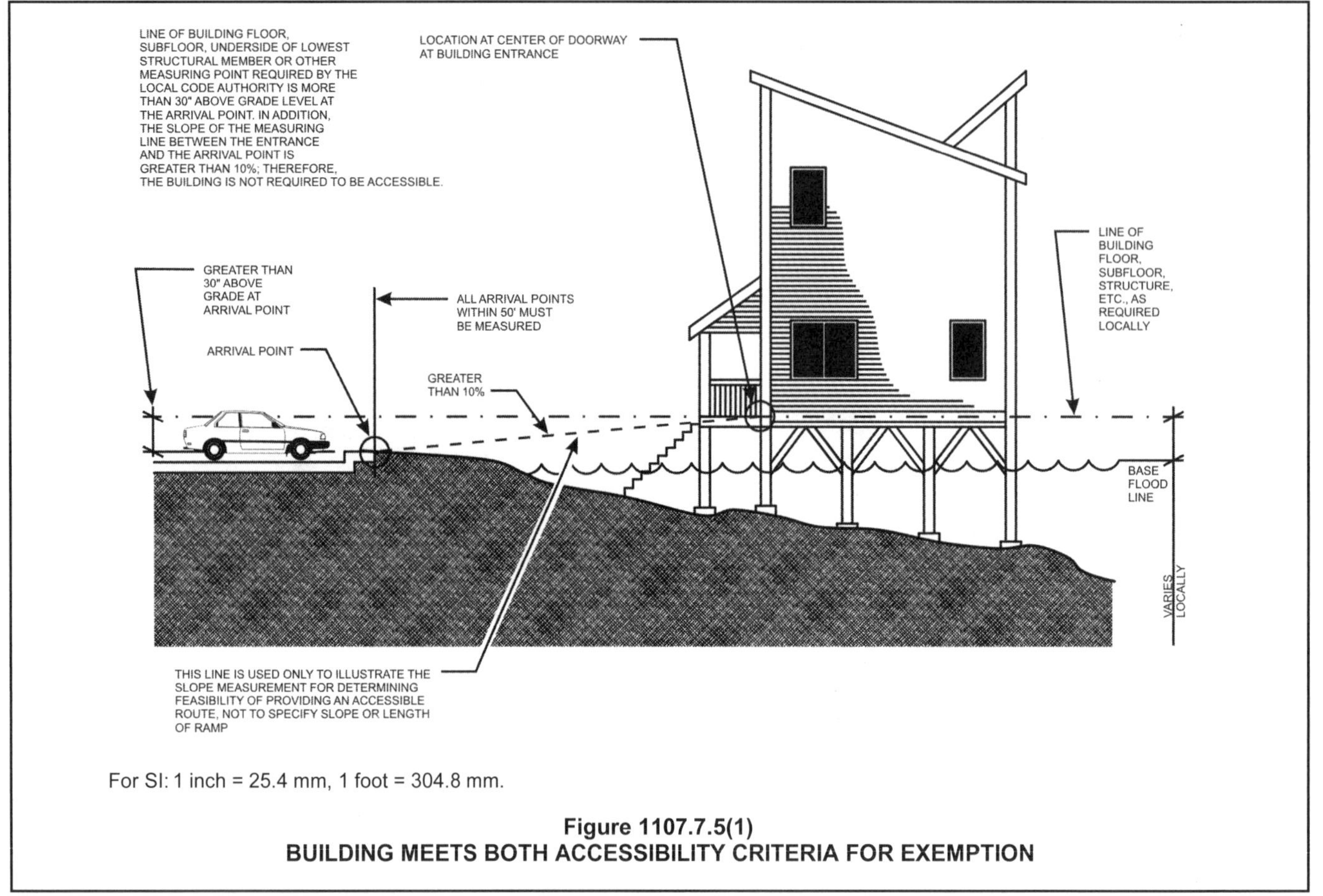

For SI: 1 inch = 25.4 mm, 1 foot = 304.8 mm.

Figure 1107.7.5(1)
BUILDING MEETS BOTH ACCESSIBILITY CRITERIA FOR EXEMPTION

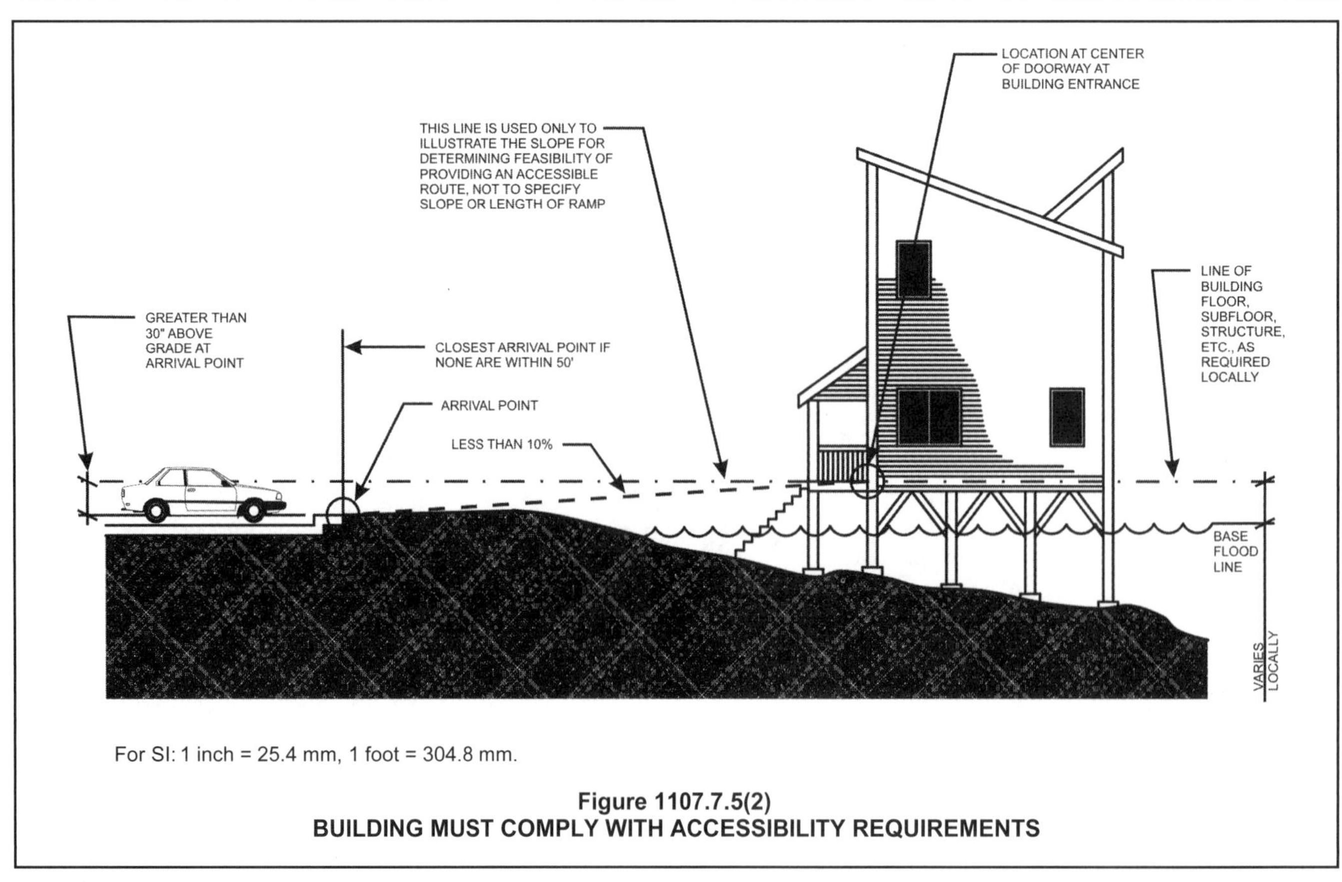

For SI: 1 inch = 25.4 mm, 1 foot = 304.8 mm.

Figure 1107.7.5(2)
BUILDING MUST COMPLY WITH ACCESSIBILITY REQUIREMENTS

SECTION 1108
SPECIAL OCCUPANCIES

1108.1 General. In addition to the other requirements of this chapter, the requirements of Sections 1108.2 through 1108.4 shall apply to specific occupancies.

❖ The criteria provided herein are occupancy specific, and are intended to result in a reasonable level of accessibility in areas with assembly seating, self-service storage facilities and judicial facilities.

1108.2 Assembly area seating. A building, room or space used for assembly purposes with *fixed seating* shall comply with Sections 1108.2.1 through 1108.2.5. Lawn seating shall comply with Section 1108.2.6. Assistive listening systems shall comply with Section 1108.2.7. Performance areas viewed from assembly seating areas shall comply with Section 1108.2.8. Dining areas shall comply with Section 1108.2.9.

❖ Sections 1108.2.1 through 1108.2.5 specifically address facilities with fixed seating utilized for purposes of viewing an event, typically facilities and spaces of occupancies of Groups A-1, A-3, A-4 and A-5. These criteria would also be applicable in assembly-type spaces with fixed seats that are located in buildings of other occupancies. Assembly spaces with fixed seating must provide accessible viewing locations that have: access to services; the number of wheelchair spaces and their associated companion seats dispersed so that persons using wheelchair spaces have options; and designated aisle seats. Lawn seating is a unique type of seating arrangement that is typically exterior and not associated with fixed seating. When there is a stage or platform associated with the assembly space, assistive listening systems and performance areas are addressed. The requirements for seating in areas for eating or drinking, typically Group A-2 spaces or facilities, and the dispersion of this seating are addressed in Section 1108.2.9.

1108.2.1 Services. If a service or facility is provided in an area that is not *accessible*, the same service or facility shall be provided on an *accessible* level and shall be *accessible*.

❖ This section establishes an important concept. The intent is that all types of services provided by a facility must be accessible. For example, after providing and dispersing wheelchair spaces in a venue, and because of tiered seating to provide a line of sight to the event, a stadium may have many tiers or sections of seating that are not accessible. This section establishes that if a souvenir stand is located in or near that non-accessible section, one of three circumstances would apply: an additional souvenir stand must be provided in the facility on an accessible level; the souvenir stand would have to be relocated to an accessible level; or the area where the souvenir stand is located must be made accessible. This section also emphasizes that not only must that service or facility be located on an accessible level, but the service itself must also be accessible. Obviously, it would be inappropriate to locate a souvenir stand on an accessible level if the approach or entrance to that stand is inaccessible.

1108.2.2 Wheelchair spaces. In rooms and spaces used for assembly purposes with *fixed seating*, *accessible wheelchair spaces* shall be provided in accordance with Sections 1108.2.2.1 through 1108.2.2.3.

❖ The intent of this section is to provide a reasonable number of spaces in an assembly occupancy with fixed seating to accommodate persons who use wheelchairs or scooters. Demographic statistics from the National Center for Health Statistics on the number of noninstitutionalized Americans who use wheelchairs indicate that these requirements are realistic. These required wheelchair spaces consist of an open, available floor space in which the wheelchair takes the place of the fixed seat that would otherwise occupy that space. Section 1108.2.3 is consistent with ICC A117.1 in requiring companion seating adjacent to each wheelchair space.

1108.2.2.1 General seating. *Wheelchair spaces* shall be provided in accordance with Table 1108.2.2.1.

❖ The number of required wheelchair spaces is indicated in Table 1108.2.2.1 and is based on the total number of fixed seats set up to view the same event, with the exception of box seats covered in Sections 1108.2.2.2 and 1108.2.2.3. For example, the fixed seating on all levels and all types provided in a sports stadium, excluding box seats, is used to calculate the number of required wheelchair spaces for general seating. The requirements for box- or suite-type seating are calculated separately. Seating provided in different rooms, such as in a series of lecture halls in a university, must be calculated separately. Note that the percentage of wheelchair spaces is less for facilities with a capacity of more than 5,000. The industry has been able to provide statistics to show that the higher percentage of wheelchair spaces is not typically utilized in large facilities.

TABLE 1108.2.2.1
ACCESSIBLE WHEELCHAIR SPACES

CAPACITY OF SEATING IN ASSEMBLY AREAS	MINIMUM REQUIRED NUMBER OF WHEELCHAIR SPACES
4 to 25	1
26 to 50	2
51 to 100	4
101 to 300	5
301 to 500	6
501 to 5,000	6, plus 1 for each 150, or fraction thereof, between 501 through 5,000
5,001 and over	36 plus 1 for each 200, or fraction thereof, over 5,000

❖ This table sets forth the required number of wheelchair spaces based on the capacity of seating in the space containing the fixed seating. Any fixed seating with less than four seats is not required to provide wheelchair spaces. There are special criteria for wheelchair spaces in facilities with seating capacities of over 500 and over 5,000. For example, six wheelchair spaces would be required for an assembly space with a seating capacity of 500. Seven wheelchair spaces would be required for a seating capacity of 501 through 650. Both of the last two rows use the term "or fraction thereof" to designate the next step up.

1108.2.2.2 Luxury boxes, club boxes and suites. In each luxury box, club box and suite within arenas, stadiums and *grandstands*, *wheelchair spaces* shall be provided in accordance with Table 1108.2.2.1.

❖ When luxury boxes, club boxes or suites are provided, each luxury box, club box or suite must have an accessible route to that box and at least one wheelchair space and associated companion space. If the luxury box, club box or suite has more than 25 seats, the number of required wheelchair spaces is increased in accordance with Table 1108.2.2.1. For example, if a stadium has three luxury boxes with 51 seats in each, at least four wheelchair spaces and their associated companion seats must be provided in each box. While the total of all three luxury boxes (i.e., 153 occupants) would only require five total wheelchair spaces, these specific types of boxes must be calculated individually.

Luxury boxes are most commonly found in larger facilities, thus the reference to arenas, stadiums and grandstands. Boxes that are portions of balconies, such as in an opera house, or boxes separated by rails in general seating areas, such as behind home plate in baseball stadiums, are addressed in Section 1108.2.2.3.

1108.2.2.3 Other boxes. In boxes other than those required to comply with Section 1108.2.2.2, the total number of *wheelchair spaces* provided shall be determined in accordance with Table 1108.2.2.1. *Wheelchair spaces* shall be located in not less than 20 percent of all boxes provided.

❖ Examples of the boxes covered in this section are the seats separated by railings or low walls behind home plate in a baseball stadium [see Commentary Figure 1108.2.2.3(1)], or side balconies in opera houses [see Commentary Figure 1108.2.2.3(2)]. Luxury boxes are typically located in a separate level, and are often attached to some type of party room behind a group of seats.

When boxes other than luxury boxes, club boxes or suites are provided, at least 20 percent of the boxes must have an accessible route to that box and at least one wheelchair space and associated companion space in that box. The total number of box seats is utilized to calculate the total number of required wheelchair spaces in accordance with Table 1108.2.2.1. For example, if a stadium has five boxes with 51 seats in each, the total of all five boxes (i.e., 255 occupants) would require five total wheelchair spaces. The wheelchair spaces could be located in one box (i.e., 20 percent minimum) or dispersed to all five boxes. Note that there is additional dispersion criteria for assembly seating based on lines of sight in Section 802.8 of ICC A117.1, as well as dispersion by level in Section 1108.2.4.

Figure 1108.2.2.3(1)
EXAMPLE OF BOX SEATING IN A SPORTS VENUE

Figure 1108.2.2.3(2)
EXAMPLE OF BOX SEATING IN A THEATER

1108.2.3 Companion seats. At least one companion seat shall be provided for each *wheelchair space* required by Sections 1108.2.2.1 through 1108.2.2.3.

❖ A companion seat allows a friend, relative, associate, etc., to accompany a person using a wheelchair at an event and to share in the experience with the same level of companionship as others attending the event. The ICC A117.1 standard permits two wheelchair spaces to be located next to each other; however, each wheelchair space must also be adjacent to at least one companion seat. Accessible seating areas are sometimes provided with nonfixed companion chairs allowing the users to adjust the adjacency of seats and wheelchair spaces to their liking. If fixed seats are provided, the companion seat must have shoulder alignment with the person using the wheelchair space.

1108.2.4 Dispersion of wheelchair spaces in multilevel assembly seating areas. In *multilevel assembly seating* areas, *wheelchair spaces* shall be provided on the main floor level and on one of each two additional floor or *mezzanine* levels. *Wheelchair spaces* shall be provided in each luxury box, club box and suite within assembly facilities.

Exceptions:

1. In *multilevel assembly seating* areas utilized for worship services where the second floor or *mezzanine* level contains 25 percent or less of the total seating capacity, *wheelchair spaces* shall be permitted to all be located on the main level.
2. In *multilevel assembly seating* areas where the second floor or *mezzanine* level provides 25 percent or less of the total seating capacity and 300 or fewer seats, all *wheelchair spaces* shall be permitted to be located on the main level.
3. *Wheelchair spaces* in team or player seating serving areas of sport activity are not required to be dispersed.

❖ When facilities provide seating on multiple levels, wheelchair spaces must also be provided on multiple levels. For example, if a theater has a main level and a balcony level, wheelchair spaces must be provided on both. If a theater has a main level and two balcony levels, wheelchair spaces must be provided on the main level and at least one of the two balcony levels. While the term "level" is difficult to define in some assembly arrangements, the intent is to provide the wheelchair user with a choice of seating similar to what is available to all persons in the space. A "level" is not each time that there is a step-down to another row of seating. Multilevel seating facilities are typically separate floor levels, commonly referred to as balconies. Subjective judgment on the part of both the designer and the building official is required for each individual facility.

There are three specific exceptions to this multilevel dispersion requirement.

Exception 1 is applicable to assembly spaces utilized for worship services. There is typically not a cost or line-of-sight issue in this type of facility, as there is in other assembly seating areas. The balcony is not required to contain accessible seating or have an accessible route, as long as accessible seating is provided on the main level and the balcony does not contain more than 25 percent of the seating.

Exception 2 is applicable to balconies in assembly facilities with a limited number of seats in the upper level. If accessible seating is provided on the main level and the balcony contains less than 25 percent of the seating and less than 300 seats, then the balcony is not required to contain accessible seating or have an accessible route. This can be used for small theaters with up to 900 seats on the main floor and 300 seats on the balcony.

Exception 3 relieves the requirement for wheelchair space dispersion among the various team or player seating options that may be provided in a facility. However, there must be at least one wheelchair space for each team.

The dispersement requirements in the code must be combined with the additional wheelchair dispersement requirements in ICC A117.1, including side to side (horizontally) across a venue, front to back (variety of distances) within the seating, and by type (distinct service or amenities) of seating. The standard also looks at requirements for line of sight or seated or standing spectators. Wheelchair spaces and their associated companion seats may be grouped, but must, at a minimum, be distributed into the number of locations indicated in Table 802.10 of ICC A117.1.

For requirements for accessible means of egress from these wheelchair spaces, please see Section 1009.1, Exception 3 and Section 1029.8.

1108.2.5 Designated aisle seats. At least 5 percent, but not less than one, of the total number of aisle seats provided shall be designated aisle seats and shall be the aisle seats located closest to *accessible routes*.

Exception: Designated aisle seats are not required in team or player seating serving *areas of sport activity*.

❖ Designated aisle seats allow some persons with mobility impairments to be able to easily access certain seats along aisles. Due to an impairment, it may be difficult or impossible for some to move further into the narrow space between rows. It is not the intent of the designated aisle seats to be transfer locations. The person using this space can be someone using a walker, crutches or cane or anyone who may have difficulty with steps; hence, the requirement that the designated aisle seats be close to accessible routes. When seats have armrests, ICC A117.1 requires each designated aisle seat to have a folding or retractable armrest allowing easier access or better fit.

1108.2.6 Lawn seating. Lawn seating areas and exterior overflow seating areas, where fixed seats are not provided, shall connect to an *accessible route*.

❖ Lawn seating on both flat and sloped sites is fairly common at less formal venues. It may be the only type of seating provided or one of several options. Sometimes lawns are overflow areas, with or without movable chairs, located beyond the fixed seating areas. In any case, these lawn areas must be located on an accessible route. Access onto the lawn seating depends on the ability of the person and/or assistance from companions (see Commentary Figure 1108.2.6).

Figure 1108.2.6
EXAMPLE OF LAWN SEATING

1108.2.7 Assistive listening systems. Each building, room or space used for assembly purposes where audible communications are integral to the use of the space shall have an assistive listening system.

Exception: Other than in courtrooms, an assistive listening system is not required where there is no audio amplification system.

❖ This section is intended to accommodate people with a hearing impairment. Assembly seating areas, such as stadiums, theaters, auditoriums, lecture halls and similar spaces, are required to make provisions for the use of assistive listening devices in accordance with Section 1108.2.7.1 and Table 1108.2.7.1. In these assembly areas, audible communication is often integral to the use and full enjoyment of the space. This requirement offers the possibility for individuals with hearing impairments to attend functions in these facilities without having to give advance notice and without disrupting the event in order to have a portable assistive listening system set up and made ready for use.

There are three primary types of listening systems available: induction loop, AM/FM and infrared. Each type of system has certain advantages and disadvantages that the designer should take into consideration when choosing the type of system that is most appropriate for the intended application. Signage notifying the general public of the availability of these systems must be provided in accordance with Section 1111.3.

If an audio amplification system is not provided within a facility, an assistive listening system is not required. Given the essential nature of the proceedings, this exception is not applicable to courtrooms. All courtrooms must have an assistive listening system.

1108.2.7.1 Receivers. The number and type of receivers shall be provided for assistive listening systems in accordance with Table 1108.2.7.1.

Exceptions:

1. Where a building contains more than one room or space used for assembly purposes, the total number of required receivers shall be permitted to be calculated based on the total number of seats in the building, provided that all receivers are usable with all systems and if the rooms or spaces used for assembly purposes required to provide assistive listening are under one management.
2. Where all seats in a building, room or space used for assembly purposes are served by an induction loop assistive listening system, the minimum number of receivers required by Table 1108.2.7.1 to be hearing-aid compatible shall not be required.

❖ Table 1108.2.7.1 specifies the number and types of required receivers that must be available to use in an assembly area. Exception 1 states that if a facility has more than one assembly space with an audio amplification system, such as a multiplex theater, then the total seating for all the spaces may be used to determine the number of receivers required.

Exception 2 states that because induction loop technology renders an entire space accessible to the assistive listening system, hearing-aid compatible receivers are not necessary.

TABLE 1108.2.7.1. See below.

❖ Table 1108.2.7.1 specifies the number and type of required receivers in an assembly occupancy. At least 25 percent of the receivers provided (but not

TABLE 1108.2.7.1
RECEIVERS FOR ASSISTIVE LISTENING SYSTEMS

CAPACITY OF SEATING IN ASSEMBLY AREAS	MINIMUM REQUIRED NUMBER OF RECEIVERS	MINIMUM NUMBER OF RECEIVERS TO BE HEARING-AID COMPATIBLE
50 or less	2	2
51 to 200	2, plus 1 per 25 seats over 50 seats*	2
201 to 500	2, plus 1 per 25 seats over 50 seats*	1 per 4 receivers*
501 to 1,000	20, plus 1 per 33 seats over 500 seats*	1 per 4 receivers*
1,001 to 2,000	35, plus 1 per 50 seats over 1,000 seats*	1 per 4 receivers*
Over 2,000	55, plus 1 per 100 seats over 2,000 seats*	1 per 4 receivers*

Note: * = or fraction thereof

less than two receivers) must be hearing-aid compatible, as persons with hearing aids typically cannot use earpieces or headphone-equipped receivers. When determining the general number of receivers required, the table is to be applied as though it reads "2 plus 1 for each additional 25 over 50, or a fraction thereof." For example, where 51 to 75 total seats are provided, three assistive listening devices are required. Where 76 to 100 total seats are provided, four assistive listening devices are required.

The number of required receivers is based on the capacity of the assembly areas. Note that as the size of the assembly area increases, the number of required receivers also increases. However, the percentage of receivers to seats decreases. This is based on the actual usage in large assembly areas.

1108.2.7.2 Ticket windows. Where ticket windows are provided in stadiums and arenas, at least one window at each location shall have an assistive listening system.

❖ An assistive-listening system is required at one window in each location where tickets are sold. Note that this is limited to stadiums and arenas where it is assumed that the technology is already provided for inside the space (see Section 1108.2.7). This will allow a person with a hearing impairment to be able to communicate with the ticket salesperson regarding date, cost and location of tickets being purchased. The New York Yankees and Minnesota Twin stadiums are currently using an induction-loop system at their ticket windows.

1108.2.7.3 Public address systems. Where stadiums, arenas and *grandstands* have 15,000 fixed seats or more and provide audible public announcements, they shall also provide prerecorded or real-time captions of those audible public announcements.

❖ If stadiums, arenas or grandstands provide public announcements, the same information should be displayed on some type of electronic signage. With seating of 15,000 or more, most stadiums, arenas and grandstands have electronic scoreboards that are capable of displaying text messages. This is especially important when dealing with emergency evacuation situations. Prerecorded messages related to emergencies should be worked out with the fire department as part of the fire and safety evacuation plans [see Section 404 of the *International Fire Code*® (IFC®)]. Fire alarm systems for assembly occupancies (Section 907.2.1.1) with an occupant load of 1,000 or more must include an emergency voice/alarm communication system. Where captioned audible public announcements are required for stadiums by this section, the emergency voice/alarm communication system must also be captioned (see Section 907.2.1.2). Specific criteria are provided in Section 907.5.2.2.4.

ICC A117.1 has new provisions for variable message signs (VMS), which provides guidance for appropriate case, style, size, spacing, contrast and rate of change. There are special allowances that address large assembly occupancies and the distance from the viewer to the sign.

1108.2.8 Performance areas. An *accessible route* shall directly connect the performance area to the assembly seating area where a *circulation path* directly connects a performance area to an assembly seating area. An *accessible route* shall be provided from performance areas to ancillary areas or facilities used by performers.

❖ Performance areas, such as stages, orchestra pits, band platforms, choir lofts and similar spaces, must be accessible. If there is a direct route from the seating to the performance area, there must also be an accessible route. For example, if steps are provided from the assembly seating area to the stage within the theater, then an accessible route (e.g., ramp or platform lift) to the stage must also be provided within the theater. An accessible route must also be provided to any ancillary areas, such as green rooms or practice/warm-up rooms. The intent is that a person with mobility impairments could participate in the event. This could include high-school graduation with students coming from the audience onto the stage to receive their diplomas; participating with the community band; playing in the orchestra for a performance; acting in a production; or giving a speech. Technical production areas for the stage, such as catwalks, are exempted under Section 1103.2.7.

1108.2.9 Dining and drinking areas. In dining and drinking areas, all interior and exterior floor areas shall be *accessible* and be on an *accessible route*.

Exceptions:

1. An *accessible route* between *accessible* levels and stories above or below is not required where permitted by Section 1104.4, Exception 1.
2. An *accessible route* to dining and drinking areas in a *mezzanine* is not required, provided that the *mezzanine* contains less than 25 percent of the total combined area for dining and drinking and the same services, and decor are provided in the *accessible* area.
3. In sports facilities, tiered dining areas providing seating required to be *accessible* shall be required to have *accessible routes* serving at least 25 percent of the dining area, provided that *accessible routes* serve *accessible* seating and where each tier is provided with the same services.
4. Employee-only work areas shall comply with Sections 1103.2.2 and 1104.3.1.

❖ Dining and drinking areas most frequently occur in Group A-2 buildings (restaurants, cafeterias, nightclubs, dinner theaters, etc.). The provisions of this section are intended to govern such areas, rather than the criteria specified in Sections 1108.2.2 through 1108.2.5. This section requires the total floor area allotted for dining and drinking to be accessible, including areas with tables and chairs, bar seating, standing spaces, gaming areas, outdoor decks, etc.

An accessible route would be required to dining areas that are raised or lowered. The intent is that a person with a disability may not be able to get to every seat, but they will still have a choice of tables.

Exception 1, with a reference back to Exception 1 in Section 1104.4, exempts a basement, mezzanine or second floor from being served by an accessible route, provided the aggregate area of the nonaccessible levels is 3,000 square feet (278.7 m^2) or less. Keep in mind that a mezzanine is required by Section 505.2 to have a clear height below of at least 7 feet (2134 mm). Therefore, Exception 1 cannot be applied to platforms or raised or lowered floor areas.

Exception 2 is intended to acknowledge a practical and reasonable limitation in buildings with a mezzanine level that provides less than 25 percent of the total area for dining and drinking. In addition, all services and amenities provided in the mezzanine must also be available in an accessible area. Any mezzanine condition that does not meet the criteria of this exception does not qualify for the exception and must be served by an accessible route. Nonqualifying examples are as follows: a dining mezzanine that contains 25 percent or more of the total dining area; a raised or depressed dining area that is not actually a mezzanine; and a mezzanine level that contains the only area where a specific service or amenity is provided, like a bar or private party room. The area of the mezzanine is not limited by this exception to 3,000 square feet (278.7 m^2); it could be a larger area or a smaller area.

Exception 3 is for sports facilities. Some sports facilities also have accommodations for dining or picnicking while watching the event. For line-of-sight issues, the dining terraces are tiered. If the same services are available on the accessible level as any other level, only 25 percent of the dining area is required to be on an accessible route.

At this time, the code does not contain specific information on tiered dining facilities that also have issue of line of sight for viewing an event, such as a dinner theater, movie theaters that offer dining during a show, or dinner seating during a sport event such as at a racetrack. These types of spaces may want to look at the provisions for both dining (see Section 1108.2.9) and seating for viewing an event (see Section 1108.2.2) and develop a reasonable compromise.

Exception 4 is simply a reference to employee work area exceptions. Employee work areas may be within the mezzanine levels, such as the area behind a bar.

1108.2.9.1 Dining surfaces. Where dining surfaces for the consumption of food or drink are provided, at least 5 percent, but not less than one, of the dining surfaces for the seating and standing spaces shall be *accessible* and be distributed throughout the facility and located on a level accessed by an *accessible route*.

❖ A dining facility may offer spaces for people to eat or drink at tables, booths, bars and counters (see Commentary Figure 1108.2.9.1). The numbers provided at all locations are used to determine the total, and then 5 percent of the total of these spaces must be capable of allowing for a person in a wheelchair to move to and sit at that location. The accessible surfaces are also required to be distributed throughout the facility, but are not necessarily required to be dispersed by type. This requirement, in conjunction with Section 1108.2.9, should provide a variety of seating options. There are many considerations. For example, a table may be at the right height and provide knee and toe clearances, but the aisle and aisle accessway layout only allows for one or two approaches to the table. A booth may allow clearance at the end, but the location may not work out if a person sitting at that location may get bumped repeatedly by wait staff and customers trying to move in and out of the area, so this may be an operational issue for the restaurant. There are practical reasons to not require dispersment by type. You could take the booth seat off one side to provide better access to a booth, but then you have banquet seating. If you require a high table to be lowered, you no longer have a high table.

The issue of whether a portion of a bar or dining counter in a restaurant is required to be lowered to be accessible is subjective. The assumption is that if other types of seating are provided adjacent to the counter, then services provided at the counter will also be available at the adjacent seating. Therefore, if adequate accessible seating is available adjacent to the bar area, the bar is not required to be lowered. If the bar is the only eating or dining surface in a restaurant, or in a separate room in the restaurant, then a portion of the bar must be made accessible.

Figure 1108.2.9.1
EXAMPLE OF DINING SURFACES

1108.3 Self-service storage facilities. *Self-service storage facilities* shall provide *accessible* individual self-storage spaces in accordance with Table 1108.3.

❖ This section addresses facilities that provide self-storage units or spaces. These types of facilities are often storage garages located in long rows. Some facilities also provide climate-controlled storage within large multistory warehouses. The key is that the storage is moved in and out and accessed by the renter of the space. The intent is to provide access for persons with disabilities to this service without requiring the entire facility to be accessible.

TABLE 1108.3
ACCESSIBLE SELF-SERVICE STORAGE FACILITIES

TOTAL SPACES IN FACILITY	MINIMUM NUMBER OF REQUIRED ACCESSIBLE SPACES
1 to 200	5%, but not less than 1
Over 200	10, plus 2% of total number of units over 200

❖ The minimum number of accessible spaces is based on the total number of self-storage spaces available in a facility.

1108.3.1 Dispersion. *Accessible* individual self-service storage spaces shall be dispersed throughout the various classes of spaces provided. Where more classes of spaces are provided than the number of required *accessible* spaces, the number of *accessible* spaces shall not be required to exceed that required by Table 1108.3. *Accessible* spaces are permitted to be dispersed in a single building of a multibuilding facility.

❖ Self-storage facilities may offer a variety of spaces, such as heated/nonheated, different sizes, etc. If the variety offered is greater than the number of accessible spaces required, the accessible spaces should be dispersed as much as possible. For example, if a facility offers the choice of a 10-foot by 10-foot (3048 mm by 3048 mm) unit and a 10-foot by 20-foot (3048 mm by 6096 mm) unit, but only one accessible space is required, then only one unit is required to be accessible. The choice of which unit is made accessible is up to the building owner or designer. When a facility has multiple buildings, accessible spaces are permitted to be located in a single building. This is in consideration that many of these facilities do not have paved roads to access them. This allows for only a small portion of the site to be considered when looking at arrival points, accessible routes and entrances.

At this time, ICC A117.1 does not provide specific information on how to make a self-service storage unit accessible. Since overhead doors are not typically considered accessible, alternatives to be discussed between the owner/designer and code official are the possibility of man doors on the side of the unit or electric garage door openers.

1108.4 Judicial facilities. Judicial facilities shall comply with Sections 1108.4.1 and 1108.4.2.

❖ Accessibility in judicial facilities (e.g., courthouses) includes access to all public areas as well as special requirements for the courtrooms, holding cells and visitation areas. This section was added for coordination with new requirements for judicial facilities from the DOJ. For requirements addressing lawyer/client visitation spaces, see Section 1109.11.2.

1108.4.1 Courtrooms. Each courtroom shall be *accessible* and comply with Sections 1108.4.1.1 through 1108.4.1.5.

❖ All courtrooms are required to be accessible for participants at every level. This would include participation as a employee of the court, counselor, litigant, juror, witness or observer. All courtrooms must be on an accessible route from both the public side and courtroom staff side. The waiting areas, lawyer meeting rooms and vestibules that are sometimes located immediately adjacent to the courtroom entrance must be accessible. Access must be readily available to the gallery, as well as the witness stand and jury box. Members of the jury must have an accessible route to the jury deliberation room. The witness box in a courtroom is often raised to have a line of sight between the judge and witness. It is not acceptable to ask someone to testify from an area outside the box; therefore, the witness box must always be accessible. Elements would include an accessible route to all portions of the courtroom unless limited by the following subsections which identify areas and features of courtrooms that have specific accessibility requirements.

An accessible route must also be available between the courtroom and the association holding cells addressed in Section 1108.4.2.

The ICC was very proud to participate in a special Courthouse Access Advisory Committee that assisted the Access Board in developing recommendations for accessibility to courtrooms and courthouses. This report includes information on both requirements and recommended design practices. The committee was able to construct actual mock-ups to verify design configurations. The committee report can be located at http://www.access-board.gov/guidelines-and-standards/buildings-and-sites/120-ada-standards/background/courthouse-access.

1108.4.1.1 Jury box. A *wheelchair space* shall be provided within the jury box.

Exception: Adjacent companion seating is not required.

❖ Jury boxes may be configured in numerous ways, including with multiple tiers. At least one wheelchair space within the "box" is required. A wheelchair space adjacent to (or outside of) the jury box could not be used to satisfy the requirement. Since only jury members are seated in the jury box, companion seating is not required.

1108.4.1.2 Gallery seating. *Wheelchair spaces* shall be provided in accordance with Table 1108.2.2.1. Designated aisle seats shall be provided in accordance with Section 1108.2.5.

❖ Gallery seating in a courtroom is regulated similarly to other assembly seating areas with respect to wheelchair spaces, associated companion seating and designated aisle seats (see Commentary Figure 1108.4.1.2).

Figure 1108.4.1.2
GALLERY SEATING

1108.4.1.3 Assistive listening systems. An assistive listening system must be provided. Receivers shall be provided for the assistive listening system in accordance with Section 1108.2.7.1.

❖ Every courtroom must have an assistive listening system. Unlike other assembly occupancies, the requirement is not dependent on an audio amplification system being provided. The assumption is that any person in a courtroom with a hearing impairment should be provided with an enhanced ability to hear the proceedings when desired. Issues of privacy may dictate the type of system provided within the courtrooms. Signage indicating that assistive listening devices are available is required and is typically located near the public entry to the courtroom (see Section 1111.3)

1108.4.1.4 Employee work stations. The judge's bench, clerk's station, bailiff's station, deputy clerk's station and court reporter's station shall be located on an accessible route. The vertical access to elevated employee work stations within a courtroom is not required at the time of initial construction, provided a *ramp*, lift or elevator can be installed without requiring reconfiguration or extension of the courtroom or extension of the electrical system.

❖ Employee work stations in courtrooms are permitted to be constructed as adaptable for a future accessible route if and when a route is necessary for a mobility-impaired employee. Depending on the type of courtroom (e.g., panel, trial, jury, traffic), employee work stations will vary. The judge is typically raised to establish a level of authority over the courtroom and to have his or her eyes approximately level with those of the standing lawyers. Associated clerk stations are raised to ease interaction with the judge. The vertical portion of the accessible route (ramp, platform lift, elevator) can be an adapting element and is expected to be predesigned and readily incorporated into the courtroom when needed. This option is consistent with the modification for employees based on the need of the employee expressed in Section 1103.2.2. This option does not apply to the raised public areas in courtrooms, namely, witness stands, jury boxes and galleries, which are required to be constructed accessible.

1108.4.1.5 Other work stations. The litigant's and counsel stations, including the lectern, shall be *accessible*.

❖ The well of the court typically contains the area for the litigant's and counselor's tables and a lectern for lawyers to address the judge and/or jury. Depending on the courtroom, this configuration may vary. Both defendant and prosecutor tables should be constructed appropriately for a front approach for work stations (see Commentary Figure 1108.4.1.5). The lectern can be adjustable so that it can work for lawyers seated in wheelchairs as well as those standing.

Figure 1108.4.1.5
COURTROOM WELL WORK STATIONS

1108.4.2 Holding cells. Central holding cells and court-floor holding cells shall comply with Sections 1108.4.2.1 and 1108.4.2.2.

❖ Where holding cells are provided, Sections 1108.4.2.1 and 1108.4.2.2 indicate which cells shall be accessible.

1108.4.2.1 Central holding cells. Where separate central holding cells are provided for adult males, juvenile males, adult females or juvenile females, one of each type shall be *accessible*. Where central holding cells are provided and are not separated by age or sex, at least one *accessible* cell shall be provided.

❖ Some courthouse facilities have a central holding area for detainees that are in the courthouse for judicial proceedings. At least one central holding cell is required to be accessible. If separate holding cells are provided for males and females or to separate juvenile offenders from the adult offenders, at least one of each type is required to be accessible.

1108.4.2.2 Court-floor holding cells. Where separate court-floor holding cells are provided for adult males, juvenile males, adult females or juvenile females, each courtroom shall be served by one *accessible* cell of each type. Where court-floor holding cells are provided and are not separated by age or sex, courtrooms shall be served by at least one *accessible* cell. *Accessible* cells shall be permitted to serve more than one courtroom.

❖ Some courthouse facilities have a holding cell located in the area of the courtroom for detainees who are in the courthouse for judicial proceedings. For security reasons, these holding cells may be provided on each floor of the facility in order to have the detainees as close to the courtrooms they are to appear in as possible. At least one holding cell per floor is required to be accessible. One accessible holding cell could serve all the courtrooms on one floor. If separate holding cells are provided for males and females or to separate juvenile offenders from the adult offenders, at least one of each type is required to be accessible.

SECTION 1109 OTHER FEATURES AND FACILITIES

1109.1 General. *Accessible* building features and facilities shall be provided in accordance with Sections 1109.2 through 1109.15.

Exception: *Accessible units*, *Type A units* and *Type B units* shall comply with Chapter 10 of ICC A117.1.

❖ This section clarifies that the provisions of Section 1109 are applicable in addition to all other requirements of Chapter 11. For example, although Section 1109.2 sets forth accessibility requirements for toilet rooms and bathing facilities within a store, it is not intended to mean that these are the only accessibility requirements that are applicable to stores. The building features and facilities covered by this section are required to be accessible to the extent set forth herein. The exception is a reminder that Accessible Type A and Type B dwelling and sleeping units should comply with Chapter 10 of ICC A117.1 within the individual dwelling units. Common and public areas associated with Group R-2 and R-3 occupancies should comply with Section 1109.

1109.2 Toilet and bathing facilities. Each toilet room and bathing room shall be *accessible*. Where a floor level is not required to be connected by an *accessible route*, the only toilet rooms or bathing rooms provided within the facility shall not be located on the inaccessible floor. Except as provided for in Sections 1109.2.2 and 1109.2.3, at least one of each type of fixture, element, control or dispenser in each accessible toilet room and bathing room shall be *accessible*.

Exceptions:

1. Toilet rooms or bathing rooms accessed only through a private office, not for *common* or *public use* and intended for use by a single occupant, shall be permitted to comply with the specific exceptions in ICC A117.1.
2. This section is not applicable to toilet and bathing rooms that serve *dwelling units* or *sleeping units* that are not required to be *accessible* by Section 1107.
3. Where multiple single-user toilet rooms or bathing rooms are clustered at a single location, at least 50 percent but not less than one room for each use at each cluster shall be *accessible*.
4. Where no more than one urinal is provided in a toilet room or bathing room, the urinal is not required to be *accessible*.
5. Toilet rooms or bathing rooms that are part of critical care or intensive care patient sleeping rooms serving *Accessible units* are not required to be *accessible*.
6. Toilet rooms or bathing rooms designed for bariatrics patients are not required to comply with the toilet room and bathing room requirement in ICC A117.1. The *sleeping units* served by bariatrics toilet or bathing rooms shall not count toward the required number of *Accessible sleeping units.*
7. Where toilet facilities are primarily for children's use, required *accessible* water closets, toilet compartments and lavatories shall be permitted to comply with children's provision of ICC A117.1.

❖ This section generally requires all toilet rooms and bathing rooms provided to be accessible. A person using a wheelchair must be able to approach and enter the room. Within the toilet room and/or bathing room, a minimum of one of each element or fixture provided is required to be accessible. Elements and fixtures include such things as water closets, lavatories, mirrors, towel dispensers, hand dryers and any other device that is installed and intended for use by the occupants of the room. Showers and bathtubs are both considered bathing fixtures, but not as different types of bathing fixtures. If a shower and a bathtub are both provided in the same bathing room, only one is required to be accessible. The designer or owner can choose which one. Requirements for the total number of bathing and toilet rooms are in Chapter 29 of the code [which is duplicated from Section 403 of the *International Plumbing Code*® (IPC®)]. Large assembly and mercantile occupancies must also include a family or assisted use toilet room in addition to their other toilet facilities (see commentary, Section 1109.2.1). Facilities with more than 20 toilet compartments or 20 lavatories in a single toilet room should also comply with Sections 1109.2.2 and 1109.2.3.

Provisions for accessible bathrooms are required even on levels that are not accessed by an accessible route. Many people benefit from the accessibility provisions in accessible bathrooms. For example, a person who is arthritic can use the grab bars to stand up and sit down, as well as finding the lever handles on the lavatory easier to use. In addition, Title I of the

ADA might require that level to be made accessible as part of an employee modification. If the bathroom is already accessible, it does not need to be modified, thus reducing the changes that need to be made to the building.

ICC A117.1 is the document referenced for all accessible toilet room requirements. The technical requirements in ICC A117.1 are based on allowing a person in a wheelchair to perform a side transfer [see Commentary Figure 1109.2(1)]. Maintaining a clear floor space at each fixture is important [see Commentary Figure 1109.2(2) for possible configurations]. The single-occupant bathroom requirements in the 2010 ADA Standard are now consistent with the provisions in ICC A117.1.

Exception 1 addresses a condition in which a toilet room or bathing room is permitted to be adaptable rather than fully accessible. The intent is that if a toilet room is part of an individual office and serves only the occupant of that office, the adaptable toilet room can be readily modified to be fully accessible based on that individual's needs. Preplanning during construction and design will facilitate future alterations. The ICC A117.1 indicates exactly what those adaptable features can be (e.g., reversible door swing; water closet seat height; blocking for grab bars; removable cabinets under the lavatory). This is not intended to be a general exception to allow a very small bathroom that would require walls or fixtures relocated to provide accessibility at a later date. The bathrooms must be sized to be accessible.

Exception 2 is intended to correlate with the provisions of Section 1107, which establish the minimum number of facilities required to be Accessible, Type A or Type B units. Without this exception, the code would literally be requiring accessible fixtures in inaccessible dwelling and sleeping units. It is important to note that the bathrooms associated with required Accessible, Type A and Type B dwelling and sleeping units must comply with the requirements in Chapter 10 of ICC A117.1.

Exception 3 specifies where toilet rooms are clustered together (i.e., entrance doors next to each other or across the hall), not all need to be accessible. In such configurations, typically found in a doctor's office or drug test center, the requirement is reduced to a 50-percent minimum. If these toilet rooms are clustered in separate locations, such as in a multiclinic facility, the 50-percent minimum would be applied to each cluster. A single-occupant women's bathroom adjacent to a single-occupant men's bathroom is not considered a cluster since they each serve a different sex. The IPC does have an allowance that would let some small occupancies have two unisex single-occupant toilet rooms, rather than having the same rooms labeled men's and women's (see Section 2902.2.1) If these toilet rooms are clustered, they can use this 50-percent exception since they are the same type.

Exception 4—The IPC permits urinals to be substituted for water closets to a maximum of 67 percent in each toilet room (see Section 419.2 of the IPC). Exception 4 states that if only one urinal is provided within a toilet room, that urinal is not required to be accessible. This situation would typically only occur in toilet rooms that also had one water closet compartment.

Exception 5—While Exception 2 would exempt all the nonaccessible patient rooms in a hospital from accessible bathroom requirements, the intent of Exception 5 is also to exempt the Accessible units that may be provided within the critical-care or intensive-care units from requiring accessible bathrooms. In critical-care or intensive-care units, the patients are

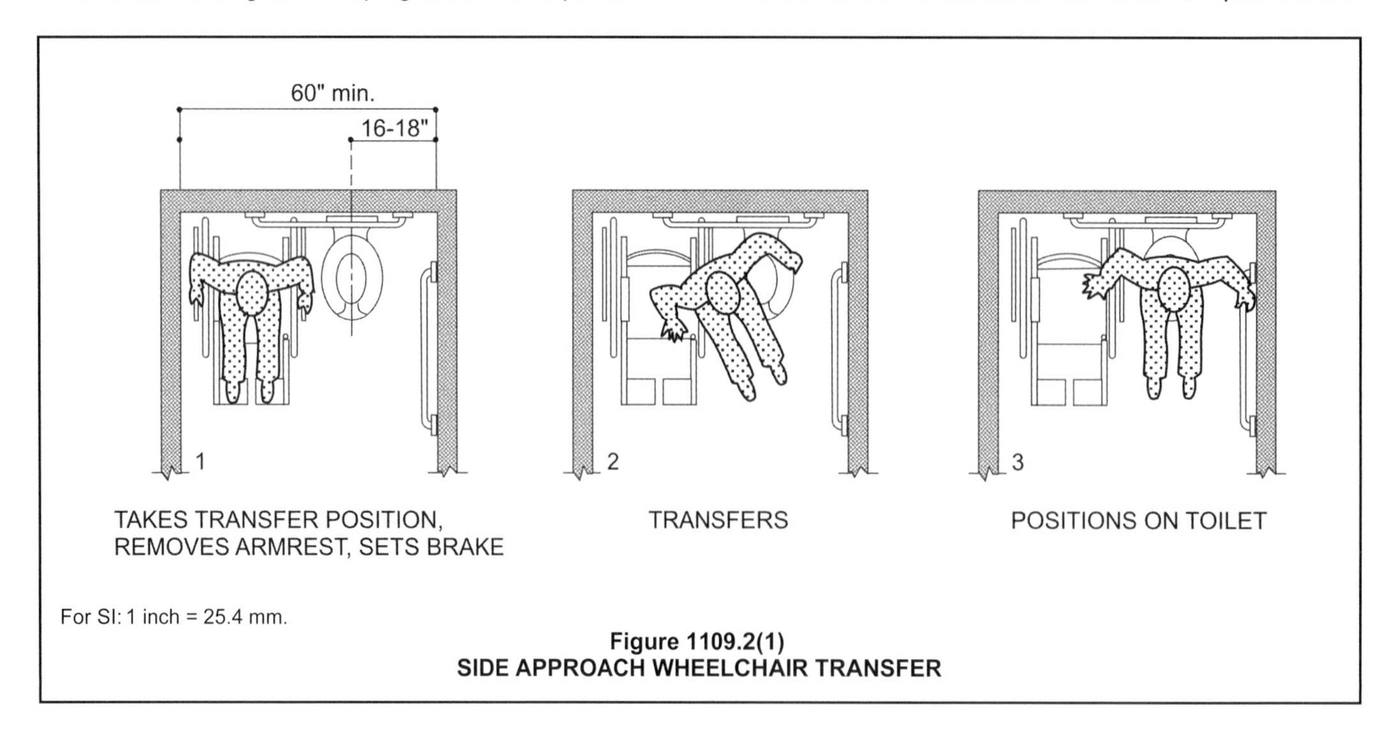

Figure 1109.2(1)
SIDE APPROACH WHEELCHAIR TRANSFER

often too ill to use the bathroom without assistance. Therefore, assistance is offered and expected in these areas to all patients. In addition, critical-care and intensive-care rooms often must be designed to maximize free space for equipment and personnel in case of emergency care situations.

Exception 6—Toilet rooms for bariatric patients must have a unique design to address the special needs for size and weight of the patient. The provisions in ICC A117.1 do not have specific requirements for bariatric toilet and bathing facilities. The patients would need higher toilets that were located farther from the wall than standard accessible toilet configurations. Designing specifically for these patients is permitted by Section 104.11 and ICC A117.1 Section 103. However, since use of these rooms would be limited to those patients, these rooms shall not count towards the number of Accessible patient sleeping units required in Section 1107.5.3.

Exception 7 acknowledges the allowances in the ICC A117.1 for a toilet room and water closet compartment specifically designed for children's use. This may be appropriate in areas such as day care facilities and certain areas of elementary schools.

1109.2.1 Family or assisted-use toilet and bathing rooms. In assembly and mercantile occupancies, an *accessible* family or assisted-use toilet room shall be provided where an aggregate of six or more male and female water closets is required. In buildings of mixed occupancy, only those water closets required for the assembly or mercantile occupancy shall be used to determine the family or assisted-use toilet room requirement. In recreational facilities where separate-sex bathing rooms are provided, an *accessible* family or assisted-

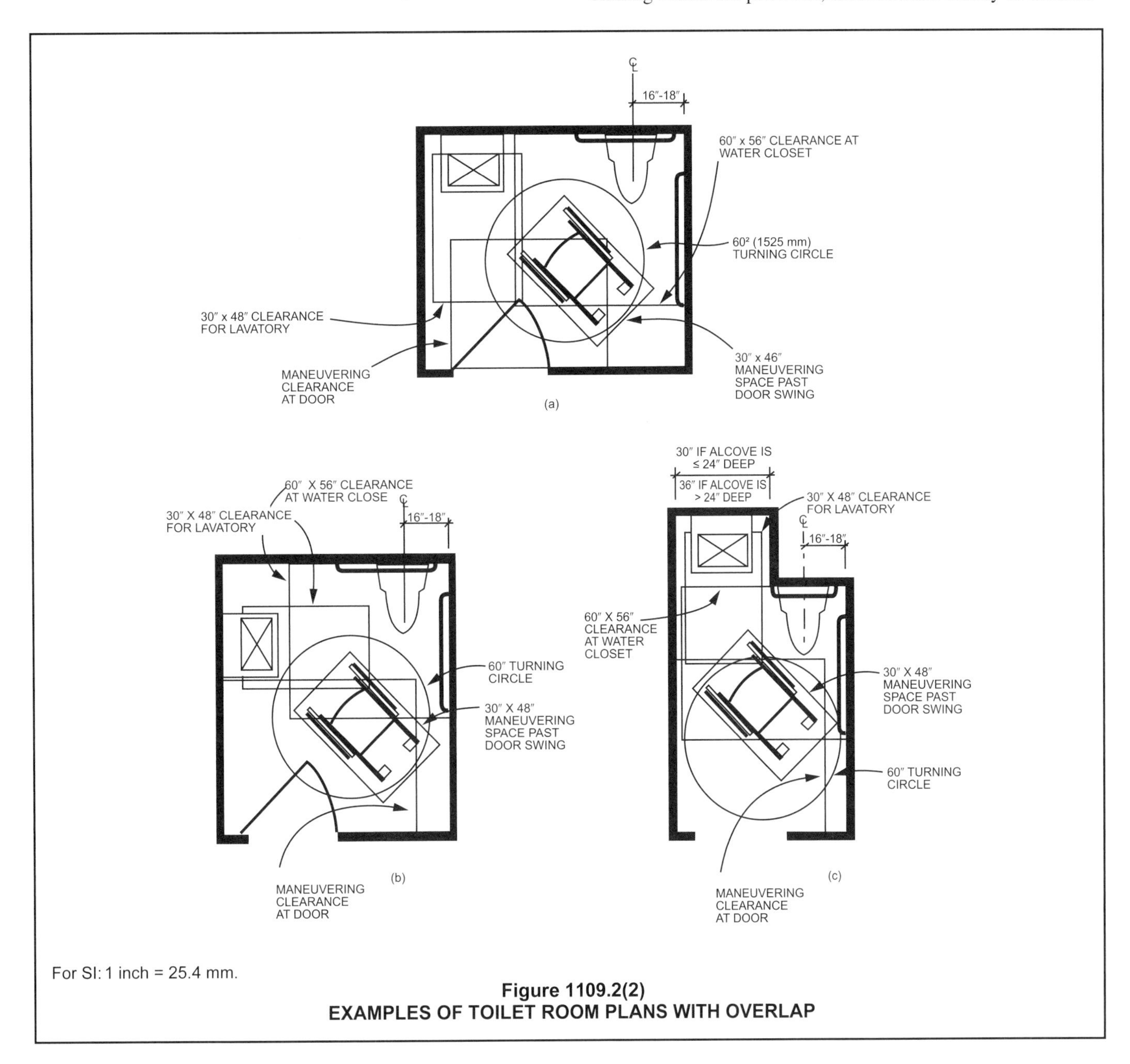

Figure 1109.2(2)
EXAMPLES OF TOILET ROOM PLANS WITH OVERLAP

use bathing room shall be provided. Fixtures located within family or assisted-use toilet and bathing rooms shall be included in determining the number of fixtures provided in an occupancy.

Exception: Where each separate-sex bathing room has only one shower or bathtub fixture, a family or assisted-use bathing room is not required.

❖ The primary issue relative to family toilet/bathing facilities is that some people with disabilities require assistance to utilize them. If that attendant is of the opposite sex, a toilet or bathing facility that can accommodate both persons is required. This person and their care giver using the facilities in multistall toilet rooms for either men or women becomes a privacy concern for everyone trying to use the toilet room. Also, it is important to note that family/assisted use bathrooms are not reserved for persons with disabilities like a parking space. This bathroom is also an advantage to any family with small children that are the opposite sex of the parent they are with. There are situations where the parent does not feel comfortable letting the child go into the opposite sex toilet room without them, but the child is too old to be comfortable going into the toilet room with the parent. This bathroom is also a benefit for persons who may need assistance because of a temporary disability, like a broken arm or leg. It is important to note that these provisions will typically not result in a substantial cost burden to the building owners, since the fixtures provided also count toward the minimum number required (see Section 2902.1.2, or Section 403.1.2 of the IPC).

The family or assisted-use bathing room requirements, although beneficial to all occupancies and building sizes, are only required in those structures that typically have large transient occupant loads or typically are occupied by large numbers of families, namely large assembly and mercantile. The section also identifies how mixed-occupancy buildings are to be addressed relative to the calculation. The fixtures provided in family toilet and bathing rooms count toward the number of required fixtures for the occupancy. The number of fixtures to be located in such rooms is limited, based on the premise that these rooms are securable (see commentary, Section 1109.2.1.2). The provision for bathing facilities is primarily geared toward recreational facilities. It is only required where the designer has chosen to provide separate-sex facilities for bathing and has provided more than one tub or shower for each sex. The exception acknowledges that smaller bathing rooms such as those described can be utilized by a disabled person with assistance without much hardship to a nondisabled person.

Many facilities have been installing family bathrooms as part of their standard bathroom layouts prior to the code requirement, viewing them as "customer friendly." For example, a person requiring assistance can also be a small child or an elderly adult. A parent shopping or attending an event with a child of the opposite sex can utilize this facility as well as an adult who may need assistance from a spouse.

Note that the requirement for family bathrooms does not exempt separate-sex bathrooms from providing accessible or ambulatory stalls. For a discussion of family bathrooms in existing buildings, see IEBC Sections 410.8.10 and 705.1.9.

1109.2.1.1 Standard. Family or assisted-use toilet and bathing rooms shall comply with Sections 1109.2.1.2 through 1109.2.1.7.

❖ Facilities are required to comply with the accessible technical provisions for toilet rooms in ICC A117.1 as well as the additional criteria established in the following subsections. In smaller tenant spaces, the IPC now permits two family or assisted-use toilet rooms in place of two single-occupant separate-sex bathrooms (see Section 2902.2.2 and Section 403.2.1 of the IPC).

1109.2.1.2 Family or assisted-use toilet rooms. Family or assisted-use toilet rooms shall include only one water closet and only one lavatory. A family or assisted-use bathing room in accordance with Section 1109.2.1.3 shall be considered a family or assisted-use toilet room.

Exception: A urinal is permitted to be provided in addition to the water closet in a family or assisted-use toilet room.

❖ A family toilet room must include a lavatory and a water closet. The exception permits a urinal to also be installed in the family toilet room if desired. If a family bathing room is provided within a facility, it can serve as the required family toilet room.

1109.2.1.3 Family or assisted-use bathing rooms. Family or assisted-use bathing rooms shall include only one shower or bathtub fixture. Family or assisted-use bathing rooms shall also include one water closet and one lavatory. Where storage facilities are provided for separate-sex bathing rooms, *accessible* storage facilities shall be provided for family or assisted-use bathing rooms.

❖ A family bathing facility is required to have one shower or tub, one water closet and one lavatory. The shower can be a transfer type, a roll-in type or a combination of the two. Accessible storage facilities, such as lockers, are also required if storage facilities are provided in the separate-sex bathing facilities. A family bathing room can also serve a dual purpose (bathing and toilet) as the required family toilet room (see commentary, Section 1109.2.1.2).

1109.2.1.4 Location. Family or assisted-use toilet and bathing rooms shall be located on an *accessible route*. Family or assisted-use toilet rooms shall be located not more than one *story* above or below separate-sex toilet rooms. The *accessible route* from any separate-sex toilet room to a family or assisted-use toilet room shall not exceed 500 feet (152 m).

❖ A one-story, 500-foot (1524 mm) limitation for access to customer toilet facilities is currently in Section 2902.3.2 (which is duplicated from Section 403.3.3 of the IPC). The distance to the customer toilet facilities is measured from the main entrance of a store or space to the toilet facility. The distance in this section is measured from the separate-sex facility to the family facility. The general travel distance requirement, in combination with this section, could result in a total travel distance for a family toilet room of two stories or 1,000 feet (3048 mm) maximum. The travel distance limitation is why there is more than one family bathroom in large facilities. An accessible route is required to provide access to the family facilities. Signage is required at both the separate-sex and family facilities in accordance with Sections 1111.1 and 1111.2.

1109.2.1.5 Prohibited location. In passenger transportation facilities and airports, the *accessible route* from separate-sex toilet rooms to a family or assisted-use toilet room shall not pass through security checkpoints.

❖ Security checkpoints in airports and similar facilities represent a potential delay, which may cause missed flights or connections. Because of security concerns, more and more facilities are adding security checkpoints. For example, many large sports facilities, courthouses and government buildings have checkpoints where the public moves into certain portions of the facilities. While not required, due to the delay in moving through a security system, a designer might want to follow this same guidance for location of the family bathrooms in other types of facilities.

1109.2.1.6 Clear floor space. Where doors swing into a family or assisted-use toilet or bathing room, a clear floor space not less than 30 inches by 48 inches (762 mm by 1219 mm) shall be provided, within the room, beyond the area of the door swing.

❖ The clear floor space provisions are intended to provide a room that is large enough to allow a person in a wheelchair to enter and close the door before utilizing the fixtures. This requirement is also in ICC A117.1 [see Commentary Figure 1109.2(2)].

1109.2.1.7 Privacy. Doors to family or assisted-use toilet and bathing rooms shall be securable from within the room.

❖ Since privacy while utilizing bathing and toilet facilities is an issue, the door to the facility must be securable from the inside. The securing mechanism must meet the operable parts requirements, including being within reach ranges and not require any tight pinching, grasping or sharp turning of the wrist to operate.

While it is not prohibited to lock bathrooms, this bathroom should be as easily accessed as the separate men's and women's bathrooms. A person who needed this facility should not have to find someone with a key for the family bathroom if the men's and women's bathrooms are readily available. This bathroom does count towards the required fixture count (see Section 2902.1.2) and as such must be available.

1109.2.2 Water closet compartment. Where water closet compartments are provided in a toilet room or bathing room, at least 5 percent of the total number of compartments shall be wheelchair *accessible*. Where the combined total water closet compartments and urinals provided in a toilet room or bathing room is six or more, at least 5 percent of the total number of compartments shall be ambulatory *accessible*, provided in addition to the wheelchair-*accessible* compartment.

❖ There are different configurations of water closet compartments that facilitate different degrees of physical disability. The provisions of ICC A117.1 establish the configuration and dimensional requirements for various types of water closet compartments. A wheelchair-accessible compartment is one in which sufficient space is provided for the wheelchair to enter completely the water closet compartment [see Commentary Figures 1109.2.2(2) and 1109.2.2(3)]. The wheelchair user then transfers from the wheelchair to the water closet in order to utilize the fixture. It is important that the required clear floor space be maintained. The configuration did not intend for a lavatory to be provided within the minimum-sized compartment. If a lavatory is located within a compartment, it must meet the same provisions as a single-occupant toilet room.

An ambulatory-accessible compartment is intended to facilitate use by a person with a mobility impairment that necessitates the use of a walking aid, such as a cane or walker [see Commentary Figure 1109.2.2(1)]. An ambulatory-accessible water closet compartment is not intended to be utilized by a person in a wheelchair. The 36-inch (914 mm) width is intended to allow standing persons to support themselves utilizing the grab bars on both sides. A wider compartment would not allow adequate bearing. Since these provisions are intended to address, within reason, the needs of both ranges of mobility impairment, this section requires a wheelchair-accessible compartment in all cases. In larger toilet rooms (i.e., those with a total of six or more water closet compartments and urinals), one ambulatory-accessible compartment is required in addition to the wheelchair-accessible compartment. In very large facilities with more than 20 toilet compartments in a single toilet room, an additional accessible and ambulatory water closet may be required to meet the 5-percent requirements.

This section is not intended to increase the required number of fixtures beyond that required by the IPC. For example, if a toilet room contains 10 water closets, eight water closet compartments may be of conventional design, one water closet compartment must be wheelchair accessible and one must be ambulatory accessible.

1109.2.3 Lavatories. Where lavatories are provided, at least 5 percent, but not less than one, shall be *accessible*. Where an *accessible* lavatory is located within the *accessible* water closet compartment at least one additional *accessible* lavatory shall be provided in the multicompartment toilet room outside the water closet compartment. Where the total lavatories provided in a toilet room or bathing facility is six or more, at least one lavatory with enhanced reach ranges shall be provided.

❖ A lavatory is a type of sink for which the primary purpose is hand washing. This section provides scoping requirements for lavatories located in spaces not specifically addressed or exempted by Section 1109.2. If a lavatory is provided in toilet and bathing rooms specifically exempted by Section 1109.2, it is not required to be accessible. Lavatories that are part of an employee work station are exempted under Section 1103.2.2. An example of a lavatory that was part of an employee's work area would be the lavatories found in the exam rooms. The lavatories are there for the nurses and doctors to wash their hands before examining a patient.

When lavatories are provided in other areas, a minimum of 5 percent, but not less than one, is required to be accessible via a front approach with appropriate knee and toe clearances.

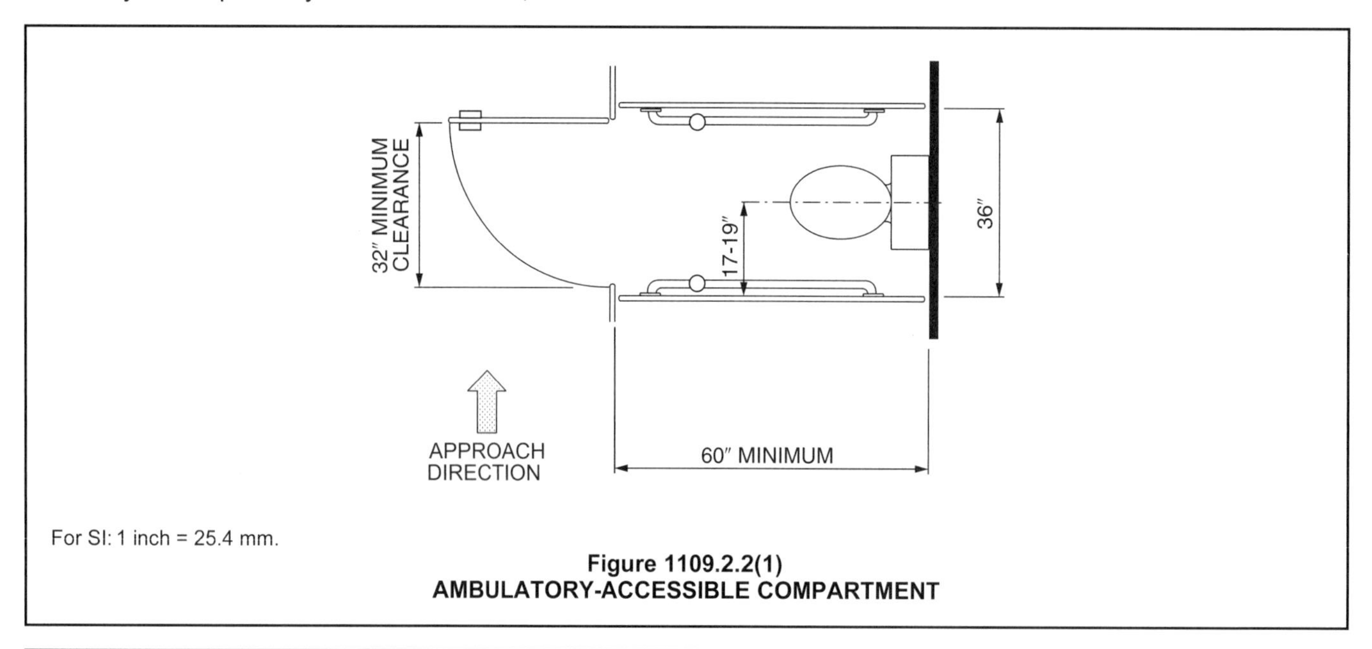

For SI: 1 inch = 25.4 mm.

Figure 1109.2.2(1)
AMBULATORY-ACCESSIBLE COMPARTMENT

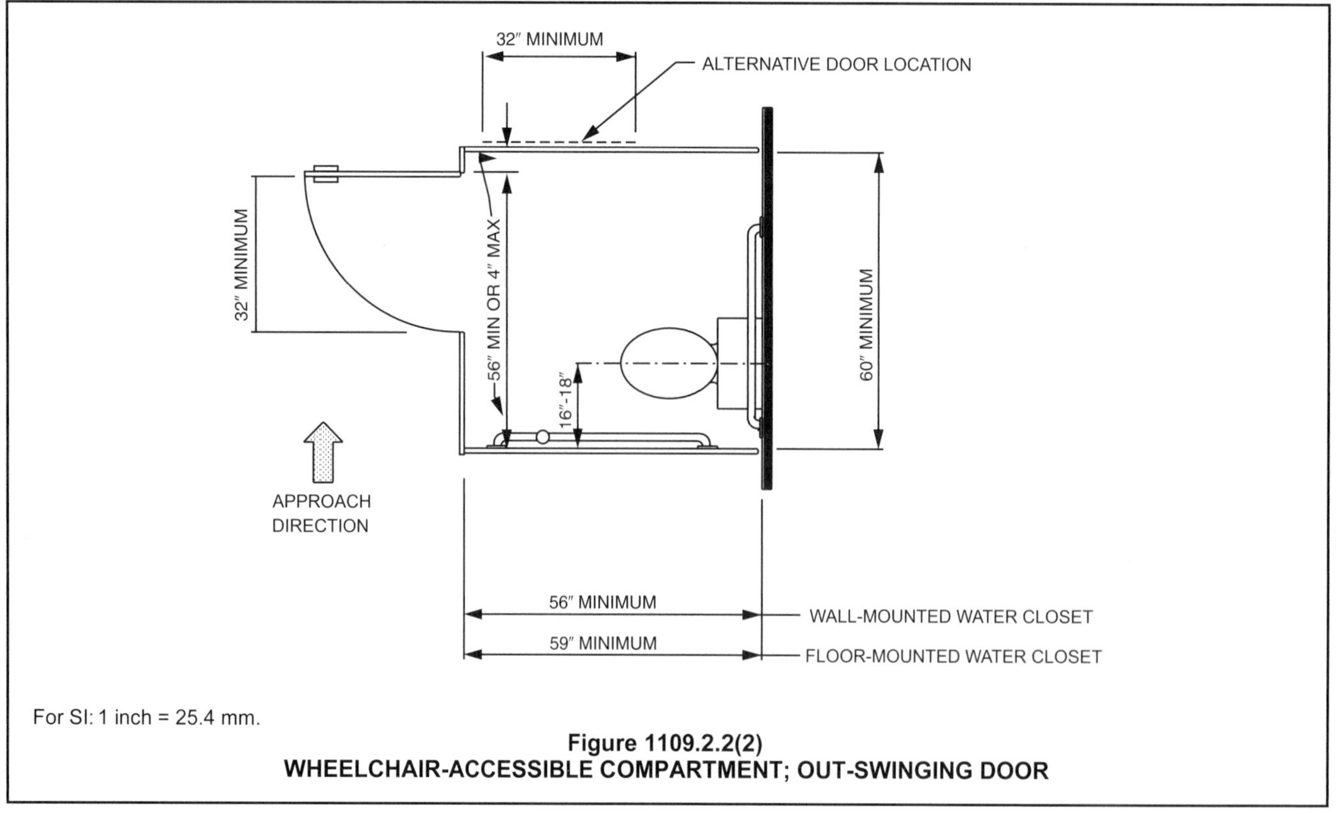

For SI: 1 inch = 25.4 mm.

Figure 1109.2.2(2)
WHEELCHAIR-ACCESSIBLE COMPARTMENT; OUT-SWINGING DOOR

A designer can choose to locate an accessible lavatory actually within the accessible toilet stall for customer convenience. If this is the case, that cannot be the only accessible lavatory within the toilet room. An additional accessible lavatory must be provided outside of the accessible stall.

In large facilities, where the designer has located six or more lavatories within one toilet facility (similar to the ambulatory stall), one of the lavatories must also be provided with enhanced reach range for the faucet controls in accordance with Section 606.5 of ICC A117.1. This would require the faucets to be located on the side of the lavatory, or provide access to the side of one of the lavatories, or to have automatic controls. This is to address the needs of persons who may have a limited reach over the lavatory where they may have difficulty reaching the controls. The lavatory with enhanced reach range could be the same lavatory as the accessible lavatory, or it could be another lavatory.

There are allowances in ICC A117.1 for lavatories specifically designed for children's use. This may be appropriate in areas such as day care facilities and certain areas of elementary schools.

1109.3 Sinks. Where sinks are provided, at least 5 percent but not less than one provided in *accessible* spaces shall be *accessible*.

Exception: Mop or service sinks are not required to be *accessible*.

❖ A sink may have numerous functions, unlike a lavatory, which is primarily provided for hand washing. Sinks in kitchens, kitchenettes and classrooms are examples that may have scoping and accessibility requirements different than those for lavatories. When sinks are provided, a minimum of 5 percent, but not less than one, is required to be accessible via a front approach with knee and toe clearances. There is a specific exception here for mop or service-type sinks. Service sinks include a wide variety of specialized sinks manufactured for multiple or specific functions. Service sinks are typically sized to be very deep and positioned for their function, which often renders them incapable of being accessible. Also, when a sink is part of an individual work station, that sink is not required to be accessible in accordance with Section 1103.2.2.

A forward approach to sinks is not always required (see commentary, Section 1109.4).

There are allowances in ICC A117.1 for sinks specifically designed for children's use. This may be appropriate in areas such as day care facilities and certain areas of elementary schools.

1109.4 Kitchens and kitchenettes. Where kitchens and kitchenettes are provided in *accessible* spaces or rooms, they shall be *accessible*.

❖ Kitchens, kitchenettes and wet bars in accessible spaces, other than within Type A and Type B dwelling units, must be accessible. An accessible kitchen would include clear floor spaces at all appliances, a sink with a front approach and a work surface with a front approach (see Section 804 of ICC A117.1 for specific details). Kitchenettes (where a cooktop or conventional range is not provided) allow a side approach to the sink (Section 606 of ICC A117.1) and do not require an accessible work surface. A common mistake is to forget that the sink must be set at a maximum height of 34 inches (864 mm) to allow for reach to the faucets (see Commentary Figure 1109.4).

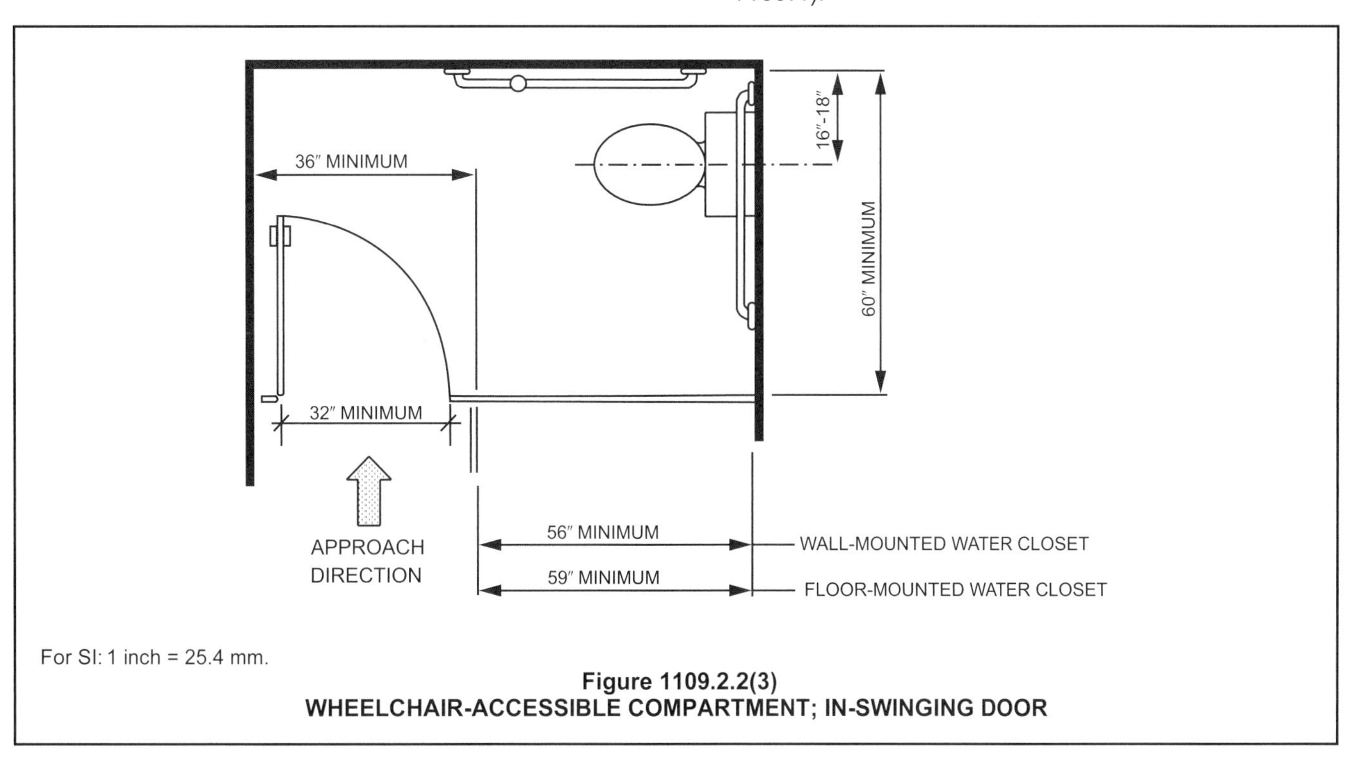

Figure 1109.2.2(3)
WHEELCHAIR-ACCESSIBLE COMPARTMENT; IN-SWINGING DOOR

Commercial kitchens are considered employee work areas. See the commentaries to Sections 1103.2.2 and 1104.3.1. Kitchenettes that are part of an employee break room are specifically excluded in the definition of employee work areas (see Chapter 2) and must be accessible.

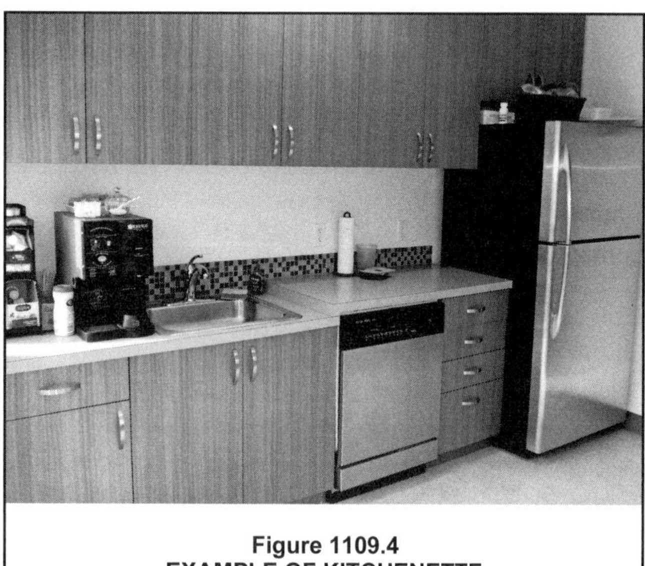

Figure 1109.4
EXAMPLE OF KITCHENETTE

1109.5 Drinking fountains. Where drinking fountains are provided on an exterior site, on a floor or within a secured area, the drinking fountains shall be provided in accordance with Sections 1109.5.1 and 1109.5.2.

❖ This section establishes a reasonable threshold for the required number of accessible drinking fountains. It should be noted that this section does not require the installation of drinking fountains where none are required or provided. Chapter 29 (which is duplicated from Section 403 of the IPC) contains criteria indicating the number of drinking fountains that are required based on occupancy. There are spaces that are not required to have drinking fountains. Examples are spaces with fewer than 15 occupants and spaces that serve water, such as a restaurant.

Current requirements for plumbing fixtures (see Table 2902 and Table 403 of the IPC) stipulate only one drinking fountain for many facilities or tenant spaces. The provisions in Section 1109.5.1 and Section 410.2 of the IPC stipulate a minimum of two drinking fountains for these facilities and spaces in order to meet the high/low criteria. Additionally, the plumbing requirements would allow for 50 percent of the drinking fountains to be substituted with water dispensers (see Section 410.3 of the IPC). While the provisions in Sections 1109.5.1 and 1109.5.2 and Section 410.2 of the IPC would not prohibit the substitution, the requirement for two drinking fountains would effectively negate this substitution unless there were at least three drinking fountains required for a floor. The IPC has defined, "drinking fountain," "water cooler" and "water dispenser" in an effort to clarify the differences between drinking fountains (including water coolers) and stand-alone water dispensers. A water dispenser cannot be considered as an alternative to either the high or low drinking fountain. However, water dispensers that are permanently installed would have to comply with Section 1109.13 for operable parts.

The IPC now includes provisions for travel distance to a drinking fountain (see Section 403.5 of the IPC), which are 500 feet (152 400 mm) and one-story maximum travel. In malls, the maximum travel is reduced to 300 feet (91 440 mm).

1109.5.1 Minimum number. No fewer than two drinking fountains shall be provided. One drinking fountain shall comply with the requirements for people who use a wheelchair and one drinking fountain shall comply with the requirements for standing persons.

Exceptions:

1. A single drinking fountain with two separate spouts that complies with the requirements for people who use a wheelchair and standing persons shall be permitted to be substituted for two separate drinking fountains.
2. Where drinking fountains are primarily for children's use, drinking fountains for people using wheelchairs shall be permitted to comply with the children's provisions in ICC A117.1 and drinking fountains for standing children shall be permitted to provide the spout at 30 inches (762 mm) minimum above the floor.

❖ Where a single drinking fountain is provided, this section mandates a minimum of two fixtures be provided: one for persons seated in a wheelchair (low) and one for standing persons (high). The high and low drinking fountains that serve a facility need not be provided at the same location in the facility. If the high and low drinking fountains are remotely located, signage is required to indicate the location of the other fountains (see Section 1111.2). Technical criteria for both high and low fountains are located in Section 602 of ICC A117.1 (see Commentary Figure 1109.5.1). The drinking fountains must be located so that they are not protruding objects (see Section 1003.3).

Exception 1 allows for a single drinking fountain to meet the requirements for both high and low heights by providing two spouts at one bowl; however, at this time there are only a few drinking fountains on the market that can meet this criterion. See the commentary to Section 1109.5.

Exception 2 allows for both the high and low drinking fountains to be designed for children. There are technical criteria in A117.1 for wheelchair-accessible drinking fountains specifically designed for children who are age 12 and younger. This may be appropriate for facilities such as day care facilities and certain areas of elementary schools.

Figure 1109.5.1
EXAMPLE OF DRINKING FOUNTAINS
FOR PERSONS USING WHEELCHAIRS (LOW) AND
STANDING PERSONS (HIGH)

1109.5.2 More than the minimum number. Where more than the minimum number of drinking fountains specified in Section 1109.5.1 is provided, 50 percent of the total number of drinking fountains provided shall comply with the requirements for persons who use a wheelchair and 50 percent of the total number of drinking fountains provided shall comply with the requirements for standing persons.

Exceptions:

1. Where 50 percent of the drinking fountains yields a fraction, 50 percent shall be permitted to be rounded up or down, provided that the total number of drinking fountains complying with this section equals 100 percent of the drinking fountains.
2. Where drinking fountains are primarily for children's use, drinking fountains for people using wheelchairs shall be permitted to comply with the children's provisions in ICC A117.1 and drinking fountains for standing children shall be permitted to provide the spout at 30 inches (762 mm) minimum above the floor.

❖ When an even number of drinking fountains is provided, half must accommodate persons seated in wheelchairs (low) and half must accommodate standing persons (high). Exception 1 addresses when an odd number of drinking fountains is provided.

An example:

- Two drinking fountains are required by Section 1109.5.1 and Section 410.2 of the IPC.
- Seven drinking fountains are provided.
- Fifty percent of seven is three and one-half.
- Rounding up yields four; rounding down yields three.
- Therefore, there are two choices:
  - Provide four low and three high fountains; or
  - Provide three low and four high fountains.
- Both choices comply since the complying fixtures total 100 percent.

This logic applies whenever an odd number of three or more drinking fountains is provided, regardless of the total quantity. This is based on the number provided—not the number required. In accordance with Section 410.3 of the IPC, a designer/owner could switch out some of the required drinking fountains for bottled water dispensers (see commentary to Section 1109.5).

Exception 2 is consistent with the same exception in Section 1109.5.2, allowing for both high and low drinking fountains to be designed for children. There are technical criteria in ICC A117.1 for wheelchair-accessible drinking fountains specifically designed for children who are age 12 and younger. This may be appropriate for day care facilities and certain areas of elementary schools.

1109.6 Saunas and steam rooms. Where provided, saunas and steam rooms shall be *accessible*.

Exception: Where saunas or steam rooms are clustered at a single location, at least 5 percent of the saunas and steam rooms, but not less than one, of each type in each cluster shall be *accessible*.

❖ The 2009 edition of the ICC A117.1 standard has included technical requirements for saunas and steam rooms. The scoping in the IBC is consistent with bathrooms and coordinates with the 2010 ADA Standard. What constitutes a "cluster" is subjective; however, typically the rooms would have to have entrances immediately adjacent or across the hall from one another to be considered "clustered."

The requirements within the standard provide for entering the room, closing the door and then transferring to the bench in the room. Removing the wheelchair from the space is not addressed, so perhaps a special chair might be utilized. Section 914 of the *International Mechanical Code*® (IMC®) addresses safety requirements for sauna rooms.

1109.7 Elevators. Passenger elevators on an *accessible route* shall be *accessible* and comply with Chapter 30.

❖ This section requires all passenger elevators on an accessible route to be accessible. The reference to Chapter 30 requires all passenger elevators to conform to general elevator safety standards as well as accessibility standards. It is not the intent of this section to prohibit service elevators, Limited Use Limited Access (LULA) elevators, private residence elevators or platform lifts (as permitted by Section 1109.8). Service, LULA and private residence elevators are recognized as types of passenger lifts by ASME A17.1/ CSA B44.

Elevators are defined in ASME A17.1 as "a hoisting and lowering mechanism, equipped with a car, that moves within guides and serves two or more landings ...." The limitations of use for the different types of elevators are controlled by the standard.

ASME limits the occupants in "freight elevators" to just the operator and the people needed to load and unload the freight. Also "construction elevators" are limited to construction purposes only. Only passenger-type elevators can serve as part of an accessible route. Therefore, freight and construction types of elevators are not required to be accessible.

The term "service elevator" is not defined in ASME A17.1. This is a slang term used by building owners to separate the "back of the house" passenger elevator used by staff (and typically designed to accommodate a heavier load) from the passenger elevators used by the general public for circulation. As part of an accessible route to employee work areas, a service elevator is required to be accessible.

LULA elevators have a maximum vertical travel distance of 25 feet (7620 mm). Private residence elevators are limited to within or serving individual dwelling units.

It should also be noted that unlike plumbing fixtures, this section does not establish that a certain number of passenger elevators are required to be accessible while other passenger elevators may be of a design that is not fully accessible. All passenger elevators that are on an accessible route are required to be accessible. If a bank of passenger elevators is provided as part of an accessible route, all elevators in that bank must be accessible. See IEBC Sections 410.8.2 and 705.1.2 for existing elevators being altered.

The reference to an accessible route (see Sections 1104.3, 1004.4 and 1109.7) is important because some areas in certain buildings may not be required to be accessible (see Sections 1103.2 and 1104.4). For example, since an elevator penthouse is not required to be accessible (see Section 1103.2.9), the passenger elevator is not required to serve that level. Keep in mind that the penthouse that contains elevator equipment must be accessed by a stairway (see Section 1011.12.1). In addition, certain types of elevators (e.g., freight elevators) may not be part of an accessible route.

Note that there are additional requirements for elevators depending on the building height:

- In buildings four stories and higher, at least one elevator must be sized in order to accommodate stretchers (see Section 3002.4).
- In buildings five stories and higher, an elevator must be available for serving as part of an accessible means of egress (see Section 1009.2.1).
- Buildings 120 feet (36 576 mm) or taller shall have two fire service access elevators, both sized to fit a stretcher (see Sections 403 and 3007).
- Occupant evacuation elevators are also an option (see Section 3008).

Note that fire service access elevators and occupant evacuation elevators can also serve as elevators that are part of an accessible means of egress.

1109.8 Lifts. Platform (wheelchair) lifts are permitted to be a part of a required *accessible route* in new construction where indicated in Items 1 through 10. Platform (wheelchair) lifts shall be installed in accordance with ASME A18.1.

1. An *accessible route* to a performing area and speaker platforms.
2. An *accessible route* to *wheelchair spaces* required to comply with the *wheelchair space* dispersion requirements of Sections 1108.2.2 through 1108.2.6.
3. An *accessible route* to spaces that are not open to the general public with an *occupant load* of not more than five.
4. An *accessible route* within an individual *dwelling unit* or *sleeping unit* required to be an *Accessible unit*, *Type A unit* or *Type B unit*.
5. An *accessible route* to jury boxes and witness stands; raised courtroom stations including judges' benches, clerks' stations, bailiffs' stations, deputy clerks' stations and court reporters' stations; and to depressed areas such as the well of the court.
6. An *accessible route* to load and unload areas serving amusement rides.
7. An *accessible route* to play components or soft contained play structures.
8. An *accessible route* to team or player seating areas serving *areas of sport activity*.
9. An *accessible route* instead of gangways serving recreational boating facilities and fishing piers and platforms.
10. An *accessible route* where existing exterior *site* constraints make use of a *ramp* or elevator infeasible.

❖ This section indicates that platform (wheelchair) lifts are only permitted to be part of a required accessible route in new construction in limited situations. If a platform lift is permitted as part of the accessible route into a space, it can also serve as part of the accessible route used for means of egress out of the space (with the exception of Item 10) if it has standby power (see Section 1007.5). There are concerns about the potential for a large number of people needing to exit the building under the allowances for Item 10; therefore, an accessible means of egress must be via another route. Platform lifts may be used more extensively in existing construction (see IEBC Sections 410.8.3 and 705.1.3). This is in recognition that circumstances in existing buildings may make it impractical to accomplish accessibility by use of an elevator, in which case a platform lift is a reasonable alternative. Accessible means of egress is not required in existing buildings undergoing alterations (see Section 1009.1, Exception 1, or IEBC Sections 410.1, Exception 2 and 705.1, Exception 2); therefore, a platform lift as part of the accessible means of egress in these situations is not an issue.

A previous edition of the technical standard (i.e., ASME A17.1-1996) required key operation to platform lifts, which unnecessarily inhibited independent access by persons with physical disabilities. The requirements for platform lifts have been removed from the elevator standard and now have their own standard, ASME A18.1, *Safety Standard for Platform Lifts and Stairway Chairlifts*. This newer standard allows push-button operation of platform lifts, thus making independent access much easier.

The listed items indicate when a platform lift can be utilized as part of a required accessible route in new construction. Items 1 and 2 allow platform lifts to be utilized for access to performing (e.g., stages, orchestra pits) and wheelchair space locations used to view an event.

Item 3 specifies that a platform lift may be used to provide access to a nonpublic area with five or less occupants, such as a projection booth.

Item 4 allows for a platform lift within an individual dwelling or sleeping unit.

Item 5 permits platform lifts to provide access to the raised or lowered areas typically found in courtroom settings. This is consistent with Section 1108.4.1.4.

Item 6 allows platform lifts for access to loading and unloading areas serving amusement rides. It does not attempt to address access to any other element of a ride. There are provisions for amusement rides in Section 1110.4.8 and Chapter 11 of the 2009 ICC A117.1.

Item 7 identifies children's play spaces, the entry to which may be raised or depressed, as an appropriate place for a platform lift. Such spaces are often associated with restaurants, shopping malls and preschools. There are provisions for children's play areas in Chapter 11 of the 2009 ICC A117.1.

Item 8 acknowledges that team or player seating may be a small depressed area (i.e., dugout) capable of being efficiently served by a platform lift.

Item 9 allows for a platform lift to serve as part of the accessible route to recreational boating and fishing piers required to be on an accessible route per Sections 1110.4.9 and 1110.4.11.

Item 10 recognizes that existing site constraints may make installation of a ramp or elevator infeasible. An example would be the situation of dealing with existing public sidewalks, easements and public ways in downtown urban areas. This situation would be most common in hilly areas where the street and sidewalk follows grade and the building's floor is level, resulting in steps up or down at entrances.

Note that a platform lift that was not part of the required accessible route could be used to facilitate access to any space. The governing factor in those situations would be the limitations of the product itself for capacity and travel distance. Item 4 permits platform lifts to be utilized within individual dwelling or sleeping units as part of a required accessible route. Single-family homes are not required to be accessible. However, persons with mobility impairments may choose to install stairway chairlifts in their private homes. Section 1011.2, Exception 3, specifies a minimum stairway width for platform lifts or chair lifts provided within dwelling units.

A platform lift is an electrically operated, mechanical device designed to transport a person who cannot use stairs over a short vertical distance. Platform lifts must be sized to accommodate a wheelchair. Platform lifts can be used by persons who use wheelchairs or scooters and persons with limited mobility. Some platform lifts are equipped with folding seats. A fold-down seat that moves up the stairway is not a platform lift, but rather a chair lift. A stairway chair lift is not suitable for a person using a wheelchair and cannot serve as part of a required accessible route. A platform lift is most suitable for changes of elevation of one story or less where the installation of a ramp is not feasible. ASME A18.1 limits the rise of platform lifts to 14 feet or less. [see Commentary Figures 1109.8(1) and 1109.8(2)].

While new technologies are expanding current options, there are two basic types of platform lifts:

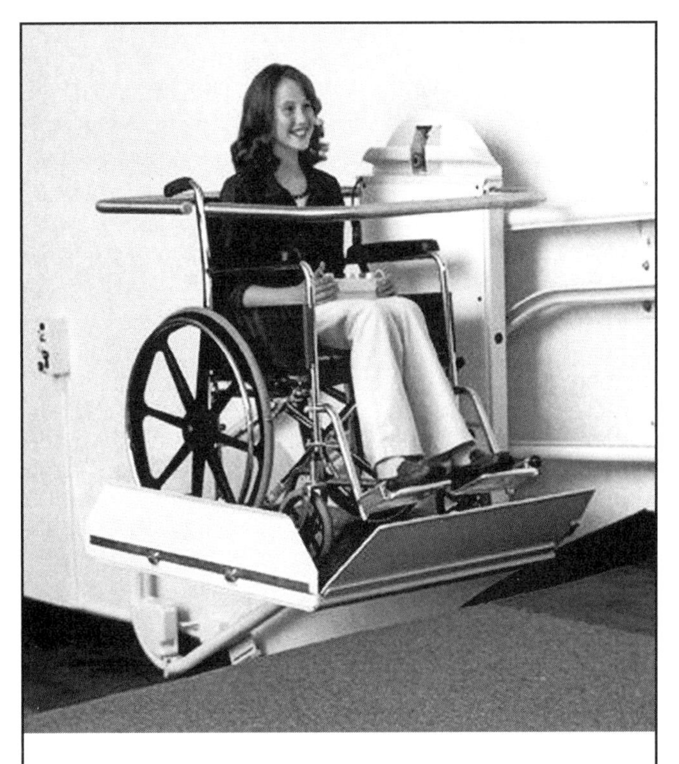

Figure 1109.8(1)
PLATFORM (WHEELCHAIR) LIFT
(Photo courtesy of Wheel-evator)

vertical lifts and inclined lifts. Vertical lifts are similar to elevators in that they travel only up and down in a fixed vertical space. Platform lifts should not be confused with Limited Access Limited Use elevators or private residence elevators (see Section 1109.7). Inclined platform lifts are usually installed in conjunction with a stairway and travel along the slope of the stairway. Inclined lifts are a design consideration for long flights of stairs where a vertical platform lift is not practical, where headroom is limited or where ceilings are low.

Figure 1109.8(2)
STAIRWAY CHAIRLIFT
(Photo courtesy of Wheel-evator)

1109.9 Storage. Where fixed or built-in storage elements such as cabinets, coat hooks, shelves, medicine cabinets, lockers, closets and drawers are provided in required *accessible* spaces, at least 5 percent, but not less than one of each type shall be *accessible*.

❖ Fixed or built-in storage elements are common in many occupancies, including both residential and nonresidential occupancies such as office buildings, recreational facilities, etc. Examples would include coat closets, storage closets, lockers, etc. The code does not require storage facilities, but when such facilities are provided, this section requires at least 5 percent, but not less than one, of each type to contain storage space in accordance with ICC A117.1 (see Commentary Figure 1109.9). This could be a portion of shelving or coat rods, a number of lockers, etc. Requirements would include a wheelchair space in front of the element, operable parts (if applicable) and the shelf, box or hook within the 15- to 48-inch (381 to 1219 mm) reach ranges. Storage within kitchens and kitchenettes is more specifically addressed in Section 804 of ICC A117.1. For the kitchen or kitchenette to be able to have both upper and lower cabinets at heights that keep the counter useable, storage elements are not required to meet the reach range. Storage elements within Accessible, Type A and Type B units are specifically addressed under Chapter 10 of ICC A117.1.

Storage elements that would be part of an employee work space, such as a lab in a doctor's office, are exempted under Section 1103.2.2.

Figure 1109.9
EXAMPLE OF LOCKER AREA

1109.9.1 Equity. *Accessible* facilities and spaces shall be provided with the same storage elements as provided in the similar nonaccessible facilities and spaces.

❖ This section is really about the issue of equal access to amenities. This requirement is consistent with the requirements for family or assisted-use bathrooms to have the same lockers, hand dryers, baby-changing stations, etc., as provided in the men's and women's bathrooms. For example, hotel rooms commonly provide closets and dressers for guests. These same storage facilities must also be available in Accessible sleeping units. They cannot be eliminated as an option to providing an accessible route or turning space within the Accessible unit. For hotel rooms, ICC A117.1 has a requirement for providing equal bathroom counter space in similar Accessible and nonaccessible guestrooms.

1109.9.2 Shelving and display units. Self-service shelves and display units shall be located on an *accessible route*. Such shelving and display units shall not be required to comply with reach-range provisions.

❖ The intent of this section is that persons who utilize wheelchairs can see items on shelves in stores and libraries, but someone may be required to assist them in reaching the items. Putting all items within the 15- to 48-inch (381 to 1219 mm) reach range would adversely affect the space allowed for products or information.

1109.10 Detectable warnings. Passenger transit platform edges bordering a drop-off and not protected by platform screens or *guards* shall have a *detectable warning*.

Exception: *Detectable warnings* are not required at bus stops.

❖ A detectable warning is a standardized feature built in or applied to walking surfaces to warn a visually impaired person of a hazard on or near his or her path of travel and which could result in injury to that person, such as moving into a street, or at a significant drop-off.

This section requires a detectable warning at the edges of passenger transit platforms where they border a drop-off (see Commentary Figure 1109.10). For example, loading platforms in both light- and heavy-rail transit stations where passengers await the arrival of trains are a serious potential hazard to a person with a vision impairment when there is no train at the station. There have been incidents where people with severe vision impairments have fallen off the transit platform and been killed or seriously injured. The presence of a detectable warning at the platform edge where it borders the drop-off would be encountered and recognized by a person with a vision impairment either through detection by a long cane or by foot contact with the detectable warning surface.

The exception for bus stops is established so as not to require a detectable warning at the curb between a sidewalk and the street. Detectable warnings at curbs are contradictory to the envisioned application of detectable warnings in the built environment. The curb itself is a recognizable and well-known cue to people with vision impairments to proceed with caution.

There is a great deal of controversy and general disagreement as to the benefit or advisability of detectable warning surfaces throughout the built environment. The use of such warnings at transit stations is, to date, the only well-documented use of detectable warnings, as discussed in studies such as "Tactile Warnings to Promote Safety in the Vicinity of Transit Platform Edges," conducted by the Urban Mass Transportation Administration, and "Pathfinder Tactile Tile Demonstration Test Project," conducted by the Metro-Dade Transit Agency. The Access Board will be looking into this issue as part of the development of the *Public Rights-of-Way Guidelines* for intersections of public sidewalks and roads. The 2010 ADA Standard has dropped the requirement for detectable warnings at curb cuts. Neither the code nor ICC A117.1 requires detectable warnings at curb cuts. ICC A117.1 only addresses the technical criteria for when someone chooses to provide detectable warnings.

ICC A117.1 prescribes the type of surface that constitutes a detectable warning in Section 705. One option for the detectable warning surface that is currently considered suitable consists of raised, truncated domes with a diameter between 0.9 inches (23 mm) and 1.4 inches (36 mm), a height of approximately 0.2 inch (5.1 mm) and center-to-center spacing of between 1.6 inches (41 mm) and 2.4 inches (61 mm). However, there is current and ongoing research into the use and application of truncated domes as a suitable detectable warning surface for various applications, including interior and exterior locations. ICC A117.1 requires that the type of detectable warning surface utilized must be standard throughout a building, facility, site or complex of buildings. If different types of detectable warning surfaces are utilized, their usefulness is diminished. The different messages they would convey to a person with a vision impairment would not be consistent and may be confusing and easily misinterpreted.

It is anticipated that future applications of detectable warnings will be considered when additional research and documentation of their usefulness and suitability is available. One issue, among many, is the durability of a detectable warning surface in an exterior application and the potential difficulty that truncated domes may present to a person in a wheelchair, as well as to nondisabled persons who may have to negotiate the surface.

Figure 1109.10
DETECTABLE WARNINGS AT TRANSIT PLATFORM EDGE

1109.11 Seating at tables, counters and work surfaces. Where seating or standing space at fixed or built-in tables, counters or work surfaces is provided in *accessible* spaces, at least 5 percent of the seating and standing spaces, but not less than one, shall be *accessible*.

> **Exception:** Check-writing surfaces at check-out aisles not required to comply with Section 1109.12.2 are not required to be *accessible*.

❖ Many occupancies, such as libraries, classrooms and other public spaces, are equipped with fixed seating, tables or both (see Commentary Figure 1109.11). The threshold of 5 percent is consistent with the requirement for accessible seating in dining areas as specified in Section 1108.2.9.1. Fixed or built-in seating that is part of an individual work station is not required to be accessible (see Section 1103.2.2). Where required, accessible tables and work surfaces are expected to include adequate knee and toe clearances as specified in ICC A117.1. For requirements for points of sales or service that do not include a work station, see Section 1109.12.3. For example, the check-writing station in the center of the bank lobby is considered a work station, whereas the teller window is typically considered a service counter. The accessible check-writing station would require knee and toe clearances, while the accessible teller window would not.

Note that the requirement addresses built-in counters, tables or work stations for standing or seating. Built-in surfaces where persons are typically standing are most often installed at a height of 36 to 48 inches (914 to 1219 mm). Depending on the overall configuration of the counters in the space, a portion of the counter for standing persons may need to be lowered to an accessible height of 28 to 34 inches (711 to 864 mm) and configured for forward approach.

The intent of Exception 1 is that if a check-writing surface is provided at a nonaccessible checkout aisle, this surface is not required to meet the counter and work surface requirements in ICC A117.1 (see Section 1109.12.2).

Figure 1109.11
EXAMPLE OF ACCESSIBLE WORK STATION

There are allowances in ICC A117.1 for work surfaces specifically designed for children's use. This may be appropriate in facilities such as day care facilities and certain areas of elementary schools, or areas frequented mostly by children, such as children's museums or the children's areas in libraries.

1109.11.1 Dispersion. *Accessible* fixed or built-in seating at tables, counters or work surfaces shall be distributed throughout the space or facility containing such elements and located on a level accessed by an *accessible route*.

❖ An appropriate distribution of seating and tables is required and is intended to be consistent with the distribution of all such seating and tables in the facility. For example, a library may contain several separate areas in which fixed seating or tables are installed. This provision would prohibit the location of all accessible seating and tables in a single specific area of the facility. However, this should not be construed to mean that every separate area in which fixed seating or tables are provided is required to contain an accessible seat or table. For example, if there are four areas in which fixed seating or tables are located, and two such seats or tables are required by this section to be accessible, the intent of this section is to require that one accessible seat or table be located in two of the four areas.

If the facility has floor levels exempted from an accessible route (see Section 1104.4), accessible work surfaces are not required to be located on the nonaccessible level.

1109.11.2 Visiting areas. Visiting areas in judicial facilities and Group I-3 shall comply with Sections 1109.11.2.1 and 1109.11.2.2.

❖ Persons appearing in court proceedings may need to talk with their lawyer, social worker, parole officer, etc. If a prisoner is serving a sentence, they may need to talk with all these individuals, as well as have visitations from family. Typically, these visitation areas include cubicles or counters in supervised areas.

1109.11.2.1 Cubicles and counters. At least 5 percent, but not less than one of the cubicles, shall be *accessible* on both the visitor and detainee sides. Where counters are provided, at least one shall be *accessible* on both the visitor and detainee sides.

> **Exception:** This requirement shall not apply to the detainee side of cubicles or counters at noncontact visiting areas not serving *Accessible unit* holding cells.

❖ If detainee/visitor counters or cubicles are provided, at least 5 percent must be accessible. Since it is not known if the person needing the accessible counter is on the visitor side or the detainee side, at least one counter location shall be accessible on both sides. In accordance with the exception, visiting areas that do not serve accessible cells would still require 5 percent of the counters or cubicles to be accessible on the visitor's side (see Commentary Figure 1109.11.2.1). This is consistent with the exception to certain areas of jails in accordance with Section 1103.2.13.

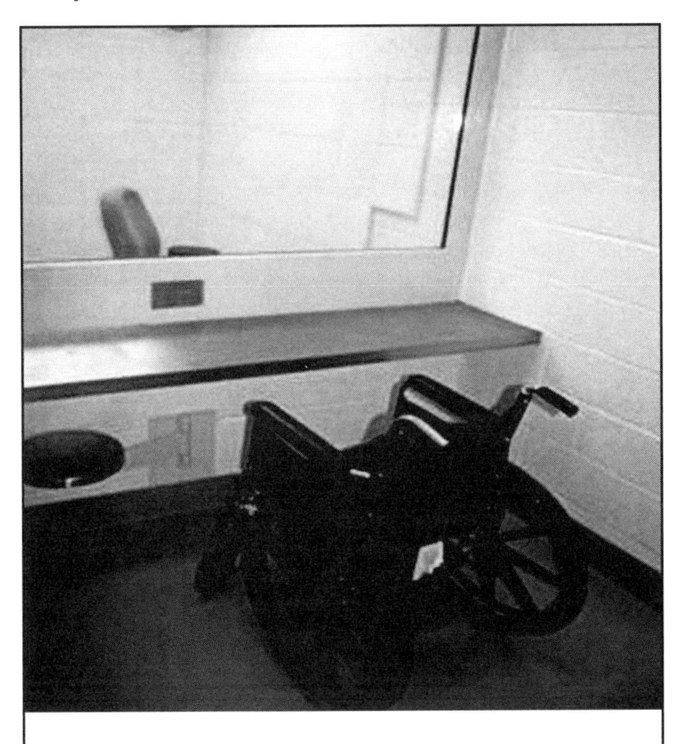

Figure 1109.11.2.1
EXAMPLE OF ACCESSIBLE VISITOR'S CUBICLE

1109.11.2.2 Partitions. Where solid partitions or security glazing separate visitors from detainees, at least one of each type of cubicle or counter partition shall be *accessible*.

❖ Section 1109.11.2.1 requires 5 percent of cubicles or counters in courthouse and jail visiting areas to be accessible in accordance with the built-in counter/work surface requirements. This section deals with the visiting situation where solid partitions of security glazing separate visitors from detainees. To facilitate communication, the system may include some type of handset, headphones or microphone. The system must consider in its design that either user may use a wheelchair or have difficulty bending or stooping. Any type of telephone or headphone system provided to address communication must have a volume control.

1109.12 Service facilities. Service facilities shall provide for *accessible* features in accordance with Sections 1109.12.1 through 1109.12.5.

❖ This section introduces five subsections that address accessibility of dressing rooms: service counters and windows; check-out aisles; food service; queues and waiting lines, etc. These types of elements are broadly termed "service facilities" and are intended to permit usage by persons with physical disabilities.

1109.12.1 Dressing, fitting and locker rooms. Where dressing rooms, fitting rooms or locker rooms are provided, at least 5 percent, but not less than one, of each type of use in each cluster provided shall be *accessible*.

❖ This section establishes a reasonable minimum threshold for accessible dressing, fitting and locker rooms. This section does not require that dressing, fitting and locker rooms be provided, but when such facilities are provided, at least 5 percent of each group of rooms is required to be accessible. This section also requires an appropriate distribution of accessible facilities based on distinct and different functions of each group of facilities provided. For example, a department store may have one group of fitting rooms serving the menswear department and a separate group of fitting rooms serving the women's wear department. This section would not permit the location of all accessible fitting rooms in one department and no accessible fitting rooms in another department. This may result in a number of accessible fitting rooms greater than 5 percent of the total number of all fitting rooms in the facility. For example, if there are two dressing rooms in each of three different departments in a retail store, a total of three accessible fitting rooms would be required: one in each of the three areas. Where amenities such as coat hooks, lockers or shelves are provided in inaccessible dressing, fitting or locker rooms, they must also be provided in the accessible rooms (see Section 1109.9). In accordance with Section 803.4 of ICC A117.1, an accessible bench must be provided in all accessible dressing, fitting and locker rooms (see Commentary Figure 1109.12.1).

1109.12.2 Check-out aisles. Where check-out aisles are provided, *accessible* check-out aisles shall be provided in accordance with Table 1109.12.2. Where check-out aisles serve different functions, accessible check-out aisles shall be provided in accordance with Table 1109.12.2 for each function. Where check-out aisles are dispersed throughout the building or facility, *accessible* check-out aisles shall also be dispersed. Traffic control devices, security devices and turnstiles located in *accessible* check-out aisles or lanes shall be *accessible*.

Exception: Where the public use area is under 5,000 square feet (465 m^2) not more than one *accessible* check-out aisle shall be required.

❖ This section applies to sales locations configured as lanes or aisles, often designed to accommodate shopping carts such as typically found in supermarkets, drugstores, discount retail stores, etc. If check-out aisles serve different functions, at least one check-out aisle that serves each function shall be accessible. If a facility offers check-out stations at a variety of locations, the accessible locations must also be dispersed. Where security devices such as turnstiles and automatically activated gates are provided, they are required to be accessible. The required number of check-out aisles for each function is set forth in Table 1109.12.2. Signage is required in accordance with Section 1111.1.

Providing a cash register at a location does not automatically create a check-out aisle. See Section 1109.12.3 for sales and service counters.

The exception allows for small retail facilities to only provide one accessible check-out counter.

TABLE 1109.12.2
ACCESSIBLE CHECK-OUT AISLES

TOTAL CHECK-OUT AISLES OF EACH FUNCTION	MINIMUM NUMBER OF ACCESSIBLE CHECK-OUT AISLES OF EACH FUNCTION
1 to 4	1
5 to 8	2
9 to 15	3
Over 15	3, plus 20% of additional aisles

1109.12.3 Point of sale and service counters. Where counters are provided for sales or distribution of goods or services, at least one of each type provided shall be *accessible*. Where such counters are dispersed throughout the building or facility, *accessible* counters shall also be dispersed.

❖ Wherever sales or service counters or windows are provided, at least one window or a portion of a counter is required to be accessible. For example, most hotel registration and check-out functions occur at a counter. The intent of this section is to require that one portion of the counter area be accessible. Accessibility is accomplished by providing a lower counter height to accommodate a person using a wheelchair and locating that counter on an accessible route. This is not necessarily intended to require that multiple lower counter heights be provided based on the different types of services that are offered. For example, this is not intended to require two separate, lower counter heights: one for hotel registration and one for check-out. One lower counter height can be provided at which both functions are accomplished. In the above-described situation, dispersion of the accessible counters would be required if the counters themselves were dispersed in separate locations.

The intent of the accessible service window is to permit the customer to interact with the service representative the same way they interact with the general public. An extra piece of counter stuck on the front of the high service counter [typically 42- to 48-inches (1067 to 1219 mm) high] may be appropriate as a temporary fix for barrier removal but is not acceptable in new construction. ICC A117.1 permits a maximum counter height of 36 inches (914 mm) at a service window (see commentary, Section 1109.11 and Commentary Figure 1109.12.3).

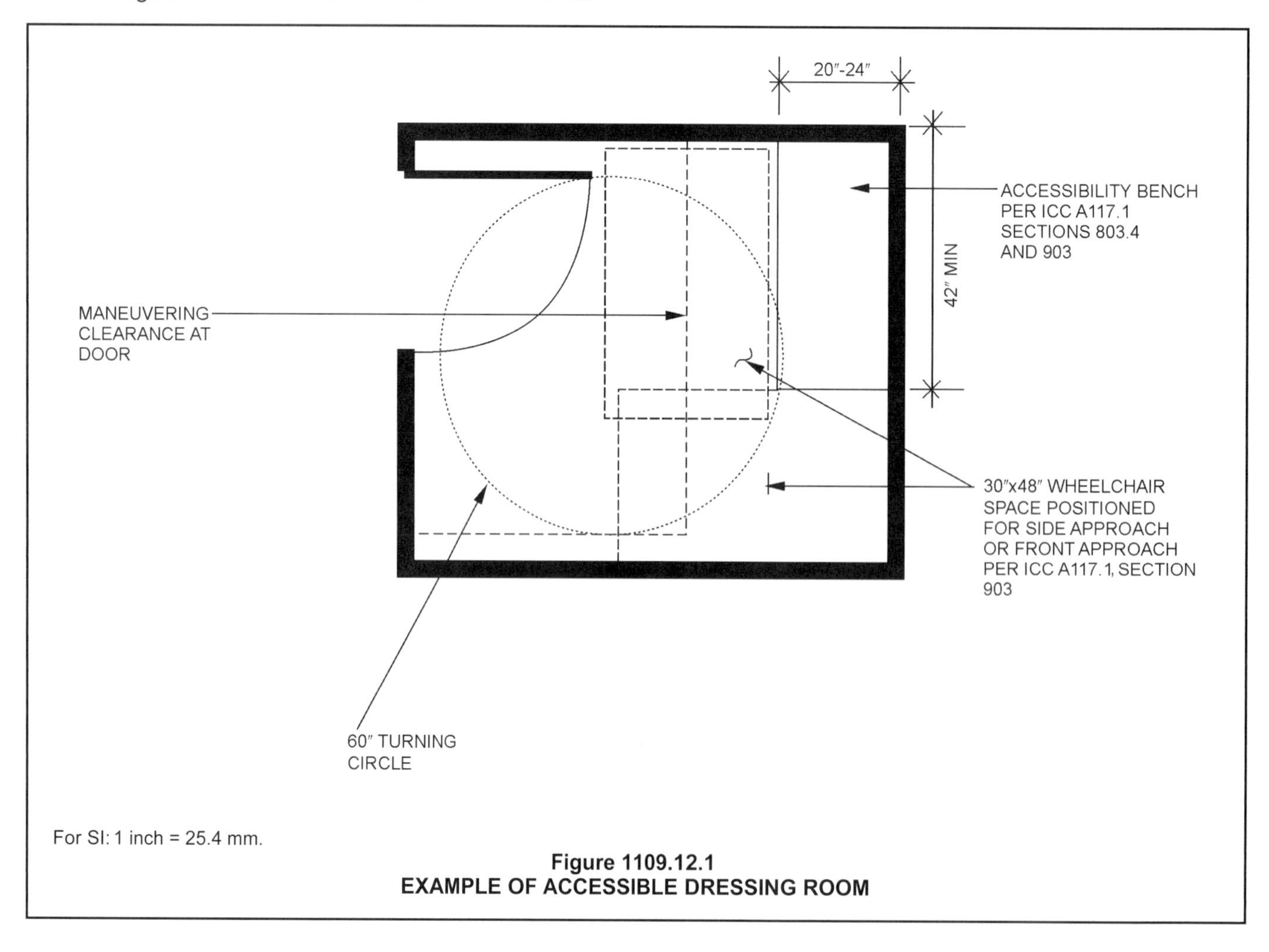

For SI: 1 inch = 25.4 mm.

Figure 1109.12.1
EXAMPLE OF ACCESSIBLE DRESSING ROOM

Figure 1109.12.3
SERVICE COUNTERS

1109.12.4 Food service lines. Food service lines shall be *accessible*. Where self-service shelves are provided, at least 50 percent, but not less than one, of each type provided shall be *accessible*.

❖ All lines that provide access to food service must meet the accessible route provisions in ICC A117.1. Where self-service shelves are utilized, such as in a cafeteria, at least half of each type must be on an accessible route and be located within reach ranges (see Commentary Figure 1109.12.4).

Figure 1109.12.4
EXAMPLE OF SELF-SERVICE SHELVING

1109.12.5 Queue and waiting lines. Queue and waiting lines servicing *accessible* counters or check-out aisles shall be *accessible*.

❖ If patrons must form a line for service, then the waiting line must meet the accessible route provisions of ICC A117.1. Accessible waiting lines are required at locations where accessible counters, windows or check-out aisles are provided, such as banks, ticket counters, fast-food establishments and similar facilities. A separate waiting line for persons with disabilities is not a viable alternative. This section does not require permanent designation of queue lines. However, where permanent crowd control barriers are provided, special attention must be paid to turns around such obstructions or dividers.

1109.13 Controls, operating mechanisms and hardware. Controls, operating mechanisms and hardware intended for operation by the occupant, including switches that control lighting and ventilation and electrical convenience outlets, in *accessible* spaces, along *accessible routes* or as parts of *accessible* elements shall be *accessible*.

Exceptions:

1. Operable parts that are intended for use only by service or maintenance personnel shall not be required to be *accessible*.
2. Electrical or communication receptacles serving a dedicated use shall not be required to be *accessible*.
3. Where two or more outlets are provided in a kitchen above a length of counter top that is uninterrupted by a sink or appliance, one outlet shall not be required to be *accessible*.
4. Floor electrical receptacles shall not be required to be *accessible*.
5. HVAC diffusers shall not be required to be *accessible*.
6. Except for light switches, where redundant controls are provided for a single element, one control in each space shall not be required to be *accessible*.
7. Access doors or gates in barrier walls and fences protecting pools, spas and hot tubs shall be permitted to comply with Section 1010.1.9.2.

❖ This section requires that any controls intended to be utilized by the occupants of the space be accessible when such controls are located in an accessible space, along accessible routes or as part of an accessible element. For example, the switch controlling the general lighting of a room that is required to be accessible also must be accessible. This is also true for most of the electrical outlets that are located in the room. Thermostats that are intended to be adjustable by the occupants must also be accessible. Another example would be the button or switch that activates an automatic hand dryer in a toilet room. If the fixture is mounted such that the outlet for the air is at an accessible height, the purpose is defeated if the switch to activate the device is inaccessible.

The exceptions listed are similar to the exceptions already located in ICC A117.1 for Accessible, Type A and Type B dwelling and sleeping units. Since the same problems exist in nonresidential facilities, the exceptions are appropriate.

Exception 1 deals with items that are not intended for use by the general occupants of the space. This would include controls such as thermostats that are intended to be adjustable only by authorized personnel. Another example would be the controls restricted to use by authorized personnel for a public address system in a meeting room.

Exception 2 deals with connections for a dedicated use. For example, the outlet behind the refrigerator, where it plugs in, is not required to be accessible.

Exception 3 is for kitchens and kitchenettes. If more than one electrical outlet is provided over a portion of the countertop, only one is required to be accessible.

Exception 4 permits floor receptacles not to be accessible, since by being in the floor they are out of the reach range.

Exception 5 is for HVAC diffusers. The typical locations for these elements are most often in the floor, high up on the walls or on the ceiling.

Exception 6 allows for redundant controls for everything but lighting. For example, a ceiling fan could be operated by a wall switch as well as by the chain on the fan itself.

Exception 7 addresses the operable parts on safety gates and doors serving pools, spas and hot tubs. The operable parts are required to be higher than normal to be above the reach of young children (see Section 1010.1.9.2).

1109.14 Fuel-dispensing systems. Fuel-dispensing systems shall be *accessible*.

❖ With the exceptions specific to fuel-dispensing systems (e.g., gas pumps) in ICC A117.1 under "operable parts," this section asks for operable parts to be within the unobstructed reach range provisions in ICC A117.1. Because of safety concerns, these operable parts do not have to meet the 5-pounds-force (22.24 N) requirement. Basically, all operable parts (e.g., pump controls, credit card readers, receipt dispensers, etc.) must be located between 15 inches and 48 inches (381 mm and 1219 mm) above the parking surface. Curbs must be considered when determining the height of controls. The intent is to address the needs of persons with limited reach range. Fuel-dispensing systems that are part of a work station are exempted by Section 1103.2.2.

1109.15 Gaming machines and gaming tables. Two percent, but not less than one, of each type of gaming table provided shall be *accessible* and provided with a front approach. Two percent of gaming machines provided shall be *accessible* and provided with a front approach. *Accessible* gaming machines shall be distributed throughout the different types of gaming machines provided.

❖ Gaming machines and tables are now found nationwide. Accessibility must be considered to allow for all people access to these services/games. Side approach access is not practical or comfortable for extended playing time. Front access allows integration with other players for equal play time and communication.

When gaming tables (e.g., blackjack, roulette, craps, poker) are provided, at least one of each type should be accessible. This may require one of the tables to be positioned at a height that would allow for all players to sit at the same height as the person using the wheelchair rather than on bar stools.

With gaming machines, there may be many different games and amounts within one facility. Distribution of 2 percent throughout the types should provide players with a variety of options.

If the variety offered is greater than the number of accessible gaming machines required, the accessible machines should be dispersed as much as possible. For example, if a facility offers 25-cent, $1 and $10 slot machines, but only two accessible machines are required, then only two of the types are required to be accessible. The choice of which machines are made accessible is up to the building owner or designer.

SECTION 1110
RECREATIONAL FACILITIES

1110.1 General. Recreational facilities shall be provided with *accessible* features in accordance with Sections 1110.2 through 1110.4.

❖ Recreational and sports facilities such as tennis courts, swimming pools, baseball fields, etc., may be part of development on a site such as a school or apartment complex. Group R-2, R-3 and R-4 facilities are addressed in Sections 1110.2 through 1110.2.3 so that equitable opportunity for use by all occupants of a multibuilding site would be evaluated. All other occupancies must comply with Section 1110.3.

The intent of the provisions is to allow for equal access for all persons to participate in the recreational activity to the best of their abilities. It is not intended to change the playing surface or how the game is played. This could be considered analogous to the approach, enter and exit requirements for employee work areas. See the commentary to Section 1110.4.1.

The Access Board has included requirements for many types of recreational facilities in the new 2010 ADA Standard for Accessible Design. The 2009 edition of ICC A117.1 includes a new chapter with technical information on accessibility for many types of recreational facilities. The new Chapter 11 in ICC A117.1 contains technical provisions for amusement rides; recreational boating facilities; exercise machinery and equipment; fishing piers and platforms; golf facilities; miniature golf facilities; play areas; swimming pools; wading pools; hot tubs; spas and shooting and firing positions.

1110.2 Facilities serving Group R-2, R-3 and R-4 occupancies. Recreational facilities that serve Group R-2, R-3 and

Group R-4 occupancies shall comply with Sections 1110.2.1 through 1110.2.3, as applicable.

❖ Group R-2, R-3 and R-4 facilities are addressed in Sections 1110.2 through 1110.2.3 so that equitable opportunity for use by all occupants of a single or multibuilding site would be evaluated.

1110.2.1 Facilities serving Accessible units. In Group R-2 and R-4 occupancies where recreational facilities serve *Accessible units*, every recreational facility of each type serving *Accessible units* shall be *accessible*.

❖ See Sections 1107.6.2.3 and 1107.6.4.1 for when Accessible units are required in Group R-2 and R-4 facilities. When a facility such as a dormitory, fraternity, sorority or group home has Accessible units, any recreational facilities that might serve the residents must also be accessible to the extent specified in Section 1110.4. Examples include an accessible route to the edge of a sand volleyball court, and an accessible route to and into a pool.

1110.2.2 Facilities serving Type A and Type B units in a single building. In Group R-2, R-3 and R-4 occupancies where recreational facilities serve a single building containing *Type A units* or *Type B units*, 25 percent, but not less than one, of each type of recreational facility shall be *accessible*. Every recreational facility of each type on a site shall be considered to determine the total number of each type that is required to be *accessible*.

❖ Many multiple-family developments include common recreational facilities, such as a swimming pool, tennis court, playground, etc., which are available for use by residents of the complex. In some cases, there may be more than one type of recreational facility located at different points within the development. This section establishes the criterion that 25 percent, but not less than one, of each type of recreational facility has to be accessible. For example, if two separate tennis court areas are provided, one containing two tennis courts and the other containing seven, then a total of three courts must be accessible. In a single building site, the three accessible tennis courts could all be in one area or divided between the two areas. This satisfies the requirement that at least 25 percent of the total number of courts be made accessible.

1110.2.3 Facilities serving Type A and Type B units in multiple buildings. In Group R-2, R-3 and R-4 occupancies on a single site where multiple buildings containing *Type A units* or *Type B units* are served by recreational facilities, 25 percent, but not less than one, of each type of recreational facility serving each building shall be *accessible*. The total number of each type of recreational facility that is required to be *accessible* shall be determined by considering every recreational facility of each type serving each building on the site.

❖ In a multibuilding site, the same criteria apply as for single building sites. However, in addition, if certain facilities only serve certain buildings, at least 25 percent of the facilities in each area must be accessible. For example, if two separate tennis court areas are provided, each serving a specific building, and one contains two tennis courts while the other contains seven courts, then 25 percent at each location, for a total of three courts, must be accessible. While the number is the same for the single building, the distribution requirement is different. In this case, one tennis court in the two-court area must be made accessible and two of the tennis courts in the seven-court area must be made accessible. This satisfies the requirement that at least 25 percent of the courts for each building are made accessible. If all tennis courts serve all of the buildings, then the distribution could be the same as discussed in Section 1110.2.3.

1110.3 Other occupancies. Recreational facilities not falling within the purview of Section 1110.2 shall be *accessible*.

❖ When tennis courts, pools, baseball diamonds, playgrounds or other recreational facilities are provided in occupancies other than the Group R-2, R-3 or R-4 occupancies discussed in Section 1110.2, all facilities must be accessible. For recreational facilities associated with Groups R-2, R-3 and R-4, see the commentary for Sections 1110.2 through 1110.2.3.

1110.4 Recreational facilities. Recreational facilities shall be *accessible* and shall be on an *accessible route* to the extent specified in this section.

❖ Accessible recreation facilities must have an accessible route to them. Depending on the type of facility, additional requirements may be indicated to provide further access, such as into a pool, or on the playing surface of a miniature golf course. It is not intended to change the nature of the game or how the game is played. The intent is to allow for someone to participate to the best of their abilities. These items are intended to harmonize with the federal 2010 ADA Standard for Accessible Design.

1110.4.1 Area of sport activity. Each *area of sport activity* shall be on an *accessible route* and shall not be required to be *accessible* except as provided for in Sections 1110.4.2 through 1110.4.14.

❖ "Area of sports activity" is defined in Chapter 2. The broad term, "area of sports activity," addresses indoor and outdoor courts, fields and other sport areas. Examples include basketball and tennis courts; practice areas for dance or gymnastics; baseball, soccer and football fields; skating rinks; running tracks; and skateboard parks. The phrase "portion...where the play or practice of a sport occurs" varies depending on the sports. Football fields include the playing field boundary lines, the end zones and the space between the boundary lines and safety border. Players may run or be pushed into this safety zone during play. In football, this safety zone is used as part of the playing field, and is therefore included in the area of sports activity.

This section requires an accessible route to the edge of a recreational facility. See Sections 1110.4.2 through 1110.4.14 for additional requirements or exceptions for specific types of recreational facilities.

1110.4.2 Team or player seating. At least one wheelchair space shall be provided in team or player seating areas serving *areas of sport activity*.

Exception: Wheelchair spaces shall not be required in team or player seating areas serving bowling lanes that are not required to be *accessible* in accordance with Section 1110.4.3.

❖ Accessible seating for sports teams and players is a logical requirement that allows players to maintain team participation when injured as well as allowing for coaches or players with mobility impairments to participate. When there is team seating provided on opposite sides for home and visitor teams, an accessible route must be provided to both locations. See Commentary Figure 1110.4.2. The exception acknowledges that only 5 percent of lanes in bowling facilities are required to be provided with an accessible route. Since teams sit at their lanes while playing, this provides a balance between player seating and access to the accessible bowling lanes (see Section 1110.4.3).

Figure 1110.4.2
EXAMPLE OF TEAM SEATING IN A BASEBALL DUGOUT

1110.4.3 Bowling lanes. An *accessible* route shall be provided to at least 5 percent, but not less than one, of each type of bowling lane.

❖ Bowling lanes are typically depressed from the main circulation routes and viewing areas in bowling facilities for sightline purposes. Since bowling lanes repeat each other, only 5 percent of bowling lanes and their associated team or player seating areas are required to be on an accessible route. See also Section 1110.4.2.

1110.4.4 Court sports. In court sports, at least one *accessible* route shall directly connect both sides of the court.

❖ If you have court sports, most games have you switch sides during the course of the game. An accessible route is required to both sides of the court without leaving the immediate area. In a multicourt facility, such as a tennis facility, best design practice would provide the accessible routes so that they would not have to go through other courts where someone else may be playing. See Commentary Figure 1110.4.4.

Figure 1110.4.4
EXAMPLE OF COURT SPORTS

1110.4.5 Raised boxing or wrestling rings. Raised boxing or wrestling rings are not required to be *accessible* or to be on an *accessible* route.

❖ Boxing and wrestling rings are raised to allow for a line a site from the audience to the ring, similar to a stage. These unique types of facilities are not required to be on an accessible route or be accessible. This exception is in step with several of the general exceptions in Section 1103.2 where small raised areas provided for specific purposes are exempted. As with any other exception, there is nothing that would prohibit the installation of specialized equipment to allow a disabled person to perform the particular function.

1110.4.6 Raised refereeing, judging and scoring areas. Raised structures used solely for refereeing, judging or scoring a sport are not required to be *accessible* or to be on an *accessible route*.

❖ In order for a judge or referee to see the entire playing surface, they may be sitting on a raised platform or high chair. Where these areas are used solely for refereeing, judging or scoring, they are not required to be accessible or be on an accessible route. See Commentary Figure 1110.4.6.

This exception is in step with several of the general exceptions in Section 1103.2 where small raised areas provided for specific purposes are exempted. As with any other exception, there is nothing that would prohibit the installation of specialized equipment to allow a disabled person to perform the particular function.

Figure 1110.4.6
EXAMPLE OF RAISED REFEREE STAND

1110.4.7 Animal containment areas. Animal containment areas that are not within public use areas are not required to be *accessible* or to be on an *accessible route*.

❖ "Public use area" is defined as "...made available to the general public." There may be areas of a facility, such as a horse riding arena, where portions of the facility are not open to the general public. Those areas are not required to be accessible or be on an accessible route. Areas such as the riding arena would have to be on an accessible route. The arena itself, as an area of sports activity, could be a dirt floor.

1110.4.8 Amusement rides. Amusement rides that move persons through a fixed course within a defined area shall comply with Sections 1110.4.8.1 through 1110.4.8.3.

Exception: Mobile or portable amusement rides shall not be required to be *accessible*.

❖ Amusement rides that move a person through a fixed course and along a specific route are required to be accessible to the extent specified. Typical examples would be a roller coaster, a Ferris wheel, a ride that moves the rider along to view different scenes (e.g., omnimover or rail transport), or pendulum or swing rides. See Commentary Figures 1110.4.8(1) and (2). There is an exception for these types of rides that are mobile or portable. For example, the portable amusement rides that come in each year for the State or County Fair are not required to meet this section. There are also many types of rides that do not have a fixed course. Rides where the route is controlled by the user, such as go-carts or bumper cars, are not required to meet this section. These rides should have the ride entrance on an accessible route, but they are not required to provide additional accessible features specified in Sections 1110.4.8.1 through 1110.4.8.3.

Figure 1110.4.8(1)
EXAMPLE OF AN
AMUSEMENT RIDE WITH A FIXED COURSE

Figure 1110.4.8(2)
EXAMPLE OF AN
AMUSEMENT RIDE WITH A FIXED COURSE

1110.4.8.1 Load and unload areas. Load and unload areas serving amusement rides shall be *accessible* and be on an *accessible route*. Where load and unload areas have more than one loading or unloading position, at least one loading and unloading position shall be on an *accessible route*.

❖ Sections 1110.4.8.1, 1110.4.8.2 and 1110.4.8.3 work together. For the types of rides scoped by Section

1110.4.8, an accessible route must be provided from the entrance to the ride to at least one of the areas where people will get on and off the ride. Since it is difficult for someone to wait on a sloped surface, Section 1109.8 allows for a platform lift to be used to provide an accessible route. In accordance with Section 1104.5, that platform lift should be in the same area as the general circulation route.

For the ride itself, there are the options of having a person transfer from his or her wheelchair to a ride seat or to physically move the wheelchair onto the ride [see Commentary Figures 1110.4.8.1(1) and (2)]. Sometimes, a type of transfer device, such as a transfer board, is necessary for someone to slide from the wheelchair over a gap to the ride seat. This transfer point must also be on an accessible route. If there is a transfer situation, best design practice may include a location for the wheelchair to be stored while the person is on the ride or a means to move the wheelchair to a different unload area.

1110.4.8.2 Wheelchair spaces, ride seats designed for transfer and transfer devices. Where amusement rides are in the load and unload position, the following shall be on an *accessible route*.

1. The position serving a wheelchair space.
2. Amusement ride seats designed for transfer.
3. Transfer devices.

❖ See the commentary to Sections 1110.4.8 and 1110.4.8.1.

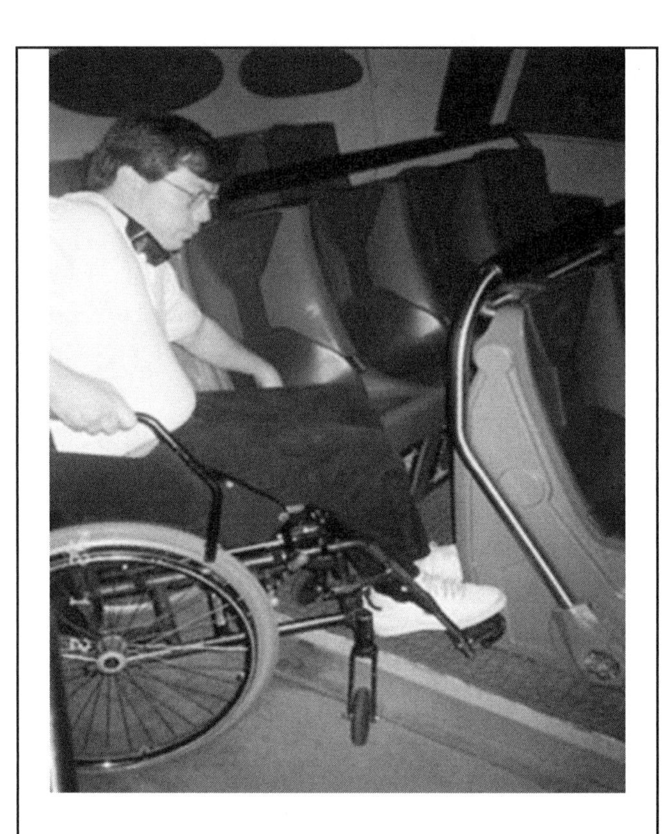

Figure 1110.4.8.1(2)
EXAMPLE OF A PERSON IN A WHEELCHAIR TRANSFERRING ONTO A RIDE SEAT

Figure 1110.4.8.1(1)
EXAMPLE OF A PERSON IN A WHEELCHAIR MOVING ONTO A RIDE

1110.4.8.3 Minimum number. Amusement rides shall provide at least one wheelchair space, amusement ride seat designed for transfer or transfer device.

Exceptions:

1. Amusement rides that are controlled or operated by the rider are not required to comply with this section.
2. Amusement rides designed primarily for children, where children are assisted on and off the ride by an adult, are not required to comply with this section.
3. Amusement rides that do not provide seats that are built-in or mechanically fastened shall not be required to comply with this section.

❖ See the commentary to Sections 1110.4.8 and 1110.4.8.1.

The exceptions apply to types of rides that need an accessible route to the load and unload areas, but do not need to have the ability to transfer directly from the wheelchair to a ride seat. If a person has limited mobility, they may be able to walk a short distance to be able to enjoy the rides. An example of Exception 1 would be bumper cars or go-carts. An example of Exception 2 would be a merry-go-round or other ride designed specifically for small children, even though an adult may choose to go on the ride with that child. An example of a ride that does not provide seats might be a water flume ride that uses inner tubes. See Commentary Figures 1110.4.8.3(1) and (2).

1110.4.9 Recreational boating facilities. Boat slips required to be *accessible* by Sections 1110.4.9.1 and 1110.4.9.2 and boarding piers at boat launch ramps required to be *accessible* by Section 1110.4.9.3 shall be on an *accessible route*.

❖ This section addresses two different types of piers that are found in a harbor or on a lake: the slips where people dock and store their boats and the piers adjacent to boat launch ramps. ICC A117.1 offers technical data on where to provide that accessible route. There are allowances for the difference in elevation between the land piers and a floating pier, as well as consideration of the different water levels caused by tides and drought.

1110.4.9.1 Boat slips. *Accessible* boat slips shall be provided in accordance with Table 1110.4.9.1. All units on the site shall be combined to determine the number of *accessible* boat slips required. Where the number of boat slips is not identified, each 40 feet (12 m) of boat slip edge provided along the perimeter of the pier shall be counted as one boat slip for the purpose of this section.

Exception: Boat slips not designed for embarking or disembarking are not required to be *accessible* or be on an *accessible route*.

❖ Marinas offer separate slips for types of boats, sometimes separating sail boats from power boats or offering slips of different lengths and widths to accommodate boats of different sizes (see Commentary Figure 1110.4.9.1). Similar to parking spaces for cars, the number of accessible spaces depends on the number of slips provided. The total number of slips provided in the arena may be added together to determine the number of slips that have to be on an accessible route.

In accordance with Section 1110.4.9.2, the accessible slips must be dispersed to the different types of slips provided. It is not the intent of the dispersion requirements to result in more accessible slips than those required by Section 1110.4.9.1.

If there are locations where slips are provided, such as areas to get gas, where it is not intended for persons to get on or off their boats, an accessible route is not required to those areas and no accessible slips are required to be dispersed to those locations.

Figure 1110.4.8.3(1)
EXAMPLE OF AN AMUSEMENT RIDE
ADDRESSED IN EXCEPTION 1

Figure 1110.4.8.3(2)
EXAMPLE OF AN AMUSEMENT RIDE
ADDRESSED IN EXCEPTION 2

TABLE 1110.4.9.1. See below.

❖ See commentary to Section 1110.4.9.1.

1110.4.9.2 Dispersion. *Accessible* boat slips shall be dispersed throughout the various types of boat slips provided. Where the minimum number of *accessible* boat slips has been met, no further dispersion shall be required.

❖ See commentary to Section 1110.4.9.1.

1110.4.9.3 Boarding piers at boat launch ramps. Where boarding piers are provided at boat launch ramps, at least 5 percent, but not less than one, of the boarding piers shall be *accessible.*

❖ Boat launch ramps are designated locations where people can move their boat between a towing trailer and the water. If the boat launch has an adjacent boarding pier to allow people to load or access their boat when it is in the water, that boarding pier is required to be accessible (see Commentary Figure 1110.4.9.3). If a boarding pier is not provided at a boat launch ramp, this provision is not intended to require that a boarding pier be provided.

1110.4.10 Exercise machines and equipment. At least one of each type of exercise machine and equipment shall be on an *accessible route.*

❖ Where exercise equipment is provided, an accessible route is required to at least one of each type of machine provided. There are no requirements to provide transfer devices, or change the nature of the equipment itself (ICC A117.1 has a specific exception for exercise equipment stating that it does not have to meet the height, grasping or force requirements for operable parts). Access to exercise equipment is necessary for persons who are in a recovery process from a temporary disability, and for persons with dis-

Figure 1110.4.9.1
EXAMPLE OF BOAT SLIPS

Figure 1110.4.9.3
EXAMPLE OF BOAT LAUNCH RAMP WITH BOARDING PIER

TABLE 1110.4.9.1
BOAT SLIPS

TOTAL NUMBER OF BOAT SLIPS PROVIDED	MINIMUM NUMBER OF REQUIRED ACCESSIBLE BOAT SLIPS
1 to 25	1
26 to 50	2
51 to 100	3
101 to 150	4
151 to 300	5
301 to 400	6
401 to 500	7
501 to 600	8
601 to 700	9
701 to 800	10
801 to 900	11
901 to 1000	12
1001 and over	12, plus 1 for every 100, or fraction thereof, over 1,000

abilities who need to maintain the muscles that they use to operate their equipment. Regarding "type," it is not the intent to require an accessible route to each of group of exercise bikes simply because they happen to be supplied by different manufacturers. Generally the muscle group exercised by a piece of equipment determines its type.

1110.4.11 Fishing piers and platforms. Fishing piers and platforms shall be *accessible* and be on an *accessible route.*

❖ Fishing piers and platforms are also required on an accessible route (see Commentary Figure 1110.4.11). In the ICC A117.1 standard there are allowances for locations where the ramp between the land elevation and the water elevation vary too greatly because of tides or topography, similar to boat piers. There is also the intent to coordinate with safety requirements due to concerns about falls. Some fishing platforms and piers do not have a barrier or guard, and the accessibility provisions do not require one. However, if a 42-inch-high (1067 mm) guard is provided for safety concerns, the requirement for a lower portion of rail to allow for a sitting person to fish is waived.

Figure 1110.4.11
EXAMPLE OF AN ACCESSIBLE FISHING PLATFORM

1110.4.12 Miniature golf facilities. Miniature golf facilities shall comply with Sections 1110.4.12.1 through 1110.4.12.3.

❖ Miniature golf has been a family pastime for decades. Undulating or sloped playing surfaces, changes in level, shooting through an object with a surprise as to where it comes out, water, sandtraps and rough turf all help make the game interesting. The provisions in the following subsections allow for access to at least portions of the course.

1110.4.12.1 Minimum number. At least 50 percent of holes on miniature golf courses shall be *accessible.*

❖ Half of the holes provided must be on an accessible route and meet the technical criteria in ICC A117.1. The first hole on the course, considered the start of play, must be accessible. The standard allows for the accessible route to be on the playing surface or adjacent where a certain reach can be maintained. Limited curbs are permitted across the route in order to keep the ball in play.

1110.4.12.2 Miniature golf course configuration. Miniature golf courses shall be configured so that the *accessible* holes are consecutive. Miniature golf courses shall provide an *accessible route* from the last *accessible* hole to the course entrance or exit without requiring travel through any other holes on the course.

Exception: One break in the sequence of consecutive holes shall be permitted provided that the last hole on the miniature golf course is the last hole in the sequence.

❖ In order that a person does not have to move through other holes that may be in play, the accessible route must not travel through nonaccessible holes. There can be a one break in the accessible holes, as long as the last hole is included in the route. Traditionally that is the hole where you can shoot to win a free game or a prize. For example, in an 18-hole course, you could make holes 1 through 5 and 15 through 18 the accessible holes. Possible routes will vary depending on the layout of the facility.

1110.4.12.3 Accessible route. Holes required to comply with Section 1110.4.12.1, including the start of play, shall be on an *accessible route.*

❖ See the commentary to Section 1110.4.12.1.

1110.4.13 Swimming pools, wading pools, hot tubs and spas. Swimming pools, wading pools, hot tubs and spas shall be *accessible* and be on an *accessible route.*

Exceptions:

1. Catch pools or a designated section of a pool used as a terminus for a water slide flume shall not be required to provide an *accessible* means of entry, provided that a portion of the catch pool edge is on an *accessible route.*
2. Where spas or hot tubs are provided in a cluster, at least 5 percent, but not less than one spa or hot tub in each cluster, shall be *accessible* and be on an *accessible route.*
3. Swimming pools, wading pools, spas and hot tubs that are required to be *accessible* by Sections 1110.2.2 and 1110.2.3 are not required to provide *accessible* means of entry into the water.

❖ Swimming pools are required to have a route to them and a route into the water. ICC A117.1 offers several different options for access into the water depending the type and size of the pool (see Commentary Figure 1110.4.13). Technical criteria are provided for sloped entries, ramps, pool lifts, transfer walls, transfer steps and pool stairs.

For Exception 1, see the commentary to Section 1110.4.13.2.

In accordance with Exception 2, when hot tubs are grouped, 5 percent must be accessible. In accordance with ICC A117.1, the accessible hot tub can use a pool or lift or a transfer wall to allow access into the water.

Exception 3 is an exception for the route into the water for swimming pools that serve Type A and Type B dwelling units in Groups R-2, R-3 and R-4. Group R-2 with Accessible units, such as dormitories, cannot use this exception.

1110.4.13.1 Raised diving boards and diving platforms. Raised diving boards and diving platforms are not required to be *accessible* or to be on an *accessible route*.

❖ Raised diving boards and diving platforms are not required to be accessible. It would be impractical to provide ramps or lifts to these very small areas.

1110.4.13.2 Water slides. Water slides are not required to be *accessible* or to be on an *accessible route*.

❖ Water slides are not required to be accessible. It would be impractical to provide ramps or lifts to these very small areas. When the swimming pool at the bottom is only for the slide, swimmers are not permitted there for safety reasons. That catch pool is not required to have a route into the water (see Section 1110.4.13, Exception 1, and Commentary Figure 1110.4.13.2).

1110.4.14 Shooting facilities with firing positions. Where shooting facilities with firing positions are designed and constructed at a site, at least 5 percent, but not less than one, of each type of firing position shall be *accessible* and be on an *accessible route*.

❖ There are a variety of shooting facilities that are used for target practice. Typically this is practice for shooting bows and arrows, handguns, rifles or shotguns. Participants aim at fixed targets, decoys or clay pigeons. These facilities can be constructed inside or outside. An accessible route is required to at least 5 percent of the firing position at each type of arrangement provided. ICC A117.1 will require a level space and a turning radius at that firing position.

SECTION 1111
SIGNAGE

1111.1 Signs. Required *accessible* elements shall be identified by the International Symbol of Accessibility at the following locations.

1. *Accessible* parking spaces required by Section 1106.1.

   Exception: Where the total number of parking spaces provided is four or less, identification of *accessible* parking spaces is not required.

2. *Accessible* parking spaces required by Section 1106.2.

   Exception: In Group I-1, R-2, R-3 and R-4 facilities, where parking spaces are assigned to specific *dwelling units* or *sleeping units*, identification of *accessible* parking spaces is not required.

3. *Accessible* passenger loading zones.
4. *Accessible* rooms where multiple single-user toilet or bathing rooms are clustered at a single location.
5. *Accessible* entrances where not all entrances are *accessible*.
6. *Accessible* check-out aisles where not all aisles are *accessible*. The sign, where provided, shall be above

POOL TYPE	SLOPED ENTRY/ RAMP	POOL LIFT	TRANSFER WALLS	TRANSFER STEPS	POOL STAIRS
Swimming pool (less than 300 linear feet of pool wall)	X	X			
Swimming pool (300 or more linear feet of pool wall) – two means of entry required	X[b]	X[b]	X	X	X
Wave action, leisure river, and other pools where user entry is limited to one area	X	X		X	
Wading pools	X				
Spas/hot tubs		X	X	X	

Notes:

a. For technical requirements for pool entry, see ICC A117.1.

b. Primary means must be by sloped entry, ramp or pool lift. The secondary means may be by any of the permitted types.

For SI: 1 foot - 304.8 mm.

Figure 1110.4.13
PERMITTED MEANS OF POOL ACCESS[a]

Figure 1110.4.13.2
EXAMPLE OF A CATCH POOL FOR A WATER SLIDE

the check-out aisle in the same location as the check-out aisle number or type of check-out identification.

7. Family or assisted-use toilet and bathing rooms.
8. *Accessible* dressing, fitting and locker rooms where not all such rooms are *accessible*.
9. *Accessible* areas of refuge in accordance with Section 1009.9.
10. Exterior areas for assisted rescue in accordance with Section 1009.9.
11. In recreational facilities, lockers that are required to be *accessible* in accordance with Section 1109.9.

❖ Identification of accessible elements can be accomplished by use of an International Symbol of Accessibility (see Commentary Figure 1111.1). These symbols are international in that they are recognized throughout the world as identifying accessibility. The signs listed here are typically visual signage. Raised letters and braille are not required.

There are 11 specific circumstances in which required accessible elements are to be identified, as indicated in Items 1 through 11 of this section. Generally, these are locations in which not all of the facilities provided are accessible and, therefore, it is necessary to identify those that are accessible so that they can be readily recognized by the intended user. For example, Section 1106.1 specifies the required number of accessible parking spaces. If these are not identified by signage, it would be difficult and unnecessarily inconvenient for one to identify their location. One of the concepts embodied by the code is to mainstream accessibility in recognition that many of the features making facilities and elements accessible are also useful and of benefit to people without disabilities. Part of this principle includes the idea that if an element is universally usable by people both with and without disabilities, there is no need for signage specifically identifying the element as being accessible. Hence, the signage requirements generally address circumstances in which not all of the elements will be accessible. Items 9 and 10 require the International Symbol of Accessibility in order to identify elements important to keep open as part of an accessible means of egress.

Item 11 is only to identify lockers in recreational facilities where use of a locker is temporary for the duration of the visit to the recreational facility. It is not intended to apply to lockers in other locations, such as lockers provided for students in school. In those situations, the identification of the assigned locker as accessible might be perceived to have negative connotations for the student assigned to the locker.

There are two exceptions specific to identification of accessible parking spaces in Items 1 and 2. Note that the spaces must be sized in accordance with Sections 1106.1 and 1106.2; the exceptions are only for the identification. Item 1 has an exception where the entire parking provided is four or fewer spaces. To require an accessible space to be reserved when it involves from 25 to 100 percent of the parking provided is an undue hardship to the building's employees and customers. With the area equal to an accessible parking space being provided, accessible parking is still available if it is needed, it is just not reserved. In Item 2, when a parking lot has assigned spaces for the people who live there, the identification to reserve an accessible space is not required. Who is assigned to that space can be handled administratively when a resident needs an accessible parking space.

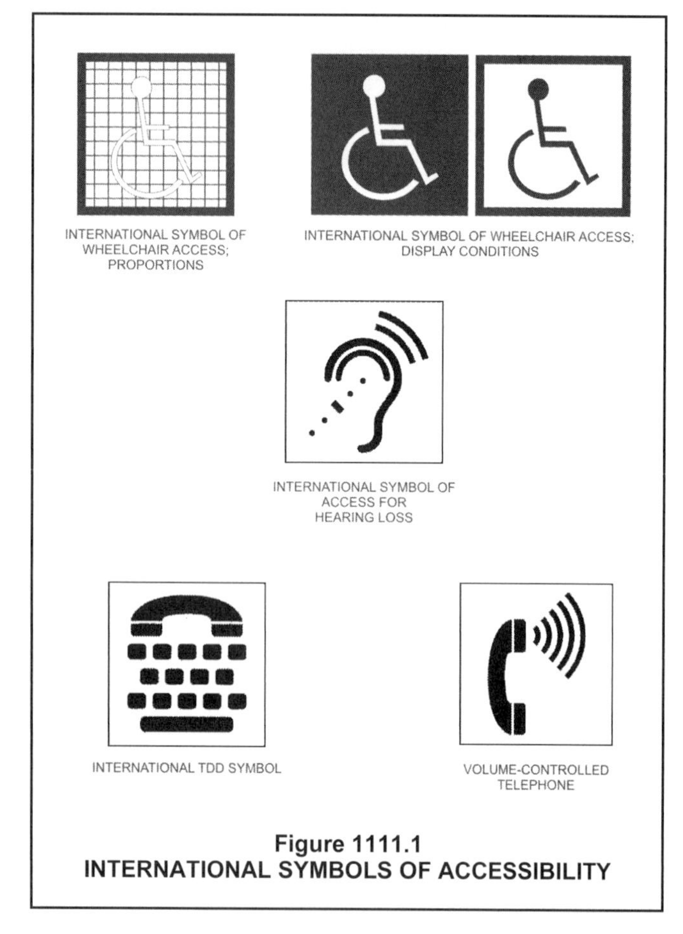

Figure 1111.1
INTERNATIONAL SYMBOLS OF ACCESSIBILITY

1111.2 Directional signage. Directional signage indicating the route to the nearest like *accessible* element shall be provided at the following locations. These directional signs shall include the International Symbol of Accessibility and sign characters shall meet the visual character requirements in accordance with ICC A117.1.

1. Inaccessible building entrances.
2. Inaccessible public toilets and bathing facilities.
3. Elevators not serving an *accessible route*.
4. At each separate-sex toilet and bathing room indicating the location of the nearest family/assisted use toilet or bathing room where provided in accordance with Section 1109.2.1.
5. At *exits* and *exit stairways* serving a required *accessible* space, but not providing an *approved* accessible *means of egress*, signage shall be provided in accordance with Section 1009.10.
6. Where drinking fountains for persons using wheelchairs and drinking fountains for standing persons are not located adjacent to each other, directional signage shall be provided indicating the location of the other drinking fountains.

❖ The direction signs listed here are typically visual signage. Raised letters and braille are not required.

There are circumstances in which it is useful and necessary to locate directional signage at certain inaccessible elements to indicate the route to the nearest like accessible element. For example, not all building entrances are required to be accessible (see Section 1105). Should a person in a wheelchair happen to approach an inaccessible building entrance at an unfamiliar facility, it is appropriate to provide direction indicating where the nearest accessible entrance is located. The same circumstance presents itself at inaccessible public toilet and bathing facilities, at elevators that do not serve an accessible route and at exits that are not part of the accessible means of egress (see Section 1009 for additional information on accessible means of egress). Directional signage is also needed to point people to specific accessible features they may need, such assembly and mercantile occupancies where family or assisted-use accessible toilet and bathing facilities are required or when high-low drinking fountains are not located adjacent to each other (Items 4 and 6).

Requiring the sign to meet visual character requirements allows for the sign to be read at the viewing distance. For example, if an entrance is not accessible and the closest a person using a wheelchair can get to the entrance is the bottom of the steps, the sign can be on the door and be large enough for a person to read from the bottom of the steps, or the sign can be smaller and placed at the bottom of the steps. Both methods would provide the information on where the accessible entrance is located.

This requirement for directional signage works in conjunction with the fact that everything is initially assumed to be accessible. Some items are not required to be identified by the International Symbols of Accessibility in the interest of mainstreaming. See the general comments to this chapter and the commentary to Section 1111.1 for a further discussion of mainstreaming.

1111.3 Other signs. Signage indicating special accessibility provisions shall be provided as shown.

1. Each assembly area required to comply with Section 1108.2.7 shall provide a sign notifying patrons of the

availability of assistive listening systems. The sign shall comply with ICC A117.1 requirements for visual characters and include the International Symbol of Access for Hearing Loss.

Exception: Where ticket offices or windows are provided, signs are not required at each assembly area provided that signs are displayed at each ticket office or window informing patrons of the availability of assistive listening systems.

2. At each door to an *area of refuge*, an exterior area for assisted rescue, an egress *stairway*, *exit passageway* and *exit discharge*, signage shall be provided in accordance with Section 1013.4.
3. At *areas of refuge*, signage shall be provided in accordance with Section 1009.11.
4. At exterior areas for assisted rescue, signage shall be provided in accordance with Section 1009.11.
5. At two-way communication systems, signage shall be provided in accordance with Section 1009.8.2.
6. In *interior exit stairways* and *ramps*, floor level signage shall be provided in accordance with Section 1023.9.
7. Signs identifying the type of access provided on amusement rides required to be *accessible* by Section 1110.4.8 shall be provided at entries to queues and waiting lines. In addition, where *accessible* unload areas also serve as *accessible* load areas, signs indicating the location of the *accessible* load and unload areas shall be provided at entries to queues and waiting lines. These directional sign characters shall meet the visual character requirements in accordance with ICC A117.1.

❖ The signs listed here are typically visual signage. Raised letters and braille are only required by Item 2.

It is considered desirable that all the requirements for accessible signage be in one location. Item 1 requires signage indicating that assistive listening systems are provided when they are required by Section 1108.2.7. The sign may be located within each of the assembly rooms, typically near the entrance, or at the ticket office. The sign should also include the International Symbol of Access for Hearing Loss. (See Commentary Figure 1111.1).

Items 2 through 6 are references to signage criteria in specific sections that have been mainstreamed in Chapter 10.

Item 7 requires one or two signs at the start of the waiting line for an amusement ride. The intent is that the rider is aware of the options before waiting in line for the ride. One sign will provide information on what transfer options are available. On most rides, for crowd control, the way in and the way out are two different paths. When the ride uses the unload area (instead of the load area) as the accessible load and unload area, directions for that alternative route must be provided.

1111.4 Variable message signs. Where provided in the locations in Sections 1111.4.1 and 1111.4.2, variable message signs shall comply with the variable message sign requirements of ICC A117.1.

❖ ICC A117.1 defines Variable Message Signs (VMS) as, "Electronic signs that have a message with the capacity to change by means of scrolling, streaming, or paging across a background." Requirements vary depending on whether the sign resolution is high or low. Requirements for high-resolution signs are consistent with printed signs. The ICC A117.1 language modifies requirements for visual signs primarily by requiring increased character height and spacing for low-resolution VMS. Even for users with unimpaired vision, there is strong research evidence that character height in low-resolution VMS, such as the LED signs that are common in the transportation environment and elsewhere, must be approximately 30 percent greater than for equivalent print signs or VMS with high resolution. The requirements for increased character height, etc., for low-resolution VMS are not applicable to high-resolution VMS, such as video (see Commentary Figure 1111.4).

The section does not "require" that VMS be installed in any facility type, but instead requires that, where provided within transportation facilities and emergency shelters, VMS comply with the standard and are usable by most of the population.

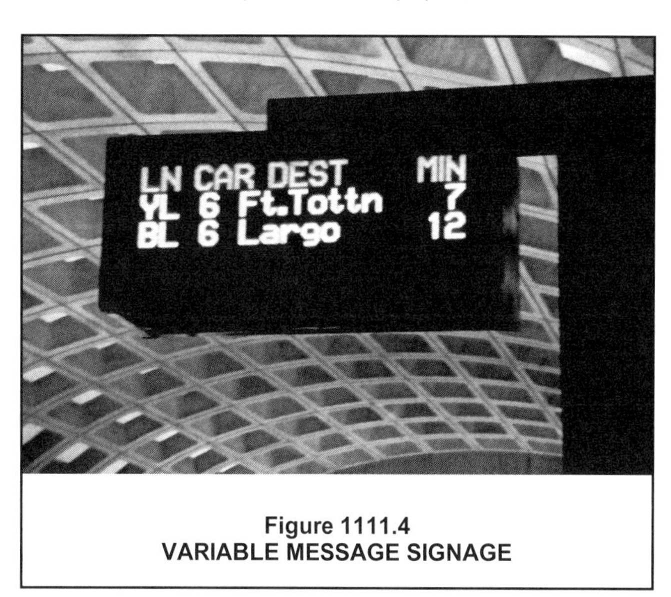

Figure 1111.4
VARIABLE MESSAGE SIGNAGE

1111.4.1 Transportation facilities. Where provided in transportation facilities, variable message signs conveying transportation-related information shall comply with Section 1111.4.

❖ The phrasing "conveying transportation-related information" was included to limit the application of the ICC A117.1 VMS requirements to signs that are necessary for the most effective use of transportation information. The intent of this text is that signs that present information that is not necessary for transportation would be exempt. For example, neither a VMS that advertised dining options in an airport nor a television set in the waiting area that had closed captioning of a news program would be required to comply with the standard's VMS provisions. However, any VMS that indicated flight departure times or gate assignments would be regulated.

Signage in transportation facilities is important for several reasons. Riders with reduced vision are especially dependent on public transportation for travel and are required by the ADA to have information that is equivalent to that provided to riders having unimpaired vision. This section coordinates the code requirements with those of the ADA. Currently, Sections E109 and E110 in Appendix E address some signage requirements within transportation facilities, stations and airports. The VMS are often used within transportation facilities (see Sections E109.2.2.2, E109.2.2.3, E109.2.7, E110.3) as a primary means of conveying information to riders. Since the ICC A117 committee has provided specific regulations for these signs, it is appropriate to reference these provisions so that designers and building officials know exactly what the technical requirements for the signs are. As noted earlier, this section does not "require" VMS, but "where provided" is intended to ensure that VMS comply with the requirements established by the ICC A117.1 committee.

The exception for equivalent audible information (see Section 1111.4.2) is not included in this section. This is done not only to coordinate with the ADA, but, also because in many such facilities, the audible information is simply not easily understood. Anyone who has stood on a subway or train platform and tried to understand an audible message at the same time that the train is pulling up to the platform will understand this exclusion. Because of the problems with hearing messages in many transportation facilities, the audible message delivery is simply not considered as being adequate to replace or eliminate the clear visual sign information.

1111.4.2 Emergency shelters. Where provided in buildings that are designated as emergency shelters, variable message signs conveying emergency-related information shall comply with Section 1111.4.

Exception: Where equivalent information is provided in an audible manner, variable message signs are not required to comply with ICC A117.1.

❖ The phrasing "conveying emergency-related information" was included in the section to limit the application of ICC A117.1 VMS requirements to signs that are necessary for the most effective use of information in emergency shelters. The intent of this text is that signs presenting information that is not necessary for the emergency use of facilities would be exempt. For example, in the case of an emergency shelter located at a school, only the signage related to the shelter would be covered. The signage dealing with the school and what school activities were occurring that week would not be expected to comply with the VMS requirements.

Emergency shelters are typically identified by a jurisdiction when they are studying emergency planning or working with the Federal Emergency Management Agency (FEMA) to develop community plans. Although any building could ultimately be pressed into service as an emergency shelter in some circumstances, the intent of this text is to apply the requirement only to those facilities that are designated in advance or during that planning stage as an intended shelter. In many communities, this may include certain schools, civic administration buildings or even large convention facilities. Emergency-related information provided by VMS in emergency shelters is most likely to be information or instructions regarding additional problems or the recovery process.

The emergency shelter provisions contain an exception that would eliminate the requirement to meet the VMS provisions of the standard. Where audible information is conveyed that is either the same or equivalent to the information provided by the sign, compliance with ICC A117.1's VMS requirements is not needed. The VMS provides the information in a visual method and ICC A117.1 provisions ensure that the information is clearly visible and legible to both the general population and to people with some level of visual impairment. Providing the equivalent information in an audible manner makes the information accessible to people with severe visual impairment or who are blind.

Bibliography

The following resource materials were used in the preparation of the commentary for this chapter o the code.

24 CFR, *Fair Housing Accessibility Guidelines* (FHAG). Washington, DC: Department of Housing and Urban Development, 1991.

36 CFR, Parts 1190 and 1191 Final Rule, *The Americans with Disabilities Act (ADA) Accessibility Guidelines; Architectural Barriers Act (ABA) Accessibility Guidelines*. Washington, DC: Architectural and Transportation Barriers Compliance Board, July 23, 2004.

42 USC 3601-88, *Fair Housing Amendments Act (FHA)*. Washington, DC: United States Code, 1988.

2010 *ADA Standard for Accessible Design*. Washington, DC: Department of Justice, September 15, 2010.

"Accessibility and Egress for People with Physical Disabilities." Falls Church, VA: CABO Board for the Coordination of the Model Codes Report, 1993.

ASME A17.1/CSA B44-07, *Safety Code for Elevators and Escalators*. New York, NY: American National Standards Institute, 2007.

ASME A18.1-05, *Safety Standard for Platform Lifts and Stairway Chairlifts*. New York, NY: American National Standards Institute, 2005.

DOJ 28 CFR, Parts 36-91, *Americans with Disabilities Act (ADA)*. Washington, DC: Department of Justice, 1991.

DOJ 28 CFR, Part 36-91 (Appendix A), *ADA Accessibility Guidelines for Building and Facilities (ADAAG)*. Washington, DC: Department of Justice, 1991.

Fair Housing Act Design Manual. Washington, DC: U.S. Department of Housing and Urban Development, Revised 1998.

FED-STD-795-88, *Uniform Federal Accessibility Standards*. Washington, DC: General Services Administration; Department of Defense; Department of Housing and Urban Development; U.S. Postal Service, 1988.

Final Report of the HUD Review of the Fair Housing Accessibility Requirements and the 2003 International Building Code (IBC). Washington, DC: February 18, 2005 (Docket No FR-4943-N-02).

Final Report of the HUD Review of the Fair Housing Accessibility Requirements and the 2006 International Building Code (IBC). Washington, DC: May 31, 2007 (Docket No FR-5136-N-01).

HUD 24 CFR, Part 100, *Fair Housing Accessibility Guidelines*. Washington, DC: U.S. Department of Housing and Urban Development, 1991.

"Pathfinder Tactile Tile Demonstration Test Project." Miami, FL: Metro-Dade Transit Agency, 1988.

"Tactile Warnings to Promote Safety in the Vicinity of Transit Platform Edges." Washington, DC: Urban Mass Transportation Administration, 1987.

Vital Health Statistics, Series 10, No. 135, Tables 1 & 2. 1977 National Health Interview Survey. Huntsville, MD: National Center for Health Statistics, 1977.

Chapter 12: Interior Environment

General Comments

Chapter 12 contains provisions governing the interior environment requirements of all buildings and structures intended for human occupancy.

Section 1201 identifies the scope of the chapter. Section 1202 provides definitions applicable to terms used in this chapter.

Section 1203 identifies special enclosed spaces where accumulated moisture must be removed by ventilation and establishes criteria (minimum openable area) for the only method of measuring compliance with the natural ventilation requirements for occupied spaces.

Section 1204 identifies the minimum space-heating requirement for interior spaces intended for human occupancy.

Section 1205 requires light for every room or space intended for human occupancy. The method of compliance is the choice of the designer, who may elect to provide artificial instead of natural light. Prescriptive requirements for stairway lighting in dwelling units are also included.

Section 1206 specifies the minimum requirements for courts and yards, including area width, accessibility for cleaning and the location of air intakes when natural light or natural ventilation is the chosen design option.

Section 1207 establishes the sound transmission control requirements for air-borne and structure-borne sound in residential buildings.

Section 1208 addresses the minimum ceiling height for all habitable and occupiable spaces along with other spaces as specified (i.e., toilet rooms and bathrooms). The minimum floor area for rooms in dwelling units is also specified.

Section 1209 provides minimal opening requirements for access to crawl spaces and attics.

Section 1210 contains requirements for toilet room surfaces and fixture surrounds.

The environmental and physiological justification for the code requirements relevant to light and ventilation of occupiable spaces is based on knowledge, technology and practices developed over centuries of building structures in which humans live and work. These design practices have been further validated by studies during the 19th and 20th centuries.

Purpose

The purpose of Chapter 12 is to establish minimum conditions for the interior environment of a building. The size of spaces, light, ventilation and noise intrusion are all addressed in order to define the acceptable conditions to which any occupant may be exposed. Design options of natural and mechanical systems are introduced and the criteria for performance are specified.

Even though it was not completely understood, the need for "fresh air" had been recognized for centuries. Designers of centuries-old adobe buildings in the southwestern United States, hide-covered Native American tepees of the plains and frame houses of early settlers in the eastern United States all relied upon the buoyancy of warm air, enabling it to rise and cooler air to flow in to replace it. Whether the design relied on solar energy, thermal mass or even wind velocity to cause the movement, it reflected a natural movement and, as a result, has been termed "natural ventilation." Only recently have we begun to recognize the reasons for ventilation and the implications of failing to provide an adequate quantity and acceptable quality of air for all occupants. The expression "sick building syndrome" has crept into our vocabulary and reflects the increased understanding of the relationship between interior environment requirements and the physiological well-being of the occupants.

The other purpose of regulating the interior environment is psychological. Merely providing adequate conditions is not sufficient if the occupant does not perceive them as adequate. Minimum space requirements (floor area, yard dimensions or ceiling height) address the need to perceive adequate light, ventilation and space to promote psychological well-being. Regulation of sound transmission also bears directly on the psychological and long-term physical well-being of the occupant.

Finally, adequate lighting from natural sources also meets the physical and psychological needs of the occupants and contributes directly to their overall safety. Safe use of any building under ordinary and emergency conditions depends greatly on proper illumination of the space. This chapter references the *International Mechanical Code*® (IMC®) as the performance standard to which ventilation must be compared and the installation standard for mechanical systems used in buildings regulated by the code.

SECTION 1201
GENERAL

1201.1 Scope. The provisions of this chapter shall govern ventilation, temperature control, lighting, *yards* and *courts*, sound transmission, room dimensions, surrounding materials and rodentproofing associated with the interior spaces of buildings.

❖ This section identifies the scope of Chapter 12. The requirements of this chapter are intended to govern and regulate the need for light, ventilation, sound transmission control, interior space dimensions and materials surrounding plumbing fixtures in all buildings. It is the intent of the code that the user must comply with these regulations for all newly constructed buildings and structures and for all buildings and structures, or portions thereof, when there is to be a change of occupancy or within any additions to a building.

SECTION 1202
DEFINITIONS

1202.1 General. The following terms are defined in Chapter 2:

❖ Definitions of terms that are associated with the content of this section are contained herein. These definitions can help in the understanding and application of the code requirements. It is important to emphasize that these terms are not exclusively related to this section but are applicable everywhere the term is used in the code. The purpose for including these definitions within this section is to provide more convenient access to them without having to refer back to Chapter 2. For convenience, these terms are also listed in Chapter 2 with a cross reference to this section. The use and application of all defined terms, including those defined herein, are set forth in Section 201.

SUNROOM.

❖ This terminology is provided in order to address separate requirements for sunrooms with regard to ventilation of adjoining spaces (see Section 1203.5.1.1). In the 2009 code the term "sunroom addition" was simplified to "sunroom." Provisions later in the chapter that still say sunroom addition. Those provisions are applicable regardless of whether the sunroom is an addition to an existing structure or is part of the original design and construction.

THERMAL ISOLATION.

❖ This terminology is required for the same reason provided in the definition of "Sunroom" (see above). Although the exact phrase "thermal isolation" is not used in the code, the phrase "thermally isolated" is found in a number of provisions. This definition provides guidance in applying this variation on the defined term.

SECTION 1203
VENTILATION

1203.1 General. Buildings shall be provided with natural ventilation in accordance with Section 1203.4, or mechanical ventilation in accordance with the *International Mechanical Code*.

Where the air infiltration rate in a *dwelling unit* is less than 5 air changes per hour when tested with a blower door at a pressure 0.2 inch w.c. (50 Pa) in accordance with Section 402.4.1.2 of the *International Energy Conservation Code—Residential Provisions*, the *dwelling unit* shall be ventilated by mechanical means in accordance with Section 403 of the *International Mechanical Code*. *Ambulatory care facilities* and Group I-2 occupancies shall be ventilated by mechanical means in accordance with Section 407 of the *International Mechanical Code*.

❖ Every room or space must be provided with ventilation. The selection of natural or mechanical ventilation on a room-by-room or space-by-space basis is the designer's prerogative. Certain conditions require mechanical exhaust even though natural ventilation systems have been selected by the designer. Section 1203.5.2 of the code, and Sections 401.6 and 403 of the IMC, direct the treatment of those special situations. Existence of these conditions, however, does not require providing mechanical ventilation other than to address the specific condition. Other rooms or spaces not affected by those conditions may be served by natural systems.

The second paragraph requires mechanical ventilation for residences when the building has been sealed very tightly against air infiltration. The measure will be 5 air changes per hour (ACH) at 50 pascals when tested in accordance with the *International Energy Conservation Code*® (IECC®), which requires air leakage testing for all new residences.

Indoor air quality has direct impact on the health of building occupants. Poor indoor air quality is listed by the Environmental Protection Agency (EPA) as being the fourth largest environmental threat to our country. A 2007 California study revealed formaldehyde exposure in most new homes is beyond limits recommended by the California Air Resources Board. Multiple studies have shown that relying on window operation to provide ventilation is not sufficient enough for the health of occupants. If unchecked, pollutants from cleaning chemicals, finishes, furniture and occupant activities can cause serious health effects on building occupants. Whole-house mechanical ventilation reduces occupant exposure to such pollutants.

Traditionally, 0.35 natural air changes per hour has been the consensus ventilation rate to provide sufficient fresh air to building occupants. This ventilation rate was typically achieved without mechanical ventilation because homes were built without an effective air barrier. As building practices have improved, homes have become tighter, and as homes become tighter, mechanical ventilation must be introduced to provide sufficient levels of ventilation.

ASHRAE 136 was developed to enable calculation of natural air changes per hour as a function of air changes at various pressures. By following the calculation procedures in this standard, it can be shown that a natural infiltration rate of 0.35 is equivalent to somewhere between 7 and 10 air changes per hour at 50 pascals (7 ACH 50 to 10 ACH 50) depending on the local climatic conditions of the home. Because most dwellings are built this tight, ASHRAE 62.2 requires mechanical ventilation for all homes, with few exceptions. However, based on ASHRAE 136, a conservative code might prescribe whole-house mechanical ventilation for any home with an infiltration leakage rate of 10 ACH 50 or less.

As a second point of reference, Chapter 6 of California's 2010 Title 24 requires that "Continuous mechanical ventilation (either exhaust or supply ventilation) must be installed when the target specific leakage area (SLA) is below 3.0." California's SLA of 3.0 is roughly equivalent to 6 ACH 50. As a third point of reference, National Association of Home Builders (NAHB) National Green Building Standard requires whole-house mechanical ventilation when the infiltration rate falls below 5.0 ACH 50. This requirement provides clear recognition from a consensus standard that whole-house mechanical ventilation should be provided for all homes that meet this threshold.

Based on the previous references, there is broad consensus across states and within consensus standards that whole-house mechanical ventilation should be required when a dwelling's infiltration falls below 5.0 ACH 50.

The section also specifies that Group I-2 and Ambulatory Care facilities be ventilated in accordance with Section 407 of the IMC. Section 407 of the IMC essentially references ASHRAE 170, which is a standard specific to the ventilation needs of health care facilities. Previously the IMC contained requirements for ventilation of certain spaces in healthcare facilities. A reference to a more specific standard was preferred. See the commentary to Section 407 of the IMC.

1203.2 Ventilation required. Enclosed *attics* and enclosed rafter spaces formed where ceilings are applied directly to the underside of roof framing members shall have cross ventilation for each separate space by ventilation openings protected against the entrance of rain and snow. Blocking and bridging shall be arranged so as not to interfere with the movement of air. An airspace of not less than 1 inch (25 mm) shall be provided between the insulation and the roof sheathing. The net free ventilating area shall be not less than $^1/_{150}$ of the area of the space ventilated. Ventilators shall be installed in accordance with manufacturer's installation instructions.

Exception: The net free cross-ventilation area shall be permitted to be reduced to $^1/_{300}$ provided both of the following conditions are met:

1. In Climate Zones 6, 7 and 8, a Class I or II vapor retarder is installed on the warm-in-winter side of the ceiling.
2. At least 40 percent and not more than 50 percent of the required venting area is provided by ventilators located in the upper portion of the *attic* or rafter space. Upper ventilators shall be located not more than 3 feet (914 mm) below the ridge or highest point of the space, measured vertically, with the balance of the *ventilation* provided by eave or cornice vents. Where the location of wall or roof framing members conflicts with the installation of upper ventilators, installation more than 3 feet (914 mm) below the ridge or highest point of the space shall be permitted.

❖ All attic spaces and each separate space formed between solid roof rafters are required to be cross ventilated where the ceiling is applied directly to the underside of the roof rafters. Care must be taken, however, to provide cross ventilation in a manner that does not introduce moisture to the attic area.

Snow infiltration can occur when the attic ventilation openings are not sufficiently protected against the entrance of snow or rain, or when more than 50 percent of the ventilation openings are located along the ridge or gable wall of the roof rather than at the eave. When the wind blows perpendicular to a roof ridge vent, a negative pressure builds up across the ridge that draws air out of the attic space through the attic vents. Cross flow of air through the attic can be achieved when outside air is drawn into the attic through the eave or cornice and exits through the ridge or gable vents. In order for this to occur, eave or cornice vents must be greater than or equal to the area of the ridge or gable vents.

Vents that permit snow or rain to infiltrate the attic are not permitted. While there is no specific test standard for this performance aspect of vents, snow infiltration through roof vents can be addressed by what is referred to as "balanced" venting. Balanced venting is providing at least 50 percent of the required ventilating area in the upper third of the space being ventilated (e.g., through ridge or gable vents). The balance of the required ventilation is provided by eave or cornice vents that are greater than or equal to the ventilation area provided by the ridge or gable vents. Most ridge vent manufacturers require slightly more ventilation area in the eaves than provided in the ridge vent to help prevent snow infiltration. If insufficient eave or cornice ventilation is provided, air will be drawn through the ridge or gable vents. Snow-laden air can enter the attic space through a ridge or gable vent. Therefore, it is preferable to have both eave and ridge vents, with the eave vent area being greater than or equal to the ridge vent area (i.e., "balanced" venting).

If an adequate amount of ventilation area in the upper portion of the space is not provided and the ventilation area is provided mainly at the eave or cornice vents, then air will enter and leave the attic space at the eave, and very little cross flow of air will occur.

A test method that has been devised for ridge vent manufacturers involves the use of a 13-foot, 6-inch-diameter (4115 mm) propeller of a 2,650-horsepower aircraft engine wind generator. Snow is simulated with fine, soft wood sawdust, added to the airstream at about 5 pounds (2 kg) per minute. Using this method, the wind speed is varied, because it is not known at what speed the most snow infiltration will occur or the factors that will determine that each wind speed was sustained for a period of 5 minutes. The entire roof system, with all vents, must be installed in the test set-up in order to get a true measure of the potential (or lack thereof) for snow infiltration.

Attic ventilation openings cannot be placed in roof areas subject to snow drifts. These roof areas are subject to greater concentrations of snow, which could increase the chances of snow entering the attic through the ventilation openings. In addition, the ventilation openings are required to be covered with corrosion-resistant mesh or similar material in accordance with Section 1203.2.1. If roof spaces are not created (e.g., solid concrete roof sections), ventilation is not required, as there is no concealed space for condensation to accumulate.

The amount of area needed for ventilating a roof space is also established in this section.

The following example illustrates the calculation of required ventilation areas for an attic space [see Commentary Figure 1203.2(1)]:

Note: The area of the attic must include the area of the eave or soffit.

Attic ventilation area = $^{1}/_{150}$ of area

Required ventilating area = $^{1,100}/_{150}$
= 7.33 sq. ft.
= 1056 sq. in.

For application of exception:

Attic ventilation area = $^{1}/_{300}$ of area

Required ventilating area = $^{1,100}/_{300}$
3.67 sq. ft.
528 sq. in.

Provide 50% by = 528 x 0.5 ridge or roof vent
= 264 sq. in.

Provide soffit ventilation = 264 sq. in. total or 264/2(50)
= 2.64 sq. in./ft.

Area of attic = 1,100 sq. ft.

For SI: 1 square inch = 0.000645 m^2,
1 square foot = 0.0929 m^2.

Common methods used to provide soffit ventilation include manufactured units, strips or soffit panels and holes or slots (with screening) that meet the criteria and are approved.

It is important to note that the distribution of ventilation openings should be uniform along the length of the soffits and ridge.

Common methods used to provide roof ridge ventilation include manufactured roof units, ridge vents and gable louvers.

Common methods used to provide soffit ventilation include manufactured units, strips or soffit panels and holes or slots (with screening) that meet the criteria and are approved.

Note that the net-free ventilation area of equipment used for soffit, roof and gable ventilation must be determined. One cannot simply calculate the ventilation area based on the opening created in the roof, wall or soffit. Products vary by manufacturer, and a review of the listing and specifications is necessary to verify actual or net-free ventilation areas.

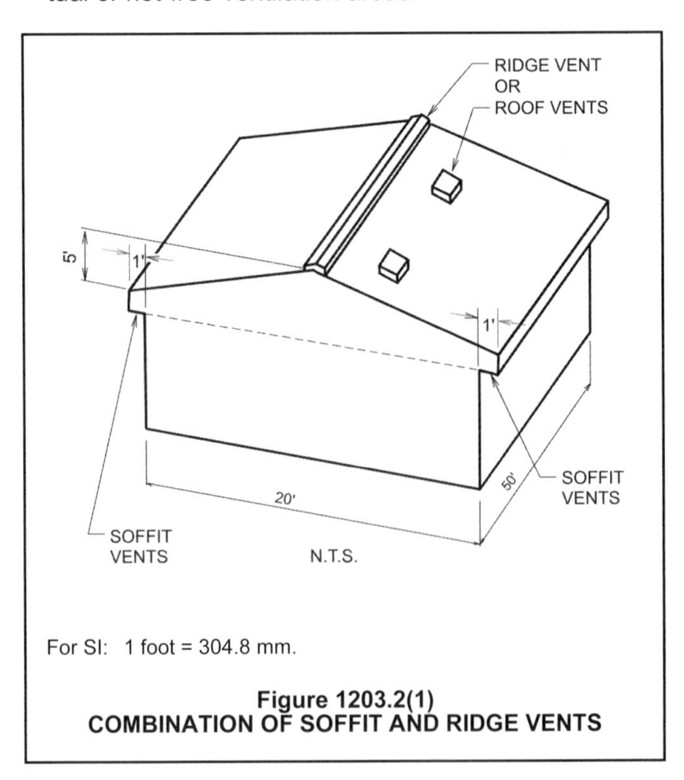

For SI: 1 foot = 304.8 mm.

Figure 1203.2(1)
COMBINATION OF SOFFIT AND RIDGE VENTS

Where an attic space is not created, but the ceiling membrane is applied directly to the bottom of the solid roof rafters, each rafter space is to be ventilated separately. In this type of installation, it is particularly important that cross ventilation is developed between each rafter space by providing vents at the ridge and eave [see Commentary Figure 1203.2(2)]. For small sections of roofs such as above dormers and for roofs that are flat or of very low slope, the 30-inch (762 mm) height difference between the highest and lowest vent locations may not be readily achievable. The designer and building official need to ensure that the ventilation openings for such roofs will result in adequate ventilation.

The exception permits a reduction in the required venting area to $^1/_{300}$ (as calculated in the example) where both conditions are met. Condition 1 recognizes that in cold climates a vapor retarder is required and permits a reduction based on the vapor retarder being installed on the warm-in-winter side of the ceiling. Condition 2 requires that proper cross ventilation be supplied. Ventilators used to apply this exception are required to be in compliance with the manufacturer's instructions to be effective.

1203.2.1 Openings into attic. Exterior openings into the *attic* space of any building intended for human occupancy shall be protected to prevent the entry of birds, squirrels, rodents, snakes and other similar creatures. Openings for ventilation having a least dimension of not less than $^1/_{16}$ inch (1.6 mm) and not more than $^1/_4$ inch (6.4 mm) shall be permitted. Openings for ventilation having a least dimension larger than $^1/_4$ inch (6.4 mm) shall be provided with corrosion-resistant wire cloth screening, hardware cloth, perforated vinyl or similar material with openings having a least dimension of not less than $^1/_{16}$ inch (1.6 mm) and not more than $^1/_4$ inch (6.4 mm). Where combustion air is obtained from an *attic* area, it shall be in accordance with Chapter 7 of the *International Mechanical Code*.

❖ Ventilation openings that would permit the entrance of small animals into the structure must be protected in accordance with this section. Hardware cloth is a particular kind of metal wire cloth screening, and perforated vinyl is a plastic screening or grid with openings of similar dimensions. Metal or vinyl are specified because of their resistance to deterioration over time. Therefore, whatever material is used, it must be nondeteriorating in addition to having the least dimension of the minimum and maximum openings of $^1/_{16}$ and $^1/_4$ inches (3.2 and 6.4 mm), respectively.

Combustion air is air supplied to the room where a fuel-burning appliance is located in order that combustion of the fuel can take place in a safe and complete manner. Chapter 7 of the IMC permits combustion air to be taken from an attic space that is ventilated by openings to the exterior, under certain conditions. Those conditions are based on the configuration of the attic space and the ventilation openings themselves (see Chapter 7 of the IMC and commentary for more information).

1203.3 Unvented attic and unvented enclosed rafter assemblies. Unvented *attics* and unvented enclosed roof framing assemblies created by ceilings applied directly to the underside of the roof framing members/rafters and the structural roof sheathing at the top of the roof framing members shall be permitted where all the following conditions are met:

1. The unvented *attic* space is completely within the *building thermal envelope*.
2. No interior Class I vapor retarders are installed on the ceiling side (*attic* floor) of the unvented *attic* assembly or on the ceiling side of the unvented enclosed roof framing assembly.

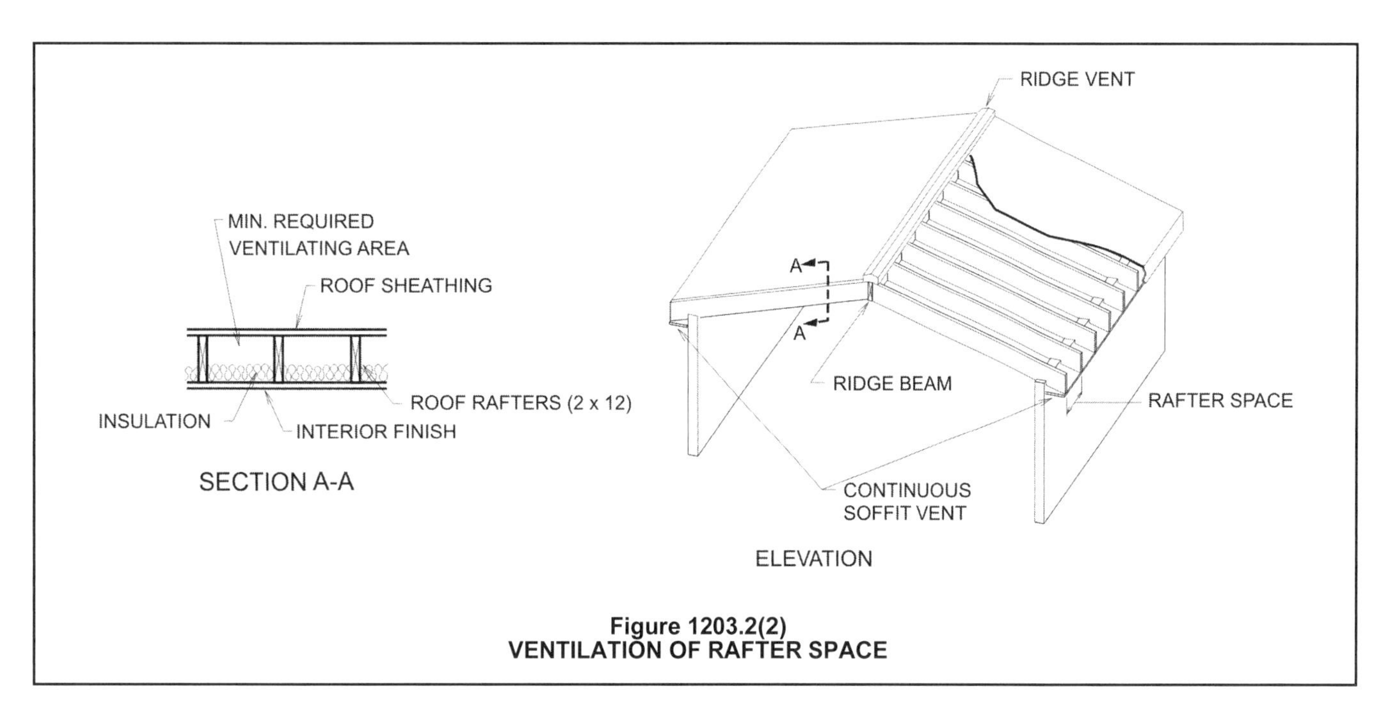

Figure 1203.2(2)
VENTILATION OF RAFTER SPACE

3. Where wood shingles or shakes are used, a minimum $^1/_4$-inch (6.4 mm) vented airspace separates the shingles or shakes and the roofing underlayment above the structural sheathing.
4. In Climate Zones 5, 6, 7 and 8, any air-impermeable insulation shall be a Class II vapor retarder or shall have a Class II vapor retarder coating or covering in direct contact with the underside of the insulation.
5. Insulation shall be located in accordance with the following:
   - 5.1. Item 5.1.1, 5.1.2, 5.1.3 or 5.1.4 shall be met, depending on the air permeability of the insulation directly under the structural roof sheathing.
     - 5.1.1. Where only air-impermeable insulation is provided, it shall be applied in direct contact with the underside of the structural roof sheathing.
     - 5.1.2. Where air-permeable insulation is provided inside the building thermal envelope, it shall be installed in accordance with Item 5.1. In addition to the air-permeable insulation installed directly below the structural sheathing, rigid board or sheet insulation shall be installed directly above the structural roof sheathing in accordance with the R values in Table 1203.3 for condensation control.
     - 5.1.3. Where both air-impermeable and air-permeable insulation are provided, the *air-impermeable insulation* shall be applied in direct contact with the underside of the structural roof sheathing in accordance with Item 5.1.1 and shall be in accordance with the R values in Table 1203.3 for condensation control. The *air-permeable insulation* shall be installed directly under the *air-impermeable insulation.*
     - 5.1.4. Alternatively, sufficient rigid board or sheet insulation shall be installed directly above the structural roof sheathing to maintain the monthly average temperature of the underside of the structural roof sheathing above 45°F (7°C). For calculation purposes, an interior air temperature of 68°F (20°C) is assumed and the exterior air temperature is assumed to be the monthly average outside air temperature of the three coldest months.
   - 5.2. Where preformed insulation board is used as the *air-permeable insulation* layer, it shall be sealed at the perimeter of each individual sheet interior surface to form a continuous layer.

Exceptions:

1. Section 1203.3 does not apply to special use structures or enclosures such as swimming pool enclosures, data processing centers, hospitals or art galleries.
2. Section 1203.3 does not apply to enclosures in Climate Zones 5 through 8 that are humidified beyond 35 percent during the three coldest months.

❖ Unvented attics are attics where the insulation and air barrier boundary are moved to be directly above the attic space, instead of on top of the ceiling. Unvented attics eliminate the extreme temperatures of the attic, thereby placing the HVAC, ducts, pipes, and anything in the attic space into a more favorable environment. Unvented attics increase energy efficiency and decrease wear and tear on equipment in the attic. This section describes attics where the insulation and air barrier are above instead of below the attic space. Moving the insulation and placing an air-impermeable barrier above the attic moderates attic conditions so they are similar to the conditions of the residential space below. The primary benefit of having the insulation and air barrier above the attic is that ducts and/or HVAC equipment in the attic are not delivering cooled air through a hot summer attic and heated air through a cold winter attic. Another benefit is to eliminate the attic vents that sometimes allow moisture to condense inside the attic, admit rain during extreme weather and possibly admit sparks in fires.

Because this space is inside the building's thermal envelope, the traditional attic ventilation required by Sections 1203.2 is not required. Unvented attics require water/moisture control. Water moves in (or out) of buildings three main ways. The greatest amount of moisture is moved as bulk water (rain or any kind of water flow). Less moisture is moved by moving moist air, such as with infiltration. The least amount of moisture is moved by moisture migration through materials. As with any attic, the roof itself is the main barrier for keeping water from entering the attic.

The provisions of this section can be applied to any attic area which is in compliance with this section. The attic is a traditional attic space, with the exception that it need not be ventilated and it will not get as hot or as cold as an attic that is open to the exterior.

It is very important that all of the five listed conditions be reviewed and considered for each building that uses the provisions of this section.

Item 1 requires that the attic space be completely contained within the building thermal envelope.

Item 2, which applies to all climate zones, prohibits the installation of a vapor retarder where it is typically installed at the ceiling level (attic floor) of a traditional ventilated attic. This assures that no barrier is installed which would separate the conditioned attic area from the remaining portion of the home. This requirement gives the attic space a limited potential to dry into the space beneath the attic so that small amounts of excess moisture can be removed from the attic. A sheet of polyurethane film or any material with a foil film facing are examples of vapor barriers that are not permitted on the attic floor.

Item 3 applies to all climate zones and contains special requirements which apply to wood shingle and shake roof coverings. Wood shakes and shingles require a vented space under them to allow the wood to dry after it gets wet from rain.

Item 4 applies only to Climate Zones 5 or higher. The air-impermeable insulation must qualify as a vapor retarder or have a vapor retarder in direct contact with the insulation on the underside (interior) face of the insulation.

Item 5 requires sufficient insulation to keep moisture from condensing on the "condensing surface" inside the attic in "average conditions." The insulation works to prevent condensation by keeping the condensing surface above the temperature where condensation will occur. Small amounts of condensation may occasionally occur at more extreme conditions; however, this is not a concern. The condensing surface is the interior side of the roof deck for air-permeable insulation and the interior of the insulation for air-impermeable insulation. The condensing surfaces differ because attic air can circulate through air-permeable insulation to contact the roof deck but can get only to the interior of air-impermeable insulation. The requirement for "air-impermeable" insulation will ensure that air and the moisture it can contain will not pass through the insulation to reach a point where it could condense because of the temperature. Item 5 specifies that air-impermeable insulation be in direct contact with the interior side of the roof deck. Air-impermeable insulation prevents the movement of moist air that comes in as infiltration through the roof into the interior of the attic. Air-impermeable insulation is defined as having an air permanence of 0.02 L/s-m^2 (at 75 Pa pressure) or less. Expanding spray foams and insulated sheathing (hard-foam sheathing board) are common types of air-impermeable insulation. When using insulated sheathing, attention to the details of completing the air sealing is required as the sheathing is installed in the roof. Fiberglass and cellulose are common types of air-permeable insulation.

Because the insulation typically used with this provision is some type of foam plastic, the requirements of Section 2603 must be applied. The provisions of Section 1203.3 do not in any way modify or eliminate the requirements for a thermal barrier (Section 2603.4). Ducts in this unvented attic construction would be considered as being inside the building thermal envelope and would not require insulation.

The provisions of this section consider the attic assembly as a "conditioned" space; there is no requirement for the space to be provided with conditioned air supply. The attic space is considered indirectly conditioned because of omission of the air barrier, insulation at the ceiling and leakage around the attic access opening. An attic assembly complying with Section 1203.3 will generally fall within the temperature ranges specified in the definition of "Conditioned space."

The key concept of this section is to move the thermal envelope (insulation) above the attic, resulting in the attic being in a conditioned (or sometimes semi-conditioned) space. Direct air supply to the attic is not required if the attic floor is not insulated; the attic temperature would be similar to interior conditioned spaces. Ducts and/or HVAC equipment in the attic also help moderate the attic conditions.

Table 1203.3. See below.

❖ This table is to be used with Section 1203.3, Items 5.1.2 and 5.1.3, in determining the correct R-value to prevent condensation inside the building thermal envelope.

1203.4 Under-floor ventilation. The space between the bottom of the floor joists and the earth under any building except spaces occupied by basements or cellars shall be provided with ventilation openings through foundation walls or *exterior walls*. Such openings shall be placed so as to provide cross ventilation of the under-floor space.

TABLE 1203.3
INSULATION FOR CONDENSATION CONTROL

CLIMATE ZONE	MINIMUM R-VALUE OF AIR-IMPERMEABLE INSULATION[a]
2B and 3B tile roof only	0 (none required)
1, 2A, 2B, 3A, 3B, 3C	R-5
4C	R-10
4A, 4B	R-15
5	R-20
6	R-25
7	R-30
8	R-35

a. Contributes to, but does not supersede, thermal resistance requirements for attic and roof assemblies in Section C402.2.1 of the *International Energy Conservation Code*.

❖ The intent of this section is to create an adequate flow of air through crawl spaces to achieve the ventilation goals of controlling temperature, humidity and accumulation of gases. The entire space must be properly ventilated by openings that are distributed to effect cross flow and include corner areas. Although the code does not specify the exact location of openings, an equal distribution of openings on at least three sides of a building, with at least one opening near each corner of the building, is typically sufficient.

Mechanical ventilating devices also can be installed to force air movement and ventilate the space, in which case the location and number of ventilation openings are less critical (see Section 1203.4.2, Exception 3). The amount of ventilation openings required can be drastically reduced if a vapor retarder is used on the ground surface in the crawl space, in accordance with Section 1203.4.2, Exception 2. Also, in accordance with Exception 3 of that section, when a vapor retarder is used, the installation of operable louvers (to close the openings in the coldest times of the year) is permitted.

1203.4.1 Openings for under-floor ventilation. The net area of ventilation openings shall be not less than 1 square foot for each 150 square feet (0.67 m^2 for each 100 m^2) of crawl-space area. Ventilation openings shall be covered for their height and width with any of the following materials, provided that the least dimension of the covering shall be not greater than $^1/_4$ inch (6.4 mm):

1. Perforated sheet metal plates not less than 0.070 inch (1.8 mm) thick.
2. Expanded sheet metal plates not less than 0.047 inch (1.2 mm) thick.
3. Cast-iron grilles or gratings.
4. Extruded load-bearing vents.
5. Hardware cloth of 0.035-inch (0.89 mm) wire or heavier.
6. Corrosion-resistant wire mesh, with the least dimension not greater than $^1/_8$ inch (3.2 mm).

❖ An example of the area calculation: A rectangular building that is 60 feet (18 288 mm) long and 20 feet (6096 mm) wide has a plan area of 1,200 square feet (111.5 m^2). The amount of ventilation opening required is $^{1,200}/_{150}$ = 8 square feet (0.74 m^2) by 144 square inches (92 903 mm^2) per square foot = 1,152 square inches (0.74 m^2). This is the total (aggregate) amount of ventilation opening that must be distributed among all the openings. This required amount of openings may be reduced by a factor of 10 if a vapor retarder is used on the ground surface in accordance with Section 1203.4.2, Exception 2.

The requirement for covering the openings with perforated plates, corrosion-resistant wire mesh or other covering is to keep small animals out. Six alternatives are given for this covering, and they all must have openings that have no dimension exceeding $^1/_4$ inch (6.4 mm).

1203.4.2 Exceptions. The following are exceptions to Sections 1203.4 and 1203.4.1:

1. Where warranted by climatic conditions, ventilation openings to the outdoors are not required if ventilation openings to the interior are provided.
2. The total area of ventilation openings is permitted to be reduced to $^1/_{1,500}$ of the under-floor area where the ground surface is covered with a Class I vapor retarder material and the required openings are placed so as to provide cross ventilation of the space. The installation of operable louvers shall not be prohibited.
3. Ventilation openings are not required where continuously operated mechanical ventilation is provided at a rate of 1.0 cubic foot per minute (cfm) for each 50 square feet (1.02 L/s for each 10 m^2) of crawl- space floor area and the ground surface is covered with a Class I vapor retarder.
4. Ventilation openings are not required where the ground surface is covered with a Class I vapor retarder, the perimeter walls are insulated and the space is conditioned in accordance with the *International Energy Conservation Code*.
5. For buildings in flood hazard areas as established in Section 1612.3, the openings for under-floor ventilation shall be deemed as meeting the flood opening requirements of ASCE 24 provided that the ventilation openings are designed and installed in accordance with ASCE 24.

❖ This section lists the locations and conditions where ventilation openings can be omitted entirely or the area of required openings can be reduced. Exception 1 could be used in extremely cold climates, where ventilation openings are a serious breach of the structure in terms of energy usage. It provides for ventilating the crawl space to the interior conditioned space of the building, which is heated and can accept moisture from the underground space without detrimental effects on the building structure.

The use of a Class I vapor retarder material on the ground surface inhibits the flow of moisture from the ground surface into the crawl space and thus reduces, if not virtually eliminates, the need for ventilation. Therefore, Exception 2 provides for a drastic reduction in the amount of ventilation openings required. While the vapor retarder may significantly reduce the moisture accumulation, ventilation openings are still required but may be equipped with manual dampers to permit them to be closed during the coldest weeks of the year in northern climates.

Exception 3 provides for the use of mechanical ventilation, such as an exhaust fan similar to a bathroom exhaust fan, to keep air moving through the crawl space. A Class I vapor retarder on the ground surface is also required when using Exception 3.

When a crawl space is provided with a Class I vapor retarder on the ground surface and is mechanically conditioned and insulated, it becomes like any other space in the conditioned structure and ventilation openings are not required in accordance with Exception 4. Requirements for insulating structures are found in the IECC.

Section 1612 of the code requires buildings located in flood hazard areas to be constructed in accordance with ASCE 24. Exception 5 is necessary to coordinate the requirements for openings in ASCE 24 with this section. In most cases, the ventilation requirements of Section 1203.4 are not satisfied by installation of the flood openings because the requirements in ASCE 24 specify that flood openings are to be no more than 1 foot (305 mm) above the adjacent grade, while most airflow vents that meet the requirements of Section 1203.3 are installed immediately below the elevated floor. If there is a sufficient number of airflow vents located within 1 foot (305 mm) of the adjacent grade, both requirements may be satisfied. In either case, both airflow ventilation and flood opening requirements must be met for buildings and structures located in flood hazard areas. For further guidance, refer to FEMA FIA-TB-1, *Openings in Foundation Walls for Buildings Located in Special Flood Hazard Areas.*

1203.5 Natural ventilation. Natural *ventilation* of an occupied space shall be through windows, doors, louvers or other openings to the outdoors. The operating mechanism for such openings shall be provided with ready access so that the openings are readily controllable by the building occupants.

❖ This section provides the standard of natural ventilation for all occupied spaces. Openings to the outdoor air, such as doors, windows and louvers, provide natural ventilation. The section does not, however, state or intend that the doors, windows or openings actually be constantly open. The intent is that they be maintained in an operable condition so that they are available for use at the discretion of the occupant.

1203.5.1 Ventilation area required. The openable area of the openings to the outdoors shall be not less than 4 percent of the floor area being ventilated.

❖ This section specifies the ratio of openable doors, windows or openings to the floor space being ventilated but does not address the distribution around the space or location of these openings. It is the designer's prerogative to distribute openings in such a manner as to accomplish the natural ventilation of the space. When inadequate natural ventilation is provided, mechanical ventilation can supplement any inadequacy (see Chapter 4 of the IMC). The plan reviewer can determine compliance with this section. For example, in Commentary Figure 1203.5.1, the combined openable area (the net-free area of a door, window, louver, vent or skylight, etc., when fully open) of double-hung windows B and C is equal to 4 percent of the floor area [300 × 0.04 = 12 square feet (1 m^2)]. The openable area of window A is not required and need not open onto a court or yard complying with Section 1206.

1203.5.1.1 Adjoining spaces. Where rooms and spaces without openings to the outdoors are ventilated through an adjoining room, the opening to the adjoining room shall be unobstructed and shall have an area of not less than 8 percent of the floor area of the interior room or space, but not less than 25 square feet (2.3 m^2). The openable area of the openings to the outdoors shall be based on the total floor area being ventilated.

Exception: Exterior openings required for *ventilation* shall be permitted to open into a sunroom with *thermal isolation* or a patio cover provided that the openable area between the sunroom addition or patio cover and the interior room shall have an area of not less than 8 percent of the floor area of the interior room or space, but not less than 20 square feet (1.86 m^2). The openable area of the openings to the outdoors shall be based on the total floor area being ventilated.

❖ Adjacent spaces with large connecting openings may share sources of light and ventilation. This section deals with the natural ventilation of connecting interior spaces, and it is the designer's obligation to place openings between rooms with exterior openings and connecting spaces without exterior openings in such a manner as to accomplish natural ventilation of the connected space. For purposes of ventilation, this section establishes a minimum openness requirement for the common wall between a room with openings to the exterior and an interior room without openings to the exterior. The minimum amount of openness required in that common wall is 8 percent of the floor area of the interior room or 25 square feet (2.33 m^2), whichever is greater. The openable area of the exterior openings in the "outer" room is required to be equal to or greater than 4 percent (in accordance with Section 1203.4.1) of the total combined floor areas served.

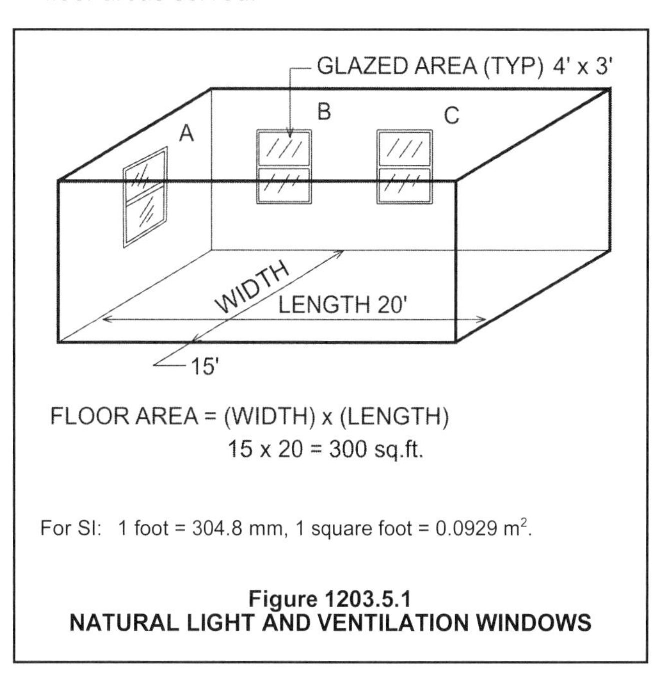

Figure 1203.5.1
NATURAL LIGHT AND VENTILATION WINDOWS

Commentary Figure 1203.5.1.1 shows a cutaway of an interior room (Space A) adjacent to a room with openings to the exterior (Space B). The openable area of exterior openings in Space B is required to be equal to or greater than 0.04 times the area of the entire space (floor area of Space A plus floor area of interior Space B). The opening in the wall between adjacent spaces must be a minimum of 25 square feet (2.33 m^2), but not less than 0.08 times the floor area of interior Space A. Since the opening between the adjacent spaces must be unobstructed in accordance with this section, a door cannot be installed in the opening.

The exception deals with a very common circumstance, especially in residential construction. As long as the sunroom is large enough and is thermally isolated, the building owner need not move ventilation openings when installing an addition that falls within the definition of "Sunroom." Sunrooms can be part of initial construction.

1203.5.1.2 Openings below grade. Where openings below grade provide required natural *ventilation*, the outside horizontal clear space measured perpendicular to the opening shall be one and one-half times the depth of the opening. The depth of the opening shall be measured from the average adjoining ground level to the bottom of the opening.

❖ This section is applicable where occupied spaces below grade are dependent on natural ventilation through structures such as window wells. In order to provide adequate ventilation, this section sets the minimum horizontal clear space adjacent to the opening used for natural ventilation. Without this minimum horizontal area, there will be inadequate air movement through the opening.

As illustrated in Commentary Figure 1203.5.1.2, the opening area required for the story below grade intended for human occupancy is:

$$A = 0.04 (L \times W)$$

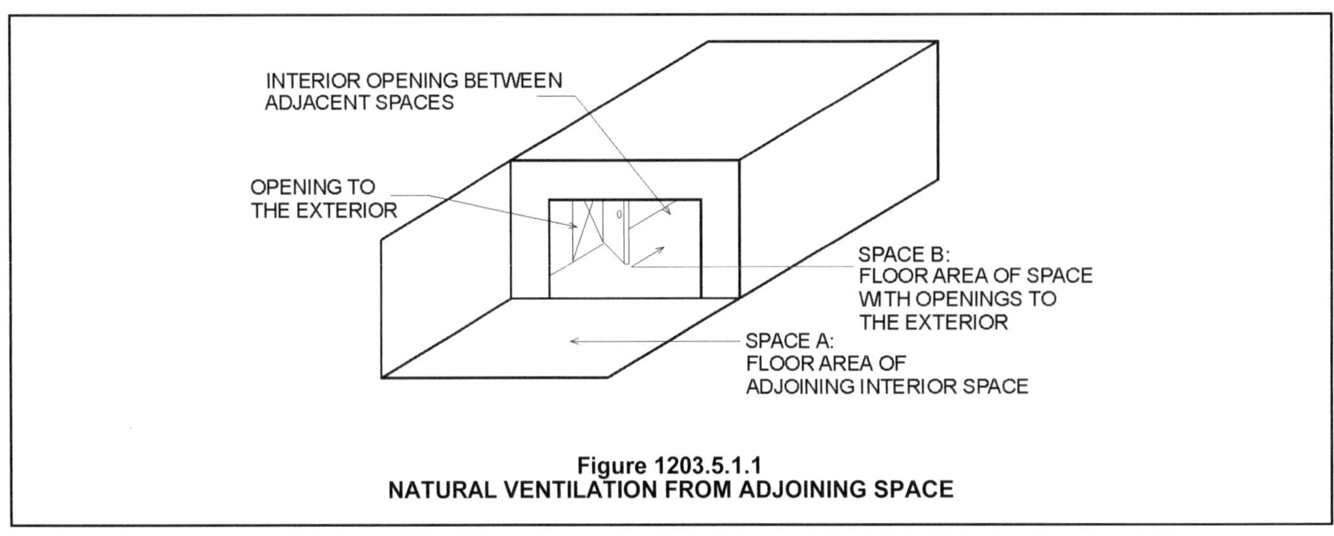

Figure 1203.5.1.1
NATURAL VENTILATION FROM ADJOINING SPACE

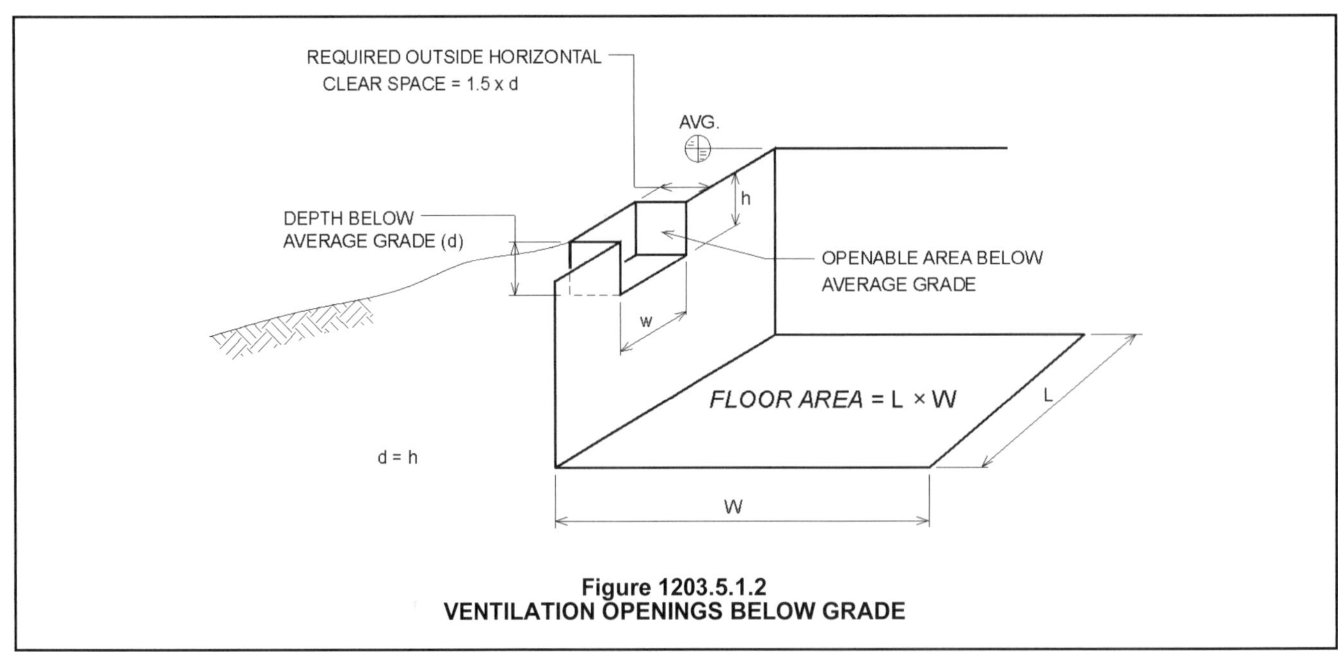

Figure 1203.5.1.2
VENTILATION OPENINGS BELOW GRADE

The area of the window in the vertical plane (w × h) must equal or exceed the required opening area. Additionally, the horizontal dimension from the window to the well wall must equal one and one-half times the depth of the openable portion of the window at the lowest point. If the story below grade is not intended for human occupancy, ventilation is required to be provided in accordance with Section 1203.4 for under-floor spaces.

1203.5.2 Contaminants exhausted. Contaminant sources in naturally ventilated spaces shall be removed in accordance with the *International Mechanical Code* and the *International Fire Code*.

❖ Contaminants in the air are to be collected and exhausted by special means. Chapters 4 and 5 of the IMC specify areas or conditions that must be separately addressed. For example, there are many operations listed in Chapter 5 of the IMC that produce contaminants that cannot be properly or safely treated by natural means. Natural ventilation only anticipates normal occupancy by people and not the extra heat loads, dust, vapors and other contaminants generated by some activities.

1203.5.2.1 Bathrooms. Rooms containing bathtubs, showers, spas and similar bathing fixtures shall be mechanically ventilated in accordance with the *International Mechanical Code*.

❖ Chapter 4 of the IMC contains provisions for bathroom ventilation, requiring mechanical exhaust without recirculation of air at specific rates that depend on the occupancy group.

1203.5.3 Openings on yards or courts. Where natural *ventilation* is to be provided by openings onto *yards* or *courts*, such *yards* or *courts* shall comply with Section 1206.

❖ To ensure that adequate air movement will be provided through openings to naturally ventilated rooms, the openings must directly connect to yards or courts with the minimum dimensions specified in Section 1206.

1203.6 Other ventilation and exhaust systems. *Ventilation* and exhaust systems for occupancies and operations involving flammable or combustible hazards or other contaminant sources as covered in the *International Mechanical Code* or the *International Fire Code* shall be provided as required by both codes.

❖ Chapter 5 of the IMC contains specific provisions for hazardous exhaust systems in occupancies such as vehicle repair garages, aircraft fueling stations, dry cleaning plants, spray-painting operations and hazardous production materials (HPM) facilities. Many of these provisions are duplicated from the *International Fire Code*® (IFC®), which also contains provisions for the handling and storage of hazardous materials.

SECTION 1204
TEMPERATURE CONTROL

1204.1 Equipment and systems. Interior spaces intended for human occupancy shall be provided with active or passive space heating systems capable of maintaining an indoor temperature of not less than 68°F (20°C) at a point 3 feet (914 mm) above the floor on the design heating day.

Exceptions: Space heating systems are not required for:

1. Interior spaces where the primary purpose of the space is not associated with human comfort.
2. Group F, H, S or U occupancies.

❖ Heating facilities are required for comfort in all new construction. The systems may be either active (such as a forced-air furnace) or passive (such as solar systems), as long as the specified performance is achieved.

The outdoor design temperatures are taken from the ASHRAE *Handbook of Fundamentals* and are listed in Appendix D of the *International Plumbing Code*® (IPC®). Outdoor design temperatures provide a baseline from which heat load calculations are made. Heating system capacity is dependent on the predicted outdoor temperatures during the heating season. As the outdoor temperature falls, the heat input to a building must increase to offset the increasing heat losses through the building envelope. Heating systems are designed to have the capacity to maintain the desired indoor temperature when the outdoor temperature is at or above the outdoor design temperature. When the outdoor temperatures are below the outdoor design temperature, the heating system will not be able to maintain a desired indoor temperature. It would be impractical, for example, to design a heating system based on the assumption that someday it might be -20°F (-29°C) outdoors if the outdoor temperature in that region rarely, if ever, dropped that low. In such a case, the heating system would be oversized and, thereby, less efficient and economical.

The winter outdoor design temperature is defined as follows: for 97.5 percent of the total hours in the northern hemisphere heating season, from December through February, the predicted outdoor temperatures will be at or above the values given in Appendix D of the IPC. It would be unreasonable to expect any heating system to maintain a desired indoor temperature when the outdoor temperature is below the design temperature. When the 97.5-percent column in Appendix D of the IPC is used, it can be assumed that the actual outdoor temperature will be at or below the design temperature for roughly 54 hours of the total of 2,160 hours in the months of December through February (2.5 percent of 2,160 hours = 54).

Exception 1 recognizes that not all interior spaces are associated with human comfort by the nature of their uses, such as a commercial cooler or freezer. These and similar spaces would not require heating systems.

Exception 2 specifically exempts Group F, H, S or U occupancies from the heating requirement in the IBC. The requirement for heating in these uses is governed by OSHA regulations (Groups F, H and S) or is not needed (Group U) or, in some cases, not desired (Group H). As an example, in the past there have been incidents of industrial bakeries being required to install heating systems in rooms that contain ovens.

SECTION 1205 LIGHTING

1205.1 General. Every space intended for human occupancy shall be provided with natural light by means of exterior glazed openings in accordance with Section 1205.2 or shall be provided with artificial light in accordance with Section 1205.3. Exterior glazed openings shall open directly onto a *public way* or onto a *yard* or *court* in accordance with Section 1206.

❖ This section establishes that an option can be exercised on a room-by-room or space-by-space basis. The option allows the designer to provide either natural light in accordance with this chapter or equivalent levels of artificial lighting.

1205.2 Natural light. The minimum net glazed area shall be not less than 8 percent of the floor area of the room served.

❖ This section establishes the minimum glazed area required based on the floor area served by the window. This is required only for spaces that are not provided with artificial light in accordance with Section 1205.3. It is the intent of the code to establish this ratio as the minimum glazed opening onto yards or courts, in accordance with Section 1205.1.

Early codes set this standard at 10 percent of the floor area served. This ratio was derived from certain architectural styles that yielded adequate light and ventilation. However, this is a more-than-adequate amount and has been reduced to the current levels because of energy conservation issues. Openings in excess of that minimum area are permitted to open onto areas other than a complying court or yard. In Commentary Figure 1203.5.1, the room dimensions are 15 feet by 20 feet (4572 mm by 6096 mm), or 300 square feet (27.9 m^2) of area. If windows B and C are double hung, with a combined glazed area of 24 square feet (2.23 m^2), they provide the minimum area required of 8 percent of the floor area (24/300 = 0.08). In this example, glazing unit A is not required for natural light; therefore, it need not face onto a required yard or court.

1205.2.1 Adjoining spaces. For the purpose of natural lighting, any room is permitted to be considered as a portion of an adjoining room where one-half of the area of the common wall is open and unobstructed and provides an opening of not less than one-tenth of the floor area of the interior room or 25 square feet (2.32 m^2), whichever is greater.

Exception: Openings required for natural light shall be permitted to open into a sunroom with *thermal isolation* or a patio cover where the common wall provides a glazed area of not less than one-tenth of the floor area of the interior room or 20 square feet (1.86 m^2), whichever is greater.

❖ In a case where a space (or room) has no glazed area open to the required courts or yards but is adjacent to one that does, it may "borrow" natural lighting from the adjacent space if: 1. The wall between the adjoining spaces is at least one-half open and unobstructed; 2. The opening equals at least 10 percent of the floor area of the interior space; and 3. The opening is not less than 25 square feet (2.33 m^2). The required glazed area facing the required court or yard must not be less than 8 percent of the total floor area of all rooms served. For example, in Commentary Figure 1205.2.1, the glazed area in Space B is required to be equal to or greater than 0.08 (floor area of Space A + floor area of Space B).

In the figure, the opening between the adjacent spaces must meet these three criteria: the wall must be at least half open and unobstructed, it must be a minimum of 25 square feet (2.33 m^2), and it must be not less than one-tenth of the floor area of Space A.

The exception deals with a very common circumstance, especially in residential construction. As long as the sunroom is large enough and is thermally isolated, the building owner need not move openings for lighting when installing an addition that falls within the definition of "Sunroom." Note that sunrooms can also be part of the initial construction of a building.

1205.2.2 Exterior openings. Exterior openings required by Section 1205.2 for natural light shall open directly onto a *public way*, *yard* or *court*, as set forth in Section 1206.

Exceptions:

1. Required exterior openings are permitted to open into a roofed porch where the porch meets all of the following criteria:
   1.1. Abuts a *public way*, *yard* or *court*.
   1.2. Has a ceiling height of not less than 7 feet (2134 mm).
   1.3. Has a longer side at least 65 percent open and unobstructed.
2. Skylights are not required to open directly onto a *public way*, *yard* or *court*.

❖ In order that enough light will be provided through openings to naturally lit rooms, the openings must open onto yards or courts with the minimum dimensions specified in Section 1206. Skylights admit light directly from above and, therefore, are not required to face a court or yard in accordance with Exception 2. Exception 1 gives the criteria by which a roofed porch may be located directly outside required openings without significantly obstructing the entrance of light to the space.

1205.3 Artificial light. Artificial light shall be provided that is adequate to provide an average illumination of 10 footcandles (107 lux) over the area of the room at a height of 30 inches (762 mm) above the floor level.

❖ The section establishes the minimum required illumination for rooms without the minimum required natural light (see Commentary Figure 1205.3). Please note that Section 1008.2 requires 1 footcandle (11 lux) of light at the walking surface of all means of egress.

1205.4 Stairway illumination. *Stairways* within *dwelling units* and *exterior stairways* serving a *dwelling unit* shall have an illumination level on tread runs of not less than 1 footcandle (11 lux). *Stairways* in other occupancies shall be governed by Chapter 10.

❖ Illumination is essential for stairway safety during normal use, as well as during egress in an emergency. The lighting must be operable by switches in the vicinity of the stairway, located as required by NFPA 70, *National Electrical Code*® (NEC). Emergency egress lighting, also referred to as "means of egress illumination," is required in occupancies other than dwelling units at a lower rate of illumination (see commentary, Sections 1008 and 1205.5).

1205.4.1 Controls. The control for activation of the required *stairway* lighting shall be in accordance with NFPA 70.

❖ The NEC provides for controls at the top and bottom of stairways within dwelling units, allowing an occupant to illuminate the stairways before traversing any stairways, regardless of the direction of travel. Illuminated switches, where required, allow an occupant to quickly find the switches when the stairways are dark.

Illumination controls for exterior stairways that are operable from the inside of a dwelling unit allow an occupant to safely egress by activating exterior stairway illumination prior to leaving the building. Exterior stairways must be provided with the minimum illumination level specified in Section 1205.4.

1205.5 Emergency egress lighting. The *means of egress* shall be illuminated in accordance with Section 1008.1.

❖ Means of egress illumination is required in all buildings to allow occupants enough light to negotiate the

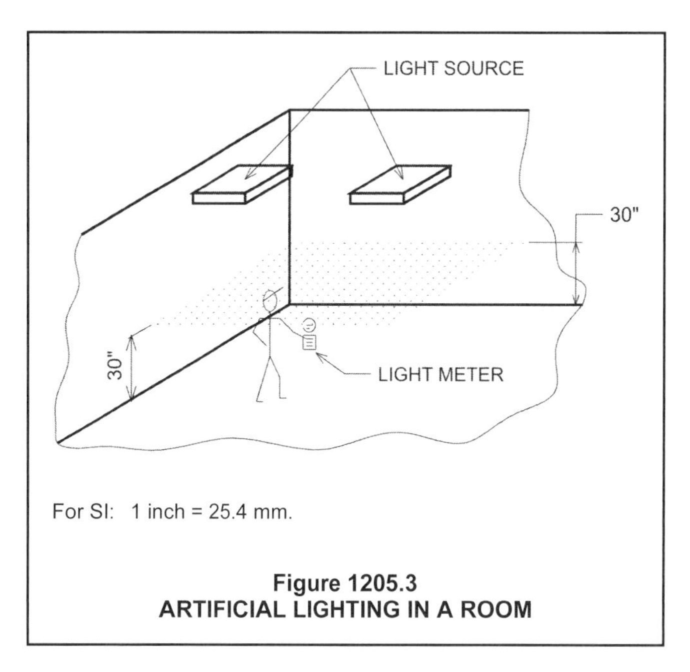

Figure 1205.3
ARTIFICIAL LIGHTING IN A ROOM

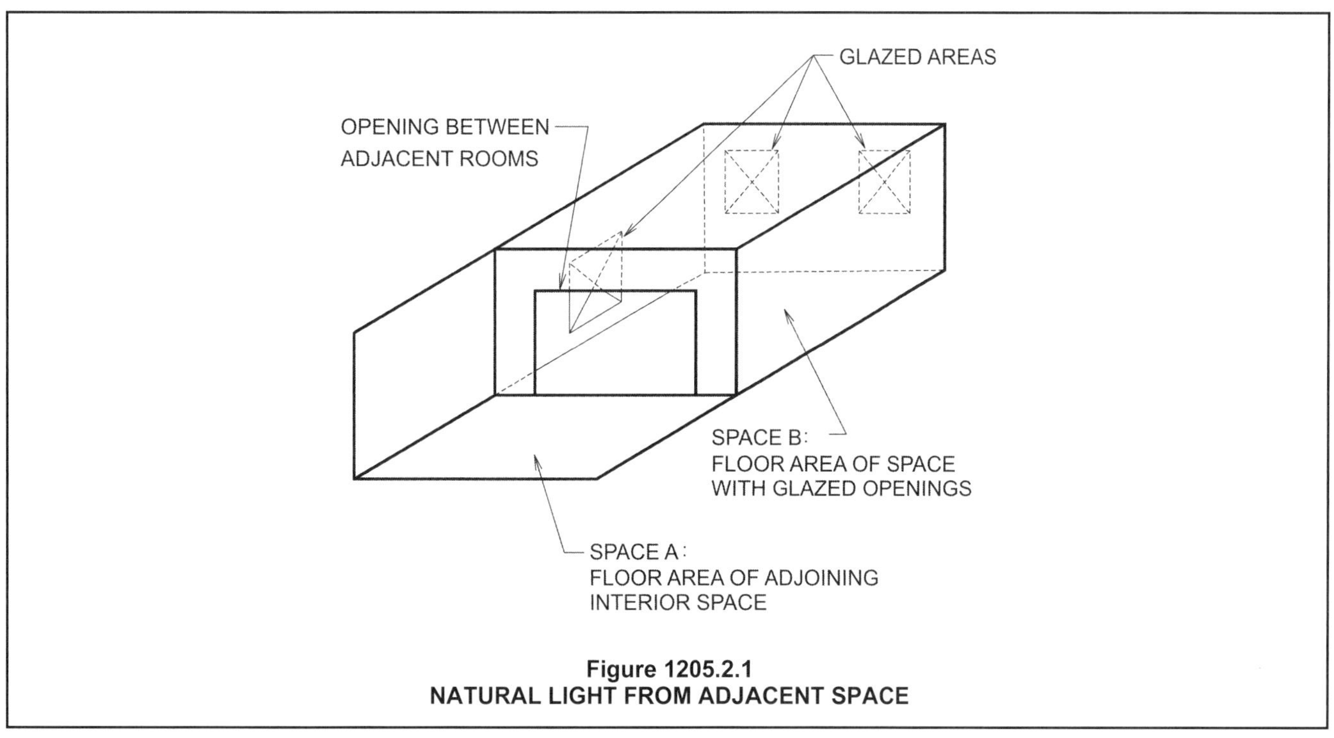

Figure 1205.2.1
NATURAL LIGHT FROM ADJACENT SPACE

exit access (such as corridors) and exits (such as enclosed stairways) at all times the building is occupied (see commentary, Section 1008.1).

SECTION 1206
YARDS OR COURTS

1206.1 General. This section shall apply to *yards* and *courts* adjacent to exterior openings that provide natural light or ventilation. Such *yards* and *courts* shall be on the same *lot* as the building.

❖ These provisions are intended to regulate those exterior areas of a building or structure that are supposed to supply required natural light or ventilation to interior spaces. These requirements are intended to increase the likelihood that the exterior walls are provided with enough adjacent open space to allow the required light and ventilating air to freely enter the exterior wall openings. These exterior areas are defined as "Courts" and "Yards." Courts and yards must be open, uncovered and on the same lot as the building. They may be either partly or wholly enclosed by the building. Requirements are provided in Section 1206, which regulates the minimum width, area, air intake and drainage of courts and yards. The requirements of Sections 1206.2 through 1206.3.3 do not apply if artificial ventilation and lighting is provided for the spaces opening onto the court or yard in accordance with Section 1203.1 or 1205.1. See Commentary Figure 1206.1 for examples of courts and yards.

1206.2 Yards. *Yards* shall be not less than 3 feet (914 mm) in width for buildings two *stories* or less above *grade plane*. For buildings more than two *stories above grade plane*, the minimum width of the *yard* shall be increased at the rate of 1 foot (305 mm) for each additional *story*. For buildings exceeding 14 *stories above grade plane*, the required width of the *yard* shall be computed on the basis of 14 *stories above grade plane*.

❖ A yard is distinguished from a court by the definitions in Chapter 2 (see commentary, Chapter 2). A court is bounded on at least three sides by exterior building walls or similar enclosing devices, whereas a yard is

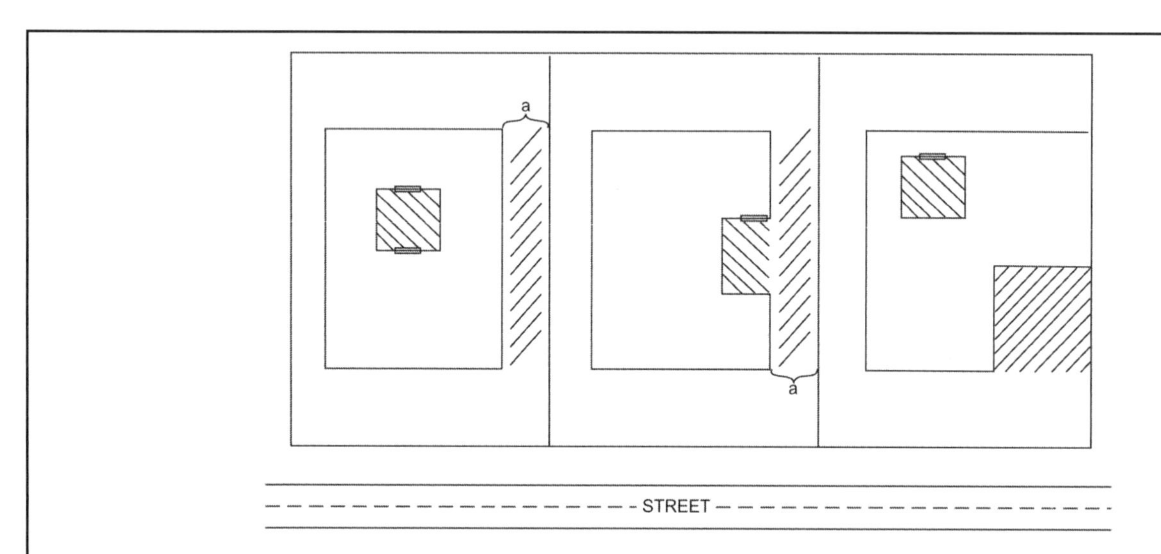

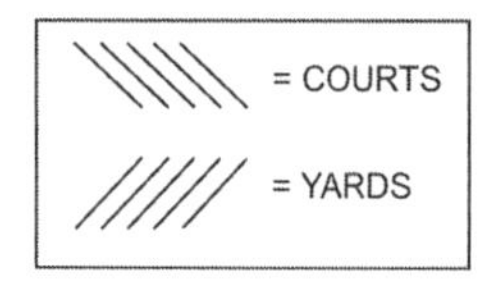

Number of Stories	Minimum width of yard in feet (a)
1 or 2	3
3	4
4	5
5	6
6	7
7	8
8	9
9	10
10	11
11	12
12	13
13	14
14	15
15 or more	15

For SI: 1 foot = 304.8 mm.

Figure 1206.1
YARDS AND COURTS

typically located between a building and a lot line and is open on at least two sides or ends (see Commentary Figure 1206.1).

The required width of a yard is measured perpendicular from the face of the wall to the opposing wall on the other side of the yard. A five-story building would be required to have a yard at least 6 feet (1829 mm) in width [3 feet (914 mm) plus 1 foot (305 mm) each for stories three through five]. A 20-story building is required to have a yard at least 15 feet (4572 mm) [3 feet (914 mm) plus 1 foot (305 mm) each for stories three through 14]. The last sentence of the section simply requires a minimum yard width of 15 feet (4572 mm) for all buildings over 14 stories in height. If the building is adjacent to a court rather than a yard, the requirements of Section 1206.3 apply. Neither Section 1206.2 nor 1206.3 is applicable if artificial lighting and ventilation is provided for spaces facing the yard or court in accordance with Sections 1203.1 and 1205.1. Note that the stories are measured above grade plane.

1206.3 Courts. *Courts* shall be not less than 3 feet (914 mm) in width. *Courts* having windows opening on opposite sides shall be not less than 6 feet (1829 mm) in width. *Courts* shall be not less than 10 feet (3048 mm) in length unless bounded on one end by a *public way* or *yard*. For buildings more than two *stories above grade plane*, the *court* shall be increased 1 foot (305 mm) in width and 2 feet (610 mm) in length for each additional *story*. For buildings exceeding 14 *stories above grade plane*, the required dimensions shall be computed on the basis of 14 *stories above grade plane*.

❖ "Courts" are defined in Chapter 2 as being bounded on no less than three sides by walls or other enclosing construction. A court adjacent to a five-story building would be required to have a width measured perpendicular from the wall facing the court of at least 6 feet (1829 mm) when required openings are on one wall and 9 feet (2743 mm) when required openings are on opposing walls [3 feet (914 mm) plus 1 foot (305 mm) each for stories three through five, or 6 feet (1829 mm) plus 1 foot (305 mm) each for stories three through five]. If the same court is bounded on all sides, the required minimum length would be 16 feet (4877 mm) [10 feet (3048 mm) plus 2 feet (610 mm) each for stories three through five]. The last sentence simply requires all buildings higher than 14 stories to have a minimum court width of 15 feet (4572 mm) without opposing required openings, a minimum width of 18 feet (5486 mm) where required openings oppose each other and a minimum length of 34 feet (10 363 mm) if bounded on all sides [width equals 3 feet (914 mm) or 6 feet (1829 mm) plus 1 foot (305 mm) each for stories three through 14, and length equals 10 feet (3048 mm) plus 2 feet (610 mm) each for stories three through 14]. The requirements of this section are not applicable if artificial lighting and ventilation are provided for spaces facing the court in accordance with Sections 1203.1 and 1205.1. Note that the stories are measured above grade plane. See Commentary Figure 1206.3 for examples of court dimensions.

1206.3.1 Court access. Access shall be provided to the bottom of *courts* for cleaning purposes.

❖ Courts must be accessed for maintenance. Clearly, a court intended to be a source of ventilation air must be maintained in a manner conducive to its purpose.

1206.3.2 Air intake. *Courts* more than two *stories* in height shall be provided with a horizontal air intake at the bottom not less than 10 square feet (0.93 m^2) in area and leading to the exterior of the building unless abutting a *yard* or *public way*.

❖ This section is applicable only to courts that are bounded on all four sides by walls or other construction. A fully bounded court takes on characteristics similar to a chimney during summer weather when the building mass is heated from the daytime sun. In order for a fully bounded court to function as an efficient source of natural ventilation, the bottom of the court must have a source of fresh air (similar to a chimney). This source of fresh air is supplied through the required opening of 10 square feet (0.93 m^2) connected directly to a street or yard. The requirements of this section are not applicable if artificial lighting and ventilation are provided for spaces facing the court in accordance with Sections 1203.1 and 1206.1.

1206.3.3 Court drainage. The bottom of every *court* shall be properly graded and drained to a public sewer or other *approved* disposal system complying with the *International Plumbing Code*.

❖ A court is an inherent water trap. A court that is not both graded and drained will accumulate water and remain in a saturated condition, which will promote an insanitary condition, including odors. Stagnant water is often a breeding ground for disease-carrying insects. Based on the design and nature of the soil after construction, paving the court may be the best solution to eliminate a problem.

SECTION 1207 SOUND TRANSMISSION

1207.1 Scope. This section shall apply to common interior walls, partitions and floor/ceiling assemblies between adjacent *dwelling units* and *sleeping units* or between *dwelling units* and *sleeping units* and adjacent public areas such as halls, *corridors*, *stairways* or *service areas*.

❖ Since noise transmission can be quantified and affects the quality of life, the code incorporates regulations that address noise transmission in multiple-family residential construction, wherein the occupants may have no control over noise. The regulated components of construction are those through which noise is primarily transmitted.

1207.2 Air-borne sound. Walls, partitions and floor/ceiling assemblies separating *dwelling units* and *sleeping units* from each other or from public or service areas shall have a sound transmission class of not less than 50, or not less than 45 if field tested, for air-borne noise when tested in accordance with ASTM E90. Penetrations or openings in construction assemblies for piping; electrical devices; recessed cabinets; bathtubs; soffits; or heating, ventilating or exhaust ducts shall be sealed, lined, insulated or otherwise treated to maintain the required ratings. This requirement shall not apply to entrance doors; however, such doors shall be tight fitting to the frame and sill.

❖ The code requires common walls between dwelling units and sleeping units, and between dwelling units and sleeping units and public areas to have a minimum sound transmission class (STC) of 50. The STC is a measure of an assembly's ability to resist sound transmission. The higher the number (rating), the higher the resistance (less sound transmission). Many standard architectural wall construction assemblies have been tested for sound transmission ratings. Construction specifications will usually say if they have been tested. Air-borne noise originates in the air, such as voice or music. For structure-borne sound, see the commentary to Section 1207.3.

As a rule, vertical assemblies meeting the requirements of this section consist of double walls or walls containing insulation similar to exterior walls.

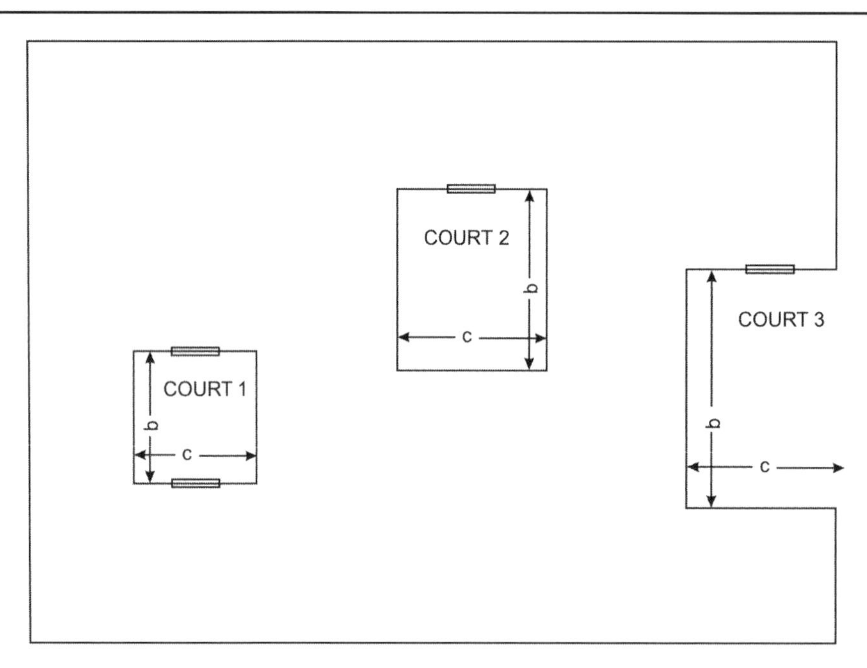

Number of Stories	Court 1		Court 2		Court 3	
	Width-b	Length-c	Width-b	Length-c	Width-b	Length-c
1 or 2	6	10	3	10	3	NA
3	7	12	4	12	4	NA
4	8	14	5	14	5	NA
5	9	16	6	16	6	NA
6	10	18	7	18	7	NA
7	11	20	8	20	8	NA
8	12	22	9	22	9	NA
9	13	24	10	24	10	NA
10	14	26	11	26	11	NA
11	15	28	12	28	12	NA
12	16	30	13	30	13	NA
13	17	32	14	32	14	NA
14	18	34	15	34	15	NA
15 or more	18	34	15	34	15	NA

N/A: Not regulated by Section 1206.3.
Note: Dimensions in feet.
For SI: 1 foot = 304.8 mm.

Figure 1206.3
DIMENSIONS OF COURTS

1207.2.1 Masonry. The sound transmission class of concrete masonry and clay masonry assemblies shall be calculated in accordance with TMS 0302 or determined through testing in accordance with ASTM E90.

❖ The STC values found in the referenced standard TMS 0302 are derived from laboratory testing of masonry assemblies in accordance with ASTM E90. This reference provides a quicker alternative for compliance with the STC ratings required by Section 1207.2 of the code for masonry and clay masonry assemblies.

1207.3 Structure-borne sound. Floor/ceiling assemblies between *dwelling units* and *sleeping units* or between a *dwelling unit* or *sleeping unit* and a public or service area within the structure shall have an impact insulation class rating of not less than 50, or not less than 45 if field tested, when tested in accordance with ASTM E492.

❖ The impact insulation class (IIC) is a measure of an assembly's ability to resist sound transmission. The higher the number (rating), the higher the resistance (less sound transmission). Floors between dwelling units and sleeping units and those floors between dwelling units and sleeping units and public areas are required to have a minimum IIC rating of 50.

Usually, floor assemblies that are carpeted meet the minimum requirement for an IIC rating of 50. Other areas with hard-surfaced finishes may require additional treatment or insulation to comply with these requirements.

There are various resource documents containing STC ratings and IIC ratings, including the following: GA 600, *Fire Resistance Design Manual*, by the Gypsum Association; NCMA TEK 69A, *New Data on Sound Reduction with Concrete Masonry Walls*, by the National Concrete Masonry Association; and BIA TN 5A, *Sound Insulation—Clay Masonry Walls*, by the Brick Institute of America. These or any other similar sources can be submitted as a basis for approval once it is demonstrated to the building official that the data are based on ASTM E90 and E492.

SECTION 1208 INTERIOR SPACE DIMENSIONS

1208.1 Minimum room widths. *Habitable spaces*, other than a kitchen, shall be not less than 7 feet (2134 mm) in any plan dimension. Kitchens shall have a clear passageway of not less than 3 feet (914 mm) between counter fronts and appliances or counter fronts and walls.

❖ This provision specifies the minimum horizontal dimensions required for all habitable rooms. Any room that functions as a living room, bedroom, dining room or any other similar habitable room must be sized such that a cylinder with a diameter of 7 feet (2134 mm) may be placed in it. Only kitchens are exempt from this requirement. This code provision allows the circulation of ventilation air through the space while maintaining reasonably sized living quarters for the occupants. "Habitable space" is defined in Chapter 2. Smaller spaces, such as bathrooms and hallways, do not need to meet the minimum room width requirements because they are not considered habitable rooms.

1208.2 Minimum ceiling heights. Occupiable spaces, *habitable spaces* and *corridors* shall have a ceiling height of not less than 7 feet 6 inches (2286 mm). Bathrooms, toilet rooms, kitchens, storage rooms and laundry rooms shall have a ceiling height of not less than 7 feet (2134 mm).

Exceptions:

1. In one- and two-family *dwellings*, beams or girders spaced not less than 4 feet (1219 mm) on center shall be permitted to project not more than 6 inches (152 mm) below the required ceiling height.
2. If any room in a building has a sloped ceiling, the prescribed ceiling height for the room is required in one-half the area thereof. Any portion of the room measuring less than 5 feet (1524 mm) from the finished floor to the ceiling shall not be included in any computation of the minimum area thereof.
3. The height of *mezzanines* and spaces below *mezzanines* shall be in accordance with Section 505.1.
4. Corridors contained within a *dwelling unit* or *sleeping unit* in a Group R occupancy shall have a ceiling height of not less than 7 feet (2134 mm).

❖ Occupiable spaces or rooms (including habitable spaces) are required to have a specific minimum ceiling height. Bathrooms, toilet rooms, kitchens, storage rooms and laundry rooms are permitted to have a lower minimum ceiling height in accordance with this section. Ceiling height is one of the variables that affects the circulation of air in a space. Additionally, there is a psychological need for spaciousness in a living space or in one of the accessory spaces.

Commentary Figure 1208.2(1) illustrates the application of Exception 1 for beams and girders spaced

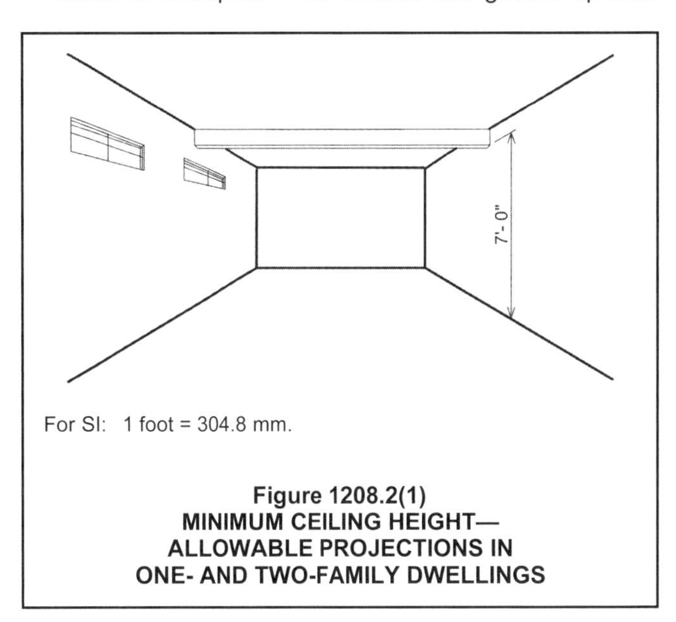

For SI: 1 foot = 304.8 mm.

Figure 1208.2(1) MINIMUM CEILING HEIGHT— ALLOWABLE PROJECTIONS IN ONE- AND TWO-FAMILY DWELLINGS

no more than 4 feet (1219 mm) on center in one- and two-family dwellings.

Commentary Figure 1208.2(2) illustrates the application of Exception 2. Rooms with sloped ceilings are required to meet two distinct conditions. First, the area of the room having a floor-to-ceiling clearance of less than 5 feet (1524 mm) does not contribute to the minimum floor area required by Section 1208.3. Second, at least one-half of the actual total area of the room must meet the minimum ceiling height requirements of Section 1208.2 [see Commentary Figure 1208.2(2)].

Exception 3 is consistent with Section 505.1, which establishes the minimum ceiling height for mezzanines at 7 feet (2134 mm). In accordance with Section 505.2, mezzanines cannot exceed one-third of the area of the room in which they are located.

Finally, Exception 4 allows corridors in Group R occupancies to be 7 feet (2134 mm). This is consistent with the IRC.

1208.2.1 Furred ceiling. Any room with a furred ceiling shall be required to have the minimum ceiling height in two-thirds of the area thereof, but in no case shall the height of the furred ceiling be less than 7 feet (2134 mm).

❖ This section only applies to rooms required to have a ceiling height of no less than 7 feet, 6 inches (2286 mm). In Commentary Figure 1208.2.1, floor area A_1 must be greater than or equal to two-thirds (length times width). Note that only those ceiling heights furred to a height of less than 7 feet, 6 inches (2286 mm) affect area A_1.

1208.3 Room area. Every *dwelling unit* shall have no fewer than one room that shall have not less than 120 square feet (13.9 m^2) of *net floor area*. Other habitable rooms shall have a *net floor area* of not less than 70 square feet (6.5 m^2).

Exception: Kitchens are not required to be of a minimum floor area.

❖ This section applies only to dwelling units. A minimum area of 120 square feet (13.9 m^2) for at least one room of each dwelling unit and a minimum of 70 square feet (6.5 m^2) for all other habitable rooms, except kitchens, are the minimum standards for habitable rooms. These minimums reflect the physiological requirements of light and ventilation and also preserve the individual's perception of space and the elements necessary for a psychological sense of well-being. The code does not regulate the minimum area of rooms and spaces in other than residential occupancies.

The exception permits smaller kitchens in dwelling units built under the code. There is no minimum area or dimension required for kitchens.

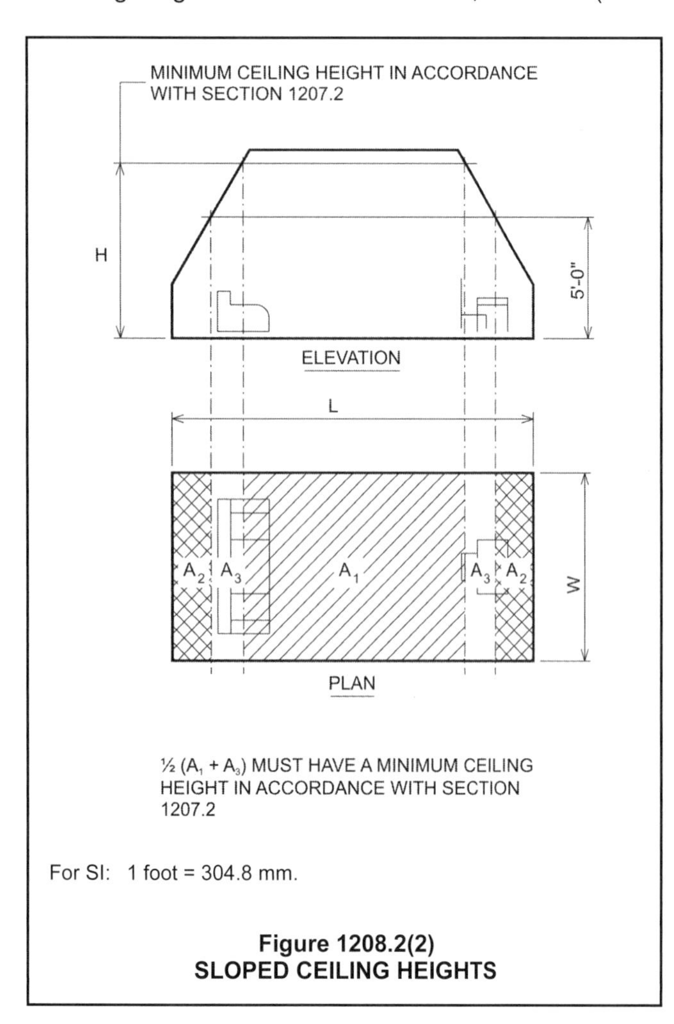

For SI: 1 foot = 304.8 mm.

Figure 1208.2(2)
SLOPED CEILING HEIGHTS

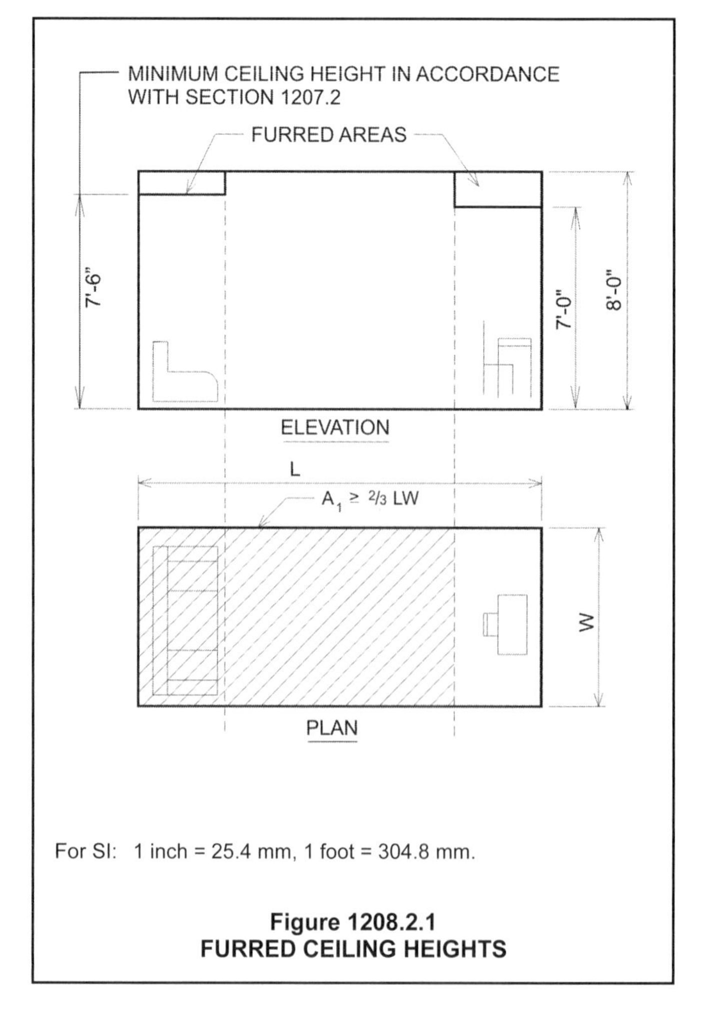

For SI: 1 inch = 25.4 mm, 1 foot = 304.8 mm.

Figure 1208.2.1
FURRED CEILING HEIGHTS

1208.4 Efficiency dwelling units. An efficiency living unit shall conform to the requirements of the code except as modified herein:

1. The unit shall have a living room of not less than 220 square feet (20.4 m^2) of floor area. An additional 100 square feet (9.3 m^2) of floor area shall be provided for each occupant of such unit in excess of two.
2. The unit shall be provided with a separate closet.
3. The unit shall be provided with a kitchen sink, cooking appliance and refrigeration facilities, each having a clear working space of not less than 30 inches (762 mm) in front. Light and *ventilation* conforming to this code shall be provided.
4. The unit shall be provided with a separate bathroom containing a water closet, lavatory and bathtub or shower.

❖ Efficiency units are very small apartments consisting of one or two rooms and a bathroom. Efficiency units that comply with this section are not required to comply with the minimum area requirements in Section 1208.3. However, the total allowable number of occupants in the dwelling is limited, depending on the area of the unit in accordance with Item 1. The purpose of both efficiency units and this section is to provide for combined use of spaces in an economical or "efficient" manner without jeopardizing health or comfort. This is possible because of the occupant limitation.

In addition to the living room, the apartment must have a separate closet and bathroom. There are no minimum area requirements for these spaces; however, the fixture clearances in the bathroom must be as required in the IPC. There is not a requirement for a separate kitchen, but if the required sink and appliances are located in the living room, they cannot encroach on the required minimum floor space.

SECTION 1209 ACCESS TO UNOCCUPIED SPACES

1209.1 Crawl spaces. Crawl spaces shall be provided with not fewer than one access opening that shall be not less than 18 inches by 24 inches (457 mm by 610 mm).

❖ The requirements of this section establish 18 inches by 24 inches (457 mm by 610 mm) as the minimum size opening for crawl spaces. If access is through a wall, the 18-inch (457 mm) minimum would be the height and the 24-inch (610 mm) minimum would be the width.

Items such as plumbing and wiring installations pass through crawl spaces at times. Required initial and periodic inspections, and repairs cannot be carried out without access to such crawl spaces.

1209.2 Attic spaces. An opening not less than 20 inches by 30 inches (559 mm by 762 mm) shall be provided to any *attic* area having a clear height of over 30 inches (762 mm). Clear headroom of not less than 30 inches (762 mm) shall be provided in the *attic* space at or above the access opening.

❖ Access to the attic provides a convenient and nondestructive means for fire department personnel to visually check for an attic fire and, if need be, gain entry to the concealed spaces and suppress a fire. Access to attic spaces can be provided through the ceiling within each compartment that is created by draftstops or through openings within the draftstops themselves. Openings located within the draftstop are required to be self-closing and the opening protective must provide structural fire integrity (the ability to remain in place) similar to the draftstop. Access is required when the attic space has a clear height greater than 30 inches (762 mm) measured from the top of the ceiling joists (or top of the floor sheathing, if present) to the underside of the roof rafters.

1209.3 Mechanical appliances. Access to mechanical appliances installed in under-floor areas, in *attic* spaces and on roofs or elevated structures shall be in accordance with the *International Mechanical Code*.

❖ Access to mechanical appliances is needed to maintain and service the equipment. See Section 306 in the IMC for detailed requirements.

SECTION 1210 TOILET AND BATHROOM REQUIREMENTS

[P] 1210.1 Required fixtures. The number and type of plumbing fixtures provided in any occupancy shall comply with Chapter 29.

❖ The minimum number of plumbing fixtures depends upon the occupant load and use of the building. Chapter 29 contains a table for determination of the number of facilities.

1210.2 Finish materials. Walls, floors and partitions in toilet and bathrooms shall comply with Sections 1210.2.1 through 1210.2.4.

1210.2.1 Floors and wall bases. In other than *dwelling units*, toilet, bathing and shower room floor finish materials shall have a smooth, hard, nonabsorbent surface. The intersections of such floors with walls shall have a smooth, hard, nonabsorbent vertical base that extends upward onto the walls not less than 4 inches (102 mm).

❖ The purpose of this requirement is to provide nonabsorbent surfaces that can be maintained in a sanitary condition. The 4-inch (152 mm) extension of the surface up the surrounding walls is so that the wall will not absorb moisture during cleaning and, thus, will be left in a clean condition. This provision does not require that the same material that is on the floor be extended up the wall. As long as the wall material is smooth, hard and nonabsorbent, and there is adequate seal between the materials to restrict moisture from getting behind the surface material and into the wall construction, the intent of the requirement should be met.

1210.2.2 Walls and partitions. Walls and partitions within 2 feet (610 mm) of service sinks, urinals and water closets shall have a smooth, hard, nonabsorbent surface, to a height of not less than 4 feet (1219 mm) above the floor, and except for structural elements, the materials used in such walls shall be of a type that is not adversely affected by moisture.

Exception: This section does not apply to the following buildings and spaces:

1. Dwelling units and sleeping units.
2. Toilet rooms that are not accessible to the public and that have not more than one water closet.

Accessories such as grab bars, towel bars, paper dispensers and soap dishes, provided on or within walls, shall be installed and sealed to protect structural elements from moisture.

❖ The walls and partitions near urinals and water closets must have the surface specified in this section since they are subject to moisture. Exception 1 recognizes that water closet facilities in dwelling units and sleeping units are not exposed to as much use as those that serve the public and, thus, are easier to maintain. Exception 2 acknowledges that toilet fixtures that do not serve the public are also subject to less use. This section requires protection of the structural supports for accessories so that they will maintain their strength.

1210.2.3 Showers. Shower compartments and walls above bathtubs with installed shower heads shall be finished with a smooth, nonabsorbent surface to a height not less than 72 inches (1829 mm) above the drain inlet.

❖ The 72-inch (1778 mm) requirement in this section is based on the height of the shower compartment and walls that are exposed to significant moisture that would cause the surface to become insanitary over a long period of time. This height is consistent with the IRC.

1210.2.4 Waterproof joints. Built-in tubs with showers shall have waterproof joints between the tub and adjacent wall.

❖ The joint between the tub and wall must be sealed to prevent moisture from getting into the supporting floor and framing. Waterproof joints are also needed to keep the concealed area of the wall in a sanitary condition.

[P] 1210.3 Privacy. Privacy at water closets and urinals shall be provided in accordance with Sections 1210.3.1 and 1210.3.2.

❖ For most people, a certain amount of privacy in public facilities is desirable. For some, it is imperative. This section provides requirements for water-closet compartments and urinal dividers.

[P] 1210.3.1 Water closet compartment. Each water closet utilized by the public or employees shall occupy a separate compartment with walls or partitions and a door enclosing the fixtures to ensure privacy.

Exceptions:

1. Water closet compartments shall not be required in a single-occupant toilet room with a lockable door.
2. Toilet rooms located in child day care facilities and containing two or more water closets shall be permitted to have one water closet without an enclosing compartment.
3. This provision is not applicable to toilet areas located within Group I-3 occupancy housing areas.

❖ Psychological studies have proven that lack of privacy places a burden on an individual's physical ability to use bathroom facilities. This is caused by uneasiness, inhibition or indignation. A partitioned compartment can provide the necessary privacy. Also, a single-occupant toilet room with a lockable door can provide the required privacy as stated in Exception 1. Exception 2, allowing an unenclosed water closet in a child care facility, recognizes the need for people to be able to assist children. Exception 3 was added in the 2009 code edition to allow for security personal to better monitor for illicit activities of inmates in Group I-3 occupancies.

[P] 1210.3.2 Urinal partitions. Each urinal utilized by the public or employees shall occupy a separate area with walls or partitions to provide privacy. The walls or partitions shall begin at a height not more than 12 inches (305 mm) from and extend not less than 60 inches (1524 mm) above the finished floor surface. The walls or partitions shall extend from the wall surface at each side of the urinal not less than 18 inches (457 mm) or to a point not less than 6 inches (152 mm) beyond the outermost front lip of the urinal measured from the finished backwall surface, whichever is greater.

Exceptions:

1. Urinal partitions shall not be required in a single-occupant or family or assisted-use toilet room with a lockable door.
2. Toilet rooms located in child day care facilities and containing two or more urinals shall be permitted to have one urinal without partitions.

❖ Users of urinals in a public toilet room need to be afforded a level of privacy for the reasons stated in the commentary for Section 1210.3.1. In addition, partitions aid in protecting adjacent users from splash. Because of the likelihood of splashing, the partition surfaces must be designed so that they can be easily cleaned and sanitized during normal restroom maintenance. Exception 1 assumes that in a single-occupant toilet room or a family/assisted-use toilet room privacy is afforded by locking the door. Exception 2 allows day care and child care centers to have one urinal out of two or more urinals in a toilet room to be without partitions. This affords care providers the ability to assist or supervise certain children with use of the urinal (e.g., those just learning to use a urinal or those with difficult clothing arrangements). Where urinal partitions are provided, occupants are more inclined to use a urinal rather than a water closet for urination. This will result in less soiling of water closet seats and surrounding surfaces. Note that where partitions are used, the distance from the center of the urinal to the face of the partition must be at least 15 inches (381 mm).The thickness of partitions cannot encroach into the required spaces for fixtures.

Bibliography

The following resource materials were used in the preparation of the commentary for this chapter of the code.

BIA TN 5A-83, *Sound Insulation—Clay Masonry Walls*. Reston, VA: Brick Institute of America, 1983.

FEMA FIA-TB 1, *Openings in Foundation Walls for Buildings Located in Special Flood Hazard Areas*. Washington, DC: Federal Emergency Management Agency, 2008.

GA 600-06, *Fire Resistance Design Manual*. Evanston, IL: Gypsum Association, 2006.

NCMA TEK 69A-92, *New Data on Sound Reduction with Concrete Masonry Walls*. Herndon, VA: National Concrete Masonry Association, 1992.

Chapter 13:
Energy Efficiency

General Comments

Chapter 13 provides for the design and construction of energy-efficient buildings and structures or portions thereof intended primarily for human occupancy by direct reference to the *International Energy Conservation Code*® (IECC®).

Purpose

The purpose of Chapter 13 is to provide minimum design requirements that will promote efficient utilization of energy in building construction. The requirements are directed toward the design of building envelopes with adequate thermal resistance and low air leakage, and toward the design and selection of mechanical, service water heating, electrical and illumination systems that will promote the effective use of depletable energy resources and encourage the use of nondepletable energy resources.

SECTION 1301
GENERAL

[E] 1301.1 Scope. This chapter governs the design and construction of buildings for energy efficiency.

❖ The scope of Chapter 13 is applicable to all buildings and structures, as well as their components and systems that are regulated by the IECC. The IECC thereby addresses the design of energy-efficient building envelopes, and the selection and installation of energy-efficient mechanical, service-water heating, electrical distribution and illumination systems and equipment for the effective use of energy in both residential and commercial buildings.

[E] 1301.1.1 Criteria. Buildings shall be designed and constructed in accordance with the *International Energy Conservation Code*.

❖ The energy conservation requirements of this chapter rely exclusively on the technical provisions of the IECC. Compliance with the IECC is the sole means of demonstrating compliance with the technical provisions of Chapter 13.

The climate basis for the 2015 IECC is derived from geographical zones that are based on multiple climate variables (so that both heating and cooling considerations are accommodated). Further, within the U.S., the zones are completely defined by political boundaries (county lines) so that code users will never have to choose from disparate climate data sources to determine local requirements. The climate zones were developed in an open process, in consultation with relevant standards committees of the American Society of Heating, Refrigerating and Air-conditioning Engineers (ASHRAE). The zones are designed to be an appropriate foundation for both residential and commercial codes, and may be useful in other contexts as well.

The IECC is designed to increase consumer awareness of a home's energy features by making baseline requirements uniform within a jurisdiction and by requiring a disclosure of each house's *R*-values, *U*-factors and heating, ventilating and air-conditioning (HVAC) efficiencies.

The IECC is designed, to the extent practicable, to incorporate aspects of the latest building science regarding energy efficiency and its effects on moisture control and durability. For example, the IECC contains provisions related to unvented crawl spaces, modifies vapor retarder requirements, requires sealing of air handlers in garages and limits worst-case glazing *U*-factors in locations where moisture condensation can be a serious problem.

Bibliography

The following resource material was used in the preparation of the commentary for this chapter of the code.

IECC-15, *International Energy Conservation Code*. Washington, DC: International Code Council, 2014.

Chapter 14: Exterior Walls

General Comments

Chapter 14 provides requirements for the materials and construction creating finished exterior surfaces of a building or structure. The chapter provides requirements resulting in minimum permitted weather resistance and fire performance.

In the past, there were relatively few materials available for application to exterior walls for protection against weather and exposure to other outside elements. With the development of new methods and materials over the years, architects and builders began to use a variety of materials for different appearances, improved insulating quality, sound transmission control and fire resistance. The code has developed prescriptive and performance regulations to control these aspects and the types and thickness of exterior wall coverings.

Purpose

The purpose of Chapter 14 is to provide the minimum requirements for exterior wall coverings. This chapter also includes the minimum regulations for materials, and the minimum thicknesses and installation requirements for exterior weather coverings and various wall veneers. Limitations on the use of combustible exterior wall finishes are also included.

SECTION 1401 GENERAL

1401.1 Scope. The provisions of this chapter shall establish the minimum requirements for exterior walls; *exterior wall* coverings; *exterior wall* openings; exterior windows and doors; architectural *trim*; balconies and similar projections; and bay and oriel windows.

❖ The requirements for exterior wall construction, including components such as openings, doors, windows, trim and balconies, are specified herein. Also specified are provisions intended to prevent damage to a building from the effects of weather and from moisture intrusion into or through the exterior walls.

SECTION 1402 DEFINITIONS

1402.1 Definitions. The following terms are defined in Chapter 2:

ADHERED MASONRY VENEER.

ANCHORED MASONRY VENEER.

BACKING.

EXTERIOR INSULATION AND FINISH SYSTEMS (EIFS).

EXTERIOR INSULATION AND FINISH SYSTEMS (EIFS) WITH DRAINAGE.

EXTERIOR WALL.

EXTERIOR WALL COVERING.

EXTERIOR WALL ENVELOPE.

FENESTRATION.

FIBER-CEMENT SIDING.

HIGH-PRESSURE DECORATIVE EXTERIOR-GRADE COMPACT LAMINATE (HPL).

HIGH-PRESSURE DECORATIVE EXTERIOR-GRADE COMPACT LAMINATE (HPL) SYSTEM.

METAL COMPOSITE MATERIAL (MCM).

METAL COMPOSITE MATERIAL (MCM) SYSTEM.

POLYPROPYLENE SIDING.

PORCELAIN TILE.

VENEER.

VINYL SIDING.

WATER-RESISTIVE BARRIER.

❖ Definitions facilitate the understanding of code provisions and minimize potential confusion. To that end, this section lists definitions of terms associated with roof assemblies, roof coverings and rooftop structures. Note that these definitions are found in Chapter 2. The use and application of defined terms, as well as undefined terms, are set forth in Section 201.

SECTION 1403 PERFORMANCE REQUIREMENTS

1403.1 General. The provisions of this section shall apply to exterior walls, wall coverings and components thereof.

❖ The exterior walls of buildings provide support for the floor and roof systems; act as barriers against the outside environment; reduce heat loss from the interior and, in some instances, provide the sole resistance to such destructive forces as earthquakes and floods. This section provides references to other applicable sections of the code that are intended to detail specific requirements to accomplish these objectives. Additionally, this chapter contains detailed requirements related to the performance of exterior wall coverings and the exterior wall envelope.

1403.2 Weather protection. Exterior walls shall provide the building with a weather-resistant *exterior wall envelope*. The *exterior wall envelope* shall include flashing, as described in Section 1405.4. The *exterior wall envelope* shall be designed and constructed in such a manner as to prevent the accumulation of water within the wall assembly by providing a *water-resistive barrier* behind the exterior veneer, as described in Section 1404.2, and a means for draining water that enters the assembly to the exterior. Protection against condensation in the *exterior wall* assembly shall be provided in accordance with Section 1405.3.

Exceptions:

1. A weather-resistant *exterior wall envelope* shall not be required over concrete or masonry walls designed in accordance with Chapters 19 and 21, respectively.
2. Compliance with the requirements for a means of drainage, and the requirements of Sections 1404.2 and 1405.4, shall not be required for an *exterior wall envelope* that has been demonstrated through testing to resist wind-driven rain, including joints, penetrations and intersections with dissimilar materials, in accordance with ASTM E331 under the following conditions:
   - 2.1. *Exterior wall envelope* test assemblies shall include at least one opening, one control joint, one wall/eave interface and one wall sill. Tested openings and penetrations shall be representative of the intended end-use configuration.
   - 2.2. *Exterior wall envelope* test assemblies shall be at least 4 feet by 8 feet (1219 mm by 2438 mm) in size.
   - 2.3. *Exterior wall envelope* assemblies shall be tested at a minimum differential pressure of 6.24 pounds per square foot (psf) (0.297 kN/m^2).
   - 2.4. *Exterior wall envelope* assemblies shall be subjected to a minimum test exposure duration of 2 hours.

   The *exterior wall envelope* design shall be considered to resist wind-driven rain where the results of testing indicate that water did not penetrate control joints in the *exterior wall* envelope, joints at the perimeter of openings or intersections of terminations with dissimilar materials.
3. Exterior insulation and finish systems (EIFS) complying with Section 1408.4.1.

❖ All exterior walls of buildings must be protected against damage caused by precipitation, wind and other weather conditions.

The main text of this section prescribes three basic components of a weather-resistive exterior wall assembly: a water-resistive barrier installed over the building substrate; flashings at penetrations and terminations of the exterior wall finish; and a means of draining moisture that may penetrate behind the finish back to the exterior.

Section 1404.2 is referenced for the requirements of the water-resistive barrier and Section 1405.3 is referenced for requirements for the flashings (see commentary, Sections 1404.2 and 1405.3). This section does not, however, contain a prescriptive requirement for the means of drainage to be provided. The method to provide the means of drainage is a performance criterion and must be evaluated based on its ability to allow moisture behind the exterior wall covering to drain back to the exterior. This may be as complicated as a rain-screen, pressure-equalized type of exterior assembly or as simple as providing discontinuities or gaps between the surface of the substrate and the back side of the finish, such as through the use of noncorrodible furring.

For common types of construction, such as vinyl siding or brick veneer, the typical practice of installing building paper and weeps will comply with the intent of this section. Discontinuities between the exterior covering and substrate must be such that they encourage the flow of moisture via gravity or capillary action to a location where the water may exit, such as at flashings and weeps. The absence of a means of drainage may result in the accumulation of moisture that becomes trapped between the finish and the substrate. Over time, extended exposure to moisture may contribute to the degradation of the finish, building substrate or even the structural elements of the exterior wall.

Exception 1 states that where the exterior wall envelope is designed and constructed of concrete or masonry materials in accordance with the requirements of Chapters 19 and 21, respectively, the water-resistive barrier and means of drainage may be omitted. This is because the penetration of moisture behind the exterior wall finish is not detrimental to concrete and masonry substrates.

Exception 2 permits the use of exterior wall finishes that do not meet the prescriptive requirements of Section 1403.2, provided that the system, with penetration details, is tested for wind-driven rain resistance. The test specimen(s) must incorporate those penetration and termination details intended for use and the system will be limited to use with those details that successfully pass the test. The minimum panel size specified represents that which is commonly used in testing to ASTM E331; however, this does not preclude the testing of larger panels if desired. The modifications to the test pressure differential and test duration are intended to represent more closely conditions that will be encountered in service. The pass/fail criteria is based upon the visual observation of moisture on the rear side of the wall assembly. In cavity-type assemblies, such as stud walls, this requires the observation of locations such as the rear face of the exterior wall sheathing and wall framing members for the presence of moisture. The test method is intended to assess the performance of the method(s) intended for use in sealing the interface between the termination of the exterior wall finish and the penetration items or abutting construction. The method is not necessarily intended to test the performance of the penetrating item.

Exception 3 relates to exterior insulation and finish systems (EIFS) constructed in accordance with Section 1408.2. This material does not have the means for draining water specified in this section.

Walls designed and constructed in accordance with this chapter must also comply with the requirements of the *International Energy Conservation Code*® (IECC®).

[BS] 1403.3 Structural. *Exterior walls*, and the associated openings, shall be designed and constructed to resist safely the superimposed loads required by Chapter 16.

❖ Exterior walls and their associated openings are required to resist all structural loads in accordance with the provisions of Chapter 16. This section is a correlative cross reference to emphasize the applicability of Chapter 16.

1403.4 Fire resistance. *Exterior walls* shall be fire-resistance rated as required by other sections of this code with opening protection as required by Chapter 7.

❖ The required fire-resistance rating of exterior walls is set forth in Tables 601 and 602. Table 602 is applicable to both load-bearing and nonload-bearing walls, since it addresses the prevention of fire spread from one building to an adjacent building. Load-bearing walls must comply with the greater of the requirements in Tables 601 and 602, based on the type of construction of the building. The size of openings in exterior walls is limited to prevent the spread of fire to other buildings. The commentary to Section 705 should also be reviewed when designing exterior walls. Trim on exterior walls is regulated by Section 1406.

The allowable size of openings in exterior walls is tabulated in Table 705.8. Section 715.4 specifies the required fire protection rating for opening protectives when required by Table 705.8. Where a fire-resistance rating is not required for the wall and unprotected openings are allowed, the glazing and the sash or frame may be of any material permitted by the code.

1403.5 Vertical and lateral flame propagation. Exterior walls on buildings of Type I, II, III or IV construction that are greater than 40 feet (12 192 mm) in height above grade plane and contain a combustible *water-resistive barrier* shall be tested in accordance with and comply with the acceptance criteria of NFPA 285. For the purposes of this section, fenestration products and flashing of fenestration products shall not be considered part of the *water-resistive barrier*.

Exceptions:

1. Walls in which the *water-resistive barrier* is the only combustible component and the *exterior wall* has a wall covering of brick, concrete, stone, terra cotta, stucco or steel with minimum thicknesses in accordance with Table 1405.2.
2. Walls in which the *water-resistive barrier* is the only combustible component and the *water-resistive barrier* has a peak heat release rate of less than 150 kW/m^2, a total heat release of less than 20 MJ/m^2 and an effective heat of combustion of less than 18 MJ/kg as determined in accordance with ASTM E1354 and has a flame spread index of 25 or less and a smoke-developed index of 450 or less as determined in accordance with ASTM E84 or UL 723. The ASTM E1354 test shall be conducted on specimens at the thickness intended for use, in the horizontal orientation and at an incident radiant heat flux of 50 kW/m^2.

❖ This section addresses the potential vertical and lateral flame spread that can occur either on or within exterior wall assemblies that contain combustible materials.

Construction practices, such as the addition of combustible water-resistant barriers, allow significant amounts of combustible materials/products (other than foam plastics) to be installed on or in exterior walls. This section requires testing in accordance with NFPA 285 for exterior walls that contain these types of combustible materials. This requirement is already in place for any exterior walls that contain foam plastic insulation or use metal composite material (MCM) exterior veneers. Small-scale testing has shown that these types of materials can provide significant amounts of combustible fuel loading to a wall assembly.

With the advent of newer exterior wall technologies, such as "rainscreen" systems, the openings in the exterior veneer will allow flames and heat to readily impact and ignite the barrier material. Due to the built-in standoffs of these systems, the barrier materials could then exhibit significant vertical or lateral flame propagation.

The testing requirements are applicable for exterior walls on Type I, II, III or IV construction since these types of construction allow either no or limited combustibles in the exterior walls. The 40-foot (12 192 mm) height limit is consistent with the code provisions for combustible exterior wall coverings.

The purpose of the last sentence is to clarify that this section does not apply to fenestration products and the flashing of fenestration products. The intent of this section is to apply to the installation of water-resistive barriers over the opaque section of exterior walls. Where water-resistive barriers are installed in a large quantity, such as over the entire opaque section of exterior walls, they can add a significant fuel load to the exterior wall. The amount of combustible material used in the flashing of fenestration products is insignificant and will contribute very little to the fuel load or spread of fire over an exterior wall.

Exception 1 recognizes that "heavy" types of noncombustible exterior wall veneers can provide protection to the water-resistive barrier, thus eliminating the need for NFPA 285 testing where the water-resistive barrier is the only combustible component in the exterior wall. Table 1405.2, which describes the allowable minimum thicknesses of brick, concrete, stone, terra cotta, stucco or steel, is referenced.

Exception 2 provides an exception for NFPA 285 testing where the water-resistive barrier is the only combustible material in any exterior wall and demonstrates low combustibility characteristics when tested in accordance with ASTM E1354 and ASTM E84. The pass criteria are based upon a proprietary test program that evaluated a number of market-available water-resistive barriers.

[BS] 1403.6 Flood resistance. For buildings in flood hazard areas as established in Section 1612.3, *exterior walls* extending below the elevation required by Section 1612 shall be constructed with flood-damage-resistant materials.

❖ The flood-damage-resistant requirements of this section apply to exterior walls enclosing areas below buildings that are elevated as required by Section 1612. Some of the properties these materials must possess include resistance to prolonged contact with water, the ability to be cleaned and disinfected after the water recedes and negligible loss of physical properties after exposure to water. Additionally, systems must possess structural strength to resist the hydrodynamic and impact forces related to flooding.

[BS] 1403.7 Flood resistance for coastal high-hazard areas and coastal A zones. For buildings in coastal high-hazard areas and coastal A zones as established in Section 1612.3, electrical, mechanical and plumbing system components shall not be mounted on or penetrate through exterior walls that are designed to break away under flood loads.

❖ ASCE 24 and Section 1612.3 require exterior walls that enclose areas below elevated buildings to break away under flood loads in coastal high-hazard areas (Zone V) and Coastal A Zones. This section prohibits the installation of penetrations for electrical, plumbing or mechanical systems through such exterior walls. The breakaway provisions are a method of protecting the remaining building frame and building system. Penetrations through these walls could defeat that purpose or reduce the ability of the breakaway system to perform as designed, thereby increasing flood damage.

SECTION 1404 MATERIALS

1404.1 General. Materials used for the construction of *exterior walls* shall comply with the provisions of this section. Materials not prescribed herein shall be permitted, provided that any such alternative has been *approved*.

❖ Section 1404 contains performance requirements and detailed installation specifications for a number of common materials used in or on exterior walls of buildings. Reference is made to the applicable chapters that contain additional requirements. These chapters should be reviewed by the reader.

It is not the intent of the code to prevent the use of any material that is equivalent in performance to those specified in Section 1404.

1404.2 Water-resistive barrier. Not fewer than one layer of No.15 asphalt felt, complying with ASTM D226 for Type 1 felt or other *approved* materials, shall be attached to the studs or sheathing, with flashing as described in Section 1405.4, in such a manner as to provide a continuous *water-resistive barrier* behind the *exterior wall* veneer.

❖ Many exterior veneers provide weather resistance but may allow either penetration of water through joints or seams or the development of condensation to occur behind the veneer. To increase the weather resistance of the wall, a layer of asphalt felt or other approved material is required to be installed over the wall backing. The felt layer and the flashing provide a water-resistive barrier behind the exterior veneer. The water-resistive membrane must be attached to the studs or sheathing in a way that will not allow the penetration of water as a result of the attachment method.

[BS] 1404.3 Wood. *Exterior walls* of wood construction shall be designed and constructed in accordance with Chapter 23.

❖ Chapter 23 contains general requirements regulating the use of wood in buildings. This section contains specific provisions for the minimum thicknesses of wood siding, wood backing surfaces and hardboard siding, as well as installation and nailing requirements for wood used as weatherboarding or a wall covering.

[BS] 1404.3.1 Basic hardboard. Basic hardboard shall conform to the requirements of AHA A135.4.

❖ Basic hardboard is required to comply with AHA A135.4. This material is not intended for use as exposed siding or as a weather covering, but it is often a component of veneered wall construction.

[BS] 1404.3.2 Hardboard siding. Hardboard siding shall conform to the requirements of AHA A135.6 and, where used structurally, shall be so identified by the *label* of an *approved* agency.

❖ Hardboard conforming to AHA A135.6 is manufactured in panels that are usually 4 feet by 8 feet (1219 mm by 2438 mm) in size or in sizes simulating drop siding where the material is to be applied horizontally over structural sheathing. When used as a panel, hardboard siding is applied directly to the wood-framing members as a backing material. Hardboard panels are not adequate to serve as a nailing base (spanning between studs) for coverings. Coverings are fastened directly at framing locations, through the panels.

[BS] 1404.4 Masonry. *Exterior walls* of masonry construction shall be designed and constructed in accordance with this section and Chapter 21. Masonry units, mortar and metal accessories used in anchored and adhered veneer shall meet the physical requirements of Chapter 21. The backing of anchored and adhered veneer shall be of concrete, masonry, steel framing or wood framing. Continuous insulation meeting the applicable requirements of this code shall be permitted between the backing and the masonry veneer.

❖ Chapter 21 contains the general requirements for the use of masonry in buildings. The material classification of masonry in this section includes materials such as concrete masonry; bricks of clay, shale or calcium silicate; clay-tile units; glazed masonry units; terra cotta; natural or cast stone; and ceramic tile. Chapter 21 and this section contain provisions for the minimum thickness of masonry units, limitations on loading, heights of veneers, backing and attachment. The last sentence allows typical continuous insulation, meeting applicable code requirements, between backing and masonry veneer. This insulation could be insulation board or other type of continuous insulations.

[BS] 1404.5 Metal. *Exterior walls* constructed of cold-formed steel, structural steel or aluminum shall be designed in accordance with Chapters 22 and 20, respectively.

❖ Chapters 20 and 22 contain the general requirements for use of metal in buildings. Chapters 20 and 22 as well as this section contain provisions for the minimum thickness of metal weather coverings, and requirements for protection from corrosion, installation and grounding.

[BS] 1404.5.1 Aluminum siding. Aluminum siding shall conform to the requirements of AAMA 1402.

❖ Although aluminum siding is addressed in Chapter 20, it is more specifically addressed here as an exterior wall covering. Aluminum siding is required to conform to the referenced standard. The thickness, installation and protection provisions of this section are also applicable (see commentary, Sections 1405.11, 1405.11.1, 1405.11.2, 1405.11.3 and 1405.11.4).

[BS] 1404.5.2 Cold-rolled copper. Copper shall conform to the requirements of ASTM B370.

❖ This material is commonly used for roofing, flashing, gutters, downspouts and general sheet metal work. ASTM B370 provides the specification for both copper sheet and copper strip materials.

[BS] 1404.5.3 Lead-coated copper. Lead-coated copper shall conform to the requirements of ASTM B101.

❖ This material is commonly used for roofing, flashing, gutters, downspouts and general sheet metal work. Among other requirements, ASTM B101 specifies that the lead must uniformly coat the surfaces, edges and ends of the copper.

[BS] 1404.6 Concrete. Exterior walls of concrete construction shall be designed and constructed in accordance with Chapter 19.

❖ Chapter 19 contains the general requirements for use of concrete in buildings, including provisions for the minimum thickness of concrete units used as veneers and wall coverings, and their installation and protection.

[BS] 1404.7 Glass-unit masonry. Exterior walls of glass-unit masonry shall be designed and constructed in accordance with Chapter 21.

❖ Structural glass block is generally regulated by the provisions of Sections 2103 and 2104. However, structural glass veneers used as wall coverings must comply with the thickness requirement of Table 1405.2.

1404.8 Plastics. Plastic panel, apron or spandrel walls as defined in this code shall not be limited in thickness, provided that such plastics and their assemblies conform to the requirements of Chapter 26 and are constructed of *approved* weather-resistant materials of adequate strength to resist the wind loads for cladding specified in Chapter 16.

❖ This section provides for the use of plastic panels for exterior walls. The code uses the term "light-transmitting plastic" to apply to plastics used for exterior wall panels and roof panels, as well as for glazing and skylights. The panels must meet the requirements of Chapter 26, as well as certain requirements of Chapter 14 for weather resistance and Chapter 16 for structural capabilities.

1404.9 Vinyl siding. Vinyl siding shall be certified and labeled as conforming to the requirements of ASTM D3679 by an *approved* quality control agency.

❖ Plastics are addressed in Chapter 26. However, polyvinyl chloride (PVC) siding is specifically addressed here as an exterior wall covering. PVC siding is required to conform to the provisions of ASTM D3679. Installation of vinyl siding is prescribed by Section 1405.14 (see the commentary to Section 1405.14 for additional discussion on the installation of vinyl siding). Note that the product must be certified and labeled, which means that the manufacturer must have regular inspections by an approved quality control agency.

1404.10 Fiber-cement siding. Fiber-cement siding shall conform to the requirements of ASTM C1186, Type A (or ISO 8336, Category A), and shall be so identified on labeling listing an *approved* quality control agency.

❖ Fiber-cement siding is also a material used for weather-resistant siding. It is manufactured from fiber-reinforced cement. The code permits its use as either panel siding or horizontal lap siding. The material standard options for compliance are ASTM C1186 and ISO 8336. Note that the product is required to bear a label, which means that the manufacturer must have regular inspections by a third-party inspection quality control agency.

1404.11 Exterior insulation and finish systems. Exterior insulation and finish systems (EIFS) and exterior insulation and finish systems (EIFS) with drainage shall comply with Section 1408.

❖ EIFS and EIFS with drainage are relatively complex exterior veneers that require great care and attention regarding the materials used and the installation details. This exterior veneer is now dealt with in Section 1408. All of the performance criteria, installation and design are covered in this section.

1404.12 Polypropylene siding. Polypropylene siding shall be certified and labeled as conforming to the requirements of ASTM D7254 and those of Section 1404.12.1 or 1404.12.2 by an approved quality control agency. Polypropylene siding shall be installed in accordance with the requirements of Section 1405.18 and in accordance with the manufacturer's instructions. Polypropylene siding shall be secured to the building so as to provide weather protection for the exterior walls of the building.

❖ The requirements for polypropylene siding are similar to the requirements for vinyl siding, except for the fire testing. The required standard for polypropylene siding is ASTM D7254. This specification addresses many of the key requirements for the material.

1404.12.1 Flame spread index. The certification of the flame spread index shall be accompanied by a test report stating that all portions of the test specimen ahead of the flame front remained in position during the test in accordance with ASTM E84 or UL 723.

❖ It is critical for exposed materials to remain in place during the flame spread testing so that the flame spread index assesses actual surface flame spread on the material surface; therefore, this section requires that the certification of compliance with ASTM D7254 indicates that the specimen remained in place during the test procedure.

1404.12.2 Fire separation distance. The fire separation distance between a building with polypropylene siding and the adjacent building shall be not less than 10 feet (3048 mm).

❖ Based on the heat-release rate exhibited by polypropylene material, a minimum distance of 10 feet (3048 mm) is required between the structure with the polypropylene siding and any adjacent structure.

1404.13 Foam plastic insulation. Foam plastic insulation used in *exterior wall covering* assemblies shall comply with Chapter 26.

❖ Foam plastic insulation is commonly included as a component in exterior wall covering assemblies for energy code compliance and is included in the current definition of exterior wall coverings. Therefore, this section appropriately references the applicable material requirements in Chapter 26.

SECTION 1405
INSTALLATION OF WALL COVERINGS

1405.1 General. *Exterior wall coverings* shall be designed and constructed in accordance with the applicable provisions of this section.

❖ Exterior wall coverings used to provide weather protection to the structure are required to comply with all the applicable provisions contained in Section 1405.

1405.2 Weather protection. *Exterior walls* shall provide weather protection for the building. The materials of the minimum nominal thickness specified in Table 1405.2 shall be acceptable as *approved* weather coverings.

❖ The exterior walls of a structure must be designed and constructed to provide for the health and safety of the occupants and to protect the structure from the detrimental effects of weather exposure.

This section introduces Table 1405.2, which is a tabulation of the minimum acceptable nominal thicknesses for a number of common weather coverings.

TABLE 1405.2. See page 14-7.

❖ This table should be used in addition to all other applicable chapters and requirements of the code for the specific material listed. Testing and experience have determined that the minimum tabulated thicknesses will be durable and protect the building against the elements for relatively long periods of time when attached and maintained as required.

Note a of Table 1405.2 allows material less than $^1/_2$-inch-thick (12.7 mm) wood siding to be used, provided that it is installed over an approved sheathing conforming to the requirements of Section 2304.6.1 (see commentary, Section 2304.6.1).

Note b of Table 1405.2 establishes that the minimum permitted nominal thickness is based on the narrowest solid thickness of material, exclusive of texture.

Section 2304.6.1 allows fiberboard and particleboard for sheathing. Although permitted as sheathing, fiberboard and particleboard materials, and other similar materials, do not provide a nailing base for weatherboarding. When used over such materials, $^1/_2$-inch (12.7 mm) wood siding is to be fastened to the wood-framing members or otherwise adequately secured.

Wood siding having a thickness of less than $^1/_2$ inch (12.7 mm) is not permitted to be fastened to foam plastic and other sheathing materials not having adequate strength as a nailing base spanning between studs or framing members.

1405.3 Vapor retarders. Vapor retarders as described in Section 1405.3.3 shall be provided in accordance with Sections 1405.3.1 and 1405.3.2, or an approved design using accepted engineering practice for hygrothermal analysis.

❖ The purpose of these requirements is to provide prescriptive methods for moisture control. The code prescribes three different vapor retarder classes, based upon the vapor permeability of the material. These are defined in Chapter 2.

1405.3.1 Class I and II vapor retarders. Class I and II vapor retarders shall not be provided on the interior side of frame walls in Zones 1 and 2. Class I vapor retarders shall not be provided on the interior side of frame walls in Zones 3 and 4. Class I or II vapor retarders shall be provided on the interior side of frame walls in Zones 5, 6, 7, 8 and Marine 4. The appropriate zone shall be selected in accordance with Chapter 3 of the *International Energy Conservation Code*—Commercial Provisions.

Exceptions:

1. Basement walls.
2. Below-grade portion of any wall.
3. Construction where moisture or its freezing will not damage the materials.
4. Conditions where Class III vapor retarders are required in Section 1405.3.2.

❖ The basic requirement is for Class I or II vapor retarders to be installed on the interior side of frame walls in Climate Zones 5, 6, 7, 8 and Marine 4. These climate zones are defined in the IECC. Figure 301.1 and Table 301.1 of the IECC provide information regarding the climate zones for all geographic locations in the United States. These are reproduced here for convenience [see Commentary Figures 1405.3.1(1) and 1405.3.1(2)]. Also, Tables 301.3(1) and 301.3(2) of the IECC define international climate zones outside of the United States. These are reprinted in this commentary as well [see Commentary Figures 1405.3.1(3) and 1405.3.1(4)]. As can be seen by studying Commentary Figure 1405.3.1(1), Climate Zones 5, 6, 7, 8 and Marine 4 are in the middle to northern portions of the continental United States and Alaska. These are areas where colder temperatures can be expected in winter months, causing moisture from the interior of the building to form condensate in the exterior walls of the building. A Class I or II vapor retarder is therefore called for on the interior side of the exterior wall.

TABLE 1405.2
MINIMUM THICKNESS OF WEATHER COVERINGS

COVERING TYPE	MINIMUM THICKNESS (inches)
Adhered masonry veneer	0.25
Aluminum siding	0.019
Anchored masonry veneer	2.625
Asbestos-cement boards	0.125
Asbestos shingles	0.156
Cold-rolled copper[d]	0.0216 nominal
Copper shingles[d]	0.0162 nominal
Exterior plywood (with sheathing)	0.313
Exterior plywood (without sheathing)	See Section 2304.6
Fiber cement lap siding	0.25[c]
Fiber cement panel siding	0.25[c]
Fiberboard siding	0.5
Glass-fiber reinforced concrete panels	0.375
Hardboard siding[c]	0.25
High-yield copper[d]	0.0162 nominal
Lead-coated copper[d]	0.0216 nominal
Lead-coated high-yield copper	0.0162 nominal
Marble slabs	1
Particleboard (with sheathing)	See Section 2304.6
Particleboard (without sheathing)	See Section 2304.6
Porcelain tile	0.25
Steel (approved corrosion resistant)	0.0149
Stone (cast artificial, anchored)	1.5
Stone (natural)	2
Structural glass	0.344
Stucco or exterior cement plaster	
Three-coat work over:	
Metal plaster base	0.875[b]
Unit masonry	0.625[b]
Cast-in-place or precast concrete	0.625[b]
Two-coat work over:	
Unit masonry	0.5[b]
Cast-in-place or precast concrete	0.375[b]
Terra cotta (anchored)	1
Terra cotta (adhered)	0.25
Vinyl siding	0.035
Wood shingles	0.375
Wood siding (without sheathing)[a]	0.5

For SI: 1 inch = 25.4 mm, 1 ounce = 28.35 g, 1 square foot = 0.093 m^2.

a. Wood siding of thicknesses less than 0.5 inch shall be placed over sheathing that conforms to Section 2304.6.

b. Exclusive of texture.

c. As measured at the bottom of decorative grooves.

d. 16 ounces per square foot for cold-rolled copper and lead-coated copper, 12 ounces per square foot for copper shingles, high-yield copper and lead-coated high-yield copper.

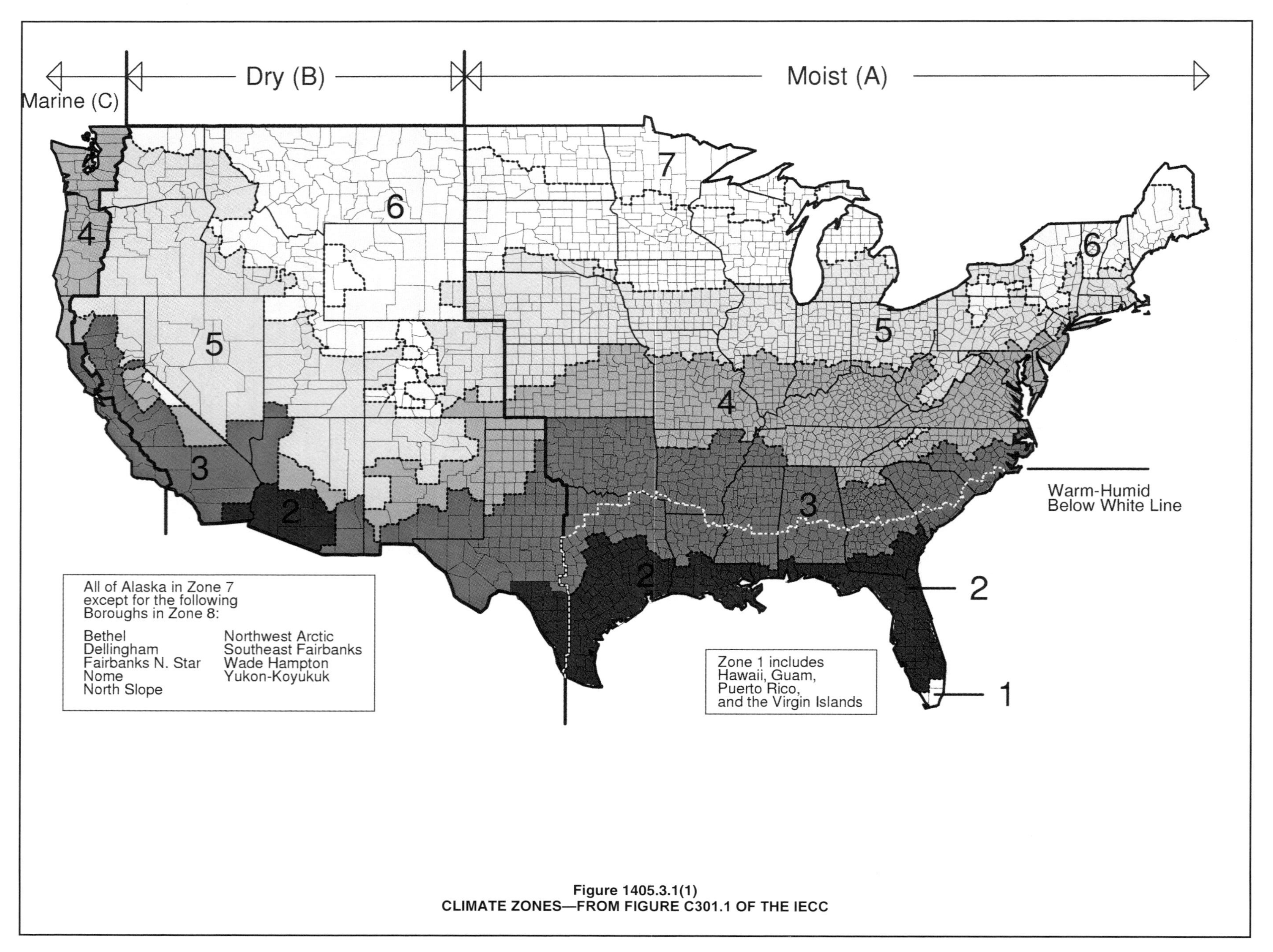

Figure 1405.3.1(1)
CLIMATE ZONES—FROM FIGURE C301.1 OF THE IECC

Key: A – Moist, B – Dry, C – Marine. Absence of moisture designation indicates moisture regime is irrelevant. Asterisk (*) indicates a warm-humid location.

US STATES

ALABAMA

3A Autauga*
2A Baldwin*
3A Barbour*
3A Bibb
3A Blount
3A Bullock*
3A Butler*
3A Calhoun
3A Chambers
3A Cherokee
3A Chilton
3A Choctaw*
3A Clarke*
3A Clay
3A Cleburne
3A Coffee*
3A Colbert
3A Conecuh*
3A Coosa
3A Covington*
3A Crenshaw*
3A Cullman
3A Dale*
3A Dallas*
3A DeKalb
3A Elmore*
3A Escambia*
3A Etowah
3A Fayette
3A Franklin
3A Geneva*
3A Greene
3A Hale
3A Henry*
3A Houston*
3A Jackson
3A Jefferson
3A Lamar
3A Lauderdale
3A Lawrence
3A Lee
3A Limestone
3A Lowndes*
3A Macon*
3A Madison
3A Marengo*
3A Marion
3A Marshall
2A Mobile*
3A Monroe*
3A Montgomery*
3A Morgan
3A Perry*
3A Pickens
3A Pike*
3A Randolph
3A Russell*
3A Shelby
3A St. Clair
3A Sumter
3A Talladega
3A Tallapoosa
3A Tuscaloosa
3A Walker
3A Washington*
3A Wilcox*
3A Winston

ALASKA

7 Aleutians East
7 Aleutians West
7 Anchorage
8 Bethel
7 Bristol Bay
7 Denali
8 Dillingham
8 Fairbanks North Star
7 Haines
7 Juneau
7 Kenai Peninsula
7 Ketchikan Gateway
7 Kodiak Island
7 Lake and Peninsula
7 Matanuska-Susitna
8 Nome
8 North Slope
8 Northwest Arctic
7 Prince of Wales Outer Ketchikan
7 Sitka
7 Skagway-Hoonah-Angoon
8 Southeast Fairbanks
7 Valdez-Cordova
8 Wade Hampton
7 Wrangell-Petersburg
7 Yakutat
8 Yukon-Koyukuk

ARIZONA

5B Apache
3B Cochise
5B Coconino
4B Gila
3B Graham
3B Greenlee
2B La Paz
2B Maricopa
3B Mohave
5B Navajo
2B Pima
2B Pinal
3B Santa Cruz
4B Yavapai
2B Yuma

ARKANSAS

3A Arkansas
3A Ashley
4A Baxter
4A Benton
4A Boone
3A Bradley
3A Calhoun
4A Carroll
3A Chicot
3A Clark
3A Clay
3A Cleburne
3A Cleveland
3A Columbia*
3A Conway
3A Craighead
3A Crawford
3A Crittenden
3A Cross
3A Dallas
3A Desha
3A Drew
3A Faulkner
3A Franklin
4A Fulton
3A Garland
3A Grant
3A Greene
3A Hempstead*
3A Hot Spring
3A Howard
3A Independence
4A Izard
3A Jackson
3A Jefferson
3A Johnson
3A Lafayette*
3A Lawrence
3A Lee
3A Lincoln
3A Little River*
3A Logan
3A Lonoke
4A Madison
4A Marion
3A Miller*
3A Mississippi
3A Monroe
3A Montgomery
3A Nevada
4A Newton
3A Ouachita
3A Perry
3A Phillips
3A Pike
3A Poinsett
3A Polk
3A Pope
3A Prairie
3A Pulaski
3A Randolph
3A Saline
3A Scott
4A Searcy
3A Sebastian
3A Sevier*
3A Sharp
3A St. Francis
4A Stone
3A Union*
3A Van Buren
4A Washington
3A White
3A Woodruff
3A Yell

CALIFORNIA

3C Alameda
6B Alpine
4B Amador
3B Butte
4B Calaveras
3B Colusa
3B Contra Costa
4C Del Norte
4B El Dorado
3B Fresno
3B Glenn

Figure 1405.3.1(2)
CLIMATE ZONES, MOISTURE REGIMES, AND WARM-HUMID DESIGNATIONS BY STATE, COUNTY AND TERRITORY—FROM TABLE C301.1 OF THE IECC

(continued)

4C Humboldt
2B Imperial
4B Inyo
3B Kern
3B Kings
4B Lake
5B Lassen
3B Los Angeles
3B Madera
3C Marin
4B Mariposa
3C Mendocino
3B Merced
5B Modoc
6B Mono
3C Monterey
3C Napa
5B Nevada
3B Orange
3B Placer
5B Plumas
3B Riverside
3B Sacramento
3C San Benito
3B San Bernardino
3B San Diego
3C San Francisco
3B San Joaquin
3C San Luis Obispo
3C San Mateo
3C Santa Barbara
3C Santa Clara
3C Santa Cruz
3B Shasta
5B Sierra
5B Siskiyou
3B Solano
3C Sonoma
3B Stanislaus
3B Sutter
3B Tehama
4B Trinity
3B Tulare
4B Tuolumne
3C Ventura
3B Yolo
3B Yuba

COLORADO

5B Adams
6B Alamosa
5B Arapahoe
6B Archuleta
4B Baca
5B Bent
5B Boulder
6B Chaffee
5B Cheyenne
7 Clear Creek
6B Conejos
6B Costilla
5B Crowley
6B Custer
5B Delta
5B Denver
6B Dolores
5B Douglas
6B Eagle
5B Elbert
5B El Paso
5B Fremont
5B Garfield
5B Gilpin
7 Grand
7 Gunnison
7 Hinsdale
5B Huerfano
7 Jackson
5B Jefferson
5B Kiowa
5B Kit Carson
7 Lake
5B La Plata
5B Larimer
4B Las Animas
5B Lincoln
5B Logan
5B Mesa
7 Mineral
6B Moffat
5B Montezuma
5B Montrose
5B Morgan
4B Otero
6B Ouray
7 Park
5B Phillips
7 Pitkin
5B Prowers
5B Pueblo
6B Rio Blanco
7 Rio Grande
7 Routt
6B Saguache
7 San Juan
6B San Miguel
5B Sedgwick
7 Summit
5B Teller
5B Washington
5B Weld
5B Yuma

CONNECTICUT

5A (all)

DELAWARE

4A (all)

DISTRICT OF COLUMBIA

4A (all)

FLORIDA

2A Alachua*
2A Baker*
2A Bay*
2A Bradford*
2A Brevard*
1A Broward*
2A Calhoun*
2A Charlotte*
2A Citrus*
2A Clay*
2A Collier*
2A Columbia*
2A DeSoto*
2A Dixie*
2A Duval*
2A Escambia*
2A Flagler*
2A Franklin*
2A Gadsden*
2A Gilchrist*
2A Glades*
2A Gulf*
2A Hamilton*
2A Hardee*
2A Hendry*
2A Hernando*
2A Highlands*
2A Hillsborough*
2A Holmes*
2A Indian River*
2A Jackson*
2A Jefferson*
2A Lafayette*
2A Lake*
2A Lee*
2A Leon*
2A Levy*
2A Liberty*
2A Madison*
2A Manatee*
2A Marion*
2A Martin*
1A Miami-Dade*
1A Monroe*
2A Nassau*
2A Okaloosa*
2A Okeechobee*
2A Orange*
2A Osceola*
2A Palm Beach*
2A Pasco*
2A Pinellas*
2A Polk*
2A Putnam*
2A Santa Rosa*
2A Sarasota*
2A Seminole*
2A St. Johns*
2A St. Lucie*
2A Sumter*
2A Suwannee*
2A Taylor*
2A Union*
2A Volusia*
2A Wakulla*
2A Walton*
2A Washington*

GEORGIA

2A Appling*
2A Atkinson*
2A Bacon*
2A Baker*
3A Baldwin
4A Banks
3A Barrow
3A Bartow
3A Ben Hill*
2A Berrien*
3A Bibb
3A Bleckley*
2A Brantley*
2A Brooks*
2A Bryan*
3A Bulloch*
3A Burke
3A Butts
3A Calhoun*
2A Camden*
3A Candler*
3A Carroll
4A Catoosa
2A Charlton*
2A Chatham*
3A Chattahoochee*
4A Chattooga
3A Cherokee
3A Clarke
3A Clay*
3A Clayton
2A Clinch*
3A Cobb
3A Coffee*
2A Colquitt*
3A Columbia
2A Cook*
3A Coweta

Figure 1405.3.1(2)
CLIMATE ZONES, MOISTURE REGIMES, AND WARM-HUMID DESIGNATIONS BY STATE, COUNTY AND TERRITORY—FROM TABLE C301.1 OF THE IECC

(continued)

3A Crawford
3A Crisp*
4A Dade
4A Dawson
2A Decatur*
3A DeKalb
3A Dodge*
3A Dooly*
3A Dougherty*
3A Douglas
3A Early*
2A Echols*
2A Effingham*
3A Elbert
3A Emanuel*
2A Evans*
4A Fannin
3A Fayette
4A Floyd
3A Forsyth
4A Franklin
3A Fulton
4A Gilmer
3A Glascock
2A Glynn*
4A Gordon
2A Grady*
3A Greene
3A Gwinnett
4A Habersham
4A Hall
3A Hancock
3A Haralson
3A Harris
3A Hart
3A Heard
3A Henry
3A Houston*
3A Irwin*
3A Jackson
3A Jasper
2A Jeff Davis*
3A Jefferson
3A Jenkins*
3A Johnson*
3A Jones
3A Lamar
2A Lanier*
3A Laurens*
3A Lee*
2A Liberty*
3A Lincoln
2A Long*
2A Lowndes*
4A Lumpkin
3A Macon*
3A Madison
3A Marion*
3A McDuffie
2A McIntosh*
3A Meriwether
2A Miller*
2A Mitchell*
3A Monroe
3A Montgomery*
3A Morgan
4A Murray
3A Muscogee
3A Newton
3A Oconee
3A Oglethorpe
3A Paulding
3A Peach*
4A Pickens
2A Pierce*
3A Pike
3A Polk
3A Pulaski*
3A Putnam
3A Quitman*
4A Rabun
3A Randolph*
3A Richmond
3A Rockdale
3A Schley*
3A Screven*
2A Seminole*
3A Spalding
4A Stephens
3A Stewart*
3A Sumter*
3A Talbot
3A Taliaferro
2A Tattnall*
3A Taylor*
3A Telfair*
3A Terrell*
2A Thomas*
3A Tift*
2A Toombs*
4A Towns
3A Treutlen*
3A Troup
3A Turner*
3A Twiggs*
4A Union
3A Upson
4A Walker
3A Walton
2A Ware*
3A Warren
3A Washington
2A Wayne*
3A Webster*
3A Wheeler*
4A White
4A Whitfield
3A Wilcox*
3A Wilkes
3A Wilkinson
3A Worth*

HAWAII

1A (all)*

IDAHO

5B Ada
6B Adams
6B Bannock
6B Bear Lake
5B Benewah
6B Bingham
6B Blaine
6B Boise
6B Bonner
6B Bonneville
6B Boundary
6B Butte
6B Camas
5B Canyon
6B Caribou
5B Cassia
6B Clark
5B Clearwater
6B Custer
5B Elmore
6B Franklin
6B Fremont
5B Gem
5B Gooding
5B Idaho
6B Jefferson
5B Jerome
5B Kootenai
5B Latah
6B Lemhi
5B Lewis
5B Lincoln
6B Madison
5B Minidoka
5B Nez Perce
6B Oneida
5B Owyhee
5B Payette
5B Power
5B Shoshone
6B Teton
5B Twin Falls
6B Valley
5B Washington

ILLINOIS

5A Adams
4A Alexander
4A Bond
5A Boone
5A Brown
5A Bureau
5A Calhoun
5A Carroll
5A Cass
5A Champaign
4A Christian
5A Clark
4A Clay
4A Clinton
5A Coles
5A Cook
4A Crawford
5A Cumberland
5A DeKalb
5A De Witt
5A Douglas
5A DuPage
5A Edgar
4A Edwards
4A Effingham
4A Fayette
5A Ford
4A Franklin
5A Fulton
4A Gallatin
5A Greene
5A Grundy
4A Hamilton
5A Hancock
4A Hardin
5A Henderson
5A Henry
5A Iroquois
4A Jackson
4A Jasper
4A Jefferson
5A Jersey
5A Jo Daviess
4A Johnson
5A Kane
5A Kankakee
5A Kendall
5A Knox
5A Lake
5A La Salle
4A Lawrence
5A Lee
5A Livingston
5A Logan
5A Macon

Figure 1405.3.1(2)
CLIMATE ZONES, MOISTURE REGIMES, AND WARM-HUMID DESIGNATIONS BY STATE, COUNTY AND TERRITORY—FROM TABLE C301.1 OF THE IECC

(continued)

4A Macoupin
4A Madison
4A Marion
5A Marshall
5A Mason
4A Massac
5A McDonough
5A McHenry
5A McLean
5A Menard
5A Mercer
4A Monroe
4A Montgomery
5A Morgan
5A Moultrie
5A Ogle
5A Peoria
4A Perry
5A Piatt
5A Pike
4A Pope
4A Pulaski
5A Putnam
4A Randolph
4A Richland
5A Rock Island
4A Saline
5A Sangamon
5A Schuyler
5A Scott
4A Shelby
5A Stark
4A St. Clair
5A Stephenson
5A Tazewell
4A Union
5A Vermilion
4A Wabash
5A Warren
4A Washington
4A Wayne
4A White
5A Whiteside
5A Will
4A Williamson
5A Winnebago
5A Woodford

INDIANA

5A Adams
5A Allen
5A Bartholomew
5A Benton
5A Blackford
5A Boone
4A Brown
5A Carroll
5A Cass
4A Clark
5A Clay
5A Clinton
4A Crawford
4A Daviess
4A Dearborn
5A Decatur
5A De Kalb
5A Delaware
4A Dubois
5A Elkhart
5A Fayette
4A Floyd
5A Fountain
5A Franklin
5A Fulton
4A Gibson
5A Grant
4A Greene
5A Hamilton
5A Hancock
4A Harrison
5A Hendricks
5A Henry
5A Howard
5A Huntington
4A Jackson
5A Jasper
5A Jay
4A Jefferson
4A Jennings
5A Johnson
4A Knox
5A Kosciusko
5A Lagrange
5A Lake
5A La Porte
4A Lawrence
5A Madison
5A Marion
5A Marshall
4A Martin
5A Miami
4A Monroe
5A Montgomery
5A Morgan
5A Newton
5A Noble
4A Ohio
4A Orange
5A Owen
5A Parke
4A Perry
4A Pike
5A Porter
4A Posey
5A Pulaski
5A Putnam
5A Randolph
4A Ripley
5A Rush
4A Scott
5A Shelby
4A Spencer
5A Starke
5A Steuben
5A St. Joseph
4A Sullivan
4A Switzerland
5A Tippecanoe
5A Tipton
5A Union
4A Vanderburgh
5A Vermillion
5A Vigo
5A Wabash
5A Warren
4A Warrick
4A Washington
5A Wayne
5A Wells
5A White
5A Whitley

IOWA

5A Adair
5A Adams
6A Allamakee
5A Appanoose
5A Audubon
5A Benton
6A Black Hawk
5A Boone
6A Bremer
6A Buchanan
6A Buena Vista
6A Butler
6A Calhoun
5A Carroll
5A Cass
5A Cedar
6A Cerro Gordo
6A Cherokee
6A Chickasaw
5A Clarke
6A Clay
6A Clayton
5A Clinton
5A Crawford
5A Dallas
5A Davis
5A Decatur
6A Delaware
5A Des Moines
6A Dickinson
5A Dubuque
6A Emmet
6A Fayette
6A Floyd
6A Franklin
5A Fremont
5A Greene
6A Grundy
5A Guthrie
6A Hamilton
6A Hancock
6A Hardin
5A Harrison
5A Henry
6A Howard
6A Humboldt
6A Ida
5A Iowa
5A Jackson
5A Jasper
5A Jefferson
5A Johnson
5A Jones
5A Keokuk
6A Kossuth
5A Lee
5A Linn
5A Louisa
5A Lucas
6A Lyon
5A Madison
5A Mahaska
5A Marion
5A Marshall
5A Mills
6A Mitchell
5A Monona
5A Monroe
5A Montgomery
5A Muscatine
6A O'Brien
6A Osceola
5A Page
6A Palo Alto
6A Plymouth
6A Pocahontas
5A Polk
5A Pottawattamie
5A Poweshiek
5A Ringgold
6A Sac
5A Scott

Figure 1405.3.1(2)
CLIMATE ZONES, MOISTURE REGIMES, AND WARM-HUMID DESIGNATIONS BY STATE, COUNTY AND TERRITORY—FROM TABLE C301.1 OF THE IECC

(continued)

5A Shelby
6A Sioux
5A Story
5A Tama
5A Taylor
5A Union
5A Van Buren
5A Wapello
5A Warren
5A Washington
5A Wayne
6A Webster
6A Winnebago
6A Winneshiek
5A Woodbury
6A Worth
6A Wright

KANSAS

4A Allen
4A Anderson
4A Atchison
4A Barber
4A Barton
4A Bourbon
4A Brown
4A Butler
4A Chase
4A Chautauqua
4A Cherokee
5A Cheyenne
4A Clark
4A Clay
5A Cloud
4A Coffey
4A Comanche
4A Cowley
4A Crawford
5A Decatur
4A Dickinson
4A Doniphan
4A Douglas
4A Edwards
4A Elk
5A Ellis
4A Ellsworth
4A Finney
4A Ford
4A Franklin
4A Geary
5A Gove
5A Graham
4A Grant
4A Gray
5A Greeley
4A Greenwood
5A Hamilton
4A Harper
4A Harvey
4A Haskell
4A Hodgeman
4A Jackson
4A Jefferson
5A Jewell
4A Johnson
4A Kearny
4A Kingman
4A Kiowa
4A Labette
5A Lane
4A Leavenworth
4A Lincoln
4A Linn
5A Logan
4A Lyon
4A Marion
4A Marshall
4A McPherson
4A Meade
4A Miami
5A Mitchell
4A Montgomery
4A Morris
4A Morton
4A Nemaha
4A Neosho
5A Ness
5A Norton
4A Osage
5A Osborne
4A Ottawa
4A Pawnee
5A Phillips
4A Pottawatomie
4A Pratt
5A Rawlins
4A Reno
5A Republic
4A Rice
4A Riley
5A Rooks
4A Rush
4A Russell
4A Saline
5A Scott
4A Sedgwick
4A Seward
4A Shawnee
5A Sheridan
5A Sherman
5A Smith
4A Stafford
4A Stanton
4A Stevens
4A Sumner
5A Thomas
5A Trego
4A Wabaunsee
5A Wallace
4A Washington
5A Wichita
4A Wilson
4A Woodson
4A Wyandotte

KENTUCKY

4A (all)

LOUISIANA

2A Acadia*
2A Allen*
2A Ascension*
2A Assumption*
2A Avoyelles*
2A Beauregard*
3A Bienville*
3A Bossier*
3A Caddo*
2A Calcasieu*
3A Caldwell*
2A Cameron*
3A Catahoula*
3A Claiborne*
3A Concordia*
3A De Soto*
2A East Baton Rouge*
3A East Carroll
2A East Feliciana*
2A Evangeline*
3A Franklin*
3A Grant*
2A Iberia*
2A Iberville*
3A Jackson*
2A Jefferson*
2A Jefferson Davis*
2A Lafayette*
2A Lafourche*
3A La Salle*
3A Lincoln*
2A Livingston*
3A Madison*
3A Morehouse
3A Natchitoches*
2A Orleans*
3A Ouachita*
2A Plaquemines*
2A Pointe Coupee*
2A Rapides*
3A Red River*
3A Richland*
3A Sabine*
2A St. Bernard*
2A St. Charles*
2A St. Helena*
2A St. James*
2A St. John the Baptist*
2A St. Landry*
2A St. Martin*
2A St. Mary*
2A St. Tammany*
2A Tangipahoa*
3A Tensas*
2A Terrebonne*
3A Union*
2A Vermilion*
3A Vernon*
2A Washington*
3A Webster*
2A West Baton Rouge*
3A West Carroll
2A West Feliciana*
3A Winn*

MAINE

6A Androscoggin
7 Aroostook
6A Cumberland
6A Franklin
6A Hancock
6A Kennebec
6A Knox
6A Lincoln
6A Oxford
6A Penobscot
6A Piscataquis
6A Sagadahoc
6A Somerset
6A Waldo
6A Washington
6A York

MARYLAND

4A Allegany
4A Anne Arundel
4A Baltimore
4A Baltimore (city)
4A Calvert
4A Caroline
4A Carroll
4A Cecil
4A Charles
4A Dorchester
4A Frederick
5A Garrett
4A Harford

Figure 1405.3.1(2)
CLIMATE ZONES, MOISTURE REGIMES, AND WARM-HUMID DESIGNATIONS BY STATE, COUNTY AND TERRITORY—FROM TABLE C301.1 OF THE IECC

(continued)

4A Howard
4A Kent
4A Montgomery
4A Prince George's
4A Queen Anne's
4A Somerset
4A St. Mary's
4A Talbot
4A Washington
4A Wicomico
4A Worcester

MASSACHSETTS

5A (all)

MICHIGAN

6A Alcona
6A Alger
5A Allegan
6A Alpena
6A Antrim
6A Arenac
7 Baraga
5A Barry
5A Bay
6A Benzie
5A Berrien
5A Branch
5A Calhoun
5A Cass
6A Charlevoix
6A Cheboygan
7 Chippewa
6A Clare
5A Clinton
6A Crawford
6A Delta
6A Dickinson
5A Eaton
6A Emmet
5A Genesee
6A Gladwin
7 Gogebic
6A Grand Traverse
5A Gratiot
5A Hillsdale
7 Houghton
6A Huron
5A Ingham
5A Ionia
6A Iosco
7 Iron
6A Isabella
5A Jackson
5A Kalamazoo
6A Kalkaska
5A Kent
7 Keweenaw
6A Lake
5A Lapeer
6A Leelanau
5A Lenawee
5A Livingston
7 Luce
7 Mackinac
5A Macomb
6A Manistee
6A Marquette
6A Mason
6A Mecosta
6A Menominee
5A Midland
6A Missaukee
5A Monroe
5A Montcalm
6A Montmorency
5A Muskegon
6A Newaygo
5A Oakland
6A Oceana
6A Ogemaw
7 Ontonagon
6A Osceola
6A Oscoda
6A Otsego
5A Ottawa
6A Presque Isle
6A Roscommon
5A Saginaw
6A Sanilac
7 Schoolcraft
5A Shiawassee
5A St. Clair
5A St. Joseph
5A Tuscola
5A Van Buren
5A Washtenaw
5A Wayne
6A Wexford

MINNESOTA

7 Aitkin
6A Anoka
7 Becker
7 Beltrami
6A Benton
6A Big Stone
6A Blue Earth
6A Brown
7 Carlton
6A Carver
7 Cass
6A Chippewa
6A Chisago
7 Clay
7 Clearwater
7 Cook
6A Cottonwood
7 Crow Wing
6A Dakota
6A Dodge
6A Douglas
6A Faribault
6A Fillmore
6A Freeborn
6A Goodhue
7 Grant
6A Hennepin
6A Houston
7 Hubbard
6A Isanti
7 Itasca
6A Jackson
7 Kanabec
6A Kandiyohi
7 Kittson
7 Koochiching
6A Lac qui Parle
7 Lake
7 Lake of the Woods
6A Le Sueur
6A Lincoln
6A Lyon
7 Mahnomen
7 Marshall
6A Martin
6A McLeod
6A Meeker
7 Mille Lacs
6A Morrison
6A Mower
6A Murray
6A Nicollet
6A Nobles
7 Norman
6A Olmsted
7 Otter Tail
7 Pennington
7 Pine
6A Pipestone
7 Polk
6A Pope
6A Ramsey
7 Red Lake
6A Redwood
6A Renville
6A Rice
6A Rock
7 Roseau
6A Scott
6A Sherburne
6A Sibley
6A Stearns
6A Steele
6A Stevens
7 St. Louis
6A Swift
6A Todd
6A Traverse
6A Wabasha
7 Wadena
6A Waseca
6A Washington
6A Watonwan
7 Wilkin
6A Winona
6A Wright
6A Yellow Medicine

MISSISSIPPI

3A Adams*
3A Alcorn
3A Amite*
3A Attala
3A Benton
3A Bolivar
3A Calhoun
3A Carroll
3A Chickasaw
3A Choctaw
3A Claiborne*
3A Clarke
3A Clay
3A Coahoma
3A Copiah*
3A Covington*
3A DeSoto
3A Forrest*
3A Franklin*
3A George*
3A Greene*
3A Grenada
2A Hancock*
2A Harrison*
3A Hinds*
3A Holmes
3A Humphreys
3A Issaquena
3A Itawamba
2A Jackson*
3A Jasper
3A Jefferson*
3A Jefferson Davis*
3A Jones*
3A Kemper

Figure 1405.3.1(2)
CLIMATE ZONES, MOISTURE REGIMES, AND WARM-HUMID DESIGNATIONS BY STATE, COUNTY AND TERRITORY—FROM TABLE C301.1 OF THE IECC

(continued)

3A Lafayette
3A Lamar*
3A Lauderdale
3A Lawrence*
3A Leake
3A Lee
3A Leflore
3A Lincoln*
3A Lowndes
3A Madison
3A Marion*
3A Marshall
3A Monroe
3A Montgomery
3A Neshoba
3A Newton
3A Noxubee
3A Oktibbeha
3A Panola
2A Pearl River*
3A Perry*
3A Pike*
3A Pontotoc
3A Prentiss
3A Quitman
3A Rankin*
3A Scott
3A Sharkey
3A Simpson*
3A Smith*
2A Stone*
3A Sunflower
3A Tallahatchie
3A Tate
3A Tippah
3A Tishomingo
3A Tunica
3A Union
3A Walthall*
3A Warren*
3A Washington
3A Wayne*
3A Webster
3A Wilkinson*
3A Winston
3A Yalobusha
3A Yazoo

MISSOURI

5A Adair
5A Andrew
5A Atchison
4A Audrain
4A Barry
4A Barton
4A Bates
4A Benton
4A Bollinger
4A Boone
5A Buchanan
4A Butler
5A Caldwell
4A Callaway
4A Camden
4A Cape Girardeau
4A Carroll
4A Carter
4A Cass
4A Cedar
5A Chariton
4A Christian
5A Clark
4A Clay
5A Clinton
4A Cole
4A Cooper
4A Crawford
4A Dade
4A Dallas
5A Daviess
5A DeKalb
4A Dent
4A Douglas
4A Dunklin
4A Franklin
4A Gasconade
5A Gentry
4A Greene
5A Grundy
5A Harrison
4A Henry
4A Hickory
5A Holt
4A Howard
4A Howell
4A Iron
4A Jackson
4A Jasper
4A Jefferson
4A Johnson
5A Knox
4A Laclede
4A Lafayette
4A Lawrence
5A Lewis
4A Lincoln
5A Linn
5A Livingston
5A Macon
4A Madison
4A Maries
5A Marion
4A McDonald
5A Mercer
4A Miller
4A Mississippi
4A Moniteau
4A Monroe
4A Montgomery
4A Morgan
4A New Madrid
4A Newton
5A Nodaway
4A Oregon
4A Osage
4A Ozark
4A Pemiscot
4A Perry
4A Pettis
4A Phelps
5A Pike
4A Platte
4A Polk
4A Pulaski
5A Putnam
5A Ralls
4A Randolph
4A Ray
4A Reynolds
4A Ripley
4A Saline
5A Schuyler
5A Scotland
4A Scott
4A Shannon
5A Shelby
4A St. Charles
4A St. Clair
4A Ste. Genevieve
4A St. Francois
4A St. Louis
4A St. Louis (city)
4A Stoddard
4A Stone
5A Sullivan
4A Taney
4A Texas
4A Vernon
4A Warren
4A Washington
4A Wayne
4A Webster
5A Worth
4A Wright

MONTANA

6B (all)

NEBRASKA

5A (all)

NEVADA

5B Carson City (city)
5B Churchill
3B Clark
5B Douglas
5B Elko
5B Esmeralda
5B Eureka
5B Humboldt
5B Lander
5B Lincoln
5B Lyon
5B Mineral
5B Nye
5B Pershing
5B Storey
5B Washoe
5B White Pine

NEW HAMPSHIRE

6A Belknap
6A Carroll
5A Cheshire
6A Coos
6A Grafton
5A Hillsborough
6A Merrimack
5A Rockingham
5A Strafford
6A Sullivan

NEW JERSEY

4A Atlantic
5A Bergen
4A Burlington
4A Camden
4A Cape May
4A Cumberland
4A Essex
4A Gloucester
4A Hudson
5A Hunterdon
5A Mercer
4A Middlesex
4A Monmouth
5A Morris
4A Ocean
5A Passaic
4A Salem
5A Somerset
5A Sussex
4A Union
5A Warren

Figure 1405.3.1(2)
CLIMATE ZONES, MOISTURE REGIMES, AND WARM-HUMID DESIGNATIONS BY STATE, COUNTY AND TERRITORY—FROM TABLE C301.1 OF THE IECC

(continued)

NEW MEXICO

4B Bernalillo
5B Catron
3B Chaves
4B Cibola
5B Colfax
4B Curry
4B DeBaca
3B Dona Ana
3B Eddy
4B Grant
4B Guadalupe
5B Harding
3B Hidalgo
3B Lea
4B Lincoln
5B Los Alamos
3B Luna
5B McKinley
5B Mora
3B Otero
4B Quay
5B Rio Arriba
4B Roosevelt
5B Sandoval
5B San Juan
5B San Miguel
5B Santa Fe
4B Sierra
4B Socorro
5B Taos
5B Torrance
4B Union
4B Valencia

NEW YORK

5A Albany
6A Allegany
4A Bronx
6A Broome
6A Cattaraugus
5A Cayuga
5A Chautauqua
5A Chemung
6A Chenango
6A Clinton
5A Columbia
5A Cortland
6A Delaware
5A Dutchess
5A Erie
6A Essex
6A Franklin
6A Fulton
5A Genesee
5A Greene
6A Hamilton
6A Herkimer
6A Jefferson
4A Kings
6A Lewis
5A Livingston
6A Madison
5A Monroe
6A Montgomery
4A Nassau
4A New York
5A Niagara
6A Oneida
5A Onondaga
5A Ontario
5A Orange
5A Orleans
5A Oswego
6A Otsego
5A Putnam
4A Queens
5A Rensselaer
4A Richmond
5A Rockland
5A Saratoga
5A Schenectady
6A Schoharie
6A Schuyler
5A Seneca
6A Steuben
6A St. Lawrence
4A Suffolk
6A Sullivan
5A Tioga
6A Tompkins
6A Ulster
6A Warren
5A Washington
5A Wayne
4A Westchester
6A Wyoming
5A Yates

NORTH CAROLINA

4A Alamance
4A Alexander
5A Alleghany
3A Anson
5A Ashe
5A Avery
3A Beaufort
4A Bertie
3A Bladen
3A Brunswick*
4A Buncombe
4A Burke
3A Cabarrus
4A Caldwell
3A Camden
3A Carteret*
4A Caswell
4A Catawba
4A Chatham
4A Cherokee
3A Chowan
4A Clay
4A Cleveland
3A Columbus*
3A Craven
3A Cumberland
3A Currituck
3A Dare
3A Davidson
4A Davie
3A Duplin
4A Durham
3A Edgecombe
4A Forsyth
4A Franklin
3A Gaston
4A Gates
4A Graham
4A Granville
3A Greene
4A Guilford
4A Halifax
4A Harnett
4A Haywood
4A Henderson
4A Hertford
3A Hoke
3A Hyde
4A Iredell
4A Jackson
3A Johnston
3A Jones
4A Lee
3A Lenoir
4A Lincoln
4A Macon
4A Madison
3A Martin
4A McDowell
3A Mecklenburg
5A Mitchell
3A Montgomery
3A Moore
4A Nash
3A New Hanover*
4A Northampton
3A Onslow*
4A Orange
3A Pamlico
3A Pasquotank
3A Pender*
3A Perquimans
4A Person
3A Pitt
4A Polk
3A Randolph
3A Richmond
3A Robeson
4A Rockingham
3A Rowan
4A Rutherford
3A Sampson
3A Scotland
3A Stanly
4A Stokes
4A Surry
4A Swain
4A Transylvania
3A Tyrrell
3A Union
4A Vance
4A Wake
4A Warren
3A Washington
5A Watauga
3A Wayne
4A Wilkes
3A Wilson
4A Yadkin
5A Yancey

NORTH DAKOTA

6A Adams
7 Barnes
7 Benson
6A Billings
7 Bottineau
6A Bowman
7 Burke
6A Burleigh
7 Cass
7 Cavalier
6A Dickey
7 Divide
6A Dunn
7 Eddy
6A Emmons
7 Foster
6A Golden Valley
7 Grand Forks
6A Grant
7 Griggs
6A Hettinger
7 Kidder

Figure 1405.3.1(2)
CLIMATE ZONES, MOISTURE REGIMES, AND WARM-HUMID DESIGNATIONS BY STATE, COUNTY AND TERRITORY—FROM TABLE C301.1 OF THE IECC

(continued)

6A LaMoure
6A Logan
7 McHenry
6A McIntosh
6A McKenzie
7 McLean
6A Mercer
6A Morton
7 Mountrail
7 Nelson
6A Oliver
7 Pembina
7 Pierce
7 Ramsey
6A Ransom
7 Renville
6A Richland
7 Rolette
6A Sargent
7 Sheridan
6A Sioux
6A Slope
6A Stark
7 Steele
7 Stutsman
7 Towner
7 Traill
7 Walsh
7 Ward
7 Wells
7 Williams

OHIO

4A Adams
5A Allen
5A Ashland
5A Ashtabula
5A Athens
5A Auglaize
5A Belmont
4A Brown
5A Butler
5A Carroll
5A Champaign
5A Clark
4A Clermont
5A Clinton
5A Columbiana
5A Coshocton
5A Crawford
5A Cuyahoga
5A Darke
5A Defiance
5A Delaware
5A Erie
5A Fairfield
5A Fayette
5A Franklin
5A Fulton
4A Gallia
5A Geauga
5A Greene
5A Guernsey
4A Hamilton
5A Hancock
5A Hardin
5A Harrison
5A Henry
5A Highland
5A Hocking
5A Holmes
5A Huron
5A Jackson
5A Jefferson
5A Knox
5A Lake
4A Lawrence
5A Licking
5A Logan
5A Lorain
5A Lucas
5A Madison
5A Mahoning
5A Marion
5A Medina
5A Meigs
5A Mercer
5A Miami
5A Monroe
5A Montgomery
5A Morgan
5A Morrow
5A Muskingum
5A Noble
5A Ottawa
5A Paulding
5A Perry
5A Pickaway
4A Pike
5A Portage
5A Preble
5A Putnam
5A Richland
5A Ross
5A Sandusky
4A Scioto
5A Seneca
5A Shelby
5A Stark
5A Summit
5A Trumbull
5A Tuscarawas
5A Union
5A Van Wert
5A Vinton
5A Warren
4A Washington
5A Wayne
5A Williams
5A Wood
5A Wyandot

OKLAHOMA

3A Adair
3A Alfalfa
3A Atoka
4B Beaver
3A Beckham
3A Blaine
3A Bryan
3A Caddo
3A Canadian
3A Carter
3A Cherokee
3A Choctaw
4B Cimarron
3A Cleveland
3A Coal
3A Comanche
3A Cotton
3A Craig
3A Creek
3A Custer
3A Delaware
3A Dewey
3A Ellis
3A Garfield
3A Garvin
3A Grady
3A Grant
3A Greer
3A Harmon
3A Harper
3A Haskell
3A Hughes
3A Jackson
3A Jefferson
3A Johnston
3A Kay
3A Kingfisher
3A Kiowa
3A Latimer
3A Le Flore
3A Lincoln
3A Logan
3A Love
3A Major
3A Marshall
3A Mayes
3A McClain
3A McCurtain
3A McIntosh
3A Murray
3A Muskogee
3A Noble
3A Nowata
3A Okfuskee
3A Oklahoma
3A Okmulgee
3A Osage
3A Ottawa
3A Pawnee
3A Payne
3A Pittsburg
3A Pontotoc
3A Pottawatomie
3A Pushmataha
3A Roger Mills
3A Rogers
3A Seminole
3A Sequoyah
3A Stephens
4B Texas
3A Tillman
3A Tulsa
3A Wagoner
3A Washington
3A Washita
3A Woods
3A Woodward

OREGON

5B Baker
4C Benton
4C Clackamas
4C Clatsop
4C Columbia
4C Coos
5B Crook
4C Curry
5B Deschutes
4C Douglas
5B Gilliam
5B Grant
5B Harney
5B Hood River
4C Jackson
5B Jefferson
4C Josephine
5B Klamath
5B Lake
4C Lane
4C Lincoln
4C Linn
5B Malheur
4C Marion

Figure 1405.3.1(2)
CLIMATE ZONES, MOISTURE REGIMES, AND WARM-HUMID DESIGNATIONS BY STATE, COUNTY AND TERRITORY—FROM TABLE C301.1 OF THE IECC

(continued)

5B Morrow
4C Multnomah
4C Polk
5B Sherman
4C Tillamook
5B Umatilla
5B Union
5B Wallowa
5B Wasco
4C Washington
5B Wheeler
4C Yamhill

PENNSYLVANIA

5A Adams
5A Allegheny
5A Armstrong
5A Beaver
5A Bedford
5A Berks
5A Blair
5A Bradford
4A Bucks
5A Butler
5A Cambria
6A Cameron
5A Carbon
5A Centre
4A Chester
5A Clarion
6A Clearfield
5A Clinton
5A Columbia
5A Crawford
5A Cumberland
5A Dauphin
4A Delaware
6A Elk
5A Erie
5A Fayette
5A Forest
5A Franklin
5A Fulton
5A Greene
5A Huntingdon
5A Indiana
5A Jefferson
5A Juniata
5A Lackawanna
5A Lancaster
5A Lawrence
5A Lebanon
5A Lehigh
5A Luzerne
5A Lycoming
6A McKean
5A Mercer
5A Mifflin
5A Monroe
4A Montgomery
5A Montour
5A Northampton
5A Northumberland
5A Perry
4A Philadelphia
5A Pike
6A Potter
5A Schuylkill
5A Snyder
5A Somerset
5A Sullivan
6A Susquehanna
6A Tioga
5A Union
5A Venango
5A Warren
5A Washington
6A Wayne
5A Westmoreland
5A Wyoming
4A York

RHODE ISLAND

5A (all)

SOUTH CAROLINA

3A Abbeville
3A Aiken
3A Allendale*
3A Anderson
3A Bamberg*
3A Barnwell*
3A Beaufort*
3A Berkeley*
3A Calhoun
3A Charleston*
3A Cherokee
3A Chester
3A Chesterfield
3A Clarendon
3A Colleton*
3A Darlington
3A Dillon
3A Dorchester*
3A Edgefield
3A Fairfield
3A Florence
3A Georgetown*
3A Greenville
3A Greenwood
3A Hampton*
3A Horry*
3A Jasper*
3A Kershaw
3A Lancaster
3A Laurens
3A Lee
3A Lexington
3A Marion
3A Marlboro
3A McCormick
3A Newberry
3A Oconee
3A Orangeburg
3A Pickens
3A Richland
3A Saluda
3A Spartanburg
3A Sumter
3A Union
3A Williamsburg
3A York

SOUTH DAKOTA

6A Aurora
6A Beadle
5A Bennett
5A Bon Homme
6A Brookings
6A Brown
6A Brule
6A Buffalo
6A Butte
6A Campbell
5A Charles Mix
6A Clark
5A Clay
6A Codington
6A Corson
6A Custer
6A Davison
6A Day
6A Deuel
6A Dewey
5A Douglas
6A Edmunds
6A Fall River
6A Faulk
6A Grant
5A Gregory
6A Haakon
6A Hamlin
6A Hand
6A Hanson
6A Harding
6A Hughes
5A Hutchinson
6A Hyde
5A Jackson
6A Jerauld
6A Jones
6A Kingsbury
6A Lake
6A Lawrence
6A Lincoln
6A Lyman
6A Marshall
6A McCook
6A McPherson
6A Meade
5A Mellette
6A Miner
6A Minnehaha
6A Moody
6A Pennington
6A Perkins
6A Potter
6A Roberts
6A Sanborn
6A Shannon
6A Spink
6A Stanley
6A Sully
5A Todd
5A Tripp
6A Turner
5A Union
6A Walworth
5A Yankton
6A Ziebach

TENNESSEE

4A Anderson
4A Bedford
4A Benton
4A Bledsoe
4A Blount
4A Bradley
4A Campbell
4A Cannon
4A Carroll
4A Carter
4A Cheatham
3A Chester
4A Claiborne
4A Clay
4A Cocke
4A Coffee
3A Crockett
4A Cumberland
4A Davidson
4A Decatur
4A DeKalb
4A Dickson
3A Dyer

Figure 1405.3.1(2)
CLIMATE ZONES, MOISTURE REGIMES, AND WARM-HUMID DESIGNATIONS BY STATE, COUNTY AND TERRITORY—FROM TABLE C301.1 OF THE IECC

(continued)

3A Fayette
4A Fentress
4A Franklin
4A Gibson
4A Giles
4A Grainger
4A Greene
4A Grundy
4A Hamblen
4A Hamilton
4A Hancock
3A Hardeman
3A Hardin
4A Hawkins
3A Haywood
3A Henderson
4A Henry
4A Hickman
4A Houston
4A Humphreys
4A Jackson
4A Jefferson
4A Johnson
4A Knox
3A Lake
3A Lauderdale
4A Lawrence
4A Lewis
4A Lincoln
4A Loudon
4A Macon
3A Madison
4A Marion
4A Marshall
4A Maury
4A McMinn
3A McNairy
4A Meigs
4A Monroe
4A Montgomery
4A Moore
4A Morgan
4A Obion
4A Overton
4A Perry
4A Pickett
4A Polk
4A Putnam
4A Rhea
4A Roane
4A Robertson
4A Rutherford
4A Scott
4A Sequatchie
4A Sevier
3A Shelby
4A Smith
4A Stewart
4A Sullivan
4A Sumner
3A Tipton
4A Trousdale
4A Unicoi
4A Union
4A Van Buren
4A Warren
4A Washington
4A Wayne
4A Weakley
4A White
4A Williamson
4A Wilson

TEXAS

2A Anderson*
3B Andrews
2A Angelina*
2A Aransas*
3A Archer
4B Armstrong
2A Atascosa*
2A Austin*
4B Bailey
2B Bandera*
2A Bastrop*
3B Baylor
2A Bee*
2A Bell*
2A Bexar*
3A Blanco*
3B Borden
2A Bosque*
3A Bowie*
2A Brazoria*
2A Brazos*
3B Brewster
4B Briscoe
2A Brooks*
3A Brown*
2A Burleson*
3A Burnet*
2A Caldwell*
2A Calhoun*
3B Callahan
2A Cameron*
3A Camp*
4B Carson
3A Cass*
4B Castro
2A Chambers*
2A Cherokee*
3B Childress
3A Clay
4B Cochran
3B Coke
3B Coleman
3A Collin*
3B Collingsworth
2A Colorado*
2A Comal*
3A Comanche*
3B Concho
3A Cooke
2A Coryell*
3B Cottle
3B Crane
3B Crockett
3B Crosby
3B Culberson
4B Dallam
3A Dallas*
3B Dawson
4B Deaf Smith
3A Delta
3A Denton*
2A DeWitt*
3B Dickens
2B Dimmit*
4B Donley
2A Duval*
3A Eastland
3B Ector
2B Edwards*
3A Ellis*
3B El Paso
3A Erath*
2A Falls*
3A Fannin
2A Fayette*
3B Fisher
4B Floyd
3B Foard
2A Fort Bend*
3A Franklin*
2A Freestone*
2B Frio*
3B Gaines
2A Galveston*
3B Garza
3A Gillespie*
3B Glasscock
2A Goliad*
2A Gonzales*
4B Gray
3A Grayson
3A Gregg*
2A Grimes*
2A Guadalupe*
4B Hale
3B Hall
3A Hamilton*
4B Hansford
3B Hardeman
2A Hardin*
2A Harris*
3A Harrison*
4B Hartley
3B Haskell
2A Hays*
3B Hemphill
3A Henderson*
2A Hidalgo*
2A Hill*
4B Hockley
3A Hood*
3A Hopkins*
2A Houston*
3B Howard
3B Hudspeth
3A Hunt*
4B Hutchinson
3B Irion
3A Jack
2A Jackson*
2A Jasper*
3B Jeff Davis
2A Jefferson*
2A Jim Hogg*
2A Jim Wells*
3A Johnson*
3B Jones
2A Karnes*
3A Kaufman*
3A Kendall*
2A Kenedy*
3B Kent
3B Kerr
3B Kimble
3B King
2B Kinney*
2A Kleberg*
3B Knox
3A Lamar*
4B Lamb
3A Lampasas*
2B La Salle*
2A Lavaca*
2A Lee*
2A Leon*
2A Liberty*
2A Limestone*
4B Lipscomb
2A Live Oak*
3A Llano*
3B Loving

Figure 1405.3.1(2)
CLIMATE ZONES, MOISTURE REGIMES, AND WARM-HUMID DESIGNATIONS BY STATE, COUNTY AND TERRITORY—FROM TABLE C301.1 OF THE IECC

(continued)

3B Lubbock
3B Lynn
2A Madison*
3A Marion*
3B Martin
3B Mason
2A Matagorda*
2B Maverick*
3B McCulloch
2A McLennan*
2A McMullen*
2B Medina*
3B Menard
3B Midland
2A Milam*
3A Mills*
3B Mitchell
3A Montague
2A Montgomery*
4B Moore
3A Morris*
3B Motley
3A Nacogdoches*
3A Navarro*
2A Newton*
3B Nolan
2A Nueces*
4B Ochiltree
4B Oldham
2A Orange*
3A Palo Pinto*
3A Panola*
3A Parker*
4B Parmer
3B Pecos
2A Polk*
4B Potter
3B Presidio
3A Rains*
4B Randall
3B Reagan
2B Real*
3A Red River*
3B Reeves
2A Refugio*
4B Roberts
2A Robertson*
3A Rockwall*
3B Runnels
3A Rusk*
3A Sabine*
3A San Augustine*
2A San Jacinto*
2A San Patricio*
3A San Saba*
3B Schleicher
3B Scurry
3B Shackelford
3A Shelby*
4B Sherman
3A Smith*
3A Somervell*
2A Starr*
3A Stephens
3B Sterling
3B Stonewall
3B Sutton
4B Swisher
3A Tarrant*
3B Taylor
3B Terrell
3B Terry
3B Throckmorton
3A Titus*
3B Tom Green
2A Travis*
2A Trinity*
2A Tyler*
3A Upshur*
3B Upton
2B Uvalde*
2B Val Verde*
3A Van Zandt*
2A Victoria*
2A Walker*
2A Waller*
3B Ward
2A Washington*
2B Webb*
2A Wharton*
3B Wheeler
3A Wichita
3B Wilbarger
2A Willacy*
2A Williamson*
2A Wilson*
3B Winkler
3A Wise
3A Wood*
4B Yoakum
3A Young
2B Zapata*
2B Zavala*

UTAH

5B Beaver
6B Box Elder
6B Cache
6B Carbon
6B Daggett
5B Davis
6B Duchesne
5B Emery
5B Garfield
5B Grand
5B Iron
5B Juab
5B Kane
5B Millard
6B Morgan
5B Piute
6B Rich
5B Salt Lake
5B San Juan
5B Sanpete
5B Sevier
6B Summit
5B Tooele
6B Uintah
5B Utah
6B Wasatch
3B Washington
5B Wayne
5B Weber

VERMONT

6A (all)

VIRGINIA

4A (all)

WASHINGTON

5B Adams
5B Asotin
5B Benton
5B Chelan
4C Clallam
4C Clark
5B Columbia
4C Cowlitz
5B Douglas
6B Ferry
5B Franklin
5B Garfield
5B Grant
4C Grays Harbor
4C Island
4C Jefferson
4C King
4C Kitsap
5B Kittitas
5B Klickitat
4C Lewis
5B Lincoln
4C Mason
6B Okanogan
4C Pacific
6B Pend Oreille
4C Pierce
4C San Juan
4C Skagit
5B Skamania
4C Snohomish
5B Spokane
6B Stevens
4C Thurston
4C Wahkiakum
5B Walla Walla
4C Whatcom
5B Whitman
5B Yakima

WEST VIRGINIA

5A Barbour
4A Berkeley
4A Boone
4A Braxton
5A Brooke
4A Cabell
4A Calhoun
4A Clay
5A Doddridge
5A Fayette
4A Gilmer
5A Grant
5A Greenbrier
5A Hampshire
5A Hancock
5A Hardy
5A Harrison
4A Jackson
4A Jefferson
4A Kanawha
5A Lewis
4A Lincoln
4A Logan
5A Marion
5A Marshall
4A Mason
4A McDowell
4A Mercer
5A Mineral
4A Mingo
5A Monongalia
4A Monroe
4A Morgan
5A Nicholas
5A Ohio
5A Pendleton
4A Pleasants
5A Pocahontas
5A Preston
4A Putnam
5A Raleigh
5A Randolph

Figure 1405.3.1(2)
CLIMATE ZONES, MOISTURE REGIMES, AND WARM-HUMID DESIGNATIONS BY STATE, COUNTY AND TERRITORY—FROM TABLE C301.1 OF THE IECC

(continued)

4A Ritchie
4A Roane
5A Summers
5A Taylor
5A Tucker
4A Tyler
5A Upshur
4A Wayne
5A Webster
5A Wetzel
4A Wirt
4A Wood
4A Wyoming

WISCONSIN

6A Adams
7 Ashland
6A Barro
7 Bayfield
6A Brown
6A Buffalo
7 Burnett
6A Calumet
6A Chippewa
6A Clark
6A Columbia
6A Crawford
6A Dane
6A Dodge
6A Door
7 Douglas
6A Dunn
6A Eau Claire
7 Florence
6A Fond du Lac
7 Forest
6A Grant
6A Green
6A Green Lake
6A Iowa
7 Iron
6A Jackson
6A Jefferson
6A Juneau
6A Kenosha
6A Kewaunee
6A La Crosse
6A Lafayette
7 Langlade
7 Lincoln
6A Manitowoc
6A Marathon
6A Marinette
6A Marquette
6A Menominee
6A Milwaukee
6A Monroe
6A Oconto
7 Oneida
6A Outagamie
6A Ozaukee
6A Pepin
6A Pierce
6A Polk
6A Portage
7 Price
6A Racine
6A Richland
6A Rock
6A Rusk
6A Sauk
7 Sawyer
6A Shawano
6A Sheboygan
6A St. Croix
7 Taylor
6A Trempealeau
6A Vernon
7 Vilas
6A Walworth
7 Washburn
6A Washington
6A Waukesha
6A Waupaca
6A Waushara
6A Winnebago
6A Wood

WYOMING

6B Albany
6B Big Horn
6B Campbell
6B Carbon
6B Converse
6B Crook
6B Fremont
5B Goshen
6B Hot Springs
6B Johnson
6B Laramie
7 Lincoln
6B Natrona
6B Niobrara
6B Park
5B Platte
6B Sheridan
7 Sublette
6B Sweetwater
7 Teton
6B Uinta
6B Washakie
6B Weston

US TERRITORIES

AMERICAN SAMOA

1A (all)*

GUAM

1A (all)*

NORTHERN MARIANA ISLANDS

1A (all)*

PUERTO RICO

1A (all)*

VIRGIN ISLANDS

1A (all)*

Figure 1405.3.1(2)
CLIMATE ZONES, MOISTURE REGIMES, AND WARM-HUMID DESIGNATIONS BY STATE, COUNTY AND TERRITORY—FROM TABLE C301.1 OF THE IECC

MAJOR CLIMATE TYPE DEFINITIONS
Marine (C) Definition—Locations meeting all four criteria: 1. Mean temperature of coldest month between -3°C (27°F) and 18°C (65°F). 2. Warmest month mean < 22°C (72°F). 3. At least four months with mean temperatures over 10°C (50°F). 4. Dry season in summer. The month with the heaviest precipitation in the cold season has at least three times as much precipitation as the month with the least precipitation in the rest of the year. The cold season is October through March in the Northern Hemisphere and April through September in the Southern Hemisphere.
Dry (B) Definition—Locations meeting the following criteria: Not marine and $P_{in} < 0.44 \times (TF - 19.5)$ [$P_{cm} < 2.0 \times (TC + 7)$ in SI units] where: P_{in} = Annual precipitation in inches (cm) T = Annual mean temperature in °F (°C)
Moist (A) Definition—Locations that are not marine and not dry.
Warm-humid Definition—Moist (A) locations where either of the following wet-bulb temperature conditions shall occur during the warmest six consecutive months of the year: 1. 67°F (19.4°C) or higher for 3,000 or more hours; or 2. 73°F (22.8°C) or higher for 1,500 or more hours.

For SI: °C = [(°F)-32]/1.8, 1 inch = 2.54 cm.

Figure 1405.3.1(3)
INTERNATIONAL CLIMATE ZONE DEFINITIONS—FROM TABLE C301.3(1) OF THE IECC

ZONE NUMBER	THERMAL CRITERIA	
	IP Units	SI Units
1	9000 < CDD50°F	5000 < CDD10°C
2	6300 < CDD50°F ≤ 9000	3500 < CDD10°C ≤ 5000
3A and 3B	4500 < CDD50°F ≤ 6300 AND HDD65°F ≤ 5400	2500 < CDD10°C ≤ 3500 AND HDD18°C ≤ 3000
4A and 4B	CDD50°F ≤ 4500 AND HDD65°F ≤ 5400	CDD10°C ≤ 2500 AND HDD18°C ≤ 3000
3C	HDD65°F ≤ 3600	HDD18°C ≤ 2000
4C	3600 < HDD65°F ≤ 5400	2000 < HDD18°C ≤ 3000
5	5400 < HDD65°F ≤ 7200	3000 < HDD18°C ≤ 4000
6	7200 < HDD65°F ≤ 9000	4000 < HDD18°C ≤ 5000
7	9000 < HDD65°F ≤ 12600	5000 < HDD18°C ≤ 7000
8	12600 < HDD65°F	7000 < HDD18°C

For SI: °C = [(°F)-32]/1.8.

Figure 1405.3.1(4)
INTERNATIONAL CLIMATE ZONE DEFINITIONS—FROM TABLE C301.3(2) OF THE IECC

1405.3.2 Class III vapor retarders. Class III vapor retarders shall be permitted where any one of the conditions in Table 1405.3.2 is met. Only Class III vapor retarders shall be used on the interior side of frame walls where foam plastic insulating sheathing with a perm rating of less than 1 is applied in accordance with Table 1405.3.2 on the exterior side of the frame wall.

❖ Wall assemblies can be designed and constructed to dry inward, outward and to both sides in all climate zones. This section allows more flexibility in the design and construction of moisture-forgiving wall systems. These requirements recognize that many common materials function to various degrees to slow the passage of moisture. In many situations, common materials such as latex paint or the kraft facing on a fiberglass batt may serve to retard moisture sufficiently. In particular, the "standard" sheet of polyethylene is usually not required as a vapor retarder in walls. This section therefore allows the use of Class III vapor retarders in lieu of the Class I or II otherwise specified in Section 1405.3.1. Classes of vapor retarders are defined in Chapter 2.

Class III vapor retarders allow more moisture vapor to pass through them. They are allowed to be used with exterior wall assemblies, described in Table 1405.3.2, that can be expected to dry. Section 1405.3.3 describes materials that are deemed to be Class III vapor retarders, and Section 1405.3.4 describes cladding deemed "vented" for the purpose of Table 1405.3.2.

TABLE 1405.3.2
CLASS III VAPOR RETARDERS

ZONE	CLASS III VAPOR RETARDERS PERMITTED FOR:[a]
Marine 4	Vented cladding over wood structural panels Vented cladding over fiberboard Vented cladding over gypsum Insulated sheathing with *R*-value ≥ R2.5 over 2 × 4 wall Insulated sheathing with *R*-value ≥ R3.75 over 2 × 6 wall
5	Vented cladding over wood structural panels Vented cladding over fiberboard Vented cladding over gypsum Insulated sheathing with *R*-value ≥ R5 over 2 × 4 wall Insulated sheathing with *R*-value ≥ R7.5 over 2 × 6 wall
6	Vented cladding over fiberboard Vented cladding over gypsum Insulated sheathing with *R*-value ≥ R7.5 over 2 × 4 wall Insulated sheathing with *R*-value ≥ R11.25 over 2 × 6 wall
7 and 8	Insulated sheathing with *R*-value ≥ R10 over 2 × 4 wall Insulated sheathing with *R*-value ≥ R15 over 2 × 6 wall

For SI: 1 pound per cubic foot = 16 kg/m^3.

a. Spray foam with a minimum density of 2 lbs/ft^3 applied to the interior cavity side of wood structural panels, fiberboard, insulating sheathing or gypsum is deemed to meet the insulating sheathing requirement where the spray foam *R*-value meets or exceeds the specified insulating sheathing *R*-value.

❖ This table describes the situations where it is permissible to use Class III vapor retarders instead of Class I or II. Class III vapor retarders allow more moisture vapor to pass through them. They are allowed to be used with exterior wall assemblies, described in Table 1405.3.2, that can be expected to dry. The table provides types of wall assemblies that can be used with Class III vapor retarders in different climate zones. The climate zones are determined based upon Chapter 3 of the IECC.

1405.3.3 Material vapor retarder class. The *vapor retarder class* shall be based on the manufacturer's certified testing or a tested assembly.

The following shall be deemed to meet the class specified:

Class I: Sheet polyethylene, nonperforated aluminum foil with a perm rating of less than or equal to 0.1.

Class II: Kraft-faced fiberglass batts or paint with a perm rating greater than 0.1 and less than or equal to 1.0.

Class III: Latex or enamel paint with a perm rating of greater than 1.0 and less than or equal to 10.0.

❖ The vapor retarder class is defined in Section 202 of the code in terms of vapor permeability. The test method used to determine perm value is ASTM E96. However, this section provides a list of materials that can be used for each class of vapor retarder and are deemed to comply with the test standard. No testing is required for these materials. All other materials are required to be tested.

1405.3.4 Minimum clear airspaces and vented openings for vented cladding. For the purposes of this section, vented cladding shall include the following minimum clear airspaces:

1. Vinyl lap or horizontal aluminum siding applied over a weather-resistive barrier as specified in this chapter.
2. Brick veneer with a clear airspace as specified in this code.
3. Other *approved* vented claddings.

❖ This section is intended to define "vented cladding" for the purposes of Table 1405.3.2. As can be seen, vented cladding is material commonly used in the code and includes vinyl siding, aluminum siding and brick with a clear airspace.

By "other approved vented claddings," the code intends materials that can be shown to allow drying of the components behind them in the same manner and at the same rate as the specific materials in Items 1 and 2.

1405.4 Flashing. Flashing shall be installed in such a manner so as to prevent moisture from entering the wall or to redirect that moisture to the exterior. Flashing shall be installed at the perimeters of exterior door and window assemblies, penetrations and terminations of *exterior wall* assemblies, *exterior wall* intersections with roofs, chimneys, porches, decks, balconies and similar projections and at built-in gutters and similar locations where moisture could enter the wall. Flashing with projecting flanges shall be installed on both sides and the ends of copings, under sills and continuously above projecting *trim*.

❖ Water that enters the exterior wall can lead to the decay of wood and the degradation of other building materials that are sensitive to moisture. Some of the most common points of moisture penetration in an exterior wall are at intersections of windows, doors, chimneys and roof lines with other framing These potential points of moisture intrusion must be closed by corrosion-resistant flashings or by other acceptable practices [see Commentary Figures 1405.4(1) through 1405.4(5)]. Materials such as asphalt-saturated felt or building paper are not acceptable for flashing at required locations, in part due to the fact that these materials have a limited ability to resist water penetration. The flashing installation locations stated in this section represent those that are most commonly encountered. As an alternative to the installation of flashings, testing of the specific penetrations, terminations and intersections of the exterior wall finish must be performed in accordance with Section 1403.2 (see commentary, Section 1403.2).

Improperly sloped flashings may enable the accumulation of moisture in the wall assembly, which may result in eventual degradation of the wall finish, building substrate or even the flashing material. Without the extension of the flashing beyond the face of the exterior wall finish, moisture may reenter the wall assembly directly below the flashing through capillary action as indicated in Commentary Figure 1405.4(5).

1405.4.1 Exterior wall pockets. In exterior walls of buildings or structures, wall pockets or crevices in which moisture can accumulate shall be avoided or protected with caps or drips, or other *approved* means shall be provided to prevent water damage.

❖ Changes in building lines and elevations, materials and tops of walls, etc., can produce wall pockets, crevices and other openings that allow the entry of rain, snow, ice and other sources of moisture. Where these openings are present, additional covers, caps, flashings, etc., must be provided to prevent the entry of moisture.

For SI: 1 inch = 25.4 mm.

Figure 1405.4(1)
FLASHING SIDING/ROOF INTERFACE

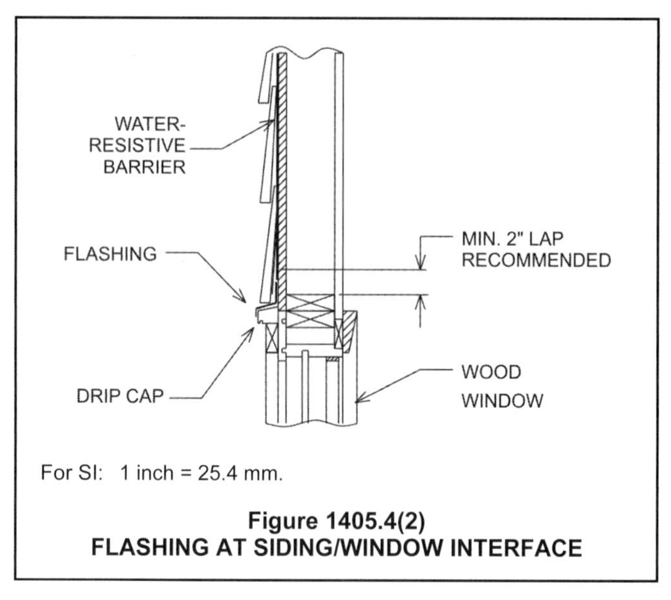

For SI: 1 inch = 25.4 mm.

Figure 1405.4(2)
FLASHING AT SIDING/WINDOW INTERFACE

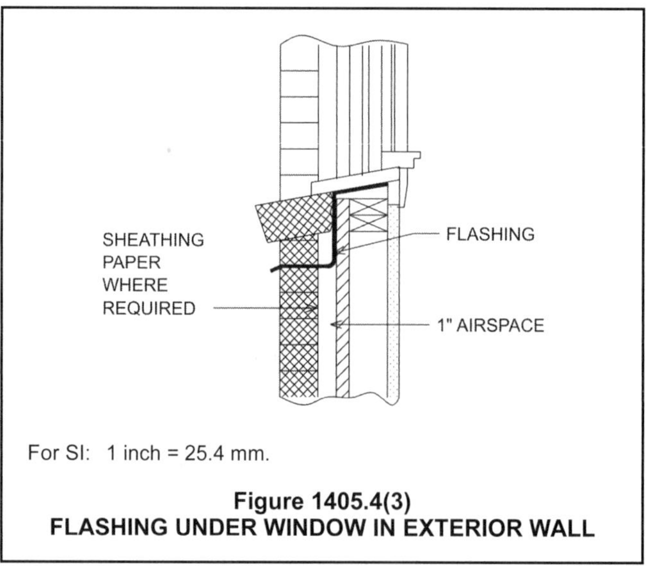

For SI: 1 inch = 25.4 mm.

Figure 1405.4(3)
FLASHING UNDER WINDOW IN EXTERIOR WALL

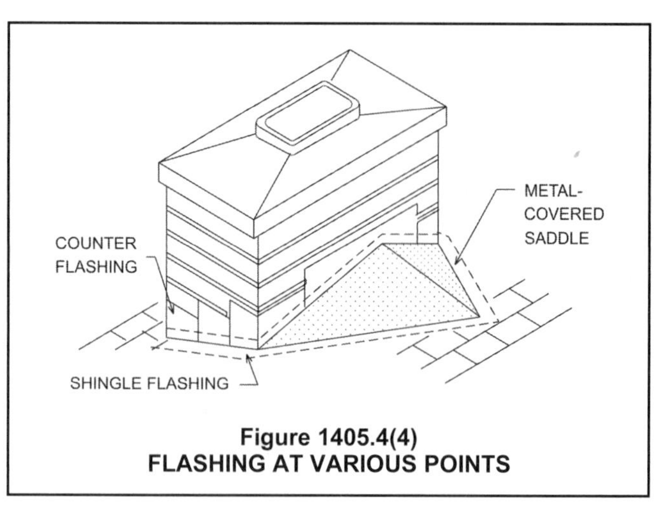

Figure 1405.4(4)
FLASHING AT VARIOUS POINTS

1405.4.2 Masonry. Flashing and weep holes in anchored veneer shall be located in the first course of masonry above finished ground level above the foundation wall or slab, and other points of support, including structural floors, shelf angles and lintels where anchored veneers are designed in accordance with Section 1405.6.

❖ Water can penetrate walls constructed of masonry materials from a number of sources. Some of the most common causes of water penetration are: the porous nature of many masonry materials; openings such as small cracks that may occur between the masonry unit and mortar joint due to expansion and contraction of the wall over time; and the interface between the masonry and penetrations or dissimilar materials.

Walls constructed of anchored masonry veneer must be designed to accommodate water penetration and to provide a means to both stop the moisture penetration and for the moisture to leave the wall. This stipulates the use of flashing to stop moisture penetration into the wall backing and small openings (weepholes) to allow the moisture to exit the wall. In order for these two components to function as intended, the weepholes must be installed immediately above the flashing so that moisture collected by the flashing can exit the wall, preventing the accumulation of water at the flashing location.

1405.5 Wood veneers. Wood veneers on exterior walls of buildings of Type I, II, III and IV construction shall be not less than 1 inch (25 mm) nominal thickness, 0.438-inch (11.1 mm) exterior hardboard siding or 0.375-inch (9.5 mm) exterior-type wood structural panels or particleboard and shall conform to the following:

1. The veneer shall not exceed 40 feet (12 190 mm) in height above grade. Where fire-retardant-treated wood is used, the height shall not exceed 60 feet (18 290 mm) in height above grade.
2. The veneer is attached to or furred from a noncombustible backing that is fire-resistance rated as required by other provisions of this code.

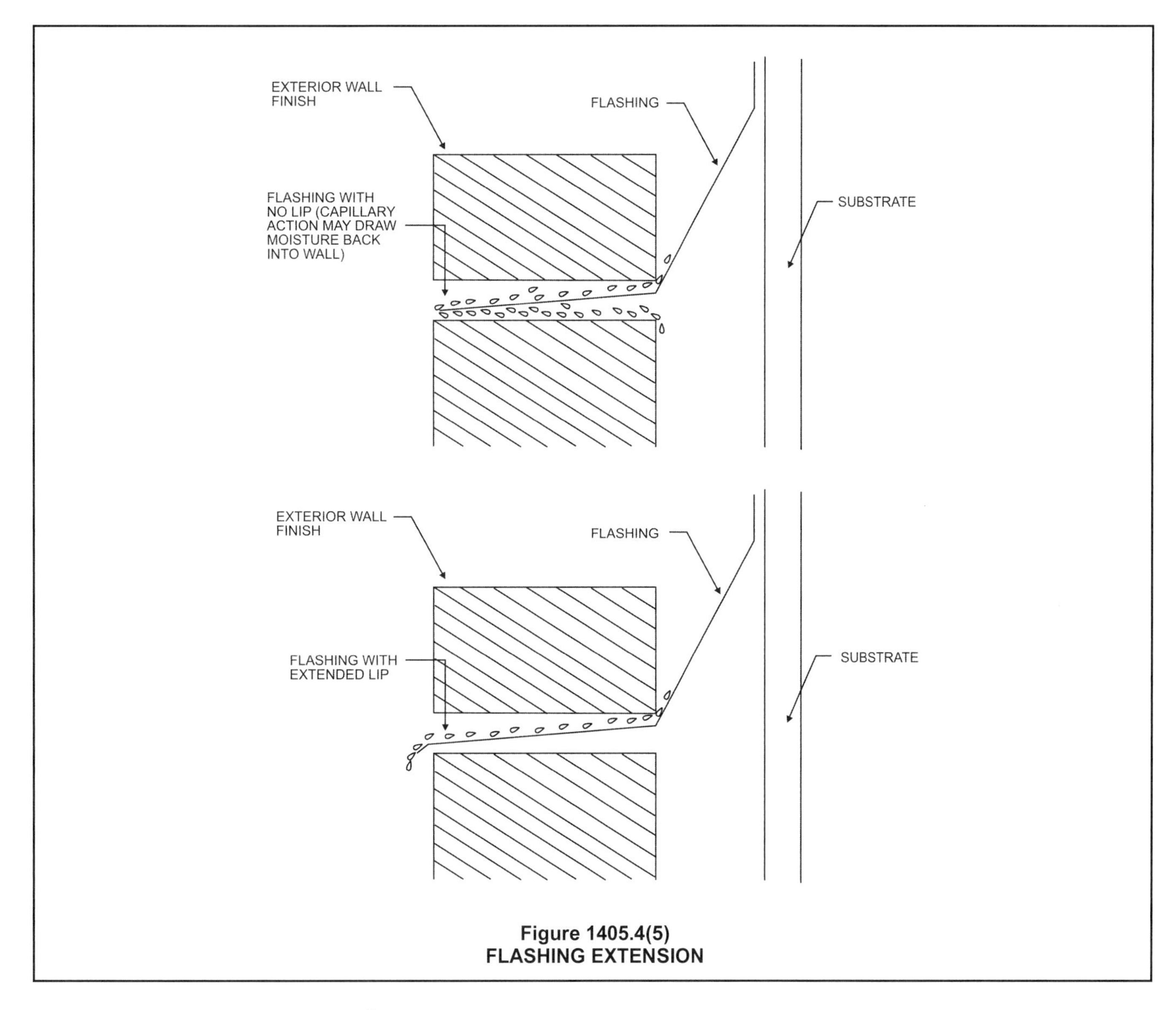

Figure 1405.4(5)
FLASHING EXTENSION

3. Where open or spaced wood veneers (without concealed spaces) are used, they shall not project more than 24 inches (610 mm) from the building wall.

❖ Wood veneers, which are combustible exterior materials, are permitted for use on buildings of other than combustible (Type V) construction where the minimum thickness requirements of this section and three construction criteria are met (see the commentary to Section 1406.2.1.1 for more discussion of wood veneers as a combustible material on exterior walls).

Item 1 limits the maximum height of the installation of the wood veneer and is coordinated with the provisions of Section 1406.2.1 (see the commentary to Section 1406.2.1 for more discussion of the limitations of use of combustible exterior materials). The increase of one additional story is permitted due to the limited combustible nature of fire-retardant-treated wood (FRTW) siding (see commentary, Section 2303.2). Item 2 requires the backing to be noncombustible, which is consistent with requirements of Chapter 6 for buildings of Type I, II, III and IV construction. The backing must also have the fire-resistance rating required by other provisions of the code, such as in Tables 601 and 602. Item 3 limits the maximum distance that wood veneers may project from the face of the building.

[BS] 1405.6 Anchored masonry veneer. Anchored masonry veneer shall comply with the provisions of Sections 1405.6, 1405.7, 1405.8 and 1405.9 and Sections 12.1 and 12.2 of TMS 402/ACI 530/ASCE 5.

❖ Sections 12.1 and 12.2 of ACI 530/ASCE 5/TMS 402 are to be used for the design and detailing requirements for anchored masonry veneers. See the commentary for ACI 530/ASCE 5/TMS 402 for an explanation of the anchored masonry veneer requirements.

[BS] 1405.6.1 Tolerances. Anchored masonry veneers in accordance with Chapter 14 are not required to meet the tolerances in Article 3.3 F1 of TMS 602/ACI 530.1/ASCE 6.

❖ The tolerances in the referenced standard apply to masonry units that are utilized in masonry wall construction and are not intended to be applied to masonry veneers.

[BS] 1405.6.2 Seismic requirements. Anchored masonry veneer located in Seismic Design Category C, D, E or F shall conform to the requirements of Section 12.2.2.10 of TMS 402/ACI 530/ASCE 5.

❖ This requires that the seismic design and detailing of anchored masonry veneer be performed in accordance with referenced standard ACI 530 for veneers located in Seismic Design Categories C, D, E and F (see Chapter 16 for more information on seismic design categories). Section 6.2.2.10 of ACI 530 contains criteria for components such as masonry anchor spacing, joint reinforcement bedding and splicing.

[BS] 1405.7 Stone veneer. Anchored stone veneer units not exceeding 10 inches (254 mm) in thickness shall be anchored directly to masonry, concrete or to stud construction by one of the following methods:

1. With concrete or masonry backing, anchor ties shall be not less than 0.1055-inch (2.68 mm) corrosion-resistant wire, or *approved* equal, formed beyond the base of the backing. The legs of the loops shall be not less than 6 inches (152 mm) in length bent at right angles and laid in the mortar joint, and spaced so that the eyes or loops are 12 inches (305 mm) maximum on center in both directions. There shall be provided not less than a 0.1055-inch (2.68 mm) corrosion-resistant wire tie, or *approved* equal, threaded through the exposed loops for every 2 square feet (0.2 m^2) of stone veneer. This tie shall be a loop having legs not less than 15 inches (381 mm) in length bent so that the tie will lie in the stone veneer mortar joint. The last 2 inches (51 mm) of each wire leg shall have a right-angle bend. One-inch (25 mm) minimum thickness of cement grout shall be placed between the backing and the stone veneer.
2. With wood stud backing, a 2-inch by 2-inch (51 by 51 mm) 0.0625-inch (1.59 mm) zinc-coated or nonmetallic coated wire mesh with two layers of water-resistive barrier in accordance with Section 1404.2 shall be applied directly to wood studs spaced not more than 16 inches (406 mm) on center. On studs, the mesh shall be attached with 2-inch-long (51 mm) corrosion-resistant steel wire furring nails at 4 inches (102 mm) on center providing a minimum 1.125-inch (29 mm) penetration into each stud and with 8d annular threaded nails at 8 inches (203 mm) on center. into top and bottom plates or with equivalent wire ties. There shall be not less than a 0.1055-inch (2.68 mm) zinc-coated or nonmetallic coated wire, or approved equal, attached to the stud with not smaller than an 8d (0.120 in. diameter) annular threaded nail for every 2 square feet (0.2 m^2) of stone veneer. This tie shall be a loop having legs not less than 15 inches (381 mm) in length, so bent that the tie will lie in the stone veneer mortar joint. The last 2 inches (51 mm) of each wire leg shall have a right-angle bend. One-inch (25 mm) minimum thickness of cement grout shall be placed between the backing and the stone veneer.
3. With cold-formed steel stud backing, a 2-inch by 2-inch (51 by 51 mm) 0.0625-inch (1.59 mm) zinc-coated or nonmetallic coated wire mesh with two layers of water-resistive barrier in accordance with Section 1404.2 shall be applied directly to steel studs spaced a not more than 16 inches (406 mm) on center. The mesh shall be attached with corrosion-resistant #8 self-drilling, tapping screws at 4 inches (102 mm) on center, and at 8 inches (203 mm) on center into top and bottom tracks or with equivalent wire ties. Screws shall extend through the steel connection not fewer than three exposed threads. There shall be not less than a 0.1055-inch (2.68 mm) corrosion-resistant wire, or approved equal, attached to the stud with not smaller than a #8 self-drilling, tapping screw extending through the steel framing not fewer than three exposed threads for every 2 square feet (0.2 m^2) of stone veneer. This tie shall be

a loop having legs not less than 15 inches (381 mm) in length, so bent that the tie will lie in the stone veneer mortar joint. The last 2 inches (51 mm) of each wire leg shall have a right-angle bend. One-inch (25 mm) minimum thickness of cement grout shall be placed between the backing and the stone veneer. The cold-formed steel framing members shall have a minimum bare steel thickness of 0.0428 inches (1.087 mm).

❖ This section permits direct anchorage of anchored stone veneers to the wall backing by one of three specified methods. It should be noted that support from below is required in accordance with Section 1405.6. Since the framing is providing only lateral anchorage, thicknesses up to 10 inches (254 mm) are permitted.

There are three methods of anchorage prescribed in this section, based upon the construction of the wall backing. The first method specifies the anchor type, size and spacing for concrete and masonry backing (see Commentary Figure 1405.7). The second method specifies the anchorage requirements for wood-frame walls. The third method specifies the anchorage requirements for cold-formed steel stud-framed walls.

All methods require that the anchorage ties be of corrosion-resistant material, such as zinc or non-metallic coated material. Corrosion-resistant ties are necessary because of the potential for corrosion caused by both water penetration and an adverse reaction of the metal with the mortar or other masonry material.

According to the first method, the specified tie spacings, in conjunction with the cement grout, are intended to achieve lateral support points of the veneer at adequate spacings. Where lateral ties are not installed at the required locations, buckling of the masonry veneer under vertical dead loads and lateral loads is possible. It should be noted that the required 1 inch (25 mm) of mortar grout creates walls that are not considered cavity wall construction.

The second and third methods require that two layers of a water-resistive barrier be installed over the wood-stud construction to provide protection from moisture penetration.

[BS] 1405.8 Slab-type veneer. Anchored slab-type veneer units not exceeding 2 inches (51 mm) in thickness shall be anchored directly to masonry, concrete or light-frame construction. For veneer units of marble, travertine, granite or other stone units of slab form, ties of corrosion-resistant dowels in drilled holes shall be located in the middle third of the

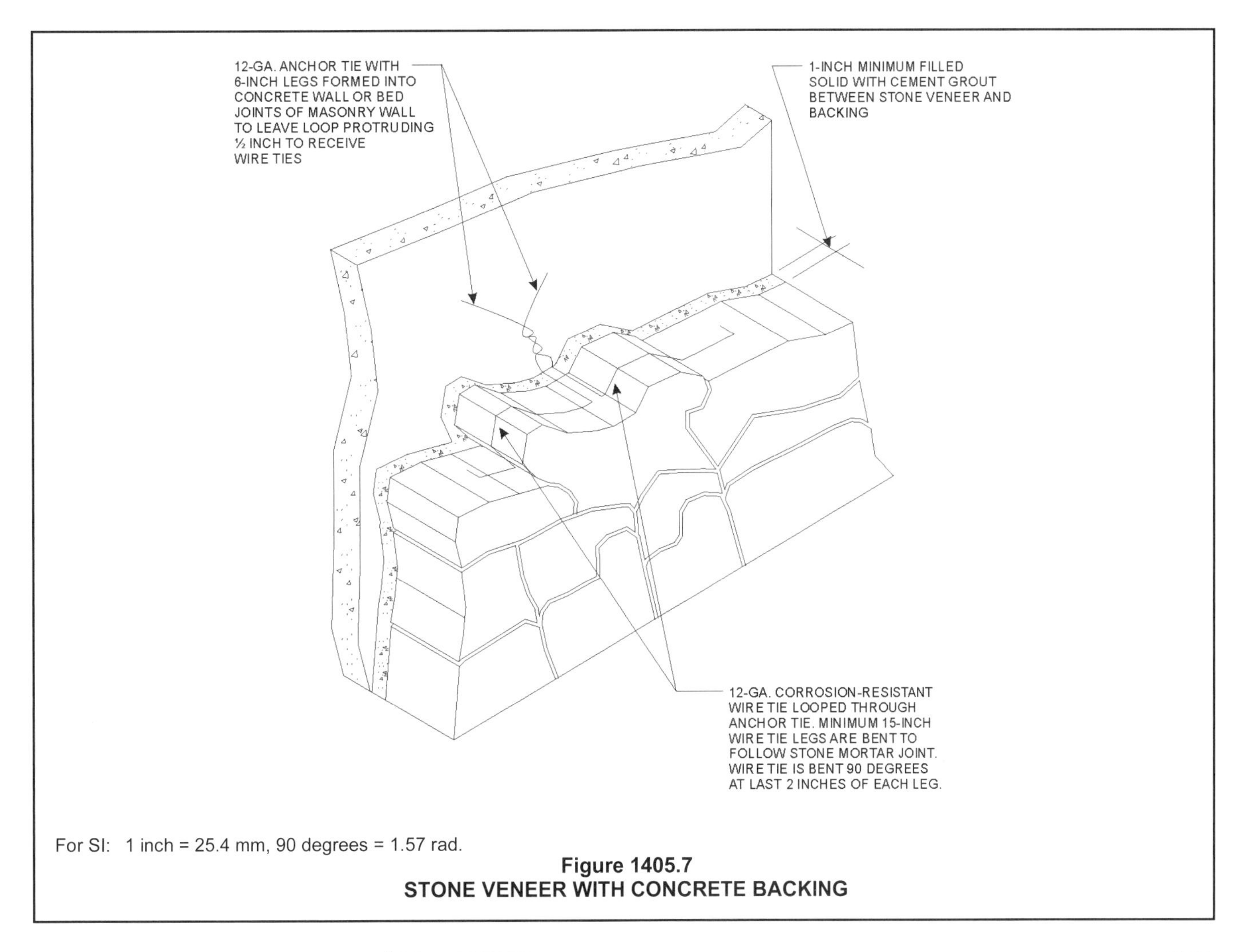

For SI: 1 inch = 25.4 mm, 90 degrees = 1.57 rad.

Figure 1405.7
STONE VENEER WITH CONCRETE BACKING

edge of the units, spaced not more than 24 inches (610 mm) apart around the periphery of each unit with not less than four ties per veneer unit. Units shall not exceed 20 square feet (1.9 m^2) in area. If the dowels are not tight fitting, the holes shall be drilled not more than 0.063 inch (1.6 mm) larger in diameter than the dowel, with the hole countersunk to a diameter and depth equal to twice the diameter of the dowel in order to provide a tight-fitting key of cement mortar at the dowel locations where the mortar in the joint has set. Veneer ties shall be corrosion-resistant metal capable of resisting, in tension or compression, a force equal to two times the weight of the attached veneer. If made of sheet metal, veneer ties shall be not smaller in area than 0.0336 by 1 inch (0.853 by 25 mm) or, if made of wire, not smaller in diameter than 0.1483-inch (3.76 mm) wire.

❖ Anchored slab-type veneer refers to thin stone masonry veneers that are individually supported. These units do not bear on the successive units below, but are anchored to and supported vertically by the backup construction with corrosion-resistant dowels located at the edges of each unit. This type of veneer anchorage applies to the use of thin stone units that are typically not set in mortar, but are made water tight at the edges through the use of sealant materials. The strength of the veneer ties is required to be adequate to support twice the weight of the veneer in tension to provide for an adequate margin of safety, such that the veneer will remain in place in case it works loose of the grout and backing.

[BS] 1405.9 Terra cotta. Anchored terra cotta or ceramic units not less than $1^5/_8$ inches (41 mm) thick shall be anchored directly to masonry, concrete or stud construction. Tied terra cotta or ceramic veneer units shall be not less than $1^5/_8$ inches (41 mm) thick with projecting dovetail webs on the back surface spaced approximately 8 inches (203 mm) on center. The facing shall be tied to the backing wall with corrosion-resistant metal anchors of not less than No. 8 gage wire installed at the top of each piece in horizontal bed joints not less than 12 inches (305 mm) nor more than 18 inches (457 mm) on center; these anchors shall be secured to $^1/_4$-inch (6.4 mm) corrosion-resistant pencil rods that pass through the vertical aligned loop anchors in the backing wall. The veneer ties shall have sufficient strength to support the full weight of the veneer in tension. The facing shall be set with not less than a 2-inch (51 mm) space from the backing wall and the space shall be filled solidly with Portland cement grout and pea gravel. Immediately prior to setting, the backing wall and the facing shall be drenched with clean water and shall be distinctly damp when the grout is poured.

❖ The minimum permitted $1^5/_8$-inch (41 mm) thickness for terra cotta units is necessary in order for each unit to have an inherent level of structural integrity. In order to increase the mechanical anchorage of the backing, dovetail webs are required and are embedded in the 2-inch (51 mm) cement grout space. The strength of the veneer ties is required to be adequate to support the full weight of the veneer in tension in order to achieve an adequate margin of safety, such that the veneer will remain in place in case it works loose of the grout and backing (see Commentary Figure 1405.9).

[BS] 1405.10 Adhered masonry veneer. Adhered masonry veneer shall comply with the applicable requirements in this section and Sections 12.1 and 12.3 of TMS 402/ACI 530/ASCE 5.

❖ Sections 12.1 and 12.3 of ACI 530/ASCE 5/TMS 402 are to be used for the design and detailing requirements for adhered masonry veneers. See the commentary for ACI 530/ASCE 5/TMS 402 for an explanation of the adhered masonry veneer requirements. Commentary Figures 1405.10(1) and 1405.10(2) illustrate two examples of an adhered masonry veneer.

[BS] 1405.10.1 Exterior adhered masonry veneer. Exterior adhered masonry veneer shall be installed in accordance with Section 1405.10 and the manufacturer's instructions.

❖ This section sets out the requirements for exterior adhered masonry veneer by reference to Section 1405.10 as a whole. Additionally, this section requires the installation to comply with the specific manufacturer's instructions.

[BS] 1405.10.1.1 Water-resistive barriers. Water-resistive barriers shall be installed as required in Section 2510.6.

❖ Requirements for the installation of water-resistive barriers are found in Section 2510.6 (stucco). Wall preparation for stucco and for exterior adhered masonry veneer is very similar, so referring to Section 2510.6 provides consistency in the code for water-resistive barrier requirements.

[BS] 1405.10.1.2 Flashing. Flashing shall comply with the applicable requirements of Section 1405.4 and the following.

❖ This section contains specific flashing requirements for exterior adhered masonry veneer at the foundation. Compliance with the general flashing requirements in Section 1405.4 is still applicable.

[BS] 1405.10.1.2.1 Flashing at foundation. A corrosion-resistant screed or flashing of a minimum 0.019-inch (0.48 mm) or 26 gage galvanized or plastic with a minimum vertical attachment flange of $3^1/_2$ inches (89 mm) shall be installed to extend not less than 1 inch (25 mm) below the foundation plate line on exterior stud walls in accordance with Section 1405.4. The water-resistive barrier shall lap over the exterior of the attachment flange of the screed or flashing.

❖ This section contains requirements for the flashing at the foundation, which are similar to the weep-screed requirements for stucco while at the same time allowing for alternatives to the stucco-specific weep screed.

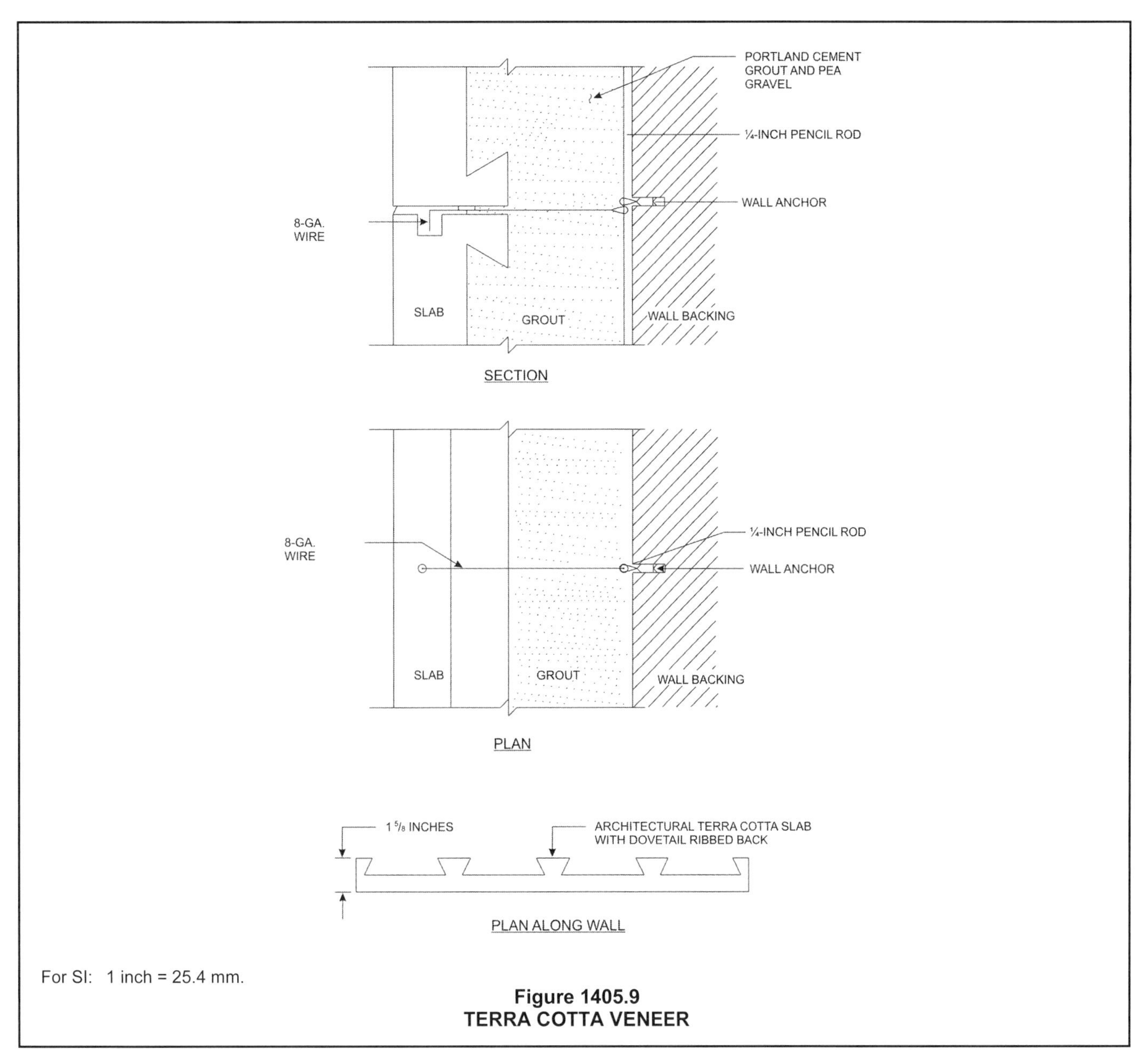

For SI: 1 inch = 25.4 mm.

Figure 1405.9
TERRA COTTA VENEER

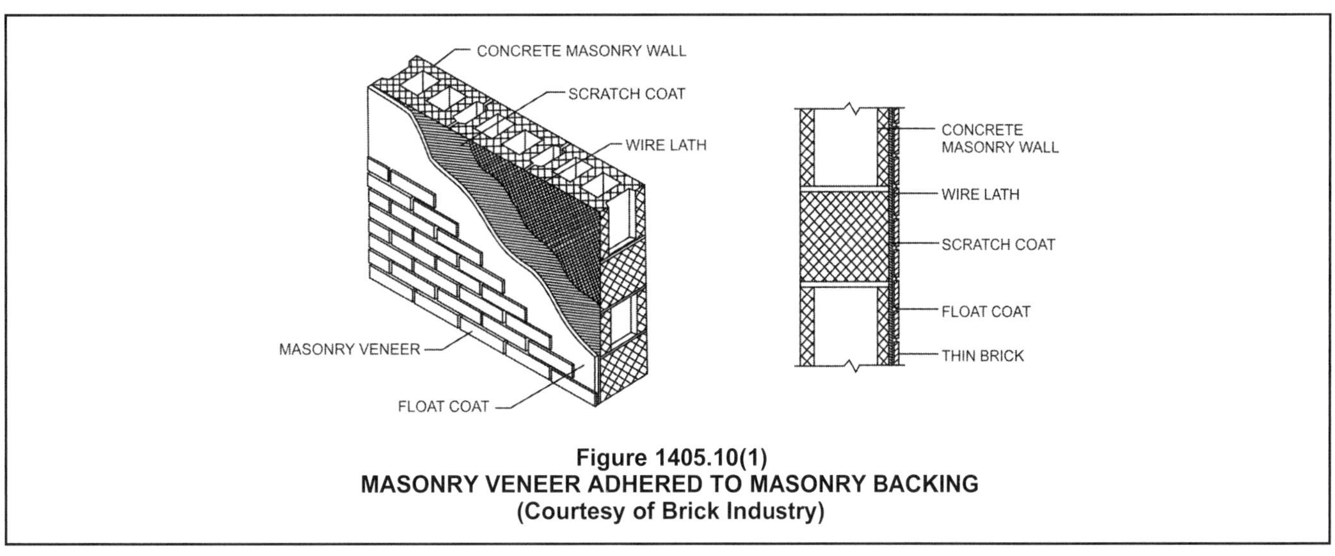

Figure 1405.10(1)
MASONRY VENEER ADHERED TO MASONRY BACKING
(Courtesy of Brick Industry)

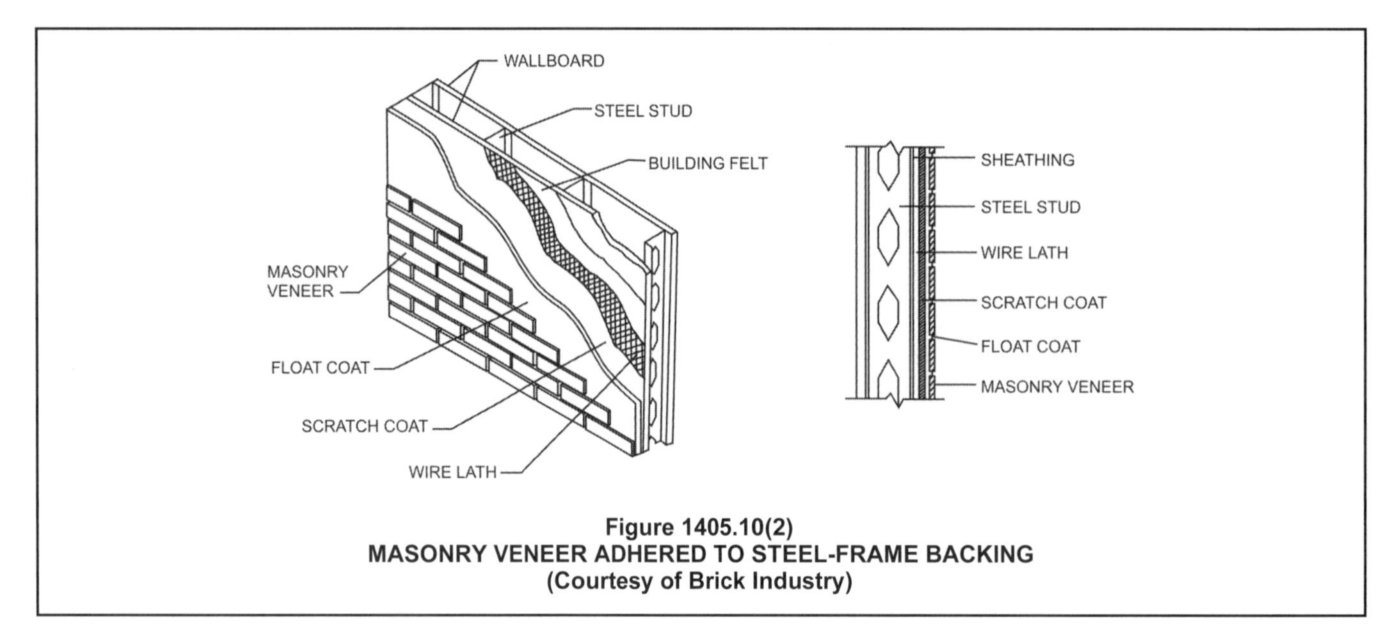

Figure 1405.10(2)
MASONRY VENEER ADHERED TO STEEL-FRAME BACKING
(Courtesy of Brick Industry)

[BS] 1405.10.1.3 Clearances. On exterior stud walls, adhered masonry veneer shall be installed not less than 4 inches (102 mm) above the earth, or not less than 2 inches (51 mm) above paved areas, or not less than $^1/_2$ inch (12.7 mm) above exterior walking surfaces that are supported by the same foundation that supports the exterior wall.

❖ The clearance requirements within this section are consistent with stucco applications. However they go one step further by specifying a minimum $^1/_2$-inch (12 mm) clearance to exterior walking surfaces that are supported by the same foundation that supports the wall to which the exterior veneer is adhered. This $^1/_2$-inch (12 mm) clearance requirement allows for architectural and aesthetic improvements in the installation of adhered masonry veneer.

[BS] 1405.10.1.4 Adhered masonry veneer installed with lath and mortar. Exterior adhered masonry veneer installed with lath and mortar shall comply with the following.

❖ The purpose of this section is to establish the minimum material and installation requirements for adhered masonry veneer installed with lath and mortar, including lathing; scratch coat; adhering veneer to scratch coat and to masonry and concrete; cold weather construction; and hot weather construction.

[BS] 1405.10.1.4.1 Lathing. Lathing shall comply with the requirements of Section 2510.

❖ Section 2510 contains material specifications and installation requirements specific to lathing for cement plaster. These requirements are applicable to adhered masonry veneer installation.

[BS] 1405.10.1.4.2 Scratch coat. A nominal $^1/_2$-inch-thick (12.7 mm) layer of mortar complying with the material requirements of Sections 2103 and 2512.2 shall be applied, encapsulating the lathing. The surface of this mortar shall be scored horizontally, resulting in a scratch coat.

❖ This section provides the minimum thickness requirement for the scratch coat to be used in the installation of the adhered veneer. The material requirements for the mortar must comply with Sections 2103.15 and 2512.2, which contain minimum mortar specifications and plasticity agent limitations.

[BS] 1405.10.1.4.3 Adhering veneer. The masonry veneer units shall be adhered to the mortar scratch coat with a nominal $^1/_2$-inch-thick (12.7 mm) setting bed of mortar complying with Sections 2103 and 2512.2 applied to create a full setting bed for the back of the masonry veneer units. The masonry veneer units shall be worked into the setting bed resulting in a nominal $^3/_8$-inch (9.5 mm) setting bed after the masonry veneer units are applied.

❖ This section provides requirements specific to adhering the masonry veneer to the scratch coat. The veneer is required to be installed in a minimum $^1/_2$-inch setting bed of mortar complying with Sections 2103.15 and 2512.2, which contain minimum mortar specifications and plasticity agent limitations.

[BS] 1405.10.1.5 Adhered masonry veneer applied directly to masonry and concrete. Adhered masonry veneer applied directly to masonry or concrete shall comply with the applicable requirements of Section 1405.10 and with the requirements of Section 1405.10.1.4 or 2510.7.

❖ Adhered masonry veneer can also be applied to masonry and concrete. This section provides minimum requirements for this application. Compliance options include application in accordance with the requirements for adhering masonry veneer with lath and mortar, bonding agents or Portland cement dash bond as described in Section 2510.7.

[BS] 1405.10.1.6 Cold weather construction. Cold weather construction of adhered masonry veneer shall comply with the requirements of Sections 2104 and 2512.4.

❖ Temperature extremes affect materials used for the installation of masonry materials. This section contains additional requirements for materials when the construction is during cold weather [ambient air temperature less than 40°F (4°C)]. Section 2104.3 references a minimum standard for construction provisions to be used. Section 2512.4 contains minimum temperature requirements for cement plaster coats.

[BS] 1405.10.1.7 Hot weather construction. Hot weather construction of adhered masonry veneer shall comply with the requirements of Section 2104.

❖ Temperature extremes affect materials used for the installation of masonry materials. This section contains additional requirements for materials when the construction is during hot weather (ambient air temperature exceeding 100°F (37.8°C)]. Section 2104.4 references a minimum standard for construction provisions to be used.

[BS] 1405.10.2 Exterior adhered masonry veneers—porcelain tile. Adhered units shall not exceed $^5/_8$ inch (15.8 mm) thickness and 24 inches (610 mm) in any face dimension nor more than 3 square feet (0.28 m^2) in total face area and shall not weigh more than 9 pounds psf (0.43 kN/m^2). Porcelain tile shall be adhered to an approved backing system.

❖ To ensure proper installation, this section limits the thickness, size and weight of porcelain tile units. These limitations are consistent with current construction practices and current adhesive capabilities.

[BS] 1405.10.3 Interior adhered masonry veneers. Interior adhered masonry veneers shall have a maximum weight of 20 psf (0.958 kg/m^2) and shall be installed in accordance with Section 1405.10. Where the interior adhered masonry veneer is supported by wood construction, the supporting members shall be designed to limit deflection to $^1/_{600}$ of the span of the supporting members.

❖ This section addresses masonry veneer that is supported on wood members. The provisions of Section 2104.1.6 are not intended to preclude masonry veneer on wood supports. The deflection limit is to minimize cracking of the veneer.

[BS] 1405.11 Metal veneers. Veneers of metal shall be fabricated from *approved* corrosion-resistant materials or shall be protected front and back with porcelain enamel, or otherwise be treated to render the metal resistant to corrosion. Such veneers shall be not less than 0.0149-inch (0.378 mm) nominal thickness sheet steel mounted on wood or metal furring strips or approved sheathing on light-frame construction.

❖ All metal veneers used as exterior weather coverings are required to be manufactured from materials that are inherently corrosion resistant or protected with other materials, such as coatings, to prevent corrosion.

For sheet steel, the minimum nominal thickness is 0.0149 inches (0.38 mm), not including the thickness of the corrosion-resistant coating. For aluminum siding, a 0.019-inch-thick (0.48 mm) value is the minimum. Aluminum siding must also conform to AAMA 1402. The minimum thicknesses for other weather coverings are obtained from Table 1405.2.

[BS] 1405.11.1 Attachment. Exterior metal veneer shall be securely attached to the supporting masonry or framing members with corrosion-resistant fastenings, metal ties or by other *approved* devices or methods. The spacing of the fastenings or ties shall not exceed 24 inches (610 mm) either vertically or horizontally, but where units exceed 4 square feet (0.4 m^2) in area there shall be not less than four attachments per unit. The metal attachments shall have a cross-sectional area not less than provided by W 1.7 wire. Such attachments and their supports shall be designed and constructed to resist the wind loads as specified in Section 1609 for components and cladding.

❖ This section prescribes minimum requirements for the attachment of metal veneers to the exterior wall construction. The attachment of the veneer may be accomplished through the use of fasteners, metal ties, clips or other means. The means of attaching panels to supporting members must be able to resist the design wind loads determined in accordance with Section 1609 as required for components and cladding.

1405.11.2 Weather protection. Metal supports for exterior metal veneer shall be protected by painting, galvanizing or by other equivalent coating or treatment. Wood studs, furring strips or other wood supports for exterior metal veneer shall be *approved* pressure-treated wood or protected as required in Section 1403.2. Joints and edges exposed to the weather shall be caulked with *approved* durable waterproofing material or by other *approved* means to prevent penetration of moisture.

❖ Supports for metal veneers are required to be protected from corrosion induced by moisture. The corrosion resistance may be provided through a surface treatment applied to the metal or through the use of materials that are inherently corrosion resistant. Where wood supports are used, they must be protected through pressure treatment or covered with a minimum of one layer of water-resistive barrier as specified in Section 1404.2. Protection of joints and edges must be provided to prevent the entrance of moisture that may result in damage to the veneer, supports or building framing.

1405.11.3 Backup. Masonry backup shall not be required for metal veneer unless required by the fire-resistance requirements of this code.

❖ The type of substrate that metal veneers may be installed over is not limited, except in cases where the wall is required to have a fire-resistance rating in accordance with other sections of the code.

1405.11.4 Grounding. Grounding of metal veneers on buildings shall comply with the requirements of Chapter 27 of this code.

❖ The need to provide electrical grounding for metal veneers has been widely discussed for more than 30 years. Examples of hazards have been demonstrated where electric services have been violated. Most metal-siding manufacturers have made provisions in their designs for grounding. This section references Chapter 27 without providing additional specifics regarding grounding of siding.

[BS] 1405.12 Glass veneer. The area of a single section of thin exterior structural glass veneer shall not exceed 10 square feet (0.93 m^2) where that section is not more than 15 feet (4572 mm) above the level of the sidewalk or grade level directly below, and shall not exceed 6 square feet (0.56 m^2) where it is more than 15 feet (4572 mm) above that level.

❖ Structural glass veneer is produced in the same manner as plate glass (see commentary, Chapter 24), with the addition of metallic oxides that are blended into the glass to provide opacity and color. To accommodate the forces due to expansion and contraction of differing materials in exterior installations, the maximum area of structural glass veneer is limited based upon the height above grade.

[BS] 1405.12.1 Length and height. The length or height of any section of thin exterior structural glass veneer shall not exceed 48 inches (1219 mm).

❖ Where Section 1405.12 regulates the maximum area of any single section of structural glass veneer, this section limits the maximum dimension of any single section to 48 inches (1219 mm) when used in exterior application.

[BS] 1405.12.2 Thickness. The thickness of thin exterior structural glass veneer shall be not less than 0.344 inch (8.7 mm).

❖ Structural glass veneers are produced in a range of thicknesses; however, the minimum thickness permitted for use as an exterior veneer is limited to $1^1/_{32}$ inches (8.7 mm).

[BS] 1405.12.3 Application. Thin exterior structural glass veneer shall be set only after backing is thoroughly dry and after application of an *approved* bond coat uniformly over the entire surface of the backing so as to effectively seal the surface. Glass shall be set in place with an *approved* mastic cement in sufficient quantity so that at least 50 percent of the area of each glass unit is directly bonded to the backing by mastic not less than $^1/_4$ inch (6.4 mm) thick and not more than $^5/_8$ inch (15.9 mm) thick. The bond coat and mastic shall be evaluated for compatibility and shall bond firmly together.

❖ Structural glass veneer is bonded to a solid backing, such as masonry, through the use of a mastic adhesive. The backing must be provided with a coating that will serve to seal it from moisture absorption, provide a uniform surface to which the mastic can adhere and provide a relatively level surface. The bond coat must be compatible for use with the backing material so that debonding will not occur over time. Some common types of bond coats are a float coat of cement mortar and portland cement stucco.

The glass veneer is attached to the bond coat by applying mastic adhesives. Mastics are characterized by their ability to remain pliable over time, rather than harden. The minimum and maximum thicknesses permitted for the mastic application are specified, since an application that is too thin would not provide an adequate bond and one that is too thick could sag over time. To minimize the possibility of debonding, it is necessary to determine that the materials intended for use as a mastic and as a bond coat are compatible with one another.

[BS] 1405.12.4 Installation at sidewalk level. Where glass extends to a sidewalk surface, each section shall rest in an *approved* metal molding, and be set at least $^1/_4$ inch (6.4 mm) above the highest point of the sidewalk. The space between the molding and the sidewalk shall be thoroughly caulked and made water tight.

❖ To protect glass veneer from breakage when installed at sidewalk level, the veneer is required to be supported by a metal molding. The molding must be protected from corrosion and be constructed of a material that is compatible for contact with the glass. The molding is to be located a minimum of $^1/_4$ inch (6.4 mm) above the sidewalk to minimize the potential for cracking of the veneer due to expansion and contraction.

[BS] 1405.12.4.1 Installation above sidewalk level. Where thin exterior structural glass veneer is installed above the level of the top of a bulkhead facing, or at a level more than 36 inches (914 mm) above the sidewalk level, the mastic cement binding shall be supplemented with *approved* nonferrous metal shelf angles located in the horizontal joints in every course. Such shelf angles shall be not less than 0.0478-inch (1.2 mm) thick and not less than 2 inches (51 mm) long and shall be spaced at *approved* intervals, with not less than two angles for each glass unit. Shelf angles shall be secured to the wall or backing with expansion bolts, toggle bolts or by other *approved* methods.

❖ Structural glass veneer installed more than 36 inches (914 mm) above sidewalk level is subject to impact from pedestrians and can pose a hazard if the veneer were to become dislodged from the backing. To minimize the potential for this occurring, it is necessary to provide additional supports that are attached to the backing. The supports are intended to supplement the mastic.

[BS] 1405.12.5 Joints. Unless otherwise specifically *approved* by the *building official*, abutting edges of thin exterior structural glass veneer shall be ground square. Mitered joints shall not be used except where specifically *approved* for wide angles. Joints shall be uniformly buttered with an *approved* jointing compound and horizontal joints shall be held to not less than 0.063 inch (1.6 mm) by an *approved* nonrigid substance or device. Where thin exterior structural glass veneer abuts nonresilient material at sides or top, expansion joints not less than $^1/_4$ inch (6.4 mm) wide shall be provided.

❖ Glass has an expansion coefficient of roughly 4.5, which is less than most metals. To minimize the potential for cracking or chipping of glass panels and the development of sharp edges due to differential movement, this section stipulates both the requirements for factory finishing of glass veneer panel edges and the minimum requirements for joints between sections.

[BS] 1405.12.6 Mechanical fastenings. Thin exterior structural glass veneer installed above the level of the heads of show windows and veneer installed more than 12 feet (3658 mm) above sidewalk level shall, in addition to the mastic cement and shelf angles, be held in place by the use of fastenings at each vertical or horizontal edge, or at the four corners of each glass unit. Fastenings shall be secured to the wall or backing with expansion bolts, toggle bolts or by other methods. Fastenings shall be so designed as to hold the glass veneer in a vertical plane independent of the mastic cement. Shelf angles providing both support and fastenings shall be permitted.

❖ While Section 1405.12.4.1 addresses the requirements for protection of pedestrians for installations of veneer that exceed 3 feet (914 mm) above sidewalk level, this section addresses installations that exceed 12 feet (3658 mm) above sidewalk level. Installations above 12 feet (3658 mm) are not subject to impact from pedestrians but are subject to wind loads. Therefore, each veneer section must be secured in place by mechanical means, in addition to the mastic and shelf angles specified in Sections 1405.12.3 and 1405.12.6, respectively.

[BS] 1405.12.7 Flashing. Exposed edges of thin exterior structural glass veneer shall be flashed with overlapping corrosion-resistant metal flashing and caulked with a waterproof compound in a manner to effectively prevent the entrance of moisture between the glass veneer and the backing.

❖ The presence of moisture behind the glass veneer can result in loss of strength of either the mastic, the bond coat or both. In addition, corrosion of the supports or fasteners can occur. This can increase the potential for failure of the structural glass veneer. This section addresses the requirements for flashing of structural glass veneers to prevent the penetration of moisture behind the veneers.

1405.13 Exterior windows and doors. Windows and doors installed in exterior walls shall conform to the testing and performance requirements of Section 1709.5.

❖ Windows and doors that are part of the exterior building envelope are to be tested for wind-load resistance in accordance with the methods specified in Section 1710.5.2 (see commentary, Section 1710.5.2).

1405.13.1 Installation. Windows and doors shall be installed in accordance with *approved* manufacturer's instructions. Fastener size and spacing shall be provided in such instructions and shall be calculated based on maximum loads and spacing used in the tests.

❖ Windows and doors that are part of the exterior envelope are to be installed in accordance with the method in which they were tested (see Section 1405.13) and the window or door manufacturer's installation instructions.

[BS] 1405.14 Vinyl siding. Vinyl siding conforming to the requirements of this section and complying with ASTM D3679 shall be permitted on exterior walls of buildings located in areas where V_{asd} as determined in accordance with Section 1609.3.1 does not exceed 100 miles per hour (45 m/s) and the *building height* is less than or equal to 40 feet (12 192 mm) in Exposure C. Where construction is located in areas where V_{asd} as determined in accordance with Section 1609.3.1 exceeds 100 miles per hour (45 m/s), or building heights are in excess of 40 feet (12 192 mm), tests or calculations indicating compliance with Chapter 16 shall be submitted. Vinyl siding shall be secured to the building so as to provide weather protection for the exterior walls of the building.

❖ The installation of vinyl siding on Type V buildings is limited based upon the basic wind speed for a given location. For areas where the basic wind speed is greater than 100 miles per hour (44 m/s), wind load resistance testing and structural calculations supporting the ability of the siding to withstand the basic wind speed must be provided.

[BS] 1405.14.1 Application. The siding shall be applied over sheathing or materials listed in Section 2304.6. Siding shall be applied to conform to the *water-resistive barrier* requirements in Section 1403. Siding and accessories shall be installed in accordance with *approved* manufacturer's instructions. Unless otherwise specified in the *approved* manufacturer's instructions, nails used to fasten the siding and accessories shall have a minimum 0.313-inch (7.9 mm) head diameter and $^1/_8$-inch (3.18 mm) shank diameter. The nails shall be corrosion resistant and shall be long enough to penetrate the studs or nailing strip at least $^3/_4$ inch (19 mm). For cold-formed steel light-frame construction, corrosion-resistant fasteners shall be used. Screw fasteners shall penetrate the cold-formed steel framing at least three exposed threads. Other fasteners shall be installed in accordance with the approved construction documents and manufacturer's instructions. Where the siding is installed horizontally, the fastener spacing shall not exceed 16 inches (406 mm) horizontally and 12 inches (305 mm) vertically. Where the siding is installed vertically, the fastener spacing shall not exceed 12 inches (305 mm) horizontally and 12 inches (305 mm) vertically.

❖ This section prescribes requirements for the installation of vinyl siding. The installation must also comply with ASTM D4756, which is referenced in ASTM D3679 (see commentary, Section 1404.9) and the manufacturer's instructions. These instructions are necessary for compliance with the performance requirements of this chapter. The prescriptive fastener requirements of this section correspond with the requirements contained in ASTM D4756. Note that other fastening methods are allowed and must be in compliance with both the manufacturer's instructions and the construction documents.

[BS] 1405.15 Cement plaster. Cement plaster applied to exterior walls shall conform to the requirements specified in Chapter 25.

❖ The materials and installation of cement plaster (sometimes referred to as "stucco") are governed by the requirements of Chapter 25. It should be noted that cement plaster is the only plaster type that is permitted for exterior use. Gypsum plaster, which is also governed by Chapter 25, is generally subject to deterioration in the presence of moisture and is therefore not permitted for use in exterior applications.

[BS] 1405.16 Fiber-cement siding. Fiber-cement siding complying with Section 1404.10 shall be permitted on exterior walls of Type I, II, III, IV and V construction for wind pressure resistance or wind speed exposures as indicated by the manufacturer's listing and *label* and *approved* installation instructions. Where specified, the siding shall be installed over sheathing or materials *listed* in Section 2304.6 and shall be installed to conform to the *water-resistive barrier* requirements in Section 1403. Siding and accessories shall be installed in accordance with *approved* manufacturer's instructions. Unless otherwise specified in the *approved* manufacturer's instructions, nails used to fasten the siding to wood studs shall be corrosion-resistant round head smooth shank and shall be long enough to penetrate the studs at least 1 inch (25 mm). For cold-formed steel light-frame construction, corrosion-resistant fasteners shall be used. Screw fasteners shall penetrate the cold-formed steel framing at least three exposed full threads. Other fasteners shall be installed in accordance with the approved construction documents and manufacturer's instructions.

❖ Fiber cement siding is also a material used for weather-resistant siding. It is manufactured from fiber-reinforced cement, and the code permits its use as either panel siding or horizontal lap siding. Sections 1405.16.1 and 1405.16.2 state that the applicable material standard is ASTM C1186. Note that the product is required to bear a label, which means that the manufacturer must have regular inspections by a third-party control agency. This section contains prescriptive fastener requirements. However, other fastening methods are allowed and must be in compliance with both the manufacturer's instructions and the construction documents.

[BS] 1405.16.1 Panel siding. Fiber-cement panels shall comply with the requirements of ASTM C1186, Type A, minimum Grade II (or ISO 8336, Category A, minimum Class 2). Panels shall be installed with the long dimension either parallel or perpendicular to framing. Vertical and horizontal joints shall occur over framing members and shall be protected with caulking, with battens or flashing, or be vertical or horizontal shiplap or otherwise designed to comply with Section 1403.2. Panel siding shall be installed with fasteners in accordance with the *approved* manufacturer's instructions.

❖ This section specifies the material standards and type of material required for fiber cement siding used in a panel configuration. In addition, details of installation necessary for fiber cement siding are specified.

[BS] 1405.16.2 Lap siding. Fiber-cement lap siding having a maximum width of 12 inches (305 mm) shall comply with the requirements of ASTM C1186, Type A, minimum Grade II (or ISO 8336, Category A, minimum Class 2). Lap siding shall be lapped a minimum of $1^1/_4$ inches (32 mm) and lap siding not having tongue-and-groove end joints shall have the ends protected with caulking, covered with an H-section joint cover, located over a strip of flashing or shall be otherwise designed to comply with Section 1403.2. Lap siding courses shall be installed with the fastener heads exposed or concealed in accordance with the *approved* manufacturer's instructions.

❖ This section specifies the material standards and type of material required for fiber cement siding used in a lap configuration. In addition, details of installation necessary for fiber cement siding are specified.

[BS] 1405.17 Fastening. Weather boarding and wall coverings shall be securely fastened with aluminum, copper, zinc, zinc-coated or other *approved* corrosion-resistant fasteners in accordance with the nailing schedule in Table 2304.10.1 or the *approved* manufacturer's instructions. Shingles and other weather coverings shall be attached with appropriate standard-shingle nails to furring strips securely nailed to studs, or with *approved* mechanically bonding nails, except where sheathing is of wood not less than 1-inch (25 mm) nominal thickness or of wood structural panels as specified in Table 2308.9.3(3).

❖ The proper attachment of exterior weatherboard or cladding is critical in order for those items to remain durable and stay in place under anticipated weather conditions. This section requires noncorrodible nails and other connectors. The requirements of Table 2304.9.1 address the attachment of wood structural panels and panels of particleboard and fiberboard. Lapped siding products of these materials are to be attached in accordance with the manufacturer's installation instructions.

The relatively narrow and variable widths of wood shingles and shakes make it necessary to provide a continuous nailing base. This is permitted to be wood sheathing, wood structural panels not less than $^5/_{16}$ inch (7.9 mm) thick or horizontal furring strips. Dimension lumber furring strips are required to be 1-inch (25 mm) minimum nominal thickness. Shingles are attached with special mechanically bonded nails.

Wood shingles and shakes are required to be fastened with at least No. 14 B&S-gage corrosion-resistant nails in accordance with Table 2304.9.1. The nail must be long enough to penetrate the framing, sheathing or furring strip at least $^3/_4$ inch (19.1 mm). In accordance with Sections 1507.8.6 (shingles) and 1507.9.7 (shakes), shingles and shakes must be fastened with not more than two fasteners per shingle or shake.

Wood siding must be attached to the sheathing or framing in accordance with Table 2304.10.1. The method of nailing for the various siding patterns is illustrated in Commentary Figure 1405.17. Wood siding, especially plain bevel siding, applied over foam plastic sheathing requires special attention to nailing. Cooperative research by the lumber and plastic industries has shown that poor performance may be expected if the nailing is inadequate. Since the foam plastic sheathing is not an adequate nail base, the siding is required to be nailed through the sheathing to the wood-framing members. Condensation can occur on the sheathing and the back of the siding, keeping the back wet. The weather face will remain dry. This differential moisture condition will cause the siding to "curl" or warp and pull improperly installed nails out of the framing members.

[BS] 1405.18 Polypropylene siding. Polypropylene siding conforming to the requirements of this section and complying with Section 1404.12 shall be limited to exterior walls of Type VB construction located in areas where the wind speed specified in Chapter 16 does not exceed 100 miles per hour (45 m/s) and the building height is less than or equal to 40 feet (12 192 mm) in Exposure C. Where construction is located in areas where the basic wind speed exceeds 100 miles per hour (45 m/s), or building heights are in excess of 40 feet (12 192 mm), tests or calculations indicating compliance with Chapter 16 shall be submitted. Polypropylene siding shall be installed in accordance with the manufacturer's instructions. Polypropylene siding shall be secured to the building so as to provide weather protection for the exterior walls of the building.

❖ The installation of polypropylene siding is limited to Type VB buildings and is further limited based upon the basic wind speed, building height and exposure category for a given location. For areas where the basic wind speed is greater than 100 miles per hour (44 m/s), wind-load-resistance testing and structural calculations supporting the ability of the siding to withstand the basic wind speed must be provided.

SECTION 1406
COMBUSTIBLE MATERIALS ON THE EXTERIOR SIDE OF EXTERIOR WALLS

1406.1 General. Section 1406 shall apply to *exterior wall coverings*; balconies and similar projections; and bay and oriel windows constructed of combustible materials.

❖ The requirements given in this section apply to exterior elements of combustible construction. The elements included are exterior wall veneers such as vinyl or wood siding, architectural trim, gutters, half-timbering, balconies, bay and oriel windows, plastic panels and foam plastics. The code requires walls to be noncombustible for buildings of construction Types I, II, III and IV, in order to limit the fuel load. Many common types of materials used for architectural enhancement of exterior walls are combustible. Therefore, the code provides limitations as well as requirements for flame spread and radiant heat exposure testing in order to mitigate the hazards associated with combustible materials as permitted elements of noncombustible construction.

1406.2 Combustible exterior wall coverings. Combustible *exterior wall coverings* shall comply with this section.

Exception: Plastics complying with Chapter 26.

❖ The specific requirements relating to the combustible exterior wall finish include resistance to ignition through radiant heat-exposure testing (see Section 1406.2.1.1) and limitations on the height and the amount of combustible materials.

The exception is provided for plastic materials that comply with the requirements of Chapter 26 for limitations on flame spread, ignition properties and amount of plastic.

1406.2.1 Type I, II, III and IV construction. On buildings of Type I, II, III and IV construction, exterior wall coverings shall be permitted to be constructed of combustible materials, complying with the following limitations:

1. Combustible exterior wall coverings shall not exceed 10 percent of an exterior wall surface area where the fire separation distance is 5 feet (1524 mm) or less.
2. Combustible exterior wall coverings shall be limited to 40 feet (12 192 mm) in height above grade plane.
3. Combustible exterior wall coverings constructed of fire-retardant-treated wood complying with Section 2303.2 for exterior installation shall not be limited in wall surface area where the fire separation distance is 5 feet (1524 mm) or less and shall be permitted up to 60 feet (18 288 mm) in height above grade plane regardless of the fire separation distance.

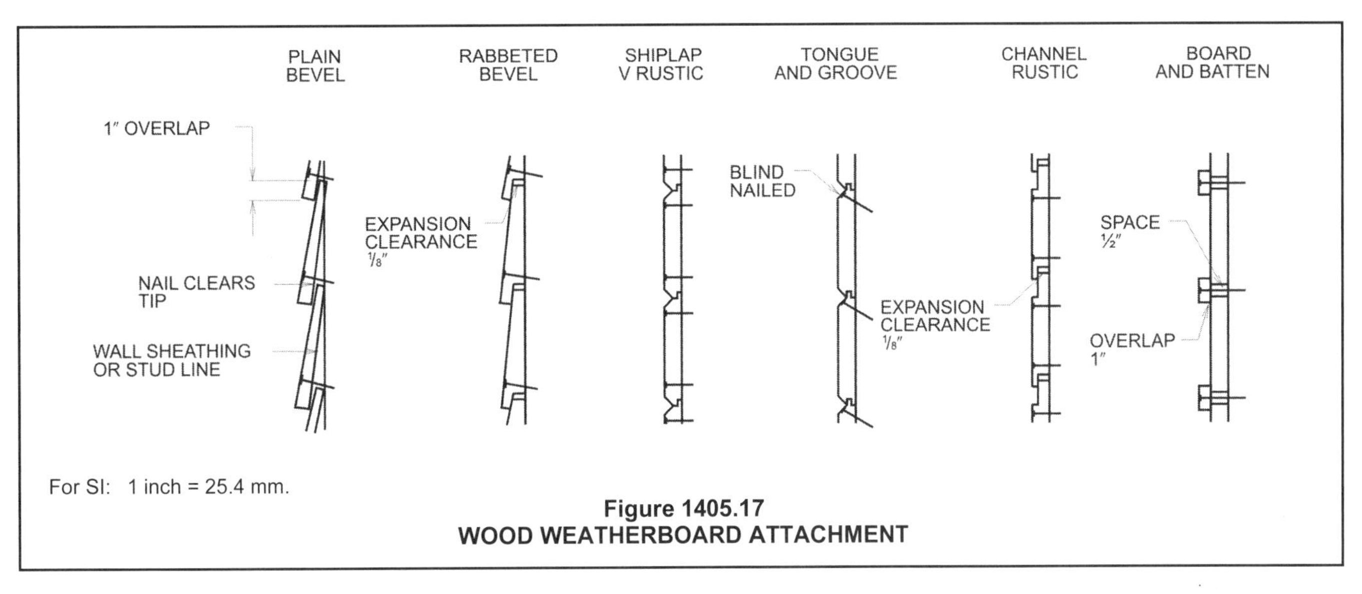

Figure 1405.17
WOOD WEATHERBOARD ATTACHMENT

4. Wood veneers shall comply with Section 1405.5.

❖ The purpose of this section is to address the hazards associated with combustible material used on exterior walls of building and structures (i.e., "Exterior wall coverings" as defined in Chapter 2). This section allows combustible exterior wall coverings under specific limitations. The first limitation addresses fire exposure from adjacent structures by limiting the area of the combustible covering when the fire separation distance is 5 feet (1524 mm) or less. When applying the second limitation, all architectural trim and exterior veneers on buildings of Type I, II, III and IV construction that are more than 40 feet (12 192 mm) above grade are required to be noncombustible. This is primarily because of the difficulties associated with fire-fighter access to higher portions of a building and to minimize potential injury from flying brands at high elevation. Combustible architectural trim and exterior veneers are permitted below the 40-foot (12 192 mm) height even when the overall building height exceeds 40 feet (12 192 mm). The third limitation addresses the FRTW. FRTW used as an exterior wall covering is unlimited in area regardless of the fire separation distance. The only qualification is that the material is limited to 60 feet (18 288 mm) in height above grade. FRTW is pressure treated with chemicals that make the wood more resistant to flame spread. Therefore, it can be expected to perform better than other combustible materials in the resistance to flame. The fourth limitation requires wood materials to comply with Section 1405.5, which contains limitations specific to wood veneers. Lastly, combustible exterior wall coverings are permitted on buildings of Type V construction without limitation. It should be noted that although Table 504.3 limits buildings of Type V construction to 40 feet (12 192 mm) in height, such buildings may be increased in height when protected with an automatic sprinkler system in accordance with Section 903.3.1.1.

1406.2.1.1 Ignition resistance. Where permitted by Section 1406.2.1, combustible exterior wall coverings shall be tested in accordance with NFPA 268.

Exceptions:

1. Wood or wood-based products.
2. Other combustible materials covered with an exterior weather covering, other than vinyl sidings, included in and complying with the thickness requirements of Table 1405.2.
3. Aluminum having a minimum thickness of 0.019 inch (0.48 mm).

❖ In order to provide a level of safety for adjoining property that is consistent with the fire-safety provisions for opening limitations in exterior walls, radiant-heat testing is required for combustible materials other than wood, using 1.25 W/cm^2 as a baseline exposure criterion.

This section requires that testing be performed in accordance with the NFPA 268 test method to measure the incident radiant heat that will cause a given exterior finish material to ignite.

There are three exceptions to the test requirements. Exception 1 is given because wood or wood-based products have been chosen as the baseline criteria for evaluation of other combustible materials. Testing in accordance with this section would be redundant since such materials are in compliance by virtue of being the baseline material.

Exception 2 relates to materials covered by one of the veneers listed in Table 1405.2, including the minimum material thickness requirements. For example, the exception would apply to the building paper under vinyl siding, not the vinyl siding itself. Again, this is an exception to eliminate any confusion that might result from the presence within a wall assembly of combustible components that are not the exposed exterior finish.

Exception 3 is provided because, although most aluminum alloys are combustible, aluminum has a higher ignition temperature than wood. Although not tested, it is anticipated that the critical radiant heat flux would therefore also be higher.

1406.2.1.1.1 Fire separation 5 feet or less. Where installed on exterior walls having a fire separation distance of 5 feet (1524 mm) or less, combustible exterior wall coverings shall not exhibit sustained flaming as defined in NFPA 268.

❖ In order to qualify for use on walls with a fire separation distance as little as 5 feet (1524 mm), the material must not flame when exposed to a 12.5 kW/m^2 exposure.

1406.2.1.1.2 Fire separation greater than 5 feet. For fire separation distances greater than 5 feet (1524 mm), any exterior wall covering shall be permitted that has been exposed to a reduced level of incident radiant heat flux in accordance with the NFPA 268 test method without exhibiting sustained flaming. The minimum fire separation distance required for the exterior wall covering shall be determined from Table 1406.2.1.1.2 based on the maximum tolerable level of incident radiant heat flux that does not cause sustained flaming of the exterior wall covering.

❖ Exterior wall coverings with a lower tolerance to radiant heat (less than 12.5 kW/m^2) can be used, but the trade-off is an increased fire separation distance to reduce the potential of ignition from radiant heat.

TABLE 1406.2.1.1.2
MINIMUM FIRE SEPARATION FOR COMBUSTIBLE EXTERIOR WALL COVERINGS

FIRE SEPARATION DISTANCE (feet)	TOLERABLE LEVEL INCIDENT RADIANT HEAT ENERGY (kW/m^2)	FIRE SEPARATION DISTANCE (feet)	TOLERABLE LEVEL INCIDENT RADIANT HEAT ENERGY (kW/m^2)
5	12.5	16	5.9
6	11.8	17	5.5
7	11.0	18	5.2
8	10.3	19	4.9
9	9.6	20	4.6
10	8.9	21	4.4
11	8.3	22	4.1
12	7.7	23	3.9
13	7.2	24	3.7
14	6.7	25	3.5
15	6.3		

For SI: 1 foot = 304.8 mm, 1 Btu/H^2 × °F = 0.0057 kW/m^2 × K.

❖ The table provides an increased minimum fire separation distance where the resistance of a material to radiant heat is less than the 12.5 kW/m^2 required for veneers on walls with a separation distance of 5 feet (1524 mm). The required resistance to radiant heat decreases with increasing separation distances. It is not anticipated that interpolation will be involved, since fire separation distances are not enforced in fractions of feet. For example, if testing indicated a tolerable level of radiant heat energy of 9.5 kW/m^2, the required minimum fire separation distance would be 10 feet (30 480 mm).

1406.2.2 Location. Combustible exterior wall coverings located along the top of exterior walls shall be completely backed up by the exterior wall and shall not extend over or above the top of the exterior wall.

❖ Combustible trim is not permitted to extend above the exterior wall to which it is attached. This is intended to limit the overall potential involvement of exterior combustible materials in a fire and to reduce the risk of fire spreading to or from such materials from the interior limits of the building. A typical example of architectural trim in this instance is a wood canopy attached to the exterior wall of a strip shopping center, with the canopy fully backed up by the wall.

1406.2.3 Fireblocking. Where the combustible exterior wall covering is furred out from the exterior wall and forms a solid surface, the distance between the back of the exterior wall covering and the exterior wall shall not exceed $1^5/_8$ inches (41 mm). The concealed space thereby created shall be fireblocked in accordance with Section 718.

Exception: The distance between the back of the exterior wall covering and the exterior wall shall be permitted to exceed $1^5/_8$ inches (41 mm) where the concealed space is not required to be fireblocked by Section 718.

❖ This section limits the area of concealed space behind combustible veneers installed on the exterior of buildings of Type I, II, III or IV construction. The intent of fireblocking is to limit the potential for fire to spread through the concealed spaces formed by combustible veneers or architectural trim. Fireblocking materials and methods are regulated by Section 718.

The exception allows for larger distances between the exterior wall covering and the exterior wall, provided that firestopping in accordance with Section 718 is not required. The applicable exception in Section 718.2.6 is Exception 2, which allows the fireblocking to be omitted when the face of the combustible exterior wall covering exposed to the concealed space is covered by the noncombustible materials listed in that exception. Furthermore, Exception 3 to Section 718.2.6 allows the omission of fireblocking in these concealed spaces when the exterior wall covering has been tested in accordance with NFPA 285 and has successfully met the acceptance criteria therein. In both those cases, there will be little opportunity for a fire in the concealed space to spread via the materials in the concealed space because of their noncombustible coverings or because they successfully passed NFPA 285 to show limited fire and flame spread over the face of, as well as within the interior of, the exterior wall system.

1406.3 Balconies and similar projections. Balconies and similar projections of combustible construction other than fire-retardant-treated wood shall be fire-resistance rated where required by Table 601 for floor construction or shall be of Type IV construction in accordance with Section 602.4. The aggregate length of the projections shall not exceed 50 percent of the building's perimeter on each floor.

Exceptions:

1. On buildings of Type I and II construction, three stories or less above *grade plane*, *fire-retardant-treated wood* shall be permitted for balconies, porches, decks and exterior stairways not used as required exits.

2. Untreated wood is permitted for pickets and rails or similar guardrail devices that are limited to 42 inches (1067 mm) in height.

3. Balconies and similar projections on buildings of Type III, IV and V construction shall be permitted to be of Type V construction, and shall not be required to have a *fire-resistance rating* where sprinkler protection is extended to these areas.

4. Where sprinkler protection is extended to the balcony areas, the aggregate length of the balcony on each floor shall not be limited.

❖ Because these elements are, in a sense, an extension of floor construction, combustible appendages are required to afford the same required fire-resistance rating as required for floor construction in Table 602, unless the appendage is of FRTW or heavy timber construction (Type IV construction). As an additional safeguard against exterior fire spread, the aggregate length of combustible appendages must not exceed 50 percent of the building perimeter on each floor. Balconies, porches, decks, supplemental exterior stairs and similar appendages in buildings of Types I and II construction are required to be constructed of noncombustible materials in order to prevent fire involvement and fire spread up or along the exterior of a noncombustible building. In buildings of Types III, IV and V construction, the use of combustible materials for these elements is permitted.

Exception 1 permits balconies and similar appendages to be constructed of FRTW where buildings of Types I and II construction do not exceed three stories in height. This is because of limited combustibility of FRTW. The three-story limitation is similar to the provisions contained in Section 1406.2.1 for combustible trim.

Exception 2 permits the use of combustible guardrails and handrails for all types of construction in an attempt to alleviate the warping and maintenance problems associated with thin lumber members used for pickets and rails. These items need not be constructed of FRTW.

Exception 3 is applicable to buildings of Types III, IV and V construction. Balconies and similar projections need not be constructed of FRTW nor have a fire-resistance rating when the appendages are protected with an automatic sprinkler system. The presence of sprinkler protection, such as a dry pendent sprinkler, will also serve to limit fire spread from floor to floor.

Balconies, porches, decks and supplemental exterior stairways that are not attached to or supported by the building are separate structures and are to be built accordingly. Although the structural support may be independent, such structures could serve to assist vertical fire spread depending on the construction of the exterior wall and the presence or lack of opening protectives.

1406.4 Bay and oriel windows. Bay and oriel windows shall conform to the type of construction required for the building to which they are attached.

Exception: *Fire-retardant-treated wood* shall be permitted on buildings three stories or less above grade plane of Type I, II, III or IV construction.

❖ In all buildings of other than Type V construction, bay windows and similar appendages are required to be constructed of noncombustible materials (see Commentary Figure 1406.4). This is consistent with the requirements for exterior walls in Type I, II, III or IV construction in Sections 602.2 through 602.4. Based upon the limited combustibility of these elements of FRTW, the exception permits their use on buildings of

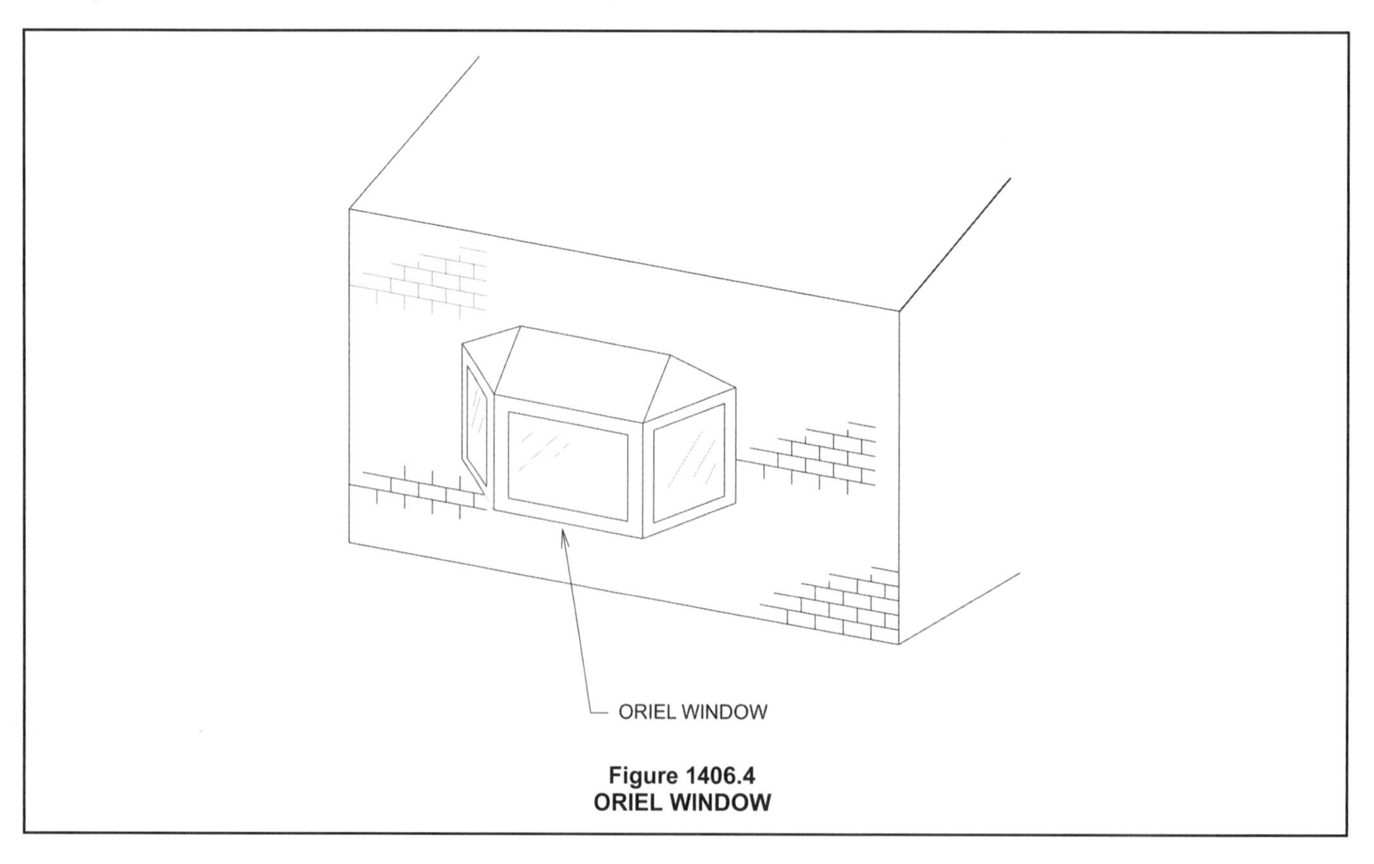

Figure 1406.4
ORIEL WINDOW

Type I, II, III or IV construction when the elements are constructed of FRTW and the building does not exceed three stories.

SECTION 1407
METAL COMPOSITE MATERIALS (MCM)

1407.1 General. The provisions of this section shall govern the materials, construction and quality of metal composite materials (MCM) for use as *exterior wall coverings* in addition to other applicable requirements of Chapters 14 and 16.

❖ MCMs are panels with a solid plastic core that is encapsulated with an aluminum facer on both sides. MCMs are typically used in exterior applications as a weather covering; however, they do not provide insulation. Section 1407 regulates MCMs that are used as exterior wall coverings. MCM used in a specific assembly, including joints and structure attachments and attachment members, are MCM systems. Commentary Figures 1407.1(1), 1407.1(2) and 1407.1(3) illustrate the details of some types of MCM systems.

1407.2 Exterior wall finish. MCM used as *exterior wall* finish or as elements of balconies and similar projections and bay and oriel windows to provide cladding or weather resistance shall comply with Sections 1407.4 through 1407.14.

❖ The plastic core of aluminum composite materials is a combustible material. This section provides a correlation with Sections 1406.2, 1406.3 and 1406.4 regulating the use of combustible materials on the exterior of buildings.

1407.3 Architectural trim and embellishments. MCM used as architectural *trim* or embellishments shall comply with Sections 1407.7 through 1407.14.

❖ In a manner similar to Section 1407.2, the provisions of this section provide a correlation with Section 1406.2.1, which regulates the use of combustible trim on the exterior of buildings.

1407.4 Structural design. MCM systems shall be designed and constructed to resist wind loads as required by Chapter 16 for components and cladding.

❖ As previously stated in numerous locations throughout this chapter, exterior weather coverings must be designed and installed to resist the design wind loads required by Chapter 16. These provisions apply to the MCM panels, the supporting elements that are part of the MCM system and the attachment of the system to the building frame.

1407.5 Approval. Results of *approved* tests or an engineering analysis shall be submitted to the *building official* to verify compliance with the requirements of Chapter 16 for wind loads.

❖ The determination that an MCM system will be able to resist a specific wind load may be accomplished through wind load testing, structural engineering calculations or a combination of both. In all instances, documentation substantiating the system's ability to resist the design wind loads is required to be submitted to the building official.

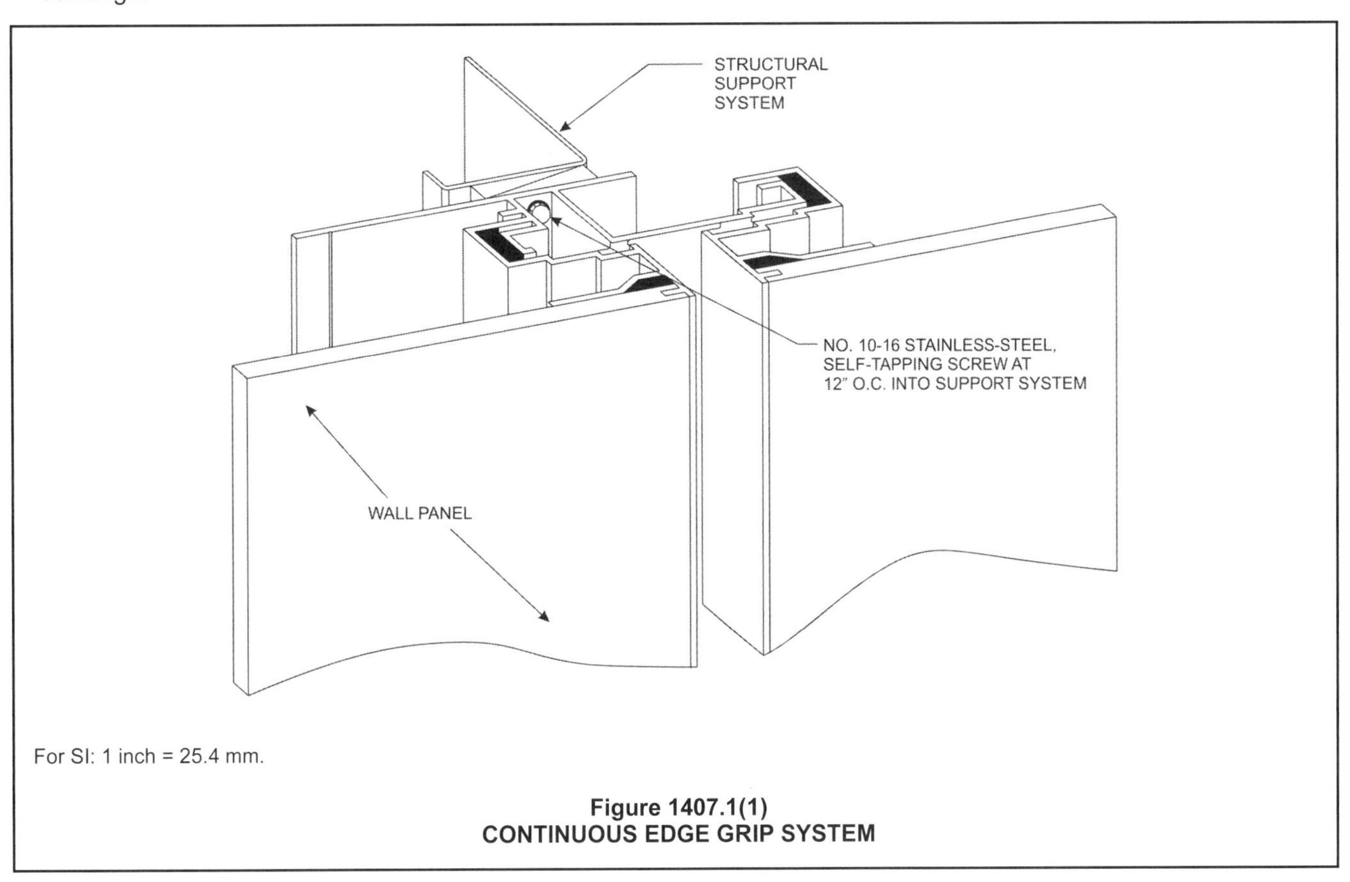

Figure 1407.1(1)
CONTINUOUS EDGE GRIP SYSTEM

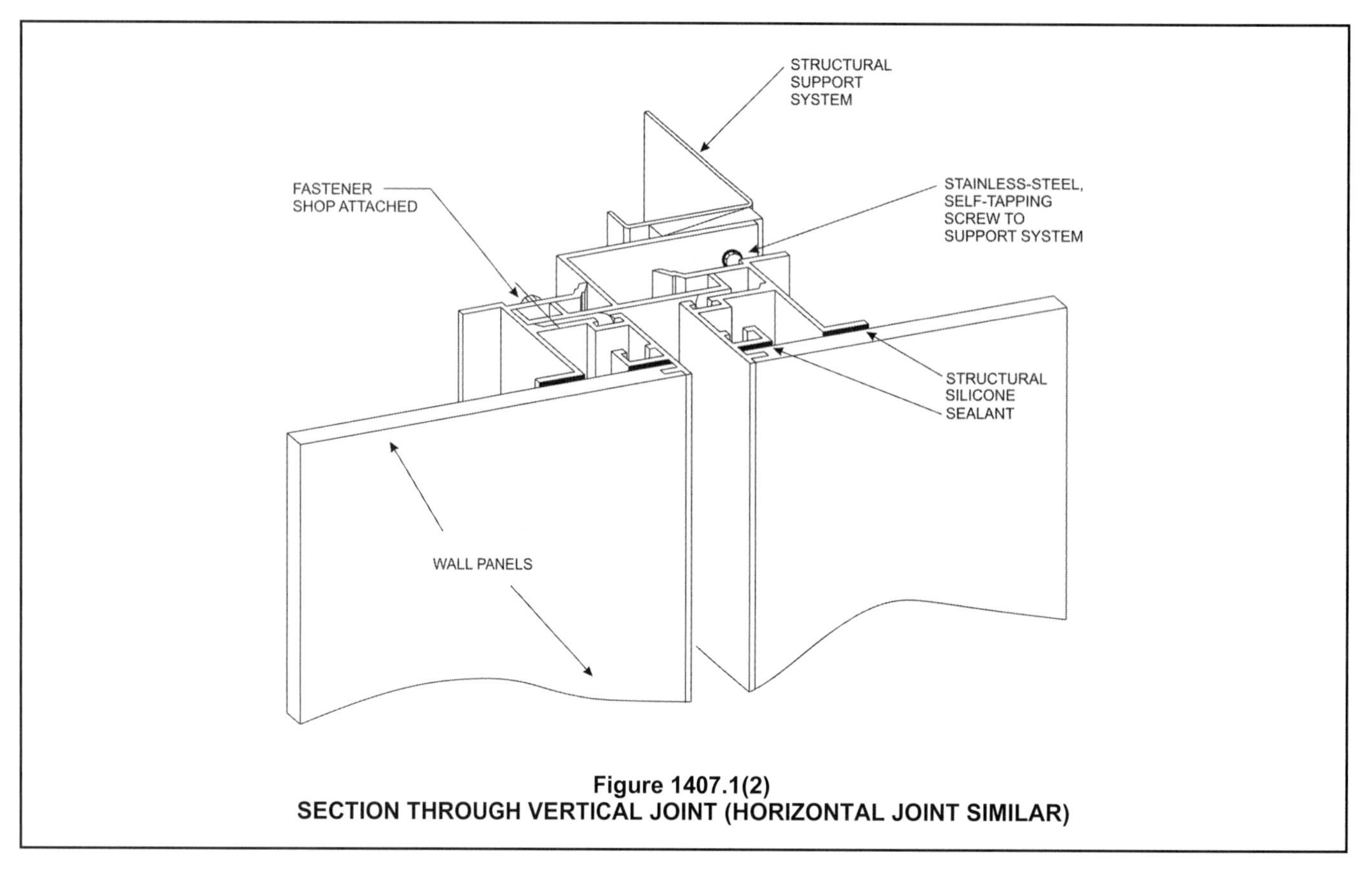

Figure 1407.1(2)
SECTION THROUGH VERTICAL JOINT (HORIZONTAL JOINT SIMILAR)

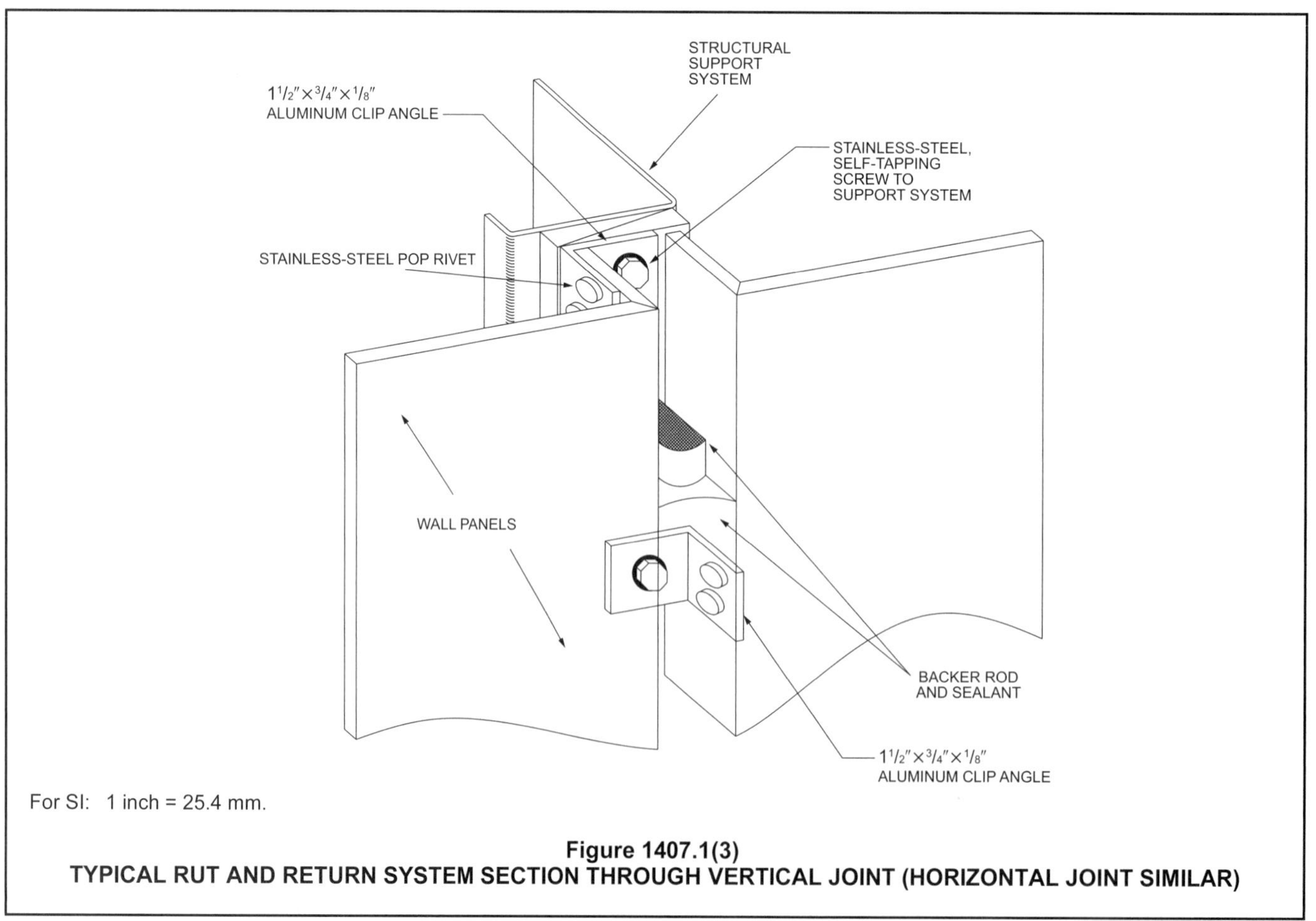

For SI: 1 inch = 25.4 mm.

Figure 1407.1(3)
TYPICAL RUT AND RETURN SYSTEM SECTION THROUGH VERTICAL JOINT (HORIZONTAL JOINT SIMILAR)

1407.6 Weather resistance. MCM systems shall comply with Section 1403 and shall be designed and constructed to resist wind and rain in accordance with this section and the manufacturer's installation instructions.

❖ As with other exterior weather coverings, MCM systems must comply with the requirements of Section 1403 to prevent the entry of moisture into the building or structure. The manufacturer's installation instructions contain details that are specific to a weathertight installation.

1407.7 Durability. MCM systems shall be constructed of *approved* materials that maintain the performance characteristics required in Section 1407 for the duration of use.

❖ All materials used as weather coverings are required to exhibit a certain level of durability to be effective in providing protection to the structure and its occupants. MCM panels are a composite construction and the bond between the aluminum facing and plastic core must be maintained for the material to maintain the performance characteristics specified in Section 1407. For example, if the bond between the panel and core is not durable, debonding could occur. This could result in detrimental performance, such as possible water intrusion.

1407.8 Fire-resistance rating. Where MCM systems are used on exterior walls required to have a *fire-resistance rating* in accordance with Section 705, evidence shall be submitted to the *building official* that the required *fire-resistance rating* is maintained.

Exception: MCM systems not containing foam plastic insulation, which are installed on the outer surface of a fire-resistance-rated *exterior wall* in a manner such that the attachments do not penetrate through the entire *exterior wall* assembly, shall not be required to comply with this section.

❖ Evidence must be submitted to confirm that the installation of MCM systems will not reduce the fire-resistance rating of exterior walls that are required by Section 704.5 to have a rating. These data may consist of testing performed in accordance with ASTM E119 of the MCM applied to the fire-resistance-rated wall assembly, as provided for in Section 703.2, or an alternative method, such as engineering analysis, as provided for in Section 703.3, respectively (see commentary, Sections 703.2 and 703.3).

1407.9 Surface-burning characteristics. Unless otherwise specified, MCM shall have a *flame spread index* of 75 or less and a smoke-developed index of 450 or less when tested in the maximum thickness intended for use in accordance with ASTM E84 or UL 723.

❖ MCMs are required to have a basic flame spread index of 75 and a smoke-developed index of 450. This is consistent with the requirements for other types of plastics intended for use in building construction, including foam plastics. As indicated in Section 1407.10.1, the flame spread rating required is more restrictive (25 or less) when MCMs are installed on buildings of Types I, II, III and IV construction.

1407.10 Type I, II, III and IV construction. Where installed on buildings of Type I, II, III and IV construction, MCM systems shall comply with Sections 1407.10.1 through 1407.10.4, or Section 1407.11.

❖ Due to the combustible core of MCM panels, the fire performance of these panels is regulated. Sections 1407.10.1 through 1407.10.4 stipulate the fire testing that is required to permit the use of these materials on buildings of other than Type V construction. It should be noted that the requirements of these sections are similar to the provisions contained in Chapter 26 for foam plastic materials. Where fire testing is specified, the panels are required to be tested in the maximum thickness intended for use. This is because the maximum thickness represents the greatest potential fuel load in a fire condition.

It should be noted that since MCM panels have an aluminum facing with a minimum thickness of 0.019 inch (0.48 mm), these materials are not required to undergo the ignition-resistance testing required by Section 1406.2.1.

1407.10.1 Surface-burning characteristics. MCM shall have a *flame spread index* of not more than 25 and a smoke-developed index of not more than 450 when tested as an assembly in the maximum thickness intended for use in accordance with ASTM E84 or UL 723.

❖ The limitations on the maximum flame spread and smoke-developed indexes correspond with the requirements of Section 2603.5.4 for foam plastic materials used on the exterior of buildings of other than Type V construction (see commentary, Section 2603.5.4).

1407.10.2 Thermal barriers. MCM shall be separated from the interior of a building by an approved thermal barrier consisting of $^1/_2$-inch (12.7 mm) gypsum wallboard or a material that is tested in accordance with and meets the acceptance criteria of both the Temperature Transmission Fire Test and the Integrity Fire Test of NFPA 275.

❖ The requirements for the installation of a thermal barrier to separate the MCM from the interior of the building are similar to the requirements contained in Section 2603.5.2. Before 1975, experience had shown that foam plastics covered with plaster or $^1/_2$-inch (12.7 mm) gypsum wallboard had performed satisfactorily in building fires. For this reason, $^1/_2$-inch (12.7 mm) gypsum wallboard was included in the code as a minimum requirement. It was recognized that specifying a single material would not be desirable in a performance code. Therefore, testing in accordance with NFPA 275 is allowed to qualify other materials. This test method was developed to specifically address the testing of materials to qualify as a thermal barrier. The test method provides specific construction samples, fire exposures and acceptance criteria to qualify a material to be a 15-minute thermal barrier. The test method addresses both the capability of the material to retard heat transfer via a fire-resistance test and to remain in place via a full-scale fire test.

1407.10.3 Thermal barrier not required. The thermal barrier specified for MCM in Section 1407.10.2 is not required where:

1. The MCM system is specifically approved based on tests conducted in accordance with NFPA 286 and with the acceptance criteria of Section 803.1.2.1, UL 1040 or UL 1715. Such testing shall be performed with the MCM in the maximum thickness intended for use. The MCM system shall include seams, joints and other typical details used in the installation and shall be tested in the manner intended for use.
2. The MCM is used as elements of balconies and similar projections, architectural *trim* or embellishments.

❖ The exemption of the installation of the thermal barrier required by Section 1407.10.2 may be accomplished in two ways. The first is through the use of large-scale fire testing of the MCM system. The system must be tested in a configuration that incorporates details that are representative of a typical installation. The second provision of this section exempts the installation of a thermal barrier where the MCM is limited to use as a component of balconies and similar projections or as architectural trim and embellishments.

1407.10.4 Full-scale tests. The MCM system shall be tested in accordance with, and comply with, the acceptance criteria of NFPA 285. Such testing shall be performed on the MCM system with the MCM in the maximum thickness intended for use.

❖ The final data required to permit the use of MCM systems on buildings of Types I, II, III and IV construction are the result of full-scale testing of the system. This testing is intended to determine that the system will not have the tendency to propagate the spread of fire over the surface of the panels or through the panel core.

1407.11 Alternate conditions. MCM and MCM systems shall not be required to comply with Sections 1407.10.1 through 1407.10.4 provided such systems comply with Section 1407.11.1, 1407.11.2, 1407.11.3 or 1407.11.4.

❖ Section 1407.11 provides an alternative to compliance with the requirements of Section 1407.10, based upon a specific set of criteria. These criteria limit the amount and height of MCM above grade on buildings of Types I, II, III and IV construction in a manner similar to the provisions of Section 1406 for combustible materials.

1407.11.1 Installations up to 40 feet in height. MCM shall not be installed more than 40 feet (12 190 mm) in height above grade where installed in accordance with Sections 1407.11.1.1 and 1407.11.1.2.

❖ The height limitation of 40 feet (12 192 mm) corresponds to the limitations of Section 1406.2.1 for architectural trim. In addition to the height limitation, the material must comply with the requirements of Sections 1407.11.1.1 and 1407.11.1.2.

1407.11.1.1 Fire separation distance of 5 feet or less. Where the *fire separation distance* is 5 feet (1524 mm) or less, the area of MCM shall not exceed 10 percent of the *exterior wall* surface.

❖ To limit the potential for the spread of fire between buildings that are in close proximity to one another, the maximum amount of combustible exterior veneer is limited in size. These limitations are identical to those of Section 1406.2.1 for combustible veneers.

1407.11.1.2 Fire separation distance greater than 5 feet. Where the *fire separation distance* is greater than 5 feet (1524 mm), there shall be no limit on the area of *exterior wall* surface coverage using MCM.

❖ Based on the decreased potential for the spread of fire between buildings with greater fire separation distances, there is no limit on the area of the MCM on such exterior walls. It is important to remember that the height of the MCM installation is also limited to 40 feet (12 190 mm). This is also a factor as it further limits the area of the MCM.

1407.11.2 Installations up to 50 feet in height. MCM shall not be installed more than 50 feet (15 240 mm) in height above grade where installed in accordance with Sections 1407.11.2.1 and 1407.11.2.2.

❖ The maximum installed height of MCM on buildings of Types I, II, III and IV construction may be increased up to 50 feet (15 240 mm) when the requirements of Sections 1407.11.2.1 and 1407.11.2.2 are complied with.

1407.11.2.1 Self-ignition temperature. MCM shall have a self-ignition temperature of 650°F (343°C) or greater when tested in accordance with ASTM D1929.

❖ Keeping with the requirements in Chapter 26 for other plastic materials, the flame spread of the MCM is not permitted to exceed 75 and the smoke-developed index may not exceed 450. In addition to the limitations on flame spread and smoke development, the MCM must be tested for self-ignition in accordance with ASTM D1929. These requirements are identical to those contained in Section 2606.4 for light-transmitting plastics (see commentary, Section 2606.4).

1407.11.2.2 Limitations. Sections of MCM shall not exceed 300 square feet (27.9 m^2) in area and shall be separated by not less than 4 feet (1219 mm) vertically.

❖ The size limitations of this section are consistent with the provisions of Section 2605.2, Item 3 and Section 2606.7.3. The area limitation is intended to minimize the amount of concentrated combustible materials on the exterior of noncombustible buildings. The separation requirement is intended to minimize the potential for extensive flame propagation over the face of the wall.

1407.11.3 Installations up to 75 feet in height (Option 1). MCM shall not be installed more than 75 feet (22 860 mm) in height above grade plane where installed in accordance with Sections 1407.11.3.1 through 1407.11.3.5.

> **Exception:** Buildings equipped throughout with an automatic sprinkler system in accordance with Section 903.3.1.1 shall be exempt from the height limitation.

❖ The maximum installed height of MCM on buildings of Type I, II, III and IV construction may be increased up to 75 feet (22 860 mm) when the requirements of Sections 1407.11.3.1 through 1407.11.3.5 are complied with. The exception allows for installations over 75 feet (22 860 mm) in recognition of the protection provided by the installation of an automatic sprinkler system throughout the building.

1407.11.3.1 Prohibited occupancies. MCM shall not be permitted on buildings classified as Group A-1, A-2, H, I-2 or I-3 occupancies.

❖ Consistent with the requirements for plastic wall panels in accordance with Section 2607.1, MCM is permitted to be installed to a maximum height of 75 feet (22 860 mm) only in occupancies classified as Groups A-3, A-4, A-5, B, E, F-1, F-2, 1-1, M, R-1, R-2, R-3, R-4, S-1, S-2 and U. MCM is not allowed for these installations in occupancies where the generated products of combustion would significantly impair the egress of the impaired, confined or crowded occupants.

1407.11.3.2 Nonfire-resistance-rated exterior walls. MCM shall not be permitted on exterior walls required to have a *fire-resistance rating* by other provisions of this code.

❖ Consistent with the requirements for plastic wall panels in accordance with Section 2607.1, MCM is not permitted where the exterior walls are required to be fire-resistance rated when installed to heights of 75 feet (22 860 mm).

1407.11.3.3 Specifications. MCM shall be required to comply with all of the following:

1. MCM shall have a self-ignition temperature of 650°F (343°C) or greater when tested in accordance with ASTM D1929.
2. MCM shall conform to one of the following combustibility classifications when tested in accordance with ASTM D635:

   Class CC1: Materials that have a burning extent of 1 inch (25 mm) or less when tested at a nominal thickness of 0.060 inch (1.5 mm) or in the thickness intended for use.

   Class CC2: Materials that have a burning rate of $2^1/_2$ inches per minute (1.06 mm/s) or less when tested at a nominal thickness of 0.060 inch (1.5 mm) or in the thickness intended for use.

❖ For installations up to 75 feet (22 860 mm) above grade plane, MCMs require a minimum self-ignition temperature of 650°F (343°C) when tested in accordance with ASTM D1929. This temperature excludes the use of easily ignited materials. For comparative purposes, untreated wood and paper ignite at about 500°F (260°C).

In order for an MCM to be approved for these installations, its smoke-development index or smoke-density rating is limited and must be evaluated by one of two test methods prescribed (smoke-developed index limited to 450 by ASTM E84 or UL 263, or a maximum average smoke-density rating of 75 by ASTM D2843). For comparative purposes, these smoke-developed values are the maximum that can be expected from some wood products tested under similar conditions. However, there is no intent to indicate a correlation between the results from the two test methods.

MCMs are also to be tested in accordance with ASTM D635 and be of combustibility Class CC1 or CC2. Class CC1 plastic materials burn at a slower rate than Class CC2 plastics. The above requirements are consistent with the requirements for plastic materials in Chapter 26.

1407.11.3.4 Area limitation and separation. The maximum area of a single MCM panel and the minimum vertical and horizontal separation requirements for MCM panels shall be as provided for in Table 1407.11.3.4. The maximum percentage of exterior wall area of any story covered with MCM panels shall not exceed that indicated in Table 1407.11.3.4 or the percentage of unprotected openings permitted by Section 705.8, whichever is smaller.

> **Exception:** In buildings provided with flame barriers complying with Section 705.8.5 and extending 30 inches (760 mm) beyond the exterior wall in the plane of the floor, a vertical separation shall not be required at the floor other than that provided by the vertical thickness of the flame barrier.

❖ Refer to Table 1407.11.3.4 for area limitations of single MCM panels and the minimum separation requirements (vertical and horizontal) between adjacent panels. The maximum percentage of wall area faced with MCM panels is limited by Table 1407.11.3.4, as well as by the percentage of unprotected openings allowed by Section 705.8. At all times, the most restrictive requirement prevails.

Vertical separation of panels (see Table 1407.11.3.4) is not required where continuous architectural projections create at least a 30-inch (762 mm) barrier (called a flame barrier) between adjacent panels. Under most conditions, this flame barrier will delay or prevent vertical fire spread from one story to the next. This horizontal barrier is seen as protection equivalent to the vertical separation typically required by Table 1407.11.3.4 (see Commentary Figure 1407.11.3.4).

TABLE 1407.11.3.4. See below.

❖ This table establishes size and separation limitations for exterior MCM panels. The requirements are determined by the combustibility class of the MCM (Class CC1 or CC2) and fire separation distance. MCM panels are only permitted where a minimum fire separation distance of 6 feet (1829 mm) is provided. The vertical separation requirements can be modified by Section 1407.11.3.4 (see Table 1407.11.3.4, Note a). The maximum area limitations of a single panel and the maximum percentage area of an exterior wall can be doubled when the structure is sprinklered throughout in accordance with NFPA 13 (see Section 1407.11.3.4).

1407.11.3.5 Automatic sprinkler system increases. Where the building is equipped throughout with an automatic sprinkler system in accordance with Section 903.3.1.1, the maximum percentage area of exterior wall of any story covered with MCM panels and the maximum square footage of a single area of MCM panels in Table 1407.11.3.4 shall be increased 100 percent. The area of MCM panels shall not exceed 50 percent of the exterior wall area of any story or the area permitted by Section 704.8 for unprotected openings, whichever is smaller.

❖ The extra protection provided by a complete NFPA 13 automatic fire sprinkler system enables the limitations placed on MCM panels to be reduced. In order for these modifications to apply, the structure must be sprinklered throughout. When this protection is achieved, the height to which plastic wall panels can be installed is not limited. Furthermore, the single panel size and exterior wall area limitations can be doubled. At no time, however, can the area of wall panels per story exceed 50 percent of the wall area of that story. The percentage limitation of Section 705.8 for unprotected openings is also applicable and the most restrictive area shall apply.

1407.11.4 Installations up to 75 feet in height (Option 2). MCM shall not be installed more than 75 feet (22 860 mm) in height above grade plane where installed in accordance with Sections 1407.11.4.1 through 1407.11.4.4.

Exception: Buildings equipped throughout with an automatic sprinkler system in accordance with Section 903.3.1.1 shall be exempt from the height limitation.

❖ The maximum installed height of MCM on buildings of Types I, II, III and IV construction may be increased

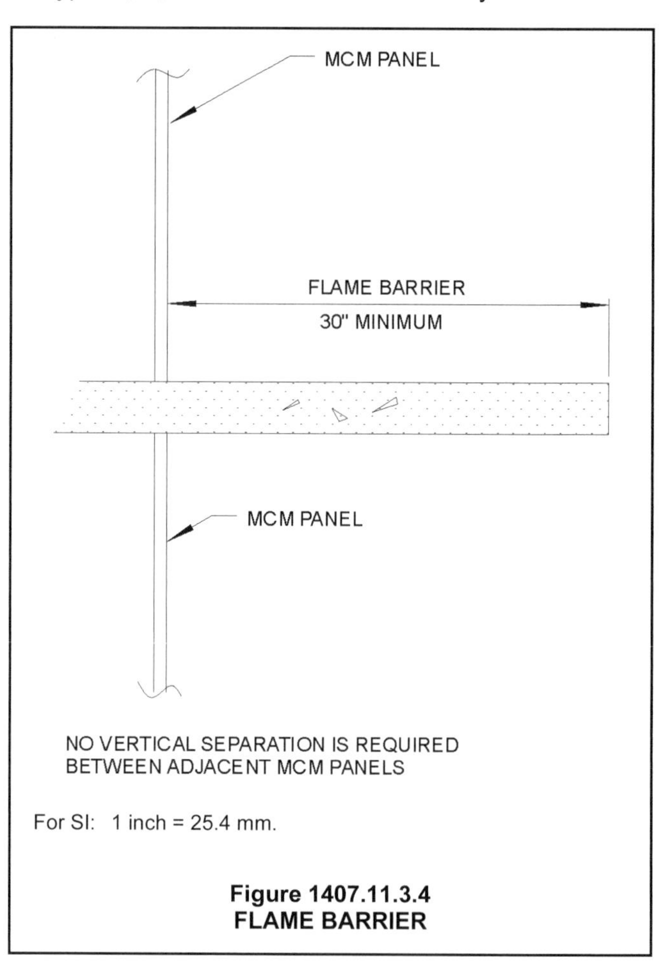

Figure 1407.11.3.4
FLAME BARRIER

TABLE 1407.11.3.4
AREA LIMITATION AND SEPARATION REQUIREMENTS FOR MCM PANELS

FIRE SEPARATION DISTANCE (feet)	COMBUSTIBILITY CLASS OF MCM	MAXIMUM PERCENTAGE AREA OF EXTERIOR WALL COVERED WITH MCM PANELS	MAXIMUM SINGLE AREA OF MCM PANELS (square feet)	MINIMUM SEPARATION OF MCM PANELS (feet)	
				Vertical	Horizontal
Less than 6	—	Not Permitted	Not Permitted	—	—
6 or more but less than 11	CC1	10	50	8	4
	CC2	Not Permitted	Not Permitted	—	—
11 or more but less than or equal to 30	CC1	25	90	6	4
	CC2	15	70	8	4
More than 30	CC1	50	Not Permitted	3[a]	0
	CC2	50	100	6[a]	3

For SI: 1 foot = 304.8 mm, 1 square foot = 0.0929 m^2.

a. For reductions in the minimum vertical separation, see Section 1407.11.3.4.

up to 75 feet (22 860 mm) when the requirements of Sections 1407.11.4.1 through 1407.11.4.4 are complied with. The exception allows for installations over 75 feet (22 860 mm) in recognition of the protection provided by the installation of an automatic sprinkler system throughout the building. These provisions are intended as an alternative to the requirements in Section 1407.11.3. This option relies on more restrictive fire separation distances and area and size limitations to achieve a level of safety equivalent to that of Section 1407.11.3. This option also allows for installation of MCM on buildings of any occupancy and on walls that are fire-resistance rated.

1407.11.4.1 Minimum fire separation distance. MCM shall not be installed on any wall with a fire separation distance less than 30 feet (9 144 mm).

Exception: Where the building is equipped throughout with an *automatic sprinkler system* in accordance with Section 903.3.1.1, the fire separation distance shall be permitted to be reduced to not less than 20 feet (6096 mm).

❖ This section requires a minimum fire separation distance of 30 feet (9144 mm), which greatly decreases the threat of fire to or from adjacent structures. The exception allows this distance to be reduced to 20 feet (6096 mm) in recognition of the protection provided by the installation of an automatic sprinkler system throughout the building.

1407.11.4.2 Specifications. MCM shall be required to comply with all of the following:

1. MCM shall have a self-ignition temperature of 650°F (343°C) or greater when tested in accordance with ASTM D1929.
2. MCM shall conform to one of the following combustibility classifications when tested in accordance with ASTM D635:

   Class CC1: Materials that have a burning extent of 1 inch (25 mm) or less when tested at a nominal thickness of 0.060 inch (1.5 mm), or in the thickness intended for use.

   Class CC2: Materials that have a burning rate of $2^1/_2$ inches per minute (1.06 mm/s) or less when tested at a nominal thickness of 0.060 inch (1.5 mm), or in the thickness intended for use.

❖ The minimum specifications for MCM used in Option 2 are identical to those required in Option 1. However, for Option 2 the combustibility classification of the MCM does not relate to differing requirements as with Option 1. These specifications, therefore, are a minimum benchmark for the MCM.

1407.11.4.3 Area and size limitations. The aggregate area of MCM panels shall not exceed 25 percent of the area of any exterior wall face of the story on which those panels are installed. The area of a single MCM panel installed above the first story above grade plane shall not exceed 16 square feet (1.5 m^2) and the vertical dimension of a single MCM panel shall not exceed 4 feet (1219 mm).

Exception: Where the building is equipped throughout with an *automatic sprinkler system* in accordance with Section 903.3.1.1, the maximum aggregate area of MCM panels shall be increased to 50 percent of the exterior wall face of the story on which those panels are installed and there shall not be a limit on the maximum dimension or area of a single MCM panel.

❖ This section sets forth the maximum area and size limitations for MCM under Option 2. These limitations are much more restrictive than those in Option 1; however, as indicated in the commentary to Section 1407.11.4, this option is applicable to buildings of any occupancy and on walls that are fire-resistance rated. The exception recognizes the extra protection provided by a complete NFPA 13 automatic fire sprinkler system and allows the limitations placed on MCM panels to be reduced. In order for these modifications to apply, the structure must be sprinklered throughout. When this protection is achieved, the single panel size (dimension and area) is not limited and exterior wall area limitations can be increased by 50 percent.

1407.11.4.4 Vertical separations. Flame barriers complying with Section 705.8 and extending 30 inches (762 mm) beyond the exterior wall or a vertical separation of not less than 4 feet (1219 mm) in height shall be provided to separate MCM panels located on the exterior walls at one-story intervals.

Exception: Buildings equipped throughout with an *automatic sprinkler system* in accordance with Section 903.3.1.1.

❖ Consistent with other provisions to limit fire spread, this section places restrictions on the size and vertical separation of MCM panels. The 4-foot (1219 mm) vertical separation is widely accepted as adequate for spandrels and other intervening construction between floors. Vertical separation of panels is not required where continuous projections create at least a 30-inch (762 mm) barrier (called a flame barrier) between adjacent panels. Under most conditions, this flame barrier will delay or prevent vertical fire spread from one story to the next. This horizontal barrier is seen as protection equivalent to the vertical separation required by this section (see Figure 1407.11.3.4).

1407.12 Type V construction. MCM shall be permitted to be installed on buildings of Type V construction.

❖ Because of the combustible nature of Type V construction, the fire testing requirements for MCM used as exterior veneers on these buildings is less severe than for other construction types. This is similar in concept to the provisions of Section 1406.

1407.13 Foam plastic insulation. MCM systems containing foam plastic insulation shall also comply with the requirements of Section 2603.

❖ The purpose of this section is to simply point the user of the code to Section 2603 for foam plastics when the MCM contains foam plastic components. The common use of foam plastic in MCM insulation would be the core of the material.

1407.14 Labeling. MCM shall be labeled in accordance with Section 1703.5.

❖ Because of the composite nature of MCMs and the proprietary nature of the panel cores, the fabrication of the panels must be subject to independent third-party verification. The verification of this inspection is evidenced by labeling of the product by the third-party agency (refer to Section 1703.5 for more discussion on the labeling of products).

SECTION 1408
EXTERIOR INSULATION AND FINISH SYSTEMS (EIFS)

1408.1 General. The provisions of this section shall govern the materials, construction and quality of exterior insulation and finish systems (EIFS) for use as *exterior wall coverings* in addition to other applicable requirements of Chapters 7, 14, 16, 17 and 26.

❖ EIFS and EIFS with drainage are relatively complex exterior veneers that require great care and attention in the materials used and the installation details. This exterior veneer is now dealt with in this separate section. All of the performance criteria, installation, and design are covered in this section. EIFS is defined in Section 1402 of this code as wall cladding systems consisting of a foam plastic insulation material covered by a plastic reinforced Portland cement base coat and a textured finish coat. The complexity of this material comes in the installation details. For the material to be able to perform as a watertight barrier, the various layers must be carefully applied to strict specifications.

1408.2 Performance characteristics. EIFS shall be constructed such that it meets the performance characteristics required in ASTM E2568.

❖ ASTM E2568 defines the necessary performance of EIFS systems. This standard contains basic test criteria for the materials and the entire system. These are generally tests that provide minimum performance levels for the system, including salt spray, accelerated weathering, ignition resistance, fire endurance and wind resistance.

[BS] 1408.3 Structural design. The underlying structural framing and substrate shall be designed and constructed to resist loads as required by Chapter 16.

❖ This section addresses the structural design of the supporting members. The section underscores the fact that Section 1408 does not obviate the need to support the EIFS with a structural support system.

1408.4 Weather resistance. EIFS shall comply with Section 1403 and shall be designed and constructed to resist wind and rain in accordance with this section and the manufacturer's application instructions.

❖ The important aspect of this section is that, like all other exterior cladding materials, EIFS must be weather resistant as well. Section 1403 provides the performance criteria for exterior cladding.

1408.4.1 EIFS with drainage. EIFS with drainage shall have an average minimum drainage efficiency of 90 percent when tested in accordance the requirements of ASTM E2273 and is required on framed walls of Type V construction, Group R1, R2, R3 and R4 occupancies.

❖ For wood-framed walls or any light-frame construction using wood sheathing or other water soluble materials, it is essential to protect the supporting substrate from moisture. The EIFS system, like all other exterior veneers, must be able to allow water that gets behind the cladding to drain.

1408.4.1.1 Water-resistive barrier. For EIFS with drainage, the *water-resistive barrier* shall comply with Section 1404.2 or ASTM E2570.

❖ The EIFS systems with drainage are required to have a barrier behind the cladding that serves as the "drainage plane" where all water penetration is stopped and directed downward and out of the wall. The section or referenced standard provide basic test criteria for the materials that would be used behind an EIFS system with drainage.

1408.5 Installation. Installation of the EIFS and EIFS with drainage shall be in accordance with the EIFS manufacturer's instructions.

❖ Manufacturers' instructions provide the proper installation parameters for these materials, including, among other information, the proper mix proportions for the cement-based substrate, the correct sequence for installation and setting times. The manufacturer's instructions form the job specifications for this material.

1408.6 Special inspections. EIFS installations shall comply with the provisions of Sections 1704.2 and 1705.16.

❖ The installation of EIFS systems require attention to detail because the performance of the system can only be assured if the system is installed correctly. This sensitivity requires that special inspections be performed to ensure that ambient conditions were within acceptable range, installation sequences were followed, and all components meet the manufacturer's requirements.

SECTION 1409 HIGH-PRESSURE DECORATIVE EXTERIOR-GRADE COMPACT LAMINATES (HPL)

1409.1 General. The provisions of this section shall govern the materials, construction and quality of High-Pressure Decorative Exterior-Grade Compact Laminates (HPL) for use as exterior wall coverings in addition to other applicable requirements of Chapters 14 and 16.

❖ High-pressure decorative exterior-grade compact laminates (HPL) are panels with a cellulose fibrous material bonded with a thermosetting resin by a high-pressure process. HPL are typically used in exterior applications as a weather covering; however, they do not provide insulation.

1409.2 Exterior wall finish. HPL used as exterior wall covering or as elements of balconies and similar projections and bay and oriel windows to provide cladding or weather resistance shall comply with Sections 1409.4 and 1409.14.

❖ The plastic core of aluminum composite material is a combustible material. This section provides a correlation with Sections 1406.2, 1406.3 and 1406.4, which regulate the use of combustible materials on the exterior of buildings.

1409.3 Architectural trim and embellishments. HPL used as architectural trim or embellishments shall comply with Sections 1409.7 through 1409.14.

❖ In a manner similar to Section 1409.2, the provisions of this section provide a correlation with Section 1406.2.1, which regulates the use of combustible trim on the exterior of buildings.

[BS] 1409.4 Structural design. HPL systems shall be designed and constructed to resist wind loads as required by Chapter 16 for components and cladding.

❖ As previously stated in numerous locations throughout this chapter, exterior weather coverings must be designed and installed to resist the design wind loads required by Chapter 16. These provisions apply to the HPL panels, the supporting elements that are part of the HPL system and the attachment of the system to the building frame.

1409.5 Approval. Results of approved tests or an engineering analysis shall be submitted to the building official to verify compliance with the requirements of Chapter 16 for wind loads.

❖ The determination that an HPL system will be able to resist a specific wind load may be accomplished through wind-load testing, structural engineering calculations or a combination of both. In all instances, documentation substantiating the system's ability to resist the design wind loads is required to be submitted to the building official.

1409.6 Weather resistance. HPL systems shall comply with Section 1403 and shall be designed and constructed to resist wind and rain in accordance with this section and the manufacturer's instructions.

❖ As with other exterior weather coverings, HPL systems must comply with the requirements of Section 1403 to prevent the entry of moisture into the building or structure. The manufacturer's installation instructions contain details that are specific to a weather-tight installation.

1409.7 Durability. HPL systems shall be constructed of approved materials that maintain the performance characteristics required in Section 1409 for the duration of use.

❖ All materials used as weather coverings are required to exhibit a certain level of durability to be effective in providing protection to the structure and its occupants.

1409.8 Fire-resistance rating. Where HPL systems are used on exterior walls required to have a *fire-resistance rating* in accordance with Section 705, evidence shall be submitted to the building official that the required *fire-resistance rating* is maintained.

> **Exception:** HPL systems not containing foam plastic insulation, which are installed on the outer surface of a fire-resistance-rated exterior wall in a manner such that the attachments do not penetrate through the entire exterior wall assembly, shall not be required to comply with this section.

❖ Evidence must be submitted to confirm that the installation of HCL systems will not reduce the fire-resistance rating of exterior walls required by Section 705.5. These data may consist of testing performed in accordance with ASTM E119 of the HCL applied to the fire-resistance-rated wall assembly, as provided for in Section 703.2, or an alternative method, such as engineering analysis, as provided for in Section 703.3 (see commentary, Sections 703.2 and 703.3).

1409.9 Surface-burning characteristics. Unless otherwise specified, HPL shall have a flame spread index of 75 or less and a smoke-developed index of 450 or less when tested in the minimum and maximum thicknesses intended for use in accordance with ASTM E84 or UL 723.

❖ HCL are required to have a basic flame spread index of 75 and a smoke-developed index of 450. This is consistent with the requirements for other types of plastics intended for use in building construction, including foam plastics. As indicated in Section 1407.10.1, the flame spread rating required is more restrictive (not more than 25) when HCL are installed on buildings of Types I, II, III and IV construction.

1409.10 Type I, II, III and IV construction. Where installed on buildings of Type I, II, III and IV construction,

HPL systems shall comply with Sections 1409.10.1 through 1409.10.4, or Section 1409.11.

❖ Due to the combustible core of HCL panels, the fire performance of these panels is regulated. Sections 1407.10.1 through 1407.10.4 stipulate the fire testing that is required to permit the use of these materials on buildings of other than Type V construction. It should be noted that the requirements of these sections are similar to the provisions contained in Chapter 26 for foam plastic materials. Where fire testing is specified, the panels are required to be tested in the maximum thickness intended for use. This is because the maximum thickness represents the greatest potential fuel load in a fire condition.

1409.10.1 Surface-burning characteristics. HPL shall have a flame spread index of not more than 25 and a smoke-developed index of not more than 450 when tested in the minimum and maximum thicknesses intended for use in accordance with ASTM E84 or UL 723.

❖ The limitations on the maximum flame spread and smoke-developed indexes of the HCL used in this application correspond with the requirements of Section 2603.5.4 for foam plastic materials used on the exterior of buildings of other than Type V construction (see commentary, Section 2603.5.4).

1409.10.2 Thermal barriers. HPL shall be separated from the interior of a building by an approved thermal barrier consisting of $^1/_2$-inch (12.7 mm) gypsum wallboard or a material that is tested in accordance with and meets the acceptance criteria of both the Temperature Transmission Fire Test and the Integrity Fire Test of NFPA 275.

❖ The requirements for the installation of a thermal barrier to separate the HPL from the interior of the building are similar to the requirements contained in Section 2603.5.2. Before 1975, experience had shown that foam plastics covered with plaster or $^1/_2$-inch (12.7 mm) gypsum wallboard had performed satisfactorily in building fires. For this reason, $^1/_2$-inch (12.7 mm) gypsum wallboard was included in the code as a minimum requirement. It was recognized that specifying a single material would not be desirable in a performance code. Therefore, testing in accordance with NFPA 275 is allowed to qualify other materials. This test method was developed to specifically address the testing of materials to qualify as a thermal barrier. The test method provides specific construction samples, fire exposures and acceptance criteria to qualify a material to be a 15-minute thermal barrier. The test method addresses both the capability of the material to retard heat transfer via a fire-resistance test and to remain in place via a full-scale fire test.

1409.10.3 Thermal barrier not required. The thermal barrier specified for HPL in Section 1409.10.2 is not required where:

1. The HPL system is specifically approved based on tests conducted in accordance with UL 1040 or UL 1715. Such testing shall be performed with the HPL in the minimum and maximum thicknesses intended for use. The HPL system shall include seams, joints and other typical details used in the installation and shall be tested in the manner intended for use.
2. The HPL is used as elements of balconies and similar projections, architectural *trim* or embellishments.

❖ The exemption of the installation of the thermal barrier required by Section 1407.10.2 may be accomplished in two ways. The first is through the use of large-scale fire testing of the HCL system. The system must be tested in a configuration that incorporates details that are representative of a typical installation. The second provision of this section exempts the installation of a thermal barrier where the use of HCL is limited to a component of balconies and similar projections or as architectural trim and embellishments.

1409.10.4 Full-scale tests. The HPL system shall be tested in accordance with, and comply with, the acceptance criteria of NFPA 285. Such testing shall be performed on the HPL system with the HPL in the minimum and maximum thicknesses intended for use.

❖ The final data required to permit the use of HCL systems on buildings of Types I, II, III and IV construction are the result of full-scale testing of the system. This testing is intended to determine that the system will not have the tendency to propagate the spread of fire over the surface of the panels or through the panel core.

1409.11 Alternate conditions. HPL and HPL systems shall not be required to comply with Sections 1409.10.1 through 1409.10.4 provided such systems comply with Section 1409.11.1 or 1409.11.2.

❖ Section 1407.11 provides an alternative to compliance with the requirements of Section 1407.10, based upon a specific set of criteria. These criteria limit the amount of material and height above grade in a manner similar to the provisions of Section 1406 for combustible materials.

1409.11.1 Installations up to 40 feet in height. HPL shall not be installed more than 40 feet (12 190 mm) in height above grade plane where installed in accordance with Sections 1409.11.1.1 and 1409.11.1.2.

❖ The height limitation of 40 feet (12 192 mm) corresponds to the limitations of Section 1406.2.2 for architectural trim. In addition to the height limitation, the material must comply with the requirements of Sections 1407.11.1 and 1407.11.2.

1409.11.1.1 Fire separation distance of 5 feet or less. Where the fire separation distance is 5 feet (1524 mm) or less, the area of HPL shall not exceed 10 percent of the exterior wall surface.

❖ To limit the potential for the spread of fire between buildings that are in close proximity to one another, the maximum amount of combustible exterior veneer is limited in size. These limitations are identical to those of Section 1406.2.1 for combustible veneers.

1409.11.1.2 Fire separation distance greater than 5 feet. Where the fire separation distance is greater than 5 feet (1524 mm), there shall be no limit on the area of exterior wall surface coverage using HPL.

❖ Based on the decreased potential for the spread of fire between buildings with greater fire separation distances, there is no limit on the area of the HCL on such exterior walls. It is important to remember that the height of the HCL installation is also limited to 40 feet (12 192 mm). This is also a factor as it further limits the area of the HCL.

1409.11.2 Installations up to 50 feet in height. HPL shall not be installed more than 50 feet (15 240 mm) in height above grade plane where installed in accordance with Sections 1409.11.2.1 and 1409.11.2.2.

❖ The maximum installed height of HCL on buildings of Type I, II, III and IV construction may be increased up to 50 feet (15 240 mm) when the requirements of Sections 1407.11.2.1 and 1407.11.2.2 are complied with.

1409.11.2.1 Self-ignition temperature. HPL shall have a self-ignition temperature of 650°F (343°C) or greater when tested in accordance with ASTM D1929.

❖ In addition to the limitations on flame spread and smoke development, the HCL must be tested for self-ignition in accordance with ASTM D1929. These requirements are identical to those contained in Section 2606.4 for light-transmitting plastics (see commentary, Section 2606.4).

1409.11.2.2 Limitations. Sections of HPL shall not exceed 300 square feet (27.9 m^2) in area and shall be separated by a minimum 4 feet (1219 mm) vertically.

❖ The size limitations of this section are consistent with the provisions of Section 2605.2, Item 3, and Section 2606.7.3. The area limitation is intended to minimize the amount of concentrated combustible materials on the exterior of noncombustible buildings. The separation requirement is intended to minimize the potential for extensive flame propagation over the face of the wall.

1409.12 Type V construction. HPL shall be permitted to be installed on buildings of Type V construction.

❖ Because of the combustible nature of Type V construction, the fire-testing requirements for HCL used as exterior veneers on these buildings is less severe than for other construction types. This is similar in concept to the provisions of Section 1406.

1409.13 Foam plastic insulation. HPL systems containing foam plastic insulation shall also comply with the requirements of Section 2603.

❖ The purpose of this section is to simply point the code user to Section 2603 for foam plastics when the HCL contains foam plastic components. The common use of foam plastic in HCL insulation would be the core of the material.

1409.14 Labeling. HPL shall be labeled in accordance with Section 1703.5.

❖ Because of the composite nature of HCL and the proprietary nature of the panel cores, the fabrication of the panels must be subject to independent third-party verification. The verification of this inspection is evidenced by labeling of the product by the third-party agency (refer to Section 1703.5 for more discussion on the labeling of products).

SECTION 1410
PLASTIC COMPOSITE DECKING

1410.1 Plastic composite decking. Exterior deck boards, stair treads, handrails and guardrail systems constructed of plastic composites, including plastic lumber, shall comply with Section 2612.

❖ Plastic composite decking is a recognized exterior deck component material. Provisions for this material are contained in Section 2612, which includes minimum requirements related to labeling, structural performance, flame spread, termite and decay resistance, and construction requirements.

Bibliography

The following resource materials were used in the preparation of the commentary for this chapter of the code.

ACI 530-13, *Building Code Requirements for Masonry Structures*. Farmington Hills, MI: American Concrete Institute, 2013.

ASCE 5-13, *Building Code Requirements for Masonry Structures*. Reston, VA: American Society of Civil Engineers, 2013.

ASTM B101-12, *Specification for Lead-coated Copper Sheet and Strip for Building Construction*. West Conshohocken, PA: ASTM International, 2012.

ASTM D3679-11, *Specification for Rigid Poly (Vinyl Chloride)(PVC) Siding*. West Conshohocken, PA: ASTM International, 2011.

ASTM D7254-07, *Standard Specification for Polypropylene (PP) Siding*. West Conshohocken, PA: ASTM International, 2007.

ASTM E96-13, *Standard Test Method for Water Vapor Transmission of Materials*. West Conshohocken, PA: ASTM International, 2013.

ASTM E119-2012A, *Test Methods for Fire Tests of Building Construction Materials*. West Conshohocken, PA: ASTM International, 2012.

ASTM E2568-09e1, *Standard Practice for Specimen Preparation and Mounting of Site-Fabricated Stretch Systems to Assess Surface Burning Characteristics*. West Conshohocken, PA: ASTM International, 2009.

NFPA 275-13, *Standard Method of Fire Tests for the Evaluation of Thermal Barriers Used Over Foam Plastic Insulation*. Quincy, MA: National Fire Protection Association, 2013.

TMS 402-2013, *Building Code for Masonry Structures*. Longmont, CO: The Masonry Society, 2013.

Chapter 15: Roof Assemblies and Rooftop Structures

General Comments

Chapter 15 regulates the materials, design, construction and quality of roofs and roof structures for all buildings and structures. Requirements for weather protection, fire classification, flashings and insulation are provided to govern the roof slab or deck and its supporting members, in addition to the covering that is applied to the roof for weather resistance, fire resistance or appearance. Specific installation requirements for certain steep-slope and low-slope roof coverings are provided.

Section 1502 lists the defined terms that are most closely associated with the provisions in this chapter.

Section 1503 provides requirements for roof coverings to protect against the effects of weather.

Section 1504 provides performance requirements including referenced standards for evaluating the performance of new and innovative roof coverings.

Section 1505 provides a classification system to determine a roof covering's effectiveness against certain fire test exposures. Specific classifications of roof coverings are required based on the type of construction classification of the structure.

Section 1506 provides requirements for the application of roof covering materials.

Section 1507 provides requirements for the installation of roof coverings.

Section 1508 establishes requirements for the use of combustible roof insulation in all types of construction.

Section 1509 provides testing and installation requirements for radiant barriers that are installed over a roof deck.

Section 1510 establishes minimum requirements for specific structures that can be constructed on and above the roof.

Section 1511 provides requirements for reroofing, including adequate structural supports, recovering of existing roof materials and replacement.

Section 1512 provides references to roof and structural roof-frame requirements necessary for solar energy systems.

Purpose

The provisions in Chapter 15 for roof construction and coverings are intended to provide a weather-protective barrier at the roof and, in most circumstances, a fire-retardant barrier to prevent flaming combustible materials, such as flying brands from nearby fires, from penetrating the roof construction. The chapter is essentially prescriptive in nature and is based on decades of experience with various traditional roofing materials. These prescriptive rules are very important for satisfactory performance of the roof covering even though the reasoning behind particular requirements may be lost. The provisions are based on preventing a repeat of past unsatisfactory performance of the various roofing materials and components.

SECTION 1501 GENERAL

1501.1 Scope. The provisions of this chapter shall govern the design, materials, construction and quality of roof assemblies, and rooftop structures.

❖ This section specifies the scope and applicability of the code for all roofs, roof assemblies and roof structures. Requirements are provided to regulate the materials, design and construction of roofs and rooftop structures such as penthouses, water tanks and other structures. See also the definition of "Roof assembly" in Section 202. Other roof assemblies addressed elsewhere in the code and not addressed in this chapter include membrane structures (see Section 3102) and light-transmitting plastics (see Section 2606).

SECTION 1502 DEFINITIONS

1502.1 Definitions. The following terms are defined in Chapter 2:

AGGREGATE.

BALLAST.

BUILDING-INTEGRATED PHOTOVOLTAIC (BIPV) PRODUCT.

BUILT-UP ROOF COVERING.

INTERLAYMENT.

MECHANICAL EQUIPMENT SCREEN.

METAL ROOF PANEL.

METAL ROOF SHINGLE.

MODIFIED BITUMEN ROOF COVERING.

PENTHOUSE.

PHOTOVOLTAIC MODULE.

PHOTOVOLTAIC PANEL.

PHOTOVOLTAIC PANEL SYSTEM.

PHOTOVOLTAIC SHINGLES.

POSITIVE ROOF DRAINAGE.

RADIANT BARRIER.

REROOFING.

ROOF ASSEMBLY.

ROOF COVERING.

ROOF COVERING SYSTEM.

ROOF DECK.

ROOF RECOVER.

ROOF REPAIR.

ROOF REPLACEMENT.

ROOF VENTILATION.

ROOFTOP STRUCTURE.

SCUPPER.

SINGLE-PLY MEMBRANE.

UNDERLAYMENT.

VEGETATIVE ROOF.

❖ Definitions facilitate the understanding of code provisions and minimize potential confusion. To that end, this section lists definitions of terms associated with roof assemblies, roof coverings and rooftop structures. Note that these definitions are found in Chapter 2. The use and application of defined terms, as well as undefined terms, are set forth in Section 201.

SECTION 1503 WEATHER PROTECTION

1503.1 General. Roof decks shall be covered with *approved* roof coverings secured to the building or structure in accordance with the provisions of this chapter. Roof coverings shall be designed and installed in accordance with this code and the *approved* manufacturer's instructions such that the roof covering shall serve to protect the building or structure.

❖ The roof covering provides protection from water intrusion into a building. The roofing system has historically been one of the most problematic areas of a building. Without a properly constructed roof assembly, water may infiltrate the building causing damage to building materials and contents. A roof assembly, for the purposes of this chapter, is defined as including the roof deck and components above that make up the roof covering. For a roof covering to perform the way it must to protect the building, it must meet certain requirements. The code provides basic requirements for the construction of roof assemblies.

The designer, in his or her roof specifications, will typically include compatibility of materials, deck type, weather conditions, roof slope, structural loads, roof drainage, roof penetrations and future reroofing.

1503.2 Flashing. Flashing shall be installed in such a manner so as to prevent moisture entering the wall and roof through joints in copings, through moisture-permeable materials and at intersections with parapet walls and other penetrations through the roof plane.

❖ Water that penetrates the building envelope can lead to the degradation of building materials that are sensitive to moisture. Flashing must be installed in a very specific manner so that moisture does not enter the building construction or get under the roof covering. As indicated, flashing is required anywhere the roof covering is interrupted or terminated.

1503.2.1 Locations. Flashing shall be installed at wall and roof intersections, at gutters, wherever there is a change in roof slope or direction and around roof openings. Where flashing is of metal, the metal shall be corrosion resistant with a thickness of not less than 0.019 inch (0.483 mm) (No. 26 galvanized sheet).

❖ Flashing consists of components that weatherproof or seal the roof system at discontinuities, such as perimeters, penetrations, walls [see Commentary Figure 1405.4(1)], expansion joints, valleys, drains and other places, where the roof covering is interrupted or terminated. For example, membrane base flashing covers the edge of the field membrane, and cap flashings or counterflashings shield the upper edges of the base flashing.

Metal flashing must be no less than the specified minimum thickness and be corrosion resistant. "Corrosion resistance" is defined in Section 202 as, "The ability of a material to withstand deterioration of its surface or its properties when exposed to its environment." This definition includes metallic and nonmetallic materials.

1503.3 Coping. Parapet walls shall be properly coped with noncombustible, weatherproof materials of a width no less than the thickness of the parapet wall.

❖ Coping is the covering piece on top of a wall, exposed to the weather and usually made of metal, masonry or stone. It is typically sloped to shed water back onto the roof.

[P] 1503.4 Roof drainage. Design and installation of roof drainage systems shall comply with Section 1503 of this code and Sections 1106 and 1108, as applicable, of the *International Plumbing Code*.

❖ The purpose of this section is to call attention to the need for providing roof drainage as well as the requirements for sizing and locating secondary drains. Positive roof drainage is required to prevent excessive ponding of water and the consequent rapid deterioration or early failure of the roof-covering material, roof structure, or the need for increased strength of the roof system to support ponding. Section 1611.2 requires the structural design calculations to consider ponding in order to prevent a progressive deflection of the structural roof members.

[P] 1503.4.1 Secondary (emergency overflow) drains or scuppers. Where roof drains are required, secondary (emergency overflow) roof drains or scuppers shall be provided where the roof perimeter construction extends above the roof in such a manner that water will be entrapped if the primary drains allow buildup for any reason. The installation and sizing of secondary emergency overflow drains, leaders and conductors shall comply with Sections 1106 and 1108, as applicable, of the *International Plumbing Code*.

❖ These requirements alert the roofer to the need for providing properly sized secondary drains or scuppers. A secondary drainage system is required where the building has parapet walls or other construction that could restrict drainage at the roof's perimeter. This section requires all buildings to have some method for preventing the accumulation of unplanned excessive rainwater.

The intent here is to limit the amount of ponding water that can accumulate on the roof due to rainfall. If the roof is constructed so water cannot pond on the roof, such as with roofs sloped toward the edge of the building, secondary drainage is not required. This criteria should be applied to all portions of the roof, so that if any portion is designed so water can pond, secondary drains would be required.

Bear in mind that a roof could be intentionally designed to allow ponding to some design depth such as for a rainwater harvesting system.

1503.4.2 Scuppers. When scuppers are used for secondary (emergency overflow) roof drainage, the quantity, size, location and inlet elevation of the scuppers shall be sized to prevent the depth of ponding water from exceeding that for which the roof was designed as determined by Section 1611.1. Scuppers shall not have an opening dimension of less than 4 inches (102 mm). The flow through the primary system shall not be considered when locating and sizing scuppers.

❖ Scuppers allow water to drain at the perimeter of a roof. These devices are commonly larger than the roof drain. This section provides performance requirements for roofs using wall scuppers as a means of secondary roof drainage. This section also provides minimum prescriptive opening dimensions for scuppers regardless of the results of the performance design. The performance design typically takes into account the flow rate, the length of the scupper opening and the water pressure (head) on the scupper.

1503.4.3 Gutters. Gutters and leaders placed on the outside of buildings, other than Group R-3, private garages and buildings of Type V construction, shall be of noncombustible material or a minimum of Schedule 40 plastic pipe.

❖ All roofs are required to be designed and built with positive drainage (see Section 1507.10.1 for the slope requirements of built-up roofs). "Positive roof drainage" is defined in Section 202. Ponding water has detrimental effects on a roof system, including roof surface and membrane deterioration due to debris accumulation; vegetation and fungal growth; deck deflections which can lead to structural damage; ice formation; tensile splitting; and the possibility of water entering the building.

1503.5 Attic and rafter ventilation. Intake and exhaust vents shall be provided in accordance with Section 1203.2 and the vent product manufacturer's installation instructions.

❖ During cold weather, condensation is deposited on cold surfaces when, for example, warm, moist air rising from the interior of a building and through the attic comes in contact with the roof deck. Ventilation minimizes condensation on the cold surfaces and, therefore, will reduce or prevent dry rot on the bottom surface of shingles or wood roof decks.

1503.6 Crickets and saddles. A cricket or saddle shall be installed on the ridge side of any chimney or penetration greater than 30 inches (762 mm) wide as measured perpendicular to the slope. Cricket or saddle coverings shall be sheet metal or of the same material as the roof covering.

Exception: Unit skylights installed in accordance with Section 2405.5 and flashed in accordance with the manufacturer's instructions shall be permitted to be installed without a cricket or saddle.

❖ A cricket or saddle is a raised roof substrate or structure constructed to channel or direct surface water around a chimney or other similar roof penetration. The exception recognizes the use of engineered skylight systems that are designed to prevent water infiltration without the use of a cricket. It applies to unit skylights that are installed in accordance with Section 2405.5.

SECTION 1504 PERFORMANCE REQUIREMENTS

1504.1 Wind resistance of roofs. Roof decks and roof coverings shall be designed for wind loads in accordance with Chapter 16 and Sections 1504.2, 1504.3 and 1504.4.

❖ It is important that roof coverings and roof decks remain intact and in place when subjected to high winds. Without an intact roof system, the building would be subjected either to water damage, which would reduce its structural stability, or to higher wind pressures for which the building is not designed. The following sections specify requirements for different roof coverings regarding wind loads. The roof decks are included in the requirements of Chapter 16.

1504.1.1 Wind resistance of asphalt shingles. Asphalt shingles shall be tested in accordance with ASTM D7158. Asphalt shingles shall meet the classification requirements of Table 1504.1.1 for the appropriate maximum basic wind speed. Asphalt shingle packaging shall bear a label to indicate compliance with ASTM D7158 and the required classification in Table 1504.1.1.

Exception: Asphalt shingles that are not included in the scope of ASTM D7158 shall be tested and labeled to indicate compliance with ASTM D3161 and the required classification in Table 1504.1.1.

❖ This section recognizes two test methods for the wind resistance of asphalt shingles: ASTM D7158, which provides a method of testing that is appropriate for sealed asphalt shingles, and ASTM D3161, which provides a method of testing that is appropriate for unsealed asphalt shingles. Each of these standards results in a shingle "classification" based on the ability of the asphalt shingle installation to resist wind uplift. Table 1504.1.1 contains minimum classification requirements based on either the mapped wind speed or the nominal design wind speed as determined from Section 1609.3.1.

TABLE 1504.1.1. See below.

❖ See the commentary to Section 1504.1.1.

1504.2 Wind resistance of clay and concrete tile. Wind loads on clay and concrete tile roof coverings shall be in accordance with Section 1609.5.

❖ To determine the minimum wind loads to use in the design and installation of clay and concrete roof tile, this section refers to Section 1609.5. When required by Table 1507.3.7, the tiles must be installed to resist the wind loads as calculated in accordance with Section 1609.5.

1504.2.1 Testing. Testing of concrete and clay roof tiles shall be in accordance with Sections 1504.2.1.1 and 1504.2.1.2.

❖ This section prescribes the test standards and criteria to be used to determine the overturning resistance and wind characteristics of concrete and clay roof tiles.

1504.2.1.1 Overturning resistance. Concrete and clay roof tiles shall be tested to determine their resistance to overturning due to wind in accordance with SBCCI SSTD 11 and Chapter 15.

❖ This section requires concrete and clay tiles to be tested to determine their overturning resistance in accordance with SBCCI SSTD 11. SBCCI SSTD 11 prescribes methods for determining the allowable overturning moment for mechanically fastened, adhesive-set and mortar-set tiles. A test procedure is also prescribed for determining the allowable uplift loads on hip/ridge tiles.

1504.2.1.2 Wind tunnel testing. Where concrete and clay roof tiles do not satisfy the limitations in Chapter 16 for rigid tile, a wind tunnel test shall be used to determine the wind characteristics of the concrete or clay tile roof covering in accordance with SBCCI SSTD 11 and Chapter 15.

❖ The wind tunnel test procedures in SBCCI SSTD 11 must be used if the roof tiles do not meet the limitations of Chapter 16 for rigid tile.

1504.3 Wind resistance of nonballasted roofs. Roof coverings installed on roofs in accordance with Section 1507 that are mechanically attached or adhered to the roof deck shall be designed to resist the design wind load pressures for components and cladding in accordance with Section 1609.

❖ This section addresses wind resistance of steep-slope systems that are mechanically attached by fasteners or adhered to the roof slab or deck by adhesives. Steep-slope roof coverings are those described in Section 1507 with a roof slope of 2:12 or greater. The basic wind speed as obtained from Chapter 16 for the building's location must be modified for the building's overall height above grade and exposure category, wind gust effect and wind importance factor. A structural design is required to demonstrate that the proposed fastening schedule will resist the wind uplift loads.

1504.3.1 Other roof systems. Built-up, modified bitumen, fully adhered or mechanically attached single-ply roof systems, metal panel roof systems applied to a solid or closely

TABLE 1504.1.1
CLASSIFICATION OF ASPHALT SHINGLES

MAXIMUM BASIC WIND SPEED, V_{ult}, FROM FIGURE 1609A, B, C OR ASCE 7	MAXIMUM BASIC WIND SPEED, V_{asd}, FROM TABLE 1609.3.1	ASTM D7158[a] CLASSIFICATION	ASTM D3161 CLASSIFICATION
110	85	D, G or H	A, D or F
116	90	D, G or H	A, D or F
129	100	G or H	A, D or F
142	110	G or H	F
155	120	G or H	F
168	130	H	F
181	140	H	F
194	150	H	F

For SI: 1 foot = 304.8 mm; 1 mph = 0.447 m/s.

a. The standard calculations contained in ASTM D7158 assume Exposure Category B or C and building height of 60 feet or less. Additional calculations are required for conditions outside of these assumptions.

fitted deck and other types of membrane roof coverings shall be tested in accordance with FM 4474, UL 580 or UL 1897.

❖ Additional requirements addressing the wind resistance of built-up, modified bitumen, fully adhered or mechanically attached single-ply roof systems, metal panel systems that are applied to a solid or closely fitted deck as well as other types of membrane roof coverings are established in this section. Besides compliance with the requirements of Section 1504.3, these roof coverings are required to be physically tested for compliance with one of three standards: FM 4474, UL 580 or UL 1897.

1504.3.2 Structural metal panel roof systems. Where the metal roof panel functions as the roof deck and roof covering and it provides both weather protection and support for loads, the structural metal panel roof system shall comply with this section. Structural standing-seam metal panel roof systems shall be tested in accordance with ASTM E1592 or FM 4474. Structural through-fastened metal panel roof systems shall be tested in accordance with FM 4474, UL 580 or ASTM E1592.

Exceptions:

1. Metal roofs constructed of cold-formed steel shall be permitted to be designed and tested in accordance with the applicable referenced structural design standard in Section 2210.1.
2. Metal roofs constructed of aluminum shall be permitted to be designed and tested in accordance with the applicable referenced structural design standard in Section 2002.1.

❖ This section specifies the test that metal panel roof systems are required to pass before installation. Standards and test methods have been established for the proper installation of metal panel roof systems whether fastened or standing seamed.

The exceptions recognize metal decks that provide both weather protection and support for structural loads, including wind and live loads. Many of these metal decks are structural elements where stresses can be calculated just like the design approach used for the roof-framing members that support the roof deck. For steel decks covered in Exception 1, the American Iron and Steel Institute (AISI) specification that is referenced in Section 2210.1 recognizes a wide variety of roofing profiles where calculations are applicable. For roofing profiles that cannot be directly calculated, the specification requires additional testing, including many of the tests currently referenced by the code for wind resistance. In addition, unlike many of the wind tests, the specification gives guidance on the selection of appropriate safety factors.

1504.4 Ballasted low-slope roof systems. Ballasted low-slope (roof slope < 2:12) single-ply roof system coverings installed in accordance with Sections 1507.12 and 1507.13 shall be designed in accordance with Section 1504.8 and ANSI/SPRI RP-4.

❖ Sections 1507.12 and 1507.13 prescribe the installation for ballasted low-slope, single-ply roof system coverings. These coverings are to be designed in accordance with Section 1504.8 and ANSI/SPRI RP-4, which is listed in Chapter 35.

1504.5 Edge securement for low-slope roofs. Low-slope built-up, modified bitumen and single-ply roof system metal edge securement, except gutters, shall be designed and installed for wind loads in accordance with Chapter 16 and tested for resistance in accordance with Test Methods RE-1, RE-2 and RE-3 of ANSI/SPRI ES-1, except V_{ult} wind speed shall be determined from Figure 1609.3(1), 1609.3(2), or 1609.3(3) as applicable.

❖ These requirements are specific to low-slope membrane roofs, namely built-up, modified bitumen and single-ply roofs. This section references the appropriate standard for low-slope roofs, which is ANSI/SPRI ES-1. This standard is important in preventing failures of the edge attachment on low-slope roof systems. ANSI/SPRI ES-1 consists of two primary parts. In the portion of ANSI/SPRI ES-1 that is included in the scope of reference, the edge metal flashings' wind resistance is determined according to RE-1, RE-2 and RE-3 test methods. By specifically referencing those test methods in ANSI/SPRI ES-1, any confusion with the determination of wind loads should be avoided since other portions of the standard can be used to determine the wind loads at a roof edge. Also, by referring the code user to code Figures 1609.3(1), 1609.3(2) and 1609.3(3), the intent that wind speeds must be determined using Chapter 16 is reinforced.

1504.6 Physical properties. Roof coverings installed on low-slope roofs (roof slope < 2:12) in accordance with Section 1507 shall demonstrate physical integrity over the working life of the roof based upon 2,000 hours of exposure to accelerated weathering tests conducted in accordance with ASTM G152, ASTM G155 or ASTM G154. Those roof coverings that are subject to cyclical flexural response due to wind loads shall not demonstrate any significant loss of tensile strength for unreinforced membranes or breaking strength for reinforced membranes when tested as herein required.

❖ This section, which addresses the effects of weather on the performance of the roof covering, requires that weather testing be performed and includes performance criteria for those roof coverings that will have to perform while subjected to wind loads.

The ASTM International (ASTM) test methods that are referenced cover exposure to carbon-arc-type or xenon-arc-type light (with or without water), fluorescent ultraviolet-condensation-type rays and concentrated natural sunlight. In conjunction with these tests, roofing materials must not show any significant signs of failure caused by cyclical wind conditions. The tensile strength in unreinforced membranes and the breaking strength in reinforced membranes must endure such conditions without any significant strength reduction.

1504.7 Impact resistance. Roof coverings installed on low-slope roofs (roof slope < 2:12) in accordance with Section 1507 shall resist impact damage based on the results of tests conducted in accordance with ASTM D3746, ASTM D4272, CGSB 37-GP-52M or the "Resistance to Foot Traffic Test" in Section 5.5 of FM 4470.

❖ All roof-covering materials must withstand impact loads, such as hail and driving rainstorms. The referenced standards set criteria for the testing of certain roofing materials, such as plastic film and bituminous roofing systems, on low-slope roofs. Each test uses similar devices to test the material. The tests consist of dropping a dart or missile, which weighs from $2^1/_2$ to 5 pounds (1.1 to 2.3 kg), approximately 4 to 5 feet (1219 to 1524 mm) and then evaluating the material for failure.

1504.8 Aggregate. Aggregate used as surfacing for roof coverings and aggregate, gravel or stone used as ballast shall not be used on the roof of a building located in a hurricane-prone region as defined in Section 202, or on any other building with a mean roof height exceeding that permitted by Table 1504.8 based on the exposure category and basic wind speed at the site.

❖ This section recognizes that aggregate, gravel and stone used on roofs can be blown off when subjected to hurricane-strength winds or higher wind effects due to the roof height on a taller building. Field assessments of damage to buildings caused by high-wind events have shown that gravel or stone blown from the roofs of buildings has exacerbated damage to neighboring buildings due to breakage of glass. Once glass is broken, higher internal pressures are created within the building, which can result in structural damage to interior walls, and to the cladding of wall and roof surfaces subjected to negative external pressures. Even where the higher internal pressure is considered and the building is designed accordingly, the breakage of windows will generally result in substantial wind and water damage to the building's interior and contents.

TABLE 1504.8. See next column.

❖ This table establishes maximum roof heights where aggregate can be used on roofs of buildings outside a hurricane-prone region. The heights are a function of basic wind speed and exposure category.

SECTION 1505 FIRE CLASSIFICATION

[BF] 1505.1 General. Roof assemblies shall be divided into the classes defined below. Class A, B and C roof assemblies and roof coverings required to be listed by this section shall be tested in accordance with ASTM E108 or UL 790. In addition, *fire-retardant-treated wood* roof coverings shall be tested in accordance with ASTM D2898. The minimum roof coverings installed on buildings shall comply with Table 1505.1 based on the type of construction of the building.

Exception: Skylights and sloped glazing that comply with Chapter 24 or Section 2610.

❖ The code designates the use of any particular classification of roof coverings based on the type of construction of the building. A minimum Class B roof covering is required for all roofs that have a minimum 1-hour fire-resistance rating in accordance with Table 602. Roofs without a required fire-resistance rating require a minimum Class C roof covering.

The exception clarifies that skylights and sloped glazing are not required to be tested for fire classification. Skylights and sloped glazing are considered to be individual assemblies that are required to meet specific requirements listed in other portions of the code. For example, the fire requirements for light-transmitting plastic skylights are covered in Section

TABLE 1504.8
MAXIMUM ALLOWABLE MEAN ROOF HEIGHT PERMITTED FOR BUILDINGS WITH AGGREGATE ON THE ROOF IN AREAS OUTSIDE A HURRICANE-PRONE REGION

NOMINAL DESIGN WIND SPEED, V_{asd} (mph)[b, d]	MAXIMUM MEAN ROOF HEIGHT (ft)[a, c]		
	Exposure category		
	B	C	D
85	170	60	30
90	110	35	15
95	75	20	NP
100	55	15	NP
105	40	NP	NP
110	30	NP	NP
115	20	NP	NP
120	15	NP	NP
Greater than 120	NP	NP	NP

For SI: 1 foot = 304.8 mm; 1 mile per hour = 0.447 m/s.

a. Mean roof height as defined in ASCE 7.

b. For intermediate values of V_{asd}, the height associated with the next higher value of V_{asd} shall be used, or direct interpolation is permitted.

c. NP = gravel and stone not permitted for any roof height.

d. V_{asd} shall be determined in accordance with Section 1609.3.1.

2610 and the material and curbing requirements for glass skylights and sloped glazing are given in Chapter 24.

TABLE 1505.1[a, b]
MINIMUM ROOF COVERING CLASSIFICATION FOR TYPES OF CONSTRUCTION

IA	IB	IIA	IIB	IIIA	IIIB	IV	VA	VB
B	B	B	C[c]	B	C[c]	B	B	C[c]

For SI: 1 foot = 304.8 mm, 1 square foot = 0.0929 m^2.

a. Unless otherwise required in accordance with the *International Wildland-Urban Interface Code* or due to the location of the building within a fire district in accordance with Appendix D.

b. Nonclassified roof coverings shall be permitted on buildings of Group R-3 and Group U occupancies, where there is a minimum fire-separation distance of 6 feet measured from the leading edge of the roof.

c. Buildings that are not more than two stories above grade plane and having not more than 6,000 square feet of projected roof area and where there is a minimum 10-foot fire-separation distance from the leading edge of the roof to a lot line on all sides of the building, except for street fronts or public ways, shall be permitted to have roofs of No. 1 cedar or redwood shakes and No. 1 shingles constructed in accordance with Section 1505.7.

❖ See the commentary to Section 1505.1.

[BF] 1505.2 Class A roof assemblies. Class A roof assemblies are those that are effective against severe fire test exposure. Class A roof assemblies and roof coverings shall be *listed* and identified as Class A by an *approved* testing agency. Class A roof assemblies shall be permitted for use in buildings or structures of all types of construction.

Exceptions:

1. Class A roof assemblies include those with coverings of brick, masonry or an exposed concrete roof deck.
2. Class A roof assemblies also include ferrous or copper shingles or sheets, metal sheets and shingles, clay or concrete roof tile or slate installed on noncombustible decks or ferrous, copper or metal sheets installed without a roof deck on noncombustible framing.
3. Class A roof assemblies include minimum 16 ounce per square foot (0.0416 kg/m^2) copper sheets installed over combustible decks.
4. Class A roof assemblies include slate installed over ASTM D226, Type II underlayment over combustible decks.

❖ Roof coverings that are effective against the severe fire-test exposure of ASTM E108 are classified as Class A. Exception 1 prescribes roof-covering materials that are identified as Class A without having to be tested in accordance with ASTM E108, based on the past performance of these materials. These include the following noncombustible coverings: brick, masonry or exposed concrete. Any assembly that is tested in accordance with ASTM E108 requirements for severe fire exposure, and that is listed and labeled by an approved testing agency as a Class A roof covering, is allowed as a roof covering on any type of construction designated in the code. Similarly, Exception 2 prescribes roof-covering materials that are identified as Class A as long as they are installed over a noncombustible deck or, in some cases, without a deck but over noncombustible framing. Exception 3 allows copper sheets over combustible decking to serve as a Class A roof covering. Exception 4 includes slate installed over a specific underlayment over a combustible deck as a Class A roof assembly based on fire tests that demonstrated the requirements of UL 790 Class A are satisfied. Lastly, note that the materials and specific installations in all four exceptions do not require listing or labeling by an approved agency.

[BF] 1505.3 Class B roof assemblies. Class B roof assemblies are those that are effective against moderate fire-test exposure. Class B roof assemblies and roof coverings shall be *listed* and identified as Class B by an *approved* testing agency.

❖ Roof coverings that are effective against the moderate fire test exposure of ASTM E108 are classified as Class B. Those roof coverings listed and identified as Class A by an approved testing agency are also allowed to be used where Class B is required. For buildings and structures of any type of construction designated in the code, Class A or B roof coverings are allowed.

[BF] 1505.4 Class C roof assemblies. Class C roof assemblies are those that are effective against light fire-test exposure. Class C roof assemblies and roof coverings shall be *listed* and identified as Class C by an *approved* testing agency.

❖ Roof coverings that are effective against the light fire test exposure of ASTM E108 are classified as Class C. There are no specific materials or products that automatically qualify as Class C roof coverings. However, note that those roof coverings listed and identified as Class A or B are allowed to be used where a Class C is required. Class C roof coverings that are listed and labeled by an approved testing agency may only be used on Types IIB, IIIB and VB construction.

[BF] 1505.5 Nonclassified roofing. Nonclassified roofing is *approved* material that is not *listed* as a Class A, B or C roof covering.

❖ Nonclassified roofing is a roofing material that is approved by the building official and is not otherwise classified (see the definition of "Approved" in Section 202).

[BF] 1505.6 Fire-retardant-treated wood shingles and shakes. *Fire-retardant-treated wood* shakes and shingles shall be treated by impregnation with chemicals by the full-cell vacuum-pressure process, in accordance with AWPA C1. Each bundle shall be marked to identify the manufactured unit and the manufacturer, and shall also be *labeled* to identify the classification of the material in accordance with the testing required in Section 1505.1, the treating company and the quality control agency.

❖ This section specifically identifies the particular pressure-impregnation method to be used to treat wood shakes and shingles. This method consists of placing bundles of the roofing materials inside a retort, which is a huge cylindrical pressure container. A vacuum is then created inside the retort, which in turn draws the air and moisture from the millions of cells per square inch in western red cedar. The fire-retardant chemical mixture is then injected at pressures reaching 150 psi (1034 kPa). The chemical is forced into every cell of the shakes and shingles—even the innermost layers. After the products are pressure treated, they are then placed in a special drying kiln where they undergo finishing with a thermal cure at temperatures of up to 2,000°F (1093°C). This section further identifies the test standard to be followed, the method of labeling and the information required per bundle of shingles in order for the building official to evaluate code compliance. This includes the testing requirements found in Section 1505.1.

[BF] 1505.7 Special purpose roofs. Special purpose wood shingle or wood shake roofing shall conform to the grading and application requirements of Section 1507.8 or 1507.9. In addition, an underlayment of $^5/_8$-inch (15.9 mm) Type X water-resistant gypsum backing board or gypsum sheathing shall be placed under minimum nominal $^1/_2$-inch-thick (12.7 mm) wood structural panel solid sheathing or 1-inch (25 mm) nominal spaced sheathing.

❖ Informal tests of special purpose roofs have shown that they provide a significant increase in protection for combustible roof coverings, and again the protection offered is considered appropriate if applied in the manner stipulated. Buildings that are permitted to utilize this roof construction are identified in Note c to Table 1505.1.

[BF] 1505.8 Building-integrated photovoltaic products. *Building-integrated photovoltaic products* installed as the roof covering shall be tested, *listed* and *labeled* for fire classification in accordance with Section 1505.1.

❖ This section requires photovoltaic products integrated into the envelope as a roof covering to comply with UL 790 or ASTM E108.

[BF] 1505.9 Photovoltaic panels and modules. Rooftop-mounted *photovoltaic panel systems* shall be tested, *listed* and identified with a fire classification in accordance with UL 1703. The fire classification shall comply with Table 1505.1 based on the type of construction of the building.

❖ This section recognizes that stand-off rack-mounted photovoltaic panels and modules are better tested in accordance with UL 1703 rather than UL 790 or ASTM E108. It also clarifies that the fire classification listed for the photovoltaic panels and modules must be consistent with the fire classification requirement for the roof covering.

[BF] 1505.10 Roof gardens and landscaped roofs. Roof gardens and landscaped roofs shall comply with Section 1507.16 and shall be installed in accordance with ANSI/SPRI VF-1.

❖ This section requires SPRI VF-1 for the fire design for roof gardens and landscaped roofs since the approach contained within the standard is more appropriate than the traditional test methods used to determine fire classification of roof coverings. This standard provides a design method that provides an acceptable level of performance of roof gardens and landscaped roofs when exposed to exterior fire sources. The general approach used in this standard is to design in fire breaks for large roof areas, around rooftop equipment and penetrations, and next to adjacent walls.

SECTION 1506 MATERIALS

1506.1 Scope. The requirements set forth in this section shall apply to the application of roof-covering materials specified herein. Roof coverings shall be applied in accordance with this chapter and the manufacturer's installation instructions. Installation of roof coverings shall comply with the applicable provisions of Section 1507.

❖ This section establishes specifications and standards for materials and installation techniques in conjunction with recognized industry standards and acceptable roofing applications. By complying with the specified material requirements, slope limitations, underlayment requirements and fastening schedules, the proposed roof covering can be expected to perform as anticipated.

1506.2 Material specifications and physical characteristics. Roof-covering materials shall conform to the applicable standards listed in this chapter.

❖ Whenever possible, this chapter references appropriate material standards for roof covering materials. Because there may be materials on the market for which the code does not currently reference a test standard, there are occasions where the building official will have to require testing by an approved lab to determine the character, quality and application limitations of the materials. This is something that should be considered as an alternative material and approached in accordance with Section 104.11.

1506.3 Product identification. Roof-covering materials shall be delivered in packages bearing the manufacturer's identifying marks and *approved* testing agency labels required in accordance with Section 1505. Bulk shipments of materials shall be accompanied with the same information issued in the form of a certificate or on a bill of lading by the manufacturer.

❖ Identification of the roofing materials is mandatory in order to verify that they comply with quality standards. In addition to bearing the manufacturer's label or identifying mark, prepared roofing and built-up roofing materials are required by the code to carry a label of an approved agency that inspects the material and finished products during manufacture.

SECTION 1507 REQUIREMENTS FOR ROOF COVERINGS

1507.1 Scope. Roof coverings shall be applied in accordance with the applicable provisions of this section and the manufacturer's installation instructions.

❖ Roof coverings serve as the membrane of the roof system, making it critical in providing weather protection. This section requires that the installation of roof coverings be in accordance with the code and the manufacturer's installation instructions. Often, other requirements must be met, such as UL or FM requirements. It is very possible that there will be times that these requirements conflict. Once adopted, the code is law. The other provisions are not law, and compliance with them does not necessarily mean that compliance with the code has been achieved. In cases where conflicts arise, the code requirements govern.

1507.2 Asphalt shingles. The installation of asphalt shingles shall comply with the provisions of this section.

❖ This section establishes the criteria for asphalt shingle roofs. Two general types of asphalt shingles specifically apply to this section: strip shingles (e.g., three-tab shingles) and individual interlocking shingles (e.g., T-lock shingles). Asphalt strip shingles, which are the most common, are produced in a variety of finished appearances, including three-tab, random or multiple-tab, no-cutout and laminated architectural.

Asphalt shingles are typically classified as one of two types: cellulose felt reinforced (e.g., organic shingles) and fiberglass mat reinforced (e.g., fiberglass shingles).

Although it is not specifically addressed in this section, the roofing industry generally recommends the attic space below asphalt shingle roofs be properly ventilated (see Section 1503.5).

Additional information regarding asphalt shingle roofs is provided in the *NRCA Roofing Manual: Steep-slope Roof Systems* published by the National Roofing Contractors Association (NRCA).

1507.2.1 Deck requirements. Asphalt shingles shall be fastened to solidly sheathed decks.

❖ Solid sheathed roof decks typically include nominal 1-inch (25 mm) solid lumber or wood structural panels. It is recommended that solid sawn lumber, not more than 8 inches (203 mm) in width, be used. Wider pieces are more likely to warp and cup. The code is specific about the types of decking allowed depending on the material used. For example, wood roof decking is addressed in Chapter 23. Manufacturers may also require the application of their roof coverings to specific roof deck materials for warranty of their products.

1507.2.2 Slope. Asphalt shingles shall only be used on roof slopes of two units vertical in 12 units horizontal (17-percent slope) or greater. For roof slopes from two units vertical in 12 units horizontal (17-percent slope) up to four units vertical in 12 units horizontal (33-percent slope), double underlayment application is required in accordance with Section 1507.2.8.

❖ Asphalt shingles are intended to be applied to a steep roof. A minimum slope is crucial in the performance of such shingles because it is a primary indicator of a roof's ability to shed water. Where the roof slope is less than four units vertical in 12 units horizontal (33-percent slope), water drainage from the roof is slower, which can cause it to back up under the roofing and cause leaks. Thus, this section also specifies double coverage of the underlayment where the slope is between two units vertical in 12 units horizontal (17-percent slope) and four units vertical in 12 units horizontal (33-percent slope). One layer of underlayment must be provided under asphalt shingle roof coverings with a slope greater than four units vertical in 12 units horizontal (33-percent slope). Also, the effect of ice dams at the eaves is more pronounced on low-sloped roofs and, as a result, special precautions are necessary for satisfactory performance of the roofing materials.

1507.2.3 Underlayment. Unless otherwise noted, required underlayment shall conform to ASTM D226, Type I, ASTM D4869, Type I, or ASTM D6757.

❖ Asphalt shingle roofs are required to be installed with a felt underlayment. Underlayment serves as secondary protection against wind-driven rain and other moisture penetration when individual shingles become damaged or dislodged. Underlayment for asphalt shingle roofs is required to comply with ASTM D226, Type I, or ASTM D4869, Type I. Products complying with the "Type I" designation of these standards are commonly called "No. 15 asphalt felt." See application requirements for underlayment in Section 1507.2.8 as well as additional restrictions on underlayment applied in high-wind areas in Section 1507.2.8.1.

Underlayment also serves as a separator sheet between the shingles and deck. As a separator, the underlayment provides some protection to the shingles from uneven roof deck edges and deck fasteners. Underlayment also helps eliminate a rectangular pattern of ridges in the membrane over insulation or deck joints.

Underlayment offers protection from resins and other chemicals contained in wood board sheathing.

Using the appropriate underlayments, decking materials and shingles will help in meeting the fire classification ratings.

1507.2.4 Self-adhering polymer modified bitumen sheet. Self-adhering polymer modified bitumen sheet shall comply with ASTM D1970.

❖ Self-adhering polymer modified bitumen sheets meeting ASTM D1970 requirements are intended for use as underlayment for ice dam protection. These underlayments have an adhesive layer that is exposed by removal of a protective sheet. The top surface of the sheet is suitable to work on during the application of the exposed roofing.

1507.2.5 Asphalt shingles. Asphalt shingles shall comply with ASTM D225 or ASTM D3462.

❖ ASTM D225 is a specification for asphalt roofing in shingle form, composed of single or multiple thicknesses of organic felt saturated and coated on both sides with asphalt and surfaced on the weather side with mineral granules. This standard is well established in the roofing industry. Generally, all organic shingles currently manufactured comply with this standard.

ASTM D3462 is a specification for asphalt roofing in shingle form, composed of single or multiple thicknesses of glass felt impregnated and coated on both sides with asphalt, and surfaced on the weather side with mineral granules. The shingles may be either locked together during installation, or have a factory-applied self-sealing adhesive. Additionally, asphalt shingles meeting this standard must pass the Class A fire exposure test requirements of ASTM E108 and the wind-resistance test requirements of ASTM D7158 (see also Section 1504.1.1).

Some "20-year warranted" shingles and "25-year warranted" shingles currently on the market may not necessarily comply with ASTM D3462. Fiberglass shingles complying with ASTM D3462 will typically be identified as such on the shingle bundle packaging or the manufacturer will supply a written certification of compliance with the standard.

1507.2.6 Fasteners. Fasteners for asphalt shingles shall be galvanized, stainless steel, aluminum or copper roofing nails, minimum 12-gage [0.105 inch (2.67 mm)] shank with a minimum $^3/_8$-inch-diameter (9.5 mm) head, of a length to penetrate through the roofing materials and a minimum of $^3/_4$ inch (19.1 mm) into the roof sheathing. Where the roof sheathing is less than $^3/_4$ inch (19.1 mm) thick, the nails shall penetrate through the sheathing. Fasteners shall comply with ASTM F1667.

❖ The fasteners allowed are either of metal that is corrosion resistant or galvanized to be corrosion resistant. It is important to use fasteners that are long enough to penetrate the roof covering, flashing, underlayment and into the deck a minimum of $^3/_4$ inch (19.1 mm). When the deck is less than $^3/_4$ inch (19.1 mm) in thickness, the fastener must penetrate through the deck. It is recommended that the nail be at least $^1/_8$ inch (3.2 mm) through the deck.

1507.2.7 Attachment. Asphalt shingles shall have the minimum number of fasteners required by the manufacturer, but not less than four fasteners per strip shingle or two fasteners per individual shingle. Where the roof slope exceeds 21 units vertical in 12 units horizontal (21:12), shingles shall be installed as required by the manufacturer.

❖ This section clarifies that the manufacturer's instructions must be consulted when installing asphalt shingles, unless the manufacturer's instructions contain fastening requirements less than those prescribed in this section. As a minimum, there needs to be four fasteners for each strip-type shingle and two fasteners for individual shingles.

For roof slopes greater than 21:12, the manufacturer should be consulted for installation requirements. Typically such an installation will include a 5-inch (127 mm) exposure, with each tab cemented in place using asphalt roofing cement compatible with the shingle.

1507.2.8 Underlayment application. For roof slopes from two units vertical in 12 units horizontal (17-percent slope) and up to four units vertical in 12 units horizontal (33-percent slope), underlayment shall be two layers applied in the following manner. Apply a minimum 19-inch-wide (483 mm) strip of underlayment felt parallel with and starting at the eaves, fastened sufficiently to hold in place. Starting at the eave, apply 36-inch-wide (914 mm) sheets of underlayment overlapping successive sheets 19 inches (483 mm) and fasten sufficiently to hold in place. Distortions in the underlayment shall not interfere with the ability of the shingles to seal. For roof slopes of four units vertical in 12 units horizontal (33-percent slope) or greater, underlayment shall be one layer applied in the following manner. Underlayment shall be applied shingle fashion, parallel to and starting from the eave and lapped 2 inches (51 mm), fastened sufficiently to hold in place. Distortions in the underlayment shall not interfere with the ability of the shingles to seal.

❖ Low-sloped roofs, 2:12 to 4:12, shed water more slowly than steeper roofs, therefore requiring greater protection from water backing up under the shingles. This is particularly important when considering wind-driven water and ice damming. For this reason, two layers of underlayment are required. For greater slopes, only one layer of underlayment is necessary, except at eaves where ice dams are potentially a problem.

Installation of underlayment material over a distorted surface can result in reduced wind and weather (moisture) resistance and poor aesthetics. Regardless of the slope, underlayment need only be fastened well enough to stay in place until the shingles are applied. The application of shingle fasteners will serve to fasten the underlayment.

1507.2.8.1 High wind attachment. Underlayment applied in areas subject to high winds [V_{asd} greater than 110 mph (49 m/s) as determined in accordance with Section 1609.3.1] shall be applied with corrosion-resistant fasteners in accordance with the manufacturer's instructions. Fasteners are to be applied along the overlap not more than 36 inches (914 mm) on center.

Underlayment installed where V_{asd}, in accordance with Section 1609.3.1, equals or exceeds 120 mph (54 m/s) shall comply with ASTM D226 Type II, ASTM D4869 Type IV, or ASTM D6757. The underlayment shall be attached in a grid pattern of 12 inches (305 mm) between side laps with a 6-inch (152 mm) spacing at the side laps. Underlayment shall be applied in accordance with Section 1507.2.8 except all laps shall be a minimum of 4 inches (102 mm). Underlayment shall be attached using metal or plastic cap nails with a head diameter of not less than 1 inch (25 mm) with a thickness of at least 32-gage [0.0134 inch (0.34 mm)] sheet metal. The cap nail shank shall be a minimum of 12 gage [0.105 inch (2.67 mm)] with a length to penetrate through the roof sheathing or a minimum of $^{3}/_{4}$ inch (19.1 mm) into the roof sheathing.

Exception: As an alternative, adhered underlayment complying with ASTM D1970 shall be permitted.

❖ These provisions are intended to improve the performance of underlayment on roofs subjected to high winds, particularly if shingles are lost due to a high-wind event. Where the nominal wind speed is greater than 110 mph (49 m/s), fasteners must be approved for asphalt shingle installation and should be flat headed and corrosion resistant.

Based on observations of roof underlayment performance both in the field following hurricanes as well as in laboratory tests, ASTM 226, Type I, underlayment is not permitted where the nominal wind speed exceeds 120 mph (54 m/s). In laboratory tests, ASTM 226, Types I and II, underlayment performed dramatically different, with the ASTM 226, Type II, material remaining in place and exhibiting fewer signs of distress. In addition, the attachment of the underlayment must be accomplished with cap nails. A 1-inch (25.4 mm) head diameter of the cap portion recognizes the type most commonly used in the field. The exception specifically permits the use of a self-adhered underlayment in accordance with Section 1507.2.4 (ASTM D1970).

1507.2.8.2 Ice barrier. In areas where there has been a history of ice forming along the eaves causing a backup of water, an ice barrier that consists of at least two layers of underlayment cemented together or of a self-adhering polymer modified bitumen sheet shall be used in lieu of normal underlayment and extend from the lowest edges of all roof surfaces to a point at least 24 inches (610 mm) inside the *exterior wall* line of the building.

Exception: Detached accessory structures that contain no conditioned floor area.

❖ Ice dams form when snow melts over the warmer parts of a roof and refreezes over the colder eaves. This ice formation acts like a dam and causes water to back up beneath the roof covering. The water will eventually leak causing damage to the structure, including the walls, ceilings and roof [see Commentary Figures 1507.8.4(1) and (2)]. In areas where this is prevalent, the installation of an ice barrier is required in accordance with this section.

The exception exempts accessory buildings from such restrictions as they are unheated structures where the need for protection against ice dams is unnecessary. The same exception is found in Sections 1507.5.4, 1507.6.4, 1507.7.4, 1507.8.4 and 1507.9.4.

1507.2.9 Flashings. Flashing for asphalt shingles shall comply with this section. Flashing shall be applied in accordance with this section and the asphalt shingle manufacturer's printed instructions.

❖ This section establishes the requirement for roof flashings for asphalt shingles to prevent leakage, and further establishes performance criteria.

1507.2.9.1 Base and cap flashing. Base and cap flashing shall be installed in accordance with the manufacturer's instructions. Base flashing shall be of either corrosion-resistant metal of minimum nominal 0.019-inch (0.483 mm) thickness or mineral-surfaced roll roofing weighing a minimum of 77 pounds per 100 square feet (3.76 kg/m^2). Cap flashing shall be corrosion-resistant metal of minimum nominal 0.019-inch (0.483 mm) thickness.

❖ Roof system edges must be weatherproofed or sealed where they intersect with other vertical components, such as walls or chimneys. Flashing is composed of two parts: the base and cap. For a general discussion of roof flashings, see the commentary to Section 1503.2.

1507.2.9.2 Valleys. Valley linings shall be installed in accordance with the manufacturer's instructions before applying shingles. Valley linings of the following types shall be permitted:

1. For open valleys (valley lining exposed) lined with metal, the valley lining shall be at least 24 inches (610 mm) wide and of any of the corrosion-resistant metals in Table 1507.2.9.2.
2. For open valleys, valley lining of two plies of mineral-surfaced roll roofing complying with ASTM D3909 or ASTM D6380 shall be permitted. The bottom layer shall be 18 inches (457 mm) and the top layer a minimum of 36 inches (914 mm) wide.
3. For closed valleys (valleys covered with shingles), valley lining of one ply of smooth roll roofing complying with ASTM D6380, and at least 36 inches (914 mm) wide or types as described in Item 1 or 2 above shall be permitted. Self-adhering polymer modified bitumen underlayment complying with ASTM D1970 shall be permitted in lieu of the lining material.

❖ Valleys are the internal angle formed by the intersection of two sloping roof planes. For asphalt shingle roofs, the valley protection is categorized as open or closed. An open valley is a method of construction in which the steep-slope roofing on both sides is trimmed along each side of the valley, exposing the valley flashing.

Closed valleys are those that are covered with shingles, and include methods known as closed-cut valleys and woven valleys.

Closed-cut valleys feature a method of valley construction in which shingles from one side of the valley extend across the valley, while shingles from the other side are trimmed back approximately 2 inches (51 mm) from the valley centerline.

Woven valley construction is a method in which shingles from both sides of the valley extend across the valley and are woven together by overlapping alternate courses as they are applied.

TABLE 1507.2.9.2. See below.

❖ These corrosion-resistant metals may be used in open-valley construction as valley lining where the lining is exposed.

1507.2.9.3 Drip edge. A drip edge shall be provided at eaves and rake edges of shingle roofs. Adjacent segments of the drip edge shall be lapped a minimum of 2 inches (51 mm). The vertical leg of drip edges shall be a minimum of $1^1/_2$ inches (38 mm) in width and shall extend a minimum of $^1/_4$ inch (6.4 mm) below sheathing. The drip edge shall extend back on the roof a minimum of 2 inches (51 mm). Underlayment shall be installed over drip edges along eaves. Drip edges shall be installed over underlayment along rake edges. Drip edges shall be mechanically fastened a maximum of 12 inches (305 mm) on center.

❖ A drip edge is a metal flashing or other overhanging component applied at the roof edge that is intended to control the direction of dripping water and help protect the underlying building components. Drip edges have an outward projecting lower edge to direct the water away from the building. It is also used to break the continuity of contact between the roof perimeter and the wall components to help prevent capillary action.

A metal drip edge is a formed metal flashing that extends back from, and bends down over, the roof edge. Along roof gables and rakes, the drip edge is applied over the underlayment. Along the eave, the underlayment is applied over the drip edge.

1507.3 Clay and concrete tile. The installation of clay and concrete tile shall comply with the provisions of this section.

❖ Additional information regarding clay and concrete tile roofs is provided in the *NRCA Roofing Manual: Steep-slope Roof Systems* published by the NRCA.

1507.3.1 Deck requirements. Concrete and clay tile shall be installed only over solid sheathing or spaced structural sheathing boards.

❖ The choice of deck material is critical to the life of the roof system. Decks must be capable of withstanding loads such as the weight of the tile and associated roof components and accessories. The weight of tile can increase significantly due to water absorption, depending on its porosity. It is important for designers to note that some types of roof tiles may develop increased porosity with age.

1507.3.2 Deck slope. Clay and concrete roof tile shall be installed on roof slopes of $2^1/_2$ units vertical in 12 units horizontal (21-percent slope) or greater. For roof slopes from $2^1/_2$ units vertical in 12 units horizontal (21-percent slope) to four units vertical in 12 units horizontal (33-percent slope), double underlayment application is required in accordance with Section 1507.3.3.

❖ Along with the manufacturer's installation specifications, this section gives specific limitations on the application of concrete and clay tile. Thus, tile is not allowed to be installed on slopes less than $2^1/_2$:12 regardless of the manufacturer's literature. Additionally, for low-sloped roofs, double underlayment is required.

1507.3.3 Underlayment. Unless otherwise noted, required underlayment shall conform to: ASTM D226, Type II; ASTM D2626 or ASTM D6380, Class M mineral-surfaced roll roofing.

❖ Because of the long service life of many tile roofs, an underlayment should be chosen that will provide protection for a comparable period of time. ASTM D226

TABLE 1507.2.9.2
VALLEY LINING MATERIAL

MATERIAL	MINIMUM THICKNESS	GAGE	WEIGHT
Aluminum	0.024 in.	—	—
Cold-rolled copper	0.0216 in.	—	ASTM B370, 16 oz. per square ft.
Copper	—	—	16 oz
Galvanized steel	0.0179 in.	26 (zinc-coated G90)	—
High-yield copper	0.0162 in.	—	ASTM B370, 12 oz. per square ft.
Lead	—	—	2.5 pounds
Lead-coated copper	0.0216 in.	—	ASTM B101, 16 oz. per square ft.
Lead-coated high-yield copper	0.0162 in.	—	ASTM B101, 12 oz. per square ft.
Painted terne	—	—	20 pounds
Stainless steel	—	28	—
Zinc alloy	0.027 in.	—	—

For SI: 1 inch = 25.4 mm, 1 pound = 0.454 kg, 1 ounce = 28.35 g, 1 square foot = 0.0929 m^2.

includes the physical requirements for Type II asphalt-saturated felt, which is commonly called No. 30 asphalt felt.

ASTM D2626 covers the felt base sheet with fine mineral surfacing on the top side, with or without perforations. It is intended that nonperforated felt be used in this application. The *NRCA Roofing Manual: Steep-slope Roof Systems* has recommendations regarding the choice of underlayment for specific roof slopes and applications.

ASTM D6380, Class M, designates mineral-surfaced roll-roofing products.

1507.3.3.1 Low-slope roofs. For roof slopes from $2^1/_2$ units vertical in 12 units horizontal (21-percent slope), up to four units vertical in 12 units horizontal (33-percent slope), underlayment shall be a minimum of two layers applied as follows:

1. Starting at the eave, a 19-inch (483 mm) strip of underlayment shall be applied parallel with the eave and fastened sufficiently in place.
2. Starting at the eave, 36-inch-wide (914 mm) strips of underlayment felt shall be applied overlapping successive sheets 19 inches (483 mm) and fastened sufficiently in place.

❖ The code requires that the underlayment be laid with two layers in a manner that provides two thicknesses at any point.

1507.3.3.2 High-slope roofs. For roof slopes of four units vertical in 12 units horizontal (33-percent slope) or greater, underlayment shall be a minimum of one layer of underlayment felt applied shingle fashion, parallel to, and starting from the eaves and lapped 2 inches (51 mm), fastened only as necessary to hold in place.

❖ A single layer of underlayment is required on all roof slopes of 4:12 or greater. Reinforced underlayment is required on all roofs with spaced sheathing in order to avoid breakthrough of the tiles.

1507.3.3.3 High wind attachment. Underlayment applied in areas subject to high wind [V_{asd} greater than 110 mph (49 m/s) as determined in accordance with Section 1609.3.1] shall be applied with corrosion-resistant fasteners in accordance with the manufacturer's installation instructions. Fasteners are to be applied along the overlap not more than 36 inches (914 mm) on center.

Underlayment installed where V_{asd}, in accordance with Section 1609.3.1, equals or exceeds 120 mph (54 m/s) shall be attached in a grid pattern of 12 inches (305 mm) between side laps with a 6-inch (152 mm) spacing at the side laps. Underlayment shall be applied in accordance with Sections 1507.3.3.1 and 1507.3.3.2 except all laps shall be a minimum of 4 inches (102 mm). Underlayment shall be attached using metal or plastic cap nails with a head diameter of not less than 1 inch (25 mm) with a thickness of at least 32-gage [0.0134 inch (0.34 mm)] sheet metal. The cap nail shank shall be a minimum of 12 gage [0.105 inch (2.67 mm)] with a length to penetrate through the roof sheathing or a minimum of $^3/_4$ inch (19.1 mm) into the roof sheathing.

Exception: As an alternative, adhered underlayment complying with ASTM D1970 shall be permitted.

❖ These provisions are intended to improve the performance of underlayment on roofs subjected to high winds, particularly if tiles are lost due to a high-wind event. Where the nominal design wind speed is greater than 110 mph (49 m/s), fasteners must be corrosion resistant in accordance with the manufacturer's instructions.

Where the nominal design wind speed exceeds 120 mph (54 m/s), the attachment of the underlayment must be accomplished with cap nails. A 1-inch (25.4 mm) head diameter of the cap portion recognizes the type most commonly used in the field. The exception specifically permits the use of a self-adhered underlayment as an alternative in accordance with ASTM D1970.

1507.3.4 Clay tile. Clay roof tile shall comply with ASTM C1167.

❖ ASTM C1167 specifies material characteristics and physical properties for clay tiles that are intended for use as a roof covering where durability and appearance are required to provide a weather-resistant surface of specified design. These tiles are made of clay, shale or other earthy substances and are heat treated at an elevated temperature. The process develops a fired bond between the particulate constituents to provide the strength and durability requirements of ASTM C1167. The tiles are classified into three grades based on their resistance to weathering. Additionally, the standard covers performance characteristics, including durability; freezing and thawing; strength; efflorescence; permeability; and reactive particulates.

1507.3.5 Concrete tile. Concrete roof tile shall comply with ASTM C1492.

❖ ASTM C1492 is the standard that covers the physical properties and material characteristics for concrete roof tiles. The water absorption requirement in ASTM C1492 for normal roof weight tiles is $12^1/_2$ percent, which is lower than the 15 percent previously listed in the code. In addition, ASTM C1492 adds clarity with regards to the water absorption requirements for medium- and lightweight roof tiles. ASTM C140 is still referenced for the testing procedures in ASTM C1492, even though it is no longer directly referenced by the code.

1507.3.6 Fasteners. Tile fasteners shall be corrosion resistant and not less than 11-gage, $^5/_{16}$-inch (8.0 mm) head, and of sufficient length to penetrate the deck a minimum of $^3/_4$ inch (19.1 mm) or through the thickness of the deck, whichever is less. Attaching wire for clay or concrete tile shall not be smaller than 0.083 inch (2.1 mm). Perimeter fastening areas include three tile courses but not less than 36 inches (914 mm) from either side of hips or ridges and edges of eaves and gable rakes.

❖ Fasteners are to be corrosion resistant, such as hot-dipped galvanized steel, aluminum or stainless steel, needle or diamond pointed, with large flat heads.

1507.3.7 Attachment. Clay and concrete roof tiles shall be fastened in accordance with Table 1507.3.7.

❖ Refer to the commentary to Section 1504.2 for other information in the code related to the wind resistance of clay and concrete roof tile.

TABLE 1507.3.7. See page 15-15.

❖ The tabulated requirements for fastener installation are based on wind speed, mean roof height and roof slope. It also covers interlocking requirements for specific areas and criteria for roofs.

1507.3.8 Application. Tile shall be applied according to the manufacturer's installation instructions, based on the following:

1. Climatic conditions.
2. Roof slope.
3. Underlayment system.
4. Type of tile being installed.

❖ Clay and concrete tiles come in two generic forms: roll tile and flat tile. Either one may be interlocking but must be applied according to the manufacturer's instructions.

1507.3.9 Flashing. At the juncture of the roof vertical surfaces, flashing and counterflashing shall be provided in accordance with the manufacturer's installation instructions, and where of metal, shall not be less than 0.019-inch (0.48 mm) (No. 26 galvanized sheet gage) corrosion-resistant metal. The valley flashing shall extend at least 11 inches (279 mm) from the centerline each way and have a splash diverter rib not less than 1 inch (25 mm) high at the flow line formed as part of the flashing. Sections of flashing shall have an end lap of not less than 4 inches (102 mm). For roof slopes of three units vertical in 12 units horizontal (25-percent slope) and over, the valley flashing shall have a 36-inch-wide (914 mm) underlayment of either one layer of Type I underlayment running the full length of the valley, or a self-adhering polymer-modified bitumen sheet complying with ASTM D1970, in addition to other required underlayment. In areas where the average daily temperature in January is 25°F (-4°C) or less or where there is a possibility of ice forming along the eaves causing a backup of water, the metal valley flashing underlayment shall be solid cemented to the roofing underlayment for slopes under seven units vertical in 12 units horizontal (58-percent slope) or self-adhering polymer-modified bitumen sheet shall be installed.

❖ Flashings are required to maintain the integrity of weather-resistant roofs (see commentary, Section 1503.2).

1507.4 Metal roof panels. The installation of metal roof panels shall comply with the provisions of this section.

❖ There are two general categories of metal roofing systems: architectural metal roofing and structural metal roofing. Architectural metal roofs are generally water-shedding roof systems. Structural metal roofs have hydrostatic (water barrier) characteristics.

Architectural metal roof systems are usually characterized by a flat pan with $^3/_4$- to $1^1/_2$-inch (19.1 to 38 mm) ribs on each side. The absence of intermediate ribs and massive side ribs gives a clean appearance, but does not provide the panels with the strength to be considered a structural panel. Architectural metal roofing systems are typically designed for steep slopes so that water will shed off the roof panels. This type of design does not require the seams to be water tight. Therefore, the slope for such nonwater-tight roof panels is required to be not less than 3:12. One exception is traditional flat seamed, soldered or welded metal roofing, which is acceptable on slopes less than 3:12. An example of roofing that has traditionally been used this way is copper. Architectural metal roofing systems require solid decking.

"Metal roof panels" are defined in Section 202 as an interlocking metal sheet having a minimum installed weather exposure of 3 square feet (0.28 m^2) per sheet. In contrast, metal roof shingles have less than 3 square feet (0.28 m^2) of exposure.

1507.4.1 Deck requirements. Metal roof panel roof coverings shall be applied to a solid or closely fitted deck, except where the roof covering is specifically designed to be applied to spaced supports.

❖ The deck for metal roofing is required to be solid or closely fitted, except where the panels are specifically designed to be applied to spaced supports. Structural standing seam metal panel roof systems possess strength characteristics that allow them to span between structural supports, and are commonly used on preengineered metal buildings. The metal panel ribs are not seamed or interlocked like a true standing seam metal roof system.

TABLE 1507.3.7
CLAY AND CONCRETE TILE ATTACHMENT[a, b, c]

<table>
<tr><th colspan="5">GENERAL - CLAY OR CONCRETE ROOF TILE</th></tr>
<tr><th>Maximum Nominal Design Wind Speed, V_{asd} [f] (mph)</th><th>Mean roof height (feet)</th><th>Roof slope < 3:12</th><th colspan="2">Roof slope 3:12 and over</th></tr>
<tr><td>85</td><td>0-60</td><td rowspan="2">One fastener per tile. Flat tile without vertical laps, two fasteners per tile.</td><td colspan="2" rowspan="2">Two fasteners per tile. Only one fastener on slopes of 7:12 and less for tiles with installed weight exceeding 7.5 lbs./sq. ft. having a width not more than 16 inches.</td></tr>
<tr><td>100</td><td>0-40</td></tr>
<tr><td>100</td><td>>40-60</td><td colspan="3">The head of all tiles shall be nailed. The nose of all eave tiles shall be fastened with approved clips. All rake tiles shall be nailed with two nails. The nose of all ridge, hip and rake tiles shall be set in a bead of roofer's mastic.</td></tr>
<tr><td>110</td><td>0-60</td><td colspan="3">The fastening system shall resist the wind forces in Section 1609.5.3.</td></tr>
<tr><td>120</td><td>0-60</td><td colspan="3">The fastening system shall resist the wind forces in Section 1609.5.3.</td></tr>
<tr><td>130</td><td>0-60</td><td colspan="3">The fastening system shall resist the wind forces in Section 1609.5.3.</td></tr>
<tr><td>All</td><td>>60</td><td colspan="3">The fastening system shall resist the wind forces in Section 1609.5.3.</td></tr>
<tr><th colspan="5">INTERLOCKING CLAY OR CONCRETE ROOF TILE WITH PROJECTING ANCHOR LUGS[d, e]
(Installations on spaced/solid sheathing with battens or spaced sheathing)</th></tr>
<tr><th>Maximum Nominal Design Wind Speed, V_{asd} [f] (mph)</th><th>Mean roof height (feet)</th><th>Roof slope < 5:12</th><th>Roof slope 5:12 < 12:12</th><th>Roof slope 12:12 and over</th></tr>
<tr><td>85</td><td>0-60</td><td rowspan="2">Fasteners are not required. Tiles with installed weight less than 9 lbs./sq. ft. require a minimum of one fastener per tile.</td><td rowspan="2">One fastener per tile every other row. All perimeter tiles require one fastener. Tiles with installed weight less than 9 lbs./sq. ft. require a minimum of one fastener per tile.</td><td rowspan="2">One fastener required for every tile. Tiles with installed weight less than 9 lbs./sq. ft. require a minimum of one fastener per tile.</td></tr>
<tr><td>100</td><td>0-40</td></tr>
<tr><td>100</td><td>>40-60</td><td colspan="3">The head of all tiles shall be nailed. The nose of all eave tiles shall be fastened with approved clips. All rake tiles shall be nailed with two nails. The nose of all ridge, hip and rake tiles shall be set in a bead of roofer's mastic.</td></tr>
<tr><td>110</td><td>0-60</td><td colspan="3">The fastening system shall resist the wind forces in Section 1609.5.3.</td></tr>
<tr><td>120</td><td>0-60</td><td colspan="3">The fastening system shall resist the wind forces in Section 1609.5.3.</td></tr>
<tr><td>130</td><td>0-60</td><td colspan="3">The fastening system shall resist the wind forces in Section 1609.5.3.</td></tr>
<tr><td>All</td><td>>60</td><td colspan="3">The fastening system shall resist the wind forces in Section 1609.5.3.</td></tr>
<tr><th colspan="5">INTERLOCKING CLAY OR CONCRETE ROOF TILE WITH PROJECTING ANCHOR LUGS
(Installations on solid sheathing without battens)</th></tr>
<tr><th>Maximum Nominal Design Wind Speed, V_{asd} [f] (mph)</th><th>Mean roof height (feet)</th><th colspan="3">All roof slopes</th></tr>
<tr><td>85</td><td>0-60</td><td colspan="3">One fastener per tile.</td></tr>
<tr><td>100</td><td>0-40</td><td colspan="3">One fastener per tile.</td></tr>
<tr><td>100</td><td>> 40-60</td><td colspan="3">The head of all tiles shall be nailed. The nose of all eave tiles shall be fastened with approved clips. All rake tiles shall be nailed with two nails. The nose of all ridge, hip and rake tiles shall be set in a bead of roofer's mastic.</td></tr>
<tr><td>110</td><td>0-60</td><td colspan="3">The fastening system shall resist the wind forces in Section 1609.5.3.</td></tr>
<tr><td>120</td><td>0-60</td><td colspan="3">The fastening system shall resist the wind forces in Section 1609.5.3.</td></tr>
<tr><td>130</td><td>0-60</td><td colspan="3">The fastening system shall resist the wind forces in Section 1609.5.3.</td></tr>
<tr><td>All</td><td>> 60</td><td colspan="3">The fastening system shall resist the wind forces in Section 1609.5.3.</td></tr>
</table>

For SI: 1 inch = 25.4 mm, 1 foot = 304.8 mm, 1 mile per hour = 0.447 m/s, 1 pound per square foot = 4.882 kg/m^2.

a. Minimum fastener size. Corrosion-resistant nails not less than No. 11 gage with $^5/_{16}$-inch head. Fasteners shall be long enough to penetrate into the sheathing $^3/_4$ inch or through the thickness of the sheathing, whichever is less. Attaching wire for clay and concrete tile shall not be smaller than 0.083 inch.

b. Snow areas. A minimum of two fasteners per tile are required or battens and one fastener.

c. Roof slopes greater than 24:12. The nose of all tiles shall be securely fastened.

d. Horizontal battens. Battens shall be not less than 1 inch by 2 inch nominal. Provisions shall be made for drainage by a minimum of $^1/_8$-inch riser at each nail or by 4-foot-long battens with at least a $^1/_2$-inch separation between battens. Horizontal battens are required for slopes over 7:12.

e. Perimeter fastening areas include three tile courses but not less than 36 inches from either side of hips or ridges and edges of eaves and gable rakes.

f. V_{asd} shall be determined in accordance with Section 1609.3.1.

1507.4.2 Deck slope. Minimum slopes for metal roof panels shall comply with the following:

1. The minimum slope for lapped, nonsoldered seam metal roof panels without applied lap sealant shall be three units vertical in 12 units horizontal (25-percent slope).
2. The minimum slope for lapped, nonsoldered seam metal roof panels with applied lap sealant shall be one-half unit vertical in 12 units horizontal (4-percent slope). Lap sealants shall be applied in accordance with the approved manufacturer's installation instructions.
3. The minimum slope for standing-seam metal roof panel systems shall be one-quarter unit vertical in 12 units horizontal (2-percent slope).

❖ Typically, structural metal roofing systems are designed to resist water at laps and seams with sealants applied in the seams. It is usually recommended that these types of roof systems have a slope of not less than $^1/_2$:12, although some manufacturers allow slopes as low as $^1/_4$:12. While the code requires a minimum slope of 3:12, a lower slope is permitted on standing-seam-type roof systems, which require only $^1/_4$:12. The term "standing seam" is used to refer to almost any roof panel with a raised vertical rib. Strictly speaking, it only indicates those metal panels that interlock or are seamed together vertically above the panel's pan.

1507.4.3 Material standards. Metal-sheet roof covering systems that incorporate supporting structural members shall be designed in accordance with Chapter 22. Metal-sheet roof coverings installed over structural decking shall comply with Table 1507.4.3(1). The materials used for metal-sheet roof coverings shall be naturally corrosion resistant or provided with corrosion resistance in accordance with the standards and minimum thicknesses shown in Table 1507.4.3(2).

❖ The requirements for metal roof coverings that incorporate supporting structural members into their design are contained in Chapter 22. Other metal-sheet roof coverings are required to comply with Table 1507.4.3(1).

Many of the materials listed in Table 1507.4.3(1) are inherently corrosion resistant, such as aluminum, copper and lead. For these corrosion-resistant materials a material standard is listed for the base material; however, for steel roofing products, coatings are added to the base material to provide the necessary corrosion resistance.

TABLE 1507.4.3(1). See page 15-17.

❖ This table provides specific guidance and appropriate metal roof coverings. The requirements vary based upon the type of metal used, such as aluminum, steel, copper, lead and zinc. In some cases, standards are referenced; in other cases, basic criteria are provided.

TABLE 1507.4.3(2). See page 15-17.

❖ Steel coating standards are located in Table 1507.4.3(2). Having this table separate from Table 1507.4.3(1) helps to clarify that these standards only address the process necessary to provide corrosion resistance for the base steel. The coating descriptions in Table 1507.4.3(2) are consistent with the applicable coating standards.

1507.4.4 Attachment. Metal roof panels shall be secured to the supports in accordance with the approved manufacturer's fasteners. In the absence of manufacturer recommendations, the following fasteners shall be used:

1. Galvanized fasteners shall be used for steel roofs.
2. Copper, brass, bronze, copper alloy or 300 series stainless-steel fasteners shall be used for copper roofs.
3. Stainless-steel fasteners are acceptable for all types of metal roofs.
4. Aluminum fasteners are acceptable for aluminum roofs attached to aluminum supports.

❖ Fasteners used to attach metal roofing to their supporting construction are to be in accordance with the roofing manufacturer's recommendations and approved for that purpose. If such approval is not available, then the code provides a default list of several fasteners that are permitted for the type of metal roof shown. The purpose of this provision is to provide compatibility with roof materials and prevent corrosion.

1507.4.5 Underlayment and high wind. Underlayment applied in areas subject to high winds [V_{asd} greater than 110 mph (49 m/s) as determined in accordance with Section 1609.3.1] shall be applied with corrosion-resistant fasteners in accordance with the manufacturer's installation instructions. Fasteners are to be applied along the overlap not more than 36 inches (914 mm) on center.

Underlayment installed where V_{asd}, in accordance with Section 1609.3.1, equals or exceeds 120 mph (54 m/s) shall comply with ASTM D226 Type II, ASTM D4869 Type IV, or ASTM D1970. The underlayment shall be attached in a grid pattern of 12 inches (305 mm) between side laps with a 6-inch (152 mm) spacing at the side laps. Underlayment shall be applied in accordance with the manufacturer's installation instructions except all laps shall be a minimum of 4 inches (102 mm). Underlayment shall be attached using metal or plastic cap nails with a head diameter of not less than 1 inch (25 mm) with a thickness of at least 32-gage [0.0134 inch (0.34 mm)] sheet metal. The cap nail shank shall be a minimum of 12 gage [0.105 inch (2.67 mm)] with a length to penetrate through the roof sheathing or a minimum of $^3/_4$ inch (19.1 mm) into the roof sheathing.

Exception: As an alternative, adhered underlayment complying with ASTM D1970 shall be permitted.

❖ These provisions are intended to improve the performance of roofs subjected to high winds, particularly if the roof covering is lost due to a high-wind event.

Where the nominal design wind speed is greater than 110 mph (49 m/s), underlayment is required to be applied with fasteners that are corrosion resistant and in accordance with the manufacturer's instructions.

Where the nominal design wind speed exceeds 120 mph (54 m/s), the attachment of the underlayment must be accomplished with cap nails. A 1-inch (25.4 mm) head diameter of the cap portion recognizes the type most commonly used in the field. The exception recognizes that a self-adhered underlayment can be used as an alternative in accordance with ASTM D1970.

1507.5 Metal roof shingles. The installation of metal roof shingles shall comply with the provisions of this section.

❖ See the commentary to Section 1507.4 for a general discussion about metal roof systems.

1507.5.1 Deck requirements. Metal roof shingles shall be applied to a solid or closely fitted deck, except where the roof covering is specifically designed to be applied to spaced sheathing.

❖ The manufacturer's instructions must be followed in addition to any restrictions required by the code.

1507.5.2 Deck slope. Metal roof shingles shall not be installed on roof slopes below three units vertical in 12 units horizontal (25-percent slope).

❖ Metal shingles cannot be installed on roof slopes less than 3:12.

1507.5.3 Underlayment. Underlayment shall comply with ASTM D226, Type I or ASTM D4869.

❖ A single layer of underlayment is the minimum requirement under all metal shingles other than flat metal shingles. The underlayment must meet the performance criteria of ASTM D226, Type I, or ASTM D4869.

1507.5.3.1 Underlayment and high wind. Underlayment applied in areas subject to high winds [V_{asd} greater than 110 mph (49 m/s) as determined in accordance with Section 1609.3.1] shall be applied with corrosion-resistant fasteners in accordance with the manufacturer's installation instructions. Fasteners are to be applied along the overlap not farther apart than 36 inches (914 mm) on center.

Underlayment installed where V_{asd}, in accordance with Section 1609.3.1, equals or exceeds 120 mph (54 m/s) shall comply with ASTM D226 Type II or ASTM D4869 Type IV. The underlayment shall be attached in a grid pattern of 12 inches (305 mm) between side laps with a 6-inch spacing (152 mm) at the side laps. Underlayment shall be applied in accordance with the manufacturer's installation instructions except all laps shall be a minimum of 4 inches (102 mm). Underlayment shall be attached using metal or plastic cap nails with a head diameter of not less than 1 inch (25 mm) with a thickness of at least 32-gage [0.0134 inch (0.34 mm)] sheet metal. The cap nail shank shall be a minimum of 12 gage [0.105 inch (2.67 mm)] with a length to penetrate

TABLE 1507.4.3(1)
METAL ROOF COVERINGS

ROOF COVERING TYPE	STANDARD APPLICATION RATE/THICKNESS
Aluminum	ASTM B209, 0.024 inch minimum thickness for roll-formed panels and 0.019 inch minimum thickness for press-formed shingles.
Aluminum-zinc alloy coated steel	ASTM A792 AZ 50
Cold-rolled copper	ASTM B370 minimum 16 oz./sq. ft. and 12 oz./sq. ft. high yield copper for metal-sheet roof covering systems: 12 oz./sq. ft. for preformed metal shingle systems.
Copper	16 oz./sq. ft. for metal-sheet roof-covering systems; 12 oz./sq. ft. for preformed metal shingle systems.
Galvanized steel	ASTM A653 G-90 zinc-coated[a].
Hard lead	2 lbs./sq. ft.
Lead-coated copper	ASTM B101
Prepainted steel	ASTM A755
Soft lead	3 lbs./sq. ft.
Stainless steel	ASTM A240, 300 Series Alloys
Steel	ASTM A924
Terne and terne-coated stainless	Terne coating of 40 lbs. per double base box, field painted where applicable in accordance with manufacturer's installation instructions.
Zinc	0.027 inch minimum thickness; 99.995% electrolytic high grade zinc with alloy additives of copper (0.08% - 0.20%), titanium (0.07% - 0.12%) and aluminum (0.015%).

For SI: 1 ounce per square foot = 0.305 kg/m^2,
1 pound per square foot = 4.882 kg/m^2,
1 inch = 25.4 mm, 1 pound = 0.454 kg.

a. For Group U buildings, the minimum coating thickness for ASTM A653 galvanized steel roofing shall be G-60.

TABLE 1507.4.3(2)
MINIMUM CORROSION RESISTANCE

55% Aluminum-zinc alloy coated steel	ASTM A792 AZ 50
5% Aluminum alloy-coated steel	ASTM A875 GF60
Aluminum-coated steel	ASTM A463 T2 65
Galvanized steel	ASTM A653 G-90
Prepainted steel	ASTM A755[a]

a. Paint systems in accordance with ASTM A755 shall be applied over steel products with corrosion-resistant coatings complying with ASTM A792, ASTM A875, ASTM A463 or ASTM A653.

through the roof sheathing or a minimum of $^3/_4$ inch (19.1 mm) into the roof sheathing.

Exception: As an alternative, adhered underlayment complying with ASTM D1970 shall be permitted.

❖ These provisions are intended to improve the performance of underlayment on roofs subjected to high winds, particularly if shingles are lost due to a high-wind event. Where the design nominal wind speed is greater than 110 mph (49 m/s), fasteners must be corrosion resistant and in accordance with the manufacturer's instructions.

Based on observations of roof underlayment performance both in the field following hurricanes as well as in laboratory tests, ASTM 226 Type I underlayment is not permitted where the nominal design wind speed exceeds 120 mph (54 m/s). In laboratory tests, ASTM 226 Types I and II underlayment performed dramatically different, with the ASTM 226 Type II material remaining in place and exhibiting fewer signs of distress. In addition, the attachment of the underlayment must be accomplished with cap nails. A 1-inch (25.4 mm) head diameter of the cap portion recognizes the type most commonly used in the field. The exception recognizes the use of a self-adhered underlayment as an alernative in accordance with ASTM D1970.

1507.5.4 Ice barrier. In areas where there has been a history of ice forming along the eaves causing a backup of water, an ice barrier that consists of at least two layers of underlayment cemented together or of a self-adhering polymer-modified bitumen sheet shall be used in lieu of normal underlayment and extend from the lowest edges of all roof surfaces to a point at least 24 inches (610 mm) inside the exterior wall line of the building.

Exception: Detached accessory structures that contain no conditioned floor area.

❖ Ice dams form when snow melts over the warmer parts of a roof and refreezes over the colder eaves. This ice formation acts like a dam and causes water to back up beneath the roof covering. The roof will eventually leak, causing damage to the structure, including the walls, ceilings and roof [see Commentary Figures 1507.8.4(1) and (2)] (see Section 1507.8.4). In areas where this is prevalent, the installation of an ice barrier is required in accordance with this section.

The exception exempts accessory buildings from such restrictions as they are unheated structures where the need for protection against ice dams is unnecessary.

1507.5.5 Material standards. Metal roof shingle roof coverings shall comply with Table 1507.4.3(1). The materials used for metal-roof shingle roof coverings shall be naturally corrosion resistant or provided with corrosion resistance in accordance with the standards and minimum thicknesses specified in the standards listed in Table 1507.4.3(2).

❖ ASTM A653 regulates steel sheet for roofing and siding that is zinc coated on continuous lines and by the cut-length method. Material of this quality is furnished flat, in coils and cut lengths, and is formed in cut lengths. Roofing and siding includes corrugated, V-crimp, roll roofing and many special patterns. Corrugated roofing and siding sheet is produced in a number of corrugations, with variations of pitch and depth. ASTM A755 covers steel sheet that is metallic coated by the hot-dipped process and prepainted by the coil-coating process with organic films for exterior-exposed building products of various qualities. Sheet material of this designation is furnished in coils, cut lengths and formed cut lengths. ASTM B101 covers lead-coated copper sheets for architectural uses.

Many of the materials listed in Table 1507.4.3(1) are inherently corrosion resistant, such as aluminum, copper and lead. For these corrosion-resistant materials, a material standard is listed for the base material. However, for steel roofing products, coatings are added to the base material to provide the necessary corrosion resistance.

Steel coating standards are located in Table 1507.4.3(2). Having this table separate from Table 1507.4.3(1) helps to clarify that these standards only address the process necessary to provide corrosion resistance for the base steel. The coating descriptions in Table 1507.4.3(2) are consistent with the applicable coating standards.

1507.5.6 Attachment. Metal roof shingles shall be secured to the roof in accordance with the *approved* manufacturer's installation instructions.

❖ Fasteners used to attach metal roofing to the roof are to have the manufacturer's approval for that purpose.

1507.5.7 Flashing. Roof valley flashing shall be of corrosion-resistant metal of the same material as the roof covering or shall comply with the standards in Table 1507.4.3(1). The valley flashing shall extend at least 8 inches (203 mm) from the centerline each way and shall have a splash diverter rib not less than $^3/_4$ inch (19.1 mm) high at the flow line formed as part of the flashing. Sections of flashing shall have an end lap of not less than 4 inches (102 mm). In areas where the average daily temperature in January is 25°F (-4°C) or less or where there is a possibility of ice forming along the eaves causing a backup of water, the metal valley flashing shall have a 36-inch-wide (914 mm) underlayment directly under it consisting of either one layer of underlayment running the full length of the valley or a self-adhering polymer-modified bitumen sheet complying with ASTM D1970, in addition to underlayment required for metal roof shingles. The metal valley flashing underlayment shall be solidly cemented to the roofing underlayment for roof slopes under seven units vertical in 12 units horizontal (58-percent slope) or self-adhering polymer-modified bitumen sheet shall be installed.

❖ See the commentary to Section 1503.2.

1507.6 Mineral-surfaced roll roofing. The installation of mineral-surfaced roll roofing shall comply with this section.

❖ Mineral-surfaced roll roofing is an asphalt roll roofing material in some cases having a selvage edge, which is a specially defined edge (lined for demarcation) designed for some special purpose, such as overlapping or seaming. It is typically 36 inches (914 mm) in width and is surfaced with coarse mineral granules. In mineral-surfaced roll roofing, the selvage is not surfaced with coarse mineral granules to allow better adhesion of the overlapping sheet.

There are two general methods for applying roll roofing: the exposed nail method and the concealed nail method. Depending on the slope and nailing method used, roll roofing can be applied parallel (downslope roof edge) or perpendicular to the eave (parallel with the rake) (see Section 1507.6.1).

1507.6.1 Deck requirements. Mineral-surfaced roll roofing shall be fastened to solidly sheathed roofs.

❖ Prior to the application of the roll roofing on wood plank or plywood decks, the deck should be inspected for delamination of plywood, warped boards and proper nailing.

For roll roofing applied parallel to the eave, a minimum roof slope of 2:12 is recommended for application by the concealed-nail method, and 6:12 is recommended for application by the exposed-nail method.

For roll roofing applied parallel to the rake, a minimum roof slope of 3:12 is recommended for application by the concealed-nail method, and 4:12 is recommended for application by the exposed-nail method.

1507.6.2 Deck slope. Mineral-surfaced roll roofing shall not be applied on roof slopes below one unit vertical in 12 units horizontal (8-percent slope).

❖ It is not recommended that asphalt roll roofing be applied to deck slopes of less than 2:12, with one exception. Roll roofing with a 19-inch (483 mm) selvage edge (commonly referred to as double coverage or split-sheet) may be applied on a deck with a slope as low as 1:12; however, it should only be used on roofs that will drain by gravity. It is not recommended for decks that will allow puddling and require evaporation to dry the roof.

1507.6.3 Underlayment. Underlayment shall comply with ASTM D226, Type I or ASTM D4869.

❖ A single layer of underlayment is the minimum requirement under mineral-surfaced rolled roofing. The underlayment must meet the performance criteria of ASTM D226 Type I, or ASTM D4869.

1507.6.3.1 Underlayment and high wind. Underlayment applied in areas subject to high winds [V_{asd} greater than 110 mph (49 m/s) as determined in accordance with Section 1609.3.1] shall be applied with corrosion-resistant fasteners in accordance with the manufacturer's installation instructions. Fasteners are to be applied along the overlap not more than 36 inches (914 mm) on center.

Underlayment installed where V_{asd}, in accordance with Section 1609.3.1, equals or exceeds 120 mph (54 m/s) shall comply with ASTM D226 Type II. The underlayment shall be attached in a grid pattern of 12 inches (305 mm) between side laps with a 6-inch (152 mm) spacing at the side laps. Underlayment shall be applied in accordance with the manufacturer's installation instructions except all laps shall be a minimum of 4 inches (102 mm). Underlayment shall be attached using metal or plastic cap nails with a head diameter of not less than 1 inch (25 mm) with a thickness of at least 32-gage [0.0134 inch (0.34 mm)] sheet metal. The cap nail shank shall be a minimum of 12 gage [0.105 inch (2.67 mm)] with a length to penetrate through the roof sheathing or a minimum of $^3/_4$ inch (19.1 mm) into the roof sheathing.

Exception: As an alternative, adhered underlayment complying with ASTM D1970 shall be permitted.

❖ These provisions are intended to improve the performance of underlayment on roofs subjected to high winds, particularly if the roof covering is lost due to a high-wind event. Where the nominal design wind speed is greater than 110 mph (49 m/s), fasteners must be corrosion resistant and in accordance with the manufacturer's instructions.

Based on observations of roof underlayment performance both in the field following hurricanes as well as in laboratory tests, ASTM 226 Type I underlayment is not permitted where the nominal design wind speed exceeds 120 mph (54 m/s). In laboratory tests, ASTM 226 Types I and II underlayment performed dramatically different, with the ASTM 226 Type II material remaining in place and exhibiting fewer signs of distress. In addition, the attachment of the underlayment must be accomplished with cap nails. A 1-inch (25.4 mm) head diameter of the cap portion recognizes the type most commonly used in the field. The exception recognizes the use of a self-adhered underlayment as an alternative in accordance with ASTM D1970.

1507.6.4 Ice barrier. In areas where there has been a history of ice forming along the eaves causing a backup of water, an ice barrier that consists of at least two layers of underlayment cemented together or of a self-adhering polymer-modified bitumen sheet shall be used in lieu of normal underlayment and extend from the lowest edges of all roof surfaces to a point at least 24 inches (610 mm) inside the exterior wall line of the building.

Exception: Detached accessory structures that contain no conditioned floor area.

❖ Ice dams form when snow melts over the warmer parts of a roof and refreezes over the colder eaves. This ice formation acts like a dam and causes water to back up beneath the roof covering. The roof will eventually leak, causing damage to the structure, including the walls, ceilings and roof [see Commentary Figures 1507.8.4(1) and (2)]. In areas where this is prevalent, the installation of an ice barrier is required in accordance with this section.

The exception exempts accessory buildings from such restrictions, as they are unheated structures where the need for protection against ice dams is unnecessary.

1507.6.5 Material standards. Mineral-surfaced roll roofing shall conform to ASTM D3909 or ASTM D6380.

❖ ASTM D6380 covers asphalt roll roofing (organic felt). ASTM D3909 covers asphalt-impregnated and coated glass felt roll roofing surfaced on the weather side with mineral granules for use as a cap sheet in the construction of built-up roofs.

1507.7 Slate shingles. The installation of slate shingles shall comply with the provisions of this section.

❖ Slate is a dense, tough, durable natural rock or stone material that is practically nonabsorbent. This natural rock has cleavage planes that allow the slate to be easily split into relatively thin layers. Slate also possesses a natural grain that usually runs perpendicular to the cleavage. Slate is usually split so the length of the shingle runs in the direction of the grain.

Slate roofing has a long service life when consideration is given to all components of the roofing system. Some grades of slate combined with proper roof deck, underlayments, fastening and accessories have a service life in excess of 75 years.

1507.7.1 Deck requirements. Slate shingles shall be fastened to solidly sheathed roofs.

❖ Due to the long service life expected of a slate roof system, careful consideration should be given to the deck material. Deck material that can be expected to last as long as the service life of the roof should be chosen. Additionally, consideration should be given to the weight of the slate when choosing the deck. Slate is available in many different thicknesses and, therefore, the weight will vary.

1507.7.2 Deck slope. Slate shingles shall only be used on slopes of four units vertical in 12 units horizontal (4:12) or greater.

❖ This type of roof covering cannot be installed on roof slopes that are less than 4:12.

1507.7.3 Underlayment. Underlayment shall comply with ASTM D226, Type II or ASTM D4869, Type III or IV.

❖ A single layer of underlayment is the minimum requirement under mineral-surfaced rolled roofing. The underlayment must meet the performance criteria of ASTM D226, Type II, or ASTM D4869, Type III and IV.

1507.7.3.1 Underlayment and high wind. Underlayment applied in areas subject to high winds [V_{asd} greater than 110 mph (49 m/s) as determined in accordance with Section 1609.3.1] shall be applied with corrosion-resistant fasteners in accordance with the manufacturer's installation instructions. Fasteners are to be applied along the overlap not more than 36 inches (914 mm) on center.

Underlayment installed where V_{asd}, in accordance with Section 1609.3.1, equals or exceeds 120 mph (54 m/s) shall comply with ASTM D226, Type II or ASTM D4869, Type IV. The underlayment shall be attached in a grid pattern of 12 inches (305 mm) between side laps with a 6-inch (152 mm) spacing at the side laps. Underlayment shall be applied in accordance with the manufacturer's installation instructions except all laps shall be a minimum of 4 inches (102 mm). Underlayment shall be attached using metal or plastic cap nails with a head diameter of not less than 1 inch (25 mm) with a thickness of at least 32-gage [0.0134 inch (0.34 mm)] sheet metal. The cap nail shank shall be a minimum of 12 gage [0.105 inch (2.67 mm)] with a length to penetrate through the roof sheathing or a minimum of $^3/_4$ inch (19.1 mm) into the roof sheathing.

Exception: As an alternative, adhered underlayment complying with ASTM D1970 shall be permitted.

❖ These provisions are intended to improve the performance of underlayment on roofs subjected to high winds, particularly if the roof covering is lost due to a high-wind event. Where the nominal design wind speed is greater than 110 mph (49 m/s), fasteners must be corrosion resistant and in accordance with the manufacturer's instructions.

Based on observations of roof underlayment performance both in the field following hurricanes as well as in laboratory tests, ASTM 226 Type I underlayment is not permitted where the nominal design wind speed exceeds 120 mph (54 m/s). In laboratory tests, ASTM 226 Types I and II underlayment performed dramatically different, with the ASTM 226 Type II material remaining in place and exhibiting fewer signs of distress. In addition, the attachment of the underlayment must be accomplished with cap nails. A 1-inch (25.4 mm) head diameter of the cap portion recognizes the type most commonly used in the field. The exception recognizes the use of a self-adhered underlayment as an alernative in accordance with ASTM D1970.

1507.7.4 Ice barrier. In areas where the average daily temperature in January is 25°F (-4°C) or less or where there is a possibility of ice forming along the eaves causing a backup of water, an ice barrier that consists of at least two layers of underlayment cemented together or of a self-adhering polymer-modified bitumen sheet shall extend from the lowest edges of all roof surfaces to a point at least 24 inches (610 mm) inside the exterior wall line of the building.

Exception: Detached accessory structures that contain no conditioned floor area.

❖ Ice dams form when snow melts over the warmer parts of a roof and refreezes over the colder eaves. This ice formation acts like a dam and causes water to back up beneath the roof covering. The roof will eventually leak, causing damage to the structure, including the walls, ceilings and roof [see Commentary Figures 1507.8.4(1) and (2)]. In areas where this is prevalent, the installation of an ice barrier is required in accordance with this section.

There is an exception to this section that exempts accessory buildings from such restrictions, as they are unheated structures where the need for protection against ice dams is unnecessary.

1507.7.5 Material standards. Slate shingles shall comply with ASTM C406.

❖ ASTM C406 covers the material characteristics, physical requirements and sampling appropriate for the selection of slate as a roofing material.

1507.7.6 Application. Minimum headlap for slate shingles shall be in accordance with Table 1507.7.6. Slate shingles shall be secured to the roof with two fasteners per slate.

❖ Slate shingles are drilled or punched with holes for fasteners with consideration for proper headlap. Fastener material should be chosen based on the expected service life. Copper slating nails, stainless steel, aluminum-alloy, bronze or cut-brass roofing nails are often specified, depending on the particular project. Unprotected black-iron, electroplated and hot-dipped galvanized fasteners are usually not recommended.

TABLE 1507.7.6
SLATE SHINGLE HEADLAP

SLOPE	HEADLAP (inches)
4:12 < slope < 8:12	4
8:12 < slope < 20:12	3
slope ≥ 20:12	2

For SI: 1 inch = 25.4 mm.

❖ Headlap is the distance of overlap measured from the uppermost ply or course to the point that it overlaps the undermost ply or course.

1507.7.7 Flashing. Flashing and counterflashing shall be made with sheet metal. Valley flashing shall be a minimum of 15 inches (381 mm) wide. Valley and flashing metal shall be a minimum uncoated thickness of 0.0179-inch (0.455 mm) zinc-coated G90. Chimneys, stucco or brick walls shall have a minimum of two plies of felt for a cap flashing consisting of a 4-inch-wide (102 mm) strip of felt set in plastic cement and extending 1 inch (25 mm) above the first felt and a top coating of plastic cement. The felt shall extend over the base flashing 2 inches (51 mm).

❖ Flashing is required at all intersections between exterior walls and roofs. These are areas where rainwater can easily penetrate the building envelope (see Section 1503.2).

1507.8 Wood shingles. The installation of wood shingles shall comply with the provisions of this section and Table 1507.8.

❖ "Wood shingles" are defined by the roofing industry as sawn wood products featuring a uniform butt thickness per individual length.

TABLE 1507.8. See page 15-22.

❖ Wood shingles cannot be installed on roof slopes below 3:12. Single-layer underlayment is required at eaves, ridges, hips, valleys and all other changes of roof slope or direction to protect the roof from leakage caused by water backup. Each shingle must be securely fastened to the roof deck with a maximum of two fasteners. Care should be taken so that the fasteners do not cause splitting of the wood shingle. Fasteners are to be specified in the manufacturer's installation instructions.

1507.8.1 Deck requirements. Wood shingles shall be installed on solid or spaced sheathing. Where spaced sheathing is used, sheathing boards shall be not less than 1-inch by 4-inch (25 mm by 102 mm) nominal dimensions and shall be spaced on centers equal to the weather exposure to coincide with the placement of fasteners.

❖ Spaced sheathing is usually of 1-inch by 4-inch (25 mm by 102 mm) or 1-inch by 6-inch (25 mm by 152 mm) softwood boards. Note that solid sheathing is required for certain locations by Section 1507.8.1.1.

1507.8.1.1 Solid sheathing required. Solid sheathing is required in areas where the average daily temperature in January is 25°F (-4°C) or less or where there is a possibility of ice forming along the eaves causing a backup of water.

❖ Solid sheathing is usually softwood panels, which provide a smooth, even base for the roofing material and help stiffen the entire roof structure. Solid sheathing provides an extra degree of protection where ice damming is a possibility.

1507.8.2 Deck slope. Wood shingles shall be installed on slopes of not less than three units vertical in 12 units horizontal (25-percent slope).

❖ Wood shingles cannot be installed on roof slopes below 3:12.

1507.8.3 Underlayment. Underlayment shall comply with ASTM D226, Type I or ASTM D4869.

❖ A single layer of underlayment is the minimum required under wood shingles. The underlayment must meet the performance criteria of ASTM D226, Type I, or ASTM D4869.

1507.8.3.1 Underlayment and high wind. Underlayment applied in areas subject to high winds [V_{asd} greater than 110 mph (49 m/s) as determined in accordance with Section 1609.3.1] shall be applied with corrosion-resistant fasteners in accordance with the manufacturer's installation instructions. Fasteners are to be applied along the overlap not more than 36 inches (914 mm) on center.

Underlayment installed where V_{asd}, in accordance with Section 1609.3.1, equals or exceeds 120 mph (54 m/s) shall comply with ASTM D226, Type II or ASTM D4869, Type IV. The underlayment shall be attached in a grid pattern of 12 inches (305 mm) between side laps with a 6-inch (152 mm) spacing at the side laps. Underlayment shall be applied in accordance with the manufacturer's installation instructions except all laps shall be a minimum of 4 inches (102 mm). Underlayment shall be attached using metal or plastic cap nails with a head diameter of not less than 1 inch (25 mm) with a thickness of at least 32-gage [0.0134 inch (0.34 mm)] sheet metal. The cap nail shank shall be a minimum of 12 gage [0.105 inch (2.67 mm)] with a length to penetrate through the roof sheathing or a minimum of $^3/_4$ inch (19.1 mm) into the roof sheathing.

Exception: As an alternative, adhered underlayment complying with ASTM D1970 shall be permitted.

❖ These provisions are intended to improve the performance of underlayment on roofs subjected to high winds, particularly if wood shingles are lost in a high-wind event. Where the nominal design wind speed is greater than 110 mph (49 m/s), fasteners must be corrosion resistant and in accordance with the manufacturer's instructions.

Based on observations of roof underlayment performance both in the field following hurricanes as well

TABLE 1507.8
WOOD SHINGLE AND SHAKE INSTALLATION

ROOF ITEM	WOOD SHINGLES	WOOD SHAKES
1. Roof slope	Wood shingles shall be installed on slopes of not less than three units vertical in 12 units horizontal (3:12).	Wood shakes shall be installed on slopes of not less than four units vertical in 12 units horizontal (4:12).
2. Deck requirement		
Temperate climate	Shingles shall be applied to roofs with solid or spaced sheathing. Where spaced sheathing is used, sheathing boards shall be not less than 1″ × 4″ nominal dimensions and shall be spaced on centers equal to the weather exposure to coincide with the placement of fasteners.	Shakes shall be applied to roofs with solid or spaced sheathing. Where spaced sheathing is used, sheathing boards shall be not less than 1″ × 4″ nominal dimensions and shall be spaced on centers equal to the weather exposure to coincide with the placement of fasteners. When 1″ × 4″ spaced sheathing is installed at 10 inches, boards must be installed between the sheathing boards.
In areas where the average daily temperature in January is 25°F or less or where there is a possibility of ice forming along the eaves causing a backup of water.	Solid sheathing is required.	Solid sheathing is required.
3. Interlayment	No requirements.	Interlayment shall comply with ASTM D226, Type 1.
4. Underlayment		
Temperate climate	Underlayment shall comply with ASTM D226, Type 1.	Underlayment shall comply with ASTM D226, Type 1.
In areas where there is a possibility of ice forming along the eaves causing a backup of water.	An ice barrier that consists of at least two layers of underlayment cemented together or of a self-adhering polymer-modified bitumen sheet shall extend from the eave's edge to a point at least 24 inches inside the exterior wall line of the building.	An ice barrier that consists of at least two layers of underlayment cemented together or of a self-adhering polymer-modified bitumen sheet shall extend from the lowest edges of all roof surfaces to a point at least 24 inches inside the exterior wall line of the building.
5. Application		
Attachment	Fasteners for wood shingles shall be hot-dipped galvanized or Type 304 (Type 316 for coastal areas) stainless steel with a minimum penetration of 0.75 inch into the sheathing. For sheathing less than 0.5 inch thick, the fasteners shall extend through the sheathing.	Fasteners for wood shakes shall be hot-dipped galvanized or Type 304 (Type 316 for coastal areas) with a minimum penetration of 0.75 inch into the sheathing. For sheathing less than 0.5 inch thick, the fasteners shall extend through the sheathing.
No. of fasteners	Two per shingle.	Two per shake.
Exposure	Weather exposures shall not exceed those set forth in Table 1507.8.7.	Weather exposures shall not exceed those set forth in Table 1507.9.8.
Method	Shingles shall be laid with a side lap of not less than 1.5 inches between joints in courses, and no two joints in any three adjacent courses shall be in direct alignment. Spacing between shingles shall be 0.25 to 0.375 inch.	Shakes shall be laid with a side lap of not less than 1.5 inches between joints in adjacent courses. Spacing between shakes shall not be less than 0.375 inch or more than 0.625 inch for shakes and taper sawn shakes of naturally durable wood and shall be 0.25 to 0.375 inch for preservative-treated taper sawn shakes.
Flashing	In accordance with Section 1507.8.8.	In accordance with Section 1507.9.9.

For SI: 1 inch = 25.4 mm, °C = [(°F) - 32]/1.8.

as in laboratory tests, ASTM 226 Type I underlayment is not permitted where the nominal design wind speed exceeds 120 mph (54 m/s). In laboratory tests, ASTM 226 Types I and II underlayment performed dramatically different, with the ASTM 226 Type II material remaining in place and exhibiting fewer signs of distress. In addition, the attachment of the underlayment must be accomplished with cap nails. A 1-inch (25.4 mm) head diameter of the cap portion recognizes the type most commonly used in the field. The exception specifically permits the use of a self-adhered underlayment as an alernative in accordance with ASTM D1970.

1507.8.4 Ice barrier. In areas where there has been a history of ice forming along the eaves causing a backup of water, an ice barrier that consists of at least two layers of underlayment cemented together or of a self-adhering polymer-modified bitumen sheet shall be used in lieu of normal underlayment and extend from the lowest edges of all roof surfaces to a point at least 24 inches (610 mm) inside the exterior wall line of the building.

Exception: Detached accessory structures that contain no conditioned floor area.

❖ Ice dams form when snow melts over the warmer parts of a roof and refreezes over the colder eaves. This ice formation acts like a dam and causes water to back up beneath the roof covering. The roof will eventually leak, causing damage to the structure, including the walls, ceilings and roof [see Commentary Figures 1507.8.4(1) and (2)]. In areas where this is prevalent, the installation of an ice barrier is required in accordance with this section.

The exception exempts accessory buildings from such restrictions, as they are unheated structures where the need for protection against ice dams is unnecessary.

1507.8.5 Material standards. Wood shingles shall be of naturally durable wood and comply with the requirements of Table 1507.8.5.

❖ Information about the installation of wood shingles is available in the *Design and Application Manual for New Roof Construction* published by the Cedar Shake and Shingle Bureau (CSSB). Additional information regarding wood shingle roofs is provided in the wood roofing section of the *NRCA Roofing Manual: Steep-slope Roof Systems* published by the NRCA.

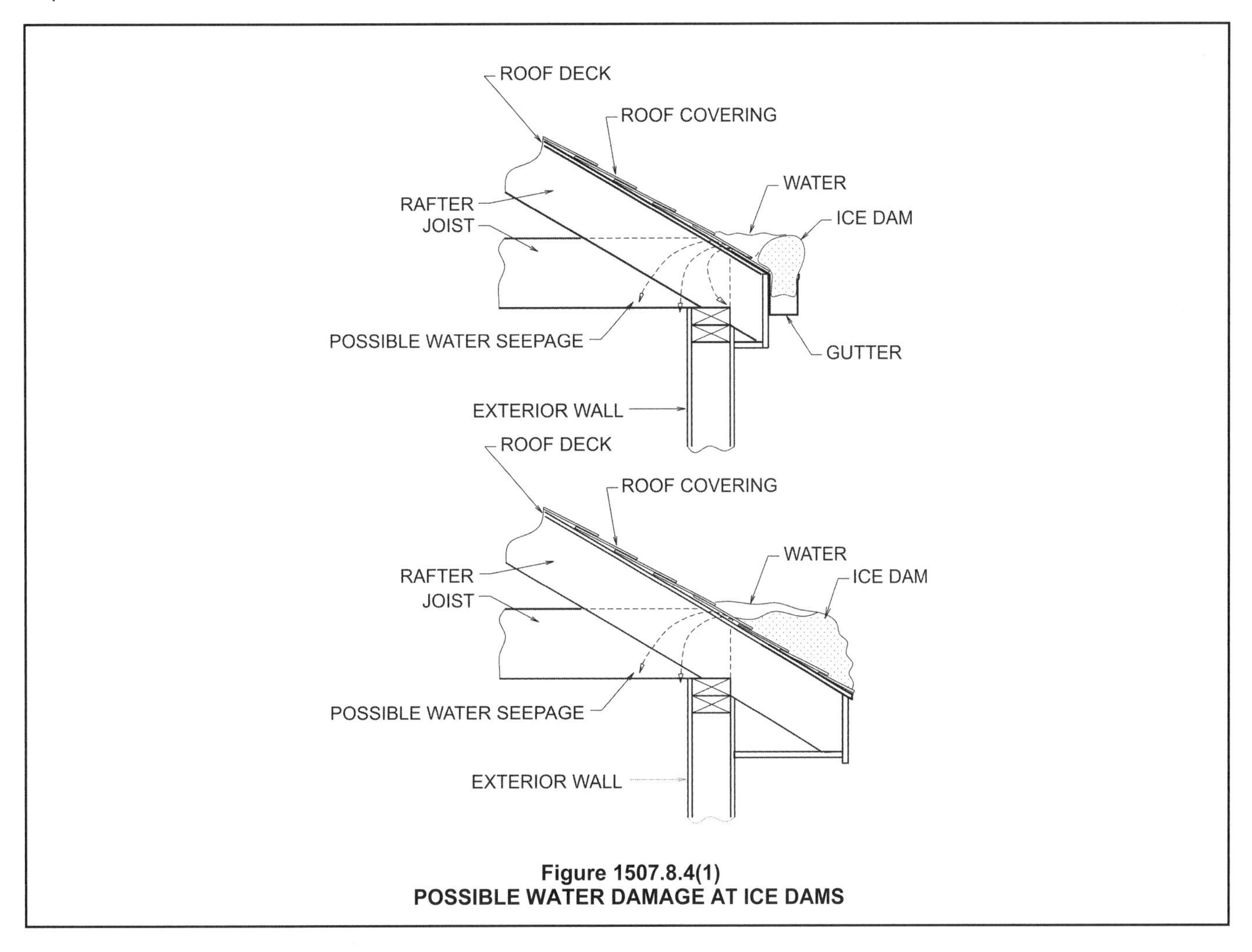

Figure 1507.8.4(1)
POSSIBLE WATER DAMAGE AT ICE DAMS

TABLE 1507.8.5
WOOD SHINGLE MATERIAL REQUIREMENTS

MATERIAL	APPLICABLE MINIMUM GRADES	GRADING RULES
Wood shingles of naturally durable wood	1, 2 or 3	CSSB

CSSB = Cedar Shake and Shingle Bureau

❖ The standards published by CSSB contain useful information on the grading rules for wood shingles. A third-party inspection agency's label is now required to document compliance of the shingle.

1507.8.6 Attachment. Fasteners for wood shingles shall be corrosion resistant with a minimum penetration of $^3/_4$ inch (19.1 mm) into the sheathing. For sheathing less than $^1/_2$ inch (12.7 mm) in thickness, the fasteners shall extend through the sheathing. Each shingle shall be attached with a minimum of two fasteners.

❖ Each shingle must be securely fastened to the roof deck with a minimum of two fasteners. Care should be taken so that the fasteners do not cause splitting of the wood shingle. Fasteners are to be as specified in the manufacturer's installation instructions.

1507.8.7 Application. Wood shingles shall be laid with a side lap not less than $1^1/_2$ inches (38 mm) between joints in adjacent courses, and not be in direct alignment in alternate courses. Spacing between shingles shall be $^1/_4$ to $^3/_8$ inch (6.4 to 9.5 mm). Weather exposure for wood shingles shall not exceed that set in Table 1507.8.7.

❖ Wood shingle exposure is specified in Table 1507.8.7. Depending on the grade of the material, total shingle length and slope of the roof deck, the table specifies the maximum length of exposure for weathering purposes.

TABLE 1507.8.7
WOOD SHINGLE WEATHER EXPOSURE AND ROOF SLOPE

ROOFING MATERIAL	LENGTH (inches)	GRADE	EXPOSURE (inches)	
			3:12 pitch to < 4:12	4:12 pitch or steeper
Shingles of naturally durable wood	16	No. 1	3.75	5
		No. 2	3.5	4
		No. 3	3	3.5
	18	No. 1	4.25	5.5
		No. 2	4	4.5
		No. 3	3.5	4
	24	No. 1	5.75	7.5
		No. 2	5.5	6.5
		No. 3	5	5.5

For SI: 1 inch = 25.4 mm.

❖ The grade of the shingle will be listed on the label required for wood shingles. Each of the three different grades is available in three different lengths. For lower-sloped roofs (3:12 to 4:12), where wind uplift is more significant, the maximum weather exposure is always less than for steeper-sloped roofs (4:12 and greater).

1507.8.8 Flashing. At the juncture of the roof and vertical surfaces, flashing and counterflashing shall be provided in accordance with the manufacturer's installation instructions, and where of metal, shall be not less than 0.019-inch (0.48 mm) (No. 26 galvanized sheet gage) corrosion-resistant metal. The valley flashing shall extend at least 11 inches (279 mm) from the centerline each way and have a splash diverter

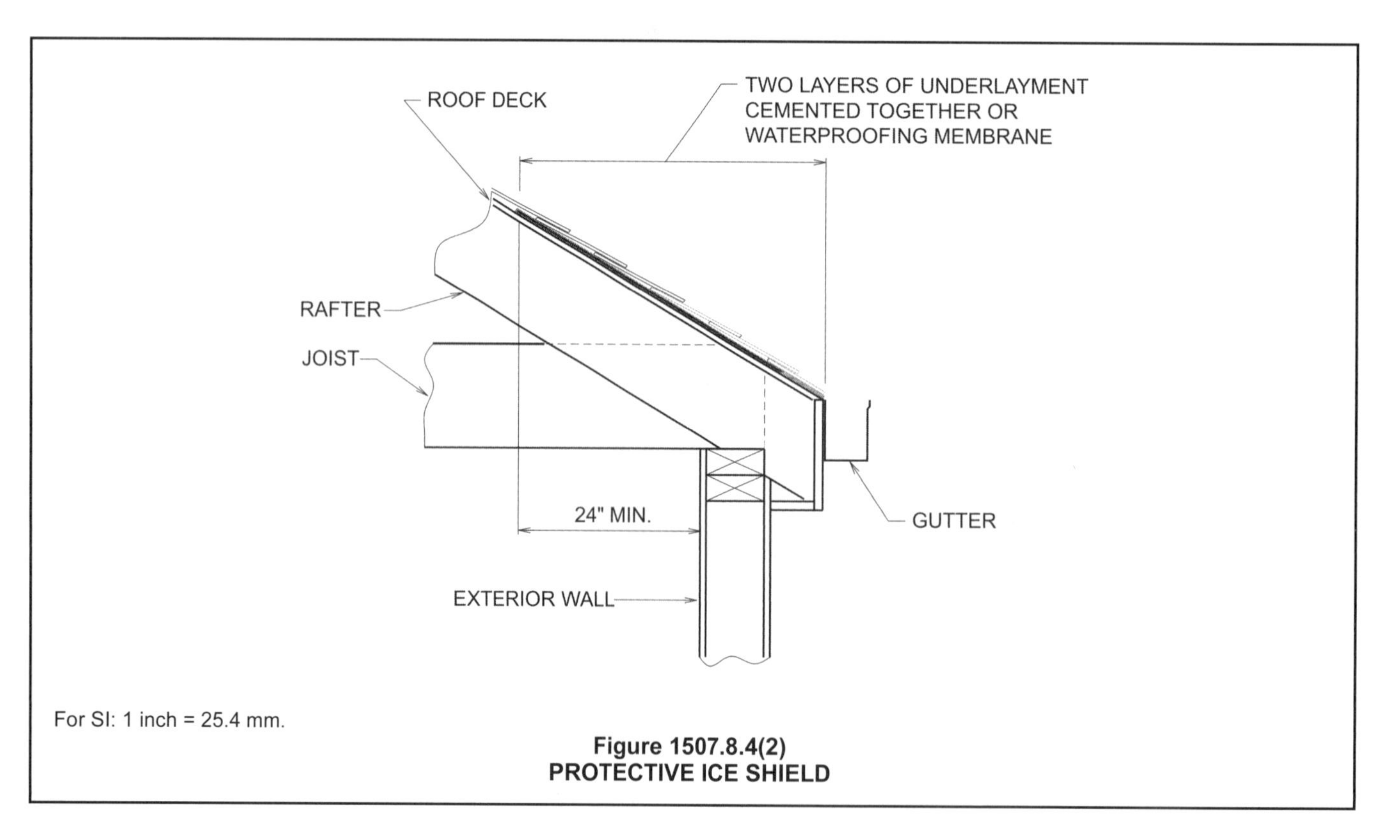

Figure 1507.8.4(2)
PROTECTIVE ICE SHIELD

rib not less than 1 inch (25 mm) high at the flow line formed as part of the flashing. Sections of flashing shall have an end lap of not less than 4 inches (102 mm). For roof slopes of three units vertical in 12 units horizontal (25-percent slope) and over, the valley flashing shall have a 36-inch-wide (914 mm) underlayment of either one layer of Type I underlayment running the full length of the valley or a self-adhering polymer-modified bitumen sheet complying with ASTM D1970, in addition to other required underlayment. In areas where the average daily temperature in January is 25°F (-4°C) or less or where there is a possibility of ice forming along the eaves causing a backup of water, the metal valley flashing underlayment shall be solidly cemented to the roofing underlayment for slopes under seven units vertical in 12 units horizontal (58-percent slope) or self-adhering polymer-modified bitumen sheet shall be installed.

❖ Improper installation of flashing is the greatest cause of failures of roof covering systems. Wherever one plane of a roof intersects another plane, flashing is required (see Section 1503.2).

1507.9 Wood shakes. The installation of wood shakes shall comply with the provisions of this section and Table 1507.8.

❖ This section establishes design and installation requirements for wood shakes, which are defined as roofing products split from logs and then shaped as required by the individual manufacturers. Wood shakes must be labeled by an approved third-party inspection agency. A quality control program is required to contain a set of grading rules.

Information about the installation of wood shakes is available in the *Design and Application Manual for New Roof Construction* published by the CSSB. Additional information is provided in the wood roofing section of the *NRCA Roofing Manual: Steep-slope Roof Systems* published by the NRCA.

1507.9.1 Deck requirements. Wood shakes shall only be used on solid or spaced sheathing. Where spaced sheathing is used, sheathing boards shall be not less than 1-inch by 4-inch (25 mm by 102 mm) nominal dimensions and shall be spaced on centers equal to the weather exposure to coincide with the placement of fasteners. Where 1-inch by 4-inch (25 mm by 102 mm) spaced sheathing is installed at 10 inches (254 mm) on center, additional 1-inch by 4-inch (25 mm by 102 mm) boards shall be installed between the sheathing boards.

❖ See the commentary to Section 1507.8.1.

1507.9.1.1 Solid sheathing required. Solid sheathing is required in areas where the average daily temperature in January is 25°F (-4°C) or less or where there is a possibility of ice forming along the eaves causing a backup of water.

❖ See the commentary to Section 1507.8.1.1.

1507.9.2 Deck slope. Wood shakes shall only be used on slopes of not less than four units vertical in 12 units horizontal (33-percent slope).

❖ Wood shakes cannot be installed on roof slopes less than 4:12 to provide for adequate drainage of the roof.

1507.9.3 Underlayment. Underlayment shall comply with ASTM D226, Type I or ASTM D4869.

❖ A single layer of underlayment is the minimum requirement under wood shakes. The underlayment must meet the performance criteria of ASTM D226, Type I, or ASTM D4869.

1507.9.3.1 Underlayment and high wind. Underlayment applied in areas subject to high winds [V_{asd} greater than 110 mph (49 m/s) as determined in accordance with Section 1609.3.1] shall be applied with corrosion-resistant fasteners in accordance with the manufacturer's installation instructions. Fasteners are to be applied along the overlap not more than 36 inches (914 mm) on center.

Underlayment installed where V_{asd}, in accordance with Section 1609.3.1, equals or exceeds 120 mph (54 m/s) shall comply with ASTM D226, Type II or ASTM D4869, Type IV. The underlayment shall be attached in a grid pattern of 12 inches (305 mm) between side laps with a 6-inch (152 mm) spacing at the side laps. Underlayment shall be applied in accordance with the manufacturer's installation instructions except all laps shall be a minimum of 4 inches (102 mm). Underlayment shall be attached using metal or plastic cap nails with a head diameter of not less than 1 inch (25 mm) with a thickness of at least 32-gage [0.0134 inch (0.34 mm)] sheet metal. The cap nail shank shall be a minimum of 12 gage [0.105 inch (2.67 mm)] with a length to penetrate through the roof sheathing or a minimum of $^3/_4$ inch (19.1 mm) into the roof sheathing.

Exception: As an alternative, adhered underlayment complying with ASTM D1970 shall be permitted.

❖ These provisions are intended to improve the performance of underlayment on roofs subjected to high winds, particularly if wood shakes are lost in a high-wind event. Where the nominal design wind speed is greater than 110 mph (49 m/s), fasteners must be corrosion resistant and in accordance with the manufacturer's instructions.

Based on observations of roof underlayment performance both in the field following hurricanes as well as in laboratory tests, ASTM 226 Type I underlayment is not permitted where the nominal design wind speed exceeds 120 mph (54 m/s). In laboratory tests ASTM 226 Types I and II underlayment performed dramatically different, with the ASTM 226 Type II material remaining in place and exhibiting fewer signs of distress. In addition, the attachment of the underlayment must be accomplished with cap nails. A 1-inch (25.4 mm) head diameter of the cap portion recognizes the type most commonly used in the field. The exception recognizes that a self-adhered underlayment can be used as an alternative in accordance with ASTM D1970.

1507.9.4 Ice barrier. In areas where there has been a history of ice forming along the eaves causing a backup of water, an ice barrier that consists of at least two layers of underlayment cemented together or of a self-adhering polymer-modified bitumen sheet shall be used in lieu of normal underlayment and extend from the lowest edges of all roof surfaces to a point at least 24 inches (610 mm) inside the exterior wall line of the building.

Exception: Detached accessory structures that contain no conditioned floor area.

❖ Ice dams form when snow melts over the warmer parts of a roof and refreezes over the colder eaves. This ice formation acts like a dam and causes water to back up beneath the roof covering. The roof will eventually leak, causing damage to the structure, including the walls, ceilings and roof [see Commentary Figures 1507.8.4(1) and (2)]. In areas where this is prevalent, the installation of an ice barrier is required in accordance with this section.

The exception exempts accessory buildings from such restrictions, as they are unheated structures where the need for protection against ice dams is unnecessary.

1507.9.5 Interlayment. Interlayment shall comply with ASTM D226, Type I.

❖ A single layer of felt interlayment must be shingled between each course on all roof slopes.

1507.9.6 Material standards. Wood shakes shall comply with the requirements of Table 1507.9.6.

❖ Information about the installation of wood shingles is available in the *Design and Application Manual for New Roof Construction* published by CSSB. Additional information is provided in the wood roofing section of the *NRCA Roofing Manual: Steep-slope Roof Systems* published by the NRCA.

TABLE 1507.9.6
WOOD SHAKE MATERIAL REQUIREMENTS

MATERIAL	MINIMUM GRADES	APPLICABLE GRADING RULES
Wood shakes of naturally durable wood	1	CSSB
Taper sawn shakes of naturally durable wood	1 or 2	CSSB
Preservative-treated shakes and shingles of naturally durable wood	1	CSSB
Fire-retardant-treated shakes and shingles of naturally durable wood	1	CSSB
Preservative-treated taper sawn shakes of Southern pine treated in accordance with AWPA U1 (Commodity Specification A, Use Category 3B and Section 5.6)	1 or 2	TFS

CSSB = Cedar Shake and Shingle Bureau.

TFS = Forest Products Laboratory of the Texas Forest Services.

❖ Wood shakes must be labeled by an approved third-party inspection agency. A quality control program is required to contain a set of grading rules. Care must be taken to completely follow the manufacturer's installation instructions for each particular product. The building official must review and approve all installations with special note of any application limitations.

1507.9.7 Attachment. Fasteners for wood shakes shall be corrosion resistant with a minimum penetration of $^3/_4$ inch (19.1 mm) into the sheathing. For sheathing less than $^1/_2$ inch (12.7 mm) in thickness, the fasteners shall extend through the sheathing. Each shake shall be attached with a minimum of two fasteners.

❖ Each shake must be secured with a minimum of two fasteners. Fasteners are to be as specified in the manufacturer's installation instructions.

1507.9.8 Application. Wood shakes shall be laid with a side lap not less than $1^1/_2$ inches (38 mm) between joints in adjacent courses. Spacing between shakes in the same course shall be $^3/_8$ to $^5/_8$ inch (9.5 to 15.9 mm) for shakes and taper sawn shakes of naturally durable wood and shall be $^1/_4$ to $^3/_8$ inch (6.4 to 9.5 mm) for preservative taper sawn shakes. Weather exposure for wood shakes shall not exceed those set in Table 1507.9.8.

❖ Care must be taken to follow completely the manufacturer's installation instructions for each particular product. The building official must review and approve all installations with special note of any application limitations.

TABLE 1507.9.8. See page 15-27.

❖ The code provides specific weather exposure limitations for two different lengths of shakes. As with wood shingles, the longer the shake, the greater the exposure allowance, since wind uplift resistance is greater for the longer shakes. The reduced weather exposure enhances the wind uplift resistance that these wood shakes will be able to provide. The exposure limitations are applicable regardless of the roof slope since they are only permitted on steep slopes roofs (4:12 or greater).

1507.9.9 Flashing. At the juncture of the roof and vertical surfaces, flashing and counterflashing shall be provided in accordance with the manufacturer's installation instructions, and where of metal, shall be not less than 0.019-inch (0.48 mm) (No. 26 galvanized sheet gage) corrosion-resistant metal. The valley flashing shall extend at least 11 inches (279 mm) from the centerline each way and have a splash diverter rib not less than 1 inch (25 mm) high at the flow line formed as part of the flashing. Sections of flashing shall have an end lap of not less than 4 inches (102 mm). For roof slopes of three units vertical in 12 units horizontal (25-percent slope) and over, the valley flashing shall have a 36-inch-wide (914 mm) underlayment of either one layer of Type I underlayment running the full length of the valley or a self-adhering polymer-modified bitumen sheet complying with ASTM D1970, in addition to other required underlayment. In areas where the average daily temperature in January is 25°F (-4°C) or less or where there is a possibility of ice forming along the eaves causing a backup of water, the metal valley flashing underlayment shall be solidly cemented to the roofing underlayment for slopes under seven units vertical in 12 units horizontal (58-percent slope) or self-adhering polymer-modified bitumen sheet shall be installed.

❖ Improper installation of flashing is the greatest cause of failure of roof covering systems. Whenever one plane of a roof intersects another plane, flashing is required where the planes intersect [see commentary, Section 1503.2 and Commentary Figures 1507.8.4(1) and (2)].

1507.10 Built-up roofs. The installation of built-up roofs shall comply with the provisions of this section.

❖ A built-up roof membrane is a continuous, semiflexible multiple-ply roof membrane, consisting of plies or layers of saturated felts, coated felts, fabrics or mats between which alternate layers of bitumen are applied. Generally, built-up roof membranes are surfaced with mineral aggregate and bitumen, a liquid-applied coating or granule-surfaced cap sheet.

1507.10.1 Slope. Built-up roofs shall have a design slope of not less than one-fourth unit vertical in 12 units horizontal (2-percent slope) for drainage, except for coal-tar built-up roofs that shall have a design slope of not less than one-eighth unit vertical in 12 units horizontal (1-percent slope).

❖ Because of their low melting points and self-healing characteristics, coal tar membranes are permitted on lower slopes of one-eighth inch per foot (1 percent).

1507.10.2 Material standards. Built-up roof covering materials shall comply with the standards in Table 1507.10.2 or UL 55A.

❖ Asphalt and coal tar are the principal bitumens used for roofing purposes. As an alternative to the ASTM material standards that are referenced, the code also permits materials that are evaluated for compliance in accordance with UL 55A.

TABLE 1507.10.2. See next column.

❖ The various types of materials permitted in typical built-up roofing must comply with applicable standards as noted in the table. The following discusses some of those standards and what they address. Note that the titles and specific editions of all the standards are located in Chapter 35 of the code.

ASTM D1863 covers the quality and grading of crushed stone, crushed slag and water-worn gravel suitable for use as aggregate surfacing on built-up roofs.

ASTM D4601 covers asphalt-impregnated and coated glass fiber base sheet, with or without perforations, for use as the first ply of the built-up roofing. When not perforated, this sheet may be used as a vapor retarder under or between roof insulation with a solid top coating of asphaltic material.

ASTM D2178 covers glass felt impregnated to varying degrees with asphalt, which may be used both with asphalts conforming to the requirements of

TABLE 1507.10.2
BUILT-UP ROOFING MATERIAL STANDARDS

MATERIAL STANDARD	STANDARD
Acrylic coatings used in roofing	ASTM D6083
Aggregate surfacing	ASTM D1863
Asphalt adhesive used in roofing	ASTM D3747
Asphalt cements used in roofing	ASTM D3019; D2822; D4586
Asphalt-coated glass fiber base sheet	ASTM D4601
Asphalt coatings used in roofing	ASTM D1227; D2823; D2824; D4479
Asphalt glass felt	ASTM D2178
Asphalt primer used in roofing	ASTM D41
Asphalt-saturated and asphalt-coated organic felt base sheet	ASTM D2626
Asphalt-saturated organic felt (perforated)	ASTM D226
Asphalt used in roofing	ASTM D312
Coal-tar cements used in roofing	ASTM D4022; D5643
Coal-tar saturated organic felt	ASTM D227
Coal-tar pitch used in roofing	ASTM D450; Type I or II
Coal-tar primer used in roofing, dampproofing and waterproofing	ASTM D43
Glass mat, coal tar	ASTM D4990
Glass mat, venting type	ASTM D4897
Mineral-surfaced inorganic cap sheet	ASTM D3909
Thermoplastic fabrics used in roofing	ASTM D5665, D5726

TABLE 1507.9.8
WOOD SHAKE WEATHER EXPOSURE AND ROOF SLOPE

ROOFING MATERIAL	LENGTH (inches)	GRADE	EXPOSURE (inches) 4:12 PITCH OR STEEPER
Shakes of naturally durable wood	18 24	No. 1 No. 1	7.5 10[a]
Preservative-treated taper sawn shakes of Southern yellow pine	18 24	No. 1 No. 1	7.5 10
	18 24	No. 2 No. 2	5.5 7.5
Taper sawn shakes of naturally durable wood	18 24	No. 1 No. 1	7.5 10
	18 24	No. 2 No. 2	5.5 7.5

For SI: 1 inch = 25.4 mm.

a. For 24-inch by 0.375-inch handsplit shakes, the maximum exposure is 7.5 inches.

ASTM D312 in the construction of built-up roofs, and asphalts conforming to the requirements of ASTM D449 in the membrane system of waterproofing.

ASTM D2626 covers the base sheet with fine mineral surfacing on the top side, with or without perforations, for use as the first ply of a built-up roof. When not perforated, this sheet may be used as a vapor retarder under roof insulation.

ASTM D226 covers asphalt-saturated organic felts, with perforations, that may be used with asphalts conforming to the requirements of ASTM D312 in the construction of built-up roofs, and with asphalts conforming to the requirements of ASTM D449 in the membrane system of waterproofing.

ASTM D312 covers four types of asphalt intended for use in built-up roof construction. The specification is intended for general classification purposes only, and does not imply restrictions on the slope at which an asphalt must be used. There are four classification types. Type I includes asphalts that are relatively susceptible to flow at roof temperatures with good adhesive and self-sealing properties. Type II includes asphalts that are moderately susceptible to flow at roof temperatures. Type III includes asphalts that are relatively nonsusceptible to flow at roof temperatures for use in built-up roof construction on slope inclines from 8.3 to 25 percent. Type IV includes asphalts that are generally nonsusceptible to flow at roof temperatures for use in built-up roof construction on slope inclines from approximately 16.7 to 50 percent.

ASTM D227 covers coal-tar saturated organic felt that may be used with coal-tar pitches conforming to the appropriate requirements of ASTM D450 in the construction of built-up roofs and in the membrane system of waterproofing.

ASTM D450 covers three types of coal-tar pitch suitable for use in the construction of built-up roofing, dampproofing and membrane waterproofing systems. Only Type I or III can be used in code-recognized roofing systems. Type I is suitable for use in built-up roofing systems with felts conforming to the requirements of ASTM D227 or as specified by the manufacturer. Type III is suitable for use in built-up roofing systems, but has less-volatile components than Type I.

ASTM D3909 covers asphalt-impregnated and coated glass felt roll roofing surfaced on the weather side with mineral granules for use as a cap sheet in the construction of built-up roofs.

As noted, this list is not all-inclusive when compared to Table 1507.10.2, but includes materials commonly used as built-up roofing materials.

1507.11 Modified bitumen roofing. The installation of modified bitumen roofing shall comply with the provisions of this section.

❖ Modified bitumen roofing is a membrane-type roofing made up of composite sheets consisting of polymer-modified bitumen often reinforced and sometimes surfaced with various types of mats, films, foils and mineral granules. The bitumen is modified through inclusion of one or more polymers. There are two general types of polymer-modified bitumen membranes: those with the principal modifier being atactic polypropylene (APP), and those with bitumen modified with styrene butadiene styrene (SBS). They differ in characteristics, as well as application. For more information on the application of polymer-modified bitumen roofing membranes, refer to the *NRCA Roofing Manual: Membrane Roof Systems*.

1507.11.1 Slope. Modified bitumen membrane roofs shall have a design slope of not less than one-fourth unit vertical in 12 units horizontal (2-percent slope) for drainage.

❖ According to this section, all roofs must have a minimum slope of $^{1}/_{4}$:12 so that there is positive drainage of storm water to gutters, roof drains and other components of an approved storm drainage system in order to divert water away from the building or structure.

1507.11.2 Material standards. Modified bitumen roof coverings shall comply with CGSB 37-GP-56M, ASTM D6162, ASTM D6163, ASTM D6164, ASTM D6222, ASTM D6223, ASTM D6298 or ASTM D6509.

❖ The materials and installation of modified bitumen roof coverings must be in accordance with the listed standards, which regulate modified bituminous roofing membranes, either prefabricated or reinforced. The standards are: CGSB 37-GP-56M, which regulates membrane, modified, bituminous, prefabricated and reinforced for roofing; ASTM D6162, which regulates SBS modified bituminous sheet materials; ASTM D6163, which regulates SBS using glass fiber reinforcements; ASTM D6164, which regulates SBS using polyester reinforcements; ASTM D6222, which regulates APP modified bituminous sheet materials; ASTM D6223, which regulates APP using polyester and glass fiber reinforcements; ASTM D6298, which regulates SBS with a metal surface; and ASTM D6509, which regulates APP using glass fiber reinforcements.

1507.12 Thermoset single-ply roofing. The installation of thermoset single-ply roofing shall comply with the provisions of this section.

❖ Thermoset is a material that solidifies or "sets" irreversibly when heated. Thinners evaporate during the curing process. Thermoset materials are those whose polymers are chemically cross-linked, commonly referred to as being "cured." Once they are fully cured, they can only be bonded to like material with an adhesive, as new molecular links may not be formed. There are four common subcategories of thermoset roof membranes: neoprene, chlorosulfonated polyethylene (CSPE), epichlorohydrin (DCH) and ethylene-propylene-diene monomer (or terpolymer) (EPDM).

1507.12.1 Slope. Thermoset single-ply membrane roofs shall have a design slope of not less than one-fourth unit vertical in 12 units horizontal (2-percent slope) for drainage.

❖ All roofs must have a minimum slope of $^1/_4$:12 so that there is positive drainage of storm water to gutters, roof drains and other components of an approved storm drainage system in order to divert water away from the building or structure.

1507.12.2 Material standards. Thermoset single-ply roof coverings shall comply with ASTM D4637, ASTM D5019 or CGSB 37-GP-52M.

❖ This section describes the standards to be used for materials and installation. The standards are: 1. ASTM D4637, which regulates unreinforced and fabric-reinforced vulcanized rubber sheets made from EPDM or polychloroprene, and is intended for use in single-ply roof membranes exposed to weather; 2. ASTM D5019, which regulates reinforced nonvulcanized polymeric sheets made from CSPE and polyisobutylene used in single-ply roof membranes; and 3. CGSB 37-GP-52M, which regulates sheet-applied elastomeric roofing membranes.

1507.12.3 Ballasted thermoset low-slope roofs. Ballasted thermoset low-slope roofs (roof slope < 2:12) shall be installed in accordance with this section and Section 1504.4. Stone used as ballast shall comply with ASTM D448 or ASTM D7655.

❖ The purpose of this section is to identify that this type of roof covering is allowed and to provide installation requirements. Further, this section specifies the minimum requirements for stone ballast to be used with this type of roofing system.

1507.13 Thermoplastic single-ply roofing. The installation of thermoplastic single-ply roofing shall comply with the provisions of this section.

❖ Thermoplastics are materials that soften when heated and harden when cooled. This process can be repeated, provided the material is not heated above the point at which decomposition occurs. Thermoplastic materials are distinguished from thermosets in that there is no chemical cross linking. Because of the materials' chemical nature, some thermoplastic membranes may be seamed by either heat (hot air) or solvent welding.

1507.13.1 Slope. Thermoplastic single-ply membrane roofs shall have a design slope of not less than one-fourth unit vertical in 12 units horizontal (2-percent slope).

❖ All roofs must have a minimum slope of $^1/_4$:12 so that there is positive drainage of storm water to gutters, roof drains and other components of an approved storm drainage system in order to divert water away from the building or structure.

1507.13.2 Material standards. Thermoplastic single-ply roof coverings shall comply with ASTM D4434, ASTM D6754, ASTM D6878 or CGSB CAN/CGSB 37-54.

❖ This section establishes standards and performance criteria for the installation of thermoplastic single-ply roof coverings. Four standards are recognized: ASTM D4434, ASTM D6754, ASTM D6878 and CGSB 37-GP-54M. These specifications cover flexible sheet made from poly (vinyl chloride) resin, ketone ethylene ester-based sheet roofing and thermoplastic polyolefin-based sheet roofing.

1507.13.3 Ballasted thermoplastic low-slope roofs. Ballasted thermoplastic low-slope roofs (roof slope < 2:12) shall be installed in accordance with this section and Section 1504.4. Stone used as ballast shall comply with ASTM D448 or ASTM D7655.

❖ The purpose of this section is to identify that this type of roof covering is allowed and to provide installation requirements. Further, this section specifies the minimum requirements for stone ballast to be used with this type of roofing system.

1507.14 Sprayed polyurethane foam roofing. The installation of sprayed polyurethane foam roofing shall comply with the provisions of this section.

❖ Sprayed polyurethane foam is a foamed plastic material constructed by mixing a two-part liquid that is spray applied to form the base of an adhered roof system. The two-part component mixture reacts chemically and immediately expands when applied through special metering equipment to form a closed-cell foam. As the foam rises, it sets into a solid and a skin forms on the surface. The foam provides a thermal insulation and the skin provides a water-resistant surface.

1507.14.1 Slope. Sprayed polyurethane foam roofs shall have a design slope of not less than one-fourth unit vertical in 12 units horizontal (2-percent slope) for drainage.

❖ All roofs must have a minimum slope of $^1/_4$:12 so that there is positive drainage of storm water to gutters, roof drains and other components of an approved storm drainage system in order to divert water away from the building or structure.

1507.14.2 Material standards. Spray-applied polyurethane foam insulation shall comply with Type III or IV as defined in ASTM C1029.

❖ ASTM C1029 is the standard that regulates spray-applied rigid cellular polyurethane thermally insulated roof coverings. This specification covers the types and physical properties of spray-applied rigid cellular polyurethane intended for use as thermal insulation. Type III or IV foam insulation is required based on their higher compressive strengths. The operating temperatures of the surfaces to which the insulation is applied cannot be lower than -22°F (-30°C) or greater than 225°F (107°C).

1507.14.3 Application. Foamed-in-place roof insulation shall be installed in accordance with the manufacturer's instructions. A liquid-applied protective coating that complies with Table 1507.14.3 shall be applied no less than 2 hours nor more than 72 hours following the application of the foam.

❖ The roof system must be protected from ultraviolet light and degradation through the use of surfacing. Liquid-applied protected elastomeric coatings are the recommended surfacing for sprayed polyurethane foam roofing systems.

TABLE 1507.14.3
PROTECTIVE COATING MATERIAL STANDARDS

MATERIAL	STANDARD
Acrylic coating	ASTM D6083
Silicone coating	ASTM D6694
Moisture-cured polyurethane coating	ASTM D6947

❖ This table identifies the applicable material standards for protective coating materials that are suitable for sprayed polyurethane foam roof systems.

1507.14.4 Foam plastics. Foam plastic materials and installation shall comply with Chapter 26.

❖ Chapter 26 addresses foam plastic materials and installation.

1507.15 Liquid-applied roofing. The installation of liquid-applied roofing shall comply with the provisions of this section.

❖ Liquid-applied coatings are nonsynthetic roofing materials that are termed "cold applied" because hot bitumen is not employed in their application.

1507.15.1 Slope. Liquid-applied roofing shall have a design slope of not less than one-fourth unit vertical in 12 units horizontal (2-percent slope).

❖ All roofs must have a minimum slope of $^1/_4$:12 so that there is positive drainage of storm water to gutters, roof drains and other components of an approved storm drainage system in order to divert water away from the building or structure.

1507.15.2 Material standards. Liquid-applied roofing shall comply with ASTM C836, ASTM C957, ASTM D1227 or ASTM D3468, ASTM D6083, ASTM D6694 or ASTM D6947.

❖ Liquid-applied roof coatings must be installed in accordance with the manufacturer's instructions. The coatings must comply with ASTM C836, ASTM C957, ASTM D1227, D3468 or ASTM D6694. ASTM C836 describes the required properties and test methods for a cold liquid-applied elastomeric membrane, one or two components, for waterproofing building decks subject to hydrostatic pressure in building areas to be occupied by personnel, vehicles or equipment. This specification only applies to a membrane system above which a separate wearing or traffic course will be applied. ASTM C957 describes the required properties and test methods for a cold liquid-applied elastomeric membrane for waterproofing building decks not subject to hydrostatic pressure. The specification applies only to a membrane system that has an integral wearing surface. It does not include specific requirements for skid resistance or fire retardance, although both may be important in specific applications. ASTM D1227 describes emulsified asphalt suitable for use as a protective coating for built-up roofs and other exposed surfaces with inclines of not less than 4 percent. ASTM D3468 describes liquid-applied neoprene and CSPE synthetic rubber solutions suitable for use in roofing and waterproofing. ASTM D6694 describes liquid-applied silicone coatings used in spray on polyurethane foam roofing. ASTM D6947 describes liquid-applied moisture cured polyurethane coatings used in spray polyurethane foam roofing systems.

1507.16 Vegetative roofs, roof gardens and landscaped roofs. *Vegetative roofs*, roof gardens and landscaped roofs shall comply with the requirements of this chapter, Sections 1607.12.3 and 1607.12.3.1 and the *International Fire Code.*

❖ The purpose of this section is to require vegetative roofs, roof gardens and landscaped roofs to comply with the requirements for other roof systems contained in this chapter. Further, this section requires these types of roofs to meet the requirements for special purpose roofs as contained in Chapters 15 and 16 as well as providing a reference to the *International Fire Code*® (IFC®).

[BF] 1507.16.1 Structural fire resistance. The structural frame and roof construction supporting the load imposed upon the roof by the *vegetative roof*, roof gardens or landscaped roofs shall comply with the requirements of Table 601.

❖ This section is a reminder to verify compliance with Table 601 for the portion of the structure that supports the additional load of a landscaped roof.

1507.17 Photovoltaic shingles. The installation of *photovoltaic shingles* shall comply with the provisions of this section.

❖ This section provides guidance for the installation of photovoltaic shingles. These shingles are integrated with the building and provide both a roof covering and source of electrical power.

1507.17.1 Deck requirements. *Photovoltaic shingles* shall be applied to a solid or closely fitted deck, except where the shingles are specifically designed to be applied over spaced sheathing.

❖ These provisions provide requirements and limitations for roof decks, roof deck slope, underlayment, underlayment application, underlayment attachment in high wind regions, ice barriers and fasteners that are considered appropriate for the installation of photovoltaic shingles. The specific requirements included here have been adapted from and are intended to be consistent with similar attributes for other steep-slope, shingle-type roof coverings. Sections 1507.17.1 and 1507.17.2 are adapted from Sections 1507.5.1 and 1507.5.2, respectively; Sections 1507.17.3 and 1507.17.4 are adapted from Sections 1507.2.3 and 1507.2.8, respectively; and Section 1507.17.5 is adapted from Section 1507.2.6.

1507.17.2 Deck slope. *Photovoltaic shingles* shall not be installed on roof slopes less than three units vertical in 12 units horizontal (25-percent slope).

❖ See the commentary to Section 1507.17.1.

1507.17.3 Underlayment. Unless otherwise noted, required underlayment shall conform to ASTM D226, ASTM D4869 or ASTM D6757.

❖ See the commentary to Section 1507.17.1.

1507.17.4 Underlayment application. Underlayment shall be applied shingle fashion, parallel to and starting from the eave, lapped 2 inches (51 mm) and fastened sufficiently to hold in place.

❖ See the commentary to Section 1507.17.1.

1507.17.4.1 High wind attachment. Underlayment applied in areas subject to high winds [V_{asd} greater than 110 mph (49 m/s) as determined in accordance with Section 1609.3.1] shall be applied with corrosion-resistant fasteners in accordance with the manufacturer's instructions. Fasteners shall be applied along the overlap at not more than 36 inches (914 mm) on center. Underlayment installed where V_{asd} is not less than 120 mph (54 m/s) shall comply with ASTM D226, Type II, ASTM D4869, Type IV or ASTM D6757. The underlayment shall be attached in a grid pattern of 12 inches (305 mm) between side laps with a 6-inch (152 mm) spacing at the side laps. Underlayment shall be applied in accordance with Section 1507.2.8 except all laps shall be a minimum of 4 inches (102 mm). Underlayment shall be attached using metal or plastic cap nails with a head diameter of not less than 1 inch (25 mm) with a thickness of not less than 32-gage [0.0134 inch (0.34 mm)] sheet metal. The cap nail shank shall be a minimum of 12 gage [0.105 inch (2.67 mm)] with a length to penetrate through the roof sheathing or a minimum of $^3/_4$ inch (19.1 mm) into the roof sheathing.

Exception: As an alternative, adhered underlayment complying with ASTM D1970 shall be permitted.

❖ See the commentary to Section 1507.17.1.

1507.17.4.2 Ice barrier. In areas where there has been a history of ice forming along the eaves causing a backup of water, an ice barrier that consists of at least two layers of underlayment cemented together or of a self-adhering polymer modified bitumen sheet shall be used instead of normal underlayment and extend from the lowest edges of all roof surfaces to a point not less than 24 inches (610 mm) inside the *exterior wall* line of the building.

Exception: Detached accessory structures that contain no conditioned floor area.

❖ See the commentary to Section 1507.17.1.

1507.17.5 Fasteners. Fasteners for photovoltaic shingles shall be galvanized, stainless steel, aluminum or copper roofing nails, minimum 12-gage [0.105 inch (2.67 mm)] shank with a minimum $^3/_8$-inch-diameter (9.5 mm) head, of a length to penetrate through the roofing materials and a minimum of $^3/_4$ inch (19.1 mm) into the roof sheathing. Where the roof sheathing is less than $^3/_4$ inch (19.1 mm) thick, the nails shall penetrate through the sheathing. Fasteners shall comply with ASTM F1667.

❖ See the commentary to Section 1507.17.1.

1507.17.6 Material standards. *Photovoltaic shingles* shall be *listed* and labeled in accordance with UL 1703.

❖ This section references UL 1703 as the standard used for determining code compliance of photovoltaic shingles.

1507.17.7 Attachment. *Photovoltaic shingles* shall be attached in accordance with the manufacturer's installation instructions.

❖ Referring to the manufacturer's instructions is appropriate since the required slope and fastening of the photovoltaic shingles are different for each manufacturer's product.

1507.17.8 Wind resistance. *Photovoltaic shingles* shall be tested in accordance with procedures and acceptance criteria in ASTM D3161. *Photovoltaic shingles* shall comply with the classification requirements of Table 1504.1.1 for the appropriate maximum nominal design wind speed. *Photovoltaic shingle* packaging shall bear a *label* to indicate compliance with the procedures in ASTM D3161 and the required classification from Table 1504.1.1.

❖ For wind resistance, the procedures and acceptance criteria used in ASTM D3161 for asphalt shingles are appropriate when adapted for photovoltaic shingles.

SECTION 1508 ROOF INSULATION

[BF] 1508.1 General. The use of above-deck thermal insulation shall be permitted provided such insulation is covered with an approved roof covering and passes the tests of NFPA 276 or UL 1256 when tested as an assembly.

Exceptions:

1. Foam plastic roof insulation shall conform to the material and installation requirements of Chapter 26.
2. Where a concrete roof deck is used and the above-deck thermal insulation is covered with an approved roof covering.

❖ This section addresses two requirements for the use of above-deck thermal insulation. The first requirement is that an approved roof covering be used. The second requirement is that an under-deck flame spread test must be conducted and passed using either NFPA 276 or the UL 1256 test methods. These tests were developed to evaluate the contribution of above-deck thermal insulations and their installation when exposed to a fire from below the deck. The primary construction for which these tests have been developed are roof assemblies that utilize a steel roof deck with above-deck insulation materials.

In addition to serving as thermal insulation, roof insulation may be used as a substrate to which the roof membrane is applied. In protected roof membrane systems, the insulation is applied over the membrane and is not expected to act as a substrate. The insulation should provide adequate support for the membrane and other associated rooftop materials and stand up to limited rooftop traffic, such as for regular roof inspections and maintenance.

Rigid board roof insulations currently among the most commonly used for low-slope roofs are: cellular glass, glass fiber, mineral fiber, perlite, phenolic foam, polyisocyanurate foam, polystyrene foam, polyurethane foam, wood fiberboard and composite board.

Nonrigid roof insulations are generally used in steep-slope and some metal-roof assembly construction. Commonly made from cellulose, glass fiber and mineral fiber, these nonrigid insulations may be available in batts, blankets and loose-fiber forms.

Exception 2 recognizes that if a concrete deck is employed as part of the assembly it will provide protection to the above-deck insulation materials. This exception will apply to nonfoam plastic roof insulation materials, since foam plastic roof insulations are covered under Exception 1 and Chapter 26 of the code.

[BF] 1508.1.1 Cellulosic fiberboard. Cellulosic fiberboard roof insulation shall conform to the material and installation requirements of Chapter 23.

❖ Chapter 23 addresses cellulosic fiberboard roof installation.

[BF] 1508.2 Material standards. Above-deck thermal insulation board shall comply with the standards in Table 1508.2.

❖ The referenced material standards provide additional guidance on the physical properties of roof insulation when utilized as above-deck components of roof assemblies.

[BF] TABLE 1508.2
MATERIAL STANDARDS FOR ROOF INSULATION

Cellular glass board	ASTM C552
Composite boards	ASTM C1289, Type III, IV, V or VI
Expanded polystyrene	ASTM C578
Extruded polystyrene	ASTM C578
Fiber-reinforced gypsum board	ASTM C1278
Glass-faced gypsum board	ASTM C1177
Mineral fiber insulation board	ASTM C726
Perlite board	ASTM C728
Polyisocyanurate board	ASTM C1289, Type I or II
Wood fiberboard	ASTM C208

❖ This table incorporates industry-recognized material standards into the code for materials commonly used in above-roof-deck insulation practices. For instance, the table references ASTM C726, which specifies the composition and physical properties for mineral fiber insulation board.

SECTION 1509
RADIANT BARRIERS INSTALLED ABOVE DECK

[BF] 1509.1 General. A *radiant barrier* installed above a deck shall comply with Sections 1509.2 through 1509.4.

❖ This section provides requirements for fire testing, proper installation, and the appropriate material standard for a radiant barrier installed above a roof deck. The definition of the term "Radiant barrier" in Chapter 2 is derived from the definition for radiant barrier in ASTM C1313.

[BF] 1509.2 Fire testing. *Radiant barriers* shall be permitted for use above decks where the *radiant barrier* is covered with an approved roof covering and the system consisting of the *radiant barrier* and the roof covering complies with the requirements of either FM 4550 or UL 1256.

❖ The key issue addressed in this section is how the fire testing of the system utilizing a radiant barrier is to be done. It clarifies that the testing must be done using the combination of the radiant barrier and the approved roof covering, and that the system must pass the fire test.

[BF] 1509.3 Installation. The low emittance surface of the *radiant barrier* shall face the continuous airspace between the *radiant barrier* and the roof covering.

❖ A radiant barrier product that is installed above the roof deck is typically located between the deck and the felt. This section clarifies orientation of the radiant barrier and air gap for above-deck radiant barriers.

[BF] 1509.4 Material standards. A *radiant barrier* installed above a deck shall comply with ASTM C1313/1313M.

❖ The standard that is referenced for radiant barriers that are used in buildings, ASTM C1313, Standard Specification for Sheet Radiant Barriers for Building Construction Applications, helps to differentiate these products from reflective insulations (see ASTM C1224, Standard Specification for Reflective Insulation for Building Applications). The abstract of ASTM C1313 reads as follows. "...The scope is specifically limited to requirements for radiant barrier sheet materials that consist of at least one surface, such as metallic foils or metallic deposits mounted or unmounted on substrates. Sheet radiant barrier materials shall consist of low emittance surface(s) that may be in combination with any substrates and adhesives required to meet the specified physical material properties."

SECTION 1510 ROOFTOP STRUCTURES

[BG] 1510.1 General. The provisions of this section shall govern the construction of rooftop structures.

❖ This section identifies and establishes the criteria used in evaluating penthouse-type roof structures. Other rooftop structures, such as water tanks, cooling towers, towers, spires, domes, cupolas, etc., are not to be considered as penthouses since the code has specific provisions for these structures (see Sections 1510.3 through 1510.6).

[BG] 1510.2 Penthouses. Penthouses in compliance with Sections 1510.2.1 through 1510.2.5 shall be considered as a portion of the story directly below the roof deck on which such penthouses are located. All other penthouses shall be considered as an additional story of the building.

❖ This section introduces the requirements necessary for a penthouse to be considered as a portion of the story directly below. Section 1510.2.1 contains height limitations, Section 1510.2.2 contains area limitations and Section 1510.2.3 contains use limitations.

[BG] 1510.2.1 Height above roof deck. Penthouses constructed on buildings of other than Type I construction shall not exceed 18 feet (5486 mm) in height above the roof deck as measured to the average height of the roof of the penthouse.

Exceptions:

1. Where used to enclose tanks or elevators that travel to the roof level, penthouses shall be permitted to have a maximum height of 28 feet (8534 mm) above the roof deck.
2. Penthouses located on the roof of buildings of Type I construction shall not be limited in height.

❖ The height of roof structures is not limited on buildings of Type I construction (see Exception 2). For other types of construction, the roof structure is limited to 18 feet (3658 mm) in height unless it encloses tanks or elevators that extend to the roof (see Exception 1), for which the maximum height is 28 feet (8534 mm). If the height exceeds that permitted, the roof structure should be counted as, and meet all the requirements for, an additional story. This section also addresses how the height of a penthouse is measured.

[BG] 1510.2.2 Area limitation. The aggregate area of penthouses and other enclosed rooftop structures shall not exceed one-third the area of the supporting roof deck. Such penthouses and other enclosed rooftop structures shall not be required to be included in determining the building area or number of stories as regulated by Section 503.1. The area of such penthouses shall not be included in determining the fire area specified in Section 901.7.

❖ The aggregate area of penthouses and any other enclosed rooftop structure is limited to one-third of the supporting roof deck. If the area of the penthouse exceeds this threshold, regardless of height, then the penthouse must be considered an additional story.

[BG] 1510.2.3 Use limitations. Penthouses shall not be used for purposes other than the shelter of mechanical or electrical equipment, tanks, or vertical shaft openings in the roof assembly.

❖ The use of the penthouse is also limited. Regardless of height or area, roof structures used for anything except sheltering electrical or mechanical equipment or vertical shaft openings must be considered as an additional story.

[BG] 1510.2.4 Weather protection. Provisions such as louvers, louver blades or flashing shall be made to protect the mechanical and electrical equipment and the building interior from the elements.

❖ Since the penthouse is a shelter for equipment or vertical shaft openings, it is often necessary to incorporate air intake and ventilation openings in the penthouse construction. In order to protect the equipment and the building from the detrimental effects of weather exposure, this section requires that measures be taken to minimize exposure to the elements. In so doing, this section essentially reiterates other provisions that address the integrity of the building envelope (see Sections 1405.2 and 1503).

[BG] 1510.2.5 Type of construction. Penthouses shall be constructed with walls, floors and roofs as required for the type of construction of the building on which such penthouses are built.

Exceptions:

1. On buildings of Type I construction, the exterior walls and roofs of penthouses with a *fire separation distance* greater than 5 feet (1524 mm) and less than 20 feet (6096 mm) shall be permitted to have not less than a 1-hour fire-resistance rating. The exterior walls and roofs of penthouses with a fire separation distance of 20 feet (6096 mm) or greater shall not be required to have a fire-resistance rating.
2. On buildings of Type I construction two stories or less in height above grade plane or of Type II construction, the exterior walls and roofs of penthouses with a *fire separation distance* greater than 5 feet (1524 mm) and less than 20 feet (6096 mm) shall be permitted to have not less than a 1-hour fire-resistance rating or a lesser fire-resistance rating as required by Table 602 and be constructed of fire-retardant-treated wood. The exterior walls and roofs of penthouses with a *fire separation distance* of 20 feet (6096 mm) or greater shall be permitted to be constructed of fire-retardant-treated wood and shall not be required to have a fire-resistance rating. Interior framing and walls shall be permitted to be constructed of fire-retardant-treated wood.
3. On buildings of Type III, IV or V construction, the exterior walls of penthouses with a fire separation distance greater than 5 feet (1524 mm) and less than

20 feet (6096 mm) shall be permitted to have not less than a 1-hour fire-resistance rating or a lesser fire-resistance rating as required by Table 602. On buildings of Type III, IV or VA construction, the exterior walls of penthouses with a fire separation distance of 20 feet (6096 mm) or greater shall be permitted to be of Type IV or noncombustible construction or fire-retardant-treated wood and shall not be required to have a fire-resistance rating.

❖ The general premise of this section requires a penthouse to be treated no differently than any other portion of the building, as it is constructed of exterior walls, a floor and a roof. Where the exterior walls of the penthouse are located within 5 feet (1524 mm) of the building's lot line, they are required to be of the same fire-resistance rating as the exterior wall of the story immediately below the penthouse. However, the exceptions recognize the reduced exposure where the exterior wall of a penthouse is set back from the exterior wall line of the building itself. Thus, the exceptions are based on two thresholds of fire separation distance: the first is greater than 5 feet (1524 mm) and up to 20 feet (6096 mm); the second is greater than 20 feet (6096 mm) (see Section 202 for the definition of "Fire separation distance").

If a penthouse is set back by a fire separation distance of 5 feet (1524 mm) or more from the lot line, the risk of exposure to adjacent property from a fire within the penthouse, as well as fire exposure to the penthouse from the adjacent lot, is reduced. Accordingly, if a penthouse is set back 5 feet (1524 mm) or more from the building's lot line, the penthouse construction may be able to take advantage of one of the exceptions, typically based on the type of construction, resulting in a reduction in rating of the exterior wall of the penthouse. It should be noted that, in many cases, penthouses located at least 20 feet (6096 mm) from the lot line are allowed further reductions.

[BG] 1510.3 Tanks. Tanks having a capacity of more than 500 gallons (1893 L) located on the roof deck of a building shall be supported on masonry, reinforced concrete, steel or Type IV construction provided that, where such supports are located in the building above the lowest *story*, the support shall be fire-resistance rated as required for Type IA construction.

❖ This section establishes the criteria used in evaluating rooftop tanks, such as water tanks.

[BG] 1510.3.1 Valve and drain. In the bottom or on the side near the bottom of the tank, a pipe or outlet, fitted with a suitable quick-opening valve for discharging the contents into a drain in an emergency shall be provided.

❖ In the event of an emergency, there must be a means to drain the tank. This not only involves a quick-opening valve on the tank but also drains that are capable of handling the discharge. The drains must be sized and installed in accordance with the *International Plumbing Code*® (IPC®).

[BG] 1510.3.2 Location. Tanks shall not be placed over or near a stairway or an elevator shaft, unless there is a solid roof or floor underneath the tank.

❖ Tanks are not permitted to be located above a stairway or elevator enclosure unless a solid roof or floor deck is provided underneath the tank. This is intended to prevent discharges from a leaking tank from affecting a shaft.

[BG] 1510.3.3 Tank cover. Unenclosed roof tanks shall have covers sloping toward the perimeter of the tanks.

❖ Roof tanks must be covered to prevent the accumulation of rain, ice and snow, which may cause an uncovered tank to overflow.

[BG] 1510.4 Cooling towers. Cooling towers located on the roof deck of a building and greater than 250 square feet (23.2 m^2) in base area or greater than 15 feet (4572 mm) in height above the roof deck, as measured to the highest point on the cooling tower, where the roof is greater than 50 feet (15 240 mm) in height above grade plane shall be constructed of noncombustible materials. The base area of cooling towers shall not exceed one-third the area of the supporting roof deck.

Exception: Drip boards and the enclosing construction shall be permitted to be of wood not less than 1 inch (25 mm) nominal thickness, provided the wood is covered on the exterior of the tower with noncombustible material.

❖ This section identifies and establishes criteria that require rooftop cooling towers to be of noncombustible construction.

[BG] 1510.5 Towers, spires, domes and cupolas. Towers, spires, domes and cupolas shall be of a type of construction having fire-resistance ratings not less than required for the building on top of which such tower, spire, dome or cupola is built. Towers, spires, domes and cupolas greater than 85 feet (25 908 mm) in height above grade plane as measured to the highest point on such structures, and either greater than 200 square feet (18.6 m^2) in horizontal area or used for any purpose other than a belfry or an architectural embellishment, shall be constructed of and supported on Type I or II construction.

❖ This section identifies the criteria used in evaluating structures such as towers, spires, domes and cupolas. When any of those structures exceed either 85 feet (25 908 mm) in height above grade plane or are greater than 200 square feet (18.6 m^2) in area, the construction type is limited to Type I or II. Section 1510.5.1 further restricts construction to noncombustible when the heights become more excessive (see commentary, Section 1510.5.1). These restrictions are related to the difficultly posed in fighting fires in such structures and also the danger to surrounding people, emergency responders and other buildings if such elements should fail during a fire.

[BG] 1510.5.1 Noncombustible construction required. Towers, spires, domes and cupolas greater than 60 feet (18 288 mm) in height above the highest point at which such structure contacts the roof as measured to the highest point on such structure, or that exceeds 200 square feet (18.6 m^2) in area at any horizontal section, or which is intended to be used for any purpose other than a belfry or architectural embellishment, or is located on the top of a building greater than 50 feet (1524 mm) in building height shall be constructed of and supported by noncombustible materials and shall be separated from the building below by construction having a fire-resistance rating of not less than 1.5 hours with openings protected in accordance with Section 712. Such structures located on the top of a building greater than 50 feet (15 240 mm) in building height shall be supported by noncombustible construction.

❖ These rooftop structures include communication towers, spires, cupolas and similar structures. Such structures must generally be constructed of the same type of construction as the building to which they are attached. Because of the difficulty in fighting fires on tall buildings and structures and the hazards associated with locating combustible materials on roofs of taller buildings, when the height of such structures exceeds 60 feet (18 288 mm) above the highest point at which it comes in contact with the roof or is in excess of 200 square feet (19 m^2) in area at any horizontal section, the structure and its supports must be of noncombustible construction. In addition, they must be separated from the building with construction that has a minimum 1.5-hour fire-resistance rating. Any opening in that construction must be protected in accordance with Section 712, which addresses the protection of openings in horizontal assemblies. While openings in horizontal assemblies are typically protected with shaft enclosures, there are floor fire door assemblies that could be used in this situation.

For similar reasons, most roof structures of any height or area placed on tall buildings [greater than 50 feet (15 240 mm) in height] are required to be constructed of noncombustible materials and be supported by noncombustible construction.

[BG] 1510.5.2 Towers and spires. Enclosed towers and spires shall have exterior walls constructed as required for the building on top of which such towers and spires are built. The roof covering of spires shall be not less than the same class of roof covering required for the building on top of which the spire is located.

❖ As with penthouses and roof structures, the code intends to obtain construction and fire resistance for enclosed towers and spires that is consistent with that of the building on which they are supported.

[BG] 1510.6 Mechanical equipment screens. *Mechanical equipment screens* shall be constructed of the materials specified for the exterior walls in accordance with the type of construction of the building. Where the fire separation distance is greater than 5 feet (1524 mm), *mechanical equipment screens* shall not be required to comply with the fire-resistance rating requirements.

❖ As the definition for "Mechanical equipment screen" makes clear, it is a rooftop structure that is not covered by a roof. This section conservatively requires that mechanical equipment screens must be constructed of the same materials as required for exterior walls, based on the type of construction of the building on which they are located. However, it does exempt them from any fire-resistance-rating requirement once the fire separation distance is greater than 5 feet (1524 mm) since, without a roof, they do not fully enclose a space and they pose a lesser exposure hazard than a penthouse.

[BG] 1510.6.1 Height limitations. *Mechanical equipment screens* shall not exceed 18 feet (5486 mm) in height above the roof deck, as measured to the highest point on the mechanical equipment screen.

Exception: Where located on buildings of Type IA construction, the height of *mechanical equipment screens* shall not be limited.

❖ The height limits specified in this section are based on those required for penthouses in Section 1510.2.1. The exception for mechanical equipment screens located on buildings of Type IA construction is for consistency, since Table 503 does not limit the building height.

[BG] 1510.6.2 Type I, II, III and IV construction. Regardless of the requirements in Section 1510.6, *mechanical equipment screens* that are located on the roof decks of buildings of Type I, II, III or IV construction shall be permitted to be constructed of combustible materials in accordance with any one of the following limitations:

1. The fire separation distance shall be not less than 20 feet (6096 mm) and the height of the *mechanical equipment screen* above the roof deck shall not exceed 4 feet (1219 mm) as measured to the highest point on the *mechanical equipment screen*.
2. The fire separation distance shall be not less than 20 feet (6096 mm) and the *mechanical equipment screen* shall be constructed of fire-retardant-treated wood complying with Section 2303.2 for exterior installation.
3. Where exterior wall covering panels are used, the panels shall have a flame spread index of 25 or less when tested in the minimum and maximum thicknesses intended for use, with each face tested independently in accordance with ASTM E84 or UL 723. The panels shall be tested in the minimum and maximum thicknesses intended for use in accordance with, and shall comply with the acceptance criteria of, NFPA 285 and shall be installed as tested. Where the panels are tested as part of an exterior wall assembly in accordance with NFPA 285, the panels shall be installed on the face of the *mechanical equipment screen* supporting structure in the same manner as they were installed on the tested exterior wall assembly.

❖ This section is an exception to the requirements in Section 1510.6 for the types of construction that typically require the exterior walls to be constructed of noncombustible materials. The three itemized limitations in this section allow for combustible materials to be used for the construction of mechanical equipment screens when the stated conditions are met.

Item 1 recognizes that the hazard posed by a combustible mechanical equipment screen located on the roof of a Type I, II, III or IV building with a fire separation distance of at least 20 feet (6096 mm) and with the height of the mechanical equipment screen limited to 4 feet (1219 mm) above the roof deck is not significant.

Item 2 is based on Exceptions 2 and 3 for penthouses in Section 1510.2.5. The two-story limit in Exception 2 for Type I buildings is omitted. This is considered reasonable since the hazard doesn't justify limiting the Type I buildings to two stories in height where fire-retardant-treated wood is used to construct an unenclosed mechanical equipment screen. The main difference between this limitation and Item 1 is that it does not include the 4-foot (1219 mm) height limit on the mechanical equipment screen. This is because it must be constructed of fire-retardant-treated wood as compared to any combustible material allowed by the code. The height of the mechanical equipment screen would still be limited to 18 feet (5486 mm) by Section 1510.6.1. It is also limited to the maximum building height, based on the type of construction of the building in accordance with Section 1510.6.1.

Item 3 applies where the combustible materials used to build the mechanical equipment screen are limited to a maximum flame spread index of 25 (similar to fire-retardant-treated wood) and the materials are successfully tested in accordance with NFPA 285, *Standard Method of Test for the Evaluation of Flammability Characteristics of Exterior Nonload-Bearing Wall Assemblies Containing Combustible Components*. This is the test method that is used to validate the use of foam plastic insulations in exterior walls of Types I, II, III and IV construction, as well as for the use of metal composite materials in accordance with Section 1407.10. Although the material would be tested as the outer face of the exterior wall in the NFPA 285 test as part of an exterior wall assembly, the test assesses the surface flame spread resistance of the materials constituting the outer face, as well as to a certain degree, the inner face where it is exposed to any open cavities in the wall assembly. The NFPA 285 test is conducted for a full 30 minutes under severe fire-exposure conditions to both the inside of the wall assembly and the outside of the wall assembly with an exterior window burner replicating a fire that has gone to post-flashover and has broken out a window, exposing the outside face of the exterior wall finish. Since the materials used to construct the mechanical equipment screen do not comprise a completely enclosed wall assembly, the maximum flame spread index of 25 has been proposed as a conservative limitation for the backside face of the material, which may not have been directly exposed to the exterior window burner flame in the NFPA 285 test. Since the NFPA 285 test is used to qualify combustible materials for use where noncombustible exterior walls are required, it seems reasonable to allow panels tested as wall assemblies to be installed as equipment screen panels, utilizing the materials tested on the exterior face of the wall system in accordance with NFPA 285.

[BS] 1510.6.3 Type V construction. The height of mechanical equipment screens located on the roof decks of buildings of Type V construction, as measured from grade plane to the highest point on the mechanical equipment screen, shall be permitted to exceed the maximum building height allowed for the building by other provisions of this code where complying with any one of the following limitations, provided the fire separation distance is greater than 5 feet (1524 mm):

1. Where the fire separation distance is not less than 20 feet (6096 mm), the height above grade plane of the mechanical equipment screen shall not exceed 4 feet (1219 mm) more than the maximum building height allowed;
2. The *mechanical equipment screen* shall be constructed of noncombustible materials;
3. The *mechanical equipment screen* shall be constructed of fire-retardant-treated wood complying with Section 2303.2 for exterior installation; or
4. Where the fire separation distance is not less than 20 feet (6096 mm), the *mechanical equipment screen* shall be constructed of materials having a flame spread index of 25 or less when tested in the minimum and maximum thicknesses intended for use with each face tested independently in accordance with ASTM E84 or UL 723.

❖ This section is an exception to the height limitations in Section 1510.6.1. It is specific for Type V buildings, which are allowed to be constructed entirely of combustible materials. The one condition that must be met for all four options in this section is that the minimum fire separation distance must be greater than 5 feet (1524 mm). That restriction is based on Section 1406.

Item 1 is based on the belief that a 4-foot (1219 mm) height increase due to the presence of a mechanical equipment screen represents a relatively minor hazard on a building of Type V construction, particularly if the fire separation distance is at least 20 feet (6096 mm).

Item 2 allows these mechanical equipment screens to be taller when they are constructed of noncombustible materials where combustible materials would otherwise be permitted. Noncombustible materials pose no additional fire load or fire exposure to the building.

Item 3 allows the use of fire-retardant-treated wood which, although combustible, does not pose a significant fire hazard when constructed as a mechanical equipment screen where there is a minimum 5-foot (1524 mm) fire separation distance.

Item 4 criteria follows from Item 3 since fire-retardant-treated wood is required to have a maximum flame spread index of 25. There is an additional requirement that the fire separation distance is at least 20 feet (6096 mm). The lesser fire hazard associated with this type of installation should allow the greater heights for the mechanical equipment screens installed on these Type V buildings.

[BS] 1510.7 Photovoltaic panels and modules. Rooftop-mounted *photovoltaic panels* and *modules* shall be designed in accordance with this section.

❖ Rooftop-mounted photovoltaic panels and modules need to comply with building code requirements similar to other rooftop structures.

[BS] 1510.7.1 Wind resistance. Rooftop-mounted *photovoltaic panels* and *modules* shall be designed for component and cladding wind loads in accordance with Chapter 16 using an effective wind area based on the dimensions of a single unit frame.

❖ This section clarifies that rooftop-mounted photovoltaic panels and modules need to resist component and cladding wind loads and specifies that a single unit must be used to establish the effective wind area.

[BS] 1510.7.2 Fire classification. Rooftop-mounted *photovoltaic panels* and *modules* shall have the fire classification in accordance with Section 1505.9.

❖ The minimum requirements set forth here are intended for the rooftop-mounted photovoltaic panels and modules to comply with the same minimum requirements as the underlying roof assembly.

[BS] 1510.7.3 Installation. Rooftop-mounted *photovoltaic panels* and *modules* shall be installed in accordance with the manufacturer's instructions.

❖ Rooftop-mounted photovoltaic panels and modules need to be installed in accordance with the manufacturer's instructions.

[BS] 1510.7.4 Photovoltaic panels and modules. Rooftop-mounted *photovoltaic panels* and *modules* shall be *listed* and labeled in accordance with UL 1703 and shall be installed in accordance with the manufacturer's instructions.

❖ This section addresses the safety of photovoltaic panels and modules by requiring these products to comply with UL 1703 and to be installed in accordance with the manufacturer's instructions. UL 1703 is a standard used to investigate photovoltaic modules and panels and includes construction and performance requirements that address potential safety hazards.

[BS] 1510.8 Other rooftop structures. Rooftop structures not regulated by Sections 1510.2 through 1510.7 shall comply with Sections 1510.8.1 through 1510.8.5, as applicable.

❖ This section addresses other rooftop structures that are not specifically regulated by Sections 1510.2 through 1510.7. These include rooftop structures such as aerial supports (Section 1510.8.1), bulkheads (Section 1510.8.2), dormers (Section 1510.8.3), fences (Section 1510.8.4) and flagpoles (Section 1510.8.5).

[BS] 1510.8.1 Aerial supports. Aerial supports shall be constructed of noncombustible materials.

Exception: Aerial supports not greater than 12 feet (3658 mm) in height as measured from the roof deck to the highest point on the aerial supports shall be permitted to be constructed of combustible materials.

❖ This section addresses construction of aerial supports.

[BS] 1510.8.2 Bulkheads. Bulkheads used for the shelter of mechanical or electrical equipment or vertical shaft openings in the roof assembly shall comply with Section 1510.2 as penthouses. Bulkheads used for any other purpose shall be considered as an additional story of the building.

❖ This provision treats bulkheads like penthouses. A definition of the term "bulkhead" that is closest to the context of Section 1510 is "a projecting framework with a sloping door giving access to a cellar stairway or a shaft."

[BS] 1510.8.3 Dormers. Dormers shall be of the same type of construction as required for the roof in which such dormers are located or the exterior walls of the building.

❖ This section recognizes typical dormer construction and requires dormers to be the same type of construction as the roof or exterior wall in which the dormer is located.

[BS] 1510.8.4 Fences. Fences and similar structures shall comply with Section 1510.6 as *mechanical equipment screens*.

❖ This section treats fences similar to mechanical equipment screens since they are similar structures and can be considered to have a similar fire hazard.

1510.8.5 Flagpoles. Flagpoles and similar structures shall not be required to be constructed of noncombustible materials and shall not be limited in height or number.

❖ This section exempts flagpoles from the rooftop height and construction limitations that apply to most other rooftop structures.

[BS] 1510.9 Structural fire resistance. The structural frame and roof construction supporting imposed loads upon the roof by any rooftop structure shall comply with the requirements of Table 601. The fire-resistance reduction permitted by Table 601, Note a, shall not apply to roofs containing rooftop structures.

❖ This section is a reminder to verify compliance with Table 601 for roofs that support the additional load of a rooftop structure. Furthermore, it clarifies that the reduction permitted in Table 601, Note a, is not applicable where the roof supports a rooftop structure.

SECTION 1511
REROOFING

1511.1 General. Materials and methods of application used for recovering or replacing an existing roof covering shall comply with the requirements of Chapter 15.

Exceptions:

1. *Roof replacement* or *roof recover* of existing low-slope roof coverings shall not be required to meet the minimum design slope requirement of one-quarter unit vertical in 12 units horizontal (2-percent slope) in Section 1507 for roofs that provide positive roof drainage.
2. Recovering or replacing an existing roof covering shall not be required to meet the requirement for secondary (emergency overflow) drains or scuppers in Section 1503.4 for roofs that provide for positive roof drainage. For the purposes of this exception, existing secondary drainage or scupper systems required in accordance with this code shall not be removed unless they are replaced by secondary drains or scuppers designed and installed in accordance with Section 1503.4.

❖ This section simply states that when a roof is replaced or recovered it must comply with Chapter 15 for the materials and methods used with exceptions for low-sloped roofs as well as a waiver of the secondary drainage provision when reroofing an existing building where the roof drains properly. This section does not mandate that the entire roof be replaced but simply that the portion being replaced complies with Chapter 15.

For low-sloped roofs, the exception indicates that reroofing (i.e., recovering or replacement) is not required to meet the $^1/_4$:12 minimum slope requirement of Section 1507, provided that the roof has positive drainage. The term "positive drainage" is defined as the drainage condition in which consideration has been made for all loading deflections of the roof deck, and additional roof slope has been provided to ensure drainage of the roof area within 48 hours of rainfall.

1511.2 Structural and construction loads. Structural roof components shall be capable of supporting the roof-covering system and the material and equipment loads that will be encountered during installation of the system.

❖ The structural integrity of the roof must be maintained during reroofing operations, which can significantly contribute to the loading of the roof due to workers and material being present during this period of time. The roof support system must be able to structurally support all additional layers of roof covering material.

1511.3 Roof replacement. *Roof replacement* shall include the removal of all existing layers of roof coverings down to the roof deck.

Exception: Where the existing roof assembly includes an ice barrier membrane that is adhered to the roof deck, the existing ice barrier membrane shall be permitted to remain in place and covered with an additional layer of ice barrier membrane in accordance with Section 1507.

❖ The term "Roof replacement" is defined in Chapter 2. When removing roof coverings, it can be difficult, if not impossible, to remove an existing layer of adhered ice barrier membrane without damaging or replacing the roof deck. The exception accounts for this condition by allowing an existing adhered ice barrier membrane to remain in place and be covered with a new ice barrier membrane as required in Section 1507, followed by the installation of the new primary roof-covering material.

1511.3.1 Roof recover. The installation of a new roof covering over an existing roof covering shall be permitted where any of the following conditions occur:

1. Where the new roof covering is installed in accordance with the roof covering manufacturer's approved instructions.
2. Complete and separate roofing systems, such as standing-seam metal roof panel systems, that are designed to transmit the roof loads directly to the building's structural system and that do not rely on existing roofs and roof coverings for support, shall not require the removal of existing roof coverings.
3. Metal panel, metal shingle and concrete and clay tile roof coverings shall be permitted to be installed over existing wood shake roofs when applied in accordance with Section 1511.4.
4. The application of a new protective coating over an existing spray polyurethane foam roofing system shall be permitted without tear off of existing roof coverings.

❖ The term "Roof recover" is defined in Chapter 2 and this section identifies conditions where a recover is permitted. Item 2 states that new roofing systems that are designed to transmit all roof loads directly to the structural supports of the building do not necessitate that the existing roofing system be removed. Item 3 allows certain roof covering, including metal, concrete panel and clay, to be placed over wood shingle and shake roofs only if any concealed combustible spaces are properly addressed in accordance with Section 1511.4 (see commentary, Section 1511.4). Item 4 refers to the practice of "recoating" in which a new protective coating is placed over an existing spray polyurethane foam roofing system. Recoating can add many years to the effective life of a spray polyurethane foam roofing system without the downside of adding significant weight associated with other reroofing-type systems.

1511.3.1.1 Exceptions. A *roof recover* shall not be permitted where any of the following conditions occur:

1. Where the existing roof or roof covering is water soaked or has deteriorated to the point that the existing roof or roof covering is not adequate as a base for additional roofing.
2. Where the existing roof covering is slate, clay, cement or asbestos-cement tile.

3. Where the existing roof has two or more applications of any type of roof covering.

❖ This section identifies the conditions where a recover is not permitted, and all layers of previously installed roof covering systems must be removed prior to the installation of the new roof covering system.

When the existing roof or roof covering is water soaked, it must be allowed to dry completely so as not to trap moisture beneath the new layer of covering. This could cause a rapid deterioration of the new covering material, as well as the existing sheathing. The existing covering is required to be removed if it cannot adequately dry out or if its physical properties have been permanently altered.

Existing roof coverings such as wood shake, slate, clay, cement or asbestos-cement tile historically do not provide an adequate base for new roof coverings and could prevent the new covering from achieving a weather-tight seal. They could also allow penetration of water, snow, etc. These types of existing coverings must always be removed.

When the existing roof has two or more layers of any type of covering system, all layers need to be removed to enable the inspector and contractor to verify that the existing sheathing is not water damaged and still capable of providing an adequate nailing base.

1511.4 Roof recovering. Where the application of a new roof covering over wood shingle or shake roofs creates a combustible concealed space, the entire existing surface shall be covered with gypsum board, mineral fiber, glass fiber or other *approved* materials securely fastened in place.

❖ "Roof recovering" is defined as the process of installing an additional roof covering over a prepared existing roof covering without removing the existing roof covering. Where recovering over wood shingles or shakes creates a combustible concealed space, the code requires that the entire surface of the wood shakes and shingles be covered with a material that will reduce the possibility of such materials adding fuel to a fire in such a space.

1511.5 Reinstallation of materials. Existing slate, clay or cement tile shall be permitted for reinstallation, except that damaged, cracked or broken slate or tile shall not be reinstalled. Existing vent flashing, metal edgings, drain outlets, collars and metal counterflashings shall not be reinstalled where rusted, damaged or deteriorated. Aggregate surfacing materials shall not be reinstalled.

❖ This section places conditions on the reuse of roof-covering materials. Historically, various types of materials have been removable without substantially damaging the material. Materials such as wood shingles and shakes, roll roofing and asphalt shingles are usually torn or cracked and cannot be reused. Fastener holes also violate the integrity of the material. Before reuse is allowed, materials such as slate, clay or cement tile should be examined thoroughly for cracks and deterioration.

1511.6 Flashings. Flashings shall be reconstructed in accordance with *approved* manufacturer's installation instructions. Metal flashing to which bituminous materials are to be adhered shall be primed prior to installation.

❖ Flashings to be reused or reconstructed must be in accordance with the manufacturer's installation instructions. Metal flashings that are to be reused for bituminous materials must be primed in accordance with the manufacturer's instructions.

SECTION 1512
PHOTOVOLTAIC PANELS AND MODULES

1512.1 Photovoltaic panels and modules. *Photovoltaic panels* and *modules* installed upon a roof or as an integral part of a roof assembly shall comply with the requirements of this code and the *International Fire Code*.

❖ Roof-mounted photovoltaic panels and modules should comply with roofing requirements, such as supporting the load of these systems on roofs and structural framing. This section also makes a needed reference to relevant requirements in the IFC.

Bibliography

The following resource materials were used in the preparation of the commentary for this chapter of the code.

ANSI/SPRI ES-1-11, *Wind Design Standard for Edge Systems Used with Low-Slope Roofing Systems*. Waltham, MA: Single-Ply Roofing Institute, 2011.

ANSI/SPRI VF-1-10, *External Fire Design Standard for Vegetative Roofs*. Waltham, MA: Single-Ply Roofing Institute, 2010.

ASTM C1167-11, *Specification for Clay Roof Tiles*. West Conshohocken, PA: ASTM International, 2011.

ASTM C1313/C 1313M-12, *Standard Specification for Sheet Radiant Barriers for Building Construction Applications*. West Conshohocken, PA: ASTM International, 2012.

ASTM D312-00 (2006), *Specification for Asphalt Used in Roofing*. West Conshohocken, PA: ASTM International, 2006.

ASTM E108-11, *Standard Test Method for Fire Tests of Roof Coverings*. West Conshohocken, PA: ASTM International, 2011.

CSSB, *New Roof Construction Manual*. Sumas, WA: Cedar Shake and Shingle Bureau, 2010.

NFPA 285-12, *Standard Fire Test Method for the Evaluation of Fire Propagation Characteristics of Exterior Nonload-bearing Wall Assemblies Containing Combustible Components.* Quincy, MA: National Fire Protection Association, 2012.

NRCA Roofing Manual: Membrane Roof Systems-2007. Rosemont, IL: National Roofing Contractors Association (NRCA), 2007.

NRCA Roofing Manual: Steep-slope Roof Systems-2009. Rosemont, IL: National Roofing Contractors Association (NRCA), 2009.

SBCCI SSTD 11-97, *Standard for Determining Wind Resistance of Concrete or Clay Roof Tiles*. Birmingham, AL: Southern Building Code Congress International, 1997.

INDEX

A

B

C

D

E

G

H

I

J

K

L

M

R

S

T

U

Y